FLAVONOIDS

Chemistry, Biochemistry and Applications

FLAVONOIDS

Chemistry, Biochemistry and Applications

Edited by
Øyvind M. Andersen
Kenneth R. Markham

CRC **Taylor & Francis**
Taylor & Francis Group
Boca Raton   London   New York

A CRC title, part of the Taylor & Francis imprint, a member of the
Taylor & Francis Group, the academic division of T&F Informa plc.

Published in 2006 by
CRC Press
Taylor & Francis Group
6000 Broken Sound Parkway NW, Suite 300
Boca Raton, FL 33487-2742

International Standard Book Number-10: 0-8493-2021-6 (Hardcover)
International Standard Book Number-13: 978-0-8493-2021-7 (Hardcover)
Library of Congress Card Number 2005048626

The front cover depicts the newly discovered (K.R. Markham et al., Phytochemistry, 55 (2000) 327-336 anthocyanic vacuolar inclusions (AVIs) in epidermal cells of blue-grey Dianthus caryophyllus (carnation). AVIs are composed of high levels of structually specific anthocyanins bound to a protein matrix. Their presence leads to flower color intensification with a marked blue shift.

Library of Congress Cataloging-in-Publication Data

Flavonoids : chemistry, biochemistry, and applications / edited by Øyvind M. Andersen and Kenneth R. Markham.
    p. cm.
   Includes bibliographical references and index.
   ISBN 0-8493-2021-6 (alk. paper)
   1. Flavonoids. I. Andersen, Øyvind M. II. Markham, Kenneth R.

QP671.F52F53 2005
612'.01528--dc22   2005048626

Visit the Taylor & Francis Web site at
http://www.taylorandfrancis.com

and the CRC Press Web site at
http://www.crcpress.com

Editors

Øyvind M. Andersen, a full professor of chemistry since 1993, has specialized in the chemistry of natural compounds. Since 2002, he has been head of the Department of Chemistry, University of Bergen, Norway. He received his Ph.D. degree in 1988 from the University of Bergen. He is author or coauthor of 100 journal articles, seven book chapters, and five patents, and has supervised over 40 M.Sc. and Ph.D. students in the fields of anthocyanins and other flavonoids. The activities of Dr. Andersen's research group have led to the establishment of several flavonoid-based companies. Most of his research projects concentrate on structural elucidation of new compounds, while others are methodology based and relate to NMR spectroscopy and various chromatographic techniques. Some of the projects focus on anthocyanin properties with the objective of exploring their pharmaceutical potential and their use as colorants in food.

Kenneth R. Markham (Ken) has recently retired from his position as group leader and distinguished scientist at Industrial Research Ltd., a New Zealand Crown Research Institute. Following completion of his B.Sc. and M.Sc. (Hons) at Victoria University, Wellington, New Zealand, and his Ph.D. in chemistry (University of Melbourne, Australia, 1963), he returned to Chemistry Division, D.S.I.R., in Lower Hutt, New Zealand, to work on xanthone chemistry. His interest in flavonoids began during a 2-year postdoctoral with Professor Tom Mabry at the University of Texas at Austin (1965–1967). The coauthored book *The Systematic Identification of Flavonoids* resulted from this interaction. On his return to New Zealand, Dr. Markham established and led the "Natural Products" section, predominantly devoted to the study of flavonoids with particular emphasis on the chemotaxonomy of New Zealand bryophytes, ferns, and gymnosperms. This work was carried out in conjunction with both local botanists and colleagues at the University of Saarbrucken (Professors Zinsmeister, Mues, and Geiger). In 1979, he worked as a visiting scientist with Professor Jeffrey Harborne at the University of Reading. This again resulted in the publication of a popular book, *Techniques of Flavonoid Identification*. Dr. Markham's flavonoid research continued at Industrial Research Ltd. and has led him into studies as diverse as plant chemotaxonomy, plant evolution, plant UV protection, historical Antarctic ozone levels, propolis and bee pollen bioactives, G.E. modification of flower color, and subcellular chemistry, to name but a few. His work has been reported in some 280 publications, including 18 invited chapters and two books, and has been recognized over the years through awards such as the Chemical Society's Easterfield Medal, Fellowship of the Royal Society of New Zealand, a Ministerial Award for Excellence in Scientific Research, and the 1999 Pergamon Phytochemical Prize.

Preface

It is with great pleasure that we accepted the offer by CRC Press to assemble and edit this compilation of reviews on flavonoids and their properties and functions for the present volume. We considered the volume timely in that the last book of this general type, *The Flavonoids — Advances in Research Since 1986* (edited by Jeffrey B. Harborne), appeared over a decade ago. Since then, advances in the flavonoid field have been nothing short of spectacular. These advances are particularly evident in the contributed chapters that cover: the discovery of a variety of new flavonoids; the application of advanced analytical techniques; genetic manipulation of the flavonoid pathway; improved understanding of flavonoid structures and physiological functions in plants and animals; and, perhaps most importantly, the significance of flavonoids to human health.

Whilst the updating aspect of the chapters is seen as the prime contribution of this book, an effort also has been made to include a summary of previous knowledge in the field to enable the reader to place new advances in this context. Chapters 1 and 2 review the application of contemporary isolation, quantification, and spectroscopic techniques in flavonoid analysis, while Chapter 3 is devoted to molecular biology and biotechnology of flavonoid biosynthesis. Individual chapters address the flavonoids in food (Chapter 4) and wine (Chapter 5), and the impact of flavonoids and other phenolics on human health (Chapter 6 and, in part, Chapter 16). Chapter 8 reviews newly discovered flavonoid functions in plants, while Chapter 9 is the first review of flavonoid–protein interactions. Chapters 10 to 17 discuss the chemistry and distribution of the various flavonoid classes including new structures reported during 1993 to 2004. A complete listing of all known flavonoids within the various flavonoid classes are found in these later chapters and the Appendix, and to date a total of above 8150 different flavonoids has been reported.

It is difficult to overstate the importance of recent advances in research on flavonoids, and we are sure that the information contained within this book will prove to be invaluable to a wide range of researchers, professionals, and advanced students in both the academic and industrial sectors.

We are greatly indebted to our authors, and are delighted that so many of the world's leading researchers in a variety of flavonoid-related fields have been willing, so generously, to share their knowledge and experience with others through their contribution to this volume. We are also very grateful to Lindsey Hofmeister, Erika Dery, Jill Jurgensen, and Tanya Gordon at Taylor and Francis, and Balaji Krishnasamy at SPI Publisher Services for their support and interest throughout the preparation of this book.

Øyvind M. Andersen and Kenneth R. Markham

Historical Advances in the Flavonoid Field — A Personal Perspective
Having been associated with flavonoid research for the past 40 years, and having witnessed the spellbinding changes that have taken place in the field during this time, it is too tempting by far not to take this opportunity to document for future researchers a brief personal perspective on developments in the flavonoid field over this period. I emphasize that this is

but a personal perspective on progress, and as such will surely exhibit some bias and deficiencies. To the aggrieved I offer my apologies.

In the early 1960s, flavonoids were widely viewed as metabolic waste products that were stored in the plant vacuole. Whilst there was interest at that time in their function as flower colorants, and in their distribution between plant taxa, the earliest investigations of their biosynthesis had just begun. In this respect it is informative to note that Tom Geissman's 1962 compilation of reviews in *The Chemistry of Flavonoid Compounds* includes nothing at all on biological function, and details only paper chromatography and absorption spectroscopy as analytical tools. At this time too, information on flavonoid distribution within the plant kingdom was still incomplete. For example, even as late as 1969 Bate-Smith wrote (in *Chemical Plant Taxonomy* edited by T. Swain) that flavonoids are rarely found in any but vascular plants. But within a few years of this statement, Markham, Porter, and others reported the widespread presence of flavonoids in mosses and liverworts and even their occurrence in an alga, *Nitella hookeri*. This, incidentally, remains the sole example of the occurrence of flavonoids in algae. To date, flavonoids have been found in all major categories of green plants except for the Anthocerotae.

By 1967 little had changed with regard to the application of physical techniques to flavonoid structure determination, with NMR and GLC yet to make an impact on the field. About this time, however, there was an upsurge in the application of flavonoid distribution to the emerging field of chemotaxonomy. Leading these researches were groups at the University of Texas at Austin (led by Tom Mabry, Ralph Alston, and Billy Turner) and at the University of Reading (led by Jeffrey Harborne). Alston and Turner's pioneering work on the tracking of plant hybridization through flavonoid analyses is still quoted today, as also is much of the anthocyanin structure and flavonoid distributional work detailed by Harborne in his 1967 book, *Comparative Biochemistry of the Flavonoids*.

Rapid advances in the application of physical techniques to flavonoid structure analysis began appearing in the literature in the mid- to late-1960s, and these were well documented in a series of books beginning in 1970 with Mabry, Markham, and Thomas' *The Systematic Identification of Flavonoids*, which reviewed, for the first time, the considerable advances made in the application of shift reagents in UV–visible absorption spectroscopy, the use of GLC for sugar analyses, and the application of ^{1}H NMR spectroscopy to flavonoid structure analysis. At this time, CCl_4-soluble TMS ether derivatives were widely used for NMR studies in the absence of readily obtainable deuterated solvents. These same derivatives, together with permethylated derivatives, were commonly also used to make flavonoid glycosides sufficiently volatile for early applications of electron impact mass spectrometry to flavonoid structure analysis (first summarized in the 1974 "Advances" book, *The Flavonoids*, Chapter 3). Contemporaneously, and detailed in the above volume, rapid advances were being reported in the structure analysis of *C*-glycosides (Chopin and Brouillant, Chapter 12) and in the biosynthesis of flavonoids (Hahlbrock and Grisebach, Chapter 16). A growing awareness of the physiological, metabolic, and evolutionary value of flavonoids was also beginning to emerge. Tony Swain, for example, was in the initial stages of formulating his innovative interpretations of the evolution of flavonoids, and the part that their "chemoecology" played in the evolution of plants (e.g., see Chapter 20 in *The Flavonoids*, 1974).

The next major advance in flavonoid structural techniques was the application of ^{13}C NMR spectroscopy. This has arguably had the greatest impact on flavonoid structure analysis since the invention of paper chromatography around 1900. For the first time, complete flavonoid structures, including flavonoid aglycones together with sugar types and linkages, could be determined using a single technique. Admittedly, the development of advanced two-dimensional techniques has further revolutionized structure analysis since the earliest applications of this technique to flavonoids in the mid-1970s, and the appearance of the first review article by Markham and Chari in 1982 (*The Flavonoids — Advances in Research*, Chapter 2). Advances in technology have diminished the sample size required for spectral analysis by more than 100 times for both ^{13}C and ^1H NMR techniques. In the early 1960s, 100-mg samples were required for proton work and in the late 1970s the same sized samples were required for carbon-13 studies.

Modern flavonoid researchers will also be aware of the impact that the more recently developed mass spectrometry techniques such as FAB, MALDI-TOF, and electrospray have had on the ability of researchers to elucidate complex flavonoid glycosidic structures through the ready determination of accurate molecular weights and limited fragmentation patterns. Similarly, the development of advanced methods of separation such as capillary electrophoresis, HPLC, and, latterly, HPLC–MS, has recently revolutionized the qualitative and quantitative analysis of flavonoid mixtures. Chemotaxonomic studies involving comparative distributional data have accordingly been vastly facilitated. Techniques such as those referred to above have enabled previously intractable flavonoid structural problems to be solved. Particularly good examples of this are to be found amongst the many complex structures currently being reported for "blue" flower pigments based on anthocyanins elaborated in an often intricate manner with large numbers of sugars, acyl groups, other flavonoids, and occasionally including metal ions.

Returning once again to the questions of function and uses, the old concept of flavonoids being merely the by-products of cellular metabolism, which are simply compartmentalized in solution in the cell vacuole, is well and truly past its use-by date. For a start, studies have revealed that flavonoids are also commonly found on the outer surfaces of leaves and flowers, albeit only the aglycone form. Additionally, flavonoids have been shown over the past few years to be found in the cell wall, the cytoplasm, in oil bodies, and associated with the nucleus and cell proteins, as well as in the vacuole. Even in the vacuole, flavonoids are not necessarily found free in solution. For example, protein-bound flavonoids have been isolated from lisianthus and other flowers in which a structurally specific binding has been identified (in anthocyanic vacuolar inclusions). It is probable that flavonoid location and specific protein binding properties will ultimately prove to relate directly to their function in plants.

Amongst the many functions now known to be performed by plant flavonoids are those of UV protection, oxidant or free radical protection, modulation of enzymic activity, allelopathy, insect attraction or repulsion, nectar guides, probing stimulants, viral, fungal, and bacterial protection, nodulation in leguminous plants, pollen germination, etc., and it is likely that this is only the tip of the iceberg. Flavonoids, it would seem, have been vital components of plants, ever since their (purported) development at the time plant life emerged from the aquatic environment, and needed protection from UV light in an atmosphere lacking today's protective ozone layer. The continued widespread accumulation of flavonoids by virtually all land-based green plants lends support to this view.

Intriguingly, it is now possible to exert some precise control over plant flavonoid composition. Manipulation of the flavonoid biosynthetic pathway in plants via genetic engineering has progressed rapidly in recent years. This has been expedited by the extensive information made available through the earlier studies of flavonoid biosynthesis pioneered in the 1960s and 1970s (see above). Genetic manipulation of the flavonoid pathway in plants has enormous potential to, for example, produce new flower colors, enhance the nutritional value of crops, and improve crop protection from UV light, microorganisms, insects, and browsing animals. Indeed, much of this work has been underway for some time and shows great promise.

Plant flavonoids have been shown in recent years to be of vital significance to mankind as well as to plants. They have been strongly implicated as active contributors to the health benefits of beverages such as tea and wine, foods such as fruit and vegetables, and even, recently, chocolate. The widely lauded "Mediterranean diet," for example, is thought to owe much of its benefits to the presence of flavonoids in the food and beverages. In the early 1990s, Hertog published aspects of the "Zutphen Elderly Study," and, in so doing, provided for the first time a sound epidemiological correlation between high food flavonoid intake and a lowering in the risk of coronary heart disease. This study also produced the first reliable estimates of average daily flavonoid intake at around 23 mg, a figure much lower than the 1000 mg that had been proposed in the 1970s. The major sources of flavonoids (in the Dutch population) were found to be tea, onions, and apples.

Other potential health benefits of dietary flavonoids are too numerous to mention here. Suffice it to say that our understanding of the importance of flavonoids in the human diet is continuing to advance rapidly. One suspects that much of the physiological activity associated with flavonoids can be attributed to (i) their proven effectiveness as antioxidants and free radical scavengers, (ii) to their metal complexing capabilities (a capability that drove early advances in absorption spectroscopy and NMR studies), and (iii) to their ability to bind with a high degree of specificity to proteins.

Because of the incredible advances that have taken place, my involvement in flavonoid studies over the past 40 years has been exciting and stimulating. I feel privileged to have been part of the discovery process. During this period flavonoids as a natural product group have risen from relative obscurity (at least in the popular media) to such prominence that educated people in the West are now not only aware of the name, but also aware of the publicized health benefits associated with their consumption.

At an academic level too, although flavonoid structure elucidation is rapidly becoming a mature science thanks to technological advances, studies of their bioavailability and physiological activity in both animals and plants is likely to become the new frontier. Exciting advances in the understanding of this physiological activity will undoubtedly lead to the more widespread application of flavonoids in the improvement of human health and in crop quality. A major influence, especially in the latter, is likely to be brought about through skilful genetic manipulation of the flavonoid biosynthetic pathway. We await this progress with eager anticipation.

Kenneth R. Markham
Industrial Research Ltd
Lower Hutt, New Zealand

List of Contributors

Jonathan E. Brown
School of Biomedical & Molecular Sciences
University of Surrey
Guildford, Surrey, U.K.

Véronique Cheynier
INRA Unite de Recherche Biopolymeres
et Aromes
Montpellier, France

Mike Clifford
School of Biomedical & Molecular Sciences
University of Surrey
Guildford, Surrey, U.K.

Olivier Dangles
UMR A 408 Safety and Quality of
Food Products
Avignon, France

Kevin M. Davies
Crop & Food Research
Palmerston North, New Zealand

Claire Dufour
UMR A 408 Safety and Quality of
Food Products
Avignon, France

Garry G. Duthie
Rowett Research Institute
Aberdeen, Scotland, U.K.

Daneel Ferreira
Natural Products Center
University of Mississippi
Oxford, Mississippi

Torgils Fossen
Department of Chemistry
University of Bergen
Bergen, Norway

Kevin S. Gould
School of Biological Sciences
University of Auckland
Auckland, New Zealand

Renée J. Grayer
Jodrell Laboratory
Royal Botanic Gardens, Kew
Richmond, Surrey, U.K.

Kurt Hostettmann
Laboratoire de Pharmacognosie
et Phytochimie
Université de Genève
Genève, Switzerland

Maurice Jay
Université Claude Bernard – Lyon I
Villeurbanne, France

Monica Jordheim
Department of Chemistry
University of Bergen
Bergen, Norway

Janet A.M. Kyle
Rowett Research Institute
Aberdeen, Scotland, U.K.

Carolyn Lister
Crop & Food Research
Christchurch, New Zealand

Jannie P.J. Marais
Natural Products Center
University of Mississippi
Oxford, Mississippi

Andrew Marston
Laboratoire de Pharmacognosie
et de Phytochimie
Université de Genève
Genève, Switzerland

Kathy E. Schwinn
Crop & Food Research
Palmerston North, New Zealand

Desmond Slade
Natural Products Center
University of Mississippi
Oxford, Mississippi

K.M. Valant-Vetschera
Institut fuer Botanik der Universitaet Wien
Wien, Austria

Nigel C. Veitch
Jodrell Laboratory
Royal Botanic Gardens, Kew
Richmond, Surrey, U.K.

Christine A. Williams
School of Plant Sciences
The University, Whiteknights
Reading, U.K.

Helen Wiseman
Department of Nutrition
  and Dietetics
King's College London
London, U.K.

Eckhard Wollenweber
Institut für Botanik
TU Darmstadt
Darmstadt, Germany

Contents

1 Separation and Quantification of Flavonoids

Andrew Marston and Kurt Hostettmann

CONTENTS

1.1 INTRODUCTION

Essential to the study of flavonoids is having the means available for their separation (analytical and preparative) and isolation. The importance of this aspect of flavonoid research can be seen in the number of review articles that refer to their chromatography.[1–6] However, the information is usually spread out over chapters in books or occurs in isolated sections devoted to individual classes of these polyphenols.

This chapter, therefore, aims to present a brief unified summary of general techniques, with reference to the different categories of structure: flavones and flavonols (and their glycosides), isoflavones, flavanones, chalcones, anthocyanins, and proanthocyanidins.

In earlier times, thin-layer chromatography (TLC), polyamide chromatography, and paper electrophoresis were the major separation techniques for phenolics. Of these methods, TLC is still the workhorse of flavonoid analysis. It is used as a rapid, simple, and versatile method for following polyphenolics in plant extracts and in fractionation work. However, the majority of published work now refers to qualitative and quantitative applications of high-performance liquid chromatography (HPLC) for analysis. Flavonoids can be separated, quantified, and identified in one operation by coupling HPLC to ultraviolet (UV), mass, or nuclear magnetic resonance (NMR) detectors. Recently, the technique of capillary electrophoresis (CE) has been gaining attention.

One feature that is of immense benefit for flavonoid analysis is the presence of the phenyl ring. This excellent chromophore is, of course, UV active and provides the reason why flavonoids are so easy to detect. Their UV spectra are particularly informative, providing considerable structural information that can distinguish the type of phenol and the oxidation pattern.

A number of techniques have been used for the preparative separation of flavonoids. These include HPLC, Diaion, Amberlite XAD-2 and XAD-7, and Fractogel TSK/Toyopearl HW-40 resins, gel filtration on Sephadex, and centrifugal partition chromatography (CPC).[7] The choice of methods and strategies varies from research group to research group and depends often on the class of flavonoid studied.

1.2   EXTRACTION

Flavonoids (particularly glycosides) can be degraded by enzyme action when collected plant material is fresh or nondried. It is thus advisable to use dry, lyophilized, or frozen samples. When dry plant material is used, it is generally ground into a powder. For extraction, the solvent is chosen as a function of the type of flavonoid required. Polarity is an important consideration here. Less polar flavonoids (e.g., isoflavones, flavanones, methylated flavones, and flavonols) are extracted with chloroform, dichloromethane, diethyl ether, or ethyl acetate, while flavonoid glycosides and more polar aglycones are extracted with alcohols or alcohol–water mixtures. Glycosides have increased water solubility and aqueous alcoholic solutions are suitable. The bulk of extractions of flavonoid-containing material are still performed by simple direct solvent extraction.

Powdered plant material can also be extracted in a Soxhlet apparatus, first with hexane, for example, to remove lipids and then with ethyl acetate or ethanol to obtain phenolics. This approach is not suitable for heat-sensitive compounds.

A convenient and frequently used procedure is sequential solvent extraction. A first step, with dichloromethane, for example, will extract flavonoid aglycones and less polar material. A subsequent step with an alcohol will extract flavonoid glycosides and polar constituents.

Certain flavanone and chalcone glycosides are difficult to dissolve in methanol, ethanol, or alcohol–water mixtures. Flavanone solubility depends on the pH of water-containing solutions.

Flavan-3-ols (catechins, proanthocyanidins, and condensed tannins) can often be extracted directly with water. However, the composition of the extract does vary with the solvent — whether water, methanol, ethanol, acetone, or ethyl acetate. For example, it is claimed that methanol is the best solvent for catechins and 70% acetone for procyanidins.[8]

Anthocyanins are extracted with cold acidified methanol. The acid employed is usually acetic acid (about 7%) or trifluoroacetic acid (TFA) (about 3%). The use of mineral acid can lead to the loss of attached acyl groups.

Extraction is typically performed with magnetic stirring or shaking but other methods have recently been introduced to increase the efficiency and speed of the extraction procedure. The first of these is called pressurized liquid extraction (PLE). By this method, extraction is accelerated by using high temperature and high pressure. There is enhanced diffusivity of the solvent and, at the same time, there is the possibility of working under an inert atmosphere and with protection from light. Commercially available instruments have extraction vessels with volumes up to about 100 ml. In a study involving medicinal plants, solvent use was reduced by a factor of two.[9] The optimization of rutin and isoquercitrin recovery from older (*Sambucus nigra*, Caprifoliaceae) flowers has been described. Application of PLE gave better results than maceration — and shorter extraction times and smaller amounts of solvent were required.[10] PLE of grape seeds and skins from winemaking wastes proved to be an efficient procedure for obtaining catechin and epicatechin with little decomposition, provided the temperature was kept below 130°C.[11]

As its name suggests, supercritical fluid extraction (SFE) relies on the solubilizing properties of supercritical fluids. The lower viscosities and higher diffusion rates of supercritical fluids, when compared with those of liquids, make them ideal for the extraction of diffusion-controlled matrices, such as plant tissues. Advantages of the method are lower solvent consumption, controllable selectivity, and less thermal or chemical degradation than methods such as Soxhlet extraction. Numerous applications in the extraction of natural products have been reported, with supercritical carbon dioxide being the most widely used extraction solvent.[12,13] However, to allow for the extraction of polar compounds such as flavonoids, polar solvents (like methanol) have to be added as modifiers. There is consequently a substantial reduction in selectivity. This explains why there are relatively few applications to polyphenols in the literature. Even with pressures of up to 689 bar and 20% modifier (usually methanol) in the extraction fluid, yields of polyphenolic compounds remain low, as shown for marigold (*Calendula officinalis*, Asteraceae) and chamomile (*Matricaria recutita*, Asteraceae).[14]

Ultrasound-assisted extraction is a rapid technique that can also be used with mixtures of immiscible solvents: hexane with methanol–water (9:1), for example, is a system used for the Brazilian plant *Lychnophora ericoides* (Asteraceae). The hexane phase concentrated less polar sesquiterpene lactones and hydrocarbons, while the aqueous alcohol phase concentrated flavonoids and more polar sesquiterpene lactones.[15]

Microwave-assisted extraction (MAE) has been described for the extraction of various compounds from different matrices.[16] It is a simple technique that can be completed in a few minutes. Microwave energy is applied to the sample suspended in solvent, either in a closed vessel or in an open cell. The latter allows larger amounts of sample to be extracted. A certain degree of heating is involved.[17]

1.3   PREPARATIVE SEPARATION

1.3.1   Preliminary Purification

Once a suitably polar plant extract is obtained, a preliminary cleanup is advantageous. The classical method of separating phenolics from plant extracts is to precipitate with lead acetate or extract into alkali or carbonate, followed by acidification. The lead acetate procedure is often unsatisfactory since some phenolics do not precipitate; other compounds may co-precipitate and it is not always easy to remove the lead salts.

Alternatively, solvent partition or countercurrent techniques may be applied. In order to obtain an isoflavonoid-rich fraction from *Erythrina* species (Leguminosae) for further purification work, an organic solvent extract was dissolved in 90% methanol and first partitioned with hexane. The residual methanol part was adjusted with water to 30% and partitioned with *t*-butyl methyl ether–hexane (9:1). This latter mixture was then chromatographed to obtain pure compounds.[18]

A short polyamide column, a Sephadex LH-20 column, or an ion exchange resin can be used. Absorption of crude extracts onto Diaion HP-20 or Amberlite XAD-2 (or XAD-7) columns, followed by elution with a methanol–water gradient, is an excellent way of preparing flavonoid-rich fractions.

1.3.2 PREPARATIVE METHODS

One of the major problems with the preparative separation of flavonoids is their sparing solubility in solvents employed in chromatography. Moreover, the flavonoids become less soluble as their purification proceeds. Poor solubility in the mobile phase used for a chromatographic separation can induce precipitation at the head of the column, leading to poor resolution, decrease in solvent flow, or even blockage of the column.

Other complications can also arise. For example, in the separation of anthocyanins and anthocyanin-rich fractions, it is advisable to avoid acetonitrile and formic acid — acetonitrile is difficult to evaporate and there is a risk of ester formation with formic acid.

There is no single isolation strategy for the separation of flavonoids and one or many steps may be necessary for their isolation. The choice of method depends on the polarity of the compounds and the quantity of sample available. Most of the preparative methods available are described in a volume by Hostettmann et al.[7]

Conventional open-column chromatography is still widely used because of its simplicity and its value as an initial separation step. Preparative work on large quantities of flavonoids from crude plant extracts is also possible. Support materials include polyamide, cellulose, silica gel, Sephadex LH-20, and Sephadex G-10, G-25, and G-50. Sephadex LH-20 is recommended for the separation of proanthocyanidins. For Sephadex gels, as well as size exclusion, adsorption and partition mechanisms operate in the presence of organic solvents. Although methanol and ethanol can be used as eluents for proanthocyanidins, acetone is better for displacing the high molecular weight polyphenols. Slow flow rates are also recommended. Open-column chromatography with certain supports (silica gel, polyamide) suffers from a certain degree of irreversible adsorption of the solute on the column.

Modifications of the method (dry-column chromatography, vacuum liquid chromatography, VLC, for example) are also of practical use for the rapid fractionation of plant extracts. VLC with a polyamide support has been reported for the separation of flavonol glycosides.[19]

Preparative TLC is a separation method that requires the least financial outlay and the most basic equipment. It is normally employed for milligram quantities of sample, although gram quantities are also handled if the mixture is not too complex. Preparative TLC in conjunction with open-column chromatography remains a straightforward means of purifying natural products, although variants of planar chromatography, such as centrifugal TLC,[7] have found application in the separation of flavonoids.

Other combinations are, of course, possible, depending on the particular separation problem. Combining gel filtration or liquid–liquid partition with liquid chromatography (LC) is one solution. Inclusion of chromatography on polymeric supports[7] can also provide additional means of solving a difficult separation.

Several preparative pressure liquid chromatographic methods are available. These can be classified according to the pressure employed for the separation:

- High-pressure (or high-performance) LC (>20 bar/300 psi)
- Medium-pressure LC (5 to 20 bar/75 to 300 psi)
- Low-pressure LC (<5 bar/75 psi)
- Flash chromatography (ca. 2 bar/30 psi)

1.3.2.1 High-Performance Liquid Chromatography

HPLC is becoming by far the most popular technique for the separation of flavonoids, both on preparative and analytical scales. Improvements in instrumentation, packing materials, and column technology are being introduced all the time, making the technique more and more attractive.

The difference between the analytical and preparative methodologies is that analytical HPLC does not rely on the recovery of a sample, while preparative HPLC is a purification process and aims at the isolation of a pure substance from a mixture.

Semipreparative HPLC separations (for 1 to 100 mg sample sizes) use columns of internal diameter 8 to 20 mm, often packed with 10 μm (or smaller) particles. Large samples can be separated by preparative (or even process-scale) installations but costs become correspondingly higher.

Optimization can be performed on analytical HPLC columns before transposition to a semipreparative scale.

The aim of this chapter is not a detailed description of the technique and instrumentation but to show applications of HPLC in the preparative separation of flavonoids. Some representative examples are given in Table 1.1. In a 1982 review of isolation techniques for flavonoids,[3] preparative HPLC had at that time not been fully exploited. However, the situation is now very different and 80% of all flavonoid isolations contain a HPLC step. Approximately 95% of reported HPLC applications are on octadecylsilyl phases. Both isocratic and gradient conditions are employed.

1.3.2.2 Medium-Pressure Liquid Chromatography

The term "medium-pressure liquid chromatography" (MPLC) covers a wide range of column diameters, different granulometry packing materials, different pressures, and a number of

---

TABLE 1.1
Preparative Separations of Flavonoids by HPLC

Sample	Column	Eluent	Reference
Phenolics from *Picea abies*	Nucleosil 100–7C$_{18}$ 250 × 21 mm	MeOH–H$_2$O, gradient	21
Chalcones from *Myrica serrata*	LiChrosorb Diol 7 μm, 250 × 16 mm	MeOH–H$_2$O, 55:45	20
	Nucleosil 100–7C$_{18}$ 250 × 21 mm	MeOH–H$_2$O, 76:24	
Flavones from *Tanacetum parthenium*	LiChrospher RP-18 250 × 25 mm	CH$_3$CN–H$_2$O, 3:7	22
Flavone glycosides from *Lysionotus pauciflorus*	LiChrosorb RP-18 250 × 10 mm	CH$_3$CN–H$_2$O, 1:4	23
Flavonoid glucuronides from *Malva sylvestris*	Spherisorb ODS-2 5 μm, 250 × 16 mm	CH$_3$CN–H$_2$O–THF–HOAc, 205:718:62:15	24
Flavonol galloyl-glycosides from *Acacia confusa*	Hyperprep ODS 250 × 10 mm	CH$_3$CN–H$_2$O, gradient	25
Flavanones from *Greigia sphacelata*	LiChrospher Diol 5 μm, 250 × 4.6 mm	Hexane–EtOAc, 7:3	26
Prenylated flavonoids from *Anaxagorea luzonensis*	Asahipack ODP-90 10 μm, 300 × 28 mm	CH$_3$CN–H$_2$O, 45:55	27
Prenylated isoflavonoids from *Erythrina vogelii*	μBondapak C$_{18}$10 μm, 100 × 25 mm	MeOH–H$_2$O, isocratic	28
Biflavones from *Cupressocyparis leylandii*	LiChrospher RP-18 7 μm, 250 × 10 mm	MeOH–H$_2$O, 72:28	29
Anthocyanin glycosides	Spherisorb ODS-2 10 μm, 250 × 10 mm	MeOH–5% HCOOH	30
Proanthocyanidins and flavans from *Prunus prostrata*	Eurospher 80 RP-18 7 μm, 250 × 16 mm	CH$_3$CN–H$_2$O (+0.1% TFA), 1:4, 3:17	31

commercially available systems. In its simplest form, MPLC is a closed column (generally glass) connected to a compressed air source or a reciprocating pump. It fulfills the requirement for a simple alternative method to open-column chromatography or flash chromatography, with both higher resolution and shorter separation times. MPLC columns have a high loading capacity — up to a 1:25 sample-to-packing-material ratio[32] — and are ideal for the separation of flavonoids.

In MPLC, the columns are generally filled by the user. Particle sizes of 25 to 200 μm are usually advocated (15 to 25, 25 to 40, or 43 to 60 μm are the most common ranges) and either slurry packing or dry packing is possible. Resolution is increased for a long column of small internal diameter when compared with a shorter column of larger internal diameter (with the same amount of stationary phase).[33] Choice of solvent systems can be efficiently performed by TLC[34] or by analytical HPLC. Transposition to MPLC is straightforward and direct.[35]

Some applications of MPLC to the separation of flavonoids are shown in Table 1.2.

1.3.2.3 Centrifugal Partition Chromatography

Various countercurrent chromatographic techniques have been successfully employed for the separation of flavonoids.[7] Countercurrent chromatography is a separation technique that relies on the partition of a sample between two immiscible solvents, the relative proportions of solute passing into each of the two phases determined by the partition coefficients of the components of the solute. It is an all-liquid method that is characterized by the absence of a solid support, and thus has the following advantages over other chromatographic techniques:

- No irreversible adsorption of the sample
- Quantitative recovery of the introduced sample
- Greatly reduced risk of sample denaturation

TABLE 1.2
Separation of Flavonoids by Medium-Pressure Liquid Chromatography

Sample	Column	Eluent	Reference
Chalcones from *Piper aduncum*	Silica gel 800 × 36 mm	Hexane–TBME–CH$_2$Cl$_2$–EtOH, 99:0.4:0.3:0.3	36
Flavonoids from *Sophora moorcroftiana*	RP-18 20 μm	MeOH–H$_2$O, 3:1	37
Flavonol glycosides from *Epilobium* species	RP-18 15–25 μm 460 × 26 mm	MeOH–H$_2$O, 35:65	38
Dihydroflavonoid glycosides from *Calluna vulgaris*	Polyamide SC-6 460 × 26 mm RP-18 20–40 μm 460 × 15 mm	Toluene–MeOH MeOH–H$_2$O	39
Prenylflavonoid glycosides from *Epimedium koreanum*	RP-8 460 × 26 mm	MeOH–H$_2$O, 2:3	40
Prenylated isoflavonoids from *Erythrina vogelii*	RP-18 15–25 μm 500 × 40 mm	MeOH–H$_2$O, 58:42, 60:40	41
Biflavonoids from *Wikstoemia indica*	RP-18 300 × 35 mm	MeOH–H$_2$O, 55:45 → 95:5	42

- Low solvent consumption
- Favorable economics

It is obvious, therefore, that such a technique is ideal for flavonoids, which often suffer from problems of retention on solid supports such as silica gel and polyamide.

Countercurrent distribution, droplet countercurrent chromatography, and rotation locular countercurrent chromatography are now seldom used but CPC, also known as centrifugal countercurrent chromatography, finds extensive application for the preparative separation of flavonoids. In CPC, the liquid stationary phase is retained by centrifugal force instead of a solid support (in column chromatography). Basically, two alternative designs of apparatus are on the market[43]: (a) rotating coil instruments; (b) disk or cartridge instruments.

Although most CPC separations are on a preparative scale, analytical instruments do exist.[44] However, these are mostly used to find suitable separation conditions for scale-up.

There are numerous examples of preparative separations of flavonoids[7,45] and some are listed in Table 1.3.

An example of the separation of flavonoid glycosides by CPC is shown in Figure 1.1. The leaves of the African plant *Tephrosia vogelii* (Leguminosae) were first extracted with dichloromethane and then with methanol. Methanol extract (500 mg) was injected in a mixture of the two phases of the solvent system and elution of the three major glycosides was achieved within 3 h.[58]

The technique of CPC was also employed as a key step in the purification of 26 phenolic compounds from the needles of Norway spruce (*Picea abies*, Pinaceae). An aqueous extract of needles (5.45 g) was separated with the solvent system $CHCl_3$–MeOH–i-PrOH–H_2O (5:6:1:4), initially with the lower phase as mobile phase and then subsequently switching to the upper phase as mobile phase. Final purification of the constituent flavonol glycosides, stilbenes, and catechins was by gel filtration and semipreparative HPLC.[20]

TABLE 1.3
Separation of Flavonoids by Centrifugal Partition Chromatography

Sample	Solvent System	Reference
Flavonoids from *Hippophae rhamnoides*	$CHCl_3$–MeOH–H_2O, 4:3:2	46
Flavonol glycosides from *Vernonia galamensis*	$CHCl_3$–MeOH–n-BuOH–H_2O, 7:6:3:4	47
Flavonol glycosides from *Picea abies*	$CHCl_3$–MeOH–i-PrOH–H_2O, 5:6:1:4	20
Flavonol glycosides from *Polypodium decumanum*	n-BuOH–EtOH–H_2O, 4:1.5:5 $CHCl_3$–MeOH–n-BuOH–H_2O, 10:10:1:6	48
Flavone *C*-glycosides from *Cecropia lyratiloba*	$CHCl_3$–MeOH–H_2O, 46:25:29 EtOAc–n-BuOH–MeOH–H_2O, 35:10:11:44	49
Biflavonoids from *Garcinia kola*	n-Hexane–EtOAc–MeOH–H_2O, 2:8:5:5	50
Isoflavones from *Astragalus membranaceus*	EtOAc–EtOH–n-BuOH–H_2O, 15:5:3:25 EtOAc–EtOH–H_2O, 5:1:5	51
Isoflavones from *Glycine max*	$CHCl_3$–MeOH–H_2O, 4:3:2 $CHCl_3$–MeOH–n-BuOH–H_2O, 8:6:1:4	52
Anthocyanidins from *Ricciocarpos natans*	n-Hexane–EtOAc–n-BuOH–HOAc–HCl 1%, 2:1:3:1:5	53
Proanthocyanidins from *Stryphnodendron adstringens*	EtOAc–n-PrOH–H_2O, 35:2:2	54
Proanthocyanidins from *Cassipourea gummiflua*	n-Hexane–EtOAc–MeOH–H_2O, 8:16:7:10	55
Anthocyanins from plants	n-BuOH–TBME–CH_3CN–H_2O, 2:2:1:5	56
Polyphenols from tea	n-Hexane–EtOAc–MeOH–H_2O, 3:10:3:10	57

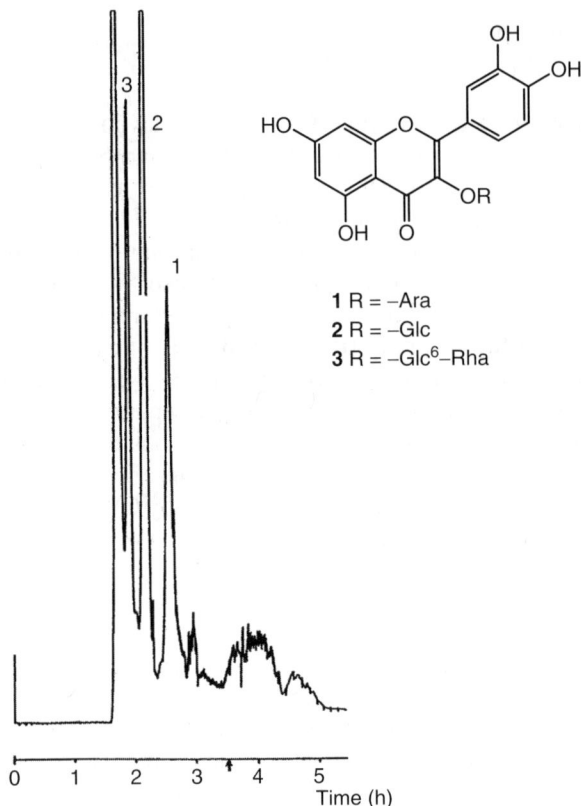

FIGURE 1.1 Separation of flavonol glycosides from *Tephrosia vogelii* (Leguminosae) with a Quattro CPC instrument. Solvent system, $CHCl_3$–MeOH–EtOH–H_2O (5:3:3:4); mobile phase, upper phase; flow-rate, 3 ml/min; detection, 254 nm; sample, 500 mg MeOH extract. (From Sutherland, I.A., Brown, L., Forbes, S., Games, G., Hawes, D., Hostettmann, K., McKerrell, E.H., Marston, A., Wheatley, D., and Wood, P., *J. Liq. Chrom. Relat. Technol.*, 21, 279, 1998. With permission.)

Four pure isoflavones were obtained from a crude soybean extract after CPC with the solvent system $CHCl_3$–MeOH–H_2O (4:3:2) (Figure 1.2). The isoflavones were isolated in amounts of 5 to 10 mg after the introduction of 150 mg of sample.[52]

A combination of gel filtration, CPC, and semipreparative HPLC was reported for the isolation of eight dimeric proanthocyanidins of general structure **1** from the stem bark of *Stryphnodendron adstringens* (Leguminosae). The CPC step involved separation with the upper layer of EtOAc–*n*-PrOH–H_2O (35:2:2) as mobile phase.[54]

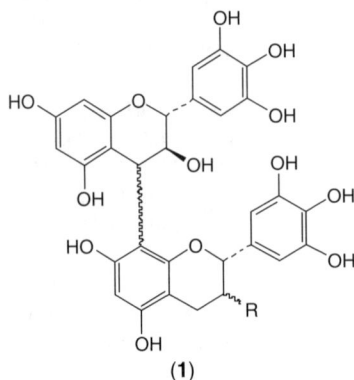

(1)

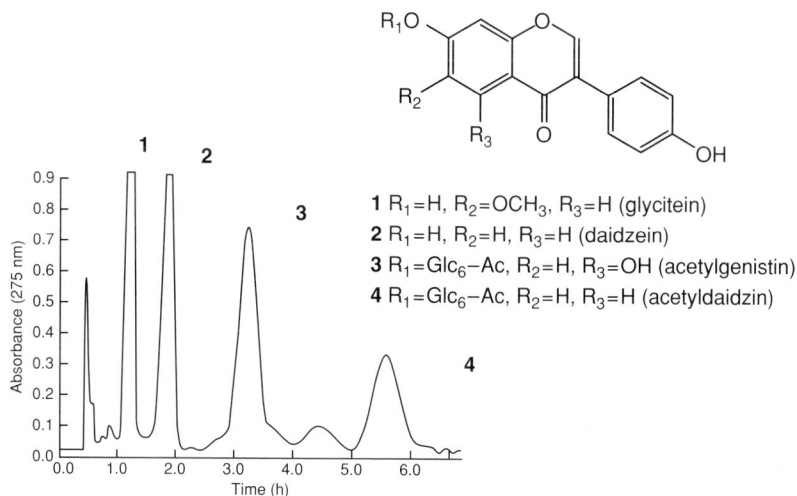

FIGURE 1.2 Separation of a crude soybean extract with a multilayer CPC instrument. Solvent system, CHCl$_3$–MeOH–H$_2$O (4:3:2); mobile phase, lower phase; flow rate, 2 ml/min; detection, 275 nm; sample, 150 mg. (From Yang, F., Ma, Y., and Ito, Y., *J. Chromatogr.*, 928, 163, 2001. With permission.)

1.4 ANALYTICAL METHODS

A herbal product contains multiple constituents that might be responsible for its therapeutic effects. It is thus necessary to define as many of the constituents as possible in order to understand and explain the bioactivity. The concept of "phytoequivalence" has been introduced in Germany to ensure consistency of phytotherapeuticals.[59] According to this concept, a chemical profile for a herbal product is constructed and compared with the profile of a clinically proven reference product. Since many of these preparations contain flavonoids, it is essential to have adequate analytical techniques at hand for this class of natural product.

Knowledge of the flavonoid content of plant-based foods is paramount to understanding their role in plant physiology and human health. Analytical methods are also important to identify adulteration of beverages, for example. And flavonoids are indispensable markers for chemotaxonomic purposes.

Various analytical methods exist for flavonoids. These range from TLC to CE. With the introduction of hyphenated HPLC techniques, the analytical potential has been dramatically extended. Gas chromatography (GC) is generally impractical, due to the low volatility of many flavonoid compounds and the necessity of preparing derivatives. However, Schmidt et al.[60] have reported the separation of flavones, flavonols, flavanones, and chalcones (with frequent substitution by methyl groups) by GC.

Quantification aspects are discussed under individual techniques.

1.4.1 SAMPLE PREPARATION

Sample preparation is included in sample handling[61] and is rapidly becoming a science in itself. The initial treatment of the sample is a critical step in chemical and biochemical analyses; it is usually the slowest step in the analysis. In the case of food and plant samples, the number and diversity of analytes is very high and efficient pretreatment is required to obtain enriched phenolic fractions.

Sample preparation methods should:[62]

- Remove possible interferents (for either the separation or detection stages) from the sample, thereby increasing the selectivity of the analytical method.
- Increase the concentration of the analyte and hence the sensitivity of the assay.
- Convert the analyte into a more suitable form for detection or separation (if needed).
- Provide robust and reproducible methods that are independent of variations in the sample matrix.

The aim of sample preparation is that the components of interest should be extracted from complex matrices with the least time and energy consumption but with highest efficiency and reproducibility. Conditions should be mild enough to avoid oxidation, thermal degradation, and other chemical and biochemical changes. Some procedures — CE, for example — necessitate more rigorous sample pretreatment than others. On the other hand, TLC requires an absolute minimum of sample preparation.

As well as typical sample preparation methods such as filtration and liquid–liquid extraction,[61] newer developments are now extensively used. The first of these is solid-phase extraction (SPE). This is a rapid, economical, and sensitive technique that uses several different types of cartridges and disks, with a variety of sorbents. Sample preparation and concentration can be achieved in a single step. Interfering sugars can be eluted with aqueous methanol on reversed-phase columns prior to elution of flavonoids with methanol.

Among the numerous applications of SPE are separations of phenolic acids and flavonoids from wines and fruit juices. Sep-Pak C_{18} cartridges have been used for the fractionation of flavonol glycosides and phenolic compounds from cranberry juice into neutral and acidic parts before HPLC analysis.[63] Antimutagenic flavonoids were identified in aqueous extracts of dry spinach after removal of lipophilic compounds by SPE.[64]

Anthocyanins were recovered from wine (after removal of ethanol) by elution from a C_{18} cartridge with an aqueous eluent of low pH.[65]

Different extracts of *Citrus* were subjected to SPE on C_{18} cartridges to remove polar components. The retained flavonoids (mainly flavanones) were eluted with methanol–dimethyl sulfoxide, which enhanced solubility of hesperidin, diosmin, and diosmetin. Recoveries of eriocitrin, naringin, hesperidin, and tangeretin from spiked samples of mesocarp tissue exceeded 96%. Flavones were relatively abundant in the leaves.[66]

SFE has long been of industrial importance but has only recently been introduced on a laboratory scale. Few applications have been reported for polyphenols but simpler phenolics have been extracted by this method, albeit with addition of methanol to the supercritical fluid.[13] Some potential may be found for online SFE, since very clean extracts (but at low extraction efficiency for phenolic compounds) can be obtained.[67]

Other innovations include PLE, MAE[68] (see Section 1.3.1), and solid-phase microextraction (SPME). SPME is a sampling method applied to GC, HPLC, and CE. It is based on adsorbent- or adsorbent-type fibers and lends itself well to miniaturization.[61]

1.4.2 THIN-LAYER CHROMATOGRAPHY

Paper chromatography and paper electrophoresis were once extensively used for the analysis of flavonoids,[5] but now the method of choice for simple and inexpensive analytical runs is TLC. The advantages of this technique are well known: short separation times, amenability to detection reagents, and the possibility of running several samples at the same time. TLC is also ideally suited for the preliminary screening of plant extracts before HPLC analysis. An excellent general text on TLC methodology has been written by Jork et al.[69] A good discussion presented by Markham in one of the earlier volumes on flavonoids describes TLC on silica gel and also two other supports, cellulose and polyamide (which now find less application).[70] In his chapter, solvent systems and spray reagents are described.

Many different solvent systems have been employed for the separation of flavonoids using TLC. Table 1.4 shows a selection for different classes of these polyphenols. Some solvent systems cited by Markham[70] are reproduced here because they still find application in the separation of flavonoids. Highly methylated or acetylated flavones and flavonols require nonpolar solvents such as chloroform–methanol (15:1). Widely distributed flavonoid aglycones, such as apigenin, luteolin, and quercetin, can be separated in chloroform–methanol (96:4) and similar polarity solvents. One system that is of widespread application for flavonoid glycosides is ethyl acetate–formic acid–glacial acetic acid–water (100:11:11:26). By the addition of ethyl methyl ketone (ethyl acetate–ethyl methyl ketone–formic acid–glacial acetic acid–water, 50:30:7:3:10), rutin and vitexin-2″-O-rhamnoside can be separated.[71] Careful choice of solvent system also allows separation of flavonoid glucosides from their galactosidic analogs.[72] This is especially important for the distinction of C-glucosides from C-galactosides. As an illustration, 8-C-glucosylapigenin (vitexin) can be separated from 8-C-galactosylapigenin with the solvent ethyl acetate–formic acid–water (50:4:10).[72]

With regard to detection, brief exposure of the TLC plate to iodine vapor produces yellow-brown spots against a white background. And, as stated by Markham,[70] flavonoids appear as dark spots against a fluorescent green background when observed in UV light (254 nm) on plates containing a UV-fluorescent indicator (such as silica gel F_{254}). In 365 nm UV light, depending on the structural type, flavonoids show dark yellow, green, or blue fluorescence, which is intensified and changed by the use of spray reagents. One of the most important of these is the "natural products reagent," which produces an intense fluorescence under 365 nm UV light after spraying with a 1% solution of diphenylboric acid-β-ethylamino ester (diphenylboryloxyethylamine) in methanol. Subsequent spraying with a 5% solution of polyethylene glycol-4000 (PEG) in ethanol lowers the detection limit from 10 μg (the average TLC detection limit for flavonoids) to about 2.5 μg, intensifying the fluorescence behavior. The colors observed in 365 nm UV light are as follows:

- Quercetin, myricetin, and their 3- and 7-O-glycosides: orange-yellow
- Kaempferol, isorhamnetin, and their 3- and 7-O-glycosides: yellow-green
- Luteolin and its 7-O-glycoside: orange
- Apigenin and its 7-O-glycoside: yellow-green

Further details about the use of the "natural products reagent" can be found in an article by Brasseur and Angenot.[73]

Aqueous or methanolic ferric chloride is a general spray reagent for phenolic compounds and gives a blue-black coloration with flavonoids. Similarly, Fast Blue Salt B forms blue or blue-violet azo dyes.

For quantitative analyses, scanning of the TLC plate with a densitometer provides good results. The flavonoids, both aglycones and glycosides, in *Vaccinium myrtillus* and *V. vitis-idaea* (Ericaceae) were determined after TLC and densitometry at 254 nm.[74] With suitable spray reagents, detection limits of 20 ng can be achieved by densitometry.[75]

Better resolution is obtained by chromatographing flavonoids on high-performance TLC (HPTLC) plates. Silica gel $60F_{254}$, RP-18, or, less frequently, Diol HPTLC plates are used for separation purposes. Methanol–water eluents are indicated for HPTLC on RP-18 chemically bonded silica gel but some acid is generally added to avoid tailing. Polar glycosides require eluents containing a high percentage of water. Special HPTLC plates have been designed for

TABLE 1.4
Solvent Systems for Thin-Layer Chromatography of Flavonoids on Silica Gel

Sample	Eluent
Flavonoid aglycones	EtOAc–*i*-PrOH–H$_2$O, 100:17:13
	EtOAc–CHCl$_3$, 60:40
	CHCl$_3$–MeOH, 96:4
	Toluene–CHCl$_3$–MeCOMe, 8:5:7
	Toluene–HCOOEt–HCOOH, 5:4:1
	Toluene–EtOAc–HCOOH, 10:4:1
	Toluene–EtOAc–HCOOH, 58:33:9
	Toluene–EtCOMe–HCOOH, 18:5:1
	Toluene–dioxane–HOAc, 90:25:4
Flavonoid glycosides	*n*-BuOH–HOAc–H$_2$O, 65:15:25
	n-BuOH–HOAc–H$_2$O, 3:1:1
	EtOAc–MeOH–H$_2$O, 50:3:10
	EtOAc–MeOH–HCOOH–H$_2$O, 50:2:3:6
	EtOAc–EtOH–HCOOH–H$_2$O, 100:11:11:26
	EtOAc–HCOOH–H$_2$O, 9:1:1
	EtOAc–HCOOH–H$_2$O, 6:1:1
	EtOAc–HCOOH–H$_2$O, 50:4:10
	EtOAc–HCOOH–HOAc–H$_2$O, 100:11:11:26
	EtOAc–HCOOH–HOAc–H$_2$O, 25:2:2:4
	THF–toluene–HCOOH–H$_2$O, 16:8:2:1
	CHCl$_3$–MeCOMe–HCOOH, 50:33:17
	CHCl$_3$–EtOAc–MeCOMe, 5:1:4
	CHCl$_3$–MeOH–H$_2$O, 65:45:12
	CHCl$_3$–MeOH–H$_2$O, 40:10:1
	MeCOMe–butanone–HCOOH, 10:7:1
	MeOH–butanone–H$_2$O, 8:1:1
Flavonoid glucuronides	EtOAc–Et$_2$O–dioxane–HCOOH–H$_2$O, 30:50:15:3:2
	EtOAc–EtCOMe–HCOOH–H$_2$O, 60:35:3:2
Flavanone aglycones	CH$_2$Cl$_2$–HOAc–H$_2$O, 2:1:1
Flavanone glycosides	CHCl$_3$–HOAc, 100:4
	CHCl$_3$–MeOH–HOAc, 90:5:5
	n-BuOH–HOAc–H$_2$O, 4:1:5 (upper layer)
Chalcones	EtOAc–hexane, 1:1
Isoflavones	CHCl$_3$–MeOH, 92:8
	CHCl$_3$–MeOH, 3:1
Isoflavone glycosides	*n*-BuOH–HOAc–H$_2$O, 4:1:5 (upper layer)
Dihydroflavonols	CHCl$_3$–MeOH–HOAc, 7:1:1
Biflavonoids	CHCl$_3$–MeCOMe–HCOOH, 75:16.5:8.5
	Toluene–HCOOEt–HCOOH, 5:4:1
Anthocyanidins and anthocyanins	EtOAc–HCOOH–2 *M* HCl, 85:6:9
	n-BuOH–HOAc–H$_2$O, 4:1:2
	EtCOMe–HCOOEt–HCOOH–H$_2$O, 4:3:1:2
	EtOAc–butanone–HCOOH–H$_2$O, 6:3:1:1
Proanthocyanidins	EtOAc–MeOH–H$_2$O, 79:11:10
	EtOAc–HCOOH–HOAc–H$_2$O, 30:1.2:0.8:8

this purpose, since normal plates can only accommodate aqueous methanol mixtures with up to about 40% water.

The European Pharmacopoeia stipulates TLC fingerprint analysis for the identification of plant drugs. This can be used, for example, in the case of hawthorn extracts (*Crataegus monogyna* and *C. laevigata*, Rosaceae), which contain flavone *O*-glycosides and flavone *C*-glycosides or passion flower extracts (*Passiflora incarnata*, Passifloraceae), which contain only flavone *C*-glycosides.

1.4.3   HIGH-PERFORMANCE LIQUID CHROMATOGRAPHY

The method of choice for the qualitative and quantitative analysis of flavonoids is HPLC. Since its introduction in the 1970s, HPLC has been used for all classes of flavonoids and hundreds of applications have been published. Numerous reviews have also appeared, such as those by Hostettmann and Hostettmann,[3] Merken and Beecher,[76] He,[77] and Cimpan and Gocan.[78]

It is not the purpose of this chapter to go into the theory of HPLC, which is adequately covered in other texts, but to describe the applications of the method. This section will concentrate on analytical applications because semipreparative HPLC has been described in Section 1.3.2. Analytical HPLC finds use in the quantitative determination of plant constituents, in the purity control of natural products, and in chemotaxonomic investigations.

For the analytical HPLC of a given subclass of flavonoids (flavones, flavonols, isoflavones, anthocyanins, etc.), the stationary phase, solvent, and gradient have to be optimized.

A very high proportion of separations are run on octadecylsilyl bonded (ODS, RP-18, or C_{18}) phases. Some reported analyses use octasilyl bonded (RP-8 or C_8) phases but these are becoming increasingly rare. Flavonoid glycosides are eluted before aglycones with these phases, and flavonoids possessing more hydroxyl groups are eluted before the less substituted analogs. As solvents for application, acetonitrile–water or methanol–water mixtures, with or without small amounts of acid, are very common. These are compatible with gradients and UV detection. Occasionally, other solvents such as tetrahydrofuran, isopropanol, or *n*-propanol are used. Acid modifiers are necessary to suppress the ionization of phenolic hydroxyl groups, giving sharper peaks with less tailing. A study has shown that there are large differences in the effectiveness of C_{18} columns for the separation of flavonoid aglycones and glycosides. While some columns give good results, others produce substantial band broadening and peak asymmetry.[79]

Octadecylsilyl stationary phases with hydrophilic endcapping have been developed for the separation of very polar analytes, which are not sufficiently retained on conventional reversed-phase columns. Among numerous other applications, they have been demonstrated to be suitable for the separation of flavonol and xanthone glycosides from mango (*Mangifera indica*, Anacardiaceae) peels.[80]

Normal phases (unmodified silica gel) are rarely employed, except for the occasional separation of weakly polar flavonoid aglycones, polymethoxylated flavones, flavanones, or isoflavones. The polymethoxylated flavones present in citrus fruits can, for example, be separated on silica gel columns. The big drawback is that solvent gradients cannot normally be run with normal phases.

Flavone *C*-glycosides generally elute with shorter retention times than the corresponding *O*-glycosides. Thus, vitexin (8-*C*-glucosylapigenin) elutes with a shorter retention time than apigenin 7-*O*-glucoside. Furthermore, 8-*C*-glycosylflavones elute with shorter retention times than the corresponding 6-*C*-glycosylflavones. Thus, apigenin 8-*C*-glucoside elutes earlier than apigenin 6-*C*-glucoside.

Flavanones elute before their corresponding flavones due to the effect of unsaturation between positions 2 and 3.

Isoflavones, chiefly found in the Leguminosae (such as soy, *Medicago sativa*, and red clover, *Trifolium pratense*) in the plant kingdom, are also successfully analyzed by HPLC on C_{18} columns.[76]

The anthocyanins exist in solution as various structural forms in equilibrium, depending on the pH and temperature. In order to obtain reproducible results in HPLC, it is essential to control the pH of the mobile phase and to work with thermostatically controlled columns. For the best resolution, anthocyanin equilibria have to be displaced toward their flavylium forms — peak tailing is thus minimized and peak sharpness improved. Flavylium cations are colored and can be selectively detected in the visible region at about 520 nm, avoiding the interference of other phenolics and flavonoids that may be present in the same extracts. Typically, the pH of elution should be lower than 2. A comparison of reversed-phase columns (C_{18}, C_{12}, and phenyl-bonded) for the separation of 20 wine anthocyanins, including mono-glucosides, diglucosides, and acylated derivatives was made by Berente et al.[81] It was found that the best results were obtained with a C_{12} 4 μm column, with acetonitrile–phosphate buffer as mobile phase, at pH 1.6 and 50°C.

Applications of HPLC to the analysis of flavonoids in medicinal and other plants are summarized by Cimpan and Gocan.[78] From the methods listed, it is noteworthy that 90% of the separations use C_{18} columns. The importance of flavonoids in foods (fruits, vegetables, and grains) means that it is indispensable to have suitable means of determining their content. The review by Merken and Beecher[76] gives an excellent summary (including full details of separation conditions) of applications of HPLC to the determination of flavones, flavonols, flavanones, isoflavones, anthocyanidins, catechins, and their respective glycosides in foods. Here again, virtually all separations are performed on RP-18 columns, with column lengths between 100 and 300 mm and with internal diameters between 2 and 5 mm. Granulometries vary from 3 to 10 μm, with most being 5 μm. Separation runs are generally up to 1 h in duration. For aglycones and glycosides of isoflavones, certain reported separations of soy-bean products (e.g., the work of Barnes et al.[82]) have employed C_8 packings, but these are rare. Some applications are given in which two or more subclasses are analyzed simultaneously, such as flavanones, flavones, and flavonols in honey, and anthocyanins, catechins, and flavonols in fruit and wines.

In general, though, there is not a single HPLC method that can solve all flavonoid separation problems. However, Sakakibara et al.[83] claim to have found a method capable of quantifying every polyphenol in vegetables, fruits, and teas. For this purpose, they used a Capcell pak C18 UG120 (250 × 4.6 mm, S-5, 5 μm) column at 35°C. Gradient elution at a flow rate of 1 ml/min was performed over 95 min with solution A (50 mM sodium phosphate [pH 3.3] and 10% methanol) and solution B (70% methanol) as follows: initially 100% of solution A; for the next 15 min, 70% A; for 30 min, 65% A; for 20 min, 60% A; for 5 min, 50% A; and finally 100% B for 25 min. Vegetable material was extracted with 90% methanol containing 0.5% acetic acid. A typical HPLC profile for 28 reference polyphenols is shown in Figure 1.3. The method allowed the determination of aglycones separately from glycosides. Information could also be obtained about simple polyphenols in the presence of more complex polycyclic polyphenols. Quantitative determination was achieved for a total of 63 different food samples.

Within the domain of medicinal plants, preparations of *Ginkgo biloba* (Ginkgoaceae) are the most widely sold phytomedicines, with sales of over US\$ 1 billion in 1998.[84] These principally involve special extracts of the leaves. Flavonoids are, at least in part, responsible for the beneficial effects of Ginkgo extracts. Generally, enriched ginkgo extracts for the preparation of ginkgo products are standardized to contain 24% flavonoids and 6% terpene

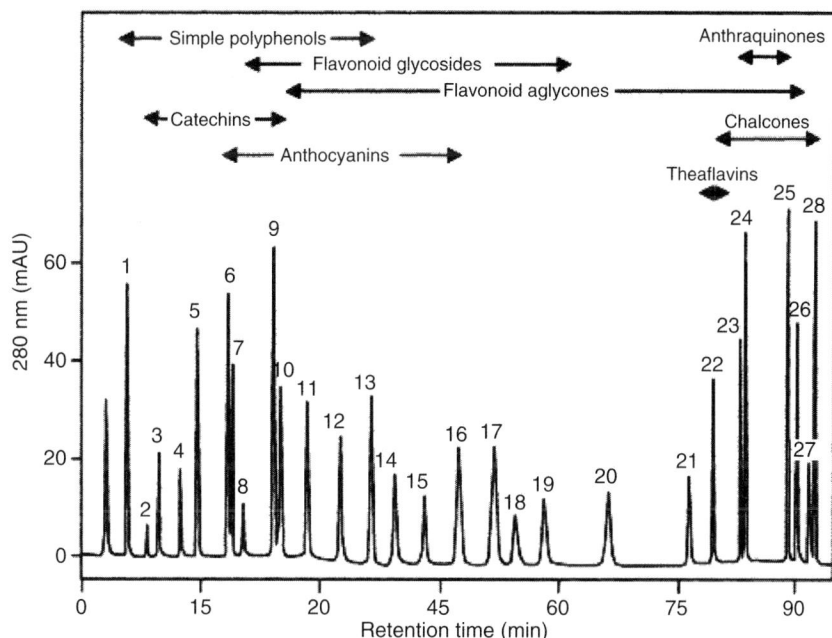

FIGURE 1.3 HPLC profile for 28 different polyphenols on a C_{18} column. Classes of compound are shown in the upper part of the chromatogram. (From Sakakibara, H., Honda, Y., Nakagawa, S., Ashida, H., and Kanazawa, K., *J. Agric. Food Chem.*, 51, 571, 2003. With permission.)

lactones and, therefore, the content of flavonoids and terpene lactones is one of the important parameters to assess the quality of ginkgo products. Fingerprint analysis for the quality control of Ginkgo preparations has shown that it is possible to separate six flavonoid aglycones, 22 flavonoid glycosides, and five biflavonoids in one run (Figure 1.4). This was achieved on a 100 × 4 mm Nucleosil 100–C_{18} 3 μm column. In order to complete the run in 30 min, a ternary mobile phase was used, consisting of isopropanol–THF (25:65) (solvent A), acetonitrile (B), and 0.5% orthophosphoric acid (solvent C). A three-pump system was required to produce a complex elution gradient (1 ml/min, detection at 350 nm) starting with 15.0% A, 1.5% B, 83.5% C and ending with 0% A, 78.0% B, 22% C.[85]

As few flavonoid glycosides are commercially available for reference purposes, their direct quantitative analysis is often impractical. It is thus common practice when investigating plant extracts to hydrolyze the glycosides and identify and quantify the released aglycones. This was the procedure adopted for the analysis of white onions (*Allium cepa*, Liliaceae) and white celery stalks (*Apium graveolens*, Apiaceae). Lyophilized plant material was extracted with 60% aqueous methanol and hydrolyzed with 1.2 *M* HCl before HPLC analysis on a Waters C_{18} Symmetry 150 × 3.9 mm (5 μm) column. As the mobile phase, a gradient of 15 to 35% acetonitrile in water adjusted to pH 2.5 with TFA was used. In nonhydrolyzed white onion extract, two major, nonidentified, flavonoid glycoside peaks were present. When the extract was hydrolyzed, these two peaks were replaced by a major peak corresponding to quercetin. Kaempferol was used as internal standard. In the analysis of nonhydrolyzed celery extract (isorhamnetin as internal standard), several peaks were observed. After hydrolysis the major peaks were due to apigenin, luteolin, and an unknown component. Quantification of the aglycones in fresh plant material was achieved by extrapolation. The limit of detection for endogenous quercetin and other flavonoid aglycones was ca. 3 μg/g fresh mass.[79]

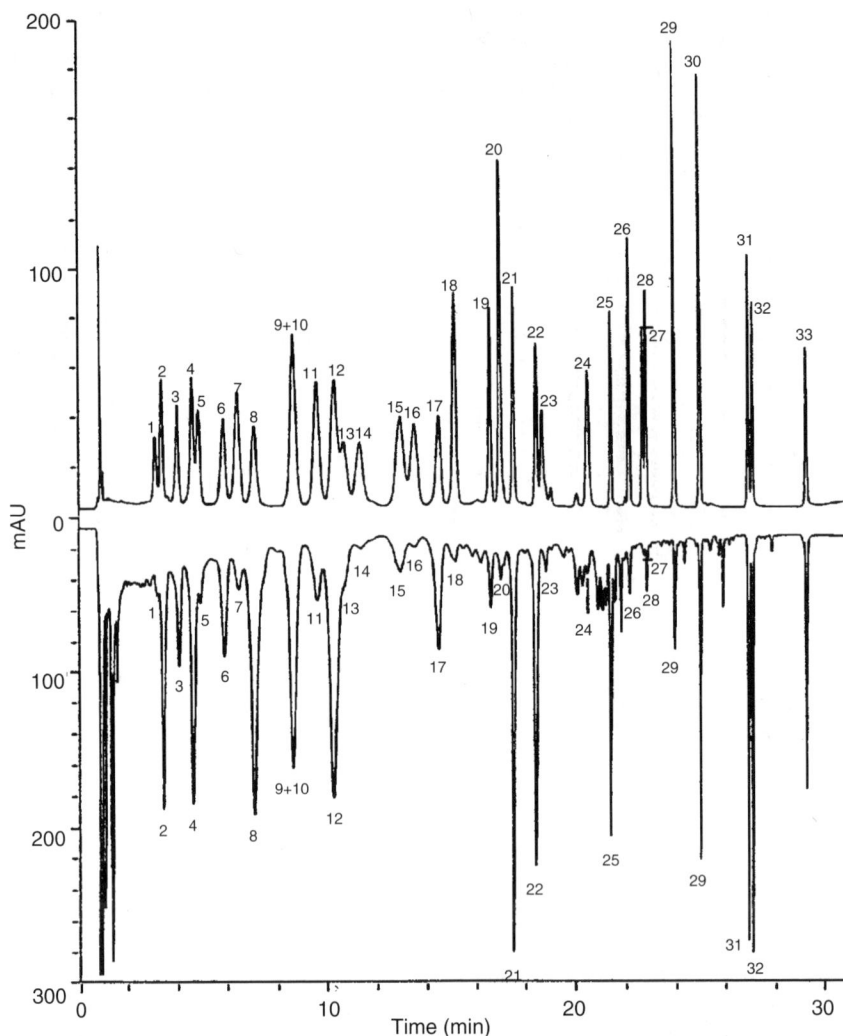

FIGURE 1.4 HPLC fingerprint of a therapeutically used dry extract of *Ginkgo biloba* leaves. Top chromatogram, pure reference flavonoids; bottom chromatogram, leaf extract. Identified flavonoids are numbered 1 to 33. (From Hasler, A., Sticher, O., and Meier, B., *J. Chromatogr.*, 605, 41, 1992. With permission.)

1.4.4 HIGH-PERFORMANCE LIQUID CHROMATOGRAPHY–ULTRAVIOLET SPECTROPHOTOMETRY

The most frequently used detection method for HPLC is UV spectrophotometry. Routine detection in HPLC is typically based on measurement of UV absorption, or visible absorption in the case of anthocyanins. No single wavelength is ideal for all classes of flavonoids since they display absorbance maxima at distinctly different wavelengths. The most commonly used wavelength for routine detection has been 280 nm, which represents a suitable compromise.

With the introduction of diode-array technology in the 1980s, a further dimension is now possible because coupled LC–UV with diode array detection (DAD) allows the chromatographic eluent to be scanned for UV–visible spectral data, which are stored and can later be compared with a library for peak identification.[86] This increases the power of HPLC analysis because with the information from the UV spectrum, it may be possible to identify the compound subclass or perhaps even the compound itself. UV spectral data of 175 flavonoids in several solvents can be found, for example, in a book by Mabry et al.[1] LC–UV with DAD enables simultaneous recording of chromatograms at different wavelengths. This improves the possibilities of quantification because detection can be performed at the wavelength maximum of the compound in question. These are typically to be found[76] at 270 and 330 to 365 nm for flavones and flavonols, at 290 nm for flavanones, at 236 or 260 nm for isoflavones, at 340 to 360 nm for chalcones, at 280 nm for dihydrochalcones, at 502 or 520 nm for anthocyanins, and at 210 or 280 nm for catechins.

Peak purity can also be determined. The spectra of eluting peaks obtained at the apex and both inflexion points of the peak can be compared in order to obtain a measure of the purity of the particular component of the sample.

LC–UV is valuable for the identification of isoflavones since their spectra differ in absorption properties from most of the other flavonoids. They have a C2–C3 double bond, with the B-ring at C3, which prevents conjugation of the phenyl group with the pyrone carbonyl group. This reduces the contribution of the B-ring to the UV spectrum and results in a peak of very low intensity in the 300 to 330 nm range (band I).

The analysis of catechins and proanthocyanidins by LC–UV presents certain problems. In general, only monomers and oligomers up to tetramers can be separated and detected as defined peaks. Polymeric forms, which may constitute the bulk of proanthocyanidins in many plant materials, are not well resolved. They give place to a drift in the baseline and the formation of characteristic humps in HPLC chromatograms. Furthermore, the spectral characteristics of these compounds do not allow easy detection and identification. Flavan-3-ols give absorption maxima at nonspecific wavelengths (270 to 290 nm) and they have lower extinction coefficients than other accompanying phenolics. Their quantification is thus not easy. The lack of reference proanthocyanidins implies that results have to be expressed with respect to other reference substances, normally catechin or epicatechin. This causes concomitant errors of quantification caused by the different extinctions shown by the individual flavan-3-ols. For reverse-phase HPLC of proanthocyanidin oligomers, the percentage of methanol or acetonitrile usually does not exceed 20%.

The coupling of HPLC with DAD allows online quantification of flavonoids in samples analyzed. Justesen et al.[87] have quantified flavonols, flavones, and flavanones in fruits, vegetables, and beverages in this fashion. The food material was extracted and then hydrolyzed to produce the corresponding flavonoid aglycones. These were analyzed on a Phenomenex C_{18} column (250 × 4.6 mm, 5 μm) using a mobile phase of methanol–water (30:70) with 1% formic acid (solvent A) and 100% methanol (solvent B). The gradient was 25 to 86% B in 50 min at a flow rate of 1 ml/min. UV spectra were recorded from 220 to 450 nm. For each compound, peak areas were determined at the wavelength providing maximal UV absorbance. Quantification was performed based on external standards. A mixture of standards of known concentrations was analyzed in duplicate before and after the batch of samples. Peak areas were used to calculate the hydrolyzed food sample flavonoid aglycone content. Method validation indicated good day-to-day variability (reproducibility) and recoveries in the range of 68 to 103%. There was low recovery of myricetin standard, presumably because of degradation during hydrolysis. Detection by online mass spectrometry (MS) was also included to check possible interferences between flavonoids eluting at similar retention times.

While identification of the peaks in a LC–UV chromatogram is possible by comparing retention times and UV spectra with authentic samples or a databank, this might not be possible for compounds with closely related structures, and wrong conclusions might be drawn. It has been established that in order to complete the characterization of phenolic compounds, reagents inducing a shift of the UV absorption maxima can be used.[1]

A postcolumn derivatization procedure, based on this technique, is possible by adding suitably modified shift reagents to the eluate leaving a HPLC column.[88] Direct information is provided about the flavonoid oxidation pattern and position of free phenolic hydroxyl groups. In the analysis of *Gentiana* (Gentianaceae) extracts, best results were obtained on a reversed-phase column with a methanol–water eluent at a pH of around 3.5, to avoid peak tailing. Classical shift reagents were adapted in order to be compatible with these conditions: sodium monohydrogenphosphate and potassium hydroxide were used as the weak and strong bases, respectively, instead of sodium acetate and sodium methanolate. In order to form a complex with the keto function, an aqueous solution of aluminum chloride was passed with the eluate through a reaction coil at 60°C. The presence of *ortho*-hydroxyl groups was shown with boric acid–sodium acetate. These shift reagents gave identical results to those obtained with classical shift reagents. The small amount of material required (50 to 100 μg of crude plant extract) in LC–UV postcolumn derivatization allows the analysis of very rare and small species, as well as single plant parts of herbarium samples.[88]

To illustrate this approach, the online identification of flavonol glycosides in *Epilobium* (Onagraceae) species will be described. Certain of these willow-herbs have important implications in the treatment of benign prostatic hyperplasia and knowledge of their flavonoid content is an aid to their identification. The aerial parts of 13 different species were extracted first with dichloromethane and then with methanol. The flavonoid-containing methanol extracts were partitioned between *n*-butanol and water; the *n*-butanol fractions were then analyzed on a Novapak C_{18} column (150×3.9 mm, 4 μm) with an acetonitrile–water gradient. TFA was added to give a pH of 3. Photodiode-array detection allowed the online recording of UV spectra (200 to 500 nm), all typical of flavonoids (Figure 1.5), except for ellagic acid (**X** in Figure 1.6). The HPLC chromatogram for *Epilobium angustifolium* is shown

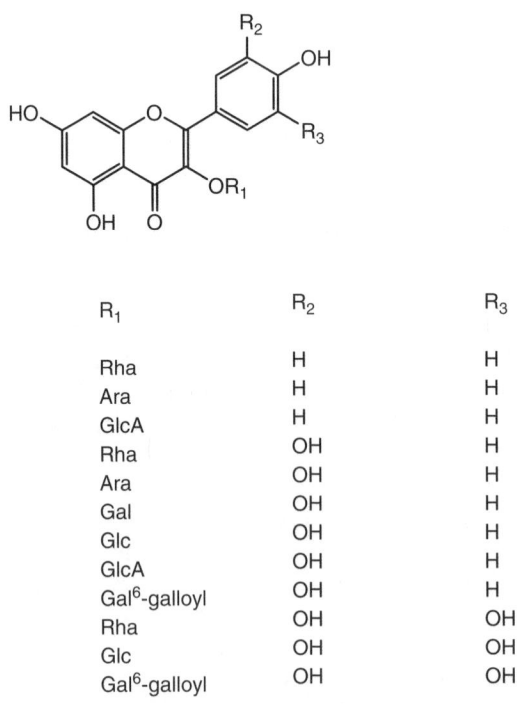

	R_1	R_2	R_3
1	Rha	H	H
2	Ara	H	H
3	GlcA	H	H
4	Rha	OH	H
5	Ara	OH	H
6	Gal	OH	H
7	Glc	OH	H
8	GlcA	OH	H
9	Gal⁶-galloyl	OH	H
10	Rha	OH	OH
12	Glc	OH	OH
13	Gal⁶-galloyl	OH	OH

FIGURE 1.5 Flavonoid glycosides isolated from the aerial parts of *Epilobium angustifolium*.

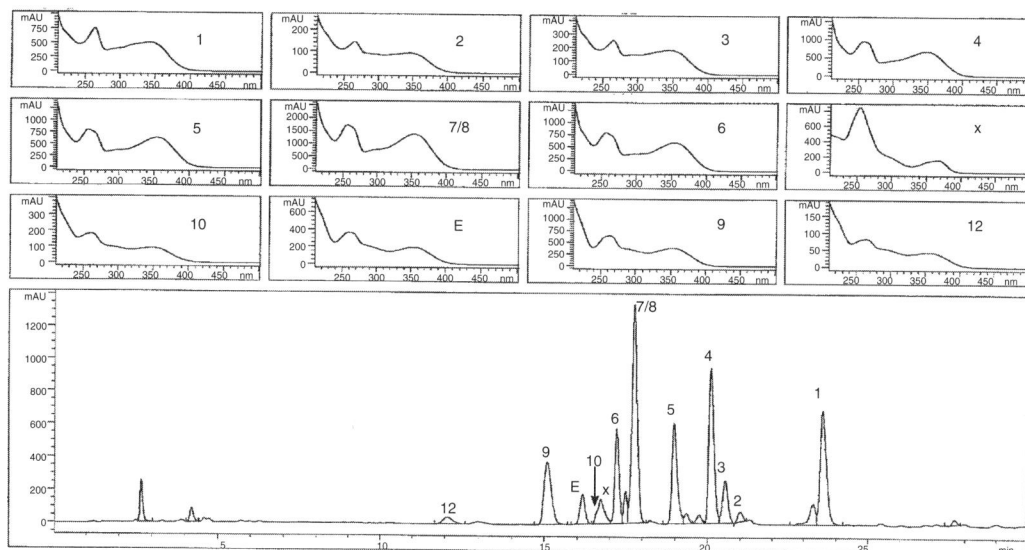

FIGURE 1.6 LC–UV chromatogram of a methanolic extract of the aerial parts of *Epilobium angustifolium* enriched by *n*-BuOH–H$_2$O partition. Column Novapak C$_{18}$ (150 × 3.9 mm, 4 μm); step gradient of CH$_3$CN–H$_2$O (containing 0.05% TFA): 0 min 10% CH$_3$CN, 4 min 12%, 12 min 12%, 16 min 18%, 30 min 25%; flow-rate 1 ml/min; chromatogram recorded at 350 nm. (From Ducrey, B., Wolfender, J.L., Marston, A., and Hostettmann, K., *Phytochemistry*, 38, 129, 1995. With permission.)

in Figure 1.6. Structure elucidation of flavonoid **E** was incomplete. In combination with shift reagents added postcolumn, LC–UV allowed determination of the hydroxylation pattern of flavonols and the position of the sugars on the aglycones. Figure 1.7 shows the UV spectra obtained online for **4** (quercitrin) after the addition of five different shift reagents. The shift of 11 nm of band II with weak base, 0.1 *M* Na$_2$HPO$_4$, was characteristic for a nonsubstituted 7-hydroxyl group. A 15 nm shift with boric acid reagent was typical for *ortho*-dihydroxyl groups on the B-ring. The shift of 42 nm of band I obtained for aluminum chloride without neutralization of the eluate was specific for a 5-hydroxyl substituent. Addition of aluminum chloride after neutralization gave a 56 nm shift of band I. This was due to a combination of an

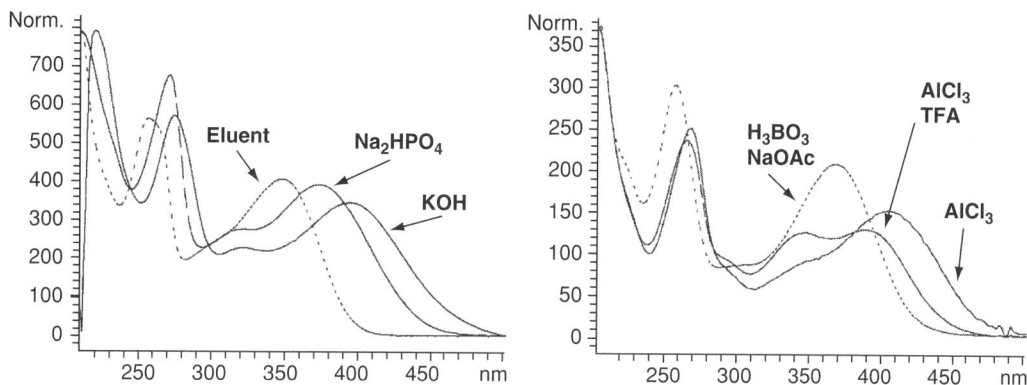

FIGURE 1.7. Online UV spectra of **4** obtained after postcolumn addition of different shift reagents. (From Ducrey, B., Wolfender, J.L., Marston, A., and Hostettmann, K., *Phytochemistry*, 38, 129, 1995. With permission.)

ortho-dihydroxyl group (C-3′ and C-4′) and to complex formation with the C-4 keto function and the 5-hydroxyl group. These data confirmed the aglycone to be quercetin. A similar procedure was adopted for the identification of the other flavonol glycosides — showing the presence of three different aglycones: kaempferol, quercetin, and myricetin. Thermospray LC–MS provided additional information on the molecular weight of the flavonol glycosides and their aglycones.[38]

1.4.5 HIGH-PERFORMANCE LIQUID CHROMATOGRAPHY–MASS SPECTROMETRY

Coupled HPLC–MS is one of the most important techniques of the last decade of the 20th century. The combination offers the possibility of taking advantage of chromatography as a separation method and MS as an identification tool. The amazing number of applications and the rapid drop in price (and size) of MS instruments has meant that the use of LC–MS is now extremely widespread. MS is one of the most sensitive methods of molecular analysis. Due to its high power of mass separation, very good selectivities can be obtained. However, the coupling between HPLC and MS has not been straightforward since the normal operating conditions of a mass spectrometer (high vacuum, high temperature, gas-phase operation, and low flow rates) are diametrically opposed to those used in HPLC, namely liquid-phase operation, high pressures, high flow rates, and relatively low temperatures.

In LC–MS, there are three general problems: the amount of column effluent that has to be introduced in the MS vacuum system, the composition of the eluent, and the type of compounds to be analyzed. Many interfaces have been developed in order to cope with these factors.[89] The interfaces must accomplish nebulization and vaporization of the liquid, ionization of the sample, removal of excess solvent vapor, and extraction of the ions into the mass analyzer. To date, no real universal interface has been constructed; each interface has characteristics that are strongly dependent on the nature of the compounds for which they are used. In LC–MS, the same rules that govern the ionization of pure compounds in the direct insertion mode are roughly preserved. Most interfaces work with reversed-phase HPLC systems, with a number of them suitable for the analysis of plant secondary metabolites. These include thermospray (TSP), continuous-flow fast-atom bombardment (CF-FAB), and electrospray (ES).[90] They cover the ionization of relatively small nonpolar products (aglycones, MW 200) to highly polar molecules (glycosides, MW 2000). Contrary to TSP or CF-FAB, where the source is in the vacuum region of the mass spectrometer, in ES the ion source is at atmospheric pressure. Atmospheric pressure ionization (API) has rendered LC–MS more sensitive and easy to handle. An API interface or source consists of five parts: (a) the liquid introduction device or spray probe; (b) the actual atmospheric pressure ion source region, where the ions are generated by means of electrospray ionization (ESI), atmospheric pressure chemical ionization (APCI), or by other means; (c) an ion sampling aperture; (d) an atmospheric pressure-to-vacuum interface; (e) an ion optical system, where the ions are subsequently transported into the mass analyser.[89] ESI and APCI are soft ionization techniques that generate mainly molecular ions for relatively small plant metabolites such as flavonoids.

Mass spectral data provide structural information on flavonoids and are used to determine molecular masses and to establish the distribution of substituents between the A- and B-rings. A careful study of fragmentation patterns can also be of particular value in the determination of the nature and site of attachment of the sugars in *O*- and *C*-glycosides. For flavonoid aglycones and glycosides with a limited number of sugar units (not more than three), TSP LC–MS analysis leads to a soft ionization, providing only intense $[M + H]^+$ ions for the aglycones and weak $[M + H]^+$ ions of glycosides (mono- or disaccharide), together with intense fragment ions due to the loss of the saccharide units, leading to the aglycone moiety $[A + H]^+$.

For example, the LC–UV–MS analysis of a methanol root extract of *Gentianella cabrerae* (Gentianaceae) gave a peak in the HPLC chromatogram (Figure 1.8), which had a UV spectrum characteristic of a xanthone (compound **2**) and one which had a UV spectrum typical for a flavonoid (compound **1**). In the TSP mass spectrum, the latter exhibited a protonated molecular ion $[M + H]^+$ at *m/z* 463 and fragments $[M + H - 90]^+$ and $[M + H - 120]^+$ at *m/z* 373 and 343, respectively, which were characteristic for the cleavage of flavone *C*-glycosides. According to these data, **1** was most probably a flavone *C*-glycoside substituted by three hydroxyl groups and one methoxyl group. Two isomeric flavones in the Gentianaceae, isoscoparin and swertiajaponin, fit with such data.[91]

In general, HPLC coupled with diode array and mass spectrometric detection provides an efficient method of rapid identification of flavonoids in a mixture. This technique now finds widespread application. By this means, 14 xanthone and flavonol glycosides were characterized online in a prepurified extract of mango (*Mangifera indica*, Anacardiaceae) peels, using ESI MS.[80] LC–UV–MS profiles of several medicinal plant extracts, including red clover (isoflavonoids), sour orange (flavanones), and astragalus (isoflavonoids and isoflavans), have been described by He.[77]

Both ES and TSP are soft ionization methods and do not typically produce many fragments. This is useful in quantitative analysis or molecular mass determination but is of little use in structure elucidation. In this case, collision-induced dissociation (CID) or collision-activated dissociation methods can be employed.[92] Fragmentation is induced in one of the high-pressure regions of the ion passageway from source to mass analyzer. Fragment ions produced by CID are very efficiently transported into the mass analyzer, providing a simple MS–MS method. With LC–MS, one analysis without CID and one with CID can be performed to obtain fragments of all components. CID is carried out to enhance fragmentation of the analytes either at the ES source (in-source CID) or in conjunction with tandem MS. In tandem MS, the first operation is to isolate a parent ion and the second is to determine the mass-to-charge ratio of the product ions formed after CID of the parent ion. The sequence of ion isolation and CID can be repeated many times in MS^n. Tandem MS and in-source CID give very similar product-ion mass spectra.

Since molecular weight information alone is insufficient for online structural determination of natural products, the fragmentation pathways of flavones and flavonols by fast-atom bombardment CID MS–MS have been documented.[93–96]

The CID MS–MS and MS^n spectra of flavonoids have been systematically studied using hybrid quadrupole time of flight (Q-TOF) and ion trap (IT) mass analyzers, under various energy conditions, to generate fragment ions.[97] These two instruments were chosen because the CID process in beam and trap systems is not generated in the same way. The results demonstrated that, if for hydroxylated flavonoids the CID MS–MS spectra generated on both instruments were similar, for partially methoxylated derivatives, there were important differences. This is a hindrance to the creation of MS–MS databases exchangeable between instruments. Generally, fragments issued from C-ring cleavage were easier to observe on a Q-TOF instrument, while losses of small molecules were favored in IT-MS. MS–MS recorded in the positive ion mode were more informative than those obtained from negative ions. Online accurate mass measurements of all MS–MS fragments were obtained on the Q-TOF instrument, while the multiple-stage MS^n capability of the IT was used to prove fragmentation pathways. Molecular formulae with an accuracy of 1.8 ppm could be produced for isovitexin on the Q-TOF instrument.[97] It is worthy of note that high-resolution MS allows molecular formulae of compounds to be assessed directly; these can then be cross-checked with spectral libraries to provide identification of unknown components.

The application of tandem MS (LC–TSP MS–MS) can be illustrated for the online characterization of flavonoids from *Gentianella cabrerae*. The LC–UV–MS of the methanol

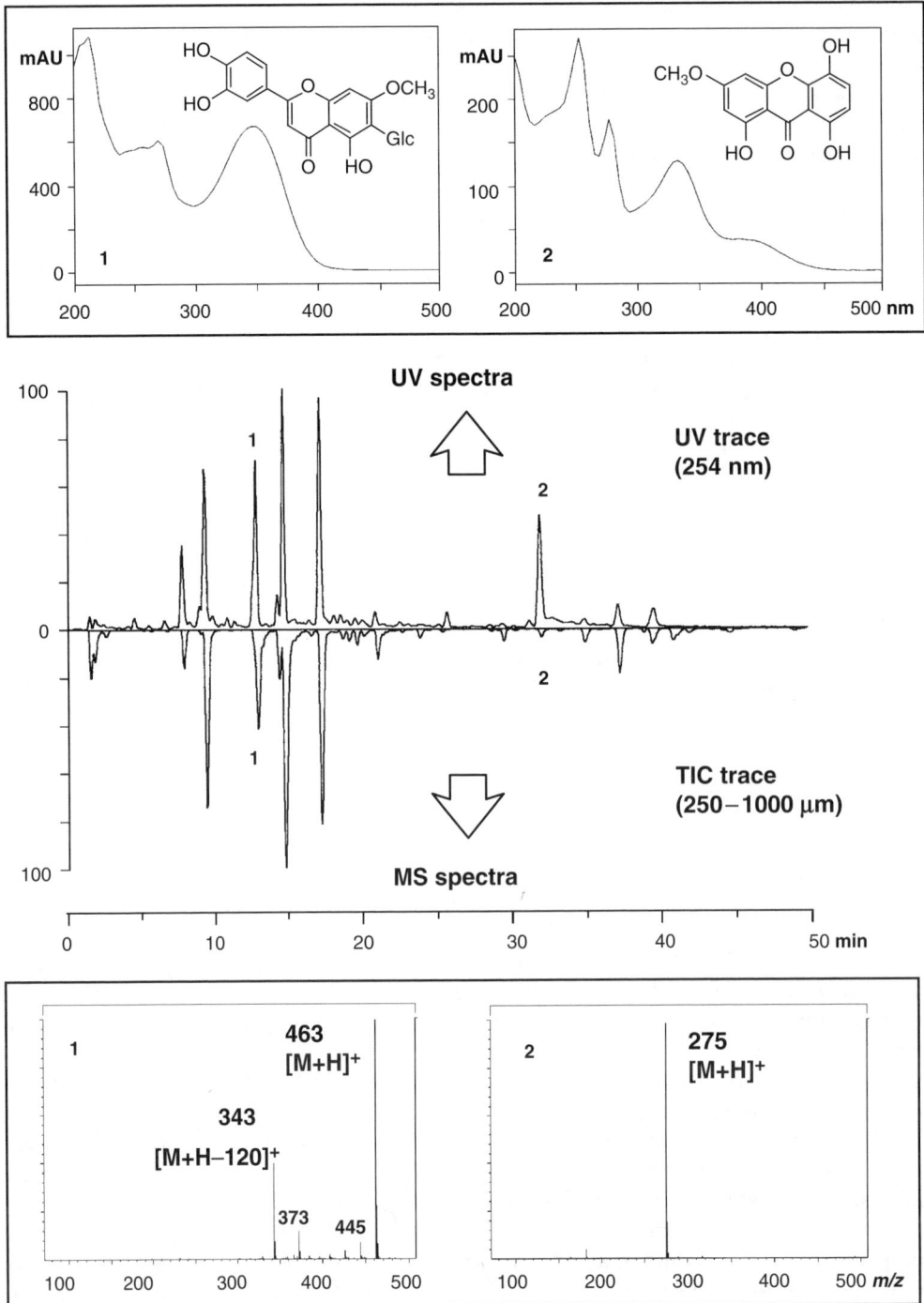

FIGURE 1.8 LC–UV–MS analysis of a methanolic extract of *Gentianella cabrerae*. The UV trace was recorded at 254 nm and the UV spectra from 200 to 500 nm. The LC–TSP-MS trace was recorded from 250 to 1000 μm. HPLC: Column Novapak C_{18} (150 × 3.9 mm, 4 μm); gradient of CH_3CN–H_2O (containing 0.1% TFA) 5:95 → 65:35 in 50 min; flow-rate 1 ml/min; 0.5 M NH_4OAc (0.2 ml/min) postcolumn. TSP: positive ion mode; filament off; vaporiser 100°C; source 280°C. (From Hostettmann, K., Wolfender, J.L., and Rodriguez, S., *Planta Med.*, 63, 2, 1997. With permission.)

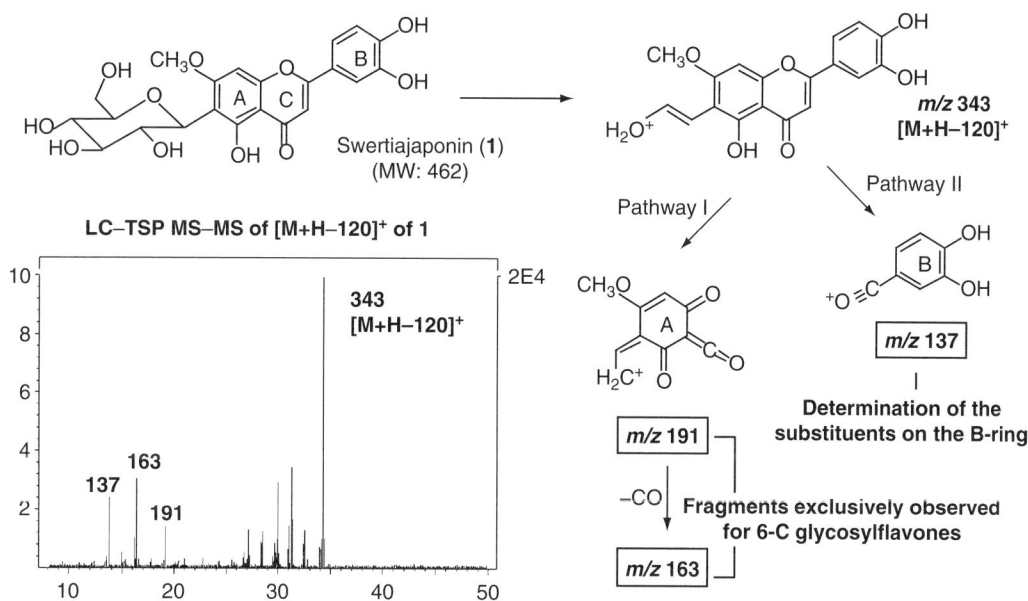

FIGURE 1.9 Online LC–TSP MS–MS of swertiajaponin (**1**) from the methanol extract of *Gentianella cabrerae*. The CID of the [M + H − 120]$^+$ ion at *m/z* 343 exhibited characteristic fragments for the substitution of the B-ring and for the glycosylation at C-6. (From Hostettmann, K., Wolfender, J.L., and Rodriguez, S., *Planta Med.*, 63, 2, 1997. With permission.)

extract indicated the presence of different flavonoid *C*-glycosides. LC–TSP MS data were insufficient to determine if compound **1** (Figure 1.9) was isoscoparin or swertiajaponin, two isomeric glycosides having a methoxyl group either at position C-3′ or at C-7, respectively. In order to complete structure determination, LC–TSP MS–MS of **1** was performed on the intense [M + H − 120]$^+$ ion (Figure 1.9). The experiment was run on a triple quadrupole instrument, in which the first quadrupole was set in order to filter the ion [M + H − 120]$^+$. This ion was then selectively fragmented by CID with argon in the collision chamber (the second quadrupole). Finally, the spectrum was recorded by scanning the third quadrupole. A classic fragmentation of the aglycone moiety of the flavonoid gave an ion at *m/z* 137. This fragment was indicative of a B-ring substituted with two hydroxyl groups, confirming the localization of the methoxyl group of **1** on the A-ring. Ions at *m/z* 163 and 191 were specific for flavone *C*-glycosides having their saccharide moiety at C-6. Thus, the flavonoid could be identified as swertiajaponin.[91]

In a study of the Guinean medicinal plant *Dissotis rotundifolia* (Melastomataceae) by hyphenated HPLC techniques, online data showed the presence of isomeric pairs of *C*-glycosylflavones in the alcoholic or hydroalcoholic extracts but these could not be distinguished either by TSP LC–MS or their UV spectra. Figure 1.10 shows the TSP LC–MS analysis of a methanol extract of *D. rotundifolia*, with the corresponding UV spectra of the four separated glycosides (**1–4**). However, tandem MS (TSP LC–MS–MS) provided a means of differentiating the isomers. There were well-defined [M + H]$^+$ pseudomolecular ions and [M + H − 120]$^+$ ion fragments for all four *C*-glycosylflavones but only isoorientin and isovitexin, the 6*C*-isomers, gave daughter ions at *m/z* 177 and *m/z* 149 from the [M + H − 120]$^+$ parent ion in the CID spectra (Figure 1.11) and hence could be distinguished from the 8*C*-isomers. The ions at *m/z* 177 and *m/z* 149 are produced after a retro-Diels–Alder fragmentation, which only occurs for isoorientin and isovitexin (Figure 1.12). By performing two parent

1 Isoorientin, R=OH   **2** Orientin, R=OH

3 Isovitexin, R=H   **4** Vitexin, R=H

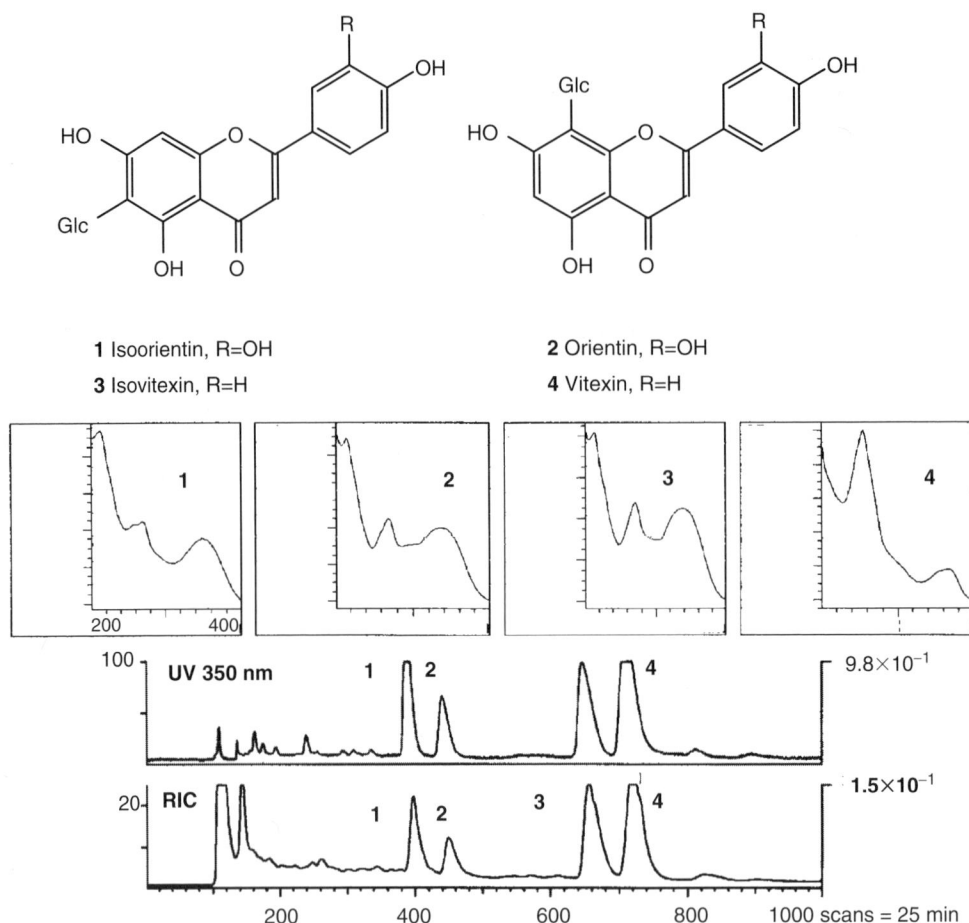

FIGURE 1.10 TSP LC–MS analysis of a methanol extract of *Dissotis rotundifolia*. Column Novapak C$_{18}$ (300 × 3.9 mm, 4 μm); CH$_3$CN–H$_2$O (14:86) (containing 0.05% TFA); flow rate 0.8 ml/min; detection 350 nm; RIC reconstructed ion current.

ion scan experiments with an isoorientin reference solution, it was possible to confirm that *m/z* 177 and *m/z* 149 were daughter ions of *m/z* 329. This is a rapid and simple means of distinguishing *C*-glycosylflavonoid isomers in plant extracts.[98]

While typical flow rates for the HPLC analyses of flavonoids lie in the 1.0 to 1.5 ml/min range, the introduction of short columns containing stationary phases with smaller pore sizes (allowing narrower peaks to be obtained in shorter separation times) means that considerably lower flow rates are the trend. Not only is there a decrease in solvent consumption but coupling to mass spectrometers or NMR instruments is facilitated.

1.4.6 HIGH-PERFORMANCE LIQUID CHROMATOGRAPHY–NUCLEAR MAGNETIC RESONANCE

Careful and critical use of the hyphenated techniques LC–UV–MS and LC–MS–MS can provide sufficient online information for the identification of small molecules such as flavonoids. However, in many cases, more data are required for an in-depth structural investigation and this can be supplied by the addition of an LC–NMR analytical capability (Figure 1.13). For practical purposes, LC–UV–MS and LC–UV–NMR are generally run as separate

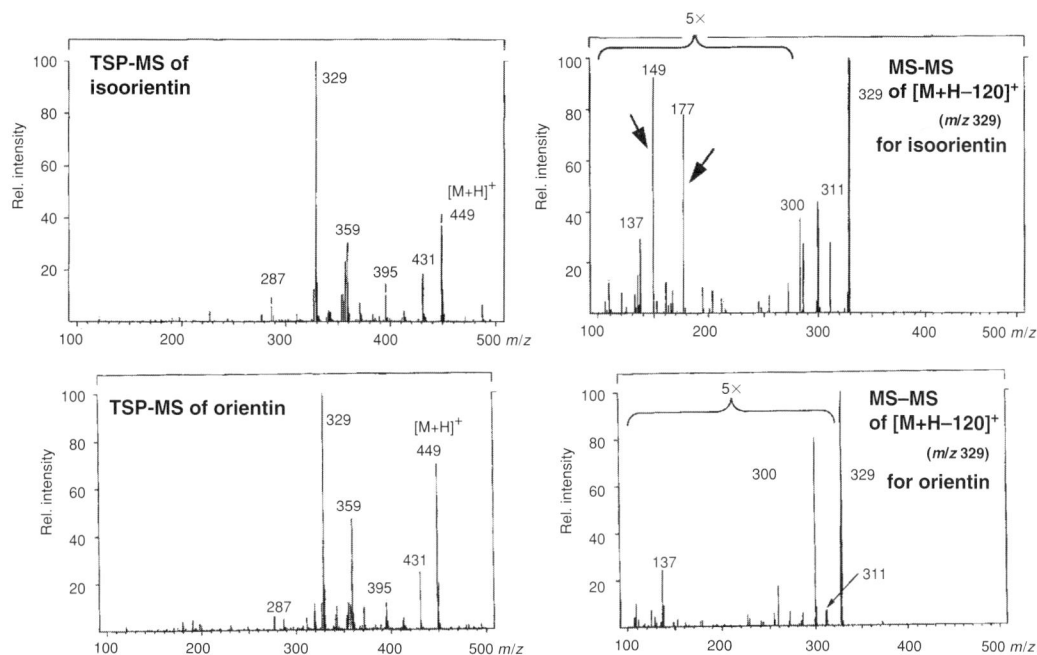

FIGURE 1.11 Online mass spectra of isoorientin (**1**) and orientin (**2**) in the TSP mode. The respective TSP MS–MS (CID) analyses of the parent $[M + H - 120]^+$ (m/z 329) ions are also shown.

operations. The coupling of HPLC with NMR spectroscopy, introduced around 1978, is one of the most powerful methods for the combined separation and structural elucidation of unknown compounds in mixtures.[99,100]

At first, LC–NMR was little used because of its lack of sensitivity. However, recent progress in pulse field gradients and solvent suppression, improvement in probe technology, and the construction of high-field magnets have given a new impulse to the technique. While HPLC–NMR coupling is relatively straightforward (the samples flow in a nonrotating 60 to 180 μl glass tube connected at both ends with HPLC tubing) compared to LC–MS, the main problem of LC–NMR is the difficulty of observing analyte resonances in the presence of the much larger resonances of the mobile phase. This problem is magnified under typical reversed-phase HPLC operating conditions, where more than one protonated solvent is used and where the resonances change frequencies during analysis in the gradient mode. Furthermore, the continuous flow of sample in the detector coil complicates solvent suppression. These problems have now been overcome by the development of fast, reliable, and powerful solvent suppression techniques, such as WET (water suppression enhanced through T_1 effects),[101] which produce high-quality spectra in both on-flow and stopped-flow modes. These techniques consist of a combination of pulsed-field gradients, shaped radiofrequency pulses, shifted laminar pulses, and selective ^{13}C decoupling, and are much faster than classical presaturation techniques previously used in the field. Thus, for typical reversed-phase HPLC analyses, nondeuterated solvents, such as methanol and acetonitrile, can be used, while water is replaced by D_2O.

The information provided by LC–NMR consists mainly of ^1H NMR spectra or $^1H-^1H$ correlation experiments. Access to ^{13}C NMR is possible but is restricted only to a very limited number of cases where the concentration of the LC peak of interest is high and ^{13}C NMR data can be deduced indirectly from inverse detection experiments. Due to the low natural

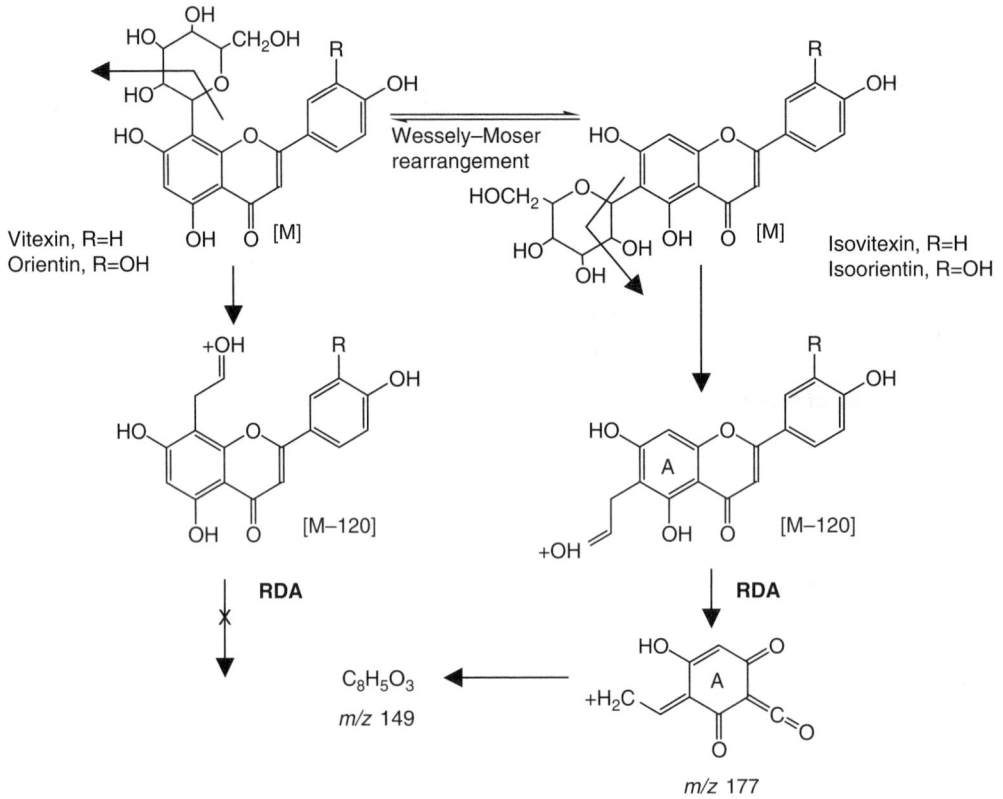

FIGURE 1.12 Specific fragmentations of the C-glycosylflavones isoorientin and isovitexin.

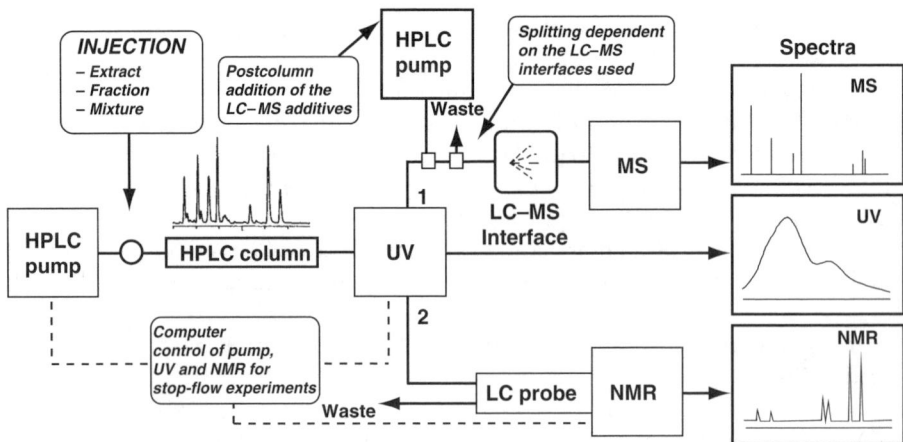

FIGURE 1.13 Schematic representation of the instrumentation used for LC–UV–MS (1) and LC–UV–NMR (2) analyses.

abundance of the ^{13}C isotope (1.1%), the sensitivity for direct measurement in the LC–NMR mode is insufficient.

LC–NMR can be operated in two different modes: on-flow and stopped-flow. In the on-flow mode, LC–NMR spectra are acquired continuously during the separation. The data are processed as a two-dimensional (2D) NMR experiment. The main drawback is the inherent low sensitivity. The detection limit with a 60 μl cell in a 500 MHz instrument for a compound with a molecular weight around 400 amu is 20 μg. Thus, on-flow LC–NMR runs are mainly restricted to the direct measurement of the main constituents of a crude extract and this is often under overloaded HPLC conditions. Typically, 1 to 5 mg of crude plant extract will have to be injected on-column.[102] In the stopped-flow mode, the flow of solvent after HPLC separation is stopped for a certain length of time when the required peak reaches the NMR flow cell. This makes it possible to acquire a large number of transients for a given LC peak and improves the detection limit. In this mode, various 2D correlation experiments (COSY, NOESY, HSQC, HMBC) are possible.

The combination of HPLC with online UV, MS, and NMR detection has proved to be a very valuable tool for the analysis of natural products in extracts or mixtures.[102,103] The field of flavonoids is no exception. The LC–NMR information obtained comes from the ^{1}H NMR spectra of selected peaks in the HPLC chromatogram. From LC–MS, A- or B-ring substitution can be deduced from the fragmentation pattern but the exact location of the substituent cannot be determined. However, for a flavonoid like apigenin, where only one hydroxyl group is located on the B-ring, ^{1}H NMR will give the substitution position because each of the three possibilities of localization of the hydroxyl group will give a unique splitting pattern. Much information can be derived about the nature and linkage positions of sugars. However, since D$_2$O is present in the eluent, exchangeable signals are not observed in the NMR spectrum.

An example of the LC–NMR stop-flow procedure is provided in the analysis of polyphenolics from the Chilean plant *Gentiana ottonis* (Gentianaceae).[104] Preliminary LC–UV analysis of a methanol extract of the roots showed the presence of several xanthones (**2, 4, 6–8**; Figure 1.14), an iridoid (**1**), and two compounds (**3, 5**) with UV spectra typical of flavonoids. TSP LC–MS provided the molecular weights of the latter two compounds and gave fragments characteristic for *C*-glycosides (losses of 90 and 120 amu). According to their UV spectra, **3** and **5** (MW 448 and 446) were, respectively, tri- and tetra-oxygenated flavone *C*-glycosides. In order to obtain further information for characterization of the polyphenols in the extract, LC–NMR was performed under the same conditions used for LC–UV–MS. However, water was replaced by D$_2$O and the amount injected was increased to 0.4 mg, which did not cause a noticeable loss in resolution. LC–^{1}H NMR spectra were recorded for all the main peaks in the stop-flow mode and the number of transients for a good signal-to-noise ratio varied between 128 and 2048. For flavone *C*-glycoside **5** (MW 446), the LC–^{1}H NMR spectrum (Figure 1.15) gave signals for six aromatic protons (δ 6.8 to 8.1), one methoxyl group (δ 4.0), and the *C*-glycoside moiety (δ 3.5 to 5.0). A pair of symmetric *ortho*-coupled protons (2H, δ 7.06, *J* = 8.3, H-3′,5′ and 2H, δ 8.00, *J* = 8.3, H-2′,6′) was characteristic for a B-ring substituted at C-4′. The singlet at δ 6.8 was attributed to H-3. A singlet at δ 6.9 was due to a proton either at position C-6 or C-8 on the A-ring. Thus, LC–UV–MS and stop-flow LC–NMR were not sufficient to fully ascertain the structure of **5**. In order to ascertain the position of *C*-glycosylation, an LC–MS–MS experiment was performed by choosing [M + H − 120]$^+$ as parent ion. This gave fragments at *m/z* 191 and 163, characteristic for 6-*C*-glycosylated flavones (described earlier). The fragment at *m/z* 121 indicated a monohydroxylated B-ring, confirming the methoxyl group to be on the A-ring. This combination of data allowed identification of **5** as 5,4′-dihydroxy-7-methoxy-6-*C*-glucosylflavone (swertisin).[104]

If full metabolite profiling of a plant extract has to be performed, LC–NMR can be run in the on-flow mode. In order to obtain adequate NMR spectra of all constituents, the amount

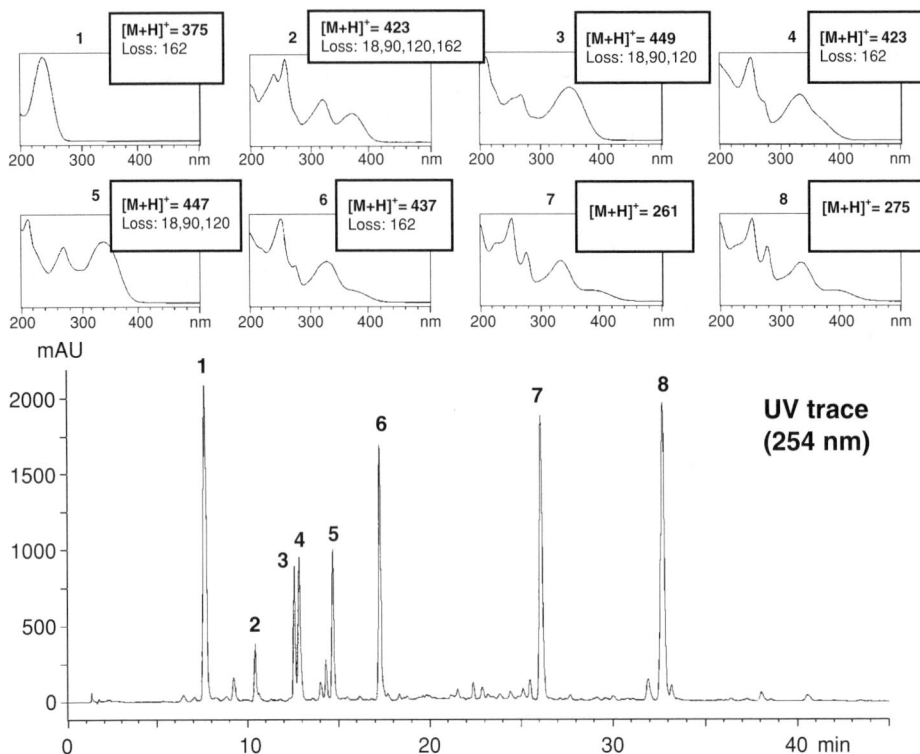

FIGURE 1.14 Online LC–UV of a methanol extract of *Gentiana ottonis*, with protonated molecular ions obtained for the main constituents by TSP LC–MS. Column Novapak C_{18} (150 × 3.9 mm, 4 μm); gradient CH_3CN-H_2O (containing 0.05% TFA) 5:95 → 65:35 in 50 min; flow rate 0.8 ml/min. (From Wolfender, J.-L., Ndjoko, K., and Hostettmann, K., in *Methods in Polyphenol Analysis*, Santos-Buelga, C. and Williamson, G., Eds., Royal Society of Chemistry, Cambridge, 2003. With permission.)

of sample injected has to be increased — this produces overloading when compared with normal analytical HPLC conditions but gives the possibility of testing for biological activity (in conjunction with a microfractionation procedure). This was the approach adopted for the investigation of new antifungal constituents from *Erythrina vogelii* (Leguminosae), a medicinal plant of the Ivory Coast.[105] In order to rapidly identify the active principles from the antifungal dichloromethane extract of the roots, preliminary analysis by LC–UV and Q-TOF LC–MS was performed. Approximately 12 major peaks were observed in the HPLC chromatogram and from UV, MS, and MS–MS online data, these were shown to be prenylated isoflavones and isoflavanones. In order to obtain more information, on-flow LC–^{1}H NMR was performed by injecting 10 mg of crude extract onto an 8 mm C_{18} radial compression column connected to the NMR instrument. At a low flow rate (0.1 ml/min), acquisition of ten LC–NMR spectra was possible. Of these ten peaks, five were found by simultaneous HPLC microfractionation to be associated with the antifungal activity of the extract. Interpretation of all online data, with emphasis on LC–NMR, allowed the identification of eight flavonoids, including a known isoflavone with antifungal activity and two putative new isoflavanones, also with antifungal activity. This dereplication procedure allowed the targeted isolation of the new antifungal compounds.[105]

Applications of LC–NMR for the online identification of flavonoids are still few and far between, one reason probably being the high cost of the apparatus. However, several other

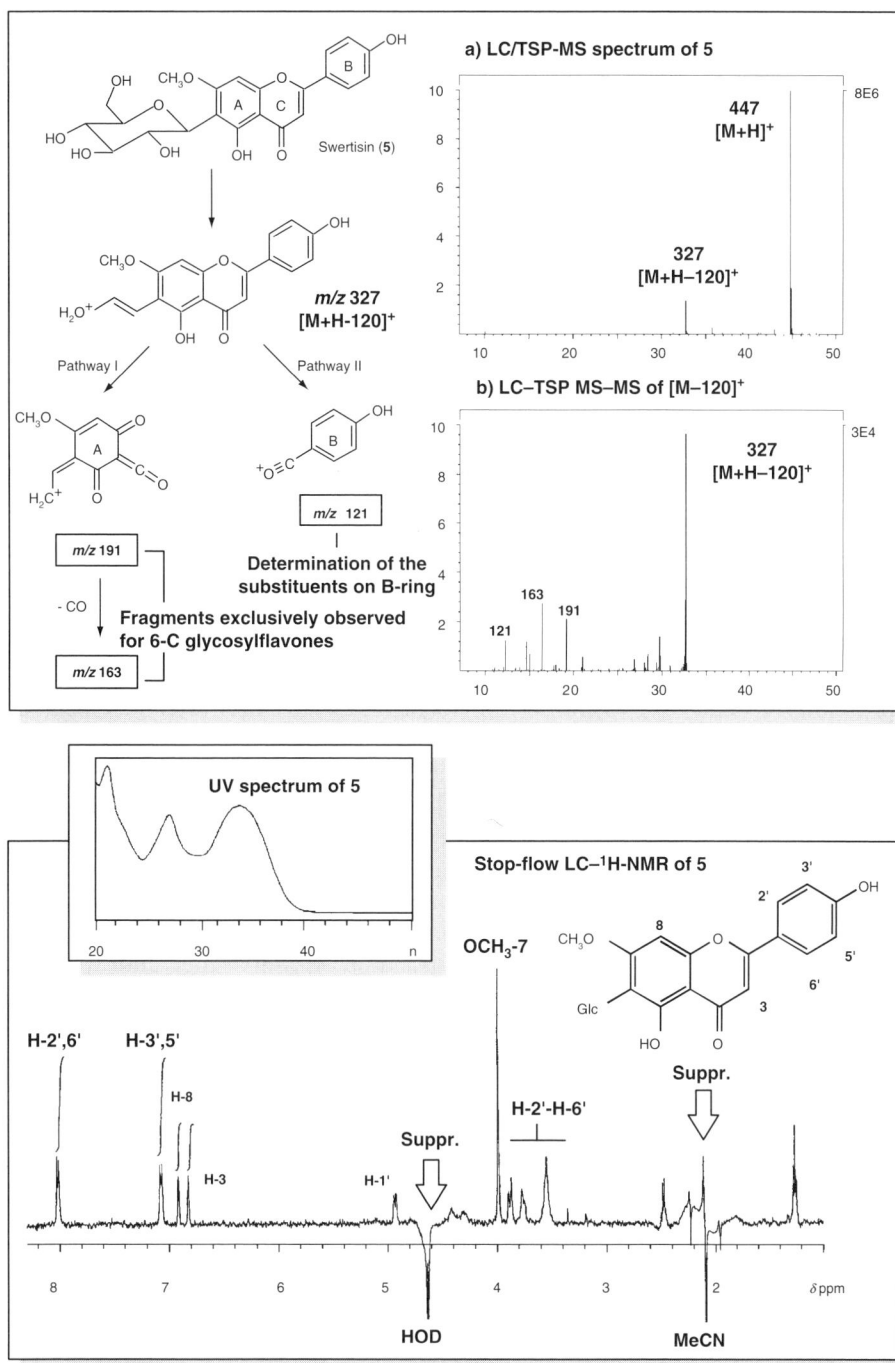

FIGURE 1.15 Stop-flow LC–^{1}H NMR spectrum of swertisin (**5**) from the methanol extract of the roots of *Gentiana ottonis*, together with the TSP LC–MS (a) and TSP LC–MS–MS (b) spectra. The LC–MS–MS analysis was performed using the fragment [M + H − 120]$^+$ (spectrum a) as parent ion. Characteristic daughter ions at *m/z* 121, 163, and 191 were observed, indicating the substitution on the A- and B-ring of the *C*-glycosylflavone. (From Wolfender, J.-L., Ndjoko, K., and Hostettmann, K., in *Methods in Polyphenol Analysis*, Santos-Buelga, C. and Williamson, G., Eds., Royal Society of Chemistry, Cambridge, 2003. With permission.)

examples do exist, in addition to those mentioned above. The technique has been successfully applied to the analysis of *Hypericum perforatum* (Guttiferae). Online identification of quercetin, several of its glycosides, and the biflavonoid I5,II8-biapigenin in an extract was possible.[106]

1.4.7 CAPILLARY ELECTROPHORESIS

CE is an analytical technique that provides high separation efficiency and short run times. When compared to HPLC, however, CE generally exhibits much lower sensitivity, a tendency to overload with samples, and less reproducible quantitative data. In contrast to HPLC, method development is more time consuming in CE — involving investigation of types, pH and concentrations of electrolytes, types and concentrations of surfactants and organic modifiers, temperatures, and applied voltages. Several modes of CE are available: (a) capillary zone electrophoresis (CZE), (b) micellar electrokinetic chromatography (MEKC), (c) capillary gel electrophoresis (CGE), (d) capillary isoelectric focusing, (e) capillary isotachophoresis, (f) capillary electrochromatography (CEC), and (g) nonaqueous CE. The simplest and most versatile CE mode is CZE, in which the separation is based on differences in the charge-to-mass ratio and analytes migrate into discrete zones at different velocities.[107] Anions and cations are separated in CZE by electrophoretic migration and electro-osmotic flow (EOF), while neutral species coelute with the EOF. In MEKC, surfactants are added to the electrolyte to form micelles. During MEKC separations, nonpolar portions of neutral solutes are incorporated into the micelles and migrate at the same velocity as the micelles, while the polar portions are free and migrate at the EOF velocity.

Applications of CE for the analysis of phytochemicals have been well documented.[108,109] CE is especially suitable for the separation of flavonoids as they are negatively charged at higher pH values.[108,110] Suntornsuk[110] has reviewed quantitative aspects and method validation of CE for flavonoids. Compared with HPLC, CE can provide an alternative analytical method when higher efficiency or higher resolution is required. For example, while TLC and HPLC analyses of passion flower do not provide adequate separation of all identified flavonoids, CE can fulfill the necessary requirements.[111] Separations of *Passiflora incarnata* (Passifloraceae) flavonoid glycosides were performed on a 50 μm internal diameter uncoated fused-silica capillary with 25 m*M* sodium borate buffer with 20% methanol (pH 9.5). The voltage was 30 kV and the temperature of the capillary maintained at 35°C. The CE instrument was equipped with a diode array detector. Twelve glycosides were satisfactorily separated within 13 min (Figure 1.16). For quantification, quercetin 3-*O*-arabinoside was used as internal standard. Calibration curves for internal standardization were established. The method was applied to the analysis of ten commercial samples of Passiflorae herba. They showed similar flavonoid patterns but differed quantitatively in individual flavonoid glycosides. Reproducibility was good, with a coefficient of variation (CV) of 2.83% for interday precision and a mean CV of 1.26% for migration time.[111]

Other applications include the online coupling of capillary isotachophoresis and CZE for the quantitative determination of flavonoids in *Hypericum perforatum* (Guttiferae) leaves and flowers. This method involved the concentration and preseparation of the flavonoid fraction before introduction into the CZE capillary. The limit of detection for quercetin 3-*O*-glycosides was 100 ng/ml.[112]

1.5 OUTLOOK

Preparative separations still present a challenge. There is no general, simple, straightforward strategy for the isolation of natural products, even if certain compounds are readily accessible by modern chromatographic techniques. Each particular separation problem has to be

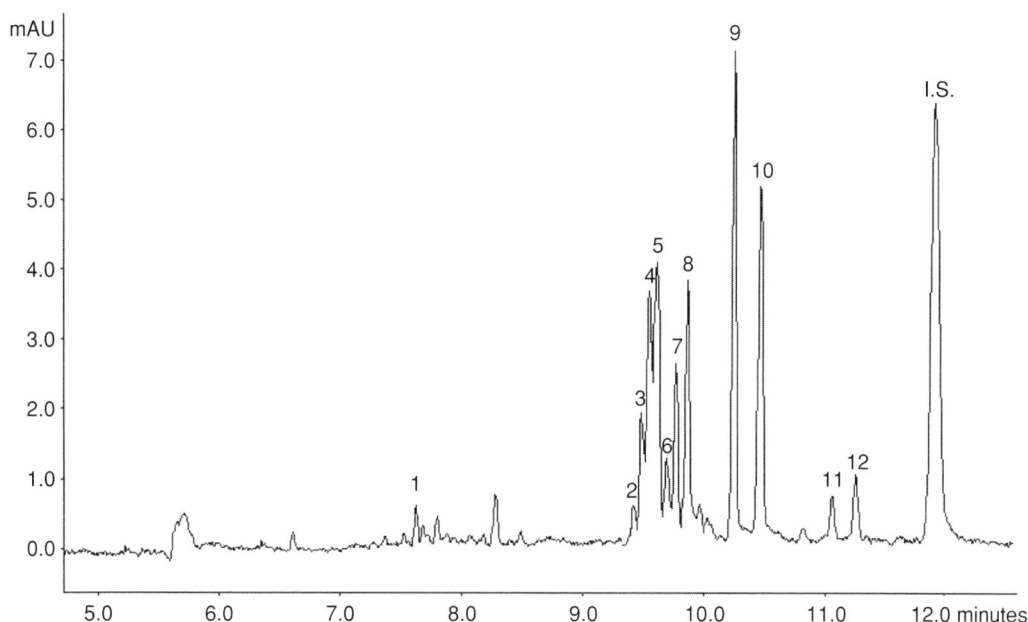

FIGURE 1.16 Electropherogram of a Passiflorae herba methanol extract. Capillary temperature 35°C; voltage 30 kV; electrolyte buffer 25 mM sodium tetraborate containing 20% MeOH (pH 9.5); UV detection at 275 nm; IS internal standard quercetin 3-O-arabinoside. (From Marchart, E., Krenn, L., and Kopp, B., *Planta Med.*, 69, 452, 2003. With permission.)

considered on its own and a suitable procedure has to be developed. In this respect, the flavonoids are no exception.

However, analytical separations of flavonoids are now routine. In quantitative measurements, the amounts of the individual components within a particular class of constituent need to be determined. Nowadays, this can easily be achieved through the use of GC, HPLC, and hyphenated techniques.

In HPLC, microbore operation is becoming popular, especially for LC–MS applications, because it allows smaller samples, faster separation times, and lower solvent consumption.

The trend is toward multiple hyphenation techniques like HPLC–UV–MS and HPLC–UV–NMR. These have an enormous potential for the rapid investigation of plant extracts.[100] Multiple hyphenation in a single system provides a better means of identification of compounds in a complex matrix.

Applications of LC–NMR are still scarce but the technique will become more widely used. The main effort for efficient exploitation of LC–NMR needs to be made on the chromatographic side, where strategies involving efficient preconcentration, high loading, stop-flow, time slicing, or low flow procedures have to be developed. Microbore columns or capillary separation methods, such as capillary LC–NMR, CE–NMR, and CEC–NMR, will find increased application, one reason being that the low solvent consumption will allow the use of fully deuterated solvents.

Other online HPLC techniques (such as LC–CD or LC–IR) are likely to be exploited. For example, a mixture of diastereoisomeric biflavonoids from the African plant *Gnidia involucrata* (Thymelaeaceae) could not be separated on a preparative scale by HPLC or crystallization. However, their analytical separation on a C_{18} column was sufficient to run an online LC–CD investigation and provide stereochemical information about the individual isomers.[113]

REFERENCES

1. Mabry, T.J., Markham, K.R., and Thomas, M.B., *The Systematic Identification of Flavonoids*, Springer-Verlag, New York, 1970.
2. Markham, K., Isolation techniques for flavonoids, in *The Flavonoids*, Harborne, J.B., Mabry, T.J., and Mabry, H., Eds., Academic Press, New York, 1975.
3. Hostettmann, K. and Hostettmann, M., Isolation techniques for flavonoids, in *The Flavonoids: Advances in Research*, Harborne, J.B. and Mabry, T.J., Eds., Chapman & Hall, London, 1982, chap. 1.
4. Harborne, J.B., Ed., *The Flavonoids: Advances in Research Since 1980*, Chapman & Hall, London, 1988.
5. Harborne, J.B., General procedures and measurement of total phenolics, in *Methods in Plant Biochemistry, Vol. 1, Plant Phenolics*, Harborne, J.B., Ed., Academic Press, London, 1989, chap. 1.
6. Santos-Buelga, C. and Williamson, G., Eds., *Methods in Polyphenol Analysis*, The Royal Society of Chemistry, Cambridge, 2003.
7. Hostettmann, K., Marston, A., and Hostettmann, M., *Preparative Chromatography Techniques: Applications in Natural Product Isolation*, 2nd ed., Springer-Verlag, Berlin, 1998.
8. Hussein, L., Fattah, M.A., and Salem, E., Characterization of pure proanthocyanidins isolated from the hulls of faba beans, *J. Agric. Food Chem.*, 38, 95, 1990.
9. Benthin, B., Danz, H., and Hamburger, M., Pressurized liquid extraction of medicinal plants, *J. Chromatogr. A*, 837, 211, 1999.
10. Dawidowicz, A.L. et al., Optimization of ASE conditions for the HPLC determination of rutin and isoquercitrin in *Sambucus nigra* L., *J. Liq. Chromatogr. Relat. Technol.*, 26, 2381, 2003.
11. Pineiro, Z., Palma, M., and Barroso, C.G., Determination of catechins by means of extraction with pressurized liquids, *J. Chromatogr. A*, 1026, 19, 2004.
12. Bevan, C.D. and Marshall, P.S., The use of supercritical fluids in the isolation of natural products, *Nat. Prod. Rep.*, 11, 451, 1994.
13. Jarvis, A.P. and Morgan, E.D., Isolation of plant products by supercritical fluid extraction, *Phytochem. Anal.*, 8, 217, 1997.
14. Hamburger, M., Baumann, D., and Adler, S., Supercritical carbon dioxide extraction of selected medicinal plants — effects of high pressure and added ethanol on yield of extracted substances, *Phytochem. Anal.*, 15, 46, 2004.
15. Sargenti, S.R. and Vichnewski, W., Sonication and liquid chromatography as a rapid technique for extraction and fractionation of plant material, *Phytochem. Anal.*, 11, 69, 2000.
16. Ganzler, K., Szinai, I., and Salgo, A., Effective sample preparation method for extracting biologically active compounds from different matrices by a microwave technique, *J. Chromatogr.*, 520, 257, 1990.
17. Kaufmann, B. and Christen, P., Recent extraction techniques for natural products: microwave-assisted extraction and pressurized solvent extraction, *Phytochem. Anal.*, 13, 105, 2002.
18. McKee, T.C. et al., Isolation and characterization of new anti-HIV and cytotoxic leads from plants, marine, and microbial organisms, *J. Nat. Prod.*, 60, 431, 1997.
19. Carlton, R.R. et al., Kaempferol-3-(2,3-diacetoxy-4-*p*-coumaroyl)rhamnoside from leaves of *Myrica gale*, *Phytochemistry*, 29, 2369, 1990.
20. Gafner, S. et al., Antifungal and antibacterial chalcones from *Myrica serrata*, *Planta Med.*, 62, 67, 1996.
21. Slimestad, R., Marston, A., and Hostettmann, K., Preparative separation of phenolic compounds from *Picea abies* by high-speed countercurrent chromatography, *J. Chromatogr. A*, 719, 438, 1996.
22. Long, C. et al., Bioactive flavonoids of *Tanacetum parthenium* revisited, *Phytochemistry*, 64, 567, 2003.
23. Liu, Y., Wagner, H., and Bauer, R., Nevadensin glycosides from *Lysionotus pauciflorus*, *Phytochemistry*, 42, 1203, 1996.
24. Billeter, M., Meier, B., and Sticher, O., 8-Hydroxyflavonoid glucuronides from *Malva sylvestris*, *Phytochemistry*, 30, 987, 1991.

25. Lee, T.-H. et al., Three new flavonol galloylglycosides from leaves of *Acacia confusa*, *J. Nat. Prod.*, 63, 710, 2000.
26. Flagg, M.L. et al., Two novel flavanones from *Greigia sphacelata*, *J. Nat. Prod.*, 63, 1689, 2000.
27. Kitaoka, M. et al., Prenylflavonoids: a new class of non-steroidal phytoestrogen (part 1). Isolation of 8-isopentenylnaringenin and an initial study on its structure–activity relationship, *Planta Med.*, 64, 511, 1998.
28. Queiroz, E.F. et al., Prenylated isoflavonoids from the root bark of *Erythrina vogelii*, *J. Nat. Prod.*, 65, 403, 2002.
29. Krauze-Baranowska, M. et al., Antifungal biflavones from *Cupressocyparis leylandii*, *Planta Med.*, 65, 572, 1999.
30. Fiorini, M., Preparative high-performance liquid chromatography for the purification of natural anthocyanins, *J. Chromatogr. A*, 692, 213, 1995.
31. Bilia, A.R. et al., Flavans and A-type proanthocyanidins from *Prunus prostrata*, *Phytochemistry*, 43, 887, 1996.
32. Leutert, T. and von Arx, E., Präparative Mitteldruck-flüssigkeitschromatographie (Preparative medium-pressure liquid chromatography), *J. Chromatogr.*, 292, 333, 1984.
33. Zogg, G.C., Nyiredy, Sz., and Sticher, O., Operating conditions in preparative medium pressure liquid chromatography (MPLC). II. Influence of solvent strength and flow rate of the mobile phase, capacity and dimensions of the column, *J. Liq. Chromatogr.*, 12, 2049, 1989.
34. Nyiredy, S., Dallenbach-Tölke, K., and Sticher, O., The "PRISMA" optimization system in planar chromatography, *J. Planar Chromatogr.*, 1, 336, 1988.
35. Schaufelberger, D. and Hostettmann, K., Analytical and preparative reversed-phase liquid chromatography of secoiridoid glycosides, *J. Chromatogr.*, 346, 396, 1985.
36. Orjala, J. et al., New monoterpene-substituted dihydrochalcones from *Piper aduncum*, *Helv. Chim. Acta*, 76, 1481, 1993.
37. Shirataki, Y. et al., Sophoraflavones H, I and J, flavonostilbenes from *Sophora moorcroftiana*, *Chem. Pharm. Bull.*, 39, 1568, 1991.
38. Ducrey, B. et al., Analysis of flavonol glycosides of thirteen *Epilobium* species (Onagraceae) by LC–UV and thermospray LC–MS, *Phytochemistry*, 38, 129, 1995.
39. Allais, D.P. et al., 3-Desoxycallunin and 2″-acetylcallunin, two minor 2,3-dihydroflavonoid glucosides from *Calluna vulgaris*, *Phytochemistry*, 39, 427, 1995.
40. Pachaly, P., Schönherr-Weissbarth, C., and Sin, K.-S., Neue Prenylflavonoid-Glykoside aus *Epimedium koreanum*, *Planta Med.*, 56, 277, 1990.
41. Atindehou, K.K. et al., Three new prenylated isoflavonoids from the root bark of *Erythrina vogelii*, *Planta Med.*, 68, 181, 2001.
42. Nunome, S. et al., *In vitro* antimalarial activity of biflavonoids from *Wikstroemia indica*, *Planta Med.*, 70, 76, 2004.
43. Conway, W.D., *Countercurrent Chromatography: Apparatus, Theory and Application*, VCH Publishers, Inc., New York, 1990.
44. Schaufelberger, D.E., Applications of analytical high-speed counter-current chromatography in natural products chemistry, *J. Chromatogr.*, 538, 45, 1991.
45. Marston, A., Slacanin, I., and Hostettmann, K. Centrifugal partition chromatography in the separation of natural products, *Phytochem. Anal.*, 1, 3, 1990.
46. Zhang, T.Y. et al., Separation of flavonoids in crude extract from sea buckthorn by countercurrent chromatography with two types of coil planet centrifuge, *J. Liq. Chromatogr.*, 11, 233, 1988.
47. Miserez, F. et al., Flavonol glycosides from *Vernonia galamensis* ssp. *nairobiensis*, *Phytochemistry*, 43, 283, 1996.
48. Vasänge, M. et al., The flavonoid constituents of two *Polypodium* species (Calaguala) and their effect on the elastase release in human neutrophils, *Planta Med.*, 63, 511, 1997.
49. Oliveira, R.R. et al., High-speed countercurrent chromatography as a valuable tool to isolate *C*-glycosylflavones from *Cecropia lyratiloba* Miquel., *Phytochem. Anal.*, 14, 96, 2003.
50. Kapadia, G.J., Oguntimein, B., and Shukla, Y.N., High-speed countercurrent chromatographic separation of biflavonoids from *Garcinia kola* seeds, *J. Chromatogr. A*, 673, 142, 1994.

51. Ma, X. et al., Preparative isolation and purification of two isoflavones from *Astragalus membranaceus* Bge. var. *mongholicus* (Bge.) Hsiao by high-speed countercurrent chromatography, *J. Chromatogr. A*, 992, 193, 2003.
52. Yang, F., Ma, Y., and Ito, Y., Separation and purification of isoflavones from a crude soybean extract by high-speed countercurrent chromatography, *J. Chromatogr. A*, 928, 163, 2001.
53. Kunz, S., Burkhardt, G., and Becker, H., Riccionidins A and B, anthocyanidins from the cell walls of the liverwort *Ricciocarpos natans*, *Phytochemistry*, 35, 233, 1994.
54. De Mello, J.P., Petereit, F., and Nahrstedt, A., Prorobinetinidins from *Stryphnodendron adstringens*, *Phytochemistry*, 42, 857, 1996.
55. Drewes, S.E. and Taylor, C.W., Methylated A-type proanthocyanidins and related metabolites from *Cassipourea gummiflua*, *Phytochemistry*, 37, 551, 1994.
56. Schwarz, M. et al., Application of high-speed countercurrent chromatography to the large-scale isolation of anthocyanins, *Biochem. Eng. J.*, 14, 179, 2003.
57. Degenhardt, A. et al., Preparative separation of polyphenols from tea by high-speed countercurrent chromatography, *J. Agric. Food Chem.*, 48, 3425, 2000.
58. Sutherland, I.A. et al., Countercurrent chromatography (CCC) and its versatile application as an industrial purification and production process, *J. Liq. Chromatogr. Relat. Technol.*, 21, 279, 1998.
59. Gaedcke, F., Phytoäquivalenz: was steckt dahinter, *Deutsch. Apoth. Ztg.*, 135, 311, 1995.
60. Schmidt, T.J., Merfort, I., and Matthiesen, U., Resolution of complex mixtures of flavonoid aglycones by analysis of gas chromatographic–mass spectrometric data, *J. Chromatogr.*, 634, 350, 1993.
61. Tura, D. and Robards, K., Sample handling strategies for the determination of biophenols in food and plants, *J. Chromatogr. A*, 975, 71, 2002.
62. Smith, R.M., Before the injection — modern methods of sample preparation for separation techniques, *J. Chromatogr. A*, 1000, 3, 2003.
63. Chen, H., Zuo, Y., and Deng, Y., Separation and determination of flavonoids and other phenolic compounds in cranberry juice by high-performance liquid chromatography, *J. Chromatogr. A*, 913, 387, 2001.
64. Edenharder, R. et al., Isolation and characterization of structurally novel antimutagenic flavonoids from spinach (*Spinacia oleracea*), *J. Agric. Food Chem.*, 49, 2767, 2001.
65. Wang, J. and Sporns, P., Analysis of anthocyanins in red wine and fruit juice using MALDI-MS, *J. Agric. Food Chem.*, 47, 2009, 1999.
66. Nogata, Y. et al., High-performance liquid chromatographic determination of naturally occurring flavonoids in *Citrus* with a photodiode-array detector, *J. Chromatogr.*, 667, 59, 1994.
67. Modey, W.K., Mulholland, D.A., and Raynor, M.W., Analytical supercritical fluid extraction of natural products, *Phytochem. Anal.*, 7, 1, 1996.
68. Eskilsson, C.S. and Björklund, E., Analytical-scale microwave-assisted extraction, *J. Chromatogr. A*, 902, 227, 2000.
69. Jork, H., Funk, W., Fischer, W., and Wimmer, H., *Thin-Layer Chromatography: Reagents and Detection Methods*, 2nd ed., VCH Verlagsgesellschaft, Weinheim, 1994.
70. Markham, K., Isolation techniques for flavonoids, in *The Flavonoids*, Harborne, J.B., Mabry, T.J., and Mabry, H., Eds., Academic Press, New York, 1975, chap. 1.
71. Wagner, H. and Bladt, S., *Plant Drug Analysis: A Thin Layer Chromatography Atlas*, 2nd ed., Springer-Verlag, Berlin, 1996.
72. Budzianowski, J., Separation of flavonoid glucosides from their galactosidic analogues by thin-layer chromatography, *J. Chromatogr.*, 540, 469, 1991.
73. Brasseur, T. and Angenot, L., Le mélange diphénylborate d'aminoéthanol-PEG 400 — un intéressant réactif de revelation des flavonoïdes, *J. Chromatogr.*, 351, 351, 1986.
74. Smolarz, H.D., Matysik, G., and Wojciak-Kosior, M., High-performance thin-layer chromatographic and densitometric determination of flavonoids in *Vaccinium myrtillus* L. and *Vaccinium vitis-idaea* L., *J. Planar Chromatogr.*, 13, 101, 2000.
75. Hiermann, A. and Bucar, F., Diphenyltin chloride as a chromogenic reagent for the detection of flavonoids on thin-layer plates, *J. Chromatogr.*, 675, 276, 1994.

76. Merken, H.M. and Beecher, G.R., Measurement of food flavonoids by high-performance liquid chromatography: a review, *J. Agric. Food Chem.*, 48, 577, 2000.

77. He, X.-G., On-line identification of phytochemical constituents in botanical extracts by combined high-performance liquid chromatographic-diode array detection-mass spectrometric techniques, *J. Chromatogr. A*, 880, 203, 2000.

78. Cimpan, G. and Gocan, S., Analysis of medicinal plants by HPLC: recent approaches, *J. Liq. Chromatogr. Relat. Technol.*, 25, 2225, 2002.

79. Crozier, A. et al., Quantitative analysis of flavonoids by reversed-phase high performance liquid chromatography, *J. Chromatogr. A*, 761, 315, 1997.

80. Schieber, A., Berardini, N., and Carle, R., Identification of flavonol and xanthone glycosides from mango (*Mangifera indica*) peels by high-performance liquid chromatography–electrospray ionization mass spectrometry, *J. Agric. Food Chem.*, 51, 5006, 2003.

81. Berente, B., Reichenbächer, M., and Danzer, K., Improvement of the HPLC analysis of anthocyanins in red wines by use of recently developed columns, *Fresenius J. Anal. Chem.*, 371, 68, 2001.

82. Barnes, S. et al., HPLC–mass spectrometry analysis of flavonoids, *Proc. Soc. Exp. Biol. Med.*, 217, 254, 1998.

83. Sakakibara, H. et al., Simultaneous determination of all polyphenols in vegetables, fruits and teas, *J. Agric. Food Chem.*, 51, 571, 2003.

84. Van Beek, T.A., Chemical analysis of *Ginkgo biloba* leaves and extracts, *J. Chromatogr. A*, 967, 21, 2002.

85. Hasler, A., Sticher, O., and Meier, B., Identification and determination of the flavonoids from *Ginkgo biloba* by high-performance liquid chromatography, *J. Chromatogr.*, 605, 41, 1992.

86. George, S. and Maute, A., A photodiode array detection system: design, concept and implementation, *Chromatographia*, 15, 419, 1982.

87. Justesen, U., Knuthsen, P., and Leth, T., Quantitative analysis of flavonols, flavones, and flavanones in fruits, vegetables and beverages by high-performance liquid chromatography with photodiode array and mass spectrometric detection, *J. Chromatogr. A*, 799, 101, 1998.

88. Hostettmann, K. et al., On-line high-performance liquid chromatography: ultraviolet–visible spectroscopy of phenolic compounds in plant extracts using post-column derivatization, *J. Chromatogr.*, 283, 137, 1984.

89. Niessen, W.M.A., State-of-the-art in liquid chromatography–mass spectrometry, *J. Chromatogr. A*, 856, 179, 1999.

90. Wolfender, J.-L. et al., Comparison of liquid chromatography/electrospray, atmospheric pressure chemical ionization, thermospray and continuous-flow fast atom bombardment mass spectrometry for the determination of secondary metabolites in crude plant extracts, *J. Mass Spectrom. Rapid Commun. Mass Spectrom.*, S35, 1995.

91. Hostettmann, K. et al., Rapid detection and subsequent isolation of bioactive constituents of crude plant extract, *Planta Med.*, 63, 2, 1997.

92. Cole, R.B., Ed., *Electrospray Ionization Mass Spectrometry: Fundamentals, Instrumentation and Applications*, Wiley, New York, 1997.

93. Lin, Y.Y., Ng, K.J., and Yang, S., Characterization of flavonoids by liquid chromatography–tandem mass spectrometry, *J. Chromatogr.*, 629, 389, 1993.

94. Ma, Y.L. et al., Characterization of flavone and flavonol aglycones by collision-induced dissociation tandem mass spectrometry, *Rapid Commun. Mass Spectrom.*, 11, 1357, 1997.

95. Li, Q.M. and Claeys, M., Characterisation and differentiation of diglycosyl flavonoids by positive ion fast atom bombardment and tandem mass spectrometry, *Biol. Mass Spectrom.*, 23, 406, 1994.

96. Grayer, R. et al., The application of atmospheric pressure chemical ionisation liquid chromatography–mass spectrometry in the chemotaxonomic study of flavonoids: characterisation of flavonoids from *Ocimum gratissimum* var. *gratissimum*, *Phytochem. Anal.*, 11, 257, 2000.

97. J.-L. Wolfender et al., Evaluation of Q-TOF-MS/MS and multiple stage IT-MS^n for the dereplication of flavonoids and related compounds in crude plant extracts, *Analusis*, 28, 895, 2000.

98. G. Rath et al., Characterization of *C*-glycosylflavones from *Dissotis rotundifolia* by liquid chromatography–UV diode array detection-tandem mass spectrometry, *Chromatographia*, 41, 332, 1995.

99. Albert, K., Liquid chromatography–nuclear magnetic resonance spectroscopy, *J. Chromatogr. A*, 856, 199, 1999.

100. Wilson, I.D., Multiple hyphenation of liquid chromatography with nuclear magnetic resonance spectroscopy, mass spectrometry and beyond, *J. Chromatogr. A*, 892, 315, 2000.

101. Smallcombe, S.H., Patt, S.L., and Keifer, P.A., WET solvent suppression and its applications to LC NMR and high-resolution NMR spectroscopy, *J. Magn. Reson. Ser. A*, 117, 295, 1995.

102. Wolfender, J.-L., Ndjoko, K., and Hostettmann, K., LC/NMR in natural products chemistry, *Curr. Org. Chem.*, 2, 575, 1998.

103. Wolfender, J.-L., Rodriguez, S., and Hostettmann, K., Liquid chromatography coupled to mass spectrometry and nuclear magnetic resonance spectroscopy for the screening of plant constituents, *J. Chromatogr. A*, 794, 299, 1998.

104. Wolfender, J.-L., Ndjoko, K., and Hostettmann, K., Application of LC–NMR in the structure elucidation of polyphenols, in *Methods in Polyphenol Analysis*, Santos-Buelga, C. and Williamson, G., Eds., Royal Society of Chemistry, Cambridge, 2003, chap. 6.

105. Queiroz, E.F. et al., On-line identification of the antifungal constituents of *Erythrina vogelii* by liquid chromatography with tandem mass spectrometry, ultraviolet absorbance detection and nuclear magnetic resonance spectrometry combined with liquid chromatographic micro-fractionation, *J. Chromatogr. A*, 972, 123, 2002.

106. Hansen, S.H. et al., High-performance liquid chromatography on-line coupled to high-field NMR and mass spectrometry for structure elucidation of constituents of *Hypericum perforatum* L., *Anal. Chem.*, 71, 5235, 1999.

107. Weston, A. and Brown, P.R., *HPLC and CE: Principles and Practice*, Academic Press, New York, 1997.

108. Tomas-Barberan, F.A., Capillary electrophoresis: a new technique in the analysis of plant secondary metabolites, *Phytochem. Anal.*, 6, 177, 1995.

109. Issaq, H., Capillary electrophoresis of natural products — II, *Electrophoresis*, 20, 3190, 1999.

110. Suntornsuk, L., Capillary electrophoresis of phytochemical substances, *J. Pharm. Biomed. Anal.*, 27, 679, 2002.

111. Marchart, E., Krenn, L., and Kopp, B., Quantification of the flavonoid glycosides in *Passiflora incarnata* by capillary electrophoresis, *Planta Med.*, 69, 452, 2003.

112. Urbanek, M. et al., On-line coupling of capillary isotachophoresis and capillary zone electrophoresis for the determination of flavonoids in methanolic extracts of *Hypericum perforatum* leaves or flowers, *J. Chromatogr. A*, 958, 261, 2002.

113. Ferrari, J. et al., Isolation and on-line LC/CD analysis of 3,8″-linked biflavonoids from *Gnidia involucrata*, *Helv. Chim. Acta*, 86, 2768, 2003.

2 Spectroscopic Techniques Applied to Flavonoids

Torgils Fossen and Øyvind M. Andersen

CONTENTS

2.1 INTRODUCTION

The purpose of this chapter is mainly to review the different spectroscopic techniques used for flavonoid analysis during the last decade. A typical analysis involving spectroscopic techniques embraces structural elucidation including determination of stereochemical attributes. However, it may also be aimed at tracing specific compounds and presenting quantitative aspects (see Chapter 1), or revealing color depiction. More than 7000 structures in various flavonoid classes have been reported in this book. Nearly half of them have been reported after 1993, which reflect that continual improvements in methods and instrumentation used for separation and structural elucidation have made it easier to use smaller flavonoid quantities to achieve results at increasing levels of precision. Recent attention regarding the variety of flavonoid structures (Chapters 10–17) and their potential properties (Chapters 4–9) has highlighted the need for understanding the physiological functions of individual flavonoids in plants and animals, and their importance to human health. Deciphering biological functions, including pharmaceutical functions, from structural flavonoid information is of increasing importance in our society.

From an analytical point of view, flavonoids may be grouped into various types of monomeric aglycones, bi-, tri-, and oligo-flavonoids including proanthocyanidins, C-alkylated flavonoids, flavonoids with different levels of O-methylation, and flavonoids with one or more saccharide units, which may include various types of acyl substituents (Chapters 10–17). The flavonoids under investigation may be part of complexes, may occur in complex matrices, and some flavonoids like the anthocyanins may exist on a variety of equilibrium forms. A successful characterization will thus follow a specific analytical route designed for the type of flavonoids under investigation, and the sort of information that is looked for. Without reference compounds the characterization of a novel compound will normally require more spectroscopic data than in the determination of a flavonoid that has been reported earlier. The amounts of flavonoids present in most plant tissues are relatively small even though the visual impression is quite striking. Methods for the characterization of individual flavonoids have traditionally reflected the lack of available material, and sensitive chromatographic and spectroscopic techniques have achieved prominence in the characterization of flavonoids.[1–3] Thus, the coupling of instruments performing chromatographic separations to those providing structural data (hyphenated methods), in particular high-performance liquid chromatography (HPLC) coupled to a diode-array detector, and a mass spectrometer or, more recently, a nuclear magnetic resonance (NMR) instrument, has had an enormous impact on structural studies involving flavonoids (see also Chapter 1).

Before a species is analyzed with respect to its flavonoid content, knowledge about earlier reports on the chemistry and flavonoid distribution within the genus and related species may be of value. The most exhaustive source for such information is *Chemical Abstracts*, and excellent reviews on structures and distribution of flavanoids have been compiled regularly.[4–12] Several reviews have recently addressed the general field of flavonoid analysis.[13–19] Among the earlier reviews in the field, we will particularly recommend consulting *Techniques of Flavonoid Identification* by Markham[2] and *Plant Phenolics* by Harborne.[3] References to review articles on specific spectroscopic techniques applied on flavonoids will be cited under the various spectroscopic methods covered in this chapter. Spectroscopic information of importance is also presented in several other chapters in this book.

In this chapter, examples of the usefulness and recent applications of the different spectroscopic techniques applied on various flavonoids will be presented. Developments in NMR instrumentation including higher fields, high-temperature superconducting probes, low-temperature coils, better radiofrequency technology, as well as improvements of techniques and computing power have made NMR spectroscopy (Section 2.2) the most important tool for structural elucidation of flavonoids when these compounds are isolated in the milligram scale. Special effort has been made in this chapter to present assigned ^{1}H and ^{13}C chemical shifts characteristic for the various flavonoid classes (Table 2.1–Table 2.6), and we present the first report of a 3D heteronuclear single-quantum coherence–total correlation spectroscopy (HSQC–TOCSY) spectrum applied to a flavonoid (Figure 2.6). Advances in mass spectrometry (MS) methodology have been shown to be extremely valuable for flavonoid analysis during the last two decades, especially the use of mild ionization techniques, which have improved the possibility of recording molecular ions and suppressed the detection limits by several orders of magnitude (Section 2.3). When flavonoid standards are not available, detailed structural information can be obtained by resorting to cone voltage fragmentation (by collision-induced dissociation (CID), tandem MS, etc.) and use of various types of mass analyzers.

Although vibrational spectroscopy (Section 2.4), infrared (IR) spectroscopy and Raman spectroscopy, is not routinely used in most flavonoids studies, the range of potential uses for these methods have been extended considerably by the development of microspectrometers with laser excitation, linked techniques, e.g., liquid chromatography (LC)–Fourier transform IR (FTIR), and two-dimensional (2D) correlation IR. Near-IR (NIR) spectroscopy has been shown to be an effective alternative method to conventional quantitative analysis of flavonoids in food, plant extracts, and pharmaceutical remedies. Absorption spectroscopy (ultraviolet, UV, or UV–Vis) (Section 2.5) will normally form part of any particular flavonoid analysis during the initial analytical stages; however, during the period of this review only minor advances in methodology were reported. In the flavonoid field, absorption spectroscopy provides most structural information about anthocyanins. Color measurements using CIE (Commission Internationale de l'Eclairage) specifications applied to pure anthocyanins and anthocyanins in plants and products derived thereof, determination of absolute configuration of flavonoid stereocenters by circular dichroism (CD) spectroscopy, and x-ray diffraction studies on solid flavonoid structures have been treated separately in Sections 2.6–2.8, respectively. Abbreviations are listed in Chapter 1.

2.2 NMR SPECTROSCOPY

2.2.1 INTRODUCTION

NMR spectroscopy is an extremely powerful analytical technique for the determination of flavonoid structures,[20–23] but it is limited by poor sensitivity, slow throughput, and difficulties in analysis of mixtures. Recent developments have, however, made NMR arguably the most important tool for complete structure elucidation of flavonoids. Today, it is possible to make complete assignments of all proton and carbon signals in NMR spectra of most flavonoids isolated in the low milligram range. These assignments are based on chemical shifts (δ) and coupling constants (J) observed in 1D ^{1}H and ^{13}C NMR spectra combined with correlations observed as crosspeaks in homo- and heteronuclear 2D NMR experiments. Other nuclei like ^{17}O NMR spectroscopy has been used to study flavonoids only in a few cases. Natural abundance ^{17}O NMR spectra have been recorded for 11 methoxyflavones,[24] and ^{17}O NMR data for some 3-arylidenechromanones and flavanones have recently been discussed in terms of mesomeric and steric substituent interactions.[25] ^{17}O NMR spectroscopy has also been used to study the effect of sugar on anthocyanin degradation and water mobility in a roselle anthocyanin model system.[26]

Advances in computing power have been an important factor for the success of more advanced NMR techniques. Running many scans and accumulating data may enhance weak signals since baseline noise, which is random, tends to cancel out. One of the main advantages to be gained from signal averaging combined with the use of Fourier transform methods and high-field superconducting magnets (up to 21 T) is the ability to routinely obtain ^{13}C NMR spectra. This carbon isotope exists in low abundance (1.108%) compared to the essentially 100% abundance of ^{1}H. NMR sensitivity also depends on magnetogyric constants, which for ^{13}C is only one-fourth of the value of ^{1}H. Thus, the sample amount required for ^{13}C NMR spectra is about ten times that for ^{1}H NMR spectra, and the number of scans is normally much higher. Progress in NMR instrumentation has been considerable during the last decade, and the recent development of cryogenic probe technology may further increase the signal-to-noise ratio by a factor of 3 to 4, as compared to conventional probes, leading to a possible reduction in experiment time of up to 16 or a reduction in required sample concentration by a factor of up to 4.[27] NMR instruments equipped with cryogenic probes have hitherto very rarely been used in structural elucidation of flavonoids.[28]

Excellent compilation of NMR data on individual flavonoids has previously been presented,[20–22] and some useful reviews in this field have also been published recently.[23,29,30] Based on a database containing 700 ^{13}C spectra of flavonoids obtained from the literature, pattern recognition has been used to assemble compatible substructures according to related spectra.[31] Some recent publications reporting flavonoid coupling constants include: NMR studies on flavones after the incorporation of ^{13}C at the carbonyl group, which allowed the measurement of two- and three-bond carbon–carbon coupling constants, ranging from 1.4 to 3.5 Hz, and the measurement of two-, three-, and four-bond carbon–hydrogen coupling constants, which ranged from 0.3 to 3.8 Hz;[32] complete assignment of the ^{1}H and ^{13}C NMR spectra of several flavones and their proton–proton and carbon–proton coupling constants, including the extreme seven-bond long-range coupling between H-7 and H-3 in 6-hydroxyflavone (0.52 Hz) and flavone (0.27 Hz).[33] Typical one-bond ^{1}H–^{13}C coupling constants of monosaccharides in anthocyanins have been observed within magnitudes of 125 and 175 Hz.[34]

Mainly during the last decade structural information regarding flavonoids associated with other molecules has been reported. High-resolution ^{1}H magic angle spinning (MAS) NMR spectroscopy has been used to investigate the structural basis for the antioxidative effects of five flavones and flavonols on the lipid molecules of cellular membranes.[35] A structural model of the solution complex between the flavonol kaempferol 7-neohesperidoside and a DNA dodecamer containing the *E. coli* wild-type lac promoter sequence (TATGTT) was obtained by simulated annealing for refinement based on distance constraints derived from nuclear Overhauser enhancement spectroscopy (NOESY) spectra.[36] The saturation transfer difference NMR technique has been used to investigate the binding of luteolin and its 7-*O*-β-D-glucopyranoside to a multi-drug-resistance transporter protein,[37] and NMR (TOCSY and NOESY spectra) has demonstrated molecular interaction of human salivary histatins with a flavan-3-ol ester, epigallocatechin gallate.[38] A review of NMR studies on the conformation of polyflavanoids and their association with proteins has been reported by Hemingway et al.[39]

The use of NMR spectroscopy in the chemical analysis of food and pharmaceutical products is very advantageous because it is nondestructive, selective, and capable of simultaneous detection of a great number of low-molecular-mass components in complex mixtures. Conventional 1D and 2D NMR and high-resolution diffusion-ordered spectroscopy have recently been used for the characterization of selected Port wine samples of different ages.[40] NMR analysis of anthocyanins and amino acids has been used to differentiate wines according to the vine variety, geographic origin, and year of production.[41] Multivariate statistical analysis of 2D NMR data of polyphenol extracts has been applied to differentiate grapevine cultivars and clones,[42] and a ^{1}H NMR method has been utilized for quality control analyses of *Ginkgo* constituents, including flavonols.[43] A very interesting metabolomic approach based on chemometric analysis of ^{1}H NMR spectra of blood plasma has been used to investigate metabolic changes following dietary intervention with soy isoflavones in healthy premenopausal women under controlled environmental conditions.[44] With respect to future *in vivo* studies, the production of ^{13}C-labeled anthocyanins in cell cultures is promising.[45]

In this chapter, NMR solvents useful for flavonoid analysis are presented (Section 2.2.2), followed by an overview of various NMR techniques, including improved 2D and 3D NMR techniques with potential in structural elucidation of flavonoids (Section 2.2.3). The application of solid-state NMR and the coupled technique LC–NMR as tools for flavonoid analysis have been highlighted in separate sections (Sections 2.2.4 and 2.2.5). Tabulated compilations of recently published NMR data (both chemical shifts and coupling constants) on flavonoids belonging to various subclasses, including individually assigned data for rotameric flavone *C*-glycoside conformers (Table 2.1 and Table 2.2), complete assignments of ^{1}H and ^{13}C chemical shifts of various flavonoid glycosyl (Table 2.3 and Table 2.4) and acyl (Table 2.5 and Table 2.6) moieties, reveal the importance of NMR for flavonoid structure elucidations. The names of the flavonoids listed in Table 2.1–Table 2.6 are collated in Table 2.7 and their structures are shown in Figure 2.8–Figure 2.16.

2.2.2 NMR SOLVENTS

The most frequently used NMR solvents for flavonoid analyses are hexadeuterodimethylsulfoxide (DMSO-*d*6) and tetradeuteromethanol (CD$_3$OD). Anthocyanins require the addition of an acid to ensure conversion to the flavylium form. For the analysis of relatively nonpolar flavonoids, solvents such as hexadeuteroacetone (acetone-*d*6), deuterochloroform (CDCl$_3$), carbontetrachloride (CCl$_4$), and pentadeuteropyridine have found some application. The choice of NMR solvent may depend on the solubility of the analyte, the temperature of the NMR experiments, solvent viscosity, and how easily the flavonoid can be recovered from the solvent after analysis. In recent years, the problem of overlap of solvent signals with key portions of the NMR spectrum has been reduced by solvent suppression and the application of improved 2D and 3D NMR techniques.

Most flavonoids in the milligram scale are dissolved relatively easily in DMSO-$d6$. The solvent peak (39.6 ppm) and the residual solvent signal (2.49 ppm) are used as secondary references for ^{13}C and ^{1}H, respectively. DMSO is viscous at room temperature (m.p. 18°C), and may be the solvent of choice for recording of optimum nuclear Overhauser effects (NOEs) in 2D NOESY experiments at room temperatures.[46] However, the relatively high melting point (m.p. 18°C) makes low-temperature studies in this solvent impossible. Alternatively, the low freezing point of CD$_3$OD (m.p. −98°C) implies that low-temperature experiments can be performed with this solvent. It is also easy to recover the flavonoid after NMR analysis by evaporating the solvent (b.p. 65°C); however, in contrast to DMSO the relatively low boiling point of CD$_3$OD limits the possibilities of high-temperature experiments like those performed to study the equilibrium of rotameric conformers of flavone C-glycosides.[47] The solubility of several types of flavonoids is more restricted in CD$_3$OD than in DMSO; however, CD$_3$OD combined with various proportions (2 to 20%) of deuterotrifluoroacetic acid, CF$_3$COOD, is at present the most common NMR solvent used for anthocyanins. However, the application of acidified NMR solvents may cause the exchange of exchangeable protons such as the methylene protons of malonyl residues and the anthocyanidin H-6 and H-8 with deuterium, thus also preventing the detection of ^{13}C correlations of these signals in heteronuclear correlation experiments.[23] Other disadvantages with acidified solvents include (1) flavonoid glycoside hydrolysis and (2) anthocyanins acylated with dicarboxylic acids may be esterified by the NMR solvent during analysis.[23] Recent NMR investigations and theoretical calculations have compared the effect of acetone-$d6$, DMSO-$d6$, and CDCl$_3$ on the conformations of 4′,7-di-hydroxy-8-prenylflavan.[48]

2.2.3 NMR Experiments

The purpose of a standard ^{1}H NMR experiment is to record chemical shifts, spin–spin couplings, and integration data, thus providing information about the relative number of hydrogen atoms. Applied to a flavonoid, this information may help in identifying the aglycone and acyl groups, the number of monosaccharides, and the anomeric configuration of the monosaccharides. However, for most flavonoids the information provided by a standard ^{1}H NMR experiment is insufficient for complete structural elucidation. Thus, ^{13}C NMR experiments (spin–echo Fourier transform, SEFT, compensated attached proton test, CAPT, etc.) combined with various 2D NMR experiments, especially those using gradient techniques that imply increased sensitivity, have to be used for assignments of all ^{1}H and ^{13}C NMR signals. For the NMR experiments described below, we recommend Braun et al.[49] for descriptions of the pulse programs and important acquisition and processing parameters. A protocol treating experimental details of modern NMR techniques for anthocyanin analysis has recently been published.[23]

Two-dimensional NMR spectra are mainly produced as contour maps. These maps may be best imagined as looking down on a forest where all the trees (representing peaks in the spectrum) have been chopped off at the same fixed height. Two-dimensional NMR spectra are produced by homonuclear and heteronuclear experiments. Homonuclear ^{1}H–^{1}H correlated NMR experiments like the double-quantum filtered COSY (^{1}H–^{1}H DQF-COSY) and ^{1}H–^{1}H TOCSY experiments generate NMR spectra in which ^{1}H chemical shifts along two axes are correlated with each other. Values on the diagonal in these spectra correspond to chemical shifts that would have been shown in a 1D ^{1}H NMR experiment. It is the off-diagonal "spots," called crosspeaks, which present information that is new. These crosspeaks arise from coupling interactions between different ^{1}H nuclei. A crosspeak observed above the diagonal will normally also be found below the diagonal, thus producing a nearly symmetrical spectrum. The 1D ^{1}H NMR spectra may be placed as projections along the top and left parts of the 2D NMR spectra. ^{1}H–^{1}H NOESY (Figure 2.3) and rotating frame Overhaüser effect spectroscopy (^{1}H–^{1}H ROESY) (Figure 2.4) experiments are other examples of homonuclear NMR experiments. Heteronuclear NMR experiments are represented by ^{1}H–^{13}C HSQC (Figure 2.1) and heteronuclear multiple bond correlation (^{1}H–^{13}C HMBC) (Figure 2.2) experiments. The ^{13}C NMR spectrum (or ^{13}C projection) is displayed along one axis and the ^{1}H NMR spectrum (or ^{1}H projection) along the other. The ^{1}H–^{13}C correlations are shown as crosspeaks in the spectrum. Contrary to homonuclear NMR experiments, there exists no diagonal and only one crosspeak is present for each correlation.

2.2.3.1 COSY and TOCSY

Two-dimensional ^{1}H–^{1}H COSY (correlation spectroscopy) experiments allow determination of the protons that are spin–spin coupled, and the spectrum shows couplings between neighboring protons ($^2J_{HH}$, $^3J_{HH}$, and $^4J_{HH}$) revealed as crosspeaks in the spectrum. The ^{1}H–^{1}H DQF-COSY experiment is a modification of the standard ^{1}H–^{1}H COSY experiment. The main advantage of the DQF technique is that noncoupled proton signals are eliminated. The DQF technique eliminates the strong solvent signal and the often very strong H_2O signal associated therewith, which may overlap with flavonoid sugar signals. The DQF-COSY experiment is routinely used in flavonoid analysis to assign all the sugar protons. The use of a "sequential walk" approach may provide information on the relative positions of individual proton signals along a spin system.

The 2D homonuclear ^{1}H–^{1}H TOCSY (Total Correlation SpectroscopY) experiment identifies protons belonging to the same spin system. As long as successive protons are coupled with coupling constants larger than 5 Hz, magnetization is transferred successively over up to five or six bonds. The presence of heteroatoms, such as oxygen, usually disrupts TOCSY transfer. Since each sugar ring contains a discrete spin system separated by oxygen, this experiment is especially useful for assignments of overlapped flavonoid sugar protons in the 1D ^{1}H NMR spectrum. It must be understood that the crosspeak intensity is not an indicator of the distance between the protons involved, and that all expected correlations may not appear in a TOCSY spectrum. To avoid this latter problem it may be helpful to record a second spectrum with another mixing time.

Most of the sugar proton signals found in flavonoids occurs in the narrow spectral region of 4.5 to 3.0 ppm. Thus, for complex flavonoids containing several sugar units, extensive overlap occurs in this part of the spectrum. ^{1}H–^{1}H sugar coupling constants for such compounds can, however, be accessible by using the selective 1D TOCSY experiment, also known as the HOHAHA (homonuclear Hartman–Hahn) experiment. In the 1D TOCSY experiment, the resonances of one proton are selected and the signal formed is transferred in a stepwise process to all protons that are *J*-coupled to this proton. Instead of crosspeaks, magnetization transfer is seen as increased multiplet intensity. Thus, this 1D TOCSY spectrum looks like a normal ^{1}H NMR spectrum including only the protons that belong to the same spin system as the chosen proton.

TOCSY experiments have together with other NMR experiments been used for the structural elucidation of flavonols from, for instance, the Indian spice *Mammea longifolia*,[50] from *Erythrina abyssinica*,[51] red onion,[52] *Polygonum viscosum*,[53] *Morina nepalensis* var. *alba*,[54] *Centaurium spicatum*,[55] *Pisum sativum* (cv Solara) shoots,[56] leaves of *Canthium dicoccum*,[57] pericarps of *Sophora japonica*,[58] flavonol and chalcone glycosides from *Bidens andicola*,[59] a new flavonol glycoside and a new pterocarpan glucoside from *Ononis vaginalis*,[60] and a flavone and three iridoids from *Stachys spinosa*.[61] TOCSY experiments have also been used for structural elucidation of anthocyanins produced in petals of genetically transformed lisianthus,[62] from blue berries of *Vaccinium padifolium*,[63] and flowers of *Ipomoea asarifolia*,[64,65] etc.

2.2.3.2 ^{1}H–^{13}C Heteronuclear NMR Experiments

The HSQC experiment is a 2D one that correlates ^{13}C nuclei with ^{1}H nuclei within a molecule by means of the one-bond coupling between them (Figure 2.1). In HSQC, the ^{1}H magnetization is directly detected and the ^{13}C magnetization is indirectly detected, unlike the HETCOR experiment, in which ^{1}H magnetization is indirectly detected (F_1) and converted to ^{13}C magnetization, which is directly detected (F_2). There are several advantages in the manner in which magnetization is detected in the HSQC experiment, including increased sensitivity (<1.0 mg flavonoid sample is sufficient) and the ability to see long-range interactions between ^{13}C and ^{1}H nuclei using a variant called HMBC (heteronuclear multiple bond correlation). After the assignment of proton signals by a combination of ^{1}H–^{1}H COSY and ^{1}H–^{1}H TOCSY experiments, the one-bond ^{1}H–^{13}C correlations observed in the HSQC spectrum allow the assignment of the corresponding ^{13}C signals. HSQC spectra have been recorded routinely for a variety of flavonoids during the last decade. A gradient-assisted heteronuclear multiple quantum correlation (GRASP-HMQC) optimized for the detection of $^1J_{CH}$ (147 Hz) and a long-range version thereof (GRASP-HMQC-LR) optimized for the detection of scalar coupling $^{2,3}J_{CH}$ (8, 5, or 3 Hz) used for the structural elucidation of an acylated flavonol glycoside[66] are alternatives to the HSQC and HMBC experiments.

The HMBC correlates proton nuclei with carbon nuclei that are separated by more than one bond (Figure 2.2). The experiment is normally optimized for $^3J_{CH}$ and $^2J_{CH}$ couplings; however, the intensity of the crosspeaks generated by this experiment

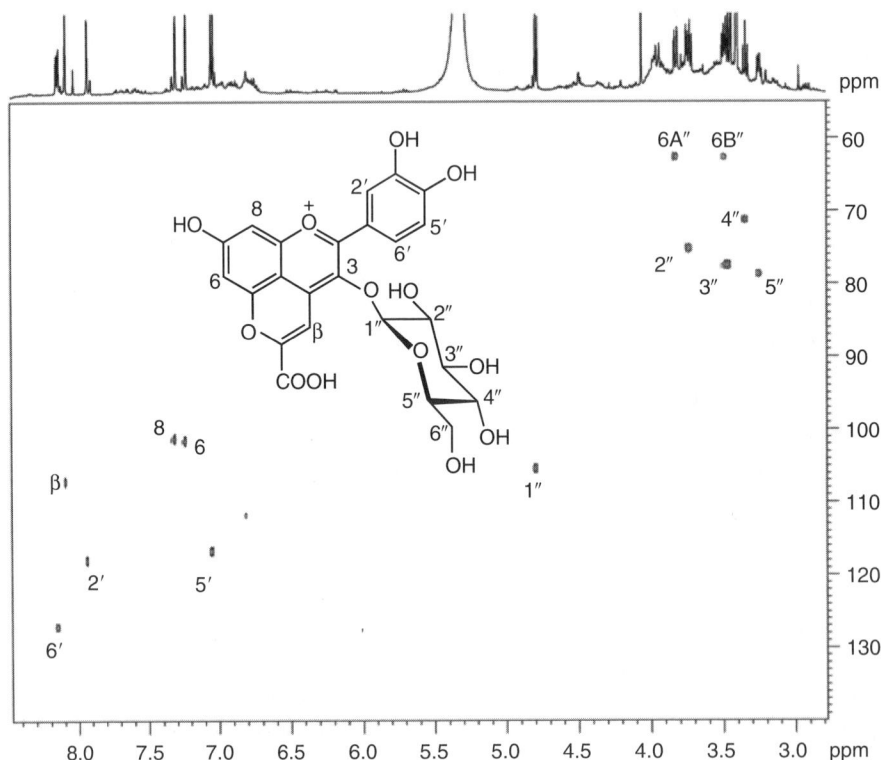

FIGURE 2.1 ^{1}H–^{13}C heteronuclear single-quantum coherence spectrum of 5-carboxypyranocyanidin 3-glucoside, showing one-bond ^{1}H–^{13}C correlation crosspeaks. The 1D ^{1}H NMR spectrum is included as a projection in the proton dimension.[176]

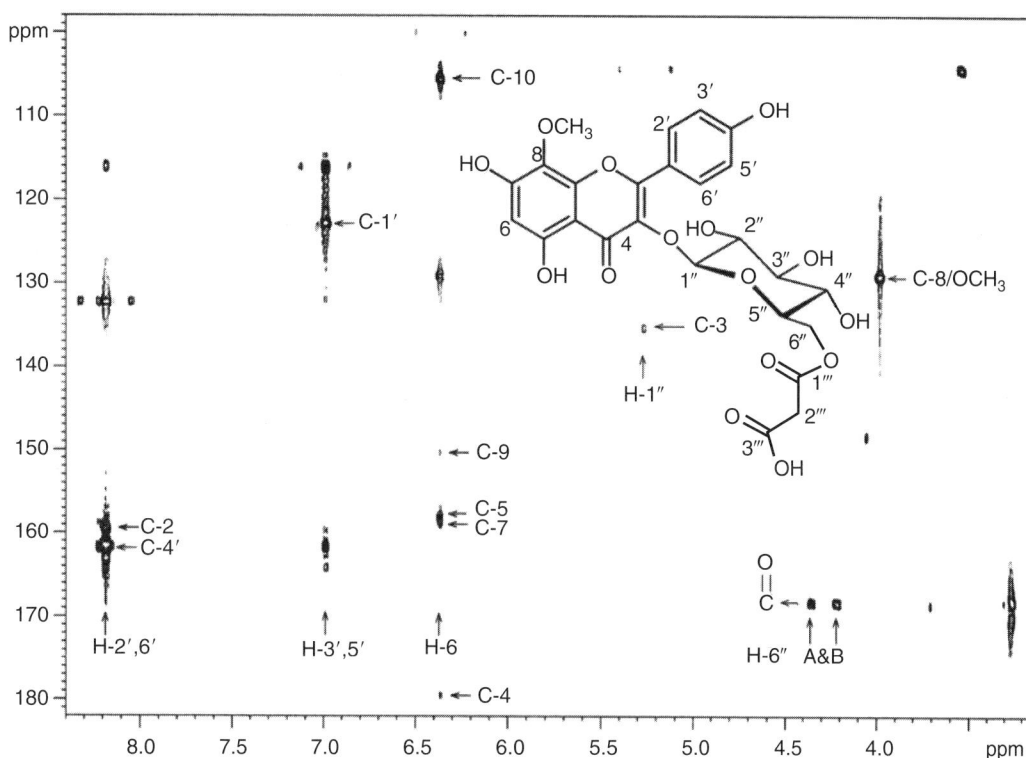

FIGURE 2.2 $^1H–^{13}C$ heteronuclear multiple bond correlation spectrum of 8-methoxykaempferol 3-O-(6″-malonyl-β-glucopyranoside) (in CD_3OD) showing multiple bond correlations important for determination of linkages between subunits and assignments of aglycone ^{13}C resonances, respectively. Horizontal arrows indicate carbon assignments, while vertical arrows show proton assignments.[176]

may sometimes be unexpected; some $^1J_{CH}$ couplings may occur as symmetrical doublets, while in rare cases crosspeaks caused by $^{n>3}J_{CH}$ couplings may be observed. In the aromatic region of the spectra some $^2J_{CH}$ couplings may be too small to be detected as crosspeaks.

Major applications of the HMBC experiment related to flavonoids include the assignment of resonances of nonprotonated carbon nuclei of the aglycone and potential acyl group(s) (Figure 2.2). Since long-range correlation of protonated carbon resonances also occurs with carbon nuclei that are separated by nonprotonated carbons or other heteronuclei like oxygen, the experiment is often used to determine the linkage points of the flavonoid building blocks (aglycone, sugar unit[s] and acyl moieties). The HMBC experiment has been successfully used for ^1H and ^{13}C NMR assignments and structural elucidations of a range of flavonoids including green tea flavonoids,[67] four new types of chalcone dimers isolated from *Myracrodruon urundeuva*,[68] diprenylated chalcones from the twigs of *Dorstenia barteri* var. *subtriangularis*,[69] flavonol and chalcone glycosides from *Bidens andicola*,[59] two unusual macrocyclic flavonols,[70] rare diastereoisomeric flavonolignans,[71] three epicatechin glucuronides isolated from plasma and urine after oral ingestion of epicatechin,[72] proanthocyanidin dimers,[73–75] a symmetrical glycosylated methylene bisflavonoid *Blutaparon portulacoides*,[76] several peracetylated proanthocyanidin trimers,[77,78] some flavones[79,80] and isoflavones,[81] 3-deoxyanthocyanins,[82] many anthocyanins,[65,83–87] some anthocyanin–flavonol complexes from flowers of chive,[88] etc.

The ^1H and ^{13}C NMR spectra of flavones and aurones are rather similar.[20,21] However, these flavonoid subgroups may be distinguished by the long-range correlations found in their HMBC spectra.[89] The ^1H and ^{13}C NMR data of the (E) and (Z) isomeric pair of 4,6,3′,4′-tetramethoxyaurone are included in Table 2.1 and Table 2.2. Gradient-enhanced (ge) HSQC and ge-HMBC studies of the flavonols quercetin and kaempferol and the flavone luteolin have been used to demonstrate that the strong intramolecular hydrogen bond of the −CO(4) and −OH(5) moieties persists over a wide range of aqueous solvent mixtures.[90] The use of reference deconvolution to suppress t_1 noise due to imperfections in spectrometer reproducibility has been described with particular emphasis on the use of HMBC applied on the proanthocyanidin, ent-epiafzelechin(4α → 8;2α → O → 7)-epicatechin.[91]

2.2.3.3 Improved Versions of the HMBC Experiment

One-bond correlations observed in the HMBC spectrum may, in some cases, provide useful structural information. On the other hand, incomplete suppression of the $^1J_{CH}$ correlations due to poor low-pass filter quality, which is often the case in the gradient-selected (gs) HMBC experiment, may complicate the spectrum considerably. Another problem associated with the gs-HMBC experiment is that the experiment may not be adjusted in a uniform manner to the wide range of multiple bond coupling constants (ranging from 1 to 25 Hz) of the compound. Thus, crosspeaks vital for the structural elucidation may be weak or totally absent from the spectrum.[92] To overcome these problems, improved HMBC experiments including 3D HMBC, in which the third dimension is used to scan the whole range of $^nJ_{CH}$ coupling constants,[93] and the 2D experiments ACCORD-HMBC,[94] IMPEACH-MBC,[95] and CIGAR-HMBC,[96] have been developed. Only the last of these has been applied to flavonoids.[97] Recently, a $^2J,^3J$-HMBC experiment has been developed that allowed differentiation between $^2J_{CH}$ and $^3J_{CH}$ correlations.[92,98]

2.2.3.4 Nuclear Overhauser Enhancement Spectroscopy

Protons that are close to each other in space may be observed as crosspeaks in a NOESY spectrum (Figure 2.3). Thus, the NOESY experiment proves to be a more sensitive alternative technique to HMBC for the determination of some linkages within a flavonoid. For instance, when sugars are attached to the 5 and 3′ hydroxyls of the aglycone, the anomeric protons show crosspeaks to H-6 and H-2′, respectively, while the anomeric proton of a sugar attached to the 7 position will exhibit crosspeaks to both H-6 and H-8. NOESY spectra have been used for unequivocal assignments of methoxyl group resonances and observation of restricted rotation of the B-ring of 2′-methoxyflavones,[99] and for the structural elucidation of two unusual macrocyclic flavonols.[70] Crosspeaks observed in a NOESY spectrum may reveal both intra- and intermolecular distances between flavonoid protons.[100] This type of information has been used for depiction of association mechanisms involving anthocyanins,[101–104] and to show that some synthesized 8-C-glucosylflavones in DMSO-$d6$ adopted conformations in which the H-2″ and H-4″ protons in the glucose moiety were oriented toward the B-ring in the flavone structure.[105] Based on NMR data and molecular dynamic simulations, a folded conformation was elucidated for the flavonol kaempferol 3-O-(2″(6‴-p-coumaroylglucosyl)-rhamnoside) in solution, implying a hydrophobic interaction between the aromatic nuclei of the aglycone and the acyl group.[106]

The NOESY experiment has also been very useful for revealing the presence of rotational conformers of dimeric flavonoids and flavone C-glycosides (Figure 2.3).[107–109] Strong exchange crosspeaks between equivalent protons of each conformer revealed the rotational equilibriums. This NOE phenomenon was first noted by Hatano et al.[110] in two conformers of procyanidin dimers.[110] The volume of the NOESY crosspeaks is related to the distance

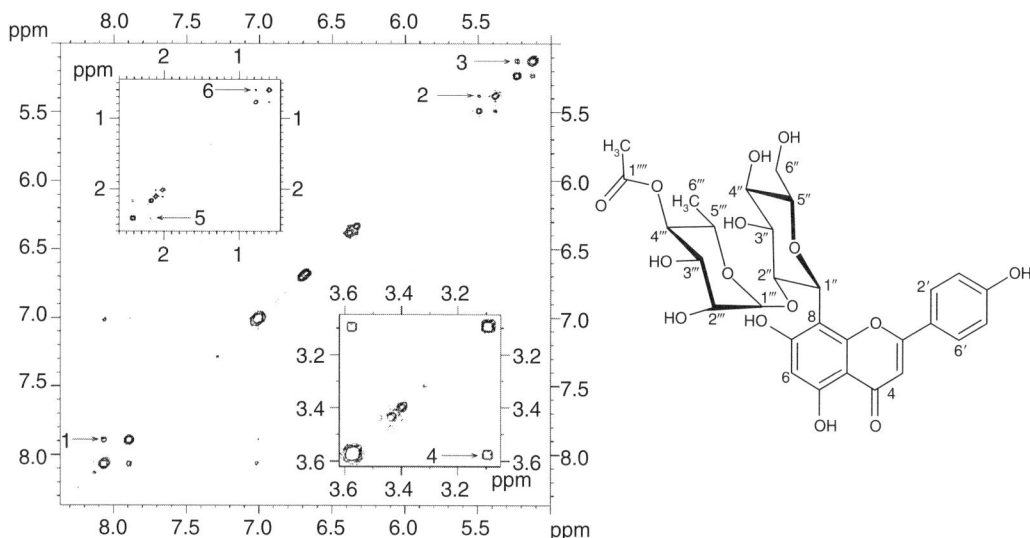

FIGURE 2.3 Expanded region of a NOESY spectrum of kaempferol 8-C-(2″-(4‴-acetylrhamnosyl)-glucoside) recorded in CD_3OD showing the exchange crosspeaks between equivalent protons of two rotameric conformers (denoted as A and B) due to rotational equilibrium. Arrows labeled by 1 to 6 show the exchange crosspeaks between each pair of the equivalent protons of H-2′,6′A/H-2′,6′B; H-1‴A/H-1‴B; H-1″A/H-1″B; H-3‴A/H-3‴B; H-5‴A/H-5‴B; and H-6‴A/H-6‴B, respectively. The relative proportions of A and B were 1.00:0.58 when dissolved in CD_3OD.[176]

between the nuclei, and 3D distance information can be estimated based on the integration of the volume of the crosspeaks. Here, crosspeaks corresponding to two protons with a known distance are used as references for the calculation of distance. Thus, intermolecular association of two anthocyanin molecules[101] and a structural model of the solution complex between a flavonol and a DNA dodecamer have been evidenced by NOESY NMR experiments and distance geometry calculations.[36]

2.2.3.5 Rotating Frame Overhauser Effect Spectroscopy

Similar to the ^{1}H–^{1}H NOESY experiment, the ^{1}H–^{1}H ROESY experiment is useful for the determination of signals arising from protons that are close in space but not necessarily connected by chemical bonds. A ROESY spectrum yields through-space correlations via the rotating frame (also called rotational nuclear) Overhauser Effect (ROE). When one multiplet is irradiated, the intensities of multiplets arising from nearby nuclei are affected. The ROESY experiment (Figure 2.4) is especially useful for cases where the NOESY signals are weak because they are near the transition between negative and positive, which may be the case for flavonoids with masses of 800 to 2000 amu. The ROESY crosspeaks are negative; however, the ROESY (and NOESY) experiment may also yield crosspeaks arising from chemical exchange. These exchange peaks are always positive.

The ROESY experiment has together with other NMR experiments been used for structural elucidation of several new flavonoids including a pentaflavonoid, ochnachalcone, isolated from the stem bark of *Ochna calodendron*,[111] some flavones,[79,80] some isoflavones from soybeans,[112] some flavonols from red onion,[52] a flavonol glycoside and a pterocarpan glucoside from *Ononis vaginalis*,[60] a dimeric proanthocyanidin,[113] some triacylated anthocyanins from *Ajuga reptans* flowers and cell cultures,[84] a rare anthocyanin from blue flowers of

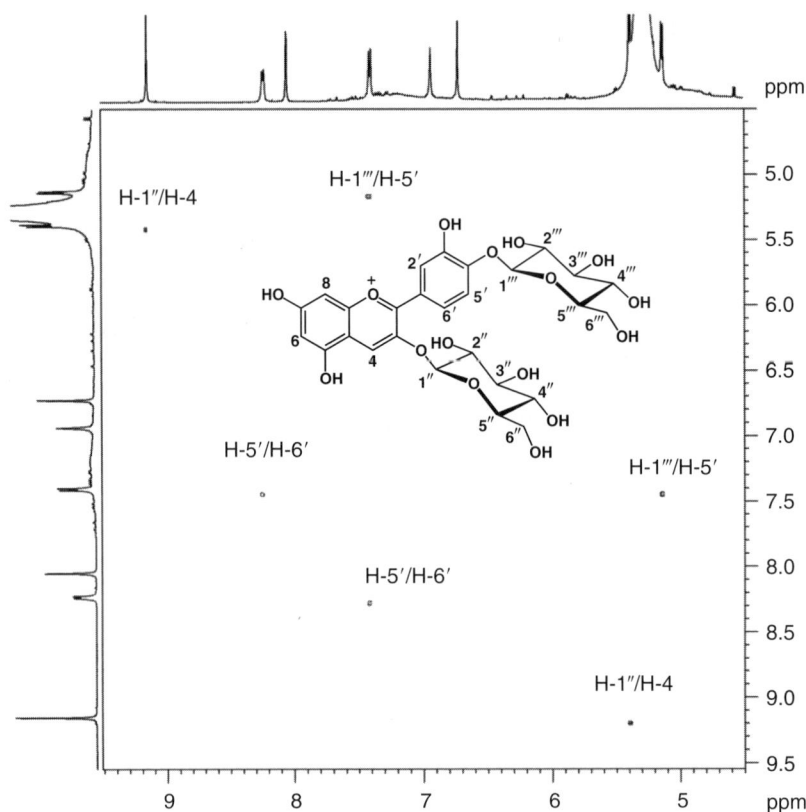

FIGURE 2.4 Expanded region of the ^{1}H–^{1}H ROESY spectrum of cyanidin 3,4′-diglucoside, recorded in CD$_3$OD–CF$_3$COOD (95:5, v/v) showing through-space correlations, which are important for determination of the linkages between the aglycone and the sugar units. Projections of the 1D ^{1}H NMR spectrum are included in both dimensions.[582]

Nymphaéa caerulea,[114] for seven natural anthocyanins stabilizing a DNA triplex,[115] etc. Sequential analysis of the oligosaccharide structures of the flavonol tamarixetin-7-*O*-rutinoside has been performed by 1D multistep-relayed COSY–ROESY experiments.[116] Selective excitation was performed by Gaussian-shaped soft pulses.

The ROESY experiment has often been used for revealing stereochemical aspects. The relative stereochemistry of two new biflavonoids isolated from the leaves of *Calycopteris floribunda* was deduced through NOE and ROESY experiments, comparative CD experiments, and optical rotation evaluations.[117] A ROESY experiment was needed to confirm the H-2, H-3-*trans* relationship of the C- and F-rings of a procyanidin trimer, since these protons showed some rather unexpected coupling constants.[77] The experiment has recently been used in the room-temperature conformational analysis of a biflavanoid and a polyhydroxylated flavanone-chromone isolated from *Cratoxylum neriifolium*.[118] Both compounds showed rotameric behavior due to the presence of a single bond between the highly substituted flavanone and flavanonol parts and the flavanone and chromone parts, respectively. Transverse-ROESY experiments in conjunction with theoretical (MM2) calculations have been used to support the proposal that the two rotamers of spinosin, a flavone *C*-glycoside, interchanged via rotation about the C6–C1″ bond.[119] The structure of β-cyclodextrin (β-CD) inclusion complexes of naringin, naringin dihydrochalcone, and the aglycone of naringin dihydrochalcones has been determined from 2D ROESY and 1D ROE experiments.[120] A quenched molecular dynamics (QMD)-ROSY study of a series of semibiosynthetically monoacylated anthocyanins produced in tissue cultures of *Daucus carota* has recently led to the identification of families of conformers of these flexible molecules that are of interest in work toward determining the mechanism for stabilization of color among these compounds in solution.[121]

2.2.3.6 Two- and Three-Dimensional HSQC-TOCSY, HSQC-NOESY, HSQC-ROESY, HMQC-NOESY, HMQC-ROESY

In recent years, several new 2D and 3D NMR techniques with significant potential in structural elucidation of flavonoids have been evolved.[122] Taking advantage of the excellent resolution in the ^{13}C dimension, interpretable NOESY, ROESY, or TOCSY data can be achieved from spectral regions with considerable overlap in the ordinary 2D NOESY, ROESY, or TOCSY spectra, respectively. The ^{13}C chemical shift values of the individual monosaccharides in polyglycosylated flavonoids can, for instance, be identified and assigned by the 2D and 3D HSQC-TOCSY experiments due to the fact that each anomeric proton exhibits crosspeaks to all ^{13}C resonances in the same spin system. The 2D HSQC-TOCSY, HSQC-NOESY, HSQC-ROESY, HMQC-NOESY, and HMQC-ROESY experiments are relatively sensitive, being only slightly less sensitive than the corresponding 2D HSQC and HMQC experiments, respectively. Despite their potential, as far as we know only 2D HSQC-TOCSY[118,123,124] has, among these techniques, been applied in the structural elucidation of flavonoids. The sugar region of the 2D ^{1}H–^{13}C gs-HSQC-TOCSY spectrum of malvidin 3-*O*-(6″-*O*-α-rhamnopyranosyl-β-glucopyranoside)-5-*O*-β-glucopyranoside, showing crosspeaks of most of the sugar signals, is presented in Figure 2.5. Here, we present the first report of a 3D HSQC-TOCSY spectrum applied to a flavonoid (Figure 2.6). The spectrum of an anthocyanin, pelargonidin 3-(6″-(4‴″-*E*-*p*-coumaroyl-α-rhamnosyl)-β-glucoside)-5-β-glucoside, is recorded in CD$_3$OD–CF$_3$COOD (95:5; v/v). To visualize the potential of this experiment, a 2D plane derived from the 3D HSQC-TOCSY spectrum shows the ^{1}H–^{13}C HSQC spectrum of all signals exhibiting TOCSY correlations to the 3-glucosyl anomeric proton, i.e., a ^{1}H–^{13}C HSQC spectrum representing no more than the correlations of the 3-glucosyl unit.

2.2.4 SOLID-STATE NMR

Solid-state NMR is a fast and nondestructive method for identification of flavonoids in the solid state, provided that a sufficient amount of the sample is available (ca. 10 mg or more).[125] The development of high-resolution ^{1}H MAS NMR spectroscopy has had a substantial impact on the ability of researchers to analyze intact tissues. Rapid spinning of the sample relative to the applied magnetic field serves to reduce line broadening. Thus, it is possible to obtain very high quality NMR spectra of whole tissue samples with no sample pretreatment. Geometric data on solids, including determination of hydrogen positions, are usually obtained by x-ray analysis of crystals. However, according to the Cambridge Structural Database, x-ray data are available only for a limited number of flavonoids, most probably due to difficulties with growing single flavonoid crystals. Solid-state ^{13}C NMR data of powder samples have been used to characterize the solid-state conformation of the flavones chrysin, apigenin, luteolin, and acacetin, the flavanones naringenin and hesperetin, and the flavonols galangin, kaempferol, quercetin, and myricetin (Figure 2.7).[125,126] Based on cross-polarization magic angle spinning (CP-MAS) solid-state ^{13}C NMR spectra, it was found that the locked conformations of the OH and OCH$_3$ groups in the solids resulted in increased shielding of carbon atoms adjacent to C–OH or C–OCH$_3$ hydrogens, thus enabling the determination of the orientations of these groups. The C5–OH (and C3–OH) hydroxyl

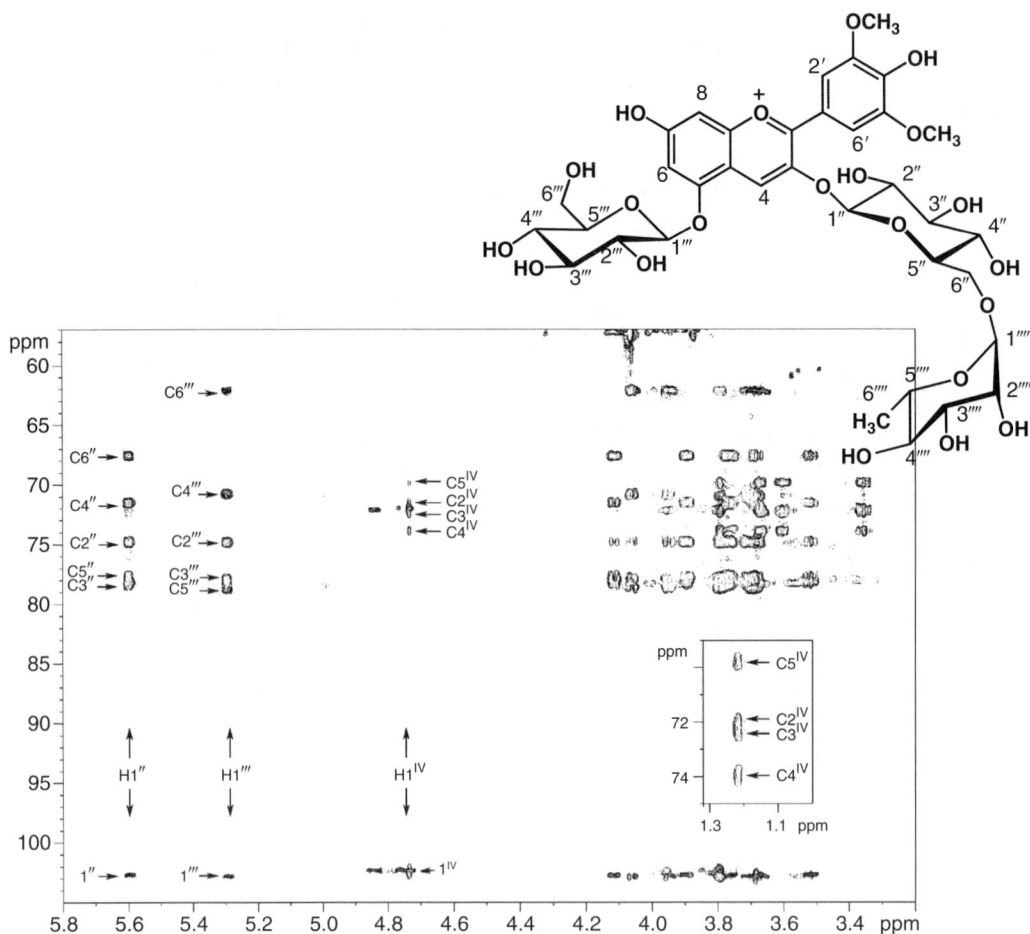

FIGURE 2.5 Region of the 2D ^{1}H–^{13}C gs-HSQC-TOCSY spectrum of malvidin 3-*O*-(6″-*O*-α-rhamno-pyranosyl-β-glucopyranoside)-5-*O*-β-glucopyranoside showing crosspeaks of most of the sugar signals. In this spectrum, each proton/carbon correlation shows crosspeaks to all ^{1}H and ^{13}C resonances belonging to the same spin system. The expanded region (boxed) reveals the correlation between H-6^{IV} and the carbons of the terminal rhamnose unit.[176]

pointed toward oxygen in the carbonyl group, forming intramolecular hydrogen bonds. Considerations of the C2′ and C6′ carbon shielding suggested that the flavonol B-ring was not coplanar with the benzopyran fragment in kaempferol, acacetin, and myricetin. More recently, assignments of ^{13}C CP-MAS NMR spectra of green and black tea, respectively, based on model compound spectra and literature data on various compounds including flavonoids, have made it possible to differentiate between commercial samples of these tea types.[127] Solid-state ^{13}C NMR and x-ray studies have also been applied on the complexes between the isoflavone genistein and various amines.[128]

2.2.5 LIQUID CHROMATOGRAPHY–NMR

The coupling of chromatographic techniques such as HPLC with NMR, LC–NMR can, in principle, provide the molecular structures of compounds in mixtures (extracts) in just one online experiment. The use of LC–NMR in the flavonoid field has been reviewed by

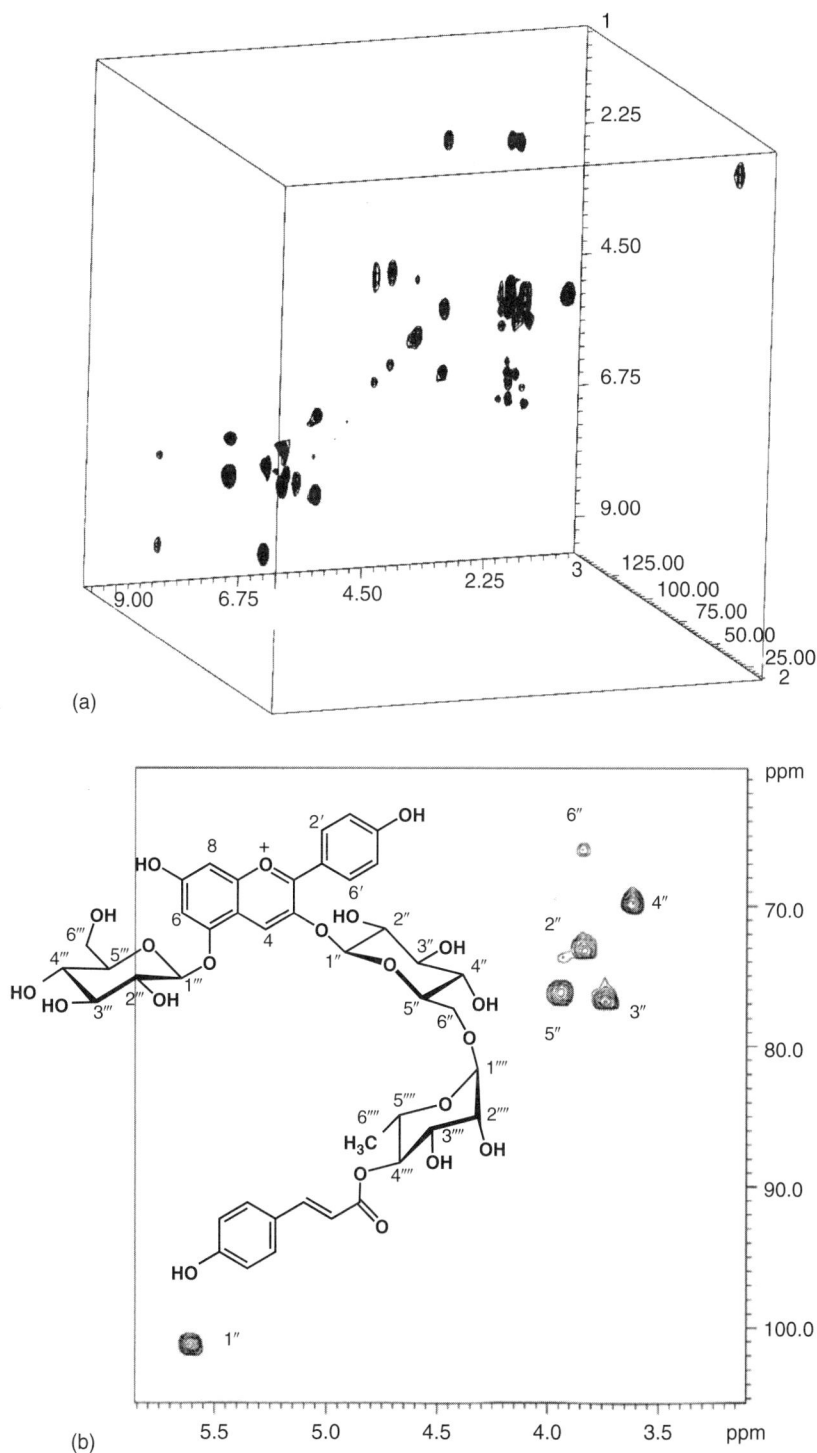

FIGURE 2.6 (a) 3D HSQC-TOCSY spectrum of pelargonidin 3-(6″-(4⁗-*E*-*p*-coumaroyl-α-rhamnosyl)-β-glucoside)-5-β-glucoside recorded in CD_3OD–CF_3COOD (95:5; v/v) at 298 K. (b) 2D plane derived from the 3D HSQC-TOCSY spectrum showing the ^1H–^{13}C HSQC spectrum of all signals exhibiting TOCSY correlations to the 3-glucosyl anomeric proton, i.e., the signals belonging to the 3-glucosyl unit.[583]

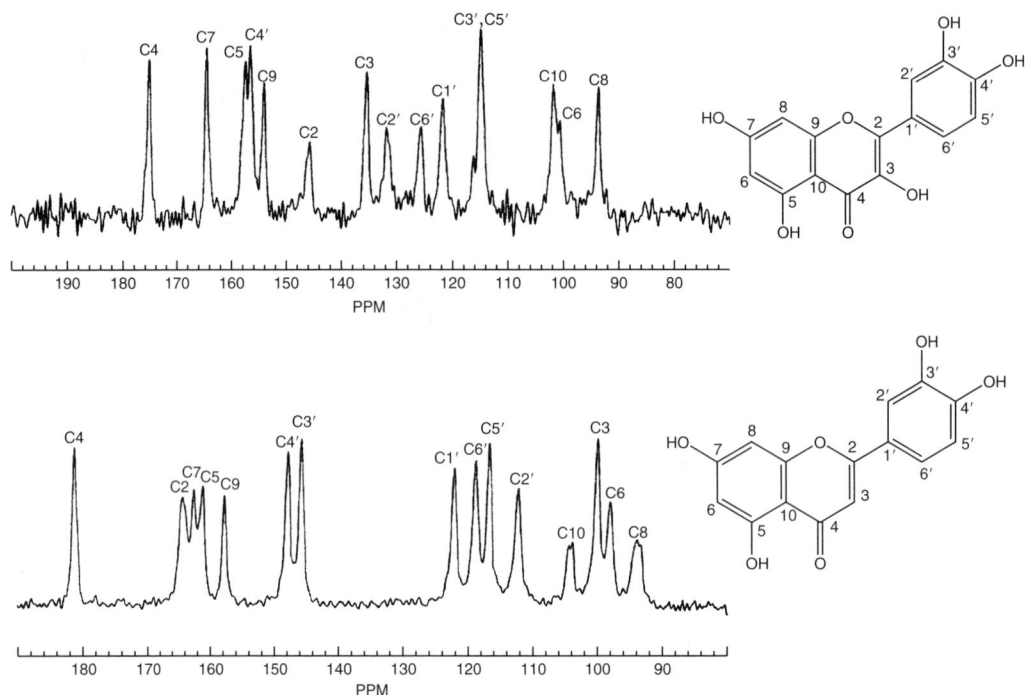

FIGURE 2.7 ^{13}C CP/MAS NMR spectra of the flavonol kaempferol (top) and the flavone luteolin (bottom). (Reprinted from Wawer, I. and Zielinska, A., *Magn. Reson. Chem.*, 39, 374, 2001. Copyright 2001 John Wiley & Sons, Ltd. With permission.)

Wolfender et al.,[129–132] and this coupled technique is treated more comprehensively in Chapter 1. The LC–NMR technique is by nature rather insensitive; however, high-field magnets and recent improvements in solvent suppression, pulse field gradients, and probe technology have made it possible to achieve useful results on various flavonoid structures.[28,133–141] The detection limit with a 60 μl cell in a 500 MHz instrument for a compound with a molecular weight of around 400 amu may typically be around 20 μg, and the information provided is hitherto mainly based on ^{1}H NMR spectra or [1]H–[1]H correlation experiments. However, a more recent system seems to be promising, affording increased sensitivity and thus shorter NMR acquisition time compared to conventional LC–NMR systems:[28] After LC separation of flavonoids in Greek oregano using ordinary nondeuterated solvents, solid-phase extraction was used for peak storage prior to the NMR analysis, and fully deuterated solvents were then used for flushing the trapped compounds into the NMR probe. Thus, the application of expensive deuterated solvents during HPLC separation and solvent suppression during NMR analysis are no longer necessary. Increased sensitivity was achieved using a newly developed cryogenic flow probe. Combining the data from the UV, MS, and NMR spectra, the flavonoids taxifolin, aromadendrin, eriodictyol, naringenin, and apigenin were identified.

2.2.6 NMR DATA ON FLAVONOID CLASSES

Assigned ^{1}H and ^{13}C chemical shifts characteristic for the various flavonoid classes are shown in Table 2.1–Table to 2.6. The names of the flavonoids listed in Table 2.1–Table 2.6 are given in Table 2.7. The structures of the flavonoids listed in Table 2.1–Table 2.7 are shown in Figure 2.8–Figure 2.16.

TABLE 2.1
¹H NMR Spectral Data of Flavonoids Recorded in Various Solvents*

No.	Flavones	3	5	6	7	8	2'	3'	4'	5'	6'	Ref.
1	6-OH-Luteolin 4'-OMe-7-(2''-α-rha-6''-ac-β-glc) (D)	6.76 s	12.69 s OH			6.94 s	7.44 d 2.3		3.88 s OMe	7.11 d 8.7	7.54 dd 8.6, 2.3	142
2	6-OH-Luteolin 7-(6''-(E)-caf)-β-glc (D)	6.61 s	12.74 br s OH			6.91 s	7.38 d 2.3			6.83 d 8.8	7.38 dd 8.8, 2.3	142
3	Isoscutellarein 7-(2''-(6''-ac)-β-all)-β-glc (D)	6.80 s	12.33 s OH	6.70 s			7.97 'd' 8.9	6.95 'd' 8.9		6.95 'd' 8.9	7.97 'd' 8.9	142
4	Isoscutellarein 4'-OMe-7-(2''-(6''-ac)-β-all)-β-glc (D)	6.89 s	12.28 s OH	6.71 s			8.09 'd' 8.9	7.14 'd' 8.9	3.88 s OMe	7.14 'd' 8.9	8.09 'd' 8.9	142
5	Apigenin 4'-(2''-(2'''-fer-glu)-glu) (M)	6.87 s		6.19 d 1.9		6.51 d 1.9	7.98 'd' 8.7	7.10 'd' 8.7		7.10 'd' 8.7	7.98 'd' 8.7	143
6	Apigenin 7-glu-4'-(2''-(2'''-fer-glu)-glu) (M)	6.61 s		6.48 d 2.0		6.79 d 2.0	7.84 'd' 9.0	7.15 'd' 9.0		7.15 'd' 9.0	7.84 'd' 9.0	143
7	Apigenin 7-glu-4'-(2''-(2'''-E-cou-glu)-glu) (M)	6.61 s		6.46 d 2.0		6.78 d 2.0	7.83 'd' 9.0	7.14 'd' 9.0		7.14 'd' 9.0	7.83 'd' 9.0	143
8	Luteolin 3'-β-glc-4'-(2''-α-rha-β-glc) (M)	6.71 s		6.23 d 1.3		6.52 d 1.3	7.98 d 1.7			7.34 d 8.0	7.70 dd 8.0, 1.7	144
9	Luteolin 3',4'-di-β-glc (M)	6.80 s		6.13 d 1.5		6.46 d 1.5	7.93 d 1.7			7.30 d 8.0	7.67 dd 8.0, 1.7	144
10	5,7,4'-tri-OH-3'-OMe-Flavone 8-C-(2''-O-β-glc-β-xyl) (M)	6.53 s		6.22 s		C-(2''-O-β-glc-β-xyl)	7.50 d 1.2	3.96 OMe		6.88 d 7.2	7.42 dd 1.2, 7.2	145
11	5,7-di-OH-3'-OMe-4'-Acetoxyflavone 8-C-(2''-O-β-glc-β-xyl) (M)	6.54 s		6.18 s		C-(2''-O-β-glc-β-xyl)	7.62 d 1.4	4.00 OMe	1.94 acetoxy	6.92 d 8.4	7.47 dd 1.4, 8.4	145
12	Iso-orientin 3'-OMe (D)	6.55 s		C-glc		6.91 s	7.56 s	3.90 s OMe		6.96 d 8.6	7.57 d 8.6	146
13	8-C-p-OH-Benz-isovitexin 4'-glc (D)	6.92 s		C-glc		C-p-OH-benz 2/6ᵃ	7.95 'd' 6.8	7.19 'd' 6.8		7.19 'd' 6.8	7.95 'd' 6.8	146
14	Apigenin 8-C-(2''-(4'''-ac-rha)-glc)† (M)	6.68 s, 6.72 s		6.39 s, 6.34 s			8.07 'd' 9.0, 7.89 'd' 9.0	7.02 'd' 9.0, 7.00 'd' 9.0		7.02 'd' 9.0, 7.00 'd' 9.0	8.07 dd 9.0, 7.89 'd' 9.0	109
15	Spinosin (D)	6.83, 6.84 s			3.89 s OMe	6.67, 6.80 s	7.97 'd' 8.7	6.95 'd' 8.7		6.95 'd' 8.7	7.97 'd' 8.7	147
16	6'''-Fer-spinosin (D)	6.85, 6.83 s			3.84 s OMe	6.77, 6.67 s	7.81 'd' 7.9	6.89 'd' 7.9		6.89 'd' 7.9	7.81 'd' 7.9	147
17	Isoscoparin 7-glc (D)	6.58 s				6.47 s	7.56 br s	3.90 s OMe		7.38 br s	6.94 m	47
18	Carlinoside (D)	6.63 s					7.45 s			6.90 d 8.0	7.56 br s	47
Flavonols												
19	Kaempferol 3-(6''-α-ara-glc) (D)			6.14 d 1.9		6.34 d 1.9	8.01 'd' 8.8	6.89 'd' 8.8		6.89 'd' 8.8	8.01 'd' 8.8	148
20	Kaempferol 3-(6''-α-ara-glc)-7-glc (D)			6.44 br s		6.76 br s	8.04 'd' 8.8	6.88 'd' 8.8		6.88 'd' 8.8	8.04 'd' 8.8	148

continued

TABLE 2.1
^{1}H NMR Spectral Data of Flavonoids Recorded in Various Solvents* — *continued*

No.	Flavones	3	5	6	7	8	2'	3'	4'	5'	6'	Ref.
21	Kaempferol 3-(2''-rha-6''-mal-glc) (M)			6.18 d 2.1		6.38 d 2.1	7.99 'd' 8.9	6.88 'd' 8.9		6.88 'd' 8.9	7.99 'd' 8.9	149
22	Kaempferol 3-glc-7-(2''-(6'''-p-cou-glc-glc)) (M)			6.31 d 1.8		6.33 d 1.8	8.07 'd' 8.0	6.91 'd' 8.0		6.91 'd' 8.0	8.07 'd' 8.0	150
23	8-OMe-Kaempferol 3-(6''-mal-glc) (M)			6.37 s		3.99 s OMe	8.18 'd' 8.9	6.99 'd' 8.9		6.99 'd' 8.9	8.18 'd' 8.9	109
24	Quercetin (M)			6.27		6.47	7.82			6.97	7.72	52
25	Quercetin 4'-glc (M)			6.25		6.44	7.81			7.35	7.75	52
26	Quercetin 3'-xyl (M)			6.27		6.47	8.17			7.06	7.91	151
28	Quercetin 3,4'-diglc (D)			6.14		6.35	7.64			7.20	7.57	52
29	Quercetin 3,7,4'-triglc (D)			6.43		6.80	7.67			7.21	7.62	52
30	Isorhamnetin 3-rut (M)			6.28		6.48	8.04			7.00	7.72	153
31	Isorhamnetin 3,7-diglc (M)			6.48 d 2		6.77 d 2	7.93 d 2	4.04 OMe		6.88 d 8	7.62 dd 8, 2	154
32	Myricetin 3-(2''-rha-glc) (M)			6.17 d 2.0		6.35 d 2.0	7.23 s	3.93 s OMe			7.23 s	149
33	Myricetin 3'-(6''-p-cou-glc) (P)			6.70 d 2		6.81 d 2	8.44 d 2				8.52 d 2	70
34	Myricetin 7-(6''-gall-glc) (D)			6.42 d 2.5		6.72 d 2.5	7.28 s				7.28 s	155
27	Myricetin 3-(2''-ac-rha) (M)			6.32		6.48	7.08				7.08	152
35	Laricitrin 3-α-ara-furanoside (D)			6.21 d 2.0		6.43 d 2.0	7.31 d 1.9	3.85 s OMe		OH	7.17 d 1.9	156
36	Laricitrin 3-glc (M)			6.27 d		6.45 d	7.62 d	4.02 OMe			7.38	153
37	Syringetin 3-(5''-glc-α-ara-furanoside) (D)			6.22 d 2.0		6.50 d 2.0	7.34 s	3.86 s OMe		3.86 s OMe	7.34 s	156
38	Syringetin 3-(6''-ac-glc) (M)			6.24 d		6.44 d	7.61 s	4.03 s OMe		4.03 s OMe	7.61 s	153
39	Syringetin 3-(6''-rha-gal) (M)			6.23 d 2.0		6.46 d 2.0	7.60 s	3.95 s OMe		3.95 s OMe	7.60 s	157
40	Syringetin 6-C-glc (D)	9.51 s OH	13.02 s OH		10.56 s OH	6.53 s	7.48 s	3.83 s OMe	9.18 s OH	3.83 s OMe	7.48 s	158

No.	Aurones	α	4	5	6	7	2'	3'	4'	5'	6'	Ref.
41	6,3'-di-OH-4,4'-di-OMe-5-Me-Aurone (D)	6.59 s	4.05 s OMe	1.98 s Me		6.61 s	7.48 d 1.5		3.83 s OMe	7.02 d 8.4	7.31 dd 1.5, 8.4	159
42	4,6,3',4'-tetra-OMe-Aurone (Z-form) (D)	6.69 s	3.92 s OMe	6.34 d 1.7	3.89 s OMe	6.72 d 1.7	7.55 d 1.7	3.83 s OMe	3.82 s OMe	7.07 d 7.9	7.54 dd 1.7, 8.0	159
43	4,6,3',4'-tetra-OMe-Aurone (E-form) (D)	7.00 s	3.89 s OMe	6.29 d 1.7	3.88 s OMe	6.51 d 1.7	8.48 d 1.8	3.84 s OMe	3.82 s OMe	7.01 d 7.9	7.54 dd 1.7, 8.0	159
44	6,3',4'-tri-OH-4-OMe-5-Me-Aurone (D)	6.54 s	4.03 s OMe	1.90 s Me		6.54 s	7.42 d 2.0			6.82 d 6.9	7.19 dd 2.0, 6.9	160
45	Maesopsin (P)	3.06 s		5.76 d 1.5	5.73 d 1.5		7.00 'd' 8.4	6.57 'd' 8.4		6.57 'd' 8.4	7.00 'd' 8.4	161

No.	Compound											Ref
46	Maesopsin 6-O-glc (two diastereoisomers) (P)	3.08 s	5.95 br s	6.05 br s		6.98 'd' 8.3	6.57 'd' 8.3, 6.56 'd' 8.3			6.57 'd' 8.3, 6.56 'd' 8.3	6.98 'd' 8.3	161
47[b]	Licoagroaurone (A)	7.46 d 8.5	6.83 d 8.5				7.61 d 2.5			6.96 d 8.5	7.42 dd 8.5, 2.5	89
	Chalcones and dihydrochalcones	2	3	4	5	6	2'	3'	4'	5'	6'	
48[c]	3'-formyl-4',6'di-OH-2'-OMe-5-Me-Chalcone (C)	7.43 m	7.53 m	7.53 m	7.53 m	7.43 m	3.90 s OMe	10.10 s CHO	12.60 s OH	2.08 s Me	13.90 s OH	162
49[d]	Chalcononaringenin 2',4'-diglc (P)	7.96 'd' 8.6	7.06 'd' 8.6		7.06 'd' 8.6	7.96 'd' 8.6		7.19 d 2.1		6.66 d 2.1		163
50[e]	2',4'-di-OH-4-OMe-6'-glc Dihydrochalcone (D)	7.05 'd' 8.7	6.60 'd' 8.7	9.10 s OH		7.05 'd' 8.7	13.36 s OH	6.15 d 2.0	3.80 s OMe	6.30 d 2.0	14.46 s OH	164
51[f]	2'-OH-3',4',6'-tri-OMe-Dihydrochalcone (A)	7.28 m	7.28 m	7.28 m	7.19 m	7.28 m	13.69 s OH	3.68 s OMe	3.95 s OMe	6.29 s	3.96 s OMe	165
	Anthocyanins	3/β	4	6	8	2'	3'	4'	5'	6'		
52	Pg 3-glc-5-(6'''-ac-glc) (M/T5)		9.19 s	7.12 d 1.7	7.01 d 2.0	8.66 'd' 9.0	7.60 'd' 9.0		7.60 'd' 9.0	8.66 'd' 9.0		166
53	Pg 3-(6''-fer-glc)-5-(6'''-mal-glc) (M/T10)		8.92 s	6.94 br s	6.92 br s	8.50 'd' 8.7	7.00 'd' 8.7		7.00 'd' 8.7	8.50 'd' 8.7		167
54	Cy 3-(6''-mal-glc) (M/T5)		9.04 d 1.0	6.76 d 1.9	6.99 dd 1.9, 1.0	8.11 d 2.2			7.11 d 8.7	8.37 dd 8.7, 2.2		168
55	Cy 3-rut (M/T5)		9.01 d 0.8	6.76 d 1.9	6.97 dd 1.9, 0.8	8.10 d 2.2			7.10 d 8.7	8.34 dd 8.7, 2.2		168
56	Cy 3-(2'',3''-digall-glc) (M/T5)		9.16 s	6.75 d 2.0	6.94 d 2.0	7.86 d 2.3			6.83 d 8.4	8.04 dd 8.4, 2.3		169
57	Cy 3,4-diglc (M/T5)		9.26 d 0.9	6.78 d 1.9	7.04 dd 1.9, 0.9	8.16 d 2.2			7.50 d 8.7	8.35 dd 8.7, 2.2		170
58	Dp 3-(6''-ac-gal) (M/T5)		9.00 d 0.8	6.75 d 1.9	6.96 dd 1.9, 0.8	7.64 s				7.64 s		171
59	Dp-3'-(2''gall-6''-ac-gal) (M/T5)		9.19 s	6.78 d 1.9	6.96 d 1.9	7.66 d 2.2				7.40 d 2.2		114
60	Pn 3-rut (M/T5)		9.08 d 0.8	6.78 d 1.9	7.03 dd 1.9, 0.8	8.32 d 2.3	4.11 s OMe		7.16 d 8.8	8.37 dd 8.8, 2.3		168
61	Pt 3,7-diglc (M/T10)		8.97 br s	6.85 d 1.8	7.28 d 1.8	8.00 br s	3.98 br s OMe			7.86 d 2.4		172
62	Pt 3-(6''-E-p-cou-glc)-5-(6'''-mal-glc) (M/T10)		8.92 s	6.99 d 1.8	6.95 d 1.8	7.93 d 2.2	3.98 s OMe			7.77 d 2.2		173
63	Mv 3-(6''-E-p-cou-glc)-5-glc (M/T5)		9.07 d 0.8	7.10 d 1.9	7.07 dd 1.9, 0.8	8.03 s	4.08 s OMe		4.08 s OMe	8.03 s		174
64	Mv 3-(6''-Z-p-cou-glc)-5-glc (M/T5)		8.81 d 0.7	7.07 d 2.0	7.00 dd 2.0, 0.7	8.03 s	4.12 s OMe		4.12 s OMe	8.03 s		174
65	Mv 3-rut-5-glc (M/T5)		8.96 s	7.08 d 1.9	7.13 d 1.9	7.96 s	4.05 s OMe		4.05 s OMe	7.96 s		174
66	Mv 3-(6''-4(''''-mal-rha)-glc)-5-(6'''-mal-glc) (M/T5)		9.13 d 0.8	7.11 d 1.9	7.26 dd 1.9, 0.8	8.14 s	4.11 s OMe		4.11 s OMe	8.14 s		174
67	Apigeninidin 5-glc (M/T1)	8.16 d 8.7	9.24 d 8.7	7.06 d 1.6	7.14 d 1.6	8.38 'd' 8.7	7.10 'd' 8.7		7.10 'd' 8.7	8.38 'd' 8.7		82
68	Luteolinidin 5-glc (M/T1)	8.26 d 8.8	9.21 d 8.8	6.98 d, <1	7.12 d, <1	7.85 d 2.0	7.11 d 8.6		7.11 d 8.6	7.99 dd 8.6, 2.0		82
69	CarboxypyrPg 3-glc (M/T5)		8.11 s	7.27[b] d 1.9	7.3[b] d 1.9	8.52 'd' 8.9	7.10 'd' 8.9		7.10 'd' 8.9	8.52 'd' 8.9		175
70	CarboxypyrCy 3-glc (M/T5)		8.08 s	7.23 d 2.0	7.31 d 2.0	7.92 d 2.4			7.04 d 8.7	8.14 dd 8.7, 2.4		176
71	CarboxypyrCy 3-(6''-mal-glc) (D/T20)		7.92 s	7.16 d 1.8	7.25 d 1.8	7.88 d 2.0			6.92 d 8.7	8.00 dd 8.7, 2.0		177
72	CarboxypyrMv 3-glc (D/T10)		7.96 s	7.20 m	7.42 m	7.80 s				7.80 s		178

continued

TABLE 2.1
^{1}H NMR Spectral Data of Flavonoids Recorded in Various Solvents* — continued

Isoflavonoids

No.	Flavonoid	2	3	5	6	7	8	2'	3'	4'	5'	6'	Ref.
73[g]	Judaicin 7-(6''-ac-glc) (D)	4.88 s	6.60 s	7.05 d 8.2	6.59 dd 8.2, 2.1		6.50 d 2.1	3.74 s OMe	6.82 s			6.90 s	179
74	Tectorigenin 4'-(6''-glc-glc) (P)	8.0 s		7.5 'd' 8.8	4.0 OMe		6.7 s	7.7 'd' 8.8	7.5 'd' 8.8		7.5 'd' 8.8	7.7 'd' 8.8	180
75[h]	7-OH-6'-OMe-3',4'-methylenedioxy-Isoflavone 7-glc (D)	8.02 s		7.98 d 8.9	7.17 dd 8.5, 2.6		7.10 d 2.3	6.87 s	6.82 s		6.82 s	3.70 s OMe	181
76	Irisjaponin A (C)	7.90 s		3.96 OMe		4.04 s OMe	6.54 s	3.84 s OMe	3.93 OMe	3.93 OMe	3.75 s OMe	6.63 s	182
77	Irisjaponin B (C)	7.85 s		3.91 OMe		4.04 s OMe	6.53 s	3.84 s OMe	3.89 s OMe	3.91 s OMe	6.74 d 8.6	7.02 d 8.6	182
78	Junipegenin B (C)	7.89 s		3.93 OMe		4.04 s OMe	6.53 s	7.10 d 2.0	3.92 s OMe	3.93 s OMe	6.94 d 8.3	7.03 dd 8.3, 2.0	182

Flavanones

No.	Flavonoid	2	3eq	3ax	5	6	7	8	2'	3'	4'	5'	6'	Ref.
79[i]	Matteucinol 7-(6''-apiofuranosyl-β-glc) (P)	5.42 dd 13.0, 3.0	2.89 dd 17.0, 3.0	3.25 dd 17.0, 13.0	12.6 s OH	2.65 s Me		2.57 s Me	7.58 'd' 7.0	7.02 'd' 7.0	3.66 s OMe			183
80[j]	Hesperitin 7-(2''-gal-6''-rha-glc) (D)	5.45 m	2.78 dd 10.5, 2.0	3.15 m	12.0 s OH	6.17 d 2.0		6.15 d 2.0	6.80 m	9.15 br s OH	3.75 s OMe			184
81[k]	5,3'-di-OH-7,4'-di-OMe-Flavanone (D)	5.33 dd	2.81 d	3.04 dd	OH 12.0 s	6.08 d	3.95 b s OMe	6.05 d	7.25	5.77 s OH	3.85 b s OMe			185
82[l]	Naringenin 7-glc (M)	5.46 dd 2.9, 13.0	2.83 dd 2.9, 17.3	3.25 dd 13.0, 17.3		6.30 d 2.2		6.27 d 2.2	7.40 'd' 8.6	6.90 'd' 8.6		6.90 'd' 8.6	7.40 'd' 8.6	109
83[m]	Naringenin 7-(6''-gall-glc) (M)	5.50 dd 3.0, 12.5	2.72 dd 17.0, 3.0	3.30 dd 17.0, 12.5		6.20 d 2.5		6.18 d 2.5	7.30 'd' 7.5	6.80 'd' 7.5		6.80 'd' 7.5	7.30 'd' 7.5	155

Dihydroflavonols and dihydroflavones

No.	Flavonoid	2	3a	3b	6	8	2'	3'	5'	6'	Ref.
84	Taxifolin 4'-glc (M)	5.08	4.59		6.02[b]	5.99[b]	7.14		7.33	7.06	52
85	Aromadendrin 7-glc (M)	5.11 d 11.7	4.68 d 11.7		6.32 d 2.2	6.30 d 2.2	7.44 'd' 8.6	6.91 'd' 8.5	6.91 'd' 8.5	7.44 'd' 8.6	186
86	Ampelopsin 7-glc (M)	4.98 d 11.3	4.61 d 11.3		6.32 d 2.2	6.30 d 2.4	6.62 s	6.62 s	6.62 s	6.62 s	186
87	2''-Ac-callunin (M)	4.63 d 11.4	5.03 d 11.4		5.93 s	6.15 s	7.39 'd' 8.6	6.81 'd' 8.6	6.81 'd' 8.6	7.39 'd' 8.6	187
88	2R,3R-trans-Aromadendrin 7-(6''-(4'''-OH-2'''-methylenebutanoyl)-glc) (D)	5.10 d 11.6	4.70 dd 11.6, 6.25	5.83 d 6.25, 5-OH; 11.80 s 5-OH; 9.60 s 4'-OH		6.15 s	7.34 'd' 8.5	6.81 'd' 8.5	6.81 'd' 8.5	7.34 'd' 8.5	188
89	(2R,3S)-(+)-3',5-di-OH-4,7-di-OMe-Dihydroflavonol (D)	6.03 d 2.2	4.80 d 2.2	11.47 s 5-OH; 3.80 s 7-OMe; 3.91 s 4'-OMe	6.16 d 2.3	6.29 d 2.3	6.77 d 2.0	9.28 s OH	6.95 d 8.3	6.67 d 8.3, 2.0	189
90	3-Desoxycallunin (M)	5.42 dd 11.8, 3.0	3.04 dd 17.2, 11.8	2.75 dd 17.2, 3.0	5.91 s	6.29 d 2.3	7.37 'd' 8.5	6.78 'd' 8.5	6.78 'd' 8.5	7.37 'd' 8.5	187

Flavan glycosides

No.	Compound	2	3	4α	4β	5	6	8	2′	5′	6′	Ref
91	Catechin 3-(6″-cin-glc) (A)	4.90 d 6.5	4.19 m	2.91 dd 16.5, 5.5	2.76 dd 16.5, 7.0		6.04 d 2.5	5.91 d 2.5	6.90 d 2.0	6.81 d 8.5	6.78 dd 8.5, 2.0	190
92	Catechin 3-(2″-cin-glc) (A)	4.65 d 8.0	4.21 m	2.90 dd 16.5, 5.5	2.63 dd 16.5, 5.5		6.03 d 2.5	5.85 d 2.5	6.90 d 2.0	6.81 d 8.0	6.71 dd 8.0, 2.0	190
93	Catechin 3-(2,″6″-dicin-glc) (A)	4.68 d 7.8	4.17 m	3.04 dd 16.5, 5.5	2.67 dd 16.5, 8.0		6.02 d 2.5	5.83 d 2.5	6.90 d 2.0	6.78 d 8.0	6.73 d 8.0, 2.0	190
94	Anadanthoside (M)	4.97 d 5.9	4.15 m	2.87 dd 15.6, 4.8	2.82 dd 15.6, 6.2	6.85 d 8.3	6.36 dd 2.0, 8.3	6.33 d 2.0	6.82 d 2.0	6.76 dd 8.3	6.72 dd 8.3, 2.0	191

Prenylated flavonoids

No.	Compound	2	3a	3b	4a	4b	5	6	7	8	2′	Ref
95	Cajanin (D)	8.22 s					12.98 s OH; 3.89 s OMe		3.85 s OMe		9.39 s OH	192
96	Indicanine C (D)	7.76 s						6.28 d 2.2		6.60 s	7.26 'd' 8.7	192
97	6-(1,1-di-Me-allyl)-7,4′-di-OH-Flavan (C)	4.97 m	2.87 m	3.11 m	2.89 m	3.16 m	7.03 brs			5.85 s OH	7.15 br d 8.5	193
98	5‡ (C)						OMe 3.73 s				7.10 'd' 8.9	194

Pterocarpans and rotenoids

No.	Compound	1	2	4	6α	6β	6a	7	10	11	11a	Ref
99	Maackianin 3-(6″-mal-glc) (D)	7.38	6.71	6.54		4.27	3.58–3.61	6.97	6.52		5.55	179
100	3,4,8,9-dimethylenedioxy-Pterocarpan (C)	6.96	6.55	3.64		4.23	3.45	6.66	6.37		5.43	195
101	Usararotenoid C (C)	7.70		6.41	4.45	4.37	4.60		6.71	7.89		196
102	12a-Epimillettosin (C)	7.67		6.37	4.44	4.36	4.59		6.54	7.74		197
103	(+)-Usararotenoid-B (C)	8.29		6.70	4.94	4.63	4.94		6.79	8.05		196

Biflavonoids

No.	Compound	2 / 2″	3 / 3″	4 / 4′α 4″β	6/5 6″/5″	8/7 8″/7″	2′ / 2‴	3′ / 3″	5′ / 5″	6′ / 6‴	Ref
104	[Catechin 3-glc-(4α → 8)-Catechin 3-(2-cin-glc)] (biflavanol) (A)	4.27 d 9.5 / 4.50 d 8.5	4.47 m / 3.98 m	4.45 m / 2.96 dd 16.5, 6.0 / 2.63 dd 16.5, 7.0	5.76 d 2.0 / 6.06 s	5.87 d 2.0	6.68 d 1.5 / 6.62 m		6.64 d 8.5 / 6.71 d 8.5	6.33 dd 8.5, 1.5 / 6.39 br d 8.5	190
105	[catechin 3-glc-(4α → 8)-Epicatechin 3-(6-cin-glc)] (biflavanol) (A)	4.38 d 10.0 / 5.23 d 4.5	4.69 dd 10.0, / 6.5 / 4.25 m	4.56 dd 6.5 / 2.50 dd 17.0, 4.6 / 2.71 dd 17.0, 4.6	5.75 d 2.0 / 6.05 s	5.87 d 2.0	6.96 d 1.5 / 6.47 br s		6.76 d 8.0 / 6.53 br d 8.0	6.70 dd 8.0, 1.5 / 6.04 m	190
106	Amentoflavone (biflavone) (D)		6.81 s / 6.76 s		6.18 d 2.5 / 6.33 s	6.43 d 2.5	7.98 d 2.5 / 7.58 'd' 8.5		7.09 d 8.5 / 6.69 'd' 8.5	7.97 dd 8.5, 2.5 / 7.58 'd' 8.5	198
107	Aulacomnium-biaureusidin (biaurone) (D)				6.06 d	6.15 d / 6.36 s	7.41 d / 7.44 d	6.45 s / 6.50 s	6.83 d	7.03 d / 7.20 dd	199
108	Cupressuflavone 7,7″-di-OMe (biflavone) (D)		6.86 s / 6.86 s		6.78 s / 6.78 s	6.78 s / 6.78 s	7.46 'd' 8.6 / 7.46 'd' 8.6	OMe; 3.82 s 3.82 s	6.77 'd' 8.6 / 6.77 'd' 8.6	7.46 'd' 8.6 / 7.46 'd' 8.6	200
109	4,4′,6-tri-O-methyl-2-Deoxymaesopsin-(2 → 7)-2,4,4′,6-tetra-O-methylmaesopsin (bibenzofuranoid) (C)	3.24 s OMe		4-OMe/6-OMe 3.70 s / 4″-OMe 3.72 s	5.93 d 2.0 / 5.88 s	5.79 d 2.0	7.13 'd' 9.0 / 7.20 'd' 9.0	3.60 d 14.0 / 4.05 d 14.0 / 3.16 d 14.0 / 3.04 d 14.0	6.63 'd' 9.0 / 6.77 'd' 9.0	7.13 'd' 9.0 / 7.20 'd' 9.0	201

continued

TABLE 2.1
^{1}H NMR Spectral Data of Flavonoids Recorded in Various Solvents* — continued

No.	Flavones	3	5	6	7	8	2'	3'	4'	5'	6'	Ref.
110	Catechin-(4α→8)-pelargonidin 3-glc (anthocyanin-flavanol)† (M/T5)	4.81 d 9.2	4.35 t 9.2	9.14 s	6.90 s		ND	8.30 'd' 9.1	8.30 'd' 9.1	7.06 'd' 9.1	7.06 'd' 9.1	108
		4.62 d 9.3	4.61 m	9.21 s	6.79 s			8.77 'd' 9	8.77 'd' 9.2	7.17 'd' 9.2	7.17 'd' 9.2	
		2/6	*3/5*	5.02 d 9.2	6.89 d 2.0			6.92 d 2.0		6.76 d 8.3	6.82 dd 8.3, 2.0	
		2'/6'	*3'''/5'''*	5.01 m				7.07 d 2.0			6.95 dd 8.3, 2.0	
111	2',4',2'''-tri-OH-4,4'''-di-OMe-4-*O*-5''' -Bichalcone (C) (Rhuschalcone 1) (bichalcone)	*2/6* 7.70 'd' 8.8	*3/5* 6.90 'd' 8.8	*3'* 6.45 d 2.4	*5'* 6.53 dd 9.0, 2.5	*6'* 8.04 d 9.1	*α* 7.71 d 15.4	*β* 7.83 d 15.4	*4-OMe* 3.85 s			202
		2'/6' 7.61 'd' 8.6	*3'''/5'''* 6.80 'd' 8.7	*3'''* 6.69 s	*5'''* 9.0, 2.5	*6'''* 7.93 s	*α'* 7.58 d 15.5	*β'* 7.85 d 15.5	*4''-OMe* 3.82 s			

Notes: pg, pelargonidin; cy, cyanidin; dp, delphinidin; pn, peonidin; pt, petunidin; mv, malvidin; all, allopyranose; ara, α-arabinopyranose; gal, galactopyranose; glc, glucopyranose; glu, glucuronic acid; rha, rhamnopyranose; xyl, xylopyranose; ac, acetyl; benz, benzoyl; caf, caffeoyl; cin, cinnamoyl; cou, *p*-coumaroyl; fer, feruloyl; gall, galloyl; mal, malonyl; sinap, sinapoyl; ND, not detected.

[a] *C-p*-OH-benz 2/6 7.07 'd' 7.7 3/5 7.90 'd' 7.7, CH$_2$ bridge 4.06 s. [b] Assignments may be reversed.

* A, acetone-*d6*; C, CDCl$_3$; M, CD$_3$OD; M/T1, CD$_3$OD-CF$_3$COOD 99:1; M/T2, CD$_3$OD-CF$_3$COOD 98:2; M/T5, CD$_3$OD-CF$_3$COOD 95:5; M/T10, CD$_3$OD-CF$_3$COOD 90:10; M/T17, CD$_3$OD-CF$_3$COOD 5:1; D, DMSO-*d6*; D/T10, DMSO-*d6*-CF$_3$COOD 90:10; D/T20, DMSO-*d6*-CF$_3$COOD 80:20; P, pyridine-*d5*; T, CF$_3$COOD; ^{13}C chemical shift values of OMe and Me groups are given in brackets.

† Rotamers: Top — chemical shift values of major rotamer.

‡5 = 3-(4'-Hydroxyphenyl)-5-methoxy-6-(3,3-dimethylallyl)-2'',2''-dimethylchromene-(5'':8,7)-3-(propyl-2-one)-4*H*-1-benzo-2,3-dihydropyran-2,4-dione.

[b] 47: 1'' (3.56 d 7.0); 2'' (5.41 m); 4'' (1.88 d 0.5 Me); 5'' (1.69 d 0.5 Me); [c] 48: β (7.96 d 16); α (7.86 d 16); [d] 49: β (8.17 d 15.3); α (8.66 d 15.3); [e] 50: β (2.80 t 7.5); α (3.12 t 7.5); [f] 51: β (2.96 t 7.4); α (3.33 t 7.4); [g] 73: OCH$_2$O (5.99 s); [h] 75: OCH$_2$O (6.00 s); [i] 79: 5' (7.02 'd' 7.0); 6' (7.58 'd' 7.0); [j] 80: 5' (6.90 m); 6' (6.80 m); [k] 81: 5' (6.92 m); 6' (7.05 d); [l] 82: 5' (6.90 'd' 8.6); 6' (6.82 'd' 8.7); [m] 83: 5' (6.80 'd' 7.5); 6' (7.30 'd' 7.5); [n] 95: 3' (6.63 d 2.3); 4' (9.31 s OH); 5' (6.36 dd 2.3, 8.3); 6' (6.96 d 8.3); [o] 96: 3' (6.82 'd' 8.7); 4' (7.19 s OH); 5' (6.82 'd' 8.7); [p] 97: 3' (6.80 br d 8.5); 5' (6.80 br d 8.5); 6' (7.15 br d 8.5); 2'' (6.18 dd 10.5, 17.7); 3''a (5.36 dd 1.0, 17.7); 3''b (5.30 dd 1.0, 10.5); 4' (1.42 s); 5'' (1.42 s). [q] 98: 3' (6.70 'd' 8.9); 5' (6.70 'd' 8.9); 6' (7.10 'd' 8.9); 7' (3.27 dd 13.9, 7.2; 3.21 dd 13.9, 7.2); 8' (5.07 td 7.2, 1.2); 10' (1.75 s); 11' (1.64 s); 2'' (1.46 s, 1.45 s Me); 3'' (5.65 d 10.1); 4'' (6.74 d 10.1); CH$_2$COCH$_3$ (3.87 d 18.1; 3.53 d 18.1); CH$_2$COCH$_3$ (2.20 s); [r] 99: 8,9 OCH$_2$O (5.92) [s] 100: 3,4 OCH$_2$O (5.84); 8,9 OCH$_2$O (5.93); [t] 101: 2,3 OCH$_2$O (5.94); 1' (3.37); 2' (5.17); 4' (1.68); 5' (1.77); OMe (3.91), [u] 102: 2,3 OCH$_2$O (5.91); 4' (5.62); 5' (6.62); 2'-Me 1 (1.45); 2'-Me (2 1.48), [v] 103: 2,3 OCH$_2$O (5.93, 5.95); OMe (3.92, 3.79).

TABLE 2.2
^{13}C NMR Spectral Data of Flavonoids Recorded in Various Solvents*

No.	Flavones	2	3	4	5	6	7	8	9	10	1'	2'	3'	4'	5'	6'	OMe	Ref.
1	6-OH-Luteolin 4'-OMe-7-(2''-α-rha-6''-ac-β-glc) (D)	163.8	103.1	182.1	146.8	130.7	151.2	93.8	149.0	105.5	123.1	113.0	146.7	151.1	112.1	118.5	55.7	142
2	6-OH-Luteolin 7-(6''-(E)-caf)-β-glc (D)	164.3	101.8	182.0	ND	130.3	151.1	93.2	148.9	105.7	121.0	112.7	145.9	150.3	115.5	118.5		142
3	Isoscutellarein 7-(2''-(6'''-ac)-β-all)-β-glc (D)	164.0	102.6	182.2	152.1	99.5	150.4	127.5	143.7	105.6	121.1	128.5	115.9	161.3	115.9	128.5	55.5	142
4	Isoscutellarein 4'-OMe-7-(2''-(6''-ac)-β-all)-β-glc (D)	163.6	103.3	182.3	152.1	99.5	150.5	127.5	143.7	105.6	122.8	128.3	114.5	162.4	114.5	128.3		143
5	Apigenin 4'-(2''-(2''-fer-glu)-glu) (M)	166.1	104.8	183.9	163.2	100.2	165.5	95.1	159.4	105.4	26.1	129.1	117.9	161.3	117.9	129.1		143
6	Apigenin 7-glu4'-(2''-(2''-(2''-fer-glu)-glu) (M)	165.9	104.9	183.8	162.9	101.3	164.5	96.0	158.9	107.3	25.8	129.2	117.9	161.4	117.9	129.2		143
7	Apigenin 7-glu4'-(2''-(2''-E-cou-glu)-glu) (M)	166.0	105.0	184.0	162.8	101.2	164.5	95.9	158.9	107.3	25.8	129.3	117.9	161.4	117.9	129.3		143
8	Luteolin 3'-β-glc-4'-(2''-α-rha-β-glc) (M)	165.1	105.2	183.8	163.2	100.1	166.6	95.1	159.4	106.9	127.1	118.1	149.2	151.4	119.2	123.3		144
9	Luteolin 3',4'-di-β-glc (M)	165.4	105.5	183.5	163.0	100.0	166.0	95.0	159.6	106.7	126.8	117.8	149.1	151.0	118.9	123.0		144
10	5,7,4'-tri-OH-3'-OMe-Flavone-8-C-β-xyl-2''-O-glc (M)	166.00	104.11	184.12	162.69	100.75	164.66	104.04	156.84	105.43	123.97	110.81	149.30	151.73	116.72	121.78	55.84	145
11	5,7-di-OH-3'-OMe-4'-Acetoxyflavone-8-C-(2''-O-glc-β-xyl) (M)	166.22	104.23	184.15	162.61	100.88	164.66	105.41	156.80	105.60	124.26	111.19	149.41	151.79	116.73	121.81		145
12	Isoorientin 3'-OMe (D)	163.03	103.03	181.85	160.48	108.72	163.64	93.64	156.21	103.29	121.32	109.99	147.90	150.59	115.68	120.23	55.84	146
13	8-C-p-OH-Benz-isovitexin 4'-O-glc (D)	162.80	103.45	182.12	157.25	107.73	160.76	106.76	153.69	103.56	123.86	127.90	116.47	160.08	116.47	127.90		146
14	Apigenin 8-C-(2''-(4''-ac-rha)-glc)† (M)	166.66, 165.95	103.58, 104.21	184.11	162.82, 163.20	99.77, 101.32	164.12, 164.34	105.75, 105.61	157.80, 156.34	105.98, 104.74	123.36, 123.32	130.13, 129.52	117.01, 117.15	162.79	117.01, 117.15	130.13, 129.52		109
15	Spinosin† (D)	163.66	102.90, 102.99	181.82, 182.15	159.57, 160.42	108.58	163.72, 164.94	90.20, 90.67	156.85, 156.96	104.09, 104.36	120.92	128.34	115.59, 115.89	160.78, 161.22	115.59, 115.89	128.34	55.99, 56.42	147
16	6'''-Fer-spinosin† (D)	163.6, 163.8	102.9, 103.1	181.7, 182.2	159.4, 160.7	108.7	165.1, 164.1	89.9, 90.6	156.8, 157.0	103.9, 104.4	121.1, 121.2	128.4, 128.5	115.70, 115.75	161.1	115.70, 115.75	128.4, 128.5	56.3, 56.0	147
17	Isoscoparin 7-glc (D)	164.3	104.7	181.7	159.2	110.2	162.2	93.4	156.2	102.9	121.2	121.2	149.3	160.1	115.3	118.8	60.2	147
	Flavonols	**2**	**3**	**4**	**5**	**6**	**7**	**8**	**9**	**10**	**1'**	**2'**	**3'**	**4'**	**5'**	**6'**		
19	Kaempferol 3-(6''-α-ara-glc) (D)	156.0	133.1	177.0	161.0	98.9	165.4	93.7	156.4	103.5	120.8	130.7	115.0	159.8	115.0	130.7		148
20	Kaempferol 3-(6''-α-ara-glc)-7-glc (D)	157.0	133.3	177.4	160.8	99.2	162.8	94.5	155.9	105.6	120.0	130.8	115.3	160.8	115.3	130.8		148
21	Kaempferol 3-(2''-rha-6''-mal-glc) (M)	161.24	134.27	179.24	163.16	99.79	165.67	94.71	158.44	105.91	123.18	132.10	116.08	159.13	116.08	132.10		149
22	Kaempferol 3-glc-7-(2''-(6'''-p-cou-glc)-glc) (M)	157.8	135.9	179.7	162.7	100.0	163.1	95.7	157.8	107.5	122.9	132.5	116.1	161.7	116.1	132.5		150
23	Quercetin (M)	148.00	137.21	177.33	162.50	99.25	165.34	94.40	158.22	104.52	124.15	115.99	148.75	146.21	116.22	121.67		52
24	8-OMe-Kaempferol 3-(6''-mal-glc) (M)	159.33	135.33	179.57	157.97	100.14	158.57	129.14 (62.01)	150.39	105.59	122.80	132.30	116.15	161.68	116.15	132.30		109
25	Quercetin 4'-glc (M)	148.04	137.89	177.36	162.47	99.31	165.66	94.47	158.19	104.54	127.60	116.49	147.83	146.79	117.62	121.26		52
26	Quercetin 3'-xyl (M)	147.17	137.47	177.36	162.53	99.30	165.65	94.42	158.17	104.55	124.31	118.54	146.32	150.49	117.19	124.92		151
28	Quercetin 3,4'-diglc (D)	155.96	134.26	174.82	161.64	99.16	164.65	94.08	156.83	104.53	124.92	115.86	147.67	146.52	116.97	121.42		52
29	Quercetin 3,7,4'-triglc (D)	156.27	134.20	177.62	160.90	99.48	162.99	94.49	156.12	105.61	124.42	116.73	147.72	146.22	115.52	120.40		52

continued

TABLE 2.2
^{13}C NMR Spectral Data of Flavonoids Recorded in Various Solvents* — continued

No.	Flavones	2	3	4	5	6	7	8	9	10	1'	2'	3'	4'	5'	6'	OMe	Ref.
30	Isorhamnetin 3-rut	158.91	135.70	179.50	ND	100.61	163.28	95.43	159.03	105.67	123.28	114.78	151.19 (57.02)	148.62	116.40	124.25		153
31	Isorhamnetin 3,7-diglc (M)	160.2	136.2	180.3	163.6	101.6	165.5	96.5	158.8	108.4	123.0	115.1	149.5 (57.5)	153.0	117.1	124.9		154
32	Myricetin 3-(2''-rha-glc) (M)	158.48	134.71	179.31	163.20	99.65	165.62	94.45	158.34	105.97	122.43	109.87	146.46	137.74	146.46	109.87		149
33	Myricetin 3'-(6''-p-couglc) (P)	147.9	138.6	177.7	104.9	162.9	99.8	166.1	94.9	157.4	123.5	110.2	148.5	140.5	148.2	113.0		70
34	Myricetin 7-(6''-gall-glc) (D)	148.2	138.2	176.3	160.3	98.8	162.5	94.6	156.0	105.0	121.1	107.9	146.1	136.2	146.1	107.9		155
27	Myricetin 3-(2''-ac-rha) (M)	158.51	135.60	179.43	163.25	99.77	165.89	94.74	159.44	105.85	121.77	109.55	146.95	137.99	146.95	109.55		152
35	Laricitrin 3-α-ara-furanoside (D)	156.4	133.4	177.7	161.3	98.6	164.2	93.6	156.9	103.9	119.8	109.5	147.8 (55.8)	137.4	145.7	105.6		156
37	Syringetin 3-(5''-glc-α-ara-furanoside) (D)	156.4	133.3	177.6	161.2	98.7	164.2	94.0	157.1	104.0	119.7	106.6	147.5 (56.0)	138.6	147.5 (56.0)	106.6		156
39	Syringetin 3-(6''-rha-gal) (M)	158.6	135.6	179.4	163.1	100.1	166.0	95.1	158.5	105.9	121.9	108.4	149.0 (57.5)	140.1	149.0 (57.5)	108.4		157
40	Syringetin 6-C-glc (D)	146.2	135.8	176.0	159.6	108.1	163.0	93.3	155.0	102.7	120.7	106.0	147.7 (56.2)	138.2	147.7 (56.2)	106.0		158
	Aurones	2	3	4	5	6	7	8	9	α	1''	2/2''	3/3''	4/4''	5/5''	6'		
41	6,3'-di-OH-4,4'-di-OMe-5-Me-Aurone (D)	145.96a	178.75	156.87 (61.39)	117.04 (8.20)	165.68b	93.30	165.30b	105.31	110.29	124.94	111.41	146.61a	149.34	112.16	123.95		159
42	4,6,3',4'-tetraOMe-Aurone (Z-form) (D)	146.15	178.90	168.04 (55.59)a	89.89	168.74 (56.12)a	94.38	158.82	104.21	110.29	124.94b	111.89	148.75 (56.42)a	150.27 (56.54)a	113.99	124.79b		159
44	6,3',4'-tri-OH-4-OMe-5-Me-Aurone (D)	145.52a	178.76	156.86	116.05	165.12b	93.26	165.63b	105.50	110.94	123.59	111.40	145.55a	147.80	117.66	124.3		159
45	Maesopsin (P)	107.4	196.8	173.8	91.1	171.6	96.8	159.8	103.1	42.1	125.9	132.6	115.7	157.2	115.7	132.6		160
46a	Maesopsin 6-O-glc diastereoisomer 1 (P)	107.7	197.0	174.6	93.3	171.4	97.6	158.4	103.6	42.1	125.6	132.5	115.8	157.2	115.8	132.5		161
46b	Maesopsin 6-O-glc diastereoisomer 2 (P)	107.6	196.9	174.5	93.2	171.4	97.3	158.3	103.4	41.9	125.6	132.5	115.7	157.2	115.7	132.5		161
47	Licoagroaurone (A)	147.4	182.9	123.4	112.9	167.7	113.1	163.6	114.9	112.0	125.7/22.6	118.8/122.1	146.2/132.8	148.1/18.1	116.5/25.8	125.5		89
	Chalcones and dihydrochalcones																	
48	3'-formyl-4',6'di-OH-2'-OMe-5-Me-Chalcone (C)	147.0	125.0	193.0	108.2	167.1 (67.0)	108.5 (192.5)	165.7 (8.1)	109.3	169.0	136.0	128.7	129.2	132.1	129.2	132.1		162
49	Chalcononaringenin 2',4'-diglc (P)	144.4	125.2	193.9	108.4	164.2	98.9	166.7	95.7	161.6	127.3	131.7	116.9	161.0	116.9	131.7		163
50	2',4'-di-OH-4-OMe-6'-glc Dihydrochalcone (D)	30.0	46.3	205.4	107.3	166.1	96.3	166.3 (56.7)	94.6	161.3	132.5	130.3	116.0	156.3	116.0	130.3		164
51	2'-OH-3',4',6'-tri-OMe-Dihydrochalcone (A)	31.3	46.7	206.1	106.6	159.6	131.6	159.9	88.3	159.9	142.7	129.3a	129.2a	126.7	129.2a	129.3a		165

Anthocyanins

		2	3	4	5	6	7	8	9	10	1'	2'	3'	4'	5'	6'	
53	Pg 3-(6″-fer-glc)-5-(6‴-mal-glc) (M/T10)	165.0	145.8	135.8	156.6	106.2	169.8	97.5	157.0	113.3	120.5	136.1	118.1	167.4	118.1	136.1	167
54	Cy 3-(6″-mal-glc) (M/T5)	164.58	145.59	136.90	159.2	103.64	170.41	95.25	157.76	113.28	121.24	118.43	147.50	155.89	117.39	128.48	168
55	Cy 3-rut (M/T5)	164.17	145.62	136.21	159.06	103.48	170.45	95.23	157.63	113.24	121.20	118.40	147.44	155.88	117.43	128.42	168
56	Cy 3-(2,3″-digall-glc) (M/T5)	164.42	145.10	136.76	159.37	103.62	170.65	94.93	157.78	113.15	120.33	117.71	147.58	155.88	117.31	128.50	169
58	Dp 3-(6″-ac-gal) (M/T5)	164.38	145.98a	135.68	159.38b	103.30	170.24	95.13	157.64b	113.87	120.91	112.71	147.57	144.91a	147.57	112.71	171
62	Pt 3-(6″-E-p-cou-glc)-5-(6‴-mal-glc) (M/T10)	164.7	146.2	134.9	156.6	106.2	169.6	97.4	157.1	113.3	119.3	109.6	149.9 (57.2)	146.3	147.7	114.2	173
65	Mv 3-rut-5-glc (M/T5)	163.98	146.05	134.56	156.65	105.55	169.88	97.68	157.14	113.14	119.50	110.87	149.82c (57.31)	147.27	149.82 (57.31)	110.87	174
63	Mv 3-(6″-E-p-cou-glc)-5-glc (M/T5)	164.89	146.00	136.38	157.34	106.11	170.01	97.70	156.87	113.47	119.58	110.95	149.84 (57.26)	147.21	149.84 (57.26)	110.95	174
66	Mv 3-(6″-(4‴-mal-rha)-glc)-5-(6‴-mal-glc) (M/T5)	164.92	146.40	135.46	156.58	106.03	169.8	97.65	157.47	113.47	119.65	111.07	149.93 (57.31)	147.37	149.93 (57.31)	111.07	174
67	Apigeninidin 5-glc (M/T1)	173.8	112.0	149.9	158.6	105.5	172.1	98.5	160.2	114.5	121.6	134.1	118.9	168.3	118.9	134.1	82
68	Luteolinidin 5-glc (M/T1)	173.1	113.0	149.7	158.2	105.3	171.1	98.5	159.6	113.7	121.9	117.0	148.7	157.2	118.8	126.6	82
69e	CarboxypyrPg 3-glc (M/T5)	166.46	135.95	150.44	154.48	101.77a	169.33	101.53a	154.48	111.00	120.95	135.04	117.40	165.75	117.40	135.04	175
70d	CarboxypyrCy 3-glc (M/T5)	166.09	136.02	149.84	154.31	101.78	169.44	101.45	154.32	110.71	121.31	118.21	147.09	154.59	117.02	127.27	176
71e	CarboxypyrCy 3-(6″-mal-glc) (D/T20)	165.01	135.25	109.79a	153.07	101.30	168.40	101.05	153.20	109.67a	120.33	118.5	146.14	153.89	116.89	126.52	177
72f	CarboxypyrMv 3-glc (D/T10)	163.4	134.8	109.3	152.7	100.8	168.2	101.1	152.7	109.5	118.4	109.4	148.2 (56.5)	143.9	148.2 (56.5)	109.4	178

Isoflavonoids

		2	3	4	4a/10	5	6	7	8	8a/9	1'	2'	3'	4'	5'	6'	OCH₂O	
73	Judaicin 7-(6″-ac-glc) (D)	67.6	129.1	120.8	117.6	127.2	109.4	157.6	103.4	153.8	119.1	152.5 (56.5)	95.5	147.7	141.0	107.5	101.2	179
74	Tectorigenin 4'-(6″-glc-glc) (P)	153.6	122.5	180.9	105.0	153.9	132.9 (60.3)	154.3	95.1	153.9	125.5	130.8	117.1	158.4	117.1	130.8		180
112	Puerarin (daidzein 8-C-glc) (D)	152.6	122.5	174.9	114.9	126.2	114.9	161.2	112.6	157.1	123.0	130.0	114.9	157.1	114.9	130.0		203
113	Calycosin (C)	153.0	123.3	174.5	116.2	127.3	115.1	162.5	102.2	157.3	124.7	116.4	146.0	147.5 (55.6)	111.9	119.7		204
75	7-OH-6-OMe-3',4'-methylene-dioxyiso-Flavone 7-glc (D)	154.7	121.7	174.2	118.3	126.8	115.3	161.3	103.4	157.0	122.6	110.9	140.3	147.9	95.5	152.8	101.1	181
76	Irisjaponin A (C)	152.5	120.0	181.2	106.5	154.8	130.4 (56.2)	155.2	93.3	153.5	118.6	149.4 (60.9)	147.3 (61.2)	143.5 (61.3)	145.9 (61.2)	108.9		182
77	Irisjaponin B (C)	152.5	120.3	181.3	106.5	154.3	130.3 (56.1)	155.0	93.2	153.5	117.0	151.5 (60.8)	142.3 (60.9)	154.3 (61.2)	107.2	125.8		182
78	Junipegenin B (C)	152.6	121.3	181.3	106.5	154.0	130.4 (56.0)	155.2	93.2	153.4	121.3	112.3	148.9 (60.9)	149.3 (60.9)	111.2	123.3		182

continued

TABLE 2.2
^{13}C NMR Spectral Data of Flavonoids Recorded in Various Solvents* — *continued*

No.	Flavonoids	2	3	4	5	6	7	8	9	10	1'	2'	3'	4'	5'	6'	OMe	Ref.
	Flavanones	2	3	4	5	6	7	8	9	10	1'	2'	3'	4'	5'	6'		
79	Matteucinol 7-(6'' apio-furanosyl-β-glc) (P)	78.8	43.6	198.4	159.5	112.3 (9.4)	162.6	111.3 (10.0)	158.3	105.8	131.7	128.3	114.5	160.3 (55.2)	114.5	128.3		183
80	Hesperitin 7-(2''-gal-6''-rha-glc) (D)	78.49	42.13	197.17	163.10	96.48	165.17	95.65	163.15	103.42	130.96	114.24	148.61	146.52 (55.76)	112.06	118.11		184
81	5,3'-di-OH-7,4'-di-OMe-Flavanone (D)	79.0	41.9	196.7	167.4	95.3	146.2 (55.6)	93.7	162.7	102.5	120.5	113.7	163.1	149.1 (55.8)	112.1	130.6		185
82	Naringenin 7-glc (M)	80.62	44.13	198.52	159.07	96.92	167.00	97.99	164.93	104.91	130.85	129.09	116.33	164.59	116.33	129.09		109
83	Naringenin 7-(6''-gall-glc) (M)	78.8	42.1	197.3	163.0	96.4	165.1	95.3	163.0	103.4	128.6	128.7	115.3	157.8	115.3	128.7		155
	Dihydroflavonols and dihydroflavones	2	3	4	5	6	7	8	9	10	1'	2'	3'	4'	5'	6'		
84	Taxifolin 4'-glc (M)	85.55	73.70	198.22	168.69	97.39a	165.72	96.32a	164.35	102.19	133.96	116.57	147.22	148.37	118.51	120.70		52
85	Aromadendrin 7-glc (M)	85.2	73.8	199.5	164.3	98.3	167.3	97.0	164.9	103.2	129.0	130.4	116.1	159.3	116.1	130.4		186
87	2''-Ac-callunin (M)	85.2	73.0	198.0	159.5	98.1	162.2	127.8	155.0	101.7	129.3	130.9	116.3	161.3	116.3	130.9		187
88	2R,3R-trans-Aromadendrin 7-(6''-(4'''-OH-2'''-methylenebutanoyl)-glc) (D)	84.0	72.5	199.7	163.5	97.6	166.0	96.1	163.4	103.0	128.3	130.5	115.8	158.7	115.8	130.5		188
89	(2R,3S)(+)-3',5-di-OH-4',7-di-OMe-Dihydroflavonol (D)	82.2	77.8	191.6	163.3	95.8	170.3 (57.0a)	95.0	165.4	103.6	129.3	119.4	147.8	149.3 (56.9a)	112.7	114.8		189
90	3-Desoxycallunin (M)	80.6	44.0	197.7	158.9	97.4	162.0	127.5	155.4	103.2	129.2	128.8	116.4	161.5	116.4	128.8		187
	Flavan glycosides	2	3	4	5	6	7	8	9	10	1'	2'	3'	4'	5'	6'		
91	Catechin 3-(6''-cin-glc) (A)	79.6	76.3	26.7	152.7	96.3	157.7	95.5	156.5	100.3	132.1	114.2	145.8	145.7	115.9	119.4		190
92	Catechin 3-(2''-cin-glc) (A)	80.4	74.8	27.7	157.1	96.3	157.8	95.5	156.7	100.5	131.9	115.4	145.8	145.8	115.8	120.1		190
93	Catechin 3-(2,6''dicinglc) (A)	80.2	75.6	24.7	157.0	96.3	157.7	95.4	156.6	100.4	131.7	115.3	145.7a	145.6a	115.8	120.1		190
94	Anadanthoside (M)	80.7	76.9	31.0	131.5	109.6	158.0	103.5	155.9	112.4	132.2	114.8	146.3	146.4	116.3	119.6		191
	Prenylated flavonoids	2	3	4	5	6	7	8	8a/9	10/4a	1'	2'	3'	4'	5'			
95g	Cajanin (D)	155.6	120.6	180.6	161.6	97.9	165.1 (56.1)	92.3	158.6	105.4	108.4	156.4	102.6	157.5	106.2			192
96h	Indicanine C (D)	150.7	125.9	175.8	158.2 (62.9)	113.4	158.7	100.7	155.7	113.1	123.4	130.4	115.8	156.4	115.8			192
97i	6-(1,1-di-Me-allyl)-7,4'-di-OH-Flavan (C)	85.0	41.5	34.9	122.4	124.3	155.3	99.8	159.8	118.7	130.0	130.9	115.8	154.7	115.8			193
98j	5‡ (C)	169.1	61.4	187.6	159.3 (62.1)	106.4	158.3	120.8	148.9	105.7	126.0	127.8	116.2	155.9	116.2			194
	Pterocarpans and rotenoids	1	1a	2	3	4	4a	5	6	6a	6b	7	7a	8	9	10	10a	
100k	3,4:8,9-dimethylene-dioxy-Pterocarpan (C)	123.5	102.1		143.0	143.3	166.8		39.6		117.0	104.1	148.3	148.3	93.3	160.0		195
114o	Isoneorautenol (C)		66.5		39.4	76.5	150.7	132.2	109.8			160.2a	103.7	157.2a	112.4b			205
115p	Erybraedin A (C)		66.8		39.8	78.8	150.8	129.3	109.7			158.4a	110.3	155.7a	112.6b			205
101l	Usararotenoid C (C)	109.9	110.5	142.3	149.4	98.5	150.7	61.7	76.5			158.1	117.3	163.3	105.9			196
102m	12a-Epimillettosin (C)	109.6		142.5	149.5	98.6	150.8	61.9	76.8			155.8	109.1	159.8	112.3			197
103n	(+)-Usararotenoid-B (C)	111.2		142.9	149.8	99.2	151.5	62.9	78.2			155.2	137.6	159.5	107.4			197
	Biflavonoids	2	3	4	5	6	7	8	9	10/α	1'	2'	3'	4'	5'	6'		
		2''	3''	4''	5''	6''	7''	8''	9''	10''/α''	1''	2'''	3'''	4'''	5''	6''		
104	[Catechin 3-glc(4α → 8)-catechin 3-(2-cin-glc)] (A)	82.5	81.2	37.5	158.7	97.0	157.2	97.5	157.2	107.2	132.2	116.9	145.8	146.5	116.6	120.9		190
		81.8	76.3	29.6	155.2	96.6	155.6	109.3	155.3	103.0	131.4	116.8	145.9	146.4	116.2	121.0		

No.	Compound																Ref.
105	[Catechin 3-glc-(4α → 8)-	82.7	82.2	37.5	159.0	97.0	157.5	97.7	157.2	107.3	132.5	116.6	146.7	146.1	116.4	121.2	190
	epicatechin 3-(2-cin-glc)] (A)	77.7	74.2	25.3	155.8	96.4	154.9	109.5	153.9	101.7	131.8	114.1	145.7	145.9	116.4	119.0	
106	Amentoflavone (biflavone) (D)	163.7	102.9	181.7	161.4	98.7	164.1	94.0	ND	103.6	123.9	131.4	120.0	ND	116.3	127.8	198
		163.8	102.6	182.2	161.5	98.9	160.5	104.0	159.7	103.7	121.4	128.2	115.8	161.0	115.8	128.2	
107	Aulacomniumbiaureusidin (biaurone) (D)	145.5	179.0	166.8	97.6	167.4	90.2	158.1	102.9	109.9	122.4	115.8	145.7	147.5	120.8	127.3	199
		145.9	180.1	165.8	108.7	165.2	90.5	158.1	102.9	109.5	123.7	117.5	145.5	147.5	115.8	124.0	
108	Cupressuflavone 7, 7''-di-OMe (D)	163.0	102.5	182.2	161.7	95.5	163.8 (56.5)	98.8	153.8	104.2	120.9	127.9	116.0	161.8	116.0	127.9	200
		161.2	102.5	182.2	161.7	95.5	163.8 (56.5)	99.0	153.8	104.2	120.9	127.9	116.0	161.8	116.0	127.9	
109	4,4',6-tri-O-methyl-2-Deoxymaesopsin- (2 → 7)-2,4,4', 6-tetra-O-methylmaesopsin* (bibenzofuranoid) (C)	91.3	197.5	158.5ᵃ	88.3	169.2	174.5	106.2	41.6	127.2	132.2ᵇ	113.4	158.4§	113.4	132.2ᵇ	201
		109.7	194.4	160.1§	103.4	169.4§	171.6	105.3	40.8	125.6	132.1ᵇ	113.9	158.9	113.9	132.1ᵇ	
110	Catechin-(4α → 8)-pelargonidin 3-glc (anthocyanin-flavanol)† (M/T5)	163.9	144.5	138.3	157.9	102.7	170.1	111.4	154.5	113.9	121.0	135.7	117.5	165.8	117.5	135.7	108
		163.6	144.6	39.0	158.2	103.4	169.9	ND	155.5	114.1	121.1	135.3	117.9	165.8	117.9	135.3	
		83.5	72.8	39.2	157.7	ND	ND		154.0	104.3	151.6	115.5	146.0	115.5	120.3		
		84.0	72.6						154.1			116.2	146.1		115.5	121.1	

		β	α	C=O	1'	2'	3'	4'	5'	6'	1	2	3	4	5	6	
		β'	α'	C=O	1''	2''	3''	4''	5''	6''	1''	2''	3''	4''	5''	6''	
111	2',4,(2''-tri-OH-4',4'''-di-OMe-4-O-5''- Bichalcone (Rhuschalcone 1) (bichalcone) (C)	143.8	118.6	191.5	113.9	166.6	101.1	166.1 (56.2)	107.7	130.9	130.7	115.8	158.6	115.8	130.7	202
		144.9	117.4	191.5	113.0	164.6	101.5	158.8 (55.6)	135.1	122.8	128.9	116.0	160.9	116.0	130.3	

Notes: pg, pelargonidin; cy, cyanidin; dp, delphinidin; pn, peonidin; pt, petunidin; mv, malvidin; all, allopyranose; ara, α-arabinopyranose; gal, galactopyranose; glc, glucopyranose; glu, glucuronic acid; rha, rhamnopyranose; xyl, xylopyranose; ac, acetyl; benz, benzoyl; caf, caffeoyl; cin, cinnamoyl; cou, p-coumaroyl; fer, feruloyl; gall, galloyl; mal, malonyl; sinap, sinapoyl; ND, not detected.
ᵃ,ᵇ,§ Assignments may be reversed.

*A, acetone-d6; C, CDCl₃; M, CD₃OD; M/T1, CD₃OD-CF₃COOD 99:1; M/T2, CD₃OD-CF₃COOD 98:2; M/T5, CD₃OD-CF₃COOD 95:5; M/T10, CD₃OD-CF₃COOD 90:10; M/T17, CD₃OD-CF₃COOD 5:1; D, DMSO-d6; D/T10, DMSO-d6-CF₃COOD 90:10; D/T20, DMSO-d6-CF₃COOD 80:20; P, pyridine-d5; T, CF₃COOD; ^{13}C chemical shift values of OMe and Me groups are given in brackets.

†Rotamers: Top — chemical shift values of major rotamer.

‡5 = 3-(4'-Hydroxyphenyl)-5-methoxy-6-(3,3-dimethylallyl)-2'',2''-dimethylchromene-(5'',6'':8,7)-3-(propyl-2-one)-4H-1-benzo-2,3-dihydropyran-2,4-dione.

ᶜ69: COOH (161.17); α (154.55); β (107.48). ᵈ70: COOH (161.44); α (155.7); β (107.4). ᵉ71: COOH (160.52); α (154.46); β (106.8). ᶠ72: COOH (160.2); α (154.5); β (106.2). ᵍ95: 6' (132.2). ʰ96: 6' (130.4); 2'' (77.8); 3'' (130.8); 4'' (116.0); 1'' (28.3). ⁱ97: 6' (130.9); 1'' (40.3); 2'' (148.7); 3'' (113.7); 4''/5'' (27.5). ʲ98: 6' (127.8); 7' (22.0); 8' (122.2); 9' (131.6); 10' (17.9); 2'' (78.3); 3'' (115.3); 4''/5'' (129.1); CH₂COCH₃ (205.9); CH₂COCH₃ (51.7); CH₂COCH₃ (28.9); 2''-Me (28.5, 28.4). ᵏ100: 11a (77.6); 11b (114.8); 3,4 OCH₂O (100.7); 8,9 OCH₂O (101.2). ˡ101: 11 (128.3); 11a (114.0); 12 (187.6); 12a (66.5); 2,3 OCH₂O (101.5); 1' (22.1); 2' (121.5); 3' (132.0); 4' (25.8); 5' (17.8); OMe (56.0). ᵐ102: 11 (129.9); 11a (110.8); 12 (189.9); 12a (66.6); 12b (113.7); 2,3 OCH₂O (101.8); 2' (78.0); 3' (129.4); 4' (115.6); 2'-Me (28.6, 28.4). ⁿ103: 11 (125.1); 11a (116.9); 12 (189.9); 12a (67.3); 12b (113.7); 2,3 OCH₂O (102.4); OMe (56.7); (61.2). ᵒ114: 1' (119.4ᵇ); 2' (122.0); 3' (114.9ᵇ); 4' (156.6ᵃ); 5' (99.4); 6' (154.4ᵇ); 2'' (78.4); 3'' (127.6); ∠'' (122.1); 5'' (26.9). ᵖ115: 1' (118.8ᵇ); 2' (122.3); 3' (108.0); 4' (153.9ᵃ); 5' (114.9); 6' (155.5ᵃ); 1'' (22.0); 2'' (121.7); 3'' (134.3); 4'' (25.0). 5'' (17.8); 1'' (23.1); 2'' (121.4); 3'' (134.9); 4'' (25.3); 5'' (17.8).

TABLE 2.3
¹H NMR Spectral Data of Glycosyl Moieties of Flavonoid Glycosides Recorded in Various Solvents*

Glycosyl Moiety	Chemical Shift (ppm)/Coupling Constants (Hz)						Ref.
	H-1	H-2	H-3	H-4	H-5	H-6	
Glucosides							
3-O-β-glc (D)	5.51	3.19	3.22	3.03	3.07	3.54 / 3.30	52
3-(6''-mal-glc) (M)	5.26 d 7.3	3.56 m	3.54 m	3.42 m	3.49 m	4.35 dd 2.1, 11.8 / 4.21 dd 5.7, 11.8	109
7-O-β-glc (D)	4.87	3.23	3.26	3.12	3.37	3.69 / 3.46	52
7-O-β-(3-p-cou-glc) (D)	5.25 d 7.8	3.51 dd 7.8, 9.4	5.1 t 9.4	3.47 t ca. 9	3.62 m	3.73 dd 1.5, 11.3 / 3.53 dd 5.4, 11.3	206
4'-O-β-glc (D)	5.09	3.31	3.26	3.16	3.41	3.69 / 3.46	52
4'-O-β-glc (M)	5.00	3.61	3.60	3.53	3.57	4.02 / 3.85	52
6-C-β-glc (D)	4.57 d 9.7	4.03 t 9.4	3.19 t 8.6	3.11 t 9.2	3.15 m	3.68 dd 1.8, 12.0 / 3.40 dd 6.3, 12.0	109
6-C-β-glc (M)	4.99 d 9.9	4.26 t 9.9	3.57 m	3.57 m	3.50 m	3.98 m / 3.84 m	109
8-C-β-glc† (D)	4.67 d 9.8 / 4.82 d 9.8	3.82 t 9.6 / 3.86 m	3.24 m / 3.31 m	3.37 t 9.4 / 3.24 m	3.21 m / 3.34 m	3.75 dd 1.4, 11.6 / 3.51 dd 6.7, 11.6 / 3.68 m / 3.48 m	109
8-C-β-glc† (M)	5.07 d 9.7 / 5.16 d 9.7	4.20 t 9.5	3.62 m / 3.67 m	3.78 t 9.2 / 3.63 m	3.56 m / 3.63 m	4.06 m / 3.94 dd 5.3, 12.0 / 3.99 m / 3.87 m	109
8-C-β-(2,(6''-di-ac-glc) (D)	4.92 d 10	5.37 t 10	3.2–3.9 m			4.47 d 12 / 4.12 dd 12, 5	207
8-C-(2''-(4'''-ac-α-rha)-**β-glc**† (M)	5.13 d 9.9 / 5.24 d 9.9	4.31 dd 9.9, 8.6 / 4.18 dd 9.8, 8.9	3.77 dd 8.6, 9.2 / 3.84 m	3.72 t 9.2 / 3.65 m	3.57 dd (b) 5.9, 9.2 / 3.67 m	4.09 dd 1.4, 12.3 / 3.91 m / 4.04 dd 1.4, 12.3 / 3.91 m	109
3-O-β-glc (M/T5)	5.40 d 7.8	3.75 dd 7.8, 9.2	3.61 t 9.2	3.51 m	3.64 m	4.00 dd 2.0, 2.0 / 3.79 dd 12.0, 6.1	170
3-O-(6''-mal-glc) (M/T5)	5.36 d 7.9	3.76 dd 7.9, 9.1	3.63 t 9.1	3.50 dd 9.1, 9.5	3.90 ddd 9.5, 7.3, 1.9	4.64 dd 12.0, 1.9 / 4.37 dd 12.0, 7.3	168
3-O-(6''-mal-glc) (D/T10)	5.41 d 7.0	3.53 t 8.5	3.43 t 9.1	3.25 t 9.0	3.48 dt 9.3	4.47 d 10.5	225
3-O-(6-rha-**glc**) (M/T5)	5.37 d 7.7	3.75 dd 7.7, 9.1	3.64 t 9.1	3.51 dd 9.1, 9.4	3.81 m	4.16 dd 11.3, 1.6 / 3.69 dd 11.3, 5.0	168
3-(2''-xyl-6''-mal-glc) (D/T10)	5.67 d 7.7	4.02 t 8.1	3.67 t 9.0	3.38 t 7.7	3.95 t 9.6	4.41 d 11.6 / 4.17 dd 11.6, 7.7	224
3-glc in Ternatin A3 (D/T10)	4.97 d 8	3.60 t 7	3.43 t 7	3.20–3.70 m	3.70–3.90 m	4.10–4.30 m / 4.50–4.70 m	221
3-glc (M/T5)	4.82 d 7.7	3.77 dd 7.7, 9.3	3.49 dd 8.9, 9.3	3.37 dd 8.9, 9.8	3.27 ddd 2.3, 6.8, 9.8	3.85 dd 11.7, 2.3 / 3.62 dd 11.7, 6.8	176

Compound							Ref.
3-(6''-rha-**glc**)-5-glc (M/T5)	5.60 d 7.7	3.77 m	3.69 m	3.52 t 9.4	3.90 ddd 2.1, 6.5, 9.0	3.90 ddd 2.1, 6.5, 9.0	4.12 dd 12.0, 2.1 / 3.77 m	174
3-(6''-rha-glc)-5-**glc** (M/T5)	5.30 d 7.7	3.79 m	3.68 m	3.68 m	3.71 m	4.06 m	174
5-O-β-glc (M/T5)	5.25 d 7.9	3.76 dd 7.9, 9.3	3.64 m	3.54 t 9.3	3.65 m	3.95 dd 12.3, 5.0 / 4.04 dd 12.0, 2.2 / 3.83 dd 12.0, 5.8	170
5-O-β-glc (D/T10)	5.10	3.45–3.55	3.37	3.24	3.37	3.74 / 3.50	220
5-O-(6''-ac-β-glc) (D/T10)	5.19 d 7.6	3.50 t 7.7	3.40 t 8.9	3.23 t 9.3	3.90 ddd 1.5, 9.0, 10.0	4.01 dd 12.0, 7.8 / 4.37 d 10.5	219
7-O-β-(3''-glc-6''-mal-**glc**) (M/T5)	5.53 d 7.7	3.55 m	3.69 m	3.52 m	3.67 m	4.49 dd 12.1, 1.8 / 4.35 dd 12.1, 5.9	170
3-[2''-(2'''-(E-caf)-glc)-6''-mal-gal]-7-[6'''-(E-caf)-**glc**]-3'-glu (D/T10)	5.35 d 7.0	3.42 m	3.36–3.42	3.36–3.42	3.90 m	4.30 m / 4.50 m	222
3'-glc in Ternatin A3 (D/T10)	5.33 d 7	3.52 t 7	3.39 t 7	3.25 t 8	3.70–3.90 m	4.10–4.30 m / 4.50–4.70 m	221
3,**4'**-di-O-β-glc (M/T5)	5.17 d 7.8	3.69 dd 7.8, 9.2	3.62 t 9.2	3.53 t 9.2	3.64 m	4.04 dd 12.0, 2.1 / 3.84 dd 12.0, 5.7	170
5'-glc in Ternatin A3 (D/T10)	5.33 d 7	3.52 t 7	3.39 t 7	3.25 t 8	3.70–3.90 m	4.50–4.70 m / 4.50–4.70 m	221
Galactosides							
3-O-β-gal (D)	5.38 d 7.7	3.58 dd 9.6, 7.7	3.38 dd 9.6, 3.3	3.66 br dd 3.3	3.34 dd 6.0, 6.3	3.46 dd 6.0, 10.6 / 3.38 dd 6.3, 10.6	109
3-O-β-(2''-O-glc-**gal**) (D)	5.68 d 7.7	3.82 dd 9.4, 3.4	3.64 dd 9.4, 3.4	3.70 dd 3.4, <1	3.29–3.44 m	3.84 dd 5.9, 10.2 / 3.50 dd 6.8, 10.2	215
3-O-β-(6''-O-rha-**gal**) (M)	5.16 d 7.8	3.92 dd 9.8, 7.8	3.66 dd 9.8, 3.5	3.90 ddd 3.5, 1.0	3.74 't' 6.3		109
3-O-β-gal (M/T5)	5.36 d 7.8	4.11 dd 7.8, 9.7	3.78 dd 9.7, 3.7	4.05 d 3.7	3.85 m	3.86 m / 3.86 m	171
3-O-(6''-ac-gal) (M/T5)	5.34 d 7.7	4.11 dd 7.7, 9.6	3.78 dd 9.6, 3.4	4.04 dd 3.4, 1.0	4.15 ddd 1.0, 3.7, 8.3	4.43 dd 8.3, 11.8 / 4.37 dd 3.7, 11.8	171
3-[2''-(2''(-(E-caf)-glc)-6''-mal-**gal**]-7-[6'''-(E-caf)-glc]-3'-glu (D/T10)	5.49 d 7.6	4.24 m	3.69 m	3.74 br s	4.14 m	4.23 m / 4.26 m	222
3'-(2''-gall-gal) (M/T5)	5.55 d 8.0	5.69 dd 8.0, 10.0	4.11 dd 10.0, 3.5	4.15 d 3.5	4.04 m	4.00 dd 11.4, 7.4 / 3.93 dd 11.4, 4.6	114
Glucuronides							
7-O-β-glu (D)	5.18 d 7.5	3.28 dd 8.5, 7.5	3.34 dd 9.4, 8.5	3.41 t 9.4	4.02 d 9.4		214
3-[2''-(2'''-(E-caf)-glc)-6''-mal-gal]-7-[6'''-(E-caf)-glc]-3'-**glu** (D/T10)	5.21 d 7.5	3.40 m	3.45 m	3.53 m	4.09 d 9.8		222
Rhamnosides							
8-C-(2''-(4'''-ac-α-**rha**)-β-glc)† (M)	5.39 d 1.8 / 5.50 d 1.7	3.91 m / 3.80 dd 1.7, 3.3	3.57 dd 3.3, 9.8 / 3.10 dd 3.3, 9.8	4.70 t 9.8 / 4.63 t 9.8	2.41 m / 2.17 m	0.77 d 6.3 / 0.60 d 6.3	109
3-O-α-rha (M)	5.44 d 1.5	4.35 dd 1.5, 3.3	3.92 dd 3.3, 9.5	3.47 t 9.5	3.64 dd 9.5, 6.2	1.08 d 6.2	152
3-O-α-(2''-ac-rha) (M)	5.49 d 1.5	5.53 dd 1.5, 3.4	3.97 dd 3.4, 9.5	3.37 m	3.48 dd 9.6, 6.2	1.05 d 6.3	152

continued

TABLE 2.3
¹H NMR Spectral Data of Glycosyl Moieties of Flavonoid Glycosides Recorded in Various Solvents* — continued

Glycosyl Moiety	Chemical Shift (ppm)/Coupling Constants (Hz)						Ref.
	H-1	H-2	H-3	H-4	H-5	H-6	
3-O-α-(4''-ac-rha) (D)	5.20 d 1	4.04 dd 2, 5	3.75 dd 2, 10	4.73 t 10	3–3.7 m	0.72 d 7	208
7-O-α-rha (D)	5.64 br s	3.92 br s	3.64 m	3.27–3.30 m	3.43 m	1.21 d 6	209
3-rha (M/T5)	6.00	4.37	4.02	3.65	3.65	1.34	223
3-(6''-**rha**-glc)-5-glc (M/T5)	4.74 d 1.5	3.79 m	3.67 dd 3.5, 9.5	3.36 t 9.5	3.60 dd 9.5, 6.3	1.22 d 6.3	174
3-O-(6''-**rha**-glc) (M/T5)	4.75 d 1.6	3.89 dd 1.6, 3.1	3.71 m	3.41 m	3.65 m	1.25 d 6.3	168
Xylosides							
3-O-β-xyl (D)	5.31 d 7.4	3.34	3.14 t 8.7	3.31	2.91 / 3.60		210
3-O-α-xyl (D)	5.20	3.17	3.14	3.21	3.54 / 2.89		210
3'-O-β-xyl (M)	4.90 d 7.6	3.63 dd 7.6, 9.0	3.56 t 8.9	3.71 m	4.10 dd 11.4, 5.3 / 3.48 dd 11.1, 10.5		151
8-C-(2''-glc-β-**xyl**) (M)	5.02 d (10)	4.32	3.90	4.08	4.28 / 3.88		145
(2''-**xyl**-glc) (C + 1 dr D)	4.52 d 7.7	3.26 m	3.30 m	3.48 m	3.65 m		211
3-(2''-**xyl**-6''-mal-glc) (D/T10)	4.59 d 7.7	2.96 t 7.7	3.08 t 8.6	2.73 t 10.7	3.14 m / 3.22 dd 4.7, 10.7		224
Arabinosides							
3'-O-α-ara-furanoside (A + D₂O)	4.12 s	3.78 br s	3.78 br s	4.20 m	3.62 br d 2.6		212
3-O-α-ara (P)	6.09 d 5.2	3.8–5.0 m	3.8–5.0 m	3.8–5.0 m	3.8–5.0 m		236
t-α-ara (D)	3.96 d 7.0	3.17 dd 8.6, 7.0	2.98 dd 8.6, 3.5	3.45 br s	3.52 dd 10.0, 4.0 / 2.94 d 10.0		213
3-O-α-ara (D)	5.22	3.73	3.48 dd 6.7, 3.1	3.62	3.18 / 3.57		210
3-O-α-ara-furanoside (D)	5.53	4.11	3.67	3.51	3.25		210
8-C-β-ara (D) (spectra recorded at 70°C)	4.81 d 9.5	4.04 br t	3.54 dd 9.0, 2.5	3.87 br s	3.93 dd 12.0, 1.5 / 3.67 d 12.0		47
Other sugars							
t-O-β-(6''-ac-all) (D)	4.93 d 8.0	3.27 dd 8.0, 2.6	3.93 't' 2.6	3.43 dd 10.0, 2.8	3.88 ddd 10.0, 4.8, 2.3	4.10 dd 12.0, 2.3 / 4.04 dd 12.0, 4.8	142
6-C-β-L-boivinoside (D)	5.33 dd 12.3, 3.1	1.50 ddd 4.3	3.86 ddd 3.4	3.25 ddd 4.3	4.04 q 6.0	1.17 d 6.0	216
6-C-β-D-olioside (D)	5.01 dd 11.7, 3.2	2.05 q 11.7 / 1.60 ddd 11.9, 5.2, 3.2	3.77 ddd 12.7, 5.2, 2.4	3.46 dd 2.4	3.65 q 6.4	1.18 d 6.4	217
6-C-chinovoside (D)	4.91, 9.4	4.28, 9.2	3.46, 9.1	3.22, 9.1	3.48, 6.2	1.36	218
6-C-fucoside (D)	4.92, 9.6	4.18, 9.4	3.64, 3.0	3.80 (small)	3.84, 6.4	1.34	218

Notes: all, allopyranose; ara, arabinopyranose; gal, galactopyranose; glc, glucopyranose; glu, glucuronic acid; rha, rhamnopyranose; xyl, xylopyranose; ac, acetyl; caf, caffeoyl; cou, p-coumaroyl; gall, galloyl; mal, malonyl; t, terminal.

* A, acetone-*d6*; C, CDCl₃; M, CD₃OD; M/T5, CD₃OD-CF₃COOD 95:5; D, DMSO-*d6*; D/T10, DMSO-*d6*-CF₃COOD 90:10; T, CF₃COOD; P, pyridine-*d5*.

†Rotamers: Top — chemical shift values of major rotamer. The parts in bold indicate the monosaccharide involved.

TABLE 2.4
^{13}C NMR Spectral Data of Glycosyl Moieties of Flavonoid Glycosides Recorded in Various Solvents*

Glycosyl Moiety	Chemical Shift (ppm)						
	C-1	C-2	C-3	C-4	C-5	C-6	Ref.
Glucosides							
3-O-β-glc (D)	100.69	74.19	76.49	70.19	77.76	61.46	52
3-(6″-mal-glc) (M)	104.36	75.61	77.83	71.15	75.53	64.89	109
7-O-β-glc (D)	99.73	73.15	76.50	69.83	77.83	60.79	52
4′-O-β-glc (D)	101.56	73.55	76.49	69.88	77.19	61.04	52
4′-O-β-glc (M)	103.42	74.82	77.54	71.32	78.36	62.44	52
6-C-β-glc (D)	73.12	70.27	79.02	70.70	81.65	61.57	109
6-C-β-glc (M)	75.28	72.60	80.12	71.79	82.62	62.86	109
8-C-β-glc[†] (D)	73.45	70.89	78.70	70.25	81.74	61.27	109
	74.35	71.09	78.72	70.60	81.93	61.35	
8-C-β-glc[†] (M)	75.36	72.85	80.33	72.30	82.93	63.23	109
	76.28	72.68	79.84	71.29	82.62	63.51	
8-C-(2″-(4‴-ac-rha)-**glc**)[†] (M)	73.77	76.09	81.80	72.51	82.99	63.13	109
	75.30	76.29	81.45	71.50	83.00	62.45	
3-O-β-glc (M/T5)	103.72	74.79	78.14	71.08	78.77	62.36	170
3-O-(6″-mal-glc) (M/T5)	103.62	74.65	77.92	71.33	75.94	65.45	168
3-**glc** (M/T5)	105.59	75.39	77.57	71.44	78.91	62.80	176
3-(6″-rha-**glc**) (M/T5)	103.53	74.69	78.02	71.22	77.44	67.79	168
3-(6′-rha-**glc**)-5-glc (M/T5)	102.59	74.74	78.34	71.47	77.74	67.51	174
3-(6″-rha-glc)-5-**glc** (M/T5)	102.75	74.74	77.89	70.75	78.67	62.08	174
5-O-β-glc (M/T5)	102.7	74.6	78.1	71.2	78.6	62.6	170
5-O-(6″-ac-β-glc) (D/T10)	101.3	73.1	75.8	70.1	74.4	63.6	219
7-O-β-(3″-glc-6″-mal-**glc**) (M/T5)	95.3	74.9	87.3	71.2	75.0	65.4	170
3,4′-di-O-β-glc (M/T5)	102.3	74.5	77.8	71.1	78.3	62.4	170
Galactosides							
3-O-β-gal (D)	102.00	71.38	73.30	68.11	75.99	60.33	109
3-O-β-(6″-O-rha-**gal**) (M)	105.98	73.14	75.08	70.18	75.28	67.32	109
3-O-β-gal (M/T5)	104.63	72.16	74.87	70.14	77.80	62.35	171
3-O-(6″-ac-gal) (M/T5)	104.06	71.89	74.60	70.31	75.15	65.20	171
3′-(2″-gall-6″-ac-gal) (M/T5)	101.69	72.67	71.96	70.24	75.34	65.13	114
Glucuronide							
7-O-β-glu (M)	102.3	74.3	77.1	72.9	76.4	173.0	72
Rhamnosides							
3-O-α-rha (M)	103.64	71.90	72.15	73.37	72.07	17.67	152
3-O-α-(2″-ac-rha) (M)	100.36	73.39	70.49	73.50	72.11	17.70	152
8-C-(2″-(4‴-ac-**rha**)-glc)[†] (M)	101.15	72.09	70.02	75.16	67.27	17.84	109
	101.19	71.80			67.03	17.95	
3-rha (M/T5)	101.99	71.40	72.30	73.28	72.21	18.04	223
3-(6″-**rha**-glc) (M/T5)	102.19	71.87	72.44	73.92	69.77	17.87	168
3-(6″-**rha**-glc)-5-glc (M/T5)	102.21	71.93	72.28	73.86	69.77	17.94	174
Xylosides							
3-O-β-xyl (D)	102.5	74.3	77.0	70.2	66.9		210
3′-O-α-xyl (D)	97.5	69.5	71.3	65.1	61.7		210

continued

TABLE 2.4
^{13}C NMR Spectral Data of Glycosyl Moieties of Flavonoid Glycosides Recorded in Various Solvents* — continued

Glycosyl Moiety	Chemical Shift (ppm)						
	C-1	C-2	C-3	C-4	C-5	C-6	Ref.
3'-O-β-xyl (M)	104.95	74.65	77.41	71.04	67.10		151
(2''-**xyl**-glc) (C + 1 drop D)	102.6	71.8	71.2	67.5	62.8		211
8-C-(2''-glc-β-**xyl**) (M)	76.04	81.34	77.04	70.17	72.10		145
Arabinosides							
3'-O-α-ara-furanoside (A + D$_2$O)	108.0	80.8	79.4	90.1	63.2		212
3-O-α-ara (P)	104.5	72.8	74.1	68.2	66.4		163
t-α-ara (D)	103.1	70.6	72.6	67.5	65.1		213
3-O-α-ara (D)	102.3	71.7	72.6	66.9	65.2		210
3-O-α-ara-furanoside (D)	108.6	83.0	77.6	86.8	61.6		210
Other sugars							
t-O-β-(6''-ac-all) (D)	102.5	71.4	70.7	66.8	71.5	63.5	142
6-C-β-boivinoside (M)	67.3	31.5	69.6	71.3	72.5	17.7	237
6-C-β-D-olioside (D)							217
6-C-chinovoside (D)							218
6-C-fucoside (D)	73.4	68.6	75.0	71.6	73.9	17.0	238

Notes: all, allopyranose; ara, arabinopyranose; gal, galactopyranose; glc, glucopyranose; glu, glucuronic acid; rha, rhamnopyranose; xyl, xylopyranose; ac, acetyl; mal, malonyl; t, terminal.

*A, acetone-*d6*; C, CDCl$_3$; M, CD$_3$OD; M/T5, CD$_3$OD-CF$_3$COOD 95:5; D, DMSO-*d6*; D/T10, DMSO-*d6*-CF$_3$COOD 90:10; T, CF$_3$COOD; P, pyridine-*d5*.

†Rotamers: Top — chemical shift values of major rotamer. The parts in bold indicate the monosaccharide involved.

2.3 MASS SPECTROMETRY

Modern mass spectrometric techniques are very well suited for the analysis of flavonoids isolated from plants and foodstuffs and in their *in vivo* metabolite forms (Table 2.8). Progress during the last two decades has made MS the most sensitive method for molecular analysis of flavonoids. MS has the potential to yield information on the exact molecular mass, as well as on the structure and quantity of compounds with the nature and within the mass range of flavonoids. Furthermore, due to the high power of mass separation, very good selectivities can also be obtained.

The purpose of the MS techniques is to detect charged molecular ions and fragments separated according to their molecular masses. Most flavonoid glycosides are polar, non-volatile, and often thermally labile. Conventional MS ionization methods like electron impact (EI) and chemical ionization (CI) have not been suitable for MS analyses of these compounds because they require the flavonoid to be in the gas phase for ionization. To increase volatility, derivatization of the flavonoids may be performed. However, derivatization often leads to difficulties with respect to interpretation of the fragmentation patterns. Analysis of flavonoid glycosides without derivatization became possible with the introduction of desorption ionization techniques. Field desorption, which was the first technique employed for the direct analysis of polar flavonoid glycosides, has provided molecular mass data and little structural information.[239] The technique has, however, been described as "notorious for the transient

TABLE 2.5
Typical ^{1}H NMR Chemical Shift Values of Acyl Moieties of Acylated Flavonoid Glycosides Recorded in DMSO-$d6$ (D) or CD$_3$OD (M) with Various Proportions of CF$_3$COOD (T) or Acetone (A)

	2	3	4	5	6	α	β	4-OH	Ref.
Acetyl (M/T10)	1.99 s								167
Malonyl (M/T10)	3.41 s								167
Succinyl (M/T2)	2.60 m	2.46 m							226
Malyl (M/T1)	2.80 dd 17, 10 4.40 dd 10, 2 2.50 dd 17, 2								227
4-OH-2-Methylenebutanoyl		2.34 t 6.5	3.45 q 6.5	6.06 s 5.44 s				4.4 t 5.4	188
Tartaryl (D/T10)	5.22 d 2.4	4.54 d 2.4							228
p-OH-Benzoyl (D/T10)	7.93 'd' 8.6				7.15 'd' 8.6				229
Galloyl (M/T5)	7.07 s				7.07 s				171
E-Cinnamoyl (A)	7.75 ddd 8.0, 1.5, 0.5	7.45 m	7.45 m	7.45 m	7.75 ddd 8.0, 1.5, 0.5	6.56 d 16.5	7.66 d 16.5		190
E-p-Coumaroyl (M/T17)	7.10 'd' 8.6	6.65 'd' 8.6		6.65 'd' 8.6	7.10 'd' 8.6	6.23 d 15.7	7.48 d 15.7		65
Z-p-Coumaroyl (D/T10)	7.09 'd' 8.6	6.31 'd' 8.6		6.31 'd' 8.6	7.09 'd' 8.6	6.37 d 13	5.72 d 13		173
E-Caffeoyl (M/T17)	6.95 d 2.0			7.12 d 8.6	6.71 dd 2.0, 8.6	5.94 d 15.7	7.18 d 15.7		65
Z-Caffeoyl (M/T5)	7.96 d 2.0			6.70 d 8.5	7.03 dd 2.0, 8.5	5.81 d 13	6.77 d 13		230
E-3,5-di-OH-Cinnamoyl (M/T17)	6.51 s		6.81 s		6.51 s	6.16 d 16.0	7.39 d 16.0		65
E-Feruloyl (D/T10)	7.08 d 2.0	OCH$_3$ 3.75 s		6.76 d 8.5	6.97 dd 2.0, 8.5	6.29 d 16.0	7.37 d 16.0		231
E-Sinapoyl (M/T5)	6.60 s	OCH$_3$ 3.80 s		OCH$_3$ 3.80 s	6.60 s	6.22 d 15.9	7.34 d 15.9		232

Notes: A, acetone-$d6$; M, CD$_3$OD: M/T1, CD$_3$OD-CF$_3$COOD 99:1; M/T5, CD$_3$OD-CF$_3$COOD 95:5; M/T10, CD$_3$OD-CF$_3$COOD 90:10; M/T17, CD$_3$OD-CF$_3$COOD 5:1; D, DMSO-$d6$; D/T10, DMSO-$d6$-CF$_3$COOD 90:10; T, CF$_3$COOD.

TABLE 2.6

Typical ^{13}C NMR Chemical Shift Values of Acyl Moieties of Acylated Flavonoid Glycosides Recorded in DMSO-d6 (D) or CD$_3$OD (M) with Various Proportions of CF$_3$COOD (T) or Acetone (A)

	1	2	3	4	5	6	α	β	C = O	Ref.
Acetyl (M/T10)	172.9	20.7								167
Malonyl (M/T10)	168.5	41.9	170.5							167
Oxalyl	169.95[a]	168.53[a]								233
Dioxalyl (M/T5)	158.1	172.4	172.4	174.8						234
Malyl (M/T1)	174.4	40.0	69.8	171.5						227
4-OH-2-Methylenebutanoyl	167.1	137.8	35.8	60.3	128					188
Tartaryl (D/T10)	171.7	74.2	69.7	167.9						228
p-OH-Benzoyl (D/T10)	120.4	131.3	115.3	162.0	115.3	131.3			165.3	235
Galloyl (M/T5)	120.83	110.48	146.33	140.14	146.33	110.48			168.11	171
E-Cinnamoyl (A)	135.7	129.2	129.9	131.0	129.9	129.2	119.6	145.1	166.3	190
E-p-Coumaroyl (M/T17)	125.0	130.1	115.8	159.8	115.8	130.1	115.5	144.8	166.2	65
Z-p-Coumaroyl (D/T10)	127.0	133.3	115.5	159.5	115.5	133.3	115.8	143.9	168.9	173
E-Caffeoyl (M/T17)	127.5	115.6	144.6	147.2	115.9	120.6	114.2	146.2	166.3	65
E-3,5-di-OH-Cinnamoyl (M/T17)	127.6	122.8	146.6	139.0	146.6	122.8	115.1	146.8	168.9	65
E-Feruloyl (D/T10)	125.6	111.7	148.0 (55.8)	149.5	115.7	123.1	114.1	145.6	166.7	231
E-Sinapoyl (M/T5)	125.80	106.51	148.92 (56.56)	139.44	148.92 (56.56)	106.51	115.00	146.94	168.42	232

Notes: A, acetone-d6; M, CD$_3$OD; M/T1, CD$_3$OD-CF$_3$COOD 99:1; M/T5, CD$_3$OD-CF$_3$COOD 95:5; M/T10, CD$_3$OD-CF$_3$COOD 90:10; M/T17, CD$_3$OD-CF$_3$COOD 5:1; D, DMSO-d6; D/T10, DMSO-d6-CF$_3$COOD 90:10; T, CF$_3$COOD; ^{13}C chemical shift values of OMe and Me groups are given in brackets.

nature of the spectra,"[240] and drawbacks related to the preparation of the MS samples have restricted application of this technique. Another method, desorption chemical ionization (DCI), provides rapid heating of the analyte and overcomes the problem of thermal decomposition inherent in conventional CI. The combined use of positive- and negative-ion DCI-MS has been shown to be an alternative approach for the structural characterization of flavonoid glycosides;[241] however, this method has been applied infrequently to flavonoid analysis in recent years.[242,243] Plasma desorption mass spectrometry (PD-MS) is another MS method used for flavonoid analysis; however, its application has in recent years been limited to some papers on anthocyanins including deoxyanthocyanidins.[244–246] Fast atom bombardment (FAB) MS is still popular for flavonoid analysis (Section 2.3.1.2). In this technique, the flavonoid is solubilized in a nonvolatile polar matrix and deposited on a copper target, which is bombarded with fast neutral energized particles such as xenon or argon and thereby inducing the desorption and ionization.

In parallel with these developments, other techniques have been introduced that were especially applicable to the combination of liquid chromatography with MS. The most interesting, from the point of view of structural studies of flavonoid glycosides, are thermospray (TSP) and atmospheric pressure ionization (API) methods, which include electrospray ionization (ESI) and atmospheric pressure chemical ionization (APCI). TSP was the first

TABLE 2.7

Compounds Included in the NMR Tables. Individual Pigment Structures Are Shown in Figure 2.8–Figure 2.16

1. 6-OH-Luteolin 4′-methyl ether-7-(2″-α-rhamnoside-6‴-acetyl-β-glucoside)
2. 6-OH-Luteolin 7-(6″-(E)-caffeoyl)-β-glucoside
3. Isoscutellarein 7-(2″-(6‴-acetyl)-β-allosyl)-β-glucoside
4. Isoscutellarein 4′-methyl ether-7-(2″-(6‴-acetyl)-β-allosyl)-β-glucoside
5. Apigenin 4′-(2″-(2‴-feruloyl-glucuronyl)-glucuronide)
6. Apigenin 7-glucuronide-4′-(2″-(2‴-feruloyl-glucuronyl)-glucuronide)
7. Apigenin 7-glucuronyl-4′-(2″-(2‴-E-p-coumaroyl-glucuronyl)-glucuronide)
8. Luteolin 3′-β-glucoside-4′-(2″-α-rhamnosyl-β-glucoside)
9. Luteolin 3′,4′-di-β-glucoside
10. 5,7,4′-tri-OH-3′-OMe-Flavone 8-C-(2″-O-β-glucosyl-β-xyloside)
11. 5,7-di-OH-3′-OMe-4′-Acetoxyflavone 8-C-(2″-O-β-glucosyl-β-xyloside)
12. Iso-orientin 3′-methyl ether
13. 8-C-p-OH-Benzoyl-isovitexin 4′-glucoside
14. Apigenin 8-C-(2″-(4‴-acetyl-rhamnosyl)-glucoside)
15. Spinosin
16. 6‴-Feruloyl-spinosin
17. Isoscoparin 7-glucoside
18. Carlinoside
19. Kaempferol 3-(6″-α-arabinosyl-glucoside)
20. Kaempferol 3-(6″-α-arabinosyl-glucoside)-7-glucoside
21. Kaempferol 3-(2″-rhamnosyl-6″-malonyl-glucoside)
22. Kaempferol 3-glucoside-7-(2″-(6‴-p-coumaroyl-glucosyl)-glucoside)
23. 8-OMe-Kaempferol 3-(6″-malonyl-glucoside)
24. Quercetin
25. Quercetin 4′-glucoside
26. Quercetin 3′-xyloside
27. Myricetin 3-(2″-acetyl-rhamnoside)
28. Quercetin 3,4′-diglucoside
29. Isorhamnetin 3-rutinoside
30. Quercetin 3,7,4′-triglucoside
31. Isorhamnetin 3,7-diglucoside
32. Myricetin 3-(2″-rhamnosyl-glucoside)
33. Myricetin 3′-(6″-p-coumaroyl-glucoside)
34. Myricetin 7-(6″-galloyl-glucoside)
35. Laricitrin 3-α-arabinofuranoside
36. Laricitrin 3-glucoside
37. Syringetin 3-(5″-glucosyl-α-arabinofuranoside)
38. Syringetin 3-(6″-acetyl-glucoside)
39. Syringetin 3-robinobioside
40. Syringetin 6-C-glucoside
41. 6,3′-di-OH-4,4′-di-OMe-5-Me-Aurone
42. 4,6,3′,4′-tetra-OMe-Aurone (Z-form)
43. 4,6,3′,4′-tetra-OMe-Aurone (E-form)
44. 6,3′,4′-tri-OH-4-OMe-5-Me-Aurone
45. Maesopsin
46. Maesopsin 6-O-glucoside (two diastereoisomers)
47. Licoagroaurone

continued

TABLE 2.7
Compounds Included in the NMR Tables. Individual Pigment Structures Are Shown in Figure 2.8–Figure 2.16 — *continued*

48. 3'-formyl-4',6'-di-OH-2'-OMe-5-Me-Chalcone
49. Chalcononaringenin 2',4'-diglucoside
50. 2',4'-diOH-4'-OMe-6'-glucoside Dihydrochalcone
51. 2'-OH-3',4',6'-tri-OMe-Dihydrochalcone
52. Pelargonidin 3-glucoside-5-(6'''-acetyl-glucoside)
53. Pelargonidin 3-(6''-feruloyl-glucoside)-5-(6'''-malonyl-glucoside)
54. Cyanidin 3-(6''-malonyl-glucoside)
55. Cyanidin 3-rutinoside
56. Cyanidin 3-(2'',3''-digalloyl-glucoside)
57. Cyanidin 3,4'-diglucoside
58. Delphinidin 3-(6''-acetyl-galactoside)
59. Delphinidin 3'-(2''-galloyl-6''-acetyl-galactoside)
60. Peonidin 3-rutinoside
61. Petunidin 3,7-diglucoside
62. Petunidin 3-(6''-*E-p*-coumaroyl-glucoside)-5-(6'''-malonyl-glucoside)
63. Malvidin 3-(6''-*E-p*-coumaroyl-glucoside)-5-glucoside
64. Malvidin 3-(6''-*Z-p*-coumaroyl-glucoside)-5-glucoside
65. Malvidin 3-rutinoside-5-glucoside
66. Malvidin 3-(6''-(4''''-malonyl-rhamnosyl)-glucoside)-5-(6'''-malonyl-glucoside)
67. Apigeninidin 5-glucoside
68. Luteolinidin 5-glucoside
69. CarboxypyranoPelargonidin 3-glucoside
70. CarboxypyranoCyanidin 3-glucoside
71. CarboxypyranoCyanidin 3-(6''-malonyl-glucoside)
72. CarboxypyranoMalvidin 3-glucoside
73. Judaicin 7-(6''-acetyl-glucoside)
74. Tectorigenin 4'-(6''-glucosyl-glucoside)
75. 7-OH-6'-OMe-3',4'-methylenedioxyisoflavone 7-glucoside
76. Irisjaponin A
77. Irisjaponin B
78. Junipegenin B
79. Matteucinol 7-(6''-apiofuranosyl-β-glucoside)
80. Hesperitin 7-(2''-galactosyl-6''-rhamnosyl-glucoside)
81. Persicogenin 5,3'-di-OH-7,4'-di-OMe-flavanone
82. Naringenin 7-glucoside
83. Naringenin 7-(6''-galloyl-glucoside)
84. Taxifolin 4'-glucoside
85. Aromadendrin 7-glucoside
86. Ampelopsin 7-glucoside
87. 2''-Accallunin
88. 2R,3R-*trans*-aromadendrin 7-(6''-(4'''-OH-2'''-methylenebutanoyl)-glucoside)
89. (2R,3S)-(+)-3',5-di-OH-4',7-di-OMe-Dihydroflavonol
90. 3-Desoxycallunin
91. Catechin 3-(6''-cinnamoyl-glucoside)
92. Catechin 3-(2''-cinnamoyl-glucoside)
93. Catechin 3-(2'',6''-dicinnamoyl-glucoside)
94. Anadanthoside

TABLE 2.7
Compounds Included in the NMR Tables. Individual Pigment Structures Are Shown in Figure 2.8–Figure 2.16 — *continued*

95. Cajanin
96. Indicanine C
97. 6-(1,1-di-Me-allyl)-7,4'-di-OH-Flavan
98. 3-(4'-hydroxyphenyl)-5-methoxy-6-(3,3-dimethylallyl)-2'',2''-dimethylchromene-(5'',6'':8,7)-3-(propyl-2-one)-4H-1-benzo-2,3-Dihydropyran-2,4-dione
99. Maackianin 3-(6''-malonyl-glucoside)
100. 3,4:8,9-Dimethylenedioxy-pterocarpan
101. Usararotenoid C
102. 12a-Epimillettosin
103. (+)-Usararotenoid-B
104. [Catechin 3-glucoside-(4α → 8)-catechin 3-(2''-cinnamoyl-glucoside)]
105. [Catechin 3-glucoside-(4α → 8)-epicatechin 3-(6''-cinnamoyl-glucoside)]
106. Amentoflavone
107. Aulacomnium–biaureusidin
108. Cupressuflavone 7,7''-dimethyl ether
109. 4,4',6-tri-O-methyl-2-deoxymaesopsin-(2 → 7)-2,4,4',6-tetra-O-Methylmaesopsin
110. Catechin-(4α → 8)-pelargonidin 3-glucoside
111. 2',2'',2'''-tri-OH-4',4'''-di-OMe-4-O-5'''-bichalcone (Rhuschalcone 1)
112. Puerarin (Daidzein 8-C-glucoside)
113. Calycosin
114. Isoneorautenol
115. Erybraedin A

method to combine true LC–MS compatibility with the ability to determine nonvolatile thermally labile compounds. The method has, for instance, enabled the analysis of mixtures of polar flavonoid glycosides,[247,248] and allowed the detection of monomeric flavan-3-ols and dimeric proanthocyanidins.[249] However, the application of this technique has some limitations related to the thermal stability of the compounds studied. This is connected to the high temperature in the TSP ion source, which is necessary for the efficient ionization of the molecules to be analyzed. In addition, the efficiency of ion production varies widely with compound type, and the flow rate and temperature of the inlet tube must be optimized for each type of compound. LC–TSP-MS has provided for the characterization of catechins and flavonoids from their CID spectra of the quasimolecular ion.[250] In this study, flavonoids exhibited three types of ring cleavage in the pyran ring, and the differentiation among flavanone, flavone, and flavonol was possible. LC–TSP-MS has also been applied for the detection and identification of a wide range of other flavonoids (see Section 1.4.5). The technique has, however, been gradually phased out by ESI and APCI, which in recent years seem to be the most useful ionization techniques for the characterization of flavonoids (see Sections 1.4.5, 2.3.1.4, 2.3.2.2, 2.3.3, and 2.3.4). These also include matrix-assisted laser desorption ionization (MALDI) MS coupled with a time-of flight (TOF) mass analyzer, which is another soft ionization technique that allows the analysis of small quantities of flavonoids (see Sections 2.3.1.3, 2.3.3, and 2.3.4). The major advantages of most of the soft ionization techniques are that they include those of minute sample sizes, and the possibility of coupling MS with different chromatographic techniques, e.g., gas chromatography (GC–MS), capillary electrophoresis (CE–MS), and, in particular, liquid chromatography (LC–MS) (see Sections 1.4.5 and 2.3.2.2).

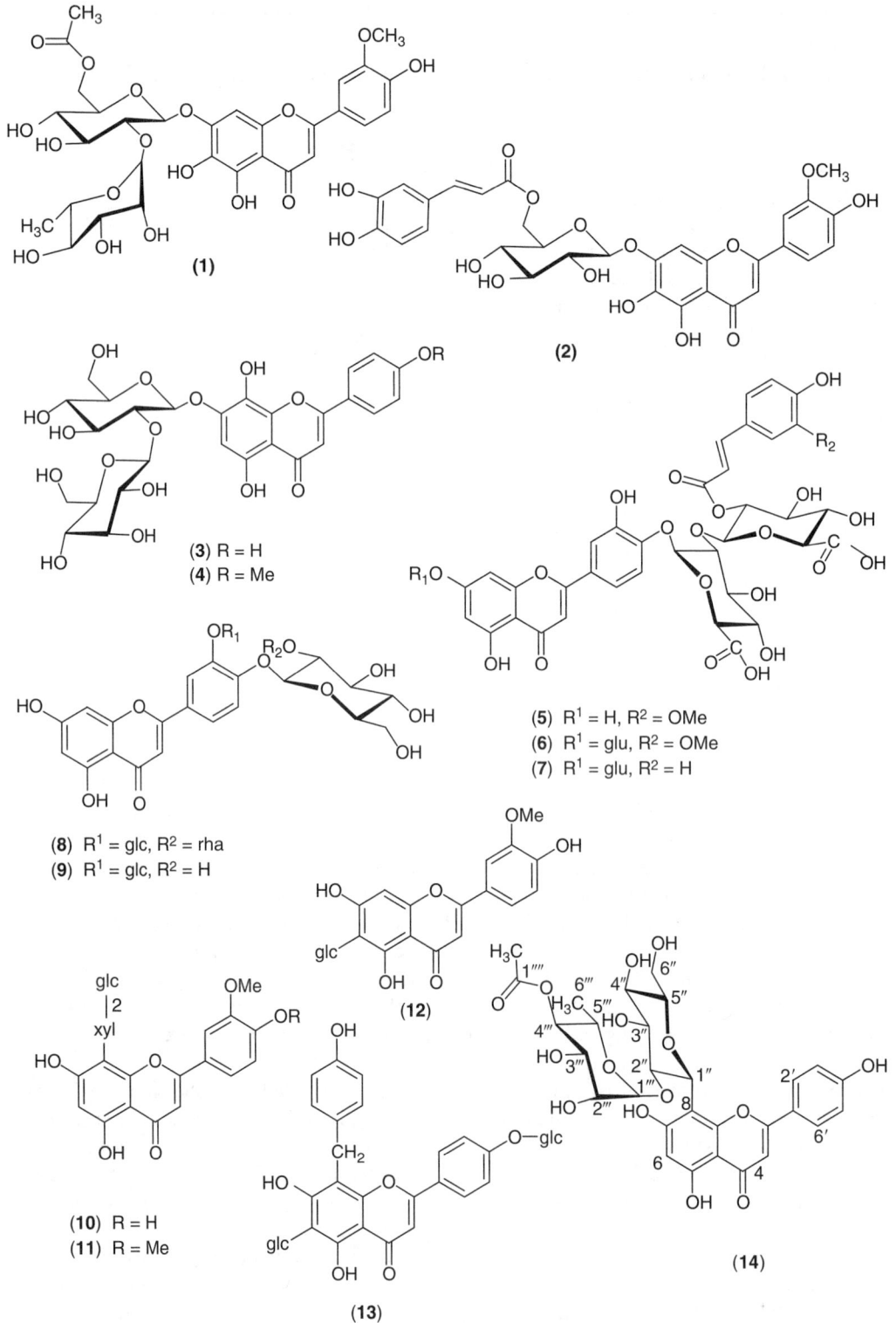

FIGURE 2.8 Structures of compounds **1**–**14**. See Table 2.7 for names.

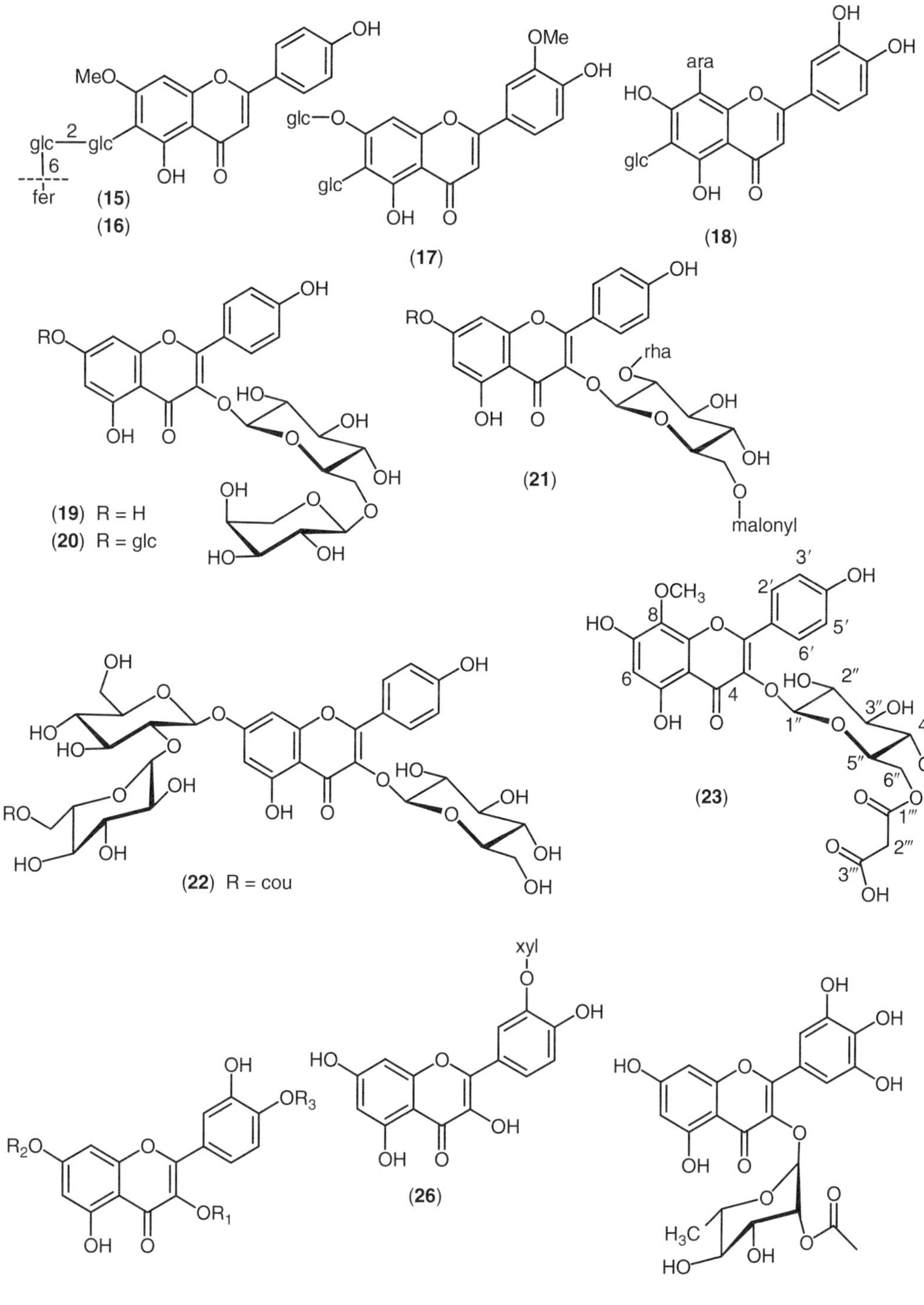

FIGURE 2.9 Structures of compounds **15–29**. See Table 2.7 for names.

(**19**) R = H
(**20**) R = glc

(**22**) R = cou

(**24**) R^1 = H, R^2 = H, R^3 = H
(**25**) R^1 = H, R^2 = H, R^3 = glc
(**28**) R^1 = glc, R^2 = H, R^3 = glc
(**29**) R^1 = glc, R^2 = glc, R^3 = glc

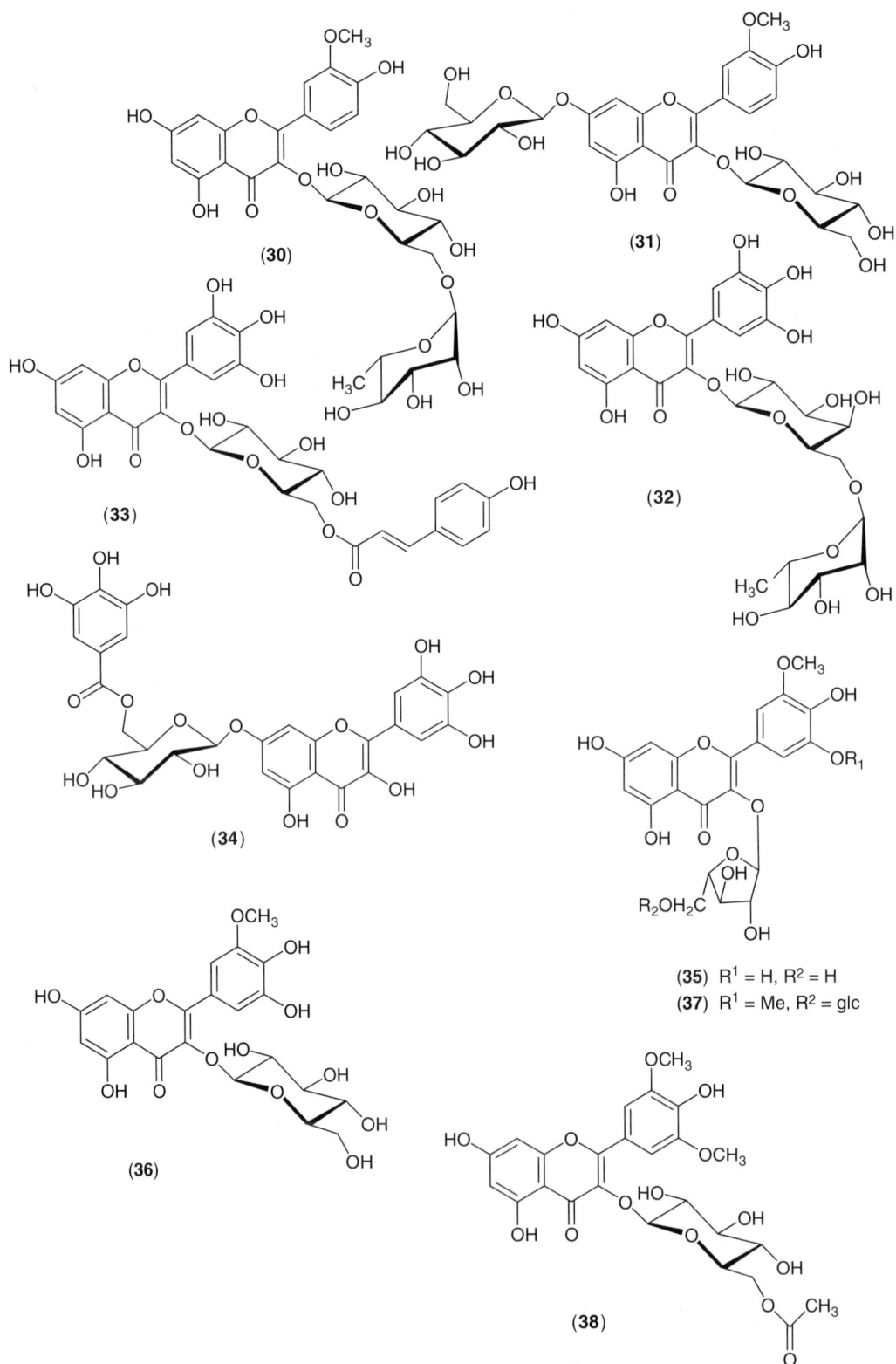

FIGURE 2.10 Structures of compounds **30**–**38**. See Table 2.7 for names.

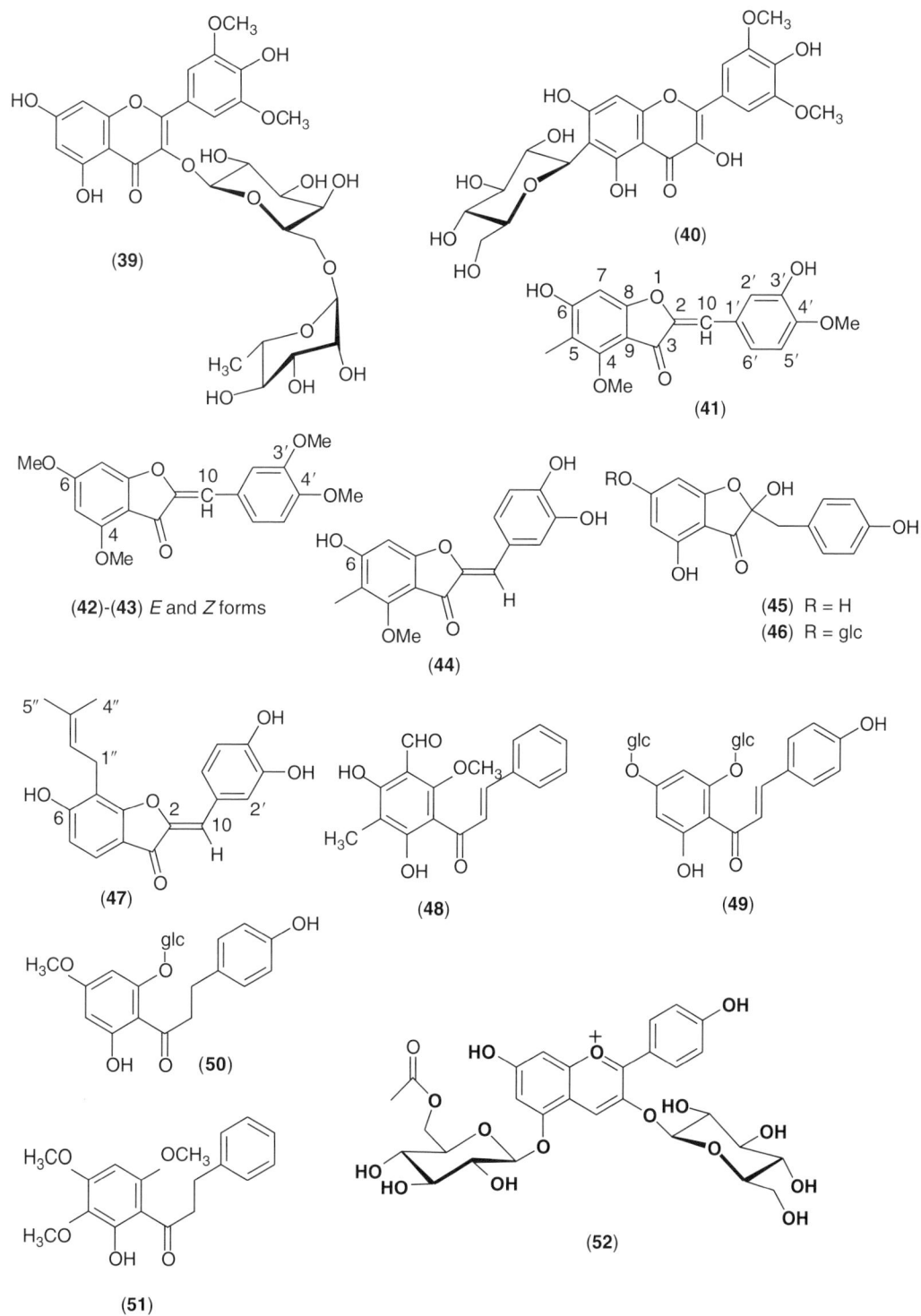

FIGURE 2.11 Structures of compounds **39–52**. See Table 2.7 for names.

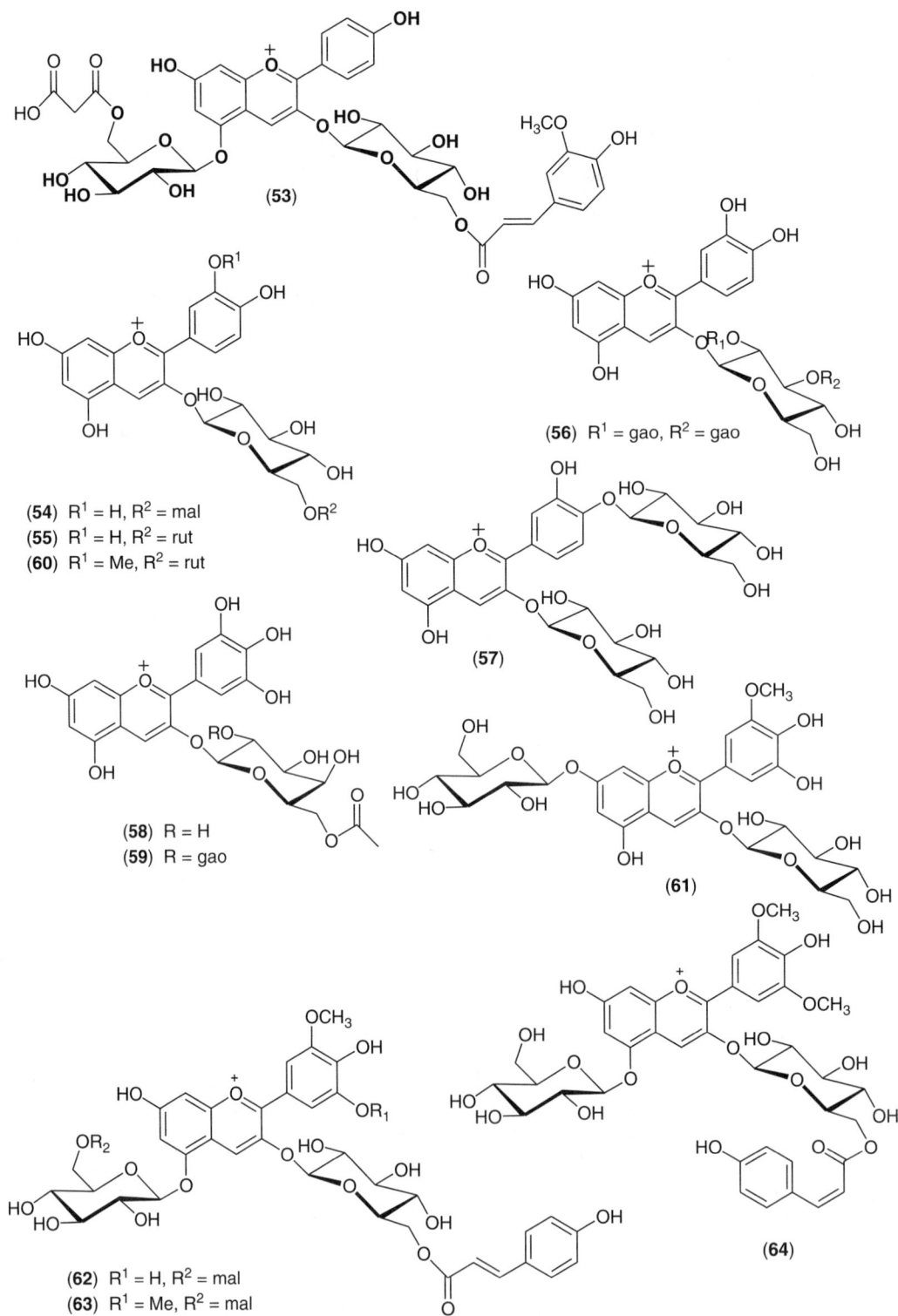

FIGURE 2.12 Structures of compounds **53–64**. See Table 2.7 for names.

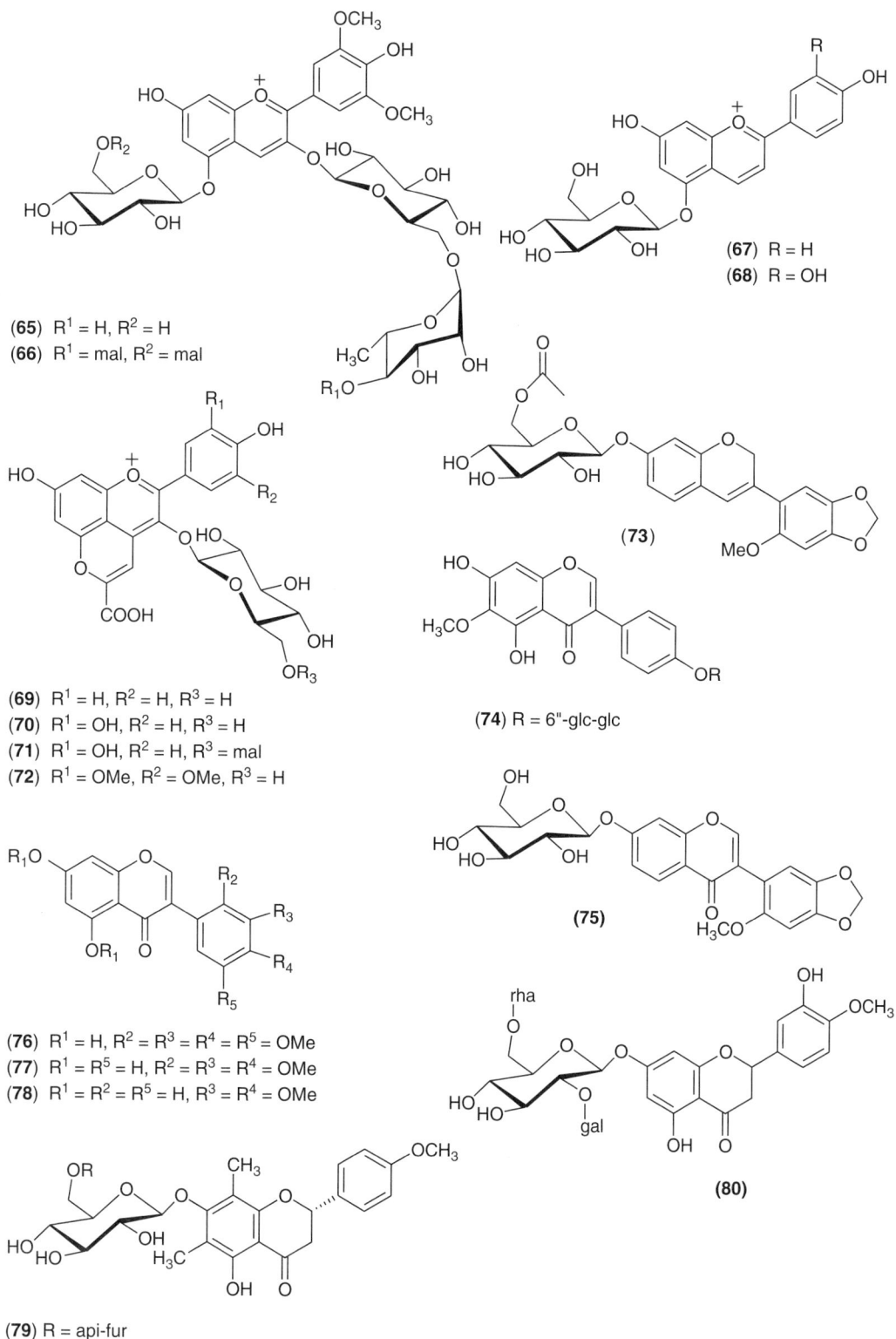

(65) R^1 = H, R^2 = H
(66) R^1 = mal, R^2 = mal

(67) R = H
(68) R = OH

(69) R^1 = H, R^2 = H, R^3 = H
(70) R^1 = OH, R^2 = H, R^3 = H
(71) R^1 = OH, R^2 = H, R^3 = mal
(72) R^1 = OMe, R^2 = OMe, R^3 = H

(73)

(74) R = 6''-glc-glc

(75)

(76) R^1 = H, R^2 = R^3 = R^4 = R^5 = OMe
(77) R^1 = R^5 = H, R^2 = R^3 = R^4 = OMe
(78) R^1 = R^2 = R^5 = H, R^3 = R^4 = OMe

(80)

(79) R = api-fur

FIGURE 2.13 Structures of compounds **65–80**. See Table 2.7 for names.

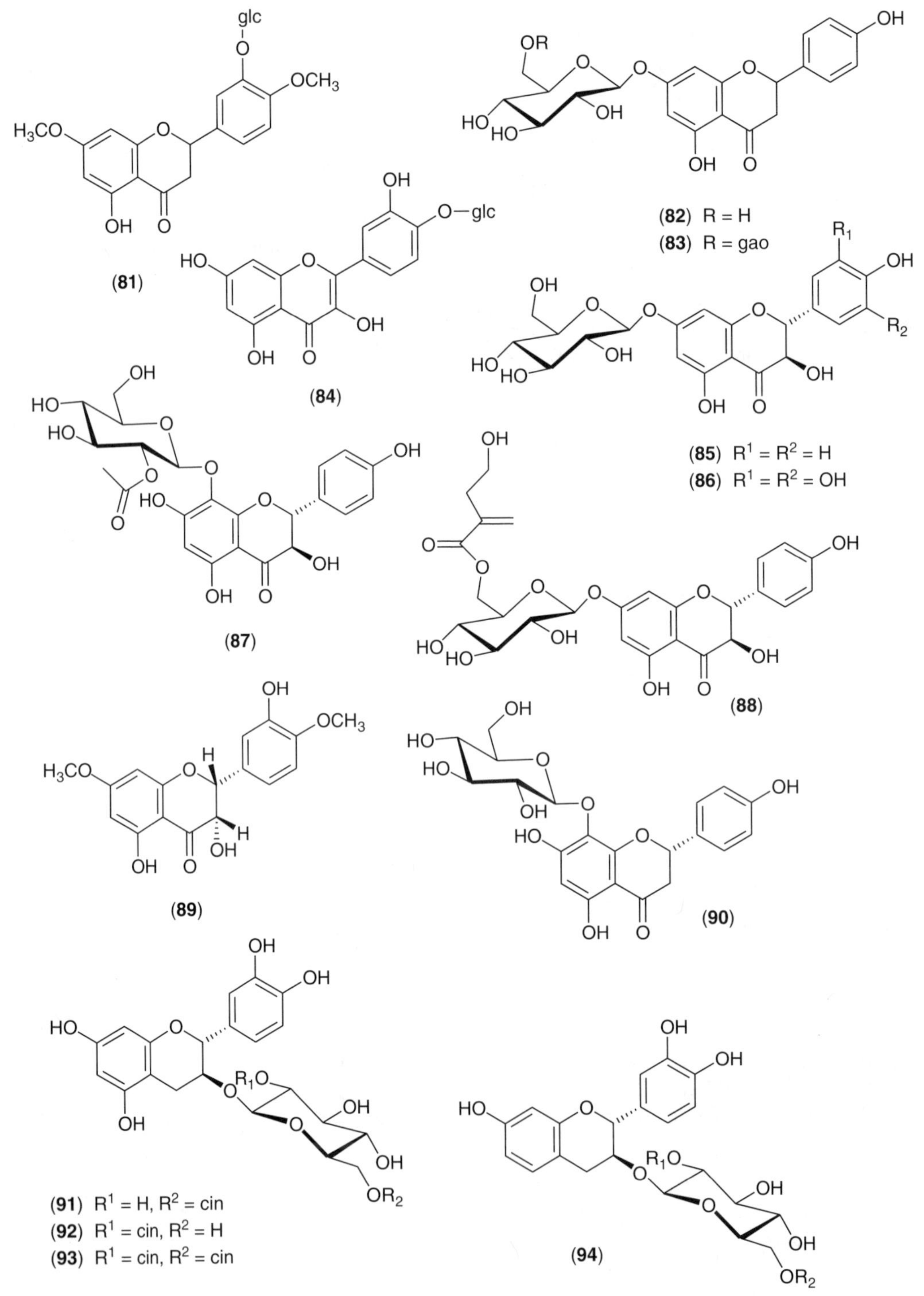

FIGURE 2.14 Structures of compounds **81–94**. See Table 2.7 for names.

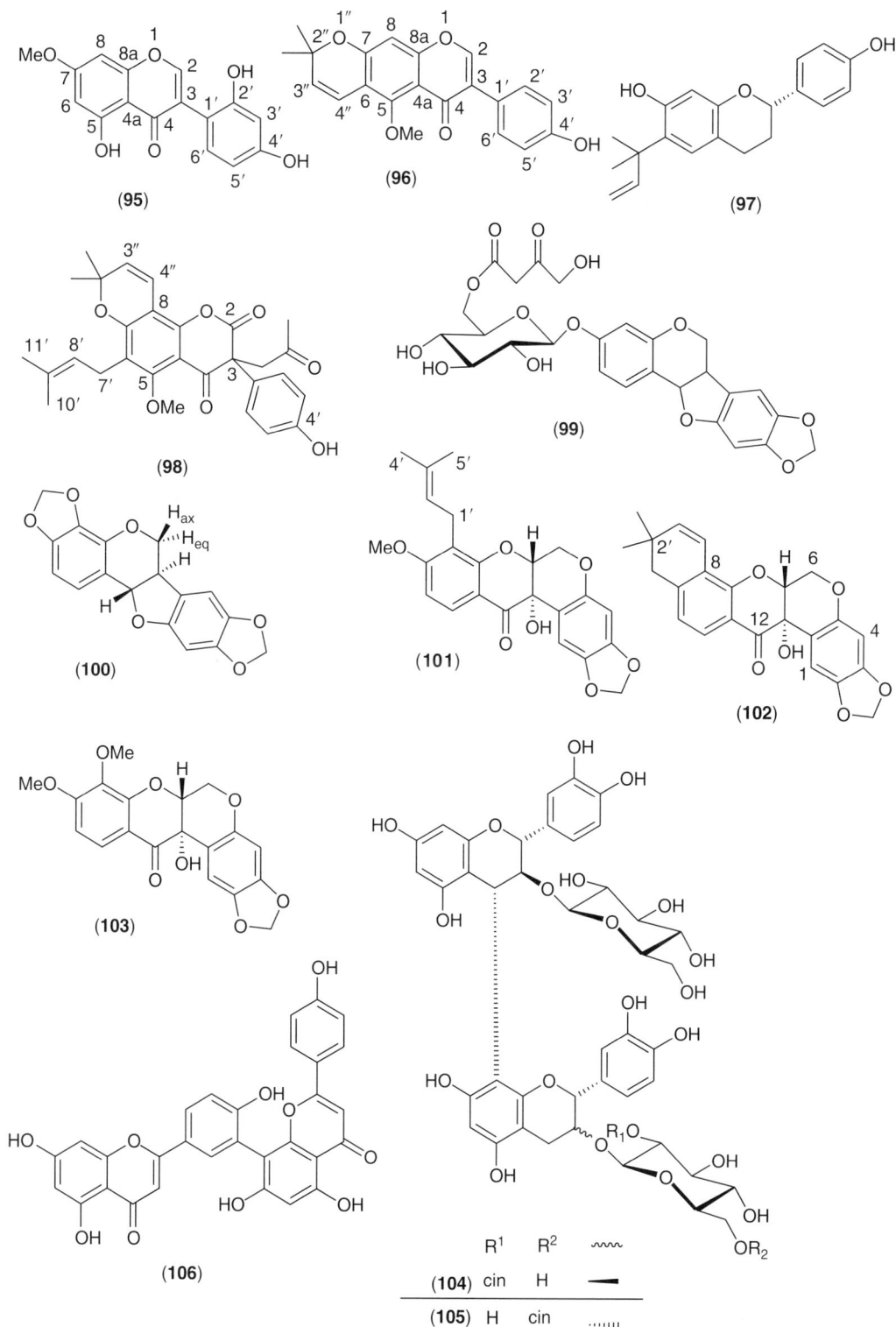

FIGURE 2.15 Structures of compounds **95–106**. See Table 2.7 for names.

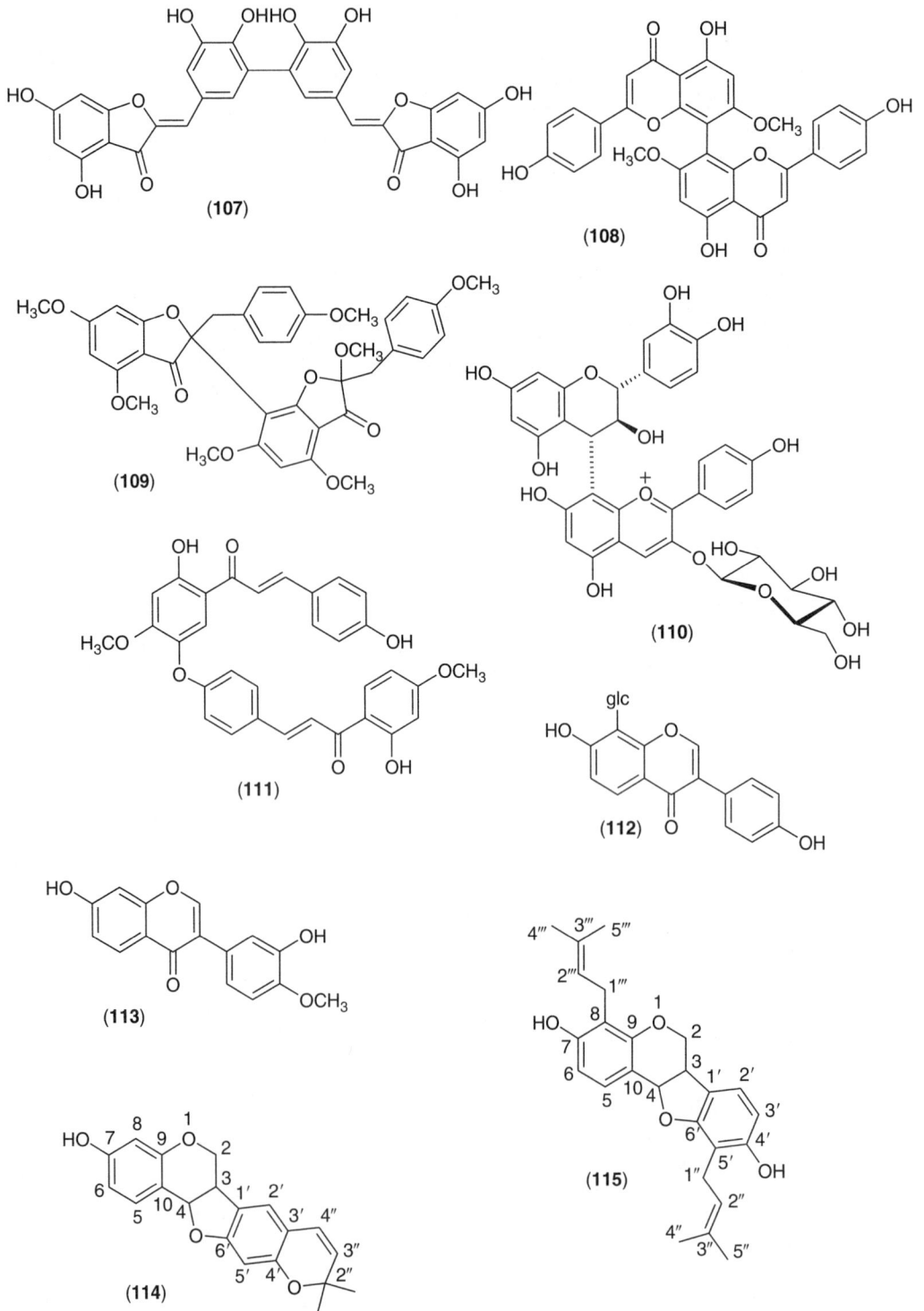

FIGURE 2.16 Structures of compounds **107–115**. See Table 2.7 for names.

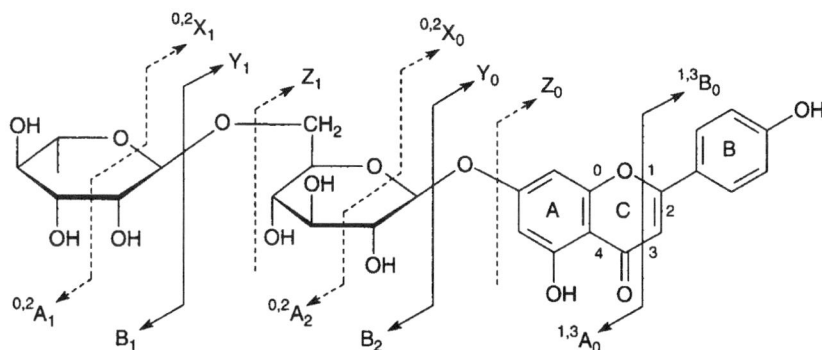

FIGURE 2.17 Ion nomenclature used for flavonoid glycosides (illustrated for apigenin 7-*O*-rutinoside). (Reprinted from Cuyckens, F. and Claeys, M., *J. Mass Spectrom.*, 39, 1, 2004. Copyright 2004 John Wiley & Sons, Ltd. With permission.)

In addition to giving accurate molecular masses of molecular ions, fragmentation patterns revealed by some MS methods may provide (a) structural information about the nature of the aglycone and substituents (sugars, acyl groups, etc.), (b) interglycosidic linkages and aglycone substitution positions, and (c) even some stereochemical information (Figure 2.17). The amount of structural information obtained for flavonoids from a mass spectrum depends on the ionization method used. The highest energy transfer occurs during EI of flavonoid aglycones, and in these cases fragmentation of molecular ions is normally seen. When soft ionization methods like ESI and APCI are applied on flavonoids in LC–MS and CE–MS systems, fragmentation of the flavonoids is not commonly seen in the spectra. However, the use of collision-induced dissociation tandem mass spectrometry (CID-MS–MS) allows detection of fragment ions.[251] During CID-MS–MS analysis precursor ions extracted in the first analyzer collide with atoms of an inert gas in the collision cell. The ionized fragments created are separated in the second analyzer. In another recent achievement, the fragment ions created can be further studied using an additional MS^n stage in a multistage tandem MS instrument with an ion trap (IT) analyzer (see Section 2.3.1.5).

At present, MS spectra alone rarely provide sufficient information for complete structural elucidation of novel complex flavonoids. Normally, MS techniques give little information about the configuration of the glycosidic linkage, and are reckoned not to be capable of distinguishing between diasteromeric sugar units. However, both FAB and ESI in combination with CID have recently been used for direct stereochemical assignment of hexose and pentose residues in acetylated flavonoid *O*-glycosides. The differentiation between a glucose residue and a galactose residue was, for instance, made by employing the [m/z 127]/[m/z 109] peak intensity ratios and the relative abundance of a fragment ion at m/z 271.[252] Combined with data obtained by other spectroscopic techniques (especially NMR and UV), MS has proved to be an invaluable tool for the identification of novel flavonoids.

Exact mass measurement at high resolution is an important tool along with other spectroscopic methods to help confirm the structure of novel flavonoids. It is used as structural proof when elemental analysis is not possible, e.g., when studying minor components. When EI-MS can be used, 1 to 10 pmol samples are required for one measurement; however, when FAB-MS is used, 0.1 to 1 nmol is normally required. The use of ESI on a double focusing mass spectrometers and MALDI-TOF-MS requires smaller amounts of sample, and subpicomole amounts of flavonoids may be adequate.

MS techniques applied to flavonoids have been covered by several recent reviews. An excellent paper by Cuyckens and Claeys[253] covers sample preparation, LC–MS analysis, and tandem mass spectrometric procedures for the characterization of flavonoid aglycones, *O*-glycosides, *C*-glycosides, and acylated glycosides in a tutorial manner. Various aglycone fragmentation patterns obtained under EI conditions can be designated according to the nomenclature proposed by Ma et al.[254] Similarly, ions formed from flavonoid glycosides can be denoted according to the nomenclature introduced by Domon and Costello (Figure 2.17).[255] The different mass spectrometric techniques applied for the analysis of flavonoid glycosides have been reviewed by Stobiecki.[256] The MS analysis of anthocyanidins and derived pigments, catechins and proanthocyanidins, and flavonols and flavonol glucuronide conjugates has recently been illustrated with case studies.[257] Mass spectrometric methods for the determination of flavonoids in biological samples have been reviewed by Prasain et al.,[258] and excellent reviews on tandem mass spectral approaches to the structural characterization of flavonoids have recently been reported.[259,260] Several general reviews covering studies on flavonoids have included MS analysis applied to flavonoids,[19,139,261,262] and important applications of the various MS techniques are covered in other chapters of this book.

The consecutive introduction and success of FAB, MALDI, APCI, ESI, new mass analyzers, as well as exciting coupled MS techniques, which progressively have drawn the main attention of mass spectroscopists for analysis of flavonoids in recent years, are described in more detail in the following sections. Some representative applications of these techniques for flavonoid analysis are presented in Table 2.9–Table 2.11.

2.3.1 MS INSTRUMENTATION AND TECHNIQUES

The three major parts of a mass spectrometer are the device for the introduction of the sample into the ionization chamber where the ionization takes place, the mass analyzer where the ions are separated according to their mass-to-charge ratios, and the detector, which can measure the quantity of negative or positive ions. The ionization methods used for flavonoid analysis can be classified as gas-phase methods including EI and CI, desorption methods including FAB and MALDI, and spray methods including ESI and APCI (Table 2.8). The different types of mass analyzers are double focusing magnetic sectors instruments, quadrupole mass filters (Q), quadrupole ion traps (Q-IT), TOF analyzers, and Fourier-transform ion-cyclotron resonance (FT-ICR).

2.3.1.1 Electron Impact and Chemical Ionization

The first conventional mode of MS involves EI ionization, in which the neutral flavonoid is impacted in the gas phase with an electron beam of 70 to 100 eV. Resulting mass spectra of the flavonoid aglycones are characterized by molecular ion peaks ($M^{+\cdot}$), and fragment ions from both the A and B rings. The use of a reactant gas in the ionization chamber, CI, normally results in the production of a more abundant molecular ion and simpler fragmentation patterns. General information about mass spectra of flavonoids recorded by these methods has been published by several authors.[2,15,256,263] More specific mass spectra analyses of 39 polymethoxylated flavones have been obtained by GC–MS.[264] In addition to the common fragmentation behavior of flavones under EI, such as retro-Diels–Alder reactions, which give characteristic fragments from the phenyl group of the flavone skeleton, new fragmentation pathways have been identified and proposed.[264]

EI and CI are normally unsuitable for most flavonoid glycosides because of their polarity and thermolability. Preparation of permethylated or trimethylsilylated derivatives may overcome this problem. However, derivatization often produces mixtures of partially derivatized compounds and may involve rearrangements,[256,265] and usually only weak molecular ion signals is observed when permethylated compounds are studied under EI conditions.

TABLE 2.8
Common Mass Spectrometry (MS) Methods Used in Recent Years for Flavonoid Analyses

Ionization Source/ Sample Introduction	Means of Ionization	Particularities	Commonly Associated Mass Analyzers
Fast atom bombardment (FAB) Direct insertion probe or LC–MS	The flavonoid is dissolved in a liquid matrix. The sample is bombarded with a fast atom beam that desorbs molecular ions and fragments from the analyte. The spectrum often contains peaks from the matrix	Mass range up to 7000 Da. Exact mass measurements are usually done by peak matching. The accuracy of the mass is the same as obtained in EI, CI. Relatively low sensitivity. Molecular ions often absent	Quadrupole
Matrix assisted laser desorption ionization (MALDI) Direct insertion probe or continuous-flow introduction. Not easily compatible with LC–MS	The flavonoid is dissolved in a solution containing an excess of a matrix that has a chromophore that absorbs at the laser wavelength. The matrix absorbs the energy from the laser pulse and produces plasma that results in vaporization and ionization of the analyte. Some structural information can be obtained in a "postsource decay" mode, or by collisional activation	High mass range. Sample amount very low (picomoles or less). Mass accuracy (0.1 to 0.01%) is normally not as high as for other mass spectrometry methods. Recent developments in delayed extraction TOF allow higher resolving power and mass accuracy. The analysis is relatively insensitive to contaminants. MS–MS difficult	Requires a mass analyzer that is compatible with pulsed ionization techniques Time of flight Fourier-transformed ion-cyclotron resonance
Electrospray ionization (ESI) Flow injection or LC–MS or CE–MS	The sample must be soluble, polar, and relatively clean. The sample solution is sprayed from a needle held at high voltage to form charged liquid droplets from which ions are desolvated. Multiply charged ions are usually produced	High mass range. Best method for analyzing multiply charged compounds. Very low chemical background leads to excellent detection limits (femtomole to picomole). Can control presence or absence of fragmentation by controlling the interface lens potentials. Compatible with MS–MS methods	Quadrupole Ion trap Time of flight Fourier-transformed ion-cyclotron resonance
Atmospheric pressure chemical ionization (APCI) Flow injection or LC–MS or CE–MS	Similar interface to that used for ESI. In APCI, a corona discharge is used to ionize the analyte in the atmospheric pressure region. Ions are formed by charge transfer from the solvent as the solution passes through a heated nebulizer into the APCI source	Mass range up to 2000 Da. The gas-phase ionization in APCI is more effective than ESI for analyzing less-polar species. Sensitivity may be high (femtomole). Compatible with MS–MS methods	Quadrupole Ion trap Time of flight

Notes: MS techniques are usually designated by the ionization source producing the ions (e.g., FAB, MALDI, ESI, APCI) and by the mass analyzer (e.g., TOF, IT) used to sort them according to their m/z values (e.g., MALDI–TOF, ESI–TOF, ESI–IT). Two analyzers are used in series in tandem MS (MS–MS) techniques (e.g., ESI–Q–TOF).

2.3.1.2 Fast Atom Bombardment

FAB-MS (Table 2.8) has been widely used for the characterization of flavonoids solubilized in a variety of matrices, and normally involve the use of xenon or argon atoms for bombardment (Table 2.9). The matrix signals may complicate interpretation of the spectra. Nevertheless, when combined with CID of positive ions and tandem mass spectrometric techniques, FAB-MS can provide information on the aglycone moiety, the carbohydrate sequence, and

TABLE 2.9
Selected Papers in Mass Spectrometry Applied to Flavonoid Analysis with Fast Atom Bombardment Ionization

Analytes	Sample	Ionization Mode	Matrix	Ref.
Anthocyanins	Pure compounds	FAB (+)	a	273
Anthocyanins	Pure compounds	FAB (+)	*m*-Nitrobenzyl alcohol	84
Anthocyanins	*Malva silvestris,* purified extract	FAB (+)	Glycerol	274
Anthocyanin–flavanol dimers	Pure compounds	FAB (+)		275
Pyranoanthocyanins	Pure compounds	FAB (+)	Glycerol	276
Anthocyanins, flavanols Flavonols, proanthocyanindins Chalcones, flavanones,	Red wine	FAB (+)	Glycerol	277
Flavonols	Pure compounds	FAB (+)	Glycerol	163
Flavanones	Pure compounds	FAB (+)	1 *N* HCl–glycerol	278
Flavones	Pure compounds	FAB (−)	Triethanolamine	279
Flavones	Pure compounds	FAB (+)	Nitrobenzyl alcohol	280
Flavone *C*-glycosides	Pure compounds	FAB (+)	1 *N* HCl–glycerol	281
Flavone *C*-glycosides	Pure compounds	FAB (+)	Glycerol and *m*-nitrobenzyl alcohol	145
Flavone *C*-glycosides Flavone *O*-glycoside,	Pure compounds	FAB (+)	Glycerol	282
Flavonols	Pure compounds	FAB (+)	Glycerol	252
Flavones, flavonols	Pure compounds	FAB (+)	Glycerol	254
Flavonols	Pure compounds	FAB (−)	2-Hydroxyethyl disulfide	283
Flavonols	Pure compound	FAB (+)	Nitrobenzene	284
Flavonols	Alkaline-earth metal complexes	FAB (±)	*m*-Nitrobenzyl alcohol	285
Flavonols	Pure compounds	FAB (+)	*m*-Nitrobenzyl alcohol	286
Flavonols	Pure compounds	FAB (+)	Lactic acid	287
Flavonols	Pure compounds	FAB (+)	Thioglycerol + NaI	288
Flavonols	Pure compounds	FAB (+)	Lactic acid	289
Flavonol sulfates	Pure compounds	FAB (+)		290
Flavonoids	Pure compounds	FAB (−)	Glycerol–thioglycerol	291
Flavonoids	Pure compounds	FAB (+)	*m*-Nitrobenzyl alcohol	292
Isoflavenes	Pure compounds	FAB (+)	*m*-Nitrobenzyl alcohol	293
Isoflavone phosphates	Pure compounds	FAB (−)	*p*-Nitrobenzyl alcohol	294
Isoflavonoids, triflavonoids	Pure compounds	FAB (+)		295
Proanthocyanidins	Rat metabolites	FAB (−)		296

ᵃDissolved in methanol and formic acid with subsequent addition of a 1:1 mixture of dithioerythritol and dithiothreitol.

the glycosylation position of glycosides,[266–269] and even stereochemical assignment of hexose and pentose residues in flavonoid *O*-glycosides.[252] An MS method based on the combined use of FAB and CID tandem MS has been used for the analysis of the fragmentation behavior of protonated 3-methoxyflavones.[268] It was shown that several diagnostic ions allowed unambiguous localization of the functional groups in the A and B rings, and isomeric 3-methoxyflavones could be differentiated using this methodology. The principles of the technique together with the processes involved in ion formation, the effect of the liquid matrix, and the optimization of FAB conditions for flavonoids have been described in several studies.[270–272] The value of FAB-MS in flavonoid analysis is demonstrated by some recent applications using both positive- and negative-ion modes (Table 2.9).

2.3.1.3 Matrix-Assisted Laser Desorption Ionization

MALDI-MS is considered a sensitive and powerful tool for the analysis of nonvolatile molecules (Table 2.8). It has greatly expanded the use of MS toward large molecules, and has revealed itself to be a powerful method for the characterization of both monomeric flavonoids as well as proanthocyanidins (Table 2.10). With this technique, fragmentation of the analyte molecules upon laser irradiation can be substantially reduced by embedding them in a light-absorbing matrix. As a result, intact analyte molecules are desorbed and ionized along with the matrix, and can be analyzed in a mass spectrometer. This soft ionization technique is mostly combined with TOF mass analyzers. A crucial factor that influences the quality of MALDI-TOF mass spectra is the crystallization of the analyte during sample preparation and the behavior of the matrix during laser irradiation. MALDI-MS can measure the mass of almost any molecule (masses up to 10^6 Da). The analysis can be performed in the linear or reflectron mode. Mass accuracy (0.1 to 0.01%) is not as high as for other MS methods; however, the analysis is relatively insensitive to contaminants. The amount of sample needed is very low (picomoles or less), and may involve only 1 to 2 μl of sample solution.[297] The technique is fast to handle, often taking less than a minute for the actual analysis after sample preparation. The use of the MALDI technique has helped in obtaining vital information in many recent flavonoid analyses (Table 2.10).

Recent developments in delayed extraction TOF allow higher resolving power and mass accuracy, and this method in the reflector mode has been used for accurate measurement of the mass of several compounds including two prenylated flavonoids.[298] However, the performance of the MALDI-TOF instrument was not better than those of the FAB and FT-ICR MS instruments, and insufficient to give acceptable accuracy for literature reporting.

The MALDI–MS technique has been extended with the so-called postsource decay method (PSD) by Spengler and coworkers.[299] This technique, which allows the determination of the fragment ions formed from the decomposition of the precursor ions of high internal energy, has been used to study the fragmentation and the fragmentation mechanisms of the flavonol glycoside rutin cationized with different alkali metal ions.[300] The technique permits the selection of a precursor molecule in a distinct mass window and the subsequent analysis of its fragments. The precursor ions passing the mass window can spontaneously fragment on their way to the detector due to the application of higher laser energies. All ions with lower and higher masses are deflected by the electrostatic device. The PSD-MALDI mass spectra of the cationized rutin molecules showed, depending on the cation, different fragmentation patterns with respect to both quality and quantity of the fragment ions formed.[300] For a more specific sequential elucidation of individual proanthocyanidin chains, MALDI-TOF-MS has been extended to PSD fragmentation.[301] Recently, Keki et al.[302] have used PSD-MALDI MS–MS to deal with the fragmentation and the fragmentation mechanisms of peracetylated isoflavone glycosides cationized with proton and various metal ions. In a very

TABLE 2.10
Papers on Matrix-Assisted Laser Desorption Ionization Mass Spectrometry with Time of Flight Mass Analyzer Applied to Flavonoid Analysis

Analyte	Sample	Ionization Mode	Matrix	Ref.
Proanthocyanidins	American cranberry (*Vaccinium macrocarpon*) fruit	(+)	DHB	304
Proanthocyanidins	Mimosa (*Acacia mearnsii*) bark tannin, Quebracho (*Schinopsois balansae*) wood tannin	Linear (+)	DHB	305
Proanthocyanidins	American cranberry (*Vaccinium macrocarpon*) concentrate juice powder	Reflectron (+)	IAA	306
Proanthocyanidins	Ruby Red sorghum (*Sorghum bicolor*) grain	Reflectron (+)	IAA	307
Proanthocyanidins	Apple (*Malus pumila*)	Reflectron ($\pm$) Linear ($\pm$)	IAA/Ag$^+$	308
Proanthocyanidins	Coffee pulp (Arabica variety)	Linear	DHB	309
Proanthocyanidins	Grape seed	Reflectron (+) Linear (+)	IAA	310
Proanthocyanidins	Leaves or needles of willow (*Salix alba*), spruce (*Picea abies*), beech (*Fagus sylvatica*), lime (*Tilia cordata*)	linear (+)	DHB, dithranol, IAA	301
Proanthocyanidins	Grape seed	Reflectron (+) Linear (+)	DHB, IAA	311
Proanthocyanidins	Grape berries (variety Gamay)	(+)	DHB	312
Proanthocyanidins	Grapes (seeds, skins and stems), Quebracho (*Schinopsis balansae*) heartwood	Reflectron (+)	DHB	313
Proanthocyanidins	Grape seed	Reflectron (+)	DHB	314
Theaflavins, thearubigins	Black tea	Linear (+)	DHB, CHCA	303
Flavonol	Rutin	Reflectron (+)	DHB	300
Anthocyanins, 3-deoxyanthocyanidins	Sorghum (*Sorghum bicolor*) plant tissue	Linear (+)	CHCA	315
Anthocyanins	Highbush blueberries (*Vaccinium corymbosum*)	Linear (+)	THAP	316
Anthocyanins	Red wines, fruit juices	Linear (+)	THAP	317
Isoflavonoids	Soybeans, tofu, isoflavone supplements	Linear (+)	THAP, DHB	318
Flavonols	Yellow onion, green tea	Linear ($\pm$)	THAP, IAA	319
Flavonols	Almond (*Prunus dulcis*) seedcoat	Linear (+)	THAP	320
Flavonols	Almond (*Prunus amygdalus*) seedcoat	Linear (+)	THAP	321

Notes: CHCA, α-cyano-4-hydroxycinnamic acid; DHB, 2,5-dihydroxybenzoic acid; IAA, *trans*-3-indoleacrylic acid; THAP, 2,4,6-trihydroxyacetophenone monohydrate.

interesting paper, the structures of theaflavins (TFs) and thearubigins (TRs) from black tea have been revealed by the use of delayed pulsed ion extraction of ions generated via MALDI-TOF-MS.[303] Spectra of standard TFs showed not only pseudomolecular ions but also ions resulting from fragmentation.

2.3.1.4 Electrospray Ionization and Atmospheric Pressure Chemical Ionization

ESI-MS (Table 2.8) was introduced by Yamashita and Fenn in 1984,[322] and this invention was recognized by The Royal Swedish Academy of Sciences with the award of The Nobel

Prize in Chemistry for 2002 partly to John B. Fenn for his pioneering work in ESI-MS. The mechanism of the transformation of ions in solution to ions in the gas phase prior to their mass analysis in a mass spectrometer together with instrumentation and applications of ESI have been reviewed by Cole.[323]

ESI is at present the most common technique used to analyze polar and nonvolatile flavonoids (from anthocyanins to condensed tannins), mainly because of the ease with which it can ionize polar and nonvolatile compounds (Table 2.11). The technique permits the detection of the molecular ion, either as a protonated molecule, $[M + H]^+$, adduct, $[M + Na]^+$, or as a deprotonated molecule, $[M - H]^-$, and causes only moderate fragmentation of the molecule as occurs with other higher-energy types of ionization techniques. MS with ESI ionization has been used for analysis of flavan-3-ols in plant extracts and in plasma samples with results achieving levels of detection of 20 ng.[324] Furthermore, this technique provides structural information for highly polymerized compounds from the interpretation of the fragmentation profiles and from the level of charge the formed ions, which can either be monocharged or appear with multiple charges.[325] There exists a very useful review on the principles, signal acquisition, and interpretation of proanthocyanidin spectra obtained by ESI-MS.[326] In a recent study, Oliveira et al.[327] described the use of ESI-MS, in combination with CID and tandem MS, for the structural characterization of anthocyanidins and anthocyanins.[327] This technique has also been used in a fragmentation study of an flavone triglycoside, kaempferol-3-O-robinoside-7-O-rhamnoside,[328] and for high mass resolution studies of a isoflavone glycoside, genistein-7-O-glucoside.[329] The ability of ES to work with liquid sample introduction techniques has made it one of the most important detectors for HPLC and capillary zone electrophoresis (Section 2.3.2).

The APCI source (Table 2.8) has been used for the analysis of various flavonoids, especially flavonols, flavones, flavanones, and chalcones (Table 2.11). APCI is based on gaseous-phase ionization, and is most suitable for compounds that are partially volatile and have a medium polarity. Thus, the application of APCI with respect to analysis of condensed tannins and anthocyanins is more limited.[257] Compared with ESI, APCI produces more fragment ions in the spectrum due to the harsher vaporization and ionization processes. More information about ESI and APCI can be found in Section 1.4.5.

2.3.1.5 Tandem (MS–MS) and Multiple (MS^n) Mass Spectrometry

In order to obtain fragmentation of ions produced by, for instance, ESI or APCI, these ions are accelerated inside the mass spectrometer so as to collide with molecules of the bath gas, usually helium.[139] Such CID of ions can be performed on all the ions emerging from the source, but this produces mixed CID spectra when more than one compound enters the source at the same time, as frequently occurs in LC–MS. To obtain pure CID spectra, the ion of interest (the precursor ion) needs to be isolated. Initially, the quadrupole mass filter was the ideal instrument to do this since its radiofrequency voltage can be set at a given value to only allow the selected precursor ion to pass through. This ion can then be accelerated into the bath gas in a collision cell (also of quadrupole design) and the products can be recorded using a third quadrupole operated in normal scanning mode. Therefore, MS–MS in space requires three quadrupoles.[139]

Different MS–MS experiments of product ion scan, precursor ion scan, and neutral loss scan modes of selected flavonoids can be carried out in order to confirm the structure of flavonoids previously detected by the full-scan mode. In the product ion scan experiments, MS–MS product ions can be produced by CID of selected precursor ions in the collision cell of the triple-quadrupole mass spectrometer (Q2) and mass analyzed using the second analyzer of the instrument (Q3). However, in the precursor ion scan experiments, Q1 scans over all possible precursors of the selected ion in Q3 of the triple quadrupole. Finally, in neutral loss scan experiments, both quadrupoles scan for a pair of ions that differ by a characteristic mass difference (neutral mass). The ESI-MS–MS experiments of product ion scan, neutral loss scan, and precursor ion scan modes have, for instance, enabled structural determination of the acylated flavonoid-*O*-glycosides and methoxylated flavonoids occurring in *Tagetes maxima*.[330] In another example, MS–MS with positive CI has been applied successfully to problems involving trace analysis of citrus flavanones and metabolite identification.[331] Positive CI-MS–MS was compared to EI-MS–MS and found to be more advantageous in searching for the common daughter ion for flavanones (*m/z* 153) in complex matrixes.

The two main methods for investigation of flavonoids using MS methods are direct infusion using a syringe, and flow injection either with or without chromatographic separation.[260] The first method allows a long and thorough sample investigation, including the acquisition of data for several consecutive fragmentation steps (MS^n experiments). This method, which is used for structure characterization, requires a relatively large amount of purified flavonoid as the sample is normally infused with a flow rate of about 3 to 10 µl/min. Flow injections allow only a short investigation time for each signal, which may be too short for MS–MS experiments. However, the Q-IT has the potential to perform MS–MS in time within one analyzer.[139] Ions of a given mass-to-charge ratio can be isolated within a Q-IT and then excited such that they collide with bath gas and the resulting product ions are trapped and scanned out to the detector. Indeed, rather than being scanned out, the cycle of ion isolation and fragmentation can be repeated a number of times to achieve multistage MS (MS^n). The development of new, more powerful mass spectrometers has thus allowed one to obtain MS–MS spectra corresponding to the fragmentation of the molecular ion previously isolated in the ionization chamber in a selective way, and, moreover, to obtain MS^n spectra up to $n = 10$,[257] which facilitates possibilities with respect to structural elucidation of unknown compounds. The application of ESI-MS^n in the analysis of the noncovalent complexes of cyclodextrins with quercetin 3-rutinoside (rutin) and quercetin has provided information about binding stoichiometry, as well as the relative stabilities and binding sites of the cyclodextrin–rutin complexes studied.[332] The diagnostic fragmentation pattern suggested that the specific inclusion complexes between rutin and the cyclodextrins could be confirmed by ESI-MS–MS alone without the need for solution-phase studies. The number of papers including MS–MS for flavonoid analysis has increased considerably in recent years, and some excellent reviews on tandem mass spectral approaches to the structural characterization of flavonoids have been reported.[259,260] Some references to other applications of MS–MS and MS^n techniques are found elsewhere in this section.

2.3.1.6 Mass Analyzers

Quadrupole mass filters, Qs, which are still widely used for flavonoid analysis, isolate ions of a selected *m/z* ratio.[330,333–336] They are mainly able to perform low-resolution mass analyses, and have a limited mass-to-charge range, typically up to *m/z* 4000, which, however, is appropriate for most flavonoid analyses. The ESI interfaces are most often used in combination with quadrupoles; on the other hand, in LC–ESI-MS, the quadrupole detector can be replaced by an IT or a TOF detector. In a Q-IT, ions are trapped in a cavity formed by three electrodes and are ejected through them by application of potentials, as a function of *m/z* values. The IT has the advantage that it can carry out sequential fragmentation first of the parent molecular ion and then of the daughter ions. Thus, it provides MS^n spectra by successive fragmentation of selected ions. The uses of IT analyzers has been described in several excellent papers.[333,334,336–340]

The TOF analyzer separates ions by virtue of their different flight times over a known distance, depending on their *m/z* value.[298,301,305,307,311,315,316,318,321,333,341,342] It supplies accurate mass determination, and has theoretically an unlimited mass range. Hybrid instruments take advantage of easy creation and isolation of molecular ions of flavonoids. The quadrupole orthogonal time-of-flight (Q-TOF) mass spectrometer is related to triple-quadrupole instruments. Ions are generated with ESI or MALDI, selected in the first quadrupole, and fragmented by collision with argon gas, and the fragments accelerated orthogonally and injected into a TOF analyzer. The advantage of the TOF detector is its higher sensitivity and better mass accuracy (at least 20 ppm) than the quadrupole detector in a triple-quadrupole instrument.

FT-ICR mass spectrometers take advantage of ion-cyclotron resonance to select and detect ions. This analyzer can be used with both ESI and MALDI interfaces. Their particular advantages are their sensitivity, extreme mass resolution, and mass accuracy. The latter allows for the determination of the empirical formulae of compounds under 1000 Da. As far as we know, this analyzer has not been applied to flavonoids.

2.3.2 COUPLED TECHNIQUES INVOLVING MASS SPECTROMETRY

Complex plant extracts and biological fluids often require very effective and sensitive separation techniques to allow the identification of the various flavonoids in the samples. The coupling of instruments performing chromatographic separations, particularly HPLC, to those providing mass structural data has in recent years had an enormous impact in flavonoid chemistry. These coupled techniques are, above all, adept at targeted analyses; i.e., determining whether a specific component is present in a plant extract or a biological fluid. They are, for instance, ideally suited to studies in systematic phytochemistry in which the occurrences of, e.g., specific flavonoids are surveyed in taxonomic groups.[139] In this section, recent papers on high-performance LC–MS, GC–MS, and CE–MS applied in the flavonoid field are considered. The usefulness of LC–MS has been thoroughly covered in other chapters of this book (e.g., Sections 1.4.5 and 5.2).

2.3.2.1 Gas Chromatography Coupled to Mass Spectrometry

GC–MS is established as a routine technique for the analysis of flavonoid aglycones and is carried out with either EI or CI sources (see Section 2.3.1.1). Because of limited volatility, analysis of flavonoid glycosides by GC–MS has not generally found favor; however, improvements in GC column technology have increased the range of flavonoids amenable to GC–MS as underivatized compounds. Schmidt et al.[343] analyzed 49 flavones, flavonols, flavanones, and chalcones without derivatization by GC and GC–MS (EI mode) using an OV-1 capillary column.[343] Recently, lipophilic and thermolabile flavonoids in various plant extracts have been characterized directly by high-temperature high-resolution GC with cold on-column injection coupled with MS.[344,345]

Employing chemical derivatization to increase volatility may extend the range of flavonoids that can be analyzed by GC–MS. However, derivatization may lead to the formation of more than one derivative from a single flavonoid. Frequently used derivatization methods are to silylate or methylate the hydroxyl groups of flavonoids. An in-vial simple and fast method for the combined methylation and extraction of phenolic acids and flavonoids in various plant extracts, followed by direct determination with GC–MS, includes the use of phase-transfer catalysis.[346] Another GC–MS method has been developed for the determination of some flavonoid aglycones and phenolic acids in human plasma.[347] The procedure involved extraction with ethyl acetate, followed by the derivatization with N,O-bis(trimethylsilyl)trifluoroacetamide + trimethylchlorosilane reagent. The trimethylsilyl derivatives formed were separated and quantified using GC–MS (EI). The average recovery was 79.3%, and the method may be used in different matrices such as serum, urine, and tissues. An isotope dilution GC–MS method has been used for the identification and quantitative determination of unconjugated lignans and isoflavonoids in human feces.[348] Following the formation of trimethylsilyl ethers, the samples were analyzed by combined capillary column GC–MS in the single-ion monitoring (SIM) mode, including corrections for all losses during the procedure using the deuterated internal standards.

2.3.2.2 High-Performance Liquid Chromatography Coupled to Mass Spectrometry

During the last decade, research efforts in the field of LC–MS have changed considerably. Technological problems in interfacing appear to be solved, and a number of interfaces have been found suitable for the analysis of flavonoids. These include TSP, continuous-flow fast-atom bombardment (CF-FAB), ESI, and APCI. LC–MS is frequently used to determine the occurrence of previously identified compounds or to target the isolation of new compounds (Table 2.11). LC–MS is rarely used for complete structural characterization, but it provides the molecular mass of the different constituents in a sample. Then, further structural characterization can be performed by LC–MS–MS and MS–MS analysis. In recent years, the combination of HPLC coupled simultaneously to a diode-array (UV–Vis) detector and to a mass spectrometer equipped with an ESI or APCI source has been the method of choice for the determination of flavonoid masses. Applications of LC–MS (and LC–MS–MS) in flavonoid analysis has recently been described in several excellent reviews,[130,139,256,257,326,336,349–353] and the usefulness of these techniques have also been thoroughly covered in other chapters of this book (e.g., Chapters 1 and 5).

Selected papers on flavonoid analyses by LC coupled to positive- or negative-mode APCI or ESI are listed in Table 2.11. These two techniques are based on API, and their operational principle is that the column effluent from the LC is nebulized into an atmospheric-pressure ion-source region. Nebulization is performed pneumatically in a heated nebulizer (APCI), by means of the action of a strong electrical field (ESI), or by a combination of both. The ions produced from the evaporating droplets are, together with solvent vapor and nitrogen, sampled in an ion-sampling aperture by supersonically expanding into this low-pressure region before transportation to the mass analyzer. The mobile phase used contains easily ionized components (e.g., trifluoroacetic acid), from which a charge may be transferred to the flavonoid, $[M+H]^+$. Both sodium $[M+23]^+$ and potassium $[M+39]^+$ adducts may be seen in these spectra. Depending on the energy of the ion source, sugar moieties may fragment off the flavonoid. Mobile-phase flow in API interfaces may differ from nanoliters (nanoelectrospray) to milliliter per minute. The temperature control of the APCI desolvation process is far less critical than in TSP-MS. In this way, a wide range of flavonoids may be analyzed under the same conditions maintained at the APCI interface. In many cases, splitting of the eluate from the LC column is necessary in order to decrease the volume of solution entering the API source.

2.3.2.3 Capillary Electrophoresis Coupled to Mass Spectrometry

The first detection of ionic species in aqueous solutions by capillary zone electrophoresis combined with ESI-MS was applied with a quadrupole mass spectrometer.[384] For the analysis of charged molecules, the high voltage applied to the electrospray needle is ideal in creating both the electrospray effect and closing the CE circuit. For uncharged molecules, though, the modified CE technique of capillary electrokinetic chromatography is necessary to achieve separation, which may create incompatibilities with ES.[385]

The CE–MS combination may provide valuable, structure-selective information about flavonoids in plant extracts; however, this coupled technique has hitherto found only very

TABLE 2.11
Selected Papers on LC–MS with Atmospheric Pressure Ionization Applied to Flavonoid Analysis

Analytes	Sample	LC Eluent	Ionization Mode	Mass Analyzer	Ref.
Anthocyanins	*Solanum stenotomum* tubers	ACN–H$_2$O, 0.1% TFA	ESI (+)	Q	354
Anthocyanins	Raspberry fruits	ACN–H$_2$O, 1% FA	APCI (+)	Q	335
Anthocyanins	Boysenberries	MeOH–H$_2$O, 5% FA	ESI (+)	Q	355
Anthocyanins	Grape juices	ACN–H$_2$O, 10% AA	ESI (+)	IT	356
Anthocyanins	Port wines	ACN–H$_2$O, 0.1% TFA	ESI (+)	Q	357
Anthocyanins	Purple corn	ACN–H$_2$O, 0.1% TFA	ESI (+)	Q	358
Flavones	Chamomile	ACN–H$_2$O, 0.1% TFA	ESI (±)	QqQ	359
Flavonols	Tomatoes	ACN–H$_2$O, 1% FA	APCI (−)	Q	360
Isoflavones and flavones	*Genista tinctoria*	ACN–H$_2$O, 0.1% AA	ESI (−)	Q	361
Isoflavones	Soy foods	ACN–H$_2$O, 0.1% TFA or 0.1% AA	APCI (±) and IS (±)	QqQ	362
Isoflavones	*Trifolium pretense*	ACN–H$_2$O, 0.2% AA	ESI (+)	Q	363
Isoflavones	*Trifolium pretense*	MeOH–10 m*M* ammonium formate buffer, pH 4.0	APCI (±)	IT and Q	364
Oligomeric anthocyanins	Grape skins	FA–H$_2$O–ACN	ESI (+)	QqQ	365
Flavonoid aglycones	Pure compounds	MeOH–H$_2$O, 0.1% FA	ESI (−)	IT	339
C-Glycosidic flavonoids	Pure compounds	ACN–H$_2$O, 0.5% AA	APCI (±) and ESI (±)	IT and Q-TOF	334
O- and *C*-Glycosidic flavonoids	*Sechium edule*	ACN–H$_2$O, 0.05% AA	ESI (−)	IT and Q	366
Flavonoids	Onion, blossom, and St. John's wort	ACN–H$_2$O, 20 m*M* TFA	ESI (+)	IT	367
Flavonoids	Pure compounds	MeOH–H$_2$O or ACN–H$_2$O, 0.1–0.4% FA or 10 m*M* AAc or 0.1% AH, 0.05% TFA	IS (±), APPI (±), and APCI (±)	QqQ	368
Flavonoids	Apples	MeOH–H$_2$O, 5% FA	ESI (+)	Q	369
Flavonoids	*Azima tetracantha*	ACN–H$_2$O–THF, 0.1% TFA	ESI (±)	QqQ	370
Flavonoids	*Oroxylum* seeds	ACN–H$_2$O, 0.2% FA	ESI (+)	IT and Q	371
Flavonoids	Tomatoes	ACN–H$_2$O–THF, 0.1% TFA	ESI (+)	QqQ	372
Flavonoids	Cocoa	ACN–H$_2$O, 0.1% FA	ESI (−)	QqQ	373
Flavonoids	Rooibos tea	ACN–H$_2$O, 0.1% AA	ESI (±)	Q	374
Flavonoids	Blood plasma and urine	ACN–H$_2$O, 2% AA	ESI (±)	QqQ	375
Flavonoids	Citrus	ACN–H$_2$O–ammonium acetate	ESI (+)	Q	376
Flavonoids	Barley	ACN–H$_2$O–MeOH, 1% AA	APCI (+)	Q	377
Flavonoids	Fresh herbs	MeOH–H$_2$O, 1% FA	APCI (−)	Q	378
Flavonoids	Red clover	ACN–H$_2$O, 0.25% AA	ESI (±)	Q	379
Flavonoids	Wood pulp, waste water	MeOH–H$_2$O, 0.5% AA	ESI (±)	QqQ	380
Flavonoids	Urine	MeOH–ACN–H$_2$O, 0.5% FA	APCI (−)	Q	381
Phenolic compounds	Soy, onions	ACN–H$_2$O, 10% FA	ESI (−)	Q	382
Phenolic compounds	Olives	MeOH–H$_2$O, 1% AA	ESI (±)	QqQ	383

Notes: AA, acetic acid; AAc, ammonium acetate; ACN, acetonitrile; AH, ammonium hydroxide; FA, formic acid; MeOH, methanol; TFA, trifluoroacetic acid; THF, Tetrahydrofuran; APCI, atmospheric pressure chemical ionization; APPI, atmospheric pressure photoionization; ESI, electrospray ionization; IS, ion spray; IT, ion trap; Q, single quadrupole; QqQ, triple quadrupole; TOF, time of flight.

limited use in flavonoid analysis. Various isoflavones have been separated on an uncoated fused-SiO$_2$ CE column (110 cm × 75 mm i.d.) using 25 mM NH$_4$OAc buffer and UV and ESI-MS detection.[386] The ESI-MS allowed recognition of the molecular masses of the isoflavones, as well as the presence of various functional groups according to observed losses from the [M – H]$^-$ ion during CID by adjusting some MS parameters. Recently, a similar CE method has been established for the analysis of a flavonoid mixture obtained from plant extracts.[387] This method used a fused-silica capillary and a buffer system consisting of 40 mM NH$_4$OAc, 15% MeCN (pH 9.5). After validation of the CE method in combination with a quadrupole mass spectrometer (with an electrospray interface and 0.1% triethylamine in 2-propanol–water [80:20 v/v] as sheath liquid in the negative-ion mode), MS detection showed sensitivity for hesperetin and naringenin similar to that of UV detection (0.4 to 0.6 mg/l). Employing external calibration allowed the reliable quantification of naringenin in a phytomedicine containing five different herbal drugs.

2.3.3 STRUCTURAL INFORMATION

The application of MS in structural elucidations of flavonoids has increased dramatically with recent developments related to soft ionization techniques, mass analyzers, and coupled MS techniques (see Sections 2.3.1 and 2.3.2). Based on relatively small flavonoid quantities, reports have provided the molecular mass in addition to structural information about the flavonoid skeleton,[252,254,333,339,340,388–392] aglycone attachment points of glycosidic residues,[392] the types of carbohydrates present,[252,388] and the types of interglycosidic linkages.[262,340,390,391] The exact location of acyl groups in the glycosidic residue is, however, difficult to define on the basis of MS data. The numerous papers on structural information about flavonoids, predominently focus on the use of CID in triple quadruple or ion trap mass spectrometers, which allows generation and analysis of accurate daughter fragments, and on the usefulness of MALDI-MS. The reader will also find references to many other important papers in the field other than in Sections 2.3.3 and 2.3.4.

A considerable amount of information has been accumulated during the review period with respect to fragmentation studies of flavonoid aglycones and their glycosides using ionization techniques such as EI and CID (Figure 2.17).[253] Tandem mass spectrometry with soft ionization methods such as FAB, ESI, and APCI have been used for the structural characterization of a variety of flavonoids, and both deprotonation[339,340,378,380,393–396] and protonation[252,254,340,380,390,397] modes combined with CID have been used (Table 2.9–Table 2.13). Due to their acidic nature, flavonoids usually give higher ion abundances upon deprotonation in the negative ESI mode than via protonation in the positive mode.

Fragmentation pathways of protonated and deprotonated molecules of *C*-glycosidic flavonoids obtained with CID tandem MS techniques have enabled important structural information to be obtained about different substitution patterns of sugars in this class of compounds.[266,334,392] Fragment ions formed by the loss of water were more pronounced for 6-*C*-glycosyl flavonoids than for the corresponding 8-*C*-glycosyl flavonoids, due to the hydrogen bonds existing between the 2″-hydroxyl of the 6-*C*-sugar unit and either the 5- or 7-hydroxyl of the aglycone (Figure 2.18). Differentiation between *O*-glycosides, *C*-glycosides, and *O,C*-diglycosides have been achieved by examining the fragmentation patterns in their first-order positive-ion spectra or low-energy CID spectra (Figure 2.17).[253] Diagnostic fragmentation patterns of flavonoids have also been reported based on a metal complexation mode in conjunction with CID.[395,399–402] EI-MS of solutions containing a flavonoid, a transition metal salt, and an auxiliary ligand has resulted in differentiation between flavonoid isomers, as well as determination of the position of glycosylation. The CID patterns can be "tuned" by changing the auxiliary ligand;[401] however, little is known about the specific

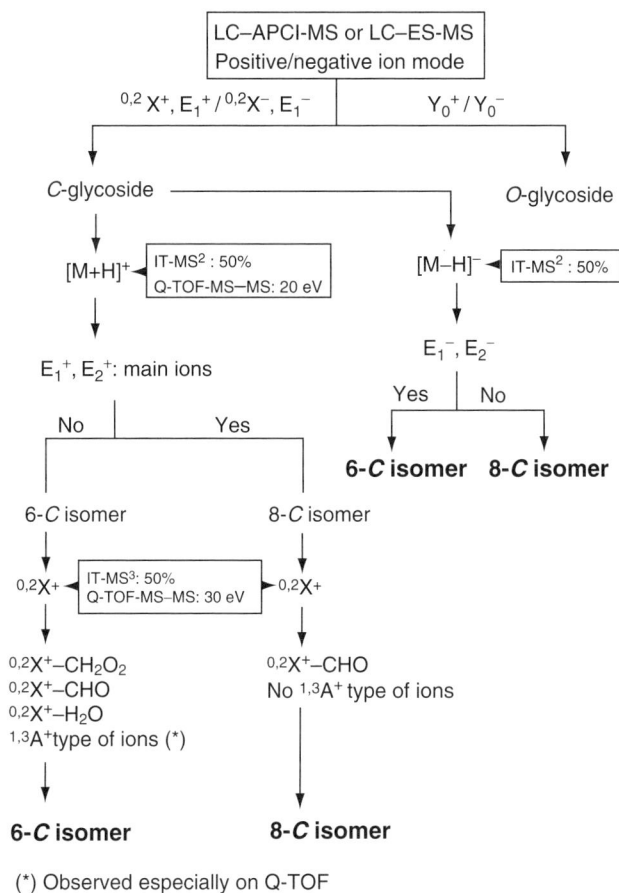

FIGURE 2.18 Identification of the 6-C and 8-C isomers of mono-C-glycosylflavones, respectively, based on diagnostic fragment ions observed in low-energy collision-induced dissociation tandem MS spectra. (Reprinted from Waridel, P. et al., *J. Chromatogr. A*, 926, 29, 2001. Copyright 2001 Elsevier Science B.V. With permission.)

structures of the metal complexes. Common auxiliary ligands used include 2,2′-bipyridine and 4,7-diphenyl-1,10-phenanthroline.

Many studies have dealt with the analysis of acylated flavonoid glycosides. MS analyses have mainly been used to obtain molecular mass information, but structure-specific information about the acyl group can be provided by neutral losses that are characteristic of the acyl group or the acylated glycosyl residue.[253] Characteristic acyl-related product ions can be observed in the $[M + H]^+$ and $[M + Na]^+$ low-energy CID spectra and radical acid-related product ions at high-energy CID conditions, which provide information on the presence and identity of the acyl group and its position on the flavonoid backbone structure.[403]

MALDI-TOF-MS has been used to identify and quantify other anthocyanins in foods.[317,404] When the anthocyanin content of highbush blueberries at different stages of anthocyanin formation were analyzed by both HPLC and MALDI-TOF-MS, it was found that both techniques provided comparable quantitative anthocyanin profiles.[317] While HPLC could distinguish anthocyanin isomers, MALDI-TOF-MS proved to be more rapid. MALDI-TOF-MS has also been used to identify the isoflavones in soy samples.[318] In a comparison of several matrices, 2′,4′,6′-trihydroxyacetophenone (THAP) and 2,5-dihydroxybenzoic acid were found most suitable. Isoflavones were predominantly ionized in a protonated form with a very small amount of sodium or potassium adduct ions. Fragmentation occurred only through loss of glycosidic residues. The same authors have used the technique to identify flavonol glycosides in yellow onion bulbs and green tea.[319] THAP was chosen as the best matrix because it worked for crude sample extracts and ionized flavonol glycosides in both positive and negative modes. In the positive mode, multiple ion forms were observed for flavonol glycosides, including $[M + H]^+$, $[M + Na]^+$, $[M + K]^+$, and $[M - H + Na + K]^+$, with further fragmentation through loss of glycosidic residues. The negative mode for all flavonol glycosides resulted in $[M - H]^-$ ion formation without detectable fragmentation. MALDI-TOF-MS has been used for structural elucidation of some flavones,[405] and a symmetrical glycosylated methylene bisflavonoid.[406]

Although proanthocyanidins are present as the second most abundant class of natural phenolic compounds after lignin,[407] relatively few MS studies appear in the literature before 1993, due to the structural complexity of these compounds. During the review period considerable progress has been achieved through studies of proanthocyanidins in various foods and beverages.[304,307–310] Using tandem MS coupled to reversed-phase HPLC (RP-HPLC), the proanthocyanidins of cocoa (*Theobroma cacao*) (Table 2.12),[373] green tea (*Camellia sinensis*),[408] and wine[409] were identified. The negative-ion mode was found to be more sensitive and selective in the studies of proanthocyanidins than the positive-ion mode.[409] Using LC–ESI-MS–MS analysis in the negative-ion mode, several new heterogeneous B-type proanthocyanidins containing (epi)afzelechin as subunits, including tetramers and pentamers, were identified in extracts of pinto beans, plums, and cinnamon (Table 2.13).[410] In MALDI analyses of Ruby Red sorghum,[307] deionization of the proanthocyanidin fractions with the Dowex 50 × 8–400 cation-exchange resin and subsequent addition of cesium trifluoroacetate (^{133}Cs) allowed the detection of exclusively $[M + Cs]^+$ ions in the spectra.

MALDI-TOF-MS has been used to characterize the molecular masses of condensed tannins with varying degrees of polymerization in unripe apples.[308] The technique has provided evidence for a catechin pentadecamer using *trans*-3-indoleacrylic acid as matrix in the presence of silver ion. Even in the absence of silver ion, the dodecamer and undecamer were observed in the positive- and negative-ion modes, respectively. The technique has also been employed to determine molecular sizes of oligomeric proanthocyanidins in coffee pulp,[309] and to characterize the polygalloyl polyflavan-3-ols (PGPF) in grape seed extracts.[310,311] Masses corresponding to a series of PGPF units inclusive of nonamers were observed in the positive-ion reflectron mode, while masses of PGPF inclusive of undecamers were observed in the positive-ion linear mode, providing the first known evidence of PGPF of this size.[310] In another study, the MALDI-TOF mass spectra of the condensed tannins of leaves and needles from willow (*Salix alba*), spruce (*Picea abies*), beech (*Fagus sylvatica*), and lime (*Tilia cordata*) in dihydroxybenzoic acid as matrix have shown signals of polymers of up to undecamers.[301] Supporting observations from NMR spectroscopy, the mass spectra of the willow, and lime leaf condensed tannins were identified as polymers with mainly procyanidin units, while the polymers of the spruce needle and beech leaves exhibit varying procyanidin–prodelphinidin ratios.

Because of their complex nature, a complete characterization of grape proanthocyanidins has so far eluded analytical chemists despite the effort devoted to it. A reliable method for total proanthocyanidin quantification and for supplying information regarding the molecular weight distribution of the most complex proanthocyanidins is still lacking. However, the potential role of the MALDI-TOF-MS technique for proanthocyanidin differentiation and as a quantification tool is promising.[305,306,311–313,411] In a recent paper, an offline coupling of size-exclusion chromatography and MALDI has been carried out to measure differences between polystyrenes and procyanidins.[314] Polystyrenes are used as standards because no

TABLE 2.12
Liquid Chromatography–Electrospray Ionization Tandem Mass Spectrometric Study of the Flavonoids of Cocoa (*Theobroma cacao*) with Negative Ion Detection

Compound	MW	t_r (min)	MS–MS ions m/z (Relative Abundance,%)	DP (V)	CE (V)
Catechin	290	8.5	289 [M − H]⁻ (40), 245 (100)	−50	−20
Epicatechin	290	11.8	289 [M − H]⁻ (40), 245 (100)	−50	−20
Luteolin 6-C-glucoside (isoorientin)	448	15.4	447 [M − H]⁻ (65), 429 (65), 357 (100), 327 (100), 285 [A − H]⁻ (50)	−60	−30
Luteolin 8-C-glucoside (orientin)	448	15.7	447 [M − H]⁻ (30), 357 (70), 327 (100), 285 [A − H]⁻ (20)	−60	−30
Apigenin 8-C-glucoside (vitexin)	432	17.8	431 [M − H]⁻ (35), 341 (30), 311 (100), 269 [A − H]⁻ (<5)	−55	−30
Apigenin 6-C-glucoside (isovitexin)	432	18.1	431 [M − H]⁻ (15), 353 (<5), 341 (40), 311 (100), 269 [A − H]⁻ (<5)	−55	−30
Quercetin 3-rutinoside (rutin)	610	18.2	609 [M − H]⁻ (100), 301 [A − H]⁻ (40)	−60	−30
Quercetin 3-galactoside (hyperoside)	464	18.4	463 [M − H]⁻ (5), 301 [A − H]⁻ (100)	−60	−38
Quercetin 3-glucoside (isoquercitrin)	464	19.0	463 [M − H]⁻ (35), 301 [A − H]⁻ (100)	−60	−32
Luteolin 7-glucoside	448	19.2	447 [M − H]⁻ (100), 285 [A − H]⁻ (100)	−60	−30
Kaempferol 3-rutinoside	594	21.3	593 [M − H]⁻ (70), 285 [A − H]⁻ (100)	−65	−45
Apigenin 7-rutinoside (isorhoifolin)	578	22.0	577 [M − H]⁻ (20), 269 [A − H]⁻ (100)	−60	−40
Naringenin 7-glucoside (prunin)	434	22.4	433 [M − H]⁻ (70), 271 [A − H]⁻ (100)	−60	−20
Kaempferol 3-glucoside	448	22.4	447 [M − H]⁻ (90), 285 [A − H]⁻ (100)	−60	−30
Quercetin 3-rhamnoside (quercitrin)	448	22.6	447 [M − H]⁻ (25), 301 [A − H]⁻ (100)	−60	−30
Naringenin 7-neohesperidoside (naringin)	580	22.6	579 [M − H]⁻ (100), 459 (20), 271 [A − H]⁻ (40)	−80	−35
Kaempferol 7-neohesperidoside	594	23.1	593 [M − H]⁻ (20), 327 (5), 285 [A − H]⁻ (100)	−70	−40
Apigenin 7-glucoside	432	23.3	431 [M − H]⁻ (100), 269 [A − H]⁻ (75)	−60	−35
Quercetin	302	33.0	301 [M − H]⁻ (60), 151 (100)	−60	−35
Luteolin	286	33.1	285 [M − H]⁻ (100), 217 (10), 199 (20), 175 (20), 151 (85), 133 (50), 107 (10)	−60	−35
Naringenin	272	34.5	271 [M − H]⁻ (25), 177 (20), 151 (100), 119 (75), 107 (35), 93 (15), 83 (10)	−60	−30
Apigenin	270	34.8	269 [M − H]⁻ (60), 151 (100)	−60	−35
Kaempferol	286	35.1	285 [M − H]⁻ (100), 217 (40), 151 (85), 133 (75)	−60	−35
Isorhamnetin	316	35.4	315 [M − H]⁻ (60), 300 (100), 151 (10)	−60	−30
Amentoflavone	538	36.3	537 [M − H]⁻ (100), 375 (65)	−60	−40

Notes: DP, declustering potential; CE, collision energy. The CE values were optimized in such a way that the sensitivity of the multiple reaction monitoring signal was at the maximum.

Source: From Sanchez-Rabaneda, F. et al., *J. Mass Spectrom.*, 38, 35, 2003. Copyright 2003 John Wiley & Sons, Ltd. With permission.

TABLE 2.13
Liquid Chromatography–Electrospray Ionization Mass Spectrometric Characteristics ([M − H]⁻ and Product Ions) of Selected Proanthocyanidin Tetramers and Pentamers from Pinto Beans, Plums, and Cinnamon, Respectively

Procyanidin Connection Sequence	[M − H]⁻	Product Ions
(Epi)Afz–(Epi)Afz–(Epi)Cat–(Epi)Cat	1121.3	849.0, 831.1, 577.2, 543.1
(Epi)Afz–(Epi)Cat–(Epi)Afz–(Epi)Cat	1121.2	849.0, 831.1, 561.1
(Epi)Afz–(Epi)Cat–(Epi)Cat–(Epi)Cat	1137.3	865.3, 577.2, 559.0
(Epi)Cat–(Epi)Afz–(Epi)Cat–(Epi)Cat	1137.3	849.3, 847.3, 577.2, 559.1
(Epi)Afz–(Epi)Afz–(Epi)Cat–(Epi)Cat–(Epi)Cat	1409.3	1119.3, 865.0, 831.1, 577.1
(Epi)Afz–(Epi)Cat–(Epi)Cat–(Epi)Cat–(Epi)Cat	1425.3	1153.3, 1135.7, 865.1, 577.1
(Epi)Cat–(Epi)Cat–A–(Epi)Cat–A–(Epi)Cat	1149.3	861.4
(Epi)Cat–(Epi)Cat–(Epi)Cat–A–(Epi)Cat	1151.2	863.5, 575.1
(Epi)Cat–(Epi)Cat–(Epi)Cat–(Epi)Cat–A–(Epi)Cat	1439.3	1151.3, 863.4, 575.2
(Epi)Afz–(Epi)Cat–A–(Epi)Cat–(Epi)Cat	1135.2	863.2, 847.2, 573.0
(Epi)Cat–A–(Epi)Cat–A–(Epi)Cat–(Epi)Cat	1149.3	859.2
(Epi)Cat–(Epi)Cat–A–(Epi)Cat–(Epi)Cat	1151.2	863.4, 861.2, 573.2
(Epi)Cat–A–(Epi)Cat–(Epi)Cat–(Epi)Cat	1151.2	861.3, 573.1
(Epi)Cat–(Epi)Cat–A–(Epi)Cat–A–(Epi)Cat–(Epi)Cat	1437.4	1147.4, 859.5
(Epi)Cat–(Epi)Cat–(Epi)Cat–A–(Epi)Cat–(Epi)Cat	1439.4	1151.2, 1149.4, 863.2, 575.1, 573.1
(Epi)Cat–(Epi)Cat–A–(Epi)Cat–(Epi)Cat–(Epi)Cat	1439.4	1151.3, 1149.4, 861.5, 577.2, 575.2, 573.1

Notes: The chirality of C-3 on the flavan-3-ols cannot be differentiated by MS. (Epi)afzelechin represents either afzelechin or epiafzelechin. Afz, afzelechin; Cat, catechin; A, A-type binding between the flavanol units, i.e., flavanols doubly linked by an additional ether bond between C-2 and O-7 in addition to the C4–C8 (or more rarely C4–C6) bond.
Source: From Gu, L. et al., *J. Mass Spectrom.*, 38, 1272, 2003. Copyright 2003 John Wiley & Sons, Ltd. With permission.

commercial procyanidin standards are available. Between 1000 and 8000 Da, there was good correlation between the MALDI and PS calibration curves. In this range, the PS calibration was correct and enabled true mass determination.

2.3.4 QUANTITATIVE CONSIDERATIONS

Recently, a method has been developed for faster evaluation of the total flavonoid content in plants and foodstuffs.[341] A TOF instrument has been used to acquire an *m/z* range of 220 to 700 with a generic gradient HPLC run, detecting both positive negative ions in alternating spectral acquisitions, and producing exact mass data during the whole run by using a Lock-Spray ESI source. Traditionally, most of the quantitative LC–MS methods utilize linear quadrupole or ion trap mass spectrometers, due to their good linear dynamic range. However, if a large number of compounds with different molecular weights is to be detected simultaneously in an LC–MS experiment using quadrupole instruments, the demand for scanning over the wide mass range decreases the sensitivity. On the other hand, if SIM is used to increase the sensitivity, the chromatographic resolution may become a problem. In view of the above considerations together with the moderately poor ionization efficiency of the flavonoids, higher detection limits with LC–MS than with LC–UV methods are usually observed. When using a TOF instrument, the necessity to compromise between sensitivity and chromatographic resolution was considerably reduced when data were acquired over a wide range of m/z ratios. The compounds were quantified using quercetin, quercitrin, rutin, and kuromanine as external standards and dextromethorphan as an internal standard. The detection limits ranged from 0.01 to 0.04 mg/ml, while the quantification ranges obtained were 0.2 to 10 mg/ml for anthocyanins and 0.2 to 4 mg/ml for the other flavonoids.[341]

The analytical performance of four modes of LC–MS, multiple MS (MS^n), and tandem MS operation (APCI and ESI with positive and negative ionizations) has been compared for two mass spectrometers, a triple-quadrupole and an ion-trap instrument.[336] With 15 flavonoids as test compounds, the use of APCI in the negative-ion mode gave the best response, with the signal intensities and the mass-spectral characteristics not differing significantly between the two instruments. Under optimum conditions, full-scan limits of detection of 0.1 to 30 mg/l were achieved in the negative APCI mode. The main fragmentations observed in the MS^n spectra on the ion trap, or the tandem MS spectra on the triple quadrupole, were generally the same. The advantage of the former approach was the added possibility to ascertain precursor–product ion relations. The best results were obtained when methanol–ammonium formate (pH 4.0) was used as LC eluent.[336] In another comparison based on five flavones or flavonols, the effect of nine different eluent compositions on the ionization efficiency has been studied using ion spray (IS), APCI, and atmospheric pressure photoionization (APPI) in positive- and negative-ion modes.[368] It was shown that the eluent composition had a major effect on the ionization efficiency, and the optimal ionization conditions were achieved in positive-ion IS and APCI using 0.4% formic acid (pH 2.3) as a buffer, and in negative-ion IS and APCI using ammonium acetate buffer adjusted to pH 4.0. For APPI work, the eluent of choice appeared to be a mixture of organic solvent and 5 mM aqueous ammonium acetate. The limits of detection (LODs) were determined in scan mode, and it was shown that negative-ion IS with an eluent system consisting of acidic ammonium acetate buffer provided the best conditions for detection of flavonoids in MS mode, their LODs ranging between 0.8 and 13 μM for an injection volume of 20 μl.[368]

A rather sensitive RP-HPLC method combined with UV (270 nm) and ESI-MS detection has been established for the determination of flavonoids and other phenolic compounds in various biological matrices.[382] LODs based on UV data of flavonoids in onion and soybean were 6 to 42 pmol injected, which corresponded with analyte concentrations of 0.08 to 0.63 mg/l. It has also been reported that 12 dietary flavonoid glycosides and aglycones in human urine have been identified and quantified by LC–MS using MeOH–ACN–formic acid as eluent, and APCI in negative mode.[381] Calibration graphs were prepared for urine, and good linearity was achieved over a dynamic range of 2.5 to 1000 ng/ml. Selected ion monitoring offered a considerable gain in selectivity as well as sensitivity, and LODs were determined to be 0.25 to 2.5 ng/ml.[381] Wogonin metabolites (flavones) have been identified in rat plasma with an LC–ESI/IT method in multistage full-scan mode.[412] On basis of this a sensitive LC–triple-quadrupole MS method using APCI in the selected reaction monitoring mode has been used to determine the concentration of wogonin and its major metabolite in rat plasma. The method had a lower limit of quantification of 0.25 ng/ml for wogonin, and has been successfully applied to a preclinical pharmacokinetic study after an oral administration of 5 mg/kg wogonin to rats. The quantitative method was validated with respect to linearity, precision, and accuracy.

Based on an in-vial derivatization method, the mass spectra of methylated flavonoids and other phenolics have been obtained via EI-MS at 70 eV.[346] Detection was performed in the selective ion monitoring mode and peaks were identified and quantified using target ions. The detection limits ranged between 2 and 40 ng/ml, whereas the limits of quantitation fall in the range of 5 to 118 ng/ml, with flavonoids accounting for the lowest sensitivity due to their multiple reaction behavior.

Probably most important for plant extract analyses, MALDI-MS is remarkably tolerant of impurities making the direct analysis of crude extracts possible. Through its capability for analyzing very small quantities of these compounds in unpurified samples, MALDI provides a sensitive means for the detection of flavonoid pigments in plant tissues. By analyzing a mixture of 3-deoxyanthocyanidins using MALDI-MS, sensitivities to the level of 15 pmol/μl have been attained for 3-deoxyanthocyanidins present in crude extracts from sorghum plant tissue, and as low as 5 pmol/μl for pure samples containing the anthocyanidin, pelargonidin, and the anthocyanin, malvin.[315]

2.4 VIBRATIONAL SPECTROSCOPY (IR AND RAMAN)

Two different types of spectroscopic techniques are most frequently used to view the fundamental modes of molecular vibrations, namely mid-IR spectroscopy and Raman spectroscopy.[413] The first method measures the absorption, transmission, or reflection of IR radiation with wavelengths in the range of 2.5 to 25 μm. The Raman method irradiates the sample with radiation of much shorter wavelengths and measures the fraction of scattered radiation for which the energy of the photon has changed. The vibrational spectra may serve as fingerprints of structure, composition, interactions, and dynamics. The reciprocal of wavelength, wavenumber (cm^{-1}), is commonly used to characterize the energy in the field of vibrational spectroscopy.

Systematic vibrational spectroscopy studies on flavonoids have occurred since the early 1950s, and most of them have been limited to a discussion on the hydroxyl and carbonyl absorption frequencies.[414] However, with the technical advances of the last two decades, the application of vibrational spectroscopy has become much more relevant in the field of flavonoid analysis.[415,416] The implementation of FTIR spectroscopy has significantly enhanced the sensitivity, and Raman spectroscopy has benefited from the availability of holographic notch filters, which efficiently suppress the strong signal from elastically scattered (Rayleigh) radiation while maintaining the Raman-shifted intensity with minimal attenuation. Furthermore, high-powered NIR semiconductor lasers and sensitive charge-coupled devices have replaced inconvenient gas lasers and light-detection technologies. Ordinary Raman spectroscopy has drawbacks in that it requires high compound concentrations, and the recorded spectrum will correspond to all molecules present in the sample. In resonance Raman spectroscopy, this is overcome through the use of laser light with a frequency corresponding to the absorption maximum of the compound to be characterized. Finally, the increase in the computing power of standard computers has facilitated more sophisticated data evaluation of both IR and Raman spectra.

The following sections describe the applications of IR, Raman, and NIR spectroscopic techniques applied to the field of flavonoids in recent years.

2.4.1 IR AND RAMAN SPECTROSCOPIC TECHNIQUES IN STUDIES OF FLAVONOID STRUCTURES

IR and Raman spectroscopic techniques have been extensively used by Merlin, Cornard, and their coworkers to achieve structural information about flavonoid geometry.[417–420] These investigations have usually been accompanied by UV–Vis spectroscopic and x-ray crystallographic analysis, as well as quantum chemical calculations. The main focus has been on the effects of position and nature of substituent (hydroxyl or methoxyl groups) on the molecular structure, including investigations of the dihedral angle between the phenyl ring and the chromone part of the molecule. The vibrational spectra of various simple flavonoids in solid state have been compared with those obtained in solutions, and differences between the solid-state spectra and solution-state spectra were explained by the possibility of the formation of intramolecular hydrogen bonds present in the solid state and under specific solution conditions, or formation of intermolecular hydrogen bonds with the solvent (CH_3OH). The Raman spectra were preferred in most of these studies because they were considerably less complex than the corresponding IR spectra. The structures of a variety of flavonoid–aluminum ion complexes, including the complexes of aluminum(III) with 3-hydroxyflavone,[421] 5-hydroxy-flavone,[422] 3′,4′-dihydroxyflavone,[423] quercetin,[424] and quercetin 3-glucoside,[425] have also been examined by this research group. The influence of pH and Al^{3+} concentration on the complex formation was considered, and molecular conformations of both the free and complexed flavonoids were proposed. Recently, these flavonoids have been used as model compounds for the study of the behavior of humic substances toward Al(III) complexation.[426] Other complexes between flavonoids and metal ions investigated by IR spectroscopy include an alumina–(+)-catechin solution system,[427] and some prepared organotin(IV) complexes with the flavonoid glycosides, rutin, hesperidin, and 2′,4′,3-trihydroxy-5′,4-dimethoxychalcone 4-rutinoside, and with the aglycones, quercetin, morin, hesperitin, and some flavones.[428] The FTIR spectra of these latter complexes were consistent with the presence of Sn–O (phenol or carbohydrate) vibrations in the compounds, and the structures of the complexes were measured by Mössbauer spectroscopy.

FTIR spectroscopy has been used for the reexamination of the carbonyl stretching frequency of some simple hydroxyflavones in argon and methanol–argon matrices,[429] and IR spectra have been recorded for some simple flavonoids including sulfonic acid derivatives.[430] The Raman spectra of six common anthocyanidins and some of their glycosylated derivatives in acidic aqueous solutions have been compared.[431] Despite great similarity between these spectra, the anthocyanin substitution pattern could easily be recognized and the effect of glycosylation was clearly visible; 5-glycosylation seemed to cause a greater perturbation in the vibrational properties than that of 3-glycosylation. A more recent Raman spectroscopy study of the structure of anthocyanins in aqueous solutions as a function of temperature has also been presented.[432]

Among the new vibrational spectroscopic techniques applied to flavonoids, the hydrophilic extracts of Scots pinewood and two model compounds, the flavone chrysin and the stilbene pinosylvin, have been characterized using UV resonance Raman spectroscopy.[433] Pinosylvin and chrysin were resonance enhanced by UV excitation. Both compounds showed very intense UV Raman bands due to alkene and aromatic structures at 1649 to 1635 cm^{-1} and 1605 to 1600 cm^{-1}, respectively. In addition, aromatic and unsaturated structures of pinosylvin and chrysin showed bands at 1582 and 1549 cm^{-1}, respectively. A very useful multichannel spectrometer with microprobe and laser excitation either in the near-UV or visible range has been used for flavonoid analysis.[415] It combines high-detection sensitivity with rapid recording of spectra. The instrument is designed to record fluorescence emission or Raman scattering spectra of samples examined in the microscope. The study of anthocyanins in *Zebrina pendula* leaves has illustrated the possibility of recording absorption, fluorescence, and Raman spectra from the same living cell.[415]

The frequencies of many vibrational normal modes do not depend only on the molecular structure. IR studies of 5-methyl-7-methoxy-isoflavone in 20 different organic solvents have been used to examine the solvent–solute interactions and to correlate solvent properties with the IR band shift.[434] It was found that no linear relationship existed between the wavenumber of the C=O stretching band and the Kirkwood–Bauer–Magat solvent parameters. However, good correlations were observed between the wavenumbers of the C=O stretching band and the solvent acceptor number, and even better correlations between the wavenumbers of the C=O stretching band and the linear solvation energy relationships. A new chemistry model within density functional theory (called CHIH-DFT) has recently been used to predict the IR and UV–Vis spectra of quercetin.[435] The predicted spectra were in good agreement with the previously reported experimental UV and IR spectra, and assignments of the principal peaks have been achieved.

2.4.2 IR AND RAMAN SPECTROSCOPIC TECHNIQUES IN STUDIES OF COMPLEXES INVOLVING FLAVONOIDS

In recent years, IR and Raman spectroscopic techniques have been applied for the characterization of flavonoid-containing systems with rather complex composition. A rapid analytical method involving attenuated total reflection (ATR) mid-IR spectroscopy and UV–Vis spectroscopy, combined with multivariate data analysis, has been applied for the discrimination of Austrian red wines.[436] By analyzing phenolic extracts (obtained by C18 solid-phase extractions followed by elution with acidified methanol) of the various wines by mid-IR spectroscopy, almost complete discrimination of all samples was achieved. Furthermore, it was possible to establish class models for five different wine cultivars and to classify the test samples correctly. In another study, the Raman spectrum of *Artocarpus heterophyllus* heartwood has shown to exhibit two characteristic bands at 1247 and 745 cm^{-1}.[437] Based on Raman measurements of pure flavones and related compounds, it was predicted that the Raman band at 1247 cm^{-1} was attributed to flavonoid-type compounds. In this case, no vibrational band corresponding to the characteristic Raman bands was observed by diffuse reflectance IR spectroscopy.[437] By using solid-state FTIR and Raman spectroscopies an inclusion complex between 2′,6′-dimethoxyflavone and formic acid has been identified.[438] The broad and intense IR absorption observed in the range 3400 to 1900 cm^{-1}, assigned to the hydrogen-bonded OH-group stretching vibration, exhibited the characteristic ABC structure of strong hydrogen-bonded complexes in good agreement with previous x-ray data showing that *cis*-formic acid was strongly hydrogen bonded to 2′,6′-dimethoxyflavone. The inclusion complex was quite unstable, and the IR spectrum clearly showed that formic acid disappeared after a period of a few months. The formation of some β-cyclodextrin inclusion complexes involving various flavanones has been investigated by FTIR and other methods.[439] Changes in the characteristic IR bands of pure substances confirmed the existence of β-cyclodextrin–flavanon complexes as new compounds with different spectroscopic bands.

In an interesting application, the interaction of polyphenols with proline-rich proteins was studied using an automated flow injection system with FTIR detection to gain insight into chemical aspects related to astringency.[416] Agarose beads carrying the proline-rich protein were placed in the IR flow cell in such a way that the beads were probed by the IR beam. By using an automated flow system, the samples were flushed over the proteins in a highly reproducible manner. Simultaneously, any retardation due to polyphenol–protein interactions taking place inside the flow cell was monitored by IR spectroscopy.[416] Recent data obtained by FTIR experiments, fluorescence spectroscopy, CD experiments, and molecular modeling have suggested that the flavone scutellarein can strongly bind to the human serum albumin.[440]

2.4.3 TWO-DIMENSIONAL IR ANALYSIS

The introduction of generalized 2D correlation IR spectroscopy, 2D-IR, by Noda[441] has extended markedly the potential of using vibrational spectroscopy for flavonoid analysis. Analyses have been performed using ordinary FTIR spectrometers, and the coordinates of the two dimensions in these spectra both use frequency or wavenumber as units (Figure 2.19). Peaks in 2D-IR spectra might show the sensitivity for each IR band or each functional group and the correlation between the functional groups, even the order of the influence when the system is subjected to a given perturbation.[441] 2D-IR spectroscopy can simplify complex spectra consisting of many overlapping bands and enhance spectral resolution by spreading peaks along the second dimension, thus enabling extraction of information that cannot be obtained straightforwardly from 1D spectra. The method has hitherto not been applied to

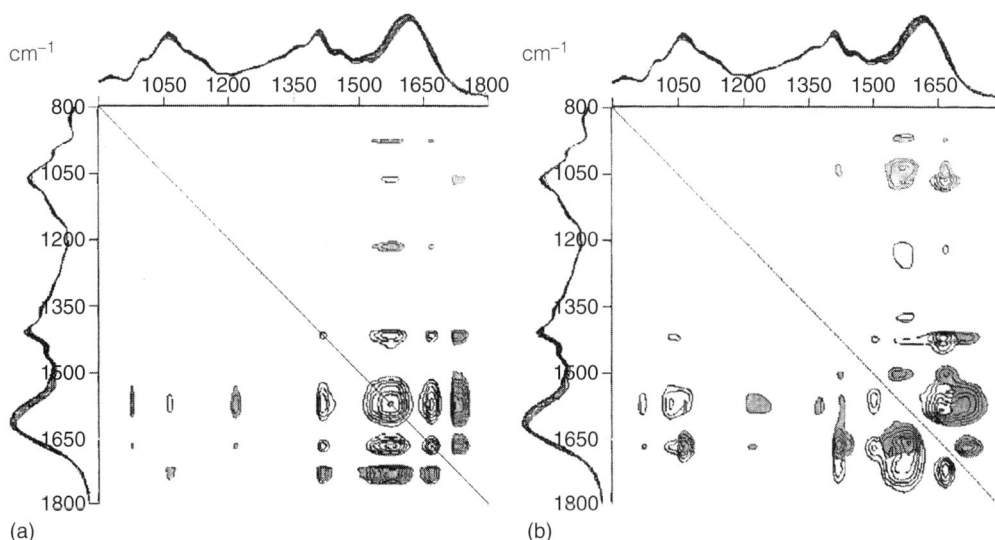

FIGURE 2.19 2D IR correlation spectra of the flavonoid-containing Chinese medicine "Quing Kai Ling." (a) Synchronous IR spectrum; (b) asynchronous IR spectrum. In analogy to the corresponding 1D IR spectra, the units of the axis in both these 2D spectra are given in cm^{-1}. (Reprinted from Zuo, L. et al., *J. Pharm. Biomed. Appl.*, 30, 1491, 2003. Copyright 2003 Elsevier Science B.V. With permission.)

pure flavonoids; however, the traditional Chinese medicine "Qing Kai Ling" injection after deterioration has been distinguished from the original formulation using FTIR and 2D-IR (Figure 2.19).[442] It has been very difficult to distinguish IR spectra of injections before and after deterioration by using the conventional 1D approach. However, higher spectral resolution and more structural information provided by 2D-IR analyses have made this differentiation possible. According to the 2D correlation analysis, the band at 1611 cm^{-1} in the conventional IR spectra, in fact, consisted of the overlap of three bands at 1572, 1667, and 1729 cm^{-1}, which were assigned, respectively, to the alkaloids, flavone derivatives, and carbonyl compounds in the injection.[442]

2.4.4 COUPLED TECHNIQUES INVOLVING VIBRATIONAL SPECTROSCOPY

Trimethylsilyl derivatives of ten hydroxy- and methoxyhydroxyflavonoids have been studied by the GC–FTIR technique.[443] The correlation found between retention and gas-phase IR data was used in structural identification of compounds having very similar chromatographic behavior. The shift of the carbonyl frequency gave information on the presence of substitution. Some hydroxy- and methoxy-substituted flavones have been studied following carbon dioxide supercritical fluid chromatography on polymethylsiloxane capillary columns using flame ionization and FTIR detection.[444]

The HPLC–FTIR technique has recently been used to identify six catechins and two methyl-xanthines present in green tea extracts.[445] A reversed-phase separation of the compounds was performed on a C-18 column equilibrated at 30°C using an isocratic mobile phase of acetonitrile–0.1% formic acid (15:85), prior to introduction to the deposition interface linked to the FTIR detector. The solvent was evaporated at 130°C and spectra were collected every 6 sec during the run. Two distinct designs for HPLC–FTIR interfaces have been developed: flow cells and solvent elimination systems. Flow cell systems acquired spectra of the eluent in the solvent matrix through IR transparent, nonhydroscopic windows. The spectrum of the solvent was then subtracted from the sample spectrum, with the result being the spectrum of the analyte of interest. Solvent elimination systems nebulized the HPLC eluent and removed the solvent before depositing the dried solutes onto an IR transparent surface. Due to the strong IR absorption of water and other solvent, solvent elimination systems are generally more sensitive than flow cell arrangements. Acid washing the drift tube was recognized as being essential to obtain reproducible catechin depositions.[445]

2.4.5 NEAR-INFRARED SPECTROSCOPY

NIR spectroscopy involves the measurement of wavelengths (800 nm to 2.5 μm) and intensity of the absorption of NIR light by a sample. This light span is energetic enough to excite overtones and combinations of molecular vibrations to higher energy levels. Several recent studies emphasize the potential of NIRS as a nondestructive and effective alternative method to conventional quantitative analysis of food, plant extracts, pharmaceutical remedies, etc. In one application, Huck et al.[446] have reported the use of NIR for quantitative analysis of the water content, ethanol, and the flavone 3′,4′,5′-trimethoxyflavone in *Primulae veris* Flos extracts. First, a calibration set using a reference method was established, e.g., HPLC for quantification of the flavone. The values obtained were correlated by the use of special software (NIRCal) to analyze the NIR spectral data. This calibration set could then be used to quantify other new samples without the need for HPLC. An NIR reflectance spectroscopy technique has been developed for the prediction of procyanidins in cocoa beans (*Theobroma cacao*),[447] for the estimation of the nasunin (anthocyanin) content in the skin of eggplant,[448] to predict the content of the dihydrochalcone aspalathin in unfermented rooibos,[449] condensed tannins concentrations in *Lotus uliginosus*,[450] tannins and total phenolics content in forage legumes,[451] and for the simultaneous determination of alkaloids and phenolic substances including flavonoids in green tea.[452] NIRS has been used for the prediction of resistance in sugarcane to stalk borer *Eldana saccharina*.[453] NIR spectra of 60 sugarcane clones varying in resistance to *E. saccharina* indicated that chlorogenates and flavonoids might be involved in the interaction between the insect and sugarcane. NIRS has also been used to estimate plant pigment content (anthocyanins, carotenoids, chlorophylls) in higher plant leaves.[454] A rapid quantitative NIRS method was established for the determination of the constituents, including the biflavone I3,II8-biapigenin, in St. John's wort extracts,[455] and for simultaneous determination of total flavones and total lactones in *Ginkgo* extracts.[456] Recent developments of multichannel dispersive Raman microprobes using NIR excitation beyond 1000 nm and linear array detectors offering good sensitivity in the NIR region of the spectrum between 1000 and 1600 nm have been applied to various samples including flavone.[457]

2.5 ULTRAVIOLET–VISIBLE ABSORPTION SPECTROSCOPY

The application of standardized UV (or UV–Vis) spectroscopy has for years been used in analyses of flavonoids. These polyphenolic compounds reveal two characteristic UV absorption bands with maxima in the 240 to 285 and 300 to 550 nm range. The various flavonoid classes can be recognized by their UV spectra,[2] and UV-spectral characteristics of individual flavonoids including the effects of the number of aglycone hydroxyl groups, glycosidic substitution pattern, and nature of aromatic acyl groups have been reviewed in several excellent books.[1,2,458]

Today, the major use of UV–Vis spectroscopy applied to flavonoids is in quantitative analyses, and the value of this method for some structural analyses is diminishing compared to the level of information gained by other modern spectroscopic techniques like NMR and

MS. This section will rather briefly concentrate on some of the more recent applications of UV absorption spectroscopy in the flavonoid field. It will mainly cover online UV absorption spectroscopy in chromatography (Section 2.5.1). Because of the current importance of UV–Vis in the study of anthocyanins, some more UV–Vis spectral details have been included related to this pigment group (Section 2.5.2). Section 2.5.3 indicates the recent use of this technique in studies of flavonoids interacting with other compounds.

2.5.1 ONLINE UV ABSORPTION SPECTROSCOPY IN CHROMATOGRAPHY

The combination of HPLC equipped with a UV–Vis DAD (see Section 1.4.4) has for the two last decades been the standard method for the detection of flavonoids in mixtures. This type of detector allows the simultaneous recording of chromatograms at different wavelengths. The HPLC–DAD (alternatively called LC–UV) method has, during the period of this review, been used for isolation, identification, screening, measurement of peak purity, or quantitative determinations of flavonoids in numerous studies, and there exist several excellent recent reviews in the field.[459–461] In the absence of standards, the method offers spectral information about individual flavonoids by recording UV–Vis spectra wherein each peak is revealed in the chromatogram. However, during elution the mobile-phase composition may vary considerably, and the various LC methods may involve different solvents and solvent compositions. There may, for instance, be a 15 nm shift toward shorter wavelengths when water is substituted for methanol. Thus, the resulting spectra of the same flavonoid may be obtained in different solvents, aggravating precise identification based on agreement with literature data obtained on pure flavonoids in a standardized solvent. When it comes to alternative quantitative methods, various catechins have been separated with gradient RP-HPLC and quantified by UV (270 nm) and fluorescence (280/310 nm, excitation/emission) detection in series.[462] The combination of HPLC with online UV, MS, and NMR detection in LC–UV–MS$^{(n)}$ (Sections 1.4.5 and 2.3.2.2) and LC–UV–NMR (Sections 1.4.6 and 2.2.5) has proved to be a very valuable tool for the analysis of natural products especially in extracts or mixtures.

The use of UV shift reagents such as $AlCl_3$ (5% in methanol)–HCl (20% aqueous), NaOMe (2.5% in methanol), and NaOAc (3 mg)–H_3BO_3 has proven to be very useful as guidelines for substitution patterns of many flavonoids; however, the use of these reagents has mainly been applicable for purified flavonoids.[1,2] By adding suitably modified shift reagents to the eluate leaving a HPLC column, similar shifts of the UV absorption maxima of flavonoids in the eluate have been induced.[463] Recently, Wolfender and his collaborators have introduced UV shift reagents by postcolumn addition in hyphenated LC–UV–MS analysis of flavonoid-containing crude extracts (Figure 2.20).[131,141,464] The shifts observed were interpreted according to the rules previously established for the analysis of pure polyphenols,[2] and permitted the localization of the position of the hydroxyl groups on most of the compounds detected in the crude extracts that were analyzed. A screening method that allowed the determination of the flavonoid composition of plant and food extracts has been based on double online detection, first at 280 nm and then at 640 nm, after derivatization with p-dimethylaminocinnamaldehyde (1% in 1.5 M sulfuric acid in methanol).[465] The colored adducts showed maximum absorption between 632 and 640 nm, thus preventing the interference of other colored compounds in the same extracts.

2.5.2 UV–VIS ABSORPTION SPECTROSCOPY ON ANTHOCYANINS

The UV–Vis spectral data on anthocyanins (typically measured in methanol containing 0.01% HCl) give important information about the nature of the aglycone and aromatic acyl

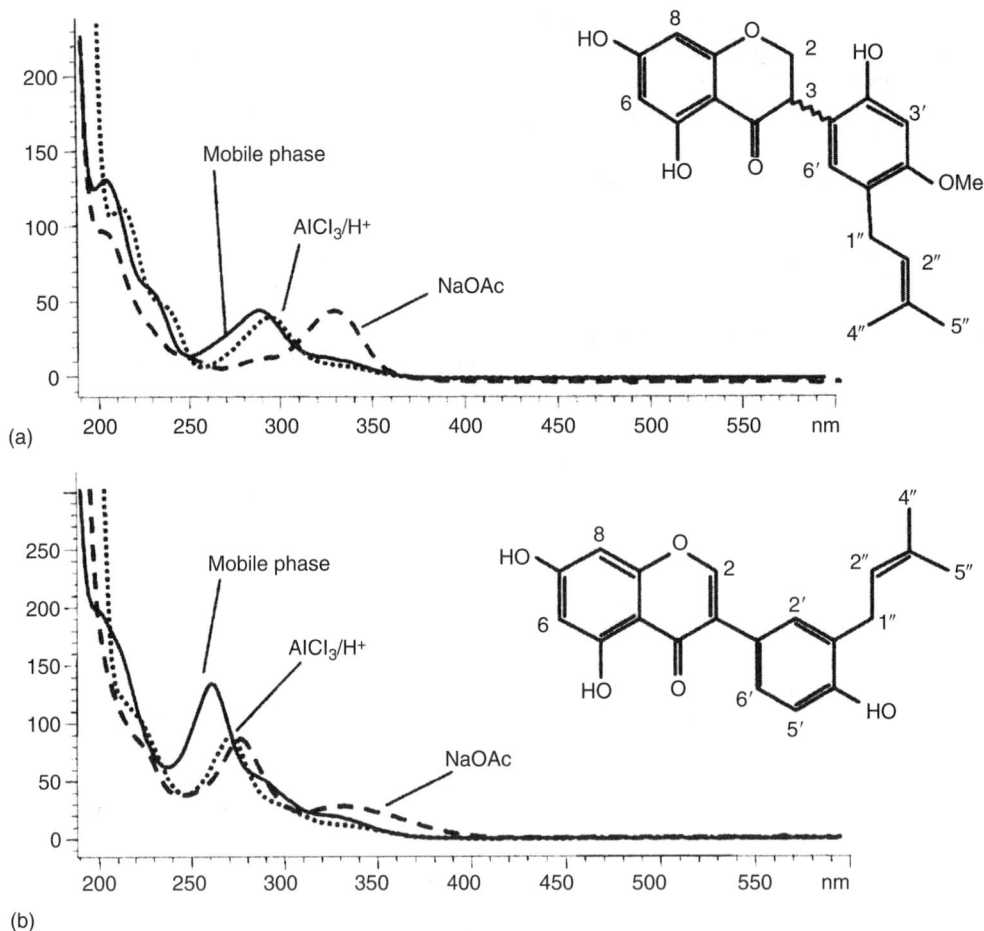

FIGURE 2.20 Complementary UV-DAD and shifted UV-DAD spectra with postcolumn addition of shift reagents of an isoflavanone (a) and an isoflavone (b) recorded online. The weak base NaOAc and acidic $AlCl_3$, respectively, were used as shift reagents. The shifted UV spectra are superimposed on the original spectra for each compound. The observed shifts provide information about the flavonoid substitution in accordance to the rules established for pure compounds. (Modified from Wolfender, J.-L., Ndjoko, K., and Hostettmann, K., *J. Chromatogr. A*, 1000, 437, 2003. Copyright 2003 Elsevier Science B.V. With permission.)

groups.[460,466] Anthocyanins with 4'-*O*-glycosylation have recently been identified,[170] and now UV–Vis spectral data for anthocyanins having glycosyl moieties connected to all the hydroxyl positions have been reported.

Some diagnostic information about the glycosidic substitution pattern may thus be revealed: anthocyanins with sugar unit(s) on the B-ring connected to the 3'-, 4'-, or 5'-hydroxyls seem to have their visible λ_{max} at shorter wavelengths (4 to 14 nm) than the corresponding anthocyanin 3-glycosides. A hypsochromic shift (12 nm) was observed for λ_{max} in the UV–Vis spectrum of cyanidin 3,4'-*O*-diglucoside, compared to the corresponding value of cyanidin 3-*O*-glucoside.[170] Also, λ_{max} in the spectra of the 3,7-*O*-diglucoside, 3,7,3'-*O*-triglucoside, and 3,7,3',5'-*O*-tetraglucoside of delphinidin have been observed at 537, 525, and 521 nm, respectively.[467,468] Anthocyanin 5-*O*- and 7-*O*-glucosides seem to exhibit their

λ_{max} at slightly longer wavelengths (5 to 9 nm) than the corresponding anthocyanin 3-O-glycosides; however, no similar effect on λ_{max} has been observed for anthocyanin 3,5-O-diglycosides and anthocyanin 3,7-O-diglycosides. It is well known that absorption spectra of pelargonidin 3-O-glycosides show higher absorbances at wavelengths around 440 nm than found in the corresponding spectra of the other common anthocyanidin 3-O-glycosides. Similarly with the spectra of pelargonidin 3-O-glycosides, a shoulder has been observed around 440 nm in the UV–Vis spectra of cyanidin and delphinidin derivatives with O-glycosyl moieties on their B-rings.[114,170] The relationship between color and substitution patterns in anthocyanidins has been investigated with the aim of developing quantitative structure–color models; and experimental data for the lowest UV transition in 20 substituted anthocyanidins have been reviewed.[469] While hypsochromic effects from hydroxyl and methoxyl moieties at positions 6 and 8 were reported, it is interesting to note that a C-glycosyl moiety in the 8-position has the opposite effect producing a more bluish color.[470]

The anthocyanins change their forms and colors depending on pH, concentration, copigmentation, and metal ions. Various inter- and intramolecular association mechanisms may be involved. Many of the studies in the field of anthocyanin and flower color, which were covered excellently by Brouillard and Dangles,[471] involve UV–Vis absorption spectroscopy. The presence of acylation with cinnamic acids can be deduced by the appearance of a peak or shoulder in the 310 to 330 nm region, while this peak is not observed in the case of benzoic acids, which have their maximum absorption between 270 to 290 nm. The $A_{\lambda max}$(acyl)–$A_{\lambda max}$(visible) ratio may be a measure of the number of aromatic acids present in the anthocyanin.[466] In the literature, there exist many recent examples on how aromatic acyl substituents of anthocyanins have caused bathochromic shifts in the UV–Vis spectra due to intra- or intermolecular 'π–π' stacking with the anthocyanidins.[472–475] In these cases, either cinnamoyl moieties or polyacylation with benzoic acids are involved. However, no significant bathochromic effects are observed on λ_{max} for cyanidin 3-(2″-galloylglucoside) and cyanidin 3-(2″,3″-digalloylglucoside), indicating the absence of intramolecular copigmentation for these pigments.[169] Dangles et al.[476] have indeed reported that two sugars are necessary as spacers to allow folding of the acyl moiety leading to higher chromophore integrity. During the review period several covalent complexes between anthocyanins and other flavonoids have been identified.[88,226,477–479] Significant bathochromic shifts of λ_{max} (12 to 20 nm) in spectra of these complexes may reveal intramolecular 'π–π' stacking of the anthocyanidin with the flavone or flavonol moiety.

Absorption spectra have also been used in the reexamination of pH-dependent color and structural transformations in aqueous solutions of some nonacylated anthocyanins and synthetic flavylium salts.[480] In a recent study, the UV–Vis spectra of flower extracts of *Hibiscus rosasinensis* have been measured between 240 and 748 nm at pH values ranging from 1.1 to 13.0.[481] Deconvolution of these spectra using the parallel factor analysis (PARAFAC) model permitted the study of anthocyanin systems without isolation and purification of the individual species (Figure 2.21). The model allowed identification of seven anthocyanin equilibrium forms, namely the flavylium cation, carbinol, quinoidal base, and E- and Z-chalcone and their ionized forms, as well as their relative concentrations as a function of pH. The spectral profiles recovered were in agreement with previous models of equilibrium forms reported in literature, based on studies of pure pigments.

Pigment stability measurements of anthocyanin-containing extracts and pure pigments is another application area of UV–Vis spectroscopy.[482–485] Color and stability studies of the 3-glucosides of the six common anthocyanidins and petunidin 3-[6-(4-(p-coumaryl)rhamnosyl)glucoside]-5-glucoside in aqueous solutions, during several months of storage, have revealed large variation between the pigments in particular at slightly acidic to slightly alkaline pH values.[485,486]

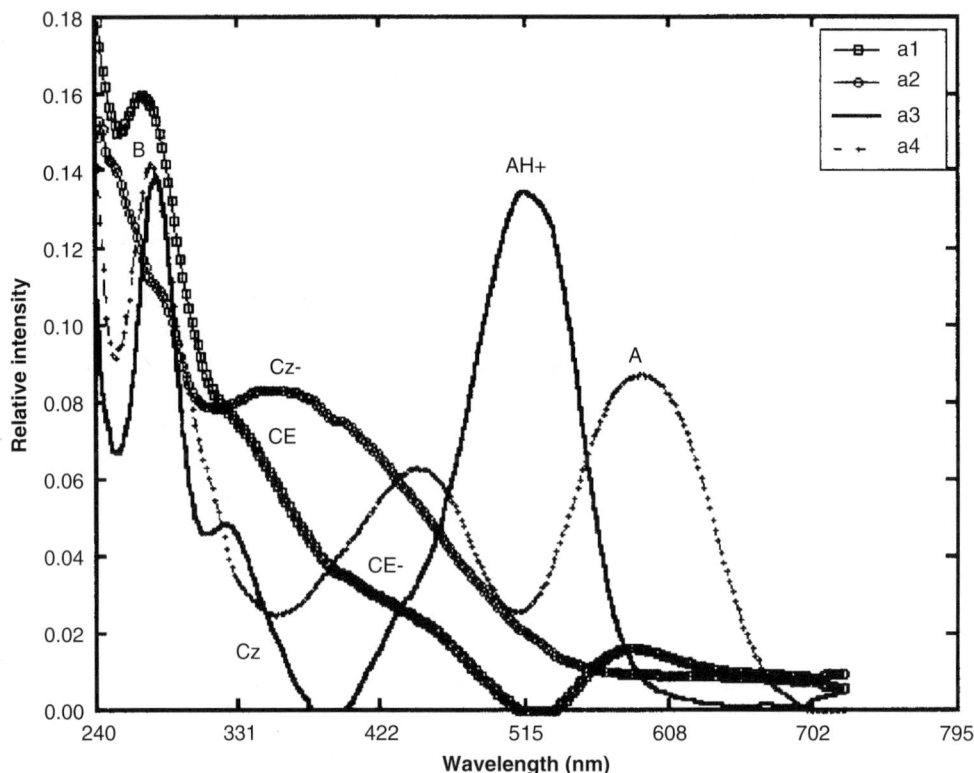

FIGURE 2.21 Spectral profiles recovered by the PARAFAC model at different pH values based on the deconvolution of UV–Vis absorption spectra, featuring the various anthocyanin secondary monomeric forms. (Reprinted from Levi, M.A.B. et al., *Talanta*, 62, 299, 2004. Copyright 2004 Elsevier Science B.V. With permission.)

Useful for quantification of anthocyanins, the molar absorption coefficients of several anthocyanins have been reviewed.[460,488] However, these compilations reveal lack of uniformity between the reported values, most probably due to the unavailability of pure anthocyanins in sufficient quantities to allow reliable weighing under optimal conditions, and the lack of standardization of anthocyanin solvent used for measurements.

2.5.3 UV–Vis Absorption Spectroscopy Involving Flavonoids in Complexes

In a recent paper, the interaction of various simple flavonoids with an anionic surfactant, sodium dodecyl sulfate (SDS) in aqueous solution, has been studied through absorption spectroscopy as a function of the concentration of the surfactant above and below the critical micelle concentration.[489] The approximate number of additive molecules (flavonoids) incorporated per micelle was estimated at a particular concentration of SDS. Incorporation of flavonoids in micelles shifted the UV absorption bands toward higher wavelengths, and the bathochromic shifts also depended upon the nature of the surfactant head group.

UV–Vis linear dichroism and (mid-)IR ATR IR analysis have been used to explain the possible association between DNA and the flavonols quercetin, rutin (quercetin-3-rutinoside), and morin (3,5,7,2′,4′-pentahydroxyflavone) in solution.[490] These nucleophilic flavonoids were shown to bind DNA by intercalation with an interaction having similar nature and geometry; however, under comparable conditions, quercetin exhibited a greater number of intercalated chromophores. The sugar part of rutin, which was arranged out of the intercalation site, did not seem to represent any steric hindrance for the interaction with DNA. This is in accordance with the findings of Nerdal et al.,[36] who determined the detailed structure of the complex between the flavonol kaempferol 7-*O*-neohesperidoside and a DNA dodecamer containing the *E. coli* wild-type *lac* promoter sequence (TATGTT) using 2D NOESY NMR. DNA, which is a target for free radicals and reactive electrophilic groups, may thus be protected by the potential close relationship with flavonols and similar compounds.

The petals of a number of flowers contain similar intensely colored intravacuolar bodies referred to as anthocyanic vacuolar inclusions (AVIs), and the presence of AVIs has been shown to have a major influence on the color of flowers by enhancing both intensity and blueness.[491] In these studies, the anthocyanin–flavonol ratios in clarified extracts of blue-gray carnation have been determined by absorption spectroscopy. The levels of anthocyanin and flavonols were calculated from the absorption at 508 and 350 nm, respectively, using molar extinction coefficients of 36,000 and 15,000, respectively. Absorbance or reflectance spectra of inner and outer petal zones of purple lisianthus, measured with an integrating sphere connected to a spectroradiometer, indicated a distinct bluing of color in the AVI-rich inner petal. This bluing was evidenced by enhanced absorbance in the longer wavelength bands at 625 nm. Thus, in the outer petal, absorbance at 625 nm was 79% of the intensity of the 545 and 575 nm peaks whilst in the inner petal it was 95%. Reflectance and transmittance were determined at 2 nm intervals over the 400 to 1100 nm waveband.[491] Flowers of the rose cultivar Rhapsody in Blue display unusual colors, changing as they age, from a vivid red-purple to a lighter and duller purple.[492] Unexpectedly, the chemical basis of these colors is among the simplest, featuring cyanidin 3,5-diglucoside as the sole pigment and quercetin and kaempferol glycosides as copigments at a relatively low copigment–pigment ratio (about 3:1), which usually produces magenta or red shades in roses. It has been revealed that the color shift to bluer shades was coupled with the progressive accumulation of cyanidin 3,5-diglucoside into AVIs, the occurrence of which increased as the petals grew older. In these studies, spectral reflectance curves between 380 and 780 nm were recorded on circular petal areas (6 mm diameter) by a spectrocolorimeter. Each color was then numerically specified in the CIELAB scale (see Section 2.6). Spectroscopic measurements of live petals were based on transmission curves between 380 and 780 nm recorded using a spectrophotometer on petals fixed on a glass microscope slide (1 mm thickness), while spectroscopic curves of portions of individual epidermal cells were recorded between 400 and 700 nm at a 2.5 nm interval (bandwidth 10 nm) using a single-beam microphotometer with a continuous interferential filter.[492]

2.6 COLOR MEASUREMENTS USING COMMISSION INTERNATIONALE DE L'ECLAIRAGE SPECIFICATIONS

Color is a complex phenomenon, and in the evaluation of color the sample must be illuminated. In the light–sample interaction different physical phenomena are observed: transmission, absorption, scattering, refraction, etc. One way of describing sample color is to use numerical terms, which can be converted to CIE (Commission Internationale de l'Eclairage) color specifications. Using the CIELAB system, the principal attributes of sample colors are lightness (L^*), saturation (C^*), and hue (h_{ab}) (Figure 2.22). L^* considers color as a source of reflected light ranging from black ($L^* = 0$) to white ($L^* = 100$). C^* describes the chroma, which correlates to the degree of gray tone of the color. The higher the C^* value, the more saturated a color is. A very high L^* value (approximately 95 or more) combined with very low C^* (approximately 4 or less) corresponds to a virtually achromatic stimulus (i.e., white). The third parameter, namely h_{ab}, defines the tonality that we normally identify with the name of a

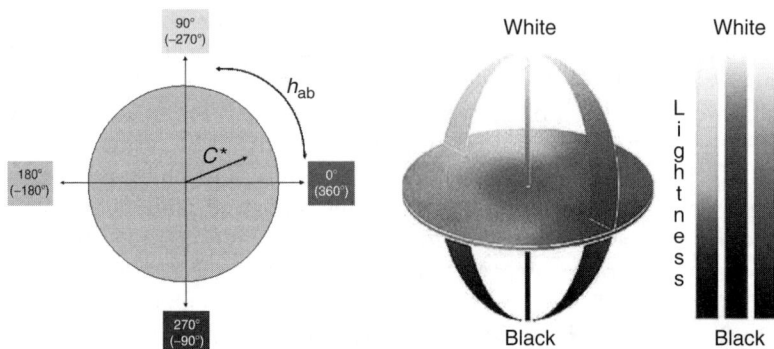

FIGURE 2.22 The three-dimensional CIELAB colour system. Left: h_{ab} (hue) describing the tonality of the color, where $0°$ corresponds to red, $90°$ corresponds to yellow, $\pm 180°$ corresponds to green, and $270°$ (or $-90°$) corresponds to blue, C^* represents the chroma (or saturation), where 0 is a gray tone, while 100 is a pure hue. Right: the lightness (L^*) goes from black (0) to white (100).

color (where $0°$ corresponds to red color, $90°$ corresponds to yellow, $180°$ [or $-180°$] corresponds to green, and $270°$ [or $-90°$] corresponds to blue). From this model, it is easy to predict that, for instance, $45°$ corresponds to orange, whereas $315°$ (or $-45°$) corresponds to purple. Furthermore, determinations of colors also depend on the illuminant (e.g., D65) and observer conditions (e.g., $10°$) under which the colors have been measured.[493–496] During the review period, color measurements using the CIE system have been performed on anthocyanin-containing samples such as intact plant parts and products derived thereof, including red wines and fruit juices, as well as on pure anthocyanins. These analyses have included stability studies, factors influencing colors, and quantitative analyses. The effects of copigmentation on the color of anthocyanins have been extensively studied by Gonnet using the CIELAB parameters.[495–497]

2.6.1 COLORIMETRIC STUDIES ON PURE ANTHOCYANINS

While most colorimetric analyses involving flavonoids have been performed on samples containing anthocyanins in mixtures with other compounds, some studies have been aimed at color analyses on pure pigments. First of all, it is obvious that the color characteristics of individual anthocyanins strongly depend on the anthocyanin concentration (Table 2.14). At pH values below 3.5, when the examined anthocyanins occur mainly in their flavylium forms, the hues of the same anthocyanin increase and the L^* values decrease considerably with increasing concentration. However, a remarkable effect was observed at high concentrations of cyanidin 3,5-diglucoside in buffer solutions at pH 2.5: the red color of the solution at 10^{-5} M (h_{ab} $358.4°$) shifted to an orange hue (h_{ab} $48.3°$) at 10^{-3} M and then back toward more reddish colors (h_{ab} $29.4°$) at even higher concentrations (5×10^{-3} M), although the $\lambda_{vis-max}$ values of all these solutions remained at similar values (510 to 511 nm).[495] Decreasing L^* values caused by increasing pigment concentration were also observed at pH values 3.5, 4.5, and 5.5; however, the pattern for the hue variation varied at these pH values,[495] most probably due to different impact of the various equilibrium forms.

The 3-glucosides of delphinidin, cyanidin, petunidin, peonidin, and malvidin, isolated from red grape skins, have been subjected to colorimetric analysis in model solutions imitating wine in the pH range 1.5 to 7.0.[498] It was revealed that both increasing number of O-substituents and degree of methylation on the B-ring of anthocyanidins led to a color shift toward more purple hues at pH 1.5 (Table 2.14).[498] The chroma of these pigments decreased

TABLE 2.14
Influence of Concentration, pH, and Solvent on Colors of Pure Anthocyanins Using CIELAB Parameters[a]

Anthocyanin[b]	pH[c]	Conc. (mM)	L*	C*	h_{ab}	Ref.
Pg	0.1% HClM	0.0154	87.1	22.1	357.3	488
Pg 3-glc	0.1% HClM	0.0176	88.1	20.2	17.6	488
Pg 3-(2-glcglc)-5-glc	0.1% HClM	0.0185	91.8	13.5	15.5	488
Pg 3-(2-glcglc)-5-glc	0.1% HClM	0.0298	78.4	39.2	53.4	488
Pg 3-(2-glcglc)-5-glc + cou	0.1% HClM	0.0173	85.3	30.3	16.4	488
Pg 3-(2-glcglc)-5-glc + fer	0.1% HClM	0.0276	82.8	36.5	19.5	488
Pg 3-(2-glcglc)-5-glc + cou + mal	0.1% HClM	0.0246	83.4	35.4	18.3	488
Pg 3-(2-glcglc)-5-glc + fer + mal	0.1% HClM	0.0270	82.6	37.0	20.5	488
Pg 3-(6-rhaglc)-5-glc + cou	0.1% HClM	0.0194	87.2	25.7	11.0	488
Pg	BpH 1.0	0.0154	90.0	16.7	22.7	488
Pg 3-glc	BpH 1.0	0.0176	90.3	17.6	44.0	488
Pg 3-(2-glcglc)-5-glc	BpH 1.0	0.0185	90.0	20.3	41.0	488
Pg 3-(2-glcglc)-5-glc	BpH 1.0	0.0298	81.8	56.0	53.5	488
Pg 3-(2-glcglc)-5-glc + cou	BpH 1.0	0.0173	86.4	26.8	23.3	488
Pg 3-(2-glcglc)-5-glc + fer	BpH 1.0	0.0276	83.4	33.7	24.1	488
Pg 3-(2-glcglc)-5-glc + cou + mal	BpH 1.0	0.0246	86.8	25.9	21.7	488
Pg 3-(2-glcglc)-5-glc + fer + mal	BpH 1.0	0.0270	83.1	33.9	22.1	488
Pg 3-(6-rhaglc)-5-glc + cou	BpH 1.0	0.0194	87.3	24.8	23.1	488
Pg 3-glc	BpH 1.1	0.10	76.6	86.9	58.7	175
5-CarboxypyranoPg 3-glc	BpH 1.1	0.10	87.5	48.7	61.1	175
Cy 3-glc	BpH 1.1	0.05	81.1	45.9	20.8	232
Cy 3-glc	BpH 1.1	0.15	63.7	81.9	42.4	232
Cy 3-(2-glcglc)-5-glc	BpH 1.1	0.05	81.5	46.7	12.8	232
Cy 3-(2-glcglc)-5-glc	BpH 1.1	0.15	69.5	78.6	38.0	232
Cy 3-(2-(2-singlc)-6-singlc)-5-glc	BpH 1.1	0.05	66.0	69.5	−14.8	232
Cy 3-(2-(2-singlc)-6-singlc)-5-glc	BpH 1.1	0.15	56.7	79.6	−9.7	232
Dp 3-glc	BpH 1.5	0.10	63.85	73.66	21.48	498
Cy 3-glc	BpH 1.5	0.10	70.4	69.0	29.5	498
Pt 3-glc	BpH 1.5	0.10	63.2	68.4	16.1	498
Pn 3-glc	BpH 1.5	0.10	73.2	61.1	24.5	498
Mv 3-glc	BpH 1.5	0.10	67.9	66.3	9.9	498
Cy 3,5-diglc	BpH 2.5	0.01	96.3	8.67	358.4	495
Cy 3,5-diglc	BpH 2.5	0.025	92.7	17.89	1.4	495
Cy 3,5-diglc	BpH 2.5	0.05	87.5	30.99	4.8	495
Cy 3,5-diglc	BpH 2.5	0.10	80.0	47.87	11.0	495
Cy 3,5-diglc	BpH 2.5	0.25	69.2	69.69	26.9	495
Cy 3,5-diglc	BpH 2.5	0.50	60.1	90.0	41.2	495
Cy 3,5-diglc	BpH 2.5	1.0	48.3	106.86	48.3	495
Cy 3,5-diglc	BpH 2.5	2.5	31.7	85.26	39.7	495
Cy 3,5-diglc	BpH 2.5	5.0	16.6	57.89	29.4	495
Cy 3-glc	BpH 3.0	0.05	82.3	40.5	5.7	232
Cy 3-glc	BpH 3.0	0.15	63.0	70.2	25.6	232
Cy 3-(2-glcglc)-5-glc	BpH 3.0	0.05	90.5	23.1	−2.8	232
Cy 3-(2-glcglc)-5-glc	BpH 3.0	0.15	76.9	52.9	4.4	232
Cy 3-(2-(2-singlc)-6-singlc)-5-glc	BpH 3.0	0.05	66.5	67.9	−17.4	232
Cy 3-(2-(2-singlc)-6-singlc)-5-glc	BpH 3.0	0.15	48.1	80.6	−5.2	232

continied

TABLE 2.14
Influence of Concentration, pH, and Solvent on Colors of Pure Anthocyanins Using CIELAB Parameters[a] — *continued*

Anthocyanin[b]	pH[c]	Conc. (mM)	L*	C*	h_{ab}	Ref.
Dp 3-glc	BpH 3.5	0.10	88.2	20.6	−5.3	498
Cy 3-glc	BpH 3.5	0.10	88.2	22.9	8.1	498
Pt 3-glc	BpH 3.5	0.10	90.2	16.6	−4.0	498
Pn 3-glc	BpH 3.5	0.10	92.1	13.9	7.5	498
Mv 3-glc	BpH 3.5	0.10	93.3	11.8	−6.9	498
Cy 3,5-diglc	BpH 3.5	0.01	98.8	1.23	1.4	495
Cy 3,5-diglc	BpH 3.5	0.025	97.9	3.25	1.3	495
Cy 3,5-diglc	BpH 3.5	0.05	96.5	6.49	1.0	495
Cy 3,5-diglc	BpH 3.5	0.10	93.9	12.54	1.5	495
Cy 3,5-diglc	BpH 3.5	0.25	86.8	27.99	2.9	495
Cy 3,5-diglc	BpH 3.5	0.50	76.8	46.95	5.9	495
Cy 3,5-diglc	BpH 3.5	1.0	62.1	66.25	17.7	495
Cy 3,5-diglc	BpH 3.5	2.5	42.2	87.89	38.2	495
Cy 3,5-diglc	BpH 3.5	5.0	24.5	71.91	35.8	495
Cy 3-glc	BpH 4.1	0.05	89.5	22.1	−2.5	232
Cy 3-glc	BpH 4.1	0.15	69.6	49.8	5.0	232
Cy 3-(2-glcglc)-5-glc	BpH 4.1	0.05	96.5	6.3	−6.3	232
Cy 3-(2-glcglc)-5-glc	BpH 4.1	0.15	89.4	20.9	−8.0	232
Cy 3-(2-(2-singlc)-6-singlc)-5-glc	BpH 4.1	0.05	72.5	50.5	−25.8	232
Cy 3-(2-(2-singlc)-6-singlc)-5-glc	BpH 4.1	0.15	50.3	79.0	−22.2	232
Cy 3,5-diglc	BpH 4.5	0.01	99.0	0.33	16.8	495
Cy 3,5-diglc	BpH 4.5	0.025	98.9	0.65	14.7	495
Cy 3,5-diglc	BpH 4.5	0.05	98.4	1.33	9.4	495
Cy 3,5-diglc	BpH 4.5	0.10	97.4	2.70	5.9	495
Cy 3,5-diglc	BpH 4.5	0.25	94.7	7.03	3.1	495
Cy 3,5-diglc	BpH 4.5	0.50	89.7	14.28	2.3	495
Cy 3,5-diglc	BpH 4.5	1.0	79.9	28.73	0.5	495
Cy 3,5-diglc	BpH 4.5	2.5	58.8	54.10	3.4	495
Cy 3,5-diglc	BpH 4.5	5.0	35.9	67.79	17.2	495
Cy 3-glc	BpH 5.1	0.05	94.9	5.0	−16.8	232
Cy 3-glc	BpH 5.1	0.15	65.5	22.6	−50.0	232
Cy 3-(2-glcglc)-5-glc	BpH 5.1	0.05	98.2	1.5	−1.1	232
Cy 3-(2-glcglc)-5-glc	BpH 5.1	0.15	94.5	6.1	−13.6	232
Cy 3-(2-(2-singlc)-6-singlc)-5-glc	BpH 5.1	0.05	82.7	24.7	−44.3	232
Cy 3-(2-(2-singlc)-6-singlc)-5-glc	BpH 5.1	0.15	52.5	65.4	−41.2	232
Cy 3,5-diglc	BpH 5.5	0.01	99.4	0.19	26.8	495
Cy 3,5-diglc	BpH 5.5	0.025	99.1	0.45	22.9	495
Cy 3,5-diglc	BpH 5.5	0.05	98.7	0.98	23.1	495
Cy 3,5-diglc	BpH 5.5	0.10	97.6	2.11	24.1	495
Cy 3,5-diglc	BpH 5.5	0.25	94.6	5.39	22.5	495
Cy 3,5-diglc	BpH 5.5	0.50	90.3	10.41	12.1	495
Cy 3,5-diglc	BpH 5.5	1.0	80.6	20.97	7.9	495
Cy 3,5-diglc	BpH 5.5	2.5	57.2	45.79	9.1	495
Cy 3,5-diglc	BpH 5.5	5.0	31.6	62.39	16.9	495
Cy 3-glc	BpH 6.0	0.05	82.6	15.4	−21.2	232
Cy 3-glc	BpH 6.0	0.15	39.9	32.5	−47.1	232

TABLE 2.14
Influence of Concentration, pH, and Solvent on Colors of Pure Anthocyanins Using CIELAB Parameters[a] — *continued*

Anthocyanin[b]	pH[c]	Conc. (mM)	L^*	C^*	h_{ab}	Ref.
Cy 3-(2-glcglc)-5-glc	BpH 6.0	0.05	96.6	4.0	−26.0	232
Cy 3-(2-glcglc)-5-glc	BpH 6.0	0.15	89.2	13.1	−28.9	232
Cy 3-(2-(2-singlc)-6-singlc)-5-glc	BpH 6.0	0.05	72.1	38.1	−49.9	232
Cy 3-(2-(2-singlc)-6-singlc)-5-glc	BpH 6.0	0.15	35.4	75.8	−43.0	232
Cy 3-glc	BpH 6.6	0.05	75.8	15.2	−17.1	232
Cy 3-glc	BpH 6.6	0.15	33.5	27.7	−50.7	232
Cy 3-(2-glcglc)-5-glc	BpH 6.6	0.05	80.7	25.9	−46.8	232
Cy 3-(2-glcglc)-5-glc	BpH 6.6	0.15	49.9	64.7	−42.3	232
Cy 3-(2-(2-singlc)-6-singlc)-5-glc	BpH 6.6	0.05	76.0	29.4	−63.3	232
Cy 3-(2-(2-singlc)-6-singlc)-5-glc	BpH 6.6	0.15	42.3	64.4	−58.2	232
Cy 3-glc	BpH 6.8	0.05	75.2	13.6	−14.8	232
Cy 3-glc	BpH 6.8	0.15	35.3	22.9	−46.8	232
Cy 3-(2-glcglc)-5-glc	BpH 6.8	0.05	73.2	34.7	−53.7	232
Cy 3-(2-glcglc)-5-glc	BpH 6.8	0.15	36.9	75.7	−46.3	232
Cy 3-(2-(2-singlc)-6-singlc)-5-glc	BpH 6.8	0.05	75.3	28.4	−75.4	232
Cy 3-(2-(2-singlc)-6-singlc)-5-glc	BpH 6.8	0.15	33.8	57.8	−67.4	232
Pg 3-glc	BpH 6.9	0.10	58.0	43.09	37.3	175
5-Carboxypyranopg 3-glc	BpH 6.9	0.10	81.2	37.5	34.8	175
Cy 3-glc	BpH 6.9	0.05	74.3	12.2	−13.8	232
Cy 3-glc	BpH 6.9	0.15	33.3	22.3	−50.0	232
Cy 3-(2-glcglc)-5-glc	BpH 6.9	0.05	69.3	38.3	−62.6	232
Cy 3-(2-glcglc)-5-glc	BpH 6.9	0.15	31.3	74.6	−51.5	232
Cy 3-(2-(2-singlc)-6-singlc)-5-glc	BpH 6.9	0.05	72.2	31.2	−88.5	232
Cy 3-(2-(2-singlc)-6-singlc)-5-glc	BpH 6.9	0.15	41.8	59.7	−72.1	232
Cy 3-glc	BpH 7.2	0.05	73.7	11.6	−15.8	232
Cy 3-glc	BpH 7.2	0.15	32.2	21.9	−47.0	232
Cy 3-(2-glcglc)-5-glc	BpH 7.2	0.05	65.4	42.2	−70.2	232
Cy 3-(2-glcglc)-5-glc	BpH 7.2	0.15	29.1	73.4	−56.1	232
Cy 3-(2-(2-singlc)-6-singlc)-5-glc	BpH 7.2	0.05	69.3	34.7	−97.7	232
Cy 3-(2-(2-singlc)-6-singlc)-5-glc	BpH 7.2	0.15	53.6	48.9	−87.7	232
Cy 3-glc	BpH 7.3	0.05	72.5	11.8	−24.2	232
Cy 3-glc	BpH 7.3	0.15	31.1	24.0	−39.7	232
Cy 3-(2-glcglc)-5-glc	BpH 7.3	0.05	64.4	43.2	−77.8	232
Cy 3-(2-glcglc)-5-glc	BpH 7.3	0.15	27.6	71.5	−59.8	232
Cy 3-(2-(2-singlc)-6-singlc)-5-glc	BpH 7.3	0.05	63.2	41.4	−100.4	232
Cy 3-(2-(2-singlc)-6-singlc)-5-glc	BpH 7.3	0.15	57.3	46.6	−97.4	232
Cy 3-glc	BpH 8.0	0.05	74.8	21.5	23.9	232
Cy 3-glc	BpH 8.0	0.15	38.7	47.2	24.1	232
Cy 3-(2-glcglc)-5-glc	BpH 8.0	0.05	67.1	52.8	−40.9	232
Cy 3-(2-glcglc)-5-glc	BpH 8.0	0.15	35.6	81.2	−34.2	232
Cy 3-(2-(2-singlc)-6-singlc)-5-glc	BpH 8.0	0.05	57.8	43.4	−87.5	232
Cy 3-(2-(2-singlc)-6-singlc)-5-glc	BpH 8.0	0.15	28.2	56.1	−77.4	232
Cy 3-glc	BpH 8.9	0.05	75.0	21.2	27.9	232
Cy 3-glc	BpH 8.9	0.15	38.7	46.9	26.9	232
Cy 3-(2-glcglc)-5-glc	BpH 8.9	0.05	67.3	49.1	−43.8	232
Cy 3-(2-glcglc)-5-glc	BpH 8.9	0.15	35.1	77.7	−34.8	232

TABLE 2.14

Influence of Concentration, pH, and Solvent on Colors of Pure Anthocyanins Using CIELAB Parameters[a] — *continued*

Anthocyanin[b]	pH[c]	Conc. (m*M*)	*L**	*C**	h_{ab}	Ref.
Cy 3-(2-(2-singlc)-6-singlc)-5-glc	BpH 8.9	0.05	56.3	35.0	−66.5	232
Cy 3-(2-(2-singlc)-6-singlc)-5-glc	BpH 8.9	0.15	22.7	38.8	−79.4	232
Cy 3-glc	BpH 9.9	0.05	73.2	16.1	17.6	232
Cy 3-glc	BpH 9.9	0.15	39.3	40.6	23.4	232
Cy 3-(2-glcglc)-5-glc	BpH 9.9	0.05	80.6	17.2	−61.2	232
Cy 3-(2-glcglc)-5-glc	BpH 9.9	0.15	52.3	332.5	−46.9	232
Cy 3-(2-(2-singlc)-6-singlc)-5-glc	BpH 9.9	0.05	55.3	20.2	−70.6	232
Cy 3-(2-(2-singlc)-6-singlc)-5-glc	BpH 9.9	0.15	19.6	24.3	−73.9	232
Cy 3-glc	BpH 10.5	0.05	62.0	28.5	−62.8	232
Cy 3-glc	BpH 10.5	0.15	24.3	49.7	−53.6	232
Cy 3-(2-glcglc)-5-glc	BpH 10.5	0.05	83.3	14.4	−155.5	232
Cy 3-(2-glcglc)-5-glc	BpH 10.5	0.15	55.5	35.6	−171.4	232
Cy 3-(2-(2-singlc)-6-singlc)-5-glc	BpH 10.5	0.05	57.2	47.2	−155.4	232
Cy 3-(2-(2-singlc)-6-singlc)-5-glc	BpH 10.5	0.15	27.6	42.3	−153.0	232

[a] Some of the values need to be regarded as transient because some pigments degrade fast at some pH values. This is especially profound for nonacylated pigments in weak acid and neutral solutions.

[b] Pg, pelargonidin; Cy, cyanidin; Dp, delphinidin; Pn, peonidin; Pt, petunidin; Mv, malvidin; glc, glucose; gal, galactose; xyl, xylose; rha, rhamnose; cou, *p*-coumaric acid; fer, ferulic acid; sin, sinapic acid; mal, malonic.

with methoxylation of the hydroxyl groups on the B-ring. Changes in lightness (*L**) corresponding to color loss and a color shift toward purple/blue were observed when increasing the pH toward 7.

In a comparative analysis of pelargonidin and several pelargonidin glycosides with different glycosylation and acylation patterns dissolved in MeOH–HCl (99.9:0.1, v/v) and aqueous buffer solution (pH 1.0), hyperchromic effects leading to higher chroma, as well as lower hue values (more reddish colors), were observed for all pelargonidin derivatives when dissolved in acidified methanolic solutions (Table 2.14).[488] The glycosylation pattern, the nature of the glycosyl substituents, as well as aromatic acylation showed significant impact on the colors of the solution. 3-Glycosylated pigments showed lower hues than the corresponding 3,5-diglycosylated pigments. The presence of cinnamoyl moieties lowered the hue value and increased the chroma, compared to similar nonacylated pigments. Contrary to this, aliphatic acylation with malonic acid had little effects on chroma and lightness, and these pigments exhibited lesser difference between the hue values obtained in aqueous and methanolic solutions, respectively.[488] Similarly, the color properties of several cyanidin-based anthocyanins, all containing a 3-glycosyl residue with different overall substitution patterns, have been analyzed in various solutions with pH ranging from 0.45 to 6.00.[499] Compared with cyanidin 3-glycosides without acylation, both acylation with cinnamic acids and glycosidic substitution at the 5-position shifted color tonalities (hue) toward purple. A small increase in color strength through acylation was also confirmed; however, it was proved that slightly different acyl moieties also affected color appearance.

An important conclusion from the colorimetric analysis of newly discovered pyranoanthocyanins is that each of them seems to retain the original color over the whole pH region

1 to 7, which implies that they, in contrast to analogous common anthocyanins, remain in the flavyllium cationic equilibrium form to a significant extent even up to pH 7.[175,500,501] The pyruvic acid pyrano-anthocyanin adducts adopt a more yellowish color than the corresponding pigments lacking the extra ring.[175,500] In contrast to this, pyrano-anthocyanins formed from the reaction between cinnamic acids and anthocyanins achieved a more bluish color.[501]

Studies of color and color stability of some anthocyanins, representing the structural variation of the common pigments isolated from many fruits and vegetables, have been performed at various pH values covering the pH region 1.0 to 10.5 and at two different concentrations (Table 2.14).[232] The hue angle shift toward bluish tones in freshly made samples of anthocyanins with 5-glucosidic substitution was amplified with aromatic acylation throughout the entire pH range except at pH 10.5. Of potential interest for weakly acidic food products, one of the pigments involved, namely cyanidin 3-(2″-(2‴-sinapoylglucosyl)-6″-sinapoylglucoside)-5-glucoside, maintained nearly the same h_{ab}, C^*, and L^* values during the whole period (98 days), in contrast to similar anthocyanins lacking the aromatic acyl moieties.

2.6.2 ANTHOCYANIN-BASED COLORS OF PLANTS AND PRODUCTS DERIVED THEREFROM

Colorimetric measurements have been applied to various analyses of red wines and model wine solutions.[500,502–507] Accurate definitions of the wines have been achieved by the L^* (lightness) and a^* (redness) values, while the representation of ΔL^* against ΔC^* revealed the color differences between various wines.[505] The color stability of wine and model wine solutions toward storage time and bleaching by sulfur dioxide has been extensively studied by Bakker et al.,[502,504,507] while the color stability of a range of anthocyanin-containing extracts, including fruit juices, has been the subject of other colorimetric studies.[508–515] It has, for instance, been shown that black carrot anthocyanins are applicable as food colorants in the acidic to weakly acidic pH region.[516]

Colorimetric analyses performed in the reflection mode have been used to determine colors of various flowers[492,517–521] as well as the colors and color development of fruits and berries.[511,522–533] In an interesting study, the "cyclamen" red (or pink) colors of some carnation cultivars have been found to be based on the macrocyclic anthocyanin, pelargonidin 3,5-diglucoside (6″,6‴-malyl diester).[534] The CIELAB coordinates revealed that these flowers showed similar colors as some rose cultivars, which, however, were mainly based on a very different pigment, cyanidin 3,5-diglucoside.

2.7 CIRCULAR DICHROISM SPECTROSCOPY

When polarized light is passed through a substance containing chiral molecules, the direction of polarization can be changed. This phenomenon is called optical activity. Optical activity of flavonoids is basically measured by two spectroscopic methods, optical rotation and dichroism. The first method observes a sample's effect on the *velocities* of right and left polarized light beams. Dichroism is defined as the *differential absorption* of radiation polarized in the two directions as a function of frequency. When applied to plane polarized light, it is called linear dichroism (LD) and for circularly polarized light, circular dichroism, CD. CD spectroscopy has during the last few decades been the most commonly used spectroscopic technique for extracting stereochemical information about flavonoids. CD measurements are simpler to perform than optical rotation measurements because they require only the detection of intensity variations, whereas rotation experiments require measurements of very small changes in electron polarization. Another benefit with CD spectroscopy is that this method, unlike optical rotation, is confined to the narrow absorption band of each chromophore involved. Thus, it is easier to determine the contribution of individual chromophores, information vital to structural analysis. Derivatization of phenolic functionalities of flavonoids may also influence the sign of the optical rotation.

2.7.1 DETERMINATION OF ABSOLUTE FLAVONOID CONFIGURATION

The utilization of CD spectroscopy in the field of flavonoids has been reviewed by Lèvai[535] and Slade et al.[536] The value of the method has primarily been related to establishing the absolute configuration of flavonoids and proanthocyanidins with stereogenic centers (chiral molecules), including the configuration of flavanones, dihydroflavonols, flavanols, flavans, isoflavans, isoflavanones, pterocarpans, neoflavonoids, 4-arylflavan-3-ols, proanthocyanidins, various classes of biflavonoids, and auronols.[107,117,535–545] These determinations have mainly been based on the sign of the observed CD bands (Figure 2.23) compared to reference compounds with defined stereochemistry. In general, Cotton effects observed in the 240 to 360 nm spectral region are similar for analogous flavonoid structures with the same configuration and independent of the substituents. Although optically active biflavonoids have been known since 1968, the first determination of the absolute configuration of optically pure biflavonoids was performed in 1995 (Figure 2.23),[537] on the basis of CD spectroscopy and single-crystal x-ray crystallography.

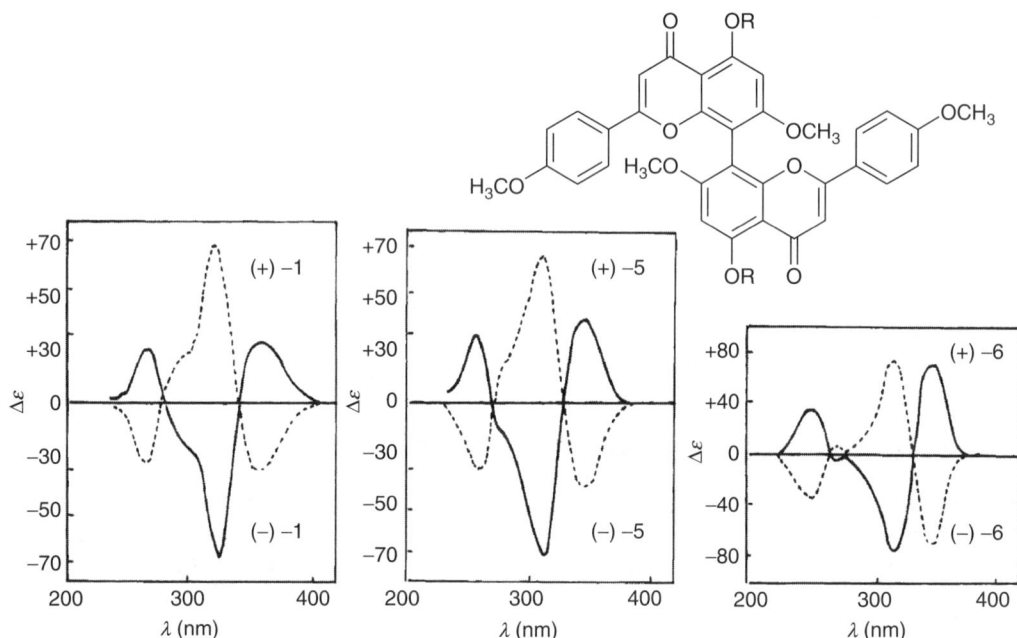

FIGURE 2.23 CD spectra of optically pure 8,8″-biflavones 5,5″-dihydroxy-4,4‴,7,7″-tetramethoxy-8,8″-biflavone (1; R = H), its camphorsulphonate derivative (6; R = camphorsulphonate), and its dimethyl derivative, namely 5,5″-dihydroxy-4,4‴,7,7″,5,5″-hexamethoxy-8,8″-biflavone (5; R = Me) (c 8 μg/ml, ethanol). These biflavones are optically active due to the atropisomerism of the biflavone moiety. The dashed lines refer to (+)-1, (+)-5, and (+)-6. The solid lines refer to (−)-1, (−)-5, and (−)-6. Cotton effects observed in the 240 to 360 nm spectral region are similar for the 8,8″-biflavones with the same configuration and independent on the substituents. (Reprinted from Zhang, F.J., Lin, G.Q., and Huang, Q.C., *J. Org. Chem.*, 60, 6427, 1995. Copyright 1995 American Chemical Society. With permission.)

During the review period, the benefits of coupled HPLC–CD as an analytical tool have been demonstrated in the flavonoid field. Based on HPLC–CD analysis, Antus et al.[546] determined the absolute configuration of several *cis*-pterocarpans with different substitution patterns. Using HPLC with a chiral stationary-phase column, coupled to a CD instrument, the separation of the C-2 diastereomers of naringin from *Citrus* juice and determination of their stereochemistry have also been achieved.[547] In another interesting report, the absolute configuration of two new 3,8''-biflavonoid diastereomers from roots of *Gnidia involucrata* have been determined by online LC–CD measurements, which also implied revision of the absolute configurations of some known biflavanones.[548]

2.7.2 CIRCULAR DICHROISM IN STUDIES OF MOLECULAR FLAVONOID INTERACTION

The term chirality has a broader sense than involving just enantiomers and diastereoisomers. Chirality may arise from spatial isomerism resulting from the lack of free rotation around single or double bonds (chiral axis). Dichroism spectra have thus been valuable tools by providing conformational-related information about molecular shape, size, and electronic properties, binding parameters in molecular complexes, and the orientation of the chromophores.[549] CD spectroscopy has, for instance, been used to gain insight into molecular association properties of anthocyanins involving aromatic stacking.[550,551] More recent applications of dichroism spectroscopy have been related to evaluation of the interaction between the flavonol rutin with cyclodextrins,[552] interactions of the flavonoids quercetin, morin, rutin, and naringin with DNA,[490,553–556] and studies of complexes between quercetin and human serum albumin, the most abundant carrier protein in blood.[557] In this last case, it was revealed that when quercetin was bound to the asymmetric albumin environment, at least three induced CD bands appeared in the spectra. The induced CD bands were then utilized for calculation of the association constant and to probe the ligand binding site.

2.8 X-RAY CRYSTALLOGRAPHY

X-ray crystallography is the most accurate method for structural elucidation of flavonoids in the solid state. The method can only be applied to crystallized compounds, which has limited the number of flavonoid crystal structures reported.

2.8.1 X-RAY STUDIES ON FLAVONOID STRUCTURES

During the period of this review, the structure of several flavonoid aglycones including flavanonols,[558,559] a flavanone,[538] a prenylated chalcone,[560] flavones,[418,420,561,562] the desoxyanthocyanidin carajurin,[563] and the conformation of the biflavonoid sciadopitysin in the solid state[564] have been determined by x-ray analysis. The constitution and configuration of procyanidin B1 was proved by the x-ray analysis of its decaacetyl derivative.[565] The structures of two flavones, two biflavonoids, and two flavanones have recently been investigated by single-crystal x-ray analysis.[566] Crystallization of the two biflavonoids as their methanol and pyridine solvates, respectively, highlighted the role of the solvent molecules in stabilizing a crystal lattice. Intermolecular 'π–π' interactions were observed in the crystal structures of the flavones and biflavonoids with centroid–centroid distances ranging from 3.70 to 3.81 Å and displacement angles ranging from 2.7 to 9.9°, but there was no 'π–π' interaction in the flavanones.[566] The flavonol mikanin-3-*O*-sulfate, in the form of its potassium salt, has been isolated from *Mikania micrantha*.[567] The crystal structures of the 1:1 complex of potassium mikanin-3-*O*-sulfate (m-3-s) with methanol showed that the potassium ions in K(m-3-s)CH$_3$OH were bridged by O5, O7, and O8 to form a chain of face-sharing KO8 coordination polyhedra, from which the aglycone units were outstretched to form a polymeric molecular column. Adjacent molecular columns were linked by 'π–π' stacking between parallel, intercalating aglycon units to form layers, which were further interconnected into a 3D supramolecular assembly.[567]

Recent literature reports on the crystal structures of flavonoid glycosides are scarce; however, the x-ray structures of the flavonol quercetin 3-methyl ether 3'-glucoside[568] and the flavone 6-hydroxyluteolin 7-rhamnoside[569] have been reported. For the latter compound, crystals of the hepta-O-acetyl derivative were used. Cartormin, a rare semi-quinone chalcone sharing a pyrrole ring C-glycoside, has been isolated from *Carthamus tinctorius*, and its structure was established from various spectral data and a single-crystal x-ray analysis.[570] Quercetin is known as a potent, competitive inhibitor of lipoxygenase LOX. Structural analysis has revealed that quercetin entrapped within this enzyme underwent degradation, and the resulting compound was identified by x-ray analysis as 3,4-dihydroxybenzoic acid positioned near the iron site.[571]

The molecular structures of various synthesized flavonoid derivatives have been examined by x-ray crystallography in recent years,[572–575] including a 3-methyl-substituted flavylium salt with photochromic properties[576] and some phosphorylated flavones.[577]

2.8.2   X-Ray Studies on Complexes Involving Flavonoids

Flavocommelin is a flavonoid component of the blue pigment commelin, which has been isolated from the petals of *Commelina communis*. Commelin is composed of six molecules each of the anthocyanin malonylawobanin and flavocommelin, and two atoms of magnesium.[578] The crystal structure of the octaacetate derivative of flavocommelin has been determined by x-ray diffraction.[579] In the crystal, the molecules were arranged parallel to each other according to the periodicity of the crystal lattice. However, intermolecular stacking of the flavanone skeletons was not observed, which suggested that the hydrophilicity of the glucose moieties was one of the important factors governing the self-association. X-ray diffraction data collected at low temperature (130 K), using synchrotron radiation, have recently been used for determination of crystal architecture and conformational properties of the inclusion complex, neohesperidin dihydrochalcone–β-cyclodextrin.[580] The complex was characterized by one aromatic part of neohesperidin dihydrochalcone deeply inserted into the hydrophobic cavity of β-cyclodextrin through the primary OH rim. The formation of other β-cyclodextrin inclusion complexes involving various flavanones,[439] as well as the inclusion behavior of both 2-hydroxypropyl β-cyclodextrin and β-cyclodextrin in solution and solid-state toward quercetin,[581] have also been subjected to x-ray diffractometric analysis. Genistein and its amine complexes with morpholine and piperazine have been studied in the solid and liquid states by x-ray crystallography and ^{13}C and ^{15}N NMR spectroscopy.[128]

REFERENCES

1. Mabry, T.J., Markham, K.R., and Thomas, M.B., *The Systematic Identification of Flavonoids*, Springer-Verlag, New York, 1970.
2. Markham, K.R., *Techniques of Flavonoid Identification*, Academic Press, London, 1982.
3. Harborne, J.B., Ed., *Methods in Plant Biochemistry, Vol. 1, Plant Phenolics*, Academic Press, London, 1989.
4. Harborne, J.B., Mabry, T.J., and Mabry, H., Eds., *The Flavonoids*, Chapman & Hall, London, 1975.
5. Harborne, J.B. and Mabry, T.J., Eds., *Flavonoids: Advances in Research*, Chapman & Hall, London, 1982.

6. Harborne, J.B., Ed., *Flavonoids: Advances in Research Since 1980*, Chapman & Hall, London, 1988.
7. Harborne, J.B., Ed., *Flavonoids: Advances in Research Since 1986*, Chapman & Hall, London, 1994.
8. Harborne, J.B. and Baxter, H., *The Handbook of Natural Flavonoids*, John Wiley, Chichester, 1999.
9. Harborne, J.B. and Williams, C.A., Anthocyanins and other flavonoids, *Nat. Prod. Rep.*, 12, 639, 1995.
10. Harborne, J.B. and Williams, C.A., Anthocyanins and other flavonoids, *Nat. Prod. Rep.*, 15, 631, 1998.
11. Harborne, J.B. and Williams, C.A., Anthocyanins and other flavonoids, *Nat. Prod. Rep.*, 18, 310, 2001.
12. Harborne, J.B. and Williams, C.A., Anthocyanins and other flavonoids, *Nat. Prod. Rep.*, 18, 310, 2004.
13. Waterman, P.G. and Mole, S., *Analysis of Phenolic Plant Metabolites*, Blackwell, Oxford, 1994.
14. Markham, K.R. and Bloor, S.J., Analysis and identification of flavonoids in practise, in *Flavonoids in Health and Disease*, Rice-Evans, C.A. and Packer, L., Eds., Marcel Dekker, 1, 1998.
15. Bohm, B.A., *Introduction to Flavonoids*, Harwood Academic Publisher, Singapore, 1998.
16. Sivam, G., Analysis of flavonoids, in *Methods of Analysis for Functional Foods and Nutraceuticals*, Hurst, W.J., Ed., CRC Press, Boca Raton, 363, 2002.
17. Santos-Buelga, C. and Williamson, G., Eds., *Methods in Polyphenol Analysis*, Royal Society of Chemistry, Cambridge, 2003.
18. Andersen, Ø.M. and Francis, G.W., Techniques of pigment identification, in *Plant Pigments and their Manipulation*, Davies, K., Ed., Blackwell Publishing, London, Chap. 10, 293, 2004.
19. Mazza, G., Cacace, J.E., and Kay, C.D., Methods of analysis for anthocyanins in plants and biological fluids. *J. AOAC Int.*, 87, 129, 2004.
20. Markham, K.R. and Chari, V.M., Carbon-13 NMR spectroscopy of flavonoids, in *The Flavonoids: Advances in Research*, Harborne, J.B. and Mabry, T.J., Eds., Chapman & Hall, London, 19, 1982.
21. Markham, K.R. and Geiger, H., 1H nuclear magnetic resonance spectroscopy of flavonoids and their glycosides in hexadeuterodimethylsulfoxide, in *The Flavonoids: Advances in Research Since 1986*, Harborne, J.B., Ed., Chapman & Hall, London, 441, 1993.
22. Agrawal, P.K., *Carbon-13 NMR of Flavonoids*, Elsevier, Amsterdam, 1989.
23. Andersen, Ø.M. and Fossen, T., Characterization of anthocyanins by NMR, in *Current Protocols in Food Analytical Chemistry*, Wrolstad, R.E., Ed., John Wiley, New York, 1, 2003.
24. Wallet, J.C., Gaydou, M., and Cung, M.T., Oxygen-17 NMR study of steric hindrance in methoxylated flavones. Substitution on the phenyl ring and the heterocycle, *Magn. Reson. Chem.*, 28, 557, 1990.
25. Toth, G. et al., 17O NMR studies on (*E*)-3-arylidenechromanone and -flavanone derivatives, *Magn. Reson. Chem.*, 39, 463, 2001.
26. Tsai, P.-J., Hsieh, Y.-Y., and Huang, T.-C., Effect of sugar on anthocyanin degradation and water mobility in a roselle anthocyanin model system using 17O NMR, *J. Agric. Food Chem.*, 52, 3097, 2004.
27. Keun, H.C. et al., Cryogenic probe 13C NMR spectroscopy of urine for metabonomic studies, *Anal. Chem.*, 74, 4588, 2002.
28. Exarchou, V. et al., LC–UV–solid-phase extraction–NMR–MS combined with a cryogenic flow probe and its application to the identification of compounds present in Greek oregano, *Anal. Chem.*, 75, 6288, 2003.
29. Horie, T. et al., 13C NMR spectral assignment of the A-ring of polyoxygenated flavones, *Phytochemistry*, 47, 865, 1998.
30. Fukai, T. and Nomura, T., New NMR structure determination methods for prenylated phenols, *Basic Life Sciences*, 66 (Plant Polyphenols 2), 259, 1999.
31. De P. Emerenciano, V. et al., Application of artificial intelligence in organic chemistry. Part XIX. Pattern recognition and structural determination of flavonoids using 13C-NMR spectra, *Spectroscopy*, 13, 181, 1997.

32. Garcia-Martinez, C. et al., Further nuclear magnetic resonance studies of flavone, *Spectroscopy*, 12, 85, 1994.
33. Aksnes, D.W., Standnes, A., and Andersen, Ø.M., Complete assignment of the 1H and 13C NMR spectra of flavone and its A-ring hydroxyl derivatives, *Magn. Reson. Chem.*, 34, 820, 1996.
34. Pedersen, A.T. et al., Anomeric sugar configuration of anthocyanin *O*-pyranosides determined from heteronuclear one-bond coupling constants, *Phytochem. Anal.*, 6, 313, 1995.
35. Scheidt, H.A. et al., Investigation of the membrane localization and distribution of flavonoids by high-resolution magic angle spinning NMR spectroscopy, *Biochim. Biophys. Acta*, 1663, 97, 2004.
36. Nerdal, W., Andersen, Ø.M., and Sletten, E., NMR studies of a plant flavonoid — DNA oligonucleotide complex, *Acta Chem. Scand.*, 47, 658, 1993.
37. Nissler, L., Gebhardt, R., and Berger, S., Flavonoid binding to a multi-drug-resistance transporter protein: an STD–NMR study, *Anal. Bioanal. Chem.*, 379, 1045, 2004.
38. Wroblewski, K. et al., The molecular interaction of human salivary histatins with polyphenolic compounds, *Eur. J. Biochem.*, 268, 4384, 2001.
39. Hemingway, R.W. et al., NMR studies on the conformation of polyflavanoids and their association with proteins, in *Advances in Lignocellulosics Characterization*, Argyropoulos, D.S., Ed., TAPPI Press, Atlanta, GA, 157, 1999.
40. Nilsson, M. et al., high-resolution NMR and diffusion-ordered spectroscopy of Port wine, *J. Agric. Food Chem.*, 52, 3736, 2004.
41. Vorveille, L., Vercauteren, J., and Rutledge, D.N., Multivariate statistical analysis of two-dimensional NMR data to differentiate grapevine cultivars and clones, *Food Chem.*, 57, 441, 1996.
42. Kosir, I.J. and Kidric, J., Use of modern nuclear magnetic resonance spectroscopy in wine analysis: determination of minor compounds, *Anal. Chim. Acta*, 458, 77, 2002.
43. Li, C.-Y. et al., Efficient 1H nuclear magnetic resonance method for improved quality control analyses of ginkgo constituents, *J. Agric. Food Chem.*, 52, 3721, 2004.
44. Solanky, K.S. et al., Application of biofluid 1H nuclear magnetic resonance-based metabonomic techniques for the analysis of the biochemical effects of dietary isoflavones on human plasma profile, *Anal. Biochem.*, 323, 197, 2003.
45. Krisa, S. et al., Production of 13C-labelled anthocyanins by *Vitis vinifera* cell suspension cultures, *Phytochemistry*, 51, 651, 1999.
46. Kondo, T. et al., Heavenly blue anthocyanin. IV. Structure determination of heavenly blue anthocyanin, a complex monomeric anthocyanin from the morning glory *Ipomoea tricolor*, by means of the negative NOE method, *Tetrahedron Lett.*, 28, 2273, 1987.
47. Nørbæk, R., Brandt, K., and Kondo, T., Identification of flavone C-glycosides including a new flavonoid chromophore from barley leaves (*Hordeum vulgare* L.) by improved NMR techniques, *J. Agric. Food Chem.*, 48, 1703, 2000.
48. Alcantara, A.F. de C. et al., NMR investigation and theoretical calculations of the effect of solvent on the conformational analysis of 4′,7-di-hydroxy-8-prenylflavan, *Quim. Nova*, 27, 371, 2004.
49. Braun, S., Kalinowski, H.-O., and Berger, S., *150 and More Basic NMR Experiments*, Wiley–VCH, Weinheim, 1998.
50. Rao, L.J.M. et al., Acylated and non-acylated flavonol monoglycosides from the Indian minor spice Nagkesar (*Mammea longifolia*), *J. Agric. Food Chem.*, 50, 3143, 2002.
51. Kamusiime, H. et al., Kaempferol 3-*O*-(2-*O*-β-glucopyranosyl-6-*O*-α-rhamnopyranosyl-β-glucopyranoside) from the African plant *Erythrina abyssinica*, *Int. J. Pharmacognosy*, 34, 370, 1996.
52. Fossen, T., Pedersen, A.T., and Andersen, Ø.M., Flavonoids from red onion (*Allium cepa*), *Phytochemistry*, 47, 281, 1998.
53. Datta, B.K. et al., A sesquiterpene acid and flavonoids from *Polygonum viscosum*, *Phytochemistry*, 54, 201, 2000.
54. Teng, R. et al., Two new acylated flavonoid glycosides from *Morina nepalensis* var. *alba* Hand.-Mazz, *Magn. Reson. Chem.*, 40, 415, 2002.
55. Shahat, A.A. et al., Anticomplement and antioxidant activities of new acetylated flavonoid glycosides from *Centaurium spicatum*, *Planta Med.*, 69, 1153, 2003.

56. Ferreres, F. et al., Acylated flavonol sophorotriosides from pea shoots, *Phytochemistry*, 39, 1443, 1995.

57. Gunasegaran, R. et al., 7-*O*-(6-*O*-benzoyl-beta-ᴅ-glucopyranosyl)-rutin from leaves of *Canthium dicoccum*, *Fitoterapia*, 72, 201, 2001.

58. Tang, Y.-P. et al., Isolation and identification of antioxidants from *Sophora japonica*, *J. Asian Nat. Prod. Res.*, 4, 123, 2002.

59. De Tommasi, N., Piacente, S., and Pizza, C., Flavonol and chalcone ester glycosides from *Bidens andicola*, *J. Nat. Prod.*, 61, 973, 1998.

60. Shaker, K.H. et al., A new triterpenoid saponin from *Ononis spinosa* and two new flavonoid glycosides from *Ononis vaginalis*, *Z. Naturforsch. B*, 59, 124, 2004.

61. Kotsos, M. et al., Chemistry of plants from Crete: stachyspinoside, a new flavonoid glycoside and iridoids from *Stachys spinosa*, *Nat. Prod. Lett.*, 15, 377, 2001.

62. Markham, K.R., Novel anthocyanins produced in petals of genetically-transformed lisianthus, *Phytochemistry*, 42, 1035, 1996.

63. Cabrita, L. and Andersen, Ø.M., Anthocyanins in blue berries of *Vaccinium padifolium*, *Phytochemistry*, 52, 1693, 1999.

64. Pale, E. et al., Acylated anthocyanins from the flowers of *Ipomoea asarifolia*, *Phytochemistry*, 48, 1433, 1998.

65. Pale, E. et al., Two triacylated and tetraglucosylated anthocyanins from *Ipomoea asarifolia* flowers, *Phytochemistry*, 64, 1395, 2003.

66. Brunner, G. et al., A novel acylated flavonol glycoside isolated from *Brunfelsia grandiflora* ssp. *Grandiflora*. Structure elucidation by gradient accelerated NMR spectroscopy at 14T, *Phytochem. Anal.*, 11, 29, 2000.

67. Davis, A.L. et al., 1H and 13C NMR assignments of some green tea polyphenols, *Magn. Reson. Chem.*, 34, 887, 1996.

68. Bandeira, M.A.M., Matos, F.J. de A., and Brazfilho, R., Structural elucidation and total assignment of the 1H and 13C NMR spectra of new chalcone dimmers, *Magn. Reson. Chem.*, 41, 1009, 2003.

69. Ngameni, B. et al., Diprenylated chalcones and other constituents from the twigs of *Dorstenia barteri* var. *subtriangularis*, *Phytochemistry*, 65, 427, 2004.

70. Elegami, A.A. et al., Two very unusual macrocyclic flavonoids from the water lily *Nymphaea lotus*, *Phytochemistry*, 63, 727, 2003.

71. Bouaziz, M. et al., Flavonolignans from *Hyparrhenia hirta*, *Phytochemistry*, 60, 515, 2002.

72. Natsume, M. et al., Structures of (−)-epicatechin glucuronide identified from plasma and urine after oral ingestion of (−)-epicatechin: differences between human and rat, *Free Radical Biol. Med.*, 34, 840, 2003.

73. Franck, U., Neszmelyi, A., and Wagner, H., 2D-NMR structure elucidation of a procyanidin from *Musanga cecropioides*, *ACH — Models Chem.*, 136, 511, 1999.

74. De Mello, J.C.P., Petereit, F., and Nahrstedt, A., A dimeric proanthocyanidin from *Stryphnodendron adstringens*, *Phytochemistry*, 51, 1105, 1999.

75. Prasad, D., Two A-type proanthocyanidins from *Prunus armeniaca* roots, *Fitoterapia*, 71, 245, 2000.

76. De Oliveira, D.B. et al., First isolation of a symmetrical glycosylated methylene bisflavonoid, *Planta Med.*, 69, 382, 2003.

77. Balas, L., Vercauteren, J., and Laguerre, M., 2D NMR structure elucidation of proanthocyanidins: the special case of the catechin-(4α-8)-catechin-(4α-8)-catechin trimer, *Magn. Reson. Chem.*, 33, 85, 1995.

78. Qa'dan, F., Petereit, F., and Nahrstedt, A., Prodelphinidin trimers and characterization of a proanthocyanidin oligomer from *Cistus albidus*, *Pharmazie*, 58, 416, 2003.

79. Damu, A.G. et al., Two acylated flavone glucosides from *Andrographis serpyllifolia*, *Phytochemistry*, 52, 147, 1999.

80. Jayakrishna, G. et al., Two new 2′-oxygenated flavones from *Andrographis elongata*, *Chem. Pharm. Bull.*, 49, 1555, 1999.

81. Manoj, C. et al., Complete 13C and 1H NMR spectral assignments of two isoflavones from the roots of *Dalbergia horrida, Magn. Reson. Chem.*, 41, 227, 2003.

82. Swinny, E., Bloor, S.J., and Wong, H., 1H and 13C NMR assignments for the 3-deoxyanthocyanins, luteolinidin-5-glucoside and apigeninidin-5-glucoside, *Magn. Reson. Chem.*, 38, 1031, 2000.

83. Kouda-Bonafos, M., Nacro, M., and Ancian, B., Total assignment of 1H and 13C NMR chemical shifts of a natural anthocyanidin, apigeninidin, using two-dimensional COLOC and HMBC techniques, *Magn. Reson. Chem.*, 34, 389, 1996.

84. Terahara, N. et al., Triacylated anthocyanins from *Ajuga reptans* flowers and cell cultures, *Phytochemistry*, 42, 199, 1996.

85. Slimestad, R., Aaberg, A., and Andersen, Ø.M., Acylated anthocyanins from petunia flowers, *Phytochemistry*, 50, 1081, 1999.

86. Torskangerpoll, K., Fossen, T., and Andersen, Ø.M., Anthocyanin pigments in tulips, *Phytochemistry*, 52, 1687, 1999.

87. Cabrita, L., Frøystein, N.Å., and Andersen, Ø.M., Anthocyanin trisaccharides in blue berries of *Vaccinium padifolium, Food Chem.*, 69, 33, 2000.

88. Fossen, T. et al., Covalent anthocyanin–flavonol complexes from flowers of chive, *Allium schoenoprasum, Phytochemistry*, 54, 317, 2000.

89. Li, W., Asada, Y., and Yoshikawa, T., Flavonoid constituents from *Glycyrrhiza glabra* hairy root cultures, *Phytochemistry*, 55, 447, 2000.

90. Exarchou, V. et al., Do strong intramolecular hydrogen bonds persist in aqueous solution? Variable temperature gradient 1H, 1H–13C GE-HSQC and GE-HMBC NMR studies of flavonols and flavones in organic and aqueous mixtures, *Tetrahedron*, 58, 7423, 2002.

91. Gibbs, A. et al., Suppression of t1 noise in 2D NMR spectroscopy by reference deconvolution, *J. Magn. Reson., Ser. A*, 101, 351, 1993.

92. Burger, R., Schorn, C., and Bigler, P., HMSC: simultaneously detected heteronuclear shift correlation through multiple and single bonds, *J. Magn. Reson.*, 148, 88, 2001.

93. Furihata, K. and Seto, H., 3D-HMBC, a new NMR technique useful for structural studies of complicated molecules, *Tetrahedron Lett.*, 37, 8901, 1996.

94. Wagner, R. and Berger, S., ACCORD-HMBC: a superior technique for structural elucidation, *Magn. Reson. Chem.*, 36, S44, 1998.

95. Hadden, C.E., Martin, G.E., and Krishnamurthy, V.V., Improved performance accordion heteronuclear multiple-bond correlation spectroscopy — IMPEACH-MBC, *J. Magn. Reson.*, 140, 274, 1999.

96. Hadden, C.E., Martin, G.E., and Krishnamurthy, V.V., Constant time inverse-detection gradient accordion rescaled heteronuclear multiple bond correlation spectroscopy: CIGAR-HMBC, *Magn. Reson. Chem.*, 38, 143, 2000.

97. Mustafa, K., Perry, N.B., and Weavers, R.T., 2-Hydroxyflavanones from *Leptospermum polygalifolium* subsp. *polygalifolium*. Equilibrating sets of hemiacetal isomers, *Phytochemistry*, 64, 1285, 2003.

98. Krishnamurthy, V.V. et al., 2J,3J-HMBC: a new long-range heteronuclear shift correlation technique capable of differentiating 2JCH from 3JCH correlations to protonated carbons, *J. Magn. Reson.*, 146, 232, 2000.

99. Banerji, A. and Luthria, D.L., Assignment of methoxy resonances and observation of restricted rotation of B-ring of 2'-methoxyflavones by 2D-NOESY in conjunction with ASIS, *Spectrosc. Lett.*, 24, 1047, 1991.

100. Nerdal, W., Pedersen, A.T., and Andersen, Ø.M., Two-dimensional nuclear Overhauser enhancement NMR experiments on pelargonidin-3-glucopyranoside, an anthocyanin of low molecular mass, *Acta Chem. Scand.*, 46, 872, 1992.

101. Nerdal, W. and Andersen, Ø.M., Intermolecular aromatic acid association of an anthocyanin (petanin) evidenced by two-dimensional nuclear Overhauser enhancement nuclear magnetic resonance experiments and distance geometry calculations, *Phytochem. Anal.*, 3, 182, 1992.

102. Gakh, E.G., Dougall, D.K., and Baker, D.C., Proton nuclear magnetic resonance studies of monoacylated anthocyanins from the wild carrot: Part 1. Inter- and intramolecular interactions in solution, *Phytochem. Anal.*, 9, 28, 1998.

103. Houbiers, C. et al., Color Stabilization of malvidin 3-glucoside: self-aggregation of the flavylium cation and copigmentation with the Z-chalcone form, *J. Phys. Chem. B*, 102, 3578, 1998.

104. Giusti, M.M., Ghanadan, H., and Wrolstad, R.E., Elucidation of the structure and conformation of red radish (*Raphanus sativus*) anthocyanins using one- and two-dimensional nuclear magnetic resonance techniques., *J. Agric. Food Chem.*, 46, 4858, 1998.

105. Kumazawa, T. et al., Synthesis of 8-C-glucosylflavones, *Carbohydr. Res.*, 334, 183, 2001.

106. Gao, J. et al., Further NMR investigation and conformational analysis of an acylated flavonol glucorhamnoside, *Magn. Reson. Chem.*, 34, 249, 1996.

107. Li, X.-C. et al., Absolute configuration, conformation, and chiral properties of flavanone-(3 → 8″)-flavone biflavonoids from *Rheedia acuminata*, *Tetrahedron*, 58, 8709, 2002.

108. Fossen,T., Rayyan, S., and Andersen, Ø.M., Dimeric anthocyanins from strawberry (*Fragaria ananassa*) consisting of pelargonidin 3-glucoside covalently linked to four flavan-3-ols, *Phytochemistry*, 65, 1421, 2004.

109. Rayyan, S. et al., Isolation and identification of flavonoids including flavone rotamers from the herbal drug "Crataegi folium cum flore," hawthorn, *Phytochem. Anal.*, 2004, in press.

110. Hatano, T. and Hemingway, R.W., Conformational isomerism of phenolic procyanidins: preferred conformations in organic solvents and water, *J. Chem. Soc. Perkin Transactions*, 2, 1035, 1997.

111. Messanga, B.B. et al., Isolation and structural elucidation of a new pentaflavonoid from *Ochna calodendron*, *New J. Chem.*, 25, 1098, 2001.

112. Hosny, M. and Rosazza, J.P.N., New isoflavone and triterpene glycosides from soybeans, *J. Nat. Prod.*, 65, 805, 2002.

113. Messanga, B.B. et al., Calodenin C. A new guibourtinidol-(4a → 8)-afzelechin from *Ochna calodendron*, *Planta Med.*, 64, 760, 2004.

114. Fossen, T. and Andersen, Ø.M., Delphinidin 3′-galloylgalactosides from blue flowers of *Nymphaéa caerulea*, *Phytochemistry*, 50, 1185, 1999.

115. Mas, T. et al., DNA triplex stabilization property of natural anthocyanins, *Phytochemistry*, 53, 679, 2000.

116. Schroeder, H. and Haslinger, E., Sequential analysis of oligosaccharide structures by selective NMR techniques. Part II. 1D multistep-relayed COSY–ROESY, *Liebigs Ann. Chem.*, 7, 751, 1993.

117. Mayer, R., Calycopterones and calyflorenones, novel biflavonoids from *Calycopteris floribunda*, *J. Nat. Prod.*, 62, 1274, 1999.

118. Kumar, V., Brecht, V., and Frahm, A.W., Conformational analysis of the biflavanoid GB2 and a polyhydroxylated flavanone-chromone of *Cratoxylum neriifolium*, *Planta Med.*, 70, 646, 2004.

119. Lewis, K.C. et al., Room-temperature (1H, 13C) and variable-temperature (1H) NMR studies on spinosin, *Magn. Reson. Chem.*, 38, 771, 2004.

120. Fronza, G. et al., Structural features of the β-CD complexes with naringin and its dihydrochalcone and aglycone derivatives by 1H NMR, *J. Inclusion Phenom. Macrocyclic Chem.*, 44, 225, 2002.

121. Whittemore, N.A. et al., A quenched molecular dynamics-rotating frame Overhauser spectroscopy study of a series of semibiosynthetically monoacylated anthocyanins, *J. Org. Chem.*, 69, 1663, 2004.

122. Braun, S., Kalinowski, H.-O., and Berger, S., *150 and More Basic NMR Experiments*, Wiley–VCH, Weinheim, 1998.

123. Abdel-Shafeek, K.A. et al., A new acylated flavonol triglycoside from *Carrichtera annua*, *J. Nat. Prod.*, 63, 845, 2000.

124. Andersen, Ø.M. et al., Chive, a specialised flavonoid producer in *Biologically-Active Phytochemicals in Food*, Special publication — Royal Society of Chemistry, 269, 171, 2001.

125. Wawer, I. and Zielinska, A., ^{13}C CP/MAS NMR studies of flavonoids, *Magn. Reson. Chem.*, 39, 374, 2001.

126. Wawer, I. and Zielinska, A., ^{13}C CP-MAS NMR studies of flavonoids. I. Solid-state conformation of quercetin, quercetin 5′-sulphonic acid and some simple polyphenols, *Solid State Nucl. Mag.*, 10, 33, 1997.

127. Martínez-Richa, A. and Joseph-Nathan, P., Carbon-13 CP-MAS nuclear magnetic resonance studies of teas, *Solid State Nucl. Mag.*, 23, 119, 2003.

128. Kozerski, L. et al., Solution and solid state 13C NMR and x-ray studies of genistein complexes with amines. Potential biological function of the C-7, C-5, and C-4′-OH groups, *Org. Biomol. Chem.*, 1, 3578, 2003.

129. Wolfender, J.-L., Ndjoko, K., and Hostettmann, K., LC/NMR in natural products chemistry, *Curr. Org. Chem.*, 2, 575, 1998.

130. Wolfender, J.-L., Rodriguez, S., and Hostettmann, K., Liquid chromatography coupled to mass spectrometry and nuclear magnetic resonance spectroscopy for the screening of plant constituents, *J. Chromatogr. A*, 794, 299, 1998.

131. Wolfender, J.-L., Ndjoko, K., and Hostettmann, K., Liquid chromatography with ultraviolet absorbance–mass spectrometric detection and with nuclear magnetic resonance spectrometry: a powerful combination for the on-line structural investigation of plant metabolites, *J. Chromatogr. A*, 1000, 437, 2003.

132. Wolfender, J.-L., Ndjoko, K., and Hostettmann, K., Application of LC–NMR in the structure elucidation of polyphenols, in *Methods in Polyphenol Analysis*, Santos-Buelga, C., and Williamson, G., Eds., Royal Society of Chemistry, Cambridge, Chap. 6, 2003.

133. Garo, E. et al., Prenylated flavanones from *Monotes engleri*. Online structure elucidation by LC/UV/NMR, *Helv. Chim. Acta*, 81, 754, 1998.

134. Hansen, S.H. et al., High-performance liquid chromatography on-line coupled to high-field NMR and mass spectrometry for structure elucidation of constituents of *Hypericum perforatum* L., *Anal. Chem.*, 71, 5235, 1999.

135. Santos, L.C. et al., Application of HPLC–NMR coupling using C30 phase in the separation and identification of flavonoids of taxonomic relevance, *Fresen. J. Anal. Chem.*, 368, 540, 2000.

136. Vilegas, W. et al., Application of on-line C30 RP-HPLC–NMR for the analysis of flavonoids from leaf extract of *Maytenus aquifolium*, *Phytochem. Anal.*, 11, 317, 2000.

137. Andrade, F.D.P. et al., Use of on-line liquid chromatography–nuclear magnetic resonance spectroscopy for the rapid investigation of flavonoids from *Sorocea bomplandii*, *J. Chromatogr. A*, 953, 287, 2002.

138. Queiroz, E.F. et al., On-line identification of the antifungal constituents of *Erythrina vogelii* by liquid chromatography with tandem mass spectrometry, ultraviolet absorbance detection and nuclear magnetic resonance spectrometry combined with liquid chromatographic micro-fractionation, *J. Chromatogr. A*, 972, 123, 2002.

139. Kite, G.C. et al., The use of hyphenated techniques in comparative phytochemical studies of legumes, *Biochem. Syst. Ecol.*, 31, 813, 2003.

140. Le Gall, G. et al., Characterization and content of flavonoid glycosides in genetically modified tomato (*Lycopersicon esculentum*) fruits, *J. Agric. Food Chem.*, 51, 2438, 2003.

141. Waridel, P. et al., Identification of the polar constituents of *Potamogeton* species by HPLC–UV with post-column derivatization, HPLC–MSn and HPLC–NMR, and isolation of a new ent-labdane diglycoside, *Phytochemistry*, 65, 2401, 2004.

142. Albach, D.C. et al., Acylated flavone glycosides from *Veronica*, *Phytochemistry*, 64, 1295, 2003.

143. Stochmal, A. et al., Acylated apigein glycosides from alfalfa (*Medicago sativa* L.) var. Artal, *Phytochemistry*, 57, 1223, 2001.

144. Bader, A. et al., Further constituents from *Caralluma negevensis*, *Phytochemistry*, 62, 1277, 2003.

145. Yayli, N., Seymen, H., and Baltaci, C., Flavone C-glycosides from *Scleranthus uncinatus*, *Phytochemistry*, 58, 607, 2001.

146. Maatooq, G.T. et al., C-p-hydroxybenzoylglycoflavones from *Citrullus colocynthis*, *Phytochemistry*, 44, 187, 1997.

147. Cheng, G. et al., Flavonoids from *Ziziphus jujuba Mill* var. Spinosa, *Tetrahedron*, 56, 8915, 2000.

148. Xie, C. et al., Flavonoid glycosides and isoquinolinone alkaloids from *Corydalis bungeana*, *Phytochemistry*, 65, 3041, 2004.

149. Kazuma, K., Noda, N., and Suzuki, M., Malonylated flavonol glycosides from the petals of *Clitoria ternatea*, *Phytochemistry*, 62, 229, 2003.

150. Fico, G. et al., Flavonoids from *Aconitum napellus* subsp. *Neomontanum*, *Phytochemistry*, 57, 543, 2001.

151. Fossen, T., Frøystein, N.Å., and Andersen, Ø.M., Myricetin 3-rhamnosyl(1 → 6)galactoside from *Nymphaéa × marliacea*, *Phytochemistry*, 49, 1997, 1998.

152. Fossen, T. et al., Flavonoids from blue flowers of *Nymphaèa caerulea*, *Phytochemistry*, 51, 1133, 1999.

153. Slimestad, R. et al., Syringetin 3-*O*-(6?-acetyl)-β-glucopyranoside and other flavonols from needles of Norway spruce, *Picea abies*, *Phytochemistry*, 40, 1537, 1995.

154. Sasaki, K. and Takahashi, T., A flavonoid from *Brassica rapa* flower as the UV-absorbing nectar guide, *Phytochemistry*, 61, 339, 2002.

155. Barakat, H.H. et al., Flavonoid galloyl glucosides from the pods of *Acacia farnesiana*, *Phytochemistry*, 51, 139, 1999.

156. Guo, J. et al., Flavonol glycosides from *Lysimachia congestiflora*, *Phytochemistry*, 48, 1445, 1998.

157. Brun, G. et al., A new flavonol glycoside from *Catharanthus roseus*, *Phytochemistry*, 50, 167, 1999.

158. Wu, J.-B. et al., A flavonol *C*-glycoside from *Moghania macrophylla*, *Phytochemistry*, 45, 1727, 1997.

159. Seabra, R.M. et al., Methoxylated aurones from *Cyperus capitatus*, *Phytochemistry*, 45, 839, 1997.

160. Seabra, R.M. et al., 6,3′,4′-Trihydroxy-4-methoxy-5-methylaurone from *Cyperus capitatus*, *Phytochemistry*, 40, 1579, 1995.

161. Li, X.-C., Cai, L., and Wu, C.D., Antimicrobial compounds from *Ceanothus americanus* against oral pathogens, *Phytochemistry*, 46, 97, 1997.

162. Ye, C.-L., Lu, Y.-H., and Wei, D.-Z., Flavonoids from *Cleistocalyx operculatus*, *Phytochemistry*, 65, 445, 2004.

163. Iwashina, T. and Kitajima, J., Chalcone and flavonol glycosides from *Asarum canadense* (Aristolochiaceae), *Phytochemistry*, 55, 971, 2000.

164. Fuendjiep, V. et al., Chalconoid and stilbenoid glycosides from *Guibourtia tessmanii*, *Phytochemistry*, 60, 803, 2002.

165. Lien, T.P. et al., Chalconoids from *Fissistigma bracteolatum*, *Phytochemistry*, 53, 991, 2000.

166. Mitchell, K.A., Markham, K.R., and Boase, M.R., Pigment chemistry and colour of *Pelargonium* flowers, *Phytochemistry*, 47, 355, 1998.

167. Hosokawa, K. et al., Seven acylated anthocyanins in the blue flowers of *Hyacinthus orientalis*, *Phytochemistry*, 38, 1293, 1995.

168. Fossen, T. and Øvstedal, D.O., Anthocyanins from flowers of the orchids *Dracula chimaera* and *D. cordobae*, *Phytochemistry*, 63, 783, 2003.

169. Fossen, T. and Andersen, Ø.M., Cyanidin 3-(2″,3″-digalloylglucoside) from red leaves of *Acer platanoides*, *Phytochemistry*, 52, 1697, 1999.

170. Fossen, T., Slimestad, R., and Andersen, Ø.M., Anthocyanins with 4′-glucosidation from red onion, *Allium cepa*, *Phytochemistry*, 64, 1367, 2003.

171. Fossen, T., Larsen, Å., and Andersen Ø.M., Anthocyanins from flowers and leaves of *Nymphaéa × marliacea* cultivars, *Phytochemistry*, 48, 823, 1998.

172. Nørbæk, R. and Kondo, T., Further anthocyanins from flowers of *Crocus antalyensis* (Iridaceae), *Phytochemistry*, 50, 325, 1999.

173. Hosokawa, K. et al., Five acylated pelargonidin glucosides in the red flowers of *Hyacinthus orientalis*, *Phytochemistry*, 40, 567, 1995.

174. Fossen, T. et al., Acylated anthocyanins from *Oxalis triangularis*, *Phytochemistry*, 66, 1133, 2004.

175. Andersen, Ø.M. et al., Anthocyanin from strawberry (*Fragaria ananassa*) with the novel aglycone, 5-carboxypyranopelargonidin, *Phytochemistry*, 65, 405, 2004.

176. Rayyan, S., Fossen, T., and Andersen, Ø.M., Unpublished results, 2004.

177. Fossen, T. and Andersen Ø.M., Anthocyanins from red onion, *Allium cepa*, with novel aglycone, *Phytochemistry*, 62, 1217, 2003.

178. Fulcrand, H. et al., A new class of wine pigments generated by reaction between pyruvic acid and grape anthocyanins, *Phytochemistry*, 47, 1401, 1998.

179. Stevenson, P.C. and Veitch, N.C., Isoflavenes from the roots of *Cicer judaicum*, *Phytochemistry*, 43, 695, 1996.

180. Farag, S.F. et al., Isoflavonoids and flavone glycosides from rhizomes of *Iris carthaliniae*, *Phytochemistry*, 50, 1407, 1999.

181. López Lázaro, M. et al., An isoflavone glucoside from *Retama sphaerocarpa* boissier, *Phytochemistry*, 48, 401, 1998.

182. Minami, H. et al., Highly oxygenated isoflavones from *Iris japonica*, *Phytochemistry*, 41, 1219, 1996.

183. Takahashi, H. et al., Triterpene and flavanone glycoside from *Rhododendron simsii*, *Phytochemistry*, 56, 875, 2001.

184. Singh, V.P., Yadav, B., and Pandey, V.B., Flavanone glycosides from *Alhagi pseudalhagi*, *Phytochemistry*, 51, 587, 1999.

185. Rawat, M.S.M. et al., A persicogenin 3′-glucoside from the stem bark of *Prunus amygdalus*, *Phytochemistry*, 38, 1519, 1995.

186. Slimestad, R., Andersen, Ø.M., and Francis, G.W., Ampelopsin 7-glucoside and other dihydroflavonol 7-glucosides from needles of *Picea abies*, *Phytochemistry*, 35, 550, 1994.

187. Allais, D.P. et al., 3-Desoxycallunin and 2″-acetylcallunin, two minor 2,3-dihydroflavonoid glucosides from *Calluna vulgaris*, *Phytochemistry*, 39, 427, 1995.

188. Binutu, O.A. and Cordell, G.A., Constituents of *Afzelia bella* stem bark, *Phytochemistry*, 56, 827, 2001.

189. Islam, M.T. and Tahara, S., Dihydroflavonols from *Lannea coromandelica*, *Phytochemistry*, 54, 901, 2000.

190. Lokvam, J., Coley, P.D., and Kursar, T.A., Cinnamoyl glucosides of catechin and dimeric procyanidins from young leaves of *Inga umbellifera* (Fabaceae), *Phytochemistry*, 65, 351, 2004.

191. Piacente, S. et al., Anadanthoside: a flavanol-3-*O*-β-D-xylopyranoside from *Anadenanthera macrocarpa*, *Phytochemistry*, 51, 709, 1999.

192. Waffo, A.K. et al., Indicanines B and C, two isoflavonoid derivatives from the root bark of *Erythrina indica*, *Phytochemistry*, 53, 981, 2000.

193. Abegaz, B.M. et al., Prenylated flavonoids, monoterpenoid furanocoumarins and other constituents from the twigs of *Dorstenia elliptica* (Moraceae), *Phytochemistry*, 65, 221, 2004.

194. Magalhães, A.F. et al., Prenylated flavonoids from *Deguelia hatschbachii* and their systematic significance in *Deguelia*, *Phytochemistry*, 57, 77, 2001.

195. Tarus, P.K. et al., Flavonoids from *Tephrosia aequilata*, *Phytochemistry*, 60, 375, 2002.

196. Yenesew, A. et al., Anti-plasmodial activities and x-ray crystal structures of rotenoids from *Millettia usaramensis* subspecies *usaramensis*, *Phytochemistry*, 64, 773, 2003.

197. Yenesew, A., Midiwo, J.O., and Waterman, P.G., Rotenoids, isoflavones and chalcones from the stem bark of *Millettia usaramensis* subspecies *usaramensis*, *Phytochemistry*, 47, 295, 1998.

198. Sannomiya, M. et al., Application of preparative high-speed counter-current chromatography for the separation of flavonoids from the leaves of *Byrsonima crassa* Niedenzu (IK), *J. Chromatogr. A*, 1035, 47, 2004.

199. Hahn, H. et al., The first biaurone, a triflavone and biflavonoids from two *Aulacomnium* species, *Phytochemistry*, 40, 573, 1995.

200. Ofman, D.J. et al., Flavonoid profiles of New Zealand kauri and other species of *Agathis*, *Phytochemistry*, 38, 1223, 1995.

201. Bekker, R. et al., Biflavonoids. Part 5: structure and stereochemistry of the first bibenzofuranoids, *Tetrahedron*, 56, 5297, 2000.

202. Masesane, I.B. et al., A bichalcone from the twigs of *Rhus pyroides*, *Phytochemistry*, 53, 1005, 2000.

203. Li, D. et al., *In vitro* enzymatic modification of puerarin to puerarin glycosides by maltogenic amylase, *Carbohydr. Res.*, 339, 2789, 2004.

204. Kamnaing, P. et al., An isoflavan–quinone and a flavonol from *Millettia laurentii*, *Phytochemistry*, 51, 829, 1999.

205. Nkengfack, A.E. et al., Prenylated isoflavanone from *Erythrina eriotricha*, *Phytochemistry*, 40, 1803, 1995.
206. Thomas, F. et al., Flavonoids from *Phlomis lychnitys*, *Phytochemistry*, 25, 1253, 1986.
207. Kato, T. and Morita, Y., C-glycosylflavones with acetyl substitution from *Rumex acetosa* L., *Chem. Pharm. Bull.*, 38, 2277, 1990.
208. Liu, K.C.S. et al., Flavonol glycosides with acetyl substitution from *Kalanchoe gracilis*, *Phytochemistry*, 28, 2813, 1989.
209. Mizuno, M. et al., Mearnsetin 3,7-dirhamnoside from *Asplenium antiquum*, *Phytochemistry*, 30, 2817, 1991.
210. Vvedenskaya, I.O. et al., Characterization of flavonols in cranberry (*Vaccinium macrocarpon*) powder, *J. Agric. Food Chem.*, 52, 188, 2004.
211. Manguro, L.O.A. et al., Flavonol glycosides of *Warburgia ugandensis* leaves, *Phytochemistry*, 64, 891, 2003.
212. Ishimaru, K. et al., Taxifolin 3-arabinoside from *Fragaria × ananassa*, *Phytochemistry*, 40, 345, 1995.
213. Takemura, M. et al., Acylated flavonol glycosides as probing stimulants of a bean aphid, *Megoura crassicauda*, from *Vicia angustifolia*, *Phytochemistry*, 61, 135, 2002.
214. Cui, C.B. et al., Constituents of a fern, *Davallia-Marinesii Moore*. 1. Isolation and structures of davallialactone and a new flavanone glucuronide, *Chem. Pharm. Bull.*, 38, 3218, 1990.
215. Adell, J., Barbera, O., and Marco, J.A., Flavonoid glycosides from *Anthyllis sericea*, *Phytochemistry*, 27, 2967, 1988.
216. Zhou, B.-N., Blaskò, G., and Cordell, G.A., Iternanthin, a C-glycosylated flavonoid from *Alternanthera philoxeroides*, *Phytochemistry*, 27, 3633, 1988.
217. Kitanaka, S., Ogata, K., and Takido, M., Studies on the constituents of the leaves of *Cassia torosa* Cav. 1. The structures of two new *C*-glycosylflavones, *Chem. Pharm. Bull.*, 37, 2441, 1989.
218. Mareck, U. et al., The 6-*C*-chinovoside and 6-*C*-fucoside of luteolin from *Passiflora edulis*, *Phytochemistry*, 30, 3486, 1991.
219. Toki, K. et al., Acylated anthocyanins in *Verbena* flowers, *Phytochemistry*, 38, 515, 1995.
220. Lu, T.S. et al., An acylated peonidin glycoside in the violet-blue flowers of *Pharbitis nil*, *Phytochemistry*, 30, 2387, 1991.
221. Terahara, N. et al., Five new anthocyanins, ternatins A3, B4, B3, B2, and D2, from *Clitoria ternatea* flowers, *J. Nat. Prod.*, 59, 139, 1996.
222. Toki, K. et al., Acylated cyanidin glycosides from the purple-red flowers of *Anemone coronaria*, *Heterocycles*, 60, 345, 2003.
223. Catalano, G., Fossen, T., and Andersen, Ø.M., Petunidin 3-*O*-alpha-rhamnopyranoside-5-*O*-beta-glucopyranoside and other anthocyanins from flowers of *Vicia villosa*, *J. Agric. Food Chem.*, 46, 4568, 1998.
224. Toki, K. et al., Two malonylated anthocyanidin glycosides in *Ranunculus asiaticus*, *Phytochemistry*, 42, 1055, 1996.
225. Saito, N. et al., Cyanidin 3-malonylglucuronylglucoside in *Bellis* and cyanidin 3-malonylglucoside in Dendranthema, *Phytochemistry*, 27, 2963, 1988.
226. Bloor, S.J. and Falshaw, R., Covalently linked anthocyanin-flavonol pigments from blue *Agapanthus* flowers, *Phytochemistry*, 53, 575, 2000.
227. Bloor, S.J., A macrocyclic anthocyanin from red/mauve carnation flowers, *Phytochemistry*, 49, 225, 1998.
228. Saito, N. et al., Acylated anthocyanins from the blue-violet flowers of *Anemone coronaria*, *Phytochemistry*, 60, 365, 2002.
229. Saito, N. et al., Acylated pelargonidin 3,7-glycosides from red flowers of *Delphinium hybridum*, *Phytochemistry*, 49, 881, 1998.
230. Yoshida, K. et al., A UV-B resistant polyacylated anthocyanin, HBA, from blue petals of morning glory, *Tetrahedron Lett.*, 44, 7875, 2003.
231. Otsuki, T. et al., Acylated anthocyanins from red radish (*Raphanus sativus* L.), *Phytochemistry*, 60, 79, 2002.

232. Torskangerpoll, K. and Andersen, Ø.M., Colour stability of anthocyanins in aqueous solutions at various pH values, *Food Chem.*, 89, 427, 2005.

233. Strack, D. et al., Cyanidin 3-oxalylglucoside in orchids. *Z. Naturforsch. C*, 41, 707, 1986.

234. Hillebrand, S., Schwarz, M., and Winterhalter, P., Characterization of anthocyanins and pyranoanthocyanins from blood orange [*Citrus sinensis* (L.) Osbeck] juice, *J. Agric. Food Chem.*, 52, 7331, 2004.

235. Terahara, N. et al., Six diacylated anthocyanins from the storage roots of purple sweet potato, *Ipomoea batatas*, *Biosci. Biotech. Biochem.*, 63, 1420, 1999.

236. Iwashina, T. et al., Flavonoids in translucent bracts of the Himalayan *Rheum nobile* (Polygonaceae) as ultraviolet shields, *J. Plant Res.*, 117, 101, 2004.

237. Suzuki, R., Okada, Y., and Okuyama, T., Two flavone *C*-glycosides from the style of *Zea mays* with glycation inhibitory activity, *J. Nat. Prod.*, 66, 564, 2003.

238. Suzuki, R., Okada, Y., and Okuyama, T., A new flavone *C*-glycosides from the style of *Zea mays* L. with glycation inhibitory activity, *Chem. Pharm. Bull.*, 51, 1186, 2003.

239. Nakanishi, T. et al., Flavonoid glycosides of the roots of *Glycyrrhiza uralensis*, *Phytochemistry*, 24, 339, 1985.

240. Wolfender, J.L. et al., Mass spectrometry of underivatized naturally occurring glycosides, *Phytochem. Anal.*, 3, 193, 1992.

241. Sakushima, A., Nishibe, S., and Brandenberger, H., Negative ion desorption chemical ionization mass spectrometry of flavonoid glycosides, *Biomed. Environ. Mass Spectrom.*, 18, 809, 1989.

242. Lee, M.S. et al., Rapid screening of fermentation broths for flavones using tandem mass spectrometry, *Biol. Mass Spectrom.*, 22, 84, 1993.

243. Rocha, L. et al., Antibacterial phloroglucinols and flavonoids from *Hypericum brasiliense*, *Phytochemistry*, 40, 1447, 1995.

244. Wood, K.V. et al., Analysis of anthocyanins and 3-deoxyanthocyanidins by plasma desorption mass spectrometry, *Phytochemistry*, 37, 557, 1994.

245. Hipskind, J., Wood, K., and Nicholson, R.L., Localized stimulation of anthocyanin accumulation and delineation of pathogen ingress in maize genetically resistant to *Bipolaris maydis* race O, *Physiol. Mol. Plant Pathol.*, 49, 247, 1996.

246. Phippen, W.B. and Simon, J.E., Anthocyanins in Basil (*Ocimum basilicum* L.), *J. Agric. Food Chem.*, 46, 1734, 1998.

247. Ducrey, B. et al., Analysis of flavonol glycosides of thirteen *Epilobium* species (Onagraceae) by LC–UV and thermospray LC–MS, *Phytochemistry*, 38, 129, 1995.

248. Slimestad, R. and Hostettmann, K., Characterization of phenolic constituents from juvenile and mature needles of Norway spruce by means of high performance liquid chromatography–mass spectrometry, *Phytochem. Anal.*, 7, 42, 1996.

249. Gabetta, B. et al., Characterization of proanthocyanidins from grape seeds, *Fitoterapia*, 71, 162, 2000.

250. Pietta, P. et al., Thermospray liquid chromatography–mass spectrometry of flavonol glycosides from medicinal plants, *J. Chromatogr. A*, 661, 121, 1994.

251. Hutton, T. and Major, H.J., Characterizing biomolecules by electrospray ionization-mass spectrometry coupled to liquid chromatography and capillary electrophoresis, *Biochem. Soc. Trans.*, 23, 924, 1995.

252. Cuyckens, F. et al., Direct stereochemical assignment of hexose and pentose residues in flavonoid *O*-glycosides by fast atom bombardment and electrospray ionization mass spectrometry, *J. Mass Spectrom.*, 37, 1272, 2002.

253. Cuyckens, F. and Claeys, M., Mass spectrometry in the structural analysis of flavonoids, *J. Mass Spectrom.*, 39, 1, 2004.

254. Ma, Y.L. et al., Characterization of flavone and flavonol aglycones by collision-induced dissociation tandem mass spectrometry, *Rapid Commun. Mass Spectrom.*, 11, 1357, 1997.

255. Domon, B. and Costello, C.E., A systematic nomenclature for carbohydrate fragmentations in FAB-MS spectra of glycoconjugates, *Glycoconjugate J.*, 5, 397, 1988.

256. Stobiecki, M., Application of mass spectrometry for identification and structural studies of flavonoid glycosides, *Phytochemistry*, 54, 237, 2000.

257. De Pascual-Teresa, S. and Rivas-Gonzalo, J.C., Application of LC–MS for the identification of polyphenols, in *Methods in Polyphenol Analysis*, Santos-Buelga, C. and Williamson, G., Eds., The Royal Society of Chemistry, Cambridge, Chap. 3, 2003.

258. Prasain, J.K., Wang, C.-C., and Barnes, S., Mass spectrometric methods for the determination of flavonoids in biological samples, *Free Radical Biol. Med.*, 37, 1324, 2004.

259. Cuyckens, F. et al., Tandem mass spectral strategies for the structural characterization of flavonoid glycosides, *Analusis*, 28, 888, 2001.

260. Kuhnle, G.G.C., Investigation of flavonoid and their *in vivo* metabolite forms using tandem mass spectrometry, *Oxidative Stress Dis.*, 9, 145, 2003.

261. Pietta, P. and Mauri, P., Analysis of flavonoids in medicinal plants, *Meth. Enzymol.*, 335, 26, 2001.

262. Careri, M., Bianchi, F., and Corradini, C., Recent advances in the application of mass spectrometry in food-related analysis, *J. Chromatogr. A*, 970, 3, 2002.

263. Markham, K.R., Flavones, flavonols and their glycosides, *Meth. Plant Biochem.*, 1, 197, 1989.

264. Berahia, T. et al., Mass spectrometry of polymethoxylated flavones, *J. Agric. Food Chem.*, 42, 1697, 1994.

265. Hedin, P.A. and Phillips, V.A., Chemical ionization (methane) mass spectrometry of sugars and their derivatives, *J. Agric. Food Chem.*, 39, 1106, 1991.

266. Li, Q.M. and Claeys, M., Characterization and differentiation of diglycosyl flavonoids by positive ion fast atom bombardment and tandem mass spectrometry, *Biol. Mass Spectrom.*, 23, 406, 1994.

267. Baracco, A. et al., A comparison of the combination of fast-atom bombardment with tandem mass spectrometry and of gas chromatography with mass spectrometry in the analysis of a mixture of kaempferol, kaempferide, luteolin and oleouropein, *Rapid Commun. Mass Spectrom.*, 9, 427, 1995.

268. Ma, Y.L., Van den Heuvel, H., and Claeys, M., Characterization of 3-methoxyflavones using fast-atom bombardment and collision-induced dissociation tandem mass spectrometry, *Rapid Commun. Mass Spectrom.*, 13, 1932, 1999.

269. Borges, C. et al., Characterisation of flavonoids extracted of *Genista tenera* by FAB MS/MS, *Adv. Mass Spectrom.*, 15, 847, 2001.

270. Self, R., Fast-atom-bombardment mass spectrometry in food science, *Appl. Mass Spectrom. Food Sci.*, 239, 1987.

271. Takayama, M. et al., Mass spectrometry of prenylated flavonoids, *Heterocycles*, 33, 405, 1992.

272. Claeys, M., Mass spectrometry in studies of flavonoid glycosides, *NATO ASI Ser., Ser. C: Math. Phys. Sci.*, 521, 165, 1999.

273. Nørbæk, R. et al., Anthocyanins in chilean species of *Alstroemeria*, *Phytochemistry*, 42, 97, 1996.

274. Farina, A. et al., HPTLC and reflectance mode densitometry of anthocyanins in *Malva Silvestris* L.: a comparison with gradient-elution reversed-phase HPLC, *J. Pharm. Biomed. Anal.*, 14, 203, 1995.

275. Shoji, T. et al., Characterization and structures of anthocyanin pigments generated in rosé cider during vinification, *Phytochemistry*, 59, 183, 2002.

276. Bakker, J. et al., Identification of an anthocyanin occurring in some red wines, *Phytochemistry*, 44, 1375, 1997.

277. Kovács, Z. and Dinya, Z., Examination of non-volatile organic compounds in red wines made in Eger, *Microchem. J.*, 67, 57, 2000.

278. Miyake, Y. et al., New potent antioxidative hydroxyflavanones produced with *Aspergillus saitoi* from flavanone glycoside in citrus fruit, *Biosci. Biotechnol. Biochem.*, 67, 1443, 2003.

279. Sharif, M. et al., Four flavonoid glycosides from *Peganum harmala*, *Phytochemistry*, 44, 533, 1997.

280. Iwashina, T., Kamenosono, T., and Ueno, T., Hispidulin and nepetin 4′-glucosides from *Cirsium oligophyllum*, *Phytochemistry*, 51, 1109, 1999.

281. Miyake, Y. et al., Isolation of *C*-glucosylflavone from lemon peel and antioxidative activity of flavonoid compounds in lemon fruit, *J. Agric. Food Chem.*, 45, 4619, 1997.

282. Yayli, N. et al., Phenolic and flavone *C*-glycosides from *Scleranthus uncinatus*, *Pharm. Biol.*, 40, 369, 2002.

283. Nielsen, J.K., Nørbæk, R., and Olsen, C.E., Kaempferol tetraglucosides from cabbage leaves, *Phytochemistry*, 49, 2171, 1998.
284. Mihara, R. et al., A novel acylated quercetin tetraglycoside from oolong tea (*Camelia sinensis*) extracts, *Tetrahedron Lett.*, 45(26), 5077, 2004.
285. Munoz, A. et al., Identification and structural assessment of alkaline-earth metal complexes with flavonols by FAB mass spectrometry, *Russ. J. Gen. Chem.*, 74, 438, 2004. Translated from *Zh. Obshch. Khim.*, 74, 482, 2004.
286. Harput, Ü.S., Saracoglu, I., and Ogihara, Y., Methoxyflavonoids from *Pinaropappus roseus*, *Turk. J. Chem.*, 28, 761, 2004.
287. Hasan, A. et al., Flavonoid glycosides and an anthraquinone from *Rumex chalepensis*, *Phytochemistry*, 39, 1211, 1995.
288. Martín-Cordero, C. et al., Novel flavonol glycoside from *Retama sphaerocarpa* Boissier, *Phytochemistry*, 51, 1129, 1999.
289. Hasan, A. et al., Two flavonol triglycosides from flowers of *Indigofera hebepetal*, *Phytochemistry*, 43, 1115, 1996.
290. Ogundipe, O.O., Moody, J.O., and Houghton, P.J., Occurrence of flavonol sulphates in *Alchornea laxiflora*, *Pharm. Biol.*, 39, 421, 2001.
291. Senatore, F., D'Agostino, M., and Dini, I., Flavonoid glycosides of *Barbarea vulgaris* L. (Brassicaceae), *J. Agric. Food Chem.*, 48, 2659, 2000.
292. Kweon, M.H. et al., Structural characterization of a flavonoid compound scavenging superoxide anion radical isolated from *Capsella bursa-pastoris*, *J. Biochem. Mol. Biol.*, 29, 423, 1996.
293. Veitch, N.C. and Stevenson, P.C., 2-Methoxyjudaicin, an isoflavene from the roots of *Cicer bijugum*, *Phytochemistry*, 44, 1587, 1997.
294. Kanakubo, A. and Isobe, M., Differentiation of sulfate and phosphate by H/D exchange mass spectrometry: application to isoflavone, *J. Mass Spectrom.*, 39, 1260, 2004.
295. Messanga, B.B. et al., Triflavonoids of *Ochna calodendron*, *Phytochemistry*, 59, 435, 2002.
296. Kida, K. et al., Identification of biliary metabolites of (−)-epigallocatechin gallate in rats, *J. Agric. Food Chem.*, 48, 4151, 2000.
297. Sporns, P. and Wang, J., Exploring new frontiers in food analysis using MALDI-MS, *Food Res. Int.*, 31, 181, 1998.
298. Fukai, T., Kuroda, J., and Nomura, T., Accurate mass measurement of low molecular weight compounds by matrix-assisted laser desorption/ionization time-of-flight mass spectrometry, *J. Am. Soc. Mass Spectrom.*, 11, 458, 2000.
299. Spengler, B., Kirsch, D., and Kaufmann, R., Metastable decay of peptides and proteins in matrix-assisted laser-desorption mass spectrometry, *Rapid Commun. Mass Spectrom.*, 5, 198, 1991.
300. Keki, S., Deak, G., and Zsuga, M., Fragmentation study of rutin, a naturally occurring flavone glycoside cationized with different alkali metal ions, using post-source decay matrix-assisted laser desorption/ionization mass spectrometry, *J. Mass Spectrom.*, 36, 1312, 2001.
301. Behrens, A. et al., MALDI-TOF mass spectrometry and PSD fragmentation as means for the analysis of condensed tannins in plant leaves and needles, *Phytochemistry*, 62, 1159, 2003.
302. Keki, S. et al., Post-source decay matrix-assisted laser desorption/ionization mass spectrometric study of peracetylated isoflavone glycosides cationized by protonation and with various metal ions, *J. Mass Spectrom.*, 38, 1207, 2003.
303. Menet, M.C. et al., Analysis of theaflavins and thearubigins from black tea extract by MALDI-TOF mass spectrometry, *J. Agric. Food Chem.*, 52, 2455, 2004.
304. Foo, L.Y. et al., The structure of cranberry proanthocyanidins which inhibit adherence of uropathogenic P-fimbriated *Escherichia coli in vitro*, *Phytochemistry*, 54, 173, 2000.
305. Pasch, H., Pizzi, A., and Rode, K., MALDI-TOF mass spectrometry of polyflavonoid tannins, *Polymer*, 42, 7531, 2001.
306. Porter, M.L. et al., Cranberry proanthocyanidins associate with low-density lipoprotein and inhibit *in vitro* Cu^{2+}-induced oxidation, *J. Sci. Food Agric.*, 81, 1306, 2001.
307. Krueger, C.G., Vestling, M.M., and Reed, J.D., Matrix-assisted laser desorption/ionization time-of-flight mass spectrometry of heteropolyflavan-3-ols and glucosylated heteropolyflavans in sorghum [*Sorghum bicolor* (L.) Moench], *J. Agric. Food Chem.*, 51, 538, 2003.

308. Ohnishi-Kameyama, M. et al., Identification of catechin oligomers from apple (*Malus pumila* cv. Fuji) in matrix-assisted laser desorption/ionization time-of-flight mass spectrometry and fast-atom bombardment mass spectrometry, *Rapid Commun. Mass Spectrom.*, 11, 31, 1997.

309. Ramirez-Coronel, M.A. et al., Characterization and estimation of proanthocyanidins and other phenolics in coffee pulp (*Coffea arabica*) by thiolysis-high-performance liquid chromatography, *J. Agric. Food Chem.*, 52, 1344, 2004.

310. Krueger, C.G. et al. Matrix-assisted laser desorption/ionization time-of-flight mass spectrometry of polygalloyl polyflavan-3-ols in grape seed extract, *J. Agric. Food Chem.*, 48, 1663, 2000.

311. Yang, Y. and Chien, M., Characterization of grape procyanidins using high-performance liquid chromatography/mass spectrometry and matrix-assisted laser desorption/ionization time-of-flight mass spectrometry, *J. Agric. Food Chem.*, 48, 3990, 2000.

312. Perret, C., Pezet, R., and Tabacchi, R., Fractionation of grape tannins and analysis by matrix-assisted laser desorption/ionisation time-of-flight mass spectrometry, *Phytochem. Anal.*, 14, 202, 2003.

313. Vivas, N. et al., Differentiation of proanthocyanidin tannins from seeds, skins and stems of grapes (*Vitis vinifera*) and heartwood of Quebracho (*Schinopsis balansae*) by matrix-assisted laser desorption/ionization time-of-flight mass spectrometry and thioacidolysis/liquid chromatography/electrospray ionization mass spectrometry, *Anal. Chim. Acta*, 513, 247, 2004.

314. Nonier, M.F. et al., Application of off-line size-exclusion chromatographic fractionation-matrix assisted laser desorption ionization time of flight mass spectrometry for proanthocyanidin characterization, *J. Chromatogr. A*, 1033, 291, 2004.

315. Sugui, J.A. et al., MALDI-TOF analysis of mixtures of 3-deoxyanthocyanidins and anthocyanin, *Phytochemistry*, 48, 1063, 1998.

316. Wang, J., Kalt, W., and Sporns, P., Comparison between HPLC and MALDI-TOF MS analysis of anthocyanins in highbush blueberries, *J. Agric. Food Chem.*, 48, 3330, 2000.

317. Wang, J. and Sporns, P., Analysis of anthocyanins in red wine and fruit juice using MALDI-MS, *J. Agric. Food Chem.*, 47, 2009, 1999.

318. Wang, J. and Sporns, P., MALDI-TOF MS analysis of isoflavones in soy products, *J. Agric. Food Chem.*, 48, 5887, 2000.

319. Wang, J. and Sporns, P., MALDI-TOF MS analysis of food flavonol glycosides, *J. Agric. Food Chem.*, 48, 1657, 2000.

320. Frison-Norrie, S. and Sporns, P., Identification and quantification of flavonol glycosides in almond seed coats using MALDI-TOF MS, *J. Agric. Food Chem.*, 50, 2782, 2002.

321. Frison, S. and Sporns, P., Variation in the flavonol glycoside composition of almond seed coats as determined by MALDI-TOF mass spectrometry, *J. Agric. Food Chem.*, 50, 6818, 2002.

322. Yamashita, M. and Fenn, J.B., Negative ion production with the electrospray ion source, *J. Phys. Chem. — US*, 88, 4671, 1984.

323. Cole, R.B., *Electrospray Ionisation Mass Spectrometry*, Wiley, New York, 1997.

324. Dalluge, J.J. et al., Capillary liquid chromatography electrospray mass spectrometry for the separation and detection of catechins in green tea and human plasma, *Rapid Commun. Mass Spectrom.*, 11, 1753, 1997.

325. Guyot, S. et al., Characterization of highly polymerized procyanidins in cider apple (*Malus sylvestris* var *kermerrien*) skin and pulp, *Phytochemistry* 44, 351, 1997.

326. Fulcrand, H. et al., Electrospray contribution to structural analysis of condensed tannin oligomers and polymers, *Basic Life Sci.*, 66, 223, 1999.

327. Oliveira, M.C., Esperanca, P., and Ferreira, M.A.A., Characterisation of anthocyanidins by electrospray ionisation and collision-induced dissociation tandem mass spectrometry, *Rapid Commun. Mass Spectrom.*, 15, 1525, 2001.

328. March, R.E., Miao, X.-S., and Metcalfe, C.D., A fragmentation study of a flavone triglycoside, kaempferol-3-*O*-robinoside-7-*O*-rhamnoside, *Rapid Commun. Mass Spectrom.*, 18, 931, 2004.

329. March, R.E. et al., A fragmentation study of an isoflavone glycoside, genistein-7-*O*-glucoside, using electrospray quadrupole time-of-flight mass spectrometry at high mass resolution, *Int. J. Mass Spectrom.*, 232, 171, 2004.

330. Parejo, I. et al., Characterization of acylated flavonoid-O-glycosides and methoxylated flavonoids from *Tagetes Maxima* by liquid chromatography coupled to electrospray ionization tandem mass spectrometry, *Rapid Commun. Mass Spectrom.*, 18, 2801, 2004.

331. Weintraub, R.A. et al., Trace determination of naringenin and hesperitin by tandem mass spectrometry, *J. Agric. Food Chem.*, 43, 1966, 1995.

332. Guo, M., Characterization of non-covalent complexes of rutin with cyclodextrins by electrospray ionization tandem mass spectrometry, *J. Mass Spectrom.*, 39, 594, 2004.

333. Wolfender, J.-L. et al., Evaluation of Q-TOF-MS/MS and multiple stage IT-MSn for the dereplication of flavonoids and related compounds in crude plant extracts, *Analusis*, 28, 895, 2001.

334. Waridel, P. et al., Evaluation of quadrupole time-of-flight tandem mass spectrometry and ion-trap multiple-stage mass spectrometry for the differentiation of C-glycosidic flavonoid isomers, *J. Chromatogr. A*, 926, 29, 2001.

335. Mullen, W., Lean, M.E.J., and Crozier, A., Rapid characterization of anthocyanins in red raspberry fruit by high-performance liquid chromatography coupled to single quadrupole mass spectrometry, *J. Chromatogr. A*, 966, 63, 2002.

336. De Rijke, E. et al., Liquid chromatography with atmospheric pressure chemical ionization and electrospray ionization mass spectrometry of flavonoids with triple-quadrupole and ion-trap instruments, *J. Chromatogr. A*, 984, 45, 2003.

337. Piovan, A., Fillipini, R., and Favretto, D., Characterization of the anthocyanins of *Caranthus roseus* (L.) G. Don *in vivo* and *in vitro* by electrospray ionization ion trap mass spectrometry, *Rapid Commun. Mass Spectrom.*, 12, 361, 1998.

338. Jungbluth, G. and Ternes, W., HPLC separation of flavonols, flavones and oxidized flavonols with UV-, DAD-, electrochemical and ESI-ion trap MS detection, *Fresenius J. Anal. Chem.*, 367, 661, 2000.

339. Fabre, N. et al., Determination of flavone, flavonol, and flavanone aglycones by negative ion liquid chromatography electrospray ion trap mass spectrometry, *J. Am. Soc. Mass Spectrom.*, 12, 707, 2001.

340. Cuyckens, F. et al., Structure characterization of flavonoid O-diglycosides by positive and negative nano-electrospray ionization ion trap mass spectrometry, *J. Mass Spectrom.*, 36, 1203, 2001.

341. Tolonen, A. and Uusitalo, J., Fast screening method for the analysis of total flavonoid content in plants and foodstuffs by high-performance liquid chromatography/electrospray ionization time-of-flight mass spectrometry with polarity switching, *Rapid Commun. Mass Spectrom.*, 18, 3113, 2004.

342. Krueger, C.G., Vestling, M.M., and Reed, J.D., Matrix-assisted laser desorption–ionization time-of-flight mass spectrometry of anthocyanin-polyflavan-3-ol oligomers in cranberry fruit (*Vaccinium macrocarpon*, Ait.) and spray-dried cranberry juice, *ACS Symp. Ser.*, 886, 232, 2004.

343. Schmidt, T.J., Merfort, I., and Matthiesen, U., Resolution of complex mixtures of flavonoid aglycones by analysis of gas chromatographic–mass spectrometric data, *J. Chromatogr.*, 634, 350, 1993.

344. Dos Santos Pereira, A. et al., Analysis and quantitation of rotenoids and flavonoids in *Derris* (*Lonchocarpus urucu*) by high-temperature high-resolution gas chromatography, *J. Chromatogr. Sci.*, 38, 174, 2000.

345. Branco, A.P. et al., Further lipophilic flavonols in *Vellozia graminifolia* (Velloziaceae) by high temperature gas chromatography: quick detection of new compounds, *Phytochem. Anal.*, 12, 266, 2001.

346. Fiamegos, Y.C. et al., Analytical procedure for the in-vial derivatization-extraction of phenolic acids and flavonoids in methanolic and aqueous plant extracts followed by gas chromatography with mass-selective detection, *J. Chromatogr. A*, 1041, 11, 2004.

347. Zhang, K. and Zuo, Y., GC–MS determination of flavonoids and phenolic and benzoic acids in human plasma after consumption of cranberry juice, *J. Agric. Food Chem.*, 52, 222, 2004.

348. Adlercreutz, H. et al., Isotope dilution gas chromatographic–mass spectrometric method for the determination of unconjugated lignans and isoflavonoids in human feces, with preliminary results in omnivorous and vegetarian women. *Anal. Biochem.*, 225, 101, 1995.

349. DaCosta, C.T., Horton, D., and Margolis, S.A., Analysis of anthocyanins in foods by liquid chromatography, liquid chromatography–mass spectrometry and capillary electrophoresis, *J. Chromatogr. A*, 881, 403, 2000.

350. Lazarus, S.A. et al., High-performance liquid chromatography/mass spectrometry analysis of proanthocyanidins in food and beverages, *Meth. Enzymol.*, 335, 46, 2001.

351. Ferreira, D. and Slade, D., Oligomeric proanthocyanidins: naturally occurring *O*-heterocycles, *Nat. Prod. Rep.*, 19, 517, 2002.

352. Flamini, R., Mass spectrometry in grape and wine chemistry. Part I: polyphenols. *Mass Spectrom. Rev.*, 22, 218, 2003.

353. Stobiecki, M. and Makkar, H.P.S., Recent advances in analytical methods for identification and quantification of phenolic compounds, *EAAP Publ.*, 110, 11–28, 2004.

354. Alcalde-Eon, C. et al., Identification of anthocyanins of pinta boca (*Solanum stenotomum*) tubers, *Food Chem.*, 86, 441, 2004.

355. Cooney, J.M., Jensen, D.J., and McGhie, T.K., LC–MS identification of anthocyanins in boysenberry extract and anthocyanin metabolites in human urine following dosing, *J. Sci. Food Agric.*, 84, 237, 2004.

356. Wang, H.B., Race, E.J., and Shrikhande, A.J., Characterization of anthocyanins in grape juices by ion trap liquid chromatography–mass spectrometry, *J. Agric. Food Chem.*, 51, 1839, 2003.

357. Mateus, N. et al., Structural diversity of anthocyanin-derived pigments in port wines, *Food Chem.*, 76, 335, 2002.

358. De Pascual-Teresa, S., Santos-Buelga, C., and Rivas-Gonzalo, J.C., LC–MS analysis of anthocyanins from purple corn cob, *J. Sci. Food Agric.*, 82, 1003, 2002.

359. Švehli'ková, V. et al., Isolation, identification and stability of acylated derivatives of apigenin 7-*O*-glucoside from chamomile (*Chamomilla recutita* [L.] Rauschert), *Phytochemistry*, 65, 2323, 2004.

360. Stewart, A.J. et al., Occurrence of flavonols in tomatoes and tomato-based products, *J. Agric. Food Chem.*, 48, 2663, 2000.

361. Luczkiewcz, M. et al., LC–DAD UV and LC–MS for the analysis of isoflavones and flavones from *in vitro* and *in vivo* biomass of *Genista tinctoria* L., *Chromatographia*, 60, 179, 2004.

362. Barnes, S., Kirk, M., and Coward, L., Isoflavones and their conjugates in soy foods: extraction conditions and analysis by HPLC–mass spectrometry, *J. Agric. Food Chem.*, 42, 2466, 1994.

363. Klejdus, B., Vitamvasova-Sterbova, D., and Kuban, V., Identification of isoflavone conjugates in red clover (*Trifolium pratense*) by liquid chromatography–mass spectrometry after two-dimensional solid-phase extraction, *Anal. Chim. Acta*, 450, 81, 2001.

364. De Rijke, E. et al., Liquid chromatography coupled to nuclear magnetic resonance spectroscopy for the identification of isoflavone glucoside malonates in *T. pratense* L. leaves, *J. Sep. Sci.*, 27, 1061, 2004.

365. Vidal, S. et al., Mass spectrometric evidence for the existence of oligomeric anthocyanins in grape skins, *J. Agric. Food Chem.*, 52, 7144, 2004.

366. Siciliano, T. et al., Study of flavonoids of *Sechium edule* (Jacq) *swartz* (Cucurbitaceae) different edible organs by liquid chromatography photodiode array mass spectrometry, *J. Agric. Food Chem.*, 52, 6510, 2004.

367. Huck, C.W., Buchmeiser, M.R., and Bonn, G.K., Fast analysis of flavonoids in plant extracts by liquid chromatography–ultraviolet absorbance detection on poly(carboxylic acid)-coated silica and electrospray ionization tandem mass spectrometric detection, *J. Chromatogr. A*, 943, 33, 2002.

368. Rauha, J.-P., Vuorela, H., and Kostiainen, R., Effect of eluent on the ionization efficiency of flavonoids by ion spray, atmospheric pressure chemical ionization, and atmospheric pressure photoionization mass spectrometry, *J. Mass Spectrom.*, 36, 1269, 2001.

369. Vrhovsek, U. et al., Quantitation of polyphenols in different apple varieties, *J. Agric. Food Chem.*, 52, 6532, 2004.

370. Bennett, R.N. et al., Profiling glucosinolates, flavonoids, alkaloids, and other secondary metabolites in tissues of *Azima tetracantha* L. (Salvadoraceae), *J. Agric. Food Chem.*, 52, 5856, 2004.

371. Chen, L.J. et al., Separation and identification of flavonoids in an extract from the seeds of *Oroxylum indicum* by CCC, *J. Liq. Chromatogr. Relat. Technol.*, 26, 1623, 2003.

372. Le Gall, G. et al., Characterization and content of flavonoid glycosides in genetically modified tomato (*Lycopersicon esculentum*) fruits, *J. Agric. Food Chem.*, 51, 2438, 2003.

373. Sanchez-Rabaneda, F. et al., Liquid chromatographic/electrospray ionization tandem mass spectrometric study of the phenolic composition of cocoa (*Theobroma cacao*), *J. Mass Spectrom.*, 38, 35, 2003.

374. Bramati, L. et al., Quantitative characterization of flavonoid compounds in Rooibos tea (*Aspalathus linearis*) by LC–UV/DAD, *J. Agric. Food Chem.*, 50, 5513, 2002.

375. Mulder, T.P.J. et al., Analysis of theaflavins in biological fluids using liquid chromatography–electrospray mass spectrometry, *J. Chromatogr. B*, 760, 271, 2001.

376. Robards, K. et al., Characterisation of citrus by chromatographic analysis of flavonoids, *J. Sci. Food Agric.*, 75, 87, 1997.

377. Maillard, M.-N., Giampaoli, P., and Cuvelier, M.-E., Atmospheric pressure chemical ionization (APCI) liquid chromatography–mass spectrometry: characterization of natural antioxidants, *Talanta*, 43, 339, 1996.

378. Justesen, U., Negative atmospheric pressure chemical ionization low-energy collision activation mass spectrometry for the characterization of flavonoids in extracts of fresh herbs, *J. Chromatogr. A*, 902, 369, 2000.

379. Lin, L.-Z. et al., LC–ESI-MS study of the flavonoid glycoside malonates of red clover (*Trifolium pratense*), *J. Agric. Food Chem.*, 48, 354, 2000.

380. Hughes, R.J. et al., A tandem mass spectrometric study of selected characteristic flavonoids, *Int. J. Mass Spectrom.*, 210/211, 371, 2001.

381. Nielsen, S.E. et al., Identification and quantification of flavonoids in human urine samples by column-switching liquid chromatography coupled to atmospheric pressure chemical ionization mass spectrometry, *Anal. Chem.*, 72, 1503, 2000.

382. Andlauer, W., Martena, M.J., and Furst, P., Determination of selected phytochemicals by reversed-phase high-performance liquid chromatography combined with ultraviolet and mass spectrometric detection, *J. Chromatogr. A*, 849, 341, 1999.

383. Ryan, D. et al., Liquid chromatography with electrospray ionisation mass spectrometric detection of phenolic compounds from *Olea europaea*, *J. Chromatogr. A*, 855, 529, 1999.

384. Olivares, J.A., et al. On-line mass spectrometric detection for capillary zone electrophoresis, *Anal. Chem.*, 59, 1230, 1987.

385. Luedtke, S. and Unger, K.K., Capillary electrochromatography — challenges and opportunities for coupling with mass spectrometry, *Chimia*, 53, 498, 1999.

386. Aramendia, M.A. et al., Determination of isoflavones by capillary electrophoresis/electrospray ionization mass spectrometry, *J. Mass Spectrom.*, S153, 1995.

387. Huck, C.W. et al., Analysis of three flavonoids by CE–UV and CE–ESI-MS. Determination of naringenin from a phytomedicine, *J. Sep. Sci.*, 25, 904, 2002.

388. Hakkinen, S. and Auriola, S., High-performance liquid chromatography with electrospray ionization mass spectrometry and diode array ultraviolet detection in the identification of flavonol aglycones and glycosides in berries, *J. Chromatogr. A*, 829, 91, 1998.

389. Careri, M., Elviri, L., and Mangia, A., Validation of a liquid chromatography ion spray mass spectrometry method for the analysis of flavanones, flavones and flavonols, *Rapid Commun. Mass Spectrom.*, 13, 2399, 1999.

390. Ma, Y.-L. et al., Mass spectrometric methods for the characterization and differentiation of isomeric *O*-diglycosyl flavonoids, *Phytochem. Anal.*, 12, 159, 2001.

391. Satterfield, M., Black, D.M., and Brodbelt, J.S., Detection of the isoflavone aglycones genistein and daidzein in urine using solid-phase microextraction-high-performance liquid chromatography–electrospray ionization mass spectrometry, *J. Chromatogr. B*, 759, 33, 2001.

392. Bylka, W., Franski, R., and Stobiecki, M., Differentiation between isomeric acacetin-6-*C*-(6″-*O*-malonyl)glucoside and acacetin-8-*C*-(6″-*O*-malonyl)glucoside by using low-energy CID mass spectra, *J. Mass Spectrom.*, 37, 648, 2002.

393. Justesen, U., Collision-induced fragmentation of deprotonated methoxylated flavonoids, obtained by electrospray ionization mass spectrometry, *J. Mass Spectrom.*, 36, 169, 2001.

394. Hvattum, E. and Ekeberg, D., Study of the collision-induced radical cleavage of flavonoid glycosides using negative electrospray ionization tandem quadrupole mass spectrometry, *J. Mass Spectrom.*, 38, 43, 2003.

395. Zhang, J. and Brodbelt, J.S., Structural characterization and isomer differentiation of chalcones by electrospray ionization tandem mass spectrometry, *J. Mass Spectrom.*, 38, 555, 2003.

396. Zhang, J. et al., Structural characterization and improved detection of kale flavonoids by electrospray ionization mass spectrometry, *Anal. Chem.*, 75, 6401, 2003.

397. Franski, R. et al., Differentiation of interglycosidic linkages in permethylated flavonoids from linked-scan mass spectra (B/E), *J. Agric. Food Chem.*, 50, 976, 2002.

398. Ma, Y.L. et al., Internal glucose residue loss in protonated *O*-diglycosyl flavonoids upon low-energy collision-induced dissociation, *J. Am. Soc. Mass Spectrom.*, 11, 136, 2000.

399. Deng, H. and Van Berkel, G.J., Electrospray mass spectrometry and UV/visible spectrophotometry studies of aluminium(III)–flavonoid complexes, *J. Mass Spectrom.*, 33, 1080, 1998.

400. Satterfield, M. and Brodbelt, J.S., Structural characterization of flavonoid glycosides by collisionally activated dissociation of metal complexes, *J. Am. Chem. Soc. Mass Spectrom.*, 12, 537, 2001.

401. Pikulski, M. and Brodbelt, J.S., Differentiation of flavonoid glycoside isomers by using metal complexation and electrospray ionization mass spectrometry, *J. Am. Chem. Soc. Mass Spectrom.*, 14, 1437, 2003.

402. Davis, B.D. and Brodbelt, J.S., Determination of the glycosylation site of flavonoid monoglucosides by metal complexation and tandem mass spectrometry, *J. Am. Chem. Soc. Mass Spectrom.*, 15, 1287, 2004.

403. Cuyckens F. et al., The application of liquid chromatography–electrospray ionization mass spectrometry and collision-induced dissociation in the structural characterization of acylated flavonol *O*-glycosides from the seeds of *Carrichtera annua*, *Eur. J. Mass Spectrom.*, 9, 409, 2003.

404. Sugui, J.A. et al., Matrix-assisted laser desorption ionization mass spectrometry analysis of grape anthocyanins, *Am. J. Enol. Vitic.*, 50, 199, 1999.

405. Nørbæk, R. et al., Flavone *C*-glycoside, phenolic acid, and nitrogen contents in leaves of barley subject to organic fertilization treatments, *J. Agric. Food Chem.*, 51, 809, 2003.

406. De Oliveira, D.B. et al., First isolation of a symmetrical glycosylated methylene bisflavonoid, *Planta Med.*, 69, 382, 2003.

407. Hemingway, R.W. and Karchesy, J.J., Chemistry and significance of condensed tannins, in *Proceedings of the First North American Tannin Conference*, Plenum Press, New York, 553, 1989.

408. Miketova, P. et al., Tandem mass spectrometry studies of green tea catechins. Identification of three minor components in the polyphenolic extract of green tea, *J. Mass Spectrom.*, 35, 860, 2000.

409. Sun, W. and Miller, J.M., Tandem mass spectrometry of the B-type procyanidins in wine and B-type dehydrodicatechins in an autoxidation mixture of (+)-catechin and (−)-epicatechin, *J. Mass Spectrom.*, 38, 438, 2003.

410. Gu, L. et al., Liquid chromatographic/electrospray ionization mass spectrometric studies of proanthocyanidins in foods, *J. Mass Spectrom.*, 38, 1272, 2003.

411. Qa'dan, F., Petereit, F., and Nahrstedt, A., Prodelphinidin trimers and characterization of a proanthocyanidin oligomer from *Cistus albidus*, *Pharmazie*, 58, 416, 2003.

412. Chen, X. et al., Quantitation of the flavonoid wogonin and its major metabolite wogonin-7β-D-glucuronide in rat plasma by liquid chromatography–tandem mass spectrometry, *J. Chromatogr. B*, 775, 169, 2002.

413. Mirabella, F.M., Ed., *Modern Techniques in Applied Molecular Spectroscopy*, John Wiley, New York, 1998.

414. Briggs, L. and Colebrook, L., Infrared spectra of flavanones and flavones. Carbonyl and hydroxyl stretching and CH out-of-plane bending absorption, *Spectrochim. Acta*, 18, 939, 1962.

415. Cornard, J.-P., Barbillat, J., and Merlin, J.-C., Performance of a versatile multichannel microprobe in Raman, fluorescence, and absorption measurements, *Microbeam Anal.*, 3, 13, 1994.

416. Edelmann, A. and Lendl, B., Toward the optical tongue: flow-through sensing of tannin–protein interactions based on FTIR spectroscopy, *J. Am. Chem. Soc.*, 124, 14741, 2002.
417. Vrielynck, L. et al., Semi-empirical and vibrational studies of flavone and some deuterated analogs, *Spectrochim. Acta A*, 50, 2177, 1994.
418. Cornard, J.-P., Vrielynck, L., and Merlin, J.-C., Structural and vibrational study of 3-hydroxy-flavone and 3-methoxyflavone, *Spectrochim. Acta A*, 51, 913, 1995.
419. Cornard, J.-P. et al., Structural study of quercetin by vibrational and electronic spectroscopies combined with semiempirical calculations, *Biospectroscopy*, 3, 183, 1997.
420. Cornard, J.-P. and Merlin, J.-C., Structural and spectroscopic investigation of 2′-methoxyflavone, *Asian J. Spectrosc.*, 3, 97, 1999.
421. Boudet, A.-C., Cornard, J.-P., and Merlin, J.-C., Conformational and spectroscopic investigation of 3-hydroxyflavone-aluminium chelates, *Spectrochim. Acta A*, 56, 829, 2000.
422. Cornard, J.-P. and Merlin, J.-C., Structural and spectroscopic investigation of 5-hydroxyflavone and its complex with aluminium, *J. Mol. Struct.*, 569, 129, 2001.
423. Cornard, J.-P., Boudet, A.-C., and Merlin, J.-C., Complexes of Al(III) with 3′,4′-dihydroxy-flavone: characterization, theoretical and spectroscopic study, *Spectrochim. Acta A*, 57, 591, 2001.
424. Cornard, J.-P. and Merlin, J.-C., Spectroscopic and structural study of complexes of quercetin with Al(III), *J. Inorg. Biochem.*, 92, 19, 2002.
425. Cornard, J.-P. and Merlin, J.-C., Complexes of aluminium(III) with isoquercitrin: spectroscopic characterization and quantum chemical calculations, *Polyhedron*, 21, 2801, 2002.
426. Cornard, J.-P. and Merlin, J.-C., Comparison of the chelating power of hydroxyflavones, *J. Mol. Struct.*, 651, 381, 2003.
427. Ramos-Tejada, M.M. et al., Investigation of alumina/(+)-catechin system properties. Part I: a study of the system by FTIR–UV–Vis spectroscopy, *Colloids Surf. B*, 24, 297, 2002.
428. Nagy, L. et al., Preparation and structural studies on organotin(IV) complexes with flavonoids, *J. Radioanal. Nucl. Chem.*, 227, 89, 1998.
429. Petroski, J.M. et al., FTIR spectroscopy of flavonols in argon and methanol/argon matrixes at 10 K. Reexamination of the carbonyl stretch frequency of 3-hydroxyflavone, *J. Phys. Chem. A*, 106, 11714, 2002.
430. Heneczkowski, M. et al., Infrared spectrum analysis of some flavonoids, *Acta Pol. Pharm.*, 58, 415, 2001.
431. Merlin, J.-C. et al., Resonance Raman-spectroscopic studies of anthocyanins and anthocyanidins in aqueous-solutions, *Phytochemistry*, 35, 227, 1994.
432. Quagliano, L.G. et al., Raman study of anthocyanins as a function of the temperature. In *Spectroscopy of Biological Molecules: New Directions*, 8th European Conference on the Spectroscopy of Biological Molecules, Enschede, Netherlands, 187, 1999.
433. Nuopponen, M. et al., A UV resonance Raman (UVRR) spectroscopic study on the extractable compounds in Scots pine (*Pinus sylvestris*) wood Part II. Hydrophilic compounds, *Spectrochim. Acta A*, 60, 2963, 2004.
434. Liu, Q., Sang, W., and Xu, X., Solvent effects on infrared spectra of 5-methyl-7-methoxy-iso-flavone in single solvent systems, *J. Mol. Struct.*, 608, 253, 2002.
435. Mendoza-Wilson, A.M. and Glossman-Mitnik, D., CHIH-DFT determination of the molecular structure, infrared and ultraviolet spectra of the flavonoid quercetin, *J. Mol. Struct. (Theochem.)*, 681, 71, 2004.
436. Edelmann, A. et al., Rapid method for the discrimination of red wine cultivars based on the mid-infrared spectroscopy of phenolic wine extracts, *J. Agric. Food Chem.*, 49, 1139, 2001.
437. Yamauchi, S., Shibutani, S., and Doi, S., Characteristic Raman bands for *Artocarpus heterophyllus* heartwood, *J. Wood Sci.*, 49, 466, 2003.
438. Vrielynck, L., Wallet, J.-C., and Merlin, J.-C., Vibrational and theoretical study of the 2′,6′-dimethoxyflavone *cis*-formic acid inclusion compound, *Spectrochim. Acta A*, 56, 2439, 2000.
439. Ficarra, R. et al., Study of flavonoids/b-cyclodextrins inclusion complexes by NMR, FT-IR, DSC, x-ray investigation, *J. Pharm. Biomed. Appl.*, 29, 1005, 2002.

440. Tian, J. et al., Probing the binding of scutellarin to human serum albumin by circular dichroism, fluorescence spectroscopy, FTIR, and molecular modeling method, *Biomacromolecules*, 5, 1956, 2004.
441. Noda, I., Generalized 2-dimensional correlation method applicable to infrared, Raman and other types of spectroscopy, *Appl. Spectrosc.*, 47, 1329, 1993.
442. Zuo, L. et al., 2D-IR correlation analysis of deteriorative process of traditional Chinese medicine "Quing Kai Ling" injection, *J. Pharm. Biomed. Appl.*, 30, 1491, 2003.
443. Horvath, E. et al., Correlation between retention behavior and GC–FTIR data in the study of flavonoids, *Talanta*, 42, 979, 1995.
444. Hadj-Mahammed, M., Badjah-Hadj-Ahmed, Y., and Meklati, B.Y., Behavior of polymethoxy-lated and polyhydroxylated flavones by carbon dioxide supercritical fluid chromatography with flame ionization and Fourier transform infrared detectors, *Phytochem. Anal.*, 4, 275, 1993.
445. Robb, C.S. et al., Analysis of green tea constituents by HPLC–FT-IR, *J. Liq. Chromatogr. Relat. Technol.*, 25, 787, 2002.
446. Huck, C. et al., Quantitative Fourier transform near infrared reflectance spectroscopy (NIRS) compared to high performance liquid chromatography (HPLC) of a flavone in *Primulae veris* Flos extracts, *Pharm. Pharmacol. Lett.*, 9, 26, 1999.
447. Whitacre, E. et al., Predictive analysis of cocoa procyanidins using near-infrared spectroscopy techniques, *J. Food Sci.*, 68, 2618, 2003.
448. Kitsuda, K. et al., Estimation of Nasunin content in the skin of eggplant "Mizunasu" fruits by nondestructive and rapid method, *J. Jpn. Soc. Food Sci.*, 50, 261, 2003.
449. Schulz, H., Joubert, E., and Schutze, W., Quantification of quality parameters for reliable evalu-ation of green rooibos (*Aspalathus linearis*), *Eur. Food Res. Technol.*, 216, 539, 2003.
450. Smith, K.F. and Kelman, W.M., Predicting condensed tannin concentrations in *Lotus uliginosus* Schkuhr using near-infrared reflectance spectroscopy, *J. Sci. Food Agric.*, 75, 263, 1997.
451. Goodchild, A.V., El Haramein, F.J., and Abd El Moneim, A., Prediction of phenolics and tannins in forage legumes by near infrared reflectance, *J. Near Infrared Spectrosc.*, 6, 175, 1998.
452. Schulz, H. et al., Application of near-infrared reflectance spectroscopy to the simultaneous prediction of alkaloids and phenolic substances in green tea leaves, *J. Agric. Food Chem.*, 47, 5064, 1999.
453. Rutherford, R.S., Prediction of resistance in sugarcane to stalk borer *Eldana saccharina* by near-infrared spectroscopy on crude budscale extracts: involvement of chlorogenates and flavonoids, *J. Chem. Ecol.*, 24, 1447, 1998.
454. Merzlyak, M.N. et al., Application of reflectance spectroscopy for analysis of higher plant pigments, *Russ. J. Plant Physiol.*, 50, 704, 2003.
455. Rager, I. et al., Rapid quantification of constituents in St. John's wort extracts by NIR spectros-copy, *J. Pharm. Biomed. Appl.*, 28, 439, 2002.
456. Hu, G. et al., Simultaneous determination of total flavones and total lactones in ginkgo extracts by near infrared spectroscopy, *Fenxi Huaxue*, 32, 1061, 2004.
457. Barbillat, J. and Da Silva, E., Near infrared Raman spectroscopy with dispersive instruments and multichannel detection, *Spectrochim. Acta A*, 53, 2411, 1997.
458. Harborne, J.B., *Comparative Biochemistry of the Flavonoids*, Academic Press, New York, 1967.
459. Merken, H.M. and Beecher, G.R., Measurement of food flavonoids by high-performance liquid chromatography: a review, *J. Agric. Food Chem.*, 48, 577, 2000.
460. Giusti, M.M. and Wrolstad R.E., Characterization and measurement of anthocyanins by UV–visible spectroscopy, in *Current Protocols in Food Analytical Chemistry*, Wrolstad, R.E. and Schwartz, R.E., Eds., John Wiley, New York, 2001.
461. Santos-Buelga, C., Garcia-Viguera, C., and Tomas-Barberan, F.A., On-line identification of flavonoids by HPLC coupled to diode array detection, in *Methods in Polyphenol Analysis*, Santos-Buelga, C. and Williamson, G., Eds., Royal Society of Chemistry, Cambridge, Chap. 5, 2003.

462. Arts, I.C.W. and Hollman, P.C.H., Optimization of a quantitative method for the determination of catechins in fruits and legumes, *J. Agric. Food Chem.*, 46, 5156, 1998.

463. Hostettmann, K. et al., On-line high-performance liquid chromatography: ultraviolet–visible spectroscopy of phenolic compounds in plant extracts using post-column derivatization, *J. Chromatogr.*, 283, 137, 1984.

464. Wolfender, J.-L. and Hostettmann, K., Liquid chromatographic–UV detection and liquid chromatographic–thermospray mass spectrometric analysis of *Chironia* (Gentianaceae) species. A rapid method for the screening of polyphenols in crude plant extracts, *J. Chromatogr. A*, 647, 191, 1993.

465. De Pascual-Teresa, S. et al., Analysis of flavanols in beverages by high-performance liquid chromatography with chemical reaction detection, *J. Agric. Food Chem.*, 46, 4209, 1998.

466. Harborne, J.B., Spectral methods of characterizing anthocyanins, *Biochem. J.*, 70, 22, 1958.

467. Yoshitama, K. and Abe, K., Chromatographic and spectral characterization of 3′-glycosylation in anthocyanidins, *Phytochemistry*, 16, 591, 1977.

468. Bloor, S., Deep blue anthocyanins from blue *Dianella* berries, *Phytochemistry*, 58, 923.

469. Torskangerpoll, K. et al., Color and substitution pattern in anthocyanidins. A combined quantum chemical–chemometrical study, *Spectrochim. Acta A*, 55, 761, 1999.

470. Saito, N. et al., The first isolation of *C*-glycosylanthocyanin from the flowers of *Tricyrtis formosana*, *Tetrahedron Lett.*, 44, 6821, 2003.

471. Brouillard, R. and Dangles, O., Flavonoids and flower colour, in *The Flavonoids, Advances in Research Since 1986*, Harborne, J. B., Ed., Chapman & Hall, London, Chap. 13, 1994.

472. Yoshida, K. et al., Contribution of each caffeoyl residue of the pigment molecule of gentiodelphin to blue color development, *Phytochemistry*, 54, 85, 2000.

473. Honda, T. and Saito, N., Recent progress in the chemistry of polyacylated anthocyanins as flower color pigments, *Heterocycles*, 56, 633, 2002.

474. Yoshida, K. et al., Influence of *E,Z*-isomerization and stability of acylated anthocyanins under the UV irradiation, *Biochem. Eng. J.*, 14, 163, 2003.

475. Oszmianski, J., Bakowska, A., and Piacente, S., Thermodynamic characteristics of copigmentation reaction of acylated anthocyanin isolated from blue flowers of *Scutellaria baicalensis* Georgi with copigments, *J. Sci. Food Agric.*, 84, 1500, 2004.

476. Dangles, O., Saito, N., and Brouillard, R., Kinetic and thermodynamic control of flavylium hydration in the pelargonidin–cinnamic acid complexation. Origin of the extraordinary flower color diversity of *Pharbitis nil*, *J. Am. Chem. Soc.*, 115, 3125, 1993.

477. Toki, K. et al., (Delphinidin 3-gentiobiosyl) (apigenin 7-glucosyl) malonate from the flowers of *Eichhornia crassipes*, *Phytochemistry*, 36, 1181, 1994.

478. Figueiredo, P. et al., New aspects of anthocyanin complexation. Intramolecular copigmentation as a means for colour loss? *Phytochemistry*, 41, 301, 1996.

479. Toki, K. et al., (Delphinidin 3-gentiobiosyl) (luteorin 7-glucosyl) malonate from the flowers of *Eichhornia crassipes*, *Heterocycles*, 63, 899, 2004.

480. Pina, F. Thermodynamics and kinetics of flavylium salts — Malvin revisited, *J. Chem. Soc., Faraday Trans.*, 94, 2109–2116, 1998.

481. Levi, M.A.B. et al., Three-way chemometric method study and UV–Vis absorbance for the study of simultaneous degradation of anthocyanins in flowers of the *Hibiscus rosa-sinensys* species, *Talanta*, 62, 299, 2004.

482. Dougall, D.K. et al., Studies on the stability and conformation of monoacylated anthocyanins. 2. Anthocyanins from wild carrot suspension cultures acylated with supplied carboxylic acids, *Carbohydr. Res.*, 310, 177, 1998.

483. Delgado-Vargas, F., Jimenez, A.R., and Paredes-Lopez, O., Natural pigments: carotenoids, anthocyanins, and betalains — characteristics, biosynthesis, processing, and stability, *CRC Crit. Rev. Food Sci.*, 40, 173, 2000.

484. Eiro, M.J. and Heinonen, M., Anthocyanin color behavior and stability during storage: effect of intermolecular copigmentation, *J. Agric. Food Chem.*, 50, 7461, 2002.

485. Giusti, M.M. and Wrolstad, R.E., Acylated anthocyanins from edible sources and their applications in food systems, *Biochem. Eng. J.*, 14, 217, 2003.

486. Fossen, T., Cabrita, L., and Andersen, Ø.M., Colour and stability of pure anthocyanins influenced by pH including the alkaline region, *Food Chem.*, 63, 435, 1998.

487. Cabrita, L., Fossen, T., and Andersen, Ø.M., Colour and stability of the six common anthocyanidin 3-glucosides in aqueous solutions, *Food Chem.*, 68, 101, 2000.

488. Giusti, M.M., Rodríguez-Saona, L.E., and Wrolstad, R.E., Molar absorptivity and color characteristics of acylated and non-acylated pelargonidin-based anthocyanins, *J. Agric. Food Chem.*, 47, 4631, 1999.

489. Naseem, B. et al., Interaction of flavonoids within organized molecular assemblies of anionic surfactant, *Colloids Surf. B*, 35, 7, 2004.

490. Solimani, R., The flavonols quercetin, rutin and morin in DNA solution: UV–vis dichroic (and mid-infrared) analysis explain the possible association between the biopolymer and a nucleophilic vegetable-dye, *Biochim. Biophys. Acta,* 1336, 281, 1997.

491. Markham, K.R. et al., Anthocyanic vacuolar inclusions — their nature and significance in flower colouration, *Phytochemistry,* 55, 327, 2000.

492. Gonnet, J.-F., Origin of the color of cv. Rhapsody in Blue Rose and some other so-called "blue" roses, *J. Agric. Food Chem.*, 51, 4990, 2003.

493. CIE, Commision Internationale de l'Eclairage, *Colorimetry*, 2nd edition, pub. 15.2, 1986.

494. Hunter, R.S. and Harold, R.W., *The Measurement of Appearance*, John Wiley, New York, Chap. 7, 1987.

495. Gonnet, J.-F., Colour effects of co-pigmentation of anthocyanins revisited — 1. A colorimetric definition using the CIELAB scale, *Food Chem.*, 63, 409, 1998.

496. Gonnet, J.-F., Colour effects of co-pigmentation of anthocyanins revisited — 2. A colorimetric look at the solutions of cyanin co-pigmented by rutin using the CIELAB scale, *Food Chem.*, 66, 387, 1999.

497. Gonnet, J.-F., Colour effects of co-pigmentation of anthocyanin revisited — 3. A further description using CIELAB differences and assessment of matched colours using the CMC model, *Food Chem.*, 75, 473, 2001.

498. Heredia, F.J. et al., Chromatic characterization of anthocyanins from red grapes — I. pH effect, *Food Chem.*, 63, 491, 1998.

499. Stintzing, F. et al., Color and antioxidant properties of cyanidin-based anthocyanin pigments, *J. Agric. Food Chem.*, 50, 6172, 2002.

500. Bakker, J. and Timberlake, C.F., Isolation, identification and characterization of new colour-stable anthocyanins occurring in some red wines, *J. Agric. Food Chem.*, 45, 35, 1997.

501. Schwartz, M. and Winterhalter, P., A novel synthetic route to substituted pyranoanthocyanins with unique colour properties, *Tetrahedron Lett.*, 44, 7583, 2003.

502. Bakker, J., Picinelli, A., and Bridle, P., Modell wine solutions: colour and composition changes during ageing, *Vitis*, 32, 111, 1993.

503. Heredia, F.J. and Guzmanchozas, M., The colour of wine — a historical-perspective. 2. Trichromatic methods, *J. Food Quality*, 16, 439, 1993.

504. Picinelli, A., Bakker, J., and Bridle, P., Modell wine solutions: effect of sulphur dioxide on colour and composition during ageing, *Vitis*, 33, 31, 1994.

505. Almela, L. et al., Comparison between the tristimulus measurements *Yxy* and *L*a*b** to evaluate the colour of young red wines, *Food Chem.*, 53, 321, 1995.

506. Almela, L. et al., Varietal classification of young red wines in terms of chemical and colour parameters, *J. Sci. Food Agric.*, 70, 173, 1996.

507. Bakker, J. et al., Effect of sulphur dioxide and must extraction on colour, phenolic composition and sensory quality of red table wine, *J. Sci. Food Agric.*, 78, 297, 1998.

508. Giusti, M.M. and Wrolstad, R.E., Radish anthocyanins as a natural red colorant for maraschino cherries, *J. Food Sci.*, 61, 688, 1996.

509. Malien-Aubert, C., Dangles, O., and Amiot, M.J., Color stability of commercial anthocyanin-based extracts in relation to the phenolic composition. Protective effects by intra- and intermolecular copigmentation, *J. Agric. Food Chem.*, 49, 170, 2001.

510. Pazmiño-Durán et al., Anthocyanins from *Oxalis triangularis* as potential food colorants, *Food Chem.*, 75, 211, 2001.

511. Yoshida, Y., Koyama, N., and Tamura, H., Color and anthocyanin composition of strawberry fruit: changes during fruit development and differences among cultivars, with special reference to the occurrence of pelargonidin 3-malonylglucoside, *J. Jpn. Soc. Hortic. Sci.*, 71, 355, 2002.

512. Tsukui, A. et al., Effect of alcoholic fermentation on the stability of purple sweet potato anthocyanins, *Food Sci. Technol. Res.*, 8, 4, 2002.

513. Choi, M.H., Kim, G.H., and Lee, H.S., Effects of ascorbic acid retention on juice color and pigment stability in blood orange (*Citrus sinensis*) juice during refrigerated storage, *Food Res. Int.*, 35, 753, 2002.

514. Cevallos-Casals, B.A. and Cisneros-Zevallos, L., Stability of anthocyanin-based aqueous extracts of Andean purple corn and red-fleshed sweet potato compared to synthetic and natural colorants, *Food Chem.*, 86, 69, 2004.

515. Rein, M.J. and Heinonen, M., Stability and enhancement of berry juice color, *J. Agric. Food Chem.*, 52, 3106, 2004.

516. Kammerer, D. and Carle, R., Quantification of anthocyanins in black carrot extracts (*Daucus carota* ssp. *Sativus* var. *atrirubens* Alef.) and evaluation of their colour properties, *Eur. Food Res. Technol.*, 219, 479, 2004.

517. Fujioka, M. et al., Anthocyanidin composition of petals in *Pelargonium* × *domesticum* Bailey, *J. Jpn. Soc. Hortic. Sci.*, 59, 823, 1991.

518. Biolley, J.P. and Jay, M., Anthocyanins in modern roses — chemical and colorimetric features in relation to the color range, *J. Exp. Bot.*, 44, 1725, 1993.

519. Tourjee, K.R., Harding, J., and Byrne, T.G., Colorimetric analysis of *Gerbera* flowers, *Hortscience*, 28, 735, 1993.

520. Toki, K. and Katsuyama, N., Pigments and color variation in flowers of *Lagerstroemia indica*, *J. Jpn. Soc. Hortic. Sci.*, 63, 853, 1995.

521. Uddin, A.F.M.J. et al., Seasonal variation in pigmentation and anthocyanidin phenetics in commercial *Eustoma* flowers, *Sci. Hortic. — Amsterdam*, 100, 103, 2004.

522. Singha, S. et al., Anthocyanin distribution in delicious apples and the relationship between anthocyanin concentration and chromaticity values, *J. Am. Soc. Hort. Sci.*, 116, 497, 1991.

523. Gonnet, J.-F. and Hieu, H., *In situ* micro-spectrophotometric and micro-spectrocolorimetric investigation of vascular pigments in flowers and cultivars of carnation (*Dianthus caryophyllus*), *J. Hort. Sci.*, 67, 663, 1992.

524. Dussi, M.C., Sugar, D., and Wrolstad, R.E., Characterizing and quantifying anthocyanins in red pears and the effect of light quality on fruit color, *J. Am. Soc. Hort. Sci.*, 120, 785, 1995.

525. Barth, M.M. et al., Ozone storage effects on anthocyanin content and fungal growth in blackberries, *J. Food Sci.*, 60, 1286, 1995.

526. Riaz, M.N. and Bushway, A.A., Effect of cultivars and weather on Hunter "L", hue angle and chroma values of red raspberry grown in Maine, *Fruit Var. J.*, 50, 131, 1996.

527. Lancaster, J.E. et al., Influence of pigment composition on skin color in a wide range of fruit and vegetables, *J. Am. Soc. Hort. Sci.*, 122, 594, 1997.

528. Fernández-Lopéz et al., Dependence between colour and individual anthocyanin content in ripening grapes, *Food Res. Int.*, 31, 667, 1999.

529. Iglesias, I. et al., Differences in fruit color development, anthocyanin content, yield and quality of seven "delicious" apple strains, *Fruit Var. J.*, 53, 133, 1999.

530. Nunes, M.C.N. et al., Fruit maturity and storage temperature influence response of strawberries to controlled atmospheres, *J. Am. Soc. Hort. Sci.*, 127, 836, 2002.

531. Shishehgarha, F., Makhlouf, J., and Ratti, C., Freeze-drying characteristics of strawberries, *Dry. Technol.*, 20, 131, 2002.

532. Mozetic, B. et al., Changes of anthocyanins and hydroxycinnamic acids affecting the skin colour during maturation of sweet cherries (*Prunus avium* L.), *Lebensm.-Wiss. u.-Technol.*, 37, 123, 2004.

533. Skupien, K. and Oszmianski, J., Comparison of six cultivars of strawberries (*Fragaria* x *ananassa* Duch.) grown in northwest Poland, *Eur. Food Res. Technol.*, 219, 66, 2004.

534. Gonnet, J.-F. and Fenet, B., "Cyclamen Red" colors based on macrocyclic anthocyanin in carnation flowers, *J. Agric. Food Chem.*, 48, 22, 2000.

535. Lèvai, A., Utilization of the chiroptical spectroscopies for the structure elucidation of flavonoids and related benzopyran derivatives, *Acta Chim. Slov.*, 45, 267, 1998.

536. Slade, D., Ferreira, D., and Marais, J.P., Circular dichroism, a powerful tool for the assessment of absolute configuration of flavonoids, in *Abstracts of Papers, 228th ACS National Meeting*, Philadelphia, 2004.

537. Zhang, F.J., Lin, G.Q., and Huang, Q.C., Synthesis, resolution and absolute configuration of optically pure 5,5″-dihydroxy-4′,4‴,7,7″-tetramethoxy–8,8″-biflavone and its derivatives, *J. Org. Chem.*, 60, 6427, 1995.

538. Kojima, K. et al., Flavanones from *Iris tenuifolia*, *Phytochemistry*, 44, 711, 1997.

539. Van Rensburg, H. et al., Circular dichroic properties of flavan-3-ols, *J. Chem. Res. Synop.* (7), 450, 1999.

540. Tseng, M.-H. et al., Allelophatic prenylflavanones from the fallen leaves of *Macaranga tanarius*, *J. Nat. Prod.*, 64, 827, 2001.

541. Friedrich, W. and Galensa, R., Identification of a new flavanol glucoside from barley (*Hordeum vulgare* L.) and malt, *Eur. Food Res. Technol.*, 214, 388, 2002.

542. Kiss, L. et al., Chiroptical properties and synthesis of enantiopure *cis* and *trans* pterocarpan skeleton, *Chirality*, 15, 558, 2003.

543. Lou, H. et al., Polyphenols from peanut skins and their free radical-scavenging effects, *Phytochemistry*, 65, 2391, 2004.

544. Mayer, R., Five biflavonoids from *Calycopteris floribunda* (Combretaceae), *Phytochemistry*, 65, 593, 2004.

545. Ferreira, D. et al., Circular dichroic properties of flavan-3,4-diols, *J. Nat. Prod.*, 67, 174, 2004.

546. Antus, S. et al., Chiroptical properties of 2,3-dihydrobenzo[*b*]furan and chromane chromophores in naturally occurring *O*-heterocycles, *Chirality*, 13, 493, 2001.

547. Caccamese, S., Manna, L., and Scivoli, G., Chiral HPLC separation and CD spectra of the C-2 diastereomers of naringin in grapefruit during maturation, *Chirality*, 15, 661, 2003.

548. Ferrari, J. et al., Isolation and on-line LC/CD analysis of 3,8″-linked biflavonoids from *Gnidia involucrata*, *Helv. Chim. Acta*, 86, 2768, 2003.

549. Bloemendal, M. and van Grondelle, R., Linear-dichroism spectroscopy for the study of structural properties of proteins, *Mol. Biol. Rep.*, 18, 49, 1993.

550. Goto, T., Tamura, H., and Kondo, T., Chiral stacking of cyanin and pelargonin — soluble and insoluble aggregates as determined by means of circular dichroism, *Tetrahedron Lett.*, 28, 5907, 1987.

551. Hoshino, T., Self-association of flavylium cations of anthocyanidin 3,5-diglucosides studied by circular dichroism and ^{1}H NMR, *Phytochemistry*, 31, 647, 1992.

552. Miyake, K. et al., Improvement of solubility and oral bioavailability of rutin by complexation with 2-hydroxypropyl-beta-cyclodextrin, *Pharm. Dev. Technol.*, 5, 399, 2000.

553. Solimani, R. et al., Flavonoid–DNA interaction studied with flow linear dichroism technique, *J. Agric. Food Chem.*, 43, 876, 1995.

554. Solimani, R., Quercetin and DNA in solution: analysis of the dynamics of their interaction with a linear dichroism study, *Int. J. Biol. Macromol.*, 18, 287, 1996.

555. Tseng, M.-H. et al., Allelophatic prenylflavanones from the fallen leaves of *Macaranga tanarius*, *J. Nat. Prod.*, 64, 827, 2001.

556. Zhu, Z., Li, C., and Li, N.-Q., Electrochemical studies of quercetin interacting with DNA, *Microchem. J.*, 71, 57, 2002.

557. Zsila, F., Bikádi, Z., and Simonyi, M., Probing the binding of the flavonoid, quercetin to human serum albumin by circular dichroism, electronic absorption spectroscopy and molecular modelling methods, *Biochem. Pharmacol.*, 65, 447, 2003.

558. Hufford, C.D. et al., Antimicrobial compounds from *Petalostemum purpureum*, *J. Nat. Prod.*, 56, 1878, 1993.

559. Selivanova, I.A. et al., Study of the crystalline structure of dihydroquercetin, *Pharm. Chem. J. — USSR*, 33, 222, 1999.

560. Yang, S.-W. et al., Munchiwarin, a prenylated chalcone from *Crotalaria trifoliastrum*, *J. Nat. Prod.*, 61, 1274, 1998.

561. Paula, V.F. et al., Chemical constituents from *Bombacopsis glabra* (Pasq.) A. Robyns: complete 1H and 13C NMR assignments and X ray structure of 5-hydroxy-3,6,7,8,4′-pentamethoxyflavone, *J. Brazil Chem. Soc.*, 13, 276, 2002.

562. Wu, J.H. et al., Desmosdumotin B: a new special flavone from *Desmos dumosus*, *Chinese Chem. Lett.*, 12, 49, 2001.

563. Devia, B. et al., New 3-deoxyanthocyanidins from leaves of *Arrabidae chica*, *Phytochem. Anal.*, 13, 114, 2002.

564. Konda, Y. et al., Conformational analysis of C3′–C8 connected biflavones, *J. Heterocyclic Chem.*, 32, 1531, 1995.

565. Weinges, K., Schick, H., and Rominger, F., x-Ray structure analysis of procyanidin B1, *Tetrahedron*, 57, 2327, 2001.

566. Jiang, R.-W. et al., Molecular structures and π–π interactions of some flavonoids and biflavonoids, *J. Mol. Struct.*, 642, 77, 2002.

567. Jiang, R.-W. et al., A novel 1:1 complex of potassium mikanin-3-*O*-sulfate with methanol, *Chem. Pharm. Bull.*, 49, 1166, 2001.

568. Zhou, Y. et al., A new flavonol glycoside from *Anaphalis sinica* Hance, *Chem. Res. Chin. Univ.*, 17, 48, 2001.

569. Bringmann, G. et al., 6-Hydroxyluteolin-7-*O*-(1″-α-rhamnoside) from *Vriesea sanguinolenta* Cogn. and Marchal (Bromeliaceae), *Phytochemistry*, 53, 965, 2000.

570. Yin, H.-B. and He, Z.-S., A novel semi-quinone chalcone sharing a pyrrole ring *C*-glycoside from *Carthamus tinctorius*, *Tetrahedron Lett.*, 41, 1955, 2000.

571. Borbulevych, O.Y. et al., Lipoxygenase interactions with natural flavonoid, quercetin, reveal a complex with protocatechuic acid in its x-ray structure at 2.1 A resolution, *Proteins*, 54, 13, 2004.

572. Uchida, T. et al., Structural study on chalcone derivatives, *Synth. Met.*, 71, 1705, 1995.

573. Rurack, K. et al., Chalcone-analogue dyes emitting in the near-infrared (NIR): influence of donor–acceptor substitution and cation complexation on their spectroscopic properties and x-ray structure, *J. Phys. Chem. A*, 104, 3087, 2000.

574. Artali, R. et al., Synthesis, x-ray crystal structure and biological properties of acetylenic flavone derivatives, *Farmaco*, 58, 875, 2003.

575. Cotelle, N. et al., Synthesis, x-ray structure and spectroscopic and electronic properties of two new synthesized flavones, *J. Phys. Org. Chem.*, 17, 226, 2004.

576. Roque, A. et al., Photochromic properties of 3-methyl-substituted flavylium salts, *Eur. J. Org. Chem.*, 16, 2699, 2002.

577. Chen, X.-L. et al., Direct observation of a series of non-covalent complexes formed by phosphorylated flavonoid–protein interactions through electrospray ionization tandem mass spectroscopy, *Anal. Chim. Acta*, 511, 175, 2004.

578. Kondo, T. et al., Structural basis of blue-color development in flower petals from *Commelina communis*, *Nature*, 358, 515, 1992.

579. Ohsawa, Y. et al., Flavocommelin octaacetate, *Acta Crystallogr. C*, 50, 645, 1994.

580. Malpezzi, L. et al., Crystal architecture and conformational properties of the inclusion complex, neohesperidin dihydrochalcone–cyclomaltoheptaose (β-cyclodextrin), by x-ray diffraction, *Carbohydr. Res.*, 339, 2117, 2004.

581. Pralhad, T. and Rajendrakumar, K., Study of freeze-dried quercetin–cyclodextrin binary systems by DSC, FT-IR, x-ray diffraction and SEM analysis, *J. Pharm. Biomed. Appl.*, 34, 333, 2004.

582. Fossen, T. and Andersen, Ø.M., Unpublished results, 2004.

583. Frøystein, N.Å., Fossen, T. and Andersen, Ø.M., Unpublished results, 2004.

3 Molecular Biology and Biotechnology of Flavonoid Biosynthesis

Kevin M. Davies and Kathy E. Schwinn

CONTENTS

3.1   THE SCOPE OF THE REVIEW

Flavonoid biosynthesis is probably the best characterized of all the secondary metabolic pathways. Therefore, this chapter cannot hope to cover all the available information in detail as well as address recent results. Since 1994, tremendous progress has been made in the identification and analysis of genes or cDNAs for flavonoid biosynthetic enzymes and regulatory factors, in the analysis of enzyme structure and function, and the targeted manipulation of flavonoid production in transgenic plants. Thus, while a general overview of the biosynthetic pathway is presented in this chapter, the focus is on these recent advances. For further information on the biochemistry and classical genetics of flavonoids, the reader is referred to previous reviews as appropriate. In particular, the biochemistry and genetics of flavonoids were covered in detail in two editions of "The Flavonoids,"[1–3] and these excellent chapters are used as general references for much of the earlier literature. Also, Bohm[4] provides an extensive review of flavonoid biosynthesis, including the biochemistry of enzyme steps for which no cDNA clones are yet available. There are also several recent reviews on individual families of enzymes, such as the acyltransferases, to which reference is given. Furthermore, much biochemical information is available on the Internet, in particular the IUBMB (http://www.chem.qmw.ac.uk/iubmb/enzyme) and BRENDA (http://www.brenda.uni-koeln.de) databases. The EC numbers listed in Table 3.1 can be used to locate appropriate entries. Previous reviews in "The Flavonoids" have featured listings of genetic mutants affecting flavonoid biosynthesis for the classical model species. Much recent progress has been made with *Arabidopsis thaliana* (thale cress), and a listing of defined loci affecting flavonoid biosynthesis in this species is included in this chapter.

The advent of functional genomics has made it possible to analyze how genes or enzymes fit as part of larger families. Within the flavonoid pathway, important biosynthetic gene families include various transferases (for acyl, glycosyl, or methyl groups), reductases, and those for oxidative reactions — membrane-bound cytochrome P450-dependent mono-oxygenases (P450s) and soluble nonheme dioxygenases. Furthermore, the regulatory factors are also typically members of large gene families. Rather than present molecular phylogenies or a discussion of flavonoid gene evolution here, the reader is referred to other publications. In particular, Tanner[5] presents recent molecular phylogenies for most of the flavonoid biosynthetic enzymes, and additional references are given for those gene families not included in that review.

3.2 OVERVIEW OF FLAVONOID BIOSYNTHESIS

The flavonoid pathway is part of the larger phenylpropanoid pathway, which produces a range of other secondary metabolites, such as phenolic acids, lignins, lignans, and stilbenes. The key flavonoid precursors are phenylalanine, obtained via the shikimate and arogenate pathways, and malonyl-CoA, derived from citrate produced by the TCA cycle. Most flavonoid biosynthetic enzymes characterized to date are thought to operate in enzyme complexes located in the cytosol. Flavonoid end products are transported to various subcellular or extracellular locations, with those flavonoids involved in pigmentation generally being transported into the vacuole.

There are many branches to the flavonoid biosynthetic pathways, with the best characterized being those leading to the colored anthocyanins and proanthocyanidins (PAs) and the generally colorless flavones, flavonols, and isoflavonoids. Genes or cDNAs have now been identified for all the core steps leading to anthocyanin, flavone, and flavonol formation, as well as many steps of the isoflavonoid branch, allowing extensive analysis of the encoded enzymes (Table 3.1). In addition, several DNA sequences are available for the modification enzymes that produce the variety of structures known within each class of compound.

Significant recent advances in our understanding of flavonoid biosynthesis include characterization of the formation of anthocyanidin 3-*O*-glucoside from leucoanthocyanidin, clarification of PA monomer formation, progress toward elucidating aurone and 3-deoxyflavonoid formation, the molecular characterization of several genes encoding enzymes that modify the flavonoid core structures, analysis of enzyme function, the determination of enzyme structures by x-ray crystallography and homology modeling, and the identification of several different groups of transcription factors regulating anthocyanin and PA biosynthesis. Data are also starting to emerge on the subcellular organization of the biosynthetic enzymes within the cytosol (and the role this may play in metabolic channeling), and transport mechanisms for the flavonoids within the cell. However, there are still major areas where data are lacking. Tertiary structures are available for only a few of the biosynthetic enzymes, little is known about the turnover or degradation of flavonoids, and details of post-transcriptional regulatory mechanisms are limited. Furthermore, the range of genes

TABLE 3.1
Flavonoid Biosynthetic Enzymes for which DNA Sequences Have Been Obtained

Enzyme	Abbreviation	EC number	Protein family	Ref.[1a]
Flavonoid precursors				
Acetyl-CoA carboxylase (cytosolic)	ACC	6.4.1.2	Biotin-containing carboxylases	8, 9
Phenylalanine ammonia-lyase	PAL	4.3.1.5	Ammonia-lyases	10
Cinnamate 4-hydroxylase	C4H	1.14.13.11	CytP450 (CYP73A)	15–17
4-Coumarate:CoA ligase	4CL	6.2.1.12	Adenylate-forming enzymes	10
Phenolic ester 3β-hydroxylase	CYP93A3	1.14.13.–	CytP450 (CYP93A)	41
The pathway to anthocyanins				
Chalcone synthase	CHS	2.3.1.74	Polyketide synthase	44, 45
Chalcone isomerase	CHI	5.5.1.6	No named family	361
Flavanone 3β-hydroxylase	F3H (FHT)	1.14.11.9	2-Oxoglutarate-dependent dioxygenase (2OGD)	64
Flavanone 4-reductase	FNR	1.1.1.234	NADPH reductase (RED)	69[b]
Dihydroflavonol 4-reductase	DFR	1.1.1.219	RED	68, 69
Anthocyanidin synthase (leucoanthocyanidin dioxygenase)	ANS (LDOX)	1.14.11.19	2OGD	64, 77
UDP-Glc:anthocyanidin 3-O-glucosyltransferase/UDP-Glc:flavonol 3-O-glucosyltransferase[c]	F3GT	2.4.1.115/ 2.4.1.91	UDP-O-Glycosyltransferase (UGT)	362
UDP-Glc:anthocyanin 5-O-glucosyltransferase[d]	A5GT	2.4.1.–	UGT	88, 102
UDP-Glc:anthocyanin 3'-O-glucosyltransferase	A3'GT	2.4.1.–	UGT	113
UDP-Rha:anthocyanidin 3-O-glucoside 6''-O-rhamnosyltransferase	A3RT	2.4.1.–	UGT	111, 112
SAM:anthocyanin 3'-O-methyltransferase	A3'OMT	2.1.1.–	SAM O-Methyltransferase (OMT)	Patent application WO03/062428
SAM:anthocyanin 3',5'-O-methyltransferase	A3'5'OMT	2.1.1.–	OMT	Patent application WO03/062428
Hydroxycinnamoyl-CoA:anthocyanin 5-O-glucoside-6'''-O-hydroxycinnamoyltransferase[e]	A5AT (Gt5AT)	2.3.1.153	Versatile acyltransferase (VAT)	121
Hydroxycinnamoyl-CoA:anthocyanidin 3-O-glucoside-6''-O-hydroxycinnamoyltransferase	A3AT (Pf3AT)	2.3.1.–	VAT	122
Malonyl-CoA:anthocyanin 5-O-glucoside-6'''-O-malonyltransferase	A5MT (Ss5MaT1)	2.3.1.–	VAT	126
Malonyl-CoA:anthocyanidin 3-O-glucoside-6''-O-malonyltransferase	A3MT (Sc3MaT, Dm3MaT1, Dv3MaT)	2.3.1.–	VAT	123–125

Malonyl-CoA:anthocyanin 5-O-glucoside-4'''-O-malonyltransferase	A5MT (Ss5MaT2)	2.3.1.–	VAT	406
Malonyl-CoA:anthocyanidin 3-O-glucoside-3'',6''-O-dimalonyltransferase	A3diMT (Dm3MaT2)	2.3.1.–	VAT	125
Flavonoid 3'-hydroxylase	F3'H	1.14.13.21	CytP450 (CYP75B)	105
Flavonoid 3',5'-hydroxylase	F3',5'H	1.14.13.–	CytP450 (CYP75A)	103

Flavones, flavonols, and flavanones

Flavonol synthase	FLS	1.14.11.–	2OGD	146
Flavone synthase I	FNSI	1.14.11.–	2OGD	145
Flavone synthase II	FNSII	1.14.13.–	CytP450 (CYP93B)	143, 144
(2S)-Flavanone 2-hydroxylase	F2H	1.14.13.–	CytP450 (CYP93B)	223
UDP-Gal:flavonoid 3-O-galactosyltransferase	F3GalT	2.4.1.–	UGT	156
UDP-Glc:flavonoid 7-O-glucosyltransferase[f]	F7GT	2.4.1.81/ 2.4.1.185	UGT	152, 153, 155
UDP-Glc:flavonoid 3,7,4'-O-glucosyltransferase[g]	UGT73G1	2.4.1.–	UGT	154
UDP-Rha:flavonol 3-O-rhamnosyltransferase	F3RT	2.4.1.159	UGT	153
UDP-Gal:flavonoid-3-O-galactosyltransferase/UDP-Gal:flavonol-3-O-galactosyltransferase	F3GalT/ PhF3GalT	2.4.1.–	UGT	156, 158
UDP-Rha:flavanone 7-O-glucoside-2''-O-rhamnosyltransferase[h]	F7RT	2.4.1.–	UGT	407
SAM:flavonoid 7-O-methyltransferase	F7OMT	2.1.1.–	OMT	160
SAM:flavonol/dihydroflavonol 3',5'-O-methyltransferase	F3'5'OMT(CrOMT2)	2.1.1.149	OMT	167
SAM:flavonoid/HCA 3'-O-methyltransferase	F3OMT (AtOMT1, CaOMT1)	2.1.1.–	OMT	164
SAM: 3'-O-methylflavonoid 4'-O-methyltransferase	F4OMT (CrOMT6)	2.1.1.75	OMT	87
SAM:trimethylflavonol 3',5'-O-methyltransferase	F3'5'OMT (CaFOMT3')	2.1.1.–	OMT	161
SAM:flavonoid O-methyltransferase[i]	FOMT (PFOMT)	2.1.1.–	OMT	116
Flavonol 3-O-sulfotransferase	F3ST	2.8.2.25	Sulfotransferase	169
Flavonol 4'-O-sulfotransferase	F4'ST	2.8.2.27	Sulfotransferase	169
Flavonoid 7-O-sulfotransferase	F7ST	2.8.2.28	Sulfotransferase	168

Chalcones

Polyketide reductase	PKR (CHR, CHKR)	1.1.1.–	Aldo/keto-reductase	177
UDP-Glc:chalcone 2'-O-glucosyltransferase	C2'GT	2.4.1.–	UGT	Patent application WO03/18682
SAM:isoliquiritigenin/licodione 2'-O-methyltransferase	C2'OMT	2.1.1.65	OMT	221, 222

continued

TABLE 3.1
Flavonoid Biosynthetic Enzymes for which DNA Sequences Have Been Obtained — continued

Enzyme	Abbreviation	EC number	Protein family	Ref.[a]
Aurones				
Aureusidin synthase	AUS	1.21.3.6	Polyphenol oxidase	212
UDP-Glc:aureusidin 7-O-glucosyltransferase	AU7GT	2.4.1.–	UGT	Patent application WO00/49155
Isoflavonoids				
2-Hydroxyisoflavanone synthase (isoflavone synthase)	2HIS (IFS)	5.4.99.–	CytP450 (CYP93C)	180–182
Isoflavone reductase	IFR	1.3.1.45	RED	195, 196
Isoflavone 2'-hydroxylase	I2'H	1.14.13.53	CytP450 (CYP81E)	193
Isoflavone 3'-hydroxylase	I3'H	1.14.13.52	CytP450 (CYP81E)	194
SAM:isoflavone 7-O-methyltransferase	I7OMT	2.1.1.150	OMT	190
SAM:2,7,4'-trihydroxyisoflavanone 4'-O-methyltransferase	HI4'OMT	2.1.1.46	OMT	187
Vestitone reductase	VR	1.1.1.246	RED	200
Pterocarpan 6a-hydroxylase (3,9-dihydroxypterocarpan 6a-hydroxylase)	P6aH (DH6aH)	1.14.13.28	CytP450 (CYP93A)	202
SAM:pterocarpan 3-O-methyltransferase	P3OMT (HM3OMT)	2.1.1.–	OMT	203
Flavonoid 6-hydroxylase	F6H	1.14.13.–	CytP450 (CYP71D)	210
UDP-Glc:formononetin 7-O-glucosyltransferase	I7GT	2.4.1.–	UGT (UGT73F1)	209
Proanthocyanidins				
Leucoanthocyanidin reductase	LAR (LCR)	1.17.1.3	RED	133
Anthocyanidin reductase	ANR	1.1.1.–	RED	136, 137

[a] Reference is given only to the first publications on the isolation and characterization of the corresponding cDNA/gene.
[b] Some DFRs show FNR activity also, and reference is given to the first example of isolation of a cDNA for such a DFR.
[c] Some 3GTs analyzed will accept other flavonoids in addition to anthocyanidins as substrates, and so the term F3GT is used.
[d] Recombinant A5GT proteins show varying degrees of anthocyanin substrate specificity.
[e] The nomenclature for the AATs follows Nakayama et al.[120] The positional numbering of the sugar hydroxyl that is modified is given followed by prime symbols to indicate which sugar is affected. The double and triple primes indicate the 3-O-glycosyl and 5-O-glycosyl, respectively. Recombinant AATs show varying degrees of substrate specificity.
[f] A number of flavonoid-7-O-GT cDNAs have been isolated, many encoding recombinant proteins with wide substrate acceptance.
[g] UGT73G1 can produce mono- or diglucosides with a wide range of flavonoid substrates, adding glucose to one or two of the indicated hydroxyls.
[h] The *Citrus* enzyme catalyzes 2''-O-rhamnosylation of the 7-O-glucoside of flavanones and flavones but not flavonols.
[i] The recombinant product is able to catalyze methylation at a variety of flavonoid hydroxyl positions.

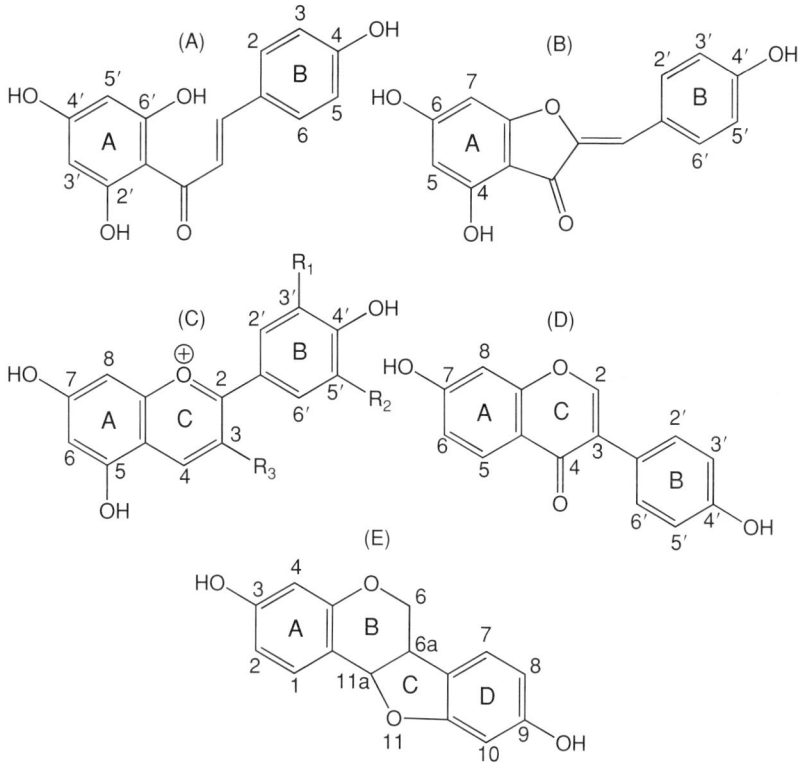

FIGURE 3.1 Base structures of a chalcone (A), an aurone (B), the main anthocyanidins (C), an isoflavonoid (D), and a pterocarpan (E). The lettering of the carbon rings is shown, as well as the numbering of the key carbons. Note that the numbering is different for each type of compound. For the majority of flavonoid types the numbering is as for the anthocyanidins, and for the cinnamic acids it is as for the chalcones. R_1, R_2, and R_3 substitutions determine the various common anthocyanidins. The common 3-hydroxyanthocyanidins ($R_3 = OH$) are pelargonidin (R_1 and $R_2 = H$), cyanidin ($R_1 = OH$ and $R_2 = H$), delphinidin (R_1 and $R_2 = OH$), peonidin ($R_1 = OCH_3$ and $R_2 = H$), petunidin ($R_1 = OCH_3$ and $R_2 = OH$), and malvidin (R_1 and $R_2 = OCH_3$). The rare 3-deoxyanthocyanidins ($R_3 = H$) are apigeninidin (R_1 and $R_2 = H$), luteolinidin ($R_1 = OH$ and $R_2 = H$), and tricetinidin ($R_1 = OH$ and $R_2 = OH$).

encoding secondary modification enzymes that have been characterized is still limited compared to the great array of known flavonoid structures. Nor have cDNAs or genes been published for some of the enzymes carrying out hydroxylation of the core flavonoid structure, such as at the C-8 and C-2′ positions.

The key enzymes involved in the formation of the hydroxycinnamic acids (HCAs) from phenylalanine and malonyl-CoA are now discussed in detail, while later sections address the branches of the flavonoid pathway leading to anthocyanins, aurones, flavones, flavonols, PAs, and isoflavonoids. This is followed by brief reviews of the regulation of flavonoid biosynthesis and the use of flavonoid genes in plant biotechnology. To assist the reader, Figure 3.1 presents the carbon numbering for the various flavonoid types discussed.

3.3 BIOSYNTHESIS OF FLAVONOID PRECURSORS

The first flavonoids, the chalcones, are formed from HCA-CoA esters, usually 4-coumaroyl-CoA (Figure 3.2), in three sequential reactions involving the "extender" molecule

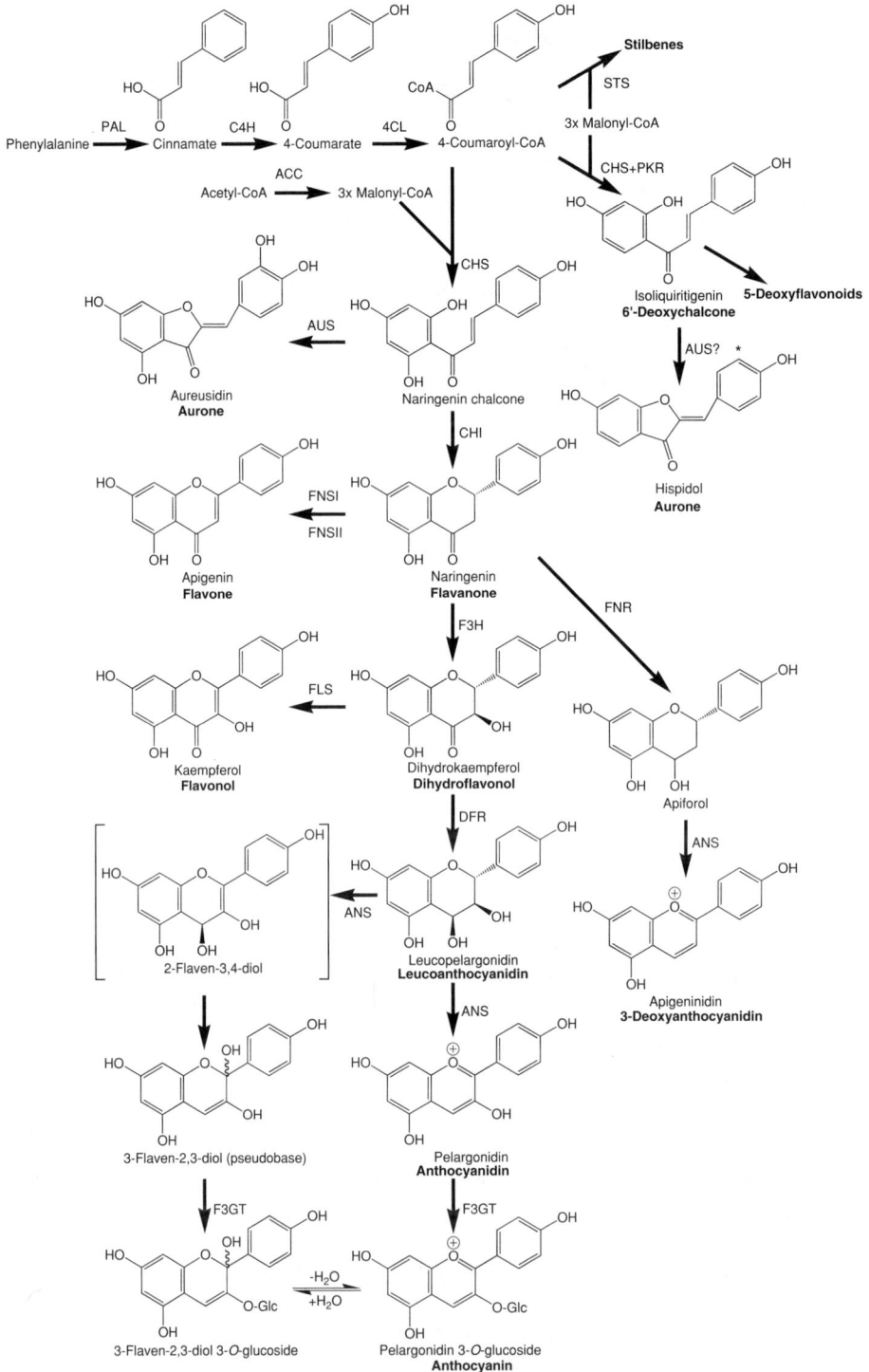

FIGURE 3.2 General phenylpropanoid and flavonoid biosynthetic pathways. The B-ring hydroxylation steps are not shown. For formation of anthocyanins from leucoanthocyanidins two routes are represented: a simplified scheme via the anthocyanidin (pelargonidin) and the likely *in vivo* route via the pseudobase. Enzyme abbreviations are defined in the text and in Table 3.1.

malonyl-CoA. In a few species, caffeoyl-CoA and feruloyl-CoA may also be used as substrates for chalcone formation.

4-Coumaroyl-CoA is produced from the amino acid phenylalanine by what has been termed the general phenylpropanoid pathway, through three enzymatic conversions catalyzed by phenylalanine ammonia-lyase (PAL), cinnamate 4-hydroxylase (C4H), and 4-coumarate: CoA ligase (4CL). Malonyl-CoA is formed from acetyl-CoA by acetyl-CoA carboxylase (ACC) (Figure 3.2). Acetyl-CoA may be produced in mitochondria, plastids, peroxisomes, and the cytosol by a variety of routes. It is the cytosolic acetyl-CoA that is used for flavonoid biosynthesis, and it is produced by the multiple subunit enzyme ATP-citrate lyase that converts citrate, ATP, and Co-A to acetyl-CoA, oxaloacetate, ADP, and inorganic phosphate.[6]

Many other compounds are involved in flavonoid biosynthesis in some species, for example, as donors for methylation or aromatic or aliphatic acylation. For intact plants, these are generally accepted to be available in the cell for the reaction to proceed if the appropriate modification activity is present.

3.3.1 ACETYL-COA CARBOXYLASE

ACC (also termed the acetyl-CoA:carbon-dioxide ligase [ADP-forming]) catalyzes the ATP-dependent carboxylation of acetyl-CoA, with Mg^{2+} as a cofactor, to form malonyl-CoA. ACC activity is found in both the plastid, where a heteromeric enzyme supports fatty acid biosynthesis, and the cytoplasm, in which a homodimeric enzyme of around 250 kDa supplies malonyl-CoA for the synthesis of a range of compounds that includes the flavonoids.[7] Full-length cDNAs for the plant cytoplasmic isoenzyme were first identified from *Medicago sativa* (alfalfa) using a probe from the human protein disulfide isomerase,[8] and from *A. thaliana* using polymerase chain reaction (PCR) with degenerate oligonucleotide sequences based on nonplant ACC sequences.[9] Sufficient amino acid similarity was found to the rate ACC to enable a full protein line-up and identification of putative binding sites for ATP, acetyl-CoA, and carboxybiotin. Sequences from several species are now available, and studies on the regulation of the gene are well advanced. ACC activity is induced in response to stimuli that increase flavonoid biosynthesis, such as ultraviolet (UV) light and fungal elicitors.

3.3.2 PHENYLALANINE AMMONIA LYASE

PAL is one of the best-characterized enzymes of plant secondary metabolism. It converts L-phenylalanine into *trans*-cinnamate (*E*-cinnamate) by the *trans*-elimination of ammonia and the pro-3*S* proton (see Ref. 4 for a full reaction discussion). The enzyme, which requires no cofactor, is a tetramer of 310–340 kDa.[3] A cDNA for PAL was first isolated from *Petroselinum crispum* (parsley),[10] and others have subsequently been isolated from numerous species. Often PAL is produced from a multigene family and is present in a variety of isoenzyme forms.

Enzymatic preparations for PAL from monocotyledonous species (monocots) can show a similar activity against tyrosine (tyrosine ammonia lyase, TAL), and TAL enzymatic preparations also show PAL activity. That a single enzyme may account for the observed co-occurring TAL and PAL activities was confirmed by Rösler et al.,[11] who showed the recombinant *Zea mays* (maize) PAL converted tyrosine to 4-coumarate directly, thus removing the requirement for the usual 4-hydroxylation step in phenylpropanoid biosynthesis.

Considerable progress has been made with elucidating the functional aspects of the PAL protein, assisted by the availability of the crystal structure for histidine ammonia lyase (HAL), which catalyzes a reaction similar to that of PAL, the conversion of L-histidine to

E-urocanic acid. Röther et al.[12] used the HAL structure to generate a homology-based model of PAL and to identify specific amino acids to target in PAL that may have an equivalent role in the active site and reaction mechanism. The same group has proposed a reaction mechanism involving electrophilic attack at the phenyl ring by a prosthetic group.[13] K_m values in the order of 0.1 mM have been reported for recombinant PAL proteins.[12] PAL activity *in vivo* appears to be modulated by signals generated by the levels of cinnamic acid, altering both gene transcription and enzyme activity (reviewed in Ref. 14).

3.3.3 Cinnamate 4-Hydroxylase

C4H catalyzes the hydroxylation of *trans*-cinnamate to *trans*-4-coumarate in the initial oxygenation step of phenylpropanoid biosynthesis, which introduces the 4′-hydroxyl that is common to most flavonoids. The isolation of a cDNA for C4H was first reported from *Helianthus tuberosus* (Jerusalem artichoke), *M. sativa*, and *Vigna radiata* (mung bean), based on purification and amino acid sequencing of the protein[15,16] or similarity to other members of the same enzyme superfamily.[17] Genes or cDNAs for C4H have subsequently been cloned from many species, with Hübner et al.[18] listing over ten species and Gravot et al.[19] noting 41 different C4H cDNAs in public databases.

Based on the amino acid sequences, two classes of C4H have been described for some species, with around 60% sequence similarity between the groups. The first sequence for the class II type was reported from *Phaseolus vulgaris* (French bean),[20] but they have also been found in other species (see, e.g., Ref. 21). The two C4H types differ at both terminal domains and in three internal domains, and it has been suggested that one type may be involved in stress responses and the other in vascular differentiation.[20,21]

Enzyme characteristics have been examined for recombinant C4H proteins from several species, including those from *P. crispum*, *P. vulgaris*, *Ammi majus*, *H. tuberosus*, and *Ruta graveolens*.[18–22] Similar K_m values toward cinnamate (2 to 10 μM) are reported, and consistently high substrate specificity (although the V_{max} values vary between studies). Only 4-coumarate is found as the *in vitro* product, with no detectable 2- or 3-coumarate production.[22]

C4H was one of the first plant enzymes of the P450 enzyme superfamily (EC 1.14.13.X) to be characterized at the encoding DNA sequence level. Several members of this family are involved in flavonoid biosynthesis, and it is worth mentioning some of the general features of this remarkable group of enzymes. P450s are heme-dependent mixed function mono-oxygenases that require O_2 and NADPH for activity. All those involved in flavonoid biosynthesis that have been characterized to date are A-type P450s localized in the microsomal fraction. P450s of plants require a second enzyme system as an additional redox partner for the transfer of electrons from NADPH to oxygen via the heme iron. This is typically the flavoprotein NADPH-P450 reductase, but sometimes NADH-dependent Cyt b_5 reductase is used in conjunction with Cyt b_5. Plant P450s are typically 50 to 60 kDa in size, and C4H is usually around 60 kDa. The general characteristics of P450s of plants are reviewed in more detail in several articles (such as in Refs. 23, 24).

P450s are grouped according to molecular phylogeny (rather than function) into families ($\geq$40% amino acid positional identity), which for plants are given a number from CYP71 to CYP99 then CYP701 onwards, and subfamilies ($\geq$55% amino acid identity) designated by a letter. Thus, C4H is in subfamily A of family 73 and termed CYP73A. Individual genes, either from different loci in a species or from different species, may then be numbered; e.g., the flavonoid 3′,5′-hydroxylase (F3′,5′H) cDNAs for the *Hf1* and *Hf2* loci of *Petunia hybrida* were named CYP75A1 and CYP75A3, and that from *Eustoma grandiflorum* (lisianthus) CYP75A5. Sequences from the same species with >97% identities are assumed to be allelic variants unless otherwise demonstrated.

P450 amino acid sequences contain several well-conserved regions, such as the heme binding domain (generally FxxGxxxCxG) and the proline rich region (PPxP) that forms a hinge between the membrane anchored N terminal and the rest of the protein. Rupasinghe et al.[25] have modeled the structure of C4H and three other *A. thaliana* P450s involved in phenylpropanoid biosynthesis, based on the amino acid sequences and crystal structures of other P450s. In addition to C4H these were the flavonoid 3′-hydroxylase (F3′H) and two enzymes involved in the monolignol pathway. The analysis showed that, despite low amino acid sequence identities in some cases, the enzymes displayed significant conservation in terms of structure and substrate recognition.

Linear furanocoumarins (psoralens) inhibit P450s as mechanism-based inactivators (suicide inhibitors). Thus, species that produce psoralens may have evolved C4H enzymes with enhanced tolerance to these compounds.[18,19] Recombinant C4H from the psoralen-producing species *R. graveolens* showed markedly slower inhibition kinetics with psoralens, and possibly biologically significant tolerance, compared to C4H from a species that does not produce the compounds (*H. tuberosus*).[19]

3.3.4   4-Coumarate:CoA Ligase

4CL activates the HCAs for entry into the later branches of phenylpropanoid biosynthesis through formation of the corresponding CoA thiol esters. 4-Coumarate, the product of C4H, is key for flavonoid biosynthesis, but 4CL will commonly accept other HCAs as substrates. Generally, 4-coumarate and caffeate are the preferred substrates, followed by ferulate and 5-hydroxyferulate, with low activity against cinnamate and none with sinapate. However, different isoenzymes have been identified that exhibit distinct substrate specificities, including within the same species. These include isoforms with a variant substrate-binding pocket that will accept sinapate,[26,27] those that will not accept ferulate,[28] and isoforms that differ in their ability to accept 5-hydroxyferulate.[29] It is thought that the different isoenzymes may have distinct roles in metabolic channeling in flavonoid or lignin biosynthesis (see, e.g., Refs. 26, 30). Supporting an important role for the variant isoforms is the observation that 4CL is encoded by gene families in all species examined to date.[31] The evolution of the *4CL* gene family of plants is discussed extensively in Refs. 27, 31.

The formation of the HCA-CoA esters proceeds through a two-step reaction, with Mg^{2+} as a cofactor. In the first step, 4-coumarate and ATP form a coumaroyl–adenylate intermediate, with the simultaneous release of pyrophosphate. In the second step, the coumaroyl group is transferred to the sulfhydryl group of CoA, with the release of AMP. The reaction mechanism of 4CL is discussed in detail in Pietrowska-Borek et al.,[32] including the newly discovered ability of recombinant *A. thaliana* 4CL2 (At4CL2) to synthesize mono- and di-adenosine polyphosphates. Based on the presence of a highly conserved peptide motif, 4CL has been placed in the adenylate-forming superfamily that also includes acetyl-CoA synthetases, long-chain fatty acyl-CoA synthetases, luciferases, and peptide synthetases.[28,31,33] 4CL contains amino acid motifs conserved either among the superfamily or just among 4CL sequences. Mutagenesis, domain swapping, and homology modeling analyses have shown the functional importance of some of these regions and identified amino acids important in the specificity shown toward the different substrates.[28,34,35] This information has been used to modify the At4CL2 isoform to accept ferulate and sinapate,[34,35] and to change the substrate specificities of the *Glycine max* (soybean) Gm4CL2 and Gm4CL3 isoforms.[33]

The route to formation of flavonoids lacking 4′-hydroxylation of the B-ring has not been elucidated. However, one possible route is the direct use of cinnamate as a substrate by 4CL. Activity on cinnamate has been shown at low levels for some of the recombinant 4CL proteins, and a separate cinnamoyl:CoA ligase with specific activity on cinnamate and not 4-coumarate has been characterized in some species.[36] Data with regard to subsequent enzymes accepting the alternative substrates are limited, but cinnamoyl-CoA is used by recombinant chalcone synthase (CHS) from several species, as well as CHS-like enzymes such as pinosylvin synthase (EC 2.3.1.146) (see, e.g., Refs. 37, 38). Indeed, the recombinant CHS from *Cassia alata* (ringworm bush) can use a wide range of substrates to make various products, including 4-deoxychalcones.[39] However, definitive evidence showing this route to 4′-deoxyflavonoids *in vivo*, from plant studies for example, has not been published.

3.3.5 MODIFICATION OF HYDROXYCINNAMIC ACID-CoA ESTERS

4-Coumaroyl-CoA is the major substrate for entry into flavonoid biosynthesis through the action of CHS. However, some CHS enzymes may also use caffeoyl-CoA or feruloyl-CoA;[2] and various HCA-CoA esters are used by the aromatic acyltransferases that modify the flavonoid glycosides. Although it might be expected that caffeoyl-CoA would be formed by direct 3-hydroxylation of 4-coumaroyl-CoA, and feruloyl-CoA by subsequent 3-*O*-methylation, varying biosynthetic routes may exist.[3,40] In particular, recombinant CYP98A3 of *A. thaliana*, which encodes an enzyme that can add a hydroxyl at the C-3 position of 4-cinnamate derivatives, shows no activity toward 4-coumaroyl-CoA, and only low-level activity toward 4-coumarate, the preferred substrates being the 5-*O*-shikimate and 5-*O*-quinate esters.[41–43] However, analysis of the corresponding mutant *ref8* supports an *in vivo* role for CYP98A3 in phenylpropanoid 3-hydroxylation. Similarly, the ferulate 5-hydroxylase (F5H, CYP84A1 of *A. thaliana*) shows greater preference toward coniferalde-hyde and coniferyl alcohol than ferulate or feruloyl-CoA. In addition, *O*-methylation can occur both on caffeoyl-CoA and 5-hydroxyferuloyl-CoA and their respective aldehyde forms. Thus, a metabolic grid seems to prevail generally for the reactions on the HCAs, particularly with respect to formation of the monolignols. Some of the downstream flavonoid biosynthesis enzymes have been studied for acceptance of *O*-methylated substrates, in particular CHS (see Section 3.4.1), and some of the oxidoreductases (see Section 3.4.5).

3.4 FORMATION OF ANTHOCYANINS

The flavonoid pathway starts with the formation of the C_{15} backbone by CHS. Chalcones are then generally directly or indirectly converted to a range of other flavonoids in a pathway of intersecting branches, with intermediate compounds being involved in the formation of more than one type of end product (Figure 3.2). This section discusses the genes and enzymes involved in the formation of the simplest common anthocyanins, 3-hydroxyanthocyanidin 3-*O*-glycosides, which require a minimum of five enzymatic steps subsequent to the formation of chalcone (Figure 3.2).

3.4.1 CHALCONE SYNTHASE

CHS carries out a series of sequential decarboxylation and condensation reactions, using 4-courmaroyl-CoA (in most species) and three molecules of malonyl-CoA, to produce a poly-ketide intermediate that then undergoes cyclization and aromatization reactions that form the A-ring and the resultant chalcone structure. The chalcone formed from 4-courmaroyl-CoA is naringenin chalcone. However, enzyme preparations and recombinant CHS proteins from some species have been shown to accept other HCA-CoA esters as substrates, such as cinnamoyl-CoA (see, e.g., Ref. 37). In particular, the *Hordeum vulgare* (barley) *CHS2* cDNA encodes a CHS protein that converts feruloyl-CoA and caffeoyl-CoA at the highest rate, and cinnamoyl-CoA and 4-courmaroyl-CoA at lower rates.[38]

The key role of CHS in flavonoid biosynthesis has made it a focus of research for many years, and it is now very well characterized. The isolation of a cDNA for CHS represented the first gene cloned for a flavonoid enzyme.[44,45] *CHS* sequences, and a series of *CHS*-like sequences, have now been characterized from many species, and Austin and Noel[46] have identified close to 650 *CHS*-like sequences in public databases.

Native CHS is a homodimer with subunits of 40 to 44 kDa. The structure of the protein produced from the *CHS2* cDNA of *M. sativa* has been determined and the residues of the active site defined.[47] It belongs to the polyketide synthase (PKS) group of enzymes that occur in bacteria, fungi, and plants, and is a type III PKS. All the reactions are carried out at a single active site without the need for cofactors.[47–49]

PKSs are characterized by their ability to catalyze the formation of polyketide chains from the sequential condensation of acetate units from malonate thioesters. In plants they produce a range of natural products with varied *in vivo* and pharmacological properties. PKSs of particular note include acridone synthase, bibenzyl synthase, 2-pyrone synthase, and stilbene synthase (STS).[46] STS forms resveratrol, a plant defense compound of much interest with regard to human health. STS shares high sequence identity with CHS, and is considered to have evolved from CHS more than once.[50] Knowledge of the molecular structure of the CHS-like enzymes has allowed direct engineering of CHS and STS to alter their catalytic activities, including the number of condensations carried out (reviewed in Refs. 46, 51, 52). These reviews also present extensive, and superbly illustrated, discussions of CHS enzyme structure and reaction mechanism.

3.4.2 CHALCONE ISOMERASE

In a reaction that establishes the flavonoid heterocyclic C-ring, chalcone isomerase (CHI) catalyzes the stereospecific isomerization of chalcones to their corresponding (2*S*)-flavanones, via an acid base catalysis mechanism.[53,54] Almost 40 years ago, the first flavonoid enzyme to be described was CHI (in the adopted hometown of the authors of this chapter).[55] Since then CHI has been analyzed in great detail, and surprisingly, it shows little similarity to other known protein sequences,[54] although CHI-like sequences have recently been reported from plants and other organisms.[56]

CHI has a deduced molecular weight (MW) of 27 to 29 kDa, although possible *in vivo* modifications have been reported.[57] Two types of CHI have been identified, the more common CHI-I type, which can use only 6′-hydroxychalcone substrates, and the CHI-II type, which can catalyze isomerization of both 6′-hydroxy- and 6′-deoxychalcones. CHI-II is prevalent in legumes, although sequence analysis and recent transgenic results[58] suggest the activity also occurs in nonlegumes. Sequences from different species for the same type of CHI show amino acid identities of >70%, while between type I and type II identities of about 50% are found. Genes for both types of CHI occur as a cluster in *Lotus japonicus*,[59] and the presence of tandem gene copies suggests an origin for *CHI-II* by gene duplication and divergence from *CHI-I*.

The structure of the recombinant *M. sativa* CHI-II protein has been determined to 1.85 Å resolution. The progress of the reaction in the reactive site cleft has been elucidated, and a full reaction sequence proposed.[52–54] However, the basis for the specificities toward 6′-hydroxy- and 6′-deoxychalcones was not resolved, although potentially key amino acid residues were identified.

With 6′-hydroxychalcones, such as naringenin chalcone, the isomerization reaction can readily occur nonenzymically to form racemic (2*R*,2*S*) flavanone. This occurs easily *in vitro* and has been reported to occur *in vivo* to the extent that moderate levels of anthocyanin can be formed.[3] However, 6′-deoxychalcones are stable under physiological conditions, due to an intramolecular hydrogen bond between the 2'-hydroxyl and the carbonyl group, and CHI-II is required to convert them to flavanones. CHI accelerates ring closure to a 10^7-fold acceleration over the spontaneous reaction rate, but with significantly slower kinetics for the 6'-deoxychalcones than 6'-hydroxychalcones; and ensures formation of the biosynthetically required (2S)-flavanones.[53,54,60] The requirement for CHI for significant flavonoid biosynthesis in some plants is well illustrated by the acyanic phenotype of the *transparent testa5* (*tt5*) *CHI* mutation of *A. thaliana*, and the increased levels of flavonols in *CaMV35S:CHI* transgenic plants of *Lycopersicon esculentum* (tomato).[61]

3.4.3 FLAVANONE 3β-HYDROXYLASE

(2S)-Flavanones are converted stereospecifically to the respective (2R,3R)-dihydroflavonols (DHFs) by flavanone 3β-hydroxylase. Stereospecificity for (2S)-flavanones has been confirmed by analysis of the recombinant protein.[62] Flavanone 3β-hydroxylase is commonly abbreviated to F3H, which is what has been used in this chapter, but FHT is also used in the literature, which agrees with the nomenclature for phenylpropanoid biosynthesis proposed in Heller and Forkmann.[3]

F3H is a soluble nonheme dioxygenase dependent on Fe^{2+}, O_2, and 2-oxoglutarate (2OG), and thus belongs to the family of 2OG-dependent dioxygenases (2OGDs). 2OGDs have been characterized from mammalian, microbial, and plant sources, and they all use four electrons generated from oxoglutarate decarboxylation to split di-oxygen and create reactive enzyme–iron species. The protein family is well represented in flavonoid biosynthesis, as can be seen from Table 3.1, and this will be discussed later. Further details on the 2OGD family are given in Refs. 62, 63.

A cDNA for F3H was first isolated from *Antirrhinum majus* (snapdragon),[64] and since then genes and cDNAs have been isolated from over a dozen other species.[5,65] The native protein is a monomer of 41 to 42 kDa, although proteolysis during purification gave values of 34 to 39 kDa in early studies of the enzyme. Using sequence comparison and analysis of recombinant proteins good progress has been made in elucidating the tertiary structure of the enzyme, the nature of the active site, and the binding of 2OG and Fe^{2+}. Britsch et al.[65] identified 14 amino acids strictly conserved among F3H sequences from several species, including histidines with a putative role in Fe^{2+} binding. Mutation analysis of recombinant *P. hybrida* F3H showed that His220, Asp222, and His278 are indeed involved in Fe^{2+} binding in the active site, and that Arg288 and Ser290 are required for 2OG binding.[66,67] Full characterization awaits determination of the crystal structure for the enzyme.

3.4.4 DIHYDROFLAVONOL 4-REDUCTASE

Dihydroflavonol 4-reductase (DFR) catalyzes the stereospecific conversion of (2R,3R)-*trans*-DHFs to the respective (2R,3S,4S)-flavan-2,3-*trans*-3,4-*cis*-diols (leucoanthocyanidins) through a NADPH-dependent reduction at the 4-carbonyl. DNA sequences for DFR were first identified from *A. majus* and *Z. mays*,[68,69] and the identity of the *Z. mays* sequence confirmed by *in vitro* transcription and translation of the cDNA and assay of the resultant protein.[70] DNA sequences have now been cloned from many species, with the size of the predicted protein averaging about 38 kDa. Stereospecificity to (2R,3R)-dihydroquercetin (DHQ) has been shown for some recombinant DFR proteins.[71]

DFR belongs to the single-domain-reductase/epimerase/dehydrogenase (RED) protein family, which has also been termed the short chain dehydrogenase/reductase (SDR) superfamily. This contains other flavonoid biosynthetic enzymes, in particular the anthocyanidin reductase (ANR), leucoanthocyanidin reductase (LAR), isoflavone reductase (IFR), and vestitone reductase (VR), as well as mammalian, bacterial, and other plant enzymes.[72,73]

The preference shown by DFR toward the three common DHFs varies markedly between species, with some enzymes showing little or no activity against dihydrokaempferol (DHK) and others showing preference toward dihydromyricetin (DHM). In particular, DFR in *Cymbidium hybrida* (cymbidium orchids), *L. esculentum*, *Petunia*, and *Vaccinium macrocarpon* (cranberry) cannot efficiently reduce DHK,[3,74,75] so that pelargonidin-based anthocyanins rarely accumulate in these species. However, DFR enzymes of many species accept all three DHFs, and DHM can be used by *Dendranthema* (chrysanthemum), *Dahlia variabilis*, *Dianthus caryophyllus* (carnation), *Matthiola*, and *Nicotiana* (tobacco) flowers even though delphinidin derivatives do not naturally occur in these ornamentals.[3,76]

Some species contain a closely related enzyme activity to DFR that can act on flavanones, termed the flavanone 4-reductase (FNR), which may represent a variant DFR form. This is discussed in more detail in Section 3.4.7. 5-Deoxyleucoanthocyanidin compounds are known to occur in legumes, and analysis of two recombinant DFR proteins (MtDFR1 and MtDFR2) from *Medicago truncatula* (barrel medic) has found activity on the 5-deoxyDHF substrates fustin and dihydrorobinetin.[71] Indeed, fustin was the preferred substrate of both recombinant enzymes. MtDFR1 and MtDFR2 showed distinct enzyme characteristics, and overexpression of MtDFR1 but not MtDFR2 promoted anthocyanin biosynthesis in flowers of *N. tabacum*.

Substrate specificity between DHK, DHQ, and DHM appears, based on chimeric DFR proteins formed using the *P. hybrida* and *Gerbera hybrida* sequences, to locate to a 26 amino acid region that may be the binding pocket for the B-ring, and as little as one amino acid change in this region can alter the specificity of the enzyme.[73]

3.4.5 ANTHOCYANIDIN SYNTHASE

The role of anthocyanidin synthase (ANS) in the biosynthetic pathway is to catalyze reduction of the leucoanthocyanidins to the corresponding anthocyanidins. However, *in vivo* it is anthocyanidins in pseudobase form that are formed, as is described below. In this chapter, use of anthocyanidin should be taken to include the pseudobase form. Furthermore, although the name ANS is commonly used, the enzyme is also referred to in the literature as leucoanthocyanidin dioxygenase (LDOX), reflecting the reaction type.

Much of the information on ANS has come not from studies on enzyme extracts but from analysis of DNA sequences and recombinant proteins. Sequences for the ANS were first isolated using transposon generated mutant lines of *A. majus* and *Z. mays*.[64,77] They encoded proteins of 40 to 41 kDa that were found to have similarity to 2OGDs, during a study on a nonflavonoid enzyme.[78] This sequence-based identification was confirmed by the *in vitro* assay of the recombinant *Perilla frutescens* protein,[79] and subsequent assays on recombinant ANS from a range of species that confirmed the requirement for Fe^{2+}, 2OG, and ascorbate.[80,81] Sequence comparisons show that ANS is more closely related to flavonol synthase (FLS), another 2OGD, than to F3H.

From extensive analysis of recombinant proteins, and the crystal structure of *A. thaliana* protein, detailed reaction mechanisms have been proposed.[80–85] The ANS reaction likely proceeds via stereospecific hydroxylation of the leucoanthocyanidin (flavan-3,4-*cis*-diol) at the C-3 to give a flavan-3,3,4-triol, which spontaneously 2,3-dehydrates and isomerizes to 2-flaven-3,4-diol, which then spontaneously isomerizes to a thermodynamically more stable anthocyanidin pseudobase, 3-flaven-2,3-diol (Figure 3.2). The formation of 3-flaven-2,3-diol via the 2-flaven-3,4-diol was previously hypothesized by Heller and Forkmann.[3] The reaction sequence, and the subsequent formation of the anthocyanidin 3-*O*-glycoside, does not require activity of a separate dehydratase, which was once postulated. Recombinant ANS and uridine diphosphate (UDP)-glucose:flavonoid 3-*O*-glucosyltransferase (F3GT, sometimes abbreviated in the literature as UF3GT, UFGT, or 3GT) are sufficient for the formation of cyanidin 3-O-glucoside from leucocyanidin, at least under mildly acid conditions, that are to be expected in a vacuole.[80]

Turnbull et al.[81,83,86] used a range of substrates to study recombinant *A. thaliana* ANS activity. *Trans*-DHQ, a potential substrate that would occur *in vivo*, was converted to quercetin in a reaction equivalent to that of the FLS. Incubation with the physiological substrate (2R,3S,4S)-3,4-*cis*-leucocyanidin resulted in the formation of *cis*-DHQ, *trans*-DHQ, quercetin, and cyanidin, with cyanidin being a minor product. The acceptance of multiple substrates and generation of a variety of products fits with the proposed 3-hydroxylation mechanism of ANS and the suggested relatively large active site cavity.[85] The overlapping *in vitro* activities of 2OGDs of flavonoid biosynthesis are discussed further in Sections 3.6 and 3.14.

Some of the 2OGDs of flavonoid biosynthesis (F3H, FNSI, FLS, and ANS) have been studied for acceptance of O-methylated substrates.[87] Substrates methylated at the 3'-hydroxyl were accepted, while methylation at the 4'- or 7-hydroxyls reduced activities to varying degrees. Multiple methylation or methylation at other positions prevented acceptance of the substrate.

For 3-deoxyanthocyanin biosynthesis the 3-hydroxyl is, of course, lacking from the ANS substrates (e.g., apiforol). Whether a specific ANS is thus involved in 3-deoxyanthocyanin biosynthesis is not clear. However, it has been postulated that the reaction may still proceed through 3-hydroxylation, and initial results suggest recombinant ANS from species that do not produce 3-deoxyanthocyanins may still use apiforol as a substrate to produce apigeninidin (results of J-I. Nakajima and K. Saito of Chiba University, Japan, with the authors' coworkers in New Zealand).

3.4.6 FORMATION OF ANTHOCYANIDIN 3-O-GLYCOSIDE

The anthocyanidin pseudobases are thought to be too unstable to accumulate in the cell, and are converted to the stable anthocyanins in what might be regarded as the final essential biosynthetic step, O-glycosylation. In the majority of plants, the initial reaction is the transfer of a glucose residue from the energy-rich nucleotide sugar (UDP-glucose) to the 3-hydroxyl of the proposed pseudobase by F3GT. As mentioned in Section 3.4.5, the action of ANS and F3GT has been demonstrated to be sufficient to convert the leucoanthocyanidins to colored anthocyanins (in an acidic environment).[80] Although the addition of glucose at the 3-hydroxyl is the most common initial activity, 3-O-galactosylation is the first reaction in some species. No cDNAs for anthocyanidin 3-O-galactosyltransferases have been published, although such sequences have been lodged with the GenBank database (accession BAD06514). 3-O-Glycosylation is often only the first of multiple sugar additions, either at other positions of the anthocyanin core structure (both A- and B-rings) or on to previously added sugars, and other glycosylations are discussed in Section 3.4.9.1.

Commonly, the F3GT is designated a flavonoid GT, as enzyme preparations from several species,[2] the recombinant *Forsythia × intermedia* and *P. hybrida* enzymes,[88,89] and the *A. majus* cDNA expressed *in vivo*,[90] can conjugate both anthocyanidin and flavonol substrates with high efficiency. However, F3GTs of some species, such as *Gentiana trifolia* and *Vitis vinifera* (grape),[91–93] may specifically or primarily act on anthocyanidin substrate, and should be termed UDP-glucose:anthocyanidin 3-O-glucosyltransferases (A3GTs) in reflection of this. Indeed, within the EC system separate classifications are given for A3GT (EC 2.4.1.115) and UDP-glucose:flavonol 3-O-glucosyltransferase (EC 2.4.1.91).

The F3GT is part of the UDPG-glycosyltransferase (UGT) family, which is family 1 of the glycosyltransferase superfamily (EC 2.4.1.X).[94,95] UGTs have a central role in detoxifying or regulating the bioactivities of a wide range of endogenous and exogenous low molecular weight compounds in both plants and mammals. In plants, they are generally monomeric, soluble enzymes of 50 to 60 kDa catalyzing O-glycoside formation, although cases of C-glycoside formation are known. Affinity for the sugar acceptor is usually high and that for the sugar donor typically lower. Conserved amino acids occur across the UGT family, in particular several histidine residues, and significantly conserved regions are illustrated in the alignment of the Plant Secondary Product Glycosyltransferase motif (PSPG-box) of 44 sequences in Ref. 95. The PSPG-box is thought to be common to UGTs involved in plant secondary metabolism, and may define a monophyletic group of genes. A nomenclature system for the UGTs has been suggested, similar to that for P450s, with groups of plant origin sharing >45% amino acid sequence identity numbered 71 to 100, subgroups with >60% identity given a letter, and the individual gene a number; e.g., the F3GT of *G. trifolia* is named UGT78B1 under this system (for details, see Refs. 95, 96). In general, the catalytic mechanism of UGTs is not well characterized, and no crystal structure of a plant enzyme has been published.

There are few reports of specific F3GT mutations, and none giving a phenotype in petal tissue, perhaps reflecting common redundancy in UGT activity. However, the first reported cloning of a plant UGT DNA sequence was based on the *F3GT bronze1* (*bz1*) mutation of *Z. mays*, in which glycosylation does not occur and a brown pigment is formed in the kernels by the condensation of the anthocyanidins in the cytosol.[4]

3.4.7 Formation of 3-Deoxyanthocyanin

The biosynthesis of 3-deoxyanthocyanins has only been studied in any detail for two grass species, *Z. mays* and *Sorghum bicolor*, and one member of the Gesneriaceae, *S. cardinalis*. It is thought that 3-deoxyanthocyanin biosynthesis occurs through the activity of FNR, so that flavan-4-ols are formed. The flavan-4-ols are then converted to anthocyanins through the action of ANS and a A5GT. The evidence to date is that, in addition to FNR activity, 3-deoxyanthocyanin biosynthesis also requires a marked reduction in the potentially competitive F3H activity.

FNR is most likely a variant form of DFR that has dual DFR/FNR activity. The recombinant DFR enzymes of *Malus domestica* (apple), *Pyrus communis* (pear), and *Z. mays*, all species that can produce 3-deoxyflavonoids under some circumstances, show both DFR and FNR activity.[97,98] In all three cases, DHFs were preferred as substrates over flavanones, supporting the need for a mechanism promoting flavanone production, i.e., a reduction in F3H activity. Enzymology studies for 3-deoxyanthocyanin-producing flower silks of *Z. mays* also show FNR activity occurs with only low levels of F3H activity.[98] The FNR was initially characterized in detail from the flowers of the Gesneriads *S. cardinalis* and *Columnea hybrida*, in which 3-deoxyanthocyanins are found in great excess to 3-hydroxyanthocyanins.[3] Recent studies on the recombinant *S. cardinalis* FNR show that this also has both DFR and FNR activity.[99]

Gene expression studies in relation to 3-deoxyflavonoid biosynthesis are limited to *S. cardinalis*, *S. bicolor*, and *Z. mays*, but all are in general agreement with the proposed biosynthetic mechanism. In petals of *S. cardinalis* transcript abundance is very high for the *FNR* and very low for *F3H*, with that for *ANS* intermediate between the two.[99] A similar pattern is found for phlobaphene production in *Z. mays* kernels, but without expression of *ANS* (see Section 3.15.3). In *S. bicolor* a similar pattern of *DFR/FNR* and *ANS* expression coincident with low *F3H* transcript levels is seen during production of 3-deoxyflavonoids in response to fungal inoculation,[100] although it has been suggested that some biosynthetic variations occur.[101]

The 3-deoxyanthocyanidins are initially glucosylated at the 5-hydroxyl. Clones for UDP-glucose:anthocyanin 5-*O*-glucosyltransferases (A5GT) have been isolated from a range of species.[88,102] However, all of these were found to require, at a minimum, prior 3-*O*-glucosylation of the substrate (see Section 3.4.9.1). This suggests that for 3-deoxyanthocyanin formation a specific A5GT may have evolved to accept the aglycone.

3.4.8 FLAVONOID 3′-HYDROXYLASE AND FLAVONOID 3′,5′-HYDROXYLASE

In a few species, the B-ring hydroxylation pattern is thought to be fully or partially determined through the HCA-CoA ester used by CHS (see Section 3.4.1). However, most commonly hydroxylation at the C-3′ and C-5′ positions is determined at the C_{15} level by the activity of two P450s, the F3′H and F3′,5′H. Genes and cDNAs for both enzymes, sometimes informally referred to as the red and blue genes because of their impact on flower color, were first cloned and characterized from *Petunia*[103–105] and subsequently from several other species (listed in Ref. 5). Based on sequence analysis, the F3′H and F3′,5′H proteins are 56 to 58 kDa in size.

The F3′H recombinant proteins of *A. thaliana*, *P. hybrida,* and *P. frutescens* accept flavanones, flavones, and DHFs as substrates,[105–107] as do enzyme preparations from plant tissues.[3] Indeed, recombinant *P. frutescens* F3′H showed a similar K_m (about $20\,\mu M$) for naringenin, apigenin, and DHK.[107] The *A. thaliana* F3′H amino acid sequence has been used to generate a model of the enzyme and examine the active site architecture and substrate recognition, as discussed in Section 3.3.3.[25] Recombinant F3′,5′H also accepts a range of substrates to carry out stepwise 3′- and 5′-hydroxylation, in particular converting naringenin or eriodictyol to pentahydroxyflavanone, DHK or DHQ to DHM, apigenin or luteolin to tricetin, and kaempferol or quercetin to myricetin.[91,103,108,109] Recombinant F3′,5′H from *P. hybrida* and *Catharanthus roseus* (Madagascar periwinkle) showed greatest activity with naringenin (K_m $7\,\mu M$) and apigenin, and decreasing activity against kaempferol and DHQ, with a preference for the 4′-hydroxylated substrates over the 3′4′-hydroxylated ones.[108,109] F3′,5′H has been shown to be able to use the alternative electron donor Cyt b_5, based on analysis of a knockout line of *P. hybrida* for the $Cytb_5$ gene (*difF*), which showed it is required for full activity of F3′,5′H but not F3′H in flowers.[110]

3.4.9 ANTHOCYANIN MODIFICATION ENZYMES

Knowledge of the genetics and molecular biology of the genes encoding the enzymes that carry out modifications of the core anthocyanin structure has advanced greatly in recent years, based on research involving a wide variety of plant species. Of the classic model species, anthocyanin modification enzymes have been studied in detail only for *P. hybrida*, regarding the production of methylated and acylated anthocyanidin 3,5-*O*-glycosides. However, 99 UGT-like sequences have been identified in the genome of *A. thaliana*, and the research currently underway to characterize them by transgenic approaches and expression *in vitro* may identify other flavonoid GTs.[96]

3.4.9.1 Anthocyanin Glycosyltransferases

Glycosylation at the 3-hydroxyl, or for 3-deoxyanthocyanidins at the 5-hydroxyl, has been discussed earlier as part of the core anthocyanin biosynthetic pathway. In this section, molecular data on the genes encoding enzymes that carry out subsequent glycosylations is reviewed. A great variety of anthocyanin glycosides occur, but at the time of the review by Heller and Forkmann[3] only a few anthocyanin GTs had been characterized biochemically, and DNA sequences were only available for the F3GT. Since then genes or cDNAs have been isolated for the UDP-rhamnose:anthocyanidin 3-*O*-glucoside *O*-rhamnosyltransferase (A3RT) from *P. hybrida*,[111,112] A5GTs,[88,102] and a UDP-glucose:anthocyanin 3'-*O*-glucosyltransferase (A3'GT).[113]

The recombinant A5GT from *P. hybrida* accepts only delphinidin 3-*O*-(4-coumaroyl)-rutinoside, consistent with the prior biochemical data.[88] In contrast, the recombinant A5GTs from *P. frutescens* and *Verbena hybrida* accept a range of 3-*O*-glycosides and acyl-glycosides, although prior 3-*O*-glycosylation is required.[102] Phylogenetic alignment of UGT sequences with proven functionality shows that the A3GT and A5GT sequences form two distinct groups, with the A3RT falling well separated from either group.[88,114] The separation of the A3RT is due to a distinctive PSPG-box, for the binding of UDP-rhamnose rather than UDP-glucose.

3.4.9.2   Anthocyanin Methyltransferases

There is a wide range of methylated flavonoids that is formed through the action of members of the *S*-adenosyl-L-methionine (SAM)-dependent methyltransferase (MT) family (EC 2.1.1.X). This large family includes enzymes targeting *O*, *C*, *N*, and *S* atoms in the methyl acceptor molecule. Those characterized to date for flavonoid biosynthesis are generally class II *O*-MTs (OMTs), which do not require Mg^{2+} and have MWs typically of 38 to 43 kDa, although one of the smaller MW class I Mg^{2+}-dependent OMT types has also been reported.[115] Although *C*-methylated flavonoids are known, there are no reports on the characterization of the enzymes. Excellent reviews of the plant OMT family are available in Refs. 116 and 117, which include presentations of molecular phylogenies, as does the review of Schröder et al.[87] Structural information on plant OMTs has been gathered from amino acid sequence analysis and the crystal structures for various animal OMTs and one plant OMT, whose crystal structure was solved to a 1.8 Å resolution.[118] Significant conserved regions include motifs for SAM (LExGxGxG) and Mg^{2+} binding (KGTVL).[117]

The well-characterized anthocyanin-related OMTs are those acting at the 3'- and 5'-hydroxyls, encoded by the genes *Mt1*, *Mt2*, *Mf1*, and *Mf2* in *P. hybrida*.[1] Anthocyanins with methylation at the 5- or 7-hydroxyls are also known. The isolation of cDNAs for anthocyanin OMTs has been reported only in the patent literature to date (International Patent Application WO03/062428), although some cDNAs were mentioned in brief in Ref. 119. The patent describes *P. hybrida*, *Fuchsia*, *Plumbago*, and *Torenia* cDNAs encoding OMTs that act on the 3'- or 3',5'-hydroxyls. The recombinant proteins act on delphinidin 3-*O*-glucoside or delphinidin 3-*O*-rutinoside to produce the 3'- or 3',5'-*O*-methylated derivatives. Interestingly, the sequences are closer to those of class I OMTs than class II.

3.4.9.3   Anthocyanin Acyltransferases

Anthocyanin acyltransferases (AATs) catalyze transfer of either aromatic or aliphatic acyl groups from a CoA-donor molecule to hydroxyl residues of anthocyanin sugar moieties; and are part of the general group of acyltransferase enzymes (EC 2.3.1.X). A wide range of activities have been characterized, including enzymes using acetyl-CoA, caffeoyl-CoA, 4-coumaroyl-CoA, malonyl-CoA, and succinyl-CoA. Nakayama et al.[120] list six aromatic and 14 aliphatic AATs, and others, such as those of *Dendranthema* × *morifolium*, have been reported since. Following the publication of the first AAT cDNA in 1998,[121] cDNA clones for two aromatic and six aliphatic AATs have been characterized at the molecular level (Table 3.1). Examples of reactions carried out by these enzymes are shown in Figure 3.3.

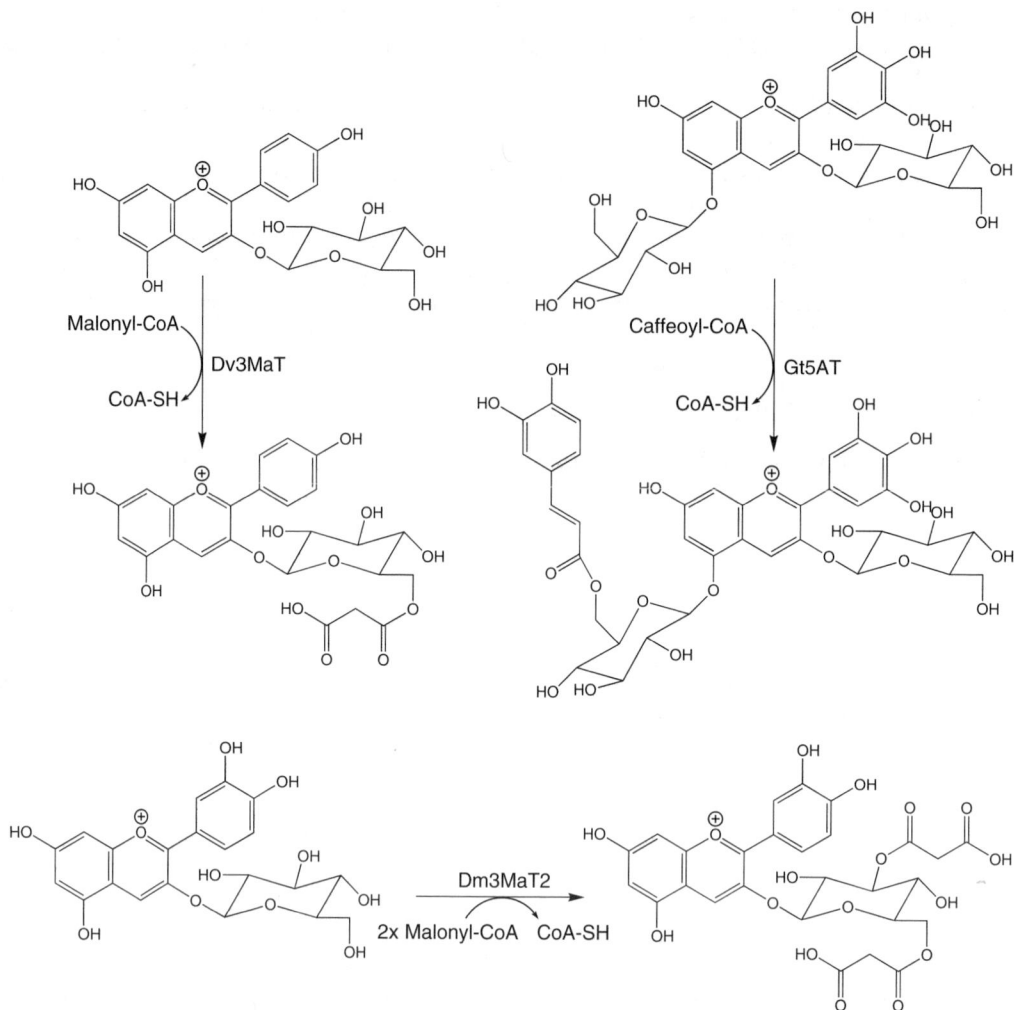

FIGURE 3.3 Examples of aliphatic and aromatic acylation of anthocyanins, as carried out by the malonyl-CoA:anthocyanidin 3-*O*-glucoside-6″-*O*-malonyltransferase (Dv3MaT from *D. variabilis*), malonyl-CoA:anthocyanidin 3-*O*-glucoside-3″,6″-*O*-dimalonyltransferase (Dm3MaT2 from *D.* X*morifolium*), and hydroxycinnamoyl-CoA:anthocyanin 5-*O*-glucoside-6‴-*O*-hydroxycinnamoyltransferase (Gt5AT from *G. triflora*).

Aromatic AAT cDNAs have been isolated from *G. triflora*[121] and *P. frutescens*.[122] The recombinant *G. triflora* protein (Gt5AT) can use either caffeoyl-CoA or 4-coumaroyl-CoA to introduce the HCA group to the glucose at the C-5 position of anthocyanidin 3,5-di-*O*-glucoside. Although specificity is shown with regard to the anthocyanin glycosylation and acylation pattern, the enzyme can accept pelargonidin, cyanidin, or delphinidin derivatives. That the number of hydroxyl groups on the B-ring does not affect reactivity is typical of AATs studied to date. The *P. frutescens* recombinant product (Pf3AT) was identified as a hydroxycinnamoyl-CoA:anthocyanidin 3-*O*-glucoside-6″-*O*-hydroxycinnamoyltransferase, utilizing cyanidin 3-*O*-glucoside and cyanidin 3,5-di-*O*-glucoside (the putative substrates *in vivo*) as well as other anthocyanins. K_m values of 6 to 227 and 113 to 777 μM have been reported for various anthocyanin substrates for the Pf3AT and Gt5AT, respectively, and 24 to 190 μM toward the HCA-CoA esters for a range of aromatic AATs.[120] The isolation and analysis of aromatic AAT cDNAs from additional species, including *P. hybrida* and *Lavandula angustifolia*, has been reported in the patent literature (International Patent Application WO96/25500).

Of the aliphatic AAT cDNAs characterized, three encode malonyl-CoA:anthocyanidin 3-*O*-glucoside-6″-*O*-malonyltransferases (A3MT) from *D. variabilis*, *Senecio cruentus* (cineraria), and *D. × morifolium*.[123–125] The *D. variabilis* and *S. cruentus* recombinant proteins (Dv3MaT and Sc3MaT) accept pelargonidin-, cyanidin-, or delphinidin 3-*O*-glucosides, but do not use anthocyanidin diglycosides. The *D. × morifolium* recombinant protein (Dm3MaT1) also does not accept diglycosides as substrates, but will use a wide range of flavonoid monoglucosides, including anthocyanidin 3-*O*-glucosides and quercetin 3-*O*-glucoside. With regard to acyl donors, Dm3MaT1 has highest activity with malonyl-CoA but also shows significant activity with succinyl-CoA. Other acyl-CoA compounds are either accepted weakly or not at all. K_m values of 19 to 57 μM toward malonyl-CoA have been reported for the aliphatic AATs.[120]

Other aliphatic AAT cDNAs characterized include ones encoding a malonyl-CoA:anthocyanin 5-*O*-glucoside-6‴-*O*-malonyltransferase (Ss5MaT) from *Salvia splendens* (scarlet sage) and a malonyl-CoA:anthocyanidin 3-*O*-glucoside-3″,6″-*O*-dimalonyltransferase (Dm3MaT2) from *D. × morifolium*.[125,126] Ss5MaT shows high specificity, accepting only "bisdemalonyl-salvianin" (a pelargonidin 3,5-di-*O*-glucoside derivative), the endogenous substrate. Recombinant Dm3MaT2 produced two products from cyanidin 3-*O*-glucoside, cyanidin 3-*O*-(6″-*O*-malonylglucoside) and cyanidin 3-*O*-(3″,6″-*O*-dimalonylglucoside), suggesting it carries out sequential acylations at the glucose 6″- and 3″-hydroxyl groups. Both Dm3MaT2 and Ss5MaT show strong preference for malonyl-CoA as the acyl donor, but Ss5MaT also shows significant usage of succinyl-CoA.

All AATs characterized to date are monomers of approximately 50 to 52 kDa in size, and are members of the large versatile acyltransferase (VAT or VPAT) family of enzymes (also known as the BAHD superfamily) involved in many primary and secondary metabolite pathways of plants and fungi.[120,127] Nakayama et al.[120] present detailed sequence comparisons among AATs and a molecular phylogeny for the VAT family. The three A3MT amino acid sequences share >70% amino acid identity. However, although the other AATs, both aliphatic and aromatic, show significant amino acid sequence identity, it is at a relatively low level. For example, Dm3MaT1 has 40% identity to Ss5MaT1, 38% to Gt5AT, and 37% to Pf3AT. Of three significant amino acid motifs identified for VATs, Motifs 1 (HxxxD) and 3 (DFGWG) are thought to be key for catalytic activity, while Motif 2 (YxGNC) might relate to recognition of the anthocyanin, as it is present only in AATs to date. The putative reaction mechanism and the possible role of conserved amino acids, based principally on mutagenesis of recombinant Ss5MaT1, are discussed in detail in Refs. 120, 126, 128. No three-dimensional structure has been published for a member of the VAT family.

The evidence suggests that AATs characterized to date are likely to be localized to the cytosol, with their deduced amino acid sequences lacking known vacuolar targeting sequences. However, it cannot be ruled out that some flavonoid ATs occur in the vacuole. Some serine carboxypeptidase-like (SCPL) enzymes, which use 1-*O*-acylglucosides as acyl group donors and catalyze acylation reactions in plant secondary metabolism, are located in the vacuole, e.g., sinapoylglucose:malate sinapoyltransferase.[129,130] It has previously been proposed that anthocyanin malonylation may occur in the vacuole, and that an SCPL-like activity is involved in the acylation of cyanidin *O*-glycosides in *Daucus carota* (wild carrot) (discussed in Ref. 131).

3.5 FORMATION OF PROANTHOCYANIDINS AND PHLOBAPHENES

The PAs, or condensed tannins, are polymers synthesized from flavan-3-ol monomer units. The phlobaphenes are 3-deoxy-PAs formed from flavan-4-ol monomers. The biosynthesis of both types of PAs follows the biosynthetic route of anthocyanins from chalcones through to the branch points to flavan-3-ol and flavan-4-ol formation. In this section, the specific enzymes forming the monomers are discussed, along with a discussion on the polymerization process. Although the chemistry of tannins is described in detail elsewhere in this book, it is useful to briefly mention the nature of the monomer subunit types and the polymer forms.

Although many variant polymer forms have been reported, the most common ones consist of linear C-4 to C-8 linkages, with the linking bond at the C-4 being *trans* with respect to the 3-hydroxyl. The subunits are usually flavan-3-ol epimers for the 3-hydroxyl, being either 2,3-*trans* or 2,3-*cis*, with the latter being prefixed with "epi." Most commonly these are (2*R*,3*S*)-*trans* or (2*R*,3*R*)-*cis*, although 2*S*-enantiomers do occur, being indicated by the "ent" prefix. Intermediates in the flavonoid pathway up to and including leucoanthocyanidins are 2,3-*trans* in stereochemistry. Flavan-3-ols may be formed by two biosynthetic routes, from either leucoanthocyanidins or anthocyanidins (Figure 3.4). The 2,3-*trans*-flavan-3-ols are produced from the leucoanthocyanidins by LAR, while the 2,3-*cis*-flavan-3-ols are produced from the anthocyanidins by ANR.

The common subunits are those with 3',4'-dihydroxylation of the B-ring (catechin and, most commonly, epicatechin) or 3',4',5'-trihydroxylation (gallocatechin and epigallocatechin). Monohydroxylation of the B-ring (at C-4'), producing the subunits of propelargonidin, is rare, as is the occurrence of subunits with hydroxyl patterns on the A- and C-rings varying from the common one of 3,5,7-hydroxylation. As the subunits for PA biosynthesis are formed

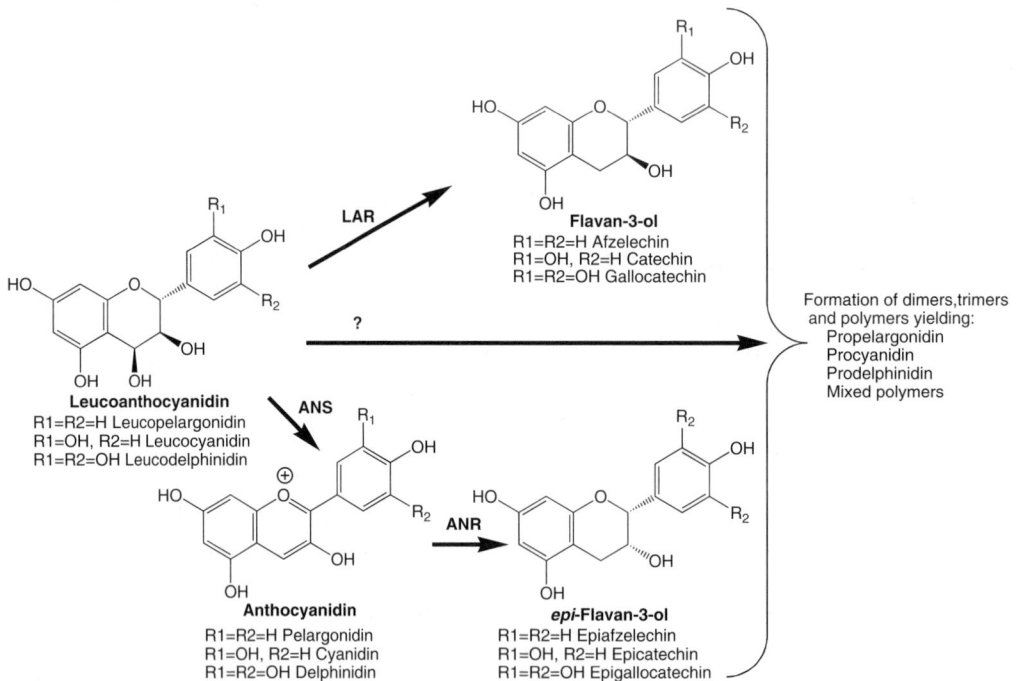

FIGURE 3.4 Biosynthetic route to proanthocyanidins from leucoanthocyanidins. The product of the ANS is given in the flavylium cation form. Enzyme abbreviations are defined in the text and in Table 3.1.

later in the pathway than flavanones, the hydroxylation status of the B-ring of the flavan-3-ols is likely to be determined by the action of the F3′H and F3′,5′H on the precursors.

3.5.1 LEUCOANTHOCYANIDIN REDUCTASE

LAR removes the 4-hydroxyl from leucoanthocyanidins to produce the corresponding 2,3-*trans*-flavan-3-ols, e.g., catechin from leucocyanidin.[5,132] Despite early biochemical characterization, it is only recently that a *LAR* cDNA was isolated and the encoded activity characterized in detail. Tanner et al.[133] purified LAR to homogeneity from *Desmodium uncinatum* (silverleaf desmodium), and used a partial amino acid sequence to isolate a *LAR* cDNA. The cDNA was expressed in *E. coli*, *N. tabacum*, and *Trifolium repens* (white clover), with the transgenic plants showing significantly higher levels of LAR activity than nontransformed plants.

LAR of *D. uncinatum* is a NAPDH-dependent reductase with closest similarity to IFR (a reductase involved in isoflavonoid biosynthesis), and like IFR, DFR, and VR (another isoflavonoid reductase), LAR belongs to the RED protein family. It shares several conserved amino acid sequence motifs with other RED proteins, but has a C terminal extension of approximately 65 amino acids of unknown function. The protein is a monomer with a predicted MW of 42.7 kDa. The preferred substrate is 3,4-*cis*-leucocyanidin, although 3,4-*cis*-leucodelphinidin and 3,4-*cis*-leucopelargonidin are also accepted. Although 2,3-*cis*-3,4-*trans*-leucoanthocyanidins have not been shown to exist as substrates *in vivo*, if LAR accepted them it would raise the possibility of a route to the 2,3-*cis*-flavan-3-ols in addition to the route through ANR. The difficulty in synthesizing 2,3-*cis*-3,4-*trans*-leucoanthocyanidins *in vitro* has prevented a definitive test of LAR activity on these substrates. However, product inhibition is about 100 times greater with catechin than epicatechin, supporting ANR as the main route. There appears to be no LAR sequence or enzyme activity in *A. thaliana*, consistent with the lack of catechin derivatives in that species.[134]

3.5.2 ANTHOCYANIDIN REDUCTASE

ANR converts anthocyanidins (presumably the pseudobase forms) to the corresponding 2,3-*cis*-flavan-3-ols, e.g., cyanidin to epicatechin. For many years little information was available on the formation of 2,3-*cis*-flavan-3-ols. However, the solution emerged from studies on the *banyuls* mutant of *A. thaliana*, which displays precocious accumulation of anthocyanins in the seed coat endothelium.[135] Initially thought to represent an anthocyanin regulatory gene, it was found that the locus encoded a DFR-like protein of 38 kDa, and as *banyuls* mutant plants also lack accumulation of PAs in the seed coat endothelium, it was suggested that the gene might encode LAR.[136] However, recombinant BANYULS from *A. thaliana* and *M. truncatula* did not show LAR activity, but rather reduced pelargonidin, cyanidin, and delphinidin to the corresponding 2,3-*cis*-flavan-3-ols.[137] Low-level production of *ent*-epiafzelechin, *ent*-epicatechin, and *ent*-epigallocatechin was also seen, although these were suspected to be artifacts of the analysis generated by chemical epimerization.[138]

Like LAR, ANR is a member of the RED protein family. Full details of the protein structure and reaction mechanism are yet to be published. However, Xie et al.[138] compared the *A. thaliana* (AtANR) and *M. truncatula* (MtANR) ANR amino acid sequences and recombinant protein activities, and made suggestions on the possible reaction series. The two recombinant proteins showed significantly different kinetic properties, substrate specificities, and cofactor requirements. Although AtANR and MtANR share only 60% sequence identity, some well-conserved domains are evident, in particular the Rossmann dinucleotide-binding domain (GxxGxxG) near the N-termini. However, two amino acid variations did occur in this domain, and this may account for the ability of MtANR to use either NADH or NAPDH, while AtANR will use only NAPDH. Both reduce pelargonidin, cyanidin, and delphinidin, with MtANR preferring cyanidin > pelargonidin > delphinidin and AtANR showing the reverse preference. With regard to *in vivo* activities, the low level of PAs (6% of wild-type levels) in the *F3'H* mutant of *A. thaliana* (*tt7*) suggests that AtANR, or other PA biosynthetic enzymes of *A. thaliana*, have a limited activity against pelargonidin.[134]

The ANR reaction involves a double reduction at the C-2 and C-3 of the anthocyanidin, allowing the inversion of C-3 stereochemistry. Xie et al.[138] postulate four possible reaction mechanisms, proceeding via either flav-3-en-ol or flav-2-en-ol intermediates. The proposed reaction mechanisms are based on anthocyanidins (the flavylium cation forms) as the starting molecules; however, as the authors acknowledge, other forms of the anthocyanidin may exist *in vivo*. In particular, the 3-flaven-2,3-diol pseudobase is thought to be the more likely *in vivo* product of the ANS.

3.5.3   Proanthocyanidin Polymerization

Nonenzymatic studies of polymerization *in vitro* suggest that it starts by the attack of the C-4 of an electrophilic leucoanthocyanidin on the C-8 of the nucleophilic flavan-3-ol initiating subunit to form the initial dimer (the chemistry of the polymerization is reviewed in Refs. 4, 5, 132). Polymerization then continues by addition of leucoanthocyanidin extension subunits in the same C4–C8 route. This suggests that the great majority of the substrate for polymerization *in planta* may not need to pass through either the LAR or ANR biosynthetic steps. However, in most cases, the extension units are 2,3-*cis*, and leucoanthocyanins are 2,3-*trans*. Thus, it may not be leucoanthocyanidins that are the extension units *in planta*, but leucoanthocyanidin derivatives. Thus, intermediates proposed for polymerization include reactive (2*R*,3*S*) compounds derived from leucoanthocyanidins, such as quinone methide and carbocation products. Another, perhaps more likely, source for the common 2,3-*cis* extension units is intermediates of the ANS and ANR biosynthetic route, possibly also as quinone methide and carbocation products. Indeed, the *tannin-deficient seed4* (*tds4*) mutant of *A. thaliana*, which is for the *ANS* gene, accumulates unidentified possible PA intermediate compounds in the developing seed.[139] These seem to fail to be transported into the vacuole as usual and accumulate in the cytoplasm, triggering the formation of multiple small vacuoles. These studies and those of Grotewold et al.[140] on *Z. mays* cell lines suggest that the final steps of PA biosynthesis may be linked to developmental changes in the cell structure. Any enzymatic polymerization process would need to evolve a mechanism for avoidance of inhibitory protein–PA interactions. A membrane-bound biosynthetic complex associated directly with the accumulation of the polymers in the vacuole might achieve this.

The recent molecular characterization of the LAR and ANR has been a major advance in the understanding of PA biosynthesis. However, it is still not known whether the polymerization of PAs occurs spontaneously in all tissues or is enzyme catalyzed in some or all cases. Flavan-3-ols and leucoanthocyanidins will polymerize spontaneously *in vitro* (reviewed in Ref. 5), and no activities responsible for the formation of PA polymers have been described. Differences in PA composition and polymer length do occur between different tissues and during tissue development, and variations from the C4–C8 linkage pattern also occur. However, these variations are not necessarily evidence of an enzymatic mechanism of polymerization, as changes in the availability or reactivity of initiating and extension units might also generate such differences. Nevertheless, a nonenzymatic polymerization process seems unlikely, given that PAs occur in some species with specific arrangements of catechin and epicatechin units. Perhaps the best evidence for biosynthetic steps after LAR and ANR are the mutant lines that have been identified that prevent PA accumulation and appear to occur after monomer formation. At least four PA mutations of *H. vulgare* (*ant25*, *ant26*, *ant27*, and *ant28*) appear to act after LAR, with evidence for *ant26* encoding a polymerization enzyme.[141] In *A. thaliana*, *TDS1*, *TDS2*, *TDS3 TDS5*, and *TDS6* act after *Banyuls*,[139] and *TT9*, *TT10*, *TT11*, *TT13*, and *TT14* act after *TT12*, which encodes a vacuolar transporter for PAs.[142] Although some of the *TDS* mutants may be allelic to the *TT* mutants, this still indicates a number of genetic steps postmonomer formation, which may include polymerization enzymes, regulatory factors, transport activities, and dirigent proteins that control stereospecificity of polymerization.

One other line of evidence supporting enzymatic polymerization comes from recent studies overexpressing ANR in transgenic plants. *N. tabacum* ANR transgenics are reported to have increased levels of at least four compounds that react with the PA stain dimethyla-minocinnamaldehyde (DMACA) but no PA polymers, supporting the requirement for further enzymatic steps.[5,133] However, lack of PA polymers may also be due to conditions in the transgenic tissues studied being unsuitable for spontaneous polymerization. Also, Xie et al.[137] reported results differing to these, as they found PA polymer-like DMACA staining compounds in leaves or petals of *A. thaliana* or *N. tabacum CaMV35S:ANR* transgenics, respectively, although it has been commented that more extensive analysis than what was presented is often required to confirm the PA polymer status.[132]

3.6 FORMATION OF FLAVONES AND FLAVONOLS

A desaturation reaction forming a double bond between C-2 and C-3 of the C-ring is involved in the formation of both flavones and flavonols, and the respective substrates involved, (2*S*)-flavanones and (2*R*,3*R*)-DHFs, differ only in the presence or absence of the 3-hydroxyl (Figure 3.2).

Two distinct FNS activities have been characterized that convert flavanones to flavones. In most plants FNS is a P450 enzyme (FNSII), but species in the Apiaceae have been found to contain the 2OGD FNSI. FNSII cDNAs were first isolated from *G. hybrida*, based on a differential display technique focusing on the conserved P450 heme-binding site,[143] and from *A. majus* and *T. fournieri* using another P450 cDNA as a probe.[144] Isolation of a cDNA for FNSI was first reported from *P. crispum*.[145] Although FNSI and FLS catalyze the equivalent 2,3-desaturation reaction, it is thought that FNSI is most likely to have evolved from the F3H in the Apiaceae.

Flavonols are formed from DHFs by the FLS. A cDNA for FLS was first isolated from *P. hybrida* using degenerate PCR primers for conserved 2OGD sequences,[146] and indeed all FLS cDNAs identified to date encode 2OGD enzymes. However, the *Torenia* FNSII shows FLS activity, raising the suggestion that, analogous to flavone formation, there are two types of FLS.[144] The recombinant *Citrus unshiu* (Satsuma mandarin) FLS had K_m values of 45 and 272 μM for DHK and DHQ, respectively.

A similar reaction mechanism has been proposed for F3H, ANS, FLS, and FNSI, involving *cis*-hydroxylation at C-3 followed by dehydration, with (2*R*,3*S*)-*cis*-DHFs as possible intermediates.[81,84] Akashi et al.[144] have suggested that the FNSII reaction likewise involves a C-2 hydroxylation step followed by dehydration. However, the studies of Martens et al.[143,145,147] suggest direct 2,3-desaturation of flavanones by FNSI and FNSII, as previously proposed from biochemical studies of FNSI.[148] A comparison of 59 2OGD amino acid sequences, including those for ANS, F3H, and FLS, identified three regions of high similarity and eight absolutely conserved amino acids.[62] These include residues with proposed functions in Fe^{2+} and 2OG binding, and two others of unknown function that are required for enzyme activity.[62,67]

Recent studies of members of the flavonoid 2OGD family show overlapping substrate and product selectivities *in vitro*. For example, the *C. unshiu* FLS has been termed a bifunctional enzyme, as the recombinant protein has both FLS and F3H activity.[62] Two groupings of flavonoid 2OGDs have been proposed, FLS/ANS and F3H/FNSI, with the former having wider substrate selectivity than the latter.[62,81,147] The overlapping activities are discussed further in Section 3.14.

3.7 FLAVONE AND FLAVONOL MODIFICATION ENZYMES

Flavones and flavonols are the substrates for a range of modification reactions, including glycosylation, methylation, acylation, and sulfation. To date, genes and cDNAs have been cloned that represent activities specific to flavones or flavonols for all of these modifications except acylation. There are also several cDNAs isolated, the encoded proteins of which accept a range of flavonoid, and even nonflavonoid, substrates. However, *in vitro* activities of recombinant proteins may not reflect their *in vivo* role. Factors such as the abundance of the protein (temporally or spatially) in relation to the potential substrate, and the involvement of metabolic channeling (see also Section 3.14), affect *in vivo* activity. However, in various transgenic experiments endogenous flavonoid GTs have been shown to accept substrates that are not normally present in the recipient species, such as 6′-deoxychalcones and isoflavonoids,[149–151] supporting broad substrate acceptance of some modification enzymes.

3.7.1 FLAVONE AND FLAVONOL GLYCOSYLTRANSFERASES

Clones have now been isolated for several types of UGT with *O*-glycosylation activity on flavones and flavonols. The recombinant proteins show wide substrate acceptance, including some that will accept both flavonoids and some of the biosynthetically unrelated betalain pigments. In general, the UGTs characterized in flavonoid biosynthesis show high regiospecificity but broad substrate acceptance, although there are exceptions.

A number of cDNAs encoding UGT activities against the 7-hydroxyl have been identified. NtGT2 from *N. tabacum* showed activity against several types of phenolic compounds.[152] Despite having closest sequence identity with A5GT sequences, no significant activity with anthocyanins was found but it catalyzed the transfer of glucose to the 7-hydroxyl of flavonols, with a K_m of 6.5 μM for the aglycone kaempferol. The recombinant protein from the *UGT73C6* gene of *A. thaliana* also transfers glucose to the 7-hydroxyl of a range of flavonols and flavones, as well as the 6-hydroxyl of the unnatural 6-hydroxyflavone substrate. However, its *in vivo* activity is likely as a UDP-Glc:flavonol-3-*O*-glycoside-7-*O*-glucosyltransferase.[153] The recombinant *Allium cepa* (onion) UGT73G1 protein also showed wide regiospecificity, adding glucose to the 3-, 7-, and 4′-hydroxyls of a wide range of flavonoids, including chalcones, flavanones, flavones, flavonols, and isoflavones, producing both mono- and diglucosides.[154] The lack of triglucoside products suggests UGT73G1 accepts only aglycone or monoglucoside substrates. Recombinant protein from a UGT cDNA from the Chinese medicinal herb *Scutellaria baicalensis* also showed activity toward the 7-hydroxyl of flavonoids, among other substrates.[155] In contrast to these activities, recombinant protein from *A. cepa* UGT73J1 showed both high regiospecificity and tight substrate specificity, adding glucose at the 7-hydroxyl of only quercetin 3-*O*-glucoside (a flavonol) and genistein (an isoflavonoid) out of many flavonoid substrates tested.[154]

A cDNA from *Vigna mungo* (black gram) seedlings encodes a protein with UDP-galactose:flavonoid 3-*O*-galactosyltransferase (F3GalT) activity.[156] A 20-fold preference for UDP-galactose over UDP-glucose was found with kaempferol as a substrate. It would accept DHFs and flavones at a lower efficiency (anthocyanins were not tested). Average amino acid sequence identity is around 35 to 45% with F3GTs and 23% to the A3RT of *P. hybrida*.

The *UGT78D1* gene of *A. thaliana* encodes a specific UDP-rhamnose:flavonol-3-*O*-rhamnosyl-transferase with activity only on kaempferol and quercetin, out of various flavonoids tested.[153]

Like many flavonoid UGTs, the UGTs reported to be involved in betalain biosynthesis in *Dorotheanthus bellidiformis* (Livingstone daisy) show precise regiospecificity but surprisingly wide substrate acceptance. Recombinant betanidin 5-*O*-glucosyltransferase will add a glucose to the 4'- and 7-hydroxyls of luteolin (a flavone) or quercetin, and the 4'-hydroxyl of cyanidin with lower efficiency;[157] betanidin 6-*O*-glucosyltransferase (B6GT) will add a glucose to the 3-hydroxyl of quercetin and cyanidin.[114] B6GT shows marked divergence at the amino acid level from the previously characterized F3GTs (for a molecular phylogeny, see Ref. 114).

P. hybrida pollen accumulates kaempferol and quercetin 3-*O*-(2″-*O*-glucopyranosyl)-galactopyranosides, which are not prevalent elsewhere in the plant, by the action of flavonol 3-*O*-galactosyltransferase (PhF3GalT) and flavonol 3-*O*-galactoside-2″-*O*-glucosyltransfer-ase (F2″GT). Miller et al.[158] isolated a cDNA for a pollen-specific gene from *P. hybrida* whose recombinant protein showed the same activity profile as the previously characterized PhF3GalT.[159] Unlike most of the GTs discussed previously, the PhF3GalT showed strong preference and high catalytic efficiency to kaempferol and quercetin, with other lower activities being limited to a range of flavonol aglycones. Notably, the PhF3GalT also catalyzed the reverse reaction, a deglycosylation. The enzyme, therefore, could be involved in modulating the abundance of a biologically active aglycone.

3.7.2 FLAVONE AND FLAVONOL METHYLTRANSFERASES

Methylation has been reported at all available hydroxyls of flavones and flavonols (the C-5, -6, -7, -8, -2', -3', -4', and -5' positions), and it can occur on both aglycone and glycoside substrates. Many of the corresponding enzyme activities have been described in detail, and typically show strong preferences with regard to substrate type and the position methylated.[3] Recently, cDNA sequences have been identified for several flavonoid OMTs, allowing sequence-based analysis and examination of recombinant protein activities. All are members of the SAM-MT family described in Section 3.4.9.2.

Induced in leaves of *H. vulgare* in response to pathogen attack is an mRNA encoding a flavonoid 7-OMT with activity against flavanone, flavone, and flavonol aglycones, with the flavone apigenin the preferred substrate.[160] Caffeic acid or glycosylated flavonoids were not accepted as substrates. Gauthier et al.[161,162] have characterized cDNA clones for two distinct enzyme activities of the semiaquatic freshwater plant *Chrysosplenium americanum* that methy-late the 3' and 5' hydroxyls of flavonoids. Recombinant F3'OMT specifically methylated the 3' or 5'-hydroxyls of 3,7,4'-trimethoxyquercetin, accepting neither quercetin nor mono- or dimethylquercetin as substrates.[161] However, in contrast, recombinant proteins from the two highly similar clones *CaOMT1* and *CaOMT2* showed 3'-OMT activity against luteolin and quercetin, and lower 3- or 5-OMT activities on caffeic and 5-hydroxyferulic acids, respect-ively.[162] An *A. thaliana* cDNA, *AtOMT1*, was originally identified as encoding a HCA OMT,[163] and the deduced amino acid sequences from *CaOMT1*, *CaOMT2*, and *AtOMT1* show high sequence identity (around 85%), and even higher identity across putative sequence motifs relating to substrate specificity and binding.[117]. Recombinant protein from *AtOMT1* showed flavonol 3'-OMT activity, using quercetin and myricetin (flavonol aglycones) effi-ciently; however, it had much lower activity with luteolin and did not accept HCAs.[164] The initial identification of *AtOMT1* as a HCA OMT based on sequence similarity illustrates the potential problems in using sequence alone to predict function (discussed in detail for OMTs in Ref. 165).

A cDNA for an OMT (PFOMT) with wide substrate acceptance that includes flavonols and HCA derivatives has been identified from *Mesembryanthemum crystallinum* (ice plant).[115]

In contrast to the other flavonol OMTs reported, the deduced amino acid sequence of PFOMT has most similarity to caffeoyl-CoA OMTs (CCoAOMTs), and it is a class I OMT with a MW of 27 kDa and a requirement for a bivalent cation such as Mg^{2+}. A wide range of flavonols, including 6-hydroxykaempferol, quercetin, 6-hydroxyquercetin (quercetagetin), 8-hydroxyquercetin (gossypetin), myricetin, and quercetin 3-O-glucoside, were accepted as substrates by the recombinant protein, as were some flavones, flavanones, and HCA-CoA esters and glucosides. Generally, potential substrates with two free hydroxyls were accepted while those with a single free hydroxyl were not. The reaction product for quercetin was shown to be isorhamnetin (3'-methoxyquercetin), while with quercetagetin five different products with 5-O-, 6-O-, 3'-O-, 5,3'-O-, or 6,3'-O-methylation were generated. This range of substrate choice and products with the recombinant protein is wider than for the purified native enzyme, the major products of which are only the 6-O- and 6,3'-O-methyl ethers. This difference has been shown to be due to the N terminal region of the protein, as a recombinant protein with the first 11 N terminal amino acids removed shows the same enzyme character-istics as the native enzyme.[166] The dual methylation reaction suggests a large and flexible active site, which is rare for the OMTs characterized to date.

Ibdah et al.[115] also examined the wider substrate acceptance of four recombinant CCoAOMTs; one each from *Stellaria longipes* (chickweed) and *A. thaliana* with high se-quence similarity to PFOMT, and one each from *N. tabacum* and *V. vinifera* with lower sequence similarity. The CCoAOMT from *S. longipes* showed the same range of substrate acceptance as PFOMT, and the same range of products from quercetagetin. The *A. thaliana* protein efficiently accepted a similar wide range of substrates, but produced only the 6-O- and 6,3'-O-methyl ethers of quercetagetin. In contrast, the *N. tabacum* and *V. vinifera* enzymes showed strong preference for caffeoyl-CoA, although they would accept other flavonoids with much lower efficiencies. The *in vitro* activity pattern is reflected by phylogenetic analysis, which groups the *M. crystallinum* and *S. longipes* sequences separately to a large group of class I CCoAOMTs (which include the *N. tabacum* and *V. vinifera* sequences).

A cDNA, *CrOMT2*, encoding a flavonoid 3',5'-OMT has been identified from *C. roseus* (during a study of alkaloid biosynthesis). The encoded protein could sequentially methylate the 3'- and 5'-hydroxyls of both myricetin and DHM, and showed weaker activity against the 3'-hydroxyl of DHQ.[167] Recombinant F3H, FNS, FLS, and ANS were all able to use the 3'-O-methylated substrates.[87] 3',5' O-methylation is characteristic of both flavonol and antho-cyanin glycosides of *C. roseus* (hirsutidin and malvidin glycosides occur), so it is possible that this represents the *in vivo* activity. Whether a separate anthocyanin 3',5'-OMT exists in *C. roseus*, or whether the *CrOMT2*-derived enzyme also methylates anthocyanins is not known. Schröder et al.[87] attempted to isolate a cDNA for anthocyanin 7-OMT from *C. roseus*, and in the process identified a clone (*CrOMT6*) encoding an OMT that specifically accepted 3'-O-methylated flavonoids as substrates, in particular flavanones, flavones, and flavonols, to produce the 3',4'-O-methylated derivatives. In a molecular phylogeny the flavonoid-related *CrOMT* sequences form a separate cluster from the *CaOMT* and *AtOMT1* sequences.

3.7.3   FLAVONE AND FLAVONOL SULFOTRANSFERASES

Flavonoids esterified with sulfate groups have been reported to occur in many plant species, in particular mono- to tetrasulfate esters of flavonols and flavones, and their methylated or glycosylated derivatives. These are likely generated by soluble sulfotransferases (STs), which transfer a sulfonate group from 3'-phosphoadenosine 5'-phosphosulfate (PAPS).[168] Two subgroups of STs have been reported. The first contains enzymes with generally wide substrate acceptance that are typically involved in detoxification of small metabolites. Enzymes of the second subgroup, which includes the flavonoid STs, show high specificity, and in animals are involved in processes such as steroid transport and inactivation. The first plant STs characterized at the molecular level were the flavonol 3-*O*- and 4′-*O*-STs (F3ST and F4′ST) from *Flaveria chloraefolia*,[169] followed by a second F3ST, BFST3, from *Flaveria bidentis*.[170] These enzymes form part of a group of flavonol STs that act sequentially to generate the range of flavonol polysulfates found in this genus. Strict specificities are shown to the 3-hydroxyl of flavonol aglycones (F3ST), or the 3′-, 4′- (F4′ST), or 7-hydroxyls of flavonol 3,3′ or 3,4′-disulfates.[171] Analysis of the recombinant proteins found the *F. chloraefolia* F3ST used only flavonol aglycones as substrates (both kaempferol and quercetin, and the methylated rhamnetin and isorhamnetin), and the F4′ST used only the flavonol 3-*O*-sulfates. BFST3 recombinant protein showed similar activity to the *F. chloraefolia* F3ST, except that kaempferol was not accepted as a substrate.

A range of ST cDNAs has been identified from functional genomics studies of *A. thaliana*,[169] including one (*AtST3*) that has been shown to encode a flavonoid 7-ST. Unlike BFST3, AtST3 recombinant protein accepts a number of flavonol and flavone aglycones, as well as their 3-*O*-monosulfated derivatives.[169] However, strict specificity to the 7-hydroxyl was found.

The plant soluble STs have around 25 to 30% amino acid identity with mammalian soluble STs, and are of a similar size.[169] Comparisons between *F. chloraefolia* F3ST and F4′ST, combined with mutational analysis and data from the crystal structure of mouse estrogen ST, have defined amino acid residues important for PAPS binding, substrate binding and catalysis, and the mechanism of sulfonate transfer.[172–175] Sequence relatedness has been used to divide the STs into families and subfamilies in a similar manner as for P450s.[169]

3.8 BIOSYNTHESIS OF 5-DEOXYFLAVONOIDS

A characteristic of legumes is the biosynthesis of 6′-deoxychalcones (chalcones lacking a hydroxyl at the C-6′ position), which are the substrates for the production of 5-deoxyflavonoids. The formation of 6′-deoxychalcones requires the activity of polyketide reductase (PKR) (also known as chalcone reductase or chalcone ketide reductase) in conjunction with CHS. It is thought that CoA-linked polyketide intermediates diffuse in and out of the CHS active site, and while unbound are reduced to alcohols by PKR.[46] The resultant hydroxyl groups are then removed from the PKR products in the final cyclization and aromatization steps catalyzed by CHS.

PKR is a NADPH-dependent monomeric enzyme of 34 to 35 kDa belonging to the aldo- and keto-reductase superfamily.[176] The first isolation and characterization of a PKR cDNA was from *G. max*.[177] The *G. max* cDNA, and cDNAs from other species, have been used to confirm the PKR activity of the recombinant protein, and to produce larger protein amounts for structural analysis.[177–179] Studies of the recombinant protein, and analysis of *35SCaMV:PKR* transgenic plants,[58,149] have also shown that PKR is able to function with CHS proteins from nonlegume species that synthesize only the common 5-hydroxyflavonoids.

3.9 BIOSYNTHESIS OF ISOFLAVONOIDS AND THEIR DERIVATIVES

Principally found in legumes, isoflavonoids are a group of compounds that originate from flavanones. The factor differentiating isoflavonoids from other flavonoids is the linking of the B-ring to the C-3 rather than the C-2 position of the C-ring. Subsequent modifications can result in a wide range of structural variation, including the formation of additional heterocyclic rings. The additional rings are principally methylenedioxy or dimethylchromene types, formed from cyclization between vicinal hydroxyl and methoxyl or monoprenyl groups, respectively. The initial steps of isoflavonoid biosynthesis are now well characterized at the molecular level, but there is limited progress on the later enzymatic steps that produce the wide range of complex derivatives found in different legume species. A general scheme for their biosynthesis is presented in Figure 3.5 and Figure 3.6.

3.9.1 2-Hydroxyisoflavanone Synthase

The entry into the isoflavonoid branch of the pathway occurs through the action of 2-hydroxyisoflavanone synthase (2HIS, also known as isoflavone synthase, IFS). 2HIS catalyzes both C-2 to C-3 aryl migration and hydroxylation of the C-2 of (2S)-flavanones to yield (2R,3S)-2-hydroxyisoflavanones. Dehydration of the 2-hydroxyisoflavanones, either spontaneously or through the action of the isoflavone dehydratase (IFD), then forms the isoflavones. The isoflavones formed will be either 5-hydroxy or 5-deoxy compounds, depending on whether they originate from the 6'-hydroxy- or the 6'-deoxychalcone pathways, respectively (e.g., genistein from naringenin and daidzein from liquiritigenin).

2HIS cDNAs were first isolated from *G. echinata* and *G. max* by a variety of functional genomics approaches.[180–182] These cDNAs were used to identify additional sequences from many legumes and some nonlegume species (see, e.g., Ref. 181), and at the time of this review there were 2HIS sequences from 14 species in public databases. The cDNAs encode P450s that have been classified as part of the CYP93 family (CYP93C) that includes FNSII (CYP93B), flavanone 2-hydroxylase (F2H, CYP93B1), and pterocarpan 6a-hydroxylase (P6aH, CYP93A). Recombinant *G. max* 2HIS expressed in insect cells was able to form isoflavones without measurable accumulation of 2-hydroxyisoflavanone intermediates.[182] This suggests that the dehydration either occurs spontaneously or as part of the 2HIS reaction, without the need for the previously proposed, and partially characterized,[183] IFD activity. However, when *G. echinata*, *G. max*, or *L. japonicus* cDNAs were expressed in yeast, 2-hydroxyisoflavanones could be identified.[180,181] Liquiritigenin was converted at a greater efficiency than naringenin, although the extent of the variation in efficiency was different for recombinant 2HIS prepared from insect or yeast systems.

2HIS sequences from different species generally show high amino acid identity scores. Some of the amino acid sequences key to the reaction have been identified by computer analysis and mutational analysis of recombinant proteins.[184] Changing Ser310 of CYP93C2 to Thr gave increased formation of 3-hydroxyflavanone (i.e., DHF), usually a minor product of the 2HIS reaction. Furthermore, replacing Lys375 with Thr gave F3H-like activity producing only DHFs. Other residues are also important for the reaction, as introduction of the defined Ser and Lys residues into the F2H sequence did not confer the ability to carry out aryl migration or 3-hydroxylation. The results support the suggestion of Hashim et al.[185] that the reaction proceeds by a radical generation at C-3 followed by migration of the aryl group from C-2 to C-3, leaving the hydroxyl at C-2.

3.9.2 4'-O-Methylation

Subsequent to the 2HIS step, a series of reactions lead to a range of plant defense compounds whose exact structures vary between species, in particular pterocarpans — such as glyceollins in *G. max* and phaseollins in *P. vulgaris* (Figure 3.5–Figure 3.7). The glyceollins and phaseollins have a free 4'-hydroxyl, but in some legume species, such as *Cicer arietinum* (chickpea), the 4'-O-methylated versions of daidzein and genistein occur, named formononetin and biochanin A, respectively.

Akashi et al.[186] suggested that the 2-hydroxyisoflavanone product of 2HIS (e.g., 2,7,4'-trihydroxyisoflavanone that dehydrates to daidzein) might be the substrate for the 4'-O-methylation reaction *in vivo*, rather than daidzein itself. This biosynthetic route is

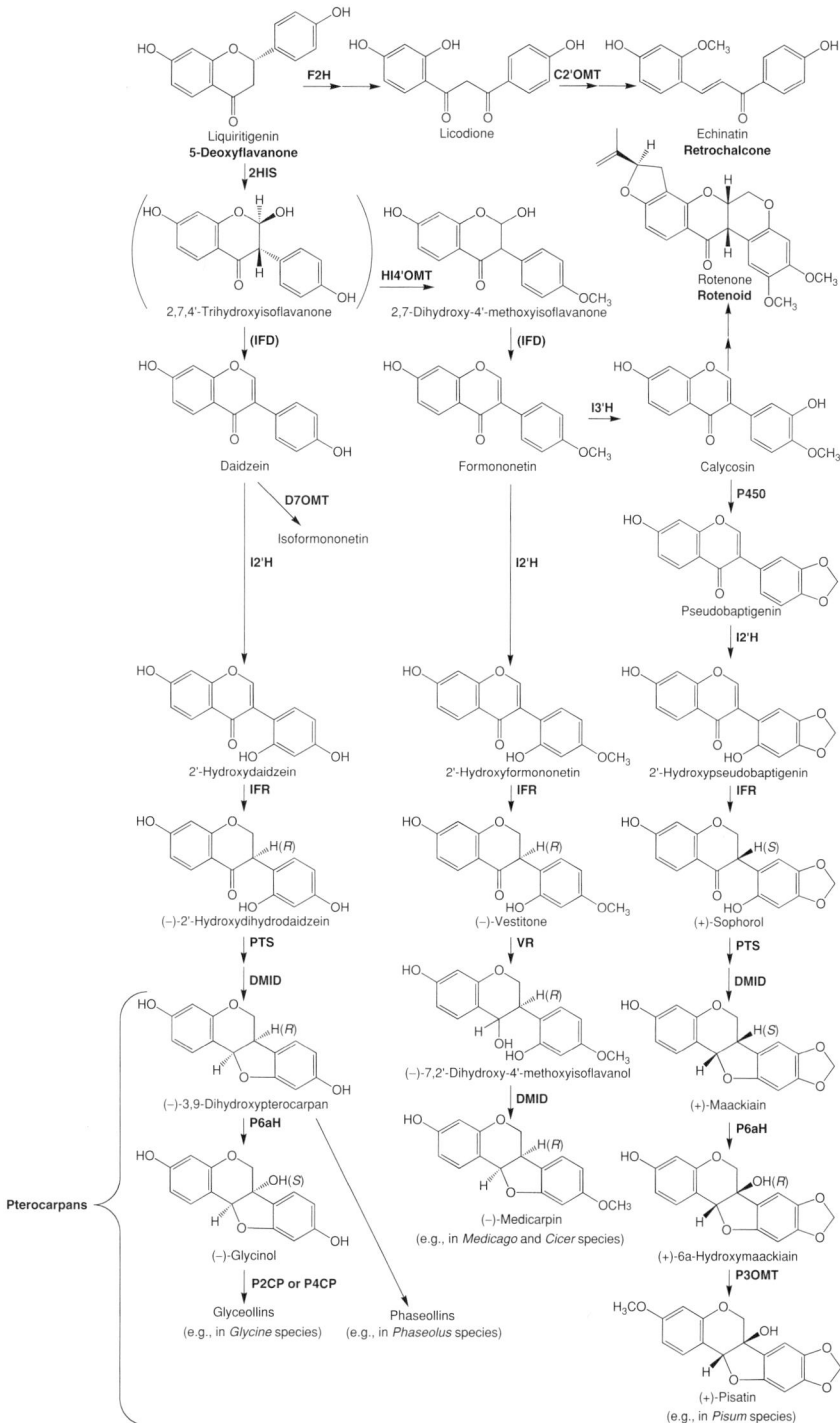

FIGURE 3.5 Biosynthetic route to isoflavonoids (and some derivatives) from the 5-deoxyflavanone liquiritigenin. A possible route to the retrochalcone echinatin is also shown. Unlabelled arrows indicate biosynthetic steps for which the enzyme(s) have not been characterized. Enzyme abbreviations are defined in the text and in Table 3.1, except for P2CP, pterocarpan 2-*C*-prenyltransferase; P4CP, pterocarpan 4-*C*-prenyltransferase.

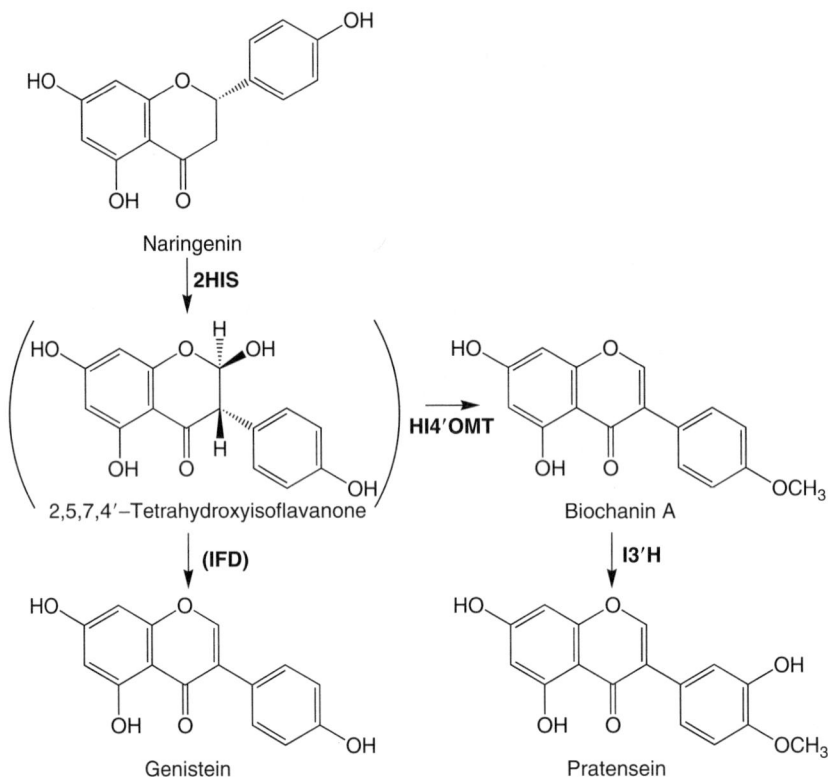

FIGURE 3.6 Biosynthetic route to isoflavones from the 5-hydroxyflavanone naringenin. Enzyme abbreviations are defined in the text.

supported by the isolation of a cDNA from *G. echinata* whose recombinant protein has 4'-OMT activity against 2,7,4'-trihydroxyisoflavanone but not daidzein, thus indicating it is a hydroxyisoflavanone 4'-OMT (HI4'OMT).[187]

An alternative biosynthetic route for the introduction of the 4'-methoxyl has been suggested from studies of *M. sativa*. Initial studies of one of the four isoflavone OMTs (IOMTs) of *M. sativa* showed it had 7-*O*-methylation activity to the A-ring of daidzein, thus forming isoformononetin from daidzein. However, overexpression of this IOMT in transgenic *M. sativa* plants enhanced the biosynthesis of 4'-*O*-methylated isoflavonoids.[188–190] The crystal structure of the IOMT[118] indicates the enzyme could accept 2,7,4'-trihydroxyisoflavanone in addition to daidzein, with the SAM methyl donor arranged close to the 7- or 4'-hydroxyl of the respective possible substrates, supporting the observed *in vitro* and *in vivo* reactions. Furthermore, IOMT8 (one of the other *M. sativa* IOMTs) was shown to localize to the endomembranes, with which 2HIS is associated, after induction of isoflavonoid biosynthesis in transgenic *M. sativa* plants.[191] Liu and Dixon[191] have proposed that close physical association of 2HIS and the IOMT causes metabolic channeling of 2,7,4'-trihydroxyisoflavanone, ensuring its 4'-*O*-methylation and the subsequent formation of formononetin, rather than its dehydration to daidzein and subsequent 7-*O*-methylation.

HI4'OMT from *G. echinata* is thought to be distinct from the *M. sativa* IOMT, because a separate daidzein 7-OMT is present in *G. echinata*, prompting the suggestion that the IOMT be renamed D7OMT.[187] The HI4'OMT amino acid sequence is closely related to that of the SAM:(+)-6a-hydroxymaackiain 3-*O*-methyltransferase (HM3OMT), which carries out a similar reaction in (+)-pisatin biosynthesis in *Pisum sativum* (pea) (see Section 3.9.7). The

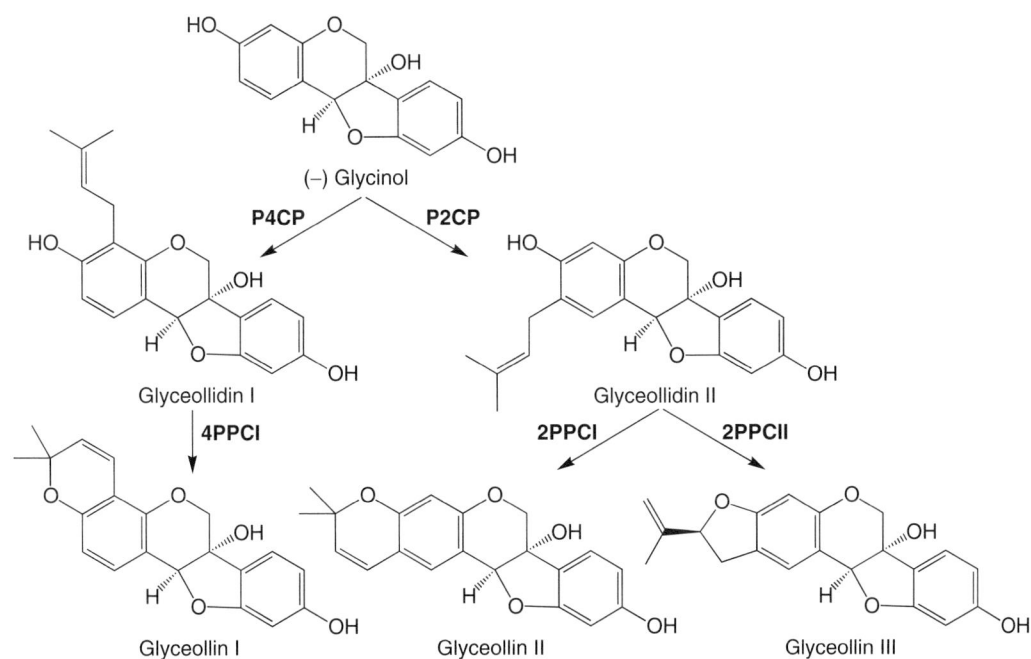

FIGURE 3.7 Biosynthetic route to complex glyceollin phytoalexins from glycinol. Enzyme abbreviations are P2CP, pterocarpan 2-*C*-prenyltransferase; P4CP, pterocarpan 4-*C*-prenyltransferase; 2PPCI/2PPCII, prenylpterocarpan cyclases acting at the C-2 prenyl group; 4PPCI, prenylpterocarpan cyclase acting at the C-4 prenyl group.

HI4′OMT protein showed activity against the 3-hydroxyl of a compound related to (+)-6a-hydroxymaackiain, (±)-medicarpin, suggesting HI4′OMT may be functionally identical to HM3OMT. However, the HM3OMT substrate is only found in species making (+)-pisatin. The *G. echinata* HI4′OMT cDNA was used to isolate HI4′OMT cDNAs from *L. japonicus*, *M. truncatula,* and other legumes. Both HI4′OMT and IOMT may be involved in formono-netin biosynthesis, perhaps in the same tissues, and the formation of heterodimers of similar OMTs has been reported.[192]

3.9.3 Isoflavone 2′- and 3′-Hydroxylase

Two P450s of the CYP81E subfamily, isoflavone 2′-hydroxylase (I2′H) and isoflavone 3′-hydroxylase (I3′H), catalyze key steps in the formation of the more complex isoflavonoids. Hydroxylation of formononetin at the C-3′ position produces calycosin, a precursor for subsequent methylenedioxy bridge formation (yielding pseudobaptigenin) as part of the branches leading to maackiain- and pisatin-type phytoalexin end products, as well as to the rotenoids and other complex derivatives. Hydroxylation at the C-2′ position of daidzein, formononetin, or pseudobaptigenin provides the hydroxyl required for C–O–C bridge for-mation that defines the pterocarpans (e.g., glycinol).

A cDNA encoding the I2′H was first identified among a group of cDNAs from *G. echinata* cell lines representing elicitor-induced mRNAs for P450s.[193] The recombinant protein catalyzed the 2′-hydroxylation of formononetin and daidzein, and had no activity with HCAs or flavanones. Subsequently, cDNA clones have been isolated from *M. truncatula* (*CYP81E7*) and other species.[194] *CYP81E7* expressed in transgenic *A. thaliana* conferred the ability to form 2′-hydroxyformononetin when formononetin was supplied. A closely related cDNA to *CYP81E7*, *CYP81E9*, was found to encode the I3′H from *M. truncatula*.[194] Recombinant MtI2′H and MtI3′H showed preference to the 5-hydroxylated substrate biochanin A over the 5-deoxy equivalent, formononetin. MtI2′H also showed the expected activity with pseudobaptigenin, daidzein, and genistein, but with reducing preference in each case. For daidzein and genistein, a lesser 3′-hydroxylase activity was also found. The preferred substrates for recombinant MtI3′H were, in descending order, biochanin A, formononetin, and 2′-hydroxyformononetin. K_m values of 67 and 50 μM toward formononetin were found for MtI2′H and MtI3′H, respectively. The I2′H and I3′H genes are differentially expressed in *M. truncatula*, in terms of spatial patterns and response to biotic and abiotic stresses.[194] The data suggest a role for I3′H in the formation of compounds such as rotenoids in response to insect attack, and I2′H in biosynthesis of antimicrobial isoflavonoids.

3.9.4 ISOFLAVONE REDUCTASE

The 2′-hydroxyisoflavones are reduced to the corresponding isoflavanones by a NADPH-dependent isoflavone reductase (IFR). The isoflavanones are the final isoflavonoid intermediates of pterocarpan biosynthesis. Variant IFR activities between species are thought to contribute to the stereochemistry of the pterocarpans produced, in particular, (+)-maackiain in *P. sativum*, (−)-maackiain in *C. arietinum*, (−)-3,9-dihydroxypterocarpan in *G. max,* and (−)-medicarpin in *M. sativa*. The (−) indicates 6a*R*11a*R* stereochemistry.

Clones corresponding to IFR were first isolated from *M. sativa*, using an antibody raised against the *P. sativum* protein to screen an expression cDNA library,[195] from *C. arietinum* based on protein purification and sequencing,[196] and subsequently from *P. sativum*.[197] IFR has a calculated MW of 34 to 35 kDa, and is a member of the RED protein family. The recombinant *M. sativa* enzyme converted 2′-hydroxyformononetin to the expected (3*R*) isomer of vestitone, and would accept 2′-hydroxypseudobaptigenin, but not formononetin, pseudobaptigenin, or several other flavonoids tested. These are similar substrate preferences to those of the purified *C. arietinum* enzyme. Surprisingly, the recombinant *P. sativum* enzyme produced a (3*R*)-isoflavanone product, (−)-sophorol, from 2′-hydroxypseudobaptigenin, rather than the (3*S*)-stereoisomer, (+)-sophorol, that would be expected from the accumulation of (+)-maackiain in this species (Figure 3.5). Thus, it is possible that an epimerase is also involved in the biosynthetic pathway to the (+)-isoflavonoids, at least in *P. sativum*.

Using the crystal structures of two related RED enzymes of lignan biosynthesis, a provisional molecular model has been produced for the *M. sativa* IFR.[198] A smaller binding pocket in the protein, in comparison to the other enzymes, is suggested to account for the specific enantiomer binding and processing of IFR.

3.9.5 VESTITONE REDUCTASE

Based on analysis of enzyme preparations, the conversion of isoflavanones to pterocarpans was thought initially to be catalyzed by a single NADPH-dependent enzyme, termed the pterocarpan synthase (PTS). However, it was subsequently shown that in *M. sativa* the conversion of vestitone to medicarpin involves two enzymes, VR and 7,2′-dihydroxy-4′-methoxyisoflavanol (DMI) dehydratase (DMID).[199] The reaction series from vestitone to the pterocarpan is thought to proceed by the VR-catalyzed reduction of vestitone to DMI, followed by the loss of water and formation of the C–O–C bridge between the heterocycle and the B-ring, catalyzed by DMID.

Guo and Paiva[200] used the amino acid sequence of purified VR to isolate a cDNA clone from *M. sativa*, which was then analyzed by expression in *E. coli*. VR, a monomeric 38 kDa protein, is another member of the RED family. It converts (3*R*)-vestitone, but not (3*S*)-vestitone, to DMI *in vitro*, although the stereospecificity may vary with VR enzymes from other species.[200] Little has been published on other VR cDNAs, although the *G. max* VR sequence is available in US patent #6,617,493. The isolation of DNA sequences for DMID has not been reported.

In some species, such as *C. arietinum* and *M. sativa*, the products of VR and DMID (maackiain and medicarpin) are the main pterocarpan phytoalexins. They are typically glucosylated and malonylated and stored in the vacuole.[201] In species such as *G. max*, *P. sativum*, and *P. vulgaris*, the pterocarpans are further converted by a series of reactions to species-specific compounds. For *G. max* and *P. sativum*, the initial reaction is a hydroxylation catalyzed by P6aH.

3.9.6 PTEROCARPAN 6A-HYDROXYLASE

The main phytoalexins in *G. max* (glyceollins) and *P. sativum* (pisatin) are pterocarpans hydroxylated at position C-6a, in a reaction carried out by the P450 P6aH (at least in *G. max*). Recombinant CYP93A1, from a cDNA isolated from elicitor-induced *G. max* cells, carries out the stereospecific and regioselective hydroxylation at the 6a position of (6a*R*,11a*R*)-3,9-dihydroxypterocarpan, with a K_m of 0.1 µ*M*, to yield 3,6a,9-trihydroxypterocarpan.[202] Given the specificity of the reaction, the encoded protein was termed the dihydroxypterocarpan 6a-hydroxylase (D6aH). P6aH activities are present in other species, but analysis of cDNA clones has not been published.

3.9.7 SAM:6A-HYDROXYMAACKIAIN 3-*O*-METHYLTRANSFERASE

Methylation of the 3-hydroxyl of (+)-6a-hydroxymaackiain by HM3OMT produces the major phytoalexin of *P. sativum*, (+)-pisatin. Two closely related cDNAs for HM3OMT were isolated from a cDNA library prepared from pathogen-induced *P. sativum* mRNA using antibodies prepared to the purified enzyme.[203] The recombinant proteins had highest activity with (+)-6a-hydroxymaackiain, a lower activity with (+)-medicarpin, and low or no activity with (−)-6a-hydroxymaackian, (−)-medicarpin, (−)-maackiain, isoliquiritigenin, daidzein, or formononetin. One of the HM3OMT proteins also had significant activity with (+)-maackiain although this is unlikely to be an *in vivo* substrate, as 3-*O*-methylmaackiain is not observed in *P. sativum* tissues.

3.9.8 PTEROCARPAN PRENYLTRANSFERASES

The formation of phytoalexins such as glyceollins and phaseollins requires *C*-prenylation by a range of pterocarpan prenyltransferase (PTP) activities, with dimethylallyl pyrophosphate (DMAPP) as the prenyl donor. For glyceollins and phaseollins, prenylation occurs at position C-2 or C-4 of glycinol or C-10 of 3,9-dihydroxypterocarpan.[204,205] However, there are differing activities in other species. For example, in *Lupinus albus* (white lupin) a prenyltransferase acting at the C-6, -8, and -3′ positions of isoflavones has been identified.[206] PTPs have also been characterized in detail for the formation of prenylated flavanones in *Sophora flavescens* (see, e.g., Ref. 207). However, no cDNA clones for flavonoid-related prenyltransferases have been published to date.

3.9.9 PRENYLPTEROCARPAN CYCLASES

The final step of glyceollin and phaseollin formation is the cyclization of the prenyl residues of glyceollidins and phaseollidins, carried out by P450 prenylcyclases (Figure 3.7). These activities have been studied in detail for the formation of three glyceollins (I, II, and III) from glyceollidin I and II, and it is thought that specific activities are involved in each reaction.[208] However, no corresponding cDNA sequences have been published to date.

3.9.10 Isoflavonoid Glucosyltransferases and Malonyltransferases

As mentioned previously, in some species the major phytoalexins are glycosylated and acylated. The final product of the isoflavonoid phytoalexin pathway in *M. sativa* is medicarpin 3-*O*-glucoside-6″-*O*-malonate, and a range of isoflavone 7-*O*-glucosides and their malonylated versions accumulate in *G. max*. There are few reports on molecular characterization of DNA sequences for the enzymes carrying out glycosylation or acylation of isoflavonoids, although some cloned GTs with wide substrate acceptance have been shown to act on isoflavones (see Section 3.7.1). However, a cDNA for one isoflavonoid-specific GT has been isolated from *G. echinata* using the *S. baicalensis* F7GT cDNA as a probe.[209] The *UGT73F1* cDNA encodes a putative UDP-glucose:formononetin 7-*O*-glucosyltransferase. The recombinant protein accepted both formononetin and daidzein efficiently, and had little activity on other flavonoids tested. Glycosylation occurred at only one, as yet unassigned, position of daidzein, which is hydroxylated at C-7 and C-4′. Since the likely *in vivo* substrate is formononetin, which has only the 7-hydroxyl free, the enzyme was termed a formononetin 7GT. In a molecular phylogeny the *UGT73F1* sequence is located in a cluster with other stress-induced GTs.

3.9.11 Flavonoid 6-Hydroxylase

The hydroxyl groups at C-5 and C-7 of the A-ring are introduced during the formation of chalcones by CHS (or at the C-7 alone if PKR is coactive). However, flavonoids also occur with C-6 and C-8 hydroxylation, including 6- and 8-hydroxyanthocyanins, flavonols, and isoflavonoids. Latunde-Dada et al.[210] have identified a cDNA representing an elicitor-induced P450 (CYP71D9) with flavonoid 6-hydroxylase (F6H) activity, which may be involved in the biosynthesis of isoflavonoids with 6,7-dihydroxylation of the A-ring. The recombinant protein did not act on the isoflavonoids or pterocarpans directly, but rather accepted flavanone and DHF substrates, including liquiritigenin, suggesting hydroxylation occurs prior to aryl migration of the B-ring. In support of this route, 2HIS was found to be able to use 6,7,4′-trihydroxyflavanone.

The F6H showed low activity against flavones and little action on the flavonol kaempferol. Nevertheless, it is possible that a similar biosynthetic route through hydroxylation of precursors, perhaps by the same F6H, is involved in the production of other 6-hydroxyflavonoids. However, an alternative activity of the 2OGD type has been characterized at the biochemical level from *C. americanum* that catalyzes the 6-hydroxylation of partially methylated flavonoids.[211]

3.10 FORMATION OF AURONES

Although it has long been thought, based on genetic mutant and biochemical evidence, that aurones are derived from chalcones, the biosynthetic mechanism has only recently been clarified, and some aspects of the enzymatic process still await *in vivo* proof. An mRNA from *A. majus*, specifically expressed in the petal epidermal cells, has been shown to encode a recombinant protein with aureusidin synthase (AUS) activity.[212–214] AUS is a variant polyphenol oxidase (PPO) that can catalyze conversion of either 2′,4′,6′, 4-etrahydroxychalcone (naringenin chalcone) or 2′,4′,6′,3,4-pentahydroxychalcone to aureusidin (3′,4′-hydroxylated) or bracteatin (3′,4′,5′-hydroxylated), respectively[212] (note that carbon numbering differs between chalcones and aurones — Figure 3.1). Thus, it carries out both B-ring hydroxylation and formation of the C-ring, and, indeed, the enzyme studies suggest 3,4,2′-hydroxylation of the chalcone substrate may be a requirement for the formation of the aurone.[214] How this observation relates to the biosynthesis of those aurones with no B-ring hydroxylation is not clear. The only studies to date published on the enzymology of B-ring deoxyaurone formation, in cell cultures of *Cephalocereus senilis* (old man cactus), did not address this step.[215] For 4-deoxyaurones, it is likely the 6′-deoxychalcones are acted on by AUS, as the compounds commonly co-occur and AUS preparations show significant activity on the 6′-deoxychalcones.[214]

PPOs are typically plastid located copper-containing glycoproteins that have activity on a wide range of phenylpropanoids. It has been known for many years that PPOs can form aurones from chalcones, and this has been demonstrated recently with a recombinant tyrosinase from *Neurospora crassa*.[212] However, AUS represents a specific variant to previously defined PPOs. In particular, it lacks a typical N terminal plastid localization signal, and it has been suggested that it may be located to the vacuole.[212] How AUS would then compete with CHI for chalcone substrate, as may happen in species such as *A. majus* that can produce aurones and anthocyanins in the same petal cells, is not clear. However, recombinant AUS can use chalcone 4′-O-glucosides as substrates,[214] raising the possibility that aurone formation *in vivo* occurs in the vacuole from glucosylated substrates. Confirmation of the role of AUS awaits characterization of specific mutants, or the production of knockout or gain of function transgenic plants, and determination of the protein localization. Bifunctional PPOs may also be involved in the biosynthesis of betalains.[216]

Aurones are commonly glycosylated at the 4- or 6-hydroxyl. International Patent Application WO00/49155 reports that the recombinant *S. baicalensis* F7GT[155] can glucosylate aureusidin at the equivalent 6-hydroxyl, and also details the cloning of similar 7-O-glucosyltransferase cDNAs from *A. majus* and *P. hybrida*. However, it is not clear how these activities relate to the aurone or chalcone 6-O-glucosyltransferase activity characterized from *Coreopsis grandiflora* that accepts both 4-deoxyaurones and 6′-deoxychalcones.[217]

To date, in addition to *CHS* mutants, no mutants have been fully characterized that abolish aurone biosynthesis, although a commercial white-flowered line of *A. majus* was shown to lack *AUS* transcript in Nakayama et al.[212] and there is a preliminary report of an EMS-generated aurone-specific mutant of *A. majus*.[218]

3.11 OTHER ACTIVITIES OF FLAVONOID BIOSYNTHESIS

There are a few well-characterized modification enzyme activities that have not been detailed in the preceding sections on flavonols, flavones, anthocyanins, and isoflavonoids, and these are discussed here. Chalcones may be modified by hydroxylation, glycosylation, or methylation. The common 6′-deoxychalcone butein is hydroxylated at the C-3′ position by a P450 enzyme thought to be distinct to the F3′H.[219] Methylation of the 2′-hydroxyl of isoliquiritigenin in legumes such as *M. sativa* and *P. sativum* forms one of the most active flavonoids for signaling to *Rhizobium* symbiots.[220] As methylation prevents formation of the flavonoid C-ring, the OMT activity will help determine the balance between nodulation-related and defense-related flavonoids. Maxwell et al.[221] isolated a cDNA from *M. sativa* for a SAM:isoliquiritigenin 2′-OMT and demonstrated the encoded activity by assaying recombinant protein from *E. coli*. As would be expected from the symbiosis function, the gene is expressed primarily in developing roots. Haga et al.[222] have presented a short report on a cDNA for a similar OMT from *G. echinata*, which in addition to acting on isoliquiritigenin, may be involved in the methylation of licodione to form the retrochalcone echinatin (Figure 3.5).

F2H was postulated to be involved in the formation of the common flavones upon its identification from *G. echinata*, as the recombinant F2H catalyzed the formation of 2-hydroxynaringenin from (2*S*)-naringenin, which yielded apigenin upon acid treatment.[180,223] Based on the subsequent identification of FNS sequences from *G. echinata* and other species, it now seems unlikely that F2H is involved in the formation of the common flavones. Rather, its *in vivo* role in *G. echinata* may be the 2-hydroxylation of liquiritigenin, which can then yield licodione upon hemiacetal opening, as part of the biosynthetic route to echinatin (see Figure 3.5). Thus, F2H could be the licodione synthase. The pathway to echinatin may continue by the methylation of licodione by the OMT described by Haga et al.[222] Alternatively, the F2H may be part of the biosynthetic pathway to flavone *C*-glycosides. 2-Hydroxyflavanones have been proposed to be direct precursors for flavone *C*-glycosides and the substrates for *C*-UGTs, thus giving a biosynthetic scheme in which glycosylation occurs prior to the formation of the flavone structure.[143,224]

The flavonoid modification enzymes for which DNA sequences are available, and which we have already discussed in this chapter, represent only a few of the expected activities. Heller and Forkmann[3] tabulated 20 to 30 characterized enzymes, and given the variety of flavonoid structures identified, this must still be only a small portion of the existing activities.

3.12   VACUOLAR IMPORTATION OF FLAVONOIDS

Although flavonoids are found in many cellular compartments, it is only the mechanisms for vacuolar import that have been characterized in any detail. Alternative import mechanisms have been found that involve direct uptake, carrier proteins, or secondary modifications triggering importation. While commonalities are found for the import of anthocyanins, flavones, flavonols, and PAs, differences have also been observed for the different types of flavonoid.

The best-characterized mechanism shares elements with general xenobiotic detoxification processes, which typically involve the addition of glycosyl, malonyl, or glutathione residues to form stable water-soluble conjugates, and the sequestration of these conjugates by ATP-binding cassette (ABC) transmembrane transporters.[225,226] In particular, glutathione-*S*-transferase (GST, EC 2.5.1.18) activities have been shown to be required for the transport of anthocyanins and PAs in some species, mostly based on studies of mutants affected in *GST* genes. In the *bronze2* (*bz2*) mutant of *Z. mays* anthocyanins undergo oxidation and condensation in the cytosol, causing bronze kernel pigmentation.[227] The *an9* mutant of *P. hybrida* has acyanic petals,[228] while the *tt19* mutant of *A. thaliana* has reduced anthocyanin and PA accumulation in seedlings and seed, respectively.[229] Although AN9 and BZ2 only share 12% amino acid identity and belong to different classes of GST (type-I and type-III, respectively), they show functional homology, being able to reciprocally complement the mutations and, furthermore, complement the mutant floral phenotype of the *flavonoid3* mutant of *D. caryophyllus*.[230] AN9 also complemented the anthocyanin phenotype of *tt19*, but it did not overcome the loss of PA production. Another indication that anthocyanin sequestration is linked to general detoxification processes is that alternatively spliced *Bz2* mRNAs accumulate in response to various stresses.[231]

Interestingly, recombinant AN9 does not glutathionate cyanidin 3-*O*-glucoside or other flavonoids *in vitro*, and no anthocyanin–glutathione conjugates have been observed *in vivo*.[232] It has been suggested, therefore, that GST directly binds anthocyanins and escorts them to the vacuole, without glutathione addition being required.[232] However, the recombinant proteins of the *A. thaliana* ABC transporter proteins AtMRP1 and AtMRP2 can transport glutathionated anthocyanin *in vitro*, as well as other glutathione *S*-conjugates and chlorophyll catabolites, supporting a role for glutathionation and subsequent import by ABC transporters.[233]

Vacuolar transport of PAs differs from that of anthocyanins in some species. In *A. thaliana*, the *TT12* locus encodes a transporter of the multidrug and toxic compound extrusion (MATE) type with 12 membrane-spanning domains, and mutations preventing TT12 function prevent accumulation of PAs.[142] Interestingly, a transcript with high sequence identity to *TT12*, *MTP77*, is upregulated in *L. esculentum* transgenics over-producing anthocyanin due to activation of an anthocyanin regulatory gene, suggesting a link between MATE transporters and anthocyanins in this species.[234]

There is also a preliminary report of a *P. frutescens* cDNA encoding a membrane protein of unknown function (8R6) that promotes anthocyanin uptake into protoplasts and anthocyanin accumulation when overexpressed in *A. thaliana* transgenics.[235] Within the vacuole of some plants (but not *A. thaliana*) the anthocyanins may occur in protein containing bodies, termed anthocyanic vacuolar inclusions (AVIs), whose function is as yet unknown but may relate to transport activities.[236,237]

An additional aspect of flavonoid transport to the vacuole is the coordination of the localization process with vacuole biogenesis. The major vacuole in a cell may grow by small pro-vacuolar vesicles that have budded off the plasma membrane, endoplasmic reticulum, or Golgi fusing with the tonoplast. Interestingly, the *tds4* mutation of *A. thaliana* (for ANS, preventing epicatechin production) prevents normal vacuole development and causes accumulation of small vesicles.[139] Other *tt* mutations that prevent PA accumulation but not epicatechin formation do not interfere with vacuole development, implying a link between epicatechin biosynthesis and vacuole development.[139]

There are few studies on vacuolar importation of flavonoids other than anthocyanins and PAs. Klein et al. reported uptake of flavone glycosides by isolated *H. vulgare* primary leaf vacuoles via a vacuolar H^+-ATPase linked mechanism,[238] and by vacuoles from *Secale cereale* (rye) mesophyll via a possible ABC transporter mechanism.[239] Li et al.[240] found medicarpin conjugated to glutathione was also sequestered by an ABC transporter mechanism.

3.13 ENZYME COMPLEXES AND METABOLIC CHANNELING

It has long been thought that biosynthetic enzymes of plants are organized into macromolecular complexes in specific subcellular locations. It was suggested in the 1970s and 1980s that for flavonoid biosynthesis a multienzyme complex might exist, loosely associated with the endoplasmic reticulum and perhaps anchored through P450 enzymes such as C4H or F3'H (for coverage of the earlier literature, see Refs. 14, 57). This proposal has received support from more recent research, in particular affinity chromatography with flavonoid enzymes, yeast two-hybrid analysis, subcellular localization studies, and transgenic plants lacking individual phenylpropanoid enzymes.[14,57,241,242] Activity studies also support direct enzyme association. For example, PKR is thought to act on an intermediate of the CHS reaction, suggesting close association of the two enzymes in the cell.[46,57] Additionally, data from transgenic experiments support the occurrence of metabolic channeling and feedback loops in the early steps of phenylpropanoid biosynthesis, in particular for PAL and C4H.[14,243,244]

A role for enzyme complexes in the metabolic channeling of substrate has been proposed, so that competition between alternate enzyme pathways can be managed and the production of a specific product from a range of possibilities is favored. Furthermore, isoforms of the different enzymes may assemble in particular complexes dedicated to specific classes of flavonoid.

3.14   ASSIGNING ENZYME FUNCTION FROM DNA SEQUENCE OR RECOMBINANT PROTEINS

As molecular studies of flavonoid biosynthesis have progressed, particular issues in assigning enzyme function from DNA sequence or recombinant protein activity have arisen that are worth commenting on briefly. Recent studies with recombinant flavonoid 2OGDs have shown overlapping substrate usage and product formation *in vitro*. For example, recombinant ANS shows overlapping activities with F3H and FLS, and recombinant FLS shows F3H and FNS activity.[62,63,84–86] However, genetic mutants suggest that these *in vitro* activities may be less prevalent *in vivo*. For example, the still active *ANS* or *FLS* genes do not complement the acyanic phenotypes of *F3H* mutants of *A. majus* or *A. thaliana*, although it is possible (but unlikely) that the genes do not have the appropriate expression patterns either temporally or spatially. The UGT enzymes also may display broader *in vitro* activities than those *in vivo*.[95,153] The reasons for the differences between the *in vitro* and *in vivo* observations are not clear. The *in vitro* assay conditions may not accurately replicate the *in vivo* environment, so that differential efficiencies with regard to the alternative substrates and isomers might not be correctly represented. Also, recombinant proteins may have small but important structural differences to the native enzyme, as well illustrated by studies of PFOMT[166] (see Section 3.7.2). Alternatively, the formation of the enzymes into complexes and the effects of metabolic channeling may control the different activities. For example, the action of F3GT on an intermediate of the ANS reaction may direct anthocyanin formation over flavonol formation, even though flavonols are the favored ANS product *in vitro*.

The second area worthy of comment is with regard to assignment of enzyme function based on amino acid sequence alone. In some cases sequence comparisons can be used to determine if a new cDNA encodes one of the well-known flavonoid biosynthetic enzymes. However, in some cases, sequence similarity is insufficient for reliable prediction of the encoded function. For example, the OMTs may have amino acid identities of above 85% but different activities.[165] For transcription factors such sequence-based assignments of function are significantly more difficult, as specific roles for a given factor may have arisen in particular species. Thus, analysis of genetic mutants and transgenic overexpression or "knockout" lines is still preferable for confirmation of the encoded activities.

3.15   REGULATION OF FLAVONOID GENE TRANSCRIPTION

The phenylpropanoid pathway involves many biosynthetic genes and several alternative branches from common precursors leading to different flavonoid types and other compounds. Such complexity requires fine-tuned control, allowing the alteration of flux as conditions vary. This control is often achieved by the coordinate regulation of multiple genes, with the groupings varying with not only the end product that is to be made, but also with respect to the species and the type and developmental stage of a tissue. Furthermore, individual biosynthetic genes may be regulated in response to a number of developmental and environmental signals. For example, in flowers the biosynthetic genes change their activities as a consequence of light and spatio-temporal developmental factors for the production of anthocyanins in the petal epidermal cells, coincident with flower fertility.

Studies to date have shown that increases in gene transcription rates generally precede flavonoid production. This observation, in conjunction with studies on flavonoid-related transcription factors (TFs), suggests that gene transcription is the key point for biosynthetic gene regulation, rather than translation or post-translational steps. Post-transcriptional regulation has been reported for flavonoid biosynthesis (reviewed in Ref. 14 and recently featured in Ref. 245), and the extent of its role in controlling the pathway may come to light as methods for studying these aspects improve. In combination with metabolic channeling, post-transcriptional regulation would provide a means of fine control on the metabolic flux through the pathway to the accumulation of end products. Nevertheless, the overall data support the idea that it is changes in biosynthetic gene transcription rate that underpin major changes in flavonoid biosynthetic activity in most situations.

The transcription rate of a given gene is principally determined by interactions of TFs specific for that gene with the RNA polymerase II-containing holoenzyme and other components of the basal transcription machinery. Gene-specific TFs bind in a sequence-specific manner to motifs (*cis*-elements) within genes, usually in gene promoters, and increase (as activators) or decrease (as repressors) the rate of transcriptional initiation.[246–248] TF activity may be modulated by a range of mechanisms, in particular competition with other TFs, direct interaction with coactivator or corepressor proteins (which themselves do not bind DNA), and reversible phosphorylation.

Research on the regulation of the flavonoid biosynthetic genes is at the forefront of general plant transcriptional regulation studies. There are data from a number of different plant species concerning the specific *cis*-elements and TFs involved, and some of the functional interactions between the different types of TF have been elucidated. In this section, the literature on regulation of flavonoid biosynthesis is reviewed in brief only, and the reader is referred to more detailed review articles where possible. For additional information on the aspects covered, as well as detailed discussion of the TF families, the reader is referred to other reviews, including a comprehensive review of the genomics of transcriptional regulation in *A. thaliana*.[52,248–255]

3.15.1 ENVIRONMENTAL REGULATION OF EARLY BIOSYNTHETIC STEPS

Studies have predominantly focused on the regulation of *CHS*, presumably because CHS activity is a key flux point for flavonoid biosynthesis. *CHS* is commonly regulated coordinately with general phenylpropanoid genes in response to abiotic and biotic stresses. Much data are available for the *cis*-regulatory elements and associated TFs for regulation of *PAL*, *C4H*, *4CL*, and *CHS* in response to UV light and pathogen-related elicitors, and cross-talk between the different stress signaling pathways is being uncovered. Furthermore, the control of flavonoid gene expression during general photomorphogenesis is being characterized as part of studies in *A. thaliana* on the constitutive photomorphogenic (COP) regulatory system.

The COP system is composed of the COP9 signalosome, a multiprotein complex encoded by several *COP*, *De-etiolated* (*DET*), and *Fusca* genes, and at least four related proteins that act outside the COP9 complex — COP1, COP10, COP1 Interacting Protein (CIP), and DET1 (reviewed in Refs. 256–258). It is thought to act via ubiquitin-mediated, light-dependent degradation of downstream signaling components of many photoreceptors, as COP1 is a putative E3 ubiquitin ligase that interacts directly with several TFs. Affected proteins encoding phenylpropanoid regulators include AtMYB21 and HY5.[259,260] *COP10* and *CIP* probably encode other components of the ubiquitin degradation pathway.

From promoter analysis and transgenic studies the light-responsive unit (LRU) of the *CHS* genes of *P. crispum* and *A. thaliana* has been shown to be necessary and sufficient for induction of gene expression in response to UV-A and blue light or UV-B (see, e.g., Refs. 261–264). The LRU contains an ACGT-containing element (ACE) and an MYB-recognition element (MRE). A variety of ACEs have been identified that are named from the last nucleotide in the recognition motif, including the A-box (TACGTA)-, C-box (GACGTC)-, G-box (CACGTG)-, and T-box (AACGTT)-type elements. Many basic region/leucine zipper (bZIP) proteins have been shown to recognize and bind ACEs present in a range of promoters (reviewed in Refs. 265, 266). In *P. crispum*, at least seven bZIP proteins (referred to as common plant regulatory factors — CPRFs) bind the LRU ACE with varying affinities. In particular, PcCPRF1 is likely to be vital to the UV-induction process, as *PcCPRF1* transcript levels increase in abundance in response to UV light and precede the increase in *CHS* transcript abundance. In *A. thaliana*, the bZIP HY5 regulates a number of pathways, including that of the phenylpropanoids, and interacts directly with the ACE of light-responsive promoters.[267] Homo- and heterodimers of CPRFs occur, and dimerization, along with phosphorylation, may be important to the subcellular localization and function of bZIP proteins.[268,269] A number of two-repeat R2R3 MYB proteins (including AmMYB305 or AmMYB340 described in Section 3.15.4) and an unusual single-repeat MYB from *P. crispum*, PcMYB1,[263] have been shown to activate phenylpropanoid gene promoters through interaction with the LRU MRE or related MREs (e.g., the P-box). The MREs are often not only involved in the response to environmental stimuli but also in the developmental and spatial control of gene expression.

MYB proteins with a repressive function are involved in the regulation of phenylpropanoid biosynthetic genes. The first indication of such a role came from overexpression of *A. majus AmMYB308* or *AmMYB330* in *N. tabacum*, which caused dramatic reductions in the levels of lignin.[270] Analysis of an *A. thaliana* mutant for the orthologous gene *AtMYB4* showed increased *C4H* transcript levels, elevated levels of sinapate esters (HCA derivatives), and enhanced tolerance to UV-B exposure.[271] In *CaMV35S:AtMyb4* plants, the *C4H*, *4CL*, and *CHS* genes were all downregulated. Although white light is required for *AtMyb4* gene expression, transcript levels fall markedly within 24 h of exposure to UV-B. Deletion analysis and creation of fusion proteins indicate a role for the C terminal domain of the AtMYB4 protein in the repression function, perhaps through an "active" repressive effect on the basal transcription machinery.[271]

The response to pathogen infection involves not only some of the *cis*-elements characterized for the UV-light response but also additional regulatory elements related to pathogen-associated stimuli. These include the H-box and a TGAC (W-box) sequence that interacts specifically with a WRKY TF.[249,272,273] H-boxes have been associated with stress induction and tissue specificity of *PAL*, *4CL*, and *CHS* genes of a number of species. For example, the multiple H-boxes of the *P. vulgaris CHS15* gene promoter, along with the neighboring G-box and ACE, contribute to both tissue specificity of gene expression and induction in response to pathogen elicitation.[272,273] Although the H-box (CCTACC) closely resembles MREs such as the P-box (e.g., CCACCTACCCC), the interacting TFs characterized to date are not MYB proteins. The KAP-1 and KAP-2 proteins of *P. vulgaris*, which interact specifically with this sequence in the *CHS15* promoter, have sequence similarity to the mammalian Ku autoantigen protein involved in control of DNA recombination and transcription.[274] KAP-2 cDNAs have been isolated from *P. vulgaris* and *M. truncatula*, and the recombinant protein has been shown to activate H-box-containing promoters *in vitro*. KAP-2 transcript is constitutively present in *P. vulgaris* tissues, suggesting post-translational control of its activity, perhaps related to the previously observed elicitor-induced phosphorylation of the protein.[275]

Pathogen elicitation can also downregulate gene transcription, as evidenced by studies on the UV-light-inducible *CHS* and *ACC* genes of *P. crispum*.[276] When plants are placed under both pathogen and UV-light stresses, the pathogen repression signal to these genes is dominant to the UV-light induction signal. For *ACC*, both signals converge on two very similar ACEs, suggesting the switch from activation to repression might be achieved through replacement of activating TFs with a repressor protein. UV-induced increases in PcCPRF1 transcript levels are prevented by elicitor treatment, while PcCPRF2 transcript levels are decreased by UV light and increased by elicitation, indicating such a regulatory system.

3.15.2 REGULATION OF ANTHOCYANIN BIOSYNTHESIS

Regulation of anthocyanin production involves transcriptional activators of the R2R3 MYB and the basic helix–loop–helix (bHLH) (or MYC) types (Table 3.2). This was first revealed by studies of the monocot *Z. mays*. It was found that the anthocyanin pathway is turned on in this species through the combined action of one member of the COLORED ALEURONE1 (C1)/PURPLE PLANT (PL) MYB family and one member of the RED1 (R)/BOOSTER1 (B) bHLH family.[14,52] The members of the MYB and bHLH families are functionally redundant, and their specific expression patterns enable spatial and temporal control of anthocyanin biosynthesis.

Activation of the biosynthetic genes is dependent on direct interaction between the MYB and bHLH TFs within the transcriptional activation complex.[14,52] The complex binds DNA through discrete *cis*-elements in the target gene promoters, one that is recognized specifically by the MYB member and one (the ARE) that is recognized by an as yet unidentified protein.[277,278] The bHLH member functions in part through the ARE, and may be the protein that binds directly to it, or alternatively, interacts with a different protein that binds to it.[278]

Other species for which the elucidation of regulatory mechanisms controlling anthocyanin pigmentation is well underway are the eudicotyledon (dicot) species *A. majus* and *P. hybrida* (primarily for floral pigmentation) and *A. thaliana* (for vegetative pigmentation) (see Table 3.2). While TF genes equivalent to those in *Z. mays* are involved, the mechanism of regulation varies. Based on mutant analyses, the encoded TFs in these species control a subset of the anthocyanin biosynthetic genes, the late biosynthetic genes (LBGs). The early biosynthetic genes (EBGs) are under independent control, by as yet unidentified TFs. Partitioning of the pathway into separately regulated units allows for independent control of the production of other flavonoid types. In insect pollinated flowers, flavonoids have roles beyond pigmentation, e.g., flavones and flavonols are frequently the basis of nectar guides. For the floral models studied, the step in the pathway at which control by the defined TFs starts is at *F3H* or *DFR*, depending on whether there is predominate coproduction with anthocyanins of flavones or flavonols, respectively.[64,89,119,279] In one of the few studies on fruit, the *A* regulatory gene of *Capsicum annuum* (bell pepper) was found to affect only transcript levels of LBGs.[280] In *A. thaliana*, the LBGs are coregulated during induction of anthocyanins in seedlings exposed to white light and for PA production.[281–283]

Modular control of the pathway is not a universal trait in dicot species. For anthocyanin production in vegetative tissues of *P. frutescens*, the biosynthetic genes are regulated as a single group, from *CHS* to *GSTs*, as occurs in *Z. mays*.[235,284,285] Furthermore, single genes may be important points of control. In *Viola cornuta ANS* may be the main regulatory target,[286] and in *V. vinifera* berries *A3GT* is regulated separately and may be the key step for triggering anthocyanin production in berries during ripening.[287,288] Modular control has also been found in the monocot *Anthurium andraeanum*; in both spathe and spadix *DFR* is regulated separately from the other genes.[289]

As in *Z. mays*, families of MYB and bHLH TFs regulating anthocyanin production have been commonly found in the other species investigated to date. Through studies of these families it has become apparent that highly similar family members may vary subtly in their regulatory activity or act at different points in a regulatory hierarchy. In *A. majus*, differential activities of the family members give variations in floral pigmentation patterns, including a striking venation pattern determined by the *Myb* gene *Venosa*.[250] Within the MYB and bHLH families in *P. hybrida*, there are indications that some exert transcriptional control over others,[290] a complexity that has not been found in *Z. mays*[291] or reported for related TFs in other species. It remains to be determined whether these *P. hybrida* TFs also function as direct regulators of the anthocyanin biosynthetic genes.

TABLE 3.2
Transcription Factors that Regulate Flavonoid Biosynthetic Genes

Species	Metabolite	Protein Type	Name	Ref.
Antirrhinum majus	Phenylpropanoids (including flavonoids)	MYB	AmMYB305	319
		MYB	AmMYB308	270
		MYB	AmMYB330	270
		MYB	AmMYB340	319
	Anthocyanins	MYB	ROSEA1	250
		MYB	ROSEA2	250
		MYB	VENOSA	250
		bHLH	DELILA	363
		bHLH	MUTABILIS	250
Arabidopsis thaliana	Phenylpropanoids	MYB	AtMYB4 (repressor)	271
	Anthocyanins and proanthocyanidins	MYB	PAP1 (AtMYB75)	364
		MYB	PAP2 (AtMYB90)	364
		MYB	TT2 (AtMYB123)	283
		bHLH	TT8	282
		bHLH	GL3/EGL3 (AtMYC-2, AtMYC-146)	365, 366, 295, 367
		WD40[a]	TTG1	293
		WRKY	TTG2	308
		MADS box	TT16	310
		WIP	TT1	311
		HD-GLABRA2	ANL2	300
		VP1-like	ABI3	303
Capsicum annuum	Anthocyanins	MYB	A	280
Fragaria × ananasa	Anthocyanins and flavonols	MYB	FaMYB1 (repressor)	297
Gerbera hybrida	Anthocyanins	MYB	GMYB10	368
		bHLH	GMYC1	369
Ipomoea tricolor	Anthocyanins	bHLH	IVORY SEED	408
Lycopersicon esculentum	Anthocyanins	MYB	ANT1	234
Perilla frutescens	Anthocyanins	MYB	MYB-P1	370
		bHLH	MYC-RP/GP	371
		WD40	PFWD	294
Petunia	Anthocyanins	MYB	AN2	372, 373
		MYB	AN4	372, 373
		bHLH	AN1	290
		bHLH	JAF13	290
		WD40	AN11	292
Sorghum bicolor	Phlobaphenes	MYB	Y	314
Vitis vinifera	Anthocyanins	MYB	MYBA	288
Zea mays	Anthocyanins	MYB	C1	374, 375
		MYB	PL	376
		bHLH	B	377
		bHLH	LC	378
		bHLH	R	379
		bHLH	SN	380
		bHLH	IN (repressor)	299
		VP1	VP1	301, 302
		WD40	PAC1	291

TABLE 3.2
Transcription Factors that Regulate Flavonoid Biosynthetic Genes — *continued*

Species	Metabolite	Protein Type	Name	Ref.
	Phlobaphenes/flavone	MYB	P1 (P)	312, 317, 381
	C-glycosides	MYB	P2	317

Note: The functions of the transcription factors have been confirmed (or indicated) by genetic mutant or plant transgenic studies. They are activators of transcription unless stated otherwise. Some are direct activators of the flavonoid biosynthetic genes, while others may encode factors upstream in a regulatory cascade.
[a]These WD40 proteins interact with the TFs for the regulation of the biosynthetic genes.

Another type of protein shown to be involved in the regulation of anthocyanin (and PA) synthesis is the WD repeat (WD40) protein.[291–293] The WD40 proteins confirmed to date as being involved in flavonoid biosynthesis (Table 3.2) have relatively high sequence identity of around 60%, and form a distinct group within the WD40 family (a molecular phylogeny of WD40 sequences is presented in Ref. 291). Based on gene expression or genetic interaction studies, the WD40 proteins do not function through direct transcriptional control of the *MYB* and *bHLH* genes, even though overproduction of MYB or bHLH factors can partially overcome some of the phenotypes of lines mutated in the WD40 gene.[291,292] Rather, they likely function as part of the MYB-bHLH transcriptional complex, perhaps providing a stabilizing influence.[285,294–296]

It remains to be determined whether other types of TFs or regulatory proteins are also directly involved in the control of anthocyanin pigmentation. As mentioned previously, the TF binding the ARE *cis*-element has not been identified, although it potentially is the bHLH factor. Furthermore, although some R2R3 MYB and bHLH TFs with a repressive effect on transcription have been identified for anthocyanin biosynthesis, their role in the overall regulatory system, as well as that of repressor TFs in general, needs further characterization. FaMYB1 likely plays a role in the regulation of anthocyanin and flavonol production in *Fragaria Xananasa* (strawberry) fruit, and has structural features in common with the repressor TF AtMYB4, suggesting it operates through a direct repression mechanism.[297] *C1-I* is a dominant negative allele of *C1* that lacks the C-terminal activation domain normally present, and probably represses anthocyanin biosynthetic genes "passively" through competition with activators for target promoter binding sites.[298] Also in *Z. mays* is *Intensifier1* (*In1*), which encodes a bHLH protein similar to R.[299] IN1 is suggested to have a repressive activity, as recessive mutations in the gene increase anthocyanin levels. It is worth noting that one of the suggested roles of the bHLH TFs is to relieve the MYB coactivators from the effect of an inhibitory factor, which is perhaps of the protein types described above.[278]

With regard to the regulation of the anthocyanin regulatory genes, progress has been made in dissecting signaling to phenylpropanoid-related TF genes during photomorphogenesis and flavonoid production in vegetative tissues and seeds. *ANTHOCYANINLESS2* (*Anl2*) of *A. thaliana* encodes a homeobox protein of the HD-GLABRA2 group that is required for anthocyanin production in the subepidermal cells of vegetative tissues.[300] Unlike the *tt* mutants, the *anl2* mutant is not altered in pigmentation of the seed coat, but it does have aberrant cellular organization in the roots. Given that homeodomain proteins are often involved in cell specification and pattern formation, *anl2* may encode an upstream regulator rather than a direct activator of the flavonoid-related *MYB* and *bHLH* genes. *Viviparous 1* (*Vp1*) of *Z. mays* encodes a distinct type of TF that has been identified subsequently in a number of species, including *A. thaliana* (*ABI3*).[301–303] In *Z. mays Vp1* has multiple regulatory roles in seed development, including both up- and downregulation of gene expression. It is required for production of anthocyanins in the aleurone and upregulates the *C1* gene directly.[304] Regulation of *C1* occurs through the *Sph cis*-element in the promoter, which is recognized by the B3 domain of VP1. VP1 likely acts as part of a complex of TFs, which may include 14-3-3 proteins.[305]

On a final note, it is being increasingly found that the anthocyanin-related factors regulate other processes. In this regard, the best characterized is the WD40 protein TRANSPARENT TESTA GLABRA1 (TTG1) of *A. thaliana*. In addition to anthocyanin pigmentation, a number of other processes occurring in epidermal cells are also dependent on TTG1, including trichome, seed mucilage, and PA production.[293,306] In a similar vein, the WD40 ANTHOCYANIN11 (AN11) of *P. hybrida* also influences seed coat development, in conjunction with the activity of the bHLH anthocyanin regulator ANTHOCYANIN1 (AN1).[307] These *P. hybrida* regulators, along with the MYB factor ANTHOCYANIN2 (AN2), also regulate vacuolar pH in the petal cells,[307] a role that appears to be played by some of the anthocyanin MYB TFs of *A. majus* as well (K. Schwinn, unpublished data).

3.15.3 REGULATION OF PROANTHOCYANIDIN BIOSYNTHESIS

Although initial progress in understanding regulation of PA biosynthesis was made with *Z. mays*, it is now best characterized in *A. thaliana*, in which PAs accumulate in the testa of the seeds. There are at least six classes of regulatory proteins that have been shown to control transcription of the genes for the PA pathway: bHLH, MADS box, R2R3 MYB, WD40, WIP, and WRKY (Table 3.3).

The bHLH and MYB factors are thought to be direct activators of PA production. Knockout and gain of function experiments have demonstrated that TT2 (MYB) and TT8 (bHLH) interact to upregulate LBGs, but not EBGs such as *CHS*.[137,282,283] TT2 may be a key determinant of the pattern of PA biosynthesis, as its transcript abundance shows much greater spatial and temporal variation than that of TT8. TT2 is also involved in regulating anthocyanin production during seedling development.

The WD40 protein is *TTG1* (described in the previous section). The WRKY TF is TTG2, which may function downstream of *TTG1*. Mutant lines for *ttg2* have reduced production of PAs and mucilage, and have fewer, less branched trichomes, although they are wild type for anthocyanin production in vegetative tissues.[308] However, *ttg2* does not affect *ANR* gene expression, suggesting a late role in PA biosynthesis or a post-transcriptional regulatory role.[309] *TT1*, encoding a member of the WIP subfamily of zinc finger proteins, is required for endothelium development, as *tt1* mutants have reduced PA production and altered seed coat morphology.[310] The MADS box protein TT16 (identical to BSISTER) also may be involved in endothelium development. Lines mutant for *tt16* have altered cell shape and reduced PA production and expression of *ANR*, but only in a specific region of the seed coat, and ectopic expression of *TT16* leads to ectopic PA accumulation.[311] Overexpression of *TT2* can partially complement the *tt16* phenotype, restoring PA biosynthesis, suggesting *TT2* acts parallel or downstream of *TT16*.

The production of 3-deoxyflavonoids, in particular the 3-deoxy-PAs, in *Z. mays*, also involves R2R3 MYB transcription factors.[312–315] In *Z. mays*, for the production of phlobaphenes, the MYB protein P1 activates *CHS* and *DFR*, and, presumably, the gene for the flavan-3-ol biosynthetic enzyme, but does not upregulate *F3H*.[312,313] The regulatory activity of P1 does not require a bHLH coactivator, even though P1 recognizes the same promoter elements of the *DFR* gene as C1, the MYB TF regulating anthocyanin synthesis. Grotewold et al.[316] compared the amino acid sequence of P1 and C1 and were

TABLE 3.3
Genetic Loci of *A. thaliana* Involved in Flavonoid Biosynthesis for which Genetic Mutants and the Encoded Product Have Been Characterized

Locus	Gene Product	Ref.[a]
Biosynthetic enzymes		
TT3	DFR	382
TT4	CHS	383
TT5	CHI	382
TT6	F3H	384
TT7	F3′H	106
TDS4/TT18	ANS	134
Banyuls	ANR	136, 137
Transporter activities		
TT12	MATE transporter	142
TT19	GST	229
Regulatory factors		
TT1	WIP	311
TT2	MYB	259
TT8	bHLH	258
TT16	MADS box	310
TTG1	WD40	293
TTG2	WRKY	308
ANL2	HD-GLABRA2	300

[a]Reference to the first publication reporting detailed characterization of the corresponding cDNA or gene.

able to identify which amino acid residues in the C1 protein defined the interaction with the *Z. mays* bHLH proteins.

3.15.4 REGULATION OF THE PRODUCTION OF OTHER FLAVONOIDS

There are much less data on the regulation of flavonoids such as the isoflavonoids, flavones, and flavonols. Some of the genes regulating the EBGs in response to environmental stimuli, as discussed in Section 3.15.1, may be involved in controlling flavonol or flavone production. However, their specific role in relation to these compounds, and the regulation of genes encoding FLS and FNS in general, has not been studied widely.

The *P1* gene of *Z. mays*, which controls phlobaphene production, along with a closely related second gene *P2*, controls production of flavone *C*-glycosides in *Z. mays* flower silks by upregulating genes required for flavanone, and possibly flavone, biosynthesis but not the subsequent genes that are required for anthocyanin production.[317] Furthermore, *P1* or *P2* driven by the *CaMV35S* promoter induces production of flavone *C*-glycosides in transgenic cell lines of *Z. mays*.[140,317] Flavonol production in *Z. mays* may be under separate regulatory control to anthocyanins or flavones.[317,318] Two *A. majus* MYB proteins with relatively high sequence identity, AmMYB305 and AmMYB340, may be involved in regulation of the EBGs required for flavonol biosynthesis.[319] AmMYB305 has been shown to activate the promoters of *CHI* and *F3H*, and AmMYB340 to regulate *CHI* and bind the "P-box" MYB recognition element that is linked to petal-enhanced expression of phenylpropanoid genes.[319,320] Like P1 of *Z. mays*, the proteins did not require a bHLH partner for their binding or activating activities.

In *P. hybrida*, genetic and biochemical evidence suggests that *An1* and *An2*, which regulate the LBGs for anthocyanin production, also positively regulate *F3',5'H* and *Cytb5*, but do not affect *F3'H* expression.[105,110] However, in leaves of *CaMV35S:Lc* transgenic *P. hybrida* both *F3'H* and *F3',5'H* are upregulated, but not *FLS*.[321] The *an4* mutation, which lacks activity of an anthocyanin regulator related to *An2* and results in acyanic pollen, does not effect the expression of *F3GalT*.[322] In *A. thaliana*, some of the genes for PA biosynthesis, such as *TTG1*, that are required for LBG expression do not affect expression of *FLS* or *F3'H*.[283]

The induction of isoflavonoid production in response to biotic signals is extensively characterized,[323] and analysis of gene promoter *cis*-elements is well advanced for both the EBGs (see Section 3.15.1 and Ref. 323) and isoflavonoid-specific genes (see, e.g., Ref. 324). However, there is no information on the TFs factors involved specifically in isoflavonoid gene regulation. Introduction of a transgene (*CRC*) for a chimeric protein with activity of both the C1 and R *Z. mays* anthocyanin-related TFs into *G. max* did not alter the transcript levels for the isoflavonoid-specific genes *2HIS*, *IFR*, and *IOMT*, although *PKR*, *FLS*, and other phenylpropanoid genes were upregulated.[325]

3.16   GENETIC MODIFICATION OF FLAVONOID BIOSYNTHESIS

In addition to being carried out as part of fundamental studies, genetic modification (GM) of flavonoid production has been used to extend existing flower color ranges or induce vegetative anthocyanin pigmentation in ornamental crops, to modify production of plant-defense flavonoids, and to increase levels of flavonoids related to human and animal health. Indeed, the flavonoid pathway has been the target of more biotechnology research than probably any other plant secondary metabolite pathway. The major approaches have been to prevent or inhibit flavonoid production, redirect substrate within the pathway, introduce new flavonoid biosynthetic activities, and modulate pathway regulation. There are numerous examples of the GM of flavonoid biosynthesis using these approaches, and due to space limitations we discuss only some of them here. An extensive listing of published examples is provided in Table 3.4–Table 3.6. An emerging area in flavonoid biotechnology is the introduction of flavonoid biosynthesis into microorganisms (see, e.g., Ref. 326), but the focus of the following sections will be on GM approaches in plants.

3.16.1   PREVENTING FLAVONOID PRODUCTION

A reduction in, or the prevention of, flavonoid biosynthesis has been demonstrated many times, and in several species, by inhibiting production of a single flavonoid biosynthetic enzyme. This approach is reviewed in Refs. 76, 327, and examples are listed in Table 3.4. It is worth noting that the first published accounts of antisense or sense RNA inhibition of plant gene expression involved *CHS* in *P. hybrida*.[328–330]

Commonly the target phenotype has been flower color. In addition to the expected white-flower phenotypes, in some species both ordered and erratic pigmentation patterns have been obtained (Table 3.4). Patterning only seems to occur in species that naturally have patterned varieties. Furthermore, some of the patterns show instability, not only within a particular plant but also in their inheritance (see, e.g., Ref. 331), which may limit the commercial usefulness of some of the more dramatic phenotypes.[332]

Approaches to inhibit anthocyanin production that target CHS can cause plant sterility, as flavonols play a role in fertility in some species.[333–335] It is possible to inhibit anthocyanin production by targeting an enzyme such as DFR, which still allows the formation of flavonols and flavones. Sense or antisense DFR transgenes have been used to reduce or prevent anthocyanin production in several species (Table 3.4), with results similar to those for CHS

TABLE 3.4
Genetic Modification of Flavonoid Production Using Inhibition of Flavonoid Biosynthetic Gene Activity by Sense or Antisense RNA

Transgene	Species Modified	Phenotype Change	Ref.
CHS[a]			
Sense and antisense	*Dendranthema*	Flower color changed from pink to white	385
Sense	*Dianthus caryophyllus*	Flower color changed from pink to white	386
Antisense	*Eustoma grandiflorum*	Flower color changed from purple to white or patterns	387
Antisense	*Gerbera hybrida*	Flower color changed from red to pink or cream	388
Antisense	*Juglans nigra* × *J. regia*	Enhanced adventitious root formation	389
Antisense	*Lotus corniculatus*	Decreased flavonoids, enhanced PA production in root cultures	336
Antisense	*Petunia hybrida*	Flower color changed from red to white or patterns	328
Antisense	*Petunia hybrida*	Flower color changed from purple to pale purple or white	327
Sense	*Petunia hybrida*	Flower color changed from purple to white or patterns	329, 330
Sense	*Rosa hybrida*	Flower color changed from red to pale red	390
Sense and antisense	*Torenia fournieri*	Flower color changed from blue to pale blue or patterns	391, 392
Sense	*Torenia hybrida*	Flower color changed from blue to white or patterns	393
F3H			
Antisense	*Dianthus caryophyllus*	Flower color changed from orange to white	394
DFR			
Sense	*Petunia hybrida*	Flower color changed from purple to white or patterns	335
Antisense	*Solanum tuberosum*	Decreased anthocyanin levels	343
Sense or antisense	*Lotus corniculatus*	Decreased PA levels	337–339
Sense and antisense	*Torenia fournieri*	Flower color changed from blue to pale blue or patterns	391, 392
Sense	*Torenia hybrida*	Flower color changed from blue to white or patterns	393
F3′,5′H			
Sense	*Petunia hybrida*	Flower color changed from dark blue to pale blue or pink	335, 395
Sense	*Torenia hybrida*	Flower color changed from blue to pink	393
FLS			
Antisense	*Eustoma grandiflorum*	Reddening effect on flower color	341
Antisense	*Petunia*	Flower color changed from purple to red	146
Antisense	*Petunia*	Flower color changed from white to pale pink	342
FNSII			
Antisense	*Torenia*	Paler flower color	350
3RT			
Antisense	*Petunia hybrida*	Flower color changed from purple to pink or patterns	111
IFR			
Sense or antisense	*Pisum sativum*	Reduced pisatin production	396
HM3OMT			
Sense or antisense	*Pisum sativum*	Reduced pisatin production and reduced pathogen resistance	396

[a]Many examples of antisense or sense suppression of *CHS* gene activity in *P. hybrida* have been published, many using it as a phenotypic marker for studying the silencing process rather than through an interest in the effect on flavonoid biosynthesis or function. Only the first reports are referenced here.

TABLE 3.5

Genetic Modification of Flavonoid Production in Plants by the Introduction of Novel Flavonoid Biosynthetic Activities or by Increasing Endogenous Activities (All "Sense" Transgenes). Examples of Complementation of Genetic Mutants Are Not Featured

Transgene	Species Modified	Phenotype[a]	Ref.
CHR	*Nicotiana tabacum*	Flower color changed from pink to pale pink	58
	Petunia	Flower color changed from white to pale yellow	149
STS	*Nicotiana tabacum*	Flower color changed from pink to pale pink	334
AUS	*Arabidopsis thaliana*	Seed color changed[b]	213
CHI	*Arabidopsis thaliana*	Increased flavonol levels	151
	Lycopersicon esculentum	Increased flavonol levels	61
DFR	*Forsythia Xintermedia*	Vegetative anthocyanin pigmentation increased	344
	Lotus corniculatus	PA types altered in cell cultures	338
	Nicotiana tabacum	Flower color changed from pink to dark pink	71, 75
	Petunia	Flower color changed from white to pink	342
	Petunia	Flower color changed from pale pink to orange or red	73, 345, 397, 398
	Solanum tuberosum	Increased anthocyanin levels	343
DFR and ANS	*Forsythia Xintermedia*	Vegetative and flower anthocyanin pigmentation increased or induced	89
DFR and F3',5'H	*Dianthus caryophyllus*	Flower color changed from pink or white to blue-purple	348
F3'H	*Petunia*	Flower color changed from lilac to pink	105
	Torenia	Reddening effect on flower color	350
F3',5'H	*Dianthus caryophyllus*	Flower color changed from pink to blue-purple	348
	Nicotiana tabacum	Change in pink shade of flowers	109, 349
	Petunia	Flower color changed from pale pink-red to magenta-deep red or patterns	146, 395, 399
F3GT	*Eustoma grandiflorum*	No change in visible phenotype[c]	90

A3'GT and F5GT	Petunia	No change in visible phenotype[d]	113
Dv3MaT	Petunia	No change in visible phenotype[e]	123
ANR	Nicotiana tabacum	Flower color changed from pink to white; PA-like compounds produced	137
2HIS	Arabidopsis thaliana	Genistein produced	151, 182, 346
	Glycine max	Isoflavone levels changed	347
	Nicotiana tabacum	Genistein produced	346
	Zea mays	No change in phenotype	346
2HIS and R/C1	Zea mays	Genistein produced in cell lines	346
2HIS CHR, and R/C1	Zea mays	Daidzein and genistein produced in cell lines	346
2HIS and CHI	Arabidopsis thaliana	Genistein produced	151
2HIS and CHI	Arabidopsis thaliana (tt6/tt3)	Genistein produced	151
2HIS, CHI, and Pap1	Arabidopsis thaliana	Genistein produced	151
IFR	Nicotiana tabacum	No change in visible phenotype[f]	150
I2'H	Arabidopsis thaliana	No change in visible phenotype[g]	194
I7OMT	Medicago sativa	Enhanced production 4'-O-methylated isoflavonoid	189

[a]Only a general indication of the phenotype is given.
[b]Seed color was restored in the tt5 mutant.
[c]A change in flavonoid glycosylation and acylation occurred.
[d]One new anthocyanin type, delphinidin 3,5,3'-tri-O-glucoside, was found.
[e]Anthocyanins with novel malonylation were formed.
[f]Cell lines were able to biotransform exogenously supplied isoflavonoids.
[g]Transgenic plants were able to convert exogenously supplied formononetin to 2-hydroxyformononetin.

TABLE 3.6
Genetic Modification of Flavonoid Production in Stably Transformed Plants or Cell Lines by Introduction of Genes Encoding Transcription Factors that Regulate Flavonoid Biosynthetic Genes (All "Sense" Transgenes)

TF Type and Transgene[a]	Species Modified	Effect on Flavonoid Production[b]	Ref.
MYB, *Ant1*	*Lycopersicon esculentum*	Anthocyanins increased	234
	Nicotiana tabacum	Anthocyanins increased	234
MYB, *C1*	*Arabidopsis thaliana*	No change in visible phenotype	355
	Medicago sativa	No change in visible phenotype	360
	Nicotiana tabacum	No change in visible phenotype	355
	Trifolium repens	Anthocyanins increased	400
MYB, *Gmyb10*	*Nicotiana tabacum*	Anthocyanins increased under high light conditions	369
MYB, *Myb.Ph2*	*Trifolium repens*	Anthocyanins increased	400
MYB, *Pap1* or *Pap2*	*Arabidopsis thaliana*	Anthocyanins and other phenylpropanoids increased	364
	Nicotiana tabacum	Anthocyanins increased	364
MYB, *Pap1* with EAR-motif repression domain	*Arabidopsis thaliana*	Anthocyanin and PA production inhibited	357, 358
MYB, *P1* or *P2*	*Zea mays* cell lines	3-Deoxyflavonoids, *C*-glycosylflavones, other phenylpropanoids and fluorescent compounds increased	140, 317
MYB, *Rosea1*	*Eustoma grandiflorum*	Anthocyanins increased	401
	Petunia	Anthocyanins increased	401
bHLH, *B-Peru*	*Medicago sativa*	No change in visible phenotype	360
	Trifolium repens	Anthocyanins increased	400
bHLH, *Delila*	*Lycopersicon esculentum*	Anthocyanins increased	402
	Nicotiana tabacum	Anthocyanins increased	402

bHLH, *Lc*	*Arabidopsis thaliana*	Anthocyanins increased	355
	Eustoma grandiflorum	No change in visible phenotype	359
	Medicago sativa	Anthocyanins and PAs increased and flavones decreased under stress conditions	360
	Lycopersicon esculentum	Increased under high light levels	403
	Nicotiana tabacum	Anthocyanins increased	355
	Pelargonium	No change in visible phenotype	359
	Petunia	Anthocyanins increased	321
bHLH, *Myc-rp* and *Myc-gp*	*Lycopersicon esculentum*	Anthocyanins increased	371
	Nicotiana tabacum	Anthocyanins increased	371
bHLH, *Sn*	*Lotus corniculatus*	Anthocyanins and proanthocyanidins increased.	356
		PAs increased in roots (and decreased in leaves of some lines)	354
bHLH and transposon, *R* and *Tag1*	*Nicotiana tabacum*	Anthocyanins increased (variegated patterns)	404
bHLH and MYB, *Cl* and *Delila*	*Arabidopsis thaliana*	No change in visible phenotype	402
bHLH and MYB, *Cl* and *R*	*Glycine max*	Ratio of genistein to daidzein reduced	325
	Zea mays cell lines	Anthocyanins, phlobaphenes, and C-glycosylflavones increased	140, 405
bHLH and MYB, *Cl* and *R* and suppressed *F3H*	*Glycine max*	Isoflavonoids increased	325
bHLH and MYB, *Lc* and *Cl*	*Arabidopsis thaliana*	Anthocyanins increased	355

[a]Except for MYB, Ph2, P1, and P2, the other TFs normally function as regulators of anthocyanin biosynthesis.

[b]Only a general indication of phenotype is given, and the full changes identified may include production of anthocyanin earlier in flower development than normal, increased anthocyanin production only under stress conditions, small increases in flavonoid levels in tissues already producing flavonoids, ectopic flavonoid production, changes in levels of nonflavonoid phenylpropanoids.

inhibition. Sense or antisense transgenes for CHS or DFR have also been used to reduce PA levels in transgenic root cultures of *Lotus corniculatus* (bird's foot trefoil).[336–339]

An alternative approach to RNA suppression for controlled reduction of flavonoid enzyme activity is the expression of single-chain antibody fragments targeted to a key enzyme or TF. Early attempts with enzymes such as DFR have been inconclusive as to the effectiveness of the technology.[340]

3.16.2 REDIRECTING SUBSTRATE IN THE FLAVONOID PATHWAY

The flavonoid pathway contains many branch points at which enzymes may compete for substrate, depending on the spatial and temporal occurrences of the enzymes and any metabolite channeling effects. Altering the balance of the competing activities may alter the levels of the different enzyme products and their derivatives. Alternatively, when a potential substrate is accumulating in tissues, a rate-limiting step may be overcome by increasing levels of the required enzyme.

Changes in the production rates of different branches of the pathway have been achieved by altering the competing activities of FLS and DFR, CHS and STS or PKR, and ANR and F3GT. Introduction of an antisense *FLS* transgene into *E. grandiflorum*,[341] *N. tabacum*,[146] or specific *Petunia* lines[342] reduced flavonol production and increased anthocyanin levels. *CaMV35S:DFR* transgenes increased anthocyanin content in flowers of *N. tabacum* and *Petunia* and tubers of *S. tuberosum*, and altered the type of PAs in *L. corniculatus* cell cultures.[75,338,342,343] In *Forsythia*, anthocyanins accumulate in some vegetative tissues, but only flavonol glycosides are found in petals. Introduction of a *CaMV35S:DFR* transgene into *F. × intermedia* increased levels of anthocyanins in tissues which normally produce them, demonstrating that DFR is rate limiting for anthocyanin biosynthesis in these cells.[344] The addition of a second sense transgene, for *M. incana ANS*, extended anthocyanin pigmentation to the flowers of the double transgenic lines.[89] 2HIS and F3H also potentially compete for substrate, and reduction of *F3H* transcript levels in transgenic *G. max* lines enhanced isoflavonoid accumulation (see Section 3.16.3.3).

STS from *V. vinifera* uses the same substrates as CHS, so that a *CaMV35S:STS* transgene introduced into *N. tabacum* reduced anthocyanin levels in the flowers of the transgenics, so that flowers were near white rather than the usual dark pink.[334] Similarly, the introduction of *CaMV35S:PKR* transgenes diverted substrate into 6′-deoxychalcone production in transgenic *N. tabacum* and *Petunia,* significantly reducing floral anthocyanin biosynthesis and resulting in pale flower colors.[58,149] A *CaMV35S:ANR* transgene introduced into *N. tabacum* changed the flower color of some transgenic lines from pink to white, presumably through competition with F3GT (see also Section 3.5.3).[137] Thus, the introduction of *STS, PKR,* or *ANR* transgenes offers a route for reduction of flavonoid biosynthesis, and may produce transgenics with more stable phenotypes than antisense RNA or sense-inhibition approaches.

The overproduction of an enzyme at a rate-limiting step to increase levels of specific flavonoids is well illustrated by the *Forsythia* ANS example described earlier, and also by the use of *CaMV35S:CHI* to increase flavonol levels in *A. thaliana* and *L. esculentum*.[61,151] In particular, fruit of *L. esculentum* plants expressing *P. hybrida CHI* had up to a 78-fold increase in flavonol content, principally rutin, in fruit peel.

3.16.3 INTRODUCING NOVEL FLAVONOID COMPOUNDS

3.16.3.1 Chalcones, Aurones, and Flavonols

The biosynthesis of the yellow flavonoids has been targeted for biotechnology applications in commercial ornamental crops, such as *Cyclamen, E. grandiflorum, Impatiens*, and *Pelargonium*, which currently lack yellow-flowered varieties. The compounds targeted to date are the aurones, which provide strong yellow colors, and the chalcones, which can provide yellow colors in certain circumstances. Aurones should offer excellent biotechnology prospects for generating yellow colors, as their direct precursors, chalcones, are ubiquitous intermediates in flavonoid biosynthesis, and cDNAs are available for the key biosynthetic activity, AUS. However, no experiments showing the use of the *AUS* cDNA to introduce aurone production in transgenic plants have been published.

A successful biotechnology route for directing chalcone accumulation has been the use of PKR to generate 6'-deoxychalcones, which are not accepted as substrates by CHI of many species. Introducing *CaMV35S:PKR* into a white-flowered line of *Petunia* generated transgenic lines that accumulated up to 50% of their petal (and pollen) flavonoids as 6'-deoxychalcones, changing the flower color from white to pale yellow.[149] Hydroxylation and glucosylation of the novel chalcones occurred at the C-3 and C-3' positions. In *N. tabacum*, the same approach resulted in the accumulation of the flavanone liquiritigenin in the petals at the expense of anthocyanin accumulation, indicating the endogenous CHI could accept the 6'-deoxychalcone substrate.[58]

3.16.3.2 Altering Dihydroflavonol 4-Reductase Activity

As discussed in Section 3.4.4, in some species DFR can show strong preference to DHF substrate with di- or trihydroxylated B-rings, limiting the production of pelargonidin-based anthocyanins. Meyer et al.,[345] in the first published case of GM of flower color, introduced a *Z. mays CaMV35S:DFR* transgene into a *P. hybrida* mutant that accumulated DHK, enabling reduction of DHK, and production of pelargonidin-based anthocyanins in the petals. Subsequent crosses of the transgenics to commercial *P. hybrida* cultivars led to F_2 lines with flower colors, including orange, novel to this species. Similar results have been obtained with other *DFR* transgenes in *P. hybrida* (Table 3.5). *DFR* transgenes have also been used to alter the type of PA accumulated[338] and to encourage accumulation of delphinidin-derived anthocyanins (see Section 3.16.3.4).

3.16.3.3 Isoflavonoids

Modification of isoflavonoid biosynthesis may have a wide range of applications for improving not only plant defense characters but also the health benefits of food for humans. With the exception of IFD (which may not be required *in vivo*), cDNA clones are available for all of the enzymes needed for the production of the isoflavonoid vestitone. Furthermore, as VR cDNAs have been cloned, only clones for the DMID are lacking for the biosynthetic branch of the antifungal pterocarpans. To date, experiments have focused on 2HIS, but there has also been success using IFR, I2'H, and I7OMT.

Overexpression of 2HIS genes has not only successfully introduced isoflavonoid production into species that lack this branch of the flavonoid pathway, but also altered levels of these compounds in isoflavonoid-producing species (Table 3.5). The first reported GM experiment used *CaMV35S:G.max2HIS* to produce genistein in *A. thaliana*, although only at levels of around 2 to 4 µg g^{-1} fresh weight.[182] The level of genistein could be raised approximately threefold in the transgenics by UV-B induction of the general phenylpropanoid pathway,[346] or up to 30-fold (50 µg g^{-1}) by using a mutant line (*tt3* or *tt6*) lacking *DFR* and *F3H* gene activity as the recipient of *CaMV35S:2HIS* and *CaMV35S:CHI-II* transgenes.[151] However, these levels are still low compared to, for example, isoflavonoid levels of >4 mg g^{-1} fresh weight of *G. max* seeds.[325] The experiments of Liu et al.[151] suggest competing activities within flavonoid biosynthesis and the effects of metabolic channeling may significantly affect the ability to engineer isoflavonoid biosynthesis in nonlegumes. Several enzymes may use flavanones as substrates in addition to the 2HIS, including F2H, F3H, F6H, FNR, FNS, F3'H, F3',5'H, and various GTs and OMTs. Thus, understanding the relationship between these activities will be important for future biotechnology approaches.

The *CaMV35S:G.max2HIS* transgene was also introduced into *N. tabacum* transgenic plants, resulting in genistein being produced in the petals (at 2 µg g^{-1} fresh weight) but not leaves.[346] Cell lines of *Z. mays* only produced genistein when the *2HIS* transgene was cointroduced with the chimeric *CRC* transgene, which induced general flavonoid biosynthesis.[346] *CRC* encodes a fusion protein comprised of two *Z. mays* TFs that regulate anthocyanin biosynthesis. By introducing a third transgene, *CaMV35S:G.maxPKR*, low levels of daidzein could be produced along with genistein. Analysis in most experiments has included a hydrolysis to yield the aglycone (genistein or daidzein). However, when extraction without hydrolysis was performed, it was found that most isoflavonoids in *A. thaliana*, *N. tabacum*, and *Z. mays* transgenics were conjugated forms. Genistin and malonyl-genistin were found in *N. tabacum* petals[346] and genistin, genistin-*O*-glucoside, genistin-*O*-rhamnoside, and a genistin glucose, rhamnose di-*O*-glycoside, in *A. thaliana*.[151]

A *CaMV35S:M.sativaIFR* transgene was used to generate transgenic *N. tabacum* cell lines, which were then analyzed for their ability to metabolize exogenously supplied isoflavonoids.[150] The transgenic cells could take up 2'-hydroxyformononetin and convert it to vestitone. Both 2'-hydroxyformononetin and vestitone accumulated as 2'- or 7-*O*-glucosides, demonstrating the activity of endogenous GTs on these compounds.

There are fewer publications on changing levels of isoflavonoids in isoflavonoid-producing species. Yu et al.[325] introduced a *CRC* transgene into *G. max*, under the control of a seed-specific promoter. Overall levels of seed isoflavonoids were slightly raised, and there was a reduction in the ratio of genistein to daidzein. Accompanying the changes in isoflavonoid levels was an accumulation of isoliquiritigenin and liquiritigenin malonyl-glucose conjugates, as well as flavonols and PA-like compounds. There was also an association of the strong phenotypes with a visual color change in some seed. Suppression of *F3H* gene activity further raised isoflavonoid levels in the *CRC* transgenics. There is also a preliminary report that the introduction of a *CaMV35S:G.max2HIS* transgene into *G. max* resulted in plants with either reduced or enhanced levels of isoflavones.[347] In addition, the role of the D7OMT in isoflavonoid biosynthesis has been studied by expression of a *CaMV35S:D7OMT* transgene in *M. sativa*.[189] Some transgenics showed higher transcript abundance for flavonoid biosynthesis enzymes after infection with *Phoma medicaginis*, as well as increased isoflavonoid levels and improved pathogen resistance.

3.16.3.4 Altering B-Ring Hydroxylation

An obvious approach for flavonoid biotechnology is to direct the accumulation of pelargonidin- or delphinidin-derived anthocyanins in species that may lack them by inhibiting the activity of the F3'H and/or F3'5'H or introducing the F3',5'H. Thus, it is no surprise that *F3',5'H* transgenes have been introduced into several species that lack the activity, including the leading ornamental crops *Dendranthema*, *Dianthus*, and *Rosa hybrida* (Table 3.5). Indeed, transgenic *Dianthus* with novel colors based on delphinidin-derived anthocyanins are now commercially available in Australia, Japan, and the United States.[348] The amount of delphinidin produced in some of the first *F3',5'H* transgenics was relatively low, due to competition from the F3'H and the substrate preference of the endogenous DFR for DHK or DHQ. Much higher levels of delphinidin-derived anthocyanins were produced by introducing both a *F3',5'H* transgene and a transgene for a *DFR* that prefers DHM to DHK as substrate (such as that from *P. hybrida*) into a plant background that accumulates DHK (a *DFR-F3'H* double mutant) (International Patent Application WO96/36716). An alternative approach has been demonstrated by Okinaka et al.,[349] who generated transgenic *Nicotiana* in which over 99% of the anthocyanins accumulated were delphinidin based by using a *Campanula medium* cDNA for a *F3',5'H* that seems particularly efficient at generating delphinidin precursors. However, formation of blue flower colors involves many factors in addition to production of delphinidin-derived anthocyanins, and many of these are still out of reach of the current gene technology. In a similar approach to that for increasing delphinidin-derived anthocyanin levels, a *F3'H* transgene has been used to increase the amount of cyanidin-based anthocyanins in *T. hybrida*.[350]

Inhibition of the activities of F3'H or F3',5'H should enable production of pelargonidin-based anthocyanins in species that lack them. *Dendranthema* is one candidate for such an approach.[351] Of course, the subsequent enzymes would need to be able to use the substrates with only monohydroxylation of the B-ring, and in particular, DFR specificity would be a consideration for some species (e.g., *Cymbidium*). Reports of transgenic plants demonstrating this approach have yet to be published.

3.16.3.5 Secondary Modifications

Although there is a wide range of cDNAs available for secondary modifications of flavonoids, biotechnology applications to date have primarily focused on modifying anthocyanin production.

Glycosylation can affect the anthocyanin-based color of flowers through both direct effects on anthocyanin chemistry and indirect effects in which the change in glycosylation has ramifications on other modifications. This later effect is nicely illustrated by experiments with the A3RT and F3GT. Inhibition of *A3RT* gene activity in *P. hybrida* produced transgenics with altered flower color. The flowers had reduced levels of malvidin derivatives and increased levels of pigments based on delphinidin and petunidin, presumably due to the anthocyanin OMT being unable to use the nonrhamnosylated substrate.[111] Introduction of a *CaMV35S:A.majusF3GT* transgene into *E. grandiflorum* resulted in a change from flavonoid 3-*O*-galactosylation to 3-*O*-glucosylation, and an associated reduction in acylation of the flavonols, but no change in flower color.[90]

Much of the focus on glycosylation activities for biotechnology has been with regard to generating blue flower colors in leading ornamental species, although published transgenic experiments to date have concentrated on model species. The A3'GT is involved in the formation of the blue anthocyanin gentiodelphin in the source species *G. triflora*,[113] and thus is a good prospect for generating anthocyanins more suited to blue flower colors. Expression of *CaMV35S:A3'GT* and *CaMV35S:A5GT* transgenes together in *P. hybrida* resulted in the formation of a new anthocyanin type, delphinidin 3,5,3'-tri-*O*-glucoside, but only at low levels (2 to 6% of total anthocyanins). No change in flower color occurred.[113]

To date, published results are available for only one of the AAT cDNAs, the *Dv3MaT*, which was overexpressed in *P. hybrida*. Although the transgenics produced malonylated anthocyanins in the petals that were novel to *P. hybrida*, no change in flower color occurred.[123] Modification of the activity of anthocyanin OMTs has been reported only in the patent literature (International Patent Application WO03/062428).

3.16.4 MODULATING PATHWAY REGULATION

TFs have been used to modify not only the amount of flavonoids in transgenic plants and cell lines of several species but also their temporal and spatial distributions, including effective use in heterologous species. A key advantage of targeting TF activity is that several biosynthetic genes may be coordinately regulated, overcoming the need to introduce multiple transgenes for biosynthetic enzymes or to identify a rate-limiting step.

Anthocyanin production has been increased in transgenic plants of several species by introduction of anthocyanin-related *Myb* or *bHLH* cDNAs under the control of the *CaMV35S* promoter. Most experiments to date have used *C1* and *R/B* family transgenes from *Z. mays*; however, increasingly TF genes from other species are proving effective (Table 3.6). The *Z. mays* TF transgenes also increased the levels of flavonols in *L. esculentum*, PAs in *L. corniculatus* and various flavonoids, and other metabolites in *Z. mays* cell lines.[140,352–354] Cointroducing transgenes for MYB and bHLH TFs that normally interact can enable higher levels of anthocyanin production, or novel spatial distribution, compared to the introduction of either transgene alone. The first example of this was for *A. thaliana*, in which plants with both *CaMV35S:C1* and *CaMV35S:Lc* transgenes had increased anthocyanin levels and novel distribution patterns not seen in plants with either transgene on its own.[355] As would be expected from the role of some TFs in repressing gene expression, appropriate transgenes can also be used to downregulate sections of the phenylpropanoid pathway.[270,271,297,356] Furthermore, TFs that normally increase transcription can be altered so that they repress it, through the addition of repression domains from other proteins, such as the ERF-associated amphiphilic repression motif.[357,358]

For reasons that are not clear, the overexpression of anthocyanin-related TFs can enhance production of other flavonoids in addition to anthocyanins. It may be that the TFs have an uncharacterized role in regulating the other pathways as part of their usual function. Alternatively, they may increase flux into, or alter flux within, the flavonoid pathway so that substrate is fed into production of the nonanthocyanin end products. A further possibility is that the effects seen are due to the nonphysiological levels of TF protein produced through the use of strong promoters, such as *CaMV35S*. Excess levels of TF protein may promote activity on gene promoter sites additional to the usual targets, as demonstrated in *CaMV35-S:AtMYB4* transgenic *A. thaliana*,[271] or interfere with the endogenous regulatory environment.

Although there has been much success with flavonoid-related TF transgenes, the technology has also proven unpredictable. This is well illustrated by the fact that different bHLH TFs, despite all being regulators of anthocyanin production, generate distinct phenotypes in recipient species (Table 3.6). Thus, one transgene may produce a phenotype in a given species while one with a similar sequence may not (see, e.g., Ref. 354), or one recipient species may develop a strong phenotype while the same transgene gives no increased pigmentation in other species (see, e.g., Refs. 321, 359). Such diverse responses likely reflect differences in the binding or activation characteristics of the introduced TFs, combined with variation in the type and activity of interacting endogenous TFs and the presence or absence of target promoter *cis*-sequences in the recipient species. The role of endogenous factors is demonstrated by the markedly different phenotypes of *M. sativa* and *P. hybrida CaMV35S:Lc* transgenics under varying environmental conditions.[254,360]

3.17   CONCLUDING COMMENTS

Since 1994, remarkable progress has been made in our understanding of flavonoid biosynthesis. DNA sequences are now available for most of the biosynthetic enzymes required to form the common flavonoid types, such as anthocyanins, flavonols, flavones, PAs, isoflavonoids, and pterocarpans. These have allowed three-dimensional structures to be established for some of the enzymes using homology modeling or x-ray crystallography. Furthermore, significant progress has been made in understanding some of the activities involved in transport of anthocyanins and PAs to the vacuole. The magnitude of the advances in molecular knowledge of the pathway is illustrated by the listing in Table 3.1 of over 50 flavonoid biosynthetic enzymes for which cDNAs or genes are available, compared to the eight that were listed in the review by Forkmann,[1] which covered the literature up to 1991. Likewise, the same review listed cloned regulatory genes from only two species, *A. majus* and *Z. mays*, while Table 3.2 lists cloned genes from 12 species.

Biotechnology of the flavonoid pathway was in its infancy at the start of the 1990s, but is currently being widely applied to develop improved cultivars of ornamental, pasture, and food crops. In particular, the recent identification of the PA monomer biosynthetic genes should be of significant benefit to agriculture in the coming decade. As knowledge of flavonoids and human health progresses, one may anticipate that gene technology will be increasingly applied to improve the health characteristics of crop plants. Important factors in the future success of flavonoid biotechnology in achieving precision engineering of the pathway will be not only continuing advances in the identification of regulatory and biosynthetic genes, but also an improved understanding of metabolic channeling. This may apply particularly to the introduction of new branches of the pathway, such as that for the pterocarpans, to target crops.

3.18 ACKNOWLEDGMENTS

It is a pleasure to thank Dr Rick Dixon and Dr Bettina Deavours for commenting on the isoflavonoid section, and Associate Professor Kevin Gould for general reviewing of the manuscript.

REFERENCES

1. Forkmann, G., Genetics of flavonoids, in *The Flavonoids: Advances in Research Since 1980*, Harborne, J.B., Ed., Chapman & Hall, London, 1994, 538.
2. Heller, W. and Forkmann, G., Biosynthesis of flavonoids, in *The Flavonoids: Advances in Research Since 1980*, Harborne, J.B., Ed., Chapman & Hall, London, 1988, 399.
3. Heller, W. and Forkmann, G., Biosynthesis of flavonoids, in *The Flavonoids: Advances in Research Since 1986*, Harborne, J.B., Ed., Chapman & Hall, London, 1994, 499.
4. Bohm, B.A., *Introduction to Flavonoids*, Harwood Academic Publishers, Amsterdam, 1998.
5. Tanner, G.J., Condensed tannins, in *Plant Pigments and their Manipulation*, Davies, K.M., Ed., Annual Plant Reviews, Sheffield Academic Press, Sheffield, 2004, 150.
6. Fatland, B.L. et al., Molecular characterization of a heteromeric ATP-citrate lyase that generates cytosolic acetyl-coenzyme A in *Arabidopsis. Plant Physiol.*, 130, 740, 2002.
7. Nikolau, B.J., Ohlrogge, J.B., and Wurtele, E.S., Plant biotin-containing carboxylases. *Arch. Biochem. Biophys.*, 414, 211, 2003.
8. Shorrosh, B.S., Dixon, R.A., and Ohlrogge, J.B., Molecular cloning, characterization, and elicitation of acetyl-CoA carboxylase from alfalfa. *Proc. Natl. Acad. Sci. USA*, 91, 4323, 1994.
9. Roesler, K.R., Shorrosh, B.S., and Ohlrogge, J.B., Structure and expression of an *Arabidopsis* acetyl-coenzyme A carboxylase gene. *Plant Physiol.*, 105, 611, 1994.
10. Kuhn, D.N. et al., Induction of phenylalanine ammonia-lyase and 4-coumarate: CoA ligase mRNAs in cultured plant cells by UV light or fungal elicitor. *Proc. Natl. Acad. Sci. USA*, 81, 1102, 1984.
11. Rösler, J. et al., Maize phenylalanine ammonia-lyase has tyrosine ammonia-lyase activity. *Plant Physiol.*, 113, 175, 1997.
12. Röther, D. et al., An active site homology model of phenylalanine ammonia-lyase from *Petroselinum crispum. Eur. J. Biochem.*, 269, 3065, 2002.
13. Schuster, B. and Rétey, J., The mechanism of action of phenylalanine ammonia-lyase: the role of prosthetic dehydroalanine. *Proc. Natl. Acad. Sci. USA*, 92, 8433, 1995.
14. Winkel-Shirley, B., Molecular genetics and control of anthocyanin expression. *Adv. Bot. Res.*, 37, 75, 2002.

15. Mizutani, M. et al., Molecular cloning and sequencing of a cDNA encoding mung bean cyto-chrome P450 (P450$_{C4H}$) possessing cinnamate 4-hydroxylase activity. *Biochem. Biophys. Res. Commun.*, 190, 875, 1993.

16. Teutsch, H.G. et al., Isolation and sequence of a cDNA encoding the Jerusalem artichoke cinnamate 4-hydroxylase, a major plant cytochrome P450 involved in the general phenylpropanoid pathway. *Proc. Natl. Acad. Sci. USA*, 90, 4102, 1993.

17. Fahrendorf, T. and Dixon, R.A., Stress responses in alfalfa (*Medicago sativa* L.) XVIII: molecular cloning and expression of the elicitor-inducible cinnamic acid 4-hydroxylase cytochrome P450. *Arch. Biochem. Biophys.*, 305, 509, 1993.

18. Hübner, S. et al., Functional expression of cinnamate 4-hydroxylase from *Ammi majus* L. *Phyto-chemistry*, 64, 445, 2003.

19. Gravot, A. et al., Cinnamic acid 4-hydroxylase mechanism-based inactivation by psoralen deriva-tives: cloning and characterization of a C4H from a psoralen producing plant — *Ruta graveolens* — exhibiting low sensitivity to psoralen inactivation. *Arch. Biochem. Biophys.*, 422, 71, 2004.

20. Nedelkina, S. et al., Novel characteristics and regulation of a divergent cinnamate 4-hydroxylase (CYP73A15) from French bean: engineering expression in yeast. *Plant Mol. Biol.*, 39, 1079, 1999.

21. Betz, C., McCollum, T.G., and Mayer, R.T., Differential expression of two cinnamate 4-hydroxylase genes in "Valencia" orange (*Citrus sinensis* Osbeck). *Plant Mol. Biol.*, 46, 741, 2001.

22. Koopmann, E., Logemann, E., and Hahlbrock, K., Regulation and functional expression of cinnamate 4-hydroxylase from parsley. *Plant Physiol.*, 119, 49, 1999.

23. Chapple, C., Molecular-genetic analysis of plant cytochrome P450-dependent monooxygenases. *Ann. Rev. Plant Physiol. Plant Mol. Biol.*, 49, 311, 1998.

24. Werck-Reichhart, D., Bak, S., and Paquette, S., Cytochromes P450, in *The Arabidopsis Book*, Somerville, C.R. and Meyerowitz, E.M., Eds., American Society of Plant Biologists, Rockville, MD, doi/10.1199/tab.0028, http://www.aspb.org/publications/arabidopsis/, 2002.

25. Rupasinghe, S., Baudry, J., and Schuler, M.A., Common active site architecture and binding strategy of four phenylpropanoid P450s from *Arabidopsis thaliana* as revealed by molecular modeling. *Prot. Eng.*, 16, 721, 2003.

26. Lindermayr, C. et al., Divergent members of a soybean (*Glycine max* L.) 4-coumarate:coenzyme A ligase gene family — primary structures, catalytic properties, and differential expression. *Eur. J. Biochem.*, 269, 1304, 2002.

27. Hamberger, B. and Hahlbrock, K., The 4-coumarate:CoA ligase gene family in *Arabidopsis thaliana* comprises one rare, sinapate-activating and three commonly occurring isoenzymes. *Proc. Natl. Acad. Sci. USA*, 101, 2209, 2004.

28. Ehlting, J., Shin, J.J.K., and Douglas, C.J., Identification of 4-coumarate:coenzyme A ligase (4CL) substrate recognition domains. *Plant J.*, 27, 455, 2001.

29. Hu, W.-J. et al., Compartmentalized expression of two structurally and functionally distinct 4-coumarate:CoA ligase genes in aspen (*Populus tremuloides*). *Proc. Natl. Acad. Sci. USA*, 95, 5407, 1998.

30. Harding, S.A. et al., Differential substrate inhibition couples kinetically distinct 4-coumarate: coenzyme A ligases with spatially distinct metabolic roles in quaking aspen. *Plant Physiol.*, 128, 428, 2002.

31. Cukovic, D. et al., Structure and evolution of 4-coumarate: coenzyme A ligase (4CL) gene families. *Biol. Chem.*, 382, 645, 2001.

32. Pietrowska-Borek, M. et al., 4-Coumarate:coenzyme A ligase has the catalytic capacity to synthe-size and reuse various (di)adenosine polyphosphates. *Plant Physiol.*, 131, 1401, 2003.

33. Lindermayr, C., Fliegmann, J., and Ebel, J., Deletion of a single amino acid residue from different 4-coumarate:CoA ligases from soybean results in the generation of new substrate specificities. *J. Biol. Chem.*, 278, 2781, 2003.

34. Stuible, H.P. and Kombrink, E., Identification of the substrate specificity-conferring amino acid residues of 4-coumarate:coenzyme A ligase allows the rational design of mutant enzymes with new catalytic properties. *J. Biol. Chem.*, 276, 26893, 2001.

35. Schneider, K. et al., The substrate specificity-determining amino acid code of 4-coumarate:CoA ligase. *Proc. Natl. Acad. Sci. USA*, 100, 8601, 2003.

36. Abd El-Mawla, A.M. and Beerhues, L., Benzoic acid biosynthesis in cell cultures of *Hypericum androsaemum. Planta*, 214, 727, 2002.

37. Fliegmann, J. et al., Molecular analysis of chalcone and dihydropinosylvin synthase from Scots pine (*Pinus sylvestris*), and differential regulation of these and related enzyme activities in stressed plants. *Plant Mol. Biol.*, 18, 489, 1992.

38. Christensen, A.B. et al., A chalcone synthase with an unusual substrate preference is expressed in barley leaves in response to UV light and pathogen attack. *Plant Mol. Biol.*, 37, 849, 1998.

39. Samappito, S. et al., Molecular characterization of root-specific chalcone synthases from *Cassia alata. Planta*, 216, 64, 2002.

40. Petersen, M., Strack, D., and Matern, U., Biosynthesis of phenylpropanoids and related compounds, in *Biochemistry of Plant Secondary Metabolism*, Wink, M., Ed., Sheffield Academic Press, Sheffield, 1999, 151.

41. Schoch, G. et al., CYP98A3 from *Arabidopsis thaliana* is a 3-hydroxylase of phenolic esters, a missing link in the phenylpropanoid pathway. *J. Biol. Chem.*, 276, 36566, 2001.

42. Franke, R. et al., The *Arabidopsis REF8* gene encodes the 3-hydroxylase of phenylpropanoid metabolism. *Plant J.*, 30, 33, 2002.

43. Nair, R.B. et al., *Arabidopsis CYP98A3 mediating aromatic 3-hydroxylation. Developmental regulation of the gene, and expression in yeast. Plant Physiol.*, 130, 210, 2002.

44. Kreuzaler, F. et al., UV-induction of chalcone synthase mRNA in cell suspension cultures of *Petroselinum hortense. Proc. Natl. Acad. Sci. USA*, 80, 2591, 1983.

45. Reimold, U. et al., Coding and 3' non-coding nucleotide sequence of chalcone synthase mRNA and assignment of amino acid sequence of the enzyme. *EMBO J.*, 2, 1801, 1983.

46. Austin, M.B. and Noel, J.P., The chalcone synthase superfamily of type III polyketide synthases. *Nat. Prod. Rep.*, 20, 79, 2003.

47. Ferrer, J.-L. et al., Structure of chalcone synthase and the molecular basis of plant polyketide biosynthesis. *Nat. Struct. Biol.*, 6, 775, 1999.

48. Tropf, S. et al., Reaction mechanisms of homodimeric plant polyketide synthases (stilbenes and chalcone synthase): a single active site for the condensing reaction is sufficient for synthesis of stilbenes, chalcones, and 6'-deoxychalcones. *J. Biol. Chem.*, 270, 7922, 1995.

49. Preisig-Mueller, R. et al., Plant polyketide synthases leading to stilbenoids have a domain catalyzing malonyl-CoA:CO_2 exchange, malonyl-CoA decarboxylation, and covalent enzyme modification and a site for chain lengthening. *Biochemistry*, 36, 8349, 1997.

50. Tropf, S. et al., Evidence that stilbene synthases have developed from chalcone synthases several times in the course of evolution. *J. Mol. Evol.*, 38, 610, 1994.

51. Noel, J.P. et al., Structurally guided alteration of biosynthesis in plant type III polyketide synthases, in *Phytochemistry in the Genomics and Post-Genomics Eras*, Romeo, J.T. and Dixon, R.A., Eds., Pergamon, New York, 2002, 197.

52. Springob, K. et al., Recent advances in the biosynthesis and accumulation of anthocyanins. *Nat. Prod. Rep.*, 20, 288, 2003.

53. Jez, J.M. and Noel, J.P., Reaction mechanism of chalcone isomerase. *J. Biol. Chem.*, 277, 1361, 2002.

54. Jez, J.M. et al., Structure and mechanism of the evolutionarily unique plant enzyme chalcone isomerase. *Nat. Struct. Biol.*, 7, 786, 2000.

55. Moustafa, E. and Wong, E., Purification and properties of chalcone–flavanone isomerase from soya bean seed. *Phytochemistry*, 6, 625, 1967.

56. Gensheimer, M. and Mushegian, A., Chalcone isomerase family and fold: no longer unique to plants. *Prot. Sci.*, 13, 540, 2004.

57. Burbulis, I.E. and Winkel-Shirley, B. Interactions among enzymes of the *Arabidopsis* flavonoid biosynthetic pathway. *Proc. Natl. Acad. Sci. USA*, 96, 12929, 1999.

58. Joung, J.Y. et al., An overexpression of chalcone reductase of *Pueraria montana* var. lobata alters biosynthesis of anthocyanin and 5'-deoxyflavonoids in transgenic tobacco. *Biochem. Biophys. Res. Commun.*, 303, 326, 2003.

59. Shimada, N. et al., A cluster of genes encodes the two types of chalcone isomerase involved in the biosynthesis of general flavonoids and legume-specific 5-deoxy(iso)flavonoids in *Lotus japonicus. Plant Physiol.*, 131, 941, 2003.

60. Bednar, R.A. and Hadcock, J.R., Purification and characterization of chalcone isomerase from soybeans. *J. Biol. Chem.*, 263, 9582, 1988.

61. Muir, S.R. et al., Overexpression of petunia chalcone isomerase in tomato results in fruit containing increased levels of flavonols. *Nat. Biotechnol.*, 19, 470, 2001.

62. Lukacin, R. et al., Flavonol synthase from *Citrus unshiu* is a bifunctional dioxygenase. *Phytochemistry*, 62, 287, 2003.

63. Prescott, A.G., Two-oxoacid-dependant dioxygenases: inefficient enzymes or evolutionary driving force, in *Evolution of Metabolic Pathways*, Romeo, J.T., Ibrahim, R., Varin, L., and De Luca, V., Eds., Elsevier Science, Oxford, 2000, 249.

64. Martin, C. et al., Control of anthocyanin biosynthesis in flowers of *Antirrhinum majus*. *Plant J.*, 1, 1991, 37.

65. Britsch, L. et al., Molecular characterization of flavanone 3β-hydroxylase: consensus sequence, comparison with related enzymes and the role of conserved histidines. *Eur. J. Biochem.*, 217, 745, 1993.

66. Lukacin, R. and Britsch, L., Identification of strictly conserved histidine and arginine residues as part of the active site in *Petunia hybrida* flavanone 3β-hydroxylase. *Eur. J. Biochem.*, 249, 748, 1997.

67. Lukacin, R., et al., Site-directed mutagenesis of the active site serine290 in flavanone 3β-hydroxylase from *Petunia hybrida*. *Eur. J. Biochem.*, 267, 853, 2000.

68. Martin, C. et al., Molecular analysis of instability in flower pigmentation of *Antirrhinum majus*, following isolation of the *pallida* locus by transposon tagging. *EMBO J.*, 4, 1625, 1985.

69. O'Reilly, C. et al., Molecular cloning of the *A1* locus of *Zea mays* using the transposable elements *En* and *Mul*. *EMBO J.*, 4, 877, 1985.

70. Reddy, A.R. et al., The *A1* (*anthocyanin-1*) locus in *Zea mays* encodes dihydroquercetin reductase. *Plant Sci.*, 52, 7, 1987.

71. Xie, D.-Y. et al., Molecular and biochemical analysis of two cDNA clones encoding dihydroflavonol-4-reductase from *Medicago truncatula*. *Plant Physiol.*, 134, 1, 2004.

72. Labesse, G. et al., Structural comparisons lead to the definition of a new superfamily of NAD(P)(H)-accepting oxidoreductases: the single-domain reductases/epimerases/dehydrogenases (the "RED" family). *Biochem. J.*, 304, 95, 1994.

73. Johnson, E.T. et al., Alteration of a single amino acid changes the substrate specificity of dihydroflavonol 4-reductase. *Plant J.*, 25, 325, 2001.

74. Johnson, E.T. et al., *Cymbidium hybrida* dihydroflavonol 4-reductase does not efficiently reduce dihydrokaempferol to produce orange pelargonidin-type anthocyanins. *Plant J.*, 19, 81, 1999.

75. Polashock, J.J. et al., Cloning of a cDNA encoding the cranberry dihydroflavonol-4-reductase (DFR) and expression in transgenic tobacco. *Plant Sci.*, 163, 241, 2002.

76. Davies, K.M. and Schwinn, K.E., Flower colour, in *Biotechnology of Ornamental Plants*, Geneve, R.L., Preece J.E., and Merkle, S.A., Eds., CAB International, Wallingford, 1997, 259.

77. Menssen, A. et al., The *En/Spm* transposable element of *Zea mays* contains splice sites at the termini generating a novel intron from a *dSpm* element in the *A2* gene. *EMBO J.*, 9, 3051, 1990.

78. Matsuda, J. et al., Molecular cloning of hyoscyamine 6β-hydroxylase, a 2-oxoglutarate-dependent dioxygenase, from cultured roots of *Hyoscyamus niger*. *J. Biol. Chem.*, 266, 9460, 1991.

79. Saito, K. et al., Direct evidence for anthocyanidin synthase as a 2-oxoglutarate-dependent oxygenase: molecular cloning and functional expression of cDNA from a red forma of *Perilla frutescens*. *Plant J.*, 17, 181, 1999.

80. Nakajima, J.-I. et al., Reaction mechanism from leucoanthocyanidin to anthocyanidin 3-glucoside, a key reaction for coloring in anthocyanin biosynthesis. *J. Biol. Chem.*, 276, 25797, 2001.

81. Turnbull, J.J. et al., Mechanistic studies on three 2-oxoglutarate-dependent oxygenases of flavonoid biosynthesis — anthocyanidin synthase, flavonol synthase, and flavanone 3 beta-hydroxylase. *J. Biol. Chem.*, 279, 1206, 2004.

82. Turnbull, J.J. et al., Purification, crystallization and preliminary x-ray diffraction of anthocyanidin synthase from *Arabidopsis thaliana*. *Acta Crystallogr. D*, 57, 425, 2001.

83. Turnbull, J.J. et al., The C-4 stereochemistry of leucocyanidin substrates for anthocyanidin synthase affects product selectivity. *Bioorg. Med. Chem. Lett.*, 13, 3853, 2003.

84. Welford, R.W.D. et al., Evidence for oxidation at C-3 of the flavonoid C-ring during anthocyanin biosynthesis. *Chem. Commun.*, 1828, 2001.

85. Wilmouth, R.C. et al., Structure and mechanism of anthocyanidin synthase from *Arabidopsis thaliana*. *Structure*, 10, 93, 2002.

86. Turnbull, J.J. et al., Are anthocyanidins the immediate products of anthocyanidin synthase? *Chem. Commun.*, 2473, 2000.

87. Schröder, G. et al., Flavonoid methylation: a novel 4'-*O*-methyltransferase from *Catharanthus roseus*, and evidence that partially methylated flavanones are substrates of four different flavonoid dioxygenases. *Phytochemistry*, 65, 1085, 2003.

88. Yamazaki, M. et al., Two flavonoid glucosyltransferases from *Petunia hybrida*: molecular cloning, biochemical properties and developmentally regulated expression. *Plant Mol. Biol.*, 48, 401, 2002.

89. Rosati, C. et al., Engineering of flower colour in forsythia by expression of two independently-transformed dihydroflavonol 4-reductase and anthocyanidin synthase genes of flavonoid pathway. *Mol. Breed.*, 12, 197, 2003.

90. Schwinn, K.E. et al., Expression of an *Antirrhinum majus* UDP-glucose:flavonoid-3-*O*-glucosyltransferase transgene alters flavonoid glycosylation and acylation in lisianthus (*Eustoma grandiflorum* Grise.). *Plant Sci.*, 125, 53, 1997.

91. Tanaka, Y. et al., Molecular and biochemical characterization of three anthocyanin synthetic enzymes from *Gentiana triflora*. *Plant Cell Physiol.*, 37, 711, 1996.

92. Do, C.B., Cormier, F., and Nicholas, Y., Isolation and characterization of a UDP-glucose:cyanidin 3-*O*-glucosyltransferase from grape cell suspension cultures (*Vitis vinifera* L.). *Plant Sci.*, 112, 43, 1995.

93. Ford, C.M., Boss, P.K., and Høj, P.B., Cloning and characterization of *Vitis vinifera* UDP-glucose:flavonoid 3-*O*-glucosyltransferase, a homologue of the enzyme encoded by the maize *Bronze-1* locus that may primarily serve to glucosylate anthocyanidins *in vivo*. *J. Biol. Chem.*, 273, 9224, 1998.

94. Vogt, T., Glycosyltransferases involved in plant secondary metabolism, in *Evolution of Metabolic Pathways*, Romeo, J.T., Ibrahim, R., Varin, L., and De Luca, V., Eds., Elsevier Science, Oxford, 2002, 317.

95. Paquette, S., Møller, B.L., and Bak, S., On the origin of family 1 plant glycosyltransferases. *Phytochemistry*, 62, 399, 2003.

96. Li, Y., Baldauf, S., Lim, E.-K., and Bowles, D.J., Phylogenetic analysis of the UDP-glycosyltransferase multigene family of *Arabidopsis thaliana*. *J. Biol. Chem.*, 276, 4338, 2001.

97. Fischer, T.C. et al., Molecular cloning, substrate specificity of the functionally expressed dihydroflavonol 4-reductases from *Malus domestica* and *Pyrus communis* cultivars and the consequences for flavonoid metabolism *Arch. Biochem. Biophys.*, 412, 223, 2003.

98. Halbwirth, H. et al., Biochemical formation of anthocyanins in silk tissue of *Zea mays*. *Plant Sci.*, 164, 489, 2003.

99. Winefield, C.S. et al., Investigation of the biosynthesis of 3-deoxyanthocyanins in *Sinningia cardinalis*. *Physiol. Plant.*, 124, 419, 2005.

100. Lo, S.C.C. and Nicholson, R.L., Reduction of light-induced anthocyanin accumulation in inoculated sorghum mesocotyls. Implications for a compensatory role in the defense response. *Plant Physiol.*, 116, 979, 1998.

101. Lo, C., Coolbaugh, R.C., and Nicholson, R.L., Molecular characterization and *in silico* expression analysis of a chalcone synthase gene family in *Sorghum bicolor*. *Physiol. Mol. Plant Pathol.*, 61, 179, 2002.

102. Yamazaki, M. et al., Molecular cloning and biochemical characterization of a novel anthocyanin 5-*O*-glucosyltransferase by mRNA differential display for plant forms regarding anthocyanin. *J. Biol. Chem.*, 274, 7405, 1999.

103. Holton, T.A. et al., Cloning and expression of cytochrome P450 genes controlling flower colour. *Nature*, 366, 276, 1993.

104. Toguri, T., Azuma, M., and Ohtani, T., The cloning and characterization of a cDNA encoding a cytochrome P450 from the flowers of *Petunia hybrida*. *Plant Sci.*, 94, 119, 1993.

105. Brugliera, F. et al., Isolation and characterization of a flavonoid 3'-hydroxylase cDNA clone corresponding to the *Ht1* locus of *Petunia hybrida*. *Plant J.*, 19, 441, 1999.
106. Schoenbohm, C. et al., Identification of the *Arabidopsis thaliana* flavonoid 3'-hydroxylase gene and functional expression of the encoded P450 enzyme. *Biol. Chem.*, 381, 749, 2000.
107. Kitada, C. et al., Differential expression of two cytochrome P450s involved in the biosynthesis of flavones and anthocyanins in chemo-varietal forms of *Perilla frutescens*. *Plant Cell Physiol.*, 42, 1338, 2001.
108. Kaltenbach, M. et al., Flavonoid hydroxylase from *Catharanthus roseus*: cDNA, heterologous expression, enzyme properties and cell-type specific expression in plants. *Plant J.*, 19, 183, 1999.
109. Shimada, Y. et al., Expression of chimeric P450 genes encoding flavonoid-3',5'-hydroxylase in transgenic tobacco and petunia plants. *FEBS Lett.*, 461, 241, 1999.
110. de Vetten, N. et al., A cytochrome b_5 is required for full activity of flavonoid 3',5'-hydroxylase, a cytochrome P450 involved in the formation of blue flower colours. *Proc. Natl. Acad. Sci. USA*, 96, 778, 1999.
111. Brugliera, F. et al., Isolation and characterization of a cDNA clone corresponding to the *Rt* locus of *Petunia hybrida*. *Plant J.*, 5, 81, 1994.
112. Kroon, J. et al., Cloning and structural analysis of the anthocyanin pigmentation locus *Rt* of *Petunia hybrida*: characterization of insertion sequences in two mutant alleles. *Plant J.*, 5, 69, 1994.
113. Fukuchi-Mizutani, M. et al., Biochemical and molecular characterization of a novel UDP-glucose:anthocyanin 3'-O-glucosyltransferase, a key enzyme for blue anthocyanin biosynthesis, from gentian. *Plant Physiol.*, 132, 1652, 2003.
114. Vogt, T., Substrate specificity and sequence analysis define a polyphyletic origin of betanidin 5- and 6-O-glucosyltransferase from *Dorotheanthus bellidiformis*. *Planta*, 214, 492, 2002.
115. Ibdah, M. et al., A novel Mg(2+)-dependent O-methyltransferase in the phenylpropanoid metabolism of *Mesembryanthemum crystallinum*. *J. Biol. Chem.*, 278, 43961, 2003.
116. Ibrahim, R.K., Bruneau, A., and Bantignies, B., Plant O-methyltransferases: molecular analysis, common signature and classification. *Plant Mol. Biol.*, 36, 1, 1998.
117. Ibrahim, R.K. and Muzac, I., The methyltransferase gene superfamily: a tree with multiple branches, in *Evolution of Metabolic Pathways*, Romeo, J.T., Ibrahim, R., Varin, L., and De Luca, V., Eds., Elsevier Science Ltd, Oxford, 2000, 349.
118. Zubieta, C. et al., Structures of two natural product methyltransferases reveal the basis for substrate specificity in plant O-methyltransferases. *Nat. Stuct. Biol.*, 8, 271, 2001.
119. Quattrocchio, F. et al., Regulatory genes controlling anthocyanin pigmentation are functionally conserved among plant species and have distinct sets of target genes. *Plant Cell*, 5, 1497, 1993.
120. Nakayama, T., Suzuki, H., and Nishino, T., Anthocyanin acyltransferases: specificities, mechanism, phylogenetics, and applications. *J. Mol. Catal.*, 23, 117, 2003.
121. Fujiwara, H. et al., cDNA cloning, gene expression and subcellular localization of anthocyanin 5-aromatic acyltransferase from *Gentiana triflora*. *Plant J.*, 16, 421, 1998.
122. Yonekura-Sakakibara, K. et al., Molecular and biochemical characterization of a novel hydroxycinnamoyl-CoA: anthocyanin 3-O-glucoside-6''-O-acyltransferase from *Perilla frutescens*. *Plant Cell Physiol.*, 41, 495, 2000.
123. Suzuki, H. et al., cDNA cloning, heterologous expressions, and functional characterization of malonyl-coenzyme A: anthocyanidin 3-O-glucoside-6''-O-malonyltransferase from dahlia flowers. *Plant Physiol.*, 130, 2142, 2002.
124. Suzuki, H. et al., Identification of a cDNA encoding malonyl-coenzyme A: anthocyanidin 3-O-glucoside-6''-O-malonyltransferase from cineraria (*Senecio cruentus*) flowers. *Plant Biotechnol.*, 20, 229, 2003.
125. Suzuki, H. et al., cDNA cloning and characterization of two *Dendranthema* × *morifolium* anthocyanin malonyltransferases with different functional activities. *Plant Sci.*, 166, 89, 2004.
126. Suzuki, H. et al., Malonyl-CoA: anthocyanin 5-O-glucoside-6'-O-malonyltransferase from scarlet sage (*Salvia splendens*) flowers. *J. Biol. Chem.*, 276, 49013, 2001.
127. St-Pierre, B. and De Luca, V., Evolution of acyltransferase genes: origin and diversification of the BAHD superfamily of acyltransferases involved in secondary metabolism, in *Evolution of Meta-*

bolic Pathways, Romeo, J.T., Ibrahim, R., Varin, L., and De Luca, V., Eds., Elsevier Science, Oxford, 2000, 285.

128. Suzuki, H., Nakayama, T., and Nishino, T., Proposed mechanism and functional amino acid residues of malonyl-CoA: anthocyanin 5-*O*-glucoside-6′′′-*O*-malonyltransferase from flowers of *Salvia splendens*, a member of the versatile plant acyltransferase family. *Biochemistry*, 42, 1764, 2003.

129. Hause, B. et al., Immunolocalization of 1-*O*-sinapoylglucose:malate sinapoyltransferase in *Arabidopsis thaliana. Planta*, 215, 26, 2002.

130. Milkowski, C. and Strack, D., Serine carboxypeptidase-like acyltransferases. *Phytochemistry*, 65, 517, 2004.

131. Lehfeldt, C. et al., Cloning of the SNG1 gene of *Arabidopsis* reveals a role for a serine carboxypeptidase-like protein as an acyltransferase in secondary metabolism. *Plant Cell*, 12, 1295, 2000.

132. Marles, M.A., Ray, H., and Gruber, M.Y., New perspectives on proanthocyanidin biochemistry and molecular regulation. *Phytochemistry*, 64, 367, 2003.

133. Tanner, G.J. et al., Proanthocyanidin biosynthesis in plants. Purification of legume leucoanthocyanidin reductase and molecular cloning of its cDNA. *J. Biol. Chem.*, 278, 31647, 2003.

134. Abrahams, S. et al., Identification and biochemical characterization of mutants in the proanthocyanidin pathway in *Arabidopsis. Plant Physiol.*, 130, 561, 2002.

135. Albert, S., Delseny, M., and Devic, M., *Banyuls*, a novel negative regulator of flavonoid biosynthesis in the *Arabidopsis* seed coat. *Plant J.*, 11, 289, 1997.

136. Devic, M. et al., The BANYULS gene encodes a DFR-like protein and is a marker of early seed coat development. *Plant J.*, 9, 387, 1999.

137. Xie, D.-Y. et al., Role of anthocyanidin reductase, encoded by *BANYULS* in plant flavonoid biosynthesis. *Science*, 299, 396, 2003.

138. Xie, D.-Y., Sharma, S.B., and Dixon, R.A., Anthocyanidin reductases from *Medicago truncatula* and *Arabidopsis thaliana. Arch. Biochem. Biophys.*, 422, 91, 2004.

139. Abrahams, S. et al., The *Arabidopsis TDS4* gene encodes leucoanthocyanidin dioxygenase (LDOX) and is essential for proanthocyanidin synthesis and vacuole development. *Plant Cell*, 35, 624, 2003.

140. Grotewold, E. et al., Engineering secondary metabolism in maize cells by ectopic expression of transcription factors. *Plant Cell*, 10, 721, 1998.

141. Jende-Strid, B., Genetic control of flavonoid biosynthesis in barley. *Hereditas*, 119, 187, 1993.

142. Debeaujon, I. et al., The *TRANSPARENT TESTA 12* gene of *Arabidopsis* encodes a multidrug secondary transporter-like protein required for flavonoid sequestration in vacuoles of the seed coat endothelium. *Plant Cell*, 13, 853, 2001.

143. Martens, S. and Forkmann, G., Cloning and expression of flavone synthase II from *Gerbera* hybrids. *Plant J.*, 20, 611, 1999.

144. Akashi, T. et al., Molecular cloning and biochemical characterization of a novel cytochrome P450, flavone synthase II, that catalyzes direct conversion of flavanones to flavones. *Plant Cell Physiol.*, 40, 1182, 1999.

145. Martens, S. et al., Cloning of parsley flavone synthase I. *Phytochemistry*, 58, 43, 2001.

146. Holton, T.A., Cornish, E.C., and Tanaka, Y., Cloning and expression of flavonol synthase from *Petunia hybrida. Plant J.*, 4, 1003, 1993.

147. Martens, S. et al., Divergent evolution of flavonoid 2-oxoglutarate-dependent dioxygenases in parsley. *FEBS Lett.*, 544, 93, 2003.

148. Britsch, L., Purification and characterization of flavone synthase I, a 2-oxoglutarate-dependent desaturase. *Arch. Biochem. Biophys.*, 282, 152, 1990.

149. Davies, K.M. et al., Production of yellow colour in flowers: redirection of flavonoid biosynthesis in *Petunia. Plant J.*, 13, 259, 1998.

150. Cooper, J.D., Qiu, F., and Paiva, N.L., Biotransformation of an exogenously supplied isoflavonoid by transgenic tobacco cells expressing alfalfa isoflavone reductase. *Plant Cell Rep.*, 20, 876, 2002.

151. Liu, C.J. et al., Bottlenecks for metabolic engineering of isoflavone glycoconjugates in *Arabidopsis. Proc. Natl. Acad. Sci. USA*, 99, 14578, 2002.

152. Taguchi, G. et al., Cloning and characterization of a glucosyltransferase that reacts on 7-hydroxyl group of flavonol and 3-hydroxyl group of coumarin from tobacco cells. *Arch. Biochem. Biophys.*, 420, 95, 2003.

153. Jones, P. et al., UGT73C6 and UGT78D1 — glycosyltransferases involved in flavonol glycoside biosynthesis in *Arabidopsis thaliana*. *J. Biol. Chem.*, 278, 43910, 2003.

154. Kramer, C.M. et al., Cloning and regiospecificity studies of two flavonoid glucosyltransferases from *Allium cepa*. *Phytochemistry*, 64, 1069, 2003.

155. Hirotani, M. et al., Cloning and expression of UDP-glucose: flavonoid 7-*O*-glucosyltransferase from hairy root cultures of *Scutelliaria baicalensis*. *Planta*, 210, 1006, 2000.

156. Mato, M. et al., Isolation and characterization of a cDNA clone of UDP-galactose: flavonoid 3-*O*-galactosyltransferase (UF3GaT) expressed in *Vigna mungo* seedlings. *Plant Cell Physiol.*, 39, 1145, 1998.

157. Vogt, T., Grimm, R., and Strack, D., Cloning and expression of a cDNA encoding betanidin 5-*O*-glucosyltransferase, a betanidin- and flavonoid-specific enzyme with high homology to inducible glucosyltransferases from the Solanaceae. *Plant J.*, 19, 509, 1999.

158. Miller, K.D. et al., Purification, cloning, and heterologous expression of a catalytically efficient flavonol-3-*O*-galactosyltransferase expressed in the male gametophyte of *Petunia hybrida*. *J. Biol. Chem.*, 274, 34011, 1999.

159. Vogt, T. and Taylor, L.P., Flavonol 3-*O*-Glycosyltransferases associated with petunia pollen produce gametophyte-specific flavonol diglycosides. *Plant Physiol.*, 108, 903, 1995.

160. Christensen, A.B. et al., A flavonoid 7-*O*-methyltransferase is expressed in barley leaves in response to pathogen attack. *Plant Mol. Biol.*, 36, 219, 1998.

161. Gauthier, A., Gulick, P.J., and Ibrahim, R.K., cDNA cloning and characterization of a 3′/5′-*O*-methyltransferase for partially methylated flavonols from *Chrysosplenium americanum*. *Plant Mol. Biol.*, 32, 1163, 1996.

162. Gauthier, A., Gulick, P.J., and Ibrahim, R.K., Characterization of two cDNA clones which encode *O*-methyltransferases for the methylation of both flavonoid and phenylpropanoid compounds. *Arch. Biochem. Biophys.*, 351, 243, 1998.

163. Zhang, H., Wang, J., and Goodman, H.M., An *Arabidopsis* gene encoding a putative 14-3-3-interacting protein, caffeic acid/5-hydroxyferulic acid *O*-methyltransferase. *Biochim. Biophys. Acta*, 1353, 199, 1997.

164. Muzac, I. et al., Functional expression of an *Arabidopsis* cDNA clone encoding a flavonol 3′-*O*-methyltransferase and characterization of the gene product. *Arch. Biochem. Biophys.*, 375, 385, 2000.

165. Schröder, G., Wehinger, E., and Schröder, J., Predicting the substrates of cloned plant *O*-methyltransferases. *Phytochemistry*, 59, 1, 2002.

166. Vogt, T., Regiospecificity and kinetic properties of a plant natural product *O*-methyltransferase are determined by its N-terminal domain. *FEBS Lett.*, 561, 159, 2004.

167. Cacace, S. et al., A flavonol *O*-methyltransferase from *Catharanthus roseus* performing two sequential methylations. *Phytochemistry*, 62, 127, 2003.

168. Marsolais, F. et al., Plant soluble sulfotransferases: structural and functional similarity with mammalian enzymes, in *Evolution of Metabolic Pathways*, Romeo, J.T., Ibrahim, R., Varin, L., and De Luca, V., Eds., Elsevier Science, Oxford, 2000, 433.

169. Varin, L. et al., Molecular characterization of two plant flavonol sulfotransferases. *Proc. Natl. Acad. Sci. USA*, 89, 1286, 1992.

170. Ananvoranich, S. et al., Cloning and regulation of flavonol 3-sulfotransferase in cell-suspension cultures of *Flaveria bidentis*. *Plant Physiol.*, 106, 485, 1994.

171. Varin, L., Flavonoid sulfation: phytochemistry, enzymology and molecular biology. *Rec. Adv. Phytochem.*, 26, 233, 1992.

172. Marsolais, F. and Varin, L., Identification of amino acid residues critical for catalysis and cosubstrate binding in the flavonol 3-sulfotransferase. *J. Biol. Chem.*, 270, 30458, 1995.

173. Marsolais, F. and Varin, L., Mutational analysis of domain II of flavonol 3-sulfotransferase. *Eur. J. Biochem.*, 247, 1056, 1997.

174. Varin, L., Marsolais, F., and Brisson, N., Chimeric flavonol sulfotransferases define a domain responsible for substrate and position specificities *J. Biol. Chem.*, 270, 12498, 1995.

175. Marsolais, F. et al., 3′-Phosphoadenosine 5′-phosphosulfate binding site of flavonol 3-sulfotransferase studied by affinity chromatography and P-31 NMR. *Biochemistry*, 38, 4066, 1999.

176. Jez, J.M., Flynn, T.G., and Penning, T.M., A new nomenclature for the aldo-keto reductase superfamily. *Biochem. Pharmacol.*, 54, 639, 1997.

177. Welle, R. et al., Induced plant responses to pathogen attack. Analysis and heterologous expression of the key enzyme in the biosynthesis of phytoalexins in soybean (*Glycine max* L. Merr. cv. Harosoy 63). *Eur. J. Biochem.*, 196, 423, 1991.

178. Welle, R. and Schröder, J., Expression cloning in *Escherichia coli* and preparative isolation of the reductase coacting with chalcone synthase during the key step in the biosynthesis of soybean phytoalexins. *Arch. Biochem. Biophys.*, 293, 377, 1992.

179. Welle, R. and Grisebach, H., Phytoalexin synthesis in soybean cells: elicitor induction of reductase involved in biosynthesis of 6′-deoxychalcone. *Arch. Biochem. Biophys.*, 272, 97, 1989.

180. Akashi, T., Aoki, T., and Ayabe, S., Cloning and functional expression of a cytochrome P450 cDNA encoding 2-hydroxyisoflavanone synthase involved in biosynthesis of the isoflavonoid skeleton in licorice. *Plant Physiol.*, 21, 821, 1999.

181. Jung, W. et al., Identification and expression of isoflavone synthase, the key enzyme for biosynthesis of isoflavones in legumes. *Nat. Biotechnol.*, 18, 208, 2000.

182. Steele, C.L. et al., Molecular characterization of the enzyme catalyzing the aryl migration reaction of isoflavonoid biosynthesis in soybean. *Arch. Biochem. Biophys.*, 367, 146, 1999.

183. Hakamatsuka, T. et al., Purification of 2-hydroxyisoflavanone dehydratase from the cell cultures of *Pueraria lobata*. *Phytochemistry*, 49, 497, 1998.

184. Sawada, Y. et al., Key amino acid residues required for aryl migration catalysed by the cytochrome P450 2-hydroxyisoflavanone synthase. *Plant J.*, 31, 555, 2002.

185. Hashim, M.F. et al., Reaction mechanism of oxidative rearrangement of flavanone in isoflavone biosynthesis. *FEBS Lett.*, 271, 219, 1990.

186. Akashi, T. et al., New scheme of the biosynthesis of formononetin involving 2,7,4′-trihydroxyisoflavanone but not daidzein as the methyl acceptor. *Biosci. Biotechnol. Biochem.*, 64, 2276, 2000.

187. Akashi, T. et al., cDNA cloning and biochemical characterization of S-adenosyl-L-methionine: 2,7,4′-trihydroxyisoflavanone 4′-*O*-methyltransferase, a critical enzyme of the legume isoflavonoid phytoalexin pathway. *Plant Cell Physiol.*, 44, 103, 2003.

188. He, X.Z. and Dixon, R.A., Affinity chromatography, substrate/product specificity, and amino acid sequence analysis of an isoflavone *O*-methyltransferase from alfalfa (*Medicago sativa* L.). *Arch. Biochem. Biophys.*, 336, 121, 1996.

189. He, X.Z. and Dixon, R.A., Genetic manipulation of isoflavone 7-*O*-methyltransferase enhances biosynthesis of 4′-*O*-methylated isoflavonoid phytoalexins and disease resistance in alfalfa. *Plant Cell*, 12, 1689, 2000.

190. He, X.Z., Reddy, J.T., and Dixon, R.A., Stress responses in alfalfa (*Medicago sativa* L). XXII. cDNA cloning and characterization of an elicitor-inducible isoflavone 7-*O*-methyltransferase. *Plant Mol. Biol.*, 36, 43, 1998.

191. Liu, C.J. and Dixon, R.A., Elicitor-induced association of isoflavone *O*-methyltransferase with endomembranes prevents the formation and 7-*O*-methylation of daidzein during isoflavonoid phytoalexin biosynthesis. *Plant Cell*, 13, 2643, 2001.

192. Frick, S. and Kutchan, T.M., Molecular cloning and functional expression of *O*-methyltransferases common to isoquinoline alkaloid and phenylpropanoid biosynthesis. *Plant J.*, 17, 329, 1999.

193. Akashi, T., Aoki, T., and Ayabe, S., CYP81E1, a cytochrome P450 cDNA of licorice (*Glycyrrhiza echinata* L.), encodes isoflavone 2′-hydroxylase. *Biochem. Biophys. Res. Commun.*, 251, 67, 1998.

194. Liu, C.J. et al., Regiospecific hydroxylation of isoflavones by cytochrome P450 81E enzymes from *Medicago truncatula*. *Plant J.*, 36, 471, 2003.

195. Paiva, N.L. et al., Stress responses in alfalfa (*Medicago sativa* L.) 11. Molecular cloning and expression of alfalfa isoflavone reductase, a key enzyme of isoflavonoid phytoalexin biosynthesis. *Plant Mol. Biol.*, 17, 653, 1991.

196. Tiemann, K. et al., Pterocarpan phytoalexin biosynthesis in elicitor-challenged chickpea (*Cicer arietinum* L.) cell cultures. Purification, characterization and cDNA cloning of NADPH:isoflavone oxidoreductase. *Eur. J. Biochem.*, 200, 751, 1991.

197. Paiva, N.L. et al., Molecular cloning of isoflavone reductase from pea (*Pisum sativum* L.): evidence for a 3R-isoflavanone intermediate in (+)-pisatin biosynthesis. *Arch. Biochem. Biophys.*, 312, 501, 1994.

198. Min, T. et al., Crystal structures of pinoresinol-lariciresinol and phenylcoumaran benzylic ether reductases and their relationship to isoflavone reductases. *J. Biol. Chem.*, 278, 50714, 2003.

199. Guo, L., Dixon, R.A., and Paiva, N.L., Conversion of vestitone to medicarpin in alfalfa (*Medicago sativa* L.) is catalyzed by two independent enzymes. Identification, purification, and characterization of vestitone reductase and 7,2'-dihydroxy-4'-methoxyisoflavanol dehydratase. *J. Biol. Chem.*, 269, 22372, 1994.

200. Guo, L. and Paiva, N.L., Molecular cloning and expression of alfalfa (*Medicago sativa* L.) vestitone reductase, the penultimate enzyme in medicarpin biosynthesis. *Arch. Biochem. Biophys.*, 320, 353, 1995.

201. Weidemann, C. et al., Medicarpin and maackiain 3-*O*-glucoside-6'-*O*-malonate conjugates are constitutive compounds in chickpea (*Cicer arietinum* L.) cell cultures. *Plant Cell Rep.*, 10, 371, 1991.

202. Schopfer, C.R. et al., Molecular characterization and functional expression of dihydroxypterocarpan 6a-hydroxylase, an enzyme specific for pterocarpanoid phytoalexin biosynthesis in soybean (*Glycine max* L.). *FEBS Lett.*, 432, 182, 1998.

203. Wu, Q.D., Preisig, C.L., and VanEtten, H.D., Isolation of the cDNAs encoding (+)6a-hydroxymaackiain 3-*O*-methyltransferase, the terminal step for the synthesis of the phytoalexin pisatin in *Pisum sativum*. *Plant Mol. Biol.*, 35, 551, 1997.

204. Welle, R. and Grisebach, H., Properties and solubilization of the prenyltransferase of isoflavonoid phytoalexin biosynthesis in soybean. *Phytochemistry*, 30, 479, 1991.

205. Biggs, D.R., Welle, R., and Grisebach, H., Intracellular localization of prenyltransferases of isoflavonoid phytoalexin biosynthesis in bean and soybean. *Planta*, 181, 244, 1990.

206. Laflamme, P. et al., Enzymatic prenylation of isoflavones in white lupin. *Phytochemistry*, 34, 147, 1993.

207. Zhao, P. et al., Characterization of leachianone G 2''-dimethylallyltransferase, a novel prenyl side-chain elongation enzyme for the formation of the lavandulyl group of sophoraflavanone G in *Sophora flavescens* Ait. cell suspension cultures. *Plant Physiol.*, 133, 1306, 2003.

208. Welle, R. and Grisebach, H., Induction of phytoalexin synthesis in soybean: enzymatic cyclization of prenylated pterocarpans to glyceollin isomers. *Arch. Biochem. Biophys.*, 263, 191, 1988.

209. Nagashima, S. et al., cDNA cloning and expression of isoflavonoid-specific glucosyltransferase from *Glycyrrhiza echinata* cell-suspension cultures. *Planta*, 218, 456, 2004.

210. Latunde-Dada, A.O. et al., Flavonoid 6-hydroxylase from soybean (*Glycine max* L.), a novel plant P-450 monooxygenase. *J. Biol. Chem.*, 276, 1688, 2001.

211. Anzelotti, D. and Ibrahim, R.K., Novel flavonol 2-oxoglutarate dependent dioxygenase: affinity purification, characterization, and kinetic properties. *Arch. Biochem. Biophys.*, 382, 161, 2000.

212. Nakayama, T. et al., Aureusidin synthase: a polyphenol oxidase homolog responsible for flower coloration. *Science*, 290, 1163, 2000.

213. Davies, K.M. et al., Genetic engineering of yellow flower colours. *Acta Hort.*, 560, 39, 2001.

214. Nakayama, T. et al., Specificity analysis and mechanism of aurone synthesis catalyzed by aureusidin synthase, a polyphenol oxidase homolog responsible for flower coloration. *FEBS Lett.*, 499, 107, 2001.

215. Liu, Q. et al., Enzymes of B-ring-deoxy flavonoid biosynthesis in elicited cell cultures of "old man" cactus (*Cephalocereus senilis*). *Arch. Biochem. Biophys.*, 321, 397, 1995.

216. Strack, D., Vogt, T., and Schliemann, W., Recent advances in betalain research. *Phytochemistry*, 62, 247, 2003.

217. Halbwirth, H. et al., Enzymatic glucosylation of 4-deoxyaurones and 6'-deoxychalcones with enzyme extracts of *Coreopsis grandiflora*, Nutt. I. *Plant Sci.*, 122, 125, 1997.

218. Schwinn, K. et al., Elucidating aurone biosynthesis for the production of yellow flower colours. *Polyphenols Commun.*, 1, 65, 2002.
219. Wimmer, G. et al., Enzymatic hydroxylation of 6′-deoxychalcones with protein preparations from petals of *Dahlia variabilis*, *Phytochemistry*, 47, 1013, 1998.
220. Maxwell, C.A. et al., A chalcone and two related flavonoids released from alfalfa roots induce nod genes of *Rhizobium meliloti*. *Plant Physiol.*, 91, 842, 1989.
221. Maxwell, C.A. et al., Molecular characterization and expression of isoliquiritigenin 2′-*O*-methyltransferase, an enzyme specifically involved in the biosynthesis of an inducer of *Rhizobium meliloti* nodulation genes. *Plant J.*, 4, 971, 1993.
222. Haga, M. et al., A cDNA for *S*-adenosyl-L-methionine: isoliquiritigenin/licodione 2′-*O*-methyltransferase (accession no. D88742) from cultured licorice (*Glycyrrhiza echinata* L.) cells. *Plant Physiol.*, 113, 663, 1997.
223. Akashi, T., Aoki, T., and Ayabe, S., Identification of a cytochrome P450 cDNA encoding (2*S*)-flavanone 2-hydroxylase of licorice (*Glycyrrhiza echinata* L.; Fabaceae) which represents licodione synthase and flavone synthase II. *FEBS Lett.*, 431, 287, 1998.
224. Kerscher, F. and Franz, G., Biosynthesis of vitexin and isovitexin: enzymatic synthesis of the C-glucosylflavones vitexin and isovitexin with an enzyme preparation from *Fagopyrum esculentum* M. seedlings. *Z. Naturforsch.*, 42c, 519, 1987.
225. Marrs, K.A., The functions and regulation of glutathione S-transferases in plants. *Annu. Rev. Plant Physiol. Plant Mol. Biol.*, 47, 127, 1996.
226. Winefield, C., The final steps in anthocyanin formation: a story of modification and sequestration. *Adv. Bot. Res.*, 37, 55, 2002.
227. Marrs, K.A. et al., A glutathione *S*-transferase involved in vacuolar transfer encoded by the maize gene *Bronze-2. Nature*, 375, 397, 1995.
228. Alfenito, M.R. et al., Functional complementation of anthocyanin sequestration in the vacuole by widely divergent glutathione S-transferases. *Plant Cell*, 10, 1135, 1998.
229. Kitamura, S., Shikazono, N., and Tanaka, A., TRANSPARENT TESTA 19 is involved in the accumulation of both anthocyanins and proanthocyanidins in *Arabidopsis. Plant J.*, 37, 104, 2004.
230. Larsen, E.S. et al., A carnation anthocyanin mutant is complemented by the glutathione S-transferases encoded by maize *Bz2* and petunia *An9. Plant Cell Rep.*, 21, 900, 2003.
231. Marrs, K.A. and Walbot, V., Expression and RNA splicing of the maize glutathione S-transferase *Bronze2* gene is regulated by cadmium and other stresses. *Plant Physiol.*, 113, 93, 1997.
232. Mueller, L.A. et al., AN9, a petunia glutathione S-transferase required for anthocyanin sequestration, is a flavonoid-binding protein. *Plant Physiol.*, 123, 1561, 2000.
233. Lu, Y.-P. et al., AtMRP2, an *Arabidopsis* ATP binding cassette transporter able to transport glutathione S-conjugates and chlorophyll catabolites: functional comparisons with AtMRP1. *Plant Cell*, 10, 267, 1998.
234. Mathews, H. et al., Activation tagging in tomato identifies a transcriptional regulator of anthocyanin biosynthesis, modification, and transport. *Plant Cell*, 15, 1689, 2003.
235. Saito, K. and Yamazaki, M., Biochemistry and molecular biology of the late-stage of biosynthesis of anthocyanin: lessons from *Perilla frutescens* as a model plant. *New Phytol.*, 155, 9, 2002.
236. Markham, K.R. et al., Anthocyanic vacuolar inclusions — their nature and significance in flower colouration. *Phytochemistry*, 55, 327, 2000.
237. Nozue, M. et al., VP24 found in anthocyanic vacuolar inclusions (AVIs) of sweet potato cells is a member of a metalloprotease family. *Biochem. Eng. J.*, 14, 199, 2003.
238. Klein, M. et al., Different energization mechanisms drive the vacuolar uptake of a flavonoid glucoside and a herbicide glucoside. *J. Biol. Chem.*, 271, 29666, 1996.
239. Klein, M. et al., A membrane-potential dependant ABC-like transporter mediates the vacuolar uptake of rye flavone glucuronides: regulation of glucuronide uptake by glutathione and its conjugates. *Plant J.*, 21, 289, 2000.
240. Li, Z.-S. et al., Vacuolar uptake of the phytoalexin medicarpin by the glutathione conjugate pump. *Phytochemistry*, 45, 689, 1997.
241. Saslowsky, D.E. and Winkel-Shirley, B., Localisation of flavonoid enzymes in *Arabidopsis* roots. *Plant J.*, 27, 37, 2001.

242. Winkel-Shirley, B., A mutational approach to dissection of flavonoid biosynthesis in *Arabidopsis*, in *Phytochemistry in the Genomics and Post-Genomics Eras*, Romeo, J.T. and Dixon, R.A., Eds., Pergamon, New York, 2002, 95.

243. Rasmussen, S. and Dixon, R.A., Transgene-mediated and elicitor-induced perturbation of metabolic channeling at the entry point into the phenylpropanoid pathway. *Plant Cell*, 11, 1537, 1999.

244. Blount, J.W. et al., Altering expression of cinnamic acid 4-hydroxylase in transgenic plants provides evidence for a feedback loop at the entry point into the phenylpropanoid pathway. *Plant Physiol.*, 122, 107–116, 2000.

245. Pairoba, C.F. and Walbot, V., Post-transcriptional regulation of expression of the *Bronze2* gene of *Zea mays* L. *Plant Mol. Biol.*, 53, 75, 2003.

246. HannaRose, W. and Hansen, U., Active repression mechanisms of eukaryotic repressors. *Trends Genet.*, 12, 229, 1996.

247. Ranish, J.A. and Hahn, S., Transcription: basal factors and activation. *Curr. Opin. Genes Dev.*, 6, 151, 1996.

248. Schwechheimer, C. and Bevan, M., The regulation of transcription factor activity in plants. *Trends Plant Sci.*, 3, 378, 1998.

249. Eulgem, T. et al., The WRKY superfamily of plant transcription factors. *Trends Plant Sci.*, 5, 199, 2000.

250. Martin, C., Jin, H., and Schwinn, K., Mechanisms and applications of transcriptional control of phenylpropanoid metabolism, in *Regulation of Phytochemicals by Molecular Techniques*, Romeo, J.T., Saunders, J.A., and Matthews, B.F., Eds., Elsevier Science, Oxford, 2001, 55.

251. Vom Endt, D., Kijne, J.W., and Memelink, J., Transcription factors controlling plant secondary metabolism: what regulates the regulators? *Phytochemistry*, 61, 107, 2002.

252. Petroni, K., Tonelli, C., and Paz-Ares, J., The MYB transcription factor family: from maize to *Arabidopsis*. *Maydica*, 47, 213, 2002.

253. Heim, M.A., et al., The basic helix–loop–helix transcription factor family in plants: a genome-wide study of protein structure and functional diversity. *Mol. Biol. Evol.*, 20, 735, 2003.

254. Davies, K.M. and Schwinn, K.E., Transcriptional regulation of secondary metabolism. *Funct. Plant Biol.*, 30, 913, 2003.

255. Riechmann, J.L., Transcriptional regulation, in *The Arabidopsis Book*, Somerville, C.R. and Meyerowitz, E.M., Eds., American Society of Plant Biologists, Rockville, MD, doi/10.1199/tab.0085, http://www.aspb.org/publications/arabidopsis/, 2002.

256. Hardtke, C.S. and Deng, X.-W., The cell biology of the COP/DET/FUS proteins. Regulating proteolysis in photomorphogenesis and beyond? *Plant Physiol.*, 124, 1548, 2000.

257. Schwechheimer, C. and Deng, X.W., COP9 signalosome revisited: a novel mediator of protein degradation. *Trends Cell Biol.*, 11, 420, 2001.

258. Hardtke, C.S. et al., Biochemical evidence for ubiquitin ligase activity of the *Arabidopsis* COP1 interacting protein 8 (CIP8). *Plant J.*, 30, 385, 2002.

259. Shin, B. et al., *AtMYB21*, a gene encoding a flower-specific transcription factor, is regulated by COP1. *Plant J.*, 30, 23, 2002.

260. Osterlund, M.T. et al., Targeted destabilization of HY5 during light-regulated development of *Arabidopsis*. *Nature*, 405, 462, 2000.

261. Schulze-Lefert, P. et al., Inducible *in vivo* DNA footprints define sequences necessary for UV light activation of the parsley chalcone synthase gene. *EMBO J.*, 8, 651, 1989.

262. Weisshaar, B. et al., Light-inducible and constitutively expressed DNA-binding proteins recognising a plant promoter element with functional relevance in light responsiveness. *EMBO J.*, 10, 1777, 1991.

263. Feldbrügge, M. et al., *PcMYB1*, a novel plant protein containing a DNA-binding domain with one MYB repeat, interacts *in vivo* with a light regulatory promoter unit. *Plant J.*, 11, 1079, 1997.

264. Hartmann, U. et al., Identification of UV/blue light-response elements in the *Arabidopsis thaliana* chalcone synthase promoter using a homologous protoplast transient expression system. *Plant Mol. Biol.*, 36, 741, 1998.

265. Menkens, A.E., Schindler, U., and Cashmore, A.R., The G-box — a ubiquitous regulatory DNA element in plants bound by the GBF family of bZIP proteins. *Trends Biochem. Sci.*, 20, 506, 1995.

266. Jakoby, M. et al., bZIP transcription factors in *Arabidopsis. Trends Plant Sci.*, 7, 106, 2002.
267. Chattopadhyay, S. et al., *Arabidopsis* bZIP protein HY5 directly interacts with light-responsive promoters in mediating light control of gene expression. *Plant Cell*, 10, 673, 1998.
268. Terzaghi, W.B., Bertekap, R.L., and Cashmore, A.R., Intracellular localization of GBF proteins and blue light-induced import of GBF2 fusion proteins into the nucleus of cultured arabidopsis and soybean cells. *Plant J.*, 11, 967, 1997.
269. Wellmer, F., Schafer, E., and Harter, K., The DNA binding properties of the parsley bZIP transcription factor CPRF4a are regulated by light. *J. Biol. Chem.*, 276, 6274, 2001.
270. Tamagnone, L. et al., The AmMYB308 and AmMYB330 transcription factors from *Antirrhinum* regulate phenylpropanoid and lignin biosynthesis in transgenic tobacco. *Plant Cell*, 10, 135, 1998.
271. Jin, H. et al., Transcriptional repression by AtMYB4 controls production of UV-protecting sunscreens in *Arabidopsis. EMBO J.*, 19, 6150, 2000.
272. Faktor, O. et al., Differential utilization of regulatory *cis*-elements for stress-induced and tissue-specific activity of a French bean chalcone synthase promoter. *Plant Sci.*, 124, 175, 1997.
273. Faktor, O. et al., The G-box and H-box in a 39 bp region of a French bean chalcone synthase promoter constitute a tissue-specific regulatory element. *Plant J.*, 11, 1105, 1997.
274. Lindsay, W.P. et al., KAP-2, a protein that binds to the H-box in a bean chalcone synthase promoter, is a novel plant transcription factor with sequence identity to the large subunit of human Ku autoantigen. *Plant Mol. Biol.*, 49, 503, 2002.
275. Yu, L.M., Lamb, C.J., and Dixon, R.A., Purification and biochemical characterization of proteins which bind to the H-box *cis*-element implicated in transcriptional activation of plant defense genes. *Plant J.*, 3, 805, 1993.
276. Logemann, E. and Hahlbrock, K., Crosstalk among stress responses in plants: pathogen defense overrides UV protection through an inversely regulated ACE/ACE type of light-responsive gene promoter unit. *Proc. Natl. Acad. Sci. USA*, 99, 2428, 2002.
277. Sainz, M.B., Grotewold, E., and Chandler, V.L., Evidence for direct activation of an anthocyanin promoter by the maize C1 protein and comparison of DNA binding by related Myb domain proteins. *Plant Cell*, 9, 611, 1997.
278. Hernandez, J.M. et al., Different mechanisms participate in the R-dependent activity of the R2R3 MYB transcription factor C1. *J. Biol. Chem.*, 279, 48205, 2004.
279. Davies, K.M. et al., Flavonoid biosynthesis in flower petals of five lines of lisianthus (*Eustoma grandiflorum* Grise.). *Plant Sci.*, 95, 67, 1993.
280. Borovsky, Y. et al., The *A* locus that controls anthocyanin accumulation in pepper encodes a MYB transcription factor homologous to *Anthocyanin2* of *Petunia. Theor. Appl. Genet.*, 109, 23, 2004.
281. Pelletier, M.K., Murrell, J., and Shirley, B.W., Characterization of flavonol synthase and leucoanthocyanidin dioxygenase genes in *Arabidopsis*. Further evidence for differential regulation of "early" and "late" genes. *Plant Physiol.*, 113, 1437, 1997.
282. Nesi, N. et al., The *TT8* gene encodes a basic helix–loop–helix domain protein required for expression of *DFR* and *BAN* genes in *Arabidopsis* siliques. *Plant Cell*, 12, 1863, 2000.
283. Nesi, N. et al., The *Arabidopsis TT2* gene encodes an R2R3 MYB domain protein that acts as a key determinant for proanthocyanidin accumulation in developing seed. *Plant Cell*, 13, 2099, 2001.
284. Dooner, H.K., Robbins, T.P., and Jorgensen, R.A., Genetic and developmental control of anthocyanin biosynthesis. *Annu. Rev. Genet.*, 25, 173, 1991.
285. Yamazaki, M. et al., Regulatory mechanisms for anthocyanin biosynthesis in chemotypes of *Perilla frutescens* var. crispa. *Biochem. Eng. J.*, 14, 191, 2003.
286. Farzad, M. et al., Differential expression of three key anthocyanin biosynthetic genes in a color-changing flower, *Viola cornuta* cv. Yesterday, Today and Tomorrow. *Plant Sci.*, 165, 1333, 2003.
287. Boss, P.K., Davies, C., and Robinson, S.P., Analysis of the expression of anthocyanin pathway genes in developing *Vitis vinifera* L. cv. Shiraz grape berries and the implications for pathway regulation. *Plant Physiol.*, 111, 1059, 1996.
288. Kobayashi, S. et al., *Myb*-related genes of the Kyoho grape (*Vitis labruscana*) regulate anthocyanin biosynthesis. *Planta*, 215, 924, 2002.

289. Collette, V.E. et al., Temporal and spatial expression of flavonoid biosynthetic genes in flowers of *Anthurium andraeanum*. *Physiol. Plant.*, 122, 297, 2004.
290. Spelt, C. et al., *Anthocyanin1* of petunia encodes a basic helix–loop–helix protein that directly activates transcription of structural anthocyanin genes. *Plant Cell*, 12, 1619, 2000.
291. Carey, C.C. et al., Mutations in the *pale aleurone color1* regulatory gene of the *Zea mays* anthocyanin pathway have distinct phenotypes relative to the functionally similar *TRANSPARENT TESTA GLABRA1* gene in *Arabidopsis thaliana*. *Plant Cell*, 16, 450, 2004.
292. de Vetten, N. et al., The *an11* locus controlling flower pigmentation in petunia encodes a novel WD-repeat protein conserved in yeast, plants, and animals. *Genes Dev.*, 11, 1422, 1997.
293. Walker, A.R. et al., The *TRANSPARENT TESTA GLABRA1* locus, which regulates trichome differentiation and anthocyanin biosynthesis in *Arabidopsis*, encodes a WD40 repeat protein. *Plant Cell*, 11, 1337, 1999.
294. Sompornpailin, K. et al., A WD-repeat-containing putative regulatory protein in anthocyanin biosynthesis in *Perilla frutescens*. *Plant Mol. Biol.*, 50, 485, 2002.
295. Payne, C.T. et al., *GL3* encodes a bHLH protein that regulates trichome development in *Arabidopsis* through interaction with GLI and TTG1. *Genetics*, 156, 1349, 2000.
296. Baudry, A. et al., TT2, TT8, and TTG1 synergistically specify the expression of *BANYULS* and proanthocyanidin biosynthesis in *Arabidopsis thaliana*. *Plant J.*, 39, 366, 2004.
297. Aharoni, A. et al., The strawberry FaMYB1 transcription factor suppresses anthocyanin and flavonol accumulation in transgenic tobacco. *Plant J.*, 28, 319, 2001.
298. Paz–Ares, J., Peterson, P.A., and Saedler, H., Molecular analysis of the *C1–I* allele from *Zea mays*: a dominant mutant of the regulatory *C1* locus. *EMBO J.*, 9, 315, 1990.
299. Burr, F.A. et al., The maize repressor-like gene *intensifier1* shares homology with the *r1/b1* multigene family of transcription factors and exhibits missplicing. *Plant Cell*, 8, 1249, 1996.
300. Kubo, H. et al., *ANTHOCYANINLESS2*, a homeobox gene affecting anthocyanin distribution and root development in *Arabidopsis*. *Plant Cell*, 11, 1217, 1999.
301. McCarty, D.R. et al., The *Viviparous-1* developmental gene of maize encodes a novel transcriptional activator. *Cell*, 66, 895, 1991.
302. Hattori, T. et al., The *Viviparous-1* gene and abscisic acid activate the *C1* regulatory gene for anthocyanin biosynthesis during seed maturation in maize. *Genes Dev.*, 6, 609, 1992.
303. Giraudat, J. et al., Isolation of the *Arabidopsis ABI3* gene by positional cloning. *Plant Cell*, 4, 1251, 1992.
304. Suzuki, M., Kao, C.Y., and McCarty, D.R., The conserved B3 domain of VIVIPAROUS1 has a cooperative DNA binding activity. *Plant Cell*, 9, 799, 1997.
305. Schultz, T.F. et al., 14-3-3 proteins are part of an abscisic acid–VIVIPAROUS1 (VP1) response complex in the *Em* promoter and interact with VP1 and EmBP1. *Plant Cell*, 10, 837, 1998.
306. Western, T.L. et al., Isolation and characterization of mutants defective in seed coat mucilage secretory cell development in *Arabidopsis*. *Plant Physiol.*, 127, 998, 2001.
307. Spelt, C. et al., ANTHOCYANIN1 of petunia controls pigment synthesis, vacuolar pH, and seed coat development by genetically distinct mechanisms. *Plant Cell*, 14, 2121, 2002.
308. Johnson, C.S., Kolevski, B., and Smyth, D.R., *TRANSPARENT TESTA GLABRA2*, a trichome and seed coat development gene of *Arabidopsis*, encodes a WRKY transcription factor. *Plant Cell*, 14, 1359, 2002.
309. Debeaujon, I. et al., Proanthocyanidin-accumulating cells in Arabidopsis testa: regulation of differentiation and role in seed development. *Plant Cell*, 15, 2514, 2003.
310. Sagasser, M. et al., *A. thaliana TRANSPARENT TESTA 1* is involved in seed coat development and defines the WIP subfamily of plant zinc finger proteins. *Genes Dev.*, 16, 138, 2002.
311. Nesi, N. et al., The *TRANSPARENT TESTA16* locus encodes the ARABIDOPSIS BSISTER MADS domain protein and is required for proper development and pigmentation of the seed coat. *Plant Cell*, 14, 2463, 2002.
312. Lechelt, C. et al., Isolation and molecular analysis of the maize *P* locus. *Mol. Gen. Genet.*, 219, 225, 1989.
313. Grotewold, E. et al., The *myb*-homologous *P* gene controls phlobaphene pigmentation in maize floral organs by directly activating a flavonoid biosynthetic gene subset. *Cell*, 76, 543, 1994.

314. Chopra, S. et al., Molecular characterization of a mutable pigmentation phenotype and isolation of the first active transposable element from *Sorghum bicolor*. *Proc. Natl. Acad. Sci. USA*, 96, 5330, 1999.

315. Chopra, S. et al., Excision of the *Candystripe1* transposon from a hyper-mutable *Y1-cs* allele shows that the sorghum *Y1* gene controls the biosynthesis of both 3-deoxyanthocyanidin phytoalexins and phlobaphene pigments. *Physiol. Mol. Plant Pathol.*, 60, 321, 2002.

316. Grotewold, E. et al., Identification of the residues in the Myb domain of maize C1 that specify the interaction with the bHLH cofactor R. *Proc. Natl. Acad. Sci. USA*, 97, 13579, 2000.

317. Zhang, P.F. et al., A maize QTL for silk maysin levels contains duplicated *Myb*-homologous genes which jointly regulate flavone biosynthesis. *Plant Mol. Biol.*, 52, 1, 2003.

318. Deboo, G.B., Albertsen, M.C., and Taylor, L.P., Flavanone 3-hydroxylase transcripts and flavonol accumulation are temporally coordinate in maize anthers. *Plant J.*, 7, 703, 1995.

319. Moyano, E., Martinez-Garcia, J.F., and Martin, C., Apparent redundancy in *myb* gene function provides gearing for the control of flavonoid biosynthesis in *Antirrhinum* flowers. *Plant Cell*, 8, 1519, 1996.

320. Sablowski, R.W.M. et al., A flower-specific Myb protein activates transcription of phenylpropanoid biosynthetic genes. *EMBO J.*, 13, 128, 1994.

321. Bradley, J.M. et al., The maize *Lc* regulatory gene up-regulates the flavonoid biosynthetic pathway of *Petunia*. *Plant J.*, 13, 381, 1998.

322. Miller, K.D., Strommer, J., and Taylor, L.P., Conservation in divergent solanaceous species of the unique gene structure and enzyme activity of a gametophytically-expressed flavonol 3-*O*-galactosyltransferase. *Plant Mol. Biol.*, 48, 233, 2002.

323. Dixon, R.A., Harrison, M.J., and Paiva, N.L., The isoflavonoid phytoalexin pathway: from enzymes to genes to transcription factors. *Physiol. Plant.*, 93, 385, 1995.

324. Subramanian, S. et al., The promoters of the isoflavone synthase genes respond differentially to nodulation and defense signals in transgenic soybean roots. *Plant Mol. Biol.*, 54, 623, 2004.

325. Yu, O. et al., Metabolic engineering to increase isoflavone biosynthesis in soybean seed. *Phytochemistry*, 63, 753, 2003.

326. Hwang, E.I. et al., Production of plant-specific flavanones by *Escherichia coli* containing an artificial gene cluster. *Appl. Environ. Microbiol.*, 69, 2699, 2003.

327. Tanaka, Y., Tsuda, S., and Kusumi, T., Metabolic engineering to modify flower colour. *Plant Cell Physiol.*, 39, 1119, 1998.

328. van der Krol, A.R. et al., An anti-sense chalcone synthase gene in transgenic plants inhibits flower pigmentation. *Nature*, 333, 866, 1988.

329. Napoli, C., Lemieux, C., and Jorgensen, R., Introduction of a chimeric chalcone synthase gene into *Petunia* results in reversible co-suppression of homologous genes *in trans*. *Plant Cell*, 2, 279, 1990.

330. van der Krol, A.R. et al., Flavonoid genes in petunia: addition of a limited number of gene copies may lead to suppression of gene expression. *Plant Cell*, 2, 291, 1990.

331. Jorgensen, R., Cosuppression, flower color patterns, and metastable gene expression states. *Science*, 268, 686, 1995.

332. Bradley, J.M. et al., Flower pattern stability in genetically modified lisianthus (*Eustoma grandiflorum*) under commercial growing conditions. *N.Z. J. Crop Hort. Sci.*, 28, 175, 2000.

333. Taylor, L.P. and Jorgensen, R., Conditional male fertility in chalcone synthase-deficient petunia. *J. Hered.*, 83, 11, 1992.

334. Fischer, R., Budde, I., and Hain, R., Stilbene synthase gene expression causes changes in flower colour and male sterility in tobacco. *Plant J.*, 11, 489, 1997.

335. Jorgensen, R.A., Que, Q.D., and Napoli, C.A., Maternally-controlled ovule abortion results from cosuppression of dihydroflavonol-4-reductase or flavonoid-3′,5′-hydroxylase genes in *Petunia hybrida*. *Funct. Plant Biol.*, 29, 1501, 2002.

336. Colliver, S.P., Morris, P., and Robbins, M.P., Differential modification of flavonoid and isoflavonoid biosynthesis with an antisense chalcone synthase construct in transgenic *Lotus corniculatus*. *Plant Mol. Biol.*, 35, 509, 1997.

337. Carron, T.R., Robbins, M.P., and Morris, P., Genetic modification of condensed tannin biosynthesis in *Lotus corniculatus*. 1. Heterologous antisense dihydroflavonol reductase down-regulates tannin accumulation in 'hairy root' cultures. *Theor. Appl. Genet.*, 87, 153, 1994.

338. Bavage, A.D. et al., Expression of an *Antirrhinum* dihydroflavonol reductase gene results in changes in condensed tannin structure and accumulation in root cultures of *Lotus corniculatus* (bird's foot trefoil). *Plant Mol. Biol.*, 35, 443, 1997.

339. Robbins, M.P. et al., Genetic manipulation of condensed tannins in higher plants. II. Analysis of birdsfoot trefoil plants harboring antisense dihydroflavonol reductase constructs. *Plant Physiol.*, 116, 1133, 1998.

340. De Jaeger, G. et al., High level accumulation of single-chain variable fragments in the cytosol of transgenic *Petunia hybrida*. *Eur. J. Biochem.*, 259, 426, 1999.

341. Nielsen, K.M. et al., Antisense flavonol synthase alters copigmentation and flower color in lisianthus. *Mol. Breed.*, 9, 217, 2002.

342. Davies, K.M. et al., Enhancing anthocyanin production by altering competition for substrate between flavonol synthase and dihydroflavonol 4-reductase. *Euphytica*, 131, 259, 2003.

343. Stobiecki, M. et al., Monitoring changes in anthocyanin and steroid alkaloid glycoside content in lines of transgenic potato plants using liquid chromatography/mass spectrometry. *Phytochemistry*, 62, 959, 2003.

344. Rosati, C. et al., Molecular cloning and expression analysis of dihydroflavonol 4-reductase gene in flower-organs of *Forsythia x intermedia*. *Plant Mol. Biol.*, 35, 303, 1997.

345. Meyer, P. et al., A new petunia flower colour generated by transformation of a mutant with a maize gene. *Nature*, 330, 677, 1987.

346. Yu, O. et al., Production of the isoflavones genistein and daidzein in non-legume dicot and monocot tissues. *Plant Physiol.*, 124, 781, 2000.

347. Jung, W.S., Chung, I.-M., and Heo, H.-Y., Manipulating isoflavone levels in plants. *J. Plant Biotechnol.*, 5, 149, 2003.

348. Lu, C. et al., Florigene flowers: from laboratory to market, in *Plant Biotechnology 2002 and Beyond*, Vasil, I.K., Ed., Kluwer Academic Publishers, Dordrecht, 2003, 333.

349. Okinaka, Y. et al., Selective accumulation of delphinidin derivatives in tobacco using a putative flavonoid 3',5'-hydroxylase cDNA from *Campanula medium*. *Biosci. Biotechnol. Biochem.*, 67, 161, 2003.

350. Ueyama, Y. et al., Molecular and biochemical characterization of torenia flavonoid 3'-hydroxylase and flavone synthase II and modification of flower color by modulating the expression of these genes. *Plant Sci.*, 163, 253, 2002.

351. Schwinn, K.E., Markham, K.E., and Given, N.K., Floral flavonoids and the potential for pelargonidin biosynthesis in commercial chrysanthemum cultivars. *Phytochemistry*, 35, 145, 1994.

352. Bovy, A. et al., High-flavonol tomatoes resulting from the heterologous expression of the maize transcription factor genes *LC* and *C1*. *Plant Cell*, 14, 2509, 2002.

353. Le Gall, G. et al., Characterization and content of flavonoid glycosides in genetically modified tomato (*Lycopersicon esculentum*) fruits. *J. Agric. Food Chem.*, 51, 2438, 2003.

354. Robbins, M.P. et al., *Sn*, a maize bHLH gene, modulates anthocyanin and condensed tannin pathways in *Lotus corniculatus*. *J. Exp. Bot.*, 54, 239, 2003.

355. Lloyd, A.M., Walbot, V., and Davis, R.W., *Arabidopsis* and *Nicotiana* anthocyanin production activated by maize regulators *R* and *C1*. *Science*, 258, 1773, 1992.

356. Damiani, F. et al., The maize transcription factor *Sn* alters proanthocyanidin synthesis in transgenic *Lotus corniculatus* plants. *Aust. J. Plant Phys.*, 26, 159, 1999.

357. Hiratsu, K. et al., Dominant repression of target genes by chimeric repressors that include the EAR motif, a repression domain, in *Arabidopsis*. *Plant J.*, 34, 733, 2003.

358. Matsui, K., Tanaka, H., and Ohme-Takagi, M., Suppression of the biosynthesis of proanthocyanidin in *Arabidopsis* by a chimeric PAP1 repressor. *Plant Biotech. J.*, 2, 487, 2004.

359. Bradley, J.M. et al., Variation in the ability of the maize *Lc* regulatory gene to upregulate flavonoid biosynthesis in heterologous systems. *Plant Sci.*, 140, 37, 1999.

360. Ray, H. et al., Expression of anthocyanins and proanthocyanidins after transformation of alfalfa with maize *Lc*. *Plant Physiol.*, 132, 1448, 2003.

361. Mehdy, M.C. and Lamb, C.J., Chalcone isomerase cDNA cloning and mRNA induction by elicitor, wounding and infection. *EMBO J.*, 6, 1527, 1987.

362. Fedoroff, N., Furtek, D., and Nelson, O.E., Cloning of the *bronze* locus in maize by a simple and generalizable procedure using the transposable element *Activator* (*Ac*). *Proc. Natl. Acad. Sci. USA*, 81, 3825, 1984.

363. Goodrich, J., Carpenter, R., and Coen, E.S., A common gene regulates pigmentation pattern in diverse plant species. *Cell*, 68, 955, 1992.

364. Borevitz, J.O. et al., Activation tagging identifies a conserved MYB regulator of phenylpropanoid biosynthesis. *Plant Cell*, 12, 2383, 2000.

365. Ramsay, N.A. et al., Two basic helix–loop–helix genes (*MYC-146* and *GL3*) from *Arabidopsis* can activate anthocyanin biosynthesis in a white-flowered *Matthiola incana* mutant. *Plant Mol. Biol.*, 52, 679, 2003.

366. Zhang, F. et al., A network of redundant bHLH proteins functions in all TTG1-dependent pathways of *Arabidopsis*. *Development*, 130, 4859, 2003.

367. Bate, N.J. and Rothstein, S.J., An Arabidopsis *Myc*-like gene (MYC-146) with homology to the anthocyanin regulatory gene Delila (accession no. AF013465). *Plant Physiol.*, 115, 315, 1997.

368. Elomaa, P. et al., Activation of anthocyanin biosynthesis in *Gerbera hybrida* (Asteraceae) suggests conserved protein–protein and protein–promoter interactions between the anciently diverged monocots and eudicots. *Plant Physiol.*, 133, 1831, 2003.

369. Elomaa, P. et al., A bHLH transcription factor mediates organ, region and flower type specific signals on dihydroflavonol-4-reductase (DFR) gene expression in the inflorescence of *Gerbera hybrida* (Asteraceae). *Plant J.*, 16, 93, 1998.

370. Gong, Z.Z., Yamazaki, M., and Saito, K., A light-inducible *Myb*-like gene that is specifically expressed in red *Perilla frutescens* and presumably acts as a determining factor of the anthocyanin forma. *Mol. Gen. Genet.*, 262, 65, 1999.

371. Gong, Z.Z. et al., A constitutively expressed *Myc*-like gene involved in anthocyanin biosynthesis from *Perilla frutescens*: molecular characterization, heterologous expression in transgenic plants and transactivation in yeast cells. *Plant Mol. Biol.*, 41, 33, 1999.

372. Quattrocchio, F. et al., Analysis of bHLH and MYB-domain proteins: species-specific regulatory differences are caused by divergent evolution of target anthocyanin genes. *Plant J.*, 13, 475, 1998.

373. Quattrocchio, F. et al., Molecular analysis of the *anthocyanin2* gene of petunia and its role in the evolution of flower color. *Plant Cell*, 11, 1433, 1999.

374. Cone, K.C., Burr, F.A., and Burr, B., Molecular analysis of the maize anthocyanin regulatory locus *C1*. *Proc. Natl. Acad. Sci. USA*, 83, 9631, 1986.

375. Paz–Ares, J. et al., Molecular cloning of the *c* locus of *Zea mays*: a locus regulating the anthocyanin pathway. *EMBO J.*, 5, 829, 1986.

376. Cone, K.C. et al., Maize anthocyanin regulatory gene *pl* is a duplicate of *c1* that functions in the plant. *Plant Cell*, 5, 1795, 1993.

377. Chandler, V.R. et al., Two regulatory genes of the maize anthocyanin pathway are homologous: isolation of *B* using *R* genomic sequences. *Plant Cell*, 1, 1175, 1989.

378. Ludwig, S.R. et al., *Lc*, a member of the maize *R* gene family responsible for tissue-specific anthocyanin production, encodes a protein similar to transcriptional activators and contains the *myc*-homology region. *Proc. Natl. Acad. Sci. USA*, 86, 7092, 1989.

379. Perrot, G.H. and Cone, K.C., Nucleotide sequence of the maize *R-S* gene. *Nucleic Acids Res.*, 17, 8003, 1989.

380. Tonelli, C. et al., Genetic and molecular analysis of *Sn*, a light-inducible, tissue specific regulatory gene in maize. *Mol. Gen. Genet.*, 225, 401, 1991.

381. Grotewold, E., Athma, P., and Peterson, T., Alternatively spliced products of the maize *P* gene encode proteins with homology to the DNA-binding domain of Myb-like transcription factors. *Proc. Natl. Acad. Sci. USA*, 88, 4587, 1991.

382. Shirley, B.W., Hanley, S., and Goodman, H.M., Effects of ionizing radiation on a plant genome: analysis of two *Arabidopsis transparent testa* mutations. *Plant Cell*, 4, 333, 1992.

383. Feinbaum, R.L. and Ausubel, F.M., Transcriptional regulation of the *Arabidopsis thaliana* chalcone synthase gene. *Mol. Cell. Biol.*, 8, 1985, 1988.

384. Wisman, E. et al., Knock-out mutants from an *En-1* mutagenized *Arabidopsis thaliana* population generate phenylpropanoid biosynthesis phenotypes. *Proc. Natl. Acad. Sci. USA*, 95, 12432, 1998.

385. Courtney-Gutterson, N. et al., Modification of flower color in florists chrysanthemum-production of a white-flowering variety through molecular-genetics. *Biotechnology*, 12, 268, 1994.
386. Gutterson, N., Anthocyanin biosynthetic genes and their application to flower colour modification through sense suppression. *Hort. Sci.*, 30, 964, 1995.
387. Deroles, S.C. et al., An antisense chalcone synthase gene leads to novel flower patterns in lisianthus (*Eustoma grandiflorum*). *Mol Breed.*, 4, 59, 1998.
388. Elomaa, P. et al., *Agrobacterium*-mediated transfer of antisense chalcone synthase cDNA to *Gerbera hybrida* inhibits flower pigmentation. *Biotechnology*, 11, 508, 1993.
389. El Euch, C. et al., Expression of antisense chalcone synthase RNA in transgenic hybrid walnut microcuttings. Effect on flavonoid content and rooting ability. *Plant Mol. Biol.*, 38, 467, 1998.
390. Firoozabady, E. et al., Regeneration of transgenic rose (*Rosa hybrida*) plants from embryogenic tissue. *Biotechnology*, 12, 609, 1994.
391. Aida, R. et al., Modification of flower colour in torenia (*Torenia fournieri* Lind.) by genetic transformation. *Plant Sci.*, 153, 33, 2000.
392. Aida, R. et al., Copigmentation gives bluer flowers on transgenic torenia plants with the antisense dihydroflavonol-4-reductase gene. *Plant Sci.*, 160, 49, 2000.
393. Suzuki, K. et al., Flower color modifications of *Torenia hybrida* by cosuppression of anthocyanin biosynthesis genes. *Mol. Breed.*, 6, 239, 2000.
394. Zuker, A. et al., Modification of flower colour and fragrance by antisense suppression of the flavanone 3-hydroxylase gene. *Mol. Breed.*, 9, 33, 2002.
395. Shimada, Y. et al., Genetic engineering of the anthocyanin biosynthetic pathway with flavonoid-3',5'-hydroxylase: specific switching of the pathway in petunia. *Plant Cell Rep.*, 20, 456, 2001.
396. Wu, Q. and VanEtten, H.D., Introduction of plant and fungal genes into pea (*Pisum sativum* L.) hairy roots reduces their ability to produce pisatin and affects their response to a fungal pathogen. *Mol. Plant Microbe Interact.*, 17, 798, 2004.
397. Helariutta, Y. et al., Cloning of a cDNA coding for dihydroflavonol-4-reductase (DFR) and characterization of *dfr* expression in corrollas of *Gerbera hybrida* var. Regina (Compositae). *Plant Mol. Biol.*, 22, 183, 1993.
398. Tanaka, Y. et al., Molecular cloning and characterization of *Rosa hybrida* dihydroflavonol 4-reductase gene. *Plant Cell Physiol.*, 36, 1023, 1995.
399. Mori, S. et al., Heterologous expression of the flavonoid 3',5'-hydroxylase gene of *Vinca major* alters flower color in transgenic *Petunia hybrida*. *Plant Cell Rep.*, 22, 415, 2004.
400. de Majnik, J. et al., Anthocyanin regulatory gene expression in transgenic white clover can result in an altered pattern of pigmentation. *Aus. J. Plant Physiol.*, 27, 659, 2000.
401. Schwinn, K. et al., Regulation of anthocyanin biosynthesis in *Antirrhinum*. *Acta Hort.*, 560, 201, 2001.
402. Mooney, M. et al., Altered regulation of tomato and tobacco pigmentation genes caused by the *delila* gene of *Antirrhinum*. *Plant J.*, 7, 333, 1995.
403. Goldsbrough, A.P., Tong, Y., and Yodder, J.I., *Lc* as a non-destructive visual reporter and transposon excision marker gene for tomato. *Plant J.*, 9, 927, 1996.
404. Liu, D., Galli, M., and Crawford, N.M., Engineering variegated floral patterns in tobacco plants using the arabidopsis transposable element *Tag1*. *Plant Cell Physiol.*, 42, 419, 2001.
405. Bruce, W. et al., Expression profiling of the maize flavonoid pathway genes controlled by estradiol-inducible transcription factors CRC and P. *Plant Cell*, 12, 65, 2000.
406. Suzuki, H. et al., Identification and characterization of a novel anthocyanin malonyltransferase from scarlet sage (*Salvia splendens*) flowers: an enzyme that is phylogenetically separated from other anthocyanin acyltransferases. *Plant J.*, 38, 994, 2004.
407. Frydman, A. et al., Citrus fruit bitter flavours: isolation and functional characterization of the gene *Cm1,2RhaT* encoding a 1,2 rhamnosyltransferase, a key enzyme in the biosynthesis of the bitter flavonoids of citrus. *Plant J.*, 40, 88, 2004.
408. Park, K.-I. et al., An intragenic tandem duplication in a transcriptional regulatory gene for anthocyanin biosynthesis confers pale-colored flowers and seeds with fine spots in *Ipomoea tricolor*. *Plant J.*, 38, 840, 2004.

4 Flavonoids in Foods

Janet A.M. Kyle and Garry G. Duthie

CONTENTS

4.1   INTRODUCTION

A "poor" diet is a major contributing factor to the etiology of chronic diseases such as heart disease and cancer.[1,2] However, defining what constitutes a "healthy" diet remains contentious, as it is difficult to definitively ascribe beneficial and detrimental properties to the diverse components of the many foods we consume. Nevertheless, considerable evidence indicates that adequate fruit and vegetable consumption has a role in maintaining health and preventing disease.[3–9] Some of these protective effects may be due to flavonoids, which are widely distributed in plant-based foods at varying levels.[10–13] For example, numerous *in vitro* investigations have demonstrated potent effects of flavonoids in mammalian systems that are potentially anticarcinogenic and antiatherogenic.[14–16] These include antioxidant protection of DNA and low-density lipoprotein, modulation of inflammation, inhibition of platelet aggregation, estrogenic effects, and modulation of adhesion receptor expression.[14–18] The role of flavonoids in health is extensively covered in Chapter 6.

From a nutritional perspective, the actual importance of flavonoids to health and disease remains unclear. Unlike the recognized micronutrients that can be obtained from plant-based diets, such as vitamin E and vitamin C, a lack of dietary flavonoids does not result in obvious deficiency syndromes. Consequently, the initial classification of some citrus flavonoids as "vitamin P"[19] was later revoked.[10] In addition, epidemiological studies relating intake of flavonoids to disease incidence or risk do not give consistent results. For example, the average combined intake of flavonols and flavones in a cross-cultural correlation study composed of 16 cohorts followed up for 25 years after initial baseline measurements collected around 1960[20] was found to be inversely associated with coronary heart disease mortality, statistically explaining 25% of the variability in rates across the cohorts. In contrast, increased risk with increasing flavonoid intake has also been observed (Table 4.1). Similarly, although some studies have found positive inverse relationships between flavonoid intake and several cancers, others have failed to demonstrate significant statistical associations (Table 4.2).

Early analysis and identification of flavonoids in plant materials and products led to estimated intakes of up to 1 g/day in the United States.[10] This approximation included flavanones, flavonols, and flavones (160 to 175 mg/day), anthocyanins (180 to 215 mg/day), catechins (220 mg/day), and biflavans (460 mg/day). More recent estimates focusing on flavonol and flavone intake indicate that these early intake levels may be too high. One explanation for this is that flavonoid analysis initially employed semiquantitative spectrophotometric measurement.[21–24] Analytical methodology has since progressed with the development of more sensitive and specific techniques. Optimized and better-validated sample preparation and hydrolysis techniques are now commonly used.[25–30] For example, Hertog et al.[25] optimized and tested the completeness of acid hydrolysis and solvent extraction of flavonol glycosides to their free (aglycone) form before quantifying flavonol and flavone concentrations of freeze-dried fruit and vegetable samples. Reversed-phase high-pressure liquid chromatography (RP-HPLC) with ultraviolet (UV) detection[31] has improved isolation and separation of compounds superseding thin layer chromatography isolation and quantification by measurement of UV spectral shifts in response to addition of colorimetric reagents.

Implementing such improved methodology on a selection of nine fruits, 28 vegetables, and several different beverages commonly consumed in the Netherlands[32] generally provided lower values compared with earlier determinations.[3,33] When flavonol and flavone intake was recalculated, levels of intake in the United States fell to 13 mg/day for this subgroup of flavonoids, which is about one tenth that of earlier estimates.[20] These Dutch compositional data, with the addition of local food preferences such as berries, have since been used in several studies to assess dietary intake and potential associations with disease incidence.[34–38] Application of this limited dataset to epidemiological studies relating flavonoid intake to disease incidence has produced conflicting results (Table 4.1 and Table 4.2), probably reflecting, in part, the paucity of composition data for many foods.[17,39] In addition,

TABLE 4.1
Epidemiological Studies Investigating Dietary Flavonoid Intake and Coronary Heart Disease Incidence and Mortality

Study	Flavonoid Class	Follow-up (Years)	No. of Cases	RR (95% CI)[a]	Ref.
Cohort					
Dutch, 693 men 805 men	Flavonols[b]/flavones[c]	5	Incidence (n = 38) Mortality (n = 43)	0.52 (0.22–1.23) **0.32 (0.15–0.71)**	134
Finnish, 2,748 men and 2,385 women	Flavonols/flavones	26	Mortality, men (n = 324) Mortality, women (n = 149)	0.73 (0.41–1.32) **0.67 (0.44–1.00)**	34
US, 34,789 men	Flavonols/flavones	6	Incidence (n = 486) Mortality (n = 105)	1.08 (0.81–1.43) 0.63 (0.33–1.20)	35
Welsh, 1,900 men	Flavonols	14	Incidence (n = 186) Mortality (n = 131)	1.0 (0.9–2.9) 1.6 (0.9–2.9)	135
Dutch, 4,807 men and women	Flavonols	5.6	Incidence (n = 146) Mortality (n = 30)	0.76 (0.49–1.18) **0.35 (0.13–0.98)**	136
Finnish, 9,131 men and women	Flavonols/flavanones[d] [Quercetin]	15	Incidence (n = 806) Mortality (n = 681)	**0.79 (0.64–0.98)** [0.86 (0.70–1.05)] 0.93 (0.74–1.17) **[0.79 (0.63–0.99)]**	129
US, 34,492 women	Flavonols/flavones	10	Mortality (n = 438)	**0.62 (0.44–0.87)**	36
Dutch, 693 men	Catechins[e]	10	Incidence (n = 90)	0.70 (0.39–1.26)	137
Dutch, 805 men	Catechins		Mortality (n = 90)	**0.49 (0.27–0.88)**	
USA, 34,492 women	Catechins Catechin/epicatechin Gallated catechins[f]	13	Mortality (n = 767)	0.85 (0.67–1.07) 0.76 (0.58–1.03) 1.00 (0.77–1.29)	138

Note: Values in bold indicate statistical significance.

[a] Multivariate adjusted relative risk (95% confidence intervals) highest *vs.* lowest quintile of intake.

[b] Quercetin, kaempferol, and myricetin.

[c] Apigenin and luteolin.

[d] Hesperetin and naringenin.

[e] Catechin, epicatechin, gallocatechin, epigallocatechin, epicatechingallate, and epigallocatechingallate.

[f] Gallocatechin, epigallocatechin, epicatechingallate, epigallocatechingallate.

TABLE 4.2
Epidemiological Studies Investigating Dietary Flavonoid Intake and Cancer Incidence

Study	Flavonoid Class	Follow-up (Years)	Site of Cancer	RR (95% CI)	Ref.
Cohort					
US, 34,651 women	Catechins	13	All causes	0.97 (0.88–1.06)	139
			Bronchus and lung ($n = 549$)	0.94 (0.72–1.23)	
			Colon ($n = 635$)	1.10 (0.85–1.44)	
Dutch, 728 men	Catechins	10	Lung ($n = 12$)	0.92 (0.41–2.07)	140
			Epithelial ($n = 30$)	0.94 (0.56–1.59)	
Dutch, 246 men	Flavonol/flavone	5	All causes ($n = 27$)	1.21 (0.66–2.21)	140
			Alimentary and respiratory ($n = 19$)	1.02 (0.51–2.04)	
Finnish, 27,110 male Smokers	Flavonols/flavones	6	Lung ($n = 791$)	0.6 (0.4–0.7)	141
			Urothelial ($n = 156$)	1.2 (0.7–1.8)	
			Renal cell ($n = 92$)	0.6 (0.4–1.1)	
			Prostate ($n = 226$)	1.3 (0.9–1.8)	
			Stomach ($n = 111$)	1.2 (0.7–1.9)	
			Colorectal ($n = 133$)	1.7 (1.0–2.7)	
Finnish, 9,959 men and women	Flavonols/flavones	20	All sites ($n = 997$)	0.87 (0.70–1.09)	142
			Lung ($n = 151$)	0.53 (0.29–0.97)	
Case–control				OR (95% CI)	
Hawaii, men and women	Flavonols/flavanones [Quercetin]		Lung (582 cases and 582 controls)	0.8 (0.5–1.4) [0.7 (0.4–1.1)]	143
Spanish, women	Flavonols		Lung (103 cases and 206 controls)	0.98 (0.44–2.19)	144

concentrations of flavonoids in foods can vary by many orders of magnitude due to the influence of numerous factors such as species, variety, climate, degree of ripeness, and postharvest storage.[10,12,40]

Consequently, the aims in this chapter are to critically examine the available literature on the flavonoid composition of foods and to establish a food flavonoid database, which can be continually expanded as more information becomes available. By using predetermined selection criteria to ensure critical assessment of data quality, the intention is to provide researchers with an improved resource for use in studies exploring the relationships between flavonoid intake and health as well as highlighting important food groups where flavonoid content data are currently lacking.

4.2 DATABASE DEVELOPMENT

4.2.1 DATABASE DEVELOPMENT

To ensure comprehensive coverage of foods and relevant flavonoids, compilation of the flavonoid composition database followed a preset development profile (Figure 4.1). This was a multistage process that evolved from a review of two major food composition databases[41–43] and from other early stage nutrient bases such as those for vitamin K[44] and

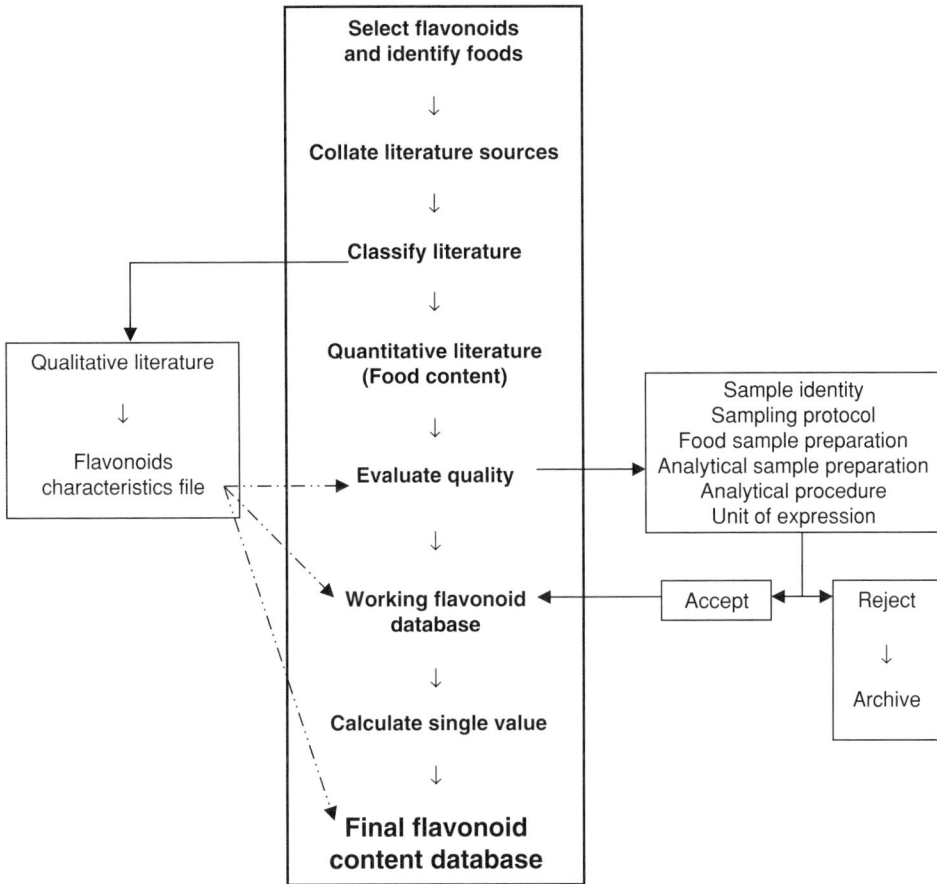

FIGURE 4.1 Schematic of selection and evaluation processes used to compile the flavonoid database.

carotenoids.[42,45] Essentially, the database development profile consists of the selection of commonly consumed flavonoid-rich foods, collation and evaluation of available literature, and, finally, acceptance or rejection of flavonoid content values (Figure 4.1).

4.2.2 SELECTION OF FOODS AND FLAVONOID CLASSES

As all foods of plant origin potentially contain flavonoids[3,10] and over 4000 individual compounds have previously been identified,[11] the development of a comprehensive flavonoid database is a huge task to undertake. Therefore, in the first instance, it was decided to focus on foods commonly available and eaten in Britain and to initially consider those flavonoid classes that have attracted most attention in relation to potential health benefits namely flavonols, flavones, catechins, flavanones, and anthocyan(id)ins.[17] Table 4.3 outlines the compounds selected to represent the five main subclasses of flavonoids. Consumer dietary surveys of the UK population were initially consulted.[46,47] Although a general indication of commonly consumed foods was found in these surveys, fruits and vegetables were presented as generalized groups such as "other fresh fruit." The fruits assigned to this group were not clearly identified. Visiting local British supermarkets proved to be the most informative method of identifying products readily available to the consumer and proved a good starting point for the literature search for flavonoid composition data (Table 4.4).

TABLE 4.3
Compounds Included in Flavonoid Database

Flavonoid Subclass	Compound

Flavonols

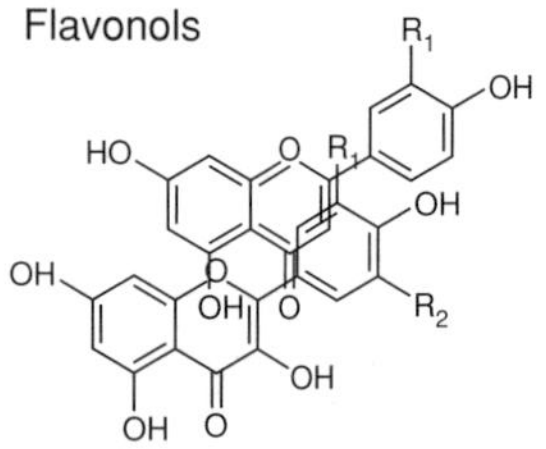

Quercetin ($R_1 = OH$, $R_2 = H$)
Kaempferol ($R_1 = R_2 = H$)
Myricetin ($R_1 = R_2 = OH$)

Flavones

Catechins

Apigenin ($R_1 = H$)
Luteolin ($R_1 = OH$)
(+)-Catechin (C) ($R_1 = H$, $R_2 = OH$)
(−)-Epicatechin (EC) ($R_1 = H$, $R_2 = H$)
(−)-Epigallocatechin (EGC) ($R_1 = OH$, $R_2 = OH$)
(−)-Epicatechingallate (ECG) ($R_1 = H$, $R_2 = -OC-Ph(OH)_3$)
(−)-Epigallocatechingallate (EGCG) ($R_1 = OH$, $R_2 = -OC-Ph(OH)_3$)
(−)-Gallocatechin (GC) ($R_1 = -OC-Ph(OH)_3$, $R_2 = H$)

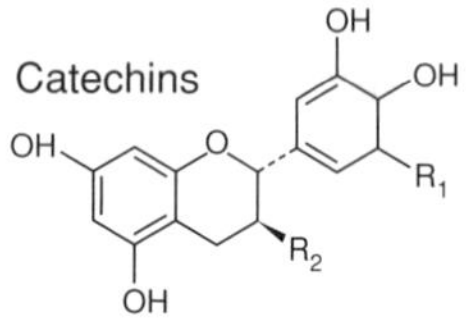

Flavanones

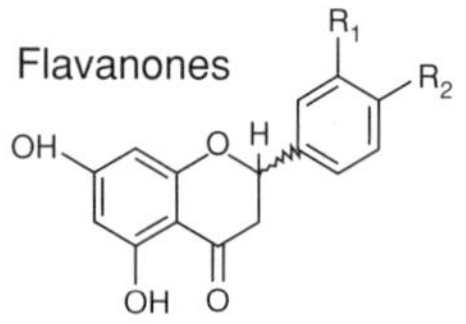

Hesperetin ($R_1 = OH$, $R_2 = OMe$)
Naringenin ($R_1 = H$, $R_2 = OH$)

Anthocyanidins

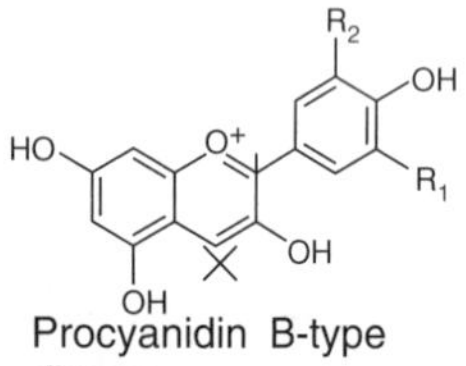

Delphinidin ($R_1 = R_2 = OH$)
Cyanidin ($R_1 = OH$, $R_2 = H$)
Pelargonidin ($R_1 = R_2 = H$)
Petunidin ($R_1 = OMe$, $R_2 = OH$)
Peonidin ($R_1 = OMe$, $R_2 = H$)
Malvidin ($R_1 = R_2 = OMe$)

Procyanidin B-type
dimers

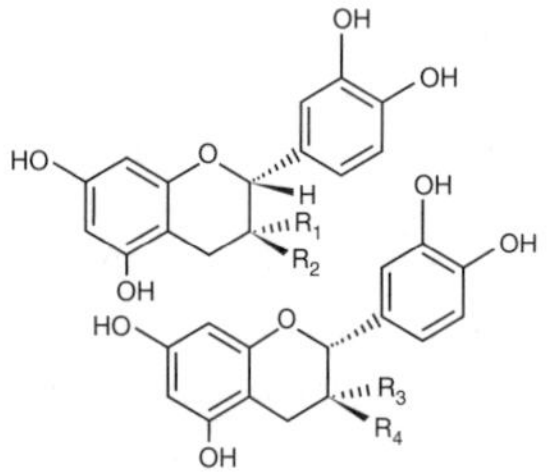

B1: R1 = OH; R2 = H; R3 = H; R4 = OH
B2: R1 = OH; R2 = H; R3 = OH; R4 = H
B3: R1 = H; R2 = OH; R3 = H; R4 = OH
B4: R1 = H; R2 = OH; R3 = OH; R4 = H

4.2.3 COLLATION AND EVALUATION OF LITERATURE SOURCES

To ensure coverage of a broad spectrum of journals containing both original and review articles, bibliography databases, such as CAB abstracts, BIDS ISI and Embase, Medline and Current Contents Agriculture and Food Citation index, were searched. Terms used for all database searches were kept simple to limit the risk of missing any original publications (Table 4.5). Initially, each flavonoid subclass was entered and then cross-searched with the

TABLE 4.4
Foods of Plant Origin Commonly Consumed in the United Kingdom and Available in Local Supermarkets[a,b]

Food Type	Food Item
Beverages	Tea, coffee, fruit and herbal drinks
	Cocoa or drinking chocolate, malted drinks
	Fruit juice concentrates, ready to drink fruit juices (still and carbonated)
	Beer, lager, wine, spirits, liqueurs, alco pops
Fresh fruits	
Normally available	Apples, bananas, oranges, grapefruit, pears, melons, grapes, lemon, strawberries
Seasonal	Satsumas, tangerines, clementines, peaches, pineapples, nectarines, plums, mangoes, raspberries, blueberries, blackcurrants, blackberries, rhubarbs
Exotic fruit section	Custard apple, passion fruit, pomegranate, sharon fruit, lychee, figs, cranberry, gooseberry
Fresh vegetables	
Normally available	Broccoli, carrots, onions, potatoes, courgette, mushrooms
	Lettuce (different varieties), tomatoes (different types), celery, cucumber, sweet pepper (red and green)
Seasonal	Spinach, winter greens, leeks, suede, turnip, Brussels sprouts, red cabbage, parsnips, corn-on-cob, red radish, mange tout, French beans, sweet potatoes, shallots, asparagus, chives
Specialist	Artichoke, squash, yam, celeric, fennel bulb, okra, oyster mushrooms
Fresh herbs	
Normally available	Oregano, basil, thyme, mint, coriander, rosemary, sage, dill

[a]Supermarkets were visited at least once per month over 4 years. Also from Gregory et al.[46] and the National Household Consumption Survey.[47]

[b]Processed food stuffs not included.

terms "composition" or "content." Common and botanical names of foods (Appendix 1) were then individually added to the search profile. Food Science and Nutrition Journal Archives were also hand searched to ensure complete coverage of journals dating back to 1976. Pre-1976 literature was referred to for flavonoid characterization purposes only.

To ensure compatibility with the Royal Society of Chemistry's food composition tables, predetermined screening procedures were used (Table 4.6), which were derived from those outlined for the nutrient tables.[41] All publications and reports on flavonoid content of foods were subsequently evaluated employing the screening procedures (Table 4.6). In brief, inclusion criteria were (a) randomly selected food items purchased from various commercial outlets during different seasons of the year, (b) food samples prepared using normal domestic

TABLE 4.5
Search Strategy for Locating Flavonoid Publications

Search Term	Publications		
	Medline	Web of Science	CAB
(Flavonols or flavones or catechins or flavanols or anthocyanins or flavanones)	13,315	4,373	6,617
and (composition or content)	833	711	4,061
and (Malus or apple or apples)	17	44	190
and (Prunus or apricot or apricots)	2	17	158
and (Allium or onion or onions)	14	19	43
and (repeated for all foods outlined in Table 4.4)			

TABLE 4.6
Selection Criteria for Addition to Database

Criteria	Outlined in Manuscript
Sample identity	Foods common or varietal name
	Country of origin and locality with details of growing conditions if possible; e.g., British tomatoes grown in greenhouses
Sampling protocol	Place and time of sample collection; e.g., purchased from local market or commercial grower in September 1999
	Number of samples collected and how these were obtained; e.g., 1 kg of fruit purchased from three shops
Food sample preparation	State of sample; e.g., raw, cooked, processed (canned or frozen), pickled
	Preparative treatment outlined; e.g., chopping or shredding
Analytical sample preparation	Storage conditions before analysis outlined; e.g., freeze dried and stored at $-20°C$
	Nature of sample analyzed; e.g., edible portion only
	Sample extraction and hydrolysis details; e.g., solvent extraction after freeze drying, with optimized acid or enzymatic hydrolysis
	Preparation of flavonoid standards and use of internal standards
Analytical procedure	Chromatographic separation and detection method used, ideally RP-HPLC with UV or fluorescent detection
	Outline of quality assurance procedures employed
Unit of expression	Method of calculating content value; e.g., based on retention time of standards, mean of duplicate samples, adjustment to recovery of internal standard
	Presentation of final value; e.g., mean $\pm$ SD as fresh weight or dry weight with original moisture content

practices such as discarding nonedible portions of onions, (c) optimized sample extraction and hydrolysis conditions clearly outlined or cited, and (d) flavonoid separation and determination conducted using standard modern techniques with validation and quality assurance measures summarized.[25,31] A worked example of the selection criteria as applied to four different onion composition studies is presented in Table 4.7. Acceptable papers were included into the working database (see Appendix 2), while manuscripts not meeting the criteria were archived and reasons for the exclusion noted.

The evaluation criteria applied during database development highlighted a lack of acceptable anthocyanin food content literature. Values were often presented as percentage of the total anthocyanin content.[48,49] In addition, test samples were frequently gathered from noncommercial sources, such as horticultural research stations.[50,51] Moreover, analytical procedures often employed spectral pH differential methodology rather than HPLC to estimate anthocyanin content.[52,53] Consequently, although there is a substantial amount of characterization information with crude estimates of total anthocyanin content (Table 4.8), anthocyanins had to be excluded from the final database. Similarly, the flavanone eriodictyol was also excluded from the final database due to a lack of rigorously analyzed quantitative information.

The evaluation process also illustrated that food content data are increasingly presented as total glycosides of flavonoids such as quercetin-3-glucoside. If a study met all the criteria except presenting aglycone data, the values were converted to aglycone format using molecular weights, an approach validated by Price et al.[54] Additionally, values expressed in parts per million or milligrams per kilogram were converted to mg/100 g. Content levels presented as dry weight were only adjusted to fresh weight concentrations if the original moisture content was available. During the process of finalizing the database, to ensure all available flavonoid composition data had been gathered, literature determining proanthocyanidin content of foods was identified. Subsequently, content values for B-type procyanidin dimers

TABLE 4.7
Assessment of Four Onion Flavonol Composition Studies Showing Acceptance or Rejection for Database Inclusion

Criteria	Hertog et al.[32]	Leighton et al.[79]	Bilyk and Sapers[89]	Crozier et al.[80]
Sample identity				
Common/varietal name	Identified	Identified	Identified	Identified
Number of varieties	Not reported	20	7	2
Country of origin	Netherlands	USA and Mexico	Not reported	UK
Growing conditions	Not reported	Identified	Not reported	Identified
Sampling protocol				
Place of purchase	Supermarket, grocer and local supermarket	Commercial growers and local suppliers	Not reported	Local supermarket
Year/season of purchase	Four times during 1991–1992	Not reported	Not reported	August 1994 and July 1995
Number of samples collected	One kilogram or three units from three locations	Not reported	Not reported	~250 g randomly selected
Food sample preparation				
State of sample	Raw	Raw	Raw	Raw and cooked
Preparative treatment	Fresh, chopped	Fresh, chopped; Cooking methods — boiled and fried	Fresh onion divided into rings, then shredded	Fresh, sliced; Cooking methods — boiled, fried, and microwaved
Analytical sample preparation				
Storage conditions prior to analysis	Freeze dried, stored at −20°C for <4 months	Analyzed fresh?	Analyzed fresh?	Freeze dried, stored at −20°C
Nature of sample analyzed	Edible portion only	Edible portion only	Both edible and inedible sections	Edible portion only
Sample extraction and hydrolysis	1.2 M HCl in 50% methanol, hydrolyzed for 2 h at 90°C	Butanol phase partitioning	Methanol, followed by acid hydrolysis	1.2 M HCl in 50% methanol, hydrolyzed for 2 h at 90°C
Use of external and internal standards	External standards	External standards	External standards	External standards
Analytical procedure				
Isolation method	C18 column	XAD-2 column	Thin layer chromatography	C18 column
Method of analysis	RP-HPLC	HPLC	HPLC	RP-HPLC
Method of detection	UV–visible	Diode array	UV–visible	UV–visible
Quality assurance	Spiked standards recovery reported, identification on aglycone RP-HPLC retention times	Mass spectrometry confirming peak identity, quantitation based on retention time	Comparison of external standard retention times	Spiked standards recovery reported, identification on aglycone RP-HPLC retention times
Unit of expression	mg/kg fresh weight	mg/kg fresh weight	g/kg fresh weight	mg/kg fresh weight
Status	**Accept**	**Accept**	**Reject**	**Accept**

TABLE 4.8
Anthocyanin Content of Some Plant-Based Food Items

Food	Content[a] (mg/100 g)	Anthocyanin[b]	Glycosides[c]
Fruits			
Red apple	0.1–0.2	Cyanidin	Glu; gala; arab
Apricot		Cyanidin	Glu
Bilberry	3.7	Malvidin; Dp; Cy; Pt; Pn	Glu; gala; arab
Blackberry	0.3–1.1	Cyanidin; Pg	Glu; gala; rut; arab; xyl
Blueberry	0.8–2.8	Malvidin; Cy; Dp; Pn; Pt	Arab; gala; glu; 6-ace-3-gly, samb; soph
Cherry	2.4	Cyanidin; Pn	Glu; rut; glu-rut; soph
Cranberry	0.1–3.6	Peonidin; Cy	Gala; glu; arab
Blackcurrants	1.3–4.0	Cyanidin; Dp	Glu; rut
Elderberry	2.0–10.0	Cyanidin	Samb; glu; samb-5-glu, 3,5-diglu, *p*-cou-glu-5-glu
Grapes (red/black)	0.7–1.1	Malvidin; Dp; Pt; Pn; Cy	Glu; glu-ace; glu-*p*-cou; rut
Litchi	0.5	Cyanidin; Mv	Glu; rut; ace-glu
Olive	—	Cyanidin	Glu; rut; caf-rut; glu-rut
Orange, blood	2	Cyanidin; Dp	Glu; mal-glu
Peach	0–0.1	Cyanidin	Glu; rut
Pear	—	Peonidin; Cy	Glu; rut
Plum	0.02–0.3	Cyanidin	Glu; rut; 3-ace-glu; gala
Pomegranate	—	Delphinidin; Cy; Pg	3,5-Diglu; glu
Raspberry	0.3–1.2	Cyanidin; Dp; Mv; Pg	3-Glu; soph, rut; 3-glu-rut
Strawberry	0.1–3.8	Pelargonidin; Cy	Rut; 3-ace-gly; glu
Vegetables			
Asparagus	—	Cyanidin; Pn	Glu; rut; glu-rut; 3,5-diglu; 3,5-caf-rham-glu
Red cabbage	0.7–0.9	Cyanidin	3,5-Diglu; soph-5-glu acylated with *p*-cou, fer, sin
Red lettuce	—	Cyanidin	Mal-glu; glu
Red onion	0.2	Cyanidin; Pn	Mal-glu; di-mal-lam; mal-lam; glu; mal-3″glu; arab
Red radish	1.5 (skin)	Pelargonidin; Cy	3,5-Diglu; soph-5-glu acylated with *p*-cou, fer, caf
Black beans	2.1	Delphinidin; Pt; Mv	Glu
Cocoa beans	0–1.0	Cyanidin	Arab; gala; arab-glu
Purple basil	0.2	Cyanidin, Pn	Glu; 3,5-glu; *p*-cou; mal; *p*-cou-glu-5-glu
Miscellaneous			
Red wine	0.01–0.5	Malvidin; Dp; Pt; Pn; Cy	Glu; ace-glu; *p*-cou-glu; 3,5-glu
Blackcurrant juice	0.02–0.1	Cyanidin	Glu; rut

[a]Crude total anthocyanin content of edible portion of plant.
[b]Cy, cyanidin; Dp, delphinidin; Mv, malvidin; Pn, peonidin; Pg, pelaronidin.
[c]Glu, glucoside; gala, galactoside; arab, arabinoside; rut, rutinoside; xyl, xyloside; rham, rhamnoside; soph, sophoroside; mal, malonyl; lam, laminaribioside; samb, sambubioside; *p*-cou, *p*-coumaroyl; ace, acetyl; caf, caffeoyl; fer, feruloyl; sin, sinapic.

(Table 4.3) — polymers of catechin and catechin esters — were found to meet the selection criteria and included in the database.

4.3 DATABASE OF FLAVONOIDS IN FOODS

The database (Table 4.9–Table 4.12) currently contains entries for 35 types of fruits, 31 vegetables, 26 beverages, eight different jams, three types of chocolate, and 12 herbs. Data are presented as average, minimum, and maximum values (mg/100 g) for each of the flavonoid

TABLE 4.9
Flavonoid Content of Fruits

	Flavonols			Flavones		Procyanidins	Catechins						Flavanones	
	Qu	K	My	Lut	Apig	B1–4	EGC	C	EC	EGCG	ECG	GC	Nar	Hesp
Apple														
Eating	3.5[a] (2.0–7.2)[b]	0.2 (Tr–0.5)	Tr	—	—	9.2 (3.8–15.4)	Tr (Tr–0.1)	0.8	6.3	—	0.03	C	—	—
Peeled	Tr	Tr	Tr	—	—	7.7	0.02	0.9	6.2	—	—	—	—	—
Sauce	2.0	Tr	—	—	—	—	—	C	6.2	—	—	—	—	—
Apricots	2.6 (2.5–2.6)	C	—	—	—	0.1	—	2.6 (0.3–5.0)	3.0	—	—	—	—	—
Avocado	—	—	—	—	—	0.02	—	—	0.3 (0.1–0.6)	—	—	—	—	—
Banana	C	C	—	—	—	C	C (Tr–0.01)	C	0.3 (Tr–0.03)	—	C	C	—	—
Bilberry	2.9 (1.7–4.1)	1.6 (Tr–4.7)	1.0 (Tr–1.8)	—	—	C	C	C	C	—	C	C	—	—
Blackberry	1.0 (0.2–1.9)	0.1 (Tr–0.2)	Tr	—	—	1.4 (0.8–2.0)	0.4 (0.0–0.7)	0.7 (0.5–1.0)	10.4 (2.4–18.1)	—	C	C	—	—
Blackcurrants	2.9 (3.7–7.9)	1.6 (Tr–1.7)	1.0 (Tr–13.1)	—	—	C	C	0.7	0.5	—	—	C	—	—
Blueberries	4.0 (2.1–7.3)	Tr	1.3 (Tr–2.5)	—	—	1.4 (0.6–2.2)	0.4 (0.2–0.6)	0.4 (0.4–1.3)	0.7 (0.1–1.1)	—	—	0.6	—	—
Cherries	1.3 (1.0–1.5)	Tr	—	—	—	2.6	0.1	1.4 (0.6–2.2)	7.5 (2.5–9.5)	—	0.2	—	—	—
Cranberries	13.0 (7.3–17.2)	Tr (Tr–0.2)	10.1 (0.4–14.2)	—	—	C	C	C	4.2	—	—	—	—	—
Custard apple	—	—	—	—	—	11.1	—	0.6	5.6	—	0.04	—	—	—
Elderberries	17.1	—	—	—	—	C	—	C	—	—	—	—	—	—

continued

TABLE 4.9
Flavonoid Content of Fruits — continued

	Flavonols			Flavones		Procyanidins	Catechins						Flavanones	
	Qu	K	My	Lut	Apig	B1–4	EGC	C	EC	EGCG	ECG	GC	Nar	Hesp
Early fig	—	—	—	—	—	0.03	—	—	—	—	—	—	—	—
Gooseberry	2.0	1.8	—	—	—	—	—	0.1	0.1	—	—	0.4	—	—
Grapefruit	0.7 (Tr–1.4)	—	—	—	C	—	—	1.7	—	—	—	—	39.2 (25.4–53.0)	1.4 (1.3–1.5)
Grapes Black	2.6 (1.5–3.7)	—	0.2 (Tr–0.5)	—	—	1.2	0.1	4.9 (0.8–8.9)	4.7 (0.7–8.6)	—	1.5 (0.2–2.8)	—	—	—
Grapes Green	0.7 (0.2–1.2)	—	—	—	—	0.9	0.04	1.4 (0.4–2.5)	0.6 (0.1–1.0)	—	—	0.03	—	—
Kiwi fruit	N	N	—	—	—	0.1	—	—	0.4 (0.3–0.5)	—	—	—	C	C
Lemon	1.3 (Tr–2.8)	C	C	0.5 (Tr–1.5)	C	—	—	—	—	—	—	—	0.8 (0.5–1.3)	18.3 (17.0–20.6)
Lime	0.4	—	—	—	—	—	—	—	—	—	—	—	3.4	43.0
Lingonberry	12.6 (10.0–16.9)	—	—	—	—	C	C	C	C	C	C	C	—	—
Mango	C	C	C	—	—	C	C	1.7	—	C	C	C	C	—
Nectarine	N	C	C	—	—	C	C	2.8	N	C	C	C	—	C
Olives	12.4 (6.0–12.2)	—	—	12.4 (6.0–12.2)	4.6 (2.0–7.1)	—	—	—	—	—	—	—	C	—
Orange	1.4 (Tr–4.3)	—	—	C	—	—	—	—	—	—	—	—	12.5 (11.0–14.5)	38.5 (31.0–43.2)

Peaches	0.5 (Tr–0.9)	C	C	—	—	—	—	3.2	—	—	1.4 (0.5–2.3)	0.7	—	0.01	—	—
Pears	0.5 (0.5–0.6)	C	—	—	—	—	—	0.7	—	—	0.1	1.5 (0.3–3.4)	—	—	—	—
Peeled	C	C	—	—	—	—	—	C	—	—	(Tr–0.1)	1.1	—	—	—	—
Persimmon	—	—	—	—	—	—	—	0.1	—	—	0.6	—	—	—	0.2	—
Plums, blue	1.2 (0.9–1.5)	—	—	—	—	—	—	16.1	—	—	4.3 (3.4–5.2)	3.6 (2.8–4.4)	—	—	—	—
Pomegranate	—	—	—	—	—	—	—	0.3	0.2	—	0.4	0.1	—	—	0.2	—
Raisins	3.6	3.2	—	—	—	—	—	—	C	—	3.0	0.7	C	C	C	C
Raspberries	0.7 (0.5–5.1)	C	—	—	—	—	—	2.8 (2.1–3.5)	1.1 (0.9–1.2)	0.5 (0.0–1.0)	5.0 (1.4–8.3)	—	—	—	—	—
Red currants	0.9 (0.6–1.3)	0.04 (Tr–0.1)	—	—	—	—	—	0.2	0.4	1.3	0.1	—	—	1.3 (1.0–1.7)	—	—
Rhubarb	C	C	—	—	C	—	—	C	C	2.2 (1.1–1.6)	0.5 (Tr–0.2)	C	0.6	C	—	—
Strawberry	0.6 (0.4–0.9)	0.8 (0.5–1.2)	—	—	—	C	—	1.9	0.2	3.0	—	0.7	0.1	—	—	—

Notes: Tr, below limit of detection; C, not quantified; N, known to be present; —, not known to be present; see Appendix 2 for references included in database. Qu, quercetin; K, kaempferol; My, myricetin; Lut, luteolin; Apig, apigenin; B1–4, B-procyanidins; EGC, epigallocatechin gallate; C, catechin; EC, epicatechin; EGCG, epigallocatechin gallate; ECG, epicatechin gallate; GC, gallocatechin; Nar, naringenin; Hesp, hesperitin.

[a] Aglycone mg/100 g fresh weight.

[b] Minimum to maximum values included in database (values given in parentheses).

TABLE 4.10
Flavonoid Content of Vegetables

	Flavonols			Flavones		Procyanidins	Catechins						Flavanones	
	Qu	K	My	Lut	Apig	B1–4	EGC	C	EC	EGCG	ECG	GC	Nar	Hesp
Broad beans	2.0[a] (Tr–2.0)[b]	0.4	2.6	Tr	—	50.0	15.7 (14.0–17.4)	14.5 (12.8–16.2)	30.0 (22.5–37.6)	Tr	Tr	4.8 (Tr–9.7)	—	—
Broad beans, boiled	0.6	Tr	0.7	Tr	—	28.1	4.7	8.2	7.8	Tr	Tr	2.7	—	—
Broad beans, canned	Tr	Tr	Tr	Tr	—	Tr	Tr	Tr	Tr	Tr	Tr	Tr	—	—
Broccoli	2.9 (1.5–4.3)	6.5 (4.1–9.4)	C	C	—	—	—	—	—	—	—	—	—	—
Broccoli, boiled	0.5	1.2			—								—	—
Brussels sprouts	0.3 (Tr–0.6)	0.8 (0.7–0.9)	—	C	—								—	—
Celery stalks	Tr	Tr	Tr	1.5 (0.5–2.2)	5.9 (1.6–10.8)								—	—
Chilli pepper, green	11.4	—		2.7	—								—	—
Chilli pepper, red	4.5	—		1.7	—								—	—
Chives	0.9	5.5	C	C	C								—	—
Cress	C	13.0	C										—	—
Endive	Tr	9.0 (1.8–13.4)		C	C								—	—
Green beans	2.6 (1.3–3.9)	0.9 (0.2–2.0)					C		0.9 (Tr–2.6)	1.5 (Tr–4.5)			—	—
Green beans, boiled	2.0	0.8					C		0.7	1.3			—	—
Green beans, processed	1.7 (1.6–1.7)	0.6 (Tr–0.6)					C		0.1 (Tr–0.1)	0.3 (Tr–0.5)			—	—
Kale	11.5 (1.4–12.0)	34.1 (2.2–47.0)											—	—
Kale, processed	4.6	29.6											—	—
Kidney beans, canned	C	C	C	C						0.4			—	—
Leeks	0.1	2.7	C						1.7				—	—
Leeks, boiled	0.03	1.1											—	—
Lettuce	3.0 (0.9–8.4)	Tr (Tr–0.3)	C	0.1	C								—	—
Lettuce, red	10.6	Tr	C	1.6	C								—	—
Onions, red	64.4 (15.6–191.7)	76.6 (15.4–190.0)	C	C	—								—	—

Food (mg/100 g fresh weight)	Qu	My	K	Apig	Lut										
Onions, white	0.03 (Tr–0.05)	0.02 (Tr–0.04)	C	C	—	—	—	—	—	—	—	—	—	—	—
Onions, yellow															
Raw	40.7 (12.4–118.7)	0.2 (Tr–1.0)	C	C	—	—	—	—	—	—	—	—	—	—	—
Fried	24.6	0.1	—	—	—	—	—	—	—	—	—	—	—	—	—
Microwaved	21.4	0.1	—	—	—	—	—	—	—	—	—	—	—	—	—
Boiled	10.8	0.1	—	—	—	—	—	—	—	—	—	—	—	—	—
Red cabbage	0.3 (0.2–0.5)	Tr	C	C	C	—	—	—	—	—	—	—	—	—	—
Red radish	Tr	0.5 (0.4–0.6)	—	—	—	—	—	—	—	—	—	—	—	—	—
	—	—													
Shallots	94.9	C	—	—	—	—	—	—	—	—	—	—	—	—	—
Spring onions	18.0	0.6	C	—	—	—	—	—	—	—	—	—	—	—	—
Sweet pepper, green	0.5	C	0.5	—	—	—	—	—	—	—	—	—	—	—	—
Sweet pepper, red	Tr	C	0.6 (0.1–1.1)	—	—	—	—	—	—	—	—	—	—	—	—
Sweet pepper, yellow	0.6 (Tr–1.3)	C	0.6 (0.2–1.1)	—	—	—	—	—	—	—	—	—	—	—	—
Tomatoes	1.1 (0.6–2.1)	0.03 (Tr–0.1)	—	—	—	—	—	—	—	—	—	—	1.1 (0.6–1.5)	C
Tomatoes, beef	0.3 (0.2–0.4)	0.02 (Tr–0.03)	—	—	—	—	—	—	—	—	—	—	C	C
Tomatoes, cherry	3.1 (1.2–6.3)	0.1 (Tr–0.12)	—	—	—	—	—	—	—	—	—	—	C	C
Tomatoes, cherry, canned	0.2	0.01	—	—	—	—	—	—	—	—	—	—	C	C
Tomatoes, plum, canned	0.3	Tr	—	—	—	—	—	—	—	—	—	—	C	C
Tomatoes, yellow	0.2	0.04	—	—	—	—	—	—	—	—	—	—	C	C
White cabbage	Tr	0.2	C	—	—	—	—	—	—	—	—	—	—	—	—
Watercress	C	C	C	—	C	—	C	—	—	—	—	—	—	—	—
Yellow beans	3.4	0.5	—	—	—	—	—	—	—	—	—	—	—	—	—

Notes: Tr, below limit of detection; C, characterized but not quantified; N, known to be present; see Appendix 2 for references included in database. Qu, quercetin; K, kaempferol; My, myricetin; Lut, luteolin; Apig, apigenin; B1–4, B-procyanidins; EGC, epigallocatechin gallate; C, catechin; EC, epicatechin; EGCG, epigallocatechin gallate; ECG, epicatechin gallate; GC, gallocatechin; Nar, naringenin; Hesp, hesperitin.

[a] Aglycone mg/100 g fresh weight.

[b] Minimum to maximum values included in database (values given in parentheses).

TABLE 4.11
Flavonoid Content of Beverages

	Flavonols			Flavones		Procyanidins	Catechins						Flavanones	
	Qu	K	My	Lut	Apig	B1–4	EGC	C	EC	EGCG	ECG	GC	Nar	Hesp
Apple juice	1.1[a] (0.1–2.9)[b]	—	—	—	—	0.05	C	1.7 (Tr–5.1)	6.5 (Tr–19.3)	—	—	C	0.8	9
Black current juice fresh	1.6 (0.7–2.5)	C	2.1 (2.1–3.2)	—	—	C	C	C	C	—	—	C	—	—
Cacao drink	C	—	Tr	—	—	0.4	Tr	0.7	0.6	Tr	Tr	Tr	—	—
Chocolate milk	0.1	Tr	Tr	—	—	C	Tr	0.9	0.3	Tr	Tr	Tr	—	—
Coffee	Tr	—	Tr	—	—	—	Tr	—	Tr	—	—	—	—	—
Cranberry juice, canned	1.2	—	2.9	—	—	C	—	0.2	C	—	—	—	—	—
Cranberry juice, fresh	17.5	—	4.7	—	—	C	—	1.0	C	—	—	—	—	—
Grape juice, black	0.5 (0.4–0.5)	—	0.6	—	—	—	—	0.4 (0.2–0.9)	Tr (Tr–0.01)	—	—	—	—	—
Grape juice, white	0.5 (0.4–0.5)	—	0.6	—	—	0.4	—	0.4 (0.2–0.9)	0.01 (Tr–0.01)	—	—	—	—	—
Grapefruit juice	0.3 (0.1–0.5)	C	—	—	C	—	—	—	—	—	—	—	23.0 (16.8–31.2)	0.5 (Tr–1.3)
Lemon juice	1.1	—	Tr	—	C	—	—	—	—	—	—	—	0.3	6.7
Lime juice	—	—	—	—	C	—	—	—	—	—	—	—	0.2	6.9
Orange juice (diluting)	0.6	—	—	—	C	—	—	—	—	—	—	—	3.3 (3.1–3.4)	21.2 (1.1–25.4)
Orange juice (fresh)	0.5 (0.1–1.0)	—	—	—	C	—	—	—	—	—	—	—	2.3 (0.8–2.9)	21.2 (9.0–38.2)

Tomato juice	1.5 (1.3–1.6)	0.1	—	—	—	—	—	—	—	—	—	—	—	—
Tea, black	2.1 (Tr–3.5)	1.5 (1.1–1.6)	0.3 (0.2–0.4)	—	5.4	—	3.9 (0.3–6.0)	0.8 (0.4–1.3)	3.7 (0.7–6.6)	6.0 (1.4–12.0)	5.9 (2.0–11.0)	1.9 (1.7–2.1)	—	—
Tea, decaffeinated	5.2	2.4	0.1	—	C	—	C	C	C	C	C	C	—	—
Tea, Earl Grey	1.7	1.7	0.3	—	C	—	0.5	0.6	1.4	4.8	7.5	0.7	—	—
Tea, green	2.1	1.3	0.8	—	3.5	—	3.3 (2.0–4.0)	0.8	4.1 (1.0–10.6)	12.1 (4.6–28.6)	5.9 (3.0–8.6)	3.1	—	—
Tea, Forest fruit	C	C	C	—	C	—	0.9	0.5	1.8	4.3	5.9	0.9	—	—
Tea, freeze dried	5.7	3.4	Tr	—	C	—	N	N	N	N	N	N	—	—
Tea, Oolong	1.3	0.9	0.5	—	—	—	6.0 (5.0–7.0)	—	2.5 (2.0–3.0)	6.5 (5.0–8.0)	4.0 (3.0–5.0)	—	—	—
Beer	0.1 (Tr–0.1)	0.8 (Tr–1.6)	—	—	1.0	—	Tr	0.4 (Tr–0.7)	0.1 (Tr–0.2)	Tr	Tr	0.05 (Tr–0.1)	—	
Cider	C	C	—	—	C	—	—	—	—	—	—	—	—	
Sherry	C	C	—	—	0.5	—	Tr	0.6	0.2	—	—	—	—	—
Wine, red	0.9 (0.4–1.3)	0.1 (Tr–0.1)	0.6 (Tr–0.6)	—	15.1 (2.7–27.4)	—	0.3	6.5 (1.8–19.1)	3.9 (1.0–8.2)	Tr	Tr	0.4	1.8	0.5
Wine, rose	N	N	—	—	0.6	—	0.1	0.7	0.4	Tr	Tr	0.2	—	
Wine, white	0.1 (Tr–0.2)	—	Tr	—	0.04	—	—	1.2 (0.1–3.5)	0.7 (0.1–2.1)	—	—	—	0.4	0.1

Note: Tr, below limit of detection; C, not quantified; N, known to be present; —, not known to be present; see Appendix 2 for references included in database.

[a] Aglycone mg/100 g fresh weight.

[b] Minimum to maximum values included in database (values given in parentheses).

TABLE 4.12
Miscellaneous Foods

	Flavonols			Flavones		Procyanidins	Catechins						Flavanones	
	Qu	K	My	Lut	Apig	B1–4	EGC	C	EC	EGCG	ECG	GC	Nar	Hesp
Jam														
Apricot	0.8[a]	0.1	—	—	—	C	—	0.5	0.5	—	—	—	—	—
Cherry	C	C	—	—	—	C	C	0.2	0.9	—	C	—	—	—
Forest fruit	—	—	—	—	—	—	—	0.1	1.6	—	—	—	C	—
Honey	0.2 (Tr–0.4)[b]	0.1 (Tr–0.1)	0.1 (0.1–0.2)	0.3 (Tr–0.7)	0.03	—	—	—	—	—	—	—	—	—
Peach	0.4	0.3	C	—	—	—	—	C	—	—	—	—	—	—
Plum	0.7	C	—	—	—	C	—	C	C	—	—	—	—	—
Raspberry	3.3 (3.0–3.7)	0.5 (0.4–0.5)	—	—	—	C	—	C	C	—	—	—	—	—
Strawberry	0.5 (0.4–0.6)	0.3 (0.1–0.4)	—	—	—	C	—	0.9	0	—	—	—	—	—
Confectionary														
Chocolate, dark	C	—	—	—	—	2.1	C	6.6 (1.3–12.0)	21.8 (2.2–41.5)	—	—	C	—	—
Chocolate, fancy and filled	C	—	—	—	—	C	C	2.2	6.3	—	—	C	—	—
Chocolate, milk	C	—	—	—	—	2.1	C	2.3 (1.3–3.3)	7.4 (2.2–12.5)	—	—	C	—	—

Herbs

	Qu	K	My	Lut	Apig	B1–4	EGC	C	EC	EGCG	ECG	GC	Nar	Hesp
Basil, sweet	5.0	—	—	C	C	—	—	—	—	—	—	—	—	—
Coriander	C	Tr	—	—	—	—	—	—	—	—	—	—	—	—
Dill	(48–110)	C	(16–24)	—	—	—	—	—	—	—	—	—	—	—
Lemon balm	C	—	—	C	—	—	—	—	—	—	—	—	—	—
Lovage	C	C	—	—	—	—	—	—	—	—	—	—	—	—
Marjorum	—	—	—	C	3.5	—	—	—	—	—	—	—	—	—
Mint	—	—	—	C (11–42)	C (18–99)	—	—	—	—	—	—	—	—	Tr
Oregano	—	—	—	25.1	C	—	—	—	—	—	—	—	—	—
Parsley	C	0.6	(Tr–1.1)	2.4 (1.1–3.7)	217.9 (185–250)	—	—	—	—	—	—	—	—	—
Rosemary	—	—	—	C	1.1	—	—	—	—	—	—	—	—	Tr
Sage	—	—	—	33.4	2.4	—	—	—	—	—	—	—	—	—
Thyme	—	—	—	39.5	C	—	—	—	—	—	—	—	—	—

Notes: Tr, below limit of detection; C, not quantified: N, known to be present; —, not known to be present; see Appendix 2 for references included in database. Qu, quercetin; K, kaempferol; My, myricetin; Lut, luteolin; Apig, apigenin; B1–4, B-procyanidins; EGC, epigallocatechin gallate; C, catechin; EC, epicatechin; EGCG, epigallocatechin gallate; ECG, epicatechin gallate; GC, gallocatechin; Nar, naringenin; Hesp, hesperitin.

[a] Aglycone mg/100 g fresh weight.

[b] Minimum to maximum values included in database (values given in parentheses).

subclasses; i.e., flavonols, flavones, proanthocyanidins, catechins, and flavanones. All the flavonol, flavone, and flavanone content data have been determined after optimized hydrolysis of their respective glycosides to the aglycone. Catechin monomers such as epicatechin gallate occur naturally in their free form and are presented as mg catechin/100 g. Procyanidins are presented as their monomeric unit (−)-epicatechin.

4.3.1 Fruits

Catechins, flavonols, and proanthocyanidins are abundant in fruits. In contrast, flavanones and flavones are restricted to citrus varieties such as oranges and lemons (Table 4.9). In some fruits (e.g., apples), flavonols are principally present in the skin and hence peeling significantly reduces levels unlike catechins which are found in the flesh of fruits. Overall, catechins were the most abundant flavonoid, (+)-catechin (C) and (−)-epicatechin (EC) being particularly prevalent. Black grapes (4.9 and 4.7 mg/100 g, C and EC, respectively) are one of the richest fruit sources of catechins followed by apples (0.8 and 6.3 mg/100 g, C and EC, respectively). Catechins are also relatively abundant in stone fruits, such as blue plums (4.3 and 3.6 mg/100 g, C and EC, respectively) and apricots (2.6 and 3.0 mg/100 g, C and EC, respectively). The gallic acid esters of catechin, (−)-epigallocatechin, (−)-epigallocatechin gallate, (−)-epicatechin gallate, and (+)-gallocatechin, are relatively uncommon in fruits, with only berries, currants, and grapes containing small amounts. Strawberries were found to contain the most complex mixture of catechins, comprising catechin (75% of total catechins), ECG (18% of total catechins), EGC (5% of total catechins), and GC (3% of total catechins). Catechin esters have been characterized, but not measured, in some fruits such as nectarines and mangos. Type B procyanidins are also present in fruits, with apples (3.8 to 15.4 mg/100 g), plums (16.1 mg/100 g), and peaches (3.2 mg/100 g) containing the highest concentrations. However, citrus fruits do not appear to contain detectable levels of catechins or type B procyanidins.

Quercetin is the most common flavonol in fruits, elderberries (17.0 mg/100 g), lingonberries (12.6 mg/100 g), and cranberries (13.0 mg/100 g) being particularly rich sources. Berries and currants are also the fruits containing most kaempferol and myricetin. For example, these two flavonols account for 29 and 18%, respectively, of the total flavonol content of the bilberry. Although kaempferol and myricetin have also been identified in fruits such as peaches and pears, concentrations are generally too low to be readily quantified in the whole fruit. The skin of these fruits contains these flavonols in significant amounts; however, their flesh, which constitutes >70% of the fresh weight, does not. Consequently, when analyzed as normally eaten only trace levels are present.

Often termed the citrus flavonoids, flavanones are only found in citrus fruits such as oranges, grapefruit, and lemons. Although naringenin is present at greater concentrations than hesperetin in grapefruit (39.2 and 1.4 mg/100 g, respectively), the latter is the dominant form in oranges, lemons, and limes. Although citrus fruit also contains low levels of flavones, the olive is by far the richest source of luteolin and apigenin (12.4 and 4.6 mg/100 g, respectively).

4.3.2 Vegetables

Allium (e.g., onions), *Brassica* (e.g., broccoli and kale), and *Lactuca* (e.g., lettuce) varieties of vegetables and tomatoes (*Lycopersicon* species) are abundant sources of flavonols, primarily quercetin and kaempferol (Table 4.10). Flavones are also found in some vegetables such as celery, sweet peppers, and lettuce. Catechins and type B procyanidins, however, have not been found in leafy green or root vegetables but have been detected in legumes such as broad and green beans. The tomato is the only vegetable (although taxonomically a fruit) to possibly contain the flavanones naringenin and hesperetin.

Of the *Allium* species, shallots and red onions represent the richest potential source of quercetin containing 95 and 64 mg/100 g, respectively. *Brassica* vegetables including broccoli, kale, cabbage, and brussels sprouts tend to contain complex mixtures of flavonols, with significant quantities of kaempferol and myricetin glycosides present in addition to quercetin conjugates. Kale is a good example of this with mean levels of 11.5 mg quercetin/100 g and 34.1 mg kaempferol/100 g. Legumes such as green and broad beans also contain complex mixtures, mainly of flavonols and catechins. For example, broad beans contain (−)-epicatechin (30.0 mg/100 g), (+)-catechin (14.5 mg/100 g), (−)-gallocatechin (4.8 mg/100 g), and quercetin, myricetin, and kaempferol at concentrations below 3.0 mg/100 g.

Green chilli pepper is one of the few vegetables to contain both flavonols (quercetin, 11.39 mg/100 g) and flavones (luteolin, 2.7 mg/100 g) at detectable levels. Celery and sweet ball peppers are the main food sources of flavones independent of flavonols.

4.3.3 BEVERAGES

Catechins are often the most common flavonoids in beverages such as fruit juice, tea, and wine (Table 4.11). These tend to contain complex mixtures of simple catechins and their gallated esters. Type B procyanidins have frequently been characterized in beverages such as fruit juices; however, reliable quantitative data are limited. Flavonols are also present in most beverages while flavanones are again restricted to citrus juices such as grapefruit and orange. The presence of flavones in beverages is not well described with only some characterization information available in the literature.

Fruit juice contains both catechins and flavonols. Apple juice is one of the richest juice sources of catechins (containing 6.3 mg (−)-epicatechin/100 ml and 0.8 mg (+)-catechin/100 ml) whereas cranberry juice contains the most flavonols, mainly in the form of quercetin and myricetin (17.5 mg/100 ml and 4.7 mg/100 ml, respectively).

Tea is the only analyzed beverage to contain (−)-epigallocatechingallate (EGCG) in quantifiable amounts. EGCG and (−)-epicatechingallate (ECG) are the most abundant forms, each contributing 27% to the total catechin content (22.2 mg/100 ml) of black tea. Three flavonols (quercetin, kaempferol, and myricetin) are also found in tea. For example, 100 g of decaffeinated tea contains 5.2 mg quercetin, 2.4 mg kaempferol, and 0.1 mg myricetin.

Wine also contains a complex mix of catechins, flavonols, procyanidins, and flavanones. Red wine contains higher flavonoid levels than white or rosé wines. Procyanidins usually represent 50% of the flavonoids found in red wine, followed by catechins (37%). A similar profile is observed with beer where again procyanidins dominate accounting for 42% of total flavonoid content.

4.3.4 MISCELLANEOUS FOODS

Jam, confectionery, and herb compositional data are presented in Table 4.12. Honey contains low levels of both flavonols and flavones, and the presence of the flavanone naringenin has also been documented. Fruit jams also contain low levels of flavonols and catechins, which generally reflect the flavonoid profile of the whole fruit.

Quantitative data for chocolate are limited, but the available literature demonstrates that it is a good potential source of (+)-catechin, (−)-epicatechin, and type B procyanidins. Dark chocolate, for example, contains 6.6 mg, 21.8 mg, and 2.1 mg/100 g of catechin epicatechin, and procyanidins, respectively.

Compositional analysis data for herbs are also limited; however, these plants may be rich sources of flavones. For example, parsley is the major source of apigenin (217.9 mg/100 g) in the whole database, while sage and thyme are rich in luteolin (33.4 and 39.5 mg/100 g, respectively).

4.4 QUALITY AND COMPLETENESS OF FLAVONOID CONTENT DATA

The data in Table 4.9–Table 4.12 are comprehensive estimates of five classes of flavonoids in commonly available foods in the United Kingdom. Moreover, these estimates are derived from critically assessed published sources and the evaluation procedures adopted ensured the inclusion of content values for edible parts of plant materials available to the UK consumer. A USDA compiled database (http://www.nal.usda.gov/fnic/foodcomp/Data/Flav/flav.pdf) aimed primarily at the North American diet is also available. These databases are in contrast to another literature-derived database that is available for flavonoids[55] where data quality was not formally assessed and flavonoid values determined using semiquantitative methods were also included.

Nevertheless, the current database also has unavoidable limitations as the selection of appropriate food items and flavonoid values still required an element of operator judgment. For example, potentially useful information was excluded because (a) only experimental plants rather than commercially available varieties were analyzed (e.g., Ref. 56), (b) the country of origin did not supply the UK market (e.g., Ref. 57), (c) values were expressed only on a dry weight basis (e.g., Ref. 58), (d) final figures were presented as percentage of the total content (e.g., Ref. 59), and (e) the flavonoid contribution from the edible portions were difficult to separate from the whole fruit (e.g., Ref. 60). In addition, flavonoid data for several commonly consumed items are likely to be missing. For example, no flavanone data are available for satsumas, tangerines, and clementines, although these are seasonally abundant in UK supermarkets. There is also an overall lack of information on commonly consumed items such as herbs, fruit tea, and beer.[61] Additionally, bias may be introduced due to the relative number of compositional studies relating to each of the different flavonoid subclasses. For example, there are five large studies comprehensively identifying flavonols and flavones in several foods and beverages,[25,29,32,62–66] but fewer for catechins,[67–69] procyanidins,[69] and flavanones.[29]

4.4.1 Factors Affecting Flavonoid Content of Food

Another potential source of error in the database relates to the possibility that the flavonoid content of fruits and vegetables analyzed in a particular study do not reflect "normal" levels in the products. Such regional differences are frequently cited in order to explain the apparent lack of association between dietary components and disease.[2,70] The present database attempts to minimize this effect by including flavonoid values of products from countries known to export to the United Kingdom as over 50% of fresh produce consumed in this country is imported from the global market to ensure a year round supply.[71] Apples, for example, are imported from 24 different countries including France, Argentina, New Zealand, and Canada with British varieties being available for 9 months of the year (freshinfo. com).

However, other factors affecting flavonoid levels such as analytical variations, environmental factors, species characteristics, and the effects of processing and storage are more difficult to take into account when compiling the database.

4.4.1.1 Analytical Variations

The methods of extraction and analysis can markedly affect the determination of flavonoids in foods.[12,13,31,72–76] Rigorous application of the selection criteria (Table 4.6) may minimize this confounding effect. Overall, sample preparation and extraction techniques along the lines described by Merken and Beecher[31] were considered acceptable. These included freeze-drying, extraction either with aqueous methanol containing an antioxidant such as BHT or by solid-phase columns, filtration and the reduction of flavonol conjugates to the "free" aglycone by acid hydrolysis, enzyme digestion, or alkaline hydrolysis. Acceptable separation methodology normally involved RP-HPLC with UV, diode array, or electrochemical detection. Fluorescence detection, capillary zone electrophoresis, and micellar electrokinetic capillary chromatography were also included if identification and confirmation of eluted peaks was based on comparison with external standards, or if mass spectroscopy or nuclear magnetic resonance was used to confirm structural identity.

4.4.1.2 Environmental Factors

The flavonoid content of plant foods may be affected by growing conditions.[3,10] For example, red wine produced in the warm, dry, and sunny conditions prevalent in the New World tend to contain more quercetin and myricetin (but less catechin) than the wines produced in the cooler and damper regions of Northern Europe.[77,78] Similar regional and climatic effects on flavonoid content have been observed for many different fruits and vegetables.[65,79–81] Concentrations of flavonol and flavanone monoglycosides, for example, are greatest on the surface of plants grown in or originating from arid and semiarid habitats.[11,79,82] However, flavonoid profiles are also influenced by irrigation, which, for example, modifies concentrations and types of anthocyanins and catechins in berries.[82,83]

Marked differences in flavonoid content can even occur within a single variety depending on numerous factors such as maturity at harvesting, storage, use of glass and polythene, and organic cultivation methods.[32,65,68,84–86] This latter factor may be one reason why the flavonoid content of foods from Hungary are much higher than those from Western Europe, the enhanced levels reflecting the role of flavonoids in plants as insecticides and antimicrobial and antifungal agents.[11,85] Interestingly, Hungary is one of the main suppliers of organic vegetables to UK supermarkets.

Such environmental influences may account for why, for example, quercetin levels in Spanish cherry tomatoes range from 3.8 mg/100 g to 20.0 mg/100 g during a single year[80] and why produce grown in polythene tunnels with reduced UV exposure contain 98% less flavonoid glycosides than when grown in the open air.[10,11,40,82] The degree to which such environmental factors decrease the accuracy of the database is impossible to quantify but is likely to be minimized by the computation of average content values from a wide range of sources. This, in turn, is likely to reflect the average intake of a population exposed to a diverse range of products over the longer term.

4.4.1.3 Species Characteristics

Computation of a single value from a wide range of sources will also minimize analogous confounding effects of varietal differences as flavonoid subclasses can vary widely between different cultivars of fruits and vegetables.[11,32,87,88] Examples of such differences include flavonols in berries,[89] flavones in honey[90] and olives,[91] catechins in pears[92] and apples,[86] procyanidins in apples[93] and blueberries,[94] and flavanones in citrus fruit[95] and grapefruit juice.[96] Typically, for example, the flavonol content of 12 chilli-pepper varieties ranges from 0 to 85 mg/100 g.[97] Such differences can be ascribed to: (a) genetic mutations influencing the synthesis and accumulation of flavonoids in tissue[11,79]; (b) the degree of pigmentation,[56,79,89,98–100] particularly in berries[101] and onions[3] (although this has been recently disputed[54,56,79,80] as original determinations may have included the nonedible and anthocyanin-rich skin of the onion); (c) the stage of maturity,[13,102] quercetin levels, for example, tending to decline as fresh peppers ripen[103] whereas fresh young tea leafs contain more catechin derivatives than older ones.[104]

In addition, there are varietal effects on the degree and type of glycolsylation of flavonoids.[105] For example, quercetin rhamnoside is the most abundant glycoside in Jonagold and Golden Delicious apples whereas in Cox's orange and Elstar varieties, quercetin galactoside and arabinoside dominate.[86] As the nature of the sugar attachment may influence the bioavailability of a particular flavonoid,[106,107] it may ultimately be preferable for the database to show flavonoid content data as glycosides. At present, such information is lacking as the majority of studies employ hydrolysis to liberate the aglycone from the food matrix. Alternatively, once the biological significance of different glycosides have been determined, it may be possible to calculate conversion factors for flavonoids that are analogous to the current means of expressing tocopherol homologs as vitamin E equivalents.[41]

4.4.1.4   Processing and Storage

In general, industrially produced products such as tea, red wine, and fruit juice have significantly different flavonoid levels and profiles than the original fresh product.[108–112] Processing and preservation can expose fresh products to increased risk of oxidative damage and the activation of oxidative enzymes such as polyphenol oxidase.[92,113] In addition, procedures such as solvent extraction, sulfur dioxide treatment, pasteurization, enzymic clarification, heating, canning, irradiation, drying, and fermentation have been reported to affect procyanidin and catechin concentrations in fruit juice,[108–110,114–117] quercetin glucosides, catechins, and procyanidins in grapes,[118] procyanidin and flavonol levels in tomatoes and related sauces,[119] and quercetin concentrations in berries.[81]

Domestic preparation procedures such as chopping, shredding, peeling, and cooking may also affect flavonoid content accounting, for example, for 21 to 54% losses of flavonols in onions[54,120,121] as well as inducing glucosidase-mediated formation of monoglucosides and free quercetin from diglucosides.[56] Boiling is reported to lead to reduced flavonol contents of onions,[54,80,120] broccoli,[122] tomatoes,[80] asparagus, and green beans[120] although the effects of microwave cooking and frying may be less marked[80,120] due to decreased leaching of flavonoids from the foods during these cooking procedures.[54] Therefore, to minimize the confounding effects of such procedures, where possible values obtained from cooked and processed products were also included in the database.

4.5   ESTIMATED DIETARY FLAVONOID INTAKE

4.5.1   Estimation of Dietary Flavonoid Intake

Investigation of the relationship between diet and the development of chronic diseases requires an assessment tool that provides a valid estimate of "usual intake."[123] There is a wide variety of techniques available to assess dietary intake either prospectively or retrospectively. These range from those that provide relatively precise measurements of individual diet such as weighed intake records or duplicate diets to those that broadly rank intake in large cohort studies into high, medium, and low categories such as diet history or food frequency questionnaires. The advantages and disadvantages of these measurements have been extensively discussed and reviewed by several key workers in the field of dietary assessment.[2,123–125] Dietary assessment workers consistently agree that after selection of the most appropriate measurement tool, accurate and representative nutrient data on food composition are required.

4.5.2   Previous Estimations of Dietary Flavonoid Intake

An initial estimate of flavonoid intake of 1000 mg/day was calculated in the United States during the 1970s using semiquantitative food composition data (Table 4.13).[10] This estimate was not questioned until the 1990s with the calculation of dietary flavonol and flavone

TABLE 4.13
Average Daily Flavonoid Intake in the United States During 1971

Flavonoid	mg/day
Flavanones, flavones, flavonols	160–175 (110–121)[a]
Anthocyanins	180–215 (124–148)
Catechins	220 (152)
Biflavans	460 (317)
Total flavonoids	1020–1070 (703–738)

[a]Expressed as quercetin-3-rhamnoside (converted to quercetin aglycone).
Source: Reproduced from Kuhnau, J., *World Rev. Nutr. Diet.*, 24, 117, 1976. With permission.

aglycone (including glycosides hydrolyzed to their free form) intake from Dutch composition data for 28 vegetables, nine fruits, and several beverages.[32,126] Intake estimates for cohorts from seven countries ranged from 3 mg/day in Finland to 65 mg/day in Japan.[20] When the data of Kuhnau,[10] originally expressed as quercetin-3-rhamnoside equivalents, are converted to quercetin aglycones (703 to 738 mg/day), flavanones, flavonols, and flavones contribute 110 to 121 mg/day (Table 4.13). This suggests that the earlier study somewhat overestimated dietary flavonoid intake.

Several dietary flavonoid intake studies have now been completed using the Dutch composition data often with additional estimates of flavonoid content of local food preferences such as berries (Table 4.14). Comparison of these intake studies indicates that quercetin is consistently the main contributor to flavonol and flavone intake. In the Netherlands, for example, quercetin accounts for 70% of the 23 mg/day total flavonol and flavone intake followed by kaempferol (17%), myricetin (6%), luteolin (4%), and apigenin (3%).[127]

TABLE 4.14
Estimated Daily Dietary Flavonoid Intakes by Different Countries

Flavonoid	Country	mg/day[a]	Main Dietary Sources	Ref.
Flavonols/flavones				
	Netherlands	23	Tea (48%), onions (29%), apples (7%)	27
	United States	20–24	Tea (26%), onions (24%), apples (8%)	145
	United Kingdom (Wales)	26	Tea (82%), onions (10%)	135
	Finland	4	Apples and onions	34
	Spain	5	Tea (26%), onions (23%), apples (8%)	144
	Japan	16	Onions (46%), molokheya (10%), apples (7%), green tea (5%)	146
Catechins				
	Netherlands	50	Tea (83%), chocolate (6%), apples and pears (6%)	130
	United States	25	Tea (59%), apples and pears (26%)	139
Flavanones				
	Finland	20	Orange and grapefruit	129

Different countries obtain flavonoids from differing sources with, for example, green tea being the main contributor to intake in Japan, red wine in Italy, and apples in Finland.[20] However, the original dietary data for these investigations were collected during the early 1960s and may be outdated, as dietary patterns have changed during the last 40 years. For example, green tea consumption in Japan accounted for 80% of flavonol intake in 1960 but only 5% in 1997.[57] Table 4.14 outlines recent dietary flavonoid estimates and main dietary sources using data gathered after 1985. Interestingly, Japanese flavonol and flavone intake is reported to have fallen from 65 to 17 mg/day with onions, not green tea, now being the main dietary source. This may reflect increasing westernization of the Japanese diet.[128]

4.5.3 ESTIMATION OF SCOTTISH DIETARY FLAVONOID INTAKE USING THE NEW COMPREHENSIVE FLAVONOID DATABASE

The flavonoid database described in this chapter was applied to 4-day weighed food records obtained from healthy Scottish men ($n = 41$) and women ($n = 52$) to provide a provisional estimate of flavonoid intake in Scotland. All subjects consumed foods containing flavonols, procyanidins, and catechins, dietary intakes of which are given in Table 4.15. The main flavonol consumed was quercetin, accounting for 66 and 63% of the total flavonol intake of 18.8 mg/day. Primary sources of flavonols were from black tea (42.7%), onions (14.3%), apples (10.2%), and lager (7.2%) (Table 4.16).

Procyanidins and catechins were primarily obtained from black tea (procyanidins >47.6% and catechins >57.6%) and apples (procyanidins >15.8% and catechins >7.5%) (Table 4.16). The main catechins consumed were EGCG (23%), ECG (22%), and EC (25%).

Flavones and flavanones were less frequently consumed during the 4-day collection period. Flavones were not consumed at all by 38 participants, while 29 people did not consume any citrus flavonoids — flavanones. The interquartile range of intake of flavones was relatively limited, ranging from 0.0 to 2.0 mg/day. Flavone consumption was not normally distributed and was negatively skewed toward a lack of consumption of foods rich in flavones such as olives and lettuce. Likewise, flavanone intake was also not normally distributed with a mean flavanone intake of 1 mg/day compared to the median intake of 1.2 mg/day. This is accounted for by the fact that the range of flavanone intakes was very wide (0 to 239 mg/day), 36% of participants not consuming any flavanone-rich foods. The main dietary

---

TABLE 4.15
Estimated Dietary Flavonoid Intake by UK (Scottish) Population ($n = 81$)

Flavonoid Subclass	Daily intake (mg) (Median [Range])
Anthocyanins	
Flavonols (quercetin + kaempferol + myricetin)	18.8 (1.9–51.3)
Flavones (apigenin + luteolin)	0.1 (0–6.7)
Proanthocyanidins (procyanidin B1 + B2 + B3 + B4)	22.5 (0–144.5)
Catechins (C + EC + EGC + ECG + EGCG + GC)[a]	59.0 (1.8–263.3)
Flavanones (hesperidin + naringenin)	1.2 (0–238.6)

[a]C, catechin; EC, epicatechin; EGC, epigallocatechin; ECG, epicatechin gallate; EGCG, epigallocatechin gallate; GC, gallocatechin.

TABLE 4.16
Main Dietary Sources of Flavonoids from UK (Scottish) Diet[a]

Flavonols		Flavones		Procyanidins		Catechins		Flavanones	
Food Item	**%[b]**	**Food Item**	**%**	**Food Item**	**%**	**Food Item**	**%**	**Food Item**	**%**
Black tea	42.7	Sweet peppers[c]	24.4	Black tea	49.5	Black tea	63.6	Orange juice	37.8
Onion	14.3	Lettuce	18.1	Apples	20.6	Apples	11.2	White wine	19.2
Apples	10.2	Cheese and tom. pizza	11.9	Red wine	8.9	Red wine	4.9	Red wine	19.0
Lager	7.2	Vegetable soup	7.6	Lager	8.0	White wine	4.9	Orange	11.6
Tomato	2.4	Honey	5.7	Chocolate milk	4.4	Chocolate milk	4.1	Vegetable samosas	4.8
Baked beans	2.1	Scotch broth	3.7	Strawberries	0.9	Lager	3.9	Grapefruit juice	3.4
Red wine	2.8	Chilli cone carne	3.6	Peaches	0.7	Apple juice	2.7	Lemon juice	2.4
Orange Juice	2.0	Celery	2.8	Grapes	0.4	Grapes	1.8	Lemon sorbet	0.9

[a] Data collected from 40 males and 41 females using 4-day weighed intake records.
[b] Percentage contribution of the top eight dietary flavonoid sources.
[c] Sweet peppers — total contribution of red, green, and yellow peppers.

sources of flavanones were orange juice (37.8%) and wine (red wine 19.0% and white wine 19.2%) (Table 4.16).

4.5.4 COMPARISON WITH OTHER ESTIMATES OF DIETARY FLAVONOID INTAKE

Several dietary flavonoid intake studies have now been completed using the Dutch composition data often with additional estimates of flavonoid content of local food preferences such as berries (Table 4.14). The intake data reported above are the first estimation of dietary intake of different types of flavonoids by a UK population. Flavonol, flavone, and catechin intakes were comparable with Dutch literature[127,130] and the only other UK dietary flavonol investigation[131] (Table 4.14). The proportional contribution by individual flavonols and catechins to total flavonol and total catechin intakes, respectively, were analogous with Dutch intake and tea was their main dietary source. The flavanone intake of 1.2 mg/day is less than Finnish estimates of 20.2 ± 27.6 mg/day (mean $\pm$ SD) for men and women combined.[129] This may reflect differences in dietary preferences although it should also be noted that the Finnish data were calculated from a food frequency questionnaire, a dietary assessment method known to overestimate intake compared with weighed records. Procyanidin intake has not previously been estimated, as there has been a lack of reliable content values. However, Santo-Buelga and Scalbert[75] noted that a rough estimated intake of proanthocyanidins in Spain might range from 10 to >100 mg. If this estimate is correct, then the procyanidins type B1-4 measured here represents a small fraction of the total.

Identification of black tea as the primary source of flavonols (42.7%), procyanidins (49.5%), and catechins (63.6%) is again consistent with published literature for tea-drinking nations.[130,131] For example, a study of 1900 Welsh men also observed that tea was the main dietary source of flavonols.[131] Interestingly, 5.4 ± 3.0 cups of tea were consumed by these subjects per day between 1979 and 1983 whereas the Scottish participants reported consuming only 2.8 ± 2.4 cups of tea per day. This may reflect the current downward trend in tea consumption in the United Kingdom especially by adults under 50 years[61] and also suggests that flavonols are obtained from other food sources in the Scottish diet. Hertog,[131] for example, did not report lager as a major source of flavonols in the Welsh population despite its widespread consumption.[46] This is possibly because lager was not included in the Dutch database.[25,32,126]

4.5.5 FUTURE REQUIREMENTS TO IMPROVE DATABASE

The compilation of the database has further emphasized the diversity of potential dietary sources of flavonoids. It has also been used to give the first estimation of dietary intake of different types of flavonoids by a UK population. However, continual update is required to accommodate the increasingly varied purchasing patterns of foods in the United Kingdom[61] as well as for use in countries with markedly different dietary habits. There is a current lack of directly analyzed flavonoid measurements for composite meals. Therefore, recipe calculations would have to include assumptions for cooking losses or gains, which could subsequently introduce bias when calculating dietary intake.[132] Moreover, it is also important to confirm the flavonoid profile of composite dishes as they may contain unusual complex mixtures. For example, bolognaise sauce contains all five flavonoid subclasses (Table 4.17) whereas no individual food in the database exhibits such a profile. Despite these reservations, calculating the flavonoid content of recipes may be reasonably robust. Fjelkner-Modig et al.,[133] found close agreement between the directly analyzed and calculated quercetin levels in moussaka. However, it is possible that flavonoid values for retail products such as canned soups and preprepared meals may be underestimated as recipes and proportions of ingredients for such products are usually unavailable. Similarly, there is only limited compositional analysis for fruit juices and herbal tea despite their growing popularity.[61] Also, currently lacking are data on flavonoid contents of satsumas, tangerines, and clementines (which were consumed by 25% of subjects during the 4-day record collection period) and herbs commonly added to recipe dishes. Both are potentially good sources of flavonoids.

Despite these caveats, the database constructed in the present study is a comprehensive assessment of the currently available data on the flavonoid contents of foods.

4.6 CONCLUSIONS

A comprehensive and critical review of food flavonoid literature has led to the development of a food composition database for flavonols, flavones, procyanidins, catechins, and flavanones. This database can now be used and continuously updated to estimate flavonoid intake of populations, to identify dietary sources of flavonoids, and to assess associations between flavonoid intake and disease. However, there is a need for better food composition data for flavones, procyanidins, and flavanones as current literature is sparse particularly for citrus fruits, fruit juices, and herbs. In addition, anthocyanin food composition data are lacking although validated methods of determination are becoming available.

4.7 ACKNOWLEDGMENTS

Compilation of this flavonoid database could not have taken place without funding from the Scottish Executive Environment and Rural Affairs Department (SEERAD), the Ministry of Agriculture Fisheries and Food, UK, and the EU Framework V Programme. We are also grateful to Dr Geraldine McNeil for much needed help and advice.

TABLE 4.17
Estimated Flavonoid Content of Bolognaise Sauce Using Standard Recipe

Ingredients	Wt (g)	Flavonols			Flavones		Procyanidin	Catechins						Flavanones	
		Qu	K	Myr	Lut	Apig	B1–4	EGC	C	EC	EGCG	ECG	GC	Nar	Hesp
Clove garlic, crushed	0.8	Tr	—	—	—	—	—	—	—	—	—	—	—	—	—
Onions chopped	60.0	24.4	0.1	C[a]	—	C	—	—	—	—	—	—	—	—	—
Minced beef	500.0	—	—	—	—	—	—	—	—	—	—	—	—	—	—
Vegetable oil	10.0	—	—	—	—	—	—	—	—	—	—	—	—	—	—
Carrots, chopped	40.0	C	C	—	—	—	—	—	—	—	—	—	—	—	—
Celery chopped	30.0	Tr[b]	Tr	Tr	0.5	1.8	—	—	—	—	—	—	—	—	—
Tomato puree	10.0	0.4	Tr	—	—	—	—	—	—	—	—	—	—	—	—
Canned tomatoes	397.0	1.2	Tr	—	—	—	—	—	—	—	—	—	—	—	—
Stock	125.0	—	—	—	—	—	—	—	—	—	—	—	—	—	—
Red wine	125.0	1.2	0.1	0.7	—	—	18.8	0.4	8.2	4.8	Tr	Tr	0.5	2.2	0.7
Salt	2.5	—	—	—	—	—	—	—	—	—	—	—	—	—	—
Pepper	1.8	—	—	—	—	—	—	—	—	—	—	—	—	—	—
Dried mixed herbs	1.8	—	—	—	—	—	—	—	—	—	—	—	—	—	—
Total raw	1303.8	27.2	0.2	0.7	0.5	1.8	18.8	0.4	8.2	4.8	0.0	0.0	0.5	2.2	0.7
Cooked dish	886.6[c]														
Flavonoid content (mg/cooked dish)[d]		13.6	0.1	0.4	0.2	0.9	9.4	0.2	4.1	2.4	0.0	0.0	0.3	1.1	0.3
Flavonoid content (mg/100 g)		1.5	Tr	Tr	Tr	0.1	1.1	Tr	0.5	0.3	—	—	Tr	0.1	Tr

Source: Ingredients — recipe taken from Holland, B. et al. The composition of foods, In: *McCance and Widdowson's The Composition of Foods*. Cambridge: Royal Society of Chemistry & MAFF, 1991. With permission.

[a] C — Characterized but not quantified.

[b] Tr — below limit of detection (<0.1 mg/100).

[c] Cooked dish weight (g) = total raw weight − 32% weight loss during cooking.[41]

[d] Flavonoid content (mg/cooked dish) = total raw flavonoid content − (total raw flavonoid content × % nutrient cooking losses).[41]

REFERENCES

1. Doll, R. and Peto, R., eds., *The Causes of Cancer: Quantitative Estimates of the Avoidable Risks of Cancer in the United States Today*. New York: Oxford University Press, 1981.
2. Willett, W.C., ed., *Nutritional Epidemiology*. London: Oxford University Press, 1998.
3. Herrmann, K., Flavonols and flavones in food plants: a review, *J. Food Technol.*, 11, 433, 1976.
4. Steinmetz, K.A. and Potter, J.D., Vegetables, fruit, and cancer. 1. Epidemiology, *Cancer Causes Control*, 2, 325, 1991.
5. Block, G., Patterson, B., and Subar, A., Fruit, vegetables, and cancer prevention: a review of the epidemiological evidence, *Nutr. Cancer*, 18, 1, 1992.
6. Tavani, A. and LaVecchia, C., Fruit and vegetable consumption and cancer risk in a Mediterranean population, *Am. J. Clin. Nutr.*, 61, S1374, 1995.
7. Steinmetz, K.A. and Potter, J.D., Vegetables, fruit, and cancer prevention: a review, *J. Am. Diet. Assoc.*, 96, 1027, 1996.
8. Ness, A.R. and Powles, J.W., Fruit and vegetables, and cardiovascular disease: a review, *Int. J. Epidemiol.*, 26, 1, 1997.
9. Block, G., Patterson, B., and Subar, A., Fruit, vegetables, and cancer prevention: a review of the epidemiological evidence, *Nutr. Cancer*, 18, 1, 1992.
10. Kuhnau, J., The flavonoids. A class of semi-essential food components: their role in human nutrition, *World Rev. Nutr. Diet.*, 24, 117, 1976.
11. Harborne, J.B., ed., *The Flavonoids: Advances in Research Since 1986*. London: Chapman & Hall, 1994.
12. Robards, K. and Antolovich, M., Analytical chemistry of fruit bioflavonoids — a review, *Analyst*, 122, R11, 1997.
13. Escarpa, A. and Gonzalez, M.C., An overview of analytical chemistry of phenolic compounds in foods, *Crit. Rev. Anal. Chem.*, 31, 57, 2001.
14. Morton, L.W. et al., Chemistry and biological effects of dietary phenolic compounds: relevance to cardiovascular disease, *Clin. Exp. Pharmacol. Physiol.*, 27, 152, 2000.
15. Birt, D.F., Hendrich, S., and Wang, W.Q., Dietary agents in cancer prevention: flavonoids and isoflavonoids, *Pharmacol. Ther.*, 90, 157, 2001.
16. Nijveldt, R.J. et al., Flavonoids: a review of probable mechanisms of action and potential applications, *Am. J. Clin. Nutr.*, 74, 418, 2001.
17. Duthie, G.G., Duthie, S.J., and Kyle, J.A.M., Plant polyphenols in cancer and heart disease: implications as nutritional antioxidants, *Nutr. Res. Rev.*, 13, 79, 2000.
18. Manthey, J.A., Biological properties of flavonoids pertaining to inflammation, *Microcirculation*, 7, S29, 2000.
19. Benthsath, A., Rusznyak, I., and Szent-Gyorgyi, A., Vitamin P, *Nature*, 139, 326, 1937.
20. Hertog, M.G.L. et al., Flavonoid intake and long-term risk of coronary heart disease and cancer in the seven countries study, *Arch. Intern. Med.*, 155, 381, 1995.
21. Herrmann, K., Uber das Vokommon und die bedeutung der anthocyanine in lebensmitteln, *Z. Lebensm. Unters. Forsch.*, 148, 290, 1972.
22. Herrmann, K., Die phenolischen inhaltsstoffe des obstes. I. Bisherige Kenntnisse über Vorkommen, Gehalte sowie Veranderungen wahrend des Fruchtwachstums, *Z. Lebensm. Unters. Forsch.*, 151, 41, 1973.
23. Wildanger, W. and Herrmann, K., Flavonole und flavone der Germüsearten. I. Flavonole der Kohlarten, *Z. Lebensm. Unters. Forsch.*, 152, 134, 1973.
24. Mosel, H.D. and Herrmann, K., Changes in catechins and hydroxycinnamic acid derivatives during development of apples and pears, *J. Sci. Food Agric.*, 25, 251, 1974.
25. Hertog, M.G.L., Hollman, P.C.H., and Venema, D.P., Optimization of a quantitative HPLC determination of potentially anticarcinogenic flavonoids in vegetables and fruits, *J. Agric. Food Chem.*, 40, 1591, 1992.
26. Escarpa, A. and Gonzalez, M.C., High-performance liquid chromatography with diode-array detection for the determination of phenolic compounds in peel and pulp from different apple varieties, *J. Chromatogr. A*, 823, 331, 1998.

27. Escarpa, A., Perez-Cabrera, C., and Gonzalez, M.C., Optimization and validation of a fast liquid gradient for determination of prominent flavan-3-ols and flavonols in fresh vegetables, *J. High Resolut. Chromatogr.*, 23, 637, 2000.

28. Pietta, P.G. et al., Optimization of separation selectivity in capillary electrophoresis of flavonoids, *J. Chromatogr. A*, 680, 175, 1994.

29. Justesen, U., Knuthsen, P., and Leth, T., Quantitative analysis of flavonols, flavones, and flavanones in fruits, vegetables and beverages by high-performance liquid chromatography with photodiode array and mass spectrometric detection, *J. Chromatogr. A*, 799, 101, 1998.

30. Arts, I.C.W. and Hollman, P.C.H., Optimization of a quantitative method for the determination of catechins in fruits and legumes, *J. Agric. Food Chem.*, 46, 5156, 1998.

31. Merken, H.M. and Beecher, G.R., Measurement of food flavonoids by high-performance liquid chromatography: a review, *J. Agric. Food Chem.*, 48, 577, 2000.

32. Hertog, M.G.L., Hollman, P.C.H., and Katan, M.B., Content of potentially anticinogenic flavonoids of 28 vegetables and 9 fruits commonly consumed in the Netherlands, *J. Agric. Food Chem.*, 40, 2379, 1992.

33. Starke, H. and Herrmann, K., Flavonols and flavones of vegetables. 7. Flavonols of leeks, chives and garlic, *Z. Lebensm. Unters. Forsch.*, 161, 25, 1976.

34. Knekt, P., Reunanen, A., and Maatela, J., Flavonoid intake and coronary mortality in Finland: a cohort study, *Br. J. Nutr.*, 312, 478, 1996.

35. Rimm, E.B. et al., Relation between intake of flavonoids and risk for coronary heart disease in male health professionals, *Ann. Intern. Med.*, 125, 384, 1996.

36. Yochum, L. et al., Dietary flavonoid intake and risk of cardiovascular disease in postmenopausal women, *Am. J. Epidemiol.*, 149, 943, 1999.

37. Hirvonen, T. et al., Intake of flavonols and flavones and risk of coronary heart disease in male smokers, *Epidemiology*, 12, 62, 2001.

38. Tabak, C. et al., Chronic obstructive pulmonary disease and intake of catechins, flavonols, and flavones — the MORGEN Study, *Am. J. Resp. Crit. Care Med.*, 164, 61, 2001.

39. Cook, N.C. and Samman, S., Flavonoids — chemistry, metabolism, cardioprotective effects, and dietary sources, *J. Nutr. Biochem.*, 7, 66, 1996.

40. Herrmann, K., On the occurrence of flavone glycosides in vegetables, *Z. Lebensm. Unters. Forsch.*, 186, 1, 1988.

41. Holland, B. et al., The composition of foods, in: *McCance and Widdowson's The Composition of Foods*. Cambridge: Royal Society of Chemistry & MAFF, 1991, p. 1.

42. Mangels, A.R. et al., Carotenoid content of fruits and vegetables: an evaluation of analytic data, *J. Am. Diet. Assoc.*, 3, 284, 1993.

43. Greenfield, H., ed., *Quality and Accessibility of Food Related Data*. New York: Marcel Dekker, 1995.

44. Bolton-Smith, C. et al., Compilation of a provisional UK database for the phylloquinone (vitamin K-1) content of foods, *Br. J. Nutr.*, 83, 389, 2000.

45. Granado, F. et al., Variability in the intercomparison of food carotenoid content data: a user's point of view, *Crit. Rev. Food Sci. Nutr.*, 37, 621, 1997.

46. Gregory, J. et al., *The Dietary and Nutritional Survey of British Adults*. London: HMSO, 1990.

47. MAFF (Ministry of Agriculture, Food and Fisheries), *Household Food Consumption and Expenditure*. Annual Report of the National Food Survey Committee. London: HMSO, 1996.

48. Mazza, G., Anthocyanins in grapes and grape products, *Crit. Rev. Food Sci. Nutr.*, 35, 341, 1995.

49. Goiffon, J.P., Mouly, P.P., and Gaydou, E.M., Anthocyanic pigment determination in red fruit juices, concentrated juices and syrups using liquid chromatography, *Anal. Chim. Acta*, 382, 39, 1999.

50. Kalt, W. et al., Anthocyanin content and profile within and among blueberry species, *Can. J. Plant Sci.*, 79, 617, 1999.

51. Ordaz-Galindo, A. et al., Purification and identification of Capulin (*Prunus serotina*) anthocyanins, *Food Chem.*, 65, 201, 1999.

52. Gao, L. and Mazza, G., Quantitation and distribution of simple and acylated anthocyanins and other phenolics in blueberries, *J. Food Sci.*, 59, 1057, 1994.

53. de Ancos, B., Gonzalez, E., and Cano, M.P., Differentiation of raspberry varieties according to anthocyanin composition, *Z. Lebensm. Unters. Forsch.*, 208, 33, 1999.

54. Price, K.R., Bacon, J.R., and Rhodes, M.J.C., Effect of storage and domestic processing on the content and composition of flavonol glucosides in onion (*Allium cepa*), *J. Agric. Food Chem.*, 45, 938, 1997.

55. Linseisen, J., Radtke, J., and Wolfram, G., Flavonoid intake of adults in a Bavarian subgroup of the national food consumption survey, *Z. Ernahrungswiss.*, 36, 403, 1997.

56. Rhodes, M.J.C. and Price, K.R., Analytical problems in the study of flavonoid compounds in onions, *Food Chem.*, 57, 113, 1996.

57. Arai, Y. et al., Dietary intakes of flavonols, flavones and isoflavones by Japanese women and the inverse correlation between quercetin intake and plasma LDL cholesterol concentration, *J. Nutr.*, 130, 2243, 2000.

58. Awad, M.A. and de Jager, A., Flavonoid and chlorogenic acid concentrations in skin of "Jona-gold" and "Elstar" apples during and after regular and ultra low oxygen storage, *Postharvest Biol. Technol.*, 20, 15, 2000.

59. Ferreres, F. et al., An HPLC technique for flavonoid analysis in honey, *J. Sci. Food Agric.*, 56, 49, 1991.

60. Tomas-Barberan, F.A. et al., HPLC–DAD–ESIMS analysis of phenolic compounds in nectarines, peaches, and plums, *J. Agric. Food Chem.*, 49, 4748, 2001.

61. MAFF (Ministry of Agriculture, Food and Fisheries), *Household Food Consumption And Expenditure*. Annual Report of the National Food Survey Committee. London: HMSO, 2000.

62. Pietta, P.G. et al., Identification of flavonoid metabolites after oral administration to rats of a *Ginkgo biloba* extract, *J. Chromatogr. B — Biomed. Appl.*, 673, 75, 1995.

63. Lugasi, A. and Hovari, J., Flavonoid aglycons in foods of plant origin I. Vegetables, *Acta Alimen.*, 29, 345, 2000.

64. Lugasi, A. and Hovari, J., Flavonoid aglycons in foods of plant origin II. Fresh and dried fruits, *Acta Alimen.*, 31, 63, 2002.

65. Hakkinen, S.H. and Torronen, A.R., Content of flavonols and selected phenolic acids in strawberries and Vaccinium species: influence of cultivar, cultivation site and technique, *Food Res. Int.*, 33, 517, 2000.

66. Torronen, R. et al., Flavonoids and phenolic acids in selected berries, *Cancer Lett.*, 114, 191, 1997.

67. Arts, I.C.W., van de Putte, B., and Hollman, P.C.H., Catechin contents of foods commonly consumed in The Netherlands. 2. Tea, wine, fruit juices, and chocolate milk, *J. Agric. Food Chem.*, 48, 1752, 2000.

68. Arts, I.C.W., van de Putte, B., and Hollman, P.C.H., Catechin contents of foods commonly consumed in The Netherlands. 1. Fruits, vegetables, staple foods, and processed foods, *J. Agric. Food Chem.*, 48, 1746, 2000.

69. De Pascual-Teresa, S., Santos-Buelga, C., and Rivas-Gonzalo, J.C., Quantitative analysis of flavan-3-ols in Spanish foodstuffs and beverages, *J. Agric. Food Chem.*, 48, 5331, 2000.

70. WCRF (World Cancer Research Fund), *Nutrition and the Prevention of Cancer: A Global Perspective*. Washington, DC: American Institute for Cancer Research, 1997.

71. HMSO, DEFRA Import Statistics, 2004.

72. Harborne, J.B., Nature, distribution and function of plant flavonoids, *Prog. Clin. Biol. Res.*, 213, 15, 1986.

73. Rice-Evans, C.A. and Packer, L., *Flavonoids in Health and Disease*. New York: Marcel Dekker, 1998.

74. Antolovich, M., Prenzler, P., Robards, K., and Ryan, D., Sample preparation in the determination of phenolic compounds in fruit, *Analyst*, 125, 989, 2000.

75. Santos-Buelga, C. and Scalbert, A., Proanthocyanidins and tannin-like compounds — nature, occurrence, dietary intake and effects on nutrition and health, *J. Sci. Food Agric.*, 80, 1094, 2000.

76. Nuutila, A.M., Kammiovirta, K., and Oksman-Caldentey, K.M., Comparison of methods for the hydrolysis of flavonoids and phenolic acids from onion and spinach for HPLC analysis, *Food Chem.*, 76, 519, 2002.

77. Goldberg, D.M. et al., Quercetin and p-coumaric acid concentrations in commercial wines, *Am. J. Enol. Vitic.*, 49, 142, 1998.

78. McDonald, M.S. et al., Survey of the free and conjugated myricetin and quercetin content of red wines of different geographical origins, *J. Agric. Food Chem.*, 46, 368, 1998.

79. Leighton, T. et al., Molecular characterization of quercetin and quercetin glycosides in *Allium* vegetables: their effects on malignant cell transformation. In: Huang, M.T., Ho, C.T., and Lee, C.Y., eds., *Phenolic Compounds in Food and their Effects on Health II. Antioxidants and Cancer Prevention*. New York: American Chemical Society, 1992, p. 220.

80. Crozier, A. et al., Quantitative analysis of the flavonoid content of commercial tomatoes, onions, lettuce, and celery, *J. Agric. Food Chem.*, 45, 590, 1997.

81. Hakkinen, S.H. et al., Influence of domestic processing and storage on flavonol contents in berries, *J. Agric. Food Chem.*, 48, 2960, 2000.

82. Cuadra, P. and Harborne, J.B., Changes in epicuticular flavonoids and photosynthetic pigments as a plant response to UV-B radiation, *Z. Naturforsch. C*, 51, 671, 1996.

83. Esteban, M.A., Villanueva, M.J., and Lissarrague, J.R., Effect of irrigation on changes in the anthocyanin composition of the skin of cv Tempranillo (*Vitis vinifera* L) grape berries during ripening, *J. Sci. Food Agric.*, 81, 409, 2001.

84. Hayashi, H. et al., Flavonoid variation in the leaves of *Glycyrrhiza glabra*, *Phytochemistry*, 42, 701, 1996.

85. Ren, H.F. et al., Antioxidative and antimicrobial activities and flavonoid contents of organically cultivated vegetables, *J. Jap. Soc. Food Sci. Technol.*, 48, 246, 2001.

86. van der Sluis, A.A. et al., Activity and concentration of polyphenolic antioxidants in apple juice. 1. Effect of existing production methods, *J. Agric. Food Chem.*, 50, 7211, 2002.

87. Glusker, J.P. and Rossi, M., Molecular aspects of chemical carcinogens and bioflavonoids. In: Cody, V., Middleton, E., Jr., and Harborne, J.B., eds., *Plant Flavonoids in Biology and Medicine: Biochemical, Pharmacological, and Structure–Activity Relationships*. New York: Alan, R. Liss, 1986, p. 395.

88. Hempel, J. and Bohm, H., Quality and quantity of prevailing flavonoid glycosides of yellow and green French beans (*Phaseolus vulgaris* L), *J. Agric. Food Chem.*, 44, 2114, 1996.

89. Bilyk, A. and Sapers, G.M., Distribution of quercetin and kaempferol in lettuce, kale, chive, garlic chive, leek, horseradish, red radish, and red cabbage tissues, *J. Agric. Food Chem.*, 33, 226, 1985.

90. Martos, I. et al., Flavonoids in monospecific eucalyptus honeys from Australia, *J. Agric. Food Chem.*, 48, 4744, 2000.

91. Romani, A., Polyphenolic content in five Tuscany cultivars of *Olea europaea* L., *J. Agric. Food Chem.*, 47, 964, 1999.

92. Amiot, M.J. et al., Influence of cultivar, maturity stage and storage conditions on phenolic composition and enzymatic browning of pear fruits, *J. Agric. Food Chem.*, 43, 1132, 1995.

93. Hammerstone, J.F., Lazarus, S.A., and Schmitz, H.H., Procyanidin content and variation in some commonly consumed foods, *J. Nutr.*, 130, 2086S, 2000.

94. Prior, R.L. et al., Identification of procyanidins and anthocyanins in blueberries and cranberries (*Vaccinium* spp.) using high-performance liquid chromatography/mass spectrometry, *J. Agric. Food Chem.*, 49, 1270, 2001.

95. Rouseff, R.L., Martin, S.F., and Youtsey, C.O., Quantitative survey of narirutin, naringin, hesperidin and neohesperidin in *Citrus*, *J. Agric. Food Chem.*, 35, 1027, 1987.

96. Ross, S.A. et al., Variance of common flavonoids by brand of grapefruit juice, *Fitoterapia*, 71, 154, 2000.

97. Lee, Y., Howard, L.R., and Villalon, B., Flavonoids and antioxidant activity of fresh pepper (*Capsicum annuum*) cultivars, *J. Food Sci.*, 60, 473, 1995.

98. Fieschi, M., Codignola, A., and Luppi Mosa, A.M., Mutagenic flavonol aglycones in infusions and in fresh and picked vegetables, *J. Food Sci.*, 54, 1492, 1989.

99. Rhodes, M.J.C., Physiologically-active compounds in plant foods: an overview, *Proc. Nutr. Soc.*, 55, 371, 1996.

100. Ferreres, F. et al., Evaluation of pollen as a source of kaempferol in rosemary honey, *J. Sci. Food Agric.*, 77, 506, 1998.

101. Bilyk, A. and Sapers, G.M., Varietal differences in the quercetin, kaempferol, and myricetin contents of highbush blueberry, cranberry, and thornless blackberry fruits, *J. Agric. Food Chem.*, 34, 585, 1986.

102. Aherne, S.A. and O'Brien, N.M., Dietary flavonols: chemistry, food content, and metabolism, *Nutrition*, 18, 75, 2002.

103. Howard, L.R. et al., Changes in phytochemical and antioxidant activity of selected pepper cultivars (*Capsicum* species) as influenced by maturity, *J. Agric. Food Chem.*, 48, 1713, 2000.

104. Lin, Y.L. et al., Composition of polyphenols in fresh tea leaves and associations of their oxygen-radical-absorbing capacity with antiproliferative actions in fibroblast cells, *J. Agric. Food Chem.*, 44, 1387, 1996.

105. DuPont, M.S. et al., Effect of variety, processing, and storage on the flavonoid glycoside content and composition of lettuce and endive, *J. Agric. Food Chem.*, 48, 3957, 2000.

106. Hollman, P.C.H. and Katan, M.B., Bioavailability and health effects of dietary flavonols in man, *Arch. Toxicol. Suppl.*, 20, 237, 1998.

107. Day, A.J., Bao, Y., Morgan, M.R., and Williamson, G., Conjugation position of quercetin glucuronides and effect on biological activity, *Free Radical Biol. Med.*, 29, 1234, 2000.

108. Spanos, G.A. and Wrolstad, R.E., Influence of processing and storage on the phenolic composition of Thompson seedless grape juice, *J. Agric. Food Chem.*, 38, 1565, 1990.

109. Spanos, G.A., Wrolstad, R.E., and Heatherbell, D.A., Influence of processing and storage on the phenolic composition of apple juice, *J. Agric. Food Chem.*, 38, 1572, 1990.

110. Spanos, G.A. and Wrolstad, R.E., Influence of variety, maturity, processing and storage on the phenolic composition of pear juice, *J. Agric. Food Chem.*, 38, 817, 1990.

111. Balentine, D.A., Wiseman, S.A., and Bouwens, L.C.M., The chemistry of tea flavonoids, *Crit. Rev. Food Sci. Nutr.*, 37, 693, 1997.

112. Soleas, G.J., Diamandis, E.P., and Goldberg, D.M., Wine as a biological fluid: history, production, and role in disease prevention, *J. Clin. Lab. Anal.*, 11, 287, 1997.

113. Lee, C.Y., Kagan, V., Jaworski, A.W., and Brown, S.K., Enzymatic browning in relation to phenolic compounds and polyphenoloxidase activity among various peach cultivars, *J. Agric. Food Chem.*, 38, 99, 1990.

114. Lu, Y.R. and Foo, L.Y., Identification and quantification of major polyphenols in apple pomace, *Food Chem.*, 59, 187, 1997.

115. Bengoechea, M.L. et al., Phenolic composition of industrially manufactured purees and concentrates from peach and apple fruits, *J. Agric. Food Chem.*, 45, 4071, 1997.

116. Robards, K. et al., Phenolic compounds and their role in oxidative processes in fruits, *Food Chem.*, 66, 401, 1999.

117. Chen, H., Zuo, Y.G., and Deng, Y.W., Separation and determination of flavonoids and other phenolic compounds in cranberry juice by high-performance liquid chromatography, *J. Chromatogr. A*, 913, 387, 2001.

118. Karadeniz, F., Durst, R.W., and Wrolstad, R.E., Polyphenolic composition of raisins, *J. Agric. Food Chem.*, 48, 5343, 2000.

119. Re, R., Bramley, P.M., and Rice-Evans, C., Effects of food processing on flavonoids and lycopene status in a Mediterranean tomato variety, *Free Radical Res.*, 36, 803, 2002.

120. Ewald, C., Effect of processing on major flavonoids in processed onions, green beans, and peas, *Food Chem.*, 64, 231, 1999.

121. Gennaro, L., Flavonoid and carbohydrate contents in Tropea red onions: effects of homelike peeling and storage, *J. Agric. Food Chem.*, 50, 1904, 2002.

122. Price, K.R. et al., Composition and content of flavonol glycosides in broccoli florets (*Brassica oleracea*) and their fate during cooking, *J. Sci. Food Agric.*, 77, 468, 1998.

123. Bingham, S.A. et al., Comparison of dietary assessment methods in nutritional epidemiology — weighed records v 24-h recalls, food frequency questionnaires and estimated diet records, *Br. J. Nutr.*, 72, 619, 1994.

124. Bingham, S.A. et al., Nutritional methods in the European prospective investigation of cancer in Norfolk, *Public Health Nutr.*, 4, 847, 2001.

125. Nelson, M., The validation of dietary assessment, in: Margetts, B.M. and Nelson, M., eds., *Design Concepts in Nutritional Epidemiology.* Oxford: Oxford University Press, 1997, p. 241.

126. Hertog, M.G.L., Hollman, P.C.H., and van de Putte, B., Content of potentially anticarcinogenic flavonoids of tea infusions, wines, and fruit juices, *J. Agric. Food Chem.*, 41, 1242, 1993.

127. Hertog, M.G.L. et al., Intake of potentially anticarcinogenic flavonoids and their determinants in adults in the Netherlands, *Nutr. Cancer*, 20, 21, 1993.

128. Kimira, M. et al., Japanese intake of flavonoids and isoflavonoids from foods, *J. Epidemiol.*, 8, 168, 1998.

129. Knekt, P. et al., Flavonoid intake and risk of chronic diseases, *Am. J. Clin. Nutr.*, 76, 560, 2002.

130. Arts, I.C.W. et al., Catechin intake and associated dietary and lifestyle factors in a representative sample of Dutch men and women, *Eur. J. Clin. Nutr.*, 55, 76, 2001.

131. Hertog, M.G.L., Antioxidant flavonols and ischemic heart disease in a Welsh population of men: The Caerphilly Study, *Am. J. Clin. Nutr.*, 65, 1489, 1997.

132. Flegal, K.M., Evaluating epidemiologic evidence of the effects of food and nutrient exposures, *Am. J. Clin. Nutr.*, 69, 1339S, 1999.

133. Fjelkner-Modig, S. et al., Flavonoids in vegetables — the influence of processing, Proceedings of the 2nd International Conference on Natural Antioxidants and Anticarcinogens in Nutrition, Health and Disease, 1998.

134. Hertog, M.G.L. et al., Dietary antioxidant flavonoids and risk of coronary heart disease: the Zutphen Elderly Study, *Lancet*, 342, 1007, 1993.

135. Hertog, M.G.L. et al., Antioxidant flavonols and ischemic heart disease in a Welsh population of men: The Caerphilly Study, *Am. J. Clin. Nutr.*, 65, 1489, 1997.

136. Geleijnse, J.M. et al., Inverse association of tea and flavonoid intakes with incident myocardial infarction: the Rotterdam Study, *Am. J. Clin. Nutr.*, 75, 880, 2002.

137. Arts, I.C.W. et al., Catechin intake might explain the inverse relation between tea consumption and ischemic heart disease: the Zutphen Elderly Study, *Am. J. Clin. Nutr.*, 74, 227, 2001.

138. Arts, I.C.W. et al., Dietary catechins in relation to coronary heart disease death among postmenopausal women, *Epidemiology*, 12, 668, 2001.

139. Arts, I.C.W. et al., Dietary catechins and cancer incidence among postmenopausal women: the Iowa Women's Health Study (United States), *Cancer Causes Control*, 13, 373, 2002.

140. Arts, I.C.W. et al., Dietary catechins and epithelial cancer incedence: the Zutphen Elderly Study, *Int. J. Cancer*, 92, 298, 2001.

141. Hirvonen, T. et al., Flavonol and flavone intake and the risk of cancer in male smokers (Finland), *Cancer Causes Control*, 12, 789, 2001.

142. Knekt, P. et al., Dietary flavonoids and the risk of lung cancer and other malignant neoplasms, *Am. J. Epidemiol.*, 146, 223, 1997.

143. Le Marchand, L. et al., Intake of flavonoids and lung cancer, *J. Nat. Cancer Inst.*, 92, 154, 2000.

144. Garcia-Closas, R. et al., Intake of specific carotenoids and flavonoids and the risk of lung cancer in women in Barcelona, Spain, *Nutr. Cancer — Int. J.*, 32, 154, 1998.

145. Sampson, L. et al., Flavonol and flavone intakes in US health professionals, *J. Am. Diet. Assoc.*, 102, 1414, 2002.

146. Arai, Y. et al., Dietary intakes of flavonols, flavones and isoflavones by Japanese women and the inverse correlation between quercetin intake and plasma LDL cholesterol concentration, *J. Nutr.*, 130, 2243, 2000.

147. Lugasi, A. and Hovari, J., Flavonoid aglycons in foods of plant origin I. Vegetables, *Acta Aliment.*, 29, 345, 2000.

148. Mattila, P. Astola, J, and Kumpulainen, J., Determination of flavonoids in plant material by HPLC with diode-array and electro-array detections, *J. Agric. Food Chem.*, 48, 5834, 2000.

149. Valles, B.S. et al., High-performance liquid chromatography of the neutral phenolic compounds of low molecular weight in apple juice, *J. Agric. Food Chem.*, 42, 2732, 1994.

150. Hakkinen, S.H. et al., Content of the flavonols quercetin, myricetin, and kaempferol in 25 edible berries, *J. Agric. Food Chem.*, 47, 2274, 1999.

151. Mikkonen, T.P. et al., Flavonol content varies among black currant cultivars, *J. Agric. Food Chem.*, 49, 3274, 2001.
152. Gorinstein, S. et al., Comparison of some biochemical characteristics of different citrus fruits, *Food Chem.*, 74, 309, 2001.
153. Esti, M., Cinquanta, L., and LaNotte, E., Phenolic compounds in different olive varieties, *J. Agric. Food Chem.*, 46, 32, 1998.
154. Vlahov, G., Flavonoids in three olive (*Olea europaea*) fruit varieties during maturation, *J. Sci. Food Agric.*, 58, 157, 1992.
155. Schieber, A., Keller, P., and Carle, R., Determination of phenolic acids and flavonoids of apple and pear by high-performance liquid chromatography, *J. Chromatogr. A*, 910, 265, 2001.
156. Makris, D.P. and Rossiter, J.T., Domestic processing of onion bulbs (*Allium cepa*) and asparagus spears (*Asparagus officinalis*): effect on flavonol content and antioxidant status, *J. Agric. Food Chem.*, 49, 3216, 2001.
157. Price, K.R. et al., Composition and content of flavonol glycosides in broccoli florets (*Brassica olearacea*) and their fate during cooking, *J. Sci. Food Agric.*, 77, 468, 1998.
158. Justesen, U. and Knuthsen, P., Composition of flavonoids in fresh herbs and calculation of flavonoid intake by use of herbs in traditional Danish dishes, *Food Chem.*, 73, 245, 2001.
159. Ferreres, F., Gil, M.I., and Tomas-Barberan, F.A., Anthocyanins and flavonoids from shredded red onion and changes during storage in perforated films, *Food Res. Int.*, 29, 389, 1996.
160. Patil, B.S., Pike, L.M., and You, K.S., Variation in the quercetin content in different coloured onions (*Allium cepa* L.), *J. Am. Soc. Hort. Sci.*, 120, 909, 1995.
161. Martinez-Valverde, I. et al., Phenolic compounds, lycopene and antioxidant activity in commercial varieties of tomato (*Lycopersicum esculentum*), *J. Sci. Food Agric.*, 82, 323, 2002.
162. Stewart, A.J. et al., Occurrence of flavonols in tomatoes and tomato-based products, *J. Agric. Food Chem.*, 48, 2663, 2000.
163. Hertog, M.G.L., Hollman, P.C.H., and van de Putte, B., Content of potentially anticarcinogenic flavonoids of tea infusions, wines, and fruit juices, *J. Agric. Food Chem.*, 41, 1242, 1993.
164. Bronner, W.E. and Beecher, G.R., Extraction and measurement of prominent flavonoids in orange and grapefruit juice concentrates, *J. Chromatogr. A*, 705, 247, 1995.
165. Mouly, P.P. et al., Differentiation of citrus juices by factorial discriminant analysis using liquid chromatography of flavanone glycosides, *J. Agric. Food Chem.*, 42, 70, 1994.
166. Achilli, G. et al., Identification and determination of phenolic constituents in natural beverages and plant extracts by means of a coulometric electrode array system, *J. Chromatogr.*, 632, 111, 1993.
167. Bremner, P.D. et al., Comparison of the phenolic composition of fruit juices by single step gradient HPLC analysis of multiple components versus multiple chromatographic runs optimised for individual families, *Free Radical Res.*, 32, 549, 2000.
168. Leuzzi, U. et al., Flavonoids in pigmented orange juice and second-pressure extracts, *J. Agric. Food Chem.*, 48, 5501, 200.
169. Marini, D. and Balestrieri, F., Multivariate analysis of flavanone glycosides in citrus juices, *Ital. J. Food Sci.*, 7, 255, 1995.
170. Ooghe, W.C. et al., Characterization of orange juice (*Citrus sinensis*) by flavanone glycosides, *J. Agric. Food Chem.*, 42, 2183, 1994.
171. Khokhar, S. and Magnusdottir, S.G.M., Total phenol, catechin, and caffeine contents of teas commonly consumed in the United Kingdom, *J. Agric. Food Chem.*, 50, 565, 2002.
172. Price, K.R., Rhodes, M.J.C., and Barnes, K.A., Flavonol glycoside content and composition of tea infusions made from commercially available teas and tea products, *J. Agric. Food Chem.*, 46, 2517, 1998.
173. Bronner, W.E. and Beecher, G.R., Method for determining the content of catechins in tea infusions by high-performance liquid chromatography, *J. Chromatogr. A*, 805, 137, 1998.
174. de Colmenares, N.G. et al., Isolation, characterisation and determination of biological activity of coffee proanthocyanidins, *J. Sci. Food Agric.*, 77, 368, 1998.
175. Fabios, M. et al., Phenolic compounds and browning in sherry wines subjected to oxidative and biological aging, *J. Agric. Food Chem.*, 48, 2155, 2000.

176. Guillen, D.A., Barroso, C.G., and Perez-Bustamante, J.A., Automation of sample preparation as a preliminary stage in the high-performance liquid chromatographic determination of polyphenolic compounds in sherry wines, *J. Chromatogr. A*, 730, 39, 1996.

177. Simonetti, P., Pietta, P., and Testolin, G., Polyphenol content and total antioxidant potential of selected Italian wines, *J. Agric. Food Chem.*, 45, 1152, 1997.

178. Frankel, E.N., Waterhouse, A.L., and Teissedre, P.L., Principal phenolic phytochemicals in selected California wines and their antioxidant activity in inhibiting oxidation of human low density lipoproteins, *J. Agric. Food Chem.*, 43, 890, 1995.

179. Salagoity-Auguste, M.H. and Bertrand, A., Wine phenolics — analysis of low molecular weight components by high performance liquid chromatography, *J. Sci. Food Agric.*, 35, 1241, 1984.

180. Carando, S. and Teissedre, P.L., Catechin and procyanidin levels in French wines: contribution to dietary intake, *Basic. Life Sci.*, 66, 725, 1999.

181. Carando, S. et al., Levels of flavan-3-ols in French wines, *J. Agric. Food Chem.*, 47, 4161, 1999.

182. Tomas-Lorente, F. et al., Phenolic compounds analysis in the determination of fruit jam genuineness, *J. Agric. Food Chem.*, 40, 1800, 1992.

183. Martos, I. et al., Flavonoid composition of Tunisian honeys and propolis, *J. Agric. Food Chem.*, 45, 2824, 1997.

184. Zafrilla, P., Ferreres, F., and Tomas-Barberan, F.A., Effect of processing and storage on the antioxidant ellagic acid derivatives and flavonoids of red raspberry (*Rubus idaeus*) jams, *J. Agric. Food Chem.*, 49, 3651, 2001.

185. Adamson, G.E. et al., HPLC method for the quantification of procyanidins in cocoa and chocolate samples and correlation to total antioxidant capacity, *J. Agric. Food Chem.*, 47, 4184, 1999.

186. Zheng, W. and Wang, S.Y. Antioxidant activity and phenolic compounds in selected herbs, *J. Agric. Food Chem.*, 49, 5165, 2001.

187. Justesen, U. and Knuthsen, P. Composition of flavonoids in fresh herbs and calculation of flavonoid intake by use of herbs in traditional Danish dishes, *Food Chem.*, 73, 245, 2001.

APPENDIX 1

BOTANICAL NAMES OF FRUITS

Common Name	Botanical Name
Apple	*Malus silvestris*
Apricot	*Prunus armeniaca*
Avocado	*Persea armeniaca*
Banana	*Musa acuminata*
Bilberry	*Vaccinium myrtillus*
Blackberry, European	*Rubus fruiticosus*
Blackberry, western trailing	*Rubus ursinus*
Blueberry, high-bush	*Vaccinium corymbosum*
Blueberry, low-bush	*Vaccinium angustifolium*
Brambleberry	*Rubus arcticus*
Cherry, sour	*Prunus cerasus*
Cherry, sweet	*Prunus avium*
Cranberry	*Vaccinium macrocarpum*
Currants, black	*Ribes nigrum*
Currants, red	*Ribes rubrum, sativum*
Custard apple	*Annona acuminata*
Eggplant/aubergine	*Solanum melongena*
Elderberry	*Sambucus nigra*
Fig	*Ficus carica*
Gooseberry	*Ribes uva crispa, grossularia*
Grapefruit	*Citrus paradisi*
Grapes, European	*Vitris vinifera*
Lemon	*Citrus lemon*
Lime	*Citrus aurantifolia*
Mandarin	*Citrus reticulata*
Mango	*Mangifera indica*
Olive	*Olea europaea*
Orange, Seville	*Citrus aurantium*
Orange, Valencia	*Citrus sinensis*
Passion fruit	*Passiflora edulis*
Peach	*Prunus persica*
Pear	*Pirus communis*
Pepper, red	*Capsicum annuum*
Persimmon	*Diospyros kaki, virginiana*
Plum, blue	*Prunus domestica*
Plum, yellow	*Prunus salicina*
Pomegranate	*Punica granatum*
Raspberry, black	*Rubus occidentalis*
Raspberry, red	*Rubus idaeus, strigosus*
Rhubarb	*Rheum rhaponticum, tataricum*
Strawberry	*Fragaria vesca, ananassa*
Tangerine	*Citrus deliciosa*
Tomato	*Lycopersicum esculentum*

BOTANICAL NAMES OF VEGETABLES

Common Name	Botanical Name
Asparagus	*Asparagus officinalis*
Artichoke	*Cynara scolymus*
Broad bean	*Vicia faba*
Broccoli	*Brassica oleracea*
Kale	*Brassica oleracea* var. *botrytis*
Brussels sprouts	*Brassica oleraces* var. *gemmifera*
Endive	*Cichorium intybus*
Spinach	*Spinacea oleracea*
Fennel	*Foeniculum vulgare*
Leek	*Allium porrum*
Lettuce	*Lactuca sativa*
Parsley	*Petroselinum crispum, sativum, hortense*
Red cabbage	*Brassica oleracea capitata rubra*
Salad greens	*Valerianella olitoria*
Turnip	*Brassica rapa var. perviridis*
Carrot	*Daucus carota*
Celery	*Apium graveolens*
Onion	*Allium cepa*
Parsnip	*Pastinaca sativa*
Potato	*Solanum tuberosum*
Radish, red	*Raphanus sativus*
Sweet potato	*Ipomoea batatas*

BOTANICAL NAMES OF HERBS AND SPICES

Common Name	Botanical Name
Basil, sweet	*Ocimum basilicum*
Caper	*Capparis ovata, spinosa*
Caraway	*Carum carvi*
Chamomile	*Anthemis nobilis*
Chervil	*Torilis tenella, nodosa*
Coriander	*Coriandrum sativum*
Dill	*Anethum graveolens*
Hops	*Humulus lupulus*
Lemon balm	*Melissa officinalis*
Marjoram	*Majorana hortensis*
Mint	*Mentha var.*
Mustard	*Sinapis alba, nigra*
Oregano	*Origanum vulgare*
Parsley	*Petroselinum crispum*
Peppermint	*Metha spicata, piperita*
Rosemary	*Rosmarinus officinalis*
Sage	*Salvia officinalis*
Tarragon	*Artemisia dranunculus*
Thyme	*Thymus vulgaris, collinus, nummularis*
Watercress	*Nasturtium officinale*

BOTANICAL NAMES OF LEGUMES

Common Name	Botanical Name
Broad bean	*Vicia faba*
Chick pea	*Cicer arietinum*
French bean	*Vicia faba*
Kidney bean	*Phaseolus vulgaris, lunatus, radiatus*
Lima bean	*Phaseolus vulgaris, lunatus, radiatus*
Mung bean	*Phaseolus vulgaris, lunatus, radiatus*
Pea	*Pisum sativum*
Sesame	*Sesamum indicum*
Soyabean	*Glycine max*

APPENDIX 2

References Used for Database Compilation

TABLE 4.9
Flavonoid Content of Fruits

	Flavonols	Flavones	Procyanidins	Catechins	Flavanones
Apple					
Eating	32, 29		69	67, 69, 86	29
Peeled	32			67	
Sauce	32			67	
Apricots	32, 29		69	67, 69	
Avocado			69	67, 69	
Banana			69	67, 69	
Bilberry	65, 81, 150		69	67, 69	
Blackberries	101		69	67, 69	
Blackcurrants	29, 81, 150, 151		67	67	
Blueberries	29, 81, 101, 150		67, 69	67, 69	
Cherries	29, 32		67, 69	67, 69	
Cranberries	29, 32, 148, 150		67, 69	67, 69	
Custard apple			69	69	
Elderberries	150		69	69	
Early fig			67	67	
Gooseberry	150		67	67	
Grapefruit	29				29
Grapes					
Black	29, 32		67, 69	67, 69	
Green	29, 32		67, 69	67, 69	
Kiwi fruit			67, 69	67, 69	
Lemon	29, 148	29, 148, 152			29, 148
Lime	29				29
Lingonberry	81, 148, 150				
Mango				67	
Nectarine				67	
Olives	91, 153, 154	91, 153, 154			
Orange	29, 148				29, 148
Peaches	32, 60		67, 69	67, 69	
Pears	29, 32		67, 69	67, 69, 155	
Peeled	32		32	32	
Persimmon			67, 69	67, 69	
Plums, blue	29, 32		67, 69	67, 69	
Pomegranate			69	69	
Raisins	118			67	
Raspberries	29, 81, 150		67, 69	67, 69	
Red currants	29, 32		67, 69	67, 69	
Rhubarb			67	67	
Strawberry	32, 65, 81, 150		67, 69	67, 69	

TABLE 4.10
Flavonoid Content of Vegetables

	Flavonols	Flavones	Procyanidins	Catechins	Flavanones
Asparagus	156				
Asparagus, boiled	156				
Broad beans	32		67, 69	67, 69	
Broad beans, boiled	32		67	67	
Broad beans, canned	32				
Broccoli	29, 32, 63, 148, 157				
Broccoli, boiled	157				
Brussels sprouts	29, 32				
Celery stalks	32, 80				
Chilli pepper, green	97, 103				
Chilli pepper, red	97, 103				
Chives	89				
Cress	158				
Endive	32, 105			67, 69	
Green beans	27, 29, 88, 120			67, 69	
Green beans, boiled	120				
Green beans, processed	27, 32			67	
Kale	29, 32, 89				
Kale, processed	32				
Kidney beans, canned				67	
Leeks	29, 32, 89				
Leeks, boiled					
Lettuce	32, 80, 89, 105				
Lettuce, red	32, 105				
Onions, red	29, 54, 79, 148, 159, 160				
Onions, white	79, 160				
Onions, yellow	25, 29, 32, 54, 79, 80, 120, 148, 159, 160				
Fried	80				
Microwaved	80				
Boiled	54, 80				
Red cabbage	32, 89				
Red radish	89				
Shallots	79				
Spring onions	29				
Sweet pepper, green	29, 32	29, 32			
Sweet pepper, red		29, 32			
Sweet pepper, yellow	29, 32, 103	29, 32, 103			
Tomatoes	29, 32, 80, 161, 162				29, 161
Tomatoes, beef	80, 162				
Tomatoes, cherry	80, 162				
Tomatoes, cherry, canned	162				
Tomatoes, plum, canned	162				
Tomatoes, yellow	162				
White cabbage	32				
Watercress	158				
Yellow beans	27				

TABLE 4.11
Flavonoid Content of Beverages

	Flavonols	Flavones	Procyanidins	Catechins	Flavanones
Apple juice	109, 149, 163		109	67, 109, 149	29
Black currant juice fresh	81				
Cranberry juice, canned	117				
Cranberry juice, fresh	117				
Grape juice, black	163			67	
Grape juice, white				67	
Grapefruit juice	163				96, 164, 165
Lemon juice	163, 166				165, 166
Lime juice					165
Orange juice (diluting)	163				164, 169
Orange juice (fresh)	29, 163				29, 168, 169, 170
Pear juice	110		110	110	
Tomato juice	162, 163, 167				167
Tea, black	29, 148, 163, 171, 172		67	67, 69, 148	
Tea, decaffeinated	172				
Tea, Earl Grey	163			67	
Tea, green	148, 163		67	67, 148, 173	
Tea, forest fruit				67	
Tea, freeze dried	172				
Tea, Oolong	163			148	
Coffee	148		67, 69, 174	67, 69, 148	
Cacao drink			69	69	
Chocolate milk	163		67	67	
Beer	163, 166		69	67, 69, 166	
Cider			69	69	
Sherry			175	175, 176	
Wine, red	29, 77, 78, 148, 166, 177, 179		69, 180	67, 69, 148, 166, 178, 179, 181	
Wine, rose			69	69	
Wine, white	163, 166, 177, 178		69	67, 69, 166	166

TABLE 4.12
Miscellaneous Foods

	Flavonols	Flavones	Procyanidins	Catechins	Flavanones
Jam					
Apricot	182		67, 69	67, 69	
Cherry			67, 69	67, 69	
Forest fruit				67	
Honey	59, 60, 90, 183	60, 183			60
Peach	182			67, 69	
Plum	182		69	69	
Raspberry	184		69	69	
Strawberry	65, 182		67, 69	67, 69	
Confectionary					
Chocolate, dark			67, 69, 185	67, 69, 185	
Chocolate, fancy and filled			67	67	
Chocolate, milk			67, 69, 185	67, 69, 185	
Herbs					
Basil, sweet		186			
Coriander	187				
Dill	187				
Lemon balm		187			
Lovage	187				
Marjorum		186			
Mint	187	187			
Oregano		186			
Parsley	29, 148, 187	29, 148, 187			
Rosemary		186, 187			186
Sage		186			
Tarrogon	187	187			
Thyme	187	187			

5 Flavonoids in Wine

Véronique Cheynier

CONTENTS

5.1 INTRODUCTION

Flavonoids (*sensu largo*, i.e., including flavanoids) are important components of grapes and essential to wine quality. They are responsible for the color and astringency of red wines as well as for the yellow hue of oxidized white wines, and are also involved in the development of haze and precipitates, and other technological problems (e.g., clogging of filtration membranes, adsorption on tank surface). They have nutritional and pharmacological properties, as discussed in detail in Chapter 6, and may play a part in the health benefits attributed to moderate wine consumption.

Grape flavonoid composition has been extensively studied since the pioneering work of Ribéreau-Gayon.[1,2] It consists primarily of anthocyanins (in red varieties) and flavanols, along with smaller amounts of flavonols and dihydroflavonols. The main representatives of each of these classes are well known but a few additional minor compounds have been formally identified in recent papers. However, major advances in the last decade concern structural determination of grape proanthocyanidins. Such progress is largely due to the development of new analytical methods, including in particular coupling of high-performance liquid chromatography with mass spectrometric detection (HPLC–MS).

Wine composition depends not only on the type of grape used as raw material, which is influenced by varietal and agricultural factors, but also on the wine-making process, which determines extraction of flavonoids into the liquid phase and their subsequent reactions. The reactions of anthocyanins and proanthocyanidins play a major role in organoleptic changes taking place during wine aging. Conversion of grape anthocyanins to other pigments and its role in color changes, from the purple nuance of young wines toward the red-brown tint of matured wine, was described by Somers in 1971.[3] Also, the decrease of astringency has long been ascribed to reactions of proanthocyanidins, based on their characteristic C–C bond-breaking and bond-making processes and on oxidation mechanisms.[4] However, some of the usually acknowledged reaction products, namely direct anthocyanin–flavanol adducts[3] and adducts in which the anthocyanin and flavanol moieties are linked through a methylmethine bridge, often called ethyl bridge,[5] have only recently been demonstrated to occur in wine,[6,7] whereas others such as the xanthylium ions arising from anthocyanin and flavanol reactions[3] have not yet been confirmed. In addition, several so far unsuspected mechanisms have been unraveled in the last few years and the resulting products have been identified.[6,8–15]

The determination of organoleptic properties of grape and wine flavonoids is another extremely active research area. The color properties of new wine pigments have been studied and compared to those of their anthocyanin precursors.[16–19] The influence of proanthocyanidin structures on their taste has also been investigated. It was thus demonstrated that higher molecular weight proanthocyanidins are both water soluble and highly astringent,[20] ruling out earlier assumptions.[21]

This chapter will provide a detailed account of the newly acquired data on flavonoid composition and distribution in grapes, and on flavonoid reactions in wine and structures of the resulting products. It will first present recent advances in analytical procedures that have rendered such progress possible and, for many of them, have been developed on grape extracts before being more widely applied to other plant material. Finally, it will review briefly current knowledge on the properties associated with flavonoids and their derivatives in wine.

5.2 ANALYTICAL PROCEDURES

5.2.1 EXTRACTION AND FRACTIONATION OF FLAVONOIDS FROM GRAPE AND WINE

Extraction of flavonoids from grapes is classically carried out with organic solvents, starting from fresh, frozen, or freeze-dried material. The most commonly used solvents are methanol, ethanol, and acetone, which can be used pure or mixed with water. Extraction of anthocyanins is commonly achieved at low temperatures with acidified methanol. The use of acid maintains the anthocyanins in the most stable flavylium forms but may cause degradation of acylated anthocyanins. The use of 1% HCl in methanol may induce partial hydrolysis of acylated grape anthocyanins[22] whereas that of 0.1% HCl in methanol causes no degradation.[23] In our experience, 60% acetone in water was as efficient as acidified methanol for the extraction of anthocyanins from grape skins, as reported earlier for strawberry anthocyanins.[24] Proanthocyanidins and especially larger molecular weight polymers are also extracted from grape seeds,[25] skins,[26] and stems[27] with aqueous acetone (60 to 70%). Selective extraction of oligomers can be performed by using ethyl acetate–water (90:10, v/v)[28] but higher yields are obtained with methanol–water (60:40, v/v).[29]

Fractionation of flavonoids from grape extracts or wines can be achieved by liquid–liquid extraction or solid-phase extraction (SPE) procedures. Liquid–liquid extraction using water and diethyl ether enables the recovery of flavanol monomers and flavonol aglycones in the organic phase, whereas anthocyanins, flavonol glycosides, and proanthocyanidins remain in the aqueous phase. Ethyl acetate extracts flavanol monomers and oligomers along with flavonols. SPE on tC18 Sep-Pak cartridges permits similar separations with significant reduction of solvent volumes and of fractionation time.[30] Successive elutions with diethyl ether, ethyl acetate, and methanol allow the separation of flavanol monomers and oligomers that are recovered, respectively, in diethyl ether and ethyl acetate, whereas proanthocyanidin polymers and anthocyanins are eluted together with methanol. However, the application of this procedure to wine flavonoids may lead to contamination of the ethyl acetate phase with *p*-coumaroylated anthocyanins and derived pigments present in red wines.[31] The latter have been separated from genuine anthocyanins by extraction with isoamyl alcohol.[3,7]

Liquid–liquid extraction has been developed further in counter current chromatography (CCC) procedures. CCC has been first used to separate flavanol monomers, dimers, oligomers (i.e., trimers and tetramers), and polymers from white wines using the partition between water and ethyl acetate.[32] Recent refinements of the technique are based on the introduction of a centrifugal force in different types of equipment, namely multilayer coil counter current chromatography (MLCCC), high-speed counter current chromatography, and centrifugal partition chromatography. CCC facilitates the separation of various types of pigments from red wine.[33] The technique has been successfully upgraded for the isolation of anthocyanins at the preparative scale.[34] Its association with gradient elution using increasing percentage of acetonitrile in the organic phase much improved its resolution.[35]

Fractionation of grape seed flavanols on a molecular weight basis has been achieved by low-pressure chromatography on Sephadex LH-20[36] and Fractogel (*syn* Toyopearl) TSK HW-40 (F),[37,38] or TSK HW-50 (F),[26] using methanol to elute monomers and oligomers, up to the tetramers, and acetone–water (70:30, v/v) to recover larger molecular weight procyanidins with no further separation. Elution of procyanidins in increasing molecular weight order indicates that separation of these molecules on Sephadex and Toyopearl phases actually relies primarily upon adsorption rather than exclusion processes. Gel permeation chromatography protocols using dimethylformamide–3 M aqueous ammonium formate (95.5:0.5, v/v) solvent on a polystyrene–divinylbenzene column[39] and acetone–8 M urea (60:40, v/v) adjusted to pH 2 on a Toyopearl column[40] have been proposed but the resolution was very poor when tested on procyanidins from grape seeds. The application of the latter method to red wine allowed the recovery of early eluting pigments that were claimed to be tannin-derived pigments.[41]

Flavanol monomers and oligomers from grape seeds[25,42] and skins[26] have also been separated on a molecular weight basis using normal-phase HPLC methods adapted from a thin layer chromatography (TLC) procedure proposed for apple procyanidins.[43] However, no relationship between the chain length and the retention time could be established when comparing both extracts, due to differences in proanthocyanidin structures.[44] In addition, the presence of other ultraviolet (UV) absorbing material (e.g., anthocyanins, which elute much later than procyanidin dimers under these chromatographic conditions) may lead to misinterpretation of the molecular weight distribution. Finally, proanthocyanidin fractions with different degrees of polymerization were obtained from grape seed and skin extracts by precipitation with chloroform–methanol (75:25, v/v) and gradual dissolution with increasing amounts of methanol in the solvent, but they showed important overlapping.[45,46]

5.2.2 ANALYSIS OF LOWER MOLECULAR WEIGHT FLAVONOIDS

Flavonoids from grape extracts and wines are usually analyzed by reverse-phase HPLC coupled with diode array detection (DAD), enabling the distinction of the various classes of flavonoid compounds on the basis of their characteristic UV–visible spectrum.[47] For instance, grape anthocyanins (structures given in Figure 5.1) present absorption maxima in the range 510 to 528 nm as well as in the UV region (around 280 nm). Anthocyanins having two (cyanidin and peonidin; Figure 5.1, $R_2 = H$) or three (delphinidin, petunidin, malvidin; Figure 5.1, $R_2 = OH$, OCH_3) substituents on the B-ring can be distinguished as the former have absorption maxima about 10 nm lower than the latter in the aglycone ($R = R' = OH$),

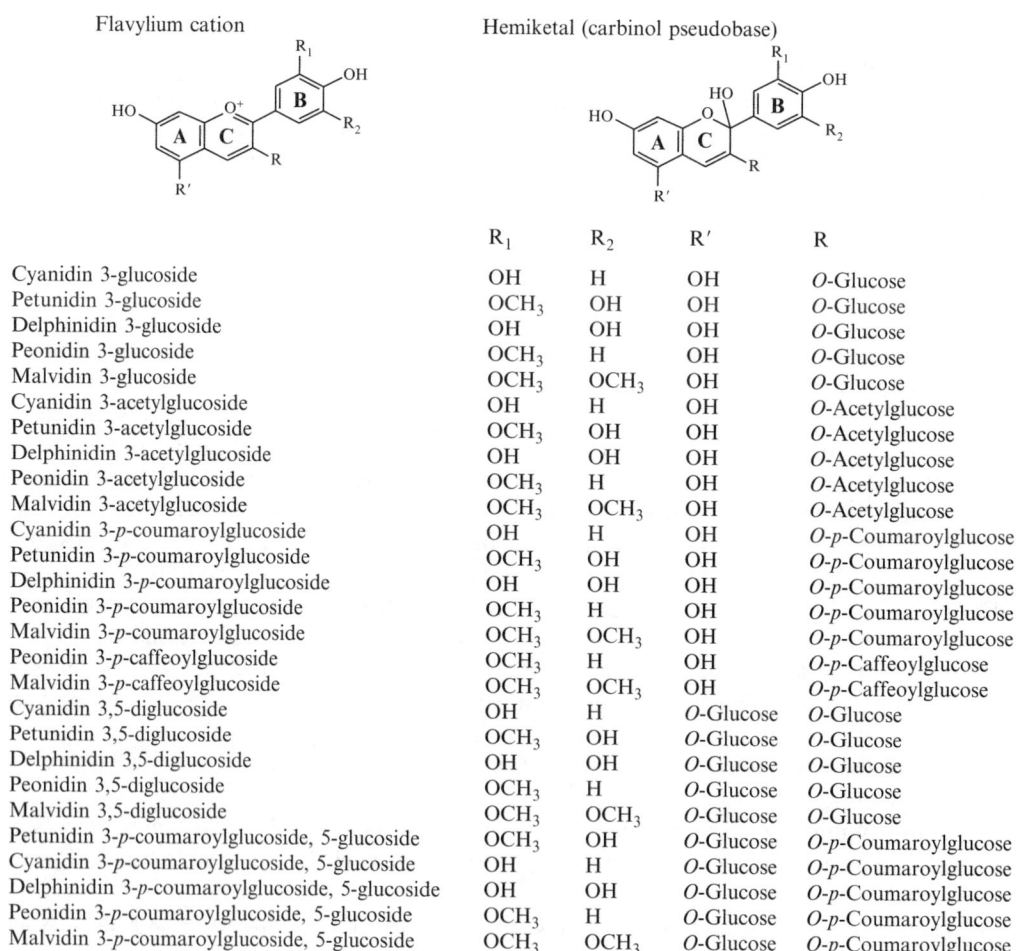

	R_1	R_2	R'	R
Cyanidin 3-glucoside	OH	H	OH	O-Glucose
Petunidin 3-glucoside	OCH$_3$	OH	OH	O-Glucose
Delphinidin 3-glucoside	OH	OH	OH	O-Glucose
Peonidin 3-glucoside	OCH$_3$	H	OH	O-Glucose
Malvidin 3-glucoside	OCH$_3$	OCH$_3$	OH	O-Glucose
Cyanidin 3-acetylglucoside	OH	H	OH	O-Acetylglucose
Petunidin 3-acetylglucoside	OCH$_3$	OH	OH	O-Acetylglucose
Delphinidin 3-acetylglucoside	OH	OH	OH	O-Acetylglucose
Peonidin 3-acetylglucoside	OCH$_3$	H	OH	O-Acetylglucose
Malvidin 3-acetylglucoside	OCH$_3$	OCH$_3$	OH	O-Acetylglucose
Cyanidin 3-p-coumaroylglucoside	OH	H	OH	O-p-Coumaroylglucose
Petunidin 3-p-coumaroylglucoside	OCH$_3$	OH	OH	O-p-Coumaroylglucose
Delphinidin 3-p-coumaroylglucoside	OH	OH	OH	O-p-Coumaroylglucose
Peonidin 3-p-coumaroylglucoside	OCH$_3$	H	OH	O-p-Coumaroylglucose
Malvidin 3-p-coumaroylglucoside	OCH$_3$	OCH$_3$	OH	O-p-Coumaroylglucose
Peonidin 3-p-caffeoylglucoside	OCH$_3$	H	OH	O-p-Caffeoylglucose
Malvidin 3-p-caffeoylglucoside	OCH$_3$	OCH$_3$	OH	O-p-Caffeoylglucose
Cyanidin 3,5-diglucoside	OH	H	O-Glucose	O-Glucose
Petunidin 3,5-diglucoside	OCH$_3$	OH	O-Glucose	O-Glucose
Delphinidin 3,5-diglucoside	OH	OH	O-Glucose	O-Glucose
Peonidin 3,5-diglucoside	OCH$_3$	H	O-Glucose	O-Glucose
Malvidin 3,5-diglucoside	OCH$_3$	OCH$_3$	O-Glucose	O-Glucose
Petunidin 3-p-coumaroylglucoside, 5-glucoside	OCH$_3$	OH	O-Glucose	O-p-Coumaroylglucose
Cyanidin 3-p-coumaroylglucoside, 5-glucoside	OH	H	O-Glucose	O-p-Coumaroylglucose
Delphinidin 3-p-coumaroylglucoside, 5-glucoside	OH	OH	O-Glucose	O-p-Coumaroylglucose
Peonidin 3-p-coumaroylglucoside, 5-glucoside	OCH$_3$	H	O-Glucose	O-p-Coumaroylglucose
Malvidin 3-p-coumaroylglucoside, 5-glucoside	OCH$_3$	OCH$_3$	O-Glucose	O-p-Coumaroylglucose

FIGURE 5.1 Structures of grape anthocyanins.

3-monoglucoside (R = O-glucose, R' = OH), and 3,5-diglucoside (R = R' = O-glucose) series. Glycosylation induces a hypsochromic shift of the absorption maximum while acylation with hydroxycinnamic acids is associated with a characteristic shoulder around 310 nm for p-coumaric acid (R = O-p-coumaroylglucose) and 325 nm for caffeic acid (R = O-caffeoyl-glucose). Coupling of HPLC with MS equipped with mild ionization techniques such as electrospray ionization (ESI-MS) or atmospheric pressure chemical ionization (APCI-MS) has become increasingly popular, as discussed below. Coupling of liquid chromatography with nuclear magnetic resonance (LC–NMR) certainly appears promising,[48] but has not yet to our knowledge been applied to grape or wine flavonoids.

5.2.2.1 HPLC Coupled to Mass Spectrometry

The first application of HPLC–ESI-MS to grape polyphenols was published in 1995.[49] Operation in the positive ion mode made it possible to detect 17 anthocyanins (namely the 3-glucoside, 3-acetylglucoside, and 3-p-coumaroylglucoside of cyanidin, delphinidin, petuni-din, peonidin, and malvidin and the 3-caffeoylglucosides of the last two anthocyanidins) in a grape extract and characterize them through their typical fragmentation patterns obtained after increasing the orifice voltage, corresponding to the loss of sugar or acylated sugar residues. The reported mass signals were improperly interpreted as $[M + H]^+$ ions that could then arise from the anthocyanin quinonoidal bases but in fact corresponded to the flavylium cations (M^+) that predominate under the acidic conditions used in HPLC elution (7% HCOOH).

A year later, the first application of HPLC–ESI-MS analysis to a red wine allowed the detection of two new pigments at $m/z = 609$ and $m/z = 755$.[50] Their fragmentation patterns yielded the same aglycone (at $m/z = 447$) after the loss of glucose (-162) and p-coumaroyl-glucose (-308) moieties, respectively, showing that they were, respectively, the glucoside and p-coumaroylglucoside of the same anthocyanidin cation. Further studies (see Section 5.5.3.3) showed that they were phenylpyranomalvidin 3-glucoside and its p-coumaroylated derivative,[8] confirming this hypothesis. Since then, a number of other anthocyanin-derived pigments have then been similarly detected and characterized in wine (see Section 5.5)[6,7,14,15,51–53] or wine-like model systems.[8,10,13,54–58] Performances of APCI and ESI sources operated in the negative and positive ion modes were compared on a series of standard compounds.[51] The best detection of anthocyanins, as their cationic forms, was achieved with ESI in the positive ion mode, in acidified medium (10% formic acid).

The negative ion mode is usually preferred for uncharged flavonoids, including flava-nols,[27,59–61] flavonols,[62] dihydroflavonols,[27] and flavanol reaction products,[9,11,63,64] which are then detected as the deprotonated $[M – H]^-$ species. It also proved more sensitive for the detection of some new anthocyanin-derived pigments bearing a carboxylic group[10] and for anthocyanin–flavanol adducts in which the anthocyanin moiety is not in the cationic form.[7]

Tandem mass spectrometry (MS–MS) and ion trap mass spectrometry (IT-MS, MS^n), which permit fragmentation patterns to be obtained on selected individual ions, are progres-sively replacing the classical ESI-quadrupole mass spectrometers in HPLC–MS coupling. MS–MS and MS^n have been successfully applied to identify anthocyanins in the juice of concord grape (*Vitis labrusca*),[65] and anthocyanins and flavonol glycosides in table grapes.[62] Fragmentation patterns also provided insight on the sequences of flavanol units in proantho-cyanidin oligomers[66] and of anthocyanin and flavanol units in flavanol–anthocyanin ad-ducts.[58]

Classical fragmentation patterns of flavonoid glycosides, based on the consecutive losses of sugar and acylglycoside residues, enable the determination of the nature of the aglycone and the mass and sequence of its substituents. Fragmentation patterns obtained from

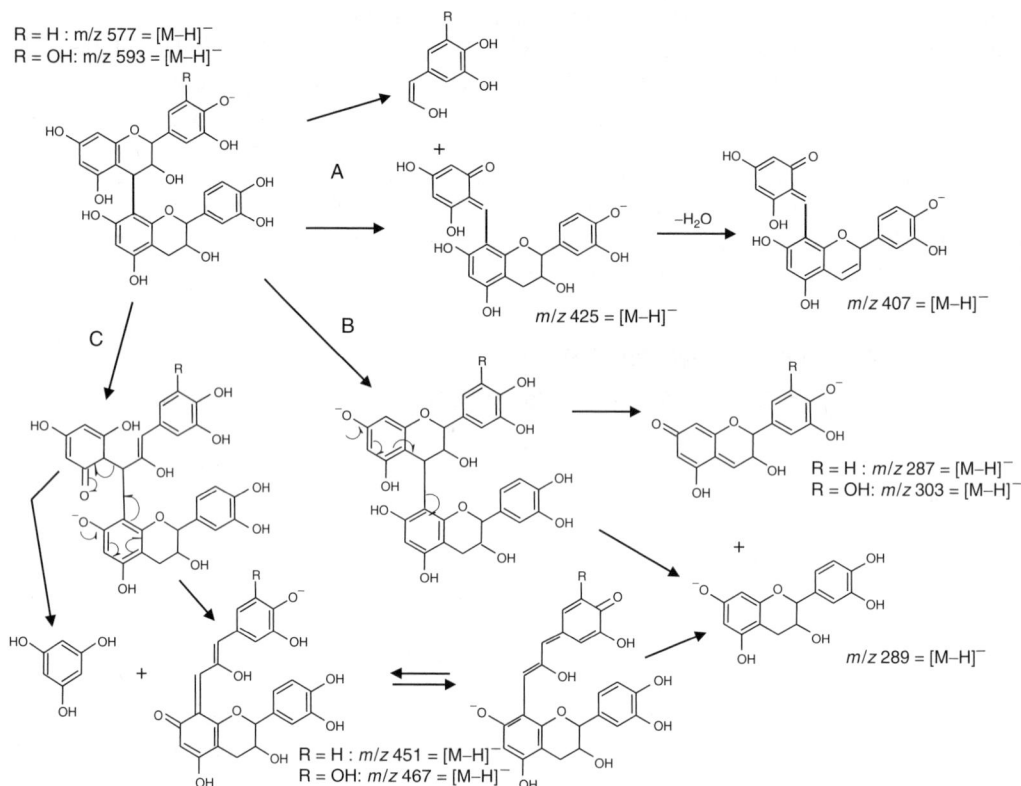

FIGURE 5.2 Mass fragmentation patterns of flavanol oligomers.

proanthocyanidins in the ESI-quadrupole apparatus are the same as in fast atom bombardment mass spectrometry (FAB-MS) operated in the negative ion mode, with one ion peak corresponding to the retro-Diels–Alder (RDA) fission (Figure 5.2, **A**) and two fragments generated by breakage of the interflavanic linkage, following the quinone methide process (Figure 5.2, **B**).[67] Fragmentation experiments performed with an ion trap mass analyzer generate additional fragments corresponding to the loss of a phloroglucinol moiety $(C_6H_6O_3)$ from the upper unit (−126 a.m.u.) (Figure 5.2, **C**).[68]

Analysis of dimeric proanthocyanidins with constitutive units showing different molecular weights demonstrates that RDA fission takes place specifically on the upper unit (substituted only in C4) and thus can be used to determine the sequence.[66–68] Cleavage of the interflavanic linkage also yields different ion species, namely the quinone methide, detected as $[M − 3H]^-$ from the upper unit, and the pseudo-molecular ion $[M − H]^-$ from the lower unit, allowing distinction between both units as illustrated in Figure 5.3. Other useful fragmentations for proanthocyanidin characterization include the loss of galloyl moieties (−152 a.m.u.) in galloylated proanthocyanidins and that of the benzylthioether substituent (−124 a.m.u.) in derivatives obtained after thiolysis (see Section 5.2.2.2).

5.2.2.2   Analysis of Flavonoid Constitutive Moieties

However, MS does not allow to distinguish between isomers such as glucose and galactose in flavonoid glycosides, or (−)-epicatechin and (+)-catechin units in procyanidins. Sugars in glycosides can be identified by using specific glycosidases (e.g., β-glucosidase, α-rhamnosi-

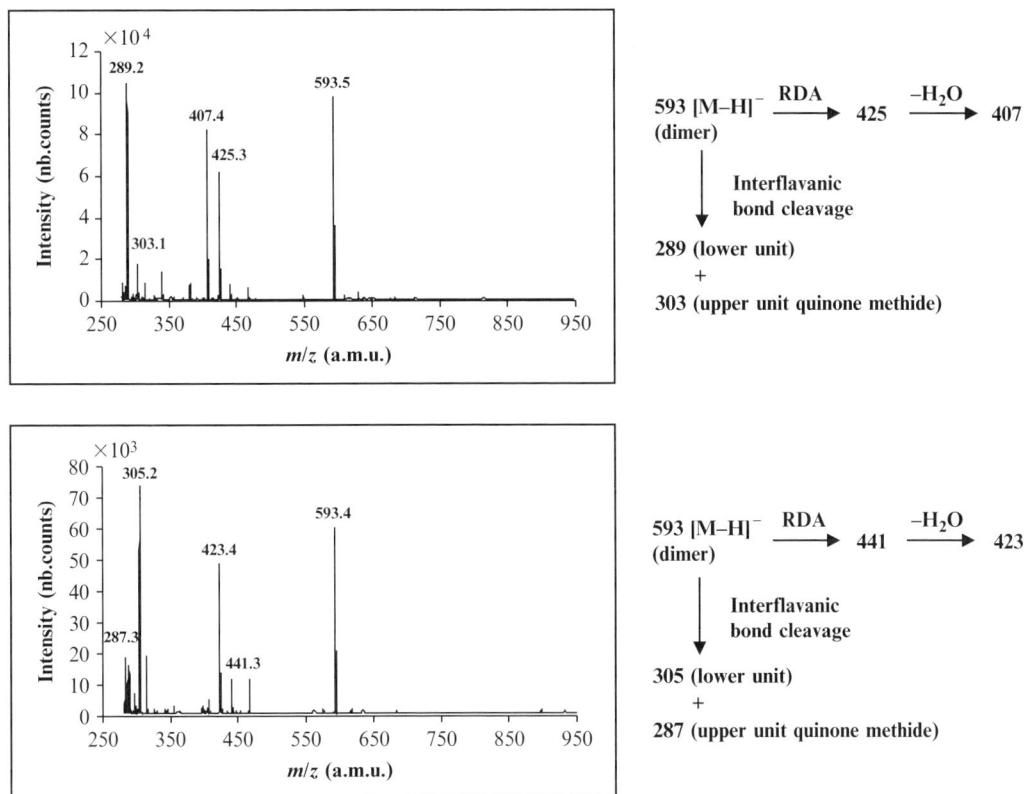

FIGURE 5.3 ESI-Q mass spectra obtained from two proanthocyanidin dimers.

dase) or by analyzing the sugar moiety released after acid-catalyzed cleavage of the glycosidic linkage. This has been achieved by comparison with reference compounds in TLC and more recently by gas chromatography (GC) analysis of the alditol acetate derivatives.[50] Further information on the linkage position on the sugar moiety was obtained by methylation of the sugar-free hydroxyl groups followed by acid hydrolysis, conversion of the partly methylated sugars to alditol acetates, and GC–MS analysis. Identification of *p*-coumaric acid in the *p*-coumaroylated anthocyanin derivative was carried out by alkaline hydrolysis followed by HPLC analysis of the released compounds.[50]

Proanthocyanidin constitutive units can be determined by acid-catalyzed cleavage in the presence of a nucleophilic agent. Commonly used nucleophiles include benzylhydrosulfide (*syn* phenylmethanethiol, toluene-α-thiol), the method then referred to as thiolysis,[69] and 1,3,5-trihydroxybenzene (*syn.* phloroglucinol).[70,71] Rupture of the interflavanoid bond in acidified methanol yields a carbocation from the upper and extension units of the molecule (initially substituted in C4) whereas the lower part (nonsubstituted in C4) is released as such. The carbocation then reacts with the nucleophiles to give a stable adduct. HPLC analysis of the reaction mixture allows the separation of lower units from derivatized units and, within each group, the distinction between isomers on the basis of their retention times.[38,72,73] However, some epimerization, especially from (−)-epicatechin (2,3 *cis*) to its *trans* isomer, may take place when the reaction is carried out at high temperature. In addition, no information is available on the linkage position in dimeric species. Nevertheless, starting from the trimers, cleavage under mild conditions followed by HPLC analysis gives access to constitutive dimers and thus to complete identification, provided that the linkage position in the released dimeric fragments can be unambiguously established.[73] Acylation with gallic acid is maintained under the mild acidic conditions used in thiolysis and phloroglucinolysis so that galloylated units present either in upper or in terminal positions can be determined. Tannase, i.e., galloyl acyl hydrolase, which cleaves the ester bond in galloylated flavanols, can also be used for identification purposes.[38,74]

5.2.2.3 Nuclear Magnetic Resonance

Formal identification of proanthocyanidin dimers, including determination of the linkage position, can best be achieved by using two-dimensional NMR techniques.[75] Proton–proton and proton–carbon correlations enable to distinguish between the different flavanol moieties and to establish their sequence. In particular, attribution of the residual proton of the lower unit A-ring (H6 or H8) is based on long distance carbon–proton correlations established by heteronuclear multiple bond correlation (HMBC) experiments. The principle is as follows: the residual proton correlates with two carbons around 150 ppm, which are C5 and C7 in the case of H6 and C7 and C8a in the case of H8. C8a can be attributed and distinguished from C5 through its correlation with the H2 proton through the oxygen atom of the heterocyclic ring, as illustrated in Figure 5.4. It should, however, be pointed out that H2–H3 coupling constants cannot be relied upon to distinguish between 2–3 *cis* and 2–3 *trans* units within an oligomer.[76] In this case, less sophisticated methods such as thiolysis may prove complementary to NMR to unequivocally establish structures.

The same two-dimensional NMR approach has enabled the identification of tannin-like derivatives.[8–10,77–80] Attribution of C8a, as described above, is not possible in the case of anthocyanins, due to the lack of H2 proton. Nuclear Overhauser effect spectroscopy experiments have been used as an alternative strategy to overcome this problem. For instance, observation of through space couplings between B-ring protons (H2′, H6′, OCH$_3$) and the protons of the methylmethine bond in a methylmethine-linked anthocyanin dimer is only possible in the case of a C8 linkage and thus allows one to rule out the C6-linkage position.[13]

5.2.3 ANALYSIS OF HIGHER MOLECULAR WEIGHT FLAVONOIDS

HPLC separation, as described above, is restricted to rather simple compounds that represent only a small proportion of flavonoids. Indeed, proanthocyanidin analysis becomes increasingly difficult as their molecular weight increases, due to the larger number of possible structures, smaller amounts of each individual compound, and poorer resolution of the chromatographic profiles. This is especially true of grape proanthocyanidins, which, unlike those of apple or cacao consisting only of epicatechin units, are based on four major

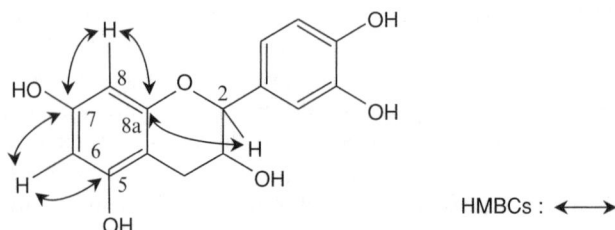

FIGURE 5.4 NMR HMBCs used to attribute flavanol H6 and H8 protons.

constitutive units and are thus extremely heterogeneous. Also, reactions of grape flavonoids in wine yield a large diversity of compounds that appear as an unresolved hump in the chromatographic profiles. To overcome this problem, analyses have been performed directly on grape or wine extracts, without prior HPLC separation.

5.2.3.1 Acid-Catalyzed Cleavage in the Presence of Nucleophilic Agents

The average composition of proanthocyanidin extracts can be determined by HPLC after acid-catalyzed cleavage in the presence of a nucleophilic agent, as described above. The most commonly used nucleophiles are benzylhydrosulfide[7,25,26,81] and phloroglucinol,[82,83] but cysteamine has recently been proposed.[84] Provided each unit can be individually quantified, after isolation and calibration, such methods allow one to determine proanthocyanidin content in a given tannin fraction, by summing the concentrations of all released units, to calculate their mean DP (mDP = ([upper and extension units]$_M$ + [end units]$_M$)/[end units]$_M$) and give insight to the nature and average proportions of their constitutive units, including galloylated units.

Thiolysis also proved useful for the analysis of derived tannins. Methylmethine-linked tannin-like compounds resulting from acetaldehyde-mediated condensation of flavanols (see Section 5.5.3.2) yielded several adducts when submitted to acid-catalyzed cleavage in the presence of ethanethiol, providing information on the position of linkages in the original ethyl-linked compounds.[64,85] Thiolysis of red wine extracts released benzylthioether derivatives of several anthocyanin–flavanol adducts,[7,52] indicating that such structures were initially linked to proanthocyanidins. However, some of the flavonoid derivatives present in wine (e.g., flavanol–anthocyanins[86]) are resistant to thiolysis, while others (e.g., flavanol–ethyl anthocyanins) were only partly cleaved.[87] Thiolysis, thus, appears as a rather simple, sensitive, and powerful tool for quantification and characterization of proanthocyanidins, but provides mostly qualitative data for their reaction products.

5.2.3.2 Mass Spectrometry

Mass spectrometry methods based on soft ionization techniques, ESI[6,59,61,88,89] and matrix-assisted laser desorption ionization/time-of-flight (MALDI-TOF),[90,91] have been successfully applied for the direct analysis of grape and wine extracts and for monitoring flavonoid reactions in model solution studies.[13,59,63] They give access to the molecular weights of the different species present in a fraction or extract and, through fragmentation patterns, provide important information on their constitutive units. Description of the various MS techniques can be found in Chapters 1 and 2.

Analysis of a grape seed procyanidin extract by direct ESI-MS in the negative ion mode[6] enabled to detect signals corresponding to not only $[M - H]^-$ ions of procyanidin oligomers (up to the pentamer) but also to doubly charged ($[M - 2H]^{2-}$) ions for higher molecular weight species (from DP5 to DP9). Arguments to distinguish between moncharged and doubly charged signals rely on the distance between the isotopic peaks due to the natural occurrence of carbon 13 (1%), as the spacing between the isotopic peaks is equal to one mass unit for monocharged ions and 0.5 mass units for doubly charged ions. The interpretation with higher charge states or several charge states in a single mass peak is limited by the resolution of the mass spectrometer. Moncharged signals of DPn species overlap with doubly charged signals of DP2n species but those corresponding to doubly charged ions of species with uneven DP can be unambiguously attributed. ESI-MS analysis of a grape seed extract thus allowed detecting a complete series of oligomers, from DP2 to DP9, as well as the corresponding monogalloylated and digalloylated species. Analysis of grape seed tannins by

ESI-MS in the positive ion mode enabled to detect $[M + H]^+$ ions up to the galloylated pentamer.[61] Analysis of wine proanthocyanidins showed the presence of additional series of species with 16 mass unit differences to signals given by grape seed procyanidins. These were attributed to the presence of (epi)gallocatechin units in mixed procyanidin–prodelphinidin oligomers.[59]

The analysis of proanthocyanidins has also been achieved by MALDI-TOF MS. This technique has a higher mass range than other MS techniques and produces only single ionization, unlike ESI-MS,[91] but it requires incorporation of the sample in a matrix. Only *trans*-3-indolacetic acid (IAA) and 2,5 dihydroxybenzoic acid (DHB) have proven to be suitable matrices for proanthocyanidin analysis. It has been postulated that the intensity of each peak signal in the mass spectrum can be used as a measure of relative abundance but no calibration has been performed, because of the lack of available standards, to confirm this hypothesis. In addition, the charge density decreases as the molecular weight increases so that larger molecular weight tannins are more difficult to detect.[91]

ESI-MS was also applied to investigate acetaldehyde-induced condensation reactions in wine-like model solutions, enabling one to detect ethyl-linked catechin oligomers up to the hexamer as the monocharged $[M - H]^-$ ions[60,92] and signals corresponding to dimeric, trimeric, and tetrameric species containing one anthocyanin and one, two, or three flavanol units as the M^+ flavylium ions.[6,60,93] A tetrameric species containing two malvidin 3-glucoside units and two flavanol units was also found as the doubly charged M^{2+} ion at $m/z = 822$ ions.[6,60] Similarly, ESI-MS analysis in the positive ion mode of a polymeric fraction arising from reaction of malvidin 3-glucoside with acetaldehyde allowed the detection of singly charged and doubly charged ions corresponding to methylmethine-linked malvidin 3-glucoside trimers and tetramers in which the constitutive anthocyanin moieties were present as different forms (i.e., flavylium ions, quinonoidal bases, hemiketal).[13]

Two ESI-MS studies aiming at profiling anthocyanin pigments in grape and wine extracts have been published.[88,94] The first one used ESI-MS^n spectra in the positive ion mode to identify anthocyanins in grape extracts. MS^n was effective for differentiating most isobaric compounds but it did not allow one to distinguish between malvidin 3,5-diglucoside and malvidin 3-caffeoylglucoside, as they are based on the same aglycone. This was achieved by deuterium exchange experiments, leading to different mass shifts in agreement with the respective numbers of exchangeable protons in the molecules.[94] The second study used nanoelectrospray tandem mass spectrometry (nano-ESI-MS–MS) to screen for potential anthocyanin-derived pigments in red wine polyphenol extracts purified by SPE. Replacement of conventional electrospray by nanoelectrospray gives access to greater sensitivity while allowing one to work with smaller sample size. Neutral loss scanning for specific elimination masses was used for detection of particular derivatives (e.g., 162 for glucosyl derivatives).[88] Characteristic elimination masses provided by MS–MS analysis of anthocyanin aglycones served as signatures for the precursor anthocyanidin derivatives.

A recent study aiming at establishing wine fingerprints without any prior purification step compared positive ion and negative ion ESI combined with Fourier transformed (FT) ion cyclotron resonance mass spectrometry.[95] FT-MS provides mass resolving power and accurate mass determination, making it theoretically possible to assign molecular formula unambiguously. The positive ion mass spectra of red wines were dominated by anthocyanins but also showed proanthocyanidin dimers, flavonols, and dihydroflavonols (as their protonated forms and potassium adducts). Most of these compounds were previously known wine constituents but identification of some methoxylated flavanol and flavonol derivatives that have been reported for the first time requires confirmation. The negative ion spectra provided no further information on the flavonoid composition but exhibited greater variations among the wines and thus appear more promising for fingerprinting purposes.

5.3 GRAPE FLAVONOIDS

5.3.1 STRUCTURE AND DISTRIBUTION IN GRAPE

5.3.1.1 Anthocyanins

Grape anthocyanins (Figure 5.1) are based on five anthocyanidins, namely cyanidin, peonidin, petunidin, delphinidin, and malvidin. Those of *Vitis vinifera* are mostly 3-monoglucosides, whereas non-*vinifera* species also contain substantial amounts of 3,5-diglucosides.[1] The 3-acetylglucoside, 3-*p*-coumaroylglucoside, and 3-caffeoylglucoside of these anthocyanidins are also present in most grape varieties. The occurrence of malvidin 3-caffeoylglucoside has been known for many years, but that of other caffeoylglucosides has only recently been confirmed. Peonidin caffeoylglucoside has been detected in *V. vinifera* grapes by HPLC–MS.[49] The caffeoylglucosides of cyanidin, delphinidin, and petunidin are present only in trace amounts, but were identified in fractions obtained by MLCCC separation of a grape skin extract.[35] The 3,5-diglucosides and the 3-(*p*-coumaroylglucoside),5-glucosides of the five anthocyanidins were identified by ESI-MS–MS in Concord (*Vitis labrusca*) grape juice[65] and in hybrid grape varieties.[94] Anthocyanidin 3,5-diglucosides were also detected by MS in a red wine made from *V. vinifera* var. Dornfelder[96] and in grape skin extracts.[35] Additional signals found in the Dornfelder wine after fractionation were attributed to their acetyl and *p*-coumaroyl derivatives, the former reported for the first time in grape.[96]

Anthocyanins are localized in skins, except in a few varieties, referred to as teinturier (e.g., *V. vinifera* var. Alicante Bouschet, var. Gamay Fréaux), that also contain anthocyanins in their pulp. In the skin, they are present in the first external layers of the hypodermal tissue,[97] and exclusively in the vacuoles.[98,99] Light microscope observations have shown the presence of organites containing higher concentrations of anthocyanins in the vacuoles of the inner side of skins from ripe Pinot noir berries.[100] Resonance Raman microscopy spectra indicated that anthocyanins are mainly in the neutral quinonoidal form within these structures whereas they are present as flavylium cations in the outer cell vacuoles.[100] These intravacuolar bodies are similar to those described in red cabbage,[101] and more recently in flower petals.[102] They have been referred to as anthocyanoplasts,[100,101] but this nomenclature is improper as it implies a membrane boundary that could not be detected by electron microscopy.[102,103]

5.3.1.2 Flavanols

Flavan-3-ols are encountered in grape as monomers, oligomers, and polymers. Within the grape berry, they are localized mostly in seeds and skins although trace amounts of monomers and dimers have been detected in pulp, especially in the teinturier variety Alicante Bouchet.[104] Major monomers are (+)-catechin, (−)-epicatechin, and (−)-epicatechin 3-gallate.[105] Gallocatechin[106] in *V. vinifera* and catechin 3-gallate[107] and gallocatechin 3-gallate[108] in non-*vinifera* grapes have also been reported.

Grape seed proanthocyanidins are based on catechin, epicatechin, and epicatechin 3-gallate units and thus partly galloylated procyanidins.[25,38,109] A number of B-type procyanidin dimers and trimers, including some galloylated derivatives, have been identified in grape[38,109] in addition to the C4–C8-linked dimers (B1–B4).[110] The presence of procyanidin A2 has also been reported,[111,112] but its identification was based only on HPLC coelution with procyanidin A2 isolated from horse chestnut and has not been confirmed.[112] Finally, catechin–catechin 3-gallate was identified in non-*vinifera* varieties.[108,113] Additional dimers based on both prodelphinidin and procyanidin units, presumably arising from grape skins, were also found in wine. Gallocatechin–gallocatechin, gallocatechin–catechin, and catechin–gallocatechin were tentatively identified in wine on the basis of their mass spectra and relative retention times in HPLC[66] whereas epigallocatechin–catechin, epicatechin–gallocatechin, and epicatechin–epigallocatechin were characterized by mass fragmentation and thiolysis.[87] Flavanol monomers and oligomers identified in grape and wine extracts are listed in Table 5.1.

These lower molecular weight compounds make up only a relatively small proportion of grape proanthocyanidins, which consist mostly of higher oligomers and polymers, as in most other plant species.[25,26,114,115] Heterogeneity of proanthocyanidins increases with their chain length, due to the diversity of constitutive units, linkage positions, and possible sequences.

TABLE 5.1
Flavanols Identified in Grapes and Wines

	Source	Ref.
Monomers		
(+)-Catechin	Berries	105
(−)-Epicatechin	Berries	105
(−)-Epicatechin gallate	Berries	105
(+)-Catechin gallate	Berries	107
(+)-Gallocatechin	Berries	106
(+)-Gallocatechin gallate	Berries	108
Additional monomeric units in oligomers and polymers		
(−)-Epigallocatechin	Skins, Stems	26
		27
(−)-Epigallocatechin gallate	Skins, Stems	26
		27
Dimers		
(−)-Epicatechin–(4β-8)–(+)-catechin (B1)	Berries	110
(−)-Epicatechin–(4β-8)–(−)-epicatechin (B2)	Berries	110
(+)-Catechin–(4α-8)–(+)-catechin (B3)	Berries	110
(+)-Catechin–(4α-8)–(−)-epicatechin (B4)	Berries	110
(−)-Epicatechin–(4β-6)–(−)-epicatechin (B5)	Seeds	38
(+)-Catechin–(4α-6)–(+)-catechin (B6)	Seeds	38
(−)-Epicatechin–(4β-6)–(+)-catechin (B7)	Seeds	38
(+)-Catechin–(4α-6)–(−)-epicatechin (B8)	Seeds	38
(−)-Epigallocatechin–(+)-catechin	Wine	87
(−)-Epicatechin–(+)-gallocatechin	Wine	87
(−)-Epicatechin–(+)-epigallocatechin	Wine	87
(−)-Epicatechin 3-gallate–(4β-8)–(+)-catechin (B1 3-gallate)	Seeds	38
(−)-Epicatechin 3-gallate–(4β-8)–(−)-epicatechin (B2 3-gallate)	Seeds	38
(−)-Epicatechin–(4β-8)–(−)-epicatechin 3-gallate (B2 3′-gallate)	Seeds	109
(+)-Catechin–(4α-8)–(−)-epicatechin 3-gallate (B4 3-gallate)	Seeds	38
(−)-Epicatechin 3-gallate–(4β-8)–(−)-epicatechin 3-gallate (B2 3,3′-digallate)	Seeds	38
Trimers		
(−)-Epicatechin–(4β-8)–(−)-epicatechin–(4β-8)–(−)-epicatechin (C1)	Wine	32
(−)-Epicatechin–(4β-8)–(−)-epicatechin–(4β-8)–(+)-catechin	Wine	32
(−)-Epicatechin–(4β-8)–(−)-epicatechin–(4β-6)–(+)-catechin	Seeds	38
(−)-Epicatechin–(4β-6)–(−)-epicatechin–(4β-8)–(−)-epicatechin	Seeds	38
(−)-Epicatechin–(4β-8)–(−)-epicatechin–(4β-6)–(−)-epicatechin	Seeds	38
(−)-Epicatechin–(4β-6)–(−)-epicatechin–(4β-8)–(+)-catechin	Seeds	38
(−)-Epicatechin–(4β-8)–(−)-epicatechin 3-gallate–(4β-8)–(+)-catechin	Seeds	38

This results in poorer resolution in all separation methods and renders isolation and formal identification of individual compounds almost impossible beyond the tetramer.

Application of thiolysis to grape proanthocyanidin polymers showed that those extracted from seeds are partly galloylated procyanidins[25] whereas those of skins[26] and stems[27] consist of both procyanidins and prodelphinidins, confirming earlier results obtained by ^{13}C NMR.[116] The major constitutive units of grape skin proanthocyanidins are epicatechin and epigallocatechin. Their 3-gallates are also encountered as extension units whereas catechin and gallocatechin are relatively more abundant in the terminal positions.[26] Much higher average degrees of polymerization were calculated in skins (about 30)[26] than in seeds[25] and stems[27] (around 10). The proportions of galloylated units are also quite different in skins (5%), stems (15%), and seeds (30%).

As discussed above, the development of mild MS techniques has led to further progress in the determination of proanthocyanidin size distribution. In particular, ESI-MS studies have demonstrated that prodelphinidin and procyanidin units coexist within the polymers, where they seem distributed at random.[60] A list of mass signals attributed to proanthocyanidins detected in grape or wine extracts is given in Table 5.2.

Flavanol content is higher in seeds than in skins but the contribution of the latter to the entire grape composition may exceed that of the former in some varieties showing small berry size. Microscopic observations have shown the presence of tannin aggregates in the vacuoles of skin cells. Most epidermal cells but only some hypodermal cells, which are more abundant in the external layers, contain tannins.[117] In addition, tannins associated with vacuole membranes and to the cell walls were described.[118] Whether these associations result from artifacts in preparation or extraction procedures or exist *in planta* is unknown and the structure of tannins present in the vacuole has not been compared to that of tannins linked to membrane material. Nevertheless, monitoring of tannin composition during grape maturation has shown that thiolysis yields gradually decrease as the berry ripens.[119] Analysis of flavonoid extracts from unripe and ripe red berries by ^{13}C NMR and ESI-MS suggested that proanthocyanidins in the ripe berry skins are associated with some polysaccharides and also with small amounts of anthocyanins.

5.3.1.3 Flavonols, Flavones, and Dihydroflavonols

Flavonols are found in grape skins as 3-glycosides of quercetin, kampferol, isorhamnetin, and myricetin.[2,120,121] Small amounts of the corresponding aglycones have also been reported. The major flavonols in grape are quercetin 3-glucoside and quercetin 3-glucuronide.[2,121–123] Other grape flavonols include myricetin 3-glucoside, kampferol 3-glucoside,[2] isorhamnetin glucoside, myricetin glucuronide, kampferol glucuronide, kampferol galactoside, kampferol glucosylarabinoside, quercetin glucosylgalactoside, quercetin glucosylxyloside,[121] and quercetin 3-rhamnosylglucoside (i.e., rutin).[120,124] Rutin identification was initially based only on comparison of HPLC retention time and UV–visible spectrum with those of a commercial standard and has therefore been considered questionable but it has recently been confirmed by MS–MS analysis in the skins of table grape[62] and by ^{1}H NMR on the compound isolated from vine leaves.[125] The latter authors have also identified quercetin 3-galactosylrhamnoside and two flavone glycosides, namely luteolin 7-glucoside and apigenin 7-glucoside in vine leaves. Finally, dihydroquercetin 3-rhamnoside (astilbin) and dihydrokampferol 3-rhamnoside (engeletin) have been identified in grape skins[126] and stems.[27]

5.3.2 Impact of Varietal and Environmental Factors on Grape Flavonoids

Anthocyanin composition varies greatly within species and cultivars. The total content (from 500 mg/kg up to 3 g/kg) and the proportions of each anthocyanin are varietal characteristics

TABLE 5.2
Proanthocyanidin Signals Detected in Grape and Wine Extracts by ESI-MS in the Negative Ion Mode

DP	(epi)Catechin	Galloyl	(epi)Gallocatechin	Mass Signals	Ref.
2	2			$[M - H]^- = 577$	6
2	2	1		$[M - H]^- = 729$	6
2	1		1	$[M - H]^- = 593$	60
2			2	$[M - H]^- = 609$	60
3	3			$[M - H]^- = 865$	6
3	3	1		$[M - H]^- = 1017$	6
3	2		1	$[M - H]^- = 881$	60
3	1		2	$[M - H]^- = 897$	60
4	4			$[M - H]^- = 1153$	6
4	4	1		$[M - H]^- = 1305$	6
4	3		1	$[M - H]^- = 1169$	60
4	3	1	1	$[M - H]^- = 1321$	60
4	2		2	$[M - H]^- = 1185$	60
4	1		3	$[M - H]^- = 1201$	60
4			4	$[M - H]^- = 1217$	60
5	5			$[M - H]^- = 1441, [M - 2H]^{2-} = 720$	6
5	4		1	$[M - H]^- = 1457, [M - 2H]^{2-} = 728$	60
5	3		2	$[M - H]^- = 1473, [M - 2H]^{2-} = 736$	60
5	2		3	$[M - H]^- = 1489$	60
5	5	1		$[M - H]^- = 1595$	87
				$[M - 2H]^{2-} = 796$	6
5	5	2		$[M - 2H]^{2-} = 873$	6
6	6			$[M - H]^- = 1731$	87
6	5		1	$[M - H]^- = 1747$	87
6	4		2	$[M - H]^- = 1763$	87
6	6	1		$[M - 2H]^{2-} = 940$	6
6	6	3		$[M - 2H]^{2-} = 1093$	6
7	7			$[M - 2H]^{2-} = 1008$	6
7	7	1		$[M - 2H]^{2-} = 1084$	6
7	7	2		$[M - 2H]^{2-} = 1161$	6
7	7	3		$[M - 2H]^{2-} = 1237$	6
8	8	1		$[M - 2H]^{2-} = 1229$	6
8	8	3		$[M - 2H]^{2-} = 1381$	6
9	9			$[M - 2H]^{2-} = 1297$	6
9	9	1		$[M - 2H]^{2-} = 1373$	6
9	9	2		$[M - 2H]^{2-} = 1449$	6

and may be used to some extent as chemotaxonomy criteria.[127,128] For instance, Pinot noir grapes contain no acylated anthocyanin. Detection of anthocyanin 3,5-diglucosides in wine is considered to attest the use of non-*vinifera* varieties in wine but improvement of the analytical methods has enabled to demonstrate that anthocyanin 3,5-diglucosides are also present in small amounts in *V. vinifera* varieties. Red and white varieties also present large differences in flavonol composition. Among some studied wine cultivars,[129] quercetin derivatives always predominate but myricetin derivatives and isorhamnetin 3-glucoside appeared restricted to red cultivars. No myricetin derivative was detected in a series of table grape varieties while isorhamnetin glucoside was found only in the white Superior seedless and Moscatel

Italia cultivars.[62] Grape proanthocyanidin composition varies only slightly between varieties and no particular pattern was observed for red and white cultivars. The data are still too scarce to conclude whether it can serve for authentication purposes.

Grape flavonoid composition also varies with environmental conditions[122,130,131] and changes throughout ripening. Anthocyanin biosynthesis starts rather late in grape maturation and shows maximum activity in the first weeks immediately after veraison (i.e., the beginning of berry softening and color change). Anthocyanin accumulation is regulated by both light exposure and temperature.[132] It usually continues until harvest but a drop may be observed in the latest phase of maturation, especially in hot climates.[132,133]

Treatment of grape berries with 2-phloroethylphosphonic acid, a compound known to enhance ethylene production, increased expression of genes encoding for major enzymes in the anthocyanin biosynthetic pathway, namely chalcone synthase (CHS), flavanone 3-hydroxylase (F3H), leucoanthocyanin dioxygenase (LDOX), and UDPglucose-flavonoid 3-O-glucosyltrasferase (UFGT), and anthocyanin accumulation.[134] The transcript accumulation was sustained for 20 days after veraison while anthocyanin levels determined by HPLC analysis decreased, suggesting that modulation of anthocyanins may involve mechanisms other than transcriptional. In addition, red pigments determined by measuring absorbance values at 520 nm of acidified methanolic extracts increased over the same period of time. This may mean that anthocyanins were, in fact, converted to other pigment species that were not taken into account in the HPLC analysis but contributed to absorbance in the red region. Incorporation of anthocyanins in polymeric structures has been postulated to take place during grape maturation.[135] Anthocyanin oligomers were recently detected in grape skin extract by ESI-MS.[136]

The concentration of proanthocyanidins is highest a few weeks after veraison both in skins and in seeds.[83] Accumulation of flavanol monomers is concomitant with that of proanthocyanidins in seeds. In skins, their maximum concentration is reached earlier and begins to decrease as proanthocyanidins accumulate. The decrease in proanthocyanidin concentration observed in the latest stages of grape maturation has been attributed to reduced extractability resulting from interactions with other cell constituents such as polysaccharides or proteins.[83,119] Results regarding changes in proanthocyanidin composition during ripening appear contradictory: the average chain length has been reported to decrease,[109] increase,[119] or remain constant[83,137] from veraison to harvest, depending on the variety.[138] A temporary increase in epigallocatechin extension units was also observed in the skins of the red variety Shiraz immediately before veraison. Since these units are trihydroxylated on the B-ring like the majority of grape anthocyanins, a coordinated process between tannin and anthocyanin accumulation was suggested.[83]

Flavonol synthesis occurs in two main periods; the first one around flowering and the second after the main period of anthocyanin biosynthesis.[123] In the latter phase, flavonol accumulation is highly dependent on environmental factors and, in particular, much increased by sun exposure of the berries.[122,132]

Based on the currently available data, grape anthocyanin and flavonol profiles are determined by genetic characteristics whereas their content varies with the vine growing conditions. Available information still appears too limited and contradictory to draw any conclusion on the impact of genetic and environmental factors on grape proanthocyanidin composition.

5.4 EXTRACTION OF GRAPE FLAVONOIDS INTO THE WINE

Since flavonoids are localized almost exclusively in the solid parts of the cluster (skins, seeds, stems), their transfer into the must and wine is primarily determined by the extent of maceration allowed in the wine-making process. Thus, white wines are usually obtained by direct pressing of red or white grapes whereas fermentation on skins, enabling extraction of anthocyanins from these tissues, is required to make red wine. Diffusion starts after crushing of the grape and continues until the wine is separated from the solid residue (marc) by racking or pressing. Diffusion kinetics depends both on the solubility of the compounds and on their localization in the berry. It is further modulated by other factors, including the concentration of alcohol and sulfur dioxide in the liquid phase, the temperature, and the extent of must homogenization.

White musts and wines made without maceration contain very low amounts of flavonoids. However, when making white wine from white grapes, skin contact at low temperature is sometimes performed before pressing and fermentation to increase extraction of volatile compounds and aroma precursors. After 4 h of skin contact, the concentration of flavanol monomers and dimers in must was increased threefold.[139] Delays between harvest and pressing, especially if sulfur dioxide is added to prevent oxidation, as well as thorough pressing, similarly result in increased concentrations of flavonoids in white musts and wines.[140,141]

In red wine making, anthocyanins diffuse rapidly. Their concentration reaches a maximum after a few days and then steadily decreases as they are involved in various reactions.[115] Monitoring of proanthocyanidin composition in the fermenting musts by HPLC after thiolysis demonstrated that proanthocyanidins from skins diffuse faster than those from seeds, due either to their localization or to the higher water solubility of prodelphinidins compared to galloylated procyanidins.[115] Extraction of proanthocyanidins from seeds starts after a lag phase, when the level of alcohol increases.[142] Polymers with highest molecular weight also diffuse slower than those of lower molecular weight so that the average chain length gradually increases.[115]

Maceration at low temperature before fermentation starts enhanced extraction of anthocyanins and proanthocyanidins from skins whereas postfermentation maceration increased that of procyanidins from seeds. The levels of anthocyanins and proanthocyanidins recovered in wine at the end of fermentation represent about 40 and 20% of their amounts in grape, respectively.[143] Anthocyanin recoveries were lower after 3-week additional maceration, representing only 20% of the grape initial content. Extraction of the pomace hardly increased the recovery yield, indicating that a major proportion of flavonoids had been converted to other species or had been irreversibly adsorbed on the solid material during fermentation. This provides good evidence that anthocyanin and tannin reactions that have been reported to take place during aging actually start very early in the wine-making process. Degradation of anthocyanins and tannins is even faster when the wine is made with carbonic maceration, a process consisting in maintaining the grape under a carbon dioxide atmosphere for a few days as performed, for instance, in Beaujolais.[142]

5.5 REACTIONS OF FLAVONOIDS IN MUSTS AND WINES

Changes in flavonoid composition involve both enzymatic and chemical processes. The former is restricted to the early stages of wine-making whereas the latter rapidly becomes prevalent as the enzymes become inactivated, and continues throughout aging. Whether they are biochemical or chemical, these processes rely primarily upon the reactivity of phenolic compounds, which is based on the reactivity of the phenol hydroxyl group itself but can be modulated by the presence of substituents. Additional reactions involve substituents or substitution bonds (e.g., enzymatic or acid-catalyzed hydrolysis of the glycosidic or ester linkages). A list of flavonoid derivatives formed by these reactions processes identified in wine or in wine-like model solutions is given in Table 5.3.

TABLE 5.3
Flavonoid Reaction Products Formed in Wine and Wine-Like Model Systems

Compound	Source	m/z (Parent and Fragment Ions)	λ_{max}	Identification Methods	Ref.
Anthocyanin reaction products					
Caffeoyltartaric acid–malvidin 3-*O*-glc adducts (e.g., **5 1**, Figure 5.4)					
Flavylium form (five isomers)	Model solution	801 ([M − 2H]⁻)/803 ([M]⁺)	285, 327 (sh), 540	HPLC-DAD–ESI-MS	55
Hemiketal form (six isomers)	Model solution	819 ([M − H]⁻)	285, 325	HPLC-DAD–ESI-MS	55
Flavylium form (four isomers)	Oxidized grape (gamay beaujolais)	803 ([M]⁺)		HPLC-DAD–ESI-MS	142
Caffeoyltartaric acid–peonidin 3-*O*-glc adducts (e.g., **5 2**, Figure 5.4)					
Flavylium form (four isomers)	Oxidized grape (gamay beaujolais)	773 ([M]⁺)	285, 327 (sh), 540	HPLC-DAD–ESI-MS	142
F–A⁺ or A⁺–F adducts					
(epi)Catechin–malvidin 3-*O*-glc	Red wines, model solutions	781 ([M]⁺; 331)	280, 531	HPLC-DAD–ESI-MS	189
(epi)Catechin–peonidin 3-*O*-glc	Red wine extract[a]	767 ([M]⁺)		HPLC-ESI-Q-MS	87
(epi)Catechin–petunidin 3-*O*-glc	Red wine extract[a]	751 ([M]⁺)		HPLC-ESI-Q-MS	87
(epi)Catechin–malvidin 3-*O*-glc	Red wine extract[a]	781 ([M]⁺; 619)		HPLC-ESI-Q-MS	87
(epi)Catechin–malvidin 3-*O*-(6-*O*-acetyl)glc	Red wine extract[a]	823 ([M]⁺)		HPLC-ESI-Q-MS	87
(epi)Catechin–malvidin 3-*O*-(6-*O*-*p*-coumaroyl)glc	Red wine extract[a]	927 ([M]⁺)		HPLC-ESI-Q-MS	87
(epi)Catechin–delphinidin 3-*O*-glc	Red wine extract[a]	781 ([M]⁺)		HPLC-ESI-Q-MS	96
(epi)Catechin–cyanidin 3-*O*-glc	Red wine extract[a]	765 ([M]⁺)			96
(epi)Catechin–petunidin 3-*O*-glc	Red wine extract[a]	7795 ([M]⁺)			96
(epi)Catechin–peonidin 3-*O*-glc	Red wine extract[a]	779 ([M]⁺)			96
F–A⁺ adducts					
(epi)Catechin–malvidin 3-*O*-glc (two isomers)	Red wine extract[b]	781 ([M]⁺; 763, 619, 601, 493, 467)	282, 535	HPLC-DAD–ESI-MS	86
(epi)Catechin–malvidin 3-*O*-glc (two isomers)	Model solution	781 ([M]⁺; 763, 619, 601, 493, 467)	282, 353, 535	HPLC-DAD–ESI-MS	58, 86
Epicatechin–malvidin 3-*O*-glc (**5 3**)	Synthesis	781 ([M]⁺; 763, 619, 601, 493, 467)		HPLC-ESI-IT-MS	86

continued

TABLE 5.3
Flavonoid Reaction Products Formed in Wine and Wine-Like Model Systems — continued

Compound	Source	m/z (Parent and Fragment Ions)	λ$_{max}$	Identification Methods	Ref.
Catechin–malvidin 3-O-glc (two isomers) (5 4)	Synthesis	781 ([M]$^+$; 763, 619, 601, 493, 467)		HPLC–ESI-IT-MS	86
Epicatechin–epicatechin–malvidin 3-O-glc (5 5)	Model solution	1069 ([M]$^+$)		HPLC–ESI-IT-MS	58
A$^+$–F adducts					
Malvidin 3-O-glc–epicatechin (4–8)-epicatechin 3'-O-gallate (5 6)	Model solution	1221 ([M + H]$^+$; 1203, 1059, 1041, 907, 755, 617)			58
Epicatechin (4–8)-(malvidin 3-O-glc (4–6)-epicatechin 3'-O-gallate) (5 7)		1221 ([M + H]$^+$; 1203, 1059, 1041, 771)		HPLC–ESI-IT-MS	58
A-Type anthocyanin–flavanol adducts					
A-Type malvidin 3-glc-(epi)catechin (two isomers)	Red wine extract[a]	783 ([M + H]$^+$; 621, 631, 469)/781 ([M − H]$^−$		HPLC–ESI-Q-MS	7
A-Type malvidin 3-glc-catechin (two isomers)	Model solution	783 ([M + H]$^+$; 621, 631, 469)		HPLC–ESI-Q-MS	7
Malvidin 3,5-di-glc-(2-O-7, 4–8)-catechin (5 8)	Model solution			^{1}H, ^{13}C NMR	179
Malvidin 3-glc-(2-O-7, 4–8)-epicatechin (5 9)	Model solution	783 ([M + H]$^+$; 765, 747, 657, 631, 621, 495, 469, 451)/781 ([M − H]$^−$; 655, 629, 601, 583, 493)		HPLC–ESI-IT-MS, NMR	193
A-Type petunidin 3-glc-(epi)catechin (5 10)	Red wine extract[a]	769 ([M + H]$^+$)		HPLC–ESI-Q-MS	87
A-Type malvidin 3-acetylglc-(epi)catechin (5 11)	Red wine extract[a]	825 ([M + H]$^+$)		HPLC–ESI-Q-MS	87
A-Type malvidin 3-glc-procyanidin dimer (5 12)	Red wine extract[a]	1071 ([M + H]$^+$)		HPLC–ESI-Q-MS	87
Benzylthioether of A-type malvidin 3-glc-(epi)catechin (5 13)	Red wine extract[a] after thiolysis	903 ([M − H]$^−$; 779, 491, 287)		HPLC–ESI-Q-MS	87
Anthocyanin dimers					
Malvidin 3-glc dimer (A$^+$–AOH)	Model solution	1071 ([M + H]$^+$)		HPLC–ESI-IT-MS	58
Malvidin–malvidin 3-glc (A$^+$–AOH)	Model solution	839 ([M + H]$^+$)		HPLC–ESI-IT-MS	58
Flavanol oxidation products					
B-Type dehydrodicatechins					
Catechin-(6–8)-catechin	Model solutions	577 ([M − H]$^−$)		FAB-MS, NMR	171
Catechin-(3'-O-8 or 4'-O-8)-catechin (two isomers)	Model solutions	577 ([M − H]$^−$)		FAB-MS, NMR	171
B-Type dehydrodicatechins	Model solutions	577 ([M − H]$^−$; 559, 533, 439, 425, 393)		ESI-MS–MS	172
A-Type dehydrodicatechins (two isomers)		575 ([M − H]$^−$)		FAB-MS, NMR	171

Methylmethine-bridged adducts

Compound	Source	m/z	λ_{max} (nm)	Method	Ref.
Epicatechin–CH–CH₃–malvidin 3-glc (two isomers)	Model solutions	809 ($[M]^+$)	280, 536, 280, 542	HPLC-DAD, FAB-MS	196
Catechin–CH–CH₃–malvidin 3-glc (two isomers)	Model solutions	809 ($[M]^+$)	280, 540	HPLC-DAD, FAB-MS	187
(epi)Catechin–CH–CH₃–malvidin 3-glc	Red wine	809 ($[M]^+$)		HPLC-DAD–ESI-MS	6
Epicatechin–(8-(CH–CH₃)-8)–malvidin 3-glc (**5 14**)	Model solutions	809 ($[M]^+$)	280, 540	HPLC-DAD–ESI-MS, NMR	199
Catechin–(8-(CH–CH₃)-8)–malvidin 3-glc (**5 15**)	Red wine^b	781 ($[M]^+$)		HPLC–ESI-MS	52
(epi)Catechin–(CH–CH₃)–delphinidin 3-glc (**5 16**)	Red wine^b	795 ($[M]^+$)		HPLC–ESI-MS	52
(epi)Catechin–(CH–CH₃)–petunidin 3-glc (**5 17**)	Red wine^b	779 ($[M]^+$)		HPLC–ESI-MS	52
(epi)Catechin–(CH–CH₃)–peonidin 3-glc (**5 18**)	Red wine^b			HPLC–ESI-MS	52
(epi) (Catechin)$_2$–(CH–CH₃)–malvidin 3-glc (**5 19**)	Red wine^b	1097 ($[M]^+$)		HPLC–ESI-MS	52
Malvidin 3-glc–(CH–CH₃–epicatechin)$_2$	Model solutions	1125 ($[M]^+$)		HPLC–ESI-MS	56
(CH–CH₃–Malvidin 3-glc)$_2$–epicatechin	Model solutions	664 ($[M]^{2+}$)		HPLC–ESI-MS	56
Malvidin 3-glc–(CH–CH₃–epicatechin)$_3$	Model solutions	1441 ($[M]^+$)		HPLC–ESI-MS	56
Malvidin 3-glc–(CH–CH₃–epicatechin)$_2$–CH–CH₃–malvidin 3-glc	Model solutions	822 ($[M]^{2+}$)		HPLC–ESI-MS	56
(epi)Catechin–CH–CH₃–(epi)catechin	Model solutions	605 ($[M-H]^-$)		HPLC–ESI-MS	92
Epicatechin–CH–CH₃–epicatechin (four isomers : 6–6, 8–8, 6–8 R, S)	Model solutions	607 ($[M+H]^+$)		HPLC–ESI-MS	56
Catechin–CH–CH₃–catechin (four isomers : 6–6, 8–8, 6–8 R, S)	Model solutions	605 ($[M-H]^-$)		NMR	77
(epi)Catechin–(CH–CH₃)–(epi)catechin	Red wine	605 ($[M-H]^-$)		HPLC–ESI-MS	6, 11
(epi)Catechin–(CH–CH₃)–(epi)catechin)$_2$	Model solutions	921 ($[M-H]^-$)		HPLC–ESI-MS	92
(epi)Catechin–(CH–CH₃)–(epi)catechin)$_2$	Red wine	921 ($[M-H]^-$)		HPLC–ESI-MS	6
(epi)Catechin–(CH–CH₃)–(epi)catechin)$_3$	Model solutions	1237 ($[M-H]^-$)		HPLC–ESI-MS	92
(epi)Catechin–(CH–CH₃)–(epi)catechin)$_4$	Model solutions	1553 ($[M-H]^-$)		HPLC–ESI-MS	92
(epi)Catechin–(CH–CH₃)–(epi)catechin)$_5$	Model solutions	1869 ($[M-H]^-$)		HPLC–ESI-MS	92
Malvidin 3-glc–(8-(CH–CH₃)-8)–malvidin 3-glc (**5 20**)	Model solution, wine	1029 ($[M]^+$) : A^+–AOH 1011 ($[M]^+$) : A^+–A 506 ($[M]^{2+}$) : A^+–A^+	280, 460 (sh), 536	HPLC-DAD–ESI-MS	13

continued

TABLE 5.3
Flavonoid Reaction Products Formed in Wine and Wine-Like Model Systems — *continued*

Compound	Source	m/z (Parent and Fragment Ions)	λ_{max}	Identification Methods	Ref.
Malvidin 3-glc–(CH–CH$_3$–malvidin 3-glc)$_2$	Model solution	1529 ([M]$^+$) : A$^+$–A$_2$ 765 ([M]$^{2+}$) : (A$^+$)$_2$–A		HPLC–ESI-MS	13
Malvidin 3-glc–(CH–CH$_3$–malvidin 3-glc)$_3$	Model solution	683 ([M]$^{3+}$) : (A$^+$)$_3$–A 1024 ([M]$^{2+}$) : (A$^+$)$_2$–A$_2$ 1042 ([M]$^{2+}$) : (A$^+$)$_2$–AOH$_2$		HPLC–ESI-MS	13, 19
Malvidin 3-glc–(CH–CH$_3$–malvidin 3-glc)$_4$	Model solution	865 ([M]$^{3+}$) : (A$^+$)$_3$–A$_2$ 1283 ([M]$^{2+}$) : (A$^+$)$_2$–A$_3$		HPLC–ESI-MS	19
Pyranoanthocyanins					
Pyranomalvidin 3-glc (**5 21**)	Wine	517 ([M]$^+$; 355)	270, 355, 498	FAB-MS, NMR	17
	Model solution		280–490	HPLC–DAD–ESI-MS	57, 149
Pyranopetunidin 3-glc	Model solution	503 ([M]$^+$)		HPLC–DAD–ESI-MS	57
Pyranopeonidin 3-glc	Model solution	487 ([M]$^+$)		HPLC–DAD–ESI-MS	57
Pyranocyanidin 3-glc	Model solution	473 ([M]$^+$)		HPLC–DAD–ESI-MS	57
Pyranodelphinidin 3-glc	Model solution	489 ([M]$^+$)		HPLC–DAD–ESI-MS	57
Pyranomalvidin 3-acetylglc (**5 22**)	Wine	559 ([M]$^+$; 355)	270, 355, 503	FAB-MS, NMR	17
(epi)Catechin pyranomalvidin 3-glc	Wine, marc	805 ([M]$^+$)		HPLC–ESI-MS	202
8-Catechin pyranomalvidin 3-glc (**5 23**)	Port wine	805 ([M]$^+$; 643, 491)	503	HPLC–DAD–ESI-MS, NMR	203
8-Epicatechin pyranomalvidin 3-glc (**5 24**)	Port wine	805 ([M]$^+$; 643, 491)	503	HPLC–DAD–ESI-MS, NMR	203
(epi)Catechin pyranomalvidin 3-acetylglc	Marc	847 ([M]$^+$)		HPLC–ESI-MS	202
(epi)Catechin pyranomalvidin 3-*p*-coumaroylglucoside	Wine, marc	951 ([M]$^+$)		HPLC–ESI-MS	202
8-Catechin pyranomalvidin 3-*p*-coumaroylglc (**5 26**)	Port wine	951 ([M]$^+$; 643)	280, 313, 503	HPLC–DAD–ESI-MS, NMR	53
8-Epicatechin pyranomalvidin 3-*p*-coumaroylglc (**5 27**)	Port wine	951 ([M]$^+$; 643)	280, 313, 503	HPLC–DAD–ESI-MS, NMR	53
Di(epi)catechin pyranomalvidin 3-glucoside	Wine, marc	1093 ([M]$^+$)		HPLC–ESI-MS	202
Epicatechin–(4→8)-epicatechin pyranomalvidin 3-glc	Model solution	1093 ([M]$^+$)		HPLC–ESI-MS	54
8-Catechin-(4→8)-catechin pyranomalvidin 3-glc (**5 25**)	Port wine	1093 ([M]$^+$)	520	HPLC–DAD–ESI-MS, NMR	203
(Epi)catechin)$_2$ pyranomalvidin 3-acetylglc	Marc	1135 ([M]$^+$)		HPLC–ESI-MS	202
(Epi)catechin)$_2$ pyranomalvidin 3-*p*-coumaroylglc	Marc	1239 ([M]$^+$)		HPLC–ESI-MS	202

Compound	Source	m/z	λmax	Technique	Ref.
8-Epicatechin-(4-8)-catechin-pyranomalvidin 3-p-coumaroylglc (5 28)	Port wine	1239 ([M]$^+$; 931)	280, 313, 512	HPLC-DAD-ESI-MS, NMR	53
((Epi)catechin)$_3$ pyranomalvidin 3-glc	Marc	1381 ([M]$^+$)		HPLC-ESI-MS	202
((Epi)catechin)$_3$ pyranomalvidin 3-acetylglc	Marc	1423 ([M]$^+$)		HPLC-ESI-MS	202
((Epi)catechin)$_3$ pyranomalvidin 3-p-coumaroylglc	Marc	1527 ([M]$^+$)		HPLC-ESI-MS	202
((Epi)catechin)$_4$ pyranomalvidin 3-glc	Marc	1669 ([M]$^+$)		HPLC-ESI-MS	202
4-Hydroxyphenyl pyranomalvidin 3-glc (5 29)	Wine, model solution	609 ([M]$^+$; 447)	280, 507	HPLC-DAD-MS / ^{1}H NMR	50 / 8
4-Hydroxyphenyl pyranomalvidin 3-p-coumaroylglc (5 30)	Wine, model solution	755 ([M]$^+$; 447)	280, 313 (sh), 507	HPLC-DAD-MS / ^{1}H NMR	50 / 8
4-Hydroxyphenyl pyranopetunidin 3-glc (5 31)	Wine	595 ([M]$^+$; 433)		ESI-MS-MS	88
4-Hydroxyphenyl pyranopeonidin 3-glc (5 32)	Wine	579 ([M]$^+$; 417)		ESI-MS-MS	88
4-Hydroxyphenyl pyranopetunidin 3-acetylglc (5 33)	Wine	637 ([M]$^+$; 433)		ESI-MS-MS	88
4-Hydroxyphenyl pyranopeonidin 3-acetylglc (5 34)	Wine	621 ([M]$^+$; 417)		ESI-MS-MS	88
4-Hydroxyphenyl pyranomalvidin 3-acetylglc (5 35)	Wine	651 ([M]$^+$; 447)		ESI-MS-MS	88
4-Hydroxyphenyl pyranopetunidin 3-p-coumaroylglc (5 36)	Wine	741 ([M]$^+$; 433)		ESI-MS-MS	88
4-Hydroxyphenyl pyranopeonidin 3-p-coumaroylglc (5 37)	Wine	725 ([M]$^+$; 417)		ESI-MS-MS	88
Catechyl pyranomalvidin 3-glc (5 38)	Wine	625 ([M]$^+$; 463)		ESI-MS-MS	88
Catechyl pyranomalvidin 3-acetylglc (5 39)	Wine	667 ([M]$^+$; 463)		ESI-MS-MS	88
Catechyl pyranomalvidin 3-p-coumaroylglc (5 40)	Wine	771 ([M]$^+$; 463)		ESI-MS-MS	88
Guaiacyl pyranomalvidin 3-glc (5 41)	Wine	639 ([M]$^+$; 477)		ESI-MS-MS	88
Guaiacyl pyranomalvidin 3-acetylglc (5 42)	Wine	681 ([M]$^+$; 477)		ESI-MS-MS	88
Guaiacyl pyranomalvidin 3-p-coumaroylglc (5 43)	Wine	785 ([M]$^+$; 477)		ESI-MS-MS	88
Syringyl pyranomalvidin 3-glc (5 44)	Wine	669 ([M]$^+$; 507)		ESI-MS-MS	88
Syringyl pyranomalvidin 3-acetylglc (5 45)	Wine	711 ([M]$^+$; 507)		ESI-MS-MS	88
Carboxy pyranomalvidin 3-glc (5 46)	Grape marc	559 ([M$^+$ − 2H$^+$]; 515, 353)	507	HPLC-ESI-MS	10
Carboxy pyranopeonidin 3-glc	Grape marc	529 ([M$^+$ − 2H$^+$]; 485, 323)	507	HPLC-ESI-MS	10
Carboxy pyranopetunidin 3-glc	Grape marc	547 ([M$^+$ − 2H$^+$]; 501, 339)	507	HPLC-ESI-MS	10
Carboxy pyranodelphinidin 3-glc	Grape marc	531 ([M$^+$ − 2H$^+$]; 487)	507	HPLC-ESI-MS	10
Carboxy pyranomalvidin 3-acetylglc (5 47)	Grape marc	601 ([M$^+$ − 2H$^+$]; 556, 353)	507	HPLC-ESI-MS	10
Carboxy pyranomalvidin 3-p-acetylglc (5 47)	Wine	603 ([M]$^+$; 561)	503, 370	HPLC-DAD-ESI-MS	293
Carboxy pyranomalvidin 3-caffeoylglc	Model solution	723 ([M$^+$ − 2H$^+$]; 556, 353)	514	HPLC-DAD-ESI-MS	57

continued

TABLE 5.3
Flavonoid Reaction Products Formed in Wine and Wine-Like Model Systems — *continued*

Compound	Source	m/z (Parent and Fragment Ions)	λ_{max}	Identification Methods	Ref.
Carboxy pyranomalvidin 3-*p*-coumaroylglc	Model solution	705 ([M$^+$ − 2H$^+$]$^-$; 556, 353)	514	HPLC-DAD–ESI-MS	57
Carboxy pyranomalvidin 3-*p*-coumaroylglc	Model solution	707 ([M]$^+$; 561)	264, 306, 520	HPLC-DAD, FAB-MS	219
Carboxy pyranomalvidin 3-*p*-coumaroylglc (**5 48**)	Wine	707 ([M]$^+$; 561)	503, 370	HPLC-DAD–ESI-MS, NMR	293
Vinyl-epicatechin–(epi)catechin pyranomalvidin 3-glc (**5 49**)	Port wine	1119 ([M]$^+$; 957, 829, 667)	575	HPLC-DAD–ESI-MS, NMR	15
Vinyl-epicatechin–(epi)catechin pyranomalvidin 3-*p*-coumaroylglc (**5 50**)	Port wine	1265 ([M]$^+$; 957)	575	HPLC-DAD–ESI-MS, NMR	15
Vinylcatechin pyranomalvidin 3-*p*-coumaroylglc	Model solution	977 ([M]$^+$; 669)	575	HPLC-DAD–ESI-MS, NMR	15
Products of condensation with other aldehydes					
Catechin-(8-(CH−COOH)-8)–catechin (**5 51**)	Model solution	635 ([M − H]$^-$)	280	HPLC-DAD–ESI-MS, NMR	9
Catechin-(8-(CH−COO−CH$_2$−CH$_3$)-8)–catechin	Model solution	663 ([M − H]$^-$)	280	HPLC-DAD–ESI-MS	222
Catechin-(6-(CH−COOH)-6)–catechin	Model solution	635 ([M − H]$^-$)	280	HPLC-DAD–ESI-MS, NMR	79
Catechin-(6-(CH−COOH)-8)–catechin (two isomers, *R* and *S*)	Model solution	635 ([M − H]$^-$)	280	HPLC-DAD–ESI-MS, NMR	79
8-Formylcatechin	Model solution	635 ([M − H]$^-$)	295, 340	HPLC-DAD–ESI-MS, NMR	213
6-Formylcatechin	Model solution	635 ([M − H]$^-$)	295, 340	HPLC-DAD–ESI-MS, NMR	213
6,8-Diformylcatechin	Model solution	635 ([M − H]$^-$)	295, 340	HPLC-DAD–ESI-MS, NMR	213
Catechin-(carboxymethine–catechin)$_2$	Model solution	981 ([M − H]$^-$)	280	HPLC-DAD–ESI-MS, NMR	9
Catechin-(furfurylmethine)–catechin (four isomers)	Model solutions	657 ([M − H]$^-$; 505, 367, 289)	280	HPLC-DAD–ESI-MS	214
Catechin(5-hydroxymethylfurfuryl)–catechin (four isomers)	Model solutions	687 ([M − H]$^-$; 289)	280	HPLC-DAD–ESI-MS	214
Catechin-(furfurylmethine)–malvidin 3-glucoside (two isomers)	Model solution	859 ([M$^+$ − 2H$^+$]$^-$)	280, 450, 545	HPLC-DAD–ESI-MS	214
Catechin(5-hydroxymethylfurfurylmethine)–malvidin 3-glucoside (two isomers)	Model solution	889 ([M$^+$ − 2H$^+$]$^-$)	280, 450, 545	HPLC-DAD–ESI-MS	214

Catechin–(CH_2)–malvidin 3-glucoside	Model solution	593 (M^+; 303)	280, 531	HPLC-DAD–ESI-MS	215
Catechin–(CH_2–CH_3)–malvidin 3-glucoside	Model solution	621 (M^+; 331)	280, 541	HPLC-DAD–ESI-MS	215
Catechin–(CH–CH_3)$_2$)–malvidin 3-glucoside	Model solution	634 (M^+; 344)	280, 541	HPLC-DAD–ESI-MS	215
Catechin–(CH_2–CH–(CH_3)$_2$)–malvidin 3-glucoside	Model solution	649 (M^+; 359)	280, 541	HPLC-DAD–ESI-MS	215
Catechin–($CHCH_3$–CH_2CH_3)–malvidin 3-glucoside	Model solution	649 (M^+; 359)	280, 541	HPLC-DAD–ESI-MS	215
Catechin–(-benzyl)–malvidin 3-glucoside	Model solution	669 (M^+; 379)	280, 541	HPLC-DAD–ESI-MS	215

Xanthylium salts

Carboxy dicatechin xanthylium (**5 52**)	Model solution	617 (M^+); 615 ($[M^+ - 2H^+]^-$; 571, 463, 419)	273, 308 (sh), 444	HPLC-DAD–ESI-MS, NMR	217
Carboxy-dicatechin xanthylium (four isomers: 6-6, 6-8 *R*, *S*, 8-8)	Model solution	615 ($[M^+ - 2H^+]^-$; 571, 463, 419)	273, 308 (sh), 444	HPLC-DAD–ESI-MS	12
Carboxy dicatechin xanthene (**5 53**)	Model solution	617 ($[M - H]^-$)		HPLC-DAD–ESI-MS	217
Ethylcarboxy dicatechin xanthylium (**5 54**) (four isomers)	Model solution	643 ($[M^+ - 2H^+]^-$)	277, 308 (sh), 459	HPLC-DAD–ESI-MS, NMR	222
Ethylcarboxy dicatechin xanthene (**5 55**)	Model solution	645 ($[M - H]^-$)		HPLC-DAD–ESI-MS	222
Hydroxy dicatechin xanthylium (three isomers)	Model solution	587 ($[M^+ - 2H^+]^-$; 435)	280, 450	HPLC-DAD–ESI-MS, NMR	294
Furfuryl-dicatechin xanthylium	Model solution	637 ($[M^+ - 2H^+]^-$)	280, 440	HPLC-DAD–ESI-MS	214
Furfuryl dicatechin xanthene	Model solution	639 ($[M - H]^-$)	280	HPLC-DAD–ESI-MS	214
Hydroxymethylfurfuryl-dicatechin xanthylium	Model solution	667 ($[M^+ - 2H^+]^-$)	280, 440	HPLC-DAD–ESI-MS	214
Hydroxymethylfurfuryl dicatechin xanthene	Model solution	669 ($[M - H]^-$)	280	HPLC-DAD–ESI-MS	214

5.5.1 Reactivity of Flavonoid Compounds

The reactivity of polyphenolic compounds is due, on the one hand, to the acidity of their phenolic hydroxyl groups and, on the other hand, to the resonance between the free electron pair on the phenolic oxygen and the benzene ring, which increases electron delocalization and confers the position of substitution adjacent to the hydroxyl group a partial negative charge and thus a nucleophilic character. The A-ring shared by all grape flavonoids possesses two nucleophilic sites, in C8 and C6 positions, due to activation by the hydroxyl groups of its phloroglucinol (1,3,5-trihydroxy)-type structure.

Anthocyanins are usually represented as the red flavylium cations (Figure 5.1, left). However, this form is predominant only in very acidic solvents (pH < 2) such as those used for HPLC analysis. In mildly acidic media, the flavylium cations undergo proton transfer and hydration reactions, respectively, generating the quinonoidal base and the hemiketal (*syn* carbinol) form (Figure 5.1, right) that can tautomerize to the chalcone.[144,145] Thus, at wine pH, malvidin 3-glucoside occurs mostly as the colorless hemiketal (75%), the red flavylium cation, yellow chalcone, and blue quinonoidal base being only minor species.

The phloroglucinol A-ring of the anthocyanin hemiketal is nucleophilic, as described more generally for grape flavonoids, whereas the C-ring of the flavylium form, bearing a cationic charge in C2 or C4, reacts as an electrophile. Classical examples of nucleophilic addition reactions onto the flavylium cation are those of water and bisulfite that have long been known to result in anthocyanin bleaching. NMR studies demonstrated that addition of water occurs mostly in C2 position, the 4-carbinol being only a minor product,[146] whereas addition of sulfur dioxide yields the two C4-sulfonate adducts.[147] Addition with other nucleophiles normally occurs in C4 position but C2 adducts resulting from reaction of anthocyanins with acetone were isolated and characterized.[148]

Another reaction of the flavylium cation has recently been demonstrated.[8,10,57,149–151] It involves concerted addition of compounds possessing a polarizable double bond on the electron-deficient site C–4 and the oxygen of the 5-hydroxyl group of the anthocyanin. The new pigments thus formed, showing a second pyran ring, have been referred to as vitisins,[152] but the term pyranoanthocyanins proposed by Lu and Foo[151] is preferred.

Flavanols also react both as nucleophiles, through their A-ring, and as elecrophiles, through the carbocations formed after acid-catalyzed cleavage of the interflavanoid linkages. The latter reaction, restricted to oligomers and polymers, was shown to occur spontaneously at wine pH values.[4,153]

The basic forms of phenols (phenolate anions) are easily oxidized to semiquinone radicals through electron transfer. These radicals can then react with another radical to form an adduct through radical coupling or, in the case of *o*-diphenols, undergo a second oxidation step yielding *o*-quinones that are electrophiles as well as oxidants.[154,155] Oxidation reactions are very slow in wine, due to the low proportion of phenolate ions at wine pH values, but take place extremely rapidly when oxidative enzymes are involved (see Section 5.5.2.2).

5.5.2 Enzymatic Reactions

Major enzymes catalyzing flavonoid reactions are oxidative enzymes (i.e., polyphenoloxidases and peroxidases) arising from grape but also from molds contaminating them. Various hydrolytic enzymes (glycosidases, esterases), excreted by the fermentation yeasts or fungi or present in preparations added for technological purposes (e.g., pectinases), are also encountered in wine.

5.5.2.1 Hydrolytic Enzymes

Pectinases and β-glucanases are the only enzymes allowed in wine-making by European legislation. They are used as clarification and filtration agents and also to release aroma compounds that are mostly present in grape as nonvolatile glycosidic precursors.[156] Pectolytic enzymes are also reported to increase extraction of phenolic compounds and wine color intensity.[157]

Such rather crude commercial preparations often contain other activities, such as cinnamoylesterases, tannin-acyl hydrolase (tannase), and glycosidases. Hydrolysis of flavonol glucosides catalyzed by β-glucosidase has been reported to induce haze formation, due to precipitation of insoluble flavonol aglycones.[158] Deglucosylation of anthocyanins results in discoloration, owing to instability of the resultant anthocyanidins. This can be further enhanced in the presence of cinnamoylesterase hydrolyzing the linkage between *p*-coumaric acid and glucose in *p*-coumaroylated anthocyanins.[159] However, β-glucosidases from *Aspergillus niger* show different substrate affinity patterns so that activities needed to hydrolyze aroma precursors and plant cell wall polysaccharides can be separated from undesirable anthocyanase activities.[160]

Tannase activity (tannin acyl hydrolase, EC 3.1.1.20) has been reported in numerous fungi, including *A. niger* and *Botrytis cinerea*.[161] It catalyzes cleavage of the ester bond in galloylated flavanols. Its action in wine is rather weak, probably due to inhibition by high levels of phenolic compounds. In addition, some galloyl groups in procyanidin polymers are resistant to tannase hydrolysis, because of steric hindrance or other molecular interactions.[162]

5.5.2.2 Enzymatic Oxidation and Subsequent Reactions

The major enzyme responsible for polyphenol oxidation in wine-making is grape polyphenoloxidase (PPO, *syn* catecholoxidase, EC 1.10.3.1), showing both cresolase (hydroxylating monophenols to *o*-diphenols (1)) and catecholase (oxidizing *o*-diphenols to *o*-quinones (2)) activities, using molecular oxygen as co-substrate

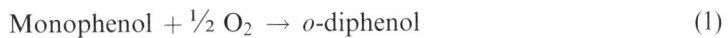

$$\text{Monophenol} + \tfrac{1}{2}\,O_2 \rightarrow o\text{-diphenol} \tag{1}$$

$$o\text{-Diphenol} + \tfrac{1}{2}\,O_2 \rightarrow o\text{-quinone} + H_2O \tag{2}$$

Laccase (EC 1.10.3.2), arising from fungal contamination by *B. cinerea*, which is the agent of gray and noble rot, also catalyzes reaction 2 and accepts a wider range of substrates, including, in particular, glycosylated flavonoids such as anthocyanins and also *p*-diphenols (oxidized to *p*-quinones).[163]

Peroxidase (POD) catalyzes oxidation of a wide range of *o*-diphenolic substrates to *o*-quinones, using hydrogen peroxide as a co-substrate (3):

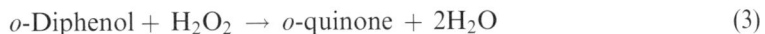

$$o\text{-Diphenol} + H_2O_2 \rightarrow o\text{-quinone} + 2H_2O \tag{3}$$

This activity is present in grape but has never been reported in musts where PPO is extremely active and the availability of hydrogen peroxide is limited. Studies performed on pear[164] suggest that POD may participate in enzymatic oxidation together with PPO that generate hydrogen peroxide through coupled oxidation processes.[165] It may also be involved in anthocyanin degradation occurring in postharvest storage. Decay of catalase activity, which catalyzes conversion of hydrogen peroxide to water, concomitant with anthocyanin degradation and induction of POD activity, has been observed in grapes submitted to carbonic maceration.[142]

Flavonoids are not usually directly involved in enzymatic oxidation since they are very poor substrates for grape polyphenoloxidase compared to caffeoyltartaric acid,[166] which is also very abundant in grape musts. *ortho*-Diphenolic flavonoid aglycones and, in particular, flavanol monomers can be oxidized by grape PPO. Glycosylated flavonoids (anthocyanins, flavonol glycosides) and proanthocyanidins are poor substrates for grape PPO but may be oxidized by laccase and by peroxidase. Moreover, they can participate in oxidation reactions through coupled oxidation and nucleophilic addition processes with enzymatically generated quinones. In particular, kinetic studies showed that the quinone arising from enzymatic oxidation of caffeoyltartaric acid is a strong oxidant, capable of oxidizing most flavonoid *o*-diphenols, including catechins,[167] epicatechin gallate,[168] procyanidins,[168,169] and cyanidin 3-glucoside[170] to the corresponding secondary *o*-quinones. Coupled oxidation of cyanidin 3-glucoside is illustrated in Figure 5.5(**A**). Such coupled oxidation reactions result in the recycling of caffeoyltartaric acid that is regenerated by reduction of its quinone and can then be reoxidized by PPO as long as enzyme activity and oxygen are available.

The electrophilic primary and secondary quinones undergo addition of nucleophiles, including flavonoids. For instance, nucleophilic addition of catechin to its enzymatically generated quinone yielded a catechin dimer in which the catechin moieties are linked through a C6′–C8 biphenyl linkage.[171] This B-type dehydrodicatechin further oxidized to yellow pigments. Additional dehydrodicatechins arise from radical coupling of the catechin semi-quinones formed by retro-disproportionation, in which the catechin moieties are linked

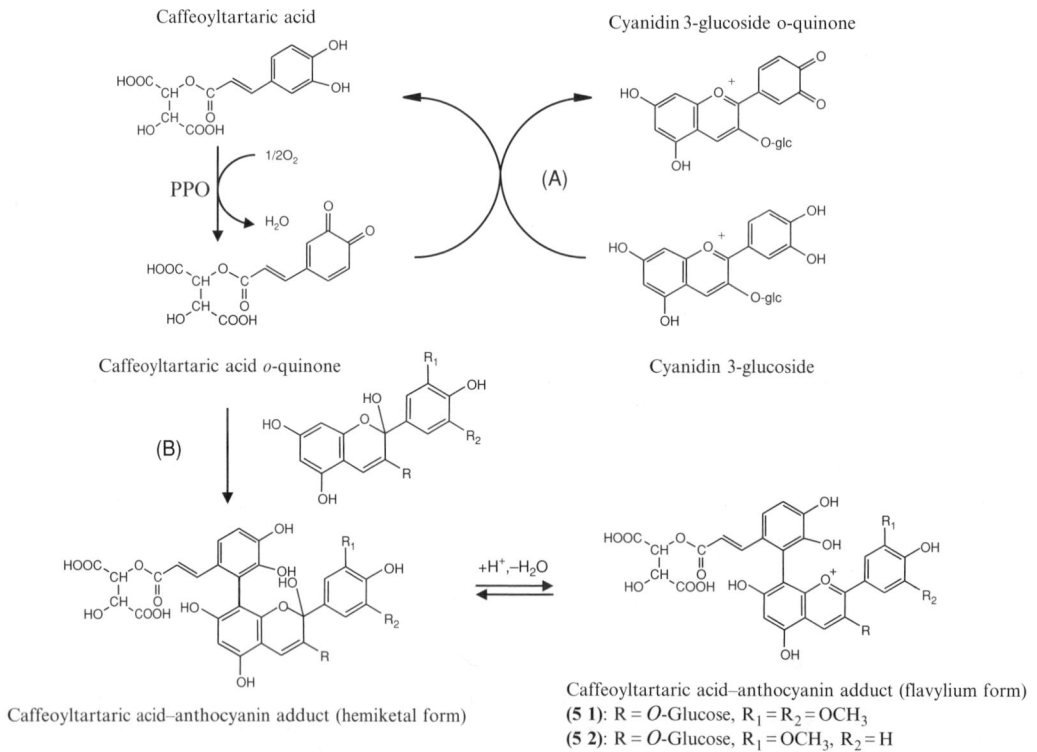

FIGURE 5.5 Anthocyanin reactions with caffeoyltartaric acid *o*-quinone coupled oxidation of cyanidin 3-glucoside (A) and nucleophilic addition of anthocyanins (B). Anthocyanin–caffeoyltartaric acid adducts are represented as one of the possible isomers, namely 2-(8-anthocyanin) caffeoyltartaric acid (**5 1**).

through phenyl–ether bonds (C3′–O–C8) or (C4′–O–C8), especially in solutions incubated at lower pH values. Catechin dimers resulting from oxidation can be distinguished from their procyanidin isomers on the basis of specific mass fragments obtained by ESI-MS, at m/z 451 for procyanidin dimers (cf. Figure 5.2) and at m/z 393 for dehydrodicatechins.[172]

Addition of anthocyanins onto the caffeoyltartaric acid quinone formed by PPO oxidation was similarly demonstrated.[55,170] Although malvidin 3-glucoside does not possess an o-diphenolic structure and is thus not susceptible to coupled oxidation, it was rapidly degraded when incubated with PPO and caffeoyltartaric acid. The apparent rate of caffeoyltartaric acid oxidation was not modified by the presence of malvidin 3-glucoside but the concentration of quinones was much lower, suggesting that they were trapped by nucleophilic addition of the anthocyanin (Figure 5.5, **B**).[170] LC–MS analysis of the oxidized solution allowed the detection of six malvidin 3-glucoside–caffeoyltartaric acid adducts both under the flavylium form ([M]$^+$ at $m/z = 803$ in the positive ion mode, [M − 2H]$^-$ at $m/z = 801$ in the negative ion mode) and the hydrated form ([M − H]$^-$ at $m/z = 819$ in the negative ion mode).[55] Comparison of the reaction rates at pH 1.7 and pH 3.4 confirmed that the addition involved the nucleophilic anthocyanin hemiketal, in agreement with the postulated mechanism. Unfortunately, no adduct was formed in sufficient amount to be isolated and identified. However, on the basis of quinone and anthocyanin reactivities, the linkages can be reasonably assumed to be in C8 and possibly C6 positions on the anthocyanin. Nucleophilic addition in positions 2 and 5 of caffeoyltartaric acid quinones has been observed earlier with glutatione.[173,174] Nucleophilic addition in position 6,[175] as well as radical coupling leading to phenyl ether bonds in position 3 or 4,[171] as demonstrated with catechin, are also possible. Finally, the side chain double bond of caffeoyltartaric acid can be the site of nucleophilic attack, as shown with caffeic acid.[176] However, the UV–visible spectra of all caffeoyltartaric acid–malvidin 3-glucoside adducts showed absorbance at 328 nm characteristic of the side chain conjugation of caffeic acid derivatives, ruling out this hypothesis. One of the possible isomers (2-(8-anthocyanin)caffeoyltartaric acid) (**5 1**) is shown in Figure 5.5. Four caffeoyltartaric acid–malvidin 3-glucoside adducts and four additional compounds detected at $m/z = 773$ in the positive ion mode, presumably corresponding to caffeoyltartaric acid–peonidin 3-glucoside adducts (**5 2**), were also detected after oxidation of Gamay beaujolais grapes.[142]

5.5.3 CHEMICAL REACTIONS

Color and taste changes taking place during wine aging have long been ascribed to conversion of grape anthocyanins to polymeric pigments through addition reactions with flavanols.[3] Three mechanisms have been postulated.

The first one involves nucleophilic addition of flavanols (in C8 or C6) onto the C4 position of the anthocyanin flavylium ion yielding 4-flavanyl-anthocyanins, which are also referred to as anthocyanin–flavanol (A–F) or anthocyanin–tannin (A–T) adducts (Figure 5.6). The formation of 4-phloroglucinylflavene and 4-flavanylflavene by addition of phloroglucinol and catechin, respectively, onto a flavylium was first demonstrated by Jurd.[177,178] The latter was oxidized to the corresponding 4-flavanylflavylium by a second molecule of the flavylium salt whereas rearrangement of the former yielded a structure similar to A-type proanthocyanidins, in which the anthocyanin and phloroglucinol units are linked through both C–C and ether linkages. This rearrangement was also observed for the flavanylflavene formed from malvidin 3,5-diglucoside and catechin.[179] Finally, xanthylium salt structures were proposed for yellow-orange products formed by reaction of malvidin 3,5-diglucoside with catechin,[5] phloroglucinol, epicatechin, and catechin-3-O-gallate[180] but no structural characterization was provided.

The second mechanism is based on nucleophilic addition onto the carbonium ion formed in acid solutions from flavan 3,4-diols[181] or by cleavage of procyanidins.[69,182] A condensation

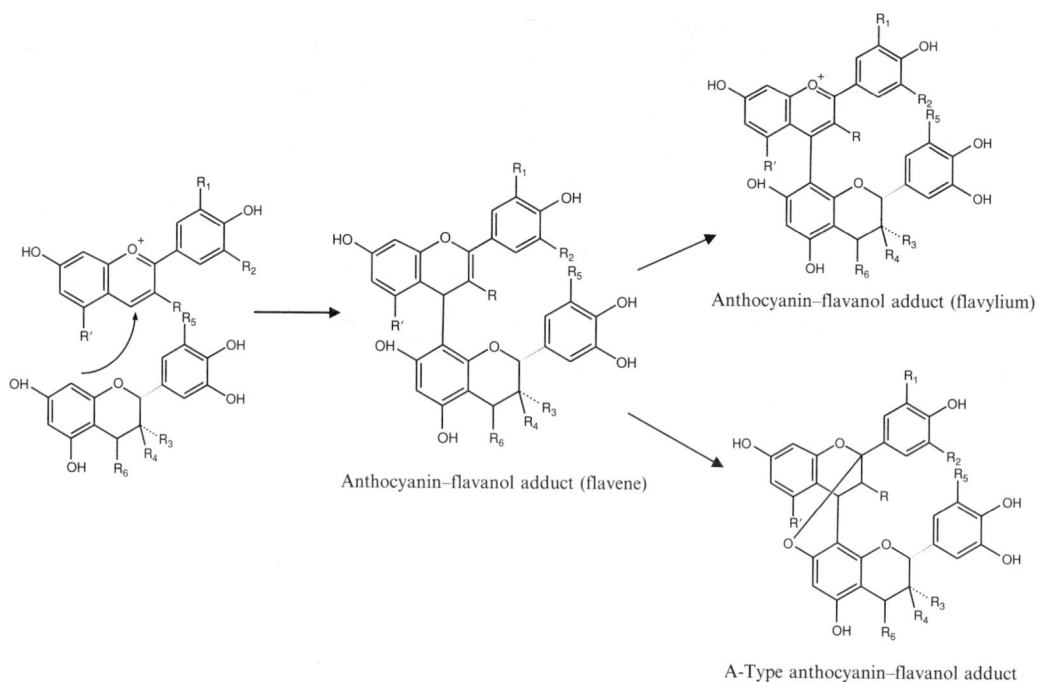

FIGURE 5.6 Postulated mechanism for formation of anthocyanin–flavanol adducts (flavanyl–antho-cyanins). R, R′, R_1, R_2 as in Figure 5.1; R_3 = H, OH, galloyl; R_4, R_5 = H, OH; R_6 = H, flavanyl.

product of 5,7,3′,4′-tetramethoxyflavan 3,4-diol and phloroglucinol was prepared as a model proanthocyanidin and identified by ^{1}H NMR after methylation of the phloroglucinol hydroxyls, confirming the postulated addition mechanism.[183] Yellow products formed after acid treatment of this adduct were postulated to be xanthylium salts, resulting from oxidation of an intermediate xanthene.[184] According to these authors, a similar mechanism may explain the formation of so-called phlobaphene pigments from proanthocyanidins. Acid-catalyzed degradation of proanthocyanidins was also shown to take place at wine pH values.[4,153] In the presence of large amounts of flavanol monomers, proanthocyanidin losses were much reduced and oligomeric species gradually replaced higher molecular weight polymers as the monomers added to the intermediate carbocation released by acid-catalyzed cleavage. Finally, addition of anthocyanins, either in the flavene form[181] or in the hemiketal form,[185] onto the carbonium ion, leading to F–A adducts was described. The resulting flavene (F–A) or hemiketal (F–AOH) adducts both generate the corresponding flavylium (F–A$^+$), through oxidation or dehydration reactions, respectively.

Reactions of anthocyanins and flavanols take place much faster in the presence of acetaldehyde[5,186,187] that is present in wine as a result of yeast metabolism and can also be produced through ethanol oxidation, especially in the presence of phenolic compounds,[188] or introduced by addition of spirit in Port wine technology. The third mechanism proposed involves nucleophilic addition of the flavanol onto protonated acetaldehyde, followed by protonation and dehydration of the resulting adduct and nucleophilic addition of a second flavonoid onto the carbocation thus formed.[5] The resulting products are anthocyanin–flavanol adducts in which the flavonoid units are linked in C6 or C8 position through a methylmethine bond, often incorrectly called ethyl-link in the literature.

Earlier investigations relied upon model solution studies starting with grape components or related molecules and comparison of color characteristics of the reaction products with those of wine pigments, but none of these structures or reactions had been formally demonstrated in wine until recently. The development of more sensitive and selective analytical techniques, such as HPLC coupled to diode array detector and MS, has enabled the characterization of various wine components and to postulate the reaction mechanisms generating them. These mechanisms can then be investigated in model solution studies and characteristics of the resulting products compared with those of wine constituents. Conversely, products obtained in wine-like solutions serve to develop specific analytical tools as well as chromatographic and spectral data that are used for determination of new products in wine. Recent studies based on these complementary approaches have confirmed the occurrence of various expected structures, and discovered numerous others, as developed below.

5.5.3.1 Products of Direct Anthocyanin–Flavanol Addition

Pigments resulting from direct reaction of anthocyanins with flavanols were first detected in wine in 1999.[87,189] HPLC–ESI-MS analysis of the aqueous phase recovered after isoamyl alcohol extraction of a 2-year-old cabernet sauvignon wine showed a series of signals at mass values corresponding to those of covalent adducts of flavanol monomers with the major grape anthocyanins as their flavylium form,[87] but did not allow to distinguish between F–A$^+$ (Figure 5.7A) and A$^+$–F (Figure 5.7B) derivatives. MS^n experiments performed on the major compound, detected at m/z 781, which was postulated to correspond to the flavylium form of an (epi)catechin–malvidin 3-glucoside adduct,[19,86] indicated that the flavanol unit is in the upper position. To confirm this hypothesis and further characterize the structures, flavanol–anthocyanin adducts (F–A$^+$) were prepared by hemisynthesis using two different approaches. Epicatechin–malvidin 3-glucoside (**5 3**) was obtained by incubating procyanidin B2 3′-O-gallate (epicatechin 3-gallate (4–8)-epicatechin) and malvidin 3-glucoside at pH 2.[58] Catechin–malvidin 3-glucoside (**5 4**) was synthesized by using a protocol adapted from the synthesis of procyanidin dimers,[190] involving nucleophilic addition of malvidin 3-glucoside onto the flavan 3,4-diol resulting from reduction of taxifolin.[86]

These experiments demonstrated the occurrence in wine of F–A$^+$ adducts arising from cleavage of procyanidins followed by nucleophilic addition of anthocyanins. Detection of epicatechin–epicatechin–malvidin 3-glucoside (**5 5**) (m/z 1069), resulting from addition of epicatechin–malvidin 3-glucoside onto the intermediate epicatechin carbonium ion, in the model solution containing B2 3′-O-gallate and malvidin 3-glucoside, is another argument in favor of this mechanism. F–A$^+$ adducts were not detected at pH 3.8, meaning that their formation was limited by the rate of proanthocyanidin acid-catalyzed cleavage.[58] At this pH value, reaction of malvidin 3-glucoside with procyanidin B2 3′-O-gallate yielded two A$^+$–F adducts (m/z 1221 in the positive ion mode) resulting from nucleophilic addition of the procyanidin dimer onto malvidin 3-glucoside. MS^n fragmentation experiments (Figure 5.8) established that malvidin 3-glucoside was linked to the epicatechin unit in one of them and to the epicatechin 3-O-gallate unit in the other.[58] The anthocyanin–procyanidin adducts were no longer detected after thiolysis. Two additional ions at m/z 903 and 933 were attributed to the benzylthioether of malvidin 3-glucoside–epicatechin and to malvidin 3-glucoside–epicatechin 3-O-gallate, confirming the postulated structures (**5 6, 5 7**) as shown in Figure 5.9.

Additional compounds corresponding to anthocyanin dimers in which one of the anthocyanins is in the flavylium form and the other in the hydrated form were detected in the solutions incubated at pH 3.8. Such products arise from nucleophilic addition of the hemiketal onto the flavylium, confirming that, at this pH value, anthocyanins exist and react under both forms, as expected from their hydration equilibrium.

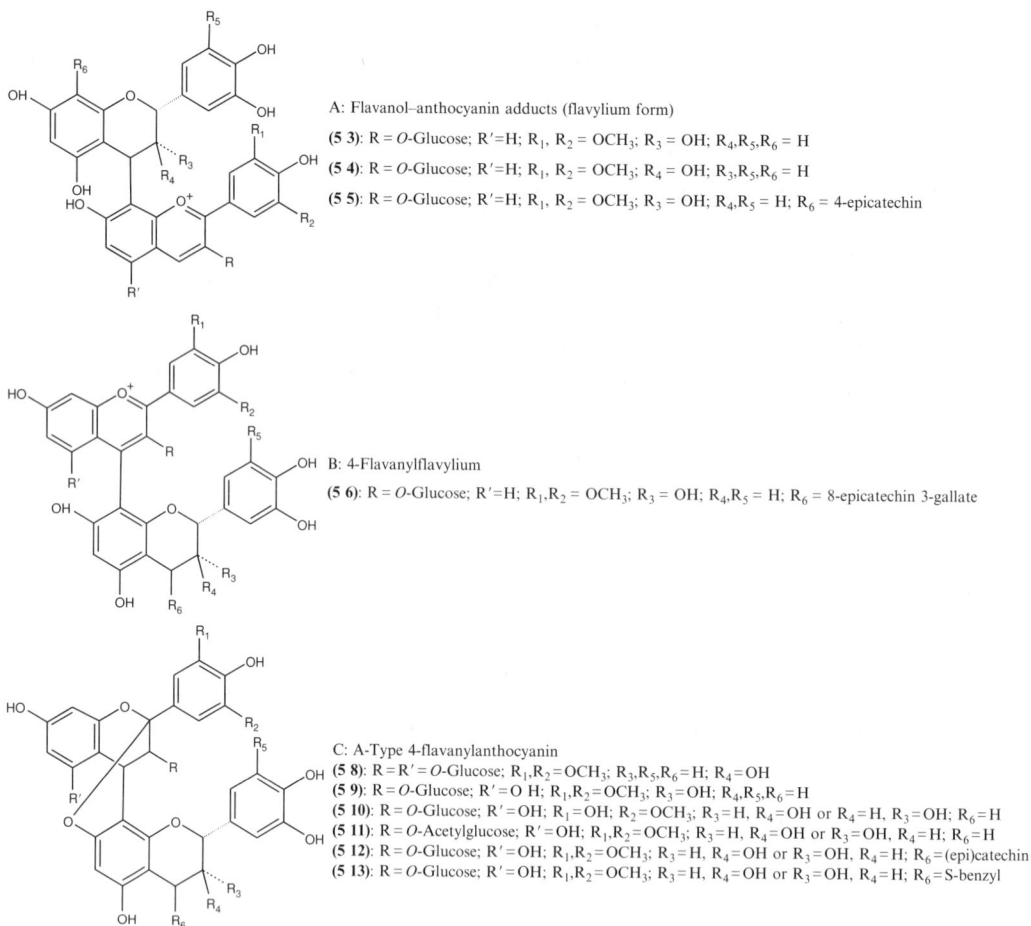

A: Flavanol–anthocyanin adducts (flavylium form)

(5 3): R = O-Glucose; R′=H; R_1, R_2 = OCH_3; R_3 = OH; R_4,R_5,R_6 = H

(5 4): R = O-Glucose; R′=H; R_1, R_2 = OCH_3; R_4 = OH; R_3,R_5,R_6 = H

(5 5): R = O-Glucose; R′=H; R_1, R_2 = OCH_3; R_3 = OH; R_4,R_5 = H; R_6 = 4-epicatechin

B: 4-Flavanylflavylium

(5 6): R = O-Glucose; R′=H; R_1,R_2 = OCH_3; R_3 = OH; R_4,R_5 = H; R_6 = 8-epicatechin 3-gallate

C: A-Type 4-flavanylanthocyanin

(5 8): R = R′ = O-Glucose; R_1,R_2 = OCH_3; R_3,R_5,R_6 = H; R_4 = OH

(5 9): R = O-Glucose; R′ = O H; R_1,R_2 = OCH_3; R_3 = OH; R_4,R_5,R_6 = H

(5 10): R = O-Glucose; R′ = OH; R_1 = OH; R_2 = OCH_3; R_3 = H, R_4 = OH or R_4 = H, R_3 = OH; R_6 = H

(5 11): R = O-Acetylglucose; R′ = OH; R_1,R_2 = OCH_3; R_3 = H, R_4 = OH or R_3 = OH, R_4 = H; R_6 = H

(5 12): R = O-Glucose; R′ = OH; R_1,R_2 = OCH_3; R_3 = H, R_4 = OH or R_3 = OH, R_4 = H; R_6 = (epi)catechin

(5 13): R = O-Glucose; R′ = OH; R_1,R_2 = OCH_3; R_3 = H, R_4 = OH or R_3 = OH, R_4 = H; R_6 = S-benzyl

FIGURE 5.7 Structures of anthocyanin–flavanol and flavanol–anthocyanin adducts.

HPLC–MS analysis of the above-mentioned cabernet sauvignon wine extract[7] also showed the presence of two compounds with the mass signals expected from the flavene forms of malvidin 3-glucoside–(epi)catechin adducts (MW = 782). This flavene structure (cf. Figure 5.6) has been postulated for a colorless product formed in wine-like model solutions containing catechin and malvidin 3-glucoside.[191,192] Incubation of malvidin 3-glucoside and catechin in ethanol yielded two colorless compounds showing the same mass spectra and eluting at the same retention times as those found in the wine extract.[7] Their resistance to thiolysis ruled out the postulated flavene structure and led to the proposal of a structure similar to A-type proanthocyanidins (Figure 5.7, **C**), as described earlier for a malvidin 3,5-diglucoside–catechin adduct.[179] Two-dimensional NMR spectrometry of the major adduct isolated from the model solution confirmed the postulated malvidin 3-glucoside (C2–O–C7, C4–C8) epicatechin structure (5 9).[193]

Other A-type anthocyanin–flavanol adducts, namely petunidin 3-glucoside–(epi)catechin (5 10), malvidin 3-acetylglucoside–(epi)catechin (5 11), and malvidin 3-glucoside–procyanidin dimer (5 12), were similarly detected in the red wine extract.[87] Finally, a signal was detected at m/z 903 in the negative ion mode after thiolysis of the wine extract. Its characteristic fragments allowed to attribute it to the benzylthioether of A-type malvidin 3-glucoside–(epi)catechin (5 13) arising from thiolysis of A-type malvidin 3-glucoside–procyanidins.

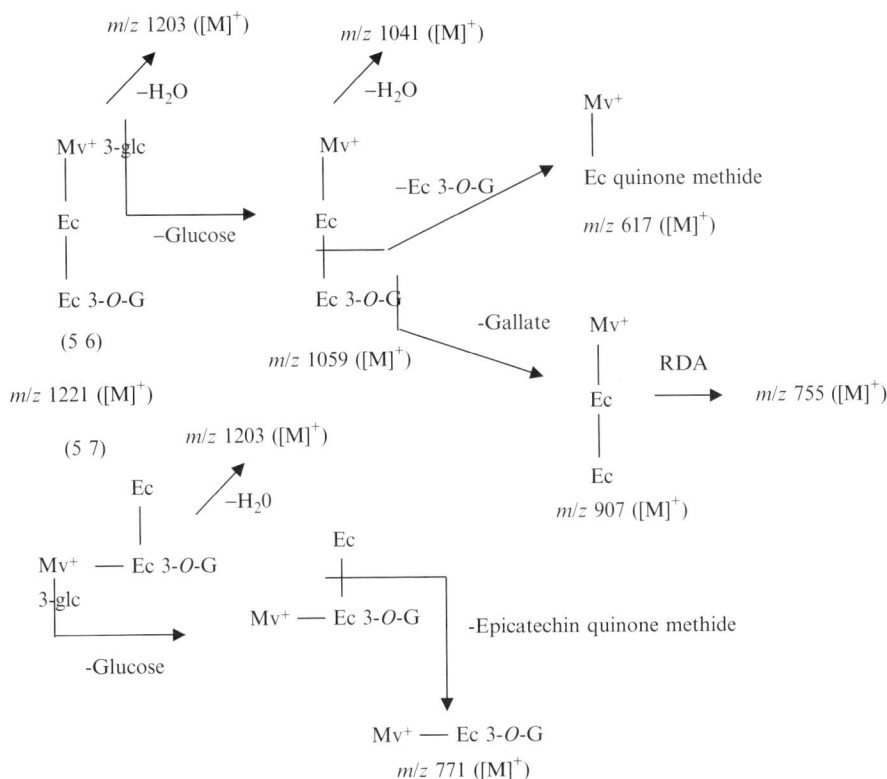

FIGURE 5.8 Mass fragmentation patterns of malvidin 3-glucoside–procyanidin B2 3'-gallate adducts.

5.5.3.2 Products of Acetaldehyde-Mediated Reactions

The effect of acetaldehyde on the resistance of wine pigments to pH-induced changes was first reported by Singleton and coworkers,[194] who postulated that acetaldehyde reacts with anthocyanins by an acid-catalyzed reaction to form polymeric pigments. Acetaldehyde was shown not only to increase color intensity but also to promote color loss and formation of colorless anthocyanin derivatives.[195] Model solution studies confirmed that reactions of anthocyanins with flavanols are much faster in the presence of acetaldehyde, causing a rapid increase in color intensity and a light absorption shift toward purple.[5] This was attributed to the formation of products consisting of anthocyanin and flavanol moieties linked through CH–CH₃ bridges on the C8 summits (Figure 5.10A). Incubation of malvidin 3-glucoside with epicatechin[196] or with catechin[187,197] yielded two products exhibiting maximum absorbance around 540 nm in the visible region and molecular ions (M⁺) at m/z 809,[187,198] in agreement with the proposed structure. The methylmethine bridge was postulated to be linked on the C8 of the anthocyanin unit and on C6 and C8 on the flavanol, respectively, for each isomer. C8–(CH–CH₃)–C8-linked isomers differing by the configuration of the asymmetric methine carbon were also envisaged.[197] NMR analysis of the four methylmethine (epi)catechin–malvidin 3-glucoside adducts was recently published.[199] It was concluded that isomers within each pair (**5 14, 5 15**) differ by the configuration of the methine carbone (*R* or *S*) of their C8–(CH–CH₃)–C8 bridge.

After longer incubation times, the chromatographic profiles of solutions containing flavanol monomers, malvidin 3-glucoside, and acetaldehyde became more complex and

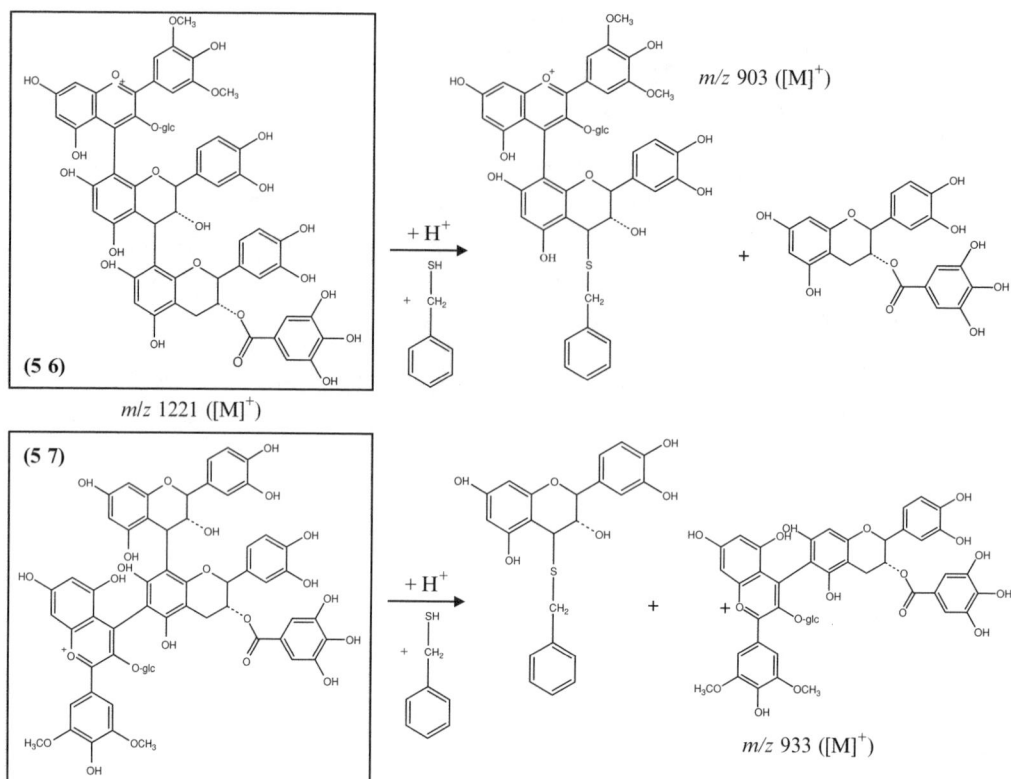

FIGURE 5.9 Hypothetical structures and thiolysis reaction of malvidin 3-glucoside–procyanidin B2 3′-gallate adducts.

unresolved late eluting compounds accumulated.[56] Their UV–visible spectra showed further a bathochromic shift in the visible region (λ_{max} 555 nm) compared to those of the dimers (λ_{max} 540 nm). Detection of mass signals corresponding to methylmethine-linked trimers containing two epicatechin and one malvidin 3-glucoside, one epicatechin and two malvidin 3-glucoside, and tetramers with three epicatechin and one malvidin 3-glucoside, or two epicatechin and two malvidin 3-glucoside, confirmed that the polymerization reaction continued.[56] Similar dimeric and trimeric adducts were formed by incubation of cyanidin 3-glucoside with (epi)catechin.[93] The presence of these various compounds and that of ethyl-linked flavanol dimers (Figure 5.10B) indicate that anthocyanin and flavanol units compete in the nucleophilic addition reaction. In addition, formation of trimers implies that both positions 6 and 8 are reactive, at least in flavanol units. Acetaldehyde-mediated reactions of procyanidins and anthocyanins were also demonstrated in model solutions, their rate increasing with the flavanol chain length.[200]

Monitoring of acetaldehyde-induced polymerization of catechin and epicatechin by HPLC–MS[63] demonstrated the formation of several methylmethine-linked flavanol dimers, trimers, and tetramers. Detection of the intermediate ethanol adducts confirmed the mechanism postulated by Timberlake and Bridle,[5] which involves protonation of acetaldehyde in the acidic medium, followed by nucleophilic attack of the resulting carbocation by the flavan unit. The ethanol adduct then loses a water molecule and gives a new carbocation that undergoes nucleophilic attack by another flavanol molecule. Four dimers (C6–C6, C8–C8, and C6–C8, *R* and *S*) were formed from each monomeric flavanol.[64,77] When both epicatechin and catechin units were present, additional isomers containing both types of units were

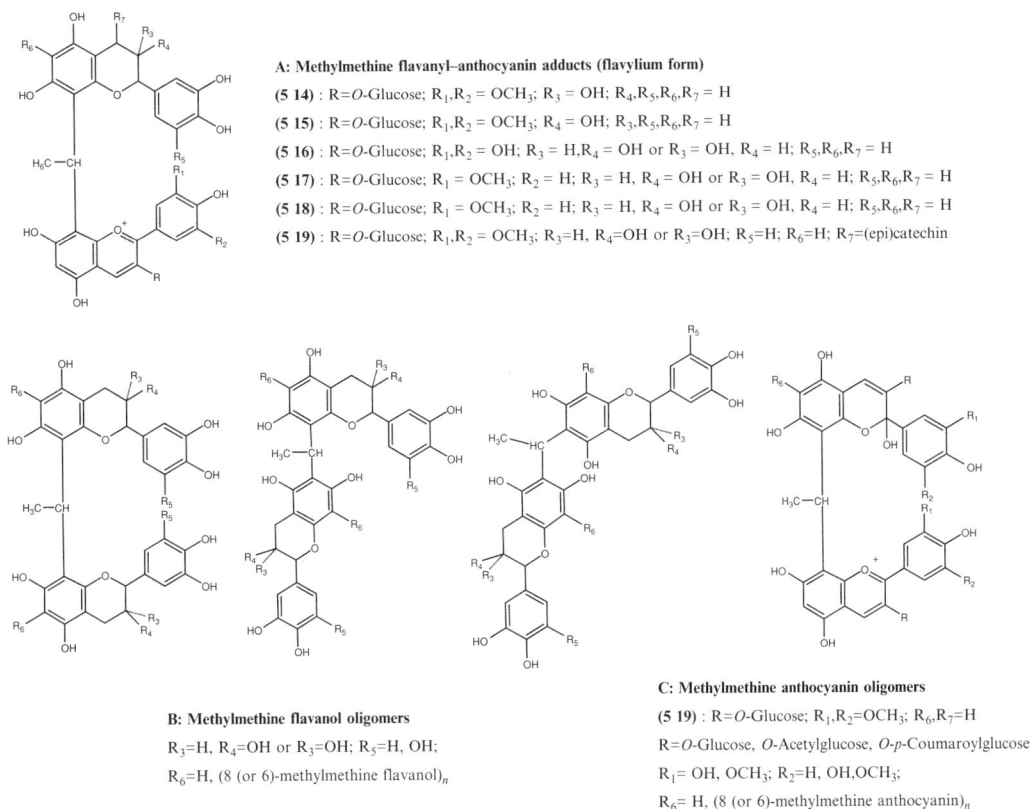

A: Methylmethine flavanyl–anthocyanin adducts (flavylium form)

(5 14) : R=*O*-Glucose; R_1,R_2 = OCH$_3$; R_3 = OH; R_4,R_5,R_6,R_7 = H

(5 15) : R=*O*-Glucose; R_1,R_2 = OCH$_3$; R_4 = OH; R_3,R_5,R_6,R_7 = H

(5 16) : R=*O*-Glucose; R_1,R_2 = OH; R_3 = H,R_4 = OH or R_3 = OH, R_4 = H; R_5,R_6,R_7 = H

(5 17) : R=*O*-Glucose; R_1 = OCH$_3$; R_2 = H; R_3 = H, R_4 = OH or R_3 = OH, R_4 = H; R_5,R_6,R_7 = H

(5 18) : R=*O*-Glucose; R_1 = OCH$_3$; R_2 = H; R_3 = H, R_4 = OH or R_3 = OH, R_4 = H; R_5,R_6,R_7 = H

(5 19) : R=*O*-Glucose; R_1,R_2 = OCH$_3$; R_3=H, R_4=OH or R_3=OH; R_5=H; R_6=H; R_7=(epi)catechin

B: Methylmethine flavanol oligomers

R_3=H, R_4=OH or R_3=OH; R_5=H, OH;

R_6=H, (8 (or 6)-methylmethine flavanol)$_n$

C: Methylmethine anthocyanin oligomers

(5 19) : R=*O*-Glucose; R_1,R_2=OCH$_3$; R_6,R_7=H

R=*O*-Glucose, *O*-Acetylglucose, *O*-*p*-Coumaroylglucose

R_1= OH, OCH$_3$; R_2=H, OH,OCH$_3$;

R_6= H, (8 (or 6)-methylmethine anthocyanin)$_n$

FIGURE 5.10 Structures of methylmethine-linked flavonoid derivatives.

formed. As the reaction was allowed to proceed, successive condensations led to numerous oligomers and polymers.[63]

The methylmethine-bridged oligomers are rather unstable in wine-like media, due to the susceptibility of the methylmethine–flavanol bonds to acid-catalyzed cleavage.[64,201] Methylmethine-linked anthocyanin–flavanol adducts are much more resistant than flavanol oligomers under the very acidic conditions used for thiolysis. However, when incubated at higher pH values (2.5 and 5), they underwent spontaneous cleavage and released malvidin 3-glucoside.[18] The effect of pH is probably related to the higher proportions of adducts under the hemiketal form that is more favorable to acid-catalyzed cleavage than the flavylium form, as mentioned above.

Finally, the occurrence of methylmethine-linked malvidin 3-glucoside oligomers (Figure 5.10C) was also reported.[13] HPLC–MS analysis of the major pigment formed after incubation of malvidin 3-glucoside and acetaldehyde in wine-like solutions at pH 3.2 gave signals at m/z 1029, 1011, and 506, which could be attributed to different forms of a methylmethine-linked malvidin 3-glucoside dimer, in which the anthocyanin units exist as one hydrated form and one flavylium form (as shown in Figure 5.10), one quinonoidal base and one flavylium, and two flavylium cations, respectively. NMR spectroscopic analysis of the dimer performed in 10% TFA in deuterated DMSO to ensure that it was in the bis-flavylium form demonstrated that it was a 8,8-methylmethine-linked malvidin 3-glucoside dimer (**5 20**). Other mass signals detected in the solution were attributed to methylmethine-linked malvidin 3-glucoside trimers, tetramers, and pentamers in which constitutive anthocyanin units were present as different forms, namely flavylium (A^+), quinonoidal base (A), or hemiketal (AOH).[13,19] NMR analysis of the oligomeric fraction confirmed that even in strongly acidic medium, the anthocyanin units in the oligomers were present as these different forms. As stated above for flavanol derivatives, the occurrence of polymeric species indicates that the C6 position can also participate in nucleophilic addition reactions although it is less reactive than the C8 position. This reactivity is attributable to the hemiketal units that are present in methyl-methine anthocyanin polymers. In contrast, anthocyanin units in methylmethine-linked anthocyanin–flavanol adducts are mostly under the flavylium form and thus act as end points in the condensation process.

HPLC–ESI-MS analysis of red wines demonstrated the presence of (epi)catechin–(CH–CH_3)–malvidin 3-glucoside under the flavylium form (**5 14, 5 15**),[6] malvidin 3-glucoside–(CH–CH_3)–malvidin 3-glucoside under the hemiketal–flavylium form (**5 20**),[13] and methyl-methine-linked (epi)catechin dimers[6,11] and trimers.[6] Other signals detected were attributed to the flavylium forms of (epi)catechin–(CH–CH_3)–delphinidin 3-glucoside (**5 16**), (epi)cate-chin–(CH–CH_3)–petunidin 3-glucoside (**5 17**), (epi)catechin–(CH–CH_3)–peonidin 3-glucoside (**5 18**), and procyanidin dimer–(CH–CH_3)–malvidin 3-glucoside (**5 19**).[52]

In addition to the purple methylmethine-linked adducts, pigments with maximum absorb-ance in the range 500 to 510 nm were observed in solutions containing acetaldehyde and anthocyanins.[197] A pyranomalvidin 3-glucoside structure (Figure 5.11, **5 21**) formed through an addition mechanism involving, on the one hand, the electronic deficient site C-4 and the 5-hydroxyl group of the malvidin 3-glucoside flavylium cation, and, on the other hand, the polarizable double of the acetaldehyde enolic form, followed by dehydration and rearomati-zation, was proposed for the major one on the basis of its mass and UV–visible spectra.[150] Simultaneously, the same product and its acetyl derivative (**5 22**) were isolated from wine and called vitisin B and acetylvitisin B.[17] Their ^{1}H NMR data confirmed the postulated

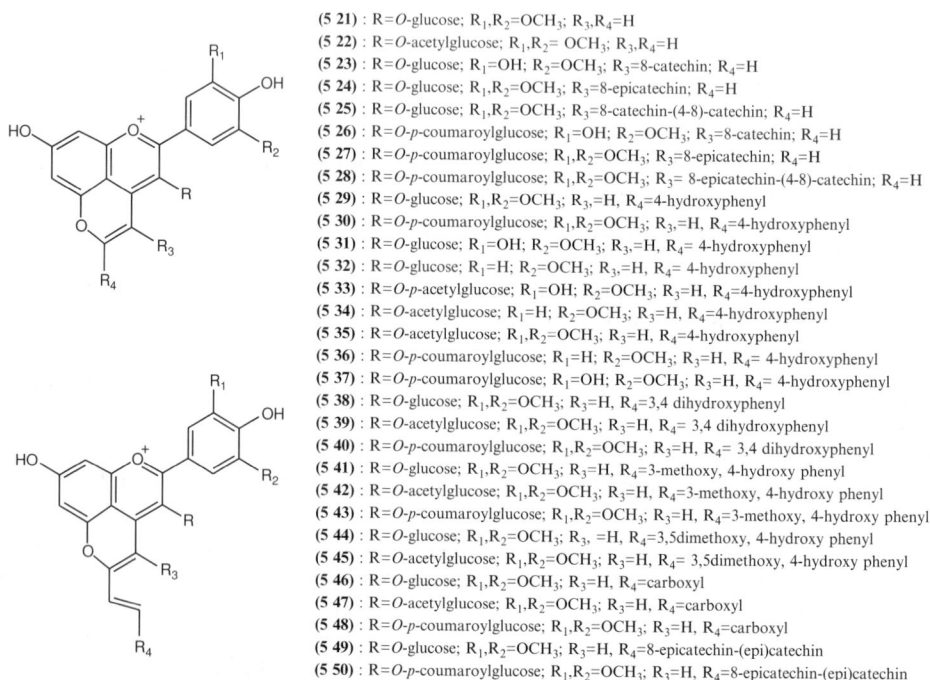

(**5 21**) : R=O-glucose; R_1,R_2=OCH_3; R_3,R_4=H
(**5 22**) : R=O-acetylglucose; R_1,R_2= OCH_3; R_3,R_4=H
(**5 23**) : R=O-glucose; R_1=OH; R_2=OCH_3; R_3=8-catechin; R_4=H
(**5 24**) : R=O-glucose; R_1,R_2=OCH_3; R_3=8-epicatechin; R_4=H
(**5 25**) : R=O-glucose; R_1,R_2=OCH_3; R_3=8-catechin-(4-8)-catechin; R_4=H
(**5 26**) : R=O-p-coumaroylglucose; R_1=OH; R_2=OCH_3; R_3=8-catechin; R_4=H
(**5 27**) : R=O-p-coumaroylglucose; R_1,R_2=OCH_3; R_3=8-epicatechin; R_4=H
(**5 28**) : R=O-p-coumaroylglucose; R_1,R_2=OCH_3; R_3= 8-epicatechin-(4-8)-catechin; R_4=H
(**5 29**) : R=O-glucose; R_1,R_2=OCH_3; R_3,=H, R_4=4-hydroxyphenyl
(**5 30**) : R=O-p-coumaroylglucose; R_1,R_2=OCH_3; R_3,=H, R_4=4-hydroxyphenyl
(**5 31**) : R=O-glucose; R_1=OH; R_2=OCH_3; R_3,=H, R_4= 4-hydroxyphenyl
(**5 32**) : R=O-glucose; R_1=H; R_2=OCH_3; R_3,=H, R_4= 4-hydroxyphenyl
(**5 33**) : R=O-p-acetylglucose; R_1=OH; R_2=OCH_3; R_3=H, R_4=4-hydroxyphenyl
(**5 34**) : R=O-acetylglucose; R_1=H; R_2=OCH_3; R_3=H, R_4=4-hydroxyphenyl
(**5 35**) : R=O-acetylglucose; R_1,R_2=OCH_3; R_3=H, R_4=4-hydroxyphenyl
(**5 36**) : R=O-p-coumaroylglucose; R_1=H; R_2=OCH_3; R_3=H, R_4= 4-hydroxyphenyl
(**5 37**) : R=O-p-coumaroylglucose; R_1=OH; R_2=OCH_3; R_3=H, R_4= 4-hydroxyphenyl
(**5 38**) : R=O-glucose; R_1,R_2=OCH_3; R_3=H, R_4=3,4 dihydroxyphenyl
(**5 39**) : R=O-acetylglucose; R_1,R_2=OCH_3; R_3=H, R_4= 3,4 dihydroxyphenyl
(**5 40**) : R=O-p-coumaroylglucose; R_1,R_2=OCH_3; R_3=H, R_4= 3,4 dihydroxyphenyl
(**5 41**) : R=O-glucose; R_1,R_2=OCH_3; R_3=H, R_4=3-methoxy, 4-hydroxy phenyl
(**5 42**) : R=O-acetylglucose; R_1,R_2=OCH_3; R_3=H, R_4=3-methoxy, 4-hydroxy phenyl
(**5 43**) : R=O-p-coumaroylglucose; R_1,R_2=OCH_3; R_3=H, R_4=3-methoxy, 4-hydroxy phenyl
(**5 44**) : R=O-glucose; R_1,R_2=OCH_3; R_3, =H, R_4=3,5dimethoxy, 4-hydroxy phenyl
(**5 45**) : R=O-acetylglucose; R_1,R_2=OCH_3; R_3=H, R_4= 3,5dimethoxy, 4-hydroxy phenyl
(**5 46**) : R=O-glucose; R_1,R_2=OCH_3; R_3=H, R_4=carboxyl
(**5 47**) : R=O-acetylglucose; R_1,R_2=OCH_3; R_3=H, R_4=carboxyl
(**5 48**) : R=O-p-coumaroylglucose; R_1,R_2=OCH_3; R_3=H, R_4=carboxyl
(**5 49**) : R=O-glucose; R_1,R_2=OCH_3; R_3=H, R_4=8-epicatechin-(epi)catechin
(**5 50**) : R=O-p-coumaroylglucose; R_1,R_2=OCH_3; R_3=H, R_4=8-epicatechin-(epi)catechin

FIGURE 5.11 Structures of pyranoanthocyanins.

structure.[17] Reaction of grape anthocyanin 3-glucosides with acetaldehyde yielded the corresponding series of pyranoanthocyanins that were also formed after yeast fermentation of a synthetic must containing anthocyanins.[57]

Another product obtained by incubation of procyanidin dimer B2, malvidin 3-glucoside, and acetaldehyde in a model wine solution was tentatively identified to a pyranoanthocyanin–procyanidin adduct from its mass and UV–visible spectrum.[54] Other flavanyl-pyranoanthocyanins based on malvidin 3-glucoside or its acetyl and *p*-coumaroyl esters and (epi)catechin monomer through tetramer were detected in wine or marc extracts.[202] Complete structural identification was provided by mass spectrometry and two-dimensional NMR spectroscopy for the (+)-catechin, (−)-epicatechin, and procyanidin B3 (catechin–(4α-8)–catechin) derivatives of pyranomalvidin 3-glucoside (**5 23, 5 24, 5 25**),[203] and for the (+)-catechin, (−)-epicatechin, and procyanidin B1 (epicatechin–(4β-8)–catechin) derivatives of pyranomalvidin 3-*p*-coumaroyl-glucoside (**5 26, 5 27, 5 28**)[53] isolated from Port wine. Flavanyl-pyranoanthocyanins may arise from nucleophilic addition of the flavanol onto the pyranoanthocyanin or from addition of the double bond of a vinylflavanol onto the anthocyanin.[80,201]

5.5.3.3 Other Pyranoanthocyanins

The first two pyranoanthocyanin structures were detected in wine HPLC profiles in 1996.[50] Their UV–visible spectra showed a visible absorbance band with a more pointed shape and hypsochromically shifted compared to that of genuine anthocyanins, but differed from each other by the presence of a shoulder around 313 nm (of the least polar one), characteristic of *p*-coumaroylated derivatives. Their mass spectra (see Section 5.2.2.1) indicated that they were, respectively, the glucoside and *p*-coumaroylglucoside of the same anthocyanin-derived aglycone, as confirmed by hydrolysis experiments. Identical products were obtained by reaction of malvidin 3-glucoside and malvidin 3-*p*-coumaroylglucoside, respectively, with vinylphenol. ^{1}H NMR analysis demonstrated that they were phenylpyranomalvidin 3-glucoside (**5 29**) and phenylpyranomalvidin 3-*p*-couramoylglucoside (**5 30**),[8] formed by the addition of the vinylphenol double bond onto the flavylium followed by an oxidation step. The vinylphenol precursor is known to be present in wine as a result of decarboxylation of *p*-coumaric acid, catalyzed by a side yeast enzymatic activity.[204,205] Similar products were obtained by reaction of vinylphenol with other grape anthocyanins[16] and by reaction of other vinylphenol derivatives (e.g., vinylsyringol) with malvidin 3-glucoside.[206]

Analysis of anthocyanin derivatives by neutral loss scanning for precursor ions with elimination masses corresponding to glucosyl, acetylglucoside, and *p*-coumaroylglucoside residues detected guaiacyl (3-methoxy, 4-hydroxyphenyl), catechyl (3,4-dihydroxyphenyl), and syringyl (3,5-dimethoxy, 4-hydroxyphenyl) pyranoanthocyanins (**5 38–5 45**) in addition to the already reported vinylphenol adducts (**5 32–5 37**) in extracts from red grape skin and red wine.[88] These derivatives are produced by reaction of anthocyanins with vinylguaiacol, vinylcatechol, and vinylsyringol, which may be formed by decarboxylation of the corresponding hydroxycinnamic acids, i.e., ferulic acid, caffeic acid, and sinapic acid, known to be present in wine.[207] Catechyl- and guaiacyl-pyranomalvidin 3-glucoside were actually obtained by this reaction and characterized by NMR.[208] Nevertheless, the slow accumulation of catechyl pyranomalvidin 3-glucoside (called Pinotin A) observed during aging of Pinotage wine[209] did not seem compatible with the fast rate reported for addition of vinylphenols onto anthocyanins.[16] An alternative pathway involving reaction of anthocyanins with *p*-hydroxycinnamic acids was demonstrated.[209,210]

Shortly after the discovery of phenyl pyranoanthocyanins, another series of anthocyanin derivatives showing similar UV–visible spectra, suggesting that they were also derived from a pyranoanthocyanin chromophore, and masses differing from those of grape anthocyanins by an excess of 68 units was detected in marc.[10,149,150,206] MS analysis of all these pigments gave characteristic fragments at −44, corresponding to the loss of a carboxylic group, leading to the suggestion of a general carboxypyranoanthocyanin structure. Signals from NMR analysis of the major product were in agreement with this hypothesis. The same malvidin 3-glucoside and malvidin 3-acetylglucoside derivatives were simultaneously detected in wine,[17] but an alternative structure was proposed.[152] Finally, the same product was obtained by reaction of pyruvic acid and malvidin 3-glucoside, definitively confirming the carboxypyranoanthocyanin structure (5 46).[10] Its formation mechanism involves addition of the double bond of the pyruvic acid enol form onto the flavylium cation followed by dehydration and rearomatization, as described above for acetaldehyde addition.

This mechanism can be extrapolated to other enolizable precursors potentially present in wine, including yeast metabolites such as α-ketoglutatic acid and 2-hydroxybutan-2-one,[57] but also to acetone, which can react with anthocyanins during solvent extraction procedures.[57,151] The resulting products are pyranoanthocyanins as presented in Figure 5.11, with $R_3 = CH_2-COOH$, $R_4 = COOH$; $R_3 = H$, $R_4 = CHOH-CH_3$; $R_3 = H$, $R_4 = CH_3$.

Finally, other pyranoanthocyanin derivatives showing maximum absorption in the visible range at 575 nm were recently isolated from Port wine.[15] The characterization of these blue pigments by ESI-MS and NMR showed that they consist of pyranomalvidin 3-glucoside (5 49) and its p-coumaroylated derivative (5 50) linked to a flavanol dimer through a vinyl bridge. Similar pigments were obtained by incubating catechin and carboxypyranomalvidin 3-p-coumaroylglucoside in ethanol–water (20:80, v/v) acidified at pH 2 under oxidative conditions. A mechanism involving nucleophilic addition of a vinylflavanol double bond onto the C10 carbon of carboxypyranoanthocyanin followed by decarboxylation and oxidation was proposed.

5.5.3.4 Condensation Reactions with Other Aldehydes

The implication of aldehydes other than acetaldehyde in flavanol polymerization was first demonstrated in wine-like hydroalcoholic solutions containing catechin and tartaric acid, the major organic acid in wine, oxidized in the presence of catalytic amounts of iron.[9] Replacement of tartaric acid and ethanol by other acids and solvents showed that tartaric acid was essential for this reaction. Reaction products showing identical retention times, UV and mass spectra were obtained in higher rates by incubating glyoxylic acid (COOH–CHO) with catechin. Analysis of the major one by two-dimensional NMR showed that it is a catechin dimer in which both catechin units are linked through a carboxymethine bond, in their C8 positions (5 51), as shown in Figure 5.12 (left). A mechanism involving oxidation of tartaric acid to glyoxylic acid, catalyzed by ferric ions, and condensation of glyoxylic acid with catechin, as described above for acetaldehyde-mediated condensation reactions, was thus postulated. Detection of the intermediate glyoxylic acid adducts and carboxymethine-linked trimers confirmed this hypothesis. Three other flavanol dimers in which the flavanol moieties are linked through C6–C6 or C6–C8 (R or S) carboxylmethine bonds were also identified by NMR spectroscopy.[79] Copper ions can replace iron in metal catalysis of tartaric acid oxidation to glyoxylic acid.[211,212] Other colorless products exhibiting characteristic UV spectra with absorption maxima at 295 nm and a shoulder at 340 nm were formed in the solution containing glyoxylic acid and catechin. Mass spectrometry and NMR analysis showed that they are 6- or 8-formylcatechin derivatives, probably arising from decarboxylation of the intermediate glyoxylic acid catechin adducts.[213]

Several aldehydes, namely furfural, 5-hydroxymethylfurfural,[214] isovaleraldehyde, benzaldehyde, propionaldehyde, isobutyraldehyde, formaldehyde, and 2-methylbutyraledehyde,[215] were shown to react with catechin and malvidin 3-glucoside in the same way as acetaldehyde or

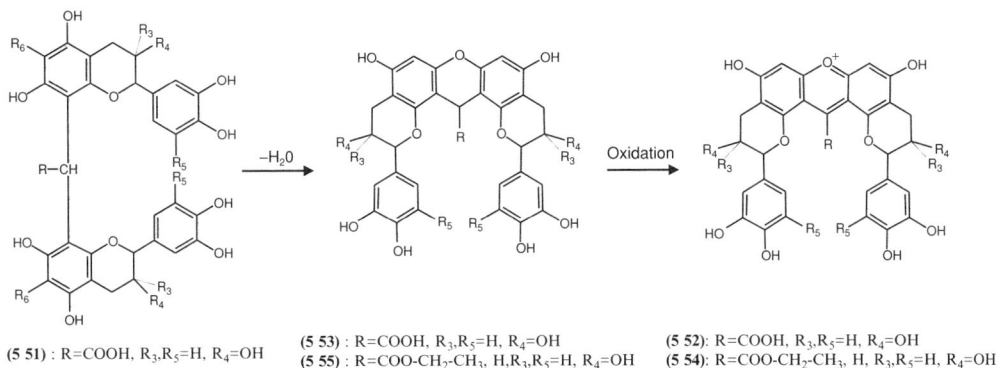

(5 51) : R=COOH, R₃,R₅=H, R₄=OH

(5 53) : R=COOH, R₃,R₅=H, R₄=OH
(5 55) : R=COO-CH₂-CH₃, H,R₃,R₅=H, R₄=OH

(5 52): R=COOH, R₃,R₅=H, R₄=OH
(5 54): R=COO-CH₂-CH₃, H, R₃,R₅=H, R₄=OH

FIGURE 5.12 Structure of catechin–(8-(CH–COOH)-8)–catechin) (**5 51**) and formation of its xanthene (**5 53**) and xanthylium (**5 52**) derivatives.

glyoxylic acid. Furfural and 5-hydroxymethylfurfural are sugar degradation products that may be formed during toasting of oak barrels and present in barrel-aged wines, whereas all the others are constituents of spirits used in the production of fortified wines such as Port wine.

The carboxymethine-linked catechin dimers are unstable and proceed to yellow products, showing absorbance maxima around 440 and 460 nm.[10,216] The major compound formed by incubating the (8–8) linked dimer in hydroalcoholic solution was isolated.[217] Its UV–visible spectrum in acidic medium showed two maxima at 273 and 444 nm that were bathochromically shifted after addition of sodium hydroxide, suggesting a xanthylium structure. Mass spectrometry data and complete assignment of the protons and carbons from two-dimensional NMR experiments led to the proposal of the xanthylium structure (**5 52**). This molecule may result from cyclization of the carboxymethine dimer followed by oxidation of the resulting xanthene (**5 53**), which was also detected in the solution. The xanthene was then obtained by reduction of the xanthylium and its structure confirmed by MS and NMR analysis.[217] The other pigment formed in hydroalcoholic solution was shown to be the ethyl ester of the carboxy xanthylium (**5 54**). Incubation of the other carboxymethine bis-catechin isomers yielded other isomers of the carboxy and ethylcarboxy xanthylium salts, through 5–7 or 7–7 dehydration.[12] Mass signals corresponding to the xanthene and xanthylium arising from dehydration of the carboxymethine (epi)catechin dimers were found in wine stored with no special care to avoid oxidation,[12] demonstrating the occurrence of such reactions in wine. Furfuryl- and hydroxymethylfurfuryl-bridged catechin dimers also proceeded to the corresponding xanthene and xanthylium salts[214] (structures as shown in Figure 5.12, with R = furfuryl, hydroxymethylfurfuryl).

Other xanthylium salt structures in which one of the catechin A-rings was substituted with an hydroxyl (in 6 or 8 position) or an ethylcarboxy group (in C8) were proposed on the basis of their NMR and MS data and postulated to result from condensation of the formylcatechin derivatives.[213] Finally, carboxymethine-linked trimeric structures containing xanthylium and quinonoidal moieties were postulated for pigments showing absorbance maxima at 560 nm, on the basis of their mass signals at *m/z* 959 and 961 in the positive ion mode.[218]

5.5.4 FACTORS CONTROLLING FLAVONOID REACTIONS IN WINE

As discussed above, studies performed in wine-like model systems and identification of flavonoid-derived species in grape, musts, wines, or marcs have confirmed a number of reaction mechanisms and shown that they actually occur under conditions encountered in wine. Major processes thus demonstrated are based on

- Nucleophilic addition of anthocyanins and flavanols on electrophiles such as quinones, flavylium cations, protonated aldehydes, and carbocations resulting from acid-catalyzed cleavage of proanthocyanidins.
- Acid-catalyzed cleavage of proanthocyanidins and methylmethine-linked species, generating flavanol carbocations and vinylflavanols, respectively.
- Formation of pyranoanthocyanins through reaction of flavylium cations with compounds possessing a polarizable double bond, namely vinylphenol derivatives (including vinylflavanols and hydroxycinnamic acids) and enolizable aldehydes and ketones (e.g., acetaldehyde and pyruvic acid).

It should be emphasized that all known flavonoid derivatives are present in wines only in small amounts so that they account together for a minor proportion of wine polyphenol composition. However, detection of series of derivatives based on all grape anthocyanins or flavanols with different chain lengths suggests that each of the rather simple molecules identified can be considered as a marker of a whole group of similar structures. In addition, most of the primary structures formed are also highly reactive and can undergo acid-catalyzed cleavage or nucleophilic addition reactions similar to those of their precursors, leading to increased structural diversity.

As these reactions occur simultaneously in wine, flavonoid composition depends on the availability of the different precursors involved, the relative reaction rates, and product stability. The respective levels of anthocyanins and flavanols in wine are primarily determined by grape composition, itself influenced by the variety and degree of ripening (see Section 5.3.2), but can be modulated by extraction processes (see Section 5.4). Since both species compete in nucleophilic addition processes, high levels of anthocyanins should result in increased amounts of anthocyanin polymers and anthocyanin–flavanol adducts while high levels of flavanols favor the formation of flavanol polymers and their xanthylium derivatives. Other important precursors include yeast metabolites (e.g., aldehydes, pyruvic acid, vinylphenols) and oxidation products of ethanol (i.e., acetaldehyde) and tartaric acid (i.e., glyoxylic acid). Specific wine-making processes such as addition of wine spirits as performed in Port wine technology also increase the level of acetaldehyde, and possibly of other aldehydes, which explains why many of the flavanyl-pyranoanthocyanins[53,203] as well as the vinylflavanyl-pyranoanthocyanins[15] were first identified in these wines.

Finally, reactions of flavonoid and nonflavonoid precursors are affected by other parameters like pH, temperature, presence of metal catalysts, etc. In particular, pH values determine the relative nucleophilic and electrophilic characters of both anthocyanins and flavanols. Studies performed in model solutions showed that acetaldehyde-mediated condensation is faster at pH 2.2 than at pH 4 and limited by the rate of aldehyde protonation.[64] The formation of flavanol–anthocyanin adducts was also limited by the rate of proanthocyanidin cleavage, which was shown to take place at pH 3.2,[153] but not at pH 3.8.[58] Nucleophilic addition of anthocyanins was faster at pH 3.4 than at pH 1.7,[55] but still took place at pH values much lower than those encountered in wine, as evidenced by the formation of anthocyanin–caffeoyltartaric acid adducts,[55] methylmethine anthocyanin–flavanol adducts,[18,56] and flavanol–anthocyanin adducts.[58] The formation of pyranoanthocyanins requiring the flavylium cation was faster under more acidic conditions, as expected, but took place in the whole wine pH range. Thus, the availability of either the flavylium or the hemiketal form does not seem to limit any of the anthocyanin reactions.

Methylmethine flavanol oligomers are more susceptible to acid-catalyzed cleavage than proanthocyanidins. C–C bonds between the methylmethine bridge and the anthocyanin unit are extremely resistant in strongly acidic medium (e.g., thiolysis conditions) where the anthocyanin is in the flavylium form[56] whereas they are rather labile at wine pH,[18] probably due to the higher proportion of anthocyanins in the hemiketal form. Cleavage of the methylmethine bonds results in rearrangement to other unstable ethyl-linked species[64] or to flavanyl-pyranoanthocyanins[19,201] and vinylflavanyl-pyranoanthocyanins.[15]

Pyranoanthocyanins are extremely stable compared to anthocyanins and methylmethine-linked adducts but a decrease in their concentration was observed in model solutions,[219,220] possibly due to their involvement in more complex structures. Higher temperatures speed up not only the formation of carboxypyranoanthocyanins but also their degradation, so that higher concentrations were reached at temperatures 10 to 15°C, which are usual for wine storage.[220] The rate of anthocyanin hemiketal isomerization to the unstable chalcone form is also highly influenced by temperature.[221]

Oxidation is another important factor for the wine-aging process. Major oxidation reactions taking place in wine following oxygen exposure actually involve other wine constituents that are primarily ethanol and, in the presence of metal ions, tartaric acid[9] rather than flavonoids, although phenolic compounds have been shown to participate in oxidation of ethanol to acetaldehyde.[188]

Wine is exposed to oxygen in transfer operations such as racking or bottling. Some oxygen may also dissolve at the wine surface during storage, especially in barrels. A process referred to as micro-oxygenation, in which oxygen is supplied continuously in quantities small enough to avoid any accumulation, has been proposed to mimic oxidation conditions encountered in barrel aging. Monitoring of flavonoid composition in control and micro-oxygenated wines showed that concentration of anthocyanins and flavanols decreased during aging.[52] Oxygenated wines contained significantly lower levels of anthocyanins and flavanols than the control wines. The concentration of methylmethine-linked species decreased in control wines, confirming the lability of these pigments,[18] and increased in oxygenated wines as a result of acetaldehyde accumulation. Other derived pigments including pyranoanthocyanins and flavanol–anthocyanin adducts accumulated during storage and were significantly more abundant in the oxygenated wines, possibly because their formation mechanisms require an oxidation step to recover the flavylium moiety. Oxidation reactions are also responsible for the formation of yellow pigments from flavanols. Catechin auto-oxidation in wine-like medium yields the same products as enzymatic oxidation although much slower.[216] In the presence of iron or copper ions, oxidation of tartaric acid takes over, leading to carboxy-methine-linked flavanol oligomers[9] and xanthylium salts derived from them after oxidation of the intermediate xanthene,[78,222] which are also formed by ascorbic acid-induced oxidation.[223,224]

5.6 PHYSICOCHEMICAL AND ORGANOLEPTIC PROPERTIES OF GRAPE AND WINE FLAVONOIDS

5.6.1 Impact of Flavonoid Reactions on Wine Color

The color of an anthocyanin solution is determined by the proportions of the different anthocyanin forms, namely red flavylium cation, violet quinonoidal bases, colorless water or sulfite adducts, and, finally, yellow chalcones. At wine pH, the C2-water adduct (hemiketal or carbinol and its open-chain *cis*-retrochalcone isomer) is actually the predominant form of malvidin 3-glucoside and other grape anthocyanin monoglucosides.[144] These species do not contribute red color. In addition, the chalcone is unstable. Cleavage of this open-chain form generates 2,4,6-trihydroxybenzaldehyde from the A-ring[225] and hydroxycinnamic acids from the B-ring (e.g., syringic acid in the case of malvidin 3-glucoside).[144]

Thus, the intense red wine color and its preservation over years require some pigment stabilizing mechanisms to take place. Such stabilization is achieved, on the one hand, through complexation of the anthocyanin chromophores with other species and, on the other hand, through conversion of labile anthocyanins to more stable derived pigments. The former mechanism may be the first step leading to the latter.[226]

Molecules involved in association with anthocyanins can be an identical anthocyanin molecule, an aromatic acyl substituent in the anthocyanin itself, or another molecule, the processes referred to as self-association, intramolecular copigmentation, and intermolecular copigmentation, respectively. Their mechanisms have been thoroughly investigated and are described in detail in excellent reviews.[227,228] The major driving force is hydrophobic vertical stacking to form $\pi-\pi$ complexes from which water is excluded. Both the flavylium cations and quinonoidal bases but not the hemiketal form are planar hydrophobic structures that can stack to protect themselves from the water environment. The enhanced color intensity resulting from self-association or copigmentation is due to a shift of the hydration balance toward the pigment forms involved in these stable complexes. It can thus be expected to be particularly important in the wine pH range where hydrated forms normally predominate. The bathochromic effect often associated with copigmentation is attributed to the larger amount of quinonoidal base formed by deprotonation of the flavylium. The role of copigmentation in wine color can be estimated by comparing wine absorbance values in the visible range before and after disruption of copigmentation complexes by dilution in a wine-like buffer.[229] Copigmentation has been reported to account for 30 to 50% of the color of young red wines, on the basis of such measurements.

The conversion of anthocyanins to the various pigments mentioned in the earlier sections increases the range of available colors. Moreover, substitutions of the C-ring as encountered in some of the derivatives impede nucleophilic addition of sulfite[230] or water,[231,232] thus increasing color stability.

Pyranoanthocyanins are orange pigments[10,16,50,54,57,202] but further substitution with vinylbenzyl derivatives yield blue colors.[15,233] These pigments are remarkably resistant to sulfite bleaching and hydration compared to anthocyanins.[16,17] Color intensity was only reduced by half over the pH range 2.2 to 7.2, whereas a malvidin 3-glucoside solution is almost colorless at pH 4 to 5.5. Above pH 6, a bathochromic shift of the absorbance maximum in the visible region and the appearance of another maximum around 600 nm indicated the presence of neutral and anionic quinonoidal bases.[17] Pyranoanthocyanins are also more stable over time than anthocyanins themselves,[16] so that their contribution to wine color is expected to increase during aging. Carboxypyranoanthocyanins were actually the major pigments detected in the HPLC profile of a grape marc extract[10] and could serve as markers of changes taking place during wine aging.[52]

The UV–visible spectra of anthocyanin oligomers and anthocyanin–flavanol adducts resulting from condensation with aldehydes are bathochromically shifted compared to those of their precursors[5] (10 nm for linear substituents, 20 nm for branched substituents[215]). The molar extinction coefficient of methylmethine-linked catechin–malvidin 3-glucoside adduct in 10% ethanol solution adjusted to pH 0.5 with hydrochloric acid (17,100)[18] is slightly lower than that of malvidin 3-glucoside (20,200).[234] The methylmethine-catechin derivative is much more resistant to discoloration through hydration and sulfite bleaching than genuine grape anthocyanins.[18] Since the C-ring of the anthocyanin moiety in the dimer is not substituted, its greater protection against nucleophilic attack of water (and sulfites) may be due to stabilization through sandwich-type stacking as demonstrated for similar products obtained from a synthetic anthocyanin.[235]

The hydration and protonation reactions of the methylmethine-linked malvidin 3-glucoside dimer can be summarized as follows, with AH^+, AOH, and A representing the flavylium, hemiketal, and quinonoidal base forms, respectively:

$$AH^+-CH-CH_3-AH^+ \xrightleftharpoons[K_{h1}]{+H_2O/-H^+} AH^+-CH-CH_3-AOH \xrightleftharpoons[K_{h2}]{+H_2O/-H^+} AOH-CH-CH_3-AOH \quad (4)$$

$$fAH^+-CH-CH_3-AOH \xrightleftharpoons[K]{} A-CH-CH_3-AOH + H^+ \quad (5)$$

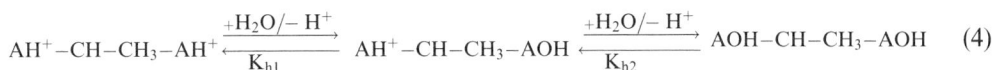

Spectrophotometric studies were conducted from pH 0.1 to 5.7. The thermodynamic constants calculated, assuming that the absorbance of AH^+–AH^+ was twice that of AH^+–AOH and that the second hydration constant (pK_{h2}) was equal to the proton transfer constant (pK), were 1.8 (pK_{h1}) for the first hydration reaction and 4.6 ($pK_{h2} \approx pK$) for the second,[19,236] whereas the hydration and proton transfer constants calculated for malvidin 3-glucoside are 2.6 and 4.25, respectively.[144] Based on these hydration constants, the only significant form of the methylmethine dimer at wine pH is AH^+–AOH, in which one of the anthocyanin moieties is under the red flavylium form and the other one is hydrated, as predicted from mass spectrometry data. Thus, conversion of grape anthocyanins (75 to 80% colorless AOH, 20 to 25% red AH^+ in wine pH range) to the methylmethine bis-anthocyanin (50% AOH, 50% AH^+) may be responsible not only for a shift toward a more purple tint but also for a twofold increase in color intensity.

The flavylium ions of direct flavanol–anthocyanin adducts (i.e., A^+–F and F–A^+) and anthocyanin dimers (A^+–AOH) have the same UV–visible spectra as anthocyanins. Species in which the anthocyanin is substituted in the 4-position (A^+–F, A^+–AOH) are expected to be resistant to sulfite bleaching and hydration whereas F–A^+ is as susceptible to water addition as their anthocyanin precursor.[237]

Color changes were monitored in solutions containing malvidin 3-glucoside alone or with procyanidin B2 3′-gallate incubated at pH 2 and pH 3.8.[58] Absorbance values at 520 nm of the pH 2 solutions decreased over time and were highly correlated with the amount of malvidin 3-glucoside. However, the 520 nm absorbance values of the same solutions measured after adjusting the pH at 3.8 remained constant throughout the incubation period, meaning that formation of new pigments that are less susceptible to hydration and mostly present in colored forms at pH 3.8 compensated for the loss of malvidin 3-glucoside. Increased resistance to sulfite bleaching was also observed, especially in the solution containing both the pigment and the flavanol. Similar trends, with slightly higher proportions of sulfite bleaching resistant pigments, were observed in the solutions incubated at pH 3.8, containing A^+–F and A^+–AOH adducts. Browning (estimated by the 420 nm absorbance values) occurred in the solution containing B2 3′-gallate incubated at pH 2 and in both solutions incubated at pH 3.8. Absorbance values at 620 nm also increased over time, especially in solutions incubated at pH 3.8, suggesting that some of the derived pigments were under the quinonoidal form. Although the dimeric and trimeric reaction products detected in these solutions are present in too low amounts to explain their color properties, they may serve as markers of reaction processes leading to a whole range of pigments based on similar structures.

The influence of controlled oxygenation on color characteristics of red wine was studied and correlated with changes in flavonoid composition over a 7-month period.[52] Pigments formed during aging were less red and more yellow and showed higher resistance to sulfite bleaching than their anthocyanin precursors, as described earlier,[3] whereas those resulting from oxygenation were more purple. Higher levels of pyranoanthocyanins and methylmethine-linked pigments were associated with aging and oxidation, respectively, suggesting that both types of derivatives play a part in the observed color changes.

Browning of white wines was shown to be correlated to their flavanol content.[139,238] Flavanol auto-oxidation and glyoxylic acid-mediated condensation resulting from oxidation of tartaric acid may contribute to the browning process. The latter mechanism yields much more intense xanthylium yellow pigments[12,218] and may also be involved in pinking of white wine,[239,240] since some of the products resulting from glyxoxylic acid-mediated reactions are purple pigments.[213]

5.6.2 IMPACT OF FLAVONOID REACTIONS ON WINE TASTE PROPERTIES

The major organoleptic character associated with flavonoids is astringency although the lower molecular weight flavanols have also been reported to contribute bitterness.[21,241,242] The physiological grounds of astringency that is described as drying, roughing, or puckering of the mouth mucosa are still obscure. However, it is generally accepted that it is not a taste perceived through recognition by taste receptors, but a tactile sensation.[243–247]

Astringency of tannins results from their interactions with salivary proteins and glycoproteins, in particular proline rich proteins, causing a loss in the lubricating power of the saliva, or with the glycoproteins of the mouth epithelium. Flavonoid protein interactions are reviewed in Chapter 8. Briefly, the affinity of polyphenols for proteins depends primarily on the number of phenolic moieties, which are the major interactions sites in the molecule,[248,249] the presence of several phenolic rings in a tannin molecule enabling it to build bridges between the proteins[250] or with other polyphenols.[251] All flavonoids can precipitate proteins if present in sufficient amounts[250] but precipitation increases with the degree of polymerization and the number of galloyl units in the polyphenol structure.[162,252–256] Nevertheless, precipitation does not necessarily reflect astringency that might also be related to conformational changes in the protein structure induced by formation of soluble complexes with tannins.[257]

Spectroscopic methods such as NMR,[251,258–262] MS,[263] and light scattering[262,264–266] have been used to study auto-association of flavonoids and their complexation with peptides in solution. Mechanisms involving hydrophobic interactions and hydrogen bonding were thus proposed. In addition, colloidal particles derived from flavanol aggregation might play an important role in tannin associations with macromolecules.[264,267]

Within a series of flavanol monomers and dimers, self-association and formation of soluble complexes with peptides, detected by MS, increased with the chain length and with the presence of galloyl substituents.[263] Aggregation of lower molecular weight flavanols increased with their molecular weight but particle size decreased with larger polymers.[264] Methylmethine-linked catechin dimers also formed colloidal particles.[268] Aggregation is strongly influenced by ethanol concentration and ionic strength. Moreover, the presence of polysaccharides was shown to modify flavanol aggregation[264,265] as well as precipitation of pentagalloylglucose[260] and wine proanthocyanidins[269] by gelatins.

Proanthocyanidin astringency has been reported to increase with chain length, up to the decamer level, and to decrease beyond this value, as the polymers become insoluble.[21] However, higher molecular weight proanthocyanidins (mDP > 20) were shown to be present in a red wine and selectively precipitated by proteins used as fining agents,[270,271] meaning that they were soluble and presumably astringent.

Assessment of taste is achieved by sensory analysis, from very simple experiments such as triangular tests aiming at determining detection thresholds to complex descriptive analysis approaches. A method referred to as time–intensity that consists in recording continuously the intensity of a given sensation over time under standardized conditions has been applied to study flavonoid bitterness and astringency properties.[247,272–279]

Recent studies performed using this method have shown that flavanol bitterness decreases from monomer to trimer.[242] Epicatechin was perceived more bitter than catechin and the C4–C6-linked catechin dimer more bitter than other procyanidin dimers with C4–C6 linkages. This may be due to the higher lipophilic character of these molecules facilitating their diffusion to the gustatory receptor.[21] Bitterness of procyanidin fractions in 5% ethanol decreased with their mean degree of polymerization (3, 10, 70).[280] No bitterness was detected when tasting the same fractions in a wine-like solution containing 13% ethanol and acidified with tartaric acid.[20]

Astringency was classically considered to increase with flavanol chain length and decrease beyond the octamer level,[21,281] but, to our knowledge, this had never been confirmed experimentally as higher molecular weight fractions were not available. In fact, recent sensory studies performed on a series of proanthocyanidin fractions isolated from grape or apple and differing in chain length, galloylation rate, and content of epigallocatechin units showed that polymeric fractions (mDP 30, 70) were by far the most astringent.[20] Larger molecular weight proanthocyanidins extracted from apple (mDP 70), grape seeds (mDP 10, 20% galloylated units), and grape skins (mDP 20, 5% galloylated units, 20% prodelphinidin units) exhibited similar astringency when tasted at the same concentration (0.5 g/l) in a model wine medium, in spite of their large composition differences determined by thiolysis.[20] This confirmed earlier results obtained in citric acid solutions and in white wine.[282] The higher percentage of galloylation in the seed proanthocyanidins actually compensated for their lower molecular weight since the same fraction after degalloylation with tannase was scored similar to the mDP12 fraction from grape skins.

The decrease of astringency occurring during wine aging is usually ascribed to the conversion of proanthocyanidins to less astringent and eventually insoluble derivatives through polymerization reactions. However, the recent findings developed above suggest that this assumption has to be at least partly revised. On the one hand, astringency of flavonoid derivatives increases with their molecular weight so that reactions leading to higher molecular weight species may result in enhanced rather than decreased astringency. On the other hand, flavonoid reactions in wine yield not only larger molecules, through acetaldehyde-induced polymerization and formation of anthocyanin–flavanol adducts, but also lower molecular weight compounds through acid-catalyzed cleavage, especially if large amounts of monomeric flavonoids such as anthocyanins, are present. An average size reduction of wine molecules might be regarded as a possible alternative explanation for the loss of astringency associated with wine aging.

Very little is known of the sensory properties of the various tannin and anthocyanin-derived species identified. Conversion of proanthocyanidins to methylmethine-linked oligomers has been shown to occur during persimon ripening and postulated to participate in astringency reduction.[85,283] However, a mDP5 methylmethine-linked catechin fraction obtained by acetaldehyde-mediated polymerization of catechin was equally astringent as equivalent chain length procyanidins.[284]

The astringency of wine tannin fractions appears to be correlated to the content of flavanol units released after thiolysis regardless of their environment in the original molecules.[282] Anthocyanins contributed neither bitterness nor astringency.[285] Whether incorporation of anthocyanin moieties in tannin-derived structures affects their interactions with proteins and taste properties remains to be investigated.

Taste perception of flavanols is also greatly affected by other constituents of the medium. In particular, lowering of pH leads to a significant increase in astringency whereas increasing the level of ethanol enhances bitterness.[286,287] The gustatory perception of tannins may also be altered by the presence of polysaccharides and proteins. A mechanism involving interaction of tannins with soluble pectins released during ripening, impeding their binding to salivary protein, has been proposed to explain changes occurring during fruit maturation.[249,288,289] The formation of soluble and colloidal polysaccharide–tannin complexes in wine-like model systems was demonstrated by light scattering.[264] Polysaccharides isolated from wine inhibited aggregation of flavanols, except type II rhamnogalaturoanan dimer, which enhanced it.[264] Carbohydrates of different origins also solubilized flavanol–protein complexes, ionic polysaccharides being more effective.[290] Similarly, analyses of the wines before and after protein fining suggested that the reduction of astringency induced by fining was due to the presence of soluble tannin–protein complexes, along with removal of highly polymerized and highly galloylated tannins.[271,291]

A sensory study based on an incomplete factorial design allowed to demonstrate that astringency of procyanidins was reduced in the presence of rhamnogalaturonan II added at levels encountered in wine but was modified neither by anthocyanins nor by the other wine polysaccharides (mannoproteins and arabinogalactan proteins).[292] Increase in ethanol level resulted in higher bitterness perception but had no effect on astringency.

The role of colloidal phenomena in astringency perception cannot be easily inferred from the available data. Aggregation kinetics and particle size as well as astringency intensity increased with the concentration of flavanols and with their chain length. Factors decreasing flavanol particle size such as presence of ethanol or of manoproteins and arabinogalactan proteins had no effect on astringency perception. In contrast, the presence of RGII and proteins, both of which increase particle size, reduced astringency perception, possibly because the flavanol involved in these aggregates could no longer interact with salivary or mouth proteins.

REFERENCES

1. Ribéreau-Gayon, P., Recherches sur les anthocyanes des végétaux, Application au genre Vitis, Thesis Faculté des Sciences de Bordeaux, Bordeaux, 1959.
2. Ribéreau-Gayon, P., Les composés phénoliques du raisin et du vin II. Les flavonosides et les anthocyanosides. *Ann. Physiol. Vég. 6*, 211, 1964.
3. Somers, T.C., The polymeric nature of wine pigments. *Phytochemistry 10*, 2175, 1971.
4. Haslam, E., In vino veritas: oligomeric procyanidins and the ageing of red wines. *Phytochemistry 19*, 2577, 1980.
5. Timberlake, C.F. and Bridle, P., Interactions between anthocyanins, phenolic compounds, and acetaldehyde and their significance in red wines. *Am. J. Enol. Vitic. 27*, 97, 1976.
6. Cheynier, V. et al., ESI-MS analysis of polyphenolic oligomers and polymers. *Analusis 25*, M32, 1997.
7. Remy, S. et al., First confirmation in red wine of products resulting from direct anthocyanin–tannin reactions. *J. Sci. Food Agric. 80*, 745, 2000.
8. Fulcrand, H. et al., Structure of new anthocyanin-derived wine pigments. *J. Chem. Soc. Perkin Trans. 1 7*, 735, 1996.
9. Fulcrand, H. et al., An oxidized tartaric acid residue as a new bridge potentially competing with acetaldehyde in flavan-3-ol condensation. *Phytochemistry 46*, 223, 1997.
10. Fulcrand, H. et al., A new class of wine pigments yielded by reactions between pyruvic acid and grape anthocyanins. *Phytochemistry 47*, 1401, 1998.
11. Saucier, C., Little, D., and Glories, Y., First evidence of acetaldehyde–flavanol condensation products in red wine. *Am. J. Enol. Vitic. 48*, 370, 1997.
12. Es-Safi, N.E. et al., Xanthylium salts formation involved in wine colour changes. *Int. J. Food Sci. Technol. 35*, 63, 2000.
13. Atanasova, V. et al., Structure of a new dimeric acetaldehyde malvidin 3-glucoside condensation product. *Tetrahedron Lett. 43*, 6151, 2002.
14. Mateus, N. et al., Structural diversity of anthocyanin-derived pigments in port wines. *Food Chem. 76*, 335, 2002.
15. Mateus, N. et al., A new class of blue anthocyanin-derived pigments isolated from red wines. *J. Agric. Food Chem. 51*, 1919, 2003.
16. Sarni-Manchado, P. et al., Stability and color of unreported wine anthocyanin-derived pigments. *J. Food Sci. 61*, 938, 1996.
17. Bakker, J. and Timberlake, C.F., Isolation, identification, and characterization of new color-stable anthocyanins occurring in some red wines. *J. Agric. Food Chem. 45*, 35, 1997.

18. Escribano-Bailon, T. et al., Color and stability of pigments derived from the acetaldehyde-mediated condensation between malvidin-3-O-glucoside and (+)-catechin. *J. Agric. Food Chem.* *49*, 1213, 2001.

19. Atanasova, V., Réactions des composés phénoliques induites dans les vins rouges par la technique de micro-oxygénation. Caractérisation de nouveaux produits de condensation des anthocyanes avec l'acétaldéhyde, Thesis ENSAM, Montpellier, 2003.

20. Vidal, S. et al., The mouth-feel properties of grape and apple proanthocyanidins in a wine-like medium. *J. Sci. Food Agric. 83*, 564, 2003.

21. Lea, A.G.H., Bitterness and astringency: the procyanidins of fermented apple ciders. In *Bitterness in Foods and Beverages.* Developments in Food Science 25 (ed. R.L. Rouseff), Elsevier, Amsterdam, 1990, p. 123.

22. Revilla, E., Ryan, J.M., and Martin Ortega, G., Comparison of several procedures used for the extraction of anthocyanins from red grapes. *J. Agric. Food Chem. 46*, 4592, 1998.

23. Escribano-Bailon, M.T. and Santos-Buelga, C., Polyphenol extraction from food. In *Methods in Polyphenol Analysis* (eds C. Santos-Buelga and G. Williamson), The Royal Society of Chemistry, Cambridge, 2003, p. 1.

24. Garcia-Viguera, C., Zafrilla, P., and Tomas-Barberan, F.A., The use of acetone as an extraction solvent for anthocyanins from strawberry fruit. *Phytochemistry 9*, 274, 1998.

25. Prieur, C. et al., Oligomeric and polymeric procyanidins from grape seeds. *Phytochemistry 35*, 781, 1994.

26. Souquet, J.-M. et al., Polymeric proanthocyanidins from grape skins. *Phytochemistry 43*, 509, 1996.

27. Souquet, J.-M. et al., Phenolic composition of grape stems. *J. Agric. Food Chem. 48*, 1076, 2000.

28. Pekic, B. et al., Study of the extraction of proanthocyanidins from grape seeds. *Food Chem. 61*, 201, 1998.

29. Sun, B., Spranger, M., and Ricardo da Silva, J., Extraction of grape seed procyanidins using different organic solvents. In *Polyphenols Communications 96* (eds J. Vercauteren, C. Cheze, M. Dumon, and J. Weber), Groupe Polyphenols, Bordeaux, 1996, p. 169.

30. Sun, B. et al., Separation of grape and wine proanthocyanidins according to their degree of polymerization. *J. Agric. Food Chem. 46*, 1390, 1998.

31. Cheynier, V. and Fulcrand, H., Analysis of proanthocyanidins and complex polyphenols. In *Methods in Polyphenol Analysis* (eds C. Santos-Buelga and G. Williamson), Royal Society of Chemistry, London, 2003, p. 282.

32. Lea, A.G.H. et al., The procyanidins of white grapes and wines. *Am. J. Enol. Vitic. 30*, 289, 1979.

33. Degenhardt, A. et al., Preparative isolation of anthocyanins by high-speed countercurrent chromatography and application of the color activity concept to red wine. *J. Agric. Food Chem. 48*, 5812, 2000.

34. Schwarz, M. et al., Application of high-speed countercurrent chromatography to the large-scale isolation of anthocyanins. *Biochem. Eng. J. 14*, 179, 2003.

35. Vidal, S. et al., Fractionation of grape anthocyanin classes using multilayer coil countercurrent chromatography with step gradient elution. *J. Agric. Food Chem. 52*, 713, 2004.

36. Lea, A.G.H. and Timberlake, C.F., The phenolics of ciders. 1. Procyanidins. *J. Sci. Food Agric. 25*, 1537, 1974.

37. Derdelinckx, G. and Jerumanis, J., Separation of malt hop proanthocyanidins on Fractogel TSK HW-40 (S). *J. Chromatogr. 285*, 231, 1984.

38. Ricardo da Silva, J.M. et al., Procyanidin dimers and trimers from grape seeds. *Phytochemistry 30*, 1259, 1991.

39. Bae, Y.S., Foo, L.Y., and Karchesy, J.J., GPC of natural procyanidin oligomers and polymers. *Holzforschung 48*, 4, 1994.

40. Yanagida, A. et al., Fractionation of apple procyanidins by size-exclusion chromatography. *J. Chromatogr. 855*, 181, 1999.

41. Shoji, T., Yanagida, A., and Kanda, T., Gel permeation chromatography of anthocyanin pigments from rose cider and red wine. *J. Agric. Food Chem. 47*, 2885, 1999.

42. Rigaud, J. et al., Normal-phase high-performance liquid chromatographic separation of procyanidins from cacao beans and grape seeds. *J. Chromatogr. A 654*, 255, 1993.

43. Lea, A., The phenolics of cider: oligomeric and polymeric procyanidins. *J. Sci. Food Agric. 29*, 471, 1978.

44. Cheynier, V. et al., Size separation of condensed tannins by normal-phase high-performance liquid chromatography. In *Methods in Enzymology, Volume 299. Oxidants and Antioxidants. Part A.* (ed. L. Packer), Academic Press, San Diego, 1999, p. 178.

45. Labarbe, B. et al., Quantitative fractionation of grape proanthocyanidins according to their degree of polymerization. *J. Agric. Food. Chem. 47*, 2719, 1999.

46. Saucier, C. et al., Rapid fractionation of grape seed proanthocyanidins. *J. Agric. Food Chem. 49*, 5732, 2001.

47. Santos-Buelga, C. and Williamson, G., *Methods in Polyphenolics Analysis*. The Royal Society of Chemistry, Cambridge, 2003, p. 383.

48. Wolfender, J.L., Ndjoko, K., and Hostettmann, K., The potential of LC–NMR in phytochemical analysis. *Phytochem. Anal. 12*, 2, 2001.

49. Baldi, A. et al., HPLC/MS application to anthocyanins of *Vitis vinifera* L. *J. Agric. Food Chem. 43*, 2104, 1995.

50. Cameira dos Santos, J.P. et al., Detection and partial characterization of new anthocyanin-derived pigments in wine. *J. Sci. Food Agric. 70*, 204, 1996.

51. Revilla, I. et al., Identification of anthocyanin derivatives in grape skin extracts and red wines by liquid chromatography with diode array and mass spectrometric detection. *J. Chromatogr. A 847*, 83, 1999.

52. Atanasova, V. et al., Effect of oxygenation on polyphenol changes occurring in the course of wine making. *Anal. Chim. Acta 458*, 15, 2002.

53. Mateus, N. et al., Isolation and structural characterization of new acylated anthocyanin-vinyl-flavanol pigments occurring in aging red wines. *J. Agric. Food Chem. 51*, 277, 2003.

54. Francia-Aricha, E.M. et al., New anthocyanin pigments formed after condensation with flavanols. *J. Agric. Food Chem. 45*, 2262, 1997.

55. Sarni-Manchado, P., Cheynier, V., and Moutounet, M., Reaction of enzymically generated quinones with malvidin-3-glucoside. *Phytochemistry 45*, 1365, 1997.

56. Es-Safi, N.E. et al., Studies on the acetaldehyde-induced condensation of (−)-epicatechin and malvidin 3-O-glucoside in a model solution system. *J. Agric. Food Chem. 47*, 2096, 1999.

57. BenAbdeljalil, C. et al., Mise en évidence de nouveaux pigments formés par réaction des anthocyanes avec des métabolites de levures. *Sci. Aliments 20*, 203, 2000.

58. Salas, E. et al., Reactions of anthocyanins and tannins in model solutions. *J. Agric. Food Chem. 51*, 7951, 2003.

59. Fulcrand, H. et al., Electrospray contribution to structural analysis of condensed tannin oligomers and polymers. In *Plant Polyphenols 2. Chemistry, Biology, Pharmacology, Ecology* (eds G.G. Gross, R.W. Hemingway, T. Yoshida, and S.J. Branham), Kluwer Academic/Plenum Publisher, New York, 1999, p. 223.

60. Fulcrand, H. et al., Study of wine tannin oligomers by on-line liquid chromatography electrospray ionisation mass spectrometry. *J. Agric. Food Chem. 47*, 1023, 1999.

61. Gabetta, B. et al., Characterization of proanthocyanidins from grape seeds. *Fitoterapia 71*, 162, 2000.

62. Cantos, E., Espin, J., and Tomas-Barberan, F., Varietal differences among the polyphenol profiles of seven table grape cultivars studied by LC–DAD–MS–MS. *J. Agric. Food Chem. 50*, 5691, 2002.

63. Fulcrand, H. et al., Study of the acetaldehyde induced polymerisation of flavan-3-ols by liquid chromatography ion spray mass spectrometry. *J. Chromatogr. 752*, 85, 1996.

64. Es-Safi, N.E. et al., Competition between (+)-catechin and (−)-epicatechin in acetaldehyde-induced polymerization of flavanols. *J. Agric. Food Chem. 47*, 2088, 1999.

65. Giusti, M.M. et al., Electrospray and tandem mass spectroscopy as tools for anthocyanin characterization. *J. Agric. Food Chem. 47*, 4657, 1999.

66. de Pascual-Teresa, S., Rivas-Gonzalo, J.C., and Santos-Buelga, C., Prodelphinidins and related flavanols in wine. *Int. J. Food Sci. Technol. 35*, 33, 2000.

67. Barofsky, D., FAB-MS applications in the elucidation of proanthocyanidin structure. In *Chemistry and Significance of Condensed Tannins* (eds R. Hemingway and J. Karchesy), Plenum Press, New York, 1988, p. 175.

68. Friedrich, W., Eberhardt, A., and Galensa, R., Investigation of proanthocyanidins by HPLC with electrospray ionization mass spectrometry. *Eur. Food Res. Technol. 211*, 56, 2000.

69. Thompson, R.S. et al., Plant proanthocyanidins. Part. I. Introduction: the isolation, structure, and distribution in nature of plant procyanidins. *J. Chem. Soc. Perkin Trans. I* 1387, 1972.

70. Foo, L. and Porter, L., Prodelphinidin polymers: definition of structural units. *J. Chem. Soc. Perkin Trans. I* 1186, 1978.

71. Matsuo, T., Tamaru, K., and Itoo, S., Chemical degradation of condensed tannin with phloroglucinol in acidic solvents. *Agric. Biol. Chem. 48*, 1199, 1984.

72. Shen, Z. et al., Procyanidins and polyphenols of *Larix gmelini* bark. *Phytochemistry 25*, 2629, 1986.

73. Rigaud, J. et al., Micro method for the identification of proanthocyanidin using thiolysis monitored by high-performance liquid chromatography. *J. Chromatogr. 540*, 401, 1991.

74. Boukharta, M., Girardin, M., and Metche, M., Procyanidines galloylées du sarment de vigne (*Vitis vinifera*) separation et identification par chromatographie liquide haute performance et chromatographie en phase gazeuse. *J. Chromatogr. 455*, 406, 1988.

75. Balas, L. and Vercauteren, J., Extensive high-resolution reverse 2D NMR analysis for the structural elucidation of procyanidin oligomers. *Magn. Reson. Chem. 32*, 386, 1994.

76. Balas, L., Vercauteren, J., and Laguerre, M., 2D NMR structure elucidation of proanthocyanidins: the special case of the catechin-(4a-8)-catechin-(4a-8)-catechin trimer. *Magn. Reson. Chem. 33*, 85, 1995.

77. Saucier, C. et al., NMR and molecular modeling: application to wine ageing. *J. Chim. Phys. 95*, 357, 1998.

78. Es-Safi, N.E. et al., Structure of a new xanthylium salt derivative. *Tetrahedron Lett. 40*, 5869, 1999.

79. Es-Safi, N.E. et al., 2D NMR analysis for unambiguous structural elucidation of phenolic compounds formed through reaction between (+)-catechin and glyoxylic acid. *Magn. Reson. Chem. 40*, 693, 2002.

80. Mateus, N. et al., Identification of anthocyanin–flavanol pigments in red wines by NMR and mass spectrometry. *J. Agric. Food Chem. 50*, 2110, 2002.

81. Souquet, J.-M., Cheynier, V., and Moutounet, M., Phenolic composition of grape stems. In *XIX emes journées internationales d'études des polyphénols* (eds F. Charbonnier, J.M. Delacotte and C. Ronaldo), France, 1998.

82. Kennedy, J. and Jones, G.P., Analysis of proanthocyanidin cleavage products following acid-catalysis in the presence of excess phloroglucinol. *J. Agric. Food Chem. 49*, 1740, 2001.

83. Downey, M., Harvey, J., and Robinson, S., Analysis of tannins in seeds and skins of Shiraz grapes throughout berry development. *Aust. J. Grape Wine Res. 9*, 15, 2003.

84. Torres, J. and Lozano, C., Chromatographic characterization of proanthocyanidins after thiolysis with cysteamine. *Chromatographia 54*, 523, 2001.

85. Tanaka, T. et al., Chemical evidence for the de-astringency (insolubilisation of tannins) of persimmon fruit. *J. Chem. Soc. Perkin Trans. I* 3013, 1994.

86. Salas, E. et al., Demonstration of the occurrence of flavanol–anthocyanin adducts in wine and in model solutions. *Anal. Chim. Acta 513*, 325, 2004.

87. Remy, S., Caractérisation des composés phénoliques polymériques des vins rouges, Thèse de doctorat INAP-G, Paris, France, 1999.

88. Hayasaka, Y. and Asenstorfer, R.E., Screening for potential pigments derived from anthocyanins in red wine using nanoelectrospray tandem mass spectrometry. *J. Agric. Food Chem. 50*, 756, 2002.

89. Hayasaka, Y. et al., Characterization of proanthocyanidins in grape seeds using electrospray mass spectrometry. *Rapid Commun. Mass Spectrom. 17*, 9, 2003.

90. Krueger, C.G. et al., Matrix-assisted laser desorption/ionization time-of-flight mass spectrometry of polygalloyl polyflavan-3-ols in grape seed extract. *J. Agric. Food Chem. 48*, 1663, 2000.

91. Yang, Y. and Chien, M., Characterization of grape procyanidins using high-performance liquid chromatography/mass spectrometry and matrix-assisted laser desorption/ionization time-of-flight mass spectrometry. *J. Agric. Food Chem. 48*, 3990, 2000.

92. Fulcrand, H. et al., LC-MS study of acetaldehyde induced polymerisation of flavan-3-ols. In *18th International Conference on Polyphenols* (eds J. Vercauteren, C. Chèze, M. Dumon, and J. Weber), Bordeaux, France, 1996, p. 203.

93. Guerra, C. et al., Partial characterization of coloured polymers of flavan-3-ols-anthocyanins by mass spectrometry. *Proc. In vino Anal. Sci.* 124, 1997.

94. Favretto, D. and Flamini, R., Application of electrospray ionization mass spectrometry to the study of grape anthocyanins. *Am. J. Enol. Vitic. 51*, 55, 2000.

95. Cooper, J.J. and Marshall, A.G., Electrospray ionisation Fourier transform mass spectrometric analysis of wine. *J. Agric. Food Chem. 49*, 5710, 2001.

96. Heier, A. et al., Anthocyanin analysis by HPLC/ESI-MS. *Am. J. Enol. Vitic. 53*, 78, 2002.

97. Ros Barcelo, A. et al., The histochemical localization of anthocyanins in seeded and seedless grapes (*Vitis vinifera*). *Sci. Hortic. 57*, 265, 1994.

98. Garcia Florenciano, E. et al., Pattern of anthocyanin deposition in vacuoles of suspension cultured grapevine cells. *Int. J. Exp. Bot. 53*, 47, 1992.

99. Amrani-Joutei, K. and Glories, Y., Etude de la localisation et de l'extractibilité des tanins et des anthocyanes de la pellicule de raisin. In *Oenologie 95, 5ème Symposium International d'Oenologie* (ed. A. Lonvaud-Funel), Lavoisier Tec et Doc, Paris, 1995, p. 119.

100. Merlin, J., Statoua, A., and Brouillard, R., Investigation of the *in vivo* organization of anthocyanins using resonance Raman microspectrometry. *Phytochemistry 24*, 1575, 1985.

101. Pecket, R.C. and Small, C.J., Occurrence, location and development of anthocyanoplasts. *Phytochemistry 19*, 2571, 1980.

102. Markham, K.R. et al., Anthocyanic vacuolar inclusions — their nature and significance in flower colouration. *Phytochemistry 55*, 327, 2000.

103. Amrani-Joutei, K., Localisation des anthocyanes et des tanins dans le raisin. Etude de leur extractibilité. Université Bordeaux II, Bordeaux, 1993.

104. Bourzeix, M., Weyland, D., and Heredia, N., Etude des catéchines et des procyanidols de la grappe de raisin, du vin et d'autres dérivés de la vigne. *Bull. l'OIV 669–670*, 1171, 1986.

105. Su, C. and Singleton, V., Identification of three flavan-3-ols from grapes. *Phytochemistry 8*, 1553, 1969.

106. Piretti, M.V., Ghedini, M., and Serrazanetti, G., Isolation and identification of the polyphenolic and terpenoid constituents of *Vitis vinifera*. v. Trebbiano variety. *Ann. Chim. 66*, 429, 1976.

107. Lee, C.Y. and Jaworski, A.W., Phenolic compounds in white grapes grown in New York. *Am. J. Enol. Vitic. 38*, 277, 1987.

108. Lee, C.Y. and Jaworski, A.W., Identification of some phenolics in white grapes. *Am. J. Enol. Vitic. 41*, 87, 1990.

109. Czochanska, Z., Foo, L., and Porter, L., Compositional changes in lower molecular weight flavans during grape maturation. *Phytochemistry 18*, 1819, 1979.

110. Weinges, K. and Piretti, M.V., Isolierung des C30H26O12-procyanidins B1 aus Wientrauben. *Liebigs Ann. Chem. 748*, 218, 1971.

111. Salagoity-Auguste, M.-H. and Bertrand, A., Wine phenolics — analysis of low molecular weight components by high performance liquid chromatography. *J. Sci. Food Agric. 35*, 1241, 1984.

112. Glories, Y. et al., Identification et dosage de la procyanidine A2 dans les raisin et les vins de *Vitis vinifera* L.C. V. merlot noir, cabernet sauvignon et cabernet franc. In *Polyphenols Communications 96* (eds J. Vercauteren, C. Chèze, M. Dumon, and J.-F. Weber), Groupe Polyphenols, Bordeaux, 1996, p. 153.

113. Lee, C.Y. and Jaworski, A.W., Major phenolic compounds in ripening white grapes. *Am. J. Enol. Vitic. 40*, 43, 1989.

114. Czochanska, Z. et al., Polymeric proanthocyanidins. Stereochemistry, structural units and molecular weight. *J. Chem. Soc. Perkin Trans. I* 2278, 1980.

115. Cheynier, V. et al., The structures of tannins in grapes and wines and their interactions with proteins. In *Wine. Nutritional and Therapeutic Benefits* (ed. T.R. Watkins), American Chemical Society, Washington, DC, 1997, p. 81.

116. Czochanska, Z. et al., Direct proof of a homogeneous polyflavan-3-ol structure for polymeric proanthocyanidins. *J. Chem. Soc. Chem. Commun.* 375, 1979.

117. Park, H.S., Fougere-Rifot, M., and Bouard, J. Les tanins vacuolaires de la baie de Vitis vinifera L var. Merlot à maturité. In *Oenologie 95* (ed. A. Lonvaud-Funel), Lavoisier, Paris, 1995, p. 115.

118. Amrani-Joutei, K., Glories, Y., and Mercier, M., Localisation des tanins dans la pellicule de baie de raisin. *Vitis 33*, 133, 1994.

119. Kennedy, J.A. et al., Composition of grape skin proanthocyanidins at different stages of berry development. *J. Agric. Food Chem. 49*, 5348, 2001.

120. Wulf, L.W. and Nagel, C.W., Analysis of phenolic acids and flavonoids by high-pressure liquid chromatography. *J. Chromatogr. 116*, 271, 1976.

121. Cheynier, V. and Rigaud, J., HPLC separation and characterization of flavonols in the skins of *Vitis vinifera* var. Cinsault. *Am. J. Enol. Vitic. 37*, 248, 1986.

122. Price, S.F. et al., Cluster sun exposure and quercetin in pinot noir grapes and wine. *Am. J. Enol. Vitic. 46*, 187, 1995.

123. Downey, M., Harvey, J., and Robinson, S., Synthesis of flavonols and expression of flavonol synthase genes in the developing grape berries of Shiraz and Chardonnay (*Vitis vinifera* L.). *Aust. Grape Wine Res. 9*, 110, 2003.

124. Lamuela-Raventos, R. and Waterhouse, A.L., A direct HPLC separation of wine phenolics. *Am. J. Enol. Vitic. 45*, 1, 1994.

125. Hmamouchi, M. et al., Flavones and flavonols in leaves of some Moroccan *Vitis vinifera* cultivars. *J. Agric. Food Chem. 47*, 186, 1996.

126. Trousdale, E. and Singleton, V.L., Astilbin and engeletin in grapes and wines. *Phytochemistry 22*, 619, 1983.

127. Mazza, G. and Miniati, E., Grapes. In *Anthocyanins in Fruits, Vegetables and Grains* (eds G. Mazza and E. Miniati), CRC Press, Boca Raton, FL, 1993, p. 149.

128. Roggero, J.P. et al., Composition anthocyanique des cépages. I: Essai de classification par analyse en composantes principales et par analyse factorielle discriminante. *Rev. Fr. Oenol. 112*, 41, 1988.

129. Cheynier, V. and Rigaud, J., Identification et dosage de flavonols du raisin. *Bulletin Liaision Groupe, Polyphenols*, 13, 442, 1986.

130. Larice, J.-L. et al., Composition anthocyanique des cépages. II — Essai de classification sur trois ans par analyse en composantes principales et étude des variations annuelles de cépages de même provenance. *Rev. Fr. Oenol. 121*, 7, 1989.

131. Rigaud, J. et al., Caractérisation des flavonoïdes de la baie de raisin. Application à une étude terroir. In *Oenologie 95* (ed. A. Lonvaud-Funel), Lavoisier, Paris, 1996, p. 137.

132. Haselgrove, L. et al., Canopy microclimate and berry composiiton; the effect of bunch exposure on the phenolic composition of *Vitis vinifera* L cv Shiraz grape berries. *Aust. Grape Wine Res. 6*, 141, 2000.

133. Roggero, J.P., Coen, S., and Ragonnet, B., High performance liquid chromatography survey on changes in pigment content in ripening grapes of Syrah. An approach to anthocyanin metabolism. *Am. J. Enol. Vitic. 37*, 77, 1986.

134. El-Kereamy, A. et al., Exogenous ethylene stimulates the long term expression of genes related to the anthocyanin synthesis in grape berries. *Physiol. Plant 119*, 1, 2003.

135. Kennedy, J.A., Matthews, M.A., and Waterhouse, A.L., Effect of maturity and vine water status on grape skin and wine flavonoids. *Am. J. Enol. Vitic. 53*, 268, 2002.

136. Vidal, S. et al., Mass spectrometric evidence for the existence of oligomeric anthocyanins in grape skins. *J. Agric. Food Chem. 52*, 7144, 2004.

137. Brossaud, F., Analyse des constituants phénoliques du cabernet franc et effet terroir en Anjou-Saumur, Thèse de doctorat ENSA Rennes, France, 1999.

138. de Freitas, V.A.P. and Glories, Y., Concentration and compositional changes of procyanidins in grape seeds and skin of white *Vitis vinifera* varieties. *J. Sci. Food Agric. 79*, 1601, 1999.

139. Cheynier, V. et al., Effect of pomace contact and hyperoxidation on the phenolic composition and quality of Grenache and Chardonnay wines. *Am. J. Enol. Vitic. 40*, 36, 1989.

140. Somers, T.C. and Pocock, K.F., Phenolic assessment of white musts: varietal differences in free-run juices and pressings. *Vitis 30*, 189, 1991.

141. Yokotstuka, K., Effect of press design and pressing pressures on grape juice components. *J. Ferment. Bioeng. 70*, 15, 1990.

142. Labarbe, B., Le potentiel polyphénolique de la grappe de Vitis vinifera var. Gamay noir et son devenir en vinification beaujolaise, ENSA-M, UMI, UMII, Montpellier, France, 2000.

143. Augustin, M., Etude de l'influence de certains facteurs sur les composés phénoliques du raisin et du vin, Université de Bordeaux II, 1996.

144. Brouillard, R. and Delaporte, B., Chemistry of anthocyanin pigments. 2. Kinetic and thermo-dynamic study of proton transfer, hydration, and tautomeric reactions of malvidin 3-glucoside. *J. Am. Chem. Soc.* 99, 8461, 1977.

145. Brouillard, R. and Dubois, J.-E., Mechanism of the structural transformations of anthocyanins in acidic media. *J. Am. Chem. Soc.* 99, 1359, 1977.

146. Cheminat, A. and Brouillard, R., NMR investigation of 3-O(β-D-glucosyl)malvidin structural transformations in aqueous solutions. *Tetrahedron Lett.* 27, 4457, 1986.

147. Berke, B. et al., Bisulfite addition to anthocyanins: revisited structures of colourless adducts. *Tetrahedron Lett.* 39, 5771, 1998.

148. Lu, Y., Foo, L.Y., and Wong, H., Isolation of the first C-2 addition products of anthocyanins. *Tetrahedron Lett.* 43, 6621, 2002.

149. Cheynier, V. et al., Reactivity of phenolic compounds in wine: diversity of mechanisms and resulting products. In *In Vino Analytica Scientia*, Bordeaux, 1997, p. 143.

150. Cheynier, V. et al., Application des techniques analytiques à l'étude des composés phénoliques et de leurs réactions au cours de la vinification. *Analusis 25*, M14, 1997.

151. Lu, Y. and Foo, Y., Unusual anthocyanin reaction with acetone leading to pyranoanthocyanin formation. *Tetrahedron Lett.* 42, 1371, 2001.

152. Bakker, J. et al., Identification of an anthocyanin occuring in some red wines. *Phytochemistry 44*, 1375, 1997.

153. Vidal, S. et al., Changes in proanthocyanidin chain-length in wine-like model solutions. *J. Agric. Food Chem.* 50, 2261, 2002.

154. Lunte, C.E., Wheeler, J.F., and Heineman, W.R., Determination of selected phenolic acids in beer extract by liquid chromatography with voltametric–amperometric detection. *Analyst 113*, 95, 1988.

155. Hapiot, P. et al., Oxidation of caffeic acid and related hydroxycinnamic acids. *J. Electroanal. Chem.* 405, 169, 1996.

156. Crouzet, J. et al., Les enzymes en oenologie. In *Oenologie. Fondements scientifiques et technologiques* (ed. C. Flanzy), Lavoisier, Paris, 1998, p. 361.

157. Fernandez-Zurbano, P. et al., Effects of maceration time and pectolytic enzymes added during maceration on the phenolic composition of must. *J. Food Sci. Technol. Int.* 5, 319, 1999.

158. Somers, T.C. and Ziemelis, G., Flavonol haze in white wines. *Vitis 24*, 43, 1985.

159. Dugelay, I. et al., Role of cinnamoyl esterase activities from enzyme preparations on the formation of volatile phenols during winemaking. *J. Agric. Food Chem.* 41, 2092, 1993.

160. Le Traon-Masson, M.-P. and Pellerin, P., Purification and characterization of two β-D-glucosidases from an *Aspergillus niger* enzyme preparation: affinity and specificity toward glucosylated compounds characteristic of the processing of fruits. *Enzyme Microb. Technol.* 22, 374, 1998.

161. Okamura, S. and Watanabe, M., Purification and properties of hydroxycinnamic acid ester hydrolase from *Aspergillus japonicus. Agric. Biol. Chem.* 46, 1839, 1982.

162. Prieur-Delorme, C., Caractérisation chimique des procyanidines de pépins de raisin *Vitis vinifera*. Application à l'étude des propriétés organoleptiques des vins, Thesis Université Montpellier II, Montpellier, 1994.

163. Dubernet, M., Recherches sur la tyrosinase de Vitis vinifera et la laccase de Botrytis cinerea. Applications technologiques, Université de Bordeaux II, 1974.

164. Richard-Forget, F. and Gauillard, F., A possible involvement of peroxidase in enzymatic browning of pear. In *Plant Peroxidases: Biochemistry and Physiology* (eds C. Obinger, U. Burner, R. Ebermann, C. Penel, and H. Greppin), Université de Genève, Genève, 1996, p. 264.

165. Jiang, Y. and Miles, P.W., Generation of H_2O_2 during enzymic oxidation of catechin. *Phytochemistry 33*, 29, 1993.

166. Gunata, Y.Z., Sapis, J.C., and Moutounet, M., Substrates and carboxylic acid inhibitors of grape polyphenoloxidases. *Phytochemistry 26*, 1573, 1987.

167. Cheynier, V., Basire, N., and Rigaud, J., Mechanism of *trans*-caffeoyl tartaric acid and catechin oxidation in model solutions containing grape polyphenoloxidase. *J. Agric. Food Chem. 37*, 1069, 1989.

168. Cheynier, V., Osse, C., and Rigaud, J., Oxidation of grape juice phenolic compounds in model solutions. *J. Food Sci. 53*, 1729, 1988.

169. Cheynier, V. and Ricardo Da Silva, J.M., Oxidation of grape procyanidins in model solutions containing *trans*-caffeoyl tartaric acid and grape polyphenoloxidase. *J. Agric. Food Chem. 39*, 1047, 1991.

170. Sarni, P. et al., Mechanisms of anthocyanin degradation in grape must-like model solutions. *J. Sci. Food Agric. 69*, 385, 1995.

171. Guyot, S., Vercauteren, J., and Cheynier, V., Colourless and yellow dimers resulting from (+)-catechin oxidative coupling catalysed by grape polyphenoloxidase. *Phytochemistry 42*, 1279, 1996.

172. Sun, W. and Miller, J.M., Tandem mass spectrometry of the B-type procyanidins in wine and B-type dehydrodicatechins in an autoxidation mixture of (+)-catechin and (−)-epicatechin. *J. Mass Spectrom. 38*, 438, 2003.

173. Cheynier, V. et al., Characterization of 2-*S*-glutathionylcaftaric acid and its hydrolysis in relation to grape wines. *J. Agric. Food Chem. 34*, 217, 1986.

174. Salgues, M., Cheynier, V., and Gunata, Z., Oxidation of grape juice 2-*S*-glutathionyl caffeoyl tartaric acid by *Botrytis cinerea* laccase and characterization of a new substance: 2,5 di-*S*-glutathionyl caffeoyl tartaric acid. *J. Food Sci. 5*, 1191, 1986.

175. Piretti, M.V., Serrazanetti, G.P., and Paglione, G., The enzymatic oxidation of (+)catechin in the presence of sodium benzene sulphinate. *Ann. Chim. 67*, 395, 1977.

176. Fulcrand, H. et al., Characterization of compounds obtained by chemical oxidation of caffeic acid in acidic conditions. *Phytochemistry 35*, 499, 1994.

177. Jurd, L. and Waiss, A.C.Jr, Anthocyanins and related compounds — VI. Flavylium salt–phloroglucinol condensation product. *Tetrahedron 21*, 1471, 1965.

178. Jurd, L., Anthocyanidins and related compounds — XI. Catechin–flavylium salt condensation reactions. *Tetrahedron 23*, 1057, 1967.

179. Bishop, P.B. and Nagel, C.W., Characterization of the condensation product of malvidin 3,5-diglucoside and catechin. *J. Agric. Food Chem. 32*, 1022, 1984.

180. Liao, H., Cai, Y., and Haslam, E., Polyphenol interactions. Anthocyanins: co-pigmentation and colour changes in red wines. *J. Sci. Food Agric. 59*, 299, 1992.

181. Jurd, L., Review of polyphenol condensation reactions and their possible occurrence in the aging of wines. *Am. J. Enol. Vitic. 20*, 197, 1969.

182. Geissman, T. and Dittamr, H., A proanthocyanidin from avocado seed. *Phytochemistry 4*, 359, 1965.

183. Jurd, L. and Lundin, R., Anthocyanidins and related compounds — XII. Tetramethylleucocyanidin-phloroglucinol and resorcinol condensation products. *Tetrahedron 24*, 2653, 1968.

184. Jurd, L. and Somers, T.C., The formation of xanthylium salts from proanthocyanidins. *Phytochemistry 9*, 419, 1970.

185. Ribéreau-Gayon, P., The anthocyanins of grapes and wines. In *Anthocyanins as Food Colors* (ed. P. Markakis), Academic Press, New York, 1982, p. 209.

186. Baranowski, J.D. and Nagel, C.W., Kinetic of malvidin-3-glucoside condensation in wine model systems. *J. Food Sci. 48*, 419, 1983.

187. Bakker, J., Pinicelli, A., and Bridle, P., Model wine solutions: colour and composition changes during ageing. *Vitis 32*, 111, 1993.

188. Wildenradt, H.L. and Singleton, V.L., The production of acetaldehyde as a result of oxidation of phenolic compounds and its relation to wine aging. *Am. J. Enol. Vitic. 25*, 119, 1974.

189. Vivar-Quintana, A.M. et al., Formation of anthocyanin-derived pigments in experimental red wines. *Food Sci. Technol. Int. 5*, 347, 1999.

190. Delcour, J.A., Ferreira, D., and Roux, D.G., Synthesis of condensed tannins part 9. The condensation sequence of leucocyanidin with (+)-catechin and with the resultant procyanidins. *J. Chem. Soc. Perkin Trans. I* 1711, 1983.

191. Mirabel, M. et al., Copigmentation in model wine solutions: occurrence and relation to wine aging. *Am. J. Enol. Vitic. 50*, 211, 1999.

192. Santos-Buelga, C. et al., Contribution to the identification of the pigments responsible for the browning of anthocyanin–flavanol solutions. *Z. Lebensm. Unters. Forsch. 209*, 411, 1999.

193. Remy-Tanneau, S. et al., Characterization of a colorless anthocyanin-flavan-3-ol dimer containing both carbon–carbon and ether interflavanoid linkages by NMR and mass spectrometries. *J. Agric. Food Chem. 51*, 3592, 2003.

194. Singleton, V., Berg, H., and Guyomon, J., Anthocyanin color level in port-type wine as affected by the use of wine spirits containing aldehydes. *Am. J. Enol. Vitic. 15*, 75, 1964.

195. Berg, H. and Akiyoshi, M., On the nature of reactions responsible for color behavior in red wine: a hypothesis. *Am. J. Enol. Vitic. 26*, 134, 1975.

196. Roggero, J.P., Coen, S., and Archier, P., Etude par CLHP de la réaction glucoside de malvidine-acétaldéhyde-composés phénoliques. *Connaissance Vigne Vin 21*, 1987.

197. Rivas-Gonzalo, J.C., Bravo-Haro, S., and Santos-Buelga, C., Detection of compounds formed through the reaction of malvidin-3-monoglucoside in the presence of acetaldehyde. *J. Agric. Food Chem. 43*, 1444, 1995.

198. Archier, P., Etude analytique et interprétation de la composition polyphénolique des produits de *Vitis vinifera*, Thèse Université d'Aix-Marseille, 1992.

199. Lee, D.F., Swinny, E.E., and Jones, G.P., NMR identification of ethyl-linked anthocyanin-flavanol pigments formed in model wine ferments. *Tetrahedron Lett. 45*, 1671, 2004.

200. Dallas, C., Ricardo-da-Silva, J.M., and Laureano, O., Interactions of oligomeric procyanidins in model wine solutions containing malvidin-3-glucoside and acetaldehyde. *J. Sci. Food Agric. 70*, 493, 1996.

201. Cheynier, V., Es-Safi, N.-E., and Fulcrand, H., Structure and colour properties of anthocyanins and related pigments. In *International Congress on Pigments in Food and Technology* (eds M.I.M. Mosquera, M.J. Galan, and D.H. Mendez), Sevilla, Spain, 1999, p. 23.

202. Asenstorfer, R.E., Hayasaka, Y., and Jones, G.P., Isolation and structures of oligomeric wine pigments by bisulfite-mediated ion-exchange chromatography. *J. Agric. Food Chem. 49*, 5957, 2001.

203. Mateus, N. et al., Identification of anthocyanin–flavanol pigments in red wines by NMR and mass spectrometry. *J. Agric. Food Chem. 50*, 2110, 2002.

204. Etievant, P.X., Volatile phenol determination in wine. *J. Agric. Food Chem. 29*, 65, 1981.

205. Chatonnet, P. et al., Synthesis of volatile phenols by *Saccharomyces cerevisiae* in wines. *J. Sci. Food Agric. 62*, 191, 1993.

206. Fulcrand, H. et al., New-anthocyanin derived wine pigments. In *18th International Conference on Polyphenols* (eds J. Vercauteren, C. Chèze, M. Dumon, and J. Weber), Bordeaux, France, 1996, p. 259.

207. Baderschneider, B. and Winterhalter, P., Isolation and characterization of novel benzoates, cinnamates, flavonoids, and lignans from riesling wine and screening for antioxidant activity. *J. Agric. Food Chem. 49*, 2788, 2001.

208. Hakansson, A.E. et al., Structures and colour properties of new red wine pigments. *Tetrahedron Lett. 44*, 4887, 2003.

209. Schwarz, M., Wabnitz, T.C., and Winterhalter, P., Pathway leading to the formation of anthocyanin–vinylphenol adducts and related pigments in red wines. *J. Agric. Food Chem. 51*, 3682, 2003.

210. Schwarz, M. and Winterhalter, P., A novel synthetic route to substituted pyranoanthocyanins with unique colour properties. *Tetrahedron Lett. 44*, 7583, 2003.

211. Es-Safi, N.E., Cheynier, V., and Moutounet, M., Effect of copper on oxidation of (+)-catechin in a model solution system. *Int. J. Food Sci. Technol. 38*, 153, 2003.

212. Clark, A.C., Prenzler, P.D., and Scollary, G.R., The role of copper(II) in the bridging reactions of (+)-catechin by glyoxylic acid in a model white wine. *J. Agric. Food Chem. 51*, 6204, 2003.

213. Es-Safi, N.E. et al., New phenolic compounds obtained by evolution of (+)-catechin and glyoxylic acid in hydroalcoholic medium. *Tetrahedron Lett. 41*, 1917, 2000.

214. Es-Safi, N.E., Cheynier, V., and Moutounet, M., Study of the reactions between (+)-catechin and furfural derivatives in the presence or absence of anthocyanins and their implication in food color change. *J. Agric. Food Chem. 48*, 5946, 2000.

215. Pissara, J. et al., Reaction between malvidin 3-glucoside and (+)-catechin in model solutions containing different aldehydes. *J. Food Sci. 68*, 476, 2003.

216. Oszmianski, J., Cheynier, C., and Moutounet, M., Iron-catalyzed oxidation of (+)-catechin in wine-like model solutions. *J. Agric. Food Chem. 44*, 1972, 1996.

217. Es-Safi, N.E. et al., Structure of a new xanthylium salt derivative. *Tetrahedron Lett. 40*, 5869, 1999.

218. Es-Safi, N.E. et al., New phenolic compounds formed by evolution of (+)-catechin and glyoxylic acid in hydroalcoholic solution and their implication in color changes of grape-derived food. *J. Agric. Food Chem. 48*, 4233, 2000.

219. Romero, C. and Bakker, J., Interactions between grape anthocyanins and pyruvic acid, with effects of pH and acid concentration on anthocyanin composition and color in model solutions. *J. Agric. Food Chem. 47*, 3130, 1999.

220. Romero, C. and Bakker, J., Effect of storage temperature and pyruvate on kinetics of anthocyanin degradation, vitisin A formation and color characteristics of model solution. *J. Agric. Food Chem. 48*, 2135, 2000.

221. Brouillard, R. and Lang, J., The hemiacetal-cis-chalcone equilibrium of malvin, a natural anthocyanin. *Can. J. Chem. 68*, 755, 1990.

222. Es-Safi, N.E. et al., New polyphenolic compounds with xanthylium skeletons formed through reaction between (+)-catechin and glyoxylic acid. *J. Agric. Food Chem. 47*, 5211, 1999.

223. Bradshaw, M.P., Prenzler, P.D., and Scollary, G.R., Ascorbic acid-induced browning of (+)-catechin in a model wine system. *J. Agric. Food Chem. 49*, 934, 2001.

224. Bradshaw, M.P. et al., Defining the ascorbic acid crossover from anti-oxidant to pro-oxidant in a model wine matrix containing (+)-catechin. *J. Agric. Food Chem. 51*, 4126, 2003.

225. Piffaut, B. et al., Comparative degradation pathways of malvidin3,5-diglucoside after enzymatic and thermal treatments. *Food Chem. 50*, 115, 1994.

226. Brouillard, R. and Dangles, O., Anthocyanin molecular interactions: the first step in the formation of new pigments during wine aging. *Food Chem. 51*, 365, 1994.

227. Goto, T. and Kondo, T., Structure and molecular stacking of anthocyanins. Flower color variation. *Angew. Chem. Int. Ed. Engl. 30*, 17, 1991.

228. Brouillard, R. and Dangles, O., Flavonoids and flower colour. In *The Flavonoids. Advances in Research Since 1986* (ed. J.B. Harborne), Chapman & Hall, London, 1993, p. 565.

229. Boulton, R., The copigmentation of anthocyanins and its role in the color of red wine: a critical review. *Am. J. Enol. Vitic. 52*, 67, 2001.

230. Timberlake, C.F. and Bridle, P., Flavylium salts resistant to sulphur dioxide. *Chem. Ind.* 1489, 1968.

231. Brouillard, R., Iabucci, G.A., and Sweeny, J.G., Chemistry of anthocyanin pigments. 9. UV–visible spectrophotometric determination of the acidity constants of apigeninidin and three related 3-deoxyflavylium salts. *J. Am. Chem. Soc. 104*, 7585, 1982.

232. Mazza, G. and Brouillard, R., Color stability and structural transformations of cyanidin 3,5-diglucoside and four deoxyanthocyanins in aqueous solutions. *J. Agric. Food Chem. 35*, 1987.

233. Roehri-Stoeckel, C. et al., Synthetic dyes: simple and original ways to 4-substituted flavylium salts and their corresponding vitisin derivatives. *Can. J. Chem. 79*, 1173, 2001.

234. Heredia, F.J. et al., Chromatic characterization of anthocyanins from red grapes — I. pH effect. *Food Chem. 63*, 491, 1998.

235. Escribano-Bailon, M., Dangles, O., and Brouillard, R., Coupling reactions between flavylium ions and catechin. *Phytochemistry 41*, 1583, 1996.

236. Atanosova, V. et al., First evidence of acetaldehyde-induced anthocyanin polymerisation. In *Polyphenol Communications 2002* (ed. I. Hadrami), Marrakech, 2002, p. 417.

237. Salas, E. et al., Structure determination and color properties of a newly synthesized direct-linked flavanol–anthocyanin dimer. *Tetrahedron Lett. 45*, 8725, 2004.

238. Simpson, R.F., Factors affecting oxidative browning of white wine. *Vitis 21*, 233, 1982.

239. Simpson, R.F., Pinking in Australian white table wines. *Aust. Wine Brew. Spirit Rev.* 56, 1977.
240. Simpson, R.F., Oxidative pinking in white wines. *Vitis 16*, 286, 1977.
241. Noble, A., Bitterness and astringency in wine. In *Bitterness in Foods and Beverages* (ed. R. Rousseff), Elsevier, Amsterdam, 1990, p. 145.
242. Gacon, K., Peleg, H., and Noble, A.C., Bitterness and astringency of flavan-3-ol monomers, dimers and trimers. *Food Qual. Prefer. 7*, 343, 1996.
243. Bate-Smith, E., Astringency in foods. *Food 23*, 124, 1954.
244. Breslin, P.A. et al., Psychophysical evidence that oral astringency is a tactile sensation. *Chem. Senses 18*, 405, 1993.
245. Green, B.G., Oral astringency: a tactile component of flavor. *Acta Psychol. 84*, 119, 1993.
246. Clifford, M.N., Astringency. In *Phytochemistry of Fruits and Vegetables* (eds F. Tomas-Barberan and R. Robins), Clarendon Press, Oxford, 1997, p. 87.
247. Lee, C.B. and Lawless, H.T., Time-course of astringent sensations. *Chem. Senses 16*, 225, 1991.
248. Haslam, E., Polyphenol–protein interactions. *Biochem. J. 139*, 285, 1974.
249. Haslam, E. and Lilley, T.H., Natural astringency in foodstuffs. A molecular interpretation. *Crit. Rev. Food Sci. Nutr. 27*, 1, 1988.
250. McManus, J.P. et al., Polyphenol interactions. Part 1. Introduction; some observations on the reversible complexation of polyphenols with proteins and polysaccharides. *J. Chem. Soc. Perkin Trans. II* 1429, 1985.
251. Baxter, N.J. et al., Multiple interactions between polyphenols and a salivary proline-rich protein repeat result in complexation and precipitation. *Biochemistry 36*, 5566, 1997.
252. Okuda, T., Mori, K., and Hatano, T., Relationships of the structures of tannins to the binding activities with hemoglobin and methylene blue. *Chem. Pharm. Bull. 33*, 1424, 1985.
253. Porter, L.J. and Woodruffe, J., Haemanalysis: the relative astringency of proanthocyanidin polymers. *Phytochemistry 23*, 1255, 1984.
254. Ricardo da Silva, J.M. et al., Interaction of grape seed procyanidins with various proteins in relation to wine fining. *J. Sci. Food Agric. 57*, 111, 1991.
255. Sarni-Manchado, P., Cheynier, V., and Moutounet, M., Interactions of grape seed tannins with salivary proteins. *J. Agric. Food Chem. 47*, 42, 1999.
256. Maury, C. et al., Influence of fining with different molecular weight gelatins on proanthocyanidin composition and perception of wines. *Am. J. Enol. Vitic. 52*, 140, 2001.
257. Bate-Smith, E.C., Haemanalysis of tannins: the concept of relative astringency. *Phytochemistry 12*, 907, 1973.
258. Hatano, T. and Hemingway, R.W., Association of (+)-catechin and catechin-(4α → 8)-catechin with oligopeptides. *J. Chem. Soc. Chem. Commun.* 2537, 1996.
259. Hatano, T., Yoshida, T., and Hemingway, R.W., Interaction of flavonoids with peptides and proteins and conformations of dimeric flavonoids in solution. In *Plant Polyphenols 2. Chemistry, Biology, Pharmacology, Ecology* (eds G.G. Gross, R.W. Hemingway, T. Yoshida, and S.J. Branham), Kluwer Academic/Plenum Publisher, New York, 1999, p. 509.
260. Luck, G. et al., Polyphenols, astringency and prolin-rich proteins. *Phytochemistry 37*, 357, 1994.
261. Charlton, A.J., Haslam, E., and Williamson, M.P., Multiple conformations of the proline-rich-protein/epigallocatechin gallate complex determined by time-averaged nuclear Overhauser effects. *J. Am. Chem. Soc. 124*, 9899, 2002.
262. Charlton, A. et al., Polyphenol/peptide binding and precipitation. *J. Agric. Food Chem. 50*, 1593, 2002.
263. Sarni-Manchado, P. and Cheynier, V., Study of noncovalent complexation between catechin derivatives and peptide by electrospray ionization-mass spectrometry (ESI-MS). *J. Mass Spectrom. 37*, 609, 2002.
264. Riou, V. et al., Aggregation of grape seed tannins in model — effect of wine polysaccharides. *Food Hydrocoll. 16*, 17, 2002.
265. Poncet-Legrand, C. et al., Flavan-3-ol aggregation in model ethanolic solutions: incidence of polyphenol structure, concentration, ethanol content and ionic strength. *Langmuir 19*, 10563, 2003.

266. Edelmann, A. et al., Dynamic light scattering study of the complexation between proline rich proteins and flavan-3-ols: influence of molecular structure, polyphenol/protein ratio and ionic strength. In *Oenologie 2003* (eds A. Lonvaud-Funel, G. De Revel, and P. Darriet), Tec et Doc, Arcachon, 2003, p. 408.

267. Saucier, C., Glories, Y., and Roux, D., Interactions tanins-colloides: nouvelles avancées concernant la notion de "bons" et de "mauvais" tanins. *Rev. Oenol. 94*, 9, 2000.

268. Saucier, C. et al., Characterization of (+)-catechin-acetaldehyde polymers: a model for colloidal state of wine polyphenols. *J. Agric. Food Chem. 45*, 1045, 1997.

269. Maury, C., Etude des phénomènes impliqués dans les collages protéiques en oenologie, Thèse Doctorat Montpellier II-ENSAM, Montpellier, 2001.

270. Sarni-Manchado, P. et al., Analysis and characterization of wine condensed tannins precipitated by protein used as fining agent in enology. *Am. J. Enol. Vitic. 50*, 81, 1999.

271. Maury, C. et al., Influence of fining with different molecular weight gelatins on proanthocyanidin composition and perception of wines. *Am. J. Enol. Vitic. 52*, 140, 2001.

272. Fischer, U. and Noble, A.C., The effect of ethanol, catechin concentration, and pH on sourness and bitterness of wine. *Am. J. Enol. Vitic. 45*, 6, 1994.

273. Guinard, J.-X., Pangborn, R.M., and Lewis, M.J., The time-course of astringency in wine upon repeated ingestion. *Am. J. Enol. Vitic. 37*, 184, 1986.

274. Kallithraka, S., Bakker, J., and Clifford, M.N., Red wine and model wine astringency as affected by malic and lactic acid. *J. Food Sci. 62*, 416, 1997.

275. Kallithraka, S., Bakker, J., and Clifford, M.N., Effect of pH on astringency in model solutions and wines. *J. Agric. Food Chem. 45*, 2211, 1997.

276. Kallithraka, S., Bakker, J., and Clifford, M.N., Evaluation of bitterness and astringency of (+)-catechin and (−)-epicatechin in red wine and in model solutions. *J. Sens. Stud. 12*, 25, 1997.

277. Naish, M., Clifford, M.N., and Birch, G.G., Sensory astringency of 5-O-caffeoylquinic acid, tannic acid and grape seed tannin by a time-intensity procedure. *J. Sci. Food Agric. 61*, 57, 1993.

278. Robichaud, J.L. and Noble, A.C., Astringency and bitterness of selected phenolics in wine. *J. Sci. Food Agric. 53*, 343, 1990.

279. Valentova, H. et al., Time-intensity studies of astringent taste. *Food Chem. 78*, 29, 2002.

280. Vidal, S. et al., Effect of tannin composition and wine carbohydrates on astringency and bitterness. In *5th Pangborn Sensory Science Symposium*, Boston, 2003.

281. Goldstein, J.L. and Swain, T., The inhibition of enzymes by tannins. *Phytochemistry 4*, 185, 1965.

282. Brossaud, F., Cheynier, V., and Noble, A., Bitterness and astringency of grape and wine polyphenols. *Aust. J. Grape Wine Res. 7*, 33, 2001.

283. Matsuo, T. and Itoo, S., A model experiment for de-astringency of persimmon fruit with high carbon dioxide treatment: *in vitro* gelation of kaki-tannin by reacting with acetaldehyde. *Agric. Biol. Chem. 46*, 683, 1982.

284. Vidal, S. et al., Taste and mouth-feel properties of different types of tannin-like polyphenolic compounds and anthocyanins in wine. *Anal. Chim. Acta 513*, 57, 2004.

285. Vidal, S. et al., The mouth feel properties of polysaccharides and anthocyanins in a wine-like medium. *Food Chem. 85*, 519, 2004.

286. Noble, A., Why do wines taste bitter and feel astringent? In *Wine Flavor* (eds A.L. Waterhouse and S. Eberler), American Chemical Society, Washington, DC, 1998, p. 156.

287. Noble, A.C., Astringency and bitterness of flavonoid phenols. In *Chemistry of Taste: Mechanisms, Behaviors, and Mimics* (eds P. Given and D. Paredes), American Chemical Society, Washington, DC, 2002, p. 192.

288. Ozawa, T., Lilley, T.H., and Haslam, E., Polyphenol interactions: astringency and the loss of astringency in ripening fruit. *Phytochemistry 26*, 2937, 1987.

289. Taira, S., Ono, M., and Matsumoto, N., Reduction of persimmon astringency by complex formation between pectin and tannins. *Postharvest Biol. Technol. 12*, 265, 1998.

290. de Freitas, V., Carvalho, E., and Mateus, N., Study of carbohydrate influence on protein–tannin aggregation by nephelometry. *Food Chem. 81*, 503, 2003.

291. Maury, C. et al., Influence of fining with plant proteins on proanthocyanidin composition of red wines. *Am. J. Enol. Vitic. 54*, 105, 2003.
292. Vidal, S. et al., Use of an experimental design approach for evaluation of key wine components on mouth-feel perception. *Food Qual. Prefer. 15*, 209, 2004.
293. Mateus, N. et al., Occurrence of anthocyanin-derived pigments in red wines. *J. Agric. Food Chem. 49*, 4836, 2001.
294. Es-Safi, N.E., Cheynier, V., and Moutounet, M., Implication of phenolic reactions in food organoleptic properties. *J. Food Compost. Anal. 16*, 535, 2003.

6 Dietary Flavonoids and Health — Broadening the Perspective

Mike Clifford and J.E. Brown

CONTENTS

6.1   INTRODUCTION

Toward the end of the 20th century, epidemiological studies and associated meta-analyses suggested strongly that long-term consumption of diets rich in plant foods offered some protection against chronic diseases, especially cancer.[1-7] Because uncontrolled production of free radicals was thought to be significantly implicated in the etiology of cancer,[8-11] these observations focused attention on the possible role of radical scavenging and radical suppressing nutrients and nonnutrients in explaining the apparent benefit of such diets.[12-15] The realization that free radicals were similarly implicated in the etiology of many other chronic diseases,[16,17] along with the recognition of the "French Paradox"[18] and the seminal papers from Hertog et al.,[19,20] immediately focused attention on flavonoids and the foods and beverages rich therein. An unfortunate, but unintended side effect of these papers was the tendency of many investigators to think of dietary phenols, polyphenols, and tannins (PPT) as encompassing only the flavonoids, and the flavonoids per se to encompass only the three flavonols and two flavones that featured in those studies,[19,20] but this is misleading and was never intended. This particular combination of events almost certainly resulted in many subsequent investigations adopting a too narrow focus.[21]

Subsequent epidemiological studies have supported the association between better health and long-term consumption of diets rich in foods of plant origin.[22-29] However, whether this is because such diets minimize exposure to deleterious substances (e.g., oxidized cholesterol, pyrolysis mutagens, salt, saturated fat, etc.), or maximize intake of certain beneficial nutrients (e.g., isothiocyanates and other sulfur-containing plant constituents, mono-unsaturated fatty acids, and poly-unsaturated fatty acids, PPT, polyacetylenes, selenium, terpenes, etc.) or some combination as advocated in the "Polymeal" concept, remains unknown.[30,31] An in vitro study indicates that there may be mechanistic basis for true synergy between PPT and isothiocyanates.[32]

In contrast, more recent studies seeking to assess the suggested link between the consumption of flavonols and flavones, or other flavonoids, have given much less consistent results. Some studies have suggested a possible protective effect of flavonoids against vascular diseases[33-37] or certain (but not all) cancers,[38-43] whereas other studies have suggested no protective effect or even an increased risk in certain populations.[44-50] Interestingly, an investigation of the relationship between the consumption of broccoli and other cruciferous vegetables and the risk of breast cancer in premenopausal women attributed the beneficial effects to isothiocyanates and not to the phenolic components,[51] although these crops are good sources of dietary phenols[52,53] including flavonoids,[54-56] and a potential for synergy in vivo has been demonstrated.[32,57]

In the same time period, various studies have suggested beneficial effects associated with raised consumption of other classes of dietary phenols. For example, increased coffee consumption has been linked with reduced incidence of type II diabetes.[58-63] Similarly, increased consumption of lignans (or at least greater plasma concentrations of their metabolites) has been linked with reduced incidence of estrogen-related cancers in some[64-66] but not all studies,[67,68] and a prospective study was equivocal.[69] It has been suggested that this inconsistency might have a genetic basis.[70] Increased consumption of isoflavones has also been associated with decreased risk of estrogen-related cancers and vascular diseases.[40,71]

This brief introduction demonstrates that the relationships between diet and health are far from simple and most certainly far from fully understood, but for a critical and detailed review of epidemiological data the reader is referred to an excellent paper by Arts and Hollman.[72] The objective of the review that follows is to record recent changes in the perceived role of flavonoids as health-promoting dietary antioxidants and place these observations in a broader context embracing other dietary phenols, and mechanisms other than simple radical scavenging and radical suppression.

6.2 THE DIVERSITY OF DIETARY PPT

PPT may be classified in several ways; for example, by biosynthetic origin, occurrence, function or effect, or structure.[73–75] A classification based on structure and function will be used in this chapter.[76] Simple phenols are substances containing only one aromatic ring and bearing at least one phenolic hydroxyl group and possibly other substituents, whereas polyphenols contain more than one such aromatic ring. Phenols and polyphenols may occur as unconjugated aglycones or, as conjugates, frequently with sugars or organic acids, but also with amino acids, lipids, etc.[77] The commonest simple phenols are cinnamates that have a C_6–C_3 structure[78,79] accompanied by C_6–C_2 and C_6–C_1 compounds, and a few unsubstituted phenols.[80–82] In general, these occur as conjugates. Flavonoids are the most extensively studied polyphenols, all characterized by a C_6–C_3–C_6 structure, subdivided by the nature of the C_3 element into anthocyanins, chalcones, dihydrochalcones, flavanols, flavanones, flavones, flavonols, isoflavones, and proanthocyanidins. The flavanols and proanthocyanidins generally occur unconjugated but the others normally occur as glycosides. Since the seminal paper of Hertog et al.,[83] there has been a tendency to think of dietary PPT as encompassing only the flavonoids, and the flavonoids *per se* to consist only of the three flavonols and two flavones that featured in that study, but this is misleading and was never intended. It is not possible to say with precision just how many individual PPT occur regularly in human diets, but on present evidence a figure in excess of 200 seems reasonable.[77]

The term "tannin" refers historically to crude plant preparations that are capable of converting hides to leather[84] and such preparations are not consumed as human food. However, the functional polyphenols contained therein at high concentration may also occur in certain foods and beverages but at comparatively low concentrations that would render them totally ineffective for producing leather. These polyphenols may be subdivided into the flavonoid-derived proanthocyanidins (condensed tannins)[85,86] and the gallic acid-derived and ellagic acid-derived hydrolyzable tannins, this latter subgroup of more restricted occurrence in human food (but commoner in some animal feeds).[87] The phloroglucinol-derived phlorotannins, while never used for preparing leather, also have a limited occurrence in human food.[80] The more recent term "phytoestrogen" refers to substances with estrogenic or antiandrogenic activity at least *in vitro*, and encompasses some isoflavones, some stilbenes, some lignans, and some coumarins.[88] The lignans are not estrogen-active until transformed by the gut microflora.[88,89] "Antioxidants" is a third function-based term much used to describe PPT, but individual compounds differ markedly in their ability to scavenge reactive oxygen species and reactive nitrogen species, and inhibit oxidative enzymes. Mammalian metabolites of PPT do not necessarily retain fully the antioxidant ability of the PPT found in plants and especially not that of their aglycones as commonly tested *in vitro*.[90,91]

The PPT discussed above are substances found in healthy and intact plant tissues, and mainly have known structures. However, many traditional foods and beverages as consumed have been produced by more or less extensive processing of such plant tissues, resulting in biochemical or chemical transformations of the naturally occurring PPT. In some cases, for example, black tea, matured red wines, and coffee beverage, these transformations may be substantial, generating large quantities of substances not found in the original plant material. There have been significant advances in the last decade in the chemistry of both red wine[92–97] and black tea.[98–106] This includes the characterization of the first large-mass thearubigin derived from four flavanol monomers and containing three benztropolone moieties,[107] and evidence that peroxidase-like reactions are involved in their production.[101] The derived polyphenols of matured wines are discussed eloquently by Cheynier in Chapter 5. However, the structures of the majority of these novel compounds have yet to be elucidated. Although often described as tannins, the beverages containing these substances are not functional tanning agents, and these substances should be referred to collectively as derived polyphenols (rather than tannins) until such time as their full structural characterization permits a more precise nomenclature.[76]

6.3 THE INTAKE OF PPT

There have been several attempts to estimate the quantities of PPT consumed, either by using diet diaries or food frequency questionnaires and data on the typical composition of individual commodities[33,36,41,47,50,86,108–113] or by diet analysis.[42,46,83] In comparison with the comprehensive databases providing the content in the diet of the established micro- and macronutrients, data for the contents of PPT are much more limited. Those data that are available for PPT content have been obtained by many different methods of analysis, rarely take account of the effects of agricultural practice, season, cooking, or commercial processing, are not necessarily just for the edible portion, and may be for varieties of fruits and vegetables different from those consumed in a particular diet under investigation.[71,110,111]

These are potentially serious limitations since quantitatively cultivars may differ substantially in composition, and the nonedible parts of fruits and vegetables may differ greatly both quantitatively and qualitatively, compared with the flesh or juice.[114] Thermal processing of tea beverages results in significant flavanol epimerization and some food products present the consumer with epimers that do not occur naturally.[115,116] In addition, domestic cooking and commercial processing may, in some cases, cause extensive leaching and destruction,[56,117–122] although such data as are available sometimes are not completely in agreement[56,122] and much remains to be investigated.[123]

Data based on analysis of particular diets avoid these limitations but are usually restricted to a few PPT because of the difficulties and cost associated with quantifying so many individual compounds of known structure, to say nothing of the serious difficulties associated with quantifying the uncharacterized derived polyphenols.[124] When such data are available, they are usually for PPT as aglycones released by hydrolysis (to simplify the analysis still further) and generally for the flavonols and flavones first studied by Hertog et al.[83] since these are amongst the easiest to determine.[33,34,36,38,39,41,46,47,71,125–127] There are more limited data for flavanones[47,126] and isoflavones after hydrolysis,[47,71] and flavanols, and proanthocyanidins[86,128–130] (which occur as aglycones).

The lack of comprehensive and reliable food composition tables that encompass the PPT (and other nonnutrients) in commodities *as consumed* seriously inhibits the demonstration of statistical relationships between intake and health or disease and may be a factor contributing to the apparent inconsistency in the outcome of epidemiological studies (discussed above). This lack also impedes proper risk assessment of botanicals,[131] as discussed below. A significant development is the creation of three online free-access databases. One provides data for contents in 205 commodities for most classes of flavonoids including the flavanol-derived theaflavins and thearubigins of black tea, but excluding isoflavones and dihydrochalcones.[132] The second provides data for isoflavones in 128 commodities,[133] and the third for flavanols and proanthocyanidins in 225 commodities.[134] A similar development at the University of Surrey in the United Kingdom covers an even wider range of dietary phenols (including chlorogenic acids, benzoic acids, phenyl alcohols, lignans, and derived polyphenols) in some 80 commodities, but is not yet available online. Developed as part of an EU research program, the Vegetal Estrogens in Nutrition and the Skeleton (VENUS) database constructed in Microsoft Access 2000 contains the daidzein and genistein contents of 791 foods, more limited data for coumestrol, formomonetin, and biochanin A, plus levels for the lignans matairesinol and secoisolariciresinol in 158 foods.[135]

A commercial database covers cinnamates, flavonoids, isoflavones, and lignans plus many nonphenolic plant constituents.[136] Chapter 4 in this volume describes the development of a UK-focused database covering primarily anthocyanidins, flavanols, flavanones, flavones, and flavonols.

While recognizing the limitations (discussed above) of such an approach to estimating diet composition and the intake of PPT, using the University of Surrey database in conjunction with diet diaries available from their other studies[137–139] has produced interesting data (Table 6.1) and insights. From Table 6.1 it is clear that PPT intakes may vary substantially, and that the flavones and flavonols, upon which most emphasis has so far been placed,[33,34,36,38,39,41,46,47,71,83,125–127] form a comparatively small part of the total intake for the populations studied. The relatively low consumption of chalcones and dihydrochalcones, isoflavones, anthocyanins, and stilbenes reflects the comparatively low consumption of apples and ciders, soya products, dark berries, and red wines by these populations. The significant contributions made by the hydroxycinnamates (in these populations primarily reflecting coffee consumption[78,79]) and derived polyphenols (in these populations primarily reflecting black tea consumption[140–142]) are striking. In this context, "black tea" refers to the beverage prepared from the fermented leaf (as distinct from green tea) and not to the addition

TABLE 6.1
Mean Dietary Intakes of 14 Classes of PPT as Determined from Diet-Diaries and a Food Composition Database

PPT	103 UK Females Aged 20–30 Years[a]		50 UK Males Aged 27–57 Years[b]	
	Estimated as Conjugates	Estimated as Aglycones[c]	Estimated as Conjugates	Estimated as Aglycones[c]
Total, range	100–2300		30–2200	
Total, mean	780	451	1058	598
Hydroxybenzoates	15		23	
Hydroxycinnamates	353	176	670	335
Total flavonoids	210	105	205	103
Anthocyanins	5		9	
Chalcones and dihydrochalcones	0.7			
Flavanols	64		58	
Flavanones	22		89	
Flavones	72		17	
Flavonols	35		26	
Isoflavones	9		0.13	
Proanthocyanidins	7		6	
Ellagitannins	23			
Derived polyphenols	170	170	160	160
Stilbenes	9			
Lignans	0.04			

[a]From Ref. 108.
[b]From Ref. 113.
[c]Aglycones are estimated approximately by taking rutin as a representative flavonoid and 5-caffeoylquinic acid as a representative hydroxycinnamate and adjusting for the relevant molecular masses.
Source: Reprinted from Clifford, M.N., *Planta Med.,* 12, 1103, 2004. With permission.

or otherwise of milk to the beverage prior to consumption. This domination by PPT from black tea and coffee indicates the importance also of considering the hydroxycinnamates and derived polyphenols whenever assessing the dietary significance of PPT, and clearly shows the limitations of looking only at flavonols and flavones after hydrolysis no matter how precise *per se* the data for these aglycones might be.

It is important to stress that data for the composition of black tea and coffee beverage reflect exactly what is consumed (with the exception of the dregs left in the cup) since all transformations associated with processing and domestic preparation have already taken place. Moreover, NEODIET publications[77] are replete with analytical data from numerous sources for the composition of these beverages (thus better avoiding extreme values associated with any peculiarity of the material analyzed or method of analysis) compared with data for many fruits and vegetables. Accordingly, the estimated consumption figures obtained using this University of Surrey database are likely to be more accurate than would have been the case if solid foods were the major sources of PPT, and data were for raw foods rather than after cooking or processing. This argument applies also to the data for PPT delivered by wines and juices. Using this approach has led us to estimate typical mean intakes of PPT for the two populations so far studied to be in the range 450 to 600 mg calculated as aglycones.

6.4 ABSORPTION OF PPT AND THE NATURE OF THE PLASMA METABOLITES

6.4.1 INTRODUCTION

There have been comparatively few human studies using synthetic labeled forms of dietary PPT. In the absence of the comprehensive data that such studies provide, it is difficult to estimate the precise levels of absorption. Generally, estimates have been based on the urinary excretion of recognizable aglycones released after deconjugation with commercial sulfatase/β-glucuronidase. A review of 97 bioavailability studies where original data have been recalculated to 50 mg aglycone equivalents found that excretion in urine ranged from 0.3 to 43% of the dose.[143] In some but not all studies this would include measurement also of the methylated forms, but data on amino acid conjugates are generally conspicuous by their absence. Generally, there would be no account of material that had been absorbed and excreted in bile, or absorbed after transformation to phenolic acids by the gut microflora, and even when phenolic acids have been fed there has rarely been any attempt to quantify the excretion of glycine or glutamine conjugates. Accordingly such data as are available, and hence used by Manach et al.,[143] will be underestimates of the true absorption, but it is impossible to judge by what amount. On present data,[143] gallic acid and isoflavones are the most well absorbed dietary PPT, though neither are major components of the European diet. Flavanols, flavanones, and flavonol glycosides are intermediate, whereas proanthocyanidins, flavanol gallates, and anthocyanins are least well absorbed. Data for other classes are either nonexistent or inadequate — lack of data for chlorogenic acids and derived polyphenols, major contributors to total PPT intake, is a serious gap in our knowledge. The number of volunteers studied has generally been small (rarely more than ten), thus giving little insight into the between-person variation (whether phenotypic or genotypic), and very few studies have investigated the effects of repeat dosing at typical dietary levels which is how many PPT-rich commodities are consumed. All of these shortcomings must be addressed in future studies.

Although little studied, it is clear that the absorption of dietary PPT may be influenced by the matrix in which they are consumed, with enhanced excretion in urine of easily recognized mammalian conjugates observed when presented in foods with a higher fat content.[144–148] In contrast, addition of milk to tea does not significantly affect absorption of either flavanols[149] or flavonols[150] despite suggestions that it might.[46] Alcohol seems not to affect the absorption of (+)-catechin,[151–153] resveratrol or quercetin aglycones,[152,153] or malvidin-3-glucoside from red wine,[152] but hastens (+)-catechin clearance from the plasma compartment either by more rapid excretion or more extensive methylation.[151]

Extensive studies in humans and animals have indicated that some PPT can be absorbed in the small intestine, for example, certain cinnamate conjugates,[154,155] flavanols[156] (that occur naturally as aglycones), and quercetin-3-glucoside and quercetin-4'-glucoside.[157–159] In contrast, quercetin, quercetin-3-galactoside, quercetin-3-rutinoside (rutin), naringenin-7-glucoside, genistein-7-glucoside, and cyanidin-3,5-diglucoside seem not to be.[159,160] Mechanisms of absorption have not been completely elucidated but involve *inter alia* interaction of certain glucosides with the active sugar transporter (SGLT1) and lumenal lactase–phloridzin hydrolase, passive diffusion of the more hydrophobic aglycones, or absorption of the glucoside and interaction with cytosolic β-glucosidase. Although varying with PPT subclass and matrix, when expressed relative to the total intake of PPT, only some 5 to 10% of the amount consumed is absorbed at this site. The major part of that absorbed in the duodenum (not less than 90 to 95% for every substance so far studied) enters the circulation as mammalian conjugates produced by a combination of methylation, sulfate conjugation, glucuronide conjugation plus glycine conjugation in the case of phenolic acids.[91] Only a very small amount of the total PPT consumed, maximally 5 to 10%, enters the plasma as unchanged plant phenols.

The 90 to 95% of the total PPT ingested, plus any mammalian glucuronides excreted through the bile, pass to the colon where they are metabolized by the gut microflora. Transformations may be extensive, and include the removal of sugars, removal of phenolic hydroxyls, fission of aromatic rings, hydrogenation, and metabolism to carbon dioxide, possibly via oxaloacetate.[161] A substantial range of microbial metabolites has been identified, including phenols and aromatic acids, phenolic acids, or lactones possessing 0, 1, or 2 phenolic hydroxyls and up to five carbons in the side chain.[162–189] Many of these metabolites arise from flavonoids and nonflavonoids, requiring a broader approach. Certain *Eubacterium* spp. and *Peptostreptococcus* spp. are able to convert plant lignans to mammalian lignans.[190] *Eubacterium* is of particular interest since this species not only metabolizes dietary (poly)-phenols,[183–185,187,190–194] but also produces butyrate,[195] a preferred energy source for colonic epithelial cells thought to play an important role in maintaining colon health in humans. Butyrate encourages differentiation of cultured colon cells and through PPARγ activation decreases absorption by the paracellular route.[196] *Clostridium orbiscindens*, a somewhat atypical member of the genus, is also of interest for its ability to metabolize flavonoids.[186,197] The yield of phenolic or aromatic acids is variable (up to ten times) between individuals, and for some individuals can vary with substrate,[198] but can be substantial (up to 50%) relative to the intake of PPT substrates.[167,169,177–180,188,199,200]

There is evidence from cell culture studies that some of the aromatic or phenolic acids, e.g., benzoic, salicylic, *m*-coumaric, *p*-coumaric, ferulic, 3-hydroxyphenylpropionic, and 3,4-dihydroxyphenylpropionic, are transported actively by the monocarboxylate transporter MCT1,[201–206] but gallic acid and intact 5-caffeoylquinic acid are not. These latter acids may enter by the paracellular route,[207,208] but absorption by this route is thought to be inhibited by butyrate enhancing PPARγ activation.[196] *In vitro*, green tea flavanols inhibit the active transport of phenolic acids by MCT1, but the significance *in vivo* of this observation is uncertain.[209] Although these acids occur in the plasma primarily as mammalian conjugates some reports suggest that a variable portion may be present in the free form.[155,210] Table 6.2 summarizes in a semiquantitative manner, so far as current knowledge allows, the fate of a "typical" daily consumption of some 450 to 600 mg of PPT (calculated as aglycones) previously defined in Table 6.1.

There have been significant advances in our knowledge of the PPT-derived metabolites that occur in plasma, i.e., identity of conjugating species, position(s) of conjugation, and

TABLE 6.2
Fate of Ingested PPT

	Aglycones (mg)
Estimated mean daily consumption (mg/day)(from Table 6.1)	450–600
• ~5 to 10% of intake absorbed in duodenum and excreted in urine. Of this	22–60
• 5 to 10% unchanged plant (poly)phenols, and	<6
• 90 to 95% mammalian conjugates	20–55
• ~90 to 95% fermented in colon (unabsorbed PPT + enteric and entero-hepatic cycling of glucuronides, etc.)	400–570
• ≤ Poorly defined and very variable portion (5 to 50%?) absorbed depending on individual's flora and substrates. Mainly mammalian conjugates of microbial metabolites	20–285

Source: Reprinted from Clifford, M.N., *Planta Med.*, 12, 1103, 2004. With permission.

concentrations achieved. It is now clear that there are some significant differences between humans and laboratory animals in this regard.[144,211,212]

6.4.2 FLAVANOLS, FLAVANOL GALLATES, AND PROANTHOCYANIDINS

Flavanols are unique amongst the flavonoids because they occur in the diet as aglycones.[55,85] Epimerization at C2 occurs between consumption and appearance in the plasma[213] but the chirality of the metabolites has not been measured routinely and it must be assumed that the data that follow have, unwittingly, been obtained on mixtures of isomers. The biological significance of the epimerization is unknown, although in rats, the novel epimers ((−)-catechin, (−)-gallocatechin, and their gallates) are more effective in inhibiting cholesterol absorption.[214] Nonconjugated (−)-epigallocatechin (up to 0.08 μM),[156] (−)-epigallocatechin gallate (EGCG) (up to 0.34 μM),[215,216] and traces of (+)-catechin (<2 nM)[217] have been found in human plasma but free (−)-epicatechin has not.[156] Three (−)-epicatechin metabolites, (−)-epicatechin-3′-*O*-glucuronide, 4′-*O*-methyl-(−)-epicatechin-3′-*O*-glucuronide, and 4′-*O*-methyl-(−)-epicatechin-5 or 7-*O*-glucuronide have been purified from human urine,[211] whereas the exact fate of (+)-catechin is not known although there is evidence for the formation of (+)-catechin sulfates, sulfo-glucuronides, and 4′-methylated conjugates in plasma and urine.[147,218] In contrast, (−)-epicatechin gallate (ECG) and EGCG appear to be excreted in bile.[149,156,219,220] ECG is extensively methylated by human liver catechol *O*-methyl transferase (COMT) at the 4′ position and to a lesser extent at the 3′ position.[221,222] EGCG is metabolized first to the 4″-methyl ether and then to the 4′,4″-dimethyl ether.[222] Glucuronidation on the B-ring or the D-ring (gallate ester) of EGCG greatly inhibited the methylation on the same ring, but glucuronidation on the A-ring of EGCG or EGC did not affect their methylation. Only a small proportion of the methylflavanols occurs unconjugated to glucuronide or sulfate, and total MeEGC in plasma exceeds total EGC by a factor of ×4 to ×6.[221]

Traces of procyanidins B1 and B2 have been detected in plasma after hydrolysis of glucuronide and sulfate conjugates.[223,224]

6.4.3 FLAVONOL GLYCOSIDES

In general, flavonols occur in the diet as glycosides.[55] Small amounts of unconjugated aglycones may be found in some red wines.[225] Flavonol glycosides and quercetin aglycone have not been convincingly demonstrated in plasma,[212,226,227] although kaempferol aglycone has been observed.[228] At least five different quercetin glucuronides[229] as well as some sulfates and mixed conjugates, and four isorhamnetin (3′-methylquercetin) conjugates have been found in human plasma after consumption of foods containing quercetin glycosides. In these studies, isorhamnetin glucuronides accounted for ~30% of the total glucuronides, but only quercetin occurred in a sulfated form. Mixed conjugates accounted for some 11% of the total quercetin. Overall, one fifth of the absorbed quercetin was isorhamnetin and one third was sulfated, leaving some 45% as quercetin glucuronides or mixed conjugates. Some studies have failed to detect flavonol sulfates, but this may be due in part to their destruction by the use of acetone in sample preparation rather than nonformation.

The main kaempferol metabolite in human plasma is the 3-glucuronide.[228] The three major metabolites of quercetin are quercetin-3-glucuronide, quercetin-3′-sulfate, and iso-rhamnetin-3-glucuronide (found at 0.1 to 1 μM). These are accompanied by lesser amounts of the 4′-glucuronides of quercetin and isorhamnetin, and several uncharacterized metabolites.[226] Quercetin-7-glucuronide was not detected[226] in human plasma, although it is a major metabolite in rats. At low quercetin doses (up to 10 mg), ~100% conversion to isorhamnetin has been reported,[229,230] whereas at higher doses a quercetin–isorhamnetin ratio of 5:1 has been observed.[226]

Using human liver slices, it has now been shown that quercetin-3-glucuronide (a major human conjugate) and quercetin-7-glucuronide (a minor human but major rat conjugate) can both be converted to quercetin-3′-sulfate suggesting that there will be local and transient release of quercetin aglycone wherever β-glucuronidase is active,[231] and this might explain the detection of trace amounts of [2-^{14}C]quercetin in liver.[232] The extent and rate of this transformation were increased when COMT was inhibited.[231]

6.4.4 FLAVONES AND POLYMETHYLFLAVONES

Flavones occur widely in the diet as *O*-glycosides and *C*-glycosides, and in citrus fruits as unconjugated polymethyl-flavones.[55] Apigenin glucuronides have been detected in urine after volunteers consumed parsley.[233] Luteolin aglycone administered to volunteers has been detected in plasma as a monoglucuronide accompanied by a trace of unconjugated luteolin.[234,235] Chrysin is transformed primarily to the 7-glucuronide with much smaller yields of the 7-sulfate.[236] Apigenin has been shown *in vitro* and *in vivo* to behave synergistically with sulforophane, an isothiocyanate from cruciferae.[32,57]

6.4.5 ISOFLAVONES

Isoflavone glycosides (of which a portion is acylated) are characteristic of legumes such as soya. Fermented soya products predominantly contain the aglycones whereas nonfermented products retain the β-glycosides.[237] Soya derivatives are an important component of many processed foods, including human milk-replacers, and such usage increases the intake in western societies.[238] In many dietary supplements and extracts, the contents of isoflavone aglycones as a percentage of total isoflavones may be significant (~2 to 15%) and a few may contain much larger proportions (>85%).[239] The aglycones are more rapidly absorbed than the glycosides.[240]

Genistein-7-glucoside and daidzein-7-glucoside have not been found in human plasma[241] but the aglycones have been observed.[239] Unconjugated plasma genistein may reach ~0.4 μM in males given an aglycone dose of 16 mg/kg body weight, but this is unlikely to be achieved from dietary sources even in Asian populations,[242] and concentrations of 50 to 100 nM are unlikely to be exceeded.[243] Human metabolism of isoflavone glycosides produces genistein and daidzein 7-glucuronides/7-sulfates and 4′,7-diconjugates (including diglucuronides and mixed conjugates), with monoglucuronides predominant.[244,245] Daidzein-4',7-diconjugates would not have an unconjugated phenolic hydroxyl so could not function as antioxidants. It has been suggested on the basis of studies *in vitro* that genistein may be converted to fatty acyl esters in human plasma,[246] thus facilitating its incorporation into low-density lipoproteins (LDL) in a form not detected by routine methods of analysis.

Mean plasma isoflavone concentrations (after deconjugation) some 5 to 6 h after consumption of soya-based foods have been reported as follows: daidzein 0.5 to 3.1 μM; genistein 0.3 to 4.1 μM; glycitein 0.20 to 0.85 μM.[247–250] Some data for the individual conjugates of daidzein and genistein have been reported. The average pattern was ~54% 7-glucuronide, 25% 4'-glucuronide, 13% monosulfates, 7% free daidzein, 0.9% sulfoglucuronides, 0.4% diglucuronide, and <0.1% disulfate.[251] A study by Shelnutt et al.[252] confirmed in plasma the greater prevalence of genistein monoglucuronides compared with genistein monosulfates (mean ratio ~5:1) whereas for daidzein the equivalent mean ratio was ~1.3:1. Also, for both aglycones, mixed conjugates appeared to account for some 45% of total conjugates. The gut flora metabolism is discussed below and more extensively in Chapter 4.

6.4.6 FLAVANONES

Flavanones occur as glycosides in the diet.[114] Studies where volunteers consumed orange juice, grapefruit juice, or a meal containing cooked tomato paste have led to the detection in plasma of hesperetin and naringenin as mammalian conjugates. C_{max} values, after deconjugation, of up to 6 μM naringenin and up to 2.2 μM hesperetin have been recorded. Hesperetin occurred as glucuronides (87%) and sulfo-glucuronides (13%), and naringenin as glucuronides. Neither aglycone was found. The uptake of naringenin from tomato paste, relative to intake, was greater than from the citrus juices.[146,253–257] Grapefruit juice, rich in naringenin, can inhibit clearance of CYP 3A4-metabolized drugs.[258]

6.4.7 CHALCONES, DIHDYROCHALCONES, AND RETROCHALCONES

The various chalcones have a comparatively restricted dietary occurrence, with apple and apple products as the major source.[114] After consumption of alcoholic cider (1.1 l) a trace of phloridzin has been detected in the plasma of one out of six volunteers,[230] but all six urines yielded phloretin after enzymic hydrolysis of mammalian conjugates. Studies *in vitro* with human microsomes indicate that the chalcone xanthohumol from hops is monoglucuronidated at C4 and C4',[259] and some additional transformations of the aglycone have been observed using rat liver microsomes.[260] Three glucuronides of licochalcone A, including the 4-glucuronide (*E* isomer) and 4'-glucuronide (*E* and *Z* isomers), have been detected in plasma and urine.[261] The glucuronide of a hydroxylated metabolite and mercapturic acid conjugates have been produced using rabbit and pig liver microsomes.[262]

6.4.8 ANTHOCYANINS

Anthocyanins occur in plants as (acylated) glycosides with the C3 hydroxyl always occupied, and sometimes the C5 hydroxyl also.[263] It is now clear that anthocyanins are found in human plasma as the intact glucosides, rutinosides, sambubiosides, sophorosides, and caffeic acid conjugates of sophorosides.[152,264–274] However, only some 0.01 to 0.2% of the dose is excreted in urine and maximal plasma concentrations for the six glycosides so far reported range from 1 to 129 nM with a mean total in one study of ≈150 nM for several glycosides consumed simultaneously. Alcohol appears to have little effect on the absorption of malvidin-3-glucoside, but plasma concentration appears to be a function of the dose. There is a pronounced interindividual variation.[152] More recently, evidence for methylation has also been presented, and a range of mammalian conjugates, anthocyanidin glucuronides, anthocyanidin sulfo-glucuronides, and anthocyanin glucuronides, have been characterized, demonstrating that some 2% of the dose may be absorbed.[264,268] The relatively unstable free aglycones have not been detected. At plasma pH values, anthocyanins will be present as the pseudo-base or quinoidal base rather than the cation characteristic of acidic plant tissues.[263] Since anthocyanins can exist in a retrochalcone form it is possible that some metabolism involves this chemical species (see above).

6.4.9 STILBENES

Resveratrol (3,5,4′-trihydroxystilbene) occurs in a limited number of foods and beverages (e.g., grapes, wines, and peanuts) either as the aglycone or the 3-glucoside (piceid). *Cis* and *trans* isomers are encountered. Some plant materials contain "oligomers" but these materials are rarely significant dietary components.[88] Resveratrol is sulfated and glucuronidated and very little free resveratrol is found even after large oral doses.[153,275] Resveratrol-3-sulfate inhibits CYP3A4 *in vitro* (IC_{50} 1 μM).[276]

6.4.10 HYDROXYBENZOIC ACIDS, HYDROXYCINNAMIC ACIDS, AND ASSOCIATED CONJUGATES

Hydroxybenzoic acids are comparatively minor, but widespread components of the diet.[82] The hydroxycinnamic acids, especially the chlorogenic acids of coffee, are a major contributor to the total dietary intake of PPT.[79]

Chlorogenic acid (5-caffeoylquinic acid) is absorbed, apparently by the paracellular route,[208] and can be detected unchanged in plasma.[277] The concentrations are low (19 to 45 nM) and some studies have failed to detect this compound,[155] possibly because in the individuals studied absorption by the paracellular route was more limited,[196] or because it was hydrolyzed by commercial β-glucuronidases during sample work-up.[143] Mammalian conjugates of ferulic (max 200 nM) and sinapic acid (max 40 nM) have been detected in volunteers' plasma after consumption of cereals containing cinnamate-esterified arabinoxylans.[278] The bioavailability of chlorogenic acid and other cinnamic acid conjugates is thus largely dependent on gut microflora metabolism.[154,155,169,278–280]

Gut flora metabolism involves dehydroxylation, hydrogenation, and shortening of the cinnamate side chain, followed by mammalian methylation, and sulfate, glucuronide, and glycine conjugation. Recognized metabolites include monoglucuronides of caffeic, ferulic, isoferulic, and vanillic acids after consumption of coffee, artichoke, red wine, or cider.[179,230,277,281–284] In one study,[155] free caffeic acid accounted for some 15 to 30% of the total (~506 nM). In contrast, after consumption of beer (500 ml), plasma was reported to contain *unconjugated* caffeic acid (0.03 to 0.30 μM), vanillic acid (0.07 to 0.09 μM), and syringic acid (up to 0.05 μM).

Curcumin (diferuloylmethane) has very low oral bioavailability, but is rapidly absorbed and low nanomolar levels of the parent compound and its glucuronide and sulfate conjugates can be detected in human plasma and portal circulation after very high (nondietary) intakes (3.6 g/day for 1 week). Metabolic reduction occurs in the liver,[285,286] and glutathione adducts have been observed *in vitro*.[287]

Gallic acid mono- and dimethyl ethers have been found as conjugates in human plasma after the consumption of either gallic acid, or beverages containing gallic acid or flavanol gallates.[284,288,289] After consumption of 1.1 l of red wine, plasma treated with deconjugation enzymes contained gallic acid 4-*O*-methyl ether at concentrations up to 0.2 μM.[284]

6.4.11 OLEUROPEIN, TYROSOL, AND HYDROXYTYROSOL

Oleuropein, a conjugate of hydroxytyrosol (3,4-dihydroxybenzyl alcohol), is a characteristic but very variable component of olives and olive oil.[81] After consumption of 25 ml virgin olive oil, hydroxytyrosol, 3-O-methylhydroxytyrosol (homovanillyl alcohol), and homovanillic acid increase in plasma, as conjugates, predominantly glucuronide.[290,291] Oleuropein may be deconjugated by the gut microflora.[144]

6.4.12 HYDROLYZABLE TANNINS

Hydrolyzable tannins are comparatively restricted in the human diet[87] and there are no human metabolic data. Studies in rats have indicated that some 63% of a dose of 1 g/kg commercial tannic acid is excreted unchanged in the feces accompanied by small amounts of gallic acid, pyrogallol, and resorcinol. Plasma after enzymic hydrolysis was found to contain 4-O-methylgallic acid, pyrogallol, and resorcinol. Urine also contained a small amount of gallic acid after enzymic hydrolysis. The most notable observation from this study is the failure of the gut microflora to metabolize the galloylglucoses efficiently, at least at this substantial dose. The viability or composition of the gut microflora was not reported.[292]

Punicalagin is absorbed by rats, and ~3 to 6% of the dose has been characterized as metabolites in urine and feces.[293] Plasma metabolites included punicalagin at ~30 μg/ml (~15 nM) and glucuronides of methylether derivatives of ellagic acid. The major urine metabolites were 6H-dibenzo[b,d]pyran-6-one derivatives as aglycones or glucuronides. Punicalin, ellagic acid, and gallagic acid were reported in feces, along with 3,8-dihydroxy-6H-dibenzo[b,d]pyran-6-one. This metabolite has previously been reported in the feces of species (e.g., beaver) consuming large amounts of ellagitannins and is considered to be a hyaluronidase inhibitor. Metabolite production was biphasic. In the first 20 days the main metabolites in biological fluids were derived from punicalagin by hydrolysis and further conjugation (methyl ethers or glucuronic acid derivatives). Beyond 20 days the microflora metabolites start to appear in feces and their mammalian conjugates become the main metabolites in plasma and urine. This dramatic change could be explained by changes in the composition or biochemical competence of the gut microflora.

6.4.13 LIGNINS AND LIGNANS

Lignans are chiral cinnamate-derived glycosides of interest primarily because some are converted by the gut microflora (deglycosylations, meta-demethylations, and para-dehydroxylations without enantiomeric inversion) to the so-called mammalian lignans[190,294–297] that after absorption are glucuronidated or sulfated[244,298] and subject to phase II hydroxylations.[297] The aglycones of enterodiol and enterolactone exhibit estrogenic activity[299] and increased excretion of mammalian lignans has been associated with a decreased incidence of breast cancer.[64,300]

The plant precursors of mammalian lignans include secoisolaricinol, laricresinol, matairesinol, 7-hydroxymatairesinol, pinoresinol, and lignin.[301,302] Flaxseed[303] and whole cereal grains[297,304,305] are considered the most important dietary sources, but many others are known, for example, strawberry achenes, berries, coffee beans, tea leaves, etc.[298,305–308]

6.4.14 DERIVED POLYPHENOLS

There have been very few studies on the absorption and metabolism of derived polyphenols despite them forming a very significant proportion of the PPT intake. Theaflavin absorption is unexpectedly rapid, but extremely limited, with a maximal plasma concentration, after deconjugation, of 1 ng/ml 2 h after a dose of 700 mg.[309] Extensive gut flora metabolism of derived polyphenols in black tea has been demonstrated by the detection in urine of substantial amounts of hippuric acid.[154,167,198] Theasinensins A and D, B-ring linked dimers of EGCG, can be absorbed by mice, and have been found in mouse plasma.[310,311]

6.4.15 PERSON-TO-PERSON VARIATION

The human data available are still comparatively limited. Most human studies show a considerable interindividual variation. Occasionally, a much greater variation is suggested, as, for example, the detection of phloridzin in the plasma of only one out of six volunteers,[230] and the detection of curcumin sulfate in the feces of only one volunteer out of 15.[312] The limited nature of the data available make it difficult to generalize with regard to either efficacy or safety.

6.5 PPT METABOLITES IN TISSUES

6.5.1 PSEUDO-PHARMACOKINETIC AND REDOX PROPERTIES

Since for the majority of dietary PPT neither the conjugates consumed nor their free aglycones are detectable in plasma, it is rarely possible to perform true pharmacokinetic analyses. Most so-called pharmacokinetic data that have been published relate to the concentrations of aglycones released after hydrolysis of mammalian conjugates in plasma or urine with commercial β-glucuronidase or sulfatase, and the data so obtained are better referred to as pseudo-pharmacokinetics. It must be noted that some glucuronides are insensitive to some commercial β-glucuronidases (although some can hydrolyze chlorogenic acids[143]), and thus misleading data may be obtained. In addition, there are no convenient sources of enzymes to hydrolyze glycine or glutamine conjugates. These can be cleaved by 6 M HCl, but in some cases this process destroys the phenolic moiety.[313]

Published human data are summarized in Table 6.3. Although the maximum concentration achieved transiently varies to some extent with PPT subclass and matrix in which it is consumed, it is unlikely that plasma metabolite concentrations will routinely exceed 10 μM in total, and ~1 μM for total aglycones. The reported T_{max} values range from 1 to 2.5 h for substances absorbed in the duodenum,[147,151,157,158,160,216,223,314–317] up to 5 to 12 h when microbial metabolism is a prerequisite.[160,177,255] Published elimination half-lives are very variable, ranging from as low as 1 h[221,318] to values in excess of 20 h.[157,158,160] The very low values may be artifacts of observation periods less than the true half-life, whereas the very high values may be exaggerated because of a biphasic elimination reflecting significant enterohepatic circulation of glucuronides. Indeed, mammalian conjugates produced in the gastrointestinal epithelium do not necessarily enter the circulation — a significant portion is returned to the gut lumen[231,319,320] where they may be deconjugated and further degraded by the gut microflora.

The effect of repeat dosing has rarely been studied. Repeated consumption of green or black tea produced only slight day-on-day accumulation of flavanols in plasma,[149,321] and modest increase in the intrinsic resistance of isolated LDL to oxidation *ex vivo*,[322] suggesting that, in general, significant elimination occurs in a time period shorter than the interval between repeat doses.

Table 6.4 summarizes the concentrations of a range of endogenous (i.e., nondietary) simple phenols, including α-tocopherol, and ascorbate in plasma from healthy individuals. The total simple phenol and ascorbate concentration is between 159 and 380 μM. The maximum additional concentration that is likely to be achieved from dietary sources, 3 to 22 μM, is marginal by comparison adding only between 0.3 and 5% if it is assumed, quite reasonably, that the "typical" mean intake is taken over three equal meals. Many people consume a much smaller quantity of dietary PPT and even those consuming double the

TABLE 6.3
Plasma Pseudo-Pharmacokinetics After Consumption of Normal Portions of Rich Sources

PPT Subclass	C_{max} (nM) Unchanged[a]	C_{max} (nM) Mammalian Conjugates	Urine Excretion %
Anthocyanins	10–150	Traces	ND–0.1
Flavanols, low fat	40–140	1000–2000	0.5–4.0
Flavanols, high fat	150–220	Up to 6200	25–30
Flavonol glycosides	Minute traces	ND	0.5–2.5
Flavonol aglycones	Minute traces	350–1100	
Flavanone glycosides	Minute traces	120–1500	4–10
Isoflavone glycosides	Minute traces	900–4000	20–55
Isoflavone aglycones	10–150	500–6500	
Cinnamates and chlorogenic acids	Up to 120	Up to 500	1–2
Resveratrol aglycone	Minute traces	No quantitative data	64–70
Oleuropein	ND	Up to 60	55–60
Phenolic gut flora metabolites		20–60	Up to 50
Hypothetical total if all consumed in one meal	250–780	2890–21,720	

[a] C_{max} = maximum concentration achieved transiently in plasma.
Note: ND, not detected.
Source: Reprinted from Clifford, M.N., *Planta Med.,* 12, 1103, 2004. With permission.

TABLE 6.4
Plasma Concentrations (μM) of Endogenous (Nondietary) Phenols and Other Plasma Antioxidants

	Plasma Concentration in Healthy Individuals
Homogentisic acid	0.014–0.070[a]
p-Hydroxyphenyl lactate	40–90[b]
p-Hydroxyphenyl pyruvate	14–60[b]
Tyrosine	60–130[b,c]
Ascorbate	40–70[d,e]
α-Tocopherol	5–30[f]
Total endogenous phenols and antioxidants	159–380
Hypothetical total diet-derived phenols	3.1–22.5[g]
Averaged over three meals gives a transient increase of between 0.3 and 5%.	~1–7.5[g]
Many people consume much less	

[a] From Ref. 523.
[b] From Ref. 524.
[c] From Ref. 525.
[d] From Ref. 526.
[e] From Ref. 527.
[f] From Ref. 528.
[g] From Table 6.3.
Source: Reprinted from Clifford, M.N., *Planta Med.,* 12, 1103, 2004. With permission.

average amount (450 to 600 mg calculated as aglycones) adopted in this review will only achieve a transient 5 to 10% increase in total plasma antioxidant content.

Many investigators have attempted to demonstrate increases in plasma antioxidant capability following the consumption of foods, beverages, or supplements rich in PPT. Table 6.5 summarizes the outcomes of 37 such studies.[210,216,266,284,317,323–348] The test substances included a range of fruit and vegetable products, including juices, alcoholic beverages, tea, and chocolate. In view of the calculations presented in Table 6.3 and Table 6.4, it is perhaps not surprising that increases in plasma antioxidant capacity were often undetectable, and at best, small and transient. Moreover, in four studies that produced increases in plasma antioxidant capability it could be attributed, at least in part, to increased plasma ascorbate.[336,343,346]

In view of these observations, it is instructive also to consider the redox potentials of PPT-derived mammalian metabolites that are known to reach plasma, and to compare these with the corresponding values for the endogenous plasma antioxidants. The polyphenols with the lowest redox potentials are flavonoids with vicinal hydroxyls in the B-ring, and conjugation extending to the A-ring, e.g., quercetin aglycone (330 mV at pH 7).[349] If the conjugation does not extend beyond the B-ring, then the redox potential is significantly higher even for (−) epigallocatechin gallate (480 mV at pH 7)[350] with three vicinal hydroxyls. The value rises again when there are only two vicinal hydroxyls (e.g., (+)-catechin 570 mV[351] or caffeic acid 540 mV[351]), a single *para*-hydroxyl (e.g., hesperidin 720 mV[351]), or isolated (*meta*) hydroxyls (e.g., resorcinol 810 mV[352]). These comparisons are extended to the endogenous (nondietary) plasma antioxidants in Table 6.6. Figure 6.1 illustrates the marked effects of mammalian and microbial metabolism on the redox potential of PPT aglycones that are frequently examined in *in vitro* systems designed to demonstrate their potent antioxidant properties.

Table 6.6 indicates that the diet-derived PPT metabolites are able thermodynamically to scavenge some or all of the damaging radicals should they come into contact. However, these metabolites are so hydrophilic, e.g., quercetin-3-glucuronide ($K = 0.008$)[353,354] compared with quercetin ($K = 66$)[353,354] and α-tocopherol ($K = 550$),[355] that it is unlikely they will encounter lipid-derived radicals. However, any phenoxyl radicals generated will have to be removed either by transfer of the unpaired electron to an endogenous scavenger such as α-tocopherol, ascorbate, glutathione, or serum albumin, or by dismutation or disproportionation although these latter mechanisms seem somewhat unlikely *in vivo* because of the relatively low phenoxyl radical concentrations. The implied demand for α-tocopherol and ascorbate is particularly interesting, since two of the supplementation studies (Table 6.5) and a study in which rats were fed secoisolariciresinol produced reductions in plasma ascorbate or α-tocopherol,[337,356] and the major sources of dietary PPT (coffee and black tea) supply neither. Moreover, it is known that for ~14% of the over-65 population subgroup in the United Kingdom the mean plasma ascorbate value is below 11 μM,[357] indicating biochemical depletion,[358] suggesting that for heavy consumers of black tea or coffee within this population subgroup the transient concentration of PPT metabolites may approach or even exceed plasma ascorbate.

From the data assembled, it is difficult to envisage how diet-derived PPT metabolites can make a major contribution to radical scavenging in plasma compared with the contribution to be expected from the endogenous antioxidants in healthy individuals replete in ascorbate.[21,359] An independent, but contemporaneous review of 93 intervention studies[360] reached a similar conclusion with regard to foods and beverages and *in vivo* antioxidant effects. Herbal remedies and dietary supplements were sometimes more effective, reflecting PPT doses substantially above those achieved by diet alone, and frequently manifest through endpoints not directly associated with antioxidant effects. Contrary to the view expressed over the last decade there can now be little doubt, that if diets rich in fruits and vegetables are advantageous, at least in part by virtue of their content of PPT, then mechanisms other than antioxidant ability are implicated.[21,361–363]

TABLE 6.5
The Outcome of 37 Studies[a] in Which Volunteers Were Given Foods, Beverages, or Supplements Rich in PPT and Plasma was Analyzed for Total Antioxidant Activity

Thirty-seven studies, (poly)phenol-rich diet compared with control
- Thirteen studies (three high and three very high doses) showed no change in plasma antioxidant status *ex vivo*
- Twenty-four studies (ten low and seven moderate doses) showed small and transient increases in plasma antioxidant status *ex vivo*
- One showed reduction in plasma vitamin E
- One showed reduction in plasma ascorbate and glutathione

[a]From Refs. 210, 216, 221, 255, 266, 284, 317, 318, 323–348.
Source: Reprinted from Clifford, M.N., *Planta Med.*, 12, 1103, 2004. With permission.

6.5.2 BINDING TO PLASMA PROTEINS

Data for the metabolites in plasma are generally for the unbound forms, but there is ample evidence that PPT bind noncovalently to proteins.[364] Most studies on PPT–protein interaction have focused on protein utilization or astringency[365–368] but a few studies have addressed binding to plasma proteins and lipoproteins.[149,342,369–371] Strongest binding has been associated with 1,2-dihydroxyphenols and proline-rich proteins such as those characteristic of human saliva and structure–activity relationships have been reported.[372,373]

Evidence has been presented from studies *in vitro* that unconjugated flavanols and proanthocyanidins bind preferentially to histidine-rich glycoprotein[374] or to Apo-A1, the major protein in human LDL, but preferentially to transferrin in rat plasma.[371] However, although unconjugated flavanols may occur in human plasma, the majority of the phenolic metabolites, as discussed above, do not have a 1,2-dihydroxyphenol moiety and are relatively hydrophilic, and weak associations are therefore to be expected, possibly explaining why dialysis removes such metabolites from isolated LDL.[323] Only traces of flavonoid-like substances have been recovered from plasma LDL of unsupplemented individuals. Tentatively, rutin (not confirmed by LC–MS and probably a misidentification) and quercetin-3-glucuronide were detected at 93 and 73 pmol/mg protein, respectively.[369] Following green tea consumption (eight cups per day for 3 days), flavanols are associated primarily with the plasma proteins ($\sim 0.47\,\mu M$) and high-density lipoproteins ($\sim 0.17\,\mu M$), with lesser amounts in the LDL ($0.077 \pm 0.021\,\mu M$) and least in the very low-density lipoproteins ($\sim 0.08\,\mu M$). Feeding isoflavones in burgers or soya bars to provide doses ranging from ~ 0.03 to 1 mg/kg body weight resulted in less than 1% of total plasma isoflavones recovered from LDL proteins,[342] with genistein, daidzein, and equal concentrations of ~ 10, ~ 3, and up to 0.2 pmol/mg protein, respectively.[375] Generally, the levels of PPT metabolites incorporated in LDL have been insufficient to increase its intrinsic resistance to oxidation *ex vivo*,[149,323,337,376] although there have been exceptions.[322,375]

6.5.3 EFFECTS ON THE VASCULAR SYSTEM

In the context of vascular disease, numerous studies have focused on the ability of phenolic compounds, as pure aglycones and as glycosides, to delay the oxidation of LDL *in vitro*.[377–379] This work has been paralleled by studies investigating the propensity with which the consumption of PPT-rich foods and beverages reduce the oxidation of LDL *ex vivo*.[149,322,337,341,348] The results of these studies, employing realistic PPT intakes, have

TABLE 6.6
A Summary of Published Data for Transient Maximal Plasma Concentrations of Diet-Derived (Poly)Phenols, Typical Plasma Concentrations of Endogenous Phenols and Antioxidants, and Associated Redox Potentials (pH 7)

Mammalian Metabolite Hydroxylation Pattern	Maximal Transient Concentration (μM)	Redox Potential (mV) at pH 7
1,2,3-*vic*	0.14[a]	400–600[d,e,f]
1,2-*vic*	0.8[a]	500–650[d,g,h]
Single *para* or isolated *meta* hydroxyls	10[a]	700–1050[d,e,g,i,j,k]
Blocked/inactive	?	Inactive

Damaging radicals		Redox Potential (mV) at pH 7
Hydroxyl radical		2310[h]
Superoxide radical anion		1800[h]
Alkoxyl radical		1600[l]
Alkyl-peroxyl radical		1000 $\pm$ 60[h,l,m]
PUFA (bis-allylic) radical		600[h]

Endogenous phenols and antioxidants in plasma	Typical plasma concentration (μM)	Redox Potential (mV) at pH 7
Endogenous *p*-phenols	114–280[a]	≈ 700[d,e,g,i,j,k]
α-Tocopherol	5–30[b]	≈ 500[d,h]
Ascorbate (depleted)	50–70[b] (≤ 11)[c]	≈ 280[e,h]
Glutathione		-276[n]

[a]From Table 6.3.
[b]From Table 6.4.
[c]From Ref. 357.
[d]From Ref. 350.
[e]From Ref. 352.
[f]From Ref. 529.
[g]From Ref. 351.
[h]From Ref. 530.
[i]From Ref. 531.
[j]From Ref. 532.
[k]From Ref. 533.
[l]From Ref. 534.
[m]From Ref. 535.
[n]From Ref. 536.
Source: Reprinted from Clifford, M.N., *Planta Med.*, 12, 1103, 2004. With permission.

shown either a small decrease in LDL oxidizability or no change at all. However, it does not necessarily follow that the reductions in cardiovascular disease (CVD) associated with PPT intake, shown in epidemiological studies, will relate solely to the ability of phenols to modify LDL oxidizability. CVD engages a variety of cell types, including endothelial cells, vascular smooth muscle cells (VSMC), leukocytes, and platelets.[380] Indeed, a key early event in atherogenesis is the adhesion of leukocytes to the arterial wall and their subsequent movement into the subendothelial space.[380] This process is mediated via the expression of adhesion molecules on the surface of endothelial cells that are expressed constitutively but can be significantly induced by proinflammatory mediators, such as tumor necrosis factor-α (TNF-

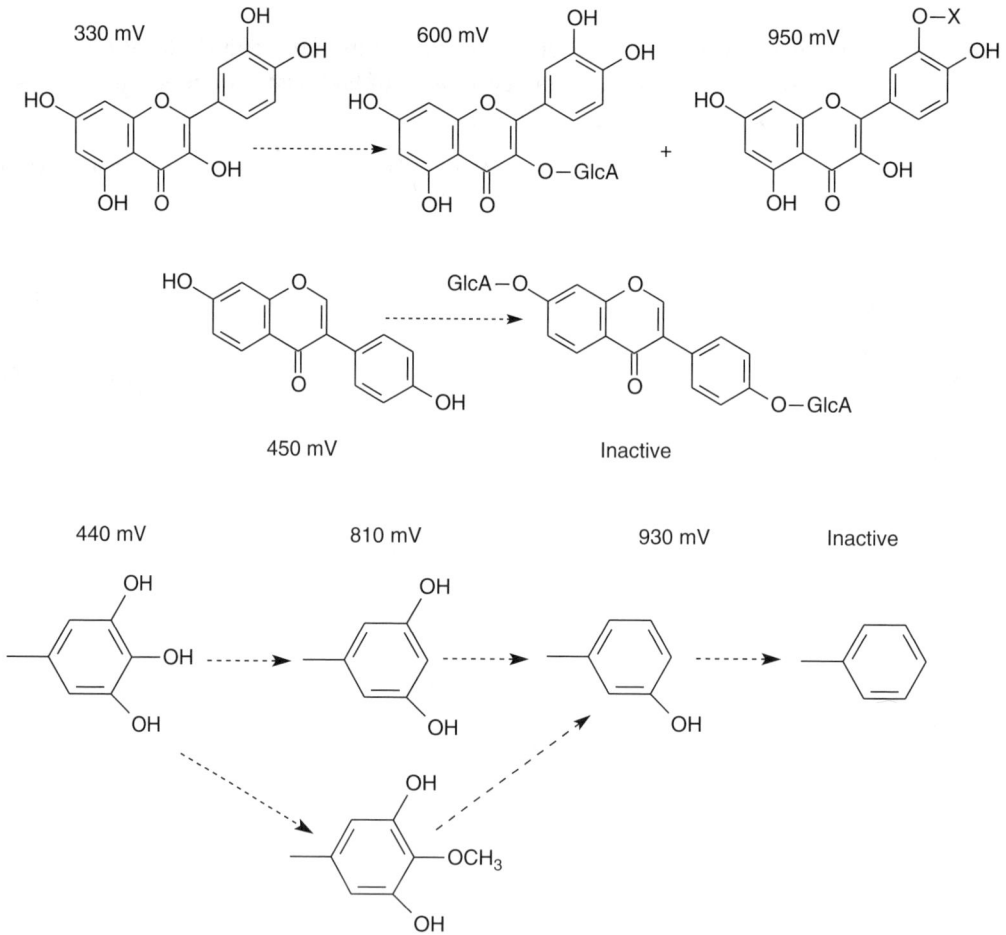

FIGURE 6.1 Illustration of the effects of mammalian metabolism and microbial metabolism on the redox potential of (poly)phenols found in plasma compared with their precursors in the diet and the aglycones commonly used in studies *in vitro*. (Reprinted from Clifford, M.N., *Planta Med.*, 12, 1103, 2004. With permission.)

α), interleukin-1β (IL-1β), and other stimuli. The redox-sensitive transcription factor, nuclear factor-κB (NF-κB), under the control of IκB, is considered essential in mediating the response to TNF-α as many relevant genes contain NF-κB binding sites including vascular cell adhesion molecule-1 (VCAM-1), intercellular adhesion molecule-1 (ICAM-1), and monocyte chemotactic protein (MCP).[381] Mitogen-activated protein (MAP) kinases, activated via receptor tyrosine kinases, are also important in vascular gene regulation. Three main groups of MAP kinases exist and include extracellular signal regulated kinase 1 and 2 (ERK1/2), c-Jun N-terminal kinase (JNK), and p38. These cell-signaling pathways are considered targets for the action of PPT and their effect on atherosclerotic and thrombotic pathways may be important. The modulation of these and other pathways by PPT will be discussed in the context of the vasculature and their potential impact on CVD.

Red wine consumption in humans reduces TNF-α-induced adhesion of monocytes to endothelial cells *ex vivo*[382] and is associated with the downregulation of monocyte adhesion molecules, in particular very late activation antigen-4 (VLA-4). Estruch et al.[383] reported similar findings for VLA-4 and also showed that levels of MCP-1, VCAM-1, and ICAM-1 were decreased after red wine consumption. Interestingly, control studies with gin revealed no effect on these variables indicating that red wine phenolic compounds were likely to be responsible. Other studies have shown that red wine prevents the activation of NF-κB in human peripheral blood mononuclear cells induced by a fat-rich meal.[384] Control studies with vodka highlighted the potential role of the phenolic component of red wine.[384] Studies with tea have revealed little effect on markers of vascular inflammation[385] although clear beneficial effects have been demonstrated in terms of other indices of vascular function.[386] Grape seed proanthocyanidins reduce adhesion molecule expression in systemic sclerosis *in vivo*[387] although the effect *in vitro* was only evident on VCAM-1 expression.[388] Indeed, many PPT attenuate the endothelial production of VCAM-1 induced by either TNF-α or IL-1β *in vitro*, including EGCG and ECG,[389] oleuropein and hydroxytyrosol,[390] apigenin and luteolin,[391] resveratrol,[390] genistein,[392,393] and a variety of gallates.[394] Furthermore, TNF-α-induced MCP-1 secretion is inhibited by genistein,[393] daidzein,[392] petunidin-3-glucoside, and delphinidin-3-glucoside[395] as well as anthocyanin-rich berry extracts.[396,397] These effects overall appear to reduce the ability of monocytes to adhere to TNF-α-activated endothelial cells.[389,398] However, it is evident that while some PPT operate by preventing NF-κB activation,[384,394,396,398] others perhaps operate via a different system.[389] Indeed, quercetin reduces ICAM-1 expression by preventing TNF-α activation of activator protein-1 (AP-1) via the JNK pathway.[399] A common theme with many of the experiments discussed above is a lack of consideration for either the form of the PPT in the incubation or the concentration used. Many studies utilize concentrations far in excess of those attainable *in vivo*, for example, quercetin at $100\,\mu M$[384] and potentially dangerous if it were achieved since $20\,\mu M$ quercetin arising from intravenous infusion caused liver and kidney damage in cancer patients.[400] Such *in vitro–in vivo* disparity highlights the need to evaluate the properties of mammalian metabolites at realistic concentrations. Some workers have tried to overcome this by assessing the effects of PPT metabolites produced by intragastric administration to rats.[401] Interestingly, the metabolites of (+)-catechin were more effective at reducing monocyte adhesion than the parent compound. However, the concentrations and identities of the metabolites used are not clearly stated (~6 μM for catechin) and appear to be higher than those normally present in plasma after a catechin-rich meal. Metabolites of the other flavanols, for which such concentrations might more easily be approached (see above), were not evaluated in this system, but quercetin metabolites were inactive.

The accumulation and proliferation of VSMC within the arterial wall is another key aspect of atherosclerosis in which PPT may also have a role. VSMC proliferation involves the activation of MAP kinases that regulate downstream targets related to cell cycle, proliferation, and migration.[402] Quercetin inhibits serum-induced VSMC proliferation, migration of VSMC from arterial explants, and platelet-derived growth factor (PDGF)-induced phosphorylation of p38 MAP kinase.[403] Quercetin also reduces TNF-α-induced activation of matrix metalloproteinase-9 expression via ERK1/2, AP-1, and NF-κB inhibition.[404] EGCG reduces PDGF-BB-induced activation of MAP kinases in VSMC although no effect was observed with angiotensin II induction.[405] Red wine polyphenols reduce vascular endothelial growth factor (VEGF) release in response to PDGF, an effect that also involves inhibition of the p38 MAP kinase pathway.[406] A red wine extract and resveratrol have also been shown to reduce VSMC proliferation.[407] Physiologically attainable concentrations of EGCG and EGC are effective at reducing VSMC proliferation[408]; however, in general the concentrations employed in these studies are greater than those that can be attained normally. Furthermore, these studies focus on aglycones rather than the metabolites present *in vivo*. It is interesting to note, however, that quercetin-3-glucuronide (10 μM) inhibits angiotensin II JNK activation reducing AP-1 binding and a decrease in VSMC hypertrophy.[409,410] On certain diets quercetin-3-glucuronide might reach a transient 1 μM but it is not known whether long-term exposure to such a concentration might be modestly effective, or whether other quercetin metabolites might also contribute. Studies of this kind are required to assess the effects of PPT *in vivo*.

The effect with which PPT modulate the response of platelets is also pertinent to vascular disease, in particular, thrombosis. Resting platelets inhibit the respiratory burst of neutrophils whereas thrombin-activated platelets increase the respiratory burst. Quercetin and resveratrol at picomolar concentrations attenuate this response by preserving endothelial CD39/ATP-dase,[411] and on present evidence (see above) such concentrations might be achieved locally following deglucuronidation at a site of inflammation.

The modulation of nitric oxide (NO) production by endothelial cells is a further route through which PPT may be important owing to the potent vasoprotective properties of NO. Polyphenol-rich beverages and extracts such as red wine,[412–414] black tea,[415] cocoa,[416] and artichoke[417] can increase the expression and activity of endothelial NO synthase (eNOS) and consequently NO formation. Furthermore, EGCG,[418] resveratrol,[419] cyanidin-3-glucoside,[420,421] genistein,[422] and luteolin and its 7-glucoside[417] have been shown to modulate NO production. Some phenols appear to operate via activation of the phosphatidylinositol-3 OH kinase/Akt pathway and an increase in intracellular calcium[423] while others do not.[422] An upstream regulator may include p38 MAP kinase.[415] These effects have generally been demonstrated only at concentrations greater than $10 \mu M$, although cyanidin-3-glucoside was effective at $0.1 \mu M$,[424] a concentration that might reasonably be achieved *in vivo*. Genistein was effective *in vitro* at $1 \mu M$,[422] a concentration that might conceivably be approached following supplementation and local deglucuronidation, but the situation is far from straightforward. Deglucuronidation would be consequent upon inflammation and likely accompanied by superoxide radical anion leading to peroxynitrite formation from NO. Peroxynitrite is capable of initiating lipid peroxidation, and of nitrosating tyrosine residues in proteins, thus potentially interfering with cell signaling.[425] Quercetin-3-glucuronide, one of the three major human conjugates of dietary quercetin glycosides, and quercetin that contemporaneously could be produced from it by deglucuronidation, have been shown *in vitro* to protect the vascular endothelium[353,409,426,427] from the damaging effects of peroxynitrite and suppress peroxynitrite-induced consumption of lipophilic antioxidants in human LDL.[426] A closer focus on plasma metabolites employed at realistic *in vivo* concentrations is required to disentangle these complexities and assess properly the importance of dietary PPT on NO production, cell signaling, and vascular tone.

Another human metabolite, quercetin-4′-glucuronide, inhibits xanthine oxidase *in vitro* at a concentration in plasma that on normal diets can realistically be approached $(K_i = 0.25 \mu M)$.[428] Various mammalian conjugates of quercetin suppress the formation of conjugated dienes,[429] and some daidzein and genistein glucuronides bind to estrogen receptors and may occur *in vivo* at a sufficient concentration to exert a modest estrogenic effect.[430]

Although classically mammalian conjugates of drugs are viewed as significantly less active than the parent drug and therefore irrelevant physiologically, as illustrated by the examples above, this is not inevitably the case when PPT are considered. However, it should be noted that even when studies *in vitro* show an effect of a plasma metabolite at a concentration that might reasonably be expected *in vivo*, protein binding may impede or prevent this effect.

6.5.4 TRANSFORMATION OF PPT METABOLITES AFTER ABSORPTION

Although with few exceptions (some flavanols, isoflavones) PPT aglycones do not normally occur in tissues, it is now recognized that transient and local deconjugation could occur in plasma, liver, and probably other tissues. Studies using human liver slices have demonstrated the potential for hydrolysis of quercetin glucuronides and the transient release of quercetin.[431] β-Glucuronidase is released to the plasma from the liver[432] and from activated neutrophils and eosinophils under oxidative challenge.[433,434] β-Glucuronidase is able *in vitro* to release luteolin from luteolin glucuronide.[431] The activity of β-glucuronidase is significantly raised in the plasma of hemodialysis patients compared with healthy controls,[433] and the lower plasma pH value associated with a site of inflammation is optimal for this enzyme.[235] Collectively these data suggest that aglycones might be released from glucuronides *in vivo* raising the possibility of biologically significant aglycone-mediated interactions at sites of inflammation, but possibly not in healthy tissues or diseased tissues where inflammation has not occurred.[435] It has been reported that estrogen-3-sulfates can be deconjugated *in vivo*,[436] but whether this enzyme is able similarly to deconjugate sulfated PPT metabolites is unknown. *In vitro*, human CYP 1A2 and CYP 2C9 demethylate certain flavonoids,[437] and hepatic demethylation of methoxyestradiol has been observed in hamsters,[438] but it is not known whether the human isoforms can demethylate PPT *in vivo*. If so, then these enzymes may also generate the aglycones locally.

Gut microflora metabolites may also be important. As discussed above, the mammalian lignan aglycones, enterodiol and enterolactone, are estrogenic,[299] and equol is more estrogenic than its dietary precursor, the isoflavone daidzein.[439]

Some C_6–C_3, C_6–C_2, and C_6–C_1 metabolites produced by the gut microflora from a wide range of dietary PPT inhibit platelet aggregation *in vitro*.[440,441] Animal and studies *in vitro* also suggest that some C_6 C_2,[442] and especially C_6–C_3[443] metabolites interfere with various enzymes in the mevalonate pathway including HMG-CoA reductase, the rate-limiting enzyme in hepatic cholesterol biosynthesis, albeit at concentrations unlikely to occur in plasma. However, these observations are of interest since commodities rich in PPT that would yield such metabolites, and the metabolites when given *per os*, have been shown to inhibit platelet aggregation[441] or to have cholesterol-lowering activity in animal[443–448] and human studies,[449] and such gut flora metabolites may have contributed to the *in vivo* effect. Interference in the mevalonate pathway, particularly HMG-CoA reductase inhibition, may have broader human significance.[450] Individual flora differ extensively in their yields of such metabolites[167,180,183,184] or hippuric acid[154,167,198] and this may be an important factor in interindividual response to diet that goes largely unrecognized in epidemiological studies. Modulation of the flora by dietary PPT, i.e., prebiotic effects, is discussed below.

Nonenzymic transformations should also be considered. Studies *in vitro* indicate that epigallocatechin gallate is transformed to B-ring linked dimers (theasinensins A and D and an oolongtheanin) in alkaline media such as plasma and intestinal fluid.[311] The interaction of several dietary phenols and peroxyl radicals[451–453] has been studied *in vitro* and several EGCG[453] and genistein[451] transformation products have been identified. Although EGCG and genistein have been detected in plasma as aglycones it is not known whether such interactions occur *in vivo* where EGCG and genistein concentrations are comparatively low and other plasma constituents (e.g., α-tocopherol, ascorbate, and serum proteins) may interfere.

6.6 PPT PRIOR TO ABSORPTION

6.6.1 INTERACTION WITH TISSUES AND NUTRIENTS PRIOR TO ABSORPTION

Over the past decade, much emphasis has been placed on the role of dietary PPT on vascular health and disease. However, on the basis of the foregoing arguments it is clear that the major part of the plant PPT consumed never reach the plasma and systemic circulation. It is equally clear that the tissues most exposed to unchanged plant PPT are those of the oro-gastrointestinal tract, and hence the potential for biologically significant effects may be considerably greater here.

For example, the mouth is exposed to a substantial, albeit transient and spasmodic, flux of PPT. It has been reported that quercetin-4′-glucoside, quercetin-3-glucoside, phloretin-2′-glucoside, and genistein-7-glucoside can be rapidly hydrolyzed, and quercetin-3-rutinoside slowly hydrolyzed, by mammalian or microbial enzymes in saliva. The rate of hydrolysis varied 20-fold across 17 volunteers.[454,455] Quercetin thus released is a substrate for salivary peroxidase, transformed to a 2,3,5,7,3',4'-hexahydroxyflavanone-like compound and two "dimeric" species,[456] but the significance of these transformations, for health or taste, is unknown.

Several black tea derived polyphenols, including theaflavin-3,3'-digallate, can inhibit *in vitro* the growth of human esophageal squamous carcinoma cells at concentrations near $20 \mu M$,[107] and theaflavin monogallates have *in vitro* inhibited growth of human colon cancer cell lines at $3 \mu M$.[457] The concentration of the four major theaflavins in a typical brew of black tea will individually exceed $3 \mu M$ and approach or exceed $20 \mu M$ and these *in vitro* observations are of interest especially with reference to the mouth and esophagus where tea beverage has not yet been diluted. Whether beneficial effects might accrue *in vivo* is uncertain, but evidence from animal studies suggests that diets rich in PPT that reach the colon may protect rodents from carcinogens,[458–460] although the human epidemiological evidence for protection against cancer, as discussed above, is mixed.[39,41,43] However, it must be noted that consumption of PPT-rich beverages at temperatures able to scald the esophagus and adjacent tissues can cause damage that predisposes to the development of cancer.[461] Fortunately, these damaging effects are not seen when the beverage is consumed at more modest temperatures,[462–464] but these observations do raise the question of what might happen at preexisting sites of ulceration, especially as many dietary PPT would still retain their free *vic*-dihydroxyphenyl moieties and thus the ability to redox cycle.

PPT having 1,2-dihydroxyphenyl or α-hydroxy-keto (e.g., benztropolone) or 1,3-diketo (e.g., curcumin)[465–467] moieties are strong binders of divalent and trivalent metal ions, and have been blamed for impaired absorption of iron.[468,469] While this may be significant for those who are already iron-deficient and dependent on poorly utilized inorganic iron supplements, for those on balanced western diets this may be of less significance than once thought,[470] although still a cause for concern[131] as the major sources of PPT are poor sources of ascorbate. Such metal binding might be one mechanism contributing to protection against carcinogenesis.[460]

Tannin–protein binding has long been known to have an economic impact on feed utilization by domestic animals, but adverse effects in humans have only rarely been demonstrated,[131] although they may have been overlooked during famines. These interactions in the gastrointestinal tract were once thought to be comparatively nonspecific but evidence is accumulating that there may also be more specific interactions.[471]

6.6.2 MODULATION OF THE GUT MICROFLORA — PREBIOTIC EFFECTS

As discussed above, the transformation of dietary PPT, including the generation of phytoestrogens, by the gut microflora has been known for many years. In contrast, the potential for prebiotic effects has only recently been recognized. Changes over 20 days in the metabolites produced from punicalagin have been interpreted as due to changes in the composition or biochemical competence of the gut microflora.[293] Several studies indicate that regular consumption of green tea (poly)phenols influences the composition of the gut microflora in humans, pigs, and sheep; for example, lowering the colonic pH value, suppressing *Clostridium perfringens*, and increasing the proportion of bifidobacteria without inhibiting lactic acid bacteria,[472–474] but the mechanisms are uncertain.[475] However, assuming the colon contents are some 200 g and the daily intake of total PPT is 1 g, then if 90% passes to the colon concentrations in excess of 4 mg/g are plausible and such concentrations might influence which species grow efficiently. For example, black tea theaflavins are antibacterial against a wide range of organisms.[476]

Although the precise yield of aromatic and phenolic acids produced during the gut flora transformations of PPT are unknown, these metabolites may have a prebiotic or antimicrobial capability. Volunteer studies[167,178] have demonstrated the production endogenously of up to 1100 mg benzoic acid per day, i.e., implying colon concentrations in excess of 5 mg/g (or 7 mg/g if expressed as 3-phenylpropionic acid, the putative immediate precursor). Six of 28 phenolic acids tested *in vitro* produced 50% inhibition of *Listeria monocytogenes* at concentrations in the range 1 to 2 mg/ml, and a further six at concentrations in the range 3 to 5 mg/ml, when applied in a medium at pH 6.2 and thus representative of the human large bowel where pH ranges from 5.7 in the cecum to 6.7 in the rectum.[477] The properties of gut flora metabolites after absorption have been discussed above.

6.6.3 Modulation of the Postprandial Surge in Plasma Glucose — Glycemic Index

There is a growing body of evidence suggesting that diets rich in PPT may influence the absorption and metabolism of glucose, resulting in a lower glycemic index[478] than would otherwise be expected. Red wine,[479] coffee,[480] and apple juice[481] have all been shown in controlled volunteer studies to slow glucose absorption and reduce the postprandial surge in plasma glucose, an event known to be an independent risk factor for coronary heart disease.[482] Studies in which volunteers consumed normal portions of PPT-rich foods[480] have also produced reductions in the postprandial concentrations of plasma insulin and glucose-dependent insulinotropic polypeptide (GIP) and elevation in the concentration of glucagon-like polypeptide-1 (GLP-1), and a polyphenol-enriched diet has been reported to reduce the incidence and severity of nephropathy in type II diabetics.[483]

A prospective study of 17,000 people suggested that the mean relative risk of developing type II diabetes was only 0.5 (0.35–0.72) in those individuals habitually consuming six or more cups of coffee per day compared with those consuming two or less ($p = 0.0002$).[58] The results of subsequent epidemiological studies on coffee consumption[58–63] have been in good agreement.

The reduced glycemic index has been attributed to PPT-mediated inhibition of α-amylase,[484,485] maltase,[486] or α-glucosidase (sucrase),[485,487] but the inhibition of these enzymes is not relevant when volunteers have been given glucose *per se*.[480,481] These observations are more conveniently explained by an effect on the active glucose transporter (SGLT1) in the duodenum. Phloridzin, a dihydrochalcone glucoside characteristic of apples and apple products,[114] but now known to be more widely distributed,[488] competes for the active site both *in vitro* and *in vivo* when given intraperitoneally.[489–492] Other dietary PPT (EGCG, EGC, and 5-caffeoylquinic acid) have been shown *in vitro* to dissipate the sodium gradient essential to the operation of SGLT1,[493,494] and several quercetin glucosides have been shown to interact with it and thus to have the potential to interfere in glucose transport.[495–501] While these effects on glucose absorption and the associated hormones (insulin, GIP, and GLP-1) are modest, they have been achieved in volunteers consuming sensible quantities of common dietary components (as distinct from effects seen only *in vitro* with high levels of PPT aglycones never seen in the diet). Such effects repeated daily, or even several times daily for say 30 years, might in part explain the reduced incidence of chronic disease, especially type II diabetes and the metabolic syndrome, in later life associated with diets rich in fruits and vegetables.

6.7 SAFETY ASSESSMENT OF DIETARY PPT

Normal dietary exposure over many years has highlighted two areas of concern arising through PPT binding of nutrients in the gastrointestinal tract. Impaired absorption of trace metals and impaired protein utilization have been referred to above.

However, in recent years PPT consumption has begun to change. The public has come to believe that diets rich in PPT are health promoting, and industry has increasingly marketed products and supplements rich in PPT. Such products range from traditional foods or beverages with a long history of safe consumption to botanical extracts in tablets or capsules marketed sometimes in an uncontrolled manner on the web.[131] There is probably little cause for concern when some individuals increase their consumption of PPT-rich commodities such as wine, tea, cranberries, soya, etc., provided that they do not exceed levels that other populations have long been exposed to without evidence of harm. Even so, it is important that the possibility of genetic polymorphisms rendering some individuals more susceptible to adverse effects is not ignored,[502] and pharmaceutical-like preparations (tablets, capsules, drops) might present particular difficulties. Neonates lack many important detoxification systems, e.g., CYP 1A2, CYP 3A4, glucuronidation, glycine conjugation, and renal excretion, and are much more vulnerable.[502] CYP 3A4 is moderately variable in healthy adults with more significant ethnic variation and a reduced activity in the elderly, increasing the risk of an adverse reaction between PPT that inhibit this enzyme and drugs that require it for clearance, as discussed below.

While some PPT-rich botanicals are derived from conventional foods, beverages, or herbs, e.g., soya, rosemary, or green tea, others may be derived from materials that have no significant history of use as food or food ingredients despite a history of usage as herbal medicines in some area of the world, e.g., *Ginkgo biloba*.[503] In some cases the supposed active principle(s) of such botanicals have been more or less purified and marketed in a form more concentrated than could be obtained from foods, with correspondingly increased bioavailability and bioefficacy.[360] Mennen et al.[131] have drawn attention to the availability on the web of "tablets or capsules containing 300 mg quercetin, 1 g citrus flavonoids or 20 mg resveratrol with suggested use of 1 to 6 tablets or capsules per day," and point out that if such an exhortation is followed, then intakes of particular PPT could be some 100-fold higher than normally achieved on typical western diets.[131] Nonculinary extraction processes may alter the composition compared with normal domestic practices, and an aqueous alcoholic extract of tea buds sold as a slimming aid had to be withdrawn from the market because of cases of severe liver toxicity that have not been observed following the consumption of conventional green tea brews.[131,504] The toxic principle has not been identified,[505] but it has been shown that EGCG, 4″-O-methyl-EGCG, and 4′,4″-di-O-methyl-EGCG are potent inhibitors (IC$_{50}$ ~0.2 μM) of COMT, and thus might interfere with drug clearance requiring this enzyme.[222] It is well known that some flavonoid-rich commodities, e.g., grapefruit juice, can impair clearance of clinical drugs that are metabolized by CYP 3A4. This has been attributed variously to inhibition by naringenin, 6′,7′-dihydroxybergammotin, and other undefined substances.[258,506–510] Resveratrol-3-sulfate also inhibits CYP3A4 *in vitro* (IC$_{50}$ 1 μM).[276]

High doses of PPT with a *vic*-dihydroxy structure (e.g., protocatechuic acid, caffeic acid, quercetin) given orally to animals have produced forestomach and kidney tumors, and chronic nephropathy. Such PPT have relatively low redox potentials and redox cycle forming quinones or quinone-methides. In tissues low in glutathione, e.g., plasma, these quinones may be scavenged by ascorbate, but glutathione is able *in vitro* to out-compete ascorbate and in glutathione-rich tissues glutathionyl adducts form preferentially.[511] If cellular glutathione is depleted, protein sulfydryls are arylated. Candidate proteins identified include glutathione-*S*-transferase P1-1 (25% inhibition *in vitro* at 1 μM quercetin)[512] and calcium ATPase,[511] and their inhibition further depletes the endogenous cellular defenses against electrophiles, including ultimate carcinogens.[511,513–519] Normally, as discussed above, only small amounts of unchanged PPT or PPT metabolites with unconjugated *vic*-dihydroxy structures reach the plasma but when the protection normally afforded by the alimentary tract is bypassed, as in cancer patients given quercetin intravenously at a dose of 1700 mg/m², plasma quercetin reached 20 µM and some nephrotoxicity was observed.[400]

Diets rich in millet have been associated with endemic goiter in parts of West Africa where millet is a staple. The damage has been attributed to vitexin, a *C*-glycosyl flavone, that in rats has antithyroid activity and that *in vitro* inhibits thyroid peroxidase and the free radical iodination step in thyroid hormone biosynthesis.[520] Isoflavones have produced similar antithyroid effects in rats, but clinical studies in adults have not.[243] However, this remains a possible concern in infants fed soya-based milk-replacers, especially if iodine supply is compromised.

From the phytoestrogen standpoint (see also Chapter 4), isoflavone intakes of 0.2 to 5 mg/day, typical of western diets,[238,243,521] and 20 to 120 mg/day, typical of Asian diets, appear to be safe, but there is concern that higher intakes can have adverse (antiandrogenic) effects on male and female fertility and sexual development *in utero* and *postpartum*.[131] However, since genistein also inhibits tyrosine kinases its estrogenic effect is weaker than might otherwise have been expected.[522] The greatest cause for concern, however, is the potential for an antiluteinizing hormone effect in baby boys aged up to 6 months who on a body weight basis receive very high levels of isoflavones in soya-based infant formula.[131] Although clinical evidence has not been produced to substantiate that this occurred, manufacturers have reformulated their soya infant formulas with low-isoflavone soya protein preparations.[243] Nevertheless, the risk remains if ill-informed parents supplement their infants inappropriately.

Clearly, care must be exercised to ensure that dietary manipulation and supplementation do not produce oral loads able to swamp the body's defenses. For the reasons outlined above, a decision tree has been developed to assist in the risk assessment of botanical products.[503]

6.8 CONCLUSIONS AND FUTURE RESEARCH REQUIREMENTS

Although it is widely accepted that diets rich in fruits and vegetables are beneficial to health, the explanation(s) remain obscure. That dietary flavonoids are key drivers has achieved the status of dogma, but proof is lacking. It is clear that very little of the plant PPT ever reaches the tissues unchanged, and it is essential that the properties of the metabolites are properly addressed. In this regard, it is very interesting that the importance of the gut microflora in transforming dietary PPT has been rediscovered. It is now clear that cinnamates and derived polyphenols are the major dietary PPT for many populations, and these PPT share many gut flora metabolites with the flavonoids. For these two reasons these PPT deserve more attention in the future, and it would be illogical to consider flavonoids in isolation. Gut flora and mammalian metabolism combine to eradicate the powerful antioxidant capability shown *in vitro* by many aglycones, and coupled with the weak or zero antioxidant activity of the metabolites suggests that radical scavenging mechanisms are, at most, a minor part of the *in vivo* story, possibly restricted to sites of inflammation where localized deconjugation might occur. Nevertheless, there is good evidence that extensive initial conjugation is desirable since grossly elevated levels of powerful PPT-derived antioxidants could cause more harm than good. Supplementation must be approached with caution.

If PPT are beneficial to health, then explanations other than antioxidant effects or radical scavenging must be sought, and effects prior to absorption certainly deserve greater consideration since it is the tissues of the oro-gastrointestinal tract that see the greatest dose. Although as yet far from proven to be biologically significant, prebiotic effects, effects on the glycemic index, incretin hormones, and postprandial surge in glucose, and the effects of gut flora or mammalian metabolites on cell signaling systems, are beginning to receive serious consideration and systematic investigation. The preliminary results suggest that beneficial effects might be achieved at realistic dietary intakes. Even if unequivocally established, these effects are likely to be modest, but with the potential possibly to yield health benefit over years or decades — fully in keeping with the epidemiology — they are unlikely to be curative. Other as yet unrecognized mechanisms may also operate at normal dietary levels.

The task remaining is massive. If significant progress is to be made cost-effectively greater coordination of efforts and resources will be necessary. For improved epidemiological data and more reliable risk assessment it is essential that more precise estimates are obtained of what actually reaches the mouth. The quantitative effects of commercial operations and culinary practice on the content of PPT must be better defined and the associated structural transformations associated with traditional processing or cooking need to be better understood. Attention must be focused on effects in humans, but animal studies with ^{14}C-labeled test compounds are essential in determining the fate of ingested material. Since polyphenols are fragmented by the gut, microflora studies may have to be performed with more than one position of labeling to determine the identity and disposition of these metabolites. The results will have to be confirmed in volunteers, perhaps using ^{13}C-labeled materials.

Volunteer studies must employ larger groups that are genetically defined and representative, and studies must be longer term, better to reflect long-term dietary practice. When ethically acceptable, volunteer studies comparing clinically defined at-risk groups with matched healthy controls could be informative. Eventually, studies must address the possibility of synergy, not only between classes of PPT, but also with other classes of nonnutrients, and multiple endpoints must be assessed so as to encompass as many disease states or mechanisms as possible.

REFERENCES

1. Lipkin, M., Uehara, K., Winawer, S., Sanchez, A., Bauer, C., Phillips, R., Lynch, H.T., Blattner, W.A., and Fraumeni, J.F., Seventh-Day Adventist vegetarians have a quiescent proliferative activity in colonic mucosa, *Cancer Lett.*, 26, 139, 1985.
2. Steinmetz, K.A. and Potter, J.D., Vegetables, fruit and cancer. I. Epidemiology, *Cancer Causes Control*, 5, 325, 1991.
3. Block, G., Patterson, B., and Subar, A., Fruit, vegetables and cancer prevention: a review of the epidemiological evidence, *Nutr. Cancer*, 18, 1, 1992.
4. Hertog, M.G., Bueno-de-Mesquita, H.B., Fehily, A.M., Sweetnam, P.M., Elwood, P.C., and Kromhout, D., Fruit and vegetable consumption and cancer mortality in the Caerphilly Study, *Cancer Epidemiol. Biomarkers Prev.*, 5, 673, 1996.
5. World Cancer Research Fund, *Food, Nutrition and the Prevention of Cancer: A Global Perspective*, American Institute for Cancer Research, Washington DC, 1997.
6. Department of Health, *Nutritional Aspects of the Development of Cancer*, HMSO, London, 1998.
7. Wallstrom, P., Wirfalt, E., Janzon, L., Mattisson, I., Elmstahl, S., Johansson, U., and Berglund, G., Fruit and vegetable consumption in relation to risk factors for cancer: a report from the Malmo Diet and Cancer Study, *Public Health Nutr.*, 3, 263, 2000.
8. Babbs, C.F., Free radicals and the etiology of colon cancer, *Free Radical Biol. Med.*, 8, 191, 1990.
9. Guyton, K.Z. and Kensler, T.W., Oxidative mechanisms in carcinogenesis, *Br. Med. Bull.*, 49, 523, 1993.
10. Goldstein, B.D. and Witz, G., Free radicals in carcinogenesis, *Free Radical Res. Commun.*, 11, 3, 1990.
11. Steinmetz, K.A. and Potter, J.D., Vegetables, fruit and cancer. II. Mechanisms, *Cancer Causes Control*, 5, 427, 1991.
12. Wattenberg, L.W., Chemoprevention of cancer, *Cancer Res.*, 45, 1, 1985.
13. Wattenberg, L.W., Inhibition of carcinogenesis by minor anutrient constituents of the diet., *Proc. Nutr. Soc.*, 49, 173, 1990.
14. Weisburger, J.H., Nutritional approach to cancer prevention with emphasis on vitamins, antioxidants and carotenoids, *Am. J. Clin. Nutr.*, 53, 226S, 1991.

15. Wattenberg, L.W., Inhibition of carcinogenesis by minor dietary constituents, *Cancer Res.*, 52, 2085, 1992.
16. Kehrer, J.P., Free radicals as mediators of tissue injury and disease, *Crit. Rev. Toxicol.*, 23, 21, 1993.
17. Stohs, S.J., The role of free radicals in toxicity and disease, *J. Basic Clin. Physiol. Pharmacol.*, 6, 205, 1995.
18. Renaud, S. and de Lorgeril, M., Wine, alcohol, platelets and the French paradox for coronary heart disease, *Lancet*, 339, 1523, 1992.
19. Hertog, M.G.L., Feskens, E.J.M., Hollman, P.C.H., Katan, M.B., and Kromhout, D., Dietary antioxidant flavonoids and risk of coronary heart disease: the Zutphen Study, *Lancet*, 342, 1007, 1993.
20. Hertog, M.G., Kromhout, D., and Aravanis, C., Flavonoid intake and long term risk of coronary heart disease and cancer in the seven countries study, *Arch. Intern. Med.*, 155, 381, 1995.
21. Clifford, M.N., Diet-derived phenols in plasma and tissues and their implications for health, *Planta Med.*, 12, 1103, 2004.
22. Joshipura, K.J., Ascherio, A., Manson, J.E., Stampfer, M.J., Rimm, E.B., Speizer, F.E., Hennekens, C.H., Spiegelman, D., and Willett, W.C., Fruit and vegetable intake in relation to risk of ischemic stroke, *J. Am. Med. Assoc.*, 282, 1233, 1999.
23. Bazzano, L.A., He, J., Ogden, L.G., Loria, C.M., Vupputuri, S., Myers, L., and Whelton, P.K., Fruit and vegetable intake and risk of cardiovascular disease in US adults: the first National Health and Nutrition Examination Survey Epidemiologic Follow-up Study, *Am. J. Clin. Nutr.*, 76, 93, 2002.
24. Truswell, A.S., Cereal grains and coronary heart disease, *Eur. J. Clin. Nutr.*, 56, 1, 2002.
25. Rissanen, T.H., Voutilainen, S., Virtanen, J.K., Venho, B., Vanharanta, M., Mursu, J., and Salonen, J.T., Low intake of fruits, berries and vegetables is associated with excess mortality in men: the Kuopio Ischaemic Heart Disease Risk Factor (KIHD) Study, *J. Nutr.*, 133, 199, 2003.
26. Vanharanta, M., Voutilainen, S., Rissanen, T.H., Adlercreutz, H., and Salonen, J.T., Risk of cardiovascular disease-related and all-cause death according to serum concentrations of enterolactone: Kuopio Ischaemic Heart Disease Risk Factor Study, *Arch. Intern. Med.*, 163, 1099, 2003.
27. Johnsen, N.F., Hausner, H., Olsen, A., Tetens, I., Christensen, J., Knudsen, K.E., Overvad, K., and Tjonneland, A., Intake of whole grains and vegetables determines the plasma enterolactone concentration of Danish women, *J. Nutr.*, 134, 2691, 2004.
28. Hung, H.C., Joshipura, K.J., Jiang, R., Hu, F.B., Hunter, D., Smith-Warner, S.A., Colditz, G.A., Rosner, B., Spiegelman, D., and Willett, W.C., Fruit and vegetable intake and risk of major chronic disease, *J. Natl. Cancer Inst.*, 96, 1577, 2004.
29. Jansen, M.C., McKenna, D., Bueno-de-Mesquita, H.B., Feskens, E.J., Streppel, M.T., Kok, F.J., and Kromhout, D., Reports: quantity and variety of fruit and vegetable consumption and cancer risk, *Nutr. Cancer*, 48, 142, 2004.
30. Franco, O.H., Bonneux, L., Peeters, A., and Steyerberg, E.W., The Polymeal: a more natural, safer, and probably tastier (than the Polypill) strategy to reduce cardiovascular disease by more than 75%, *Br. Med. J.*, 329, 18, 2004.
31. Johnson, I.T., New approaches to the role of diet in the prevention of cancers of the alimentary tract, *Mutat. Res.*, 551, 9, 2004.
32. Svehlikova, V., Wang, S., Jakubikova, J., Williamson, G., Mithen, R., and Bao, Y., Interactions between sulforaphane and apigenin in the induction of UGT1A1 and GSTA1 in CaCo-2 cells, *Carcinogenesis*, 25, 1629, 2004.
33. Knekt, P., Jarvinen, R., Reunanen, A., and Maatela, J., Flavonoid intake and coronary mortality in Finland: a cohort study, *Br. Med. J.*, 312, 478, 1996.
34. Keli, S.O., Hertog, M.G., Feskens, E.J., and Kromhout, D., Dietary flavonoids, antioxidant vitamins, and incidence of stroke: the Zutphen Study, *Arch. Intern. Med.*, 156, 637, 1996.
35. Yochum, L., Kushi, L.H., Meyer, K., and Folsom, A.R., Dietary flavonoid intake and risk of cardiovascular disease in postmenopausal women, *Am. J. Epidemiol.*, 149, 943, 1999.

36. Hirvonen, T., Pietinen, P., Virtanen, M., Ovaskainen, M.L., Hakkinen, S., Albanes, D., and Virtamo, J., Intake of flavonols and flavones and risk of coronary heart disease in male smokers, *Epidemiology,* 12, 62, 2001.

37. Mennen, L.I., Sapinho, D., de Bree, A., Arnault, N., Bertrais, S., Galan, P., and Hercberg, S., Consumption of foods rich in flavonoids is related to a decreased cardiovascular risk in apparently healthy French women, *J. Nutr.,* 134, 923, 2004.

38. Knekt, P., Jarvinen, R., Seppanen, R., Hellovaara, M., Teppo, L., Pukkala, E., and Aromaa, A., Dietary flavonoids and the risk of lung cancer and other malignant neoplasms, *Am. J. Epidemiol.,* 146, 223, 1997.

39. Garcia-Closas, R., Gonzalez, C.A., Agudo, A., and Riboli, E., Intake of specific carotenoids and flavonoids and the risk of gastric cancer in Spain, *Cancer Causes Control,* 10, 71, 1999.

40. Birt, D.F., Hendrich, S., and Wang, W., Dietary agents in cancer prevention: flavonoids and isoflavonoids, *Pharmacol. Ther.,* 90, 157, 2001.

41. Hirvonen, T., Virtamo, J., Korhonen, P., Albanes, D., and Pietinen, P., Flavonol and flavone intake and the risk of cancer in male smokers (Finland), *Cancer Causes Control,* 12, 789, 2001.

42. Arts, I.C., Jacobs, D.R., Gross, M., Harnack, L.J., and Folsom, A.R., Dietary catechins and cancer incidence among postmenopausal women: the Iowa Women's Health Study (United States), *Cancer Causes Control,* 13, 373, 2002.

43. Sun, C.L., Yuan, J.M., Lee, M.J., Yang, C.S., Gao, Y.T., Ross, R.K., and Yu, M.C., Urinary tea polyphenols in relation to gastric and esophageal cancers: a prospective study of men in Shanghai, China, *Carcinogenesis,* 23, 1497, 2002.

44. Rimm, E.B., Katan, M.B., Ascherio, A., Stampfer, M.J., and Willett, W.C., Relation between intake of flavonoids and risk for coronary heart disease in male health professionals, *Ann. Intern. Med.,* 125, 384, 1996.

45. Goldbohm, R.A., Hertog, M.G., Brants, H.A., van Poppel, G., and van den Brandt, P.A., Consumption of black tea and cancer risk: a prospective cohort study, *J. Natl. Cancer Inst.,* 88, 93, 1996.

46. Hertog, M.G.L., Sweetman, P.M., Fehily, A.M., Elwood, P.C., and Kromhout, D., Antioxidant flavonols and ischaemic heart disease in a Welsh population of men: the Caerphilly Study, *Am. J. Clin. Nutr.,* 65, 1489, 1997.

47. Kimira, M., Arai, Y., Shimoi, K., and Watanabe, S., Japanese intake of flavonoids and isoflavonoids from foods, *J. Epidemiol.,* 8, 168, 1998.

48. Skibola, C.F. and Smith, M.T., Potential health impacts of excessive flavonoid intake, *Free Radical Biol. Med.,* 29, 375, 2000.

49. Peters, U., Poole, C., and Arab, L., Does tea affect cardiovascular disease? A meta-analysis, *Am. J. Epidemiol.,* 154, 495, 2001.

50. Sesso, H.D., Gaziano, J.M., Liu, S., and Buring, J.E., Flavonoid intake and the risk of cardiovascular disease in women, *Am. J. Clin. Nutr.,* 77, 1400, 2003.

51. Ambrosone, C.B., McCann, S.E., Freudenheim, J.L., Marshall, J.R., Zhang, Y., and Shields, P.G., Breast cancer risk in premenopausal women is inversely associated with consumption of broccoli, a source of isothiocyanates, but is not modified by GST genotype, *J. Nutr.,* 134, 1134, 2004.

52. Plumb, G.W., Price, K.R., Rhodes, M.J.C., and Williamson, G., Antioxidant properties of the major polyphenolic compoumds in broccoli, *Free Radical Res.,* 27, 429, 1997.

53. Price, K.R., Causcelli, F., Colquhoun, I.J., and Rhodes, M.J.C., Hydroxycinnamic acid esters from broccoli florets, *Phytochemistry,* 45, 1683, 1997.

54. Hollman, P.C.H., Hertog, M.G.L., and Katan, M.B., Analysis and health effects of flavonoids, *Food Chem.,* 57, 43, 1996.

55. Hollman, P.C.H. and Arts, I.C.W., Flavonols, flavones and flavanols — nature, occurrence and dietary burden, *J. Sci. Food Agric.,* 80, 1081, 2000.

56. Price, K.R., Casuscelli, F., Colquhoun, I.J., and Rhodes, M.J.C., Composition and content of flavonol glycosides in broccoli floreets (*Brassica oleracea*) and their fate during cooking, *J. Sci. Food Agric.,* 77, 468, 1998.

57. Petri, N., Tannergren, C., Holst, B., Mellon, F.A., Bao, Y., Plumb, G.W., Bacon, J., O'Leary, K.A., Kroon, P.A., Knutson, L., Forsell, P., Eriksson, T., Lennernas, H., and Williamson, G.,

Absorption/metabolism of sulforaphane and quercetin, and regulation of phase II enzymes, in human jejunum *in vivo, Drug Metab. Dispos.,* 31, 805, 2003.

58. van Dam, R.M. and Feskens, E.J., Coffee consumption and risk of type 2 diabetes mellitus, *Lancet,* 360, 1477, 2002.

59. Saremi, A., Tulloch-Reid, M., and Knowler, W.C., Coffee consumption and the incidence of type 2 diabetes, *Diabetes Care,* 26, 2211, 2003.

60. Agardh, E.E., Carlsson, S., Ahlbom, A., Efendic, S., Grill, V., Hammar, N., Hilding, A., and Ostenson, C.G., Coffee consumption, type 2 diabetes and impaired glucose tolerance in Swedish men and women, *J. Intern. Med.,* 255, 645, 2004.

61. Rosengren, A., Dotevall, A., Wilhelmsen, L., Thelle, D., and Johansson, S., Coffee and incidence of diabetes in Swedish women: a prospective 18-year follow-up study, *J. Intern. Med.,* 255, 89, 2004.

62. Salazar-Martinez, E., Willett, W.C., Ascherio, A., Manson, J.E., Leitzmann, M.F., Stampfer, M.J., and Hu, F.B., Coffee consumption and risk for type 2 diabetes mellitus, *Ann. Intern. Med.,* 140, 1, 2004.

63. Tuomilehto, J., Hu, G., Bidel, S., Lindstrom, J., and Jousilahti, P., Coffee consumption and risk of type 2 diabetes mellitus among middle-aged Finnish men and women, *J. Am. Med. Assoc.,* 291, 1213, 2004.

64. Pietinen, P., Stumpf, K., Mannisto, S., Kataja, V., Uusitupa, M., and Adlercreutz, H., Serum enterolactone and risk of breast cancer: a case–control study in eastern Finland, *Cancer Epidemiol. Biomarkers Prev.,* 10, 339, 2001.

65. Boccardo, F., Lunardi, G., Guglielmini, P., Parodi, M., Murialdo, R., Schettini, G., and Ruba-gotti, A., Serum enterolactone levels and the risk of breast cancer in women with palpable cysts, *Eur. J. Cancer,* 40, 84, 2004.

66. McCann, S.E., Muti, P., Vito, D., Edge, S.B., Trevisan, M., and Freudenheim, J.L., Dietary lignan intakes and risk of pre- and postmenopausal breast cancer, *Int J. Cancer,* 111, 440, 2004.

67. Kilkkinen, A., Virtamo, J., Vartiainen, E., Sankila, R., Virtanen, M.J., Adlercreutz, H., and Pietinen, P., Serum enterolactone concentration is not associated with breast cancer risk in a nested case–control study, *Int. J. Cancer,* 108, 277, 2004.

68. Zeleniuch-Jacquotte, A., Adlercreutz, H., Shore, R.E., Koenig, K.L., Kato, I., Arslan, A.A., and Toniolo, P., Circulating enterolactone and risk of breast cancer: a prospective study in New York, *Br. J. Cancer,* 91, 99, 2004.

69. den Tonkelaar, I., Keinan-Boker, L., Veer, P.V., Arts, C.J., Adlercreutz, H., Thijssen, J.H., and Peeters, P.H., Urinary phytoestrogens and postmenopausal breast cancer risk, *Cancer Epidemiol. Biomarkers Prev.,* 10, 223, 2001.

70. McCann, S.E., Moysich, K.B., Freudenheim, J.L., Ambrosone, C.B., and Shields, P.G., The risk of breast cancer associated with dietary lignans differs by CYP17 genotype in women, *J. Nutr.,* 132, 3036, 2002.

71. Arai, Y., Watanabe, S., Kimira, M., Shimoi, K., Mochizuki, R., and Kinae, N., Dietary intakes of flavonols, flavones and isoflavones by Japanese women and the inverse correlation between quercetin intake and plasma LDL cholesterol concentration, *J. Nutr.,* 130, 2243, 2000.

72. Arts, I.C. and Hollman, P.C., Polyphenols and disease risk in epidemiologic studies, *Am. J. Clin. Nutr.,* 81, 317S, 2005.

73. Ribereau-Gayon, P., *Plant Phenolics,* Hafner Publishing Company, New York, 1972.

74. Shahidi, F. and Naczk, M., *Food Phenolics: Sources; Chemistry; Effects; Applications,* Technomic Publishing Inc., Lancaster, PA, 1995.

75. Parr, A.J. and Bolwell, G.P., Phenols in the plant and in man. The potential for possible nutritional enhancement of the diet by modifying phenolic content or composition, *J. Sci. Food Agric.,* 80, 985, 2000.

76. Clifford, M.N., The health effects of tea and tea components. Appendix 1. A nomenclature for phenols with special reference to tea, *CRC Crit. Rev. Food Sci. Nutr.,* 41, 393, 2001.

77. Lindsay, D.G. and Clifford, M.N., Special issue devoted to critical reviews produced within the EU Concerted Action "Nutritional Enhancement of Plant-based Food in European Trade" (NEODIET), *J. Sci. Food Agric.,* 80, 793, 2000.

78. Clifford, M.N., Chlorogenic acids and other cinnamates — nature, occurrence and dietary burden, *J. Sci. Food Agric.*, 79, 362, 1999.

79. Clifford, M.N., Chlorogenic acids and other cinnamates — nature, occurrence, dietary burden, absorption and metabolism, *J. Sci. Food Agric.*, 80, 1033, 2000.

80. Clifford, M.N., Miscellaneous phenols in foods and beverages — nature, occurrence and dietary burden, *J. Sci. Food Agric.*, 80, 1126, 2000.

81. Soler-Rivas, C., Espin, J.C., and Wichers, H.J., Oleuropein and related compounds, *J. Sci. Food Agric.*, 80, 1013, 2000.

82. Tomás-Barberán, F.A. and Clifford, M.N., Dietary hydroxybenzoic acid derivatives — nature, occurrence and dietary burden, *J. Sci. Food Agric.*, 80, 1024, 2000.

83. Hertog, M.G.L., Hollman, P.C.H., and Katan, M.B., Content of potentially anticarcinogenic flavonoids of 28 vegetables and 9 fruits consumed in The Netherlands, *J. Agric. Food Chem.*, 40, 2379, 1992.

84. Haslam, E., *Practical Polyphenolics. From Structure to Molecular Recognition and Physiological Action*, Cambridge University Press, Cambridge, 1998.

85. Santos-Buelga, C. and Scalbert, A., Proanthocyanidins and tannin-like compounds — nature, occurrence, dietary intake and effects on nutrition and health, *J. Sci. Food Agric.*, 80, 1094, 2000.

86. Gu, L., Kelm, M.A., Hammerstone, J.F., Beecher, G., Holden, J., Haytowitz, D., Gebhardt, S., and Prior, R.L., Concentrations of proanthocyanidins in common foods and estimations of normal consumption, *J. Nutr.*, 134, 613, 2004.

87. Clifford, M.N. and Scalbert, A., Ellagitannins — nature, occurrence and dietary burden, *J. Sci. Food Agric.*, 80, 1118, 2000.

88. Cassidy, A., Hanley, B., and Lamuela-Raventós, R.M., Isoflavones, lignans and stilbenes — origins, metabolism and potential importance to human health, *J. Sci. Food Agric.*, 80, 1044, 2000.

89. Setchell, K.D.R., Lawson, A.M., Borriello, S.P., Harkness, R., Gordon, H., Morgan, D.M.L., Kirk, D.N., Adlercreutz, H., and Anderson, L.C., Lignan formation in man — microbial involvement and possible roles in relation to cancer, *Lancet*, ii, 4, 1981.

90. Diplock, A.T., Defence against reactive oxygen species, *Free Radical Res.*, 29, 463, 1999.

91. Kroon, P.A., Clifford, M.N., Crozier, A., Day, A.J., Donovan, J.L., Manach, C., and Williamson, G., How should we assess the effects of exposure to dietary polyphenols *in vitro*? *Am. J. Clin. Nutr.*, 80, 15, 2004.

92. Bakker, J., Bridle, P., Honda, P., Kuwano, H., Saito, N., Terahara, N., and Timberlake, C.F., Identification of an anthocyanin occurring in some red wines, *Phytochemistry*, 44, 1375, 1997.

93. Bakker, J. and Timberlake, C.F., Isolation, identification and characterization of new color-stable anthocyanins occurring in some red wines, *J. Agric. Food Chem.*, 45, 35, 1997.

94. Cameira-dos-Santos, P.-J., Brillouet, J.-M., Cheynier, V., and Moutounet, M., Detection and partial characterisation of new anthocyanin-derived pigments in wine, *J. Sci. Food Agric.*, 70, 204, 1996.

95. Fulcrand, H., Benabdeljalil, C., Rigaud, J., Cheynier, V., and Moutounet, M., A new class of wine pigments yielded by reaction between pyruvic acid and the grape anthocyanins, *Phytochemistry*, 47, 1401, 1998.

96. Fulcrand, H., Cameira-dos-Santos, P.-J., Sarni-Manchado, P., Cheynier, V., and Favre-Bonvin, J., Structure of new anthocyanin-derived wine pigments, *J. Chem. Soc. Perkin Trans. 1*, 735, 1996.

97. Vivar-Quintana, A.M., Santos-Buelga, C., Francia-Aricha, E., and Rivas-Gonzalo, J.C., Formation of anthocyanin-derived pigments in experimental red wines, *Food Sci. Technol. Int.*, 5, 347, 1999.

98. Davies, A.P., Goodsall, C., Cai, Y., Davis, A.L., Lewis, J.R., Wilkins, J., Wan, X., Clifford, M.N., Powell, C., Thiru, A., Safford, R., and Nursten, H.E., Black tea dimeric and oligomeric pigments — structures and formation, in *Plant Polyphenols 2. Chemistry, Biology, Pharmacology, Ecology*, Gross, G.G., Hemingway, R.W., and Yoshida, T., Eds., Kluwer Academic, Dordrecht, 1998/1999.

99. Davis, A.L., Lewis, J.R., Cai, Y., Powell, C., Davies, A.P., Wilkins, J.P.G., Pudney, P., and Clifford, M.N., A polyphenolic pigment from black tea, *Phytochemistry*, 46, 1397, 1997.

100. Menet, M.C., Sang, S., Yang, C.S., Ho, C.T., and Rosen, R.T., Analysis of theaflavins and thearubigins from black tea extract by MALDI-TOF mass spectrometry, *J. Agric. Food Chem.,* 52, 2455, 2004.

101. Sang, S., Tian, S., Stark, R.E., Yang, C.S., and Ho, C.T., New dibenzotropolone derivatives characterized from black tea using LC/MS/MS, *Bioorg. Med. Chem.,* 12, 3009, 2004.

102. Sang, S., Tian, W., Meng, X., Stark, R.E., Rosen, R.T., Yang, C.S., and Ho, C.-T., Theadi-benztropolone A, a new type pigment from enzymatic oxidation of (−)-epicatechin and (−)-epigallocatechin gallate and characterized from black tea using LC/MS/MS, *Tetrahedron Lett.,* 43, 7129, 2002.

103. Tanaka, T., Mine, C., and Kuono, I., Structures of two new oxidation products of green tea polyphenols generated by model tea fermentation, *Tetrahedron,* 58, 8851, 2002.

104. Tanaka, T., Mine, C., Inoue, K., Matsuda, M., and Kouno, I., Synthesis of theaflavin from epicatechin and epigallocatechin by plant homogenates and role of epicatechin quinone in the synthesis and degradation of theaflavin, *J. Agric. Food Chem.,* 50, 2142, 2002.

105. Tanaka, T., Betsumiya, Y., Mine, C., and Kouno, I., Theanaphthoquinone, a novel pigment oxidatively derived from theaflavin during tea fermentation, *Chem. Commun.,* 1365, 2000.

106. Tanaka, T., Inoue, K., Betsumiya, Y., Mine, C., and Kouno, I., Two types of oxidative dimerization of the black tea polyphenol theaflavin, *J. Agric. Food Chem.,* 49, 5785, 2001.

107. Sang, S., Lambert, J.D., Tian, S., Hong, J., Hou, Z., Ryu, J.H., Stark, R.E., Rosen, R.T., Huang, M.T., Yang, C.S., and Ho, C.T., Enzymatic synthesis of tea theaflavin derivatives and their anti-inflammatory and cytotoxic activities, *Bioorg. Med. Chem.,* 12, 459, 2004.

108. Gosnay, S.L., Bishop, J.A., New, S.A., Catterick, J., and Clifford, M.N., Estimation of the mean intakes of fourteen classes of dietary phenolics in a population of young British women aged 20–30 years, *Proc. Nutr. Soc.,* 61, 125A, 2002.

109. Kühnau, J., The flavonoids. A class of semi-essential food components: their role in human nutrition, *World Rev. Nutr. Diet.,* 24, 117, 1976.

110. Linseisen, J., Radtke, J., and Wolfram, G., Flavonoid intake of adults in a Bavarian subgroup of the national food consumption survey, *Z. Ernährungswiss.,* 364, 403, 1998.

111. Radtke, J., Linseisen, J., and Wolfram, G., Phenolic acid intake of adults in a Bavarian subgroup of the national food consumption survey, *Z. Ernährungswiss.,* 37, 190, 1998.

112. Sampson, L., Rimm, E., Hollman, P.C., de Vries, J.H., and Katan, M.B., Flavonol and flavone intakes in US health professionals, *J. Am. Diet. Assoc.,* 102, 1414, 2002.

113. Woods, E., Clifford, M.N., Gibbs, M., Hampton, S., Arendt, J., and Morgan, L., Estimation of mean intakes of 14 classes of dietary phenols in a population of male shift workers, *Proc. Nutr. Soc.,* 62, 60A, 2003.

114. Tomás-Barberán, F.A. and Clifford, M.N., Flavanones, chalcones and dihydrochalcones — nature, occurrence and dietary burden, *J. Sci. Food Agric.,* 80, 1073, 2000.

115. Suzuki, M., Sano, M., Yoshida, R., Degawa, M., Miyase, T., and Maeda-Yamamoto, M., Epimerization of tea catechins and *O*-methylated derivatives of (−)-epigallocatechin-3-*O*-gallate: relationship between epimerization and chemical structure, *J. Agric. Food Chem.,* 51, 510, 2003.

116. Seto, R., Nakamura, H., Nanjo, F., and Hara, Y., Preparation of epimers of tea catechins by heat treatment, *Biosci. Biotechnol. Biochem.,* 61, 1434, 1997.

117. Price, K.R., Bacon, J.R., and Rhodes, M.J.C., Effect of storage and domestic processing on the content and composition of flavonol glucosides in onion (*Allium cepa*), *J. Agric. Food Chem.,* 45, 938, 1997.

118. Ewald, C., Fjelkner-Modig, S., Johanssen, K., Sjöholma, I., and Åkesson, B., Effect of processing on major flavonoids in processed onions, green beans, and peas, *Food Chem.,* 64, 231, 1999.

119. Chuda, Y., Suzuki, M., Nagata, T., and Tsushida, T., Contents and cooking loss of three quinic acid derivatives from Garland (*Chrysanthemum coronarium* L.), *J. Agric. Food Chem.,* 46, 1437, 1998.

120. Ioku, K., Aoyama, Y., Tokuno, A., Terao, J., Nakatani, N., and Takei, Y., Various cooking methods and the flavonoid content in onion, *J. Nutr. Sci. Vitaminol. (Tokyo),* 47, 78, 2001.

121. Brenes, M., Garcia, A., Dobarganes, M.C., Velasco, J., and Romero, C., Influence of thermal treatments simulating cooking processes on the polyphenol content in virgin olive oil, *J. Agric. Food Chem.*, 50, 5962, 2002.

122. Vallejo, F., Tom, F.A., and Garc, C., Phenolic compound contents in edible parts of broccoli inflorescences after domestic cooking, *J. Sci. Food Agric.*, 83, 1511, 2003.

123. Cheynier, V., Polyphenols in foods are more complex than often thought, *Am. J. Clin. Nutr.*, 81, 223S, 2005.

124. Bond, T.A., Lewis, J.R., Davis, A., and Davies, A.P., Analysis and purification of catechins and their transformation products, in *Methods in Polyphenol Analysis*, Santos-Buelga, C. and Williamson, G., Eds., Royal Society of Chemistry, Cambridge, 2003, p. 11.

125. Justesen, U. and Knuthsen, P., Composition of flavonoids in fresh herbs and calculation of flavonoid intake by use of herbs in traditional Danish dishes, *Food Chem.*, 73, 245, 2001.

126. Justesen, U., Knuthsen, P., Andersen, N.L., and Leth, T., Estimation of daily intake distribution of flavonols and flavanones in Denmark, *Scand. J. Nutr.*, 44, 158, 2000.

127. Knekt, P., Isotupa, S., Rissanen, H., Heliovaara, M., Jarvinen, R., Hakkinen, S., Aromaa, A., and Reunanen, A., Quercetin intake and the incidence of cerebrovascular disease, *Eur. J. Clin. Nutr.*, 54, 415, 2000.

128. Arts, I.C.W., Catechin contents of foods commonly consumed in the Netherlands. 2. Tea, wine, fruit juices, and chocolate milk, *J. Agric. Food Chem.*, 48, 1752, 2000.

129. Arts, I.C.W., van de Putte, B., and Hollman, P.C.H., Catechin contents of foods commonly consumed in the Netherlands. 1. Fruits, vegetables, staple foods, and processed foods, *J. Agric. Food Chem.*, 48, 1746, 2000.

130. Pascual-Teresa, S., Santos-Buelga, C., and Rivas-Gonzalo, J.C., Quantitative analysis of flavan-3-ols in Spanish foodstuffs and beverages, *J. Agric. Food Chem.*, 48, 5331, 2000.

131. Mennen, L.I., Walker, R., Bennetau-Pelissero, C., and Scalbert, A., Risks and safety of polyphenol consumption, *Am. J. Clin. Nutr.*, 81, 326S, 2005.

132. USDA, Flavonoid Content of Selected Foods, 2004, http://www.nal.usda.gov/fnic/foodcomp/Data/Flav/flav.html.

133. USDA–Iowa State University, USDA–Iowa State University Database on the Isoflavone Content of Foods, Release 1.3, 2002, http://www.nal.usda.gov/fnic/foodcomp/Data/isoflav/isoflav.html.

134. USDA, Proanthocyanidin Content of Selected Foods, 2004, http://www.nal.usda.gov/fnic/foodcomp/Data/PA/PA.html.

135. Kiely, M., Faughnan, M., Wahala, K., Brants, H., and Mulligan, A., Phyto-oestrogen levels in foods: the design and construction of the VENUS database, *Br. J. Nutr.*, 89 (Suppl. 1), S19, 2003.

136. BBSRC Institute of Food Research, NOTISPLUS: A Database of Bioactive Compounds Found in Food Plants, 2005, http://www.ifr.bbsrc.ac.uk/NOTIS/.

137. Porteous, L., Nutritional and Dietary Risk Factors for Osteoporosis: An Investigation of Association Between Diet and Indices of Bone Health in Young British Women Aged 25–30 Years, BSc Thesis, University of Surrey, 2001.

138. Bennett, G., Food Choices, Dietary Patterns and Circulation Adaptation in Shift Workers: An Investigation into the Possible Impact on Coronary Heart Disease Risk, BSc Thesis, University of Surrey, 2001.

139. Paul, N., Do Eating Patterns and Post-Prandial Hormonal and Metabolic Markers Vary Significantly According to Shift Type and Circadian Status in Swing Shift Workers in the Offshore Oil Industry, BSc Thesis, University of Surrey, 2001.

140. Balentine, D.A., Wiseman, S.A., and Bouwens, C.M., The chemistry of tea flavonoids, *CRC Crit. Rev. Food Sci. Nutr.*, 37, 693, 1997.

141. Clifford, M.N., Tea, in *Food Industries Manual*, Ranken, M.D., Kill, R.C., and Baker, C.J.G., Eds., Blackie, Glasgow, 1997, p. 10, Hot beverages.

142. Harbowy, M.E. and Balentine, D.A., Tea chemistry, *CRC Crit. Rev. Plant Sci.*, 16, 415, 1997.

143. Manach, C., Williamson, G., Morand, C., Scalbert, A., and Remesy, C., Bioavailability and bioefficacy of polyphenols in humans. I. Review of 97 bioavailability studies, *Am. J. Clin. Nutr.*, 81, 230S, 2005.

144. Visioli, F., Galli, C., Grande, S., Colonnelli, K., Patelli, C., Galli, G., and Caruso, D., Hydroxytyrosol excretion differs between rats and humans and depends on the vehicle of administration, *J. Nutr.*, 133, 2612, 2003.

145. Azuma, K., Ippoushi, K., Ito, H., Higashio, H., and Terao, J., Combination of lipids and emulsifiers enhances the absorption of orally administered quercetin in rats, *J. Agric. Food Chem.*, 50, 1706, 2002.

146. Bugianesi, R., Catasta, G., Spigno, P., D'Uva, A., and Maiani, G., Naringenin from cooked tomato paste is bioavailable in men, *J. Nutr.*, 132, 3349, 2002.

147. Baba, S., Osakabe, N., Yasuda, A., Natsume, M., Takizawa, T., Nakamura, T., and Terao, J., Bioavailability of (−)-epicatechin upon intake of chocolate and cocoa in human volunteers, *Free Radical Res.*, 33, 635, 2000.

148. Piskula, M.K. and Terao, J., Quercetin's solubility affects its accumulation in rat plasma after oral administration, *J. Agric. Food Chem.*, 46, 4313, 1998.

149. van het Hof, K., Wiseman, S.A., Yang, C.S., and Tijburg, L.B., Plasma and lipoprotein levels of tea catechins following repeated tea consumption, *Proc. Soc. Exp. Biol. Med.*, 220, 203, 1999.

150. van het Hof, K.H., Kivits, G.A., Weststrate, J.A., and Tijburg, L.B., Bioavailability of catechins from tea: the effect of milk, *Eur. J. Clin. Nutr.*, 52, 356, 1998.

151. Bell, J.R.C., Donovan, J.L., Wong, R., Waterhouse, A.L., German, J.B., Walzem, R.L., and Kasim-Karakas, S.E., (+)-Catechin in human plasma after ingestion of a single serving of reconstituted red wine, *Am. J. Clin. Nutr.*, 71, 103, 2000.

152. Bub, A., Watzl, B., Heeb, D., Rechkemmer, G., and Briviba, K., Malvidin-3-glucoside bioavailability in humans after ingestion of red wine, dealcoholized red wine and red grape juice, *Eur. J. Nutr.*, 40, 113, 2001.

153. Goldberg, D.M., Yan, J., and Soleas, G.J., Absorption of three wine-related polyphenols in three different matrices by healthy subjects, *Clin. Biochem.*, 36, 79, 2003.

154. Olthof, M.R., Hollman, P.C., Buijsman, M.N., van Amelsvoort, J.M., and Katan, M.B., Chlorogenic acid, quercetin-3-rutinoside and black tea phenols are extensively metabolized in humans, *J. Nutr.*, 133, 1806, 2003.

155. Nardini, M., Cirillo, E., Natella, F., and Scaccini, C., Absorption of phenolic acids in humans after coffee consumption, *J. Agric. Food Chem.*, 50, 5735, 2002.

156. Lee, M.-J., Wang, Z.-Y., Li, H., Chen, L., Sun, Y., Gobbo, S., Balentine, D.A., and Yang, C.S., Analysis of plasma and urinary tea polyphenols in human subjects, *Cancer Epidemiol. Biomarkers Prev.*, 4, 393, 1995.

157. Hollman, P.C. and Katan, M.B., Health effects and bioavailability of dietary flavonols, *Free Radical Res.*, 31 (Suppl.), S75, 1999.

158. Olthof, M.R., Hollman, P.C., Vree, T.B., and Katan, M.B., Bioavailabilities of quercetin-3-glucoside and quercetin-4'-glucoside do not differ in humans, *J. Nutr.*, 130, 1200, 2000.

159. Cermak, R., Landgraf, S., and Wolffram, S., Quercetin glucosides inhibit glucose uptake into brush-border-membrane vesicles of porcine jejunum, *Br. J. Nutr.*, 91, 849, 2004.

160. Hollman, P.C.H., van Trijp, J.M., Buysman, M.N., van der Gaag, M.S., Mengelers, M.J., de Vries, J.H.M., and Katan, M.B., Relative bioavailability of the antioxidant quercetin from various foods in man, *FEBS Lett.*, 418, 152, 1997.

161. Walle, T., Walle, U.K., and Halushka, P.V., Carbon dioxide is the major metabolite of quercetin in humans, *J. Nutr.*, 131, 2648, 2001.

162. Aura, A.M., Martin-Lopez, P., O'Leary, K.A., Williamson, G., Oksman-Caldentey, K.M., Poutanen, K., and Santos-Buelga, C., *In vitro* metabolism of anthocyanins by human gut microflora, *Eur. J. Nutr.*, 44, 133–142, 2005.

163. Aura, A.M., O'Leary, K.A., Williamson, G., Ojala, M., Bailey, M., Puupponen-Pimia, R., Nuutila, A.M., Oksman-Caldentey, K.M., and Poutanen, K., Quercetin derivatives are deconjugated and converted to hydroxyphenylacetic acids but not methylated by human fecal flora *in vitro*, *J. Agric. Food Chem.*, 50, 1725, 2002.

164. Blum, A.L., Haemerli, U.P., and Lorenz-Meyer, H., Is phloridzin or its aglycon the inhibitor of intestinal glucose transport? A study in normal and lactase-deficient men, *Eur. J. Clin. Invest.*, 5, 285, 1975.

165. Bokkenheuser, V.D., Shackleton, C.H., and Winter, J., Hydrolysis of dietary flavonoid glycosides by strains of intestinal Bacteroides from humans, *Biochem. J.*, 248, 953, 1987.

166. Bokkenheuser, V.D. and Winter, J., Hydrolysis of flavonoids by human intestinal bacteria, *Prog. Clin. Biol. Res.*, 280, 143, 1988.

167. Clifford, M.N., Copeland, E.L., Bloxsidge, J.P., and Mitchell, L.A., Hippuric acid is a major excretion product associated with black tea consumption, *Xenobiotica*, 30, 317, 2000.

168. Coldham, N.G., King, L.J., Macpherson, D.D., and Sauer, M.J., Biotransformation of genistein in the rat: elucidation of metabolite structure by product ion mass fragmentology, *J. Steroid Biochem. Mol. Biol.*, 70, 169, 1999.

169. Couteau, D., McCartney, A.L., Gibson, G.R., Williamson, G., and Faulds, C.B., Isolation and characterization of human colonic bacteria able to hydrolyse chlorogenic acid, *J. Appl. Microbiol.*, 90, 873, 2001.

170. Deprez, S., Brezillon, C., Rabot, S., Philippe, C., Mila, I., Lapierre, C., and Scalbert, A., Polymeric proanthocyanidins are catabolized by human colonic microflora into low-molecular-weight phenolic acids, *J. Nutr.*, 130, 2733, 2000.

171. Griffiths, L.A., Mammalian metabolism of flavonoids, in *The Flavonoids: Advances in Research*, Harborne, J.B. and Mabry, T.J., Eds., Chapman & Hall, London, 1982.

172. Groenewoud, G. and Hundt, H.K.L., The microbial metabolism of (+)-catechins to two novel diarylpropan-2-ol metabolites *in vitro*, *Xenobiotica*, 14, 711, 1984.

173. Jenkins, D.J., Wolever, T.M., Taylor, R.H., Barker, H., Fielden, H., Baldwin, J.M., Bowling, A.C., and Newman, H.C., Glycaemic index of foods: a physiological basis for carbohydrate exchange, *Am. J. Clin. Nutr.*, 34, 362, 1981.

174. Justesen, U., Arrigoni, E., Larsen, B.R., and Amado, R., Degradation of flavonoid glycosides and aglycones during *in vitro* fermentation with human faecal flora, *Lebensmitt. Wissensch. Technol.*, 33, 424, 2000.

175. Knight, S., Metabolism of Dietary Polyphenols by Gut Flora, PhD Thesis, University of Surrey, 2004.

176. Krishnamurty, H.G., Cheng, K.J., Jones, G.A., Simpson, F.J., and Watkin, J.E., Identification of products produced by the anaerobic degradation of rutin and related flavonoids by *Butyrivibrio* sp. C3, *Can. J. Microbiol.*, 16, 759, 1970.

177. Li, C., Lee, M.-J., Sheng, S., Meng, X., Prabhu, S., Winnik, B., Huang, B., Chung, J.Y., Yan, S., Ho, C.-T., and Yang, C.S., Structural identification of two metabolites of catechins and their kinetics in human urine and blood after tea ingestion, *Chem. Res. Toxicol.*, 13, 177, 2000.

178. Olthof, M.R., Bioavailability of Flavonoids and Cinnamic Acids and their Effect on Plasma Homocysteine in Humans, PhD Thesis, Wageningen University, 2001.

179. Olthof, M.R., Hollman, P.C.H., and Katan, M.B., Chlorogenic acid and caffeic acid are absorbed in humans, *J. Nutr.*, 131, 66, 2001.

180. Sawai, Y., Kohsaka, K., Nishiyama, Y., and Ando, K., Serum concentrations of rutoside metabolites after oral administration of a rutoside formulation to humans, *Arzneimittel-Forsch.*, 37, 729, 1987.

181. Scheline, R.R., *Mammalian Metabolism of Plant Xenobiotics*, Academic Press, London, 1978.

182. Scheline, R.R., The metabolism of (+)-catechin to hydroxyphenylvaleric acids by the intestinal microflora, *Biochim. Biophys. Acta*, 222, 228, 1970.

183. Schneider, H. and Blaut, M., Anaerobic degradation of flavonoids by *Eubacterium ramulus*, *Arch. Microbiol.*, 173, 71, 2000.

184. Schneider, H., Schwiertz, A., Collins, M.D., and Blaut, M., Anaerobic transformation of quercetin-3-glucoside by bacteria from the human intestinal tract, *Arch. Microbiol.*, 171, 81, 1999.

185. Schneider, H., Simmering, R., Hartmann, L., and Blaut, M., Degradation of quercetin-3-glucoside in gnotobiotic rats associated with human intestinal bacteria, *J. Appl. Microbiol.*, 89, 1027, 2000.

186. Schoefer, L., Mohan, R., Schwiertz, A., Braune, A., and Blaut, M., Anaerobic degradation of flavonoids by *Clostridium orbiscindens*, *Appl. Environ. Microbiol.*, 69, 5849, 2003.

187. Simmering, R., Kleessen, B., and Blaut, M., Quantification of the flavonoid-degrading bacterium *Eubacterium ramulus* in human fecal samples with a species-specific oligonucleotide hybridization probe, *Appl. Environ. Microbiol.*, 65, 3705, 1999.

188. Ward, N.C., Croft, K.D., Puddey, I.B., and Hodgson, J.M., Supplementation with grape seed polyphenols results in increased urinary excretion of 3-hydroxyphenylpropionic acid, an important metabolite of proanthocyanidins in humans, *J. Agric. Food Chem.,* 52, 5545, 2004.

189. Winter, J., Moore, L.H., Dowell, V.R., and Bokkenheuser, V.D., C-ring cleavage of flavonoids by human intestinal bacteria, *Appl. Environ. Microbiol.,* 55, 1203, 1989.

190. Wang, L.Q., Meselhy, M.R., Li, Y., Qin, G.W., and Hattori, M., Human intestinal bacteria capable of transforming secoisolariciresinol diglucoside to mammalian lignans, enterodiol and enterolactone, *Chem. Pharm. Bull.,* 48, 1606, 2000.

191. Schoefer, L., Mohan, R., Braune, A., Birringer, M., and Blaut, M., Anaerobic *C*-ring cleavage of genistein and daidzein by *Eubacterium ramulus, FEMS Microbiol. Lett.,* 208, 197, 2002.

192. Braune, A., Gutschow, M., Engst, W., and Blaut, M., Degradation of quercetin and luteolin by *Eubacterium ramulus, Appl. Environ. Microbiol.,* 67, 5558, 2001.

193. Wang, L.Q., Meselhy, M.R., Li, Y., Nakamura, N., Min, B.S., Qin, G.W., and Hattori, M., The heterocyclic ring fission and dehydroxylation of catechins and related compounds by *Eubacterium* sp. strain SDG-2, a human intestinal bacterium, *Chem. Pharm. Bull.,* 49, 1640, 2001.

194. Hur, H. and Rafii, F., Biotransformation of the isoflavonoids biochanin A, formononetin, and glycitein by *Eubacterium limosum, FEMS Microbiol. Lett.,* 192, 21, 2000.

195. Barcenilla, A., Pryde, S.E., Martin, J.C., Duncan, S.H., Stewart, C.S., Henderson, C., and Flint, H.J., Phylogenetic relationships of butyrate-producing bacteria from the human gut, *Appl. Environ. Microbiol.,* 66, 1654, 2000.

196. Kinoshita, M., Suzuki, Y., and Saito, Y., Butyrate reduces colonic paracellular permeability by enhancing PPARgamma activation, *Biochem. Biophys. Res. Commun.,* 293, 827, 2002.

197. Winter, J., Popoff, M.R., Grimont, P., and Bokkenheuser, V.D., *Clostridium orbiscindens* sp. nov., a human intestinal bacterium capable of cleaving the flavonoid C-ring, *Int. J. Syst. Bacteriol.,* 41, 355, 1991.

198. Mulder, T.P., Rietveld, A.G., and van Amelsvoort, J.M., Consumption of both black tea and green tea results in an increase in the excretion of hippuric acid into urine, *Am. J. Clin. Nutr.,* 81, 256S, 2005.

199. Hoey, L., Rowland, I.R., Lloyd, A.S., Clarke, D.B., and Wiseman, H., Influence of soya-based infant formula consumption on isoflavone and gut microflora metabolite concentrations in urine and on faecal microflora composition and metabolic activity in infants and children, *Br. J. Nutr,* 91, 607, 2004.

200. Gonthier, M.P., Cheynier, V., Donovan, J.L., Manach, C., Morand, C., Mila, I., Lapierre, C., Remesy, C., and Scalbert, A., Microbial aromatic acid metabolites formed in the gut account for a major fraction of the polyphenols excreted in urine of rats fed red wine polyphenols, *J. Nutr.,* 133, 461, 2003.

201. Takanaga, H., Tamai, I., and Tsuji, A., pH-dependent and carrier-mediated transport of salicylic acid across Caco-2 cells, *J. Pharm. Pharmacol.,* 46, 567, 1994.

202. Tsuji, A., Takanaga, H., Tamai, I., and Terasaki, T., Transcellular transport of benzoic acid across Caco-2 cells by a pH-dependent and carrier-mediated transport mechanism, *Pharm. Res.,* 11, 30, 1994.

203. Tamai, I., Takanaga, H., Maeda, H., Yabuuchi, H., Sai, Y., Suzuki, Y., and Tsuji, A., Intestinal brush-border membrane transport of monocarboxylic acids mediated by proton-coupled transport and anion antiport mechanisms, *J. Pharm. Pharmacol.,* 49, 108, 1997.

204. Konishi, Y. and Shimizu, M., Transepithelial transport of ferulic acid by monocarboxylic acid transporter in Caco-2 cell monolayers, *Biosci. Biotechnol. Biochem.,* 67, 856, 2003.

205. Konishi, Y., Kobayashi, S., and Shimizu, M., Transepithelial transport of *p*-coumaric acid and gallic acid in Caco-2 cell monolayers, *Biosci. Biotechnol. Biochem.,* 67, 2317, 2003.

206. Konishi, Y. and Kobayashi, S., Microbial metabolites of ingested caffeic acid are absorbed by the monocarboxylic acid transporter (MCT) in intestinal Caco-2 cell monolayers, *J. Agric. Food Chem.,* 52, 6418, 2004.

207. Konishi, Y., Hitomi, Y., and Yoshioka, E., Intestinal absorption of *p*-coumaric and gallic acids in rats after oral administration, *J. Agric. Food Chem.,* 52, 2527, 2004.

208. Konishi, Y. and Kobayashi, S., Transepithelial transport of chlorogenic acid, caffeic acid, and their colonic metabolites in intestinal caco-2 cell monolayers, *J. Agric. Food Chem.*, 52, 2518, 2004.

209. Konishi, Y., Kobayashi, S., and Shimizu, M., Tea polyphenols inhibit the transport of dietary phenolic acids mediated by the monocarboxylic acid transporter (MCT) in intestinal Caco-2 cell monolayers, *J. Agric. Food Chem.*, 51, 7296, 2003.

210. Ghiselli, A., Natella, F., Guidi, A., Montanari, L., Fantozzi, P., and Scaccini, C., Beer increases plasma antioxidant capacity in humans, *J. Nutr. Biochem.*, 11, 76, 2000.

211. Natsume, M., Osakabe, N., Oyama, M., Sasaki, M., Baba, S., Nakamura, Y., Osawa, T., and Terao, J., Structures of (−)-epicatechin glucuronide identified from plasma and urine after oral ingestion of (−)-epicatechin: differences between human and rat, *Free Radical Biol. Med.*, 34, 840, 2003.

212. Walle, T., Otake, Y., Walle, U.K., and Wilson, F.A., Quercetin glucosides are completely hydrolyzed in ileostomy patients before absorption, *J. Nutr.*, 130, 2658, 2000.

213. Yang, B., Arai, K., and Kusu, F., Determination of catechins in human urine subsequent to tea ingestion by high-performance liquid chromatography with electrochemical detection, *Anal. Biochem.*, 283, 77, 2000.

214. Ikeda, I., Kobayashi, M., Hamada, T., Tsuda, K., Goto, H., Imaizumi, K., Nozawa, A., Sugimoto, A., and Kakuda, T., Heat-epimerized tea catechins rich in gallocatechin gallate and catechin gallate are more effective to inhibit cholesterol absorption than tea catechins rich in epigallocatechin gallate and epicatechin gallate, *J. Agric. Food Chem.*, 51, 7303, 2003.

215. Nakagawa, K., Okuda, S., and Miyazawa, T., Dose-dependent incorporation of tea catechins, (−)-epigallocatechin-3-gallate and (−)-epigallocatechin, into human plasma, *Biosci. Biotechnol. Biochem.*, 61, 1981, 1997.

216. Kimura, M., Umegaki, K., Kasuya, Y., Sugisawa, A., and Higuchi, M., The relation between single/double or repeated tea catechin ingestions and plasma antioxidant activity in humans, *Eur. J. Clin. Nutr.*, 56, 1186, 2002.

217. Donovan, J.L., Bell, J.R., Kasim-Karakas, S., German, J.B., Walzem, R.L., Hansen, R.J., and Waterhouse, A.L., Catechin is present as metabolites in human plasma after consumption of red wine, *J. Nutr.*, 129, 1662, 1999.

218. Donovan, J.L., Kasim-Karakas, S., German, J.B., and Waterhouse, A.L., Urinary excretion of catechin metabolites by human subjects after red wine consumption, *Br. J. Nutr.*, 87, 31, 2002.

219. Yang, C.S., Chen, L., Lee, M.J., Balentine, D.A., Kuo, M.C., and Schantz, S.P., Blood and urine levels of tea catechins after ingestion of different amounts of green tea by human volunteers, *Cancer Epidemiol. Biomarkers Prev.*, 7, 351, 1998.

220. van Amelsvoort, J.M., Van Hof, K.H., Mathot, J.N., Mulder, T.P., Wiersma, A., and Tijburg, L.B., Plasma concentrations of individual tea catechins after a single oral dose in humans, *Xenobiotica*, 31, 891, 2001.

221. Meng, X., Lee, M.J., Li, C., Sheng, S., Zhu, N., Sang, S., Ho, C.T., and Yang, C.S., Formation and identification of 4′-O-methyl-(−)-epigallocatechin in humans, *Drug Metab. Dispos.*, 29, 789, 2001.

222. Lu, H., Meng, X., and Yang, C.S., Enzymology of methylation of tea catechins and inhibition of catechol-*O*-methyltransferase by (−)-epigallocatechin gallate, *Drug Metab. Dispos.*, 31, 572, 2003.

223. Holt, R.R., Lazarus, S.A., Sullards, M.C., Zhu, Q.Y., Schramm, D.D., Hammerstone, J.F., Fraga, C.G., Schmitz, H.H., and Keen, C.L., Procyanidin dimer B2 [epicatechin-(4beta-8)-epicatechin] in human plasma after the consumption of a flavanol-rich cocoa, *Am. J. Clin. Nutr.*, 76, 798, 2002.

224. Sano, A., Yamakoshi, J., Tokutake, S., Tobe, K., Kubota, Y., and Kikuchi, M., Procyanidin B1 is detected in human serum after intake of proanthocyanidin-rich grape seed extract, *Biosci. Biotechnol. Biochem.*, 67, 1140, 2003.

225. Mcdonald, M.S., Hughes, M., Burns, J., Lean, M.E.J., Matthews, D., and Crozier, A., Survey of the free and conjugated myricetin and quercetin content of red wines of different geographical origins, *J. Agric. Fdoo Chem.*, 46, 368, 1998.

226. Day, A.J., Mellon, F., Barron, D., Sarrazin, G., Morgan, M.R.A., and Williamson, G., Human metabolism of flavonoids: identification of plasma metabolites of quercetin, *Free Radical. Res.*, 35, 941, 2001.

227. Graefe, E.U., Wittig, J., Mueller, S., Riethling, A.K., Uehleke, B., Drewelow, B., Pforte, H., Jacobasch, G., Derendorf, H., and Veit, M., Pharmacokinetics and bioavailability of quercetin glycosides in humans, *J. Clin. Pharmacol.,* 41, 492, 2001.

228. DuPont, M.S., Day, A.J., Bennett, R.N., Mellon, F.A., and Kroon, P.A., Absorption of kaempferol from endive, a source of kaempferol-3-glucuronide, in humans, *Eur. J. Clin. Nutr.,* 58, 947, 2004.

229. Wittig, J., Herderich, M., Graefe, E.U., and Veit, M., Identification of quercetin glucuronides in human plasma by high-performance liquid chromatography–tandem mass spectrometry, *J. Chromatogr. B: Biomed. Sci. Appl.,* 753, 237, 2001.

230. DuPont, M.S., Bennett, R.N., Mellon, F.A., and Williamson, G., Polyphenols from alcoholic apple cider are absorbed, metabolized and excreted by humans, *J. Nutr.,* 132, 172, 2002.

231. O'Leary, K.A., Day, A.J., Needs, P.W., Mellon, F.A., O'Brien, N.M., and Williamson, G., Metabolism of quercetin-7- and quercetin-3-glucuronides by an *in vitro* hepatic model: the role of human β-glucuronidase, sulfotransferase, catechol-*O*-methyltransferase and multi-resistant protein 2 (MRP2) in flavonoid metabolism, *Biochem. Pharmacol.,* 65, 479, 2003.

232. Mullen, W., Graf, B.A., Caldwell, S.T., Hartley, R.C., Duthie, G.G., Edwards, C.A., Lean, M.E., and Crozier, A., Determination of flavonol metabolites in plasma and tissues of rats by HPLC-radiocounting and tandem mass spectrometry following oral ingestion of [2-^{14}C]quercetin-4'-glucoside, *J. Agric. Food Chem.,* 50, 6902, 2002.

233. Nielsen, S.E., Young, J.F., Daneshvar, B., Lauridsen, S.T., Knuthsen, P., Sandström, B., and Dragsted, L.O., Effect of parsley intake on urinary apigenin excretion, blood antioxidant enzymes and biomarkers for oxidative stress in humans, in *Natural Antioxidants and Anticarcinogens in Nutrition, Health and Disease,* Kumpulainen, J.T., and Salonen, J.T., Eds., Royal Society of Chemistry, Cambridge, 1999.

234. Shimoi, K., Okada, H., Furugori, M., Goda, T., Takase, S., Suzuki, M., Hara, Y., Yamamoto, H., and Kinae, N., Intestinal absorption of luteolin and luteolin 7-*O*-β-glucoside in rats and humans, *FEBS Lett.,* 438, 220, 1998.

235. Shimoi, K., Saka, N., Kaji, K., Nozawa, R., and Kinae, N., Metabolic fate of luteolin and its functional activity at focal site, *Biofactors,* 12, 181, 2000.

236. Walle, T., Otake, Y., Brubaker, J.A., Walle, U.K., and Halushka, P.V., Disposition and metabolism of the flavonoid chrysin in normal volunteers, *Br. J. Clin. Pharmacol.,* 51, 143, 2001.

237. Coward, L., Barnes, N.C., Setchell, K.D.R., and Barnes, S., Genistein, daidzein, and their β-glycoside conjugates: antitumor isoflavones in soybean foods from American and Asian diets, *J. Agric. Food Chem.,* 41, 1961, 1993.

238. Clarke, D.B. and Lloyd, A.S., Dietary exposure estimates of isoflavones from the 1998 UK Total Diet Study, *Food Addit. Contam.,* 21, 305, 2004.

239. Setchell, K.D., Brown, N.M., Desai, P., Zimmer-Nechemias, L., Wolfe, B.E., Brashear, W.T., Kirschner, A.S., Cassidy, A., and Heubi, J.E., Bioavailability of pure isoflavones in healthy humans and analysis of commercial soy isoflavone supplements, *J. Nutr.,* 131, 1362S, 2001.

240. Izumi, T., Piskula, M.K., Osawa, S., Obata, A., Tobe, K., Saito, M., Kataoka, S., Kubota, Y., and Kikuchi, M., Soy isoflavone aglycones are absorbed faster and in higher amounts than their glucosides in humans, *J. Nutr.,* 130, 1695, 2000.

241. Setchell, K.D., Brown, N.M., Zimmer-Nechemias, L., Brashear, W.T., Wolfe, B.E., Kirschner, A.S., and Heubi, J.E., Evidence for lack of absorption of soy isoflavone glycosides in humans, supporting the crucial role of intestinal metabolism for bioavailability, *Am. J. Clin. Nutr.,* 76, 447, 2002.

242. Busby, M.G., Jeffcoat, A.R., Bloedon, L.T., Koch, M.A., Black, T., Dix, K.J., Heizer, W.D., Thomas, B.F., Hill, J.M., Crowell, J.A., and Zeisel, S.H., Clinical characteristics and pharmacokinetics of purified soy isoflavones: single-dose administration to healthy men, *Am. J. Clin. Nutr.,* 75, 126, 2002.

243. Barnes, S., Phyto-oestrogens and osteoporosis: what is a safe dose? *Br. J. Nutr.,* 89 (Suppl. 1), S101, 2003.

244. Adlercreutz, H., van der, W.J., Kinzel, J., Attalla, H., Wahala, K., Makela, T., Hase, T., and Fotsis, T., Lignan and isoflavonoid conjugates in human urine, *J. Steroid Biochem. Mol. Biol.,* 52, 97, 1995.

245. Doerge, D.R., Chang, H.C., Churchwell, M.I., and Holder, C.L., Analysis of soy isoflavone conjugation *in vitro* and in human blood using liquid chromatography–mass spectrometry, *Drug Metab. Dispos.*, 28, 298, 2000.

246. Kaamanen, M., Adlercreutz, H., Jauhiainen, M., and Tikkanen, M.J., Accumulation of genistein and lipophilic genistein derivatives in lipoproteins during incubation with human plasma *in vitro*, *Biochim. Biophys. Acta*, 1631, 147, 2003.

247. King, R.A., Daidzein conjugates are more bioavailable than genistein conjugates in rats, *Am. J. Clin. Nutr.*, 68, 1496S, 1998.

248. Watanabe, S., Yamaguchi, M., Sobue, T., Takahashi, T., Miura, T., Arai, Y., Mazur, W., Wahala, K., and Adlercreutz, H., Pharmacokinetics of soybean isoflavones in plasma, urine and feces of men after ingestion of 60 g baked soybean powder (kinako), *J. Nutr.*, 128, 1710, 1998.

249. Zhang, Y., Wang, G.J., Song, T.T., Murphy, P.A., and Hendrich, S., Urinary disposition of the soybean isoflavones daidzein, genistein and glycitein differs among humans with moderate fecal isoflavone degradation activity, *J. Nutr.*, 129, 957, 1999.

250. Coward, L., Kirk, M., Albin, N., and Barnes, S., Analysis of plasma isoflavones by reversed-phase HPLC-multiple reaction ion monitoring-mass spectrometry, *Clin. Chim. Acta*, 247, 121, 1996.

251. Clarke, D.B., Lloyd, A.S., Botting, N.P., Oldfield, M.F., Needs, P.W., and Wiseman, H., Measurement of intact sulfate and glucuronide phytoestrogen conjugates in human urine using isotope dilution liquid chromatography–tandem mass spectrometry with [13C(3)]isoflavone internal standards, *Anal. Biochem.*, 309, 158, 2002.

252. Shelnutt, S.R., Cimino, C.O., Wiggins, P.A., Ronis, M.J., and Badger, T.M., Pharmacokinetics of the glucuronide and sulfate conjugates of genistein and daidzein in men and women after consumption of a soy beverage, *Am. J. Clin. Nutr.*, 76, 588, 2002.

253. Erlund, I., Silaste, M.L., Alfthan, G., Rantala, M., Kesaniemi, Y.A., and Aro, A., Plasma concentrations of the flavonoids hesperetin, naringenin and quercetin in human subjects following their habitual diets, and diets high or low in fruit and vegetables, *Eur. J. Clin. Nutr.*, 56, 891, 2002.

254. Erlund, I., Meririnne, E., Alfthan, G., and Aro, A., Plasma kinetics and urinary excretion of the flavanones naringenin and hesperetin in humans after ingestion of orange juice and grapefruit juice, *J. Nutr.*, 131, 235, 2001.

255. Manach, C., Morand, C., Gil-Izquierdo, A., Bouteloup-Demange, C., and Remesy, C., Bioavailability in humans of the flavanones hesperidin and narirutin after the ingestion of two doses of orange juice, *Eur. J. Clin. Nutr.*, 57, 235, 2003.

256. Lee, Y.S. and Reidenberg, M.M., A method for measuring naringenin in biological fluids and its disposition from grapefruit juice by man, *Pharmacology*, 56, 314, 1998.

257. Fuhr, U. and Kummert, A.L., The fate of naringin in humans: a key to grapefruit juice–drug interactions, *Clin. Pharm. Ther.*, 58, 365, 1995.

258. Bailey, D.G., Malcolm, J., Arnold, O., and Spence, J.D., Grapefruit juice–drug interactions, *Br. J. Clin. Pharmacol.*, 46, 101, 1998.

259. Yilmazer, M., Stevens, J.F., and Buhler, D.R., *In vitro* glucuronidation of xanthohumol, a flavonoid in hop and beer, by rat and human liver microsomes, *FEBS Lett.*, 491, 252, 2001.

260. Yilmazer, M., Stevens, J.F., Deinzer, M.L., and Buhler, D.R., *In vitro* biotransformation of xanthohumol, a flavonoid from hops (*Humulus lupulus*), by rat liver microsomes, *Drug Metab. Dispos.*, 29, 223, 2001.

261. Nadelmann, L., Tjornelund, J., Christensen, E., and Hansen, S.H., High-performance liquid chromatographic determination of licochalcone A and its metabolites in biological fluids, *J. Chromatogr., B: Biomed. Sci. Appl.*, 695, 389, 1997.

262. Nadelmann, L., Tjornelund, J., Hansen, S.H., Cornett, C., Sidelmann, U.G., Braumann, Christensen, E., and Christensen, S.B., Synthesis, isolation and identification of glucuronides and mercapturic acids of a novel antiparasitic agent, licochalcone A, *Xenobiotica*, 27, 667, 1997.

263. Clifford, M.N., Anthocyanins — nature, occurrence and dietary burden, *J. Sci. Food Agric.*, 80, 1063, 2000.

264. Felgines, C., Talavera, S., Gonthier, M.P., Texier, O., Scalbert, A., Lamaison, J.L., and Remesy, C., Strawberry anthocyanins are recovered in urine as glucuro- and sulfoconjugates in humans, *J. Nutr.*, 133, 1296, 2003.

265. Nielsen, I.L., Dragsted, L.O., Ravn-Haren, G., Freese, R., and Rasmussen, S.E., Absorption and excretion of blackcurrant anthocyanins in humans and watanabe heritable hyperlipidemic rabbits, *J. Agric. Food Chem.,* 51, 2813, 2003.

266. Mazza, G., Kay, C.D., Cottrell, T., and Holub, B.J., Absorption of anthocyanins from blueberries and serum antioxidant status in human subjects, *J. Agric. Food Chem.,* 50, 7731, 2002.

267. Rechner, A.R., Kuhnle, G., Hu, H., Roedig-Penman, A., van den Braak, M.H., and Rice-Evans, C.A., The metabolism of dietary polyphenols and the relevance to circulating levels of conjugated metabolites, *Free Radical Res.,* 36, 1229, 2002.

268. Wu, X., Cao, G., and Prior, R.L., Absorption and metabolism of anthocyanins in elderly women after consumption of elderberry or blueberry, *J. Nutr.,* 132, 1865, 2002.

269. Cao, G., Muccitelli, H.U., Sanchez-Moreno, C., and Prior, R.L., Anthocyanins are absorbed in glycated forms in elderly women: a pharmacokinetic study, *Am. J. Clin. Nutr.,* 73, 920, 2001.

270. Matsumoto, H., Inaba, H., Kishi, M., Tominaga, S., Hirayama, M., and Tsuda, T., Orally administered delphinidin 3-rutinoside and cyanidin 3-rutinoside are directly absorbed in rats and humans and appear in the blood as the intact forms, *J. Agric. Food Chem.,* 49, 1546, 2001.

271. Netzel, M., Strass, G., Janssen, M., Bitsch, I., and Bitsch, R., Bioactive anthocyanins detected in human urine after ingestion of blackcurrant juice, *J. Environ. Pathol. Toxicol. Oncol.,* 20, 89, 2001.

272. Miyazawa, T., Nakagawa, K., Kudo, M., Muraishi, K., and Someya, K., Direct intestinal absorption of red fruit anthocyanins, cyanidin-3-glucoside and cyanidin-3,5-diglucoside, into rats and humans, *J. Agric. Food Chem.,* 47, 1083, 1999.

273. Bitsch, I., Janssen, M., Netzel, M., Strass, G., and Frank, T., Bioavailability of anthocyanidin-3-glycosides following consumption of elderberry extract and blackcurrant juice, *Int J. Clin. Pharmacol. Ther.,* 42, 293, 2004.

274. Harada, K., Kano, M., Takayanagi, T., Yamakawa, O., and Ishikawa, F., Absorption of acylated anthocyanins in rats and humans after ingesting an extract of *Ipomoea batatas* purple sweet potato tuber, *Biosci. Biotechnol. Biochem.,* 68, 1500, 2004.

275. Walle, T., Hsieh, F., DeLegge, M.H., Oatis, J.E., and Walle, U.K., High absorption but very low bioavailability of oral resveratrol in humans, *Drug Metab. Dispos.,* 32, 1377–1382, 2004.

276. Yu, C., Shin, Y.G., Kosmeder, J.W., Pezzuto, J.M., and van Breemen, R.B., Liquid chromatography/tandem mass spectrometric determination of inhibition of human cytochrome P450 isozymes by resveratrol and resveratrol-3-sulfate, *Rapid Commun. Mass Spectrom.,* 17, 307, 2003.

277. Cremin, P., Kasim-Karakas, S., and Waterhouse, A.L., LC/ES–MS detection of hydroxycinnamates in human plasma and urine, *J. Agric. Food Chem.,* 49, 1747, 2001.

278. Kern, S.M., Bennett, R.N., Mellon, F.A., Kroon, P.A., and Garcia-Conesa, M.T., Absorption of hydroxycinnamates in humans after high-bran cereal consumption, *J. Agric. Food Chem.,* 51, 6050, 2003.

279. Gonthier, M.P., Verny, M.A., Besson, C., Remesy, C., and Scalbert, A., Chlorogenic acid bioavailability largely depends on its metabolism by the gut microflora in rats, *J. Nutr.,* 133, 1853, 2003.

280. Plumb, G.W., García-Conesa, M.T., Kroon, P.A., and Williamson, G., Metabolism of chlorogenic acid by human plasma, liver, intestine and gut microflora, *J. Sci. Food Agric.,* 79, 390, 1999.

281. Jacobson, E.A., Newmark, H., Baptista, J., and Bruce, W.R., A preliminary investigation of the metabolism of dietary phenolics in humans, *Nutr. Rep. Int.,* 28, 1409, 1983.

282. Rechner, A.R., Spencer, J.P., Kuhnle, G., Hahn, U., and Rice-Evans, C.A., Novel biomarkers of the metabolism of caffeic acid derivatives *in vivo, Free Radical Biol. Med.,* 30, 1213, 2001.

283. Rechner, A.R., Pannala, A.S., and Rice-Evans, C.A., Caffeic acid derivatives in artichoke extract are metabolised to phenolic acids *in vivo, Free Radical Res.,* 35, 195, 2001.

284. Caccetta, R.A., Croft, K.D., Beilin, L.J., and Puddey, I.B., Ingestion of red wine significantly increases plasma phenolic acid concentrations but does not acutely affect *ex vivo* lipoprotein oxidizability, *Am. J. Clin. Nutr.,* 71, 67, 2000.

285. Gupta, B. and Ghosh, B., *Curcuma longa* inhibits TNF-alpha induced expression of adhesion molecules on human umbilical vein endothelial cells, *Int. J. Immunopharmacol.,* 21, 745, 1999.

286. Garcea, G., Jones, D.J., Singh, R., Dennison, A.R., Farmer, P.B., Sharma, R.A., Steward, W.P., Gescher, A.J., and Berry, D.P., Detection of curcumin and its metabolites in hepatic tissue and portal blood of patients following oral administration, *Br. J. Cancer,* 90, 1011, 2004.

287. Wortelboer, H.M., Usta, M., van der Velde, A.E., Boersma, M.G., Spenkelink, B., van Zanden, J.J., Rietjens, I.M., van Bladeren, P.J., and Cnubben, N.H., Interplay between MRP inhibition and metabolism of MRP inhibitors: the case of curcumin, *Chem. Res. Toxicol.*, 16, 1642, 2003.

288. Shahrzad, S. and Bitsch, I., Determination of gallic acid and its metabolites in human plasma and urine by high-performance liquid chromatography, *J. Chromatogr. B: Biomed. Sci. Appl.*, 705, 87, 1998.

289. Hodgson, J.M., Morton, L.W., Puddey, I.B., Beilin, L.J., and Croft, K.D., Gallic acid metabolites are markers of black tea intake in humans, *J. Agric. Food Chem.*, 48, 2276, 2000.

290. Caruso, D., Visioli, F., Patelli, R., Galli, C., and Galli, G., Urinary excretion of olive oil phenols and their metabolites in humans, *Metabolism*, 50, 1426, 2001.

291. Miro-Casas, E., Covas, M.I., Farre, M., Fito, M., Ortuno, J., Weinbrenner, T., Roset, P., and de la Torre, T.R., Hydroxytyrosol disposition in humans, *Clin. Chem.*, 49, 945, 2003.

292. Nakamura, Y., Tsuji, S., and Tonogai, Y., Method for analysis of tannic acid and its metabolites in biological samples: application to tannic acid metabolism in the rat, *J. Agric. Food Chem.*, 51, 331, 2003.

293. Cerda, B., Llorach, R., Ceron, J.J., Espin, J.C., and Tomas-Barberan, F.A., Evaluation of the bioavailability and metabolism in the rat of punicalagin, an antioxidant polyphenol from pomegranate juice, *Eur. J. Nutr.*, 42, 18, 2003.

294. Axelson, M., Sjovall, J., Gustafsson, B.E., and Setchell, K.D., Origin of lignans in mammals and identification of a precursor from plants, *Nature*, 298, 659, 1982.

295. Borriello, S.P., Setchell, K.D., Axelson, M., and Lawson, A.M., Production and metabolism of lignans by the human faecal flora, *J. Appl Bacteriol.*, 58, 37, 1985.

296. Saarinen, N.M., Smeds, A., Makela, S.I., Ammala, J., Hakala, K., Pihlava, J.M., Ryhanen, E.L., Sjoholm, R., and Santti, R., Structural determinants of plant lignans for the formation of enterolactone *in vivo*, *J. Chromatogr. B: Anal. Technol. Biomed. Life Sci.*, 777, 311, 2002.

297. Jacobs, E., Kulling, S.E., and Metzler, M., Novel metabolites of the mammalian lignans enterolactone and enterodiol in human urine, *J. Steroid Biochem. Mol. Biol.*, 68, 211, 1999.

298. Dean, B., Chang, S., Doss, G.A., King, C., and Thomas, P.E., Glucuronidation, oxidative metabolism, and bioactivation of enterolactone in rhesus monkeys, *Arch. Biochem. Biophys.*, 429, 244, 2004.

299. Tou, J.C., Chen, J., and Thompson, L.U., Flaxseed and its lignan precursor, secoisolariciresinol diglycoside, affect pregnancy outcome and reproductive development in rats, *J. Nutr.*, 128, 1861, 1998.

300. Brooks, J.D., Ward, W.E., Lewis, J.E., Hilditch, J., Nickell, L., Wong, E., and Thompson, L.U., Supplementation with flaxseed alters estrogen metabolism in postmenopausal women to a greater extent than does supplementation with an equal amount of soy, *Am. J. Clin. Nutr.*, 79, 318, 2004.

301. Heinonen, S., Nurmi, T., Liukkonen, K., Poutanen, K., Wahala, K., Deyama, T., Nishibe, S., and Adlercreutz, H., *In vitro* metabolism of plant lignans: new precursors of mammalian lignans enterolactone and enterodiol, *J. Agric. Food Chem.*, 49, 3178, 2001.

302. Begum, A.N., Nicolle, C., Mila, I., Lapierre, C., Nagano, K., Fukushima, K., Heinonen, S.M., Adlercreutz, H., Remesy, C., and Scalbert, A., Dietary lignins are precursors of mammalian lignans in rats, *J. Nutr.*, 134, 120, 2004.

303. Nesbitt, P.D., Lam, Y., and Thompson, L.U., Human metabolism of mammalian lignan precursors in raw and processed flaxseed, *Am. J. Clin. Nutr.*, 69, 549, 1999.

304. Juntunen, K.S., Mazur, W.M., Liukkonen, K.H., Uehara, M., Poutanen, K.S., Adlercreutz, H.C., and Mykkanen, H.M., Consumption of wholemeal rye bread increases serum concentrations and urinary excretion of enterolactone compared with consumption of white wheat bread in healthy Finnish men and women, *Br. J. Nutr.*, 84, 839, 2000.

305. Glitso, L.V., Mazur, W.M., Adlercreutz, H., Wahala, K., Makela, T., Sandstrom, B., and Bach Knudsen, K.E., Intestinal metabolism of rye lignans in pigs, *Br. J. Nutr.*, 84, 429, 2000.

306. Mazur, W.M., Uehara, M., Wahala, K., and Adlercreutz, H., Phyto-oestrogen content of berries, and plasma concentrations and urinary excretion of enterolactone after a single strawberry-meal in human subjects, *Br. J. Nutr.*, 83, 381, 2000.

307. Nurmi, T., Heinonen, S., Mazur, W., Deyama, T., Nishibe, S., and Adlercreutz, H., Lignans in selected wines, *Food Chem.,* 83, 303, 2003.

308. Mazur, W.M., Wahala, K., Rasku, S., Salakka, A., Hase, T., and Adlercreutz, H., Lignan and isoflavonoid concentrations in tea and coffee, *Br. J. Nutr.,* 79, 37, 1998.

309. Mulder, T.P., van Platerink, C.J., Wijnand Schuyl, P.J., and van Amelsvoort, J.M., Analysis of theaflavins in biological fluids using liquid chromatography–electrospray mass spectrometry, *J. Chromatogr. B: Biomed. Sci. Appl.,* 760, 271, 2001.

310. Tomita, I., Sano, M., Sasaki, K., and Miyase, T., Tea catechin (EGCG) and its metabolites as antioxidants, in *Functional Foods for Disease Prevention,* 213th Meeting of the American Chemical Society, Shibamoto, T., Terao, T., and Osawa, T., Eds., American Chemical Society, Washington, DC, 1998, p. 21.

311. Yoshino, K., Suzuki, M., Sasaki, K., Miyase, T., and Sano, M., Formation of antioxidants from (−)-epigallocatechin gallate in mild alkaline fluids, such as authentic intestinal juice and mouse plasma, *J. Nutr. Biochem.,* 10, 223, 1999.

312. Sharma, R.A., McLelland, H.R., Hill, K.A., Ireson, C.R., Euden, S.A., Manson, M.M., Pirmohamed, M., Marnett, L.J., Gescher, A.J., and Steward, W.P., Pharmacodynamic and pharmacokinetic study of oral Curcuma extract in patients with colorectal cancer, *Clin Cancer Res.,* 7, 1894, 2001.

313. Clifford, M.N., Kellard, B., and Ah-Sing, E., Caffeoyl-tyrosine from green robusta coffee beans, *Phytochemistry,* 28, 1989, 1989.

314. Hollman, P.C.H., van der Gaag, M.S., Mengelers, M.J.B., van Trijp, J.M.P., de Vries, J.H.M., and Katan, M.B., Absorption and disposition kinetics of the dietary antioxidant quercetin in man, *Free Radical Biol. Med.,* 21, 703, 1996.

315. McAnlis, G.T., McEneny, J., Pearce, J., and Young, I.S., Absorption and antioxidant effects of quercetin from onions, in man, *Eur. J. Clin. Nutr.,* 53, 92, 1999.

316. van het Hof, K., Wiseman, S.A., de Boer, H., Weststrate, N.L.J., and Tijburg, L., Bioavailability and antioxidant activity of tea polyphenols in humans, in *COST 916: Bioactive Plant Cell Wall Components in Nutrition and Health. Polyphenols in Food,* Amadó, R., Andersson, H., Bardócz, S., and Serra, F., Eds., EU, Luxembourg, 1998, p. 17.

317. Wang, J.F., Schramm, D.D., Holt, R.R., Ensunsa, J.L., Fraga, C.G., Schmitz, H.H., and Keen, C.L., A dose–response effect from chocolate consumption on plasma epicatechin and oxidative damage, *J. Nutr.,* 130, 2115S, 2000.

318. Yang, C.S., Lee, M.J., and Chen, L., Human salivary tea catechin levels and catechin esterase activities: implication in human cancer prevention studies, *Cancer Epidemiol. Biomarkers Prev.,* 8, 83, 1999.

319. Bock, K.W., Eckle, T., Ouzzine, M., and Fournel Gigleux, S., Coordinate induction by antioxidants of UDP-glucuronosyltransferase UGT1A6 and the apical conjugate export pump MRP2 (multidrug resistance protein 2) in Caco-2 cells, *Biochem. Pharmacol.,* 59, 467, 2000.

320. Hong, J., Lu, H., Meng, X., Ryu, J.H., Hara, Y., and Yang, C.S., Stability, cellular uptake, biotransformation, and efflux of tea polyphenol (−)-epigallocatechin-3-gallate in HT-29 human colon adenocarcinoma cells, *Cancer Res.,* 62, 7241, 2002.

321. Widlansky, M.E., Duffy, S.J., Hamburg, N.M., Gokce, N., Warden, B.A., Wiseman, S., Keaney, J.F., Jr., Frei, B., and Vita, J.A., Effects of black tea consumption on plasma catechins and markers of oxidative stress and inflammation in patients with coronary artery disease, *Free Radical Biol. Med.,* 38, 499, 2005.

322. Ishikawa, T., Suzukawa, M., Ito, T., Yoshida, H., Ayaori, M., Nishiwaki, M., Yonemura, A., Hara, Y., and Nakamura, H., Effect of tea flavonoid supplementation on the susceptibility of low-density lipoprotein to oxidative modification, *Am. J. Clin. Nutr.,* 66, 261, 1997.

323. Carbonneau, M.A., Leger, C.L., Monnier, L., Bonnet, C., Michel, F., Fouret, G., Dedieu, F., and Descomps, B., Supplementation with wine phenolic compounds increases the antioxidant capacity of plasma and vitamin E of low-density lipoprotein without changing the lipoprotein Cu(2+)-oxidizability: possible explanation by phenolic location, *Eur. J. Clin. Nutr.,* 51, 682, 1997.

324. Conquer, J.A., Maiani, G., Azzini, E., Raguzzini, A., and Holub, B.J., Supplementation with quercetin markedly increases plasma quercetin concentration without effect on selected risk factors for heart disease in healthy subjects, *J. Nutr.,* 128, 593, 1998.

325. Eccleston, C., Baoru, Y., Tahvonen, R., Kallio, H., Rimbach, G.H., and Minihane, A.M., Effects of an antioxidant-rich juice (sea buckthorn) on risk factors for coronary heart disease in humans, *J. Nutr. Biochem.*, 13, 346, 2002.

326. Hodgson, J.M., Puddey, I.B., Croft, K.D., Burke, V., Mori, T.A., Caccetta, R.A., and Beilin, L.J., Acute effects of ingestion of black and green tea on lipoprotein oxidation, *Am. J. Clin. Nutr.*, 71, 1103, 2000.

327. Hodgson, J.M., Puddey, I.B., Croft, K.D., Mori, T.A., Rivera, J., and Beilin, L.J., Isoflavonoids do not inhibit *in vivo* lipid peroxidation in subjects with high-normal blood pressure, *Atherosclerosis*, 145, 167, 1999.

328. Leenen, R., Roodenburg, A.J., Tijburg, L.B., and Wiseman, S.A., A single dose of tea with or without milk increases plasma antioxidant activity in humans, *Eur. J. Clin. Nutr.*, 54, 87, 2000.

329. Lotito, S.B. and Frei, B., Relevance of apple polyphenols as antioxidants in human plasma: contrasting *in vitro* and *in vivo* effects, *Free Radical Biol. Med.*, 36, 201, 2004.

330. Mathur, S., Devaraj, S., Grundy, S.M., and Jialal, I., Cocoa products decrease low density lipoprotein oxidative susceptibility but do not affect biomarkers of inflammation in humans, *J. Nutr.*, 132, 3663, 2002.

331. Mitchell, J.H. and Collins, A.R., Effects of a soy milk supplement on plasma cholesterol levels and oxidative DNA damage in men — a pilot study, *Eur. J. Nutr.*, 38, 143, 1999.

332. Nakagawa, K., Ninomiya, M., Okubo, T., Aoi, N., Juneja, L.R., Kim, M., Yamanaka, K., and Miyazawa, T., Tea catechin supplementation increases antioxidant capacity and prevents phospholipid hydroperoxidation in plasma of humans, *J. Agric. Food Chem.*, 47, 3967, 1999.

333. Natella, F., Belelli, F., Gentili, V., Ursini, F., and Scaccini, C., Grape seed proanthocyanidins prevent plasma postprandial oxidative stress in humans, *J. Agric. Food Chem.*, 50, 7720, 2002.

334. Nigdikar, S.V., Williams, N.R., Griffin, B.A., and Howard, A.N., Consumption of red wine polyphenols reduces the susceptibility of low-density lipoproteins to oxidation *in vivo*, *Am. J. Clin. Nutr.*, 68, 258, 1998.

335. O'Byrne, D.J., Devaraj, S., Grundy, S.M., and Jialal, I., Comparison of the antioxidant effects of Concord grape juice flavonoids alpha-tocopherol on markers of oxidative stress in healthy adults, *Am. J. Clin. Nutr.*, 76, 1367, 2002.

336. Pedersen, C.B., Kyle, J., Jenkinson, A.M., Gardner, P.T., McPhail, D.B., and Duthie, G.G., Effects of blueberry and cranberry juice consumption on the plasma antioxidant capacity of healthy female volunteers, *Eur. J. Clin. Nutr.*, 54, 405, 2000.

337. Princen, H.M., van Duyvenvoorde, W., Buytenhek, R., Blonk, C., Tijburg, L.B., Langius, J.A., Meinders, A.E., and Pijl, H., No effect of consumption of green and black tea on plasma lipid and antioxidant levels and on LDL oxidation in smokers, *Arterioscler. Thromb. Vasc. Biol.*, 18, 833, 1998.

338. Rein, D., Lotito, S., Holt, R.R., Keen, C.L., Schmitz, H.H., and Fraga, C.G., Epicatechin in human plasma: *in vivo* determination and effect of chocolate consumption on plasma oxidation status, *J. Nutr.*, 130, 2109S, 2000.

339. Serafini, M., Maiani, G., and Ferro-Luzzi, A., Alcohol-free red wine enhances plasma antioxidant capacity in humans, *J. Nutr.*, 128, 1003, 1998.

340. Simonetti, P., Ciappellano, S., Gardana, C., Bramati, L., and Pietta, P., Procyanidins from *Vitis vinifera* seeds: *in vivo* effects on oxidative stress, *J. Agric. Food Chem.*, 50, 6217, 2002.

341. Stein, J.H., Keevil, J.G., Wiebe, D.A., Aeschlimann, S., and Folts, J.D., Purple grape juice improves endothelial function and reduces the susceptibility of LDL cholesterol to oxidation in patients with coronary artery disease, *Circulation*, 100, 1050, 1999.

342. Tikkanen, M.J., Wahala, K., Ojala, S., Vihma, V., and Adlercreutz, H., Effect of soybean phytoestrogen intake on low density lipoprotein oxidation resistance, *Proc. Natl. Acad. Sci. USA*, 95, 3106, 1998.

343. van den Berg, R., van Vliet, T., Broekmans, W.M.R., Cnubben, N.H.P., Vaes, W.H.J., Roza, L., Haenen, G.R.M.M., Bast, A., and van den Berg, H., A vegetable/fruit concentrate with high antioxidant capacity has no effect on biomarkers of antioxidant status in male smokers, *J. Nutr.*, 131, 1714, 2001.

344. Wan, Y., Vinson, J.A., Etherton, T.D., Proch, J., Lazarus, S.A., and Kris-Etherton, P.M., Effects of cocoa powder and dark chocolate on LDL oxidative susceptibility and prostaglandin concentrations in humans, *Am. J. Clin. Nutr.*, 74, 596, 2001.

345. Whitehead, T.P., Robinson, D., Allaway, S., Syms, J., and Hale, A., Effect of red wine ingestion on the antioxidant capacity of serum (see comments), *Clin. Chem.*, 41, 32, 1995.

346. Young, J.F., Nielsen, S.E., Haraldsdottir, J., Daneshvar, B., Lauridsen, S.T., Knuthsen, P., Crozier, A., Sandstrom, B., and Dragsted, L.O., Effect of fruit juice intake on urinary quercetin excretion and biomarkers of antioxidative status, *Am. J. Clin. Nutr.*, 69, 87, 1999.

347. Maxwell, S. and Thorpe, G., Tea flavonoids have little short term impact on serum antioxidant activity, *Br. Med. J.*, 313, 229, 1996.

348. Fuhrman, B., Lavy, A., and Aviram, M., Consumption of red wine with meals reduces the susceptibility of human plasma and low-density lipoprotein to lipid peroxidation, *Am. J. Clin. Nutr.*, 61, 549, 1995.

349. Jovanovic, S.V., Steenken, S., Hara, Y., and Simic, M.G., Reduction potentials of flavonoid and model phenoxyl radicals. Which ring in flavonoids is responsible for antioxidant activity? *J. Chem. Soc. Perkin Trans. 2*, 2497, 1996.

350. Jovanovic, S.V., Hara, Y., Steenken, S., and Simic, M.G., Antioxidant potential of gallocatechins. A pulse radiolysis and laser photolysis study, *J. Am. Chem. Soc.*, 117, 9881, 1995.

351. Jovanovic, S.V., Steenken, S., Tosic, M., Marjanovic, B., and Simic, M.G., Flavonoids as antioxidants, *J. Am. Chem. Soc.*, 116, 4846, 1994.

352. Steenken, S. and Neta, P., One-electron redox potentials of phenols. Hydroxy and amino-phenols and related compounds of biological interest, *J. Phys. Chem.*, 3661, 1982.

353. Shirai, M., Moon, J.H., Tsushida, T., and Terao, J., Inhibitory effect of a quercetin metabolite, quercetin 3-*O*-β-D-glucuronide, on lipid peroxidation in liposomal membranes, *J. Agric. Food Chem.*, 49, 5602, 2001.

354. Terao, J., Murota, K., and Moon, J.-H., Quercetin glucosides as dietary antioxidants in blood plasma: modulation of their function by metabolic conbversion, in *Free Radicals in Chemistry, Biology and Medicine*, Yoshikawa, T., Toyokuni, Y., Yamamoto, Y., and Naito, Y., Eds., OIAC International, London, 2000, p. 50.

355. Liao, K. and Yin, M., Individual and combined antioxidant effects of seven phenolic agents in human erythrocyte membrane ghosts and phosphatidylcholine liposome systems: importance of the partition coefficient, *J. Agric. Food Chem.*, 48, 2266, 2000.

356. Frank, J., Eliasson, C., Leroy-Nivard, D., Budek, A., Lundh, T., Vessby, B., Aman, P., and Kamal-Eldin, A., Dietary secoisolariciresinol diglucoside and its oligomers with 3-hydroxy-3-methyl glutaric acid decrease vitamin E levels in rats, *Br. J. Nutr.*, 92, 169, 2004.

357. Finch, S., Doyle, W., Lowe, C., Bates, C.J., Prentice, A., Smithers, G., and Clarke, P.C., *National Diet and Nutrition Survey. People Aged 65 Years and Over*, The Stationery Office, London, 1998.

358. Sauberlich, H.E., Human requirements and needs. Vitamin C status: methods and findings, *Ann. N.Y. Acad. Sci.*, 258, 438, 1975.

359. Ferroni, F., Maccaglia, A., Pietraforte, D., Turco, L., and Minetti, M., Phenolic antioxidants and the protection of low density lipoprotein from peroxynitrite-mediated oxidations at physiologic CO_2, *J. Agric. Food Chem.*, 52, 2866, 2004.

360. Williamson, G. and Manach, C., Bioavailability and bioefficacy of polyphenols in humans. II. Review of 93 intervention studies, *Am. J. Clin. Nutr.*, 81, 243S, 2005.

361. Halliwell, B., Rafter, J., and Jenner, A., Health promotion by flavonoids, tocopherols, tocotrienols, and other phenols: direct or indirect effects? Antioxidant or not? *Am. J. Clin. Nutr.*, 81, 268S, 2005.

362. Collins, A.R., Assays for oxidative stress and antioxidant status: applications to research into the biological effectiveness of polyphenols, *Am. J. Clin. Nutr.*, 81, 261S, 2005.

363. Scalbert, A., Johnson, I.T., and Saltmarsh, M., Polyphenols: antioxidants and beyond, *Am. J. Clin. Nutr.*, 81, 215S, 2005.

364. Sarni-Manchado, P. and Cheynier, V., Study of non-covalent complexation between catechin derivatives and peptides by electrospray ionization mass spectrometry, *J. Mass Spectrom.*, 37, 609, 2002.

365. Haslam, E. and Lilley, T.H., Natural astringency in foodstuffs — a molecular interpretation, *CRC Crit. Rev. Food Sci. Nutr.*, 27, 1, 1988.

366. Noble, A.C., Factors influencing perception of bitterness and astringency in foods and beverages, in *Polyphenols 98. XVIXe Journeés Internationales Groupe Polyphénols*, INRA Editions, France, 1998,

367. Clifford, M.N., Astringency, in *Phytochemistry of Fruits and Vegetables*, Tomás-Barberán, F.A. and Robins, R.J., Eds., Clarendon Press, London, 1997, p. 5.

368. Lesschaeve, I. and Noble, A.C., Polyphenols: factors influencing their sensory properties and their effects on food and beverage preferences, *Am. J. Clin. Nutr.*, 81, 330S, 2005.

369. Lamuela-Raventos, R.M., Covas, M.I., Fito, M., Marrugat, J., and de la Torre-Boronat, M.C., Detection of dietary antioxidant phenolic compounds in human LDL, *Clin. Chem.*, 45, 1870, 1999.

370. Manach, C., Morand, C., Texier, O., Favier, M.-L., Agullo, G., Demigné, C., Régérat, F., and Rémésy, C., Quercetin metabolites in plasma of rats fed diets containing rutin or quercetin, *J. Nutr.*, 125, 1911, 1995.

371. Brunet, M.J., Blade, C., Salvado, M.J., and Arola, L., Human apo A-I and rat transferrin are the principal plasma proteins that bind wine catechins, *J. Agric. Food Chem.*, 50, 2708, 2002.

372. Bacon, J.R. and Rhodes, M.J., Binding affinity of hydrolyzable tannins to parotid saliva and to proline-rich proteins derived from it, *J. Agric. Food Chem.*, 48, 838, 2000.

373. Bacon, J.R. and Rhodes, M.J.C., Development of a competition assay for the evaluation of binding of human parotid salivary proteins to dietary complex phenols and tannins using a peroxidase-labeled tannin, *J. Agric. Food Chem.*, 46, 5083, 1998.

374. Sazuka, M., Itoi, T., Suzuki, Y., Odani, S., Koide, T., and Isemura, M., Evidence for the interaction between (−)-epigallocatechin gallate and human plasma proteins fibronectin, fibrinogen, and histidine-rich glycoprotein, *Biosci. Biotechnol. Biochem.*, 60, 1317, 1996.

375. Wiseman, H., O'Reilly, J.D., Adlercreutz, H., Mallet, A.I., Bowey, E.A., Rowland, I.R., and Sanders, T.A., Isoflavone phytoestrogens consumed in soy decrease F(2)-isoprostane concentrations and increase resistance of low-density lipoprotein to oxidation in humans, *Am. J. Clin. Nutr.*, 72, 395, 2000.

376. McAnlis, G.T., McEneny, J., Pearce, J., and Young, I.S., Black tea consumption does not protect low density lipoprotein from oxidative modification, *Eur. J. Clin. Nutr.*, 52, 202, 1998.

377. Salah, N., Miller, N.J., Paganga, G., Tijburg, L., Bolwell, G.P., and Rice-Evans, C.A., Polyphenolic flavanols as scavengers of aqueous phase radicals and as chain-breaking antioxidants, *Arch. Biochem. Biophys.*, 322, 339, 1995.

378. Kahkonen, M. and Heinonen, M., Antioxidant activity of anthocyanins and their aglycons, *J. Agric. Food Chem.*, 51, 628, 2005.

379. Brown, J.E., Khodr, H., Hider, R.C., and Rice-Evans, C.A., Structural dependence of flavonoid interactions with Cu^{2+} ions: implications for their antioxidant properties, *Biochem. J.*, 330, 1173, 1998.

380. Ross, R., Atherosclerosis — an inflammatory disease, *N. Engl. J. Med.*, 340, 115, 1999.

381. Lu, L., Chen, S.S., Zhang, J.Q., Ramires, F.J., and Sun, Y., Activation of nuclear factor-κB and its proinflammatory mediator cascade in the infarcted rat heart, *Biochem. Biophys. Res. Commun.*, 321, 879, 2004.

382. Badia, E., Sacanella, E., Fernandez-Sola, J., Nicolas, J.M., Antunez, E., Rotilio, D., De Gaetano, G., Urbano-Marquez, A., and Estruch, R., Decreased tumor necrosis factor-induced adhesion of human monocytes to endothelial cells after moderate alcohol consumption, *Am. J. Clin. Nutr.*, 80, 225, 2004.

383. Estruch, R., Sacanella, E., Badia, E., Antunez, E., Nicolas, J.M., Fernandez-Sola, J., Rotilio, D., De Gaetano, G., Rubin, E., and Urbano-Marquez, A., Different effects of red wine and gin consumption on inflammatory biomarkers of atherosclerosis: a prospective randomized crossover trial: effects of wine on inflammatory markers, *Atherosclerosis*, 175, 117, 2004.

384. Blanco-Colio, L.M., Valderrama, M., Alvarez-Sala, L.A., Bustos, C., Ortego, M., Hernandez-Presa, M.A., Cancelas, P., Gomez-Gerique, J., Millan, J., and Egido, J., Red wine intake prevents nuclear factor-κB activation in peripheral blood mononuclear cells of healthy volunteers during postprandial lipemia, *Circulation*, 102, 1020, 2000.

385. de Maat, M.P.M., Pijl, H., Kluft, C., and Princen, H.M.G., Consumption of black or green tea has no effect on inflammation, haemostasis and endothelial markers in smoking healthy individuals, *Eur. J. Clin. Nutr.,* 54, 757, 2000.

386. Hodgson, J.M., Puddey, I.B., Burke, V., Watts, G.F., and Beilin, L.J., Regular ingestion of black tea improves brachial artery vasodilator function, *Clin. Sci. (Lond.),* 102, 195, 2002.

387. Kalfin, R., Righi, A., Del Rosso, A., Bagchi, D., Gererini, D., Cerinic, M., and Das, D., Activin, a grape seed-derived proanthocyanidin extract, reduces plasma levels of oxidative stress and adhesion molecules (ICAM-1, VCAM-1 and E-selectin) in systemic sclerosis, *Free Radical Res.,* 36, 819, 2002.

388. Sen, C. and Bagchi, D., Regulation of inducible adhesion molecule expression in human endotheila cells by grape seed proanthocyanidin extract, *Mol. Cell. Biochem.,* 216, 1, 2001.

389. Ludwig, A., Lorenz, M., Grimbo, N., Steinle, F., Meiners, S., Bartsch, C., Stangl, K., Baumann, G., and Stangl, V., The tea flavonoid epigallocatechin-3-gallate reduces cytokine-induced VCAM-1 expression and monocyte adhesion to endothelial cells, *Biochem. Biophys. Res. Commun.,* 316, 659, 2004.

390. Carluccio, M.A., Siculella, L., Ancora, M.A., Massaro, M., Scoditti, E., Storelli, C., Visioli, F., Distante, A., and De Caterina, R., Olive oil and red wine antioxidant polyphenols inhibit endothelial activation: antiatherogenic properties of mediterranean diet phytochemicals, *Arterioscler. Thromb. Vasc. Biol.,* 23, 622, 2003.

391. Choi, J.S., Choi, Y.J., Park, S.H., Kang, J.S., and Kang, Y.H., Flavones mitigate tumor necrosis factor-α-induced adhesion molecule upregulation in cultured human endothelial cells: role of nuclear factor-κB, *J. Nutr.,* 134, 1013, 2004.

392. Gottstein, N., Ewins, B., Eccleston, C., Hibbard, G., Kavanagh, I., Minihane, A.-M., Weiberg, P., and Rimbach, G., Effect of genistein and daidzein on platelet aggregation and monocyte and endothelial function, *Br. J. Nutr.,* 89, 607, 2003.

393. Rimbach, G., Weinberg, P.D., Pascual-Teresa, S., Alonso, M.G., Ewins, B.A., Turner, R., Minihane, A.M., Botting, N., Fairley, B., and Matsugo, S., Sulfation of genistein alters its antioxidant properties and its effect on platelet aggregation and monocyte and endothelial function, *Biochim. Biophys. Acta — Gen. Subjects,* 1670, 229, 2004.

394. Murase, T., Kume, N., Hase, T., Shibuya, Y., Nishizawa, Y., Tokimitsu, I., and Kita, T., Gallates inhibit cytokine-induced nuclear translocation of NF-kappaB and expression of leukocyte adhesion molecules in vascular endothelial cells, *Arterioscler. Thromb. Vasc. Biol.,* 19, 1412, 1999.

395. Garcia-Alonso, M., Rimbach, G., Rivas-Gonzalo, J.C., and de Pascual-Teresa, S., Antioxidant and cellular actvities of anthocyanins and their corresponding vitisins — studies in platelets, monocytes and human endothelial cells, *J. Agric. Food Chem.,* 52, 3378, 2004.

396. Atalay, M., Gordillo, G., Roy, S., Rovin, B., Bagchi, D., Bagchi, M., and Sen, C.K., Anti-angiogenic property of edible berry in a model of hemangioma, *FEBS Lett.,* 544, 252, 2003.

397. Youdim, K.A., McDonald, J., Kalt, W., and Joseph, J.A., Potential role of dietary flavonoids in reducing microvascular endothelium vulnerability to oxidative and inflammatory insults, *J. Nutr. Biochem.,* 13, 282, 2002.

398. Kumar, A., Dhawan, S., Hardegen, N.J., and Aggarwal, B.B., Curcumin (diferuloylmethane) inhibition of tumor necrosis factor (TNF)-mediated adhesion of monocytes to endothelial cells by suppression of cell surface expression of adhesion molecules and of nuclear factor-κB activation, *Biochem. Pharmacol.,* 55, 775, 1998.

399. Kobuchi, H., Roy, S., Sen, C.K., Nguyen, H.G., and Packer, L., Quercetin inhibits inducible ICAM-1 expression in human endothelial cells through the JNK pathway, *Am. J. Physiol.,* 277, C403, 1999.

400. Ferry, D.R., Smith, A., Malkhandi, J., Fyfe, D.W., deTakats, P.G., Anderson, D., Baker, J., and Kerr, D.J., Phase I clinical trial of the flavonoid quercetin: pharmacokinetics and evidence for *in vivo* tyrosine kinase inhibition, *Clin. Cancer Res.,* 2, 659, 1996.

401. Koga, T. and Meydani, M., Effect of plasma metabolites of (+)-catechin and quercetin on monocyte adhesion to human aortic endothelial cells, *Am. J. Clin. Nutr.,* 73, 941, 2001.

402. Hedges, J.C., Dechert, M.A., Yamboliev, I.A., Martin, J.L., Hickey, E., Weber, L.A., and Gerthoffer, W.T., A role for p38MAPK/HSP27 pathway in smooth muscle cell migration, *J. Biol. Chem.,* 274, 24211, 1999.

403. Alcocer, F., Whitley, D., Salazar-Gonzalez, J.F., Jordan, W.D., Sellers, M.T., Eckhoff, D.E., Suzuki, K., MacRae, C., and Bland, K.L., Quercetin inhibits human vascular smooth muscle cell proliferation and migration, *Surgery,* 131, 198, 2002.

404. Moon, S.K., Cho, G.O., Jung, S.Y., Gal, S.W., Kwon, T.K., Lee, Y.C., Madamanchi, N.R., and Kim, C.H., Quercetin exerts multiple inhibitory effects on vascular smooth muscle cells: role of ERK1/2, cell-cycle regulation, and matrix metalloproteinase-9, *Biochem. Biophys. Res. Commun.,* 301, 1069, 2003.

405. Ahn, H.Y., Hadizadeh, K.R., Seul, C., Yun, Y.P., Vetter, H., and Sachinidis, A., Epigallocatechin-3-gallate selectively Inhibits the PDGF-BB-induced intracellular signaling transduction pathway in vascular smooth muscle cells and inhibits transformation of *cis*-transfected NIH 3T3 fibroblasts and human glioblastoma cells (A172), *Mol. Biol. Cell,* 10, 1093, 1999.

406. Oak, M.H., Chataigneau, M., Keravis, T., Chataigneau, T., Beretz, A., Andriantsitohaina, R., Stoclet, J.C., Chang, S.J., and Schini-Kerth, V.B., Red wine polyphenolic compounds inhibit vascular endothelial growth factor expression in vascular smooth muscle cells by preventing the activation of the p38 mitogen-activated protein kinase pathway, *Arterioscler. Thromb. Vasc. Biol.,* 23, 1001, 2003.

407. Araim, O., Ballantyne, J., Waterhouse, A.L., and Sumpio, B.E., Inhibition of vascular smooth muscle cell proliferation with red wine and red wine polyphenols, *J. Vasc. Surg.,* 35, 1226, 2002.

408. Locher, R., Emmanuele, L., Suter, P.M., Vetter, W., and Barton, M., Green tea polyphenols inhibit human vascular smooth muscle cell proliferation stimulated by native low-density lipoprotein, *Eur. J. Pharmacol.,* 434, 1, 2002.

409. Yoshizumi, M., Tsuchiya, K., Suzaki, Y., Kirima, K., Kyaw, M., Moon, J.H., Terao, J., and Tamaki, T., Quercetin glucuronide prevents VSMC hypertrophy by angiotensin II via the inhibition of JNK and AP-1 signaling pathway, *Biochem. Biophys. Res. Commun.,* 293, 1458, 2002.

410. Kyaw, M., Yoshizumi, M., Tsuchiya, K., Izawa, Y., Kanematsu, Y., and Tamaki, T., Atheroprotective effects of antioxidants through inhibition of mitogen-activated protein kinases, *Acta Pharmacol. Sin.,* 25, 977, 2004.

411. Kaneider, N.C., Mosheimer, B., Reinisch, N., Patsch, J.R., and Wiedermann, C.J., Inhibition of thrombin-induced signaling by resveratrol and quercetin: effects on adenosine nucleotide metabolism in endothelial cells and platelet-neutrophil interactions, *Thromb. Res.,* 114, 185, 2004.

412. Benito, S., Lopez, D., Saiz, M.P., Buxaderas, S., Sanchez, J., Puig-Parellada, P., and Mitjavila, M.T., A flavonoid-rich diet increases nitric oxide production in rat aorta, *Br. J. Pharmacol.,* 135, 910, 2002.

413. Leikert, J.F., Rathel, T.R., Wohlfart, P., Cheynier, V., Vollmar, A.M., and Dirsch, V.M., Red wine polyphenols enhance endothelial nitric oxide synthase expression and subsequent nitric oxide release from endothelial cells, *Circulation,* 106, 1614, 2002.

414. Wallerath, T., Poleo, D., Li, H., and Forstermann, U., Red wine increases the expression of human endothelial nitric oxide synthase: a mechanism that may contribute to its beneficial cardiovascular effects, *J. Am. Coll. Cardiol.,* 41, 471, 2003.

415. Anter, E., Thomas, S.R., Schulz, E., Shapira, O.M., Vita, J.A., and Keaney, J.F., Activation of endothelial nitric-oxide synthase by the p38 MAPK in response to black tea polyphenols, *J. Biol. Chem.,* 279, 46637, 2004.

416. Karim, M., McCormick, K., and Kappagoda, C.T., Effect of cocoa extracts on endothelium-dependent relaxation, *J. Nutr.,* 130, 2105S, 2000.

417. Li, H., Xia, N., Brausch, I., Yao, Y., and Forstermann, U., Flavonoids from artichoke (*Cynara scolymus* L.) up-regulate endothelial-type nitric-oxide synthase gene expression in human endothelial cells, *J. Pharmacol. Exp. Ther.,* 310, 926, 2004.

418. Lorenz, M., Wessler, S., Follmann, E., Michaelis, W., Dusterhoft, T., Baumann, G., Stangl, K., and Stangl, V., A constituent of green tea, epigallocatechin-3-gallate, activates rndothelial nitric oxide synthase by a phosphatidylinositol-3-OH-kinase-, cAMP-dependent protein kinase-, and Akt-dependent pathway and leads to endothelial-dependent vasorelaxation, *J. Biol. Chem.,* 279, 6190, 2004.

419. Wallerath, T., Deckert, G., Ternes, T., Anderson, H., Li, H., Witte, K., and Forstermann, U., Resveratrol, a polyphenolic phytoalexin present in red wine, enhances expression and activity of endothelial nitric oxide synthase, *Circulation,* 106, 1652, 2002.

420. Xu, J.W., Ikeda, K., and Yamori, Y., Cyanidin-3-glucoside regulates phosphorylation of endothelial nitric oxide synthase, *FEBS Lett.,* 574, 176, 2004.

421. Xu, J.W., Ikeda, K., and Yamori, Y., Upregulation of endothelial nitric oxide synthase by cyanidin-3-glucoside, a yypical anthocyanin pigment, *Hypertension,* 44, 217, 2004.

422. Liu, D., Homan, L.L., and Dillon, J.S., Genistein acutely stimulates nitric oxide synthesis in vascular endothelial cells by a cyclic adenosine 5′-monophosphate-dependent mechanism, *Endocrinology,* 145, 5532, 2004.

423. Lorenz, M., Wessler, S., Follmann, E., Michaelis, W., Dusterhoft, T., Baumann, G., Stangl, K., and Stangl, V., A constituent of green tea, epigallocatechin-3-gallate, activates rndothelial nitric oxide synthase by a phosphatidylinositol-3-OH-kinase-, cAMP-dependent protein kinase-, and Akt-dependent pathway and leads to endothelial-dependent vasorelaxation, *J. Biol. Chem.,* 279, 6190, 2004.

424. Xu, J.W., Ikeda, K., and Yamori, Y., Upregulation of endothelial nitric oxide synthase by cyanidin-3-glucoside, a typical anthocyanin pigment, *Hypertension,* 44, 217, 2004.

425. Diplock, A.T., Charleux, J.-L., Crozier-Willi, G., Bung, P., Kok, F.J., Rice-Evans, C.A., Roberfroid, M., Stahl, W., and Viña-Ribes, J., Functional food science and defence against reactive oxidative species, *Br. J. Nutr.,* 80, S77, 1998.

426. Terao, J., Yamaguchi, S., Shirai, M., Miyoshi, M., Moon, J.H., Oshima, S., Inakuma, T., Tsushida, T., and Kato, Y., Protection by quercetin and quercetin 3-*O*-β-D-glucuronide of peroxynitrite-induced antioxidant consumption in human plasma low-density lipoprotein, *Free Radic. Res.,* 35, 925, 2001.

427. Nees, S., Weiss, D.R., Reichenbach-Klinke, E., Rampp, F., Heilmeier, B., Kanbach, J., and Esperester, A., Protective effects of flavonoids contained in the red vine leaf on venular endothelium against the attack of activated blood components *in vitro, Arzneimittel-Forsch.,* 53, 330, 2003.

428. Day, A.J., Bao, Y., Morgan, M.R.A., and Williamson, G., Conjugation position of quercetin glucuronides and effect on biological activity, *Free Radical Biol. Med.,* 29, 1234, 2000.

429. Manach, C., Morand, C., Crespy, V., Demigné, C., Texier, O., Régérat, F., and Remesy, C., Quercetin is recovered in human plasma as conjugated derivatives which retain antioxidant properties, *FEBS Lett.,* 426, 331, 1998.

430. Zhang, Y., Song, T.T., Cunnick, J.E., Murphy, P.A., and Hendrich, S., Daidzein and genistein glucuronides *in vitro* are weakly estrogenic and activate human natural killer cells at nutritionally relevant concentrations, *J. Nutr.,* 129, 399, 1999.

431. O'Leary, K.A., Day, A.J., Needs, P.W., Sly, W.S., O'Brien, N.M., and Williamson, G., Flavonoid glucuronides are substrates for human liver β-glucuronidase, *FEBS Lett.,* 503, 103, 2001.

432. Lampe, J.W., Li, S.S., Potter, J.D., and King, I.B., Serum β-glucuronidase activity is inversely associated with plant-food intakes in humans, *J. Nutr.,* 132, 1341, 2002.

433. Shimoi, K., Saka, N., Nozawa, R., Sato, M., Amano, I., Nakayama, T., and Kinae, N., Deglucuronidation of a flavonoid, luteolin monoglucuronide, during inflammation, *Drug Metab. Dispos.,* 29, 1521, 2001.

434. Marshall, T., Shult, P., and Busse, W.W., Release of lysosomal enzyme β-glucuronidase from isolated human eosinophils, *J. Allergy Clin. Immunol.,* 82, 550, 1988.

435. Spencer, J.P., Schroeter, H., Crossthwaithe, A.J., Kuhnle, G., Williams, R.J., and Rice-Evans, C., Contrasting influences of glucuronidation and *O*-methylation of epicatechin on hydrogen peroxide-induced cell death in neurons and fibroblasts, *Free Radical Biol. Med.,* 31, 1139, 2001.

436. Pasqualini, J.R., Gelly, C., Nguyen, B.L., and Vella, C., Importance of estrogen sulfates in breast cancer, *J. Steroid Biochem.,* 34, 155, 1989.

437. Breinholt, V.M., Offord, E.A., Brouwer, C., Nielsen, S.E., Brosen, K., and Friedberg, T., *In vitro* investigation of cytochrome P450-mediated metabolism of dietary flavonoids, *Food Chem. Toxicol.,* 40, 609, 2002.

438. Zhu, B.T., Evaristus, E.N., Antoniak, S.K., Sarabia, S.F., Ricci, M.J., and Liehr, J.G., Metabolic deglucuronidation and demethylation of estrogen conjugates as a source of parent estrogens and

catecholestrogen metabolites in Syrian hamster kidney, a target organ of estrogen-induced tumor-igenesis, *Toxicol. Appl. Pharmacol.*, 136, 186, 1996.

439. Shutt, D.A. and Cox, R.I., Steroid and phyto-oestrogen binding to sheep uterine receptors *in vitro*, *J. Endocrinol.*, 52, 299, 1972.

440. Kim, D.H., Jung, E.A., Sohng, I.S., Han, J.A., Kim, T.H., and Han, M.J., Intestinal bacterial metabolism of flavonoids and its relation to some biological activities, *Arch. Pharm. Res.*, 21, 17, 1998.

441. Yasuda, T., Takasawa, A., Nakazawa, T., Ueda, J., and Ohsawa, K., Inhibitory effects of urinary metabolites on platelet aggregation after orally administering Shimotsu-To, a traditional Chinese medicine, to rats, *J. Pharm. Pharmacol.*, 55, 239, 2003.

442. Bhat, C.S. and Ramasarma, T., Effect of phenyl and phenolic acids on mevalonate-5-phosphate kinase and mevalonate-5-pyrophosphate decarboxylase of the rat brain, *J. Neurochem.*, 32, 1531, 1979.

443. Lee, J.S., Choi, M.S., Jeon, S.M., Jeong, T.S., Park, Y.B., Lee, M.K., and Bok, S.H., Lipid-lowering and antioxidative activities of 3,4-di(OH)-cinnamate and 3,4-di(OH)-hydrocinnamate in cholesterol-fed rats, *Clin Chim. Acta*, 314, 221, 2001.

444. Bok, S.H., Lee, S.H., Park, Y.B., Bae, K.H., Son, K.H., Jeong, T.S., and Choi, M.S., Plasma and hepatic cholesterol and hepatic activities of 3-hydroxy-3-methyl-glutaryl-CoA reductase and acyl CoA: cholesterol transferase are lower in rats fed citrus peel extract or a mixture of citrus bioflavonoids, *J. Nutr.*, 129, 1182, 1999.

445. Kim, H.K., Jeong, T.S., Lee, M.K., Park, Y.B., and Choi, M.S., Lipid-lowering efficacy of hesperetin metabolites in high-cholesterol fed rats, *Clin. Chim. Acta*, 327, 129, 2003.

446. Lee, C.H., Jeong, T.S., Choi, Y.K., Hyun, B.H., Oh, G.T., Kim, E.H., Kim, J.R., Han, J.I., and Bok, S.H., Anti-atherogenic effect of citrus flavonoids, naringin and naringenin, associated with hepatic ACAT and aortic VCAM-1 and MCP-1 in high cholesterol-fed rabbits, *Biochem. Biophys. Res. Commun.*, 284, 681, 2001.

447. Matsumoto, N., Okushio, K., and Hara, Y., Effect of black tea polyphenols on plasma lipids in cholesterol-fed rats, *J. Nutr. Sci. Vitaminol.*, 44, 337, 1998.

448. Yamakoshi, J., Kataoka, S., Koga, T., and Ariga, T., Proanthocyanidin-rich extract from grape seeds attenuates the development of aortic atherosclerosis in cholesterol-fed rabbits, *Atheroscler-osis*, 142, 139, 1999.

449. Kurowska, E.M., Spence, J.D., Jordan, J., Wetmore, S., Freeman, D.J., Piche, L.A., and Serra-tore, P., HDL-cholesterol-raising effect of orange juice in subjects with hypercholesterolemia, *Am. J. Clin. Nutr.*, 72, 1095, 2000.

450. Mo, H., and Elson, C.E., Studies of the isoprenoid-mediated inhibition of mevalonate synthesis applied to cancer chemotherapy and chemoprevention, *Exp. Biol. Med. (Maywood)*, 229, 567, 2004.

451. Arora, A., Valcic, S., Cornejo, S., Nair, M.G., Timmermann, B.N., and Liebler, D.C., Reactions of genistein with alkylperoxyl radicals, *Chem. Res. Toxicol.*, 13, 638, 2000.

452. Sang, S., Tian, S., Wang, H., Stark, R.E., Rosen, R.T., Yang, C.S., and Ho, C.T., Chemical studies of the antioxidant mechanism of tea catechins: radical reaction products of epicatechin with peroxyl radicals, *Bioorg. Med. Chem.*, 11, 3371, 2003.

453. Valcic, S., Burr, J.A., Liebler, D.C., and Timmermann, B.N., Antioxidant chemistry of green tea catechins. Identification of products of the reaction of (−)-pigallocatechin gallate and (−)-epigallocatechin from their reactions with peroxyl radicals, *Chem. Res. Toxicol.*, 13, 801, 2000.

454. Hirota, S., Nishioka, T., Shimoda, T., Miura, K., Ansal, T., and Takahama, U., Quercetin gluco-sides are hydrolysed to quercetin in human oral cavity to participate in peroxidase-dependent scavenging of hydrogen peroxide, *Food Sci. Technol. Res.*, 7, 239, 2001.

455. Walle, T., Browning, A.M., Steed, L.L., Reed, S.G., and Walle, U.K., Flavonoid glucosides are hydrolyzed and thus activated in the oral cavity in humans, *J. Nutr.*, 135, 48, 2005.

456. Takahama, U., Hirota, S., Nishioka, T., and Yoshitama, K., Oxidation of quercetin by salivary components 1. Salivary peroxidase-dependent oxidation of quercetin and characterization of the oxidation products, *Food Sci. Technol. Res.*, 8, 148, 2002.

457. Lu, J., Ho, C.T., Ghai, G., and Chen, K.Y., Differential effects of theaflavin monogallates on cell growth, apoptosis, and Cox-2 gene expression in cancerous versus normal cells, *Cancer Res.*, 60, 6465, 2000.

458. Caderni, G., Filippo, C.d., Luceri, C., Salvadori, M., Giannini, A., Biggeri, A., Remy, S., Cheynier, V., and Dolara, P., Effects of black tea, green tea and wine extracts on intestinal carcinogenesis induced by azoxymethane in F344 rats, *Carcinogenesis*, 21, 1965, 2000.

459. Giovannelli, L., Testa, G., de Filippo, C., Cheynier, V., Clifford, M.N., and Dolara, P., Effect of complex polyphenols and tannins from red wine on DNA oxidative damage of rat colon mucosa *in vivo*, *Eur. J. Nutr.*, 39, 207, 2000.

460. Casalini, C., Lodovici, M., Briani, C., Paganelli, G., Remy, S., Cheynier, V., and Dolara, P., Effect of complex polyphenols and tannins from red wine (WCPT) on chemically induced oxidative DNA damage in the rat, *Eur. J. Nutr.*, 38, 190, 1999.

461. IARC, *IARC Monographs on the Evaluation of Carcinogenic Risks to Humans. Volume 51: Coffee, Tea, Maté, Methylxanthines and Glyoxal*, IARC, Lyons, 1991.

462. Terry, P., Lagergren, J., Wolk, A., and Nyren, O., Drinking hot beverages is not associated with risk of oesophageal cancers in a Western population, *Br. J. Cancer*, 84, 120, 2001.

463. Sewram, V., De Stefani, E., Brennan, P., and Boffetta, P., Maté consumption and the risk of squamous cell esophageal cancer in Uruguay, *Cancer Epidemiol. Biomarkers Prev.*, 12, 508, 2003.

464. Castellsague, X., Munoz, N., De Stefani, E., Victora, C.G., Castelletto, R., and Rolon, P.A., Influence of maté drinking, hot beverages and diet on esophageal cancer risk in South America, *Int. J. Cancer*, 88, 658, 2000.

465. O'Coinceanainn, M., Astill, C., and Baderschneider, B., Coordination of aluminium with purpurogallin and theaflavin digallate, *J. Inorg. Biochem.*, 96, 463, 2003.

466. O'Coinceanainn, M., Bonnely, S., Baderschneider, B., and Hynes, M.J., Reaction of iron(III) with theaflavin: complexation and oxidative products, *J. Inorg. Biochem.*, 98, 657, 2004.

467. Tonnesen, H.H., Chemistry of curcumin and curcuminoids, in *Phenolic Compounds in Foods and their Effects on Health. 1 Analysis, Occurrence and Chemistry*, Ho, C.-T., Lee, C.Y., and Huang, M.-T., Eds., American Chemical Society, Washington, DC, 1992.

468. Cook, J.D., Reddy, M.B., and Hurrell, R.F., The effect of red and white wines on nonheme-iron absorption in humans, *Am. J. Clin. Nutr.*, 61, 800, 1995.

469. Hurrell, R.F., Reddy, M., and Cook, J.D., Inhibition of non-haem iron absorption in man by polyphenolic-containing beverages, *Br. J. Nutr.*, 81, 289, 1999.

470. Temme, E.H. and Van Hoydonck, P.G., Tea consumption and iron status, *Eur. J. Clin. Nutr.*, 56, 379, 2002.

471. Frazier, R.A., Papadopoulou, A., Mueller-Harvey, I., Kissoon, D., and Green, R.J., Probing protein–tannin interactions by isothermal titration microcalorimetry, *J. Agric. Food Chem.*, 51, 5189, 2003.

472. Goto, K., Kanaya, S., Nishikawa, T., Hara, H., Terada, A., Ishigami, T., and Hara, Y., The influence of tea catechins on fecal flora of elderly residents in long-term care facilities, *Ann. Long-Term Care*, 6, 43, 1998.

473. Hara, Y., Influence of tea catechins on the digestive tract, *J. Cell. Biochem.*, 27 (Suppl.), 52, 1997.

474. Okubo, T., Ishihara, N., Oura, A., Serit, M., Kim, M., Yamamoto, T., and Mitsuoka, T., *In vivo* effect of tea polyphenol intake on human intestinal microflora and metabolism, *Biosci. Biotechnol. Biochem.*, 56, 588, 1992.

475. Min, B.R., Attwood, G.T., Reilly, K., Sun, W., Peters, J.S., Barry, T.N., and McNabb, W.C., *Lotus corniculatus* condensed tannins decrease *in vivo* populations of proteolytic bacteria and affect nitrogen metabolism in the rumen of sheep, *Can. J. Microbiol.*, 48, 911, 2002.

476. Yam, T.S., Shah, S., and Hamilton-Miller, J.M.T., Microbiological activity of whole and fractionated crude extracts of tea (*Camellia sinensis*), and of tea components, *FEMS Microbiol. Lett.*, 152, 169, 1997.

477. Fallingborg, J., Intraluminal pH of the human gastrointestinal tract, *Dan. Med. Bull.*, 46, 183, 1999.

478. Thompson, L.U., Yoon, J.H., Jenkins, D.J., Wolever, T.M., and Jenkins, A.L., Relationship between polyphenol intake and blood glucose response of normal and diabetic individuals, *Am. J. Clin. Nutr.*, 39, 745, 1984.

479. Gin, H., Rigalleau, V., Caubet, O., Masquelier, J., and Aubertin, J., Effects of red wine, tannic acid, or ethanol on glucose tolerance in non-insulin-dependent diabetic patients and on starch digestibility *in vitro*, *Metabolism*, 48, 1179, 1999.

480. Johnston, K.L., Clifford, M.N., and Morgan, L.M., Coffee acutely modifies gastrointestinal hormone secretion and glucose tolerance in humans: glycemic effects of chlorogenic acid and caffeine, *Am. J. Clin. Nutr.*, 78, 728, 2003.

481. Johnston, K.L., Clifford, M.N., and Morgan, L.M., Possible role for apple juice phenolic compounds in the acute modification of glucose tolerance and gastrointestinal hormone secretion in humans, *J. Sci. Food Agric.*, 82, 1800, 2002.

482. Coutinho, M., Gerstein, H.C., Wang, Y., and Yusef, S., The relationship between glucose and incident cardiovascular events: a metaregression analysis of published data from 20 studies of 95,783 individuals followed for 12.4 years, *Diabetes Care*, 22, 233, 1999.

483. Facchini, F.S. and Saylor, K.L., A low-iron-available, polyphenol-enriched, carbohydrate-restricted diet to slow progression of diabetic nephropathy, *Diabetes*, 52, 1204, 2003.

484. Hara, Y. and Honda, M., The inhibition of α-amylase by tea polyphenols, *Agric. Biol. Chem.*, 54, 1939, 1990.

485. Matsumoto, N., Ishigaki, F., Ishigaki, A., Iwashina, H., and Hara, Y., Reduction of blood-glucose levels by tea catechin, *Biosci. Biotechnol. Biochem.*, 57, 525, 1993.

486. Matsui, T., Ebuchi, S., Kobayashi, M., Fukui, K., Sugita, K., Terahara, N., and Matsumoto, K., Anti-hyperglycemic effect of diacylated anthocyanin derived from *Ipomoea batatas* cultivar Ayamurasaki can be achieved through the α-glucosidase inhibitory action, *J. Agric. Food Chem.*, 50, 7244, 2002.

487. Kawabata, J., Mizuhata, K., Sato, E., Nishioka, T., Aoyama, Y., and Kasai, T., 6-hydroxyflavonoids as α-glucosidase inhibitors from marjoram (*Origanum majorana*) leaves, *Biosci. Biotechnol. Biochem.*, 67, 445, 2003.

488. Hilt, P., Schieber, A., Yildirim, C., Arnold, G., Klaiber, I., Conrad, J., Beifuss, U., and Carle, R., Detection of phloridzin in strawberries (*Fragaria x ananassa* Duch.) by HPLC–PDA–MS/MS and NMR spectroscopy, *J. Agric. Food Chem.*, 51, 2896, 2003.

489. Nakazawa, F., Influence of phloridzin on intestinal absorption, *Tohokv J. Exp. Med.*, 3, 288, 1922.

490. Alvarado, F., Hypothesis for the interaction of phlorizin and phloretin with membrane carriers for sugars, *Biochim. Biophys. Acta*, 135, 483, 1967.

491. Alvarado, F. and Crane, R.K., Phlorizin as a competitve inhibitor of the active transport of sugars by hamster small intestine, *Biochim. Biophys. Acta*, 56, 170, 1962.

492. Crane, R.K., Intestinal absorption of sugars, *Physiol. Rev.*, 40, 789, 1960.

493. Welsch, C.A., Lachance, P.A., and Wasserman, B.P., Dietary phenolic compounds: inhibition of Na^+-dependent D-glucose uptake in rat intestinal brush border membrane vesicles, *J. Nutr.*, 119, 1698, 1989.

494. Kobayashi, Y., Suzuki, M., Satsu, H., Arai, S., Hara, Y., Suzuki, K., Miyamoto, Y., and Shimizu, M., Green tea polyphenols inhibit the sodium-dependent glucose transporter of intestinal epithelial cells by a competitive mechanism, *J. Agric. Food Chem.*, 48, 5618, 2000.

495. Noteborn, H.P.J.M., Jansen, E., Benito, S., and Mengelers, M.J.B., Oral absorption and metabolism of quercetin and sugar-conjugated derivatives in specific transport systems, *Cancer Lett.*, 114, 175, 1997.

496. Walgren, R.A., Walle, U.K., and Walle, T., Transport of quercetin and its glucosides across human intestinal epithelial Caco-2 cells, *Biochem. Pharmacol.*, 55, 1721, 1998.

497. Gee, J.M., DuPont, M.S., Rhodes, M.J.C., and Johnson, I.T., Quercetin glucosides interact with the intestinal glucose transport pathway, *Free Radic. Biol. Med.*, 25, 19, 1998.

498. Walgren, R.A., Lin, J.T., Kinne, R.K., and Walle, T., Cellular uptake of dietary flavonoid quercetin 4'-β-glucoside by sodium-dependent glucose transporter SGLT1, *J. Pharmacol. Exp. Ther.*, 294, 837, 2000.

499. Walgren, R.A., Karnaky, K.J., Jr., Lindenmayer, G.E., and Walle, T., Efflux of dietary flavonoid quercetin 4'-β-glucoside across human intestinal Caco-2 cell monolayers by apical multidrug resistance-associated protein-2, *J. Pharmacol. Exp. Ther.*, 294, 830, 2000.

500. Song, J., Kwon, O., Chen, S., Daruwala, R., Eck, P., Park, J.B., and Levine, M., Flavonoid inhibition of sodium-dependent vitamin C transporter 1 (SVCT1) and glucose transporter iso-form 2 (GLUT2), intestinal transporters for vitamin C and glucose, *J. Biol. Chem.,* 277, 15252, 2002.

501. Wolffram, S., Block, M., and Ader, P., Quercetin-3-glucoside is transported by the glucose carrier SGLT1 across the brush border membrane of rat small intestine, *J. Nutr.,* 132, 630, 2002.

502. Dorne, J.L.C.M., Walton, K., and Renwick, A.G., Human variability in xenobiotic metabolism and pathway-related uncertainty factors for chemical risk assessment: a review, *Food Chem. Toxicol.,* 43, 203, 2005.

503. Walker, R., Criteria for risk assessment of botanical food supplements, *Toxicol. Lett.,* 149, 187, 2004.

504. Agence Française de Securite Sanitaire des Produite de Sante, Tea extract, 2005, http://agmed. sante.gouv.fr/htm/10/filcoprs/030401.htm.

505. Schmidt, M., Schmitz, H.J., Baumgart, A., Guedon, D., Netsch, M.I., Kreuter, M.H., Schmidlin, C.B., and Schrenk, D., Toxicity of green tea extracts and their constituents in rat hepatocytes in primary culture, *Food Chem. Toxicol.,* 43, 307, 2005.

506. Bailey, D.G., Kreeft, J.H., Munoz, C., Freeman, D.J., and Bend, J.R., Grapefruit juice–felodipine interaction: effect of naringin and 6′,7′-dihydroxybergamottin in humans, *Clin. Pharm. Ther.,* 64, 248, 1998.

507. Fuhr, U., Maier-Bruggemann, A., Blume, H., Muck, W., Unger, S., Kuhlmann, J., Huschka, C., Zaigler, M., Rietbrock, S., and Staib, A.H., Grapefruit juice increases oral nimodipine bioavail-ability, *Int. J. Clin. Pharmacol. Ther.,* 36, 126, 1998.

508. He, K., Iyer, K.R., Hayes, R.N., Sinz, M.W., Woolf, T.F., and Hollenberg, P.F., Inactivation of cytochrome P450 3A4 by bergamottin, a component of grapefruit juice, *Chem. Res. Toxicol.,* 11, 252, 1998.

509. Schmiedlin-Ren, P., Edwards, D.J., Fitzsimmons, M.E., He, K., Lown, K.S., Woster, P.M., Rahman, A., Thummel, K.E., Fisher, J.M., Hollenberg, P.F., and Watkins, P.B., Mechanisms of enhanced oral availability of CYP3A4 substrates by grapefruit constituents. Decreased enterocyte CYP3A4 concentration and mechanism-based inactivation by furanocoumarins, *Drug Metab. Dispos.,* 25, 1228, 1997.

510. Edwards, D.J., Bellevue, F.H., and Woster, P.M., Identification of 6′,7′-dihydroxybergamottin, a cytochrome P450 inhibitor, in grapefruit juice, *Drug Metab. Dispos.,* 24, 1287, 1996.

511. Boots, A.W., Haenen, G.R., den Hartog, G.J., and Bast, A., Oxidative damage shifts from lipid peroxidation to thiol arylation by catechol-containing antioxidants, *Biochim. Biophys. Acta,* 1583, 279, 2002.

512. van Zanden, J.J., Ben Hamman, O., van Iersel, M.L., Boeren, S., Cnubben, N.H., Lo, B.M., Vervoort, J., van Bladeren, P.J., and Rietjens, I.M., Inhibition of human glutathione *S*-transferase P1-1 by the flavonoid quercetin, *Chem. Biol. Interact.,* 145, 139, 2003.

513. Monks, T.J. and Jones, D.C., The metabolism and toxicity of quinones, quinonimines, quinone methides, and quinone-thioethers, *Curr. Drug Metab.,* 3, 425, 2002.

514. Monks, T.J., Hanzlik, R.P., Cohen, G.M., Ross, D., and Graham, D.G., Contemporary issues in toxicology. Quinone chemistry and toxicity, *Toxicol. Appl. Pharmacol.,* 112, 2, 1992.

515. Hagiwara, A., Hirose, M., Takahashi, S., Ogawa, K., Shirai, T., and Ito, N., Forestomach and kidney carcinogenicity of caffeic acid in F344 rats and C57BL/6N × C3H/HeN F_1 mice, *Cancer Res.,* 51, 5655, 1991.

516. Nakamura, Y., Torikai, K., and Ohigashi, H., Toxic dose of a simple phenolic antioxidant, protocatechuic acid, attenuates the glutathione level in ICR mouse liver and kidney, *J. Agric. Food Chem.,* 49, 5674, 2001.

517. Awad, H.M., Boersma, M.G., Boeren, S., van Bladeren, P.J., Vervoort, J., and Rietjens, I.M., Structure–activity study on the quinone/quinone methide chemistry of flavonoids, *Chem. Res. Toxicol.,* 14, 398, 2001.

518. Awad, H.M., Boersma, M.G., Boeren, S., van der, W.H., van Zanden, J., van Bladeren, P.J., Vervoort, J., and Rietjens, I.M., Identification of *o*-quinone/quinone methide metabolites of quercetin in a cellular *in vitro* system, *FEBS Lett.,* 520, 30, 2002.

519. Boots, A.W., Kubben, N., Haenen, G.R., and Bast, A., Oxidized quercetin reacts with thiols rather than with ascorbate: implication for quercetin supplementation, *Biochem. Biophys. Res. Commun.*, 308, 560, 2003.

520. Gaitan, E., Lindsay, R.H., Reichert, R.D., Ingbar, S.H., Cooksey, R.C., Legan, J., Meydrech, E.F., Hill, J., and Kubota, K., Antithyroid and goitrogenic effects of millet: role of *C*-glycosyl-flavones, *J. Clin. Endocrinol. Metab.*, 68, 707, 1989.

521. van Erp-Baart, M.A., Brants, H.A., Kiely, M., Mulligan, A., Turrini, A., Sermoneta, C., Kilkkinen, A., and Valsta, L.M., Isoflavone intake in four different European countries: the VENUS approach, *Br. J. Nutr.*, 89 (Suppl. 1), S25, 2003.

522. Akiyama, T., Ishida, J., Nakagawa, S., Ogawara, H., Watanabe, S., Itoh, N., Shibuya, M., and Fukami, Y., Genistein, a specific inhibitor of tyrosine-specific protein kinases, *J. Biol. Chem.*, 262, 5592, 1987.

523. Deutsch, J.C. and SanthoshKumar, C.R., Dehydroascorbic acid undergoes hydrolysis on solubilization which can be reversed with mercaptoethanol, *J. Chromatogr. A*, 746, 300, 1996.

524. Deutsch, J.C., Determination of *p*-hydroxyphenylpyruvate, *p*-hydroxyphenyllactate and tyrosine in normal human plasma by gas chromatography–mass spectrometry isotope-dilution assay, *J. Chromatogr. B*, 690, 1, 1997.

525. Heinecke, J.W., Tyrosyl radical production by myeloperoxidase: a phagocyte pathway for lipid peroxidation and dityrosine cross-linking of proteins, *Toxicology*, 177, 11, 2002.

526. Gregory, J.R., Collins, D.L., Davies, P.S.W., Hughes, J.M., and Clarke, P.C., *National Diet and Nutrition Survey. Children Aged One and a Half Years to Four and a Half Years*, HMSO, London, 1995.

527. Gregory, J., Lowe, S., Bates, C.J., Prentice, A., Jackson, L.V., Smithers, G., Wenlock, R., and Farron, M., *National Diet and Nutrition Survey. Young People Aged 4 to 18 Years*, The Stationery Office, London, 2000.

528. Halliwell, B. and Gutteridge, J.M., *Free Radicals in Biology and Medicine*, 3th edition, Oxford University Press, Oxford, 1999.

529. Jovanovic, S.V., Hara, Y., Steenken, S., and Simic, M.G., Antioxidant potential of theaflavins. A pulse radiolysis study, *J. Am. Chem. Soc.*, 119, 5337, 1997.

530. Buettner, G.R., The pecking order of free radicals and antioxidants: lipid peroxidation, alpha-tocopherol, and ascorbate, *Arch. Biochem. Biophys.*, 300, 535, 1993.

531. Jovanovic, S.V., Steenken, S., Boone, C.W., and Simic, M.G., H-atom transfer is a preferred antioxidant mechanism of curcumin, *J. Am. Chem. Soc.*, 121, 9677, 1999.

532. Steenken, S. and Neta, P., Electron transfer rates and equilibria between substituted phenoxide ions and phenoxyl radicals, *J. Phys. Chem.*, 83, 1134, 1979.

533. Jovanovic, S.V., Tosic, M., and Simic, M.G., Use of Hammett correlation and σ^+ for calculation of one-elctron redox potentials of antioxidants, *J. Phys. Chem.*, 95, 10824, 1991.

534. Koppenol, W.H., Oxyradical reactions: from bond-dissociation energies to reduction potentials, *FEBS Lett.*, 264, 165, 1990.

535. Jovanovic, S.V., Jankovic, I., and Josimovic, L., Electron transfer reactions of alkyl peroxy radicals, *J. Am. Chem. Soc.*, 114, 9018, 1992.

536. Jones, D.P., Carlson, J.L., Mody, V.C., Cai, J., Lynn, M.J., and Sternberg, P., Redox state of glutathione in human plasma, *Free Radical Biol. Med.*, 28, 625, 2000.

7 Isoflavonoids and Human Health

Helen Wiseman

CONTENTS

7.1 INTRODUCTION

Isoflavonoid phytoestrogens such as the soy isoflavones genistein and daidzein are plant-derived nonsteroidal estrogen mimics, often referred to as phytoestrogens (other phytoestrogens include lignans such as secoisolariciresinol, coumestans such as coumestrol, and prenyl-flavonoids such as 8-prenylnaringenin; see Figure 7.1), that are extensively investigated to determine their potential, particularly in the protection of human health.[1–9]

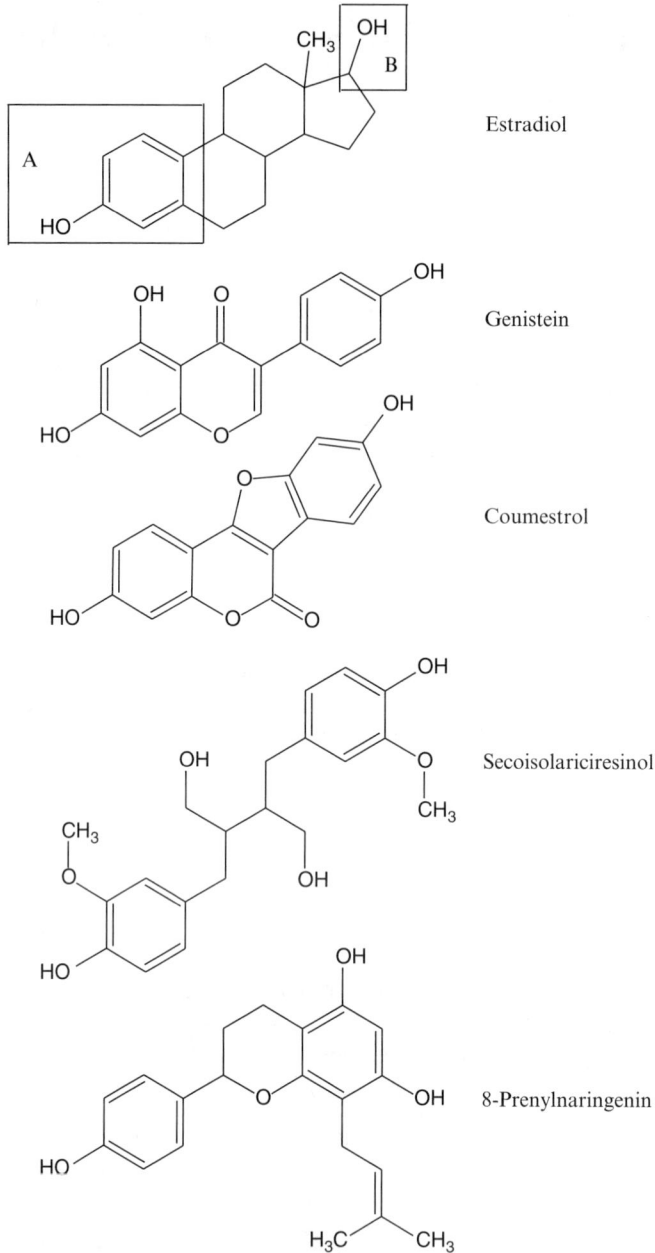

FIGURE 7.1 The structural relationship between phytoestrogens and 17β-estradiol.

7.2 ISOFLAVONOIDS: DIETARY SOURCES AND INTAKES, METABOLISM, AND BIOAVAILABILITY

7.2.1 DIETARY SOURCES AND INTAKES

Isoflavonoids include the isoflavones genistein and daidzein, which occur mainly as the glycosides genistin and daidzin (see Figure 7.2), respectively, in soybeans and consequently in a wide range of soy-derived foods and to a lesser extent in other legumes.[10–12] Traditional

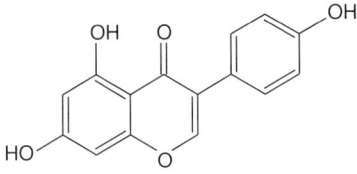

Genistein

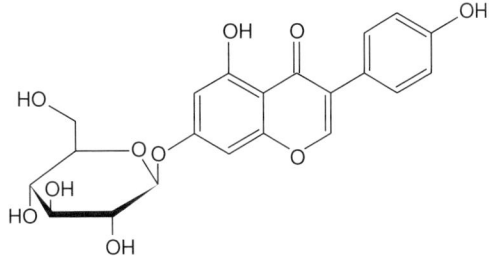

Genistin (genistein glucoside)

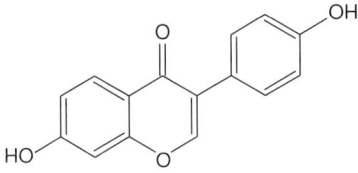

Daidzein

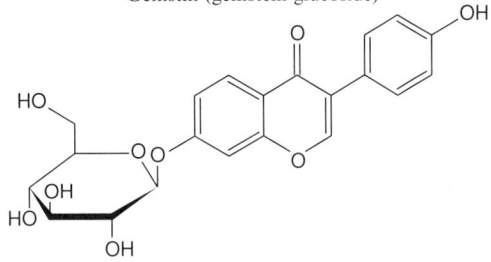

Daidzin (daidzein glucoside)

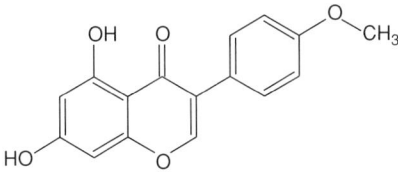

Biochanin A

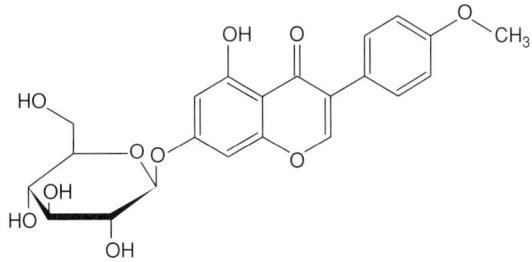

Ononin (biochanin A glucoside)

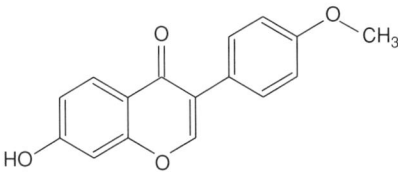

Formononetin

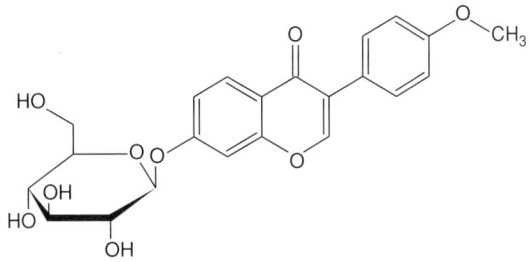

Sissotrin (formononetin glucoside)

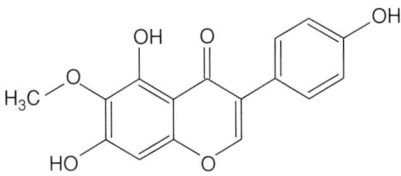

Glycitein

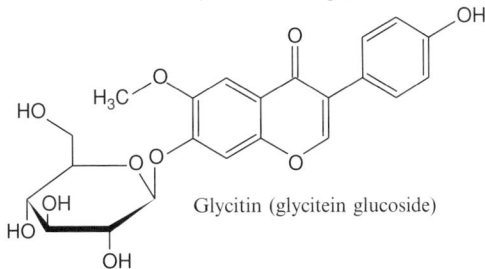

Glycitin (glycitein glucoside)

FIGURE 7.2 Chemical structures of the main soy (genistein and daidzein) and red clover (biochanin A and formononetin) isoflavonoids (aglycones and glucosides).

soy foods are made from soy beans and include both fermented and nonfermented foods. Nonfermented soy foods contain isoflavones mostly present as β-glucosides, some of which are esterified with malonic acid or acetic acid. Fermented soy foods such as miso or tempeh contain mostly unconjugated isoflavones.[13] Some alcoholic beverages such as beer contain significant amounts of isoflavones.[14] The isoflavone aglycone and glucoconjugate content of high- and low-soy UK foods used in nutritional studies has been reported.[11] Soybeans (774 mg/kg isoflavones) and soybean-containing foods had the highest isoflavone content of the foods examined. The low-soy foods all contained very low concentrations (<8 mg/kg) of the isoflavone aglycones and glucoconjugates.[11]

Dietary exposure estimates of isoflavones have been made from the 1998 UK Total Diet Study.[15] Each Total Diet Study group consisted of composite samples, one for each of the 20 designated food groups. The composite samples were taken to represent the average consumption of all the individual food elements in each group, processed in the form that they were consumed in. In the Total Diet Survey, individual composites of the bread, processed meat, and fish food groups contained >5 mg/kg of the individual isoflavones, daidzein, genistein, and glycitein. In addition, individual composites from the groups, miscellaneous cereals, other vegetables, fruit products, and nuts contained >1 mg/kg of isoflavones. The Total Diet Survey sample collection model for the average adult consumer, following weighting for average consumption of food from each Total Diet Survey group, gave an estimated daily intake of 3 mg/day of combined isoflavone aglycones.[15] This indicates that the UK dietary intake of isoflavone phytoestrogens is higher than previously estimated and this is likely to be due in part to the extensive use of soy products in processed foods.[15] An investigation into the intake of dietary phytoestrogens by Dutch women indicates that pea, nuts, grain products, and soy products were the main sources of isoflavones; and also, despite their low concentrations of isoflavones compared to soy foods, coffee and tea, because of their high levels of consumption.[16] The main sources of lignans were grain products, fruit and alcoholic beverages, and in this population the phytoestrogen intake consisted largely of lignans.[16] Generally, in the diets of Europeans, food sources of lignans are more widely consumed than those of isoflavones and lignans may be the more important dietary source of phytoestrogens.[17]

Estimating dietary intakes of isoflavone phytoestrogens can be difficult because of inadequate information regarding the phytoestrogen contents of foods. The validation of a newly constructed phytoestrogen database containing more than 600 values given to foods, by using duplicate diet analysis together with measurements of isoflavone concentrations in urine and plasma, has shown that the 24-h urinary excretion and timed plasma concentrations of genistein and daidzein can be used as biomarkers of intake.[18] However, it should be noted that the considerable variation found in the isoflavone content of isolated soy proteins used in food manufacture and in commercial milks exposes the limitations of using food databases for estimating daily isoflavone intakes.[19]

7.2.2 METABOLISM AND BIOAVAILABILITY

The metabolism and bioavailability of isoflavonoids is likely to be of crucial importance to their ability to help protect human health against disease.[3,20,21] Many studies have been published on the metabolism and bioavailability of isoflavones in adults.[22–25] The metabolism of isoflavones is of particular interest because the potency of isoflavone metabolites differs from that of the parent compounds.[26] The daidzein metabolite equol is three times as potent as is daidzein in an endometrial tumor line. Equol is also a more potent antioxidant *in vitro* (see Sections 7.3.5 and 7.4.2)[27–30] and the clinical significance of the ability to form equol has been considered in depth.[20]

Daidzin and genistin (and to a lesser extent glycitin) and their acetylglucosides and malonylglucosides are the predominant isoflavone forms in soy foods.[11] After ingestion isoflavones are hydrolyzed by mammalian lactase phlorizin hydrolase,[21] which releases the aglycones daidzein, genistein, and glycitein. These may be absorbed or further metabolized by the gut microflora to metabolites, including the conversion of daidzein to the isoflavan equol or O-desmethylangolensin (O-DMA), and the conversion of genistein to p-ethyl phenol. More recently, 4-hydroxyphenyl-2-propionic acid was identified from the metabolism of genistein in rats.[31] Studies have shown that particular bacterial groups are involved in the metabolism of the isoflavone glycosides.[32]

Unconjugated isoflavones are absorbed quickly from the upper small intestine in rats,[33] whereas the glycoside conjugates are absorbed more slowly, which is consistent with their hydrolysis at more distal sites in the intestine to the unconjugated isoflavones.[34] Isoflavonoids, once absorbed, are rapidly converted to their β-glucuronides[33] and sulfate ester conjugates.[35] These biological conjugates circulate in the plasma and are excreted in the urine and feces. Indeed, intact sulfate and glucuronide isoflavone conjugates have been measured in human urine, following soy consumption.[36] Concentrations of the free aglycones of up to 22% of genistein and 18% of daidzein were observed, and the average pattern of daidzein conjugates was 54% 7-glucuronide, 25% 4′-glucuronide, 13% monosulfates, 7% free daidzein, 0.9% sulfoglucuronides, 0.4% diglucuronide, and 0.1% disulfate.[36]

Plasma isoflavonoid levels in Japanese and Finnish men have been measured and the means of the total daidzein, genistein, O-DMA, and equol levels were approximately 17-, 44-, 33-, and 55-fold higher for the Japanese subjects compared to the Finnish ones.[37] In postmenopausal Australian women following consumption of soy flour, mean plasma levels of daidzein and equol of 68 and 31 ng/ml, respectively, were observed.[38] Interestingly, only 33% of subjects were able to metabolize daidzein to equol.[38] Increased concentrations of both genistein and daidzein were observed in male subjects following ingestion of a cake containing soy flour and cracked linseed, within 30 min of consumption, with maximum concentrations reached by 5.5 to 8.5 h.[39]

The importance of the gut microflora in the metabolism of isoflavones has been demonstrated. Antibiotic administration blocks isoflavone metabolism and germfree animals do not excrete metabolites.[40] Moreover, only germfree rats colonized with microflora from a good equol producer excrete equol when fed soy.[41]

Interindividual variation in ability to metabolize daidzein to equol (more estrogenic and a more potent antioxidant than daidzein) could thus influence the potential health protective effects of soy isoflavones.[20] The extent of gut microflora metabolism in humans is variable, approximately 35% of a Western population can produce equol.[23,38,42]

A variable metabolic response to isoflavones has been shown for subjects following consumption of soy flour; urinary excretion concentrations of genistein, daidzein, equol, and O-DMA were increased 8-, 4-, 45-, and 66-fold, respectively, compared to baseline.[43] Considerable interindividual variation in metabolic response was reported with the peak levels of equol showing the most variation.[43]

In healthy young adults, when diets high or low in isoflavones (textured soy protein product containing 56 or 2 mg/day) were each consumed for 14 days separated by a 25-day washout period, considerable interindividual variation in metabolic response was found.[23] In addition, the good equol excretors (36% of subjects) consumed significantly less fat and more carbohydrate (also greater amounts of nonstarch polysaccharide, NSP) compared to the poor equol excretors.[23] Female equol excretors have been reported to consume a higher percentage of energy as carbohydrate and also greater amounts of plant protein and NSP than non-equol excretors.[42]

Dietary modification, such as feeding wheat bran or soy protein, has been unsuccessful at changing equol-producing capability,[44] which indicates that the intestinal microflora of an individual is relatively stable and resistant to change.

The bioavailability of soy isoflavones has been shown to depend on the gut microflora.[45] Metabolism by the gut microflora is an important factor influencing the disposition of chemicals in the gut and can result in activation of substances to more biologically active products. The presence of different populations of microflora in the human gut may influence the bioavailability of isoflavones. The identification of the bacterial species involved in the conversion of daidzein to equol is of considerable importance and very challenging given the large number of bacteria present in the colon and small intestine. In a study that identified equol producers by culturing fecal flora from healthy Japanese adults after they consumed 70 g of tofu, three strains of bacteria were reported to convert pure daidzein to equol *in vitro*: the gram-negative *Bacteroides ovatus* spp., the gram-positive *Streptococcus intermedius* spp., and *Ruminococcus productus* spp.[46]

Chronic consumption of a high-soy diet (104 mg isoflavones/day) has recently been compared with a low-soy diet (0.54 mg/day) in 76 healthy young adults; after the 10-week diet period, concentrations of the isoflavonoids in plasma, urine, and feces were significantly higher in the high-soy group than in the low-soy group.[47] Although interindividual variation in isoflavone metabolism was high (34% of subjects were good equol producers), intraindividual variation (assessed by comparing midpoint with endpoint results) in metabolism was low. Only concentrations of *O*-DMA in plasma and urine appeared to be influenced by sex, with men having significantly higher concentrations than women.[47] Other studies suggested that men and women respond differently to chronic exposure to isoflavones,[48–50] although there are considerable inconsistencies in the data. Furthermore, chronic soy isoflavone consumption did not appear to induce many significant changes to the gut metabolism of isoflavones, other than effects on β-glucosidase activity (this was significantly higher in subjects who consumed the high-soy diet than in those who consumed the low-soy diet), suggesting that the gut bacteria and enzymes responsible for equol or *O*-DMA production are not inducible.[47]

Further to the possible contribution that soy isoflavonoids may make to adult human health, the possible health consequences of early life soy exposure are also attracting attention.[51,52] Although there have been many studies on the metabolism and bioavailability of isoflavonoids in adults, there is little information available in infants and children. The gut microflora in early childhood is very different to that in adulthood; therefore, it is important to characterize developmental changes in isoflavone metabolism in early life. The development of the microflora occurs gradually and it can take several years before an adult-type flora is established. This has important implications for isoflavone metabolism.

The urinary excretion of isoflavonoids has recently been investigated in 60 infants and children (aged 4 months to 7 years, divided into four age groups). The study compared infants and children who had been fed soy-based formulas in early infancy, to control (cows'-milk formula-fed) infants and children.[53] Infants aged 4 to 6 months, fed soy-based infant formulas (but not those fed cows'-milk formulas) were found to excrete considerable amounts of genistein, daidzein, and glycitein in urine, indicating that these compounds are well absorbed and that the required glucosidase activity has developed.[53] The majority of the soy-based infant formula-fed infants and about one half of the cows'-milk formula-fed group (after a soy challenge) were capable of converting daidzein to *O*-DMA. However, conversion of daidzein to equol was observed in very few children, even in the oldest age group (3 to 7 years). These findings indicate that there appears to be no lasting effect of early-life isoflavone exposure on isoflavone metabolism and have important developmental implications for isoflavonoid bioavailability.[53]

7.3 ISOFLAVONOIDS AND CANCER PREVENTION

7.3.1 HORMONE-DEPENDENT CANCER PREVENTION BY ISOFLAVONOIDS

The possible role of isoflavonoids in the prevention of cancer and in particular hormone-dependent cancers such as breast and prostate cancer is currently extensively investigated.[6,8,54–57] In addition, consumption of soy foods rich in isoflavones has been weakly associated with reduced colon cancer.[54,58] Colon cancer risk is influenced by estrogen exposure; although the mechanism of action has not been fully elucidated, studies with estrogen receptor α (see Section 7.3.3) knockout mice indicate that it may be independent of estrogen receptor α.[58]

Breast and prostate cancer is much less prevalent in Far Eastern countries, where there is an abundance of soy phytoestrogens in the diet, compared to Western ones. Emigration of people from Pacific Rim countries to the United States has been shown to increase their risk of breast and prostate cancer. The increase in prostate cancer risk in men occurs in the same generation, whereas for women the increase in breast cancer risk is observed in the next generation.[59] These changes in breast and prostate cancer risk have been mostly attributed to changes in diet, in particular the switch to a low-soy Western diet: in countries such as Japan, Korea, China, and Taiwan, the mean daily intake of soy products has been estimated to be in the range 10 to 50 g compared to only 1 to 3 g in the United States.[60] Increased soy intake has been associated with a lowered risk of breast cancer in two out of four epidemiological studies that examined a wide range of dietary components in relation to breast cancer risk: no significant effect was observed in the other two studies.[1,60]

High urinary excretion of both equol and enterolactone (mammalian metabolite of plant lignans) has been found to be associated with a significant decrease in breast cancer risk in an epidemiological case–control study in breast cancer patients.[61] Although this could suggest the possible importance of isoflavonoid and lignan metabolism in decreased breast cancer risk, the phytoestrogen excretion observed may just be a marker of dietary differences.[1]

The possible protective effect of isoflavonoids against prostate cancer has recently been reviewed[56] and it is of particular interest that equol may be a novel antiandrogen that inhibits prostate growth and hormone feedback in rat studies.[62] The role of isoflavonoids in the prevention of breast cancer is, however, the main focus of the next section of this chapter.

7.3.2 ESTROGENS AND RISK OF BREAST CANCER

Breast cancer is still a major cause of death for women in Western countries.[63] Breast cancer is thought to have many causes, ranging from gene profile to diet and lifestyle and mutations in particular tumor suppressor genes such as BRCA1, BRCA2, and p53 are likely to be of particular importance. A ribonucleotide reductase gene (p53R2) has been shown to be directly involved in the p53-dependent cell cycle checkpoint for DNA damage, thus clarifying the relationship between a ribonucleotide reductase activity involved in repair of damaged DNA and tumor suppression by p53.[64] Furthermore, the structural basis has been established for the recognition and repair by 8-hydroxyguanine DNA glycosylase of the oxidative DNA base damage product (and endogenous mutagen) 8-hydroxyguanine.[65]

The role of endogenous estrogens in breast cancer risk is widely recognized. Different forms of estrogen metabolism result in the formation of mitogenic endogenous estrogens or the metabolic activation of estrogens that can result in carcinogenic free-radical-mediated DNA damage. Pregnancy appears to be important in breast cancer risk: women who have never been through pregnancy have the greatest risk of breast cancer. For those women who have been through pregnancy, multipregnancies are of no greater benefit than a single pregnancy and a pregnancy earlier in life is more protective than one later in life. This is because of the differentiation of breast epithelial cells into milk-producing cells that occurs in the breast during pregnancy and the apoptosis of breast cells that occurs in the breast following pregnancy, thus providing a chance for elimination of mutated epithelial cells.[1]

7.3.3 Estrogen Receptor Mediated Events

Estrogens play a vital role in the growth, development, and homeostasis of estrogen responsive tissues. The estrogen receptor mediates the biological activity of estrogens and is a ligand-inducible nuclear transcription factor: estrogen binds to the ligand-binding domain of the estrogen receptor resulting in either the activation or repression of target genes.[66,67] The selective estrogen receptor antagonist raloxifene, structurally related to the anticancer drug tamoxifen,[63] can inhibit the mitogenic effects of estrogen in reproductive tissues, while maintaining the beneficial effects of estrogen in other tissues. The crystal structures of the ligand-binding domain of the estrogen receptor complexed to either 17β-estradiol or to raloxifene have been reported,[68] thus providing structural evidence for the mechanisms of estrogen receptor agonism and antagonism. A combination of specific polar and nonpolar interactions enables the estrogen receptor to selectively recognize and bind 17β-estradiol with great affinity. The estrogen receptor is the only steroid receptor able to additionally interact with a large number of nonsteroidal compounds, which frequently show a structural similarity to the steroid nucleus of estrogen, including phytoestrogens, and drug and environmental xenestrogens such as dioxins.[67] In particular, a phenolic ring analogous to ring A in estradiol is required (see Figure 7.1) and these structural features enable them to bind to estrogen receptors to elicit responses ranging from agonism to antagonism of the endogenous hormone ligand.[69]

Originally, it was accepted that only one estrogen receptor existed (the classical estrogen receptor, ERα). This is in contrast to other members of the nuclear receptor superfamily where multiple forms have been found. A separate subtype, termed estrogen receptor β (ERβ), has subsequently been identified in cDNA libraries from rat prostate and ovary tissues.[70] ERβ shows a different tissue distribution from ERα. ERβ was first reported to be strongly expressed in ovary, uterus, brain, bladder, testis, prostate, and lung.[71] Expression of ERβ appears to occur at different sites in the brain from ERα.[71] Evidence has since been found, using reverse-transcription polymerase chain reaction, for the presence of ERβ in normal human breast tissue[72] and ERβ has been shown to be highly expressed in rat breast tissue using specific antibodies.[73] ERβ has also been found to be expressed in both bone[74–76] and the cardiovascular system.[77]

7.3.4 Animal Models

Studies using animal models provided the initial experimental evidence that soy can prevent breast cancer.[1,60] Results from 26 animal studies of experimental carcinogenesis have shown that in 17 of these studies (65%) protective effects were reported: the risk of cancer (incidence, latency, or tumor number) was greatly reduced, and no studies reported that soy intake increased tumor development.[60] In a rat model of breast cancer (7,12-dimethylbenz[a]anthracene (DMBA) induced), genistein administered in high doses by injection to young animals suppressed the number of mammary tumors observed over a 6-month period by 50% and delayed the appearance of the tumors,[78] indicating the likely importance of the timing of exposure to the protective components of soy. In later studies, similar levels of protection were achieved by adding genistein to the feed (0.25 g/kg) given to the mother such that the offspring were exposed to dietary genistein from conception to day 21 postpartum.[79] By contrast, when pregnant female rats were treated daily with subcutaneous injections of genistein (doses given were 20, 100, or 300 μg/day) between days 15 and 20 of gestation, this *in utero* exposure was found to dose-dependently increase the incidence of DMBA-induced mammary tumors in female offspring.[80] However, when prepubertal rats were treated with 20 μg of genistein (~1 mg/kg body weight) between postnatal days 7 and 20, this greatly reduced the multiplicity but not the incidence of DMBA-induced mammary tumors and 60% of the tumors that did occur were not malignant offspring.[81] Furthermore, injection of prepubertal rats with genistein (500 μg/g body weight) or estradiol benzoate (500 ng/g body weight) on days 16, 18, and 20 showed that both treatments resulted in significantly increased mammary gland terminal end buds and increased ductal branching compared to controls, indicating an ER-dependent action of genistein in mammary gland proliferation and differentiation, which could be protective against mammary cancer.[82] Overall, these results indicate that genistein has very complex effects on carcinogen-induced mammary cancer in the rat model and great care is required in interpreting these results and drawing parallels with human breast cancer.

Biochanin A, found in certain subterranean clovers, including red clover, and converted to genistein by demethylation in the liver in addition to in the breast, was a good anticancer agent when administered following the carcinogen,[83] suggesting the benefits of isoflavones other than genistein may not be solely restricted to early life exposure. The fermented soy food miso (contains mostly unconjugated isoflavones) and tamoxifen acted together to cause an additive reduction in the number of mammary tumors in the rat model[84] and this may be of considerable importance for women on standard tamoxifen therapy.[63]

In the mouse model of breast cancer (tamoxifen is estrogenic in this model), maternal genistein exposure (pregnant mice injected with 20 μg/day between days 15 and 20 of gestation) resulted in similar effects to that of estrogen on mammary gland development,[85] and further studies are needed to determine whether these estrogenic changes could lead to an increased risk of breast tumors. In addition, when human breast cancer cells (MCF-7) are grown orthotopically in ovariectomized rats, addition of genistein to their diets resulted in an increase in the growth of the cancer cells.[86] This could have important implications (in relation to tumor reoccurrence) for the consumption of phytoestrogens including isoflavonoids by women who have had their ovaries removed following a diagnosis of breast cancer. But as the removal of the ovaries will greatly reduce the growth of breast cancer cells, then any increase in risk caused by dietary phytoestrogens would probably be less than if the ovaries remained intact.[1]

7.3.5 MECHANISMS OF ANTICANCER ACTION OF ISOFLAVONOIDS

Genistein is a potent and specific *in vitro* inhibitor of tyrosine kinase action in the autophosphorylation of the epidermal growth factor (EGF) receptor[87] and is thus frequently used as a pharmacological tool. The EGF receptor is overexpressed in many cancers, in particular those with the greatest ability for metastasis[88] and it has therefore often been assumed that some of the anticancer effects of genistein are mediated via inhibition of tyrosine kinase activity; however, this is likely to be an oversimplification of the true *in vivo* situation.[1]

Isoflavonoids have biphasic effects on the proliferation of breast cancer cells in culture; at concentrations greater than $5\,\mu M$, genistein exhibits a concentration-dependent ability to inhibit both growth factor-stimulated and estrogen-stimulated (reversed by 17β-estradiol) cell proliferation.[89] Genistein at low concentrations can, in the absence of any estrogens, stimulate the growth of estrogen receptor-positive MCF-7 cells.[90,91] Genistein does not, however, stimulate the growth of estrogen receptor-negative breast cancer cells,[92,93] it only inhibits cell proliferation in these cell lines.[92] Equol is a much more potent stimulator than daidzein of the expression of estrogen-specific genes.[94] It is of great interest that phytoestrogen-responsive genes (PE-13.1 and pRDA-D) have been identified and characterized from MCF-7 cells and it may be possible to use these as molecular markers in elucidating the role phytoestrogens, including isoflavonoids, play in cancer prevention.[95]

Although genistein is a much better ligand for ERβ than for the ERα (20-fold higher binding affinity),[71] it can also act as an estrogen agonist via both ERα and ERβ in some test systems.[96,97] However, genistein also behaves as a partial estrogen agonist in human kidney cells transiently expressing ERβ, suggesting that it may be a partial estrogen antagonist in some cells expressing ERβ.[98] Furthermore, although genistein binds to the ligand binding domain of ERβ in a manner similar to that observed for17β-estradiol, in the ERβ–genistein complex the AF-2 helix (H12) does not adopt the normal agonist type position, but instead takes up a similar orientation to that induced by ER antagonists such as raloxifene.[99] This suboptimal alignment of the transactivation helix is in keeping with the reported partial agonist activity of genistein via ERβ.[98]

Mechanisms other than those involving estrogen receptors are also likely to be involved in the inhibition of cell proliferation by genistein. This is because genistein inhibits both the EGF-stimulated as well as the 17β-estradiol-stimulated growth of MCF-7 cells.[87] It has been suggested that the inhibitory action of genistein on cell proliferation involved effects on the autophosphorylation of the EGF receptor in membranes and demonstrated in membranes isolated from cells.[87] Although studies have shown that exposure to genistein can reduce the tyrosine phosphorylation of cell proteins in whole cell lysates, studies using cultured human breast and prostate cancer cells, have not, however, confirmed that genistein has a direct effect on the autophosphorylation of the EGF receptor.[100] Furthermore, *in vivo* studies in male rats have shown that genistein decreases the amount of EGF receptor present in the prostate, indicating that the observed decrease in tyrosine phosphorylation may be only a secondary effect of the influence of genistein on the expression or turnover of EGF receptor.[1,101]

Many other mechanisms of action for isoflavonoids and genistein in particular have been suggested. These include inhibition of DNA topoisomerases,[102] inhibition of cell cycle progression,[103] inhibition of angiogenesis,[104,105] tumor invasiveness,[106] inhibition of enzymes involved in estrogen biosynthesis,[107] effects on the expression of DNA transcription factors c-fos and c-jun[108] and on transforming growth factor-β (TGF-β).[103,109] Effects have also been reported on reactive oxygen species,[28,110] oxidative membrane damage,[30] membrane rigidity,[111] similar to those found previously to contribute to the antioxidant action of tamoxifen,[63,112,113] and oxidative damage *in vivo*.[114]

Antioxidant properties have been reported for isoflavones both *in vitro* and *in vivo*.[28,30,110,114] Equol, in model membrane systems, was a more effective antioxidant than genistein or the parent compound daidzein[30] and shows structural similarity to the tocopherols.[1] Daidzein and geinstein showed antioxidant action in primary and cancer lymphocytes (Jurkat cells), both isoflavones increased DNA protection against oxidative damage and decreased lipid peroxidation.[115] Moreover, a protective effect was achieved at concentrations that can be achieved in plasma following soy consumption. An important aspect of cancer risk is the involvement of the inflammatory response, which involves the production of cytokines and proinflammatory oxidants such as the hypochlorous acid produced by neutrophils and peroxynitrite by macrophages, which react with phenolic tyrosine residues on proteins to form chloro- and nitrotyrosine.[116] It has been reported that neutrophil myeloperoxidase chlorinates and nitrates isoflavones and enhances their antioxidant properties, thus soy isoflavones may have potentially protective benefits at sites of inflammation.[116,117] Antioxidant action could also contribute to anticancer ability because reactive oxygen species could initiate signal transduction through the mitogen activated protein (MAP) kinases.[1,118]

There have been a number of reports relating to the possible antioxidant effects of isoflavone consumption. Soy isoflavone consumption as a soy protein burger (56 mg isoflavones/day for 17 days) decreased plasma F_2-isoprostane concentrations in healthy young adults.[114] Consumption of a soy isoflavone supplement (50 mg isoflavones, twice a day for 3 weeks) decreased a biomarker of DNA oxidative damage (white cell 5-hydroxy-methyl-2′-deoxyuridine concentrations) but did not alter plasma F_2-isoprostane concentrations.[119] Furthermore, consumption of soy protein (110 mg isoflavones/day for 4 weeks) decreased plasma peroxide concentrations and increased total antioxidant status but did not effect a biomarker of oxidative DNA damage (urinary 8-hydroxy-2-deoxyguanosine concentrations).[120] It is of considerable interest that widely differing effects in relation to the potential benefits to human health are frequently reported for isoflavones consumed within the food matrix in soy foods, compared to those consumed in capsule or tablet form as dietary supplements (see Section 7.4.2).

Angiogenesis, the formation of new blood vessels, is normally an important process involved in productive function, development, and wound repair. Disease states, however, often involve persistent and unregulated angiogenesis. The growth and metastasis of tumors is dependent on angiogenesis. Genistein is a potent inhibitor of angiogenesis *in vitro*[104] and thus could have therapeutic applications in the treatment of chronic neovascular diseases including solid tumor growth[121] and inhibition of neovascularization of the eye by genistein has been reported.[105] Recently, novel molecular targets for the inhibition of angiogenesis by genistein have been discovered including tissue factor, endostatin, and angiostatin.[122]

Genistein may enhance the action of transforming growth factor-β (TGFβ).[103,109] This action may be a link between the effects of genistein in a variety of chronic diseases,[1] including atherosclerosis and hereditary hemorrhagic telangiectasia (the Osler–Weber–Rendu syndrome) in which defects in TGFβ have been characterized.[123]

7.3.6 CLINICAL STUDIES

Few studies have yet been reported on use of phytoestrogens as preventative agents for breast cancer. In one study, an isolated soy protein beverage (42 mg genistein and 27 mg daidzein) was administered daily for 6 months to healthy pre- and postmenopausal women and breast cancer risk factors were measured in nipple aspirate fluid (NAF).[124] Although no change in NAF was observed in postmenopausal women, premenopausal women showed an increase in NAF volume, which persisted even after treatment ended and indicates the isoflavones were having an undesirable estrogenic effect in the premenopausal women.[124] This provides some cause for concern that risk of premenopausal breast cancer may actually be enhanced by phytoestrogens,[1] although further studies are needed. Furthermore, in a study of 84 normal premenopausal women, consumption of a soy supplement (60 g soy, 45 mg total isoflavones) for 14 days resulted in a weak estrogenic response in the breast: nipple aspirate levels of apolipoprotein D were significantly lowered and pS2 levels were significantly raised.[125]

Mammographic breast density has been consistently associated with risk for breast cancer. A review of case–control studies showed odds ratios for breast cancer in women with the highest versus the lowest mammographic breast density ranged from 2.1 to 6.0.[126] Although the reasons for this are not fully understood it is possible that breast density acts as a biomarker for the past and current reproductive and hormonal events that influence breast cancer risk.[127] Mammographic breast density can thus also be used as biomarker of estrogenic or antiestrogenic effects of a particular treatment on breast tissue.[128] Consumption of a dietary supplement that provided red clover-derived isoflavones (26 mg biochanin A, 16 mg formononetin, 1 mg genistein, and 0.5 mg daidzein) for 12 months did not increase mammographic breast density in postmenopausal women, suggesting neither estrogenic nor antiestrogenic effects, of this supplement at the dose given, on the breast.[128]

7.4 PROTECTION BY ISOFLAVONOIDS AGAINST CARDIOVASCULAR DISEASE

7.4.1 CHOLESTEROL-LOWERING AND ISOFLAVONOIDS

Estrogen administration in postmenopausal women has been observed to produce cardio-protective benefits. The exact biomolecular mechanisms for this cardioprotection are unclear but it is likely that actions mediated both through the estrogen receptors, such as the beneficial alteration in lipid profiles and upregulation of the low-density lipoprotein (LDL) receptor, and independently of the estrogen receptors, such as antioxidant action, contribute to the observed cardioprotective effects of estrogens.

Lower incidence of heart disease has also been reported in populations consuming large amounts of soy products. Lowering of cholesterol is probably the best-documented cardioprotective effect of soy.[129,130] Soy protein incorporated into a low-fat diet can reduce cholesterol and LDL-cholesterol concentrations and the soy isoflavones are likely to contribute to these effects.[131] Soy isoflavones have been reported to improve cardiovascular risk factors in peripubertal rhesus monkeys,[132,133] and inflammatory markers in atherosclerotic, ovariectomized monkeys.[134] The potential role of phytoestrogens, including isoflavonoids, as cardioprotective agents has been extensively reviewed.[4,5,135]

A recent meta-analysis of eight randomized controlled trials of soy protein consumption in humans has found that with identical soy protein intake, high isoflavone intake led to significantly greater decreases in serum LDL cholesterol than low isoflavone intake, indicating that isoflavones have LDL-cholesterol-lowering effects that are independent of the soy protein.[136]

Consumption of soy protein (40 g/day providing either 56 mg isoflavones/day or 90 mg isoflavones/day) or cesin and nonfat dry milk (40 g/day) by postmenopausal women for 6 months showed a significant decrease in non-high-density lipoprotein (HDL) cholesterol and a significant increase in mononuclear cell LDL receptor mRNA and HDL cholesterol in both of the soy isoflavone groups compared to the control group.[137] Indeed, consumption of soy protein (20 g/day containing 80 mg isoflavones/day for 5 weeks) in high-risk middle-aged men (45 to 59 years of age) in Scotland significantly decreased non-HDL cholesterol and blood pressure, compared to the control treatment.[138]

However, studies in hypercholesterolemic subjects, using soy protein depleted of isoflavones have shown that soy protein independently of isoflavones can favorably affect LDL size, LDL particle distribution was shifted to a less atherogenic pattern,[139] and can decrease triglyceride concentrations, triglyceride fatty acid fractional synthesis rate, and cholesterol concentrations.[140]

By contrast, a meta-analysis of randomized controlled trials indicates that consumption of isolated isoflavones did not appear to have any significant effect on serum cholesterol, suggesting further studies investigating possible interactions of isoflavones with other components of soy protein are needed.[141] Indeed, a 12-month intervention with red clover-derived isoflavones (26 mg biochanin A, 16 mg formononetin, 1 mg genistein, and 0.5 mg daidzein) administered in the form of a dietary supplement found only modest protective benefits (decreases in triglycerides and plasminogen activator inhibitor type 1) in perimenopausal women.[142] This could, however, relate to the use of isolated isoflavones consumed as a dietary supplement rather than in soy protein — see Section 7.4.2. This study also found interactions between the apolipoprotein E (apoE) genotype and treatment tended to be significant for changes in total and LDL cholesterol in 49- to 65-year-old women, with isoflavone treatment potentially beneficial.[142] ApoE is an important factor influencing blood lipid profiles and the women were genotyped for polymorphisms in the gene encoding apoE to determine potential gene-treatment interactions.[142]

It is of related interest that a cross-sectional study in 301 postmenopausal women (60 to 75 years of age) living in the Netherlands, where isoflavone and lignan intakes were assessed with a food frequency questionnaire covering habitual diet during the year prior to the study, reported that high intakes of isoflavones were associated with lower levels of the atherogenic lipoprotein Lp(a) but had little effect on plasma lipids (total cholesterol, LDL, and HDL cholesterol and triglycerides), suggesting that at low levels of intake dietary isoflavones have a limited effect on plasma lipids.[143]

7.4.2 ANTIOXIDANT ACTION

Antioxidant action is one of the mechanisms that may contribute to the cardiovascular protective effects of soy and soy isoflavones. Antioxidant properties have been reported for isoflavones both *in vitro* and *in vivo* (see Section 7.3.5).

The oxidation hypothesis of atherosclerosis states that the oxidative modification of LDL (or other lipoproteins) is important and possibly obligatory in the pathogenesis of the atherosclerotic lesion; thus, it has been suggested that inhibiting the oxidation of LDL will decrease or prevent atherosclerosis and clinical sequelae.[144] LDL oxidation also has important implications for vascular health function. High concentrations of LDL may inhibit arterial function in terms of the release of nitric oxide from the endothelium and many of these effects are mediated by lipid oxidation products.[145] Furthermore, oxidized LDL inhibits endothelium-dependent nitric oxide-mediated relaxations in isolated rabbit coronary arteries.[146] Oxidized LDL induces apoptosis in vascular cells including macrophages and this is prevented by nitric oxide.[147]

Isoprostanes are a relatively new class of lipids and are produced *in vivo* principally by a free radical-catalyzed peroxidation of polyunsaturated fatty acids.[148–150] Isoprostanes are isomers of the conventional enzymatically derived prostaglandins. F_2-isoprostanes are the most studied species and are isomers of the enzyme-derived prostaglandin $F_{2\alpha}$. F_2-isoprostanes are considered to provide a reliable biomarker for oxidative stress and resultant oxidative lipid damage *in vivo* because of their mechanism of formation, chemical stability, and specific structural features that enable them to be distinguished from other free radical-generated products. Increased concentrations of F_2-isoprostanes have been consistently reported in association with cardiovascular risk factors such as chronic cigarette smoking, diabetes mellitus, and hypercholesterolemia. Furthermore, some F_2-isoprostanes possess potent biological activities indicating that they may also act as mediators of the cellular effects of oxidative stress. Oxidative stress may also lead to raised blood pressure, another cardiovascular risk factor, possibly via effects on arterial function.[148–150]

The effect of dietary soy isoflavones on the F_2-isoprostane 8-*epi*-prostaglandin $F_{2\alpha}$ (8-*epi*-$PGF_{2\alpha}$ biomarker for *in vivo* lipid peroxidation) and on resistance of LDL to oxidation has been reported.[114] In a randomized crossover study in 24 young healthy male and female subjects, consuming diets that were rich in soy that was high (56 mg isoflavones/day: 35 mg genistein and 21 mg daidzein) or low in isoflavones (2 mg isoflavones/day), each for 2 weeks, plasma concentrations of the F_2-isoprostane 8-*epi*-$PGF_{2\alpha}$ were significantly lower after the high-isoflavone dietary treatment than after the low-isoflavone dietary treatment (326 ± 32 and 405 ± 50 ng/l, respectively, $P = 0.028$).[114] The lag time for copper-ion-induced LDL oxidation was longer (48 ± 2.4 and 44 ± 1.9 min, respectively, $P = 0.017$).[114]

This increased resistance of LDL to oxidation is in agreement with the findings of a number of studies with dietary soy, including the increase in lag time in a study in six young healthy male and female subjects who consumed three soy bars per day (providing 57 mg total isoflavones/day: 36 mg genistein and 21 mg daidzein) for 2 weeks.[151] Although HDL-derived 17β-estradiol fatty acid esters have been shown to accumulate in LDL *ex vivo*[152] and esterified isoflavones can also be incorporated into LDL *ex vivo*,[153] it has not yet been shown that isoflavones can be esterified to LDL *in vivo*. Increased resistance to LDL oxidation has also been reported in a 12-week single open-group dietary intervention with soy foods (60 mg isoflavones/day) in 42 normal postmenopausal women.[154] A randomized crossover study in 25 hyperlipidemic male and female subjects, consuming soy-based breakfast cereals (168 mg total isoflavones/day) and control breakfast cereals, each for 3 weeks, reported decreased oxidized LDL (total conjugated diene content) following consumption of the soy-based breakfast cereal compared to the control.[155]

In contrast to these clear antioxidant effects of dietary soy, which is likely to be mediated by soy isoflavones, when isoflavones are extracted from soy and used to make supplements, this appears to reduce their antioxidant efficacy, as indicated by a wide range of studies. A 4-month randomized crossover placebo-controlled study in 14 healthy premenopausal women using an isoflavone supplement to deliver 86 mg/day for two menstrual cycles reported no change in LDL lag time.[156] Furthermore, a number of studies investigating effects of soy isoflavone supplements on oxidative stress found no effect on F_2-isoprostane concentrations. Consumption of 55 mg isoflavones/day for 8 weeks by 59 male and female subjects (35 to 69 years of age) with high-normal blood pressure,[157] consumption of 50 mg isoflavones, twice a day for 3 weeks by 12 male and female subjects (22 to 56 years of age),[119] and consumption of 30 mg isoflavones twice a day for 12 weeks by 36 postmenopausal subjects (H. Wiseman et al., unpublished results) all had no effect on F_2-isoprostane concentrations. To try and understand more fully how the form in which isoflavones are consumed influences their ability to protect human health, we have recently carried out a crossover study in healthy young women, comparing the effects of isoflavones within the food matrix with those in supplements, on biomakers of oxidative stress, including F_2-isoprostanes (H. Wiseman et al., unpublished results).

7.4.3 ARTERIAL FUNCTION

Arterial function is vital to the prevention of ischemic changes in the organs that the arteries deliver blood to, and is particularly relevant to ischemic heart disease. A recent population-based study (the Rotterdam study) has shown arterial stiffness (or compliance) to be strongly associated with atherosclerosis at various sites in the vasculature (aorta and carotid artery).[158] Mechanisms of soy-mediated vascular protection may include effects on arterial function, including flow-mediated endothelium-dependent vasodilation (reflecting endothelial function) and systemic arterial compliance (reflecting arterial elasticity) and these have been measured in a number of studies. A randomized double-blind study administering either soy protein isolate (118 mg isoflavones/day) or caesin placebo for 3 months to 213 healthy male and postmenopausal subjects (50 to 75 years of age) showed a significant improvement in peripheral pulse wave velocity (reflecting peripheral vascular resistance and one component, together with systemic arterial compliance, of vascular function) but worsened flow-mediated vasodilation in the men and had no significant effect on the flow-mediated vasodilation in the postmenopausal women.[159] Furthermore, consumption of soy protein with isoflavones (107 mg isoflavones/day for 6 weeks) in a randomized, crossover study had favorable effects on the endothelium (postocclusion peak flow velocity of the brachial artery was significantly lower, consistent with a vasodilatory response) in healthy menopausal women.[160]

In a placebo-controlled, randomized, crossover study with 21 peri- and postmenopausal women treated for 5 weeks with a supplement delivering 80 mg total soy isoflavones/day, a significant improvement was reported in systemic arterial compliance, but had no effect on flow-mediated vasodilation.[161] This lack of an effect on flow-mediated vasodilation is in agreement with a study of similar design, in 20 postmenopausal women and again using a supplement to provide 80 mg total soy isoflavones/day.[162]

It is noteworthy that a cross-sectional study in 301 postmenopausal women (60 to 75 years of age) in the Netherlands, where isoflavone and lignan intakes were assessed with a food frequency questionnaire covering habitual diet during the year prior to the study, reported no associations between isoflavone intake and vascular function, including endothelial function, blood pressure, and hypertension, and this is in contrast to the observed protective effect of dietary lignan intake on blood pressure and hypertension.[163]

7.4.4 CELLULAR EFFECTS

Vascular protection could also be conferred by the ability of genistein to inhibit proliferation of vascular endothelial cells and smooth muscle cells and to increase levels of growth factors,[164] including the cytokine transforming growth factor β (TGF-β). Phytoestrogens including the soy isoflavones genistein and daidzein and the daidzein metabolite equol have all been reported to inhibit growth and MAP kinase activity in human aortic smooth muscle cells[165] and thus may confer protective effects on the cardiovascular system by inhibiting vascular remodeling and neointima formation. TGF-β helps maintain normal vessel wall structure and promotes smooth muscle cell differentiation, while preventing their migration and proliferation. Genistein has been shown to increase TGF-β secretion by cells in culture[166] and, as previously suggested for tamoxifen,[167] increased TGF-β production may be a mediator of some of the cardioprotective effects of soy.[166] However, we have recently found no effect on plasma TGF-β concentrations following consumption of soy either high (56 mg/day) or low (2 mg/day) in isoflavones for 2 weeks in a randomized crossover study in young healthy subjects.[168]

7.5 PROTECTION BY ISOFLAVONOIDS AGAINST OSTEOPOROSIS, COGNITIVE DECLINE, AND MENOPAUSAL SYMPTOMS?

7.5.1 OSTEOPOROSIS

Osteoporosis is a chronic disease in which the bones become brittle and break more easily. Postmenopausal women may suffer hip fractures caused by osteoporosis, which develops primarily as a consequence of the low estrogen levels that occur after the menopause. Premenopausal women are, therefore, protected by their estrogen levels against osteoporosis. Although calcium supplementation is important before the menopause, on its own it cannot stop bone loss in perimenopausal and postmenopausal women. Hormone replacement therapy (HRT) can be very effective, 0.625 mg/day of conjugated estrogens has been reported to prevent bone loss[169] and HRT is osteoprotective if taken postmenopausally for more than 24 months.[170] The drug ipriflavone (a synthetic isoflavone derivative) at a dose of 600 mg/day can prevent the increase in bone turnover and the decrease bone density in postmenopausal women.[171] The protective effects of HRT, together with the finding that ERβ is highly expressed in bone and appears to mediate a distinct mechanism of estrogen action,[74,75] suggests that phytoestrogens may thus protect women against postmenopausal bone loss.[172] The potential skeletal benefits of soy isoflavones has recently been reviewed.[173]

Consumption by postmenopausal women (6-month parallel-group design) of soy protein (40 g/day providing either 56 mg isoflavones/day or 90 mg isoflavones/day) compared to cesin and nonfat dry milk (40 g/day) produced significant increases in bone mineral content and density in the lumbar spine (but not in any other parts of the body), but only in the higher isoflavone (90 mg/day) group compared to the control group.[137] Daily intake for 2 years of two glasses of soymilk containing 76 mg of isoflavones has been reported to prevent lumbar spine bone loss in postmenopausal women.[174] Moreover, consumption of a red clover-derived isoflavone supplement (43.5 mg/day isoflavones) for 1 year significantly decreased loss of lumbar spine bone mineral content and bone mineral density and increased concentrations of the bone formation markers.[175] Similarly, consumption of a soy isoflavone supplement (80 mg/day isoflavones) for 1 year were found to have a beneficial effect on hip bone mineral content in postmenopausal Chinese women with a low initial bone mass.[176]

Consumption of soy foods (providing 60 mg/day isoflavones) for 12 weeks by postmenopausal women has been found to significantly decrease clinical risk factors for osteoporosis (short-term markers of bone turnover) including decreased urinary N-telopeptide excretion (bone resorption marker) and increased serum osteocalcin (bone formation marker).[154] Furthermore, consumption of a soy isoflavone supplement containing 61.8 mg of isoflavones for 4 weeks by postmenopausal Japanese women significantly decreased excretion of bone resorption markers.[177]

A study in 500 Australian women (aged 40 to 80 years) has shown that higher isoflavone intakes are associated with higher concentrations of bone alkaline phosphatase, a short-term marker of bone formation and turnover.[178]

Treatment of postmenopausal women with osteoporosis with raloxifene (60 mg/day or 120 mg/day for 36 months) was found to significantly increase bone mineral density in the spine and femoral neck and decrease the risk of vertebral fracture compared to the placebo treatment.[179] Treatment with raloxifene increased the risk of venous thromboembolism compared to the placebo group and was also associated with a lower risk of breast cancer and did not cause breast pain or vaginal bleeding.[179]

7.5.2 Menopausal Symptoms and Cognitive Decline

The estrogenic properties of phytoestrogens may also help with menopausal symptoms such as hot flushes and vaginitis.[180] An improvement in hot flushes with dietary supplementation with 45 g of raw soy flour per day has been reported; however, an improvement was also seen with white wheat flour (contains very little in the way of phytoestrogen content).[181] Furthermore, consumption of a red clover-derived isoflavone supplement (80 mg/day isoflavones) has been reported to significantly decrease menopausal hot flush symptoms compared with placebo.[182] However, two recent systematic reviews have reached differing conclusions; while the first review concludes that there is some evidence to support the efficacy of soy and soy isoflavone preparation for perimenopausal symptoms,[183] the second one concludes that isoflavone phytoestrogens available as soy foods, soy extracts, and red clover extracts do not improve hot flushes or other menopausal symptoms,[184] suggesting that further studies are needed.

Consumption of soy foods for 10 weeks (100 mg/day isoflavones) has been reported to improve human memory in young healthy adults[185,186] and consumption of a soy isoflavone supplement for 12 weeks (60 mg/day isoflavones) to improve cognitive function in postmenopausal women.[187] By contrast, consumption of soy protein (99 mg isoflavones/day) for 12 months failed to improve cognitive function in postmenopausal women,[188,189] suggesting further clinical trials are required to fully determine the possible health beneficial effects of isoflavones against cognitive decline.

7.6 ISOFLAVONOIDS: POTENTIAL RISKS

Phytoestrogens can cause infertility in some animals and thus concerns have been raised over their consumption by human infants. The isoflavones found in a subterranean clover species (in Western Australia) have been identified as the agents responsible for an infertility syndrome in sheep.[190] Soy isoflavones in the diets of cheetahs in captivity has been shown to lead to their infertility.[191] Most animals that are bred commercially and domestic animals, however, are fed diets containing soy (up to 20% by weight) without any apparent reproductive problems.[1] No reproductive abnormalities have been found in peripubertal rhesus monkeys[132] nor in people living in countries where soy consumption is high. Indeed, the finding that dietary isoflavones are excreted into breast milk by soy-consuming mothers suggests that in cultures where the consumption of soy products is the normal dietary practice, breast-fed infants are exposed to high levels again without any adverse effects.[192] Isoflavone exposure shortly after birth at a critical developmental period through breastfeeding may protect against cancer and may be more important to the observation of lower cancer rates in populations in the Far East than adult dietary exposure to isoflavones.[192]

There have been some concerns expressed regarding the possible health consequences in adulthood (endocrinological and reproductive outcomes) of early-life isoflavone exposure from soy-based infant formula.[193] The daily exposure of infants to isoflavones in soy-based infant formulas is 6- to 11-fold higher on a body weight basis than the dose that has hormonal effects in adults consuming soy foods.[194] However, evidence from adult and infant populations indicates that dietary isoflavones in soy-based infant formulas do not adversely affect human growth, development, or reproduction.[51,52,193,195–197]

Although toxicity from isoflavones may arise from their action as alternative substrates for the enzyme thyroid peroxidase[198] and people in southeast Asia would be protected by the dietary inclusion of iodine-rich seaweed products, a recent study has shown that isoflavone supplements do not affect thyroid function in iodine-replete postmenopausal women.[199] Considerations of the safety of soy isoflavones is an area of great interest in relation to their potential benefits to human health and has recently been comprehensively reviewed.[200]

7.7   ISOFLAVONOIDS AND HUMAN HEALTH: CONCLUSIONS

The important question of whether isoflavonoids should be used to protect human health clearly requires much more information to be provided by appropriate studies. Factors such as age and biological responsiveness to the different potential protective or even harmful effects of isoflavonoids (these will change with age) appear to play an important role.

Postmenopausal women may well benefit in terms of protection against heart disease and osteoporosis, from estrogen replacement therapeutics strategies that utilize isoflavonoids. Older men may also benefit from protection against problems such as prostate cancer and cardiovascular disease.

In 1999, the US FDA allowed health claims (on food labels) on the association between soy protein and reduced risk of coronary heart disease for foods containing ≥6.25 g of soy protein, assuming either four servings, or that a total of 25 g of soy protein are consumed daily. Furthermore, in 2002, the UK Joint Health Claims Initiative approved a health claim on the association between soy protein and cholesterol reduction, "the inclusion of at least 25 g of soy protein per day, as part of a diet low in saturated fat, can help reduce blood cholesterol levels" and it is important to note that this claim relates to soy protein that has retained its naturally occurring isoflavones.[201]

An important *caveat* is that the epidemiological evidence suggesting protection from hormone-dependent cancer by isoflavone phytoestrogens is based on foods rather than isoflavone extracts, which in the future could include isoflavone-containing therapeutics. Indeed, the preparation of isoflavone extracts from soy protein could well result in the loss of important components that act synergistically with the isoflavones. This approach could also result in daily isoflavonoid intake increased too far above normal dietary levels such that toxicity occurs.

The results of further extensive studies are thus clearly needed before further decisions can be made regarding the future of isoflavonoids in human health.

REFERENCES

1. Barnes S. Phytoestrogens and breast cancer. *Ballieres Clin. Endocrinol. Metab.* 12, 559–579, 1998.
2. Bingham SA, Atkinson C, Liggins J, Bluck L, Coward A. Plant oestrogens: where are we now. *Br. J. Nutr.* 79, 393–406, 1998.
3. Wiseman H. The bioavailability of non-nutrient plant factors: dietary flavonoids and phyto-oestrogens. *Proc. Nutr. Soc.* 58, 139–146, 1999.
4. Wiseman H. The therapeutic potential of phytoestrogens. *Exp. Opin. Invest. Drugs* 9, 1829–1840, 2000.
5. Wiseman H. Dietary phytoestrogens, oestrogens and tamoxifen: mechanisms of action in modulation of breast cancer risk and in heart disease prevention. In *Biomolecular Free Radical Toxicity: Causes and Prevention* (H Wiseman, P Goldfarb, TJ Ridgway, A Wiseman, editors) John Wiley, Chichester, pp. 170–208, 2000.
6. Cornwell T, Cohick W, Raskin I. Dietary phytoestrogens and health. *Phytochemistry* 65, 995–1016, 2004.
7. Stevens JF and Page JE. Xanthohumol and related prenylflavonoids from hops and beer: to your good health. *Phytochemistry* 65, 1317–1330, 2004.
8. Dixon RA. Phytoestrogens. *Annu. Rev. Plant Biol.* 55, 225–261, 2004.
9. Wiseman H. Phytochemicals (b) Epidemiological factors. In *Encyclopedia of Human Nutrition* (B Caballero, L Allen, A Prentice, editors). Academic Press, London, 2005, 497–508.
10. Reinli K, Block G. Phytoestrogen content of foods — a compendium of literature values. *Nutr. Cancer* 26, 123–148, 1996.
11. Wiseman H, Casey K, Clarke DB, Barnes KA, Bowey E. Isoflavone aglycone and glucoconjugate content of high- and low-soy U.K. foods used in nutritional studies. *J. Agric. Food Chem.* 50, 1404–1410, 2002.
12. Messina MJ. Legumes and soybeans: overview of their nutritional profiles and health effects. *Am. J. Clin. Nutr.* 70, 439S–450S, 1999.
13. Coward L, Barnes NC, Setchell KDR, Barnes S. Genistein, daidzein and their β-glycoside conjugates: antitumor isoflavones in soybean foods from American and Asian diets. *J. Agric. Food Chem.* 41, 1961–1967, 1993.
14. Lapcik O, Hill M, Hampl R, Wahala K, Adlercreutz H. Identification of isoflavonoids in beer. *Steroids* 63, 14–20, 1998.
15. Clarke DB, Lloyd AS. Dietary exposure estimates of isoflavones from the 1998 UK Total Diet Study. *Food Addit. Contam.* 21, 305–316, 2004.
16. Boker LK, Van der Schouw YT, De Kleijn MJ et al. Intake of dietary phytoestrogens by Dutch women. *J. Nutr.* 132, 1319–1328, 2002.
17. Fletcher RJ. Food sources of phyto-oestrogens and their precursors in Europe. *Br. J. Nutr.* 89 (Suppl. 1), S39–S43, 2003.
18. Ritchie MR, Morton MS, Deighton N, Blake A, Cummings JH. Plasma and urinary phyto-oestrogens as biomarkers of intake: validity by duplicate diet analysis. *Br. J. Nutr.* 91, 447–457, 2004.
19. Setchell KD, Cole SJ. Variations in isoflavone levels in soy foods and soy protein isolates and issues related to isoflavone databases and food labelling. *J. Agric. Food Chem.* 51 (14), 4146–4155, 2003.
20. Setchell DR, Brown NM, Lydeking-Olsen E. The clinical importance of the metabolite equol — a clue to the effectiveness of soy and its isoflavones. *J. Nutr.* 132, 3577–3584, 2002.

21. Rowland I, Faughnan M, Hoey L et al. Bioavailability of phyto-oestrogens. *Br. J. Nutr.* 89 (Suppl. 1), S45–S58, 2003.

22. Wantanabe S, Yamaguchi M, Sobue T et al. Pharmacokinetics of soyabean isoflavones in plasma, urine and feces of men after ingestion of 60 g baked soybean powder (Kinako). *J. Nutr.* 128, 1710–1715, 1998.

23. Rowland IR, Wiseman H, Sanders TAB, Adlercreutz H, Bowey EA. Interindividual variation in metabolism of soy isoflavones and lignans: influence of habitual diet on equol production by the gut microflora. *Nutr. Cancer* 36, 27–32, 2000.

24. Setchell KD, Faughnan MS, Avades T et al. Comparing the pharmacokinetics of daidzein and genistein with the use of ^{13}C-labelled tracers in premenopausal women. *Am. J. Clin. Nutr.* 77, 411–419, 2003.

25. Setchell KDR, Brown NM, Deai PB et al. Bioavailability, disposition and dose–response effects of soy isoflavones when consumed by healthy women at physiologically typical dietary intakes. *J. Nutr.* 133, 1027–1035, 2003.

26. Markiewicz L, Garey J, Aldercreutz H, Gurpide E. *In-vitro* bioassays of non-steroidal phytoestrogens. *J. Steroid Biochem.* 45, 399–405, 1993.

27. Hodgson JM, Croft KD, Puddey IB, Mori TA, Beilin LJ. Soybean isoflavonoids and their metabolic products inhibit *in vitro* lipoprotein oxidation in serum. *J. Nutr. Biochem.* 7, 664–669, 1996.

28. Mitchell JH, Gardner PT, Mcphail DB et al. Antioxidant efficacy of phytoestrogens in chemical and biological model systems. *Arch. Biochem. Biophys.* 360, 142–148, 1998.

29. Arora A, Nair NG, Strasburg GM. Antioxidant activities of isoflavones and their biological metabolites in a liposomal system. *Arch. Biochem. Biophys.* 356, 133–141, 1998.

30. Wiseman H, O'Reilly J, Lim P et al. Antioxidant properties of the isoflavone phytoestrogen functional ingredient in soy products. In *Functional Foods, the Consumer, the Products and the Evidence* (M Sadler, M Saltmarsh, editors), Royal Society of Chemistry, Cambridge, pp. 80–86, 1998.

31. Coldham NG, Darby C, Hows M et al. Comparative metabolism of genistin by human and rat gut microflora: detection and identification of the end-products of metabolism. *Xenobiotica* 32, 45–62, 2002.

32. Hur H-G, Lay JL, Beger RD, Freeman JP, Rafii F. Isolation of human intestinal bacteria metabolising the natural isoflavone glycosides daidzin and genistin. *Arch. Microbiol.* 174, 422–428, 2000.

33. Sfakianos J, Coward L, Kirk M, Barnes S. Intestinal uptake and biliary excretion of the isoflavone genistein in the rat. *J. Nutr.* 127, 1260–1268, 1997.

34. King RA, Bursill DB. Plasma and urinary kinetics of the isoflavones daidzein and genistein after a single soy meal in humans. *Am. J. Clin. Nutr.* 67, 867–872, 1998.

35. Yasuda T, Mizunuma S, Kano Y et al. Urinary and biliary metabolites of genistein in rats. *Biol. Pharm. Bull.* 19, 413–417, 1996.

36. Clarke DB, Lloyd AS, Botting NP et al. Measurement of intact sulfate and glucuronide phytoestrogen conjugates in human urine using isotope dilution liquid chromatography–tandem mass spectrometry with [^{13}C$_3$] isoflavone internal standards. *Anal. Biochem.* 309, 158–172, 2002.

37. Adlercreutz H, Markkanen H, Watanabe S. Plasma concentrations of phytoestrogens in Japanese men. *Lancet* 342, 1209–1210, 1993.

38. Morton MS, Wilcox G, Wahlqvist, ML, Griffiths K. Determination of lignans and isoflavonoids in human female plasma following dietary supplementation. *J. Endocrinol.* 142, 251–259, 1994.

39. Morton MS, Matos-Ferreira A, Abranches-Monteiro L. et al. Measurement and metabolism of isoflavonoids and lignans in the human male. *Cancer Lett.* 114, 145–151, 1997.

40. Setchell KD, Cassidy A. Dietary isoflavones: biological effects and relevance to human health. *J. Nutr.* 129, 758S–767S, 1999.

41. Bowey E, Aldercreutz A, Rowland I. Metabolism of isoflavones and lignans by the gut microflora: a study in germ-free and human flora associated rats. *Food Chem. Toxicol.* 41, 631–636, 2003.

42. Lampe JW, Karr SC, Hutchins AM, Slavin JL. Urinary equol excretion with a soy challenge: influence of habitual diet. *Proc. Soc. Exp. Biol. Med.* 217, 335–339, 1998.

43. Kelly GE, Joannou GE, Reeder AY, Nelson C, Waring MA. The variable metabolic response to dietary isoflavones in humans. *Proc. Soc. Exp. Biol. Med.* 208, 40–43, 1995.
44. Lampe JW, Skor HE, Li S et al. Wheat bran and soy protein do not alter urinary excretion of the isoflavan equol in premonopausal women. *J. Nutr.* 131, 740–744, 2001.
45. Xu X, Harris KS, Wang H-J, Murphy PA, Hendrich S. Bioavailability of soybean isoflavones depends upon gut microflora in women. *J. Nutr.* 125, 2307–2315, 1995.
46. Ueno T, Uchiyama S. Identification of the specific intestinal bacteria capable of metabolizing soy isoflavone to equol. *Am. Nutr. Metab.* 45, 114 (abstract), 2001.
47. Wiseman H, Casey K, Bowey EA et al. Influence of 10 wk of soy consumption on plasma concentrations and excretion of isoflavonoids and on gut microflora metabolism in healthy adults. *Am. J. Clin. Nutr.* 80, 692–699, 2004.
48. Lu LJ, Grady JJ, Marshall MV, Ramanujam VM, Anderson KE. Altered time course of urinary daidzein and genistein excretion during chronic soya diet in healthy male subjects. *Nutr. Cancer* 24, 311–323, 1995.
49. Lu LJ, Lin SN, Grady JJ, Nagamani M, Anderson KE. Altered kinetics and extent of urinary daidzein and genestein excretion in women during chronic soy exposure. *Nutr. Cancer* 26, 289–302, 1996.
50. Lu LJ, Anderson KE. Sex and long-term diets affect the metabolism and excretion of soy isoflavones in humans. *Am. J. Clin. Nutr.* 68 (Suppl.), 1500S–1504S, 1998.
51. Badger TM, Ronis MJ, Hakkak R, Rowlands JC, Korourian S. The health consequences of early soya consumption. *J. Nutr.* 132, 559S–565S, 2002.
52. Mendez MA, Anthony MS, Arab L. Soy-based formulae and infant growth and development. *J. Nutr.* 132, 2127–2130, 2002.
53. Hoey L, Rowland IR, Lloyd AS, Clarke DB, Wiseman H. Influence of soya-based infant formula consumption on isoflavone and gut microflora metabolite concentrations in urine and on faecal microflora composition and metabolic activity in infants and children. *Br. J. Nutr.* 91, 607–616, 2004.
54. Adlercreutz H. Phyto-oestrogens and cancer. *Lancet Oncol.* 3, 32–40, 2002.
55. Greenwald P. Clinical trials in cancer prevention: current results and perspectives for the future. *J. Nutr.* 134 (12 Suppl.), 3507S–3512S, 2004.
56. Holzbeierlein JM, McIntosh, Thrasher JB. The role of soy phytoestrogens in prostate cancer. *Curr. Opin. Urol.* 15, 17–22, 2005.
57. Magee PJ, Rowland IR. Phyto-oestrogens, their mechanism of action: current evidence for a role in breast and prostate cancer. *Br. J. Nutr.* 91, 513–531, 2004.
58. Guo JY, Li X, Browning JD Jr et al. Dietary soy isoflavones and estrone protect ovariectomized ERalphaKO and wild-type mice from carcinogen-induced colon cancer. *J. Nutr.* 134, 179–182, 2004.
59. Shimizu H, Ross RK, Bernstein L et al. Cancers of the prostate and breast among Japanese and white immigrants in Los Angeles County. *Br. J. Cancer* 63, 963–966, 1991.
60. Messina M, Persky V, Setchell KDR, Barnes S. Soy intake and cancer risk: a review of the *in vitro* and *in vivo* data. *Nutr. Cancer* 21, 113–131, 1994.
61. Ingram D, Sanders K, Kolybaba M, Lopez D. Case control study of phyto-oestrogens and breast cancer. *Lancet* 350, 990–994, 1997.
62. Lund TD, Munson DJ, Haldy ME et al. Equol is a novel anti-androgen that inhibits prostate growth and hormone feedback. *Biol. Reprod.* 70, 1188–1195, 2004.
63. Wiseman H. *Tamoxifen: Molecular Basis of use in Cancer Treatment and Prevention*, John Wiley, Chichester, 1994.
64. Tanaka H, Arakawa H, Yamaguchi T et al. A ribonucleotide reductase gene involved in a p53-dependent cell-cycle checkpoint for DNA damage. *Nature* 404, 42–49, 2000.
65. Bruner SD, Norman DPG, Verdine GL. Structural basis for recognition and repair of the endogenous mutagen 8-oxoguanine in DNA. *Nature* 403, 859–866, 2000.
66. Katzenellenbogen JA, O'Malley BW, Katzenellenbogen BS. Tripartite steroid hormone receptor pharmacology: interaction with multiple effector sites as a basis for the cell- and promoter-specific action of these hormones. *Mol. Endocrinol.* 10, 119–131, 1996.

67. Bronsens JJ, Parker MG. Oestrogen receptor hijacked. *Nature* 423, 487–488, 2003.

68. Brzozowski AM, Pike ACW, Dauter Z et al. Molecular basis of agonism and antagonism in the oestrogen receptor. *Nature* 389, 753–758, 1997.

69. Miksicek RJ. Estrogenic flavonoids. Structural requirements for biological activity. *Proc. Soc. Exp. Biol. Med.* 208, 44–50, 1995.

70. Kuiper GGJM, Enmark E, Pelto-Huikko M, Nilsson S, Gustafsson J-A. Cloning of a novel estrogen receptor expressed in rat prostate and ovary. *Proc. Natl. Acad. Sci. USA* 93, 5925–5930, 1996.

71. Kuiper GGJM, Carlsson B, Grandien K et al. Comparison of the ligand binding specificity and transcript tissue distribution of estrogen receptors α and β. *Endocrinology* 138, 863–870, 1997.

72. Crandall DL, Busler DE, Novak TJ, Weber, Kral JG. Identification of estrogen receptor beta RNA in human breast and abdominal Subcutaneous adipose tissue, *Biochem. Biophys. Res. Commun.* 248, 523–526, 1998.

73. Saji S, Jensen EV, Nilsson S, Rylander T, Warner M, Gustafsson J-A. Estrogen receptors α and β in the rodent mammary gland. *Proc. Natl. Acad. Sci. USA* 97, 337–342, 2000.

74. Onoe, Y, Miyaura C, Ohta H, Nozawa S, Suda T. Expression of estrogen receptor β in rat bone. *Endocrinology* 138, 4509–4512, 1997.

75. Arts J, Kuiper GGJM, Janssen, JMMF et al. Differential expression of estrogen receptors α and β mRNA during differentiation of human osteoblast SV-HFO cells. *Endocrinology* 138, 5067–5070, 1997.

76. Stossi F, Barnett DH, Frasor J et al. Transcriptional profiling of oestrogen-regulated gene expression via estrogen receptor (ER) α or ERβ in human osteosarcoma cells: distinct and common target genes for these receptors. *Endocrinology* 145, 3473–3486, 2004.

77. Makela S, Savolainen H, Aavik E et al. Differentiation between vasculoprotective and uterotropic effects of ligands with different binding affinities to estrogen receptors α and β. *Proc. Natl. Acad. Sci. USA* 96, 7077–7082, 1999.

78. Lamartiniere CA, Moore JB, Brown NA et al. Genistein suppresses mammary cancer in rats. *Carcinogenesis* 16, 2833–2840, 1995.

79. Fritz, W, Wang, J, Coward L, Lamartiniere CA. Dietary genistein: perinatal mammary cancer prevention, bioavailability and toxicity testing in the rat. *Carcinogenesis* 19, 2151–2158, 1998.

80. Hilakivi-Clarke L, Cho E, Onojafe I, Raygada M, Clarke R. Maternal exposure to genistein during pregnancy increases carcinogen-induced mammary tumorigenesis in female rat offspring. *Oncol. Rep.* 6, 1089–1095, 1999.

81. Hilakivi-Clarke L, Onojafe I, Raygada M et al. Prepubertal exposure to zearalenone or genistein reduces mammary tumorigenesis. *Br. J. Cancer* 80, 1682–1688, 1999.

82. Cotroneo MS, Wang J, Fritz WA, Eltoum IE, Lamartiniere CA. Genistein action in the prepubertal mammary gland in a chemoprevention model. *Carcinogenesis* 23, 1467–1474, 2002.

83. Gotoh T, Yamada K, Yin H et al. Chemoprevention of *N*-nitroso-*N*-methylurea-induced rat mammary carcinogenesis by soy foods or biochanin A. *Jpn. J. Cancer Res.* 89, 137–142, 1998.

84. Gotoh T, Yamada K, Ito A et al. Chemoprevention of *N*-nitroso-*N*-methylurea-induced rat mammary cancer by miso and tamoxifen, alone and in combination. *Jpn. J. Cancer Res.* 89, 487–495, 1998.

85. Hilakivi-Clarke L, Cho E, Clarke R. Maternal genistein exposure mimics the effects of oestrogen on mammary gland development in female mouse offspring. *Oncol. Rep.* 5, 609–615, 1998.

86. Hsieh CY, Santell RC, Haslam SZ, Helferich WG. Estrogenic effects of genistein on the growth of estrogen receptor-positive human breast cancer (MCF-7) cells *in vitro* and *in vivo*. *Cancer Res.* 58, 3833–3838, 1998.

87. Akiyama T, Ishida J, Nakagawa S et al. Genistein, a specific inhibitor of tyrosine-specific protein kinases. *J. Biol. Chem.* 262, 5592–5595, 1987.

88. Kim JW, Kim YT, Kim DK et al. Expression of epidermal growth factor receptor in carcinoma of the cervix. *Gynecol. Oncol.* 60, 283–287, 1996.

89. So FV, Guthrie N, Chambers AF, Carroll KK. Inhibition of proliferation of estrogen receptor-positive MCF-7 human breast cancer cells by flavonoids in the presence and absence of excess estrogen. *Cancer Lett.* 112, 127–133, 1997.

90. Wang C, Kurzer MS. Phytoestrogen concentration determines effects on DNA synthesis in human breast cancer cells. *Nutr. Cancer* 28, 236–247, 1997.

91. Zava DT, Duwe G. Estrogenic and antiproliferative properties of genistein and other isoflavonoids in human breast cancer cells *in vitro*. *Nutr. Cancer* 27, 31–40, 1997.

92. Peterson TG, Barnes S. Genistein inhibition of the growth of human breast cancer cells: independence from estrogen receptors and the multi-drug resistance gene. *Biochem. Biophys. Res. Commun.* 179, 661–667, 1991.

93. Wang TTY, Sathymoorthy N, Phang JM. Molecular effects of genistein on estrogen receptor-mediated pathways. *Carcinogenesis* 17, 271–275, 1996.

94. Sathyamoorthy N, Wang TT. Differential effects of dietary phyto-oestrogens daidzein and equol on human breast cancer MCF-7 cells. *Eur. J. Cancer* 33, 2384–2389, 1997.

95. Ramanathan L, Gray WG. Identification and characterization of a phytoestrogen-specific gene from the MCF-7 human breast cancer cell. *Toxicol. Appl. Pharm.* 191, 107–117, 2003.

96. Kuiper GGJM, Lemmen JG, Carlsson BO et al. Interaction of estrogenic chemicals and phytoestrogens with estrogen receptor β. *Endocrinology* 139, 4252–4263, 1998.

97. Mueller SO, Simon S, Chae K, Metzler M, Korach KS. Phytoestrogens and their human metabolites show distinct agonistic and antagonistic properties on estrogen receptor (alpha) and ER(beta) in human cells. *Toxicol. Sci.* 81, 530–531, 2004.

98. Barkhem T, Carlsson B, Nilsson Y et al. Differential response of estrogen receptor α and estrogen receptor β to partial estrogen agonists/antagonists. *Mol. Pharmacol.* 54, 105–112, 1998.

99. Pike ACW, Brzozowski AM, Hubbard RE et al. Structure of the ligand-binding domain of oestrogen receptor beta in the presence of a partial agonist and a full antagonist. *EMBO J.* 18, 4608–4618, 1999.

100. Peterson TG, Barnes S. Genistein inhibits both estrogen and growth factor stimulated proliferation of human breast cancer cells. *Cell Growth Differ.* 7, 1345–1351, 1996.

101. Dalu A, Haskell JF, Coward L, Lamartiniere CA. Genistein, a component of soy, inhibits the expression of the EGF and ErbB/Neu receptors in the rat dorsolateral prostate. *Prostate* 37, 36–43, 1998.

102. Kondo K, Tsuneizumi K, Watanabe T, Oishi M. Induction of *in vivo* differentiation of mouse embryonal carcinoma (F9) cells by inhibitors of topoisomerases. *Cancer Res.* 50, 5398–5404, 1991.

103. Kim H, Peterson TG, Barnes S. Mechanisms of action of the soy isoflavone genistein: emerging role of its effects through transforming growth factor beta signaling pathways. *Am. J. Clin. Nutr.* 68, 1418S–1425S, 1998.

104. Fotsis T, Pepper M, Adlercreutz H et al. Genistein, a dietary-derived inhibitor of *in vitro* angiogenesis. *Proc. Natl. Acad. Sci. USA* 90, 2690–2694, 1993.

105. Kruse FE, Joussen AM, Fotsis T et al. Inhibition of neovacularization of the eye by dietary factors exemplified by isoflavonoids. *Ophthalmologie* 94, 152–156, 1997.

106. Yan CH, Han R. Genistein suppresses adhesion-induced protein tyrosine phosphorylation and invasion of B16–B16 melanoma cells. *Cancer Lett.* 129, 117–124, 1998.

107. Makela S, Poutanen M, Kostlan ML et al. Inhibition of 17 beta-hydroxysteroid oxidoreductase by flavonoids in breast and prostate cancer cells. *Proc. Soc. Exp. Biol. Med.* 217, 310–316, 1998.

108. Wei H, Barnes S, Wang Y. Inhibitory effect of genistein on a tumor promoter-induced c-fos and c-jun expression in mouse skin. *Oncol. Rep.* 3, 125–128, 1996.

109. Sathyamoorthy N, Gilsdorf JF, Wang TTY. Differential effects of genistein on transforming growth factor beta-1 expression in normal and malignant mammary epithelial cells. *Anticancer Res.* 18, 2449–2453, 1998.

110. Wei H, Bowen R, Cai Q et al. Antioxidant and antipromotional effects of the soybean isoflavone genistein. *Proc. Soc. Exp. Biol. Med.* 208, 124–130, 1995.

111. Tsuchiya H, Nagayama M, Tanaka T et al. Membrane-rigidifying effects of anti-cancer dietary factors. *Biofactors* 16 (3–4), 45–56, 2002.

112. Wiseman H, Cannon M, Arnstein HRV, Halliwell B. Mechanism of inhibition of lipid peroxidation by tamoxifen and 4-hydroxytamoxifen introduced into liposomes. Similarity to cholesterol and ergosterol. *FEBS lett.* 274, 107–110, 1990.

113. Wiseman H, Quinn P, Halliwell B. Tamoxifen and related compounds decrease membrane fluidity in liposomes. Mechanism for the antioxidant action of tamoxifen and relevance to its anticancer and cardioprotective actions? *FEBS Lett.* 330, 53–56, 1993.

114. Wiseman H., O'Reilly JD, Adlercreutz H et al. Isoflavone phytoestrogens consumed in soy decrease F_2-isoprostane concentrations and increase resistance of low-density lipoprotein to oxidation in humans. *Am. J. Clin. Nutr.* 72, 395–400, 2000.

115. Foti P, Erba D, Riso P et al. Comparison between daidzein and genistein antioxidant activity in primary and cancer lymphocytes. *Arch. Biochem. Biophys.* 433, 431–427, 2005.

116. D'Alessandro T, Prasain J, Benton MR et al. Polyphenols, inflammatory response, and cancer prevention: chlorination of isoflavones by human neutrophils. *J. Nutr.* 133, 3773S–3777S, 2003.

117. Boersma BJ, D'Alessandro T, Benton MR et al. Neutrophil myeloperoxidase chlorinates and nitrates soy isoflavones and enhances their antioxidant properties. *Free Radical Biol. Med.* 35, 1417–1430, 2003.

118. Wiseman H, Halliwell B. Damage to DNA by reactive oxygen and nitrogen species: role in inflammatory disease and progression to cancer. *Biochem. J.* 313, 17–29, 1996.

119. Djuric Z, Chen G, Doerge DR, Heilbrun LK, Kucuk O. Effect of soy isoflavone supplementation on markers of oxidative stress in men and women. *Cancer Lett.* 172, 1–6, 2001.

120. Bazzoli DL, Hill S, DiSilvestro RA. Soy protein antioxidant actions in active, young adult women. *Nutr. Res.* 22, 807–815, 2002.

121. Fotsis T, Pepper M, Adlercreutz H et al. Genistein, a dietary ingested isoflavonoid, inhibits cell proliferation and *in vitro* angiogenesis. *J. Nutr.* 125, 790S–797S, 1995.

122. Su SJ, Yeh TM, Chuang WJ et al. The novel targets for anti-angiogenesis of genistein on human cells. *Biochem. Pharmacol.* 69, 307–318, 2005.

123. Johnson DW, Berg JN, Baldwin MA et al. Mutations in the activin receptor-like kinase I gene in hereditary haemorrhagic telangiectasia type 2. *Nat. Genet.* 13, 189–195, 1996.

124. Petrakis N, Barnes S, King EB et al. Stimulatory influence of soy protein isolate on breast secretion in pre- and postmenopausal women. *Cancer Epidemiol. Biomarkers Prev.* 5, 785–794, 1996.

125. Hargreaves DF, Potten, CS, Harding, C et al. Two-week soy supplementation has estrogenic effect on normal premenopausal breast. *J. Clin. Endocrinol. Metab.* 84, 4017–4024, 1999.

126. Boyd NF, Lockwood GA, Byng JW, Tritchler DL, Yaffe MJ. Mammographic densities and breast cancer risk. *Cancer Epidemiol. Biomarkers Prev.* 7, 1133–1144, 1998.

127. Boyd NF, Lockwood GA, Martin LJ et al. Mammographic density as a marker of susceptibility to breast cancer: a hypothesis. *IARC Sci. Publ.* 154, 163–169, 2001.

128. Atkinson C, Warren RML, Sala E et al. Red clover-derived isoflavones and mammographic breast density: a double-blind, randomized, placebo-controlled trial. *Breast Cancer Res.* 6, R170–R179, 2004.

129. Sirtori CR, Lovati MR. Soy proteins and cardiovascular disease. *Curr. Atheroscler. Rep.* 3, 47–53, 2001.

130. Clarkson, TB. Soy, phytoestrogens and cardiovascular disease. *J. Nutr.* 132, 566S–569S, 2002.

131. Anderson JW, Johnstone BM, Cook-Newell ME. Meta-analysis of the effects of soy protein intake on serum lipids. *N. Engl. J. Med.* 333, 276–282, 1995.

132. Anthony MS, Clarkson TB, Hughs CL Jr, Morgan TM, Burke GL. Soybean isoflavones improve cardiovascular risk factors without affecting the reproductive system of peripubertal rhesus monkeys. *J. Nutr.* 126, 43–50, 1996.

133. Clarkson TB, Anthony MS, Williams JK, Honore EK, Cline JM. The potential of soybean phytoestrogens for postmenopausal hormone replacement therapy. *Proc. Soc. Exp. Biol. Med.* 217, 365–368, 1998.

134. Register TC, Cann JA, Kaplan JR et al. Effects of soy isoflavones and conjugated equine oestrogens on inflammatory markers in atherosclerotic, ovariectomized monkeys. *J. Clin. Endocrinol. Metab.* 90, 1734–1740, 2005.

135. Park D, Huang T, Frishman WH. Phytoestrogens as cardioprotective agents. *Cardiol. Rev.* 13, 13–17, 2005.

136. Zhuo X-G, Melby MK, Wantanabe S. Soy isoflavone intake lowers serum LDL cholesterol: a meta-analysis of 8 randomized controlled trials in humans. *J. Nutr.* 134, 2395–2400, 2004.

137. Potter SM, Baum JA, Teng H et al. Soy protein and isoflavones: their effects on blood lipids and bone density in postmenopausal women. *Am. J. Clin. Nutr.* 68 (Suppl.), 1375S–1379S, 1998.

138. Sagara M, Kanda T, Njelekera M et al. Effects of dietary intake of soy protein and isoflavones on cardiovascular disease risk factors in high risk, middle-aged men in Scotland. *J. Am. Coll. Nutr.* 23, 85–91, 2004.

139. Desroches S, Mauger JF, Ausman LM, Lichtenstein AH, Lamarche B. Soy protein favourably affects LDL size independently of isoflavones in hypercholesterlemic men and women. *J. Nutr.* 134, 574–579, 2004.

140. Wang Y, Jones PJ, Ausman LM, Lichtenstein AH. Soy protein reduces triglyceride levels and triglyceride fatty acid fractional synthesis rate in hypercholesterolemic subjects. *Atherosclerosis* 173, 269–275, 2004.

141. Yeung J, Yu TF. Effects of isoflavones (soy phyto-estrogens) on serum lipids: analysis of randomized controlled trials. *Nutr. J.* 2, 15, 2003.

142. Atkinson C, Oosthuizen W, Scollen et al. Modest protective effects of isoflavones from a red clover-derived dietary supplement on cardiovascular disease risk factors in perimenopausal women, and evidence of an interaction with ApoE genotype in 49–65 year-old women. *J. Nutr.* 134, 1759–1764, 2004.

143. Kreijkamp-Kaspers S, Kok L, Bots ML, Grobbee DE, Van Der Schouwy T. Dietary phytoestrogens and plasma lipids in Dutch postmenopausal women; a cross-sectional study. *Atherosclerosis* 178, 95–100, 2005.

144. Witzum JL. The oxidation hypothesis of atherosclerosis. *Lancet* 344, 793–795, 1994.

145. Bruckdorfer KR. Antioxidants, lipoprotein oxidation, and arterial function. *Lipids* 31, S83–S85, 1996.

146. Buckley, C, Bund SJ, McTaggart F et al. Oxidized low-density lipoproteins inhibit endothelium-dependent relaxations in isolated large and small rabbit coronary arteries. *J. Auton. Pharmacol.* 16, 261–267, 1996.

147. Heinloth A, Brune B, Fischer B, Galle J. Nitric oxide prevents oxidised LDL-induced p53 accumulation, cytochrome c translocation and apoptosis in macrophages via guanylate cyclase stimulation. *Atherosclerosis* 162, 93–101, 2002.

148. Pratico D, Lawson, JA, Rokach, J, Fitzgerald GA. The isoprostanes in biology and medicine. *Trends Endocrinol. Metab.* 12, 243–247, 2001.

149. Fam SS, Morrow JD. The isoprostanes: unique products of arachidonic acid oxidation a review. *Curr. Med. Chem.* 10, 1723–1740, 2003.

150. Halliwell B, Whiteman M. Measuring reactive species and oxidative damage *in vivo* and in cell culture: how should you do it and what do the results mean? *Br. J. Pharm.* 142, 231–155, 2004.

151. Tikkanen MJ, Wahala K, Ojala S, Vihma V, Adlercreutz H. Effect of soybean phytoestrogen intake on low density lipoprotein oxidation resistance. *Proc. Natl. Acad. Sci. USA* 95, 3106–3110, 1998.

152. Helisten H, Hockerstedt A, Wahala K et al. Accumulation of high-density lipoprotein-derived estradiol-17β fatty acid esters in low-density lipoprotein particles. *J. Clin. Endocrinol. Metab.* 86, 1294–1300, 2001.

153. Meng QH, Hockerstedt A, Heinonen S et al. Antioxidant protection of lipoproteins containing estrogens: *in vitro* evidence for low- and high-density lipoproteins as estrogen carriers. *Biochim. Biophys. Acta* 1439, 331–340, 1999.

154. Scheiber MD, Liu, JH, Subbiah, MTR, Rebar RW, Setchell KDR. Dietary inclusion of whole soy foods results in significant reductions in clinical risk factors for osteoporosis and cardiovascular disease in normal postmenopausal women. *Menopause* 8, 384–392, 2001.

155. Jenkins DJ, Kendall, CW, Vidgen E et al. Effect of soy-based breakfast cereal on blood lipids and oxidized low-density lipoprotein. *Metabolism* 49, 1496–1500, 2000.

156. Samman S, Lyons-Wall PM, Chan GS, Smith SJ, Petocz P. The effect of supplementation with isoflavones on plasma lipids and oxidisability of low density lipoprotein in premenopausal women. *Atherosclerosis* 147, 277–283, 1999.

157. Hodgson JM, Puddey IB, Croft KD et al. Isoflavonoids do not inhibit *in vivo* lipid peroxidation in subjects with high-normal blood pressure. *Atherosclerosis* 145, 167–172, 1999.

158. Van Popele NM, Grobbee DE, Bots ML et al. Association between arterial stiffness and atherosclerosis: the Rotterdam study. *Stroke* 32, 454–460, 2001.

159. Teede HJ, Dalais FS, Kotsopulos D et al. Dietary soy has both beneficial and potentially adverse cardiovascular effects: a placebo-controlled study in men and postmenopausal women. *J. Clin. Endocrinol. Metab.* 86, 3053–3060, 2001.

160. Steinberg FM, Guthrie NL, Villablanca AC, Kumar K, Murray MJ. Soy protein with isoflavones has favourable effects on endothelia function that are independent of lipid and antioxidant effect in health postmenopausal women. *Am. J. Clin. Nutr.* 78, 123–130, 2003.

161. Nestel, PJ, Yamashita T, Sasahara T, Pomeroy S, Dart A, Komesaroff P, Owen A, Abbey M. Soy isoflavones improve systemic arterial compliance but not plasma lipid in menopausal and perimenopausal women. *Arterioscler. Thromb. Vasc. Biol.* 17, 3392–3398, 1997.

162. Simons LA, von Konigsmark M, Simons J, Celermajer DS. Phytoestrogens do not influence lipoprotein levels or endothelial function in healthy, postmenopausal women. *Am. J. Cardiol.* 85, 1297–1301, 2000.

163. Kreijkamp-Kaspers S, Kok L, Bots ML, Grobbee DE, Van Der Schouwy T. Dietary phytoestrogens and vascular function in postmenopausal women: a cross-sectional study. *J. Hypertens.* 22, 1381–1388, 2004.

164. Raines EW, Ross R. Biology of atherosclerotic plaque formation: possible role of growth factors in lesion development and the potential impact of soy. *J. Nutr.* 125, 624S–630S, 1995.

165. Dubey RK, Gillespie DG, Imthurn B et al. Phytoestrogens inhibit growth and MAP kinase activity in human aortic smooth muscle cells. *Hypertension* 33, 177–182, 1999.

166. Kim H, Xu J, Su Y et al. Actions of the soy phytoestrogen genistein in models of human chronic disease: potential involvement of transforming growth factor β. *Biochem. Soc. Trans.* 29, 216–222, 2001.

167. Grainger DJ, Metcalfe JC. Tamoxifen: teaching an old drug new tricks. *Nat. Med.* 2, 381–385, 1996.

168. Sanders TAB, Dean TS, Grainger D, Miller GJ, Wiseman H. Moderate intakes of soy isoflavones increase HDL but do not influence transforming growth factor β concentrations and haemostatic risk factors for coronary heart disease in healthy subjects. *Am. J. Clin. Nutr.* 76, 373–377, 2002.

169. Thorneycroft IH. The role of oestrogen replacement therapy in the prevention of osteoporosis. *Am. J. Obstet. Gynecol.* 160, 1306–1310, 1989.

170. Fentiman IS, Wang DY, Allen DS et al. Bone density of normal women in relation to endogenous and exogenous oestrogens. *Br. J. Rheumatol.* 33, 808–815, 1994.

171. Gambacciani M, Ciaponi M, Cappagli B, Piaggesi L, Genazzani AR. Effects on combined low dose of the isoflavone derivative ipriflavone and estrogen replacement on bone mineral density and metabolism in postmenopausal women. *Maturitas* 28, 75–81, 1997.

172. Anderson JJ, Garner SC. Phytoestrogens and bone. *Ballieres Clin. Endocrinol. Metab.* 12, 543–557, 1998.

173. Messina M, Ho S, Alekel DL. Skeletal benefits of soy isoflavones: a review of the clinical trial epidemiologic data. *Curr. Opin. Clin. Nutr. Metab. Care* 7, 649–658, 2004.

174. Lydeking-Olsen E, Beck-Jensen JE, Setchell KD, Holm-Jensen T. Soymilk or progesterone for prevention of bone loss: a 2 year randomised, placebo-controlled trial. *Eur. J. Nutr.* 43, 246–257, 2004.

175. Atkinson C, Compston JE, Day NE, Dowsett M, Bingham SA. The effects of phytoestrogen isoflavones on bone density in women: a double-blind, randomised, placebo-controlled trial. *Am. J. Clin. Nutr.* 79, 326–333, 2004.

176. Chen Y-M, Ho SC, Lam SSH et al. Soy isoflavones have a favourable effect on bone loss in Chinese postmenopausal women with lower bone mass: a double-blind, randomized, controlled trial. *J. Clin. Endocrinol. Metab.* 88, 4740–4747, 2003.

177. Uesugi T, Fukui Y, Yamori Y. Beneficial effects of soybean isoflavone supplementation on bone metabolism and serum lipids in postmenopausal Japanese women: a four-week study. *J. Am. Coll. Nutr.* 21, 97–102, 2002.

178. Hanna K, Wong J, Patterson C, O'Neill S, Lyons-Wall P. Phytoestrogen intake, excretion and markers of bone health in Australian women. *Asia Pacific J. Clin. Nutr.* 13 (Suppl.), S74, 2004.

179. Ettinger B, Black DM, Mitlak BH et al. Reduction of vertebral risk in postmenopausal women with osteoporosis treated with raloxifene. Results from a 3-year randomized clinical trial. *J. Am. Med. Assoc.* 282, 637–645, 1999.

180. Eden J. Phytoestrogens and the menopause. *Ballieres Clin. Endocrinol. Metab.* 12, 581–587, 1998.

181. Murkies AL, Lombard C, Strauss BIG et al. Dietary flour supplementation decreases postmenopausal hot flushes: effect of soy and wheat. *Maturitas* 21, 189–195, 1995.

182. Van Der Weijer P, Barentsen R. Isoflavones from red clover (Promensil(R)) significantly reduce menopausal hot flush symptoms compared with placebo. *Maturitas* 42, 187, 2002.

183. Huntley AL, Ernst E. Soy for the treatment of perimenopausal symptoms — a systematic review. *Maturitas* 47, 1–9, 2004.

184. Krebs EE, Ensrud KE, MacDonald R, Wilt TJ. Phytoestrogens for treatment of menopausal symptoms: a systematic review. *Obtophystet. Gynecol.* 104, 824–836, 2004.

185. File SE, Jarrett N, Fluck E et al. Eating soya improves human memory. *Psychopharmacology* 157, 430–436, 2001.

186. File SE, Duffy R, Wiseman H. Soya improves human memory. In *Soy and Health, Clinical Evidence and Dietetic Applications* (CK Descheemaeter, I Debruyre, editors), Garant Apeldoorn-Antwerp, pp. 167–173, 2002.

187. Duffy R, Wiseman H, File S. Improved cognitive function in postmenopausal women after 12 weeks of consumption of a soy extract containing isoflavones. *Pharmacol. Biochem. Behav.* 75, 721–729, 2003.

188. Kreijkamp-Kaspers S, Kok L, Grobbee DE et al. Effect of soy protein containing isoflavones on cognitive function, bone mineral density and plasma lipids in postmenopausal women. *J. Am. Med. Assoc.* 292, 65–74, 2004.

189. Kreijkamp-Kaspers S, Kok L, Grobbee DE et al. Effect of soy protein containing isoflavones on cognitive function, bone mineral density and plasma lipids in postmenopausal women. *Obstet. Gynecol. Surv.* 60, 41–43, 2005.

190. Adams NR. Detection of the effects of phytoestrogens on sheep and cattle. *J. Anim. Sci.* 73, 1509–1515, 1995.

191. Setchell KD, Gosselin SJ, Welsh MB et al. Dietary oestrogens — a probable cause of infertility and liver disease in captive cheetahs. *Gastroenterology* 93, 225–233, 1987.

192. Franke AA, Yu MC, Maskarinec G et al. Phytoestrogens in human biomatrices including breast milk. *Biochem. Soc. Trans.* 27, 308–318, 1999.

193. Strom BL, Schinnar R, Ziegler EE et al. Exposure to soy-based formula in infancy and endocrinological and reproductive outcomes in young adulthood. *J. Am. Med. Assoc.* 286, 807–814, 2001.

194. Setchell KD, Zimmer-Nechemias L, Cai J, Heubi JE. Exposure of infants to phyto-oestrogens from soy-based infant formula. *Lancet* 350, 23–27, 1997.

195. Chen A, Rogan WJ. Isoflavones in soy infant formula: a review of evidence for endocrine and other activity in infants. *Annu. Rev. Nutr.* 24, 33–54, 2004.

196. Merritt RJ, Jenks BH. Safety of soy-based infant formulas containing isoflavones: the clinical evidence. *J. Nutr.* 134, 1220S–1224S, 2004.

197. Giampietro PG, Bruno G, Furcolo G et al. Soy protein formulas in children: no hormonal effects in long term feeding. *J. Pediatr. Endocrinol. Metab.* 17, 191–196, 2004.

198. Divi RL, Chang HC, Doerge DR. Anti-thyroid isoflavones from soybean: isolation, characterization and mechanisms of action. *Biochem. Pharmacol.* 54, 1087–1096, 1997.

199. Bruce B, Messina M, Spiller GA. Isoflavone supplements do not affect thyroid function in iodine replete postmenopausal women. *J. Med. Food* 6, 309–316, 2003.

200. Munro IC, Harwood M, Hlywka JJ et al. Soy Isoflavones: a safety review. *Nutr. Rev.* 61, 1–33, 2003.

201. British Nutrition Foundation, Briefing paper, 2002.

8 Flavonoid Functions in Plants

Kevin S. Gould and Carolyn Lister

CONTENTS

8.1 INTRODUCTION

The flavonoids are a remarkable group of plant metabolites. No other class of secondary product has been credited with so many — or such diverse — key functions in plant growth and development. Many of these tasks are critical for survival, such as attraction of animal vectors for pollination and seed dispersal, stimulation of *Rhizobium* bacteria for nitrogen fixation, promotion of pollen tube growth, and the resorption of mineral nutrients from senescing leaves. Others provide a competitive edge to plants that grow under suboptimal environments. Flavonoids, for example, are known to enhance tolerance to a variety of abiotic stressors, they are employed as agents of defense against herbivores and pathogens, and they form the basis for allelopathic interactions with other plant species. The flavonoids are evidently extremely useful to plants, and it is not surprising, therefore, that species from all orders of the plant kingdom, from the basal liverworts to the most advanced angiosperms, invest significant amounts of metabolic energy into the production of these compounds.

The past decade has witnessed resurgence in research activity on the functions of flavonoids in plants. There are several reasons for this. First, advances in molecular biology, coupled with an improved knowledge of the pathway for flavonoid biosynthesis, have led to the production of plant mutants that are deficient or superabundant in one or more flavonoid pigments. Comparisons of mutant and wild-type phenotypes have permitted hypotheses for flavonoid function to be tested directly. Second, improvements in analytical techniques (e.g., high-performance liquid chromatography, liquid chromatography–mass spectrometry, and nuclear magnetic resonance spectroscopy) for flavonoid compounds have stimulated the search for novel compounds useful for the manipulation of flower color. These, in turn, prompted the discovery of hitherto unknown functions of flavonoids in plant reproduction. Third, concerns about the enlarging ozone hole and the increased exposure of biota to ultraviolet (UV) radiation led to the quest for sunscreens — and to the knowledge that some flavonoids play an important role in protecting plants from harmful UV-B levels. Fourth, there has been an explosive interest in flavonoids, particularly the anthocyanins, as potential nutritional supplements for humans. This contributed to the discovery of their antioxidant roles *in planta*. Finally, advances in field-portable instrumentation have enabled hypotheses for flavonoid function to be tested directly in the field.

In this chapter, we review experimental and theoretical evidence for the main hypotheses for flavonoid functions in plants. Our discussion distinguishes between the functions of the red and blue flavonoids (anthocyanins and 3-deoxyanthocyanins) and those of the colorless (or yellow) remainder. Recent evidence indicates that these two subsets differ markedly both in range and type of functions and, for many species, in their cellular location within the plant tissue. The property of anthocyanin molecules to absorb green light, for example, affords unique capabilities, such as the protection of chloroplasts from the damaging effects of strong irradiance, and as a visible cue to some animals. Flavonols and flavones, on the other hand, do not directly affect photosynthesis, but they can act as chemical signals or UV guides to attract or deter insects, and are highly effective UV filters. Thus, the comparison of the "colorful" versus "colorless" flavonoids provides an instructive insight into the divergent evolution of the roles of flavonoids from a common biosynthetic pathway.

8.2 ANTHOCYANINS AND 3-DEOXYANTHOCYANINS

8.2.1 DEFENSE

One of the longest-standing hypotheses for the presence of anthocyanins in leaves — that they serve to protect plants against herbivory or pathogenic attacks — has received strong theoretical support in recent years, yet still lacks compelling direct evidence. Anthocyanins have been implicated as aposematic colorants, as insect feeding-deterrents, and as antifungal agents, but there is as yet no good reason to believe that defense is the unifying, or indeed even a primary reason for the production of these pigments in vegetative tissues.[1,2]

The anthocyanins, in contrast to certain other phenolic compounds (see Section 8.3.4), are not toxic to higher animal species.[3,4] For the invertebrates, too, anthocyanin toxicity seems to be the exception rather than the rule. The most abundant foliar anthocyanin, cyanidin-3-O-glucoside, has been reported to inhibit the growth of larvae of the tobacco budworm, *Heliothis virescens*, an important pest of cotton and other crops.[5] However, anthocyanin-rich extracts did not influence the feeding behavior or survival rate of aphids, nor inhibit larval growth of the fruitworm.[6–8] Fecundity was reduced in aphids that had been fed on preinfested, red portions of *Sorghum* leaves; the response could not be causally attributed to anthocyanins, however, because leaves that had turned red from water stress, rather than preinfestation, had no effect on fecundity.[9] It seems unlikely, therefore, that anthocyanin toxicity plays any major role in the defense against insect grazers.

There is, nonetheless, evidence that some insects preferentially avoid eating red-leafed plants. California maple aphids *Periphyllus californiensis*, for example, have been observed to colonize the yellow-orange leaves of Japanese maples, yet almost entirely avoid red-leafed individuals.[10,11] Hamilton and Brown[12] attributed such discriminatory behavior to the effects of aposematic coloration. The red colors in autumn leaves, they reasoned, serve as an honest signal of defensive commitment against insect herbivores. In their survey of the literature on 262 north-temperate tree species, autumn foliage coloration was noted to be strongest in those species that face a high diversity of damaging specialist aphids. The authors suggested that the red light reflected from anthocyanic leaves supplies a "pick on someone else" warning signal to the aphids, which are known to use color cues in host selection. One problem with this hypothesis, however, is that red light is thought to lie beyond the perception of most aphid species[13]; the red warning is being issued to a blind audience! An alternative explanation could be that the red coloration renders leaves unattractive to the insects. Aphids are most attracted to green or yellow light, the very waveband that is deficient in the spectrum that is reflected from anthocyanic leaves.[14] Thus, red foliage may simply appear unpalatable to potential herbivores. The costs to the plant associated with the biosynthesis of anthocyanin pigment may be more than offset by the gains to be had from herbivore deterrence.

Similar hypotheses have been proposed to explain the presence of anthocyanins in the young leaves of growth flushes in many tropical, and some temperate plant species. When such leaves lack chlorophyll they are particularly conspicuous, at least to the human eye, varying from scarlet to red, crimson, mauve, or blue.[15–18] The absence of chlorophyll means that flushing red leaves hold less nitrogen and other minerals than do young green leaves, and this may make them a less attractive meal for potential herbivores. Moreover, when chlorophyll is eventually synthesized, the leaves are already fully expanded and possess tough cellular features that would impair their digestibility.[19] The red colors might, in addition, serve to undermine the visual camouflage of invertebrates; herbivory would be reduced by exposing the browsing insects to their predators.[20]

For many species, however, leaf flushes contain appreciable amounts of chlorophyll in addition to the anthocyanins,[21] and the pigment combination generates brown or almost black colors if in sufficient concentration. Dark colors can camouflage leaves against the exposed soil and litter of forest floors.[22,23] They may also serve to mimic dead leaves.[24] Indeed, even the brilliant red and crimson flushing leaves might appear dark or dead to a potential herbivore, since most nonmammalian folivores lack red light receptors.[25] Juniper[18] noted that flushing leaves often appear flaccid and hang down vertically from the branches "like wet facial tissue." He argued that the color, texture, and orientation of such leaves would deter predation at this vulnerable stage in development by obscuring the characteristic "leaf cues" as perceived by an insect. Anthocyanin biosynthesis was considered a significant part of this complex strategy to ensure leaf longevity.

An alternative postulate for red coloration in tropical young leaves is that the anthocyanins protect against fungal infection. Such protection may be critical during the vulnerable period of leaf expansion before the formation of a protective cuticle and lignified cell walls. In support of this hypothesis, Coley and Aide[26] showed that the leaf-cutting ant *Atta columbica* preferentially avoids red-leafed species. The ant cultivates fungus on acquired leaf pieces, and it is the fungus, rather than the leaves, that serves as the sole food source for larvae and much of the diet for workers and queen. The ant avoids cutting leaves that contain antifungal compounds, perhaps because of its dependence on the fungus. In feeding trials using leaf discs from 20 plant species, ant preference decreased significantly as the anthocyanin content increased. Moreover, when oat flakes impregnated with cyanidin chloride were supplied as the only food source, the ants again showed dosage-dependant avoidance. The authors suggested that even low concentrations of anthocyanins might be detrimental to fungal colonies.

There is not, however, very much *direct* evidence for fungicidal properties of the anthocyanin pigments. This is in sharp contrast to certain 3-deoxyanthocyanidins, which have been shown to be highly effective agents against both fungi and bacteria. Apigeninidin significantly inhibited the growth of *Fusarium oxysporum*, *Gibberella zeae*, *Gliocladium roseum*, *Alternaria solani*, and *Phytophthora infestans* on agar plates.[27] It also inhibited the growth of certain gram-positive bacteria (*Bacillus cereus*, *Staphylococcus aureus*, *Staphylococcus epidermidis*, and *Streptococcus faecalis*) and, to a lesser extent, gram-negative bacteria (*Escherichia coli*, *Serratia marcens*, and *Shigella flexneri*) on agar.[28] Antifungal properties of apigeninidin have been postulated to play a role in mold resistance in *Sorghum* seeds, where the 3-deoxyanthocyanidin is abundant.

Although anthocyanins do not seem to have mycocidal properties *per se*, they may still function to protect some plants from invading pathogens. Page and Towers[29] recently demonstrated a novel mechanism whereby a mix of cyanidin-3-*O*-glucoside and cyanidin-3-*O*-(6-*O*-malonylglucoside) could confer protection to silver beachwood (*Ambrosia chamissonis*). This plant holds large amounts of thiarubrine pigments in its stems, petioles, and roots. The thiarubrines, which are toxic to a variety of organisms including insects, bacteria, and fungi, contain a 1,2-dithiin chromophore that is highly unstable in light.[30–33] Even short exposures to visible light or UV radiation convert the red thiarubrine A to 2,6-dithiabicyclo[3.1.0]hexane intermediates, and then to the colorless, inactive, thiophene A (Figure 8.1). *A. chamissonis* naturally grows at sunny locations along the Pacific coast of North America, and requires, therefore, a mechanism to prevent the decomposition of these photolabile defense compounds. In the stems and petioles, thiarubrine A is held in subepidermal laticifers surrounded by a sheath of anthocyanin-containing cells. The absorbance spectra for cyanidin-3-*O*-glucoside and thiarubrine A are strikingly similar in the visible region of the light spectrum. Thus, anthocyanins could contribute to defensive armory by intercepting quanta that would otherwise degrade the potent, toxic thiarubrines. Consistent with this hypothesis, surgical removal of the anthocyanic sheath led to rapid destruction of thiarubrine A under the light of a microscope.[29] Furthermore, thiarubrines in the roots, which would not normally experience high irradiances, are not enshrouded by anthocyanins. The metabolic investment into anthocyanin biosynthesis is evidently part of a coordinated strategy to ameliorate defense responses, rather than an extravagant byproduct of a saturated flavonoid pathway. It seems highly likely that other photolabile protectants could also benefit from an anthocyanin shield. The light-screening hypothesis for anthocyanin function as a defense agent presents an exciting new avenue for future research.

8.2.2 PROTECTION FROM SOLAR ULTRAVIOLET

The anthocyanins have often been included alongside other flavonoids and the hydroxycinnamic acids as potential UV protectants. Most anthocyanins absorb UV radiation between

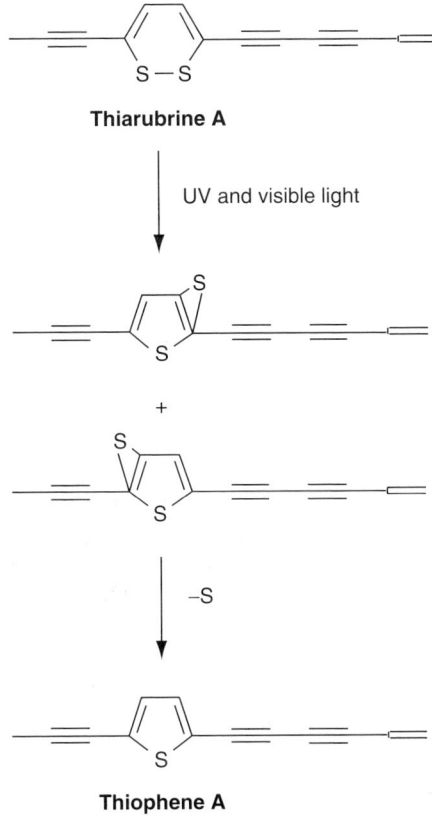

FIGURE 8.1 Conversion of the antimicrobial agent, thiarubrine A, to its photoproduct, thiophene A, by exposure to ultraviolet and visible radiation. (Redrawn from Page, J.E. and Towers, G.H.N., *Planta*, 215, 478, 2002. With permission.)

270 and 290 nm, and the acylated anthocyanins, in particular, are strong UV-B absorbers.[34,35] Red leaves often reflect significantly less solar UV than do green leaves as a consequence of their higher UV absorbance.[36] If the energy of the absorbed UV is not transmitted to cellular organelles, then anthocyanins could serve to moderate the damage to DNA, proteins, and membranes in plants that grow naturally in high UV-B environments.

Consistent with a protective role against solar UV, anthocyanin biosynthesis has been observed many times to be induced or upregulated in plants following exposure to supplementary UV radiation,[37–52] although exceptions have been noted.[53–57] There are also reports describing more direct evidence for anthocyanin involvement in UV protection. For example, in cultured cells of the cornflower *Centaurea cyanus*, the presence of cyanidin 3-*O*-(6-*O*-malonyl) glucoside appreciably reduced the extent of damage to DNA following UV-B or UV-C irradiation.[58] Similarly, an anthocyanin-rich extract from red apple skins protected against UV-B-induced damage to plasmid DNA,[59] and flavonoids from a red-leafed maize mutant significantly impeded the dimerization of adjacent pyrimidines on the same strand of DNA, one of the key indicators of DNA damage (Figure 8.2), in tissues irradiated with UV-B or UV-C.[60] Mutants of *Arabidopsis thaliana* that are deficient in flavonoids, including the anthocyanins, but which have normal levels of sinapate esters, are more sensitive to UV-B radiation than is the wild type when grown under high irradiance.[61] Finally, red-leafed *Coleus*

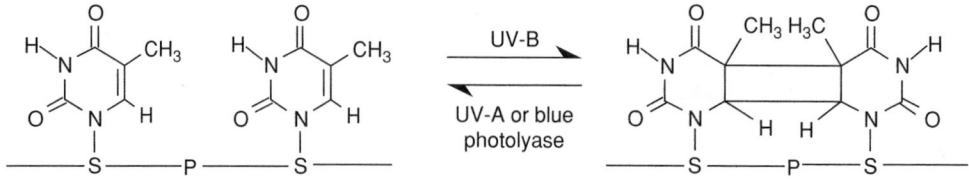

FIGURE 8.2 Formation of a thymine–thymine dimer by UV-B radiation, and repair by UV-A or blue light-activated photolyase.

varieties that had been pretreated with UV-B or UV-C retained greater photosynthetic efficiencies under strong light than did green-leafed varieties, indicating that their acylated anthocyanins effectively shielded the proteins associated with photosynthetic electron transport.[62]

Notwithstanding these data, a substantial body of recent empirical evidence argues strongly against any major involvement of anthocyanins in the protection from UV. First, in most red-leafed species that have been examined, the anthocyanins do not reside at the most suitable cellular location for UV screening. An effective UV filter needs to remove the short-wavelength UV component before it reaches the chloroplasts, at the same time providing minimal attenuation of the photosynthetically active radiation.[63] Because chloroplasts are generally located in the subepidermal tissues of vegetative shoots, the outermost layer, known as the dermal system, is considered to be the most appropriate site for UV interception.[64] UV-screening pigments must be present in epidermal cell walls as well as in the vacuole if they are to prevent significant transmission of UV light to subjacent chlorenchyma.[65–67] Colorless phenolic compounds, including certain flavonoids and hydroxycinnamic acids, have indeed been found in epidermal cell vacuoles, walls, cuticles, and trichomes in the leaves of many species (Table 8.1).[68–80] The anthocyanins, too, have been found in epidermal cell vacuoles of roots,[81,82] various floral organs,[67,83,84] and some leaves, including *Arabidopsis*.[85] Cell-wall-bound anthocyanins and "anthocyanin-like pigments" have been observed in the leaves of certain liverwort and moss gametophytes.[86–88] In most red-leafed species, however, the anthocyanins are present as solutions inside the vacuoles of palisade and spongy mesophyll cells, the very cells that require protection from the effects of excess UV radiation. This seems to be equally true of senescing leaves[21,89] as of expanding and mature leaves.[90–92] Thus, although the red cell vacuoles might well capture stray rays of UV light, their location within the internal tissues of the leaf precludes any major role as a UV filter.

Several studies involving plant mutants have also concluded that factors other than anthocyanin pigmentation contribute to UV protection. For example, Lois and Buchanan[103] identified in *Arabidopsis* a mutation that led to dramatic increases in sensitivity to UV radiation. The mutant was found to be deficient in certain flavonoids, particularly a rhamnosylated derivative of kaempferol, yet it held normal amounts of anthocyanin. Similarly, Klaper et al.[104] found that neither a deficiency nor a surfeit of anthocyanin influenced the growth responses of *Brassica rapa* mutants to supplementary UV-B treatment. The authors suggested that flavonoids other than the anthocyanins might respond more rapidly and to a greater extent to UV-B exposure. Red-leafed plants have been noted to contain high levels of colorless phenolic compounds which contribute significantly more than anthocyanins to the total absorbance of UV-A and UV-B radiation.[105]

There is, moreover, evidence that for some species the presence of anthocyanins can impair, rather than enhance plant performance under UV radiation. Ryan et al.[106] noted that the transgenic *leaf color* line of *Petunia*, which had elevated levels of anthocyanins resulting from

TABLE 8.1
Examples of Plant Species from which UV-B Protective Flavonoids Have Been Identified

Plant Species	Flavonoid Location	Protective Flavonoids	Ref.
Arabidopsis thaliana	Epidermal cells	Kaempferol-3-gentiobioside-7-rhamnoside; kaempferol-3,7-dirhamnoside	94
Hordeum vulgare	Epidermal cells	Saponarin; lutonarin	95
Brassica oleracea	Epidermal cells	Cyanidin glycosides	96
Zea mays	Epidermal cells	Anthocyanins	60
Gnaphalium luteo-album	Leaf wax	Calycopterin; 3'-methoxycalycopterin	97
Gnaphalium vira-vira	Leaf wax	7-*O*-Methylaraneol	98
Marchantia polymorpha	Thallus cells	Luteolin-7-glucuronide; luteolin-3,4'-diglucuronide	99
Sinapus alba	Epidermal cells	Anthocyanin glycosides; quercetin glycosides	54
Oryza sativa	Epidermal cells	Iso-orientin acylated glucosides	100
Pinus sylvestris	Epidermal cells	3'',6''-di-*p*-Coumarylkaempferol-3-glucoside; 3'',6''-di-*p*-coumarylquercetin-3-glucoside	78
Brassica napus	Epidermal cells	Quercetin-3-sophoroside-7-glucoside; quercetin-3-sinapylsophoroside-7-glucoside	101
Quercus ilex	Leaf hairs	Acylated kaempferol glycosides	102

Source: Adapted from Harborne, J.B. and Williams, C.A., *Phytochemistry*, 55, 481, 2000. Copyright (2000), With permission from Elsevier.

the action of a maize flavonoid regulatory gene, grew slower than wild-type plants in a UV-B enriched environment. Likewise, the anthocyanic seedlings of *Impatiens capensis* accumulated less biomass and produced fewer flowers and fruits than did green seedlings after UV stress,[107] and purple-leaf mutants of rice grew less, and had more evidence of DNA damage after UV treatment than did green-leafed lines with a similar genetic background.[46,108]

To understand the apparent detrimental effect of anthocyanins on the growth of UV-irradiated plants, Hada et al.[109] compared the rates of synthesis of cyclobutane pyrimidine dimers (CPDs) and of DNA repair in the near-isogenic lines of purple- and green-leafed rice. Short-term exposures to UV-B caused greater CPD production in the green than in the purple line. Under continuous UV exposure, however, the trend reversed; the purple line accumulated significantly more CPDs than did the green line. The difference was attributable to rates of CPD repair, which ran substantially faster in the green line. To repair UV-damaged DNA, plants employ photolyase, an enzyme that uses blue or UV-A light to re-monomerize the pyrimidine dimers (Figure 8.2). The anthocyanins in purple rice absorbed the energy of the incident UV-A, effectively preventing the photoactivation of photolyase. Thus, any short-term benefit to be gained from the absorption of UV-B by anthocyanins is offset by their property to limit the rate of DNA repair. Protection from the effects of solar UV cannot be the primary function of anthocyanins in shoots, although other flavonoids assume this role (see Section 8.3.1.1).

8.2.3 PHOTOPROTECTION

When plants receive more light energy than can be used for photochemistry, they show a characteristic decline in the quantum efficiency of photosynthesis, termed photoinhibition (reviewed in Ref. 110). Most plants experience photoinhibition at some point over the course of their lives, often on a regular or even a daily basis. The excess quanta from photoinhibitory light fluxes can usually be channeled into quenching mechanisms such as xanthophyll pigments associated with the chloroplasts, which ultimately release the energy as heat. When the excess energy can be contained in this way, photoinhibition is said to be *dynamic*; it is reversible, usually short lived, and causes no long-term damage to the photosynthetic apparatus. However, prolonged exposures to strong light, particularly when combined with other environmental stressors such as cold or heat, can saturate the quenching mechanisms and lead to *chronic* photoinhibition. Reactive oxygen species are then generated, with the potential to destroy thylakoid membranes and to denature proteins associated with photosynthetic electron transport, particularly those of photosystem II. Characterized by chlorophyll bleaching or necrotic lesions, chronic photoinhibition can lead to long-term damage or even the death of the plant.

A substantial body of empirical evidence indicates that anthocyanins in leaves can reduce the severity of photoinhibitory damage to plants under stress. A photoprotective role for anthocyanins is hardly a new idea; as early as the 19th century, the German physiologist Pringsheim suggested that anthocyanins might protect the photosynthetic machinery by screening out the most damaging wavelengths (cited in Ref. 111). Only recently, however, have technological advances in field-portable instrumentation permitted the hypothesis to be tested directly. Foremost among these are the pulse-modulated chlorophyll fluorometers that are used to dissect photosynthetic function in the intact plant under field conditions. The kinetics of the generation and decay of chlorophyll fluorescence provide information on quantum efficiency, rates of electron transport, and energy quenching processes.[112,113] Recent data suggest that chlorophyll fluorescence parameters can differ markedly between red and green leaves under light stress.

Central to the photoprotection hypothesis is the property of anthocyanins to absorb visible radiation. Solutions of anthocyanins in the vacuoles of living cells typically show strong absorbance in the 500 to 600 nm waveband (Figure 8.3). Consequently, red leaves absorb substantially more yellow-green light than do acyanic (green) leaves of comparable age and structure.[14,62,92,114–120] The amount of extra light that is absorbed (often more than 20% greater than green leaves) is a direct function of the amount of anthocyanin present, irrespective of the location of the red cells within the leaf tissues. Interestingly, the amount of red light that is reflected from red leaves often only poorly correlates to anthocyanin content;

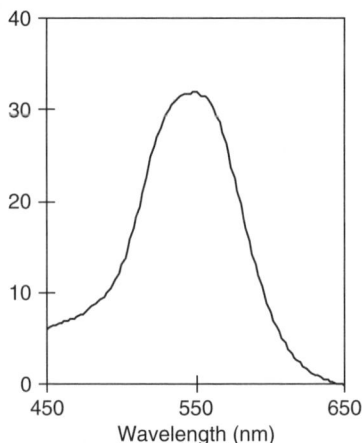

FIGURE 8.3 *In vivo* absorptance spectrum for a single anthocyanic cell vacuole in a leaf of *Aristotelia serrata*. (K.S. Gould, unpublished data.)

chlorophyll content and morphological and anatomical features of the leaf surface evidently influence red reflectance to the greater extent.

The fate of these extra absorbed green quanta is not known. It is very clear, however, that the energy is not transferable to the chloroplasts for use in photosynthesis. Profiles of the absorption of red, blue, and green light within the various leaf tissues have been compared recently for red and green leaves of the New Zealand tree *Quintinia serrata*.[121] The presence of anthocyanin in a cell vacuole dramatically reduced the amount of green light that could penetrate down to the lowermost tissues in the leaf. When anthocyanins were present in the upper epidermal layer, green light utilization was restricted to an extremely narrow band of palisade mesophyll at the top of the leaf. Similarly, when anthocyanins were located in the palisade mesophyll layers, the subjacent spongy mesophyll cells were deprived of green light. Thus, although red leaves absorb the more green light in total, their photosynthetic tissues actually receive less green light than do those of green leaves. Consistent with previous observations of grape leaves,[122] anthocyanins did not affect the distribution of blue light within the leaf, and they modified the gradient of red light only slightly.

Green light is an important contributor to photosynthesis, especially in the lower spongy mesophyll tissues.[123,124] A deficit of green light should, therefore, reduce the overall photosynthetic productivity. Accordingly, red leaves have been found to have a lower quantum efficiency of photosynthesis, and a higher threshold irradiance at which saturation of photosynthesis is achieved, relative to green leaves. Red leaves can also display some physiological characteristics of shade plants, possibly as a result of having chloroplasts develop under the anthocyanin optical filter.[121,125] The differences between photosynthesis of red and green leaves are usually small, however, and are often statistically insignificant,[120,126–129] although more substantial differences have been noted for certain species.[62]

Although anthocyanins seem to have only a modest effect on photosynthesis under nonsaturating light, they can nonetheless impact dramatically on the propensity for photoinhibition under conditions of strong light. By absorbing the green, high-energy photons that would otherwise excite chlorophyll *b*, the anthocyanins have the potential to reduce the frequency of dynamic photoinhibition, moderate the severity of chronic photoinhibition, and expedite photosynthetic recovery. Photoprotective advantages of red versus green leaves have been demonstrated many times in disparate plant species, as evidenced from the kinetics of chlorophyll fluorescence during and following light stress.[115,129–134] For *Cornus stolonifera*, for example, 30-min exposure to white light at $1500\ \mu mol\ m^{-2}\ sec^{-1}$ caused a 60% reduction in the quantum yield of red leaves, but an almost 100% reduction in acyanic leaves (Figure 8.4). When the light was turned off, quantum yields of the red leaves recovered to their maximum value after only 80 min, yet the acyanic leaves did not reach their pretreatment state even after 6 h.

Foliar anthocyanins are evidently an effective device for photoprotection under high light environments. They do not, however, substitute for the protection conferred by xanthophyll-cycle pigments. Recent experiments using flavonoid- or xanthophyll-deficient mutants of *A. thaliana* indicate that the xanthophylls have a greater role in the protection of plants from short-term light stress, but the flavonoids, including the anthocyanins, are the more effective photoprotectants in the long term.[135]

The photoprotection hypothesis for anthocyanins is attractive because it can explain why red-leafed plants occur across such diverse environments. In the tropics, for example, anthocyanins in flushing red leaves may provide a critical photoprotective role to the nascent chloroplasts until adequate levels of xanthophylls have been synthesized.[136] In the understorey, anthocyanins might protect shade-acclimated plants from the effects of sunflecks, which can be 2000-fold brighter than the usual light level, or canopy gaps caused by tree blowdown.[131,137] A photoprotective function of anthocyanins would also benefit leaves in

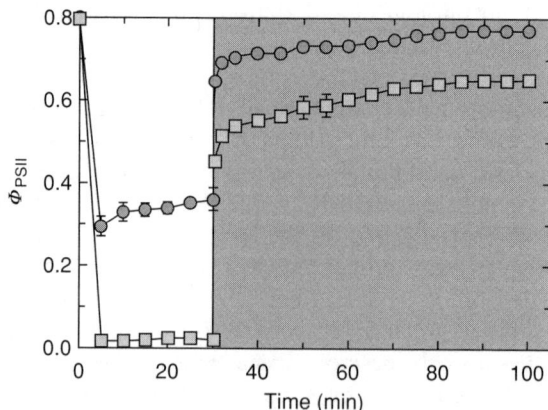

FIGURE 8.4 Changes in the quantum yield of photosystem II (ϕ_{PSII}) following irradiation of red (circles) and acyanic (squares) leaves of *Cornus stolonifera*. The strong, white light ($1500\,\mu mol\ m^{-2}\ sec^{-1}$) was turned off after 30 min, as indicated by the shaded box. (Redrawn from Feild, T.S., Lee, D.W., and Holbrook, N.M., *Plant Physiol.*, 127, 566, 2001. With permission.)

sunny, cold climates where, because of reduced Calvin cycle activity, the absorbed light energy is in excess of the capacity to utilize ATP and NADPH in photosynthesis.[132] The photoprotection hypothesis has even been used to explain the *de novo* synthesis of anthocyanins in the autumn foliage of deciduous trees. Autumn senescence involves the rapid liberation of nitrogen from the degradation of stroma proteins and thylakoid membranes, and its resorption into the branches. If degrading chlorophyll is not protected from light, however, reactive oxygen is formed that could jeopardize the resorptive process. As an optical screen, anthocyanins may provide photoprotection during the dismantling of the photosynthetic apparatus.[115,138–142] In support of this hypothesis, the efficiency of nitrogen resorption from senescing leaves has been found to be significantly greater in wild type than in anthocyanin-deficient mutants of three deciduous woody species subjected to high light and low temperatures.[143]

The photoprotection hypothesis is not without its critics,[62] and it certainly does not account for every occurrence of anthocyanins, such as in the leaf veins, in stems and petioles, and in underground organs such as roots and tubers. Nonetheless, there is now a compelling body of empirical data consistent with a photoprotective function in the photosynthetic cells of leaves. Photoprotection might well be the primary role of anthocyanin in the vegetative shoots of many species.

8.2.4 ANTIOXIDANT ACTIVITY

Over the past decade, several hundred research articles have been published highlighting the antioxidant potency of anthocyanin-rich extracts from fruits and vegetables. It is evident from *in vitro* studies that the anthocyanins are exceptionally strong antioxidants; they scavenge almost all species of reactive oxygen and nitrogen with an efficiency up to four times greater than those of ascorbic acid and α-tocopherol.[144–148] In experiments involving rats or animal cell cultures subjected to oxidative stress, anthocyanins have been found to reduce significantly the extent of DNA damage, lipid peroxidation, and low-density lipoprotein oxidation.[149–153] It is reasonable, therefore, to ask whether anthocyanins might serve a similar function in plants — to ameliorate defense responses against oxidative stress caused by growing in unfavorable environments.

A growing body of experimental evidence does indeed indicate that anthocyanins contribute to the control of levels of reactive oxygen in plant cells. In apple fruit, for example, chlorophyll bleaching (a symptom of photooxidative damage) was considerably lower in red than in green regions after exposure to strong light.[154] Similarly, the combination of strong light and low temperatures caused significantly more lipid peroxidation in anthocyanin-deficient mutants of *Arabidopsis* than in wild type.[135] Gamma irradiation triggered the biosynthesis of anthocyanin in wild-type *Arabidopsis* plants, which were subsequently better scavengers of superoxide radicals than were nonirradiated or anthocyanin-deficient mutants.[155] Moreover, only those plants that contained both anthocyanin and ascorbic acid were able to grow normally and flower after gamma irradiation. Anthocyanins, therefore, can hold the key to survival under severe conditions of oxidative stress.

Anthocyanins have the potential to moderate the total oxidative load via three mechanisms. First, they can chelate to copper and iron, thereby decreasing the possibility of hydroxyl radical production from Haber–Weiss reactions.[156–159] These chelates might also protect other low molecular weight antioxidants (LMWAs), such as ascorbate and α-tocopherol, from autoxidation by transition metals.[160] Anthocyanin–transition metal chelation has been demonstrated *in vitro* many times,[161,162] but is unlikely to feature significantly *in planta*.

The property of anthocyanins to attenuate light presents a second possible mechanism for reducing the oxidative load. Chloroplasts, mitochondria, and peroxisomes are the chief sources of reactive oxygen in plant cells. Under conditions of excess light, (i) chlorophyll excitation leads to the production of singlet oxygen, (ii) electrons leak from reduced ferredoxin in photosystem I to molecular oxygen, forming superoxide radicals, and (iii) hydrogen peroxide may be produced in the peroxisomes by the C2 oxidative photosynthesis cycle.[163–165] In theory, therefore, any process that serves to reduce the amount of light incident on chloroplasts, such as the optical masking of chlorophyll by anthocyanins, should also diminish the numbers of reactive oxygen species produced by photooxidation.[115,166] The effectiveness of photoabatement by anthocyanins was demonstrated recently using a model system comprising suspensions of chloroplasts isolated from lettuce leaves.[119] When used to shield the illuminated chloroplasts, a red cellulose filter whose optical properties approximated that of anthocyanin at pH 1 effected a 33% decline in numbers of superoxide radicals, and a corresponding 37% reduction in chlorophyll oxidation, as compared to unshielded chloroplasts. By absorbing green light and yet permitting the transmission of red and blue wavebands to the chloroplasts, anthocyanins are an efficient device to reduce the oxidative burden without seriously compromising photosynthesis.

The third mechanism for antioxidant activity of anthocyanins is through the direct scavenging or quenching of reactive oxygen and nitrogen. Cyanidin-3-*O*-glucoside, which is believed to be the most common anthocyanin in vegetative shoots, heads the antioxidant league for these pigments. Its high scavenging activity has been ascribed to the reducing power of the two ortho-related hydroxyl groups on the B ring, the oxonium ion in the C ring and the numbers and positions of free hydroxyl groups on the anthocyanin molecule.[156,167] Anthocyanins that donate protons to free radicals become aryloxyl radicals themselves. These may be stabilized through electron delocalization or via self-association,[161,168] or else be regenerated to anthocyanins using other antioxidants such as ascorbic acid.[169]

It could be argued, however, that the vacuolar location of the pigmented anthocyanins precludes any major role in free-radical scavenging *in planta*. Almost all free radicals originate from organelles, the plasma membrane, and the apoplasm of plant cells. Cytoplasmic antioxidants, including ascorbate, α-tocopherol, and the extremely efficient enzyme superoxide dismutase, are therefore more optimally located to scavenge organelle-derived reactive oxygen. Nonetheless, a significant antioxidant advantage of anthocyanic cells over acyanic cells has been demonstrated *in vivo*. In microscopic examinations of mechanically injured leaf tissues, red pigmented cells were found to eliminate H_2O_2 at rates substantially faster than those of green cells.[170] Similarly, slices of sweet potato tubers perfused with H_2O_2 showed rapid inactivation of the oxidant exclusively in those cells that held anthocyanin.[171] In both of those studies, the responses were attributable to anthocyanins *per se*, rather than to other, possibly colorless phenolics associated with the red cells. It is significant that the fluorescent probes used to visualize H_2O_2 did not penetrate the tonoplast of the plant cells; therefore, changes in fluorescence associated with H_2O_2 scavenging were the result of cytoplasmic rather than vacuolar influences. Although anthocyanins are prominent as red solutions inside the cell vacuole, they are thought to be synthesized in the cytoplasm as an equilibrium of colorless tautomers.[172,173] Hence, there is a real possibility that the scavenging of H_2O_2 is undertaken by colorless, cytosolic anthocyanins prior to their transport into the vacuole. Both the colorless and red forms of anthocyanin have strong antioxidant potentials, as measured by lipid peroxidation assays and by cyclic voltammetry.[174–176] Colorless tautomers of cyanidin 3-O-(6-O-malonyl) glucoside at pH 7, the approximate pH of the cytosol, were shown to remove up to 17% of the superoxide radicals generated by irradiated suspensions of chloro-plasts isolated from lettuce.[119] Thus, the cytosolic anthocyanins have the potential to ameli-orate a plant's defense system under conditions of oxidative stress.

Plant tissues normally hold a suite of enzymatic and LMWAs, any combination of which serves to protect membranes and DNA from the effects of reactive oxygen.[177] How import-ant, therefore, is the contribution of anthocyanin to this antioxidant pool? From the limited data available, it appears that reliance on anthocyanins for antioxidant protection varies considerably among species and even across ecotypes. For example, in the shade plant *Elatostema rugosum*, anthocyanins contribute to the LMWA pool more than any other constituent phenolic, and total scavenging activity correlates linearly with anthocyanin con-tent.[175] In contrast, green- and red-leafed morphs of the sun plant *Q. serrata* show similar ranges of scavenging capacities, and use hydroxycinnamic acids as the most active LMWA component.[176] For wild-type *Arabidopsis,* the contribution of anthocyanin to the total super-oxide radical-scavenging activity was almost twice as great in the Landsberg *erecta* ecotype as in the Columbia ecotype.[155] Thus, it would seem that anthocyanin biosynthesis can enhance, but is not usually a prerequisite for protection from oxidative stress. It is likely to be of greatest use under harsh environments where the capacities for other methods of energy dissipation (e.g., via xanthophylls) and free radical scavenging (e.g., by enzymes) have been exhausted.

8.2.5 ANTHOCYANINS AS A GENERALIZED STRESS RESPONSE

There is a strong association between anthocyanin biosynthesis and plant stress. Almost all biotic and abiotic stressors   including herbivory, fungal and viral pathogens, wounding, temperature extremes, high light, UV radiation, mineral nutrient imbalances, drought, salin-ity, anoxia, ozone exposure, and herbicides — can cause anthocyanin levels to increase in vegetative shoots and roots.[178–180] There is evidence that anthocyanins offer protection against many of these stressors. Anthocyanins, for example, have been associated with enhanced tolerance to chilling and freezing temperatures,[181–186] to heavy metals,[187–191] and to water stress.[192–194] Along with certain other flavonoids and compounds such as abscisic acid, the jasmonates, and ethylene, it seems that the anthocyanins may function to mitigate the effects of general stress, and are therefore a useful component of the *general adaptation syndrome*.[195]

There have been several attempts to provide a unifying hypothesis for the mechanism by which anthocyanins provide protection against such a diverse assortment of environmental stressors. Steyn et al.[166] argued that because light can become toxic to green tissues under various conditions of abiotic and biotic stress, the photoprotective properties of anthocyanins are paramount. Chalker-Scott[178,179] provided a compelling case for an osmoregulatory role of anthocyanins in plant cells, since most kinds of suboptimal environments induce water stress, either directly or indirectly. Antioxidant protection is a third potential candidate for the all-encompassing explanation for anthocyanin function,[118] given that supernumerary reactive oxygen and nitrogen species are generated especially by plant tissues under stress. All of these hypotheses warrant further examination in plants from disparate taxonomic groups across diverse ecosystems. It seems equally possible, however, that there is no preeminent function of anthocyanins in vegetative tissues. Anthocyanins and 3-deoxyanthocyanins have been identified among various orders of bryophytes, the most primitive group of land plants with an estimated 470 million years of evolution.[196] It is likely, therefore, that these compounds have been "hijacked" over the course of evolution to perform an array of protective tasks that contribute in different ways and to different degrees to the physiology of plants.[197] Anthocyanins offer the potential for multifaceted, versatile, and effective protection to plants under stress.

8.3 COLORLESS FLAVONOIDS

8.3.1 STRESS PROTECTION

In common with the anthocyanins, the colorless and yellow flavonoids are also inducible by numerous, disparate stressors. The best known of these is probably exposure to UV radiation, although flavonoids also accumulate in response to wounding, pathogen infection, high light, chilling, ozone, or nutrient deficiency. Antioxidant protection provides a possible common thread linking all of these different responses.

8.3.1.1 Ultraviolet Radiation

Unlike the anthocyanins, the colorless flavonoids are thought to be primarily involved in the protection against UV radiation. UV light is a potential hazard to plants because it can damage DNA and impair various physiological processes.[198] Plants often respond to UV light by the activation of flavonoid biosynthetic genes.[199–201] The flavones and flavonols are strongly UV absorbing and accumulate mainly in epidermal cells after UV induction, indicating that they may function as a shield for protecting photosynthetic tissues.[200,202,203] Analyses of mutants defective in flavonoid biosynthesis have indicated the importance of flavonoids for UV tolerance.[103,203] When *Arabidopsis* mutants lacking in flavonols were placed under UV light, growth was retarded.[61] In rye (*Secale cereale*) seedlings, flavonoids have been shown to protect against the damaging effects of shortwave UV on the photosynthetic apparatus.[80]

Fluorescence microscopy has shown that UV-B attenuating compounds are localized mainly in the upper epidermis (e.g., in *Brassica napus*).[101] Interestingly, the smaller amounts of flavonoids present in mesophyll tissue and in the lower epidermis are not inducible by UV-B, but rather represent the constitutive fraction. Changes in the levels of flavonols have been observed not only inside epidermal cells, but also in hairs and epicuticular wax (Table 8.1). Flavonoids are generally present as water-soluble glycosides in cell vacuoles, but when present in the epicuticular wax on the leaf surface, flavonoids are nonglycosylated, are very often *O*-methylated, and are lipophylic. *O*-Methylation tends to shift the UV absorption properties to shorter wavelengths so that they typically absorb significantly in the 250 to 320 nm region. Thus, *O*-methylated flavonoids are better able to protect plant leaves from UV-B damage, as demonstrated in two *Gnaphalium* species.[98] One further site of flavonoid synthesis is in the leaf hairs or trichomes on leaves. Pubescence enriched with acylated flavonol glycosides has been shown to provide UV protection in leaves of *Quercus ilex*.[102]

Most studies on UV protection have focused on leaves; relatively few have studied other plant organs, such as fruit. Because the anatomy, histology, and physiology differ markedly between fruit and leaves, the two may also differ in UV shielding mechanisms. Studies on apple (*Malus pumila*) skins showed that flavonoids (mainly quercetin glycosides) accumulate during acclimation to strong sunlight and can serve as efficient UV-B screens.[57] Similarly, for berries of a white grape cultivar (*Vitis vinifera* L. cv. Bacchus), flavonols and hydroxycinnamic acids were identified as the main groups of UV-absorbing phenolics.[204] The flavonols alone provided efficient shielding from UV-A, but the combination of flavonols and hydroxycinnamic acids was required for tolerance to UV-B. These results confirm that parts of the plant as dissimilar as fruits and leaves can share similar mechanisms of UV protection.

Many studies have shown that the flavonoids, along with certain other phenolics, increase in UV-B-treated plants, but until recently it has been difficult to assign a specific role for these compounds in UV-B resistance. Moreover, it has remained uncertain whether any one group of flavonoids was more important than others. Jaakola et al.[205] examined the activation of flavonoid biosynthesis by solar radiation in bilberry (*Vaccinium myrtillus*) leaves. The flavonol quercetin, various anthocyanins glycosides, and the hydroxycinnamic acids were shown collectively to play a predominant role in the defense against high solar radiation; all of these compounds increased markedly with greater exposure to sunlight. In contrast, levels of polymeric procyanidin decreased, indicating they do not participate in protecting leaf tissues from excess light. In soybean (*Glycine max*), the isoflavonoids are highly responsive to those wavelengths that are most affected by variations in ozone levels, suggesting that they function as UV protectants in the field.[206]

Recent work with several species (e.g., *Petunia*,[207] *Arabidopsis*,[208] *Vicia faba*,[209] *B. napus*,[101] *Oryza sativa*[100]) has provided evidence that UV light can induce the preferential synthesis of those flavonols that have the higher levels of hydroxylation (e.g., quercetin instead of kaempferol, and isoorientin instead of isovitexin), rather than a general increase in all flavonols. For *B. napus*, quercetin glycosides are UV-B inducible and kaempferol glycosides form a constitutive shield.[101] An increase in dihydroxylated compounds as a UV-B response has also been shown to apply to different populations within a species; in clover (*Trifolium repens*), for example, large differences between populations in both productivity and tolerance to UV-B stress were correlated to differences in amounts of quercetin.[210] The effect of UV-B on the levels of flavonol glycosides was synergistically enhanced by water stress. These changes were more pronounced for the *ortho*-dihydroxylated quercetin, rather than for the monohydroxylated kaempferol glycosides.[211] Similarly, a *Petunia* mutant for which kaempferol was the predominant flavonol grew at a significantly slower rate under UV-B treatment than did the wild-type plants, which accumulated mainly quercetin.[207] These observations suggest that quercetin confers a protective advantage that is not matched in the mutant, even though it has the higher overall flavonol levels. Hydroxylation does not affect the UV-absorbing properties *per se*, but it does affect the antioxidant capacities of these compounds. Thus, the selective advantage of quercetin under UV-B stress is likely to be related to its higher capacity to scavenge UV-generated free radicals.[208,212] Interestingly, in Scots pine (*Pinus sylvestris*) both kaempferol and quercetin-3-glucoside increase after UV-B treatment, with kaempferol derivatives increasing in primary needles and quercetin derivatives increasing in cotyledonary needles.[78] The reason for these differences is not immediately apparent.

Some studies have shown no changes in flavonoid levels in response to UV. Originally, only the flavonoids were thought to serve as UV-screening pigments.[63] It is now clear, however, that other phenolics, such as hydroxycinnamic acids and their esters, are also involved. The importance of each group of compounds seems to vary across plant species as well as between leaves of different developmental stages. One study has shown that for *Arabidopsis* the hydroxycinnamates are more effective UV-B protectants than flavonoids.[213] In contrast, Li et al.[61] demonstrated for *Arabidopsis* that flavonoids are required in addition to other phenolic compounds *in vivo*. In rye (*S. cereale*), hydroxycinnamates in epidermal cells are the dominant UV-B protective compounds at the early stages of primary leaf development.[69] However, during subsequent leaf development and acclimation this function is increasingly replaced by epidermal flavonoids (e.g., isovitexin 2″-*O*-glycosides).

8.3.1.2 Temperature Stress

In addition to the induction of anthocyanin biosynthesis, chilling stress has also been shown to promote the formation of colorless flavonoids. Cold treatments (and drought stress) caused increases in levels of (−)-epicatechin and hyperoside (quercetin 3-galactoside) in two species of hawthorn, *Crataegus laevigata* and *C. monogyna*. Such treatments also enhanced the antioxidant capacity of the shoot extracts, and this may be the primary function of these cold-inducible flavonoids.[214]

The flavonoids are also believed to play a role in the responses to heat stress. Recently, Coberly and Rausher[215] reported that flavonoids can mitigate the adverse effects of heat stress on fertilization and early seed maturation. Two possible mechanisms were suggested: (i) flavonoids promote the overall well-being of plants under heat stress (e.g., by scavenging free radicals); or (ii) heat stress inhibits one or more process directly involved in fertilization or ovule maturation, and flavonoids ameliorate these effects. As discussed in Section 8.3.2, flavonoids have been found to play key roles in plant fertility.

High temperatures have been shown to influence flavonoid gene expression.[216] There are also synergistic responses to the effects of high temperatures and UV-B. In cucumber (*Cucumis sativus* L.) seedlings, synthesis of a quercetin-like compound was preferentially enhanced after UV-B exposure.[217] When plants irradiated with low and medium doses of UV-B were subsequently heat stressed (46°C for 1 h), survival improved by 112 and 82% and shoot elongation increased by 35 and 40%, respectively, relative to the controls that received no UV-B. A synergism between UV-B tolerance and heat tolerance could be used by growers to precondition seedlings in UV-B-deficient greenhouses, and may benefit plants under the predicted global warming scenario.

8.3.1.3 Heavy Metal Tolerance

There is accumulating evidence that flavonoids participate in the resistance to high levels of metals in soils. Roots of maize (*Zea mays*) plants that had been exposed to aluminum have been found to exude high levels of phenolic compounds.[218,219] The observation is consistent with the metal-binding activity of many flavonoids. It has been argued, however, that phenolics, including flavonoids, are unlikely to be effective in binding metals in an acid environment because H^+ ions compete for complex formation. Kidd et al.[218] provided empirical evidence that certain flavonoids can form complexes with aluminum in the root. Indeed, morin, a pentahydroxyflavone, complexes with aluminum and is routinely used to stain for aluminum in the root apoplast.[220] Catechin also forms stable aluminum complexes,[218] and green tea, which contains high levels of catechin, accumulates and tolerates high tissue levels of aluminum.

The distribution of aluminum in the root tips of the aluminum-tolerant forage legume, *Lotus pedunculatus* Cav., a species that also accumulates condensed tannin (proanthocyanidin), has also been investigated.[221] In osmium-fixed samples from high and low aluminum concentrations, aluminum was generally found in association with osmium-binding vacuoles. Because of the high affinity of osmium for condensed tannin, it was hypothesized that condensed tannins possibly bind and detoxify aluminum in the root apices. Alternatively, in common with other abiotic stressors, high concentrations of metals such as copper and aluminum could result in the production of reactive oxygen species.[222] Plants with high flavonoid production may be able to combat this oxidative stress. It has also been suggested, however, that Al^{3+} has the potential to induce oxidative stress in plants by stimulating the prooxidant nature of endogenous phenolic compounds.[223]

8.3.1.4 Oxidative Stress

Antioxidant capacities are a feature of plant phenolics in general[145,224] and like the antho-cyanins, many of the colorless flavonoids are also thought to be key players in the abrogation of oxidative stress. Although they are localized largely in vacuoles,[225] smaller pools of flavonoids are thought to occur in the cytoplasm.[101] This has recently been demonstrated using mature petals of lisianthus (*Eustoma grandiflorum*), for which ~14% of the flavonoids reside within the cytoplasm, 30% in the cell wall, and 56% in the vacuole.[226] Cytoplasmic flavonoids are better sited than vacuolar flavonoids to interact with reactive oxygen species generated by organelles,[227] and they have a demonstrable antioxidant potential at the neutral pH of the cytosol.[228] A vacuolar location would not, however, impede a second putative function of flavonoids, that of dissipating excess absorbed UV energy; indeed, it can be argued that an epidermal vacuolar location is ideal for that purpose.[207]

Under high-stress conditions, more H_2O_2 is produced by the chloroplasts, mitochondria, and peroxisomes than can be accommodated by peroxidase or catalase, and the oxidant can diffuse across the tonoplast into the cell vacuole. It has been proposed that vacuolar flavo-noids in conjunction with peroxidase serve to scavenge excess H_2O_2 (Figure 8.5). Studies with the leaves of the tropical tree *Schefflera arboricola* have shown that quercetin and kaempferol glycosides, in particular, can act in this manner,[169] although all classes of flavonoids can act as radical scavengers to a greater or lesser extent.[145,229] Since the oxidation of flavonols by H_2O_2 requires peroxidases, it might therefore be an advantage for the plant to have a colocalization of flavonoids and peroxidase in the vacuole.

8.3.2 Reproduction and Early Plant Development

Flavonoids often participate in plant reproduction, in the protection of reproductive tissues and seeds, and in seedling development. This may, in part, be due to their role in UV light shielding (thereby protecting DNA) and antioxidant properties, but other functions are also important.

8.3.2.1 Chemical Signals as Attractants for Pollination and Seed Dispersal

Plants that are insect pollinated generally have flowers with large, brightly colored petals, in contrast to most wind-pollinated plants for which flowers are small, dull, and often apetalous (e.g., *Petunia* versus maize).[202] Pigmentation presumably acts as a signal to attract pollinating insects or birds. The anthocyanins make the most obvious contribution to flower color,[230] though other flavonoids can also assist in this. There are a few flavonoids that are yellow, e.g., chalcones and aurones, but these pigments are restricted to relatively few plant species. Yellow color can also be produced by flavonols following methylation, certain types of glycosylation, or certain A-ring hydroxylation patterns.[231] Yellow petal colors have also been shown to result from an aggregation of (colorless) flavonol glycosides on a protein matrix in the cytoplasm of epidermal cells.[226] Various flavonols and flavones act as

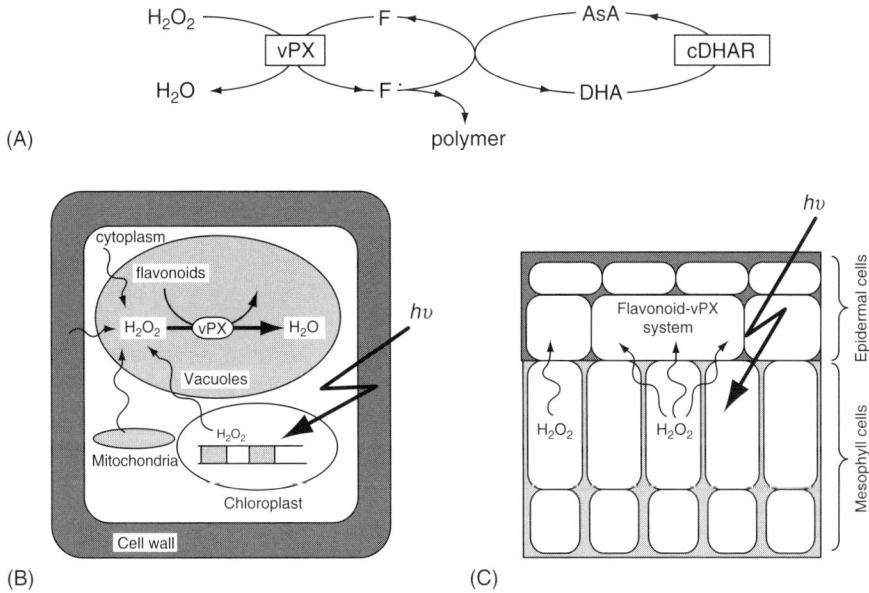

(A)

(B)									(C)

FIGURE 8.5 A proposed diagram for the protective function of flavonoids during stress (A). Scheme of the H_2O_2-scavenging mechanism by flavonoids. vPX, vacuolar peroxidase; F, flavonoid; F$^\bullet$, flavonoid radical; AsA, ascorbic acid; DHA, dehydroascorbic acid; $h\upsilon$, light energy; and cDHAR, cytosolic dehydroascorbic acid reductase. The diffusive nature of H_2O_2 enables vPX to scavenge it in vacuoles, even if the generating site is other than a vacuole (B). This concept can be expanded to cell–cell interaction. The photoproduced H_2O_2 may leak out from mesophyll cells and be scavenged in epidermal cells that have high flavonoid content (C). (Redrawn from Yamasaki, H., Sakihama, Y., and Ikehara, N., *Plant Physiol.*, 115, 1405, 1997. Copyright the American Society of Plant Biologists and reprinted with permission.)

copigments with anthocyanins leading to an intensification of flower color.[231] Colorless flavonoids may also provide "body" to flowers, to give a cream or ivory appearance in petals that would otherwise be translucent.[231]

Petals and other floral organs often show intricate patterns under UV light. They are especially common among yellow flowered species, for which the principal yellow pigment is usually a carotenoid. However, the UV-absorbing regions that give rise to these patterns are colorless or yellow flavonol glycosides and chalcones, which quench the energy of light reflected from carotenoids in the chromoplasts.[232] UV patterning can also be provided by colorless flavonol glycosides and chalcones, e.g., in *Coronilla* and *Potentilla*.[233] Flavonoids can determine which animal vector is attracted to effect pollination. In *Petunia integrifolia*, for example, flowers are pink and are pollinated by bees, while in the related species, *Petunia axillaris,* flowers are white and are pollinated by moths.[234] It is assumed that these moths are attracted by the high amounts of fluorescent flavonols. Similarly, the corolla of *B. rapa* has a UV-absorbing zone in its center, known as a nectar guide, for attracting pollinating insects.[235] The pigment responsible, identified as isorhamnetin 3,7-*O*-di-β-D-glucopyranoside, is present at 13-fold greater amounts in the basal parts of the petals than in the apical regions. This difference in flavonoid content is presumed to contribute to the visual attractiveness of *B. rapa* flowers to insect pollinators.

Flavonoids in fruit probably serve to attract frugivores that assist in seed dispersal. This is especially important for larger plants such as trees, for which seeds need to be transported some critical distance away from the parent to ensure germination.

8.3.2.2 Flavonoids in Pollen, and Their Role in Plant Sexual Reproduction

Anthocyanins, flavonols, or chalcones often accumulate in both the male and female sex organs in a flower, including the pollen grains.[192] Insects can detect light from the UV to orange waveband,[236] and are possibly attracted to pollen whose color contrasts against petals due to UV reflective or absorptive flavonoids. However, if pollen pigments were present primarily as insect attractants then these pigments should be redundant, and therefore unlikely to be present in wind-pollinated (anemophilous) species. Nevertheless, many anemophilous species do contain flavonoids.[237] Indeed, there are relatively few plant species that only produce nonpigmented (white) pollen.

Alternative hypotheses for pollen pigments hold that flavonoids are involved in pollen-style incompatibility relationships via pollen germination, or else serve as growth regulators for pollen tube development. Certain flavonoids have also been postulated to ameliorate the effects of heat stress during fertilization.[215]

In order to understand the role of flavonoids in pollen, flavonoid-deficient mutants have been produced through artificial methods, such as cosuppression, antisense expression, chemical mutagenesis, and ionizing radiation. Examples include *Petunia*[238–241] and *A. thaliana*.[242] Most mutants produce white pollen that is sterile in self-pollinations (e.g., maize[243]). In *Petunia*, white pollen failed to produce functional pollen tubes both in an *in vitro* pollen germination assay and on stigmas from the same mutant.[240] However, on wild-type stigmas the white pollen was functional and extracts from wild-types stigmas could "rescue" *in vitro* pollen tube growth, indicating that the stigma contains other factors that can complement the mutation.[244] The active compound(s) in the stigma were shown to be flavonols. Similarly, flavonols (quercetin, kaempferol, myricetin), though no other class of flavonoid, have been shown to strongly promote pollen germination frequency and pollen tube growth of tobacco (*Nicotiana tabacum*) *in vitro*.[245] In contrast to *Petunia*, the flavonoid-deficient mutant in *Arabidopsis* had fertile pollen. Ylstra et al.[246] suggested that other phenylpropanoids or else compounds such as the (brassino-)phytosteroids might compensate for the absence of flavonoids. Song et al.[247] showed that sinapate esters, rather than flavonoids, may have a physiological role in pollen tube growth in *Arabidopsis*.

Notwithstanding these studies, the precise function of flavonoids in fertilization and pollen tube growth remains unclear. It has been speculated that the flavonols form a gradient along which the growing pollen tubes are guided to their target, the ovule.[192] However, there has been no direct experimental evidence to support this idea. It is interesting to note that certain mycorrhizal fungi demonstrate enhanced hyphal growth on a medium enriched in the same flavonols that promote pollen tube growth, and it is possible that there is a common mechanism of action, e.g., a regulatory effect on cell elongation.[248] It has been postulated that flavonols play a structural role in the membranes of rapidly growing pollen tubes and that these may not be essential for pollen germinating with a short transmitting tract.[249] However, in *Petunia* the absence of flavonoids interferes with the ability of pollen to penetrate the style. Derksen et al.[250] noted that in flavonol-deficient pollen tubes of *Petunia hybrida*, the structure of the primary wall at the tip dramatically changed before it disrupted. It was concluded that flavonols act on precursors of the pollen tube wall and interfere with a cross-linking system in the wall, possibly via extensions.

The observations that flavonols are not involved in the fertilization process in certain species, and that this function can be completed using other compounds, suggest that flavonols only affect fertility indirectly.[239] There are various examples of cross-talk between branch pathways of phenylpropanoid metabolism,[61,251] or the shikimate pathway.[252] The absence of flavonols in maize and *Petunia* could affect the accumulation of other compounds that are more specifically required for male fertility. Thus, differences between species in terms of flavonoid requirement for male fertility could relate to differences in the integration of metabolic pathways.[242] Nonetheless, virtually all plants accumulate flavonoids in pollen,[253] including *Arabidopsis*, which suggests that they are of fundamental importance to the male gametophyte.

Another hypothesis for the role of pollen flavonoids is that they shield the vulnerable haploid genome from the mutagenic effects of UV light or, indeed, any auto- or photo-oxidative stress. There is only limited experimental evidence in support of this hypothesis for pollen. Indirect evidence has come from studies comparing flavonoid levels in pollen exposed to different UV light regimes. Pollen collected from UV-irradiated maize and Californian poppy (*Eschscholzia californica*) contained more flavonoids than did control pollen.[254,255] However, in experiments comparing white pollen to pigmented pollen of *Ipomoea purpurea*, light was shown to have little discernible overall effect on fertilization success, with no effect on either pollen donor or pollen recipient.[215]

8.3.2.3 Seeds and Seedling Development

The roles of flavonoids in seeds and seedling development have been reviewed previously.[256,257] Ndakidemi and Dakora[256] specifically discussed the role of flavonoids in legume seed development, viz: (i) nodulation; (ii) arbuscular mycorrhizal fungal development; (iii) resistance to microbial pathogens; (iv) defense against insect pests; (v) controlling the parasitic plant *Striga*; and (vi) as allelochemicals in controlling weeds in cropping systems. Many of these functions also apply to other plant parts, but there are also examples of flavonoid-enriched seed coats for which specific roles have been ascribed.

Other putative functions of flavonoids in seeds have received less attention. It was thought that flavonoids are necessary for embryo development, yet cosuppresion of chalcone synthase (*Chs*) gene has been found not to cause female sterility.[258] Experiments by Jorgensen et al.[258] with *P. hybrida* led to the hypothesis that accumulation of dihydroflavonols in the seed coat inhibits embryo growth, either directly or indirectly. However, there appears to be no evidence of exactly how this may happen *in vivo*. There is some evidence that proanthocyanidins assist in the reinforcement of plant tissues; seeds of snap bean (*Phaseolus vulgaris*), which lack proanthocyanidins, are more sensitive to mechanical stress and water stress than are wild-type seeds.[259] The flavonoids, particularly proanthocyanidins, in the seed coat contribute to the maintenance of seed dormancy as well as increasing seed longevity in storage.[257,260,261] Testa pigmentation also appeared to confer resistance to solute leakage, imbibition damage, and attack by soil-borne fungi, thereby improving seed vigor and germination in legumes.[262,263] High concentrations of phenolics in emerging seedlings have been associated with protection against UV-B damage at this critical stage of development.[264]

Kaempferol can increase seed production in *Petunia*, whether produced as a result of pollination or by wounding.[265] It has been suggested that the reproductive function of flavonols may have evolved from a defensive role. Allelopathic flavonoids in the stigma may prevent introduction of pathogens into the pistil.

Seed and root exudates can affect growth of neighboring plants, including weeds in cropping systems. Various groups of compounds have been suggested as possible allelochemicals, including the flavonoids, although evidence for this is very limited. Nonetheless, the isoflavonoids medicarpin, 4-methoxymedicarpin, sativan, and 5-methoxysativan from alfalfa (*Medicago sativa*) seeds can inhibit seedling development in other species when released into the soil.[266,267]

8.3.3 SIGNALING — A ROLE AS CHEMICAL MESSENGERS

It has been suggested that the synthesis of flavones, flavanones, and flavonols could have evolved primarily as chemical messengers, such as defense molecules.[268] Indeed, flavonoids have recently been described as a novel class of hormones.[269] Interestingly, in this regard, the flavonoids share some structural and chemical features with the steroids, retinoids, thyroid hormone, prostaglandins, and fatty acids. A signaling capacity would explain many of the functions of flavonoids, both in plants and in humans.

Flavonoids are thought to be particularly useful as signals in plant–rhizosphere interactions. Roots are exposed to a variety of soil microorganisms that use plants as a food source or niche habitat. These relationships can ultimately be detrimental or beneficial to the plant, yet both types involve flavonoid synthesis. The stages of communication between interacting plants and microorganisms typically involve signal exchange and perception, followed by the invasion of the microbe and structural changes in the plant. The involvement of flavonoids as regulatory signals in such interactions has been reviewed previously.[270–276] The main beneficial roles are in attracting microorganisms for nodulation and mycorrhizal association. The growth of other beneficial soil microflora has also been demonstrated; for example, luteolin and quercetin from alfalfa seed exudates promote growth of.[277]

8.3.3.1 Defense Against Pathogenic Microbes

The isoflavonoids, in particular, act as effective phytoalexins, which can be defined as small molecular weight antimicrobial compounds or biological stress metabolites. They can be constitutive, or else are inducible by wounding or biological attack. The constitutive versus inducible response varies between different species, and can also vary within the plant depending on age or environment. Isoflavonoids, especially the pterocarpans, isoflavans, isoflavones, and isoflavanones, are extremely toxic to fungal pathogens. These flavonoids inhibit fungal spore germination, germ tube elongation, and hyphal growth through causing damage to membrane systems.[278,279] Other compounds such as kievitone and phaseollin isoflavonoids lead to mycelial leakage of metabolites and shrinkage of hyphal tip protoplasts.[280] Antimicrobial activity has also been noted in the flavans, flavanones, 3-hydroxyflavanones, and flavonols.

Antimicrobial flavonoids are particularly rich in seed coats and in the sapwood of trees.[281] Various studies have shown that the seeds of grain legumes (e.g., *G. max*, *P. vulgaris*, *Pisum sativum*, *Canavalia ensiformis*, *Arachis hypogea* and *Cicer arietinum*) store antifungal isoflavonoids for defense against pathogen infection prior to germination.[256] The isoflavonoid pterocarpans maackianin and pisatin (Figure 8.6) act as classical phytoalexins in the interaction between garden pea (*P. sativum*) and the fungal pathogen *Nectria haematococca*. Enzymes that detoxify maackianin and pisatin have been identified as fungal virulence factors.[282,283] Pisatin induces a protein that appears to control the transcription of a fungal detoxification enzyme.[284] In addition to serving as a cue for phytoalexin detoxification, pisatin, along with other flavonoid compounds, apparently stimulates the germination of spores of *N. haematococca*.[285] *Sorghum* seeds that contain the higher concentrations of proanthocyanidins show enhanced tolerance to aflatoxin-producing *Aspergilllus* fungus.[286] The soybean isoflavones daidzein and genistein (Figure 8.6) stimulate chemotropic behavior in hyphal germination of the oomycete pathogen *Phytophthora sojae*, suggesting that they might help the hyphal tips of zoospores encysted adjacent to roots to locate their host.[287]

8.3.3.2 Roles in *Agrobacterium* Infection

Flavonoids from pollen and stigmas of *P. hybrida* (kaempferol 3-glucosylgalactoside, quercetin 3-glucosylgalactoside, rutin, myricetin 3-galactoside, narcissin, and apigenin 7-glucosides) have all been found to induce the *vir* region of the *Agrobacterium tumefaciens* Ti plasmid, which is responsible for virulence.[288] Along with other *vir* inducing factors like cinnamic acids, these flavonoids might play an even more important role in the natural

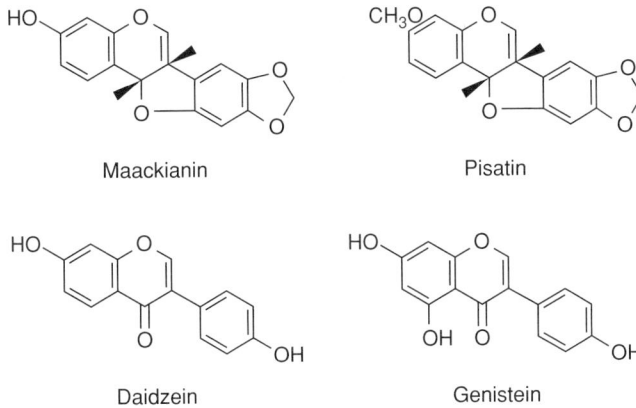

FIGURE 8.6 Structure of the phytoalexin isoflavonoid pterocarpans, maackianin, and pisatin from garden pea, and the isoflavones daidzein and genistein from soybean.

infection of plants by *Agrobacterium* than does acetosyringone, which until now has been found only in tobacco.

Agrobacterium tumefaciens-induced plant tumors accumulate considerable amounts of free auxin. The overproduction of auxin is required for tumor growth, and it appears that both the accumulation and maintenance of high auxin levels are dependent on flavonoids. To determine possible mechanisms by which high auxin concentrations are maintained, Schwalm et al.[289] examined for associations between the distribution patterns of auxins and flavonoids, both in tumors from *T. repens* and in *A. thaliana* shoots. Tumor-specific flavones, isoflavones, and pterocarpans were detected, namely 7,4'-dihydroxyflavone (DHF), formononetin, and medicarpin. DHF was also the dominant flavone in high free-auxin regions of *Arabidopsis* leaf primordia. Moreover, these flavonoids were localized at the sites of strongest auxin-inducible chalcone synthase (CHS) genes. It was concluded that CHS-dependent flavonoid aglycones are possibly endogenous regulators of the basipetal auxin flux, thereby leading to free-auxin accumulation in *A. tumefaciens*-induced tumors. This, in turn, triggers vigorous proliferation and vascularization of the tumor tissues and suppresses their subsequent differentiation. The role of flavonoids in auxin regulation is discussed in more detail below.

8.3.3.3 Legume Nodulation

Flavonoid levels in plants can be affected by their nutritional status. Low nitrate concentrations, for example, induce flavonoid accumulation, which then serve as chemoattractants to nitrogen-fixing bacteria. Nodulation then restores the nitrogen economy of the plant.[290] In *L. pedunculatus,* proanthocyanidins are not formed under high-nitrogen conditions,[291,292] presumably because the requirement for nodulation is not so critical.

Flavonoids act as signal molecules in the early stage of legume–*Rhizobium* symbiosis (as reviewed in Refs. 273, 276). In legumes, flavonoid biosynthetic genes are active in young root cap cells and in the zones where root hairs emerge.[293] Flavonoids are released into the soil, which then induce various *nod* genes in soil-borne rhizobia.[294–296] *Nod* genes are responsible for the synthesis of lipochitin-oligosaccharides (LCOs), also called Nod factors, which are species-specific signaling molecules that initiate root nodule development. Genetic evidence indicates that a protein, termed NodD, functions as the receptor for the flavonoid signal.[297] NodD protein from different bacteria differ in their response to distinct flavonoids excreted by their corresponding host plant, thus helping determine host specificity.[298] The *nod* gene inducers so far identified from seeds and seed coats of legumes include flavones,[294,299]

TABLE 8.2
Some *nod* Gene Inducers Released by Different Legumes

Source	Inducers	Ref.
Medicago sativa		
Root exudates	7,4'-Dihydroxyflavone; 4,4'-dihydroxy-3'-methoxychalcone; liquiritigenin; formononetin; formononetin-7-*O*-(6'-*O*-malonylglucoside); formononetin-7-*O*-glycoside	303, 306–308
Seed exudates	Luteolin; chrysoeriol; 5,3'-Dimethoxyluteolin; 5-Methoxyluteolin; Trigonelline; Stachydrine	294, 307, 309
Phaseolus vulgaris		
Root exudates	Genistein; genistein-7-*O*-glycoside; eriodictyol; naringenin; daidzein; coumestrol; liquiritigenin; isoliquiritigenin	304, 306, 310
Seed exudates	Delphinidin-3-*O*-glycosides; petunidin-3-*O*-glycosides; malvidin-3-*O*-glycosides; quercetin-3-*O*-glycosides; kaempferol-3-*O*-glycosides	304
Pisum sativum		
Seed rinse	Eriodictyol; apigenin-7-*O*-glucoside	311
Glycine max		
Root exudates	Isoliquiritigenin	312
Seed exudates	Genistein; genistein-7-*O*-(6'-*O*-malonylglucoside); daidzein-7-*O*-(6'-*O*-malonylglucoside); daidzein-2-*O*-glucoside with unidentified acylation	301, 313
Trifolium repens		
Seedling extract	7,4'-Dihydroxyflavone; geraldone (7,4'-dihydroxy-3'-methoxyflavone); 4'-hydroxy-7-methoxyflavone	295
Vicia sativa		
Root exudate	3,5,7,3'-Tetrahydroxy-4'-methoxyflavanone; 7,3'-dihydroxy-4'-methoxyflavanone; 4,2',4'-trihydroxychalcone; 4,4'-dihydroxy-2'-methoxychalcone; naringenin; liquiritigenin; 7,4'-dihydroxy-3'-methoxyflavanone; 5,7,4'-trihydroxy-3'-methoxyflavanone; 5,7,3'-trihydroxy-4'-methoxyflavanone	296, 300

Source: From Jain, V. and Nainawatee, H.S., *J. Plant Biochem. Biotechnol.*, 11, 1, 2002. With permission.

flavonones,[300] isoflavones,[301,302] flavonols,[300,303] and anthocyanins.[304] Details of specific flavonoids involved in *nod* gene induction are given in Table 8.2. Siqueira et al.[305] also list a large number of examples of flavonoids involved in nodulation.

Certain flavonoids also act as inhibitors of *nod* gene expression.[311,314,315] Examples of such compounds, which are also referred to as anti-inducers, are given in Table 8.3. The mechanism of action is often by competitive inhibition; chemical structures of these inhibitors are similar to those of *nod* inducers, and inhibition can be overcome by increasing the concentration of inducers.[314] Inhibitors are sometimes strain specific, acting on a few strains belonging to the same cross-inoculation group.[316] The isoflavone daidzein induces *nod* gene expression in *Bradyrhizobium japonicum* (nodulates soybeans)[301] but is an inhibitor in *Rhizobium trifolii* (nodulates clovers)[315] and *Rhizobium leguminosarum* (nodulates peas).[311] These inhibitors may contribute to host specificity. The balance between stimulatory and inhibitory flavonoids in roots and root exudates may contribute to the regulation of nodulation. The environment may also influence whether compounds act as inducers or inhibitors of *nod* induction.[315] Spatiotemporal distribution of flavonoids in the rhizosphere and at the root surface is likely to determine the levels of induction of rhizobial *nod* genes.

TABLE 8.3
Some Major Antagonists of Inducers of Rhizobial and Bradyrhizobial *nod* Genes

Symbiotic Association	Antagonist Compounds	Ref.
Medicago sativa and *Rhizobium meliloti*	Umbelliferone, morin, quercetin	314
Trifolium spp. and *Rhizobium. leguminosarum* bv. *Trifolii*	Umbelliferone, formononetin	315
Phaseolus vulgaris and *Rhizobium leguminosarum* bv. *Phaseoli*	Umbelliferone, acetosyringone	316
Pisum sativum and *Rhizobium leguminosarum* bv. *Viciae*	Daidzein, genistein, kaempferol, acetovanollone, syringaldehyde	311
Glycine max and *Bradyrhizobium japonicum*	7-Hydroxy-5-methyl flavone, flavone, kaempferol, chrysin	316

Source: From Jain, V. and Nainawatee, H.S., *J. Plant Biochem. Biotechnol.*, 11, 1, 2002. With permission.

Some flavonoids may also play a broader role in *Rhizobium* infection, affecting other microbial processes such as by promoting growth,[277,317] affecting metabolism[317,318] and positive chemotaxis.[319]

Nodule organogenesis is thought to involve auxin and possibly also cytokinin in the stimulation of cell divisions and regulation of root differentiation.[320] Mathesius[321] tested whether those flavonoids that preferentially accumulate in cells undergoing early nodule organogenesis could affect peroxidase-driven auxin turnover, thereby explaining local changes in auxin distribution during nodule formation in white clover (*T. repens* cv. Haifa). A derivative of 7,4′-dihydroxyflavone (DHF), as well as free DHF, strongly inhibited auxin breakdown by peroxidase at concentrations estimated in the root tissue. Formononetin, an isoflavonoid accumulating in nodule primordia, accelerated auxin breakdown by peroxidase at concentrations estimated to be present in the roots. The data indicated that local changes in flavonoid accumulation could indeed regulate local auxin levels during nodule organogenesis. A model for the interaction of flavonoids with peroxidases has been proposed to explain changes in auxin during nodule development (Figure 8.7). A similar mechanism could be involved in lateral root and root gall development, and the wider role of flavonoids in auxin regulation is discussed below.

8.3.3.4 Mycorrhizal Fungi

The capacity for nitrogen fixation via nodulating bacteria is limited to relatively few plant species. In contrast, arbuscular mycorrhizal fungi (AMF) associations with roots occur in about 80% of plant species. The mutualistic association is important for improving the nutritional status of plants in soil where nutrients such as phosphate are limited. As with nodulation, exudates from seeds and seedlings affect plant infection by AMF, and flavonoids are one group of compounds present in such exudates. Nodulation and mycorrhizal formation on the same plant mutually increase each other's establishment, probably because of the involvement of flavonoids in both processes.[322] This may be due to flavonoids being stimulators of both. There are numerous reports of the effects of specific flavonoids on mycorrhizae in a wide range of plant species (Table 8.4). Effects of flavonoids have been demonstrated on spore germination, hyphal growth, and root colonization (reviewed in Refs. 271, 323). It has been commented that AMF might have genus- or even species-specific signal requirements during the AMF symbiosis, yet some flavonoids exhibit a general stimulatory effect on different AMF genera. For example, quercetin greatly stimulates hyphal development of

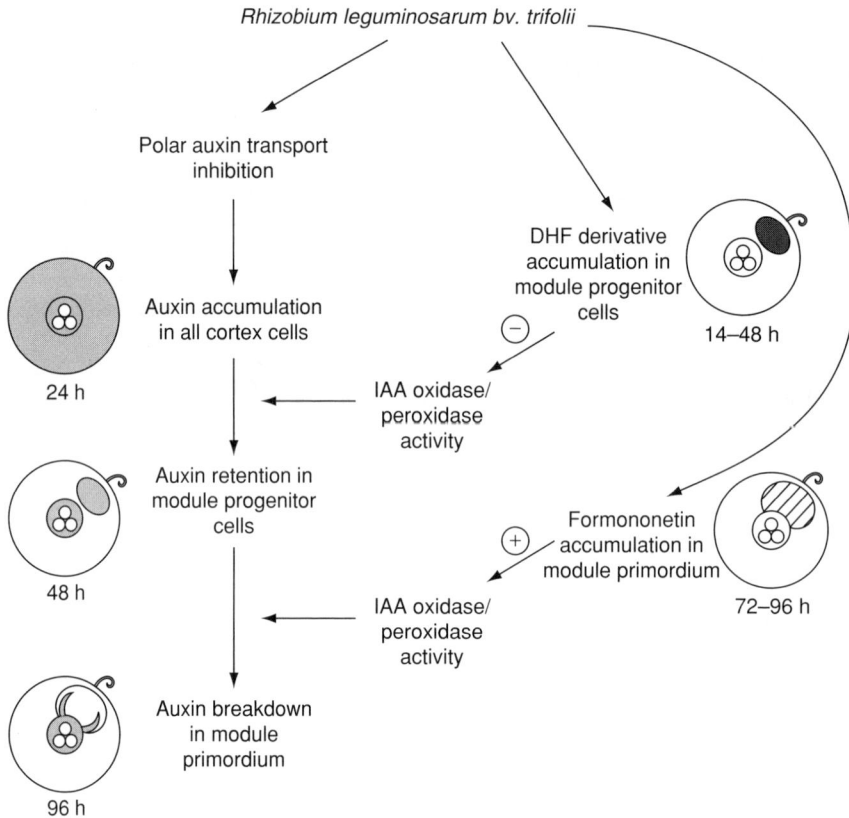

FIGURE 8.7 Model for the role of *Rhizobium*-induced flavonoids in the regulation of auxin balance during nodule formation in white clover. (Redrawn from Mathesius, U., *J. Exp. Bot.*, 52, 419, 2001. With permission of Oxford University Press.)

various species of fungi, e.g., *Gigaspora* and *Glomus*.[324–326] However, studies with biochanin A showed stimulation of *Glomus*[271,327] but not *Gigaspora* species.[324,325]

Flavonoid levels or compositions change in the root from the early stages of appressoria formation through to the later stages of symbiosis when the AMF is well established. Thus, flavonoids have been suggested as signaling compounds during root colonization[271] although this has been debated considerably[248]. Larose et al.[331] presented various data that point

TABLE 8.4
Some Flavonoids Involved in Mycorrhizal Associations

Symbiotic Association	Flavonoids	Effect	Ref.
Medicago sativa and *Glomus* spp.	Quercetin and quercetin-3-*O*-galactoside	Promotion of spore germination, hyphal growth, and branching	277, 328
Trifolium repens and *Glomus* spp.	Formononetin and biochanin A	Promotion of root colonization, stimulation fungal growth	305, 327
Eucalyptus globulus and *Pisolithus* spp.	Rutin	Stimulation of hyphal growth	329
Pinus sylvestris and *Suillas* spp.	Root exudates	Enhanced spore germination	330

strongly toward the action of AMF-derived signal(s) on plant roots before root colonization. Two recent reports show a role of flavonoids as regulatory signals for the susceptibility of roots to AMF at the beginning of the formation of the symbiosis.[332,333] Guenoune et al.[332] demonstrated that the flavonoid medicarpin accumulates in roots with high phosphate levels. Medicarpin exhibits a strong inhibitory effect on hyphal growth of *Glomus intraradices*, thereby preventing roots with a high phosphate status from being colonized by AMF. These data support the hypothesis that during early stages of colonization by *G. intraradices*, suppression of defense-related properties is associated with the successful establishment of AMF symbiosis. Akiyama et al.[333] investigated melon (*Cucumis melo*) roots inoculated with or without the fungus *Glomus caledonium* under low phosphate conditions. Accumulation of a *C*-glycosylflavone, isovitexin 2″-*O*-beta-glucoside, occurred in phosphate-deficient, nonmycorrhizal roots but not in mycorrhizal roots, nor in phosphate-supplemented roots. This experiment indicated that the accumulation of the compound was caused by a phosphate deficiency. Treatment of roots with isovitexin 2″-*O*-beta-glucoside increased root colonization under both low and high phosphate conditions. These findings suggest that the phosphate deficiency-induced *C*-glycosylflavonoid is involved in the regulation of AMF colonization in melon roots.

Flavonoid accumulation during mycorrhizal associations has been shown both to vary over time and to be AMF specific. Larose et al.[331] studied the accumulation of flavonoids in roots of *M. sativa* after the exposure to *Glomus mosseae*, *G. intraradices,* or *Gigaspora rosea*. It was shown that flavonoid accumulation in *M. sativa* roots (i) is induced before root colonization, pointing toward the presence of a fungal-derived signal, (ii) varies according to the developmental stage of the symbiosis, and (iii) depends on the species of root-colonizing fungus.

Flavonoids are not always essential signal molecules in mycorrhizal symbioses, however; various other phenolic acids can elicit similar effects.[248] Compounds that actively promote the growth of AMF (i.e., kaempferol, quercetin, and myricetin)[324,325] are also required for pollen germination and growth of the germ tube.[244,245] Thus, these compounds may have some general effect on cell elongation.[248] It has been concluded that flavonoids stimulate, but are not essential for mycorrhiza formation.[334]

8.3.3.5 Parasitic Plants

Parasitic plants often use chemicals released by their host plant to stimulate seed germination, to locate the host, or for haustorial development. Many different compounds are involved, including strigolactones, quinones, coumarins, flavonoids, and other phenolics. Flavonoids contribute to signaling in some species but not others. Haustorial development in *Triphysaria versicolor* can be induced *in vitro* by the anthocyanidins petunidin, cyanidin, pelargonidin, delphinidin, as well as their glycosides obtained from the host plant.[335,336] Anthocyanins are not usually found in root exudates, however, and thus the mechanism by which they affect natural signals for parasitic plants in the soil is not clear.

Flavonoids in the bark of the host *Malosma laurina* induce stem coiling in the parasitic plant *Cuscuta sublinclusa*.[337] Similarly, xenognosins, the 2′-formononetin derivatives from *Astragalus* spp., attract a parasitic angiosperm *Agalinis purpurea* to the roots and subsequently initiate growth of the haustorium that allows the parasite to attach itself to its host.[338] However, flavonoids are not always essential for parasitism, as demonstrated using *Orabanche aegyptiaca* on *Arabidopsis*.[339] It has been argued that flavonoids have to be oxidized to quinines before they can actively induce haustorial development.[335,340,341] Thus, the compelling data from the *in vitro* experiments do not necessarily recapitulate what happens *in vivo*.

Certain flavonoids are used by plants to protect them from invasion by parasites. For example, poplar (*Populus* spp.) cultivars produce a chemical barrier to parasitization by mistletoe (*Viscum album*).[342] Resistant poplar cultivars were significantly higher in flavonols and flavones compared to susceptible cultivars. Likewise, in *Streblus asper* the bark and wood of trees that are resistant to the parasite *Cuscuta reflexa* hold higher levels of flavonoids, as well as steroids and alkaloids.[343]

8.3.3.6 Regulation of Auxin

Auxins are implicated in many developmental and physiological responses, including regulation of the rate of organ elongation, phototropism, and gravitropism. The hormone might also assist plant stress responses through its involvement in stomatal aperture[344] and by reallocating resources under poor growth conditions.[345] Auxin moves from cell to cell in a polar fashion, exhibiting a basipetal polarity in stems and a more complex polarity in roots.[346] Polar auxin transport is controlled by several types of proteins, including auxin influx and efflux carriers, which pump auxin into and out of plant cells, respectively. In the 1960s, it was shown that monohydroxy B-ring flavonoids were involved in degradation of indole acetic acid (IAA), whereas dihydroxy B-ring flavonoids inhibited IAA-degrading activity.[347,348] There is now accumulating evidence for a role of flavonoids as endogenous regulators of auxin transport. However, they are not the sole regulators of auxin movement.

The development of nodules, lateral roots, and galls may all be mediated by plant flavonoids through a perturbence of the root auxin balance.[349] Flavonoids are induced in root cortical cells before and during their division during the formation of nodules[350] and galls.[351,352] Flavonoids have been found to affect the activity of a peroxidase that regulates auxin turnover,[321] and they are inhibitors of auxin transport.[349,353] Quercetin, apigenin, and kaempferol can outcompete auxins for binding sites on plasma membranes.[349] Flavonoids may affect auxin distribution and local concentrations and thus modulate auxin-mediated processes that range from gene expression and ion transport to cell and organ differentiation. However, glycosylation interferes with binding and thus only aglycones are active inhibitors of auxin transport. Since the glycosidic form of flavonoids is most abundant in plant cells, their role remains unclear.

Evidence that flavonoids regulate auxin accumulation *in vivo* was obtained using the flavonoid-deficient mutant *tt4* of *Arabidopsis*.[354] In whole seedling [14C]indole-3-acetic acid transport studies, the pattern of auxin distribution in this mutant was shown to be altered relative to that of wild-type plants. The defect appeared to be in auxin accumulation, as a considerable amount of auxin escaped from the roots through leakage from the root tip. Treatment of the *tt4* mutant with the missing intermediate naringenin restored normal auxin distribution and accumulation by the root. The ability of flavonoids to prevent auxin leakage from the tip provides compelling evidence that endogenous flavonoids are required for normal auxin accumulation by plant cells. Brown et al.[353] also demonstrated that flavonoids act as endogenous regulators of auxin transport. Auxin transport measurements in the inflorescence and the hypocotyls of two different *tt4* mutants, which block flavonoid synthesis, indicate that auxin transport is elevated in the absence of endogenous flavonoids. Growth of plants on naringenin leads to growth and gravity inhibition consistent with inhibition of auxin transport.

Further studies are required to fully elucidate the role of flavonoids in auxin regulation *in vivo* to determine, for example, whether changes in the synthesis or deposition of specific flavonoids within the cell act to change the rate or direction of auxin transport.[355] There is the question of how such different organs or developmental outcomes as nodules, lateral roots, arbuscules, and root galls can involve such similar responses in the plant. How these differences in organogenesis are established remains largely unknown. One possibility is that gene duplication and formation of multigene families has allowed temporal and spatial differences in expression of different members of gene families from previous developmental pathways to symbiotic or pathogenic ones. These cell- and tissue-specific differences may confer specificity to different organs while maintaining gene function.

8.3.4 PROTECTION FROM INSECT AND MAMMALIAN HERBIVORY

Flavonoids, along with other phenolics, help protect plants from herbivory by both insects and mammals. Although one of the most studied groups of flavonoids in regard to this function is the isoflavonoids, other classes of flavonoids are apparently involved, including anthocyanins, flavones, flavonols, and proanthocyanidins (Table 8.5). Although most research has used leaves, protective flavonoids have also been found in other plant parts, such as the roots and seed coats. Insecticidal activity of flavonoids is achieved through various mechanisms including their effects as feeding deterrents,[356] digestion inhibitors,[357] and direct toxicity.[358–360] Lipid-soluble flavonoids in leaves form phenolic resins that deter feeding by insects and can bind irreversibly with plant proteins to form flavonoid-based tannins that are unpalatable to herbivores.[3] One of the best-known and commercially valuable flavonoid insecticide is the family of rotenoid isoflavonoids, in particular rotenone. These compounds are present in roots and aerial parts of many tropical species of Fabaceae including the genera *Derris*, *Lonchcarpus*, *Mundulea*, and *Tephrosia*.[361] Rotenone is potent against a wide range of pests including leaf-chewing beetles, caterpillars, flea beetles, and aphids. The metabolite acts by specifically inhibiting the NADH-dependent dehydrogenase step of the mitochondrial respiratory chain, thus impairing O_2 uptake by insects.[361]

In addition to their being feeding deterrents, flavonol glycosides can also function as phagostimulants to insects. In some cases, the taste of the flavonoids may be associated with attraction or repellence of herbivores. Quercetin-3-glucoside, which occurs in the pollen of sunflower (*Helianthus annuus*), is phagoactive for the western corn rootworm (*Diabrotica virgifera*), which feeds on this pollen.[379] However, it must be noted that flavonoids are not the only phagostimulant present in the pollen and there are some that are more active than the flavonol. There are cases where a feeding attractant becomes a deterrent as concentration increases. For example, in studies with clover (*Trifolium subterraneum*) and the red-legged earth mite (*Halotydeus destructor*), genistein showed 93% deterrence when supplied at a concentration of 0.08%, and 68% deterrence at 0.045%, yet was found to be an attractant to the mite at 0.01%.[380]

Some flavonoids, such as dicoumerol, can also serve as herbivore deterrents for mammals. Higher oligomeric forms of proanthocyanidins are feeding deterrents, or else they impair digestion due to their ability to precipitate proteins. This has been reviewed previously.[381,382] Some mammals (various herbivores, though not carnivores) have adapted to a diet containing condensed tannins by the production of proline-rich proteins in the saliva.[93] These proteins have a strong affinity for tannins and bind them in the mouth so that the hydrogen-bonded complex passes through the stomach without causing any damage. There are examples where the tannin-binding capacity is restricted to condensed tannins and the reaction does not occur with hydrolyzable tannins.[383] Scandinavian and North American moose can feed on twigs and bark from a range of trees and shrubs but they cannot eat tissue of *Rubus* and *Alnus*, which contain both classes of tannin. North American deer, on the other hand, can eat more widely as their salivary proteins can bind both types of tannins.[384] Estrogenic isoflavonoids affect lambing rates in sheep[385] and plants may have evolved these compounds to control the reproductive capacity of their foragers.

TABLE 8.5
Examples of Flavonoids Acting as Feeding Deterrents for Insects

Plant	Compound(s)	Insect Pest	Ref.
Trifolium repens Roots	Medicarpin (isoflavone)	Feeding deterrent to the beetle *Costelytra zealandica*	362
Gossypium sp. Buds	Cyanidin-3-glucoside (anthocyanin); gossypetin 8-*O*-rhamnoside and gossypetin 8-*O*-glucoside (flavonols)	Tobacco budworm, *Heliothis viriscens*	5, 363
Vigna unguiculate Seed coats	Proanthocyanins	Resistance to cowpea weevil *Callosobruchus maculatus*	364
Ulex europaeus Root bark	Ulexones A (isoflavone)	Feeding deterrent for larvae of *Costelytra zealandica*	365
Arachis hypogaea Leaf bud petioles	Procyanidin	Affects fecundity of groundnut aphid (*Aphis craccivora*)	366
Lonchocarpus castilloi Heartwood	Castillen D (aurone) and castillen E (dihydrochalcone)	Termite *Cryptotermes brevis* feeding deterrence, but were not toxic	367
Lotus pedunculatus Roots, leaves	Vestitol (isoflavonoid)	*Heteronychus aratoo*	368
Zea mays Silks	Maysin (*C*-glycosyl flavone)	Corn earworm, *Helicoverpa zea*	369
Melicope subunifoliolata Leaves	Meliternatin (3,5-dimethoxy-3',4',6,7-bismethylendioxyflavone) and six other minor polyoxygenated flavones	Strong feeding deterrent activity against *Sitophilus zeamais;* larvicidal activity against *Aedes aegypti*	370
Nothofagus spp. (Chile and New Zealand) Leaves	Galangin (flavonol) and the stilbene pinosylvin (appear to act in concert)	Deter feeding by leafrollers (*Ctenopsteustis obliquana, Epiphyas postvittana*)	371
Pinus banksiana Needles	Rutin and quercetin-3-glucoside (flavonol glycosides)	Reduced growth and increased mortality of gypsy moth (*Lymantria dispar*)	372
Oryza sativa Leaves	Schaftoside, isoschaftoside, neoschaftoside (glycoflavones)	Sucking deterrent to the brown plant hopper (*Niloparvata lugens*)	373
	Sakuranetin (flavanone) in conjunction with chlorogenic acid	Resistance to stem nematode (*Ditylenchus angustus*) attack	374
Glycine max Leaves	Coumestrol, phaseollin, afrormosin (isoflavonoids)	Soybean looper (*Pseudoplusia includens*); larvae of *Pectinophora gossypiella*	375, 376
	Daidzin and genistin (isoflavones)	Stink bug (*Nezara viridula*)	377
	Rutin and quercetin-3-glucosylgalactoside (flavonols) and genistein (isoflavone)	Cabbage looper (*Trichoplusia ni*)	Cited in 93
Mucuna spp. Seeds	Tannins	Bruchids	359
Triticum spp. Stems and leaves	Tricin (flavone) and unidentified *C*-glycosyl-flavones	Aphids — *Schizaphis graminum* and *Myzus persicae*	378

8.4 CONCLUSIONS

Despite the resurgence in research activity on flavonoid function, many questions remain unanswered. Some functions are only partially understood, and there are probably many others not yet uncovered. For example, there have been several intriguing reports that describe correlations between flavonoid content and morphology. In *Antirrhinum*, the intensity of anthocyanin pigmentation in the flowers depends upon the shape of cells in the petals.[386] In maize, mutant endosperm cells show both an abnormality in shape and a blockage in anthocyanin biosynthesis, indicating a possible connection between flavonoid precursors and cell morphology.[387] Although these may perhaps be explained by some of the activities described above (e.g., regulation of auxin), Tamagnone et al.[388] have postulated a more direct function for flavonoid intermediates in tissue development.

Investigations into the medicinal properties of flavonoids have also revealed novel mechanisms of action; for example, in mediating nucleic acid strand scission, and the inhibition or induction of certain enzymes. It is unclear, however, if these functions have any physiological significance in the plant. Woo et al.[389] have suggested that flavonoids in plants may affect gene expression by acting on a putative hormone receptor in the nuclear membrane, or else they could change the activity of regulatory proteins, such as tyrosine kinase, that are involved in cell division. This is being investigated further.

It is fascinating that this one class of secondary compounds has such a diversity of functions. Multigene families in the flavonoid pathway have presumably lead to specialization of flavonoid gene members in processes as disparate as signaling, defense, development, flower pigmentation, and cell wall modification.[274,390] Different plant species may use different mechanisms to distribute flavonoids among subcellular compartments, and multiple mechanisms are used in individual species.[391] In addition to the flavonoids as a group displaying a diversity of functions, individual compounds also show multifarious functions. The different functions often share common mechanisms. For example, the ability of flavonoids to act as antioxidants is behind their role in combating many different types of stresses. Similarly, the role of flavonoids in regulation of auxin distribution influences plant responses to nodulating bacteria, mycorrhizal fungi, and *Agrobacterium*.

Future applications of analytical methodology and molecular biology techniques are likely to reveal much more about flavonoid function in plants in the coming decades. A more complete understanding of flavonoid function would provide the foundation for further manipulating plants to cope with environmental stress.

REFERENCES

1. Lee, D.W. and Gould, K.S., Anthocyanins in leaves and other vegetative organs: an introduction, in *Anthocyanins in Leaves, Advances in Botanical Research*, vol. 37, Gould, K.S. and Lee, D.W., Eds., Academic Press, Amsterdam, 2002, 1.
2. Simmonds, M.S.J., Flavonoid–insect interactions: recent advances in our knowledge, *Phytochemistry*, 64, 21, 2003.
3. Harborne, J.B., Flavonoid pigments, in *Herbivores: Their Interaction with Secondary Plant Metabolites*, Rosenthal, G.A. and Janzen, D.H., Eds., Academic Press, New York, 1979, 619.
4. Harborne, J.B., *Introduction to Ecological Biochemistry*, 4th ed., Academic Press, New York, 1997.
5. Hedin, P.A. et al., Multiple factors in cotton contributing to resistance to the tobacco budworm *Heliothis virescens* F., in *Plant Resistance to Insects*, Heden, P.A., Ed., American Chemical Society, Washington, 1983, 347.
6. Quiros, C.F. et al., Resistance in tomato to the pink form of the potato aphid *Macrosiphum euphorbiae*. The role of anatomy, epidermal hairs and foliage composition, *J. Am. Soc. Hortic. Sci.*, 102, 166, 1977.
7. Isman, M.B. and Duffy, S.S., Toxicity of tomato phenolic compounds to the fruitworm *Heliothis zea*, *Entomol. Exp. Appl.*, 31, 370, 1982.
8. Gonzales, W.L. et al., Host plant changes produced by the aphid *Sipha flava*: consequences for aphid feeding behaviour and growth, *Entomol. Exp. Appl.*, 103, 107, 2002.
9. Costa-Arbulú, C. et al., Feeding by the aphid *Sipha flava* produces a reddish spot on leaves of *Sorghum halepense*: an induced defense? *J. Chem. Ecol.*, 23, 273, 2001.

10. Furuta, K., Early budding of *Acer palmatum* caused by the shade; intraspecific heterogeneity of the host for the maple aphid, *Bull. Tokyo Univ. For.*, 82, 137, 1990.
11. Furuta, K., Host preference and population dynamics in an autumnal population of the maple aphid, *Periphyllus californiensis* Shinji (Homoptera, Aphididae), *J. Appl. Entomol.*, 102, 93, 1986.
12. Hamilton, W.D. and Brown, S.P., Autumn tree colours as a handicap signal, *Proc. R. Soc. Lond. Ser. B*, 268, 1489, 2001.
13. Prokopy, R.J. and Owens, E.D., Visual detection of plants by herbivorous insects. *Annu. Rev. Entomol.*, 28, 337, 1983.
14. Neill, S.O. and Gould, K.S., Optical properties of leaves in relation to anthocyanin concentration and distribution, *Can. J. Bot.*, 77, 1777, 1999.
15. Richards, P.W. *The Tropical Rainforest*, Cambridge University Press, Cambridge, 1952.
16. Lee, D.W., Brammeier, S., and Smith, A.P., The selective advantages of anthocyanins in developing leaves of mango and cacao, *Biotropica*, 19, 40, 1987.
17. Tuohy, J.M. and Choinski, J.S., Comparative photosynthesis in developing leaves of *Brachystegia spiciformis* Benth., *J. Exp. Bot.*, 41, 919, 1990.
18. Juniper, B.E., Flamboyant flushes: a reinterpretation of non-green flush colours in leaves, *Int. Dendrol. Soc. Yearb.*, 49, 1993.
19. Kursar, T.A. and Coley, P.D., Delayed greening in tropical leaves: an antiherbivore defense? *Biotropica*, 24, 256, 1992.
20. Lev-Yadun, S. et al., Plant coloration undermines herbivorous insect camouflage, *BioEssays* 26, 1126, 2004.
21. Lee, D.W. and Collins, T.H., Phylogenetic and ontogenetic influences on the distribution of anthocyanins and betacyanins in leaves of tropical plants, *Int. J. Plant Sci.*, 162, 1141, 2001.
22. Blanc, P., Biologie des plantes des sous-bois tropicaux, Thèse de Doctorat d'Etat, Université Pierre et Marie Curie (Paris 6), Paris, 1989.
23. Givnish, T.J., Leaf mottling: relation to growth form and leaf phenology and possible role as camouflage, *Funct. Ecol.*, 4, 463, 1990.
24. Stone, B.C. Protective colouration of young leaves in certain Malaysian palms. *Biotropica*, 11, 126, 1979.
25. Dominy, N.D. et al., Why are young leaves red? *Oikos*, 98, 163, 2002.
26. Coley, P.D. and Aide, T.M., Red coloration of tropical young leaves: a possible antifungal defence? *J. Trop. Ecol.*, 5, 293, 1989.
27. Schutt, C. and Netzly, D., Effect of apiforol and apigeninidin on growth of selected fungi, *J. Chem. Ecol.*, 17, 2261, 1991.
28. Stonecipher, L.L., Hurley, P.S., and Netzley, D.H., Effect of apigeninidin on the growth of selected bacteria, *J. Chem. Ecol.*, 19, 1021, 1993.
29. Page, J.E. and Towers, G.H.N., Anthocyanins protect light-sensitive thiarubrine phototoxins, *Planta*, 215, 478, 2002.
30. Guillet, G. et al., Multiple modes of insecticidal action of three classes of polyacetylene derivatives from *Rudbeckia hirta* (Asteraceae), *Phytochemistry*, 46, 495, 1997.
31. Block, E. et al., The photochemistry of thiarubrine A and other 1,2-dithiins: formation of 2,6-dithiabicyclo[3.1.0]hex-3-enes, *J. Am. Chem. Soc.*, 118, 4719, 1996.
32. Page, J.E., Block, E., and Towers, G.H.N., Visible light photochemistry and phototoxicity of thiarubrines, *Photochem. Photobiol.*, 170,159, 1999.
33. Towers, G.H.N. et al., Antibiotic properties of thiarubrine A, a naturally occurring dithiacyclohexadiene polyine, *Planta Med.*, 51, 225, 1985
34. Markham, K.R., *Techniques of Flavonoid Identification*, Academic Press, London, 1982.
35. Giusti, M.M., Rodriguez-Saona, L.E., and Wrolstad, R.E., Molar absorptivity and color characteristics of acylated and non-acylated pelargonidin-based anthocyanins, *J. Agric. Food Chem.*, 47, 4631, 1999.
36. Lee, D.W. and Lowry, J.B., Young-leaf anthocyanin and solar ultraviolet, *Biotropica*, 12, 75, 1980.
37. Alexieva, V. et al., The effect of drought and ultraviolet radiation on growth and stress markers in pea and wheat, *Plant Cell Environ.*, 24, 1337, 2001.

38. Balakumar, T., Babu, V.H.B., and Paliwal K., On the interaction of UV-B radiation (280–315 nm) with water stress in crop plants, *Physiol. Plant.,* 87, 217, 1993.

39. Brandt, K., Giannini, A., and Lercari, B., Photomorphogenic responses to UV radiation III: a comparative study of UVB effects on anthocyanin and flavonoid accumulation in wild-type and aurea mutant of tomato (*Lycopersicon esculentum* Mill.), *Photochem. Photobiol.,* 62, 1081, 1995.

40. Chimphango, S.G.M., Musil, C.F., and Dakora, F.D., Response of purely symbiotic and NO$_3$-fed nodulated plants of *Lupinus luteus* and *Vicia atropurpurea* to ultraviolet radiation, *J. Exp. Bot.,* 54, 1771, 2003.

41. Hashimoto, T., Shichijo, C., and Yatsuhashi, H., Ultraviolet action spectra for the induction and inhibition of anthocyanin synthesis in broom sorghum seedlings, *J. Photochem. Photobiol. B,* 11, 353, 1991.

42. Jain, V.K. and Guruprasad, K.N., Differences in the synthesis of anthocyanin in *Sorghum bicolor* after exposure to ultraviolet-A, blue, red, far-red and white light, *Plant Physiol. Biochem.,* 17, 23, 1990.

43. Lancaster, J.E. et al., Induction of flavonoids and phenolic acids in apple by UV-B and temperature, *J. Hortic. Sci. Biotechnol.,* 75, 142, 2000.

44. Khare, M. and Guruprasad, K.N., UV-B-induced anthocyanin synthesis in maize regulation by FMN and inhibitors of FMN photoreactions, *Plant Sci.,* 91, 1, 1993.

45. Mendez, M., Jones, D.G., and Manetas, Y., Enhanced UV-B radiation under field conditions increases anthocyanin and reduces the risk of photoinhibition but does not affect growth in the carnivorous plant *Pinguicula vulgaris*, *New Phytol.,* 144, 275, 1999.

46. Maekawa, M. et al., Differential responses to UV-B irradiation of three near isogenic lines carrying different purple leaf genes for anthocyanin accumulation in rice (*Oryza sativa* L.), *Breeding Sci.,* 51, 27, 2001.

47. Ravindran, K.C. et al., Influence of UV-B supplemental radiation on growth and pigment content in *Suaeda maritima* L., *Biol. Plant.,* 44, 467, 2001.

48. Reddy, V.S. et al., Ultraviolet-B-responsive anthocyanin production in a rice cultivar is associated with a specific phase of phenylalanine ammonia lyase biosynthesis. *Plant Physiol.,* 105, 1059, 1994.

49. Singh, A., Tamil, S.M., and Sharma, R., Sunlight-induced anthocyanin pigmentation in maize vegetative tissues, *J. Exp. Bot.,* 50, 1619, 1999.

50. Takeda, J. and Abe, S., Light-induced synthesis of anthocyanin in carrot cells in suspension. IV. The action spectrum, *Photochem. Photobiol.,* 56, 69, 1992.

51. Thangeswaran, A., Jeyachandran, R., and George, V.K., A preliminary investigation on UV-filters in two cultivars of mungean, *Environ. Ecol.,* 9,1038, 1991.

52. Yakimchuk, R. and Hoddinott, J., The influence of ultraviolet-B light and carbon dioxide enrichment on the growth and physiology of seedlings of three conifer species, *Can. J. For. Res.,* 24, 1, 1994.

53. Bacci, L. et al., UV-B radiation causes early ripening and reduction in size of fruits in two lines of tomato (*Lycopersicon esculentum* Mill.), *Glob. Change Biol.,* 5, 635, 1999.

54. Buchholz, G., Ehmann, B., and Wellmann, E., Ultraviolet light inhibition of phytochrome-induced flavonoid biosynthesis and DNA photolyase formation in mustard cotyledons (*Sinapis alba* L.), *Plant Physiol.,* 108, 227, 1995.

55. Jordan, B.R. et al., The effect of ultraviolet-B radiation on gene expression and pigment composition in etiolated and green pea leaf tissue: UV-B-induced changes are gene-specific and dependent upon the developmental stage, *Plant Cell Environ.,* 17, 45, 1994.

56. Nissim-Levi, A. et al., Effects of temperature, UV-light and magnesium on anthocyanin pigmentation in cocoplum leaves, *J. Hortic. Sci. Biotechnol.,* 78, 61, 2003.

57. Solovchenko, A. and Schmitz-Eiberger, M., Significance of skin flavonoids for UV-B-protection in apple fruits, *J. Exp. Bot.,* 54, 1977, 2003.

58. Takahashi, A., Takeda, K., and Ohnishi, T., Light-induced anthocyanin reduces the extent of damage to DNA in UV-irradiated *Centaurea cyanus* cells in culture, *Plant Cell Physiol.,* 32, 541, 1991.

59. Kootstra, A., Protection from UV-B-induced DNA damage by flavonoids, *Plant Mol. Biol.*, 26, 771, 1994.

60. Stapleton, A.E. and Walbot, V., Flavonoids can protect maize DNA from the induction of ultraviolet radiation damage, *Plant Physiol.*, 105, 881, 1994.

61. Li, J. et al., *Arabidopsis* flavonoid mutants are hypersensitive to UV-B radiation, *Plant Cell*, 5, 171, 1993.

62. Burger, J. and Edwards, G.E., Photosynthetic efficiency, and photodamage by UV and visible radiation, in red versus green leaf coleus varieties, *Plant Cell Physiol.*, 37, 395, 1996.

63. Caldwell, M.M., Robberecht, R., and Flint, S.D., Internal filters: prospects for UV-acclimation in higher plants, *Physiol. Plant.*, 58, 445, 1983.

64. Day, T.A., Vogelmann, T.C., DeLucia, E.H., Are some plant life forms more effective than others in screening out ultraviolet-B radiation? *Oecologia*, 92, 513, 1992.

65. Ålenius, C.M., Vogemann, T.C., and Bornman, J.F., A three-dimensional representation of the relationship between penetration of u.v.-B radiation and u.v.-screening pigments in leaves of *Brassica napus*, *New Phytol.*, 131, 297, 1995.

66. Bornman, J.F. and Vogelmann, T.C., Effect of UV-B radiation on leaf optical properties measured with fibre optics, *J. Exp. Bot.*, 42, 547, 1991.

67. Gorton, H.L. and Vogelmann, T.C., Effects of epidermal cell shape and pigmentation on optical properties of *Antirrhinum* petals at visible and ultraviolet wavelengths, *Plant Physiol.*, 112, 879, 1996.

68. Brown, J. and Tevini, M., Regulation of UV-protective pigment synthesis in the epidermal layer of rye seedlings (*Secale cereale* L. Cv. Kustro), *Photochem. Photobiol.*, 57, 318, 1993.

69. Burchard, P., Bilger, W., and Weissenbock, G., Contribution of hydroxycinnamates and flavonoids to epidermal shielding of UV-A and UV-B radiation in developing rye primary leaves as assessed by ultraviolet-induced chlorophyll fluorescence measurements, *Plant Cell Environ.*, 23, 1373, 2000.

70. Day, T.A., Relating UV-B radiation screening effectiveness of foliage to absorbing-compound concentration and anatomical characteristics in a diverse group of plants, *Oecologia*, 95, 542, 1993.

71. Hoque, E. and Remus, G., Natural UV-screening mechanisms of Norway spruce (*Picea abies* [L.] Karst.) needles, *Photochem. Photobiol.*, 69, 177, 1999.

72. Hutzler, P. et al., Tissue localization of phenolic compounds in plants by confocal laser scanning microscopy, *J. Exp. Bot.*, 49, 953, 1998.

73. Karabourniotis, G. and Fasseas, C., The dense indumentum with its polyphenol content may replace the protective role of the epidermis in some young xeromorphic leaves, *Can. J. Bot.*, 74, 347, 1996.

74. Kolb, C.A. et al., Effects of natural intensities of visible and ultraviolet radiation on epidermal ultraviolet screening and photosynthesis in grape leaves, *Plant Physiol.*, 127, 863, 2001.

75. Laakso, K., Sullivan, J.H., and Huttunen, S., The effects of UV-B radiation on epidermal anatomy in loblolly pine (*Pinus taeda* L.) and Scots pine (*Pinus sylvestris* L.), *Plant Cell Environ.*, 23, 461, 2000.

76. Olsson, L.C, Veit, M., and Bornman, J.F., Epidermal transmittance and phenolic composition in leaves of atrazine-tolerant and atrazine-sensitive cultivars of *Brassica napus* grown under enhanced UV-B radiation, *Physiol. Plant.*, 107, 259, 1999.

77. Schmitz-Hoerner, R. and Weissenboeck, G., Contribution of phenolic compounds to the UV-B screening capacity of developing barley primary leaves in relation to DNA damage and repair under elevated UV-B levels, *Phytochemistry*, 64, 243, 2003.

78. Schnitzler, J.-P. et al., UV-B screening pigments and chalcone synthase mRNA in needles of Scots pine seedlings, *New Phytol.*, 132, 247, 1996.

79. Sullivan, J.H. et al., Changes in leaf expansion and epidermal screening effectiveness in *Liquidambar styraciflua* and *Pinus taeda* in response to UV-B radiation, *Physiol. Plant.*, 98, 349, 1996.

80. Tevini, M., Braun, J., and Fieser, G., The protective function of the epidermal layer of rye seedlings against ultraviolet-B radiation, *Photochem. Photobiol.*, 53, 329, 1991.

81. Kubo, H. et al., ANTHOCYANINLESS2, a homeobox gene affecting anthocyanin distribution and root development in *Arabidopsis, Plant Cell,* 11, 1217, 1999.
82. Solangaarachchi, S.M. and Gould, K.S., Anthocyanin pigmentation in the adventitious roots of *Metrosideros excelsa, N.Z.J. Bot.,* 39, 161, 2001.
83. Markham, K.R. et al., Anthocyanic vacuolar inclusions — their nature and significance in flower colouration, *Phytochemistry,* 55, 327, 2000.
84. Wannakrairoj, S. and Kamemoto, H., Histological distribution of anthocyanins in *Anthurium* spathes, *Hortscience,* 25, 809, 1990.
85. Lloyd, A.M. et al., Epidermal cell fate determination in *Arabidopsis*: patterns defined by a steroid-inducible regulator, *Science,* 266, 436, 1994.
86. Kunz, Z., Burkhardt, G., and Becker, H., Riccionidins A and B, anthocyanidins from the cell walls of the liverwort *Ricciocarpos natans, Phytochemistry,* 35, 233, 1994.
87. Post, A., Photoprotective pigment as an adaptive strategy in the Antarctic moss *Ceratodon purpureus, Polar Biol.,* 10, 241, 1990.
88. Post, A. and Vesk, M., Photosynthesis, pigments, and chloroplast ultrastructure of an Antarctic liverwort from sun-exposed and shaded sites, *Can. J. Bot.,* 70, 2259, 1992.
89. Lee, D.W., Anthocyanins in leaves: distribution, phylogeny and development, in *Anthocyanins in Leaves. Advances in Botanical Research*, vol. 37, Gould, K.S. and Lee, D.W., Eds., Academic Press, Amsterdam, 2002, 37.
90. Gould, K.S. and Quinn, B.D., Do anthocyanins protect leaves of New Zealand native species from UV-B? *N.Z. J. Bot.,* 37, 175, 1999.
91. Gould, K.S. et al., Functional role of anthocyanins in the leaves of *Quintinia serrata, J. Exp. Bot.,* 51, 1107, 2000.
92. Woodall, S.S., Dodd, I.C., and Stewart, G.R., Contrasting leaf development within the genus *Syzygium, J. Exp. Bot.,* 49, 79, 1998.
93. Harborne, J.B. and Williams, C.A., Advances in flavonoid research since 1992, *Phytochemistry*, 55, 481, 2000.
94. Ormrod, D.P., Landry, L.G., and Conklin, P.L., Short-term UV-B radiation and ozone exposure effects on aromatic secondary metabolite accumulation and shoot growth of flavonoid-deficient *Arabidopsis* mutants, *Physiol. Plant.*, 93, 602, 1995.
95. Reuber, S., Bornman, J.F., and Weissenbock, G., A flavonoid mutant of barley (*Hordeum vulgare* L.) exhibits increased sensitivity to UV-B radiation in the primary leaf, *Plant Cell Environ.*, 19, 593, 1996.
96. Gitz, D.C., Liu, L., and McClure, J.W., Phenolic metabolism, growth, and UV-B tolerance in phenylalanine ammonia-lyase-inhibited red cabbage seedlings, *Phytochemistry*, 49, 377, 1998.
97. Cuadra, P., Harborne, J.B., and Waterman, P.G., Increases in surface flavonols and photosynthetic pigments in *Gnaphalium luteo-album* in response to UV-B radiation, *Phytochemistry*, 45, 1377, 1997.
98. Cuadra, P. and Harborne, J.B., Changes in epicuticular flavonoids and photosynthetic pigments as a polant response to UV-B radiation, *Z. Naturforsch.*, 51c, 671, 1996.
99. Markham, K.R. et al., An increase in the luteolin:apigenin ratio in *Marchantia polymorpha* on UV-B enhancement, *Phytochemistry*, 48, 791, 1998.
100. Markham, K.R. et al., Possible protective role for 3′,4′-dihydroxyflavones induced by enhanced UV-B in a UV-tolerant rice cultivar, *Phytochemistry*, 49, 1913, 1998.
101. Olsson, L.C. et al., Differential flavonoid response to enhanced UV-B radiation in *Brassica napus, Phytochemistry,* 49, 1021, 1998.
102. Skaltsa, H. et al., UV-B protective potential and flavonoid content of leaf hairs of *Quercus ilex, Phytochemistry*, 37, 987, 1994.
103. Lois, R. and Buchanan, B.B., Severe sensitivity to ultraviolet radiation in an *Arabidopsis* mutant deficient in flavonoid accumulation. II. Mechanisms of UV-resistance in Arabidopsis, *Planta*, 194, 504, 1994.
104. Klaper, R., Frankel, S., and Berenbaum, M.R., Anthocyanin content and UVB sensitivity in *Brassica rapa, Photochem. Photobiol.,* 63, 811, 1996.

105. Woodall, G.S. and Stewart, G.R., Do anthocyanins play a role in UV protection of the red juvenile leaves of *Syzygium*? *J. Exp. Bot.*, 49, 1447, 1998.

106. Ryan, K.G. et al., UVB radiation induced increase in quercetin:kaempferol ratio in wild-type and transgenic lines of *Petunia*, *Photochem. Photobiol.*, 68, 323, 1998.

107. Dixon, P., Weinig, C., and Schmitt, J., Susceptibility to UV damage in *Impatiens capensis* (Balsaminaceae): testing for opportunity costs to shade-avoidance and population differentiation, *Am. J. Bot.*, 88, 1401, 2001.

108. Kang, H.-S., Hidema, J., and Kumagai, T., Effects of light environment during culture on UV-induced cyclobutyl pyrimidine dimers and their photorepair in rice (*Oryza sativa* L.), *Photochem. Photobiol.*, 68, 71, 1998.

109. Hada, H. et al., Higher amounts of anthocyanins and UV-absorbing compounds effectively lowered CPD photorepair in purple rice (*Oryza sativa* L.), *Plant Cell Environ.*, 26, 1691, 2003.

110. Long, S.P., Humphries, S., and Falkowski, P.G., Photoinhibition of photosynthesis in nature, *Annu. Rev. Plant Physiol. Plant Mol. Biol.*, 45, 633, 1994.

111. Wheldale, M. *The Anthocyanin Pigments of Plants*, Cambridge University Press, Cambridge, 1916.

112. Krause, G.H. and Weis, E., Chlorophyll fluorescence and photosynthesis: the basics, *Annu. Rev. Plant Physiol. Plant Mol. Biol.*, 42, 313, 1991.

113. Maxwell, K. and Johnson, G.N., Chlorophyll fluorescence — a practical guide, *J. Exp. Bot.*, 51, 659, 2000.

114. Eller, B.M., Glättli, R., and Flach, B., Optical properties and pigments of sun and shade leaves of the beech (*Fagus silvatica* L.) and the copper-beech (*Fagus silvatica* cv. Atropunicea), *Flora,* 171, 170, 1981.

115. Feild, T.S., Lee, D.W., and Holbrook, N.M., Why leaves turn red in autumn. The role of anthocyanins in senescing leaves of red-osier dogwood, *Plant Physiol.*, 127, 566, 2001.

116. Gausman, H.W., Visible light reflectance, transmittance, and absorptance of differently pigmented cotton leaves, *Remote Sens. Environ.*, 13, 233, 1982.

117. Gitelson, A.A., Merzlyak, M.N., and Chivkunova, O.B., Optical properties and nendestructive estimation of anthocyanin content in plant leaves, *Photochem. Photobiol.*, 74, 38, 2001.

118. Gould, K.S., Neill, S.O., and Vogelmann, T.C., A unified explanation for anthocyanins in leaves? in *Anthocyanins in Leaves. Advances in Botanical Research*, vol. 37, Gould, K.S. and Lee, D.W., Eds., Academic Press, Amsterdam, 2002, 167.

119. Neill, S.O. and Gould, K.S., Anthocyanins in leaves: light attenuators or antioxidants? *Funct. Plant Biol.*, 30, 865, 2003.

120. Pietrini, F. and Massacci, A., Leaf anthocyanin content changes in *Zea mays* L. grown at low temperature: significance for the relationship between the quantum yield of PSII and the apparent quantum yield of CO_2 assimilation, *Photosynth. Res.*, 58, 213, 1998.

121. Gould, K.S. et al., Profiles of photosynthesis within red and green leaves of *Quintinia serrata* A. Cunn., *Physiol. Plant.*, 116, 127, 2002.

122. Karabourniotis, G., Bornman, J.F., and Liakoura, V., Different leaf surface characteristics of three grape cultivars affect leaf optical properties as measured with fibre optics: possible implication in stress tolerance, *Aust. J. Plant Physiol.*, 26, 47, 1999.

123. Sun, J., Nishio, J.N., and Vogelmann, T.C., Green light drives CO_2 fixation deep within leaves, *Plant Cell Environ.*, 39, 1020, 1998.

124. Nishio, J.N., Why are higher plants green? Evolution of the higher plant photosynthetic pigment complement, *Plant Cell Environ.*, 23, 539, 2000.

125. Manetas, Y. et al., Exposed red anthocyanic leaves of *Quercus coccifera* display shade characteristics, *Funct. Plant Biol.*, 30, 265, 2003.

126. Barker, D.H., Seaton, G.G.R., and Robinson, S.A., Internal and external photoprotection in developing leaves of the CAM plant *Cotyledon orbiculata*, *Plant Cell Environ.*, 20, 617, 1997.

127. Choinski, J.S. Jr. and Johnson, J.M., Changes in photosynthesis and water status of developing leaves of *Brachystegia spiciformis* Benth., *Tree Physiol.*, 13, 17, 1993.

128. Marini, R.P., Do net gas exchange rates of green and red peach leaves differ? *Hortscience*, 21, 118, 1986.

129. Pietrini F., Ianelli, M.A., and Massacci, A., Anthocyanin accumulation in the illuminated surface of maize leaves enhances protection from photo-inhibitory risks at low temperature, without further limitation to photosynthesis, *Plant Cell Environ.,* 25, 1251, 2002.

130. Dodd, I.C. et al., Photoinhibition in differently coloured juvenile leaves of *Syzygium* species, *J. Exp. Bot.,* 49, 1437, 1998.

131. Gould, K.S. et al., Why leaves are sometimes red, *Nature,* 378, 241, 1995.

132. Krol, M. et al., Low-temperature stress and photoperiod affect an increased tolerance to photo-inhibition in *Pinus banksiana* seedlings, *Can. J. Bot.,* 73, 1119, 1995.

133. Manetas, Y., Drinia, A., and Petropoulou, Y., High contents of anthocyanins in young leaves are correlated with low pools of xanthophyll cycle components and low risk of photoinhibition, *Photosynthetica,* 40, 349, 2002.

134. Smillie, R.M. and Hetherington, S.E., Photoabatement by anthocyanin shields photosynthetic systems from light stress, *Photosynthetica,* 36, 451, 1999.

135. Havaux, M. and Kloppstech, K., The protective functions of carotenoid and flavonoids pigments against excess visible radiation at chilling temperature investigated in *Arabidopsis npq* and *tt* mutants, *Planta,* 213, 953, 2001.

136. Gamon, J.A. and Surfus, J.S., Assessing leaf pigment content and activity with a reflectometer, *New Phytol.,* 143, 105, 1999.

137. Hackett, W.P., Differential expression and functional significance of anthocyanins in relation to phasic development in *Hedera helix* L., in *Anthocyanins in Leaves. Advances in Botanical Research,* vol. 37, Gould, K.S. and Lee, D.W., Eds., Academic Press, Amsterdam, 2002, 95.

138. Garcia-Plazaola, J.I., Hernandez, A., and Becerril, J.M., Antioxidant and pigment composition during autumnal leaf senescence in woody deciduous species differing in their ecological traits, *Plant Biol.,* 5, 557, 2003.

139. Hoch, W.A., Zeldin, E.L., and McCown, B.H., Physiological significance of anthocyanins during autumnal leaf senescence, *Tree Physiol.,* 21, 1, 2001.

140. Lee, D.W. Anthocyanins in autumn leaf senescence, in *Anthocyanins in Leaves. Advances in Botanical Research,* vol. 37, Gould, K.S. and Lee, D.W., Eds., Academic Press, Amsterdam, 2002, 147.

141. Lee, D.W. et al., Pigment dynamics and autumn leaf senescence in a New England deciduous forest, eastern USA, *Ecol. Res.,* 18, 677, 2003.

142. Schaberg, P.G. et al., Factors influencing red expression in autumn foliage of sugar maple trees, *Tree Physiol.,* 23, 325, 2003.

143. Hoch, W.A., Singsaas, E.L., and McCown, B.H., Resorption protection. Anthocyanins facilitate nutrient recovery in autumn by shielding leaves from potentially damaging light levels, *Plant Physiol.,* 133, 1, 2003.

144. Bors, W., Michel C., and Saran, M., Flavonoid antioxidants: rate constants for reactions with oxygen radicals, *Methods Enzymol.,* 234, 420, 1994.

145. Rice-Evans, C.A., Miller, N.J., and Paganga, G., Antioxidant properties of phenolic compounds, *Trends Plant Sci.,* 2, 152, 1997.

146. Tsuda, T., Kato, Y., and Osawa, T., Mechanism for the peroxynitrite scavenging activity by anthocyanins, *FEBS Lett.,* 484, 207, 2000.

147. Wang, H., Cao, G., and Prior, R.L., Oxygen radical absorbing capacity of anthocyanins, *J. Agric. Food Chem.,* 45, 304, 1997.

148. Yamasaki, H., Uefuji, H., and Sakihama, Y., Bleaching of the red anthocyanin induced by super-oxide radical, *Arch. Biochem. Biophys.,* 332, 183, 1996.

149. Ferguson, L.R., Role of plant polyphenols in genomic stability, *Mutat. Res.,* 475, 89, 2001.

150. Frankel, E.F., Waterhouse, A.L., and Teissedre, P.L., Principal phenolic phytochemicals in selected California wines and their antioxidant activity in inhibiting oxidation of human low-density lipoproteins, *J. Agric. Food Chem.,* 43, 890, 1995.

151. Hu, C. et al., Black rice (*Oryza sativa* L. indica) pigmented fraction suppresses both reactive oxygen species and nitric oxide in chemical and biological model systems, *J. Agric. Food Chem.,* 51, 5271, 2003.

152. Lazze, M.C. et al., Anthocyanins protect against DNA damage induced by *tert*-butyl-hydroperoxide in rat smooth muscle and hepatoma cells, *Mutat. Res.*, 535, 103, 2003.

153. Ramirez-Tortosa, C. et al., Anthocyanin-rich extract decreases indices of lipid peroxidation and DNA damage in vitamin E-depleted rats, *Free Radical Biol. Med.*, 31, 1033, 2001.

154. Merzlyak, M.N. and Chivkunova, O.B., Light-stress-induced pigment change and evidence for anthocyanin photoprotection in apples, *J. Photochem. Photobiol. B*, 55, 154, 2000.

155. Nagata, T. et al., Levels of active oxygen species are controlled by ascorbic acid and anthocyanin in *Arabidopsis, J. Agric. Food Chem.*, 51, 2992, 2003.

156. Van Acker, S.A.B.E. et al., Structural aspects of antioxidant activity in flavonoids, *Free Radical Biol. Med.*, 20, 331, 1996.

157. Brown, J.E. et al., Structural dependence of flavonoid interactions with Cu^{2+} ions: implications for their antioxidant properties, *Biochem. J.*, 330, 1173, 1998.

158. Sarma, A.D. and Sharma, R., Anthocyanin–DNA copigmentation complex: mutual protection against oxidative damage, *Phytochemistry*, 52, 1313, 1999.

159. Yoshino, M. and Murakami, K., Interaction of iron with polyphenolic compounds: application to antioxidant characterization, *Anal. Biochem.*, 257, 40, 1998.

160. Halliwell, B. and Gutteridge, J.M.C., *Free Radicals in Biology and Medicine*, 3rd ed., Oxford University Press, Oxford, 1999.

161. Bors, W. et al., Flavonoids as antioxidants: determination of radical scavenging efficiencies, *Methods Enzymol.*, 186, 343, 1990.

162. George, F., Figueiredo, P., and Brouillard, R. Malvin Z-chalcone: an unexpected new open cavity for the ferric cation, *Phytochemistry*, 50, 1391, 1999.

163. Alscher, R.G., Donahue, J.L., and Cramer, C.L., Reactive oxygen species and antioxidants: relationships in green cells, *Physiol. Plant.*, 100, 224, 1997.

164. Foyer, C.H., The contribution of photosynthetic oxygen metabolism to oxidative stress in plants, in *Oxidative Stress in Plants*, Inzé, D. and van Montagu, M., Eds., Taylor & Francis, London, 2002, 33.

165. Foyer, C.H., Lelandais, M., and Kunert, K.H., Photooxidative stress in plants, *Physiol. Plant.*, 92, 696, 1994.

166. Steyn, W.J. et al., Anthocyanins in vegetative tissues: a proposed unified function in photoprotection, *New Phytol.*, 155, 349, 2002.

167. Rice-Evans, C.A., Miller, N.J., and Paganga, G., Structure–antioxidant activity relationships of flavonoids and phenolic acids, *Free Radical Biol. Med.*, 20, 933, 1996. Erratum in *Free Radical Biol. Med.*, 21, 417, 1996.

168. Larson, R., *Naturally Occurring Antioxidants*, Lewis Publisher, Boca Raton, 1997.

169. Yamasaki, H., Sakihama, Y., and Ikehara, N., Flavonoid-peroxidase reaction as a detoxification mechanism of plant cells against H_2O_2, *Plant Physiol.*, 115, 1405, 1997.

170. Gould, K.S., McKelvie, J., and Markham, K.R., Do anthocyanins function as antioxidants in leaves? Imaging of H_2O_2 in red and green leaves after mechanical injury, *Plant Cell Environ.*, 25, 1261, 2002.

171. Philpott, M. et al., *In situ* and *in vitro* antioxidant activity of sweetpotato anthocyanins, *J. Agric. Food Chem.*, 52, 1511, 2004.

172. Brouillard, R. and Dangles, O., Flavonoids and flower colour. In *The Flavonoids — Advances in Research Since 1986*, Harborne, J.B., Ed., Chapman & Hall, London, 1994, 565.

173. Marrs, K.A. et al., A glutathione S-transferase involved in vacuolar transfer encoded by the maize gene *Bronze-2, Nature*, 375, 397, 1995.

174. Lapidot, T. et al., pH-dependent forms of red wine anthocyanins as antioxidants, *J. Agric. Food Chem.*, 47, 67, 1999.

175. Neill, S.O. et al., Antioxidant activities of red versus green leaves in *Elatostema rugosum, Plant Cell Environ.*, 25, 539, 2002.

176. Neill, S.O. et al., Antioxidant capacities of green and cyanic leaves in the sun species, *Quintinia serrata, Funct. Plant Biol.*, 29, 1437, 2002.

177. Polle, A., Defense against photooxidative damage in plants, in *Oxidative Stress and the Molecular Biology of Antioxidant Defenses*, Scandalios, J.G., Ed., Cold Spring Harbor Laboratory Press, New York, 1997, 623.

178. Chalker-Scott, L., Environmental significance of anthocyanins in plant stress responses, *Photochem. Photobiol.*, 70, 1, 1999.

179. Chalker-Scott, L., Do anthocyanins function as osmoregulators in leaf tissues, in *Anthocyanins in Leaves. Advances in Botanical Research*, vol. 37, Gould, K.S. and Lee, D.W., Eds., Academic Press, Amsterdam, 2002, 103.

180. Close, D.C. and Beadle, C.L., The ecophysiology of foliar anthocyanin, *Bot. Rev.*, 69, 149, 2003.

181. Christie, P.J., Alfenito, M.R., and Walbot, V., Impact of low-temperature stress on general phenylpropanoid and anthocyanin pathways: enhancement of transcript abundance and anthocyanin pigmentation in maize seedlings, *Planta*, 194, 541, 1994.

182. Leng, P. and Qi, J.X., Effect of anthocyanin on David peach (*Prunus davidiana* Franch) under low temperature stress, *Sci. Hortic.*, 97, 27, 2003.

183. McKown, R., Kuroki, G., and Warren, G., Cold responses of *Arabidopsis* mutants impaired in freezing tolerance, *J. Exp. Bot.*, 47, 1919, 1996.

184. Nozzolillo, C. et al., Anthocyanins of jack pine (*Pinus banksiana*) seedlings, *Can. J. Bot.*, 80, 796, 2002.

185. Oberbauer, S.F. and Starr, G., The role of anthocyanins for photosynthesis of Alaskan arctic evergreens during snowmelt, in *Anthocyanins in Leaves. Advances in Botanical Research*, vol. 37, Gould, K.S. and Lee, D.W., Eds., Academic Press, Amsterdam, 2002, 129.

186. Solecka, D. and Kacperska, A., Phenylpropanoid deficiency affects the course of plant acclimation to cold, *Physiol. Plant.*, 119, 253, 2003.

187. Hale, K.L. et al., Anthocyanins facilitate tungsten accumulation in *Brassica*, *Physiol. Plant.*, 116, 351, 2002.

188. Hale, K.L. et al., Molybdenum sequestration in *Brassica* species. A role for anthocyanins? *Plant Physiol.*, 126, 1391, 2001.

189. Krupa, Z., Baranowska, M., and Orzol, D., Can anthocyanins be considered as heavy metal stress indicator in higher plants? *Acta Physiol. Plant.*, 18, 147, 1996.

190. Marrs, K.A. and Walbot, V., Expression and RNA splicing of the maize glutathione S-transferase Bronze2 gene is regulated by cadmium and other stresses, *Plant Physiol.*, 113, 93, 1997.

191. Suvarnalatha, G.., Rajendran, L., and Ravishankar, G.A., Elicitation of anthocyanin production in cell cultures of carrot (*Daucus carrota* L.) by using elicitors and abiotic stress, *Biotechnol. Lett.*, 16, 1275, 1994.

192. Farrant, J.M., A comparison of mechanisms of desiccation tolerance among three angiosperm resurrection plant species, *Plant Ecol.*, 151, 29, 2000.

193. Farrant, J.M. et al., An investigation into the role of light during desiccation of three angiosperm resurrection plants, *Plant Cell Environ*, 26, 1275, 2003.

194. Sherwin, H.W. and Farrant, J.M., Protection mechanisms against excess light in the resurrection plants *Craterostigma wilmsii* and *Xerophyta viscose*, *Plant Growth Regul.*, 24, 203, 1998.

195. Leshem, Y.Y. and Kuiper, P.J.C., Is there a GAS (general adaptation syndrome) response to various types of environmental stress? *Biol. Plant.*, 38, 1, 1996.

196. Qiu, Y.-L. et al., The gain of three mitochondrial introns identifies liverworts as the earliest land plants, *Nature*, 394, 671, 1998.

197. Gould, K.S., Nature's Swiss army knife: the diverse protective roles of anthocyanins in leaves, *J. Biomed. Biotechnol.*, 5, 314, 2004.

198. Stapleton, A.E., Ultraviolet radiation and plants: burning questions, *Plant Cell*, 4, 1353, 1992.

199. Kubasek, W.L. et al., Regulation of flavonoid biosynthetic genes in germinating *Arabidopsis* seedlings, *Plant Cell*, 4, 1229, 1992.

200. Schmelzer, E., Jahnen, W., and Hahlbrock, K., *In situ* localization of light-induced chalcone sythase mRNA, chalcone sythase, and flavonoid endproducts in epidermal cells of parsley leaves, *Proc. Natl. Acad. Sci. USA*, 85, 2989, 1988.

201. Van Tunen, A.J. et al., Cloning of the two chalcone flavanone isomerase genes from *Petunia hybrida*: coordinate, light-regulated and differential expression of flavonoid genes, *EMBO J.*, 7, 1257, 1988.
202. Koes, R.E., Quattrocchio, F., and Mol, J.N.M., The flavonoid biosynthetic pathway in plants: function and evolution, *BioEssays*, 16, 123, 1994.
203. Shirley, B.W., Flavonoid biosynthesis: "new" functions for an "old" pathway, *Trends Plant Sci.*, 1, 377, 1996.
204. Kolb, C.A. et al., UV screening by phenolics in berries of grapevine (*Vitis vinifera*), *Funct. Plant Biol.*, 30, 1177, 2003.
205. Jaakola, L. et al., Activation of flavonoid biosynthesis by solar radiation in bilberry (*Vaccinium myrtillus* L.) leaves, *Planta*, 218, 721, 2004.
206. Mazza, C.A. et al., Functional significance and induction by solar radiation of ultraviolet-absorbing sunscreens in field-grown soybean crops, *Plant Physiol.*, 122, 117, 2000.
207. Ryan, K.G. et al., Flavonoid gene expression and UV photoprotection in transgenic and mutant *Petunia* leaves, *Phytochemistry*, 59, 23, 2002.
208. Ryan, K.G. et al., Flavonoids and UV photoprotection in *Arabidopsis* mutants, *Z. Naturforsch. C*, 56, 745, 2001.
209. Rozema, J. et al., The role of UV-B radiation in aquatic and terrestrial ecosystems — an experimental and functional analysis of the evolution of UV-absorbing compounds, *J. Photochem. Photobiol. B*, 66, 2, 2002.
210. Hofmann, R.W. et al., Responses of nine *Trifolium repens* L. populations to ultraviolet-B radiation: differential flavonol glycoside accumulation and biomass production, *Ann. Bot.*, 86, 527, 2000.
211. Hofmann, R.W. et al., Responses to UV-B radiation in *Trifolium repens* L. — physiological links to plant productivity and water availability, *Plant Cell Environ.*, 26, 603, 2003.
212. Smith, G.J. and Markham, K.R., Tautomerism of flavonol glucosides: relevance to plant UV protection and flower colour, *J. Photochem. Photobiol. A*, 118, 99, 1998.
213. Landry, L.G., Chapple, C.C.S., and Last, R.L., *Arabidopsis* mutants lacking phenolic sunscreens exhibit enhanced ultraviolet-B injury and oxidative damage, *Plant Physiol.*, 109, 1159, 1995.
214. Kirakosyan, A. et al., Antioxidant capacity of polyphenolic extracts from leaves of *Crataegus laevigata* and *Crataegus monogyna* (Hawthorn) subjected to drought and cold stress, *J. Agric. Food Chem.*, 51, 3973, 2003.
215. Coberly, L.C. and Rausher, M.D., Analysis of a chalcone synthase mutant in *Ipomoea purpurea* reveals a novel function for flavonoids: amelioration of heat stress, *Mol. Ecol.*, 12, 1113, 2003.
216. Oren-Shamir, M. and Levi-Nissim, A., Temperature effects on the leaf pigmentation of *Cotinus coggygria* "Royal Purple", *J. Hortic. Sci.*, 72, 425, 1997.
217. Teklemariam, T. and Blake, T.J., Effects of UVB preconditioning on heat tolerance of cucumber (*Cucumis sativus* L.), *Environ. Exp. Bot.*, 50, 169, 2003.
218. Kidd, P.S. et al., The role of root exudates in aluminium resistance and silicon-induced amelioration of aluminium toxicity in three varieties of maize (*Zea mays* L.), *J. Exp. Bot.*, 52, 1339, 2001.
219. Cocker, K.M., Evans, D.E., and Hobson, M.J., The amelioration of aluminium toxicity by silicon in higher plants: solution chemistry or an *in planta* mechanism? *Physiol. Plant.*, 104, 608, 1998.
220. Gunse, B., Poschenrieder, C., and Barcelo, J., The role of ethylene metabolism in the short-term responses to aluminium by roots of two maize cultivars different in Al-resistance, *Environ. Exp. Bot.*, 43, 73, 2000.
221. Stoutjesdijk, P.A., Sale, P.W., and Larkin, P.J., Possible involvement of condensed tannins in aluminium tolerance of *Lotus pedunculatus*, *Aust. J. Plant Physiol.*, 28, 1063, 2001.
222. Babu, T.S. et al., Similar stress responses are elicited by copper and ultraviolet radiation in the aquatic plant *Lemna gibba*: implication of reactive oxygen species as common signals, *Plant Cell Physiol.*, 44, 1320, 2003.
223. Sakihama, Y. and Yamasaki, H., Lipid peroxidation induced by phenolics in conjunction with aluminum ions, *Biol. Plant.*, 45, 249, 2002.
224. Takahama, U., Oxidation of flavonols by hydrogen peroxide in epidermal and guard cells of *Vicia faba* L., *Plant Cell Physiol.*, 29, 433, 1988.

225. Charriere-Ladreix, Y. and Tissut, M., Foliar flavonoid distribution during *Spinacia* chloroplast isolation, *Planta*, 151, 309, 1981.

226. Markham, K.R., Gould, K.S., and Ryan, K.G., Cytoplasmic accumulation of flavonoids in flower petals and its relevance to yellow flower colouration, *Phytochemistry*, 58, 403, 2001.

227. Winkel-Shirley, B., Evidence for enzyme complexes in the phenylpropanoid and flavonoid pathways, *Physiol. Plant.*, 107, 142, 1999.

228. Davies, J.M., The bioenergetics of vacuolar H+ pumps, in: *The Plant Vacuole*, Leigh, R.A., Sanders, D., and Callow, J.A., Eds., Academic Press, New York, 1997, 339.

229. Shahidi, F. and Wanasundara, P.K., Phenolic antioxidants, *Crit. Rev. Food Sci. Nutr.*, 32, 67, 1992.

230. Kong, J.-M. et al., Analysis and biological activities of anthocyanins, *Phytochemistry*, 64, 923, 2003.

231. Forkmann, G., Flavonoids as flower pigments: the formation of the natural spectrum and its extension by genetic engineering, *Plant Breeding*, 106, 1, 1991.

232. Harborne, J.B., Functions of flavonoids in plants, in *Chemistry and Biochemistry of Plant Pigments*, Goodwin, T.W., Ed., vol. 1, Academic Press, London, 1976, 736.

233. Harborne, J.B., The flavonoids: recent advances, in *Plant Pigments*, Goodwin, T.W., Ed., Academic Press, London, 1988, 299.

234. Wijsman, H.J.W., On the interrelationships of certain species of *Petunia* II. Experimental data: crosses between different taxa, *Acta Bot. Neerl.*, 32, 97, 1983.

235. Sasaki, K. and Takahashi, T., A flavonoid from *Brassica rapa* flower as the UV-absorbing nectar guide, *Phytochemistry*, 61, 339, 2002.

236. Primack, R.B., Ultraviolet patterns in flowers, or flowers as viewed by insects, *Arnoldia*, 42, 139, 1982.

237. Lubliner-Mianowska, K., The pigments of pollen grains, *Acta Soc. Bot. Pol.*, 24, 609, 1955.

238. Van Tunen, A.J., Hartman, S.A., and Mur, L.A., Regulation of chalcone flavanone isomerase (CHI) gene expression in *Petunia hybrida*: the use of alternative promoters in corolla, anthers and pollen, *Plant Mol. Biol.*, 12, 539, 1989.

239. Van der Meer, I.M. et al., Antisense inhibition of flavonoid biosynthesis in *Petunia* anthers results in male sterility, *Plant Cell*, 4, 253, 1992.

240. Taylor, L.P. and Jorgensen, R., Conditional male fertility in chalcone sythase-deficient *Petunia*, *J. Hered.*, 83, 11, 1992.

241. Napoli, C.A. et al., *White anther*: A *Petunia* mutant that abolishes pollen flavonol accumulation, induces male sterility, and is complemented by a chalcone synthase transgene, *Plant Physiol.*, 120, 615, 1999.

242. Burbulis, I.E., Iacobucci, M., and Shirley, B.W., A null mutation in the first enzyme of flavonoid biosynthesis does not affect male fertility in *Arabidopsis, Plant Cell*, 1013, 1996.

243. Coe, E.H., McCormick, S., and Modena, S.A., White pollen in maize, *J. Hered.*, 83, 11, 1981.

244. Mo, Y., Nagel, C., and Taylor, L.P., Biochemical complementation of chalcone synthase mutants defines a role for flavonols in functional pollen, *Proc. Nat. Acad. Sci. USA*, 89, 7213, 1992.

245. Ylstra, B. et al., Flavonols stimulate development, germination, and tube growth of tobacco pollen, *Plant Physiol.*, 100, 902, 1992.

246. Ylstra, B., Muskens, M., and van Tunen, A.T., Flavonols are not essential for fertilization in *Arabidopsis thaliana*, *Plant Mol. Biol.*, 32, 1155, 1996.

247. Song, K.S. et al., Phenolic function in pollen growth in *Arabidopsis thaliana*, *Acta Hortic.*, 447, 223, 1997.

248. Becard, G. et al., Flavonoids are not necessary plant signal compounds in arbuscular mycorrhizal symbioses, *Mol. Plant–Microbe Interact.*, 8, 252, 1995.

249. Ylstra, B. et al., Flavonols and fertilization in *Petunia hybrida*: localization and mode of action during pollen tube growth, *Plant J.*, 6, 201, 1994.

250. Derksen, J. et al., Pollen tubes of flavonol-deficient *Petunia* show striking alterations in wall structure leading to tube disruption, *Planta*, 207, 575, 1999.

251. Loake, G.J. et al., Phenylpropanoid pathway intermediates regulate transient expression of a chalcone synthase gene promoter, *Plant Cell*, 3, 829, 1991.

252. Yao, K.N., de Luca, V., and Brisson, N., Creation of a metabolic sink for tryptophan alters the phenylpropanoid pathway and the susceptibility of potato to *Phytophthora infestans*, *Plant Cell*, 7, 1787, 1995.

253. Wiermann, R. and Vieth, K., Outer pollen wall, an important accumulation site for flavonoids, *Protoplasma*, 118, 230, 1983.

254. Santos, A. et al., Biochemical and ultrastructural changes in pollen of *Zea mays* L. grown under enhanced UV-B radiation, *Ann. Bot.*, 82, 641, 1998.

255. Wakelin, A.M., Biochemistry and Inheritance of Pollen/Petal Colour Mutants in Californian Poppy — Investigating the Role of Carotenoid Pigments in Pollen, PhD thesis, Lincoln University, New Zealand, 2001.

256. Ndakidemi, P.A. and Dakora, F.D., Legume seed flavonoids and nitrogenous metabolites as signals and protectants in early seedling development, *Funct. Plant Biol.*, 30, 729, 2003.

257. Winkel-Shirley, B., Flavonoids in seeds and grains: physiological function, agronomic importance and the genetics of biosynthesis, *Seed Sci. Res.*, 8, 415, 1998.

258. Jorgensen, R.A., Que, Q.D., and Napoli, C.A., Maternally-controlled ovule abortion results from cosuppression of dihydroflavonol-4-reductase or flavonoid-3',5'-hydroxylase genes in *Petunia hybrida*, *Funct. Plant Biol.*, 29, 1501, 2002.

259. Moore, R.P., Effects of mechanical injuries on viability, in *Viability of Seeds*, Roberts, E.H., Ed., Chapman & Hall, London, 1972, 100.

260. Debeaujon, I., Leon-Kloosterziel, K.M., and Koornneef, M., Influence of the testa on seed dormancy, germination, and longevity in *Arabidopsis, Plant Physiol.*, 122, 403, 2000.

261. Debeaujon, I. et al., Proanthocyanidin-accumulating cells in *Arabidopsis* testa: regulation of differentiation and role in seed development, *Plant Cell*, 15, 2514, 2003.

262. Powell, A.A., The importance of genetically determined seed coat characteristics to seed quality in grain legumes, *Ann. Bot.*, 63, 169, 1989.

263. Kantar, F., Pilbeam, C.J., and Hebblethwaite, P.D., Effect of tannin content opf faba bean (*Vicia faba*) seed on seed vigour, germination and field emergence, *Ann. Appl. Biol.*, 128, 85, 1996.

264. Zavala, J.A. and Botto, F., Impact of solar UV-B on seedling emergence, chlorophyll fluorescence, and growth and yielf of radish (*Rhaphanus sativus*), *Funct. Plant Biol.*, 29, 797, 2002.

265. Vogt, T. et al., Pollination- or wound-induced kaempferol accumulation in *Petunia* stigmas enhances seed production, *Plant Cell*, 6, 11, 1994.

266. Dornbos, D.L., Spencer, G.F. Jr., and Miller, R.W., Medicarpin delays alfalfa seed germination and seedling growth, *Crop Sci.*, 30, 162, 1990.

267. Guenzi, W.D., Kehr, W.R., and McCalla, T.M., Water-soluble phytotoxic substances in alfalfa forage: variation with variety, cutting, year and stage of growth, *Agron. J.*, 56, 499, 1964.

268. Stafford, H.A., Flavonoid evolution: an enzymic approach, *Plant Physiol.*, 96, 680, 1991.

269. Baker, M.E., Flavonoids as hormones — a perspective from an analysis of molecular fossils, *Adv. Exp. Med. Biol.*, 439, 249, 1998.

270. Phillips, D.A., Flavonoids: plant signals to soil microbes, in *Phenolic Metabolism in Plants*, Stafford, H.A. and Ibrahim, R.K., Eds., Plenum Press, New York, 1992, 201.

271. Vierheilig, H. et al., Flavonoids and arbuscular-mycorrhizal fungi, *Adv. Exp. Med. Biol.*, 439, 9, 1998.

272. Straney, D. et al., Host recognition by pathogenic fungi through plant flavonoids, in *Flavonoids in Cell Function*, Buslig, B. and Manthy, J., Eds., Kluwer Academic/Plenum Publishers, New York, 2002, 9.

273. Sundaravarathan, S. and Kannaiyan, S., Role of plant flavonoids as signal molecules to *Rhizobium, Biotechnol. Biofertil.*, 144, 2002.

274. Cohen, M.F. et al., Roles of plant flavonoids in interactions with microbes: from protection against pathogens to the mediation of mutualism, *Rec. Res. Dev. Plant Physiol.*, 2, 157, 2001.

275. Stafford, H.A., Roles of flavonoids in symbiotic and defense functions in legume roots, *Bot. Rev.*, 63, 27, 1997.

276. Jain, V. and Nainawatee, H.S., Plant flavonoids: signals to legume nodulation and soil micro-organisms, *J. Plant Biochem. Biotechnol.*, 11, 1, 2002.

277. Hartwig, U.A., Joseph, C.M., and Phillips, D.A., Flavonoids released naturally from alfalfa seeds enhance growth rate of *Rhizobium meliloti*, *Plant Physiol.*, 95, 797, 1991.

278. Skipp, R.A. and Bailey, J.A., The fungitoxicity of isoflavonoid phytoalexins measured using different types of bioassay, *Physiol. Plant Pathol.*, 11, 101, 1977.

279. Higgins, V.J., The effect of some pterocarpanoid phytoalexins on germ tube elongation of *Stemphylium botryosum*, *Phytopathology*, 78, 339, 1978.

280. Van Etten, H.D. and Bateman, D.F., Studies on the mode of action of the phytoalexin phaseollin, *Phytopathology*, 62, 1363, 1971.

281. Kemp, M.S. and Burden, R.S., Phytoalexins and stress metabolites in the sapwood of trees, *Phytochemistry*, 25, 1261, 1986.

282. Enkerli, J., Bhatt, G., and Covert, S.F., Maackiain detoxification contributes to the virulence of *Nectria haematococca* MP VI on chickpea, *Mol. Plant–Microbe Interact.*, 11, 317, 1998.

283. Wasmann, C.C. and Van Etten, H.D., Transformation-mediated chromosome loss and disruption of a gene for pisatin demethylase decrease the virulence of *Nectria haematococca* on pea, *Mol. Plant–Microbe Interact.*, 9, 793, 1996.

284. He, J., Ruan, Y., and Straney, D., Analysis of determinants of binding and transcriptional activation of the pisatin-responsive DNA binding factor of *Nectria haematococca*, *Mol. Plant–Microbe Interact.*, 9, 171, 1996.

285. Ruan, Y., Kotraiah, V., and Straney, D.C., Flavonoids stimulate spore germination in *Fusarium solani* pathogenic on legumes in a manner sensitive to inhibitors of cAMP-dependent protein kinase, *Mol. Plant–Microbe Interact.*, 8, 929, 1995.

286. Waniska, R.D., Structure, phenolic compounds and antifungal proteins of *Sorghum* caryopses, in *Proceedings of International Consultation*, Chandrashekar, A., Bandyopadhyay, R., and Hall, A.J., Eds., ICRISTAT, India, 2000, 72.

287. Morris, P.F., Bone, E., and Tyler, B.M., Chemotropic and contact responses of *Phytophthora sojae* hyphae to soybean isoflavonoids and artificial substrates, *Plant Physiol.*, 117, 1171, 1998.

288. Zerback, R., Dressler, K., and Hess, D., Flavonoid compounds from pollen and stigma of *Petunia hybrida*: inducers of the *vir* region of the *Agrobacterium tumefaciens* Ti plasmid, *Plant Sci.*, 62, 83, 1989.

289. Schwalm, K. et al., Flavonoid-related regulation of auxin accumulation in *Agrobacterium tumefaciens*-induced plant tumors, *Planta*, 218, 163, 2003.

290. Dixon, R.A. and Paiva, N.L., Stress-induced phenylpropanoid metabolism, *Plant Cell*, 7, 1085, 1995.

291. Pankhurst, C.E., Craig, A.S., and Jones, W.T., Effectiveness of Lotus root nodules. 1. Morphology and flavolan content of nodules formed on *Lotus pedunculatus* by fast-growing *Lotus rhizobia*, *J. Exp. Bot.*, 30, 1085, 1979.

292. Pankhurst, C.E. and Jones, W.T., Effectiveness of *Lotus* root nodules. 3. Effect of combined nitrogen on nodule effectiveness and flavolan synthesis in plant roots, *J. Exp. Bot.*, 30, 1109, 1979.

293. Yang, W.C. et al., *In situ* localization of chalcone synthase mRNA in pea root nodule development, *Plant J.*, 2, 143, 1992.

294. Peters, N.K., Frost, J.W., and Long, S.R., A plant flavone, luteolin, induces expression of *Rhizobium meliloti* nodulation genes, *Science*, 233, 917, 1986.

295. Redmond, J.W. et al., Flavones induce expression of nodulation genes in *Rhizobium*, *Nature*, 323, 632, 1986.

296. Zaat, S.A.J. et al., Induction of a nodA promoter of the *Rhizobium leguminosarum* Sym plasmid PRL1JI by plant flavanones and flavones, *J. Bacteriol.*, 169, 198, 1987.

297. Fisher, R.F. and Long, S.R., Rhizobium–plant signal exchange, *Nature*, 357, 655, 1992.

298. Spaink, H.P. et al., *Rhizobium* nodulation gene nodD as a determinant of host specificity, *Nature*, 328, 337, 1987.

299. Sadowsky, M.J., Cregan, P.B., and Keyser, H.H., Nodulation and nitrogen fixation efficacy of *Rhizobium fredii* with *Phaseolus vulgaris* genotypes, *Appl. Environ. Microbiol.*, 54, 1907, 1988.

300. Recourt, K. et al., Inoculation of *Vicia sativa* subsp. nigra roots with *Rhizobium leguminosarum* biovar viciae results in release of nod gene activating flavanones and chalcones, *Plant Mol. Biol.*, 16, 841, 1991.

301. Kosslak, R.M. et al., Induction of *Bradyrhizobium japonicum* common nod genes by isoflavones isolated from *Glycine max*, *Proc. Natl. Acad. Sci. USA*, 84, 7428, 1987.
302. Pueppke, S.G. et al., Release of flavonoids by the soybean cultivars McCall and Peking and their perception as signals by the nitrogen-fixing symbiont *Sinorhizobium fredii*, *Plant Physiol.*, 117, 599, 1998.
303. Maxwell, C.A. et al., A chalcone and two related flavonoids released from alfalfa roots induce nod genes of *Rhizobium meliloti*, *Plant Physiol.*, 91, 842, 1989.
304. Hungria, M., Joseph, C.M., and Phillips, D.A., Rhizobium nod gene inducers exuded naturally from roots of common bean (*Phaseolus vulgaris* L.), *Plant Physiol.*, 97, 759, 1991.
305. Siqueira, J.O. et al., Significance of phenolic compounds in plant soil microbial systems, *Crit. Rev. Plant Sci.*, 10, 63, 1991.
306. Dakora, F.D., Joseph, C.M., and Phillips, D.A., Alfalfa (*Medicago sativa* L.) root exudates contain isoflavonoids in the presence of *Rhizobium meliloti*, *Plant Physiol.*, 101, 819, 1993.
307. Phillips, D.A., Joseph, C.M., and Maxwell, C.A., Trigonelline and stachydrine released from alfalfa seeds activate NodD2 protein in *Rhizobium meliloti*, *Plant Physiol.*, 99, 1526, 1992.
308. Leon-Barrios, M. et al., Isolation of *Rhizobium meliloti nod* gene inducers from alfalfa rhizosphere soil, *Appl. Environ. Microbiol.*, 59, 636, 1993.
309. Hartwig, U.A. and Phillips D.A., Release and modification of nod-gene-inducing flavonoids from alfalfa seeds, *Plant Physiol.*, 95, 804, 1991.
310. Bolanos-Vasquez, M.C. and Werner, D., Effects of *Rhizobium tropici*, *R. etli*, and *R. legumino-sarum* bv. *phaseoli* on nod gene-inducing flavonoids in root exudates of *Phaseolus vulgaris*, *Mol. Plant–Microbe Interact.*, 10, 339, 1997.
311. Firmin, J.L. et al., Flavonoid activation of nodulation genes in *Rhizobium* reversed by other compounds present in plants, *Nature*, 324, 90, 1986.
312. Kape, R. et al., Isoliquiritigenin, a strong nod gene- and glyceollin resistance-inducing flavonoid from soybean root exudate, *Appl. Environ. Microbiol.*, 58, 1705, 1992.
313. Smit, G. et al., *Bradyrhizobium japonicum nodD*$_1$ can be specifically induced by soybean flavonoids that do not induce the *nodYABCSUIJ* operon, *J. Biol. Chem.*, 267, 310, 1992.
314. Peters, N.K. and Long, S.R., *Rhizobium meliloti* nodulation gene inducers and inhibitors, *Plant Physiol.*, 88, 396, 1988.
315. Djordjevic, M.A. et al., Clovers secrete compounds which either stimulate or repress nod gene function in *Rhizobium trifolii*, *EMBO J.*, 6, 1173, 1987.
316. Kosslak, R.M. et al., Strain-specific inhibition of nod gene induction in *Bradyrhizobium japonicum* by flavonoid compounds, *Appl. Environ. Microbiol.*, 56, 1333, 1990.
317. Jain, V. and Nainawatee, H.S., Flavonoids influence growth and saccharide metabolism of *Rhizobium meliloti*, *Folia Microbiol.*, 44, 311, 1999.
318. Jain, V., Garg, N., and Nainawatee, H.S., Effect of *nod* regulators on the ammonia assimilation enzymes of *Rhizobium* sp *vigna* and *Rhizobium meliloti*, *Natl. Acad. Sci. Lett.*, 15, 345, 1992.
319. Caetano-Anolles, G., Christ-Estes, D.K., and Bauer, W.D., Chemotaxis of *Rhizobium meliloti* to the plant flavone luteolin requires functional nodulation genes, *J. Bacteriol.*, 170, 3164, 1988.
320. Hirsch, A.M. and Fang, Y., Plant hormones and nodulation: what's the connection, *Plant Mol. Biol.*, 26, 5, 1994.
321. Mathesius, U., Flavonoids induced in cells undergoing nodule organogenesis in white clover are regulators of auxin breakdown by peroxidase, *J. Exp. Bot.*, 52, 419, 2001.
322. Xie Z.P. et al., Rhizobial nodulation factors stimulate mycorrhizal colonization of nodulating and nonnodulating soybeans, *Plant Physiol.*, 108, 1519, 1995.
323. Morandi, D., Occurrence of phytoalexins and phenolic compounds in endomycorrhizal inter-actions, and their potential role in biological control, *Plant Soil*, 185, 241, 1996.
324. Becard, G., Douds, D.D., and Pfeffer, P.E., Extensive *in vitro* hyphal growth of vesicular-arbuscular mycorrhizal fungi in the presence of CO_2 and flavonols, *Appl. Environ. Microbiol.*, 58, 821, 1992.
325. Chabot, S. et al., Hyphal growth promotion *in vitro* of the VA mycorrhizal fungus, *Gigaspora margarita* Becker and Hall, by the activity of structurally specific flavonoid compounds under CO_2-enriched conditions, *New Phytol.*, 122, 461, 1992.

326. Rhlid, R.B. et al., Isolation and identification of flavonoids from Ri T-DNA-transformed roots (*Daucus carota*) and their significance in vesicular-arbuscular mycorrhiza, *Phytochemistry*, 33, 1369, 1993.

327. Nair, M.G., Safir, G.R., and Siqueira, J.O., Isolation and identification of vesicular-arbuscular mycorrhiza-stimulatory compounds from clover (*Trifolium repens*) roots, *Appl. Environ. Microbiol.*, 57, 434, 1991.

328. Tsai, S.M. and Phillips, D.A., Flavonoids released naturally from alfalfa promote development of symbiotic Glomus spores *in vitro, Appl. Environ. Microbiol.*, 57, 1485, 1991.

329. Lagrange, H., Jay-Allgmand, C., and Lapeyrie, F., Rutin, the phenolglycoside from eucalyptus root exudates, stimulates *Pisolithus* hyphal growth at picomolar concentrations, *New Phytol.*, 149, 349, 2001.

330. Fries, N. et al., Abietic acid, an activator of basidiospore germination in ectomycorrhizal species of the genus *Suillus* (Boletaceae), *Expt. Mycol.*, 11, 360, 1987.

331. Larose, G. et al., Flavonoid levels in roots of *Medicago sativa* are modulated by the developmental stage of the symbiosis and the root colonizing arbuscular mycorrhizal fungus, *J. Plant Physiol.*, 159, 1329, 2002.

332. Guenoune, D. et al., The defense response elicited by the pathogen *Rhizoctonia solani* is suppressed by colonization of the AM-fungus *Glomus intraradices*, *Plant Sci.*, 160, 925, 2001.

333. Akiyama, K., Matsuoka, H., and Hayashi, H., Isolation and identification of a phosphate deficiency-induced C-glycosylflavonoid that stimulates arbuscular mycorrhiza formation in melon roots, *Mol. Plant–Microbe Interact.*, 15, 334, 2002.

334. Xie, Z.P. et al., Nod factors and tri-iodobenzoic acid stimulate mycorrhizal colonization and affect carbohydrate partitioning in mycorrhizal roots of *Lablab purpureus*, *New Phytol.*, 139, 361, 1998.

335. Lynn, D.G. and Chang, M., Phenolic signals in cohabitation: implications for plant development, *Annu. Rev. Plant Physiol. Plant Mol. Biol.*, 41, 497, 1990.

336. Albrecht, H., Yoder, J.I., and Phillips, D.A., Flavonoids promote haustoria formation in the root parasite *Triphysaria versicolor*, *Plant Physiol.*, 119, 585, 1999.

337. Kelly, C.K., Plant foraging: a marginal value model and coiling response in *Cuscuta subinclusa*, *Ecology*, 71, 1916, 1990.

338. Steffens, J.C. et al., Molecular specificity of haustorial induction in *Agalinis purpurea* (L.) Raf. (Schrophulariaceae), *Ann. Bot.*, 50, 1, 1982.

339. Westwood, J.H., Characterization of the *Orobanche-Arabidopsis* system for studying parasite–host interactions, *Weed Sci.*, 48, 742, 2000.

340. Kim, D. et al., On becoming a parasite. Evaluating the role of wall oxidases in parasitic plant development, *Chemico-Biol. Interact.*, 5, 103, 1998.

341. Matvienko, M. et al., Quinone oxidoreductase message levels are differentially regulated in parasitic and non-parasitic plants exposed to allelopathic quinines, *Plant J.*, 25, 375, 2001.

342. Salle, G.C., Hariri, E.B., and Andary, C., Polyphenols and resistance of poplar (*Populus* spp.) to mistletoe (*Viscum album* L.), *Acta Hortic.*, 381, 756, 1994.

343. Julie, S. and Daniel, M., Chemical changes induced by *Cuscuta reflexa* on *Streblus asper* Lour., *Natl. Acad. Sci. Lett.*, 19, 185, 1996.

344. Dietrich, P., Sanders, D., and Hedrich, R., The role of ion channels in light-dependent stomatal opening, *J. Exp. Bot.*, 52, 1959, 2001.

345. Palme, K. and Galweiler, L., PIN-pointing the molecular basis of auxin transport, *Curr. Opin. Plant Biol.*, 2, 375, 1999.

346. Lomax, T.L., Muday, G.K., and Rubery, P., Auxin transport, in *Plant Hormones: Physiology, Biochemistry, and Molecular Biology*, Davies, P.J., Ed., Kluwer Academic Press, Norwell, The Netherlands, 1995, 509.

347. Furuya, M., Galston, A.W., and Stowe, B.B., Isolation from peas of cofactors and inhibitors of indoyl-3-acetic acid oxidase, *Nature*, 193, 456, 1962.

348. Galston, A.W., Flavonoids and photomorophogenesis in peas, in *Perspectives in Phytochemistry*, Harborne, J.B. and Swain, T., Eds., Academic Press, New York, 1969, 193.

349. Jacobs, M. and Rubery, P.H., Naturally occurring auxin transport regulators, *Science*, 241, 346, 1988.

350. Mathesius, U. et al., Flavonoids synthesized in cortical cells during nodule initiation are early developmental markers in white clover, *Mol. Plant–Microbe Interact.*, 11, 1223, 1998.

351. Hartley, S.E., Are gall insects large rhizobia? *Oikos*, 84, 333, 1999.

352. Hutangura, P. et al., Auxin induction is a trigger for root gall formation caused by root-knot nematodes in white clover and is associated with the activation of the flavonoid pathway, *Aust. J. Plant Physiol.*, 26, 221, 1999.

353. Brown, D.E. et al., Flavonoids act as negative regulators of auxin transport *in vivo* in *Arabidopsis*, *Plant Physiol.*, 126, 524, 2001.

354. Murphy, A., Peer, W.A., and Taiz, L., Regulation of auxin transport by aminopeptidases and endogenous flavonoids, *Planta*, 211, 315, 2000.

355. Mathesius, U., Conservation and divergence of signalling pathways between roots and soil microbes — the *Rhizobium*-legume symbiosis compared to the development of lateral roots, mycorrhizal interactions and nematode-induced galls, *Plant Soil*, 255, 105, 2003.

356. Hedin, P.A. and Waage, S.K., Roles of flavonoids in plant resistance to insects, in *Plant Flavonoids in Biology and Medicine: Biochemical, Pharmacological and Structure Activity Relationship*, Cody, V., Middleton, E., and Harborne, J.B., Eds., Alan Liss, New York, 1986, 87.

357. Fahey, G.C. Jr. and Jung, H.G., Phenolic compounds in forages and fibrous feedstuffs, in *Toxicants of Plant Origin, Vol. IV. Phenolics*, Cheeke, P.R., Ed., CRC Press, Boca Raton, FL, 1989, 123.

358. Freeland, W.J., Calcott, P.H., and Anderson, L.R., Tannins and saponin: interaction in herbivore diets, *Biochem. Syst. Ecol.*, 13, 189, 1985.

359. Rajaram, N. and Jandardhanan, K., The biochemical composition and nutritional potential of the tribal pulse, *Mucuna gigantean* (Willd.) DC., *Plant Food Hum. Nutr.*, 44, 45, 1991.

360. Regnault-Roger, C. et al., Isoflavonoids involvement in the non-adaptability of *Acanthoscelides obtectus* SAY (Bruchidae, Coleoptera) to soybean (*Glycine max*) seeds, in *Proceedings of the 16th Annual Meeting of the International Society of Chemical Ecology*, International Society of Chemical Ecology, Marseilles, France, 1999, 110.

361. Dakora, F.D., Plant flavonoids: biological molecules for useful exploitation, *Aust. J Plant Physiol.*, 22, 87, 1995.

362. Sutherland, O.R.W. et al., Insect feeding deterrent activity of phytoalexin isoflavonoids, *Biochem. Syst. Ecol.*, 8, 73, 1980.

363. Hedin, P.A., Jenkins, J.N., and Parrott, W.L., Evaluation of flavonoids in *Gossypium arboreum* (L.) cottons as potential source of resistance to tobacco budworm, *J. Chem. Ecol.*, 18, 105, 1992.

364. Oigianbe, N.O. and Onigbinde, A.O., The association between some physiochemical characteristics and susceptibility of cowpea [*Vigna unguiculata* (L.) Walp] to *Callosobruchus maculates* (F), *J. Stored Prod. Res., 32*, 7, 1996.

365. Russell, G.B., Sirat, H.M., and Sutherland, O.R.W., Isoflavones from root bark of gorse, *Phytochemistry*, 29, 1287, 1990.

366. Grayer, R.J. et al., Condensed tannin levels and resistance of groundnuts against *Aphis craccivora*, *Phytochemistry*, 31, 3795, 1992.

367. Reyes-Chilpa, R. et al., Antitermitic activity of *Lonchocarpus castilloi* flavonoids and heartwood extracts, *J. Chem. Ecol.*, 21, 455, 1995.

368. Russell, G.B. et al., Vestitol: a phytoalexin with insect feeding-deterrent activity, *J. Chem. Ecol.*, 4, 571, 1978.

369. Byrne, P.F. et al., Quantitative trait loci and metabolic pathways: genetic control of the concentration of maysin, a corn earworm resistance factor, in maize silks, *P. Natl. Acad. Sci. USA*, 93, 8820, 1996.

370. Ho, S.H. et al., Meliternatin: a feeding deterrent and larvicidal polyoxygenated flavone from *Melicope subunifoliolata*, *Phytochemistry*, 62, 1121, 2003.

371. Russell, G.B. et al., Patterns of bioactivity and herbivory on *Nothofagus* species from Chile and New Zealand, *J. Chem. Ecol.*, 26, 41, 2000.

372. Beninger, C.W. and Abou-Zaid, M.M., Flavonol glycosides from four pine species that inhibit early instar gypsy moth (Lepidoptera: *Lymantriidae*) development, *Biochem. Syst. Ecol.*, 25, 505, 1997.

373. Grayer, R.J. et al., Phenolics in rice phloem sap as sucking deterrents to the brown planthopper, *Nilaparvata lugens*, *Acta Hortic.*, 381, 691, 1994.

374. Plowright, R.A. et al., The induction of phenolic compounds in rice after infection by the stem nematode *Ditylenchus angustus*, *Nematologica*, 42, 564, 1996.

375. Caballero, P. and Smith, C.M., Isoflavones from an insect-resistant variety of soybean and the molecular structure of afrormosin, *J. Nat. Prod.*, 49, 1126, 1986.

376. Schoonhoven, L.M., Secondary plant substances and insects, *Rec. Adv. Phytochem.*, 5, 197, 1972.

377. Piubelli, G.C. et al., Flavonoid increase in soybean as a response to *Nezara viridula* injury and its effect on insect-feeding preference, *J. Chem. Ecol.,* 29, 1223, 2003.

378. Dreyer, D.L. and Jones, K.C., Feeding deterency of flavonoids and related phenolics towards *Schizaphis graminum* and *Myzus persicae*: aphid feeding deterrents in wheat, *Phytochemistry*, 20, 2489, 1981.

379. Lin, S. and Mullin, C.A., Lipid, polyamide and flavonol phagostimulants for the adult Western corn rootworm from sunflower pollen, *J. Agric. Food Chem.*, 47, 1223, 1999.

380. Wang, S.F., Ridsdill-Smith, T.J., and Ghisalberti, E.L., Role of isoflavonoids in resistance of subterranean clover trifoliates to the redlegged earth mite *Halotydeus destructor*, *J. Chem. Ecol.*, 24, 2089, 1998.

381. Harborne, J.B., Plant polyphenols and their role in plant defence mechanisms, in *Polyphenols 1994*, INRA, Paris, 1994, 19.

382. Harborne J.B., Recent advances in chemical ecology, *Nat. Prod. Rep.*, 16, 509, 1999.

383. Juntheikki, M.R., Comparison of tannin-binding proteins in saliva of Scandinavian and North American moose (*Alces alces*), *Biochem. Syst. Ecol.*, 24, 595, 1996.

384. Hagerman, A.E. and Robbins, C.T., Specificity of tannin-binding salivary proteins relative to diet selection by mammals, *Can. J. Zool.*, 71, 628, 1993.

385. Shutt, D.A., The effects of plant oestrogens on animal reproduction, *Endeavour*, 35, 110, 1976.

386. Noda, K. et al., Flower colour intensity depends on specialized cell shape controlled by a MYB-related transcription factor, *Nature*, 369, 661, 1997.

387. Gavazzi, G. et al., *Dap* (Defective *a*leurone *p*igmentation) mutations affect maize aleurone development, *Mol. Gen. Genet.*, 8, 1985, 1997.

388. Tamagnone, L. et al., Inhibition of phenolic acid metabolism results in precocious cell death and altered cell morphology in leaves of transgenic tobacco plants, *Plant Cell*, 10, 1801, 1998.

389. Woo, H.H. et al., Flavonoids: signal molecules in plant development, *Adv. Exp. Med. Biol.*, 505, 51, 2002.

390. Lawson, C.G.R. et al., *Rhizobium* inoculation and physical wounding result in the rapid induction of the same chalcone synthase copy in *Trifolium subterraneum*, *Mol. Plant–Microbe Interact.*, 7, 498, 1994.

391. Winkel-Shirley, B., Biosynthesis of flavonoids and effects of stress, *Curr. Opin. Plant Biol.*, 5, 218, 2002.

9 Flavonoid–Protein Interactions

Olivier Dangles and Claire Dufour

CONTENTS

9.1 GENERAL CONSIDERATIONS

Naturally occurring flavonoids are ubiquitous in all parts of plants where they display a huge structural diversity based on different C rings (Figure 9.1), aglycone substitutions by OH and OMe groups, and glycosidation and acylation patterns. In addition to their important biological roles in plants and plant–insect interactions, flavonoids have been thoroughly investigated during the last two decades because of their possible health effects in man via a diet rich in plant products.[1,2] Indeed, flavonoids, especially flavanols, flavonols, and anthocyanins, are relatively abundant in human diet, partially bioavailable, and possibly involved in still incompletely understood mechanisms related to the prevention of cancers, cardiovascular diseases, and neurodegenerescence. However, whatever these mechanisms may be, most of them must be related to at least one of the two fundamental properties of flavonoids: their reducing ability (antioxidant properties by electron or H-atom donation) and their ability to interact with proteins.[3] Flavonoid–protein interaction in plants is an important issue, for instance, in relation to flavonoid biosynthesis and flavonoid-mediated chemical defense mechanisms. However, this chapter will essentially deal with flavonoid–protein interactions in man and their possible implications for human health. Given the wealth of

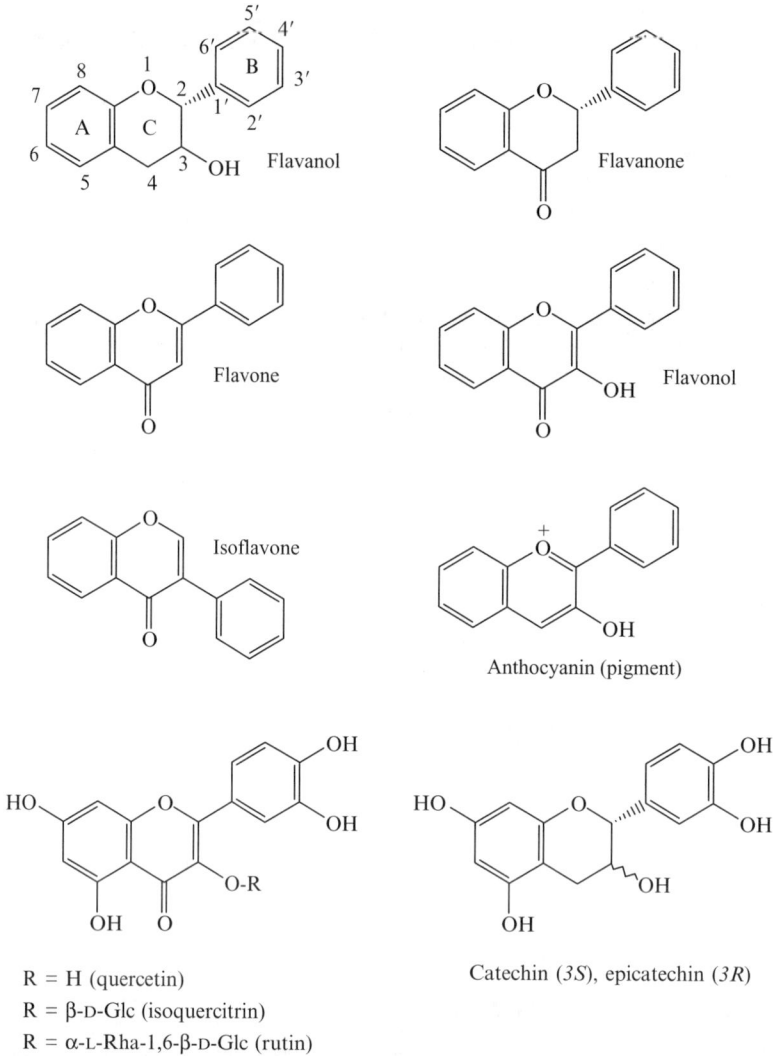

FIGURE 9.1 The main flavonoid classes and examples of common dietary flavonoids.

literature data in this important research field, this review is only aimed at providing a wide overview but by no means an exhaustive report.

9.1.1 THE BIOLOGICAL SIGNIFICANCE OF FLAVONOID–PROTEIN INTERACTIONS IN MAN

Biochemical studies devoted to the possible health effects of flavonoids try to assess either their nutritional value in the prevention of degenerative diseases or their therapeutic value as potential drugs. In the latter case, structure–activity relationships are typically established that include not only naturally occurring flavonoids but also synthetic analogs. In the former case, only dietary flavonoids are considered and bioavailability data must be kept in mind when it comes to building biologically relevant models. Fortunately, flavonoid bioavailability is a fast developing field and a lot of valuable information on the mechanisms by which flavonoids are absorbed from the intestine and metabolized (microbial catabolism, conjugation in liver and enterocytes) has emerged during the last 10 years.[4,5] In fact, flavonoid–protein interactions involved in flavonoid bioavailability, in addition to interactions taking place before intestinal absorption (e.g., interactions with salivary proteins), are currently the sole binding processes with clear *in vivo* biological significance. However, little is known so far about the possible delivery of flavonoid conjugates to specific tissues. On the other hand, flavonoids have been very often pointed out as *in vitro* enzyme inhibitors and ligands of receptors involved in signal transduction.[6,7] However, the biological relevance of these flavonoid–protein binding processes to the field of human nutrition still awaits *in vivo* validation. In particular, the crucial point would be the demonstration that the flavonoid in its bioavailable (typically conjugated) form can accumulate near the target protein in concentrations high enough for the interaction to take place. During the last decade, cell effects distinct from the antioxidant activity by radical scavenging and metal chelation have been frequently evoked to interpret flavonoid-mediated health effects. Indeed, the rather low circulating concentrations of dietary flavonoids (typically in the range 0.1 to 1 μM), and the demonstration that phenolic OH groups critical to the radical scavenging activity can be conjugated in the circulating forms, do not argue in favor of *in vivo* effects dominated by the antioxidant activity except, possibly, in the gastrointestinal tract where high concentrations of native flavonoids can accumulate after a meal. Hence, it can be speculated that most of the flavonoid-mediated health effects involve interactions of flavonoids with specific biological targets, mainly proteins. It is important to underline that the term flavonoid–protein interactions in the literature does not always refer to a direct molecular contact (complexation) but may point to regulation by flavonoids of gene expression for specific proteins. Cytochrome P450 enzymes are a good example of proteins whose function can be regulated by flavonoids via such diverse mechanisms.[8]

9.1.2 MOLECULAR INTERACTIONS RESPONSIBLE FOR FLAVONOID–PROTEIN COMPLEXATION

Intrinsically, the phenolic nucleus is a structural unit that is favorable to molecular (non-covalent) interactions with proteins. These interactions can be divided into two classes[3]:

- *Van der Waals interactions*: the nonpolar polarizable aromatic ring can develop strong dispersion interactions with amino acid residues displaying similar properties. These interactions are strengthened by the partial desolvation experienced by the two surfaces coming into contact and the simultaneous release of water molecules from the solvation shells where they are highly ordered to the bulk solvent where they develop more hydrogen bonds with other water molecules (hydrophobic effect). These two distinct processes (ligand–protein dispersion interactions and the hydrophobic effect) are often (erroneously) referred to collectively as "hydrophobic interactions." The importance of dispersion interactions in the stability of the flavonoid–protein complexes could be reflected by the rather general trend emerging from structure–affinity relationships that the flat more polarizable (iso)flavones and flavonols (electron delocalization spread over the three rings) are generally better ligands than flavanones and flavanols (see below).
- *Electrostatic interactions*: in the case of phenols, hydrogen bonding is probably the most important interaction falling in this category. Indeed, the OH group can act as a hydrogen bond donor (via its acidic proton) and a hydrogen bond acceptor (via its nonconjugated lone pair lying in the plane of the phenolic nucleus) toward polar amino acid residues and peptide bonds. In addition, flavonoids having an electron-withdrawing 4-keto group possess a fairly acidic 7-OH group because of a strong electron delocalization in the corresponding phenolate anions. In the case of flavones and flavonols, a similar electronic effect also raises the acidity of the 4'-OH group.

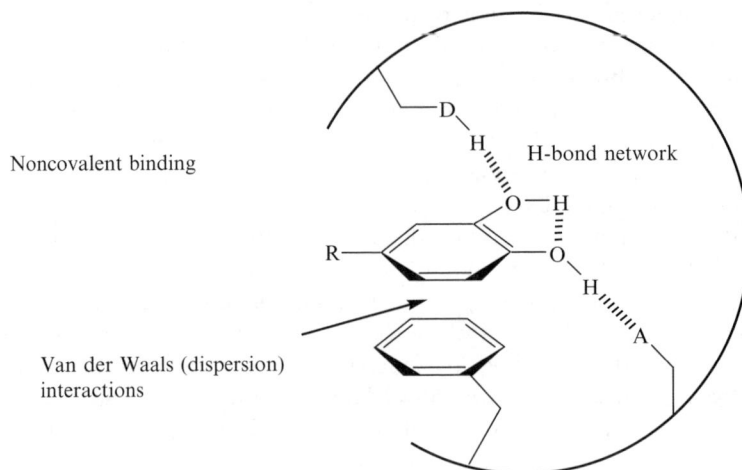

FIGURE 9.2 A schematic representation of molecular interactions at work in flavonoid–protein binding.

Hence, 4-keto flavonoids can be partially dissociated at neutral pH and eventually involved in attractive coulombic interactions with positively charged amino acid residues.

The main molecular interactions involved in phenol–protein interactions are arbitrarily represented in Figure 9.2. A real case of flavonoid–protein hydrogen bond network is depicted in Figure 9.3. It refers to one of the rare examples of a flavonoid–protein complex

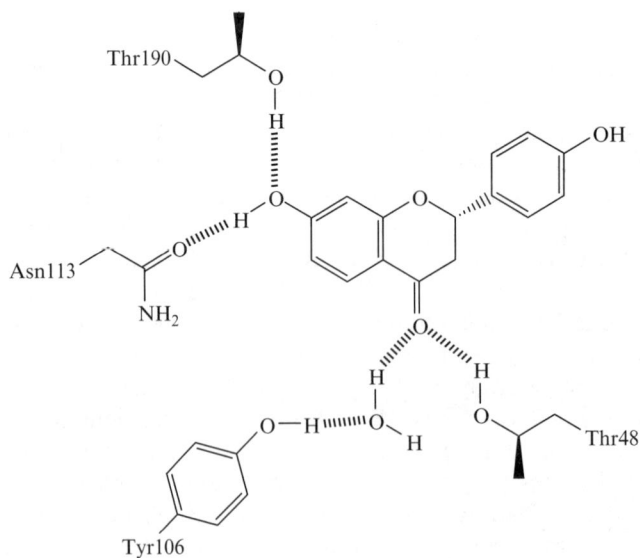

FIGURE 9.3 The hydrogen bond network in the (S)-4′,7-dihydroxyflavanone–chalcone isomerase complex. (Adapted from Jez, J.M., Bowman, M.E., and Noel, J.P., *Biochemistry*, 41, 5168, 2002. With permission.)

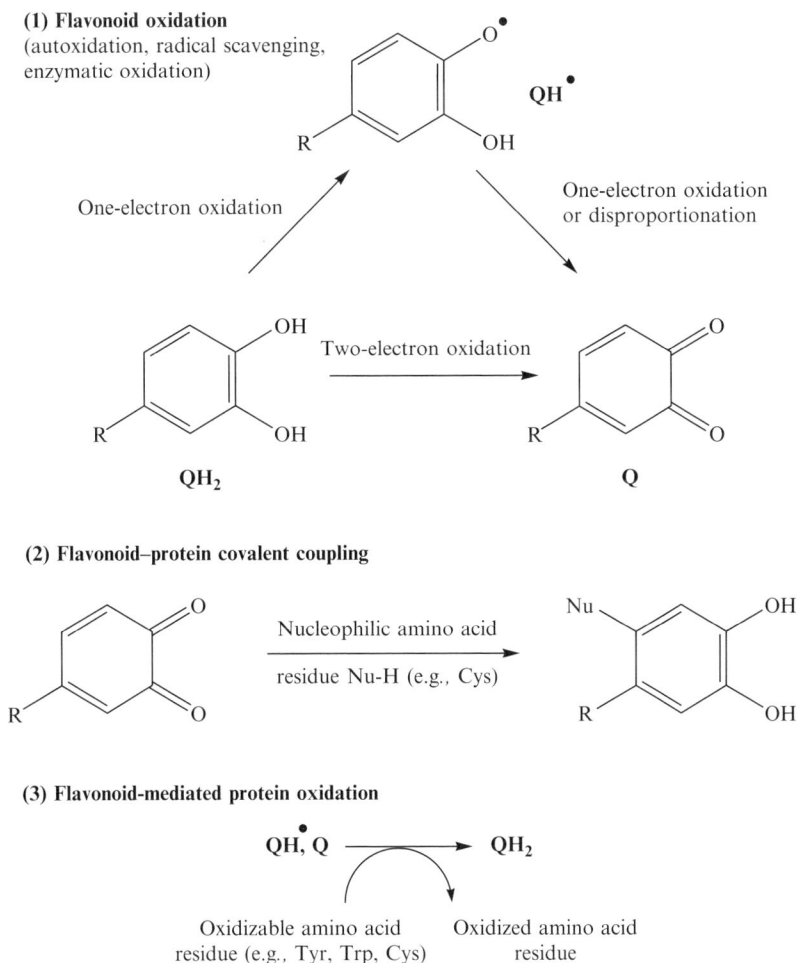

(1) Flavonoid oxidation
(autoxidation, radical scavenging, enzymatic oxidation)

QH$^\bullet$

One-electron oxidation

One-electron oxidation or disproportionation

Two-electron oxidation

QH$_2$

Q

(2) Flavonoid–protein covalent coupling

Nucleophilic amino acid residue Nu-H (e.g., Cys)

(3) Flavonoid-mediated protein oxidation

QH$^\bullet$, Q $\longrightarrow$ QH$_2$

Oxidizable amino acid residue (e.g., Tyr, Trp, Cys)

Oxidized amino acid residue

FIGURE 9.4 Flavonoid oxidation and protein modifications.

whose structure has been elucidated by x-ray crystallography, the complex of the plant enzyme chalcone isomerase with its product (S)-4′,7-dihydroxyflavanone.[9]

Finally, in addition to these noncovalent and reversible interactions, flavonoid–protein redox reactions and oxidative covalent coupling may result from one- or two-electron oxidation of the flavonoid brought about by such mechanisms as: autoxidation (oxidation by dioxygen catalyzed by metal ion traces), scavenging of reactive oxygen species (ROS) (antioxidant activity) eventually produced by the protein itself, and enzymatic oxidation. Being both electrophilic and oxidizing, flavonoid oxidation products (aryloxyl radicals, quinones, and quinonoid compounds) may react with nucleophilic or oxidizable amino acid residues, thereby irreversibly modifying the protein by covalent coupling or oxidation (Figure 9.4).

9.1.3 SPECIFICITY OF FLAVONOID–PROTEIN INTERACTIONS

Given the intrinsic propensity of the phenolic nucleus for developing molecular interactions, it is no surprise that examples of flavonoid–protein complexation are numerous and concern a wide variety of proteins. However, the question of their specificity deserves examination.

In the case of conformationally open proteins (random coils) with multiple binding sites for polyphenols such as proline-rich salivary proteins, binding constants are quite low for small polyphenols (gallates, catechin) but increase sharply when the number of polyphenolic nuclei increases (flavanol-3-O-gallates, oligomeric procyanidins, polygalloylglucose), thus allowing multiple molecular contacts along the protein chain with a preference for the hydrophobic proline residues.[10,11] Similarly, oxidative polymerization of catechin or poly-condensation with aldehydes produces a strong increase in the affinity for xanthine oxidase.[12,13] Such trends reflect the approximately additive character of hydrogen bonding and Van der Waals interactions and suggest rather unspecific binding along an extended protein chain or at the surface of globular proteins.[3,14] By contrast, structure–affinity relationships with various globular proteins having well-defined binding cavities (enzymes, receptors) clearly point to specific interactions with properly substituted flavonoids (generally, flavone, isoflavone, or flavonol aglycones) reaching dissociation constants in the range 1 nM to 1 μM. Interestingly, a screening of 20 tannins including 10 flavanols for 16 brain receptors has shown a rather high specificity in the displacement of specific radioligands.[15] For instance, only two receptors, 5HT1 and the β-adrenoceptor, are significantly inhibited (>60%) by more than one flavanol at the concentration of 10 μM. In addition, whereas the β-adrenoceptor is inhibited by epicatechin and its 3'-deoxy analog, 5HT1 appears more sensitive to 4,8-procyanidin dimers. The origin of specific flavonoid–protein interactions may be traced to the structural resemblance between some properly substituted flavonoids and a variety of bioactive compounds including coenzymes (e.g., nucleotides), steroid hormones, and neurotransmitters (e.g., catecholamines) (Figure 9.5). An unexpected case of specific flavonoid–protein binding occurring *in vivo* can be found in the vacuoles of some flower

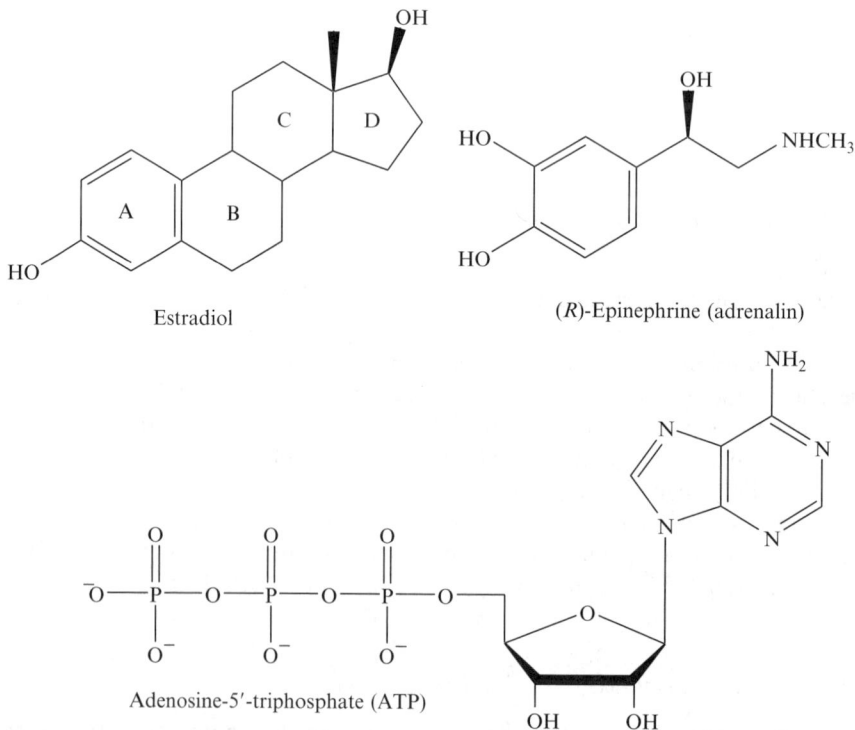

Estradiol

(*R*)-Epinephrine (adrenalin)

Adenosine-5'-triphosphate (ATP)

FIGURE 9.5 Structures of biologically important molecules with a formal structural resemblance to flavonoids.

petals (e.g., lisianthus and carnation) where specific anthocyanins are sequestered within protein matrices.[16] In purple lisianthus, for instance, four minor acylated anthocyanin-3,5-diglycosides are tightly bound (noncovalent association) to a protein aggregate including three major components (MW 34 to 50 kDa) whereas their more abundant triglycoside analogs remain in solution.

9.2 EXAMPLES OF FLAVONOID–PROTEIN INTERACTIONS IN MAN

9.2.1 INTERACTIONS PRIOR TO INTESTINAL ABSORPTION

9.2.1.1 Interactions with Salivary Proteins and Astringency

The interaction of tannins (oligomeric procyanidins) with salivary proteins is a key step in the development of the oral sensation of dryness and roughness called astringency, which is typical of tannin-rich food. Basic salivary proteins are proline-rich proteins (25 to 42%) that can bind and precipitate polyphenols. They comprise a typical 19-residue sequence that is essentially constituted of proline, glutamine, and glycine residues. This sequence is repeated 5 to 15 times to form proteins of ~150 residues that display extended conformations with multiple binding sites for polyphenols.[10] Polyphenol–salivary protein interactions are especially significant with procyanidins (condensed tannins) and galloylated derivatives of D-glucose (hydrolyzable tannins) that possess several 1,2-dihydroxy- or 1,2,3-trihydroxybenzene nuclei allowing multiple interactions with the protein. When the polyphenol concentration is increased to provide sufficient coating of the protein, interactions of the polyphenol molecules (probably under their self-associated form) with a second protein molecule lead to extensive aggregation to form colloidal particles of different sizes. Ultimately, precipitation occurs.

Detailed investigations by nuclear magnetic resonance[4,5,10,11,17] using the typical proline-rich sequences (~20 residues including seven to eight Pro residues) have provided an outline of the main characteristics of the binding:

- The proline residues themselves are privileged sites for interactions with the phenolic nuclei. For instance, β-1,2,3,4,6-penta-O-galloyl-D-glucose (PGG) is able to stack one of its galloyl units on the most accessible face of the pyrrolidine heterocycle (that presenting the Cα–H group). The interaction is essentially driven by dispersion forces and the hydrophobic effect with the possible participation of H bonds between some phenolic OH groups (donors) and the O atom of the tertiary amide group linking the Pro residue to the preceding residue.
- The size of the binding site is restricted to a proline residue for small polyphenols (propylgallate, flavanol monomers) only. As the polyphenol size increases, the number of binding sites is reduced. For instance, up to seven to eight epigallocatechin-3-O-gallate (EGCG) molecules (one per Pro residue) can be simultaneously bound to the 19-mer peptide (pH 3.8, 27°C, 10% DMSO). By contrast, the stoichiometry of the peptide–PGG complex is 1:2.
- Multidentate binding of the polyphenol is important for the formation of stable complexes. Indeed, taking into account polyphenol self-association and assuming identical peptide binding sites (probably, a crude approximation), dissociation constants can be estimated from binding isotherms expressing the chemical shift changes of sensitive peptide proton signals as a function of the polyphenol concentration. K_d values are 0.05, 2.4, 5.0, and 0.05 mM for PGG, EGCG, epicatechin, and its 4,8-β dimer (procyanidin B2), respectively (pH 3.8, 3°C, 10% dimethyl sulfoxide [DMSO], except for epicatechin [no DMSO]).

– Cooperativity between peptide binding sites may be important. Thus, the affinity of EGCG for the 19-mer peptide (eight Pro residues) is ~50 times as strong as for a 7-mer peptide with only one to two binding site(s) for EGCG. Similarly, complexation with a full-length basic salivary protein is also much stronger than with the peptide models.[18] Hence, it may be proposed that salivary proteins wrap around the polyphenol molecules to allow multiple interactions. However, since proline residues provide rigidity to the protein backbone, this process is expected to take place with little loss in rotational freedom and little change in the extended conformation of the peptide.

– From the concentration dependence of the polyphenol chemical shifts as a function of the polyphenol concentration in the absence and presence of peptide, it is clear that polyphenol self-association is only weakly affected by the peptide. Thus, peptide–polyphenol binding probably involves noncovalent polyphenol oligomers that are more effective than the corresponding monomers at developing interactions with a second peptide molecule to trigger precipitation.

– Polyphenol–salivary protein interaction is essentially pH independent in the range 3.8 to 6.0 whereas the extent of precipitation is higher and the particle sizes larger at higher pH. Indeed, at higher pH, the net charge of the peptide moieties must be lower, thus favoring aggregation. On the other hand, at the highest pH values, partial autoxidation of EGCG may take place during the 24-h incubation of the polyphenol with the protein, thus providing electrophilic o-quinone intermediates that may couple to the nucleophilic residues of the protein. Hence, it cannot be excluded that the formation of covalent polyphenol–protein linkages is partially responsible for the increased aggregation at higher pH.

The influence of procyanidin structure on the extent of salivary protein precipitation has also been studied by light scattering techniques.[19] Factors favoring precipitation are: catechin moieties rather than epicatechin moieties; the presence of 3-O-galloyl residues; a higher degree of polymerization; and 4,8-interflavan linkages rather than 4,6-interflavan linkages. However, it must be kept in mind that the extent of precipitation may not be a faithful reflection of the polyphenol–protein affinity in solution (see above).

Despite their very short sequence (7 to 38 amino acid residues for the 12 histatins identified so far), the histidine-rich salivary protein histatins have also been reported to precipitate tannins, eventually more efficiently than proline-rich proteins, especially at neutral pH and high tannin concentration.[20,21] A detailed NMR analysis of the binding between EGCG and histatin 5, a 24-mer that is very rich in basic His, Lys, and Arg residues (~60%) and devoid of secondary structure, has revealed noncooperative binding of six to seven flavanol molecules with a dissociation constant of 1 mM (pH 3.0, 25°C).[22]

9.2.1.2 Interactions in the Gastrointestinal Tract

The formation of insoluble tannin–dietary protein complexes and inhibition of digestive enzymes by tannins have been investigated for a long time for their implications in the low digestibility of nutrients in animals.[23] However, data on the structure and stability of the corresponding complexes seem scarce. In addition, the interactions of flavonoids other than tannins with dietary proteins and their possible consequences on flavonoid bioavailability and antioxidant activity in the gastrointestinal tract are essentially not documented. Regarding digestive enzymes, some properly hydroxylated flavone and flavonol aglycones (e.g., quercetin) have been reported to inhibit the protease trypsin with IC_{50} values in the range 10 to 60 μM.[24]

9.2.2 INTERACTIONS INVOLVED IN FLAVONOID BIOAVAILABILITY

9.2.2.1 Intestinal Absorption

After tannins, flavonoid glycosides are by far the most common dietary sources of flavonoids. However, intestinal absorption requires prior deglycosylation. This step, which does not take place in the gastric compartment, occurs in the large intestine as a result of microbial metabolism.[25] In addition, some flavonoid glucosides can be absorbed directly from the small intestine upon deglucosylation by the enzyme lactase phlorizin hydrolase (LPH) and subsequent passive diffusion of the aglycones through the enterocyte layer (route 1). However, a minor route (route 2) could also consist in active transport of the glucosides through the sodium-dependent intestinal D-glucose carrier SGLT1 and subsequent deglucosylation by cytosolic β-glucosidase. Using inhibitors of LPH and SGLT1, it was demonstrated that route 1 is the sole absorption mechanism for quercetin-3-glucoside (Q3Glc) whereas route 2 could contribute to the absorption of quercetin-4′-glucoside (Q4′Glc).[26] Hydrolysis of flavonoid glucosides by LPH has been studied in detail. This enzyme, which is present on the luminal side of the brush border of the small intestine, primarily catalyzes the hydrolysis of lactose. However, LPH also displays a second site for the hydrolysis of more lipophilic substrates such as the dihydrochalcone glucoside phlorizin. Selective inhibition of the phlorizin site only marginally affects the efficiency of LPH at hydrolyzing Q3Glc and Q4′Glc, thus demonstrating that deglucosylation mainly occurs at the lactase site.[27] Q3Glc and Q4′Glc have similar K_m (40 to 50 μM) and k_{cat} values and are more rapidly hydrolyzed than isoflavone glucosides. The ability of extracts of human small intestine and liver to hydrolyze flavonoid glucosides (likely due to the cytosolic broad-specificity β-glucoside) was also investigated.[28] The flavonol Q4′Glc and the isoflavone genistein 7-glucoside are hydrolyzed with similar efficiencies by both tissues with apparent K_m values of 32 and 14 μM, respectively. By contrast, Q3Glc, quercetin-3,4′-diglucoside, kaempferol-3-glucoside, and rutin (Q-3-β-D-Glc-1,6-α-L-Rha) are not hydrolyzed (with the exception of a small proportion of Q3Glc with the small intestine extract). Hence, in the flavonol family, a Glc moiety at the 4′ position and a free OH group at the 3 position seem required for activity.

The fraction of flavonoids that reaches the colon can be extensively metabolized by microflora enzymes. This may be an important step in flavonoid bioavailability, especially for flavonoids that are essentially not absorbed from the small intestine such as the oligomeric procyanidins.[29] Flavonoid glycosides (native forms) and glucuronides (excreted in the bile) can be hydrolyzed to the corresponding aglycones and further processed into phenolic acids.[25,30,31] Both the flavonoid aglycones and their phenolic acid derivatives can then be absorbed from the colon. In fact, phenolic acids and their glucuronides are detected in much larger concentration in blood and urine than the conjugates of the flavonoid aglycones themselves, thus raising the possibility that the cleavage products are relevant active forms *in vivo.*[31,32] However, no detailed biochemical information (enzyme identification and isolation for the determination of structure–activity relationships, K_m values, etc.) is currently available about the degradation of flavonoids by microflora enzymes.

9.2.2.2 Conjugation Enzymes

The detection of sulfates and glucuronides of quercetin and 3′-O-methylquercetin in the plasma of volunteers having absorbed a flavonol-rich meal[33] has prompted detailed investigations of quercetin metabolism in the enterocyte and liver. By recording the formation of the different quercetin glucuronides (QGlcU) from quercetin and UDP-glucuronic acid in the presence of human liver cell-free extract, it was shown that conjugation on the B ring (positions 3′ and 4′) takes place with low K_m values (0.5 to 1 μM), which suggest a high affinity of quercetin for the corresponding isoforms of liver UDP-glucuronyl transferase.[34] Although quercetin has a lower affinity for isoforms catalyzing conjugation at the 7 position (together with marginal conjugation at the 3 position), the V_{max} value is higher. Hence, at high quercetin concentration, glucuronidation at the 7 position may be predominant. Alternatively, quercetin and other flavonols have been shown to be substrates of purified isoforms of UDP-glucuronyl transferase.[35,36] Using human liver cell-free extracts, it was also possible to demonstrate that quercetin glucuronides undergo further metabolism in the liver.[37] For instance, deconjugation to free quercetin takes place rapidly and is as efficient for each of the four glucuronides tested (positions 3, 3', 4', and 7). These results are in agreement with a preliminary investigation with human cell extracts from liver, small intestine, and neutrophils.[38] In addition, a detailed kinetic analysis with pure β-glucuronidase showed that Q4'GlcU ($K_m = 48$ μM) displays more affinity for the enzyme than its 3 and 7 regioisomers ($K_m = 167$ and 237 μM, respectively) although deconjugation at the 3 position is faster based on the k_{cat}/K_m values. Hence, net glucuronidation of dietary flavonols in the liver could be the result of a balance between the activity of UDP-glucuronyl transferase and β-glucuronidase that would be normally shifted toward the former. Since conjugation dramatically alters the antioxidant activity of flavonoids and their interactions with proteins,[33,34] their eventual deconjugation could be of great importance in the development of flavonoid-mediated health effects. Using intact hepatocytes, methylation of Q3GlcU and Q7GlcU at the 3' and 4' positions as a result of catalysis by catechol O-methyltransferase and formation of quercetin 3'-sulfate (following initial deglucuronidation) were also observed.[37] By contrast, Q4'GlcU is not metabolized under the same conditions.

9.2.2.3   Plasma Proteins

The absorption bands of flavonol conjugates detected in the plasma of rats fed quercetin or rutin are bathochromically shifted in comparison to typical flavonol absorption bands.[39] A similar phenomenon can be reproduced by adding serum albumin (SA) to flavonol solutions, thus suggesting that the circulating flavonols are bound to SA. Since then, several studies have investigated flavonol–albumin complexation. For instance, the human SA (HSA)–quercetin binding constant (high-affinity site) has been evaluated at physiological pH by circular dichroism ($14.6 \pm 2.1 \times 10^3 M^{-1}$, 37°C), ultracentrifugation ($267 \pm 33 \times 10^3 M^{-1}$, 20°C), and fluorescence ($50 \pm 8 \times 10^3 M^{-1}$, 25°C).[40–43] These values reflect the moderate affinity of quercetin for albumin in line with similar observations for a variety of drugs and xenobiotics.[44] On the other hand, the sole flavonols that can be detected in plasma after a meal rich in quercetin or rutin are glucuronides and sulfoglucuronides of quercetin and 3'-O-methylquercetin.[33] Hence, investigating the binding of quercetin conjugates (or models of quercetin conjugates), rather than that of quercetin itself, is more biologically relevant. To that purpose, the influence of quercetin methylation, glycosidation, and sulfation on albumin binding has been assessed.[42,43] (Table 9.1). Quercetin-7-O-sulfate retains a high affinity for BSA and HSA whereas additional sulfation of 4'-OH markedly weakens the binding to both albumins. Glycosidation and sulfation of 3-OH, methylation of 3'-OH (isorhamnetin), and deletion of 3'-OH (kaempferol) all significantly lower the affinity to BSA, the binding to HSA being much less affected. These observations outline the importance of a free OH at position 3' for a strong binding to BSA. Luteolin (no 3-OH) is as efficient as quercetin in binding BSA but not HSA. The influence of glycosyl and acyl substituents is also reflected in the large differences in affinity to BSA displayed by chrysin (strong binding competitively to quercetin), chrysin-7-O-β-D-glucoside and p-methoxycinnamic acid methyl ester (weak binding), and chrysin-7-O-(6-O-p-methoxycinnamoyl)-β-D-glucoside (strong binding noncompetitively to quercetin).[45]

TABLE 9.1

Binding Constants (K) for Flavonoid–Serum Albumin Complexes (1:1 Stoichiometry), pH 7.4 Phosphate–NaCl Buffer, 25°C

Flavonoid[a]	$K \pm$ SD ($\times 10^3\ M^{-1}$)	
	BSA	HSA
Quercetin	134 ± 6^b	50 ± 8^c
Quercetin-7-O-sulfate	143 ± 29	87 ± 23
Quercetin-3-O-sulfate	21 ± 2	58 ± 5
Quercetin-4′,7-O-disulfate	25 ± 2	6 ± 1^d
3′-O-Methylquercetin	24 ± 1^d	N/A
Quercetin-3-O-β-D-Glc	13 ± 4^d	16 ± 1
Rutin	11 ± 7^d	—
Kaempferol	12 ± 1^d	129 ± 13^d
Luteolin	143 ± 35	14 ± 3
Luteolin-7-O-β-D-Glc	N/A	N/A
Flavonoid[e]		
4′-O-Methylquercetin	51 ± 2^d	75 ± 3^d
Diosmetin	66 ± 2	53 ± 1
Baicalein	148 ± 2^d	74 ± 4
Baicalein-7-O-β-D-GlcU	78 ± 1^d	117 ± 4
Genistein	30 ± 1	11 ± 1
Naringenin	49 ± 1	3 ± 1
Catechin	N/A	N/A

[a]Flavonoids forming fluorescent complexes.
[b]SD: standard deviation from four experiments.
[c]From triplicates.
[d]From duplicates. Otherwise SD from curve fitting.
[e]Assuming competitive binding with quercetin (fluorescent indicator).
N/A, not applicable.

   While nonfluorescent in its free state, SA-bound quercetin is strongly fluorescent in the visible range. Hence, quercetin can be used as a fluorescent probe for investigating the binding of flavonoids that are nonfluorescent in both their free and SA-bound states such as flavones and flavonols lacking a free 4′-OH, flavanones, and isoflavones[42,43] (Table 9.1). Assuming competitive binding, it can be noticed that methylation of 4′-OH (quercetin vs. tamarixetin, luteolin vs. diosmetin) decreases the BSA binding constants by a factor 2 to 3. By contrast, the affinity for HSA of tamarixetin (4′-O-methylquercetin), identified as a quercetin metabolite in bile and urine, remains high. Baicalein (5,6,7-trihydroxyflavone) and its 7-O-β-D-glucuronide (baicalin) are tightly bound to both albumins. Thus, the relatively bulky glucuronyl moiety that is typical of most flavonoid conjugates does not hamper the binding in this case. The isoflavone genistein appears as a poor albumin ligand. The lack of conjugation or a non-planar C ring in the flavanone naringenin may be responsible for its relatively low affinity for both albumins, especially HSA. These structural characteristics are even more pronounced in the flavanol catechin, which actually does not affect the fluorescence of albumin-bound quercetin, thus suggesting either no affinity for albumin or binding to a totally independent site.

   Despite its moderate intensity (binding constants K in the range 10^4 to $10^5\ M^{-1}$), the interaction between albumin and flavones and flavonols is strong enough to ensure a quasi-total complexation of the circulating forms, owing to the respective concentrations of the partners (0.1 to 1 μM for the circulating forms, 0.6 mM for serum albumin). Hence, it may be reasonably proposed that flavonols and flavones circulate as albumin complexes rather than in their free form.

Additional investigations aimed at demonstrating specific binding to a particular SA site have been carried out. Displacement studies with known site markers (dansylasparagine and warfarin for subdomain IIA, ibuprofen and diazepam for subdomain IIIA)[43] are in agreement with preliminary results[41]: the quercetin high-affinity binding site appears to be located in subdomain IIA for both HSA and BSA. In the case of HSA, the quercetin binding cavity in subdomain IIA is large enough to accommodate additional ligands such as salicylate[40] and warfarin.[41,43] Furthermore, this site, which is known to bind a large variety of xenobiotics, is lined by positively charged basic residues (Lys, Arg) and nonpolar residues (Tyr, Trp, Leu), in agreement with a complexation driven by dispersion interactions, the hydrophobic effect, and, possibly, attractive coulombic interactions involving partially dissociated flavonol molecules. Finally, flavonoid–albumin binding is noncompetitive to that of fatty acids.[40]

Interactions with other plasma proteins are still poorly documented. Weak binding of quercetin with α1-glycoprotein has been reported.[41] Moreover, when a sample of human plasma is eluted on an EGCG-linked sepharose column, the sole proteins that are retained are fibronectin, fibrinogen, and histidine-rich glycoprotein.[46] Binding of EGCG to each of these plasma proteins was confirmed using dot binding assays on nitrocellulose sheets. In addition, the binding was shown to be restricted to flavanols having a 3-O-galloyl moiety. Interaction of catechin with the protein component (Apo A-1) of high-density lipoproteins has also been proposed from gel electrophoresis experiments and comparison with molecular mass markers.[47]

9.2.3 Interactions Potentially Involved in Cellular Health Effects

9.2.3.1 ATP-Binding Proteins

Flavonoids are known to inhibit the function of many ATP-binding proteins,[6,7] such as mitochondrial ATPase, myosin, Na/K and Ca plasma membrane ATPases, protein kinases, topoisomerase II, and multidrug resistance (MDR) proteins. In general, inhibition takes place through binding of the flavonoids to the ATP-binding site. Only two cases relevant to the inhibition of carcinogenesis by flavonoids[48,49] will be discussed in detail.

9.2.3.1.1 Kinases

Phosphorylation of proteins at OH groups of serine, threonine, and tyrosine residues is an important mechanism of intracellular signal transduction involved in various cellular responses including the regulation of cell growth and proliferation.[49,50] The reaction makes use of ATP as a phosphate donor and is catalyzed by protein kinases. For instance, growth factor hormones bind to extracellular domains of large transmembrane receptors that display a tyrosine kinase moiety in their intracellular portion. As a consequence of hormone–receptor binding, the receptor dimerizes and becomes active in the phosphorylation of proteins close to the membrane, thereby triggering a large number of signaling pathways themselves involving other PKs, such as PKC, a Ser/Thr PK, and mitogen-activated PKs (MAPKs). On the other hand, each phase of the cell cycle, during which the DNA is replicated and the chromosomes built and then separated, is characterized by intense bursts of phosphorylation controlled by highly regulated kinases called cyclin-dependent kinases (CDKs).

A possible mechanism for the potential anticarcinogenic effects of flavonoids could be their ability to inhibit various PKs (demonstrated with purified enzymes or cell extracts). For instance, the isoflavone genistein has been shown to inhibit the epidermal growth factor (EGF) receptor in the submicromolar range by competing with ATP for its binding site.[51] Similarly, butein (2′,3,4,4′-tetrahydroxychalcone) appears as a specific inhibitor of tyrosine kinases (IC_{50} for EGF receptor $= 65$ μM) acting competitively to ATP and noncompetitively to the phosphate acceptor and having no affinity for Ser/Thr PKs such as PKC and the cAMP-dependent PKA.[52] However, structural changes may affect both the selectivity toward different kinases and the binding site. For instance, silymarin, a flavanonol analog, was suggested to bind to the hormone-binding site in competition with EGF.[53] Recently, PKC was shown to be efficiently inhibited by flavones and flavonols having a 3′,4′-dihydroxy substitution on the B ring (efficient concentrations 50 in the range 1 to 10 μM).[54,55] Hydrogen bonding between these two OH groups and the enzyme seems a key determinant of the complexation that takes place in the ATP-binding site, in competition with the cofactor. Similar structure–activity relationships were also established in the inhibition by flavonoids of phosphoinositide 3-kinase (PI3-K), a lipid kinase catalyzing phosphorylation of inositol lipids at the D3 position of the inositol ring to form new intracellular lipid second messengers.[55] Flavonoids were also demonstrated to inhibit CDKs.[56] Consistently, studies with intact cells have shown that various flavonoids can cause cell cycle arrest in correlation to their ability to inhibit CDKs.[53,57]

Flavonoids can also modulate the activity of MAPKs as a possible mechanism for their potential antineurodegenerative action[58,59] and protection against autoimmune, allergic, and cardiovascular diseases.[60,61] For instance, investigations on intact antigen-presenting dendritic cells have shown that the MAP kinases involved in cell maturation (ERK, p38 kinase, JNK) can be activated by bacterial lipopolysaccharide and that this activation is strongly inhibited by pretreatment of the cells by EGCG.[61] However, no evidence is provided that the mechanism actually proceeds via direct EGCG–MAPK inhibition. Interestingly, quercetin-3-O-glucuronide (Q3GlcU), a quercetin metabolite, is more specific than quercetin in inhibiting the activity of MAP kinases in vascular smooth muscle cells.[60] Indeed, pretreatment of the cells with Q3GlcU selectively prevents the activation of c-Jun N terminal kinase (JNK) by angiotensin II. Since JNK is responsible for c-Jun phosphorylation that produces a component of transcription factor AP-1, its inhibition blocks AP-1-mediated gene expression, which is involved in the growth of vascular smooth muscle cells. Once more, the molecular mechanism underlying the effect of the flavonoid remains unclear. Indeed, angiotensin-II-mediated JNK activation may be due to ROS produced by NADH/NADPH oxidase. Hence, its inhibition may be a combination of antioxidant action (electron transfer to ROS) and flavonoid–protein interaction (e.g., direct inhibition of JNK or NADH/NADPH oxidase). As a recent example of progress made in elucidating the molecular mechanisms by which flavonoids regulate gene expression, it has been shown that the potent inhibition by luteolin of the lipopolysaccharide-activated transcriptional activity of the nuclear factor κB in rat fibroblasts does not proceed by inhibition of the release of NF−κB from its cytoplasmic complex with inhibitor IκB nor by its translocation into the nucleus and subsequent binding to DNA.[62] In fact, luteolin activates the PKA pathway, thereby stimulating the production of phosphorylated proteins that will compete with NF−κB for coactivator CBP. The mechanism of the luteolin-mediated PKA activation is not elucidated yet although modulation of cAMP level via inhibition of cAMP phosphodiesterase by luteolin is suggested.

Pretreatment of neurons by flavonoids (epicatechin and its 3′-O-methylether, kaempferol) strongly inhibits cell death induced by oxidized low-density lipoproteins (ox-LDL) without reduction of ox-LDL uptake or intracellular oxidative stress.[59] Cell protection is selectively correlated to inactivation of JNK, thus suggesting that, irrespective of their H-atom donating activity, flavonoids can selectively attenuate a pro-apoptotic signaling cascade involving MAPKs.

9.2.3.1.2 Multidrug Resistance Proteins

Cancer cells typically overexpress ATP-dependent transmembrane transporters capable of expelling a wide variety of chemically unrelated drugs used in cancer therapy. This phenomenon is known as multidrug resistance (MDR). Inhibition of MDR proteins, such as the P-glycoprotein (Pgp), to prevent drug efflux during cancer therapy has thus potential clinical value.

Quercetin was shown to efficiently inhibit the Pgp-mediated drug efflux by inhibiting the ATPase activity required for transport.[63] From investigations using a soluble cytosolic portion of mouse Pgp, which includes the nucleotide- and drug-binding domains, it was possible to monitor flavonoid binding by fluorescence as well as its influence on ATP binding and the efflux of the anticancer steroid drug RU 486.[64] Flavones (aglycones) bearing OH groups at positions 3 and 5 come up as efficient mouse Pgp ligands with apparent dissociation constants lower than 10 μM. By contrast, the quercetin glycoside rutin, the flavanone naringenin, and the isoflavone genistein have low affinity for Pgp. Interestingly, flavones and flavonols behave as bifunctional inhibitors whose binding site overlap the vicinal binding sites for both ATP and RU 486. Those trends were confirmed using a cytosolic portion of Pgp from the parasite *Leishmania tropica*.[65] In the presence of the most efficient Pgp inhibitors, the drug daunomycin was shown to accumulate in resistant parasites and inhibit their growth. Interestingly, the presence of a 1,1-dimethylallyl substituent at position 8 of the flavone nucleus markedly increases the affinity for Pgp (a factor ~20 for apigenin and kaempferide) with apparent dissociation constants of ~1 μM. These observations prompted the search for more active amphiphilic flavonoids bearing saturated or unsaturated (prenyl, geranyl) hydrocarbon chains.[66,67] For instance, 4-*n*-octyloxy-2′,4′,6′-trihydroxychalcone displays an optimized affinity for mouse Pgp with a dissociation constant of 20 nM.

The interaction of flavonoids with MDR proteins is of interest not only in the field of cancer prevention and therapy but also in the field of flavonoid bioavailability. Indeed, using specific inhibitors of MDR proteins, it was shown that multiresistant protein 2 (MRP2), but not Pgp, is involved in the efflux of quercetin conjugates from human hepatic cells.[37] Similar observations were made with the cell line Caco-2 (a popular model for human intestinal absorption) where MRP2 was found responsible for the efficient efflux of chrysin after its conjugation into a mixture of glucuronide and sulfate.[68]

9.2.3.2 Ligand–Receptor Interactions

9.2.3.2.1 Estrogen Receptors

Estrogen hormones influence the growth, differentiation, and functioning of many tissues from the male and female reproductive systems. They also have cardioprotective effects. Some dietary flavonoids, especially the isoflavone genistein, belong to the phytoestrogen family as they tightly bind both estrogen receptors (ER) α and β and trigger gene activation as full agonists.[69] This effect could provide a basis for interpreting the inverse relation between the risk of prostate and breast cancers and the intake of isoflavone-rich soy foods that has been put forward by epidemiological studies.

The affinity of genistein (4′,5,7-trihydroxyisoflavone) for ERα and ERβ is 0.7 and 13% of that for the endogenous ligand 17β-estradiol, respectively.[69] Hence, the corresponding dissociation constants can be calculated to be as low as 7 nM (ERα) and 0.6 nM (ERβ). The high affinity of isoflavones for ERα and ERβ can be interpreted by the isoflavones A and C rings mimicking the estrogen A and B rings. The binding of apigenin (4′,5,7-trihydroxyflavone) and kaempferol (3,4′,5,7-tetrahydroxyflavone) to ERβ is six to seven times weaker than that of genistein. Moreover, deletion of the 5-OH phenolic group of genistein to give daidzein weakens the binding by a factor 3 to 4 and 13 for ERα and ERβ, respectively. In comparison to their aglycones, daidzein and genistein glucuronides have an affinity for the ERβ that is reduced by a factor 10 and 40, respectively.[70] Nonetheless, given the relatively high peak serum concentrations of total daidzein and genistein (0.5 to 1 μM) reached after consumption of soy food, isoflavone glucuronide–ERβ binding appears strong enough to be of biological significance. Like 17β-estradiol binding, genistein binding promotes dimerization of the receptor and subsequent binding to DNA at the estrogen receptor element,[71] thereby inducing gene activation. The estrogenic potency of genistein on ERα and ERβ is 0.025 and 0.8% that of 17β-estradiol, respectively.[69]

9.2.3.2.2 GABA-A Receptor

γ-Aminobutyric acid (GABA) is the major inhibitory neurotransmitter in the central nervous system. GABA-A receptors are pentameric *trans*-membrane proteins having a central chloride ion channel. The chloride flux can be regulated by a variety of neuroactive ligands including flavones that are able to bind to the benzodiazepine (BDZ) binding site at the interface between α and γ subunits. Flavones typically behave as partial agonists potentiating the GABA-activated ion current in a sub-maximal manner.[72] As such, they are potential anxiolytic agents. Naturally occurring flavones typically bind to the BDZ receptor with moderate affinity (K_i in the range 1 to 100 μM) and have been selected as leads for structure–affinity relationship studies including numerous synthetic flavonoid analogs. Hence, the introduction of methyl and nitro groups as well as halogen atoms on the A and B rings has emerged as a powerful way to increase the flavonoid–receptor interaction. For instance, 6-methyl-3′,5-dinitroflavone binds with a record K_i value of 1.9 nM.[73] Moreover, recent investigations[74] have highlighted that some naturally occurring 2′-hydroxyflavones can approach such an affinity ($K_i = 6.1$ nM for 5,7,2′-trihydroxy-6,8-dimethoxyflavone extracted from a Chinese medicinal herb). It has been proposed that the critical interactions between a flavonoid ligand and the BDZ receptor include hydrogen bonding at the C ring oxygen atoms and hydrophobic interactions (possibly π-stacking interactions involving aromatic amino acid residues) developed by the A and B rings.[72,73]

Since flavonoids are able to bind to the BDZ binding site of the GABA-A receptor, they might well interact with the BDZ binding site of the mitochondrial permeability transition protein, thereby modulating oxidative stress-induced apoptosis.[58]

9.2.3.2.3 Adenosine Receptors

Adenosine receptors (subtypes A_1, A_{2A}, A_{2B}, and A_3) play a role in brain and heart protection as well as in the modulation of the immune and inflammatory systems. Hence, adenosine antagonists have potential pharmacological value. Interestingly, several flavone and flavonol aglycones have rather high affinities for adenosine receptors.[75] In addition, varying the substitution of the flavone nucleus by OH and OR groups (R = Me, Et, nPr) can produce fairly selective A_3 ligands with K_i values of the order of 1 μM (close to that of the natural ligand adenosine). Alkylation of the OH groups typically increases both the affinity and selectivity.[76] In a functional A_3 receptor assay, the most potent ligands were also shown to reverse inhibition of adenylyl cyclase, thus demonstrating antagonism.

9.2.3.3 Redox Enzymes

Lipoxygenases (LOX), cycloxygenases (COXs), and xanthine oxidase (XO) are metalloenzymes whose catalytic cycle involves ROS such as lipid peroxyl radicals, superoxide, and hydrogen peroxide. LOXs and COXs catalyze important steps in the biosynthesis of leucotrienes and prostaglandins from arachidonic acid, which is an important cascade in the development of inflammatory responses. XO catalyzes the ultimate step in purine biosynthesis, the conversion of xanthine into uric acid. XO inhibition is an important issue in the treatment of gout. Flavonoids may exert part of their antioxidant and anti-inflammatory activities via direct inhibition of LOXs, COXs, and XO. Typically, interpretation of the inhibition studies are complicated because of the possible combination of distinct inhibition mechanisms: formation of noncovalent enzyme-inhibitor complexes, direct scavenging by flavonoid antioxidants of ROS inside or outside the catalytic pocket (with simultaneous oxidation of the flavonoids), chelation of the enzyme metal centers by the flavonoids, and enzyme inactivation by reactive aryloxyl radicals, quinones, or quinonoid compounds produced upon flavonoid oxidation that may eventually form covalent adducts with the enzyme.

9.2.3.3.1 Lipoxygenases and Cycloxygenases

Mammalian 15-lipoxygenase 1 (15-LOX1) has been proposed as an endogenous pro-oxidant enzyme capable of oxidizing LDLs, an early event in the development of atherosclerosis.[77] Hence, its inhibition by flavonoids is a potential mechanism for the prevention of cardiovascular diseases by these antioxidants. Recently, the inhibition by flavonoids of the peroxidation of linoleic acid catalyzed by rabbit reticulocyte 15-LOX1 has been investigated.[78] Flavone and flavonol aglycones come up as the most potent inhibitors and affect enzyme activity in three distinct ways: prolongation of the initial lag phase during which the accumulation of lipid hydroperoxides is very slow, lowering of the maximal peroxidation rate during the subsequent phase of hydroperoxide accumulation, and inactivation of the enzyme in a third phase due to the combined action of the flavonoid and intermediates of the catalytic cycle. In addition, the inhibition (assessed as IC_{50} values from the concentration dependence of the percentage of peroxidation rate decrease during the second phase) is insensitive to the presence of Fe(III) (an observation ruling out iron–flavonoid chelation as a possible inhibition mechanism) and stronger with flavones and flavonols having a catechol group (a critical determinant of radical scavenging efficiency) either on the A ring or on the B ring (most potent inhibitors with IC_{50} ~1 μM: luteolin, baicalein, fisetin). Overall, a mechanism combining direct inhibition (noncompetitively to linoleic acid) and radical scavenging can be proposed. Interestingly, QGlcU, the main circulating quercetin metabolites, retain the ability to inhibit soybean LOX.[34] For instance, Q3'GlcU ($K_d = 6.5$ μM), Q4'GlcU ($K_d = 8.4$ μM), and Q7GlcU ($K_d = 6.0$ μM) display an affinity for the enzyme that is only two to three times as low as that of quercetin ($K_d = 2.8$ μM), whereas Q3GlcU ($K_d = 60$ μM) is a much poorer inhibitor.

The inhibition of the 15-LOX-induced peroxidation of LDL by quercetin and its 3-, 4'-, and 7-monoglucosides (QGlc) has also been addressed.[79] Quercetin, Q7Glc, and Q3Glc, which all possess a catechol group in the B ring, inhibit LDL peroxidation with IC_{50} values in the range 0.3 to 0.5 μM and efficiently spare endogenous LDL-bound α-tocopherol. By contrast, Q4'Glc, which lacks a free catechol group, is less effective ($IC_{50} = 1.2$ μM) and does not spare α-tocopherol. These results suggest that the inhibition of LDL peroxidation by quercetin and its glucosides mainly proceeds via peroxyl radical scavenging and regeneration of α-tocopherol from the corresponding α-tocopheryl radical rather than by direct enzyme inhibition. Interestingly, the percentage of residual flavonol (initial concentration: 1 μM) after 6 h of incubation in the presence of 15-LOX and in the absence of LDL (pH 7.4, 20°C) is ~0, 20, 40, and 90% for Q4'Glc, quercetin, Q7Glc, and Q3Glc, respectively. By contrast, in the absence of 15-LOX and in the presence of LDL, only quercetin and Q7Glc are partially consumed (residual flavonol = 60%). This unexpected result shows that 15-LOX can catalyze the autoxidation of quercetin and its monoglucosides in a way that is highly dependent on the site of glucosidation. Finally, in the presence of both 15-LOX and LDL, the flavonols are totally consumed in less than 2 h (with the exception Q3Glc consumed in 4 h) in agreement with an inhibition dominated by redox processes.

Quercetin and a selection of naturally occurring prenylated flavonoids used as anti-inflammatory agents in Chinese medicine were also tested for their ability to inhibit the synthesis of 12- and 5-hydroxyeicosatetraenoic acids from arachidonic acid catalyzed by 12-LOX (from platelets) and 5-LOX (from polymorphonuclear leucocytes), respectively.[80] Similarly, quercetin and the prenylated flavonoids were tested for their ability to inhibit the synthesis of thromboxane B2 and prostaglandins E2 and D2 from arachidonic acid catalyzed by COX1 and COX2, respectively. Quercetin appears as a potent inhibitor of 5-LOX ($IC_{50} = 0.8$ μM), a more modest inhibitor of 12-LOX ($IC_{50} = 12$ μM) and COX1 ($IC_{50} = 8$ μM), and a very poor inhibitor of COX2 ($IC_{50} = 76$ μM). Interestingly, apigenin is inactive toward all four enzymes. In addition, some chalcones and flavanones having a lavandulyl (5-methyl-2-isopropenyl-hex-4-enyl) group at position 8 (A ring) and a 2',4'-dihydroxy substitution on the B ring are selective inhibitors of 5-LOX and COX1, and inactive toward 12-LOX and COX2.

9.2.3.3.2 Xanthine Oxidase

Regarding XO activity and the simultaneous formation of uric acid (from xanthine) and superoxide, flavonoids can act as true enzyme inhibitors (formation of enzyme-inhibitor complexes) thereby quenching both superoxide and uric acid formation or by scavenging superoxide that can be independently recorded using chemiluminescence or colorimetric methods. Hence, it is possible to rank flavonoids in distinct classes[81]: superoxide scavengers without inhibitory activity on XO, XO inhibitors without additional superoxide scavenging activity (IC_{50} for uric acid formation $\approx IC_{50}$ for superoxide scavenging), XO inhibitors with an additional superoxide scavenging activity (IC_{50} for uric acid formation $> IC_{50}$ for superoxide scavenging), XO inhibitors with an additional pro-oxidant effect in superoxide production (IC_{50} for uric acid formation $< IC_{50}$ for superoxide scavenging), weak XO inhibitors with an additional pro-oxidant effect in superoxide production, and flavonoids with no effect on XO and superoxide. A planar C ring (flavones, flavonols) seems required for XO inhibition (IC_{50} in the range 0.5 to 10 μM with the exception of 3-hydroxyflavone, which does not interact with XO). Hence, catechins have pure superoxide scavenging activity and do not interact with XO. Only flavonols with a catechol group on the B ring (quercetin, myricetin, fisetin) display additional superoxide scavenging activity. By contrast, some hydroxylated flavones (chrysin, apigenin, luteolin) show an underlying pro-oxidant activity. Interestingly, glycosidation of the flavonoid nucleus generally abolishes XO inhibition. For instance, the IC_{50} values of quercetin for XO inhibition and superoxide scavenging are 2.6 and 1.6 μM, respectively. By contrast, quercetin-3-O-rhamnoside (quercitrin) has an IC_{50} of 8.1 μM for superoxide scavenging but no longer inhibits XO ($IC_{50} > 100$ μM). Similarly, Q3GlcU and Q7GlcU are very poor XO inhibitors ($K_d \geq 100$ μM).[34] However, Q3'GlcU ($K_d = 1.4$ μM) and Q4'GlcU ($K_d = 0.25$ μM) are strong inhibitors of XO, the latter as potent as quercetin itself. Hence, while glucuronidation at the 3' or 4' position suppresses the free catechol moiety of the B ring and thereby most of the radical scavenging activity, the affinity for XO is spared as if flavonol–XO binding took place with marginal participation of the B ring. While epicatechin and its oligomers do not inhibit XO, oligomers of epicatechin-3-O-gallate (4β-8 interflavan linkage) are inhibitors whose potency increases with the number of monomer units ($IC_{50} = 7.2$ to 4.4 μM from dimer to tetramer).[82] Accordingly, a French maritime pine bark extract (pycnogenol) rich in procyanidins (75% weight, DP 2 to 7) was found to strongly reduce XO activity and retard the electrophoretic mobility of the protein under nondenaturing conditions only.[83] In addition, pure low molecular weight components of the extract are without effect. Hence, it can be concluded that XO inhibition proceeds by binding to XO of the high DP procyanidins (DP >3). Moreover, the binding is noncompetitive with respect to the substrate xanthine, abolished by polyethylene glycol or the surfactant Triton X-100 and unaffected by addition of sodium chloride or urea. Hence, it can be proposed that the binding does not take place to the xanthine binding site and primarily involves dispersion forces and the hydrophobic effect. In the same work, the activities of catalase (from human erythrocyte), horseradish peroxidase, and soybean lipoxygenase were also shown to be inhibited by the pine bark extract. For catalase at least, this may proceed by procyanidin–protein binding since the electrophoretic mobility of catalase under nondenaturing conditions is also decreased by the extract.

Similarly, catechin polymers formed upon horseradish peroxidase-catalyzed oxidation of catechin or polycondensation of catechin with aldehydes prove much more efficient than catechin (at identical monomer concentration) at inhibiting XO and superoxide formation.[12,13] A more detailed investigation with the catechin–acetaldehyde polycondensate (which is expected to form in wine because of the microbial oxidation of ethanol to acetaldehyde) shows that inhibition is noncompetitive to xanthine and likely occurs via binding to the FAD or Fe/S redox centers involved in electron transfers from the reduced molybdenum center to dioxygen with simultaneous production of superoxide.[13]

9.2.3.3.3 Peroxidases and Tyrosinases

These enzymes have been used to oxidize flavonoids for investigating the reactivity and potential toxicity of their aryloxyl radicals (one-electron oxidation) and o-quinones (two-electron oxidation). For instance, 4′-hydroxyflavonoids are quickly converted into aryloxyl radicals that can oxidize glutathione and NADH with concomitant reduction of dioxygen and ROS formation.[84–86] This process provides a possible metal-independent mechanism for the pro-oxidant activity of flavonoids. Alternatively, 3′,4′-dihydroxyflavonoids are oxidized into the corresponding semiquinone radicals, which quickly disproportionate. The resulting o-quinones can then be reduced by NADH (hydride ion transfer) or form conjugates with glutathione without dioxygen activation.[84,85,87,88] Flavonoid-derived o-quinones can eventually react with Cys residues of proteins to form (reversible) covalent adducts. The latter process provides an original mechanism for the inhibition by quercetin of glutathione S-transferase P1-1,[89] an enzyme involved in the defense against electrophiles and in MDR of tumor cells.

9.2.3.3.4 Cytochrome P450

These heme-containing monooxygenases include several isoforms (1A1, 1A2, 1B1, 3A4, etc.) with different tissue distributions and play a key role in the metabolism of endogenous substrates (e.g., steroids) and xenobiotics (food components, drugs, carcinogens, pollutants).[8] Although the metabolism of xenobiotics by cytochrome P450 (CYP, phase I enzymes) typically results in more hydrophilic compounds that are more readily excreted after eventual conjugation by phase II enzymes (e.g., UDP glucuronyltransferases, sulfotransferases), toxic reactive intermediates, including free radicals, can be formed. Indeed, CYPs are responsible for the conversion of some procarcinogens (e.g., polyaromatic hydrocarbons or PAHs) into carcinogens (e.g., PAH epoxides).

CYP–flavonoid interactions are a good example of the multiple ways flavonoids can affect enzymatic activities, i.e., from the regulation of gene expression to direct binding to the processed enzymes.[8] Flavonoids can induce, or eventually inhibit, the biosynthesis of CYP 1A1 via interactions with the aryl hydrocarbon receptor (AhR), a cytosolic protein that, once activated by a ligand, translocates to the nucleus and, in association with the AhR translocator, forms a transcription factor for CYP 1A1. For instance, in human breast cancer cells, quercetin binds to AhR as an agonist (in competition with the typical AhR ligand 2,3,7,8-tetrachlorodibenzo-p-dioxin) and stimulates gene expression for CYP 1A1 with a parallel increase in CYP 1A1-mediated O-deethylation of 7-ethoxyresorufin.[90] This process is strikingly dependent on the hydroxylation pattern of the B ring since kaempferol (3′-deoxyquercetin) binds AhR as an antagonist (no subsequent activation of CYP 1A1). It is also highly dependent on the cell type since, in hepatic cells, quercetin binds to AhR as an antagonist, thereby inhibiting gene expression for CYP 1A1 and benzo[a]pyrene activation.[91] This provides a possible mechanism for the anticancer activity of quercetin.

Flavonoids, especially flavones and flavonols, also directly bind to several CYP isoforms (1A1, 1A2, 1B1, 3A4) involved in xenobiotics metabolism and inhibit enzyme activity. Structure–activity relationships[92–95] show rather high isoform selectivities depending on the flavonoid substitution pattern and contrasted inhibition mechanisms. For instance, inhibition by flavonoids of 7-methoxyresorufin O-demethylation in microsomes enriched in CYP 1A1 and 1A2 reveals that galangin (3,5,7-trihydroxyflavone) is a mixed inhibitor of CYP 1A2 ($K_i = 8$ nM) and a five times less potent inhibitor of CYP 1A1. By contrast, 7-hydroxyflavone is a competitive inhibitor of CYP 1A1 ($K_i = 15$ nM) and a six times less potent inhibitor of CYP 1A2.[95] In addition, fairly selective inhibition of CYP 1B1 (specifically detected in cancer cells) by some flavonoids has been reported. For example, 5,7-dihydroxy-4′-methoxyflavone inhibits 1B1, 1A1, and 1A2 with IC_{50} values of 7, 80, and 80 nM, respectively.[92]

Eventually, flavonoids can be hydroxylated or demethylated by CYPs. For instance, hesperetin (4′-methoxy-3′,5,7-trihydroxyflavanone) is specifically demethylated by CYPs 1A1 and 1B1, but not by CYPs 1A2 and 3A4.[92] In addition, 3,5,7-trihydroxyflavone undergoes sequential CYP 1A1-catalyzed hydroxylation at C4′ and C3′ to finally yield quercetin.[96,97] These reactions may be relevant to flavonoid metabolism and cytotoxicity since the corresponding products are more reducing and thus more prone to autoxidation with simultaneous ROS production.

Finally, flavonoids are also able to inhibit CYP19 or aromatase, an enzyme catalyzing a three-step oxidation sequence resulting in aromatization of the A ring of male steroid hormones (androgens) to yield estrogens. Together with flavonoid–estrogen receptor binding, this process could be relevant to the prevention of hormone-dependent cancers by flavonoids. Binding studies[98,99] show that flavonoids are more efficient inhibitors than isoflavonoids with the A and C rings of the former possibly mimicking the androgen D and C rings, respectively. The simultaneous presence of a 4-keto group (possibly interacting with the heme iron center as an axial ligand in agreement with the observed high spin–low spin transition upon binding) and a 7-OH group is critical for a high affinity. By contrast, hydroxylation at positions 3 and 6 is strongly destabilizing. Flavanones bind almost as tightly as flavones but isoflavones are only poor ligands[99] ($K_i = 2.6$, 5.1, and 123 µM for 5,7-dihydroxyflavone, ($\pm$)-4′,5,7-trihydroxyflavanone, and 4′,5,7-trihydroxyisoflavone, respectively) in sharp contrast to the structural requirement for strong flavonoid–estrogen receptor binding (see above).

Interestingly, 17β-hydroxysteroid dehydrogenase, another redox enzyme involved in steroid metabolism, is also strongly inhibited by 7-hydroxyflavonoids.[100] For instance, the flavone apigenin is more potent at inhibiting 17β-hydroxysteroid dehydrogenase ($IC_{50} = 0.3$ µM) than aromatase ($IC_{50} = 2.9$ µM) and the isoflavone genistein, which is only a weak aromatase inhibitor, inhibits 17β-hydroxysteroid dehydrogenase with an IC_{50} of 1 µM.

9.2.3.4 Modulation of Antioxidant Properties and Oxidation Pathways by Binding to Proteins

In addition to enzymatic oxidation, flavonoid oxidation can take place via autoxidation (metal-catalyzed oxidation by dioxygen) and ROS scavenging. The former process can be related to flavonoid cytotoxicity (ROS production) while the latter is one of the main antioxidant mechanisms. Both processes may be modulated by flavonoid–protein binding. Although poorly documented so far, these points could be important and, for instance, albumin–flavonoid complexes with an affinity for LDL could act as the true plasma antioxidants participating in the regeneration of α-tocopherol from the α-tocopheryl radical formed

upon scavenging of LDL-bound lipid peroxyl radicals. In addition, flavonoid–protein complexation can be expected to provide protection to the protein against oxidative degradation.

In principle, addition of an oxidizing agent to a mixture of flavonoid and protein can
cause degradation of both partners. From the influence of the protein on the kinetics of
flavonoid oxidation, it can be decided whether the bound flavonoid molecule is still accessible
to the oxidizing agent, i.e., whether it is still antioxidant. On the other hand, a reliable
procedure for monitoring protein oxidation (with or without cleavage of peptide bonds) is
needed to assess the eventual protection of the protein by the flavonoid. For example,
inhibition of the enzymatic activity of butyrylcholine esterase, a contaminant of commercially
available serum albumin, is a sensitive marker of protein degradation by peroxyl radicals.[101]

The influence of serum albumin on quercetin oxidation has been investigated with
different one-electron or two-electron oxidizing agents: the peroxyl radicals formed by
thermal decomposition of diazo compound AAPH in the presence of dioxygen, sodium
periodate, and potassium nitrosodisulfonate. Rather unexpectedly, quercetin–BSA binding
does not affect the rate of quercetin oxidation by periodate[102] and even accelerates the rate
of quercetin oxidation by nitrosodisulfonate (Dufour et al., unpublished results). Hence,
SA-bound quercetin remains fully accessible to these small oxidizing agents, thus suggesting
that it retains its antioxidant activity. In the first step, the quercetin o-quinone (in fast
equilibrium with a p-quinone methide form) that is barely detectable in the absence of BSA
because of subsequent fast water addition, is strongly stabilized, a likely consequence of
ligand–protein charge transfer interactions and a low local water concentration. Serum
albumin was also demonstrated to protect the quercetin p-quinone methide–water adduct
from further degradation leading to 3,4-dihydroxybenzoic acid and 2-oxo-2-(2,4,6-trihydrox-
yphenyl)acetic acid. Investigations with other flavonoids confirm that albumin only weakly
affects the kinetics of flavonoid oxidation while leaving the product distribution essentially
unchanged. Unlike quercetin and kaempferol, whose oxidation leads to C ring cleavage,
oxidation of luteolin, isoquercitrin, and catechin, either in their free or BSA-bound form,
preferentially leads to dimers.

Under conditions where the protein is not oxidatively degraded, its influence on the
radical scavenging activity of flavonoids can be more readily assessed. For example, whereas
the reaction of BSA and gelatin with the ABTS radical cation is negligible, sorghum procyanidin (15 to 17 flavanol units) can scavenge up to six ABTS radicals per monomer at pH
7.4.[103] Of the two proteins, only gelatin slightly affects the kinetics of radical scavenging.
However, the overall stoichiometry remains unaffected. At pH 4.9, the inhibition of ABTS
scavenging by both proteins, especially gelatin, is somehow stronger but tends to vanish with
time. In the case of BSA at least, these observations could simply point to negligible protein–
flavonoid binding. However, covalent coupling between the proteins and procyanidin quinones may take place during radical scavenging. Indeed, although the amount of procyanidin–
protein precipitate is not affected by ABTS, procyanidin oxidation by ABTS leads to
precipates that cannot be resolubilized. Using the ABTS scavenging test, it was also possible
to demonstrate that quercetin and human plasma exert nonadditive antioxidant activities,[104]
thus suggesting that binding of quercetin to plasma proteins masks part of the electron-
donating activity of quercetin. The masking effect decreases in the series quercetin > rutin > catechin. The same trend emerges when plasma is replaced by serum albumin.
Interestingly, the binding affinity to serum albumin[42,43] parallels the masking effect. However, it must be noted that ABTS is much bulkier than ROS. Hence, its scavenging by
polyphenols is expected to be especially sensitive to steric hindrance brought by the protein
environment.

Since lipid peroxidation is clearly related to the onset of atherosclerosis and the impair-
ment of membrane functions, the influence of proteins on the ability of flavonoids to inhibit

lipid peroxidation deserves examination. Such investigations have been carried out with BSA and lecithin liposomes.[105] Whereas BSA alone already slows down the formation of lipid hydroperoxides and hexanal, its influence on the antiperoxidizing activity of the selected polyphenols is highly dependent on the polyphenolic structure. Hence, BSA lowers the inhibition of hydroperoxide formation by catechins and caffeic acid, enhances inhibition by malvidin and rutin, and leaves essentially unchanged inhibition by quercetin. No clear interpretation based on polyphenol–BSA binding can be given.

The influence of a protein environment on the antioxidant activity of flavonoids can be readily evaluated by monitoring the inhibition by flavonoids of the peroxidation of HSA-bound linoleic acid in plasma-mimicking conditions.[106] The AAPH-initiated peroxidation of linoleic acid leads to four isomeric hydroperoxides that further react to form the corresponding ketodienes and alcohols. As expected, the formation of the lipid peroxidation products is inhibited more efficiently by 3′,4′-dihydroxyflavonoids (quercetin, quercetin-3-β-D-glucoside > catechin) than by flavonoids having a monohydroxylated B ring (kaempferol, 3′-O-methylquercetin, quercetin-3,4′-β-D-diglucoside). More importantly, the strong binding of quercetin to HSA (noncompetitively to linoleic acid) does not alter its antiperoxidizing activity. In contrast, α-tocopherol, although much more potent than flavonoids in the absence of HSA, and ascorbate are only weakly active. Thus, in plasma, the flavonol–albumin complex could be regarded as an antioxidant species with the flavonol molecule efficiently trapping the peroxyl radicals derived from AAPH and eventually from the lipid. Similarly, HSA-bound quercetin efficiently protects the enzyme butyrylcholine esterase (a typical contaminant of commercially available HSA) from oxidative damage by AAPH-derived peroxyl radicals.[101]

Oxidation of catechols in the presence of a protein may lead to extensive catechol–protein covalent coupling (Figure 9.4) as demonstrated in the case of the chlorogenic acid–BSA couple.[107] Autoxidation of EGCG at pH 4.9 in the presence of Zn(II) cations was shown to generate semiquinone radicals (stabilized by Zn(II) binding) mainly on the B ring moiety.[108] In the presence of BSA, EGCG autoxidation is accompanied by irreversible protein precipitation suggesting covalent EGCG–BSA coupling that probably involves EGCG o-quinones in fast disproportionation equilibrium with the semiquinone radicals. Finally, incubation of Hep G2 and Caco-2 cells with [^{14}C]quercetin results in quercetin–protein covalent coupling (as much as 10% of the total cellular content of quercetin in the case of Caco-2 cells).[109] The process is insensitive to the presence of an excess ascorbate, which rules out significant autoxidation of quercetin in the buffer or cell culture medium. Hence, quercetin oxidation could involve ROS within the cells. The quercetin-derived quinone or quinone methide intermediates thus formed are then proposed to add to specific cell proteins (MW ~55–80 kDa in the case of Hep G2 cells), the major cell proteins remaining unaltered.

More generally, one-electron oxidation of protein-bound phenols to form reactive aryloxyl radicals is a possible pro-oxidant mechanism since these radicals can propagate H-atom or electron transfers within the protein. In addition to phenol–protein covalent coupling, these phenol-mediated oxidative damages to proteins could be detrimental to their function as enzymes, receptors, and membrane transporters. For instance, investigations by capillary electrophoresis have shown that quercetin in concentrations lower than 25 μM potentiates HSA degradation by AAPH-derived peroxyl radicals.[110]

9.3 CONCLUSION AND PERSPECTIVES

Flavonoids, as food components or potential drugs, interact with a wide range of proteins by distinct mechanisms: weak and rather unspecific binding of tannins to proline-rich or histidine-rich random coils leading to protein precipitation, specific enzyme inhibition, and ligand–receptor interactions mostly involving flavonoid aglycones with an unsaturated C ring, binding to transport proteins. In addition, flavonoid–protein interactions can modulate the redox properties of flavonoids that underline their antioxidant, and eventually their pro-oxidant, activity. After electrophilic activation by one-electron or two-electron oxidation, flavonoids can also form covalent bonds with proteins.

Whether these binding processes are relevant to human health is not clearly demonstrated yet. However, one of the most promising perspectives is the participation of flavonoids in the regulation of gene expression, possibly by direct interactions with specific receptors and nuclear factors. For example, quercetin has been shown to increase the intracellular glutathione level by activating the promoter of the catalytic subunit of γ-glutamylcysteine synthetase.[111] This effect is fairly specific since myricetin (5′-hydroxyquercetin) and two quercetin 3-glycosides are inactive. As a possible molecular mechanism, quercetin could help release specific nuclear factors (from inert cytosolic complexes), thereby facilitating their translocation into the nucleus. On the other hand, some flavones, isoflavones, and flavonols are also known to activate peroxisome proliferator-activated receptor-γ (PPAR-γ), thus leading to suppression of inducible COX-2 and NO synthase in mouse macrophages.[112] A likely mechanism consists in allosteric binding of the flavonoids to PPAR-γ and subsequent modification of the receptor conformation.

In conclusion, spectacular advances in the fields of flavonoid bioavailability and flavonoid-mediated cell effects in relation to the development of new biological tools (e.g., proteomic analysis, reporter genes) have been achieved during the last decade. A more coherent picture of the ways flavonoids combine their redox properties and affinity to specific proteins is emerging. This wealth of new chemical and biological information suggests that the elucidation of *in vivo* molecular mechanisms and receptors involved in flavonoid health effects is at hand.

REFERENCES

1. Harborne, J.B. and Williams, C.A., Advances in flavonoid research since 1992, *Phytochemistry*, 55, 481, 2000.
2. Parr, A.J. and Bolwell, G.P., Phenols in the plant and in man. The potential for possible nutritional enhancement of the diet by modifying the phenol content and profile, *J. Sci. Food Agric.*, 80, 985, 2000.
3. Haslam, E., Natural polyphenols (vegetable tannins) as drugs: possible modes of action, *J. Nat. Prod.*, 59, 205, 1996.
4. Hollman, P.C.H. and Katan, M.B., Dietary flavonoids: intake, health effects and bioavailability, *Food Chem. Toxicol.*, 37, 937, 1999.
5. Scalbert, A. and Williamson, G., Dietary intake and bioavailability of polyphenols, *J. Nutr.*, 130, 2073S, 2000.
6. Havsteen, B.H., The biochemistry and medical significance of flavonoids, *Pharmacol. Ther.*, 96, 67, 2002.
7. Middleton, E. and Kandaswami, C., The impact of plant flavonoids on mammalian biology: implications for immunity, inflammation and cancer, in *The Flavonoids, Advances in Research Since 1986*, Harborne, J.B., Ed., Chapman & Hall, London, 1994, 619.
8. Hodek, P., Trefil, P., and Stiborova, M., Flavonoids: potent and versatile biologically active compounds interacting with cytochromes P450, *Chem. Biol. Interact.*, 139, 1, 2002.
9. Jez, J.M., Bowman, M.E., and Noel, J.P., Role of hydrogen bonds in the reaction mechanism of chalcone isomerase, *Biochemistry*, 41, 5168, 2002.
10. Charlton, A.J. et al., Polyphenol/peptide binding and precipitation, *J. Agric. Food Chem.*, 50, 1593, 2002.

11. Baxter, N.J. et al., Multiple interactions between polyphenols and a salivary proline-rich protein repeat result in complexation and precipitation, *Biochemistry*, 36, 5566, 1997.

12. Kurisawa, M. et al., Amplification of antioxidant activity and xanthine oxidase inhibition of catechin by enzymatic polymerization, *Biomacromolecules*, 4, 469, 2003.

13. Kim, Y.J. et al., Superoxide anion scavenging and xanthine oxidase inhibition of (+)-catechin–aldehyde polycondensates. Amplification of the antioxidant property of (+)-catechin by polycondensation with aldehydes, *Biomacromolecules*, 5, 547, 2004.

14. Spencer, C.M. et al., Polyphenol complexation — some thoughts and observations. *Phytochemistry*, 27, 2397, 1988.

15. Zhu, M. et al., Plant polyphenols: biologically active compounds or non-selective binders to proteins? *Phytochemistry*, 44, 441, 1997.

16. Markham, K.R. et al., Anthocyanic vacuolar inclusions — their nature and significance in flower colouration, *Phytochemistry*, 55, 327, 2000.

17. Murray, N.J. et al., Study of the interaction between salivary proline-rich proteins and a polyphenol by 1H-NMR spectroscopy, *Eur. J. Biochem.*, 219, 923, 1994.

18. Charlton, A.J. et al., Tannin interactions with a full-length human salivary protein display a stronger affinity than with single proline-rich repeats, *FEBS Lett.*, 382, 289, 1996.

19. De Freitas, V. and Mateus, N., Structural features of procyanidin interactions with salivary proteins, *J. Agric. Food Chem.*, 49, 940, 2001.

20. Yan, Q. and Bennick, A., Identifications of histatins as tannin-binding proteins in human saliva, *Biochem. J.*, 311, 341, 1995.

21. Naurato, N. et al., Interaction of tannin with human salivary histatins, *J. Agric. Food Chem.*, 47, 2229, 1999.

22. Wroblewski, K. et al., The molecular interactions of human salivary histatins with polyphenolic compounds, *Eur. J. Biochem.*, 268, 4384, 2001.

23. Hagerman, A.E., Tannin–protein interactions, in *Phenolic Compounds in Food and their Effects on Health I. Analysis, Occurrence, and Chemistry.* ACS Symposium Series 506, Ho, C.T., Lee, C.Y., and Huang, M.T., Eds., American Chemical Society, Washington, DC, 1992, 236.

24. Maliar, T. et al., Structural aspects of flavonoids as trypsin inhibitors, *Eur. J. Med. Chem.*, 39, 241, 2004.

25. Bokkenheuser, V.D., Shackleton, C.H.L., and Winter, J., Hydrolysis of dietary flavonoid glycosides by strains of intestinal *Bacteroides* from humans, *Biochem. J.*, 248, 953, 1987.

26. Day, A.J. et al., Absorption of quercetin-3-glucoside and quercetin-4'-glucoside in the rat small intestine: the role of lactase phlorizin hydrolase and the sodium-dependent glucose transporter, *Biochem. Pharmacol.*, 65, 1199, 2003.

27. Day, A.J. et al., Dietary flavonoid and isoflavone glycosides are hydrolysed at the lactase site of lactase phlorizin hydrolase, *FEBS Lett.*, 468, 166, 2000.

28. Day, A.J. et al., Deglycosylation of flavonoid and isoflavonoid glycosides by human small intestine and liver β-glucosidase activity, *FEBS Lett.*, 436, 71, 1998.

29. Deprez, S. et al., Polymeric proanthocyanidins are catabolized by a human colonic microflora into low molecular weight phenolic acids, *J. Nutr.*, 130, 2733, 2000.

30. Aura, A.M. et al., Quercetin derivatives are deconjugated and converted to hydroxyphenylacetic acids but not methylated by human fecal floral *in vitro, J. Agric. Food Chem.*, 50, 1725, 2002.

31. Rechner, A.R. et al., The metabolic fate of dietary polyphenols in humans, *Free Radical Biol. Med.*, 33, 220, 2002.

32. Kim, D.H. et al., Intestinal bacterial metabolism of flavonoids and its relation to some biological activities, *Arch. Pharm. Res.*, 21, 17, 1998.

33. Manach, C. et al., Quercetin is recovered in human plasma as conjugated derivatives which retain antioxidant properties, *FEBS Lett.*, 426, 331, 1998.

34. Day, A.J. et al., Conjugation position of quercetin glucuronides and effect on biological activity, *Free Radical Biol. Med.*, 29, 1234, 2000.

35. King, C.D. et al., The glucuronidation of exogenous and endogenous compounds by stably expressed rat and human UDP-glucuronyltransferase, *Arch. Biochem. Biophys.*, 332, 92, 1996.

36. Oliveira, E.J. and Watson, D.G., *In vitro* glucuronidation of kaempferol and quercetin by human UGT-1A9 microsomes, *FEBS Lett.*, 471, 1, 2000.

37. O'Leary et al., Metabolism of quercetin-7- and quercetin-3-glucuronides in an *in vitro* hepatic model: the role of human β-glucuronidase, sulfotransferase, catechol O-methyltransferase and multi-resistant protein 2 (MRP2) in flavonoid metabolism, *Biochem. Pharmacol.*, 65, 479, 2003.

38. O'Leary et al., Flavonoid glucuronides are substrates for human liver β-glucuronidase, *FEBS Lett.*, 503, 103, 2001.

39. Manach, C. et al., Quercetin metabolites in plasma of rats fed diets containing rutin or quercetin, *J. Nutr.*, 125, 1911, 1995.

40. Zsila, F., Bikádi, Z., and Simonyi, M., Probing the binding of the flavonoid, quercetin to human serum albumin by circular dichroism, electronic absorption spectroscopy and molecular modelling methods, *Biochem. Pharmacol.*, 65, 447, 2003.

41. Boulton, D.W., Walle, U.K., and Walle, T., Extensive binding of the bioflavonoid quercetin to human plasma proteins, *J. Pharm. Pharmacol.*, 50, 243, 1998.

42. Dangles, O. et al., Binding of flavonoids to plasma proteins, *Methods Enzymol.*, 335, 319, 2001.

43. Dufour, C. and Dangles, O., Flavonoid–serum albumin complexation: determination of binding constants and binding sites by fluorescence spectroscopy, *Biochim. Biophys. Acta*, 1721, 164, 2005.

44. Peters, T., Ed., *All About Albumin*, Academic Press, San Diego, 1996.

45. Alluis, B. and Dangles, O., Acylated flavone glucosides: synthesis, conformational investigation and complexation properties, *Helv. Chim. Acta*, 82, 2201, 1999.

46. Sazuka, M. et al., Evidence for the interaction between (−)-epigallocatechin gallate and human plasma proteins fibronectin, fibrinogen, and histidine-rich glycoprotein, *Biosci. Biotechnol. Biochem.*, 60, 1317, 1996.

47. Brunet, M.J. et al., Human Apo A-I and rat transferrin are the principal plasma proteins that bind wine catechins, *J. Agric. Food Chem.*, 50, 2708, 2002.

48. Yang, C.S. et al., Inhibition of carcinogenesis by polyphenolic compounds, *Annu. Rev. Nutr.*, 21, 381, 2001.

49. Birt, D.F., Hendrich, S., and Wang, W., Dietary agents in cancer prevention: flavonoids and isoflavonoids, *Pharmacol. Ther.*, 90, 157, 2001.

50. Bridges A.J., Chemical inhibitors of protein kinases, *Chem. Rev.*, 101, 2541, 2001.

51. Akiyama, T. et al., Genistein, a specific inhibitor of tyrosine specific protein kinase, *J. Biol. Chem.*, 262, 5592, 1987.

52. Yang, E.B. et al., Butein, a specific protein tyrosine kinase inhibitor, *Biochem. Biophys. Res. Commun.*, 245, 435, 1998.

53. Zi, X.L. et al., A flavonoid antioxidant, silymarin, inhibits activation of erbB1 signaling and induces cyclin-dependent kinase inhibitors, G1 arrest and anticarcinogenic effects in human prostate carcinoma DU145 cells, *Cancer Res.*, 58, 1920, 1998.

54. Agullo, G. et al., Relationship between flavonoid structure and inhibition of phosphatidylinositol-3 kinase: a comparison with tyrosine kinase and protein kinase C inhibition, *Biochem. Pharmacol.*, 53, 1649, 1997.

55. Gamet-Payrastre L. et al., Flavonoids and the inhibition of PKC and PI 3-kinase, *Gen. Pharmacol.*, 32, 279, 1999.

56. De Azevedo, W.F. et al., Structural basis for specificity and potency of a flavonoid inhibitor of human CDK2, a cell cycle kinase, *Proc. Natl. Acad. Sci. USA*, 93, 2735, 1996.

57. Casagrande, F. and Darbon, J.M., Effects of structurally related flavonoids on cell cycle progression of human melanoma cells: regulation of cyclin-dependent kinases CDK2 and CDK1, *Biochem. Pharmacol.*, 61, 1205, 2001.

58. Schroeter, H. et al., MAPK signaling in neurodegeneration: influences of flavonoids and of nitric oxide, *Neurobiol. Aging*, 23, 861, 2002.

59. Schroeter, H. et al., Flavonoids protect neurons from oxidized low-density-lipoprotein-induced apoptosis involving c-Jun N-terminal kinase (JNK), c-Jun and caspase-3, *Biochem. J.*, 358, 547, 2001.

60. Yoshizumi, M. et al., Quercetin glucuronide prevents VSMC hypertrophy by angiotensin II via the inhibition of JNK and AP-1 signaling pathway, *Biochem. Biophys. Res. Commun.*, 293, 1458, 2002.

61. Ahn, S.C. et al., Epigallocatechin-3-gallate, constituent of green tea, suppresses the LPS-induced phenotypic and functional maturation of murine dendritic cells though inhibition of mitogen-activated protein kinases and NF-κB, *Biochem. Biophys. Res. Commun.*, 313, 148, 2004.

62. Kim, S.H. et al., Luteolin inhibits the nuclear factor-κB transcriptional activity in Rat-1 fibroblasts, *Biochem. Pharmacol.*, 66, 955, 2003.

63. Shapiro, A.B. and Ling, V., Effects of quercetin on Hoechst 33342 transport by purified and reconstituted P-glycoprotein, *Biochem. Pharmacol.*, 53, 587, 1997.

64. Conseil, G. et al., Flavonoids: a class of modulators with bifunctional interactions at ATP- and steroid-binding sites on mouse P-glycoprotein, *Proc. Natl. Acad. Sci. USA*, 95, 9831, 1998.

65. Perez-Victoria, J.M. et al., Correlation between the affinity of flavonoids binding to the cytosolic site of *Leishmania tropica* multidrug transporter and their efficiency to revert parasite resistance to daunomycin, *Biochemistry*, 38, 1736, 1999.

66. Maitrejean, M. et al., The flavanolignan silybin and its hemisynthetic derivatives, a novel series of potential modulators of P-glycoprotein, *Bioorg. Med. Chem.*, 10, 1, 1999.

67. Bois, F. et al., Synthesis and biological activity of 4-alkoxy chalcones: potential hydrophobic modulators of P-glycoprotein-mediated multidrug resistance, *Bioorg. Med. Chem.*, 7, 2691, 1999.

68. Walle, U.K., Galijatovic, A., and Walle, T., Transport of the flavonoid chrysin and its conjugated metabolites by the human intestinal cell line Caco-2, *Biochem. Pharmacol.*, 58, 431, 1999.

69. Kuiper, G.G. et al., Interaction of estrogenic chemicals and phytoestrogens with estrogen receptor β, *Endocrinology*, 139, 4252, 1998.

70. Zhang, Y. et al., Daidzein and genistein glucuronides *in vitro* are weakly estrogenic and activate human natural killer cells at nutritionally relevant concentrations, *J. Nutr.*, 129, 399, 1999.

71. Kostelac, D., Rechkemmer, G., and Briviba, K., Phytoestrogens modulate binding response of estrogen receptors α and β to the estrogen response element, *J. Agric. Food Chem.*, 51, 7632, 2003.

72. Huang, X. et al., 3D-QSAR model of flavonoids binding at benzodiazepine site in GABA$_A$ receptors, *J. Med. Chem.*, 44, 1883, 2001.

73. Dekermendjian, K. et al., Structure–activity relationships and molecular modeling analysis of flavonoids binding to the benzodiazepine site of the rat brain GABA$_A$ receptor complex, *J. Med. Chem.*, 42, 4343, 1999.

74. Huen, M.S.Y. et al., Naturally occurring 2'-hydroxyl-substituted flavonoids as high-affinity benzodiazepine site ligands, *Biochem. Pharmacol.*, 66, 2397, 2003.

75. Ji, X.D., Melman, N., and Jacobson, K.A., Interactions of flavonoid and other phytochemicals with adenosine receptors, *J. Med. Chem.*, 39, 781, 1996.

76. Karton, Y. et al., Synthesis and biological activities of flavonoid derivatives as adenosine receptor antagonists, *J. Med. Chem.*, 39, 2293, 1996.

77. Cathcart, M.K. and Folcik, V.A., Lipoxygenases and atherosclerosis: protection versus pathogenesis, *Free Radical Biol. Med.*, 28, 1726, 2000.

78. Sadik, C.D., Sies, H., and Schewe, T., Inhibition of 15-lipoxygenases by flavonoids: structure–activity relations and mode of action, *Biochem. Pharmacol.*, 65, 773, 2003.

79. Da Silva, E.L., Tsushida, T., and Terao, J., Inhibition of mammalian 15-lipoxygenase-dependent lipid peroxidation in low-density lipoprotein by quercetin and quercetin monoglucosides, *Arch. Biochem. Biophys.*, 349, 313, 1998.

80. Chi, Y.S. et al., Effects of naturally occurring prenylated flavonoids on enzymes metabolizing arachidonic acid: cyclooxygenases and lipoxygenases, *Biochem. Pharmacol.*, 62, 1185, 2001.

81. Cos, P. et al., Structure–activity relationship and classification of flavonoids as inhibitors of xanthine oxidase and superoxide scavengers, *J. Nat. Prod.*, 61, 71, 1998.

82. Hatano, T. et al., Effects of interactions of tannins with co-existing substances. VII. Inhibitory effects of tannins and related polyphenols on xanthine oxidase, *Chem. Pharm. Bull.*, 38, 1224, 1990.

83. Moini, H., Guo, Q., and Packer, L., Enzyme inhibition and protein-binding action of the procyanidin-rich french maritime pine bark extract, pycnogenol: effect on xanthine oxidase, *J. Agric. Food Chem.*, 48, 5630, 2000.

84. Galati, G. et al., Prooxidant activity and cellular effects of the phenoxyl radicals of dietary flavonoids and other polyphenolics, *Toxicology*, 177, 91, 2002.

85. Galati, G. et al., Peroxidative metabolism of apigenin and naringenin versus luteolin and quercetin: glutathione oxidation and conjugation, *Free Radical Biol. Med.*, 30, 370, 2001.

86. Galati, G. et al., Glutathione-dependent generation of reactive oxygen species by the peroxidase-catalysed redox cycling of flavonoids, *Chem. Res. Toxicol.*, 12, 521, 1999.

87. Moridani, M.Y. et al., Catechin metabolism: glutathione conjugate formation catalysed by tyrosinase, peroxidase and cytochrome P450, *Chem. Res. Toxicol.*, 14, 841, 2001.

88. Boersma, M.G. et al., Regioselectivity and reversibility of the glutathione conjugation of quercetin quinone methide, *Chem. Res. Toxicol.*, 13, 185, 2000.

89. Van Zanden, J.J. et al., Inhibition of human glutathione S-transferase P1-1 by the flavonoid quercetin, *Chem. Biol. Interact.*, 145, 139, 2003.

90. Ciolino, H.P., Daschner, P.J., and Yeh, G.C., Dietary flavonoids quercetin and kaempferol are ligands of aryl hydrocarbon receptor that affect CYP1A1 differentially, *Biochem. J.*, 340, 715, 1999.

91. Kang, Z.C., Tsai, S.J., and Lee, H., Quercetin inhibits benzo[*a*]pyrene-induced DNA adducts in human Hep G2 cells by altering cytochrome P-450 1A1 expression, *Nutr. Cancer*, 35, 175, 1999.

92. Doostdar, H., Burke, M.D., and Mayer, R.T., Bioflavonoids: selective substrates and inhibitors for cytochromes P450 CYP1A and CYP1B1, *Toxicology*, 144, 31, 2000.

93. Lee, H. et al., Structure-related inhibition of human hepatic caffeine N3-demethylation by naturally occurring flavonoids, *Biochem. Pharmacol.*, 55, 1369, 1998.

94. Moon, J.Y., Lee D.W., and Park, K.H., Inhibition of 7-ethoxycoumarin O-deethylase activity in rat liver microsomes by naturally occurring flavonoids: structure–activity relationships, *Xenobiotica*, 28, 117, 1998.

95. Zhai, S. et al., Comparative inhibition of human cytochromes P450 1A1 and 1A2 by flavonoids, *Drug Metab. Dispos.*, 26, 989, 1998.

96. Silva, I.D. et al., Metabolism of galangin by rat cytochromes P450: relevance to the genotoxicity of galangin, *Mutat. Res.*, 393, 247, 1997.

97. Silva, I.D. et al., Involvement of rat cytochrome 1A1 in the biotransformation of kaempferol to quercetin: relevance to the genotoxicity of kaempferol, *Mutagenesis*, 12, 383, 1997.

98. Ibrahim, A.R. and Abdul-Hajj, Y.J., Aromatase inhibition by flavonoids, *J. Steroid Biochem. Mol. Biol.*, 37, 257, 1990.

99. Kao, Y.C. et al., Molecular basis of the inhibition of human aromatase (estrogen synthetase) by flavone and isoflavone phytoestrogens: a site-directed mutagenesis study, *Environ. Health Perspect.*, 106, 85, 1998.

100. Le Bail, J.C. et al., Aromatase and 17β-hydroxysteroid dehydrogenase inhibition by flavonoids, *Cancer Lett.*, 133, 101, 1998.

101. Salvi, A. et al., Protein protection by antioxidants: development of a convenient assay and structure–activity relationships of natural polyphenols, *Helv. Chim. Acta*, 85, 867, 2002.

102. Dangles, O., Dufour, C., and Bret, S., Flavonol–serum albumin complexation. Two-electron oxidation of flavonols and their complexes with serum albumin. *J. Chem. Soc. Perkin Trans. 2*, 737, 1999.

103. Riedl, K.M. and Hagerman, A.E., Tannin–protein complexes as radical scavengers and radical sinks, *J. Agric. Food Chem.*, 49, 4917, 2001.

104. Arts, M.J.T.J. et al., Masking of antioxidant capacity by the interactions of flavonoids with protein, *Food Chem. Toxicol.*, 39, 787, 2001.

105. Heinonen, M. et al., Effect of protein on the antioxidant activity of phenolic compounds in a lecithin–liposome oxidation system, *J. Agric. Food Chem.*, 46, 917, 1998.

106. Dangles, O. and Dufour, C., Antioxidant activity of dietary flavonoids, in *Proceedings of the XIth Biennial Meeting of the Society for Free Radical Research*, Pasquier, C., Ed., 2002, 533.

107. Rawel, H.M., Rohn, S., Kruse, H.P., and Kroll, J., Structural changes induced in bovine serum albumin by covalent attachment of chlorogenic acid, *Food Chem.*, 78, 443, 2002.

108. Hagerman, A.E., Dean, R.T., and Davies, M.J., Radical chemistry of epigallocatechin gallate and its relevance to protein damage, *Arch. Biochem. Biophys.*, 414, 115, 2003.

109. Walle, T., Vincent, T.S., and Walle, U.K., Evidence of covalent binding of the dietary flavonoid quercetin to DNA and protein in human intestinal and hepatic cells, *Biochem. Pharmacol.*, 65, 1603, 2003.
110. Salvi, A. et al., Structural damage to proteins caused by free radicals: assessment, protection by antioxidants, and influence of protein binding, *Biochem. Pharmacol.*, 61, 1237, 2001.
111. Myhrstad, M.C.W. et al., Flavonoids increase the intracellular glutathione level by transactivation of the γ-glutamylcysteine synthetase catalytical subunit promoter, *Free Radical Biol. Med.*, 32, 386, 2002.
112. Liang, Y.C. et al., Suppression of inducible cyclooxygenase and nitric oxide synthase through activation of peroxisome proliferator-activated receptor-γ by flavonoids in mouse macrophages, *FEBS Lett.*, 496, 12, 2001.

10 The Anthocyanins

Øyvind M. Andersen and Monica Jordheim

CONTENTS

10.1   INTRODUCTION

The anthocyanins constitute a major flavonoid group that is responsible for cyanic colors ranging from salmon pink through red and violet to dark blue of most flowers, fruits, and leaves of angiosperms. They are sometimes present in other plant tissues such as roots, tubers, stems, bulbils, and are also found in various gymnosperms, ferns, and some bryophytes. As described below, the past decade has witnessed a renaissance in research activities on and general interests in these water-soluble pigments in several areas. When searching Chemical Abstracts/Medline for the word anthocyanin, the number of articles obtained was 790 in 2003 compared to 257 in 1993.

This chapter follows on from those of the four previous editions of *The Flavonoids*,[1–4] and the three review articles of Harborne and Williams.[5–7] It will confine its attention largely to a detailed account on anthocyanin structures reported after 1992 (Section 10.2 and Table 10.2). Special effort has been made to present a comprehensive overview of all the various anthocyanins with complete structures in the literature (Section 10.2 and Appendix A). Many anthocyanins reported in checklists of previous reviews[3,8] have been excluded from Appendix A mainly because of the lack of experimental proofs for proper determination of the linkage point(s) between one or more of the glycosidic units involved. For instance, after careful considerations of the data used as evidence for determination of the linkage position of the monosaccharides in the different anthocyanidin 5-monoglycosides presented in the various reports in the literature, we have excluded these anthocyanins apart from the deoxyanthocyanidin 5-glycosides from Appendix A.

Anthocyanin production by cell cultures (Section 10.3.1) and synthesis (Section 10.3.2), and anthocyanin localization in plant cells (Section 10.4) have been treated separately due to important progress in these fields in recent years. Motivated by the many reports on anthocyanins from various sources in the review period (Table 10.2), some chemotaxomic considerations have been included in Section 10.5. This chapter has not dealt with other articles than those written in the English or German language. Thus, this review has most probably not given the right credit, in particular, to the Japanese research groups of Tadeo Kondo and Kumi Yoshida; Norio Sait, Fumi Tatsuzawa, and Toshio Honda; and Norihiko Terahara.

10.1.1   ANTHOCYANIN STRUCTURES

The total number of different anthocyanins reported to be isolated from plants in this review is 539 (Appendix A). This number includes 277 anthocyanins that have been identified later than 1992. Several previously reported anthocyanins have for the first time received complete structural elucidation, and some structures have been revised. The majority of anthocyanins with the most complex structures and highest molecular masses have been reported in the period of this review.

Two classes of dimeric anthocyanins isolated from plants (section 10.2.6) have been identified in plants for the first time. One class includes pigments where an anthocyanin and a flavone or flavonol are linked to each end of a dicarboxylic acyl unit.[9–12] The other class includes four different catechins linked covalently to pelargonidin 3-glucoside.[13] During the last decade, seven new desoxyanthocyanidins and a novel type of anthocyanidin called pyranoanthocyanidins have been reported (Section 10.2.2). Toward the end of the 20th century, several color-stable 4-substituted anthocyanins, pyranoanthocyanins, were discovered in small amounts in red wine and grape pomace.[14–16] Recently, similar compounds have been isolated from extracts of petals of *Rosa hybrida* cv. "M'me Violet,"[17] scales of red onion,[18] and strawberries.[13,19] About 94% of the new anthocyanins in the period of this review are based on only six anthocyanidins (Table 10.2).

The first natural *C*-glycosylanthocyanin has recently been isolated from flowers of *Tricyrtis formosana*.[20] No new monosaccharide units have been identified in anthocyanins during the last decade; however, two new disaccharides[21–23] and one new trisaccharide[24,25] have been reported connected to anthocyanidins (Section 10.2.4). The two anthocyanins from blue flowers of *Nymphaéa caerulea*[26] and the two minor anthocyanins from red onions[18] are, with the exception of the desoxyanthocyanins, the only anthocyanins without a sugar in the 3-positions. Among the new anthocyanins reported after 1992, around 88% contain acyl group(s). The acyl groups, *E*-3,5-dihydroxycinnamoyl in a triacylated-tetraglucosylated cyanidin derivative from *Ipomoea asarifolia*,[27] and tartaryl in four anthocyanins isolated from flowers of *Anemone coronaria*,[28,29] have, for the first time, been identified as part of an anthocyanins (Section 10.2.5). The first anthocyanins found conjugated with sulfate, malvidin 3-glucoside-5-[2-(sulfato)glucoside] and malvidin 3-glucoside-5-[2-(sulfato)-6-(malonyl)glucoside], have been isolated from violet flowers of *Babiana stricta*.[30] Six novel anthocyanins made in transgenic plants[31,32] and four novel anthocyanins produced in plant cell cultures[33–36] have been included in Appendix A. Some interesting research on the complex metalloanthocyanins is outlined in Section 10.2.6.

10.1.2 NUTRITIONAL SUPPLEMENTS — HEALTH ASPECTS

There has been an explosive interest in anthocyanins as potential nutritional supplements for humans. Regular consumption of anthocyanins and other polyphenols in fruits, vegetables, wines, jams, and preserves is associated with probable reduced risks of chronic diseases such as cancer, cardiovascular diseases, virus inhibition, Alzheimer's disease. Anthocyanins and other flavonoids are regarded as important nutraceuticals mainly due to their antioxidant effects, which give them a potential role in prevention of the various diseases associated with oxidative stress. However, flavonoids have further been recognized to modulate the activity of a wide range of enzymes and cell receptors.[37] In spite of the voluminous literature available, Western medicine has, however, not yet used flavonoids therapeutically, even though their safety record is exceptional. Aspects related to the impact of flavonoids on human health are presented in Chapter 6. The literature on the occurrence of anthocyanins and other flavonoids in foods, their possible dietary effects, bioavaiability, metabolism, pharmacokinetic data, and safety has recently been reviewed by several authors.[33–51] The current knowledge on various molecular evidences of cancer chemoprevention by anthocyanins has been summarized by Hou.[52] He divided the mechanisms into antioxidation, the molecular mechanisms involved in anticarcinogenesis, and the molecular mechanisms involved in the apoptosis induction of tumor cells.

10.1.3 FOOD COLORANTS

There is a worldwide interest in further use of food colorants from natural sources as a consequence of perceived consumer preferences as well as legislative action in connection with synthetic dyes. Several excellent overviews of the common anthocyanin food dyes, quantitative and qualitative aspects of anthocyanins used in food products, and physicochemical properties of anthocyanins (color characteristics and stability) have been presented in the period of this review.[53–57] An impressive compilation of the anthocyanin content of a variety of fruits, vegetables, and grains has been published by Mazza and Miniati.[58] The flavonoid composition of foods is treated in Chapter 4. Different types of anthocyanin-derived pigments, including the pyranoanthocyanins originating by cycloaddition of diverse compounds at C-4 and the 5-hydroxyl of anthocyanidins, and compounds resulting from the condensation between anthocyanins and flavanols, either direct or mediated by acetaldehyde or other compounds, are generated in wine during storage.[59–61] This has led to enlightenment of the color changes that take place in red wine (Chapter 5).

10.1.4 MOLECULAR BIOLOGY, BIOSYNTHESIS, AND FUNCTIONS

Advances in molecular biology, coupled with improved knowledge of anthocyanin biosynthesis, have led to increased interests in cultivars and plant mutants with new colors and shapes. A general overview of the biosynthetic pathway leading to flavonoids and recent advances in the molecular biology and biotechnology of flavonoid biosynthesis is presented in Chapter 3. Several reviews in the field of anthocyanins have recently been reported. Springob et al.[62] covered the biochemistry, molecular biology, and regulation of anthocyanin biosynthesis, with particular emphasis on mechanistic features and late steps of anthocyanin biosynthesis, including glycosylation and vacuolar sequestration. Irani et al.[63] focused on molecular mechanisms of the regulation of anthocyanin biosynthesis, and the factors that influence the pigmentation properties of anthocyanins, while Ben-Meir et al.[64] outlined the biochemistry and genetics of the pathway leading to anthocyanin production, and provide an overview on the application of the generated knowledge toward molecular breeding of ornamentals. Other related excellent reviews of Forkmann and coworkers have covered classical versus molecular breeding of ornamentals,[65] metabolic engineering and applications of flavonoids,[66] and biosynthesis of flavonoids.[67]

Anthocyanic coloration plays a vital role in the attraction of insects and birds, leading to pollination and seed dispersal, but their appearance in young leaves and seedlings is often transient. There is increasing evidence that anthocyanins, particularly when they are located at the upper surface of the leaf or in the epidermal cells, have a role to play in the physiological survival of plants. It has been outlined that foliar anthocyanins accumulate in young, expanding foliage, in autumnal foliage of deciduous species, in response to nutrient deficiency, temperature changes, or ultraviolet (UV) radiation exposure, and in association with damage or defense against browsing herbivores or pathogenic fungal infection.[42,68–70] The functions of anthocyanins have in this context mainly been hypothesized as a compatible solute contributing to osmotic adjustment to drought and frost stress, an antioxidant, and a UV and visible light protectant. The flavonoid functions in plants are treated in Chapter 7.

10.1.5 ANALYTICAL METHODS AND INSTRUMENTATION

Continual improvements in methods and instrumentation (e.g., high-performance liquid chromatography [HPLC], liquid chromatography–mass spectrometry [LC–MS], and nuclear magnetic resonance [NMR] spectroscopy) used for separation and structural elucidation of anthocyanins (see Chapters 1 and 2) have made it easier to use smaller quantities of material, and to achieve results at increasing levels of precision. New anthocyanins regularly turn up in plant sources that already have been well investigated before (Table 10.2). Most anthocyanins show instability toward a variety of chemical and physical parameters, including oxygen, high temperatures, and most pH values.[49] The various anthocyanins have similar structures and may occur in complex mixtures, which makes them rather difficult to isolate. A routine analysis may involve just one HPLC injection (5 μl) of a crude extract of dried petal (0.5 mg). However, a typical structural elucidation of a novel anthocyanin may demand more plant material (above 100 g) subjected to extraction with acidified alcoholic solvent, followed by purification and separation using various chromatographic techniques before structural elucidation by spectroscopy and sometimes chemical degradation.[50,51] Recent MS and two-dimensional (2D) NMR techniques have, in particular, become important for the determination of many anthocyanin linkage positions and identification of aliphatic acyl groups. Analytical methods for extraction, separation, and characterization of anthocyanins have been treated in a number of recent reviews.[72–79]

10.2 ANTHOCYANIN CHEMISTRY

10.2.1 GENERAL ASPECTS AND NOMENCLATURE

The anthocyanins consist of an aglycone (anthocyanidin), sugar(s), and, in many cases, acyl group(s). The anthocyanidins are derivatives of 2-phenylbenzopyrylium (flavylium cation) (Table 10.1). The numbering of the left structure in Table 10.1 is used for most anthocyanins. The pyranoanthocyanins are based on the skeleton represented by the structure on the right in Table 10.1. A more systematic name, e.g., 5-carboxypyranopelargonidin, can be 5-carboxy-2-(4-hydroxyphenyl)-3,8-dihydroxy-pyrano[4,3,2-*de*]-1-benzopyrylium. When a given anthocyanin is dissolved in water, a series of secondary structures are formed from the flavylium cation according to different acid–base, hydration, and tautomeric reactions.[80]

While 31 monomeric anthocyanidins (Table 10.1) have been properly identified, around 90% of all anthocyanins (Appendix A) are based on only six anthocyanidins, pelargonidin (Pg), cyanidin (Cy), peonidin (Pn), delphinidin (Dp), petunidin (Pt), and malvidin (Mv). Among the 539 anthocyanins or anthocyanidins that have been identified, 97% are glycosidated (Figure 10.1). The 3-desoxyanthocyanidins, sphagnorubins and rosacyanin B (Table 10.1, Figure 10.3), are the only anthocyanidins found in their nonglycosidated form in plants. Nearly all reports on anthocyanins specifying the D or L configuration of the anthocyanin sugar moieties (monosaccharides), lack experimental evidence for this type of assignments.

10.2.2 ANTHOCYANIDINS

In addition to the 18 anthocyanidins listed previously,[4] Table 10.1 contains seven new desoxyanthocyanidins and a novel type of anthocyanidin called pyranoanthocyanidins. While 31 monomeric anthocyanidins have been properly identified, most of the anthocyanins are based on cyanidin (30%), delphinidin (22%), and pelargonidin (18%), respectively (Figure 10.2). Altogether 20% of the anthocyanins are based on the three common anthocyanidins (peonidin, malvidin, and petunidin) that are methylated. Around 3, 3, and 2% of the anthocyanins or anthocyanidins are labeled as 3-desoxyanthocyanidins, rare methylated anthocyanidins, and 6-hydroxyanthocyanidins, respectively.

In bryophytes, anthocyanins are usually based on 3-desoxyanthocyanidins located in the cell wall. A new anthocyanidin, riccionidin A (Figure 10.3), has been isolated from the liverwort *Ricciocarpos natans*.[81] It could be derived from 6,7,2′,4′,6′-pentahydroxyflavylium, having undergone ring closure of the 6′-hydroxyl at the 3-position. Its visible spectrum in methanolic HCl is at 494 nm. This pigment was accompanied by riccionidin B, which most probably is based on two molecules of riccionidin A linked via the 3′- or 5′-positions. Both pigments were also detected in the liverworts *Marchantia polymorpha*, *Riccia duplex*, and *Scapania undulata*.[81] Somewhat, unexpectedly, riccionidin A has also been isolated from adventitious root cultures of *Rhus javanica* (Anacardiaceae).[82]

A new 3-desoxyanthocyanidin, 7-*O*-methylapigeninidin, has been isolated in low concentration from grains and leaf sheaths of *Sorghum caudatum*.[83] Its UV–vis spectrum recorded in methanol with 0.1% HCl showed absorption maxima at 278.6 and 476.4 nm. The secondary-ion MS spectrum showed a strong $[M]^+$ ion at m/z 269, consistent with the $C_{16}H_{13}O_4$ molecular formula. The ^1H and ^{13}C NMR spectral data were closely related to those of apigeninidin, except for a singlet at 4.08 ppm (aromatic *O*-methyl group). This methyl group was located at C-7 after observation of DIFFNOEs between the methyl protons and both the H-8 and H-6 protons. The pigment was found in low concentration both in grains and in leaf sheaths. A similar 3-desoxyanthocyanidin has been detected in grains of *Sorghum bicolor* after incubation with the fungus *Colletotrichum sublineolum*.[84] In addition to plasma desorption MS data, bathochromic shift analyses indicated that the structure of the compound was

TABLE 10.1

The Structures of Naturally Occurring Anthocyanidin.[a] The Numbering of the Structure on the Left is Used for all Anthocyanins; the Numbering for the Pyranoanthocyanins is Given in the Structure on the Right

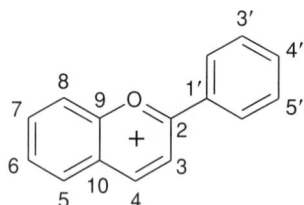

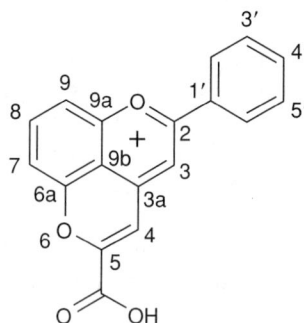

	Substitution Pattern					
	3	**5 (6a)[b]**	**6 (7)[b]**	**7 (8)[b]**	**3′**	**4′**	**5′**
Common anthocyanidins							
Pelargonidin (Pg)	OH	OH	H	OH	H	OH	H
Cyanidin (Cy)	OH	OH	H	OH	H	OH	H
Delphinidin (Dp)	OH	OH	H	OH	OH	OH	OH
Peonidin (Pn)	OH	OH	H	OH	OMe	OH	H
Petunidin (Pt)	OH	OH	H	OH	OMe	OH	OH
Malvidin (Mv)	OH	OH	H	OH	OMe	OH	OMe
Rare methylated anthocyanidins							
5-MethylCy	OH	OMe	H	OH	OH	OH	H
7-MethylPn (rosinidin)	OH	OH	H	OMe	OMe	OH	H
5-MethylDp (pulchellidin)	OH	OMe	H	OH	OH	OH	OH
5-MethylPt (europinidin)	OH	OMe	H	OH	OMe	OH	OH
5-MethylMv (capensinidin)	OH	OMe	H	OH	OMe	OH	OMe
7-MethylMv (hirsutidin)	OH	OH	H	OMe	OMe	OH	OMe
6-Hydroxylated anthocyanidins							
6-HydroxyPg	OH	OH	OH	OH	H	OH	H
6-HydroxyCy	OH	OH	OH	OH	OH	OH	H
6-HydroxyDp	OH	OH	OH	OH	OH	OH	OH
3-Desoxyanthocyanidins							
Apigeninidin (Ap)	H	OH	H	OH	H	OH	H
Luteolinidin (Lt)	H	OH	H	OH	OH	OH	H
Tricetinidin (Tr)	H	OH	H	OH	OH	OH	OH
7-MethylAp[c]	H	OH	H	OMe	H	OH	H
5-MethylLt[c]	H	OMe	H	OH	OH	OH	H
5-Methyl-6-hydroxyAp (carajurone)[c]	H	OMe	OH	OH	H	OH	H
5,4′-Dimethyl-6-hydroxyAp (carajurin)	H	OMe	OH	OH	H	OMe	H
5-Methyl-6-hydroxyLt[c]	H	OMe	OH	OH	OH	OH	H
5,4′-Dimethyl-6-hydroxyLt[c]	OH	OMe	OH	OH	OH	OMe	H
Riccionidin A[c, d]	OH	H	OH	OH	H	OH	H

TABLE 10.1
The Structures of Naturally Occurring Anthocyanidins.[a] The Numbering of the Structure on the Left is Used for all Anthocyanins; the Numbering for the Pyranoanthocyanins is Given in the Structure on the Right — *continued*

	Substitution Pattern						
	3	5 (6a)[b]	6 (7)[b]	7 (8)[b]	3'	4'	5'
Pyranoanthocyanidins							
5-CarboxypyranoPg[c]	OH	O–	H	OH	H	OH	H
5-CarboxypyranoCy[c,e]	OH	O–	H	OH	OH	OH	H

[a]Sphagnorubins A–C from peat moss, *Sphagnum*, have not been included (Figure 10.3).
[b]The numbers in parentheses correspond to the pyranoanthocyanidins.
[c]New anthocyanidins (reported between 1992 and 2004).
[d]Ring closure on the basis of ether linkage between the 3- and 6'-positions. Riccionidin A and its dimer, riccionidin B, have an additional OH-group in the 2'-position (Figure 10.3).
[e]Rosacyanin B (Figure 10.3).

consistent with that of 5-*O*-methylluteolinidin. The spectrum of this phytoalexin, which showed greater fungitoxicity than luteolinidin, revealed an absorption maximum at 495 nm.

Although a previous synthesis of the desoxyanthocyanidin carajurin, 6,7-dihydroxy-5,4'-dimethoxy-flavylium (isolated from leaves of *Arrabidaea chica*), was published in 1953,[85]

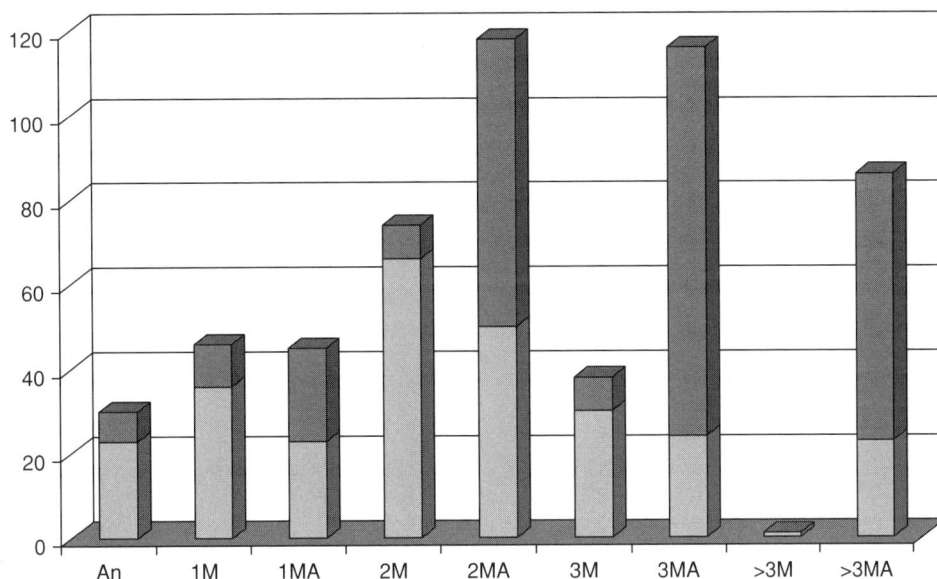

FIGURE 10.1 Anthocyanidins (An) and anthocyanins grouped according to their content of monosaccharide units and acylation. The vertical axis represents the number of pigments. The upper dark part of each bar represents the anthocyanins reported later than 1992. 1M, one monosaccharide unit; 1MA, one monosaccharide unit plus acylation; 2M, two monosaccharide units; 2MA, two monosaccharide plus acylation; 3M, three monosaccharide units; 3MA, three monosaccharide units plus acylation; >3M, more than three monosaccharide units; >3MA, more than three monosaccharide units plus acylation.

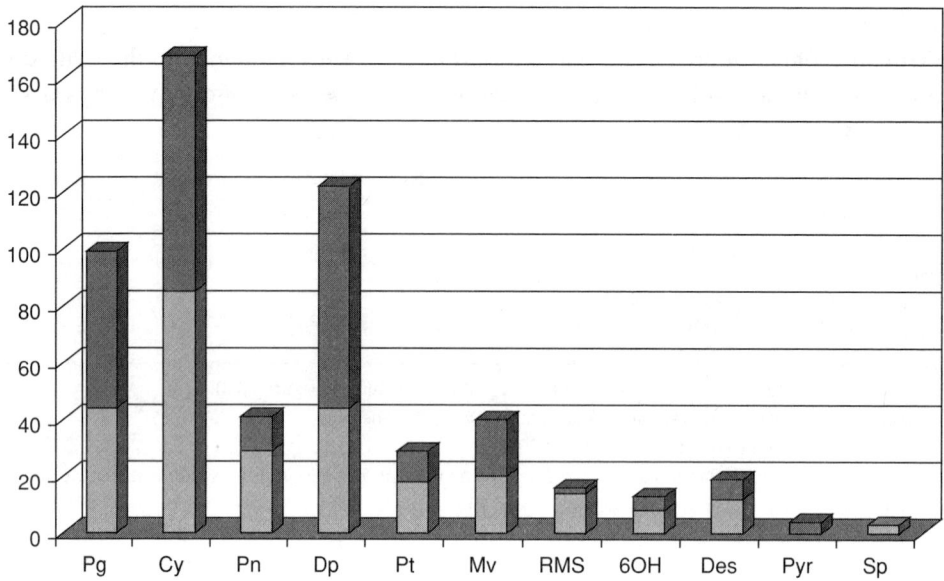

FIGURE 10.2 The number of anthocyanins based on the various anthocyanidins. The upper dark part of each bar represents the anthocyanins reported later than 1992. Pg, pelargonidin; Cy, cyanidin; Pn, peonidin, Dp, delphinidin; Pt, petunidin; Mv, malvidin; RMS, rare methylated structures; 6OH, 6-hydroxy-; Des, desoxy-; Pyr, pyrano-; Sp, sphagnorubins. See Table 10.1 for structures.

some authors have considered the structure of this pigment to be only partially described.[1,58] More recently, two groups[86,87] have nearly simultaneously confirmed the structure of carajurin — even with a crystal structure.[87] The structure of carajurone was also revised to be 6,7,4′-trihydroxy-5-methoxy-flavylium.[86] Additionally, two new 3-desoxyanthocyanidins,

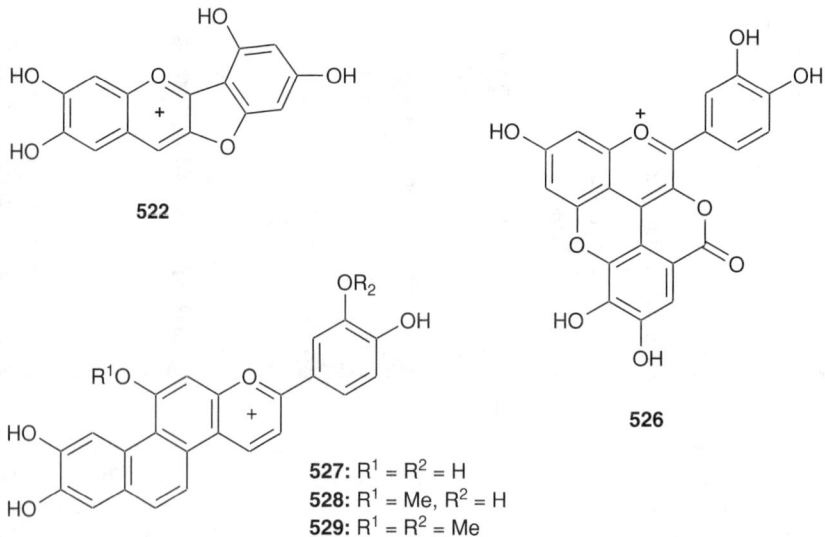

FIGURE 10.3 Some anthocyanidins with unusual structures; **522**, riccionidin A; **526**, rosacyanin B; **527**–**529**, sphagnorubins A–C.

6,7,3'-trihydroxy-5,4'-dimethoxy-flavylium[86] and 6,7,3',4'-tetrahydroxy-5-methoxy-flavy-lium,[86,87] were isolated from these leaves, which are traditionally used by some indigenous populations of South America for body painting and for dyeing fibers.

In recent years, several color-stable 4-substituted anthocyanins have been discovered in small amounts in red wine and grape pomace (see Chapter 5). Vitisin A and acetylvitisin A were identified as the 3-glucoside and the 3-acetylglucoside of malvidin containing an additional $C_3H_2O_2$ unit linking the C-4 and the C-5 hydroxyl group. Vitisin B and acetylvitisin B were identified as analogous pigments having a $CH=CH$ moiety instead of the $C_3H_2O_2$ unit.[15] The suggested structure for carboxypyranomalvidin (vitisidin A) was later slightly revised by Fulcrand et al.,[16] who proved that the $C_3H_2O_2$ unit was part of a pyran ring having a free acid group. They suggested that vitisin A was formed by cycloaddition of pyruvic acid involving both C-4 and the hydroxyl at C-5 of malvidin. Four reported methylpyranoantho-cyanins isolated from blackcurrant seeds[88] were later shown to be the oxidative cycloaddition products of the extraction solvent (acetone) and the natural anthocyanins.[89] Pyranoantho-cyanidins generated from the respective glycosides after hydrolysis were found to undergo rearrangement to form a new type of furoanthocyanidins.[90] Recently, four new pyranoantho-cyanins, namely pyranocyanin C and D and pyranodelphinidin C and D, were isolated by the same group from an extract of blackcurrant seeds.[91] These pigments were absent in fresh extracts, and their levels increased gradually with time. Their formation was likely to be from the reaction of the anthocyanins and *p*-coumaric acid in the extracts.

The first pyranoanthocyanidin isolated from intact plants received the trivial name rosa-cyanin B (Figure 10.3).[17] This violet pigment was isolated in small amounts from the petals of *Rosa hybrida* cv. "M'me Violet." Its structure was revealed mainly by high resolution fast atom bombardment MS and NMR (1D and 2D). Rosacyanin B is very stable in acidic alcoholic solutions; however, under neutral or weakly acidic aqueous conditions it is precipitated before forming the colorless pseudobase. This anthocyanidin contains no sugar units. Recently, four anthocyanins with the same aglycone, 5-carboxypyranocyanidin, have been isolated from acidified, methanolic extracts of the edible scales, as well as from the dry outer scales of red onion (*Allium cepa*).[18] Two of the structures were elucidated by 2D NMR spectroscopy and LC–MS as the 3-glucoside and 3-[6-(malonyl)glucoside] (**525**) of this 4-substituted aglycone. The two analog pigments methylated at either the terminal carboxyl group of the acyl moiety or at the aglycone carboxyl were most probably formed by esterification of **525** with the solvent (acidified methanol) during the isolation process. Another 3-glucoside (**523**) with the new 4-substituted aglycone, 5-carboxypyranopelargoni-din, has been isolated in small amounts from the acidified, methanolic extract of strawberries (*Fragaria × ananassa*).[19] By comparison of UV–vis absorption spectra, **523** showed in contrast to ordinary pelargonidin 3-glucoside (**5**) a local absorption peak around 360 nm, a hypso-chromic shift (8 nm) of the visible absorption maximum, and lack of a distinct UV absorption peak around 280 nm. The similarities between the absorption spectra of **523** in various acidic and neutral buffer solutions implied restricted formation of the instable colorless equilibrium forms, which are typical for most anthocyanins in weakly acidic solutions. The molar absorptivity of **523** varied little with pH contrary to similar values of, for instance, **5**.

Each anthocyanidin is involved in a series of equilibria giving rise to different forms, which exhibit their own properties including color.[80] One- and two-dimensional NMR have been used to characterize the various forms of malvidin 3,5-diglucoside present in aqueous solution in the pH range 0.3 to 4.5 and to determine their molar fractions as a function of pH.[92] In addition to the flavylium cation, two hemiacetal forms and both the *cis* and *trans* forms of chalcone were firmly identified. In a reexamination, the intricate pH-dependent set of chemical reactions involving synthetic flavylium compounds (e.g., 4'-hydroxyflavylium) was confirmed to be basically identical to those of natural anthocyanins (e.g., malvidin 3,5-diglucoside) in acidic and neutral media.[94] For each process, a kinetic expression was deduced allowing calculation of all the equilibrium constants and most of the rate constants in the system. In recent years, Pina et al. have performed a systematic investigation of the photochemical and thermal reactions of synthetic flavylium compounds.[94] They have shown that 4′,7-dihydroxy-flavylium (AH^+) in a water–ionic liquid biphasic system can be used as a write–read–erase system. In acid media, the *trans*-chalcone form is soluble in ionic liquids and is thermally metastable, but reacts photochemically (write) to give the yellow flavylium salt, which can be optically read without being erased. The system is prepared for a new cycle by two consecutive pH "jumps."[95] These results are very interesting since the flavylium compounds represent examples of multistate or multifunctional chemical systems that may be used for information processing at the molecular level according to principles similar to those that govern information transfer in living organisms. In particular, flavylium compounds can behave as optical memories and logic gates systems: a write–read–erase molecular switch.

10.2.3 ANTHOCYANINS NOT BASED ON THE COMMON ANTHOCYANIDINS

Although most new anthocyanins discovered during the last decade have been based on the six common anthocyanidins (Figure 10.2), some rare exceptions with limited distribution have been reported. The major 3-deoxyanthocyanin isolated from the fern *Blechnum novae-zelandiae* was determined to be luteolinidin 5-[3-(glucosyl)-2-(acetyl)glucoside] by HPLC, NMR (1D, 2D), and electrospray MS.[96] The new 3-[6-(*p*-coumaryl)glucoside] and 3-glucoside of hirsutidin together with the known corresponding petunidin and malvidin derivatives have been identified in extracts of both cell suspensions and fresh flowers of *Catharanthus roseus*.[97] The extracts were analyzed by positive-ion electrospray ionization MS, and collision experiments were performed on molecular ions by means of ion trap facilities. Purified compounds were also analyzed by thin-layer chromatography and UV–vis spectroscopy.

The 3-rutinoside and 3-glucoside of 6-hydroxycyanidin have previously been isolated from red flowers of *Alstroemeria* cultivars,[98] whereas 6-hydroxydelphinidin 3-rutinoside, occurred in pink-purple flowers of five cultivars.[99] During the period of this review, the 3-[6-(malonyl)glucoside] of 6-hydroxycyanidin and 6-hydroxydelphinidin in addition to 6-hydroxydelphinidin 3-glucoside have been identified in various *Alstroemeria* cultivars.[100–102] The position of the 6-hydroxyl of these anthocyanidins was unambiguously assigned by homo- and heteronuclear NMR techniques.[103] Flower color, hue, and color intensity of fresh tepals of 28 Chilean species and 183 interspecific hybrids have been described by CIELAB parameters.[104] Compared with flowers containing exclusively cyanidin 3-glycosides, the hues of flowers with 6-hydroxycyanidin 3-glycosides were more reddish. Substitution of the anthocyanidin A-ring with 6-hydroxyl causes a hypsochromic shift (~15 nm) in the visible spectra, which has diagnostic value.[103] The relationship between flower color and anthocyanin content has also been investigated in 45 *Alstroemeria* cultivars by Tatsuzawa et al.[105] The major anthocyanins of outer perianths were cyanidin 3-rutinoside and 6-hydroxycyanidin 3-rutinoside in cultivars with red flowers, 6-hydroxydelphinidin 3-rutinoside in those that were red-purple, and delphinidin 3-rutinoside in purple ones. Recently, the same group has isolated the 3-(glucoside) and 3-[6-(rhamnosyl)glucoside] of 6-hydroxypelargonidin (aurantinidin) from extracts of the orange-red flowers of the *Alstroemeria* cultivars "Oreiju," "Mayprista," and "Spotty-red."[106] Aurantinidin has previously been reported to occur in *Impatiens aurantiaca* (Balsaminaceae).[107]

10.2.4 GLYCOSIDES

Most anthocyanins contain two, three, or just one monosaccharide unit (Figure 10.1); however, as much as seven glucosyl units have been found in ternatin A1 (*Clitoria ternatea*)[108] and cyanodelphin (*Delphinium hybridum*).[109] Altogether 240 and 24 anthocyanins have been reported to contain a disaccharide and a trisaccharide, respectively, while no tetrasaccharide has been found yet in anthocyanins. The sugar moieties are connected to the anthocyanidins through *O*-linkages; however, recently Saito et al.[20] have isolated 8-*C*-glucosylcyanidin 3-[6-(malonyl)glucoside] from the purple flowers of *Tricyrtis formosana* cultivar Fujimusume (Liliaceae). Although *C*-glycosylation is common in other flavonoids, especially flavones (Chapter 13), this is the first report of a natural *C*-glycosylanthocyanin.

10.2.4.1 Monosaccharides

The anthocyanin monosaccharides are represented by glucose, galactose, rhamnose, arabinose, xylose, and glucuronic acid. There is no new monosaccharide attached to anthocyanidins reported in the period of this review. Glucosyl moieties have been identified in as much as 90% of the anthocyanins, while the rarest monosaccharide in anthocyanins, glucuronosyl, is limited to 11 anthocyanins (Figure 10.4). This latter monosaccharide has previously been identified in anthocyanins isolated from flowers of *Helenium autumnale* and *Bellis perennis* (Compositae),[110,111] and tentatively identified as luteolinidin 4′-glucuronide in flower extracts of *Holmskioldia sanguinea*.[112] More recently, three delphinidin and three cyanidin derivatives based on 3-[2-(2-(caffeylglucosyl)galactoside]-7-[6-(caffeyl)glucoside]-3′-glucuronoside have been isolated from flowers of *Anemone coronaria*.[28,29] Glucuronosyl units have also been found in two anthocyanin–flavonol complexes isolated from chive flowers (*Allium schoenoprasum*), however, linked to the flavonol moieties.[11] The acid function of the glucuronosyl of these latter complexes was considerably methylesterified during extraction with methanol containing as little as 1% trifluoroacetic acid.

10.2.4.2 Disaccharides

The following disaccharides have previously[4] been found linked to anthocyanidins: 2-glucosylglucose (sophorose), 6-rhamnosylglucose (rutinose), 2-xylosylglucose (sambubiose), 6-glucosylglucose (gentiobiose), 6-rhamnosylgalactose (robinobiose), 2-xylosylgalactose (lathyrose), 2-rhamnosylglucose (neohesperidose), 3-glucosylglucose (laminariobiose), 6-arabinosylglucose, 2-glucurononylglucose, 6-glucosylgalactose, and 4-arabinosylglucose (Figure 10.5). In addition, several anthocyanidin disaccharides have been detected without proper determination of the linkage points between the monosaccharides. During the period of this review, Yoshida et al.[21] have isolated delphinidin 3-[6-(*E-p*-coumaryl)glucoside]-5-[6-(malonyl)-4-(rhamnosyl)-glucoside] (muscarinin A) from purplish-blue spicate flower petals of *Muscari armeniacum* (Liliaceae), which contained an interesting 1 → 4 linkage between the rhamnose and one of the glucose units. Another new disaccharide, 2-glucosylgalactose, has been found in pelargonidin 3-[2-(2-*E*-caffeyl)glucosyl)-galactoside] isolated from sepals of *Pulsatilla cernua* (Ranunculaceae),[22] in the major anthocyanin, cyanidin 3-[2-(glucosyl)galactoside], isolated from scarlet fruits of *Cornus suecica* (Cornaceae),[23] and in seven exciting acylated cyanidin and delphinidin derivatives from flowers of *Anemone coronaria*[28,29] (Table 10.2).

Among the new anthocyanins, which have been reported after 1992, 47, 22, and 36 contain sophorose, rutinose, and sambubiose, respectively (Figure 10.5). Most of the anthocyanins containing sophorose were first isolated from species belonging to Convolvulaceae (22) and Cruciferae (13) (Table 10.2). This disaccharide has also been identified in new anthocyanins isolated from *Ajuga* (Labiatae),[36,113,114] *Consolida* (Ranunculaceae),[115] *Begonia* (Begoniaceae),[116] and in the flavonol unit of two covalent anthocyanin–flavonol complexes from *Allium* (Alliaceae).[11]

Most of the novel anthocyanins containing rutinose (20) have been isolated from species belonging to the genera *Petunia* and *Solanum* in Solanaceae (Table 10.2). However,

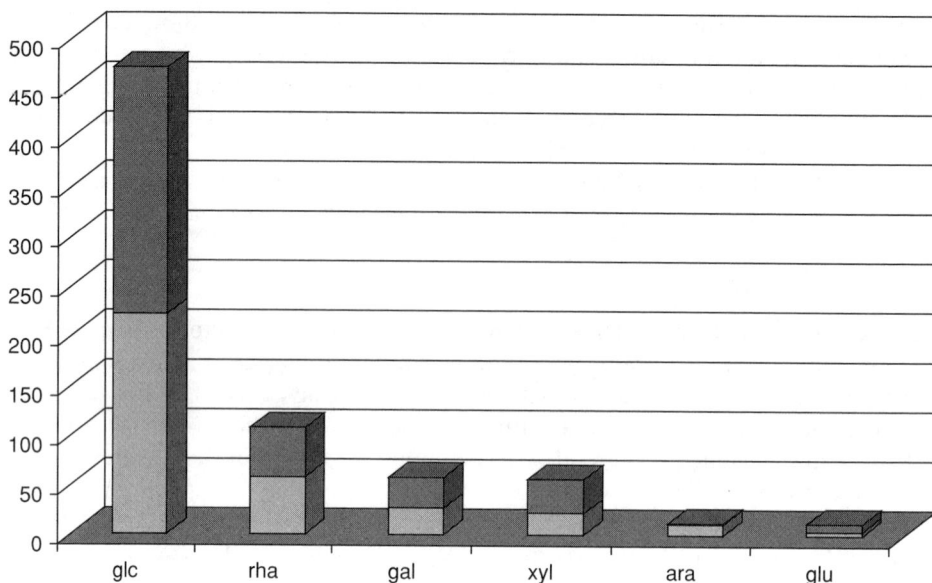

FIGURE 10.4 Numbers of anthocyanins containing the various monosaccharides identified in anthocyanins. The upper dark part of each bar represents the anthocyanins reported later than 1992. glc, glucose; rha, rhamnose; xyl, xylose; gal, galactose; ara, arabinose; glu, glucuronic acid. Some anthocyanins contain more than one type of monosaccharide.

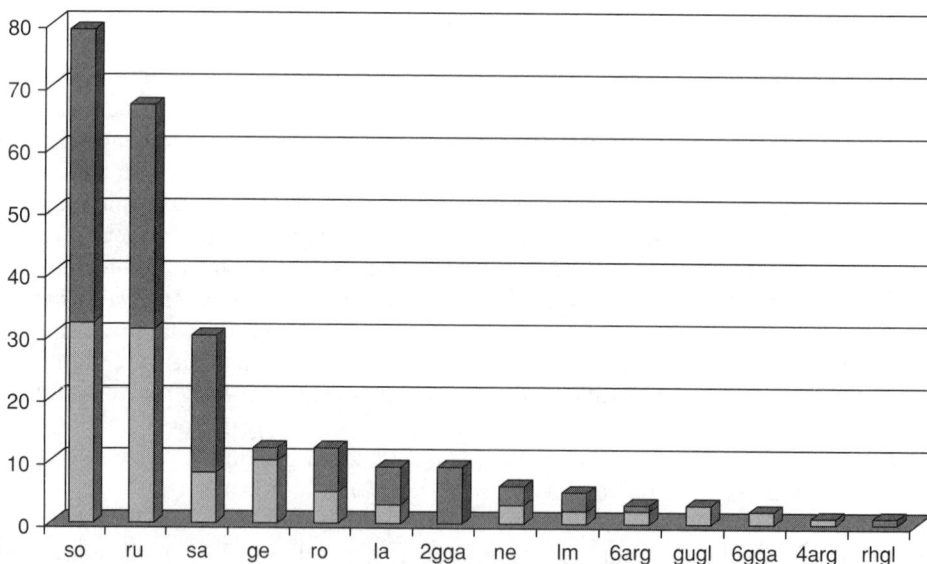

FIGURE 10.5 Numbers of anthocyanins containing the various disaccharides identified in anthocyanins. The upper dark part of each bar represents the anthocyanins reported later than 1992. so, 2-glucosylglucose; ru, 6-rhamnosylglucose; sa, 2-xylosylglucose; ge, 6-glucosylglucose; ro, 6-rhamnosylgalactose; la, 2-xylosylgalactose; 2gga, 2-glucosylgalactose; ne, 2-rhamnosylglucose; lm, 3-glucosylglucose; 6arg, 6-arabinosylglucose; gugl, 2-glucuronylglucose; 6gga, 6-glucosylgalactose; 4arg, 4-arabinosylglucose; rhgl, 4-rhamnosylglucose.

TABLE 10.2
Occurrence of Anthocyanins Reported after 1992

Taxon	Organ	Pigment[a]	References[b]
BRYOPHYTA			
BLECHNACEAE			
Blechnum novae-zelandiae	Tissue	Lt5-[3-(glc)-2-(ace)glc] (**514**)[c]	*96* (2001)
HEPATOPHYTA			
MARCHANTIACEAE			
Marchantia polymorpha	Cell walls	Riccionidin A (**522**), riccionidin B[d, e]	*81* (1994)
RICCIACEAE			
Riccia duplex	Cell walls	Riccionidin A (**522**), riccionidin B[d, e]	*81* (1994)
Ricciocarpus natans			
SCAPANIACEAE			
Scapania undulata	Cell walls	Riccionidin A (**522**), riccionidin B[d, e]	*81* (1994)
GYMNOSPERMAE			
Pinaceae			
Abies, Picea, Pinus, Pseudotsuga, Tsuga[f]			
27 spp.	Cones	Cy3-glc, Dp3-glc, Pn3-glc, Pt3-glc	*295* (2003)
Pinus banksiana	Needles	Cy3-glc, Dp3-glc, Pn3-glc, Pt3-glc, Mv3-glc	*338* (2002)
ANGIOSPERMAE			
MONOCOTYLEDONEAE			
Palmae (= Arecaceae)			
Euterpe edulis[g]	Fruits	Cy3-glc, Cy3-[6-(rha)glc]	*339* (1994)
Pinanga polymorpha[g] (Asian palm)	Fruits	Cy3-glc, Cy3-[6-(rha)glc]	*339* (1994)
Commelinaceae			
Tradescantia pallida	Leaves	Cy3-[6-(2,5-di-(fer)ara)glc]7,3'-[6-(fer)glc] (**251**)	*136* (1995)
Gramineae (= Poaceae)			
Alopecurus, Anthoxanthum, Avenula, Bothriochloa, Dactylis, Deschampsia, Elymus, Festuca, Holcus, Hordeum, Mischanthus, Molinia, Oryza, Phalaris, Phleum, Poa, Sinarundinaria, Zea,[f] 23 spp.	Grass	Cy3-[3,6-di-(mal)glc], Cy3-[6-(mal)glc], Cy3-[6-(rha)glc], Cy3-glc, Dp3-[6-(mal)glc], Dp3-glc, Pn3-[6-(mal)glc], Pn3-[di(mal)-glc][e], Pn3-glc	*296* (2001)
Panicum melinis[g]	Flowers	Cy3-[caf-6-(ara)glc], Cy3-[6-(ara)glc]	*340* (1992)
Pennisetum setaceum (Rubrum Red Riding Hood)	Leaves and flowers	Cy3-[6-(rha)glc], Cy3-glc	*341* (2004)
Phalaris arundinacea	Flower tops	Cy3-[3,6-di-(mal)glc], Cy3-[6-(mal)glc], Cy3-glc, Pn3-glc	*342* (2001)
Phragmites australis	Flowers	Cy3-[6-(mal)glc], Cy3-[6-(suc)glc] (**135**), Cy3-glc	*221* (1998)
Sorghum bicolor[g]	Grains	5MLt (**519**)	*84* (1996) *83* (1997)
Sorghum caudatum	Grains	7Map (**516**), Ap,Lt,Lt5-glc	*343* (1994) *344* (2003)

continued

TABLE 10.2

Occurrence of Anthocyanins Reported after 1992 — *continued*

Taxon	Organ	Pigment[a]	References[b]
Triticum aestivum[g] (cv. Katepwa)	Grains (purple wheat)	Cy3-gal, Cy3-glc, Pn3-glc	
Triticum aestivum[g] (cv. Purendo 38)	Grains (blue wheat)	Cy3-glc, Pn3-glc	*344* (2003)
Zea mays	Leaves and flowers	Cy3-[3,6-di-(mal)glc], Cy3-[6-(mal)glc], Cy3-glc, Pn3-glc	*342* (2001)
Iridaceae			
Babiana stricta	Violet flowers	Mv3-glc-5-[2-(S)-6-(mal)glc] (**472**), Mv3-glc-5-[2-(S)glc] (**467**), Mv3-glc-5-[6-(mal)glc] (**469**)	*30* (1994)
Crocus antalyensis	Flowers	Dp3,5-di-glc, Dp3,7-di-glc, Dp3-glc-5-[6-(mal)glc] (**338**), Pt3,5-di-glc, Pt3,7-di-glc (**427**)	*345* (1999)
Crocus chrysanthus (Skyline)	Flowers	Mv3,7-di[6-(mal)glc] (**473**), Pt3,7-di[6-(mal)glc] (**439**)	*346* (1998)
Crocus chrysanthus (Eyecatcher)	Flowers	Dp3-[6-(rha)glc], Pt3-[6-(rha)glc]	*346* (1998)
Crocus sieberi ssp. *sublimes* (Tricolor)	Flowers	Dp3,5-di-glc, Pt3,5-di-glc	*346* (1998)
Liliaceae			
Dianella nigra	Berries	Dp3-glc-7,3',5'-tri[6-(cum)glc] (**399**), Dp3,7,3',5'-tetra[6-(cum)glc] (**403**)	*146* (2001)
Dianella tasmanica	Berries	Dp3-glc-7,3',5'-tri[6-(cum)glc] (**399**), Dp3-glc-7-glc-3',5'-di[6-(cum)glc] (**391**)	*146* (2001)
Hyacinthus orientalis	Red flowers	Pg3-[6-(caf)glc]-5-[6-(mal)glc], Pg3-[6-(caf)glc]-5-glc, Pg3-[6-(cum)glc]-5-[4-(mal)glc] (**50**), Pg3-[6-(cum)glc]-5-[6-(ace)glc] (**49**), Pg3-[6-(fer)glc]-5-[6-(mal)glc] (**54**), Pg3-[6-(fer)glc]-5-glc (**46**), Pg3-glc-5-[6-(mal)glc] (**35**), Pg3-[6-(Z-cum)glc]-5-glc (**42**)	*201* (1995) *327* (1995)
Hyacinthus orientalis	Blue flowers	Cy3-[6-(cum)glc]-5-[6-(mal)glc], Dp3-[6-(caf)glc]-5-[6-(mal)glc], Dp3-[6-(cum)glc]-5-[6-(mal)glc], Dp3-[6-(cum)glc]-5-glc, Dp3-[6-(Z-cum)glc]-5-[6-(mal)glc], Pg3-[6-(cum)glc]-5-[6-(mal)glc], Pt3-[6-(cum)glc]-5-[6-(mal)glc] (**443**)	*328* (1995)
Lilium	Flowers	Cy3-[6-(rha)glc]-7-glc (**124**), Cy3-[6-(rha)glc]	*120* (1999)
Muscari armeniacum	Blue flowers	Dp3-[6-(cum)glc]-5-[6-(mal)-4-(rha)glc] (**370**), Dp3-glc, Pt3-glc, Mv3-glc	*21* (2002)
Tricyrtis formosana	Flowers	Cy3-[6-(mal)glc]8C-glc (**151**)	*20* (2003)
Tulipa (Queen Wilhelmina)	Flowers	Cy3-[6-(2-(ace)rha)glc] (**144**), Cy3-[6-(rha)glc], Pg3-[6-(2-(ace)rha)glc] (**29**)[d], Pg3-[6-(rha)glc]	*122* (1999)
Tulipa gesneriana	Anthers	Dp3-[6-(2-(ace)rha)glc] (**332**), Dp3-[6-(3-(ace)rha)glc] (**333**), Dp3-[6-(rha)glc]	*121* (1999)
Alliaceae			
Agapanthus praecox ssp. *orientalis*	Flowers	Dp3-[6-(cum)glc]-7-glc]Kae3-glc-7-xyl-4'-glc suc (**538**), Dp3-[6-(cum)glc]-7-glc]Kae3,7-di-glc-4'-glc suc (**539**)	*12* (2000)
Allium cepa	Bulbs	5-Carboxypyranocyanidin 3-glucoside (**524**), 5-Carboxypyranocyanidin 3-[6-(malonyl)glucoside] (**525**)	*18* (2003)

TABLE 10.2
Occurrence of Anthocyanins Reported after 1992 — *continued*

Taxon	Organ	Pigment[a]	References[b]
Allium cepa	Bulbs	Cy3,4′-di-glc (**113**), Cy3,5-di-glc, Cy3-[3-(glc)-6-(mal)glc] (**150**), Cy3-[3-(glc)-6-(mal)glc]4′-glc (**192**), Cy3-[3-(glc)-6-(Me-mal)glc], Cy3-[3-(glc)glc], Cy3-[6-(mal)glc], Cy3-glc, Cy4′-glc (**100**), Cy7-[3-(glc)-6-(mal)glc]4′-glc (**193**), Pn3-[6-(mal)glc], Pn3-[6-(mal)glc]-5-glc (**283**)	*135* (1994) *147* (2003)
Allium sativum	Inner scale leaves	Cy3-[(ace)glc][e], Cy3-[3-(mal)glc], Cy3-[3,6-di-(mal)glc], Cy3-[6-(mal)glc], Cy3-glc	*204* (1997)
Allium schoenoprasum	Flowers	(6″-(Cy3-glc)) (2″″-(Kae3-[2″-(glc)(glc)]-7-glu))malonate (**536**), (6″-(Cy3-[3″-(ace)glc])) (2″″-(Kae3-[2″-(glc)(glc)]-7-glu))malonate (**537**), Cy3-[3,6-di-(mal)glc], Cy3-[6-(ace)glc], Cy3-[6-(mal)glc], Cy3-glc	*11* (2000)
Allium victorialis	Stems	Cy3-[3-(mal)glc] (**133**), Cy3-[3,6-di-(mal)glc] (**141**), Cy3-[6-(mal)glc], Cy3-glc	*202* (1995)
Triteleia bridgesii	Flowers	Cy3-[cum-glc-cum-glc]-5-[mal-glc][e], Dp3-[6-(4-(glc)cum)glc]-5-[6-(mal)glc] (**372**), Dp3-[6-(cum)glc]-5-[6-(mal)glc], Dp3-[6-(cum)glc]-5-glc, Dp3-[6-(Z-cum)glc]-5-[6-(mal)glc]	*166* (1998)
Alstroemeriaceae			
Alstroemeria (Westland, Tiara)	Flowers/ tepals	Cy-6OH-3-[6-(mal)glc] (**500**), Cy-6OH-3-[6-(rha)glc], Cy-6OH-3-glc, Dp-6OH-3-[6-(mal)glc] (**504**), Dp-6OH-3-[6-(rha)glc], Dp-6OH-3-glc (**502**), Pg-6OH-3-[6-(rha)glc] (**496**), Pg-6OH-3-glc (**495**), Cy3-[6-(mal)glc], Cy3-[6-(rha)glc], Cy3-glc, Dp3-[6-(mal)glc], Dp3-[6-(rha)glc], Dp3-glc, Pg3-[6-(rha)glc]	*100* (1996) *101* (2001) *102* (2002) *122* (2003)
Musaceae			
Musa × *paradisiaca*[g]	Bracts	Cy3-[6-(rha)glc], Dp3-[6-(rha)glc], Mv3-[6-(rha)glc], Pg3-[6-(rha)glc], Pn3-[6-(rha)glc]	*347* (2001)
Orchidaceae			
Bletilla striata	Flowers	Cy3-[6-(mal)glc]-7-[6-(cum)glc]-3′-[6-(4-(6-(4-(glc)cum)glc)cum)glc] (**257**), Cy3-glc-7-[6-(cum)glc]-3′-[6-(4-(6-(4-(glc)cum)glc)cum)glc] (**254**), Cy3-[6-(mal)glc]-7-[6-(caf)glc]-3′-[6-(4-(6-(4-(glc)caf)glc)caf)glc] (**258**), Cy3-glc-7-[6-(caf)glc]-3′-[6-(4-(6-(4-(glc)caf)glc)caf)glc] (**256**)	*157* (1995)
Dendrobium (Pompadour)	Purple flowers	Cy3-[6-(mal)glc]7,3′-di[6-(sin)glc], Cy3-glc-7,3′-di[6-(sin)glc]	*181* (2002)
Dendrobium phalaenopsis (Pramot)	Red-purple flowers	Cy3-[6-(mal)glc]7,3′-di[6-(4-(glc)hba)glc] (**253**)	*189* (1994)
Dracula chimaera	Flowers	Cy3-[6-(mal)glc], Cy3-[6-(rha)glc], Cy3-glc, Pn3-[6-(rha)glc], Pn3-[6-(mal)glc]	*330* (2003)
Dracula cordobae × *Laeliocattleya* (minipurple)	Flowers	Cy3-[6-(mal)glc], Cy3-[6-(rha)glc], Cy3-glc, Pn3-[6-(rha)glc], Pn3-[6-(mal)glc]	*330* (2003)

continued

TABLE 10.2
Occurrence of Anthocyanins Reported after 1992 — *continued*

Taxon	Organ	Pigment[a]	References[b]
Laelia pumila			
Cattleya walkeriana	Flowers	Cy3-[6-(mal)glc]-7-[6-(cum)glc]-3′-[6-(4-(6-(caf)glc)cum)glc] (**245**), Cy3-[6-(mal)glc]-7-[6-(cum)glc]-3′-[6-(4-(6-(caf)glc)caf)glc] (**247**), Cy3-[6-(mal)glc]-7-[6-(caf)glc]-3′-[6-(4-(6-(caf)glc)caf)glc] (**249**), Cy3-[6-(mal)glc]-7-[6-(fer)glc]-3′-[6-(4-(6-(fer)glc)caf)glc][e], Cy3-[6-(mal)glc]-7-[6-(cum)glc]-3′-[6-(4-(6-(cum)glc)cum)glc] (**244**)	*155* (1994) *156* (1996)
Phalaenopsis equestris			
P. intermedia			
P. leucorrhoda			
P. sanderiana,			
P. schilleriana	Flowers	Cy3-[6-(mal)glc]7,3′-di[6-(sin)glc] (**230**), Cy3-glc-7,3′-di[6-(sin)glc] (**222**)	*180* (1997)
Sophronitis coccinea	Flowers	cumCy3-[mal-glc]7,3′-di-glc[e], Cy3,3′-di-glc-7-[6-(caf)glc] (**204**), Cy3-[6-(mal)glc]-7-[6-(caf)glc]-3′-glc (**210**), Cy3-[6-(mal)glc]-7-[6-(fer)glc]-3′-glc (**211**), ferCy3,7,3′-tri-glc[e]	*176* (1998)
Vanda	Flowers	Cy3-[6-(mal)glc]7,3′-di[6-(fer)glc][d], Cy3-[6-(mal)glc]7,3′-di[6-(sin)glc], Cy3-glc-7,3′-di[6-(sin)glc], Dp3-[mal-glc]7,3′-di[fer-glc][d, g], Dp3-[6-(mal)glc]7,3′-di[6-(sin)glc] (**382**), Dp3-glc-7,3′-di[6-(sin)glc] (**376**), sinferCy3-[mal-glc]7,3′-di-glc[e], sinferDp3-[mal-glc]7,3′-di-glc[d,e]	*182* (2004)
Pontederiaceae			
Eichhornia crassipes	Blue-purple flowers	(6‴-(Dp3-[6″-(glc)glc])) (6″-(Ap7-glc))malonate (**534**), (6‴-(Dp3-[6″-(glc)glc])) (6″-(Lt 7-glc))malonate (**535**)	*9* (1994) *10* (2004)
DICOTYLEDONEAE			
Aceraceae			
Acer platanoides	Leaves	Cy3-[2-(gao)-6-(rha)glc](161), Cy3-[2-(gao)glc] (**139**), Cy3-[2,3di(gao)glc] (**143**), Cy3-[6-(rha)glc], Cy3-glc	*193* (1992) *195* (1999)
Anacardiaceae			
Rhus javanica	Adventitious roots	Riccionidin A	*82* (2000)
Araliaceae			
Aralia cordata	Cultured cells	Pn3-[2-(xyl)gal] (**264**)	*133* (1994)
Fatsia japonica	Berries	Cy3-[2-(xyl)gal]	*134* (1992)
Apocyanaceae			
Catharanthus roseus[g]	Cell suspension	Hi3-[6-(cum)glc] (**493**), Hi3-glc (**492**), Mv3-[6-(cum)glc] (**463**), Mv3-glc, Pt3-[6-(cum)glc], Pt3-glc	*97* (1998) *348* (2003)
Balanophoraceae			
Cynomorium coccineum[g]	Floral tissue	Cy3-glc	*339* (1994)
Begoniaceae			
Begonia	Flowers	Cy3-[2-(xyl)-6-(caf)glc] (**158**), Cy3-[2-(xyl)-6-(Z-caf)glc] (**159**)[g], Cy3-[2-(glc)-6-(cum)glc] (**167**)[g], Cy3-[2-(glc)-6-(Z-cum)glc] (**168**)[g], Cy3-[2-(xyl)-6-(cum)glc] (**156**)[g], Cy3-[2-(xyl)-6-(Z-cum)glc] (**157**)[g]	*116* (1995)

TABLE 10.2
Occurrence of Anthocyanins Reported after 1992 — *continued*

Taxon	Organ	Pigment[a]	References[b]
Bignoniaceae			
Arrabidaea chica	Leaves	6-OH-5-Me-Ap (**517**), 6-OH-5-Me-Lt (**520**), 5,4′-Me-6-OH (**521**)	*86* (2001) *87* (2002)
Boraginaceae			
Lobostemon	Flowers	Cy3,5-di-glc, Cy3-[6-(rha)glc], Cy3-glc, Dp3,5-di-glc, Dp3-[6-(rha)glc], Dp3-glc	*299* (1997)
Burseraceae			
Dacryodes edulis[g]	Skin, pulp	Cy3-glc/gal, Pt3-glc/gal, Pn3-glc/gal	*349* (2003)
Campanulaceae			
Campanula isophylla			
C. carpatica			
C. poskarshyana[f]	Flowers	Dp3-[6-(rha)glc]-7-[6-(4-(6-(4-(glc)hba)glc)hba)glc] (**404**), Dp3-[6-(rha)glc]-7-[6-(4-(6-(hba)glc)hba)glc] (**387**), Dp3-[6-(rha)glc]-7-glc (**318**), Dp3-[6-(rha)glc]-7-[6-(4-(6-(4-(6-(hba)glc)hba)glc)hba)glc]	*117* (1993)
Caprifoliaceae			
Lonicera caerulea	Fruits	Cy3,5-di-glc, Cy3-[6-(rha)glc], Cy3-glc, Pg3-glc, Pn3-[6-(rha)glc], Pn3-glc	*350* (2004)
Sambucus canadensis	Flowers	Cy3-[6-(Z-cum)-2-(xyl)glc]-5-glc (**196**), Cy3-[6-(cum)-2-(xyl)glc]-5-glc, Cy3-[2-(xyl)glc], Cy3-[2-(xyl)glc]-5-glc, Cy3,5-di-glc, Cy3-glc	*125* (1995)
Caryophyllaceae			
Dianthus caryophyllus	Flowers	Cy3,5-glc(6″,6‴-mal diester) (**152**), Cy3-[6-(mly)glc]-5-glc, Pg3,5-glc(6″,6‴-mal diester) (**38**), Dp3,5-glc(6″,6‴-mal diester) (**339**)[h], Dp3-[6-(mly)glc]-5-glc (**340**)[h], Dp3,5-di[6-(mly)glc] (**347**)[h]	*214* (1998) *215* (2000) *216* (2000) *31* (2003)
Compositae (=Asteraceae)			
Cichorium intybus	Flowers	Dp3,5-di[6-(mal)glc] (**346**), Dp3-[6-(mal)glc]-5-glc (**337**), Dp3glc5-[6-(mal)glc], Dp3,5-di-glc	*351* (2002)
Dendranthema grandiflorum	Purple-red flowers	Cy3-[3,6-di(mal)glc]	*206* (1997)
Felicia amelloides	Flowers	Dp3-[2-(rha)glc]-7-[6-(mal)glc] (**357**)	*139* (1999)
Gynura aurantiaca cv.	Leaves	Cy3-[6-(mal)glc]-7-[6-(4-(6-(caf)glc)caf)glc]-3′-[6-(caf)glc] (**248**)	*248* (1994)
Helianthus annuus[g]	Purple sunflower seeds	Cy3-ara, Cy3-glc, Cy3-xyl, Cy3-[di(mal)-glc], Cy3-[di(mal)-xyl][d], Cy3-[mal-ara], Cy3-[mal-glc], Cy3-[mal-xyl]	*197* (1994)
Lactuca sativa	Leaves	Cy3-[6-(mal)glc]	*352* (1996)
Senecio cruentus	Flowers	Cy3-[6-(mal)glc]-3′-[6-(caf)glc] (**186**), Cy3-glc-3′-[6-(caf)glc] (**172**), Cy3,3′-glc, Pg3-[6-(mal)glc], Pg3-[6-(mal)glc]-7-[6-(4-(6-(caf)glc)caf)glc] (**85**), Pg3-[6-(mal)glc]-7-[6-(caf)glc] (**53**)	*353* (1993) *354* (1995)
Evolvulus pilosus	Blue flowers	Dp3-[6-(4-(6-(3-(glc)caf)glc)caf)glc]-5-[6-(mal)glc] (**397**), Dp3,5-di-glc, Dp3-[6-(4-(6-(3-(glc)caf)glc)caf)glc]-5-glc (**392**)	*313* (1994) *223* (1996)

continued

TABLE 10.2
Occurrence of Anthocyanins Reported after 1992 — *continued*

Taxon	Organ	Pigment[a]	References[b]
Convolvulaceae			
Ipomoea asarifolia	Flowers	Cy3-[2-(6-(caf)glc)-6-(4-(6-(hca)glc)caf)glc]-5-glc (**243**), Cy3-[2-(6-(cum)glc)-6-(4-(6-(cum)glc)caf)glc]-5-glc (**241**), Cy3-[2-(6-(caf)glc)-6-(caf)glc)]-5-glc, Cy3-[2-(6-(cum)glc)-6-(caf)glc]-5-glc (**214**)	*355* (1998) *27* (2003)
Ipomoea batatis (*batatas*)	Purple tubers	Cy3-[2-(glc)glc]-5-glc, Cy3-[2-(6-(cum)glc)glc]-5-glc (**199**), Cy3-[6-(caf)-2-(glc)glc]-5-glc (**205**), Pn3-[6-(caf)-2-(glc)glc]-5-glc, Cy3-[2-(6-(hba)glc)-6-(caf)glc]-5-glc (**212**), Cy3-[2-(6-(caf)glc)-6-(caf)glc]-5-glc, Cy3-[2-(6-(fer)glc)-6-(caf)glc]-5-glc (**217**), Pn3-[2-(6-(fer)glc)-6-(caf)glc]-5-glc, Pn3-[2-(6-(caf)glc)-6-(caf)glc]-5-glc, Pn[2-(6-(hba)glc)-6-(caf)glc]-5-glc (**291**)	*357* (1992) *356* (1997) *188* (1999) *35* (2000)
Ipomoea purpurea	Brownish-red flowers	Cy3-[2-(6-(4-(6-(3-(glc)caf)glc)caf)glc)glc] (**236**), Cy3-[2-(glc)glc], Cy3-[2-(6-(caf)glc)glc] (**174**), Cy3-[2-(glc)-6-(caf)glc] (**173**)	*358* (1998)
Ipomoea purpurea	Violet-blue flowers	Cy3-[2-(6-(3-(glc)caf)glc)-6-(4-(6-(caf)glc)caf)glc]-5-glc (**255**), Cy3-[2-(6-(3-(glc)caf)glc)-6-(caf)glc]-5-glc (**235**), Cy3-[2-(6-(caf)glc)-6-(caf)glc]-5-glc (**215**), Cy3-[2-(glc)glc]-5-glc	*305* (1995)
Ipomoea purpurea	Red-purple flowers	Pg3-[2-(6-(3-(glc)caf)glc)-6-(4-(6-(caf)glc)caf)glc]-5-glc (**93**), Pg3-[2-(6-(caf)glc)-6-(4-(6-(caf)glc)caf)glc]-5-glc (**91**), Pg3-[2-(6-(caf)glc)-6-(caf)glc]-5-glc (**76**), Pg3-[2-(6-(3-(glc)caf)glc)-6-(caf)glc]-5-glc, Pg3-[2-(glc)-6-(caf)glc]-5-glc	*306* (1996)
Pharbitis nil/Ipomoea nil	Flowers	Cy3-[2-(glc)-6-(4-(glc)caf)glc]-5-glc (**232**), Cy3-[6-(3-(glc)caf)glc]-5-glc (**206**), Pg3,5-di-glc, Pg3-[2-(6-(3-(glc)caf)glc)-6-(4-(6-(caf)glc)caf)glc]-5-glc, Pg3-[2-(6-(caf)glc)-6-(caf)glc]-5-glc, Pg3-[2-(glc)-6-(caf)glc]-5-glc (**64**), Pg3-[2-(6-(cum)glc)glc]-5-glc (**60**), Pg3-[2-(glc)-6-(4-(glc)caf)glc]-5-glc (**89**), Pg3-[6-(3-(glc)caf)glc] (**45**), Pg3-[6-(3-(glc)caf)glc]-5-glc (**65**), Pg3-[6-(caf)glc] (**28**), Pg3-glc, Pn3,5-di-glc, Pn3-[2-(glc)-2-(6-(glc)caf)glc]-5-glc, Pn3-[2-(glc)glc]-5-glc, Pn3-[6-(3-(glc)caf)glc] (**285**), Pn3-[6-(3-(glc)caf)glc]-5-glc (**290**), Pn3-glc	*312* (1993) *309* (1994) *359* (1996) *310* (2001) *360* (2001)
Coriariaceae			
Coriaria myrtifolia[g]	Fruits	Dp-, Cy-, Pt-, Pn-, Mv3-glc, Dp-, Cy-, Pt-, Pn-, Mv3-gal	*361* (2002)
Cornaceae			
Cornus suecica	Fruits	Cy3-[2-(glc)gal] (**114**), Cy3-[2-(glc)glc], Cy3-glc, Cy3-gal	*23* (1998)
Crassulaceae			
Crassula, Cotyledon, Tylecodon,[f, g] 22 spp.	Flower	Cy3-[2-(glc)glc], Cy3-[2-(xyl)glc], Cy3-glc, Dp3-[2-(xyl)glc], Pn3-glc	*298* (1995)
Cruciferae (Brassicaceae)			
Brassica campestris	Stem	Cy[2-(2-(sin)glc)-6-(fer)glc]-5-[6-(mal)glc] (**229**), Cy[2-(2-(sin)glc)-6-(cum)glc]-5-[6-(mal)glc] (**227**), Cy[2-(2-(sin)glc)-6-(fer)glc]-5-glc, Cy[2-(2-(sin)glc)-6-(cum)glc]-5-glc	*178* (1997)

TABLE 10.2
Occurrence of Anthocyanins Reported after 1992 — *continued*

Taxon	Organ	Pigment[a]	References[b]
Arabidopsis thaliana	Leaves and stems	Cy3-[6-(4-(glc)cum)-2-(2-(sin)xyl)glc]-5-[6-(mal)glc] (**239**)[d]	*167* (2002)
Matthiola incana	Flowers	Cy3-[2-(2-(sin)xyl)-6-(caf)glc]-5-[6-(mal)glc] (**225**), Cy3-[2-(2-(sin)xyl)-6-(cum)glc]-5-[6-(mal)glc] (**224**), Cy3-[2-(2-(sin)xyl)-6-(fer)glc]-5-[6-(mal)glc] (**226**), Cy3-[2-(2-(sin)xyl)-6-(fer)glc]-5-glc (**218**), Pg3-[2-(2-(fer)xyl)-6-(fer)glc]-5-[6-(mal)glc] (**83**), Pg3-[2-(2-(sin)xyl)-6-(cum)glc]-5-[6-(mal)glc] (**84**), Pg3-[2-(2-(sin)xyl)-6-(cum)glc]-5-glc (**75**), Pg3-[2-(xyl)glc]-5-glc, Pg3-[2-(xyl)-6-(cum)glc]-5-[6-(mal)glc] (**68**), Pg3-[2-(2-(sin)xyl)-6-(fer)glc]-5-[6-(mal)glc] (**86**), Pg3-[2-(2-(sin)xyl)-6-(fer)glc]-5-glc (**81**), Pg3-[2-(xyl)-6-(fer)glc]-5-[6-(mal)glc] (**70**), Pg3-glc	*177* (1995) *173* (1996)
Raphanus sativus	Callus	Cy3-[2-(6-(fer)glc)-6-(caf)glc]-5-glc, Pg3-[2-(2-(fer)glc)-6-(fer)glc]-5-glc (**80**), Pg3-[2-(6-(caf)glc)-6-(caf)glc]-5-glc, Pg3-[2-(6-(caf)glc)-6-(cum)glc]-5-glc (**73**), Pg3-[2-(6-(caf)glc)-6-(fer)glc]-5-glc (**77**), Pg3-[2-(6-(fer)glc)-6-(caf)glc]-5-glc (**78**), Pg3-[2-(6-(fer)glc)-6-(cum)glc]-5-glc (**74**), Pg3-[2-(6-(fer)glc)-6-(fer)glc]-5-glc (**79**), Pg3-[2-(6-(fer)glc)glc]-5-glc (**67**), Pg3-[2-(glc)-6-(caf)glc]-5-glc, Pg3-[2-(glc)-6-(cum)glc]-5-[6-(mal)glc] (**71**), Pg3-[2-(glc)-6-(cum)glc]-5-glc (**61**), Pg3-[2-(glc)-6-(fer)glc]-5-[6-(mal)glc] (**72**), Pg3-[2-(glc)-6-(fer)glc]-5-glc (**66**)	*315* (1998) *314* (2002)
Ericaceae			
Vaccinium padifolium	Berries	Cy3-[6-(rha)-2-(xyl)glc], Dp3-rha, Mv3-[2-(xyl)glc] (**451**), Mv3-[6-(rha)glc], Pn3-[2-(xyl)glc], Pn3-[6-(rha)-2-(xyl)glc] (**274**), Pt3-[2-(xyl)glc] (**423**), Pt3-[6-(rha)-2-(xyl)glc] (**430**)	*124* (1999) *25* (2000)
Euphorbiaceae			
Acalypha hispida	Flowers	Cy3-[(2-(gao)-6-(rha)gal] (**160**)[d], Cy3-[2-(gao)gal], Cy3-gal	*130* (2003)
Gentianaceae			
Eustoma grandiflorum, after genetical transformation	Flowers	Cy3-[6-(rha)gal]-5-[6-(cum)glc] (**180**), Cy3-gal-5-[6-(cum)glc] (**162**), Dp3-[6-(rha)gal]-5-glc (**316**), Dp3-[6-(rha)gal]-5-[6-(cum)glc] (**361**), Dp3-[6-(rha)gal]-5-[6-(Z-cum)glc] (**362**), Dp3-[6-(rha)glc]-5-[6-(cum)glc] (**363**), Dp3-[6-(rha)glc]-5-[6-(Z-cum)glc] (**364**), Dp3-[6-(rha)gal]-5-[6-(fer)glc] (**367**), Dp3-[6-(rha)gal]-5-[6-(Z-fer)glc] (**368**), Dp3-gal-5-[6-(cum)glc] (**342**), Dp3-gal-5-[6-(Z-cum)glc] (**343**), Dp3-glc-5-[6-(cum)glc]	*129* (1993) *123* (1996)
Gentiana	Blue flowers	Dp3,3′-di-glc-5-[6-(caf)glc] (**369**), Dp3,3′-di-glc-5-[6-(cum)glc] (**366**), Dp3-glc-5-,3′-di[6-(caf)glc], Dp3-glc-5-[6-(caf)glc]-3′-[6-(cum)glc] (**373**), Dp3-glc-5-[6-(cum)glc] (**344**), Dp3-glc-5-[6-(cum)glc]-3′-[6-(caf)glc] (**374**)	*154* (1997)
Gentiana	Pink flowers	Cy3-glc, Cy3-glc-5,3′-di[6-(caf)glc] (**216**), Cy3-glc-5-[6-(caf)glc] (**171**), Cy3-glc-5-[6-(cum)glc] (**165**)	*153* (1995)

continued

TABLE 10.2

Occurrence of Anthocyanins Reported after 1992 — *continued*

Taxon	Organ	Pigment[a]	References[b]
Geraniaceae			
Geranium pratense			
G. sanguineum			
G. Johnson's Blue	Flowers	Mv3,5-di-glc, Mv3-glc, Mv3-glc-5-[6-(ace)glc] (**466**), Mv5-glc	*211* (1997)
Geranium sylvaticum	Flowers	Cy3,5-di-glc, Cy3-glc, Dp3-glc, Mv3,5-di-glc, Mv3-[6-(ace)glc]-5-glc (**465**)	*210* (1995)
Pelargonium domesticum ('Dubonnet')	Flowers	Cy3-glc-5-[6-(ace)glc] (**147**), Cy3,5-di-glc, Dp3-glc-5-[6-(ace)glc] (**334**), Dp3,5-di-glc, Mv3,5-di-glc, Mv3-glc-5-[6-(ace)glc], Pg3-glc-5-[6-(ace)glc], Pg3,5-di-glc, Pn3-glc-5-[6-(ace)glc] (**282**), Pn3,5-di-glc, Pt3-glc-5-[6-(ace)glc] (**437**), Pt3,5-di-glc	*317* (1998)
Grossulariaceae			
Ribes nigrum	Berries	Cy3-[6-(rha)glc], Dp3-glc (pyranocyanidin A, pyranocyanidin B, pyranodelphinin A, and pyranodelphinin B)[d, i]	*88* (2000) *89* (2001)
		Cy3-[6-(cum)glc], Cy3-[ara], Cy3-glc, Dp3-[6-(cum)glc], Dp3-[6-(rha)glc], Mv3-[6-(rha)glc], Mv3-glc, Pg3-[6-(rha)glc], Pg3-glc, Pn3-[6-(rha)glc], Pn3-glc, Pt3-[6-(rha)glc], Pt3-glc (pyranocyanidin C, pyranocyanidin D, pyranodelphinin C, and pyranodelphinin D)[d]	*365* (2002) *91* (2002)
Labiatae (=Lamiaceae)			
Ajuga reptans	Flowers, and cell cultures	Dp3-[2-(6-(cum)glc)-6-(cum)glc]-5-[6-(mal)glc] (**379**), Cy3-[2-(6-(cum)glc)-6-(cum)glc]-5-[6-(mal)glc] (**223**), Dp3-[2-(glc)glc]-5-glc, Cy3-[2-(glc)glc]-5-glc, Dp3-[2-(6-(fer)glc)-6-(cum)glc]-5-[6-(mal)glc] (**380**)[d], Dp3-[2-(6-(fer)glc)-6-(fer)glc]-5-[6-(mal)glc] (**381**), Cy3-[2-(6-(cum)glc)-6-(cum)glc]-5-glc (**213**), Dp3-[di-fer(2glc-glc)]-5-glc[e], Cy3-[fer-cum(2glc-glc)]-5-[mal-glc][e]	*113* (1996) *114* (2001)
Lamium, Salvia, Thymus, 49 spp.	Flowers	Cy3-[6-(cum)glc]-5-[4-(mal)-6-(mal)glc] (**354**)	*200* (1992)
Ocimum basilicum[g] (Dark Opal, Holy Sacred Red, Opal, Osmin Purple, Purple Bush, Purple Ruffles, Red Rubin, Rubin)	Flowers and leaves	Cy3,5-di-glc, Cy3-glc, Cy3-[cum-glc], Cy3-[cum-glc]-5-glc, Pn3,5-di-glc, Pn3-[cum-glc]-5-glc	*320* (1998)
Salvia patens	Flowers	Dp3-[6-(cum)glc]-5-[6-(mal)glc] + apigenin7,4′-di-glc + Mg (protodelphin)	*226* (1994)
Salvia uliginosa	Flowers	Dp3-[6-(cum)glc]-5-[4-(ace)-6-(mal)glc] (**353**)	*209* (1999)
Leguminosae (=Fabaceae)			
Amphithalea, Coelidium, Hypocalyptus, Liparia,[f, g] 10 spp.	Flowers	Cy3-glc, Cy3-[6-(ace)glc], Cy3-[6-(cum)glc], Mv3-glc, Pn3-glc, Cy3-[2-(glc)glc], Pg3-[2-(glc)glc]	*326* (1995)

TABLE 10.2
Occurrence of Anthocyanins Reported after 1992 — *continued*

Taxon	Organ	Pigment[a]	References[b]
Cassia auriculata	Heartwood	Pg5-gal[d, e]	*362* (1994)
Clitoria ternatea		Dp3-[2-(rha)-6-(mal)glc] (**337**), Dp3-[2-(rha)glc],	*144* (1996)
		Dp3-[6-(mal)glc], Dp3-[6-(mal)glc]-3′-[6-(4-(6-	*145* (1998)
		(cum)glc)cum)glc]5′-[6-(cum)glc] (**409**), Dp3-[6-	*141* (2003)
		(mal)glc]-3′-[6-(4-(6-(4-(glc)cum)glc)cum)glc]5′-[6-(4-	
		(glc)cum)glc] (**415**), Dp3-[6-(mal)glc]-3′-[6-(4-(6-(4-	
		(glc)cum)glc)cum)glc]5′-glc (**407**), Dp3-[6-(mal)glc]-	
		3′-[6-(4-(6-(cum)glc)cum)glc]5′-[6-(4-(glc)cum)glc]	
		(**411**), Dp3-[6-(mal)glc]-3′-[6-(4-(6-(4-	
		(glc)cum)glc)cum)glc]5′-[6-(cum)glc] (**410**), Dp3-[6-	
		(mal)glc]-3′-[6-(4-(6-(cum)glc)cum)glc]5′-glc (**395**),	
		Dp3-[6-(mal)glc]-3′-[6-(4-(glc)cum)glc]5′-[6-(cum)glc]	
		(**396**), Dp3-[6-(mal)glc]-3′-[6-(4-(glc)cum)glc]5′-glc	
		(**388**), Dp3-[6-(mal)glc]-3′,5′-di-[6-(cum)glc] (**378**),	
		Dp3-[6-(mal)glc]-3′-[6-(cum)glc]5′-glc (**371**),	
		Dp3-[6-(mal)glc]-3′,5-di-glc (**358**), Dp3-glc,	
		Dp3-glc-3′,5′-di-[6-(4-(glc)cum)glc] (**405**),	
		Dp3-glc-3′-[6-(4-(glc)cum)glc]5′-glc (**386**)	
Glycine max	Seed coats	Cy3-glc, Dp3-glc, Pt3-glc	*363* (2001)
Lupinus (Russell hybrids)	Blue flowers	(Dp3-[6-(mal)glc]apigenin7-[6-(mal)glc]malonic residue + Fe)[d], (Dp3-[6-(mal)glc]luteolin7-[6-(mal)glc]malonic residue + Fe)[d]	*225* (1993)
Lupinus (Russell hybrids)	Pink flowers	Cy3-[6-(mal)glc], Pg3-[6-(mal)glc]	*225* (1993)
Phaseolus coccinesu	Seed coats	Dp3-glc	*322* (1996)
Phaseolus lunatus	Seed coats	Pn3-glc, Pn3-[6-(rha)glc]	*322* (1996)
Phaseolus vulgaris[g]	Seed coats	Cy3,5-di-glc, Cy3-glc, Dp3-glc, Pg3-glc, Pt3-glc, Pt3,5-di-glc, Mv3-glc, Mv3,5-di-glc	*323* (1997) *363* (2001)
Pisum spp.	Purple pod	Dp3-[2-(xyl)gal]-5-glc (**314**), Dp3-[2-(xyl)gal]-5-[6-(ace)glc] (**356**)	*131* (2000)
Podalyria,[g] 7 spp.	Flowers	Cy3-glc, Cy3-[6-(cum)glc]	*325* (1994)
Vicia villosa	Blue flowers	Dp3-[6-(rha)glc]-5-glc, Pt3-[6-(rha)glc]-5-glc, Mv3-[6″rha)glc]-5-glc	*321* (1998)
Vigna angularis	Grains	Cy	*322* (1996)
Vigna subterranean	Grains	Cy3-glc, Mv3-glc, Pt3-glc	*324* (1997)
Virgilia,[g] 2 spp.	Flowers	Cy3-glc, Pn3-glc, Cy3-[6-(ace)glc], Pn3-[6-(ace)glc], Cy3-[6-(cum)glc]	*325* (1994)
Linaceae			
Linum grandiflorum	Flowers	Cy3-[6-(rha)glc], Dp3-[2-(xyl)-6-(rha)glc] (**312**), Dp3-[6-(rha)glc]	*24* (1995)
Lobeliaceae			
Lobelia erinus	Flowers	Cy3-[6-(4-(*Z/E*-cum)rha)glc]-5-[6-(mal)glc]-3′-[6-(caf)glc] (**238**)	*118* (1995) *365* (1996)
Malvaceae			
Lavatera maritima[g]	Flowers	Mv3-[6-(mal)glc]-5-glc (**468**), Mv3-[6-(mal)glc] (**462**)	*339* (1994)
Melastomataceae			
Tibouchina urvilleana	Flowers	Mv3-[6-(cum)glc]-5-[2-(ace)xyl] (**471**)	*208* (1993)

continued

TABLE 10.2
Occurrence of Anthocyanins Reported after 1992 — *continued*

Taxon	Organ	Pigment[a]	References[b]
Myrtaceae			
Eugenia umbelliflora[g]	Berries	Cy3-glc, Dp3-glc, Mv3-glc, Pg3-glc, Pn3-glc, Pt3-glc	*366* (2003)
Nymphaéaceae			
Nymphaéa alba	Leaves	Cy3-[2-(gao)-6-(ace)gal], Cy3-[6-(ace)gal] (**129**), Cy3-gal, Dp3-[2-(gao)-6-(ace)gal], Dp3-[2-(gao)gal], Dp3-[6-(ace)gal], Dp3-gal	*192* (2001)
Nymphaéa caerulea (=*N. capensis*)	Flowers	Dp3'-[2-(gao)gal] (**325**), Dp3'-[2-(gao)-6-(ace)gal] (**331**)	*26* (1999)
Nymphaéa marliacea[f]	Leaves	Cy3-[2-(gao)-6-(ace)gal] (**142**), Dp3-gal, Dp3-[6-(ace)gal] (**322**), Dp3-[2''''(gao)gal]	*367* (1997)
	Flowers	Dp3-[2-(gao)-6-(ace)gal] (**330**)	*191* (1998)
Oxalidaceae			
Oxalis triangularis[g]	Leaves	Mv3-[6-(rha)glc]-5-glc, Mv3-mal-[6-(rha)glc]-5-glc, Mv3di-mal-[6-(rha)glc]-5-glc	*368* (2001)
Papaveraceae			
Meconopsis grandis,[g] *M. horridula,* and *M. betonicifolia*	Flowers	Cy3-[2-(xyl)-6-(mal)glc]-7-glc (**190**)	*127* (1996) *128* (2001)
Passifloraceae			
Passiflora edulis	Fruit	Cy3-[6-(mal)glc], Cy3-glc, Pg3-glc	*369* (1997)
Passiflora suberosa	Fruit	Cy3-[6-(mal)glc], Cy3-glc, Dp3-[6-(mal)glc], Dp3-glc, Pg3-[6-(mal)glc], Pg3-glc, Pt3-[6-(mal)glc], Pt3-glc	*369* (1997)
Primulaceae			
Cyclamen persicum (Bonfire)	Flowers	Pn3-[2-(rha)glc] (**268**)[d]	*140* (1999)
Cyclamen persicum (Sierra Rose)	Flowers	Pn3,5-di-glc, Cy3,5-di-glc, Mv3,5-di-glc	*140* (1999)
Ranunculaceae			
Aconitum chinense	Flowers	Dp3-[6-(rha)glc]-7-[6-(4-(6-(hba)glc)hba)glc]	*184* (1994)
Anemone coronaria	Flowers	Cy3-[2-(2-(caf)glc)-6-(3-(2-tar)mal)gal]-7-[6-(caf)glc]-3'-glu (**246**), Cy3-[2-(2-(caf)glc)-6-(mal)glc]-7-[6-(caf)glc]-3'-glu (**240**), Cy3-[2-(2-(caf)glc)glc]-7-[6-(caf)glc]-3'-glu (**234**), Dp3-[2-(2-(caf)glc)gal-(3-(2-tar)mal)gal]-7-[6-(caf)glc] (**383**), Dp3-[2-(2-(caf)glc)-6-(3-(2-tar)mal)gal]-7-[6-(caf)glc]-3'-glu (**401**), Dp3-[2-(2-(caf)glc)-6-(mal)gal]-7-[6-(caf)glc]-3'-glu (**398**), Dp3-[2-(2-(caf)glc)gal]-7-[6-(caf)glc]-3'-glu (**393**), Pg3-[2-(xyl)-6-(mal)gal] (**32**), Pg3-[2-(xyl)-6-(Me-mal)gal] (**33**), Pg3-[2-(xyl)gal], Pg3-[2-(xyl)-6-(3-(3-(4-(glc)caf)2-tar)mal)gal] (**82**)	*132* (2001) *28* (2002) *29* (2003)
Consolida armeniaca	Flowers	Dp3-[6-(mal)glc]-7-[6-(4-(6-(hba)glc)hba)glc] (**377**)[d], Dp3-[6-(mal)glc]-7-[2-(glc)-6-(4-(6-(hba)glc)hba)glc] (**394**)[d], Dp3-[6-(mal)glc]-7-[2-(6-(hba)glc)-6-(4-(6-(hba)glc)hba)glc] (**400**)[d], Dp3-[6-(mal)glc]-7-[glc-2-(6-(hba)glc)-6-(4-(6-(hba)glc)hba)glc][e]	*115* (1996)
Delphinium hybridum	Flowers	Dp3-[6-(rha)glc], Dp3-[6-(rha)glc]-7-glc, Pg3,7-di-glc, Pg3-[6-(mal)glc]-7-[6-(4-(glc)hba)glc] (**69**), Pg3-[6-(mal)glc]-7-glc (**36**), Pg3-[6-(rha)glc]-7-[6-(4-(glc)hba)glc] (**87**), Pg3-[6-(rha)glc]-7-[6-(hba)glc] (**57**), Pg3-[6-(rha)glc]-7-glc (**21**), Pg3-glc-7-[6-(4-(glc)hba)glc] (**58**)	*119* (1998) *331* (1999)

TABLE 10.2
Occurrence of Anthocyanins Reported after 1992 — *continued*

Taxon	Organ	Pigment[a]	References[b]
Pulsatilla cernua	Flowers	Pg3-[2-(2-(caf)glc)gal] (**44**)	*22* (1998)
Ranunculus asiaticus	Flower	Cy3-[2-(xyl)-6-(mal)glc] (**148**), Cy3-[2-(xyl)glc], Dp3-[2-(xyl)-6-(mal)glc] (**335**), Dp3-[2-(xyl)glc]	*126* (1996)
Rhamnaceae			
Ceanothus papillosus	Flowers	Dp3-[6-(rha)glc]7,3′-di[6-(cum)glc] (**389**), Dp3-[6-(rha)glc]-7-[6-(cum)glc]-3′-glc (**384**)	*32* (1997)
Rosaceae			
Fragaria × *ananassa*	Berries	A5-CarboxypyranoPg3-glc (**523**), Pg3-glc Afzelechin(4α → 8)Pg3-glc (**530**), Epiafzelechin (4α → 8)Pg3-glc (**531**), Catechin (4α → 8)Pg3-glc (**532**), Epicatechin (4α → 8)Pg3-glc (**533**)	*19* (2004) *13* (2004)
Prunus cerasus (Balaton, Montmorency)	Fruit	Cy3-[2-(glc)-6-(rha)glc], Cy3-[6-(rha)glc], Cy3-glc	*370* (1997)
Rosa (*Cinnamomeae*, *Chinenses*, *Gallicanae*), 44 spp.	Flowers	Cy3,5-di-glc, Cy3-[2-(glc)glc], Cy3-[6-(cum)glc], Cy3-[6-(rha)glc], Cy3-glc, Pg3,5-di-glc, Pg3-glc, Pn3,5-di-glc, Pn3-[6-(cum)glc], Pn3-[6-(rha)glc]	*300* (1995) *301* (2000)
Rosa hybrida	Flower	Rosacyanin B (**526**)	*17* (2002)
Rubus iaciniatus	Berries	Cy3-[6-di-(oxa)glc][d, g]	*57* (2002)
Rubiaceae			
Cephaelis subcoriacea[g]	Fruits	Cy3-glc	*339* (1994)
Rutaceae			
Citrus sinensis (Florida)[g]	Juice	Cy3-[6-(mal)glc], Cy3-glc	*371* (2002)
Sapindaceae			
Litchi chinensis Sonn.	Pericarp	Cy3-[6-(rha)glc], Cy3-glc, Cy3-gal, Pg3,7-di-glc	*372* (1993) *373* (2000) *374* (2004)
Scrophulariaceae			
Mimulus cardinalis[g]	Flowers	Cy3-glc, Pg3-glc	*375* (1997)
Mimulus lewisii[g]	Flowers	Cy3-glc, Pg3-glc	*375* (1997)
Solanaceae			
Petunia[f] (Baccara, Carpet, Celebrity, Fantasy, Fulcon, Madness, Prime Time), 17 spp.	Pink flowers	Pn3-[6-(4-(4-(6-(caf)glc)cum)rha)glc]-5-glc (**296**), Pn3-[6-(4-(4-(glc)cum)rha)glc]-5-glc (**295**), Pn3-[cum-6-(rha)glc]-5-glc, Pn3caf[6-(rha)glc]-5-glc, Pn3-[6-(rha)glc]-5-glc	*165* (2004)
Petunia "Mitchell" (*P. axillaries* × *P. hybrida*)[g]	Leaves	Pt3-[cum-6-(rha)glc]-5-glc, Pt3-[6-(rha)glc]-5-glc, Pt3-[caf-6-(rha)glc]-5-glc	*376* (1998)
Petunia exserta	Flowers	Cy3-[6-(rha)glc], Cy3-glc, Pg3-[6-(rha)glc], Pg3-glc	*377* (1999)
Petunia guarapuavensis	Flowers	Mv3-[6-(4-(4-(6-(caf)glc)cum)rha)glc]-5-glc (**482**), Mv3-[6-(4-(4-(6-(caf)glc)caf)rha)glc]-5-glc (**484**)	*162* (1997)
Petunia hybrida	Flowers	Mv3-[6-(4-(4-(6-(caf)glc)caf)rha)glc]-5-glc, Mv3-[6-(4-(4-(6-(caf)glc)cum)rha)glc]-5-glc, Mv3-[6-(4-(4-(6-(cum)glc)cum)rha)glc]-5-glc (**481**), Mv3-[6-(4-(4-(6-(fer)glc)cum)rha)glc]-5-glc (**483**), Mv3-[6-(4-(caf)rha]-5-glc (**478**), Mv3-[6-(4-(*Z*-cum)rha)glc]-5-glc (**477**), Pt3-[6-(4-(4-(6-(caf)glc)cum)rha)glc]-5-glc (**445**)	*159* (1998) *168* (1999) *161* (2001)

continued

TABLE 10.2
Occurrence of Anthocyanins Reported after 1992 — *continued*

Taxon	Organ	Pigment[a]	References[b]
Petunia integrifolia subsp. *inflata*	Strains	Mv3-[6-(4-(4-(6-(caf)glc)cum)rha)glc] (**480**), Mv3-[6-(4-(caf)rha)glc], Mv3-[6-(4-(cum)rha)glc]	*163* (1999)
Petunia occidentalis	Flowers	Dp3-[6-(4-(caf)rha)glc]-5-glc (**365**), Mv3-[caf-glc-rut][e], Pt3-[6-(4-(4-(glc)cum)rha)glc]-5-glc (**444**), Pt3-[caf-glc-caf-rut]-5-glc[e], Pt3-[caf-glc-cum-rut]-5-glc (**445**)[e], Pt3-[caf-glc-rut][e], Pt3-[cum-glc-rut]-5-glc[e]	*160* (1999)
Petunia reitzii	Flowers	Dp3-[6-(4-(4-(6-(caf)glc)cum)rha)glc]-5-glc (**390**), Dp3-[6-(4-(4-(glc)cum)rha)glc]-5-glc (**385**), Dp3-[6-(rha)-2-(caf)glc]-5-glc, Dp3-[6-(rha)-2-(cum)glc]-5-glc, Dp3-[6-(rha)-2-(*Z*-cum)glc]-5-glc, Dp3-[6-(rha)glc], Dp3-[6-(rha)glc]-5-glc, Pt3-[6-(rha)glc]-5-glc, Pt3-[6-(rha)-2-(caf)glc]-5-glc, Pt3-[6-(rha)-2-(cum)glc]-5-glc, Pt3-[6-(rha)-2-(*Z*-cum)glc]-5-glc	*164* (2000)
Solanum melongena	Skin	Dp3-[6-(4-(cum)rha)glc]-5-glc, Dp3-[6-(4-(*Z*-cum)rha)glc]-5-glc[d]	*378* (2001)
Solanum stenotomum[g]	Tubers	Dp3-[cum-6-(rha)glc]-5-glc, Mv3-[cum-6-(rha)glc]-5-glc, Mv3-[fer-6-(rha)glc]-5-glc, Pn3-[caf-6-(rha)glc]-5-glc, Pn3-[cum-6-(rha)glc]-5-glc, Pn3-[fer-6-(rha)glc]-5-glc, Pt3-[caf-6-(rha)glc]-5-glc, Pt3-[cum-6-(rha)glc], Pt3-[cum-6-(rha)glc]-5-glc, Pt3-[*Z*-cum-6-(rha)glc]-7-glc[e], Pt3-[fer-6-(rha)glc]-5-glc	*379* (2003)
Solanum andigena × *Solanum tuberosum*	Tubers	Pg3-[6-(4-(fer)rha)glc]-5-glc (**62**), Pg3-[6-(4-(cum)rha)glc]-5-glc	*171* (1998)
Solanum tuberosum (Congo)	Tubers and shoots	Mv3-[6-(4-(fer)rha)glc]-5-glc (**479**), Pt3-[6-(4-(fer)rha)glc]-5-glc (**443**)	*172* (2000)
Solanum tuberosum	Tubers	Pn3-[6-(4-(caf)rha)glc]-5-glc (**288**), Pn3-[6-(4-(cum)rha)glc]-5-glc, Pt3-[6-(4-(caf)rha)glc]-5-glc (**442**), Pt3-[6-(4-(cum)rha)glc]-5-glc (**441**)	*169* (2003)
Theaceae			
Camellia sinensis	Leaves	Cy3-gal, Dp3-[6-(cum)gal] (**326**), Dp3-gal	*158* (2001)
Visnea mocanera[g]	Fruits	Cy3-gal, Cy3-glc, Dp3-glc, Mv3-glc, Pn3-glc, Pt3-glc	*380* (1996)
Umbelliferae (=Apiaceae)			
Daucus carota (Nentes scarlet-104)	Cell culture	Cy3-[2-(xyl)gal], Cy3-glc	*381* (2000)
Glehnia littoralis	Petiole-derived callus cultures	Cy3-[2-(xyl)-6-(6-(fer)glc)glc] (**203**)	*34* (1998)
Verbenaceae			
Verbena hybrida	Flowers	Cy3,5-di[6-(ace)glc] (**181**), Cy3-[6-(mal)glc], Pg3,5-di[6-(ace)glc] (**47**), Pg3,5-di-glc, Pg3-[6-(ace)glc], Pg3-[6-(ace)glc]-5-glc, Pg3-[6-(mal)glc], Pg3-[6-(mal)glc]-5-[6-(ace)glc] (**48**), Pg3-glc-5-[6-(ace)glc] (**31**)	*382* (1991) *383* (1995) *384* (1995)

TABLE 10.2
Occurrence of Anthocyanins Reported after 1992 — _continued_

Taxon	Organ	Pigment[a]	References[b]
Vitaceae			
Vitis vinifera [g]	Berries	Dp-, Cy-, Pt3-[6-(ace)glc], Pn3-[6-(ace)glc] (**278**), Mv3-[6-(ace)glc] (**461**), Dp3-[6-(cum)glc] (**327**), Cy-, Pt-, Pn-, Mv3-[6-(cum)glc], Dp-, Cy-, Pt-, Pn-, Mv3-glc, Dp-, Pt3,5-di-glc, Pn3-[6-(caf)glc] (**281**), Mv3-[6-(caf)glc]	_385_ (1995) _386_ (2004)

Notes: ace, acetic acid; oxa, oxalic acid; mal, malonic acid; suc, succinic acid; mly, malic acid; hba, _p_-OH-benzoic acid; gao, gallic (tri-OH-benzoyl) acid; cum, _p_-coumaric acid; caf, caffeic acid; fer, ferulic acid; sin, sinapic acid; hca, 3,5-diOHcinnamic acid; tar, tartaric acid; ara, arabinose; xyl, xylose; rha, rhamnose; gal, galactose; glc, glucose; glu, glucuronic acid; 2-(xyl)glc, sambubiose; 2-(xyl)gal, lathyrose; 2-(rha)glc, neohesperidose; 6-(rha)gal, robinose; 6-(rha)glc, rutinose; 2-(glc)glc, sophorose; 3-(glc)glc, laminariobiose; 6-(glc)glc, gentiobiose.

[a]See Table 10.1 for anthocyanidin abbreviations and linkage positions.
[b]Numbers in brackets refer to the year of the publication.
[c]Numbers in bold represent the first report of the pigments not listed in previous editions of _The Flavonoids_.[1–4] See Appendix A.
[d]Possibly new anthocyanins.
[e]Pigments assigned tentatively.
[f]Each genus involved may include one or more samples (species or plant organs). Each sample contains one or more of the listed anthocyanins.
[g]Structures are based on TLC and HPLC or MS data.
[h]Produced by genetically modified violet carnations.
[i]Produced during experimental workup.

new anthocyanins based on rutinose have also been reported from _Campanula_ (Campanulaceae),[117] _Lobelia_ (Lobeliaceae),[118] _Ceanothus_ (Rhamnaceae),[32] _Delphinium_ (Ranunculaceae),[119] _Lilium_,[120] _Tulipa_ (Liliaceae),[121,122] and _Alstroemeria_ (Alstroemeriaceae).[106] It is interesting to observe that a transgenic _Eustoma_ (Gentianaceae) line produced by insertion of an _Antirrhinum majus_ cDNA coding for UDP-glucosyl:flavonoid-3-_O_-glucosyltransferase contained in its petals significant levels of 3-rutinoside and 3-glucoside derivatives of delphinidin, in which glucose has replaced the 3-_O_-linked galactose present in the original anthocyanins of the nontransformed plants.[123] After 1992, sambubiose units have mainly been found in new anthocyanins from _Matthiola_ and _Arabidopsis_ (Cruciferae) (Table 10.2), and in some new anthocyanins from _Begonia_ (Begoniaceae),[116] _Vaccinium_ (Ericaceae),[124] _Sambucus_ (Caprifoliaceae),[125] _Ranunculus_ (Ranunculaceae),[126] and _Meconopsis_ (Papaveraceae).[127,128]

Among the disaccharides with more limited occurrence in anthocyanins, new reports on robinobiose include six 3-[6-(rhamnosyl)galactosides-5-glycosides of delphinidin and cyanidin isolated from purple _Lisianthus_ (Gentianaceae) flowers,[129] in addition to cyanidin 3-[2-(galloyl)-6-(rhamnosyl)galactoside] from red flowers of the chenille plant _Acalypha hispida_ (Euphorbiaceae).[130] Two new anthocyanins containing lathyrose, delphinidin 3-[2-(xylosyl)-galactoside]-5-[6-(acetyl)glucoside] and its deacetylated derivative, have been isolated from purple pods of pea (_Pisum_ spp.).[131] Both pigments showed moderate stability and anti-oxidative activity in a neutral aqueous solution. Three anthocyanins with a pelargonidin 3-lathyroside skeleton acylated with malonic acid have been isolated from scarlet flowers of _Anemone coronaria_.[132] A minor anthocyanin accumulated in the cultured cells of _Aralia cordata_ (Araliaceae) has been identified as peonidin 3-lathyroside.[133] This disaccharide has previously been isolated mainly from species belonging to Umbelliferae[4] and Araliaceae.[134] Laminariobiose seems to have an even more restricted occurrence in anthocyanins, identified in anthocyanins isolated mainly from the genus *Allium*.[18,135] Quite extraordinary, this disaccharide has been found linked to the 5-position of the major 3-deoxyanthocyanin isolated from the fern *Blechnum novae-zelandiae*.[96]

The structure of the major anthocyanin tradescantin in *Tradescantia pallida* has been determined to be cyanidin 3-[6-(2,5-di-(ferulyl)arabinosyl)glucoside]-7,3′-di-[6-(ferulyl)glucoside].[136] The same disaccharide, 6-arabinofuranosyl-glucopyranose, has previously been found in zebrinin isolated from *Zebrina pendula*,[137] which belongs to the same family (Commelinaceae) as *Tradescantia*. Tradescantin displays a very high stability,[138] most probably because of its structural conformation, which allows sandwich-type complex formation. Two malonylated 3-gentiobiosyl derivatives of delphinidin, linked covalently to different flavones, have been isolated from the flowers of *Eichhornia crassipes* (Pontederiaceae).[9,10] Anthocyanins that contain neohesperidose have been isolated from petals of the blue marguerite daisy *Felicia amelloides* (Asteraceae) as delphinidin 3-neohesperidoside-7-[6-(malonyl)glucoside],[139] as peonidin 3-neohesperidoside from petals of two cultivars of *Cyclamen persicum* (Primulaceae),[140] and as the 3-[2-(rhamnosyl)-6-(malonyl)glucoside] and its deacylated form of delphinidin from petals of a mauve line of *Clitoria ternatea* (Leguminosae).[141] Neohesperidose has previously only been identified in anthocyanins from species belonging to the gymnosperm family Podocarpaceae.[142]

10.2.4.3   Trisaccharides

Altogether 19 anthocyanins based on seven trisaccharides, 2-glucosyl-6-rhamnosylglucose, 2-xylosyl-6-rhamnosylglucose, 2-xylosyl-6-glucosylgalactose, 2-xylosyl-6-glucosylglucose (new), 6-(6-glucosylglucosyl)glucose, 3-(3-glucosylglucosyl)glucose, and 3-glucosyl-6-glucosylglucose, have been identified. Among the novel anthocyanins containing a trisaccharide reported after 1992, the 3-[6-(rhamnosyl)-2-(xylosyl)glucoside] of delphinidin has been isolated as the major anthocyanin from scarlet flowers of *Linum grandiflorum*,[106] while the same triglycoside of petunidin and peonidin has been isolated in minor amounts from fruits of *Vaccinium padifolium*.[25] These latter fruits also contain three novel 3-sambubiosides of petunidin, peonidin, and malvidin,[124] in contrast to previous reports on pigments of plants in Ericaceae, which show that a variety of 3-monoglycosides are regularly present. It is interesting to observe that cyanidin 3-[6-(6-((*E*)-sinapyl)glucosyl)-2-(xylosyl)glucoside] has been reported to be produced in *Glehnia littoralis* (Umbelliferae) callus cultures,[34] and not the analogous 6-rhamnosyl-2-xylosylgalactose derivative identified in its relative *Daucus carota*.[33]

10.2.4.4   Glycosidic Linkages

Anthocyanins bear glycosidic moieties in the anthocyanidin 3-, 5-, 7-, 3′-, or 5′-position (Figure 10.6). Nearly all anthocyanins have a sugar located at the 3-position (Appendix A). The only exceptions are the 3′-[2-(galloyl)galactoside] and 3′-[2-(galloyl)-6-(acetyl)galactoside] of delphinidin isolated from blue flowers of the African water lily *Nymphaéa caerulea*[26] and the 4′-glucoside and 7-[3-(glucosyl)-6-(malonyl)glucoside]-4′-glucoside of cyanidin from red onion (*Allium cepa*).[18] The desoxyanthocyanins (Appendix A) including the new luteolinidin 5-[3-(glucosyl)-2-(acetyl)glucoside] recently isolated from the fern *Blechnum novae-zelandiae*,[96] of course, cannot have any sugar in their 3-positions. Several anthocyanidin 5-glycosides and anthocyanidin 7-glycosides without sugar in their 3-positions have previously been reported.[8] However, all of these may be classified as tentative structures due to lack of data (e.g., long-range ^{1}H–^{13}C couplings in heteronuclear NMR spectra) for absolute identification of the linkage positions.

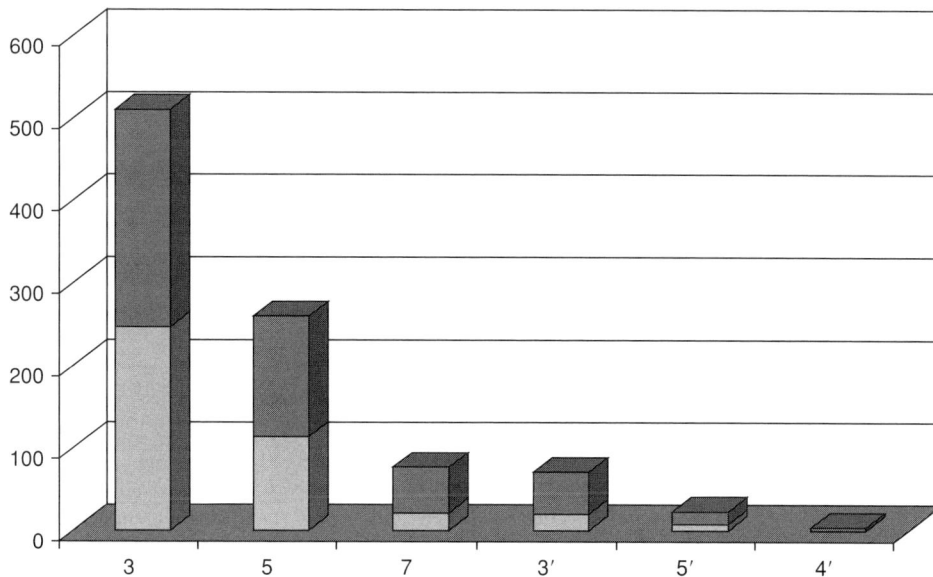

FIGURE 10.6 The number of anthocyanins with glycosyl moieties linked to the various anthocyanidin positions. The upper dark part of each bar represents the anthocyanins reported later than 1992. See Table 10.1 for structures.

About 146 and 65 novel anthocyanins reported after 1992 have a second sugar moiety located at their 5- and 7-positions, respectively (Appendix A). With the exception of two desoxyanthocyanins from the fern *Blechnum procerum*,[143] no identified anthocyanin has sugars linked to both the 5- and 7-positions. In the same period, 51 new anthocyanins with a glucosyl moiety in the 3'-position have been reported. Most of these have been isolated from species belonging to Orchidaceae (16), *Clitoria* (Leguminosae) (12), Ranunculaceae (6), and Gentianaceae (5). However, a few were found in Compositae, Liliaceae, Rhamnaceae, Nymphaeaceae, Lobeliaceae, and Commelinaceae (Table 10.2). In contrast to most anthocyanidin 3'-glycosides reported, five delphinidin and cyanidin derivatives from flowers of *Anemone coronaria* (Table 10.2) have glucuronic acid instead of glucose linked to the B-ring.[28,29] Similarly, glucose was replaced with galactose in delphinidin 3'-[2-(galloyl)-galactoside] and its acetylated derivative isolated from extracts of *Nymphaéa caerulea* flowers.[26]

In addition to the previously identified anthocyanins with a sugar in their 5'-positions from flowers of *Clitoria ternatea* (five anthocyanins) and *Lobelia erinus* (two anthocyanins), 12 further ternatins and preternatins have been isolated from *Clitoria ternatea* (Leguminosae),[144,145] and three anthocyanins from the genus *Dianella* (Liliaceae). The three polyacetylated delphinidin 3,7,3',5'-tetraglucosides from berries of two *Dianella* species[146] showed exceptional blueness at *in vivo* pH values due to effective intramolecular copigmentation involving *p*-coumarylglucose units at the 7-, 3'-, and 5'-positions of the aglycone. It has also been reported that the five new polyacylated delphinidin 3,3',5'-triglucosides from *Clitoria ternatea* flowers formed intramolecular stacking between the aglycone and the 3',5'-coumaryl-glucosyl side chains in solution.[144]

Four cyanidin 4'-glucosides have recently been isolated from pigmented scales of red onion.[147] These structures were established by extensive use of 2D NMR spectroscopy and electrospray LC–MS. The previous report on 4'-glucosyl linkages in two anthocyanins from

Hibiscus esculentus[148] was mainly based on lack of bathochromic shifts by the addition of $AlCl_3$ to a solution containing these anthocyanins, which is inadequate as proof for absolute identification of the 4′-linkages. Among the desoxyanthocyanins, Nair and Mohandoss[112] have previously indicated the occurrence of luteolinidin 4′-glucuronide from flowers *Holmskioldia sanguinea* (Verbenaceae). Compared with spectra of cyanidin 3-glycosides, the cyanidin 4′-glucosides from red onions showed hypsochromic shifts (10 nm) of visible λ_{max} and hyperchromic effects on wavelengths around 440 nm, similar to pelargonidin 3-glycosides.[147] Glycosidic substitution of the other B-ring hydroxyl groups (3′ and 5′) causes similar hypsochromic shifts of visible λ_{max}.[3]

10.2.5 ANTHOCYANINS WITH ACYLATION

More than 65% of the reported anthocyanins with properly identified structures are acylated, and anthocyanin diversity is highly associated with the nature, number, and linkage positions of the acyl groups. The aromatic acyl groups of these anthocyanins include various hydroxycinnamic acids (*p*-coumaric, caffeic, ferulic, sinapic, and 3,5-dihydroxycinnamic acids) and two hydroxybenzoic acids (*p*-hydroxybenzoic and gallic acids). Malonic acid is the most frequent aliphatic acyl group, while acetic, malic, oxalic, succinic, and tartaric acids have a more restricted distribution. As much as four different acyl groups located at four different glycosyl moieties have been identified in Lobelinin B isolated from flowers of *Lobelia erinus*.[149] The number of anthocyanins containing the different acyl moieties is presented in Figure 10.7.

During the period of this review, Pale et al.[27] have isolated a triacylated-tetraglucosylated cyanidin derivative (**243**) from flowers of *Ipomoea asarifolia*, which contained *E*-3,5-dihydroxycinnamic acid linked to the 6-position of one of the glucosyl moieties. This acyl

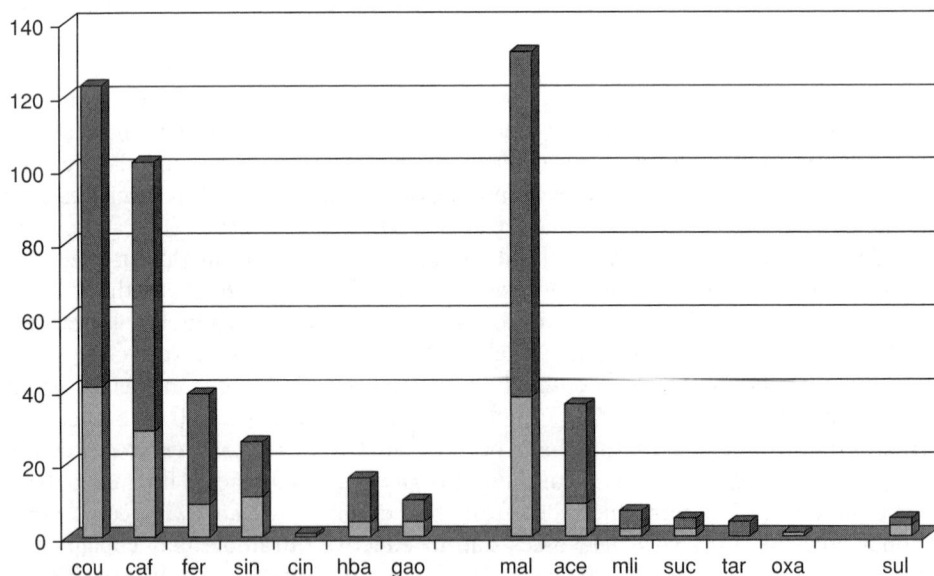

FIGURE 10.7 Numbers of anthocyanins containing the various acyl moieties identified in anthocyanins. The upper dark part of each bar represents the anthocyanins reported later than 1992. cou, *p*-coumaric acid; caf, caffeic acid; fer, ferulic acid; sin, sinapic acid; cin, 3,5-dihydroxycinnamic acid; hba, *p*-hydroxybenzoic acid; gao, gallic acid; mal, malonic acid; ace, acetic acid; mli, malic acid; suc, succinic acid; tar, tartaric acid; oxa, oxalic acid; sul, sulfate.

group has previously not been identified in any anthocyanin. Several nonnaturally occurring acids have also been identified in acylated forms in anthocyanins isolated from wild carrot suspension cultures provided with these acids in the medium.[150,151] Among the dicarboxylic acids, tartaryl has for the first time been identified in four anthocyanins (**82, 246, 383, 401**) isolated from flowers of *Anemone coronaria*.[28,132] Interestingly, the first anthocyanins found conjugated with sulfate, malvidin 3-glucoside-5-[2-(sulfato)glucoside], and malvidin 3-glucoside-5-[2-(sulfato)-6-(malonyl)glucoside], have been isolated from violet flowers of *Babiana stricta* (Iridaceae).[30] Several other flavonoid classes than anthocyanins have previously been reported to contain sulfate groups.[152]

10.2.5.1 Acylation with Phenolic Acids

Two hundred and seventy-nine anthocyanins with aromatic acylation have been identified (Appendix A), and 67% have been reported after 1992. Ninety-three anthocyanins are reported to be acylated with both aromatic and aliphatic acyl groups.

10.2.5.1.1 *p-Coumaric Acid*

More than 150 anthocyanins acylated with *p*-coumaric acid have been reported. Among these, most of the 132 anthocyanins, which have been assigned with a complete structure (Appendix A), have the *p*-coumaryl unit(s) in glucose 6-position(s) within a glycosidic moiety linked to the 3-position of the aglycone. In anthocyanins from the genera *Eustoma*[123,129] and *Gentiana*[153,154] (Gentianaceae), the *p*-coumaryl was found to be linked to the 5-glucosyl and not the 3-glucosyl. However, in Albireodelphin D (**373**), isolated from *Gentiana*,[154] this acyl moiety was connected to the 3′-glucosyl. This latter linkage position of the *p*-coumaryl has also been identified in delphinidin 3-[6-(rhamnosyl)glucoside]-7-[6-(*p*-coumaryl)glucoside]-3′-[6-(*p*-coumaryl)glucoside] isolated from flowers of *Ceanothus papillosus* (Rhamnaceae),[32] and in anthocyanins from orchids (Orchidaceae),[155–157] *Clitoria ternatea* (Leguminosae) (called ternatins),[4,144,145] and *Dianella* spp. (Liliaceae) fruits.[146] Most of these anthocyanins from the two latter genera contained another *p*-coumaryl unit located within the 5′-glycosyl moieties. In delphinidin 3,7,3′,5′-tetra[6-(*p*-coumaryl)glucoside] from *Dianella* spp., as many as four 6-*p*-coumarylglucoside units were located at different aglycone positions.[146]

In anthocyanin 3-rutinosides acylated with *p*-coumaric acid isolated from species in Solanaceae, this acyl group was found to be linked to the rhamnose 4-position (Appendix A). A similar linkage has also been reported in anthocyanins from species belonging to *Lobelia erinus* (Lobeliaceae),[118,149] and previously from *Silene dioica* (Caryophyllaceae), *Viola* spp. (Violaceae), and *Iris* spp. (Iridaceae). The identification by Terahara et al.[158] of *p*-coumaryl linked to galactose in delphinidin 3-[6-(*E-p*-coumaryl)galactoside] isolated from leaves of red flower tea, *Camellia sinensis*, is really outstanding.

In ten anthocyanins from various *Petunia* spp. (Solanaceae),[159–165] the coumaryl unit had another glucosyl moiety connected to its 4-hydroxyl group, making diverse anthocyanins with 3-glucosyl-rhamnosyl-*p*-coumaryl-glucosyl-(acyl?) moieties. Similar esterification has also been found in 14 ternatins from *Clitoria ternatea* (Leguminosae),[4,144,145] four anthocyanins from orchids (Orchidaceae),[155–157] in delphinidin 3-[6-(4-(glucosyl)coumaryl)glucoside]-5-[6-(malonyl)glucoside] from flowers of *Triteleia bridgesii* (Liliaceae),[166] in cyanidin 3-[2-(glucosyl)-6-(4-(glucosyl)coumaryl)glucoside]-5-glucoside isolated from cabbage,[137] and in cyanidin 3-[6-(4-(glucosyl)coumaryl)-2-(2-(sinapyl)xylosyl)glucoside]-5-[6-(malonyl)glucoside] from leaves and stems of *Arabidopsis thaliana*.[167] In orchids, these alternating 3-glucosyl-*p*-coumaryl-glucosyl chains were characteristically located in the aglycone 3′-positions, while most of the ternatins contained this type of chain in both the aglycone 3′- and 5′-positions. The largest monomeric anthocyanin recorded to date, ternatin A1, has been found to be built by delphinidin, seven glucosyls, four *p*-coumaryls, and one malonyl units (molecular mass: 2108.87 g mol^{-1}).[108]

10.2.5.1.2 Caffeic Acid

Most of the 100 anthocyanins that are acylated with caffeic acid have this cinnamyl moiety linked to a glucosyl 6-position (Appendix A). Some anthocyanins isolated from species belonging to Solanaceae (**288, 365, 442, 478, 484**)[159–164,168,169] and *Silene dioica* (Caryophyllaceae) (**166, 201**)[170] have a caffeyl located on the 4-position of the rhamnosyl unit of the disaccharide rutinose. This acyl group has also been shown to be attached to the 2- and 5-hydroxyl groups of arabinose in anthocyanidin 6-arabinofuranosylglucosides (**166, 201**) isolated from *Zebrina pendula* (Commelinaceae).[137] Additionally, it has been linked to the 2-hydroxyl group of glucose in pelargonidin 3-[2-(2-(*E*-caffeyl)glucosyl)galactoside] from flowers of *Pulsatilla cernua*,[22] and to the same position of the same disaccharide in several anthocyanins (**234, 240, 246, 383, 393, 398, 401**) isolated from flowers of *Anemone coronaria*,[28,29] which also belong to Ranunculaceae.

Rather exceptionally, caffeic acid is found to be esterified with a tartaryl hydroxyl in one end, and a glucosyl in the phenolic 4-hydroxyl in the other end in pelargonidin 3-[2-(xylosyl)-6-(3-(3-(4-(glucosyl)caffeyl)-2-tartaryl)malonyl)galactoside] isolated from scarlet flowers of *Anemone coronaria* "St. Brigid."[132] Several anthocyanins from species within each of the families Compositae, Orchidaceae, Solanaceae, and Convolvulaceae have a glucosyl in the phenolic 4-hydroxyl group of caffeic acid, making a chain of alternating glucosyl–caffeyl–glucosyl units (Table 10.2). In 15 anthocyanins from species in Convolvulaceae (Table 10.2), the last caffeyl unit with a glucosyl linked to the phenolic end has remarkably the sugar linked to the 3-hydroxyl of caffeic acid.

10.2.5.1.3 Ferulic Acid

More than 59 different anthocyanins have been reported to be acylated with ferulic acid; however, only 39 have been assigned with a complete structure. These anthocyanins were isolated for the first time from species belonging to the following families: Cruciferae (18), Labiatae (5), Solanaceae (4), Umbelliferae (3), Convolvulaceae (2), Gentianaceae (2), Liliaceae (2), Commelinaceae (1), Lobeliaceae (1), and Orchidaceae[4] (Table 10.2). Most of the anthocyanins have the ferulyl unit(s) in glucose 6-position(s) within a glycosidic moiety linked to the aglycone 3-position. However, this cinnamic acid has also been shown to be attached to the 4-hydroxyl group of rhamnose in three anthocyanidin 3-rutinosides (**62, 443, 479**) from red[171] and purple[172] potatoes (Solanaceae), and to the 2-hydroxyl group of xylose in pelargonidin 3-[6,2-di-(ferulyl)sambubioside]-5-[6-(malonyl)glucoside] isolated from flowers of *Matthiola incana* (Cruciferae).[173] Remarkably, this acyl group has been shown to be attached to the 2- and 5-hydroxyl groups of arabinose in cyanidin 3-[6-(2,5-di-(*E*-ferulyl)arabinosyl)-glucoside]-7, 3'-di-[6-(*E*-ferulyl)-glucoside].[136] This major anthocyanin from leaves of *Tradescantia pallida* (Commelinaceae) shows an extra absorption band at 583 nm at pH values above 4.0, which makes this pigment highly colored at these pH values.[174] The high stability[138] of this type of pigments with as much as four acyl units was assumed to be due to the difficulty of water molecule diffusion to the hydrophobic center formed by the acyl groups and the aglycone.[175]

Ferulyl has been found as part of the glycosidic moiety linked to the anthocyanidin 5-position only in delphinidin 3-[6-(rhamnosyl)galactoside]-5-[6-(ferulyl)glucoside] from flowers of *Eustoma grandiflorum* (Gentianaceae),[129] to the 7-position in cyanidin 3-[6-(malonyl)glucoside]-7-[6-(*E*-ferulyl)glucoside]-3'-glucoside from *Sophronitis coccinea* (Orchidaceae)[176] and cyanidin 3-[6-(2,5-di-(*E*-ferulyl)arabinosyl)-glucoside]-7,3'-di-[6-(*E*-ferulyl) glucoside] from *Tradescantia pallida*,[136] and to the 5'-position in Lobelinin B (**413**) from *Lobelia erinus* (Lobeliaceae).[149]

In cyanidin 3-[6-(4-(glucosyl)ferulyl)sophoroside]-5-glucoside from red cabbage[137] and malvidin 3-[6-(4-(4-(6-ferulyl)glucosyl)-*E-p*-coumaryl)rhamnosyl)glucoside]-5-glucoside from flowers *Petunia hybrida*,[161] the ferulyl unit has a glucosyl moiety linked to its phenolic 4-hydroxyl group.

10.2.5.1.4 Sinapic Acid
Most of the 26 anthocyanins that are acylated with sinapic acid have been isolated from cabbage (*Brassica* spp.), stock (*Matthiola incana*), mustard (*Sinapis alba*), and mustard weed (*Arabidopsis thaliana*) in Cruciferae, yam (*Dioscorea alata*) in Dioscoreaceae, and from the genera *Phalaenopsis*, *Dendrobium,* and *Vanda* in Orchidaceae[4] (Table 10.2). In addition, the 3-[6-(6-(sinapyl)glucosyl)-2-(xylosyl)galactoside] and 3-[6-(6-(sinapyl)glucosyl)galactoside] of cyanidin have been isolated from cell suspension cultures of an Afghan cultivar of *Daucus carota* (Umbelliferae).[33]

In anthocyanins, sinapyl is mainly connected to the 6-hydroxyl of glucose or 2-hydroxyl of xylose. This latter linkage was found in nine anthocyanins from stock, mustard, and mustard weed in Cruciferae.[167,173,177] In some anthocyanins isolated from cabbage,[178] sinapyl has also been found in the glucose 2-position (**219**, **220**, **227**, **229**), while the 3-[4-(sinapyl)-6-(glucosyl)glucosides] of cyanidin and peonidin isolated from jam[179] have this acyl group located in the glucose 4-position.

In anthocyanins isolated from orchids (**222**, **230**, **376**, **382**), the sinapyl was part of the glycosidic moieties linked to the aglycone 7- and 3'-positions,[180–182] while the other anthocyanins mentioned above had this acyl group linked to the glycoside in the aglycone 3-position.

10.2.5.1.5 p-Hydroxybenzoic Acid
Most of the 16 anthocyanins that are acylated with *p*-hydroxybenzoic acid (Appendix A) have been isolated from flowers of *Delphinium hybridum*,[109,119,183] *Aconitum chinense*,[184] *Consolida armeniaca* (Ranunculaceae),[115] *Campanula* species (Campanulaceae),[117,185,186] or roots of *Ipomoea batatas* (Convolvulaceae).[187,188] In the examined species from Ranunculacea and *Campanula,* one to as many as four *p*-hydroxybenzoyl moieties belong to the anthocyanidin 7-glycoside. In *p*-hydroxybenzoylated anthocyanins from sweet potatoes (*Ipomoea batatas*), this acyl group is located within the 3-glucoside. In addition, cyanidin 3-[6-(malonyl)glucoside]-7,3'-[6-(4-(glucosyl)-*p*-hydroxybenzoyl)glucoside] and cyanidin 3-[6-(6-(*p*-hydroxybenzoyl)glucosyl)-2-(xylosyl)galactoside] have been isolated from red-purple flowers of *Dendrobium* "Pramot" (Orchidaceae)[189] and cell suspension cultures of *Daucus carota* (Umbelliferae),[33] respectively. When determined, the *p*-hydroxybenzoyl of anthocyanins is in all cases linked to the 6-position of one glucosyl moiety.

10.2.5.1.6 Gallic Acid
Eleven anthocyanins have been reported to be galloylated. Anthocyanins with gallic acid (3,4,5-trihydroxybenzoic acid) as acyl substituent have been found in leaves of the giant water lily (*Victoria amazonica*) in Nymphaéaceae as the 3-[2-(galloyl)galactoside] of delphinidin and cyanidin.[190] These anthocyanins together with their 6''-acetylated analogs were also identified in *Nymphaéa × marliacea*[191] and *Nymphaéa alba*.[192] The 2''-galloylgalactoside and 2-(galloyl)-6-(acetyl)galactoside of delphinidin have been identified in the blue flowers of *Nymphaéa caerulea*; however, these moieties were in this latter species located in the rare 3'-position.[26]

The distribution of galloylated anthocyanins outside Nymphaéaceae seems to be restricted. They occur in leaves in Aceraceae as the 3-[6-(galloyl)glucoside], 3-[6-(galloyl) rutinoside], and 3-[2,3-di-(galloyl)glucoside] of cyanidin,[193–195] and as the 3-[2-(galloyl) galactoside] and 3-[2-(galloyl)-6-(rhamnosyl)galactoside] of cyanidin in flowers of *Acalypha hispida* (Euphorbiaceae).[130] An anthocyanin from seeds of *Abrus precatorius* (Leguminosae) has tentatively been identified as delphinidin 3-*p*-coumaryl-galloylglucoside.[196]

10.2.5.2 Acylation with Aliphatic Acids

Among the 178 anthocyanins with aliphatic acylation (Appendix A), more than 70% have been reported after 1992.

10.2.5.2.1 Malonic Acid

One or two malonyl units have been found in 132 anthocyanins from a variety of families, and 100 have been reported during the period of this review. Most of these pigments have the malonyl unit(s) in glucose 6-position(s); however, the anthocyanins containing malonyl, which have been isolated from various flowers of *Anemone coronaria* (Ranunculaceae),[28,29,132] has this dicarboxylic acyl group linked to the galactose 6-position. In another interesting report, Mazza and Gao[197] have isolated the 3-malonylxyloside, 3-dimalonylxyloside, and 3-malonylarabinoside of cyanidin from seeds of *Helianthus annuus*; however, in these cases the linkage positions have not been determined.

From species in Labiatae, the 3-[6-(*p*-coumaryl)glucoside]-5-[4,6-di-(malonyl)glucoside] of cyanidin, delphinidin, and pelargonidin, and the 3-[6-(caffeyl)glucoside]-5-[4,6-di-(malonyl)glucoside] of the two latter anthocyanidins have been reported.[203–205] The location of malonyl to the glucose 4-position has also been reported for cyanidin 3-[4-(malonyl)-2-(glucuronosyl)glucoside] and pelargonidin 3-[6-(*p*-coumaryl)glucoside]-5-[4-(malonyl) glucoside] isolated from red flowers of *Bellis perennis* (Compositae)[113] and *Hyacinthus orientalis* (Liliaceae),[201] respectively. Malonyl substitution in the sugar 3-position of anthocyanins has been found in 3-[3-(malonyl)glucoside] and 3-[3,6-di-(malonyl)glucoside] of cyanidin isolated from various organs of some *Allium* species.[202–205] This latter anthocyanin has also been identified in flowers of chrysanthemum (*Dendranthema grandiflorum*)[206] and in many grasses.[207] On mild hydrolysis, it was converted to the 3-[3-(malonyl)glucoside], indicating that the malonic acid linkage to the 3-position of glucose was more stable than the more common linkage at the 6-position of glucose.[202]

The occurrence of the diester structures of the malonic acid moiety in natural anthocyanin pigments has so far been reported in pigments from flowers of *Eichhornia crassipes*[9,10] and chive, *Allium schoenoprasum*,[11] where the anthocyanin–flavone and anthocyanin–flavonol disubstituted malonate structures were exhibited, respectively (Figure 10.8 and Figure 10.9). In some anthocyanins from flowers of *Anemone coronaria*, malonic acid is esterified with galactose in one end and tartaryl in the other end.[29,132]

Malonylation has been found in two 6-hydroxyanthocyanins from *Alstroemeria*, the 3-[6-(malonyl)glucosides] of 6-hydroxycyanidin and 6-hydroxydelphinidin,[101,102] and in 5-carboxypyranocyanidin 3-[6-(malonyl)glucoside] from red onion.[18] A pigment containing methyl-malonyl, pelargonidin 3-[2-(xylosyl)-6-(methyl-malonyl)galactoside], has been reported to have been isolated from scarlet flowers of *Anemone coronaria*.[132] Methyl esterification of the terminal carboxyl group of malonyl units can easily occur in acidified methanolic solvents during extraction and in the isolation process.[167,342]

10.2.5.2.2 Acetic Acid

Altogether 36 anthocyanins with one or two acetyl groups have been identified (Appendix A). The number given in parenthesis shows the number of anthocyanins reported as novel pigments in species in the various families: Verbenaceae (8), Liliaceae (5), Geraniaceae (6), Vitaceae (5), Nymphaeaceae (5), Alliaceae (1), Theaceae (1), Rutaceae (1), Melastomataceae (1), Labiatae (1), Leguminosae (1), and Blechnaceae (1).

The acetyl groups are, in general, linked to glucosyl 6-positions; however, in anthocyanins from *Nymphaea* (**129**, **142**, **322**, **330**, **331**) and *Tulipa* (**144**, **332**, **333**), the galactosyl and rhamnosyl moieties, respectively, are acetylated. In the latter cases, the acetyl groups are located in the rhamnosyl 2- or 3-positions. Linkage to the rare 3-position has also been reported for the acetyl group of an anthocyanin–flavonol complex (**537**) from *Allium schoe-noprasum*.[11] Two other outstanding acetylated pigments, malvidin 3-[6-(*p*-coumaryl) glucoside]-5-[2-(acetyl)xyloside] and delphinidin 3-[6-(*p*-coumaryl)glucoside]-5-[4-(acetyl)-6-(malonyl)glucoside], have been isolated from flowers of *Tibouchina urvilleana*[208] and *Salvia uliginosa*,[209] respectively. The acyl group of the two similar acetylated malvidin 3,5-diglucosides (**465, 466**), which have been isolated from *Geranium* flowers, is reported to be attached to either the 3-glucosyl[210] or the 5-glucosyl,[211] with the latter as the most probable linkage in both cases. The only report of acetylated anthocyanins outside angiosperm families is luteolinidin 5-[2-(acetyl)-3-(glucosyl)glucoside] from the fern *Blechnum novae-zelandiae*.[96]

10.2.5.2.3 Malic, Succinic, Oxalic, and Tartaric Acids

Anthocyanins acylated with the dicarboxylic malic acid seem to be restricted to carnations (*Dianthus*). Wild-type carnation petals have been found to contain the 3-[6-(malyl)glucosides], 3-[6-(malyl)glucoside]-5-glucosides, and the 3-glucoside-5-glucoside (6″,6‴-malyl diesters) of pelargonidin and cyanidin.[212–216] These two latter macrocyclic pigments[219,221] are the only anthocyanins where the same acyl group is connected to two monosaccharides, which are attached at different locations at the anthocyanidin. The orientation of the malyl groups of these anthocyanins has been determined by long-range 2D heteronuclear NMR techniques (HMBC).[216] Recently, Fukui et al.[31] have shown that marketed genetically modified violet carnations produce analogous delphinidin-type anthocyanins by heterologous flavonoid 3′, 5′-hydroxylase gene expression.

Anthocyanins acylated with succinic acid have previously been isolated from some species in the genus *Centaurea* (Compositae), and identified as the 3-[6-succinylglucoside]-5-glucosides of cyanidin and pelargonidin.[217–219] Later, the former pigment isolated from cornflower, *Centaurea cyanus*, was shown to be part of the self-assembled supramolecule protocyanin, composed of six molecules each of malonylflavone and succinylcyanin complexed with magnesium and ferric ions (Section 10.2.7).[220] Succinyl has also been reported to occur in minor amounts in flowering tops of *Phragmites australis* (Gramineae) as cyanidin 3-[6-(succinyl)-glucoside],[221] and the structures of the two major anthocyanins (Figure 10.9) in blue *Agapanthus* (Agapanthaceae) flowers have been found to be based on a *p*-coumarylated delphinidin diglycoside attached to a flavonol triglycoside via a succinic acid diester link.[12]

Oxalic acid is another dicarboxylic acid with restricted occurrence as an acyl moiety of anthocyanins. In addition to some European orchids flowers (Section 10.5.10),[222] cyanidin 3-dioxalylglucoside has been isolated from fruits of *Rubus laciniatus*.[57] Altogether six different anthocyanins acylated with oxalic acid have been reported. With the exception of cyanidin 3-[6-(oxalyl)glucoside], evidences for proper determination of the linkage position of this acyl moiety are absent.

Tartaric acid has an even more limited distribution as acylation agent of anthocyanins, identified in only four anthocyanins (**82, 246, 383,** and **401**) isolated from flowers of *Anemone coronaria*.[28,132] In these pigments, one of the tartaryl hydroxyl groups is esterified with a carboxyl group of malonic acid. In pelargonidin 3-[2-(xylosyl)-6-(3-(3-(4-(glucosyl)caffeyl)-2-tartaryl)malonyl)galactoside] (**82**), the other hydroxyl group of tartaryl is esterified with caffeyl.

10.2.6 Dimeric Flavonoids Including an Anthocyanin Unit

Most reported anthocyanins are monomeric in nature; however, new types of flavonoids consisting of an anthocyanin moiety covalently linked to another flavonoid unit have been reported in the period of this review. One class includes one anthocyanin unit and one flavone or flavonol unit attached covalently to each end of a common dicarboxylic acid. The other class involves one anthocyanin moiety covalently linked directly to a flavanol unit.

Two dimeric pigments ((6'''-(delphinidin 3-[6''-(glucosyl)glucoside])) (6''-(apigenin 7-glucosyl))malonate (**534**) and (6'''-(delphinidin 3-[6''-(glucosyl)glucoside])) (6''-(luteolin 7-glucosyl))malonate (**535**) (Figure 10.8) have been isolated from the blue-violet flowers of *Eichhornia crassipes* (Pontederiaceae) by Toki et al.[9,10] The major *Eichhornia* anthocyanin A has an apigenin 7-glucoside molecule (flavone) attached with an ester bond to one end of malonic acid, and delphinidin 3-gentiobioside linked with a similar bond to the other end. The minor *Eichhornia* anthocyanin B has a similar structure with apigenin 7-glucoside replaced by luteolin 7-glucoside. The three-dimensional structure of these pigments was suggested from the observation of negative Cotton effects at λ_{max} (535 and 547 nm,

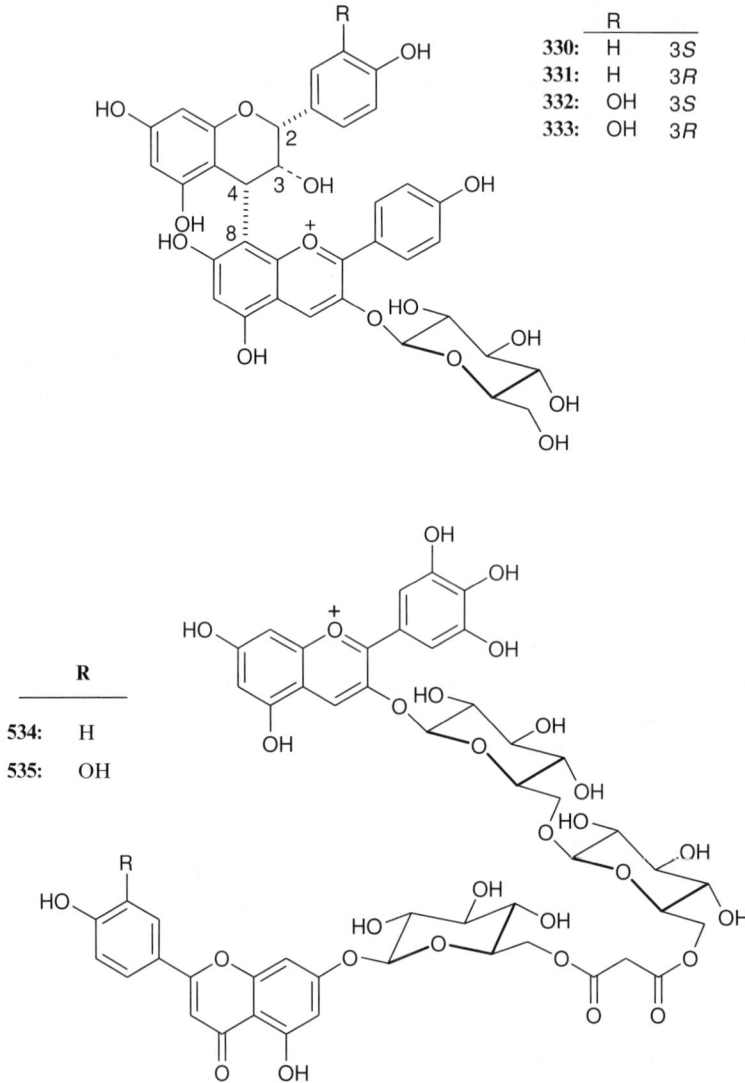

FIGURE 10.8 Afzelechin(4α → 8)Pg3glc (**530**), epiafzelechin (4α → 8)Pg3glc (**531**), catechin(4α → 8)Pg3glc (**532**), epicatechin(4α → 8)Pg3glc (**533**) isolated from extracts of strawberries, and (6'''-(delphinidin 3-[6''-(glucosyl)glucoside])) (6''-(apigenin 7-glucosyl))malonate (**534**) and (6'''-(delphinidin 3-[6''-(glucosyl)glucoside])) (6''-(luteolin 7-glucosyl))malonate (**535**) from blue-violet flowers of *Eichhornia crassipes*. Pg3glc, pelargonidin 3-glucoside.

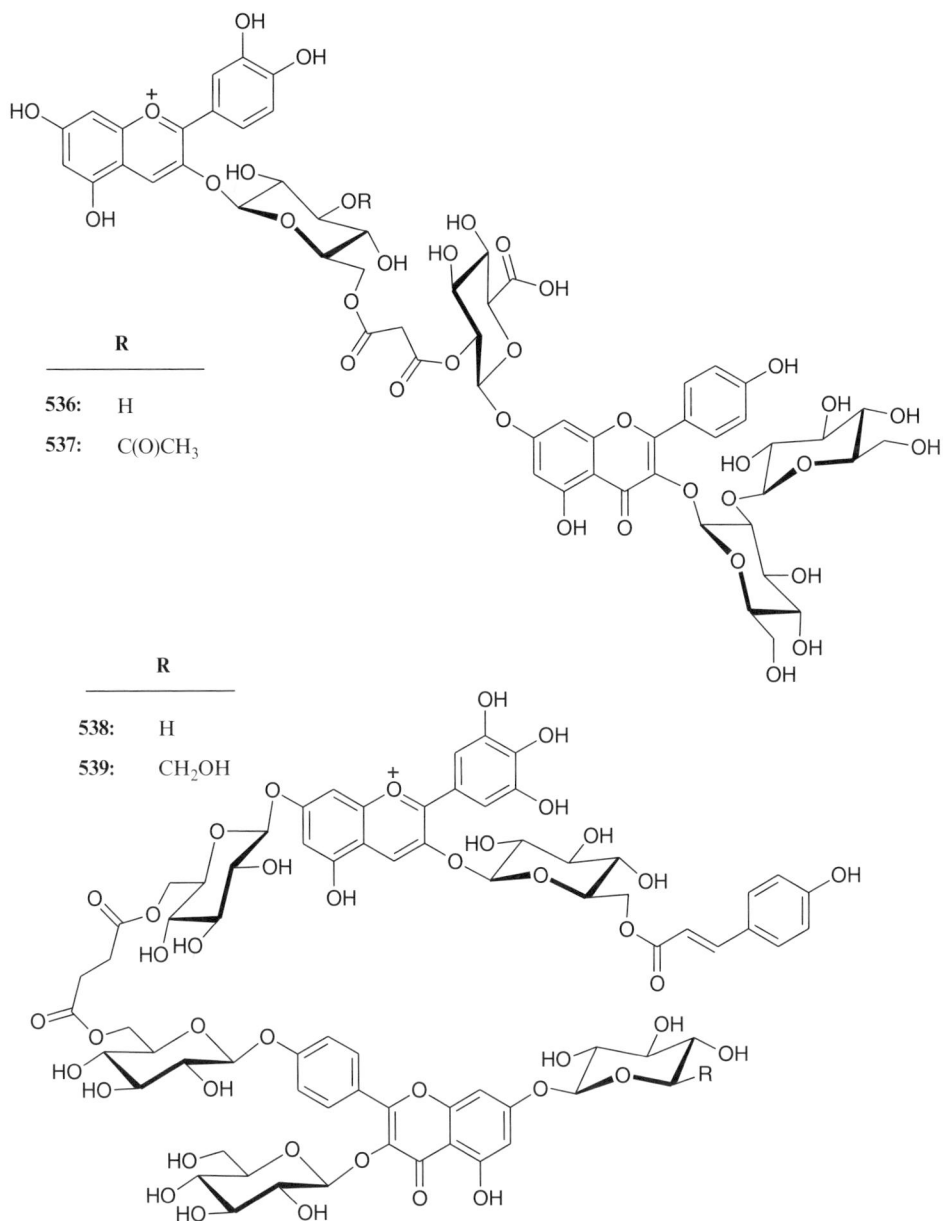

FIGURE 10.9 (6″-(cyanidin 3-glucosyl)) (2‴′-(kaempferol 3-[2″-(glucosyl)(glucoside)]-7-glucuronosyl))-malonate (**536**) and (6″-(cyanidin 3-[3″-(acetyl)glucosyl])) (2‴′-(kaempferol 3-[2″-(glucosyl)(glucoside)]-7-glucuronosyl))malonate (**537**) isolated from pale-purple flowers of chive (*Allium schoenoprasum*),[11] and (6″-(delphinidin 3-[6″-(*p*-coumaryl)glucoside]-7-glucosyl)) (6‴′-(kaempferol 3-glucoside-7-xyloside-4′-glucosyl))succinate (**538**) and (6‴-(delphinidin 3-[6″-(*p*-coumaryl)glucoside]-7-glucosyl)) (6‴′-(kaempferol 3-glucoside-7-glucoside-4′-glucosyl))succinate (**539**) from blue *Agapanthus* flowers.[12]

respectively). The chromophore (delphinidin) and the copigment (flavone) occupy a folding conformation as a binary complex.[9,10] *Eichhornia* anthocyanin A exhibited remarkable color stability in aqueous solution at mildly acidic pH values.[223] The existence of intramolecular hydrophobic interactions between the chromophoric skeleton and the flavone group was indicated by reduction in the hydration constant when compared with the parent delphinidin 3-glycoside. The existence of other anthocyanin–flavone complexes has previously been shown or indicated in flowers of orchids,[222,224] lupins,[225] and *Salvia patens*.[226]

Two anthocyanin–flavonol complexes, ((6″-(cyanidin 3-glucosyl)) (2″″-(kaempferol 3-[2″-(glucosyl)(glucoside)]-7-glucuronosyl))malonate (**536**) and (6″-(cyanidin 3-[3″-(acetyl)-glucosyl])) (2″″-(kaempferol 3-[2″-(glucosyl)(glucoside)]-7-glucuronosyl))malonate (**537**) (Figure 10.9), have been isolated from the pale-purple flowers of chive (*Allium schoenoprasum*).[11] These pigments, which were supplied with complete NMR assignments, were based on either cyanidin 3-glucoside or cyanidin 3-[3-(acetyl)glucoside] esterified to one end of malonic acid, and kaempferol 3-[2-(glucosyl)glucoside]-7-glucosiduronic acid connected to the other end. Compared to similar spectra of the same monomeric anthocyanins, the bathochromic shifts (9 nm) in the UV–vis spectra of the complexes revealed intramolecular association between the anthocyanin and flavonol units, which influenced the pigment color. The chemical shifts of the anthocyanidin H-4 in the two complexes were 0.3 ppm upfield for the same shifts of anthocyanins without connection to a flavonol moiety. Two similar complexes, (6‴-(delphinidin 3-[6″-(p-coumaryl)glucoside]-7-glucosyl)) (6″″-(kaempferol 3-glucoside-7-xyloside-4′-glucosyl))succinate (**538**) and (6‴-(delphinidin 3-[6″-(p-coumaryl)glucoside]-7-glucosyl)) (6″″-(kaempferol 3-glucoside-7-glucoside-4′-glucosyl))succinate (**539**) (Figure 10.9), have been isolated from the blue *Agapanthus* flowers (Agapanthaceae).[12] In these structures, succinate was involved instead of malonate to connect delphinidin 3-[6-(p-couma-loyl)glucoside]-7-glucoside to either kaempferol 3,4′-di-glucoside-7-xyloside or kaempferol 3,7,4′-tri-glucoside.

The other class involving one anthocyanin moiety covalently linked directly to a flavanol unit has been found as minor pigments in extracts of strawberries.[13] These purple complexes were characterized by UV–vis spectroscopy, 2D NMR techniques, and electrospray MS to be afzelechin(4α → 8)pelargonidin 3-O-β-glucopyranoside (**530**), epiafzelechin(4α → 8)pelargo-nidin 3-O-β-glucopyranoside (**531**), catechin(4α → 8)pelargonidin 3-O-β-glucopyranoside (**532**), and epicatechin(4α → 8)pelargonidin 3-O-β-glucopyranoside (**533**) (Figure 10.8). The stereochemistry at the 3- and 4-positions of the flavan-3-ols was elucidated after assumption of the R-configuration at C-2. Each of the four pigments occurred in the NMR solvent as a pair of rotamers. Other complexes based on anthocyanins or anthocyanidins directly linked to flava-nol(s) have been detected in wine mainly by LC–MS techniques[59–62] (see Chapter 5).

10.2.7 METALLOANTHOCYANINS

In the period of this review, Kondo et al. have presented several extraordinary papers in the field of metalloanthocyanins and their contribution to blue flower colors. Commelinin[227] from blue flowers of *Commelina communis* has been found to consist of six molecules of delphinidin 3-[6-(p-coumaryl)glucoside]-5-[6-(malonyl)glucoside] (malonylawobanin) copig-mented with six flavone (flavocommelin) molecules complexed with two magnesium atoms (Figure 10.10). Self-association was shown to exist between the anthocyanin moieties. Intact commelinin was also obtained by reconstruction of this supramolecule from its components. Cd-commelinin, in which Mg^{2+} was replaced with Cd^{2+}, has been studied with x-ray crystal structure. The blue flower color development and the stability of the color were explained by metal complexation of the anthocyanin and intermolecular hydrophobic association.[227] The octaacetate derivative of the flavone part of this molecule has been determined by x-ray diffraction,[228] and in the crystal the flavone molecules were arranged parallel to each other according to the periodicity of the crystal lattice. Intermolecular stacking of the flavone skeletons was, however, not observed, and the hydrophilicity of the glucose moieties was suggested as an important factor governing the self-association.

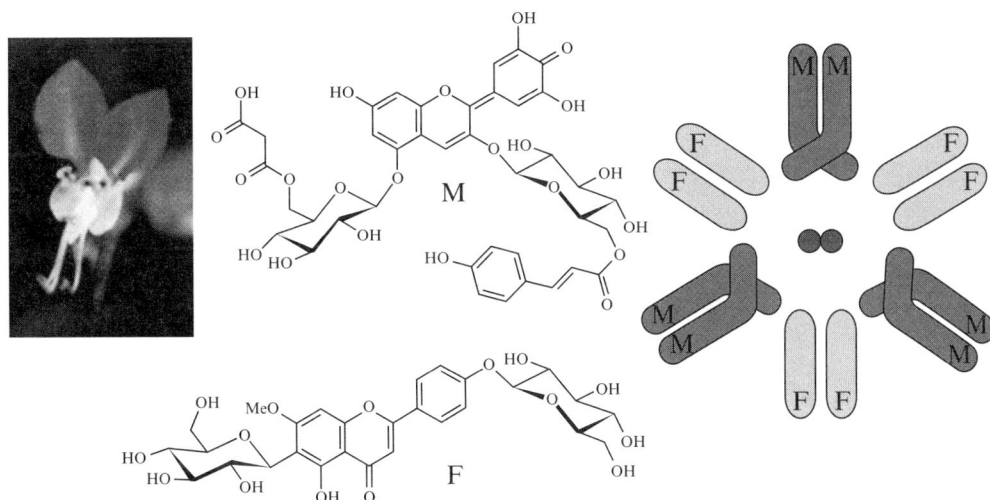

FIGURE 10.10 The illustration figure (right) indicates the mechanism known to operate for commelinin found in *Commelina communis* by Kondo et al. Commelinin consists of six molecules of delphinidin 3-[6-(*p*-coumaryl)glucoside]-5-[6-(malonyl)glucoside] (malonylawobanin) (M, purple) copigmented with six flavone molecules (F, yellow) complexed with two magnesium atoms (red). (After Kondo, T. et al., *Nature*, 358, 515, 1992. With permission.)

Another metalloanthocyanin, protodelphin, similar to commelinin and protocyanin, has been isolated from flowers of *Salvia patens*.[226] Protodelphin has been shown to be composed of six molecules of delphinidin 3-[6-(*p*-coumaryl)glucoside]-5-[6-(malonyl)glucoside], six molecules of apigenin 7,4′-diglucosides, and two magnesium ions.[229] Compared to commelinin, protodelphin includes the same anthocyanin, malonylawobanin, which is, however, another flavone. Takeda et al.[226] have been able to resynthesize the natural blue pigment *in vitro* by adding the three components together. Mg^{2+} could be substituted *in vitro* by other divalent metal cations (e.g., Co^{2+}, Ni^{2+}, Zn^{2+}, and Cd^{2+}). Using synthetic apigenin 7,4′-diglucosides derived from D- or L-glucose, Kondo et al.[229] have shown that the sugars of the three flavone molecules in the self-association complex oriented the structure into an M helix (for D-glucose) or P helix (for L-glucose). The M helix was able to intercalate with the Mg^{2+}-malonylawobanin complex to form native pigment, whereas the P helix could not. They concluded that restricted chiral and structural recognition controlled the entire self-assembly of the metalloanthocyanin, and was responsible for the blue flower color.

In 1998, Kondo et al.[220] proposed a new molecular mechanism for blue color expression with protocyanin from cornflower, *Centaurea cyanus*. The protocyanin structure is similar to that of commelinin, composed of six molecules each of apigenin 7-glucuronide-4′-[6-(malonyl)glucoside] and succinylcyanin (**153**), complexed with magnesium and ferric ions.[230] It was proposed that the molecular stacking of the aromatic units prevented hydration of the anthocyanidin nucleus. The blue color of protocyanin was found to be caused by ligand to metal charge transfer (LMCT) interaction between succinylcyanin and Fe^{3+}.[220] This color mechanism (Figure 10.11) is different from that known to operate for commelinin.

Metal complexes involving anthocyanins have previously been proposed also in blue flowers of *Hydrangea macrophylla*.[80] Later, the color change and variation of hydrangea has been suggested to be caused by free Al^{3+} complexation of those components of the complex responding to slight vacuolar pH change.[231,232]

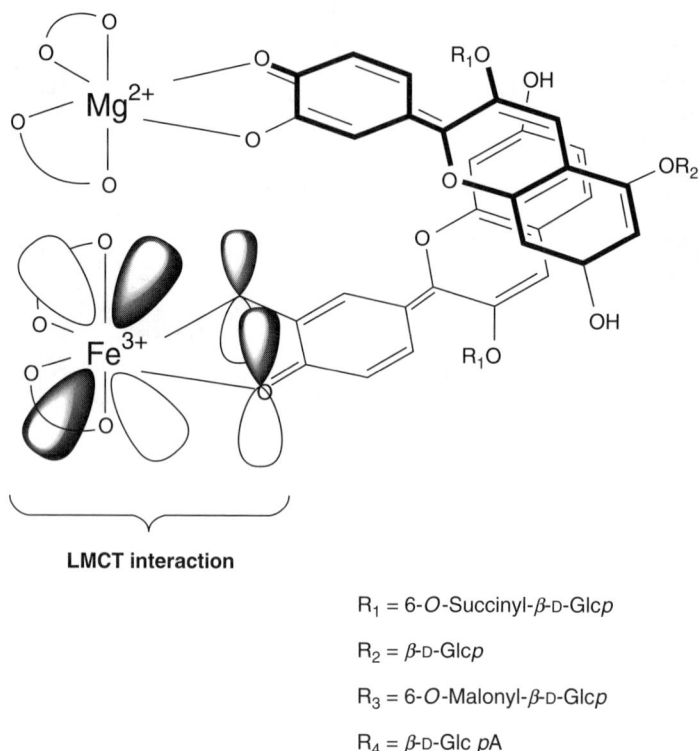

LMCT interaction

R_1 = 6-O-Succinyl-β-D-Glc$p$

R_2 = β-D-Glc$p$

R_3 = 6-O-Malonyl-β-D-Glc$p$

R_4 = β-D-Glc pA

FIGURE 10.11 Schematic representation of the LMCT interaction between the anthocyanin succinyl-cyanin and Fe^{3+} in protocyanin. (From Kondo, T. et al., *Tetrahedron Lett.*, 39, 8307, 1998. With permission.)

It is well known that anthocyanins with hydroxyl groups in *ortho*-position to each other form complexes with aluminum ion leading to bathochromic and hyperchromic shifts in their absorption spectra. Complexation of Al^{3+} with synthetic and natural anthocyanins has been investigated in aqueous solutions within the pH range 2 to 5.[233,234] As shown by UV–vis spectroscopic data, the complexes involved not only the colored forms but also the colorless forms of the pigment. ^{1}H NMR analysis in CD_3OD confirmed the conversion of the antho-cyanin from the red flavylium form to the deep-purple quinonoidal forms upon coordination to Al^{3+}.[233] From relaxation kinetics measurements (pH jump), complexation constants of Al^{3+} and several synthetic and natural anthocyanins have been calculated.[233–235]

10.2.8 ANTHOCYANIN COLORS AND STABILITY

One of the best-established functions of anthocyanins is in the production of flower color and the provision of colors attractive to plant pollinators. Considerable effort has been made to give explanations for the color variations expressed by anthocyanins in plants. Various factors including concentration and nature of the anthocyanidin, anthocyanidin equilibrium forms, the extent of anthocyanin glycosidation and acylation, the nature and concentration of copigmentation, metal complexes, intra- and intermolecular association mechanisms, and influence of external factors like pH, salts, etc. have been found to have impact on antho-cyanin colors.[80]

The role of anthocyanins and flavones in providing stable blue flower colors in the angiosperms has more recently been outlined by Harborne and Williams.[42] It was apparent from the data collected that delphinidin was the most common anthocyanidin in blue flowers, and that copigmentation with a specific flavone constituent was the most common mechanism for shifting the mauve colors of delphinidin glycosides toward blue. These anthocyanin–flavone complexes, where they exist, showed high flavone to anthocyanin ratios (e.g., 10:1), except when a metal cation was also present. The anthocyanin–flavone complexes in *Eichhornia crassipes*[9,10] are unique in that the anthocyanin and the flavones were covalently linked through a central aliphatic acyl residue (Section 10.2.6). Similar bathochromic color effects have recently also been reported for dimeric anthocyanin–flavonol complexes in blue *Agapanthus* flowers (Agapanthaceae),[12] and in purple flowers of chive (*Allium schoenoprasum*).[11] In the latter case, the anthocyanidin is cyanidin. More about metalloanthocyanins and dimeric flavonoids containing anthocyanins is found in Sections 10.2.5 and 10.2.6, respectively.

Some very interesting macrocyclic anthocyanins, the 3-glucoside-5-glucoside (6″,6‴-malyl diesters) of cyanidin and pelargonidin, have been isolated from carnations (*Dianthus*).[214,215] Gonnet and Fenet[215] have shown by CIELAB parameters that carnations with "cyclamen red" colors closely matched those of petals of some *Rosa* cultivars considered for comparison. However, these cyclamen colors in roses were based on the presence of cyanidin 3,5-diglucoside in contrast to the carnations, which contained pelargonidin derivatives acylated with malic acid. These macrocyclic anthocyanins were, however, relatively labile and underwent readily ring opening to furnish the corresponding 3-(6″-malylglucoside)-5-glucosides, and could only be extracted if neutral solvents were employed. *In vivo* these rare pigments appear to be stabilized by copigmentation with associated flavones. Wild-type carnations lack a flavonoid 3′,5′-hydroxylase gene, which implies that they cannot produce the corresponding delphinidin derivatives. Recently, Fukui et al.[31] showed that marketed genetically modified violet carnations produced analogous delphinidin-type anthocyanins by heterologous flavonoid 3′,5′-hydroxylase gene expression. They concluded that the bluish hue of the transgenic carnation flowers was accounted for by three factors: accumulation of the delphinidin-type anthocyanins, the presence of strong copigmentation with flavone derivative, and a relatively high vacuolar pH of 5.5.

The relationship between color and substitution patterns in anthocyanidins has been investigated with the aim of developing quantitative structure–color models.[236] Experimental data for the lowest UV transition in 20 substituted anthocyanidins were reviewed. While a hypsochromic effect from hydroxyl and methoxyl moieties at positions 6 and 8 was reported here,[236] it is interesting to note that a *C*-linked sugar in the 8-position has the opposite effect producing a more bluish color.[20] Color and stability studies of the 3-glucosides of the six common anthocyanidins and petanin, petunidin 3-[6-(4-(*p*-coumaryl)rhamnosyl)glucoside]-5-glucoside, in aqueous solutions during several months of storage revealed especially large variation at slightly acidic to slightly alkaline pH values.[237,238] ^{1}H NMR spectroscopy has been used to characterize the aggregation processes leading to color stabilization of malvidin 3-glucoside.[239] The concentrations of the different forms in aqueous solution were determined as a function of pH for several values of the total anthocyanin concentration. The pH-dependent color and structural transformations in aqueous solutions of some non-acylated anthocyanins and synthetic flavylium salts have been reexamined.[240,241]

It is known that polyacylated anthocyanins are very stable in neutral or weakly acidic aqueous solutions, whereas simple anthocyanins are quickly decolorized by hydration at the 2-position of the anthocyanidin nucleus. The shift to blue colors and increased anthocyanin stability are in many cases simply achieved by intramolecular copigmentation involving stacking between the anthocyanidin and the aromatic acyl groups.[233,242–244] Both intra- and intermolecular interactions have been found in aromatic monoacylated anthocyanins studied by 1D and 2D proton NMR spectroscopy.[245] These compounds formed strong intramolecular π-complexes between the cinnamic acyl group and the anthocyanidin nucleus, the double bond of the acyl group involved as well as its aromatic ring. Upon increasing the concentration of the anthocyanins or lowering the temperature of the NMR sample, the π-complexes formed multinuclear complexes as shown by the resultant negative nuclear Overhauser effect (NOE) values. In a very detailed review, recent progress in the chemistry of polyacylated anthocyanins as flower color pigments has been outlined by Honda and Saito.[246] It was recognized that both the bluing effect and stabilization of flower colors remarkably depended on the number of aromatic acids presented in the polyacylated anthocyanins. After classification of the polyacylated anthocyanins into seven types by the substitution pattern of the acyl functions, it was concluded that anthocyanins with the aromatic acyl groups in glycosyls in both the 7- and 3′-positions were considered to make the most stable colors in the flowers.

Red-purple colors in the flowers of orchids have been shown to be derived from altogether 15 cyanidin and peonidin glycosides, with aromatic acylated sugars attached at both the 7- and 3′-positions (Table 10.2).[155–157,176,180–182,189] Intramolecular associations between these planar molecules in orchids provided stable colors without the need for any copigment or metal cation.[247] Figueiredo et al.[247] proposed that the glycosyl-acyl "side chains" attached to both positions 3′ and 7 of the chromophore favored a better overlap and stronger interaction with the π-system of the central chromophore (Figure 10.12), than what was observed for other acylated anthocyanins. They supported the assessment by molecular calculations, which gave minimum energy conformation for a "sandwich" type with the 3′-chain folded "over" and the 7-chain folded "under" the chromophore (Figure 10.12). Similar acylation of glycosyls in anthocyanidin 7- and 3′-positions has also been reported for anthocyanins in Commelinaceae,[136] Compositae,[248] Liliaceae,[148] and Rhamnaceae.[32] The three acylated delphinidin 3,7,3′,5′-tetraglucosides from berries of two *Dianella* species (Liliaceae) showed exceptional blueness at *in vivo* pH values due to effective intramolecular copigmentation involving

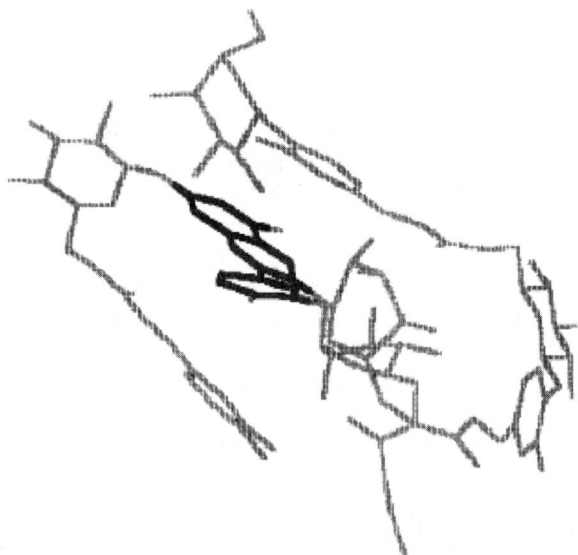

FIGURE 10.12 Structure of cyanidin 3-[6-(malonyl)glucoside]-7-[6-(caffeyl)glucoside]-3′-[6-(4-(6-(4-(glucosyl)caffeyl)glucosyl)caffeyl)glucoside] (**258**) optimized by molecular calculations. The anthocyanidin is represented in black, while substituents are drawn in gray. The acylated side chains attached to the 3′- and 7-positions of the chromophore interact favorably with the π-system of the central chromophore. (From Figueiredo, P. et al., *Phytochemistry*, 51, 125, 1999. With permission.)

p-coumaryl-glucose units (GC) at the aglycone 7-, 3'-, and 5'-positions.[148] Evidence was presented to show that the effectiveness of the copigmentation could be ranked: 3',5'-GC > 7-GC > 3-GC. The blue flower color of *Ceanothus papillosus* (Rhamnaceae) was proposed to arise from a supramolecular complex of high stoichiometry including anthocyanins and kaempferol 3-[2-(xylosyl)rhamnoside].[32] This copigmentation effect appeared to be quite specific, and did not occur to the same extent with other more common flavonols. An extraordinary, long wavelength visible absorption maximum at 680 nm was produced, which conferred additional blueness. Figueiredo et al.[247] have also explored the role of malonic acid residues that are present in many anthocyanins. These dicarboxylic acyl groups appeared to provide color stabilization, due to an increase in acidity in the vacuolar solution of the petal. The pK_a of malonic acid was 2.83 and deprotonation of the malonyl group provided protection against alkanization of the medium and hence loss of color.

According to our data, only 17 anthocyanins acylated with hydroxycinnamic acids occur in both the *cis* (*Z*) and *trans* (*E*) configurations (Appendix A). George et al.[249] have compared the pairs of 3-[6-(*E/Z-p*-coumaryl)glucoside]-5-[6-(malonyl)glucosides] of malvidin and delphinidin. They observed that the *cis* isomers exhibited ε values about 1.5 times greater than the *trans* isomers, in both pairs. It was calculated (pK'_h values) that the *cis* forms were less prone to undergo hydration reactions forming the colorless anthocyanin forms. Based on computed structures, the more coplanar arrangement allowed by the *cis* isomers was postulated as the rationale supporting the enhanced color stability.[249] When considering the color effect of this type of intramolecular copigmentation *in vivo*, one should bear in mind that the *trans* isomer seems to predominate, and that the conversion between the two isomers is rare. When Yoshida et al.[250] studied the *E,Z*-isomerization reaction and stability of several types of acylated anthocyanins under the influence of UV irradiation, their interest was focused on the reason why isomerization reaction of some acyl residues was prevented in living plant cells. They concluded that the stability of anthocyanins under irradiation highly depends on molecular stacking. They proposed that light energy absorbed by cinnamoyl residues might be transferred to the anthocyanidin nucleus and released without any isomerization reaction or degradation of pigments. Thus, the flower color may be stable for a long time under strong solar radiation.

10.3 ANTHOCYANIN PRODUCTION

10.3.1 CELL CULTURES

The interest in and demand for natural food colorants and pharmacologically interesting natural compounds have encouraged new research initiatives aimed at the development of more efficient means of harvesting anthocyanins. When growth procedures are optimized, cell culture systems have the potential of producing both higher anthocyanin concentrations within reduced time, and another selection of anthocyanins relative to production in whole plants. In a recent general review, Ramachandra Rao and Ravishankar[251] have dealt with the production of high-value secondary metabolites including anthocyanins through plant cell cultures, shoot cultures, root cultures, and transgenic roots obtained through biotechnological means. In an overview of the status and prospects in the commercial development of plant cell cultures for production of anthocyanin, Zhang and Furusaki[252] have focused on strategies for enhancement of anthocyanin biosynthesis to achieve economically viable technology.

Production of anthocyanins in plant cell and tissue cultures has been reported for more than 30 species including *Daucus carota*, *Fragaria × ananassa*, *Vaccinium* spp., *Vitis hybrida*, *Solanum tuberosum*, *Malus sylvestris*, *Aralia cordata*, *Perilla frutescens*, *Ipomoea batatas*, *Euphorbia millii*, *Strobilanthes dyeriana*, *Hibiscus sabariffa*, *Dioscorea cirrhosa*, etc.[35,252,253]

The production has shown to be influenced by a variety of environmental stimuli such as light irradiation, UV light, low temperature, oxygen level, hormones, fungal elicitors, low nutrient levels.[251–253] Increased level of oxygen supply and light irradiation have, for instance, shown independently positive influence on the production of anthocyanins in suspended cultures of *Perilla frutescens* cells in a bioreactor.[254] However, a combination of irradiation with a higher oxygen supply reduced the production. In *Vaccinium pahalae* cell cultures, anthocyanin yield was enhanced by increasing sucrose concentration in the liquid suspension medium and by manipulating the initial inoculum density.[255]

A cell culture system has the potential advantage of facilitating selective production of certain anthocyanins. The nine acylated anthocyanins produced by flowers of *Hyacinthus orientalis* regenerated *in vitro* were identical to those of field-grown flowers.[256] However, the concentration of cyanidin 3-[6-(*p*-coumaryl)glucoside]-5-[6-(malonyl)glucoside] was considerably higher in the regenerated flowers. Lower concentration of 2,4-dichlorophenoxyacetic acid in the medium used for strawberry suspension cultures has, for instance, limited cell growth and enhanced both anthocyanin production and anthocyanin methylation.[253] The ratio of peonidin-3-glucoside to the total anthocyanin content increased significantly under these conditions. A methylated anthocyanin like peonidin 3-glucoside is normally not found in intact strawberries, and although the activity of anthocyanin methyltransferase was not measured by Nakamura et al.,[253] the results indicated that lower 2,4-dichlorophenoxyacetic acid concentrations enhanced the activity of anthocyanin methyltransferase. Do and Cormier[257] have reported that increased osmotic potential in the medium resulted in a significant intracellular accumulation of peonidin-3-glucoside in *Vitis vinifera* cells. Similarly, jasmonic acid has been reported to increase the peonidin 3-glucoside content considerably, while the other major anthocyanins only experienced smaller increments.[258]

Another example of using cell cultures has been presented by Dougall et al.[151] To improve understanding of the ways in which cinnamic acid groups alter the color retention of anthocyanins, a series of anthocyanins that differed systematically in their acyl group was needed. When cinnamic acids were fed to wild carrot suspension cultures, the proportion of acylated to nonacylated anthocyanins increased.[151] With high relevance for future metabolic studies, *Vitis vinifera* cells grown in a bioreactor have been used for production of isotopically ^{13}C-labeled phenolic substances such as anthocyanins.[259,260] The enrichment of labeling (between 40 and 65%) obtained for all compounds is sufficient to investigate their absorption and metabolism in humans. Similarly, ^{14}C-L-phenylalanine has been incorporated into a range of polyphenolic compounds when fed to cell cultures.[261,262] Experiments with *Vaccinium pahalae* berries and *Vitis vinifera* suspension cultures, using [^{14}C]-sucrose as the carbon source, have demonstrated a 20 to 23% efficiency of ^{14}C incorporation into the flavonoid-rich fractions.[263]

To improve production of anthocyanins, efforts have mainly been devoted to the optimization of biosynthetic pathways by both process and genetic engineering approaches. The productivity in the cultures is, however, determined by synthetic capacity, storage capacity, and the capacity to metabolize the compounds in the transport and detoxification processes.[264] The potential of manipulation and optimization of postbiosynthetic events have recently been reviewed by Zhang et al.[265] These events, including chemical and enzymatic modifications, transport, storage or secretion, and catabolism or degradation, were outlined with anthocyanin production in plant cell cultures as case studies.

Bioreactor-based systems for mass production of anthocyanins from cultured plant cells have been described for several species.[254,260,266–270] A highly productive cell line of *Aralia cordata* obtained by continuous cell aggregate cloning has, for instance, been reported to yield anthocyanins in concentrations as high as 17% on a dry weight basis.[266] However, to date economic feasibility has not been established in part because of some unique engineering challenges inherent in mass cultivation of plant cultures.

10.3.2 SYNTHESIS

Since the early contributions of Willstatter and Robinson, several alternative approaches following mainly two routes have been considered for synthesis of anthocyanins.[271] One of the routes includes condensation reactions of 2-hydroxybenzaldehydes with acetophenones, while the other uses transformations of anthocyanidin-related compounds like flavonols, flavanones, and dihydroflavonols to yield flavylium salts. The urge for plausible sequences of biosynthetic significance has sometimes motivated this latter approach. In the period of this review, new synthetically approaches in the field have also predominantly been following the same general routes; however, some new features have been shown in synthesis of pyranoanthocyanidins.

In an interesting paper by Bernini et al.,[272] compounds with a flavonoid structure have been selectively oxyfunctionalized at the C-2 carbon atom by dimethyldioxirane (DMD). Products obtained in this way appeared to be useful starting materials to access anthocyanidins. An example of this route is presented in Scheme 10.1. Here, 2,4-*cis*-flavane-4-acetate (**A**) was oxidized by DMD at room temperature, affording the corresponding C-2 hydroxy derivative (**B**) as the only product (63% yield). Further treatment of **B** with silica gel eliminated acetic acid to give **C** quantitatively. Then **C** was easily transformed into the flavylium salt (**D**) by simple addition of a 37% solution of HCl in water.

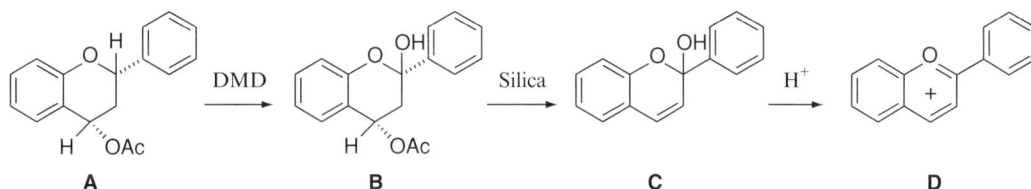

Following more traditionally routes, a recent one-step synthesis giving 3-deoxyanthocyanidins (apigeninidin and luteolinidin) in high yield involved the condensation reaction between 2,4,6-triacetoxybenzaldehyde and an acetophenone derivative in anhydrous methanolic hydrogen chloride.[273] The reduction of quercetin 3-*O*-(6-*O*-rhamnosylglucoside), rutin, by zinc amalgam in 3% absolute methanolic hydrochloric acid has provided pure cyanidin 3-*O*-(6-*O*-rhamnosylglucoside) in good yield.[274]

New 3-desoxyanthocyanidins have been prepared according to a Grignard reaction of some flavones with appropriate alkyl- and aryl-magnesium bromides.[275] The reaction of 5,7-dihydroxyflavone (chrysin) with an excess of phenylmagnesium bromide in THF under reflux conditions, followed by hydrochloric acid hydrolysis, afforded **1** in Scheme 10.2. Flavylium salts bearing a substituent at the 4-position are important compounds since they are known to be less sensitive to nucleophiles, especially water, and they give only minor amounts of the colorless pseudobases.[276]

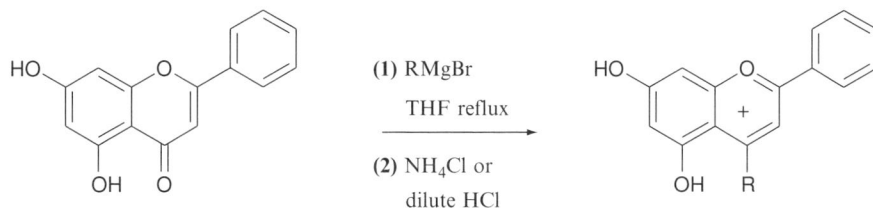

Flavylium salts with 4-methyl substitution (e.g., **2** in Scheme 10.2) might be further gently reacted with aromatic aldehydes affording anthocyanidin pigments of the pyranoanthocyanidin type.[275]

Compared to the more common flavylium salts, these latter pigments showed large bathochromic shifts (from about 100 to 200 nm), thus providing blue-violet and green colors, which are difficult to obtain with anthocyanidins in general.

The pyranoanthocyanin in red wine called vitisin A has been found by synthetic experiments to be formed by the reaction of malvidin 3-glucoside with pyruvic acid.[16] Its formation resulted from cyclization between C-4 and the hydroxyl group at C-5 of the original flavylium moiety with the double bond of the enolic form of pyruvic acid, followed by dehydration and rearomatization steps. Other pyranoanthocyanins in wine formed by the reaction of malvidin 3-glucoside with vinylphenols were first observed by Fulcrand et al.,[277] and later synthesized by nucleophilic addition of vinylphenols to malvidin 3-glucoside.[278] The former research group has also synthesized similar orange-colored anthocyanins from monoglucosylated anthocyanins and ethanol, alpha-ketoglutaric acid, acetone, or 3-hydroxybutan-2-one.[279] See Chapter 5 for further information about anthocyanins in wine.

Pyranoanthocyanins have also been formed by oxidative cycloaddition between anthocyanins from blackcurrant seeds and acetone,[90] which should be kept in mind when acetone is used as solvent during extraction and isolation of natural anthocyanins. When the same research group isolated four new pyranoanthocyanins (pyranocyanin C and D and pyranodelphinin C and D) from an extract of blackcurrant seeds, these structures were confirmed by chemical synthesis involving blackcurrant anthocyanins and p-coumaric acid.[91] Recently, pyranoanthocyanins have been obtained by a simple one-step reaction involving anthocyanins and cinnamic acids bearing at least one electron-donating substituent at the para-position.[280] Through this type of reaction with p-dimethylamino cinnamic acid, which is similar to the reaction of 4-substituted 3-desoxyanthocyanins with p-dimethylamino cinnamaldehyde,[275] synthetic malvidin- and cyanidin-based anthocyanins with violet hues have been prepared.[280]

10.4 ANTHOCYANIN LOCALIZATION IN PLANT CELLS

Anthocyanins are the most common water-soluble pigments in the plant kingdom, and are normally found dissolved uniformly in vacuolar solutions of epidermal cells. However, in cases like the Sphagnorubins,[281] the pigments are so tenaciously bound to the cell wall that they are only extracted with difficulty.

In species from more than 33 families, anthocyanins have been found located in pigmented bodies in vacuoles described as anthocyanoplasts (ACPs).[282,283] While the cells matured, these structures often disappeared. The ACPs were thought to be membranous, enclosing the site of anthocyanin biosynthesis. When the nature of ACPs from colored skin tissues of "Kyoho" grapes was investigated under a microscope, it was indicated that each ACP was surrounded by a transparent membrane, which maintained the concentration of anthocyanin within an ACP higher than that in the vacuole.[284] Later, it was, however, reported that the spherical pigmented bodies in cells of sweet potato (*Ipomoea batatas*) in suspension culture might be protein matrices,[285,286] and that ACPs possess neither a membrane boundary nor an internal structure.[286–288] A protein labeled VP24 was found to accumulate in the intravacuolar pigmented globules (cyanoplasts) and not in the tonoplast.[286] The expression of VP24 was closely correlated with the accumulation of anthocyanin in the cell lines, and this protein was suggested to play an important role in trapping of large amounts of anthocyanins that have been transported into these vacuoles. More recently, sequence analysis has indicated that matureVP24 peptide is a member of the metalloprotease family, and might be a novel vacuolar localized aminopeptidase.[289,290] Significant evidences have also been presented that enzymes involved in anthocyanin biosynthesis were associated with the cytoplasmic face of the endoplasmic reticulum.[291] Although synthesized in the cytoplasm, the anthocyanins, like many other flavonoids, may be rapidly transported across the tonoplast into the cell vacuole as indicated for anthocyanins of maize tissues.[292] Anthocyanins thus assume their distinct color after transport to the vacuole; however, very little is still known about the trafficking mechanisms of anthocyanins and their precursor in plant cells.

Not much was documented about the chemical nature and the functional significance of these inclusions in petal cells before Markham el al.[293] reported intensively colored intravascular bodies in petals of carnations (*Dianthus caryophyllus*) and lisianthus (*Eustoma grandiflorum*), which they named anthocyanic vacuolar inclusions (AVIs). The AVIs occurred predominantly in the adaxial epidermal cells, and their presence was shown to have major influence on flower color by enhancing both intensity and blueness. Electron microscopy studies on lisianthus epidermal tissue failed to detect a membrane boundary in AVI bodies, and the isolated AVIs were shown to have a protein matrix. The presence of large AVIs produced marked color intensification in the inner zone of the petal by concentrating anthocyanins above levels that would be possible in vacuolar solution. A minor subset of the total anthocyanins (four cyanidin and delphinidin acylated 3,5-diglycosides) was bound to this matrix rather than the acylated delphinidin triglycosides, which were the major forms present in the petal extracts.[293] "Trapped" anthocyanins differed from solution anthocyanins only in that they lacked a terminal rhamnose on the 3-linked galactose, revealing the specificity of the interactions. Recently, Conn et al.[294] have reported that AVIs appeared as dark red-to-purple spheres of various sizes in vacuoles in two lines of grapevine (*Vitis vinifera*) cell suspension culture due to their interaction with anthocyanins. Compared with the total anthocyanin profile, the profile of the AVI-bound anthocyanins showed an increase of ~28 to 29% in acylated (*p*-coumarylated) anthocyanins in both lines. Figure 10.13 shows AVIs from *Vitis vinifera*,[294] and AVIs from *Eustoma grandiflorum* and *Dianthus caryophyllus*.[293]

10.5   CHEMOTAXONOMIC PATTERNS

The structures of nearly 540 different anthocyanins have been elucidated, and more than half of these have been reported after 1992. In the following sections, some chemotaxonomic considerations within 11 families have been included. These families have been chosen due to the fact that most of the new anthocyanins have been isolated from species belonging to these families. Each family has been represented with a general structure including all the anthocyanins we have registered in our files as identified in one or more species belonging to this family. The chemotaxonomic considerations are mainly limited to the pattern revealed by the new anthocyanins reported in these families in the period of this review. Some additional reports of chemotaxonomic relevance are mentioned below.

The anthocyanin content of male and female cones of 27 species of *Abies*, *Picea*, *Pinus*, *Pseudotsuga*, and *Tsuga* (Pinaceae) has been determined, and only four anthocyanins were found.[295] The anthocyanin content of 23 grass species (Poaceae) belonging to five subfamilies has been resolved, and altogether 11 anthocyanins were identified.[296] A comprehensive survey of the flower pigment composition of 70 species and subspecies, 43 cultivars, and six artificial hybrids of *Crocus* (Iridaceae) used for chemotaxonomic investigations has been carried out.[297] The anthocyanin content of petals of 28 Chilean species of *Alstroemeria* (Alstroemeriaceae) and 183 interspecific hybrids has been analyzed by HPLC.[104] Flower color, hue, and intensity were measured by CIELAB parameters in fresh petals and compared with their anthocyanin content and the estimated flavonoid concentrations. The chemical basis of red flower pigmentation in species of *Cotyledon*, *Crassula*, and *Tylecodon* (Crassulaceae) has been outlined.[298] Only six major anthocyanins were found in the 10 species and 22 samples investigated, and each of the genera showed a characteristic combination of compounds. The chemical basis of flower pigmentation based on six major anthocyanins in the genus

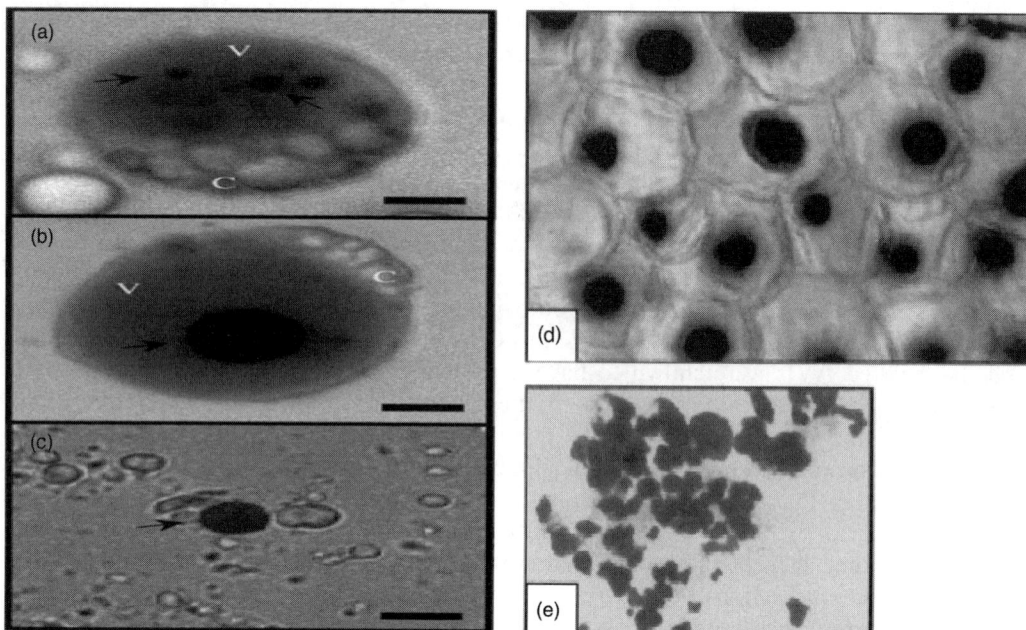

FIGURE 10.13 Left: Microscopic images of dark-grown *Vitis vinifera* protoplast of FU-2 line under bright field microscopy (a, b) before and (c) after lysis, showing a single AVI in solution. Arrows indicate AVIs, and the black bar is scaled to 10 μm. V, vacuole; C, cytoplasm. (From Conn, S. et al., *Biotechnol. Lett.*, 25, 835, 2003. With permission.) Right: (d) AVIs isolated from *Eustoma grandiflorum*, line 54, inner petal region, and (e) AVIs in an adaxial epidermal peel of *Dianthus caryophyllus*. (From Markham, K.R. et al., *Phytochemistry*, 55, 327, 2000. With permission.)

Lobostemon (Boraginaceae) has been presented.[299] During a wide survey of flower flavonoids in a variety of sections in the genus *Rosa*, altogether 11 anthocyanins were identified.[300,301] According to the anthocyanin distribution patterns in the genus, eight groups were classified chemotaxonomically.

10.5.1 ALLIACEAE

The genus *Allium* (Alliaceae) comprises important vegetables like onions and chive. Among the 14 anthocyanins, which have been identified in various *Allium* species (Figure 10.14), ten novel anthocyanins with characteristic structures have been reported from this genus after 1993. Several of the anthocyanins from species in *Allium* have either a glucosyl (laminariobioside), malonyl, or acetyl moiety linked distinctively to the glucosyl 3-position,[11,135,147, 202–205,302] which is very unusual. Most of the anthocyanins in this genus are based on cyanidin. Two covalent anthocyanin–flavonol complexes (Figure 10.9) have been isolated from the flowers of chive, *Allium schoenoprasum*.[11]

Red onion reveals one of the most characteristic anthocyanin patterns ever found. The main anthocyanins of some cultivars have been identified as the 3-[6-(malonyl)glucoside], 3-[3-(glucosyl)-6-(malonyl)glucoside], 3-[3-glucosyl)glucoside], and 3-glucoside of cyanidin.[135,203,205] Without support from NMR or MS data, the 3-arabinoside and 3-malonylarabinoside of cyanidin have been reported to be among the main anthocyanins of Spanish red onion (cultivar "Morada de Amposta").[303] Among the minor anthocyanins the 3-[3,6-(dimalonyl)glucoside] and 3-[3-(malonyl)glucoside] of cyanidin as well as the 3,5-diglucosides of cyanidin and peonidin have been found.[203] Additionally, peonidin 3-glucoside and peonidin

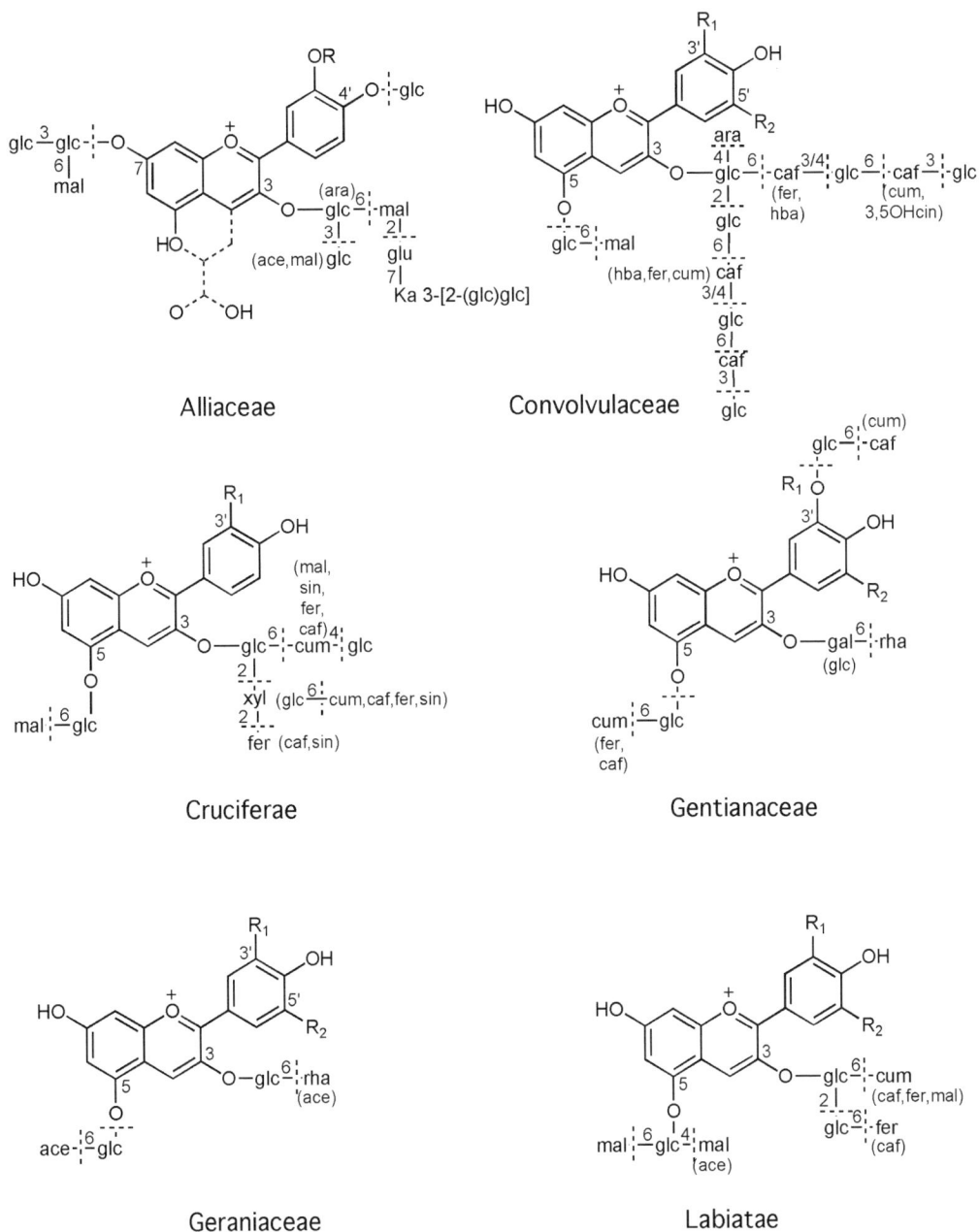

FIGURE 10.14 The structures represent a general presentation of all the anthocyanins identified in each of the families Alliaceae, Convolvulaceae, Cruciferae, Gentianaceae, Geraniaceae, and Labiatae. See Table 10.2 for abbreviations.

3-malonylglucoside have been reported, however, without determination of the linkage between the acyl group and the sugar of the latter pigment.[205] Recently, two pyranoanthocyanins, the 3-glucoside and 3-[6-(malonyl)glucoside] of 5-carboxypyranocyanidin,[18] as well as four anthocyanins having a glucosyl in the cyanidin 4′-position,[149] the 4′-glucoside, 3-glucoside-4′-glucoside, 3-[3-(glucosyl)-6-(malonyl)glucoside]-4′-glucoside, and 7-[3-(glucosyl)-6-(malonyl)glucoside]-4′-glucoside[149] have been isolated in minor amounts.

10.5.2 CONVOLVULACEAE

Altogether 43 different anthocyanins have been reported to occur in Convolvulaceae (Figure 10.14). Most of these pigments have a glucose unit in the 5-position, and either a mono- or diacylated glucose or sophorose unit in the 3-position of peonidin, pelargonidin, cyanidin, and rarely delphinidin. The acyl groups are mainly caffeyl or coumaryl; however, ferulyl, *p*-hydroxybenzoyl, and malonyl have been reported. A triacylated-tetraglucosylated cyanidin derivative (**243**), which contained the novel *E*-3,5-dihydroxycinnamic acid linked to the 6-position of one of the glucosyl moieties, has recently been isolated from flowers of *Ipomoea asarifolia*.[27] Even though some ternatins (**414–416, 418**) and cyanodelphin (**417**) actually have larger masses, the "Heavenly Blue Anthocyanin" (**297**), which is a peonidin 3-sophoro-side-5-glucoside with three caffeylglucose residues,[304] is among the largest monomeric anthocyanins that has been isolated ($[M]^+ = 1759$ amu).

The Morning Glory flowers (*Ipomoea/Pharbitis*) exist in a wide range of color forms. There is a good correlation between scarlet flower color and the occurrence of pelargonidin derivatives.[305,306] Lu et al.[307] have showed that the flower color of *Pharbitis nil* gradually shifts to the blue region with increasing numbers of caffeic acid residues in polyacylated pelargonidin glycosides. Blue and mauve flower colors, attractive to bee pollinators, are generally based on delphinidin, petunidin, or malvidin. However, exceptional cases are found, for instance, in *Ipomoea tricolor* and *Pharbitis nil* where the blue flower colors are caused by the peonidin-based "Heavenly Blue Anthocyanin."[308–310] Yoshida et al.[311] have shown that the color change of *Ipomoea tricolor* while flowering was due to vacuolar pH changes from 6.6 to 7.7, at which range the anhydrobase anion of HBA was formed and stabilized by intramolecular stacking. A different acylated peonidin glycoside, isomeric with "Heavenly Blue Anthocyanin,"[312] and related cyanidin derivatives have been obtained from violet-blue flowers of *Pharbitis nil*[308] and *Ipomoea purpurea*,[305] while related pelargonidin derivatives have been isolated from red-purple flowers.[306,308] One blue flowered species with more expectable delphinidin chromophore has been discovered in *Evolvulus pilosus*. The pigment in this plant, **397**, which contains one malonyl unit linked to the 5-glucoside, provides a stable blue color at neutral pHs maintained by intermolecular stacking of the caffeyl moieties between the flavylium chromophores.[313] None of the other anthocyanins reported from this family are malonylated.

10.5.3 CRUCIFERAE

In total, around 45 different anthocyanins have been reported to occur in various species belonging to Cruciferae (Brassicaceae), and nearly all of them share the same type of structures (Figure 10.14). They have a glucosyl at the 5-position, which in some cases is malonylated, and a disaccharide at the 3-position, which is acylated with one or two cinnamic acids. In the genera *Brassica*[178] and *Raphanus*,[314,315] this disaccharide is a sophorose, while *Arabidopsis*,[167] *Sinapis*,[316] and *Matthiola*[177,173] have a sambubioside in the same position. When one cinnamyl moiety occurs, it is attached to the 6-position of the inner hexose (glucose). The second cinnamic acid is linked to the 2-position of the outer hexose (xylose) when a sambubioside occurs, and at the 6-position of the outer hexose (glucose) when the corresponding sophoroside occurs. The cinnamic acids include coumaric acid, caffeic acid, ferulic acid, and sinapic acid. All anthocyanins are either based on cyanidin or pelargonidin, and 28 of these have been reported as novel in the period of this review.

10.5.4 GENTIANACEAE

Altogether 22 anthocyanins, including three produced by genetic engineering, have been reported to occur in flowers from the family Gentianaceae (Figure 10.14). Nineteen among these have been identified as novel compounds in the period of this review. The examined species belonging to *Gentiana* contain one or more anthocyanins with a 3,5,3'-triglycosidic substitution pattern.[153,154] The blue flowers are reported to contain delphinidin derivatives, while the pink flowers of this genus contain cyanidin derivatives.[153,154] All the monosaccharides directly linked to the aglycone are glucosyl moieties. This is in contrast to anthocyanins isolated from *Eustoma grandiflorum*, which have a galactosyl or robinobiosyl attached to the 3-position, and no glycosyl moiety linked to the 3'-position.[123,129] The new anthocyanins based on delphinidin isolated from flowers of *Eustoma grandiflorum* have been produced by genetic engineering after transformation with a UDP-glucose:flavonoid-3-*O*-glucosyltransferase cDNA from *Antirrhinum majus*. The galactose at the 3-position has been partly replaced in some anthocyanins by glucose during anthocyanin synthesis in the transformed plant.[123]

10.5.5 GERANIACEAE

Thirteen different anthocyanins from species in Geraniaceae have been identified (Figure 10.14), and seven among these are novel pigments reported after 1994. The major anthocyanins of various *Pelargonium* species and cultivars were identified as the 3,5-diglucosides and 3-glucoside-5-[6-(acetyl)glucosides] of the six common anthocyanidins.[317] The major factors responsible for color variation were shown to be the types and relative levels of pigments present. Variations in pH and copigment levels were not found to contribute significantly. Flowers with colors ranging from cream and pink through to deep purple, including salmon, orange, and red, were studied. While either flavonols or carotenoids were responsible for cream or yellow coloration, all other colors resulted from anthocyanin mixtures. Similar malvidin 3,5-diglucosides have also been reported in purplish-blue *Geranium* flowers.[210,211] Even though the acetyl in these pigments has been reported attached both to the 3-glucose and to the 5-glucose, the latter is the most probable position.[211] In addition, minor amounts of the 3,5-diglucosides of malvidin and cyanidin and the 3-glucosides of cyanidin and delphinidin have been identified in purplish-blue flowers of *Geranium sylvaticum*.[210]

10.5.6 LABIATAE

Around 40 anthocyanins have been identified in one or more species belonging to Labiatae (Figure 10.14), and during the last decade seven have been identified for the first time in this family. The Labiatae is a large, highly evolved plant family with a wide geographical distribution, which constitutes an appropriate group of plants for investigation of the relationship between anthocyanin structure and flower color. Studies on the distribution pattern of anthocyanins in species of *Salvia* and other genera have shown that the red, scarlet, and pink-colored flower varieties contained pelargonidin, the blue ones delphinidin, and the amethyst- and grape-violet-colored ones were based on cyanidin derivatives.[318,200] In a survey of 49 Labiatae species and cultivars, Saito and Harborne[200] confirm the universal occurrence of anthocyanin 3,5-diglucosides, and the widespread occurrence of both aromatic and aliphatic acylation with *p*-coumaric and caffeic acids in the majority of the species. For the more conjugated anthocyanins, the disaccharide sophorose in the 3-position is a distinct feature. Lu and Foo[319] have recently reviewed the polyphenolic content of *Salvia* including 20 anthocyanins. Phippen and Simon.[320] have identified anthocyanins from the herb *Ocimum basilicum* with a slightly simpler structures than the more substituted anthocyanins found in *Ajuga reptans*,[110,111] *Ajuga pyramidalis*,[36] and *Salvia uliginosa*.[209]

The location of an aliphatic acyl group to the glucose 4-position as found in the 3-[6-(*p*-coumaryl)glucoside]-5-[4,6-di-(malonyl)glucosides] of cyanidin, delphinidin, and pelargonidin, the 3-[6-(caffeyl)glucoside]-5-[4,6-di-(malonyl)glucoside] of the two latter anthocyanidins,[198–200] and delphinidin 3-[6-(*p*-coumaryl)glucoside]-5-[4-(acetyl)-6-(malonyl)glucoside] from flowers of *Salvia uliginosa*[209] is very characteristic for Labiatae. Protodelphin (**351**), a metalloantho-cyanin similar to commelinin and protocyanin (Section 10.2.7), has been detected in flowers of *Salvia patens*.[226]

10.5.7 LEGUMINOSAE

More than 60 different anthocyanins have been identified in species belonging to the family Leguminosae (Figure 10.15). In the period of this review, 17 novel anthocyanins have been reported from this family, of which 13 are from *Clitoria ternatea*.[141,144,145] Blue petals of *Clitoria ternatea* contain mainly ternatins (polyacylated delphinidin 3,3′,5′-triglucosides) and preternatins. The change in flower color from blue to mauve in this species is caused by lack of (polyacylated) glucosyl substitutions at both the 3′- and 5′-positions of the ternatins.[141]

The novel delphinidin 3-xylosylgalactoside 5-acetylglucoside and its deacetylated deriva-tive have been isolated from purple pods of pea (*Pisum* spp.).[129] Anthocyanins containing 3-galactoside (including 3-lathyroside) have previously been identified in the genus *Lathyrus*. The 3-rhamnoside-5-glucosides of petunidin (71%), delphinidin (12%), and malvidin (9%) have been isolated from the purple-blue flowers of *Vicia villosa*,[321] while the 3-glucosides of delphinidin, petunin, and malvidin have been mainly identified in species belonging to *Vigna*, *Phaseolus*, and *Glycine*.[322–324] Other anthocyanin patterns have been shown to be of chemo-taxonomic value in the tribes *Podalyrieae*[325] and *Liparieae*.[326] The considerable variation in flower colors in the former tribe was not reflected in the chemical variation because the flower pigments were surprisingly conservative.

10.5.8 LILIACEAE

Around 38 different anthocyanins have been reported to occur in various species belonging to Liliaceae (Figure 10.15). Seventeen have been reported after 1992, and many of these have extraordinary structures. Recently, Saito et al.[20] have isolated 8-*C*-glucosylcyanidin 3-[6-(malonyl)glucoside] from the purple flowers of *Tricyrtis formosana* cultivar Fujimusume, which is the first natural *C*-glycosylanthocyanin to be reported. Three acylated delphinidin 3,7,3′,5′-tetraglucosides have been isolated from the remarkably colored blue berries of two *Dianella* species.[146] One of these pigments had as much as four 6-*p*-coumarylglucoside units located at different aglycone positions. Acetyl groups in anthocyanins are, in general, linked to glucosyl 6-positions; however, in four anthocyanins from tulips the acetyl group is located in the rhamnosyl 2- or 3-position.[121,122] In delphinidin 3-[6-(4-(glucosyl)coumaryl)glucoside]-5-[6-(malonyl)glucoside], which is one of the major anthocyanins isolated from blue-purple flowers of *Triteleia bridgesii*, the coumaryl unit had another glucosyl moiety connected to its 4-hydroxyl group.[166] The other pigments in these flowers were delphinidin and cyanidin 3,5-diglucoside acylated with *p*-coumaryl and in some cases malonyl. Nearly 20 similar anthocyanidin 3,5-diglucosides with a cinnamic acid derivative located to the 6-position of the 3-sugar and possible malonyl or acetyl connected to the 5-sugar have been isolated from flowers of *Hyacinthus orientalis*.[201,327,328] Red *Hyacinthus* petals contain pelargonidin deriva-tives, while the blue flowers have mainly delphinidin derivatives. Anthocyanins from flowers of *Lilium* have been reported to contain the 3-rutinoside-7-glucoside and 3-rutinoside of cyanidin.[120]

10.5.9 NYMPHAEACEAE

Eleven different anthocyanins from water lilies have been identified (Figure 10.15), and six among these are novel compounds reported after 1992. One or more anthocyanins containing a galloylgalactosyl moiety have been isolated from all examined species belonging to the

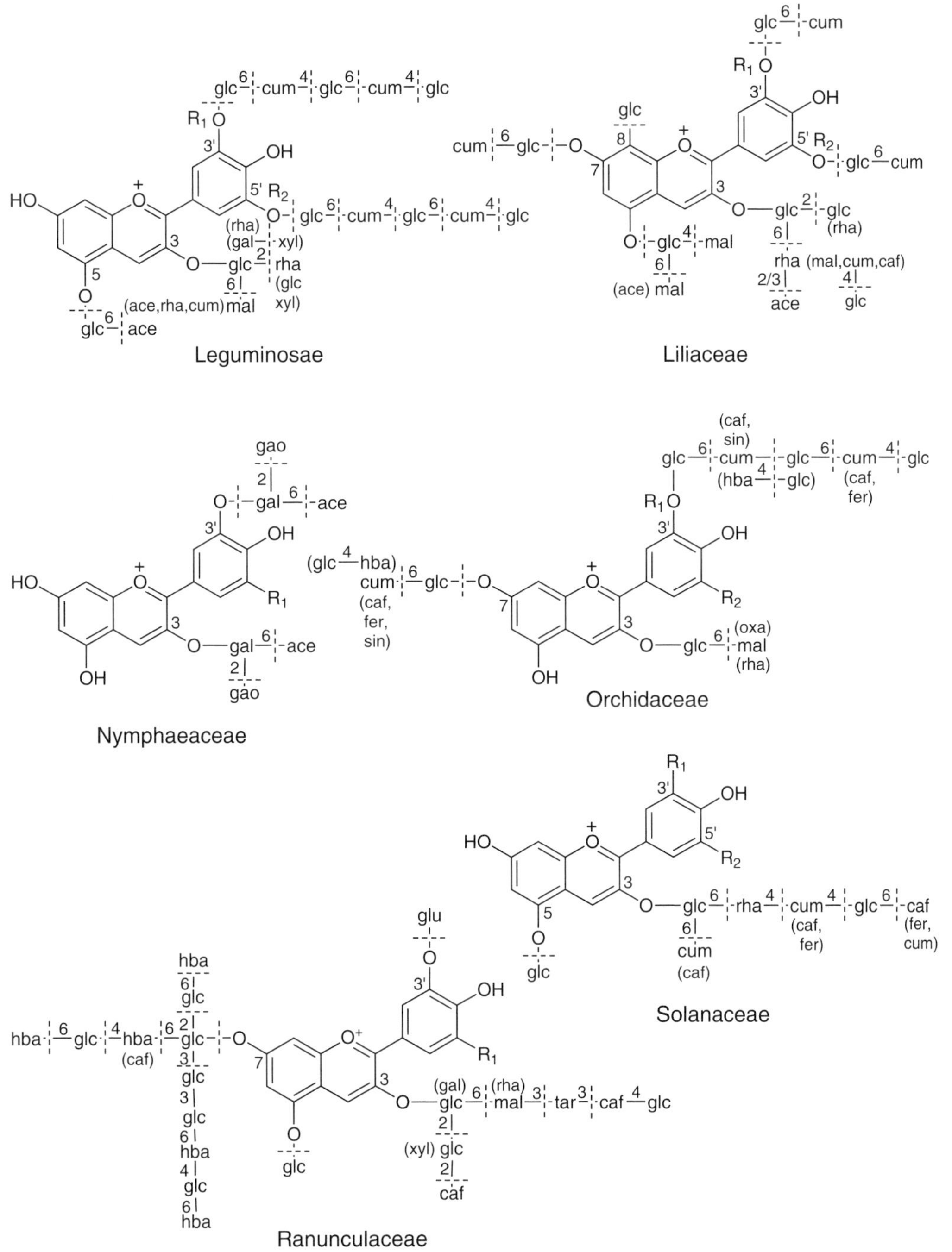

FIGURE 10.15 The structures represent a general presentation of all the anthocyanins identified in each of the families Leguminosae, Liliaceae, Nymphaeaceae, Orchidaceae, Ranunculaceae, and Solanaceae. See Table 10.2 for abbreviations.

family Nymphaeaceae.[26,190–192] This glycosidic moiety is located at the anthocyanidin 3-position in pink or reddish flowers and leaves; however, in the blue flowers of *Nymphaea caerulea* this moiety is located in the rare 3′-position. Some of these pigments are further acetylated on the galactosyl moiety. The pigments from *Nymphaea caerulea* are extraordinary due to their lack of glycosidic substitution at the 3-hydroxyl.[192] The distribution of galloylated anthocyanins outside Nymphaeaceae is restricted to *Acalypha hispida* (Euphorbiaceae)[130] and some species in Aceraceae.[193–195]

10.5.10   ORCHIDACEAE

Around 30 anthocyanins have been identified in orchids (Figure 10.15), and 16 among these are novel compounds reported after 1993. In orchids, cyanidin 3-oxalylglycosides have previously been reckoned to be remarkable taxonomic markers of certain European genera (e.g., *Dactylorhiza*, *Nigritella*, *Orchis*, and *Ophrys*).[222] Outside Orchidaceae, anthocyanins acylated with oxalic acid have only been reported to occur in Evergreen Blackberry (*Rubus laciniatus* Willd.).[57] More recent reports on anthocyanins from flowers of the genera *Dendrobium*,[181,189] *Laelia*,[155,156] *Cattleya*,[155,156] *Bletilla*,[157] *Phalaenopsis*,[180,329] *Sophronitis*,[176] and *Vanda*[179] in subfamily Epidendroideae show, however, a very different and characteristic pattern including substitution at the anthocyanidin 3,7,3′-positions. Flowers of two *Dracula* species (same subfamily) contain only various 3-glycosides of cyanidin and peonidin.[330]

10.5.11   RANUNCULACEAE

Nearly 35 different anthocyanins have been reported to occur in one or more species in the family Ranunculaceae (Figure 10.15), and 24 of these have been reported after 1992 as novel compounds. Flowers of species in the genera *Delphinium* (blue),[109] *Consolida* (blue-violet),[115] and *Aconitum* (purplish-blue)[184] contain similar anthocyanins with polyacyl substitution based on *p*-hydroxybenzoylglucose residues at the 7-hydroxyl of delphinidin, in addition to a more simple glycosyl moiety at the 3-position. Red flowers of *Delphinium hybridum*[119,331] share a similar 3,7-disubstitution pattern based on pelargonidin instead of delphinidin.

The reports on anthocyanins from *Anemone coronaria*,[28,29,132] *Pulsatilla cernua*,[22] and *Ranunculus asiaticus*[126] reveal structures that are very different from the anthocyanins described above. Many of these latter pigments are based on lathyroside or sambubioside residues located at the 3-hydroxyl of delphinidin, pelargonidin, or cyanidin (Table 10.2). Among these, four unusual anthocyanins containing the acyl moiety tartaryl have been isolated from *Anemone coronaria*.[28,29,132] Six of the anthocyanins in this plant have a glucuronoside in the 3′-position.

10.5.12   SOLANACEAE

Altogether 50 different anthocyanins (Figure 10.15) have been reported to occur in the Solanaceae family.[161,169] Reports on anthocyanins from the genus *Solanum*, including potatoes (*Solanum tuberosum*), reveal a dominance of one or more of the 3-[6-(4-*p*-coumaryl-rhamnosyl)-glucoside]-5-glucosides of malvidin, petunidin, delphinidin, peonidin, cyanidin, and pelargonidin. In some cases, the *p*-coumaryl moiety is replaced with sinapyl or caffeyl, and in other cases, the anthocyanins are reported without acylation. Similar anthocyanins, but pelargonidin derivatives, have also been found within the ornamental genus *Petunia*. In addition, diacylated 3-rutinoside-5-glucosides of malvidin, petunidin, delphinidin, and peonidin have also been isolated from flowers of various *Petunia* species and hybrids.[159–165] Only some of the oldest reports on anthocyanins from species within Solanaceae present anthocyanins containing the disaccharides gentiobiose, sophorose, and a 3,7-diglucoside substitution pattern.

The inheritance and biosynthesis of anthocyanin pigmentation in *Petunia* and *Solanum* have received immense interest throughout many decades,[332–334] and the first experiments in genetically modifying anthocyanin flower color were carried out on *Petunia hybrida*.[335]

This subject has been treated comprehensively in Chapter 3. Colorful potatoes have been suggested as potential sources for food colorants,[336,337] and the major anthocyanin of the purple-fleshed potato "Congo," petanin, shows higher color stability than simple anthocyanins at most pH values.[236] The use of anthocyanins as taxonomic markers in the genus *Petunia* discussed in relation to the flower color and possible pollination vectors has been described by Ando et al.[160] based on the presence of at least 24 anthocyanins in the flowers of 20 native taxa.

REFERENCES

1. Timberlake, C.F. and Bridle, P. Anthocyanins, *The Flavonoids*, Harborne, J.B., Mabry, T.J., and Mabry, H., Eds., Academic Press, New York, 1975, Chap. 5.
2. Hrazdina, G., Anthocyanins, *The Flavonoids: Advances in Research*, Harborne, J.B. and Mabry, T.J., Eds., Chapman & Hall, London, 1982, Chap. 3.
3. Harborne, J.B. and Grayer, R.J., The Anthocyanins, *The Flavonoids: Advances in Research Since 1980*, Harborne, J.B., Ed., Chapman & Hall, London, 1988, Chap. 1.
4. Strack, D. and Wray, V. The Anthocyanins, *The Flavonoids: Advances in Research Since 1986*, Harborne, J.B., Ed., Chapman & Hall, London 1994, Chap. 1.
5. Harborne, J.B. and Williams, C.A., Anthocyanins and other flavonoids, *Nat. Prod. Rep.*, 12, 639, 1995.
6. Harborne, J.B. and Williams, C.A., Anthocyanins and other flavonoids, *Nat. Prod. Rep.*, 15, 631, 1998.
7. Harborne, J.B. and Williams, C.A., Anthocyanins and other flavonoids, *Nat. Prod. Rep.*, 18, 310, 2001.
8. Harborne, J.B. and Baxter, H., *The Handbook of Natural Flavonoids*, John Wiley, Chichester, 1999.
9. Toki, K. et al., Delphinidin 3-gentiobiosyl)(Apigenin 7-glucosyl)malonate from the flowers of *Eichhornia crassipes*, *Phytochemistry*, 36, 1181, 1994.
10. Toki, K. et al., Delphinidin 3-gentiobiosyl) (luteolin 7-glucosyl)malonate from the flowers of *Eichhornia crassipes*, *Heterocycles*, 63, 899, 2004.
11. Fossen, T. et al., Covalent anthocyanin–flavonol complexes from flowers of chive, *Allium schoenoprasum*, *Phytochemistry*, 54, 317, 2000.
12. Bloor, S.J. and Falshaw, R., Covalently linked anthocyanin–flavonol pigments from blue *Agapanthus* flowers, *Phytochemistry*, 53, 575, 2000.
13. Fossen, T., Rayyan, S., and Andersen, Ø.M., Dimeric anthocyanins from strawberry (*Fragaria ananassa*) consisting of pelargonidin 3-glucoside covalently linked to four flavan-3-ols, *Phytochemistry*, 65, 1421, 2004.
14. Bakker, J. et al., Identification of an anthocyanin occurring in some red wines, *Phytochemistry*, 44, 1375, 1997.
15. Bakker, J. and Timberlake, C.F., Isolation, identification, and characterization of new color-stable anthocyanins occurring in some red wines, *J. Agric. Food. Chem.*, 45, 35, 1997.
16. Fulcrand, H. et al., A new class of wine pigments generated by reaction between pyruvic acid and grape anthocyanins, *Phytochemistry*, 47, 1401, 1998.
17. Fukui, Y. et al., Structure of rosacyanin B, a novel pigment from petals of *Rosa hybrida*, *Tetrahedron Lett.*, 43, 2637, 2002.
18. Fossen, T. and Andersen, Ø.M., Anthocyanins from red onion, *Allium cepa*, with novel aglycone, *Phytochemistry*, 62, 1217, 2003.
19. Andersen, Ø.M. et al., Anthocyanin from strawberry (*Fragaria ananassa*) with the novel aglycon, 5-carboxypyranopelargonidin, *Phytochemistry*, 65, 405, 2004.
20. Saito, N. et al., The first isolation of *C*-glycosylanthocyanin from the flowers of *Tricyrtis formosana*, *Tetrahedron Lett.*, 44, 6821, 2003.
21. Yoshida, K. et al., Structure of muscarinin A, and acylated anthocyanin, from purplish blue spicate flower petals of *Muscari armeniacum*, *ITE Letters on Batteries, New Technologies & Medicine*, 3, 35, 2002.

22. Yoshitama, K. et al., An acylated pelargonidin diglycoside from *Pulsatilla cernua*, *Phytochemistry*, 47, 105, 1998.

23. Slimestad, R. and Andersen, Ø.M., Cyanidin 3-(2-glucosylgalactoside) and other anthocyanins from fruits of *Cornus suecica*, *Phytochemistry*, 49, 2163, 1998.

24. Toki, K. et al., Delphinidin 3-xylosylrutinoside in petals of *Linum grandiflorum*, *Phytochemistry*, 39, 243, 1995.

25. Cabrita, L., Frøystein, N.Å., and Andersen, Ø.M., Anthocyanin trisaccharides in blue berries of *Vaccinium padifolium*, *Food Chem.*, 69, 33, 2000.

26. Fossen, T. and Andersen, Ø.M., Delphinidin 3′-galloylgalactosides from blue flowers of *Nymphaéa caerulea*, *Phytochemistry*, 50, 1185, 1999.

27. Pale, E. et al., Two triacylated and tetraglucosylated anthocyanins from *Ipomoea asarifolia* flowers, *Phytochemistry*, 64, 1395, 2003.

28. Saito, N. et al., Acylated anthocyanins from the blue-violet flowers of *Anemone coronaria*, *Phytochemistry*, 60, 365, 2002.

29. Toki, K. et al., Acylated cyanidin glycosides from the purple-red flowers of *Anemone coronaria*, *Heterocycles*, 60, 345, 2003.

30. Toki, K. et al., Malvidin 3-glucoside-5-glucoside sulphates from *Babiana stricta*, *Phytochemistry*, 37, 885, 1994.

31. Fukui, Y. et al., A rationale for the shift in colour towards blue in transgenic carnation flowers expressing the flavonoid 3′,5′-hydroxylase gene, *Phytochemistry*, 63, 15, 2003.

32. Bloor, S.J., Blue flower colour derived from flavonol–anthocyanin co-pigmentation in *Ceanothus papillosus*, *Phytochemistry*, 45, 1399, 1997.

33. Glaessgen, W.E. et al., Anthocyanins from cell suspension cultures of *Daucus carota*, *Phytochemistry*, 31, 1593, 1992.

34. Miura, H. et al., Anthocyanin production of *Glehnia littoralis* callus cultures, *Phytochemistry*, 48, 279, 1998.

35. Terahara, N. et al., Anthocyanins in callus induced from purple storage root of *Ipomoea batatas* L., *Phytochemistry*, 54, 919, 2000.

36. Madhavi, D.L. et al., Characterization of Anthocyanins from *Ajuga pyramidalis* Metallica Crispa cell cultures, *J. Agric. Food Chem.*, 44, 1170, 1996.

37. Middleton, E. Jr., Kandaswami, C., and Theoharides, T.C., The effects of plant flavonoids on mammalian cells: implications for inflammation, heart disease, and cancer, *Pharmacol. Rev.*, 52, 673, 2000.

38. Bohm, H. et al., Flavonols, flavone and anthocyanins as natural antioxidants of food and their possible role in the prevention of chronic diseases, *Z. Ernahrungsw.*, 37, 147, 1998.

39. Clifford, M.N., Anthocyanins—nature, occurrence and dietary burden, *J. Sci. Food Agric.*, 80, 1063, 2000.

40. Duthie, G.G., Duthie, S.J., and Kyle, J.A.M., Plant polyphenols in cancer and heart disease: implications as nutritional antioxidants, *Nut. Res. Rev.*, 13, 79, 2000.

41. Duthie, G.G. et al., Antioxidant activity of anthocyanins *in vitro* and *in vivo*, *ACS Symposium Series*, 871 (Nutraceutical Beverages), American Chemical Society, Washington, DC, 90, 2004.

42. Harborne, J.B. and Williams, C.A., Advances in flavonoid research since 1992, *Phytochemistry*, 55, 481, 2000.

43. Scalbert, A. and Williamson, G., Dietary intake and bioavailability of polyphenol, *J. Nutr.*, 130, 2073, 2000.

44. Murkovic, M., Physiological action of anthocyans, *Ernaehrung Medizin*, 17, 167, 2002.

45. Rice-Evans, C.A. and Packer, L., *Flavonoids in Health and Disease*, 2nd edn, Marcel Dekker, New York, 2003, 467.

46. Pietta, P., Minoggio, M., and Bramati, L., Plant polyphenols: structure, occurrence and bioactivity, *Studies in Natural Products Chemistry*, 28 (Bioactive Natural Products (Part I)), Elsevier Science Publishers, Amsterdam, 257, 2003.

47. Kong, J.M. et al., Analysis and biological activities of anthocyanins, *Phytochemistry*, 64, 923, 2003.

48. Prior, R.L., Absorption and metabolism of anthocyanins: potential health effects, *Phytochemicals: Mechanisms of Action*, 4th International Phytochemical Conference, "Phytochemicals: Mechanisms of Action", Pomona, CA (October 21–22, 2002), 2004.

49. Walle, T., Absorption and metabolism of flavonoids, *Free Radical Biol. Med.*, 36, 829, 2004.
50. Stintzing, F.C. and Carle, R., Functional properties of anthocyanins and betalains in plants, food, and in human nutrition, *Trends Food Sci. Technol.*, 15, 19, 2004.
51. Manach, C. et al., Polyphenols: food sources and bioavailability, *Am. J. Clin. Nutr.*, 79, 727, 2004.
52. Hou, D.X., Potential mechanisms of cancer chemoprevention by anthocyanins, *Curr. Mol. Med.*, 3, 149, 2003.
53. Bridle, P. and Timberlake, C.F., Anthocyanins as natural food colors-selected aspects, *Food Chem.*, 58, 103, 1996.
54. Delgado-Vargas F., Jimenez, A.R., and Paredes-Lopez, O., Natural pigments: carotenoids, anthocyanins, and betalains — characteristics, biosynthesis, processing, and stability, *Crit. Rev. Food Sci. Nutr.*, 40, 173, 2000.
55. Francis, F.J., Anthocyanins and betalains: composition and applications, *Cereal Foods World*, 45, 208, 2000.
56. Wrolstad, R.E., Anthocyanins, *IFT Basic Symposium Series*, 14 (Natural Food Colorants), 237, 2000.
57. Stintzing, F.C. et al., A novel zwitterionic anthocyanin from evergreen blackberry (*Rubus laciniatus* Wild), *J. Agric. Food Chem.*, 50, 396, 2002.
58. Mazza, G. and Miniati, E., Eds., *Anthocyanins in Fruits, Vegetables and Grains*, CRC Press, Boca Raton, 1993.
59. Brouillard, R., Chassaing, S., and Fougerousse, A., Why are grape/fresh wine anthocyanins so simple and why is it that red wine color lasts so long? *Phytochemistry*, 64, 1179, 2003.
60. Remy-Tanneau, S. et al., Characterization of a colorless anthocyanin-flavan-3-ol dimer containing both carbon–carbon and ether interflavanoid linkages by NMR and mass spectrometry, *J. Agric. Food Chem.*, 51, 3592, 2003.
61. Salas, E. et al., Demonstration of the occurrence of flavanol–anthocyanin adducts in wine and in model solutions, *Anal. Chim. Acta*, 513, 325, 2004.
62. Springob, K. et al., Recent advances in the biosynthesis and accumulation of anthocyanins, *Nat. Prod. Rep.*, 20, 288, 2003.
63. Irani, N.G., Hernandez, J.M., and Grotewold, E., Regulation of anthocyanin pigmentation, *Recent Adv. Phytochem.*, 37, 59, 2003.
64. Ben-Meir, H., Zuker, A., Weiss, D., and Vainstein, A., Molecular control of floral pigmentation: anthocyanins, *Breed. Ornamentals*, 253, 2002.
65. Forkmann, G., Classical Versus Molecular Breeding of Ornamentals, Proceedings of the International Symposium held in Muenchen, Germany (August 25–29), *Acta Hortic.*, 612, 193, 2003.
66. Forkmann, G. and Martens, S., Metabolic engineering and applications of flavonoids, *Curr. Opin. Biotechnol.*, 12, 155, 2001.
67. Forkmann, G. and Heller, W., Biosynthesis of flavonoids, *Compr. Nat. Prod. Chem.*, 1, 713, 1999.
68. Gould, K.S. and Lee, D.W., Anthocyanins in leaves, *Advances in Botanical Research*, 37, Callow, J.A., Ed., Academic Press, Amsterdam, 2002.
69. Simmonds, M.S.J., Flavonoid–insect interactions: recent advances in our knowledge, *Phytochemistry*, 64, 21, 2003.
70. Close, D.C. and Beadle, C.L., The ecophysiology of foliar anthocyanin, *Bot. Rev.*, 69, 149, 2003.
71. Francis, F.J., Food colorants: anthocyanins, *Crit. Rev. Food Sci. Nutr.*, 28, 273, 1989.
72. Takeoka, G. and Dao, L., Anthocyanins, in *Methods of Analysis for Functional Foods and Nutraceuticals*, Hurst, W.J., Ed., CRC Press, Boca Raton, 2002, 219.
73. Andersen, Ø.M. and Francis, G.W., Techniques of pigment identification, *Plant Pigments and their Manipulation*, Davies, K., Ed., Blackwell Publishing, London, 2004, Chap. 10. 293.
74. DaCosta, C.T., Horton, D., and Margolis, S.A., Analysis of anthocyanins in foods by liquid chromatography, liquid chromatography–mass spectrometry and capillary electrophoresis, *J. Chromatogr. A*, 881, 403, 2000.
75. Andersen, Ø.M., How easy is it nowadays to analyse anthocyanins, in *Polyphenols 2000*, Martens, S., Treutter, D., and Forkmann, G., Eds., Technische Universitat Munchen, Freising, 2001, 49.
76. Wrolstad, R.E. et al., Analysis of anthocyanins in nutraceuticals, *ACS Symp. Ser.*, 803, 42, 2002.

77. Rivas-Gonzalo, J.C., Analysis of anthocyanins, in *Methods in Polyphenol Analysis*, Santos-Buelga, C. and Williamson, G., Eds., Royal Society of Chemistry, Cambridge, 2003, 228.

78. Swinny, E.E. and Markham, K.R., Applications of flavonoid analysis and identification techniques: isoflavones (phytoestrogens) and 3-deoxyanthocyanidins, *Flavonoids in Health and Disease*, 2th edn, Rice-Evans, C.A. and Packer, L., Eds., Dekker/CRC Press, London, 2003, 97.

79. Andersen, Ø.M. and Fossen, T., Characterization of anthocyanins by NMR, Unit F.1.4, *Current Protocols in Food Analytical Chemistry*, Wrolstad, R., Ed., John Wiley, New York, 2003.

80. Brouillard, R. and Dangles, O., Flavonoids and flower colour, *The Flavonoids, Advances in Research Since 1986*; Harborne, J.B., Ed., Chapman & Hall, London, 1994, Chap. 13.

81. Kunz, S., Burkhardt, G., and Becker, H., Riccionidins A and B, anthocyanidins from the cell walls of the liverwort *Ricciocarpos natans*, *Phytochemistry*, 35, 233, 1994.

82. Taniguchi, S. et al., Galloylglucoses and riccionidin A in *Rhus javanica* adventitious root cultures, *Phytochemistry*, 53, 357, 2000.

83. Pale, E. et al., 7-*O*-Methylapigenidin, an anthocyanidin from *Sorghum caudatum*, *Phytochemistry*, 45, 1091, 1997.

84. Lo, S.-C. et al., Phytoalexin accumulation in sorghum: identification of a methyl ether of luteolinidin, *Physiol. Mol. Plant Pathol.*, 49, 21, 1996.

85. Ponniah, L. and Seshadri, T.R., Nuclear oxidation in flavones and related compounds, *Proc. Indian Acad. Sci. A*, 38, 77, 1953.

86. Zorn, B. et al., 3-Desoxyanthocyanidins from *Arrabidaea chica*, *Phytochemistry*, 56, 831, 2001.

87. Devia, B. et al., New 3-Deoxyanthocyanidins from Leaves of *Arrabidaea chica*, *Phytochem. Anal.*, 13, 114, 2002.

88. Lu, Y., Sun, Y., and Yeap Foo, L., Novel pyranoanthocyanins from black currant seeds, *Tetrahedron Lett.*, 41, 5975, 2000.

89. Lu, Y.R. and Foo, L.Y., Unusual anthocyanin reaction with acetone leading to pyranoanthocyanin formation, *Tetrahedron Lett.*, 42, 1371, 2001.

90. Lu, Y., Foo, L.Y., and Wong, H., Isolation of the first C-2 addition products of anthocyanins, *Tetrahedron Lett.*, 43, 6621, 2002.

91. Lu, Y., Foo, L.Y., and Sun, Y., New pyranoanthocyanins from black currant seeds, *Tetrahedron Lett.*, 43, 7341, 2002.

92. Santos, H. et al., Elucidation of the multiple equilibria of malvin in aqueous solution by one- and two-dimensional NMR, *Phytochemistry*, 33, 1227, 1993.

93. Pina, F., Thermodynamics and kinetics of flavylium salts — malvin revisited, *J. Chem. Soc., Faraday Trans.*, 94, 2109, 1998.

94. Pina, F., Maestri, M., and Balzani, V., Photochromic systems based on synthetic flavylium compounds and their potential use as molecular-level memory elements, in *Handbook of Photochemistry and Photobiology*, Abdel-Mottaleb, M.S.A. and Nalwa, H.S.Eds., American Scientific Publishers, Stevenson Ranch, 2003, 3, 411.

95. Pina, F. et al., Thermal and photochemical properties of 4′,7-dihydroxyflavylium in water-ionic liquid biphasic systems: a write–read–erase molecular switch, *Angew. Chem. Int. Ed.*, 43, 1525, 2004.

96. Swinny, E.E., A novel acetylated 3-deoxyanthocyanidin laminaribioside from the fern *Blechnum novae-zelandiae*, *Z. Naturforsch., C: Biosci.*, 56, 177, 2001.

97. Piovan, A., Filippini, R., and Favretto, D., Characterization of the anthocyanins of *Catharanthus roseus* (L.) G. Don *in vivo* and *in vitro* by electrospray ionization in trap mass spectrometry, *Rapid Commun. Mass Spectrom.*, 12, 361, 1998.

98. Saito, N. et al., Anthocyanin glycosides from the flowers *Alstroemeria*, *Phytochemistry*, 24, 2125, 1985.

99. Saito, N. et al., 6-Hydroxyanthocyanidin glycosides in the flowers of *Alstroemeria*, *Phytochemistry*, 27, 1399, 1988.

100. Nørbæk, R. et al., Anthocyanins in chilean species of *Alstroemeria*, *Phytochemistry*, 42, 97, 1996.

101. Tatsuzawa, F. et al., 6-Hydroxycyanidin 3-malonylglucoside from the flowers of *Alstroemeria* "Tiara", *Heterocycles*, 55, 1195, 2001.

102. Tatsuzawa, F. et al., Two novel 6-hydroxyanthocyanins in the flowers of *Alstroemeria* "Westland", *Heterocycles*, 57, 1787, 2002.

103. Nygard, A.M. et al., Structure determination of 6-hydroxycyanidin- and 6-hydroxydelphinidin-3-(6''-O-alpha-L-rhamnopyranosyl-beta-D-glucopyranosides) and other anthocyanins from *Alstroemeria* cultivars, *Acta Chem. Scand.*, 51, 108, 1997.
104. Nørbæk, R., Christensen, L.P., and Brandt, K., An HPLC investigation of flower color and breeding of anthocyanins in species and hybrids of *Alstroemeria*, *Plant Breed.*, 117, 63, 1998.
105. Tatsuzawa, F. et al., Flower colors and anthocyanin pigments in 45 cultivars of *Alstroemeria* L., *J. Jpn. Soc. Hortic. Sci.*, 72, 243, 2003.
106. Tatsuzawa, F. et al., 6-Hydroxypelargonidin glycosides in the orange-red flowers of *Alstroemeria*, *Phytochemistry*, 62, 1239, 2003.
107. Clevenger, S., New anthocyanidin in *Impatiens*, *Can. J. Biochem. Physiol.*, 42, 154, 1964.
108. Terahara, N. et al., Structure of ternatin A1, the largest ternatin in the major blue anthocyanins from *Clitoria ternatea* flower, *Tetrahedron Lett.*, 31, 2921, 1990.
109. Kondo, T. et al., Structure of cyanodelphin, a tetra-*p*-hydroxybenzoated anthocyanin from blue flower of *Delphinium hybridum*, *Tetrahedron Lett.*, 32, 6375, 1991.
110. Takeda, K., Harborne, J.B., and Self, R., Identification and distribution of malonated anthocyanins in plants of the Compositae, *Phytochemistry*, 25, 1337, 1986.
111. Toki, K., Saito, N., and Honda, T., 3 Cyanidin 3-glucuronylglucosides from red flowers of *Bellis perennis*, *Phytochemistry*, 30, 3769, 1991.
112. Nair, A.G.R. and Mohandoss, S., Feruloylscutellarin and other rare flavonoids from *Holmskioldia sanguinea*, *Indian J. Chem., Sect. B.*, 24, 323, 1985.
113. Terahara, N. et al., Triacylated anthocyanins from *Ajuga reptans* flowers and cell cultures, *Phytochemistry*, 42, 199, 1996.
114. Terahara, N. et al., Acylated anthocyanidin 3-sophoroside-5-glucosides from *Ajuga reptans* flowers and the corresponding cell cultures, *Phytochemistry*, 58, 493, 2001.
115. Saito, N. et al., Acylated delphinidin glycosides in the blue-violet flowers of *Consolida armeniaca*, *Phytochemistry*, 41, 1599, 1996.
116. Chirol, N. and Jay, M., Acylated anthocyanins from flowers of *Begonia*, *Phytochemistry*, 40, 275, 1995.
117. Brandt, K. et al., Structure and biosynthesis of anthocyanins in flowers of *Campanula*, *Phytochemistry*, 33, 209, 1993.
118. Saito, N. et al., Acylated cyanidin 3-rutinoside-5,3'-diglucoside from the purple-red flower of *Lobelia erinus*, *Phytochemistry*, 39, 423, 1995.
119. Saito, N. et al., Acylated pelargonidin 3,7-glycosides from red flowers of *Delphinium hybridum*, *Phytochemistry*, 49, 881, 1998.
120. Nørbæk, R. and Kondo, T., Anthocyanins from flowers of *Lilium* (Liliaceae), *Phytochemistry*, 50, 1181, 1999.
121. Nakayama, M. et al., Anthocyanins in the dark purple anthers of *Tulipa gesneriana*: identification of two novel delphinidin 3-*O*-(6-*O*-(acetyl-α-rhamnopyranosyl)-β-glucopyranosides), *Biosci. Biotechnol. Biochem.*, 63, 1509, 1999.
122. Torskangerpoll, K., Fossen, T., and Andersen, Ø.M., Anthocyanin pigments of tulips, *Phytochemistry*, 52, 1687, 1999.
123. Markham, K.R., Novel anthocyanins produced in petals of genetically transformed Lisianthus, *Phytochemistry*, 42, 1035, 1996.
124. Cabrita, L. and Andersen, Ø.M., Anthocyanins in blue berries of *Vaccinium padifolium*, *Phytochemistry*, 52, 1693, 1999.
125. Nakatani, N. et al., Acylated anthocyanins from fruits of *Sambucus canadensis*, *Phytochemistry*, 38, 755, 1995.
126. Toki, K., Two malonylated anthocyanidin glycosides in *Ranunculus asiaticus*, *Phytochemistry*, 42, 1055, 1996.
127. Takeda, K. et al., A malonylated anthocyanin and flavonols in the blue flowers of *Meconopsis*, *Phytochemistry*, 42, 863, 1996.
128. Tanaka, M. et al., A malonylated anthocyanin and flavonols in blue *Meconopsis* flowers, *Phytochemistry*, 56, 373, 2001.

129. Markham, K.R. and Ofman, D.J., Lisianthus flavonoid pigments and factors influencing their expression in flower colours, *Phytochemistry*, 34, 679, 1993.

130. Reiersen, B. et al., Anthocyanins acylated with gallic acid from chenille plant, *Acalypha hispida*, *Phytochemistry*, 64, 867, 2003.

131. Terahara, N. et al., New anthocyanins from purple pods of Pea (*Pisum* spp.), *Biosci. Biotechnol. Biochem.*, 64, 2569, 2000.

132. Toki, K. et al., Anthocyanins from the scarlet flowers of *Anemone coronaria*, *Phytochemistry*, 56, 711, 2001.

133. Asada, Y., Sakamoto, K., and Furuya, T., A minor anthocyanin from cultured cells of *Aralia cordata*, *Phytochemistry*, 35, 1471, 1994.

134. Terahara, N. et al., Cyandin 3-lathyroside from berries of *Fatsia japonica*, *Phytochemistry*, 31, 1446, 1992.

135. Terahara, N., Yamaguchi, M., and Honda, T., Malonylated anthocyanins from bulbs of Red Onion, *Allium cepa* L., *Biosci. Biotechnol. Biochem.*, 58, 1324, 1994.

136. Baublis, A.J. and Berber-Jiménez, M.D., Structural and conformational characterization of stable anthocyanin from *Tradescantia pallida*, *J. Agric. Food Chem.*, 43, 640, 1995.

137. Idaka, E. et al., Structure of zebrinin, a novel acylated anthocyanin isolated from *Zebrina pendula*, *Tetrahedron Lett.*, 28, 1901, 1987.

138. Baublis, A.J. and Berberjimenez, M.D., Structural and conformational characterization of a stable anthocyanin from *Tradescantia-pallida*, *J. Agric. Food Chem.*, 43, 640, 1994.

139. Bloor, S.J., Novel pigments and copigmentation in the blue marguerite daisy, *Phytochemistry*, 50, 1395, 1999.

140. Webby, R.F. and Boase, M.R., Peonidin 3-*O*-neohesperidoside and other flavonoids from *Cyclamen persicum* petals, *Phytochemistry*, 52, 939, 1999.

141. Kazuma, K., Noda, N., and Suzuki, M., Flavonoid composition related to petal color in different lines of *Clitoria ternatea*, *Phytochemistry*, 64, 1133, 2003.

142. Andersen, Ø.M., Delphinidin-3-neohesperidoside and cyanidin-3-neohesperidoside from receptacles of Poducarpus species, *Phythochemistry*, 28, 495, 1989.

143. Crowden, R.K. and Jarman, S.J., 3-Deoxyanthocyanins from fern *Blechnum-procerum*, *Phytochemistry*, 13, 1947, 1974.

144. Terahara, N. and Oda, M., Five new anthocyanins, ternatins A3, B4, B2, and D2, from *Clitoria ternatea* flowers, *J. Nat. Prod.*, 59, 139, 1996.

145. Terahara, N. et al., Eight new anthocyanins, ternatins C1–C5 and D3 and preternatins A3 and C4 from young *Clitoria ternatea* flowers, *J. Nat. Prod.*, 61, 1361, 1998.

146. Bloor, S.J., Deep blue anthocyanins from blue *Dianella* berries, *Phytochemistry*, 58, 923, 2001.

147. Fossen, T., Slimestad, R., and Andersen, Ø.M., Anthocyanins with 4'-glucosidation from red onion, *Allium cepa*, *Phytochemistry*, 64, 1367, 2003.

148. Hedin, P.A. et al., Isolation and structural determination of 13 flavonoid glycosides in *Hibiscus esculentus* (okra), *Am. J. Bot.*, 55, 431, 1968.

149. Kondo, T. et al., Structure of lobelinin A and B. Novel anthocyanins acylated with three and four different organic acids, respectively, *Tetrahedron Lett.*, 30, 6055, 1989.

150. Baker, D.C. et al., Effects of supplied cinnamic-acids and biosynthetic intermediates on the anthocyanins accumulated by wild carrot suspension-cultures, *Plant Cell Tis. Organ Cult.*, 39, 79, 1994.

151. Dougall, D.K. et al., Anthocyanins from wild carrot suspension cultures acylated with supplied carboxylic acids, *Carbohydr. Res.*, 310, 177, 1998.

152. Harborne, J.B., *The Flavonoids: Advances in Research Since 1986*, Chapman & Hall, London, 1994.

153. Hosokawa, K. et al., Three acylated cyanidin glucosides in pink flowers of *Gentiana*, *Phytochemistry*, 40, 941, 1995.

154. Hosokawa, K. et al., Seven acylated anthocyanins in blue flowers of *Gentiana*, *Phytochemistry*, 45, 167, 1997.

155. Tatsuzawa, F. et al., An acylated cyanidin glycoside in the red-purple flowers of x *Laeliocattleya* cv Mini Purple, *Phytochemistry*, 37, 1179, 1994.

156. Tatsuzawa, F. et al., Acylated cyanidin 3,7,3'-triglucosides in flowers of x *Laeliocattleya* cv. Mini Purple and its relatives, *Phytochemistry*, 41, 635, 1996.

157. Saito, N. et al., Acylated cyanidin glycosides in the purple-red flowers of *Bletilla striata*, *Phytochemistry*, 40, 1523, 1995.

158. Terahara, N. et al., Anthocyanins from red flower tea (Benibana-cha), *Camellia sinensis*, *Phytochemistry*, 56, 359, 2001.

159. Fukui, Y. et al., Structures of two diacylated anthocyanins from *Petunia hybrida* cv. Surfinia Violet Mini, *Phytochemistry*, 47, 1409, 1998.

160. Ando, T. et al., Floral anthocyanins in wild taxa of *Petunia*, *Biochem. Syst. Ecol.*, 27, 623, 1999.

161. Gonzalez, E., Fougerousse, A., and Brouillard, R., Two diacylated malvidin glycosides from *Petunia hybrida* flowers, *Phytochemistry*, 58, 1257, 2001.

162. Tatsuzawa, F. et al., Diacylated malvidin 3-rutinoside-5-glucosides from the flowers of *Petunia guarapuavensis*, *Heterocycles*, 45, 1197, 1997.

163. Tatsuzawa, F. et al., Acylated malvidin 3-rutinosides in dusky violet flowers of *Petunia integrifolia* subsp. *inflata*, *Phytochemistry*, 52, 351, 1999.

164. Tatsuzawa, F. et al., Acylated delphinidin 3-rutinoside-5-glucosides in the flowers of *Petunia reitzii*, *Phytochemistry*, 54, 913, 2000.

165. Tatsuzawa, F. et al., Acylated peonidin 3-rutinoside-5-glucosides from commercial Petunia cultivars with pink flowers, *Heterocycles*, 63, 509, 2004.

166. Toki, K., Saito, N., and Honda, T., Acylated anthocyanins from the blue-purple flowers of *Triteleia bridgesii*, *Phytochemistry*, 48, 729, 1998.

167. Bloor, S.J. and Abrahams, S., The structure of the major anthocyanin in *Arabidopsis thaliana*, *Phytochemistry*, 59, 343, 2002.

168. Slimestad, R., Aaberg, A., and Andersen, Ø.M., Acylated anthocyanins from petunia flowers, *Phytochemistry*, 50, 1081, 1999.

169. Fossen, T. et al., Anthocyanins from a Norwegian potato cultivar, *Food Chem.*, 81, 433, 2003.

170. Kamsteeg, J. et al., Anthocyanins isolated from petals of various genotypes of red Campion (*Silene-dioica* (L.) Clairv.), *Z. Naturforsch., C: Biosci.*, 33, 475, 1978.

171. Naito, K. et al., Acylated pelargonidin glycosides from a red potato, *Phytochemistry*, 47, 109, 1998.

172. Fossen, T. and Andersen, Ø.M., Anthocyanins from tubers and shoots of the purple potato, *Solanum tuberosum, J. Hortic. Sci. Biotechnol.*, 75, 360, 2000.

173. Saito, N. et al., Acylated pelargonidin 3-sambubioside-5-glucosides in *Matthiola incana*, *Phytochemistry*, 41, 1613, 1996.

174. Lin, M., Shi, Z., and Francis, F.J., A simple method of analysis for *Tradescantia* anthocyanins, *J. Food Sci.*, 57, 766, 1992.

175. Shi, Z., Francis, F.J., and Daun, H., Quantitative comparison of the stability of anthocyanins from *Brassica oleracea* and *Tradescantia pallida* in non-sugar drink model and protein model systems, *J. Food Sci.*, 57, 768, 1992.

176. Tatsuzawa, F. et al., Acylated cyanidin glycosides in the orange-red flowers of *Sophronitis coccinea*, *Phytochemistry*, 49, 869, 1998.

177. Saito, N. et al., Acylated cyanidin 3-sambubioside-5-glucosides in *Matthiola incana*, *Phytochemistry*, 38, 1027, 1995.

178. Suzuki, M., Nagata, T., and Terahara, N., New acylated anthocyanins from *Brassica campestris* var. *chinensis, Biosci. Biotechnol. Biochem.*, 61, 1929, 1997.

179. Yoshida, K. et al., Structures of alatanin A, B and C isolated from edible purple yam *Dioscorea alata, Tetrahedron Lett.*, 32, 5575, 1991.

180. Tatsuzawa, F. et al., Acylated cyanidin glycosides in the red-purple flowers of *Phalaenopsis*, *Phytochemistry*, 45, 173, 1997.

181. Williams, C.A. et al., Acylated anthocyanins and flavonols from purple flowers of *Dendrobium* cv. "Pompadour", *Biochem. Syst. Ecol.*, 30, 667, 2002.

182. Tatsuzawa, F. et al., Acylated anthocyanins in the flowers of *Vanda* (Orchidaceae), *Biochem. Syst. Ecol.*, 32, 651, 2004.

183. Kondo, T. et al., Structure of violdelphin, an anthocyanin from violet flower of *Delphinium hybridum*, *Chem. Lett.*, 1, 137, 1990.

184. Takeda, K. et al., The anthocyanin responsible for purplish blue flower colour of *Aconitum chinense*, *Phytochemistry*, 36, 613, 1994.

185. Asen, S., Stewart, R.N., and Norris, K.H., Pelargonidin 3-[di (*p*-hydroxybenzoyl)rutinoside]-7-glucoside from flowers of *Campanula*, *Phytochemistry*, 18, 1251, 1979.

186. Terahara, N. et al., Structures of campanin and rubrocampanin: two novel acylated anthocyanins with *p*-hydroxybenzoic acid from the flowers of bellflower, *Campanula medium* L., *J. Chem. Soc., Perkin Trans. 1*, 3327, 1990.

187. Miyazaki, T., Tsuzuki, W., and Suzuki, T., Composition and structure of anthocyanins in the periderm and flesh of sweet potatoes, *J. Jpn. Soc. Hortic. Sci.*, 60, 217, 1991.

188. Terahara, N. et al., Six diacylated anthocyanins from the storage roots of purple sweet potato, *Ipomoea batatas*, *Biosci. Biotechnol. Biochem.*, 63, 1420, 1999.

189. Saito, N. et al., An acylated cyanidin glycoside from the red-purple flowers of *Dendrobium*, *Phytochemistry*, 37, 245, 1994.

190. Strack, D. et al., Two anthocyanins acylated with gallic acid from the leaves of *Victoria amazonica*, *Phytochemistry*, 31, 989, 1992.

191. Fossen, T., Larsen, Å., and Andersen, Ø.M., Anthocyanins from flowers and leaves of *Nymphaéa* × *marliacea* cultivars, *Phytochemistry*, 48, 823, 1998.

192. Fossen, T. and Andersen, Ø.M., Cyanidin 3-(6″-acetylgalactoside) and other anthocyanins from reddish leaves of the water lily, *Nymphaéa alba*, *J. Hortic. Sci. Biotechnol.*, 76, 213, 2001.

193. Ji, S.B. et al., Galloylcyanidin glycosides from *Acer*, *Phytochemistry*, 31, 655, 1992.

194. Ji, S.B. et al., Distribution of anthocyanins in Aceraceae leaves, *Biochem. Syst. Ecol.*, 20, 771, 1992.

195. Fossen, T. and Andersen, Ø.M., Cyanidin 3-(2″,3″-digalloylglucoside) from red leaves of *Acer platanoides*, *Phytochemistry*, 52, 1697, 1999.

196. Karawya, M.S. et al., Anthocyanins from the seeds of *Abrus precatorius*, *Fitoterapia*, 52, 175, 1981.

197. Mazza, G. and Gao, L., Malonylated anthocyanins in purple sunflower seeds, *Phytochemistry*, 35, 237, 1994.

198. Kondo, T. et al., Structure of monardaein, a bis-malonylated anthocyanin isolated from golden balm, *Monarda didyma*, *Tetrahedron Lett.*, 26, 5879, 1985.

199. Kondo, T. et al., Structure of anthocyanins in scarlet, purple, and blue flowers of *Salvia*, *Tetrahedron Lett.*, 30, 6729, 1989.

200. Saito, N. and Harborne, J.B., Correlations between anthocyanin type, pollinator and flower colour in the Labiatae, *Phytochemistry*, 31, 3009, 1992.

201. Hosokawa, K. et al., Acylated anthocyanins from red *Hyacinthus orientalis*, *Phytochemistry*, 39, 1437, 1995.

202. Andersen, Ø.M. and Fossen, T., Anthocyanins with an unusual acylation pattern from stem of *Allium victorialis*, *Phytochemistry*, 40, 1809, 1995.

203. Fossen, T. et al., Characteristic anthocyanin pattern from onions and other *Allium* spp., *J. Food Sci.*, 61, 703, 1996.

204. Fossen, T. and Andersen, Ø.M., Malonated anthocyanins of garlic *Allium sativum* L., *Food Chem.*, 58, 215, 1997.

205. Donner, H., Gao, L., and Mazza, G., Separation and characterization of simple and malonylated anthocyanins in red onions, *Allium cepa* L., *Food Res. Int.* 30, 637, 1998.

206. Nakayama, M. et al., Identification of cyanidin 3-*O*-(3″,6″-beta-dimalonyl-beta-glucopyranoside) as a flower pigment of chrysanthemum (*Dendranthema grandiflorum*), *Biosci. Biotechnol. Biochem.*, 61, 1607, 1997.

207. Fossen, T. et al., Anthocyanins of grasses, *Biochem. Syst. Ecol.*, 30, 855, 2002.

208. Terahara, N. et al., A diacylated anthocyanin from *Tibouchina urvilleana* flowers, *J. Nat. Prod.*, 56, 335, 1993.

209. Ishikawa, T. et al., An acetylated anthocyanin from the blue petals of *Salvia uliginosa*, *Phytochemistry*, 52, 517, 1999.

210. Andersen, Ø.M., Viksund, R.I., and Pedersen, A.T., Malvidin 3-(6-acetylglucoside)-5-glucoside and other anthocyanins from flowers of *Geranium sylvaticum*, *Phytochemistry*, 38, 1513, 1995.

211. Markham, K.R., Mitchell, K.A., and Boase, M.R., Malvidin-3-*O*-glucoside-5-*O*-(6-acetylglucoside) and its colour manifestation in "Johnson's Blue" and other "Blue" geraniums, *Phytochemistry*, 45, 417, 1997.

212. Terahara, N. et al., Anthocyanins acylated with malic acid in *Dianthus caryophyllus* and *D. deltoids*, *Phytochemistry*, 25, 1715, 1986.

213. Terahara, N. and Yamaguchi, M., 1H NMR spectral analysis of the malylated anthocyanins from *Dianthus*, *Phytochemistry*, 25, 2906, 1986.

214. Bloor, S.J., A macrocyclic anthocyanin from red/mauve carnation flowers, *Phytochemistry*, 49, 225, 1998.

215. Gonnet, J.F. and Fenet, B., "Cyclamen red" colors based on a macrocyclic anthocyanin in carnation flowers, *J. Agric. Food Chem.*, 48, 22, 2000.

216. Nakayama, M. et al., Cyclic malyl anthocyanins in *Dianthus caryophyllus*, *Phytochemistry*, 55, 937, 2000.

217. Tamura, H. et al., Structures of a succinyl anthocyanin and a malonyl flavone, two constituents of the complex blue pigment of cornflower *Centaurea cyanus*, *Tetrahedron Lett.*, 24, 5749, 1983.

218. Sulyok, G. and Laszlo-Bencsik, A., Cyanidin 3-(6-succinyl glucoside)-5-glucoside from flowers of seven *Centaurea species*, *Phytochemistry*, 24, 1121, 1985.

219. Takeda, K. et al., Pelargonidin 3-(6″-succinyl glucoside)-5-glucoside from pink *Centaurea cyanus* flowers. *Phytochemistry*, 27, 1228, 1988.

220. Kondo, T. et al., A new molecular mechanism of blue colour development with protocyanin, a supramolecular pigment from Cornflower, *Centaurea cyanus*, *Tetrahedron Lett.*, 39, 8307, 1998.

221. Fossen, T. and Andersen, Ø.M., Cyanidin 3-*O*-(6″-succinyl-β-glucopyranoside) and other anthocyanins from *Phragmites australis*, *Phytochemistry*, 49, 1065, 1998.

222. Strack, D., Busch, E., and Klein, E., Anthocyanin patterns in European orchids and their taxonomic and phylogenetic relevance, *Phytochemistry*, 28, 2127, 1989.

223. Figueiredo, P. et al., New aspects of anthocyanin complexation. Intramolecular copigmentation as a means for colour loss? *Phytochemistry*, 41, 301, 1996.

224. Uphoff, W., Identification of European orchids by determination of the anthocyanin concentration during development of the blossoms, *Experientia*, 38, 778, 1982.

225. Takeda, K., Harborne, J.B., and Watermann, P.G., Malonylated flavonoids and blue flower colour in lupin, *Phytochemistry*, 34, 421, 1993.

226. Takeda, K. et al., A blue pigment complex in flowers of *Salvia patens*, *Phytochemistry*, 35, 1167, 1994.

227. Kondo, T. et al., Structural basis of blue-colour development in flower petals from *Commelina communis*, *Nature*, 358, 515, 1992.

228. Ohsawa, Y. et al., Flavocommelin octaacetate, *Acta Crystallogr., Sect. C: Crystal Struct. Commun.*, 50, 645, 1994.

229. Kondo, T., Oyama, K., and Yoshida, K., Chiral molecular recognition on formation of a metalloanthocyanins: a supramolecular metal complex pigment from blue flowers of *Salvia patens*, *Angew. Chem. Int. Ed. Engl.*, 40, 894, 2001.

230. Kondo, T. et al., Composition of procyanin, a self-assembled supramolecular pigment from the blue cornflower, *Centaurea-cyanus*, *Angew. Chem. Int. Ed. Engl.*, 33, 978, 1994.

231. Kondo, T. et al., Cause of flower color variation of hydrangea, *Hydrangea macrophylla*, *Tennen Yuki Kagobutsu Toronkai Koen Yoshishu*, 41, 265, 1999.

232. Yoshida, K., Oyama, K., and Kondo, T., Flower color development and nano-science, *Yuki Gosei Kagaku Kyokaishi*, 62, 490, 2004.

233. Dangles, O., Elhabiri, M., and Brouillard, R., Kinetic and thermodynamic investigation of the aluminium–anthocyanin complexation in aqueous solution, *J. Chem. Soc., Perkin Trans. 2*, 2587, 1994.

234. Elhabiri, M. et al., Anthocyanin–aluminium and –gallium complexes in aqueous solution, *J. Chem. Soc., Perkin Trans. 2*, 355, 1997.

235. Margarida, C.M. et al., Complexation of aluminium(III) by anthocyanins and synthetic flavylium salts. A source for blue and purple color, *Inorg. Chim. Acta*, 356, 51, 2003.
236. Torskangerpoll, K. et al., Color and substitution pattern in anthocyanidins. A combined quantum chemical–chemometrical study, *Spectrochim. Acta, Part A*, 55, 761, 1999.
237. Fossen, T., Cabrita, L., and Andersen, Ø.M., Color and stability of pure anthocyanins influenced by pH including the alkaline region, *Food Chem.*, 63, 435, 1998.
238. Cabrita, L., Fossen, T., and Andersen, Ø.M., Colour and stability of the six common anthocyanidin 3-glucosides in aqueous solutions, *Food Chem.*, 68, 101, 2000.
239. Houbiers, C. et al., Color stabilization of malvidin 3-glucoside: self-aggregation of the flavylium cation and copigmentation with the Z-chalcone form, *J. Phys. Chem. B*, 102, 3578, 1998.
240. Pina, F., Thermodynamics and kinetics of flavylium salts — malvin revisited, *J. Chem. Soc., Faraday Trans.*, 94, 2109, 1998.
241. Melo, M.J., Moncada, M.C., and Pina, F., On the red colour of raspberry (*Rubus idaeus*), *Tetrahedron Lett.*, 41, 1987, 2000.
242. Dangles, O., Saito, N., and Brouillard, R., Anthocyanin intramolecular copigment effect, *Phytochemistry*, 34, 119, 1993.
243. Dangles, O., Anthocyanin complexation and color expression, *Analusis*, 25, 50, 1997.
244. Redus, M., Baker, D.C., and Dougall, D.K., Rate and equilibrium constants for the dehydration and deprotonation reactions of some monoacylated and glycosylated cyanidin derivatives, *J. Agric. Food Chem.*, 47, 3449, 1999.
245. Gakh, E.G., Dougall, D.K., and Baker, D.C., Proton nuclear magnetic resonance studies of monoacylated anthocyanins from the wild carrot: part 1. Inter- and intra-molecular interactions in solution, *Phytochem. Anal.*, 9, 28, 1998.
246. Honda, T. and Saito, N., Recent progress in the chemistry of polyacylated anthocyanins as flower color pigments, *Heterocycles*, 56, 633, 2001.
247. Figueiredo, P. et al., New features of intramolecular copigmentation by acylated anthocyanins, *Phytochemistry*, 51, 125, 1999.
248. Yoshitama, K. et al., A stable reddish purple anthocyanin in the leaf of *Gynura aurantiaca* cv. "Purple Passion", *J. Plant Res.*, 107, 209, 1994.
249. George, F. et al., Influence of *trans–cis* isomerization of coumaric acid substituents on colour variance and stabilisation in anthocyanins, *Phytochemistry*, 57, 791, 2001.
250. Yoshida, K. et al., Influence of *E,Z*-isomerization and stability of acylated anthocyanins under the UV irradiation, *Biochem. Eng. J.*, 14, 163, 2003.
251. Ramachandra Rao, S. and Ravishankar, G.A., Plant cell cultures: chemical factories of secondary metabolites, *Biotechnol. Adv.*, 20, 101, 2002.
252. Zhang, W. and Furusaki, S., Production of anthocyanins by plant cell cultures, *Biotechnol. Bioprocess. Eng.*, 4, 231, 1999.
253. Nakamura, M., Seki, M., and Furusaki, S., Enhanced anthocyanin methylation by growth limitation in strawberry suspension culture, *Enzyme Microb. Technol.*, 22, 404, 1998.
254. Zhong, J.J. et al., Enhancement of anthocyanin production by *Perilla frutescens* cells in a stirred bioreactor with internal light irradiation, *J. Ferment. Bioeng.*, 75, 299, 1993.
255. Smith, M.A.L. et al., Continuous cell culture and product recovery from wild *Vaccinium pahalae* germplasm, *J. Plant Physiol.*, 150, 462, 1997.
256. Hosokawa, K. et al., Acylated anthocyanins in red flowers of *Hyacinthus orientalis* regenerated *in vitro*, *Phytochemistry*, 42, 671, 1996.
257. Do, C.B. and Cormier, F., Accumulation of peonidin 3-glucoside enhanced by osmotic stress in grape (*Vitis vinifera* L.) cell suspension, *Plant Cell Tissue Organ Culture*, 24, 49, 1991.
258. Curtin, C., Zhang, W., and Franco, C., Manipulating anthocyanin composition in *Vitis vinifera* suspension cultures by elicitation with jasmonic acid and light irradiation, *Biotechnol. Lett.*, 25, 1131, 2003.
259. Krisa, S. et al., Production of ^{13}C-labelled anthocyanins by *Vitis vinifera* cell suspension cultures, *Phytochemistry*, 51, 651, 1999.
260. Aumont, V. et al., Production of highly ^{13}C-labeled polyphenols in *Vitis vinifera* cell bioreactor cultures, *J. Biotechnol.*, 109, 287, 2004.

261. Wharton, P.S. and Nicholsen, R.L., Temporal synthesis and radiolabelling of the sorghum 3-deoxyanthocyanidin phytoalexins and the anthocyanin, cyanidin 3-dimalonyl glucoside, *New Phytol.*, 145, 457, 2000.

262. Vitrac, X. et al., Carbon-14 biolabelling of wine polyphenols in *Vitis vinifera* cell suspension cultures, *J. Biotechnol.*, 95, 49, 2002.

263. Grusak, M.A. et al., An enclosed-chamber labeling system for the safe ^{14}C-enrichment of phytochemicals in plant cell suspension cultures, *In Vitro Cell. Dev. Biol. Plant*, 40, 80, 2004.

264. Endress, R., Plant cells as producers of secondary compounds, in *Plant Cell Biotechnology*, Endres, R., Ed., Springer-Verlag, Berlin, 1994, 121.

265. Zhang, W., Curtin, C., and Franco, C., Towards manipulation of post-biosynthetic events in secondary metabolism of plant cell cultures, *Enzyme Microb. Technol.*, 30, 688, 2002.

266. Kobayashi, Y. et al., Large-scale production of anthocyanin by *Aralia cordata* cell suspension cultures, *Appl. Microbiol. Biotechnol.*, 40, 215, 1993.

267. Decendit, A. et al., Anthocyanins, catechins, condensed tannins and piceid production in *Vitis vinifera* cell bioreactor cultures, *Biotechnol. Lett.*, 18, 659, 1996.

268. Kurata, H. et al., Intermittent light irradiation with second- or hour-scale periods controls anthocyanin production by strawberry cells, *Enzyme Microb. Technol.*, 26, 621, 2000.

269. Honda, H. et al., Enhanced anthocyanin production from grape callus in an air-lift type bioreactor using a viscous additive-supplemented medium, *J. Biosci. Bioeng.*, 94, 135, 2002.

270. Meyer, J.E., Pepin, M.-F., and Smith, M.A.L., Anthocyanin production from *Vaccinium pahalae*: limitations of the physical microenvironment, *J. Biotechnol.*, 93, 45, 2002.

271. Iacobucci, G.A. and Sweeny, J.G., The chemistry of anthocyanins, anthocyanidins and related flavylium salts, *Tetrahedron Lett.*, 39, 3005, 1983.

272. Bernini, R., Mincione, E., Sanetti, A., Bovicelli, P., and Lupattelli, P., An efficient oxyfunctionalisation by dimethyldioxirane of the benzylethereal carbon of flavonoids; a general and useful way to anthocyanidins, *Tetrahedron Lett.*, 38, 4651, 1997.

273. Mas, T., A new and convenient one-step synthesis of the natural 3-deoxyanthocyanidins apigeninidin and luteolinidin chlorides from 2,4,6-triacetoxybenzaldehyde, *Synthesis (Stuttgart)*, 12, 1878, 2003.

274. Elhabiri, M. et al., A convenient method for conversion of flavonols into anthocyanins, *Tetrahedron Lett.*, 36, 4611, 1995.

275. Roehri-Stoeckel, C. et al., Synthetic dyes: simple and original ways to 4-substituted flavylium salts and their corresponding vitisin derivatives, *Can. J. Chem.*, 79, 1173, 2001.

276. Brouillard, R., The *in vivo* expression of anthocyanin colour in plants, *Phytochemistry*, 22, 1311, 1983.

277. Fulcrand, H. et al., Structure of new anthocyanin-derived wine pigments, *J. Chem. Soc., Perkin Trans. 1*, 735, 1996.

278. Håkansson, A.E. et al., Structures and color properties of new red wine pigments, *Tetrahedron Lett.*, 44, 4887, 2003.

279. Benabdeljalil, C. et al., Evidence of new pigments resulting from reaction between anthocyanins and yeast metabolites, *Sci. Aliment.*, 20, 203, 2000.

280. Schwarz, M. and Winterhalter, P., A novel synthetic route to substituted pyranoanthocyanins with unique color properties, *Tetrahedron Lett.*, 44, 7583, 2003.

281. Vowinkel, E., Cell wall pigments of peat mosses. 2. Structure of sphagnorubin, *Chem. Ber.*, 108, 1166, 1975.

282. Pecket, R.C. and Small, C.J., Occurrence, location and development of anthocyanoplasts, *Phytochemistry*, 19, 2571, 1980.

283. Nozzolillo, C. and Ishikura, N., An investigation of the intracellular site of anthocyanoplasts using isolated protoplasts and vacuoles, *Plant Cell Rep.*, 7(6), 389, 1988.

284. Nakamura, M., Anthocyanoplasts in "Kyoho" grapes, *J. Jpn. Soc. Hortic. Sci.*, 62, 353, 1993.

285. Nozue, M. et al., Detection and characterization of a vacuolar protein (VP24) in anthocyanin-producing cells of sweet potato in suspension culture, *Plant Cell Physiol.*, 36, 883, 1995.

286. Nozue, M. et al., Expression of vacuolar protein (VP24) in anthocyanin producing cells of sweet potato in suspension culture, *Plant Physiol.*, 115, 1065, 1997.

287. Nozzolillo, C., Anthocyanoplasts: organelles or inclusions? *Polyphenols Actual.*, 11, 16, 1994.
288. Cormier, F., Food colorants from plant cell cultures, *Recent Adv. Phytochem.*, 31, 201, 1997.
289. Xu, W. et al., Primary structure and expression of a 24-kDa vacuolar protein (VP24) precursor in anthocyanin-producing cells of sweet potato in suspension culture, *Plant Physiol.*, 125, 447, 2001.
290. Nozue, M. et al., VP24 found in anthocyanic vacuolar inclusions (AVIs) of sweet potato cells is a member of a metalloprotease family, *Biochem. Eng. J.*, 14, 199, 2003.
291. Hrazdina, G. and Jensen, R.A., Spatial organization of enzymes in plant metabolic pathways, *Annu. Rev. Plant Physiol. Plant Mol. Biol.*, 43, 241, 1992.
292. Marrs, K.A. et al., A glutathione S-transferase involved in vacuolar transfer encoded by the maize gene Bronze-2, *Nature*, 375, 397, 1995.
293. Markham, K.R. et al., Anthocyanic vacuolar inclusions — their nature and significance in flower colouration, *Phytochemistry*, 55, 327, 2000.
294. Conn, S., Zhang, W., and Franco, C., Anthocyanic vacuolar inclusions (AVIs) selectively bind acylated anthocyanins in *Vitis vinifera* L. (grapevine) suspension culture, *Biotechnol. Lett.*, 25, 835, 2003.
295. Griesbach, R.J. and Santamour, F.S. Jr., Anthocyanins in cones of *Abies, Picea, Pinus, Pseudotsuga* and *Tsuga* (*Pinaceae*), *Biochem. Syst. Ecol.*, 31, 261, 2003.
296. Fossen, T. et al., Anthocyanins of grasses, *Biochem. Syst. Ecol.*, 30, 855, 2002.
297. Nørbæk, R. et al., Flower pigment composition of *Crocus* species and cultivars used for a chemotaxonomic investigation, *Biochem. Syst. Ecol.*, 30, 763, 2002.
298. Van-Wyk, B.E. and Winter, P.J.D., The homology of red flower colour in *Crassula, Cotyledon* and *Tylecodon* (Crassulaceae), *Biochem. Syst. Ecol.*, 23, 291, 1995.
299. Van-Wyk, B.E., Winter, P.J.D., and Buys, M.H., The major flower anthocyanins of *Lobostemon* (Boraginaceae), *Biochem. Syst. Ecol.*, 25, 39, 1997.
300. Mikanagi, Y. et al., Flower flavonol and anthocyanin distribution in subgenus *Rosa*, *Biochem. Syst. Ecol.*, 23, 183, 1995.
301. Mikanagi, Y. et al., Anthocyanins in flowers of genus *Rosa*, sections *Cinnamomeae* (=*Rosa*), *Chinenses, Gallicanae* and some modern garden roses, *Biochem. Syst. Ecol.*, 28, 887, 2000.
302. Du, C.T., Wang, P.L., and Francis, F.J., Cyanidin-3-laminariobioside in spanish red onion (*Allium cepa*), *J. Food Sci.*, 39, 1265, 1975.
303. Ferreres, F., Gil, M.I., and Tomas-Barberan, F.A., Anthocyanins and flavonoids from shredded red onion and changes during storage in perforated films, *Food Res. Int.*, 29, 389, 1996.
304. Kondo, T. et al., Heavenly blue anthocyanin. IV. Structure determination of heavenly blue anthocyanin, a complex monomeric anthocyanin from the morning glory *Ipomoea tricolor*, by means of the negative NOE method, *Tetrahedron Lett.*, 28, 2273, 1987.
305. Saito, N. et al., Acylated cyanidin glycosides in the violet-blue flowers of *Ipomoea purpurea*, *Phytochemistry*, 40, 1283, 1995.
306. Saito, N. et al., Acylated pelargonidin glycosides in red-purple flowers of *Ipomoea purpurea*, *Phytochemistry*, 43, 1365, 1996.
307. Lu, T.S. et al., Acylated pelargonidin glycosides in the red-purple flowers of *Pharbitis nil*, *Phytochemistry*, 31, 289, 1992.
308. Lu, T.S. et al., Acylated peonidin glycosides in the violet-blue cultivars of *Pharbitis nil*, *Phytochemistry*, 31, 659, 1992.
309. Saito, N. et al., Acylated pelargonidin glucosides in the maroon flowers of *Pharbitis nil*, *Phytochemistry*, 35, 407, 1994.
310. Toki, K. et al., Acylated pelargonidin 3-sophoroside-5-glucosides from the flowers of the Japanese Morning Glory cultivar "Violet", *Heterocycles*, 55, 1241, 2001.
311. Yoshida, K. et al., Cause of blue petal color, *Nature*, 373, 291, 1995.
312. Saito, N. et al., An acylated cyanidin 3-sophoroside-5-glucoside in the violet-blue flowers of *Pharbitis nil*, *Phytochemistry*, 33, 245, 1993.
313. Toki, K. et al., An acylated delphinidin glycoside in the blue flowers of *Evolvulus pilosus*, *Phytochemistry*, 36, 609, 1994.
314. Otsuki, T. et al., Acylated anthocyanins from red radish (*Raphanus sativus* L.), *Phytochemistry*, 60, 79, 2002.

315. Giusti, M.M., Ghanadan, H., and Wrolstad, R.E., Elucidation of the structure and conformation of red radish (*Raphanus sativus*) anthocyanins using one- and two-dimensional nuclear magnetic resonance techniques, *J. Agric. Food Chem.*, 46, 4858, 1998.

316. Takeda, K., Fischer, D., and Grisebach, H., Anthocyanin composition of *Sinapis alba*, light induction of enzymes and biosynthesis, *Phytochemistry*, 27, 1351, 1988.

317. Mitchell, K.A., Markham, K.R., and Boase, M.R., Pigment chemistry and colour of *Pelargonium* flowers, *Phytochemistry*, 47, 355, 1998.

318. Haque, M.S., Ghoshal, D.N., and Ghoshal, K.K., Anthocyanins in *Salvia* — their significance in species relationship and evolution, *Proc. Indian Nat. Sci. Acad., Part B*, 47, 204, 1981.

319. Lu, Y. and Foo, L.Y., Polyphenolics of *Salvia* — a review, *Phytochemistry*, 59, 117, 2002.

320. Phippen, W.B. and Simon, J.E., Anthocyanins in Basil (*Ocimum basilicum* L.), *J. Agric. Food Chem.*, 46, 1734, 1998.

321. Catalano, G., Fossen, T., and Andersen, Ø.M., Petunidin 3-*O*-alpha-rhamnopyranoside-5-*O*-beta-glucopyranoside and other anthocyanins from flowers of *Vicia villosa*, *J. Agric. Chem.*, 46, 4568, 1998.

322. Yoshida, K. et al., Structural analysis and measurement of anthocyanins from colored seed coats of *Vigna*, *Phaseolus*, and *Glycine* legumes, *Biosci. Biotechnol. Biochem.*, 60, 589, 1996.

323. Takeoka, G.R. et al., Characterization of black bean (*Phaseolus vulgaris* L.) anthocyanins, *J. Agric. Food Chem.*, 45, 3395, 1997.

324. Pale, E. et al., Anthocyanins from Bambara groundnut (*Vigna subterranea*), *J. Agric. Food Chem.*, 45, 3359, 1997.

325. Van Wyk, B.E. and Winter, P.J.D., Chemotaxonomic value of anthocyanins in *Podalyria* and *Virgilia* (tribe *Podalyrieae*: Fabaceae), *Biochem. Syst. Ecol.*, 22, 813, 1994.

326. Van Wyk, B.E., Winter, P.J.D., and Schutte, A.L., Chemotaxonomic value of anthocyanins in the tribe *Liparieae* (Fabaceae), *Biochem. Syst. Ecol.*, 23, 295, 1995.

327. Hosokawa, K. et al., Five acylated pelargonidin glucosides in the red flowers of *Hyacinthus orientalis*, *Phytochemistry*, 40, 567, 1995.

328. Hosokawa, K. et al., Seven acylated anthocyanins in the blue flowers of *Hyacinthus orientalis*, *Phytochemistry*, 38, 1293, 1995.

329. Griesbach, R.J., Flavonoid copigments and anthocyanin of *Phalaenopsis schilleriana*, *Lindleyana*, 5, 231, 1990.

330. Fossen, T. and Øvstedal, D.O., Anthocyanins from flowers of the orchids *Dracula chimaera* and *D. cordobae*, *Phytochemistry*, 63, 783, 2003.

331. Honda, K., Tsutsui, K., and Hosokawa, K., Analysis of the flower pigments of some *Delphinium* species and their interspecific hybrids produced via ovule culture, *Sci. Hortic.*, 82, 125, 1999.

332. Winkel-Shirley, B., Molecular genetics and control of anthocyanin expression, anthocyanins in leaves, *Adv. Bot. Res.*, 37, 75, 2002.

333. Brown, C.R. et al., Breeding studies in potatoes containing high concentrations of anthocyanins, *Am. J. Potato Res.*, 80, 241, 2003.

334. De Jong, W.S. et al., Candidate gene analysis of anthocyanin pigmentation loci in the Solanaceae, *Theor. Appl. Genet.*, 108, 423, 2004.

335. Shimada, Y. et al., Genetic engineering of the anthocyanin biosynthetic pathway with flavonoid-3′,5′-hydroxylase: specific switching of the pathway in petunia, *Plant Cell Rep.*, 20, 456, 2001.

336. Opheim, S. and Andersen, Ø.M., Anthocyanins in the genus *Solanum*. *Phytochemistry* (*Life Sci. Adv.*), 11, 239, 1992.

337. Rodriguez-Saona, L.E., Giusti, M.M., and Wrolstad, R.E., Anthocyanin pigment composition of red-fleshed potatoes, *J. Food Sci.*, 63, 458, 1998.

338. Nozzolillo, C. et al., Anthocyanins of jack pine (*Pinus banksiana*) seedlings, *Can. J. Bot.*, 80, 796, 2002.

339. Harborne, J.B., Saito, N., and Detoni, C.H., Anthocyanins of *Cephaelis*, *Cynomorium*, *Euterpe*, *Lavatera* and *Pinanga*, *Biochem. Syst. Ecol.*, 22, 835, 1994.

340. Stringheta, P.C., Bobbio, P.A., and Bobbio, F.O., Stability of anthocyanic pigments from *Panicum melinis*, *Food Chem.*, 44, 37, 1992.

341. Beckwith, A.G. et al., Relationship of lich quantity and anthocyanin production in *Pennisetum setaceum* cvs. Rubrum and Red Riding Hood, *J. Agric. Food Chem.*, 52, 456, 2004.

342. Fossen, T., Slimestad, R., and Andersen, Ø.M., Anthocyanins from maize (*Zea mays*) and reed canarygrass (*Phalaris arundinacea*), *J. Agric. Food Chem.*, 49, 2318, 2001.

343. Kouda-Bonafos, M. et al., Isolation of apigeninidin from leaf sheaths of *Sorghum caudatum*, *J. Chem. Ecol.*, 20, 2123, 1994.

344. Abdel-Aal, E-S.M. and Hucl, P., Composition and stability of anthocyanins in Blue-grained wheat, *J. Agric. Food Chem.*, 51, 2174, 2003.

345. Nørbæk, R. and Kondo, T., Further anthocyanins from flowers of *Crocus antalyensis* (Iridaceae), *Phytochemistry*, 50, 325, 1999.

346. Nørbæk, R. and Kondo, T., Anthocyanins from flowers of *Crocus* (Iridaceae), *Phytochemistry*, 47, 861, 1998.

347. Pazmiño-Durán, E.A. et al., Anthocyanins from banana bracts, (*Musa* × *paradisiaca*) as potential food colorants, *Food Chem.*, 73, 327, 2001.

348. Filippini, R. et al., Production of anthocyanins by *Catharanthus roseus*, *Fitoterapia*, 74, 62, 2003.

349. Missang, C.E., Guyot, S., and Renard, M.G.C., Flavonols and anthocyanins of Bush Butter, *Dacryodes edulis* (G. Don) H.J Lam, fruit. Changes in their composition during ripening, *J. Agric. Food Chem.*, 41, 7475, 2003.

350. Chaovanalikit, A., Thompson, M.M., and Wrolstad, R.E., Characterization and quantification of anthocyanins and polyphenolics in blue honeysuckle (*Lonicera caerulea* L.), *J. Agric. Food Chem.*, 52, 848, 2004.

351. Nørbæk, R., Nielsen, K., and Kondo, T., Anthocyanins from flowers of *Cichorium intybus*, *Phytochemistry*, 60, 357, 2002.

352. Yamaguchi, M. et al., Cyanidin 3-malonylglucoside and malonyl-coenzyme A: anthocyanidin malonyltransferase in *Lactuca sativa* leaves, *Phytochemistry*, 42, 661, 1996.

353. Terahara, N., Toki, K., and Honda, T., Acylated anthocyanins from flowers of Cineraria, *Senecio cruentus*, red cultivar, *Z. Naturforsch., C: Biosci.*, 48(5–6), 430, 1993.

354. Toki, K. et al., Acylated pelargonidin 3,7-glycosides from pink flowers of *Senecio cruentus*, *Phytochemistry*, 38, 1509, 1995.

355. Pale, E. et al., Acylated anthocyanins from the flowers of *Ipomoea asarifolia*, *Phytochemistry*, 48, 1433, 1998.

356. Goda, Y. et al., Two acylated anthocyanins from purple sweet potato, *Phytochemistry*, 44, 183, 1997.

357. Odake, K. et al., Chemical structures of two anthocyanins from purple sweet potato, *Ipomoea batatas*, *Phytochemistry*, 31, 2127, 1992.

358. Saito, N. et al., Acylated cyanidin 3-sophorosides in the brownish-red flowers of *Ipomoea purpurea*, *Phytochemistry*, 49, 875, 1998.

359. Saito, N. et al., Acylated peonidin glycosides in the slate flowers of *Pharbitis nil*, *Phytochemistry*, 41, 1607, 1996.

360. Toki, K. et al., A novel acylated pelargonidin 3-sophoroside-5-glucoside from grayish-purple flowers of the Japanese Morning Glory, *Heterocycles*, 55, 2261, 2001.

361. Escribano-Bailón, M.T. et al., Anthocyanin composition of the fruit of *Coriaria myrtifolia* L., *Phytochem. Anal.*, 13, 354, 2002.

362. Rai, K.N., Singh, M.P., and Sinha, B.K., Pelargonidin-5-*O*-β-D-galactosidefrom heartwood of *Cassia auriculata* Linn., *Asian J. Chem.*, 6(3), 696, 1994.

363. Choung, M.G. et al., Isolation and determination of anthocyanins in seed coats of black soybean (*Glycine max* (L.) Merr.), *J. Agric. Food Chem.*, 49, 5848, 2001.

364. Slimestad, R. and Solheim, H., Anthocyanins from black currants (*Ribes nigrum* L.), *J. Agric. Food Chem.*, 50, 3228, 2002.

365. Corrigenda: (Saito, N. et al., *Phytochemistry*, 39, 423, 1995), *Phytochemistry*, 41, 1637, 1996.

366. Kuskoski, E.M. et al., Characterization of anthocyanins from the fruits of Baguaqu (*Eugenia umbelliflora* Berg), *J. Agric. Food Chem.*, 51, 5450, 2003.

367. Fossen, T. and Andersen, Ø.M., Acylated anthocyanins from leaves of the water lily, *Nymphaéa* × *marliacea*, *Phytochemistry*, 46, 353, 1997.

368. Pazmiño-Durán, E.A. et al., Anthocyanins from *Oxalis tringularis* as potential food colorants, *Food Chem.*, 75, 211, 2001.

369. Kidøy, L. et al., Anthocyanins in Fruits of *Passiflora edulis* and *P. suberosa*, *J. Food Compost. Anal.*, 10, 49, 1997.

370. Wang, H. et al., Quantification and characterization of anthocyanins in Balaton tart cherries, *J. Agric. Food Chem.*, 45, 2556, 1997.

371. Lee, H.S., Characterization of major anthocyanins and color of red-fleshed Budd Blood Orange (*Citrus sinensis*), *J. Agric. Food Chem.*, 50, 1243, 2002.

372. Underhill, S.J.R. and Critchley, C., Physiological, biochemical and anatomical changes in lychee (*Litchi chinensis* Sonn.) pericarp during storage, *J. Hortic. Sci.*, 68, 327, 1993.

373. Sarni-Manchado, P. et al., Phenolic composition of Litchi fruit pericarp, *J. Agric. Food Chem.*, 48, 5995, 2000.

374. Zhang, Z. et al., Purification and structural analysis of anthocyanins from litchi pericarp, *Food Chem.*, 84, 601, 2004.

375. Wilbert, S.M., Schemske, D.W., and Bradshaw, H.D. Jr., Floral anthocyanins from two Monkey-flower species with different pollinators, *Biochem. Syst. Ecol.*, 25, 437, 1997.

376. Bloor, S.J. et al., Identification of flavonol and anthocyanin metabolites in leaves of *Petunia* "Mitchell" and its *LC* transgenic, *Phytochemistry*, 49, 1427, 1998.

377. Griesbach, R.J., Stehmann, J.R., and Meyer, F., Anthocyanins in the "red" flowers of *Petunia exserta*, *Phytochemistry*, 51, 525, 1999.

378. Lin, H-.C., Liou, C-.C., and Tsai, T-.C., *Taiwan Nongye Huaxue Yu Shipin Kexue*, 39, 370, 2001 (Abstract).

379. Alcalde-Eon, C. et al., Identification of anthocyanins of pinta boca (*Solanum stenotomum*) tubers, *Food Chem.*, 86, 441, 2004.

380. Hernández-Pérez, M. et al., Phenolic composition of the "Mocán" (*Visnea mocanera* L.f.), *J. Agric. Food Chem.*, 44, 3512, 1996.

381. Narayan, M.S. and Venkataraman, L.V., Characterisation of anthocyanins derived from carrot (*Daucus carota*) cell culture, *Food Chem.*, 70, 361, 2000.

382. Toki, K. et al., Pelargonidin 3-acetylglucoside in *Verbena* flowers, *Phytochemistry*, 30, 3828, 1991.

383. Toki, K. et al., Pelargonidin 3-glucoside-5-acetylglucoside in *Verbena* flowers, *Phytochemistry*, 40, 939, 1995.

384. Toki, K. et al., Acylated anthocyanins in *Verbena* flowers, *Phytochemistry*, 38, 515, 1995.

385. Baldi, A. et al., HPLC/MS application to anthocyanins of *Vitis vinifera* L., *J. Agric. Food Chem.*, 43, 2104, 1995.

386. Núñez, V. et al., *Vitis vinifera* L. cv. Graciano grapes characterized by its anthocyanin profile, *Postharvest Biol. Technol.*, 31, 69, 2004.

APPENDIX A

Checklist of all natural anthocyanidins (in bold) and anthocyanins. Anthocyanidins found in plants without sugar are labeled with numbers. For anthocyanins found after 1992 the numbers in italic correspond to the numbers in the reference list.

A1   Pelargonidin (3,5,7,4′-tetrahydroxyflavylium)

1   3-Arabinoside
2   3-Xyloside
3   3-Rhamnoside
4   3-Galactoside
5   3-Glucoside
6   3-[2-(Xylosyl)galactoside]
7   3-[2-(Xylosyl)glucoside]
8   3-Rhamnoside-5-glucoside
9   3-[6-(Rhamnosyl)galactoside]

10   3-[2-(Rhamnosyl)glucoside]
11   3-[6-(Rhamnosyl)glucoside]
12   3-Galactoside-5-glucoside
13   3-Glucoside-5-glucoside
14   3-Glucoside-7-glucoside
15   3-[2-(Glucosyl)glucoside]
16   3-[6-(Glucosyl)glucoside]
17   3-[6-(Rhamnosyl)-2-(xylosyl)glucoside]
18   3-[2-(Xylosyl)glucoside]-5-glucoside
19   3-[6-(Rhamnosyl)galactoside]-5-glucoside
20   3-[6-(Rhamnosyl)glucoside]-5-glucoside
21   3-[6-(Rhamnosyl)glucoside]-7-glucoside (*331*)
22   3-[6-(Rhamnosyl)-2-(glucosyl)glucoside]
23   3-[2-(Glucosyl)glucoside]-5-glucoside
24   3-[2-(Glucosyl)glucoside]-7-glucoside
25   3-[6-(Acetyl)glucoside]
26   3-[6-(Malonyl)glucoside]
27   3-[6-(Malyl)glucoside]
28   3-[6-(Caffeyl)glucoside] (*360*)
29   3-[6-(2-(Acetyl)rhamnosyl)glucoside] (*122*)
30   3-[6-(Acetyl)glucoside]-5-glucoside (*384*)
31   3-Glucoside-5-[6-(acetyl)glucoside]
32   3-[2-(Xylosyl)-6-(malonyl)galactoside] (*29*)
33   3-[2-(Xylosyl)-6-(methyl-malonyl)galactoside] (*29*)
34   3-[6-(Malonyl)glucoside]-5-glucoside
35   3-Glucoside-5-[6-(malonyl)glucoside] (*327*)
36   3-[6-(Malonyl)glucoside]-7-glucoside (*331*)
37   3-[6-(Malonyl)-2-(glucosyl)glucoside]
38   3-Glucoside-5-glucoside (6″,6‴-malyl diester) (*215*)
39   3-[6-(Succinyl)glucoside]-5-glucoside
40   3-[6-(4-(*p*-Coumaryl)rhamnosyl)glucoside]
41   3-[6-(*p*-Coumaryl)glucoside]-5-glucoside
42   3-[6-(*p*-Coumaryl)glucoside]-5-glucoside *Z* (*327*)
43   3-[6-(Caffeyl)glucoside]-5-glucoside
44   3-[2-(2-(Caffeyl)glucosyl)galactoside] (*22*)
45   3-[6-(3-(Glucosyl)caffeyl)glucoside] (*360*)
46   3-[6-(Ferulyl)glucoside]-5-glucoside (*327*)
47   3-[6-(Acetyl)glucoside]-5-[6-(acetyl)glucoside] (*382*)
48   3-[6-(Malonyl)glucoside]-5-[6-(acetyl)glucoside] (*384*)
49   3-[6-(*p*-Coumaryl)glucoside]-5-[6-(acetyl)glucoside] (*327*)
50   3-[6-(*p*-Coumaryl)glucoside]-5-[4-(malonyl)glucoside] (*201*)
51   3-[6-(*p*-Coumaryl)glucoside]-5-[6-(malonyl)glucoside]
52   3-[6-(Caffeyl)glucoside]-5-[6-(malonyl)glucoside]
53   3-[6-(Malonyl)glucoside]-7-[6-(caffeyl)glucoside] (*354*)
54   3-[6-(Ferulyl)glucoside]-5-[6-(malonyl)glucoside] (*327*)
55   3-[6-(*p*-Coumaryl)glucoside]-5-[4-(malonyl)-6-(malonyl)glucoside]
56   3-[6-(Caffeyl)glucoside]-5-[4-(malonyl)-6-(malonyl)glucoside]
57   3-[6-(Rhamnosyl)glucoside]-7-[6-(*p*-hydroxybenzoyl)glucoside] (*331*)
58   3-Glucoside-7-[6-(4-(glucosyl)*p*-hydroxybenzoyl)glucoside] (*331*)
59   3-[6-(4-(*p*-Coumaryl)rhamnosyl)glucoside]-5-glucoside

60 3-[2-(6-(*p*-Coumaryl)glucosyl)glucoside]-5-glucoside (*310*)
61 3-[2-(Glucosyl)-6-(*p*-coumaryl)glucoside]-5-glucoside (*314*)
62 3-[6-(4-(Ferulyl)rhamnosyl)glucoside]-5-glucoside (*171*)
63 3-[2-(6-(Caffeyl)glucosyl)glucoside]-5-glucoside
64 3-[2-(Glucosyl)-6-(caffeyl)glucoside]-5-glucoside (*359*)
65 3-[6-(3-(Glucosyl)caffeyl)glucoside]-5-glucoside (*360*)
66 3-[2-(Glucosyl)-6-(ferulyl)glucoside]-5-glucoside (*314*)
67 3-[2-(6-(Ferulyl)glucosyl)glucoside]-5-glucoside (*314*)
68 3-[2-(Xylosyl)-6-(*p*-coumaryl)glucoside]-5-[6-(malonyl)glucoside] (*173*)
69 3-[6-(Malonyl)glucoside]-7-[6-(4-(glucosyl)*p*-hydroxybenzoyl)glucoside] (*331*)
70 3-[2-(Xylosyl)-6-(ferulyl)glucoside]-5-[6-(malonyl)glucoside] (*173*)
71 3-[2-(Glucosyl)-6-(*p*-coumaryl)glucoside]-5-[6-(malonyl)glucoside] (*314*)
72 3-[2-(Glucosyl)-6-(ferulyl)glucoside]-5-[6-(malonyl)glucoside] (*314*)
73 3-[2-(6-(Caffeyl)glucosyl)-6-(*p*-coumaryl)glucoside]-5-glucoside (*314*)
74 3-[2-(6-(Ferulyl)glucosyl)-6-(*p*-coumaryl)glucoside]-5-glucoside (*314*)
75 3-[2-(2-(Sinapyl)xylosyl)-6-(*p*-coumaryl)glucoside]-5-glucoside (*173*)
76 3-[2-(6-(Caffeyl)glucosyl)-6-(caffeyl)glucoside]-5-glucoside (*306*)
77 3-[2-(6-(Caffeyl)glucosyl)-6-(ferulyl)glucoside]-5-glucoside (*314*)
78 3-[2-(6-(Ferulyl)glucosyl)-6-(caffeyl)glucoside]-5-glucoside (*314*)
79 3-[2-(6-(Ferulyl)glucosyl)-6-(ferulyl)glucoside]-5-glucoside (*314*)
80 3-[2-(2-(Ferulyl)glucosyl)-6-(ferulyl)glucoside]-5-glucoside (*315*)
81 3-[2-(2-(Sinapyl)xylosyl)-6-(ferulyl)glucoside]-5-glucoside (*173*)
82 3-[2-(Xylosyl)-6-(3-(3-(4-(glucosyl)caffeyl)-2-tartaryl)malonyl)galactoside] (*29*)
83 3-[2-(2-(Ferulyl)xylosyl)-6-(ferulyl)glucoside]-5-[6-(malonyl)glucoside] (*173*)
84 3-[2-(2-(Sinapyl)xylosyl)-6-(*p*-coumaryl)glucoside]-5-[6-(malonyl)glucoside] (*173*)
85 3-[6-(Malonyl)glucoside]-7-[6-(4-(6-(caffeyl)glucosyl)caffeyl)glucoside] (*354*)
86 3-[2-(2-(Sinapyl)xylosyl)-6-(ferulyl)glucoside]-5-[6-(malonyl)glucoside] (*173*)
87 3-[6-(Rhamnosyl)glucoside]-7-[6-(4-(glucosyl)*p*-hydroxybenzoyl)glucoside] (*331*)
88 3-[2-(6-(3-(Glucosyl)caffeyl)glucosyl)glucoside]-5-glucoside
89 3-[2-(Glucosyl)-6-(4-(glucosyl)caffeyl)glucoside]-5-glucoside (*310*)
90 3-[2-(6-(3-(Glucosyl)caffeyl)glucosyl)-6-(caffeyl)glucoside]-5-glucoside
91 3-[2-(6-(Caffeyl)glucosyl)-6-(4-(6-(caffeyl)glucosyl)caffeyl)glucoside]-5-glucoside
    (*306*)
92 3-[6-(Rhamnosyl)glucoside]-7-[6-(4-(6-(4-(6-(*p*-hydroxybenzoyl)glucosyl)*p*-
    hydroxybenzoyl)glucosyl)*p*-hydroxybenzoyl)glucoside]
93 3-[2-(6-(3-(Glucosyl)caffeyl)glucosyl)-6-(4-(6-(caffeyl)glucosyl)caffeyl)glucoside]-
    5-glucoside (*306*)
94 3-[2-(6-(3-(Glucosyl)caffeyl)glucosyl)-6-(4-(6-(3-(glucosyl)caffeyl)glucosyl)caffeyl)
    glucoside]-5-glucoside

A2 Cyanidin (3,5,7,3′,4′-pentahydroxyflavylium)
95   3-Arabinoside
96   3-Xyloside
97   3-Rhamnoside
98   3-Galactoside
99   3-Glucoside
100 4′-Glucoside (*147*)
101 3-Arabinoside-5-glucoside
102 3-[2-(Xylosyl)galactoside]
103 3-[2-(Xylosyl)glucoside]

104   3-Rhamnoside-5-glucoside
105   3-Glucoside-7-rhamnoside
106   3-[6-(Rhamnosyl)galactoside]
107   3-[2-(Rhamnosyl)glucoside]
108   3-[6-(Rhamnosyl)glucoside]
109   3-Galactoside-5-glucoside
110   3-Glucoside-5-glucoside
111   3-Glucoside-7-glucoside
112   3-Glucoside-3'-glucoside
113   3-Glucoside-4'-glucoside (*135*)
114   3-[2-(Glucosyl)galactoside] (*23*)
115   3-[2-(Glucosyl)glucoside]
116   3-[3-(Glucosyl)glucoside]
117   3-[6-(Glucosyl)glucoside]
118   3-[2-(Glucuronosyl)glucoside]
119   3-[2-(Xylosyl)-6-(rhamnosyl)glucoside]
120   3-[2-(Xylosyl)glucoside]-5-glucoside
121   3-[2-(Xylosyl)-6-(glucosyl)galactoside]
122   3-[6-(Rhamnosyl)glucoside]-5-glucoside
123   3-[6-(Rhamnosyl)glucoside]-3'-glucoside
124   3-[6-(Rhamnosyl)glucoside]-7-glucoside (*120*)
125   3-[2-(Glucosyl)-6-(rhamnosyl)glucoside]
126   3-Glucoside-5-glucoside-3'-glucoside
127   3-Glucoside-7-glucoside-3'-glucoside
128   3-[2-(Glucosyl)glucoside]-5-glucoside
129   3-[6-(Acetyl)galactoside] (*192*)
130   3-[4-(Acetyl)glucoside]
131   3-[6-(Acetyl)glucoside]
132   3-[6-(Oxalyl)glucoside]
133   3-[3-(Malonyl)glucoside] (*202*)
134   3-[6-(Malonyl)glucoside]
135   3-[6-(Succinyl)glucoside] (*221*)
136   3-[6-(Malyl)glucoside]
137   3-[6-(*p*-Coumaryl)glucoside]
138   3-[2-(Galloyl)galactoside]
139   3-[2-(Galloyl)glucoside] (*193*)
140   3-[6-(Caffeyl)glucoside]
141   3-[3-(Malonyl)-6-(malonyl)glucoside] (*202*)
142   3-[2-(Galloyl)-6-(acetyl)galactoside] (*367*)
143   3-[2-(Galloyl)3-(galloyl)glucoside] (*193*)
144   3-[6-(2-(Acetyl)rhamnosyl)glucoside] (*122*)
145   3-[6-(4-(Acetyl)rhamnosyl)glucoside]
146   3-[6-(Acetyl)glucoside]-5-glucoside
147   3-Glucoside-5-[6-(acetyl)glucoside] (*317*)
148   3-[2-(Xylosyl)-6-(malonyl)glucoside] (*126*)
149   3-[6-(Malonyl)glucoside]-5-glucoside
150   3-[3-(Glucosyl)-6-(malonyl)glucoside] (*135*)
151   3-[6-(Malonyl)glucoside]6-*C*-[glucoside] (*20*)
152   3-Glucoside-5-glucoside (6″,6‴-malyl diester) (*214*)
153a 3-[6-(Succinyl)glucoside]-5-glucoside

153b   3-[6-(Succinyl)glucoside]-5-glucoside[a]
154   3-[4-(Malonyl)-2-(glucuronosyl)glucoside]
155   3-[6-(Malonyl)-2-(glucuronosyl)glucoside]
156   3-[2-(Xylosyl)-6-(*p*-coumaryl)glucoside] (*116*)
157   3-[2-(Xylosyl)-6-(*p*-coumaryl)glucoside] *Z* (*116*)
158   3-[2-(Xylosyl)-6-(caffeyl)glucoside] (*116*)
159   3-[2-(Xylosyl)-6-(caffeyl)glucoside] *Z* (*116*)
160   3-[2-(Galloyl)-6-(rhamnosyl)galactoside] (*130*)
161   3-[2-(Galloyl)-6-(rhamnosyl)glucoside]
162   3-Galactoside-5-[6-(*p*-coumaryl)glucoside] (*129*)
163   3-[6-(*p*-Coumaryl)glucoside]-5-glucoside
164   3-[6-(*p*-Coumaryl)glucoside]-5-glucoside *Z*
165   3-Glucoside-5-[6-(*p*-coumaryl)glucoside] (*153*)
166   3-[6-(4-(Caffeyl)rhamnosyl)glucoside]
167   3-[2-(Glucosyl)-6-(*p*-coumaryl)glucoside] (*116*)
168   3-[2-(Glucosyl)-6-(*p*-coumaryl)glucoside] *Z* (*116*)
169   3-[6-(Caffeyl)glucoside]-5-glucoside
170   3-[6-(Caffeyl)glucoside]-5-glucoside *Z*
171   3-Glucoside-5-[6-(caffeyl)glucoside] (*153*)
172   3-Glucoside-3'-[6-(caffeyl)glucoside] (*353*)
173   3-[2-(Glucosyl)-6-(caffeyl)glucoside] (*358*)
174   3-[2-(6-(Caffeyl)glucosyl)glucoside] (*358*)
175   3-[6-(Ferulyl)glucoside]-5-glucoside
176   3-[6-(6-(Ferulyl)glucosyl)galactoside]
177   3-[6-(6-(Sinapyl)glucosyl)galactoside]
178   3-[4-(Sinapyl)-6-(glucosyl)glucoside]
179   3-[6-(6-(Sinapyl)glucosyl)glucoside]
180   3-[6-(Rhamnosyl)galactoside]-5-[6-(*p*-coumaryl)glucoside] (*129*)
181   3-[6-(Acetyl)glucoside]-5-[6-(acetyl)glucoside] (*382*)
182   3-[6-(Malonyl)glucoside]-5-[6-(malonyl)glucoside]
183   3-[6-(*p*-Coumaryl)glucoside]-5-[6-(malonyl)glucoside]
184   3-[6-(*p*-Coumaryl)glucoside]-5-[6-(malonyl)glucoside] *Z*
185   3-[6-(Caffeyl)glucoside]-5-[6-(malonyl)glucoside]
186   3-[6-(Malonyl)glucoside]-3'-[6-(caffeyl)glucoside] (*353*)
187   3-[6-(Ferulyl)glucoside]-5-[6-(malonyl)glucoside]
188   3-[6-(*p*-Coumaryl)glucoside]-5-[4-(malonyl)-6-(malonyl)glucoside]
189   3-[6-(*p*-Coumaryl)glucoside]-5-[4-(malonyl)-6-(malonyl)glucoside] *Z*
190   3-[6-(Malonyl)-2-(xylosyl)glucoside]-7-glucoside (*127, 128*)
191   3-[6-(Malonyl)-2-(glucosyl)glucoside]-5-glucoside
192   3-[3-(Glucosyl)-6-(malonyl)glucoside]-4'-glucoside (*147*)
193   -7-[3-(Glucosyl)-6-(malonyl)glucoside]-4'-glucoside (*147*)
194   3-[2-(Xylosyl)-6-(6-(*p*-hydroxybenzoyl)glucosyl)galactoside]
195   3-[6-(*p*-Coumaryl)-2-(xylosyl)glucoside]-5-glucoside
196   3-[6-(*p*-Coumaryl)-2-(xylosyl)glucoside]-5-glucoside *Z* (*125*)
197   3-[2-(Xylosyl)-6-(6-(*p*-coumaryl)glucosyl)galactoside]
198   3-[6-(4-(*p*-Coumaryl)rhamnosyl)glucoside]-5-glucoside
199   3-[2-(6-(*p*-Coumaryl)glucosyl)glucoside]-5-glucoside (*357*)

[a]Component of the metalloanthocyanin protocyanin.

200   3-[6-(*p*-Coumaryl)-2-(glucosyl)glucoside]-5-glucoside
201   3-[6-(4-(Caffeyl)rhamnosyl)glucoside]-5-glucoside
202   3-[2-(Xylosyl)-6-(6-(ferulyl)glucosyl)galactoside]
203   3-[2-(Xylosyl)-6-(6-(ferulyl)glucosyl)glucoside] (*34*)
204   3-Glucoside-3'-glucoside-7-[6-(caffeyl)glucoside] (*176*)
205   3-[6-(Caffeyl)-2-(glucosyl)glucoside]-5-glucoside (*356*)
206   3-[6-(3-(Glucosyl)caffeyl)glucoside]-5-glucoside (*312*)
207   3-[6-(Ferulyl)-2-(glucosyl)glucoside]-5-glucoside
208   3-[2-(Xylosyl)-6-(6-(sinapyl)glucosyl)galactoside]
209   3-[6-(Sinapyl)-2-(glucosyl)glucoside]-5-glucoside
210   3-[6-(Malonyl)glucoside]-3'-glucoside-7-[6-(caffeyl)glucoside] (*176*)
211   3-[6-(Malonyl)glucoside]-3'-glucoside-7-[6-(ferulyl)glucoside] (*176*)
212   3-[2-(6-(*p*-Hydroxybenzoyl)glucosyl-6-(caffeyl)glucoside]-5-glucoside (*188*)
213   3-[2-(6-(*p*-Coumaryl)glucosyl)-6-(*p*-coumaryl)glucoside]-5-glucoside (*114*)
214   3-[2-(6-(*p*-Coumaryl)glucosyl)-6-(caffeyl)glucoside]-5-glucoside (*27*)
215   3-[2-(6-(Caffeyl)glucosyl)-6-(caffeyl)glucoside]-5-glucoside (*305*)
216   3-Glucoside-5-[6-(caffeyl)glucoside]-3'-[6-(caffeyl)glucoside] (*153*)
217   3-[2-(6-(Ferulyl)glucosyl)-6-(caffeyl)glucoside]-5-glucoside (*35*)
218   3-[6-(Ferulyl)-2-(2-(sinapyl)xylosyl)glucoside]-5-glucoside (*173*)
219   3-[2-(2-(Sinapyl)glucosyl)-6-(*p*-coumaryl)glucoside]-5-glucoside
220   3-[2-(2-(Sinapyl)glucosyl)-6-(ferulyl)glucoside]-5-glucoside
221   3-[2-(6-(Sinapyl)glucosyl)-6-(sinapyl)glucoside]-5-glucoside
222   3-Glucoside-7-[6-(sinapyl)glucoside]-3'-[6-(sinapyl)glucoside] (*180*)
223   3-[2-(6-(*p*-Coumaryl)glucosyl)-6-(*p*-coumaryl)glucoside]-5-[6-(malonyl)glucoside]
       (*114*)
224   3-[6-(*p*-Coumaryl)-2-(2-(sinapyl)xylosyl)glucoside]-5-[6-(malonyl)glucoside] (*173*)
225   3-[6-(Caffeyl)-2-(2-(sinapyl)xylosyl)glucoside]-5-[6-(malonyl)glucoside] (*177*)
226   3-[6-(Ferulyl)-2-(2-(sinapyl)xylosyl)glucoside]-5-[6-(malonyl)glucoside] (*173*)
227   3-[2-(2-(Sinapyl)glucosyl)-6-(*p*-coumaryl)glucoside]-5-[6-(malonyl)glucoside] (*178*)
228   3-[2-(6-(Ferulyl)glucosyl)-6-(ferulyl)glucoside]-5-[6-(malonyl)glucoside] (*36*)
229   3-[2-(2-(Sinapyl)glucosyl)-6-(ferulyl)glucoside]-5-[6-(malonyl)glucoside] (*178*)
230   3-[6-(Malonyl)glucoside]-7-[6-(sinapyl)glucoside]-3'-[6-(sinapyl)glucoside] (*180*)
231   3-[2-(Glucosyl)-6-(4-(glucosyl)coumaryl)glucoside]-5-glucoside
232   3-[2-(Glucosyl)-6-(4-(glucosyl)caffeyl)glucoside]-5-glucoside (*312*)
233   3-[2-(Glucosyl)-6-(4-(glucosyl)ferulyl)glucoside]-5-glucoside
234   3-[2-(2-(Caffeyl)glucosyl)galactoside]-7-[6-(caffeyl)glucoside]-3'-glucuronoside (*29*)
235   3-[2-(6-(3-(Glucosyl)caffeyl)glucosyl)-6-(caffeyl)glucoside]-5-glucoside (*305*)
236   3-[2-(6-(4-(6-(3-(Glucosyl)caffeyl)glucosyl)caffeyl)glucosyl)glucoside] (*358*)
237   3-[6-(6-(Sinapyl)glucosyl)glucoside]-7-[6-(sinapyl)glucoside]-3'-glucoside
238   3-[6-(4-(*p*-Coumaryl)rhamnosyl)glucoside]-5-[6-(malonyl)glucoside]-3'-[6-
       (caffeyl)glucoside] *Z, E* (*118, 365*)
239   3-[6-(4-(Glucosyl)coumaryl)-2-(2-(sinapyl)xylosyl)glucoside]-5-[6-(malonyl)glucoside]
       (*167*)
240   3-[2-(2-(Caffeyl)glucosyl)-6-(malonyl)galactoside]-7-[6-(caffeyl)glucoside]-3'-
       glucuronoside (*28*)
241   3-[2-(6-(*p*-Coumaryl)glucosyl)-6-(4-(6-(*p*-coumaryl)glucosyl)caffeyl)glucoside]-
       5-glucoside (*27*)
242   3-[6-(2-(Caffeyl)arabinosyl)glucoside]-7-[6-(caffeyl)glucoside]-3'-[6-(caffeyl)glucoside]
243   3-[2-(6-(Caffeyl)glucosyl)-6-(4-(6-(3,5-dihydroxycinnamyl)glucosyl)caffeyl)glucoside]-
       5-glucoside (*355*)

244 3-[6-(Malonyl)glucosyl)-7-[6-(*p*-coumaryl)glucoside]-3′-[6-(4-(6-(*p*-coumaryl)glucosyl)coumaryl)glucoside] (*156*)

245 3-[6-(Malonyl)glucoside]-7-[6-(*p*-coumaryl)glucoside]-3′-[6-(4-(6-(caffeyl)glucosyl)coumaryl)glucoside] (*155*)

246 3-[2-(2-(Caffeyl)glucosyl)-6-[3-(2-(tartaryl)malonyl)galactoside]-7-[6-(caffeyl)glucoside]-3′-gulucuronoisde] (*132*)

247 3-[6-(Malonyl)glucoside]-7-[6-(*p*-coumaryl)glucoside]-3′-[6-(4-(6-(caffeyl)glucosyl)caffeyl)glucoside] (*156*)

248 3-[6-(Malonyl)glucoside]-7-[6-(4-(6-(caffeyl)glucosyl)caffeyl)glucoside]-3′-[6-(caffeyl)glucoside] (*248*)

249 3-[6-(Malonyl)glucoside]-7-[6-(caffeyl)glucoside]-3′-[6-(4-(6-(caffeyl)glucosyl)caffeyl)glucoside] (*156*)

250 3-[6-(2-(Caffeyl)-5-(caffeyl)arabinosyl)glucoside]-7-[6-(caffeyl)glucoside]-3′-[6-(caffeyl)glucoside]

251 3-[6-(2-(Ferulyl)-5-(ferulyl)arabinosyl)glucoside]-7-[6-(ferulyl)glucoside]-3′-[6-(ferulyl)glucoside] (*136*)

252 3-[6-(3-(Glucosyl)-6-(sinapyl)glucosyl)glucoside]-7-[6-(sinapyl)glucoside]-3′-glucoside

253 3-[6-(Malonyl)glucoside]-7-[6-(4-(glucosyl)*p*-hydroxybenzoyl)glucoside]-3′-[6-(4-(glucosyl)*p*-hydroxybenzoyl)glucoside] (*189*)

254 3-Glucoside-7-[6-(*p*-coumaryl)glucoside]-3′-[6-(4-(6-(4-(glucosyl)coumaryl)glucosyl)coumaryl)glucoside] (*157*)

255 3-[2-(6-(3-(Glucosyl)caffeyl)glucosyl)-6-(4-(6-(caffeyl)glucosyl)caffeyl)glucoside]-5-glucoside (*305*)

256 3-Glucoside-7-[6-(caffeyl)glucoside]-3′-[6-(4-(6-(4-(glucosyl)caffeyl)glucosyl)caffeyl)glucoside] (*157*)

257 3-[6-(Malonyl)glucoside]-7-[6-(*p*-coumaryl)glucoside]-3′-[6-(4-(6-(4-(glucosyl)coumaryl)glucosyl)coumaryl)glucoside] (*157*)

258 3-[6-(Malonyl)glucoside]-7-[6-(caffeyl)glucoside]-3′-[6-(4-(6-(4-(glucosyl)caffeyl)glucosyl)caffeyl)glucoside] (*157*)

A3   Peonidin (3′-methoxy-3,5,7,4′-tetrahydroxyflavylium)

259 3-Arabinoside

260 3-Rhamnoside

261 3-Galactoside

262 3-Glucoside

263 3-Arabinoside-5-glucoside

264 3-[2-(Xylosyl)galactoside] (*133*)

265 3-[2-(Xylosyl)glucoside]

266 3-[4-(Arabinosyl)glucoside]

267 3-Rhamnoside-5-glucoside

268 3-[2-(Rhamnosyl)glucoside] (*140*)

269 3-[6-(Rhamnosyl)glucoside]

270 3-Galactoside-5-glucoside

271 3-Glucoside-5-glucoside

272 3-[2-(Glucosyl)glucoside]

273 3-[6-(Glucosyl)glucoside] (*25*)

274 3-[6-(Rhamnosyl)-2-(xylosyl)glucoside]

275 3-[6-(Rhamnosyl)glucoside]-5-glucoside

276 3-[2-(Glucosyl)glucoside]-5-glucoside

277 3-[6-(6-(Glucosyl)glucosyl)glucoside]

278   3-[6-(Acetyl)glucoside] (*385*)
279   3-[6-(Malonyl)glucoside]
280   3-[6-(*p*-Coumaryl)glucoside]
281   3-[6-(Caffeyl)glucoside] (*386*)
282   3-Glucoside-5-[6-(acetyl)glucoside] (*317*)
283   3-[6-(Malonyl)glucoside]-5-glucoside (*147*)
284   3-[6-(*p*-Coumaryl)glucoside]-5-glucoside
285   3-[6-(3-(Glucosyl)caffeyl)glucoside] (*360*)
286   3-[4-(Sinapyl)-6-(glucosyl)glucoside]
287   3-[6-(4-(Coumaryl)rhamnosyl)glucoside]-5-glucoside
288   3-[6-(4-(Caffeyl)rhamnosyl)glucoside]-5-glucoside (*169*)
289   3-[6-(Caffeyl)-2-(glucosyl)glucoside]-5-glucoside
290   3-[6-(3-(Glucosyl)caffeyl)glucoside]-5-glucoside (*360*)
291   3-[2-(6-(*p*-Hydroxybenzoyl)glucosyl)-6-(caffeyl)glucoside]-5-glucoside (*35*)
292   3-[2-(6-(Caffeyl)glucosyl)-6-(caffeyl)glucoside]-5-glucoside
293   3-[2-(6-(Caffeyl)glucosyl)-6-(ferulyl)glucoside]-5-glucoside
294   3-[6-(4-(Glucosyl)caffeyl)-2-(glucosyl)glucoside]-5-glucoside
295   3-[6-(4-(4-(Glucosyl)coumaryl)rhamnosyl)glucoside]-5-glucoside (*165*)
296   3-[6-(4-(4-(6-(Caffeyl)glucosyl)coumaryl)rhamnosyl)glucoside]-5-glucoside (*165*)
297   3-[2-(6-(3-(Glucosyl)caffeyl)glucosyl)-6-(4-(6-(3-
       (glucosyl)caffeyl)glucosyl)caffeyl)glucoside]-5-glucoside

A4   Delphinidin (3,5,7,3′,4′,5′-hexahydroxyflavylium)
298   3-Arabinoside
299   3-Rhamnoside
300   3-Galactoside
301   3-Glucoside
302   3-[2-(Xylosyl)galactoside]
303   3-[2-(Xylosyl)glucoside]
304   3-Rhamnoside-5-glucoside
305   3-[2-(Rhamnosyl)glucoside]
306   3-[6-(Rhamnosyl)galactoside]
307   3-[6-(Rhamnosyl)glucoside]
308   3-Glucoside-5-glucoside
309   3-Glucoside-7-glucoside
310   3-[2-(Glucosyl)glucoside]
311   3-[6-(Glucosyl)glucoside]
312   3-[2-(Xylosyl)-6-(rhamnosyl)glucoside] (*24*)
313   3-[6-(Rhamnosyl)galactoside]-5-rhamnoside
314   3-[2-(Xylosyl)galactoside]-5-glucoside (*131*)
315   3-[2-(Xylosyl)glucoside]-5-glucoside
316   3-[6-(Rhamnosyl)galactoside]-5-glucoside (*123*)
317   3-[6-(Rhamnosyl)glucoside]-5-glucoside
318   3-[6-(Rhamnosyl)glucoside]-7-glucoside (*117*)
319   3-Glucoside-7-glucoside-3′-glucoside
320   3-Glucoside-3′-glucoside-5′-glucoside
321   3-[2-(Glucosyl)glucoside]-5-glucoside
322   3-[6-(Acetyl)galactoside] (*367*)
323   3-[6-(Acetyl)glucoside]
324   3-[6-(Malonyl)glucoside]

325   3′-[2-(Galloyl)galactoside] (*26*)
326   3-[6-(*p*-Coumaryl)galactoside] (*158*)
327   3-[6-(*p*-Coumaryl)glucoside] (*386*)
328   3-[6-(*p*-Coumaryl)glucoside] *Z*
329   3-[2-(Galloyl)galactoside]
330   3-[2-(Galloyl)-6-(acetyl)galactoside] (*191*)
331   3′-[2-(Galloyl)-6-(acetyl)galactoside] (*26*)
332   3-[6-(2-(Acetyl)rhamnosyl)glucoside] (*121*)
333   3-[6-(3-(Acetyl)rhamnosyl)glucoside] (*121*)
334   3-Glucoside-5-[6-(acetyl)glucoside] (*317*)
335   3-[2-(Xylosyl)-6-(malonyl)glucoside] (*126*)
336   3-[2-(Rhamnosyl)-6-(malonyl)glucoside]
337   3-[6-(Malonyl)glucoside]-5-glucoside (*144*)
338   3-Glucoside-5-[6-(malonyl)glucoside] (*345*)
339   3-Glucoside-5-glucoside (6″,6‴-malyl diester) (*215*)
340   3-[6-(Malyl)glucoside]-5-glucoside (*31*)
341   3-[6-(*p*-Coumaryl)glucoside]-5-glucoside
342   3-Galactoside-5-[6-(*p*-coumaryl)glucoside] (*123*)
343   3-Galactoside-5-[6-(*p*-coumaryl)glucoside] *Z* (*123*)
344   3-Glucoside-5-[6-(*p*-coumaryl)glucoside] (*154*)
345   3-[6-(Acetyl)glucoside]-5-[6-(acetyl)glucoside]
346   3-[6-(Malonyl)glucoside]-5-[6-(malonyl)glucoside] (*351*)
347   3-[6-(Malyl)glucoside]-5-[6-(malyl)glucoside] (*31*)
348   3-[6-(*p*-Coumaryl)glucoside]-5-[6-(malonyl)glucoside]
349   3-[6-(*p*-Coumaryl)glucoside]-5-[6-(malonyl)glucoside] *Z*
350   3-[6-(*p*-Coumaryl)glucoside]-5-[6-(malonyl)glucoside][b]
351   3-[6-(*p*-Coumaryl)glucoside]-5-[6-(malonyl)glucoside][c]
352   3-[6-(Caffeyl)glucoside]-5-[6-(malonyl)glucoside]
353   3-[6-(*p*-Coumaryl)glucoside]-5-[4-(acetyl)-6-(malonyl)glucoside] (*209*)
354   3-[6-(*p*-Coumaryl)glucoside]-5-[4-(malonyl)-6-(malonyl)glucoside] (*200*)
355   3-[6-(Caffeyl)glucoside]-5-[4-(malonyl)-6-(malonyl)glucoside]
356   3-[2-(Xylosyl)galactoside]-5-[6-(acetyl)glucoside] (*131*)
357   3-[2-(Rhamnosyl)glucoside]-7-[6-(malonyl)glucoside] (*139*)
358   3-[6-(Malonyl)glucoside]-3′-glucoside-5′-glucoside (*141*)
359   3-[6-(4-(*p*-Coumaryl)rhamnosyl)glucoside]-5-glucoside
360   3-[6-(4-(*p*-Coumaryl)rhamnosyl)glucoside]-5-glucoside *Z*
361   3-[6-(Rhamnosyl)galactoside]-5-[6-(*p*-coumaryl)glucoside] (*123*)
362   3-[6-(Rhamnosyl)galactoside]-5-[6-(*p*-coumaryl)glucoside] *Z* (*123*)
363   3-[6-(Rhamnosyl)glucoside]-5-[6-(*p*-coumaryl)glucoside] (*123*)
364   3-[6-(Rhamnosyl)glucoside]-5-[6-(*p*-coumaryl)glucoside] *Z* (*123*)
365   3-[6-(4-(Caffeyl)rhamnosyl)glucoside]-5-glucoside (*160*)
366   3-Glucoside-5-[6-(*p*-coumaryl)glucoside]-3′-glucoside (*154*)
367   3-[6-(Rhamnosyl)galactoside]-5-[6-(ferulyl)glucoside] (*123*)
368   3-[6-(Rhamnosyl)galactoside]-5-[6-(ferulyl)glucoside] *Z* (*123*)
369   3-Glucoside-5-[6-(caffeyl)glucoside]-3′-glucoside (*154*)
370   3-[6-(*p*-Coumaryl)glucoside]-5-[4-(rhamnosyl)-6-(malonyl)glucoside]

---

[b]Component of the metalloanthocyanin commelinin.
[c]Component of the metalloanthocyanin protodelphin.

371 3-[6-(Malonyl)glucoside]-3′-[6-(*p*-coumaryl)glucoside]-5′-glucoside (*141*)
372 3-[6-(4-(Glucosyl)coumaryl)glucoside]-5-[6-(malonyl)glucoside] (*166*)
373 3-Glucoside-5-[6-(caffeyl)glucoside]-3′-[6-(*p*-coumaryl)glucoside] (*154*)
374 3-Glucoside-5-[6-(*p*-coumaryl)glucoside]-3′-[6-(caffeyl)glucoside] (*154*)
375 3-Glucoside-5-[6-(caffeyl)glucoside]-3′-[6-(caffeyl)glucoside]
376 3-Glucoside-7-[6-(sinapyl)glucoside]-3′-[6-(sinapyl)glucoside] (*182*)
377 3-[6-(Malonyl)glucoside]-7-[6-(4-(6-(*p*-hydroxybenzoyl)glucosyl)*p*-hydroxybenzoyl)glucoside] (*115*)
378 3-[6-(Malonyl)glucoside]-3′-[6-(*p*-coumaryl)glucoside]-5′-[6-(*p*-coumaryl)glucoside] (*141*)
379 3-[2-(6-(*p*-Coumaryl)glucosyl)-6-(*p*-coumaryl)glucoside]-5-[6-(malonyl)glucoside] (*113*)
380 3-[2-(6-(Ferulyl)glucosyl)-6-(*p*-coumaryl)glucoside]-5-[6-(malonyl)glucoside] (*114*)
381 3-[2-(6-(Ferulyl)glucosyl)-6-(ferulyl)glucoside]-5-[6-(malonyl)glucoside] (*114*)
382 3-[6-(Malonyl)glucoside]-7-[6-(sinapyl)glucoside]-3′-[6-(sinapyl)glucoside] (*182*)
383 3-[2-(2-(Caffeyl)glucosyl)-6-(3-(2-tartaryl)malonyl)galactoside]-7-[6-(caffeyl)glucoside] (*29*)
384 3-[6-(Rhamnosyl)glucoside]-7-[6-(*p*-coumaryl)glucoside]-3′-glucoside (*32*)
385 3-[6-(4-(4-(Glucosyl)coumaryl)rhamnosyl)glucoside]-5-glucoside (*164*)
386 3-Glucoside-3′-[6-(4-(glucosyl)coumaryl)glucoside]-5′-glucoside (*141*)
387 3-[6-(Rhamnosyl)glucoside]-7-[6-(4-(6-(*p*-hydroxybenzoyl)glucosyl)*p*-hydroxybenzoyl)glucoside] (*117*)
388 3-[6-(Malonyl)glucoside]-3′-[6-(4-(glucosyl)coumaryl)glucoside]-5′-glucoside (*141*)
389 3-[6-(Rhamnosyl)glucoside]-7-[6-(*p*-coumaryl)glucoside]-3′-[6-(*p*-coumaryl)glucoside] (*32*)
390 3-[6-(4-(4-(6-(Caffeyl)glucosyl)coumaryl)rhamnosyl)glucoside]-5-glucoside (*164*)
391 3-Glucoside-7-glucoside-3′-[6-(*p*-coumaryl)glucoside]-5′-[6-(*p*-coumaryl)glucoside] (*146*)
392 3-[6-(4-(6-(3-(Glucosyl)caffeyl)glucosyl)caffeyl)glucoside]-5-glucoside (*223*)
393 3-[2-(2-(Caffeyl)glucosyl)galactoside]-7-[6-(caffeyl)glucoside]-3′-glucuronoside (*29*)
394 3-[6-(Malonyl)glucoside]-7-[2-(glucosyl)-6-(4-(6-(*p*-hydroxybenzoyl)glucosyl)*p*-hydroxybenzoyl)glucoside] (*115*)
395 3-[6-(Malonyl)glucoside]-3′-[6-(4-(6-(*p*-coumaryl)glucosyl)coumaryl)glucoside]-5′-glucoside (*141*)
396 3-[6-(Malonyl)glucoside]-3′-[6-(4-(glucosyl)coumaryl)glucoside]-5′-[6-(*p*-coumaryl)glucoside] (*141*)
397 3-[6-(4-(6-(3-(Glucosyl)caffeyl)glucosyl)caffeyl)glucoside]-5-[6-(malonyl)glucoside] (*313*)
398 3-[2-(2-(Caffeyl)glucosyl)-6-(malonyl)galactoside]-7-[6-(caffeyl)glucoside]-3′-glucuronoside (*29*)
399 3-Glucoside-7-[6-(*p*-coumaryl)glucoside]-3′-[6-(*p*-coumaryl)glucoside]-5′-[6-(*p*-coumaryl)glucoside] (*146*)
400 3-[6-(Malonyl)glucoside]-7-[2-(6-(*p*-hydroxybenzoyl)glucosyl)-6-(4-(6-(*p*-hydroxybenzoyl)glucosyl)*p*-hydroxybenzoyl)glucoside] (*115*)
401 3-[2-(2-(Caffeyl)glucosyl)-6-(3-(2-tartaryl)malonyl)galactoside]-7-[6-(caffeyl)glucoside]-3′-glucuronoside (*29*)
402 3-[6-(Malonyl)glucoside]-7-[6-(4-(6-(caffeyl)glucosyl)caffeyl)glucoside]-3′-[6-(caffeyl)glucoside]
403 3-[6-(*p*-Coumaryl)glucoside]-7-[6-(*p*-coumaryl)glucoside]-3′-[6-(*p*-coumaryl)glucoside]-5′-[6-(*p*-coumaryl)glucoside] (*146*)

404 3-[6-(Rhamnosyl)glucoside]-7-[6-(4-(6-(4-(glucosyl)*p*-hydroxybenzoyl)glucosyl)*p*-hydroxybenzoyl)glucoside] (*117*)

405 3-Glucoside-3′-[6-(4-(glucosyl)coumaryl)glucoside]-5′-[6-(4-(glucosyl)coumaryl)glucoside] (*141*)

406 3-[6-(Rhamnosyl)glucoside]-7-[6-(4-(6-(4-(6-(*p*-hydroxybenzoyl)glucosyl)*p*-hydroxybenzoyl)glucosyl)*p*-hydroxybenzoyl)glucoside]

407 3-[6-(Malonyl)glucoside]-3′-[6-(4-(6-(4-(glucosyl)coumaryl)glucosyl)coumaryl)glucoside]-5′-glucoside (*141*)

408 3-[6-(Malonyl)glucoside]-3′-[6-(4-(glucosyl)coumaryl)glucoside]-5′-[6-(4-(glucosyl)coumaryl)glucoside]

409 3-[6-(Malonyl)glucoside]-3′-[6-(4-(6-(*p*-coumaryl)glucosyl)coumaryl)glucoside]-5′-[6-(*p*-coumaryl)glucoside] (*145*)

410 3-[6-(Malonyl)glucoside]-3′-[6-(4-(6-(4-(glucosyl)coumaryl)glucosyl)coumaryl)glucoside]-5′-[6-(*p*-coumaryl)glucoside] (*141*)

411 3-[6-(Malonyl)glucoside]-3′-[6-(4-(6-(*p*-coumaryl)glucosyl)coumaryl)glucoside]-5′-[6-(4-(glucosyl)coumaryl)glucoside] (*141*)

412 3-[6-(4-(*p*-Coumaryl)rhamnosyl)glucoside]-5-[6-(malonyl)glucoside]-3′-[6-(caffeyl)glucoside]-5′-[6-(caffeyl)glucoside]

413 3-[6-(4-(*p*-Coumaryl)rhamnosyl)glucoside]-5-[6-(malonyl)glucoside]-3′-[6-(caffeyl)glucoside]-5′-[6-(ferulyl)glucoside]

414 3-[6-(Malonyl)glucoside]-3′-[6-(4-(6-(*p*-coumaryl)glucosyl)coumaryl)glucoside]-5′-[6-(4-(6-(*p*-coumaryl)glucosyl)coumaryl)glucoside]

415 3-[6-(Malonyl)glucoside]-3′-[6-(4-(6-(4-(glucosyl)coumaryl)glucosyl)coumaryl)glucoside]-5′-[6-(4-(glucosyl)coumaryl)glucoside] (*141*)

416 3-[6-(Malonyl)glucoside]-3′-[6-(4-(6-(4-(glucosyl)coumaryl)glucosyl)coumaryl)glucoside]-5′-[6-(4-(6-(*p*-coumaryl)glucosyl)coumaryl)glucoside]

417 3-[6-(Rhamnosyl)glucoside]-7-[3-(3-(6-(4-(6-(*p*-hydroxybenzoyl)glucosyl)*p*-hydroxybenzoyl)glucosyl)glucosyl)-6-(4-(6-(*p*-hydroxybenzoyl)glucosyl)*p*-hydroxybenzoyl)glucoside]

418 3-[6-(Malonyl)glucoside]-3′-[6-(4-(6-(4-(glucosyl)coumaryl)glucosyl)coumaryl)glucoside]-5′-[6-(4-(6-(4-(glucosyl)coumaryl)glucosyl)coumaryl)glucoside]

A5 Petunidin (3′-methoxy-3,5,7,4′,5′-pentahydroxyflavylium)
419 3-Arabinoside
420 3-Rhamnoside
421 3-Galactoside
422 3-Glucoside
423 3-[2-(Xylosyl)glucoside] (*25*)
424 3-Rhamnoside-5-glucoside
425 3-[6-(Rhamnosyl)glucoside]
426 3-Glucoside-5-glucoside
427 3-Glucoside-7-glucoside (*345*)
428 3-[2-(Glucosyl)glucoside]
429 3-[6-(Glucosyl)glucoside]
430 3-[6-(Rhamnosyl)-2-(xylosyl)glucoside] (*25*)
431 3-[6-(Rhamnosyl)glucoside]-5-glucoside
432 3-[6-(6-(Glucosyl)glucosyl)glucoside]
433 3-[2-(Glucosyl)-6-(rhamnosyl)glucoside]-3′-glucoside
434 3-[6-(Acetyl)glucoside]

435   3-[6-(Malonyl)glucoside]
436   3-[6-(*p*-Coumaryl)glucoside]
437   3-Glucoside-5-[6-(acetyl)glucoside] (*317*)
438   3-[6-(*p*-Coumaryl)glucoside]-5-glucoside
439   3-[6-(Malonyl)glucoside]-7-[6-(malonyl)glucoside] (*346*)
440   3-[6-(*p*-Coumaryl)glucoside]-5-[6-(malonyl)glucoside]
441   3-[6-(4-(*p*-Coumaryl)rhamnosyl)glucoside]-5-glucoside (*169*)
442   3-[6-(4-(Caffeyl)rhamnosyl)glucoside]-5-glucoside (*169*)
443   3-[6-(4-(Ferulyl)rhamnosyl)glucoside]-5-glucoside (*172, 328*)
444   3-[6-(4-(4-(Glucosyl)coumaryl)rhamnosyl)glucoside]-5-glucoside (*160*)
445   3-[6-(4-(4-(6-(Caffeyl)glucosyl)coumaryl)rhamnosyl)glucoside]-5-glucoside (*161*)

A6   **Malvidin (3′,5′-dimethoxy-3,5,7,4′-tetrahydroxyflavylium)**
446   3-Arabinoside
447   3-Rhamnoside
448   3-Galactoside
449   3-Glucoside
450   3-Xyloside-5-glucoside
451   3-[2-(Xylosyl)glucoside]
452   3-Rhamnoside-5-glucoside
453   3-[6-(Rhamnosyl)glucoside]
454   3-Glucoside-5-glucoside
455   3-Glucoside-7-glucoside
456   3-[3-(Glucosyl)glucoside]
457   3-[6-(Glucosyl)glucoside]
458   3-[6-(Rhamnosyl)glucoside]-5-glucoside
459   3-[2-(Glucosyl)glucoside]-5-glucoside
460   3-[6-(6-(Glucosyl)glucosyl)glucoside]
461   3-[6-(Acetyl)glucoside] (*385*)
462   3-[6-(Malonyl)glucoside] (*339*)
463   3-[6-(*p*-Coumaryl)glucoside] (*97*)
464   3-[6-(Caffeyl)glucoside]
465   3-[6-(Acetyl)glucoside]-5-glucoside (*210*)
466   3-Glucoside-5-[6-(acetyl)glucoside] (*211*)
467   3-Glucoside-5-[2-(sulfato)glucoside] (*30*)
468   3-[6-(Malonyl)glucoside]-5-glucoside (*339*)
469   3-Glucoside-5-[6-(malonyl)glucoside] (*30*)
470   3-[6-(*p*-Coumaryl)glucoside]-5-glucoside
471   3-[6-(*p*-Coumaryl)glucoside]-5-[2-(acetyl)xyloside] (*208*)
472   3-Glucoside-5-[2-(sulfato)-6-(malonyl)glucoside] (*30*)
473   3-[6-(Malonyl)glucoside]-7-[6-(malonyl)glucoside] (*346*)
474   3-[6-(*p*-Coumaryl)glucoside]-5-[6-(malonyl)glucoside]
475   3-[6-(*p*-Coumaryl)glucoside]-5-[6-(malonyl)glucoside] *Z*
476   3-[6-(4-(*p*-Coumaryl)rhamnosyl)glucoside]-5-glucoside
477   3-[6-(4-(*p*-Coumaryl)rhamnosyl)glucoside]-5-glucoside *Z* (*161*)
478   3-[6-(4-(Caffeyl)rhamnosyl)glucoside]-5-glucoside (*161*)
479   3-[6-(4-(Ferulyl)rhamnosyl)glucoside]-5-glucoside (*172*)
480   3-[6-(4-(4-(6-(Caffeyl)glucosyl)coumaryl)rhamnosyl)glucoside] (*163*)
481   3-[6-(4-(4-(6-(*p*-Coumaryl)glucosyl)coumaryl)rhamnosyl)glucoside]-5-glucoside
        (*168*)

482   3-[6-(4-(4-(6-(Caffeyl)glucosyl)coumaryl)rhamnosyl)glucoside]-5-glucoside (*162*)
483   3-[6-(4-(4-(6-(Ferulyl)glucosyl)coumaryl)rhamnosyl)glucoside]-5-glucoside (*161*)
484   3-[6-(4-(4-(6-(Caffeyl)glucosyl)caffeyl)rhamnosyl)glucoside]-5-glucoside (*162*)

A7   5-Methylcyanidin (5-methoxy-3,7,3′,4′-tetrahydroxyflavylium)
485   3-Glucoside

A8   7-Methylpeonidin = Rosinidin (7,3′-dimethoxy-3,5,4′-trihydroxyflavylium)
486   3-Glucoside-5-glucoside

A9   5-Methyldelphinidin = Pulchellidin (5-methoxy-3,7,3′,4′,5′-pentahydroxyflavylium)
487   3-Rhamnoside
488   3-Glucoside

A10   5-Methylpetunidin = Europinidin (5,3′-dimethoxy-3,7,4′,5′-tetrahydroxyflavylium)
489   3-Galactoside
490   3-Glucoside

A11   5-Methylmalvidin = Capensinidin (3,7,4′-trihydroxy-5,3′,5′-trimethoxyflavylium)
491   3-Rhamnoside

A12   7-Methylmalvidin = Hirsutidin (3,5,4′-trihydroxy-7,3′,5′-trimethoxyflavylium)
492   3-Glucoside (*97*)
493   3-[6-(*p*-Coumaryl)glucoside] (*97*)
494   3-Glucoside-5-glucoside

A13   6-Hydroxypelargonidin (3,5,6,7,4′-pentahydroxyflavylium)
495   3-Glucoside (*102*)
496   3-[6-(Rhamnosyl)glucoside] (*101*)
497   3-Glucoside-5-glucoside
498   3-[2-(Glucosyl)glucoside]

A14   6-Hydroxycyanidin (3,5,6,7,3′,4′-hexahydroxyflavylium)
499   3-Glucoside
500   3-[6-(Malonyl)glucoside] (*100*)
501   3-[6-(Rhamnosyl)glucoside]

A15   6-Hydroxydelphinidin (3,5,6,7,3′,4′,5′-heptahydroxyflavylium)
502   3-Glucoside (*101*)
503   3-[6-(Rhamnosyl)glucoside]
504   3-[6-(Malonyl)glucoside] (*101*)

A16   Apigeninidin (5,7,4′-trihydroxyflavylium)
505   Apigeninidin
506   5-Glucoside
507   7-Glucoside
508   5-Glucoside-7-glucoside
509   5-[5-(Caffeyl)glucoside]

A17 Luteolinidin (5,7,3′,4′-tetrahydroxyflavylium)
510   Luteolinidin
511   5-Glucoside
512   7-Glucoside
513   5-Glucoside-7-glucoside
514   5-[3-(Glucosyl)-2-(acetyl)glucoside] (*96*)

A18 Tricetinidin (5,7,3′,4′,5′-pentahydroxyflavylium)
515   Tricetinidin

A19 7-Methylapigeninidin (5,4′-dihydroxy-7-methoxyflavylium)
516   7-Methylapigeninidin (*83*)

A20 6-Hydroxy-5-methylapigeninidin = Carajuron (5-methoxy-6,7,4′-trihydroxyflavylium)
517   6-Hydroxy-5-methylapigeninidin (*86*)

**A21 5,4′-Dimethyl-6-hydroxyapigeninidin = Carajurin (6,7-dihydroxy-5,4′-
        dimethoxyflavylium)**
518   5,4′-Dimethyl-6-hydroxyapigeninidin

A22 5-Methylluteolinidin (5-methoxy-7,3′,4′-trihydroxyflavylium)
519   5-Methylluteolinidin (*84*)

A23 6-Hydroxy-5-methylluteolinidin (5-methoxy-6,7,3′,4′-tetrahydroxyflavylium)
520   6-Hydroxy-5-methylluteolinidin (*86*)

A24 5,4′-Dimethyl-6-hydroxyluteolinidin (5,4′-dimethoxy-6,7,3′,4′-tetrahydroxyflavylium)
521   5,4′-Dimethyl-6-hydroxyluteolinidin (*86*)

A25 Riccionidin A (Figure 10.3)
522   Riccionidin A (*96*)

A26 5-Carboxypyranopelargonidin (Table 10.1)
523   5-Carboxypyranopelargonidin 3-glucoside (*13*)

A27 5-Carboxypyranocyanidin 3-glucoside (Table 10.1)
524   5-Carboxypyranocyanidin 3-glucoside (Table 10.1) (*18*)
525   5-Carboxypyranocyanidin 3-[6-(malonyl)glucoside] (Table 10.1) (*18*)

A28 Rosacyanin B (Figure 10.3)
526   Rosacyanin B (*17*)

A29 Sphagnorubin A (Figure 10.3)
527   Sphagnorubin A

A30 Sphagnorubin B (Figure 10.3)
528   Sphagnorubin B

A31 Sphagnorubin C (Figure 10.3)
529   Sphagnorubin C

Dimeric flavonoids including anthocyanidin

530 Afzelechin(4α → 8)pelargonidin 3-glucoside (Figure 10.8) (*13*)

531 Epiafzelechin(4α → 8)pelargonidin 3-glucoside (Figure 10.8) (*13*)

532 Catechin(4α → 8)pelargonidin 3-glucoside (Figure 10.8) (*13*)

533 Epicatechin(4α → 8)pelargonidin 3-glucoside (Figure 10.8) (*13*)

534 (6‴-(Delphinidin 3-[6″-(glucosyl)glucoside])) (6″-(apigenin 7-glucosyl))malonate (Figure 10.8) (*9*)

535 (6‴-(Delphinidin 3-[6″-(glucosyl)glucoside])) (6″-(luteolin 7-glucosyl))malonate (Figure 10.8) (*10*)

536 (6″-(Cyanidin 3-glucosyl)) (2⁗-(kaempferol 3-[2″-(glucosyl)(glucoside)]-7-glucuronosyl))malonate (Figure 10.9) (*11*)

537 (6″-(Cyanidin 3-[3″-(acetyl)glucosyl])) (2⁗-(kaempferol 3-[2″-(glucosyl) (glucoside)]-7-glucuronosyl))malonate (Figure 10.9) (*11*)

538 (6‴-(Delphinidin 3-[6″-(*p*-coumaryl)glucoside]-7-glucosyl)) (6‴-(kaempferol 3-glucoside-7-xyloside-4′-glucosyl))succinate (Figure 10.9) (*12*)

539 (6‴-(Delphinidin 3-[6″-(*p*-coumaryl)glucoside]-7-glucosyl)) (6‴-(kaempferol 3 glucoside-7-glucoside-4′-glucosyl))succinate (Figure 10.9) (*12*)

11 Flavans and Proanthocyanidins

Daneel Ferreira, Desmond Slade, and Jannie P.J. Marais

CONTENTS

11.1 INTRODUCTION

In order to ensure continuity of the excellent coverage of this topic, first by Haslam[1,2] and later by Porter,[3,4] this chapter will follow broadly the format of the 1988[3] and 1994[4] contributions. Complementary to these are reviews by Hemingway,[5] Porter,[6] and Ferreira et al.[7–12] Owing to the comprehensiveness of the Porter reviews[3,4] and in view of the large number of compounds covered therein, the current contribution is focused on developments in the post-1992 period.

The proanthocyanidins represent a major group of phenolic compounds that occur ubiquitously in woody and some herbaceous plants.[3–5,8,11,12] Leucocyanidins are defined as monomeric flavonoids (flavan-3,4-diols and flavan-4-ols), which produce anthocyanidins (**1**) by cleavage of a C—O bond on heating with mineral acid.[2,3] The proanthocyanidins are flavan-3-ol oligomers that produce anthocyanidins by cleavage of a C—C interflavanyl bond under strongly acidic conditions. This classification is somewhat arbitrary since several biflavonoids with C—O—C interflavanyl bonds and a few trimeric analogs containing both C—C and C—O—C interflavanyl bonds have recently been identified (see Table 11.10 and Table 11.11).

Together with the bi- and triflavonoids, the proanthocyanidins constitute the two major classes of complex C_6—C_3—C_6 secondary metabolites. The bi- and triflavonoids[5,13] represent products of oxidative coupling of flavones, flavonols, dihydroflavonols, flavanones, isoflavones, aurones, auronols, and chalcones and thus consistently possess a carbonyl group at C-4 or its equivalent in every constituent flavonoid unit (see Chapter 17). The proanthocyanidins, on the contrary, are usually thought to originate by ionic coupling at C-4 (C-ring) of an electrophilic flavanyl unit, generated from a flavan-4-ol[4] or a flavan-3,4-diol,[5] to a nucleophilic flavanyl moiety, often a flavan-3-ol. However, the limits between the bi- or triflavonoids and proanthocyanidins are of random nature in view of the growing number of "mixed" dimers, e.g., flavan-3-ol → dihydroflavonol, and "nonproanthocyanidins" comprising oxidatively coupled flavan-3-ols.

The past 20 to 25 years have witnessed remarkable growth in our understanding of the basic structures and physicochemical properties of the proanthocyanidins. When taken in conjunction with a growing realization of their biological significance (described in, e.g., Chapters 4 to 6), such a comprehension of their chemical characteristics has highlighted the importance of this area of natural products chemistry.

11.2   NOMENCLATURE

The system of nomenclature proposed by Hemingway et al.[14] and extended by Porter[3,4] is applied consistently and is briefly summarized as follows:

1. The names of the basic flavan units are given in Table 11.1. All flavans and flavan-3-ols in this list possess 2S and 2R,3S absolute configuration, respectively, for example, catechin (2). Those with a 2R,3R configuration are prefixed with "epi," e.g., epicatechin (3). Units possessing a 2S configuration are differentiated by the enantio (ent) prefix, e.g., ent-epicatechin (4) exhibiting 2S,3S absolute configuration (Figure 11.1).
2. The flavanoid skeleton is drawn and numbered in the way as illustrated for catechin (2).
3. The location of the interflavanyl bond in dimers and oligomers is denoted within parentheses as in carbohydrates. The orientation of the interflavanyl bond at C-4 is denoted as α or β as in the IUPAC rules. Thus, the familiar procyanidin B-1 (5) is named epicatechin-(4β → 8)-catechin, the analogous prodelphinidin (6) is named epigallocatechin-(4β → 8)-catechin, and the 2S analog (7) is named ent-catechin-(4β → 6)-ent-epiafzelechin.
4. A-type proanthocyanidins are often incorrectly named due to the fact that the DEF moiety in, e.g., trimeric analogs, is rotated through 180°. The proposed system[3,4,15] cognizant of this aspect will thus be used. Proanthocyanidin A-2 (8) is thus named epicatechin-(2β → 7, 4β → 8)-epicatechin. The proper name for the trimeric analog 9 is epicatechin-(2β → 7, 4β → 8)-epicatechin-(4β → 8)-epicatechin.
5. The rules apply equally well to the nomenclature of ether-linked proanthocyanidins, e.g., compound 10 is epioritin-(4β → 3)-epioritin-4β-ol.[16]
6. The constituent units of proanthocyanidin oligomers may be differentiated by using ABC, DEF, etc., designations as indicated in structure 9, or alternatively using T, M, or B designations for top, middle, or bottom units, respectively, as was proposed by Porter.[4] These systems have the advantage that the same numbering system may be used for each monomeric unit.

With reference to Table 11.1, butiniflavan (11) is named from three proanthocyanidin dimers based on 4-subtituted 2S-7,3',4'-trihydroxyflavan (2S-flavans unsubstituted at C-3 possess the same orientation of substituents at C-2 as 2R-flavan-3-ols) isolated from *Cassia petersiana*.[17] The name is derived from the close structural relationship between flavan (11)

TABLE 11.1
Proanthocyanidin Nomenclature: Proanthocyanidin Type and Names of (2R,3S) Monomer Units (2S-Flavans)

Proanthocyanidin	Monomer	Hydroxylation Pattern						
		3	5	7	8	3'	4'	5'
Procassinidin	Cassiaflavan	H	H	OH	H	H	OH	H
Probutinidin	Butiniflavan	H	H	OH	H	OH	OH	H
Proapigeninidin	Apigeniflavan	H	OH	OH	H	H	OH	H
Proluteolinidin	Luteoliflavan	H	OH	OH	H	OH	OH	H
Protricetinidin	Tricetiflavan	H	OH	OH	H	OH	OH	OH
Prodistenidin	Distenin	OH	OH	OH	H	H	H	H
Propelargonidin	Afzelechin	OH	OH	OH	H	H	OH	H
Procyanidin	Catechin	OH	OH	OH	H	OH	OH	H
Prodelphinidin	Gallocatechin	OH	OH	OH	H	OH	OH	OH
Proguibourtinidin	Guibourtinidol	OH	H	OH	H	H	OH	H
Profisetinidin	Fisetinidol	OH	H	OH	H	OH	OH	H
Prorobinetinidin	Robinetinidol	OH	H	OH	H	OH	OH	OH
Proteracacinidin	Oritin	OH	H	OH	OH	H	OH	H
Promelacacinidin	Mesquitol	OH	H	OH	OH	OH	OH	H
Propeltogynidin	Peltogynane	OCH$_2$-	H	OH	H	H	OH	OH
Promopanidin	Mopanane	OCH$_2$-	H	OH	H	OH	OH	H

and the 2S-7,3',4'-trihydroxyflavanone, butin. Proanthocyanidins with a butiniflavan moiety substituted at C-4 are then classified as probutinidins (see Table 11.14, Figure 11.2).

11.3   STRUCTURES AND DISTRIBUTION

The naturally occurring compounds in the flavan, flavan-3-ol, flavan-4-ol, flavan-3,4-diol, and proanthocyanidin classes, together with their plant sources, are listed in Table 11.2–Table 11.17. The lists are confined to new compounds reported in the post-1992 period or those that have been overlooked in the 1994 review, and therefore must be considered in conjunction with the corresponding tables of the Porter reviews to be comprehensive. Since many of the monomeric analogs have been published under trivial names these will be retained to facilitate electronic literature searches. Unfortunately, a considerable number of these potentially chiral compounds have been reported without assignment of absolute configuration, and are hence presented as such.

11.3.1   Flavans, Flavan-3-ols, Flavan-4-ols, and Flavan-3,4-diols

Owing to the purported role of the flavans and flavan-3-ols as nucleophilic chain-terminating units, and of the flavan-4-ols and flavan-3,4-diols (leucoanthocyanidins) as electrophilic chain-extension units in the biosynthesis of the proanthocyanidins,[4] the chemistry of these four classes of compounds is intimately linked to that of the proanthocyanidins.

The most important features of the flavans and flavan-3-ols pertaining to the chemistry of the proanthocyanidins are the nucleophilicity of their A-rings, the aptitude of the heterocyclic ring of flavan-3-ols to cleavage and subsequent rearrangements, the susceptibility of analogs with pyrocatechol- or pyrogallol-type B-rings to phenol oxidative

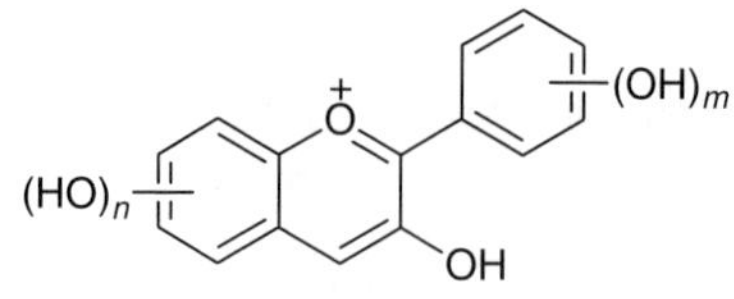

(1)

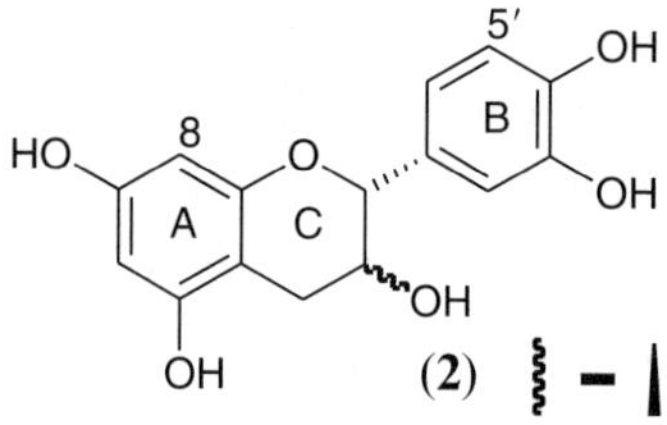

(2)

(3)

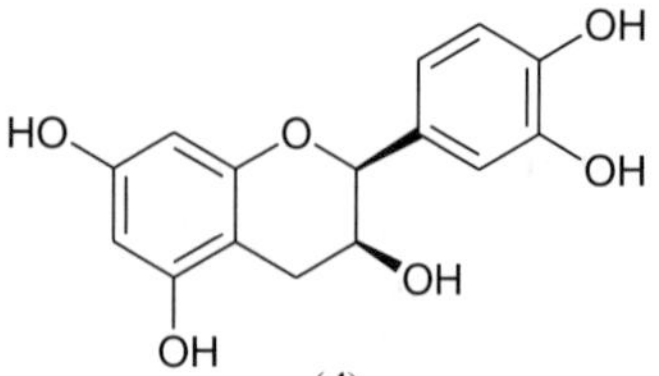

(4)

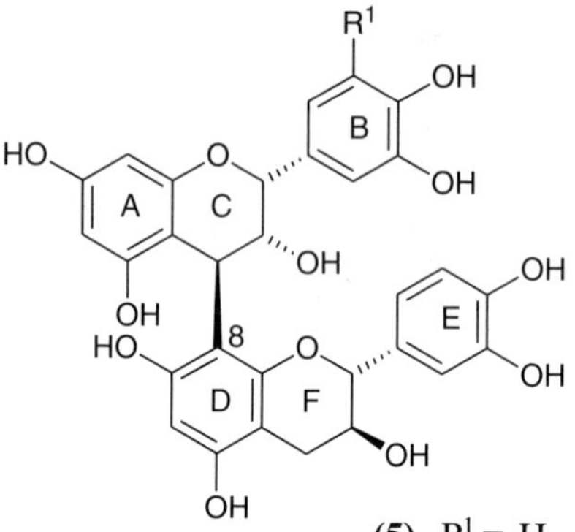

(5) R[1] = H
(6) R[1] = OH

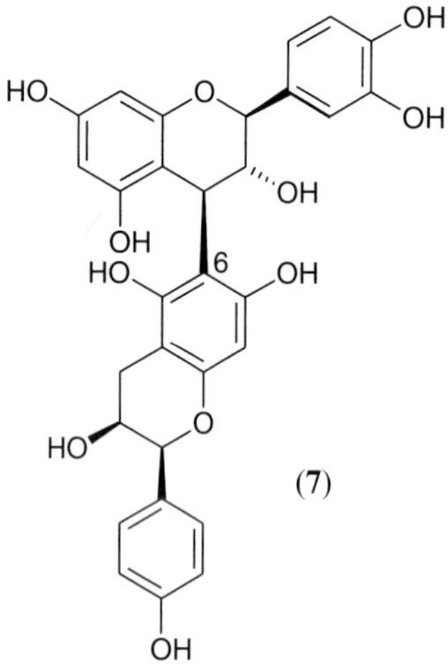

(7)

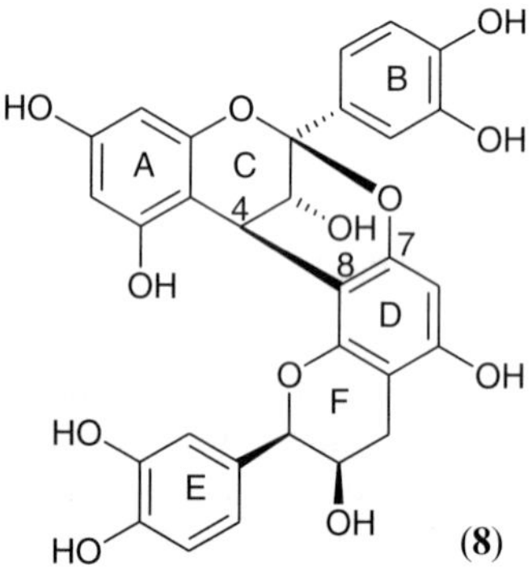

(8)

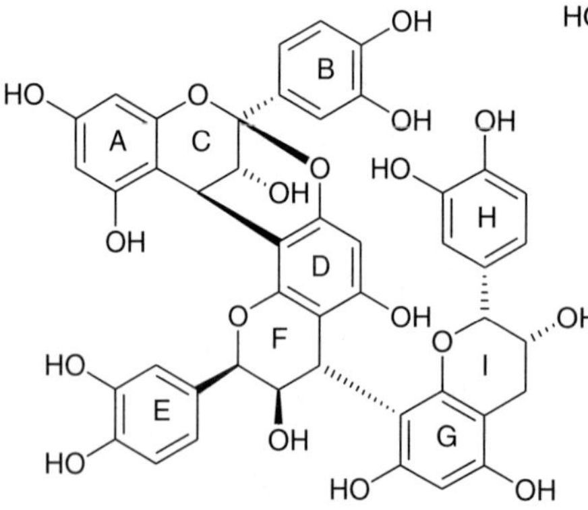

(9)

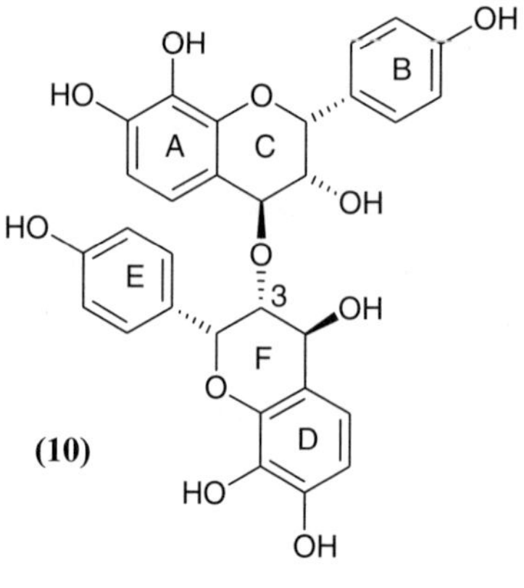

(10)

FIGURE 11.1 Structures of compounds **1**–**10**.

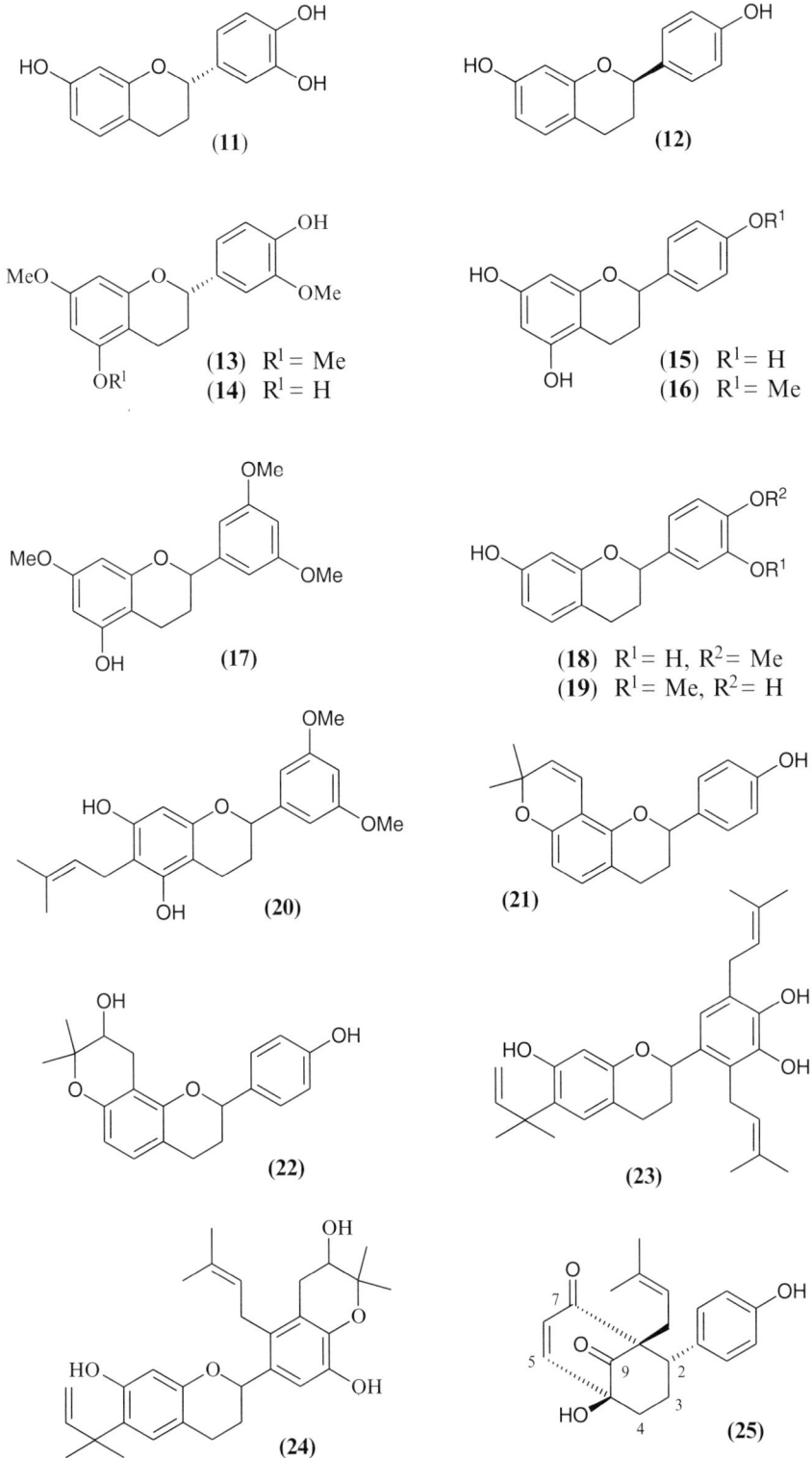

FIGURE 11.2 Structures of compounds **11**–**25**.

TABLE 11.2
The Natural Flavans

Class	Compound	Structure	Source	Ref.
1. Simple flavans	Tupichinol C	(12)	*Tupistra chinensis*	18
	5,7,3'-Tri-OMe-4-OH	(13)	*Mariscus psilostachys*	19
	5,4'-Di-OH-7,3'-di-OH	(14)		
	5,7,4'-Tri-OH	(15)	*Faramea guianensis*	20
	5,7-Di-OH-4'-OMe	(16)		
	5-OH-7,3',5'-tri-OMe	(17)	*Cyperus conglomerates*	19
	7,3'-Di-OH-4'-OMe	(18)	*Terminalea argentea*	22
	7,4'-Di-OH-3'-OMe	(19)		
2. Prenylated		(20)	*Cyperus conglomerates*	19
		(21)	*Brosimum acutifolium*	23
		(22)		
	Kazinol Q	(23)	*Broussonetia kazinoki*	24
	Kazinol R	(24)		
	Acutifolin A	(25)	*Brosimum acutifolium*	25
	Acutifolin B	(26)		
	Acutifolin C	(27)		
	Acutifolin D/E	(28)/(29)		
	Acutifolin F	(30)		

coupling, and the conformational mobility of their benzopyran rings. The predominant feature of the flavan-4-ols and flavan-3,4-ols in the same context is their role as precursors of flavan-4-yl carbocations or A-ring quinone methide electrophiles. The stability of the carbocations and hence the ease and rate of their formation from the benzylic alcohol precursors is dependent on the degree of delocalization of the positive charge over the A-ring, and on the potential of the B-ring to contribute toward stabilization via an A-conformation.[10]

11.3.1.1 Flavans

In contrast to the ubiquitous distribution of flavonoids hydroxylated at C-3 or C-4 of their heterocyclic rings, the unsubstituted flavans (2-phenylchromans) are more rarely found. The flavans co-occur with chalcones, flavanones, flavan-3,4-diols, flavonols, and 1,3-diphenyl-propanes.[10]

Several new simple flavans, including tupichinol C (**12**)[18] with its rare (2*R*) absolute configuration, were identified. Among the remaining analogs the absolute configuration at C-2 has been established for flavans **13** and **14** only.[19] Flavan **13** had 2*S* absolute configuration while **14** was obtained as a racemate.[19] For the remaining simple flavans, **15**, **16**,[20] **17**,[21] and **18** and **19**,[22] as well as the prenylated analogs **20**,[21] **21**, **22**,[23] kazinols Q (**23**) and R (**24**),[24] and acutifolins A–F (**25–30**),[25] (Figure 11.2 and Figure 11.3) the configurational issue has simply been ignored. This is an unfortunate situation since simple techniques like measurement of optical rotation and circular dichroic (CD) data readily facilitate unequivocal assignment of absolute configuration. For example, simple flavans with 2*S* absolute configuration usually exhibit negative optical rotations,[26] and vice versa for 2*R* configuration.[27] The more sensitive CD technique exhibits a negative Cotton effect for the ${}^1L_b$ electronic transition of the aromatic chromophore in the 270 to 290 nm region of the CD spectra of 2*S*-flavans and a positive one for 2R-flavans.[28,29] Since derivatization of phenolic functionalities may sometimes influence the sign of the optical rotation, the CD method should be routinely employed to unequivocally establish the absolute configuration of nonracemic flavans.

Biogenetically, acutifolin A (**25**) may be derived from the co-occurring brosimine B (**31**) via base-catalyzed pyran-ring cleavage and oxygenation of the A-ring (Scheme 11.1). Subsequent recyclization of the B-ring quinone methide intermediate (**32**) would then lead to the formation of the unique bicyclo[3.3.1]non-3-ene-2,9-dione ring system of acutifolin A (**25**).[25]

The flavans exhibit a wide range of biological activities. Details of these may be found in the appropriate references.

11.3.1.2   Flavan-3-ols

Two flavan-3-ols with new hydroxylation patterns were reported (Table 11.3). (2R,3R)-5,7,2′,5′-Tetrahydroxyflavan-3-ol (**33**) was obtained from *Prunus prostrata*[30] and (+)-5,6,7,8,3′,4′-hexahydroxyflavan-3-ol (elephantorrhizol) (**34**) from *Elephantorrhiza goetzei*.[31] The absolute configuration of **33** was established by comparison of CD data with those of epicatechin, and that of **34** based on the positive optical rotation. The new epirobinetinidol (**35**) was isolated from commercial wattle (*Acacia mearnsii*) bark exstract,[32] *ent*-mesquitol (**36**) from *Dichrostachys cinerea*,[33] and the first guibourtinidol, (2R,3S)-guibourtinol (**37**), from *Cassia abbreviata*.[34] The absolute configuration of these compounds was again properly assessed via the CD for **35** and **37**, and sign of the optical rotation for **36** (Figure 11.4).

A large number and variety of new derivatives of known flavan-3-ols have be reported (Table 11.3). It should again be emphasized that in many instances the issue of absolute configuration assessment has simply been ignored. As for the flavans, this may be conveniently done by CD. The CD curves of flavan-3-ols exhibit two Cotton effects for the ^1L_a and ^1L_b electronic transitions of the A-ring aromatic chromophore in the 240 and 280 nm regions, respectively.[35–37] Analogs with 2R and 2S absolute configurations gave negative and positive Cotton effects, respectively, in the 280 nm (^1L_b transition) region. The sign of the Cotton effect of the ^1L_a transition at ~240 nm is consistently opposite to that at longer wavelength.

In addition to identification of flavan-3-ols and derivatives from natural sources (Table 11.3, Figure 11.3–Figure 11.5, Figure 11.7, and Figure 11.8), several synthetic studies and efforts at establishing absolute configuration have been reported. The modified Mosher method has been successfully applied to configurational definition of the flavan-3-ols and 4-arylflavan-3-ols,[69] and the A-type proanthocyanidins.[70] The first stereoselective synthesis of a series of flavan-3-ol permethylaryl ethers as well as the free phenolic forms was recently developed.[34,37,71] Substantial efforts were also devoted toward the synthesis of [13]C- and other labeled flavan-3-ols.[72–74] The 4β-carboxymethylepicatechin, dryopteric acid (**46**), was recently synthesized via nucleophilic substitution at C-4 of 4β-acetoxyepicatechin.[47]

The synthetic protocol (Scheme 11.2) toward the flavan-3-ol permethylaryl ethers is based upon the transformation of *retro*-chalcones into 1,3-diarylpropenes. These compounds are then subjected to asymmetric dihydroxylation to give diarylpropan-1,2-diols that are used as chirons for essentially enantiopure flavan-3-ols. The protocol is demonstrated in Scheme 11.2 for the synthesis of the tetra-*O*-methyl-3-*O*-acetyl derivatives **61a**, **61b**, **62a**, and **62b** of (+)-catechin (**2**), (−)-*ent*-catechin, (−)-epicatechin (**3**), and (+)-*ent*-epicatechin (**4**).[71]

The (*E*)-*retro*-2-methoxymethylchalcone methyl ether (**57**) was transformed by consecutive reduction (H$_2$–Pd and NaBH$_4$) and elimination [SOCl$_2$ and 1,8-diazabicyclo[5.4.0]undec-7-ene(1,8-DBU)] of the ensuing alcohol (**58**), exclusively affording the (*E*)-1,3-diarylpropene

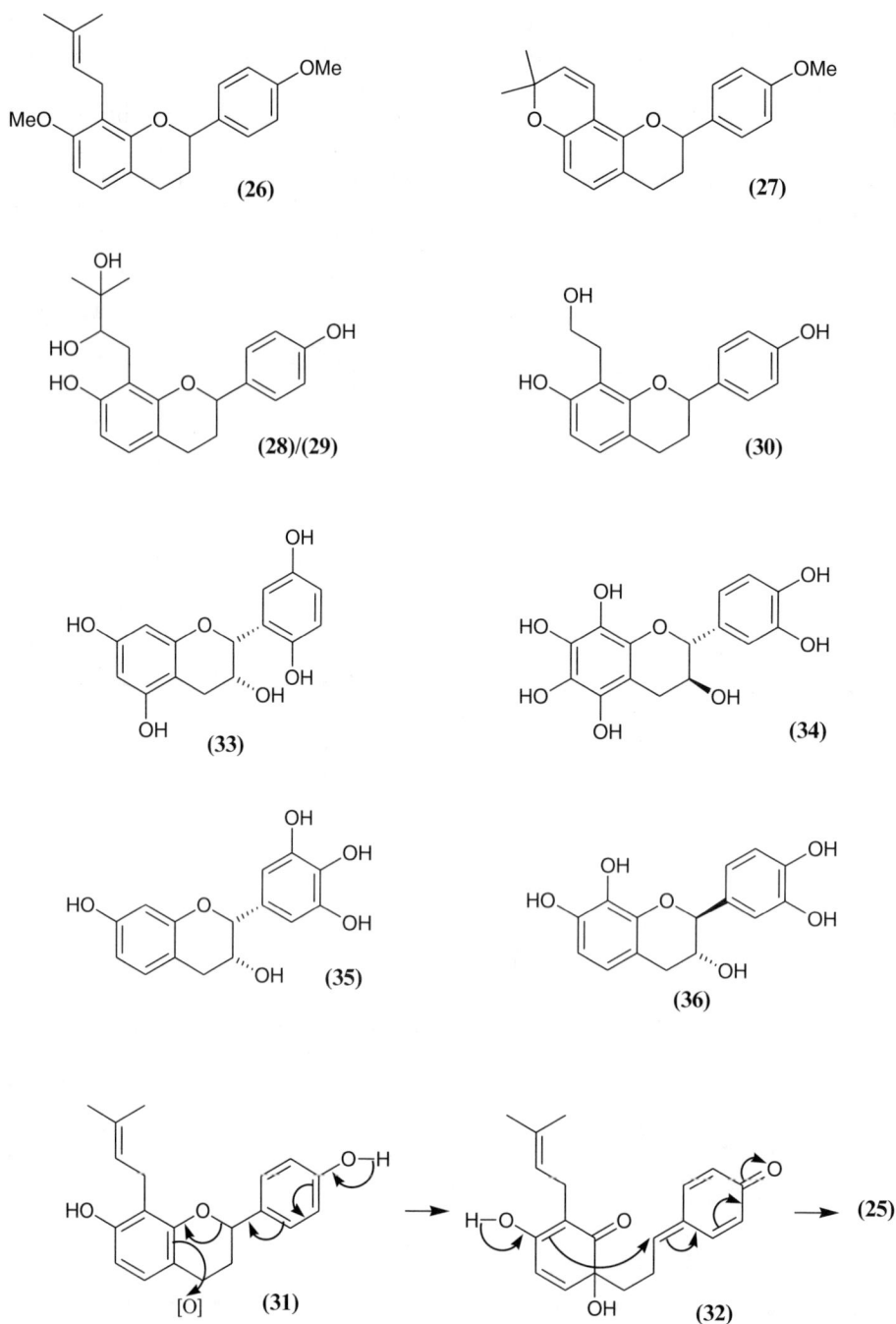

FIGURE 11.3 Structures of compounds **26–36** including Scheme 11.1: Proposed biogenetic conversion of brosimine B (**31**) into acutifolin A (**25**).

(deoxodihydrochalcone) (**59**). Treatment of this (E)-propene at 0°C with AD-mix-α (see Ref. 71 for the relevant K.B. Sharpless references) in the two-phase system Bu^{t}OH–H$_2$O (1:1) afforded the (+)-(1S,2S)-*syn*-diol (**60a**) in 80% yield and high optical purity (99% ee). The (−)-(1R,2R)-*syn*-diol (**60b**) was similarly formed by using AD-mix-β in the same two-phase

TABLE 11.3
Naturally Occurring Flavan-3-ols and Related Compounds

Class	Compound	Structure	Source	Ref.
1. Free phenolic	(2R,3R)-5,7,2',5'-Tetra-OH	(33)	*Prunus prostrata*	30
	Elephantorrhizol	(34)	*Elephantorrhiza goetzei*	31
	Epirobinetinidol	(35)	Commercial wattle bark extract	32
	Ent-mesquitol	(36)	*Dichrostachys cinerea*	33
	Guibourtinidol	(37)	*Cassia abbreviata*	34
2. *O*-Glycosides	Catechin-3-*O*-β-D-Glp	(38)	*Quercus marilandica* Muenchh.	38
	Catechin-7-*O*-β-D-Glp-3'-Me	(39)	*Picea abies*	39
	Afzelechin-3-*O*-α-L-Rhp	(44)	*Cassipourea gerrardii*	45
	Afzelechin-4'-*O*-β-D-Glp	(45)	*Selliguea feei*	46
	Epiafzelechin-3-*O*-β-D-Alp	(47)	*Drynaria propinqua*	48
	Epicatechin-5-*O*-β-D-Glp	(48)	*Davallia mariesii* Moore	49
	Davalliosides A and B	(50), (51)		51
	Epicatechin-5-*O*-β-D-Xylp	(66)	*Brosimopsis acutifolium*	58
	Anadanthoside	(69)	*Anadenanthera macrocarpa*	60
	Epicatechin-5-*O*-β-D-Glp-3-benzoate	(71)	*Celastrus orbiculatus*	62
	Barbatoflavan	(72)	*Campanula barbata*	63
	Ent-afzelechin-7-*O*-β-D-Glp	(73)	*Daphniphyllum oldhamii*	64
	Catechin-5-*O*-β-D-Glp-4'-Me	(76)	*Celastrus angulatus*	65
	Catechin-7-*O*-β-D-Glp	(78)	*Hordeum vulgare* L.	67
	Catechin-3-*O*-β-D-Glp (2-cinnamoyl)	(79)	*Inga umbellifera*	68
	Catechin-3-*O*-β-D-Glp (6-cinnamoyl)	(80)		
	Catechin-3-*O*-β-D-Glp (2,6-bis-cinnamoyl)	(81)		
3. *C*-Glycoside	Epicatechin-8-C-β-D-Galp	(77)	*Theobroma cacao* L.	66
4. Simple esters	Epigallocatechin-3-*O*-Ga	(40)	*Cistus incanus*	40
	Epigallocatechin-3-*O*-Ga-5,3',5-tri-OMe	(41)	*Sedum sediforum*	41
	Epigallocatechin-3-*O*-(4-OH-Bz)	(42)	*Cistus salvifolius*	42
	Gallocatechin-3'-*O*-Ga	(53)	*Pithecellobium lobatum*	52
	Gallocatechin-4'-*O*-Ga	(54)		
	Gallocatechin-7,3'-di-*O*-Ga	(55)		
	Gallocatechin-7,4'-di-*O*-Ga	(56)		
	Epigallocatechin-3-*O*-(3,5-di-Me)-Ga	(67)	*Stryphnodendron adstringens*	59
	Epigallocatechin-3-*O*-(4-OH-3-OMe)-benzoate	(68)		
	Amurensisin (see Section 11.3.2.1.1)	(120)	*Vitus amurensis*	104
5. Simple ethers	*Ent*-gallocatechin-4'-Me	(43)	*Cassia trachypus*	43
	Gallocatechin-4'-Me		*Panda oleosa*	44
	Epicatechin-5,7,3'-tri-Me	(49)	*Cinnamomum camphora*	50
	Epicatechin-5,7-di-Me-3', 4'-methylenedioxy	(52)		
	Peltogynoid analog	(63)	*Cassine papillosa*	33
	Catechin-7-Me	(64)	*Prunus prostrata*	30
	Catechin-3'-Me	(65)	*Pinus sylvestris*	57
	Guibourtinidol-7-Me	(70)	*Crinum bulbispermum*	61
	Tupichinol A	(74)	*Tupistra chinensis*	18
	Tupichinol B	(75)		
6. C-4 Alkylated	Dryopteric acid	(46)	*Selliguea feei*	46, 47

continued

TABLE 11.3
Naturally Occurring Flavan-3-ols and Related Compounds — *continued*

Class	Compound	Structure	Source	Ref.
7. A-ring substituted	Pyranochromene	(83)	*Lupinus angostifolius*	75
	Pyranochromene	(84)		
	Cinchonain	(85)	*Castanopsis hystrix*	76
	Cinchonain	(86)		
	Apocynin A	(87)	*Apocynum venetum*	77
	Apocynin B	(88)		
	Cinchonain	(89)	*C. hystrix*	26
	Cinchonain	(90)		
	Apocynin C	(91)	*A. venetum*	77
	Apocynin D	(92)		
	Shanciol	(93)/(94)	*Pleione bulbocodioides*	78
	Shanciol A	(95)		80, 81
	Shanciol B	(96)		
	Shanciol C	(97)		

system (87% yield, 99% ee). These conversions proceeded slowly with reaction times in the range 12 to 48 h. The enantiomeric purity of the diols was determined by observing the H-1 and H-2 spin systems of the corresponding mono- or bis-MTPA esters, which show different chemical shifts in the ^{1}H nuclear magnetic resonance (NMR) spectra of the two diastereoisomers. The absolute configuration was tentatively assigned according to the Sharpless model (see Ref. 71 for appropriate references) for AD-mix (Figure 11.6, Scheme 11.3).

Simultaneous deprotection and cyclization of diols **60a** and **60b** with 3 *M* HCl in MeOH followed by acetylation yielded the 2,3-*trans*- (~50%) (**61a** and **61b**) and for the first time 2,3-*cis*-flavan-3-ol methylether acetate derivatives (~20%) (**62a** and **62b**) in excellent enantiomeric excesses (>99%). The optical purity was assessed by ^{1}H NMR using [Eu(hfc)$_3$] as chiral shift reagent. The absolute configuration of the derivatives of the *trans*- and *cis*-flavan-3-ol derivatives was assigned by comparison of CD data with those of authentic samples in the catechin or epicatechin series flavan-3-ols.[35,36] Thus, the absolute configuration of the flavan-3-ol methyl ether acetates confirms the assigned configuration of the diols as derived from the Sharpless model.

The protocol was subsequently adapted to also provide synthetic access to a variety of flavan-3-ol diastereoisomers in their free phenolic forms.[34,37]

The co-occurrence of 4′-*O*-methylepigallocatechin and peltogynoid-type analog (**63**) in *Cassine papillosa* was demonstrated[53] and their biogenetic relationship established by transformation of the flavan-3-ol into **63** using acetone and toluene-*p*-sulfonic acid. This reaction occurs readily at room temperature, proceeds with retention of the C-2 and C-3 configurations, and is general for a variety of flavan-3-ols.[54] Twelve years after the latter paper was published, Fukuhara et al.[55,56] confirmed these novel observations using a similar process to form the "planar" catechin analog (**82**) (Figure 11.7).

The rare series of naturally occurring flavan-3-ols with an additional pyran ring[4] was extended by identification of the two epicatechin analogs **83** and **84**,[75] as well as four phenylpropanoid-substituted catechins, **85**, **86**, **89**, and **90**,[76] of the cinchonain type (Figure 11.8). The structure of the latter four compounds, including their absolute configurations, were elucidated on the basis of spectroscopic evidence and also by the synthesis of analogs **89**

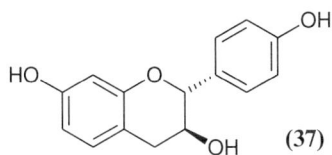

(37)

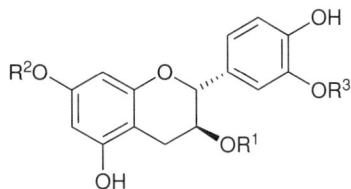

(38) R^1 = β-D-glucopyranosyl-*O*-α-L-rhamnopyranosyl, $R^2 = R^3$ = H
(39) R^1 = H, R^2 = β-D-glucopyranosyl, R^3 = Me

(40) R^1 = 3,4,5-tri-OH-benzoyl(galloyl), $R^2 = R^3$ = H
(41) R^1 = galloyl, $R^2 = R^3$ = Me
(42) R^1 = 4-OH-benzoyl, $R^2 = R^3$ = H

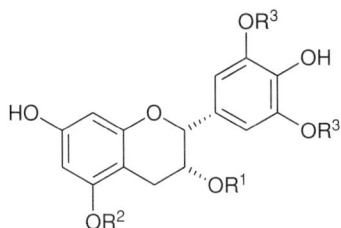

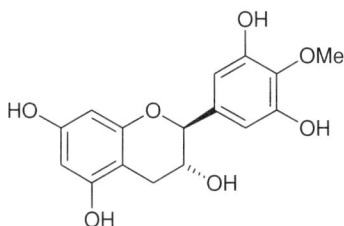

(43)

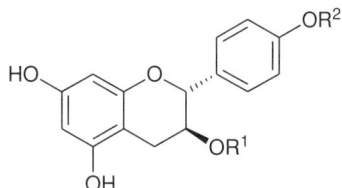

(44) R^1 = α-L-rhamnopyranosyl, R^2 = H
(45) R^1 = H, R^2 = β-D-glucopyranosyl

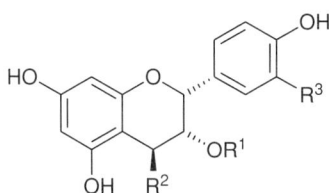

(46) R^1 = H, R^2 = CH$_2$CO$_2$H, R^3 = OH
(47) R^1 = β-D-allopyranosyl, R^2 = H, R^3 = H

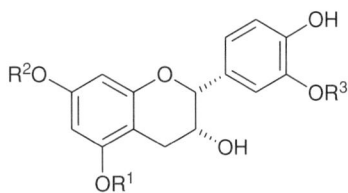

(48) R^1 = β-D-glucopyranosyl, $R^2 = R^3$ = H
(49) $R^1 = R^2 = R^3$ = Me

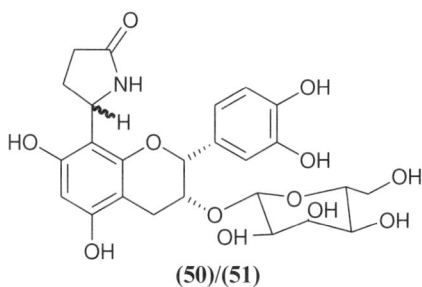

(50)/(51)

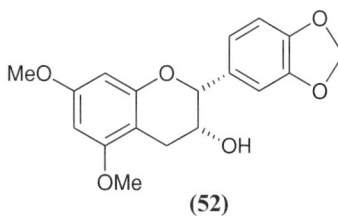

(52)

FIGURE 11.4 Structures of compounds **37**–**52**.

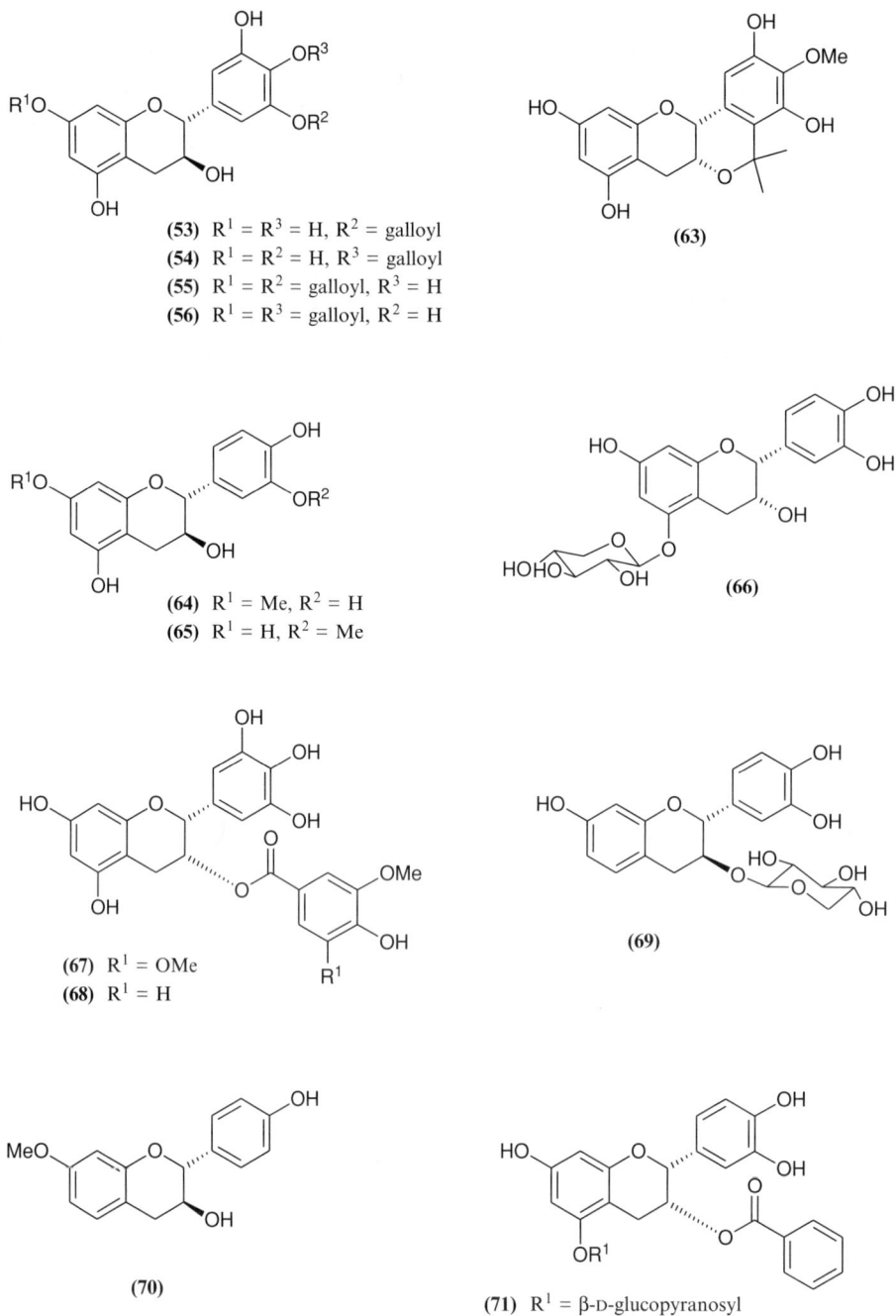

FIGURE 11.5 Structures of compounds **53–56** and **63–71**.

and **90** via acid-catalyzed condensation of catechin and caffeic acid (see also Ref. 77). Comparison of the CD data of compounds **85**, **86**, **89**, and **90** with those of cinchonains 1a–1d has led to extensive revision of previous structures (see Ref. 76). The gallocatechin- and epigallocatechin-type analogs, apocynins A–D (**87**, **88**, **91**, and **92**) were identified in *Apocynum venetum*.[78]

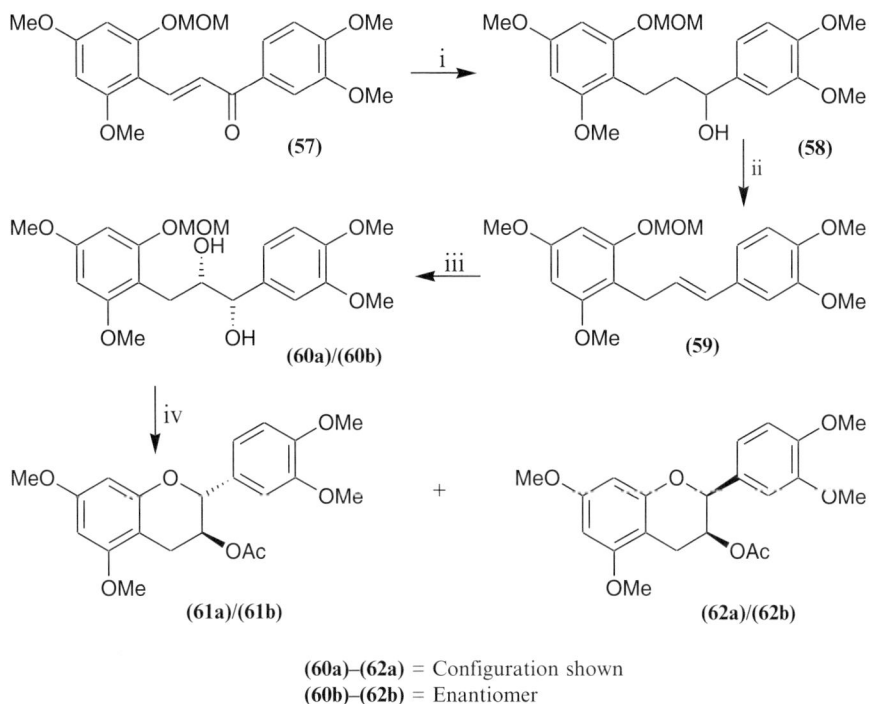

Scheme 11.2

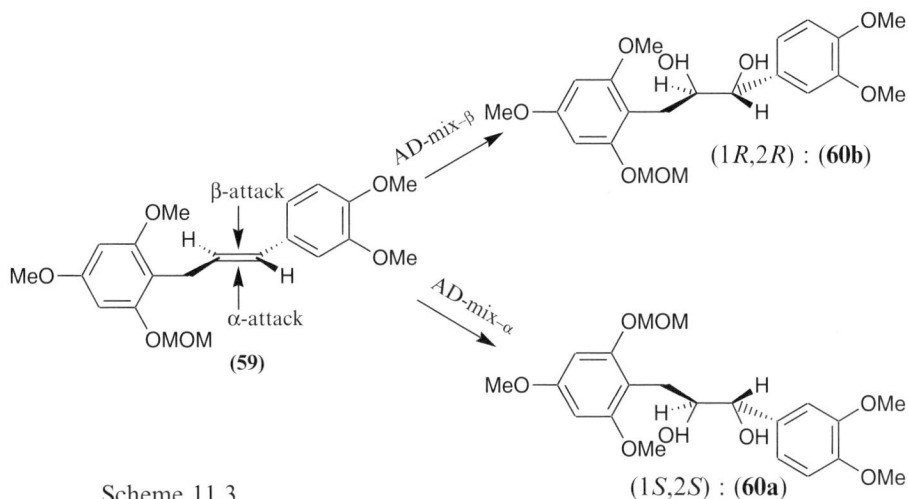

Scheme 11.3

FIGURE 11.6 Structures of compounds **57–62** including Scheme 11.2: Reagents and conditions: i, Pd–H$_2$, EtOH, then NaBH$_4$, EtOH; ii, SOCl$_2$, CH$_2$Cl$_2$, then 1,8-DBU, CH$_2$Cl$_2$, reflux; iii, AD-mix-α or AD-mix-β, Bu^{t}OH–H$_2$O (1:1, v/v), MeSO$_2$NH$_2$, 0°C; iv, 3 M HCl, MeOH–H$_2$O (3:1, v/v), then Ac$_2$O, pyridine, and Scheme 11.3: Sharpless model for assigning absolute configuration of *syn*-diols.

An interesting new group of flavan-3-ol analogs containing an additional C$_6$–C$_2$ unit at C-5 (A-ring) has been isolated from the tubes of *Pleione bulbocodioides*.[79–81] Shanciol, called a dihydrophenanthropyran, was obtained in partially racemized form (**93**)/(**94**), and is accompanied by shanciol A (**95**), shanciol B (**96**), and shanciol E (**97**). These compounds are

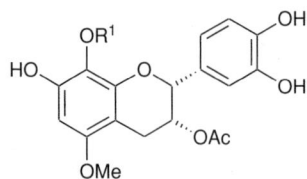

(72) R^1 = β-D-glucopyranosyl

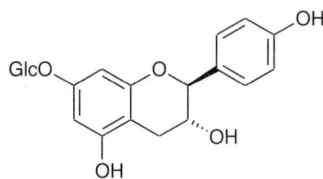

(73)

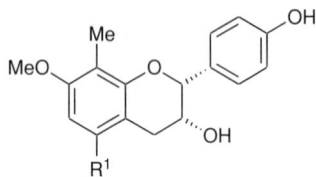

(74) R^1 = H
(75) R^1 = OMe

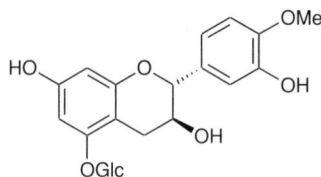

(76)

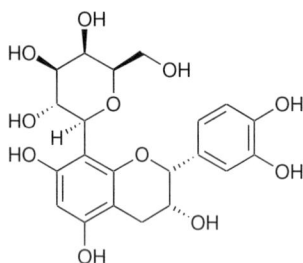

(77)

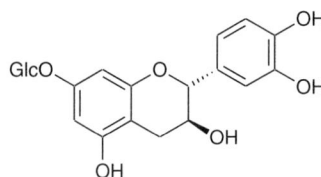

(78)

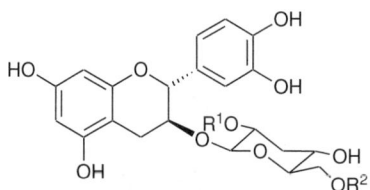

(79) R^1 = cinnamoyl, R^2 = H
(80) R^1 = H, R^2 = cinnamoyl
(81) R^1 = R^2 = cinnamoyl

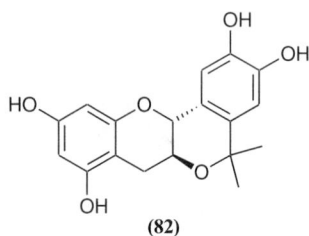

(82)

FIGURE 11.7 Structures of compounds **72–81**.

conspicuously accompanied by their isoflavan-4-ol isomers, e.g., shanciol C (**98**).[80] Flavan-3-ols with an A-ring phenethyl group, e.g., **95**, were implied as possible biogenetic precursors to these dihydrophenanthropyrans (Figure 11.8 and Figure 11.9).[80]

Nonproanthocyanidins with flavan or flavan-3-ol constituent units as well as "complex tannins," i.e., polyphenols in which a flavan-3-ol unit is connected to a hydrolyzable tannin through a C—C linkage are discussed in Sections 11.3.3 and 11.3.4, respectively.

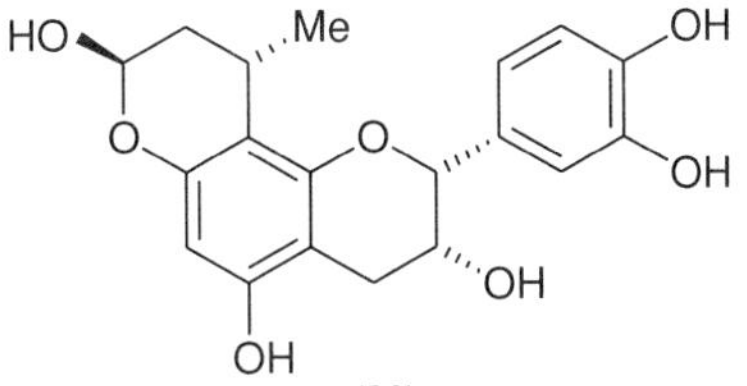

FIGURE 11.8 Structures of compounds **82–97**.

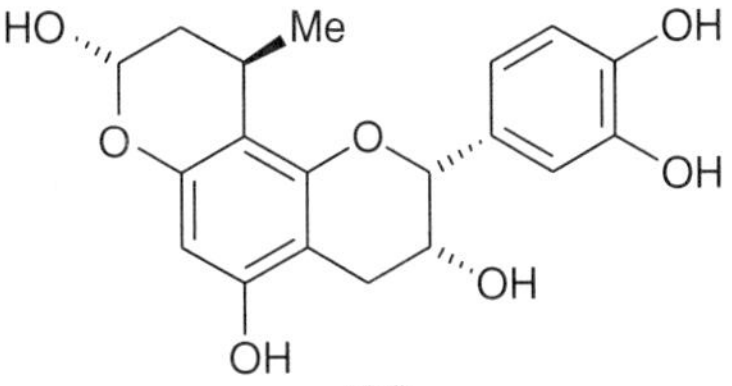

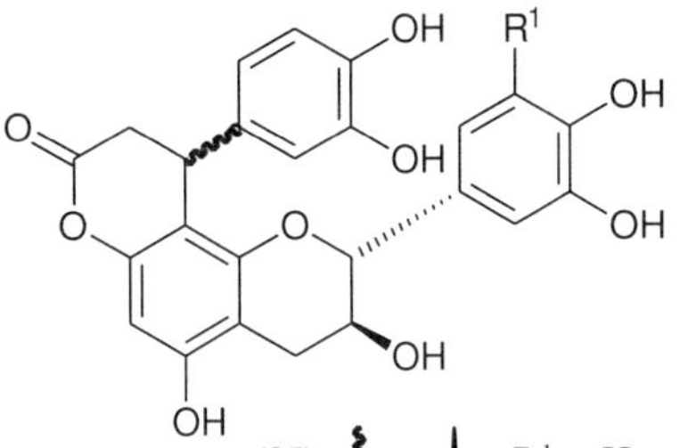

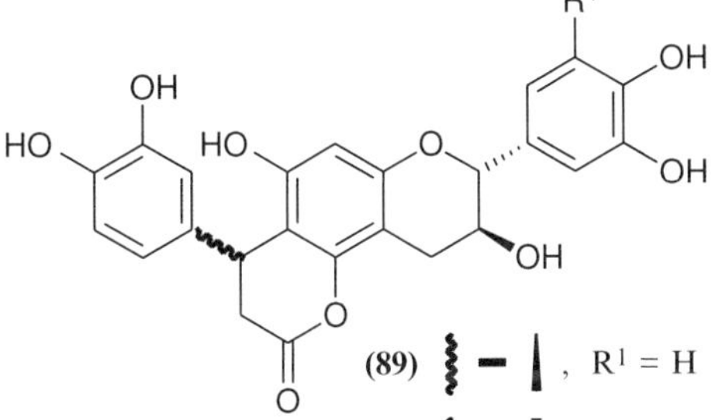

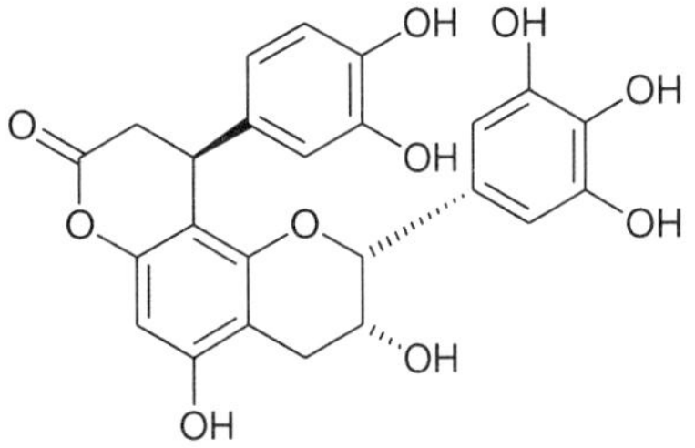

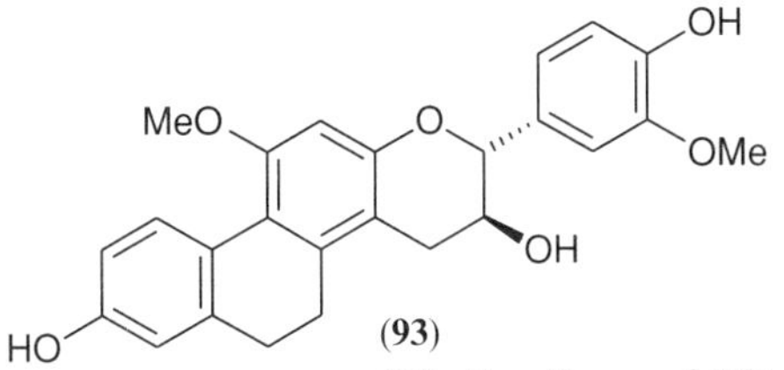

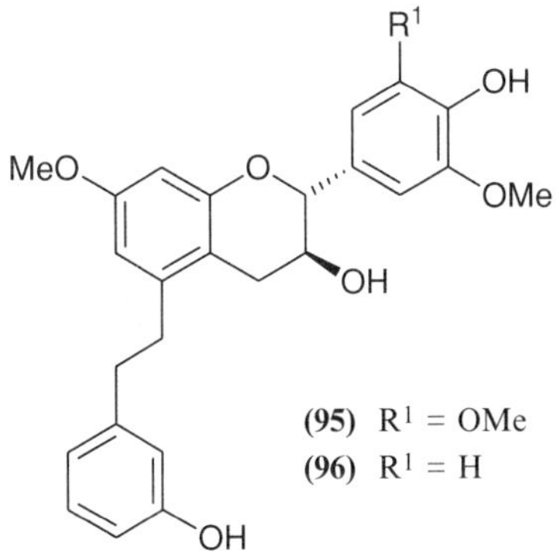

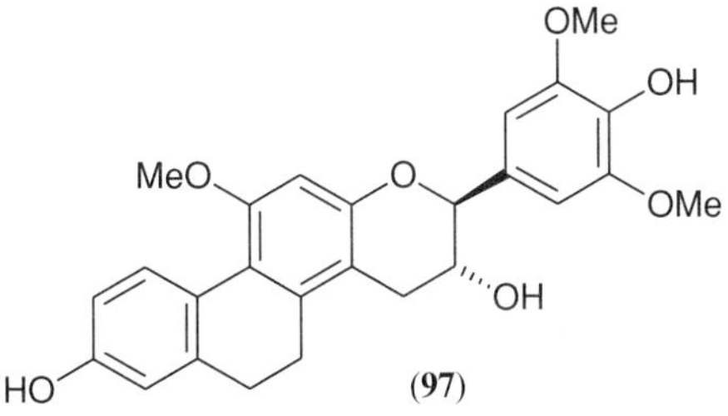

11.3.1.3   Flavan-4-ols

A limited number of new flavan-4-ols has been reported since 1992 (Figure 11.9). Additions to this class of compounds comprise the 2,4-*cis*-flavan-4-ols (**99** and **100**),[82] pneumatopterins A–D (**101–104**),[83] and xuulanins (**105** and **106**).[84] (2S)-5,7,4′-Trihydroxyflavan-4-ol (**107**) was claimed to have been identified from natural sources also.[85] This may be unlikely in view of the high reactivity of 5-oxyflavan-3,4-diols as electrophiles in weakly acidic conditions.[7,10] The absolute configuration of compound **100** was established as 2S,4S by reference to the ORD data of its likely flavanone precursor and using the relative configuration as established by ^{1}H NMR coupling constants of the heterocyclic protons.[82]

A series of flavan-4-ols, e.g., **108**, was conveniently prepared by metal hydride reduction of the corresponding flavanone.[86] The flavan-4-ols were converted into the 4-methoxyflavans, e.g., **109**, by acid-catalyzed solvolysis in methanol. Both these classes of compounds are currently evaluated as anticancer drugs. Enantiomerically enriched *cis*-flavan-4-ols have been prepared by lipase-catalyzed kinetic resolution of racemic counterparts.[87]

11.3.1.4   Flavan-3,4-diols

The four flavan-3,4-diols (**110–113**) (Figure 11.9) were isolated from the seeds of *Musa sapientum*.[88] These were claimed to be the leucoguibourtinidins *ent*-epiguibourtinidol-4α-ol (**110**), *ent*-guibourtinidol-4α-ol (**111**), (−)-(2S,3R,4R)-2,3-*trans*-flavan-3,4-diol (**112**), the first flavan-3,4-diol devoid of A-ring hydroxylation, and the leucopelargonidin, *ent*-epiafzelechin-4α-ol (**113**). The relative 2,3-*trans*-3,4-*cis* configuration of compounds **111** and **112** ($^3J_{2,3} = 10.5$, $^3J_{3,4} = 3.1$ Hz and $^3J_{2,3} = 10.5$, $^3J_{3,4} = 2.5$ Hz, respectively) was convincingly deduced from ^{1}H NMR data of the per-*O*-acetyl derivatives. However, the proposed 2,3-*cis*-3,4-*trans* configuration of the *O*-acetyl derivatives of **110** and **113** did not appear to be supported by coupling constants ($^3J_{2,3} = 2.5$, $^3J_{3,4} = 10.5$ Hz and $^3J_{2,3} = 3.5$, $^3J_{3,4} = 9.5$ Hz, respectively) and suggests that a reinterpretation of data is now required. A further feature that casts doubt on the validity of the structural claims is the apparent stability of leucopelargonidin (**113**) in contrast to the well-established reactivity of flavan-3,4-diols with 5,7-dihydroxy A-rings.[10]

The 8-*O*-methylepioritin-4α-ol (**114**)[89] complements the rare series of flavan-3,4-diols where *O*-methylation had occurred at one of the phenolic hydroxyl functions (Figure 11.10). Compounds **115–122** were obtained from various Lonchocarpus species.[82,84,90] The enzymatic conversion of a variety of dihydroflavonols via stereospecific reduction to *cis*-flavan-3,4-diols using flower extracts of *Dianthus caryoplyllus* L. was also demonstrated.[91] In addition, the chiroptical data of the eight diastereomeric flavan-3,4-diols have recently been reported.[92] These data permit unambiguous assignment of the absolute configuration at C-2 via the sign of the ^1L_b transition of the aromatic A-ring chromophore near 280 nm. A negative Cotton effect in this region invariably reflects 2R configuration and a positive one 2S configuration. The configuration at the C-3 and C-4 stereocenters then follows from the relative configurational assignment via ^{1}H NMR coupling constants.

For additional information regarding the chemistry of the flavans, flavan-3-ols, flavan-4-ols, and flavan-3,4-diols the reader is referred to Refs. 7–12.

11.3.2   Proanthocyanidins

11.3.2.1   B-Type Proanthocyanidins

Proanthocyanidins of the B-type are characterized by singly linked flavanyl units, usually between C-4 of the chain-extension unit and C-6 or C-8 of the chain-terminating moiety. They are classified according to the hydroxylation pattern of the chain-extension units (see Table 11.1).

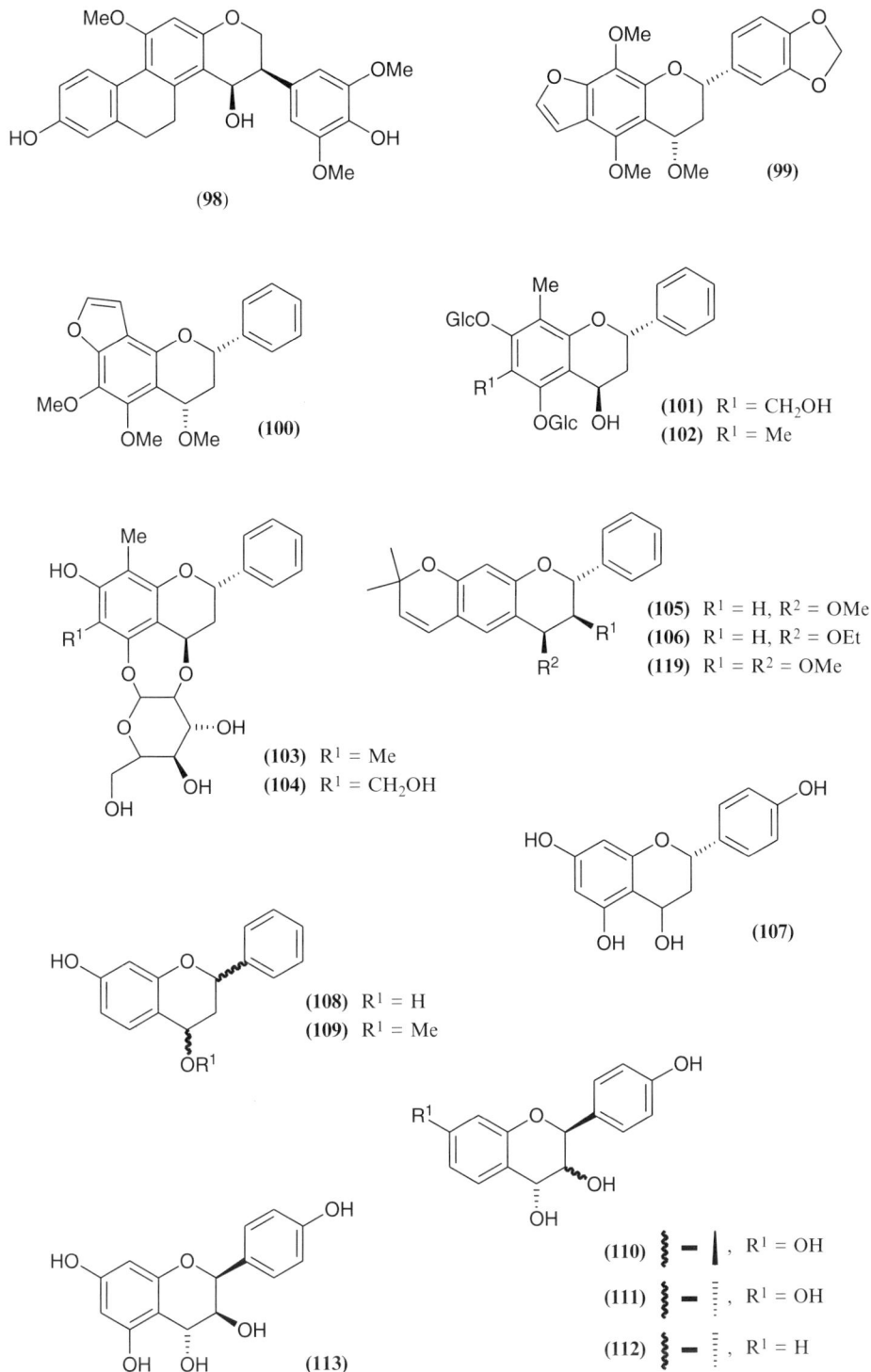

FIGURE 11.9 Structures of compounds 98–113 and 119.

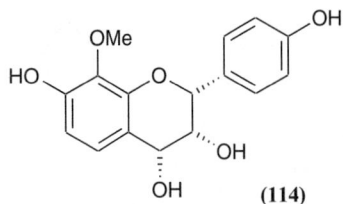

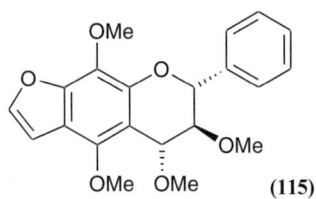

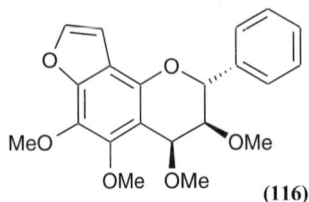

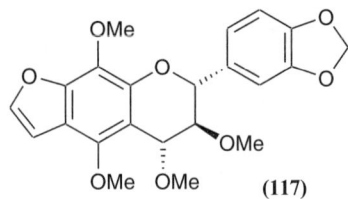

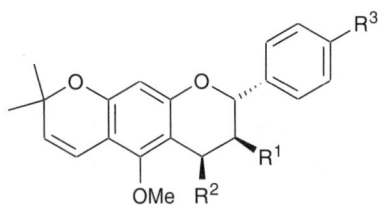

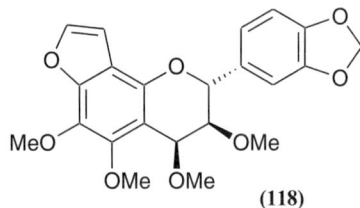

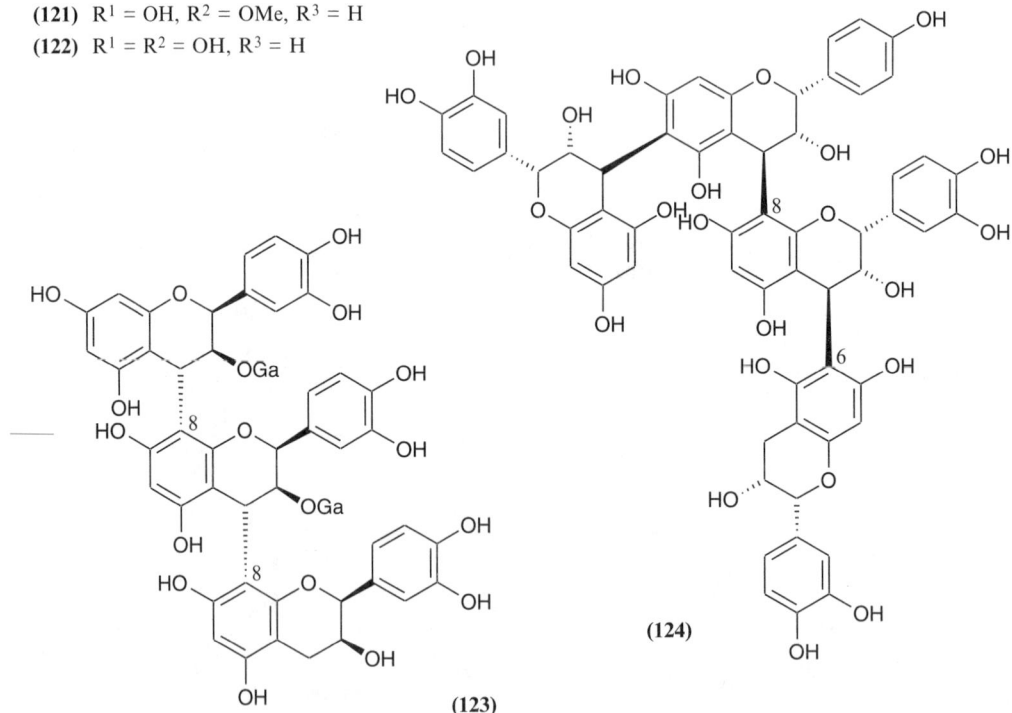

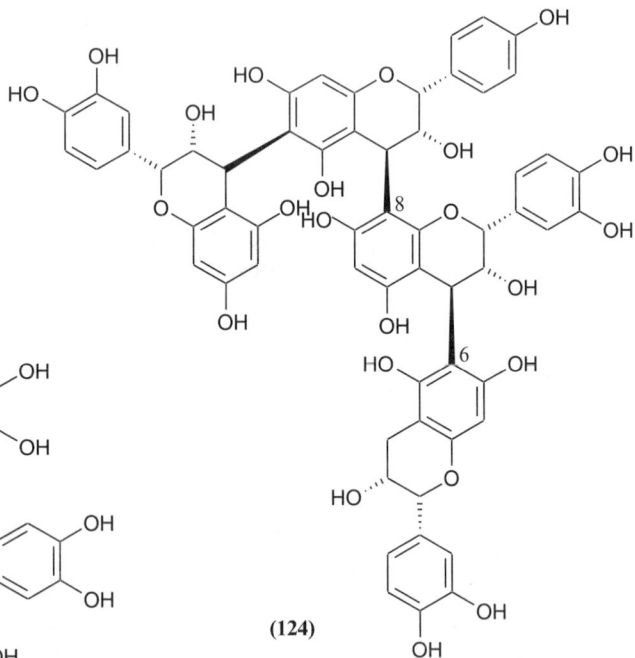

FIGURE 11.10 Structures of compounds **114–118** and **120–124**.

(120) $R^1 = R^2 = OMe$, $R^3 = OH$
(121) $R^1 = OH$, $R^2 = OMe$, $R^3 = H$
(122) $R^1 = R^2 = OH$, $R^3 = H$

A considerable number of new compounds have been added since 1992, such a sustained research effort motivated by the growing realization of the importance of these secondary metabolites in health, food, agriculture, and various other industrial applications. Information in this regard may be found in the cited references. A significant number of the new entries are derivatized via, e.g., galloylation and glycosylation, hence stressing the relevance of the flavan-3-ols exhibiting similar derivatization (Table 11.3).

11.3.2.1.1 Procyanidins (3,5,7,3′,4′-Pentahydroxylation)

The procyanidins maintained their position as a dominant and ubiquitous group of naturally occurring proanthocyanidins. A limited number of structures will be shown, especially with a view to demonstrate perusal of the nomenclature system at the tri- and tetrameric levels, e.g., trimer (123) and tetramer (124) (Figure 11.10). New structures for di-, tri-, and tetrameric procyanidins are listed in Table 11.4. The roots of *Polygonum coriarium* afforded a "mixed" procyanidin–prodelphinidin glucosylated tetra- and hexamer, both carrying various galloyl groups.[102] Possible structures were also proposed for the "mixed" procyanidin and prodelphinidin tri- and tetramers from *Geranium sanguineum*,[103] and for oligomeric proanthocyanidins D_{14}–D_{19} from *Quercus robur*.[100]

A number of "mixed" procyanidins or prodelphinidin analogs with exceptionally complex structures have been identified from the roots of *Clementsia semenovii*[105] and *Rhodiala pamiroalaica*.[106,107] Owing to the space requirements for the naming of these macromolecules, these are listed in the text only. In addition, the authors stated that indicated configurations are relative; hence, it is unclear whether the proanthocyanidin community should indeed consider the structures of these compounds as well as those cited in Refs. 100–103, 105–107 as sufficiently defined. The analogs from *C. semenovii* are: CS-3, 7-*O*-(6-*O*-galloyl-β-D-Glcp → 6-*O*-β-D-Glcp → 6-*O*-β-D-Glcp → 6-*O*-β-D-Glcp → 6-*O*-β-D-Glcp)-(+)-catechin-(4α → 8)-(−)-epigallocatechin-(4β → 8)-(+)-catechin-(4α → 8)-(−)-epigallocatechin-(4β → 8)-(−)-epigallocatechin-(4β → 8)-(−)-epigallocatechin, and CS-4, 3-*O*-galloyl-7-*O*-[6-*O*-galloyl-β-D-Glcp → 6-*O*-β-D-Glcp → 6-*O*-β-D-Glcp]-(+)-gallocatechin-(4α → 8)-[(+)-catechin-(4α → 8)-3-*O*-galloyl-(−)-epigallocatechin]₂-(4β → 8)-(−)-epigallocatechin.

The compounds from *R. pamiroalaica* are: RP-1, 7-*O*-[6-*O*-galloyl-β-D-Glcp → *O*-β-D-Glcp → *O*-β-D-Glcp]-(+)-gallocatechin-(4α → 8)-(−)-epicatechin-(4β → 8)-(−)-epicatechin-(4β → 8)-(+)-catechin-(4α → 8)-5-*O*-[6-*O*-galloyl-β-D-Glcp → *O*-β-D-Glcp → *O*-β-D-Glcp]-(+)-catechin, RP-2, 7-*O*-[*O*-β-D-Glcp → *O*-β-D-Glcp]-(−)-epicatechin-(4β → 6)-7-*O*-β-D-Glcp-(−)-epicatechin-(4β → 6)-3-*O*-galloyl-(−)-epigallocatechin-(4β → 6)-3-*O*-galloyl-(−)-epigallocatechin-(4β → 6)-3-*O*-galloyl-5-*O*-β-D-Glcp-(−)-epicatechin, RP-3, 7-*O*-(6-*O*-galloyl-β-D-Glcp)-3-*O*-galloyl-(−)-epigallocatechin-(4β → 8)-[(−)-epicatechin-(4β → 8)-(3-*O*-galloyl-(−)-epigallocatechin)]₂-(4β → 8)-5-*O*-(β-D-Glcp-6-*O*-β-D-Glcp)-(+)-catechin, and RP-4, 7-*O*-(6-*O*-galloyl-β-D-Glcp)-3-*O*-galloyl-(−)-epigallocatechin-(4β → 8)-[3-*O*-galloyl-(−)-epicatechin]-(4β → 8)-[3-*O*-galloyl-(−)-epicatechin]-(4β → 8)-[3-*O*-galloyl-(−)-epicatechin]-(4β → 8)-[3-*O*-galloyl-5-(β-D-Glcp → 6-*O*-galloyl-β-D-Glcp)]-(−)-epigallocatechin.

We also need to point out the often improper use of proanthocyanidin nomenclature. In Ref. 104, both vitisinol (125) and amurensisin (126) were classified as procyanidins; per definition they do not belong to this class of compounds (Figure 11.11). Vitisinol (125) is rather a member of the nonproanthocyanidin class with flavan or flavan-3-ol constituent units (see Section 11.3.3), while amurensisin (126) is simply a gallic acid derivative of epicatechin (see Section 11.3.1.2).

In addition to the contributions dealing with the isolation and structural elucidation of procyanidins, several excellent papers describing the synthesis and chemical manipulation of the proanthocyanidins in general have been published. These may be listed as follows:

TABLE 11.4
Naturally Occurring Flavan-4-ols and Flavan-3,4-diols

Class	Compound	Structure	Source	Ref.
1. Flavan-4-ol	Furanoflavan	(99)	*Lonchocarpus subglaucescens*	82
	Furanoflavan	(100)		
	Pneumatopterin A	(101)	*Pneumatopteris pennigera*	83
	Pneumatopterin B	(102)		
	Pneumatopterin C	(103)		
	Pneumatopterin D	(104)		
	Xuulanin	(105)	*Lonchocarpus xuul*	84
	4β-Demethylxuulanin-4β-ethyl ether	(106)		
	(2S)-5,7,4′-Tri-OH-flavan-4-ol	(107)	*Streptococcus sobrinus*	85
2. Flavan-3,4-diol	*Ent*-epiguibourtinidol-4α-ol	(110)	*Musa sapientum*	88
	Ent-guibourtinidol-4α-ol	(111)		
	2,3-*trans*-3,4-*cis*-Flavan-3,4-diol	(112)		
	Ent-epiafzelechin-4α-ol	(113)		
	8-*O*-methylepioritin-4α-ol	(114)	*Acacia caffra*	89
	Compound	(115)	*Lonchocarpus subglaucescens*	82
	Compound	(116)		
	Compound	(117)		
	Compound	(118)		
	Compound	(119)	*L. xuul*	84
	Compound	(120)	*L. xuul/L. yucatanensis*	90
	Compound	(121)		
	Compound	(122)		

1. Continuation of the crucial role of acid-catalyzed depolymerization of proanthocyanidin polymers and subsequent capture of incipient flavanyl-4-carbocations by sulfur and oxygen nucleophiles, toward structural elucidation.[108–113] This approach was successfully applied to demonstrate the presence of covalently bonded glycoside moieties in the chain-extension units of mangrove (*Bruguiera gymnorrhiza*) bark proanthocyanidins,[108] and the presence of 3-*O*-galloyl groups in the extension units of polymeric proanthocyanidins of *Hamamelis virginiana*.[111] It also highlighted the risk of using extended thiolysis to provide meaningful estimates of the molecular weight of polymeric proanthocyanidins, as well as the use of thiolysis as a means of obtaining "quantitative" information on the composition of "mixed" polymers.[109] The latter work was also extended to the acid-catalyzed phloroglucinolysis of Pecan (*Caraya illinoensis*) nut pith proanthocyanidins.[113] A conspicuous feature of the degradation studies on the mangrove and pecan proanthocyanidins is the demonstration of the susceptibility of the C_4–C_{10} bond of both the C-4 thio- and phloroglucinol adducts to cleavage under acidic conditions. This is demonstrated in Figure 11.11, Scheme 11.4 for an intermediate thioether (127). Protonation of the electron-rich A-ring leads to the weakening and subsequent cleavage of the C_4–C_{10} bond in 128 under influence of the electron-donating benzyl sulfanyl (or phloroglucinol[113]) moiety. The sulfonium species (129) or its phloroglucinol equivalent[113] is then susceptible to various rearrangements. Note that this process represents the equivalent of the cleaving of the interflavanyl

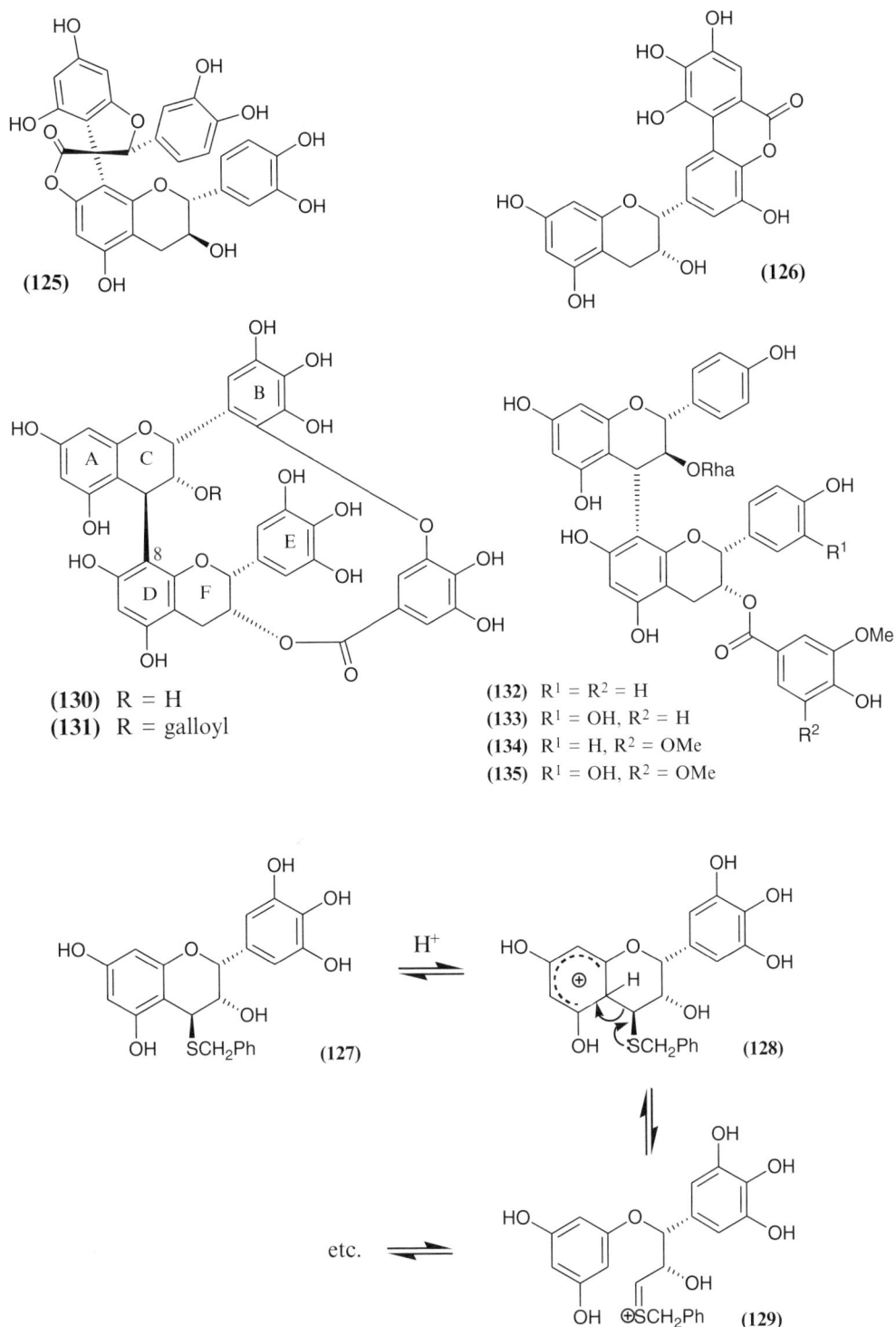

(125)

(126)

(130) R = H
(131) R = galloyl

(132) R¹ = R² = H
(133) R¹ = OH, R² = H
(134) R¹ = H, R² = OMe
(135) R¹ = OH, R² = OMe

H⁺

(127)

(128)

etc.

(129)

FIGURE 11.11 Structures of compounds 125–135 including Scheme 11.4: Proposed route to the acid-catalyzed cleavage of the C_4–C_{10} bond in C-4 thiobenzylflavan-3-ols.

bond under acidic conditions but under the influence of an external electron-donating sulfur or phloroglucinol nucleophile.

2. Synthesis as a means of unequivocal structure elucidation. Synthetic studies continued to make important contributions toward the unambiguous structure elucidation of the naturally occurring proanthocyandins. Prevalent amongst these is the synthesis of a series of di- and trimeric profisetinidins exhibiting both 2,3-*trans*- and *cis*-relative configuration of the constituent fisetinidol units,[114] the development of a protocol to generate the interflavanyl bond of procyanidins under neutral conditions,[115] the synthesis of ^{2}H-,[116] ^{13}C-, and ^{14}C-labeled catechins and procyanidins,[117–121] the synthesis of a high molecular mass condensed tannin,[122] synthesis of ether-linked proteracacinidins or promelacacinidins,[123] and a highlight of the review period, a series of papers dealing with synthesis of the proanthocyanidins found in cacao,[124] unequivocal proof of the 4β-configuration in procyanidins B-2,[125] stereoselective synthesis of an unnatural procyanidin diastereoisomer, epicatechin-(4α → 8)-epicatechin,[126] and a process to synthesize (6 → 6)-, (6 → 8)-, and (8 → 8)-linked catechin and epicatechin dimers as well as their 3,3-di-*O*-gallate esters.[127] In addition, a stereoselective route to octa-*O*-benzylprocyanidin B-3 was also described.[128]

3. X-ray and CD analysis. The structure of procyanidin B-1 was unequivocally confirmed by x-ray analysis of its deca-*O*-acetyl derivative by Weinges,[129] one of the pioneers in the field of proanthocyanidin chemistry. One of the most powerful methods to establish the absolute configuration at C-4 of the T-unit in dimeric A- and B-type proanthocyanidins remains the chiroptical method via application of the aromatic quadrant rule.[130] This has been repeatedly demonstrated by the author's own work and several other contributions listed in Refs. 7–12.

11.3.2.1.2 Prodelphinidins (3,5,7,3′,4′,5′-Hexahydroxylation)

In addition to the "mixed" procyanidin–prodelphinidin tri- and tetramers listed in Table 11.5 and Section 11.3.2.1.1,[105–107] a large number of new prodelphinidins have been reported. These are collated in Table 11.6. The roots of *C. semenovii* afforded a complex series of prodelphinidin oligomers.[138,139] The structures were deduced by chemical and enzymatic degradation, as well as spectroscopic data. These are compounds CS-1, 7-*O*-[6-*O*-galloyl-β-D-Glcp-6 → *O*-β-D-Glcp-6 → *O*-β-D-Glcp-6 → *O*-β-D-Glcp]-(+)-gallocatechin-(4α → 8)-(+)-gallocatechin-(−)-epigallocatechin-(4β → 8)-(−)-epigallocatechin-(4β → 8)-(−)-epigallocatechin-(4β → 8)-(+)-catechin, CS-2, 3-*O*-galloyl-7-*O*-(β-D-Glcp-6 → *O*-β-D-Glcp-(−)-epigallocatechin-(4α → 8)-[3-*O*-galloyl-(−)-epicatechin]-(4β → 8)-[3-*O*-galloyl-(−)-epigallocatechin]-(4β → 8)-[3-*O*-galloyl-5-*O*-(6-*O*-galloyl-*O*-β-D-Glcp)]-(−)-epicatechin, rhodichimoside, 7-*O*-[β-D-Glcp-*O*-β-D-Glcp]$_2$-3-*O*-galloyl-(−)-epicatechin]-(4β → 8)-[3-*O*-galloyl-(−)-epigallocatechin]-(4β → 8)-3-*O*-galloyl-(−)-epigallocatechin, rhodichin, 7-*O*-β-D-Glcp-3-*O*-galloyl-(−)-epigallocatechin-(4β → 8)-[(−)-epigallocatechin]$_2$-(4β → 8)-epigallocatechin-(4β → 6)-3-*O*-galloyl-(−)-epigallocatechin.

A "mixed" propelargonidin–prodelphinidin–procyanidin pentamer, PZ-5, was identified in *Ziziphus jujuba*.[140] Prodelphinidin polymers of undefined structure were also obtained from white clover (*Trifolium repens* L.) flowers[141] and *Onobrychis viciifolia* (sainfoin).[142]

The two unique dimeric compounds, samarangenins A (**130**) and B (**131**),[133] comprise two epigallocatechin moieties possessing the characteristic C → D-ring linkage of B-type proanthocyanidins as well as a C–O bond between a B-ring carbon and an oxygen function of the pyrogallyl-type ring of the 3-*O*-galloyl moiety at the DEF unit (Figure 11.11). Accordingly, they represent the first doubly linked proanthocyanidins that possess interflavanyl bonds originating from both two-electron (the 4 → 8 bond) and one-electron (the carbon → oxygen) bond processes.

TABLE 11.5
The Natural Procyanidins

Class	Compound	Structure	Source	Ref.
1. Dimers	Catechin-(4α → 8)-gallocatechin (tentative)		*Cistus incanus*	40
	Catechin-(4α → 8)-epigallocatechin		*Croton lechleri*	93
	Catechin-(4α → 8)-catechin-3-O-β-D-Glp		*Quercus marilandica* Muenchh.	38
	3-O-Ga-*ent*-epicatechin-(4α → 8)-*ent*-epicatechin		*Byrsonima crassifolia*	96
	Ent-epicatechin-(4α → 8)-*ent*-epicatechin-3-O-Ga			
	3-O-Ga-*ent*-epicatechin-(4α → 8)-*ent*-epicatechin-3-O-Ga			
	Ent-epicatechin-(4α → 6)-*ent*-epicatechin			
	Epicatechin-(4β → 8)-catechin-3-O-(4-OH)benzoate		*Hamamelis virginiana*	97
	Epicatechin-(4β → 8)-epigallocatechin		*Ziziphus jujuba*	98
	Epicatechin-(4β → 8)-gallocatechin		*Alhagi sparsifolia*	99
	3-O-Ga-catechin-(4β → 8)-catechin-3-O-Ga		*Quercus robur*	100
	Epicatechin-(4β → 6)-epicatechin-3-O-Ga		*Vitus amurensis*	104
	Catechin-3-O-β-D-Glp-(4α → 8)-catechin-3-O-β-D-Gl(2-cinnamoyl)pyranoside		*Inga umbellifera*	68
	Catechin-3-O-β-D-Glp-(4α → 8)-epicatechin-3-O-β-D-Gl(6-cinnamoyl)pyranoside			
2. Trimers	Epicatechin-(4β → 8)-epicatechin-(4β → 6)-epicatechin	**(123)**	*Craetagus laevigata/Malus pumila*	164, 165
	3-O-Ga-*ent*-epicatechin-(4α → 8)-3-O-Ga-*ent*-epicatechin-(4α → 8)-*ent*-epicatechin		*Byrsonima crassifolia*	96
	3-O-Ga-epicatechin-(4β → 8)-3-O-Ga-*ent*-epicatechin-(4α → 8)-*ent*-epicatechin			
3. "Mixed" trimers	Catechin-(4α → 8)-gallocatechin-(4α → 8)-gallocatechin		*Lonchocarpus* spp.	90
	Epiafzelechin-(4β → 8)-epicatechin-(4β → 8)-catechin		*Ziziphus jujuba*	98
4. "Mixed" tetramers	Epicatechin-(4β → 6)-epiafzelechin-(4β → 8)-epicatechin-(4β → 6)-epicatechin (davallin)	**(124)**	*Davalia mariesii*	94, 95
	3-O-Ga-7-O-[O-(6-O-Ga)-β-D-Glp]-epigallocatechin-(4β → 8)-epicatechin-(4β → 8)-epicatechin-(4β → 8)-epigallocatechin		*Polygonum coriarium*	101
	Epicatechin-(4β → 8)-3-O-Ga-epigallocatechin-(4β → 8)-epicatechin-(4β → 8)-epicatechin			

TABLE 11.6
The Natural Prodelphinidins

Class	Compound	Structure	Source	Ref.
1. Dimers	Gallocatechin-(4α → 6)-gallocatechin		*Cistus incanus*	40
	Gallocatechin-(4α → 6)-epigallocatechin		*Croton lechleri*	93
	Gallocatechin-(4α → 8)-epicatechin			
	Epigallocatechin-(4β → 8)-epigallocatechin		*Cistus salvifolius*	42
	Epigallocatechin-(4β → 8)-gallocatechin (prodelphinidin B1)		*Psidium guajava/*	131,
			Pithecellobium	132
			lobatum	
	Samarangenin A	**(130)**	*Syzygium*	133
			samarangens/	
			S. aqueum	
	Samarangenin B	**(131)**		
	Epigallocatechin-(4β → 6)-epigallocatechin		*Stryphnodendron*	59
			adstringens	
	4-O-Methylgallocatechin-(4α → 8)-4′-O-methylgallocatechin			134
	3-O-Ga-Epigallocatechin-(4β → 6)-gallocatechin		*Cistus incanus*	135
	3-O-Ga-Epigallocatechin-(4β → 8)-gallocatechin			
	3-O-Ga-Epigallocatechin-(4β → 8)-epigallocatechin		*C. salvifolius*	42
	Epigallocatechin-(4β → 6)-epigallocatechin-3-O-Ga			
	Epigallocatechin-(4β → 8)-epigallocatechin-3-O-(4-OH)-benzoate		*S. adstringens*	59
	Gallocatechin-(4α → 8)-epigallocatechin-3-O-Ga			
	Gallocatechin-(4α → 8)-epigallocatechin-3-O-(4-OH)-benzoate			
2. Trimers	Gallocatechin-(4α → 8)-gallocatechin-(4α → 8)-epigallocatechin		*Croton lechleri*	93
	Gallocatechin-(4α → 8)-gallocatechin-(4α → 8)-gallocatechin		*Ribes nigrum*	136
	Epigallocatechin-(4β → 8)-gallocatechin-(4α → 8)-catechin		*Cistus albidus*	137
	Epigallocatechin-(4β → 8)-gallocatechin-(4α → 8)-gallocatechin			

11.3.2.1.3 Propelargonidins (3,5,7,4′-Tetrahydroxylation)

The limited numbers of new propelargonidins are listed in Table 11.7. This is in addition to the "mixed" procyanidin–propelargonidin tetramer, davallin (Table 11.5),[94,95] the "mixed" propelargonidin–procyanidin trimer, epiafzelechin-(4β → 8)-epicatechin-(4β → 8)-catechin (Table 11.5),[98] and the "mixed" propelargonidin–prodelphinidin–procyanidin pentamer PZ-2.[140]

11.3.2.1.4 Profisetinidins (3,7,3′,4′-Tetrahydroxylation)

The profisetinidins are the most important proanthocyanidins of commerce, making up the major constituents of wattle and quebracho tannins. New entries are collated in Table 11.8.

The structures of the di- and trimeric profisetinidins from *Pithecellobium dulce* (Guamúchil) were rigorously corroborated via synthesis.[114] The synthetic approach was additionally motivated by the precariousness of unequivocally differentiating between 2,3-*cis*-3,4-*trans*- and 2,3-*cis*-3,4-*cis*-configurations of the chain-extension units on the basis of ^{1}H NMR coupling constants.[150,151] Furthermore, the powerful nuclear Overhauser effect (NOE) method for differentiating between 2,4-*cis*- and 2,4-*trans*-substitution[152] is less useful at the di- and trimeric levels due to the adverse effects of dynamic rotational isomerism about the interflavanyl bond(s) on ^{1}H NMR spectra at ambient temperatures.

TABLE 11.7
The Natural Propelargonidins

Class	Compound	Structure	Source	Ref.
1. Dimers	3-O-(α-L-Rhamnopyranosyl)afzelechin-(4α → 8)-epiafzelechin-3-O-vanillate	(132)	*Joannesia princeps*	143
	3-O-(α-L-Rhamnopyranosyl)afzelechin-(4α → 8)-epicatechin-3-O-vanillate	(133)		
	3-O-(α-L-Rhamnopyranosyl)afzelechin-(4α → 8)-epiafzelechin-3-O-syringate	(134)		
	3-O-(α-L-Rhamnopyranosyl)afzelechin-(4α → 8)-epicatechin-3-O-syringate	(135)		
	3-O-Ga-epiafzelechin-(4β → 8)-epicatechin-3-O-Ga[a]		Green tea	144
	3-O-Ga-epiafzelechin-(4β → 8)-epicatechin-3-O-Ga		Green tea	145
	3-O-Ga-epiafzelechin-(4β→6)-epicatechin-3-O-Ga			
	Epiafzelechin-(4β → 8)-4β-carboxymethylepiafzelechin Me ester		*Drynaria fortunei*	146
2. Trimers	Epiafzelechin-(4β → 8)-epiafzelechin-(4β → 8)-4-Me-epigallocatechin		*Heisteria pallida*	147
	Afzelechin-(4α → 8)-afzelechin-(4α → 8)-afzelechin (C-glucosylated?)		*Aegle marmelos*	148

[a] (4β → 8) bond tentatively assigned.

The principles that govern the base-catalyzed C-ring isomerization of dimeric profisetinidins[152–154] also control a series of similar rearrangements in the trimeric bis-fisetinidol-catechins (**136–139**)[155–158] and related bis-fisetinidol-epicatechins (Figure 11.12).[159] Analogs possessing constituent chain-extension units with 3,4-*cis*-configuration (ABC unit in, e.g., **137**) are subject to extensive 1,3-migrations in intermediate B-ring quinone methides and hence to the formation of exceptionally complex reaction mixtures.[156–159] Compound **139** afforded no less than 16 conversion products, the structures of which were unequivocally established.[157] This has subsequently led to the development of a more controlled synthesis that is based upon the repetitive formation of the interflavanyl bond and pyran ring rearrangement of the chain-extension unit under mild basic conditions.[160] Thus, in contrast to the unrestrained course of the base-catalyzed C-ring rearrangement reactions of profisetinidin triflavonoids possessing 2,3-*trans*–3,4-*cis*-flavanyl constituent units, the stepwise construction of the dipyranochromene framework via sequential interflavanyl bond formation and pyran ring rearrangement permitted concise synthetic access to phlobatannins at the trimeric level, e.g., the hexahydrodipyrano[2,3-*f*:2′,3′-*h*]chromene (**140**).[160]

The readily occurring cleavage of the interflavanyl bond in proanthocyanidins that exhibit C-5 oxygenation of the A-ring of their chain-extension units, with sulfur or oxygen nucleophiles under acid catalysis, plays a crucial role in the structural elucidation of this complex group of natural products. In the 5-deoxy series of compounds, e.g., the fisetinidol-(4α → 8)-catechin (**141**), this C(sp^3)–C(sp^2) bond is remarkably stable under a variety of conditions and has resisted efforts at cleavage in a controlled fashion. Such a stable interflavanyl bond has adversely affected both the structural investigation of the polymeric proanthocyanidins in black wattle bark and of those from other commercial sources, as well as the establishment of the absolute configuration of the chain-terminating flavan-3-ol moieties in the 5-deoxyoligoflavanoids. It has recently been demonstrated that the interflavanyl bond

TABLE 11.8
The Natural Profisetinidins

Class	Compound	Structure	Source	Ref.
1. Dimers	Fisetinidol-(4α → 8)-6-Me-catechin		Commercial wattle bark extract	32
	Epifisetinidol-(4β → 8)-catechin		*Pithecellobium dulce*	114
	Epifisetinidol-(4β → 8)-epicatechin			
2. Trimers	Bis-epifisetinidol-(4β → 6:4β → 8)-catechin		*P. dulce*	114
	Bis-epifisetinidol-(4β → 6:4β → 8)-epicatechin			
	Fisetinidol-(4α → 8)-catechin-(6 → 4β)-epifisetinidol			
	Fisetinidol-(4α → 8)-epicatechin-(6 → 4β)-epifisetinidol			
	Bis-fisetinidol-(4α → 6:4α → 8)-catechin-3-*O*-Ga		*Burkea africana*	149
3. Phlobatannin related	Compound	**(145)**	*Guibourtia coleosperma*	155
	Compound	**(146)**	*Colophospermum mopane*	155
	Compound	**(147)**	*Baikiaea plurijuga*	156
	Compound	**(148)**		
	Compound	**(149)**	*C. mopane*	156
	Compound	**(150)**	*B. plurijuga*	156
	Compound	**(151)**	*C. mopane/B. plurijuga*	159
	Compound	**(152)**	*B. plurijuga*	

in the proanthocyanidins, including the 5-deoxy analogs and their permethylaryl ethers, is readily subject to reductive cleavage with sodium cyanoborohydride [Na(CN)BH₃] in trifluoroacetic acid (TFA) at 0°C.[161–163] Thus, under acidic conditions profisetinidin (**141**) is protonated at C-8 (D-ring) and the interflavanyl bond in intermediate **142** cleaved by delivery of the equivalent of hydride ion either at C-4(C) to give the constituent flavan-3-ol moieties **2** and **143**, or at C-2(C) to give the 1,3-diarylpropan-2-ol (**144**) and catechin (**2**) (see Ref. 163) (Figure 11.13, Scheme 11.5). The protocol is as effective for the permethylaryl ether of **141**, for procyanidin B-3 and B-4, and also for A-type proanthocyanidins[163] (*vide infra*).

11.3.2.1.5 Prorobinetinidins (3,7,3′,4′,5′-Pentahydroxylation)
The prorobinetinidins also feature prominently in wattle bark extract.[3] In the period under review, new entries, listed in Table 11.9, have been identified from wattle bark,[32] *Robinia pseudacacia*,[166] and *Stryphnodendron adstringens*.[167] The recent investigation of commercial wattle bark extract[32] has led to the identification of robinetinidol-(4β → 8)-catechin (**153**) and robinetinidol-(4β → 8)-gallocatechin (**154**), the first prorobinetinidins with a 3,4-*cis* C-ring configuration, as well as a prorobinetinidin-type tetrahydropyrano[2,3-*h*]chromene isomerization intermediate (**155**) (Figure 11.14). Compounds of the class **155** as well as dimeric analogs possessing the DEF-GHI core of **155** readily form from the "conventional" tri- and biflavanoids, respectively, by rearrangement of the pyran heterocycle(s) under mild basic conditions. The mechanisms explaining their intricate genesis were extensively reviewed.[7,9,10] Such a susceptibility of the constituent flavanyl units of proanthocyanidins to intramolecular rearrangement via B-ring quinone methide intermediates under basic conditions was also demonstrated in an unusual dimerization–rearrangement reaction of catechin at pH 12 and 40°C.[168]

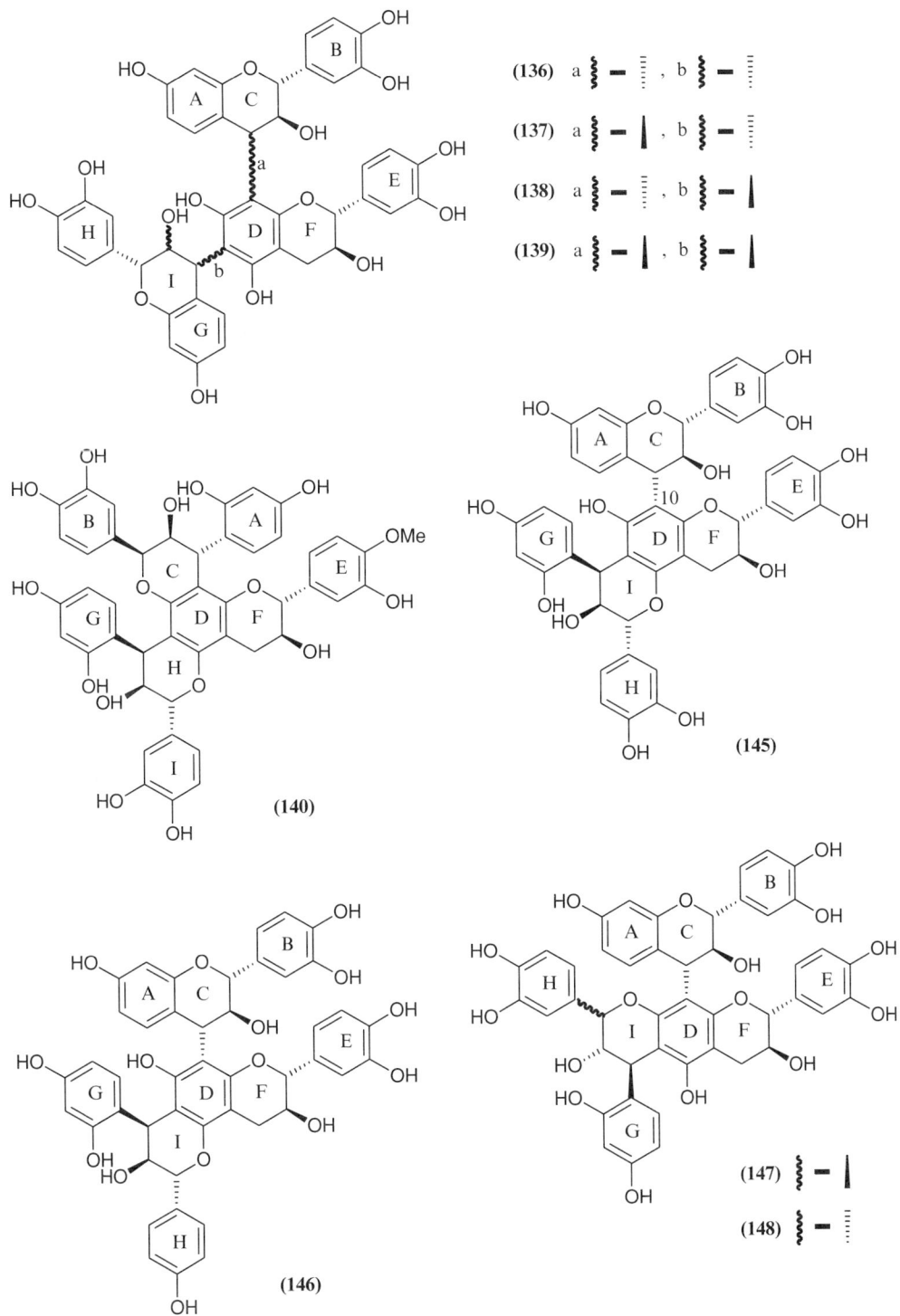

FIGURE 11.12 Structures of compounds **136–140** and **145–148**.

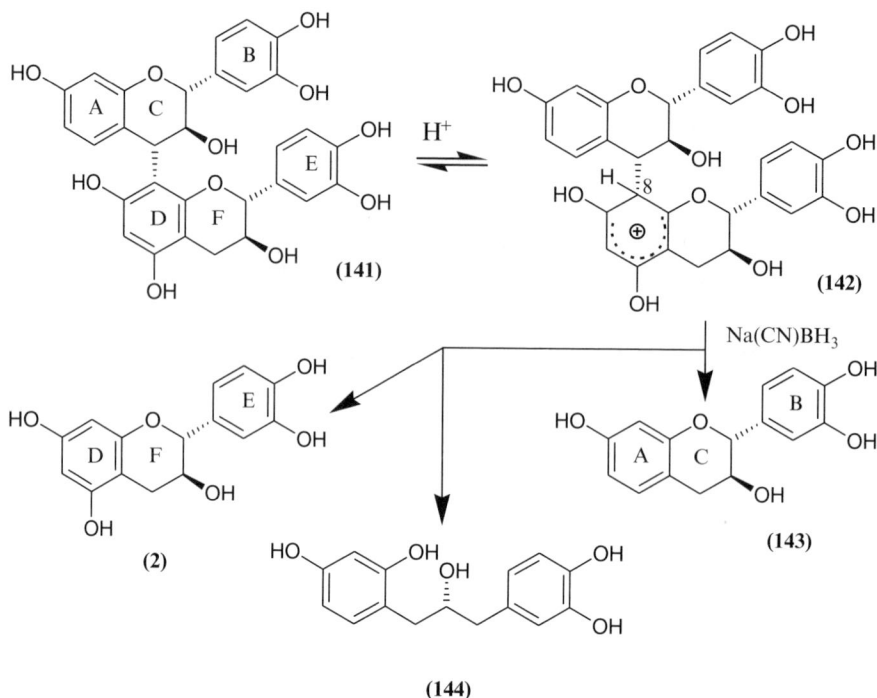

FIGURE 11.13 Structures of compounds **141–144** including Scheme 11.5: Proposed route to the reductive cleavage of the interflavanyl bond in proanthocyanidins.

In the heartwood of *R. pseudacacia*, the flavan-3,4-diol, leucorobinetinidin (robinetinidol-4α-ol), as the incipient electrophile for prorobinetinidin biosynthesis, coexists with a variety of monomeric flavonoids invariably possessing C-4 oxygenation,[166] hence reducing the nucleophilicity of their A-rings compared to that of the corresponding functionality in the C-4 deoxy compounds, e.g., catechin. *R. pseudacacia* therefore represents a rare metabolic pool where oligomer formation has to occur via the action of a very potent electrophile on chain-terminating units apparently lacking the nucleophilicity that is associated with natural sources in which proanthocyanidin formation is paramount. The diversity of the oxidation level of the chain-terminating moieties suggests that the biflavanoids in *R. pseudacacia* may be interrelated via oxidation–reduction of these units.

11.3.2.1.6 Proteracacinidins (3,7,8,4′-Tetrahydroxylation)
In contrast to the large number of oligomeric proanthocyanidins with resorcinol- or phloroglucinol-type A-rings of the chain-extension units,[3,4] those possessing pyrogallol-type A-rings (7,8-dihydroxylation) are rare. The first proteracacinidin analogs were identified as recently as 1994. Since then a considerable number (Table 11.10) of these compounds have been isolated from two members of the Leguminosae, *Acacia caffra* and *A. galpinii*.[16,169–177] A conspicuous feature of the compounds listed in Table 11.10 is the heterogeneity of the interflavanyl bond in this class of naturally occurring proanthocyanidins. Both carbon and oxygen centers participate in interflavanyl bond formation and it would appear as if both one- and two-electron processes play a prominent role in establishing the interflavanyl linkage(s). This presumably reflects the poor nucleophilicity of the pyrogallol-type A-ring of the monomeric flavan-3,4-diol precursors, hence permitting alternative centers to participate in the interflavanyl bond forming process.

TABLE 11.9
The Natural Prorobinetinidins

Class	Compound	Structure	Source	Ref.
1. Dimers	Robinetinidol-(4β → 8)-catechin	(153)	Commercial	32
	Robinetinidol-(4β → 8)-gallocatechin	(154)	black wattle bark	
	Robinetinidol-(4β → 6)-robinetinidol-4β-ol		*Robinia*	166
	Robinetinidol-(4β → 6)-robinetinidol-4α-ol		*pseudacacia*	
	Robinetinidol-(4α → 2′)-robinetinidol-4β-ol			
	Robinetinidol-(4α → 2′)-robinetinidol-4α-ol			
	Robinetinidol-(4α → 2′)-dihydrorobinetin			
	Robinetinidol-(4β → 6)-dihydrorobinetin[a]			
	Robinetinidol-(4α → 8)-dihydrorobinetin			
	Robinetinidol-(4α → 2′)-robinetin[a]			
	Robinetinidol-(4β → 2′)-7,3′,4′,5′-tetra-OH-flavone			
	Robinetinidol-(4β → 8)-epigallocatechin		*Stryphnodendron*	167
	Robinetinidol-(4α → 8)-epigallocatechin		*adstringens*	
	Robinetinidol-(4β → 8)-epigallocatechin-3-*O*-Ga			
	Robinetinidol-(4α → 8)-epigallocatechin-3-*O*-Ga			
	Robinetinidol-(4α → 6)-gallocatechin			
	Robinetinidol-(4α → 6)-epigallocatechin			
2. Phlobatannin related	Compound	(155)	Commercial black wattle bark	32

[a] Dihydrorobinetin is (2*R*,3*R*)-2,3-*trans*-7,3′,4′,5′-tetrahydroxydihydroflavonol and robinetin the corresponding flavonol.

Considerable effort has been devoted to confirm the constitution and absolute configuration of both the carbon–carbon and carbon–oxygen analogs via biomimetic syntheses.[16,170–172,176] In addition, the compounds possessing carbon–oxygen linkages are susceptible to ready reductive cleavage under acidic conditions, hence permitting unequivocal structural elucidation of the constituent flavanyl moieties.[174] Definition of the axial chirality of analogs possessing biphenyl linkages, e.g., the four compounds from *A. galpinii*[173] and compound **156**, was unambiguously done by circular dichroism (Figure 11.15).

11.3.2.1.7 Promelacacinidins (3,7,8,3′,4′-Pentahydroxylation)

The natural occurrence of promelacacinidins was until recently restricted to the heartwoods of *Prosopis glandulosa*[3] and *Acacia melanoxylon*.[4] The investigations of the heartwood constituents of *A. caffra* and *A. galpinii* also revealed the presence of a limited number of promelacacinidins (Table 11.11) as well as "mixed" di- and trimeric proteracacinidin or promelacacinidins (see Table 11.10).

Notable from Table 11.10 and Table 11.11 is the considerable number of proteracacinidins and promelacacinidins at both the di- and trimeric levels possessing 2,3-*cis*-3,4-*cis*-flavan-3,4-diol "terminating" moieties. It has been demonstrated that melacacidin (**158**) is basically inert toward solvolysis or epimerization at C-4.[179,180] This was recently confirmed during the thiolysis of 4β-chloroepioritin derivatives.[181] Such stability may be ascribed to hydrogen bonding between the axial C-3(OH) and the heterocyclic oxygen, which locks the C-ring in a half-chair conformation with C_{2eq}, C_{3ax}, and C_{4eq} substituents.[179,180] In this conformation, the appropriate C_4 σ* orbital is at an angle of ~45° above the plane of the

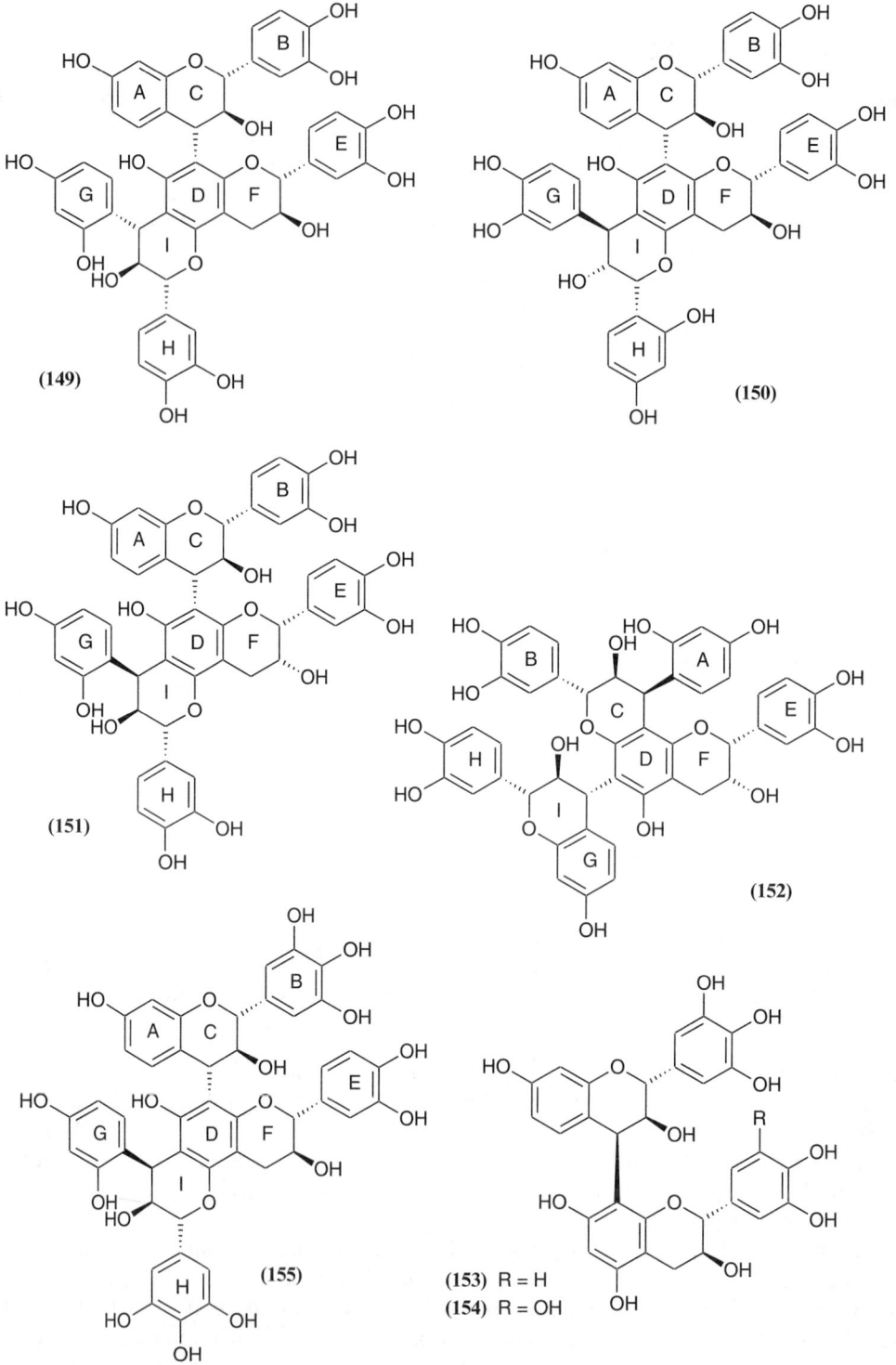

FIGURE 11.14 Structures of compounds **149–155**.

TABLE 11.10
The Natural Proteracacinidins

Class	Compound	Structure	Source	Ref.
1. Dimers	*Ent*-oritin-(4β → 7:5 → 6)-epioritin-4α-ol	**(156)**	*Acacia caffra*	169
	Ent-oritin-(4β → 5)-epioritin-4β-ol			89
	Epioritin-(4β → 3)-epioritin-4β-ol	**(10)**		16
	Oritin-(4α → 7:5 → 6)-epioritin-4α-ol		*A. galpinii*	173
	Oritin-(4β → 7:5 → 6)-epioritin-4α-ol			
	Epioritin-(4β → 7:5 → 6)-epioritin-4α-ol			
	Epioritin-(4β → 7:5 → 6)-oritin-4α-ol			
	Epioritin-(4β → 5:3 → 4)-oritin-4α-ol	**(157)**	*A. caffra*	173
	Oritin-(4α → 5)-epioritin-4β-ol		*A. galpinii*	175
	Ent-epioritin-(4α → 5)-epioritin-4β-ol		*A. caffra*	
	Epioritin-(4β → 5)-epioritin-4α-ol		*A. galpinii*	
	Ent-oritin-(4β → 5)-epioritin-4α-ol			
	Epioritin-(4β → 6)-oritin-4α-ol		*A. caffra*	176
	Epioritin-(4β → 6)-*ent*-oritin-4α-ol		*A. galpinii*	
	Ent-oritin-(4β → 6)-epioritin-4α-ol			
	Ent-oritin-(4β → 5)-oritin-4α-ol			
	Ent-oritin-(4α → 6)-epioritin-4α-ol		*A. caffra*	
	Ent-oritin-(4α → 6)-oritin-4α-ol		*A. galpinii*	
	Ent-oritin-(4α → 6)-epioritin-4β-ol		*A. caffra*	
	Epioritin-4α-ol-(6 → 6)-epioritin-4β-ol		*A. galpinii*	177
	(2*S*)-7,8,4′-Trihydroxyflavan-(4β → 6)-epioritin-4α-ol		*A. caffra*	182
2. "Mixed" dimers	Epimesquitol-(4β → 4)-epioritin-4β-ol		*A. caffra*	16
	Epioritin-(4β → 6)-epimesquitol-4α-ol		*A. caffra*	176
	Epioritin-(4β → 6)-epimesquitol-4β-ol			
	Epimesquitol-(4β → 6)-epioritin-4α-ol			
3. Trimers	Epioritin-(4β → 6)-epioritin-(4α → 4)-epioritin-4α-ol		*A. caffra*	174
	Epioritin-(4β → 3)-epioritin-(4β → 6)-epioritin-4β-ol			
	Epioritin-(4β → 6)-oritin-(4α → 6)-epioritin-4α-ol		*A. caffra*	177
	Oritin-(4β → 6)-oritin-(4α → 6)-epioritin-4α-ol		*A. galpinii*	
	Epioritin-(4β → 6)-epioritin-(4β → 6)-epioritin-4α-ol		*A. caffra*	177
4. "Mixed" trimers	Epioritin-(4β → 3)-epioritin-(4β → 6)-epimesquitol-4α-ol		*A. caffra*	174

A-ring and "buried" in the heterocyclic ring that screens its overlap by an external nucleophile. Since a C-4 antibonding orbital orthogonal to the A-ring would permit the most effective delocalization of A-ring electron density or stabilization of electron deficiency at C-4, it is clear why an all-*cis* C-ring configuration is more common for flavan-3,4-diols with 7,8-dihydroxylated A-rings. These compounds, no doubt, will have a reduced need for delocalization of the aromatic A-ring electron density than their counterparts with more electron-rich resorcinol- and phloroglucinol-type A-rings. It may then also explain the stability and abundance of the flavan-3,4-diol, teracacidin, as well as the growing number of di- and trimers with 2,3-*cis*-3,4-*cis*-flavanyl constituent units all possessing 7,8-dihydroxy A-rings and axial C-3 hydroxyl groups.[4,89,171,172,174,177]

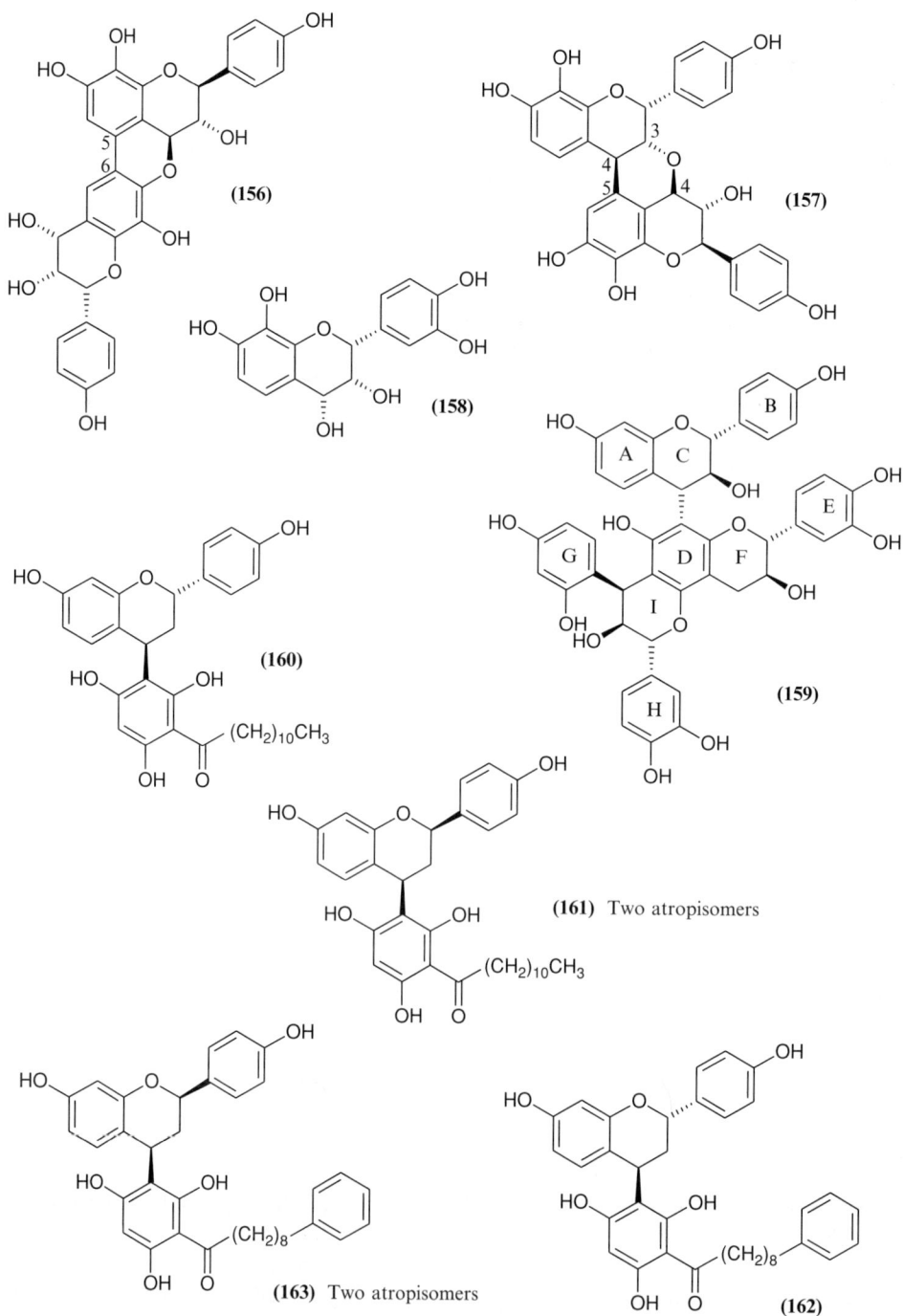

FIGURE 11.15 Structures of compounds **156–163**.

11.3.2.1.8 Proguibourtinidins (3,7,4′-Trihydroxylation)

The pro- and leucoguibourtinidins with their 7,4′-dihyroxy phenolic functionality represent a rare group of proanthocyanidins. New analogs are listed in Table 11.12. The structures of the dimers from *C. abbreviata* were rigorously established using ¹H NMR and CD techniques, as

TABLE 11.11
The Natural Promelacacinidins

Class	Compound	Source	Ref.
1. Dimers	Epimesquitol-(4β → 6)-epimesquitol-4β-ol	*Acacia caffra*	170
	Mesquitol-(4β → 6)-3,7,8,3′,4′-pentahydroxyflavone	*A. nigrescens*	178
	Epimesquitol-(4β → 5)-3,7,8,3′,4′-pentahydroxyflavone		
2. "Mixed" dimers	See Table 11.10		
3. "Mixed" trimers	See Table 11.10		

well as semisynthesis from starting materials with known absolute configuration.[183] The structure of guibourtinidol-(4α → 8)-afzelechin from *Ochna calodendron* was claimed on the basis of ^{1}H NMR evidence only,[184] while the known guibourtinidol-(4α → 8)-epiafzelechin was also identified in the stems of *Cassia biflora*.[185]

11.3.2.1.9   Procassinidins (7,4′-Dihydroxylation)
The procassinidins with their 7,4′-dihydroxyflavan chain-extension units represent a rare group of naturally occurring proanthocyanidins and only four compounds were reported in the previous review.[4] Seven new analogs were identified in the bark of *C. petersiana*[186] and are listed in Table 11.13. Synthesis as the permethylaryl acetate derivatives was done by reduction of the flavanone, including the optically pure (2S)-di-O-methylliquirtigenin, to the flavan-4-ol, which then served as electrophile in the Lewis acid (TiCl$_4$) catalyzed coupling with the appropriate flavan-3-ol permethylaryl ether, e.g., penta-O-methylepi- or -gallocatechin.[186] The cassiflavan-type myristinins A–F (**160–163**) (Figure 11.15) were identified in *Myristica cinnamomea*.[214]

11.3.2.1.10   Probutinidins (7,3′,4′-Trihydroxylation)
The probutinidins (see Section 11.2) represent a second class of proanthocyanidins with flavan chain-extension units. Only five members of this class of compounds have been identified (Table 11.14). Their structures and absolute configurations were also confirmed by synthesis via reduction of the flavanone, butin, followed by acid-catalyzed condensation with the appropriate flavan-3-ol.[17,187] A notable feature of the synthetic studies was the apparent preference for (4 → 8) bond formation reported by both groups of authors.

TABLE 11.12
The Natural Proguibourtinidins

Class	Compound	Structure	Source	Ref.
1. Dimers	Guibourtinidol-(4β → 8)-epiafzelechin		*Cassia abbreviata*	183
	Guibourtinidol-(4β → 8)-epicatechin			
	Guibourtinidol-(4β → 8)-afzelechin			
	Guibourtinidol-(4α → 6)-afzelechin			
	Ent-guibourtinidol-(4β → 8)-epicatechin			
	Guibourtinidol-(4α → 8)-afzelechin		*Ochna calodendron*	184
2. Phlobatannin related	Compound	(**159**)	*Colophospermum mopane*	155

TABLE 11.13
The Natural Procassinidins

Class	Compound	Structure	Source	Ref.
Dimers	Cassiaflavan-(4α → 8)-epicatechin		*Cassia petersiana*	186
	Cassiaflavan-(4α → 8)-epigallocatechin			
	Cassiaflavan-(4β → 8)-epicatechin			
	Cassiaflavan-(4β → 8)-epigallocatechin			
	Cassiaflavan-(4β → 8)-gallocatechin			
	Ent-cassiaflavan-(4β → 8)-epicatechin			
	Cassiaflavan-(4α → 6)-epicatechin			
	Cassiaflavan (myristinin A)	**(160)**	*Myristica cinnamomea*	214
	Cassiaflavan (myristinin D)	**(162)**		
	Ent-cassiaflavan atropisomers (myristinin B and C)	**(161)**		
	Ent-cassiaflavan atropisomers (myristinin E and F)	**(163)**		

11.3.2.2 A-Type Proanthocyanidins

In contrast to proanthocyanidins of the B-type, where the constituent flavanyl units are linked *via* only one bond, analogs of the A-class possess an unusual second ether linkage to C-2 of the T-unit. This feature introduces a high degree of conformational stability at the interflavanyl bonding axes of dimeric analogs that culminates in high quality and unequivocal NMR spectra, conspicuously free of the effects of dynamic rotational isomerism. Compounds of this class are readily recognizable from the characteristic AB-doublet ($^3J_{3,4} = 3.4$ Hz) for both 3,4-*trans*- and 3,4-*cis*-C-ring protons[32] in the heterocyclic region of their ^{1}H NMR spectra, and may possess either (2α,4α)- or (2β,4β)-double interflavanyl linkages. These alternatives are readily differentiated via CD data, which show negative and positive Cotton effects in the 220 to 240 nm region for (2α,4α)- and (2β,4β)-configurations, respectively.[32,188] As a consequence of these favorable structural features and because of their considerable biological activity,[189,190] a substantial effort has been devoted to a continued search for new analogs. The proposed and by now well-established system of nomenclature (see Section 11.2) will be used, leading in some instances to a change of published names as far as the α,β-designations are concerned. New entries are listed in Table 11.15.

The report on aesculitannins A–G from the seed shells of *Aesculus hippocastanum*[191] demonstrates three important chemical methods to facilitate the unequivocal structural elucidation of the A-type proanthocyanidins. These protocols include thiolytic degradation using phenylmethanethiol in acidic medium, oxidative formation of the ether linkage when

TABLE 11.14
The Natural Probutinidins

Class	Compound	Source	Ref.
Dimers	Butiniflavan-(4β → 8)-catechin	*Cassia nomame*	187
	Butiniflavan-(4α → 8)-catechin		
	Butiniflavan-(4β → 8)-epicatechin	*C. petersiana*	17
	Butiniflavan-(4α → 8)-epicatechin		
	Butiniflavan-(4β → 8)-epigallocatechin		

the carbon–hydrogen bond at C-2 and the 4-flavanyl constituent are cofacial using hydrogen peroxide in sodium hydrogen carbonate solution, and transformation of the thermodynamically less stable 2,3-*cis*-flavan-3-ol moieties into 2,3-*trans* units via base-catalyzed epimerization at C-2.

An interesting paper reported the conversion of B- into A-type proanthocyanidins via oxidation using 1,1-diphenyl-2-picrylhydrazyl (DPPH) radicals under neutral conditions.[212] The feasibility of the method was demonstrated by transformation of procyanidins B-1 and B-2 into proanthocyanidins A-1 and A-2, respectively. Indirect evidence for the intermediacy of *p*-quinone methides of types **164** and **165** in the oxidative conversion of B- into A-type proanthocyanidins came from the oxidation of epigallocatechin with the homogenate of banana flesh polyphenol oxidase (Figure 11.16).[213] When the anthocyanin malvidin-3-*O*-glucoside was treated with epicatechin in ethanol at 35°C in a sealed tube, the major product was identified as the promalvidin A-type compound, malvidin-3-glucoside-(2 → 7, 4 → 8)-epicatechin (**166**).[215]

The problem of assigning the absolute configuration at the stereocenters of the F-ring of the A-type proanthocyanidins was also addressed.[163] This straightforward general chemical method is based upon the consecutive reductive cleavage of the acetal functionality (cleavage a) and interflavanyl C—C bond (cleavage b). Thus, separate treatment of the hepta-*O*-methyl ethers **167** and **168** of procyanidins A-1 and A-2 with Na(CN)BH₃ in TFA at 0°C (Figure 11.17, Scheme 11.6) gave conversion into the respective monomeric units, i.e., the *ent*-catechin and catechin derivatives (**171** and **173**) from the A-1 derivative (**167**), and the *ent*-catechin and epicatechin derivatives (**171** and **172**) from the A-2 derivative (**168**).

Cleavage "a" of the carbon–oxygen bond is presumably triggered by protonation of the C-7 (D-ring) acetal oxygen and concomitant delivery of the equivalent of a hydride ion at the antibonding (σ*) orbital of the carbon–oxygen bond in a predominant S_N2 manner. Such a transfer of hydride ion apparently occurs from a complex between the reducing agent and the axial C-3 (C-ring) oxygen lone pair, the proximity of the boron–hydrogen bonds to the backside of the acetal carbon atom being a prerequisite for reduction of the acetal bond. Reduction thus leads to "inversion" of configuration at C-2 (C-ring) of both B-type procyanidin intermediates (**169** and **170**). These intermediates are prone to facile reductive cleavage of their interflavanyl bonds (cleavage b)[162] to give the *ent*-catechin derivative (**171**) from the ABC unit and, respectively, the epicatechin and catechin derivatives (**172** and **173**) from the DEF moieties. The location of the free hydroxyl group at the A-ring unambiguously defines the D-ring oxygen that is involved in the acetal functionality of the parent compounds (**167** and **168**). A similar reductive cleavage using Zn–HCl was also used to facilitate structural elucidation of the dracoflavans (Table 11.15).[203]

Despite the apparent clarity of the nomenclature rules, several papers in the area of the A-type proanthocyanidins still lack proper implementation of these rules. The reader must therefore ascertain the correctness of published names. In addition, the reader is also referred to the growing body of evidence of the physiological importance of these compounds, data of which can be found in several of the papers listed in the references.

11.3.3 Nonproanthocyanidins with Flavan or Flavan-3-ol Constituent Units

In addition to the extensive range of di- and trimeric oligoflavanoids with rearranged C-rings, dubbed phlobatannins (see Section 11.3.2.1.4, Table 11.8, Table 11.9, and Table 11.12), daphnodorins A–D and larixinol were earlier reported as rearranged biflavonoid metabolites comprising either flavan or flavan-3-ol and 4,2′,4′,6′-tetrahydroxychalcone (chalconarin-genin) constituent units.[3,4] Since then a considerable number of new analogs, collated in Table 11.16, have been reported.

TABLE 11.15
The Natural A-Type Proanthocyanidins

Class	Compound	Source	Ref.
1. Dimers	Procyanidins		
	Epicatechin-(2β → 7:4β → 6)-epicatechin (procyanidin A-6)	*Aesculus hippocastanum*	191
	Epicatechin-(2β → 5:4β → 6)-epicatechin (procyanidin A-7)	*A. hippocastanum/Theobroma cacao*	191, 192
	Epicatechin-(2β → 7:4β → 6)-catechin	*Arachis hypogea* L.	193
	Epicatechin-(2β → 7:4β → 6)-*ent*-catechin		
	Epicatechin-(2β → 7:4β → 6)-*ent*-epicatechin		
	Epicatechin-(2β → 7:4β → 8)-afzelechin (geranin B)	*Geranium niveum*	199
	Ent-epicatechin-(2α → 7:4α → 8)-catechin (pavetannin A-2)	*Pavetta owariensis*	194
	3-*O*-α-L-Arabinopyranosyl-*ent*-epicatechin-(2β → 7:4α → 8)-catechin	*Theobroma cacao* L.	66
	Propelargonidins		
	Epiafzelechin-(2β → 7:4β → 8)-*ent*-afzelechin	*Cassipourea gerrardii*	70
	7-*O*-Me-Epiafzelechin-(2β → 7:4β → 8)-epiafzelechin	*C. gummiflua*	195
	7-*O*-Me-Epiafzelechin-(2β → 7:4β → 8)-*ent*-afzelechin		
	Ent-epiafzelechin-(2α → 7:4α → 8)-quercetin	*Prunus prostrata*	30
	Ent-epiafzelechin-3-*O*-*p*-OH-benzoate-(2α → 7:4α → 8)-epiafzelechin	*P. armeniaca*	196
	Ent-epiafzelechin-(2α → 7:4α → 8)-*ent*-afzelechin		197
	Epiafzelechin-(2β → 7:4β → 8)-afzelechin (geranin A)	*Geranium niveum*	198, 199
	Epiafzelechin-(2β → 7:4β → 8)-gallocatechin (geranin C)		
	Prodelphinidins		
	Epigallocatechin-(2β → 7:4β → 8)-epicatechin	*Dioclea lasiophylla*	200
	Prorobinetinidins		
	Robinetinidol-(2β → 7:4β → 8)-catechin	*Acacia mearnsii*	32
	Proluteolinidins		
	7-*O*-β-D-Glp-luteoliflavan-(2β → 7:4β → 8)-(2R)-5,7,3′,4′-tetrahydroxyflavanone (diinsininol)	*Sarcophyte piriei*	202
	7-*O*-β-D-Glp-luteoliflavan-(2β → 7:4β → 8)-*ent*-naringenin (diinsinin)		
	Prodistenidins		
	6-Methyl-5-*O*-methyl-epidistenin-(2β → 7:4β → 8)-(2S)-7-OH-5-OMe-flavan (dracoflavan B₁)	Dragon's blood (Daemonorops)	203
	6-Methyl-5-*O*-methyl-*ent*-epidistenin-(2α → 7:4α → 8)-(2S)-7-OH-5-OMe-flavan (dracoflavan B₂)		
	Miscellaneous		
	7-Hydroxy-5-methoxy-6-methylflavan-(2β → 7:4β → 8)-(2S)-7-OH-5-OMe-flavan (dracoflavan C₁)	Dragon's blood (Daemonorops)	203
	7-Hydroxy-5-methoxy-6-methylflavan-(2α → 7:4α → 8)-(2S)-7-OH-5-OMe-flavan (dracoflavan C₂)		

5,7-Di-OH-6-Me-flavan-(2β → 7:4β → 8)-7-OH-5-OMe-flavan (dracoflavan D$_1$)		
5,7-Di-OH-6-Me-flavan-(2α → 7:4β → 8)-7-OH-5-OMe-flavan (dracoflavan D$_2$)		
2. Trimers		
Procyanidins		
Epicatechin-(4β → 8)-epicatechin-(2β → 7:4β → 8)-epicatechin (aesculitannin A)	*Aesculus hippocastanum*	191
Epicatechin-(2β → 7,4β → 8)-*ent*-catechin-(4β → 8)-epicatechin (aesculitannin B)[a]		
Epicatechin-(2β → 7,4β → 8)-epicatechin-(2β → 7,4β → 8)-epicatechin (aesculitannin C)		
Epicatechin-(2β → 7,4β → 8)-epicatechin-(2β → 7,4β → 8)-catechin (aesculitannin D)		
Epicatechin-(2β → 7,4β → 8)-epicatechin-(4β → 8)-*ent*-epicatechin (pavetannin B-1)	*Pavetta owariensis*	194
Epicatechin-(2β → 7,4β → 8)-epicatechin-(4β → 8)-epicatechin (pavetannin B-2)	*P. owariensis/Ecdysanthera utilis*	194, 209
Epicatechin-(2β → 7,4β → 8)-epicatechin-(4β → 6)-epicatechin (pavetannin B-3)	*P. owariensis*	194
Epicatechin-(2β → 7,4β → 6)-*ent*-epicatechin-(4α → 8)-epicatechin (pavetannin B-4)	*P. owariensis*	204
Epicatechin-(2β → 7,4β → 6)-catechin-(4α → 8)-epicatechin (pavetannin B-5)	*P. owariensis*	194, 204
Epicatechin-(2β → 7,4β → 8)-epicatechin-(4β → 8)-catechin (pavetannin B-6)		
Epicatechin-(2β → 7,4β → 8)-*ent*-epicatechin-(2α → 7,4α → 8)-*ent*-catechin (pavetannin B-7)	*P. owariensis*	205
Epicatechin-(2β → 7,4β → 8)-epicatechin-(2β → 7,4β → 8)-*ent*-catechin (pavetannin B-8)	*Aesculus hippocastanum*	206
Epicatechin-(2β → 7,4β → 8)-catechin-(4β → 8)-catechin		
Epicatechin-(4α → 8)-epicatechin-(2β → 7,4β → 8)-epicatechin		
Epicatechin-(2β → 7,4β → 6)-epicatechin-(2β → 7,4β → 8)-epicatechin	*Parameria laevigata* Moldenke	207
Epicatechin-(4β → 6)-epicatechin-(2β → 7,4β → 8)-epicatechin	*Vaccinium macrocarpon* Ait. (Cranberry)	189
3-*O*-Arabinopyranosylepicatechin-(2β → 7,4β → 8)-epicatechin-(4β → 8)-epicatechin (3-T-*O*-α-L-arabinopyranosylcinnamtannin B$_1$)	*Theobroma cacao* L.	66
3-*O*-Galactopyranosylepicatechin-(2β → 7,4β → 8)-epicatechin-(4β → 8)-epicatechin (3-T-*O*-β-D-galactopyranosylcinnamtannin B$_1$)		
Propelargonidins		
Epiafzelechin-(2β → 7,4β → 8)-epiafzelechin-(4β → 8)-afzelechin (selligueain A)	*Selliguea fecei*	46, 208
Epiafzelechin-(2β → 7,4β → 8)-epiafzelechin-(4β → 8)-afzelechin-4β-acetic acid methyl ester [epiafzelechin-(2β → 7,4β → 8)-epiafzelechin-(4β → 8)-3'-deoxydryopteric acid methyl ester] (selligueain B)		
Epiafzelechin-(2β → 7,4β → 8)-afzelechin-(2β → 7,4β → 8)-afzelechin (geranin D)	*Geranium niveum*	199
3. Tetramers		
Procyanidins		
Epicatechin-(4β → 6)-epicatechin-(2β → 7,4β → 8)-epicatechin-(4β → 8)-epicatechin (pavetannin C-1)	*Pavetta owariensis*	205, 209
Epicatechin-(2β → 7,4β → 8)-*ent*-epicatechin-(4α → 8)-epicatechin-(4α → 8)-epicatechin (pavetannin C-2)		

continued

TABLE 11.15
The Natural A-Type Proanthocyanidins — *continued*

Class	Compound	Source	Ref.
	Epicatechin-(2β → 7,4β → 8)-epicatechin-(4β → 8)-epicatechin (aesculitannin E)	*Aesculus hippocastanum*	192
	Epicatechin-(2β → 7,4β → 8)-*ent*-catechin-(4β → 8)-[b]-epicatechin-(4β → 8)-epicatechin (aesculitannin F)		
	Epicatechin-(2β → 7,4β → 8)-[b]-epicatechin-(2β → 7,4β → 8)-epicatechin (aesculitannin G)		
	Epicatechin-(2β → 7,4β → 6)-epicatechin-(4β → 8)-epicatechin (parameritannin A-1)	*Parameria laevigata* Moldenke	207, 211
	Epicatechin-(2β → 5,4β → 6)-epicatechin-(2β → 7,4β → 8)-epicatechin-(4β → 8)-epicatechin (parameritannin A-2)		
	Epicatechin-(2β → 7,4β → 6)-epicatechin-(2β → 7,4β → 8)-epicatechin-(4β → 8)-epicatechin (parameritannin A-3)		
"Mixed" propelargonidins/procyanidins			
	Epiafzelechin-(2β → 7,4β → 8)-epicatechin-(4β → 8)-epicatechin (pavetannin C-3)	*Pavetta owariensis*	210
	Epiafzelechin-(2β → 7,4β → 8)-*ent*-afzelechin-(4α → 8)-epicatechin-(2α → 7,4α → 8)-*ent*-catechin (pavetannin C-4)		
	Epiafzelechin-(2β → 7,4β → 8)-*ent*-catechin-(4α → 8)-epicatechin-(2α → 7,4α → 8)-*ent*-catechin (pavetannin C-5)		

[a] In the original paper aesculitannin B was named epicatechin-(2β → 7,4β → 8)-*ent*-catechin-(4α → 8)-epicatechin.
[b] Indicated as (4α → 8) in the original manuscript.

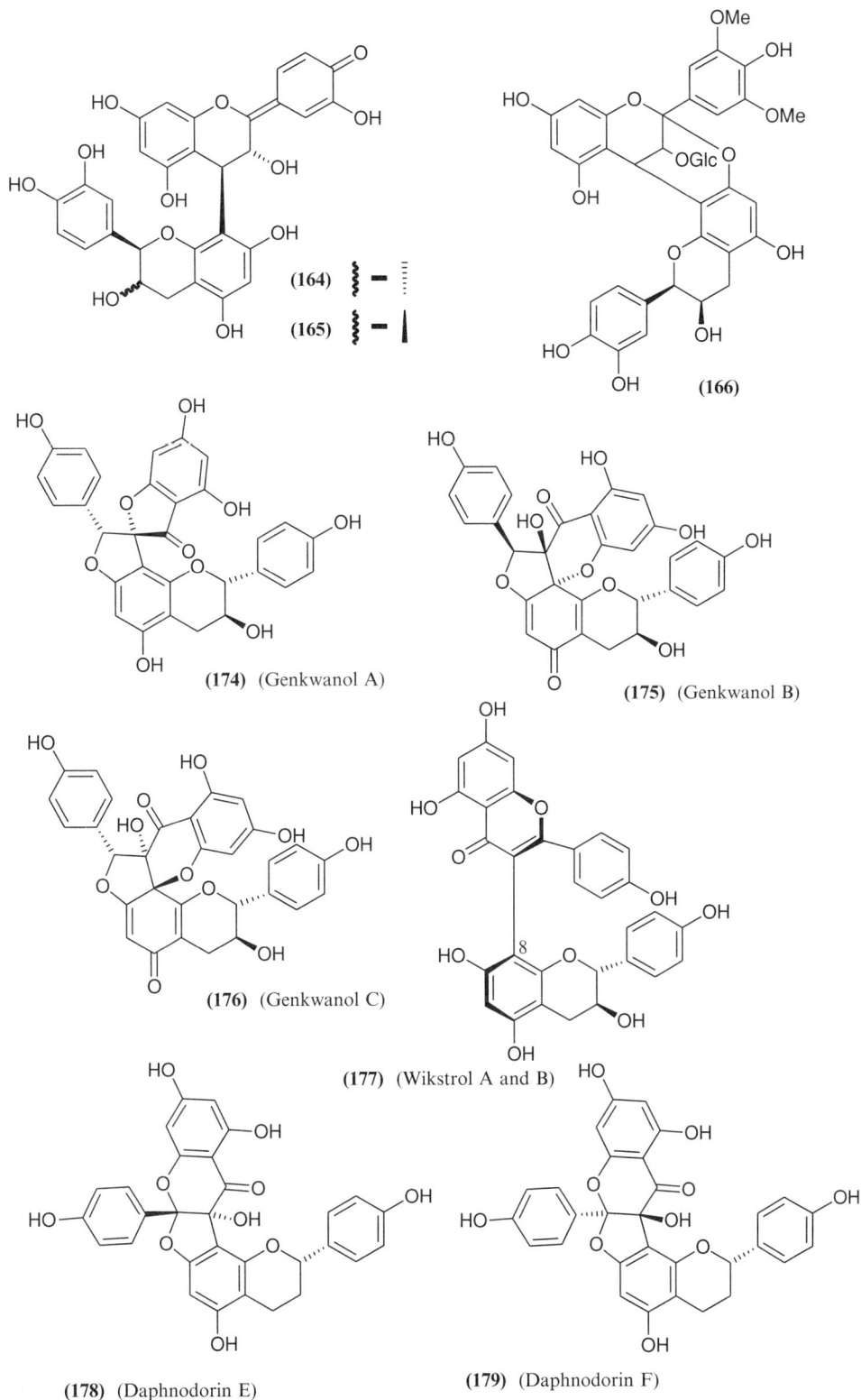

FIGURE 11.16 Structures of compounds **164–166** and **174–179**.

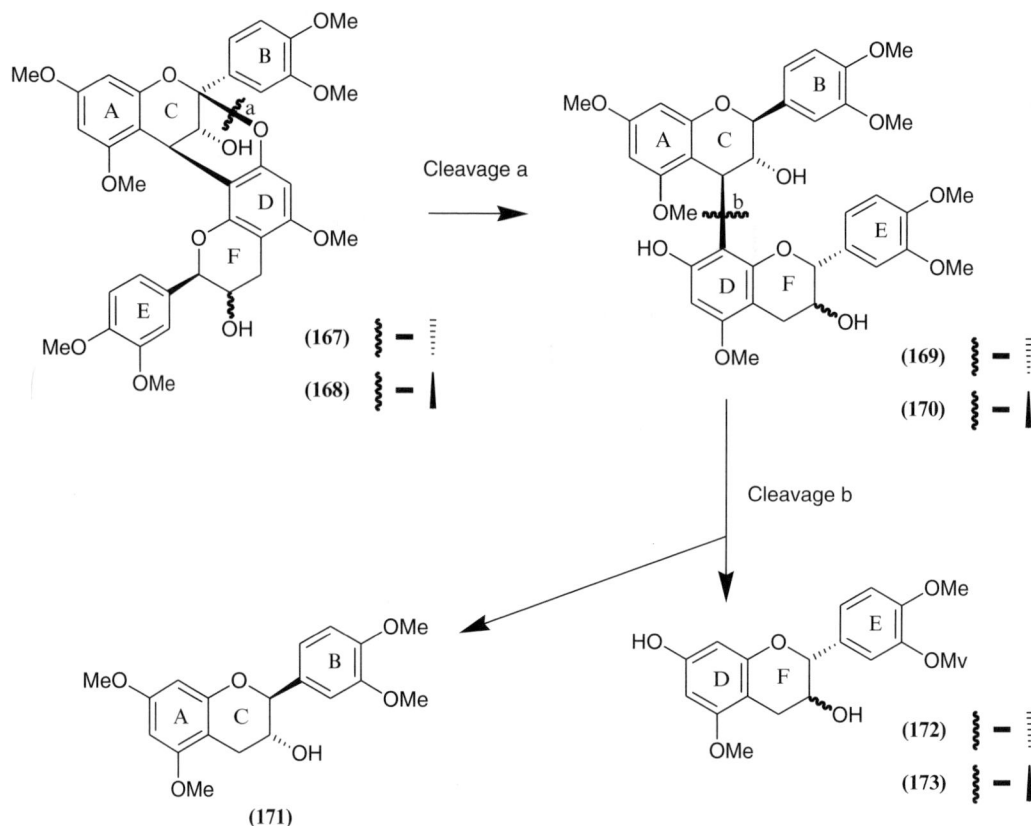

FIGURE 11.17 Scheme 11.6: Proposed route to the reductive cleavage of both interflavanyl bonds in A-type proanthocyanidins including the structures of compounds **167–173**.

The basic carbon framework of an afzelechin moiety coupled at C-8 to the α-carbon of a chalconaringenin unit is evident in the structures of the three genkwanols (**174–176**) (Figure 11.16). Their structures were meticulously corroborated by the collective utilization of ^{1}H and ^{13}C NMR data, x-ray analysis, and the modified Mosher method.[216–218] Two related but nonrearranged compounds comprising 5,7,4′-trihydroxflavone and afzelechin constituent units, the (3 → 8)-coupled atropisomeric wikstrols A and B (**177**), were shown to arise by acid-catalyzed rearrangement of daphnodorin B.[219] Additional entries are shown in Figure 11.18.

New structures or chemistry emerging from the remarkable "black tea pool" are also listed in Table 11.16. Theaflavonin (**190**) and desgalloyl theaflavonin (**191**) are B,B′-linked bisflavonoids formed presumably via oxidative coupling of the flavonol glucoside, isomyricitrin, and 3-O-galloylepigallocatechin and epigallocatechin, respectively (Figure 11.19).[225] The R absolute configuration of the atropisomeric biphenyl linkages of compounds **189–191** was established by comparison of CD data with those of theasinensis C and E possessing R and S axial chirality, respectively. Theadibenzotropolone A (**192**), the first theaflavin-type trimer in black tea, and theaflavin-3-gallate were formed by the reaction of (−)-epicatechin and (−)-epigallocatechin gallate with horseradish peroxidase in the presence of H_2O_2.[226] Its presence in black tea was confirmed by liquid chromatography–electrospray ionization tandem mass spectrometry (MS). An informative schematic representation of the enzymatic

TABLE 11.16
The Natural Nonproanthocyanidins with Flavan or Flavan-3-ol Constituent Units

Compound	Structure	Source	Ref.
Genkwanol A	(174)	*Daphne genkwa* Sieb. et Zucc.	216
Genkwanol B	(175)		217
Genkwanol C	(176)		218
Wikstrol A atropisomer	(177)	*Wikstroemia sikokiana*	219
Wikstrol B atropisomer	(177)		
Daphnodorin E	(178)	*D. odora* Thumb.	220
Daphnodorin F	(179)		
Daphnodorin G	(180)	*D. odora*	221
Daphnodorin H	(181)		
Daphnodorin I	(182)		
Daphnodorin J	(183)	*D. odora*	222
Daphnodorin K	(184)		
Daphnodorin L	(185)		
Daphnodorin M	(186)	*D. acutiloba*	223
Daphnodorin N	(187)		
Damalachawin	(188)	*Dracaena cinnabari*	224
Theogallinin	(189)	*Camellia sinensis* (black tea)	225
Theoflavonin	(190)		
Desgalloyl theaflavonin	(191)		
Theadibenzotropolone A	(192)	*C. sinensis* (black tea)	226
Theaflavate A	(193)		227
Theacitrin A	(194)		228
Theaflavate B	(195)		229
Isotheaflavin-3′-*O*-gallate	(196)		
Neotheaflavin-3-*O*-gallate	(197)		

conversion of the polyphenols in the tea leaves was also reported.[225] Theaflavate A (193) formed when epicatechin-3-*O*-gallate was oxidized with $K_3Fe(CN)_6$.[227] Compounds **195–197** were also produced via chemical oxidation of the appropriate flavan-3-ol or flavan-3-*O*-gallate precursors (Figure 11.20).[229]

11.3.4 COMPLEX TANNINS

The term, complex tannin, appears to be established as descriptor for the class of polyphenols in which a flavan-3-ol unit, representing a constituent unit of the "condensed tannins" (proanthocyanidins), is connected to a "hydrolyzable (gallo-or ellagi-) tannin" through a carbon–carbon linkage. Since the first demonstration of their natural occurrence,[230] a considerable number of these unique secondary metabolites have been reported.[3,4] New additions (Table 11.17) to this series of compounds come exclusively from the groups of Nonaka and Nishioka, and Okuda and Yoshida in Japan.

Malabathrins A (198), E (199), and F (200) (Figure 11.20 and Figure 11.21), which are composed of a C-glucosidic ellagitannin and a C—C coupled epicatechin moiety, were isolated from the leaves of *Melastoma malabatricum*.[231] The *S* chirality for both the hexahydroxydiphenoyl (HHDP) groups in malabathrin A (198) was deduced from its CD curve, which exhibited positive and negative Cotton effects at 233 and 262 nm, respectively. Its

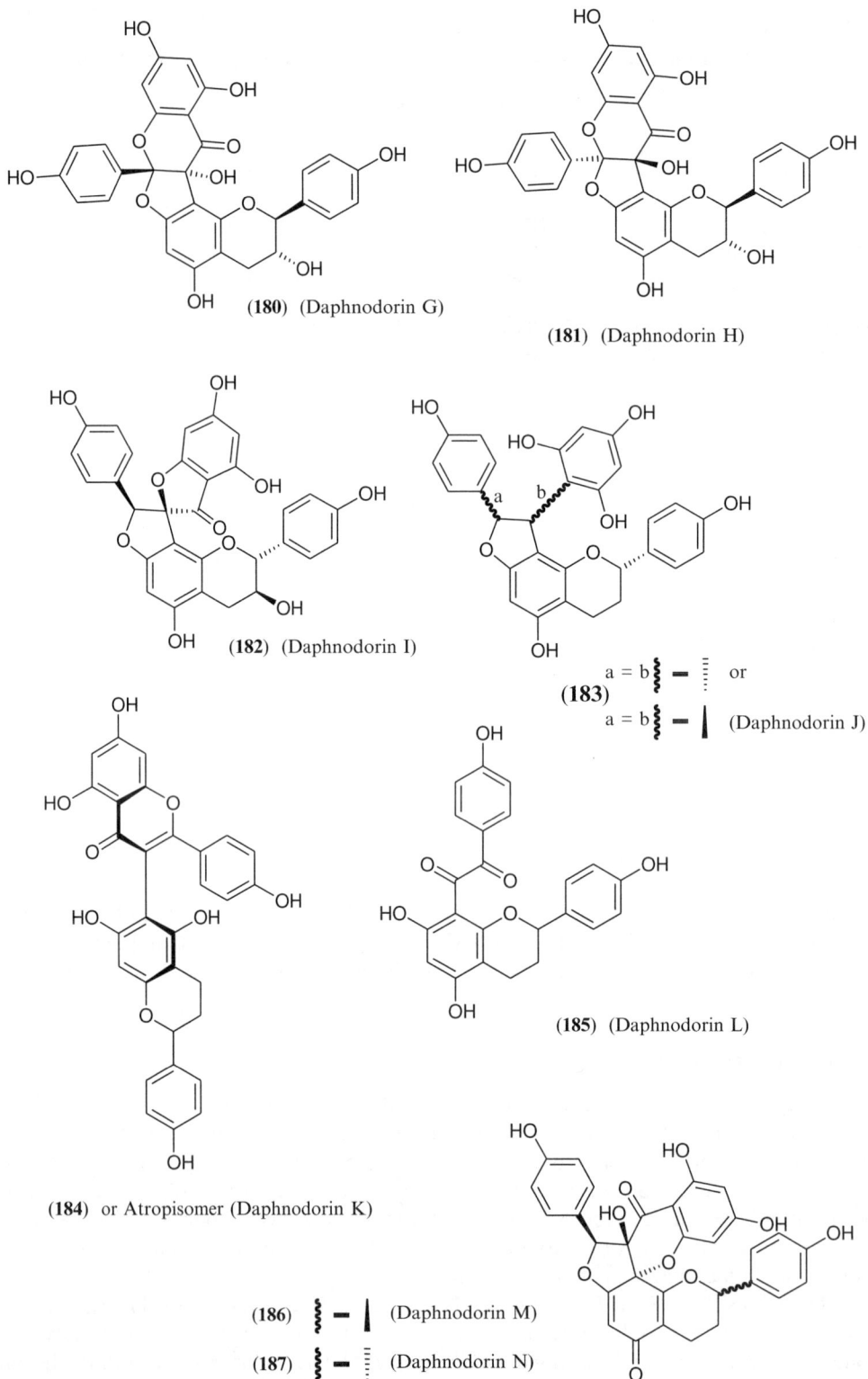

FIGURE 11.18 Structures of compounds 180–187.

(188) (Damalachawin)

(189) R = galloyl (Theogallinin)

(192)

(190) R¹ = galloyl, R² = β-D-glucopyranosyl
(Theaflavonin)
(191) R¹ = H, R² = β-D-glucopyranosyl
(Desgalloyl theaflavonin)

(193) (Theaflavate A)

(194) (Theacitrin A)

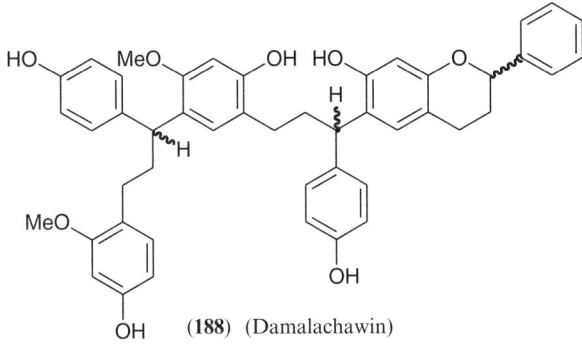

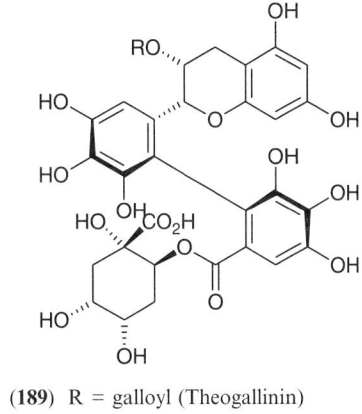

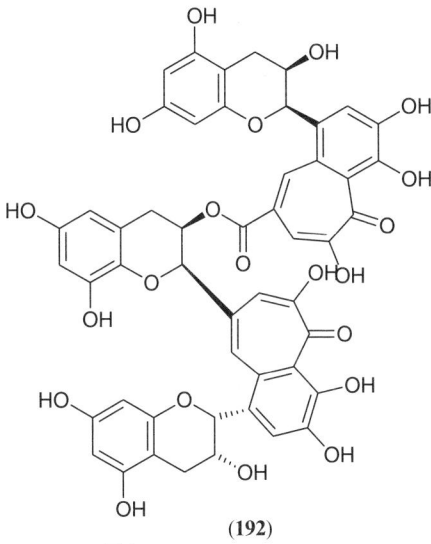

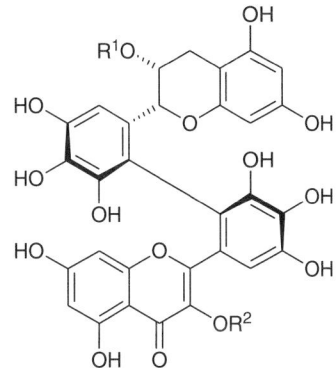

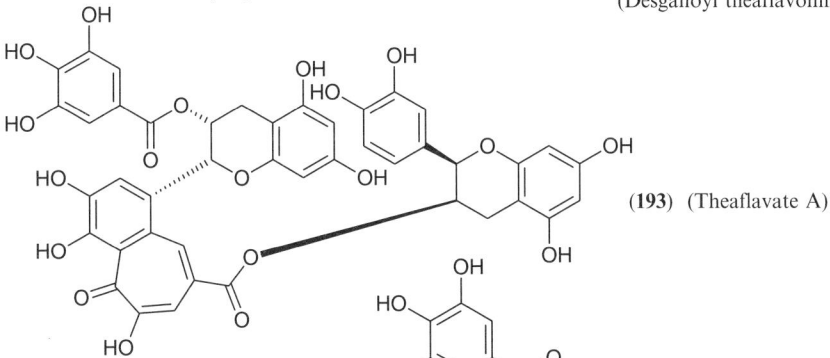

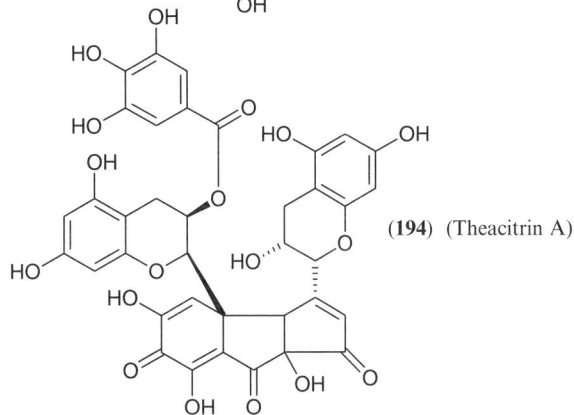

FIGURE 11.19 Structures of compounds 188–194.

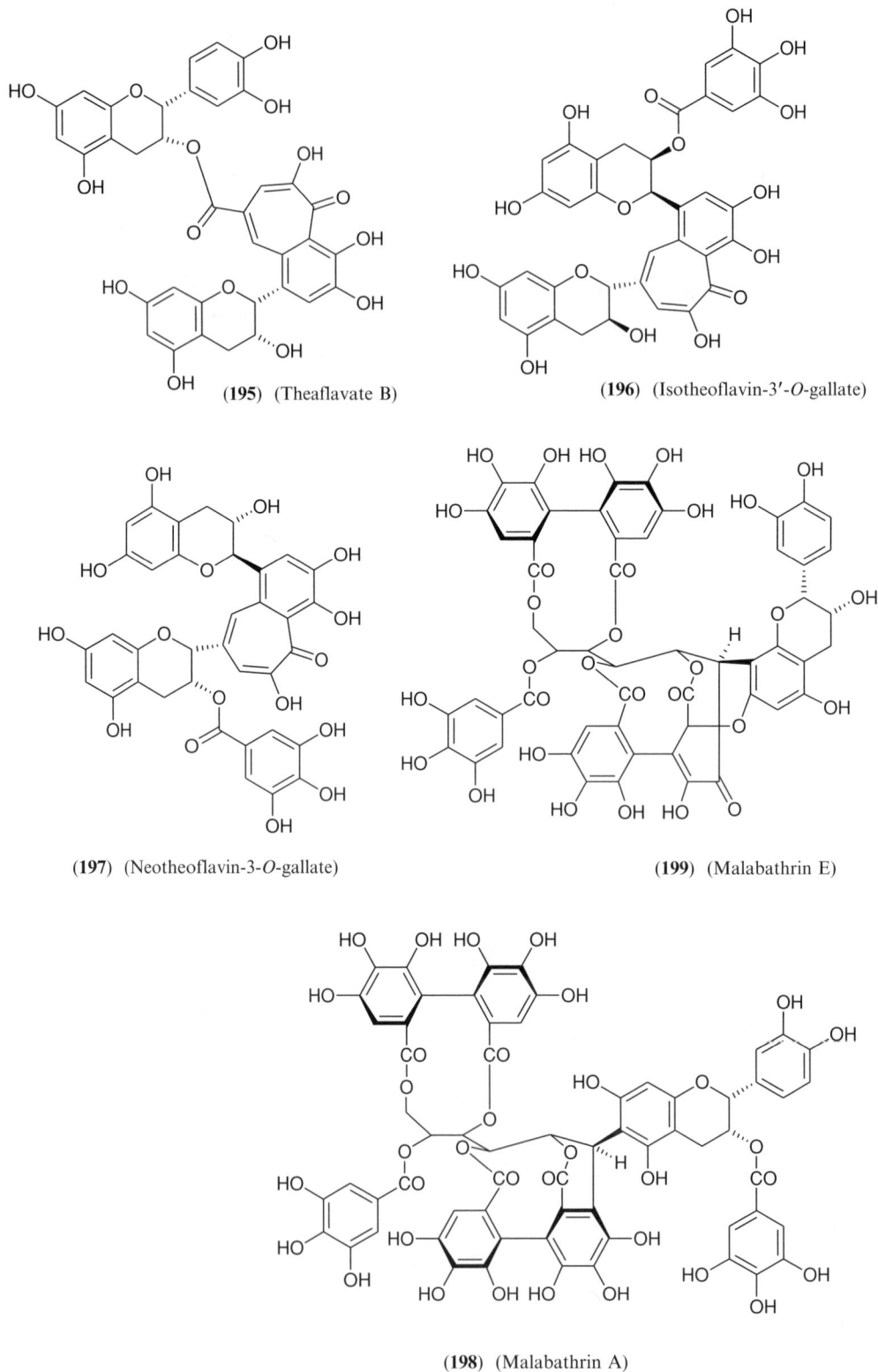

(195) (Theaflavate B)

(196) (Isotheoflavin-3′-O-gallate)

(197) (Neotheoflavin-3-O-gallate)

(199) (Malabathrin E)

(198) (Malabathrin A)

FIGURE 11.20 Structures of compounds 195–199.

TABLE 11.17
The Natural Complex Tannins

Compound	Structure	Source	Ref.
Malabathrin A	(198)	*Melastoma malabatricum*	231
Malabathrin E	(199)		
Malabathrin F	(200)		
Guajavin A	(201)	*Psidium guajava*	232
Guajavin B	(203)		
Psidinin A	(205)		
Psidinin B	(207)		
Psidinin C	(209)		
Psiguavin	(211)		
Strobilanin	(213)	*Platycarya strobilacea* Sieb. et Zucc.	233
Camelliatannin C	(214)	*Camellia japonica*	234–236
Camelliatannin D	(216)		
Camelliatannin E	(215)		
Camelliatannin F	(217)		
Camelliatannin G	(218)		
Stachyuranin A	(219)	*Stachyurus praecox* Sieb. et Zucc.	237
Stachyuranin B	(220)		
Stachyuranin C	(221)		

structure was unequivocally confirmed by acid-catalyzed condensation of the ellagitannin, casuarinin, and epicatechin-3-*O*-gallate. The cyclopentenone moiety in malabathrins E (199) and F (200) is regarded as the product of oxidative conversion of the HHDP group at $O_{(2)}$–$O_{(3)}$ of glucose in, e.g., malabathrin A (198).[231]

The ^{1}H NMR spectrum of guajavin A (201) at room temperature was complicated due to the effects of conformational isomerism, a feature that was commonly observed in complex tannins where the C-8 position of the flavan-3-ol unit is substituted.[232] Structural elucidation of guajavins A (201) and B (203) was done by comparison of their ^{13}C NMR data with those of the related catechin analogs, stenophyllanin A (202) and acutissimin B (204), respectively, and by synthesis via acid-catalyzed condensation of gallocatechin with the ellagitannins, stachyurin, and vescalagin, respectively. The ^{1}H and ^{13}C NMR spectra of psidinins A (205), B (207), and C (209) similarly resembled those of their catechin analogs, mongolicains A (206) and B (208), and stenophyllinin A (210), respectively, thus readily facilitating the structural elucidation of the former three compounds (Figure 11.22).

Psiguavin (211), in which the B-ring of the flavan-3-ol unit is extensively rearranged, is considered to be derived biosynthetically from eugenigrandin A (partial structure [212]) by successive oxidation of the pyrogallol B-ring, benzylic acid-type rearrangement, and decarboxylation, followed by oxidative coupling as is indicated in Figure 11.23 (Scheme 11.7).[232]

Camelliatannins C (214) and E (215) with their C-6 and C-8 substituted epicatechin moieties, respectively, represent the first examples of complex tannins lacking a C—C bond between C-1 of glucose and the HHDP group at O-2 of the glucose unit (Figure 11.24). These bonds could, however, be readily formed by treatment of analogs 214 and 215 with polyphosphoric acid, hence transforming them into camelliatannins B and A, respectively.[235] Camelliatannins C (214) and E (215) may thus be considered as precursors to the "normal" type of

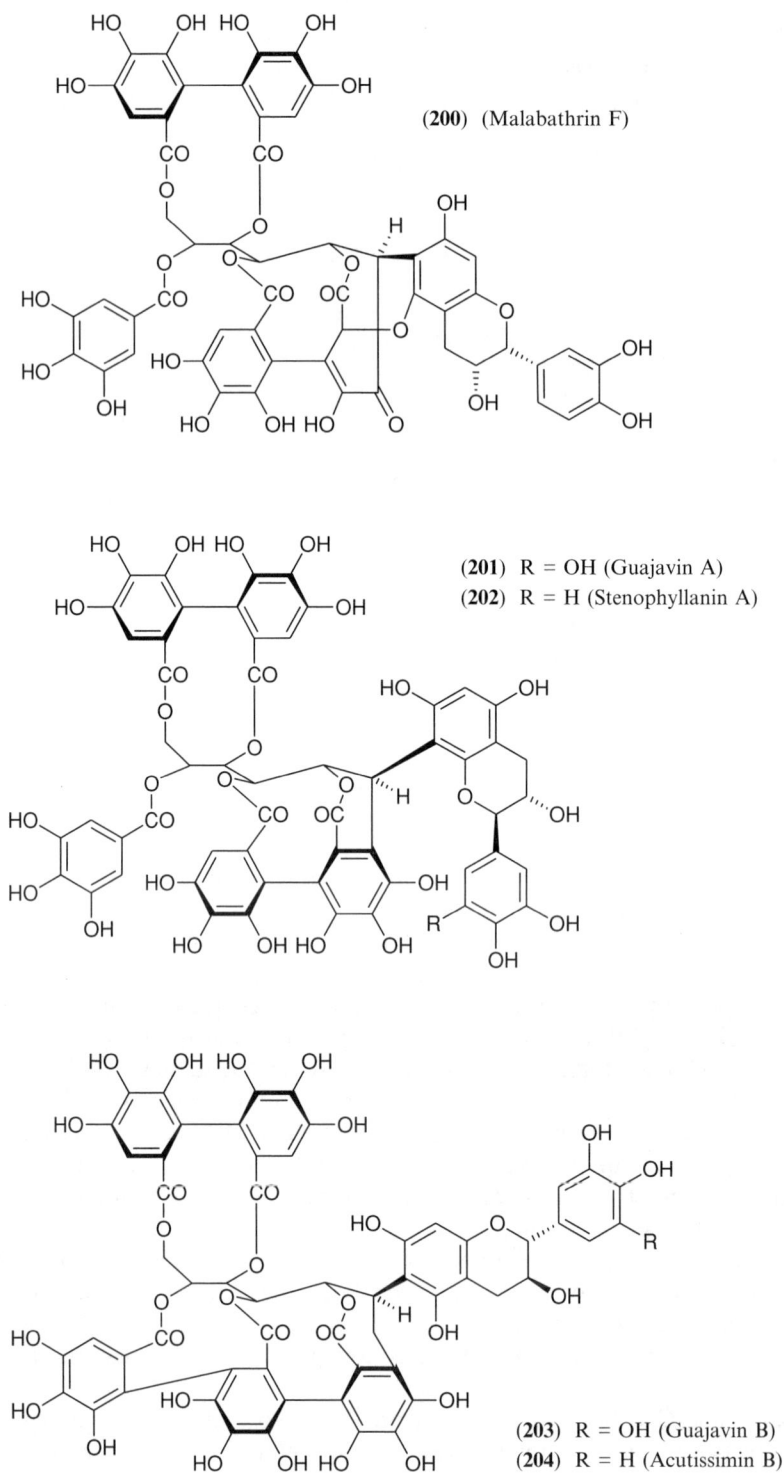

FIGURE 11.21 Structures of compounds **200**–**204**.

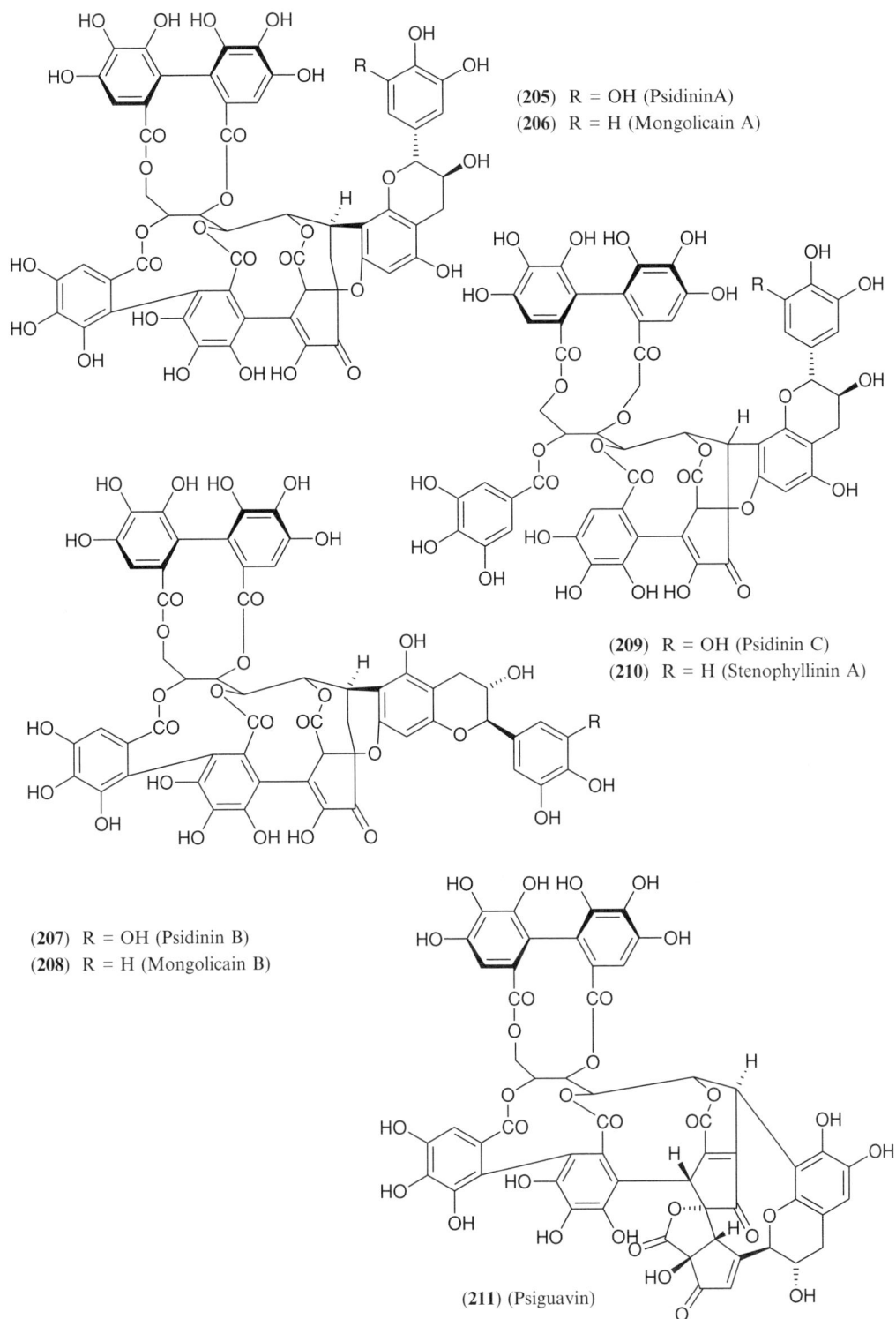

(205) R = OH (PsidininA)
(206) R = H (Mongolicain A)

(209) R = OH (Psidinin C)
(210) R = H (Stenophyllinin A)

(207) R = OH (Psidinin B)
(208) R = H (Mongolicain B)

(211) (Psiguavin)

FIGURE 11.22 Structures of compounds 205–211.

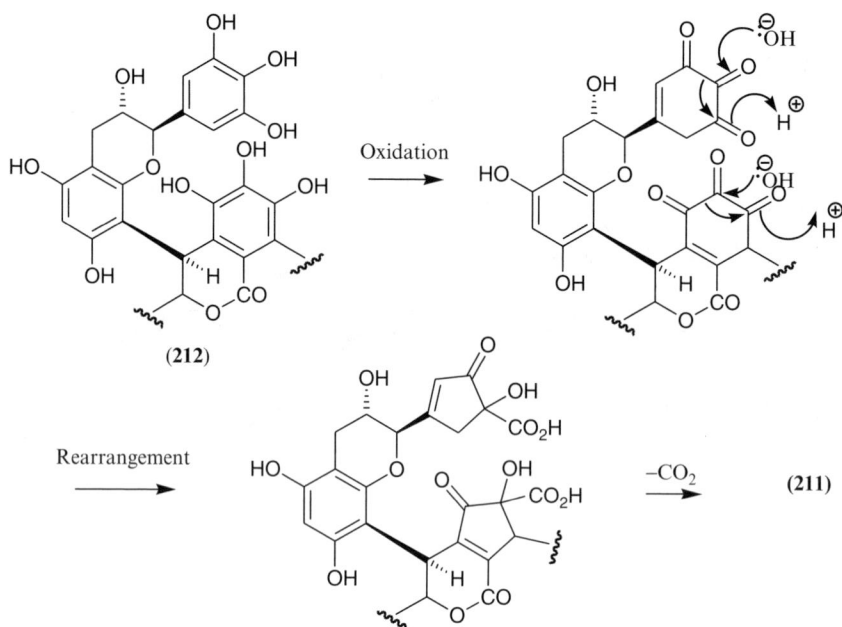

FIGURE 11.23 Scheme 11.7: Proposed route to the transformation of eugenigrandin A (**212**) into psiguavin (**211**).

complex tannins and may be anticipated to co-occur in plant sources containing the latter class of metabolites. Camelliatannin D (**216**) represents the first example composed of dimeric hydrolyzable tannin and flavan-3-ol constituent units.[234] Camelliatannin A presumably serves as the biogenetic precursor to both camelliatannins F (**217**) and G (**218**).

Stachyuranins A (**219**) and B (**220**) also lack the C—C linkage between C-1 of the glucose moiety and the aroyl group at O-2 of glucose (Figure 11.25). When dissolved in aqueous methanol at room temperature, stachyuranin A (**219**) is gradually converted into stenophyllanin A,[230] which presumably suggests that the latter compound is produced nonenzymatically from **219** in plants.

11.3.5 MISCELLANEOUS

A new series of polyphenols, named castavinols, has been isolated from Bordeaux red wines.[238] The three analogs (**222–224**) (Figure 11.25) (or their C-ring enantiomers) were speculated to form via [1,2]-addition of diacetyl, a C-4 yeast metabolite, to, e.g., malvidin-3-O-glucoside (**225**) (Figure 11.26, Scheme 11.8). The resultant intermediate (**226**) would then cyclize by addition of the vinylic double bond to the electrophilic carbonyl carbon to give a C-3 carbocation (**227**), which is reduced, presumably under enzymatic control from the yeast, to compound **222**. The remaining analogs (**223** and **224**) may similarly be derived from peonidin- and petunidin-3-glucoside, respectively.

11.3.6 NMR AND CONFORMATIONAL ANALYSIS OF PROANTHOCYANIDINS

Conformational analysis of proanthocyanidins is, in principle, concerned with the conformation of the pyran heterocycle and with the phenomenon of conformational isomerism due to

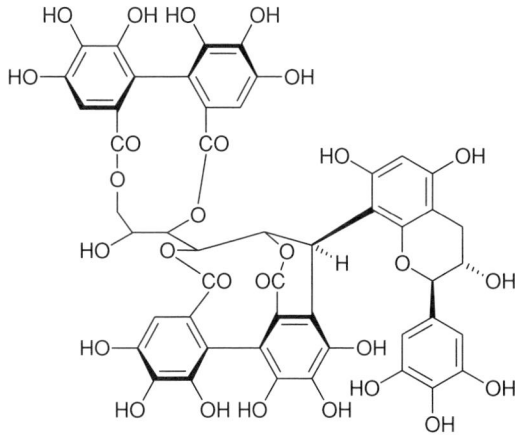

(**213**) Strobilanin

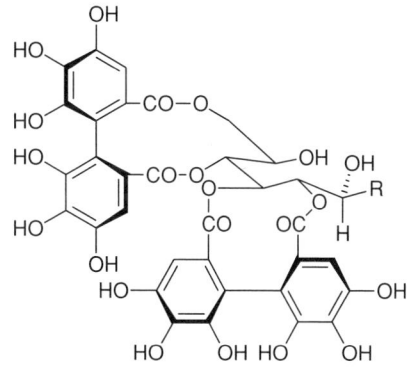

(**214**) R= 6-epicatechin (Camelliatannin C)
(**215**) R= 8-epicatechin (Camelliatannin E)

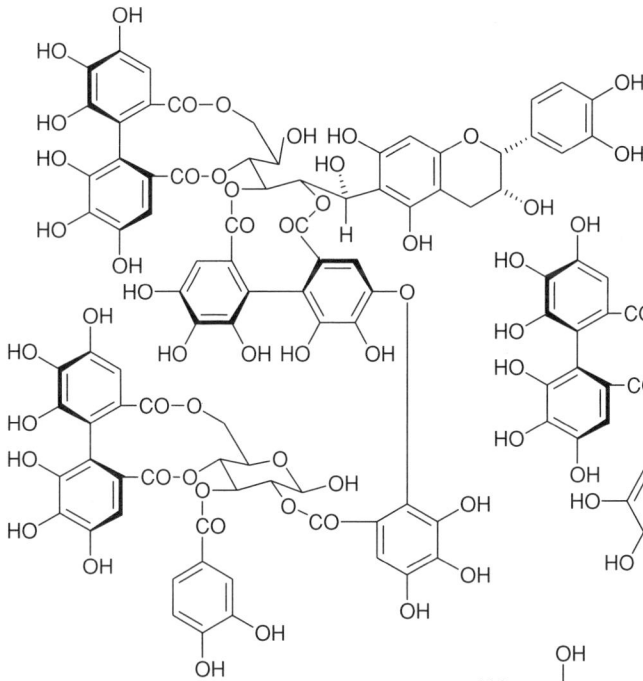

(**217**) (Camelliatannin F)

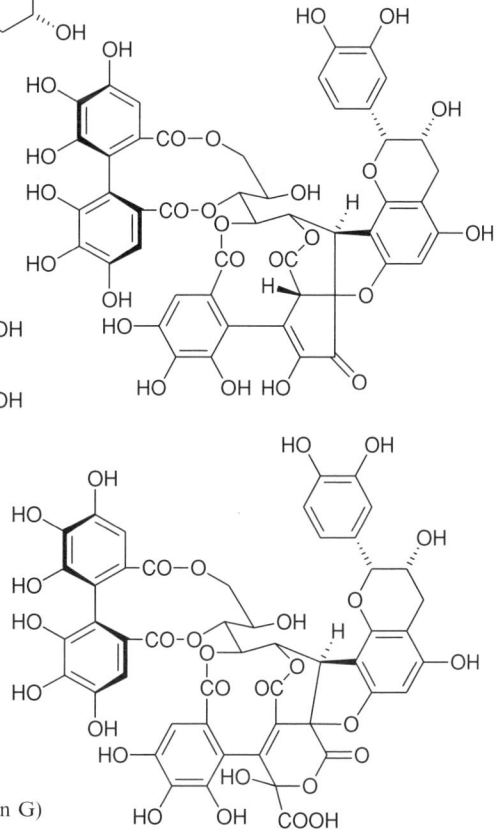

(**216**) (Camelliatannin D)

(**218**) (Camelliatannin G)

FIGURE 11.24 Structures of compounds **213**–**218**.

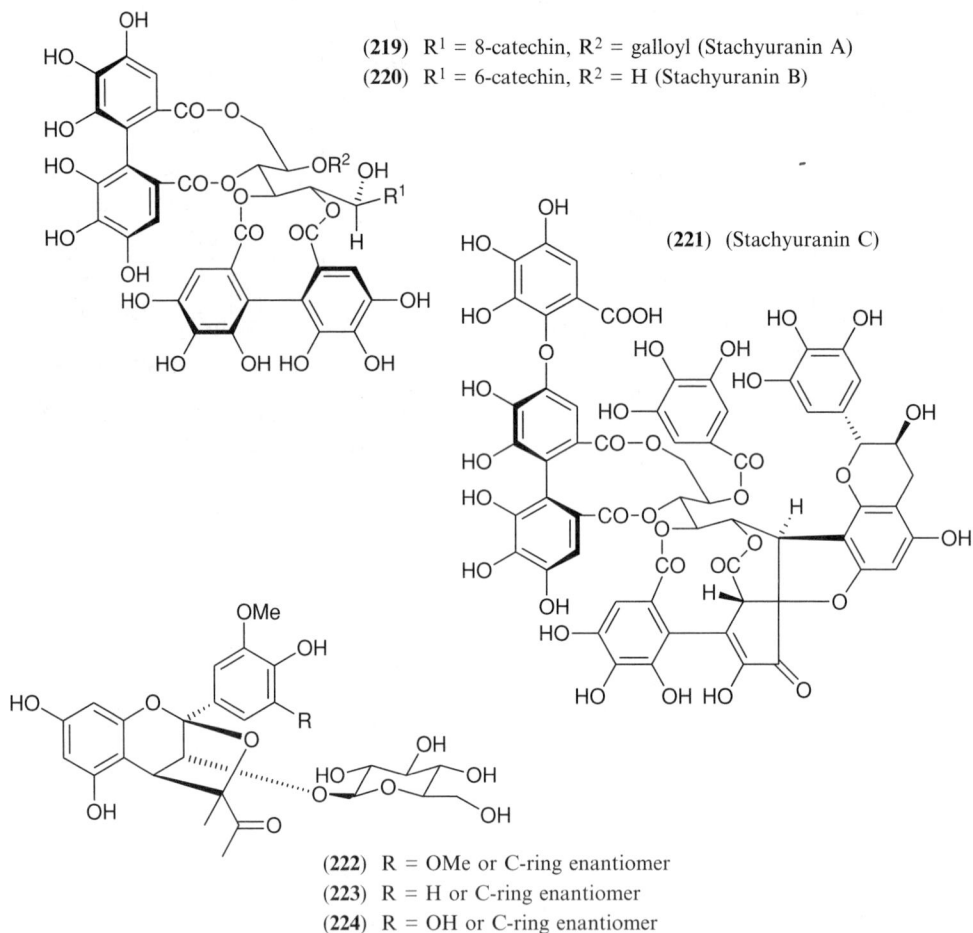

(219) R^1 = 8-catechin, R^2 = galloyl (Stachyuranin A)
(220) R^1 = 6-catechin, R^2 = H (Stachyuranin B)

(221) (Stachyuranin C)

(222) R = OMe or C-ring enantiomer
(223) R = H or C-ring enantiomer
(224) R = OH or C-ring enantiomer

FIGURE 11.25 Structures of compounds **219–224**.

restricted rotation about the interflavanyl bond axis or axes. Realization of the fact that the conformational itinerary of the heterocyclic rings involves a dynamic equilibrium between E- and A-conformers[239] had a profound impact in this field. Useful discussions dealing with various aspects of the conformational behavior of the proanthocyanidins may be extracted from Refs. 4, 7–12.

The utilization of the full array of modern ^{1}H and ^{13}C NMR methodology has led to various contributions regarding the proton and carbon assignments as well as the conformations of di- and trimeric proanthocyanidins.[240–249] Vercauteren and coworkers[241] differentiated C-4 → C-8- and C-4 → C-6-linked peracetylated procyanidins [catechin-(4α → 8)-catechin-(4α → 6)-catechin] by initial complete assignment of proton and carbon resonances of the appropriate dimers, followed by observation of differences in correlation of H-4 (C) with the quaternary carbons of the A–C- and D–F-ring junctions. The same group[242] also confirmed the C-8 substitution of the DEF and GHI units in the peracetate of procyanidin C-1 [catechin-(4α → 8)-catechin-(4α → 8)-catechin] in an HMBC experiment. De Bruyne et al.[243] subsequently provided a full analysis of the ^{1}H and ^{13}C NMR data of procyanidin B-3 [catechin-(4α → 8)-

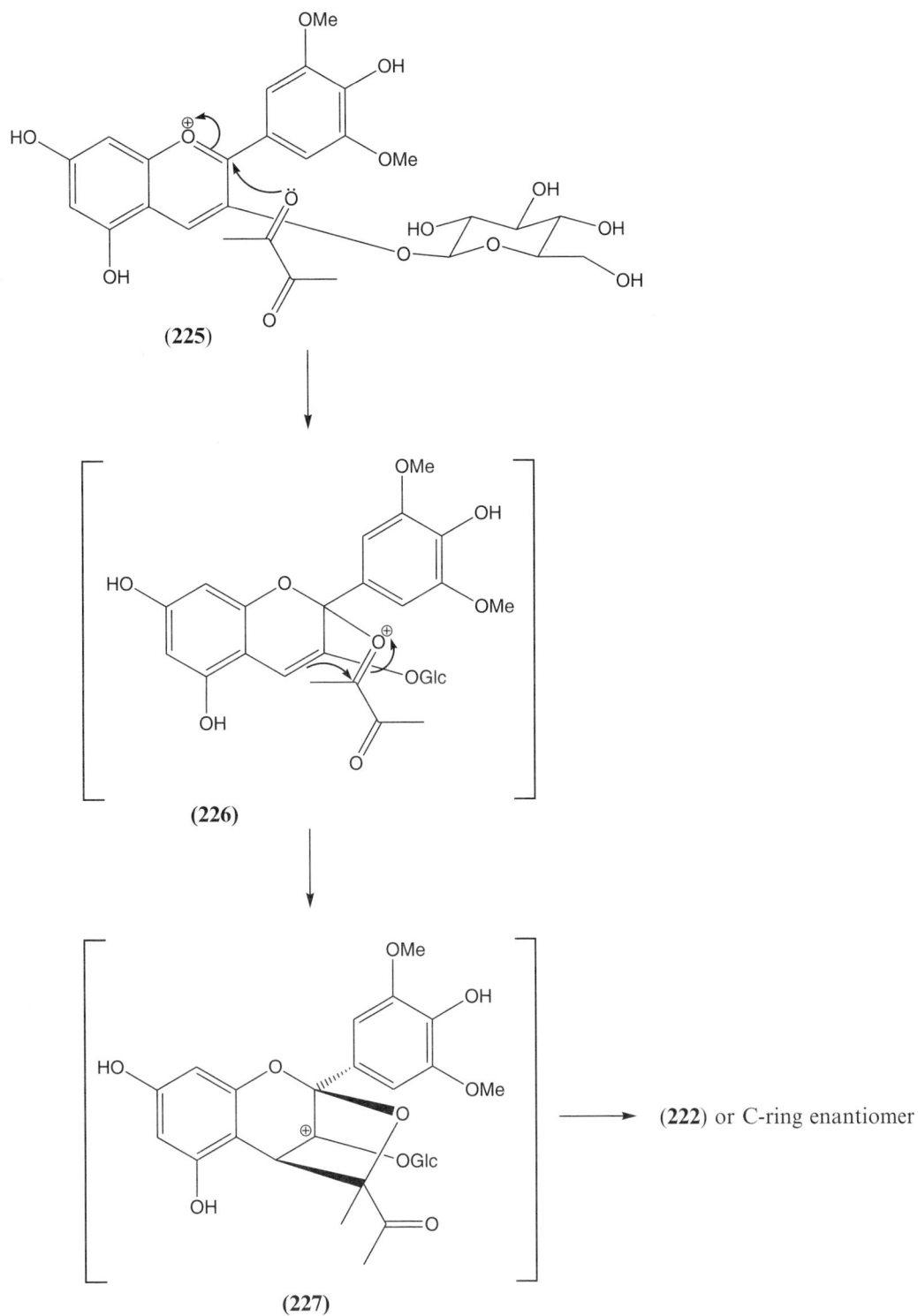

FIGURE 11.26 Scheme 11.8: Proposed route to the formation of castavinols including the structures of compounds **225**–**227**.

catechin]. Two-dimensional NMR sequences were also used to assign the proton and carbon resonances of procyanidin A-2 [epicatechin-(2β → 7, 4β → 8)-catechin].[244] Rimpler and coworkers[240] used the peracetates of tri- and tetrameric procyanidins with epicatechin constituent units to define a further diagnostic shift parameter facilitating establishment of the interflavanyl bonding position. Williamson and coworkers[245] similarly reported fully assigned ¹H and ¹³C NMR data of procyanidin B-2 [epicatechin-(4β → 8)-epicatechin] and its per-O-acetyl derivative. Hemingway et al.[246] produced evidence that both H-6 and C-6 resonances of free phenolic flavan-3-ols are downfield from H-8 and C-6, hence indicating that the chemical shifts of A-ring protons are indeed reversed compared to those commonly reported. Tobiason et al.[247] demonstrated the temperature dependence of the pyran ring proton coupling constants of catechin. Such a temperature dependence of the coupling constants was also reproduced from the Boltzmann distribution of the conformational ensemble generated by the GMMX searching program.[247] Additional useful information may be extracted from Ref. 248 as well as the preceding two papers in the series by the same authors. Reference 248 deals with the complete and unequivocal ¹H and ¹³C NMR assignments of a range of green tea polyphenols.

Investigations of the conformational properties of the flavan-3-ols and oligomeric proanthocyanidins have hitherto involved a variety of molecular mechanics and molecular orbital computations in combination with crystal structures, time-resolved fluorescence, as well as ¹H and ¹³C NMR methods. Representative references to all these techniques may be found in the papers listed in Refs. 241–247, 250. These "NMR papers" incidentally also represent the major contributions regarding the conformation of proanthocyanidins, and may be summarized in a conformational context by reference to the significant contributions of Hatano and Hemingway.[251,252]

NMR analysis of procyanidin B-1 [epicatechin-(4β → 8)-catechin] and B-3 [catechin-(4α → 8)-catechin] permitted full assignment of the H- and C-resonances for both the compact (**228**) and the more extended (**229**) conformer in the free phenolic form (Figure 11.27). In organic solvents, the more extended rotamer of procyanidin B-1 is preferred over the more compact rotamer (10:7) but in D₂O the more compact rotamer dominates (10:2). When procyanidin B-3 is dissolved in organic solvents, the more compact rotamer (**228**) is slightly preferred (8:10). With D₂O as solvent only trace proportions of the more extended rotamer (**229**) are detected. In this solvent, rotational conformation exchange is detected despite the observation of two distinct and sharp sets of signals for each rotamer. The heterocyclic ring of the ABC unit exists in an approximate half-chair conformation in each rotamer for both procyanidins B-1 and B-3. The heterocyclic ring of the ABC unit exists in an approximate half-chair conformation in each rotamer for both procyanidin B-1 and B-3. Coupling constants of the protons of the pyran rings of the DEF moieties indicate substantial axial orientation of the E-ring (see compounds **230** and **231** for E- and A-conformers of the DEF unit of procyanidin B-3). Line shape analysis of H-3 (F) indicated that the "abnormal" coupling constants of the F-ring protons were reminiscent of a comparatively high-energy skewed-boat conformation for procyanidin B-1 and between a half-chair and a skewed-boat for procyanidin B-3 rather than to E ⇋ A-conformational exchange, which has hitherto been used to explain the smaller than anticipated coupling constants.

Hatano and Hemingway[251,252] used NOE studies to assess the association of catechin and procyanidin B-3 with oligopeptides (see also relevant results in Refs. 253–256). The observed intermolecular NOEs indicating the preferred sites in the association of catechin and procyanidin B-3 with the tetrapeptide Gly–Pro–Gly–Gly are shown in **232** and **233**, respectively. The molecular shapes of both the polyphenol and polypeptide are important features as far as selectivity is concerned.

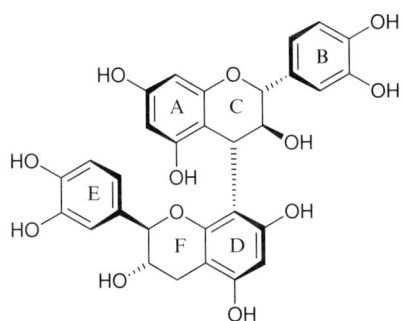

(228) (Compact rotamer of procyanidin B-3)
C_3–C_4–C_8–C_9 torsion angle, (−)

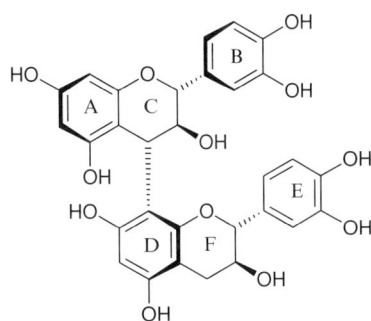

(229) (Extended rotamer of procyanidin B-3)
C_3–C_4–C_8–C_9 torsion angle, (+)

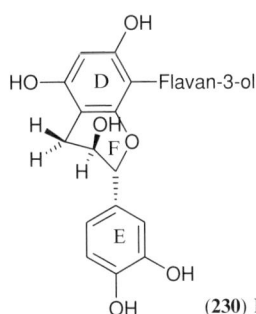

(230) E-conformer

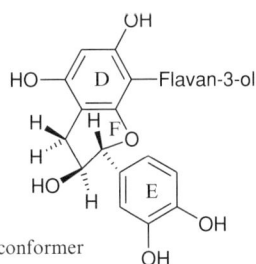

(231) A-conformer

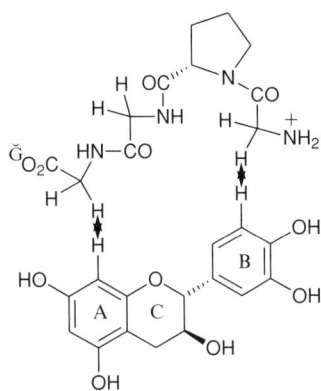

(232)

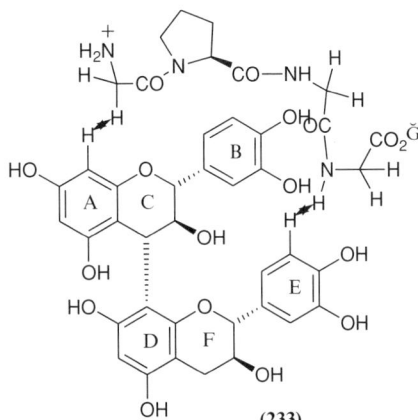

(233)

FIGURE 11.27 Structures of compounds **228–233**.

11.3.7 HPLC–MS ANALYSIS OF PROANTHOCYANIDINS

The various MS methods to determine the molecular composition of the constituent mono-meric units in proanthocyanidins oligomers are summarized in Ref. 257. Contributions focusing on proanthocyanidin analysis via the HPLC–MS protocol included a wide range of plant-derived foods and beverages, and are summarized in Refs. 12, 258–261. In addition, references to additional significant contributions in this area are readily available via several of the excellent electronic search engines that are at our disposal.

REFERENCES

1. Haslam, E., Natural proanthocyanidins, in *The Flavonoids*, Harborne, J.B., Mabry, T.J., and Mabry, H., Eds., Chapman & Hall, London, 1975, chap. 10.
2. Haslam, E., Proanthocyanidins, in *The Flavonoids — Advances in Research*, Harborne, J.B. and Mabry, T.J., Eds., Chapman & Hall, London, 1982, chap. 7.
3. Porter, L.J., Flavans and proanthocyanidins, in *The Flavonoids — Advances in Research Since 1980*, Harborne, J.B., Ed., Chapman & Hall, London, 1988, chap. 2.
4. Porter, L.J., Flavans and proanthocyanidins, in *The Flavonoids — Advances in Research Since 1986*, Harborne, J.B., Ed., Chapman & Hall, London, 1994, chap. 2.
5. Hemingway, R.W., Biflavonoids and proanthocyanidins, in *Natural Products of Woody Plants 1: Chemicals Extraneous to the Lignocellulosic Cell Wall*, Rowe, J.W., Ed., Springer Verlag, New York, 1989, chap. 7.6.
6. Porter, L.J., Condensed tannins, in *Natural Products of Woody Plants 1: Chemicals Extraneous to the Lignocellulosic Cell Wall*, Rowe, J.W., Ed., Springer Verlag, New York, 1989, chap. 7.7.
7. Ferreira, D. et al., Diversity of structure and function in oligomeric flavanoids, *Tetrahedron*, 48, 1743, 1992.
8. Ferreira, D. and Bekker, R., Oligomeric proanthocyanidins: naturally occurring *O*-heterocycles, *Nat. Prod. Rep.*, 13, 411, 1996.
9. Ferreira, D. et al., Condensed tannins, *Fortschr. Chem. Org. Naturst.*, 77, 21, 1999.
10. Ferreira, D., Nel, R.J.J., and Bekker, R., Condensed tannins, in *Comprehensive Natural Products Chemistry*, Barton, D.H.R. and Nakanishi, K., Eds., Pergamon Press (Elsevier), New York, 1999, 3, chap. 3.19.
11. Ferreira, D. and Li., X-C., Oligomeric proanthocyanidins: naturally occurring *O*-heterocycles, *Nat. Prod. Rep.*, 17, 193, 2000.
12. Ferreira, D. and Slade, D., Oligomeric proanthocyanidins: naturally occurring *O*-heterocycles, *Nat. Prod. Rep.*, 19, 517, 2002.
13. Geiger, H., Biflavonoids and triflavonoids, in *The Flavonoids — Advances in Research Since 1986*, Harborne, J.B., Ed., Chapman & Hall, London, 1994, chap. 4.
14. Hemingway, R.W., Foo, L.Y., and Porter, L.J., Linkage isomerism in trimeric and polymeric 2,3-*cis*-procyanidins, *J. Chem. Soc., Perkin Trans. 1*, 1209, 1982.
15. Kolodziej, H. et al., On the nomenclature of oligoflavanoids with an A-type unit, *J. Nat. Prod.*, 56, 1199, 1993.
16. Bennie, L. et al., Oligomeric flavanoids. Part 32. Structure and synthesis of ether-linked proteracacinidin and promelacacinidin proanthocyanidins from *Acacia caffra*, *Phytochemistry*, 53, 785, 2000.
17. Coetzee, J. et al., Oligomeric flavanoids. Part 30. Structure and synthesis of butiniflavan-epicatechin and -epigallocatechin probutinidins, *Phytochemistry*, 52, 737, 1999.
18. Pan, W-B. et al., New flavans, spirostanol sapogenins and a pregnane genin from *Tupistra chinensis* and their cytotoxicity, *J. Nat. Prod.*, 66, 161, 2003.
19. Garo, E. et al., Five flavans from *Mariscus psilostachys*, *Phytochemistry*, 43, 1265, 1996.
20. Sauvain, M. et al., Isolation of flavans from the Amazonian shrub *Faramea guianensis*, *J. Nat. Prod.*, 57, 406, 1994.
21. Abdel-Mogib, M., Baisaf, S.A., and Ezmirly, S.T., Two novel flavans from *Cyperus conglomeratus*, *Pharmazie*, 55, 693, 2000.
22. Garcez, F. et al., Triterpenoids, lignans and flavans from *Terminalea argentea*, *Biochem. Syst. Ecol.*, 31, 229, 2003.
23. Torres, S.L. et al., Two flavans from *Brosimum acutifolium*, *Phytochemistry*, 44, 347, 1997.
24. Ko, H-H. et al., Cytotoxic isoprenylated flavans of *Broussonetia kazinoki*, *J. Nat. Prod.*, 62, 164, 1999.
25. Takashima, J. and Ohsaki, A., Acutifolins A-F, a new flavan-derived constituent and five new flavans from *Brosimum acutifolium*, *J. Nat. Prod.*, 64, 1493, 2001.
26. Achenbach, H., Stoecker, M., and Constenia, M.A., Constituents of tropical medicinal plants. Part 31. Flavonoid and other constituents of *Bauhinia manca*, *Phytochemistry*, 27, 1835, 1988.

27. Ikuta, J. et al., Constituents of the cultivated mulberry tree. Part XXXII. Components of *Broussonetia kazinoki* Sieb. 1. Structures of two new isoprenylated flavans and five new isoprenylated 1,3-diarylpropane derivatives, *Chem. Pharm. Bull.*, 34, 1968, 1986.

28. Cardillo, G. et al., Constituents of Dragon's blood resin. I. Structure and absolute configuration of new optically active flavans, *J. Chem. Soc. C*, 3967, 1971.

29. Antus, S. et al., Chiroptical properties of 2,3-dihydrobenzo[*b*]furan and chromane chromophores in naturally occurring *O*-heterocycles, *Chirality*, 13, 493, 2001.

30. Bilia, A.R. et al., Flavans and A-type proanthocyanidins from *Prunus prostrata*, *Phytochemistry*, 43, 887, 1996.

31. Moijo, F., Gashe, B.A., and Mejinda, R.R.T., A new flavan from *Elephantorrhiza goetzei*, *Fitoterapia*, 70, 412, 1999.

32. Cronje, A. et al., Oligomeric flavanoids. Part 16. Novel prorobinetinidins and the first A-type proanthocyanidin with a 5-deoxy A- and a 3,4-*cis*-C-ring from the maiden investigation of commercial wattle bark extract, *J. Chem. Soc., Perkin Trans. 1*, 2467, 1993.

33. Rao, R.J. et al., Novel 3-*O*-acyl mesquitol analogues as free radical scavengers and enzyme inhibitors: synthesis, biological evaluation and structure–activity relationship, *Biorg. Med. Chem. Lett.*, 13, 2777, 2003.

34. Nel, R.J.J. et al., The novel flavan-3-ol, (2*R*,3*S*)-guibourtinidol and its diastereomers, *Phytochemistry*, 52, 1153, 1999.

35. Korver, O. and Wilkins, C.K., Circular dichroism of flavanols, *Tetrahedron*, 27, 5459, 1971.

36. Van Rensburg, H. et al., Circular dichroic properties of flavan-3-ols, *J. Chem. Res. (S)*, 450, 1999.

37. Nel, R.J.J. et al., Stereoselective synthesis of flavonoids. Part 8. Free phenolic flavan-3-ols, *J. Chem. Res. (S)*, 606, 1999 (*M*, 2610).

38. Bae, Y-S. et al., Flavan and procyanidin glycosides from the bark of blackjack oak, *Phytochemistry*, 35, 473, 1994.

39. Pan, H. and Lundgren, L.N., Phenolic extractions from root bark of *Picea abies*, *Phytochemistry*, 39, 1423, 1995.

40. Petereit, F., Kolodziej, H., and Nahrstedt, A., Flavan-3-ols and proanthocyanidins from *Cistus incanus*, *Phytochemistry*, 30, 981, 1991.

41. Sakar, M.K., Petereit, F., and Nahrstedt, A., Two phloroglucinol glucosides, flavan gallates and flavonol glycosides from *Sedum sediforum* flowers, *Phytochemistry*, 33, 171, 1993.

42. Danne, A., Petereit, F., and Nahrstedt, A., Flavan-3-ols, prodelphinidins and further polyphenols from *Cistus salvifolius*, *Phytochemistry*, 37, 533, 1994.

43. Delle Monache, F. et al., A bianthraquinone and 4'-*O*-methyl-*ent*-gallocatechin from *Cassia trachypus*, *Phytochemistry*, 31, 259, 1991.

44. Gareia, J. et al., 4'-*O*-Methylgallocatechin from *Panda oleosa*, *Phytochemistry*, 32, 1626, 1993.

45. Drewes, S.E., Taylor, C.W., and Cunningham, A.B., (+)-Afzelechin 3-rhamnoside from *Cassipourea gerrardii*, *Phytochemistry*, 31, 1073, 1992.

46. Baek, N.I. et al., Flavonoids and a proanthocyanidin from rhizomes of *Selliguea feei*, *Phytochemistry*, 37, 513, 1994.

47. Ohmori, K., Ushimaru, N., and Suzuki, K., Stereoselective substitution of flavan skeletons: synthesis of dryopteric acid, *Tetrahedron Lett.*, 43, 7753, 2002.

48. Liu, S.Q., Xiao, Z.Y., and Feng, R., A flavanol glycoside from *Drynaria propinqua*, *Phytochemistry*, 35, 1595, 1994.

49. Cui, C-B. et al., Constituents of a fern, *Davallia mariessi* Moore. IV. Isolation and structures of a novel norcarotane sesquiterpene glycoside, a chromone glucuronide, and two epicatechin glycosides, *Chem. Pharm. Bull.*, 40, 2035, 1992.

50. Mukerjee, R.K., Fujimoto, Y., and Kakinuma, K., 1-(ω-Hydroxyfattyacyl)glycerols and two flavanols from *Cinnamomum camphora*, *Phytochemistry*, 37, 1641, 1994.

51. Cui, C-B. et al., Davallioside A and B, novel flavan-3-ol derivatives with a γ-lactam, from the rhizomes of *Davallia mariesii* Moore, *Chem. Pharm. Bull.*, 38, 2620, 1990.

52. Lee, M-W. et al., Tannins and related compounds. III. Flavan-3-ol gallates and proanthocyanidins from *Pithecellobium lobatum*, *Phytochemistry*, 31, 2117, 1992.

53. Drewes, S.E. and Mashimbye, M.J., Flavanoids and triterpenoids from *Cassine papillosa* and the absolute configuration of 11,11-dimethyl-1,3,8,10-tetrahydroxy-9-methoxypeltogynan, *Phytochemistry*, 32, 1041, 1993.

54. Van der Westhuizen, J.H., Steenkamp, J.A., and Ferreira, D., An unusual reaction of flavan-3-ols with acetone of relevance to the formation of the tetracyclic ring system in peltogynoids, *Tetrahedron*, 46, 7849, 1990.

55. Fukuhara, K. et al., Enhanced radical-scavenging activity of a planar catechin analogue, *J. Am. Chem. Soc.*, 124, 5952, 2002.

56. Fukuhara, K. et al., A planar catechin analogue as a promising antioxidant with reduced prooxidant activity, *Chem. Res. Toxicol.*, 16, 81, 2003.

57. Pan, H. and Lundgren, L.N., Phenolics from inner bark of *Pinus sylvestris*, *Phytochemistry*, 42, 1185, 1996.

58. Ferrari, F., Delle Monache, F., and De Lima, R.A., (−)-Epicatechin 5-*O*-β-D-xylopyranoside from *Brosimopsis acutifolium*, *Phytochemistry*, 47, 1165, 1998.

59. De Mello, P., Petereit, F., and Nahrstedt, A., Flavan-3-ols and prodelphinidins from *Stryphnodendron adstringens*, *Phytochemistry*, 41, 807, 1996.

60. Piacente, S. et al., Anadanthoside: a flavanol-3-*O*-β-D-xylopyranoside from *Anadenanthera macrocarpa*, *Phytochemistry*, 51, 709, 1999.

61. Ramadan, M.A. et al., Minor phenolics from *Crinum bulbispermum* bulbs, *Phytochemistry*, 54, 891, 2000.

62. Hwang, B.Y. et al., Antioxidant benzoylated flavan-3-ol glycoside from *Celastrus orbiculatus*, *J. Nat. Prod.*, 64, 82, 2001.

63. Cuendet, M., Potterat, O., and Hostettmann, K., Flavonoids and phenylpropanoid derivatives from *Campanula barbata*, *Phytochemistry*, 56, 631, 2001.

64. Shao, Z-Y., Zhu, D-Y., and Guo, Y.W., A new flavan-3-ol glucoside from *Daphniphyllum oldhamii*, *Chin. Chem. Lett.*, 14, 926, 2003.

65. Liu, X., Zhao, D., and Wang, H., Phenolic compounds from *Celastrus angulatus*, *J. Ind. Chem. Soc.*, 79, 259, 2002.

66. Hatano, T. et al., Proanthocyanidin glucosides and related polyphenols from cacao liquor and their antioxidant effects, *Phytochemistry*, 59, 749, 2002.

67. Friedrich, W. and Galensa, R., Identification of a new flavanol glucoside from barley (*Hordeum vulgare* L.) and malt, *Eur. Food Res. Technol.*, 214, 388, 2002.

68. Lokvam, J., Coley, P.D., and Kursar, T.A., Cinnamoyl glucosides of catechin and dimeric procyanidins from young leaves of *Ingab umbellifera* (Fabaceae), *Phytochemistry*, 65, 351, 2004.

69. Rossouw, W. et al., Oligomeric flavanoids. Part 17. Absolute configurations of flavan-3-ols and 4-arylflavan-3-ols via a Mosher's method, *Tetrahedron*, 50, 12477, 1994.

70. Drewes, S.E. et al., Epiafzelechin-(4β → 8,2β → 0 → 7)-*ent*-afzelechin from *Cassipourea gerrardii*, *Phytochemistry*, 31, 2491, 1992.

71. Van Rensburg, H. et al., Enantioselective synthesis of flavonoids. Part 3. *Trans*- and *cis*-flavan-3-ol methyl ether acetates, *J. Chem. Soc., Perkin Trans. 1*, 3415, 1997.

72. May, B., Arnaudinaud, V., and Vercauteren, J., Gram-scale production and applications of optically pure ^{13}C-labeled (+)-catechin and (−)-epicatechin, *Eur. J. Org. Chem.*, 12, 2379, 2001.

73. Kori, T. et al., Preparation of 4-deuterio or tritioflavan-3-ols, *Jpn. Kokai, Tokyo Koho Pat. No JP200113169*, 2001.

74. Jew, S-S. et al., Enantioselective synthesis of (2*R*,3*S*)-(+)-catechin, *Tetrahedron: Asymmetry*, 13, 715, 2002.

75. Stobiecki, M. and Popenda, M., Flavan-3-ols from seeds of *Lupinus angostifolius*, *Phytochemistry*, 37, 1707, 1994.

76. Chen, H.F. et al., Tannins and related compounds. Part 121. Phenylpropanoid-substituted catechins from *Castanopsis hystrix* and structure revisions of cinchonains, *Phytochemistry*, 33, 183, 1993.

77. Awale, S. et al., Facile and regioselective synthesis of phenylpropanoid-substituted flavan-3-ols, *Org. Lett.*, 4, 1707, 2002.

78. Fan, W. et al., Apocynins A–D; new phenylpropanoid-substituted flavan-3-ols isolated from leaves of *Apocynum venetum* (Luobuma-ye), *Chem. Pharm. Bull.*, 47, 1049, 1999.

79. Bai, L. et al., Shanciol, a dihydrophenanthropyran from *Pleione bulbocodioides*, *Phytochemistry*, 41, 625, 1996.

80. Bai, L., Yamaki, M., and Tagaki, S., Flavan-3-ols and dihydrophenanthropyrans from *Pleione bulbocodioides*, *Phytochemistry*, 47, 1125, 1998.

81. Bai, L. et al., Four stilbenoids from *Pleione bulbocodioides*, *Phytochemistry*, 48, 327, 1998.

82. Magalhaes, A.F. et al., Twenty-three flavonoids from *Lonchocarpus subglaucescens*, *Phytochemistry*, 42, 1459, 1996.

83. Tanaka, N. et al., Four new flavan-4-ol glucosides from *Pneumatopteris pennigera*, *Aust. J. Chem.*, 50, 329, 1997.

84. Borges-Argaez, R., Pena-Rodriquez, L.M., and Waterman, P.G., Flavonoids from the stem bark of *Lonchocarpus xuul*, *Phytochemistry*, 54, 611, 2000.

85. Mitsunaga, T. et al., Inhibitory effects of bark proanthocyanidins on the activities of glucosyl-transferases of *Streptococcus sobrinus*, *J. Wood Chem. Technol.*, 17, 327, 1997.

86. Pouget, C. et al., Synthesis and structure of flavan-4-ols and 4-methoxyflavans as new potential anticancer drugs, *Tetrahedron*, 56, 6047, 2000.

87. Todoroki, T., Saito, A., and Tanaka, A., Lipase-catalyzed kinetic resolution of ($\pm$)-*cis*-flavan-4-ol and its acetate: synthesis of chiral 3-hydroxyflavanones, *Biosci. Biotechnol. Biochem.*, 66, 1172, 2002.

88. Ali, M. and Bhutani, K., Flavan-3,4-diols from *Musa sapientum* seeds, *Pharmazie*, 48, 455, 1993.

89. Malan, E., A (4$\beta \rightarrow$ 5)-linked proteracacinidin dimer from the heartwood of *Acacia caffra*, *Phytochemistry*, 40, 1519, 1995.

90. Borges-Argaez, R., Pena-Rodriguez, L.M., and Waterman, P.G., Flavonoids from two *Lonchocarpus* species of the Yucutan peninsula, *Phytochemistry*, 60, 533, 2002.

91. Stich, K. et al., Enzymatic conversion of dihydroflavonols to flavan-3,4-diols using extracts of *Dianthus caryophyllus* L. (carnation), *Planta*, 187, 103, 1992.

92. Ferreira, D. et al., Circular dichroic properties of flavan-3,4-diols, *J. Nat. Prod.*, 67, 174, 2004.

93. Cai, Y. et al., Biological and chemical investigation of dragon's blood from croton species of South America. Part 1. Polyphenolic compounds from *Croton lechleri*, *Phytochemistry*, 30, 2033, 1991.

94. Cui, C.B., Davallin, a new tetrameric proanthocyanidin from the rhizomes of *Davallia mariesii* Moore, *Chem. Pharm. Bull.*, 39, 2179, 1991.

95. Cui, C.B. et al., Constituents of a fern, *Davallia mariesii* Moore. V. Isolation and structure of davallin, a new tetrameric proanthocyanidin, and two new phenolic glycosides, *Chem. Pharm. Bull.*, 41, 1491, 1993.

96. Geiss, F. et al., Proanthocyanidins with (+)-epicatechin units from *Byrsonima crassifolia* bark, *Phytochemistry*, 39, 635, 1995.

97. Hartisch, C. and Kolodziej, H., Galloylhamameloses and proanthocyanidins from *Hamamelis virginiana*, *Phytochemistry*, 42, 191, 1996.

98. Malik, A. et al., Proanthocyanidins of *Ziziphus jujuba*, *Chem. Nat. Compd.*, 33, 165, 1997.

99. Malik, A. et al., Catechins and proanthocyanidins of *Alhagi sparsifolia*. I, *Chem. Nat. Compd.*, 33, 174, 1997.

100. Kuliev, Z.A. et al., Study of the catechins and proanthocyanidins from *Quercus robur*, *Chem. Nat. Compd.*, 33, 642, 1997.

101. Keneshov, B.M. et al., Proanthocyanidins of *Polygonum coriarium*. III. Structures of proantho-cyanidins T1 and T2, *Chem. Nat. Compd.*, 33, 453, 1997.

102. Keneshov, B.M. et al., Proanthocyanidins of *Polygonum coriarium*. IV. Structures of proantho-cyanidins T3 and T4, *Chem. Nat. Compd.*, 33, 548, 1997.

103. Mavlyanov, S.M. et al., Tannins of *Geranium sanguineum*, *Chem. Nat. Compd.*, 33, 179, 1997.

104. Wang, J-N. et al., Procyanidins from the seeds of *Vitis amurensis*, *Phytochemistry*, 53, 1097, 2000.

105. Matamarova, K.N. et al., Oligomeric proanthocyanidins glycosides of *Clementsia semenovii*. II, *Chem. Nat. Compd.*, 35, 39, 1999.

106. Ismailov, A.E. et al., Oligomeric proanthocyanidins glycosides of *Rhodiola pamiroalaica*. II, *Chem. Nat. Compd.*, 35, 33, 1999.

107. Ismailov, A.E. et al., Oligomeric proanthocyanidins glycosides of *Rhodiola pamiroalaica*, *Chem. Nat. Compd.*, 34, 450, 1998.

108. Achmadi, S. et al., Catechin-3-*O*-rhamnoside chain extender units in polymeric proanthocyanidins from mangrove bark, *Phytochemistry*, 35, 217, 1994.

109. McGraw, G.W., Steynberg, J.P., and Hemingway, R.W., Condensed tannins: a novel rearrangement of procyanidins and prodelphinidins in thiolytic cleavage, *Tetrahedron Lett.*, 34, 987, 1993.

110. Foo, L.Y. et al., Proanthocyanidins from *Lotus corniculatus*, *Phytochemistry*, 41, 617, 1996.

111. Dauer, A., Rimpler, H., and Hensel, A., Polymeric proanthocyanidins from the bark of *Hamamelis virginiana*, *Planta Med.*, 69, 89, 2003.

112. Foo, L.Y. et al., Proanthocyanidins from *Lotus pedunculatus*, *Phytochemistry*, 45, 1689, 1997.

113. Steynberg, P.J. et al., Acid-catalyzed rearrangements of flavan-4-phloroglucinol derivatives to novel 6-hydroxyphenyl-6a,11b-dihydro-6*H*-[1]benzofuro[2,3-*c*]chromenes and hydroxyphenyl-3,2′-spirobi[dihydro[1]benzofurans], *J. Chem. Soc., Perkin Trans. 1*, 2395, 1997.

114. Steynberg, P.J. et al., Oligomeric flavanoids. Part 26. Structure and synthesis of the first profisetinidins with epifisetinidol constituent units, *J. Chem. Soc., Perkin Trans. 1*, 1943, 1997.

115. Steynberg, P.J. et al., Oligomeric flavanoids. Part 27. Interflavanyl bond formation in procyanidins under neutral conditions, *Tetrahedron*, 54, 8153, 1998.

116. Pierre, M-C., Cheze, C., and Vercauteren, J., Deuterium labeled procyanidin synthesis, *Tetrahedron Lett.*, 38, 5639, 1997.

117. Déprez, S., Mila, I., and Scalbert, A., Carbon-14 biolabeling of (+)-catechin and proanthocyanidin oligomers in willow tree cuttings, *J. Agric. Food Chem.*, 47, 4219, 1999.

118. Nay, B. et al., Total synthesis of labeled flavonoids. 2. ^{13}C-labeled ($\pm$)-catechin from potassium[^{13}C]cyanide, *Eur. J. Org. Chem.*, 1279, 2000.

119. Nay, B., Arnaudinaud, V., and Vercauteren, J., Gram-scale production and applications of optically pure ^{13}C-labeled (+)-catechin and (−)-epicatechin, *Eur. J. Org. Chem.*, 2379, 2001.

120. Arnaudinaud, V. et al., Total synthesis of isotopically labeled flavonoids. Part 5. Gram-scale production of ^{13}C-labeled (−)-procyanidin B$_3$, *Tetrahedron Lett.*, 42, 5669, 2001.

121. Arnaudinaud, V. et al., Total synthesis of isotopically labeled flavonoids. Part 3. ^{13}C-labeled (−)-procyanidin B$_3$ from 1-[^{13}C]-acetic acid, *Tetrahedron Lett.*, 42, 1279, 2001.

122. Yoneda, S., Kawamoto, H., and Nakatsubo, F., Synthesis of high molecular mass condensed tannins by cationic polymerization of flavan-3,4-carbonate, *J. Chem. Soc., Perkin Trans. 1*, 1025, 1997.

123. Bennie, L. et al., Oligomeric flavanoids. Part 32. Structure and synthesis of ether-linked proteracacinidin and promelacacinidin proanthocyanidins from *Acacia caffra*, *Phytochemistry*, 53, 785, 2000.

124. Tückmantel, W., Kozikowski, A.P., and Romanczyk, L.J. Jr., Studies in polyphenol chemistry and bioactivity. 1. Preparation of building blocks from (+)-catechin. Procyanidin formation. Synthesis of the cancer cell growth inhibitor, 3-*O*-galloyl-(2*R*,3*R*)-epicatechin-4β,8-[3,*O*-galloyl-(2*R*,3*R*)-epicatechin], *J. Am. Chem. Soc.*, 121, 12073, 1999.

125. Kozikowski, A.P., Tückmantel, W., and George, C., Studies in polyphenol chemistry and bioactivity. 2. Establishment of interflavan linkage regio- and stereochemistry by oxidative degradation of an *O*-alkylated derivative of procyanidins B$_2$ to (*R*)-(−)-2,4-diphenylbutyric acid, *J. Org. Chem.*, 65, 5371, 2000.

126. Kozikowski, A.P., Tückmantel, W., and Hu, Y., Studies in polyphenol chemistry and bioactivity. 3. Stereocontrolled synthesis of epicatechin-4α,8-epicatechin, an unnatural isomer of the B-type procyanidins, *J. Org. Chem.*, 66, 1287, 2001.

127. Tückmantel, W., Kozikowski, A.P., and Romanczyk, L.J., Methods for the preparation of catechin and epicatechin dimers, *Int. Pub. No., WO 00/61547*, 2000.

128. Saito, A. et al., Synthetic studies of proanthocyanidins. Highly stereoselective synthesis of the catechin dimer, procyanidin B-3, *Biosci. Biotechnol. Biochem.*, 66, 1764, 2002.

129. Weinges, K., Schich, H., and Rominger, F., X-ray structure analysis of procyanidin B-1, *Tetrahedron*, 57, 2327, 2001.

130. DeAngelis, G.G. and Wildman, W.C., Circular dichroism studies. 1. A quadrant rule for the optically active aromatic chromophore in rigid polycyclic systems, *Tetrahedron*, 25, 5099, 1969.

131. Tanaka, T. et al., Tannins and related compounds. CXVI. Six new complex tannins, guiajavins, psidinins and psiguavin from the bark of *Psidium guajava* L., *Chem. Pharm. Bull.*, 40, 2092, 1992.

132. Lee, M. et al., Tannins and related compounds. III. Flavan-3-ol gallates and proanthocyanidins from *Pithecellobium lobatum*, *Phytochemistry*, 31, 2117, 1992.

133. Nonaka, G. et al., Tannins and related compounds. CXIX. Samarangenins A and B, novel proanthocyanidins with doubly bonded structures, from *Syzygium samarangens* and *S. aqueum*, *Chem. Pharm. Bull.*, 40, 2671, 1992.

134. De Mello, J.C.P., Petereit, F., and Nahrstedt, A., A dimeric proanthocyanidin from *Stryphnodendron adstringens*, *Phytochemistry*, 51, 1105, 1999.

135. Danne, A., Petereit, F., and Nahrstedt, A., Proanthocyanidins from *Cistus incanus*, *Phytochemistry*, 34, 1129, 1993.

136. Tits, M. et al., Prodelphinidins from *Ribes nigrum*, *Phytochemistry*, 31, 971, 1992.

137. Qa'dan, F., Petereit, F., and Nahrstedt, A., Prodelphinidin trimers and characterization of a proanthocyanidin oligomer from *Cistus albidus*, *Pharmazie*, 58, 416, 2003.

138. Matamarova, K.N. et al., Oligomeric proanthocyanidin glycosides of *Clementsia semenovii*, *Chem. Nat. Compd.*, 34, 676, 1999.

139. Kuliev, Z.A. et al., Oligomeric proanthocyanidin glycosides of *Clementsia semenovii* and their biological activity. III, *Chem. Nat. Compd.*, 36, 60, 2000.

140. Malik, A. et al., New oligomeric proanthocyanidins from *Ziziphus jujuba*, *Chem. Nat. Compd.*, 38, 40, 2002.

141. Foo, L.Y. et al., The phenols and prodelphinidins of white clover flowers, *Phytochemistry*, 54, 539, 2000.

142. Marais, J.P.J. et al., Polyphenols, condensed tannins and other natural products in *Onobrychus viciifolia* (sainfoin), *J. Agric. Food Chem.*, 48, 3440, 2000.

143. Achenbach, H. and Benirschke, G., Joannesilactone and other compounds from *Joannesia princeps*, *Phytochemistry*, 45, 149, 1997.

144. Kiehne, A., Lakenbrink, C., and Engelhardt, U.H., Analysis of proanthocyanidins in tea samples. Part 1. LC–MS results, *Z. Lebensm. Unters. Forsch. A*, 205, 153, 1997.

145. Lakenbrink, C., Engelhardt, U.H., and Wray, V., Identification of two novel proanthocyanidins in green tea, *J. Agric. Food Chem.*, 47, 4621, 1999.

146. Chang, E-J. et al., Proliferative effects of flavan-3-ols and propelargonidins from rhizomes of *Drynaria fortunei* on MCF-7 and osteoblastic cells, *Arch. Pharm. Res.*, 26, 620, 2003.

147. Dirsch, V., Neszmelyi, A., and Wagner, H., A trimeric propelargonidin from stem bark of *Heisteria pallida*, *Phytochemistry*, 34, 291, 1993.

148. Abeysekera, A.M. et al., An immunomodulatory C-glucosylated propelargonidin from the unripe fruit of *Aegle marmelos*, *Fitoterapia*, 67, 367, 1996.

149. Mathisen, E. et al., Antioxidants from the bark of *Burkea africana*, an African medicinal plant, *Phytother. Res.*, 16(S1), 148, 2002.

150. Clark-Lewis, J.H., Flavan derivatives. XXI. Nuclear magnetic resonance spectra, configuration and conformation of flavan derivatives, *Aust. J. Chem.*, 21, 2059, 1968.

151. Van der Westhuizen, J.H., Ferreira, D., and Roux, D.G., Synthesis of condensed tannins. Part 2. Synthesis by photolytic rearrangement, stereochemistry, and circular dichroism of the first 2,3-*cis*-3,4-*cis*-4-arylflavan-3-ols, *J. Chem. Soc., Perkin Trans. 1*, 1220, 1981.

152. Steynberg, J.P. et al., Oligomeric flavanoids. Part 4. Base-catalyzed conversions of (−)-fisetinidol-(+)-catechin profisetinidins with 2,3-*trans*-3,4-*cis*-flavan-3-ol constituent units, *J. Chem. Soc., Perkin Trans. 1*, 1331, 1988.

153. Steynberg, J.P. et al., Oligomeric flavanoids. Part 3. Structure and synthesis of phlobatannins related to (−)-fisetinidol-(4α,6)- and (4α,8)-(+)-catechin profisetinidins, *J. Chem. Soc., Perkin Trans. 1*, 3323, 1988.

154. Steynberg, J.P. et al., Oligomeric flavanoids. Part 6. Evidence supporting the inversion of absolute configuration at 3-C associated with base catalyzed A-/B-ring interchange of precursors having 2,3-*trans*-3,4-*cis*-flavan-3-ol constituent units, *Heterocycles*, 28, 923, 1989.

155. Steynberg, J.P. et al., Oligomeric flavanoids. Part 11. Structure and synthesis of the first phlobatannins related to (4α,6:4α,8)-bis-(−)-fisetinidol-(+)-catechin profisetinidin triflavanoids, *J. Chem. Soc., Perkin Trans. 1*, 235, 1990.

156. Bonnet, S.L. et al., Structure and synthesis of phlobatannins related to the (4β,6:4α,8)-bis-fisetinidol-catechin profisetinidin triflavanoid, *Phytochemistry*, 43, 215, 1996.

157. Bonnet, S.L. et al., Structure and synthesis of phlobatannins related to the (4β,6:4β,8)-bis-fisetinidol-catechin profisetinidin triflavanoid, *Phytochemistry*, 43, 229, 1996.

158. Bonnet, S.L. et al., Structure and synthesis of phlobatannins related to the (4α,6:4β,8)-bis-fisetinidol-catechin profisetinidin triflavanoid, *Phytochemistry*, 43, 241, 1996.

159. Bonnet, S.L. et al., Structure and synthesis of phlobatannins related to bis-fisetinidol-epicatechin profisetinidin triflavanoids, *Phytochemistry*, 43, 253, 1996.

160. Saunders, C.M. et al., Controlled biometric synthesis of profisetinidin triflavanoids related phlobatannins, *Tetrahedron*, 52, 6003, 1996.

161. Steynberg, P.J. et al., Cleavage of the interflavanyl bond in 5-deoxy (A-ring) proanthocyanidins, *J. Chem. Soc., Chem. Commun.*, 31, 1994.

162. Steynberg, P.J. et al., Oligomeric flavanoids. Part 19. Reductive cleavage of the interflavanyl bond in proanthocyanidins, *J. Chem. Soc., Perkin Trans. 1*, 3005, 1995.

163. Steynberg, P.J. et al., Oligomeric flavanoids. Part 25. Cleavage of the acetal functionality in A-type proanthocyanidins, *Tetrahedron*, 53, 2591, 1997.

164. Svedstrom, U. et al., Isolation and identification of oligomeric procyanidins from *Craetagus* leaves and flowers, *Phytochemistry*, 60, 821, 2002.

165. Shaji, T. et al., Isolation and structure elucidation of some procyanidins from apple by low-temperature nuclear magnetic resonance, *J. Agric. Food Chem.*, 51, 3806, 2003.

166. Coetzee, J. et al., Oligomeric flavanoids. Part 18. Dimeric prorobinetinidins from *Robinia pseudacacia*, *Tetrahedron*, 51, 2339, 1995.

167. De Mello, J.P., Petereit, F., and Nahrstedt, A., Prorobinetinidins from *Stryphnodendron adstringens*, *Phytochemistry*, 42, 857, 1996.

168. Ohara, S. and Hemingway, R.W., Condensed tannins: the formation of a diarylpropanol-catechinic acid dimer from base-catalyzed reactions of (+)-catechin, *J. Wood Chem. Technol.*, 11, 195, 1991.

169. Malan, E. et al., A novel doubly-linked proteracacinidin analog from *Acacia caffra*, *Tetrahedron Lett.*, 35, 7415, 1994.

170. Malan, E. and Sireeparsad, A., The structure and synthesis of the first dimeric proteracacinidins from *Acacia galpinii*, *Phytochemistry*, 38, 237, 1995.

171. Coetzee, J., Malan, E., and Ferreira, D., Oligomeric flavanoids. Part 28. Structure and synthesis of ether-linked (4-*O*-3)-bis-teracacinidins, a novel class of naturally occurring proanthocyanidins, *J. Chem. Res. (S)*, 526, 1998.

172. Coetzee, J., Malan, E., and Ferreira, D., Oligomeric flavanoids. Part 29. Structure and synthesis of novel ether-linked (4-*O*-4)-bis-teracacinidins, *Tetrahedron*, 54, 9153, 1998.

173. Bennie, L. et al., Oligomeric flavanoids. Part 34. Doubly-linked proteracacinidin analogues from *Acacia caffra* and *Acacia galpinii*, *Tetrahedron*, 57, 661, 2001.

174. Bennie, L. et al., Structure and stereochemistry of triflavanoids containing both ether and carbon–carbon interflavanyl bonds, *Phytochemistry*, 57, 1023, 2001.

175. Bennie, L. et al., Structure and stereochemistry of dimeric proteracacinidins possessing the rare C-4(C) → C-5(D) interflavanyl linkage, *Phytochemistry*, 59, 673, 2002.

176. Bennie, L. et al., (4 → 6)-Coupled proteracacinidins and promelacacinidins from *Acacia galpinii* and *Acacia caffra*, *Phytochemistry*, 60, 521, 2002.

177. Bennie, L. et al., Trimeric proteracacinidins and a (6 → 6)-bis-leucoteracacinidin from *Acacia galpinii* and *Acacia caffra*, *Phytochemistry*, 65, 215, 2004.

178. Howell, H. et al., Identification of two novel promelacacinidin dimers from *Acacia nigrescens*, *J. Nat. Prod.*, 65, 769, 2002.

179. Clark-Lewis, J.W. and Mortimer, P.I., Flavan derivatives. III. Melacacidin and isomelacacidin from Acacia species, *J. Chem. Soc.*, 4106, 1960.

180. Clark-Lewis, J.W. and Williams, L.R., Flavan derivatives. XVII. Epimerization of the benzylic 4-hydroxyl group in flavan-3,4-diols and the formation of 4-alkyl ethers by solvolysis, *Aust. J. Chem.*, 20, 2152, 1967.

181. Coetzee, J., Malan, E., and Ferreira, D., The formation and stability of flavans with 2,3-*cis*-3,4-*cis* configuration, *Tetrahedron*, 55, 9999, 1999.

182. Malan, E. et al., The structure and synthesis of a 7,8,4'-trihydroxyflavan-epioritin dimer from *Acacia caffra*, *Phytochemistry*, 44, 529, 1997.

183. Malan, E. et al., The structure and synthesis of proguibourtinidins from *Cassia abbreviata*, *Phytochemistry*, 41, 1209, 1996.

184. Messanga, B. et al., Calodenin C. A new guibourtinidol-(4α → 8)-afzelechin from *Ochna calodendron*, *Plant Med.*, 64, 760, 1998.

185. Rani, M. and Kalidhar, S.B., Chemical investigation of *Cassia biflora* stems, *J. Indian Chem. Soc.*, 75, 386, 1998.

186. Coetzee, J. et al., Oligomeric flavanoids. Part 31. Structure and synthesis of the first procassinidin dimers based on epicatechin, and gallo- and epigallocatechin, *Phytochemistry*, 53, 795, 2000.

187. Hatano, T. et al., Flavan dimers with lipase inhibitory activity from *Cassia nomame*, *Phytochemistry*, 46, 893, 1997.

188. Barrett, M.W. et al., Plant proanthocyanidins. Part 6. Chiroptical studies. Part 95. Circular dichroism of procyanidins, *J. Chem. Soc., Perkin Trans. 1*, 2375, 1979.

189. Foo, L.Y. et al., A-type proanthocyanidin trimers from cranberry that inhibit adherence of uropathogenic P-fimbriated *Escherichia coli*, *J. Nat. Prod.*, 63, 1225, 2000.

190. Foo, L.Y. et al., The structure of cranberry proanthocyanidins which inhibit adherence of uropathogenic P-fimbriated *Escherichia coli*, *Phytochemistry*, 54, 173, 2000.

191. Morimoto, S., Nonaka, G., and Nishioka, I., Tannins and related compounds. LIX. Aesculitannins, novel proanthocyanidins with doubly-bonded structures from *Aesculus hippocastanum* L., *Chem. Pharm. Bull.*, 35, 4717, 1987.

192. Porter, L.J., Ma, Z., and Chan, B.G., Cacao procyanidins: major flavanoids and identification of some minor metabolites, *Phytochemistry*, 30, 1657, 1991.

193. Lou, H. et al., A-type proanthocyanidins from peanut skins, *Phytochemistry*, 51, 297, 1999.

194. Balde, A.M. et al., Proanthocyanidins from stem bark of *Pavetta owariensis*. Part 2. Dimeric and trimeric proanthocyanidins possessing a doubly linked structure from *Pavetta owariensis*, *Phytochemistry*, 30, 4129, 1991.

195. Drewes, S.E. and Taylor, C.W., Methylated A-type proanthocyanidins and related metabolites from *Cassipourea gummiflua*, *Phytochemistry*, 37, 551, 1994.

196. Prasad, D. et al., An A-type proanthocyanidin from *Prunus armeniaca*, *J. Nat. Prod.*, 61, 1123, 1998.

197. Rawat, M.S.M. et al., Proanthocyanidins from *Prunus armeniaca* roots, *Phytochemistry*, 50, 321, 1999.

198. Calzada, F. et al., Geranins A and B, new antiprotozoal proanthocyanidins from *Geranium niveum*, *J. Nat. Prod.*, 62, 705, 1999.

199. Calzada, F. et al., Geranins C and D, additional new antiprotozoal A-type proanthocyanidins from *Geranium niveum*, *Planta Med.*, 67, 677, 2001.

200. Barreiros, A.L.B.S. et al., A-type proanthocyanidin antioxidant form *Dioclea lasiophylla*, *Phytochemistry*, 55, 805, 2000.

201. Ma, C-M. et al., Inhibitory effects on HIV-1 protease of constituents from the wood of *Xanthoceras sorbifolia*, *J. Nat. Prod.*, 63, 238, 2000.

202. Ogundaini, A. et al., Isolation of two new antiinflammatory biflavanoids from *Sarcophyte piriei*, *J. Nat. Prod.*, 59, 587, 1996.

203. Arnone, A. et al., Constituents of Dragon's blood. 5. Dracoflavans B_1, B_2, C_1, C_2, D_1, and D_2, new A-type deoxyproanthocyanidins, *J. Nat. Prod.*, 60, 97, 199.

204. Balde, A.M. et al., Proanthocyanidins from stem bark of *Pavetta owariensis*, 3. NMR study of acetylated trimeric proanthocyanidins possessing a doubly-linked structure, *J. Nat. Prod.*, 56, 1078, 1993.

205. Balde, A.M. et al., Oligomeric proanthocyanidins possessing a doubly linked structure from *Pavetta owariensis*, *Phytochemistry*, 38, 719, 1995.

206. Santos-Buelga, C., Kolodziej, H., and Treutter, D., Procyanidin trimers possessing a doubly linked structure from *Aesculus hippocastanum*, *Phytochemistry*, 38, 499, 1995.

207. Kamiya, K. et al., Studies on Jamu and the medicinal resources of Indonesia. Part 4. Studies on the constituents of bark of *Parameria laevigata* Moldenke, *Chem. Pharm. Bull.*, 49, 551, 2001.

208. Baek, N.I. et al., Potential sweetening agents of plant origin. 29. Studies on Indonesian medicinal plants. 6. Selligueain A, a novel highly sweet proanthocyanidin from the rhizomes of *Selliguea feei*, *J. Nat. Prod.*, 56, 1532, 1993.

209. Lin, L.C., Kuo, Y.C., and Chou, C.J., Immunomodulatory proanthocyanidins from *Ecdysanthera utilis*, *J. Nat. Prod.*, 65, 505, 2002.

210. Balde, A. et al., Tetrameric proanthocyanidins containing a double interflavanoid (A-type) linkage from *Pavetta owariensis*, *Phytochemistry*, 40, 933, 1995.

211. Kamiya, K. et al., A-type proanthocyanidins from the bark of *Parameria laevigata*, *Heterocycles*, 60, 1697, 2003.

212. Kondo, K. et al., Conversion of procyanidin B-type (catechin dimer) to A-type: evidence for abstraction of C-2 hydrogen in catechin during radical oxidation, *Tetrahedron Lett.*, 41, 485, 2000.

213. Tanaka, T., Kondou, K., and Kouno, I., Oxidation and epimerization of epigallocatechin in banana fruits, *Phytochemistry*, 53, 311, 2000.

214. Sawadjoon, S. et al., Atropisomeric myristinins: selective COX-2 inhibitors and antifungal agents from *Myristica cinnamomea*, *J. Org. Chem.*, 67, 5470, 2002.

215. Remy-Tanneau, S. et al., Characterization of a colorless anthocyanin-flavan-3-ol dimer containing both carbon-carbon and ether interflavanoid linkages by NMR and mass spectrometry, *J. Agric. Food Chem.*, 51, 3592, 2003.

216. Baba, K. et al., Structure of a new spirobiflavonoid, genkwanol A, from the root of *Daphne genkwa* Sieb. et Zucc., *Yakugaku Zasshi*, 107, 525, 1987.

217. Baba, K. et al., A spirobiflavonoid genkwanol B from *Daphne genkwa*, *Phytochemistry*, 31, 975, 1992.

218. Baba, K., Taniguchi, M., and Kozawa, M., A third spirobiflavonoid genkwanol C from *Daphne genkwa*, *Phytochemistry*, 33, 914, 1993.

219. Baba, K., Taniguchi, M., and Kozawa, M., Three biflavonoids from *Wikstroemia sikokiana*, *Phytochemistry*, 37, 879, 1994.

220. Baba, K. et al., Biflavonoids from *Daphne odora*, *Phytochemistry*, 38, 1021, 1995.

221. Taniguchi, M. and Baba, K., Three biflavonoids from *Daphne odora*, *Phytochemistry*, 42, 1447, 1996.

222. Taniguchi, M., Fujiwara, A., and Baba, K., Three flavonoids from *Daphne odora*, *Phytochemistry*, 45, 183, 1997.

223. Taniguchi, M. et al., Two biflavonoids from *Daphne acutiloba*, *Phytochemistry*, 49, 863, 1998.

224. Himmelreich, U. et al., Damalachawin, a triflavonoid of a new structural type from Dragon's blood of *Dracaena cinnabari*, *Phytochemistry*, 39, 949, 1995.

225. Hashimoto, F., Nonaka, G., and Nishioka, I., Tannins and related compounds. CXIV. Structure of novel fermentation products, theogallinin, theaflavonin and desgalloyl theaflavonin from black tea, and changes of tea leaf polyphenols during fermentation, *Chem. Pharm. Bull.*, 46, 1383, 1992.

226. Sang, S. et al., Theadibenzotropolone A, a new type pigment from enzymatic oxidation of (−)-epicatechin and (−)-epigallocatechin gallate and characterized from black tea using LC/MS/MS, *Tetrahedron Lett.*, 43, 7129, 2002.

227. Wan, X. et al., A new type of tea pigment — from the chemical oxidation of epicatechin gallate and isolated from tea, *J. Sci. Food Agric.*, 74, 401, 1997.

228. Davies, A.L. et al., A polyphenolic pigment from black tea, *Phytochemistry*, 46, 1397, 1997.

229. Lewis, J.R. et al., Theaflavate B, isotheaflavin-3′-O-gallate and neotheaflavin-3-O-gallate: three polyphenolic pigments from black tea, *Phytochemistry*, 49, 2511, 1998.

230. Nonaka, G., Nishimura, H., and Nishioka, I., Tannins and related compounds. Part 26. Isolation and structures of stenophyllanins A, B and C, novel tannins from *Quercus stenophylla*, *J. Chem. Soc., Perkin Trans. 1*, 163, 1985.

231. Yoshida, T. et al., Tannins and related polyphenols of melastomataceous plants. V. Three new complex tannins from *Melastoma malabathricum* L., *Chem. Pharm. Bull.*, 40, 1727, 1992.

232. Tanaka, T. et al., Tannins and related compounds. CXVI. Six new complex tannins from the bark of *Psidium guajava* L., *Chem. Pharm. Bull.*, 40, 2092, 1992.

233. Tanaka, T. et al., Tannins and related compounds. CXXIV. Five new ellagitannins, platycaryanins A, B, C, and D, and platycariin, and a new complex tannin, strobilanin, from the fruits and bark of *Platycarya strobilacea* Sieb. et Zucc., and biomimetic synthesis of C-glycosidic ellagitannins from glucopyranose-based ellagitannins, *Chem. Pharm. Bull.*, 41, 1708, 1993.

234. Han, L. et al., Tannins of theaceous plants. V. Camelliatannins F, G and H, three new tannins from *Camellia japonica*, *Chem. Pharm. Bull.*, 42, 1399, 1994.

235. Hatano, T. et al., Tannins and related polyphenols of theaceous plants. VIII. Camelliatannins C and E, new complex tannins from *Camellia japonica* leaves, *Chem. Pharm. Bull.*, 43, 1629, 1995.

236. Hatano, T. et al., Camelliatannin D, a new inhibitor of bone resorption from *Camellia japonica* leaves, *Chem. Pharm. Bull.*, 43, 2033, 1995.

237. Han, L. et al., Tannins of Stachyurus species. III. Stachyuranins A, B and C, three new complex tannins from *Stachyurus praecox* leaves, *Chem. Pharm. Bull.*, 43, 2109, 1995.

238. Castagnino, C. and Vercauteren, J., Castavinol, a new series of polyphenols from Bordeaux red wine, *Tetrahedron Lett.*, 37, 7739, 1996.

239. Porter, L.J. et al., Conformational analysis of flavans: proton NMR and molecular mechanical (MM_2) studies of the benzopyran ring of 3′,4′,5,7-tetrahydroxyflavan-3-ols: the crystal and the molecular structure of the procyanidin (2R,2S,4R)-3′,4′,5,7-tetramethoxy-4-(2,4,6-trimethoxyphenyl)flavan-3-ol, *J. Chem. Res. (S)*, 86 (M, 830), 1986.

240. Hör, M., Heinrich, M., and Rimpler, H., Proanthocyanidin polymers with antisecretory activity and proanthocyanidin oligomers from *Guazuma ulmifolia*, *Phytochemistry*, 42, 109, 1996.

241. Balas, L. and Vercauteren, J., Extensive high-resolution reverse 2D NMR analysis for the structural elucidation of procyanidin oligomers, *Magn. Reson. Chem.*, 32, 386, 1994.

242. Balas, L., Vercauteren, J., and Laguerre, M., 2D NMR structure elucidation of proanthocyanidins: the special case of the catechin-($4\alpha \rightarrow 8$)-catechin-($4\alpha \rightarrow 8$)-catechin trimer, *Magn. Reson. Chem.*, 33, 85, 1995.

243. De Bruyne, T. et al., Unambiguous assignments for free dimeric proanthocyanidin phenols from 2D NMR, *Phytochemistry*, 43, 265, 1996.

244. Vivas, N. et al., A complete structural and conformational investigation of procyanidin A_2 dimer, *Tetrahedron Lett.*, 37, 2015, 1996.

245. Khan, M.L., Haslam, E., and Williamson, M.P., Structure and conformation of the procyanidin B-2 dimer, *Magn. Reson. Chem.*, 35, 854, 1997.

246. Hemingway, R.W. et al., Conformation and complexation of tannins: NMR spectra and molecular search modeling of flavan-3-ols, *Magn. Reson. Chem.*, 34, 424, 1996.

247. Tobiason, F.L. et al., Temperature dependence of (+)-catechin pyran ring proton coupling constants as measured by NMR and modeled using GMMX search methodology, *Tetrahedron Lett.*, 38, 985, 1997.

248. Vdovin, A.D., Kuliev, Z.A., and Abdullaev, N.D., ^{1}H and ^{13}C NMR spectrometry in the study of flavan-3-ols, proanthocyanidins, and their derivatives. III. ^{13}C NMR spectroscopy of flavan-3-ols an proanthocyanidins, *Chem. Nat. Compd.*, 33, 417, 1997.

249. Davies, A.L. et al., ^{1}H and ^{13}C assignments of some green tea polyphenols, *Magn. Reson. Chem.*, 34, 887, 1996.

250. Laouenan, P. et al., NMR structural investigations and conformational analysis of condensed tannins. A continuing challenge due to restricted rotation about the interflavanyl linkage, *Analusis*, 25, M29, 1997.

251. Hatano, T. and Hemingway, R.W., Association of (+)-catechin and catechin-($4\alpha \rightarrow 8$)-catechin with oligopeptides, *J. Chem.. Soc., Chem. Commun.*, 2537, 1996.

252. Hatano, T. and Hemingway, R.W., Conformational isomerism of phenolic procyanidins: preferred conformations in organic solvents and water, *J. Chem. Soc., Perkin Trans. 2*, 1035, 1997.

253. Charlton, A.J. et al., Tannin interactions with a full-length human salivary proline-rich protein display a stronger affinity than with single proline-rich repeats, *FEBS Lett.*, 382, 289, 1996.

254. Baxter, N.J. et al., Multiple interactions between polyphenols and a salivary proline-rich protein repeat results in complexation and precipitation, *Biochemistry*, 36, 5566, 1997.
255. Charlton, A.J., Haslam, E., and Williamson, M.P., Multiple conformations of the proline-rich protein/epigallocatechin gallate complex determined by time-averaged nuclear Overhauser effects, *J. Am. Chem. Soc.*, 124, 9899, 2002.
256. Charlton, A.J. et al., Polyphenol/peptide binding and precipitation, *J. Agric. Food Chem.*, 50, 1593, 2002.
257. Lazarus, S.A. et al., High-performance liquid chromatography/mass spectrometry analysis of proanthocyanidins in foods and beverages, *J. Agric. Food Chem.*, 47, 3693, 1999.
258. Lazarus, S.A. et al., High-performance liquid chromatography/mass spectrometric analysis of proanthocyanidins in foods and beverages, *Methods Enzymol.*, 335, 46, 2001.
259. Lazarus, S.A. et al., Analysis and purification of proanthocyanidin oligomers, *Methods Polyphenol Anal.*, 267, 2003.
260. Gu, L. et al., Liquid chromatography/electron spray ionization mass spectrometric studies of proanthocyanidins in foods, *J. Mass Spectrom.*, 38, 1272, 2003.
261. Gu, L. et al., Screening of foods containing proanthocyanidins and their structural characterization using LC–MS/MS and thiolytic degradation, *J. Agric. Food Chem.*, 51, 7513, 2003.

12 Flavones and Flavonols

K.M. Valant-Vetschera and E. Wollenweber

CONTENTS

12.1   INTRODUCTION

Flavonoid analyses are mostly concentrating on plants which are of either pharmaceutical interest or of commercial value. In addition, flavonoids are important factors in biological interactions between living organisms. This is best illustrated by the last review "Advances in flavonoid research since 1992" focusing on these topics.[1] In contrast, mere distribution studies or chemosystematically oriented compilations are rare (e.g., on Asteraceae).[2] Naturally, the presently known distribution of flavones and flavonols in plants reflects the current scientific interests, and hence the interpretation of their chemodiversity must be made with caution.

The main part of this compilation consists of extensive tables listing the flavonoids and their plant sources, which are commented accordingly. The data originate primarily from excerpts of current literature, the use of *Chemical Abstracts* and of *Current Contents* (Life Sciences and Agriculture) databases, supported by a review on prenylated flavonoids[3] and data taken from the *Handbook of Natural Flavonoids*.[4] Whenever possible, original literature was consulted to verify structures and their sources. The use of electronically available information lead to inclusion of most recent publications, but the present compilation cannot be claimed to be complete. Apologies go to colleagues whose publications may have been overlooked, and notification on reports that escaped our attention is strongly encouraged.

For compilation and arrangement of compounds, earlier reviews and surveys were taken as the basis.[5,6] In comparison to the previously published reviews, the increasing number and complexity of structures observed is striking. Thus it became quite difficult to list all of these structures in a logical sequence, particularly prenylated derivatives with additional cyclized substituents. Substitution patterns used for grouping of the flavone and flavonol derivatives are as follows.

OH-, OMe-groups; *C*-methyl; methylenedioxy groups; *C*-prenylation; *O*-prenylation; (dihydro)furanosubstitution; pyranosubstitution; complex cyclosubstitution; aromatic substitution; esterification; chlorination. These residues may also occur combined in one flavonoid structure.

In many cases, abbreviation of substitutents could no longer be made without ending up with hardly understandable chemical nomenclature (e.g., complex-*O*-cyclosubstitution). This problem was already obvious in the publication of Barron and Ibrahim[3] who shifted to illustrations of such complex compounds. Consequently, figures of structures showing characteristic substitution patterns will complete the tabulated information provided here.

12.2   ORGANIZATION OF THE TABLES

All tables are organized along the same lines. For the numbering of the basic flavonoid molecule, we refrained to use the system being recommended by the Royal Society of Chemistry, in which primed numbers are mixed in with unprimed numbers. Instead, we still use the more commonly practiced system of most flavonoid scientists: the structures are arranged by number and position, in ascending order, of substituent at ring C being cited first, followed by the substituents in ring A, and then by those in ring B in primed numbers. We also prefer this convention for our tables, for reasons of increased structural information. Thus, for instance, cirsilineol is 5,4'-dihydroxy-6,7,3'-trimethoxyflavone, morin is 3,5,7,2',4'-pentahydroxyflavone, to give two examples. To further increase the value of information, compounds are listed according to increasing numbers of hydroxyl- and methoxyl-groups, and by increasing complexity of other substituents. Further columns inform about trivial names of the compounds, plant sources (species) and families (abbreviated), accumulation sites, and references.

Plant sources are not listed for widespread compounds such as apigenin or kaempferol and some further compounds as had been done previously.[6] Instead, they are either marked as "many records," or, where applicable, as "widespread in Asteraceae" or other larger families. Acknowledged trivial names are cited in column 3, unreasonable names such as "dechloro-chlorapigenin" are omitted. We strictly follow the priority rule when a new trivial name had first been given to a *different* structure. In some cases, widely accepted synonyms are included (e.g., cyclomulberrin/cyclomulberrochromene[3]).

The listing of accumulation sites forms an essential part of the tables (under the column head "Plant organs"). In particular, attention focuses on the presence of flavonoids in exudates if indicated by the authors (marked with the abbreviation "ext.," if not clear from a term like "bud exudate"). Critical sources are listings such as "whole plant" which could mean anything from aerial parts to inclusion of roots and flowers. Despite our efforts, we received a single reply only from the authors addressed. Hence the correctness of "whole plant" was confirmed only for two flavonols isolated from *Andrographis viscosula*.[7] Details on specific accumulation sites and specific accumulation trends will be discussed in the text relating to the respective tables.

For easier navigation through the tabulated data, some further specifics should be pointed out to the reader. In this context, citing "last edition" refers to Wollenweber.[3]

- Abbreviations such as -OH = hydroxyl, -OMe = methoxy, -Me = methyl are used throughout text and tables. Further abbreviations used to describe the flavonoid structure are explained in a footnote to Table 12.3.
- Basic OH-substituted compounds not (yet) found in nature are noted in brackets, prior to their corresponding methyl derivates.
- Compounds already listed in previous editions, but without being reported from any new source, are only cited by name.
- Products that have been reported for the first time since compilation of the previous tables are marked with an asterisk. The same applies for new trivial names.
- Names missing in previous editions have been added now (also marked by an asterisk), explaining for literature citations older than the beginning of the reporting period (e.g., Ref. 8).
- In column "Plant species," these are grouped by families and are listed alphabetically for each flavonoid. For individual compounds of single publications, a maximum of two species per genus is noted; more than two species are abbreviated by "spp."
- In the same column, flavonoids known so far only in glycosidic combination or in acylated form are marked "Glycoside only" and "Ester only," respectively. No citations or sources are specified in these cases.
- Abbreviations of family names given in the following column should be generally understandable. Rare family names are written in full.
- Synthesis is noted under the head "Plant organs." Some recently synthesized compounds that are likely to be found in nature, sooner or later, are cited in brackets, for example, the 7-methyl ether of 6-hydroxygalangin, or 8-*C*-methylapigenin. These are marked "synthesis only."
- Some previously compiled flavonoids,[6] which are still only known as synthetic products, have been omitted from the current tables.
- Revised structures are indicated referring to respective publications.

12.3 FLAVONES

A total of 309 entries on the distribution of flavones and their methyl ethers are summarized in Table 12.1. The substitution patterns range from unsubstituted flavone to octa-*O*-substituted flavones. As expected, the number of plant species accumulating these structures is

TABLE 12.1
Flavones and Their Methyl Ethers

No.	OH-Substitution	OMe-Substitution	Trivial Name	Plant Species	Family	Plant Organ	Ref.
	Unsubstituted flavone						
1			Flavone				
	Mono-O-substituted flavones						
2	5-OH		Primuletin				
3		5-OMe					
4	(6-OH)	6-OMe					
5	(7-OH)	7-OMe					
6	2'-OH						
7		2'-OMe					
8	(3'-OH)	3'-OMe					
9	(4'-OH)	4'-OMe					
	Di-O-substituted flavones						
10	(5,6-diOH)	6-OMe					
11		5,6-diOMe					
12	5,7-diOH		Chrysin	Anomianthus dulcis	Annonac.	Leaf	131
				Artemisia campestris spp. glutinosa	Asterac.	Aerial p., ext.	132
				Baccharis viminea	Asterac.	Aerial p., ext.	129
				Baccharis viminea	Asterac.	Aerial p., ext.	133
				Mikania hirsutissima	Asterac.	Aerial parts	134
				Heliotropium pycnophyllum	Boragin.	Aerial p., ext.	36
				Eriodictyon sessilifolium	Hydrophyll.	Leaf resin	135
				Mimulus moschata	Scrophul.	Aerial p., ext.	108
				Propolis from Egypt			136
				European Propolis			137
				Bees Wax			138

No.	Name	Substitution (OH)	Substitution (OMe)	Species	Family	Plant part	Ref.
13	Tectochrysin	5-OH	7-OMe	Uvaria rufas	Annonac.	Root	139
				Baccharis viminea	Asterac.	Aerial p., ext.	129
				Baccharis viminea	Asterac.	Aerial p., ext.	133
				Lychnophora markgravii	Asterac.	Aerial parts	140
				Godmania aesculifolia	Bignon.	Aerial p., ext.	141
				Heliotropium pycnophyllum	Boragin.	Aerial p., ext.	36
				Pelargonium crispum	Geraniac.	Leaf exudate	142
				Collinsonia canadensis	Lamiac.	Aerial p., ext.	143
				Hoslundia opposita	Lamiac.	Twigs	144
				Bees wax			138
14			5,7-diOMe	Leptospermum scoparium	Myrtac.	Leaf	21
15	Primetin	5,8-diOH	—				
16		5,2'-diOH	2'-OMe				
17		5-OH (6,7-diOH)	—				
18		7-OH (6,3'-diOH)	6-OMe*	Dalbergia cochinchinensis	Fabac.	Stem	145
19		(6,4'-diOH)	6,3'-diOMe				
20		6-OH (7,8-diOH)	4'-OMe	Glycoside only			
21			7,8-diOMe*	Godmania aesculifolia	Bignon.	Leaf	146
22		7,4'-diOH		Dracaena cinnabari	Agavac.	Resin	147
				Glycyrrhiza eurycarpa	Fabac.	Root	17
				Glycyrrhiza pallidiflora	Fabac.	Root	148
23	Isopratol	4'-OH	7-OMe	Trigonella spp.	Fabac.	Aerial parts	149
24		8,2'-diOH					
25		8-OH					
26		2',5'-diOH					
27		3',4'-diOH	2'-OMe				

continued

TABLE 12.1
Flavones and Their Methyl Ethers — continued

No.	OH-Substitution	OMe-Substitution	Trivial Name	Plant Species	Family	Plant Organ	Ref.
	Tri-O-substituted flavones						
28	5,6,7-triOH		Baicalein				
29	5,7-diOH	6-OMe	Oroxylin A	Gomphrena boliviana	Amaranth.	Whole plant	150
				Scutellaria seleriana	Lamiac.	Aerial parts	151
				Adenostoma sparsifolium	Rosac.	Aerial p., ext.	19
				Bupleurum scorzonerifolium	Umbellif.	Root	152
30	5,6-diOH	7-OMe	Negletein	Desmos chinensis	Anonaceae	Root	62
31	7-OH	5,6-diOMe		Gomphrena boliviana	Amaranth.	Whole plant	150
				Collinsonia canadensis	Lamiac.	Aerial p., ext.	143
32	5-OH	6,7-diOMe	Mosloflavone*	Uvaria rufas	Annonac.	Root	139
				Mosla soochouensis	Lamiac.	Whole herb	153
33		5,6,7-triOMe					
34	5,7,8-triOH		Norwogonin	Scutellaria planifolia	Lamiac.	Root	55
35	5,8-diOH	7-OMe*	Pediflavone*	Didymocarpus pedicellatus	Gesneriac.	Immat. leaf	154
36	5,7-diOH	8-OMe	Wogonin	Tetracera indica	Dilleniac.	Aerial parts	155
				Scutellaria planifolia	Lamiac.	Root	55
				Adenostoma sparsifolium	Rosac.	Aerial p., ext.	19
				Bupleurum scorzonerifolium	Umbellif.	Root	152
37	7-OH	5,8-diOMe		Polygonum senegalense	Polygon.	Leaf surface	156
38	5-OH	7,8-diOMe	7-Methyl-wogonin	Andrographis affinis	Acanthac.	Whole plant	157
				Uvaria rufas	Annonac.	Root	139
39	5-OH	7,8-diOMe	Moslosooflavone*	Nothofagus solandri	Fagac.	Aerial p., ext.	158
				Collinsonia canadensis	Lamiac.	Aerial p., ext.	143
				Mosla soochouensis	Lamiac.	Whole herb	153
40	—	5,7,8-triOMe					
41	5-OH	6,2'-diOMe					
42	(5,6,2'-triOH)	5,6,2'-triOMe					
43	(5,6,3'-triOH)	5,6,3'-triOMe					

No.	OH	OMe	Trivial name	Species	Family	Plant part	Ref.
44	5,7,2'-triOH	5-OMe*		*Scutellaria planifolia*	Lamiac.	Root	55
45	7,2'-diOH		Echioidinin	*Andrographis alata*	Acanthac.	Whole plant	159
46	5,2'-diOH	7-OMe		*Andrographis lineata*	Acanthac.	Whole plant	14
				Andrographis rothii	Acanthac.	Whole plant	13
				Andrographis viscosula	Acanthac.	Whole plant	10
47	5-OH	7,2'-diOMe*		*Andrographis rothii*	Acanthac.	Whole plant	13
48		5,7,2'-triOMe*		*Andrographis viscosula*	Acanthac.	Whole plant	10
49	5,7,4'-triOH		Apigenin	*Many records*			
50	7,4'-diOH	5-OMe	Thevetiaflavon	*Many records*			
51	5,4'-diOH	7-OMe	Genkwanin	*Many records*			
52	5,7-diOH	4'-OMe	Acacetin	*Artemisia afra*	Asterac.	Aerial parts	160
				Artemisia diffusa	Asterac.	Aerial p., ext.	161
				Baccharis trinervis	Asterac.	Leaf	162
				Calea tenuifolia	Asterac.	Aerial parts	163
				Hieracium amplexicaule	Asterac.	Aerial p., ext.	32
				Ophrysosporus charrua	Asterac.	Aerial parts	164
				Eucryphia, 6 spp.	Cunoniac.	Leaf, bud, ext.	165
				Marchesinia brachiata	Hepaticae	Thallus	166
				Monoclea gottschei	Hepaticae	Thallus	22
				Cunila angustifolia	Lamiac.	Aerial parts	167
				Dorystoechas hastata	Lamiac.	Aerial p., ext.	168
				Lycopus virginicus	Lamiac.	Aerial parts	169
				Perowskia spp.	Lamiac.	Aerial p., ext.	170
				Salvia sclarea	Lamiac.	Aerial p., ext.	168
				Salvia syriaca	Lamiac.	Aerial parts	171
				Salvia, 3 spp.	Lamiac.	Aerial p., ext.	41
				Sideritis spp.	Lamiac.	Aerial parts	172
				Teucrium marum, T. polium	Lamiac.	Aerial p., ext.	168
				Mirabilis viscosa	Nyctagin.	Aerial p., ext.	43
53	7-OH	5,4'-diOMe		*Currania robertiana*	Pteridaceae	Frond exud.	173
54	5-OH	7,4'-diOMe		*Escallonia pulverulenta*	Saxifrag.	Aer. p., res. ex.	42

continued

TABLE 12.1
Flavones and Their Methyl Ethers — continued

No.	OH-Substitution	OME-Substitution	Trivial Name	Plant Species	Family	Plant Organ	Ref.
55		5,7,4'-triOMe		*Anarrhinum forskalii*	Scrophul.	Aerial p., ext.	108
				Antirrhinum, 4 spp.	Scrophul.	Aerial p., ext.	108
56	5,8,2'-triOH						
57	7,8,4'triOH						
58	5,2',5'-triOH			Glycoside only			
	(6,2',3'-triOH)						
59	(7,2',4'triOH)	6,2',3'-triOMe					
60	7,3',4'-triOH	7,2',4'-triOMe*		*Albizia odoratissima*	Mimosac.	Root bark	174
61	7,4'-diOH	3'-OMe	Geraldone				
62	7,3'-diOH	4'-OMe*	Farnisin*	*Acacia farnesiana*	Mimosac.	Seed	175
63	3'-OH	7,4'-diOMe	Tithonine	*Albizia odoratissima*	Mimosac.	Root bark	174
64	3'-OH	7,4'-diOMe	Tithonine	*Virola michelli*	Myristic.	Leaf	176
65	7-OH	3',4'-diOMe*		*Launaea asplenifolia*	Asterac.	Whole plant?	8
66		7,3',4'-triOMe				Synthesis	176
67	(3',4',5'-triOH)	3',4',5'-triOMe*		*Primula veris*	Primul.	Flower	177
Tetra-O-substituted flavones (5,6,7,8-tetraOH)							
68	5,8-diOH	6,7-diOMe		*Fissistigma lanuginosum*	Annonac.	Leaf	178
				Betula davurica	Betulaceae	Leaf	179
						Synthesis	24
69	5,7-diOH	6,8-diOMe*		*Scutellaria repens*	Lamiac.	Root	180
						Synthesis	24
						Synthesis	24
70	8-OH	5,6,7-triOMe*	Alnetin	*Godmania aesculifolia*	Bignon.	Aerial p., ext.	141
				Nothofagus cunninghamii	Fagac.	Aerial p., ext.	158
71	5-OH	6,7,8-triOMe		*Lindera lucida*	Laurac.	Twigs	181

No.	OH	OMe	Trivial name	Species	Family	Plant part	Ref.
72		5,6,7,8-tetraOMe		*Godmania aesculifolia*	Bignon.	Leaf	146
		5,6,7,8-tetraOMe		*Nothofagus cunninghamii*	Fagac.	Aerial p., ext.	158
	(5,6,7,2'-tetraOH)						
73	5,7,2'-triOH	6-OMe		*Centaurea jacea*	Asterac.	Aerial p., ext.	182
74	5,6,7,4'-OH		Scutellarein	*Duranta plumieri*	Verben.	Stem	183
75	5,7,4'-triOH	6-OMe	Hispidulin	Many records, mostly from Asteraceae	Asterac.	Aerial p., ext.	133
				Ambrosia ambrosioides	Asterac.	Aerial parts	184
76	5,6,4'-triOH	7-OMe	Sorbifolin	*Onopordon sibthorpianum*	Asterac.	Aerial parts	185
				Mentha × piperita	Lamiac.	Aerial parts	186
				Thymus herba barona	Lamiac.	Aerial parts	
77	5,6,7-triOH	4'-OMe		Many records, mostly from Asteraceae			
78	5,4'-diOH	6,7-diOMe	Cirsimaritin	Many records, mostly from Asteraceae			
79	5,7-diOH	6,4'-diOMe	Pectolinarigenin	*Artemisia argyi*	Asterac.	Aerial parts	187
80	5,6-diOH	7,4'-diOMe	Ladanein	*Marrubium trachyticum*	Lamiac.	Aerial parts	187
				Mentha × piperita	Lamiac.	Aerial parts	185
				Micromeria albanica	Lamiac.	Aerial parts	188
				Nepeta pungens, N. saturejoides	Lamiac.	Leaf surface	189
				Ocinum, 5 spp.	Lamiac.	Leaf surface	190
				Orthosiphon stamineus	Lamiac.	Aerial parts	191
				Salvia cyanescens	Lamiac.	Aerial parts	192
				Salvia hypoleuca, S. stenophylla	Lamiac.	Aerial p., ext.	41
				Salvia syriaca	Lamiac.	Aerial p., ext.	171
81	7-OH	5,6,4'-triOMe		Glycoside only			
82	6-OH	5,7,4'-triOMe		*Orthosiphon stamineus*	Lamiac.	Aerial parts	191
83	5-OH	6,7,4'-triOMe	Salvigenin	Many records, mostly Asteraceae and Lamiaceae			
84		5,6,7,4'-tetraOMe		*Chromolaena odorata*	Asterac.	Aerial p., ext.	193
		5,6,7,4'-tetraOMe		*Citrus sinensis*	Rutac.	Oil	16
		5,6,7,4'-tetraOMe		*Orthosiphon stamineus*	Lamiac.	Aerial parts	191
	(5,7,8,2'-tetraOH)						
85	5,8,2'-triOH	7-OMe	Scutevulin				
86	5,7,2'-triOH	8-OMe		*Ficus altissima*	Morac.	Aerial parts	30

continued

TABLE 12.1
Flavones and Their Methyl Ethers — *continued*

No.	OH-Substitution	OMe-Substitution	Trivial Name	Plant Species	Family	Plant Organ	Ref.
87	5,2'-diOH	7,8-diOMe	Skullcapflavone I	*Andrographis affinis*	Acanthac.	Whole plant	157
				Andrographis elongata	Acanthac.	Whole plant	194
				Andrographis lineata	Acanthac.	Whole plant	14
				Andrographis rothii	Acanthac.	Whole plant	13
				Scutellaria planifolia	Lamiac.	Root	55
88	5,7-diOH	8,2'-diOMe		*Andrographis affinis*	Acanthac.	Whole plant	157
89	7-OH	5,8,2'-triOMe		*Andrographis echioides*	Acanthac.	Whole plant	195
90	5-OH	7,8,2'-triOMe		*Andrographis paniculata*	Acanthac.	Whole plant	11
91	5,7,8,4'-tetraOH		Isoscutellarein	*Baccharis pilularis*	Asterac.	Aerial p., ext.	196
				Odixia achlaena, O. angusta	Asterac.	Aerial p., ext.	197
				Ozothamnus scutellifolius	Asterac.	Aerial p., ext.	197
92	5,8,4'-triOH	7-OMe		*Centaurea chilensis*	Asterac.	Leaf + stem	198
93	5,7,4'-triOH	8-OMe	4'-Hydroxy-wogonin	*Chrysothamnus nauseosus*	Asterac.	Aerial p., ext.	199
				Madia, 3 spp.	Asterac.	Aerial p., ext.	200
				Zinnia acerosa	Asterac.	Aerial p., ext.	133
				Licania densiflora	Chrysobal.	Aerial parts	202
				Scutellaria repens	Lamiac.	Root	180
				Bupleurum scorzonerifolium	Umbellif.	Root	152
				Verbena littoralis	Verben.	Aerial parts	203
94	5,7,8-triOH	4'-OMe	Takakin	*Odixia achlaena, O. angusta*	Asterac.	Aerial p., ext.	197
	5,7,8-triOH	4'-OMe	Takakin	*Ozothamnus scutellifolius*	Asterac.	Aerial p., ext.	197
95	5,4'-diOH	7,8-diOMe					
96	5,8-diOH	7,4'-diOMe*	Bucegin	*Helicteres isora*	Sterculiac.	Leaf	204
97	5,7-diOH	8,4'-diOMe		*Chrysothamnus nauseosus*	Asterac.	Aerial p., ext.	199
				Madia sativa	Asterac.	Aerial p., ext.	133
				Madia, 4 spp.	Asterac.	Aerial p., ext.	201
				Zinnia acerosa	Asterac.	Aerial p., ext.	133

No.	OH	OMe	Name	Species	Family	Plant part	Ref.
98	5-OH	7,8,4'-triOMe		Eucryphia jinksi	Cunoniac.	Leaf, bud, ext.	165
				Calceolaria irazuensis	Scrophul.	Aerial p., ext.	108
				Asterella blumeana	Hepaticae	Thallus	205
				Citrus reticulata	Rutac.	Fruit peel	206
				Calceolaria, 3 spp.	Scrophul.	Aerial p., ext.	207
				Calceolaria irazuensis	Scrophul.	Aerial p., ext.	108
99		5,7,8,4'-tetraOMe		Citrus sinensis	Rutac.	Oil	16
100	5,7,2',5'-tetraOH*	7,2',5'-triOMe*		Scutellaria baicalensis	Lamiac.	Root	208
101	5-OH	5,7,2'5'-tetraOMe*		Andrographis neesiana	Acanthac.	Whole plant	12
102				Andrographis rothii	Acanthac.	Whole plant	13
103	(5,6,2',6'-tetraOH)	6,2',6'-triOMe	Zapotinin	Casimiroa tetrameria	Rutac.	Leaf	209
104	5-OH	5,6,2',6'-tetraOMe	Zapotin	Casimiroa tetrameria	Rutac.	Leaf	209
105	(5,6,3',5'-tetraOH)	5,6,3',5'-tetraOMe	Cerosillin				
106	5,7,2',3'-tetraOH		Norartocarpetin				
107	5,7,2',4'-tetraOH		Artocarpetin				
108	5,2',4'-triOH	7-OMe		Scutellaria planifolia	Lamiac.	Root	55
109	5-OH	7,2',4'-triOMe		Andrographis viscosula	Acanthac.	Whole plant	10
110		5,7,2',4'-tetraOMe		Andrographis elongata	Acanthac.	Whole plant	194
111	5,7,2',6'-tetraOH						
112	5,2',6'-triOH	7-OMe*					
113	5,7,2'-triOH	6'-OMe					
114	5-OH	7,2'6'triOMe*		Andrographis paniculata	Acanthac.	Whole plant	11
115	5,7,3',4'-tetraOH		Luteolin	Many records			
116	7,3',4'-triOH	5-OMe		Arnica longifolia	Asterac.	Flower	210
117	5,3',4'-triOH	7-OMe		Artemisia barrelieri	Asterac.	Aerial p., ext.	132
				Dabautia arborea	Asterac.	Leaf exudate	211
				Wunderlichia crulsiana	Asterac.	Aerial parts	212
				Heliotropium stenophyllum	Boragin.	Leaf exudate	213
				Nonea lutea, N. pulla	Boragin.	Aerial p., ext.	36

continued

TABLE 12.1
Flavones and Their Methyl Ethers — *continued*

No.	OH-Substitution	OMe-Substitution	Trivial Name	Plant Species	Family	Plant Organ	Ref.
118	5,7,4'-triOH	3'-OMe	Chrysoeriol	*Eucryphia lucida, E. jinksii*	Cunoniac.	Leaf, bud, ext.	165
				Salvia hypoleuca	Lamiac.	Aerial p., ext.	41
				Salvia sclarea	Lamiac.	Aerial p., ext.	168
				Sideritis spp.	Lamiac.	Aerial parts	172
				Teucrium marum	Lamiac.	Aerial p., ext.	168
				Antirrhinum, 3 spp.	Scrophul.	Aerial p., ext.	108
				Petunia parviflora	Solanac.	Aerial p., ext.	214
119	5,7,3'-triOH	4'-OMe	Diosmetin	Many records			
				Artemisia iwayomogi, A. molinieri	Asterac.	Aerial p., ext.	161
				Artemisia caerulescens	Asterac.	Aerial p., ext.	132
				Dubautia arborea	Asterac.	Leaf exudate	211
				Eupatorium altissimum	Asterac.	Aerial p., ext.	215
				Ozothamnus scutellifolius	Asterac.	Aerial p., ext.	197
				Tithonia calva	Asterac.	Aerial p., ext.	182
				Nonea rosea	Boragin.	Aerial p., ext.	36
				Aeonium glutinosum	Crassul.	Aerial p., ext.	216
				Cyperus alopecuroides	Cyperac.	Inflor.	217
				Acacia farnesiana	Mimosac.	Seed	175
120	7,4'-diOH	5,3'-diOMe		*Hypericum perforatum*	Guttiferae	Callus	218
121	7,3'-diOH	5,4'-diOMe*		*Phyllospadix japonica*	Zosterac.	Whole plant	219
122	5,4'-diOH	7,3'-diOMe	Velutin	*Artemisia caerulescen*	Asterac.	Aerial p., ext.	132
				Artemisia iwayomog	Asterac.	Aerial p., ext.	161
				Artemisia oliveriana	Asterac.	Aerial p., ext.	220
				Bahia glandulos	Asterac.	Aerial parts	221
				Bracteantha viscosa	Asterac.	Aerial p., ext.	222
				Haplopappus baylahuen	Asterac.	Aerial p., ext.	223
				Helichrysum bracteatum	Asterac.	Stem and leaf resin	32
				Heterotheca pilosa	Asterac.	Aerial p., ext.	182
				Madia sativa	Asterac.	Aerial p., ext.	133
				Senecio viscosa	Asterac.	Aerial p., ext.	32

No.	OH	OMe	Trivial name	Species	Family	Plant part	Ref.
				Nonea lutea, N. pulla	Boragin.	Aerial p., ext.	36
				Eucryphia, 4 spp.	Cunoniac.	Leaf, bud, ext.	165
				Monoclea gottschei	Hepaticae	Thallus	22
				Eriodictyon sessilifolium	Hydrophyll.	Leaf resin	135
				Salvia candidissima	Lamiac.	Aerial parts	224
				Salvia chinopeplica	Lamiac.	Leaf	225
				Salvia sclarea	Lamiac.	Aerial p., ext.	168
				Teucrium marum	Lamiac.	Aerial p., ext.	168
				Kitaibelia vitifolia	Malvac.	Aerial p., ext.	226
				Mirabilis viscosa	Nyctagin.	Aerial p., ext.	43
				Antirrhinum, 3 spp.	Scrophul.	Aerial p., ext.	108
				Petunia parviflora	Solanac.	Aerial p., ext.	214
				Salpiglossis sinuata	Solanac.	Aerial p., ext.	214
				Lethedon tannaensis	Thymelaeac.	Leaf	227
				Lantana montevidensis	Verben.	Aerial p., ext.	228
123	5,3'-diOH	7,4'-diOMe	Pilloin	*Artemisia iwayomogi*	Asterac.	Aerial p., ext.	161
				Baccharis trinervis	Asterac.	Aerial p., ext.	162
				Onoprodon laconicum	Asterac.	Leaf	184
				Godmania aesculifolia	Bignon.	Aerial parts	141
				Eucryphia milleganii	Cunoniac.	Aerial p., ext.	165
				Lycopus virginicus	Lamiac.	Leaf, bud, ext.	169
				Notholaena nivea	Pteridaceae	Aerial parts	45
				Antirrhinum braun-blanquetii, A. graniticum	Scrophul.	Frond exud.	108
124	5,7-diOH	3',4'-diOMe		*Calea tenuifolia*	Asterac.	Aerial p., ext.	163
				Chrysothamnus viscidiflorus	Asterac.	Aerial parts	199
				Monoclea gottschei, M. forsteri	Hepaticae	Aerial p., ext.	22
				Asarina barkleyana	Scrophul.	Thallus	108
125	5-OH	7,3',4'-triOMe		*Baccharis trinervis*	Asterac.	Aerial p., ext.	162
				Calea tenuifolia	Asterac.	Leaf	163
				Nonea pulla	Boragin.	Aerial parts	36
				Eucryphia lucida, E.milleganii	Cunoniac.	Aerial p., ext.	165
				Orthosiphon stamineus	Lamiac.	Leaf, bud, ext.	191
				Sideritis spp.	Lamiac.	Aerial parts	172
				Teucrium botrys	Lamiac.	Aerial parts	168

continued

TABLE 12.1

Flavones and Their Methyl Ethers — continued

No.	OH-Substitution	OMe-Substitution	Trivial Name	Plant Species	Family	Plant Organ	Ref.
126		5,7,3',4'-tetraOMe		Kitaibelia vitifolia	Malvac.	Aerial p., ext.	226
				Anarrhinum forskalii	Scrophul.	Aerial p., ext.	108
				Antirrhinum, 5 spp.	Scrophul.	Aerial p., ext.	108
				Lethedon tannaensis	Thymelaeac.	Leaf	227
				Lantana montevidensis	Verben.	Aerial p., ext.	228
127	(6,7,3',4'-tetraOH) 7,3'-diOH	6,4'-diOMe	Abrectorin				
128	(6,7,3',5'-tetraOH) 7-OH	6,3',5'-triOMe*	Grantionin*	Inula grantioides	Asterac.	Aerial parts	34
129	(7,2',4',5'-tetra-OH)	7,2',4',5'-tetraOMe*		Calliandra californica	Fabac.	Root	229
130	7,3',4',5'-tetraOH						
Penta-O-substituted flavones							
131	(5,6,7,8,2'-pentaOH) (5,8,2'-triOH)	6,7-diOMe		Scutellaria baicalensis	Lamiac.	Root	[a]
132	5,7,2'-triOH	6,8-diOMe*		Scutellaria repens	Lamiac.	Root	180
133	5,2'-diOH	6,7,8-triOMe	Tenaxin 1			Synthesis	24
134	(5,6,7,8,4'-pentaOH) 5,8,4'-triOH	6,7-diOMe	Isothymusin	Agastache barberi	Lamiac.	Leaf surface	189
				Becium grandiflorum	Lamiac.	Leaf, ext.	230
				Nepeta spp.	Lamiac.	Leaf surface	189
				Ocimum gratissimum	Lamiac.	Aerial parts	231
				Prunus cerasus	Rosac.	Fruit	232
					—	Synthesis	24
135	5,7,4'-triOH	6,8-diOMe	Desmethyl-sudachitin	Ambrosia trifida	Asterac.	Aerial p., ext.	233
				Madia capitata	Asterac.	Aerial p., Ext.	200
				Biebersteinia orphanidis	Bieberstein.	Leaf surface	234
				Scutellaria repens	Lamiac.	Root	180

No.	OH	OMe	Name	Species	Family	Part	Ref.
136	5,6,4'-triOH	7,8-diOMe	Thymusin	*Mentha × piperita*	Lamiac.	Aerial parts	185
				Nepeta assurgens	Lamiac.	Leaf surface	189
				Origanum × intercedens	Lamiac.	Glandul. hairs	235
				Thymus herba baron	Lamiac.	Aerial parts	186
137	5,7,8-triOH	6,4'-diOMe*	Pilosin *	*Ocimum americanum var. pilosum*	Lamiac.	Leaf surface	190
138	5,6,8-triOH	7,4'-diOMe					
139	5,6,7-triOH	8,4'-diOMe*					
140	5,4'-diOH	6,7,8-triOMe	Xanthomicrol	*Calycadenia truncata*	Asterac.	Leaf exudate	236
				Bracteantha viscosa	Asterac.	Aerial p., ext.	222
				Helichrysum, 8 spp.	Asterac.	Whole plant	237
				Hymenoxis scaposa	Asterac.	Aerial parts	238
				Varthemia iphionoides	Asterac.	Whole plant	239
				Ononis natrix	Fabac.	Aerial parts	240
				Cunila angustif., C. incana	Lamiac.	Aerial pars	167
				Dracunculus kotschyi	Lamiac.	Leaf surface	189
				Nepeta, 5 spp.	Lamiac.	Leaf surface	189
				Ocimum gratissimum	Lamiac.	Aerial parts	231
				Satureja montana	Lamiac.	Aerial p., ext.	241
				Thymus herba barona	Lamiac.	Aerial parts	186
141	(5,8-diOH	6,7,4'-triOMe)	"Pedunculin"	*Tithonia*, 5 spp.	Asterac.	Vegetative p.	[b]
				Ocimum, 4 spp.	Lamiac.	Leaf surface	190
142	5,7-diOH	6,8,4'-triOMe	Nevadensin	*Ambrosia trifida*	Asterac.	Aerial p., ext.	233
				Baccharis griesebachii	Asterac.	Res. exud.	242
				Madia capitata	Asterac.	Aerial p., ext.	200
				Tithonia calva	Asterac.	Aerial p., ext.	182
				Simsia cronquistii	Asterac.	Aerial parts	243
				Viguiera rosei	Asterac.	Aerial p., ext.	244
				Biebersteinia orphanidis	Bieberstein.	Leaf surface	234
				Ocimum, 5 spp.	Lamiac.	Leaf surface	190
				Rosa centifolia cv. muscosa	Rosac.	Aerial p., ext.	18
				Tamarix dioica	Tamaric.	Aerial parts	245
						Synthesis	24
143	5,6-diOH	7,8,4'-triOMe	Pebrellin*	*Mentha × piperita*	Lamiac.	Aerial parts	185
						Synthesis	24

continued

TABLE 12.1
Flavones and Their Methyl Ethers — continued

No.	OH-Substitution	OMe-Substitution	Trivial Name	Plant Species	Family	Plant Organ	Ref.
144	4'-OH	5,6,7,8-tetraOMe*		Nothofagus menziesii	Fagac.	Aerial p., ext.	158
				Citrus reticulata	Rutac.	Peel	206
145	7-OH	5,6,8,4'-tetraOMe				Only synth.	24
	6-OH	5,7,8,4'-tetraOMe*					
146	5-OH	6,7,8,4'-tetraOMe	Gardenin B	Baccharis griesebachii	Asterac.	Res. exud.	242
				Biebersteinia orphanidis	Bierberstein.	Leaf surface	234
				Godmania aesculifolia	Bignon.	Aerial p., ext.	141
				Ononis natrix	Fabac.	Aerial parts	240
				Cunila angustifolia, C. fasciculata	Lamiac.	Aerial parts	167
				Nepeta transcaucasica	Lamiac.	Aerial p., ext.	246
				Ocimum, 7 spp.	Lamiac.	Leaf surface	190
				Satureja montana	Lamiac.	Aerial p., ext.	241
				Rosa centifolia cv. muscosa	Rosac.	Aerial p., ext.	18
				Tamarix dioica	Tamaric.	Aerial parts	245
147	(5,6,7,2',4'-pentaOH)	5,6,7,8,4'-pentaOMe	Tangeretin	Citrus sinensis	Rutac.	Fruit peel oil	16
		5,6,7,8,4'-pentaOMe	Tangeretin	Citrus "Dancy tangerine"	Rutac.	Leaf	114
148	5,7,2'-triOH	6,4'-diOMe*	Tamaridone*	Tamarix dioica	Tamaric.	Aerial parts	245
149	5,6,7,3',4'-pentaOH		6-Hydroxy-luteolin	Tabebuia caraiba	Bignon.	Leaf	247
150	6,7,3',4'-tetraOH	5-OMe*	Carajuflavone*	Arrabidaea chica f. cuprea	Bignon.	Leaf	248
151	5,7,3',4'-tetraOH	6-OMe	Nepetin	Many records, mostly from Asteraceae			
152	5,6,3',4'-tetraOH	7-OMe	Pedalitin	Leiothrix flavescens	Eriocaul.	Capitula	15
				Monoclea gottschei	Hepaticae	Thallus	22
				Mentha pulegium	Lamiac.	Leaf surface	249
				Salvia blepharophylla	Lamiac.	Leaf	250
				Mentha suaveolens	Lamiac.	Leaf surface	249
153	5,6,7,4'-tetraOH	3'-OMe	Nodifloretin				
154	5,6,7,3'-tetraOH	4'-OMe					
155	7,3',4'-triOH	5,6-diOMe					
156	5,3',4'-triOH	6,7-diOMe	Cirsiliol	Many records, mostly Asterac. and Lamiac.			
157	5,7,4'-triOH	6,3'-diOMe	Jaceosidin	Many records, mostly Asterac. and Lamiac.			

No.				Species	Family	Plant part	Ref.
158	5,7,3'-triOH	6,4'-diOMe	Desmethyl-centaureidin	Arnica longifolia	Asterac.	Flower	210
				Baccharis gaudichaudiana	Asterac.	Aerial parts	251
				Sideritis spp.	Lamiac.	Aerial parts	172
				Duranta plumieri	Verben.	Stem	183
159	5,6,4'-triOH	7,3'-diOMe		Arctotis venusta	Asterac.	Aerial p., ext.	32
				Artemisia argyi	Asterac.	Aerial parts	186
				Monoclea gottschei	Hepaticae	Thallus	22
				Acinos alpinus, Ac. suaveolens	Lamiac.	Aerial parts	252
				Calamintha nepeta	Lamiac.	Aerial parts	252
				Mentha, 3 spp.	Lamiac.	Aerial parts	252
				Micromeria, 3 spp.	Lamiac.	Aerial parts	252
				Ocinum lamiifolium	Lamiac.	Leaf surface	190
				Origanum, 7 spp.	Lamiac.	Aerial parts	252
				Origanum × intercedens	Lamiac.	Glandul. hairs	235
				Thymbra capitata	Lamiac.	Aerial parts	253
				Thymbra capitata, T. spicata	Lamiac.	Aerial parts	252
160	5,6,3'-triOH	7,4'-diOMe		Ocinum lamiifolium	Lamiac.	Leaf surface	190
	5,6,3'-triOH	7,4'-diOMe		Salvia blepharophylla	Lamiac.	Leaf	250
161	3',4'-diOH	5,6,7-triOMe*	Cirsilineol	Arrabidaea brachypoda	Bignon.	Epicut. wax	254
162	5,4'-diOH	6,7,3'-triOMe		Artemisia oliveriana	Asterac.	Aerial p., ext.	220
				Centaurea macrocephala	Asterac.	Aerial p., ext.	255
				Oncosiphon grandiflorum	Asterac.	Aerial p., ext.	32
				Palafoxia spacelata	Asterac.	Aerial p., ext.	182
				Tecomella undulata	Bignon.	Leaf	256
				Hyssopus officinalis	Lamiac.	Aerial p., ext.	170
				Lycopus europaeus	Lamiac.	Aerial p., ext.	170
				Ocimum, spp.	Lamiac.	Leaf surface	190
				Teucrium alyssifolium	Lamiac.	Aerial parts	257
				Teucrium marum	Lamiac.	Aerial p., ext.	168
				Thymus herba barona	Lamiac.	Aerial parts	186
				Ziziphoras hispanica	Lamiac.	Aerial p., ext.	170
				Lippia citriodora	Verben.	Aerial parts	258
163	5,3'-diOH	6,7,4'-triMe	Eupatorin			Synthesis	259
				Dorema aucheri	Umbellif.	Aerial p., ext.	260

continued

TABLE 12.1
Flavones and Their Methyl Ethers — continued

No.	OH-Substitution	OMe-Substitution	Trivial Name	Plant Species	Family	Plant Organ	Ref.
				Achillea santolina	Asterac.	Veget. parts	261
				Arnica longifolia	Asterac.	Flower	210
				Chromolaena arnottiana	Asterac.	Aerial parts	262
				Eupatorium altissimum	Asterac.	Aerial p., ext.	215
				Mikania minima	Asterac.	Aerial parts	263
				Vernonia saligna	Asterac.	Aerial parts	264
				Mentha × piperita	Lamiac.	Aerial parts	185
				Ocimum, spp.	Lamiac.	Leaf surface	190
				Orthosiphon stamineus	Lamiac.	Aerial parts	191
				Salvia macrosiphon, S. mirzayani	Lamiac.	Aerial p., ext.	41
				Salvia syriaca	Lamiac.	Aerial parts	171
				Sideritis spp.	Lamiac.	Aerial parts	172
				Trichostema lanata	Lamiac.	Aerial p., ext.	41
164	5,7-diOH	6,3',4'-triOMe	Eupatilin	Artemisia giraldii	Asterac.	Aerial parts	265
				Artemisia ludoviciana var. mexicana	Asterac.	Aerial parts	266
				Artemisia oliveriana	Asterac.	Aerial p., ext.	220
				Artemisia mongolica	Asterac.	Aerial parts	265
				Artemisia umbelliformis	Asterac.	Gland. trich.	267
				Artemisia nitida, A. verlotiorum	Asterac.	Aerial p., ext.	132
				Baccharis gaudichaudiana	Asterac.	Aerial parts	251
				Inula brittanica	Asterac.	Aerial p., ext.	196
				Tanacetum polycephalum	Asterac.	Aerial p., ext.	220
				Monoclea gottschei	Hepaticae	Thallus	22
				Salvia sclarea	Lamiac.	Aerial p., ext.	168
				Sideritis spp.	Lamiac.	Aerial parts	172
165	5,6-diOH	7,3',4'-triOMe		Artemisia argyri	Asterac.	Aerial parts	186
				Calamintha nepeta	Lamiac.	Aerial parts	252
				Mentha × piperita	Lamiac.	Aerial parts	185
				Micromeria, 3 spp.	Lamiac.	Aerial parts	252
				Ocimum lamiifolium	Lamiac.	Leaf surfce	190

No.	Substitution	Name	Species	Family	Plant part	Ref.
166	4'-OH 5,6,7,3'-tetraOMe*		*Origanum onites*	Lamiac.	Aerial parts	252
167	3'-OH 5,6,7,4'-tetraOMe		*Salvia sclarea*	Lamiac.	Aerial p., ext.	168
168	7-OH 5,6,3',4'-tetraOMe		*Salvia syriaca*	Lamiac.	Aerial parts	171
169	6-OH 5,7,3',4'-tetraOMe		*Satureja thymbra*	Lamiac.	Aerial parts	252
			Thymbra capitata, T. spicata	Lamiac.	Aerial parts	252
170	5-OH 6,7,3',4'-tetraOMe	Ageconyflavon B*	*Ageratum conyzoides*	Asterac.	Whole plant	268
			Achillea conferta	Asterac.	Aerial parts	269
			Achillea santolina	Asterac.	Aerial parts	53
			Artemisia argyri	Asterac.	Aerial parts	186
			Artemisia austriaca	Asterac.	Aerial parts	270
			Artemisia giraldii	Asterac.	Aerial parts	265
			Artemisia sieversiana	Asterac.	Aerial parts	271
			Artemisia mongolica, A. verlotiorum	Asterac.	Aerial p., ext.	132
			Centaurea macrocephala	Asterac.	Aerial p., ext.	255
			Chromolaena arnottiana	Asterac.	Aerial parts	262
			Eupatorium altissimum	Asterac.	Aerial p., ext.	215
			Lagophylla glandulosa	Asterac.	Leaf exudate	236
			Parthenium incanum	Asterac.	Aerial p., ext.	182
			Cunila angustifolia, C. incana	Lamiac.	Aerial parts	167
			Mentha longifolium	Lamiac.	Aerial parts	272
			Mentha pulegium	Lamiac.	Leaf surface	249
			Micromeria albanica	Lamiac.	Aerial parts	188
			Ocimum americanum var. americanum	Lamiac.	Leaf surface	190
			Orthosiphon stamineus	Lamiac.	Aerial parts	191
			Salvia dominica	Lamiac.	Aerial p., ext.	168
			Salvia macrosiphon, S.mirzayani	Lamiac.	Aerial p., ext.	41
			Salvia syriaca	Lamiac.	Aerial p., ext.	168
			Sideritis spp.	Lamiac.	Aerial parts	172
			Teucrium alyssifolium	Lamiac.	Aerial parts	257
			Teucrium botrys	Lamiac.	Aerial p., ext.	168
			Teucrium pseudochamaepitys	Lamiac.	Aerial parts	273

continued

TABLE 12.1
Flavones and Their Methyl Ethers — continued

No.	OH-Substitution	OMe-Substitution	Trivial Name	Plant Species	Family	Plant Organ	Ref.
171		5,6,7,3',4'-pentaOMe	Sinensetin	Ziziphoras hispanica	Lamiac.	Aerial p., ext.	170
				Citrus "Dancy tangerine"	Rutac.	Leaf	114
				Chromolaena odorata	Asterac.	Aerial p., ext.	193
				Conoclinium coelestinum	Asterac.	Aerial p., ext.	215
				Eupatorium coelestinum	Asterac.	Aerial parts	274
				Orthosiphon stamineus	Lamiac.	Aerial parts	191
				Citrus sinensis	Rutac.	Fruit peel oil	16
				Citrus "Dancy tangerine"	Rutac.	Leaf	114
	(5,7,8,2',3'-pentaOH)						
172	5,2',3'-triOH	7,8-diOMe	(Norwightin)	Glycoside only			
173	5,3'-diOH	7,8,2'-triOMe	Wightin				
174	5-OH	7,8,2',3'-tetraOMe		Mentha longifolia	Lamiaceae	Aerial parts	272
	(5,7,8,2',4'-pentaOH)						
175	5-OH	7,8,2',4'-tetraOMe					
	(5,7,8,2',5'-pentaOH)						
176	5,2',5'-triOH	7,8-diOMe	Rehderianin I				
177	5-OH	7,8,2',5'tetraOMe*		Andrographis affinis	Acanthac.	Whole plant	157
	(5,7,8,2',6'-pentaOH)						
178	5,2',6'-triOH	7,8-diOMe					
179	5,7,2'-triOH	8,6-diOMe					
180	5,2'-diOH	7,8,6-triOMe	Rivularin	Scutellaria baicalensis	Lamiac.	Root	208
181	5,7-diOH	8,2',6'-triOMe					
182	5-OH	7,8,2',6'-tetraOMe	Altisin				
183	5,7,8,3',4'-pentaOH		Hypolaetin	Licania pyrifolia	Chrysobal.	Aerial parts	202
184	5,7,3',4'-tetraOH	8-OMe	Onopordin	Centaurea chilensis	Asterac.	Leaf+stem	198
				Madia, 4 spp.	Asterac.	Aerial p., ext.	200
				Onopordum laconicum	Asterac.	Aerial parts	184
				Viguiera decurrens	Asterac.	Aerial p., ext.	244
185	5,7,8,4'-tetraOH	3-OMe		Glycoside only			
186	5,7,8,3'-tetraOH	4-OMe		Glycoside only			

No.	OH	Name	OMe	Species	Family	Part	Ref.
187	5,7,4'-triOH		8,3'-diOMe	Conyza spp.	Asterac.	Resin. exud.	42
				Hemizonia lutescens	Asterac.	Aerial p., ext.	133
188	8,3'-diOH		5,7,4'-triOMe*	Verbena littoralis	Verben.	Aerial parts	203
189	5,7-diOH		8,3',4'-triOMe	Cowania mexicana var. Stansburiana	Rosac.	Aerial p., ext.	19
190	5-OH		7,8,3',4'-tetraOMe	Cowania mexicana var. Stansburiana	Rosac.	Aerial p., ext.	19
191	(5,7,8,3',4'-penta-OH)	Isosinensetin	5,7,8,3',4'-pentaOMe	Citrus sinensis	Rutac.	Fruit peel oil	16
				Ficus altissima	Morac.	Aerial parts	30
192	5,7-diOH		8,3',5'-triOMe*	Limnophila rugosa	Scrophul.	Aerial parts	275
					—		—
193	(5,6,2',3',4'-penta-OH)		5,6,2',3',4'-pe-OMe*	Casimiroa tetrameria	Rutac.	Leaf	209
					—		—
194	(5,6,2',3',6'-penta-OH)		5,6,2',3',6'-pe-OMe	Casimiroa tetrameria	Rutac.	Leaf	209
195	(5,6,2',4',5'-penta-OH) / 5,4',5'-triOH		6,2'-diOMe*	Teucrium quadrifarium	Lamiac.	?	276
196	(5,6,3',4',5'-pentaOH)	Cerosillin B	5,6,3',4',5'-pentaOMe				
197	(5,7,2',3',4'-penta-OH)*		5,7,2',3',4'-penta-OMe*	Andrographis lineata	Acanthac.	Whole plant	14
198	5,7,2',4',5'-pentaOH	Isoetin (Hieracin)		Glycoside only			
199	5-OH		7,2',4',5'-tetraOMe	Artemisia campestris ssp. glutinosa	Asterac.	Aerial parts	277
				Calliandra californica	Fabac.	Root	229
200	(5,7,2',4',6'-pentaOH)		5,7,2',4',6'-pentaOMe*	Andrographis viscosula	Acanthac.	Whole plant	10
201	5,7,3',4',5'-pentaOH	Tricetin		Eucalyptus globulus	Myrtac.	Pollen	20
				Kunzea ericoides	Myrtac.	Pollen	20
				Leptospermum scoparium	Myrtac.	Pollen	20
				Metrosideros excelsa	Myrtac.	Pollen	20
				Metrosideros umbellata	Myrtac.	Pollen	20
202	5,7,4',5'-tetraOH	Selgin	3'-OMe	Artemisia caerulescens	Asterac.	Aerial p., ext.	132
203	5,7,3',5'-tetraOH		4'-OMe	Nonea lutea, N. pulla	Boragin.	Aerial p., ext.	36
				Asarina procumbens	Scrophul.	Aerial p., ext.	278
204	5,7,5'-triOH	Apometzgerin	3',4'-diOMe	Nonea pulla	Boragin.	Aerial p., ext.	36
				Asarina procumbens	Scrophul.	Aerial p., ext.	278

continued

TABLE 12.1
Flavones and Their Methyl Ethers — continued

No.	OH-Substitution	OMe-Substitution	Trivial Name	Plant Species	Family	Plant Organ	Ref.
205	5,7,4'-triOH	3',5'-diOMe	Tricin	Epimedium brevicornum	Berberid.	Aerial parts	279
				Agelaea pentagyna	Connarac.	Leaf	280
				Castilleja fissifolia	Scrophul.	Aerial p., ext.	278
				Xerophyta retinervis	Velloziac.	Leaf	105
206	5,5'-diOH	7,3',4'-triOMe*	Lethedocin*	Lethedon tannaensis	Thymelaeac.	Leaf	227
207	5,4'-diOH	7,3',5'-triOMe*	7-Methyl-tricin*	Centaurea incana	Asterac.	Aerial parts	281
				Betonica officinalis (=Stachys)	Lamiac.	Aerial parts	282
				Lethedon tannaensis	Thymelaeac.	Leaf	227
208	5,7-diOH	3',4',5'-triOMe		Nonea pulla	Boragin.	Aerial p., ext.	36
				Asarina procumbens	Scrophul.	Aerial p., ext.	108
209	5-OH	7,3',4',5'-tetraOMe	Corymbosin	Centaurea incana	Asterac.	Aerial parts	281
				Walsura piscidia	Meliac.	Aerial parts	283
210		5,7,3',4',5'-pentaOMe		Ficus maxima	Morac.	Leaf	77
				Murraya paniculata	Rutac.	Leaf	284
				Neoraputia paraensis	Rutac.	Aerial parts	285
				Neoraputia paraensis	Rutac.	Fruit	286
	(5,8,3',4',5'-pentaOH)*						
211	5,5'-diOH	8,3',4'-triOMe*		Artemisia giraldii	Asterac.	Aerial parts	287
	(6,7,3',4',5'-pentaOH)						
212	6,7,4'-triOH	3',5'-diOMe*	Prosogerin E				
213	6,7-diOH	3',4',5'-triOMe	Prosogerin D	Artemisia giraldii	Asterac.	Aerial parts	287
214	7-OH	6,3',4',5'-tetraOMe	Prosogerin C				
215		6,7,3',4',5'-pentaOMe					
	Hexa-O-substituted flavones						
	(5,6,7,8,2',4'-hexaOH)						
216	5,2',4'-triOH	6,7,8-triOMe*	Tamadone*	Tamarix dioica	Tamaric.	Aerial parts	245
	(5,6,7,8,2',5'-hexaOH)						
217	5,2',5'-triOH	6,7,8-triOMe					
	(5,6,7,8,2',6'-hexaOH)						
218	5,6,2',6'-tetraOH	7,8-diOMe					

No.	Hydroxyl	Methoxyl	Trivial name	Species	Family	Plant part	15
219	5,2',6'-triOH	6,7,8-triOMe					
220	5,6,2'-triOH	7,8,6'-triOMe					
221	5,2'-diOH	6,7,8,6'-tetraOMe	Skullcapflavon II				
222	5,6,7,8,3',4'-hexaOH*			*Leiothrix flavescens*	Eriocaul.	Capitula	15
223	5,7,8,3',4'-pentaOH	6-OMe		Glycoside only			
224	5,6,7,3',4'-pentaOH	8-OMe		*Nepeta*, 6 spp.	Lamiac.	Leaf surface	189
225	5,8,3',4'-tetraOH	6,7-diOMe*		*Helichrysum*, 8 spp.	Asterac.	Whole plant	237
226	5,7,3',4'-tetraOH	6,8-diOMe		*Helichrysum*, 8 spp.	Asterac.	Whole plant	237
227	5,6,7,8-tetraOH	3',4'-diOMe					
228	5,3',4'-triOH	6,7,8-triOMe	Sideritiflavon	*Madia*, 3 spp.	Asterac.	Aerial p., ext.	200
				Mentha spicata	Lamiac.	Leaf	288
				Thymus herba barona	Lamiac.	Aerial parts	186
229	5,8,4'triOH	6,7,3'triOMe		*Mentha longifolia*	Lamiac.	Aerial parts	272
230	5,7,4'-triOH	6,8,3'-triOMe	Sudachitin	*Tithonia calva*	Asterac.	Aerial p., ext.	182
				Viguiera rosei	Asterac.	Aerial p., ext.	244
				Biebersteinia orphanidis	Bieberstein.	Leaf surface	234
231	5,7,3'-triOH	6,8,4'-triOMe	Acerosin	*Calycadenia*, 3 spp.	Asterac.	Leaf exudate	236
				Biebersteinia orphanidis	Bieberstein.	Leaf surface	234
232	5,6,4'-triOH	7,8,3'-triOMe	Thymonin	*Madia capitata*	Asterac.	Aerial p., ext.	200
				Acinos alpinus, Ac. suaveolens	Lamiac.	Aerial parts	252
				Calamintha nepeta, C. sylvatica	Lamiac.	Aerial parts	252
				Mentha, 3 spp.	Lamiac.	Aerial parts	252
				Mentha × piperita	Lamiac.	Aerial parts	185
				Mentha spicata	Lamiac.	Leaf	288
				Micromeria, 3 spp.	Lamiac.	Aerial parts	252
				Origanum, 9 spp.	Lamiac.	Aerial parts	252
				Origanum × intercedens	Lamiac.	Glandul. hairs	235
				Satureja salzmannii	Lamiac.	Aerial parts	252
				Thymus spp.	Lamiac.	Aerial parts	252
233	8,3'-diOH	5,6,7,4'-tetraOMe*		*Vernonia saligna*	Asterac.	Aerial parts	264
234	5,4'-diOH	6,7,8,3'-tetraOMe		*Calycadenia multiglandulosa*	Asterac.	Leaf exudate	236
				Madia dissitiflora	Asterac.	Aerial p., ext.	200
				Cleome droserifolia	Capparid.	Aerial parts	289

continued

TABLE 12.1
Flavones and Their Methyl Ethers — *continued*

No.	OH-Substitution	OMe-Substitution	Trivial Name	Plant Species	Family	Plant Organ	Ref.
235	5,3'-diOH	6,7,8,4'-tetraOMe	Gardenin D	*Cunila angustifolia*	Lamiac.	Aerial parts	167
236	5,7-diOH	6,8,3',4'-tetraOMe	Hymenoxin	*Micromeria albanica*	Lamiac.	Aerial parts	188
				Thymus herba barona	Lamiac.	Aerial parts	186
				Calycadenia truncata, C. villosa	Asterac.	Leaf exudate	236
				Tithonia calva	Asterac.	Aerial p., ext.	182
				Viguiera rosei	Asterac.	Aerial p., ext.	244
				Biebersteinia orphanidis	Bieberstein.	Leaf surface	234
				Ononis natrix	Fabac.	Aerial parts	240
				Ocimum × citriodorum	Lamiac.	Leaf surface	190
				Citrus "Dancy tangerine"	Rutac.	Leaf	114
237	5,6-diOH	7,8,3',4'-tetraOMe		*Mentha spica*	Lamiac.	Leaf	288
				Micromeria albanica	Lamiac.	Aerial parts	188
				Origanum onites	Lamiac.	Aerial parts	290
				Satureja salzmannii	Lamiac.	Aerial parts	290
				Thymus membranaceus	Lamiac.	Aerial parts	290
238	4'-OH	5,6,7,8,3'-pentaOMe					
239	3'-OH	5,6,7,8,4'-pentaOMe					
240	5-OH	6,7,8,3',4'-pentaOMe	5-Desmethyl-nobiletin	*Cunila incana*	Lamiac.	Aerial parts	167
				Mentha spicata	Lamiac.	Leaf	288
				Micromeria albanica	Lamiac.	Aerial parts	188
				Ocimum americanum var. americanum	Lamiac.	Leaf surface	190
				Satureja montana	Lamiac.	Aerial p., ext.	241
				Citrus sinensis	Rutac.	Fruit peel oil	16
				Citrus "Dancy tangerine"	Rutac.	Leaf	114
				Murraya paniculata	Rutac.	Leaf	291
241		5,6,7,8,3',4'-hexaOMe	Nobiletin	*Antirrhinum graniticum*	Scrophul.	Aerial p., ext.	108
				Conoclinium greggii	Asterac.	Aerial parts	292
				Ozothamnus lycopodioides	Asterac.	Aerial p., ext.	69
				Viguiera rosei	Asterac.	Aerial p., ext.	244
				Citrus sinensis	Rutac.	Fruit peel oil	16
				Citrus "Dancy tangerine"	Rutac.	Leaf	114

No.	OH	OMe	Name	Plant source	Family	Plant part	Ref.
242	5,6,7,2',4',5'-hexaOH			Glycoside only			
243	5,7,2',4'-tetraOH	6,5'-diOMe	Arcapillin			Synthesis	293
244	5,2',4'-triOH	6,7,5'-triOMe	Tabularin				
245	5,7-diOH	6,2',4',5'-tetraOMe					
246	5-OH	6,7,2',4',5'-pentaOMe	Agehoustin G	Murraya paniculata	Rutac.	Leaf	291
247	5,3',4',5'-tetraOH (5,6,7,3',4',5'-hexaOH)	6,7-diOMe		Artemisia assoana	Asterac.	Aerial parts	26
248	5,7,3',4'-tetraOH	6,5'-diOMe		Conoclinium coelestinum	Asterac.	Aerial p., ext.	215
249	5,3',5'-triOH	6,7,4'-triOMe*		Conoclinium greggii	Asterac.	Aerial parts	292
250	5,3',4'-triOH	6,7,5'-triOMe		Artemisia argyri	Asterac.	Aerial parts	186
251	5,7,4'-triOH	6,3',5'-triOMe		Neoraputia paraensis	Rutac.	Aerial parts	285
252	5,7,3'-triOH	6,4',5'-triOMe		Murraya paniculata	Rutac.	Leaf	291
253	5,4'-diOH	6,7,3',5'-tetraOMe					
254	5,3'-diOH	6,7,4',5'-tetraOMe					
255	6,7-diOH	5,3',4',5'-tetraOMe					
256	5,7-diOH	6,3',4',5'-tetraOMe					
257	5,6-diOH	7,3',4',5'-tetraOMe	Arteanoflavon			Synthesis	294
258	4'-OH	5,6,7,3',5'-pentaOMe*	Ageconyflavon C*			Synthesis	259
259	6-OH	5,7,3',4',5'-pentaOMe*				Synthesis	259
260	5-OH	6,7,3',4',5'-pentaOMe	Umuhengerin	Ageratum conyzoides	Asterac.	Whole plant	268
						Synthesis	259
261	(5,6,8,3',4',5'-hexaOH)	5,6,7,3',4',5'-hexaOMe		Chromolaena arnottiana	Asterac.	Aerial parts	262
				Murraya paniculata	Rutac.	Leaf	291
				Murraya paniculata	Rutac.	Leaf	291
				Neoraputia paraensis	Rutac.	Aerial parts	285
						Synthesis	28
262	5,6,8,3'-tetraOH (5,6,8,3',4',5'-hexaOH)	4',5'-diOMe	Zhizimin	Ageratum conyzoides	Asterac.	Aerial parts	c
263		(5,6,8,3',4',5'-hexaOMe)		Glycoside only			
264	5,7,8,2',3',4'-hexaOH			Glycoside only			
265	5,4'-diOH	7,8,2',3'-tetraOMe					
266	5-OH	7,8,2',3',4'-pentaOMe	Serpyllin	Andrographis lineata	Acanthac.	Whole plant	14
267	5,7,3',6'-tetraOH (5,7,8,2',3',6'-hexaOH)	8,2'-diOMe	Ganhuangenin				

continued

TABLE 12.1
Flavones and Their Methyl Ethers — continued

No.	OH-Substitution	OMe-Substitution	Trivial Name	Plant Species	Family	Plant Organ	Ref.
	(5,7,8,2',4',5'-hexaOH)						
268	5,7,2',4'-tetraOH	8,5'-diOMe					
	(5,7,8,2',5',6'-hexaOH)						
269	5,7,2',5'-tetraOH	8,6'-diOMe					
	(5,7,8,3',4',5'-hexaOH)						
270	5,7,3',4',5'-pentaOH	8-OMe					
271	5,7,3',5'-tetraOH	8,4'-diOMe					
272	5,7,5'-triOH	8,3',4'-triOMe					
273	5,3'-diOH	7,8,4',5'-tetraOMe					
274	5,7-diOH	8,3',4',5'-tetraOMe		*Bracteantha viscosa*	Asterac.	Aerial p., ext.	222
275		5,7,8,3',4',5'-hexaOMe					
	(5,6,2',3',4',6'-hexaOH)						
276		5,6,2',3',4',6'-hexaOMe		*Casimiroa tetrameria*	Rutac.	Leaf	209
	(5,6,2',3',5',6'-hexOH)						
277		5,6,2',3',5',6'-hexaOMe		*Casimiroa tetrameria*	Rutac.	Leaf	209
	(5,7,2',3',4',5'-hexaOH)						
278	5,7,4'-triOH	2',3',5'-triOMe*		*Psiadia punctulata*	Asterac.	Leaf	295
279	5,7,3'-triOH	2',4',5'-triOMe*		*Psiadia arabica*	Asterac.	Aerial parts	9
280	5,4'-diOH	7,2',3',5'-tetraOMe*		*Psiadia punctulata*	Asterac.	Leaf	295
281	5,7-diOH	2',3',4',5'-tetraOMe*		*Psiadia punctulata*	Asterac.	Leaf	295
282	5,3'-diOH	7,2',4',5'-tetraOMe					
283	5-OH	7,2',3',4',5'-pentaOMe		*Psiadia punctulata*	Asterac.	Leaf	295
	Hepta-O-substituted flavones						
	(5,6,7,8,2',3',6'-heptaOH)						
284	5,7,3',6'-tetraOH	6,8,2'-triOMe*		*Scutellaria planipes*	Scrophul.	Root	55
	(5,6,7,8,2',4',5'-heptaOH)						
285	5,7,2',4'-tetraOH	6,8,5'-triOMe					
286	5,2',5'-triOH	6,7,8,4'-tetraOMe					
287	5,2',4'-triOH	6,7,8,5'-tetraOMe	Agecorynin-D	*Ononis natrix*	Fabac.	Aerial p., ext.	278
288	2',4'-diOH	5,6,7,8,5'-pentaOMe					

No.	Hydroxylation	Methoxylation	Name	Plant source	Family	Part	Ref.
289	5,2'-diOH	6,7,8,4',5'-pentaOMe	Agehoustin F				
290	5-OH	6,7,8,2',4',5'-hexaOMe	Agehoustin E	*Murraya paniculata*	Rutac.	Leaf	291
291		5,6,7,8,2',4',5'-heptaOMe	Agecorynin-C	*Tamarix dioica*	Tamaric.	Aerial parts	245
	(5,6,7,8,3',4',5'-OH)						
292	5,3',5'-triOH	6,7,8,4'-tetraOMe	Gardenin E	*Cleome droserifolia*	Capparid.	Aerial parts	289
293	5,7,5'-triOH	6,8,3',4'-tetraOMe	Scaposin	*Tamarix dioica*	Tamaric.	Aerial parts	245
294	5,6,7-triOH	8,3',4',5'-tetraOMe	"Trimethyl-wogonin"	*Murraya paniculata*	Rutac.	Leaf	291
295	5,4'-diOH	6,7,8,3',5'-pentaOMe*					
296	5,3'-diOH	6,7,8,4',5'-pentaOMe	Gardenin C				
297	4'-OH	5,6,7,8,3',5'-hexaOMe					
298	3'-OH	5,6,7,8,4',5'-hexaOMe					
299	8-OH	5,6,7,3',4',5'-hexaOMe					
300	5-OH	6,7,8,3',4',5'-hexaOMe	Gardenin A			Synthesis	24
				Murraya paniculata	Rutac.	Leaf	291
				Tamarix dioica	Tamaric.	Aerial parts	245
301		5,6,7,8,3',4',5'-heptaOMe	5'-Methoxy-nobiletin	*Conoclinium greggii*	Asterac.	Aerial parts	292
				Ozothamnus lycopodioides	Asterac.	Aerial p., ext.	69
	(5,6,7,2',3',4',5'-heptaOH)						
302			Psiadiarabicin				
303	5,3'-diOH	6,7,2',4',5'-pentaOMe	Agecorynin G	*Ageratum corymbosum*	Asterac.	Aerial parts	296
304	3'-OH	5,6,7,2',4',5'-hexaOMe	Agecorynin F	*Ageratum corymbosum*	Asterac.	Aerial parts	296
305	5-OH	6,7,2',3',4',5'-hexaOMe	Agehoustin B				
306		5,6,7,2',3',4',5'-heptaOMe					
	Octa-O-substituted flavones						
	(5,6,7,8,2',3',4',5'-octaOH)						
307	5,3'-diOH	6,7,8,2',4',5'-hexaOMe	Agehoustin D				
308	3'-OH	5,6,7,8,2',4',5'-heptaOMe	Agehoustin C				
309		5,6,7,8,2',3',4',5'-octaOMe	Agehoustin A				

*For explanation, please see text.
[a]Revised to 6,8-diMe, see text.
[b]Revised to 6,8,4'-triOMe, see text.
[c]Revised to 5,6,7,3',4',5'-Me, see text.

growing. Since the last complilation,[6] some eight compounds (scut-6-Me, scut-6,7-diMe, scut-6,4′-diMe, scut-6,7,4′-triMe; lut-3′-Me, 6-OH-lut-6-Me, 6-OH-lut-6,7-diMe, and 6-OH-lut-6,3′-diMe) fall in the category of "widespread" in addition to apigenin, genkwanin (apigenin-7-Me), acacetin (ap-4′-Me), and luteolin. Thus, no specific sources have been listed for these compounds.

The number of newly described structures increased by about 50 entries during the reporting period. These include a series of 2′- and 5′-substituted flavones, which have been reported from several Asteraceae such as from leaves of *Psiadia punctulata*.[9] The genus *Andrographis* (Acanthaceae) yielded several of these more complex substituted flavones, isolated from whole plants.[10–14] In contrast to other reports on *Andrographis*,[7] the meaning of "whole plant" could not be clarified, with root tissue probably included in the analysis as well. Since all *Andrographis* species are annuals, inclusion of root tissue probably has little influence on the flavonoid composition. Capitula of *Leiothrix flavescens* (*Eriocaulaceae*) yielded a new flavone with a rare 5,6,7,8,3′,4′-hexahydroxy substitution (compound 264 in Table 12.1).[15] This is remarkable insofar as many of the listed flavones are (poly)methoxy derivatives and other hexahydroxyflavones are known as glycosides only.

Most of the source reports concern equally the families of the Asteraceae and Lamiaceae, followed by Rutaceae. However, it must be taken into consideration that the long list may rather be due to the number of species and not to the number of genera. The large number of results in both families may also be due to the research focus on these groups by the authors. In these families, flavone accumulation is mostly reported in leaves, aerial parts and in exudates. Species of the genus *Scutellaria* (Lamiaceae) form an exception, with analyses concentrating on roots since those are used pharmaceutically. According to the distribution of flavones in aerial parts and leaves in other Lamiaceae, similar results should also be expected from *Scutellaria*. Genera of the Rutaceae accumulate flavones primarily in aerial parts and leaves. Many of these compounds, however, were found in fruit peels of *Citrus*, particularly those with higher methylation patterns (e.g., compound 194, Table 12.1[16]). None of the reports, however, indicate possible external accumulation on vegetative tissue of Rutaceae.

Reports on other families are much lower in number. In Fabaceae and Mimosaceae, flavones and their methyl ethers are described from all parts of the plants, including roots of, for example, *Glycyrrhiza eurycarpa*, which primarily accumulates a series of prenylated derivatives.[17] None of the reports cited here indicate external occurrence. In Rosaceae, only few reports exist on the external accumulation of higher methylated flavone aglycones.[18,19] The occurrence of tricetin was proved for the pollen of several genera from the Myrtaceae,[20] with only one report concerning accumulation in leaves.[21] Very few reports exist on families such as Solanaceae and Moraceae. This is quite in contrast to the large number of reports on prenylated flavones accumulated particularly in the Moraceae. Of the nonflowering plants, only few reports relate to the fronds of ferns. Within the mosses, apparently only the thalli of Hepaticae yielded flavones and their methyl ethers.[22]

Some flavone structures have been revised during the reporting period. The structure of 5,8,2′-triOH-6,7-diOMe flavone (compound 131 in Table 12.1) had been ascribed to a product isolated from *Scutellaria baicalensis*.[23] After synthesis, it needs to be revised to 5,7,2′-triOH-6,8-diOMe flavone (compound 132, Table 12.1).[24] Pedunculin, earlier isolated from *Tithonia* species and claimed as 5,8-diOH-6,7,4′-triOMe-flavone (compound 141, Table 12.1),[25] needs to be revised, after synthesis, to 5,7-diOH-6,8,4′-triOMe flavone = nevadensin (compound 142 in Table 12.1).[24] In the previous review, the compound 5,6,7,4′-tetraOH-3′,5′-diOMe had erroneously been cited as a component of *Artemisia assoana*.[6] Data have now been included for the correct structure, 5,7,4′-triOH-6,3′,5′-triOMe flavone (compound 251 in Table 12.1).[26] A further flavone reported from *Ageratum conyzoides* (compound 263 in

Table 12.1) as 5,6,8,3',4',5'-hexamethoxyflavone,[27] was revised to 5,6,7,3',4',5'-hexaOMe flavone (compound 261 in Table 12.1) after synthesis.[28]

12.4 FLAVONOLS

Some 393 reports on flavonols and their distribution are listed in Table 12.2. During the reporting period, the number of new sources also increased, leading to reduction of listings in the very widespread compounds kaempferol, kaempferol-3-methyl ether, quercetin, and quercetin-3-methyl ether. About 54 compounds are reported as new structures, a number equaling that of the flavones. These include a series of polymethoxylated derivatives from species of the Asteraceae, where they are reported to occur in aerial parts as well as in leaf exudates. Species from the Rutaceae accumulate highly methoxylated flavonols in leaves as well as in fruit peels, whereas species of *Fabaceae* were found to accumulate such compounds mainly in the heartwood. A hexamethoxylated flavonol (compound 369 in Table 12.2) was isolated from *Distemonanthus benthamianus* (Fabaceae),[29] a species also known for accumulation of complex cycloflavonols (Table 12.4). Of the Moraceae, only one report concerns the genus *Ficus*, which produces another hexamethoxylated flavonol (compound 279 in Table 12.2) in the aerial parts.[30] The same applies to accumulation of flavonol aglycones in roots of *Duroia hirsuta* (Rubiaceae).[31] The number of 2'- and 5'-substituted derivatives appears to be lower than that of the corresponding flavones.

Most of the new source reports concern species from the Asteraceae, with many of the flavonols being isolated from aerial parts, where they are accumulated externally. They range from simple to more complex structures. There appears to be a tendency towards 6-methoxylation rather than towards 8-methoxylation, in addition to possible OMe-substitution of other positions of the flavonol molecule. Flavonols with 6,8-di-*O*-methylation and additional OMe-groups are also found in several genera such as *Senecio*,[32] *Psiadia*,[33] or *Inula*,[34] to cite but a few examples.

Aerial parts, fruits, flowers, and bark tissue of a series of Rutaceae species yielded a number of hexamethoxylated flavonols. Once more, the complexity of metabolic pathways in this family is demonstrated by the formation of such compounds. The number of entries for this family is the second largest following the Asteraceae, but it must be taken into account that only a few genera of this large family are concerned. The third largest group concerns *Heliotropium* species of the family Boraginaceae, where particularly leaf exudates yielded flavonols.[35,36] For species of *Alkanna*, flavonols were reported for aerial parts without indicating possible external occurrence.[37] Interestingly, almost no flavones were reported from *Heliotropium* (see Table 12.1), and species of the genus *Nonea* were so far found to accumulate flavones only in their exudates.[36] Further distribution studies will have to confirm the possible chemosystematic value of these accumulation trends.

A number of new listings concern the families of Scrophulariaceae and Solanaceae. In both cases the number of reports concerning external accumulation is also increased. Thus, further research will probably reveal that this phenomenon is more widespread in these families as is obvious from the present data. In Fabaceae, most reports concern accumulation in heartwood, with a few exceptions such as leaves of *Millettia racemosa*.[38] However, no indication to possible external accumulation is made. Similar to flavone accumulation data, pollen of Myrtaceae were also found to accumulate flavonols.[20] Very few reports exist on Gymnosperms such as *Cryptomeria* (Taxodiaceae)[39] or *Ephedra*,[40] without indication of external accumulation. So far, no new reports on flavones are known for these taxa.

In contrast to the numerous reports on flavones in Lamiaceae, only very few genera were found to accumulate flavonols in their exudates. The accumulation of 5,6-di-*O*-methylated derivatives in species of *Salvia*[41] may be of chemosystematic significance, in relation to other

TABLE 12.2
Flavonols and Their Methyl Ethers

No.	OH-Substitution	OMe-Substitution	Trivial Name	Plant Species	Family	Plant Organ	Ref.
	Di-O-substituted flavonols						
1	3,7-diOH						
2		3,7-diOMe*		Pongamia pinnata	Fabac.	Root bark	297
3	(3,4'-diOH) 3-OH	4'-OMe					
	Tri-O-substituted flavonols						
4	3,5,7-triOH		Galangin	Baccharis viminea	Asterac.	Aerial p., ext.	129
				Cassinia quinquefaria	Asterac.	Aerial p., ext.	222
				Flourensia cernua	Asterac.	Aerial p., ext.	182
				Gnaphalium microcephalum	Asterac.	Aerial p., ext.	298
				Helichrysum, 8 spec.	Asterac.	Aerial p., ext.	237
				Helichrysum aureum	Asterac.	Aerial p., ext.	32
				Helichrysum aureonitens	Asterac.	Aerial parts	299
				Heterothalamus psiadioides	Asterac.	Leaf	300
				Odixia, 6 spp.	Asterac.	Aerial p., ext.	197
				Heliotropium filifolium	Boragin.	Leaf exudate	35
				Millettia racemosa	Fabac.	Leaf	38
				Nothofagus alessandri	Fagac.	Leaf	301
				Nothofagus antarctica	Fagac.	Aerial p., ext.	302
				Nothofagus, 6 spp.	Fagac.	Aerial p., ext.	158
				Ribes viscosissimum	Grossular.	Leaf exudate	303
				Woodsia scopulina	Pteridac.	Fronds, ext.?	304
				Bees wax			138
				Propolis from Arizona			305
				European Propolis			137
				Propolis from Chile			306
				Propolis from Egypt			136
5	5,7-diOH	3-OMe		Helichrysum aureum	Asterac.	Aerial p., ext.	32
				Helichrysum picardii	Asterac.	Aerial p. (ext.)	307
				Lychnophora markgravii	Asterac.	Aerial parts	140

No.	OMe (trivial name)	OH	Species	Family	Plant part	Ref.
			Ozothamnus, 5 spp.	Asterac.	Aerial p., ext.	197
			Pseudognaphalium cheiranthifolium	Asterac.	Res. exudate	308
			Heliotropium filifolium	Boragin.	Leaf exudate	35
			Heliotropium huascuense	Boragin.	Leaf exudate	309
			Heliotropium megalanthum	Boragin.	Leaf exudate	213
			Heliotropium pycnophyllum	Boragin.	Aerial p., ext.	36
			Heliotropium sinuatum	Boragin.	Leaf exudate	310
			Heliotropium stenophyllum	Boragin.	Aerial p., ext.	36
			Nothofagus, 5 spp.	Fagac.	Aerial p., ext.	158
			Woodsia scopulina	Pteridac.	Fronds, ext.?	304
6	5-OMe	3,7-diOH	Propolis from Arizona			305
			Propolis from Chile			306
7	7-OMe (Izalpinin)	3,5-diOH	Baccharis viminea	Asterac.	Aerial p., ext.	129
			Helichrysum aureum	Asterac.	Aerial p., ext.	32
			Ozothamnus ledifolius	Asterac.	Aerial p., ext.	197
			Heliotropium pycnophyllum	Boragin.	Aerial p., ext.	36
			Capparis tweediana	Capparid.	Leaf	311
			Nothofagus antarctica	Fagac.	Aerial p., ext.	302
			Nothofagus, 6 spp.	Fagac.	Aerial p., ext.	158
			Propolis from Arizona			305
8	3,7-diOMe	5-OH	Pseudognaphalium cheiranthifolium	Asterac.	Res. exudate	308
			Heliotropium huascuense	Boragin.	Leaf exudate	309
			Heliotropium megalanthum	Boragin.	Leaf exudate	213
			Heliotropium pycnophyllum	Boragin.	Aerial p., ext.	36
			Escallonia leucantha	Saxifrag.	Resin. exud.	42
			Notholaena ekmanii	Pteridac.	Frond exud.	312
9	5,7-diOMe	3-OH				
10	3,5,7-triOMe	—				
11	8-OMe	3,7-diOH (3,7,8-triOH)				
12		3,7,4'-triOH (3,5,2'-triOH)				

continued

TABLE 12.2
Flavonols and Their Methyl Ethers — continued

No.	OH-Substitution	OMe-Substitution	Trivial Name	Plant Species	Family	Plant Organ	Ref.
	Tetra-O-substituted flavonols						
13	3,5,6,7-tetraOH		6-Hydroxy-galangin	Adenostoma sparsifolium	Rosac.	Aerial p., ext.	19
				Cassinia quinquefaria	Asterac.	Aerial p., ext.	222
				Platanus acerifolia	Platanac.	Bud	a
14	3,5,7-triOH	6-OMe	Alnusin	Anaphalis margaritacea	Asterac.	Aerial p., ext.	298
				Cassinia quinquefaria	Asterac.	Aerial p., ext.	222
				Gnaphalium microcephalum	Asterac.	Aerial p., ext.	298
				Adenostoma sparsifolium	Rosac.	Aerial p., ext.	19
15	(3,5,6-triOH	7-OMe)				Synth. only	313
16	5,7-diOH	3,6-diOMe		Gomphrena boliviana, G. martiana	Amaranth.	Whole plant	150
				Anaphalis margaritacea	Asterac.	Aerial p., ext.	298
				Gnaphalium microcephalum	Asterac.	Aerial p., ext.	298
				Helichrysum, 8 spp.	Asterac.	Aerial p., ext.	237
				Pseudognaphalium cheiranthifolium	Asterac.	Aerial parts	308
17	5,6-diOH	3,7-diOMe		Gnaphalium affine	Asterac.	Aerial parts	314
18	3,7-diOH	5,6-diOMe		Salvia columbariae	Lamiac.	Aerial p., ext.	41
	3,7-diOH	5,6-diOMe		Trichostema lanatum	Lamiac.	Aerial p., ext.	41
19	3,6-diOH	5,7-diOMe				Synthesis	47
20	3,5-diOH	6,7-diOMe					
21	5-OH	3,6,7-triOMe	Alnustin				
22		3,5,6,7-tetraOMe					
23	3,5,7,8-tetraOH		8-Hydroxy-galangin			Synthesis	47
24	5,7,8-triOH	3-OMe		Nothofagus, 3 spp.	Fagac.	Aerial p., ext.	158
	5,7,8-triOH	3-OMe		Ozothamnus, 3 spp.	Asterac.	Aerial p., ext.	197
				Ozothamnus ledifolius	Asterac.	Aerial p., ext.	197
				Helichrysum aureum	Asterac.	Aerial p., ext.	32
				Ozothamnus expansifolius	Asterac.	Aerial p., ext.	197
25	3,5,8-triOH	7-OMe		Nothofagus antarctica	Fagac.	Aerial p., ext.	302
26	3,5,7-triOH	8-OMe		Nothofagus alessandri	Fagac.	Leaf	301
				Nothofagus, 7 spp.	Fagac.	Aerial p., ext.	158

No.	OH	OMe	Name	Species	Family	Plant part	Ref.
27	5,8-diOH	3,7-diOMe	Isognaphalin	Woodsia scopulina	Pteridac.	Fronds, ext.?	304
28	5,7-diOH	3,8-diOMe	Gnaphalin	Nothofagus solandri	Fagac.	Aerial p., ext.	158
				Anaphalis margaritacea	Asterac.	Aerial p., ext.	298
				Gnaphalium luteo-album	Asterac.	Aerial p., ext.	315
				Gnaphalium microcephalum	Asterac.	Aerial p., ext.	298
				Gymnosperma glutinosum	Asterac.	Aerial p., ext.	196
				Helichrysum aureum	Asterac.	Aerial p., ext.	32
				Helichrysum bracteiferum	Asterac.	Aerial p., ext.	237
				Helichrysum picardii	Asterac.	Aerial p. (ext.)	307
				Ozothamnus ericifolius., O. purpurescens	Asterac.	Aerial p., ext.	197
29	3,5-diOH	7,8-diOMe		Nothofagus alessandri	Fagac.	Leaf	301
				Nothofagus, 5 spp.	Fagac.	Aerial p., ext.	158
				Ozothamnus ledifolius	Asterac.	Aerial p., ext.	197
				Nothofagus solandri	Fagac.	Aerial p., ext.	158
				Woodsia scopulina	Pteridac.	Fronds, ext.?	304
30	7-OH	3,5,8-triOMe		Ozothamnus expansifolius, O. hookeri	Asterac.	Aerial p., ext.	197
31	5-OH	3,7,8-triOMe	Methylgnaphalin	Nothofagus solandri, N. truncata	Fagac.	Aerial p., ext.	158
32	5-OH	3,7,8-triOMe	Methylgnaphalin	Many records			
33	3-OH	5,7,8-triOMe		Many records			
34		3,5,7,8-tetraOMe		Propolis from Arizona			305
35	3,5,7,2'-tetraOH		Datiscetin	Ambrosia trifida	Asterac.	Aerial p., ext.	233
36	3,5,2'-triOH	7-OMe	Datin	Artemisia campestris	Asterac.	Aerial p., ext.	132
37	3,5,7-triOH	2'-OMe	Ptaeroxylol	Baccharis pilularis	Asterac.	Aerial p., ext.	133
38	3,5,4'-tetraOH		Kaempferol	Madia elegans	Asterac.	Aerial p., ext.	201
39	5,7,4'-triOH	3-OMe	Isokaempferide	Madia sativa	Asterac.	Leaf exudate	236
40	3,7,4'-triOH	5-OMe		Ozothamnus expansifolius, O. scutellifolius	Asterac.	Aerial p., ext.	197
41	3,5,4'-triOH	7-OMe	Rhamnocitrin	Stevia subpubescens	Asterac.	Aerial p., ext.	182
				Heliotropium stenophyllum	Boragin.	Aerial p., ext.	36

continued

TABLE 12.2
Flavonols and Their Methyl Ethers — continued

No.	OH-Substitution	OMe-Substitution	Trivial Name	Plant Species	Family	Plant Organ	Ref.
				Heliotropium chenopodiaceum var. *ericoideum*	Boragin.	Leaf exudate	35
				Aeonium leucoblepharum, Ae.nobile	Crassul.	Aerial p., ext.	216
				Nothofagus cunninghamii	Fagac.	Aerial p., ext.	158
				Aniba sp.	Laurac.	Wood, bark	316
				Mirabilis viscosa	Nyctagin.	Aerial p., ext.	43
				Notholaena nivea	Pteridac.	Frond exud.	45
				Calceolaria irazuensis	Scrophul.	Aerial p., ext.	108
				Mimulus cardinalis	Scrophul.	Aerial p., ext.	108
				Solanum paludosum	Solanac.	Aerial parts	317
				Viscum cruciatum	Viscac.	Cuticular wax	318
42	3,5,7-triOH	4'-OMe	Kaempferide	*Baccharis pilularis*	Asterac.	Aerial p., ext.	133
				Baccharis viminea	Asterac.	Aerial p., ext.	129
				Chrysothamnus nauseosus	Asterac.	Aerial p., ext.	133
				Ozothamnus scutellifolius	Asterac.	Aerial p., ext.	197
				Eucryphia jinksii	Cunoniac.	Leaf, bud, ext.	165
				Nothofagus menziesii, N. nervosa	Fagac.	Aerial p., Ext.	158
				Eriodictyon sessilifolium	Hydrophyll.	Leaf resin	135
				Mirabilis viscosa	Nyctagin.	Aerial p., ext.	43
				Currania robertiana	Pteridac.	Frond exud.	173
				Calceolaria irazuensis	Scrophul.	Aerial p., ext.	108
				Brazilian propolis			319
				Propolis from Chile			306
43	7,4'-diOH	3,5-diOMe		*Chrysothamnus nauseosus*	Asterac.	Aerial p., ext.	199
44	5,4'-diOH	3,7-diOMe	Kumatakenin	*Achillea ageratum*	Asterac.	Aerial parts	320
				Alkanna orientalis	Boragin.	Aerial parts	37
				Heliotropium chenopodiaceum var. *ericoideum*	Boragin.	Resin. exud.	35
				Heliotropium pycnophyllum	Boragin.	Aerial p., ext.	36
				Cleome spinosa	Capparid.	Aerial p., ext.	321
				Aeonium spp.	Crassul.	Aerial p., ext.	216
				Eucryphia lucida	Cunoniac.	Leaf, bud. ext.	165

No.	OH	OMe	Name	Species	Family	Plant part	Ref.
				Nothofagus cunninghamii	Fagac.	Aerial p., ext.	158
				Pelargonium fulgidum	Geraniac.	Leaf exudate	142
				Salvia cyanescens	Lamiac.	Aerial parts	192
				Mirabilis viscosa	Nyctagin.	Aerial p., ext.	43
				Bosistoa brassii	Rutac.	Leaf	322
				Evodia merrillii	Rutac.	Fruit	323
				Calceolaria arachnoidea	Scrophul.	Aerial p., ext.	108
				Chamaesaracha sordida	Solanac.	Aerial p., ext.	214
				Solanum paludosum	Solanac.	Aerial parts	317
				Viscum album	Viscac.	Aerial p., ext.	44
				Viscum cruciatum	Viscac.	Cut. wax	318
				Amomum koenigii	Zingib.	Fruit	324
45	5,7-diOH	3,4-diOMe	Ermanin	—	Boragin.	Aerial parts	37
				Aeonium spp.	Crassul.	Aerial p., bud, ext.	216
				Eucryphia lucida	Cunoniac.	Leaf, bud, ext.	165
				Nothofagus menziesii, N. nervosa	Fagac.	Aerial p., ext.	158
				Fouquieria splendens	Fouquierac.	Aerial p., ext.	325
				Mirabilis viscosa	Nyctagin.	Aerial p., ext.	43
				Currania robertiana	Pteridac.	Frond exud.	173
				Notholaena nivea	Pteridac.	Frond exud.	45
				Barosma crenulata	Rutac.	Aerial p., ext.	326
				Petunia surfina	Solanac.	Aerial p., ext.	214
				Viscum album	Viscac.	Cut. wax	318
				Brazilian Propolis			319
46	3,7-diOH	5,4'-diOMe*		*Amomum koenigii*	Zingib.	Fruit	324
47	3,5-diOH	7,4'-diOMe		*Flourensia cernua*	Asterac.	Aerial p., ext.	182
				Haplopappus hirtellus	Asterac.	Aerial parts	327
				Madia elegans	Asterac.	Aerial p., ext.	201
				Ozothamnus scutellifolius	Asterac.	Aerial p., ext.	197
				Serratula strangulata	Asterac.	Whole plant	328
				Stevia subpubescens	Asterac.	Aerial p., ext.	182
				Heliotropium stenophyllum	Boragin.	Aerial p., ext.	36
				Cleome spinosa	Capparid.	Aerial p., ext.	321
				Aeonium sedifolium	Crassul.	Aerial p., ext.	216

continued

TABLE 12.2
Flavonols and Their Methyl Ethers — continued

No.	OH-Substitution	Trivial Name	OMe-Substitution	Plant Species	Family	Plant Organ	Ref.
48	4'-OH		3,5,7-triOMe	Salvia chinopeplica	Lamiac.	Leaf	225
				Aniba spp.	Laurac.	Wood, bark	316
				Mirabilis viscosa	Nyctagin.	Aerial p., ext.	43
				Notholaena nivea	Pteridac.	Frond exud.	45
				Calceolaria mexicana	Scrophul.	Aerial p., ext.	108
49	5-OH		3,7,4'-triOMe	Artemisia rupestris	Asterac.	Aerial parts	132
				Baccharis pilularis	Asterac.	Aerial p., ext.	129
				Grindelia nana	Asterac.	Aerial p., ext.	133
				Grindelia tenella	Asterac.	Aerial p., ext.	182
				Haplopappus hirtellus	Asterac.	Aerial parts	327
				Haplopappus sonorensis	Asterac.	Aerial parts	329
				Ozothamnus scutellifolius	Asterac.	Aerial p., ext.	197
				Senecio viscosa	Asterac.	Aerial p., ext.	32
				Stevia subpubescens	Asterac.	Aerial p., ext.	182
				Xanthocephalum gymnosp.	Asterac.	Aerial p., ext.	182
				Heliotropium pycnophyllum	Boragin.	Aerial p., ext.	36
				Cleome spinosa	Capparid.	Aerial p., ext.	321
				Aeonium goochia, Ae. Lindleyi	Crassul.	Aerial p., ext.	216
				Eucryphia lucida, E. milliganii	Cunoniac.	Leaf, bud, ext.	165
				Nothofagus cunninghamii	Fagac.	Aerial p., ext.	158
				Dorystoechas hastata	Lamiac.	Aerial parts	168
				Aniba spp.	Laurac.	Wood and bark	316
				Mirabilis viscosa	Nyctagin.	Aerial p., ext.	43
				Currania robertiana	Pteridac.	Frond exud.	173
				Calceolaria chelidonioides	Scrophul.	Aerial p., ext.	108
				Cryptomeria japonica	Taxodiac.	Leaf	39
50	3,5,8,4'-tetraOH	Pratoletin	3,5,7,4'-tetraOMe*	Amomum koenigii	Zingib.	Fruit	324
51	(3,6,7,4'-tetraOH)			Amomum koenigii	Zingib.	Fruit	324

No.	Hydroxyl	Methoxyl	Trivial name	Plant source	Family	Plant part	Ref.
52	6,7,4'-triOH	3-OMe		Graziela mollissima	Asterac.	Aerial parts	330
53	7,4'-diOH	3,6-diOMe					
54	3,7,8,4'-tetraOH						
55	7,4'-diOH	3,8-diOMe*					
56	3,7,3',4'-tetraOH		Fisetin	Parkia clappertoniana	Fabac.	Leaf	331
				Acacia montana	Mimosac.	Heartwood	332
				Dalbergia odorifera	Fabac.	Heartwood	333
				Glycyrrhiza spec.	Fabac.	Liquorice	334
				Millettia racemosa	Fabac.	Leaf	38
57	7,3',4'-triOH	3-OMe	Geraldol				
58	3,7,4'-triOH	3'-OMe					
59	3-OH	7,3',4'-triOMe					
60		3,7,3',4'-tetraOMe					
Penta-O-substituted flavonols (3,5,6,7,8-pentaOH)							
61	5,7,8-triOH	3,6-diOMe*		Cassinia arcuata	Asterac.	Aerial p., ext.	278
62	3,5,7-triOH	6,8-diOMe		Anaphalis margaritacea	Asterac.	Aerial p., ext.	298
				Helichrysum, 8 spp.	Asterac.	Aerial p., ext.	237
						Synthesis	335
63	5,8-diOH	3,6,7-triOMe		Pseudognaphalium cheiranthifolium	Asterac.	Aerial parts	308
						Synthesis	336
64	5,7-diOH	3,6,8-triOMe	Araneol	Anaphalis margaritacea	Asterac.	Aerial p., ext.	298
				Gymnosperma glutinosum	Asterac.	Aerial p., ext.	196
				Gymnosperma glutinosum	Asterac.	Aerial p., ext.	133
				Helichrysum, 8 spp.	Asterac.	Aerial p., ext.	237
						Synthesis	335
65	3,5-diOH	6,7,8-triOMe		Gnaphalium chilense, G. microcephalum	Asterac.	Aerial p., ext.	298
66	5-OH	3,6,7,8-tetraOMe		Gnaphalium affine	Asterac.	Aerial parts	314
67		3,5,6,7,8-pentaOMe		Pseudognaphalium cheiranthifolium, P. vira vira	Asterac.	Leaf trichome	337
68	3,5,6,7,4'-pentaOH*		6-Hydroxy-kaempferol	Helminthia echioides	Asterac.	Aerial parts	338
69	5,6,7,4'-tetraOH	3-OMe	Vogeletin	Glycoside only		Synthesis	47
70	3,6,7,4'-tetraOH	5-OMe				Synthesis	28

continued

TABLE 12.2
Flavonols and Their Methyl Ethers — continued

No.	OH-Substitution	OMe-Substitution	Trivial Name	Plant Species	Family	Plant Organ	Ref.
71	3,5,7,4'-tetraOH	6-OMe		*Ambrosia artemisifolia*	Asterac.	Aerial p., ext.	182
				Ageratina espinosa	Asterac.	Aerial p., ext.	133
				Carthamus tinctorius	Asterac.	Petal (nat.?)	339
				Centaurea incana	Asterac.	Aerial parts	281
				Chrysactinia mexicana	Asterac.	Aerial p., ext.	182
				Eupatorium altissimum, E. serotinum	Asterac.	Aerial p., ext.	215
				Heteranthemis viscidihirta	Asterac.	Leaf exudate	340
				Xanthium strumarium	Asterac.	Aerial p., ext.	32
				Aeonium, 3 spp.	Crassul.	Aerial p., ext.	216
				Adenostoma sparsifolium	Rosac.	Aerial p., ext.	19
				Brazilian propolis			341
72	3,5,6,4'-tetraOH	7-OMe		Glycoside only		Synthesis	342
73	3,5,6,7-tetraOH	4'-OMe				Synthesis	313
74	5,7,4'-triOH	3,6-diOMe				Synthesis	47
				Achillea micrantha	Asterac.	Aerial parts	343
				Ageratina espinosa	Asterac.	Aerial p., ext.	133
				Ambrosia chamissonis	Asterac.	Aerial p., ext.	233
				Brickellia eupatorioides	Asterac.	Aerial p., ext.	215
				Calycadenia multiglandulosa, C. villosa	Asterac.	Leaf exudate	236
				Centaurea, 4 spp.	Asterac.	Aerial p., ext.	182
				Eupatorium altissimum	Asterac.	Aerial p., ext.	215
				Flourensia cernua	Asterac.	Aerial p., ext.	182
				Grindelia robusta	Asterac.	Leaf exudate	344
				Grindelia squarrosa	Asterac.	Aerial p., ext.	182
				Heteranthemis viscidihirta	Asterac.	Leaf exudate	340
				Heterotheca villosa	Asterac.	Aerial p., ext.	182
				Oncosiphon grandiflorum	Asterac.	Aerial p., ext.	32
				Perityle lemmonii	Asterac.	Aerial p., ext.	133
				Psiadia dentata	Asterac.	Leaf	345
				Stevia berlandieri	Asterac.	Aerial p., ext.	182

No.	OH	Name	OMe	Species	Family	Part	Ref.
				Tanacetum parthenium	Asterac.	Leaf, flower	49
				Alkanna orientalis	Boragin.	Aerial parts	37
				Aeonium, 5 spp.	Crassul.	Aerial p., ext.	216
				Salvia cyanescens	Lamiac.	Aerial parts	192
				Gardenia, 5 spp.	Rubiac.	Bud exudate	346
				Barbacenia rubro-virens	Velloziac.	Leaf surface	105
75	5,6,4'-triOH		3,7-diOMe	*Parthenium incanum*	Asterac.	Aerial p., ext.	182
				Pulicaria dysenterica	Asterac.	Aerial p., ext.	182
				Pulicaria dysenterica	Asterac.	External	347
				Tanacetum parthenium	Asterac.	Aerial parts	48
76	5,6,7-triOH		3,4'-diOMe[a]	*Pulicaria dysenterica*	Asterac.	External	347
						Synthesis	28
77	3,7,4'-triOH		5,6-diOMe	*Salvia columbariae*	Lamiac.	Aerial p., ext.	41
				Trichostema lanatum	Lamiac.	Aerial p., ext.	41
78	3,5,4'-triOH	Eupalitin	6,7-diOMe	*Artemisia austriaca*	Asterac.	Aerial parts	270
				Baccharis pilularis	Asterac.	Aerial p., ext.	129
				Brickellia longifolia	Asterac.	Aerial parts	348
				Heterotheca villosa	Asterac.	Aerial p., ext.	182
				Aeonium glutinosum	Crassul.	Aerial p., ext.	216
79	3,5,7-triOH	Betuletol	6,4'-diOMe	*Gnaphalium microcephalum*	Asterac.	Aerial p., ext.	298
				Mirabilis viscosa	Nyctagin.	Aerial p., ext.	43
				Rosa centifolia cv. *muscosa*	Rosac.	Aerial p., ext.	18
				Brazilian Propolis			319
						Synthesis	342
80	3,5,6-triOH		7,4'-diOMe*	*Zanthoxylum bungeanum*	Rutac.	Pericarp	349
81	3,5,6-triOH		7,4'-diOMe			Synthesis	313
82	5,4'-diOH	Penduletin	3,6,7-triOMe	*Achillea ageratum*	Asterac.	Aerial parts	320
				Achillea nana	Asterac.	Aerial p., ext.	196
				Ageratina espinosa	Asterac.	Aerial p., ext.	133
				Baccharis pedunculata	Asterac.	Aerial parts	350
				Baccharis trinervis	Asterac.	Aerial parts	351
				Brickellia eupatorioides	Asterac.	Aerial p., ext.	215
				Flourensia cernua	Asterac.	Aerial p., ext.	182
				Grindelia tarapacana	Asterac.	Aerial p., ext.	352

continued

TABLE 12.2
Flavonols and Their Methyl Ethers — *continued*

No.	OH-Substitution	OMe-Substitution	Trivial Name	Plant Species	Family	Plant Organ	Ref.
				Grindelia robusta	Asterac.	Leaf exudate	344
				Lagophylla glandulosa	Asterac.	Leaf exudate	236
				Oncosiphon grandiflorum	Asterac.	Aerial p., ext.	32
				Tanacetum polycephalum	Asterac.	Aerial p., ext.	220
				Alkanna orientalis	Boragin.	Aerial parts	37
				Aeonium, 3 spp.	Crassul.	Aerial p., ext.	216
				Trixis vauhieri	Asterac.	Leaf	353
83	5,7-diOH	3,6,4'-triOMe	Santin	*Achillea latiloba*	Asterac.	Aerial p., ext.	354
				Achillea atrata ssp. *multifida*	Asterac.	Aerial parts	355
				Achillea multifida	Asterac.	Aerial p., ext.	196
				Ageratina espinosa	Asterac.	Aerial p., ext.	133
				Anthemis tinctoria	Asterac.	Aerial p., ext.	356
				Artemisia barrelieri	Asterac.	Aerial parts	132
				Brickellia eupatorioides	Asterac.	Aerial p., ext.	215
				Eupatorium cannabinum	Asterac.	Aerial p., ext.	216
				Grindelia tarapacana	Asterac.	Aerial p., ext.	298
				Grindelia glutinosa	Asterac.	Leaf exudate	344
				Grindelia squarrosa	Asterac.	Aerial p., ext.	182
				Perityle lemmonii	Asterac.	Aerial p., ext.	133
				Stevia berlandieri	Asterac.	Aerial p., ext.	182
				Tanacetum microphyllum	Asterac.	Aerial parts	357
				Aeonium, 3 spp.	Crassul.	Aerial p., ext.	216
84	5,6-diOH	3,7,4'-triOMe	"Tanetin"	*Drummondita hassellii*	Rutac.	Aerial parts	358
85	3,4'-diOH	5,6,7-triOMe	Candidol	*Pulicaria dysenterica*	Asterac.	External	347
86	3,5-diOH	6,7,4'-triOMe	Mikanin	*Tanacetum parthenium*	Asterac.	Aerial parts	b
87	5-OH	3,6,7,4'-tetraOMe		*Baccharis pilularis*	Asterac.	Aerial p., ext.	196
				Achillea sibirica subsp. *mongolica*	Asterac.	Aerial p., ext.	354
				Ageratina espinosa	Asterac.	Aerial p., ext.	133
				Brickellia eupatorioides	Asterac.	Aerial p., ext.	215

No.	Trivial name	Substitution	Species	Family	Plant part	Ref.
		(3,5,6,8,4'-pentaOH)	*Grindelia glutinosa*	Asterac.	Leaf exudate	344
			Parthenium incanum	Asterac.	Aerial p., ext.	182
			Aeonium, 3 spp.	Crassul.	Aerial p., ext.	216
			Drummondita hassellii	Rutac.	Aerial parts	358
88	"Candiron"	(5,4'-diOH 3,6,8-triOMe)	*Tephrosia candida*	Fabac.	Seed	c
	—				Synthesis	51
		(3,5,7,8,2'-pentaOH)	Ester only			
89		3,5,8,2'-tetraOH 7-OMe				
90		5,2'-diOH 3,7,8-triOMe				
91		5-OH 3,7,8,2'-tetraOMe				
92	Herbacetin	3,5,7,8,4'-pentaOH	*Ozothamnus hookeri*	Asterac.	Aerial p., ext.	197
93		5,7,8,4'-tetraOH 3-OMe	*Baccharis pilularis*	Asterac.	Aerial p., ext.	196
			Ozothamnus, 3 spp.	Asterac.	Aerial p., ext.	197
94	Pollenitin	3,5,8,4'-tetraOH 7-OMe	*Ephedra aphylla*	Ephedrac.	Aerial parts	40
95	Sexangularetin	3,5,7,4'-tetraOH 8-OMe	*Ozothamnus*, 3 spp.	Asterac.	Aerial p., ext.	197
96		3,5,7,8-tetraOH 4'-OMe	*Pentagramma triangularis*	Pteridac.	Frond exudate	359
97		5,8,4'-triOH 3,7-diOMe	*Ozothamnus hookeri*	Asterac.	Aerial p., ext.	197
98		5,7,4'-triOH 3,8-diOMe	*Brachyglottis cassinioides*	Asterac.	Leaf exudate	360
			Chrysothamnus nauseosus	Asterac.	Aerial p., ext.	199
			Haplopappus deserticola	Asterac.	Resin. exud.	361
			Ozothamnus, 3 spp.	Asterac.	Aerial p., ext.	197
			Cleome spinosa	Capparid.	Aerial p., ext.	321
			Calcolaria arachnoidea	Scrophul.	Aerial p., ext.	108
99		5,7,8-triOH 3,4'-diOMe*	*Pityrogramma triangularis*	Pteridac.	Frond exud.	362
100		3,5,4'-triOH 7,8-diOMe	*Ozothamnus expansifolius, O. obcordatus*	Asterac.	Aerial p., ext.	197
101		3,5,8-triOH 7,4'-diOMe			Synthesis	313
102	Prudomestin	3,5,7-triOH 8,4'-diOMe	*Calceolaria irazuensis*	Scrophul.	Aerial p., ext.	108
103		5,4'-diOH 3,7,8-triOMe	*Ozothamnus*, 4 spp.	Asterac.	Aerial p., ext.	197
			Cleome spinosa	Capparid.	Aerial p., ext.	321
			Calceolaria chelidonioides, C. tripartita	Scrophul.	Aerial p., ext.	108
104		5,8-diOH 3,7,4'triMe	*Ozothamnus obcordattus*	Asterac.	Aerial p., ext.	197
105		5,7-diOH 3,8,4'-triOMe	*Haplopappus deserticola*	Asterac.	Resin. exud.	361

continued

TABLE 12.2
Flavonols and Their Methyl Ethers — continued

No.	OH-Substitution	OMe-Substitution	Trivial Name	Plant Species	Family	Plant Organ	Ref.
				Helichrysum foetidum	Asterac.	Aerial p., ext.	182
				Nothofagus, 3 spp.	Fagac.	Aerial p., ext.	158
				Currania robertiana	Pteridac.	Frond exud.	173
				Calceolaria chelidonioides	Scrophul.	Aerial p., ext.	108
106	3,5-diOH	7,8,4'-triOMe	Tambulin	Helianthus annuus	Asterac.	Leaf	363
				Ozothamnus expansifolius, O. obcordatus	Asterac.	Aerial p., ext.	197
				Drummondita calida	Rutac.	Aerial parts	358
				Calceolaria irazuensis	Scrophul.	Aerial p., ext.	108
107	5-OH	3,7,8,4'-tetraOMe	Flindulatin	Helichrysum foetidum	Asterac.	Aerial p., ext.	182
				Ozothamnus, 4 spp.	Asterac.	Aerial p., ext.	197
				Cleome spinosa	Capparid.	Aerial p., ext.	321
				Nothofagus menziesii, N. nervosa	Fagac.	Aerial p., ext.	158
				Drummondita calida	Rutac.	Aerial parts	358
108	(3,6,7,8,4'-pentaOH)	3,5,7,8,4'-pentaOMe*	Auranetin				
109	-	3,6,7,8,4'-pentaOMe					
110	3,5,7,2',4'-pentaOH		Morin	Millettia racemosa	Fabac.	Leaf	38
111	5,7-diOH (3,5,7,2',5'-pentaOH)	3,2',4'-triOMe					
112	3,5,2'-triOH	7,5'-diOMe	Viscidulin I	Blumea balsamifera	Asterac.	Aerial parts	364
113	3,5,7,2',6'-pentaOH						
114	3,5,7,2',6'-pentaOH	3,5,7,2',6'-pentaOMe					
115	3,5,7,3',4'-pentaOH	3-OMe	Quercetin	Many records			
116	5,7,3',4'-tetraOH	5-OMe	Azaleatin	Many records			
117	3,7,3',4'-tetraOH						
118	3,5,3',4'-tetraOH	7-OMe	Rhamnetin	Artemisia campestris ssp. glutinosa	Asterac.	Aerial parts	132
				Baccharis pilularis	Asterac.	Aerial p., ext.	129
				Baccharis pilularis	Asterac.	Aerial p., ext.	133
				Cassinia vauvilliersii	Asterac.	Aerial parts	365
				Chromolaena odorata	Asterac.	Aerial p., ext.	193
				Chrysothamnus nauseosus	Asterac.	Aerial p., ext.	133

No.	Name	Substitution	Base	Species	Family	Plant part	Ref.
				Flourensia thurifera	Asterac.	Leaf resin	366
				Madia elegans	Asterac.	Aerial p., ext.	201
				Ozothamnus expansifolius, O. scutellifolius	Asterac.	Aerial p., ext.	197
				Pulicaria dysenterica	Asterac.	Aerial p., ext.	182
				Heliotropium stenophyllum	Boragin.	Aerial p., ext.	36
				Capparis tweediana	Capparid.	Leaf	311
				Aeonium, 3 spp.	Crassul.	Aerial p., ext.	216
				Nothofagus obliqua	Fagac.	Aerial p., ext.	158
				Vellozia streptophylla	Velloziac.	Leaf surface	105
				Viscum album, V. cruciatum	Viscac.	Cut. wax	318
119	Isorhammetin	3'-OMe	3,5,7,4'-tetraOH	*Ambrosia ambrosioides*	Asterac.	Aerial p., ext.	133
				Baccharis vininea	Asterac.	Aerial p., ext.	129
				Chrysothamnus nauseosus	Asterac.	Aerial p., ext.	133
				Ericameria linearifolia	Asterac.	Aerial p., ext.	133
				Eriophyllum staechadifolium	Asterac.	Aerial p., ext.	133
				Grindelia nana	Asterac.	Aerial p., ext.	133
				Haplopappus baylahuen	Asterac.	Leaf and stem resin	223
				Lychnophora diamantina	Asterac.	Aerial parts	367
				Ozothamnus, 3 spp.	Asterac.	Aerial p., ext.	197
				Pulicaria dysenterica	Asterac.	Aerial p., ext.	182
				Pulicaria gnaphalodes	Asterac.	Aerial p., ext.	220
				Siegesbeckia jorullensis	Asterac.	Aerial p., ext.	182
				Capparis tweediana	Capparid.	Leaf	311
				Aeonium decorum	Crassul.	Aerial p., ext.	216
				Nothofagus obliqua	Fagac.	Aerial p., ext.	158
				Eriodictyon sessilifolium	Hydrophyll.	Leaf resin	135
				Kitaibelia vitifolia	Malvac.	Aerial p., ext.	226
				Mirabilis viscosa	Nyctagin.	Aerial p., ext.	43
				Polygonum punctatum	Polygon.	Aerial parts	368
				Vellozia conicostigma	Velloziac.	Leaf surface	105
				Viscum album, V. cruciatum	Viscac.	Cut. wax	318
				Propolis from Arizona			305
120	Tamarixetin	4'-OMe	3,5,7,3'-tetraOH	*Chromolaena odorata*	Asterac.	Aerial p., ext.	193
				Vellozia streptophylla	Velloziac.	Leaf surface	105

continued

TABLE 12.2
Flavonols and Their Methyl Ethers — *continued*

No.	OH-Substitution	OMe-Substitution	Trivial Name	Plant Species	Family	Plant Organ	Ref.
121	7,3',4'-triOH	3,5-diOMe	Caryatin	*Artemisia* spp.	Asterac.	Aerial parts	132
122	5,3',4'-triOH	3,7-diOMe		*Baccharis pilularis*	Asterac.	Aerial p., ext.	133
				Chrysothamnus viscidiflorus	Asterac.	Aerial p., ext.	199
				Eirmocephala megaphylla	Asterac.	Aerial parts	369
				Flourensia cernua	Asterac.	Aerial p., ext.	182
				Grindelia tarapacana	Asterac.	Aerial p., ext.	298
				Haplopappus taeda	Asterac.	Stems	370
				Holocarpha, 3 spp.	Asterac.	Leaf resin	371
				Ozothamnus lycopodioides, O. scutellifolius	Asterac.	Aerial p., ext.	197
				Palafoxia sphacelata	Asterac.	Aerial p., ext.	182
				Stegesbeckia jorulensis., S. orientalis	Asterac.	Aerial p., ext.	182
				Heliotropium pycophyllum., H. stenophyllum	Boragin.	Aerial p., ext.	36
				Aeonium spp.	Crassul.	Aerial p., ext.	216
				Eucryphia milliganii, E. moorei	Cunoniac.	Leaf, bud, ext.	165
				Pelargonium fulgidum, P. quercifolium	Geraniac.	Leaf exudate	142
				Mirabilis viscosa	Nyctagin.	Aerial p., ext.	43
				Rubus phoenicolasius	Rosac.	Aerial p., ext.	18
				Calceolaria, 4 spp.	Scrophul.	Aerial p., ext.	108
				Petunia surfina	Solanac.	Aerial p., ext.	214
				Salpiglossis sinuata	Solanac.	Aerial p., ext.	214
				Lantana camara	Verbenac.	Aerial p., ext.	228
				Viscum album	Viscac.	Aerial p., ext.	44
				Viscum cruciatum	Viscac.	Cut. wax	318
123	5,7,4'-triOH	3,3'-diOMe		*Anarthria scabra*	Anarthriac.	Leaf	554
				Heliotropium sinuatum	Boragin.	Leaf exudate	310
				Heliotropium stenophyllum	Boragin.	Aerial p., ext.	36
				Cleome amplyocarpa	Capparid.	Exudate	372
				Eucryphia lucida	Cunoniac.	Leaf, bud, ext.	165
				Cyperus alopecuroides	Cyperac.	Aerial parts	217

No.	Substitution	Name	Species	Family	Part	Ref.
			Nothofagus menziesii, N. nervosa	Fagac.	Aerial p., ext.	158
			Fouquieria splendens	Fouquierac.	Aerial p., ext.	325
			Mirabilis viscosa	Nyctagin.	Aerial p., ext.	43
			Cotoneaster microphylla	Rosac.	Aerial p., ext.	373
			Rubus phoenicolasius	Rosac.	Aerial p., ext.	18
			Barosma crenulata	Rutac.	Aerial p., ext.	326
			Melicope coodeana	Rutac.	Leaf	374
			Calceolaria, 5 spp.	Scrophul.	Aerial p., ext.	108
			Mimulus lewisii	Scrophul.	Leaf exudate	375
			Viscum album	Viscac.	Aerial p., ext.	44
			Viscum cruciatum	Viscac.	Cut. wax	318
			Propolis from Arizona			305
124	5,7,3'-triOH 3,4'-diOMe		*Asteriscus graveolens*	Asterac.	Aerial parts	111
			Chrysothamnus viscidiflorus	Asterac.	Aerial p., ext.	199
			Calycadenia truncata	Asterac.	Leaf exudate	236
			Flourensia cernua	Asterac.	Aerial p., ext.	182
			Grindelia tarapacana	Asterac.	Aerial p., ext.	298
			Psiadia dentata	Asterac.	Leaf	345
			Eucryphia milliganii	Cunoniac.	Leaf, bud, ext.	165
			Cyperus alopecuroides	Cyperac.	Aerial parts	217
			Petunia surfina	Solanac.	Aerial p., ext.	214
			Vellozia streptophylla	Velloziac.	Leaf surface	105
125	3,7,4'-triOH 5,3'-diOMe		*Heliotropium stenophyllum*	Boragin.	Leaf ex.	213
126	3,7,3'-triOH 5,4'-diOMe*		*Anarthria laevis*	Anarthriac.	Leaf	554
			Rhododendron ellipticum	Ericac.	Leaf	376
127	3,5,4'-triOH 7,3'-diOMe	Rhamnazin	*Ambrosia trifida*	Asterac.	Aerial p., ext.	233
			Artemisia campestris	Asterac.	Aerial parts	132
			Grindelia nana	Asterac.	Aerial p., ext.	133
			Heterotheca villosa	Asterac.	Aerial p., ext.	182
			Madia elegans	Asterac.	Aerial p., ext.	201
			Siegesbeckia jorullensis	Asterac.	Aerial p., ext.	182
			Stevia subpucescens	Asterac.	Aerial p., ext.	182
			Wedelia biflora	Asterac.	Leaf	377
			Heliotropium stenophyllum	Boragin.	Aerial p., ext.	36

continued

TABLE 12.2
Flavonols and Their Methyl Ethers — *continued*

No.	OH-Substitution	OMe-Substitution	Trivial Name	Plant Species	Family	Plant Organ	Ref.
				Capparis tweediana	Capparid.	Leaf	311
				Aeonium, 3 spp.	Crassul.	Aerial p., ext.	216
				Eucryphia jinksii	Cunoniac.	Leaf, bud, ext.	165
				Nothofagus cunninghamii	Fagac.	Aerial p., ext.	158
				Kitaibelia vitifolia	Malvac.	Aerial p., ext.	226
				Mirabilis viscosa	Nyctagin.	Aerial p., ext.	43
				Polygonum punctatum	Polygon.	Aerial parts	368
				Notholaena nivea	Pteridaceae	Frond exud.	45
				Salpiglossis sinuata	Solanac.	Aerial p., ext.	214
				Viscum cruciatum	Viscac.	Cut. wax	318
				Propolis from Arizona			305
128	3,5,3'-triOH	7,4'-diOMe	Ombuin	*Chromolaena odorata*	Asterac.	Aerial p., ext.	193
129	3,5,7-triOH	3',4'-diOMe	Dillenetin	*Amomum koenigii*	Zingib.	Fruit	324
130	7,4'-diOH	3,5,3'-triOMe		*Chromolaena odorata*	Asterac.	Aerial p., ext.	193
131	5,4'-diOH	3,7,3'-triOMe	Pachypodol	*Chrysothamnus viscidiflorus*	Asterac.	Aerial p., ext.	199
				Flourensia cernua	Asterac.	Aerial p., ext.	182
				Grindelia robusta	Asterac.	Leaf exudate	344
				Grindelia tarapacana	Asterac.	Aerial p., ext.	352
				Grindelia squarrosa	Asterac.	Aerial p., ext.	182
				Heterotheca villosa	Asterac.	Aerial p., ext.	182
				Senecio viscosa	Asterac.	Aerial p., ext.	32
				Senecio viscosissimus	Asterac.	External	378
				Xanthocephalum gymnospermoides	Asterac.	Aerial p., ext.	182
				Heliotropium sinuatum	Boragin.	Leaf exudate	310
				Nothofagus cunninghamii	Fagac.	Aerial p., ext.	158
				Mirabilis viscosa	Nyctagin.	Aerial p., ext.	43
				Bosistoa floydii, *B. medicinalis*	Rutac.	Leaf	322
				Melicope elleryana	Rutac.	Fruit	379
				Melicope ternata	Rutac.	Bark	380

No.			Name	Species	Family	Plant part	Ref.
132	5,3'-diOH	3,7,4'-triOMe	Ayamin	Euodia merrillii	Rutac.	Fruit	323
				Euodia viticina	Rutac.	Fruit	381
				Adenosma capitatum	Scrophul.	External?	382
				Mimulus lewisii	Scrophul.	Leaf exudate	375
				Petunia surfina	Solanac.	Aerial p., ext.	214
				Salpiglossis sinuata	Solanac.	Aerial p., ext.	214
				Viscum album	Viscac.	Aerial p., ext.	44
				Viscum cruciatum	Viscac.	Cut. wax	318
133	5,7-diOH	3,3',4'-triOMe		Bahia glandulosa	Asterac.	Aerial parts	221
				Grindelia squarrosa, tenella	Asterac.	Aerial p., ext.	182
				Haplopappus hirtellus	Asterac.	Aerial parts	327
				Ozothamnus scutellifolius	Asterac.	Aerial p., ext.	197
				Pisadia dentata	Asterac.	Leaf	345
				Siegesbeckia jorullensis, orientalis	Asterac.	Aerial p., ext.	182
				Stevia subpubescens	Asterac.	Aerial p., ext.	182
				Heliotropium chenopodiaceum var. ericoideum	Boragin.	Exudate	35
				Heliotropium pycnophyllum	Boragin.	Aerial p., ext.	36
				Eucryphia lucida, E. milliganii	Cunoniac.	Leaf, bud, ext.	165
				Petunia surfina	Solanac.	Aerial p., ext.	214
				Lantana camara	Verbenac.	Aerial p., ext.	228
				Amomum koenigii	Zingib.	Fruit	324
				Flourensia cernua	Asterac.	Aerial p., ext.	182
				Grindelia nana	Asterac.	Aerial p., ext.	133
				Grindelia robusta	Asterac.	Leaf exudate	344
				Siegesbeckia jorullensis	Asterac.	Aerial p., ext.	182
				Eucryphia lucida, E. milliganii	Cunoniac.	Leaf, bud, ext.	165
				Mimulus lewisii	Scrophul.	Leaf exudate	375
				Petunia surfina	Solanac.	Aerial p., ext.	214
				Barbacenia rubro-virens	Velloziac.	Leaf surface	105
134	3,7-diOH	5,3',4'-triOMe*		Amomum koenigii	Zingib.	Fruit	324
135	3,5-diOH	7,3',4'-triOMe		Chrysothamnus viscidiflorus	Asterac.	Aerial p., ext.	199
				Chromolaena odorata	Asterac.	Aerial p., ext.	193
				Aeonium arboreum	Crassul.	Aerial p., ext.	216
				Kitaibelia vitifolia	Malvac.	Aerial p., ext.	226

continued

TABLE 12.2
Flavonols and Their Methyl Ethers — continued

No.	OH-Substitution	OMe-Substitution	Trivial Name	Plant Species	Family	Plant Organ	Ref.
136	4'-OH	3,5,7,3'-tetraOMe		Mirabilis viscosa	Nyctagin.	Aerial p., ext.	43
				Amomum koenigii	Zingib.	Fruit	324
137	5-OH	3,7,3',4'-tetraOMe	Retusin	Distemonanthus benthamianus	Fabac.	Heartwood	383
				Artemisia rupestris	Asterac.	Aerial parts	132
				Brickellia eupatorioides	Asterac.	Aerial p., ext.	215
				Grindelia nana	Asterac.	Aerial p., ext.	133
				Grindelia tenella	Asterac.	Aerial p., ext.	182
				Siegesbeckia jorullensis, S.orientalis	Asterac.	Aerial p., ext.	182
				Urolepis hecatantha	Asterac.	Aerial parts	262
				Xanthocephalum gymnospermoides	Asterac.	Aerial p., ext.	182
				Aeonium lindleyi	Crassul.	Aerial p., ext.	216
				Eucryphia lucida, E. milliganii	Cunoniac.	Leaf, bud, ext.	165
				Bridelia ferruginea	Euphorb.	Stem bark	384
				Nothofagus cunninghamii	Fagac.	Aerial p., ext.	158
				Mirabilis viscosa	Nyctagin.	Aerial p., ext.	43
				Evodia merrillii	Rutac.	Fruit	323
				Solanum plusodum	Solanac.	Aerial parts	317
				Cryptomeria japonica	Taxodiac.	Leaf	39
				Amomum koenigii	Zingib.	Fruit	324
138	3-OH	5,7,3',4'-tetraOMe					
139	(3,5,7,3',5'-pentaOH)	3,5,7,3',4'-pentaOMe		Amomum koenigii	Zingib.	Fruit	324
140	3,5,7-triOH	3',5'-diOMe	Morelosin				
141	3,6,7,3',4'-pentaOH		Rhynchosin				
142	6,7,3',4'-tetraOH	3-OMe*		Graziela mollissima	Asterac.	Aerial parts	330
143	3,6,3',4'-tetraOH	7-OMe*		Dalbergia odorifera	Fabac.	Heartwood	333
144	(7-OH)	3,6,3',4'-tetraOMe)	"Santoflavone"	Achillea santolina	Asterac.	Aerial parts	d
145	3,7,8,3',4'-pentaOH	3-OMe	Melanoxetin	Acacia karroo, A. montana	Mimosac.	Heartwood	332
146	7,8,3',4'-tetraOH	3-OMe	Transilitin	Acacia nigrescens	Mimosac.	Heartwood	385

No.	Hydroxyl	Methoxyl	Name	Species	Family	Source	Ref.
147	3,7,3',4'-tetraOH	8-OMe		Acacia karroo ssp. montana	Mimosac.	Heartwood	332
148	3,7,8,4'-tetraOH	3'-OMe*		Acacia nigrescens	Mimosac.	Heartwood	385
149	7,8,4'-triOH	3,3'-diOMe*		Acacia karroo ssp. montana	Mimosac.	Heartwood	332
150	3,7,2',3',4'-pentaOH (G)						
151	3,7,3',4',5'-pentaOH		Robinetin				
152	3,7,4'-triOH	3',5'diOMe*	Laurentinol*	Millettia laurentii	Fabac.	Heartwood	386
153	4',5'-diOH	3,7,3'-triOMe*		Duroia hirsuta	Rubiac.	Root	31
154	3,4'-diOH	7,3',5'-triOMe*		Duroia hirsuta	Rubiac.	Root	31
155	4'-OH	3,7,3',5'-tetraOMe*		Duroia hirsuta	Rubiac.	Root	31
Hexa-O-substituted flavonols							
(3,5,6,7,8,3'-hexaOH)							
156	5-OH	3,6,7,8,3'-pentaOMe	Emmaosunin	Ester only		Synthesis	335
(3,5,6,7,8,4'-hexaOH)							
157	5,7,8,4'-tetraOH	3,6-diOMe					
158	5,8,4'-triOH	3,6,7-triOMe					
159	5,7,4'-triOH	3,6,8-triOMe	Sarothrin			Synthesis	335
160	3,5,4'-triOH	6,7,8-triOMe					
161	5,7,8-triOH	3,6,4'-triOMe		Ester only			
162	5,4'-diOH	3,6,7,8-tetraOMe	Calycopterin	Trixis vauthieri	Asterac.	Leaf	353
163	5,7-diOH	3,6,8,4'-tetraOMe	Araneosol	Nothofagus, 3 spp.	Fagac.	Aerial p., ext.	158
				Rosa centifolia	Rosac.	Aerial p., ext.	18
				Drummondita calida	Rutac.	Aerial parts	358
						Synthesis	335
164	3,8-diOH	5,6,7,4'-tetraOMe	Eriostemin			Synthesis	336
165	3,5-diOH	6,7,8,4'-tetraOMe*		Helichrysum cassianum	Asterac.	Aerial p., ext.	387
						Synthesis	336
166	5-OH	3,6,7,8,4'-pentaOMe	5-Hydroxy-auranetin	Gnaphalium affine	Asterac.	Aerial parts	314
				Cleome spinosa	Capparid.	Aerial p., ext.	321
				Nothofagus menziesii, N. nervosa	Fagac.	Aerial p., ext.	158
				Rosa centifolia cv. muscosa	Rosac.	Aerial p., ext.	18
				Drummondita calida	Rutac.	Aerial parts	358
				Antirrhinum hispanicum	Scrophul.	Aerial p., ext.	108
						Synthesis	336

continued

TABLE 12.2
Flavonols and Their Methyl Ethers — continued

No.	OH-Substitution	OMe-Substitution	Trivial Name	Plant Species	Family	Plant Organ	Ref.
167	(3,5,6,7,2',3'-hexaOH)	3,5,6,7,8,4'-hexaOMe*		Drummondita calida	Rutac.	Aerial parts	358
						Synthesis	336
168	5,2',3'-triOH (3,5,6,7,2',4'-hexaOH)	3,6,7-triOMe*		Vitex rotundifolia	Verbenac.	Fruit	388
169	5,4-diOH (3,5,6,7,2',5'-hexaOH)	3,6,7,2'-tetraOMe	Chrysosplin				
170	5-OH	3,6,7,2',5'OMe*	Grantioidin				
171	3,5,6,7,3',4'-hexaOH		Quercetagetin			Synthesis	47
172	5,6,7,3',4'-pentaOH	3-OMe				Synthesis	28[e]
173	3,6,7,3',4'-pentaOH	5-OMe	"Allopatuletin"	Tagetes patula	Asterac.	Petals	389
174	3,5,7,3',4'-pentaOH	6-OMe	Patuletin	Anthemis tinctoria	Asterac.	Flower	132
				Artemisia barrelieri	Asterac.	Aerial parts	281
				Centaurea incana	Asterac.	Aerial parts	182
				Chrysactinia mexicana	Asterac.	Aerial p., ext.	133
				Eriophyllum confertum	Asterac.	Aerial p., ext.	390
				Pallenis spinosa	Asterac.	Aerial parts	342
175	3,5,6,3',4'-pentaOH	7-OMe				Synthesis	47
176	3,5,6,7,4'-pentaOH	3'-OMe		Glycoside only		Synthesis	47
177	3,5,6,7,3'-pentaOH	4'-OMe		Glycoside only		Synthesis	47
178	5,7,3',4'-tetraOH	3,6-diOMe	Axillarin	Ambrosia chamissonis	Asterac.	Aerial p., ext.	233
				Artemisia australis	Asterac.	Aerial parts	132
				Asteriscus sericeus	Asterac.	Aerial p., ext.	32
				Bahia pringlei	Asterac.	Aerial parts	391
				Calycadenia, 3 spp.	Asterac.	Leaf exudate	236
				Eriophyllum confertum	Asterac.	Aerial p., ext.	133
				Eriophyllum staechadifolium	Asterac.	Aerial p., ext.	133
				Gymnosperma glutinosa	Asterac.	Aerial p., ext.	196

No.	Substitution	OMe	Name	Species	Family	Part	Ref.
179	5,6,3',4'-tetraOH	3,7-diOMe	Tomentin	*Gymnosperma glutinosum*	Asterac.	Aerial p., ext.	133
				Helichrysum, 8 spp.	Asterac.	Aerial p., ext.	237
				Holocarpha heermannii	Asterac.	Leaf resin	371
				Inula brittanica	Asterac.	Aerial p., ext.	196
				Inula germanica	Asterac.	Aerial p., ext.	32
				Madia sativa	Asterac.	Leaf exudate	236
				Madia, 4 spp.	Asterac.	Aerial p., ext.	200
				Oncosiphon grandiflorum	Asterac.	Aerial p., ext.	32
				Ozothamnus lycopodioides	Asterac.	Aerial p., ext.	197
				Tanacetum balsamita	Asterac.	Aerial p., ext.	32
				Tanacetum parthenium, T. vulgaris	Asterac.	Leaf	49
				Xanthium strumarium	Asterac.	Aerial p., ext.	32
				Cleome amplyocarpa	Capparid.	Exudate	372
180	5,6,7,4'-tetraOH	3,3'-diOMe		*Ambrosia artemisifolia*	Asterac.	Aerial p., ext.	182
181	5,6,7,3'-tetraOH	3,4'-diOMe		*Artemisia abrotanum*	Asterac.	Aerial parts	392
				Holocarpha, 4 spp.	Asterac.	Leaf resin	371
				Inula germanica	Asterac.	Aerial p., ext.	32
				Madia sativa	Asterac.	Aerial p., ext.	133
				Tanacetum parthenium	Asterac.	Aieral parts	48
						synthesis	28
				Artemisia abrotanum	Asterac.	Aerial parts	392
						synthesis	28
182	3,5,3',4'-tetraOH	6,7-diOMe	Eupatolitin	*Pulicaria gnaphaloides*	Asterac.	Aerial p., ext.	220
183	3,5,7,4'-tetraOH	6,3'-diOMe	Spinacetin	*Anthemis tinctoria*	Asterac.	Flower	389
				Eriophyllum confertum	Asterac.	Aerial p., ext.	133
						Synthesis	342
184	3,5,7,3'-tetraOH	6,4'-diOMe	Laciniatin	*Chromolaena odorata*	Asterac.	Aerial p., ext.	193
						Synthesis	342
185	5,3',4'-triOH	3,6,7-triOMe	Chrysosplenol-D	*Achillea clusiana*	Asterac.	Aerial p., ext.	196
				Artemisia australis, A. mongolica	Asterac.	Aerial parts	132
				Brickellia eupatorioides	Asterac.	Aerial p., ext.	215
				Flourensia cernua	Asterac.	Aerial p., ext.	182
				Hemizonia lutescens	Asterac.	Aerial p., ext.	133
				Heterotheca villosa	Asterac.	Aerial p., ext.	182

continued

TABLE 12.2
Flavonols and Their Methyl Ethers — continued

No.	OH-Substitution	OMe-Substitution	Trivial Name	Plant Species	Family	Plant Organ	Ref.
				Inula brittanica	Asterac.	Aerial p., ext.	32
				Inula germanica	Asterac.	Aerial p., ext.	32
				Madia sativa	Asterac.	Leaf exudate	236
				Oncosiphon grandiflorum	Asterac.	Aerial p., ext.	32
				Pulicaria gnaphalodes	Asterac.	Aerial p., ext.	220
				Tanacetum polycephalum	Asterac.	Aerial p., ext.	220
				Rosa centifolia cv. *muscosa*	Rosac.	Aerial p., ext.	18
186	5,7,4'-triOH	3,6,3'-triOMe	Jaceidin	*Achillea clusiana*	Asterac.	Aerial p., ext.	196
				Achillea micrantha	Asterac.	Aerial parts	343
				Asteriscus graveolens	Asterac.	Aerial parts	111
				Asteriscus sericeus	Asterac.	Aerial p., ext.	32
				Bahia pringlei	Asterac.	Aerial parts	391
				Centaurea, 3 spp.	Asterac.	Aerial p., ext.	182
				Eriophyllum staechadifolium	Asterac.	Aerial p., ext.	133
				Eupatorium buniifolium	Asterac.	Aerial parts	393
				Flourensia cernua	Asterac.	Aerial p., ext.	182
				Inula brittanica	Asterac.	Aerial p., ext.	196
				Lagophylla glandulosa	Asterac.	Leaf exudate	236
				Pulicaria gnaphalodes	Asterac.	Aerial p., ext.	220
				Tanacetum parthenium	Asterac.	Leaf	49
				Alkanna orientalis	Boragin.	Aerial parts	37
				Mirabilis viscosa	Nyctagin.	Aerial p., ext.	43
				Melicope coodeana	Rutac.	Leaf	374
187	5,7,3'-triOH	3,6,4'-triOMe	Centaureidin	*Achillea atrata* ssp. *multifida*	Asterac.	Aerial parts	355
				Achillea multifida	Asterac.	Aerial p., ext.	196
				Ambrosia chamissonis	Asterac.	Aerial p., ext.	233
				Anthemis tinctoria	Asterac.	Aerial p., ext.	356
				Artemisia abrotanum	Asterac.	Aerial p., ext.	392
				Artemisia barrelieri	Asterac.	Aerial parts	132
				Baccharis saligna	Asterac.	Aerial parts	394

No.	OH	OMe	Name	Species	Family	Plant part	Ref.
188	5,6,4'-triOH	3,7,3'-triOMe	Chrysosplenol-C	Bahia xylopoda	Asterac.	Aerial parts	391
				Centaurea jacea	Asterac.	Aerial p., ext.	182
				Eriophyllum confertum	Asterac.	Aerial p., ext.	133
				Eupatorium buniifolium	Asterac.	Aerial parts	393
				Grindelia robusta	Asterac.	Leaf exudate	344
				Grindelia tarapacana	Asterac.	Aerial p., ext.	352
				Stevia berlandieri	Asterac.	Aerial p., ext.	182
				Asteriscus sericeus	Asterac.	Aerial p., ext.	32
				Tanacetum microphyllum	Asterac.	Aerial parts	395
				Aeonium spp.	Crassul.	Aerial p., ext.	216
189	5,6,3'-triOH	3,7,4'-triOMe	Oxyayanin-B	Tanacetum parthenium	Asterac.	Aerial parts	48
190	3,5,4'-triOH	6,7,3'-triOMe	Veronicafolin	Pterocaulon sphacelatum	Asterac.	Aerial parts	396
191	3,5,3'-triOH	6,7,4'-triOMe	Eupatin	Pulicaria dysenterica	Asterac.	External	347
192	3,5,7-triOH (3,5,6-triOH)	6,3',4'-triOMe 7,3',4'-triOMe)*		Pulicaria dysenterica	Asterac.	Aerial p., ext.	182
193	7,4'-diOH	3,5,6,3'-tetraOMe		Arnica longifolia	Asterac.	Flower	210
194	6,4'-diOH	3,5,7,3'-tetraOMe				Only synth.	313
195	5,4'-diOH	3,6,7,3'-tetraOMe	Chrysosplenetin	Achillea ageratum	Asterac.	Aerial parts	320
				Artemisia clusiana	Asterac.	Aerial p., ext.	196
				Artemisia mongolica	Asterac.	Aerial parts	132
				Artemisia nana	Asterac.	Aerial p., ext.	196
				Artemisia sieversii	Asterac.	Aerial parts	271
				Grindelia robusta	Asterac.	Leaf exudate	344
				Grindelia tarapacana	Asterac.	Aerial p., ext.	352
				Haplopappus bezanillanus	Asterac.	Aerial parts	327
				Inula brittanica	Asterac.	Aerial p., ext.	129
				Inula brittanica	Asterac.	Aerial p., ext.	32
				Parthenium incanum	Asterac.	Aerial p., ext.	182
				Pulicaria gnaphalodes	Asterac.	Aerial p., ext.	220
				Adenosma capitatum	Scrophul.	External?	382

continued

TABLE 12.2
Flavonols and Their Methyl Ethers — *continued*

No.	OH-Substitution	OMe-Substitution	Trivial Name	Plant Species	Family	Plant Organ	Ref.
196	5,3'-diOH	3,6,7,4'-tetraOMe	Casticin	Achillea sibirica subsp.mongolica	Asterac.	Aerial p., ext.	354
				Artemisia abrotanum	Asterac.	Aerial parts	392
				Lagophylla glandulosa	Asterac.	Leaf exudate	236
				Parthenium incanum	Asterac.	Aerial p., ext.	182
				Tanacetum polycephalum	Asterac.	Aerial parts	220
				Vitex rotundifolia	Verbenac.	Fruit	388
197	5,7-diOH	3,6,3',4'-tetraOMe	Bonanzin	Bahia xylopoda	Asterac.	Aerial parts	391
198	5,6-diOH	3,7,3',4'-tetraOMe		Pulicaria dysenterica	Asterac.	Aerial p., ext.	32
199	3,3'-diOH	5,6,7,4'-tetraOMe	Eupatoretin				
200	3,5-diOH	6,7,3',4'-tetraOMe		Artemisia annua	Asterac.	Aerial parts	397
201	3,5-diOH	6,7,3',4'-tetraOMe		Baccharis saligna	Asterac.	Aerial parts	394
				Parthenium incanum	Asterac.	Aerial p., ext.	182
202	4'-OH	3,5,6,7,3'-pentaOMe		Pallemis spinosa	Asterac.	Aerial parts	390
203	7-OH	3,5,6,3',4'-pentaOMe*		Vigua spiralis	Fabac.	Leaf and stem	398
				Citrus sinensis	Rutac.	Fruit peel oil	16
204	5-OH	3,6,7,3',4'-pentaOMe	Artemetin	Achillea conferta	Asterac.	Aerial parts	269
				Achillea sibirica subsp.mongolica	Asterac.	Aerial p., ext.	354
				Artemisia annua	Asterac.	Aerial parts	397
				Artemisia mongolica, A.verlotiorum	Asterac.	Aerial p., ext.	132
				Inula brittanica	Asterac.	Aerial p., ext.	196
				Inula brittanica	Asterac.	Aerial p., ext.	32
				Parthenium incanum	Asterac.	Aerial p., ext.	182
				Ficus altissima	Morac.	Aerial parts	30
				Adenosma capitatum	Scrophul.	External?	382
				Vitex rotundifolia	Verbenac.	Fruit	388
205	3-OH	5,6,7,3',4'-pentaOMe*	Marionol*	Chromolaena odorata	Asterac.	Aerial p., ext.	399
206	—	3,5,6,7,3',4'-hexaOMe (7-chloro-derivative)		Pallemis spinosa	Asterac.	Aerial parts	390
	(3,5,6,8,3',4'-hexaOH)						
	(3,5,7,8,2',3'-hexaOH)						
207	5,2',3'-triOH	3,7,8-triOMe					

No.	Hydroxylation	Methylation	Name / Source	Family	Plant part	Ref.
	(3,5,7,8,2',4'-hexaOH)					
208	3,5,2'-triOH	7,8,4'-triOMe				
209	5,4'-diOH	3,7,8,2'-tetraOMe				
210	5-OH	3,7,8,2',4'-pentaOMe	Ester only			
	(3,5,7,8,2',5'-hexaOH)					
211	5,2',5'-triOH	3,7,8-triOMe				
212	5,5'-diOH	3,7,8,2'-tetraOMe				
213	3,5,7,8,3',4'-hexaOH		Gossypetin			
214	5,7,8,3',4'-pentaOH	3-OMe	Glycoside only			
215	3,5,8,3',4'-pentaOH	7-OMe	Glycoside only			
216	3,5,7,3',4'-pentaOH	8-OMe	Glycoside only			
217	3,5,7,8,4'-pentaOH	3'-OMe				
218	7,8,3',4'-triOH	3,5-diOMe*	Eugenia edulis	Myrtac.	Leaf	400
219	5,8,3',4'-tetraOH	3,7-diOMe	Chorizanthe diffusa	Polygon.	Whole plant	7
220	5,7,3',4'-tetraOH	3,8-diOMe	Madia sativa	Asterac.	Aerial p., ext.	133
221	3,5,3',4'-tetraOH	7,8-diOMe	Madia, 4 spp.	Asterac.	Aerial p., ext.	200
222	5,7,8,4'-tetraOH	3,3'-diOMe	Ozothamnus hookeri	Asterac.	Aerial p., ext.	70
223	5,7,8,3'-tetraOH	3,4'-diOMe	Zanthoxylum alatum	Rutac.	Seed	401
224	3,5,8,3'-tetraOH	7,4'-diOMe	Glycoside only			
225	3,5,7,4'-tetraOH	8,3'-diOMe	Glycoside only			
226	3,5,7,3'-tetraOH	8,4'-diOMe				
227	5,3',4'-triOH	3,7,8-triOMe	Calycadenia ciliata, C. multiglandulosa	Asterac.	Leaf exudate	236
			Madia anomala	Asterac.	Aerial p., ext.	200
			Ozothamnus lycopodioides	Asterac.	Aerial p., ext.	197
			Calceolaria tenella	Scrophul.	Aerial p., ext.	108
228	5,4'-diOH	3,7,3'-triOMe	Haplopappus deserticola	Asterac.	Resin. exud.	361
229	5,7,4'-triOH	3,8,3'-triOMe	Boronia coerulescens	Rutac.	Aerial parts	95
230	3,5,4'-triOH	7,8,3'-triOMe	Calceolaria scabiosifolia	Scrophul.	Aerial p., ext.	108
231	3,5,7-triOH	8,3',4'-triOMe	Synthesis			402

continued

TABLE 12.2
Flavonols and Their Methyl Ethers — continued

No.	OH-Substitution	OMe-Substitution	Trivial Name	Plant Species	Family	Plant Organ	Ref.
232	5,4'-diOH	3,7,8,3'-tetraOMe	Ternatin	Euodia viticina	Rutac.	Fruit	403
				Melicope elleryana	Rutac.	Fruit	379
				Melicope simplex, M. ternata	Rutac.	Bark	380
				Boronia coerulescens	Rutac.	Aerial parts	95
233	5,3'-diOH	3,7,8,4'-tetraOMe		Calceolaria tenella	Scrophul.	Aerial p., ext.	108
234	5,8-diOH	3,7,3',4'-tetraOMe		Solanum paludosum	Solanac.	Aerial parts	317
235	5,7-diOH	3,8,3',4'-tetraOMe				Synthesis	402
236	7-OH	3,5,8,3',4'-pentaOMe				Synthesis	402
237	5-OH	3,7,8,3',4'-pentaOMe		Helichrysum foetidum	Asterac.	Aerial p., ext.	182
				Ozothamnus lycopodioides	Asterac.	Aerial p., ext.	197
				Murraya paniculata	Rutac.	Fruit	404
				Solanum paludosum	Solanac.	Aerial parts	317
238	—	3,5,7,8,3',4'-hexaOMe		Murraya paniculata	Rutac.	Fruit	404
239	3,6,7,8,3',4'-hexaOH						
240	3,5,7,2',3',4'-hexaOH						
241	5,2',3'-triOH	3,7,4'-triOMe	Apuleidin	Glycoside only			
242	3,5,7,2',4',5'-hexaOH		5'-Hydroxy-morin	Glycoside only			
243	5,7,4',5'-tetraOH	3,2'-diOMe					
244	5,7,2',5'-tetraOH	3,4'-diOMe					
245	5,2',5'-triOH	3,7,4'-triOMe	Oxyayanin-A	Psiadia punctulata	Asterac.	Aerial parts	33
246	3,5,7-triOH	2',4',5'-triOMe*					
247	2',5'-diOH	3,5,7,4'-tetraOMe					
248	5,2'-diOH	3,7,4',5'-tetraOMe					
249	5'-OH	3,5,7,2',4'-pentaOMe					
250	2'-OH	3,5,7,4',5'-pentaOMe					
251	5-OH	3,7,2',4',5'-pentaOMe					
252	—	3,5,7,2',4',5'-hexaOMe					
253	(3,6,7,2',4',5'-hexaOH) 2',5'-diOH	3,6,7,4'-tetraOMe					

No.	OH	OMe	Name	Species	Family	Plant part	Ref.
254	6,5'-diOH	3,7,2',4'-tetraOMe		*Bridelia ferruginea*	Euphorb.	Stem bark	384
255	3,5,7,3',4',5'-hexaOH		Myricetin	*Millettia racemosa*	Fabac.	Leaf	38
				Nothofagus antarctica	Fagac.	Aerial p., ext.	158
				Nothofagus antarctica	Fagac.	Aerial p., ext.	302
				Plinia pinnata	Myrtac.	Aerial parts	405
				Eucalyptus globulus	Myrtac.	Pollen	20
				Kunzea ericoides	Myrtac.	Pollen	20
256	5,7,3',4',5'-pentaOH	3-OMe	Annulatin	Glycoside only			
257	3,7,3',4',5'-pentaOH	5-OMe		Glycoside only			
258	3,5,3',4',5'-pentaOH	7-OMe	Europetin				
259	3,5,7,4',5'-pentaOH	3'-OMe	Laricytrin				
260	3,5,7,3',5'-pentaOH	4'-OMe	Mearnsetin				
261	5,7,3',5'-tetraOH	3,4'-diOMe					
262	5,7,3',4'-tetraOH	3,5'-diOMe					
263	3,5,3',5'-tetraOH	7,4'-diOMe					
264	3,5,7,5'-tetraOH	3',4'-diOMe					
265	3,5,7,4'-tetraOH	3',5'-diOMe	Syringetin	*Eugenia edulis*	Myrtac.	Leaf	400
266	7,4',5'-triOH	3,5,3'-triOMe		*Aeonium*, 4 spp.	Crassul.	Aerial p., ext.	216
267	5,4',5'-triOH	3,7,3'-triOMe		*Aeonium sedifolium*	Crassul.	Aerial p., ext.	216
268	5,3',5'-triOH	3,7,4'-triOMe		*Aeonium sedifolium*	Crassul.	Aerial p., ext.	216
269	5,7,5'-triOH	3,3',4'-triOMe		*Bridelia ferruginea*	Euphorb.	Stem bark	384
270	5,7,4'-triOH	3,3',5'-triOMe		*Xanthocephalum gymnospermum*	Asterac.	Aerial p., ext.	182
271	3,5,5'-triOH	7,3',4'-triOMe		*Tillandsia usneoides*	Brom.	Aerial p., ext.	406
272	5,7-diOH	3',4',5'-triOMe*	Ferrugin*	*Bridelia ferruginea*	Euphorb.	Stem bark	384
273	5,5'-diOH	3,7,3',4'-tetraOMe		*Heliotropium megalanthum*	Boragin.	Leaf exudate	378
274	5,4'-diOH	3,7,3',5'-tetraOMe		*Xanthocephalum gymnospermum*	Asterac.	Aerial p., ext.	182
275	5,7-diOH	3,3',4',5'-tetraOMe	Combretol	*Aeonium*, 5 spp.	Crassul.	Aerial p., ext.	216
276	3,5-diOH	7,3',4',5'-tetraOMe*		*Bosistoa floydii*	Rutac.	Leaf	322
277	5-OH	3,7,3',4',5'-pentaOMe		*Murraya paniculata*	Rutac.	Leaf	284
278	*3-OH*	*5,7,3',4',5'-pentaOMe*					

continued

TABLE 12.2
Flavonols and Their Methyl Ethers — continued

No.	OH-Substitution	OMe-Substitution	Trivial Name	Plant Species	Family	Plant Organ	Ref.
279		3,5,7,3',4',5'-hexaOMe*		Ficus altissima	Morac.	Aerial parts	30
280	3,5,8,3',4',5'-hexaOH (3,6,7,3',4',5'-hexaOH)			Murraya paniculata	Rutac.	Flower	61
281	3,7,4',5'-tetraOH (3,7,8,2',4',5'-hexaOH)	6,3'-diOMe*	Inucrithmin*	Inula crithmoides	Asterac.	Aerial parts	407
282	2'-OH	3,7,4',5'-pentaOMe*		Parkia clappertoniana	Fabac.	Leaf	331
Hepta-O-substituted flavonols							
	(3,5,6,7,8,2',4'-OH)						
283	5,2',4'-triOH	3,6,7,8-tetraOMe					
284	5,4'-diOH	3,6,7,8,2'-pentaOMe					
285	5,2'-diOH	3,6,7,8,4'-pentaOMe					
286	5-OH	3,6,7,8,2',4'-hexaOMe					
	(3,5,6,7,8,2',5'-OH)						
287	5-OH	3,6,7,8,2',5'-hexa-OMe*	Grantiodinin*	Inula grantioides	Asterac.	Aerial parts	34
289	3,5,6,7,8,3',4'-heptaOH						
290	5,7,8,3',4'-pentaOH	3,6-diOMe					
291	3,5,7,3',4'-pentaOH	6,8-diOMe					
292	5,7,3',4'-tetraOH	3,6,8-triOMe		Calycadenia, 3 spp. Madia, 3 spp.	Asterac. Asterac.	Synthesis Leaf exudate Aerial p., ext. Synthesis	335 236 200 335
293	5,6,3',4'-tetraOH	3,7,8-triOMe				Synthesis	335
294	3,5,3',4'-tetraOH	6,7,8-triOMe				Synthesis	335
295	3,5,7,4'-tetraOH	6,8,3'-triOMe	Limocitrol			Synthesis	335
296	3,5,7,3'-tetraOH	6,8,4'-triOMe	Isolimocitrol	Glycoside only		Synthesis	335
297	5,3',4'-triOH	3,6,7,8-tetraOMe		Helichrysum, 8 spp. Madia, 3 spp.	Asterac. Asterac.	Aerial parts, ext. Aerial p., ext.	237 200
298	5,7,4'-triOH	3,6,8,3'-tetraOMe		Rosa centifolia cv. Muscosa	Rosac.	Aerial p., ext. Synthesis	18 335
299	5,7,3'-triOH	3,6,8,4'-tetraOMe				Synthesis	335

No.	OH substitution	OMe substitution	Trivial name	Species	Plant source	Family	Ref.
300	5,6,3'triOH	3,7,8,4'tetraOME*		Calycadenia ciliata	Leaf exudate	Asterac.	236
301	3,5,4'-triOH	6,7,8,3'-tetraOMe			Synthesis		336
302	3,5,3'-triOH	6,7,8,4'-tetraOMe*		Zieridium pseudobtusifolium	Leaf	Rutac.	408
					Synthesis		335
303	3,5,8-triOH	6,7,3'4'-tetraOMe*		Murraya paniculata var. omphalocarpa		Rutac.	409
304	5,4-diOH	3,6,7,8,3'-pentaOMe		Calycadenia ciliata, C. multiglandulosa	Leaf exudate	Asterac.	236
305	5,4-diOH	3,6,7,8,3'-pentaOMe		Gnaphalium luteo-album	Leaf exudate	Asterac.	315
				Madia dissitiflora	Aerial p., ext.	Asterac.	200
				Polanisia dodecandra	Aerial parts	Capparid.	410
					Synthesis		336
306	5,3'-diOH	3,6,7,8,4'-pentaOMe		Calycadenia truncata, C. mollis	Leaf exudate	Asterac.	236
				Polanisia dodecandra	Aerial parts	Capparid.	410
				Acronychia porteri	Leaf	Rutac.	408
				Zieridium pseudobtusifolium	Leaf	Rutac.	408
307	5,7-diOH	3,6.8,3',4'-pentaOMe*		Melicope coodeana	Leaf	Rutac.	374
308	3,5-diOH	6,7,8,3',4'-pentaOMe*		Acronychia porteri	Leaf	Rutac.	408
309	7-OH	3,5,6,8,3',4'-hexaOMe*		Citrus sinensis	Fruit peel oil	Rutac.	16
310	5-OH	3,6,7,8,3',4'-hexaOMe		Acronychia porteri	Leaf	Rutac.	408
311	3-OH	5,6,7,8,3',4'-hexaOMe	Natsudaidain	Citrus hassaku, C. madurensis	Leaf	Rutac.	410
312		3,5,6,7,8,3',4'-heptaOMe	"HEPTA"	Citrus hassaku, C. madurensis	Leaf	Rutac.	410
313		3,5,6,7,8,3',4'-heptaOMe		Citrus sinensis	Fruit peel oil	Rutac.	16
314	(3,5,6,7,2',3',4'-OH) 5,6,2',3'-tetraOH	3,7,4'-triOMe	Apuleisin				
315	(3,5,6,7,2',4',5'-OH) 5,7,4',5'-tetraOH	3,6,2'-triOMe					
316	5,7,2',4'-tetraOH	3,6,3'-triOMe					
317	5,7,2',5'-tetraOH	3,6,4'-triOMe					
318	5,2',5'-triOH	3,6,7,4'-tetraOMe					
319	5,7,5'-triOH	3,6,2',4'-tetraOMe*					
320	5,7,2'-triOH	3,6,4',5'-tetraOMe					
321	5,6,5'-triOH	3,7,2',4'-tetraOMe					
322	2',5'-diOH	3,5,6,7,4'-pentaOMe	Apulein	Eupatorium buniifolium	Leaf and stem	Asterac.	412
323	6,5'-diOH	3,5,7,2',4'-pentaOMe					
324	6,2'-diOH	3,5,7,4',5'-pentaOMe					

continued

TABLE 12.2
Flavonols and Their Methyl Ethers — continued

No.	OH-Substitution	OMe-Substitution	Trivial Name	Plant Species	Family	Plant Organ	Ref.
325	5,5'-diOH	3,6,7,2',4'-pentaOMe		Glycoside only			
326	5,2'-diOH	3,6,7,4',5'-pentaOMe	Brickellin	Eupatorium buniifolium	Asterac.	Aerial parts	413
327	5,6-diOH	3,7,2',4',5'-pentaOMe					
328	5-OH	3,6,7,2'4'5'hexaOMe*		Eupatorium buniifolium	Asterac.	Aerial parts	413
329	3-OH	5,6,7,2',4',5'-hexaOMe*					
330		3,5,6,7,2',4',5'-heptaOMe					
	(3,5,6,7,3',4',5'-OH)						
331	5,7,3',4',5'-pentaOH	3,6-diOMe	6-Hydroxy-myricetin				
332	5,6,7,3',5'-pentaOH	3,4'-diOMe					
333	3,5,7,3',5'-pentaOH	6,4'-diOMe					
334	5,7,4',5'-tetraOH	3,6,3'-triOMe					
335	5,7,3',5'-tetraOH	3,6,4'-triOMe					
336	5,6,7,4'-tetraOH	3,3',5'-triOMe					
337	3,5,7,4'-tetraOH	6,3',5'-triOMe		Glycoside only			
338	5,3',5'-triOH	3,6,7,4'-tetraOMe					
339	5,7,5'-triOH	3,6,3',4'-tetraOMe					
340	5,7,4'-triOH	3,6,3',5'-tetraOMe					
341	5,6,5'-triOH	3,7,3',4'-tetraOMe	Apuleitrin				
342	7,5'-diOH	3,5,7,3',4'-pentaOMe	Apuleirin				
343	5,5'-diOH	3,6,7,3',4'-pentaOMe					
344	5,3'-diOH	3,6,7,4',5'-pentaOMe					
345	3,5-diOH	6,7,3',4',5'-pentaOMe					
346	4'-OH	3,5,6,7,3',5'-hexaOMe					
347	5-OH	3,6,7,3',4',5'-hexaOMe		Murraya paniculata	Rutac.	Fruit	404
348		3,5,6,7,3',4',5'-heptaOMe		Murraya paniculata	Rutac.	Flower	414
	(3,5,6,7,3',4',6'-heptaOH)						
349	5,6-diOH	3,6,7,3',4'-pentaOMe					
	(3,5,7,8,2',3',4'-heptaOH)						
350	5-OH	3,7,2',3',4'-pentaOMe		Ester only			
351	(3,5,7,8,2',4',5'-heptaOH)						

No.	OH substitution	OMe substitution	Common name	Species / Notes	Family	Tissue	Ref.
352	5,7,2',4'-tetraOH	3,8,5'-triOMe					
353	5,3',4'-triOH	3,7,8,5'-tetraOMe					
354	5,7,3'-triOH	3,8,4',5'-tetraOMe					
355	3,5,7,8,3',4',5'-heptaOH	8-OMe	Hibiscetin	Glycoside only			
356	3,5,7,3',4',5'hexaOH	3,7-diOMe*		Chorizanthe diffusa	Polygon.	Whole plant	7
357	5,8,3',4',5'-pentaOH	8,5'-diOMe		Glycoside only			
358	3,5,7,3'4'-pentaOH	3,7,4'-triOMe					
359	5,8,3',5'-tetraOH	3,8,4'-triOMe					
360	5,7,3',5'-tetraOH	8,3',5'-triOMe					
361	3,5,7,4'-tetraOH	3,7,8,4'-tetraOMe		Glycoside only			
362	5,3',5'-triOH	3,8,4',5'-tetraOMe					
363	5,7,3'-triOH	3,8,3',4',5'-pentaOMe					
364	5,7-diOH	7,8,3',4',5'-pentaOMe	Conyzatin				
365	3,5-diOH	3,5,7,3',4',5'-hexaOMe*					
366	8-OH	3,7,8,3',4',5'-hexaOMe*		Murraya paniculata	Rutac.	Fruit	404
367	5-OH	3,5,7,8,3',4',5'-heptaOMe		Murraya paniculata	Rutac.	Fruit	404
368							
	(3,5,7,2',3',4',6'-OH)						
369	5-OH	3,7,2',3',4',6'-hexaOMe*		Distemonanthus benthamianus	Fabac.	Heartwood	29
	Octa-O-substituted flavonols						
	(3,5,6,7,8,2',4',5'-octaOH)						
370	5,7,4'-triOH	3,6,8,2'-tetraOMe					
371	5,7,2',5'-triOH	3,6,8,4'-tetraOMe					
372	5,7,2',4'-triOH	3,6,8,5'-tetraOMe					
373	5,6,2',5'-triOH	3,7,8,4'-tetraOMe					
374	5,4',5'-triOH	3,6,7,8,2'-pentaOMe					
375	5,7,5'-triOH	3,6,8,2',4'-pentaOMe					
376	5,7,2'-triOH	3,6,8,4',5'-pentaOMe					
377	5,2'-diOH	3,6,7,8,4',5'-hexaOMe		Zieridium pseudobtusifolium	Rutac.	Leaf	408
378	5,7-OH	3,6,8,2',4',5'-hexaOMe					
379		3,5,6,7,8,2',4',5'-octaOMe					
	(3,5,6,7,8,3',4',5'-octaOH)	—	Purpurascenin (6,8-Dihydroxy-myricetin)				
380	5,7,3',4',5'-pentaOH	3,6,8-triOMe					
381	5,7,3',5'-tetraOH	3,6,8,4'-tetraOMe					

continued

TABLE 12.2
Flavonols and Their Methyl Ethers — continued

No.	OH-Substitution	OMe-Substitution	Trivial Name	Plant Species	Family	Plant Organ	Ref.
382	5,7,3',4'-tetraOH	3,6,8,5'-tetraOMe					
383	5,6,3',5'-tetraOH	3,7,8,4'-tetraOMe					
384	5,3',5'-triOH	3,6,7,8,4'-pentaOMe					
385	5,7,4'-triOH	3,6,8,3',5'-pentaOMe					
386	5,7,3'-triOH	3,6,8,4',5'-pentaOMe					
387	3,5,3'-triOH	6,7,8,4',5'-pentaOMe*	Desmethyl-digicitrin	*Zieridium pseudobtusifolium*	Rutac.	Leaf	408
388	3',5'-diOH	3,5,6,7,8,4'-hexaOMe					
389	5,3'-diOH	3,6,7,8,4',5'-hexaOMe	Digicitrin	*Zieridium pseudobtusifolium*	Rutac.	Leaf	408
390	5,7-diOH	3,6,8,3',4',5'-hexaOMe		*Gymnosperma glutinosum*	Asterac.	Aerial p., ext.	129
391	3'-OH	3,5,6,7,8,4',5'-heptaOMe					
392	5-OH	3,6,7,8,3',4',5'-heptaOMe					
393		3,5,6,7,8,3',4',5'-octaOMe	Exoticin				

*For explanation, please see text.
[a]Revised to 3,5,7,8-tetraOH, see text.
[b]Revised to 3,6,4'-triMe, see text.
[c]Revised to 3,6,7-triMe, see text.
[d]Revised to 5-OH-6,7,3'4'-tetraOMe, see text.
[e]Revised to Queg-7-Me, see text.

species of this genus. Single reports exist for families such as the Saxifragaceae[42] or Nyctaginaceae[43]. New results on Rosaceae and Viscaceae deserve special consideration. A quite complex derivative (compound 298 in Table 12.2) was isolated from several Rosaceae. By contrast, rather simple derivatives were found in the leaf wax of *Viscum* spp.[44] Particularly with the Rosaceae, more results in this direction are to be expected when more material is analyzed. Frond exudates of several ferns proved to be a rich source for various flavonol derivatives, which outnumber the few corresponding flavones.[45]

Several compounds were structurally revised. 6-Hydroxygalangin (compound 13 in Table 12.2), as reported from *Platanus* buds,[46] was revised to 8-hydroxygalangin after synthesis (compound 23, Table 12.2).[47] 6-Hydroxykaempferol-3,7,4'-triMe had been reported as "tanetin" (compound 84, Table 12.2) from *Tanacetum parthenium*.[48] Its structure was later revised to 6-hydroxykaempferol-3,6,4'-triMe = santin (compound 83, Table 12.2).[49] The name "tanetin"is hence obsolete. 5,4-diOH-3,6,8-triOMe-flavone had been isolated from *Tephrosia candida* and named "candiron" (compound 88, Table 12.2).[50] Synthesis revealed that the structure must be revised to 5,4'-diOH-3,6,7-triOMe-flavone = penduletin (compound 82, Table 12.2).[51] The name "candiron" must not be used, therefore. "Santoflavone," a compound isolated from *Achillea santolina* and claimed to be 7-OH,3,6,3',4'-tetramethoxyflavone (compound 144, Table 12.2),[52] was later revised to 5-hydroxy-6,7,3',4'-tetraOMe flavone (compound 170, Table 12.1).[53] Bhardwaj et al. had reported "allopatuletin" to be a 3,6,7,3',4'-pentahydroxy-5-methoxyflavone (compound 173, Table 12.2), from *Tagetes pendula*.[54] After synthesis, revision of this structure to quercetagetin-7-Me is required (compound 175, Table 12.2).[47] Zhang et al. reported "viscidulin III" from the roots of *Scutellaria planipes*.[55] Unfortunately, it remains dubious whether the authors used the name in its initial meaning, that is, as 3,5,7,3'-tetraOH-2',4'-diOMe flavone or as its revised structure 5,7,3',6'-tetraOH-8,2'-diOMe (for which the name ganhuangenin would apply).[56] Since the authors did not answer relevant requests, this report has not been included in our tables.

12.5 FLAVONES WITH OTHER SUBSTITUENTS

As already mentioned in Section 12.1, several biosynthetic trends are included in this section. These concern *C*-methylated as well as other *C*-substituted flavones, further methylenedioxy derivatives and structures resulting from prenylation and cyclization processes. Most of the prenyl side chains (C5, C10, or C15) are linked directly to the flavone molecule; rarely *O*-prenylation occurs. Extensive modification of the terpenoic side chain may occur by further oxidation, reduction, dehydration, and cyclization. In addition, cyclization of the terpenoid chains with an *ortho*-phenolic OH-group to give pyrano- or furano-derivatives is quite common. Apart from the flavanones, the flavones are the second most abundant class of isoprenylated flavonoids.[3] Compounds which have not been included here are those resulting from Diels–Alder reaction, forming adducts (e.g., Brosimone D) or dimeric flavones (for these structures, see Ref. 3). Most reports concern a few genera of the Moraceae and Fabaceae, exhibiting quite diverse biosynthetic capacities in terms of complex substitution patterns, a phenomenon earlier also addressed by Barron and Ibrahim[3]. Thus, flavonoid profiles of some of these genera will be discussed separately.

12.5.1 *C*-Methylflavones and *C2/C3*-Substituted Flavones

Direct methylation through *C*-bonds appears to be common in the positions 6 and 8 of the flavonoid molecule. Other positions are rarely *C*-methylated (C7, saltillin; C3, a glycoside only; compound 23, Table 12.3). Most reports concentrate on species from the family Myrtaceae, where *C*-methylflavones also occur externally.[57] *Desmos cochinchinensis* (Annonaceae) was

TABLE 12.3
Flavones with Other Substituents

C-Methyl- and C2/C3-substituted flavones

No.	OH- and OMe-Substitution	Other Substituents	Trivial Name	Plant Species	Family	Plant Organ	Ref.
1	5,7-diOH	6-Me	Strobochrysin	Matteucia orientalis	Aspid.	Rhizome	415
2	5,7-diOH	6,8-diMe*	Matteuorien*	Leptospermum scoparium	Myrtac.	Leaf	416
3	5-OH, 7-OMe	6-Me		Leptospermum scoparium	Myrtac.	Leaf	59
4	5-OH, 7-OMe	6,8-diMe*		Desmos cochinchinensis	Annon.	Leaf	58
5	5-OH, 7-OMe	6,8-diMe*	Desmosflavone	Desmos chinensis	Annon.	Seeds	60
6	5,7-diOMe	6-Me*	Saltillin	Leptospermum scoparium	Myrtac.	Leaf	417
7	5-OH, 4'-OMe	7-Me					
8	5,7-diOH, 8-OMe	6-Me					
9	5,2'-diOH,7-OMe	6,8-diMe*					
10	5,7,4'-triOH	6-Me*	6-Methyl-apigenin	Trianthema portulacastrum	Aizoaoac.	Whole plant	418
11	5,7,4'-triOH	8-Me*	8-Methyl-apigenin	Valeriana wallichii	Valerian.	Aerial parts	419
						Synthesis	420
12	5,7,4'-triOH	3-Me	Glycoside only			Synthesis	420
13	5,7,4'-triOH	6,8-diMe*	6,8-Dimethyl-apigenin Syzalterin*	Callistemon, 5 spp.	Myrtac.	External	57
				Pancratium maritimum	Amaryllid.	Bulb	421
				Syzygium alternifolium	Myrtac.	Leaf	422
14	5,4'-diOH, 7-OMe	6-Me	8-Desmethyl-sideroxylin	Callistemon, 8 spp.	Myrtac.	External	57
				Eucalyptus saligna	Myrtac.	Leaf wax	423
15	5,4'-diOH, 7-OMe	6,8-diMe	Sideroxylin			Synthesis	420
				Callistemon, 9 spp.	Myrtac.	External	57
				Eucalyptus saligna	Myrtac.	Leaf wax	423
				Leptospermum laevigatum	Myrtac.	Leaf wax	424
16	5,7-diOH, 4'-OMe	6-Me*				Synthesis only	420
17	5,7-diOH, 4'-OMe	8-Me*				Synthesis only	420
18	5-OH, 7,4'-diOMe	6-Me	8-Desmethyleucalyptin	Callistemon, 5 spp.	Myrtac.	External	57
19	5-OH, 7,4'-diOMe	6,8-diMe	Eucalyptin	Callistemon lanceolatus	Myrtac.	Leaf	425
				Callistemon, 5 spp.	Myrtac.	External	57
20	5,7,3',4'-tetraOH	3-Me	Glycoside only				
21	5,7,3',4'-tetraOH	6-Me*	6-Methylluteolin*	Salvia nemorosa	Lam.	Aerial parts	426

No.				Species	Family	Plant part	Ref.
22	5,3′,4′-triOH-7-OMe	6-Me*		*Hydrastis canadensis*	Ranunculac.	Root	427
23	5,3′,4′-triOH-7-OMe	6,8-diMe*		*Hydrastis canadensis*	Ranunculac.	Root	427
24	5-OH	6-Me, 8-diMe, 7 = O	Dasytrichone	*Dasymaschalon trichophorum*	Annonac.	Stems, leaves	63
25	5,7-diOH	8-Me, 6-CHO	Unonal	*Dasymaschalon rostratum*	Annonac.	Stems	64
				Desmos chinensis	Annonac.	Seeds	60
				Desmos chinensis	Annonac.	Root	62
				Desmos cochinchinensis	Annonac.		58
26	5,7-diOH	6-Me, 8-CHO	Isounonal	*Dasymaschalon rostratum*	Annonac.	Stem	64
				Desmos chinensis	Annonac.	Seeds	60
27	5-OH, 7-OMe	8-Me, 6-CHO	Unonal-7-Me				
28	5,3′,4′-triOH-7-OMe	6,8-diCH$_3$-5′-C5*	Muxiangrine III*	*Elsholtzia stauntonii*	Lam.	Aerial parts	65
29	5,5′-diOH, 7-OMe	6-CH$_3$, 4,3′-ODmp*	Muxiangrine II*	*Elsholtzia stauntonii*	Lam.	Aerial parts	65
30	5,3′-diOH, 7-OMe	6,8-diCH$_3$, 4,3′-ODmp*	Muxiangrine I*	*Elsholtzia stauntonii*	Lam.	Aerial parts	65
31	5,7,4′-triOH	6-C2*	Drymariatin; glycoside only				
32	5-OH, 7-OMe	6-CH$_2$CHO, see Figure 1	Hoslundal				
33	5,7,3′,4′-tetraOH	6-Acrylic acid*, see Figure 1	DeMe-torosaflavone D*	*Cassia nomane*	Fabac.	Aerial parts	68
34	5,7,3′-triOH, 4′-OMe	6-Acrylic acid *, see Figure 1	Torosaflavone D*	*Cassia torosa*	Fabac.	Leaf	67
Methylendioxy-flavones							
35	5-OH	6,7-OCH$_2$O		*Millettia erythrocalyx*	Fabac.	Leaf	428
36	7-OMe	3′,4′-OCH$_2$O*		*Millettia leucantha*	Fabac.	Stem bark	429
37	7-OMe	3′,4′-OCH$_2$O*					
38	7-OMe	5,6-/3′,4′-diOCH$_2$O					
39	5,6-diOMe	3′,4′-OCH$_2$O					
40	5-OH-7-OMe	3′,4′-OCH$_2$O*		*Millettia leucantha*	Fabac.	Stem bark	429
41	5,7-diOMe	3′,4′-OCH$_2$O*	Prosogerin-A	*Neoraputia magnifica*	Rutac.	Fruit	286
42	7-OH, 6-OMe	3′,4′-OCH$_2$O		*Millettia erythrocalyx*	Fabac.	Stem bark	428
43	6,7-diOMe	3′,4′-OCH$_2$O	Milletenin C				
44	5,4′-diOH	6,7-OCH$_2$O	Kanzakiflavon-2				
45	7,2′diOMe	4′, 5′-OCH$_2$O*	Millettocalyxin A*	*Millettia erythrocalyx*	Fabac.	Stem bark	428
46	7,8-diOMe	3′,4′-OCH$_2$O*		*Albizia odoratissima*	Fabac. (Mim.)	Root bark	174
47	5-OH, 6,8-OMe	3′,4′-OCH$_2$O*		*Limnophila indica*	Scroph.	Whole plant	430
48	5,6,7-triOMe	3′,4′-OCH$_2$O*	Ageconyflavon A*	*Ageratum conyzoides*	Ast.	Whole plant	268

continued

TABLE 12.3
Flavones with Other Substituents — continued

No.	OH- and OMe-Substitution	Other Substituents	Trivial Name	Plant Species	Family	Plant Organ	Ref.
49	5,7,4'triOMe	3',4'-OCH₂O*		Neoraputia magnifica	Rutac.	Fruit	286
50	5,7,5'-triOMe	3',4'-OCH₂O		Neoraputia paraensis	Rutac.	Aerial parts	285
51	5,8-diOH, 4'-OMe	6,7-O₂CH₂	Kanzakiflavon-1				
52	5,7-diOH, 6,8-diOMe	3',4'-OCH₂O	Linderoflavone A				
53	5,6,7,8-tetraOMe	3',4'-OCH₂O	Linderoflavone B	Ozothamnus lycopodioides	Ast.	Leaf exudate	69
54	5,6,7,5'-tetraOMe	3',4'-OCH₂O		Ficus maxima	Morac.	Leaf	77
				Neoraputia paraensis	Rutac.	Aerial parts	285
55	5,3',4',5'-tetraOMe	6,7-O₂CH₂		Ageratum conyzoides	Ast.	Aerial parts	27
56	7-OH, 5,6,8,5'-OMe	3',4'-OCH₂O		Ageratum tomentosum var. bracteatum	Ast.	Leaf + flower	431
57	5,6,7,8,5'-pentaOMe	3',4'-OCH₂O	Eupalestin	Ozothamnus lycopodioides#	Ast.	Leaf exudate	69
C-Prenylflavones							
58	5-OMe	7,8-di-C5*	5-Methoxy-7,8-diprenylflavone	Tephrosia barbigera	Fabac.	Seeds	4
59	7-OMe	8-C5-OH	trans-Lanceolatin; Lanceolatin A				
60	5,7-diOH	6-C5	6-Prenylchrysin			Synthesis	432
61	5,7-diOH	8-C5	8-Prenylchrysin			Synthesis	432
62	5-OH, 7-OMe	8-C5	Tephrinone				
63	5,7-diOMe	8-C5-OH	cis-Tephrostachin				
64	5,7-diOMe	8-C5-OH	trans-Tephrostachin				
65	5,7-diOMe	8-C5	trans-Anhydrotephr				
66	7,4'-diOH	6-C5*	Licoflavon A*	Glycyrrhiza eurycarpa	Fabac.	Root	17
67	7,4'-diOH	8-C5		Glycyrrhiza echinata	Fabac.	Cell culture	3
68	7,4'-diOH	8-C5(OH)₂*	Brosimacutin F	Brosimum acutifolium	Morac.	Bark	433
69	7,4'-diOH	3'-C5*	Kanzonol D	Glycyrrhiza eurycarpa	Fabac.	Root	17
70	7,4'-diOH	6,3'-diC5*	Licoflavone B			Synthesis	434
				Glycyrrhiza inflata	Fabac.	Root	74
71	7,4'-diOH	6,3'-diC5	(Prenyllicoflavone A)	Glycyrrhiza glabra	Fabac.	Root	75
72	5,7-OH, 6OMe	3-C5		Ehretia ovalifolia	Borag.	Leaves	78

No.	Substituent	Prenyl	Compound	Species	Family	Plant part	Ref.
73	5,7,4'-triOH	6-C5*	6-Prenyl-apigenin	Cudrania cochinchinensis	Morac.	Root	435
				Dorstenia ciliata	Morac.	Aerial parts	436
				Dorstenia kameruniana	Morac.	Leaf	437
				Maclura pomifera	Morac.	Fruit	438
74	5,7,4'-triOH	6-C5-OH*	Dinklagin C*	Dorstenia dinklagii	Morac.	Twigs	439
				Maclura pomifera	Morac.	Stem, leaf	440
75	5,7,4'-triOH	8-C5*	Licoflavone C*, 8-prenylapigenin	Cudrania cochinchinensis	Morac.	Root	435
				Dorstenia ciliata	Morac.	Twigs	436
				Dorstenia poinsettifolia	Morac.	Leaf	441
				Glycyrrhiza inflata	Fabac.	Root	74
76	5,7,4'-triOH	8-C5	8-Prenylapigenin			Synthesis	442
77	5,7,4'-OH	8-C5-OH*	Ephedroidin	Genista ephedroides	Fabac.	Aerial parts	443
78	5,7,4'-triOH	3'-C5	Yinyanghuo D	Epimedium sagittatum	Berberidac.	Leaf	444
79	5,7,4'-triOH	3'-C5-OH		Vancouveria hexandra	Berberidac.	Underground parts	553
80	5,7,4'-OH	6-C10	Albanin D, revised			Synthesis	71
81	5,7,4'-triOH	3'-C10	Kuwanon S	Morus alba	Morac.	Root bark	445
82	5,7,4'-triOH	6,2'-diC5	Isoartocarpin	Artocarpus integrifolia	Morac.	Heartwood	3
83	5,7,4'-triOH	6,3'-diC5	Gancaonin Q				
84	5,7,4'-triOH	6,3'-diC5-OH		Vancouveria hexandra	Berberidac.	Underground parts	553
85	5,7,4'-triOH	8,3'-diC5					
86	5,7,4'-triOH	3',5'-diC5	Honyucitrin	—			
87	5,7,4'-triOH	3'-C5, 5'-C5-OH*	Yinyanghuo B*	Epimedium sagittatum	Berberidac.	Leaf	444
88	5,4'-diOH, 7-OMe	8-C5	Artonin U	Artocarpus heterophyllus	Morac.	Bark	446
89	5,7-diOH, 4'-OMe	8-C5					
90	5,7-diOH, 4'-OMe	8,3'-diC5					
91	7,2',4'-triOH	3-C10	Rubraflavone A				
92	5,7,2',4'-tetraOH	3-C5	Albanin A				
93	5,7,2',4'-tetraOH	6-C5	Artocarpesin	Cudrania cochinchinensis	Morac.	Root	435
				Artocarpus elasticus	Morac.	Wood	447
				Artocarpus heterophyllus	Morac.	Heartwood	448
				Maclura pomifera	Morac.	Fruit	438
94	5,7,2',4'-tetraOH	6-C5-OH	Oxidihydroartocarpesin				
95	5,7,2',4'-tetraOH	6-C10	Albanin E, revised			Synthesis	71
96	5,7,2',4'-tetraOH	8-C15	Moralbanone*	Morus alba	Morac.	Root bark	445

continued

TABLE 12.3
Flavones with Other Substituents — continued

No.	OH- and OMe-Substitution	Other Substituents	Trivial Name	Plant Species	Family	Plant Organ	Ref.
97	5,7,2',4'-tetraOH	3,6-diC5	Cudraflavone C	*Cudrania tricuspidata*	Morac.	Root bark	449
98	5,7,2',4'-tetraOH	3,8-diC5	Mulberrin, Kuwanon C	*Morus australis*	Morac.	Root bark	450
99	5,7,2',4'-tetraOH	3-C10, 6-C5	Rubraflavone C		Morac.		
100	5,7,2',4'-tetraOH	8-C5, 3-C10*	Artocommunol CD*	*Artocarpus communis*	Morac.	Root cortex	451
101	5,7,2',4'-tetraOH	6,5'-diC5	Cudraflavone D	*Cudrania tricuspidata*	Morac.	Root bark	449
102	5,7,2',4'-tetraOH	3,3'-diC5	Kuwanon T	*Artocarpus heterophyllus*	Morac.	Root bark	452
103	5,7,2',4'-tetraOH	3,6,8-triC5*	Artelasticin*	*Artocarpus elasticus*	Morac.	Wood	447
104	5,7,2',4'-tetraOH	3,6,8-triC5	Dorsilurin D	*Dorstenia psilurus*	Morac.	Root	453
105	5,7,2',4'-tetraOH	6,8,3'-triC5*	Dorsilurin A*	*Dorstenia psilurus*	Morac.	Root	454
106	5,2',4'-triOH, 7-OMe	3-C5	Integrin		—	—	455
107	5,2',4'-triOH,7-OMe	8-C5	Artocarpetin A	*Artocarpus heterophyllus*	Morac.	Root	456
108	5,2',4'-triOH, 7-OMe	3,6-diC5	Artocarpin	*Clarisia racemosa*	Morac.	?	
109	5,2',4'-triOH, 7-OMe	3-C5, 8-C10	Brosimone H		—	—	
110	5,3',4'-triOH-7-OMe	6,8-diCH3, 5'-C5*	Muxiangrine III*	*Elsholtzia stauntonii*	Lam.	Aerial parts	65
111	5,7,2'-triOH, 4'-OMe	6,8-diC5					
112	5,4'-diOH, 7,2'-OMe	8-C5*	Artocarpetin B	*Artocarpus heterophyllus*	Morac.	Root bark	452
113	5,7,3',4'-tetraOH	6-C5*	Gancaonin O*; 6-prenylluteolin	*Glycyrrhiza uralensis*	Fabac.	Aerial parts	457
114	5,7,3',4'-tetraOH	6-C5*	8-Prenylluteolin	*Hypericum perforatum*	Hypericaceae	Callus	218
115	5,7,3',4'-tetraOH	8-C5					
116	5,7,3',4'-tetraOH	8,5'-diC5	Epimedokoreanin B	*Epimedium koreanum*	Berberidac.	Aerial parts	458
117	5,7,4'-triOH, 3'-OMe	6-C5*	6-Prenylchrysoeriol	*Dorstenia mannii*	Morac.	Aerial parts	459
118	5,7,4'-triOH, 3'-OMe	6-C10	Cannflavin A				
119	5,3'-diOH, 7,4'-diOMe	6-C5	Cannflavin B				
120	5,7,2',5'-OH-4'-OMe	3,3'-diC5*	Heteroartonin A*	*Artocarpus heterophyllus*	Morac.	Root bark	452
121	5,2',5'-OH-7,4'-OMe	3-C5*	Artoindonesianin Q*	*Artocarpus champeden*	Morac.	Heartwood	460
122	5,7,5'-OH-2',4'-OMe	3-C5*	Artoindonesianin R*	*Artocarpus champeden*	Morac.	Heartwood	460
123	5,7,3',4',5'-pentaOH	3-C5	Asplenetin				
124	5,7,4'-OH-3',5'-OMe	8-C5	Baohuosu				
	O-Prenylflavones						
125	4'-OH, 5-OMe	7-O-C5 (epoxy)		*Achyrocline flaccida*	Ast.	Aerial parts	76
126	4'-OH,7-OMe	7-O-C5 (epoxy)*		*Achyrocline flaccida*	Ast.	Aerial parts	76

No.			Name	Species	Family	Part	Ref.
127	6,7,4'-triOMe	5-O-allyl		*Ficus maxima*	Morac.	Leaf	77
128	5,6,7,3',5'-OMe	4'-O-C5*		*Millettia erythrocalyx*	Fabac.	Stem bark	461
129	7-OMe-6-OC5	4',5'-OCH₂O*	Millettocalyxin B*	*Helminthostachys zeylanica*	Ophioglossac.	Rhizome	462
130	5,4'-diOH, 6-OMe	7,8-fur	Ugonine C	*Pongamia pinnata*	Fabac.	Root	297
131	6-O-C5	7,8-fur*	Ovalifolin	*Millettia erythrocalyx*	Fabac.	Leaf	461
Pyranoflavones							
132	5-OH	7,8-ODmp*		*Dahlstedtia pentaphylla*	Fabac.	Root	463
				Lonchocarpus subglaucesc.	Fabac.	Root	464
133	5-OH	7,8-ODmp		*Tephrosia praecans*	Fabac.	Seed	465
134	5-OMe	7,6-ODmp		*Pongamia pinnata*	Fabac.	Root	297
135	5-OMe	7,8-ODmp	Isopongaflavon (Candidin)	*Tephrosia tunicata*	Fabac.	Root	466
						Synthesis	467
136	5-OMe	7,8-ODmp-diOAc, see Figure 12.2		*Pongamia pinnata* (syn. *P- glabra*)	Fabac.	Stem bark	468
137	6-OMe	7,8-ODmp*		*Lonchocarpus subglaucescens*	Fabac.	Root	464
138	2-OH	4,3/6,5/7,8-triODmp, 4'=O see Figure 12.2	Dorsilurin E	*Dorstenia psilurus*	Morac.	Root	453
139	5'-OMe	7,8-ODmp					
140	5,7-diOH	4',3'-ODmp*	Yinyanghuo C*	*Epimedium sagittatum*	Berberidac.	Leaf	444
				Vancouveria hexandra	Berberidac.	Underground parts	553
						Synthesis	434
141	5,4'-diOH	7,6-ODmp	Carpachromene*	*Lonchocarpus xuul, L. yucatanensis*	Fabac.	Leaf	469
				Dorstenia kameruniana	Morac.	Leaf	437
				Maclura pomifera	Morac.	Stem and leaf	440
				Atalantia monophylla	Rutac.	Leaf	470
142	5,4'-diOMe	7,6-ODmp-OH*	Dinklagin B*	*Dorstenia dinklagii*	Morac.	Twig	439
143	5,4'-diOMe	7,6-ODmp		*Lonchocarpus xuul, L. yucatanensis*	Fabac.	Leaf	469
144	5,4'-diOH	7,8-ODmp	Atalantoflavon				
145	5,4'-diOH	7,6-/3,6'-ODmp	Isocyclomorusin	*Artocarpus altilis*	Morac.	Stem	471
146	5,4'-diOH	7,8-/3,6'-ODmp	Cyclomorusin = cyclomulberrochromene	*Morus alba*	Morac.	Root bark	445
				Artocarpus communis	Morac.	Root cortex	451

continued

TABLE 12.3
Flavones with Other Substituents — continued

No.	OH- and OMe-Substitution	Other Substituents	Trivial Name	Plant Species	Family	Plant Organ	Ref.
147	5,7,4'-triOH	3,6'-ODmp*	Cyclocommunol*	*Artocarpus communis*	Morac.	Root bark	472
148	5,7,5'-triOH	4',3'-ODmp*	Yinyanghuo E	*Epimedium sagittatum*	Berberidac.	Leaf	444
149	5,2',4'-triOH	7,6-ODmp	Cycloartocarpesin	*Cudrania tricuspidata*	Morac.	Root bark	449
				Maclura pomifera	Morac.	Cell culture	473
150	7,4'-diOH-3'-OMe	5,6-ODmp	Ciliatin B	*Dorstenia ciliata*	Morac.	Aerial parts	436
151	5,4'-diOH-3'-OMe	7,6-ODmp		*Lonchocarpus xuul, L. yucatanensis*	Fabac.	Leaf	469
152	5,4'-diOH, 3'-OMe	7,8-ODmp	Racemoflavon	*Artocarpus champeden*	Morac.	Bark	474
153	5,7,3',4'-tetraOH	2',3'-ODmp*	Cyclochampedol*	*Neoraputia paraensis*	Rutac.	Aerial parts	475
154	5,4'-diOH,3',5'-diOMe-	7,6-ODmp		*Neoraputia paraensis*	Rutac.	Aerial parts	285
155	5-OH, 7,8,3',4'-OMe	5,6-ODmp		*Neoraputia paraensis*	Rutac.	Aerial parts	475
156	5-OH-7,3',4',5'-OMe	7,6-ODmp*		*Neoraputia paraensis*	Rutac.	Aerial parts	285
157	5,4'-diOH,8,3',5'-triOMe	7,6-ODmp*		*Neoraputia alba*	Rutac.	Leaf	476
158	5-OH, 8,3',4',5'-tetra-OMe	7,6-ODmp					
159	7,8,3',4',5'-pentaOMe	5,6-ODmp					
160	5,2',4'-triOH	7,6-ODmp-C5*	Australon A*	*Morus australis*	Morac.	Root bark	450
161	5,2',4'-triOH	7,8-ODmp-C5	Brosimone G	*Brosimopsis oblongifolia*	Morac.	Root	477
Furanoflavones							
162		7,8-fur	Lanceolatin B	*Millettia peguensis*	Fabac.	Leaf, st. bark	478
				Millettia sanagana	Fabac.	Root bark	79
163	5-OH	7,8-fur	Pongaglabol	*Pongamia pinnata*	Fabac.	Root	297
				Millettia peguensis	Fabac.	Leaf, st. bark	478
164	5-OMe	7,6-fur	Pinnatin	*Tephrosia purpurea*	Fabac.	Aerial parts	479
165	6-OMe	7,8-fur	Kanjone*	*Millettia sanagana*	Fabac.	Root bark	480
				Pongamia glabra	Fabac.	Seed	480
				Millettia erythrocalyx	Fabac.	Root	461
166	8-OMe	7,6-fur		*Millettia peguensis*	Fabac.	Leaf, st. bark	478
				Millettia sanagana	Fabac.	Root bark	79
167	2'-OMe	7,8-fur		*Pongamia glabra*	Fabac.	Leaf	480
				Pongamia glabra	Fabac.	Seed	481

No.	Substitution	Substitution	Name	Species	Family	Plant part	Ref.
168	2'-OMe	7,8-fur	Pongol-Me	*Milletia erythrocalyx*	Fabac.	Leaf	461
169	2'-OMe	7,8-fur					
170	3'-OH	7,8-fur					
171	3'-OMe	7,6-fur	Pongone	*Pongamia glabra*	Fabac.	Flower	482
172	4'-OH	7,8-fur	Isopongaglabol				482
173	4'-OMe	7,6-fur	Glabone	*Derris mollis*	Fabac.	Root	483
174	4'-OMe	7,8-fur					
175	5,4'-diOH-	7,6-dihydrofur-C3, see Figure 12.3	Ciliatin A*	*Dorstenia ciliata*	Morac.	Aerial parts	436
176	5,4'-diOH-	7,6-dihydrofur-C3-OH, see Figure 12.3	"Compound 8"	*Maclura pomifera*	Morac.	Stem and leaf	440
177	5,4'-diOH	7,6-dihydrofurODmp-OH*		*Maclura pomifera*	Morac.	Stem and leaf	440
178	5,3'-diOH	7,6-dihydrofur-2''-C3/4',5'-dihfur-OH-5''-C3-OH; see Figure 12.6	Epimedokoreanin A*	*Epimedium koreanum*	Berberidac.	Aerial parts	484
179	5,4'-diOH	7,8-dihydrofurODmp-OH, see Figure 12.6	Artelastofuran*	*Artocarpus elasticus*	Morac.	Wood	485
180	4'-OH, 6-OMe	7,8-fur	6-OMe-isopongaglabol				
181	2',5'-diOMe	7,8-fur*	Millettocalyxin C*	*Millettia erythrocalyx*	Morac.	Leaf	461
182	5,3',4'-triOH	7,6-bisfurano — see Figure 12.3	Demethyltorosaflavone C	*Cassia nomane*	Fabac.	Aerial parts	68
183	5,3'-diOH, 4'-OMe	7,6- bisfurano — see Figure 12.3	Torosaflavone C	*Cassia torosa*	Fabac.	Leaf	67
Furano- and pyranosubstitution							
184		6,5-ODmp, 7,8-fur*	Sanaganone*	*Millettia sanagana*	Fabac.	Root bark	79
185	5,3',6'-triOH	7,6-ODmp, 4',5'-dihydrofur-2''-C3	Dihydrofur-arto-biliichromene b1				
186	5,3',6'-triOH	7,6-ODmp, 4',5'-dihydrofur-2''-C3	Dihydrofur-arto-biliichromene b2				
187	5,3',4'-triOH	7,6-ODmp, 6',5'-dihydrofur-2''-C3	Dihydrofur-arto-biliichromene a				
C-prenyl- and C-pyranosubstitution							
188	5-OH	6-C5, 7,8-ODmp	Fulvinervin B	*Tephrosia fulvinervis*	Fabac.	Pods	486
189	5-OH	6-C5-OH, 7,8-ODmp	Fulvinervin C	*Tephrosia fulvinervis*	Fabac.	Seed	487
190	7-OH	6-C5, 3',4'-ODmp*	Kanzonol E*	*Glycyrrhiza eurycarpa*	Fabac.	Root	17
						Synthesis	434
191	5,7-diOH	5'-C5-OH, 4',3'-ODmp	Yinyanghuo A*	*Epimedium sagittatum*	Berberidac.	Leaf	444
192	5,2'-diOH	6-C5, 7,8/4',5'-diODmp		*Euchresta formosana*	Fabac.	Root	488
193	5,2'-diOH	8-C5, 7,6/4',5'-diODmp		*Euchresta formosana*	Fabac.	Root	488
194	5,4'-diOH	3-C5-OH, 7,8-ODmp	Artocommunol CC*	*Artocarpus communis*	Morac.	Root cortex	451
195	5,4'-diOH *	6-C5, 7,8-ODmp*	Laxifolin*	*Derris laxiflora*	Fabac.	Root	489
				Derris laxiflora	Fabac.	Root	490

continued

TABLE 12.3
Flavones with Other Substituents — continued

No.	OH- and OMe-Substitution	Other Substituents	Trivial Name	Plant Species	Family	Plant Organ	Ref.
196	5,4'-diOH	8-C5, 7,6-ODmp*	Isolaxifolin*	*Derris laxiflora*	Fabac.	Root	489
197	5,7,2'-triOH	3-C5, 3',4'-ODmp	Kuwanon B	*Derris laxiflora*	Fabac.	Root	490
198	5,7,4'-triOH	8-C5, 2',3-ODmp	Cyclomulberrin				
199	5,7-diOH	3'-C5, 4',5'-ODmp	Morusin; Mulberrochromene	*Vancouveria hexandra*	Berberidac.	Underground parts	553
200	5,2',4'-triOH	3-C5, 7,8-ODmp	Oxydihydro-morusin	*Morus australis*	Morac.	Root bark	450
201	5,2',4'-triOH	3-C5-OH, 7,8-ODmp	Rubraflavone D	*Morus alba*	Morac.	Root bark	445
202	5,2',4'-triOH	3-C10, 7,6-ODmp	Dorsilurin B*	*Dorstenia psiluris*	Morac.	Root	454
203	5,2',4'-triOH	3,6-diC5, 7,8-ODmpOH*	Morusignin L	*Morus insignis*	Morac.	Root bark	491
204	5,2',4'-triOH	7,6-ODmp, 3-C5-OH	Artonin E; KB-3	*Artocarpus communis*	Morac.	Shoot bark	492
205	5,2',4',5'-tetraOH	3-C5, 7,8-ODmp	KB-2	*Artocarpus kemando*	Morac.	Stem bark	493
206	5,2',4',5'-tetraOH	3-C5, 7,8-ODmp		*Artocarpus communis*	Morac.	Shoot bark	494
207	5,2',4',5'-tetraOH	3,8-diC5; 7,6-ODmp, see Figure 12.2	Heterophyllin	*Artocarpus heterophyllus*	Fabac.	Root bark	452
208	5,2',4',5'-tetraOH	3'-C5, 7,6-ODmp	Artobilochromene				
209	5,7,2'-triOH	3-C5, 2',3'-ODmp-C5	Sanggenon K	*Morus sp.*	Morac.	Root bark	495
210	5,7,2'-triOH	3-C5, 3',4'-ODmp-C5	Sanggenon J	*Morus sp.*	Morac.	Root bark	495
C-linked aromatic substituents							
211	5,7-diOH	6-C-cinnamyl		chinese propolis			81
212	5,7,4'-triOH	8-C-p-OH-benzyl*		*Thymus hirtus*	Lam.	Aerial parts	80
213	5,7,3',4'-tetraOH	8-C-p-OH-benzyl*	p-Hydroxybenzylluteolin	*Thymus hirtus*	Lam.	Aerial parts	80
214	5,7,3'-OH-4'-OMe	8-C-p-OH-benzyl*	p-Hydroxybenzyl-diosmetin	*Thymus hirtus*	Lam.	Aerial parts	80
C-linked ketopyrano substituents							
215	5-OH,7-OMe	6,5''-Ketopyrano...3''-OH*, see Figure 12.4	Hosloppin (3''-O-demethylhoslundin)*	*Hoslunda opposita*	Lam.	Leaf	496
216	5-OH, 7-OMe	6,5''-Ketopyrano...3''-Me, see Figure 12.4	Hoslundin				
217	5,7-diOMe	6,5''-Ketopyrano...3''-OH, see Figure 12.4	5-OMe-hoslundin	*Hoslunda opposita*	Lam.	Twigs	144

No.	Subst.	Structure	Name	Species	Family	Plant part	Ref.
218	5,7-diOMe	6,6''-Ketopyrano...3''-OH*, see Figure 12.4	Oppositin	*Bidens pilosa*	Ast.	Aerial parts	497
				Hoslunda opposita	Lam.	Twigs	144

Tephrosia flavones

No.	Subst.	Structure	Name	Species	Family	Plant part	Ref.
219		7,8-bisfur —	Glabratephrinol	*Tephrosia apollinea*	Fabac.	Seed	498
220		7,8-bisfur —	Glabratephrin	*Tephrosia semiglabra*	Fabac.	Aerial parts and root	499
221		7,8- bisfur —	Semiglabrinol	*Tephrosia semiglabra*	Fabac.	Aerial part. roots	500
222		7,8- bisfur —	Semiglabrin	*Tephrosia semiglabra*	Fabac.	Aerial part. roots	500
				Tephrosia purpurea	Fabac.	Aerial parts	479
223		7,8-bisfur —OH	Pseudosemiglabrinol	*Tephrosia apollinea*	Fabac.	Aerial parts	501
224		7,8-bisfur —Oac — -	Pseudosemiglabrin	*Tephrosia semiglabra*	Fabac.	Aerial parts	502
225	5-OMe	7,8-bisfur —*, see Figure 12.5	Enantiomultijugin	*Tephrosia viciodes*	Fabac.	Aerial parts	503
226	5-OMe	7,8-bisfur —	Multijugin	*Tephrosia multijuga*	Fabac.	Aerial parts and root	504
227	5-OMe	7,8-bisfur —	Multijuginol	*Tephrosia multijuga*	Fabac.	Aerial parts and root	504
228	5-OMe	7,8-pyr-fur (3''-oxo)	Stachyoidin	*Tephrosia polystachyoides*	Fabac.	Not mentioned	3
229	5-OMe	7,8-pyr-fur (3''-oxo-4''-OAc)	Tephrodin	*Tephrosia polystachyoides*	Fabac.	Not mentioned	3
230	5-OMe	7,8-fur-C4-diOAc	Polystachin	*Tephrosia polystachyoides*	Fabac.	Aerial parts	555
231	5,7-diOMe	8-diMe-oxo-furano*, see Figure 12.5	Hookerianin*	*Tephrosia hookeriana*	Fabac.	Root	505
232	5-OMe	7,8-oxofuryl*, see Figure 12.5	Tephrorianin*	*Tephrosia hookeriana*	Fabac.	Pods	82
233	7-OMe	8-furyl (2'',4''-diOH)	Tepurindol	*Tephrosia purpurea*	Fabac.	Root	500
234	7-OMe	8-furyl (4''-oxo)	Tephroglabrine	*Tephrosia purpurea*	Fabac.	Root	500
235	7-OMe	8-furyl (2'' = oxo), see Figure 12.5	Apollinine	*Tephrosia apollinea*	Fabac.	Seed	500
236	5,7-diOMe	8-furyl (2''-oxo)	Tachrosin	*Tephrosia polystachyoides*	Fabac.	Leaf and stem	500

***Artocarpus* flavones**

No.	Subst.	Structure	Name	Species	Family	Plant part	Ref.
237	5,7,5'-triOH,4'-OMe	3,6'-cyclo C6-C3; 6-C5	Cycloaltilisin	*Artocarpus altilis*	Morac.	Stem	471
238	5,2',5'-triOH,7,4'-diOMe	3,6'-cyclo C6-C3*	Artoindonesianin S*	*Artocarpus champeden*	Morac.	Heartwood	460
239	5,7,2',5'-tetraOH, 4'-OMe	3,6'-cyclo C6-C3*	Artoindonesianin T*	*Artocarpus champeden*	Morac.	Heartwood	460
240	5,7,2',4'-tetraOH	3,6'cyclo-C6-diMe-fur*	Artoindonesianin P*	*Artocarpus lanceifolius*	Morac.	Tree bark	506
241	5,4'-diOH, 7,2'-diOMe	3,6'-cyclo C6-5'-fur*	Artonin L*	*Artocarpus heterophyllus*	Morac.	Root bark	507
242	5,2',4'-triOH, 7-OMe	3,6'-cyclo C6-5'-fur*, see Figure 12.6	Artonin K*	*Artocarpus heterophyllus*	Morac.	Root bark	492
243	5,7,2',4'-tetraOH-	3,6'-cyclo-C6-5'-fur; 4'-C5*	Artonin J*	*Artocarpus heterophyllus*	Morac.	Root bark	507
244	5,2',4'-triOH,7-OMe	3,6'-cyclo-C6-5'-fur; 4'-C5*	Artonin T*	*Artocarpus heterophyllus*	Morac.	Root bark	446

continued

TABLE 12.3
Flavones with Other Substituents — continued

No.	OH- and OMe-Substitution	Other Substituents	Trivial Name	Plant Species	Family	Plant Organ	Ref.
245	5,2',4',5'-tetra-OH	7,8-ODmp, 3,6'-cyclo-C6	KB-1	Artocarpus communis	Morac.	Shoot bark	494
246	5-OH, 4'-OMe	7,8-ODmp; 3,6'-cyclo-C6-C3	Artomunoxanthentrione	Artocarpus communis	Morac.	Root bark	508
247	5,2',4',5'-tetraOH	7,8-ODmp; 3,6'-cyclo-C6-C3	Artobiloxanthone	Artocarpus nobilis	Morac.	Bark	509
248	5,2',5'-triOH, 4'-OMe	7,8-ODmp, 3,6'-cyclo C6	Artomunoxanthone	Artocarpus communis	Morac.	Bark	492
				Artocarpus communis	Morac.	Root bark	3
249	5,2',4';triOH	7,8-ODmp, 3,6'-cyclo-C6-diMe-fur	Cycloartobiloxanthone	Artocarpus nobilis	Morac.	Bark	509
250	5,2',4'-triOH	7,6-ODmp; 3,6'-cyclo C6-5'-fur	Artonin M	Artocarpus communis	Morac.	Bark	492
251	5,2',4'-triOH	8-C10, 7,6-ODmp-3,6' cyclo-C6-diMe-fur*	Artoindonesianin A*	Artocarpus rigida	Morac.	Bark	552
252	5,2'-diOH, 4'-OMe	7,8-ODmp, 3,6'-cyclo C6-5'-fur*	Cycloartomunoxanthone*	Artocarpus champeden	Morac.	Root	510
253	5,2'-diOH, 4'-OMe	6-C5; 7,8-ODmp, 3,6'-cyclo C6-5'-fur*	Artonin F*	Artocarpus communis	Morac.	Root bark	511
254	5,2',5'-triOH, 7-OMe	3',4'-ODmp, 3,6'-cyclo-C6-C3*, see Figure 12.6	Artonol E*	Artocarpus communis	Morac.	Bark	512
255	5,7,2',4'-tetraOH	6-C5, 3',4'-ODmp, 3,6'-cycloC6-C3*	Artonin N*	Artocarpus communis	Morac.	Bark	492
256	5,2',5'-triOH	7,8-ODmp, 3',4'-ODmp, 3,6'-cyclo-C6-C3*	Artonol C*	Artocarpus rigida	Morac.	Bark	552
257	5,7,4'-triOH, 3',6'-di-oxo	6,5'-C5; 3,6'-cycloC6-C3*	Artonin O*	Artocarpus communis	Morac.	Bark	492
				Artocarpus rigida	Morac.	Bark	552
258	5-OH; 2',5'-di-oxo	7,8-ODmp, 3',4'-ODmp, 3,6'-cyclo-C6-C3*, see Figure 12.6	Artonol D*	Artocarpus kemando	Morac.	Stem bark	493
259	5,4'-dOH, 2',5'-di-oxo	7,8-ODmp; 3,6'-cycloC6-C3, 2',5'-epoxy*	Artonin P*	Artocarpus communis	Morac.	Bark	492
				Artocarpus rigida	Morac.	Bark	552
260	5OH,7OMe	6-C5, 3,6'-cyclo O-C6-C3*	Artoindonesianin B*	Artocarpus champeden	Morac.	Root	510
261	5,4'-diOH	8-C5, 7,6-ODmp, 3,6'-cycl-OC5*	Artelastochromene*	Artocarpus elasticus	Morac.	Wood	447
262	5,4'diOH#	8-C5-7,6-fur-C3-OH, 3,6'-cyclo-O-C6-C3-OH*	Carpelastofuran*	Artocarpus elasticus	Morac.	Wood	513
263	5-OH,4'OMe	7,8-ODmp, C5-O-C5; 3,6'-cycl-O-C5*, see Figure 12.6	Artocommunol CA*	Artocarpus communis	Morac.	Root cortex	451
264	5,7,4'-triOH	6-C5, 3,6'-cycl-O-C5*	Cyclocommunin = (isocyclomulberrin*)	Artocarpus altifolius	Morac.	Stem	471
				Artocarpus communis	Morac.	Root bark	472
				Artocarpus elasticus	Morac.	Wood	485
265	5,7,4'-triOH	6-C5, 3,6'-cycl-O-C5	Brosimone I	Brosimopsis oblongifolia	Morac.	Root	477
266	5,7,4'-triOH	6,8-diC5, 3,6'-cycl-O-C5	Artelastin*	Artocarpus elasticus	Morac.	Wood	447

No.	Substitution	Structure	Compound	Species	Family	Part	Ref
267	5,7,4'-triOH	6,8-diC5, 3,6'-cyclo-O-C6-C3-OH, see Figure 12.6	Artelastocarpin	*Artocarpus elasticus*	Morac.	Wood	485
268	5,4'-diOH, 7-OMe	3,6'-cyclo O-C6*	Oxyisocyclointegrin*	*Artocarpus elasticus*	Morac.	Wood	513
269	5,4'-diOH, 7-OMe	3,6'-cyclo O-C7*	Cyclointegrin*	*Artocarpus integrifolia*	Morac.	Heartwood	3
270	5,4'-diOH, 7-OMe	6-C5, 3,6'-cyclo O-C6-C3	Artonin S	*Artocarpus integrifolia*	Morac.	Heartwood	3
271	5,4'-diOH, 7-OMe	6-C5, 3,6'-cycl-O-C5	Cycloartocarpin	*Artocarpus heterophyllus*	Morac.	Bark; shoot	446
272	5,4',5'-triOH	8-C5, 7,6-ODmp, 3,6'-cycl-O-C5	Cycloheterophyllin				
273	5,5'-diOH, 4'-OMe	7,8-ODmp, 3,6'-cycl-O-C5	Cycloartomunin	*Artocarpus communis*	Morac.	Root bark	511
274	5,3',4'-triOH,7-OMe	8-C5, 3,6'-cycl-O-C5-ODmp*	Dihydro-isocycloartomunin*	*Artocarpus communis*	Morac.	Root bark	511
Flavone–coumarin hybrids							
275	5,7,4'-triOH	6-(8''-umbelliferyl)-*, see Figure 12.7		*Gnidia soccotrana*	Thymeleac.		83
276	5,7,4'-triOH	8-(6''-umbelliferyl)-*, see Figure 12.7		*Gnidia soccotrana*	Thymeleac.		83

Notes: C5 means, e.g., Me_2CH–CH– or Me_2C = CH–CH$_2$–; C5-OH means, e.g., $Me_2C(OH)CH_2$–CH$_2$–; C10 means, e.g., geranyl- or lavandulyl-. For further possibilities see Barron and Ibrahim.[3] The pyrano ring is indicated by –Odmp for oxygen-linked dimethylallyl unit. For the meaning of C3, cyclo-C6, etc. see examples in figures, as indicated for related structures.

*For explanation, please see text.

reported as a new source for desmosflavone (5-OH-7OMe-6,8-diMe flavone),[58] a compound already known from *Leptospermum scoparium* (Myrtaceae).[59]

C-Formylflavones, being substituted at the 6 and 8 position (unonal and related compounds), have been reported for species of the genus *Desmos* (Annonaceae).[60–62] Another derivative with unusual substitution, dasytrichone, was isolated from two species of *Dasymascholon* (Annonaceae).[63,64] Thus, the accumulation of these types of compounds in Annonaceae may be of chemosystematic significance. Muxiangrines I and II, comprising a pyrano structure and *C*-methyl substitution (compounds 28, 29, Table 12.3), have been isolated along with muxiangrine III (*C*-methyl- and prenyl-substitution; compound 30, Table 12.3) from the aerial parts of *Elsholtzia stauntonii* (Lamiaceae).[65]

Drymariatin, a *C*-2-substituted derivative, is known as glycoside only (*Drymaria diandra*, Caryophyllaceae).[66] Another *C*-2-substituted flavone, hoslundal, was earlier reported from *Hoslunda* (Lamiaceae).[6] From species of *Cassia* (Fabaceae), flavones with a *C*-6-acrylic acid substituent were isolated.[67,68] This substitution pattern appears to be quite unique. Hoslundal, torosaflavone D, and its dimethyl derivative (for formulae see Figure 12.1) were all detected in either leaves or aerial parts, but not indicated as exudate compounds.

12.5.2 METHYLENEDIOXYFLAVONES

Predominantly, this substitution occurs in the 3′,4′-position of ring B, rarely between neighboring OH-groups in ring A. Such compounds were reported from Fabaceae, Rutaceae, and some Asteraceae, from all parts of the plants. Only in rare cases, their accumulation as exudate constituents was documented (*Ozothamnus*, Asteraceae).[69,70]

12.5.3 *C*-PRENYLFLAVONES

Linear substituted prenyl flavones exhibit a tendency towards prenylation at positions 3,6- and 8- of the flavonoid molecule. 3′-Prenylation of ring B occurs occasionally. The prenyl residue is mostly of the 3,3-dimethylallyl structure or the OH-equivalent of it. The prenyl residue "1,1-dimethylallyl" is rare. Geranylated flavones are also not very

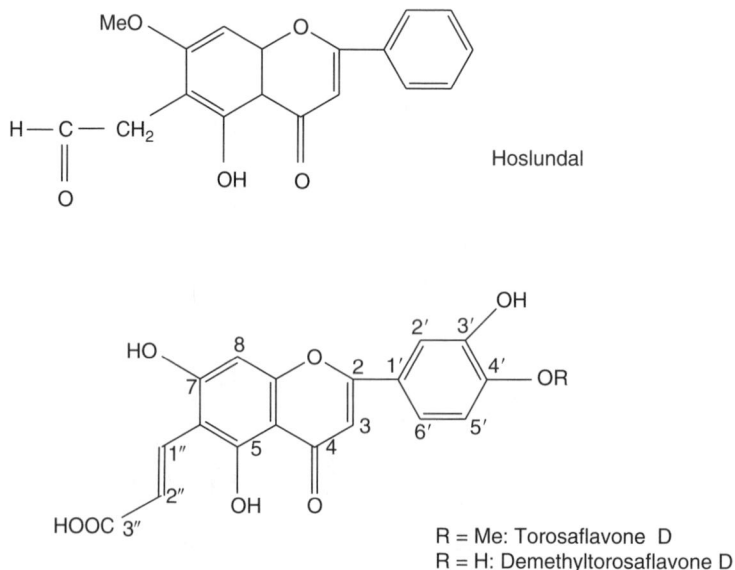

FIGURE 12.1 C2/C3-substituted flavones.

common (e.g., albanin D). For further types of substituents the reader may consult the review of Barron and Ibrahim.[3] The 7,6-chromenoflavone carpachromene affords an example of a more widespread occurrence in species of Fabaceae, Moraceae, and Rutaceae (Table 12.3). Similar results may be expected for other complex flavones in the future.

Some problematical structures and names should be mentioned here. Revised structures concern albanins D (compound 80, Table 12.3) and E (compound 95, Table 12.3) from *Morus alba,* which are not 8-geranyl-, but 6-geranyl-derivatives of 5,7,4'-triOH and 5,7,2',4'-tetraOH flavone, respectively.[71] The flavone lanceolatin A (compound 59, Table 12.3) is definitely a *C*-prenylated flavone, isolated originally from the stems of *Tephrosia lanceolata.*[72] Thus, this name must not be used for a biflavone from *Lophira lanceolata* as Pegnyemb et al. have done later.[73] The compound 7,4'-diOH-6,3'-diC$_5$ flavone, isolated from the roots of *Glycyrrhiza inflata* and named "licoflavone B" by Kajiyamam et al.,[74] was later isolated from the roots of *G. glabra* and named prenyllicoflavone A (compound 70, Table 12.3).[75] Consequently, the latter name falls into the category of synonyms.

12.5.4 *O*-PRENYLFLAVONES

Only very few flavones of this type exist, in most cases showing various other types of substituents as well. *O*-prenylation is known to occur at position 6, 7, or 4'-OH. Most of the substituents are 3,3-dimethylallyl structures. Whereas epoxyprenyl derivatives have been reported from the aerial parts of *Achyrocline flaccida* (Asteraceae),[76] a new compound with 4'-*O*-dimethylallyl substitution was later reported from the leaves of *Ficus maxima* (compound 128, Table 12.3).[77] There is no indication of external accumulation in any of the plants listed. Millettocalyxin B (compound 129, Table 12.3) from the stem bark of *Millettia erythrocalyx* represents an example of a mixed structure (methylenedioxy- and *O*-prenylsubstitution). Similarly, ovalifolin (compound 131, Table 12.3) has a furano-substitutent in addition and is being reported for two new sources of Fabaceae. The name ovalifolin (published 1974) has priority for this furanoflavone; its use for a structurally different compound from *Ehretia ovalifolia*[78] is, therefore, obsolete.

12.5.5 PYRANOFLAVONES

These types of flavones are characterized by cyclization between an OH-group with a prenyl residue to result in chromene or chromane structures, abbreviated as O-Dmp in Table 12.3. This section contains quite a large number of compounds and sources. Most cyclizations take place beween 7-OH and 6- or 8-position of ring A, particularly observed in species of Rutaceae and Fabaceae. Additional cyclization between 2'-OH and *C*-3 is mainly found in members of the Moraceae, whereas cyclization between 3'-OH or 4'-OH with the neighboring prenyl is reported from Berberidaceae (yinyanghuo C, E; compounds 140, 148, Table 12.3). Pyranoflavones are reported to occur in all parts of the plants, but there are no reports explicitly citing their external accumulation. Some special structures are illustrated in Figure 12.2, such as a flavone acetylated at the chromene residue from *Pongamia pinnata* (compound 136, Table 12.3) or the tri-ODmp-substituted dorsilurin E (compound 138, Table 12.3).

12.5.6 FURANOFLAVONES

Cyclization resulting in furano-substitution is quite frequent between 7-OH and the neighboring 6 or 8-position of ring A. Most reports concentrate on Fabaceae and Moraceae, again from all parts of the plants, except external accumulation. The number of compounds known is smaller than that of the pyranoderivatives. Furano- substitution at ring B is rarely observed (e.g. epimedokoreanin A; compound 178, Table 12.3 and in Figure 12.3) from *Epimedium*

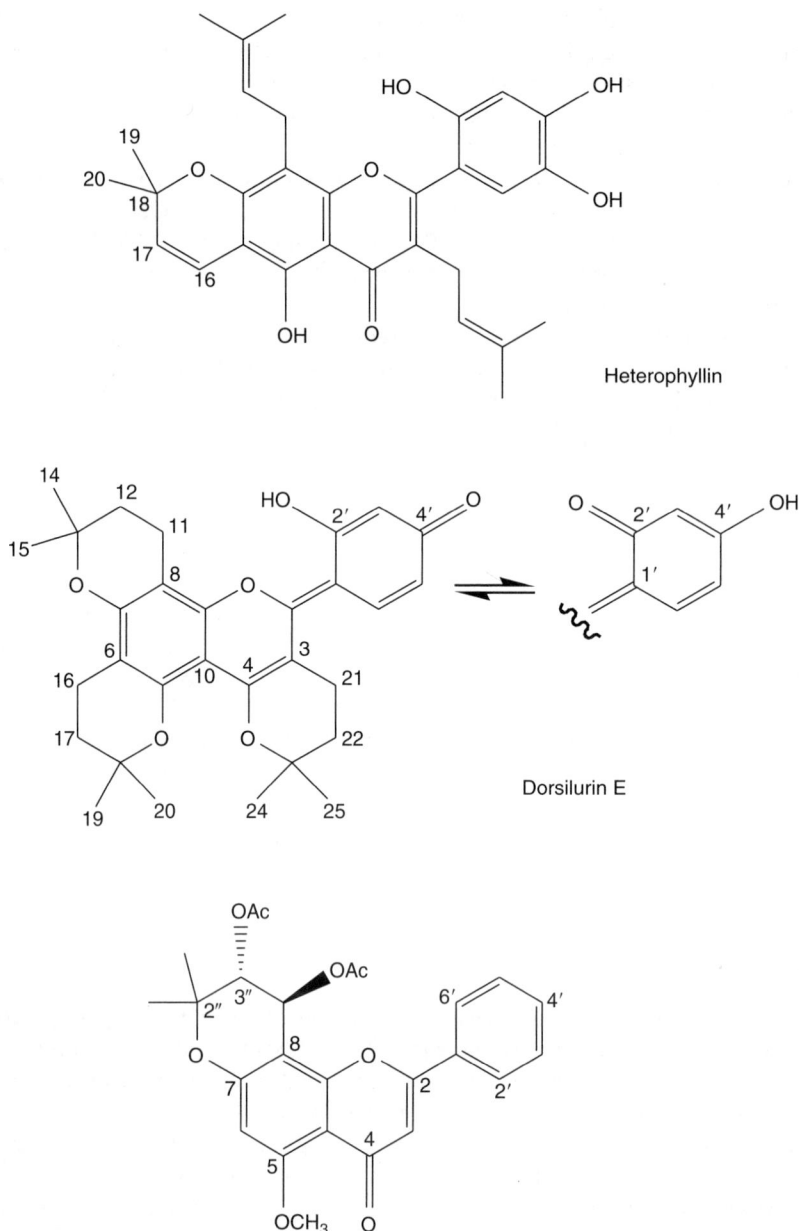

Heterophyllin

Dorsilurin E

FIGURE 12.2 (Dihydro-) pyranoflavones.

(Berberidaceae). This compound exhibits a more complex furano-substitution, similar to ciliatin A (Figure 12.3) and "compound 8" (Figure 12.3) from species of Moraceae. Torosaflavone C and its demethyl derivative, both isolated from *Cassia* species[67,68] are bisfurano-substituted (compounds 182 and 183, Table 12.3 and in Figure 12.3), similar to compounds found in *Tephrosia* species (see Section 12.5.10). In *Epimedium*, there appears to be a strong tendency towards cyclization between 3′-position and 4′-OH, both in pyrano- and furano-derivatives. Further distribution studies are needed to confirm these biosynthetic trends and their possible chemosystematic significance.

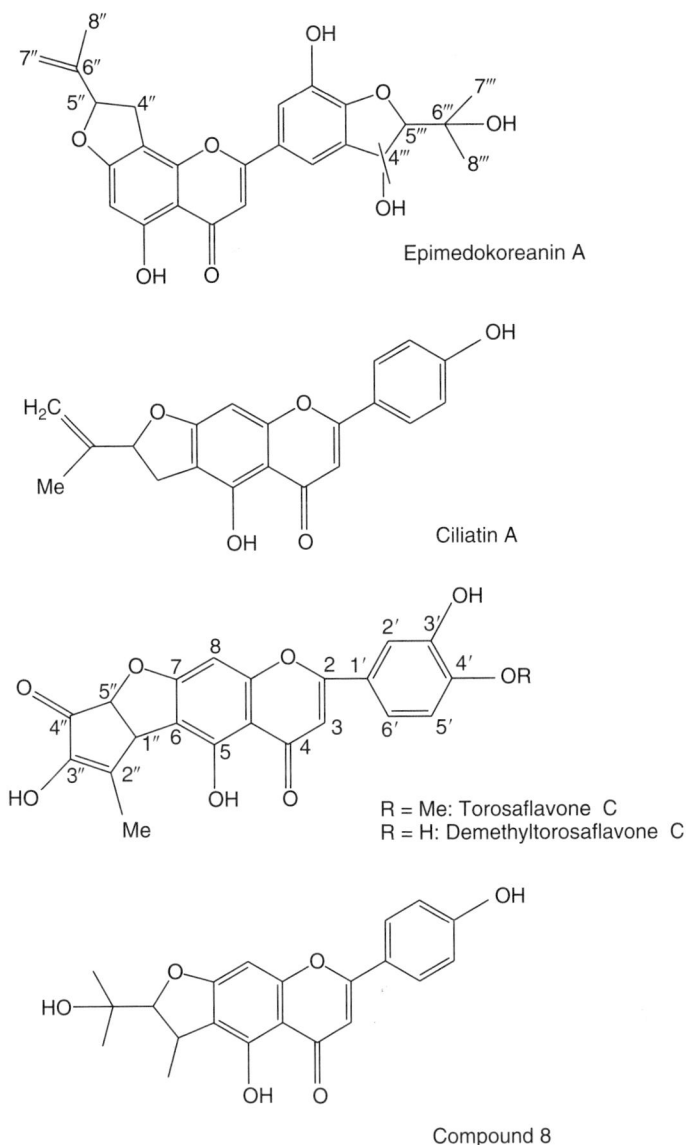

FIGURE 12.3 Furanoflavones.

12.5.7 FURANO- AND PYRANO-SUBSTITUTION

With the exception of sanaganone, all other compounds listed in this section exhibit a chromeno-structure between 7-OH and *C*-6 of ring A, whereas the furano-substitution occurs on ring B. During the reporting period, only sanaganone is a newly described compound, occurring in the root bark of *Millettia* (Fabaceae).[79]

12.5.8 *C*-PRENYL- AND PYRANO-SUBSTITUTION

This section comprises a series of flavones, mostly reported from genera of the Moraceae (*Morus*, *Dorstenia*, *Artocarpus*) and the Fabaceae (*Derris*, *Tephrosia*, *Euchresta*). Very few reports exist on *Vancouveria* and *Epimedium* (Berberidaceae). In contrast to most of the other reported structures, these Berberidaceae flavones are again B-ring-substituted only (e.g., Yinyanghuo A, compound 191, Table 12.3). The majority of the other flavones are cyclized to chromene structures between 7,6- or 7,8 of ring A, and *C*-5-substitution concentrates on positions 3, 6, and 8 (see heterophyllin, Figure 12.2). They have been reported from all parts of the plants, but there is a strong indication for roots and root barks as major sources. No reference exists on external accumulation of such compounds.

12.5.9 *C*-Linked Aromatic- and Ketopyrano-Substitution

C-Linked aromatic substituents are reported to occur in aerial parts of *Thymus hirtus*,[80] along with one report on Chinese propolis.[81] In aerial parts of some members of Asteraceae and Lamiaceae, flavones with a *C*-6-ketopyrano-substitution were reported (e.g., hosloppin, oppositin; see Figure 12.4). These aromatic substituents are positioned at the 6-, and rarely, at the 8-position of ring A. They are all reported from aerial parts, but no reference is made to their possible occurrence as exudates constituents.

12.5.10 *Tephrosia* Flavones

The genus *Tephrosia* (Fabaceae) was selected to demonstrate the biosynthetic capacity of flavone substitution. In particular, there is a strong tendency towards formation of furano-residues, linked through *C*-bonds on position 8 of the flavone nucleus (e.g., apollinine, hookerianin; Figure 12.5). The basic flavone structure is mostly 5- and 7-*O*-methylated. These compounds have been exclusively reported to occur in roots, leaf and stem as well as

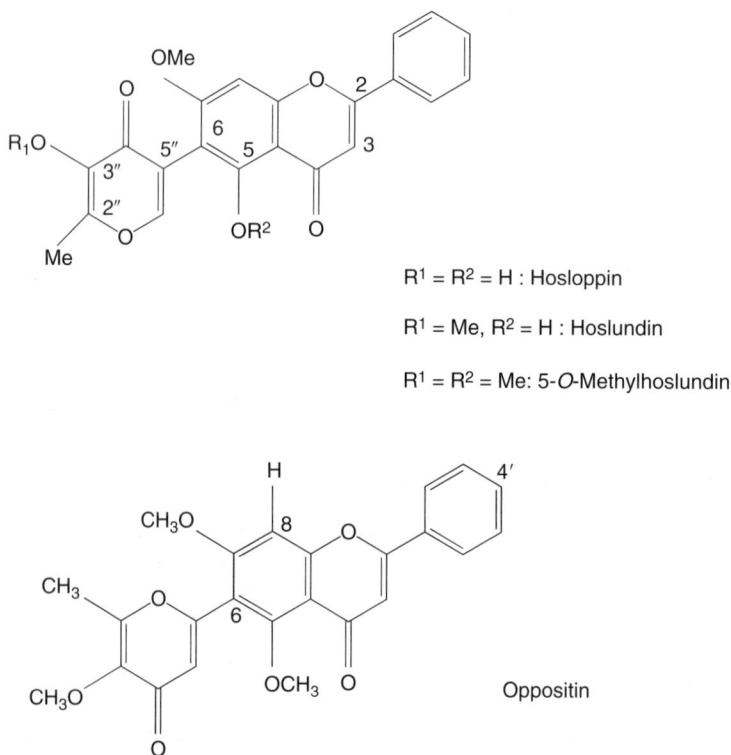

$R^1 = R^2 = H$: Hosloppin

$R^1 = Me, R^2 = H$: Hoslundin

$R^1 = R^2 = Me$: 5-*O*-Methylhoslundin

Oppositin

FIGURE 12.4 Flavones with ketopyrano substitution.

in seeds of *Tephrosia* species. Several compounds are 7,8-disubstituted (furanogroup between 7-OH and *C*-8; e.g., tephrorianin; Figure 12.5) isolated from pods.[82] Aerial parts of *Tephrosia* spp. yielded primarily bisfurano structures, which are also 7,8-disubstituted, such as semi-glabrin, multijugin, and enantiomultijugin (Figure 12.5). There is also a tendency observed towards acetylation on the bisfurano moiety. Even more complex structures arise by addition of bicyclic substituents, being combined from furano and pyrano residues (stachyoidin, tephrodin, compounds 228 and 229, Table 12.3). These structures have been known for a long time, their accumulation site, however, being not indicated.[3] Only two pyranoflavones are accumulated: isopongaflavone and 5-OMe-7,6-(2,2-dimethylchromeno)-flavone (compounds 135 and 136, Table 12.3). Further compounds include *C*-prenylpyranoflavones such

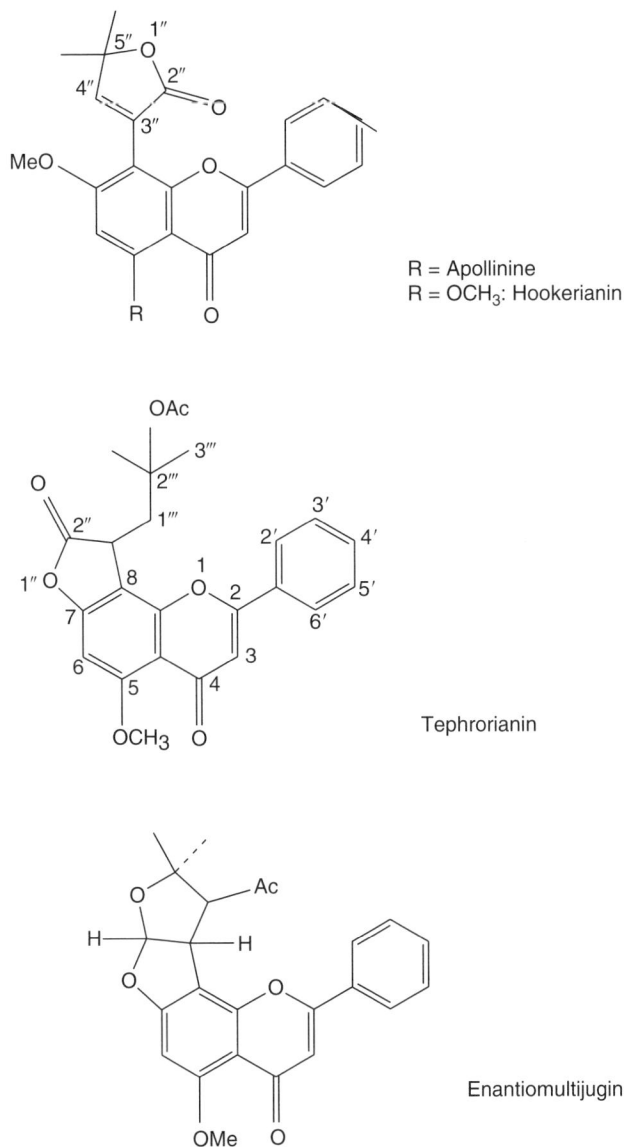

R = Apollinine
R = OCH₃: Hookerianin

Tephrorianin

Enantiomultijugin

FIGURE 12.5 Furanoflavones from *Tephrosia*.

as fulvinervins (compounds 188 and 189, Table 12.3). Generally, the trend to 7-OH-*C*-8-substitution is quite prominent in this genus. Altogether, the biosynthetic capacity of *Tephrosia* is remarkable.

12.5.11 *ARTOCARPUS* FLAVONES

The genus *Artocarpus* (Moraceae) affords a prime example of biosynthetic activities to produce complex cyclized flavone structures, being primarily accumulated in addition to *C*-prenylated and pyrano-substituted flavones (see Table 12.3; Section 12.5.2 and Section 12.5.8). Apart from the pyrano substitution also encountered in other plants (see Section 12.5.5, Table 12.3), cyclization occurs between the 6'- and 3-position of the flavone molecule to yield xanthonoid structures (e.g., artonol E, D; Figure 12.6). A further and even more prominent tendency is represented by cyclization between the 2'-OH-group and the position 6 of the flavonoid nucleus. In this case, cyclization may lead to either six- or seven-membered rings including oxygen (see structures of artocommunol CA and artelastocarpin in Figure 12.6). Artonin K serves as another example of complex cyclization (Figure 12.6).

Generally, cyclization between 7-OH and *C*-8 occurs in various flavone derivatives of *Artocarpus* (e.g., artonol D; artocommunol CA, artelastofuran; Figure 12.6.). However, *A. rigida* yielded both 7, 6- and 7, 8-cyclized compounds (artonin M, P; compounds 250 and 259, Table 12.3). In this section, compounds with pyranosubstitution at both ring A and B are listed, for example, artonol D (Figure 12.6). All of these compounds appear to be mainly accumulated in rather lignified parts of the plant, such as heartwood, bark, and shoot. It is hardly perceivable that such complex compounds could occur in exudates of aerial parts.

12.5.12 FLAVONE–COUMARIN HYBRIDS

Only two flavones of this type are known so far. They were reported from *Gnidia soccotrana* (Thymeleaceae).[83] For formula see Figure 12.7. Similarly, only a few flavonols are known as hybrid structures, but none of them with coumarins (see Table 12.4).

12.6 FLAVONOLS WITH OTHER SUBSTITUENTS

Data on this type of flavonols are summarized in Table 12.4. In contrast to the corresponding flavones, the number and complexity of derivatives is smaller. This concerns particularly the formation of furano-, pyrano- and other cyclic flavonols. There is a remarkable number of *O*-prenylated flavonols known to date, contrasting to only very few flavones exhibiting this substitution pattern (see Table 12.3). Similar trends have been earlier documented in the review of Barron and Ibrahim.[3] The occurrence of a series of glycosides based on *C*-prenylated structures is considerable.[3] This substitution trend concerns also some of the dihydroflavonols, thus indicating specific enzyme activities probably dependent on the presence of a 3-OH group.

12.6.1 *C*-METHYLFLAVONOLS

In this section, reports concentrate on genera from the families Caesalpiniaceae, Myrtaceae, and Velloziaceae. Apart from the Myrtaceae, no *C*-methylflavones have been reported to occur in the other two families as yet (see Table 12.3). In Myrtaceae, *C*-methylflavonols have been found also in exudates.[57] Most of the other species listed here accumulate these flavonols in the leaves without further specification. Fungal sources include two species of *Colletotrichum*, where *C*-methylflavonols have been found in the culture filtrate.[84] So far, no C2- or C3- *C*-linked flavonols have been reported as was the case with the flavones.

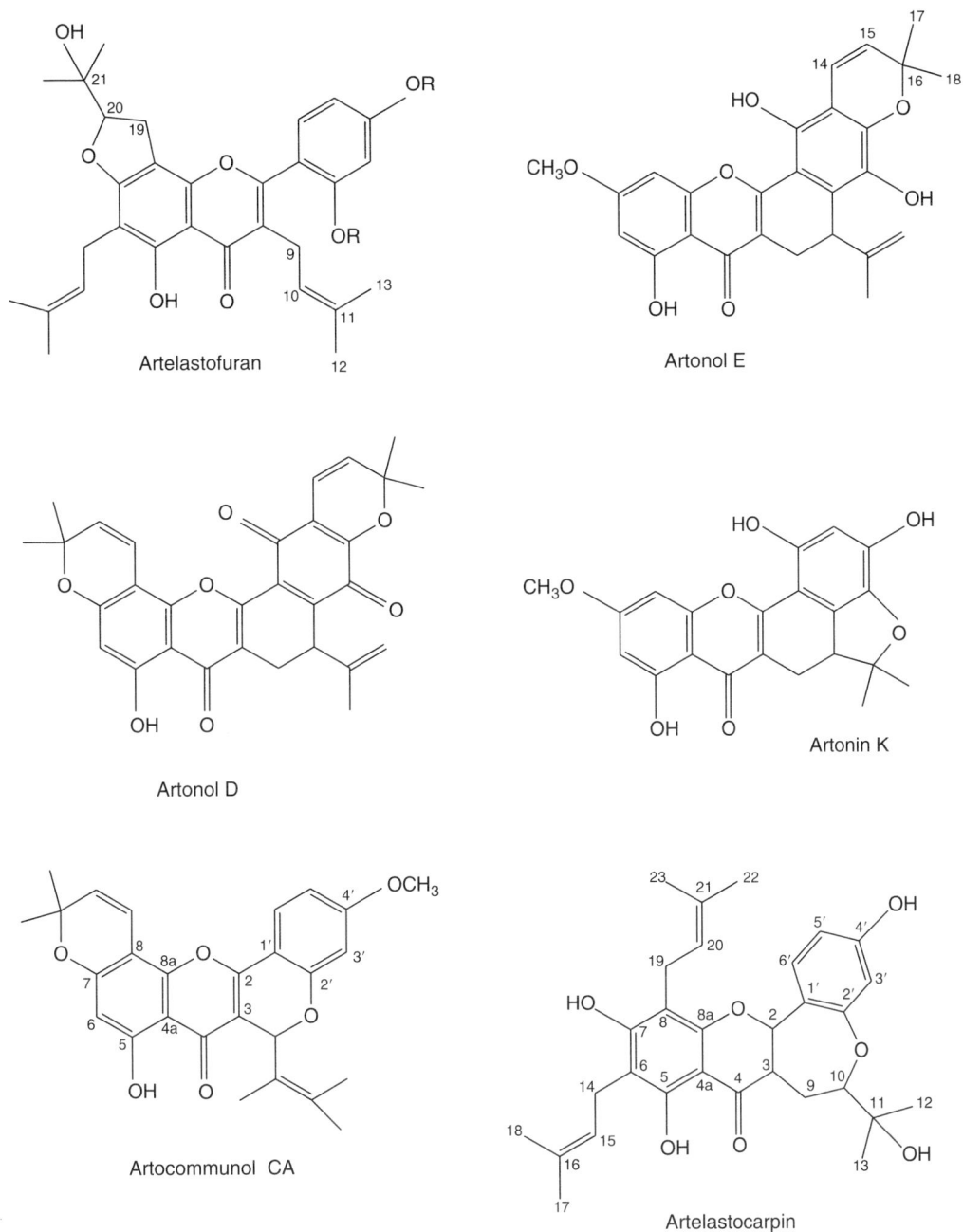

FIGURE 12.6 Flavones from *Artocarus* spp.

12.6.2 METHYLENEDIOXYFLAVONOLS

The Rutaceae appear to be a rich source of flavonols with methylenedioxy substitution. Their main accumulation site is apparently bark tissue, followed by leaves. It may be assumed that some of these leaf constitutents are accumulated externally as was also the case with the corresponding flavone derivatives of other plant families. In general, the Rutaceae exhibit a

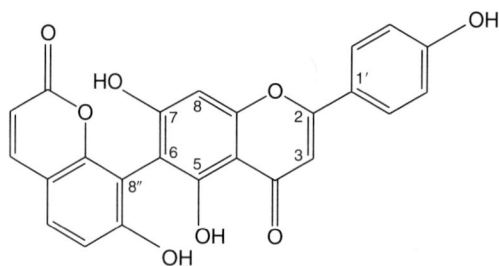

6-(8″-Umbelliferyl)-apigenin

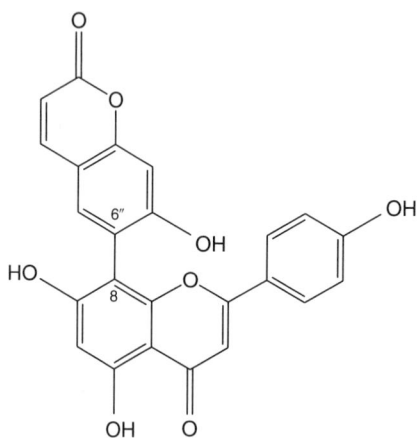

8-(6″-Umbelliferyl)-apigenin

FIGURE 12.7 Flavone–coumarin hybrids.

trend rather to produce flavonols of this type, whereas one corresponding flavone only (ageconyflavon A) is so far known from this family. Very few reports concern the genus *Millettia* (Fabaceae; stem bark)[85] and some Amaranthaceae (aerial parts, whole plants).[86]

12.6.3   *C*-PRENYLFLAVONOLS

Within the complex flavonols, *C*-5 substitution is by far the most prominent trend in terms of numbers of compounds and sources. Also, the number of structurally different *C*-5-residues is remarkable. Apart from the common "3,3-dimethylallyl" and the rarer "1,1-dimethylallyl," several hydroxylated *C*-5-residues such as in topazolin hydrate, isolated from roots of *Lupinus luteus* (Fabaceae; compound 94 in Table 12.4). Another source of such compounds is the whole plant of *Duranta repens* (Verbenaceae).[87,88] Plant organs accumulating *C*-prenylflavo-nols range from buds to leaves and from aerial parts to roots. Only the bud constituents may be considered as exudate compounds. From *Lilium candidum*, a more complex flavonol was reported (Figure 12.8).[89] This genus also accumulates N-containing flavonoid derivatives discussed earlier.[6]

Some remarks should be made regarding structures revised during the reporting period. A flavonol isolated from *Glycyrrhiza lepidota*, named glepidotin, had been ascribed the structure of 8-*C*-prenylgalangin.[90] Comparison with the synthetic product and its isomer revealed that it is, in fact, 6-*C*-prenylgalangin (compound 78, Table 12.4).[91] Noricaritin

TABLE 12.4
Flavonols with Other Substituents

No.	O-Substitution	Other Substituents	Trivial Name	Plant Species	Family	Plant Organ	Ref.
	C-Methylflavonols						
1	3,5,7-triOH	8-Me	8-Methylgalangin				
2	3,5,7-triOH	6,8-diMe					
3	5,7-diOH, 3-OMe	8-Me					
4	5,7-diOH, 3-OMe	6,8-diMe					
5	3,5-diOH, 7-OMe	8-Me					
6	3,5-diOH, 7-OMe	6,8-diMe					
7	3,5,6,7-tetraOH	8-Me	Isoplatanin				
8	3,5,7,8-tetraOH	6-Me	Platanin				
9	5,7,8-triOH, 3-OMe	6-Me					
10	3,5,7-triOH, 8-OMe	6-Me	Pityrogrammin				
11	5,6,4'-triOH, 3-OMe	8-Me	Sylpin				
12	3,5,7,4'-tetraOH	6-Me	6-Methylkaempferol				
13	5,7,4'-triOH, 3-OMe	6-Me					
14	5,7,4'-triOH, 3-OMe	6,8-diMe		*Piliostigma thonningii*	Caesalpin.	Leaf	514
15	5,4-diOH-3,7-diOMe	6-Me	8-Desmethyllatifolin	*Callistemon,* 4 spp.	Myrt.	External	57
				Leptospermum laevigatum	Myrt.	Leaf wax	424
16	5,4-diOH, 3,7-diOMe	6,8-diMe	Latifolin	*Piliostigma thinningii*	Caesalpin.	Leaf	514
17	5-OH, 3,7,4'-triOMe	6-Me	8-Desmethylkalmiatin				
18	5-OH, 3,7,4'-triOMe	6,8-diMe	Kalmiatin				
19	5,4-diOH-3,6,7-triOMe	6-Me*		*Colletotrichum dematium*	Fungus!	(cult. filtr.)	84
20	5,4-diOH-3,6,7-triOMe	8-Me*		*Vellozia laevis*	Velloz.	Leaf surface	105
				Vellozia nanuzae	Velloz.	Leaf	515
21	3,5,8,4-OH, 7-OMe	6-Me					
22	5,8,4-triOH, 3,7-OMe	6-Me					
23	5,4-diOH-3,7,8-triOMe	6-Me*		*Colletotrichum dematium*	Fungus!	(cult. filtr.)	84
24	3,5,7,3',4'-pentaOH	6-Me	Pinoquercetin				
25	(3,5,7,3',4'-pentaOH)	8-Me	Glycoside only				

continued

TABLE 12.4
Flavonols with Other Substituents — continued

No.	O-Substitution	Other Substituents	Trivial Name	Plant Species	Family	Plant Organ	Ref.
26	3,5,7,3',4'-pentaOH	6,8-diMe		Piliostigma thonningii	Caesalpin.	Leaf	514
27	5,7,3',4'-OH, 3-OMe	6-Me		Vellozia phalocarpus	Velloz.	Leaf	105
				Vellozia phalocarpus	Velloz.	Leaf	516
				Xerophyta retinervis	Velloz.	Leaf	105
28	5,7,3',4'-tetraOH-3-OMe	6,8-diMe*		Piliostigma thonningii	Caesalpin.	Leaf	514
29	5,3',4'-triOH-3,7-diOMe	6-Me*		Piliostigma thonningii	Caesalpin.	Leaf	514
30	5,3',4'-triOH-3,7-diOMe	6,8-diMe*		Piliostigma thonningii	Caesalpin.	Leaf	514
31	5,7,4'-OH, 3,3'-diOMe	6-Me					
32	5,7,4'-OH, 3,3'-diOMe	6,8-diMe					
33	5,4'-diOH, 3,7,3'-OMe	6-Me		Callistemon salignus	Myrt.	External	57
				Leptospermum laevigatum	Myrt.	Leaf wax	424
				Piliostigma thonningii	Caesalpin.	Leaf	514
34	5,4'-diOH, 3,7,3'-OMe	6,8-diMe		Leptospermum laevigatum	Myrt.	Leaf wax	424
35	5-OH-3,7,3',4'-OMe	6-Me*		Vellozia epidendroides, V.lilacina*	Velloz.	Leaf	517
36	5,3',4'-OH-3,6,7-OMe	8-Me*		Vellozia nanuzae	Velloz.	Leaf	515
37	5,7,4'-OH-3,6,3'-OMe	8-Me*		Vellozia stipitata	Velloz.	Whole plant	104
38	5,4'-OH,3,6,7,3'-OMe*	8-Me*		Vellozia nanuzae	Velloz.	Leaf	515
				Vellozia epidendroides, V.lilacina*	Velloz.	Leaf	517
				Vellozia laaevis	Velloz.	Leaf	516
				Vellozia laevis, V. phalocarpa	Velloz.	Leaf	105
39	5,3'-OH,3,6,7,4'-OMe	8-Me*	Alltaudiol				
40	3,5,7,3',4',5'-hexaOH	6-Me					
41	5,7,3',4',5'-OH, 3-OMe	6-Me	Dumosol				
42	5,7,3',4',5'-OH, 3-OMe	6,8-diMe					
43	3,5,7,3',5'-OH, 4'-OMe	6-Me					
44	3,5,7,3',5'-OH, 4'-OMe	6,8-diMe					
45	5,7,3',5'-OH, 3,4'-OMe	6-Me					
46	5,7,3',5'-OH, 3,4'-OMe	6,8-diMe					

Methylendioxyflavonols

No.	Substitution	Methylenedioxy	Trivial name	Species	Family	Part	Ref.
47	3,5-diOMe	6,7-OCH_2O		*Gomphrena martiana, G. boliviana*	Amaranth.	Whole plant	150
48	3,5-diOMe	6,7-/3',4'-diOCH_2O	Meliternatin	*Melicope simplex, M. ternata*	Rut.	Bark	380
49	3,7-diOMe	3',4'-OCH_2O		*Comptonella microcarpa*	Rut.	Leaf (ext.?)	518
50	5-OH,3,7-diOMe	3',4'-OCH_2O	Desmethylkanugin	*Millettia leucantha*	Fab.	Stem bark	85
51	3,5,7-triOMe	3',4'-OCH_2O	Isokanugin	*Melicope simplex, M. ternata*	Rut.	Bark	380
52	3,5,8-triOH	6,7-/3',4'-diOCH_2O		*Comptonella microcarpa*	Rut.	Leaf (ext.?)	518
53	3,5,8-triOMe	6,7-OCH_2O					
54	3,5,8-triOMe	3',4'-OCH_2O					
55	3,5,8-triOMe	6,7-/3',4'-diOCH_2O					
56	3,5,3'-triOMe	6,7-/4',5'-diOCH_2O					
57	3,5,4'-triOH	6,7-OCH_2O	Gomphrenol				
58	3,7,3'-triOMe	3',4'-OCH_2O	Kanugin				
59	3,7-diOH,5,6-diOMe	3',4'-OCH_2O					
60	5-OH, 3,6,7-triOMe	3',4'-OCH_2O	Melisimplin	*Melicope simplex, M. ternata*	Rut.	Bark	380
61	3,5,6,7-tetraOMe	3',4'-OCH_2O	Melisimplexin	*Melicope ternata*	Rut.	Bark	380
62	5,7-diOH-3,8-diOMe	3',4'-OCH_2O*		*Melicope coodeana*	Rut.	Leaf	374
63	5-OH, 3,7,8-triOMe	3',4'-OCH_2O	5-Desmethylmeliternin	*Melicope simplex*	Rut.	Bark	380
64	7-OH,3,5,8-triOMe	3',4'-OCH_2O					
65	3,5,7,8-tetraOMe	3',4'-OCH_2O	Meliternin	*Comptonella microcarpa*	Rut.	Leaf (ext.?)	518
66	3,5,8,3'-tetraOMe	6,7-/3',4'-diOCH_2O		*Melicope ternata*	Rut.	Bark	380
67	5,3',4'-OH, 3-OMe	6,7-OCH_2O		Glycoside only			
68	3,5,3'-triOH,4'-OMe	6,7-OCH_2O*	Wharangin	*Blutaparon portulacoides*	Amaranth.	Aerial parts	519
69	5,3',4'-triOH, 3-OMe	7,8-OCH_2O	Melinervin	Glycoside only			
70	5,4'-OH, 3,3'-diOMe	6,7-OCH_2O					
71	3,5,7-triOH, 6,8-diOMe	3',4'-OCH_2O					
72	5,7-diOH-3,6,8-triOMe	3',4'-OCH_2O*		*Melicope coodeana*	Rut.	Leaf	374
73	3,5-diOH, 6,7,8-triOMe	3',4'-OCH_2O		*Melicope triphylla*	Rut.	Leaf	520
74	7-OH-3,5,6,8-tetraOMe	3',4'-OCH_2O*		*Comptonella microcarpa*	Rut.	Leaf	518
75	5-OH, 3,6,7,8-tetraOMe	3',4'-OCH_2O	5-Desmethylmelibentin	*Melicope triphylla*	Rut.	Leaf	521
76	3,5,6,7,8-pentaOMe	3',4'-OCH_2O	Melibentin	*Melicope ternata*	Rut.	Bark	380
77	3,5,8,3',4'-pentaOMe	6,7-OCH_2O	Melicophyllin	*Melicope ternata*	Rut.	Bark	380

continued

TABLE 12.4
Flavonols with Other Substituents — continued

No.	O-Substitution	Other Substituents	Trivial Name	Plant Species	Family	Plant Organ	Ref.
	C-Prenylflavonols						
78	3,5,7-triOH	6-C5	Glepidotin A; 6-prenylgalangin; revised		Synthesis		91
79	3,5,7-triOH	8-C5	8-(1,1-dimethylallyl)-galangin	Platanus acerifolius	Platanac.	Bud	522
80	3,7,4'-triOH	8-C5-OH	Platanetin				
81	3,5,7,8-tetraOH	6-C5		Platanus acerifolia	Platanac.	Bud	46
82	3,5,7,4'-tetraOH	6-C5*	Licoflavonol*	Glycyrrhiza spp.	Fab.	Root	523
83	3,5,7,4'-tetraOH	6-C5-OH			Synthesis		91
84	3,5,7,4'-tetraOH	6-C10*	Macarangin*	Macaranga vedeliana	Euphorb.	Leaf	524
				Monotes africanus	Dipteroc.	Leaf	525
				Macaranga denticulata	Euphorb.	Leaf	526
85	3,5,7,4'-tetraOH	8-C5	Noranhydroicaritin 8-(1,1-dimethylallyl)-kaempferol	Epimedium koreanum	Berb.	Aerial parts	527
				Platanus acerifolia	Platanac.	Bud	46
86	3,5,7,4'-tetraOH	6-C5-OH	Noricaritin, revised			—	91
87	3,5,7,4'-tetraOH	8-C-(3-methyl-succinoyl), see Figure 12.8		Lilium candidum	Liliac.		89
88	3,5,7,4'-tetraOH	8-C10*	Isomacarangin*	Macaranga schweinfurthii	Euphorb.	Leaf	528
89	3,5,7,4'-tetraOH	3'-C5	Isolicoflavonol	Glycyrrhiza spec.	Fab.	Liquorice	334
90	3,5,7,4'-tetraOH	6,8-diC5	6,8-Diprenylkaempferol	Platanus acerifolia	Platanac.	Bud	46
91	3,5,7,4'-tetraOH	6,3'-diC5	Glyasperin A; 6,3'-diprenylkaempferol	Monotes africanus	Dipteroc.	Leaf	525
92	3,5,7,4'-tetraOH	8,3'-diC5	Broussoflavonol F	Glycyrrhiza aspera	Fab.	Root	529
93	5,7,4'-triOH, 3-OMe	6-C5	Topazolin	Broussonetia papyrifera	Morac.	Root bark	93
94	5,7,4'-triOH, 3-OMe	6-C5-OH	Topazolin hydrate	Lupinus luteus	Fab.	Root	530
95	3,7,4'-triOH, 5-OMe	8-C5	Sophoflavescenol	Lupinus luteus	Fab.	Root	530
96	3,5,4'-triOH, 7-OMe	8-C5	Isoanhydroicaritin	Sophora flavescens	Fab.	Roots	531
97	3,5,7-triOH, 4'-OMe	8-C5	Anhydroicaritin	Epimedium koreanum	Berb.	Aerial parts	527
98	3,5,7-triOH, 4'-OMe	8-C5-OH	Icaritin				
99	3,5,7-triOH, 4'-OMe	8-C5-OMe*	Brevicornin*	Epimedium brevicornum	Berb.	Aerial parts	279
100	5,7,4'-triOH, 3,6-diOMe	3'-C5-OH	Aliarin				

No.	Substitution	Prenyl	Name	Species	Family	Part	Ref.
101	3,7,4'-triOH, 5,6-diOMe	3'-C5-OH		Duranta repens	Verben.	Whole plant	87
102	5,7-diOH, 3,6,4'-triOMe	3'-C5	Viscosol	Duranta repens	Verben.	Whole plant	88
103	5,7-diOH,3,6,4'-triOMe	3'-C5-OH*		Duranta repens	Verben.	Whole plant	87
104	3,7-diOH,5,6,4'-triOMe	3'-C5-OH					
105	3,5,7,2',4'-pentaOH	8-C10	Kushenol C				
106	3,5,7,2',4'-pentaOH	8-C10-OH	Kushenol G				
107	3,5,7,3',4'-pentaOH	6-C5*	Gancaonin P*	Glycyrrhiza uralensis	Fab.	Aerial parts	457
				Dorstenia ciliata	Morac.	Aerial parts	436
108	3,5,7,4'-tetraOH, 3'-OMe	6-C5	Gancanonin P -3'-Me	Glycyrrhiza uralensis	Fab.		532
109	3,5,7,3',4'-pentaOH	6-C5-OH	Glycoside only		—		
110	3,5,7,3',4'-pentaOH	5'-C5	Uralenol	Glycyrrhiza uralensis	Fab.	Leaves	533
				Broussonetia papyrifera	Morac.		534
111	3,5,7,3',4'-pentaOH	6,5'-diC5*	Broussonol E*	Broussonetia kazinoki	Morac.	Leaf	98
112	3,5,7,3',4'-pentaOH	8,5'-diC5*	Broussonol D*	Broussonetia kazinoki	Morac.	Leaf	98
113	3,5,7,3',4'-penta-OH	8,2',6'-triC5	8,2',6'-Triprenylquercetin	Petalostemum purpureum	Fab.	Root, may be artifact	535
114	3,5,7,3',4'-pentaOH	8,2',3'-triC5*	Broussoflavonol G*	renamed from Broussoflavonol "E"			94
115	3,5,7,3',4'-pentaOH	8,2',6'-triC5	Broussoflavonol C				
116	5,7,3',4'-OH-3-OMe	6-C5*		Vellozia coronata, V. namuzae	Velloz.	Leaf surface	515
117	5,7,3',4'-tetraOH, 3-OMe	5'-C5	Uralenol-3-Me	Glycyrrhiza uralensis	Fab.	Leaves	536
118	5,7,3',4'-OH, 3-OMe	6,8-diC5	Broussoflavonol B				
119	3,5,7,4'-tetraOH-3'-OMe	6,8-diC5*	Dorsmanin D*	Dorstenia mannii	Morac.	Twigs	454
120	5,3',4'-OH-3,7-OMe	6-C5*		Vellozia coronata	Velloz.	Leaf surface	515
121	5,7,4'-triOH-3,8-diOMe	6-C5*		Vellozia scoparia	Velloz.	Leaf surface	105
122	3,5,3'-OH, 7,4'-diOMe	6-C5*	6-Prenylherbacetin*	Artemisia campestris glutinosa	Ast.	Aerial parts	277
123	3,5,3'-OH, 7,4'-diOMe	8-C5	Isorhynchospermin Rhynchospermin				
124	5,4'-diOH, 3,7,3'-OMe	8-C5	8-Prenylpachypodol				
125	3,6,7,3',4'-pentaOH	2'-C5	Neouralenol	Glycyrrhiza uralensis	Fab.	Leaves	533
126	5,6,4',5'-tetraOH-3OMe	2'-C5	Uralene	Glycyrrhiza uralensis	Fab.	Leaves	536
	O-Prenylflavonols						
127	5,7-diOH,3-OMe	4-O-C5*		Boronia coerulescens	Rut.	Aerial parts	95
128	5,7-diOH, 3-OMe	4-O-C5		Bosistoa brassii	Rut.	Leaf	537
129	5-OH, 3,8-diOMe	7-O-C5 (epoxy)	7-Epoxyprenylgnaphaliin				

continued

TABLE 12.4
Flavonols with Other Substituents — continued

No.	O-Substitution	Other Substituents	Trivial Name	Plant Species	Family	Plant Organ	Ref.
130	5,7-diOH,8-OMe*	3,4'-di-O-C5*		*Boronia coerulescens*	Rut.	Aerial parts	95
131	3,5,7-OH,6-OMe	4'-O-C5*		*Boronia coerulescens*	Rut.	Aerial parts	95
132	5,7-OH,3,6-OMe	4'-O-C5*		*Boronia coerulescens*	Rut.	Aerial parts	95
133	3,5,7-triOH, 8-OMe*	4'-O-C5*		*Boronia coerulescens*	Rut.	Aerial parts	95
134	5,7-diOH,3,8-diOMe*	4'-O-C5*		*Boronia coerulescens*	Rut.	Aerial parts	95
135	5,4'-diOH, 3,3'-diMe	7-O-C5		*Euodia glabra*	Rut.	Shoot bark	538
				Melicope elleryana	Rut.	Fruit	379
136	5,4'-diOH, 3,3'-diOMe	7-O-C5 (epoxy)					
137	5,7-diOH, 3,3'-diOMe	4'-O-C5		*Bosistoa medicinalis*	Rut.	Leaf	322
138	5,3'-diOH, 3,4'-diOMe	7-O-C5		*Melicope elleryana*	Rut.	Fruit	379
139	4'-OH, 3,5,3'-triOMe	7-O-C5					
140	3,5,3'-triOH-6,7-diOMe	4'-O-C10-OH	Geranioloxyalatum flavone	*Zanthoxylum alatum*	Rut.	Seed	401
141	5,7-diOH, 3,8,3'-OMe*	4'-O-C5*		*Boronia coerulescens*	Rut.	Aerial parts	95
142	3,5,4'-triOH-8,3'-diOMe*	7-O-C5*		*Melicope micrococca*	Rut.	Aerial parts	96
143	5,4'-diOH, 3,8,3'-OMe	7-O-C5*		*Boronia coerulescens*	Rut.	Aerial parts	95
144	5,4'-diOH, 3,8,3'-OMe	7-O-C5		*Melicope elleryana*	Rut.	Fruit	379
145	4'-OH, 3,5,8,3'-tetraOMe	7-O-C5*		*Melicope triphylla*	Rut.	Leaf	521
146	3,5,8,3',4'-pentaOMe	7-O-C5		*Melicope triphylla*	Rut.	Leaf + bark	539
147	5,4'-OH, 3,6,8,3'-OMe	7-O-C5		*Boronia coerulescens*	Rut.	Aerial parts	95
148	3,5,8,4'-OH-7,3'-OMe	6-O-C5		*Melicope micrococca*	Rut.	Aerial parts	96
149	5-OH,3,8-diOMe,7-O-C5	3',4'-OCH$_2$O		*Melicope triphylla*	Rut.	Leaf	521
150	3,5,8-triOMe, 7-O-C5	3',4'-OCH$_2$O		*Achyrocline flaccida*	Ast.	Aerial parts	540
				Melicope triphylla	Rut.	Leaf + bark	539
152	3,5,6,8-tetraOMe-7O-C5	3',4'-OCH$_2$O		*Comptonella microcarpa*	Rut.	Leaf (ext.?)	518
				Comptonella microcarpa	Rut.	Leaf (ext.?)	518
Pyranoflavonols							
153	3,5,4'-triOH	7,6-ODmp	Desmethyl-anhydroicaritin				91
154	3-OMe	7,8-ODmp	Karanjachromene				
155	3,5-diOMe	7,8-ODmp*		*Lonchocarpus latifolius*	Fab.	Root	541
156	3,6-diOMe	7,8-ODmp		*Lonchocarpus latifolius*	Fab.	Root	541

No.	Substitution		Name	Species	Family	Part	Ref
157	3,5,4'-triOH	7,8-ODmp	Citrusinol	Citrus reticulata*	Rut.	Fruit	542
158	4'-OH, 3,6-diOMe	7,8-ODmp*		Dorstenia poinsettifolia var. angusta	Morac.	Leaves	441
159	3,5,3',4'-tetraOH	7,8-ODmp, 6-C5*, see Figure 12.9	Poinsettifolin A*				436
160	5,3',4'-OH, 3-OMe*	7,6-ODmp*		Dorstenia ciliata	Morac.	Aerial parts	97
161	3,5,3',4'-tetraOH	7,8-ODmp-Me	Sarothranol*	Hypericum japonicum	Hyperic.	Whole plant	543
162	3,5-diOMe, 3',4'-OCH2O	7,8-ODmp		Asclepias syriaca	Asclepiad.	Leaf	541
163	3,6-diOMe, 3',4'-O2CH2	7,8-ODmp	Pongachromene	Lonchocarpus latifolius	Fab.	Root	544
164	3,5,7-triOH	4',3'-cycl-OC5-OH*	Glycyrrhiza-flavonol A*	Derris araripensis	Fab.	Root	544
				Glycyrrhiza sp.	Fab.	Liquorice	334
Furanoflavonols							
165	3-OMe	7,6-fur*	Ponganone XI*	Pongamia pinnata	Fab.	Root	297
166	3-OMe	7,8-fur	Karanjin	Dahlstedtia pentaphylla	Fab.	Root	463
167	3,4'-diOMe	7,8-fur		Derris mollis	Fab.	Root	483
168	3,5,6-triOMe	7,8-fur		Derris araripensis	Fab.	Root	544
169	3-OMe, 3',4'-OCH2O	7,8-fur	Pongapin	Lonchocarpus latifolius	Fab.	Root	541
170	3,3'-diOMe, 4',5'-OCH2O	7,8-fur	3'-OMe-pongapin				
171	3,5,6-triOMe, 3',4'-O2CH2	7,8-fur	5,6-diOMe-pongapin	Derris araripensis	Fab.	Root	544
172	3,5,3',4'-tetraOH	5'-C5; 7,8-triMe-fur	Broussonol C	Broussonetia kazinoki	Morac.	Leaf	98
173	3,5,3'triOH, 4',5'-ODmp*	7,8-triMe-fur*	Broussonol B*	Broussonetia kazinoki	Morac.	Leaf	98
Prenyl- and pyrano-substitution							
174	3,5-diOH	8-C5, 7,6-ODmp	Sericetin				
175	3,5,7-triOH	6,8-diC5-4',3'-ODMp*	Dorsilurin C*	Dorstenia psilurus	Morac.	Root	453
176	3,3',4'-triOH	7,6-ODmp, 8-C5	Macaflavon I				
177	3',4'-diOH, 3-OMe	7,6-ODmp, 8-C5	Macaflavon II				
178	3,5,7,3'tetraOH	4',5'ODmp-8-C5*	Broussonol A*	Broussonetia kazinoki	Morac.	Leaf	98
179	3,5,7,4'-tetraOH	3',2'-ODmp-C6*	Petalopurpurenol*	Petalostemon purpureus	Fab.	Root	99
180	3,5,3',4'-tetraOH	7,8-ODmp-6-C10*	Dorsmanin C*	Dorstenia mannii	Morac.	Twigs	454
181	5,3',4'-OH, 3-OMe	8-C5, 7,6-ODmp	Broussoflavonol A	Broussonetia papyrifera	Morac.	Cortex	545
182	3,5,5'-tetraOH	8,2'-diC5, 3',4'-ODmp, see Figure 12.9	Broussoflavonol D				
183	3,5,7,5'-tetraOH	8,2'-diC5, 3',4'-ODmp	Broussoflavonol E	Broussonetia papyrifera	Morac.	Root bark	546

continued

TABLE 12.4
Flavonols with Other Substituents — continued

No.	O-Substitution	Other Substituents	Trivial Name	Plant Species	Family	Plant Organ	Ref.
	Flavonols with aromatic substituents						
184	3,5,7,4'-tetraOH	8-C-p-OH-benzyl*	p-Hydroxybenzyl-kaempferol*	*Thymus hirtus*	Lam.	Aerial parts	80
185	3,5,7,3',4'-pentaOH	8-C-p-OH-benzyl*	p-Hydroxybenzyl-quercetin*	*Thymus hirtus*	Lam.	Aerial parts	80
186	5,7-diOH, 3,4'-diOMe	8-C-p-OH-phenylethyl	Haplopappin	*Haplopappus foliosus*	Ast.		100
	Various cyclo-flavonols						
187	5,7,4,5'-tetraOH	3,2-O-CH2-		*Acacia crombei, A. carnei*	Mimosac.		
188	5,7,3',4'-tetraOH-6OMe	3,2-O-CH2-	Benthamianin	*Distemonanthus benthamianus*	Fab.	Heartwood	
189	5,7,3',4'-tetraOH-6OMe	3,2-O-CH2-	Distemonanthin	*Distemonanthus benthamianus*	Fab.	Heartwood	
190	7,4,5'-triOH	3,2-O-CH2-OH, see Figure 12.9	Fasciculiferin	*Acacia fasciculifera*	Mimosac.	Heartwood	547
191	7,4,5'-triOH	3,2-O-CH2-	Peltogynin	*Acacia peuce, A. crombei, A. fasciculifera*	Mimosac.	Heartwood	548
192	7,3',4'-triOH	3,2-O-CH2-	Mopanin	*Colophospermum mopane*	Caesalpin.	Stem	549
193	5-OH, 7-OMe, 3',4'-O2CH2	3,2-O-CH2-	Pulcherrimin	*Caesalpinia pulcherrima*	Caesalpin.	Stem	550
194	5-OH, 6,7-diOMe 3',4'-O2CH2	3,2-O-CH2-	6-OMe-pulcherrimin	*Caesalpinia pulcherrima*	Caesalpin.	Stem	550
195	3,5-diOH	7,8-O-cycl-phenylethyl	Calomelanol D	*Pityrogramma calomelanos*	Pteridaceae	Frond exudate	121
196	3,5,4'-triOH	7,8-O-cycl-phenylethyl		*Pityrogramma calomelanos*	Pteridaceae	Frond exudate	122
	Hybrid structures						
197	3,5,7,4'-tetraOH	Diterpene-flavonol, see Figure 12.9	Denticulaflavonol	*Macaranga denticulata*	Euphorb.	Leaves	526
198	5'-OH, 3,5,2'-OMe	Flavono-lignoid		*Distemonanthus benthamianus*	Fab.	Heartwood	102

*For explanation, please see text.
For abbreviations see footnote to Table 12.3.

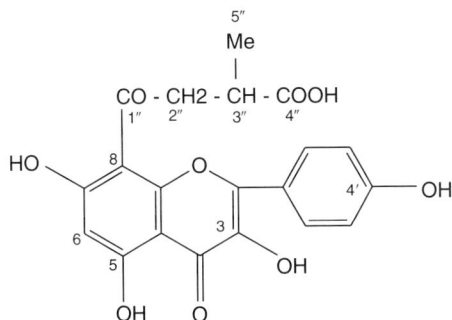

FIGURE 12.8 Flavonol from *Lilium candidum*.

(compound 86, Table 12.4), originally isolated from *Bursera leptophloeos* as 3,5,7,4'-tetraOH-8-*C*-hydroxyprenylflavone,[92] needs to be revised to the corresponding 6-*C*-derivative.[91] Finally, the structure of broussoflavonol E (3,5,7,3',4'-pentaOH-8,2',6'-triprenylflavone) from *Broussonetia papyrifera*[93] was later revised to 3,5,7,3',4'-pentaOH-8,2',3'-triprenylflavone and renamed broussoflavonol G (compound 114, Table 12.4).[94]

12.6.4 *O*-PRENYLFLAVONOLS

These compounds are almost exclusively accumulated in aerial parts and leaves from genera of the Rutaceae such as *Bosistoa*, *Boronia*, and *Melicope*, and rarely in *Euodia* and *Zanthoxylum*. There is a strong tendency towards prenylation at the 7-OH or 4'-OH group. From aerial parts of *Boronia coerulescens*, a derivative with 3,4'-*O*-prenylation was also described (compound 130, Table 12.4).[95] Similarly, only one *C*-6-*O*-derivative was reported from the aerial parts of *Melicope*.[96] Two 7-epoxyderivatives have already been listed in the previous survey.[6] Combination with methylenedioxy substitution is less frequent, concerning a few sources of Rutaceae and Asteraceae only. Probably, these constituents are partly accumulated externally.

12.6.5 PYRANOFLAVONOLS

In contrast to corresponding flavones, only a few structures are reported with a pyrano-substitution, mostly of the chromeno-type between the 7-OH and the neighboring *C*-8. Major sources are roots, aerial parts, and leaves from Rutaceae, Moraceae, and Fabaceae. The earlier reported *Asclepias syriaca* (Asclepiadaceae) affords a rare source of such structures. So far, the corresponding 7,6-chromeno structures (sarothranol) are known only from whole plants of *Hypericum japonicum* (Hypericaceae).[97]

Poinsettifolin A serves as an example of a recently isolated flavonol with a *C*-5-unit attached to the chromene structure (Figure 12.9). Desmethylanhydroicaritin, isolated from *Bursera leptophloea* as 3,5,4'-trihydroxy-7,8-pyranoflavon,[92] needs to be revised to 3,5,4'-triOH-7,6-pyranoflavone (compound 157, Table 12.4).[91]

12.6.6 FURANOFLAVONOLS

Major sources of the few flavonols with furano-substitution are roots of various Fabaceae genera. With exception of ponganone XI (7,6-furano-), all compounds listed here are furano-substituted between 7-OH and *C*-8 of ring A. Similarly, compounds exhibiting

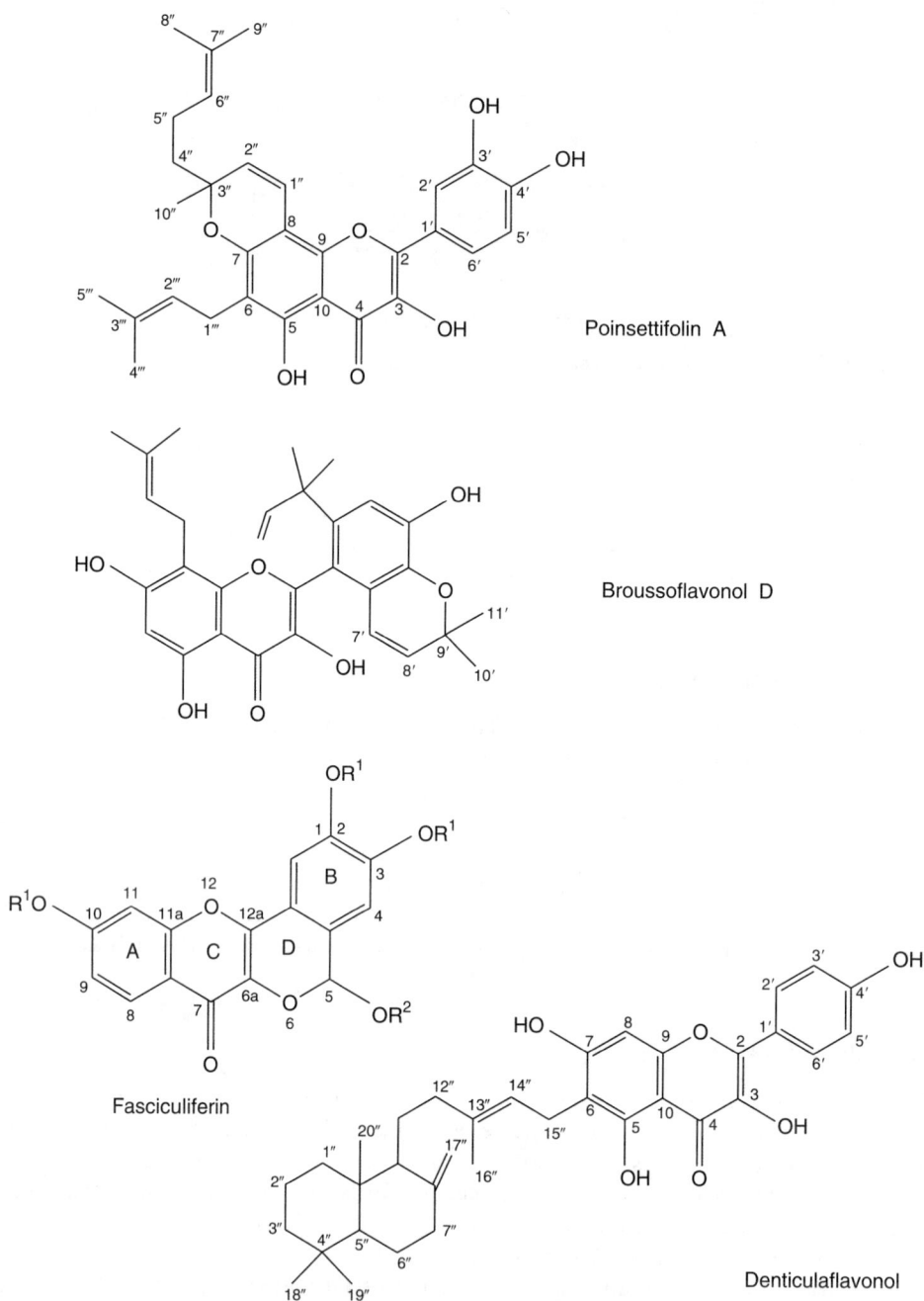

FIGURE 12.9 Complex flavonol structures.

additional substitution, such as methylenedioxy-, pyrano-, or C-prenylmoieties, are of limited number. During the reporting period, roots of *Lonchocarpus latifolius* (Fabaceae) were described as new source for pongapin, a methylendioxyfuranoflavonol. Leaves of *Broussonetia kazinoki* (Moraceae) represent a new source for mixed furanoflavonols such as broussonol C and B, respectively, with no indication as to possible external accumulation.[98]

12.6.7 C-PRENYL- AND PYRANO-SUBSTITUTION

Flavonols with this substitution pattern occur mainly in species of Moraceae. A related structure, petalopurpurenol was found in roots of *Petalostemon purpureus* (Fabaceae).[99] Cyclization into chromene structures may occur between 7-OH and *C*-6 or 7-OH and *C*-8, but substitution between 3'-OH and *C*-4' is also encountered (e.g., broussoflavonol D; Figure 12.9).

12.6.8 FLAVONOLS WITH AROMATIC SUBSTITUENTS

As with the corresponding flavones, aerial parts of *Thymus hirtus* (Lamiaceae) afforded *p*-OH-benzyl derivatives of kaempferol and quercetin, respectively.[80] Earlier, *Haplopappus foliosus* (Asteraceae) was reported to accumulate haplopappin, a phenylethyl substituted quercetin derivative.[100] Similar substituted flavones have also been found mainly to occur in members of Lamiaceae and Asteraceae (see Table 12.3), thus being probably chemosystematically significant accumulation trends.

Frond exudates of the genus *Pityrogramma* (Pteridaceae) afforded calomelanol D and a related structure (compounds 195, 196, Table 12.4), being cyclized between 7-OH and *C*-8, whereby a phenylethyl unit is further attached to the 7,8-chromene unit. A corresponding flavone derivative was also isolated.[101] By comparison with flavanone and chalcone analogues, Iinuma et al. suggested nonenzymatical processes upon which linear and angular calomelanols should be formed and hence coined the term "tertiary metabolites" for such compounds.[101] This could possibly apply also to some of the other complex flavone and flavonol derivatives presented in this chapter.

12.6.9 VARIOUS CYCLOFLAVONOLS

This section comprises a series of structurally different flavonols, with specific cyclization not falling in any of the other categories listed. These include the so-called "peltogynoids," which are cyclized between 3-OH and *C*-2'- of ring B (as in fasciculiferin, Figure 12.9). They were reported from heartwood and stems of some Fabaceae and Caesalpiniaceae, respectively. 6-Methoxypulcherrimin bears a methylenedioxy group in addition. No new source has been published during the reporting period.

12.6.10 HYBRID STRUCTURES

Only two compounds are listed here, in which a flavonol molecule is linked to a biosynthetically different product such as a terpenoid (denticulaflavonol; Figure 12.9) and a flavonol-lignoid structure isolated from heartwood of *Distemonanthus benthamianus* (Fabaceae).[102] In addition, the latter taxon accumulates derivatives with a 3-OH-2'-cyclization as mentioned before (see previous section). Earlier, similar substituted flavone-lignoid derivatives named scutellaprostins were reported from *Scutellaria* spp. (for references see Wollenweber[6]).

12.6.11 FURANOFLAVONOLS OF VELLOZIA

Species of the genus *Vellozia* have been extensively studied for their flavonoid complement in relation to chemosystematics.[103–107] In addition to a series of *C*-methylflavonols and two *C*-prenylated flavonols, derivatives of vellokaempferol and velloquercetin are accumulated in whole plants, leaves, and leaf exudates. The basic structure of these compounds is characterized by 7,6-isopropenylfurano substitution, based upon kaempferol, quercetin, and their *O*-methyl ethers. In addition, 8-*C*-methyl derivatives of these compounds were also identified from leaves of *V. stipitata*.[104] So far, species of this genus are the only reported sources of these compounds, which in parts have been proved to be accumulated externally.[104–106] Structures are exemplified by Figure 12.10.

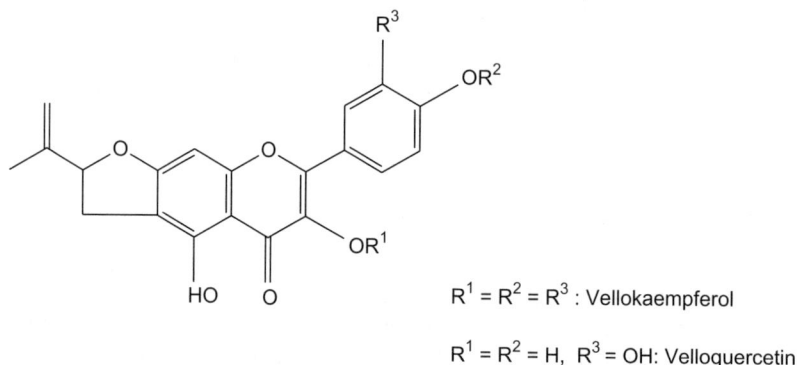

$R^1 = R^2 = R^3$: Vellokaempferol

$R^1 = R^2 = H$, $R^3 = OH$: Velloquercetin

FIGURE 12.10 Furanoflavonols from *Vellozia*.

12.7 FLAVONE AND FLAVONOL ESTERS

The occurrence of acylated flavones and flavonols still appears interesting enough to justify a short paragraph on this subject (for compilation see Table 12.5). Of the flavones, only three compounds are known to date, with one newly reported isobutyrate flavone from leaf exudates of *Asarina procumbens* (Scrophulariaceae).[108] One further compound, the 5'-benzoate of 8,2',5'-trihydroxyflavone, was isolated recently from the exudate of *Primula palinuri* (Iinuma and Wollenweber, unpublished).

By comparison, a series of mostly monoacylated flavonols is known to date and recent reports increased the number slightly. Four new products came from *Pseudognaphalium robustum* and *Tanacetum microphyllum* (both Asteraceae), and from *Adina cordifolia* (Rubiaceae). A diacetylated compound (3,5-diacetyltambulin) was recently isolated from the bark of *Zanthoxylum integrifoliolum* (Rutaceae).[109] Since most of the flavonols are monoacylated, the accumulation of quercetin tetraacetate in *Adina cordifolia*[110] is a remarkable result. Altogether, the newly reported compounds occur scattered in the plant kingdom; their occurrence is so far of little chemosystematic value. Aerial parts of *Tanacetum microphyllum* (Asteraceae) yielded a derivative, which is structurally not an ester. It is, indeed, a carbo-methoxy derivative of 6-hydroxyluteolin-4'-methyl ether (compound 34 in Table 12.5). No other flavonoid of this type is known so far.

Several of these compounds are accumulated externally as was proven for the farinose exudates of ferns, the flavonoid aglycones from *Primula* species, or those washed from the surface leaves and stems from Asteraceae. Heartwood (*Adina cordifolia*) also is a well-known accumulation site for lipophilic products. In this case, the occurrence of a natural tetraacetylflavonol is of particular interest. However, the accumulation site of the angelate of tri-*O*-methylgossypetin from *Polygonum flaccidum*[111] is unclear. Since it was isolated along with caryophyllenexpoxide, borneol, sitosterin, and stigmasterol, it might well be present in some lipophilic epicuticular material. Generally speaking, the production of acylated flavonoids still is a rare phenomenon, and so far Cheilanthoid ferns are the most important source of such products.

12.8 CHLORINATED FLAVONOIDS

These are extremely rare natural products. Among the flavones, 6-chloroapigenin is the only compound of this type occurring naturally. It had been isolated in 1980 from an *Equisetum* species.[112] The flavonol chlorflavonin (5,2'-diOH-3,7,8-triOMe-3'-chloroflavone) had been isolated from *Aspergillus candidus* already in 1969.[113] Since then, only the 7-chloroderivatives

TABLE 12.5
Flavone and Flavonol Esters

No.	OH-Substitution	OMe-Substitution	Acyl Moiety	Plant Species	Family	Plant Organ	Ref.
	Flavon-ester						
1	5-diOH		7-*O*-benzoate				
2	2'-diOH		5'-*O*-acetate				
3	5,4'-diOH	7,8-diOMe	6-*O*-isobutyrate*	*Asarina procumbens*	Scroph.	Leaf exudate	108
	Flavonol-ester						
4	5,7-diOH	3-OMe	8-*O*-butyrate*	*Pseudognaphalium robustum*	Ast.	Resinous exudate	551
5	5,7-diOH	3-OMe	8-Me-butenoate				
6	3,5-diOH	7-OMe	8-*O*-acetate				
7	3,5-diOH	7-OMe	8-*O*-butyrate				
8	5-OH	3,7-diOMe	8-*O*-acetate				
9	5-OH	3,7-diOMe	8-*O*-butyrate				
10	5,7-diOH	3,6-diOMe	8-*O*-butyrate*	*Pseudognaphalium robustum*	Ast.	Resinous exudate	551
11	3,5,2'-triOH	7-OMe	8-*O*-acetate				
12	3,5,4'-triOH	7-OMe	8-*O*-acetate				
13	3,5,4'-triOH	7-OMe	8-*O*-butyrate				
14	3,5-diOH	7,4'-diOMe	8-*O*-acetate				
15	3,5-diOH	7,4'-diOMe	8-*O*-butyrate				
16		7,8,4'-triOMe	3,5-di-*O*-acetate*	*Zanthoxylum integrifolium*	Rutac.	Fruit	109
17	5,7,3',4'-tetraOH	(Quercetin)	3-*O*-isobutyrate				
18	3,5,7,4'-tetraOH	(Quercetin)	3'-*O*-isobut.				
19	3,5,7,3'-tetraOH	(Quercetin)	4'-*O*-isobut.				
20	3'-OH	(Quercetin)	3,5,7,4'-tetra-*O*-acetate*	*Adina cordifolia*	Naucleac.	Heartwood	110
21	5,7-diOH	3,3'-diOMe	4'-*O*-Me-butyrate				
22	5,7-diOH	3,3'-diOMe	4'-*O*-isovalerate				
23	5,7-diOH	3,6,4'-triOMe	8-*O*-tigliate				
24	5,4'-diOH	3,6,7-triOMe	8-*O*-Me-butyrate				
25	3,5,6,3'-tetraOH	4'-OMe	7-*O*-acetate*	*Tanacetum microphyllum*	Ast.	Aerial parts	395
26	5,5'-diOH	3,7,8-triOMe	2'-*O*-acetate				
27	3,5,3'-triOH	7,4'-diOMe	8-*O*-butyrate				
28	3,5,3'-triOH	7,4'-diOMe	8-*O*-butyrate				
29	5,4'-diOH	3,7,3'-triOMe	8-*O*-acetate				
30	5,7-diOH	8,3',4'-triOMe	3-*O*-angelate				
31	5,7,2'-triOH	3,4'-diOMe	5'-*O*-acetate				
32	5,7,3'-triOH	3,4'-diOMe	5'-*O*-acetate				
33	5-OH	3,7,2',3',4'-pentaOMe	8-*O*-acetate				
	Carbomethoxy-flavonol						
34	3,5,3'-OH	4'-OMe	7-COOCH$_3$*	*Tanacetum microphyllum*	Ast.	Aerial parts	395

of 3,5,6,8,4'-pentamethoxyflavone and of 3,5,6,8,3',4'-hexamethoxyflavone, both isolated from leaves of *Citrus* "Dancy tangerine,"[114] were reported as new chloroflavonols. The earlier expectation that such compounds might be found in Asteraceae, which are known for the production of other chlorinated natural products, was not fulfilled.

12.9 FLAVONOIDS OF *HELMINTHOSTACHYS*

Rhizomes of *Helminthostachys zeylanica* (Ophioglossaceae) yielded a series of complex flavone and flavonol derivatives with singular structures. They were named ugonins A–I. Whereas some flavones of this medicinally used plant are known for a long time,[115] new data include also a series of flavonols.[116] The complexity of these compounds is remarkable. Some of the structures are depicted in Figure 12.11.

12.10 COMMENTS ON DISTRIBUTION AND ACCUMULATION

In a compilation such as the present one, it is tempting to interpret the data regarding substitution patterns and distribution within the plant kingdom. However, the data presented here only comprise those not included in previous editions (e.g., Wollenweber[6]), hence the distribution picture sure is somewhat distorted. The new entries concern not only several families of the Angiosperms, but include also results on ferns and mosses. Although not from a plant source in the strict sense, the *C*-methylflavonols from fungi such as *Colletotrichum dematium*[84] were also included (Table 12.4). This organism is pathogenic to *Epilobium angustifolium*. The possible uptake of these compounds by the fungus from the plant can be excluded since substitution patterns of the fungal flavonols and the flavonoids from *Epilobium* do not coincide.[117] Another fungus, *Aspergillus flavus*, proved to be able to synthesize prenylated naringenin derivatives from supplied flavanones.[118] Apparently, these fungi are able to metabolize flavonoids from precursors, but nothing is known about the basic biosynthesis of flavonoids in these organisms.

One of the hypotheses regarding evolutionary aspects of flavonoid diversification concerns the concept of flavonol accumulation in basal Angiosperms versus flavone accumulation in advanced families. Recently, some further efforts have been made towards defining the flavone/flavonols ratio in Dicotyledonae and their relation to lignification,[119] indicating an increased tendency towards flavonol accumulation in lignified plants, whereas herbaceous species tend to accumulate more of the flavones. From the presented entries, it appears that flavone derivatives are more abundant in Lamiaceae than flavonols. In the Asteraceae, however, more data concern the flavonols. Both families are more or less herbaceous and members of the more advanced Angiosperms.

It might be of more value to check the substitution patterns for their chemosystematic significance, as had been done earlier in frequency analysis.[56] According to current data, 6-substitution, both —OH and —OMe, appears to be more frequent than the corresponding 8-substitution in flavones. The number of their 6,8-diOMe derivatives is quite considerable though. By comparison, the number of the related 6,8-OH-flavones is restricted to a few compounds reported from natural sources (compounds 136, 222, 227, and 262 in Table 12.1). All of the other polyhydroxylated structures have so far not been found as natural products. A similar ratio between 6- and 8-substitution was found with the flavonols, but the number of naturally occurring 6,8-diOH flavonols is limited to two compounds only (compounds 239 and 289 in Table 12.2). Further accumulation trends of possible chemosystematic relevance have been discussed in the respective sections.

Aspects on substitution patterns of prenylated flavonoids and their derivatives including the frequency of occurrence have been discussed in detail.[3] Thus, these aspects will not be

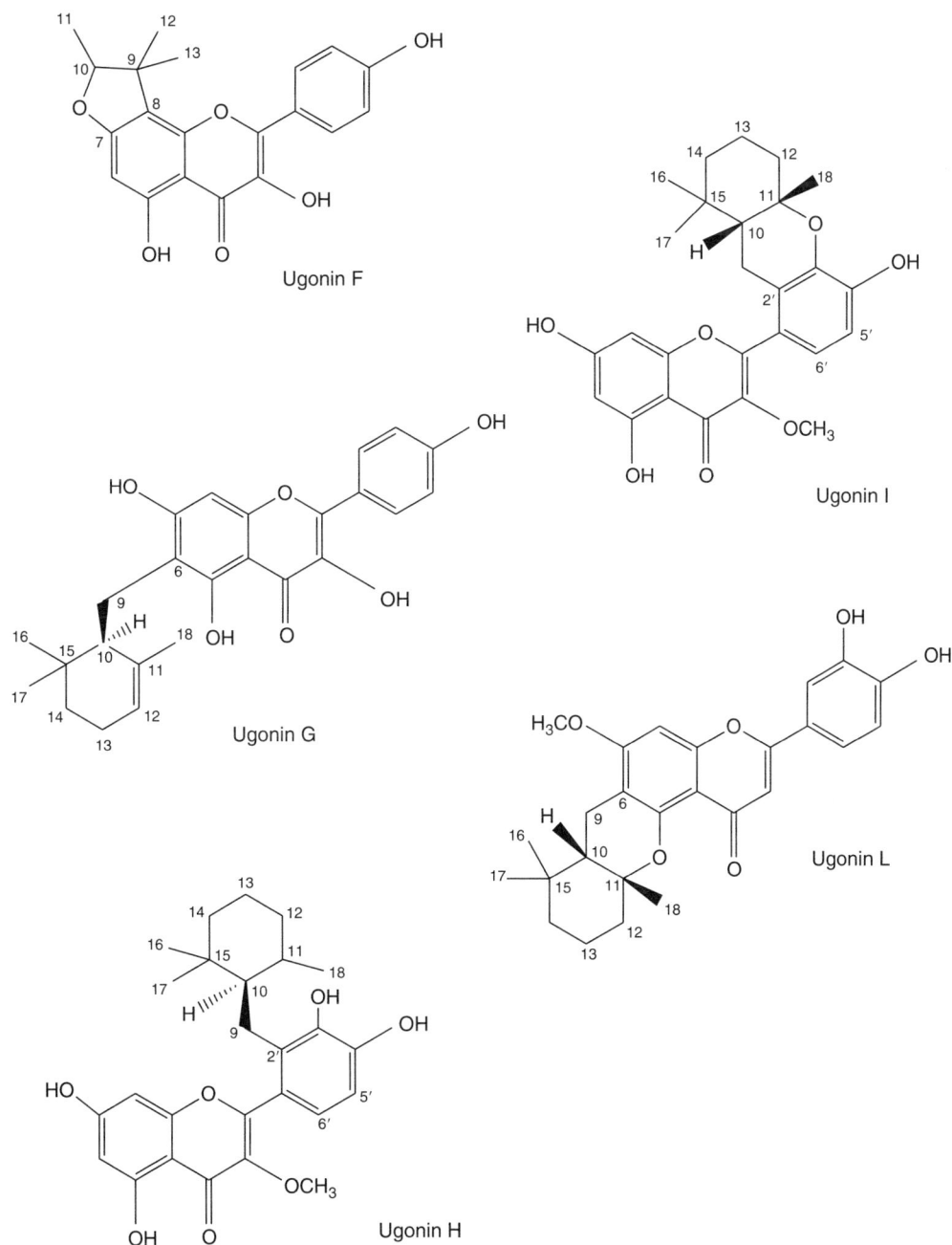

FIGURE 12.11 Flavonoids from *Helminthostachys*.

covered here. From the present data, the tendency to produce complex flavones is predominant. Such products are of limited distribution in few genera of some families only, as exemplified in the respective sections of this chapter (e.g., *Artocarpus-Moraceae*; *Tephrosia-Fabaceae*). The degree of complexity differs between flavones and flavonols. Thus, the flavones outnumber the flavonols, both in number and complexity. Interestingly, the number of *O*-prenylated flavonols is much higher than that of the corresponding flavones. The occurrence of complex flavones and flavonols in various *Pteridophyta* and *Hepaticae* is further remarkable. Apparently, various evolutionary independent groups in the plant kingdom are able to synthesize biosynthetically complex products.

Some general comments on the external accumulation and excretion of flavonoid aglycones should be made. In most cases, flavone and flavonol methyl ethers contribute to the exudates, sometimes including methylenedioxy derivatives, *C*-methylderivatives, and flavone and flavonol esters. Complex structures, however, are rarely reported to occur in exudates, as for example, *C*-prenylflavonols in Velloziaceae,[105] or 7,8-cycloflavonols from frond exudates of Pteridaceae.[120–122] The most complex structures found in *Artocarpus* spp. are not known to be excreted, for reasons unknown. Maybe the responsible enzymes cannot be compartimented in the cells of, for example, glandular hairs, which are frequently the accumulation site of lipophilic material. For flavonoid aglycones, this was established for instance in a study on *Mentha*.[123] Several other studies are similarly conclusive (e.g., Heinrich et al.[124]). The presence of the basic enzymes for flavonoid production in head cells of glands was reported for *Primula kewensis*.[125] *Primula* species are widely known for their production of flavonoid exudates. Afolayan and Meyer[126] hypothesized that exudate flavonoids of *Helichrysum aureonitens* were probably produced by the endoplasmatic reticulum in the secreting trichomes. There would be many more interesting aspects to be discussed, such as ecological significance (e.g., Tattini et al.[127]) or ontological differentiation during plant development,[128] which would be beyond the scope of this review. Ecological and other aspects not fully discussed here have been addressed in previous publications.[6,129,130] It is hoped that the aspects discussed here will stimulate further research outside the flavonoid community as well.

REFERENCES

1. Harborne, J.B. and Williams, C.A., Advances in flavonoid research since 1992. *Phytochemistry*, 55, 481, 2000.
2. Bohm, B.A. and Stuessy, T.F., *Flavonoids of the Sunflower Family (Asteraceae)*. Springer, Vienna/ New York, 2001.
3. Barron, D. and Ibrahim, R.K., Isoprenylated flavonoids — a survey, *Phytochemistry*, 43, 921, 1996.
4. Harborne, J.B. and Baxter, H., *The Handbook of Natural Flavonoids*, vol. 1. Wiley & Sons, Chichester, 1999.
5. Wollenweber, E. and Dietz, V.H., Occurrence and distribution of free flavonoid aglycones in plants, *Phytochemistry*, 20, 869, 1981.
6. Wollenweber, E., Flavones and flavonols, in *The Flavonoids — Advances in Research since 1986*, J.B. Harborne, Ed., Chapman and Hall, London, Chap. 7, 1994.
7. Chung, H.S. et al., Flavonoid constituents of *Chorizanthe diffusa* with potential cancer chemopreventive activity, *J. Agric. Food Chem.*, 47, 36, 1999.
8. Gupta, D.R., Dhiman, R.P., and Ahmed, B., Constituents of *Launaea asplenifolia*, *Pharmazie*, 40, 273, 1985.
9. Mossa, J.S. et al., A flavone and diterpene from *Psiadia arabica*, *Phytochemistry*, 31, 2863, 1992.
10. Rao, Y.K. et al., Flavones from *Andrographis viscosula*, *Phytochemistry*, 61, 927, 2002.
11. Reddy, M.K. et al., A flavone and an unusual 23-carbon terpenoid from *Andrographis paniculata*, *Phytochemistry*, 62, 1271, 2003.
12. Reddy, M.K. et al., A new chalcone and a flavone from *Andrographis neesiana*, *Chem. Pharm. Bull.*, 51, 854, 2003.
13. Reddy, M.K. et al., Two new flavonoids from *Andrographis rothii*, *Chem. Pharm. Bull.*, 51, 191, 2003.
14. Kishore, P.H. et al., Flavonoids from *Andrographis lineate*, *Phytochemistry*, 63, 457, 2003.
15. Santos, L.C. et al., Xanthones and flavonoids from *Leiothrix curvifolia* and *Leiothrix flavescens*, *Phytochemistry*, 56, 853, 2001.

16. Chen, J., Montanari, A.M., and Widmer, W.W., Two new polymethoxylated flavones, a class of compounds with potential anticancer activity, isolated from cold pressed dancy tangerine peel oil solids, *J. Agric. Food Chem.*, 45, 364, 1997.

17. Fukai, T., Nishizawa, J., and Nomura, T., Five isoprenoid-substituted flavonoids from *Glycyrrhiza eurycarpa*, *Phytochemistry*, 35, 515, 1994.

18. Wollenweber, E., Dörr, M., and Armbruster, S., Flavonoid-Aglyka als Drüsenprodukte der Moosrose (*Rosa centifolia* cv. *muscosa*) und der japanischen Weinbeere (*Rubus phoenicolasius*), *Z. Naturforsch.*, 48c, 956, 1993.

19. Wollenweber, E. et al., Lipophilic exudate constituents of some Rosaceae from the Southwestern USA, *Z. Naturforsch.*, 51c, 296, 1996.

20. Campos, M.G., Webby, R.F., and Markham, K.R., The unique occurrence of the flavone aglycon tricetin in Myrtaceae pollen, *Z. Naturforsch.*, 58c, 944, 2002.

21. Mayer, R., A β-hydroxychalcone from *Leptospermum scoparium*, *Planta Med.*, 59, 269, 1993.

22. Kraut, L., Klein, R., and Mues, R., Flavonoid diversity in the liverwort genus *Monoclea*, *Z. Naturforsch.*, 47c, 794, 1992.

23. Takagi, S., Yamaki, M., and Inoue, K., Flavone di-C-glycosides from Scutellaria baicalensis, *Phytochemistry*, 20, 2443, 1981.

24. Horie, T. et al., Synthesis of 5,8-dihydroxy-6,7-dimethoxyflavones and revised structures for some natural flavones, *Phytochemistry*, 39, 1201, 1995.

25. La Duke, J.C., Flavonoid chemistry and systematics of *Tithonia*, *Am. J. Bot.*, 69, 784, 1982.

26. Martinez, V. et al., Phenolic and acetylenic metabolites from *Artemisia assoana*, *Phytochemistry*, 26, 2619, 1987.

27. González, A.G. et al., Methoxyflavones from *Ageratum conyzoides*, *Phytochemistry*, 30, 1269, 1991.

28. Horie, T. et al., Studies of the selective O-alkylation and dealkylation of flavonoids. XIV. A convenient method for synthesizing 5,6,7,8-trihydroxy-3-methoxyflavones from 6-hydroxy-3,5, 7-trimethoxyflavones, *Bull. Chem. Soc. Jpn*, 66, 877, 1993.

29. Malan, E., A flavonol with a tetrasubstituted B-ring from *Distemonanthus benthamianus*, *Phytochemistry*, 32, 1631, 1993.

30. Sharaf, M., Abu-Gabal, N.S., and El-Ausari, M.A., Exudate flavonoids from *Ficus altissima*, *Biochem. Syst. Ecol.*, 28, 291, 2000.

31. Aquino, R. et al., New 3-methoxyflavones, an iridoid lactone and a flavonol from *Duroia hirsuta*, *J. Nat. Prod.*, 62, 560, 1999.

32. Wollenweber, E. et al., Exudate flavonoids in several Asteroideae and Cichorioideae (Asteraceae), *Z. Naturforsch.*, 52c, 137, 1997.

33. Midiwo, J.O. and Owuor, F.A.O., Epicuticular flavonoids of *Psiadia puntulata* and *Polygonum senegalense*, *5th NAPRECA Symposium on Natural Products, Antananarivo, Madagascar*, p. 81, 1992.

34. Ahmad, V.U. and Ismail, N., 5-Hydroxy-3,6,7,2′,5′-pentamethoxyflavone from *Inula grantioides*, *Phytochemistry*, 30, 1040, 1991.

35. Urzua, A. et al., Flavonoids in the resinous exudate of Chilean *Heliotropium* species from *Cochranea* section, *Biochem. Syst. Ecol.*, 21, 744, 1993.

36. Wollenweber, E. et al., On the occurrence of exudate flavonoids in the Borage family (Boraginaceae), *Z. Naturforsch.*, 57c, 445, 2002.

37. El-Sohly, H.N. et al., Antiviral flavonoids from *Alkanna orientalis*, *Planta Med.*, 63, 384, 1997.

38. Ganapaty, S., Pushpalatha, V., and Naidu, K.C., Flavonoids from *Millettia racemosa* Benth., *Indian Drugs*, 36, 635, 1999.

39. Su, W.-C., Fang, J.-M., and Cheng, Y.-S., Flavonoids and lignans from leaves of *Cryptomeria japonica*, *Phytochemistry*, 40, 563, 1995.

40. Hussein, S.A. et al., Flavonoids from *Ephedra aphylla*, *Phytochemistry*, 45, 1529, 1997.

41. Wollenweber, E. et al., Exudate flavonoids of some *Salvia* and a *Trichostoma* species, *Z. Naturforsch.*, 47c, 782, 1992.

42. Urzua, A. et al., Flavonoid aglycones in the resinous exudate of some Chilean plants, *Fitoterapia*, 62, 358, 1991.

43. Wollenweber, E. and Dörr, M., Exudate flavonoids from aerial parts of *Mirabilis viscosa* (Nyctaginaceae), *Biochem. Syst. Ecol.*, 24, 799, 1996.
44. Wollenweber, E., Wieland, A., and Haas, K., Epicuticular waxes and flavonol aglyones of the European mistletoe, *Viscum album*, *Z. Naturforsch.*, 55c, 314, 2000.
45. Wollenweber, E. et al., Flavonoid aglycones and a dihydrostilbene from the frond exudate of *Notholaena nivea*, *Phytochemistry*, 33, 611, 1993.
46. Kaouadji, M. et al., 6-Hydroxygalangin and C-prenylated kaempferol derivatives from *Platanus acerifolia* Buds., *Phytochemistry*, 31, 2131, 1992.
47. Horie, T. et al., Studies of the selective O-alkylation and dealkylation of flavonoids. XVIII. A convenient method for synthesizing 3,5,6,7-tetrahydroxyflavones, *Bull. Chem. Soc. Jpn.*, 68, 2033, 1995.
48. Williams, C.A. et al., A biologically active lipophilic flavonol from *Tanacetum parthenium*, *Phytochemistry*, 38, 267, 1995.
49. Williams, C.A. et al., The flavonoids of *Tanacetum parthenium* and *T. vulgare* and their anti-imflammatory properties, *Phytochemistry*, 51, 417, 1999.
50. Parmar, V.S. et al., Isolation of candirone: a novel pentaoxygenated pattern in a natural occurring 2-phenyl-4H-1-benzopyran-4-one from *Tephrosia candida*, *Tetrahedron*, 43, 4241, 1987.
51. Horie, T. et al., Revised structure of a natural flavone from *Tephrosia candida*, *Phytochemistry*, 37, 1189, 1994.
52. Ahmad, V.U. et al., Santoflavone, a 5-deoxyflavonoid from *Achillea santolina*, *Phytochemistry*, 38, 1305, 1995.
53. Balboul, B.A.A.A. et al., A guaianolide and a gemacranolide from *Achillea sontolina*, *Phytochemistry*, 446, 1045, 1997.
54. Bhardwaj, D.K. et al., Quercetagetin-5-methyl ether from the petals of *Tagetes patula*, *Phytochemistry*, 19, 713, 1980.
55. Zhang, Y.-Y. et al., Studies on the constituents of roots of *Scutellaria planipes*, *Planta Med.*, 63, 536, 1997.
56. Wollenweber, E. and Jay, M., Flavones and flavonols, in *The Flavonoids — Advances in Research since 1980*, J.B. Harborne, Ed., Chapman and Hall, London/New York, Chap. 7, 1988.
57. Wollenweber, E. et al., C-Methyl-flavonoids from leaf exudates of some Myrtaceae, *Phytochemistry*, 55, 965, 2000.
58. Wu, J.H. et al., Isolation and identification of flavones from *Desmos cochinchinensis*, *Acta Pharm. Sin.*, 29, 621, 1994.
59. Mayer, R., Flavonoids from *Leptospermum scoparium*, *Phytochemistry*, 29, 1340, 1990.
60. Ju, J.-H. and Tu, J.-G., Studies on chemical constituents of seeds of *Desmos chinensis*, *Zhonguo Zhongyao Zahzi*, 24, 418, 1999.
61. Wu, T.-S. et al., A flavonoid and indole alkaloid from flowers of *Murraya paniculata*, *Phytochemistry*, 37, 287, 1994.
62. Wu, J.H. et al., Chemical constituents from the root of *Desmos chinensis*, *Chin. Tradit. Herb. Drugs*, 31, 567, 2000.
63. Liu, Y.-L. et al., Dasytrichone, a novel flavone from *Dasymaschalon trichophorum* with cancer chemopreventive potential, *Nat. Prod. Lett.*, 161, 1992.
64. Zhou, L.D. et al., A-ring formylated flavonoids and oxoaporphinoid alkaloid from *Dasymaschalon rostratum*, *China J. Chin. Mat. Med.*, 26, 39, 2001.
65. Zheng, S. et al., Three new C-methylated flavones from *Elsholtzia stauntonii*, *Planta Med.*, 65, 173, 1999.
66. Ding, Z., Zhou, J., and Tan, N., A novel flavonoid glycoside from *Drymania diandra*, *Planta Med.*, 65, 578, 1999.
67. Kitanaka, S. and Takido, M., Studies on the constituents of the leaves of *Cassia torosa*: 2. The structure of two novel flavones, torosaflavone-C and torosaflavone D, *Chem. Pharm. Bull.*, 39, 3254, 1991.
68. Kitanaka, S. and Takido, M., Studies on the constituents of the leaves of *Cassia torosa*: 3. The structures of two new flavone glycosides, *Chem. Pharm. Bull.*, 40, 249, 1992.

69. Rumbero, A., Arriaga-Giner, F.J., and Wollenweber, E., A new oxyprenyl coumarin and highly methylated flavones from the exudate of *Ozothamnus lycopodioides* (Asteraceae), *Z. Naturforsch.*, 55c, 1, 2000.

70. Rumbero, A., Arriaga-Giner, F.J., and Wollenweber, E., New constituents of the leaf and stem exudates of *Ozothamnus hookeri* (Asteraceae), *Z. Naturforsch.*, 55c, 318, 2000.

71. Fukai, T. and Nomura, T., Revised structures of albanins D and E, geranylated flavones from *Morus alba*, *Heterocycles*, 32, 499, 1991.

72. Ayengar, K.N.M., Sastry, B.R., and Rangaswami, S., Structure of lanceolatin-A, *Ind. J. Chem.*, 11B, 85, 1973.

73. Pegnyemb, D.E., Ghogomu-Tih, R., and Sondengam, B.L., Minor biflavonoids of *Lophira lanceolata*, *J. Nat. Prod.*, 57, 1275, 1994.

74. Kajiyama K. et al., New prenylflavones and dibenzoylmethane from Glycyrrhiza inflata, *J. Nat. Prod.*, 55, 1197, 1992.

75. Kitagawa, I. et al., Chemical studies of Chinese licorice-roots. I. Elucidation of five new flavonoid constituents from the roots of *Glycyrrhiza glabra* collected in Xinjiang, *Chem. Pharm. Bull.*, 42, 1056, 1994.

76. Norbedo, C., Ferraro, G., and Coussio, J.D., Flavonoids from *Achyrocline flaccida*, *Phytochemistry*, 23, 2698, 1984.

77. Diaz, G. et al., Methoxyflavones from *Ficus maxima*, *Phytochemistry*, 45, 1697, 1997.

78. Khattab, A.M., Grace, M.H., and El-Khrisy, E.A., A new flavone derivative from *Ehretia ovalifolia* leaves, *Pharmazie*, 56, 661, 2001.

79. Mbafor, J.T. et al., Furanoflavones from root bark of *Millettia sanagana*, *Phytochemistry*, 40, 949, 1995.

80. Merghem, R. et al., Five 8-C-benzylated flavonoids from *Thymus hirtus* (Labiateae), *Phytochemistry*, 38, 637, 1995.

81. Usia, T. et al., Constituents of Chinese propolis and their antiproliferative activities, *J. Nat. Prod.*, 65, 673, 2002.

82. Vanangamudi, A. et al., Tephrorianin: a novel flavone from *Tephrosia hookeriana*, *Fitoterapia*, 68, 543, 1997.

83. Franke, K., Porzel, A., and Schmidt, J., Flavone–coumarin hybrids from *Gnidia socotrana*, *Phytochemistry*, 61, 873, 2002.

84. Abou-Zaid, M. et al., C-Methylflavonols from the fungus *Colletotrichum dematium epilobii*, *Phytochemistry*, 45, 957, 1997.

85. Phrutivorapongkul, A. et al., Studies on the chemical constituents of stem bark of *Millettia leucantha*: isolation of new chalcones with cytotoxic, antiherpes simplex virus and anti-inflammatory activities, *Chem. Pharm. Bull.*, 51, 187, 2003.

86. Pomilio, A.B. et al., Antimicrobial constituents of *Gomphrena martiana* and *Gomphrena boliviana*, *J. Ethnopharmacol.*, 36, 155, 1992.

87. Anis, I. et al., Thrombin inhibitory constituents from *Duranta repens*, *Helv. Chim. Acta*, 84, 649, 2001.

88. Anis, I. et al., Enzyme inhibitory constituents from *Duranta repens*, *Chem. Pharm. Bull.*, 50, 515, 2002.

89. Buckova, A. et al., A new acylated kaempferol derivative from *Lilium candidum*, *Phytochemistry*, 27, 1914, 1988.

90. Mitscher, L.A. et al, Antimicrobial agents from higher plants: prenylated flavonoids and other phenols from *Glycyrrhiza lepidota*, *Phytochemistry*, 22, 573, 1983.

91. Fukai, T. and Nomura, T., ^{1}H-NMR chemical shift of the flavonol 5-hydroxy proton as a characterization of 6- or 8-isoprenoid substitution, *Heterocycles*, 34, 1213, 1992.

92. Souza, M.P., Machado, M.I.L., and Braz-Filho, R., Six flavonoids from *Bursera leptophloeos*, *Phytochemistry*, 28, 2467, 1989.

93. Fang, S.-C. et al., Isoprenylated flavonols of Formosan *Broussonetia papyrifera*, *Phytochemistry*, 38, 535, 1995.

94. Lin, C.-N. et al., Revised structure of broussoflavonol G and the 2D-NMR spectra of some related prenylflavonoids, *Phytochemistry*, 41, 1215, 1996.

95. Ahsan, M. et al., Novel O-prenylated flavonoids from two varieties of *Boronia coerulescens*, *Phytochemistry*, 37, 259, 1994.
96. Sultana, N., Hartley, T.G., and Waterman, P.G., Two novel prenylated flavones from the aerial parts of *Melicope micrococca*, *Phytochemistry*, 50, 1249, 1999.
97. Ishiguro, K. et al., An isopentenylated flavonol from *Hypericum japonicum*, *Phytochemistry*, 32, 1583, 1993.
98. Zhang, P.-C. et al., Five new diprenylated flavonols from the leaves of *Broussonetia kazinoki*, *J. Nat. Prod.*, 64, 1206, 2001.
99. Huang, L. et al., A new prenylated flavonol from the root of *Petalostemon purpureus*, *J. Nat. Prod.*, 59, 290, 1996.
100. Tschesche, R. et al., Haplopappin, ein 8-(α-Methylbenzyl) Flavonoid aus *Haplopappus foliosus*, *Liebigs Ann. Chem.*, 2465, 1985.
101. Iinuma, M. et al., Two biflavonoids in the farinose exudate of *Pentagramma triangularis*, *Phytochemistry*, 35, 1043, 1994.
102. Malan, E., Swinny, E., and Ferreira, D., A 3-oxygenated flavonolignoid from *Distemonanthus benthamianus*, *Phytochemistry*, 37, 1771, 1994.
103. Harborne, J.B. et al., Six dihydrofuranoflavonols from the leaf surface of *Vellozia*, *Phytochemistry*, 31, 305, 1992.
104. Williams, C.A. et al., Six further lipophilic flavonols from the leaf of *Vellozia stipitata*, *Phytochemistry*, 32, 731, 1993.
105. Williams, C.A. et al., Differences in flavonoid patterns between genera within the Velloziaceae, *Phytochemistry*, 36, 931, 1994.
106. Branco, A. et al., Two monoisoprenylated flavonoids from *Vellozia graminifolia*, *Phytochemistry*, 47, 471, 1998.
107. Branco, A. et al., Further lipophilic flavonols in *Vellozia graminifolia* (Velloziaceae) by high temperature gas chromatography: quick detection of new compounds, *Phytochem. Anal.*, 12, 266, 2001.
108. Wollenweber, E., Dörr, M., and Roitman, J.N., Epicuticular flavonoids of some Scrophulariaceae, *Z. Naturforsch.*, 55c, 5, 2000.
109. Chen, I.-S. et al., Chemical constituents and biological activities of the fruit of *Zanthoxylum integrifolium*, *J. Nat. Prod.*, 62, 833, 1999.
110. Rao, M.S., Duddeck, H., and Dembinski, R., Isolation and structural elucidation of 3,4',5,7-tetraacetyl quercetin from *Adina cordifolia* (Karam ki Gaach), *Fitoterapia*, 73, 353, 2002.
111. Ahmed, A.A. et al., Flavonoids of *Asteriscus graveolens*, *J. Nat. Prod.*, 54, 1092, 1991.
112. Syrchina, A.I. et al., 6-Chloroapigenin from *Equisetum arvense*, *Khim. Prir. Soedin.*, 4, 499, 1980.
113. Bird, A.E. and Marshall, A.C., Structure of chlorflavonin, *J. Chem. Soc. C*, 2418, 1969.
114. Chen, J. and Montanari, A.M., Isolation and identification of new polymethoxyflavonoids from dancy tangerine leaves, *J. Agric. Food Chem.*, 46, 1235, 1998.
115. Murakami, T. et al., Chemische Untersuchungen über die Inhaltsstoffe von *Helminthostachys zeylanica* I, *Chem. Pharm. Bull.*, 21, 1849, 1973.
116. Huang, Y.-L. et al., Antioxidant flavonoids from the rhizomes of *Helminthostachys zeylanica*, *Phytochemistry*, 64, 1277, 2003.
117. Hiermann, A. et al., Isolierung des antiphlogistischen Wirkprinzips von *Epilobium angustifolium*, *Planta Med.*, 57, 357, 1991.
118. Tahara, S., Tanaka, M., and Barz, W., Fungal metabolism of prenylated flavonoids, *Phytochemistry*, 44, 1031, 1997.
119. Soares, G.L.G. and Kaplan, M.A.C., Analysis of flavone–flavonol ratio in Dicotyledonae, *Bot. J. Linn. Soc.*, 135, 61, 2001.
120. Iinuma, M., Tanaka, T., and Asai, F., Flavonoids in frond exudates of *Pityrogramma tartarea*, *Phytochemistry*, 36, 941, 1994.
121. Asai, F. et al., Five complex flavonoids in the farinose exudate of *Pityrogramma calomelanos*, *Phytochemistry*, 31, 2487, 1992.
122. Asai, F. et al., Two complex flavonoids in the farinose exudate of *Pityrogramma calomelanos*, *Heterocycles*, 33, 229, 1992.

123. Voirin, B., Bayet, C., and Colson, M., Demonstration that flavone aglycones accumulate in the peltate glands of *Mentha* × *piperita* leaves., *Phytochemistry*, 34, 85, 1993.

124. Heinrich, G. et al., Glandular hairs of *Sigesbeckia jorullensis* Kunth (Asteraceae): morphology, histochemistry and composition of essential oil, *Ann. Bot.*, 89, 459, 2002.

125. Schöpker, H. et al., Phenylalanin ammonia-lyase and chalcone synthase in glands of *Primula kewensis* (W. Wats.): immunofluorescence and immunogold localization, *Planta*, 196, 712, 1995.

126. Afolayan, A.J. and Meyer, J.J.M., Morphology and ultrastructure of secreting and nonsecreting foliar trichomes of *Helichrysum aureonitens* (Asteraceae)., *Int. J. Plant Sci.*, 156, 481, 1995.

127. Tattini, M. et al., Flavonoids accumulate in leaves and glandular trichomes of *Phillyrea latifolia* exposed to excess solar radiation, *New. Phytol.*, 148, 69, 2000.

128. Laitinen, M.-L., Julkunen-Tiitto, R., and Rousi, M., Foliar phenolic composition of European white birch during bud unfolding and leaf development, *Physiol. Plant.*, 114, 450, 2002.

129. Wollenweber, E. and Valant-Vetschera, K.M., New results with exudate flavonoids in Compositae, in *Compositae: Systematics*, vol. 1. *Proceedings of the International Compositae Conference*, Kew, 1994, D.J.N. Hind and H.J. Beentje, Eds., Royal Botanic Gardens, Kew, 1996.

130. Valant-Vetschera, K.M. and Wollenweber, E., Exudate flavonoid aglycones of *Achillea* sect. *Achillea* and sect. *Babounya*: a comparative study, *Biochem. Syst. Ecol.*, 22, 609, 1994.

131. Sinz, A. et al., Phenolic compounds from *Anomianthus dulcis*, *Phytochemistry*, 50, 1069, 1999.

132. Valant-Vetschera, K.M., Fischer, R., and Wollenweber, E., Exudate flavonoids in species of *Artemisia* (Asteraceae-Anthemideae): new results and chemosystematic interpretation, *Biochem. Syst. Ecol.*, 31, 487, 2003.

133. Wollenweber, E. et al., Exudate flavonoids in Asteraceae from Arizona, California, and Mexico, *Z. Naturforsch.*, 52c, 301, 1997.

134. Ohkoshi, E., Makino, M., and Fujimoto, Y., Studies on the constituents of *Mikania hirsutissima* (Compositae), *Chem. Pharm. Bull.*, 47, 1436, 1999.

135. Arriaga, J.F. et al., Three new benzoic derivatives from the glandular excretion of *Eriodictyon sessilifolium* (Hydrophyllaceae), *Z. Naturforsch.*, 43 c, 337, 1988.

136. Hegazi, A.G. and Abd El Hardy, F.K., Egyptian propolis: 1-antimicrobial activity and chemical composition of Upper Egypt propolis, *Z. Naturforsch.*, 56c, 82, 2001.

137. Bankova, V. et al., Chemical composition of European propolis: expected and unexpected results, *Z. Naturforsch.*, 57c, 530, 2002.

138. Tomás-Barberán, F.A., Ferreres, F., and Tomás-Lorente, F., Flavonoids from *Apis mellifera* beeswax, *Z. Naturforsch.*, 48c, 68, 1993.

139. Chantrapromma, K. et al., 5-Hydroxy-7-methoxy-2-phenyl-4H-1-benzopyran-4-one (tectochrysin) and 2,5-dihydroxy-7-methoxy-2-phenyl-2,3-dihydro-4H-1-benzopyran-4-one: isolation from *Uvaria rufas* and X-ray structures, *Aust. J. Chem.*, 42, 2289, 1989.

140. Sartori, F.T. et al., Phytochemical study of *Lychnophera markgravii* (Asteraceae), *Biochem. Syst. Ecol.*, 30, 609, 2002.

141. Wollenweber, E., Dörr, M., and Gómez, L.D.P., Exudate flavonoids in *Godmania aesculifolia* (Bignoniaceae), *Biochem. Syst. Ecol.*, 24, 481, 1996.

142. Williams, C.A. et al., Chrysin and other leaf exudate flavonoids in the genus *Pelargonium*, *Phytochemistry*, 46, 1349, 1997.

143. Stevens, J.F. et al., A novel 2-hydroxyflavanone from *Collinsonia canadensis*, *J. Nat. Prod.*, 62, 392, 1999.

144. Ngadjui, B.T. et al., Opposition and 5-O-methylhoslundin, pyrone-substituted flavonoids of *Hoslundia opposita*, *Phytochemistry*, 32, 1313, 1993.

145. Pathak, V. et al., Antiandrogenic phenolic constituents from *Dalbergia cochinchinensis*, *Phytochemistry*, 46, 1219, 1997.

146. Stermitz, F.R., Arslanian, R.L., and Castro, O., Flavonoids from the leaf surface of *Godmania aesculifolia*, *Biochem. Syst. Ecol.*, 20, 481, 1992.

147. Masaoud, M. et al., Flavonoids of Dragon's blood from *Dracaena cinnabari*, *Phytochemistry*, 38, 745, 1995.

148. Kajiyama, K. et al., Flavonoids and isoflavonoids of chemotaxonomic significance from *Glycyrrhiza pallidiflora* (Leguminosae), *Phytochemistry*, 31, 3229, 1993.

149. Kawashty, S.A. et al., The chemosystematics of Egyptian *Trigonella* species, *Biochem. Syst. Ecol.*, 26, 851, 1998.
150. Pomilio, A.B. et al., Antimicrobial constituents of *Gomphrena martiana* and *Gomphrena boliviana*, *J. Ethnopharmacol.*, 36, 155, 1992.
151. Esquivel, G., Calderón, J.S., and Flores, E., A neo-clerodane diterpenoid from *Scutellaria seleriana*, *Phytochemistry*, 47, 135, 1998.
152. Chang, W.-L. et al., Immunosuppresive flavones and lignans from *Bupleurum scorzonerifolium*, *Phytochemistry*, 64, 1375, 2003.
153. Wang, Q. et al., A new 2-hydroxyflavanone from *Mosla soochouensis*, *Planta Med.*, 65, 729, 1999.
154. Guha, P.K. and Bhattacharyya, A., 5,8-Dihydroxy-7-methoxyflavone from the immature leaves of *Didymocarpus pedicellata*, *Phytochemistry*, 31, 1833, 1992.
155. Harrison, L.J., Sia, G.L., and Sim, K.Y., 5,7-Dihydroxy-8-methoxyflavone from *Tetracera indica*, *Planta Med.*, 60, 493, 1994.
156. Midiwo, J.O. et al., Flavonoids of *Polygonum senegalense*. Part II. More surface and internal tissue flavonoid aglycones, *Bull. Chem. Soc. Ethiop.*, 6, 119, 1992.
157. Reddy, M.V.B. et al., New 2'-oxygenated flavonoids from *Andrographis affinis*, *J. Nat. Prod.*, 66, 295, 2003.
158. Wollenweber, E. et al., Taxonomic significance of flavonoid variation in temperate species of *Nothofagus*, *Phytochemistry*, 62, 1125, 2003.
159. Damu, A.G. et al., A flavone glycoside from *Andrographis alata*, *Phytochemistry*, 49, 1811, 1998.
160. Nkunya, M.H.H., Weenen, H., and Kinabo, L.S., Constituents of *Artemisia afra*, *Fitoterapia*, 63, 279, 1992.
161. Valant-Vetschera, K.M. and Wollenweber, E., Flavonoid aglycones from the leaf surfaces of *Artemisia* spp. (Compositae-Anthemideae), *Z. Naturforsch.*, 50c, 353, 1995.
162. Herrera, J.C. et al., Analysis of 5-hydroxy-7-methoxyflavones by normal HPLC, *J. Chromatogr. A*, 740, 201, 1996.
163. Köhler, I. et al., *In vitro* antiplasmodial investigation of medicinal plants from El Salvador, *Z. Naturforsch.*, 58c, 277, 2002.
164. Favier, L.S. et al., Diterpenoids and flavonoids from *Ophryosporus charrua*, *Phytochemistry*, 45, 1469, 1997.
165. Wollenweber, E. et al., Variation in flavonoid exudates in *Eucryphia* species from Australia and South America, *Biochem. Syst. Ecol.*, 28, 111, 2000.
166. Nagashima, F., Murakami, Y., and Asakawa, Y., Aromatic compounds from the Ecuadorian liverwort *Marchesinia brachiata*: a revision, *Phytochemistry*, 51, 1101, 1999.
167. Bordignon, S.A.L., Montanha, J.A., and Schenkel, E.P., Flavones and flavanones from South American *Cunila* species (Lamiaceae), *Biochem. Syst. Ecol.*, 31, 785, 2003.
168. Valant-Vetschera, K.M., Roitman, J.N., and Wollenweber, E., Chemodiversity of exudate flavonoids in some members of the Lamiaceae, *Biochem. Syst. Ecol.*, 31, 1279, 2003.
169. Kartnig, Th., Bucar, F., and Neuhold, S., Flavones from the above ground parts of *Lycopus virginicus*, *Planta Med.*, 59, 563, 1993.
170. Tomás-Barberán, F.A. and Wollenweber, E., Flavonoid aglycones from the leaf surfaces of some Labiatae species, *Plant. Syst. Evol.*, 173, 109, 1990.
171. Hatam, N.A.R. and Yousif, N.J., Flavonoids from *Salvia syriaca*, *Int. J. Pharm.*, 30, 109, 1992.
172. Gil, M.I. et al., Distribution of flavonoid algycones and glycosides in *Sideritis* species from the Canary Islands and Madeira, *Phytochemistry*, 34, 227, 1993.
173. Wollenweber, E. et al., Acylphloroglucinols and flavonoid aglycones produced by external glands on the leaves of two *Dryopteris* ferns and *Currania robertiana*, *Phytochemistry*, 48, 931, 1998.
174. Rao, Y.K. et al. Two new 5-deoxyflavones from *Albizia odoratissima*, *Chem. Pharm. Bull.*, 50, 1271, 2002.
175. Sahu, N.P., Achari, B., and Banerjee, S., 7,3'-Dihydroxy-4'-methoxyflavone from seeds of *Acacia farnesiana*, *Phytochemistry*, 49, 1425, 1998.
176. Carvalho, J.C.T. et al., Anti-inflammatory activity of flavone and some of its derivatives from *Virola michelli* Heckel, *J. Ethnopharmcol.*, 64, 173, 1999.

177. Huck, Ch.W. et al., Isolation and structural elucidation of 3′,4′,5′-trimethoxyflavone from the flowers of *Primula veris*, *Planta Med.*, 65, 491, 1999.

178. Alias, Y., Awang, K., and Hadi, A.H.A., An antimitotic and cytotoxic chalcone from *Fissistigma lanuginosum*, *J. Nat. Prod.*, 58, 1160, 1995.

179. Fuchino, H. et al., Chemical evaluation of *Betula* species in Japan. IV. Constituents of *Betula davurica*, *Chem. Pharm. Bull.*, 46, 166, 1998.

180. Matsuura, Y., Miyaichi, Y., and Tomimori, T., Studies on the Nepalese crude drugs. 19. On the flavonoid and phenylethanoid constituents of the root of *Scutellaria repens*, *Yakugaku Zasshi*, 114, 775, 1994.

181. Leong, Y.-W. et al., A dihydrochalcone from *Lindera lucida*, *Phytochemistry*, 47, 891, 1998.

182. Wollenweber, E. et al., Exudate flavonoids in miscellaneous Asteraceae, *Phytochem. Bull.*, 21, 19, 1989.

183. Babu, G.J., Naidu, K.C., and Ganapatyr, S., Phytoconstituents from the stem of *Duranta plumieri*, *Indian Drugs*, 35, 514, 1998.

184. Lazari, D., Skaltsa, H., and Harvala, C., Flavonoids of *Onopordum sibthorpianum* and *Onopordum laconicum*, *Biochem. Syst. Ecol.*, 26, 105, 1998.

185. Voirin, B. and Bayet, C., Developmental variations in leaf flavonoid aglycones of *Mentha* × *Piperita*, *Phytochemistry*, 31, 2299, 1992.

186. Corticchiato, M. et al., Free flavonoids aglycones from *Thymus herba barona* and its monoterpenoid chemotypes, *Phytochemistry*, 40, 115, 1995.

187. Seo, J.-M. et al., Antitumor activity of flavones isolated from *Artemisia argyi*, *Planta Med.*, 69, 218, 2003.

188. Çitoglu, G.S. and Aksit, F., Occurrence of marrubiin and ladanein in *Marrubium trachyticum* Boiss. from Turkey, *Biochem. Syst. Ecol.*, 30, 885, 2002.

189. Tomás-Barberán, F.A. et al., Flavonoids from some Yugoslavian *Micromeria* species: chemotaxonomical aspects, *Biochem. Syst. Ecol.*, 19, 697, 1991.

190. Jamzad, Z. et al., Leaf surface flavonoids in Iranian species of *Nepeta* (Lamiaceae) and some related genera, *Biochem. Syst. Ecol.*, 31, 587, 2003.

191. Grayer, R.J. et al., Distribution of 8-hydroxylated leaf-surface flavones in the genus *Ocimum*, *Phytochemistry*, 56, 559, 2001.

192. Tezuka, Y. et al., Constituents of the Vietnamese medicinal plant *Orthosiphon stamineus*, *Chem. Pharm. Bull.*, 48, 1711, 2000.

193. Gökdil, G. et al., Terpenoids and flavonoids from *Salvia cyanescens*, *Phytochemistry*, 46, 799, 1997.

194. Wollenweber, E., Dörr, M., and Muniappan, R., Exudate flavonoids in a tropical weed, *Chromolaena odorata* L., *Biochem. Syst. Ecol.*, 23, 873, 1995.

195. Jayakrishna, G. et al., Two new 2′-oxygenated flavones from *Andrographis elongata*, *Chem. Pharm. Bull.*, 49, 1555, 2001.

196. Jayaprakasam, B. et al., Dihydroechioidinin, a flavanone from *Andrographis echioides*, *Phytochemistry*, 52, 935, 1999.

197. Valant-Vetschera, K.M. and Wollenweber, E., Exudate leaf flavonoids of *Achillea clusiana* Tausch and related species, *Biochem. Syst. Ecol.*, 24, 477, 1996.

198. Wollenweber, E. et al., Rare flavonoids from *Odixia* and *Ozothamnus* spp. (Asteraceae, Gnaphalieae), *Z. Naturforsch.*, 52c, 571, 1997.

199. Sepulveda, S. et al., Constituents of *Centaurea chilensis*, *Fitoterapia*, 65, 88, 1994.

200. Stevens, J.F. et al., Leaf surface flavonoids of *Chrysothamnus*, *Phytochemistry*, 51, 771, 1999.

201. Wollenweber, E. et al., Flavonoids and terpenoids from the resinous exudates of *Madia* species (Asteraceae, Helenieae), *Z. Naturforsch.*, 58c, 153, 2003.

202. Braca, A. et al., Flavonoids and triterpenoids from *Licania heteromorpha* (Chrysobalanac.), *Biochem. Syst. Ecol.*, 27, 527, 1999.

203. Li, Y. et al., Littorachalcone, a new enhancer of NGF-mediated neurite outgrowth, from *Verbena littoralis*, *Chem. Pharm. Bull.*, 51, 872, 2003.

204. Ramesh, P. and Yuvarajan, C.R., A new flavone methyl ether from *Helicteres isora*, *J. Nat. Prod.*, 58, 1242, 1995.

205. Neves, M. et al., Three triterpenoids and one flavonoid from the liverwort *Asterella blumeana* grown *in vitro*, *Phytother. Res.*, 12, S21, 1998.

206. Iinuma, M. et al., Studies on the constituents of useful plants. 5. Multisubstituted flavones in the fruit peel of *Citrus reticulata* in their examination by GLC, *Chem. Pharm. Bull.*, 23, 717, 1980.

207. Wollenweber, E., Mann, K., and Roitman, J.N, Flavonoid aglycones excreted by three *Calceolaria* species, *Phytochemistry*, 28, 2213, 1989.

208. Zhang, Y.-Y. et al., Four flavonoids from *Scutellaria baicalensis*, *Phytochemistry*, 35, 511, 1994.

209. Heneka, B., Isolierung gastrointestinal wirksamer Inhaltsstoffe aus *Casimiroa tetrameria*, einer yukatekischen Arzneipflanze der Maya (Mexiko), Doctoral Thesis, Freiburg i. Br., 2001

210. Skibinski, A., Merfort, I., and Willuhn, G., Thirty-seven flavonoids from flowers of *Arnica longifolia*, *Phytochemistry*, 37, 1635, 1994.

211. Bohm, B.A., Major exudate flavonoids of *Dubautia arborea* (Asteraceae), *Biochem. Syst. Ecol.*, 27, 755, 1999.

212. André, A.C.G.M., Dias, D.A., and Vichnewski, W., Flavonoids of *Wunderlichia crulsiana*, *Biochem. Syst. Ecol.*, 30, 483, 2002.

213. Urzua, A. et al., External flavonoids from *Heliotropium megalanthum* and *H. huascoense* (Boraginaceae). Chemotaxonomic considerations, *Soc. Chil. Quim*, 45, 23, 2000.

214. Wollenweber, E. and Dörr, M., Exudate flavonoids in some Solanaceae, *Biochem. Syst. Ecol.*, 23, 457, 1995.

215. Wollenweber, E. et al., External flavonoids of 12 species of North American Eupatorieae (Asteraceae), *Z. Naturforsch.*, 51c, 893, 1996.

216. Stevens, J.F., Hart, H.T., and Wollenweber, E., The systematic and evolutionary significance of exudates flavonoids in *Aeonium*, *Phytochemistry*, 39, 805, 1995.

217. Nassar, M.J. et al., A benzoquinone and flavonoids from *Cyperus alopecuroides*, *Phytochemistry*, 60, 385, 2002.

218. Dias, A.C.P. et al., Unusual flavonoids produced by callus of *Hypericum perforatum*, *Phytochemistry*, 48, 1165, 1998.

219. Takagi, M. et al., Flavonoids in the sea-grass, *Phyllospadix japonica*, *Agric. Biol. Chem.*, 43, 2417, 1979.

220. Wollenweber, E. and Rustaiyan, A., Exudate flavonoids in three Persian Asteraceae species, *Biochem. Syst. Ecol.*, 19, 673, 1991.

221. Perez-Castorena, A.-L., Martinez-Vazquez, M., and de Vilar, A.R., Diterpenes of *Bahia glandulosa*, *Phytochemistry*, 46, 729, 1997.

222. Wollenweber, E. et al., Exudate flavonoids from two Australian Asteraceae, *Bracteantha viscosa* and *Cassinia quinquefaria*, *Phytochemistry*, 33, 871, 1993.

223. Nuñez-Alarcon, J. and Quinones, M., Epicuticular flavonoids from *Haplopappus baylahuen* and the hepatoprotective effect of the isolated 7-methylaromadendrin, *Biochem. Syst. Ecol.*, 23, 453, 1993.

224. Topcu, G. et al., Terpenoids and flavonoids from the aerial parts of *Salvia candidissima*, *Phytochemistry*, 40, 501, 1995.

225. Amaro-Luis, J., Herrera, J.R., and Luis, J.G., Abietane diterpenoids from *Salvia chinopeplica*, *Phytochemistry*, 47, 895, 1998.

226. Wollenweber, E. and Dörr, M., Exudate flavonoids of *Kitaibelia vitifolia* (Malvaceae), *Biochem. Syst. Ecol.*, 24, 801, 1996.

227. Zahir, A., Jossang, A., and Bodo, B., DANN topoisomerase I inhibitors: cytotoxic flavones from *Lethodon tannaensis*, *J. Nat. Prod.*, 59, 701, 1996.

228. Wollenweber, E. et al., Flavonoid aglycones and triterpenoids from the leaf exudate of *Lantana camara* and *Lantana montevidensis*, *Biochem. Syst. Ecol.*, 25, 269, 1997.

229. Encarnacion, D.E. et al., Two new flavones from *Calliandra californica*, *J. Nat. Prod.*, 57, 1307, 1994.

230. Grayer, R.J. and Veitch, N.C., An 8-hydroxylated external flavone and its 8-O-glucoside from *Becium grandiflorum*, *Phytochemistry*, 47, 779, 1998.

231. Vieira, R.F. et al., Genetic diversity of *Ocimum gratissimum* L. based on volatile oil constituents, flavonoids and RAPD markers, *Biochem. Syst. Ecol.*, 29, 287, 2001.

232. Wang, H. et al., Antioxidant polyphenols from tart cherries (*Prunus cerasus*), *J. Agric. Food Chem.*, 47, 840, 1999.

233. Wollenweber, E. et al., Exudate flavonoids in three *Ambrosia* species, *Nat. Prod. Lett.*, 7, 109, 1995.

234. Greenham, J. et al., A distinctive flavonoid density for the anomalous genus *Biebersteinia*, *Phytochemistry*, 56, 87, 2001.

235. Bosabalidis, A., Gabrieli, Ch., and Niopas, J., Flavone aglycones in glandular hairs of *Origanum* × *intercedens*, *Phytochemistry*, 49, 1549, 1998.

236. Bohm, B.A. et al., Non-polar flavonoids of *Calycadenia*, *Lagophylla* and *Madia*, *Phytochemistry*, 31, 1261, 1992.

237. Randriaminahy, M. et al., Lipophilic constituents from *Helichrysum species* endemic to Madagascar, *Z. Naturforsch.*, 47c, 10, 1992.

238. Ferracini, V.L. et al., Sesquiterpene lactones and one flavonoid from *Hymenoxis scaposa* var. *linearis*, *Biochem. Syst. Ecol.*, 22, 111, 1994.

239. Afifi, F.U. et al., Antifungal flavonoids from *Varthemia iphionoides*, *Phytother. Res.*, 5, 173, 1991.

240. Al-Khalil, S. et al., N-arachidylanthranilic acid, a new derivative from *Ononis natrix*, *J. Nat. Prod.*, 58, 760, 1995.

241. Wollenweber, E. and Valant-Vetschera, K.M., External flavonoids of *Satureja montana*, *Fitoterapia*, 62, 462, 1991.

242. Feresin, G.E. et al., Constituents of the Argentinian medicinal plant *Baccharis grisebachii* and their antimicrobial activity, *J. Ethnopharmacol.*, 89, 73, 2003.

243. Maldonado, E., Hernández, E., and Ortega, A., Amides, coumarins and other constituents from *Simsia cronquistii*, *Phytochemistry*, 31, 1413, 1992.

244. Wollenweber, E. et al., External flavonoids of three species of *Viguiera*, section Hypargyrea (Asteraceae), *Z. Naturforsch.*, 50c, 588, 1995.

245. Parmar, V.S. et al., Highly oxygenated bioactive flavones from *Tamarix*, *Phytochemistry*, 36, 507, 1994.

246. Tomás-Barberán, F.A. et al., External and vacuolar flavonoids from *Nepeta transcaucasica*, *Biochem. Syst. Ecol.*, 20, 589, 1992.

247. Blatt, C.T.T., Salatino, A., and Salatino, M.L.F., Flavonoids of *Tabebuia caraiba* (Bignoniaceae), *Biochem. Syst. Ecol.*, 24, 89, 1996.

248. Takemura, O.S. et al., A flavone from leaves of *Arrabidaea chica* f. *cuprea*, *Phytochemistry*, 38, 1299, 1995.

249. Zaidi, F. et al., Free flavonoid aglycones from leaves of *Mentha pulegium* and *Mentha suaveolens* (Labiatae), *Phytochemistry*, 48, 991, 1998.

250. Bisio, A. et al., Flavonoide und Triterpenoide aus *Salvia blepharophylla*, *Pharmazie*, 52, 330, 1997.

251. Akaike, S. et al., A new *ent*-clerodane diterpene from the aerial parts of *Baccharis gaudichaudiana*, *Chem. Pharm. Bull.*, 51, 197, 2003.

252. Tomás-Barberán, F.A., Husain, S.Z., and Gil, M.I., The distribution of methylated flavones in the Lamiaceae, *Biochem. Syst. Ecol.*, 16, 43, 1988.

253. Barberán, F.A.T., Hernández, L., and Tomás, F., A chemotaxonomic study of flavonoids in *Thymbra capitata*, *Phytochemistry*, 25, 561, 1986.

254. Alcerito, T. et al., Foliar epicuticular wax of *Arrabidaea brachypoda*: flavonoids and antifungal activity, *Biochem. System. Ecol.*, 30, 677, 2002.

255. Wollenweber, E., External leaf flavonoids of *Centaurea macrocephala*, *Fitoterapia*, 62, 364, 1991.

256. Azam, M.M. and Ghanim, A., Flavones from leaves of *Tecombella undulata* (Bignon.), *Biochem. Syst. Ecol.*, 28, 803, 2000.

257. Topcu, G. et al., A new flavanone from *Teucrium alyssifolium*, *Turk. J. Chem.*, 20, 265, 1996.

258. Valentao, P. et al., Analysis of vervain flavonoids by HPLC/diode array detector method. Its application to quality control, *J. Agric. Food Chem.*, 47, 4579, 1999.

259. Nakayama, M. et al., Separation of 5,6,7-trisubstituted flavone derivatives by HLPC, *J. Chem. Soc. Jpn.*, 1390, 1978.

260. Wollenweber, E., Dörr, M., and Rustaiyan, A., *Dorema aucheri* — the first umbelliferous plant found to produce exudate flavonoids, *Phytochemistry*, 38, 1417, 1995.

261. Elgamal, M.H.A., Hanna, A.G., and Duddeck, H., Constituents of *Achillea santolina*, *Fitoterapia*, 62, 349, 1991.

262. De Guttierez, A.N. et al., Sesquiterpene lactones, a labdane and other constituents of *Urolepis hecatantha* and *Chromolaena arnottiana*, *Phytochemistry*, 39, 795, 1995.

263. Cuenca, M. et al., Sesquiterpene lactones of *Mikania minima*, *Phytochemistry*, 34, 1509, 1993.

264. Huang, Y., Ding, Z.-H., and Liu, J.-K., A new highly oxygenated flavone from *Vernonia saligna*, *Z. Naturforsch. C*, 58, 347, 2003.

265. Tan, R.X. et al., Mono- and sesquiterpenes and antifungal constituents from *Artemisia* species, *Planta Med.*, 65, 64, 1999.

266. Ruiz-Cancino, A., Cano, A.E., and Delgado, G., Sesquiterpene lactones and flavonoids from *Artemisia ludoviciana* ssp. *mexicana*, *Phytochemistry*, 33, 1113, 1993.

267. Cappelletti, E.M., Caniato, R., and Appendino, G., Localization of the cytotoxic hydroperoxyeudesmanolide of *Artemisia umbelliformis*, *Biochem. Syst. Ecol.*, 14, 183, 1986.

268. Vyas, A.V. and Mulchandani, N.B., Polyoxygenated flavones from *Ageratum conyzoides*, *Phytochemistry*, 25, 2625, 1986.

269. Nadir, M.T. et al., The constituents of *Achillea conferta*: phytochemical and antimicrobial studies, *Int. J. Pharm.*, 29, 89, 1991.

270. Çubukçu, B. and Melikoglu, G., Flavonoids of *Artemisia austriaca*, *Planta Med.*, 61, 488, 1995.

271. Tang, H.Q. et al., Terpenoids and flavonoids from *Artemisia* species, *Planta Med.*, 66, 391, 2000.

272. Jahan, N., Malik, A., and Muhammad, P., New flavonoid from *Mentha longifolia*, *Heterocycles*, 55, 1951, 2001.

273. Savona, G., Flavones of *Teucrium pseudochamaepitys*, *Anales Quim.*, 75, 433, 1979.

274. Le-Van, N. and Pham, T.V.C., Two new flavones from *Eupatorium coelestinum*, *Phytochemistry*, 18, 1859, 1979.

275. Mukherjee, K.S. et al., A new flavonoid from *Limnophila rugosa*, *Fitoterapia*, 74, 188, 2003.

276. Xie, N. et al., Flavones from *Teucrium quadrifarium*, *Zhongguo Yaoke Daxue Xuebao*, 22, 200, 1991.

277. Tarhouni, M.R., Isolation and characterization of flavonoids from *Artemisia compestris* L. subsp. *glutinosa* plant, *J. Soc. Chim.*, 12, 891, 1996.

278. Wollenweber, E. et al., Externally accumulated flavonoids in three Mediterranean *Ononis* species, *Z. Naturforsch.*, 58c, 771, 2003.

279. Guo, B.-L. et al., Brevicornin, a flavonol from *Epimedium brevicornu*, *Phytochemistry*, 41, 991, 1996.

280. Kuwabara, H. et al., Tricin from a Malagasy connaraceous plant with potent antihistaminic activity, *J. Nat. Prod.*, 66, 1273, 2003.

281. Akkal, S. et al., Flavonoids from *Centaurea incana* (Asteraceae), *Biochem. Syst. Ecol.*, 25, 361, 1997.

282. Kobzar, A.-J. and Nikonor, G.K., Flavonoids from overground parts of *Betonica officinalis*, *Khim. Prim. Soedin.*, 636, 1986.

283. Balakrishna, K. et al., Constituents of *Walsura piscidia*, *Fitoterapia*, 66, 548, 1995.

284. Kinoshita, T. and Firman, K., Myricetin 5,7,3′,4′,5′-pentamethyl ether and other methylated flavonoids from *Murraya paniculata*, *Phytochemistry*, 45, 179, 1997.

285. Souza, J.P.I., Arruda, A.C., and Arruda, M.P.S., Highly methoxylated flavones from *Neoraputia paraensis*, *Fitoterapia*, 66, 465, 1995.

286. Tomazela, D.M. et al., Pyrano chalcones and a flavone from *Neoraputia magnifica* and their *Trypanosoma cruzi* glycosomal glyceraldehyde-3-phosphate dehydrogenase-inhibitory activities, *Phytochemistry*, 55, 643, 2000.

287. Zheng, W.F. et al., Two flavones from *Artemisia giraldii* and their antimicrobial activity, *Planta Med.*, 62, 160, 1996.

288. Yamamura, S. et al., Antihistaminic flavones and aliphatic glycosides from *Mentha spicata*, *Phytochemistry*, 48, 131, 1998.

289. Fushiya, S. et al., Flavonoids from *Cleome droserifolia* supress NO production in activated macrophages *in vitro*, *Planta Med.*, 65, 404, 1999.

290. Tomás-Barberán, F.A., Husain, S.Z., and Gil, M.I., The distribution of methylated flavones in the Lamiaceae, *Biochem. Syst. Ecol.*, 16, 43, 1988.

291. Kinoshita, T. and Firman, K., Highly oxygenated flavonoids from *Murraya paniculata*, *Phytochemistry*, 42, 1207, 1996.

292. Martinez-Vazquez, M. et al., Methylated flavones from *Conoclidium greggii*, *J. Nat. Prod.*, 56, 1410, 1993.

293. Horie, T. et al., Studies of the selective O-alkylation and dealkylation of flavonoids. 13. An improved method for synthesizing 5,6,7-trihydroxyflavones from 6-hydoxy-5,7-dimethoxyflavones. *J. Org. Chem.*, 57, 3343, 1992.

294. Horie, T., Synthesis of 5,6,7-trihydroxyflavones, *J. Chem. Soc. Jpn.*, 747, 1978.

295. Juma, B.F. et al., Flavonoids and phenylpropanoids in the surface exudate of *Psiadia punctulata*, *Phytochemistry*, 57, 571, 2001.

296. Quijano, L. et al., Flavonoids from *Ageratum corymbosum*, *Phytochemistry*, 31, 2859, 1992.

297. Tanaka, T. et al., Flavonoids in root bark of *Pongamia pinnata*, *Phytochemistry*, 31, 993, 1992.

298. Wollenweber, E. et al., Rare flavonoid aglycones from *Anaphalis margaritacea* and two *Gnaphalium* species, *Z. Naturforsch.*, 48c, 420, 1993.

299. Meyer, J.J.M. et al., Antiviral activity of galangin isolated from the aerial parts of *Helichrysum aureonitens*, *J. Ethnopharmacol.*, 56, 165, 1997.

300. Kerber, V. A., Miguel, O.G., and Moreira, E.A., Flavonoids from *Heterothalamus psiadioides*, *Fitoterapia*, LXIV, 185, 1993.

301. Russell, G.B. et al., Patterns of bioactivity and herbivory on *Nothofagus* species from Chile and New Zealand, *J. Chem. Ecol.*, 26, 41, 2000.

302. Wollenweber, E., Stüber, A., and Kraut, L., Flavonoid aglycones and flavonol glycosides in lipophilic leaf exudate of *Nothofagus antarctica*, *Phytochemistry*, 44, 1399, 1997.

303. Bohm, B.A., External and vacuolar flavonoids of *Ribes vixcosissimum*, *Biochem. Syst. Ecol.*, 21, 745, 1993.

304. Economides, C. and Adam, K.-P., Lipophilic flavonoids from the fern *Woodsia scopulina*, *Phytochemistry*, 49, 859, 1998.

305. Wollenweber, E. et al., A novel caffeic acid derivative and other constituents of *Populus* bud excretion and propolis (bee-glue), *Z. Naturforsch.*, 42c, 1030, 1987.

306. Munoz, O. et al., Phenolic compounds of propolis from central Chilean matorral, *Z. Naturforsch.*, 57c, 273, 2002.

307. Tomas-Lorente, F. et al., Antimicrobial phenolics from *Helichrysum picardii*, *Fitoterapia*, 62, 521, 1991.

308. Urzua, A. et al., Flavonoids and diterpenoids in the trichome resinous exudates from *Pseudognaphalium cheiranthifolium*, *P. heterotrichium* and *P. vira vira*, *Biochem. Syst. Ecol.*, 23, 459, 1995.

309. Villarroel, L. et al., *Heliotropium huascoense* resin exudates: chemical constituents and defensive properties, *J. Nat. Prod.*, 64, 1123, 2001.

310. Torres, R. et al., Flavonoides del exudado resinoso de *Heliotropium sinuatum*, *Bol. Soc. Quim.*, 41, 195, 1996.

311. Pelotto, J.P. and del Pero Martinez, M.A., Flavonoid aglycones from Argentinian *Capparis* species (Capparaceae), *Biochem. Syst. Ecol.*, 26, 577, 1998.

312. Wollenweber, E. and Roitman, J.N., New frond exudate flavonoids from cheilanthoid ferns, *Z. Naturforsch.*, 46c, 325, 1991.

313. Tominaga, H. and Horie, T., Studies of the selective *O*-alkylation of flavonoids. XV. A convenient synthesis of 3,5,6-trihydroxy-7-methoxyflavone and revised structures of two natural flavones, *Bull. Chem. Soc. Jpn.*, 66, 2668, 1993.

314. Morimoto, M.S., Kumeda, S., and Komai, K., Insect antifeedant flavonoids from *Gnaphalium affine* D. Don, *J. Agric. Food Chem.*, 48, 1888, 2000.

315. Cuadra, P., Harborne, J.B., and Waterman, P.G., Increases in surface flavonols and photosynthetic pigments in *Gnaphalium luteo-album* in response to UV-B radiation, *Phytochemistry*, 45, 1377, 1997.

316. Rossi, M.H., Yoshida, M., and Maia, J.G.S., Neolignans, styrylpyrones and flavonoids from an *Aniba* species, *Phytochemistry*, 45, 1263, 1997.

317. Sarmento Silva, T.M. et al., Flavonoids and an alkanide from *Solanum paludosum*, *Biochem. Syst. Ecol.*, 30, 479, 2002.

318. Haas, K., Bauer, M., and Wollenweber, E., Cuticular waxes and flavonol aglycones of mistletoes, *Z. Naturforsch.*, 58c, 464, 2003.

319. Banskota, A.H. et al., Chemical constituents of Brazilian propolis and their cytotoxic activities, *J. Nat. Prod.*, 61, 896, 1998.

320. Vieira, L. et al., Germacranes and flavonoids from *Achillea ageratum*, *Phytochemistry*, 45, 111, 1997.

321. Wollenweber, E. and Dörr, M., Flavonoid aglycones of *Cleome spinosa* (Cleomaceae), *Phytochem. Bull.*, 24, 2, 1992.

322. Auzi, A.A., Hartley, T.G., and Waterman, P.G., Distribution of flavonoids, alkaloids, acetophenones and phloroglucinols in the genus *Bosistoa* (Rutaceae), *Biochem. Syst. Ecol.*, 25, 611, 1997.

323. Chou, C.-J. et al., Novel acetophenones from fruits of *Evodia merrillii*, *J. Nat. Prod.*, 55, 795, 1992.

324. Dong, H. et al., Eicosenones and methylated flavonols from *Amomum koenigii*, *Phytochemistry*, 50, 899, 1999.

325. Wollenweber, E. and Yatskievych, G., External flavonoids of Ocotillo (*Fouquieria splendens*), *Z. Naturforsch.*, 49c, 689, 1994.

326. Wollenweber, E. and Graven, E.H., Flavonoid aglycones of oval leaf buchu, *Barosma crenulata*, *Fitoterapia*, 63, 86, 1992.

327. Maldonado, Z., Hoeneisen, M., and Silva, M., Constituents of *Haploappus bezanillanus* and *H. hirtellus*, *Bol. Soc. Chil. Quim.*, 38, 43, 1993.

328. Wang, S. et al., Identification and determination of ecdysones and flavonoids in *Serratula strangulata* by micellar electrokinetic capillary chromatography, *Planta Med.*, 68, 1029, 2002.

329. Murillo, J.J. et al., Antimicrobial flavones from *Haplopappus sonorensis*, *Fitoterapia*, 74, 226, 2003.

330. Nakashima, C. et al., Two flavones from *Graziela mollissima*, *Phytochemistry*, 37, 285, 1994.

331. Lemmich, E. et al., 5-Deoxyflavones from *Parkia clappertoniana*, *Phytochemistry*, 42, 1011, 1996.

332. Malan, E. and Swartz, P., A comparative study of the phenolic products in the heartwood of *Acacia karroo* from two different localities, *Phytochemistry*, 39, 791, 1995.

333. Chan, S.-C. et al., Three new flavonoids and antiallergic, anti-inflammatory constituents from the heartwood of *Dalbergia odorifera*, *Planta Med.*, 64, 153, 1998.

334. Hatano, T. et al., Phenolic constituents of liquorice. VII. A new chalcone with a potent radical scavenging activity and accompanying phenolics from liquorice, *Chem. Pharm. Bull.*, 45, 1485, 1997.

335. Horie, T. et al., Studies of the selective O-alkylation and dealkylation of flavonoids. XXI. A convenient method for synthesizing 3,5,7-trihydroxy-6,8-dimethoxylflavones and 5,7-dihydroxy-3,6,8-trimethoxyflavones, *Bull. Chem. Soc. Jpn.*, 69, 1033, 1996.

336. Horie, T. et al., Studies of the selective O-alkylation and dealkylation of flavonoids. XIX. A convenient method for synthesizing 3,5,6,7,8-pentaoxygenated flavones, *Chem. Pharm. Bull.*, 43, 2054, 1995.

337. Mendoza, L. and Urzua, A., Minor flavonoids and diterpenoids in the resinous trichome exudates from *Pseudognaphalium cheiranthifolium*, *P. heterotrichium*, *P. vira vira* and *P. robustum*, *Biochem. Syst. Ecol.*, 26, 469, 1998.

338. Milovanovic, M. and Djermanovic, M., Constituents of *Helminthia echoides*, *Fitoterapia*, 65, 377, 1994.

339. Hattori, M. et al., 6-Hydroxykaempferol and its glycosides from *Carthamus tinctonius* petals, *Phytochemistry*, 31, 4001, 1992.

340. Valant-Vetschera, K.M. et al., New exudate flavonoids of species from the *Chrysanthemum* complex (Asteraceae–Anthemidae), *Biochem. Syst. Ecol.*, 31, 545, 2003.

341. Tazawa, S., Warasina, T., and Noro, T., Studies on the constituents of Brazilian propolis, *Chem. Pharm. Bull.*, 47, 1388, 1999.

342. Horie, T. et al., Studies of the selective O-alkylation and dealkylation of flavonoids. XXII. A convenient method for synthesizing 3,5,7-trihydroxy-6-methoxyflavones, *Chem. Pharm. Bull.*, 45, 446, 1997.

343. Hatam, N.A.R. and Seifert, K., Flavonoids from *Achillea micrantha*, *Planta Med.*, 60, 600, 1994.

344. Timmermann, B. et al., External flavonoids in two *Grindelia* species, *Z. Naturforsch.*, 49c, 395, 1994.

345. Jakobsen, T.H. et al., 3-Methoxyflavones and a novel coumarin from *Psiadia dentata*, *Biochem. Syst. Ecol.*, 29, 963, 2001.
346. Miller, J.M. et al., Unusual flavonols from bud exudates of Fijian *Gardenia* species (Rubiaceae), *Indian J. Chem.*, 28B, 1093, 1989.
347. Williams, C.A., Harborne, J.B., and Greenham, J., Geographical variation in the surface flavonoids of *Pulicaria dysenterica*, *Biochem. System. Ecol.*, 28, 679, 2000.
348. El-Sayed, N.H. et al., Flavonoids of *Brickellia longifolia*, *Phytochemistry*, 29, 2364, 1990.
349. Xiong, Q., Shi, D., and Mizuno, M., Flavonol glucosides in pericarps of *Zanthoxylum bungeanum*, *Phytochemistry*, 39, 723, 1995.
350. Rahalison, L. et al., Antifungal principles of *Baccharis pedunculata*, *Planta Med.*, 61, 360, 1995.
351. Sharp, H. et al., 6-Oxygenated flavones from *Baccharis trinervis* (Asteraceae), *Biochem. Syst. Ecol.*, 29, 105, 2001.
352. Wollenweber, E. et al., Exudate flavonoids from *Grindelia tarapacana* from Chile, *Z. Naturforsch.*, 48c, 533, 1993.
353. Ribeiro, A. and Pilo-Veloso, D., Trypanocidal flavonoids from *Trixis vauthieri*, *J. Nat. Prod.*, 60, 836, 1997.
354. Valant-Vetschera, K.M. and Wollenweber, E., Leaf exudate flavonoids of *Achillea sibirica* subsp. *mongolica* and *A. latiloba*, *Biochem. Syst. Ecol.*, 27, 523, 1999.
355. Aljancic, I. et al., Flavones and sesquiterpenes from *Achillea atrata* subsp. *multifida*: antimicrobial activity, *J. Nat. Prod.*, 62, 909, 1999.
356. Wollenweber, E. and Mayer, K., Exudate flavonoids of *Anthemis nobilis* and *A. tinctoria*, *Fitoterapia*, 62, 365, 1991.
357. Martinez, J. et al., Isolation of two flavonoids from *Tanacetum microphyllum* as PMA-induced ear edema inhibitors, *J. Nat. Prod.*, 60, 142, 1997.
358. Rashid, M.A. et al., Alkaloids, flavonols and coumarins from *Drummondita hassellii* and *D. calida*, *Phytochemistry*, 31, 1265, 1992.
359. Wollenweber, E., Armbruster, S., and Roitman, J.N., A herbacetin methyl ether from the farinose exudate of a *Pentagramma triangularis* hybrid, *Phytochemistry*, 37, 455, 1994.
360. Reid, A.R. and Bohm, B.A., External and vacuolar flavonoids of *Brachyglottis cassinioides*, *Biochem. Syst. Ecol.*, 21, 746, 1993.
361. Tojo, E. et al., Clerodane diterpenes from *Haplopappus deserticola*, *Phytochemistry*, 52, 1531, 1999.
362. Iinuma, M. et al., Unusual biflavonoids in the farinose exudate of *Pentagramma triangularis*, *Phytochemistry*, 44, 705, 1997.
363. Macias, F.A. et al., Bioactive flavonoids from *Helianthus annuus* cultivars, *Phytochemistry*, 45, 683, 1997.
364. Barua, N.C. and Sharma, R.P., (2R, 3R)-7,5′-dimethoxy-3,5,2′-trihydroxyflavanone from *Blumea balsamifera*, *Phytochemistry*, 31, 4040, 1992.
365. Reid, A.R. and Bohm, B.A., Vacuolar and exudates flavonoids of New Zealand *Cassinia* (As teraceae: Gnaphalieae), *Biochem. Syst. Ecol.*, 22, 501, 1994.
366. Faini, F. et al., Chemistry toxicity and antifeedant activity of the resin of *Flourensia thurifera*, *Biochem. Syst. Ecol.*, 25, 189, 1997.
367. Da Costa, F.B. et al., Flavonoids and helianolides from *Lychnophora diamantina*, *Phytochemistry*, 34, 261, 1993.
368. Marin, J.C. et al., Fitoquímica y evaluación de la acción biológica de *Polygonum punctatum*, *Rev. Latinomer. Quim.*, 29, 100, 2001.
369. Borkosky, S. et al., Sesquiterpene lactones and other constituents of *Eirmocephala megaphylla* and *Cyrtocymura cincta*, *Phytochemistry*, 42, 1637, 1996.
370. Marambio O. and Silva, M., New compounds isolated from *Haplopappus taeda*, *Bol. Soc. Chil. Quim.*, 34, 105, 1989.
371. Bohm, B.A., Crins, W.J., and Wells, T.C., External flavonoids of *Holocarpha* (Asteraceae: Madiinae), *Biochem. Syst. Ecol.*, 22, 859, 1994.
372. Sharaf, M., Mansour, R.M.A., and Saleh, N.A.M., Exudate flavonoids from aerial parts of four *Cleome* species, *Biochem. Syst. Ecol.*, 20, 443, 1992.

373. Wollenweber, E., On the distribution of exudate flavonoids among Angiosperms, *Rev. Latinoamer. Quim.*, 21, 115, 1990.

374. Simonsen, H.T. et al., Methylenedioxy- and methoxyflavones from *Melicope coodeana* syn. *Euodia simplex*, *Phytochemistry*, 60, 817, 2002.

375. Bohm, B.A., Exudate flavonoids of *Mimulus lewisii*, *Biochem. Syst. Ecol.*, 20, 591, 1992.

376. Ho, L.-K. and Lin, W.-N., Quercetin 5,4'-dimethyl ether from *Rhododendron ellipticum*, *Phytochemistry*, 39, 463, 1995.

377. Miles, D.H. et al., Potential agrochemicals from leaves of *Wedelia biflora*, *Phytochemistry*, 32, 1427, 1993.

378. Urzua, A. et al., Comparative leaf surface chemistry from *Senecio cerberoanus* and *Senecio viscosissimus*, *Biochem. Syst. Ecol.*, 28, 399, 2000.

379. Wang, E. et al., The crystal structure of the dimorphic forms of a new flavonoid from the Australian tree *Melicope elleryana*, *Aust. J. Chem.*, 54, 739, 2001.

380. Cambie, R.C., Pan, Y.P., and Bowden, B.F., Flavonoids of the barks of *Melicope simplex* and *Melicope ternate*, *Biochem. Syst. Ecol.*, 24, 461, 1996.

381. Likhiwitayawuid, K., Jongbunprasert, V., and Chanmahasathien, W., Flavones from *Euodia viticina*, *Planta Med.*, 61, 590, 1995.

382. Phuong, N.M. et al., Methoxylated flavones from *Adenosoma capitatum*, *Pharmazie*, 52, 647, 1997.

383. Happi, E.N. and Mpondo, T.N., Two polymethoxylated flavones from *Distemonanthus benthamianus*, *J. Nat. Prod.*, 57, 291, 1994.

384. Cimanga, K. et al., Complement-inhibiting constituents of *Bridelia ferruginea* stem bark, *Planta Med.*, 65, 213, 1999.

385. Malan, E., 7,8,4'-Trihydroxy-3,3'-dimethoxyflavone from the heartwood of *Acacia nigrescens*, *Phytochemistry*, 33, 733, 1993.

386. Kamnaing, P. et al., An isoflavan-quinone and a flavonol from *Millettia laurentii*, *Phytochemistry*, 51, 829, 1999.

387. Wollenweber, E., Stevens, J.F., and Ivancic, M., Flavonoid aglycones and a thiophene derivative from *Helichrysum cassianum*, *Phytochemistry*, 47, 1441, 1998.

388. Ko, W.G. et al., Polymethoxyflavonoids from *Vitex rotundifolia* inhibit proliferation by inducing apoptosis in human myeloid leukemia cells, *Food Chem. Toxicol.*, 38, 861, 2000.

389. Masterova, I. et al., Phenolic substances in flowers of *Anthemis tinctoria*, *Fitoterapia*, 64, 277, 1993.

390. Ahmed, A.A., Spaller, M., and Mabry, T.J., Flavonoids in *Pallenis spinosa* (Asteraceae), *Biochem. Syst. Ecol.*, 20, 785, 1992.

391. Pérez, A.-L., Nieto, D.A., and de Vivar, A.R., Sesquiterpenoid and other metabolites from two *Bahia* species, *Phytochemistry*, 29, 901, 1990.

392. Bergendorff, O. and Sterner, O., Spasmolytic flavonols from *Artemisia abrotanum*, *Planta Med.*, 61, 370, 1995.

393. Caula, S.A. et al., Polyphenols isolated from *Eupatorium buniifolium*, *Rev. Latinoamer. Quim.*, 22, 1, 1991.

394. Quijano, L. et al., The molecular structure of maniladiol from *Baccharis salicina*, *Phytochemistry*, 49, 2065, 1998.

395. Abad, M.J., Bermejo, P., and Villar, A., Anti-inflammatory activity of two flavonoids from *Tanacetum microphyllum*, *J. Nat. Prod.*, 56, 1164, 1993.

396. Semple, S.-J. et al., Antiviral flavonoid from *Pterocaulon sphacelatum*, an Australian aboriginal medicine, *J. Ethnopharmacol.*, 68, 283, 1999.

397. Zheng, G.-Q., Cytotoxic terpenoids and flavonoids from *Artemisia annua*, *Planta Med.*, 60, 54, 1994.

398. Zallocchi, E.M. and Pomilio, A.B., Flavonoids from Vigna candida, V. spiralis and V. adenantha, *Fitoterapia*, **65**, 470, 1994.

399. Wollenweber, E. and Roitman, J.N., A novel methyl ether of quercetagetin from *Chromolaena odorata* leaf exudates, *Biochem. Syst. Ecol.*, 24, 479, 1996.

400. Hussein, S.A.M. et al., Polyoxygenated flavonoids from *Eugenia edulis*, *Phytochemistry*, 64, 883, 2003.

401. Ramidi, R. and Ali, M., Two new flavonoids from the seeds of *Zanthoxylum alatum*, *Pharmazie*, 54, 781, 1999.

402. Horie, T. et al., Studies of the selective O-alkylation and dealkylation of flavonoids. X. Selective demethylation of 7-OH-3,5,8-triOMe flavones with anhydrous aluminium halide in acetonitrile or ether, *J. Org. Chem.*, 52, 4702, 1987.

403. Likhiwitayawuid, K., Jongbunprasert, V., and Chanmahasathien, W., Flavones from *Euodia viticina*, *Planta Med.*, 61, 590, 1995.

404. Ferracin, R.J. et al., Flavonoids from the fruits of *Murraya paniculata*, *Phytochemistry*, 47, 393, 1998.

405. Mendez, J. et al., 5,7,2′,5′-Tetrahydroxyflavonol 3-rhamnoside from *Plinia pinnata*, *Phytochemistry*, 36, 1087, 1994.

406. Wollenweber, E., Mann, K., and Roitman, J.N., A myricetin tetramethyl ether from the leaf and stem surface of *Tillandsia usneoides*, *Z. Naturforsch.*, 47c, 638, 1992.

407. El-Lakany, A.M. et al., New methoxylated flavonols from *Inula crithmoides* L., *Pharmazie*, 51, 435, 1996.

408. Lichius, J.J. et al., Antimitotic and cytotoxic flavonols from *Zieridium pseudobtusifolium* and *Acronychia porteri*, *J. Nat. Prod.*, 57, 1012, 1994.

409. Kinoshita, T., A new flavonol from *Murraya paniculata* var. *omphalocarpa*: ^{13}C-NMR as a useful tool for structure elucidation of polyoxyflavones, *Heterocycles*, 50, 269, 1999.

410. Shi, Q. et al., Antitumor agents. 54. Cytotoxic and antimitotic flavonols from *Polanisia dodecandra*, *J. Nat. Prod.*, 58, 475, 1995.

411. Itoigawa, M., Takeya, K., and Furukawa, H., Cardiotonic flavonoids from *Citrus* plants (Rutaceae), *Biol. Pharm. Bull.*, 17, 1519, 1994.

412. Muschietti, L. et al., 5,7,5′-Trihydroxy-3,6,2′,4′-tetramethoxyflavone from *Eupatorium buniifolium*, *Phytochemistry*, 36, 1085, 1994.

413. Muschietti, L. et al., 2′-oxygenated flavonoids from *Eupatorium buniifolium*, *Planta Med.*, 59, suppl., 1993.

414. Lin, J.-K. and Wu, T.-S., Constituents of flowers of *Murraya paniculata*, *J. Chin. Chem. Soc.*, 41, 213, 1994.

415. Basnet, P. et al., Five new C-methyl flavonoids, the potent aldose inhibitors from *Matteucia orientalis*, *Chem. Pharm. Bull.*, 43, 1558, 1995.

416. Häberlein, H. and Tschiersch, K.-P., On the occurrence of methylated and methoxylated flavonoids in *Leptospermum scoparium*, *Biochem. Syst. Ecol.*, 26, 97, 1998.

417. Häberlein, H. and Tschiersch, K.-P., Triterpenoids and flavonoids from *Leptospermum scoparium*, *Phytochemistry*, 35, 765, 1994.

418. Kokpol, U. et al., A C-methylflavone from *Trianthema portulacastrum*, *Phytochemistry*, 44, 719, 1997.

419. Wasowski, C. et al., Isolation and identification of 6-methylapigenin, a comparative ligand for the brain GABA-A receptors, from *Valeriana wallichii*, *Planta Med.*, 68, 932, 2002.

420. Hauteville, M. et al., Synthesis of 5-hydroxy-6- and 8-methylflavones and their ultraviolet spectral differentiation, *Phytochemistry*, 48, 547, 1998.

421. Youssef, D.T.A., Ramadan, M.A., and Khalif, A.A., Acetophenones, a chalcone, a chromone and flavonoids from *Pancratium maritimum*, *Phytochemistry*, 49, 2579, 1998.

422. Rao, J.R. and Rao, R.S., Syzalterin, a new 6,8-diC-methylflavone from *Syzygium alternifolium* leaves, *Indian J. Chem.*, 30B, 66, 1991.

423. Sarker, S.D. et al., Sideroxylin and 8-demethylsideroxylin from *Eucalyptus saligna* (Myrtaceae), *Biochem. Syst. Ecol.*, 29, 759, 2001.

424. Wollenweber, E., Mann, K., and Roitman, J.R., C-methyl flavones from the leaf wax of *Leptospermum laevigatum* (Myrtaceae), *Z. Naturforsch.*, 51c, 8, 1996.

425. Huq, F. and Misra, L.N., An alkenol and C-methylated flavones from *Callistemon lanceolatus* leaves, *Planta Med.*, 63, 369, 1997.

426. Milovanovic, M. et al., Chemical constituents of *Salvia nemorosa* L. and its antioxidant effect in lard, *J. Serb. Chem. Soc.*, 61, 423, 1996.

427. Hwang, B.Y. et al., Antimicrobial constituents from goldenseal (the rhizomes of *Hydrastis canadensis*) against selected oral pathogens, *Planta Med.*, 69, 623, 2003.

428. Sritularek, B. et al., Flavonoids from the roots of *Millettia erythrocalyx*, *Phytochemistry*, 61, 943, 2002.

429. Phrutivorapongkul, A. et al., Studies on the chemical constituents of stem bark of *Millettia leucantha*: isolation of new chalcones with cytotoxic, antiherpes simplex virus and anti-inflammatory activities, *Chem. Pharm. Bull.*, 51, 187, 2003.

430. Mukherjee, K.S. et al., A methylenedioxy flavone from *Limnophila indica*, *Phytochemistry*, 49, 2533, 1998.

431. Vázquez, M.M., Amaro, A.R., and Joseph-Nathan, P., Three new flavonoids from *Ageratum tomentosum* var. *bracteatum*, *Phytochemistry*, 27, 3706, 1988.

432. Daskiewicz, J.-B., Bayet, C., and Barron, D., Regioselective syntheses of 6-(1,1-dimethylallyl)- and 8-(3,3-diimethylallyl) chrysins, *Tetrahedron*, 58, 3589, 2002.

433. Takashima, J. and Ohsaki, A., Brosimacutins A-I, nine new flavonoids from *Brosimum acutifolium*, *J. Nat. Prod.*, 65, 1843, 2002.

434. Gulácsi, K. et al., A short and facile synthetic route to prenylated flavones. Cyclodehydrogenation of prenylated 2'-hydroxychalcones by a hypervalent iodine reagent, *Tetrahedron*, 54, 13867, 1998.

435. Chang, C.H. et al., Flavonoids and a prenylated xanthone from *Cudrania cochinchinensis* var. *gerontogea*, *Phytochemistry*, 40, 945, 1995.

436. Ngadjui, B.T. et al., Prenylated and geranylated chalcones and flavones from the aerial parts of *Dorstenia ciliata*, *Bull. Chem. Soc. Ethiop.*, 16, 157, 2002.

437. Abegaz, B.M. et al., Prenylated chalcones and flavanones from the leaves of *Dorstenia kameruniana*, *Phytochemistry*, 49, 1147, 1998.

438. Delle Monache, G. et al., Two isoflavones and a flavone from the fruits of *Maclura pomifera*, *Phytochemistry*, 37, 893, 1994.

439. Ngadjui, B.T. et al., Dinklagins A, B and C: three prenylated flavonoids and other constituents from the twigs of *Dorstenia dinklagei*, *Phytochemistry*, 61, 99, 2002.

440. Lee, S.-J. et al., Prenylated flavonoids from *Maclura pomifera*, *Phytochemistry*, 49, 2573, 1998.

441. Tsopmo, A. et al., Geranylated flavonoids from *Dorstenia poinsettifolia*, *Phytochemistry*, 48, 345, 1998.

442. Raguenet, H., Barron, D., and Mariotte, A.-M., Total synthesis of 8-(1,1-dimethylallyl)-apigenin, *Heterocycles*, 43, 277, 1996.

443. Pistelli, L. et al., Flavonoids from *Genista ephedroides*, *J. Nat. Prod.*, 61, 1404, 1998.

444. Chen, C.-C. et al., New prenylflavones from the leaves of *Epimedium sagittatum*, *J. Nat. Prod.*, 59, 412, 1996.

445. Du, J. et al., Antiviral flavonoids from the root bark of *Morus alba* L., *Phytochemistry*, 62, 1235, 2003.

446. Aida, M. et al., Artonins Q, R, S, T, and U, five new isoprenylated phenols from the bark of *Artocarpus heterophyllus*, *Heterocycles*, 39, 847, 1994.

447. Kijjoa, A. et al., Prenylflavonoids from *Artocarpus elasticus*, *Phytochemistry*, 43, 691, 1996.

448. Sato, M. et al., Flavones with antibacterial activity against carcinogenic bacteria, *J. Ethnopharmacol.*, 54, 171, 1996.

449. Hano, Y. et al. Cudraflavones C and D, two new prenylflavones from the root bark of *Cudrania tricuspidata*, *Heterocycles*, 31, 1339, 1990.

450. Ko, H.-H. et al., Bioactive constiuents of *Morus australis* and *Broussonetia papyrifera*, *J. Nat. Prod.*, 60, 1008, 1997.

451. Chan, S.-C., Ko, H.-H., and Liu, C.-N., New prenylflavonoids from *Artocarpus communis*, *J. Nat. Prod.*, 66, 427, 2003.

452. Chung, M.I. et al., Prenylflavonoids of *Artocarpus heterophyllus*, *Phytochemistry*, 40, 1279, 1995.

453. Ngadjui, B.T. et al., Dorsilurins C, D and E, three prenylated flavonoids from the roots of *Dorstenia psilurus*, *Phytochemistry*, 52, 731, 1999.

454. Ngadjui, B.T. et al., Prenylated flavones and phenylpropanoid derivatives from roots of *Dorstenia psilurus*, *Phytochemistry*, 48, 733, 1998.

455. Lin, C.N., Lu, C.-M., and Huang, P.L., Flavonoids from *Artocarpus heterophyllus*, *Phytochemistry*, 39, 1447, 1995.

456. Cunha, M.P.S., Pinto, A.C., and Braz-filho, R., Two flavonoids from *Clarisia racemosa, J. Braz. Chem. Soc.*, 5, 101, 1994.

457. Fukai, T. et al., Structures of five new prenylated flavonoids, gancaonins L, M, N, O and P from aerial parts of *Glycyrrhiza uralensis, Heterocycles*, 31, 373, 1990.

458. Li, W.K., Zhang, R.Y., and Xiao, P.G., Epimedokorenin B and Epimedokoreanin C form the aerial parts of *Epimedium koreanum, Acta Pharm. Sin.*, 29, 835, 1994.

459. Ngadjui, B.T. et al., Prenylated flavonoids from the aerial parts of *Dorstenia mannii, Phytochemistry*, 55, 915, 2000.

460. Syah, Y.M. et al., Artoindonesianins Q-T, four isoprenylated flavones from *Artocarpus champeden* Spreng. (Moraceae), *Phytochemistry*, 61, 949, 2002.

461. Sritularek, B. et al., New flavones from *Millettia erythrocalyx, J. Nat. Prod.*, 65, 589, 2002.

462. Murakami, T. et al., Chemische Untersuchungen über die Inhaltsstoffe von *Helminthostachys zeylanica* II, *Chem. Pharm. Bull.*, 21, 1851, 1973.

463. Garcez, F.R. et al., Prenylated flavonoids as evolutionary indicators in the genus *Dahlstedtia, Phytochemistry*, 27, 1079, 1988.

464. Magalhães, A.F. et al., Twenty-three flavonoids from *Lonchocarpus subglaucescens, Phytochemistry*, 42, 1459, 1996.

465. Camele, G. et al., Three new flavonoids from *Thephrosia praecans, Phytochemistry*, 19, 707, 1980.

466. Andrei, C.C. et al., C-prenylflavonoids from roots of *Tephrosia tunicata, Phytochemistry*, 55, 799, 2000.

467. Prasad, K.J., Periasamy, P.A., and Vijayalakshmi, C.S., A facile synthesis of isopongaflavone, atalantoflavone dimethylether, racemoflavone dimethylether, and methylene dioxy isopongaflavone, *J. Nat. Prod.*, 56, 208, 1993.

468. Carcache-Blanco, E.J. et al., Constituents of the stem bark of *Pongamia pinnata* with the potential to induce quinone reductase, *J. Nat. Prod.*, 66, 1197, 2003.

469. Borges-Argáez, R., Pena-Rodriguez, K.M., and Waterman, P.G., Flavonoids from two *Lonchocarpus* species of the Yucatan peninsula, *Phytochemistry*, 60, 533, 2002.

470. Saraswathy, A. et al., Carpachromene from *Atalantia monophylla, Fitoterapia*, 69, 463, 1998.

471. Chen, C.-C., Huang, Y.-L., and Ou, J.-C., Three new prenylflavones from *Artocarpus altilis, J. Nat. Prod.*, 56, 1594, 1993.

472. Lin, C.-N. and Shieh, W.-L., Pyranoflavonoids from *Artocarpus communis, Phytochemistry*, 31, 2922, 1992.

473. Delle Monache, G. et al., Comparison between metabolite products in cell culture and whole plant of *Maclura pomifera, Phytochemistry*, 39, 575, 1995.

474. Achmad, S.A. et al., A new prenylated flavone from *Artocarpus champeden, J. Nat. Prod.*, 59, 878, 1996.

475. Souza, J.P.I. et al., Prenylated flavones from *Neoraputia paraensis, Phytochemistry*, 52, 1705, 1999.

476. Arruda, A.C. et al., Two pyranoflavones from *Neoraputia alba, Phytochemistry*, 30, 3157, 1991.

477. Ferrari, F., Messana, I., and De Araujo, M.d.C.M., Structures of three new flavone derivatives, Brosimones G, H, and I, from *Brosimopsis oblongifolia, Planta Med.*, 55, 70, 1989.

478. Ganapaty, S. et al., Flavonoids from *Milletti peguensis* Ali (Fabaceae), *Biochem. Syst. Ecol.*, 26, 125, 1998.

479. Ahmad, V.U. et al., Flavonoids of *Tephrosia purpurea, Fitoterapia*, 70, 443, 1999.

480. Malik, S.B., Sharma, P., and Seshadri, T., Furanoflavonoids from leaves of *Pongamia glabra, Indian J. Chem.*, 15B, 536, 1977.

481. Pathak, V.P., Saini, T.R., and Khanna, R.N., Isopongachromene A, a chromenoflavone from *Pongamia glabra* seeds, *Phytochemistry*, 22, 308, 1983.

482. Ganguly A. and Bhattacharyya, A., Pongone: a new furanoflavone from the flowers of *Pongamia glabra, Planta Med.*, 54, 90, 1988.

483. Lyra, D.A. et al., Flavonoids from *Derris mollis, Gazz. Chim. Ital.*, 109, 93, 1979.

484. Li, W. K., Xiao, P.-G., and Zhang, R.-Y., A difuranoflavone from *Epimedium koreanum, Phytochemistry*, 38, 807, 1995.

485. Kijjoa, A. et al., Further prenylated flavonoids from *Artocarpus elasticus, Phytochemistry*, 47, 875, 1998.

486. Venkata Rao, E., Venkataratnam, G., and Vilain, C., Flavonoids from *Tephrosia fulvinervis*, *Phytochemistry*, 24, 2427, 1985.

487. Venkataratnam, G., Venkata Rao, E., and Vilain, C., Fulvinervin C, a flavone from *Tephrosia fulvinervis*, *Phytochemistry*, 25, 1507, 1986.

488. Mizuno, M. et al., Four flavonoids in the roots of *Euchresta formosana*, *Phytochemistry*, 30, 3095, 1991.

489. Lin, Y.-L., Chen, Y.-L., and Kuo, Y.-H., Three new flavonoids, 3′-methoxylupinifolin, laxifolin, and isolaxifolin from the roots of *Derris laxiflora*, *Chem. Pharm. Bull.*, 39, 3132, 1991.

490. Lin, Y.-L., Chen, Y.-L., and Kuo, Y.-H., Two new flavanones and two new chalcones from the root of *Derris laxiflora*, *Chem. Pharm. Bull.*, 40, 2295, 1992.

491. Hano, Y. et al., Components of the root bark of *Morus insignins*. 3. Structures of three new isoprenylated xanthones morusignins I, J and K and an isoprenylated flavone morusignin L., *Heterocycles*, 36, 1359, 1993.

492. Aida, M. et al., Artonols A, B, C, D, and E, five new isoprenylated phenols from the bark of *Artocarpus communis*, *Heterocycles*, 45, 163, 1997.

493. Seo, E.-K. et al., Bioactive prenylated flavonoids from the stem bark of *Artocarpus kemando*, *Arch. Pharm. Res.*, 26, 124, 2003.

494. Fujimoto, Y. et al., New flavones from *Artocarpus communis*, *Chem. Pharm. Bull.*, 38, 1787, 1990.

495. Hano, Y. and Nomura T., Constituents of the Chinese crude drug "Sang-bai-pi" (*Morus* root barks), IV. Structures of four new flavonoids, Sanggenon H, I, J and K, *Heterocycles*, 20, 1971, 1983.

496. Ngadjui, B.T. et al., Hosloppin, a new pyrone-substituted flavonoid from *Hoslundia opposita*, *J. Nat. Prod.*, 58, 109, 1995.

497. Sarker, S.D. et al., 5-O-methylhoslundin: an unusual flavonoid from *Bidens pilosa* (Asteraceae), *Biochem. Syst. Ecol.*, 28, 591, 2000.

498. Waterman, P.G. and Khalid, S.A., The major flavonoids of the seed of *Tephrosia apollinea*, *Phytochemistry*, 19, 909, 1980.

499. Vleggaar, R. et al. Flavonoids from *Tephrosia* — XI. The structure of glabratephrin, *Tetrahedron*, 34, 1405, 1978.

500. Pelter, A. et al., 8-Substituted flavonoids and 3′-substituted 7-oxygenated chalcones from *Tephrosia purpurea*, *J. Chem. Soc. Perkin Trans. 1*, 2491, 1981.

501. Ahmad, S., Natural occurrence of *Tephrosia* flavones, *Phytochemistry*, 25, 955, 1986.

502. Jonathan, L.T. et al., Pseudosemiglabrin, a platelet aggregation inhibitor from *Tephrosia semiglabra*, *J. Nat. Prod.*, 53, 1572, 1990.

503. Gómez-Garibay, F. et al., Euantiomultijugin, a flavone from *Tephrosia vicioides*, *Phytochemistry*, 31, 2925, 1992.

504. Vleggaar, R., Smalberger, T.M., and Van den Berg, A.J., Flavonoids from *Tephrosia*, IX. The structure of multijugin and multijuginol, *Tetrahedron*, 31, 2571, 1975.

505. Prabhakar, P. et al., Hookerianin: a flavone from *Tephrosia hookeriana*, *Phytochemistry*, 43, 315, 1996.

506. Hakim, E.H. et al., Artoindonesianin P, a novel prenylated flavone with cytotoxic activity from *Artocarpus lanceifolius*, *Fitoterapia*, 73, 668, 2002.

507. Aida, M., Constituents of the Moraceae plants. Part 16. Artonins J, K, and L, three new isoprenylated flavones from the root bark of *Artocarpus heterophyllus* Lamk, *Heterocycles*, 36, 575, 1993.

508. Shieh, W.-L. and Lin, C.-N., A quinoid pyranobenzoxanthone and pyranodihydrobenzoxanthone from *Artocarpus communis*, *Phytochemistry*, 31, 364, 1992.

509. Sultanbawa, M.U.S. and Surendrakumar, S., Two pyronodihydrobenzoxanthones from *Artocarpus nobilis*, *Phytochemistry*, 28, 599, 1989.

510. Hakim, E.H. et al., Artoindosenianins A and B, two new prenylated flavones from the root of *Artocarpus champede*, *J. Nat. Prod.*, 62, 613, 1999.

511. Lin, C.-N. and Shieh, W.-L., Prenylflavonoids and a pyronodihydrobenzoxanthone from *Artocarpus communis*, *Phytochemistry*, 30, 1669, 1991.

512. Hano, Y. et al., Artonins E and F, two new prenylflavones from the bark of *Artocarpus communis*, *Heterocycles*, 31, 877, 1990.

513. Cidade, H.M. et al., Artelastocarpin and carpelastofuran, two new flavones, and cytotoxicities of prenyl flavonoids from *Artocarpus elasticus* against three cancer cell lines, *Planta Med.*, 67, 867, 2001.

514. Ibewuike, J.C. et al., Piliostigmin, a 2-phenoxychrome, and C-methylflavonol from *Piliostigma thonningii*, *Phytochemistry*, 43, 687, 1996.

515. Harborne, J.B. et al., Variations in the lipophilic and vacuolar flavonoids of the genus *Vellozia*, *Phytochemistry*, 35, 1475, 1994.

516. Williams, C.A. et al., Occurrence of C-methylflavonols in leaves of *Vellozia*, *Phytochemistry*, 31, 555, 1992.

517. Harborne, J.B. et al., Ten isoprenylated and C-methylated flavonoids from the leaves of three *Vellozia* species, *Phytochemistry*, 34, 219, 1993.

518. Girard, C. et al., Polyoxygenated flavones from the leaves of *Comptonella microcarpa*, *J. Nat. Prod.*, 62, 1188, 1999.

519. Ferreira, E.O. and Dias, D.A., A methylenedioxyflavonol from aerial parts of *Blutaparon portulacoides*, *Phytochemistry*, 53, 145, 2000.

520. Hou, R.-S. et al., Cytotoxic flavonoids from the leaves of *Melicope triphylla*, *Phytochemistry*, 35, 271, 1994.

521. Higa, M. et al., Flavonoid constituents of *Melicope triphylla*, *Yakugaku Zasshi*, 110, 822, 1990.

522. Kaouadji, M. and Ravanal, P., Further non-polar flavonols from *Platanus acerifolia* buds, *Phytochemistry*, 29, 1348, 1990.

523. Saitoh, T., Kinoshita, T., and Shibata, S., Flavonols of licorice root, *Chem. Pharm. Bull.*, 24, 1242, 1976.

524. Hnawia, E. et al., A geranyl substituted flavonol from *Macaranga vedeliana*, *Phytochemistry*, 29, 2367, 1990.

525. Meragelman, K.M., McKee, T.C., and Boyd, M.R., Anti HIV prenylated flavonoids from *Monotes africanus*, *J. Nat. Prod.*, 64, 546, 2001.

526. Sutthivaiyakit, S. et al., Diterpenylated and prenylated flavonoids from *Macaranga denticulate*, *Tetrahedron*, 58, 3619, 2002.

527. Sun, P. et al., Studies on the constituents of *Epimedium koreanum*, *Chem. Pharm. Bull.*, 46, 355, 1998.

528. Beutler, J.A., McCall, K.L., and Boyd, M.R., A novel geranylflavone from *Macaranga schweinfurteii*, *Nat. Prod. Lett.*, 13, 29, 1999.

529. Zeng, L. et al., Four new prenylated flavonoids, glyasperins A, B, C, and D from the roots of *Glycyrrhiza aspera*, *Heterocycles*, 34, 575, 1992.

530. Tahara, S., Hashidoko, Y., and Mizutani, J., New 3-methoxyflavones in the roots of yellow Lupin (*Lupinus luteus* cv. Topaz), *Agric. Biol. Chem.*, 51, 1039, 1987.

531. Woo, E.R. et al., A new prenylated flavonol from the roots of *Sophora flavescens*, *J. Nat. Prod.*, 61, 1552, 1998.

532. Jia, S.S. et al., Isolation and identification of Gancaonin P-3′-methylether from the leaves of *Glycyrrhiza uralensis*, *Acta Pharm. Sin.*, 28, 623, 1993.

533. Jia, S.S., Ma, C.M., and Wang, J.M., Studies on flavonoid constituents isolated from the leaves of *Glycyrrhiza uralensis*, *Acta Pharm. Sin.*, 25, 758, 1990.

534. Chen, R.M. et al., Natural PTP1B inhibitors form *Broussonetia papyrifera*, *Bioorg. Med. Chem. Lett.*, 12, 3387, 2002.

535. Hufford, C.D. et al., Antimicrocial compounds from *Petalostemum purpureum*, *J. Nat. Prod.*, 56, 1878, 1993.

536. Jia, S.S. et al., The new isoprenyl flavonoids from the leaves of *Glycyrrhiza uralensis*, *Yaoxue Xuebao*, 28, 28, 1993.

537. Parsons, I.C. et al., New triterpenes and flavonoids from the leaves of *Bosistoa brassii*, *J. Nat. Prod.*, 56, 46, 1993.

538. Fraser, A.W. and Lewis, J.R., Two flavonols from *Euodia glabra*, *Phytochemistry*, 12, 1787, 1973.

539. Higa, M. et al., Flavonoid constituents of *Melicope triphylla*, *Yakugaku Zasshi*, 107, 954, 1987.

540. Broussalis, A.M. et al., Phenolic constituents of four *Achyrocline* species, *Biochem. Syst. Ecol.*, 16, 401, 1998.

541. Magalhães, A.F. et al., Flavonoids from *Lonchocarpus latifolius* roots, *Phytochemistry*, 55, 787, 2000.

542. Saxena, V.K. and Shrivastava, P., 4′-Hydroxy-3,6-dimethoxy-6″,6″-dimethylchromeno-(7,8,2″,3″)-flavone from *Citrus reticulate* cv *blanco*, *Phytochemistry*, 36, 1039, 1994.

543. Gonnet, J.-F., Kozjek, F., and Favre-Bonvin, J., Les flavonols d′*Asclepias syriaca*, *Phytochemistry*, 12, 2773, 1973.

544. Nascimento do, MC. and Mors, W.B., Flavonoids of *Derris araripensis*, *Phytochemistry*, 20, 147, 1981.

545. Matsumoto, J. et al., Components of *Broussonetia papyrifera*. I. Structures of the two new isoprenylated flavonols and two chalcone derivatives, *Chem. Pharm. Bull.*, 33, 3250, 1985.

546. Fukai, T. and Nomura, T., Revised structures of Broussoflavonols C and D, and the structure of Broussoflavonol E, *Heterocycles*, 29, 2379, 1989.

547. Van Heerden, F.R. et al., Metabolites from the purple heartwoods of the Mimosiodeae. Part 4. *Acacia fasciculifera* F. Muell ex Benth: Fasciculiferin, fasciluliferol, and the synthesis of 7-aryl- and 7-flavanyl-peltogynoids, *J. Chem. Soc. Perkin Trans. 1*, 2483, 1981.

548. Brandt, E.V. and Roux, D.G., Metabolites from the purple heartwood of Mimosoideae I, *Acacia peuce* F. Muell: the first natural 2,3-*cis*-peltogynoids, *J. Chem. Soc. Perkins Trans. 1*, 777, 1979.

549. Drewes, S.E. and Roux, D.G., Isolation of mopanin from *Colospermum mopane* and interrelation of flavonoid components of *Peltogyne* spp., *J. Chem. Soc. C*, 1407, 1967.

550. McPherson, D.D. et al., Peltogynoids and homoisoflavonoids from *Caesalpinia pulcherrima*, *Phytochemistry*, 22, 2835, 1983.

551. Urzua, A. et al., Acylated flavonoids from *Pseudognaphalium* species, *J. Nat. Prod.*, 62, 381, 1999.

552. Hano, Y., Inami, R., and Nomura, T., Components of the bark of *Artocarpus rigida*. 2. Structures of four new isoprenylated flavone derivatives artonins M, N, O, and P, *Heterocycles*, 35, 1341, 1993.

553. Iinuma, M. et al., Five phenolic compounds in the underground parts of *Vancouveria hexandra*, *Heterocycles*, 35, 407, 1993.

554. Williams, C.A. et al., Flavonoid evidence and the classification of the Anarthriaceae within the Poales, *Phytochemistry*, 45, 1189, 1997.

555. Vleggaar, R., Smalberger, T.M., and Van Aswegen, J.L., Flavonoids from *Tephrosia*. X. The structure of polystachin, *S. Afr. J. Chem.*, 31, 47, 1978.

APPENDIX

Trivial Name

Trivial Name	Substitution
Abrectorin	7,3'-diOH, 6,4'-diOMe — flavone
Acacetin	5,7-diOH, 4'-OMe — flavone
Acerosin	5,7,3'-triOH, 6,8,4'-triOMe — flavone
Ageconyflavon A	5,6,7-triOMe, 3',4'-OCH$_2$O — flavone
Ageconyflavon B	4'-OH, 5,6,7,3'-tetraOMe — flavone
Ageconyflavon C	4'-OH, 5,6,7,3',5'-OMe — flavone
Agecorynin F	5-OH, 6,7,2',3',4',5'-hexaOMe — flavone
Agecorynin G	3'-OH, 5,6,7,2',4',5'-hexaOMe — flavone
Agecorynin-C	5,6,7,8,2',4',5'-heptaOMe — flavone
Agecorynin-D	5,2',4'-triOH, 6,7,8,5'-tetraOMe — flavone
Agehoustin A	5,6,7,8,2',3',4',5'-octaOMe — flavone
Agehoustin B	5,6,7,2',3',4',5'-heptaOMe — flavone
Agehoustin C	3'-OH, 5,6,7,8,2',4',5'-heptaOMe — flavone
Agehoustin D	5,3'-diOH, 6,7,8,2',4',5'-hexaOMe — flavone
Agehoustin E	5-OH, 6,7,8,2',4',5'-hexaOMe — flavone
Agehoustin F	5,2'-diOH, 6,7,8,4',5'-pentaOMe — flavone
Albanin A	5,7,2',4'-tetraOH, 3-C5 — flavone
Albanin D, revised	5,7,4'-OH, 6-C10 — flavone
Albanin E, revised	5,7,2',4'-tetraOH, 6-C10 — flavone
Aliarin	5,7,4'-triOH, 3,6-diOMe, 3'-C5-OH — flavonol
"Allopatuletin" — obsolete	
Alluaudiol	5,7,3',4',5'-OH, 3-OMe, 6-Me — flavonol
Alnetin	5-OH, 6,7,8-triOMe — flavone
Alnusin	3,5,7-triOH, 6-OMe — flavonol
Alnustin	5-OH, 3,6,7-triOMe — flavonol
Altisin	5-OH, 7,8,2',6'-tetraOMe — flavone
Anhydroicaritin	3,5,7-triOH, 4'-OMe, 8-C5 — flavonol
Annulatin	5,7,3',4',5'-pentaOH, 3-OMe — flavonol
Apigenin	5,7,4'-triOH — flavone
Apollinine	7-OMe, 8-furyl (2'' = oxo) — flavone, see Figure 12.5
Apometzgerin	5,7,5'-triOH, 3',4'-diOMe — flavone
Apuleidin	5,2',3'-triOH, 3,7,4'-triOMe — flavonol
Apulein	2',5'-diOH, 3,5,6,7,4'-pentaOMe — flavonol
Apuleirin	7,5'-diOH, 3,5,7,3',4'-pentaOMe — flavonol
Apuleisin	5,6,2',3'-tetraOH, 3,7,4'-triOMe — flavonol
Apuleitrin	5,6,5'-triOH, 3,7,3',4'-tetraOMe — flavonol
Araneol	5,7-diOH, 3,6,8-triOMe — flavonol
Araneosol	5,7-diOH, 3,6,8,4'-tetraOMe — flavonol
Arcapillin	5,2',5'-triOH, 6,7,5'-triOMe — flavone
Arteanoflavon	5,7-diOH, 6,3',4',5'-tetraOMe — flavone
Artelasticin	5,7,2',4'-tetraOH, 3,6,8-triC5 — flavone
Artelastin	5,7,4'-triOH, 6,8-diC5, 3,6'-cycl-OC5 — flavone
Artelastocarpin	5,7,4'-triOH, 6,8-diC5, 3,6'-cyclo-O-C6-C3-OH — flavone, see Figure 12.6
Artelastochromene	5,4'-diOH, 8-C5, 7,6-ODmp, 3,6-cycl-OC5 — flavone
Artelastofuran	5,4'-diOH, 7,8-dihydrofurODmp-OH — flavone, see Figure 12.6
Artemetin	5-OH, 3,6,7,3',4'-pentaOMe — flavonol
Artobilochromene	5,2',4',5'-tetraOH, 3'-C5, 7,6-ODmp — flavone

continued

APPENDIX

Trivial Name — *continued*

Trivial Name	Substitution
Artobiloxanthone	5,2′,4′,5′-tetraOH, 7,8-ODmp, 3,6′-cyclo-C6-C3 — flavone
Artocarpesin	5,7,2′,4′-tetraOH, 6-C5 — flavone
Artocarpetin	5,2′,4′-triOH, 7-OMe — flavone
Artocarpetin A	5,2′,4′-triOH,7-OMe, 8-C5 — flavone
Artocarpetin B	5,4′-diOH, 7,2′-OMe, 8-C5 — flavone
Artocarpin	5,2′,4′-triOH, 7-OMe, 3,6-diC5 — flavone
Artocommunol CA	5-OH,4′OMe, 7,8-ODmp, C5-O-C5; 3,6′-cycl-O-C5 — flavone, see Figure 12.6
Artocommunol CC	5,4′-diOH, 3-C5-OH, 7,8-ODmp — flavone
Artocommunol CD	5,7,2′,4′-tetraOH, 8-C5, 3-C10 — flavone
Artoindonesianin A	5,2′,4′-triOH, 8-C10, 7,6-ODmp-3,6′ cyclo-C6-diMe-fur — flavone
Artoindonesianin B	5-OH, 7OMe, 6-C5, 3,6′-cyclo O-C6-C3 — flavone
Artoindonesianin P	5,7,2′,4′-tetraOH, 3,6′cyclo-C6-diMe-fur — flavone
Artoindonesianin Q	5,2′,5′-OH, 7,4′-OMe, 3-C5 — flavone
Artoindonesianin R	5,7,5′, OH-2′,4′-OMe, 3-C5 — flavone
Artoindonesianin S	5,2′,5′-triOH, 7,4′-diOMe, 3,6-cyclo C6-C3 — flavone
Artoindonesianin T	5,7,2′,5′-tetraOH, 4′-OMe, 3,6-cyclo C6-C3 — flavone
Artomunoxanthentrione	5-OH, 4′-OMe, 7,8-ODmp; 3,6′-cycloC6-C3 — flavone
Artomunoxanthone	5,2′,5′-triOH, 4′-OMe, 7,8-ODmp, 3,6′-cyclo C6 — flavone
Artonin E ("KB-3")	5,2′4′,5′-tetraOH, 3-C5, 7,8-ODmp — flavone
Artonin F	5,2′-diOH, 4′-OMe, 6-C5;7,8-ODmp, 3,6′-cyclo C6-5′-fur — flavone
Artonin J	5,7,2′,4′-tetraOH, 3,6′-cycloC6-5′-fur; 4′-C5 — flavone
Artonin K	5,2′,4′-triOH, 7-OMe, 3,6-cyclo C6-5′-fur — flavone, see Figure 12.6
Artonin L	5,4′-diOH, 7,2′-diOMe, 3,6-cyclo C6-5′-fur — flavone
Artonin M	5,2′,4′-triOH, 7,6-ODmp; 3,6′-cyclo C6-5′-fur — flavone
Artonin N	5,7,2′,4′-tetraOH, 6-C5, 3′,4′-ODmp, 3,6′-cycloC6-C3 — flavone
Artonin O	5,7,4′-triOH, 3′,6′-di-oxo, 6,5′-C5; 3,6′-cycloC6-C3 — flavone
Artonin P	5,4′-dOH, 2′,5′-di-oxo, 7,8-ODmp; 3,6′-cycloC6-C3, 2′,5′-epoxy — flavone
Artonin S	5,4′-diOH, 7-OMe, 6-C5, 3,6-cyclo O-C6-C3 — flavone
Artonin T	5,2′,4′-triOH, 7-OMe, 3,6-cycloC6-5′-fur; 4′-C5 — flavone
Artonin U	5,4′-diOH, 7-OMe, 8-C5 — flavone
Artonol C	5,2′,5′-triOH, 7,8-ODmp, 3′,4′-ODmp, 3,6′-cyclo-C6-C3 — flavone
Artonol D	5-OH, 2′,5′-di-oxo, 7,8-ODmp, 3′,4′-ODmp, 3,6′-cyclo-C6-C3 — flavone, see Figure 12.6
Artonol E	5,2′,5′-triOH, 7-OMe, 3′,4′-ODmp, 3,6′-cyclo-C6-C3 — flavone, see Figure 12.6
Asplenetin	5,7,3′,4′,5′-pentaOH, 3-C5 — flavone
Atalantoflavon	5,4′-diOH, 7,8-ODmp — flavone
Auranetin	3,6,7,8,4′-pentaOMe — flavonol
Australon A	5,2′,4′-triOH, 7,6-ODmp-C5 — flavone
Axillarin	5,7,3′,4′-tetraOH, 3,6-diOMe — flavonol
Ayanin	5,3′-diOH, 3,7,4′-triOMe — flavonol
Azaleatin	3,7,3′,4′-tetraOH, 5-OMe — flavonol
Baicalein	5,6,7-triOH — flavone
Baohuosu	5,7,4′-OH. 3′,5′-OMe, 8-C5 — flavone
Benthamianin	5,7,3′,4′-tetraOH, 6OMe, 3,2′-O-CH2 — flavonol
Betuletol	3,5,7-triOH, 6,4′-diOMe — flavonol
Bonanzin	5,7-diOH, 3,6,3′,4′-tetraOMe — flavonol
Brevicornin	3,5,7-triOH, 4′-OMe, 8-C5-OMe — flavone
Brickellin (revised)	5,2′-diOH, 3,6,7,4′,5′-pentaOMe — flavonol
Brosimacutin F	7,4′-diOH, 8-C5(OH)2 — flavone

APPENDIX

Trivial Name — *continued*

Trivial Name	Substitution
Brosimone G	5,2',4'-triOH, 7,8-ODmp-C5 — flavone
Brosimone H	5,2',4'-triOH, 7-OMe, 3-C5, 8-C10 — flavone
Brosimone I	5,7,4'-triOH, 6-C5, 3,6'-cycl-O-C5 — flavone
Broussoflavonol A	5,3',4'-OH, 3-OMe, 8-C5, 7,6-ODmp — flavonol
Broussoflavonol B	5,7,3',4'-OH, 3-OMe, 6,8-diC5 — flavonol
Broussoflavonol C	3,5,7,3',4'-pentaOH, 8,2',6'-triC5 — flavonol
Broussoflavonol D	3,5,7,5'-tetraOH, 8,2'-diC5, 3',4'-ODmp — flavonol, see Figure 12.9
Broussoflavonol E	3,5,7,5'-tetraOH, 8,2'-diC5, 3',4'-ODmp — flavonol
Broussoflavonol F	3,5,7,4'-tetraOH, 8,3'-diC5 — flavonol
Broussoflavonol G	3,5,7,3',4'-pentaOH, 8,2',3'-triC5 — flavonol
Broussonol A	3,5,7,3'tetraOH, 4',5'ODmp-8-C5 — flavonol
Broussonol B	3,5,3'triOH, 4',5'-ODmp, 7,8-triMe-fur — flavonol
Broussonol C	3,5,3',4'-tetraOH, 5'-C5; 7,8-triMe-fur — flavonol
Broussonol D	3,5,7,3',4'-pentaOH, 8,5'-diC5 — flavonol
Broussonol E	3,5,7,3',4'-pentaOH, 6,5'-diC5 — flavonol
Bucegin	5,7-diOH, 8,4'-diOMe — flavone
Calomelanol D	3,5,4'-triOH, 7,8-O-cycl-phenylethyl — flavonol
Calycopterin	5,4'-diOH, 3,6,7,8-tetraOMe — flavonol
Candidin (syn.: Isopongaflavone)	
Candidol	3,4'-diOH, 5,6,7-triOMe — flavonol
"Candiron" — obsolete	
Cannflavin A	5,7,4'-triOH, 3'-OMe, 6-C10 — flavone
Cannflavin B	5,3'-diOH, 7,4'-diOMe, 6-C5 — flavone
Carajuflavone	6,7,3',4'-tetraOH, 5-OMe — flavone
Carpachromene	5,4'-diOH, 7,6-ODmp — flavone
Carpelastofuran	5,4'diOH, 8-C5-7,6-fur-C3-OH, 3,6'-cyclo-O-C6-C3-OH — flavone
Caryatin	7,3',4'-triOH, 3,5-diOMe — flavonol
Casticin	5,3'-diOH, 3,6,7,4'-tetraOMe — flavonol
Centaureidin	5,7,3'-triOH, 3,6,4'-triOMe — flavonol
Cerosillin	5,6,3',5'-tetraOMe — flavone
Cerosillin B	5,6,3',4',5'-pentaOMe — flavone
Chlorflavonin	5,2'-diOH, 3,7,8-triOMe, 3'-chloro — flavonol
6-Chloroapigenin	5,7,4'-triOH, 6-chloro — flavone
Chrysin	5,7-diOH — flavon
Chrysoeriol	5,7,4'-triOH, 3'-OMe — flavone
Chrysosplenetin	5,4'-diOH, 3,6,7,3'-tetraOMe — flavonol
Chrysosplenol-C	5,6,4'-triOH, 3,7,3'-triOMe — flavonol
Chrysosplenol-D	5,3',4'-triOH, 3,6,7-triOMe — flavonol
Chrysosplin	5,4'-diOH, 3,6,7,2'-tetraOMe — flavonol
Ciliatin A	5,4'-diOH, 7,6-dihydrofur-C3 — flavone, see Figure 12.3
Ciliatin B	7,4'-diOH, 3'-OMe, 5,6-ODmp — flavone
Cirsilineol	5,4'-diOH, 6,7,3'-triOMe — flavone
Cirsiliol	5,3',4'-triOH, 6,7-diOMe — flavone
Cirsimaritin	5,4'-diOH, 6,7-diOMe — flavone
Citrusinol	3,5,4'-triOH, 7,8-ODmp — flavonol
Combretol	5-OH, 3,7,3'4'5'-pentaOMe — flavonol

continued

APPENDIX

Trivial Name — *continued*

Trivial Name	Substitution
Conyzatin	5,7-diOH, 3,8,3',4',5'-pentaOMe — flavonol
Corymbosin	5-OH, 7,3',4',5'-tetraOMe — flavone
Cudraflavone C	5,7,2',4'-tetraOH, 3,6-diC5 — flavone
Cudraflavone D	5,7,2',4'-tetraOH, 6,5'-diC5 — flavone
Cycloaltilisin	5,7,5'-triOH,4'-OMe, 3,6'-cyclo C6-C3; 6-C5 — flavone
Cycloartobiloxanthone	5,2',4',triOH, 7,8-ODmp, 3,6'cyclo-C6-diMe-fur — flavone
Cycloartocarpesin	5,2',4'-triOH, 7,6-ODmp — flavone
Cycloartocarpin	5,4'-diOH, 7-OMe, 6-C5, 3,6'-cycl-O-C5 — flavone
Cycloartomunin	5,5'-diOH, 4'-OMe, 7,8-ODmp, 3,6'-cycl-O-C5 — flavone
Cycloartomunoxanthone	5,2'-diOH, 4'-OMe, 7,8-ODmp, 3,6'-cyclo C6-5'-fur — flavone
Cyclochampedol	5,7,3',4'-tetraOH, 2',3-ODmp — flavone
Cyclocommunin (syn.: Isocyclomulberrin)	5,7,4'-triOH, 6-C5, 3,6'-cycl-O-C5 — flavone
Cyclocommunol	5,7,4'-triOH, 3,6'-ODmp — flavone
Cycloheterophyllin	5,4',5'-triOH, 8-C5, 7,6-ODmp, 3,6'-cycl-O-C5 — flavone
Cyclointegrin	5,4'-diOH, 7-OMe, 3,6'-cyclo O-C7 D202 — flavone
Cyclomorusin (syn.: Cyclomulberrochromene)	5,4'-diOH, 7,8-/3,6'-ODmp — flavone
Cyclomulberrin	5,7,4'-triOH, 8-C5, 2',3-ODmp — flavon
Cyclomulberrochromene (syn.: Cyclomorusin)	
Dasytrichone	5-OH, 6-Me, 8-diMe, 7 = O — flavone
Datin	3,5,2'-triOH, 7-OMe — flavonol
Datiscetin	3,5,7,2'-tetraOH — flavonol
Demethyltorosaflavone C	5,3',4'-triOH, 7,6- bisfurano — flavone, see Figure 12.3
Demethyltorosaflavone D	5,7,3',4'-tetraOH, 6-acrylic acid — flavone, see Figure 12.1
Denticulaflavonol	3,5,7,4'-tetraOH, diterpene-flavonol — flavonol, see Figure 12.9
Desmethylcentaureidin	5,7,3'-triOH, 6,4'-diOMe — flavone
Desmethylsudachitin	5,7,4'-triOH, 6,8-diOMe — flavone
Desmethylanhydroicaritin	3,5,4'-triOH, 7,6-ODmp — flavonol
Desmethyldigicitrin	3,5,3'-triOH, 6,7,8,4',5'-pentaOMe — flavonol
8-Desmethyleucalyptin	5-OH, 7,4'-diOMe, 6-Me — flavone
8-Desmethylkalmiatin	5-OH, 3,7,4'-triOMe, 6-Me — flavonol
Desmethylkanugin	3,7-diOMe, 3',4'-OCH2O — flavonol
8-Desmethyllatifolin	5,4'-diOH, 3,7-diOMe, 6-Me — flavonol
5-Desmethylmelibentin	5-OH, 3,6,7,8-tetraOMe, 3',4'-O2CH2 — flavonol
5-Desmethylmeliternin	5-OH, 3,7,8-triOMe, 3',4'-OCH2O — flavonol
5-Desmethylnobiletin	5-OH, 6,7,8,3',4'-pentaOMe — flavone
8-Desmethylsideroxylin	5,4'-diOH, 7-OMe, 6-Me — flavone
Desmosflavone	5-OH, 7-OMe, 6,8-diMe — flavone
Digicitrin	5,3'-diOH, 3,6,7,8,4',5'-hexaOMe — flavonol
Dihydrofuranoartobilichromene a	5,3',4'-triOH, 7,6-ODmp, 6',5'-dihydrofur-2''-C3 — flavone
Dihydrofuranoartobilichromene b1	5,3',6'-triOH, 7,6-ODmp, 4',5'-dihydrofur-2''-C3 — flavone
Dihydrofuranoartobilichromene b2	5,3',6'-triOH, 7,6-ODmp, 4',5'-dihydrofur-2''-C3 — flavone
Dillenetin	3,5,7-triOH, 3',4'-diOMe — flavonol
5,6-Dimethoxypongapin	3,5,6-triOMe, 3',4'-O2CH2, 7,8-fur — flavonol
Dihydroisocycloartomunin	5,3',4'-triOH,7-OMe, 8-C5, 3,6'-cycl.-O-C5-ODmp — flavone
8-(1,1-Dimethylallyl)-galangin	3,5,7-triOH, 8-C5 — flavonol
8-(1,1-Dimethylallyl)-kaempferol	3,5,7,4'-tetraOH, 8-C5 — flavonol
6,8-Dimethylapigenin	5,7,4'-triOH, 6,8-diMe — flavone
Diosmetin	5,7,3'-triOH, 4'-OMe — flavone

APPENDIX

Trivial Name — *continued*

Trivial Name	Substitution
6,8-Diprenylkaempferol	see Glyasperin A
Dinklagin B	5,4'-diOMe, 7,6-ODmp-OH — flavone
Distemonanthin	5,7,3',4'-tetraOH, 6-OMe, 3,2'-O-CH2 — flavonol
Dorsilurin A	5,7,2',4'-tetraOH, 6,8,3'-triC5 — flavone
Dorsilurin B	5,2',4'-triOH, 3,6-diC5, 7,8-ODmpOH — flavone
Dorsilurin C	3,5,7-triOH, 6,8-diC5-4',3'-ODMp — flavonol
Dorsilurin D	5,7,2',4'-tetraOH, 3,6,8-triC5 — flavone
Dorsilurin E	2'-OH, 4,3/6,5/7,8-triODmp, 4' = O . . . — flavone, see Figure 12.2
Dorsmanin C	3,5,3',4'-tetraOH, 7,8-ODmp-6-C10 — flavonol
Dorsmanin D	3,5,7,4'-tetraOH, 3'-OMe, 6,8-diC5 — flavonol
Dumosol	3,5,7,3',5'- pentaOH, 4'-OMe, 6-Me — flavonol
Echioidinin	5,2'-diOH, 7-OMe — flavone
Emmaosunin	5-OH, 3,6,7,8,3'-pentaOMe — flavonol
Enantiomultijugin	5-OMe, 7,8-bisfur — flavone, see Figure 12.5
Ephedroidin	5,7,4'-OH, 8-C5-OH — flavone
Epimedokoreanin A	5,3'-diOH, 7,8-dihydrofur-2''-C3/4',5'-dihfur-OH-5''-C3-OH . . . — flavone, see Figure 12.3
Epimedokoreanin B	5,7,3',4'-tetraOH, 8,5'-diC5 — flavone
7-Epoxyprenylgnaphaliin	5-OH, 3,8-diOMe, 7-O-C5 (epoxy) — flavonol
Eriostemin	3,8-diOH, 5,6,7,4'-tetraOMe — flavonol
Ermanin	5,7-diOH, 3,4'-diOMe — flavonol
Eucalyptin	5-OH, 7,4'-diOMe, 6,8-diMe — flavone
Eupalestin	5,6,7,8,5'-pentaOMe, 3',4'-OCH2O — flavone
Eupalitin	3,5,4'-triOH, 6,7-diOMe — flavonol
Eupatilin	5,7-diOH, 6,3',4'-triOMe — flavone
Eupatin	3,5,3'-triOH, 6,7,4'-triOMe — flavonol
Eupatolitin	3,5,3',4'-tetraOH, 6,7-diOMe — flavonol
Eupatoretin	3,3'-diOH, 5,6,7,4'-tetraOMe — flavonol
Eupatorin	5,3'-diOH, 6,7,4'-triOMe — flavone
Europetin	3,5,3',4',5'-pentaOH, 7-OMe — flavonol
Exoticin	3,5,6,7,8,3',4',5'-octaOMe — flavonol
Farnisin	7,3'-diOH, 4'-OMe — flavone
Fasciculiferin	7,4,5'-triOH, 3,2'-O-CH2-OH — flavonol, see Figure 12.9
Ferrugin	5,7-diOH, 3',4',5'-triOMe — flavonol
Fisetin	3,7,3',4'-tetraOH — flavonol
Flindulatin	5-OH, 3,7,8,4'-tetraOMe — flavonol
Fulvinervin B	5-OH, 6-C5, 7,8-ODmp — flavone
Fulvinervin C	5-OH, 6-C5-OH, 7,8-ODmp — flavone
Galangin	3,5,7-triOH — flavonol
Gancaonin O	5,7,3',4'-tetraOH, 6-C5 — flavone
Gancaonin P	3,5,7,3',4'-pentaOH, 6-C5 — flavonol
Gancaonin Q	5,7,4'-triOH, 6,3'-diC5 — flavone
Ganhuangenin	5,7,3',6'-tetraOH, 8,2'-diOMe — flavone
Gardenin A	5-OH, 6,7,8,3',4',5'-hexaOMe — flavone
Gardenin B	5-OH, 6,7,8,4'-tetraOMe — flavone
Gardenin C	5,3'-diOH, 6,7,8,4',5'-pentaOMe — flavone
Gardenin D	5,3'-diOH, 6,7,8,4'-tetraOMe — flavone

continued

APPENDIX

Trivial Name — *continued*

Trivial Name	Substitution
Gardenin E	5,3′,5′-triOH, 6,7,8,4′-tetraOMe — flavone
Genkwanin	5,4′-diOH, 7-OMe — flavone
Geraldol	3,7,4′-triOH, 3′-OMe — flavonol
Geraldone	7,4′-diOH, 3′-OMe — flavone
Geranioloxyalatum flavone	3,5,3′-triOH-6,7-diOMe, 4′-O-C10-OH — flavonol
Glabone	4′-OMe, 7,6-fur — flavone
Glabratephrin	7,8-bisfur — flavone
Glabratephrinol	7,8-bisfur — flavone
Glepidotin A	3,5,7-triOH, 6-C5 — flavonol
Glyasperin A	3,5,7,4′-tetraOH, 6,3′-diC5 — flavonol
Glycyrrhiza-flavonol A	3,5,7-triOH, 4′,3′-cycl-OC5-OH — flavonol
Gnaphalin	5,7-diOH, 3,8-diOMe — flavonol
Gomphrenol	3,5,4′-triOH, 6,7-OCH2O — flavonol
Gossypetin	3,5,7,8,3′,4′-hexaOH — flavonol
Grantiodinin	5-OH, 3,6,7,8,2,5′-hexa-OMe — flavonol
Grantioidin	5-OH, 3,6,7,2′5′pentaOMe — flavonol
Grantionin	7-OH, 6,3′,5′-triOMe — flavone
Haplopappin	5,7-diOH, 3,4′-diOMe, 8-C-p-OH-phenylethyl — flavonol
Herbacetin	3,5,7,8,4′-pentaOH — flavonol
Heteroartonin A	5,7,2′,5′-OH-4′, OMe, 3,3′-diC5 — flavone
Heterophyllin	5,2′,4′,5′-tetraOH, 3,8-diC5; 7,6-ODmp — flavone, see Figure 12.2
Hibiscetin	3,5,7,8,3′,4′,5′-heptaOH — flavonol
Hispidulin	5,7,4′-triOH, 6-OMe — flavone
Honyucitrin	5,7,4′-triOH, 3′,5′-diC5 — flavone
Hookerianin	5,7-diOMe, 8-diMe-oxo-furano — flavone, see Figure 5
Hosloppin	5-OH,7-OMe, 6,5″-Ketopyrano....3″-OH — flavone, see Figure 12.4
Hoslundin	5-OH, 7-OMe, 6,5″-Ketopyrano....3″-Me — flavone, see Figure 12.4
5-Hydroxyauranetin	5-OH, 3,6,7,8,4′-pentaOMe — flavonol
6-Hydroxygalangin	3,5,6,7-tetraOH — flavonol
8-Hydroxygalangin	3,5,7,8-tetraOH — flavonol
5′-Hydroxymorin	3,5,7,2′,4′,5′-hexaOH — flavonol
6-Hydroxykaempferol	3,5,6,7,4′-pentaOH — flavonol
6-Hydroxyluteolin	5,6,7,3′,4′-pentaOH — flavone
6-Hydroxymyricetin	3,5,6,7,3′,4′,5′-OH — flavonol
4′-Hydroxywogonin	5,7,4′-triOH, 8-OMe — flavone
Hymenoxin	5,7-diOH, 6,8,3′,4′-tetraOMe — flavon
Hypolaetin	5,7,8,3′,4′-pentaOH, — flavone
Icaritin	3,5,7-triOH, 4′-OMe, 8-C5-OH — flavonol
Integrin	5,2′,4′-triOH, 7-OMe, 3-C5 — flavone
Inucrithmin	3,7,4′,5′-tetraOH, 6,3′-diOMe — flavonol
Isoanhydroicaritin	3,5,4′-triOH, 7-OMe, 8-C5 — flavonol
Isoartocarpin	5,7,4′-triOH, 6,2′-diC5 — flavone
Isocyclomorusin	5,4′-diOH, 7,6-/3,6′-ODmp — flavone
Isocyclomulberrin (syn.: Cyclocommunin)	
Isoetin	5,7,2′,4′,5′-pentaOH — flavone
Isognaphalin	5,8-diOH, 3,7-diOMe — flavonol
Isokaempferide	5,7,4′-triOH, 3-OMe — flavonol
Isokanugin	3,5,7-triOMe, 3′,4′-OCH2O — flavonol

APPENDIX

Trivial Name — *continued*

Trivial Name	Substitution
Isolaxifolin	5,4'-diOH, 8-C5, 7,6-ODmp — flavone
Isolicoflavonol	3,5,7,4'-tetraOH, 3'-C5 — flavonol
Isolimocitrol	3,5,7,3'-tetraOH, 6,8,4'-triOMe — flavonol
Isomacarangin	3,5,7,4'-tetraOH, 8-C10 — flavonol
Isoplatanin	3,5,6,7-tetraOH, 8-Me — flavonol
Isopongaflavone (syn.: Candidin)	5-OMe, 7,8-ODmp — flavone
Isopongaglabol	4'-OH, 7,8-fur — flavone
Isopratol	4'-OH, 7-OMe — flavone
Isorhamnetin	3,5,7,4'-tetraOH, 3'-OMe — flavonol
Isorhynchospermin	3,5,3'-OH, 7,4'-diOMe, 6-C5+D348 — flavonol
Isoscutellarein	5,7,8,4'-tetraOH — flavone
Isosinensetin	5,7,8,3',4'-pentaOMe — flavone
Isothymusin	5,8,4'-triOH, 6,7-diOMe — flavone
Isouononal	5,7-diOH, 6-Me, 8-CHO — flavone
Izalpinin	3,5-diOH, 7-OMe — flavonol
Jaceidin	5,7,4'-triOH, 3,6,3'-triOMe — flavonol
Jaceosidin	5,7,4'-triOH, 6,3'-diOMe — flavone
Kaempferide	3,5,7-triOH, 4'-OMe — flavonol
Kaempferol	3,5,7,4'-tetraOH — flavonol
Kalmiatin	5-OH, 3,7,4'-triOMe, 6,8-diMe — flavonol
Kanjone	6-OMe, 7,8-fur — flavone
Kanugin	3,7,3'-triOMe, 3',4'-OCH2O — flavonol
Kanzakiflavon-1	5,8-diOH, 4'-OMe, 6,7-O2CH2 — flavone
Kanzakiflavon-2	5,4'-diOH, 6,7-OCH2O — flavone
Kanzonol D	7,4'-diOH, 3'-C5 — flavone
Kanzonol E	7-OH, 6-C5, 3',4'-ODmp — flavone
Karanjachromene	3-OMe, 7,8-ODmp — flavonol
Karanjin	3-OMe, 7,8-fur — flavonol
KB-1	5,2'4',5'-tetra-OH, 7,8-ODmp, 3,6'-cyclo C6 — flavone
KB-2	5,2'4',5'-tetraOH, 3-C5, 7,8-ODmp — flavone
Kumatakenin	5,4'-diOH, 3,7-diOMe — flavonol
Kushenol C	3,5,7,2',4'-pentaOH, 8-C10 — flavonol
Kushenol G	3,5,7,2',4'-pentaOH, 8-C10-OH — flavonol
Kuwanon B	5,7,2'-triOH, 3-C5, 3',4'-ODmp — flavone
Kuwanon C (syn.: Mulberrin)	
Kuwanon S	5,7,4'-triOH, 3'-C10 — flavone
Kuwanon T	5,7,2',4'-tetraOH, 3,3'-diC5 — flavone
Laciniatin	3,5,7,3'-tetraOH, 6,4'-diOMe — flavonol
Ladanein	5,6-diOH, 7,4'-diOMe — flavone
Lanceolatin A	7-OMe, 8-C5-OH — flavone
Lanceolatin B	7,8-fur — flavone
Laricitrin	3,5,7,4',5'-pentaOH, 3'-OMe — flavonol
Latifolin	5,4'-diOH, 3,7-diOMe, 6,8-diMe — flavonol
Laurentinol	3,7,4'-triOH, 3',5'diOMe — flavone
Laxifolin	5,4'-diOH, 6-C5, 7,8-ODmp — flavone
Lethedocin	5,5'-diOH, 7,3',4'-triOMe — flavone

continued

APPENDIX

Trivial Name — *continued*

Trivial Name

Trivial Name	**Substitution**
Licoflavone A	7,4'-diOH, 6-C5 — flavone
Licoflavone B	7,4'-diOH, 6,3'-diC5 — flavone
Licoflavone C	5,7,4'-triOH, 8-C5 — flavone
Licoflavonol	3,5,7,4'-tetraOH, 6-C5 — flavonol
Limocitrin	3,5,7,4'-tetraOH, 8,3'-diOMe — flavonol
Limocitrol	3,5,7,4'-tetraOH, 6,8,3'-triOMe — flavonol
Linderoflavone A	5,7-diOH, 6,8-diOMe, 3',4'-OCH2O — flavone
Linderoflavone B	5,6,7,8-tetraOMe, 3',4'-OCH2O — flavone
Luteolin	5,7,3',4'-tetraOH — flavone
Macaflavone I	3,3',4'-triOH, 7,6-ODmp, 8-C5 — flavonol
Macaflavone II	3',4'-diOH, 3-OMe, 7,6-ODmp, 8-C5 — flavonol
Macarangin	3,5,7,4'-tetraOH, 6-C10 — flavonol
Marionol	3-OH, 5,6,7,3',4'-pentaOMe — flavonol
Matteuorien	5,7-diOH, 6,8-diMe — flavon
Mearnsetin	3,5,7,3',5'-pentaOH, 4'-OMe — flavonol
Melanoxetin	3,7,8,3',4'-pentaOH — flavonol
Melibentin	3,5,6,7,8-pentaOMe, 3',4'-OCH2O — flavonol
Melicophyllin	3,5,8,3',4'-pentaOMe, 6,7-OCH2O — flavonol
Melinervin	3,5,7-triOH, 6,8-diOMe, 3',4'-OCH2O — flavonol
Melisimplexin	3,5,6,7-tetraOMe, 3',4'-OCH2O — flavonol
Melisimplin	5-OH, 3,6,7-triOMe, 3',4'-OCH2O — flavonol
Meliternatin	3,5-diOMe, 6,7-/3',4'-diOCH2O — flavonol
Meliternin	3,5,7,8-tetraOMe, 3',4'-OCH2O — flavonol
6-Methylapigenin	5,7,4'-triOH, 6-Me — flavone
8-Methylapigenin	5,7,4'-triOH, 8-Me — flavone
8-Methylgalangin	3,5,7-triOH, 8-Me — flavonol
Methylgnaphalin	5-OH, 3,7,8-triOMe — flavonol
6-Methylkaempferol	3,5,7,4'-tetraOH, 6-Me — flavonol
6-Methylluteolin	5,7,3',4'-tetraOH, 6-Me — flavone
7-Methyltricin	5,4'-diOH, 7,3',5'-triOMe — flavone
7-Methylwogonin	5-OH, 7,8-diOMe — flavone
Mikanin	3,5-diOH, 6,7,4'-triOMe — flavonol
Milletenin C	6,7-diOMe, 3',4'-OCH2O — flavone
Millettocalyxin A	7,2'diOMe, 4', 5'-OCH2O — flavone
Millettocalyxin B	7-OMe-6-OC5, 4',5'-OCH2O — flavone
Millettocalyxin C	2',5'-diOMe, 7,8-fur+D426 — flavone
Mopanin	7,3',4'-triOH, 3,2'-O-CH2 — flavonol
Moralbanone	5,7,2',4'-tetraOH, 8-C15 — flavone
Morelosin	3,5,7-triOH, 3',5'-diOMe — flavonol
Morin	3,5,7,2',4'-pentaOH — flavonol
Morusignin L	5,2',4'-triOH, 7,6-ODmp, 3-C5OH — flavone
Morusin (syn.: Mulberrochromene)	5,2',4'-triOH, 3-C5, 7,8-ODmp — flavone
Mosloflavone	5-OH, 6,7-diOMe — flavone
Moslosooflavone	5-OH, 7,8-diOMe — flavone
Mulberrin (syn.: Kuwanon C)	5,7,2',4'-tetraOH, 3,8-diC5 — flavone
Mulberrochromene (syn.: Morusin)	
Multijugin	5-OMe, 7,8-bisfur — flavone
Multijuginol	5-OMe, 7,8-bisfur — flavone

APPENDIX

Trivial Name — *continued*

Trivial Name	Substitution
Murrayanol	5,4′-diOH, 3,6,7,3′,5′-pentaOMe — flavonol
Muxiangrine I	5,3′-diOH, 7-OMe, 6,8-diCH3, 4′,3′-ODmp — flavone
Muxiangrine II	5,5′-diOH, 7-OMe, 6-CH3, 4′,3′-ODmp — flavone
Muxiangrine III	5,3′,4′-triOH-7-OMe, 6,8-diCH3-5′-C5 — flavone
Myricetin	3,5,7,3′,4′,5′-hexaOH — flavonol
Natsudaidain	3-OH, 5,6,7,8,3′,4′-hexaOMe — flavonol
Negletein	5,6-diOH, 7-OMe — flavone
Neouralenol	3,6,7,3′,4′-pentaOH, 2′-C5 — flavonol
Nepetin	5,7,3′,4′-tetraOH, 6-OMe — flavone
Nevadensin	5,7-diOH, 6,8,4′triOMe — flavone
Nobiletin	6,6,7,8,3′4′-hexaOMe — flavone
Nodifloretin	5,6,7,4′-tetraOH, 3′OMe — flavone
Noranhydro-icaritin	3,5,7,4′-tetraOH, 8-C5 — flavonol
Norartocarpetin	5,7,2′4′-tetraOH — flavone
Norartocarpin (syn.: Mulberrrin)	
Noricaritin (revised)	3,5,7,4′-tetraOH, 6-C5-OH — flavonol
Norwightin	5,7,8,2′,3′-pentaOH — flavone
Norwogonin	5,7,8-triOH — flavone
Ombuin	3,5,3′-triOH, 7,4′-diOMe — flavonol
Onopordin	5,7,3′4′-tetraOH, 8-OMe — flavone
Oppositin	5,7-diOMe, 6,6″-Ketopyrano....3″-OH — flavone, see Figure 12.4
Oroxylin	5,7-diOH, 6-OMe — flavone
Ovalifolin	6-O-C5, 7,8-fur — flavone
Oxidihydroartocarpesin	5,7,2′,4′-tetraOH, 6-C5-OH — flavone
Oxyayanin-A	5,2′,5′-triOH, 3,7,4′-triOMe — flavonol
Oxyayanin-B	5,6,3′-triOH, 3,7,4′-triOMe — flavonol
Oxydihydromorusin	5,2′,4′-triOH, 3-C5-OH, 7,8-ODmp — flavone
Oxyisocyclointegrin	5,4′-diOH, 7-OMe, 3,6′-cyclo O-C6 — flavone
Pachypodol	5,4′-diOH, 3,7,3′-triOMe — flavone
Patuletin	3,5,7,3′,4′-pentaOH, 6-OMe — flavone
Pectolinarigenin	5,7-diOH, 6,4′-diOMe — flavone
Pedalitin	5,6,3′,4′-tetraOH, 7-OMe — flavone
Pediflavone	5,8-diOH, 7-OMe — flavone
"Pedunculin" — obsolete	
Peltogynin	7,4,5′-triOH, 3,2′-O-CH2 — flavonol
Penduletin	5,4′-diOH, 3,6,7-triOMe — flavone
Petalopurpurenol	3,5,7,4′-tetraOH, 3′,2′-ODmp-C6 — flavonol
p-Hydoxybenzylluteolin	5,7,3′,4′-tetraOH, 8-C-p-OH-benzyl — flavone
p-Hydroxybenzyldiosmetin	5,7,3′-OH, 4′-OMe, 8-C-p-OH-benzyl — flavone
p-Hydroxybenzylkaempferol	3,5,7,4′-tetraOH, 8-C-p-OH-benzyl — flavonol
p-Hydroxybenzylquercetin	3,5,7,3′,4′-pentaOH, 8-C-p-OH-benzyl — flavonol
Pilloin	5,3′-diOH, 7,4′-diOMe — flavone
Pilosin	5,7,8-triOH, 6,4′-diOMe — flavone
Pinnatin	5-OMe, 7,6-fur — flavone
Pinoquercetin	3,5,7,3′,4′-pentaOH, 6-Me — flavonol
Pityrogrammin	3,5,7-triOH, 8-OMe, 6-Me — flavonol

continued

APPENDIX

Trivial Name — *continued*

Trivial Name	Substitution
Platanetin	3,5,7,8-tetraOH, 6-C5 — flavonol
Platanin	3,5,7,8-tetraOH, 6-Me — flavonol
Poinsettifolin A	3,5,3′,4′-tetraOH, 7,8-ODmp, 6-C5 — flavonol, see Figure 12.9
Pollenitin	3,5,8,4′-tetraOH, 7-OMe — flavonol
Polystachin	5-OMe, 7,8-fur-C4-diOAc — flavone
Pongachromene	3,5-diOMe, 3′,4′-OCH2O, 7,8-ODmp — flavonol
Pongaglabol	5-OH, 7,8-fur — flavone
Ponganone XI	3-OMe, 7,6-fur — flavonol
Pongapin	3-OMe, 3′,4′-OCH2O, 7,8-fur — flavonol
Pongone	3′-OMe, 7,6-fur — flavone
Pratensin A	5,7-diOH, 3,6,4′-triOMe, 8-tigliat — flavonol
Pratensin B	5,4′-diOH, 3,6,7-triOMe, 8-Me-but — flavonol
Pratoletin	3,5,8,4′-tetraOH — flavonol
6-Prenylapigenin	5,7,4′-triOH, 6-C-5 — flavone
8-Prenylapigenin, see Licoflavone C	
6-Prenylchrysin	5,7-diOH, 6-C5 — flavone
8-Prenylchrysin	5,7-diOH, 8-C5 — flavone
6-Prenylchrysoeriol	5,7,4′-triOH, 3′-OMe, 6-C5 — flavone
6-Prenylgalangin, see Glepidotin A	
6-Prenylherbacetin	5,7,4′-triOH, 3,8-diOMe, 6-C5 — flavonol
Prenyllicoflavone A (syn.: licoflavone B)	
6-Prenylluteolin, see Gancaonin O	
8-Prenylluteolin	5,7,3′,4′-tetraOH, 8-C5 — flavone
8-Prenylpachypodol	5,4′-diOH, 3,7,3′-OMe, 8-C5 — flavonol
Primetin	5,8-diOH — flavone
Primuletin	5-OH — flavone
Prosogerin A	7-OH, 6-OMe, 3′,4′-OCH2O — flavone
Prosogerin C	6,7,3′,4′,5′-pentaOMe — flavone
Prosogerin D	7-OH, 6,3′,4′,5′-tetraOMe — flavone
Prosogerin E	6,7-diOH, 3′,4′,5′-triOMe — flavone
Prudomestin	3,5,7-triOH, 8,4′-diOMe — flavonol
Pseudosemiglabrin	7,8-bisfur -OAc — flavone
Pseudosemiglabrinol	7,8-bisfur -OH — flavone
Psiadiarabicin	5,3′-diOH, 6,7,2′,4′,5′-pentaOMe — flavone
Ptaeroxylol	3,5,7-triOH, 2′-OMe — flavonol
Pulcherrimin	5-OH, 7-OMe, 3′,4′-O2CH2, 3,2′-O-CH2 — flavonol
Purpurascenin	3,5,6,7,8,2′,4′,5′-octaOMe — flavonol
Quercetagetin	3,5,6,7,3′,4′-hexaOH — flavonol
Quercetin	3,5,7,3′,4′-pentaOH — flavonol
Racemoflavon	5,4′-diOH, 3′-OMe, 7,8-ODmp — flavone
Rehderianin I	5,2′,5′-triOH, 7,8-diOMe — flavone
Retusin	5-OH, 3,7,3′,4′-tetraOMe — flavonol
Rhamnazin	3,5,4′-triOH, 7,3′-diOMe — flavonol
Rhamnetin	3,5,3′,4′-tetraOH, 7-OMe — flavonol
Rhamnocitrin	3,5,4′-triOH, 7-OMe — flavonol
Rhynchosin	3,6,7,3′,4′-pentaOH — flavonol
Rhynchospermin	3,5,3′-OH, 7,4′-diOMe, 8-C5 — flavonol
Rivularin	5,2′-diOH, 7,8,6′-triOMe — flavone

APPENDIX

Trivial Name — *continued*

Trivial Name	Substitution
Robinetin	3,7,3',4',5'-pentaOH — flavonol
Rubraflavone A	7,2',4'-triOH, 3-C10 — flavone
Rubraflavone C	5,7,2',4'-tetraOH, 3-C10, 6-C5 — flavone
Rubraflavone D	5,2',4'-triOH, 3-C10, 7,6-ODmp — flavone
Saltillin	5-OH, 4'-OMe, 7-Me — flavone
Salvigenin	5-OH, 6,7,4'-triOMe — flavone
Sanaganone	6,5-ODmp, 7,8-fur — flavone
Sanggenon J	5,7,2'-triOH, 3-C5, 3',4'-ODmp-C5 — flavone
Sanggenon K	5,7,2'-triOH, 3-C5, 2',3'-ODmp-C5 — flavone
Santin	5,7-diOH, 3,6,4'-triOMe — flavonol
"Santoflavone" — obsolete	
Sarothranol	5,3',4'-OH, 3-OMe, 7,6-ODmp — flavonol
Sarothrin	5,7,4'-triOH, 3,6,8-triOMe — flavonol
Scaposin	5,7,5'-triOH, 6,8,3',4'-tetraOMe — flavone
Scutellarein	5,6,7,4'-OH — flavone
Scutevulin	5,7,2'-triOH, 8-OMe — flavone
Selgin	5,7,4',5'-tetraOH, 3'-OMe — flavone
Semiglabrin	7,8- bisfur — flavone
Semiglabrinol	7,8- bisfur — flavone
Sericetin	3,5-diOH, 8-C5, 7,6-ODmp — flavonol
Serpyllin	5-OH, 7,8,2',3',4'-pentaOMe — flavone
Sexangularetin	3,5,7,4'-tetraOH, 8-OMe — flavonol
Sideritiflavon	5,3',4'-triOH, 6,7,8-triOMe — flavone
Sideroxylin	5,4'-diOH, 7-OMe, 6,8-diMe — flavone
Sinensetin	5,6,7,3',4'-pentaOMe — flavone
Skullkapflavone I	5,2'diOH, 7,8-diOMe — flavone
Skullcapflavone II	5,2'-diOH, 6,7,8,6'-tetraOMe — flavone
Sophoflavescenol	3,7,4'-triOH, 5-OMe, 8-C5 — flavonol
Sorbifolin	5,6,4'-triOH, 7-OMe — flavone
Spinacetin	3,5,7,4'-tetraOH, 6,3'-diOMe — flavonol
Stachyoidin	5-OMe, 7,8-pyr-fur (3''-oxo) — flavone
Strobochrysin	5,7-diOH, 6-Me — flavone
Sudachitin	5,7,4'-triOH, 6,8,3'-triOMe — flavone
Sylpin	5,6,4'-triOH, 3-OMe, 8-Me — flavonol
Syringetin	3,5,7,4'-tetraOH, 3',5'-diOMe — flavonol
Syzalterin	5,7,4'triOH, 6,8-diMe — flavone
Tabularin	5,7-diOH, 6,2',4',5'-tetraOMe — flavone
Tachrosin	5,7-diOMe, 8-furyl (2''-oxo) — flavon
Takakin	5,7,8-triOH, 4'-OMe — flavone
Tamadone	5,2',4'-triOH, 6,7,8-triOMe — flavon
Tamaridone	5,7,2'-triOH, 6,4'-diOMe — flavone
Tamarixetin	3,5,7,3'-tetraOH, 4'-OMe — flavonol
Tambulin	3,5-diOH, 7,8,4'-triOMe — flavonol
"Tanetin" — obsolete	
Tangeretin	5,6,7,8,4'-pentaOMe — flavone
Tectochrysin	5-OH, 7-OMe — flavone

continued

APPENDIX

Trivial Name — *continued*

Trivial Name	Substitution
Tenaxin 1	5,2'-diOH, 6,7,8-triOMe — flavone
Tephrinone	5-OH, 7-OMe, 8-C5 — flavone
Tephrodin	5-OMe, 7,8-pyr-fur (3''-oxo-4''-OAc) — flavone
Tephroglabrine	7-OMe, 8-furyl (4''-oxo) — flavone
Tephrorianin	5-OMe, 7,8-oxofuryl — flavone, see Figure 5
cis-Tephrostachin	5,7-diOMe, 8-C5-OH — flavone
Tepurindol	7-OMe, 8-furyl (2'',4''-diOH) — flavone
Ternatin	5,4'-diOH, 3,7,8,3'-tetraOMe — flavonol
Thevetiaflavon	7,4'-diOH, 5-OMe — flavone
Thymonin	5,6,4'-triOH, 7,8,3'-triOMe — flavone
Thymusin	5,6,4'-triOH, 7,8-diOMe — flavone
Tithonine	3'-OH, 7,4'-diOMe — flavone
Tomentin	5,6,3',4'-tetraOH, 3,7-diOMe — flavonol
Topazolin	5,7,4'-triOH, 3-OMe, 6-C5 — flavonol
Topazolin hydrate	5,7,4'-triOH, 3-OMe, 6-C5-OH — flavonol
Torosaflavone C	5,3'-diOH, 4'-OMe, 7,6- bisfurano — flavone, see Figure 12.3
Torosaflavone D	5,7,3'-triOH, 4'-OMe, 6-acrylic acid — flavone, see Figure 12.1
trans-Anhydrotephronin	5,7-diOMe, 8-C5 — flavone
Transilitin	7,8,3',4'-tetraOH, 3-OMe — flavonol
trans-Lanceolatin (syn.: Lanceolatin A)	
trans-Tephrostachin	5,7-diOMe, 8-C5-OH — flavone
Tricetin	5,7,3',4',5'-pentaOH — flavone
Tricin	5,7,4'-triOH, 3',5'-diOMe — flavone
Trimethylwogonin	5,6,7-triOH, 8,3',4',5'-tetraOMe — flavone
8,2',6'-Triprenylquercetin	3,5,7,3',4'-penta-OH, 8,2',6'-triC5 — flavonol
Ugonine C	5, 4'-diOH, 6-OMe, 7,8-fur — flavone
Umhugengerin	5-OH, 6,7,3',4',5'-pentaOMe — flavone
Unonal	5,7-diOH, 8-Me, 6-CHO — flavone
Uralene	5,6,4',5'-tetraOH, 3OMe, 2'-C5 — flavonol
Uralenol	3,5,7,3',4'-pentaOH, 5'-C5 — flavonol
Velutin	5,4'-diOH, 7,3'-diOMe — flavone
Veronicafolin	3,5,4'-triOH, 6,7,3'-triOMe — flavonol
Viscidulin I	3,5,7,2',6'-pentaOH — flavonol
Viscidulin III	3,5,7,3'-tetraOH, 2',4'-diOMe — flavon
Viscosol	5,7-diOH, 3,6,4'-triOMe, 3'-C5 — flavonol
Wharangin	5,3',4'-triOH, 3-OMe, 7,8-OCH2O — flavonol
Wightin	5,3'-diOH, 7,8,2'-triOMe — flavone
Wogonin	5,7-diOH, 8-OMe — flavone
Xanthomicrol	5,4'-diOH, 6,7,8-triOMe — flavone
Yinyanghuo A	5,7-diOH, 5'-C5-OH, 4',3'-ODmp — flavone
Yinyanghuo B	5,7,4'-triOH, 3'-C5, 5'-C5-OH — flavone
Yinyanghuo C	5,7-diOH, 4',3'-ODmp — flavone
Yinyanghuo D	5,7,4'-triOH, 3'-C5 — flavone
Yinyanghuo E	5,7,5'-triOH, 4',3'-ODmp — flavone
Zapotin	5,6,2',6'-tetraOMe — flavone
Zapotinin	5-OH, 6,2',6'-triOMe — flavone

13 Flavone and Flavonol O-Glycosides

Christine A. Williams

CONTENTS

13.1 INTRODUCTION

Flavone and flavonol O-glycosides make up one of the largest classes of flavonoid constituents with over 2000 known structures. There are 279 glycosidic combinations of the most common flavonol aglycone, quercetin, and 347 kaempferol O-glycosides listed in the check list (Appendix B) for the period ending December 2003. The group includes any bound form of flavone or flavonol such as acylated and sulfated derivatives and not only those with just sugar. Thus, the number of possible combinations is enormous because of the wide structural variation, i.e., in (1) the hydroxylation and methoxylation pattern of the aglycone; (2) the number and nature of sugars and their position of attachment through hydroxyl groups to the aglycone; (3) the nature of the sugar linkage to the aglycone, different interglycosidic linkages and whether the sugars are in the pyranose or furanose form; (4) the nature, number, and position of attachment of aliphatic or aromatic acyl groups to one or more sugars or directly through a hydroxyl group to the aglycone; and (5) the presence and position of attachment of one or more sulfate groups through a hydroxyl group of the aglycone or that of a sugar. Sugars can also be attached through a carbon bond but these compounds, the C-glycosyl-flavonoids, are dealt within Chapter 14.

The monosaccharides most frequently found in O-combination with flavone and flavo-nols are glucose and rhamnose and less frequently arabinose, xylose, and glucuronic acid. These sugars are usually present in the expected pyranose form and with the appropriate linkage, i.e., β for glucose, galactose, xylose, and glucuronic acid and α for rhamnose. Arabinose can occur in either the pyranose or furanose form and with an α- or β-linkage. Seven other monosaccharides have been found occasionally linked to flavones or flavonols (see Section 13.3.1). The most usual place of sugar attachment in flavonols is at the 3-hydroxyl and at the 7-hydroxyl in flavones but sugars have been found at all the other possible positions.

In the first edition of *The Flavonoids*,[1] published in 1975, 134 flavone glycosides and 252 flavonol glycosides were listed. The original intent of this first edition was to compile an updated version of the classic monograph edited by the late Ted Geissman, entitled *The Chemistry of Flavonoid Compounds*, which was published in 1962, in which only 30 flavone and 54 flavonol glycosides were described.[2] The glycosides listed in the monograph included the best-known flavonol glycoside, quercetin 3-rhamnosyl(1 → 6)glucoside (rutin) and the common flavone glycoside, luteolin 7-glucoside (glucoluteolin), but did not include any acylated or sulfated derivatives. Thirteen years later, in the first edition of *The Flavonoids*,[1] nine of the listed 134 flavone glycosides and 18 of the 247 recorded flavonol glycosides were found to be acylated with acids such as *p*-coumaric, caffeic, ferulic, sinapic, gallic, benzoic, *p*-hydroxybenzoic, acetic, and malonic. However, the most exciting find of this review period was the discovery that flavonoid conjugates are not such a rarity in the plant kingdom as once thought and that they are in fact frequent constituents of many salt-tolerant and water-stress resilient plants. At this stage, only 11 had been fully characterized but many more had been detected and flavonoid sulfates were regularly recorded on a presence or absence basis. The second edition of *The Flavonoids*, which covered the new compounds found between 1975 and 1980,[3] showed a large increase in the number of both acylated and sulfated glycosides bringing the total number of flavone glycosides in that check list to 271 and flavonol glycosides to 486. Both the third edition, covering the years 1981 to 1985,[4] and the fourth edition[5] (1986 to 1991) of *The Flavonoids* showed a steady increase in the number of flavone and flavonol glycosides to 345 and 647 and 463 and 906, respectively, but with comparatively small increases in the number of new acylated and sulfated glycosides. No trisaccharides were reported in association with flavones or flavonols in Geissman's monograph[2] and only a small number of flavonol triosides, some with linear and some with branched sugars, were recorded in book one[1] of *The Flavonoids*. It is not until book four[5] that there is a marked increase in the number of trisaccharides with 4 new linear and 11 new branched structures recorded. Flavone triosides continued to be of rare occurrence in books three[4] and four.[5] However, the first known tetrasaccharide, [rhamnosyl(1 → 4)glucosyl]sophorose, was recorded in 1987 (book four)[5] in combination with the flavone acacetin (apigenin 4'-methyl ether) at the 7-position with an acetyl group at the 6''-position of the sophorose in leaves of *Peganum harmala* (Zygophyllaceae).[6]

The main remit of this chapter is to provide reference and plant source details of new flavone and flavonol *O*-glycosides discovered since 1991, i.e., covering the years 1992 to 2003. A checklist of all (as far as possible) known structures is also included in Appendices A and B. A series of reviews, which include most of the data on new *O*-glycosylflavones and flavonols presented here, have appeared in *Natural Product Reports* and cover the years 1992 to 1994,[7] 1995 to 1997,[8] and 1998 to 2000,[9] and with a fourth (2001 to 2003)[10] in press. Other useful sources of data are *The Phytochemical Dictionary*,[11] *The Handbook of Natural Flavo-noids*,[12] and for general background reading Jeffrey Harborne's *Comparative Biochemistry of the Flavonoids*.[13]

13.2 SEPARATION, PURIFICATION, AND IDENTIFICATION

The traditional methods of separation, purification, and identification of flavone and flavonol *O*-glycosides, including paper, column, and thin layer chromatography and UV spectral analysis of pure compounds, acid, alkaline, and enzyme hydrolysis, and identification of the resulting aglycones and sugars are well described by Mabry et al.,[14] by Markham,[15,16] and by Harborne.[17] High-performance liquid chromatography (HPLC) is now a standard technique used in all phytochemistry laboratories. HPLC with UV detection is an established method for separating and detecting flavonoid glycosides in complex mixtures, for comparing flavonoid profiles of related plant taxa and for identifying known compounds. However, it has been recently superceded by HPLC–mass spectrometry (MS) techniques such as liquid chromatography (LC)–MS, which gives the molecular weight of the glycoside and LC–MS–MS, which also gives the mass spectrum of the fragmentation ions. Atmospheric pressure chemical isolation (APCI) is especially useful for obtaining the molecular weight of the glycoside, together with the molecular ions for the aglycone and any intermediate sugars or acylated sugars from very small amounts (~0.1 mg) of pure compound. In their paper in *Phytochemical Analysis*, Grayer et al.[18] describe the application of APCI to a chemotaxonomic study of the flavonoids in the genus *Ocimum*. Fast atom bombardment MS gives a strong molecular ion, which indicates the number and type of sugar units present and useful fragmentation patterns with obvious loss of sugars and any methoxyls from the aglycone. This method has tended to be replaced by electrospray MS, which is less costly but does not give so much fragmentation data, although MS–MS on the resulting product ions can give further useful fragmentation ions. However, the techniques of choice for flavonoid glycoside identification are ^{1}H NMR, ^{13}C NMR, and two-dimensional nuclear magnetic resonance (NMR), which allow complete characterization including details of form and linkage of the sugars, the nature and position of acyl groups, and number of sulfate groups. For more detailed information on HPLC, MS, NMR, and other recent separation, purification, and identification techniques see Chapters 1 and 2.

13.3 NEW FLAVONE AND FLAVONOL *O*-GLYCOSIDES

Some 228 new flavone *O*-glycosides and over 500 new flavonol *O*-glycosides have been reported in the period 1992 to 2003. This brings the number of known structures listed in the check lists (Appendix A) to 679 flavone and 1331 flavonol glycosides. (Appendix B) The new glycosides are listed in Table 13.1[19–164] and Table 13.2,[23,107,165–479] respectively. In all these entries and in later tables the sugars are assumed to be in the pyranose form and to have the appropriate linkage, i.e., β for glucose, α for rhamnose, etc. except where otherwise stated. Reports of new monosaccharides, disaccharides, trisaccharides, tetrasaccharides, acylating agents, and sulfate conjugates will be considered first.

13.3.1 Monosaccharides

A list of all the monosaccharides that have been found in *O*-combination with flavones or flavonols are given in Table 13.3. These include five sugars, fructose, allulose, lyxose, fucose, and glucosamine, which have been recorded since 1992. α-D-Fructofuranose has been reported from leaves of *Crataegus pinnatifida* (Rosaceae)[41] linked to both the C-8 position and the 7-hydroxyl of apigenin. Here, the *C*- and *O*-glycosidic linkages to the sugar form a unique ring structure (Pinnatifinoside A, **13.1**). There has been one previous report of fructose, as a tricin fructosylglucoside in *Hyacinthus orientalis* (Liliaceae),[480] but the structure of this glycoside was never confirmed. An acetylated derivative, Pinnatifinoside B (**13.2**), and

TABLE 13.1
New Flavone Glycosides

Glycoside	Source	Family	Ref.
5,7-Dihydroxyflavone (chrysin)			
7-(4″-Acetylglucoside)	*Calicotome villosa*	Leguminosae	19
7-(6″-Acetylglucoside)	Aerial parts		
5,6,7-Trihydroxyflavone (baicalein)			
7-(6″-Malonylglucoside)	*Cephalocereus senilis* suspension cultures	Cactaceae	20
Baicalein 5-methyl ether			
7-Glucoside	*Cephalocereus senilis* chitin-treated cell suspension cultures	Cactaceae	21
Baicalein 6-methyl ether (oxoxylin A)			
5-Rhamnoside	*Trichosanthes anguina* seeds	Cucurbitaceae	22
7-Glucosyl(1 → 3)rhamnoside	*Eupatorium africanum* whole plant	Compositae	23
Baicalein 7-methyl ether (negletein)			
6-Xyloside	*Bauhinia purpurea* stems	Leguminosae	24
6-Glucoside	*Colebrookea oppositifolia* bark	Labiatae	25
6-Rhamnoside(1 → 2)fucoside	*Origanum vulgare*	Labiatae	26
5,7,8-Trihydroxyflavone (norwogonin)			
5-Glucoside	*Pyracantha coccinea* roots	Rosaceae	27
7-Galactoside	*Scutellaria ocellata* and *S. nepetoides*	Labiatae	28
7-Hydroxy-5,8-dimethoxyflavone			
7-Glucoside	*Scutellaria immaculate* aerial parts	Labiatae	29
7-Glucuronide	*Scutellaria rivularis* roots	Labiatae	30
5,7,2′-Trihydroxyflavone			
7-Glucoside	*Scutellaria ramosissima* aerial parts	Labiatae	31
2′-Glucoside			
5,2′-Dihydroxy-7-methoxyflavone (echioidinin)			
5-Glucoside	*Andrographis alata* whole plant	Acanthaceae	32
2′-(6″-Acetylglucoside)	*Andrographis affinis* whole plant	Acanthaceae	33
5,7,4′-Trihydroxyflavone (apigenin)			
7-Apiofuranosyl(1 → 6)glucoside	*Gonocaryum calleryanum* leaves	Icacinaceae	34
7-Cellobioside	*Salvia uliginosa* petals	Labiatae	35
7-Sophorotrioside	*Leptostomum macrocarpon* gametophyte	Bryales	36
7-(2^G-Rhamnosyl)rutinoside	*Ligustrum vulgare* leaves	Oleaceae	37
7-(2^G-Rhamnosyl)gentiobioside	*Lonicera gracilepes* var. *glandulosa* leaves	Caprifoliaceae	38
7-Rhamnoside-4′-glucosylrhamnoside	*Asplenium normale*	Aspleniaceae	39
7-Cellobioside-4′-glucoside	*Salvia uliginosa* petals	Labiatae	35
7-Glucosyl(1 → 2)glucuronide-4′-glucuronide	*Medicago sativa* aerial parts	Leguminosae	40
Pinnatifinoside A (**13.1**)	*Crataegus pinnatifida* var. *major* leaves	Rosaceae	41
Pinnatifinoside B (**13.2**)			
Pinnatifinoside C (**13.3**)			
Pinnatifinoside D (**13.4**)			
7-(2″-*E*-*p*-Coumaroylglucoside)	*Echinops echinatus* flowers	Compositae	42
7-(3″-*p*-Coumaroylglucoside)	*Stachys aegyptiaca* aerial parts	Labiatae	43
7-(6″-*E*-*p*-Coumarylgalactoside)	*Lagopsis supina* whole plant	Labiatae	44
7-(6″-*E*-Caffeoylglucoside)	*Bellis perennis* flowers	Compositae	45

TABLE 13.1
New Flavone Glycosides — *continued*

Glycoside	Source	Family	Ref.
7-[6″-(3-Hydroxy-3-methylglutaryl)glucoside]	*Chamaemelum nobile* flowers	Compositae	46
7-(3″,6″-Di-*E-p*-coumaroylgalactoside)	*Lagopsis supina* whole plant	Labiatae	44
7-(3″-Acetyl-6″-*E-p*-coumaroylglucoside)	*Blepharis ciliaris* aerial parts	Acanthaceae	47
7-Rhamnosyl(1 → 6)(4″-*E-p*-methoxylcinnamoylglucoside)	*Chrozophora oblongifolia* aerial parts	Euphorbiaceae	48
4′-(2″-Feruloylglucuronosyl)(1 → 2)glucuronide	*Medicago sativa* aerial parts	Leguminosae	49
7-Glucuronosyl(1 → 3)[(2″-*p*-coumaroylglucuronosyl)(1 → 2)glucuronide]	*Medicago sativa* aerial parts	Leguminosae	40
7-Glucuronosyl(1 → 3)[(2‴-feruloylglucuronosyl)(1 → 2) glucuronide]			
7-(2″-Feruloylglucuronosyl)(1 → 2) glucuronide-4′-glucuronide			
7-Glucuronide-4′-(2‴-*E-p*-coumaroylglucuronosyl)(1 → 2)glucuronide	*Medicago sativa* aerial parts	Leguminosae	49
7-Glucuronide-4′-(2‴-feruloylglucuronosyl)(1 → 2) glucuronide			
Apigenin 7-methyl ether (genkwanin)			
4′-α-L-Arabinopyranosyl(1 → 6)galactoside	*Salvia moorcroftiana* whole plant	Labiatae	50
4′-Rhamnosyl(1 → 2)[rhamnosyl(1 → 6)galactoside]			
Apigenin 4′-methyl ether (acacetin)			
7-Rhamnoside	*Peganum harmalai*	Zygophyllaceae	51
7-Glucosyl(1 → 4)xyloside	*Centratherum anthelminticum* seeds	Compositae	52
7-Apiosyl(1 → 6)glucoside	*Crotalaria podocarpa* aerial parts	Leguminosae	53
7-(2^G-Rhamnosyl)rutinoside	*Buddleia officinalis* flowers	Loganiaceae	54
7-Rhamnosyl(1 → 2)glucosyl(1 → 2)glucoside	*Peganum harmala* leaves	Zygophyllaceae	51
7-Rhamnosyl(1 → 2)glucosyl(1 → 2)glucosyl(1 → 2)glucoside	*Peganum harmala* leaves	Zygophyllaceae	55
7-(4‴-Acetylrutinoside)	*Thalictrum przewalskii* aerial parts	Ranunculaceae	56
7-Rhamnosyl(1 → 6)[2″-acetylglucosyl(1 → 2)glucoside]	*Dendranthema lavandulifolium* whole plant	Compositae	57
7-[6‴-Acetylglucosyl(1 → 2)]rhamnosyl(1 → 6)glucoside	*Calamintha glandulosa* leaves	Labiatae	58
7-Glucosyl(1 → 6)[3‴-acetylrhamnosyl(1 → 2)glucoside]	*Peganum harmala* leaves	Zygophyllaceae	51
7-(4⁗-Acetylrhamnosyl)(1 → 6) glucosyl(1 → 3)(6″-acetylglucoside)	*Thalictrum przewalskii* aerial parts	Ranunculaceae	56
6-Hydroxyapigenin (scutellarein)			
7-Xylosyl(1 → 2)xyloside	*Hebe stenophyllum* leaves	Scrophulariaceae	59
7-Xylosyl(1 → 2)glucoside			
7-Xylosyl(1 → 6)galactoside	*Semecarpus kurzii* leaves	Anacardiaceae	60
7-Glucuronosyl(1 → 2)glucuronide	*Perilla ocimoides* leaves	Labiatae	61
7-[6″-(3-Hydroxy-3-methylglutaryl)glucoside]	*Frullania muscicola* whole plant	Frullaniaceae	62
Scutellarein 6-methyl ether (hispidulin)			
7-Rhamnoside	*Picnomon acarna* aerial parts	Compositae	63
7-Methylglucuronide	*Centaurea furfuracea* aerial parts	Compositae	64

continued

TABLE 13.1
New Flavone Glycosides — *continued*

Glycoside	Source	Family	Ref.
4'-Glucoside	*Cirsium oligophyllum* leaves	Compositae	65
7-Xylosyl(1 → 2)xyloside	*Chirita fimbrisepala* roots	Gesneriaceae	66
7-Neohesperidoside	*Ipomoea purpurea* flowers	Convolvulaceae	67
7-(6''-*E*-*p*-Coumaroylglucoside)	*Eriocaulon buergerianum* capitula	Eriocaulaceae	68
Scutellarein 4'-methyl ether			
7-Rutinoside	*Teucridium parvifolium* leaves, stems, and fruits	Labiatae	69
7-(2'',6''-Diacetylalloside)	*Sideritis perfoliata*	Labiatae	70
Scutellarein 5,4'-dimethyl ether			
7-Glucoside	*Striga passargei* whole plant	Scrophulariaceae	71
7-(4^{Rha}-Acetylrutinoside)			
Scutellarein 6,7-dimethyl ether			
4'-Glucuronide	*Conyza linifolia*	Compositae	72
Scutellarein 6,4'-dimethyl ether (pectolinarigenin)			
7-(6''-Acetylglucoside)	*Lantana camara* aerial parts	Verbenaceae	73
7-(2'''-Acetylrutinoside)	*Linaria japonica* whole plant	Scrophulariaceae	74
7-(3'''-Acetylrutinoside)			
7-(4'''-Acetylrutinoside)	*Linaria haelava* whole plant	Scrophulariaceae	75
Scutellarein 7,4'-dimethyl ether			
6-Xylosyl(1 → 2)glucoside	*Gelonium multiflorum* seeds	Euphorbiaceae	76
6-Neohesperidoside			
Scutellarein 6,7,4'-trimethyl ether (salvigenin)			
5-[6''-Acetylglucosyl (1 → 3)galactoside]	*Striga aspera*	Scrophulariaceae	77
8-Hydroxyapigenin (isoscutellarein)			
7-Glucosyl(1 → 2)xyloside	*Sideritis* spp. aerial parts	Labiatae	78
8-Sophoroside	*Gratiola officinalis* leaves	Scrophulariaceae	79
8-(6''-*E*-*p*-Coumaroylglucoside)	*Stachys aegyptiaca* whole plant	Labiatae	80
8-(2''-Sulfatoglucuronide)	*Helicteres angustifolia* root bark	Sterculiaceae	81
8-(2'',4''-Disulfatoglucuronide)	*Helicteres isora* fruit	Sterculiaceae	82
Isoscutellarein 4'-methyl ether			
8-Glucoside	*Glossostemon bruguieri* roots	Sterculiaceae	83
8-(6''-*n*-butylglucuronide)	*Helicteres isora* fruit	Sterculiaceae	82
7-Allosyl(1 → 2)glucoside	*Sideritis javalambrensis* aerial parts	Labiatae	84
8-(2''-Sulfatoglucuronide)	*Helicteres angustifolia* root bark	Sterculiaceae	81
8-(2'',4''-Disulfatoglucuronide)	*Helicteres isora* fruit	Sterculiaceae	82
6,8-Dihydroxy-7,4'-dimethoxyflavone			
6-Rutinoside	*Dicliptera riparia* whole plant	Acanthaceae	85
6-(4''-Acetylrhamnosyl)(1 → 6)glucoside			
5,7,2'-Trihydroxy-6-methoxyflavone			
7-Glucoside	*Scutellaria amoena* roots	Labiatae	86
7-Methylglucuronide			
5,7,2',6'-Tetrahydroxyflavone			
2'-Glucoside	*Scutellaria baicalensis* hairy root cultures	Labiatae	87

TABLE 13.1
New Flavone Glycosides — *continued*

Glycoside	Source	Family	Ref.
5,2′,6′-Trihydroxy-7-methoxyflavone			
2′-Glucoside	*Andrographis alata* whole plant	Acanthaceae	88
5,7,8,2′-Tetrahydroxyflavone			
7-Glucuronide	*Scutellaria rivularis* roots	Labiatae	30
5,2′-Dihydroxy-7,8-dimethoxyflavone (skullcapflavone 1)			
2′-Glucoside	*Andrographis paniculata* roots	Acanthaceae	89
2′-(2″-*E*-Cinnamoylglucoside)	*Andrographis serpyllifolia* whole plant	Acanthaceae	90
2′-(3″-*E*-Cinnamoylglucoside)			
2′-(4″-*E*-Cinnamoylglucoside)	*Andrographis elongata* whole plant	Labiatae	91
5,7,3′,4′-Tetrahydroxyflavone (luteolin)			
5-Glucuronide-6″-methyl ester	*Dumortiera hirsuta* gametophytes	Hepaticae	92
5-Rutinoside	*Salvia lavandulifolia* ssp. *oxyodon* aerial parts	Labiatae	93
7-Glucosyl(1 → 4)-α-L-arabinopyranoside	*Cassia glauca* seeds	Leguminosae	94
7-Xylosyl(1 → 6)glucoside (primeveroside)	*Halenia corniculata* whole plant	Gentianaceae	95
7-Apiosyl(1 → 6)glucoside	*Phlomis nissolii* aerial parts	Labiatae	96
7-Rhamnosyl(1 → 6)galactoside (7-robinobioside)	*Pteris cretica* fronds	Adiantaceae	97
7-Sophoroside	*Pteris cretica* aerial parts	Adiantaceae	98
7-Galactosyl(1 → 6)galactoside	*Anogeissus latifolia*	Combretaceae	99
7-Galactosylglucuronide	*Andryala reguisina* aerial parts	Compositae	100
7-Glucoside-3′-glucuronide	*Melissa officinalis* leaves	Labiatae	101
3′-Xylosyl(1 → 2)glucoside	*Viburnum grandiflorum* leaves	Caprifoliaceae	102
4′-Rutinoside	*Dalbergia stipulacea* leaves	Leguminosae	103
7-Sophorotrioside	*Leptostomum macrocarpon* gametophytes	Bryales	36
7-(6″-*p*-Benzoylglucoside)	*Vitex agnus-castus* root bark	Verbenaceae	104
7-[6″-(2-Methylbutyryl)glucoside]	*Arnica chamissonis* flowers	Compositae	105
7-[6″-(3-Hydroxy-3-methylglutaryl)glucoside]	*Frullania muscicola* whole plant	Frullaniaceae	62
3′-(3″-Acetylglucuronide)	*Rosmarinus officinalis* leaves	Labiatae	106
3′-(4″-Acetylglucuronide)			
7-Glucosyl(1 → 6)(4‴-caffeoylglucoside)	*Lonicera implexa* leaves	Caprifoliaceae	107
7-Glucoside-4′-(*Z*-2-methyl-2-butenoate) (7-glucoside-4′-angelate)	*Polygonum aviculare* whole plant	Polygonaceae	108
7-Apiosyl(1 → 2)[glucosyl (1 → 4)(6-malonylglucoside)]	*Capsicum anuum* fruit	Solanaceae	109
7-(Acetylsophorotrioside)	*Leptostomum macrocarpon* gametophytes	Bryales	36
7-(6⁗-Acetylallosyl)(1 → 3)glucosyl (1 → 2)glucoside (veronicoside A)	*Veronica didyma*	Scrophulariaceae	110
7-(2″-Feruloylglucuronosyl) (1 → 2)glucuronide-4′-glucuronide	*Medicago sativa* aerial parts	Leguminosae	40
7-(2″-Sulfatoglucoside)	*Thalassia testudinum* leaves	Hydrocharitaceae	111
Luteolin 5-methyl ether			
7-Xylosyl(1 → 6)glucoside	*Dirca palustris* twigs	Thymelaeaceae	112
Luteolin 7-methyl ether			

continued

TABLE 13.1
New Flavone Glycosides — *continued*

Glycoside	Source	Family	Ref.
3'-Glucoside	*Avicennia marina* aerial parts	Avicenniaceae	113
3'-Galactoside			
Luteolin 3'-methyl ether (chrysoeriol)			
7-α-ʟ-Arabinofuranosyl(1 → 6)glucoside	*Tagetes patula* whole plant	Compositae	114
7-Apiosyl(1 → 6)glucoside	*Phlomis nissolii* aerial parts	Labiatae	96
7-Neohesperidoside	*Morinda morindoides* leaves	Rubiaceae	115
7,4'-Diglucuronide	*Medicago sativa* aerial parts	Leguminosae	116
7-(3″-*Z-p*-coumaroylglucoside)	*Ballota acetabulosa* flowering aerial parts	Labiatae	117
7-(3″,6″-di-*E-p*-coumaroylglucoside)	*Marrubium velutinum* aerial parts	Labiatae	118
7-(2‴-Feruloylglucuronosyl) (1 → 2)glucuronide	*Medicago sativa* aerial parts	Leguminosae	116
7-[Glucuronosyl(1 → 3)(2‴-feruloylglucuronosyl)] (1 → 2)glucuronide			
Luteolin 4'-methyl ether (diosmetin)			
3'-Glucoside	*Cassia torosa* leaves	Leguminosae	119
7-Arabinosyl(1 → 6)glucoside	*Galium palustre*	Rubiaceae	120
7-Xylosyl(1 → 6)glucoside (as a mixture)			
7-Xylosyl(1 → 6)glucoside	*Hebe parviflora* and *H. traversii* leaves	Scrophulariaceae	59
7-Neohesperidoside (neodiosmin)	*Citrus aurantium* leaves	Rutaceae	121
7-(2″,6″-Dirhamnosyl)glucoside	*Buddleia madagascariensis* leaves	Loganiaceae	122
7-Apiosyl(1 → 2)(6″-acetylglucoside)	*Paullinia pinnata* leaves	Sapindaceae	123
Luteolin 5,3'-dimethyl ether			
7-Glucoside	*Pyrus serotina* leaves	Rosaceae	124
4'-Glucoside			
Luteolin 5,4'-dimethyl ether			
7-Xylosyl(1 → 6)glucoside	*Dirca palustris* dried twigs	Thymelaeaceae	112
Luteolin 7,3'-dimethyl ether			
4'-Apiosyl(1 → 2)glucoside	*Viscum alniformosanae* leaves and stems	Loranthaceae	125
Luteolin 5,3',4'-trimethyl ether			
7-Xylosyl(1 → 6)glucoside	*Dirca palustris* dried twigs	Thymelaeaceae	112
7-Rutinoside			
Luteolin 7,3',4'-trimethyl ether			
5-Glucoside	*Lethedon tannaensis* leaves	Thymelaeaceae	126
5-Xylosyl(1 → 6)glucoside			
6-Hydroxyluteolin			
6-Rhamnoside	*Erythroxylum leal-costae* leaves	Erythroxylaceae	127
7-Xylosyl(1 → 2)xyloside	*Hebe stenophylla* aerial parts	Scrophulariaceae	128
7-Xylosyl(1 → 6)glucoside	*Hebe stenophylla* aerial parts	Scrophulariaceae	59
7-Sambubioside	*Hebe stricta* leaves	Scrophulariaceae	129
7-[3″-(3-Hydroxy-3-methylglutaryl)glucoside]	*Frullania muscicola* whole plant	Frullaniaceae	62
7-[4″-(3-Hydroxy-3-methylglutaryl)glucoside]			
7-[6″-(3-Hydroxy-3-methylglutaryl)glucoside]			
7-(6″-*E*-Caffeoylglucoside)	*Veronica liwanensis* and *V. longifolia* aerial parts	Scrophulariaceae	130

TABLE 13.1
New Flavone Glycosides — *continued*

Glycoside	Source	Family	Ref.
6-Glucoside-7-[6′′′-(3-hydroxy-3-methylglutaryl)glucoside]	*Frullania teneriffae*	Hepaticae	131
7-[6′′-(3-Hydroxy-3-methylglutaryl)glucoside]-3-glucuronide	*Frullania cesatiana*	Hepaticae	131
6-Methoxyluteolin (nepetin, eupafolin)			
7-Glucuronide	*Digitalis lanata* leaves	Scrophulariaceae	132
7-Methylglucuronide			
4′-Glucoside	*Cirsium oligophyllum* leaves	Compositae	65
7-Rhamnoside-3′-xyloside	*Chenopodium ambrosioides*	Chenopodiaceae	133
7-[6′′-(2-Methylbutyryl)glucoside]	*Arnica chamissonis* flowers	Compositae	105
6-Hydroxyluteolin 3′-methyl ether (nodifloretin)			
7-[6′′-(3-Hydroxy-3-methylglutaryl)glucoside]	*Frullania polysticta* whole plant	Frullaniaceae	62
6-Hydroxyluteolin 4′-methyl ether			
7-Rhamnosyl(1 → 2)(6′′-acetylglucoside)	*Veronica liwanensis* and *V. longifolia* aerial parts	Scrophulariaceae	130
6-Hydroxyluteolin 6,3′-dimethyl ether			
5-Rhamnoside	*Tridax procumbens* leaves	Compositae	134
7-Rutinoside	*Kichxia elatine* aerial parts	Scrophulariaceae	135
6-Hydroxyluteolin 5,6,3′,4′-tetramethyl ether			
7-Cellobioside	*Sphaeranthus indicus* stems	Compositae	136
8-Hydroxyluteolin (hypolaetin)			
8-Rhamnoside	*Erythroxylum leal-costae* leaves	Erythroxylaceae	127
7-Sophoroside	*Gratiola officinalis* leaves	Scrophulariaceae	79
8-Glucoside-3′-rutinoside	*Cornulaca monacantha* aerial parts	Chenopodiaceae	137
7-Sulfatoglucoside	*Leptocarpus elegans* culm	Restionaceae	138
7-Sulfatogalactoside			
7-Sulfatoglucuronide	*Meeboldina thysanantha* culm	Restionaceae	138
7-Sulfate-8-glucoside	*Hypolaena fastigiata* culm	Restionaceae	138
Hypolaetin 7-methyl ether			
3′-Sulfatogalactoside	*Leptocarpus tenax* culm	Restionaceae	138
3′-Sulfatoglucuronide	*Leptocarpus elegans* culm	Restionaceae	138
Hypolaetin 3′-methyl ether			
8-Glucuronide	*Gratiola officinalis* leaves	Scrophulariaceae	79
7-Sophoroside			
Hypolaetin 4′-methyl ether			
7-(6′′′-Acetylallosyl)(1 → 2)(6′′-acetylglucoside)	*Sideritis syriaca* and *S. scardica*	Labiatae	139
Hypolaetin 7,3′-dimethyl ether			
4′-Glucoside	*Leptocarpus elegans* culm	Restionaceae	138
5,6,4′-Trihydroxy-7,8-dimethoxy flavone (thymusin)			
6-Isobutyrate	*Asarina procumbens* aerial parts	Scrophulariaceae	140
5,8,4′-Trihydroxy-6,7-dimethoxy flavone (isothymusin)			
8-Glucoside	*Becium grandiflorum* leaves	Labiatae	141
5,7-Dihydroxy-6,8,4′-trimethoxy flavone (nevadensin)			

continued

TABLE 13.1
New Flavone Glycosides — *continued*

Glycoside	Source	Family	Ref.
5-Glucoside	*Lysionotus pauciflorus* aerial parts	Gesneriaceae	142
7-Glucoside	*Lysionotus pauciflorus* aerial parts	Gesneriaceae	143
5-Gentiobioside	*Lysionotus pauciflorus* aerial parts	Gesneriaceae	142
7-Rutinoside	*Lysionotus pauciflorus* aerial parts	Gesneriaceae	143
5,8-Dihydroxy-6,7,4′-trimethoxyflavone (8-hydroxysalvigenin)			
8-Glucoside	*Isodon erianderianus* leaves	Labiatae	144
5,6,7,8,3′,4′-Hexahydroxyflavone			
7-Glucoside	*Juniperus zeravschanica* fruits	Cupressaceae	145
5,6,7,3′,4′-Pentahydroxy-8-methoxyflavone			
7-Glucoside	*Vellozia nanuzae* leaves	Velloziaceae	146
5,6,3′,4′-Tetrahydroxy-7,8-dimethoxyflavone (pleurostimin 7-methyl ether)			
6-Glucoside	*Vellozia nanuzae* leaves	Velloziaceae	146
5,2′,6′-Trihydroxy-6,7-dimethoxyflavone			
2′-Glucoside	*Scutellaria baicalensis* roots	Labiatae	147
5,2′,6′-Trihydroxy-7,8-dimethoxyflavone			
2′-Glucuronide	*Scutellaria rivularis* roots	Labiatae	30
5,2′-Dihydroxy-7,8,6′-trimethoxyflavone			
2′-Glucuronide	*Scutellaria rivularis* roots	Labiatae	30
5,7,2′,4′,5′-Pentahydroxyflavone (isoetin)			
4′-Glucuronide	*Adonis aleppica* whole plant	Ranunculaceae	148
5,7,3′,4′,5′-Pentahydroxyflavone (tricetin)			
3′-Rhamnosyl(1 → 4)rhamnoside	*Mentha longifolia* aerial parts	Labiatae	149
3′-Glucoside-5′-rhamnoside			
7-Glucoside-3′-[6″-(3-hydroxy-3-methylglutaryl) glucoside]	*Frullania polysticta* whole plant	Frullaniaceae	62
Tricetin 7-methyl ether			
3′-Glucoside-5′-rhamnoside	*Mentha longifolia* aerial parts	Labiatae	149
Tricetin 3′-methyl ether			
7-Glucuronide	*Medicago sativa* aerial parts	Leguminosae	116
Tricetin 4′-methyl ether			
7-Apiosyl(1 → 2)(6″-acetylglucoside)	*Paullinia pinnata* leaves	Sapindaceae	123
Tricetin 3′,5′-dimethyl ether (tricin)			
7-β-D-Arabinopyranoside (setaricin)	*Setaria italica* leaves	Gramineae	150
4′-Apioside	*Salsola collina* aerial parts	Chenopodiaceae	151
7-(2″-*p*-Coumaroylglucuronosyl) (1 → 2)glucuronide	*Medicago sativa* aerial parts	Leguminosae	116
7-(2″-Feruloylglucuronosyl)(1 → 2)glucuronide			
7-(2″-Sinapoylglucuronosyl)(1 → 2)glucuronide			
7-[Glucuronosyl(1 → 3)(2‴-feruloylglucuronosyl) (1 → 2)glucuronide]			
7-[X″-(3-Hydroxy-3-methylglutaryl)glucoside]	*Frullania polysticta* whole plants	Frullaniaceae	62
Tricetin 7,3′,4′-trimethyl ether			
5-Glucoside	*Lethedon tannaensis* leaves	Thymelaeaceae	126
Tricetin 7,3′,4′,5′-tetramethyl ether			
5-Glucoside	*Lethedon tannaensis* leaves	Thymelaeaceae	126

TABLE 13.1
New Flavone Glycosides — *continued*

Glycoside	Source	Family	Ref.
5-Xylosyl(1 → 2)rhamnoside	*Bauhinia variegata* seeds	Leguminosae	152
5-Xylosyl(1 → 6)glucoside	*Lethedon tannaensis* leaves	Thymelaeaceae	126
6-Hydroxytricetin 6,3′,5′-trimethyl ether			
7-α-L-Arabinosyl(1 → 6)glucoside	*Mimosa rubicaulis* roots	Leguminosae	153
6-Hydroxytricetin 6,4′,5′-trimethyl ether			
3′-Rhamnoside	*Mimosa rubicaulis* roots	Leguminosae	154
6-Hydroxytricetin 6,7,3′,5′-tetramethyl ether			
5-Robinobioside	*Aloe barbadensis* leaves	Liliaceae	155
8-Hydroxytricetin			
5-Rhamnoside	*Argyreia speciosa* leaves	Convolvulaceae	156
5-Glucoside			
5,6,7,8,3′,4′- Hexahydroxyflavone			
7-Glucoside	*Juniperus seravschanica* fruits	Cupressaceae	157
5,7,2′,5′-Tetrahydroxy-8,6′-tetrahydroxy-8, 6′-dimethoxyflavone (viscidulin III)			
2′-Glucoside	*Scutellaria baicalensis* roots	Labiatae	158
5,2′,6′-Trihydroxy-6,7,8-trimethoxyflavone			
2′-Glucoside	*Scutellaria baicalensis* roots	Labiatae	147
5,7-Dihydroxy-6-*C*-methylflavone			
7-Xylosyl(1 → 3)xyloside	*Mosla chinensis*	Labiatae	159
5,7-Dihydroxy-6,8-di-*C*-methylflavone (matteuorien)			
7-[6″-(3-Hydroxy-3-methylglutaryl)glucoside]	*Matteuccia orientalis* rhizomes	Aspleniaceae	160
Stachysetin (**13.5**)	*Stachys aegyptiaca* aerial parts	Labiatae	161
8-Prenylapigenin			
4′-Rutinoside	*Desmodium gangeticum* stems	Leguminosae	162
3′-Prenylapigenin			
7-Rutinoside	*Pithecellobium dulce* stems	Leguminosae	163
8-*C*-Prenyl-5,7,4′-trihydroxy-3′-methoxyflavone (8-*C*-prenylchrysoeriol)			
7-Glucosyl(1 → 3)-α-L-arabinopyranoside	*Erythrina indica* seeds	Leguminosae	164

two related nonacetylated glycosides, Pinnatifinosides C and D (**13.3** and **13.4**), in which the fructose has been replaced by the sugars β-D-allulofuranose and α-D-allulofuranose, respectively, co-occurred with Pinnatifinoside A.[41] A second pentose sugar, α-D-lyxose, was found attached to the 8-hydroxyl of gossypetin (8-hydroxyquercetin) in the aerial parts of *Orostachys japonicus*,[422] a member of the Crassulaceae. This is an unexpected discovery as the 2-epimer of xylose is very rare in nature. Fucose, a characteristic constituent of algal and plant polysaccharides, has been found in combination with 5,6-dihydroxy-7-methoxyflavone (negletein) as the 6-rhamnosyl(1 → 2)fucoside in *Origanum vulgare*.[26] Glucosamine (2-amino-2-deoxyglucose) is the only amino sugar to have been found in combination with flavones or flavonols. This sugar is important in animal physiology as a component of chitin, mucoproteins, and mucopolysaccharides. The presence of α-D-glucosamine in the aerial parts of *Halocnemum strobilaceum* (Chenopodiaceae)[377] at the 7-hydroxyl of isorhamnetin is totally unexpected and should be further investigated to establish its exact location in the plant and to screen related plants for similar structures.

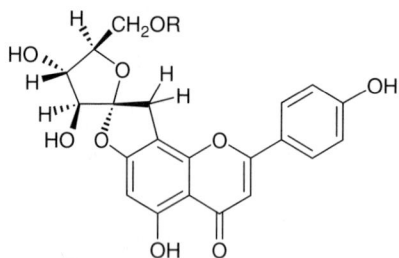

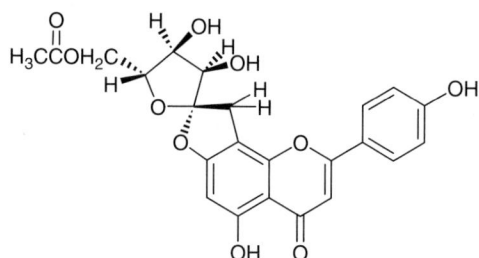

13.1 R = H
13.2 R = Acetyl

13.3

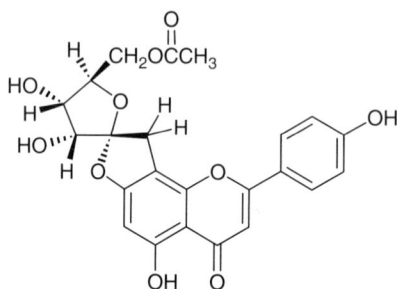

13.4

13.3.2 DISACCHARIDES

Twenty-one new disaccharides have been found in combination with flavones or flavonols since 1992. These are listed with previously known disaccharides in Table 13.4. Four of the new structures are novel sugar combinations: xylose–xylose, rhamnose–fucose, apiosyl–rhamnose, and glucose–arabinose. Both xylosyl(1 → 2)xylose and xylosyl(1 → 3)xylose have been found, the former at the 7-hydroxyl of scutellarein 6-methyl ether (patuletin) from roots of *Chirita fimbrisepala* (Gesneriaceae)[66] and its 1 → 3 isomer from *Mosla chinensis* (Labiatae)[159] at the 7-hydroxyl of 5,7,-dihydroxy-6-*C*-methylflavone. Rhamnosyl(1 → 2)fucose has already been dealt with under monosaccharides above. Apiofuranosyl(1 → 4)rhamnose has been recorded from *Chenopodium murale*[198] in combination with kaempferol at the 3-position and with rhamnose at the 7-hydroxyl. Five new glucosylarabinose isomers have been recorded during the review period. This is not surprising as arabinose can occur in either the pyranose or furanose form, with α or β linkage and with five possible linkage positions, which can all now be easily distinguished by modern NMR techniques. Glucosyl(1 → 4)-α-L-arabinopyranose was found at the 7-position of the flavone luteolin in seed of *Cassia glauca*[94] and its 1 → 5 linked furanose isomer in aerial parts of the legume, *Retama sphaerocarpa*,[402] at the 3-hydroxyl of quercetin 7,3′-dimethyl ether (rhamnazin). Glucosyl(1 → 3)-α-L-arabinopyranose occurs at the 7-hydroxyl of 8-prenylchysoeriol in seeds of *Erythrina indica*[164].The other two isomers were present only in acylated form. Thus, glucosyl(1 → 2)-β-arabinopyranose was recorded at the 3-hydroxyl of quercetin with glucose at the 7-hydroxyl and a feruloyl group at the 6-position of the glucose in a whole plant extract of *Carrichtera annua* (Cruciferae),[357] while the remaining isomer, glucosyl(1 → 2)-α-L-arabinofuranose, from *Euphorbia pachyrhiza*,[340] was found attached to the 3-hydroxyl of quercetin with a galloyl group at the 2-hydroxyl of the glucose.

TABLE 13.2
New Flavonol Glycosides

Glycoside	Source	Family	Ref.
3,5,7-Trihydroxyflavone (galangin)			
3-Glucoside-8-sulfate	*Phyllanthus virgatus* whole plant	Euphorbiaceae	165
3,7-Dihydroxy-8-methoxyflavone			
7-Rhamnoside	*Butea superba* stems	Leguminosae	166
7-Rhamnosyl(1 → 4) rhamnosyl(1 → 6)glucoside	*Shorea robusta* seeds	Dipterocarpaceae	167
5,7-Dihydroxy-3,6-dimethoxyflavone			
5-α-L-Arabinosyl(1 → 6)glucoside	*Acacia catechu* stems	Leguminosae	168
3,6,7-Trihydroxy-4′-methoxyflavone			
7-Rhamnoside	*Setaria italica* leaves	Gramineae	169
Kaempferol			
3-α-D-Arabinopyranoside	*Persea americana* leaves	Lauraceae	170
5-Glucuronide	*Leucanthemum vulgare* leaves	Compositae	171
7-Alloside	*Indigofera hebepetala* leaves	Leguminosae	172
3-Rhamnosyl(1 → 2)-α-L-arabinofuranoside (arapetaloside B)	*Artabotrys hexapetalus* leaves	Annonaceae	173
3-Xylosyl(1 → 2)glucoside	*Galium sinaicum* aerial parts	Rubiaceae	174
3-Rhamnoside(1 → 2)rhamnoside	*Cassia hirsuta* flowers	Leguminosae	175
3-Glucosyl(1 → 2)rhamnoside	*Ginkgo biloba* leaves	Ginkgoaceae	176, 177
7-Glucosyl(1 → 4)xyloside	*Crotalaria laburnifolia*	Leguminosae	178
7-Neohesperidoside	*Caralluma tuberculata*	Asclepiadaceae	179
7-Glucosyl(1 → 3)rhamnoside	*Rhodiola crenulata* roots	Crassulaceae	180
7-Sophoroside	*Crocus sativus* stamens (saffron)	Iridaceae	181
3,5-Diglucoside	*Dryopteris dickinsii* fronds	Aspleniaceae	182
3,7-Diarabinoside	*Indigofera hebepetala* leaves	Leguminosae	172
3,4′-Diglucoside	*Picea abies* needles	Pinaceae	183
7-Rhamnoside-4′-glucoside	*Pteridium aquilinum* aerial parts	Dennstaedtiaceae	184
7,4′-Diglucoside	*Cassia javanica*	Leguminosae	185
3-Xylosyl(1 → 3)rhamnosyl(1 → 6)galactoside	*Astragalus caprinus* leaves	Leguminosae	186
3-Xylosyl(1 → 6)glucosyl(1 → 2)rhamnoside	*Helicia nilagirica* leaves	Proteaceae	187
3-Rhamnosyl(1 → 3)rhamnosyl(1 → 6)glucoside	*Camellia sinensis* green tea	Theaceae	188
3-Rhamnosyl(1 → 2)glucosyl(1 → 6)galactoside	*Cassia marginata* stems	Leguminosae	189
3-Rhamnosyl(1 → 6)glucosyl(1 → 6)galactoside	*Albizia lebbeck* leaves	Leguminosae	190
3-Glucosyl(1 → 4)rhamnosyl(1 → 2)glucoside	*Allium neapolitanum* whole plant	Liliaceae	191
3-Glucosyl(1 → 2)galactosyl(1 → 2)glucoside	*Nigella sativa* seeds	Ranunculaceae	192
3-Rhamnosyl(1 → 2)[glucosyl(1 → 3)glucoside]	*Impatiens balsamina* petals	Balsaminaceae	193
3-Rhamnosyl(1 → 2)[glucosyl(1 → 4)glucoside]	*Allium neapolitanum* whole plant	Liliaceae	194
3-Glucosyl(1 → 2)[glucosyl(1 → 3)rhamnoside]	*Crocus speciosus* and *C. antalyensis* flowers	Iridaceae	195
7-(3^G-Glucosylgentiobioside)	*Brassica juncea* leaves	Cruciferae	196
3-Rhamnosyl(1 → 2)galactoside-7-α-L-arabinofuranoside			
3-Robinobioside-7-α-L-arabinofuranoside	*Indigo hebepetala* flowers	Leguminosae	197
3-Xylosyl(1 → 4)rhamnoside-7-rhamnoside	*Chenopodium murale* whole plant	Chenopodiaceae	198
3-Rhamnoside-7-xylosyl(1 → 2)rhamnoside	*Chenopodium murale* aerial parts	Chenopodiaceae	199
3-Apiosyl(1 → 4)rhamnoside-7-rhamnoside	*Chenopodium murale* whole plant	Chenopodiaceae	198

continued

TABLE 13.2
New Flavonol Glycosides — *continued*

Glycoside	Source	Family	Ref.
3-Neohesperidoside-7-rhamnoside	*Sedum telephium* subsp. *maximum* leaves	Crassulaceae	200
3-Rhamnoside-7-glucosyl(1 → 2)rhamnoside	*Siraita grosvenori* fresh fruit	Cucurbitaceae	201
3-Glucosyl(1 → 4)galactoside-7-α-L-arabinofuranoside	*Corchorus depressus* whole plant	Tiliaceae	202
3-Glucosyl(1 → 6)galactoside-7-α-L-arabinofuranoside			
3-Apioside-7-rhamnosyl(1 → 6)galactoside	*Silphium perfoliatum* leaves	Compositae	203
3-Glucosyl(1 → 2)rhamnoside-7-glucoside	*Crocus chrysanthus-biflorus* cvs "eye-catcher" and "spring pearl" flowers	Iridaceae	204
3-Gentiobioside-7-rhamnoside	*Arabidopsis thaliana* leaves	Cruciferae	205
3-Glucosyl(1 → 2)galactoside-7-glucoside	*Nicotiana* spp. flowers	Solanaceae	206
3-Sophoroside-7-glucuronide	*Allium cepa* guard cells	Alliaceae	207
3-Neohesperidoside-4′-glucoside	*Pseuderucaria clavata* aerial parts	Cruciferae	208
3-Neohesperidoside-7,4′-diglucoside	Minus flowers		
3-Gentiobioside-4′-glucoside	*Asplenium incisum* fronds	Aspleniaceae	209
3-Galactoside-3′,4′-dirhamnoside	*Astragalus tana* aerial parts	Leguminosae	210
3-Rhamnoside-7,4′-digalactoside	*Warburgia ugandensis* leaves	Rubiaceae	211
4′-Rhamnosyl(1 → 3)rhamnosyl(1 → 6)galactoside	*Rhamnus thymifolius* fruits	Rhamnaceae	212
3-Rhamnosyl(1 → 2)[xylosyl (1 → 3)rhamnosyl(1 → 6)galactoside]	*Astragalus caprinus* leaves	Leguminosae	213
3-Glucosyl(1 → 3)rhamnosyl(1 → 2) [rhamnosyl(1 → 6) galactoside]	*Maytenus aquifolium* leaves	Celestraceae	214
3-Rhamnosyl(1 → 4)rhamnosyl (1 → 6)galactoside-7-rhamnoside (3-isorhamninoside-7-rhamnoside)	*Vigna* spp. whole plant	Leguminosae	215
3-Rutinoside-7-sophoroside	*Equisetum* spp.	Equisetaceae	216
3-Sophoroside-7-cellobioside	*Brassica oleracea* leaves	Cruciferae	217
3-(2^G-Glucosylrutinoside)-7-rhamnoside	*Sophora japonica* seeds	Leguminosae	218
3-(2^G-Rhamnosylrutinoside)-7-glucoside (mauritianin 7-glucoside)	*Alangium premnifolium* leaves	Alangiaceae	219
3-Rhamnosyl(1 → 6)[rhamnosyl(1 → 2) galactoside]-7-rhamnoside	*Astragalus shikokianus* aerial parts	Leguminosae	220
3-Glucosyl(1 → 2)[rhamnosyl(1 → 6) galactoside]-7-rhamnoside	*Cephalocereus senilis* whole young plants	Cactaceae	221
3-[6″-(3-Hydroxy-3-methylglutaryl)glucoside]	*Citrus aurantifolia* callus cultures	Rutaceae	222
3-(6″-*p*-Hydroxybenzoylgalactoside)	*Persicaria lapathifolia* whole plants	Polygonaceae	223
3-(2″-Galloylarabinoside)	*Eucalyptus rostrata* leaves	Myrtaceae	224
3-(6″-Galloylgalactoside)	*Pemphis acidula* leaves	Lythraceae	225
3-(2″,6″-Digalloylglucoside)	*Loropetalum chinense* leaves	Hamamelidaceae	226
3-(2″-*E*-*p*-Coumaroyl-α-L-arabinofuranoside)	*Prunus spinosa* flowers	Rosaceae	227
3-(2″-*E*-*p*-Coumaroylrhamnoside)	*Platanus orientalis* buds	Platanaceae	228
	Platanus acerifolia buds	Platanaceae	229
3-(2″-*Z*-*p*-Coumaroylrhamnoside)	*Platanus acerifolia* buds	Platanaceae	229
3-(2″-*Z*-*p*-Coumaroylglucoside)	*Eryngium campestre* aerial parts	Umbelliferae	230
3-(4″-*p*-Coumaroylglucoside)	*Elaeagnus bockii* whole plant	Elaeagnaceae	231

TABLE 13.2
New Flavonol Glycosides — *continued*

Glycoside	Source	Family	Ref.
3-(6″-Caffeoylglucoside)	*Pteridium aquilinum* aerial parts	Dennstaedtiaceae	232
3-(5″-Feruloylapioside)	*Pteridium aquilinum* aerial parts	Dennstaedtiaceae	233
3-(6″-Feruloylglucoside)	*Polylepis incana* leaves	Rosaceae	234
3-(6″-Acetylglucoside)	*Picea abies* needles	Pinaceae	235
7-(6″-*p*-Coumaroylglucoside)	*Buddleia coriacea* aerial parts	Loganiaceae	236
3-(2″,3″-di-*E*-*p*-Coumaroylrhamnoside)	*Platanus orientalis* buds	Platanaceae	228
3-(2″,4″-di-*E*-*p*-Coumaroylrhamnoside)	*Pentachondra pumila* leaves and stems	Epacridaceae	237
3-(2″,4″-di-*Z*-*p*-Coumaroylrhamnoside)	*Laurus nobilis* leaves	Lauraceae	238
3-(2″,6″-di-*E*-*p*-Coumaroylglucoside)	*Quercus canariensis*	Fagaceae	239
3-(2′-*Z*-*p*-Coumaroyl-6″-*E*-*p*-coumaroylglucoside)			
3-(3″,6″-di-*Z*-*p*-Coumaroylglucoside) (stenopalustroside A)	*Stenochlaena palustris* leaves	Pteridaceae	240
3-(3″-*Z*-*p*-Coumaroyl-6″-*E*-feruloylglucoside) (stenopalustroside B)			
3-(3″-*Z*-*p*-Coumaroyl-6″-*E*-*p*-coumaroylglucoside) (stenopalustroside C)			
3-(3″-*E*-*p*-Coumaroyl-6″-*Z*-*p*-coumaroylglucoside) (stenopalustroside D) (isolated as a mixture)			
3-(3″-*E*-*p*-Coumaroyl-[6″-(4-*O*-(4-hydroxy-3-methoxyphenyl-1,3-dihydroxyisopropyl-feruloyl)] glucoside (stenopalustroside E)			
3-(2″-*E*-*p*-Coumaroyl-6″-acetylglucoside)	*Quercus dentata* leaves	Fagaceae	241
3-(3″-Acetyl-6″-*p*-coumaroylglucoside)	*Anaphalis aurea-punctata* whole plant	Compositae	242
3-(3″,4″-Diacetylglucoside)	*Minthostachys spicata* aerial parts	Labiatae	243
3-(6^G-Malonylneohesperidoside)	*Clitoria teratea* petals	Leguminosae	244
3-(2‴-*E*-Feruloylgalactosyl(1 → 4)glucoside)	*Allium porrum* bulbs	Alliaceae	245
3-(2‴-*E*-Feruloylgalactosyl(1 → 6)glucoside)			
3-(2^G-*E*-*p*-Coumaroylrutinoside)	*Alibertia sessilis* leaves	Rubiaceae	246
3-(6‴-Caffeoylglucosyl)(1 → 4)rhamnoside	*Rorippa indica* whole plant	Cruciferae	247
3-(6‴-*E*-Feruloylglucosyl)(1 → 2)galactoside	*Hedyotis diffusa* whole plant	Rubiaceae	248
3-(6‴-Sinapoylglucosyl)(1 → 2)galactoside	*Thevetia peruviana* leaves	Apocynaceae	249
3-(3‴-Acetyl-α-ʟ-arabinopyranosyl)(1 → 6)glucoside	*Thalictrum atriplex* aerial parts	Ranunculaceae	250
3-(6‴-Acetylglucosyl)(1 → 3)galactoside	*Ricinus communis* roots	Euphorbiaceae	251
3-(2‴-Feruloylglucosyl)(1 → 2)(6″-malonylglucoside)	*Petunia* cv "Mitchell" and its LC transgenic leaves	Solanaceae	252
3-[6″-(3-Hydroxy-3-methylglutaryl)glucoside]-7-glucoside	*Citrus aurantifolia* callus cultures	Rutaceae	222
3-(6″-Malonylglucoside)-7-glucoside	*Equisetum* spp.	Equisetaceae	216
3-(2″-*E*-*p*-Coumaroyl-α-ʟ-arabinofuranoside)-7-rhamnoside	*Prunus spinosa* leaves	Rosaceae	253
3-(3‴-*p*-Coumaroylrhamnoside)-7-rhamnoside	*Cheilanthes fragrans* aerial parts	Sinopteridaceae	254
3-(6″-*E*-*p*-Coumaroylglucoside)-7-glucoside	*Lotus polyphyllus* whole plant	Leguminosae	255
3-(2″,3″-Diacetylrhamnoside)-7-rhamnoside	*Dryopteris crassirhizoma* rhizomes	Filicales	256
3-(2″,4″-Diacetylrhamnoside)-7-rhamnoside			
3-(3″,4″-Diacetylrhamnoside)-7-rhamnoside			

continued

TABLE 13.2
New Flavonol Glycosides — *continued*

Glycoside	Source	Family	Ref.
3-(4″,6″-Diacetylglucoside)-7-rhamnoside	*Delphinium formosum* flowers	Ranunculaceae	257
3-(2‴,3‴,4‴-Triacetyl-α-L-arabinopyranosyl)(1 → 6) glucoside	*Calluna vulgaris* flowers	Ericaceae	258
3-[6″-(7⁗-Glucosyl-*p*-coumaroyl)glucosyl](1 → 2)rhamnoside	*Ginkgo biloba* leaves	Ginkgoaceae	177
3-Rhamnosyl(1 → 3)(4‴-acetylrhamnosyl)(1 → 6)glucoside	*Camellia sinensis* green tea	Theaceae	259
3-[2^{Gal}-(6‴-Feruloylglucosyl)robinobioside]	*Brunfelsia grandiflora* ssp. *grandiflora* aerial parts	Solanaceae	260
3-Rhamnosyl(1 → 3)(4‴-acetylrhamnosyl(1 → 6)galactoside)	*Rhamnus thymifolius* fruits	Rhamnaceae	212
3-Glucosyl(1 → 4)[(6‴-sinapoylglucosyl)(1 → 2)galactoside]	*Thevetia peruviana* leaves	Apocynaceae	249
3-(2⁗-Sinapoylglucosyl(1 → 4)[(6‴-sinapoylglucosyl) (1 → 2)galactoside]			
3-[2^{Gal}(4‴-Acetylrhamnosyl)robinobioside]	*Galega officinalis* aerial parts	Leguminosae	261
3-[2″-(4‴-Acetylrhamnosyl)sophoroside]	*Ammi majus* aerial parts	Umbelliferae	262
3-Neohesperidoside-7-(6″-malonylglucoside)	*Crocus chrysanthus biflorus* cvs "eye catcher" and "spring pearl" flowers	Iridaceae	204
3-Neohesperidoside-7-(6″-acetylglucoside)			
3-Neohesperidoside-7-(2″-*E*-*p*-coumaroylglucoside)	*Allium ursinum* whole plant	Liliaceae	263
3-Neohesperidoside-7-(2″-*E*-feruloylglucoside)			
3-(4‴-*p*-Coumaroylglucosyl)(1 → 2)rhamnoside-7-glucoside	*Mentha lavandulacea* aerial parts	Labiatae	264
3-(6‴-*p*-Coumaroylglucosyl)(1 → 2)rhamnoside-7-glucoside			
3-(6‴-*p*-Coumaroylglucosyl)(1 → 2)rhamnoside-7-glucoside	*Ginkgo biloba* leaves	Ginkgoaceae	265
3-Glucosyl(1 → 2)rhamnoside-7-(6″-*E*-*p*-coumaroylglucoside)	*Reseda muricata* leaves	Resedaceae	266
3-(6″-*E*-*p*-Coumaroylglucosyl)(1 → 2)glucoside-7-rhamnoside	*Aconitum napellus* ssp. *tauricum* flowers	Ranunculaceae	267
3-(6‴-*E*-Caffeoylglucosyl)(1 → 2)glucoside-7-rhamnoside			
3-Glucoside-7-(6‴-*E*-*p*-coumaroylglucosyl)(1 → 3) rhamnoside	*Aconitum napellus* ssp. *neomontanum*	Ranunculaceae	268
3-Glucoside-7-(6‴-*E*-caffeoylglucosyl)(1 → 3)rhamnoside			
3-(2‴-*E*-*p*-Coumaroylsophoroside)-7-glucoside	*Brassica oleracea* leaves	Cruciferae	269
3-(2‴-*E*-Caffeoylsophoroside)-7-glucoside			
3-(2‴-*E*-Feruloylsophoroside)-7-glucoside			
3-Apioside-7-rhamnosyl(1 → 6)(2″-*E*-caffeoylgalactoside)	*Silphium perfoliatum* leaves	Compositae	203
3-Xylosyl(1 → 2)rhamnoside-7-(4″-acetylrhamnoside)	*Kalanchoe streptantha* leaves	Crassulaceae	270
3-Glucosyl(1 → 2)(6″-acetylgalactoside)-7-glucoside	*Trigonella foenum graecum*	Leguminosae	271
3-Rhamnosyl(1 → 2)[glucosyl(1 → 3)(4‴-*p*-coumaroylrhamnosyl)(1 → 6)galactoside]	*Lysimachia capillipes* whole plant	Primulaceae	272
3-Rhamnosyl(1 → 2)[xylosyl(1 → 3)rhamnosyl(1 → 6) (3″-*p*-coumaroylgalactoside)]	*Astragalus caprinus* leaves	Leguminosae	213
3-Rhamnosyl(1 → 2)[xylosyl(1 → 3)rhamnosyl(1 → 6) (4″-*p*-coumaroylgalactoside)]			
3-Rhamnosyl(1 → 2)[xylosyl(1 → 3)rhamnosyl(1 → 6) (3″-feruloylgalactoside)]			
3-Rhamnosyl(1 → 2)[xylosyl(1 → 3)rhamnosyl(1 → 6) (4″-feruloylgalactoside)]			
3-Neohesperidoside-7-[2″-*E*-*p*-coumaroyllaminaribioside]	*Allium ursinum* whole plant	Liliaceae	263

TABLE 13.2
New Flavonol Glycosides — *continued*

Glycoside	Source	Family	Ref.
3-(2‴-*E*-Caffeoylglucosyl)(1 → 2) glucoside-7-cellobioside	*Brassica oleracea* leaves	Cruciferae	217
3-(2‴-*E*-Feruloylglucosyl)(1 → 2) glucoside-7-cellobioside			
3-(2″-*E*-Sinapoylglucosyl)(1 → 2) glucoside-7-cellobioside			
3-Glucosyl(1 → 6)[rhamnosyl(1 → 3)(2″-*E*-*p*-coumaroylglucoside)]-7-rhamnosyl(1 → 3) rhamnosyl (1 → 3)(4″-*E*-*p*-coumaroylrhamnoside)	*Planchonia grandis*	Lecythidaceae	273
3-Glucosyl(1 → 6)[rhamnosyl(1 → 3)(2″-*E*-*p*-coumaroylglucoside)]-7-rhamnosyl(1 → 3) rhamnosyl(1 → 3)(4″-*Z*-*p*-coumaroylrhamnoside)			
3-Rhamnosyl(1 → 6)[rhamnosyl(1 → 3)(2″-*E*-*p*-coumaroylglucoside)]-7-rhamnosyl(1 → 3) rhamnosyl (1 → 3)rhamnosyl(1 → 3) (4″-*E*-*p*-coumaroylrhamnoside)			
3-Sulfate-7-α-arabinopyranoside	*Atriplex hortensis* leaves	Chenopodiaceae	274
8-*C*-Sulfate	*Phyllanthus virgatus* whole plant	Euphorbiaceae	165
Kaempferol 3-methyl ether			
7-Glucuronide	*Centaurea bracteata* aerial parts	Compositae	275
Kaempferol 7-methyl ether (rhamnocitrin)			
4′-Glucoside	*Cotoneaster simonsii* leaves	Rosaceae	276
3-Xylosyl(1 → 2)[rhamnosyl(1 → 6)glucoside]	*Cestrum nocturnum* leaves	Solanaceae	277
3-Rhamnosyl(1 → 3)[apiosyl(1 → 6)glucoside]	*Mosla soochouensis* stem wood	Labiatae	278
3-Apiosyl(1 → 5)apioside-4′-glucoside	*Mosla chinensis*	Labiatae	279
3-Neohesperidoside-4′-glucoside	*Cadaba glandulosa* aerial parts	Capparidaceae	280
3-Apiosyl(1 → 5)apiosyl(1 → 2) [rhamnosyl(1 → 6)glucoside]	*Viscum angulatum* whole plant	Viscaceae	281
3-[3-Hydroxy-3-methylglutaryl(1 → 6)][apiosyl (1 → 2)galactoside]	*Astragalus caprinus* leaves	Leguminosae	186
3-[5‴-*p*-Coumaroylapiosyl(1 → 2)glucoside]	*Astragalus complanatus* seeds	Leguminosae	282
3-[5‴-Feruloylapiosyl(1 → 2)glucoside]			
3-glucoside-4′-(3‴-dihydrophaseoylglucoside)			
3-(6-*E*-3,5-Dimethoxy-4-hydroxycinnamoylglucosyl) (1 → 2)[rhamnosyl(1 → 6)glucoside]	*Cestrum nocturnum* leaves	Solanaceae	277
Kaempferol 4′-methylether (kaempferide)			
3-Rhamnoside	*Agrimonia eupatoria* aerial parts	Rosaceae	283
3-Neohesperidoside	*Costus spicatus* leaves	Costaceae	284
3-Rhamnoside-7-xyloside	*Cassia biflora* leaves	Leguminosae	285
3-(4Rha-Rhamnosylrutinoside)	*Sageretia filiformia* leaves	Rhamnaceae	286
3-(2Glc-Glucosylrutinoside)	*Dianthus caryophyllus* cv. "Novada" leafy stems and roots	Caryophyllaceae	287
3-[6‴-Acetyl(4″-α-methylsinapoylneohesperido-side)]	*Aerva tomentosa*	Amaranthaceae	288
Kaempferol 3,5-dimethyl ether			
7-Glucoside	*Nitraria tangutorum* leaves	Nitrariaceae	289

continued

TABLE 13.2
New Flavonol Glycosides — *continued*

Glycoside	Source	Family	Ref.
Kaempferol 3,7-dimethyl ether			
4′-Glucoside	*Lantana camara* leaves	Verbenaceae	290
Kaempferol 7,4′-dimethyl ether			
3-Glucoside	*Gymnotheca involucrata* whole plant	Saururaceae	291
3-Neohesperidoside	*Costus spiralis* leaves	Costaceae	292
6-Hydroxykaempferol			
3-Glucoside	*Carthamnus tinctorius* petals	Compositae	293
7-Alloside	*Tagetes erecta* flowers	Compositae	294
3-Rutinoside	*Daphniphyllum calycinum* leaves	Daphniphyllaceae	295
3,6-Diglucoside	*Carthamnus tinctorius* petals	Compositae	293
3,6,7-Triglucoside			
3-Rutinoside-6-glucoside			
7-(6″-Caffeoylglucoside)	*Eupatorium glandulosum* leaves	Compositae	296
6-Hydroxykaempferol 6-methyl ether (eupafolin)			
3-(6″-*p*-Coumaroylglucoside)	*Paepalanthus polyanthus, P. hilairei, P. robustus, P. ramosus,* and *P. denudatus* capitulae	Eriocaulaceae	297
6-Hydroxykaempferol 4′-methyl ether			
7-Glucoside	*Serratula strangulata* whole plant	Compositae	298
7-Galactoside			
6-Hydroxykaempferol 6,4′-dimethyl ether			
3-Glucoside	*Arnica montana* flowers	Compositae	299
6-Hydroxykaempferol 3,5,7,4′-tetramethyl ether			
6-Rhamnoside	*Pterocarpus marsupium* roots	Leguminosae	300
8-Hydroxykaempferol (herbacetin)			
3-β-D-Glucofuranoside	*Jungia paniculata* whole plant	Compositae	301
3-Rhamnoside-8-glucoside	*Ephedra aphylla* aerial parts	Ephedraceae	302
Herbacetin 7-methyl ether			
8-Sophoroside	*Ranunculus sardous* pollen	Ranunculaceae	303
3-(2″-*E*-Feruloylglucoside)	*Ranunculus sardous* pollen	Ranunculaceae	304
Herbacetin 8-methyl ether (sexangularetin)			
3-Neohesperidoside	*Crateagus monogyna* pollen	Rosaceae	305
Herbacetin 7,8,4′-trimethyl ether (tambulin)			
3,5-Diacetate	*Zanthoxylum integrifolium* fruits	Rutaceae	306
5,7,8-Trihydroxy-3-methoxyflavone			
8-(*E*-2-Methylbut-2-enoate)	*Pseudognaphalium robustum* and *P. cheirantifolium* whole plant	Compositae	307
5,7,8-TriOH-3,6-dimethoxyflavone			
8-(*E*-2-Methylbut-2-enoate)	*Pseudognaphalium robustum* and *P. cheirantifolium* whole plant	Compositae	307
6,8-Dihydroxykaempferol			
3-Rutinoside	*Withania somnifera* leaves	Solanaceae	308
Quercetin			
3-α-D-Arabinopyranoside	*Persea americana* leaves	Lauraceae	170
5-Glucuronide	*Leucanthemuum vulgare* leaves	Compositae	171
4′-Galactoside	*Cornulaca monacantha* aerial parts	Chenopodiaceae	309

TABLE 13.2
New Flavonol Glycosides — *continued*

Glycoside	Source	Family	Ref.
4'-Glucuronide	*Psidium guaijava* leaves	Myrtaceae	310
3-Rhamnosyl(1 → 2)-α-L-arabinofuranoside (arapetaloside A)	*Artabotrys hexapetalus* leaves	Annonaceae	173
3-α-L-Arabinofuranosyl(1 → 2)glucoside	*Prunus spinosa* leaves	Rosaceae	253
3-Xylosyl(1 → 6)glucoside	*Cistus ladanifer* pollen	Cistaceae	311
3-Rhamnosyl(1 → 2)rhamnoside	*Centaurea horrida* aerial parts	Compositae	312
3-Glucosyl(1 → 2)rhamnoside	*Ginkgo biloba* leaves	Ginkgoaceae	177
3-Galactosyl(1 → 2)rhamnoside	*Embelia schimperi* leaves	Myrsinaceae	313
3-Laminaribioside	*Pteridium aquilinum* aerial parts	Dennstaedtiaceae	314
3-Glucosyl(1 → 3)galactoside	*Filipendula formosa* aerial parts	Rosaceae	315
3-Glucosyl(1 → 4)galactoside	*Rumex chalepensis* leaves	Polygonaceae	316
3-Glucosyl(1 → 2)glucuronide	*Cordia macleodii* leaves and flowers	Boraginaceae	317
3-Rhamnoside-3'-glucoside	*Myrsine seguinii* leaves	Myrsinaceae	318
3-Xylosyl(1 → 2)rhamnosyl(1 → 6)glucoside	*Camellia saluensis* leaves	Theaceae	319
3-Xylosyl(1 → 2)rhamnoside	*Helicia nilagirica* leaves	Proteaceae	187
3-Apiosyl(1 → 2)rhamnosyl(1 → 6)glucoside	*Baccharis thesioides* aerial parts	Compositae	320
3-(6'''-Rhamnosylgentiobioside)	*Capparis spinosa* aerial parts	Capparidaceae	321
3-Rhamnosyl(1 → 2)glucosyl(1 → 6)galactoside	*Cassia marginata* stems	Leguminosae	189
3-Rhamnosyl(1 → 6)glucosyl(1 → 6)galactoside	*Albizia lebbeck* leaves	Leguminosae	190
3-Glucosyl(1 → 2)galactosyl(1 → 2)glucoside	*Nigella sativa* seeds	Ranunculaceae	192
7-(2G-Xylosylrutinoside)	*Bidens andicola* aerial parts	Compositae	322
3-Glucosyl(1 → 2)[rhamnosyl(1 → 6)galactoside]	*Thevetia peruviana* leaves	Apocynaceae	323
3-Rhamnosyl(1 → 2)-α-L-arabinopyranoside-7-glucoside	*Putoria calabrica* aerial parts	Rubiaceae	324
3-Xylosyl(1 → 2)glucoside-7-rhamnoside	*Lathyrus chrysanthus* and *L. chloranthus* flowers	Leguminosae	325
3-Galactoside-7-glucosyl(1 → 4)rhamnoside			
3-Neohesperidoside-7-rhamnoside	*Sedum telephium* subsp. *maximum* leaves	Crassulaceae	200
3-Rhamnosyl(1 → 4)rhamnoside-7-galactoside	*Maesa lanceolata* leaves	Myrsinaceae	326
3-Neohesperidoside-7-glucoside	*Nicotiana* spp. flowers	Solanaceae	206
3-Glucosyl(1 → 2)rhamnoside-7-glucoside	*Crocus chrysanthus-biflorus* cvs "eye catcher" and "spring pearl" flowers	Iridaceae	204
3-Glucosyl(1 → 2)galactoside-7-glucoside	*Trigonella foenum-graecum* stems	Leguminosae	271
3-Sophoroside-7-glucuronide	*Allium cepa* guard cells	Alliaceae	207
3-Rutinoside-3'-apioside	*Plantago ovata* and *P. psyllium* seeds	Plantaginaceae	327
3,3',4'-Triglucoside	*Eruca sativa* leaves	Cruciferae	328
3-Xylosyl(1 → 4)[xylosyl(1 → 6)glucosyl(1 → 2)rhamnoside]	*Helicia nilagirica* leaves	Protaceae	187
3-Xylosyl(1 → 3)rhamnosyl(1 → 6)[apiosyl(1 → 2)galactoside]	*Astragalus caprinus* leaves	Leguminosae	213
3-Glucosyl(1 → 4)rhamnoside-7-rutinoside	*Myrsine africana* leaves	Myrsinaceae	329
3-Rhamnosyl(1 → 6)[glucosyl(1 → 2)glucoside]-7-rhamnoside	*Warburgia ugandensis* leaves	Rubiaceae	211
7-[Xylosyl(1 → 2)rhamnosyl(1 → 2)rhamnosyl](1 → 6)glucoside	*Bidens andicola* aerial parts	Compositae	322

continued

TABLE 13.2
New Flavonol Glycosides — *continued*

Glycoside	Source	Family	Ref.
3-(4″-Malonylrhamnoside)	*Ribes alpinum* leaves	Grossulariaceae	330
3-(2″-Caffeoylglucuronide)	*Scolymus hispanicus*	Compositae	331
3-(6″-Feruloylgalactoside)	*Persicaria lapathifolia* aerial parts	Polygonaceae	332
3-(2″-Acetylrhamnoside)	*Nymphaea caerulea* blue flowers	Nymphaeaceae	333
3-(2″-Acetylgalactoside)	*Hypericum perforatum* dried crude drug	Guttiferae	334
3-(6″-*n*-Butylglucuronide) (parthenosin)	*Parthenocissus tricuspidata* leaves	Vitaceae	335
7-(6″-Galloylglucoside)	*Acacia farnesiana* pods	Leguminosae	336
7-(6″-Acetylglucoside)	*Carthamnus tinctorius* leaves	Compositae	337
4′-(6″-Galloylglucoside)	*Eucalyptus rostrata* leaves	Myrtaceae	224
3-(2″,6″-Digalloylgalactoside)	*Acer okamotanum* leaves	Aceraceae	338
3-(3″,6″-Diacetylgalactoside)	*Tagetes elliptica* aerial parts	Compositae	339
3-(2″,3″,4″-triacetylgalactoside)			
3-(2‴-Galloylglucosyl)(1 → 2)-α-L-arabinofuranoside	*Euphorbia pachyrhiza*	Euphorbiaceae	340
3-(2″-Galloylglucoside)-4′-vinylpropionate	*Psidium guaijava* seeds	Myrtaceae	341
3-(2″-Galloylrutinoside)	*Euphorbia ebractedata* aerial parts	Euphorbiaceae	342
3-(6ᴳ-Malonylneohesperidoside)	*Clitoria ternatea* petals	Leguminosae	244
3-α-L-Arabinopyranosyl(1 → 6) (2″-*E*-*p*-coumaroylglucoside)	*Vicia angustifolia* leaves and stems	Leguminosae	343
3-α-L-Arabinopyranosyl(1 → 6) (2″-*E*-*p*-coumaroylgalactoside)			
3-(2ᴳ-*E*-*p*-Coumaroylrutinoside)	*Alibertia sessilis* leaves	Rubiaceae	246
3-(2‴-*E*-Caffeoyl-α-L-arabinopyranosyl) (1 → 6)glucoside	*Morina nepalensis* var. *alba* whole plant	Morinaceae	344
3-(2‴-*E*-Caffeoyl-α-L-arabinopyranosyl) (1 → 6)galactoside			
3-(6″-Caffeoylgentiobioside)	*Lonicera implexa* leaves	Caprifoliaceae	107
3-(6″-Caffeoylsophoroside)	*Bassia muricata* aerial parts	Chenopodiaceae	345
3-(6″-Feruloylsophoroside)			
3-(2‴-Feruloylsophoroside)	*Petunia* cv "Mitchell" and its LC transgenic leaves	Solanaceae	252
3-(6‴-Sinapoylglucosyl)(1 → 2)galactoside	*Thevetia peruviana* leaves	Apocynaceae	249
3-Rhamnosyl(1 → 6)(2″-acetylglucoside)	*Prunus mume* flowers	Rosaceae	346
3-(6″-Acetylglucosyl)(1 → 3)galactoside (euphorbianin)	*Euphorbia hirta* leaves	Euphorbiaceae	347
3-(2″-Caffeoylglucoside)(1 → 2) (6″-malonylglucoside)	*Petunia* cv. "Mitchell" and its LC transgenic leaves	Solanaceae	252
3-(3″,4″-Diacetylrhamnosyl)(1 → 6)glucoside	*Tordylium apulum* aerial parts	Umbelliferae	348
3-[2‴,3‴,4‴-Triacetyl-α-L-arabinopyranosyl (1 → 6)glucoside]	*Calluna vulgaris* flowers	Ericaceae	349
3-[2‴,3‴,4‴-Triacetyl-α-L-arabinopyranosyl (1 → 6)galactoside]	*Calluna vulgaris* flowers	Ericaceae	350
3-[2‴,3‴,5‴-Triacetyl-α-L-arabinopyranosyl (1 → 6)glucoside]	*Calluna vulgaris* flowers	Ericaceae	258
3-[2″,6″-{-*p*-(7‴-Glucosyl)coumaroyl} glucosyl]rhamnoside	*Ginkgo biloba* leaves	Ginkgoaceae	177

TABLE 13.2
New Flavonol Glycosides — *continued*

Glycoside	Source	Family	Ref.
3-(6″-Malonylglucoside)-7-glucoside	*Ranunculus fluitans* leaves	Ranunculaceae	351
3-(6″-*E-p*-Coumaroylglucoside)-7-glucoside	*Lotus polyphyllus* whole plant	Leguminosae	255
3-(6″″-*p*-Coumaroylsophorotrioside)	*Pisum sativum* shoots	Leguminosae	352
3-(6″″-Caffeoylsophorotrioside)			
3-(6″″-Feruloylsophorotrioside)			
3-(6″″- Sinapoylsophorotrioside)			
3-(6″″-Feruloylglucosyl)(1 → 2) galactosyl(1 → 2)glucoside	*Nigella sativa* seeds	Ranunculaceae	192
3-Rutinoside-7-(6″-benzoylglucoside)	*Canthium dicoccum* leaves	Rubiaceae	353
3-(*p*-Coumaroylsambubioside)-7-glucoside	*Ranunculus* spp. leaves	Ranunculaceae	354
3-(6‴-*p*-Coumaroylglucosyl)(1 → 2)rhamnoside-7-glucoside	*Ginkgo biloba* leaves	Ginkgoaceae	177
3-(6″-*E-p*-Coumaroylsophoroside)-7-rhamnoside	*Aconitum napellus* ssp. *tauricum* flowers	Ranunculaceae	267
3-(*p*-Coumaroylsophoroside)-7-glucoside	*Ranunculus* spp. leaves	Ranunculaceae	354
3-(Caffeoylarabinosylglucoside)-7-glucoside			
3-(2‴-Caffeoylsambubioside)-7-glucoside			
3-(Feruloylsambubioside)-7-glucoside			
3-(4‴-Caffeoylrhamnosyl)(1 → 2)-α-L-arabinopyranoside-7-glucoside	*Putoria calabrica* aerial parts	Rubiaceae	324
3-Glucosyl-7-(6″-*E*-caffeoylglucosyl)(1 → 3) rhamnoside	*Aconitum napellus* ssp. *neomontanum* flowers	Ranunculaceae	268
3-(6″-Caffeoylsophoroside)-7-rhamnoside	*Aconitum baicalense* aerial parts	Ranunculaceae	355
3-Sophoroside-7-(6″-*trans*-caffeoylglucoside)	*Symplocarpus renifolius* leaves	Araceae	356
3-(2‴-*E*-Caffeoylsophoroside)-7-glucoside	*Brassica oleracea* leaves	Cruciferae	269
3-(2‴-*E*-Feruloylsophoroside)-7-glucoside			
3-[(6″-Feruloylglucosyl)(1 → 2)-β-arabinopyranoside]-7-glucoside	*Carrichtera annua* whole plant	Cruciferae	357
3-(6″-*E*-Sinapoylsophoroside)-7-rhamnoside	*Elaeagnus bockii* leaves	Elaeagnaceae	358
3-Caffeoylsophoroside-7-caffeoylglucoside	*Ranunculus fluitans* leaves	Ranunculaceae	351
3-Caffeoylsophoroside-7-feruloylglucoside			
3-(2″-Sinapoylglucoside)-3′-(6″-sinapoylglucoside)-4′-glucoside	*Eruca sativa* leaves	Cruciferae	328
3,4′-Diglucoside-3′-(6″-sinapoylglucoside)			
3-Rhamnosyl(1 → 6)[rhamnosyl(1 → 2)(3″-*E-p*-coumaroylgalactoside)]-7-rhamnoside	*Rhazya orientalis* aerial parts	Apocynaceae	359
3-Rhamnosyl(1 → 6)[rhamnosyl(1 → 2)(4″-*E-p*-coumaroylgalactoside)]-7-rhamnoside			
3-Rhamnosyl(1 → 2)[glucosyl(1 → 3)(4‴-*p*-coumaroylrhamnosyl)(1 → 6)galactoside]	*Lysimachia capillipes* whole plant	Primulaceae	272
Diquercetin 3-galactoside ester of tetrahydroxy-μ-truxinic acid (monochaetin, **13.6**)	*Monochaetum multiflorum* leaves	Melastomataceae	360
3-Sulfate-7-α-arabinopyranoside	*Atriplex hortensis* leaves	Chenopodiaceae	274
3-Rhamnoside-3′-sulfate	*Leea guineensis* leaves	Leeaceae	361
3-Glucoside-3′-sulfate	*Centaurea bracteata* aerial parts	Compositae	275

continued

TABLE 13.2
New Flavonol Glycosides — *continued*

Glycoside	Source	Family	Ref.
7,4′-Disulfate	*Alchornea laxiflora* whole plant	Euphorbiaceae	362
Quercetin 3-methyl ether			
5-Glucoside	*Asplenium trichomanes-ramosum* fronds	Aspleniaceae	363
7-α-L-Arabinofuranosyl(1 → 6)glucoside	*Lepisorus ussuriensis* whole plant	Polypodiaceae	364
7-Rutinoside	*Bidens leucantha* leaves	Compositae	365
7-Gentiobioside	*Lonicera implexa* leaves	Caprifoliaceae	107
7-Galactosyl(1 → 4)glucoside	*Acacia catechu* stems	Leguminosae	366
5-Glucoside-3′-sulfate	*Calorphus elongatus* culms	Restionaceae	367
Quercetin 7-methyl ether (rhamnetin)			
3-α-L-Arabinopyranosyl(1 → 3)galactoside	*Pongamia pinnata* seeds	Leguminosae	368
3-Robinobioside	*Cassia siamea* stem bark	Leguminosae	369
3-Laminaribioside	*Pteridium aquilinum* aerial parts	Dennstaedtiaceae	370
3-Gentiobioside	*Cassia fistula* roots	Leguminosae	371
3-α-L-Arabinopyranosyl(1 → 3) [galactosyl(1 → 6)galactoside]	*Pongamia pinnata* seeds	Leguminosae	368
3-[3-Hydroxy-3-methylglutaryl(1 → 6)] [apiosyl (1 → 2)galactoside]	*Astragalus caprinus* leaves	Leguminosae	186
3-(3′′′′-*p*-Coumaroylrhamnoside)	*Rhamnus petiolaris* fruit	Rhamnaceae	372
3,3′-Disulfate	*Argyreia mollis* roots	Convolvulaceae	373
3,3′,4′-Trisulfate	*Tamarix amplexicaulis* leaves	Tamaricaceae	374
Quercetin 3′-methyl ether			
3-Rhamnoside	*Oxytropis lanata* aerial parts	Leguminosae	375
5-Galactoside	*Pyrus bourgaeana* aerial parts	Rosaceae	376
7-α-D-Glucosaminopyranoside	*Halocnemum strobilaceum* aerial parts	Chenopodiaceae	377
3-α-Arabinopyranosyl(1 → 6)galactoside	*Trillium apeton* and *T. kamtechaticum* leaves	Liliaceae	378
3-Xylosyl(1 → 2)glucoside	*Lathyrus chrysanthus* and *L. chloranthus* flowers	Leguminosae	325
3-Xylosyl(1 → 6)glucoside	*Cistus ladanifer* pollen	Cistaceae	311
3-Xylosyl(1 → 2)galactoside	*Asclepias syriaca* flowers	Asclepiadaceae	379
3-Apiosyl(1 → 2)glucoside	*Vernonia galamensis* ssp. *galamensis* var. *petitiana* whole plant	Compositae	380
3-Apiosyl(1 → 2)galactoside	*Vernonia galamensis* spp. *nairobiensis* leaves	Compositae	381
3-Laminaribioside	*Pteridium aquilinum* aerial parts	Dennstaedtiaceae	314
3-Glucosyl(1 → 3)galactoside	*Achlys triphylla* underground parts	Berberidaceae	382
3-Galactoside-7-rhamnoside	*Lathyrus chrysanthus* and *L. chloranthus* flowers	Leguminosae	325
4′-Rhamnosyl(1 → 2)glucoside (crosatoside A)	*Crocus sativus* pollen	Iridaceae	383
3-Xylosyl(1 → 3)rhamnosyl(1 → 6)glucoside	*Hamada scoparia* leaves	Chenopodiaceae	384
3-Xylosylrobinobioside	*Nitraria retusa* leaves and young stems	Nitrariaceae	385
3-Apiosyl(1 → 2)[rhamnosyl(1 → 6)glucoside]	*Pituranthos tortuosus* shoots	Umbelliferae	386
3-Rhamnosyl(1 → 2)[glucosyl(1 → 6)glucoside	*Allium neapolitanum* whole plant	Liliaceae	191
3-(4^{Rha}-Galactosylrobinobioside)	*Nitraria retusa* leaves and young stems	Nitrariaceae	385
3-Galactosyl(1 → 2)[rhamnosyl(1 → 6)glucoside])	*Calotropis gigantean* aerial parts	Asclepiadaceae	387
3-Xylosyl(1 → 2)glucoside-7-rhamnoside	*Lathyrus chrysanthus* and *L. chloranthus* flowers	Leguminosae	325

TABLE 13.2
New Flavonol Glycosides — *continued*

Glycoside	Source	Family	Ref.
3-Glucosyl(1 → 6)galactoside-7-glucoside	*Heterotropa aspera* leaves	Aristolochiaceae	388
3-Rhamnosyl(1 → 2)gentiobiosyl (1 → 6)glucoside	*Allium neopolitanum* whole plant	Liliaceae	191
3-Rhamnosyl(1 → 2)gentiobioside-7-glucoside			
3-(2^G-Rhamnosylrutinoside)-7-rhamnoside	*Coleogyne ramosissima* aerial parts	Rosaceae	389
3-[6″-(2-*E*-Butenoyl)glucoside]	*Zygophyllum simplex* aerial parts	Zygophyllaceae	390
3-(2″,3″,4″-Triacetylglucoside)	*Warburgia stuhlmanii* leaves	Cannellaceae	391
7-(6″-*p*-Coumaroylglucoside)	*Buddleia coriacea* aerial parts	Loganiaceae	236
3-(6‴-*p*-Coumaroylglucosyl)(1 → 2)rhamnoside	*Ginkgo biloba* leaves	Ginkgoaceae	265
3-(3‴-Feruloylrhamnosyl)(1 → 6)galactoside	*Allium neopolitanum* whole plant	Liliaceae	191
3-(6″-*E*-Sinapoylsophoroside)	*Cassia marginata* stems	Leguminosae	189
3-(2‴-Acetyl-α-arabinopyranosyl)(1 → 6) galactoside	*Trillium apetalon* and *T. kamtschaticum* leaves	Liliaceae	378
3-Rhamnosyl(1 → 6)(2″-acetylglucoside)	*Prunus mume* flowers	Rosaceae	346
3-(6″-Acetylglucosyl)(1 → 3)galactoside	*Achlys triphylla* underground parts	Berberidaceae	382
3-(4″,6″-Diacetylglucosyl)(1 → 3)galactoside			
3-(6″-*E*-*p*-Coumaroylglucoside)-7-glucoside	*Lotus polyphyllus* whole plant	Leguminosae	255
3-[2″-(4‴-Acetylrhamnosyl)gentiobioside]	*Ammi majus* aerial parts	Umbelliferae	262
3-Rhamnosyl(1 → 6)[rhamnosyl(1 → 2) (3‴-*E*-*p*-coumaroylgalactoside]-7-rhamnoside	*Rhazya orientalis* aerial parts	Apocynaceae	359
3-Rhamnosyl(1 → 6)[rhamnosyl(1 → 2) (4″-*p*-coumarylgalactoside)]-7-rhamnoside			
3-Rhamnosyl(1 → 6)[rhamnosyl(1 → 2) (4″-*Z*-*p*-coumaroylgalactoside)]			
3-Rhamnosyl(1 → 6)[rhamnosyl(1 → 2) (4″-*E*-feruloylgalactoside)]-7-rhamnoside			
3-(4″-Sulfatorutinoside)	*Zygophyllum dumosum* aerial parts	Zygophyllaceae	392
Quercetin 4′-methyl ether (tamarixetin)			
3-Galactoside	*Cynanchum thesioides* whole plant	Asclepiadaceae	393
3-Neohesperidoside	*Costus spicatus* leaves	Costaceae	284
3-Glucosyl(1 → 2)galactoside	*Cynanchum thesioides* whole plant	Asclepiadaceae	393
3,7-Diglucoside	*Zanthoxylum bungeanum* pericarps	Rutaceae	394
3-Rutinoside-7-rhamnoside	*Cassia italica* aerial parts	Leguminosae	395
3-Glucoside-7-sulfate	*Polygonum hydropiper* leaves	Polygonaceae	396
Quercetin 3,5-dimethyl ether (caryatin)			
7-Glucoside	*Eucryphia glutinosa* twigs	Eucryphiaceae	397
Quercetin 3,7-dimethyl ether			
3′-Neohesperidoside	*Dasymaschalon sootepense* leaves	Annonaceae	398
3′-(6^G-Rhamnosylneohesperidoside)			
4′-Sulfate	*Ipomoea regnelli*	Convolulacaeae	373
Quercetin 3,3′-dimethyl ether			
7-Rutinoside	*Bidens pilosa* roots	Compositae	399
Quercetin 3,4′-dimethyl ether			
7-Glucoside	*Zanthoxylum bungeanum* pericarps	Rutaceae	394
7-α-L-Arabinofuranosyl(1 → 6)glucoside	*Punica granatum* bark	Punicaceae	400
7-Rutinoside	*Bidens pilosa* var. *radiata* aerial parts	Compositae	401
7-Rutinoside	*Bidens leucantha* leaves	Compositae	365

continued

TABLE 13.2
New Flavonol Glycosides — *continued*

Glycoside	Source	Family	Ref.
7-(2^G-Rhamnosylrutinoside)			
7-(2^G-Glucosylrutinoside)			
Quercetin 7,3′-dimethyl ether (rhamnazin)			
3-Glucosyl(1 → 5)-α-L-arabinofuranoside	*Retama sphaerocarpa* aerial parts	Leguminosae	402
3-Xylosyl(1 → 2)glucoside	*Albizzia julibrissin* seeds	Leguminosae	403
3-Glucosyl(1 → 5)[apiosyl(1 → 2)-α-L-arabinofuranoside]	*Retama sphaerocarpa* aerial parts	Leguminosae	404
Quercetin 7,4′-dimethyl ether (ombuin)			
3-Arabinofuranoside	*Coccinia indica* roots	Cucurbitaceae	405
3-Glucoside	*Gynostemma yixingense*	Cucurbitaceae	406
Quercetin 3′,4′-dimethyl ether			
3-Neohesperidoside	*Crotalaria verrucosa* stems	Leguminosae	407
3,7-Diglucoside	*Calamintha grandiflora* leaves and flowers	Labiatae	408
Quercetin 3,7,4′-trimethyl ether			
3-Sulfate	*Ipomoea regnelli*	Convolvulaceae	373
Quercetin 5,3′,4′-trimethyl ether			
3-Galactosyl(1 → 2)rhamnoside-7-rhamnoside	*Alhagi persarum* aerial parts	Leguminosae	409
Quercetin 5,7,3′,4′-tetramethyl ether			
3-Galactoside	*Sesbania aculeata*	Leguminosae	410
6-Hydroxyquercetin (quercetagetin)			
6-Glucoside	*Tagetes mandonii* aerial parts	Compositae	411
7-(6″-Isobutyrylglucoside)	*Buphthalmum salicifolium* flowers	Compositae	412
7-(6″-Isovalerylglucoside)			
7-[6″-(2-Methylbutyryl)glucoside]			
7-(6″-E-Caffeoylglucoside)	*Eupatorium glandulosum* leaves	Compositae	413
7-(6″-Acetylglucoside)			
Quercetagetin 6-methyl ether (patuletin)			
3,7-Diglucoside	*Arnica montana* flowers	Compositae	299
3-(6″-p-Coumaroylglucoside)	*Paepalanthus polyanthus,* P. hilairei, P. robustus, P. ramosus, and P. denudatus capitulae	Eriocaulaceae	297
3-(6″-E-Feruloylglucoside)	*Paepalanthus polyanthus* aerial parts	Eriocaulaceae	414
7-(6″-Isobutyrylglucoside)	*Buphthalmum salicifolium* flowers	Compositae	412
	Inula britannica flowers	Compositae	415
7-[6″-(2-Methylbutyryl)glucoside]	*Inula britannica* flowers	Compositae	415
7-(6″-Isovalerylglucoside)			
3-Rhamnoside-7-(2″-acetylrhamnoside)	*Kalanchoe brasiliensis* stems and leaves	Crassulaceae	416
3-(4″-Acetylrhamnoside)-7-rhamnoside			
3-(4″-Acetylrhamnoside)-7-(2‴-acetylrhamnoside)			
3-(2″-Feruloylglucosyl)(1 → 6)[apiosyl(1 → 2)glucoside]	*Spinacia oleracea* leaves	Chenopodiaceae	417
Quercetagetin 7-methyl ether			
6-Glucoside	*Tagetes mandonii* aerial parts	Compositae	418
4′-Glucoside	*Paepalanthus latipes* leaves and scapes	Eriocaulaceae	419

TABLE 13.2
New Flavonol Glycosides — *continued*

Glycoside	Source	Family	Ref.
3-Neohesperidoside	*Paepalanthus vellozioides* leaves and scapes	Eriocaulaceae	419
3-Cellobioside	*Paepalanthus latipes* and *P. vellozioides* leaves and scapes	Eriocaulaceae	419
3-(2‴-Caffeoylglucosyl)(1 → 2)glucuronide	*Paepalanthus latipes* leaves and scapes	Eriocaulaceae	419
Quercetagetin 3,6-dimethyl ether (axillarin)			
5-α-L-Arabinosyl(1 → 6)glucoside	*Acacia catechu* stems	Legumnosae	420
7-Sulfate	*Centaurea bracteata* roots	Compositae	421
Quercetagetin 6,3′-dimethyl ether (spinacetin)			
3-(2″-Apiosylgentiobioside)	*Spinacia oleracea* leaves	Chenopodiaceae	417
3-(2‴-Feruloylgentiobioside)			
3-(2″-*p*-Coumaroylglucosyl)(1 → 6) [apiosyl(1 → 2)glucoside]			
3-(2″-Feruloylglucosyl)(1 → 6) [apiosyl(1 → 2)glucoside]			
Quercetagetin 7,3′-dimethyl ether			
6-Glucoside	*Tagetes mandonii* aerial parts	Compositae	418
Quercetagetin 3,6,3′-trimethyl ether (jaceidin)			
5-Glucoside	*Eucryphia glutinosa* twigs	Eucryphiaceae	397
Quercetagetin 6,7,3′-trimethyl ether (veronicafolin)			
3-Glucosyl(1 → 3)galactoside	*Eupatorium africanum*	Compositae	23
Quercetagetin 6,3′,4′-trimethyl ether			
3-Glucoside	*Arnica montana*	Compositae	299
8-Hydroxyquercetin (gossypetin)			
8-α-D-Lyxopyranoside	*Orostachys japonicus* aerial parts	Crassulaceae	422
Gossypetin 8-methyl ether			
3-Xylosyl(1 → 2)rhamnoside	*Butea superba* stems	Leguminosae	423
Gossypetin 3,8-dimethyl ether			
5-Glucoside	*Eugenia edulis* leaves	Myrtaceae	424
Gossypetin 7,8-dimethyl ether			
3-Glucoside	*Erica cinerea* flowers	Ericaceae	425
4′-Glucoside			
3,3′-Disulfate	*Erica cinerea* flowers	Ericaceae	426
Gossypetin 8,3′-dimethyl ether (limocitrin)			
3-Rutinoside-7-glucoside	*Coleogyne ramosissima* aerial parts	Rosaceae	389
3,5,7,3′,4′,5′-Hexahydroxyflavone (myricetin)			
3′-Rhamnoside	*Davilla flexuosa* leaves	Dilleniaceae	427
3-Xylosyl(1 → 3)rhamnoside	*Maesa lanceolata* leaves	Myrsinaceae	326
3-Rhamnosyl(1 → 2)rhamnoside	*Licania densiflora* leaves	Chrysobalanaceae	428
3-Neohesperidoside	*Physalis angulata* leaves	Solanaceae	429
3-Robinobioside	*Nymphaéa* × *Marliacea* leaves	Nymphaceae	430
3-Rhamnoside-3′-glucoside	*Myrsine seguinii* leaves	Myrsinaceae	318
3-Galactoside-3′-rhamnoside	*Buchanania lanzan* leaves	Anacardiaceae	431
3,4′-Dirhamnoside	*Myrsine seguinii* leaves	Myrsinaceae	318
3,4′-Diglucoside	*Picea abies* needles	Pinaceae	432
3-Rhamnosyl(1 → 3)glucosyl(1 → 6)glucoside	*Oxytropis glabra*	Leguminosae	433

continued

TABLE 13.2
New Flavonol Glycosides — *continued*

Glycoside	Source	Family	Ref.
3-(2^G-Rhamnosylrutinoside)	*Clitoria ternatea* petals	Leguminosae	244
3-Glucosyl(1 → 2)rhamnoside-7-glucoside	*Crocus chrysanthus-biflorus* cvs "eye catcher" and "spring pearl" flowers	Iridaceae	204
3-(4″-Malonylrhamnoside)	*Ribes alpinum* leaves	Grossulariaceae	330
3-(2″-*p*-Hydroxybenzoylrhamnoside)	*Limonium sinense* aerial parts	Plumbaginaceae	434
3-(3″-Galloylrhamnoside)	*Myrica esculenta* bark	Myricaceae	435
3-(3″-Galloylgalactoside)			
3-(2″-Galloylglucoside)	*Geranium pratense* aerial parts	Geraniaceae	436
3-(6″-*p*-Coumaroylglucoside)	*Nymphaéa lotus* leaves	Nymphaeaceae	437
Nympholide A			

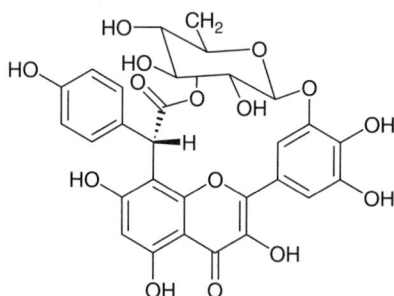

13.7

Nympholide B

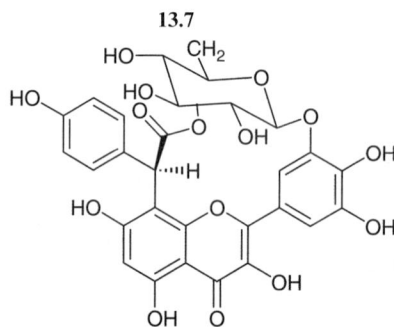

13.8

3-(2″-Acetylrhamnoside)	*Nymphaéa caerulea* blue flowers	Nymphaeaceae	333
3-(4″-Acetylrhamnoside)	*Eugenia jambola* leaves	Myrtaceae	438
7-(6″-Galloylglucoside)	*Acacia farnesiana* pods	Leguminosae	336
3-(2″,3″-Digalloylrhamnoside)	*Acacia confusa* leaves	Leguminosae	439
3-(3″,4″-Diacetylrhamnoside)	*Myrsine africana* leaves	Myrsinaceae	440
3-(2″,3″,4″-Triacetylxyloside)	*Maesa lanceolata* leaves	Myrsinaceae	326
3-(4″-Acetyl-2″-galloylrhamnoside)	*Eugenia jambolana* leaves	Myrtaceae	441
3-(3‴-6‴-Diacetylglucosyl)(1 → 4) (2″,3″-diacetylrhamnoside)	*Maesa lanceolata* leaves	Myrsinaceae	326
Myricetin 7-methyl ether			
3-(2″-Galloylrhamnoside)	*Acacia confusa* leaves	Leguminosae	439
3-(3″-Galloylrhamnoside)			
Myricetin 3′-methyl ether (larycitrin)			
3-α-ʟ-Arabinofuranoside	*Lysimachia congestiflora* whole plant	Primulaceae	442

TABLE 13.2
New Flavonol Glycosides — *continued*

Glycoside	Source	Family	Ref.
3-(4″-Malonylrhamnoside)	*Ribes alpinum* leaves	Grossulariaceae	330
Myricetin 4′-methyl ether			
3-Galactoside	*Licania heteromorpha* var.	Chrysobalanaceae	443
	heteromorpha aerial parts		
3-(4″-Acetylrhamnoside)	*Eugenia jambolana* leaves	Myrtaceae	441
Myricetin 3′,4′-dimethyl ether			
3-Rhamnoside	*Clausena excavata* aerial parts	Rutaceae	444
3-Glucoside	*Licania densiflora*	Chrysobalanaceae	429
Myricetin 3′,5′-dimethyl ether (syringetin)			
3-Rhamnosyl(1 → 5)-α-ʟ-arabinofuranoside	*Lysimachia congestiflora* whole plant	Primulaceae	442
3-Robinobioside	*Catharanthus roseus* stems	Apocynaceae	445
3-(2″,3″-Diacetylglucoside)	*Warburgia stuhlmannii* leaves	Canellaceae	391
3-(6″-Acetylglucosyl)(1 → 3)galactoside	*Achlys triphylla* underground parts	Berberidaceae	382
8-Hydroxymyricetin 8-methyl ether			
3-Rhamnoside	*Erica verticillata* aerial parts	Ericaceae	446
8-Hydroxymyricetin 8,5′-dimethyl ether			
3-Rhamnoside	*Erica verticillata* aerial parts	Ericaceae	446
8-Hydroxymyricetin 8,3′,5′-trimethyl ether			
3-Rhamnoside	*Erica verticillata* aerial parts	Ericaceae	446
3,7,3′,4′,5′-Pentalhydroxyflavone (5-deoxymyricetin, robinetin)			
7-Glucoside	*Alternanthera sessilis*	Amaranthaceae	447
3-Rutinoside	*Ateleia Herbert-smithii* leaves	Leguminosae	448
3,4′-Dihydroxy-7,3′,5′-trimethoxyflavone			
3-Galactosyl(1 → 4)xyloside	*Abrus precatorius* seeds	Leguminosae	449
5,7,8-Trihydroxy-3,6,4′-trimethoxyflavone			
8-Tiglate (pratensin A)	*Galeana pratensis* aerial parts	Compositae	450
3,5,2′-Trihydroxy-7,8,4′-trimethoxyflavone			
5-Glucosyl(1 → 2)galactoside	*Cassia occidentalis* whole plant	Leguminosae	451
3,5,6,7,8,4′-Hexahydroxy-3′-methoxyflavone			
3-Rhamnosyl(1 → 4)rhamnosyl(1 → 6)glucoside	*Eschsholtzia californica* aerial parts	Papaveraceae	452
3,5,7,4′-Tetrahydroxy-6,8,3′-trimethoxyflavone			
3-α-ʟ-Arabinopyranosyl(1 → 3)galactoside	*Pongamia pinnata* heartwood	Leguminosae	453
3-α-ʟ-Arabinopyranosyl(1 → 3)[galactosyl (1 → 6)galactoside]			
3,6,7,8,3′,4′-Hexahydroxy-5′-methoxyflavone			
7-Neohesperidoside	*Hibiscus vitifolius*	Malvaceae	454
5,7,2′,3′,4′-Pentahydroxy-3,6-dimethoxyflavone			
7-Glucoside	*Tridax procumbens* aerial parts	Compositae	455
5,2′-Dihydroxy-3,6,7,4′,5′-pentamethoxyflavone (brickellin)			
2′-Glucoside	*Chrysosplenium grayanum*	Saxifragaceae	456
3,5,7,2′,6′-Pentahydroxyflavone			
2′-Glucoside	*Scutellaria amoena* roots	Labiatae	457
5,7-Dihydroxy-3,6,8,4′-tetramethoxyflavone			
7-Glucosyl(1 → 3)galactoside	*Aspilia africana* whole plant	Compositae	458
7,4′-Dihydroxy-3,5,6,8-tetramethoxyflavone			
4′-Glucosyl(1 → 3)galactoside	*Centaurea senegalensis* whole plant	Compositae	459

continued

TABLE 13.2
New Flavonol Glycosides — *continued*

Glycoside	Source	Family	Ref.
5,8-Dihydroxy-3,6,7,4'-tetramethoxyflavone			
8-Neohesperidoside	*Peperomia pellucida*	Piperaceae	460
5,4'-Dihydroxy-6,7,8,3'-tetramethoxyflavone			
(africanutin)			
4'-Galactoside	*Eupatorium africanum*	Compositae	23
	whole plant		
3,5,7,2',3',4'-Hexahydroxyflavone	*Eupatorium sternbergianum*	Compositae	461
3-Glucoside	whole plant		
5,7,2'-Trihydroxy-3,6,4'-trimethoxyflavone			
7-Glucoside	*Tridax procumbens* whole plant	Compositae	462
5,2',4'-Trihydroxy-3,7,5'-trimethoxyflavone			
2'-Galactosyl(1 → 4)glucoside	*Albizzia procera* stems	Leguminosae	463
Methylenedioxyflavonol glycosides			
3-Methoxy-5-hydroxy-6,7-methylenedioxyflavone			
4'-Glucuronide	*Spinacia oleracea* leaves	Chenopodiaceae	464
3-Hydroxy-5,4'-dimethoxy-6,7-methylenedioxyflavone			
3-Xyloside (viviparum A)	*Polygonum viviparum*	Polygonaceae	465
3,3'-Dihydroxy-5,4'-dimethoxy-6,7-methylenedioxyflavone			
3-Xyloside (viviparum B)	*Polygonum viviparum*	Polygonaceae	465
Prenyl- and pyranoflavonol glycosides			
8-Prenylkaempferol[noranhydroicaritin, 3,5,7,4'-tetrahydroxy-8-(3'',3''-dimethylallyl)flavone]			
3,7-Diglucoside	*Vancouveria hexandra*	Berberidaceae	466
	underground and aerial parts		
8-Prenylkaempferol 7-methyl ether			
3-Rhamnosyl(1 → 3)[apiosyl(1 → 6)glucoside]	*Mosla soochouensis* stem wood	Labiatae	278
8-Prenylkaempferol 4'-methyl ether (anhydroicarinin)			
7-Glucosyl(1 → 4)glucoside (cuhuoside, 7-cellobioside)	*Epimedium acuminatum*	Berberidaceae	467
3-Rhamnosyl(1 → 6)galactoside-7-galactoside	*Sesbania grandiflora* bark	Leguminosae	468
3-Glucosyl(1 → 3)rhamnoside-7-glucoside	*Vancouveria hexandra*	Berberidaceae	466
	underground and aerial parts		
3-Rhamnosyl(1 → 2)rhamnoside-7-sophoroside (acuminatoside)	*Epimedium acuminatum* aerial parts	Berberidaceae	469
3-[4''',6'''-Diacetylglucosyl(1 → 3)-4''-acetylrhamnoside]	*Berberis dictyota* aerial parts	Berberidaceae	470
3-[2''',6'''-Diacetylglucosyl(1 → 3)-4''-acetylrhamnoside]-7-glucoside (epimedin K)	*Epimedium koreanum* aerial parts	Berberidaceae	471
3''-[4''',6'''-Diacetylglucosyl(1 → 3)-4''-acetylrhamnoside]-7-glucoside	*Epimedium koreanum* aerial parts	Berberidaceae	472
8-(3''-Hydroxy-3''-methylbutyl)kaempferol 4'-methyl ether (icaritin)			
3-Rhamnosyl(1 → 2)rhamnoside (wanepimedoside A)	*Epimedium wanshanense*	Berberidaceae	473
	whole plant		
8-(γ-Methoxy-γγ-dimethyl)propylkaempferol 4'-methyl ether			
7-Glucoside (caohuoside D)	*Epimedium koreanum* aerial parts	Berberidaceae	474
8-Prenylquercetin 4'-methyl ether			
3-Rhamnoside (caohuoside C)	*Epimedium koreanum* aerial parts	Berberidaceae	475

TABLE 13.2
New Flavonol Glycosides — *continued*

Glycoside	Source	Family	Ref.
8-Prenylquercetin 7,4′-dimethyl ether			
3-Rhamnosyl(1 → 4)rhamnoside	*Butea monosperma* stems	Leguminosae	476
6″,6″-Dimethylpyrano(2″,3″:7,8)-4′-methoxykaempferol			
3-Rhamnoside	*Epimedium acuminatum*	Berberidaceae	477
C-Methylated flavonol glycosides			
5,7-Dihydroxy-6,8-di-*C*-methyl-3-methoxyflavone			
7-Galactosyl(1 → 2)rhamnoside	*Cotula anthemoides* seeds		478
2′-*C*-Methylmyricetin			
3-Rhamnoside-5′-gallate	*Syzygium samarangense* leaves	Myrtaceae	479

All the other new disaccharides are new isomers of known sugar combinations. Among the pentose–pentose sugars are three rhamnosylarabinoses. A rhamnosyl(1 → 2)arabinose was listed in the fourth edition of *The Flavonoids*[5] but no details of form or stereochemistry were given. The new sugars include a rhamnosyl(1 → 2)-α-L-arabinopyranose, which was identified in the aerial parts of *Putoria calabrica* (Rubiaceae)[324] in combination with quercetin at the 3-hydroxyl and its furanose isomer at the 3-hydroxyl of kaempferol in leaves of *Artabotrys hexapetalus* (Annonaceae).[173] The report of myricetin 3′,5′-dimethyl ether (syringetin) 3-rhamnosyl(1 → 5)arabinofuranoside from the whole plant of *Lysimachia congestiflora*[442] provides the third new isomer. Xylosyl(1 → 4)rhamnose, found at the 3-position of kaempferol with rhamnose at the 7-hydroxyl in *Chenopodium murale*,[198] is an expectable new isomer of the known 1 → 2 and 1 → 3 linked sugars.

There are three new isomers of pentose–hexose disaccharides in Table 13.4. These include α-L-arabinopyranosyl(1 → 3)galactose found in combination with 3,5,7,4′-tetrahydroxy-6,8,3′-flavone at the 3-hydroxyl in the heartwood of the legume, *Pongamia pinnata*.[453] Lathyrose, xylosyl(1 → 6)galactose was found in a member of the Anacardiaceae, *Semecarpus kurzii*,[60] at the 7-hydroxyl of scutellarein (6-hydroxyapigenin). The third new isomer in this

TABLE 13.3
Monosaccharides of Flavone and Flavonol Glycosides

Pentoses	Hexoses	Uronic Acids
D-Apiose	D-Allose	D-Galacturonic acid[a]
L-Arabinose[b]	D-Allulose[c,d]	D-Glucuronic acid[a]
D-Fructose[c]	D-Fucose[c]	
D-Lyxose[c]	D-Galactose	
L-Rhamnose	D-Glucosamine[c]	
D-Xylose	D-Glucose	
	D-Mannose	

[a]Also reported to occur as the methyl and ethyl ethers.
[b]Known to occur in both pyranose and furanose forms; all other sugars (except apiose, fructose, and allulose) are normally in the pyranose form.
[c]Newly reported since 1992.
[d]Recorded as D-allulose but preferred name is D-psicose or D-*ribo*-2-hexulose.

group is apiosyl(1 → 6)glucose found at the 7-position of three different flavones: luteolin and its 3'-methyl ether (chrysoeriol) in *Phlomis nissolii* (Labiatae)[96] and acacetin (apigenin 4'-methyl ether) in *Crotalaria podocarpa* (Leguminosae).[53] There are two further new hexose–pentoses: glucosyl(1 → 4)xylose has been isolated at the 7-hydroxyl of kaempferol from *Crotalaria laburnifolia*[178] and galactosyl(1 → 2)rhamnose at the 3-position of quercetin from *Embelia schimperi* (Myrsinaceae).[313] Noteworthy is the first record of the hexose–hexose, cellobiose (glucosyl(1 → 4)glucose) from *Epimedium acuminatum*,[467] which was present at the 7-hydroxyl of 8-prenylkaempferol 4'-methyl ether. Cellobiose has the same sugar linkage as cellulose and has been found in a further four families in combination with two flavones and two flavonols during the review period. Thus, 6-hydroxyluteolin 5,6,3',4'-tetramethyl ether 7-cellobioside has been isolated from stems of the Composite *Sphaeranthus indicus*[136] and apigenin 7-cellobioside and 7-cellobioside-4'-glucoside from petals of *Salvia uliginosa* (Labiatae).[35] Quercetagetin 3-cellobioside has been reported in stems and scapes of *Paepalanthus latipes* and *P. vellozioides*,[419] from the monocot family Velloziaceae. Kaempferol 3-sophoroside-7-cellobioside and three acylated kaempferol 3-diglycosides, the 3-(2-*E*-caffeoylsophoroside)-7-cellobioside and the corresponding feruoyl and sinapoyl isomers, were found in leaves of *Brassica oleracea* (Cruciferae).[217] Two further glucosylgalactose isomers are listed in Table 13.4. Glucosyl(1 → 3)galactose was present in underground parts of *Achlys triphylla* (Berberidaceae),[382] in acetylated form attached to the 3-hydroxyls of both isorhamnetin and syringetin, while its 1 → 4 isomer was found at the 3-hydroxyl of quercetin in leaves of *Rumex chalepensis* (Polygonaceae).[316]

TABLE 13.4
Disaccharides of Flavone and Flavonol Glycosides

Structure	Trivial Name
Pentose–pentose	
O-β-D-Xylosyl(1 → 2)xylose[a,66]	
O-β-D-Xylosyl(1 → 3)xylose[a,159]	
O-α-L-Apiofuranosyl(1 → 2)xylose	
O-α-L-Apiofuranosyl(1 → 4)rhamnose[a,198]	
O-α-L-Rhamnosyl(1 → 5)arabinofuranose[a,442]	
O-α-L-Rhamnosyl(1 → 2)-α-L-arabinopyranose[a,324]	
O-α-L-Rhamnosyl(1 → 2)-α-L-arabinofuranose[a,b,173]	
O-α-L-Rhamnosyl(1 → 2)rhamnose	
O-α-L-Rhamnosyl(1 → 3)rhamnose	
O-α-L-Rhamnosyl(1 → 4)rhamnose	
O-α-L-Rhamnosyl(1 → 4)xylose	
O-β-D-Xylosyl(1 → 2)rhamnose	
O-β-D-Xylosyl(1 → 3)rhamnose	
O-β-D-Xylosyl(1 → 4)rhamnose[a,198]	
Pentose–hexose	
O-α-L-Arabinosyl(1 → 6)glucose	Vicianose
O-α-L-Arabinopyranosyl(1 → 3)galactose[a,453]	
O-α-D-Arabinosyl(1 → 6)galactose	
O-β-D-Xylosyl(1 → 2)glucose	Sambubiose
O-β-D-Xylosyl(1 → 6)glucose	
O-β-D-Xylosyl(1 → 2)galactose	
O-β-D-Xylosyl(1 → 6)galactose[a,60]	Lathyrose
O-β-D-Apiosyl(1 → 2)glucose	

TABLE 13.4
Disaccharides of Flavone and Flavonol Glycosides — *continued*

Structure	Trivial Name
O-β-D-Apiosyl(1 → 6)glucose[a,53,96]	
O-β-D-Apiosyl(1 → 2)galactose	
O-α-L-Rhamnosyl(1 → 2)glucose	Neohesperidose
O-α-L-Rhamnosyl(1 → 3)glucose	Rungiose
O-α-L-Rhamnosyl(1 → 6)glucose	Rutinose
O-α-L-Rhamnosyl(1 → 2)galactose	
O-α-L-Rhamnosyl(1 → 6)galactose	Robinobiose
O-α-L-Rhamnosyl(1 → 2)fucose[a,26]	

Hexose–pentose

Structure	Trivial Name
O-β-D-Glucosyl(1 → 2)-β-arabinopyranose[a,357]	
O-β-D-Glucosyl(1 → 2)-α-L-arabinofuranose[a,c,340]	
O-β-D-Glucosyl(1 → 3)-α-L-arabinopyranose[a,164]	
O-β-D-Glucosyl(1 → 4)-α-L-arabinopyranose[a,94]	
O-β-D-Glucosyl(1 → 5)-α-L-arabinofuranose[a,402]	
O-β-D-Glucosyl(1 → 2)rhamnose	
O-β-D-Glucosyl(1 → 3)rhamnose	
O-α-L-Glucosyl(1 → 4)rhamnose	
O-β-D-Glucosyl(1 → 2)xylose	
O-β-D-Glucosyl(1 → 4)xylose[a,178]	
O-β-D-Galactosyl(1 → 2)rhamnose[a,313]	
O-β-D-Galactosyl(1 → 3)rhamnose	
O-β-D-Galactosyl(1 → 4)rhamnose	

Hexose–hexose

Structure	Trivial Name
O-β-D-Glucosyl(1 → 2)glucose	Sophorose
O-β-D-Glucosyl(1 → 3)glucose	Laminaribiose
O-β-D-Glucosyl(1 → 4)glucose[a,467]	Cellobiose
O-β-D-Glucosyl(1 → 6)glucose	Gentiobiose
O-β-D-Glucosyl(1 → 2)galactose	
O-β-D-Glucosyl(1 → 3)galactose[a,382]	
O-β-D-Glucosyl(1 → 4)galactose[a,316]	
O-β-D-Galactosyl(1 → 4)glucose	Lactose
O-β-D-Galactosyl(1 → 6)glucose	
O-β-D-Galactosyl(1 → 4)galactose	
O-β-D-Galactosyl(1 → 6)galactose	
O-β-D-Allosyl(1 → 2)glucose	
O-β-D-Mannosyl(1 → 2)allose	

Pentose–uronic acid
O-α-L-Rhamnosyl(1 → 2)galacturonic acid

Uronic acid–uronic acid
O-β-D-Glucuronosyl(1 → 2)glucuronic acid

---

[a]Disaccharides newly reported since 1992 with reference number. Except where otherwise stated, sugars are assumed to be in the pyranose form and to have the appropriate linkage, i.e., β for glucosides, α for rhamnosides, etc.
[b]Reported only at the 3-hydroxyl of quercetin with glucose at the 7-position.
[c]Reported only in acylated form.

TABLE 13.5
Trisaccharides of Flavonol Glycosides

Structure	Trivial Name
Linear	
O-β-Glucosyl(1 → 4)-O-α-arabinofuranosyl(1 → 2)arabinopyranose	Primflasin
O-β-D-Xylosyl(1 → 2)-O-α-L-rhamnosyl(1 → 6)glucose[a,319]	
O-β-D-Xylosyl(1 → 3)-O-α-L-rhamnosyl(1 → 6)glucose[a,384]	
O-β-D-Xylosyl(1 → 6)-O-β-D-glucosyl(1 → 2)rhamnose[a,187]	
O-β-D-Xylosyl(1 → 3)-O-α-L-rhamnosyl(1 → 6)galactose[a,186]	
O-α-L-Rhamnosyl(1 → 3)-O-α-L-rhamnosyl(1 → 3)rhamnose[a,b,273]	
O-α-L-Rhamnosyl(1 → 2)-O-α-L-rhamnosyl(1 → 6)glucose	2'-Rhamnosylrutinoside
O-α-L-Rhamnosyl(1 → 3)-O-α-L-rhamnosyl(1 → 6)glucose[a,188]	
O-α-L-Rhamnosyl(1 → 4)-O-α-L-rhamnosyl(1 → 6)glucose[c]	
O-α-L-Rhamnosyl(1 → 2)-O-β-D-glucosyl(1 → 3)glucose	3'-Rhamnosyllaminaribiose
O-α-L-Rhamnosyl(1 → 3)-O-β-D-glucosyl(1 → 6)glucose[a,433]	
O-β-D-Glucosyl(1 → 3)-O-α-L-rhamnosyl(1 → 6)glucose	3'-Glucosylrutinoside
O-β-D-Glucosyl(1 → 4)-O-α-L-rhamnosyl(1 → 2)glucose[a,191]	
O-β-D-Glucosyl(1 → 2)-O-β-D-glucosyl(1 → 2)rhamnose	
O-β-D-Glucosyl(1 → 6)-O-β-D-glucosyl(1 → 4)rhamnose	
O-β-D-Glucosyl(1 → 2)-O-β-D-glucosyl(1 → 2)glucose	Sophorotriose
O-β-D-Glucosyl(1 → 2)-O-β-D-glucosyl(1 → 6)glucose	2'-Glucosylgentiobiose
O-β-D-Glucosyl(1 → 4)-O-β-D-glucosyl(1 → 6)glucose	6'-Maltosylglucose
O-β-D-Glucosyl(1 → 6)-O-β-D-glucosyl(1 → 4)glucose	Sorborose
O-α-L-Rhamnosyl(1 → 3)-O-α-L-rhamnosyl(1 → 6)galactose	Rhamninose
O-α-L-Rhamnosyl(1 → 4)-O-α-L-rhamnosyl(1 → 6)galactose	Isorhamninose
O-α-L-Rhamnosyl(1 → 2)-O-β-D-glucosyl(1 → 6)galactose[a,189]	
O-α-L-Rhamnosyl(1 → 6)-O-β-D-glucosyl(1 → 6)galactose[a,190]	
O-β-D-Glucosyl(1 → 3)-O-α-L-rhamnosyl(1 → 6)galactose[d]	Sugar of faralatroside
O-β-D-Glucosyl(1 → 2)-O-β-D-galactosyl(1 → 2)glucose[a,192]	
Branched	
O-α-L-Arabinopyranosyl(1 → 3)-O-[β-D-galactosyl(1 → 6)galactose][a,453]	
O-β-D-Apiosyl(1 → 2)-O-[α-L-rhamnosyl(1 → 4)glucose]	2^G-Apiosylrutinose
O-β-D-Apiosyl(1 → 2)-O-[α-L-rhamnosyl(1 → 6)galactose]	2^{Gal}-Apiosylrobinobiose
O-β-D-Glucosyl(1 → 5)-O-[β-D-apiosyl(1 → 2)-α-L-arabinofuranose][a,404]	
O-α-L-Rhamnosyl(1 → 3)-O-[β-D-apiosyl(1 → 6)glucose][a,278]	
O-β-D-Xylosyl(1 → 2)-O-[α-L-rhamnosyl(1 → 6)glucose]	2^G-Xylosylrutinose
O-α-L-Rhamnosyl(1 → 2)-O-[α-L-rhamnosyl(1 → 6)glucose]	2^G-Rhamnosylrutinose
O-α-L-Rhamnosyl(1 → 4)-O-[α-L-rhamnosyl(1 → 2)glucose]	4^G-Rhamnosylneohesperidose
O-β-D-Glucosyl(1 → 2)-O-[β-D-apiosyl(1 → 2)glucose]	
O-α-L-Rhamnosyl(1 → 6)-O-[β-D-glucosyl(1 → 2)glucose]	6^G-Rhamnosylsophorose
O-α-L-Rhamnosyl(1 → 2)-O-[β-D-glucosyl(1 → 4)glucose][a,194]	2^G-Rhamnosylcellobiose
O-α-L-Rhamnosyl(1 → 2)-O-[β-D-glucosyl(1 → 6)glucose]	2^G-Rhamnosylgentiobiose
O-β-D-Glucosyl(1 → 2)-O-[α-L-rhamnosyl(1 → 6)glucose]	2^G-Glucosylrutinose
O-β-D-Glucosyl(1 → 3)-O-[α-L-rhamnosyl(1 → 2)glucose]	3^G-Glucosylneohesperidose
O-β-D-Glucosyl(1 → 2)-O-[β-D-glucosyl(1 → 3)rhamnose][a,195]	
O-α-L-Rhamnosyl(1 → 2)-O-[β-D-glucosyl(1 → 6)galactose]	
O-β-D-Glucosyl(1 → 2)-O-[α-L-rhamnosyl(1 → 6)galactose]	2^G-Glucosylrobinobiose
O-β-D-Galactosyl(1 → 2)-O-[α-L-rhamnosyl(1 → 6)glucose]	2^G-Galactosylrutinose
O-β-D-Glucosyl(1 → 2)-O-[β-D-glucosyl(1 → 6)glucose]	2^G-Glucosylgentiobiose

TABLE 13.5
Trisaccharides of Flavonol Glycosides — *continued*

Structure	Trivial Name
O-β-D-Glucosyl(1 → 3)-*O*-[β-D-glucosyl(1 → 6)glucose][a,196]	3[G]-Glucosylgentiobiose
O-α-L-Rhamnosyl(1 → 2)-*O*-[α-L-rhamnosyl(1 → 6)galactose]	2[Gal]-Rhamnosylrobinobiose
O-α-L-Rhamnosyl(1 → 4)-*O*-[α-L-rhamnosyl(1 → 6)galactose]	4[Gal]-Rhamnosylrobinobiose
O-α-L-Rhamnosyl(1 → 6)-*O*-[α-L-rhamnosyl(1 → 2)galactose][a,220]	
O-β-D-Glucosyl(1 → 4)-*O*-[β-D-glucosyl(1 → 2)galactose][a,e,249]	

[a]Newly reported since 1992 with reference number.
[b]Present at the 7-position of kaempferol with a *p*-coumaric acid attached at the 4-hydroxyl of the first rhamnose and a known acylated branched trisaccharide at the 3-hydroxyl of the aglycone.
[c]Only present with a caffeyl or *p*-coumaryl group at the 6-hydroxyl of the second glucose.
[d]Only present with an acetyl group at the 4-hydroxyl of the rhamnose.
[e]Present at the 3-hydroxyl of kaempferol in mono- or diacylated form with sinapic acid.

13.3.3 TRISACCHARIDES

Some 11 new linear and 8 new branched trisaccharides have been discovered in combination with flavonols since 1992. These are presented in Table 13.5 together with previously known trisaccharides. Among the linear trisaccharides are seven novel sugar combinations. The most interesting are four structures with xylose as the terminal sugar. The only previously known trisaccharide containing xylose is the branched sugar, 2[G]-xylosylrutinose. The linear trisaccharide, xylosyl(1 → 2)rhamnosyl(1 → 6)glucose was found attached to the 3-hydroxyl of quercetin in leaves of *Camellia saluensis* (Theaceae),[319] while its isomer, xylosyl(1 → 3)rham-3)rhamnosyl(1 → 6)glucose, was found at the 3-position of isorhamnetin in *Hamada scoparia* (Chenopodiaceae).[384] The other two structures were both found in combination with kaempferol at the 3-position, xylosyl(1 → 6)glucosyl(1 → 2)rhamnose from *Helicia nilagirica* (Proteaceae)[187] and xylosyl(1 → 3)rhamnosyl(1 → 6)galactose from the legume *Astragalus caprinus*.[186] The former was also present in similar combination with quercetin. The first linear trirhamnose, rhamnosyl(1 → 3)rhamnosyl(1 → 3)rhamnose, has been recorded from

TABLE 13.6
Trisaccharides of Flavone Glycosides

Structure	Trivial Name
Linear	
O-α-L-Rhamnosyl(1 → 2)-*O*-β-D-glucosyl(1 → 2)glucose[a,51]	
O-β-D-Allosyl(1 → 3)-*O*-β-D-glucosyl(1 → 2)glucose[a,110]	
Branched	
O-α-L-Rhamnosyl(1 → 2)-*O*-[α-L-rhamnosyl(1 → 6)galactose][b]	2[Gal]-Rhamnosylrobinobiose
O-β-D-Glucosyl(1 → 2)-*O*-[α-L-rhamnosyl(1 → 6)glucose][b,c]	2[G]-Glucosylrutinose
O-β-D-Apiosyl(1 → 2)-*O*-[β-D-glucosyl(1 → 4)glucose][a,c,109]	2[G]-Apiosylcellobiose
O-β-D-Glucuronosyl(1 → 3)-*O*-[β-D-glucuronosyl(1 → 2)glucuronic acid][a,c,40,116]	

[a]Newly reported trisaccharides with reference number.
[b]Found previously only attached to flavonols.
[c]Present only in acylated form (see Table 13.1).

Planchonia grandis (Lecythidaceae).[273] This sugar occurred at the 7-hydroxyl of kaempferol acylated with *p*-coumaric acid at the 4-position of the first rhamnose and with a known acylated branched trisaccharide at the 3-hydroxyl. Two further acylated kaempferol glycosides with known trisaccharides at both the 3- and 7-hydroxyls were also present in this plant. This is the first record of flavonol (or flavone) glycosides containing six sugars. The three other new linear combinations were isolated as kaempferol 3-rhamnosyl(1 → 2)glucosyl (1 → 6)galactoside from *Cassia marginata*,[189] kaempferol and quercetin 3-rhamnosyl(1 → 6) glucosyl(1 → 6)galactoside from *Albizia lebbeck*[190] (both legumes), and kaempferol and quercetin 3-glucosyl(1 → 2)galactosyl(1 → 2)glucosides from seeds of *Nigella sativa* (Ranunculaceae).[192] The remaining new linear trisaccharides in Table 13.2: rhamnosyl(1 → 3) rhamnosyl(1 → 6)glucose,[192] rhamnosyl(1 → 3)glucosyl(1 → 6)glucose,[433] and glucosyl (1 → 4)rhamnosyl(1 → 2)glucose[191] are all new isomers of known structures. Details will not be given here but these sugars are marked as new in Table 13.6 and with a reference number, which relates both to the main reference list and to the reference numbers in Table 13.2.

Among the eight new branched trisaccharides are five new sugar combinations, including the first to contain arabinose. Thus, α-L-arabinopyranosyl(1 → 3)[galactosyl(1 → 6) galactose] was found at the 3-hydroxyl of 3,5,7,4'-tetrahydroxy-6,8,3'-trimethoxyflavone in the heartwood of *Pongamia pinnata* (Leguminosae)[453] and glucosyl(1 → 5)[apiosyl(1 → 2)-α-L-arabinofuranose] at the 3-hydroxyl of quercetin 7,3'-dimethyl ether in another legume, *Retama sphaerocarpa*.[404] Another novel apiose containing sugar, rhamnosyl(1 → 3)[apiosyl (1 → 6)glucose], present at the 3-hydroxyls of kaempferol 7-methyl ether and 8-prenylrhamnetin, was isolated from stemwood of the Labiate, *Mosla soochouensis*.[278] The remaining new combinations include kaempferol 3-glucosyl(1 → 2)[glucosyl(1 → 3)rhamnoside] from flowers of *Crocus speciosus* and *C. antalyensis*[195] and glucosyl(1 → 4)[glucosyl(1 → 2)galactose], which was found attached to the 3-position of kaempferol in mono- and diacylated forms with sinapic acid, in *Thevetia peruviana* (Apocynaceae).[249] Details of three new isomers of known sugar combinations, rhamnosyl(1 → 2)[glucosyl(1 → 4)glucose],[194] glucosyl(1 → 3)[glucosyl (1 → 6)glucose],[196] and rhamnosyl(1 → 6)[rhamnosyl(1 → 2)galactose][220] are given in Table 13.2 and Table 13.5.

It is of some note that trisaccharides have been found in combination with flavones for the first time. Six structures have been identified since 1992; two linear and four branched trisaccharides are listed in Table 13.6. However, flavone trioses are still of rare occurrence compared with the large number of known flavonol trisaccharide combinations. Rhamnosyl(1 → 2)glucosyl(1 → 2)glucose was found at the 7-hydroxyl of acacetin in aerial parts of *Peganum harmala* (Zygophyllaceae)[51] and allosyl(1 → 3)glucosyl(1 → 2)glucose, acetylated at the 6-position of the allose and attached to luteolin at the 7-hydroxyl, in *Veronica didyma*.[110] These are both new trisaccharides that have not been found in association with flavonols. Two of the branched sugars are also novel structures, apiosyl(1 → 2)[glucosyl(1 → 4)glucose] and glucuronosyl(1 → 3)[glucuronosyl(1 → 2)glucuronic acid]. The former was discovered in fruits of *Capsicum annum* (Solanaceae)[109] at the 7-hydroxyl of luteolin and the latter, in acylated form, attached to the 7-hydroxyls of apigenin,[40] chrysoeriol,[116] and tricin,[116] in aerial parts of *Medicago sativa* (Leguminosae). The remaining two branched trisaccharides have been found previously in combination with flavonols.

13.3.4 TETRASACCHARIDES

Only one branched tetrasaccharide was listed in the last edition of *The Flavonoids:*[5] [rhamnosyl(1 → 4)glucosyl(1 → 6)]sophorose, which was present at the 7-hydroxyl of acacetin and acetylated at the 6''-position of the sophorose in leaves of *Peganum harmala* (Zygophyllaceae).[6] Since then some eight new branched tetrasaccharides have been reported, all attached to flavonols and all with unique sugar combinations. Details of the new structures together with the known tetrasaccharide are given in Table 13.7. No linear tetrasaccharide has yet been recorded. Apiosyl(1 → 5)apiosyl[rhamnosyl(1 → 6)glucose], which was found at the 7-hydroxyl of kaempferol 7-methyl ether in the mistletoe, *Viscum angulatum*,[281] is the first report of any sugar to contain two linked apiose moieties. There are three new structures with xylose as the terminal sugar. Two were found in combination with quercetin at the 3-position. Thus, xylosyl(1 → 3)rhamnosyl(1 → 6)[apiosyl(1 → 2)galactose] was present in leaves of *Astragalus caprinus*[213] and xylosyl(1 → 4)[xylosyl(1 → 6)glucosyl(1 → 2)rhamnose] in leaves of *Helicia nilagirica*.[187] The third sugar, xylosyl(1 → 2)[rhamnosyl(1 → 2)rhamnosyl (1 → 6)glucose], was attached to the 7-hydroxyl of both quercetin 3-methyl ether and 7-methyl ether in *Bidens andicola* (Compositae).[322] Three of the remaining structures were found attached to the 3-hydroxyl of kaempferol. These are: glucosyl(1 → 3)rhamnosyl(1 → 2) [rhamnosyl(1 → 6)galactose], which was isolated from leaves of *Maytenus aquifolium*,[214] rhamnosyl(1 → 2)[xylosyl(1 → 3)rhamnosyl(1 → 6)galactose] from leaves of *Astragalus caprinus*,[213] and rhamnosyl(1 → 2)[glucosyl(1 → 3)rhamnosyl(1 → 6)galactose] present in acylated form, together with the corresponding quercetin glycoside, in *Lysimachia capillipes* (Primulaceae).[272] The eighth new sugar, rhamnosyl(1 → 2)[gentiobiosyl(1 → 6)glucose] was discovered at the 3-hydroxyl of isorhamnetin in *Allium neapolitanum*.[191]

13.3.5 Sulfate Conjugates

Only a comparatively small number of new flavonoid sulfate conjugates have been recorded between 1993 and 2003. Eleven flavone derivatives are recorded in Table 13.1 and 15 flavonol derivatives in Table 13.2, mostly from plants that grow in water-stress conditions. Amongst the flavones the most notable are six sulfate conjugates discovered in some Australian species of the monocot family, the Restionaceae. These plants are unusual in having no true leaves so that the compounds were isolated from culm tissue. Four are hypolaetin (8-hydroxyluteolin) derivatives: the 7-sulfatoglucoside and 7-sulfatoglucuronide from *Leptocarpus elegans*,[138] the 7-sulfatoglucuronide from *Meeboldina thysanantha*,[138] and the 7-sulfate-8-glucoside from *Hypolaena fastigiata*.[138] The other two flavone sulfates are hypolaetin 7-methyl ether 3'-sulfatogalactoside from *Leptocarpus tenax*[138] and the corresponding 3'-sulfatoglucuronide

---

TABLE 13.7
Tetrasaccharides of Flavone and Flavonol *O*-Glycosides

Structure	Ref.
Flavone tetrasaccharide	
[*O*-α-L-Rhamnosyl(1 → 4)-*O*-β-D-glucosyl(1 → 6)]-*O*-sophorose	6
Flavonol tetrasaccharides	
O-β-D-Apiosyl(1 → 5)-*O*-β-D-apiosyl(1 → 2)-*O*-[α-L-rhamnosyl(1 → 6)glucose][a]	281
O-β-D-Xylosyl(1 → 3)-*O*-α-L-rhamnosyl(1 → 6)-*O*-[β-D-apiosyl(1 → 2)galactose][a]	213
O-β-D-Xylosyl(1 → 4)-*O*-[β-D-xylosyl(1 → 6)-*O*-β-D-glucosyl(1 → 2)rhamnose][a]	187
O-β-D-Xylosyl(1 → 2)-*O*-[α-L-rhamnosyl(1 → 2)-*O*-α-L-rhamnosyl(1 → 6)glucose][a]	322
O-α-L-Rhamnosyl(1 → 2)-*O*-[β-D-xylosyl(1 → 3)-*O*-α-L-rhamnosyl(1 → 6)galactose][a,b]	213
O-α-L-Rhamnosyl(1 → 2)-*O*-[β-D-glucosyl(1 → 3)-*O*-α-L-rhamnosyl(1 → 6)galactose][a,b]	272
O-β-D-Glucosyl(1 → 3)-*O*-α-L-rhamnosyl(1 → 2)-*O*-[α-L-rhamnosyl(1 → 6)galactose][a]	214
O-α-L-Rhamnosyl(1 → 2)[gentiobiosyl(1 → 6)glucose][a]	191

[a]Newly reported since 1992.

[b]Present only in acylated form.

from *L. elegans*.[138] Flavonoid sulfates were found to be characteristic constituents of the Restionaceae being detected in 27% of the 115 taxa surveyed. The first report of a sulfate 2″-linked to glucose is from the water plant, *Thalassia testudinum*,[111] another monocot, where it is found in association with luteolin at the 7-position. The other flavone sulfates are 8-hydroxyapigenin (isoscutellarein) 8-(2″-sulfatoglucuronide)[81] and 8-(2″,4″-disulfatoglucuronide) and the corresponding isoscutellarein 4′-methyl ether conjugates from fruits of *Helicteres isora* (Sterculiaceae).[82]

Among the more interesting new flavonol conjugates are three sulfates linked directly to the aglycone via a carbon rather than an oxygen atom. Compounds such as these have previously only been recorded from synthesis. Thus, galangin 3-glucoside-8-*C*-sulfate and 8-*C*-sulfate and kaempferol 8-*C*-sulfate have been characterized from a whole plant extract of *Phyllanthus virgatus*,[165] a member of the Euphorbiaceae. Two new sulfated flavonols recorded from *Centaurea bracteata*[275] are quercetin 3-glucoside-3′-sulfate and quercetagetin (6-hydroxyquercetin) 3,6-dimethyl ether 7-sulfate. Another quercetin conjugate, the 3-rhamnoside-3′-sulfate, has been isolated from leaves of *Leea guineensis* (Leeaceae),[361] a monogeneric family close to the vines (Vitaceae). Quercetin 7-methyl ether 3,3′-disulfate from roots of *Argyreia mollis*[373] and quercetin 3,7-dimethyl ether 4′-sulfate and quercetin 3,7,4′-trimethyl ether 3-sulfate from aerial parts of *Ipomoea regnelli*[373] are the first sulfated flavonoids to be isolated from Convolulaceae species. Most of the other new flavonol sulfates are all single family occurrences: quercetin 7,4′-disulfate from *Alchornea laxiflora*,[362] another member of the Euphorbiaceae, and the 3,3′,4′-trisulfate from leaves of *Tamarix amplexicaulis* (Tamaricaceae),[374] isorhamnetin 3-(4″-sulfatorutinoside) from aerial parts of *Zygophyllum dumosum* (Zygophyllaceae),[392] quercetin 4′-methyl ether 3-glucoside-7-sulfate from *Polygonum hydropiper* (Polygonaceae)[396] leaves, and gossypetin 7,8-dimethyl ether 3,3′-disulfate from flowers of the bell heather, *Erica cinerea* (Ericaceae).[426] The structure of quercetin 5-glucoside-3′-

---

TABLE 13.8
Acylating Acids and Alcohol and a Lignan Found in Flavone and Flavonol Derivatives

Aliphatic Acids and Alcohol	Aromatic Acids	Sesquiterpene Acid	Lignans
Acetic	Benzoic	Dihydrophaseic	μ-Truxillic acid[360]
Malonic	*p*-Hydroxybenzoic		*p,p*-Dihydroxytruxillic acid[161]
Lactic (2-hydroxypropionic)	Gallic		
Vinylpropionic[a, 341]	Cinnamic		
Succinic	*p*-Coumaric		
Butyric	Caffeic		
Isobutyric	Ferulic		
3-Methylbutyric	Isoferulic		
Crotonic (*E*-2-butenoic)	Sinapic		
2-Methyl-2-butenoic	α-Methylsinapic[a, 288]		
n-Butanoic[a, 335]			
Isovaleric (isopentanoic)[a, 412, 417]			
Tiglic (*E*-2-methyl-2-butenoic)			
3-Hydroxy-3-methylglutaric			
Quinic			
4-Hydroxy-3-methoxyphenyl-1, 3-dihydroxypropan-2-ol[a, 240]			

[a]Newly reported since 1992 with reference numbers.

sulfate from another restionad, *Calorphus elongatus*,[367] has not been completely established. However, the possibility that it might be the corresponding 5-sulfate-3'-glucoside seems unlikely since it co-occurs with quercetin 5-glucoside.

13.3.6 ACYLATED DERIVATIVES

Some 77 new acylated flavone and 224 new acylated flavonol derivatives are included in Table 13.1 and Table 13.2, respectively. Only one new acylating acid has been found in combination with flavones and three new aliphatic acids, a new aliphatic alcohol, a lignan, and a new aromatic acid have been discovered in association with flavonols. These are listed with previously recorded structures in Table 13.8. Thus, the lignan, *p,p*-dihydroxytruxillic acid has been found linked to two molecules of apigenin 7-glucoside through the 6-positions of the two sugar moieties in the biflavone glycoside, stachysetin (**13.5**), from *Stachys aegyptiaca*.[161] μ-Truxillic acid has been identified more recently in the biflavonol glycoside, monochaetin (**13.6**). Here, two molecules of quercetin 3-galactoside are attached through the 6-hydroxyls of the two galactoses to the carboxyl groups of the μ-truxillic acid. Monochaetin was isolated from a leaf extract of the Columbian species *Monochaetum multiflorum* (Melastomataceae).[360]

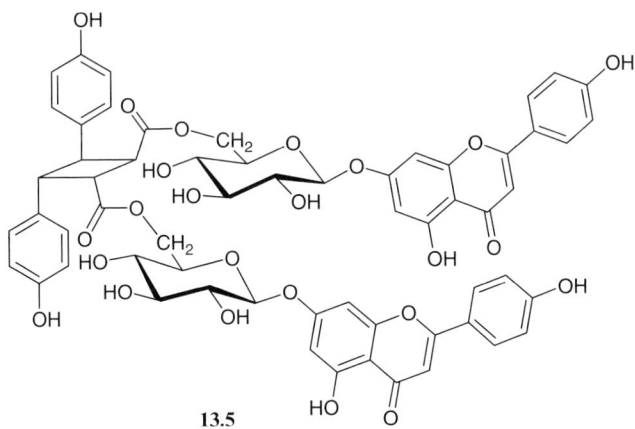

13.5

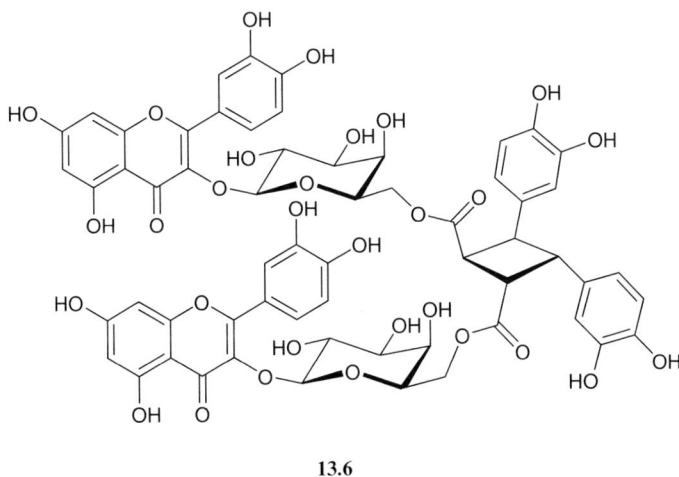

13.6

The aliphatic acid, vinylpropionic, has been found in seeds of *Psidium guaijava* (Myrtaceae)[341] directly attached to the 4'-hydroxyl of quercetin 3-(3''-galloylglucoside) and *n*-butanoic acid as quercetin 3-(6''-*n*-butylglucuronide) in leaves of the vine, *Parthenocissus tricuspidata*.[335] There are two independent reports of isovaleric (isopentanoic) acid, the first as quercetagetin 7-(6''-isovalerylglucoside) in flowers of *Buphthalmum salicifolium*[412] and the corresponding patuletin glycoside in flowers of another Composite, *Inula Britannica*.[415] The new aliphatic alcohol, 4-hydroxy-3-methoxyphenyl-1,3-dihydroxypropan-2-ol, has been found, in combination with two additional aromatic acylating acids, *p*-coumaric and ferulic, attached to kaempferol 3-glucoside (stenopalustroside E) in the fern, *Stenochlaena palustris*.[240] Three new related kaempferol glycosides, the 3-(3''-*Z*-*p*-coumaroyl-6''-feruloylglucoside) (stenopalustroside B), 3-(3-*Z*-*p*-coumaroyl-6''-*E*-*p*-coumaroylglucoside) (stenopalustroside C), and its stereoisomer, stenopalustroside D, were also present in this plant. The only new aromatic acid, α-methylsinapic, was found in *Aerva tomentosa* (Amaranthaceae),[288] attached to kaempferol 4'-methyl ether 3-(6'''-acetyl neohesperidoside) at the 4-position of the glucose moiety. Among the new reports of previously known acylating agents attached to flavonols, *p*-coumaric and acetic acids are the most common with 61 and 50 entries, respectively. Malonic acid, a frequent acylating agent in anthocyanins (see Chapter 10), is rarely found in association with flavones and flavonols. The one new flavone entry is of luteolin 7-apiosyl(1 → 2)[glucosyl(1 → 4)(malonylglucoside)] from fruits of *Capsicum annum*[109] and there are some six malonated flavonol glycosides listed in Table 13.2.[7,38,51,82,168,185]

Reports of unusual known acylating acids include 12 new entries for 3-hydroxy-3-methylglutaric acid, ten are in combination with glycosides of scutellarein, tricetin, luteolin, 6-hydroxyluteolin, and its 3-methyl ether, from four species of the liverwort genus, *Frullania*.[62,131] The other reports are of the 7-[6''-(3-hydroxy-3-methylglutaryl)glucoside] of apigenin from *Chamaemelum nobile* (Compositae)[46] and of 5,7-dihydroxy-6,8-di-*C*-methylflavone from the fern, *Matteuccia orientalis*.[160] There are also new reports of 2-methyl butyric acid and 2-methyl-butenoic (angelic) acid, the former occurs as luteolin 7-[6''-(2-methylbutryl)glucoside] in flowers of *Arnica chamissonis* (Compositae)[105] and the latter as luteolin 7-glucoside-4'-angelate in *Polygonum aviculare*.[108] Tiglic acid (*E*-2-methyl-2-butenoic acid) has been found, attached to the 8-hydroxyl of 5,7,8-trihydroxy-3,6,4'-trimethoxyflavone, in *Galeana pratensis*,[450] another Composite.

13.3.7 NEW FLAVONE GLYCOSIDES — FURTHER CONSIDERATIONS

Some 228 new flavone glycosides are listed in Table 13.1 with details of plant source and references. This increases the total of known glycosides by some third to 700 and includes 26 new apigenin, 25 new luteolin, 8 new chrysoeriol, and 2 new tricin glycosides bringing their totals to, 99, 111, 44, and 44, respectively. A complete check list of all the known flavone glycosides is given in Appendix A. There are a small number of new monoglycosides still discovered. Among the most interesting finds are three 5-glycosides of simple flavones, i.e., baicalein 6-methyl ether 5-rhamnoside from seeds of *Trichosanthes anguina* (Cucurbitaceae),[22] and the 5-glucosides of 5,7,8-trihydroxyflavone (norwogonin) and 5,2'-dihydroxy-7-methoxyflavone from *Pyracantha coccinea* (Rosaceae)[27] and *Andrographis alata* (Acanthaceae),[32] respectively. The new structures in Table 13.1 also include 32 flavone aglycones that have been found in glycosidic combination for the first time. For example, scutellarein 5,4'-dimethyl ether as the 7-glucoside and 7-(4^Rha-acetylrutinoside) from *Striga passargei* (Scrophulariaceae)[71] and four new luteolin methyl ethers, the 5,3'-dimethyl ether as the 7-glucoside and 4'-glucoside from *Pyrus serotina*,[124] the 5,4'-di- and 5,3',4'-trimethyl ether as their 7-xylosyl(1 → 6)glucosides from *Dirca palustris* (Thymelaeaceae),[112] and the 7,3',4'-trimethyl ether as the 5-glucoside and 5-xylosyl(1 → 6)glucoside in *Lethedon tannaensis*,[126] another member of the Thymelaeaceae. Among the 8-hydroxyluteolin (hypolaetin) derivatives are three glycosides from the Restionaceae: hypolaetin 7-methyl ether 3′-sulfatoglucuronide and 3′-sulfatogalactoside and hypolaetin 7,3′-dimethyl ether 4′-glucoside from three *Leptocarpus* species.[138] Four glycosides, the 5- and 7-glucosides, the 5-gentiobioside, and the 7-rutinoside, of the trimethylated flavone, nevadensin (5,7-dihydroxy-6,8,4′-trimethoxyflavone), have been reported from *Lysionotus pauciflorus* (Gesneriaceae).[142,143] 2′-Methylation and 2′-glycosylation are characteristic features of *Scutellaria* and other Labiate species. Therefore, it is not surprising to find reports of further 2′,6′-hydroxylated flavones in glycosidic combination in *Scutellaria baicalensis*[147] and *S. rivularis*.[30] Glycosides of several new methyl ethers of tricetin and 6-hydroxytricetin have also been discovered (see Table 13.1). These include further glycosides from *Lethedon tannaensis*,[126,127] namely, tricetin 7,3′,4′-trimethyl ether 5-glucoside and tricetin 7,3′,4′,5′-tetramethyl ether 5-glucoside and 5-xylosyl(1 → 6)glucoside. Two *C*-methylated flavones have been found in glycosidic combination for the first time bringing the total number of known structures to five. One of the novel glycosides, 5,7-dihydroxy-6,8-di-*C*-methylflavone 7-[6″-(3-methylglutaryl)glucoside], was isolated from the rhizomes of the fern, *Matteuccia orientalis*[160] and the other, 5,7-dihydroxy-6-*C*-methylflavone 7-xylosyl(1 → 3)xyloside, from the Labiate, *Mosla chinensis.*[159]

13.3.8 NEW FLAVONOL GLYCOSIDES — FURTHER CONSIDERATIONS

There has been a very large increase in the number of flavonol glycosides discovered, especially kaempferol derivatives. Thus, over 500 new structures have been listed in Table 13.2 bringing the total number of known flavonol glycosides to 1333. These are included with previously known flavonol glycosides in Appendix B. There are some 140 new kaempferol, 107 quercetin, and 28 new myricetin glycosides. Half of all the new flavonol glycosides are acylated, often with two or more acyl groups (see Section 13.3.6). A number of new acylated kaempferol glycosides have been isolated from ferns, for example, the 3-(6″-caffeoylglucoside) and 3-(5″-feruloyllapioside) from *Pteridium aquilinum*[232,233] and three 3-(diacetylrhamnoside)-7-rhamnoside isomers (2,′3′-, 2′,4′-, and 3′,4′-diacetyl) from *Dryopteris crasssirhizoma*.[256] Thirty-four known flavonol aglycones have been found in combination with sugars for the first time. These include three interesting new methylated 8-hydroxymyricetin derivatives, the 3-rhamnosides of the 8-mono-, 8,5′-di-, and 8,3′,5′-trimethyl ethers, from *Erica verticillata.*[446] Also two glycosides of 5-deoxymyricetin (robinetin) have been reported, the 7-glucoside from *Alternanthera sessilis* (Amaranthaceae)[447] and the 3-rutinoside from *Ateleia herbert-smithii*,[448] a member of the Leguminosae, a family rich in 5-deoxy and 5-methylated flavonoids. The first reported glycosides of gossypetin 3,8-dimethyl ether and its 7,8-isomer have been identified in *Eugenia edulis* (Myrtaceae)[424] and *Erica cinerea*,[425,426] respectively. There are six new records of 2′- or 2′,6′-hydroxylated flavonols in glycosidic combination including one from the roots of another *Scutellaria* species, *S. amoena*,[457] but most of the reports are from members of the Compositae. Two further *C*-methylated flavonols have been found in glycosidic combination, the 3-rhamnoside-5′-gallate of 2′-*C*-methyl myricetin from *Syzygium samarangense* (Myrtaceae)[479] and the 7-galactosyl(1 → 2)rhamnoside of 5,7-dihydroxy-6,8-di-*C*-methyl-3-methoxyflavone from *Cotula anthemoides* (Compositae).[478] The only previous entry was of 8-*C*-methylkaempferol 7-glucoside from roots of *Sophora leachiana* (Leguminosae).[481]

13.3.9 GLYCOSIDES OF PRENYLATED FLAVONES AND FLAVONOLS, AND OF PYRANO AND METHYLENEDIOXYFLAVONOLS

No prenylated flavone glycosides were recorded in the last edition of *The Flavonoids*. Therefore, it is significant that the present list in Table 13.1 should contain three such compounds, all isolated from members of the Leguminosae. They include 8-*C*-prenylapigenin 4′-rutinoside from *Desmodium gangeticum*,[162] 3'-*C*-prenylapigenin 7-rutinoside from *Pithecellobium dulce*[163] (both from stem tissue), and 8-*C*-prenylchrysoeriol 7-glucosyl(1 → 3)-α-L-arabino-pyranoside from seeds of *Erythrina indica*.[164]

Thirteen new prenylated flavonol glycosides have been discovered during the review period, all but three of them from species of the Berberidaceae. There are four new aglycone sugar combinations including the first prenylated quercetin glycosides, i.e., 8-prenylquercetin 4'-methyl ether 3-rhamnoside from *Epimedium koreanum* (Berberidaceae)[475] and 8-prenyl-quercetin 7,4'-dimethyl ether 3-rhamnosyl(1 → 4)rhamnoside from *Butea monosperma* (Leguminosae).[476] 6,6''-Dimethylpyrano(2'',3'':7,8)-4'-methoxykaempferol 3-rhamnoside was also present in *Epimedium acuminatum*.[477] This is only the second occurrence of a pyrano-flavonol glycoside. Three new methylenedioxyflavonol glycosides have been reported bringing the total number of known structures to four. They include two glycosides, viviparum A and B from *Polygonum viviparum*[465] and 3-methoxy-5-hydroxy-6,7-methylenedioxyflavone 4'-glucuronide from spinach, *Spinacia oleracea* (Chenopodiaceae).[464]

13.4 DISTRIBUTION PATTERNS

Flavone and flavonol *O*-glycosides are widely distributed in the angiosperms and gymno-sperms, mosses, liverworts, and ferns. However, in some monocot families they are largely replaced by or co-occur with flavone *C*-glycosides, for example, in the grasses, palms, Cyperaceae, and Iridaceae. In the dicots the more evolutionary advanced families tend to accumulate flavone *O*-glycosides and complex methylated or extra hydroxylated flavonoid aglycones, while the more primitive families produce flavonol *O*-glycosides, especially myr-icetin derivatives, together with proanthocyanins. Gymnosperms are characterized by the presence of flavonol *O*-glycosides, flavone *C*-glycosides, and biflavonoids. Flavonoid *O*- and *C*-glycosides also occur in mosses, liverworts, and ferns. Biflavonoids are important constitu-ents of the bryophytes but are rare in ferns, where flavonol and flavone *O*-glycosides, glycoflavones, and dihydroflavonols are the most frequent components. There have been no major reviews of the distribution of flavonoids in lower plants, gymnosperms, or the dicotyledons since those by Markham, Niemann, and Giannasi in the third edition of *The Flavonoids*,[4] Chapters 12 to 14, respectively. However, the chapter by Williams and Harborne on the distribution of flavonoids in the monocotyledons was updated in 1994.[482] Undoubt-edly, the current emphasis on molecular taxonomy has led to a reduction in research in chemotaxonomy with most flavonoid projects now based on a search for biologically active or medicinally useful secondary constituents. Space does not allow a complete review of flavonoid distribution here but some of the more interesting findings and the more extensive surveys will be mentioned.

Among the bryophytes, ten new flavone glycosides, all acylated with 3-hydroxy-3-methyl-glutaric acid, have been reported from three *Frullania* species.[62,131] In another liverwort, *Dumortiera hirsuta*,[92] luteolin 5-glucuronide-6''-methyl ether has been identified and apigenin and luteolin 7-sophorotrioside and luteolin 7-(acetylsophorotrioside) have been characterized from gametophytes of the moss, *Leptostomum macrocarpon*.[36] There are reports of new flavone *O*-glycosides from two ferns, luteolin 7-robinobioside[97] and 7-sophoroside[98] from *Pteris cretica* and 5,7-dihydroxy-6,8-di-*C*-methylflavone 7-[6''-(3-hydroxy-3-methylglutaryl)-glucose] from *Matteuccia orientalis*.[160] The other novel fern glycosides are all flavonol and include the unusual quercetin 3-methyl ether 5-glucoside from *Asplenium trichomanes-ramo-sum*,[363] four acylated kaempferol glycosides, stenopalustrosides B–E from *Stenochlaena palustris*,[240] which were discussed in Section 13.3.6 and quercetin 3-methyl ether 7-α-L-arabinofuranosyl(1 → 6)glucoside from *Lepisorus ussuriensis*.[364] Six further flavonol glyco-sides have been recorded for brachen, *Pteridium aquilinum*[184,232,233,314] (Table 13.2).

Reports of flavonoid *O*-glycosides from the gymnosperms include the unusual 5,6,7,8,3′,4′-hexahydroxyflavone 7-glucoside from fruits of *Juniperus zeravschanica* (Cupressaceae)[145] and kaempferol 3,4′-diglucoside from needles of the common spruce, *Picea abies*.[183] Five further glycosides have been identified in leaves of the Maidenhair tree, *Ginkgo biloba*,[176,177,265] including the unusual acylated glycosides, kaempferol and quercetin 3-[2″6‴-{-*p*-(7‴′-glucosyl)coumaroyl}glucosyl]rhamnosides,[177] in which the *p*-coumaric acid moiety is linked in a linear fashion to the 6‴- and 4‴′-hydroxyls of the two glucose molecules.

There have been three major flavonoid surveys of monocot families since 1991. Thus, 115 Restionaceae species[483] endemic to Australia were analyzed for their culm flavonoids. The data are mainly of flavonoid aglycones but a variety of new glycosides were also characterized, including six new flavone sulfate conjugates (discussed in Section 13.3.5). The aglycones were determined after acid hydrolysis and hypolaetin (found in 23 of 34 genera), luteolin (in 25 genera), flavone *C*-glycosides (in 13 genera), and sulfates (in 15 genera) were found to be the most typical flavonoid constituents. Gossypetin (in seven genera), tricin (in seven genera), and myricetin (in two genera) were relatively rare in these plants. A survey of three families related to Restionaceae, the Anarthriaceae, Ecdeiocoleaceae, and Lygineaceae,[484] which are endemic to South Western Australia, showed the regular presence of myricetin, quercetin, and isorhamnetin with only traces of kaempferol. These flavonols were present mainly as 3-*O*-glycosides but some unknown conjugates were found to be characteristic of the genus *Anarthria*. In a further flavonoid survey of Velloziaceae taxa two new unusual flavone glycosides, 5,6,7,3′.4′-pentahydroxy-8-methoxyflavone 7-glucoside and 5,6,3′,4′-tetrahydroxy-7,8-dimethoxyflavone 6-glucoside, have been identified in *Vellozia nanza*.[146] However, most genera of the Velloziaceae are characterized by the presence of lipophilic, prenyl-, pyrano-, or *C*-methylated flavonols, with some simple flavones and flavanones. Flavonol *O*-glycosides are common in *Aylthonia*, *Barbacenia*, and *Xerophyta* but most *Vellozia* species accumulate flavone *C*-glycosides. The lipophilic and vacuolar flavonoid data for *Vellozia* species are summarized by Harborne et al.[485] and the data for all the genera of the Velloziaceae by Williams et al.[486] A flavonoid survey of *Iris* species[487] showed the characteristic constituents were glycoflavones but here they co-occur with isoflavones and the xanthone mangiferin and its derivatives.

Among the dicotyledons three of the larger families will be considered, Compositae, Labiatae, and Leguminosae. The Compositae is very rich in flavonoids and there have been a number of recent surveys of tribes or genera but nearly all are confined to or concentrate on the lipophilic surface constituents. There is one useful new book by Bohm and Stuessy entitled *Flavonoids of the Sunflower Family (Asteraceae)*, published in 2001,[488] which after a general introduction to the family and to flavonoids, summarizes all the then known flavonoid data for each tribe and considers the efficacy of flavonoids at different taxonomic levels. However, it is not always reader friendly with many back references to previous sections or chapters so that access to the reference information is not as easy as it might be. The genus *Tanacetum*[489] and other members of the tribe Anthemideae[171] have been surveyed for both surface lipophilic and vacuolar flavonoids. In *Tanacetum* species, apigenin and luteolin 7-glucuronides are the characteristic vacuolar flavonoids. However, 6-hydroxyluteolin 7-glucoside was found in *T. corymbosum*, chrysoeriol 7-glucuronide in *T. parthenium*, *T. macrophyllum*, and *T. cinerariifolium*, and quercetin 7-glucuronide in *T. parthenium*, *T. corymbosum*, and *T. cinerariifolium*.[489] The lipophilic flavonoids are based mainly on 6-hydroxykaempferol 3,6,4′-trimethyl ether and quercetagetin 3,6,3′-trimethyl ether with methyl ethers of scutellarein and 6-hydroxyluteolin in some species. Both lipophilic and polar flavonoids were isolated from leaf, ray, and disc florets of other Anthemideae, *Anthemis*, *Chrysanthemum*, *Cotula*, *Ismelia*, *Leucanthemim*, and *Tripleuropermum*.[171] *Anthemis* species characteristically produced flavonol glycosides in the leaves while in the other taxa flavone O-glycosides were more usual. Two flavonol glycosides, quercetin and kaempferol 5-glucuronides, were identified in leaves of *Leucanthemum vulgare*.[171] In most of these plants ray and disc florets had noticeably different flavonoid patterns, the former based on apigenin or luteolin 7-glucoside or glucuronide and the latter having additional flavonol glycosides such as quercetin and patuletin 7-glucosides and quercetin 7-glucuronide. A similar flavonoid survey of *Pulicaria* species (tribe Inulae)[490] has also been published.

There have been a considerable number of chemotaxonomic studies of Labiate taxa since 1991. These include a survey of flavonoid aglycones and glycosides in *Sideritis* species[491] from the Canary Islands and Madeira and the discovery of a new flavone glycoside, isoscutellarein 7-glucosyl(1 → 2)xyloside, from *Sideritis luteola* and 15 other Spanish *Sideritis* species.[78] A review of the polyphenolics of the genus *Salvia*[492] includes flavonoid O-glycoside data and Tomás-Barberán et al.[493] have determined the distribution of flavonoid p-coumaroylglucoside and 8-hydroxyflavone allosylglucosides in the Labiatae. Five other studies concern leaf flavonoid glycosides as taxonomic characters in the genera *Ocimum*,[494] *Calamintha* and *Micromeria*,[495] *Teucridium* and *Tripora*,[496] *Oxera* and *Faradaya*,[497] and *Lavandula* and *Sabaudia*.[498]

The Leguminosae is phytochemically one of the most diverse families and contains a wealth of flavonoid constituents. A review of the seed polysaccharides and flavonoids, which includes flavonoid O-glycosides, provides a useful chemical overview of the family.[499] However, the definitive publications on the chemistry of the family are undoubtedly the three latest Leguminosae volumes (11a–11c) of Robert Hegnauer's *Chemotaxonomie der Pflanzen*.[500] A more recent review of the phytochemistry of the large and taxonomically difficult genus *Acacia* has been published by Seigler.[501]

REFERENCES

1. Harborne, J.B. and Williams, C.A., Flavone and flavonol glycosides, in *The Flavonoids*, Harborne, J.B., Mabry, T.J., and Mabry, H., Eds., Chapman & Hall, London, 1975, chapter 8.
2. Hattori, S., Glycosides of flavones and flavonols, in *The Chemistry of Flavonoid Compounds*, Geissman, T.A., Ed., Pergamon Press, Oxford, 1962, chapter 11.
3. Harborne, J.B. and Williams, C.A., Flavone and flavonol glycosides, in *The Flavonoids: Advances in Research*, Harborne, J.B. and Mabry, T.J., Eds, Chapman & Hall, London, 1982, chapter 5.
4. Harborne, J.B. and Williams, C.A., Flavone and flavonol glycosides, in *The Flavonoids: Advances in Research Since 1980*, Harborne, J.B., Ed., Chapman & Hall, 1988, chapter 8.
5. Harborne J.B. and Williams, C.A., Flavone and flavonol glycosides, in *The Flavonoids: Advances in Research Since 1986*, Harborne, J.B., Ed., Chapman & Hall, London, 1994, chapter 8.
6. Ahmed, A.A. and Saleh, N.A.M., Peganetin, a new branched acetylated tetraglycoside of acacetin from *Peganum harmala*, *Nat. Prod.*, 50, 256, 1987.
7. Harborne, J.B. and Williams, C.A., Anthocyanins and other flavonoids, *Nat. Prod. Rep.*, 12, 639, 1995.
8. Harborne, J.B. and Williams, C.A., Anthocyanins and other flavonoids, *Nat. Prod. Rep.*, 15, 631, 1998.
9. Harborne, J.B. and Williams, C.A., Anthocyanins and other flavonoids, *Nat. Prod. Rep.*, 18, 310, 2001.
10. Williams, C.A. and Grayer, R.J., Anthocyanins and other flavonoids, *Nat. Prod. Rep.*, 21, 539–573, 2004.
11. Harborne, J.B., Baxter, H., and Moss, G.P., Eds., *Phytochemical Dictionary*, Taylor and Francis, London, 2nd edition, 1999, chapter 37.
12. Harborne, J.B. and Baxter, J., Eds., *The Handbook of Natural Flavonoids*, Vol. 1, John Wiley and Sons, Chichester, 1999, Section 3.
13. Harborne, J.B., *Comparative Biochemistry of the Flavonoids*, Academic Press, London, 1967.

14. Mabry, T.J., Markham, K.R., and Thomas, M.B., *The Systematic Identification of Flavonoids*, Springer-Verlag, New York, 1970.

15. Markham, K.R., *Techniques of Flavonoid Identification*, Academic Press, London, 1982.

16. Markham, K.R., Flavones, flavonols and their glycosides, in *Methods in Plant Biochemistry*, Dey, P.M. and Harborne, J.B., Series Editors, Vol. 1, *Plant Phenolics*, Harborne, J.B., Ed., Academic Press, London, 1989.

17. Harborne, J.B., in *Chromatography*, Heftmann, E., Ed., Elsevier, Amsterdam, 5th edition, 1992, pp. 363–392.

18. Grayer, R.J. et al., The application of atmospheric pressure chemical ionisation liquid chromatography–mass spectrometry in the chemotaxonomic study of flavonoids: characterisation of flavonoids from *Ocimum gratissimum* var. *gratissimum, Phytochem. Anal.*, 11, 257, 2000.

19. Pistelli, L. et al., Flavonoids from *Calicotome villosa, Fitoterapia*, 74, 4171, 2003.

20. Liu, Q., Dixon, R.A., and Mabry, T.J., Additional flavonoids from elicitor-treated cell cultures of *Cephalocereus senilis, Phytochemistry*, 34, 167, 1993.

21. Liu, Q. et al., Flavonoids from elicitor-treated cell suspension cultures of *Cephalocereus senilis, Phytochemistry*, 32, 925, 1993.

22. Yadava, R.N. and Syeda, Y., A novel flavone glycoside from *Trichosanthes anguina* seeds, *Fitoterapia*, 65, 554, 1994.

23. Aqil, M., Flavonoid glycosides from *Eupatorium africanum, Ultra Sci. Phys. Sci.*, 7, 1, 1995.

24. Yadava, R.N. and Tripathi, P. A novel flavone glycoside from the stem of *Bauhinia purpurea, Fitoterapia*, 71, 88, 2000.

25. Yang, F., Li, X.-C., Wang, H.-Q., and Yang, C.-R., Flavonoid glycosides from *Colebrookea oppositifolia, Phytochemistry*, 42, 867, 1996.

26. Zheng, S. et al., Studies on the flavonoid compounds of *Origanum vulgare* L., *Indian J. Chem.*, 36B, 104, 1997.

27. Bilia, A.R. et al., Flavonoids from *Pyracantha coccinea* roots, *Phytochemistry*, 33, 1449, 1993.

28. Yuldashev, M.P. and Karimov, A., Flavonoids of *Scutellaria ocellata* and *S. nepetoides, Chem. Nat. Compd.*, 37, 431, 2002.

29. Yuldashev, M.P., Flavonoids of the aerial parts of *Scutellaria immaculata, Chem. Nat. Compd.*, 37, 428, 2002.

30. Tomimori, T., Imoto, Y., and Miyaichi, Y., Studies on the constituents of *Scutellaria* species. 13. On the flavonoid constituents of the root of *Scutellaria rivularis* Wall., *Chem. Pharm. Bull.*, 38, 3488, 1990.

31. Yuldashev, M.P., Batirov, C.K., and Malikov, V.M., New flavone glycoside from *Scutellaria ramosissima, Khim. Prir. Soedin*, 317, 1995.

32. Demu, A.G. et al., A flavone glycoside from *Andrographis alata, Phytochemistry*, 49, 1811, 1998.

33. Reddy, M.V.B. et al., New 2′-oxygenated flavonoids from *Andrographis affinis, J. Nat. Prod.*, 66, 295, 2003.

34. Taneko, T. et al., Secoiridoid and flavonoid glycosides from *Gonocaryum calleryanum, Phytochemistry*, 39, 115, 1995.

35. Veitch, N.C. et al., Flavonoid cellobiosides from *Salvia uliginosa, Phytochemistry*, 48, 389, 1998.

36. Brinkmeier, E. and Geiger, H., Flavone 7-sophorotriosides and biflavonoids from *Leptostomum macrocarpon, Z. Naturforsch. C*, 53, 1, 1998.

37. Picroni, A. and Pachaly, P., Isolation and structure elucidation of ligustroflavone, a new apigenin triglycoside from the leaves of *Ligustrum vulgare* L., *Pharmazie*, 55, 78, 2000.

38. Kikuchi, M. and Matsuda, N., Flavone glycosides from *Lonicera gracilipes* var. *glandulosa, J. Nat. Prod.*, 59, 314, 1996.

39. Iwashina, T., Matsumoto, S., and Yoshida, Y., Apigenin 7-rhamnoside-4′-glucosylrhamnoside from *Asplenium normale, Phytochemistry*, 32, 1629, 1993.

40. Stochmal, A. et al., Alfalfa (*Medicago sativa* L.) flavonoids. 1. Apigenin and luteolin glycosides from aerial parts, *J. Agric. Food Chem.*, 49, 753, 2001.

41. Zheng, P.-C. and Xu, S.-X., Flavonoid ketohexosefuranosides from the leaves of *Crataegus pinnatifida* Bge. var. *major* N.E. Br., *Phytochemistry*, 57, 1249, 2001.

42. Ram, S.N. et al., An acylflavone glucoside of *Echinops echinatus* flowers, *Planta Med.*, 62, 187, 1996.

43. El-Ansari, M.A., Nawwar, M.A., and Saleh, N.A.M., Stachysetin, a diapigenin 7-glucoside-*p*-*p*′-dihydroxytruxinate from *Stachys aegyptiaca*, *Phytochemistry*, 40, 1543, 1995.

44. Li, J. and Chen, Y., Two flavonoids from *Lagopsis supine*, *Yaoxue Xuebao*, 37, 186, 2002.

45. Gudej, J. and Nazaruk, J., Apigenin glycoside esters from flowers of *Bellis perennis* L., *Acta Pol. Pharm.*, 54, 233, 1997.

46. Tschan, G.M. et al., Chamaemeloside, a new flavonoid glycoside from *Chamaemelum nobile*, *Phytochemistry*, 41, 643, 1996.

47. Harraz, F.M. et al., Acylated flavonoids from *Blepharis ciliaris*, *Phytochemistry*, 43, 521, 1996.

48. Abdel-Rahem, S.I., Rashwan, O., and Abdel-Sattar, E., Flavonoids from *Chrozophora oblongifolia*, *Bull. Fac. Pharm. (Cairo Univ.)*, 39, 103, 2001.

49. Stochmal, A. et al., Acylated apigenin glycosides from alfalfa (*Medicago sativa* L.) var. *artal*, *Phytochemistry*, 57, 1223, 2001.

50. Zahid, M. et al., Flavonoid glycosides from *Salvia moorcroftiana*, *Carbohydr. Res.*, 337, 403, 2002.

51. Sharaf, M. et al., Four flavonoid glycosides from *Peganum harmala*, *Phytochemistry*, 44, 533, 1997.

52. Yadava, R.N. and Barsainya, D., A novel flavone glycoside from *Centratherum anthelminticum* Kuntzel, *J. Indian Chem. Soc.*, 74, 822, 1997.

53. Wanjala, C.C.W. and Majinda, R.R.T., Flavonoid glycosides from *Crotalaria*, *Phytochemistry*, 51, 705, 1999.

54. Li, J.S. et al., Separation and isolation of the flavonoids from *Buddleia officinalis* Maxim., *Yaoxue Xuebao*, 31, 849, 1996.

55. Sharaf, M., Isolation of an acacetin tetraglycoside from *Peganum harmala*, *Fitoterapia*, 67, 294, 1996.

56. Yu, S.C. et al., Flavonoid glycosides from *Thalictrum przewakskii*, *J. Asian Nat. Prod. Res.*, 1, 301, 1999.

57. Shen, Y. et al., Studies on the flavonoids from *Dendranthema lavandulifolium*, *Yaoxue Xuebao*, 32, 451, 1997.

58. Marin, P.D. et al., Acacetin glycosides as taxonomic markers in *Calamintha* and *Micromeria*, *Phytochemistry*, 58, 943, 2001.

59. Mitchell, K.A., Markham, K.R., and Bayly, M.J., Flavonoid characters contributing to the taxonomic revision of the *Hebe parviflora* complex, *Phytochemistry*, 56, 453, 2001.

60. Alam, M.S. and Jain, N., A new flavone glycoside from *Semecarpus kurzii*, *Fitoterapia*, 64, 239, 1993.

61. Yoshida, K., Kameda, K., and Kondo, T., Diglucuronoflavones from purple leaves of *Perilla ocimoides*, *Phytochemistry*, 33, 917, 1993.

62. Kraut, L., Mues, R., and Sim-Sim, M., Acylated flavone and glycerolglucosides from two *Frullania* species, *Phytochemistry*, 34, 211, 1993.

63. Laskaris, G.G., Gournelis, D.C., and Kakkalou, E., Phenolics of *Picnomon acarna*, *J. Nat. Prod.*, 58, 1248, 1995.

64. Akkal, S. et al., A new flavone glycoside from *Centaurea furfuracea*, *Fitoterapia*, 70, 368, 1999.

65. Iwashina, T., Kamenosono, K., and Ueno, T., Hispidulin and nepetin 4′-glucosides from *Cirsium ologophyllum*, *Phytochemistry*, 51, 1109, 1999.

66. Zhov, L.D., Yu, J.G., and Guo, J., Mahuangchiside, a new flavone glycoside from *Chirita fimbrisepala*, *Chin. Chem. Lett.*, 11, 131, 2000.

67. Ragunathan, V. and Sulochana, N., A new flavone glycoside from the flowers of *Ipomoea purpurea* Roth., *Indian J. Chem.*, 33B, 507, 1994.

68. Ho, J.-C. and Chen, C.M., Flavonoids from the aquatic plant *Eriocaulon buergerianum*, *Phytochemistry*, 61, 405, 2002.

69. Grayer, R.J. et al., Scutellarein 4′-methyl ether glycosides as taxonomic markers in *Teucridium* and *Tripora* (Lamiaceae, Ajugoideae), *Phytochemistry*, 60, 727, 2002.

70. Ezer, N. et al., Flavonoid glycosides and a prenylpropanoid glycoside from *Sideritis perfoliata*, *Int. J. Pharmacogn.*, 30, 61, 1992.

71. Aqil, M., Babayo, Y., and Bobboi, A., Flavone glycosides from *Striga passargei*, *U. Scientist Phil. Science*, 8, 247, 1996.

72. El-Sayed, N.H. et al., Flavonoids of *Conyza linifolia*, *Rev. Latinoam. Quim.*, 22, 89, 1991.

73. Begum, S. et al., Nematicidal constituents of the aerial parts of *Lantana camara*, *J. Nat. Prod.*, 63, 765, 2000.

74. Otsuka, H., Isolation of isolinariins A and B, new flavonoid glycosides from *Linaria japonica*, *J. Nat. Prod.*, 55, 1252, 1992.

75. Lahloub, M.F., Flavonoid, phenylpropanoid and iridoid glycosides of *Linaria haelava* (Forssk.) Dil., *Mansoura J. Pharm. Sci.*, 8, 78, 1992.

76. Das, B. and Chakravarty, A.K., Three flavone glycosides from *Gelonium multiflorum*, *Phytochemistry*, 33, 493, 1993.

77. Aqil, M. et al., A new flavone glycoside from *Striga aspera*, *Sci. Phys. Sci.*, 6, 131, 1994.

78. Palamino, O.M. et al., Isoscutellarein 7-glucosyl(1 → 2) xyloside from sixteen species of *Sideritis*, *Phytochemistry*, 42, 101, 1996.

79. Grayer-Barkmeijer, R.J. and Tomas-Barberan, F.A., 8-Hydroxylated flavone *O*-glycosides and other flavonoids in chemotypes of *Gratiola officinalis*, *Phytochemistry*, 34, 205, 1993.

80. Sharaf, M., Isoscutellarein 8-*O*-(6″-*trans*-*p*-coumaroyl)-β-D-glucoside from *Stachys aegyptiaca*, *Fitoterapia*, 69, 355, 1998.

81. Chen, Z.-T., Lee, S.-W., and Chen, C.-M., New flavonoid glycosides of *Helicteres angustifolia*, *Heterocycles*, 38, 1399, 1994.

82. Kamiya, K. et al., Flavonoid glucuronides from *Helicteres isora*, *Phytochemistry*, 57, 297, 2001.

83. Sharaf, M., El-Ansari, M.A., and Saleh, N.A.M., A new flavonoid from the roots of *Glossostemon bruguieri*, *Fitoterapia*, 69, 47, 1998.

84. Rios, J.L. et al., Antioxidant activity of flavonoids from *Sideritis javalambrensis*, *Phytochemistry*, 31, 1947, 1992.

85. Luo, Y. et al., Glycosides from *Dicliptera riparia*, *Phytochemistry*, 61, 449, 2002.

86. Zhou, Z.H., Zhang, Y.J., and Yang, C.R., New flavonoid glycosides from *Scutellaria amoena*, *Stud. Plant Sci.*, 6, 305, 1999.

87. Zhou, Y. et al., Flavonoids and phenylethanoids from hairy root cultures of *Scutellaria baicalensis*, *Phytochemistry*, 44, 83, 1997.

88. Damu, A.G., Jayaprakasam, B., and Gunasekar, D., A new flavone 2′-glucoside from *Andrographis alata*, *J. Asian Nat. Prod. Res.*, 1, 133, 1998.

89. Gupta, K.K., Taneja, S.C., and Dhar, K.L., Flavonoid glycoside of *Andrographis paniculata*, *Indian J. Chem.*, 35B, 512, 1996.

90. Damu, A.G. et al., Two acylated flavone glucosides from *Andrographis serpyllifolia*, *Phytochemistry*, 52, 147, 1999.

91. Jayakrishna, G. et al., Two new 2′-oxygenated flavones from *Andrographis elongata*, *Chem. Pharm. Bull.*, 49, 1555, 2001.

92. Kraut, L. et al., Carboxylated α-pyrone derivatives and flavonoids from the liverwort *Dumortiera hirsuta*, *Phytochemistry*, 42, 1693, 1996.

93. Zarzuelo, A. et al., Luteolin 5-rutinoside from *Salvia lavandulifolia* ssp. *oxyodon*, *Phytochemistry*, 40, 1321, 1995.

94. Salpekar, J. and Khan, S.A., Luteolin 7-*O*-β-D-glucopyranosyl(1 → 4)-*O*-α-L-arabinopyranoside from *Cassia glauca*, *Ultra Sci. Phys. Sci.*, 8, 260, 1996.

95. Rodriguez, S. et al., Xanthones, secoiridoids and flavonoids from *Halenia corniculata*, *Phytochemistry*, 40, 1265, 1995.

96. Tsopmo, A. et al., Geranylated flavonoids from *Dorstenia poinsettifolia*, *Phytochemistry*, 48, 345, 1998.

97. Imperato, F., A new flavone glycoside from the fern *Pteris cretica*, *Experientia*, 50, 1115, 1994.

98. Imperato, F. and Nazzaro, F., Luteolin 7-*O*-sophoroside from *Pteris cretica*, *Phytochemistry*, 41, 337, 1996.

99. Chaturvedi, S.K. and Saxena, V.K., Luteolin 7-*O*-β-D-galactopyranosyl(1 → 6)-*O*-β-D-galactopyranoside from the roots of *Anogeissus latifolia*, *Acta Cienc. Indica Chem., Sect. C*, 17, 155, 1991.

100. Recio, M.C. et al., Luteolin 7-galactosylglucuronide, a new flavonoid from *Andryala regusina*, *Pharmazie*, 48, 228, 1993.

101. Patora, J. and Klimek, B., Flavonoids from lemon balm (*Melissa officinalis* L., Lamiaceae), *Acta Pol. Pharm.*, 59, 139, 2002.

102. Parveen, M. et al., Luteolin 3′-xylosyl(1 → 2) glucoside from *Viburnum grandifolium*, *Phytochemistry*, 49, 2535, 1998.

103. Borai, P. and Dayal, R., A flavone glycoside from *Dalbergia stipulacea* leaves, *Phytochemistry*, 33, 731, 1993.

104. Hirobe, C. et al., Cytotoxic flavonoids from *Vitex agnus-castus*, *Phytochemistry*, 46, 521, 1997.

105. Merfort, I. and Wendisch, D., New flavonoid glycosides from Arnicae Flos DAB 9, *Planta Med.*, 58, 355, 1992.

106. Okamura, N. et al., Flavonoids in *Rosmarinus officinalis* leaves, *Phytochemistry*, 37, 1463, 1994.

107. Flamini, G. et al., Three new flavonoids and other constituents from *Lonicera implexa*, *J. Nat. Prod.*, 60, 449, 1997.

108. Sun, L. et al., The flavonoids from *Polygonum aviculare*, *Indian J. Chem.*, 41B, 1319, 2002.

109. Materska, M. et al., Isolation and structure elucidation of flavonoid and phenolic acid glycosides from pericarp of hot pepper fruit *Capsicum annum* L., *Phytochemistry*, 63, 893, 2003.

110. Wang, C.Z., Jia, Z.J., and Liao, J.C., Flavonoid and iridoid glycosides from *Veronica didyma* Tenore, *Indian J. Chem.*, 34B, 914, 1995.

111. Jensen, P.R. et al., Evidence that a new antibiotic flavone glycoside chemically defends the sea grass *Thalassia testudinum* against zoosporic fungi, *Appl. Environ. Microbiol.*, 64, 1490, 1999.

112. Ransewak, R.N. et al., Phenolic glycosides from *Dirca palustris*, *J. Nat. Prod.*, 62, 1558, 1999.

113. Sharaf, M., El-Ansari, M.A., and Saleh, N.A.M., New flavonoids from *Avicennia marina*, *Fitoterapia*, 71, 274, 2000.

114. Das, C., Tripathi, A.K., and Jogi, S.R., Chrysoeriol 7-*O*-(6-*O*-α-L-arabinopyranosyl)-β-D-glucopyranoside from *Tagetes patula*, *Orient. J. Chem.*, 12, 327, 1996.

115. Cimanga, K. et al., Flavonoid *O*-glycosides from the leaves of *Morinda morindoides*, *Phytochemistry*, 38, 1301, 1995.

116. Stochmal, A. et al., Alfalfa (*Medicago sativa* L.) flavonoids. 2. Tricin and chrysoeriol glycosides from aerial parts, *J. Agric. Food Chem.*, 49, 5310, 2001.

117. Sahpaz, S., Skaltsounis, A.-L., and Bailleul, F., Polyphenols from *Ballota acetabulosa*, *Biochem. Syst. Ecol.*, 30, 601, 2002.

118. Karioti, A. et al., Acylated flavonoid and phenylethanoid glycosides from *Marrubium velutinum*, *Phytochemistry*, 64, 655, 2003.

119. Kitanaka, S. and Takido, M., Studies on the constituents of the leaves of *Cassia torosa* Cav. 3. The structure of two flavone glycosides, *Chem. Pharm. Bull.*, 40, 249, 1992.

120. Scabra, R.M. and Alvis, E.A.C., Phenolic compounds in *Gallium palustre*, *Rev. Port. Farm.*, 45, 121, 1995.

121. Del Rio, J.A. et al., Neodiosmin, a flavone glycoside of *Citrus aurantium*, *Phytochemistry*, 31, 723, 1992.

122. Emam, A.M. et al., Two flavonoid triglycosides from *Buddleja madagascariensis*, *Phytochemistry*, 48, 739, 1998.

123. Abourashed, E.A. et al., Two new flavone glycosides from *Paullinia pinnata*, *J. Nat. Prod.*, 62, 1179, 1999.

124. Ozawa, T. et al., Identification of species-specific flavone glucosides useful as taxonomic markers in the genus *Pyrus*, *Biosci. Biotechnol. Biochem.*, 59, 2244, 1995.

125. Chou, C.J. et al., Flavonoid glycosides from *Viscum alniformosanae*, *J. Nat. Prod.*, 62, 1421, 1999.

126. Zahir, A. et al., Five new flavone 5-*O*-glycosides from *Lethedon tannaensis:* lethedosides and lethediosides, *J. Nat. Prod.*, 62, 241, 1999.

127. Chavez, J.P. et al., Flavonoids and triterpene ester derivatives from *Erythroxylum leal costae*, *Phytochemistry*, 41, 941, 1996.

128. Mitchell, K.A., Markham, K.R., and Bayly, M.J., 6-Hydroxyluteolin 7-*O*-β-D-[2-*O*-β-D-xylosylxyloside]: a novel flavone xyloxyloside from *Hebe stenophylla*, *Phytochemistry*, 52, 1165, 1999.

129. Kellam, S.J. et al., Luteolin and 6-hydroxyluteolin glycosides from *Hebe stricta*, *Phytochemistry*, 33, 867, 1993.

130. Albach, D.C. et al., Acylated flavone glycosides from *Veronica*, *Phytochemistry*, 64, 1295, 2003.

131. Kraut, L. et al., Flavonoids from some *Frullania* species (Hepaticae), *Z. Naturforsch. C*, 50, 345, 1995.

132. Hiermann, A., New flavone glycosides in the leaves of *Digitalis lanata*, *Planta Med.*, 45, 59, 1982.

133. Kamil, M., Jain, N., and Ilyas, M., A novel flavone glycoside from *Chenopodium ambrosioides*, *Fitoterapia*, 63, 230, 1992.

134. Yadava, R.N. and Saurabh, K., A new flavone glycoside, 5,7,4′-trihydroxy-6,3′-dimethoxyflavone 5-rhamnoside from leaves of *Tridax procumbens* Linn., *J. Asian Nat. Prod. Res.*, 1, 147, 1998.

135. Yuldashev, M.P., Batirov, E.E., and Malikov, V.M., Flavonoids from aerial parts of *Kichxia elatine*, *Chem. Nat. Compd.*, 32, 30, 1996.

136. Yadava, R.N. and Kumar, S., 7-Hydroxy-3′,4′,5,6-tetramethoxyflavone 7-glucosyl(1 → 4)gluco-4)glucoside: a new flavone glycoside from the stem of *Sphaeranthus indicus* Linn., *J. Inst. Chem. (India)*, 70, 164, 1998.

137. Kandil, F.E. and Husieny, H.A., A new flavonoid glycoside from *Cornulaca monacantha*, *Orient. J. Chem.*, 14, 215, 1998.

138. Williams, C.A. et al., Flavonoid patterns and the revised classification of Australian Restionaceae, *Phytochemistry*, 49, 529, 1998.

139. Abdel Sattar et al., Flavonoid glycosides from *Sideritis* species, *Fitoterapia*, 64, 278, 1993.

140. Wollenweber, E., Dorr, E., and Roitman, J.N., Epicuticular flavonoids of some Scrophulariaceae, *Z. Naturforsch. C*, 55, 5, 2000.

141. Grayer, R.J. and Veitch, N.C., An 8-hydroxylated external flavone and its 8-*O*-glucoside from *Becium grandiflorum*, *Phytochemistry*, 47, 779, 1998.

142. Liu, Y., Wagner, H., and Bauer, R., Nevadensin glycosides from *Lysionotus pauciflorus*, *Phytochemistry*, 42, 1203, 1996.

143. Wagner, Y.L.H. and Bauer, R., Phenylpropanoids and flavonoid glycosides from *Lysionotus pauciflorus*, *Phytochemistry*, 48, 339, 1998.

144. Na, Z. et al., Flavonoids from *Isodon enanderianus*, *Yunnan Zhiwu Yangiu*, 24, 121, 2002.

145. Yuldashev, M.P. and Rassulova, L.Kh., Flavonoids of *Juniperus zeravschanica*, *Chem. Nat. Compd.*, 37, 226, 2001.

146. Harborne, J.B. et al., Ten isoprenylated and *C*-methylated flavonoids from the leaves of three *Vellozia* species, *Phytochemistry*, 34, 219, 1993.

147. Ishimaru, K. et al., Two flavone 2′-glucosides from *Scutellaria baicalensis*, *Phytochemistry*, 40, 279, 1995.

148. Pauli, G.F. and Junior, P., Phenolic glycosides from *Adonis aleppica*, *Phytochemistry*, 38, 1245, 1995.

149. Sharaf, M., El-Ansari, M.A., and Saleh, N.A.M., Flavone glycosides from *Mentha longifolia*, *Fitoterapia*, 70, 478, 1999.

150. Jain, N. et al., Setaricin: a new flavone glycoside from *Setaria italica*, *Chem. Ind. (Lond.)*, 422, 1989.

151. Syrchina, A.I. et al., Tricin 4′-apioside from *Salsola collina*, *Khim. Prir. Soedin*, 439, 1992.

152. Yadava, R.N. and Reddy, V.M.S., A new flavone glycoside, 5-hydroxy-7,3′,4′,5′-tetramethoxy-flavone 5-*O*-β-D-xylopyranosyl(1 → 2)-α-L-rhamnopyranoside from *Bauhinia variegata* Linn., *J. Asian Nat. Prod. Res.*, 3, 341, 2001.

153. Yadava, R.N. and Agrawal, P.K., A new flavonoid glycoside: 5,7,4′-trihydroxy-6,3′,5′-trimethoxyflavone 7-α-L-arabinosyl(1 → 6)glucoside from roots of *Mimosa rubicaulis*, *J. Asian Nat. Prod. Res.*, 1, 15, 1998.

154. Yadava, R.N., Agrawal, P.K., and Singh, K.R., A novel flavone glycoside: 5,7,3′-trihydroxy-6,4′,5′-trimethoxyflavone 3′-*O*-α-L-rhamnopyranoside from the leaves of *Mimosa rubicaulis*, *Asian J. Chem.*, 10, 522, 1998.

155. Saxena, V.K. and Sharma, D.N., 5,4′-Dihydroxy-6,7,3′,5′-tetramethoxyflavone 5-rhamnosyl(1 → 6)galactoside from *Aloe barbadensis* (leaves), *J. Inst. Chem. (India)*, 70, 179, 1998.

156. Ahmad, M. et al., Two new flavone glycosides from the leaves of *Argyreia speciosa* (Convolvulaceae), *J. Chem. Res. Synop.*, 7, 248, 1993.

157. Yuldashev, M.P. and Rassulova, L.Kh., Flavonoids of *Juniperus zeravschanica*, *Chem. Nat. Compd.*, 37, 226, 2001.

158. Zhang, Y.-Y. et al., Four flavonoids from *Scutellaria baicalensis*, *Phytochemistry*, 35, 511, 1994.

159. Zheng, S., Sun, L., and Shen, X., Chemical constituents of *Mosla chinensis* Maxim., *Zhiwu Xuebao*, 38, 156, 1996.

160. Basnet, P. et al., Five new *C*-methylflavonoids, the potent aldose reductase inhibitors from *Matteuccia orientalis*, *Chem. Pharm. Bull.*, 43, 1558, 1995.

161. El-Ansari, M.A., Nawwar, M.A., and Saleh, N.A.M., Stachysetin, a diapigenin 7-glucoside-*p,p*′-dihydroxytruxinate from *Stachys aegyptiaca*, *Phytochemistry*, 40, 1543, 1995.

162. Yadava, R.N. and Tripathi, P., A novel flavone glycoside from the stem of *Desmodium gangeticum*, *Fitoterapia*, 69, 443, 1998.

163. Saxena, V.K. and Singhal, M., Novel prenylated flavonoid from stem of *Pithecellobium dulce*, *Fitoterapia*, 70, 98, 1999.

164. Yadava, R.N. and Reddy, K.I.S., A novel prenylated flavone glycoside from the seeds of *Erythrina indica*, *Fitoterapia*, 70, 357, 1999.

165. Huang, Y.L. et al., Tannins, flavonol sulphonates and a norlignan from *Phyllanthus virgatus*, *J. Nat. Prod.*, 61, 1194, 1998.

166. Yadava, R.N. and Reddy, K.I.S., A novel flavone glycoside from the stems of *Butea superba*, *Fitoterapia*, 69, 269, 1998.

167. Prakash, E.O. and Rao, J.T., A new flavone glycoside from the seeds of *Shorea robusta*, *Fitoterapia*, 70, 539, 1999.

168. Yadav, R.N., A new flavone glycoside from the stems of *Acacia catechu* Willd., *J. Inst. Chem. (India)*, 73, 104, 2001.

169. Jain, N. and Yadava, R.N., 3,6,7-Trihydroxy-4′-methoxyflavone glycoside from the leaves of *Setaria italica*, *J. Chem. Res.*, 265, 1995.

170. de Almeida, A.P. et al., Flavonol monoglycosides isolated from the antiviral fractions of *Persea americana* (Lauraceae) leaf infusion, *Phytother. Res.*, 12, 562, 1998.

171. Williams, C.A., Greenham, J., and Harborne, J.B., The role of lipophilic and polar flavonoids in the classification of temperate members of the Anthemideae, *Biochem. Syst. Ecol.*, 29, 929, 2001.

172. Hasan, A., Farman, M., and Ahmed, I., Flavonoid glycosides from *Indigofera hebepetala*, *Phytochemistry*, 35, 275, 1994.

173. Li, T.M., Li, W.K., and Yu, J.G., Flavonoids from *Artabotrys hexapetalus*, *Phytochemistry*, 45, 831, 1997.

174. El-Gamal, A.A. et al., Flavonol glycosides from *Galium sinaicum*, *Alex. J. Pharm. Sci.*, 13, 41, 1999.

175. Rao, K.V. et al., Flavonol glycosides from *Cassia hirsuta*, *J. Nat. Prod.*, 62, 305, 1999.

176. Markham, K.R., Geiger, H., and Jaggy, H., Kaempferol 3-*O*-glucosyl(1 → 2)rhamnoside from *Ginkgo biloba* and a reappraisal of other gluco(1 → 2, 1 → 3 and 1 → 4)rhamnoside structures, *Phytochemistry*, 31, 1009, 1992.

177. Hasler, A. et al., Complex flavonol glycosides from the leaves of *Ginkgo biloba*, *Phytochemistry*, 31, 1391, 1992.

178. Yadava, R.N. and Singh, A., A novel flavone glycoside from the seeds of *Crotalaria laburnifolia* Linn., *J. Indian Chem. Soc.*, 70, 273, 1993.

179. Rizwani, G.H. et al., Flavone glycosides of *Caralluma tuberculata* N.E. Brown, *Pak. J. Pharm. Sci.*, 3, 27, 1990.

180. Du, M. and Xie, J., Flavonol glycosides from *Rhodiola crenulata*, *Phytochemistry*, 38, 809, 1995.

181. Straubinger, M. et al., Two kaempferol sophorosides from *Crocus sativus*, *Nat. Prod. Lett.*, 10, 213, 1997.

182. Fuchino, H. et al., 5-*O*-Glucosylated kaempferols from the fern *Dryopteris dickinsii*, *Nat. Med. (Tokyo)*, 51, 537, 1997.

183. Strack, D. et al., Structures and accumulation patterns of soluble and insoluble phenolics from Norway spruce needles, *Phytochemistry*, 28, 2071, 1989.

184. Imperato, F., Kaempferol 7-*O*-rhamnoside-4′-*O*-glucoside from *Pteridium aquilinum*, *Phytochemistry*, 47, 911, 1998.

185. Pervez, M. et al., Flavonoids of *Cassia javanica* (Caesalpinaceae), *Pak. J. Sci.*, 47, 34, 1995.

186. Semmar, N. et al., Four new flavonol glycosides from the leaves of *Astragalus caprinus*, *J. Nat. Prod.*, 65, 576, 2002.

187. Wu, T., Kong, D.Y., and Li, H.T., Chemical components of *Helicia nilagirica* Beed. L., Structure of three new flavonol glycosides, *Chin. Chem. Lett.*, 13, 1071, 2002.

188. Lakenbrink, C. et al., New flavonol glycosides from tea (*Camellia sinensis*), *Nat. Prod. Lett.*, 14, 233, 2000.

189. Chauhan, D., Rai, R., and Chauhan, J.S., Two flavonoid triglycosides from *Cassia marginata*, *Indian J. Chem.*, 41B, 446, 2002.

190. El-Monsallamy, A.M.D., Leaf flavonoids of *Albizia lebbeck*, *Phytochemistry*, 48, 759, 1998.

191. Carotenuto, A. et al., The flavonoids of *Allium neapolitanum*, *Phytochemistry*, 44, 949, 1997.

192. Merfort, I. et al., Flavonol triglycosides from seeds of *Nigella sativa*, *Phytochemistry*, 46, 359, 1997.

193. Fukumoto, H. et al., Structure determination of a kaempferol 3-rhamnosyldiglucoside from *Impatiens balsamina*, *Phytochemistry*, 37, 1486, 1994.

194. Carotenuto, A. et al., The flavonoids of *Allium neapolitanum*, *Phytochemistry*, 44, 949, 1997.

195. Norbaek, R. and Kondo, T., Flavonol glycosides from flowers of *Crocus speciosus* and *C. antalyensis*, *Phytochemistry*, 51, 1113, 1999.

196. Kim et al., A new kaempferol 7-*O*-triglucoside from the leaves of *Brassica juncea* L., *Arch. Pharm. Res.*, 25, 621, 2002.

197. Hasan, A. et al., Two flavonol triglycosides from flowers of *Indigofera hebepetala*, *Phytochemistry*, 43, 1115, 1996.

198. Gohar, A.A., Maatooq, G.T., and Niwa, M., Two flavonoid glycosides from *Chenopodium murale*, *Phytochemistry*, 53, 299, 2000.

199. El-Sayed, N.H. et al., A flavonol triglycoside from *Chenopodium murale*, *Phytochemistry*, 51, 591, 1999.

200. Mulinacci, N. et al., Flavonol glycosides from *Sedum telephium* subspecies *maximum* leaves, *Phytochemistry*, 38, 531, 1995.

201. Si, J.Y. et al., Isolation and structure determination of flavonol glycosides from the fresh fruits of *Siraitia grosvenori*, *Yaoxue Xuebao*, 29, 158, 1994.

202. Zahid, M. et al., New cycloartane and flavonol glycosides from *Corchorus depressus*, *Helv. Chim. Acta*, 85, 689, 2002.

203. El-Sayed, N.H. et al., Kaempferol triosides from *Silphium perfoliatum*, *Phytochemistry*, 60, 835, 2002.

204. Norback, R., Nielsen, J.K., and Kondo, T., Flavonoids from flowers of two *Crocus chrysanthus-biflorus* cultivars: "Eye-catcher" and "Spring Pearl" (Iridaceae), *Phytochemistry*, 51, 1139, 1999.

205. Veit, M. and Pauli, G.F., Major flavonoids from *Arabidopsis thaliana*, *J. Nat. Prod.*, 62, 1301, 1999.

206. Snook, M.E. et al., The flower flavonols of *Nicotiana* species, *Phytochemistry*, 31, 1639, 1992.

207. Urushibara, S., Okuno, T., and Matsumoto, T., New flavonol glycosides, major determinants inducing the green fluorescence in the guard cells of *Allium cepa*, *Tetrahedron Lett.*, 33, 1213, 1992.

208. Shaaf, M., El-Ansari, M.A., and Saleh, N.A.M., New kaempferol tri- and tetraglycosides from *Pseuderucaria clavata*, *Fitoterapia*, 68, 62, 1997.

209. Iwashina, T. et al., Flavonol glycosides from *Asplenium foreziense* and its five related taxa and *A. incisum*, *Biochem. Syst. Ecol.*, 28, 665, 2000.

210. Alaniya, M.D. and Chkadua, N.F., Flavonoids of *Astragalus tana*, *Chem. Nat. Compd.*, 36, 537, 2000.

211. Manguro, L.O.A. et al., Flavonol glycosides of *Warburgia ugandensis* leaves, *Phytochemistry*, 64, 891, 2003.

212. Satake, T. et al., Studies on the constituents of Turkish plants. 1. Flavonol triglycosides from the fruit of *Rhamnus thymifolius*, *Chem. Pharm. Bull.*, 41, 1743, 1993.

213. Semmar, N. et al., Four new flavonol glycosides from the leaves of *Astragalus caprinus*, *Chem. Pharm. Bull.*, 50, 981, 2002.

214. Sannomiya, M. et al., A flavonoid glycoside from *Maytenus aquifolium*, *Phytochemistry*, 49, 237, 1998.
215. Zallocchi, E.M. and Pomilio, A.B., Evolution of flavonoids in the Phaseolinae, *Phytochemistry*, 37, 449, 1994.
216. Veit, M. et al., New kaempferol glycosides from *Equisetum* species, *Z. Naturforsch. B*, 48, 1398, 1993.
217. Nielsen, J.K., Norbaek, R., and Olsen, C.E., Kaempferol tetraglucosides from cabbage leaves, *Phytochemistry*, 49, 2171, 1998.
218. Wang, J.-H. et al., A flavonol tetraglycoside from *Sophora japonica* seeds, *Phytochemistry*, 63, 463, 2003.
219. Kijima, H. et al., Alangiflavoside, a new flavonol glycoside from the leaves of *Alangium premnifolium*, *J. Nat. Prod.*, 58, 1753, 1995.
220. Yahara, S., Kotjyoume, M., and Kohoda, H., Flavonoid glycosides and saponins from *Astragalus shikokianus*, *Phytochemistry*, 53, 469, 2000.
221. Liu, Q. et al., Flavonol glycosides from *Cephalocereus senilis*, *Phytochemistry*, 36, 229, 1994.
222. Berhow, M.A. et al., Acylated flavonoids in callus cultures of *Citrus aurantifolia*, *Phytochemistry*, 36, 1225, 1994.
223. Park, S.H. et al., Acylated flavonol glycosides with anti-complement activity from *Persicaria lapathifolia*, *Chem. Pharm. Bull.*, 47, 1484, 1999.
224. Okamura, H. et al., Two acylated flavonol glycosides from *Eucalyptus rostrata*, *Phytochemistry*, 33, 512, 1993.
225. Masuda, T. et al., Isolation and antioxidant activity of galloyl flavonol glycosides from the seashore plant, *Pemphis acidula*, *Biosci. Biotechnol. Biochem.*, 65, 1302, 2001.
226. Liu, Y. et al., New galloylated flavonoidal glycoside and gallotannins from leaves of *Loropetalum chinense* Oliv., *Tianran Chanwu Yanjiu Yu Kaifa*, 9, 12, 1997.
227. Olszewska, M. and Wolbis, M., Flavonoids from the flowers of *Prunus spinosa* L., *Acta Pol. Pharm.*, 58, 367, 2001.
228. Mitrokotsa, D. et al., Bioactive compounds from the buds of *Platanus orientalis* and isolation of a new kaempferol glycoside, *Planta Med.*, 59, 517, 1993.
229. Kaovadji, M., Morand, J.-M., and Garcia, J., Further acylated kaempferol rhamnosides from *Platanus acerifolia* buds, *J. Nat. Prod.*, 56, 1618, 1993.
230. Hohmann, J. et al., Flavonol acyl glycosides of the aerial parts of *Eryngium campestre*, *Planta Med.*, 63, 96, 1997.
231. Cao, S.-G. et al., Flavonol glycosides from *Elaeagnus bockii* (Elaeagnaceae), *Nat. Prod. Lett.*, 15, 1, 2001.
232. Imperato, F. and Minutiello, P., Kaempferol 3-*O*-(6″-caffeoylglucoside) from *Pteridium aquilinum*, *Phytochemistry*, 45, 199, 1997.
233. Imperato, F., Kaempferol 3-*O*-(5″-feruloylapioside) from *Pteridium aquilinum*, *Phytochemistry*, 43, 1421, 1996.
234. Catalano, S. et al., Kaempferol 3-*O*-β-D-(6″-feruloylglucoside) from *Polylepis incana*, *Phytochemistry*, 37, 1777, 1994.
235. Slimestad, R. et al., Syringetin 3-*O*-(6″-acetyl)-β-glucopyranoside and other flavonols from needles of Norway spruce, *Picea abies*, *Phytochemistry*, 40, 1537, 1995.
236. Kubo, I. and Yokokawa, Y., Two tyrosinase inhibiting flavonol glycosides from *Buddleia coriacea*, *Phytochemistry*, 31, 1075, 1992.
237. Bloor, S.J., An antimicrobial kaempferol diacylrhamnoside from *Pentachondra pumila*, *Phytochemistry*, 38, 1033, 1995.
238. Fiorini, C. et al., Acylated kaempferol glycosides from *Laurus nobilis* leaves, *Phytochemistry*, 47, 821, 1998.
239. Romussi, G., Parodi, B., and Caviglioli, G., Compounds from Cupuliferae. 15. Flavonoid glycosides from *Quercus canariensis* Willd., *Pharmazie*, 47, 877, 1992.
240. Liu, H. et al., Acylated flavonol glycosides from leaves of *Stenochlaena palustris*, *J. Nat. Prod.*, 62, 70, 1999.
241. Zhou, Y.-J. et al., Flavonoids from the leaves of *Quercus dentata*, *Indian J. Chem.*, 40B, 394, 2001.

242. Wu, Y.Q. et al., A new acylated flavonoid from *Anaphalis aureo-punctata*, *Chin. Chem. Lett.*, 14, 66, 2003.

243. Senatore, F. and De Feo, V., Flavonoid glycosides from *Minthostachys spicata* (Lamiaceae), *Biochem. Syst. Ecol.*, 23, 573, 1995.

244. Kazuma, K., Noda, N., and Suzuki, M., Malonylated flavonol glycosides from the petals of *Clitoria ternatea*, *Phytochemistry*, 62, 229, 2003.

245. Fattorusso, E. et al., The flavonoids of leek, *Allium porrum*, *Phytochemistry*, 57, 565, 2001.

246. Olea, R.S.G., Roque, N.F., and Das Bolzani, V., Acylated flavonol glycosides and terpenoids from the leaves of *Alibertia sessilis*, *J. Braz. Chem. Soc.*, 8, 257, 1997.

247. Lin, Y.L. and Kuo, Y.H., Roripanoside, a new kaempferol rhamnoside from *Rorippa indica* (L.) Hiern., *J. Chin. Chem. Soc. (Taipei)*, 42, 973, 1995.

248. Lu, C.-M. et al., A new acylated flavonol glycoside and antioxidant effects of *Hedyotis diffusa*, *Planta Med.*, 66, 374, 2000.

249. Abe, F. et al., Flavonol sinapoyl glycosides from leaves of *Thevetia peruviana*, *Phytochemistry*, 40, 577, 1995.

250. Guangyao, G. et al., A new flavonoid from the aerial part of *Thalictrum atriplex*, *Fitoterapia*, 71, 627, 2000.

251. Aquil, M. and Khan, I.Z., Three flavonol glycosides from *Ricinus communis*, *Bull. Chem. Soc. Ethiop.*, 11, 51, 1997.

252. Bloor, S.J. et al., Identification of flavonol and anthocyanin metabolites in leaves of *Petunia* "Mitchell" and its LC transgenic, *Phytochemistry*, 49, 1427, 1998.

253. Olszewska, M. and Wolbis, M., Flavonoids from the leaves of *Prunus spinosa* L., *Polish J. Chem.*, 76, 967, 2002.

254. Imperato, F., Kaempferol 3-(2″-*p*-coumaroylrhamnoside)-7-rhamnoside from *Cheilanthes fragrans*, *Phytochemistry*, 31, 3291, 1992.

255. El-Mousallami, A.M.D., Afifi, M.S., and Hussein, S.A.M., Acylated flavonol diglucosides from *Lotus polyphyllus*, *Phytochemistry*, 60, 807, 2002.

256. Min, B.-S. et al., Kaempferol acetylrhamnosides from the rhizome of *Dryopteris crassirhizoma* and their inhibitory effects on three different activities of human immunodeficiency virus-1 reverse transcriptase, *Chem. Pharm. Bull.*, 49, 546, 2001.

257. Ogden, S. et al., Acylated kaempferol glycosides from the flowers of *Delphinium formosum*, *Phytochemistry*, 49, 241, 1998.

258. Simon, J.K. et al., Two flavonol 3-[triacetylarabinosyl(1 → 6)glucosides] from *Calluna vulgaris*, *Phytochemistry*, 33, 1237, 1993.

259. Lakenbrink, C. et al., New flavonol triglycosides from tea (*Camellia sinensis*), *Nat. Prod. Lett.*, 14, 233, 2000.

260. Brunner, G. et al., A novel acylated flavonol glycoside isolated from *Brunfelsia grandiflora* ssp. *grandiflora*, *Phytochem. Anal.*, 11, 29, 2000.

261. Champavier, Y. et al., Acetylated and non-acetylated flavonol triglycosides from *Galega officinalis*, *Chem. Pharm. Bull.*, 48, 281, 2000.

262. Singab, A.N.B., Acetylated flavonol triglycosides from *Ammi majus* L., *Phytochemistry*, 49, 2177, 1998.

263. Carotenuto, A. et al., The flavonoids of *Allium ursinum*, *Phytochemistry*, 41, 531, 1996.

264. El-Desoky, S.K., El-Ansari, M.A., and El-Negoomy, S.I., Flavonol glycosides from *Mentha lavandulacea*, *Fitoterapia*, 72, 532, 2001.

265. Tang, Y. et al., Coumaroyl flavonol glycosides from the leaves of *Ginkgo biloba*, *Phytochemistry*, 58, 1251, 2001.

266. El-Sayed, N.H. et al., Kaempferol triosides from *Reseda muricata*, *Phytochemistry*, 57, 575, 2001.

267. Fico, G. et al., New flavonol glycosides from the flowers of *Aconitum napellus* ssp. *tauricum*, *Planta Med.*, 67, 287, 2001.

268. Fico, G. et al., Flavonoids from *Aconitum napellus* ssp. *neomontanum*, *Phytochemistry*, 57, 543, 2001.

269. Nielsen, J.K., Olsen, C.E., and Petersen, M.K., Acylated flavonol glycosides from cabbage leaves, *Phytochemistry*, 34, 539, 1993.

270. Costa, S.S., Jossang, A., and Bodo, B., 4''''-Acetylsagittatin A, a kaempferol triglycoside from *Kalanchoe streptantha*, *J. Nat. Prod.*, 59, 327, 1996.

271. Han, Y. et al., Flavonol glycosides from the stems of *Trigonella foenum-graecum*, *Phytochemistry*, 58, 577, 2001.

272. Xie, C. et al., Flavonol glycosides from *Lysimachia capillipes*, *J. Asian Nat. Prod. Res.*, 4, 17, 2002.

273. Crublet, M.-L. et al., Acylated flavonol glycosides from the leaves of *Planchonia grandis*, *Phytochemistry*, 64, 589, 2003.

274. Bylka, W., Stobiecki, M., and Frański, R., Sulfated flavonoid glycosides from leaves of *Atriplex hortensis*, *Acta Physiol. Plant.*, 23, 285, 2001.

275. Flamini, G., Antognoli, E., and Morelli, I., Two flavonoids and other compounds from the aerial parts of *Centaurea bracteata* from Italy, *Phytochemistry*, 57, 559, 2001.

276. Palme, E., Bilia, A.R., and Morelli, I., Flavonols and isoflavones from *Cotoneaster simonsii*, *Phytochemistry*, 42, 903, 1996.

277. Mimaki, Y. et al., Flavonol glycosides and steroidal saponins from the leaves of *Cestrum nocturnum* and their cytotoxicity, *J. Nat. Prod.*, 64, 17, 2001.

278. Zhen, S. et al., Two new flavonoids from *Mosla soochouensis* Matsuda, *Indian J. Chem.*, 37B, 1078, 1998.

279. Zheng, S. et al., Flavonoid constituents from *Mosla chinensis* Maxim., *Indian J. Chem.* 35B, 392, 1996.

280. Gohar, A.A., Flavonol glycosides from *Cadaba glandulosa*, *Z. Naturforsch. C*, 57, 216, 2002.

281. Lin, J.-H., Chiou, Y.-N., and Lin, Y.-L., Phenolic glycosides from *Viscum angulatum*, *J. Nat. Prod.*, 65, 638, 2002.

282. Cui, B. et al., Constituents of leguminose plants. XXX. Structures of three new acylated flavonol glycosides from *Astragalus complanatus* R. Br., *Chem. Pharm. Bull.*, 40, 1943, 1992.

283. Bilia, A.R. et al., A flavonol glycoside from *Agrimonia eupatoria*, *Phytochemistry*, 32, 1078, 1993.

284. da Silva, B.P., Bernardo, R.R., and Parente, J.P., Flavonol glycosides from *Costus spicatus*, *Phytochemistry*, 53, 87, 2000.

285. Jain, N. and Yadava, R.N., A novel flavonol glycoside: 4'-methoxykaempferol 3-*O*-α-L-rhamnopyranosyl 7-*O*-β-D-xylopyranoside from the leaves of *Cassia biflora*, *J. Indian Chem. Soc.* 71, 209, 1994.

286. Khar, R. and Sati, O.P., A novel flavonol glycoside from *Sageretia filiformia* leaves, *J. Indian Chem. Soc.*, 74, 506, 1997.

287. Curir, P. et al., Kaempferide triglycoside: a possible factor of resistance of carnation (*Dianthus caryophyllus*) to *Fusarium oxysporum* f. sp. *dianthi*, *Phytochemistry*, 56, 717, 2001.

288. Jasawant, B., Ragunathan, V., and Sulochana, N., A rare flavonol glycoside from *Aerva tomentosa* Forsk as antimicrobial and hepatoprotective agent, *Indian J. Chem.*, 42B, 956, 2003.

289. Duan, J. et al., The leaf flavonoids and phenolic acids of *Nitraria tangutorum* Bor., *Zhiwu Ziyuan Yu Huanjing*, 8, 6, 1996.

290. Pan, W.D. et al., Chemical constituents of the leaves of *Lantana camara*, *Yaoxue Xuebao*, 28, 35, 1993.

291. Tian, J. and Ding, L., Chemical constituents of *Gymnotheca involucrata* Pei, *Zhongguo Zhongyao Zazhi*, 26, 43, 2001.

292. da Silva, A. et al., Flavonol glycosides from leaves of *Costus spiralis*, *Fitoterapia*, 71, 507, 2000.

293. Hattori, M. et al., 6-Hydroxykaempferol and its glycosides from *Carthamus tinctorius* petals, *Phytochemistry*, 31, 4001, 1992.

294. Das, K.C. and Tripathi, A.K., 6-Hydroxykaempferol 3-*O*-β-D-alloside from *Tagetes erecta*, *Fitoterapia*, 68, 477, 1997.

295. Gamez, E.J.C. et al., Antioxidant flavonoid glycosides from *Daphniphyllum calycinum*, *J. Nat. Prod.*, 61, 706, 1998.

296. Nair, A.G.R. et al., 6-Hydroxykaempferol 7-(6''-caffeoylglucoside) from *Eupatorium glandulosum*, *Phytochemistry*, 33, 1275, 1993.

297. de Andrade, F.D.P. et al., Acyl glucosylated flavonols from *Paepalanthus* species, *Phytochemistry*, 51, 411, 1999.

298. Dai, J.-Q. et al., Two new flavone glucosides from *Serratula strangulata*, *Chem. Res. Chin. Univ.*, 17, 469, 2001.

299. Merfort, I. and Wendisch, D., New flavonoid glycosides from Arnicae Flos DAB-9, *Planta Med.*, 58, 355, 1992.

300. Yadava, R.N. and Singh, R.K.R., 6-Hydroxy-3,5,7,4′-tetramethoxyflavone 6-rhamnoside from roots of *Pterocarpus marsupium*, *Phytochemistry*, 48, 1259, 1998.

301. D'Agostino, M. et al., Flavonol glycosides from *Jungia paniculata*, *Fitoterapia*, 66, 283, 1995.

302. Hussein, S.A.M. et al., Flavonoids from *Ephedra aphylla*, *Phytochemistry*, 45, 1529, 1997.

303. Markham, K.R. and Campos, M., 7- and 8-*O*-Methylherbacetin 3-*O*-sophorosides from bee pollens and some structure/activity observations, *Phytochemistry*, 43, 763, 1996.

304. Markham, K.R., Mitchell, K.A., and Campos, M., An unusually lipophilic flavonol glycoside from *Ranunculus sardous* pollen, *Phytochemistry*, 45, 203, 1997.

305. Dauguet, J.C. et al., 8-Methoxykaempferol 3-neohesperidoside and other flavonoids from bee pollen of *Crataegus monogyna*, *Phytochemistry*, 33, 1503, 1993.

306. Chen, I.S. et al., Chemical constituents and biological activities of the fruit of *Zanthoxylum integrifolium*, *J. Nat. Prod.*, 62, 833, 1999.

307. Urzua, A. et al., Acylated flavonoids from *Pseudognaphalium* species, *J. Nat. Prod.*, 62, 381, 1999.

308. Kandil, F.E. et al., Flavonol glycosides and phenolics from *Withania somnifera*, *Phytochemistry*, 37, 1215, 1994.

309. Kandil, F.E. and Grace, M.H. Polyphenols from *Cornulaca monocantha*, *Phytochemistry*, 58, 611, 2001.

310. Kandil, F.E. et al., Flavonoids from *Psidium guaijava*, *Asian J. Chem.*, 9, 871, 1997.

311. Tomas-Lorente, F., Flavonoids from *Cistus ladanifer* bee pollen, *Phytochemistry*, 31, 2027, 1992.

312. Flamini, G. et al., A new flavonoid glycoside from *Centaurea horrida*, *J. Nat. Prod.*, 63, 662, 2000.

313. Arot, L.O.M. and Williams, L.A.D., A flavonol from *Embelia schimperi* leaves, *Phytochemistry*, 44, 1397, 1997.

314. Imperato, F., Flavonol glycosides from *Pteridium aquilinum*, *Phytochemistry*, 40, 1801, 1995.

315. Whang, W.K. et al., Flavonol glycosides from the aerial parts of *Filipendula formosa*, *Yakhak Hoechi*, 43, 5, 1999.

316. Hasan, A. et al., Flavonoid glycosides and an anthraquinone from *Rumex chalepensis*, *Phytochemistry*, 39, 1211, 1995.

317. El-Sayed, N.H. et al., Phenolics and flavonoids of *Cordia macleodii*, *Rev. Latinoam. Quim.*, 26, 30, 1998.

318. Zhang, X.N. et al., Three flavonol glycosides from leaves of *Myrsine seguinii*, *Phytochemistry*, 46, 943, 1997.

319. Zhou, Z.-H., Zhang, Y.-J., and Yang, C.-R., Saluensis, a new glycoside from *Camellia saluensis*, *Acta Bot. Yunnan.*, 22, 90, 2000.

320. Lin, Y.L. et al., A flavonol triglycoside from *Baccharis thesioides*, *Phytochemistry*, 33, 1549, 1993.

321. Sharaf, M., El-Ansari, M.A., and Saleh, N.A.M., Quercetin triglycoside from *Capparis spinosa*, *Fitoterapia*, 71, 46, 2000.

322. Tommasi, N., Piacento, S., and Pizza, C., Flavonol and chalcone ester glycosides from *Bidens andicola*, *J. Nat. Prod.*, 61, 973, 1998.

323. Tewtrakul, S. et al., Flavanone and flavonol glycosides from leaves of *Thevetia peruviana* and their HIV-1 reverse transcriptase and HIV-1 integrase inhibitory activities, *Chem. Pharm. Bull.*, 50, 630, 2002.

324. Calis, I. et al., Flavonoid, iridoid and lignan glycosides from *Putoria calabrica*, *J. Nat. Prod.*, 64, 961, 2001.

325. Markham, K.R., Hammett, K.R.W., and Ofman, D.J., Floral pigmentation in two yellow-flowered *Lathyrus* species and their hybrid, *Phytochemistry*, 31, 549, 1992.

326. Manguro, L.O.A. et al., Flavonol glycosides of *Maesa lanceolata* leaves, *Nat. Prod. Sci.*, 8, 77, 2002.

327. Nishibe, S., Kodama, A., and Noguchi, Y., Phenolic compounds from seeds of *Plantago ovata* and *P. syllium*, *Nat. Med. (Tokyo, Japan)*, 55, 258, 2001.

328. Weckerle, B. et al., Quercetin 3,3',4'-tri-*O*-β-D-glucopyranosides from leaves of *Eruca sativa* (Mill.), *Phytochemistry*, 57, 547, 2001.

329. Manguro, L.O.A., Midiwo, J.O., and Kraus, W., A new flavonol tetraglycoside from *Mysine africana* leaves, *Nat. Prod. Lett.*, 9, 121, 1996.

330. Gluchoff-Fiasson, K. et al., Three new flavonol malonylrhamnosides from *Ribes alpinum*, *Chem. Pharm. Bull.*, 49, 768, 2001.

331. Sanz, M.-J. et al., A new quercetin acyl glucuronide from *Scolymus hispanicus*, *J. Nat. Prod.*, 56, 1995, 1993.

332. Kim, Y. et al., Flavonol glycoside gallate and ferulate esters from *Persicaria lapathifolia* as inhibitors of superoxide production in human monocytes stimulated by unopsonized zymosan, *Planta Med.*, 66, 72, 2000.

333. Fossen, T. et al., Flavonoids from blue flowers of *Nymphaèa caerulea*, *Phytochemistry*, 51, 1133, 1999.

334. Jurgenliernk, G. and Nahrstedt, A., Phenolic compounds from *Hypericum perfoliatum*, *Planta Med.*, 68, 88, 2002.

335. Wang, H.K. et al., Flavonol glycosides from *Parthenocissus tricuspidata*, *Yakhak Hoechi*, 39, 289, 1995.

336. Barakat, H.H. et al., Flavonoid galloyl glucosides from the pods of *Acacia farnesiana*, *Phytochemistry*, 51, 139, 1999.

337. Lee, J.Y. et al., Antioxidative flavonoids from leaves of *Carthamnus tinctorius*, *Arch. Pharm. Res.*, 25, 313, 2002.

338. Kim, H.J. et al., A new flavonol glycoside gallate ester from *Acer okamotoanum* and its inhibitory activity against HIV-1 integrase, *J. Nat. Prod.*, 61, 145, 1998.

339. D'Agostino, M. et al., Flavonol glycosides from *Tagetes elliptica*, *Phytochemistry*, 31, 4387, 1992.

340. Aimova, M.Zh. et al., New acylated quercetin diglycoside from *Euphorbia pachyrhiza*, *Akad. Nauk. Resp. Kaz. Ser. Khim.*, 26, 1999.

341. Michael, H.N., Saliband, J.Y., and Ishek, M.S., Acylated flavonol glycoside from *Psidium guaijava*, *Pharmazie*, 57, 859, 2002.

342. Ahn, B.T. et al., A new flavonoid from *Euphorbia ebracteolata*, *Planta Med.*, 62, 383, 1996.

343. Takemura, M. et al., Acylated flavonol glycosides as probing stimulants of a bean aphid, *Megoura crassicauda*, from *Vicia angustifolia*, *Phytochemistry*, 61, 135, 2002.

344. Teng, R. et al., Two new acylated flavonoid glycosides from *Morina nepalensis* var. *alba* Hand.-Mazz., *Magn. Res. Chem.*, 40, 415, 2002.

345. Kamel, M.S. et al., Acylated flavonoid glycosides from *Bassia muricata*, *Phytochemistry*, 57, 1259, 2001.

346. Yoshikawa, M. et al., New flavonol oligoglycosides and polyacylated sucroses with inhibitory effects on aldose reductase and platelet aggregation from flowers of *Prunus mume*, *J. Nat. Prod.*, 65, 1151, 2002.

347. Aquil, M. and Khan, I.Z., Euphorbianin — a new flavonol glycoside from *Euphorbia hirta* Linn., *Global J. Pure Appl. Sci.*, 5, 371, 1999.

348. Kofinas, C. et al., Flavonoids and bioactive coumarins of *Tordylium apulum*, *Phytochemistry*, 48, 637, 1998.

349. Simon, A. et al., Further flavonoid glycosides from *Calluna vulgaris*, *Phytochemistry*, 32, 1045, 1993.

350. Simon, A. et al., Quercetin 3-[triacetylarabinosyl(1 → 6)galactoside] and chromones from *Calluna vulgaris*, *Phytochemistry*, 36, 1043, 1994.

351. Gluchoff-Fiasson, K., Fiasson, J.L., and Watson, H., Quercetin glycosides from European aquatic *Ranunculus* species of subgenus *Batrachium*, *Phytochemistry*, 45, 1063, 1997.

352. Ferreos, F. et al., Acylated flavonol sophorotriosides from pea shoots, *Phytochemistry*, 39, 1443, 1995.

353. Gunasegaran, R. et al., 7-*O*-(6-*O*-Benzyol-β-D-glucopyranosyl)rutin from leaves of *Canthium dicoccum*, *Fitoterapia*, 72, 201, 2001.

354. Gluchoff-Fiasson, K., Fiasson, J.L., and Favre-Bonvin, J., Quercetin glycosides from antartic *Ranunculus* species, *Phytochemistry*, 37, 1629, 1994.

355. Zhapova, Ts. et al., Flavonoid glycosides from *Aconitum baicalense*, *Khim. Prir. Soedin.*, 484,1992.

356. Whang, W.K. and Lee, M.T., New flavonol glycosides from the leaves of *Symplocarpus renifolius*, *Arch. Pharm. Res.*, 22, 423, 1999.

357. Abdel-Shafeck, K.A. et al., A new acylated flavonol triglycoside from *Carrichtera annua*, *J. Nat. Prod.*, 63, 845, 2000.

358. Cao, S.-G. et al., Flavonol glycosides from *Elaeagnus bockii* (Elaeagnaceae), *Nat. Prod. Lett.*, 15, 211, 2001.

359. Itoh, A. et al., Flavonoid glycosides from *Rhazya orientalis*, *J. Nat. Prod.*, 65, 352, 2002.

360. Isaza, J.-H., Ito, H., and Yoshida, T., A flavonol glycoside–lignan ester and accompanying glucosides from *Monochaetum multiflorum*, *Phytochemistry*, 58, 321, 2001.

361. Op de Beck, P. et al., Quercitrin 3′-sulphate from leaves of *Leea guineensis*, *Phytochemistry*, 47, 1171, 1998.

362. Ogundipe, O.O., Moody, J.O., and Houghton, P.J., Occurrence of flavonol sulfates in *Alchornea laxiflora*, *Pharm. Biol. (Lisse, Netherlands)*, 39, 421, 2001.

363. Iwasina, I. et al., New and rare flavonol glycosides from *Asplenium trichomanes-ramosum* as stable chemotaxonomic markers, *Biochem. Syst. Ecol.*, 23, 283, 1995.

364. Choi, Y.H. et al., A flavonoid diglycoside from *Lepisorus ussuriensis*, *Phytochemistry*, 43, 1111, 1996.

365. De Tommasi, N. and Pizza, C., Flavonol and chalcone ester glycosides from *Bidens leucantha*, *J. Nat. Prod.*, 60, 270, 1997.

366. Yadava, R.N. and Sodhi, S., A new flavone glycoside: 5,7,3′,4′-tetrahydroxy-3-methoxyflavone 7-*O*-β-D-galactopyranosyl(1 → 4)-*O*-β-D-glucopyranoside from the stem of *Acacia catechu*, *J. Asian Nat. Prod. Res.*, 4, 11, 2002.

367. Williams, C.A. et al., Flavonoid patterns and the revised classification of Australian Restionaceae, *Phytochemistry*, 49, 529, 1998.

368. Chauhan, D. and Chauhan, J.S., Flavonoid glycosides from *Pongamia pinnata*, *Pharm. Biol.*, 40, 171, 2002.

369. Tripathi, A.K. and Singh, J., A flavonoid glycoside from *Cassia siamea*, *Fitoterapia*, 64, 90, 1993.

370. Imperato, F., Rhamnetin 3-*O*-laminaribioside from *Pteridium aquilinum*, *Phytochemistry*, 45, 1729, 1997.

371. Vaishnov, M.M. and Gupta, K.R., Rhamnetin 3-*O*-gentiobioside from *Cassia fistula* roots, *Fitoterapia*, 67, 78, 1996.

372. Ozipek., M. et al., Rhamnetin 3-*p*-coumaroylrhamnoside from *Rhamnus petiolaris*, *Phytochemistry*, 37, 249, 1994.

373. Mann, P. et al., Flavonoid sulfates from the Convolvulaceae, *Phytochemistry*, 50, 267, 1999.

374. Barakat, H.H., Contribution to the phytochemical study of Egyptian Tamaricaceous plants, *Nat. Prod. Sci.*, 4, 221, 1998.

375. Iriste, V.A. and Blinova, K.F., Flavonol monosides of *Oxytropis lanata*, *Chem. Nat. Compd.*, 412, 1973.

376. Bilia, A.R. et al., Flavonol glycosides from *Pyrus bourgaena*, *Phytochemistry*, 35, 1378, 1994.

377. Miftakhova, A. F., Burasheva, G.S., and Abilov, Z.A., Flavonoids of *Halocnemum strobilaceum*, *Chem. Nat. Compd.*, 35, 100, 1999.

378. Yoshitama, K. et al., Studies on the flavonoids of the genus *Trillium* 3. Flavonol glycosides in the leaves of *T. apetalon* and *T. kamtschaticum*, *J. Plant Res.*, 110, 443, 1997.

379. Sikorska, M. and Matlawska, I., Kaempferol, isorhamnetin and their glycosides in the flowers of *Asclepias syriaca* L., *Acta Pol. Pharm.*, 58, 269, 2001.

380. Awaad, A.S. and Grace, M.H., Flavonoids and pharmacological activity of *Vernonia galamensis* ssp. *galamensis* var. *petitiana*, *Egyptian J. Pharm. Sci.*, 40, 117, 1999.

381. Miserez, F. et al., Flavonol glycosides from *Vernonia galamensis* ssp. *nairobiensis*, *Phytochemistry*, 43, 283, 1996.

382. Mizuno, M. et al., Four flavonol glycosides from *Achlys triphylla*, *Phytochemistry*, 31, 301, 1992.

383. Song, C. and Zu, R., Constituents of *Crocus sativus*. III. Structural elucidation of two new glycosides of pollen, *Huaxue Xuebao*, 49, 917, 1991.

384. Ben Salah, H. et al., Flavonol triglycosides from leaves of *Hamada scoparia*, *Chem. Pharm. Bull.*, 50, 1268, 2002.

385. Halim, A.F., Saad, H.E.A., and Hashish, N.E., Flavonol glycosides from *Nitraria retusa*, *Phytochemistry*, 40, 349, 1995.

386. Singab, A.N. et al., A new flavonoid glycoside from *Pituranthos tortuosus*, *Nat. Med. (Tokyo)*, 52, 191, 1998.

387. Sen, S., Sahu, N.P., and Mahato, S.B., Flavonol glycosides from *Calotropis gigantea*, *Phytochemistry*, 31, 2919, 1992.

388. Nishida, R., Ovipostion stimulant of a zeryntiine swallowtail butterfly, *Luehdorfia japonica*, *Phytochemistry*, 36, 873, 1994.

389. Ito, H. et al., Flavonoid and benzophenone glycosides from *Coelogyne ramosissima*, *Phytochemistry*, 54, 695, 2000.

390. Hassanean, H.A. and Desoky, E.K., An acylated isorhamnetin glucoside from *Zygophyllum simplex*, *Phytochemistry*, 31, 3293, 1992.

391. Manguro, L.O.A. et al., Flavonol and drimane-type seqiterpene glycosides of *Warburgia stuhlmannii* leaves, *Phytochemistry*, 63, 497, 2003.

392. Li, C.J. et al., A new sulfated flavonoid from *Zygophyllum dumosum*, *Nat. Prod. Lett.*, 8, 281, 1996.

393. Yuan, H.Q. and Zuo, C.X., Chemical constituents of *Cyanchum thesioides*, *Yaoxue Xuebao*, 27, 589, 1992.

394. Xiong, Q., Shi, D., and Mizuno, M., Flavonol glucosides in pericarps of *Zanthoxylum bungeanum*, *Phytochemistry*, 39, 723, 1995.

395. El-Sayed, N.H. et al., Flavonoids of *Cassia italica*, *Phytochemistry*, 31, 2187, 1992.

396. Yaga, A. et al., Antioxidative sulphated flavonoids in leaves of *Polygonum hydropiper*, *Phytochemistry*, 35, 885, 1994.

397. Sepulveda-Boza, S., Delhvi, S., and Cassels, B.K., Flavonoids from the twigs of *Eucryphia glutinosa*, *Phytochemistry*, 32, 1301, 1993.

398. Sinz, A. et al., Flavonol glycosides from *Dasymaschalon sootepense*, *Phytochemistry*, 47, 1393, 1998.

399. Brandão, M.G.L. et al., Two methoxylated flavone glycosides from *Bidens pilosa*, *Phytochemistry*, 48, 397, 1998.

400. Chauhan, D. and Chauhan, J.S., Flavonoid diglycoside from *Punica granatum*, *Pharm. Biol.*, 39, 155, 2001.

401. Wang, J. et al., Flavonoids from *Bidens pilosa* var. *radiata*, *Phytochemistry*, 46, 1275, 1997.

402. Martin-Cordero, C. et al., Flavonol glycoside from *Retama sphaerocarpa* Boissier, *Phytochemistry*, 51, 1129, 1999.

403. Yadava, R.N. and Reddy, V.M.S., A biologically active flavonol glycoside from seeds of *Albizzia julibissin* Durazz, *J. Inst. Chem. (India)*, 73, 195, 2001.

404. Martin-Cordero, C. et al., Retamatrioside, a new flavonol triglycoside from *Retama sphaerocarpa*, *J. Nat. Prod.*, 63, 248, 2000.

405. Veishnav, M.M. and Gupta, K.R., Ombuin 3-*O*-arabinofuranoside from *Coccinia indica*, *Fitoterapia*, 67, 80, 1996.

406. Si, J. et al., Isolation and identification of flavonoids from *Gynostemma yixingense*, *Zhiwu Xuebao*, 36, 239, 1994.

407. Yadava, R.N. and Mathews, S.R., A novel flavonol glycoside from *Crotolaria verrucosa*, *Fitoterapia*, 65, 340, 1994.

408. Souleles, C., Harvala, C., and Chinou, J., Flavonoids from *Calamintha grandiflora*, *Int. J. Pharmacogn.*, 29, 317, 1991.

409. Eskalieva, B.K. and Burasheva, G.Sh., Flavonoids of *Alhagi persarum*, *Chem. Nat. Compd.*, 38, 102, 2002.

410. Salpekar, J. and Khan, S.A., 5,7,3′,4′-Tetramethoxyquercetin 3-*O*-β-D-galactopyranosyl(1 → 4)-*O*-β-D-xylopyranoside from *Sesbania aculeata*, *Asian J. Chem.*, 9, 272, 1997.

411. D'Agostino, M. et al., Quercetagetin 6-*O*-β-D-glucopyranoside from *Tagetes mandonii*, *Phytochemistry*, 45, 201, 1997.

412. Hellmann, J., Muller, E., and Merfort, I., Flavonoid glucosides and dicaffeoylquinic acids from flower heads of *Buphthalmum salicifolium*, *Phytochemistry*, 51, 713, 1999.

413. Nair, A.G.R. et al., Flavonol glycosides from leaves of *Eupatorium glandulosum*, *Phytochemistry*, 40, 283, 1995.

414. Santos dos, L.C. et al., 6-Methoxyquercetin 3-*O*-(6″-*E*-feruloyl)-D-β-glucopyranoside from *Paepalanthus polyanthus* (Eriocaulaceae), *Biochem. Syst. Ecol.*, 30. 451, 2002.

415. Park, E.J., Kim, Y., and Kim, J., Acylated flavonol glycosides from the flower of *Inula britannica*, *J. Nat. Prod.*, 63, 34, 2000.

416. Costa, S.S. et al., Patuletin acetylrhamnosides from *Kalanchoe brasiliensis* as inhibitors of human lymphocyte proliferative activity, *J. Nat. Prod.*, 57, 1503, 1994.

417. Ferreres, F. et al., Acylated flavonol glycosides from spinach leaves (*Spinacia oleracea*), *Phytochemistry*, 45, 1701, 1997.

418. Sentoro, F., D'Agostino, M., and Dini, I., Two new quercetagetin *O*-glucosides from *Tagetes mandonii*, *Biochem. Syst. Ecol.*, 27, 309, 1999.

419. Vilegas, W. et al., Quercetagetin 7-methyl ether glycosides from *Paepalanthus vellozioides* and *P. latipes*, *Phytochemistry*, 51, 413, 1999.

420. Yadava, R.N., A new flavone glycoside from the stems of *Acacia catechu* Willd., *J. Inst. Chem. (India)*, 73, 104, 2001.

421. Flamini, G., Pardini, M., and Morelli, I., A flavonoid sulphate and other compounds from the roots of *Centaurea bracteata*, *Phytochemistry*, 58, 1229, 2001.

422. Sung, S.H., Jung, W.J., and Kim, Y.C., A novel flavonol lyxoside of *Orostachys japonicus* herb, *Nat. Prod. Lett.*, 16, 29, 2002.

423. Yadava, R.N. and Reddy, K.I.S., A new bioactive flavonol glycoside from the stems of *Butea superba* Roxb., *J. Asian Nat. Prod. Res.*, 1, 139, 1998.

424. Hussein, S.A.M. et al., Polyoxygenated flavonoids from *Eugenia edulis*, *Phytochemistry*, 64, 883, 2003.

425. Bennini, B. et al., The revised structure of two flavonol 3- and 4′-monoglucosides from *Erica cinerea*, *Phytochemistry*, 38, 259, 1995.

426. Chulia, A.J. et al., Two flavonol conjugates from *Erica cinerea*, *J. Nat. Prod.*, 58, 560, 1995.

427. David, J.M. et al., Flavonol glycosides from *Davilla flexuosa*, *J. Braz. Chem. Soc.*, 7, 115, 1996.

428. Braca, A. et al., Three flavonoids from *Licania densiflora*, *Phytochemistry*, 51, 1125, 1999.

429. Ismail, N. and Alam, M., A novel cytotoxic flavonoid glycoside from *Physalis angulata*, *Fitoterapia*, 72, 676, 2001.

430. Fossen, T., Froystein, N.A., and Andersen, O.M., Myricetin 3-rhamnosyl(1 → 6)galactoside from *Nymphaéa* x *marliacea*, *Phytochemistry*, 49, 1997, 1998.

431 Arya, R. et al., Myricetin 3′-rhamnoside-3-galactoside from *Buchanania lanzan* (Anacardiaceae), *Phytochemistry*, 31, 2569, 1992.

432. Slimestad, R. et al., Myricetin 3,4′-diglucoside and kaempferol derivatives from needles of Norway spruce, *Picea abies*, *Phytochemistry*, 32, 179, 1993.

433. Yu, R. et al., Chemical components of *Oxytropis glabra* DC., *Zhiwu Xuebao*, 34, 369, 1992.

434. Lin, L.-C. and Chou, C.-J., Flavonoids and phenolics from *Limonium sinense*, *Planta Med.*, 66, 382, 2000.

435. Sun, D. et al., Flavonols from *Myrica esculenta* bark, *Linchen Huaxue Yu Gongye*, 11, 251, 1991.

436. Akdemir, Z.S. et al., Polyphenolic compounds from *Geranium pratense* and their free radical scavenging activities, *Phytochemistry*, 56, 189, 2001.

437. Elegami, A.A. et al., Two very unusual macrocyclic flavonoids from the water lily, *Nymphaéa lotus*, *Phytochemistry*, 63, 727, 2003.

438. Timbola, A.K. et al., A new flavonol from leaves of *Eugenia jambolana*, *Fitoterapia*, 73, 174, 2002.

439. Lee, T.H. et al., Three new flavonol galloylglycosides from leaves of *Acacia confusa*, *J. Nat. Prod.*, 63, 710, 2000.

440. Arot, L.O.M., Midiwo, J.O., and Kraus, W.A., Flavonol glycoside from *Myrsine africana* leaves, *Phytochemistry*, 43, 1107, 1996.

441. Mahmoud, I.I. et al., Acylated flavonol glycosides from *Eugenia jambolana* leaves, *Phytochemistry*, 58, 1239, 2001.
442. Guo, J. et al., Flavonol glycosides from *Lysimachia congestiflora*, *Phytochemistry*, 48, 1445, 1998.
443. Braca, A. et al., Three flavonoids from *Licania heteromorpha*, *Phytochemistry*, 51, 1121, 1999.
444. He, H. et al., Flavonoid glycoside from *Clausena excavata*, *Yunnan Zhiwu Yanjiu*, 23, 256, 2001.
445. Brun, G., A new flavonol glycoside from *Catharanthus roseus*, *Phytochemistry*, 50, 167, 1999.
446. Gournelis, D.C., Flavonoids of *Erica verticillata*, *J. Nat. Prod.*, 58, 1065, 1995.
447. Sahu, B.R. and Chakrabarty, A.A., A flavone glycoside, robinetin 7-glucoside from *Alternanthera sessilis*, *Asian J. Chem.*, 5, 1148, 1994.
448. Veitch, N.C. et al., Six new isoflavones and a 5-deoxyflavonol glycoside from the leaves of *Ateleia herbert-smithii*, *J. Nat. Prod.*, 66, 210, 2003.
449. Yadava, R.N. and Reddy, V.M.S., A new biologically active flavonol glycoside from the seeds of *Abrus precatorius* Linn., *J. Asian Nat. Prod. Res.*, 4, 103, 2002.
450. Maldonado, E. et al., Acylated flavonols and other constituents from *Galeana pratensis*, *Phytochemistry*, 31, 1003, 1992.
451. Purwar, C. et al., New flavonoid glycosides from *Cassia occidentalis*, *Indian J. Chem.*, 42B, 434, 2003.
452. Beck, M.A. and Haberlein, H., A new flavonol 3-*O*-glycoside from *Eschscholtzia californica*, *Planta Med.*, 65, 296, 1999.
453. Agrawal, B., Hemlata, and Singh, J., Two new flavone glycosides from *Pongamia pinnata*, *Int. J. Pharmacogn.*, 31, 305, 1993.
454. Ragunathan, V. and Sulochana, N., A new flavonol bioside from the flowers of *Hibiscus vitifolius* Linn. and its hypoglycemic activity, *J. Indian Chem.*, 71, 705, 1994.
455. Ali, M., Ravinder, E., and Ramachandram, R., A new flavonoid from the aerial parts of *Tridax procumbens*, *Fitoterapia*, 72, 313, 2001.
456. Arisawa, M. et al., Novel flavonoids from *Chrysosplenium grayanum* Maxim. (Saxifragaceae), *Chem. Pharm. Bull.*, 41, 571, 1993.
457. Zhou, Z.H., Zhang, Y.J., and Yang, C.R., New flavonoid glycosides from *Scutellaria amoena*, *Stud. Plant Sci.*, 6, 305, 1999.
458. Aquil, M., Babayo, Y., and Babboi, A., Two flavonoid glycosides from *Aspilia africana*, *Ultra Sci. Phys. Sci.*, 9, 281, 1997.
459. Khan, I.Z., Kolo, B.G., and Aquil, M., Flavonol glycosides from *Centaurea senegalensis* DC., *Global J. Pure Appl. Sci.*, 4, 255, 1998.
460. Aquil, M., Rahman, F.A., and Ahmad, M.B., A new flavonol glycoside from *Peperomia pellucida*, *Ultra Sci. Phys. Sci.*, 6, 141, 1994.
461. D'Agostino, M. et al., Isolation of 3,5,7,2',3',4'-hexahydroxyflavone 3-*O*-β-D-glucopyranoside from *Eupatorium sternbergianun*, *Fitoterapia*, 65, 472, 1994.
462. Akbar, E. et al., Flavone glycosides and bergenin derivatives from *Tridax procumbens*, *Heterocycles*, 57, 733, 2002.
463. Yadava, R.N. and Tripathi, P., Chemical examination and anti-inflammatory accumulation of the extract from the stem of *Albizzia procera* Benth., *Res. J. Chem. Environ.*, 4, 57, 2000.
464. Bergman, M. et al., The antioxidant activity of aqueous spinach extract: chemical identification of active fractions, *Phytochemistry*, 58, 143, 2001.
465. Zheng, S. et al., Two new flavone glycosides from *Polygonum viviparum* L., *Indian J. Chem.*, 40B, 167, 2001.
466. Mizuno, M. et al., Two flavonol glycosides from *Vancouveria hexandra*, *Phytochemistry*, 31, 297, 1992.
467. Liang, H., Li, L., and Weimen, Y., New flavonol glycoside from *Epimedium acuminatum*, *J. Nat. Prod.*, 56, 943, 1993.
468. Das, K.C. and Tripathi, A.K., A new flavonol glycoside from *Sesbania grandiflora*, *Fitoterapia*, 69, 477, 1998.
469. Hu, B.-H., Zhou, L.D., and Liu, Y.-L., New tetrasaccharide flavonol glycoside from *Epimedium acuminatum*, *J. Nat. Prod.*, 55, 672, 1992.
470. Anam, E.A., A flavonol glycoside from *Berberis dictyota*, *Indian J. Heterocycl. Chem.*, 7, 59, 1997.

471. Sun, P. et al., A new flavonol glycoside, epimedin K, from *Epimedium koreanum, Chem. Pharm. Bull.*, 44, 446, 1996.

472. Li, W.K. et al., Flavonol glycosides from *Epimedium koreanum, Phytochemistry*, 38, 263, 1995.

473. Li, W.K., Zhang, R.Y., and Xiao, P.G., Flavonoids from *Epimedium wanshanense, Phytochemistry*, 43, 527, 1996.

474. Li, W.K. et al., Structure of a new flavonol glycoside from *Epimedium koreanum* Nakai, *Gaodeng Xuexiao Huaxue Xuebao*, 16, 1575, 1995.

475. Li, W.K. et al., Caohuoside-C from the aerial parts of *Epimedium koreanum* Nakai, *Gaodeng Xuexiao Huaxue Xuebao*, 16, 230, 1995.

476. Yadava, R.N. and Singh, R.K., A novel flavonoid glycoside from *Butea monosperma, J. Inst. Chem. (India)*, 70, 9, 1998.

477. Hu, B.H., Zhou, L.D., and Liu, Y.L., Separation and structure of acuminatin from *Epimedium acuminatum, Yaoxue Xuebao*, 27, 397, 1992.

478. Yadava, R.N. and Barsainya, B., A novel flavone glycoside from the seeds of *Cotula anthemoides, Fitoterapia*, 69, 437, 1998.

479. Nair, A.G.R. et al., New and rare flavonol glycosides from leaves of *Syzgium samarangense, Fitoterapia*, 70, 148, 1999.

480. Williams, C.A., Biosystematics of the Monocotyledonae-flavonoid patterns in leaves of the Liliaceae, *Biochem. Syst. Ecol.*, 3, 229, 1975.

481. Iinuma, M., Two *C*-methylated flavonoid glycosides from the roots of *Sophora leachiana, J. Nat. Prod.*, 54, 1144, 1991.

482. Harborne, J.B. and Williams, C.A., Recent Advances in the chemosystematics of the monocotyledons, *Phytochemistry*, 37, 3, 1994.

483. Williams, C.A. et al., Flavonoid patterns and the revised classification of Australian Restionaceae, *Phytochemistry*, 49, 529, 1998.

484. Williams, C.A. et al., Flavonoid evidence and the classification of the Anarthriaceae within the Poales, *Phytochemistry*, 45, 1189, 1997.

485. Harborne, J.B. et al., Variations in the lipophilic and vacuolar flavonoids of the genus *Vellozia, Phytochemistry*, 35, 1475, 1994.

486. Williams, C.A. et al., Differences in flavonoid patterns between genera within the Velloziaceae, *Phytochemistry*, 36, 931, 1994.

487. Williams, C.A., Harborne, J.B., and Colasante, M., Flavonoid and xanthone patterns in bearded *Iris* species and the pathway of chemical evolution, *Biochem. Syst. Ecol.*, 25, 309, 1997.

488. Bohm, B.A. and Stuessy, T.F. *Flavonoids of the Sunflower Family (Asteraceae)*, Springer-Verlag, Wien, Austria, 2001.

489. Williams, C.A., Harborne, J.B., and Eagles, J., Variations in lipophilic and polar flavonoids in the genus *Tanacetum, Phytochemistry*, 52, 1301, 1999.

490. Williams, C.A. et al., Variations in lipophilic and vacuolar flavonoids among European *Pulicaria* species, *Phytochemistry*, 64, 275, 2003.

491. Gil, M.I. et al., Distribution of flavonoid aglycones and glycosides in *Sideritis* species from the Canary Islands and Madeira, *Phytochemistry*, 34, 227, 1993.

492. Lu, Y. and Foo, L.Y., Polyphenolics of *Salvia* — a review, *Phytochemistry*, 59, 117, 2002.

493. Tomás-Barberan, F.A. et al., Flavonoid *p*-coumaroylglucosides and 8-hydroxyflavone allosylglucosides in some Labiatae, *Phytochemistry*, 31, 3097, 1992.

494. Grayer, R.J. et al., Leaf flavonoid glycosides as chemosystematic characters in *Ocimum, Biochem. System. Ecol.*, 30, 327, 2002.

495. Marin, P.D. et al., Acacetin glycosides as taxonomic markers in *Calamintha* and *Micromeria, Phytochemistry*, 58, 943, 2001.

496. Grayer, R.J. et al., Scutellarein 4'-methyl ether glycosides as taxonomic markers in *Teucridium* and *Tripora* (Lamiaceae, Ajugoideae), *Phytochemistry*, 60, 727, 2002.

497. Grayer, R.J. and Kok, R.P.J. de, Flavonoids and verbascoside as chemotaxonomic characters in the genera *Oxera* and *Faradaya* (Labiatae), *Biochem. System. Ecol.*, 26, 729, 1998.

498. Upson, T.M. et al., Leaf flavonoids as systematic characters in the genera *Lavandula* and *Sabaudia, Biochem. Syst. Ecol.*, 28, 991, 2000.

499. Hegnauer, R. and Grayer-Barkmeijer, R.J., Relevance of seed polysaccharides and flavonoids for the classification of the Leguminosae: a chemotaxonomic approach, *Phytochemistry*, 34, 3, 1993.
500. Hegnauer, R. and Hegnauer, M. *Chemotaxonomie der Pflanzen*, Vols 11a, 11b(1), and 11b(2), Birkhauser Verlag, Basel, Switzerland, 1994.
501. Seigler, D.S., Phytochemistry of *Acacia-sensu lato*, *Biochem. Syst. Ecol.*, 31, 845, 2003.

APPENDIX A

CHECKLIST OF KNOWN FLAVONE GLYCOSIDES

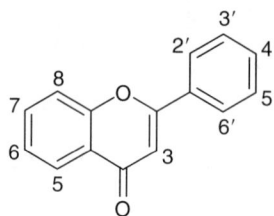

13.9

5,7-Dihydroxyflavone (chrysin)
1. 5-Xyloside
2. 5-Glucoside (toringin)
3. 7-Glucoside
4. 7-Galactoside
5. 7-Glucuronide
6. 7-Rutinoside
7. 7-Gentiobioside
8. 7-Benzoate
9. 7-(4″-Acetylglucoside)*
10. 7-(6″-Acetylglucoside)*

6-Hydroxy-4″-methoxyflavone
11. 6-Arabinoside

7,2′-Dihydroxyflavone
12. 7-Glucoside

7,4′-Dihydroxyflavone
13. 7-Glucoside
14. 4′-Glucoside
15. 7-Rutinoside

3′,4′-Dihydroxyflavone
16. 4′-Glucoside

2′,5′-Dihydroxyflavone
17. 5′-Acetate

5,6,7-Trihydroxyflavone (baicalein)
18. 6-Glucoside
19. 6-Glucuronide
20. 7-Rhamnoside
21. 7-Glucuronide
22. 7-(6″-Malonylglucoside)*

Baicalein 5-methyl ether
   23. 7-Glucoside*
Baicalein 6-methyl ether (oroxylin A)
   24. 5-Rhamnoside*
   25. 7-Glucoside
   26. 7-Glucuronide
   27. 7-Glucosyl(1 → 3)rhamnoside*
Baicalein 7-methyl ether (negletein)
   28. 5-Glucuronide
   29. 5-Glucuronosylglucoside
   30. 6-Xyloside*
   31. 6-Glucoside*
   32. 6-Rhamnosyl(1 → 2)fucoside*
Baicalein 5,6-dimethyl ether
   33. 7-Glucoside
5,7,8-Trihydroxyflavone (norwogonin)
   34. 5-Glucoside*
   35. 7-Galactoside*
   36. 7-Glucuronide
   37. 8-Glucuronide
5,7-Dihydroxy-8-methoxyflavone (wogonin)
   38. 5-Glucoside
   39. 7-Glucoside
   40. 7-Glucuronide
7-Hydroxy-5,8-dimethoxyflavone
   41. 7-Glucoside*
   42. 7-Glucuronide*
5-Hydroxy-7,8-dimethoxyflavone
   43. 5-Glucoside
7,3',4'-Trihydroxyflavone
   44. 7-Glucoside
   45. 7-Galactoside
   46. 7-Rutinoside
7,4'-Dihydroxy-3'-methoxyflavone
   47. 7-Glucoside
5,7,2'-Trihydroxyflavone
   48. 7-Glucoside*
   49. 7-Glucuronide
   50. 2'-Glucoside*
5,2'-Dihydroxy-7-methoxyflavone (echioidin)
   51. 5-Glucoside*
   52. 2'-Glucoside (echioidin)
   53. 2'-(6''-Acetylglucoside)*
7,8,4'-Trihydroxyflavone
   54. 8-Neohesperidoside
5,7,4'-Trihydroxyflavone (apigenin)
   55. 5-Glucoside
   56. 5-Galactoside
   57. 7-Arabinoside
   58. 7-Xyloside

59. 7-Rhamnoside
60. 7-Glucoside (cosmosiin)
61. 7-Galactoside
62. 7-Glucuronide
63. 7-Galacturonide
64. 7-Methylglucuronide
65. 7-Methylgalacturonide
66. 7-(6″-Ethylglucuronide)
67. 4′-Arabinoside
68. 4′-Glucoside
69. 4′-Glucuronide
70. 7-Arabinofuranosyl(1 → 6)glucoside
71. 7-Arabinopyranosyl(1 → 6)glucoside
72. 7-Xylosyl(1 → 2)glucoside
73. 7-Xylosyl(1 → 6)glucoside
74. 7-Apiofuranosyl(1 → 6)glucoside (apiin)*
75. 7-Rutinoside
76. 7-Neohesperidoside
77. 7-Rhamnosylglucuronide
78. 7-Dirhamnoside
79. 7-Glucosylrhamnoside
80. 7-Cellobioside*
81. 7-Allosyl(1 → 2)glucoside
82. 7-Galactosyl(1 → 4)mannoside
83. 7-Xylosylglucuronide
84. 7-Rhamnosylglucuronide
85. 7-Rhamnosyl(1 → 2)galacturonide
86. 7-Digalacturonide
87. 7-Galacturonylglucoside
88. 7-Glucuronosyl(1 → 2)glucuronide
89. 7,4′-Diglucoside
90. 7,4′-Dialloside
91. 7,4′-Diglucuronide
92. 7-Glucuronide-4′-rhamnoside
93. 4′-Diglucoside
94. 7-Sophorotrioside*
95. 7-(2^G-Rhamnosyl)rutinoside*
96. 7-(2^G-Rhamnosyl)gentiobioside*
97. 7-Rhamnoside-4′-glucosylrhamnoside*
98. 7-Rutinoside-4′-glucoside
99. 7-Neohesperidoside-4′-glucoside
100. 7-Cellobioside-4′-glucoside*
101. 7-Rhamnoside-4′-rutinoside
102. 7-Neohesperidoside-4′-sophoroside
103. 7-Glucosyl(1 → 2)glucuronide-4′-glucuronide*
104. 7-Digalacturonide-4′-glucoside
105. 7-Diglucuronide-4′-glucuronide
106. Pinnatifinoside A (**13.1**)
107. Pinnatifinoside B (**13.2**)
108. Pinnatifinoside C (**13.3**)

109. Pinnatifinoside D (**13.4**)
110. 7-(2″-*E*-*p*-Coumaroylglucoside)*
111. 7-(3″-*p*-Coumaroylglucoside)*
112. 7-(4″-*Z*-*p*-Coumaroylglucoside)
113. 7-(4″-*E*-*p*-Coumaroylglucoside)
114. 7-(6″-*p*-Coumaroylglucoside)
115. 7-(6″-*E*-*p*-Coumaroylgalactoside)*
116. 7-(6″-*E*-Caffeoylglucoside)*
117. 5-(6″-Malonylglucoside)
118. 7-(6″-Malonylglucoside)
119. 7-[6″-(3-Hydroxy-3-methylglutaryl)glucoside]*
120. 7-(2″-Acetylglucoside)
121. 7-(6″-Acetylglucoside)
122. 7-(6″-Crotonylglucoside)
123. 7-(2″-Acetyl-6″-methylglucuronide)
124. 7-Lactate
125. 7-(2″-Glucosyllactate)
126. 7-(2″-Glucuronosyllactate)
127. 7-Glucoside-4′-*p*-coumarate
128. 7-Glucoside-4′-caffeate
129. 7-(2″,6″-Di-*p*-coumaroylglucoside)
130. 7-(3″,6″-Di-*p*-coumaroylglucoside)
131. 7-(3″,6″-Di-*E*-*p*-coumaroylgalactoside)*
132. 7-(4″,6″-Di-*p*-coumaroylglucoside)
133. 7-(2″,3″-Diacetylglucoside)
134. 7-(3″,4″-Diacetylglucoside)
135. 7-(3″-Acetyl-6″-*E*-*p*-coumaroylglucoside)*
136. 5-Rhamnosyl(1 → 2)(6″-acetylglucoside)
137. 7-Rhamnosyl(1 → 6)(4″-*E*-*p*-methoxycinnamoylglucoside)*
138. 4′-(2″-Feruloylglucuronosyl(1 → 2)glucuronide)*
139. 7-(6‴-Acetylallosyl)(1 → 2)glucoside
140. 7-(6‴-Malonylneohesperidoside)
141. 7-(Malonylapiosyl)glucoside
142. 7-Rutinoside-4′-caffeate
143. 7-(6″-Acetylalloside)-4′-alloside
144. 7-(4″,6″-Diacetylalloside)-4′-alloside
145. 7-Glucuronide-4′-(6″-malonylglucoside)
146. 7-Glucuronosyl(1 → 3)[(2″-*p*-coumaroylglucuronosyl)(1 → 2)glucuronide]*
147. 7-Glucuronosyl(1 → 3)[(2″-feruloylglucuronosyl)(1 → 2)glucuronide]*
148. 7-(2″-Ferulylglucuronosyl)(1 → 2)glucuronide-4′-glucuronide*
149. 7-Glucuronide-4′-(2‴-*E*-*p*-coumaroylglucuronosyl)(1 → 2)glucuronide*
150. 7-Glucuronide-4′-(2‴-feruloylglucuronosyl)(1 → 2)glucuronide*
151. 7-Sulfatoglucoside
152. 7-Sulfatogalactoside
153. 7-Sulfatoglucuronide
154. 7-Sulfate

Apigenin 7-methyl ether (genkwanin)
155. 5-Glucoside
156. 4′-Glucoside
157. 5-Xylosylglucoside

158. 4'-α-L-Arabinopyranosyl(1 → 6)galactoside*
159. 4'-Glucosylrhamnoside
160. 4'-Rhamnosyl(1 → 2)[rhamnosyl(1 → 6)galactoside]*
161. 5-(6''-Malonylglucoside)

Apigenin 4'-methyl ether (acacetin)
162. 7-Rhamnoside*
163. 7-Glucoside (tilianine)
164. 7-Galactoside
165. 7-Glucuronide
166. 7-(6''-Methylglucuronide)
167. 7-Arabinosylrhamnoside
168. 7-Glucosyl(1 → 4)xyloside*
169. 7-Apiosyl(1 → 6)glucoside*
170. 7-Rutinoside (linarin)
171. 7-Neohesperidoside (fortunellin)
172. 7-Diglucoside
173. 7-Rhamnosylgalacturonide
174. 7-Glucuronosyl(1 → 2)glucuronide
175. 7-(2^G-Rhamnosylrutinoside)*
176. 7-Rhamnosyl(1 → 2)glucosyl(1 → 2)glucoside*
177. 7-Rhamnosyl(1 → 2)glucosyl(1 → 2)glucosyl(1 → 2)glucoside*
178. 7-(2''-Acetylglucoside)
179. 7-(6''-Acetylglucoside)
180. 7-(4''-Acetylrutinoside)
181. 7-(4'''-Acetylrutinoside)*
182. 7-[2'''-(2-Methylbutyryl)rutinoside]
183. 7-[3'''-(2-Methylbutyryl)rutinoside]
184. 7-Rhamnosyl(1 → 6)[2''-acetylglucosyl(1 → 2)glucoside]*
185. 7-[6'''-Acetylglucosyl(1 → 2)][rhamnosyl(1 → 6)glucoside]*
186. 7-Glucosyl(1 → 6)[3'''-acetylrhamnosyl(1 → 2)glucoside]*
187. 7-(4''''-Acetylrhamnosyl)(1 → 6)glucosyl(1 → 3)(6''-acetylglucoside)*
188. 7-[Rhamnosyl(1 → 4)glucosyl(1 → 6)](6'''-acetylsophoroside)
189. Di-6''-(acacetin-7-glucosyl)malonate

Apigenin 5,7-dimethyl ether
190. 4'-Galactoside

Apigenin 7,4'-dimethyl ether
191. 5-Xylosylglucoside

6-Hydroxyapigenin (scutellarein)
192. 5-Glucuronide
193. 6-Xyloside
194. 6-Glucoside
195. 7-Rhamnoside
196. 7-Glucoside
197. 7-Glucuronide
198. 4'-Arabinoside
199. 7-Xylosyl(1 → 2)xyloside*
200. 7-Xylosyl(1 → 4)rhamnoside
201. 7-Xylosyl(1 → 2)glucoside*
202. 7-Xylosyl(1 → 6)galactoside*
203. 7-Glucosyl(1 → 4)rhamnoside

204. 7-Rutinoside
205. 7-Neohesperidoside
206. 7-Rhamnosyl(1 → 2)galactoside
207. 7-Diglucoside
208. 7-Glucuronosyl(1 → 2)glucuronide*
209. 6-Xyloside-7-rhamnoside
210. 7,4'-Dirhamnoside
211. 7-(6''-Malonylglucoside)
212. 7-[6''-(3-Hydroxy-3-methylglutaryl)glucoside]*
213. 7-(6''-Feruloylglucuronide)
214. 7-(Sinapoylglucuronide)
Scutellarein 6-methyl ether (Hispidulin)
215. 7-Rhamnoside*
216. 7-Glucoside (homoplantaginin)
217. 7-Glucuronide
218. 7-Methylglucuronide*
219. 4'-Glucoside*
220. 7-Xylosyl(1 → 2)glucoside*
221. 7-Rutinoside
222. 7-Neohesperidoside*
223. 7-(6''-*E*-*p*-Coumaroylglucoside)*
224. 7-Sulfate
225. 4'-Sulfate
226. 7,4'-Disulfate
Scutellarein 7-methyl ether
227. 6-Glucoside
228. 6-Galactoside
229. 7-Glucoside
230. 6-Rhamnosylxyloside
Scutellarein 4'-methyl ether
231. 6-Glucoside
232. 7-Glucoside
233. 7-Glucuronide
234. 7-Rutinoside*
235. 7-Sophoroside
236. 7-(2'',6''-Diacetylalloside)*
237. 7-(*p*-Coumaroylglucosyl)(1 → 2)mannoside
Scutellarein 5,4'-dimethyl ether
238. 7-Glucoside*
239. 7-(4^{Rha}-Acetylrutinoside)*
Scutellarein 6,7-dimethyl ether
240. 4'-Glucoside
241. 4'-Glucuronide*
242. 4'-Rutinoside
Scutellarein 6,4'-dimethyl ether (pectolinarigenin)
243. 7-Rhamnoside
244. 7-Glucoside
245. 7-Glucuronide
246. 7-Glucuronic acid methyl ether
247. 7-Rutinoside

248. 7-(6''-Acetylglucoside)*
249. 7-(2'''-Acetylrutinoside)*
250. 7-(3'''-Acetylrutinoside)*
251. 7-(4'''-Acetylrutinoside)*
Scutellarein 7,4'-dimethyl ether
252. 6-Glucoside
253. 6-Xylosyl(1 → 2)glucoside*
254. 6-Neohesperidoside*
Scutellarein 6,7,4'-trimethyl ether (salvigenin)
255. 5-Glucoside
256. 5-(6''-Acetylglucosyl)(1 → 3)galactoside*
8-Hydroxyapigenin (isoscutellarein)
257. 7-Xyloside
258. 7-Glucoside
259. 7-Glucosyl(1 → 2)xyloside*
260. 8-Glucuronide
261. 7-Neohesperidoside
262. 7-Allosyl(1 → 2)glucoside
263. 8-Sophoroside*
264. 8-(6''-E-p-Coumaroylglucoside)*
265. 7-(6''-Acetylallosyl)(1 → 2)glucoside
266. 7-[6'''-Acetylallosyl(1 → 2)6''-acetylglucoside]
267. 8-(2''-Sulfatoglucuronide)*
268. 8-(2'',4''-Disulfatoglucuronide)*
8-Hydroxyapigenin 4'-methyl ether
269. 8-Glucoside*
270. 8-Glucuronide
271. 8-(6''-n-butylglucuronide)*
272. 7-Allosyl(1 → 2)glucoside*
273. 8-Xylosylglucoside
274. 7-(6'''-Acetylallosyl)(1 → 2)glucoside
275. 8-(2''-Sulfatoglucoside)
276. 8-(2''Sulfatoglucuronide)*
277. 8-(2'',4''-Disulfatoglucuronide)*
8-Hydroxyapigenin 8,4'-dimethyl ether
278. 7-Glucuronide
6,8-Dihydroxy-7,4'-dimethoxyflavone
279. 6-Rutinoside*
280. 6-(4''-Acetylrhamnosyl)(1 → 6)glucoside*
7,3',4',5'-Tetrahydroxyflavone
281. 7-Rhamnoside
282. 7-Glucoside
5,6,7,2'-Tetrahydroxyflavone
283. 7-Glucuronide
5,7,2'-Trihydroxy-6-methoxyflavone
284. 7-Glucoside*
285. 7-Methylglucuronide*
5,7,2',6'-Tetrahydroxyflavone
286. 2'-Glucoside*
5,2',6'-Trihydroxy-7-methoxyflavone

  287. 2′-Glucoside*

5,7,8,2′-Tetrahydroxyflavone

  288. 7-Glucuronide*

5,7,2′-Trihydroxy-8-methoxyflavone

  289. 7-Glucuronide

5,2′-Dihydroxy-7,8-dimethoxyflavone (skullcapflavone 1)

  290. 2′-Glucoside*

  291. 2′-(2″-*E*-Cinnamoylglucoside)*

  292. 2′-(3″-*E*-Cinnamoylglucoside)*

  293. 2′-(4″-*E*-Cinnamoylglucoside)*

5,7-Dihydroxy-8,2′-dimethoxyflavone

  294. 7-Glucuronide

5-Hydroxy-7,8,2′-trimethoxyflavone

  295. 5-Glucoside

5,7,3′,4′-Tetrahydroxyflavone (luteolin)

  296. 5-Glucoside (galuteolin)

  297. 5-Galactoside

  298. 5-Glucuronide

  299. 5-Glucuronide-6″-methyl ester*

  300. 7-Xyloside

  301. 7-Rhamnoside

  302. 7-Glucoside

  303. 7-Galactoside

  304. 7-Glucuronide

  305. 7-Galacturonide

  306. 7-Methylglucuronide

  307. 3′-Xyloside

  308. 3′-Rhamnoside

  309. 3′-Glucoside

  310. 3′-Glucuronide

  311. 3′-Galacturonide

  312. 4′-Arabinoside

  313. 4′-Glucoside

  314. 4′-Glucuronide

  315. 5-Rutinoside*

  316. 7-Dirhamnoside

  317. 7-Arabinofuranosyl(1 → 6)glucoside

  318. 7-Arabinopyranosyl(1 → 6)glucoside

  319. 7-Glucosyl(1 → 4)α-L-arabinopyranoside*

  320. 7-Xylosyl(1 → 6)glucoside (primeveroside)*

  321. 7-Apiosyl(1 → 6)glucoside*

  322. 7-Sambubioside

  323. 7-Apiosylglucoside

  324. 7-Rutinoside

  325. 7-Neohesperidoside (veronicastroside)

  326. 7-Glucosylrhamnoside

  327. 7-Robinobioside*

  328. 7-Sophoroside*

  329. 7-Gentiobioside

  330. 7-Laminaribioside

331. 7-Glucosylgalactoside
332. 7-Galactosyl(1 → 6)galactoside*
333. 7-Allosyl(1 → 2)glucoside
334. 7-Glucosylglucuronide
335. 7-Galactosylglucuronide*
336. 7-Glucuronosyl(1 → 2)glucuronide
337. 7-Glucoside-3'-xyloside
338. 7,3'-Diglucoside
339. 7-Glucoside-3'-glucuronide*
340. 7-Glucuronide-3'-glucoside
341. 7,3'-Diglucuronide
342. 7,3'-Digalacturonide
343. 7,4'-Diglucoside
344. 7-Galactoside-4'-glucoside
345. 7-Glucuronide-4'-rhamnoside
346. 7-Galacturonide-4'-glucoside
347. 7,4'-Diglucuronide
348. 3'-Xylosyl(1 → 2)glucoside*
349. 4'-Rutinoside*
350. 4'-Neohesperidoside
351. 3',4'-Diglucoside
352. 3',4'-Diglucuronide
353. 3',4'-Digalacturonide
354. 7-Rhamnosyldiglucoside
355. 7-Sophorotrioside*
356. 7-Glucosylarabinoside-4'-glucoside
357. 7-Rutinoside-3'-glucoside
358. 7-Rutinoside-4'-glucoside
359. 7-Neohesperidoside-4'-glucoside
360. 7-Glucoside-4'-neohesperidoside
361. 7-Gentiobioside-4'-glucoside
362. 7-Glucuronide-3',4'-dirhamnoside
363. 7,4'-Diglucuronide-3'-glucoside
364. 7-Glucuronosyl(1 → 2)glucuronide-4'-glucuronide
365. 7,3',4'-Triglucuronide
366. 7-Neohesperidoside-4'-sophoroside
367. 7-(6''-E-Cinnamoylglucoside)
368. 7-(2''-p-Coumaroylglucoside)
369. 7-(6''-p-Coumaroylglucoside)
370. 7-Caffeoylglucoside
371. 7-(6''-Feruloylglucoside)
372. 7-(6''-p-Benzoylglucoside)*
373. 5-(6''-Malonylglucoside)
374. 7-(6''-Malonylglucoside)
375. 7-[6''-(2-Methylbutyryl)glucoside]*
376. 7-[6''-(3-Hydroxy-3-methylglutaryl)glucoside]*
377. 3'-Acetylglucuronide
378. 7-(6''-Acetylglucoside)
379. 3'-(3''-Acetylglucuronide)*
380. 3'-(4''-Acetylglucuronide)*

381. 7-Glucosyl(1 → 6)(4‴-caffeoylglucoside)*
382. 7-Glucoside-4′-(*Z*-2-methyl-2-butenoate)*
383. 7-Glucuronide-3′-feruloylglucoside
384. 7-Neohesperidoside-6″-malonate
385. 7-(3″-Acetylapiosyl)(1 → 2)xyloside
386. 7-(6‴-Acetylallosyl)(1 → 2)glucoside
387. 7-(6‴-Acetylsophoroside)*
388. 7-Apiosyl(1 → 2)[glucosyl(1 → 4)(6-malonylglucoside)]*
389. 7-(Acetylsophorotrioside)*
390. 7-(6⁗-Acetylallosyl)(1 → 3)glucosyl(1 → 2)glucoside*
391. 7-(2″-Feruloylglucuronosyl)(1 → 2)glucuronide-4′-glucuronide*
392. 7,4′-Diglucuronide-3′-feruloylglucoside
393. 7-Lactate
394. 7-(2″-Glucosyllactate)
395. 7-(2″-Glucuronosyllactate)
396. 7-Sulfatoglucoside
397. 7-(2″-Sulfatoglucoside)*
398. 7-Sulfatoglucuronide
399. 7-Sulfate-3′-glucoside
400. 7-Sulfatorutinoside
401. 7-Sulfate-3′-rutinoside
402. 7-Sulfate
403. 3′-Sulfate
404. 4′-Sulfate
405. 7-Disulfatoglucoside
406. 7,3′-Disulfate
Luteolin 5-methyl ether
407. 7-Glucoside
408. 7-Xylosyl(1 → 6)glucoside*
Luteolin 7-methyl ether
409. 5-Glucoside
410. 3′-Glucoside*
411. 3′-Galactoside*
412. 4′-Rhamnoside
413. 5-Xylosylglucoside
414. 4′-Gentiobioside
Luteolin 3′-methyl ether (chrysoeriol)
415. 5-Glucoside
416. 7-Xyloside
417. 7-Rhamnoside
418. 7-Glucoside
419. 7-Glucuronide
420. 4′-Glucoside
421. 5-Diglucoside
422. 7-Arabinofuranosyl(1 → 2)glucoside
423. 7-α-ʟ-Arabinofuranosyl(1 → 6)galactoside*
424. 7-Apiosyl(1 → 6)glucoside*
425. 7-Rutinoside
426. 7-Neohesperidoside*
427. 7-Rhamnosylgalactoside

428. 7-Rhamnosylglucuronide
429. 7-Digalactoside
430. 7-Mannosyl(1 → 2)alloside
431. 7-Allosyl(1 → 2)glucoside
432. 7-Glucuronosyl(1 → 2)glucuronide
433. 5,4'-Diglucoside
434. 7,4'-Dixyloside
435. 7,4'-Diglucoside
436. 7-Glucuronide-4'-rhamnoside
437. 7,4'-Diglucuronide*
438. 7-Sophorotrioside
439. 7-(6''-Crotonylglucoside)
440. 7-(Malonylglucoside)
441. 7-(3''-Z-p-Coumaroylglucoside)*
442. 7-(3'',6''-Di-E-p-coumaroylglucoside)*
443. 7-p-Coumaroylglucosylglucuronide
444. 7-(2'''-Feruloylglucuronosyl(1 → 2)glucuronide)*
445. 7-(6'''-Acetylglucosyl)(1 → 2)mannoside
446. 7-[Glucuronosyl(1 → 3)(2'''-feruloylglucuronosyl)](1 → 2)glucuronide*
447. 7-Sulfatoglucoside
448. 7-Disulfatoglucoside
449. 7-Sulfate
Luteolin 4'-methyl ether (diosmetin)
450. 7-Glucoside
451. 7-Glucuronide
452. 3'-Glucoside*
453. 7-Arabinosyl(1 → 6)glucoside*
454. 7-α-Glucosyl(1 → 6)glucoside*
455. 7-β-Glucosyl(1 → 6)arabinoside
456. 7-Xylosyl(1 → 6)glucoside*
457. 7-Rutinoside (diosmin)
458. 7-Neohesperidoside (neodiosmin)*
459. 7-Diglucoside
460. 7-(2'',6''-Dirhamnosyl)glucoside*
461. 7-(6''-Malonylglucoside)
462. 7-Apiosyl(1 → 2)(6''-Acetylglucoside)*
463. 7-Sulfate
464. 3'-Sulfate
465. 7,3'-Disulfate
Luteolin 5,3'-dimethyl ether
466. 7-Glucoside*
467. 4'-Glucoside*
Luteolin 5,4'-dimethyl ether
468. 7-Xylosyl(1 → 6)glucoside*
Luteolin 7,3'-dimethyl ether
469. 5-Rhamnoside
470. 5-Glucoside
471. 4'-Glucoside
472. 4'-Apiosyl(1 → 2)glucoside*

Luteolin 7,4'-dimethyl ether
  473. 3'-Glucoside
  474. 5-Xylosylglucoside
Luteolin 3',4'-dimethyl ether
  475. 7-Rhamnoside
  476. 7-Glucuronide
Luteolin 5,3',4'-trimethyl ether
  477. 7-Xylosyl(1 → 6)glucoside*
  478. 7-Rutinoside*
Luteolin 7,3',4'-trimethyl ether
  479. 5-Glucoside*
  480. 5-Xylosyl(1 → 6)glucoside*
6-Hydroxyluteolin
  481. 5-Glucoside
  482. 6-Xyloside
  483. 6-Rhamnoside*
  484. 6-Glucoside
  485. 6-Glucuronide
  486. 7-Arabinoside
  487. 7-Xyloside
  488. 7-Rhamnoside
  489. 7-Apioside
  490. 7-Glucoside
  491. 7-Galactoside
  492. 7-Glucuronide
  493. 7-Xylosyl(1 → 2)xyloside*
  494. 7-Xylosyl(1 → 6)glucoside*
  495. 7-Rhamnosyl(1 → 4)xyloside
  496. 7-Sambubioside*
  497. 6-Glucoside-3'-rhamnoside
  498. 7-Rutinoside
  499. 7-Sophoroside
  500. 7-Gentiobioside
  501. 7-Arabinoside-4'-rhamnoside
  502. 7-(6''-Malonylglucoside)
  503. 7-[3''-(3-Hydroxy-3-methylglutaryl)glucoside]*
  504. 7-[4''-(3-Hydroxy-3-methylglutaryl)glucoside]*
  505. 7-[6''-(3-Hydroxy-3-methylglutaryl)glucoside]*
  506. 7-(6''-*E*-Caffeoylglucoside)*
  507. 7-(6'''-*p*-Coumaroylsophoroside)
  508. 7-(6'''-Caffeoylsophoroside)
  509. 6-Glucoside-7-[6'''-(3-hydroxy-3-methylglutaryl)glucoside]*
  510. 7-[6''-(3-Hydroxy-3-methylglutaryl)glucoside]-3'-glucuronide*
  511. 6-Sulfate
  512. 7-Sulfate
  513. 6,7-Disulfate
6-Methoxyluteolin
  514. 7-Glucoside
  515. 7-Glucuronide*
  516. 7-Methylglucuronide*
  517. 4'-Glucoside*

518. 7-Rutinoside
519. 7-Rhamnosyl-3′-xyloside*
520. 7-[6″-(2-Methylbutyryl)glucoside]*
521. 7-Sulfate
522. 3′,4′-Disulfate
6-Hydroxyluteolin 7-methyl ether (pedalitin)
523. 6-Glucoside (pedaliin)
524. 6-Galactoside
525. 7-Glucuronide
526. 7-Methylglucuronide
527. 6-Galactosylglucoside
6-Hydroxyluteolin 3′-methyl ether (nodifloretin)
528. 7-Diglucoside
529. 7-[6″-(3-Hydroxy-3-methylglutaryl)glucoside]*
530. 7-Sulfate
531. 6,7-Disulfate
6-Hydroxyluteolin 4′-methyl ether
532. 7-Allosyl(1 → 2)glucoside
533. 7-Rhamnosyl(1 → 2)(6″-acetylglucoside)*
534. 7-[6‴-Acetylallosyl(1 → 2)6″-acetylglucoside]
6-Hydroxyluteolin 6,7-dimethyl ether (cirsiliol)
535. 4′-Glucoside
6-Hydroxyluteolin 6,3′-dimethyl ether
536. 5-Rhamnoside*
537. 7-Rhamnoside
538. 7-Glucoside
539. 7-Rutinoside*
540. 7-Sulfate
541. 7,4-Disulfate
6-Hydroxyluteolin 6,4′-dimethyl ether
542. 7-Glucoside
543. 7-Rutinoside
6-Hydroxyluteolin 7,3′-dimethyl ether
544. 6-Glucoside
6-Hydroxyluteolin 6,7,3′-trimethyl ether (cirsilineol)
545. 4′-Glucoside
6-Hydroxyluteolin 5,6,3′,4′-tetramethyl ether
546. 7-Cellobioside*
8-Hydroxyluteolin (hypolaetin)
547. 7-Xyloside
548. 7-Glucoside
549. 8-Rhamnoside*
550. 8-Glucoside
551. 8-Glucuronide
552. 7-Sophoroside*
553. 7-Allosyl(1 → 2)glucoside
554. 8-Gentiobioside
555. 8,4′-Diglucuronide
556. 8-Glucoside-3′-rutinoside*
557. 7-(6‴-Acetylallosyl)(1 → 2)glucoside

558. 7-[6'''-Acetylallosyl(1 → 2)6''-acetylglucoside]
559. 7-[6'''-Acetylallosyl(1 → 2)3''-acetylglucoside]
560. 7-Sulfatoglucoside*
561. 7-Sulfatogalactoside*
562. 7-Sulfatoglucuronide*
563. 7-Sulfate-8-glucoside*
564. 8-Glucoside-3'-sulfate
565. 8-Sulfate

Hypolaetin 7-methyl ether
566. 3'-Sulfatogalactoside*
567. 3'-Sulfatoglucuronide*

Hypolaetin 3'-methyl ether
568. 7-Glucoside
569. 8-Glucuronide*
570. 7-Sophoroside*
571. 7-Allosyl(1 → 2)glucoside
572. 7-Mannosyl(1 → 2)glucoside

Hypolaetin 4'-methyl ether
573. 8-Glucoside
574. 8-Glucuronide
575. 7-Allosyl(1 → 2)glucoside
576. 7-(6'''-Acetylallosyl)(1 → 2)glucoside
577. 7-[6'''-Acetylallosyl(1 → 2)6''-acetylglucoside]*
578. 8-Glucoside-3'-sulfate

Hypolaetin 7,3'-dimethyl ether
579. 4'-Glucoside*

Hypolaetin 8,3'-dimethyl ether
580. 7-Glucoside

5,6,4'-Trihydroxy-7,8-dimethoxyflavone (thymusin)
581. 6-Isobutyrate*

5,8,4'-Trihydroxy-6,7-dimethoxyflavone (isothymusin)
582. 8-Glucoside*

5,7,-Dihydroxy-6,8,4'-trimethoxyflavone (nevadensin)
583. 5-Glucoside*
584. 7-Glucoside*
585. 5-Gentiobioside*
586. 7-Rutinoside*

5,8-Dihydroxy-6,7,4'-trimethoxyflavone
587. 8-Glucoside*

5,6,7,8,3',4'-Hexahydroxyflavone
588. 7-Glucoside*

5,6,7,3',4'-Pentahydroxy-8-methoxyflavone (pleurostimin)
589. 7-Apioside
590. 7-Glucoside*

5,6,3',4'-Tetrahydroxy-7,8-dimethoxyflavone (pleurostimin 7-methyl ether)
591. 6-Glucoside*

5,7,3',4'-Tetrahydroxy-6,8-dimethoxyflavone
592. 7-Glucoside

5,8,3',4'-Tetrahydroxy-6,7-dimethoxyflavone
593. 8-Glucoside

5,7,3'-Trihydroxy-6,8,4'-trimethoxyflavone (acerosin)
 594. 5-(6''-Acetylglucoside)
5,7,4'-Trihydroxy-6,8,3'-trimethoxyflavone (sudachitin)
 595. 7-Glucoside
 596. 4'-Glucoside
 597. 7-(3-Hydroxy-3-methylglutarate)-4'-glucoside
 598. 7-[6''-(3-Hydroxy-3-methylglutaryl)glucoside]
 599. 4'-[6''-(3-Hydroxy-3-methylglutaryl)glucoside]
 600. Sudachitin D
5,2',3'-Trihydroxy-7,8-dimethoxyflavone
 601. 3'-Glucoside
5,2',6'-Trihydroxy-6,7-dimethoxyflavone
 602. 2'-Glucoside*
5,2',6'-Trihydroxy-7,8-dimethoxyflavone
 603. 2'-Glucuronide*
5,2'-Dihydroxy-7,8,6'-trimethoxyflavone
 604. 2'-Glucuronide*
5-Hydroxy-7,8,2',3'-tetramethoxyflavone
 605. 5-Glucoside
5,7,2',4',5'-Pentahydroxyflavone (isoetin)
 606. 7-Glucoside
 607. 7-Arabinoside
 608. 2'-Xyloside
 609. 4'-Glucuronide*
 610. 5'-Glucoside
 611. 7-Xylosylarabinosylglucoside
 612. 7-Glucoside-2'-xyloside
 613. 2'-(4''-Acetylxyloside)
 614. 7-Glucoside-2'-(4''-acetylxyloside)
5,7,3',4',5'-Pentahydroxyflavone (tricetin)
 615. 7-Glucoside
 616. 3'-Xyloside
 617. 3'-Glucoside
 618. 3'-Rhamnosyl(1 → 4)rhamnoside*
 619. 7,3'-Diglucuronide
 620. 3'-Glucoside-5'-rhamnoside*
 621. 7-Glucoside-3'-[6''-(3-hydroxy-3-methylglutaryl)glucoside]*
 622. 7-Diglucoside
 623. 3',5'-Diglucoside
 624. 3'-Sulfate
 625. 3'-Disulfate
Tricetin 7-methyl ether
 626. 3'-Glucoside-5'-rhamnoside*
Tricetin 3'-methyl ether
 627. 7-Glucoside
 628. 7-Glucuronide*
 629. 7,5'-Diglucuronide
Tricetin 4'-methyl ether
 630. 7-Apiosyl(1 → 2)(6''-acetylglucoside)*

Tricetin 3′,4′-dimethyl ether
  631. 7-Glucuronide
Tricetin 3′,5′-dimethyl ether (Tricin)
  632. 5-Glucoside
  633. 7-β-D-Arabinopyranoside*
  634. 7-Xyloside
  635. 7-Glucoside
  636. 7-Glucuronide
  637. 4′-Apioside*
  638. 4′-Glucoside
  639. 5-Diglucoside
  640. 7-Rutinoside
  641. 7-Neohesperidoside
  642. 7-Diglucoside
  643. 7-Fructosylglucoside
  644. 7-Rhamnosylglucuronide
  645. 7-Rhamnosyl(1 → 2)galacturonide
  646. 7-Diglucuronide
  647. 5,7-Diglucoside
  648. 7-Rutinoside-4′-glucoside
  649. 7-(2″-*p*-Coumaroylglucuronosyl)(1 → 2)glucuronide*
  650. 7-(2″-Feruloylglucuronosyl)(1 → 2)glucuronide*
  651. 7-(2″-Sinapoylglucuronosyl)(1 → 2)glucuronide*
  652. 7-[Glucuronosyl(1 → 3)(2‴-feruloylglucuronosyl)](1 → 2)glucuronide*
  653. 7-[X′-(3-Hydroxy-3-methylglutaryl)glucoside]*
  654. 7-Sulfatoglucoside
  655. 7-Sulfatoglucuronide
  656. 7-Disulfatoglucuronide
3-(3-Methylbutyl)tricetin
  657. 5-Neohesperidoside
Tricetin 7,3′,4′-trimethyl ether
  658. 5-Glucoside*
Tricetin 7,3′,4′,5′-tetramethyl ether
  659. 5-Glucoside*
  660. 5-Xylosyl(1 → 2)rhamnoside*
  661. 5-Xylosyl(1 → 6)glucoside*
6-Hydroxytricetin 6,3′,5′-trimethyl ether
  662. 7-α-L-Arabinosyl(1 → 6)glucoside*
6-Hydroxytricetin 6,4′,5′-trimethyl ether
  663. 3′-Rhamnoside*
6-Hydroxytricetin 6,7,3′,5′-tetramethyl ether
  664. 5-Robinobioside*
8-Hydroxytricetin
  665. 5-Rhamnoside*
  666. 5-Glucoside*
  667. 7-Glucuronide
5,6,7,8,3′,4′-Hexahydroxyflavone
  688. 7-Glucoside*
5,7,2′,5′-Tetrahydroxy-8,6′-dimethoxyflavone (viscidulin III)
  669. 2′-Glucoside*

5,2′,6′-Trihydroxy-6,7,8-trimethoxyflavone
  670. 2′-Glucoside*
5,4′-Dihydroxy-7,8,2′,3′-tetramethoxyflavone
  671. 5-Glucoside
C-Methylflavones
5,7-Dihydroxy-6-C-methylflavone
  672. 7-Xylosyl(1 → 3)xyloside*
3-C-Methylapigenin
  673. 5-Rhamnoside
5,7,4′-Trihydroxy-3′-C-methylflavone
  674. 4′-Rhamnoside
5,7-Dihydroxy-6,8-di-C-methylflavone (matteuorien)
  675. 7-[6″-(3-Hydroxy-3-methylglutaryl)glucoside]*
3-C-Methylluteolin
  676. 5-Rhamnoside
  677. Stachysetin (**13.5**)*
Prenylated flavones
8-Prenylapigenin
  678. 4′-Rutinoside*
3′-Prenylapigenin
  679. 7-Rutinoside*
8-C-Prenyl-5,7,4′-trihydroxy-3′-methoxyflavone (8-C-Prenylchrysoeriol)
  680. 7-Glucosyl(1 → 3)-α-L-arabinopyranoside*

*Flavone glycosides newly reported since 1992.

APPENDIX B

CHECKLIST OF KNOWN FLAVONOL GLYCOSIDES

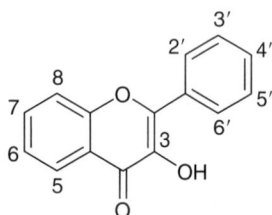

13.10

3,5,7-Trihydroxyflavone (galangin)
  1. 3-Glucoside
  2. 7-Glucoside
  3. 3-Rutinoside
  4. 3-Galactosyl(1 → 4)rhamnoside
  5. 8-Glucoside-8-sulfate*
  6. 8-Sulfate*
3,7-Dihydroxy-8-methoxyflavone
  7. 7-Rhamnoside*
  8. 7-Rhamnosyl(1 → 4)rhamnosyl(1 → 6)glucoside*
3,7,4′-Trihydroxyflavone
  9. 3-Glucoside

 10. 7-Glucoside
 11. 4′-Glucoside
 12. 7-Rutinoside
5,7-Dihydroxy-3,6-dimethoxyflavone
 13. 5-α-L-Arabinosyl(1 → 6)glucoside*
3,6,7-Trihydroxy-4′-methoxyflavone
 14. 7-Rhamnoside*
8-Hydroxygalangin 3-methyl ether
 15. 8-(*Z*-2-Methyl-2-butenoate)
 16. 8-(2-Methylbutyrate)
8-Hydroxygalangin 7-methyl ether
 17. 8-Acetate
 18. 8-Butyrate
3,7,3′,4′-Tetrahydroxyflavone (fisetin)
 19. 3-Glucoside
 20. 7-Glucoside
 21. 4′-Glucoside
 22. 7-Rutinoside
3,7,4′-Trihydroxy-3′-methoxyflavone (geraldol)
 23. 4′-Glucoside
3,5,7,4′-Tetrahydroxyflavone (kaempferol)
 24. 3-α-D-Arabinopyranoside*
 25. 3-Arabinofuranoside (juglanin)
 26. 3-Xyloside
 27. 3-Rhamnoside (afzelin)
 28. 3-Glucoside (astragalin)
 29. 3-α-D-Galactoside
 30. 3-β-D-Galactoside (trifolin)
 31. 3-Alloside (asiaticalin)
 32. 3-Glucuronide
 33. 3-(6″-Ethylglucuronide)
 34. 5-Rhamnoside
 35. 5-Glucoside
 36. 5-Glucuronide*
 37. 7-Arabinoside
 38. 7-Xyloside
 39. 7-Rhamnoside
 40. 7-Glucoside (populnin)
 41. 7-Alloside*
 42. 4′-Rhamnoside
 43. 4′-Glucoside
 44. 3-Rhamnosyl(1 → 2)-α-L-arabinofuranoside (arapetaloside B)*
 45. 3-Xylosyl(1 → 2)rhamnoside
 46. 3-Rhamnosylxyloside
 47. 3-Arabinosyl(1 → 6)galactoside
 48. 3-Xylosyl(1 → 2)glucoside*
 49. 3-Xylosyl(1 → 2)galactoside
 50. 3-Rhamnosyl(1 → 2)rhamnoside*
 51. 3-Apiosyl(1 → 2)glucoside
 52. 3-Apiosyl(1 → 2)galactoside
 53. 3-Glucosyl(1 → 2)rhamnoside*

54. 3-Glucosyl(1 → 4)rhamnoside
55. 3-Rutinoside
56. 3-Neohesperidoside
57. 3-Rhamnosyl(1 → 3)glucoside (rungioside)
58. 3-Robinobioside
59. 3-Rhamnosyl(1 → 2)galactoside
60. 3-Sambubioside
61. 3-Gentiobioside
62. 3-Sophoroside
63. 3-Glucosyl(1 → 6)galactoside
64. 3-Glucosyl(1 → 2)galactoside
65. 3-Galactosylglucoside
66. 3-Digalactoside
67. 7-Glucosyl(1 → 4)xyloside*
68. 7-Neohesperidoside*
69. 7-Glucosyl(1 → 3)rhamnoside*
70. 7-Galactosyl(1 → 4)rhamnoside
71. 7-Sophoroside
72. 3,5-Diglucoside
73. 3,5-Digalactoside
74. 3,7-Diarabinoside
75. 3-Arabinoside-7-rhamnoside
76. 3-α-L-Arabinofuranoside-7-α-L-rhamnopyranoside
77. 3-Rhamnoside-7-arabinoside
78. 3-Glucoside-7-arabinoside
79. 3-Xyloside-7-rhamnoside
80. 3-Rhamnoside-7-xyloside
81. 3-Xyloside-7-glucoside
82. 3-Glucoside-7-xyloside
83. 3,7-Dirhamnoside
84. 3-Rhamnoside-7-glucoside
85. 3-α-D-Glucoside-7-α-L-rhamnoside
86. 3-β-D-Glucoside-7-rhamnoside
87. 3-Galactoside-7-rhamnoside
88. 3-Glucoside-7-galactoside
89. 3,7-Diglucoside
90. 3-Rhamnoside-7-galacturonide
91. 3-Glucoside-7-glucuronide
92. 3-Glucuronide-7-glucoside
93. 3,4'-Dixyloside
94. 3,4'-Diglucoside*
95. 3-Rhamnoside-4'-arabinoside
96. 3-Rhamnoside-4'-xyloside
97. 3-Galactoside-4'-glucoside
98. 7,4'-Dirhamnoside
99. 7-Rhamnoside-4'-glucoside*
100. 7,4'-Diglucoside*
101. 3-Glucosyl-β-(1 → 4)arabinofuranosyl-α-(1 → 2)arabinopyranoside (primflasine)
102. 3-Xylosylrutinoside
103. 3-Xylosyl(1 → 3)rhamnosyl(1 → 6)galactoside*

104. 3-Xylosyl(1 → 6)glucosyl(1 → 2)rhamnoside*
105. 3-Rhamnosyl(1 → 2)rhamnosyl(1 → 6)glucoside
106. 3-Rhamnosyl(1 → 3)rhamnosyl(1 → 6)glucoside*
107. 3-Rhamnosyl(1 → 4)rhamnosyl(1 → 6)glucoside
108. 3-Rhamnosyl(1 → 3)rhamnosyl(1 → 6)galactoside (rhamninoside)
109. 3-Rhamnosyl(1 → 4)rhamnosyl(1 → 6)galactoside (isorhamninoside)
110. 3-Rhamnosyl(1 → 2)glucosyl(1 → 6)galactoside*
111. 3-Rhamnosyl(1 → 6)glucosyl(1 → 6)galactoside*
112. 3-Glucosyl(1 → 4)rhamnosyl(1 → 2)glucoside*
113. 3-Glucosyl(1 → 3)rhamnosyl(1 → 6)galactoside
114. 3-Glucosyl(1 → 2)gentiobioside
115. 3-Sophorotrioside
116. 3-β-Maltosyl(1 → 6)glucoside
117. 3-Glucosyl(1 → 2)galactosyl(1 → 2)glucoside*
118. 3-Xylosyl(1 → 2)[rhamnosyl(1 → 6)glucoside]
119. 3-Rhamnosyl(1 → 2)[rhamnosyl(1 → 6)glucoside] (mauritianin)
120. 3-Rhamnosyl(1 → 2)[glucosyl(1 → 3)glucoside]*
121. 3-Rhamnosyl(1 → 2)[glucosyl(1 → 4)glucoside]*
122. 3-Rhamnosyl(1 → 6)[glucosyl(1 → 2)glucoside]
123. 3-Glucosyl(1 → 2)[glucosyl(1 → 3)rhamnoside]*
124. 3-Glucosyl(1 → 2)[rhamnosyl(1 → 6)galactoside]
125. 3-Galactosyl(1 → 2)[rhamnosyl(1 → 6)glucoside]
126. 3-(2^G-Rhamnosylrutinoside)
127. 3-(2^G-Rhamnosylgentiobioside)
128. 3-(3^R-Glucosylrutinoside)
129. 3-(2^G-Glucosylrutinoside)
130. 3-(2′-Rhamnosyllaminaribioside)
131. 3-(2^G-Glucosylgentiobioside)
132. 3-Apiosyl(1 → 2)[rhamnosyl(1 → 6)galactoside]
133. 7-(3^G-Glucosylgentiobioside)*
134. 4′-Rhamnosyl(1 → 2)[rhamnosyl(1 → 6)galactoside]
135. 3-Rhamnosylarabinoside-7-rhamnoside
136. 3-Rhamnosyl(1 → 2)galactoside-7-α-L-arabinofuranoside*
137. 3-Robinobioside-7-α-L-arabinofuranoside*
138. 3-Glucosylxyloside-7-xyloside
139. 3-β-D-Apiofuranosyl(1 → 2)-α-L-arabinofuranosyl-7-α-L-rhamnoside
140. 3-Xylosyl(1 → 2)rhamnoside-7-rhamnoside (sagittatin A)
141. 3-Xylosyl(1 → 4)rhamnoside-7-rhamnoside*
142. 3-Rhamnoside-7-xylosyl(1 → 2)rhamnoside*
143. 3-Apiosyl(1 → 4)rhamnoside-7-rhamnoside*
144. 3-Rhamnosyl(1 → 4)rhamnoside-7-rhamnoside
145. 3-Rhamnosylxyloside-7-glucoside
146. 3-Neohesperidoside-7-rhamnoside*
147. 3-Glucosylrhamnoside-7-rhamnoside
148. 3-Rutinoside-7-rhamnoside
149. 3-Rhamnoside-7-rhamnosylglucoside
150. 3-Rhamnoside-7-glucosyl(1 → 2)rhamnoside*
151. 3-Glucosyl(1 → 3)rhamnoside-7-rhamnoside
152. 3-Rhamnosyl(1 → 2)galactoside-7-rhamnoside
153. 3-Robinobioside-7-rhamnoside (robinin)

154. 3-Rhamnosyl(1 → 2)galactoside-7-glucoside
155. 3-Sophoroside-7-α-ʟ-arabinoside
156. 3-Glucosyl(1 → 4)galactoside-7-α-ʟ-arabinofuranoside*
157. 3-Glucosyl(1 → 6)galactoside-7-α-ʟ-arabinofuranoside*
158. 3-Apioside-7-rhamnosyl(1 → 6)galactoside*
159. 3-Robinobioside-7-glucoside
160. 3-Lathyroside-7-rhamnoside
161. 3-Sambubioside-7-glucoside
162. 3-Rutinoside-7-glucoside
163. 3-Neohesperidoside-7-glucoside
164. 3-Glucosyl(1 → 2)rhamnoside-7-glucoside*
165. 3-Rutinoside-7-galactoside
166. 3-Rutinoside-7-glucuronide
167. 3-Sophoroside-7-rhamnoside
168. 3-Laminaribioside-7-rhamnoside
169. 3-Gentiobioside-7-rhamnoside*
170. 3-Sophoroside-7-glucoside
171. 3-Glucoside-7-sophoroside
172. 3-Gentiobioside-7-glucoside
173. 3-Glucoside-7-gentiobioside
174. 3-Glucosyl(1 → 2)galactoside-7-glucoside*
175. 3-Sophoroside-7-glucuronide*
176. 3-Gentiobioside-7-glucuronide
177. 3-Rutinoside-4'-glucoside
178. 3-Neohesperidoside-4'-glucoside*
179. 3-Neohesperidoside-7,4'-diglucoside*
180. 3-Sophoroside-4'-glucoside
181. 3-Gentiobioside-4'-glucoside*
182. 3-Glucoside-7,4'-dirhamnoside
183. 3-Galactoside-3,4'-dirhamnoside*
184. 3-Rhamnoside-7,4'-digalactoside*
185. 4'-Rhamnosyl(1 → 3)rhamnosyl(1 → 6)galactoside*
186. 3,7,4'-Triglucoside
187. 3-Rhamnosyl(1 → 2)[xylosyl(1 → 3)rhamnosyl(1 → 6)galactoside]*
188. 3-Glucosyl(1 → 3)rhamnosyl(1 → 2)[rhamnosyl(1 → 6)galactoside]*
189. 3-Xylosylrutinoside-7-glucoside
190. 3-Rhamnosyl(1 → 4)rhamnosyl(1 → 6)galactoside-7-rhamnoside*
191. 3-Sophorotrioside-7-rhamnoside
192. 3-Rutinoside-7-sophoroside*
193. 3-(2^G-Glucosylrutinoside)-7-rhamnoside*
194. 3-(2^G-Rhamnosylrutinoside)-7-glucoside* (mauritianin 7-glucoside)
195. 3-Galactosyl(1 → 6)glucoside-7-dirhamnoside (malvitin)
196. 3-Sophorotrioside-7-glucoside
197. 3-Sophoroside-7-cellobioside*
198. 3-Rhamnosyl(1 → 6)[glucosyl(1 → 2)glucoside]-7-glucoside
199. 3-(2^G-Glucosylrutinoside)-7-glucoside
200. 3-Rhamnosyl(1 → 6)[rhamnosyl(1 → 2)galactoside]-7-rhamnoside* (astrasikokioside)
201. 3-Glucosyl(1 → 2)[rhamnosyl(1 → 6)galactoside]-7-rhamnoside*
202. 3-Rutinoside-4'-diglucoside
203. 3-Gentiobioside-7,4'-bisglucoside

204. 3-*Z*/*E*-*p*-Coumarate
205. 3-[6″-(3-Hydroxy-3-methylglutaryl)glucoside]*
206. 3-(*p*-Hydroxybenzoylglucoside)
207. 3-(6″-*p*-Hydroxybenzoylgalactoside)*
208. 3-Benzoylglucoside
209. 3-(6″-Succinylglucoside)
210. 3-(6″-Malonylglucoside)
211. 3-(6″-Malonylgalactoside)
212. 3-(2″-Galloylarabinoside)*
213. 3-(2″-Galloylglucoside)
214. 3-(6″-Galloylglucoside)
215. 3-(6″-Galloylgalactoside)*
216. 3-(2″,6″-Digalloylglucoside)*
217. 3-(6″-*Z*-Cinnamoylglucoside)
218. 3-(2″-*E*-*p*-Coumaroyl-α-L-arabinofuranoside)*
219. 3-(2″-*E*-*p*-Coumaroylrhamnoside)*
220. 3-(2″-*Z*-*p*-Coumaroylrhamnoside)*
221. 3-(X″-*p*-Coumaroylglucoside) (tiliroside)
222. 3-(2″-*Z*-*p*-Coumaroylglucoside)
223. 3-(3″-*p*-Coumaroylglucoside)
224. 3-(4″-*p*-Coumaroylglucoside)*
225. 3-(6″-*p*-Coumaroylglucoside) (tribuloside)
226. 3-(6″-*p*-Coumaroylgalactoside)
227. 3-(6″-Caffeoylglucoside)*
228. 3-(5″-Feruloylapioside)*
229. 3-(6″-Feruloylglucoside)*
230. 3-(2″-Acetylrhamnoside)
231. 3-(3″-Acetylrhamnoside)
232. 3-(4″-Acetylrhamnoside)
233. 3-(6″-Acetylglucoside)*
234. 3-(3″,4″-Diacetylglucoside)*
235. 7-(6″-Succinylglucoside)
236. 7-Galloylglucoside
237. 7-(6″-*p*-Coumaroylglucoside)*
238. 3-(2″,3″-Di-*E*-*p*-coumaroylrhamnoside)*
239. 3-(2″,4″-Di-*E*-*p*-coumaroylrhamnoside)*
240. 3-(2″,4″-Di-*Z*-*p*-coumaroylrhamnoside)*
241. 3-(2″,4″-Di-*p*-coumaroylglucoside)
242. 3-(2″,6″-Di-*E*-*p*-coumaroylglucoside)
243. 3-(2″,6″-Di-*Z*-*p*-coumaroylglucoside)*
244. 3-(3″,6″-Di-*Z*-*p*-coumaroylglucoside) (stenopalustroside A)*
245. 3-(6″-*p*-Coumaroylacetylglucoside)
246. 3-(3″-*Z*-*p*-Coumaroyl-6″-feruloylglucoside) (stenopalustroside B)*
247. 3-(4″-Acetyl-6″-*p*-coumaroylglucoside)
248. 3-(3″-*Z*-*p*-Coumaroyl-6″-*E*-*p*-coumaroylglucoside) (stenopalustroside C)*
249. 3-(3″-*E*-*p*-Coumaroyl-6″-*Z*-*p*-coumaroylglucoside) (stenopalustroside D)*
250. 3-(3″-*E*-*p*-Coumaroyl-[6″-(4-*O*-{4-hydroxy-3-methoxyphenyl)-1,3-dihydroxyisopropyl-feruloyl}]glucoside (stenopalustroside E)*
251. 3-(2″-*E*-*p*-Coumaroyl-6″-acetylglucoside)*
252. 3-(3″-Acetyl-6″-*p*-coumaroylglucoside)*

253. 3-(3″,4″-Diacetylglucoside)*
254. 3-(2″,3″-Diacetyl-4′-*p*-coumaroylrhamnoside)
255. 3-(2″,3″-Diacetyl-4″-*Z*-*p*-coumaroyl)-6″-(*E*-*p*-coumaroylglucoside)
256. 3-(3″,4″-Diacetyl-2″,6″-di-*E*-*p*-coumaroylglucoside)
257. 3-Apiosylmalonylglucoside
258. 3-(6^G-Malonylneohesperidoside)*
259. 3-(2″-*E*-Feruloylgalactosyl)(1 → 4)glucoside*
260. 3-(2″-*E*-Feruloylgalactosyl)(1 → 6)glucoside*
261. 3-(2^G-*E*-*p*-Coumaroylrutinoside)*
262. 3-(4″-*E*-*p*-Coumaroylrobinobioside)
263. 3-(6‴-*p*-Coumaroylglucosyl)(1 → 2)rhamnoside
264. 3-(4″-*Z*-*p*-Coumaroylrobinobioside)
265. 3-(6″-Caffeylglucosyl)(1 → 4)rhamnoside*
266. 3-(Feruloylsophoroside) (petunoside)
267. 3-(6″-*E*-Feruloylglucosyl)(1 → 2)galactoside*
268. 3-(6‴-Sinapoylglucosyl)(1 → 2)galactoside*
269. 3-(3‴-Acetyl-α-L-arabinopyranosyl)(1 → 6)glucoside*
270. 3-(2‴-Acetylarabinosyl)(1 → 6)galactoside
271. 3-(6‴-Acetylglucosyl)(1 → 3)galactoside*
272. 3-[2‴-Feruloylglucosyl(1 → 2)6″-malonylglucoside]*
273. 3-Benzoylglucoside-7-glucoside
274. 3-*p*-Hydroxybenzoylglucoside-7-glucoside
275. 3-[6″-(3-Hydroxy-3-methylglutaryl)glucoside]-7-glucoside*
276. 3-(6″-Malonylglucoside)-7-glucoside*
277. 3-(2″-*E*-*p*-Coumaroyl-α-L-arabinofuranoside)-7-rhamnoside*
278. 3-(3″-*p*-Coumaroylrhamnoside)-7-rhamnoside*
279. 3-(6″-*E*-*p*-Coumaroylglucoside)-7-glucoside*
280. 3-Glucoside-7-(*p*-coumaroylglucoside)
281. 3-Caffeoylsophoroside
282. 3-(6‴-Caffeoylglucosyl)(1 → 2)galactoside
283. 3-(2″-Caffeoylglucoside)-7-rhamnoside
284. 3-(Caffeoylglucoside)-7-glucoside
285. 3-Feruloylglucoside-7-glucoside
286. 3-(3″-Acetylarabinofuranoside)-7-glucoside
287. 3-(4″-Acetylrhamnoside)-7-rhamnoside (sutchuenoside A)
288. 3-(6″-Acetylglucoside)-7-rhamnoside
289. 3-(6″-Acetylgalactoside)-7-rhamnoside
290. 3-(6″-Acetylglucoside)-7-glucoside
291. 3-(2″,3″-Diacetylrhamnoside)-7-rhamnoside*
292. 3-(2″,4″-Diacetylrhamnoside)-7-rhamnoside*
293. 3-(3″,4″-Diacetylrhamnoside)-7-rhamnoside*
294. 3-(4″,6″-Diacetylrhamnoside)-7-rhamnoside*
295. 3-(2‴,3‴,4‴-Triacetyl-α-L-arabinopyranosyl)(1 → 6)glucoside*
296. 3-(2‴,3‴,5‴-Triacetylarabinofuranosyl)(1 → 6)glucoside
297. 3-[6‴-(7⁗-Glucosyl-*p*-coumaroyl)glucosyl](1 → 2)rhamnoside*
298. 3-(*p*-Coumaroylsophoroside)
299. 3-(*p*-Coumarylglucoside)-4′-glucoside
300. 3-(2-Hydroxypropionylglucoside)-4′-glucoside
301. 3-[2^{Gal}-(6″-Feruloylglucosyl)robinobioside]*
302. 3-(*p*-Coumaroylsophorotrioside)

303. 3-(Feruloylsophorotrioside)
304. 3-Glucosyl(1 → 4)[(6‴-sinapoylglucosyl)(1 → 2)galactoside]*
305. 3-(2‴-Sinapoylglucosyl)(1 → 4)[(6‴-sinapoylglucosyl)(1 → 2)galactoside]*
306. 3-Rhamnosyl(1 → 4)(3‴-acetylrhamnosyl)(1 → 6)galactoside
307. 3-Rhamnosyl(1 → 3)(4‴-acetylrhamnosyl)(1 → 6)glucoside*
308. 3-Rhamnosyl(1 → 3)(2‴-acetylrhamnosyl)(1 → 6)galactoside*
309. 3-Glucosyl(1 → 3)(4‴-acetylrhamnosyl)(1 → 6)galactoside
310. 3-[2-^{Gal}-(6‴-Feruloylglucosyl)robinobioside]*
311. 3-[2^{Gal}-(4″-Acetylrhamnosyl)robinobioside]*
312. 3-[2″-(4‴-Acetylrhamnosyl)sophoroside]*
313. 3-Neohesperidoside-7-(6″-malonylglucoside)*
314. 3-(4″-*E*-*p*-Coumaroylrobinobioside)-7-rhamnoside
315. 3-(4″-*Z*-*p*-Coumaroylrobinobioside)-7-rhamnoside (variabiloside D)
316. 3-(*p*-Coumaroylrutinoside)-7-glucoside
317. 3-Neohesperidoside-7-(2″-*E*-*p*-coumaroylglucoside)*
318. 3-(4″-*p*-Coumaroylglucosyl)(1 → 2)rhamnoside-7-glucoside*
319. 3-(6″-*p*-Coumaroylglucosyl)(1 → 2)rhamnoside-7-glucoside*
320. 3-Glucosyl(1 → 2)rhamnoside-7-(6″-*E*-*p*-coumaroylglucoside)*
321. 3-(6‴-*E*-*p*-Coumaroylglucosyl)(1 → 2)glucoside-7-rhamnoside*
322. 3-Glucoside-7-(6‴-*E*-*p*-coumaroylglucosyl)(1 → 3)rhamnoside*
323. 3-(2‴-*E*-*p*-Coumaroylsophoroside)-7-glucoside*
324. 3-Apioside-7-rhamnosyl(1 → 6)(2″-*E*-caffeoylgalactoside)*
325. 3-(Caffeoylrobinobioside)-7-rhamnoside
326. 3-(6″-*E*-Caffeoylglucosyl)(1 → 2)glucoside-7-rhamnoside*
327. 3-Glucoside-7-(6″-*E*-caffeoylglucosyl)(1 → 3)rhamnoside*
328. 3-(2″-Caffeoyllaminaribioside)-7-rhamnoside
329. 3-(4″-Caffeoyllaminaribioside)-7-rhamnoside
330. 3-(2‴-*E*-*p*-Coumaroylsophoroside)-7-glucoside*
331. 3-(2‴-*E*-Caffeoylsophoroside)-7-glucoside*
332. 3-(Feruloylrobinobioside)-7-rhamnoside
333. 3-Neohesperidoside-7-(2″-E-feruloylglucoside)*
334. 3-(2‴-*E*-Feruloylsophoroside)-7-glucoside*
335. 3-Sophoroside-7-(2″-feruloylglucoside)
336. 3-(Sinapoylsophoroside)-7-glucoside
337. 3-Xylosyl(1 → 3)(4″-acetylrhamnoside)-7-rhamnoside
338. 3-Xylosyl(1 → 2)rhamnoside-7-(4⁗-acetylrhamnoside)*
339. 3-Neohesperidoside-7-(6″-acetylglucoside)*
340. 3-Glucosyl(1 → 2)(6″-acetylgalactoside)-7-glucoside*
341. 3,4′-Diglucoside-7-(2″-feruloylglucoside)
342. 3-(*p*-Coumaroylglucoside)-7,4′-diglucoside
343. 3-(2″-Feruloylglucoside)-7,4′-diglucoside
344. 3-(Sinapoylglucoside)-7-sophoroside
345. 3-(*p*-Coumaroylferuloyldiglucoside)-7-rhamnoside
346. 3-Rhamnosyl(1 → 2)[glucosyl(1 → 3)(4‴-*p*-coumaroylrhamnosyl)(1 → 6)galactoside]*
347. 3-Rhamnosyl(1 → 2)[xylosyl(1 → 3)rhamnosyl(1 → 6)(3″-*p*-coumaroylgalactoside)]*
348. 3-Rhamnosyl(1 → 2)[xylosyl(1 → 3)rhamnosyl(1 → 6)(4″-*p*-coumaroylgalactoside)]*
349. 3-Rhamnosyl(1 → 2)[xylosyl(1 → 3)rhamnosyl(1 → 6)(3″-feruloylgalactoside)]*
350. 3-Rhamnosyl(1 → 2)[xylosyl(1 → 3)rhamnosyl(1 → 6)(4″-feruloylgalactoside)]*
351. 3-Gentiobioside-7-(caffeoylarabinosylrhamnoside)
352. 3-Neohesperidoside-7-[2″-*E*-*p*-coumaroyllaminaribioside]*

353. 3-(2‴-*E*-Caffeoylglucosyl)(1 → 2)glucoside-7-cellobioside*
354. 3-(2‴-*E*-Feruloylglucosyl)(1 → 2)glucoside-7-cellobioside*
355. 3-(2‴-*E*-Sinapoylglucosyl)(1 → 2)glucoside-7-cellobioside*
356. 3-Glucosyl(1 → 6)[rhamnosyl(1 → 3)(2″-*E*-*p*-coumaroylglucoside)]-7-rhamnosyl(1 → 3)rhamnosyl(1 → 3)(4″-*E*-*p*-coumaroylrhamnoside)*
357. 3-Glucosyl(1 → 6)[rhamnosyl(1 → 3)(2″-*E*-*p*-coumaroylglucoside)]-7-rhamnosyl(1 → 3)rhamnosyl(1 → 3)(4″-*Z*-*p*-coumaroylrhamnoside)*
358. 3-Rhamnosyl(1 → 6)[rhamnosyl(1 → 3)(2″-*E*-*p*-coumaroylglucoside)]-7-rhamnosyl(1 → 3)rhamnosyl(1 → 3)(4″-*E*-*p*-coumaroylrhamnoside)*
359. 3-Sulfatorhamnoside
360. 3-α-(6″-Sulfatoglucoside)
361. 3-β-(3″-Sulfatoglucoside)
362. 3-β-(6″-Sulfatoglucoside)
363. 3-Glucuronide-7-sulfate
364. 3-Sulfatorutinoside
365. 3-(6″-Sulfatogentiobioside)
366. 3-Sulfate-7-α-arabinopyranoside*
367. 3-Sulfate
368. 7-Sulfate
369. 8-Sulfate*
370. 3,7-Disulfate
371. 3,7,4′-Trisulfate

Kaempferol 3-methyl ether
372. 7-Rhamnoside
373. 7-Glucoside
374. 7-Glucuronide*
375. 7-Rutinoside

Kaempferol 5-methyl ether
376. 3-Galactoside

Kaempferol 7-methyl ether (rhamnocitrin)
377. 3-Rhamnoside
378. 3-Alloside
379. 3-Glucoside
380. 3-Galactoside
381. 3-Glucuronide
382. 5-Glucoside
383. 4′-Glucoside*
384. 3-Rutinoside
385. 3-Neohesperidoside
386. 3-Galactoside-4′-glucoside
387. 3-Rhamnosyl(1 → 3)rhamnosyl(1 → 6)galactoside (3-rhamninoside, alaternin, catharticin)
388. 3-Rhamnosyl(1 → 4)rhamnosyl(1 → 6)galactoside (3-isorhamninoside)
389. 3-Xylosyl(1 → 2)[rhamnosyl(1 → 6)glucoside]*
390. 3-Rhamnosyl(1 → 3)[apiosyl(1 → 6)glucoside]*
391. 3-Apiosyl(1 → 5)apioside-4′-glucoside*
392. 3-Neohesperidoside-4′-glucoside*
393. 3-Apiosyl(1 → 5)apiosyl(1 → 2)[rhamnosyl(1 → 6)glucoside]*
394. 3-[3-Hydroxy-3-methylglutaryl(1 → 6)][apiosyl(1 → 2)galactoside]*
395. 3-(5‴-*p*-Coumaroylapiosyl)(1 → 2)glucoside*

396. 3-(5‴-Feruloyllapiosyl)(1 → 2)glucoside*
397. 3-(6″-*E*-Sinapoylglucosyl)(1 → 2)rutinoside*
398. 3-Glucoside-4′-(2″-dihydrophaseoylglucoside)
399. 3-Glucoside-4′-(3‴-dihydrophaseoylglucoside)*
400. 3-(6-*E*-3,5-Dimethoxy-4-hydroxycinnamoylglucosyl)(1 → 2)[rhamnosyl(1 → 6)glucoside]*
401. 3-Sulfate

Kaempferol 4′-methyl ether (kaempferide)
402. 3-Rhamnoside*
403. 3-Galactoside
404. 3-Glucuronide
405. 3-Neohesperidoside*
406. 3-Diglucoside
407. 3-Rhamnoside-7-xyloside*
408. 3,7-Dirhamnoside
409. 3-Rhamnoside-7-glucoside
410. 3-Glucoside-7-rhamnoside
411. 3,7-Diglucoside
412. 3-(4^{Rha}-Rhamnosylrutinoside)*
413. 3-(2^{Glc}-Glucosylrutinoside)*
414. 3-[6‴-Acetyl(4″-α-methylsinapoylneohesperidoside)]*
415. 3-Rhamnoside-7-(6″-succinylglucoside)
416. 3-Sulfate

Kaempferol 3,5-dimethyl ether
417. 7-Glucoside*

Kaempferol 3,7-dimethyl ether
418. 4′-Glucoside*

Kaempferol 3,4′-dimethyl ether
419. 7-Glucoside

Kaempferol 7,4′-dimethyl ether
420. 3-Glucoside*
421. 3-Neohesperidoside*
422. 3-(6″-*p*-Coumaroylglucoside)
423. 3-Sulfate

6-*C*-Methylkaempferol
424. 3-Glucoside

6-Hydroxykaempferol
425. 3-Glucoside*
426. 7-Glucoside
427. 7-Alloside*
428. 3-Rutinoside*
429. 7-Rutinoside
430. 3,6-Diglucoside*
431. 3-Rutinoside-6-glucoside*
432. 3,6,7-Triglucoside*
433. 7′-(6″-Caffeoylglucoside)*
434. 7-Acetylglucoside

6-Hydroxykaempferol 3-methyl ether
435. 6-Glucoside
436. 7-Glucoside

437. 7-Sulfate
6-Hydroxykaempferol 5-methyl ether
  438. 4'-Rhamnoside
  439. 3-Arabinosylrhamnoside
6-Hydroxykaempferol 6-methyl ether (eupafolin)
  440. 3-Rhamnoside
  441. 3-Glucoside
  442. 3-Galactoside
  443. 3-Glucuronide
  444. 7-Glucoside
  445. 4'-Rhamnoside
  446. 3-Rutinoside
  447. 3-Robinobioside
  448. 3,7-Dirhamnoside
  449. 3-(6''-*p*-Coumaroylglucoside)*
  450. 3-Rhamnoside-7-(4'''-acetylrhamnoside)
  451. 3-(3''-Acetylrhamnoside)-7-(3''-acetylrhamnoside)
  452. 3-(6''-Acetylglucoside)
  453. 3-Sulfate
6-Hydroxykaempferol 7-methyl ether
  454. 6-Rhamnosyl(1 → 4)xyloside
6-Hydroxykaempferol 4'-methyl ether
  455. 7-Glucoside*
  456. 7-Galactoside*
  457. 3,7-Dirhamnoside
6-Hydroxykaempferol 3,6-dimethyl ether
  458. 7-Glucoside
6-Hydroxykaempferol 6,7-dimethyl ether (eupalitin)
  459. 3-Rhamnoside (eupalin)
  460. 3-Galactoside
  461. 5-Rhamnoside
  462. 3-Galactosylrhamnoside
  463. 3-Diglucoside
  464. 3-Glucosylgalactoside
  465. 3-Sulfate
6-Hydroxykaempferol 6,4'-dimethyl ether
  466. 3-Glucoside*
  467. 3-Galactoside
6-Hydroxykaempferol 7,4'-dimethyl ether
  468. 3-Glucoside
  469. 3-Sulfate
6-Hydroxykaempferol 3,6,7-trimethyl ether
  470. 4'-Glucoside
6-Hydroxykaempferol 6,7,4'-trimethyl ether (mikanin)
  471. 3-Glucoside
  472. 3-Galactoside
6-Hydroxykaempferol 3,5,7,4'-tetramethyl ether
  473. 6-Rhamnoside*
8-Hydroxykaempferol (herbacetin)

474. 3-Glucoside
475. 3-β-D-Glucofuranoside*
476. 7-Arabinoside
477. 7-Rhamnoside
478. 7-Glucoside
479. 8-α-L-Arabinopyranoside
480. 8-Xyloside
481. 8-Rhamnoside
482. 8-Glucoside
483. 4′-Glucoside
484. 3-Rhamnoside-8-glucoside*
485. 3-Glucuronide-8-glucoside
486. 7-Glucosyl(1 → 3)rhamnoside
487. 8-Rutinoside
488. 8-Gentiobioside
489. 3-Glucoside-8-xyloside
490. 7-Rhamnoside-8-glucoside
491. 8,4′-Dixyloside
492. 8-Arabinoside-4′-xyloside
493. 3-Sophoroside-8-glucoside
494. 7-(6″-Quinylglucoside)
495. 8-(3″-Acetyl-α-L-arabinopyranoside)
496. 8-(3″-Acetylxyloside)
497. 8-(2″,3″-Diacetylxyloside)
498. 8-(Diacetylglucoside)
499. 8-(2″,3″,4″-Triacetylxyloside)
500. 8-Acetate
501. 8-Butyrate

Herbacetin 7-methyl ether
502. 8-Sophoroside*
503. 3-(2″-*E*-Feruloylglucoside)*
504. 8-Acetate
505. 8-Butyrate

Herbacetin 8-methyl ether (sexangularetin)
506. 3-Glucoside
507. 3-Galactoside
508. 3-Rutinoside
509. 3-Neohesperidoside*
510. 3-Sophoroside*
511. 3-Glucoside-7-rhamnoside
512. 3,7-Diglucoside
513. 3-Rhamnosylglucoside-7-rhamnoside
514. 3-Rutinoside-7-glucoside
515. 3-Glucoside-7-rutinoside

Herbacetin 7,8-dimethyl ether
516. 3-Rhamnoside

Herbacetin 7,4′-dimethyl ether
517. 8-Acetate
518. 8-Butyrate

Herbacetin 7,8,4′-trimethyl ether (tambulin)
519. 3,5-Diacetate*

5,7,8-Trihydroxy-3-methoxyflavone
  520. 8-(*E*-2-Methylbut-2-enoate)*
5,7,8-Trihydroxy-3,6-dimethoxyflavone
  521. 8-(*E*-2-Methylbut-2-enoate)*
6,8-Dihydroxykaempferol
  522. 3-Rutinoside*
3,5,7,3',4'-Pentahydroxyflavone (quercetin)
  523. 3-α-L-Arabinofuranoside (avicularin)
  524. 3-α-L-Arabinopyranoside (guaijaverin, foeniculin)
  525. 3-α-D-Arabinopyranoside*
  526. 3-β-L-Arabinoside (polystachioside)
  527. 3-Xyloside (reynoutrin)
  528. 3-Rhamnoside (quercitrin)
  529. 3-Glucoside (isoquercitrin)
  530. 3-Galactoside (hyperin)
  531. 3-Alloside
  532. 3-Glucuronide (miquelianin)
  533. 3-Galacturonide
  534. 3-(6″-Methylglucuronide)
  535. 3-(6″-Ethylglucuronide)
  536. 5-Glucoside
  537. 5-Glucuronide*
  538. 7-Arabinoside
  539. 7-Xyloside
  540. 7-Rhamnoside
  541. 7-Glucoside (quercimeritrin)
  542. 7-α-Galactoside
  543. 3'-Xyloside
  544. 3'-Glucoside
  545. 4'-Glucoside (spiraeoside)
  546. 4'-Galactoside*
  547. 4'-Glucuronide*
  548. 3-Diarabinoside
  549. 3-Arabinosylxyloside
  550. 3-Rhamnosyl(1 → 2)-α-L-arabinofuranoside (arapetaloside A)*
  551. 3-Rhamnosyl(1 → 2)arabinoside
  552. 3-α-L-Arabinofuranosyl(1 → 2)glucoside*
  553. 3-Arabinosyl(1 → 6)glucoside (vicianoside)
  554. 3-Arabinosyl(1 → 6)galactoside
  555. 3-Galactosylarabinoside
  556. 3-Dixyloside
  557. 3-Xylosyl(1 → 2)rhamnoside
  558. 3-Xylosyl(1 → 2)glucoside (3-sambubioside)
  559. 3-Xylosyl(1 → 6)glucoside*
  560. 3-Glucosyl(1 → 2)xyloside
  561. 3-Xylosyl(1 → 2)galactoside
  562. 3-Apiofuranosyl(1 → 2)arabinoside
  563. 3-Apiofuranosyl(1 → 2)xyloside
  564. 3-Rhamnosylxyloside
  565. 3-Rhamnosyl(1 → 2)rhamnoside*

566. 3-Rhamnosyl(1 → 2)galactoside
567. 3-Apiofuranosyl(1 → 2)glucoside
568. 3-Apiofuranosyl(1 → 2)galactoside
569. 3-Rutinoside (rutin)
570. 3-Neohesperidoside
571. 3-Glucosyl(1 → 2)rhamnoside*
572. 3-Glucosyl(1 → 4)rhamnoside
573. 3-Galactosyl(1 → 2)rhamnoside*
574. 3-Galactosyl(1 → 4)rhamnoside
575. 3-Rhamnosyl(1 → 6)galactoside
576. 3-Laminaribioside*
577. 3-Sophoroside
578. 3-Gentiobioside
579. 3-Galactosylglucoside
580. 3-Glucosyl(1 → 2)galactoside
581. 3-Glucosyl(1 → 3)galactoside*
582. 3-Glucosyl(1 → 4)galactoside*
583. 3-Glucosyl(1 → 6)galactoside
584. 3-Digalactoside
585. 3-Glucosylmannoside
586. 3-Glucosyl(1 → 2)glucuronide*
587. 3-Galactosylglucuronide
588. 7-Rutinoside
589. 7-Glucosylrhamnoside
590. 3,5-Digalactoside
591. 3-Arabinoside-7-glucoside
592. 3-Glucoside-7-arabinoside
593. 3-Xyloside-7-glucoside
594. 3-Galactoside-7-xyloside
595. 3,7-Dirhamnoside
596. 3-Rhamnoside-7-glucoside
597. 3-Glucoside-7-rhamnoside
598. 3-Galactoside-7-rhamnoside
599. 3,7-Diglucoside
600. 3-Galactoside-7-glucoside
601. 3-Glucoside-7-glucuronide
602. 3-Glucuronide-7-glucoside
603. 3,7-Diglucuronide
604. 3,3′-Diglucoside
605. 3-Rhamnoside-3′-glucoside*
606. 3,4′-Diglucoside
607. 7,4′-Diglucoside
608. 3-Xylosyl(1 → 2)rhamnosyl(1 → 6)glucoside*
609. 3-Xylosyl(1 → 6)glucosyl(1 → 2)rhamnoside*
610. 3-Rhamnosyl(1 → 2)rutinoside
611. 3-Rhamnosyl(1 → 4)rhamnosyl(1 → 6)glucoside
612. 3-Rhamnosyl(1 → 4)rhamnosyl(1 → 6)galactoside
613. 3-Apiosyl(1 → 2)rhamnosyl(1 → 6)glucoside*
614. 3-(6‴-Rhamnosylgentiobioside)*
615. 3-Rhamnosyl(1 → 2)glucosyl(1 → 6)galactoside*

616. 3-Rhamnosyl(1 → 6)glucosyl(1 → 6)galactoside*
617. 3-Glucosyl(1 → 4)galactosylrhamnoside
618. 3-Glucosyl(1 → 3)rhamnosyl(1 → 6)galactoside
619. 3-Sophorotrioside
620. 3-Glucosyl(1 → 2)galactosyl(1 → 2)glucoside*
621. 3-(2^G-Rhamnosylrutinoside)
622. 3-(2^G-Apiosylrutinoside)
623. 3-Rhamnosyl(1 → 2)[rhamnosyl(1 → 6)galactoside]
624. 3-Xylosyl(1 → 2)[rhamnosyl(1 → 6)glucoside]
625. 3-(2^G-Glucosylrutinoside)
626. 3-(3^R-Glucosylrutinoside)
627. 3-(2^G-Rhamnosylgentiobioside)
628. 3-Rhamnosyl(1 → 2)[glucosyl(1 → 6)galactoside]
629. 3-Glucosyl(1 → 2)[rhamnosyl(1 → 6)galactoside]*
630. 3-(2^G-Glucosylgentiobioside)
631. 7-(2^G-Xylosylrutinoside)*
632. 3-Rhamnosyl(1 → 2)-α-ʟ-arabinopyranoside-7-glucoside*
633. 3-Dixyloside-7-glucoside
634. 3-Xylosyl(1 → 2)glucoside-7-rhamnoside*
635. 3-Xyloside-7-xylosylglucoside
636. 3-Xylosyl(1 → 2)glucoside-7-glucoside (3-sambubioside-7-glucoside)
637. 3-Rutinoside-7-rhamnoside
638. 3-Neohesperidoside-7-rhamnoside*
639. 3-Glucosyl(1 → 2)rhamnoside-7-rhamnoside
640. 3-Rhamnosyl(1 → 4)rhamnoside-7-galactoside*
641. 3-Robinobioside-7-rhamnoside
642. 3-Rutinoside-7-glucoside
643. 3-Neohesperidoside-7-glucoside*
644. 3-Glucosyl(1 → 2)rhamnoside-7-glucoside*
645. 3-Glucoside-7-rutinoside
646. 3-Rhamnosyl(1 → 2)galactoside-7-glucoside
647. 3-Galactosyl-7-glucosyl(1 → 4)rhamnoside*
648. 3-Glucoside-7-neohesperidoside
649. 3-Rutinoside-7-galactoside
650. 3-Galactoside-7-neohesperidoside
651. 3-Robinobioside-7-glucoside
652. 3-Rutinoside-7-glucuronide
653. 3-Gentiobioside-7-glucoside
654. 3-Sophoroside-7-glucoside
655. 3-Glucosyl(1 → 2)galactoside-7-glucoside*
656. 3-Sophoroside-7-glucuronide*
657. 3-Gentiobioside-7-glucuronide
658. 3-Sambubioside-3′-glucoside
659. 3-Rutinoside-3′-apioside*
660. 3-Xylosyl(1 → 2)rhamnoside-4′-rhamnoside
661. 3-Rutinoside-4′-glucoside
662. 3,7,4′-Triglucoside
663. 3,3′,4′-Triglucoside*
664. 3-Xylosyl(1 → 4)[xylosyl(1 → 6)glucosyl(1 → 2)rhamnoside]*
665. 3-Xylosyl(1 → 3)rhamnosyl(1 → 6)[apiosyl(1 → 2)galactoside]*

666. 3-Rhamnosylglucoside-7-xylosylglucoside
667. 3-(2^G-Rhamnosylrutinoside)-7-glucoside
668. 3-Glucosyl(1 → 4)rhamnoside-7-rutinoside*
669. 3-Rhamnosyl(1 → 6)[glucosyl(1 → 2)glucoside]-7-rhamnoside*
670. 3-Rhamnosyldiglucoside-7-glucoside
671. 7-[Xylosyl(1 → 2)rhamnosyl(1 → 2)rhamnosyl](1 → 6)glucoside*
672. 3-Rutinoside-7,3'-bisglucoside
673. 3-Rutinoside-4'-diglucoside
674. 3-Isobutyrate
675. 3'-Isobutyrate
676. 4'-Isobutyrate
677. 3-[6''-(3-Hydroxy-3-methylglutaryl)galactoside]
678. 3-(6''-*p*-Hydroxybenzoylgalactoside)
679. 3-(4''-Malonylrhamnoside)*
680. 3-(6''-Malonylglucoside)
681. 3-(6''-Malonylgalactoside)
682. 3-(2''-Galloyl-α-arabinopyranoside)
683. 3-(2''-Galloylrhamnoside)
684. 3-(2''-Galloylglucoside)
685. 3-(6''-Galloylglucoside)
686. 3-(2''-Galloylgalactoside)
687. 3-(6''-Galloylgalactoside)
688. 3-(2''-*p*-Coumarylglucoside)
689. 3-(3''-*p*-Coumarylglucoside)
690. 3-(6''-*p*-Coumarylglucoside) (helichrysoside)
691. 3-(6''-Caffeoylgalactoside)
692. 3-(2''-Caffeoylglucuronide)*
693. 3-(6''-Feruloylgalactoside)*
694. 3-Isoferuloylglucuronide
695. 3-(2''-Acetylrhamnoside)*
696. 3-(3''-Acetylrhamnoside)
697. 3-(4''-Acetylrhamnoside)
698. 3-(6''-Acetylglucoside)
699. 3-(2''-Acetylgalactoside)*
700. 3-(3''-Acetylgalactoside)
701. 3-(6''-Acetylgalactoside)
702. 3-(6''-*n*-Butylglucuronide) (parthenosin)*
703. 7-(6''-Galloylglucoside)*
704. 7-(6''-Acetylglucoside)*
705. 4'-(6''-Galloylglucoside)*
706. 3-Diacetylglucoside
707. 7-Acetyl-3'-glucoside
708. 3-α-(2''-*p*-Hydroxybenzoyl)-4'-*p*-coumarylrhamnoside
709. 3-(2'',6''-Digalloylgalactoside)*
710. 3-(3'',6''-Di-*p*-coumarylglucoside)
711. 3-(3'',6''-Diacetylgalactoside)*
712. 3-(2'',3'',4''-Triacetylgalactoside)*
713. 3-(2'''-Galloylglucosyl)(1 → 2)-α-L-arabinofuranoside*
714. 3-(2''-Galloylrutinoside)*
715. 3-(6^G-Malonylneohesperidoside)*

716. 3-(X″-Benzoyl-X″-xylosylglucoside)
717. 3-(X″ or X‴-Benzoyl-X″-glucosylglucoside)
718. 3-α-ʟ-Arabinopyranosyl(1 → 6)(2″-*E-p*-coumaroylglucoside)*
719. 3-α-ʟ-Arabinopyranosyl(1 → 6)(2″-*E-p*-coumaroylgalactoside)*
720. 3-(2^G-*E-p*-Coumaroylrutinoside)*
721. 3-(4″-*E-p*-Coumaroylrobinobioside)
722. 3-α-(6″-*p*-Coumaroylglucosyl)(1 → 4)rhamnoside
723. 3-(2‴-*E*-Caffeoyl-α-ʟ-arabinopyranosyl)(1 → 6)glucoside)*
724. 3-(2″-*E*-Caffeoyl-α-ʟ-arabinopyranosyl)(1 → 6)galactoside)*
725. 3-(6″-Caffeoylsophoroside)*
726. 3-(6″-Caffeoylgentiobioside)*
727. 3-(Acetylrutinoside)
728. 3-(2″-Feruloylsophoroside)*
729. 3-(6″-Feruloylsophoroside)*
730. 3-(6‴-Sinapoylglucosyl)(1 → 2)galactoside*
731. 3-Rhamnosyl(1 → 6)(2″-acetylglucoside)*
732. 3-(6″-Acetylglucosyl)(1 → 3)galactoside (euphorbianin)
733. 3-(2‴-Caffeoylglucoside)(1 → 2)(6″-malonylglucoside)*
734. 3-(3,″4″-Diacetylrhamnosyl)(1 → 6)glucoside*
735. 3-[2‴,3‴,4‴-Triacetyl-α-ʟ-arabinopyranosyl(1 → 6)glucoside]*
736. 3-[2‴,3‴,4‴-Triacetyl-α-ʟ-arabinopyranosyl(1 → 6)galactoside]*
737. 3-[2‴,3‴,5‴-Triacetyl-α-ʟ-arabinopyranosyl(1 → 6)glucoside]*
738. 3-(6″-Malonylglucoside)-7-glucoside*
739. 3-(6″-*E-p*-Coumaroylglucoside)-7-glucoside*
740. 3-Feruloylglucoside-7-glucoside
741. 3-(4″-Acetylrhamnoside)-7-rhamnoside
742. 3-(6″-Acetylgalactoside)-7-rhamnoside
743. 3-(2″-Galloylglucoside)-4′-vinylpropionate*
744. 3-Malonylglucoside-4′-glucoside
745. 3-Feruloylglucoside-4′-glucoside
746. 3-[2″,6″-{-*p*-(7‴-Glucosyl)coumaroyl}glucosyl]rhamnoside*
747. 3-(6‴-*p*-Coumaroylsophorotrioside)*
748. 3-(6⁗-Caffeoylsophorotrioside)*
749. 3-(6⁗-Feruloylsophorotrioside)*
750. 3-(6⁗-Feruloylglucosyl)(1 → 2)galactosyl(1 → 2)glucoside*
751. 3-(6⁗-Sinapoylsophorotrioside)*
752. 3-Glucosyl(1 → 3)(4‴-acetylrhamnosyl)(1 → 6)-galactoside
753. 3-(3‴-Benzoylsophoroside)-7-rhamnoside
754. 3-Rutinoside-7-(6″-benzoylglucoside)*
755. 3-(*p*-Coumaroylsambubioside)-7-glucoside*
756. 3-(4″-*E-p*-Coumaroylrobinobioside)-7-rhamnoside
757. 3-(6‴-*p*-Coumaroylglucosyl)(1 → 2)rhamnoside-7-glucoside*
758. 3-(4″-*E-p*-Coumaroylrobinobioside)-7-glucoside (variabiloside A)
759. 3-(4″-*Z-p*-Coumaroylrobinobioside)-7-glucoside (variabiloside B)
760. 3-(6″-*E*-p-Coumaroylsophoroside)-7-rhamnoside*
761. 3-(*p*-Coumaroylsophoroside)-7-glucoside*
762. 3-(6″-*p*-Coumaroylgentiobioside)-7-rhamnoside
763. 3-(Caffeoylarabinosylglucoside)-7-glucoside*
764. 3-(4‴-Caffeoylrhamnosy)(1 → 2)-α-ʟ-arabinopyranoside-7-glucoside*
765. 3-(2‴-Caffeoylsambubioside)-7-glucoside*

766. 3-Glucosyl-7-(6″-*E*-caffeoylglucosyl)(1 → 3)rhamnoside*
767. 3-(6″-*E*-Caffeoylsophoroside)-7-rhamnoside*
768. 3-(2″-*E*-Caffeoylsophoroside)-7-glucoside*
769. 3-Sophoroside-7-(6″-*E*-caffeoylglucoside)*
770. 3-(6″-Feruloylglucosyl)(1 → 2)-β-arabinopyranoside-7-glucoside*
771. 3-(Feruloylsambubioside)-7-glucoside*
772. 3-(2″-*E*-Feruloylsophoroside)-7-glucoside*
773. 3-(6″-*E*-Sinapoylsophoroside)-7-rhamnoside*
774. 3-(Caffeoylsophoroside)-7-(caffeoylglucoside)*
775. 3-(Caffeoylsophoroside)-7-(feruloylglucoside)*
776. 3-Acetylsophoroside-7-rhamnoside
777. 3-Rhamnoside-7-glucoside-4′-(caffeoylgalactoside)
778. 3-Ferulylglucoside-7,4′-diglucoside
779. 3-(2″-Sinapoylglucoside)-3′-(6″-sinapoylglucoside)-4′-glucoside*
780. 3,4′-Diglucoside-3′-(6″-sinapoylglucoside)*
781. 3-Rhamnosyl(1 → 6)[rhamnosyl(1 → 2)(3″-*E*-*p*-coumaroylgalactoside)]-7-rhamnoside*
782. 3-Rhamnosyl(1 → 6)[rhamnosyl(1 → 2)(4″-*E*-*p*-coumaroylgalactoside)]-7-rhamnoside*
783. 3-Rhamnosyl(1 → 2)[glucosyl(1 → 3)(4‴-*p*-coumaroylrhamnosyl)(1 → 6)galactoside]*
784. Diquercetin 3-galactoside ester of tetrahydroxy-μ-truxinic acid (**13.6**)*
785. 3-Sulfatorhamnoside
786. 3-(3″-Sulfatoglucoside)
787. 3-Sulfate-7-α-arabinopyranoside*
788. 3-Glucuronide-7-sulfate
789. 3-Rhamnoside-3′-sulfate*
790. 3-Glucoside-3′-sulfate*
791. 3-Acetyl-7,3′,4′-trisulfate
792. 3-Sulfate
793. 3′-Sulfate
794. 3,7-Disulfate
795. 3,3′-Disulfate
796. 7,4′-Disulfate*
797. 3′,4′-Disulfate
798. 3,7,3′-Trisulfate
799. 3,7,4′-Trisulfate
800. 3,7,3′,4′-Tetrasulfate
801. 3-Sulfatoglucoside
802. 3-Glucuronide-3′-sulfate

Quercetin 3-methyl ether
803. 5-Glucoside*
804. 7-α-L-Arabinofuranosyl(1 → 6)glucoside*
805. 7-Rhamnoside
806. 7-Glucoside
807. 3′-Xyloside
808. 4′-Glucoside
809. 7-Rutinoside*
810. 7-Gentiobioside*
811. 7-Galactosyl(1 → 4)glucoside*
812. 7-Rhamnosyl-3′-xyloside
813. 7-Diglucoside-4′-glucoside
814. 5-Glucoside-3′-sulfate*

Quercetin 5-methyl ether (azaleatin)
815. 3-Arabinoside
816. 3-Rhamnoside (azalein)
817. 3-Galactoside
818. 3-Glucuronide
819. 3-Xylosylarabinoside
820. 3-Rhamnosylarabinoside
821. 3-Arabinosylgalactoside
822. 3-Rutinoside
823. 3-Diglucoside
Quercetin 7-methyl ether (rhamnetin)
824. 3-α-L-Arabinofuranoside
825. 3-α-L-Arabinopyranoside
826. 3-Rhamnoside
827. 3-Glucoside
828. 3-Galactoside
829. 5-Glucoside
830. 3'-Glucuronide
831. 3-α-Diarabinoside
832. 3-β-Diarabinoside
833. 3-α-L-Arabinopyranosyl(1 → 3)galactoside*
834. 3-Rhamnosyl(1 → 4)rhamnoside
835. 3-Rutinoside
836. 3-Neohesperidoside
837. 3-Robinobioside*
838. 3-Laminaribioside*
839. 3-Gentiobioside*
840. 3-Galactosyl(1 → 4)galactoside
841. 3-Galactosyl(1 → 6)galactoside
842. 3-Mannosyl(1 → 2)alloside
843. 3-Galactoside-3'-rhamnoside
844. 3-Rhamnosyl(1 → 3)rhamnosyl(1 → 6)galactoside (3-rhamninoside, xanthorhamnin A and B)
845. 3-α-L-Arabinopyranosyl(1 → 3)[galactosyl(1 → 6)galactoside]*
846. 3-Arabinoside-3',4'-diglucoside
847. 3,3',4'-Triglucoside
848. 3-Galactoside-3',4'-diglucoside
849. 3-Rhamnosyl(1 → 3)(4''-acetylrhamnosyl)(1 → 6)galactoside
850. 3-[3-Hydroxy-3-methylglutaryl(1 → 6)][apiosyl(1 → 2)galactoside]*
851. 3-(3''''-p-Coumaroylrhamninoside*)
852. 3-Sulfate
853. 3,3'-Disulfate*
854. 3,3',4'-Trisulfate*
855. 3,5,4'-Trisulfate-3'-glucuronide
Quercetin 3'-methyl ether (isorhamnetin)
856. 3-α-L-Arabinofuranoside
857. 3-α-L-Arabinopyranoside (distinchin)
858. 3-Xyloside
859. 3-Rhamnoside*
860. 3-Glucoside

861. 3-Galactoside
862. 3-Glucuronide
863. 5-Glucoside
864. 5-Galactoside*
865. 7-Rhamnoside
866. 7-Glucoside
867. 7-α-D-Glucosaminopyranoside*
868. 3-Arabinosyl(1 → 2)rhamnoside
869. 3-Arabinosyl(1 → 6)glucoside
870. 3-α-Arabinopyranosyl(1 → 6)galactoside*
871. 3-Xylosyl(1 → 2)glucoside*
872. 3-Xylosyl(1 → 6)glucoside*
873. 3-Xylosyl(1 → 2)galactoside*
874. 3-Apiosyl(1 → 2)glucoside*
875. 3-Apiosyl(1 → 2)galactoside*
876. 3-Rhamnosyl(1 → 2)rhamnoside
877. 3-Rutinoside (narcissin)
878. 3-Neohesperidoside
879. 3-Rhamnosyl(1 → 2)galactoside
880. 3-Robinobioside
881. 3-Laminaribioside*
882. 3-Sophoroside
883. 3-Gentiobioside
884. 3-Lactoside
885. 3-Glucosyl(1 → 2)galactoside
886. 3-Glucosyl(1 → 3)galactoside*
887. 4′-Rhamnosyl(1 → 2)glucoside (crosatoside A)*
888. 3-Arabinoside-7-rhamnoside
889. 3-Arabinoside-7-glucoside
890. 3-Glucoside-7-arabinoside
891. 3-Glucoside-7-xyloside
892. 3,7-Dirhamnoside
893. 3-Rhamnoside-7-glucoside
894. 3-Glucoside-7-rhamnoside
895. 3-Galactoside-7-rhamnoside*
896. 3,7-Diglucoside
897. 3-Galactoside-7-glucoside
898. 3-Glucoside-4′-rhamnoside
899. 3,4′-Diglucoside (dactylin)
900. 3-Galactoside-4′-glucoside
901. 7-Sophoroside
902. 3-Xylosyl(1 → 3)rhamnosyl(1 → 6)glucoside*
903. 3-Xylosylrutinoside
904. 3-Xylosylrobinobioside*
905. 3-Rhamnosyl(1 → 4)rhamnosyl(1 → 6)glucoside
906. 3-Apiosyl(1 → 2)[rhamnosyl(1 → 6)glucoside]*
907. 3-Rutinosylglucoside
908. 3-(2^G-Rhamnosylrutinoside) (typhaneoside)
909. 3-Rhamnosyl(1 → 2)[rhamnosyl(1 → 6)galactoside]
910. 3-Rhamnosyl(1 → 2)[glucosyl(1 → 6)glucoside]*

911. 3-Glucosyl(1 → 2)[rhamnosyl(1 → 6)galactoside]
912. 3-Galactosyl(1 → 2)[rhamnosyl(1 → 6)glucoside]*
913. 3-(4^{Rha}-Galactosylrobinobioside)*
914. 3-Xylosyl(1 → 2)glucoside-7-rhamnoside*
915. 3-Rutinoside-7-rhamnoside
916. 3-Rhamnosyl(1 → 2)galactoside-7-glucoside
917. 3-Robinobioside-7-rhamnoside
918. 3-Rutinoside-7-glucoside
919. 3-Sophoroside-7-rhamnoside
920. 3-Rhamnoside-7-sophoroside
921. 3-Sophoroside-7-glucoside
922. 3-Glucoside-7-gentiobioside
923. 3-Gentiobioside-7-glucoside
924. 3-Glucosyl(1 → 6)galactoside-7-glucoside*
925. 3-Rutinoside-4'-rhamnoside
926. 3-Rutinoside-4'-glucoside
927. 3-Gentiotrioside-7-glucoside
928. 3-Rhamnosyl(1 → 2)[gentiobiosyl(1 → 6)glucoside]*
929. 3-Rhamnosyl(1 → 2)[glucosyl(1 → 6)glucoside]-7-glucoside*
930. 3-(2^{G}-Rhamnosylrutinoside)-7-rhamnoside*
931. 3-Rhamnosyl(1 → 2)[rhamnosyl(1 → 6)galactoside]-7-rhamnoside
932. 3-(6''-Malonylglucoside)
933. 3-(6''-Galloylglucoside)
934. 3-(6''-p-Coumaroylglucoside)
935. 3-(2''-Acetylglucoside)
936. 3-(6''-Acetylglucoside)
937. 3-(6''-Acetylgalactoside)
938. 3-[6''-(2-E-Butenoyl)glucoside]*
939. 3-(2'',3'',4''-Triacetylglucoside)*
940. 7-(6''-p-Coumaroylglucoside)*
941. 3-(3-Methylbutyrylrutinoside)
942. 3-(3'',6''-Di-p-Coumaroylglucoside)
943. 3-(6''-p-Coumaroylglucosyl)(1 → 2)rhamnoside*
944. 3-(4'''-p-Coumaroylrobinobioside)
945. 3-(3'''-Feruloylrhamnosyl)(1 → 6)galactoside*
946. 3-(6''-E-Sinapoylsophoroside)*
947. 3-(2'''-Acetyl-α-arabinopyranosyl)(1 → 6)galactoside*
948. 3-Rhamnosyl(1 → 6)(2''-acetylglucoside)*
949. 3-(6''-Acetylglucosyl)(1 → 3)galactoside*
950. 3-(4'',6''-Diacetylglucosyl)(1 → 3)galactoside*
951. 3-(6''-E-p-Coumaroylglucoside)-7-glucoside*
952. 3-Feruloyl-7-rhamnosylglucoside
953. 3-Rhamnosyl(1 → 6)[rhamnosyl(1 → 2)(4''-Z-p-coumaroylgalactoside)]*
954. 3-[2''-(4'''-Acetylrhamnosyl)gentiobioside]*
955. 3-(p-Coumaroylrhamnosylgalactoside)-7-rhamnoside
956. 3-Rhamnosyl(1 → 6)[rhamnosyl(1 → 2)(3'''-E-p-coumaroylgalactoside)]-7-rhamnoside*
957. 3-Rhamnosyl(1 → 6)[rhamnosyl(1 → 2)(4''-p-coumaroylgalactoside)]-7-rhamnoside*
958. 3-Rhamnosyl(1 → 6)[rhamnosyl(1 → 2)(4''-E-feruloylgalactoside)]-7-rhamnoside*
959. 3-Sulfatorutinoside

960. 3-Glucuronide-7-sulfate
961. 3-Sulfate
962. 7-Sulfate
963. 3,7-Disulfate
964. 3,4'-Disulfate
965. 3,7,4'-Trisulfate
966. 3-(4''-Sulfatorutinoside)*
Quercetin 4'-methyl ether (tamarixetin)
967. 3-Rhamnoside
968. 3-Glucoside
969. 3-Galactoside*
970. 3-Neohesperidoside*
971. 3-Robinobioside
972. 3-Glucosyl(1 → 2)galactoside*
973. 3-Digalactoside
974. 3,7-Diglucoside*
975. 5-Glucoside-7-glucuronide
976. 3-Rutinoside-7-rhamnoside*
977. 3-Sulfate
978. 3-Glucoside-7-sulfate*
Quercetin 3,5-dimethyl ether (caryatin)
979. 3'-(or 4'-)Glucoside
980. 7-Glucoside*
Quercetin 3,7-dimethyl ether
981. 5-Glucoside
982. 3'-Neohesperidoside*
983. 3'-(6^G-Rhamnosylneohesperidoside)*
984. 4'-Sulfate*
Quercetin 3,3'-dimethyl ether
985. 7-Glucoside
986. 4'-Glucoside
987. 7-Rutinoside*
Quercetin 3,4'-dimethyl ether
988. 7-Glucoside*
989. 7-α-L-Arabinofuranosyl(1 → 6)glucoside*
990. 7-Rutinoside*
991. 7-(2^G-Rhamnosylrutinoside)*
992. 7-(2^G-Glucosylrutinoside)*
Quercetin 5,3'-dimethyl ether
993. 3-Glucoside
Quercetin 7,3'-dimethyl ether (rhamnazin)
994. 3-Glucoside
995. 3-Galactoside
996. 3-Rhamnoside
997. 4'-Glucoside
998. 3-Glucosyl(1 → 5)-α-L-arabinofuranoside*
999. 3-Xylosyl(1 → 2)glucoside*
1000. 3-Rutinoside
1001. 3-Neohesperidoside
1002. 3-Galactoside-4'-glucoside

1003. 3-Rhamnosyl(1 → 3)rhamnosyl(1 → 6)galactoside (xanthorhamnin C)
1004. 3-Rhamnosyl(1 → 4)rhamnosyl(1 → 6)galactoside (3-isorhamninoside)
1005. 3-Glucosyl(1 → 5)[apiosyl(1 → 2)-α-L-arabinofuranoside]*
1006. 3-Sulfate
Quercetin 7,4′-dimethyl ether (ombuin)
1007. 3-Arabinofuranoside*
1008. 3-Glucoside*
1009. 3-Galactoside
1010. 3-Rutinoside (ombuoside)
1011. 3,5-Diglucoside
1012. 3-Rutinoside-5-glucoside
1013. 3-Sulfate
Quercetin 3′,4′-dimethyl ether
1014. 3-Rutinoside
1015. 3-Neohesperidoside*
1016. 3,7-Diglucoside*
1017. 5-Glucoside-7-glucuronide
Quercetin 3,7,4′-trimethyl ether
1018. 3-Sulfate*
Quercetin 5,3′,4′-trimethyl ether
1019. 3-Galactosyl(1 → 2)rhamnoside-7-rhamnoside*
Quercetin 7,3′,4′-trimethyl ether
1020. 3-Arabinoside
1021. 3-Digalactoside
Quercetin 5,7,3′,4′-tetramethyl ether
1022. 3-Arabinoside
1023. 3-Galactoside*
1024. 3-Rutinoside
3,5,7-Trihydroxy-3′,5′-dimethoxyflavone
1025. 7-Glucoside (lagotiside)
3,5,8,5′-Tetrahydroxy-7-methoxyflavone
1026. 8-Acetate*
6-Hydroxyquercetin (quercetagetin)
1027. 3-Rhamnoside
1028. 3-Glucoside (tagetiin)
1029. 6-Glucoside*
1030. 7-Glucoside
1031. 3,7-Diglucoside
1032. 7-(6″-Isobutyrylglucoside)*
1033. 7-(6″-Isovalerylglucoside)*
1034. 7-[6″-(2-Methylbutyryl)glucoside]*
1035. 7-(6″-E-Caffeoylglucoside)*
1036. 7-(6″-Acetylglucoside)*
Quercetagetin 3-methyl ether
1037. 7-Glucoside
1038. 7-Sulfate
Quercetagetin 6-methyl ether (patuletin)
1039. 3-Xyloside
1040. 3-Rhamnoside
1041. 3-Glucoside

1042. 3-Galactoside
1043. 3-Glucuronide
1044. 5-Glucoside
1045. 7-Glucoside
1046. 3-Rutinoside
1047. 3-Robinobioside
1048. 3-Galactosylrhamnoside
1049. 3-Gentiobioside
1050. 3-Digalactoside
1051. 3,7-Dirhamnoside
1052. 3,7-Diglucoside*
1053. 3-Digalactosylrhamnoside
1054. 3-Glucosyl(1 → 6)[apiosyl(1 → 2)glucoside]
1055. 3-(6″-*p*-Coumaroylglucoside)*
1056. 3-(6″-*E*-Feruloylglucoside)*
1057. 3-(6″-Acetylglucoside)
1058. 7-(6″-Isobutyrylglucoside)*
1059. 7-[6″-(2-Methylbutyryl)glucoside]*
1060. 7-(6″-Isovalerylglucoside)*
1061. 3-Rhamnoside-7-(2‴-acetylrhamnoside)*
1062. 3-Rhamnoside-7-(3‴-acetylrhamnoside)
1063. 3-Rhamnoside-7-(4‴-acetylrhamnoside)
1064. 3-(4″-Acetylrhamnoside)-7-rhamnoside*
1065. 3-Rhamnoside-7-(3‴,4‴-diacetylrhamnoside)
1066. 3-(3″-Acetylrhamnoside)-7-(3‴-acetylrhamnoside)
1067. 3-(4″-Acetylrhamnoside)-7-(2‴-acetylrhamnoside)*
1068. 3-(4″-Acetylrhamnoside)-7-(3‴-acetylrhamnoside)
1069. 3-(4″-Acetylrhamnoside)-7-(2‴,4‴-diacetylrhamnoside)
1070. 3-(4″-Acetylrhamnoside)-7-(2‴,4‴-diacetylrhamnoside)
1071. 3-(2″-Feruloylglucosyl)(1 → 6)[apiosyl(1 → 2)glucoside]*
1072. 3-Sulfate
1073. 7-Sulfate
1074. 3,3′-Disulfate
1075. 3-Glucoside-7-sulfate
Quercetagetin 7-methyl ether
1076. 3-Glucoside
1077. 6-Glucoside*
1078. 4′-Glucoside*
1079. 3-Neohesperidoside*
1080. 3-Cellobioside*
1081. 3-(2‴-Caffeoylglucosyl)(1 → 2)glucuronide*
Quercetagetin 3′-methyl ether
1082. 3-Glucoside
1083. 3-Galactoside
1084. 7-Glucoside
Quercetagetin 4′-methyl ether
1085. 3-Arabinoside
Quercetagetin 3,6-dimethyl ether (axillarin)
1086. 7-Glucoside (axillaroside)
1087. 4′-Glucuronide

1088. 5-α-L-Arabinosyl(1 → 6)glucoside*
1089. 7-Sulfate*
Quercetagetin 3,7-dimethyl ether
1090. 6-Glucoside
1091. 6-Galactoside
1092. 4'-Glucoside
Quercetagetin 6,7-dimethyl ether
1093. 3-Rhamnoside (eupatolin)
1094. 3-Apioside
1095. 3-Glucoside
1096. 3-Galactoside
1097. 3-Glucosylgalactoside
1098. 3-Sulfate
Quercetagetin 3,3'-dimethyl ether
1099. 7-Glucoside
Quercetagetin 6,3'-dimethyl ether (spinacetin)
1100. 3-Glucoside
1101. 7-Glucoside
1102. 3-Rutinoside
1103. 3-Gentiobioside
1104. 3-(2''-Apiosylgentiobioside)*
1105. 3-(2'''-Feruloylgentiobioside)*
1106. 3-(2''-p-Coumaroylglucosyl)(1 → 6)[apiosyl(1 → 2)glucoside]*
1107. 3-(2''-Feruloylglucosyl)(1 → 6)[apiosyl(1 → 2)glucoside]*
1108. 3-Sulfate
Quercetagetin 7,3'-dimethyl ether
1109. 6-Glucoside*
Quercetagetin 3,6,7-trimethyl ether
1110. 4'-Glucoside
Quercetagetin 3,6,3'-trimethyl ether (jaceidin)
1111. 5-Glucoside*
1112. 7-Glucoside (jacein)
1113. 7-Neohesperidoside
1114. 4'-Sulfate
Quercetagetin 3,6,4'-trimethyl ether
1115. 7-Glucoside (centaurein)
Quercetagetin 3,7,4'-trimethyl ether
1116. 6-Glucoside
1117. 3'-Glucoside
Quercetagetin 6,7,3'-trimethyl ether (veronicafolin)
1118. 3-Rutinoside
1119. 3-Glucosyl(1 → 3)galactoside*
1120. 3-Digalactoside
1121. 3-Sulfate
Quercetagetin 6,7,4'-trimethyl ether (eupatin)
1122. 3-Sulfate
Quercetagetin 6,3',4'-trimethyl ether
1123. 3-Glucoside*
1124. 3-Sulfate
Quercetagetin 7,3',4'-trimethyl ether

1125. 3-Rhamnoside

Quercetagetin 3,6,7,3'-tetramethyl ether

1126. 4'-Glucoside (chrysosplenin)
1127. 4'-Galactoside (galactobuxin)

Quercetagetin 3,6,7,3',4'-pentamethyl ether (artemetin)

1128. 5-Glucosylrhamnoside

8-Hydroxyquercetin (gossypetin)

1129. 3-Glucoside
1130. 3-Galactoside
1131. 3-Glucuronide
1132. 7-Rhamnoside
1133. 7-Glucoside
1134. 8-α-D-Lyxopyranoside*
1135. 8-Rhamnoside
1136. 8-Glucoside
1137. 8-Glucuronide
1138. 3-Glucoside-8-glucuronide
1139. 3-Glucuronide-8-glucoside
1140. 7-Rhamnoside-8-glucoside
1141. 3-Gentiotrioside
1142. 3-Sophoroside-8-glucoside
1143. 8-Glucuronide-3-sulfate
1144. 3-Sulfate

Gossypetin 7-methyl ether

1145. 3-Arabinoside
1146. 3-Rhamnoside
1147. 3-Galactoside
1148. 8-Glucoside
1149. 3-Rutinoside
1150. 3-Galactoside-8-glucoside

Gossypetin 8-methyl ether (corniculatusin)

1151. 3-α-L-Arabinofuranoside
1152. 3-Glucoside
1153. 3-Galactoside
1154. 7-Glucoside
1155. 3-Xylosyl(1 → 2)rhamnoside*
1156. 3-Robinobioside

Gossypetin 3'-methyl ether

1157. 7-Glucoside
1158. 3-Rutinoside
1159. 7-Neohesperidoside (haploside F)
1160. 7-(6″-Acetylglucoside)
1161. 7-(6″-Acetylrhamnosyl)(1 → 2)glucoside

Gossypetin 3,8-dimethyl ether

1162. 5-Glucoside*

Gossypetin 7,8-dimethyl ether

1163. 3-Glucoside*
1164. 4'-Glucoside*
1165. 3,3'-Disulfate*

Gossypetin 7,4'-dimethyl ether

1166. 8-Glucoside
1167. 8-Acetate
1168. 8-Butyrate
Gossypetin 8,3′-dimethyl ether (limocitrin)
1169. 3-Rhamnoside
1170. 3-Glucoside
1171. 3-Galactoside
1172. 3-Rutinoside
1173. 7-Glucoside
1174. 7-Neohesperidoside
1175. 3-Sophoroside
1176. 3,7-Diglucoside
1177. 3-Rutinoside-7-glucoside*
1178. 7-(6″-Acetylglucoside)
1179. 7-(6″-Acetylneohesperidoside)
3,5,7,3′,4′,5′-Hexahydroxyflavone (myricetin)
1180. 3-Arabinoside
1181. 3-α-Arabinofuranoside
1182. 3-Xyloside
1183. 3-Rhamnoside (myricitrin)
1184. 3-Glucoside
1185. 3-Galactoside
1186. 7-Arabinoside
1187. 7-Glucoside
1188. 3′-Arabinoside
1189. 3′-Xyloside
1190. 3′-Rhamnoside*
1191. 3′-Glucoside
1192. 3-Dixyloside
1193. 3-Dirhamnoside
1194. 3-Xylosyl(1 → 2)rhamnoside
1195. 3-Xylosyl(1 → 3)rhamnoside*
1196. 3-Xylosylglucoside
1197. 3-Rhamnosyl(1 → 2)rhamnoside*
1198. 3-Rutinoside
1199. 3-Neohesperidoside*
1200. 3-Robinobioside*
1201. 3-Diglucoside
1202. 3-Galactosylglucoside
1203. 3-Digalactoside
1204. 3-Rhamnoside-7-glucoside
1205. 3-Rhamnoside-3′-glucoside*
1206. 3-Galactoside-3′-rhamnoside*
1207. 3,3′-Digalactoside
1208. 3,4′-Dirhamnoside*
1209. 3,4′-Diglucoside*
1210. 3-Rhamnosyl(1 → 3)glucosyl(1 → 6)glucoside*
1211. 3-(2^G-Rhamnosylrutinoside)*
1212. 3-Glucosylrutinoside
1213. 3-Triglucoside

1214. 3-Rutinoside-7-rhamnoside
1215. 3-Robinobioside-7-rhamnoside
1216. 3-Rutinoside-7-glucoside
1217. 3-Glucosyl(1 → 2)rhamnoside-7-glucoside*
1218. 3-(2″-*p*-Hydroxybenzoylrhamnoside)*
1219. 3-(4″-Malonylrhamnoside)*
1220. 3-(2″-Galloylrhamnoside)
1221. 3-(3″-Galloylrhamnoside)*
1222. 3-(4″-Galloylrhamnoside)
1223. 3-(2″-Galloylglucoside)*
1224. 3-(6″-Galloylglucoside)
1225. 3-(3″-Galloylgalactoside)*
1226. 3-(6″-Galloylgalactoside)
1227. 3-(6″-*p*-Coumarylglucoside)*
1228. Nympholide A* (**13.7**)
1229. Nympholide B* (**13.8**)
1230. 3-(2″-Acetylrhamnoside)*
1231. 3-(4″-Acetylrhamnoside)*
1232. 7-(6″-Galloylglucoside)*
1233. 3-(2″,3″-Digalloylrhamnoside)*
1234. 3-(3″,4″-Diacetylrhamnoside)*
1235. 3-(2″,3″,4″-Triacetylxyloside)*
1236. 3-(4″-Acetyl-2″-galloylrhamnoside)*
1237. 3-[3‴,6‴-Diacetylglucosyl(1 → 4)2″,3″-diacetylrhamnoside]*
1238. 3-(*p*-Coumarylrhamnosylgalactoside)
1239. 3-Glucosyl(1 → 2)(6‴-caffeylglucosyl)(1 → 2)rhamnoside-4′-
        rhamnosyl(1 → 4)xyloside (montbretin A)
1240. 3-Glucosyl(1 → 2)(6‴-*p*-coumaroylglucosyl)(1 → 2)rhamnoside-4′-
        rhamnosyl(1 → 4)xyloside (montbretin B)
1241. 3-Sulfatorhamnoside
Myricetin 3-methyl ether
1242. 7-Rhamnoside
1243. 3′-Xyloside
1244. 3′-Glucoside
1245. 7-Rhamnoside-3′-xyloside
Myricetin 5-methyl ether
1246. 3-Rhamnoside
1247. 3-Galactoside
Myricetin 7-methyl ether (europetin)
1248. 3-Rhamnoside
1249. 3-(2″-Galloylrhamnoside)*
1250. 3-(3″-Galloylrhamnoside)*
Myricetin 3′-methyl ether (larycitrin)
1251. 3-α-L-Arabinofuranoside*
1252. 3-Rhamnoside
1253. 3-Glucoside
1254. 3-Galactoside
1255. 7-Glucoside
1256. 5′-Glucoside
1257. 3-Rutinoside

1258. 3,5′-Diglucoside
1259. 3-Rhamnosylrutinoside
1260. 3-Rutinoside-7-glucoside
1261. 3,7,5′-Triglucoside
1262. 3-(4″-Malonylrhamnoside)*
1263　3-*p*-Coumarylglucoside
Myricetin 4′-methyl ether
1264. 3-Rhamnoside (mearnsitrin)
1265. 3-Galactoside*
1266. 3-Galactosyl(1 → 4)galactoside
1267. 3,7-Dirhamnoside
1268. 3-(4″-Acetylrhamnoside)*
Myricetin 3,4′-dimethyl ether
1269. 3′-Xyloside
1270. 7-Rhamnoside-3′-xyloside
Myricetin 7,4′-dimethyl ether
1271. 3-Galactoside
Myricetin 3′,4′-dimethyl ether
1272. 3-Rhamnoside*
1273. 3-Glucoside*
Myricetin 3′,5′-dimethyl ether (syringetin)
1274. 3-Arabinoside
1275. 3-Rhamnoside
1276. 3-Xyloside
1277. 3-Glucoside
1278. 3-Galactoside
1279. 3-Rhamnosyl(1 → 5)-α-L-arabinofuranoside*
1280. 3-Rutinoside
1281. 3-Robinobioside*
1282. 3-Rhamnosylrutinoside
1283. 3-Rutinoside-7-glucoside
1284. 3-(*p*-Coumarylglucoside)
1285. 3-(2″,3″-Diacetylglucoside)*
1286. 3-(6″-Acetylglucosyl)(1 → 3)galactoside*
6-Hydroxymyricetin 6,3′,5′-trimethyl ether
1287. 3-Glucoside
6-Hydroxymyricetin 3,6,3′,5′-tetramethyl ether
1288. 7-Glucoside
8-Hydroxymyricetin (hibiscetin)
1289. 3-Glucoside
1290. 8-Glucosylxyloside
8-Hydroxymyricetin 8-methyl ether
1291. 3-Rhamnoside*
8-Hydroxymyricetin 8,5-dimethyl ether
1292. 3-Rhamnoside*
8-Hydroxymyricetin 8,3′,5′-trimethyl ether
1293. 3-Rhamnoside*
3,5,7,2′-Tetrahydroxyflavone (datiscetin)
1294. 3-Glucoside
1295. 3-Rutinoside

3,7,2′,3′,4′-Pentahydroxyflavone
1296. 3-Neohesperidoside
3,7,3′,4′,5′-Pentahydroxyflavone (5-deoxymyricetin, robinetin)
1297. 7-Glucoside*
1298. 3-Rutinoside*
3,4′-Dihydroxy-7,3′,5′-trimethoxyflavone
1299. 3-Galactosyl(1 → 4)xyloside*
3,5,7,2′,6′-Pentahydroxyflavone
1300. 2′-Glucoside*
5,7-Dihydroxy-3,6,8,4′-tetramethoxyflavone
1301. 7-Glucosyl(1 → 3)galactoside*
7,4′-Dihydroxy-3,5,6,8-tetramethoxyflavone
1302. 4′-Glucosyl(1 → 3)galactoside*
5,8-Dihydroxy-3,6,7,4′-tetramethoxyflavone
1303. 8-Neohesperidoside*
5,4′-Dihydroxy-6,7,8,3′-tetramethoxyflavone (africanutin)
1304. 4′-Galactoside*
3,5,7,2′,3′,4′-Hexahydroxyflavone
1305. 3-Glucoside*
5,7,2′-Trihydroxy-3,6,4′-trimethoxyflavone
1306. 7-Glucoside*
5,2′,5′-Trihydroxy-3,7,8-trimethoxyflavone
1307. 2′-Acetate
5,2′-Dihydroxy-3,7,4′-trimethoxyflavone
1308. 2′-Glucoside
5,2′,4′-Trihydroxy-3,7,5′-trimethoxyflavone
1309. 2′-Galactosyl(1 → 4)glucoside*
5,2′-5′-Trihydroxy-3,7,4′-trimethoxyflavone
1310. 2′-Glucoside
5,6′,5′-Trihydroxy-3,7,4′-trimethoxyflavone
1311. 5′-Glucoside
5,2′-Dihydroxy-3,7,4′,5′-tetramethoxyflavone
1312. 2′-Glucoside
5,5′-Dihydroxy-3,6,7,4′-tetramethoxyflavone
1313. 5′-Glucoside
5,7,8-Trihydroxy-3,6,4′-trimethoxyflavone
1314. 8-Tiglate*
5,8,4′-Trihydroxyflavone-3,7,3′-trimethoxyflavone
1315. 8-Acetate
3,5,2′-Trihydroxy-7,8,4′-trimethoxyflavone
1316. 5-Glucosyl(1 → 2)galactoside*
3,5,6,7,8,4′-Hexahydroxy-3′-methoxyflavone
1317. 3-Rhamnosyl(1 → 4)rhamnosyl(1 → 6)glucoside*
3,5,7,3′,4′-Pentahydroxy-6,8-dimethoxyflavone
1318. 3-Arabinoside
3,5,7,4′-Tetrahydroxy-6,8,3′-trimethoxyflavone
1319. 3-α-L-Arabinopyranosyl(1 → 3)galactoside*
1320. 3-Rhamnosyl(1 → 2)glucoside
1321. 3-α-L-Arabinopyranosyl(1 → 3)[galactosyl(1 → 6)galactoside]*
3,6,7,8,3′,4′-Hexahydroxy-5′-methoxyflavone

1322. 7-Neohesperidoside*
5,7,2′,3′,4′-Pentahydroxy-3,6-dimethoxyflavone
1323. 7-Glucoside*
5,2′,5′-Trihydroxy-3,6,7,4′-tetramethoxyflavone
1324. 5′-Glucoside
5,2′-Dihydroxy-3,6,7,4′,5′-pentamethoxyflavone (brickellin)
1325. 2′-Glucoside*
5,5′-Dihydroxy-3,6,7,2′,4′-pentamethoxyflavone
1326. 5′-Glucoside
5,8-Dihydroxy-3,7,2′,3′,4′-pentamethoxyflavone
1327. 8-Acetate
5,7,3′,5′-Tetrahydroxy-3,6,8,4′-tetramethoxyflavone
1328. 3′-Glucoside
C-Methylated flavonol glycosides
5,7-Dihydroxy-6,8-di-C-methyl-3-methoxyflavone
1329. 7-Galactosyl(1 → 2)rhamnoside*
3,5,7,4′-Tetrahydroxy-8-C-methyl flavone (8-C-Methylkaempferol)
1330. 7-Glucoside
3,5,7,3′,4′,5′-Hexahydroxy-2′-C-methyl flavone (2′-C-methylmyricetin)
1331. 3-Rhamnoside-5′-gallate*
Prenylated, Pyrano and Methylenedioxy Flavonol Glycosides

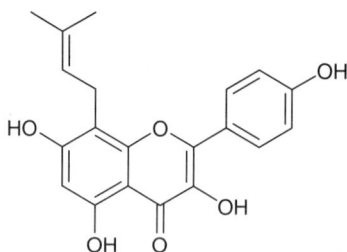

13.11

8-Prenylkaempferol (noranhydroicaritin, 3,5,7,4′-tetrahydroxy-8-(3,″3″-dimethylallyl)flavone)
    1. 3-Rhamnoside (ikaroside A)
    2. 3-Rhamnosyl(1 → 2)rhamnoside
    3. 3-Xylosyl(1 → 2)rhamnoside (ikaroside D)
    4. 3-Glucosyl(1 → 2)rhamnoside (ikaroside B)
    5. 3-Rhamnoside-7-glucoside (epimedoside A)
    6. 3,7-Diglucoside*
    7. 3-Rhamnosyl(1 → 2)xyloside-7-glucoside (epimedoside E)
    8. 3-Glucosyl(1 → 2)rhamnoside-7-glucoside (Ikaroside C, diphylloside A)
    9. 3-Rhamnosyl(1 → 2)rhamnoside-7-glucoside (diphylloside B)
    10. 3-Rhamnosyl(1 → 2)glucoside-7-glucoside
    11. 3-Glucosyl(1 → 2)rhamnoside-7-glucosyl(1 → 2)glucoside (diphylloside C)
    12. 3-Xylosyl(1 → 2)rhamnoside-7-glucosyl(1 → 2)glucoside (hexandroside C)
    13. 3-(4″-Acetylrhamnoside) (ikaroside F)
8-Prenylkaempferol 7-methyl ether
    14. 3-Rhamnosyl(1 → 3)[apiosyl(1 → 6)glucoside]*

8-Prenylkaempferol 4'-methyl ether (anhydroicaritin)
   15. 3-Rhamnoside
   16. 3-Glucoside
   17. 7-Glucoside (icariside I)
   18. 3-Rhamnosyl(1 → 2)rhamnoside*
   19. 3-Xylosyl(1 → 2)rhamnoside (sagittatoside B)
   20. 3-Rutinoside
   21. 3-Glucosyl(1 → 2)rhamnoside (sagittatoside A)
   22. 3-Rhamnoside-7-glucoside (icariin)
   23. 7-Cellobioside (cuhuoside)*
   24. 3-Xylosyl(1 → 2)rhamnoside-7-glucoside (epimedin B)
   25. 3-Xylosyl(1 → 2)rhamnoside-7-glucoside
   26. 3-Rhamnosyl(1 → 2)rhamnoside-7-glucoside (epimedin C)
   27. 3-Rhamnosyl(1 → 3)rhamnoside-7-glucoside (hexandroside D)
   28. 3-Glucosyl(1 → 2)rhamnoside-7-glucoside (epimedin A)
   29. 3-Glucosyl(1 → 3)rhamnoside-7-glucoside*
   30. 3-Galactosyl(1 → 3)rhamnoside-7-glucoside
   31. 3-Rhamnosyl(1 → 6)galactoside-7-galactoside*
   32. 3-Rhamnosyl(1 → 2)rhamnoside-7-sophoroside* (acuminatoside)
   33. 3-[3'''-Acetylxylosyl(1 → 3)4''-acetylrhamnoside] (sempervirenoside)
   34. 3-Glucosyl(1 → 2)(3''-acetylrhamnoside) (sagittatoside C)
   35. 3-Glucosyl(1 → 3)(4''-acetylrhamnoside) (epimedokoreanoside II)
   36. 3-[4''',6'''-Diacetylglucosyl(1 → 3)4''-acetylrhamnoside]*
   37. 3-[2''',6'''-Diacetylglucosyl(1 → 3)4''-acetylrhamnoside]-7-glucoside (epimedin K)
   38. 3-Xylosyl(1 → 3)(4''-acetylrhamnoside)-7-glucoside
   39. 3-[6'''-Acetylglucosyl(1 → 3)4''-acetylrhamnoside]-7-glucoside
      (epimedokoreanoside I)
   40. 3-[4''',6'''-Diacetylglucosyl(1 → 3)4''-acetylrhamnoside]-7-glucoside*
   41. 3-(6'''-Acetylgalactosyl)(1 → 3)rhamnoside-7-glucoside
   42. 3-[3'''-Acetylxylosyl(1 → 3)4''-acetylrhamnoside]-7-glucoside
8-(3''-Hydroxy-3''-methylbutyl)kaempferol 4'-methyl ether (icaritin)
   43. 3-Rhamnoside
   44. 3-Rhamnosyl(1 → 2)rhamnoside (wanepimedoside A)*
8-(γ-Methoxy-γγ-dimethyl)propylkaempferol 4'-methyl ether
   45. 7-glucoside* (caohuoside D)
8-Prenylquercetin 4'-methyl ether
   46. 3-Rhamnoside* (caohuoside C)
8-Prenylquercetin 7,4'-dimethyl ether
   47. 3-Rhamnosyl(1 → 4)rhamnoside*
6'',6''-Dimethylpyrano(2'',3'':7,8)kaempferol
   48. 3-Rhamnoside
6'',6''-Dimethylpyrano(2'',3'':7,8)-4'-methoxykaempferol
   49. 3-Rhamnoside*
3-Methoxy-5-hydroxy-6,7-methylenedioxyflavone
   50. 4'-Glucuronide*

3,5,4'-Trihydroxy-6,7-methylenedioxyflavone
   51. 3-Glucoside
3-Hydroxy-5,4'-dimethoxy-6,7-methylenedioxyflavone
   52. 3-Xyloside (viviparum A)*
3,3'-Dihydroxy-5,4'-dimethoxy-6,7-methylenedioxyflavone
   53. 3-Xyloside (viviparum B)*

*Flavonol glycosides newly reported since 1992.

14 *C*-Glycosylflavonoids

Maurice Jay, Marie-Rose Viricel, and Jean-François Gonnet

CONTENTS

14.1   NATURAL SOURCES AND SOME TAXONOMIC IMPLICATIONS

The natural sources of *C*-glycosylflavonoids reported for the last 10 years are listed in Table 14.1. The number of species possessing *C*-glycosylflavonoids is given in parentheses for each genus, except in the case where a unique species was under investigation.

14.1.1   Plant Species Rich in *C*-Glycosylflavonoids

In Bryophyta, the presence of *C*-glycosylflavones was reported for the first time in the genera *Plagiochila* and *Plagiochasma*. It is the same situation in Gymnospermae for the genera *Cycas* and *Abies*. In ferns, the confirmation of the presence of *C*-glycosylflavonoids was provided for three families: Aspleniaceae (*Asplenium* sp.), Athyriaceae, and Hymenophyllaceae (*Trichomanes* sp.). In monocots, several investigations have concerned two species, *Zea mays* and *Hordeum vulgare*, and two families, Restionaceae and Velloziaceae. In dicots, the first citations for *C*-glycosylflavones concern the families Araceae, Betulaceae, Brassicaceae, Capparaceae, Chenopodiaceae, Droseraceae, Geraniaceae, Illecebraceae, Mimosaceae, Oleaceae, Orchidaceae, Oxalidaceae, Pistaciaceae, Plumbaginaceae, Polygalaceae, Sapindaceae, Solanaceae, Stercurlariaceae, Turneraceae, Urticaceae, and Violaceae; important additions to the occurrence of *C*-glycosylflavones were given for Cucurbitaceae, Rosaceae, and Theaceae.

14.1.2   *C*-Glycosylflavonoids and Taxonomic Aspects

14.1.2.1   Ferns

The Athyriaceae and Aspleniaceae are two large families of leptosporangiate ferns. However, until the second half of the 20th century, they were united in one family because of the superficial resemblance of their sori. This similarity having been analyzed as the result of convergence rather than of close phylogenetic relationship, the two subfamilies were raised to the family rank. The investigation of Umikalsom et al.[362] provided additional chemical data for characterizing and separating the two taxa. A large collection of 15 species of *Asplenium*, four species of *Athyrium*, 12 species of *Diplazium*, and two species of *Deparia* were compared for their flavonoid contents based on proanthocyanidins, flavonol-*O*-glycosides, flavone-*O*-glycosides, and flavone-*C*-glycosides. Nearly all species representatives of these four genera showed their own specific flavonoid pattern. A large range of flavone *C*- and *O*-glycosides was found in *Asplenium*: the flavone *C*-glycosides were based on both apigenin and luteolin,

TABLE 14.1
Natural Sources of C-Glycosylflavonoids (Since 1991). The Number of Species Under Investigation Given in Parentheses

BRYOPHYTA
Frullania polysticha[186]
Frullania cesatiana[187]
Frullania tamatisci[329]
Plagiochila jamesonii[318]
Plagiochasma rupestre[318]
Plagiomnium sp.[16,399]
PTERIDOPHYTA
Aspleniaceae
Asplenium viviparum[156]
Asplenium (3)[363]
Asplenium (4 hybrids)[249]
Aspleniaceae (15)[362]
Athyriaceae
Athyriaceae (18)[362]
Hymenophyllaceae
Trichomanes (23)[379]
GYMNOSPERMAE
Ephedraceae
Ephedra aphylla[153]
Cycadaceae
Cycas panzhihuaensis[416]
Pinaceae
Abies (7)[331]
ANGIOSPERMAE MONOCOTYLEDONAE
Eriocaulaceae
Eriocaulaceae[311]
Syngonanthus (22)[309]
Gramineae
Bambusa (13)[351]
Deschampsia antarctica[383]
Hordeum vulgare[239,258,260,278,279]
Hyparrhenia hirta[37]
Sasa borealis[405]
Triticum aestivum[129,255]
Zea mays[333–336,345,346]
Iridaceae
Iris (2)[14]
Restionaceae
Restionaceae (115)[387]
Velloziacieae
Velloziaceae (4 genus)[386]
Vellozia (10)[126,127]
ANGIOSPERMAE DICOTYLEDONAE
Acanthaceae
Climacanthus nutans[352]

continued

TABLE 14.1
**Natural Sources of *C*-Glycosylflavonoids
(Since 1991). The Number of Species Under
Investigation Given in Parentheses — *continued***

Amaranthaceae
Alternanthera maritima[316]
Araceae
Arum palaestinum[8,9]
Arum dracunculus[300]
Xanthosoma violaceum[298]
Betulaceae
Betula platyphylla[204]
Bombacaceae
Bombax ceiba[99,100]
Brassicaceae
Barbarea vulgaris[324]
Capparaceae
Cleome (4), *Capparis* (3)[326]
Caryophyllaceae
Silene conoidea[15]
Stellaria media[304]
Chenopodiaceae
Beta vulgaris[109]
Combretaceae
Combretum quadrangulare[24]
Terminalia catappa[213]
Compositeae
Achillea nobilis[192,234]
Achillea setacea[235]
Achillea sp.[365]
Atractylis carduus[252]
Centaurea (2)[115–117]
Felicia amelloides[34]
Otanthus maritimus[92]
Crucifereae
Boreava orientalis[312]
Cucurbitaceae
Bryonia (2)[188,189]
Citrullus colocynthis[223]
Cucumis sativus[3,13,191,228]
Lagenaria siceraria[188,190]
Droseraceae
Drosophyllum lusitanicum[43]
Euphorbiaceae
Aleurites moluccana[253,254,268]
Glochidion zeylanica[288]
Jatropha polhiana[400]
Gentianaceae
Gentiana arisanensis[198,211]
Gentianella azurea[414]

TABLE 14.1
Natural Sources of *C*-Glycosylflavonoids
(Since 1991). The Number of Species Under
Investigation Given in Parentheses — *continued*

Tripterospermum japonicum[289]
Geraniaceae
Pelargonium reniform[200]
Pelargonium (58)[392]
Guttifereae
Clusia sandiensis[74]
Illecebraceae
Scleranthus uncinatus[403,404]
Labiateae
Faradaya (4)[120]
Ocimum (9)[121,122]
Otostegia fruticosa[2]
Oxera (20)[120]
Salvia officinalis[218]
Schnabelia tetradonta[76]
Scutellaria albida[330]
Scutellaria amoena[417]
Scutellaria baicalensis[412,413]
Scutellaria pontica[282]
Leguminoseae
Abrus precatorius[222]
Acacia saligna[89,94]
Acacia leucophloea[366]
Bocoa (2)[180]
Cassia nomame[178]
Cassia occidentalis[130]
Crotalaria thebaica[154]
Cyclopia intermedia[168]
Desmodium tortuosum[207]
Glycyrrhiza glabra[210]
Glycyrrhiza eurycarpa[216]
Herminiera elaphroxylon[93]
Lagonychium farcatum[93]
Lupinus hartwegii[169]
Lupinus luteus[406]
Mohgania macrophylla[398]
Phaseolus radiatus[165]
Phaseolineae[393]
Prosopis chilensis[88]
Rhynchosia (2)[421]
Vigna radiata[199]
Mimosaceae
Mimosa pudica[217]
Myrtaceae
Eucalyptus globulus[232]

---

continued

TABLE 14.1
Natural Sources of *C*-Glycosylflavonoids
(Since 1991). The Number of Species Under
Investigation Given in Parentheses — *continued*

Oleaceae
Ligustrum vulgare[353]
Orchidaceae
Ornithocephalinae (15)[385]
Tylostylis discolor[231]
Oxalidaceae
Biophytum sensitivum[42]
Passifloraceae
Passiflora incarnata[56,208,305,373]
Passiflora (3)[295]
Passiflora (2)[294]
Passiflora sp.[1]
Pistaciaceae
Pistacia atlantica[269]
Plumbaginaceae
Plumbago zeylanica[212]
Polygalaceae
Polygala telephioides[195]
Polygonaceae
Rumex (8)[314]
Polygonum perfoliatum[418]
Ranunculaceae
Trollius lebedouri[420]
Rhamnaceae
Ziziphus jujuba[54]
Rhamnella inaequilatera[350]
Rosaceae
Cotoneaster thymaefolia[291]
Cotoneaster wilsonii[49]
Crataegus monogyna[308]
Crataegus pinnatifida[175,411]
Crataegus sinaica[87]
Cydonia oblonga[102]
Eriobotrya japonica[159]
Rutaceae
Citrus (review)[233]
Citrus (2)[124]
Citrus sp.[23,111,259–261,361]
Feronia elephantum[85]
Raputia paraensis[22]
Sapindaceae
Allophylus edulis[143]
Saxifragaceae
Itea/Pterostemon[35]
Scrophulariaceae
Gratiola officinalis[123]

TABLE 14.1
Natural Sources of *C*-Glycosylflavonoids
(Since 1991). The Number of Species Under
Investigation Given in Parentheses — *continued*

Solanaceae
Capsicum annuum[246]
Sterculariaceae
Theobroma cacao[317]
Theaceae
Thea sp.[96,375]
Thea (90 beverages)[97]
Thymeleaceae
Daphne laureola[359]
Turneraceae
Turnera diffusa[297]
Urticaceae
Cecropia lyratifolia[285]
Verbenaceae
Verbena pinnatifida[85]
Vitex polygama[205]
Violaceae
Viola arvensis[46]
Viola yedoensis[401]
Vitidaceae
Vitis (22)[263]
Tetrastigma hemsleyanum[215]

both mono-*C*- and di-*C*-glycosylated; some of them were further *O*-glycosylated. In Athyriaceae (the other three genera), it was only apigenin-based *C*-glycosides that were found; *O*-glycosyl-*C*-glycosylflavones and *O*-glycosylflavones were absent. Accordingly, the Athyriaceae were clearly distinguished and considered as more primitive than the Aspleniaceae.

In the genus *Asplenium*, Matsumoto et al.[249] studied the flavonoid composition of four natural hybrids in natural populations where the hybrids could be collected with individuals of parental genotypes: *Asplenium normale* × *A. boreale*, *A. normale* × *A. shimurae*, *A. normale* × *A. oligophlebium*, and *A. boreale* × *A. oligophlebium*. The phenolic patterns consisted of *O*-glycosylflavones and di-*C*-glycosylflavones. Interestingly, the flavonoid composition of the hybrids was shown to be total addition of the parental attributes.

14.1.2.2 Angiospermae Monocotyledonae

14.1.2.2.1 Family Eriocaulaceae
A recent study by Salatino et al.[311] concerned the distribution of flavonoids in four genera of Eriocaulaceae: *Eriocaulon*, *Leiothrix*, *Paepalanthus*, and *Syngonanthus*. The authors compared the flavonoid patterns with the results of cladistic analyses based on 49 predominantly morphological characters. This morphometric analysis suggests that *Paepalanthus* is polyphyletic, *Eriocaulon* is closely related to some small subgroups of *Paepalanthus*, while *Leiothrix* and *Syngonanthus* appear as more advanced sister groups. Moreover, at the sectional level within *Syngonanthus*, the *Eulepis* and *Thysanocephalus* sections seemed to constitute a more advanced monophyletic group than the *Carpocephalus* and *Syngonanthus* sections.

The structural aspects of flavonoids seem to have paralleled the evolution of Eriocaulaceae. The 6-oxygenation has been superseded during evolution: in the primitive side, *Paepalanthus* and *Eriocaulon* accumulate 6-OH flavonols; in the more advanced groups, *Leiothrix*, *Syngonanthus–Carpocephalus*, *Syngonanthus–Syngonanthus*, 6-OH flavones are found; finally, in highly advanced sections, *Syngonanthus–Eulepis* and *Syngonanthus–Thysanocephalus*, 6-oxygenated derivatives are lacking while there is a prevalence of *C*-glycosylflavones.

14.1.2.2.2 Family Velloziaceae

In another group of monocots, the Velloziaceae family, chemical studies[127,386] clarified the delineation of subfamilies and genera, which had been the subject of much dispute. The flavonoid patterns were described for about 100 species representative of the subfamilies Vellozioideae (*Vellozia*, *Nanuza*, *Barbaceniopsis*, *Xerophyta*, *Talbotia*), Barbacenioideae (*Aylthonia*, *Barbacenia*, *Burlemarxia*, *Pleurostigma*). Flavone *C*-glycosides have been identified in both subfamilies; however, the Vellozioideae could be distinguished from the Barbacenioideae by the accumulation of flavone mono-*C*-glycosides rather than di-*C*-glycosides. It was apparent that *Barbaceniopsis*, *Xerophyta* (Madagascan species), and *Talbotia* differ from most other members of the subfamily in the absence of *C*-glycosylflavones, and predominance of flavonol glycosides. The species *Nanuza plicata* remains unique within Vellozioideae in the accumulation of biflavonoids. And finally the genus *Pleurostigma* was unique within the Barbacenioideae in accumulating 6-OH flavonoids instead of *C*-glycosylflavones.

14.1.2.3 Angiospermae Dicotyledonae

14.1.2.3.1 Tribe Phasaeolinae

The *C*-glycosylflavones clarify the position of the genus *Dysalobium* within the tribe Phasaeolinae (Leguminosae Fam.).[393] The Phasaeolinae is a taxonomically very complex group because of the limited number of useful morphological characters available to distinguish generic limits. The main taxa are *Phaseolus*, *Dysolobium*, *Macroptilium*, *Strophostyles*, and *Vigna*. The flavonoid profiles of 49 representative species of these five genera were compared and about 35 flavonoids were identified. A statistical procedure using binary presence–absence data for the leaf flavonoids, and based on Sneath's simple matching coefficient, indicated four main groupings, one of them restricted to *Dysalobium*, which is clearly separated from all others while it alone produces only *C*-glycosylflavones.

14.1.2.3.2 Tribe Ornithocephalinae

Ornithocephalinae,[385] a small subtribe of the family Orchidaceae, is traditionally recognized as one of the most advanced in this family. The complexity and diversity of floral and vegetative morphologies of the species of this subtribe seem to have reached a level that is apparently only paralleled by some members of the Oncidiinae usually interpreted as a putative close relative. The results of the leaf flavonoid analyses of 15 species representative of the genera, *Zygostates*, *Ornithocephalus*, *Chytroglossa*, *Phymatidium*, and *Rauhiella*, showed the presence of 16 different *C*-glycosylflavones. These taxa could be distinguished from each other by occurrence of different isomers; these were all apigenin-based structures, with a large representation of different apigenin 7,4'-dimethylether 6-*C*-glycosyl X''-*O*-glycosides, and apigenin 7-methylether 6-*C*-glycosyl-X''-*O*-glycosides (*O*-glycosidic linkages were not precisely determined). All these flavonoids are unusual and quite rare in the Angiosperms.

Methylated flavonoids, considered to be advanced chemical characters, suggested that *Zygostates*, *Ornithocephalus*, *Chytroglossa*, and *Phymatidium* might be highly evolved genera, while the presence of isovitexin and absence of methylated derivatives in *Rauhiella* clearly separated this genus from the other members.

As already mentioned, subtribe Oncidiinae has been considered as a probable sister group of the Ornithocephalinae; *Oncidium* species so far surveyed lack flavone *C*-glycosides, and possess new and unusual 6-hydroxyflavone glycosides. Thus, different flavonoid patterns did not support a close association of these two subtribes.

14.1.2.3.3   *Genus* Pelargonium

The first example concerns the chemical survey of 56 *Pelargonium* species (representative of 19 sections).[392] Their flavonoid composition revealed a large chemical diversity: flavonols, flavones, *C*-glycosylflavones, proanthocyanidins, and ellagitannins. No individual phenolic compound or group of compounds provides taxonomic markers at the sectional level. However, the data indicated which sections are homogenous in their phenolic profiles and which not and in this way could support or refute the current classification. *C*-Glycosylflavones were well represented in the sections Subsucculentia, Chorisma, Perista, Reniformia, and Jenkinsonia, for example, while they were not found in the Pelargonium, Otidia, and Cortusina sections. Interestingly, most species possessing high levels of *C*-glycosylflavones showed a strong correlation between these compounds and the presence of ellagitannins. In other sections lacking *C*-glycosylflavones, a correlation exists between proanthocyanidins and the flavonol myricetin. Thanks to these correlations, it was possible to detect in each section well placed and misplaced species. Thus in Reniformia, three species, *P. reniforme*, *P. exstipulatum*, and *P. album*, have the basic flavone-*C*-glycoside or ellagitannin profile but *P. odoratissimum* has a very different myricetin or proanthocyanidin pattern, suggesting it may need to be moved.

14.1.2.3.4   *Genus* Itea *and* Pterostemon

Itea and *Pterostemon*, sister taxa close to *Ribes* in the Saxifragaceae family, were studied[35] for chemical data in comparison to gene sequence data. Recent phylogenetic analyses of *rbc*L, 18S rDNA, *matK*, and *atp*B sequences all concur in suggesting that *Itea* and *Pterostemon* are sister taxa, close to *Ribes* in the Saxifragaceae as a part of a larger Saxifragales clade that also includes Hamamelidaceae, Crassulaceae, Penthoraceae, etc. The flavonoid profile of *Pterostemon* comprises *O*-glycosides of quercetin and *C*-glucosylflavones (vitexin, isovitexin, orientin and their *O*-glycoside derivatives, especially *X″-O*-xylosides). This *C*-glycosylflavone pattern resembles very closely that observed in *Itea*, and provides support for the closeness of their relationship recently demonstrated on the basis of gene sequence data. But this finding indicates that the flavonoid profiles of *Itea* and *Pterostemon* are dramatically different from the overall profile of a large number of species of Saxifragaceae, which is mainly based upon various *O*-glycosylflavonols but with no trace of *C*-glycosylflavones. If the DNA studies indicate that *Itea* and *Pterostemon* are sister taxa within the Saxifragaceae, the phenolic contents demonstrate that the two genera possess unusual and non-saxifragoid flavonoid chemistry. On the evolutionary sequence of flavonoid production, it appears that these genera represent a derived clade with a replacement of flavonols by *C*-glycosylflavones, and that *Pterostemon*, due to *O*-glycosylflavonols, possibly represents an intermediate form between *Itea* and the true Saxifragaceae.

14.1.2.3.5   *Genus* Centaurea

Based on glycosylflavones within blue flowering alpine cornflowers, *Centaurea montana* L. and *Centaurea triumfetti* All. (Compositae), a major discussion concerned the influence of microscale environmental conditions and the role of reproductive modes on the distribution of *C*-glycosylflavones. Twenty-one out of the 48 flavonoid glycosides detected in plants from different origins are basic 6- and 8-*C*-monoglucosides of apigenin, luteolin, and chrysoeriol, 6,8-di-*C*-glucosyl apigenin, 2″-*O*-glucosides and arabinosides of the 6-*C*-glucosides, and caffeoyl derivatives of 2″-*O*-glucosides along with some other incompletely identified *C*-glycosides. The other 27 are *O*-glycosides, mono- and di-glycosides, most of which were based on the above three aglycones.[115]

In individual plants of *Centaurea montana* originating from different restricted areas in the French southern Alps, in addition to practically pure *O*-glycosidic patterns, these molecules are arranged in many diverse assemblages of simple or complex *C*-glycoside derivatives — along with *O*-glycosides in some. This results in an extraordinary diversity of flavonoid patterns of this species, which comprise from five to more than 20 compounds.

This flavonoid variation is first correlated with the phytosociological origin of *Centaurea montana* plants: those displaying *C*-glycosidic patterns (all types) are largely predominant in meadows of *Triseto-Polygonion* while in tall grass prairies under *Larix* of *Adenostylion*, *O*-glycosidic types are the most frequent. In some meadows, the individuals with different types of *C*-glycosidic patterns are distributed according to microstational parameters as detected by aerial infrared remote sensing.

The origin of the huge chemical diversity in *Centaurea montana* was shown to result from its reproductive mode combining vegetative reproduction and strictly allogamous pollination (completely preventing autogamy and strongly limiting fertility between genetically related partners), both confirmed by flavonoid analysis of wild and experimental plants from breeding experiences. Clonal ramets are readily identifiable by identical flavonoid profiles of closely collected individuals. Interindividual diversity is also consistently observed in the progeny of most of the experimental crosses, revealing a generalized heterozygosis of wild individuals, an expectable consequence of the obligate allogamy of this species.

For instance, in a breeding experience involving two wild partners with *O*-glycosidic dominant profiles (more than 75% of the total flavonoids) and collected in a single meadow, one out the three descendants displayed a *C*-glycosidic pattern featuring 2″-*O*-glycosides of mono-*C*-glucosides (35% of the total) and their caffeoyl derivatives (45% of the total) along with the three basic *C*-glycosides (10%). By contrast, in the lineage of combinations of other individuals featuring *O*-glycosidic patterns also, only *O*-glycosidic phenotypes were observed, in which the traces of *C*-glycosides present in the parental fingerprints (10 to 20% of the total) completely disappeared. Consequently, *C*-glycosidic phenotypes seemed to proceed from a hypostatic determinism. This was extensively confirmed by many experiments with individuals displaying the two extreme phenotypes detected in *Centaurea montana*. Thus, when crossing plants with *O*-glycosidic (but *C*-glycosides — all types — remaining present, each 2 to 5% of the total) and *C*-glycosidic (blend of "complex" [85%] and "simple" [5%] types; *O*-glycosides: <10% of the total) patterns, 80% of the progeny displayed phenotypes close to that of the *O*-glycosidic parental type. Only one *C*-glycosidic pattern was observed in this progeny, in which the accumulation of caffeoyl conjugates was noticeably reduced.

In addition to a hypostatic determinism of *C*-glycosidic pathways vs. *O*-glycosidic ones, the variations observed suggest that each of the biosynthetic steps of the *C*-glycosidic metabolism is independently affected by heterogeneity. Regarding the genetic control involved, the loss or gain of molecules or their quantitative variations suggests the existence of heterozygosis of dominant or recessive characters in a polygenic system.[117]

Flavonoid glycoside variations of comparable amplitude are also observable in the individual fingerprints of sister species *Centaurea triumfetti*.[116] All the features of variation in *Centaurea triumfetti* closely compare to those observed in *Centaurea montana* and its origin probably rests on the same biological parameters, mainly obligate allogamy coupled to vegetative multiplication. The main difference between these two species is *Centaurea triumfetti*, which completely lacks the three caffeoyl derivatives of *O*-glucosyl-*C*-glucosyl flavones. Based on reports of the flavonoid chemistry of tetraploids (Mears, 1980), this feature can be regarded as a clue to the existence of relationships between these species; e.g., *Centaurea montana* ($2n = 44$) could be the tetraploid of *Centaurea triumfetti* ($2n = 22$). In this framework, the gain in the flavonoid profiles of tetraploid cytotypes of "new" compounds (complex *C*-glycosides, here) structurally based on those present in the profiles of diploid cytotypes (basic mono-*C*-glycosides and their *O*-glucosidic derivatives) is interpretable as functional (de)repression of existing structural genes (present but inactive in the diploid cytotypes) during the polyploidization process. Consequently, the large presence of simple *C*-glycosides in the fingerprints of *Centaurea triumfetti* individuals (diploids), which are the biosynthetic precursors of most of the major substituted molecules in the profiles of *Centaurea montana* (tetraploids), can be regarded as a sign of the existence of relationships between these two species, in the context of a polyploidization process.

14.2 NATURALLY OCCURRING *C*-GLYCOSYLFLAVONOIDS

Natural *C*-glycosylflavonoids are presented in four groups: the mono-*C*-glycosylflavonoids (Table 14.2), the di-*C*-glycosylflavonoids (Table 14.3), the *O*-glycosyl-*C*-glycosylflavonoids (Table 14.4), and the *O*-acyl-*C*-glycosylflavonoids (Table 14.5).

For each compound, the first line or reference indicates the first mention of this molecule in the phytochemical literature. Additional lines (when present) give information on other citations during the last 12 years. New *C*-glycosylflavonoids described since 1992 are specifically marked by an asterisk for their first mention in the plant kingdom.

14.2.1 MONO-*C*-GLYCOSYLFLAVONOIDS

For the period under review, 17 new mono-*C*-glycosylflavonoids were described. Two interesting features are evident:

- Seven new aglycones support the *C*-glycosidic bond in this class: 5,7-dihydroxyflavone (chrysin) in *Scutellaria*,[257,413] 5-hydroxy 7-methoxyflavone (tectochrysin) in *Piper*,[265] 5,7,2',4',5'-pentahydroxyflavone (isoetin) in *Hordeum*,[279] 5,7,2',3',5',6'-hexahydroxyflavone in *Polygala*,[195] 3,5,7,4'-tetrahydroxy 3',5'-dimethoxyflavone (myricitrin) in *Moghania*,[398] 5,7,2',4',5'-pentahydroxyflavonol (5'-OH morin) in *Bombax*,[99] and 5,7-dihydroxy 6,2',4',5'-tetramethoxyisoflavone in *Dalbergia*.[306]
- Three original *C*-substitutions other than sugar have been observed: 4-hydroxy-1-ethyl benzene in Cucumerin A and B isolated from *Cucumis*,[228] *p*-hydroxybenzyl in *Citrullus*,[223] and 1,5,8-trihydroxy 3-methoxyxanthone in *Swertia*[381] (Figure 14.1).

14.2.2 DI- AND TRI-*C*-GLYCOSYLFLAVONOIDS

Thirteen new compounds were found for the first time during the last 12 years:

- Two new sugars were linked by a *C*-glycosidic bond: β-D-ribopyranoside in *Passiflora*[56] and β-D-6-deoxygulopyranoside in *Viola*.[46] A 3,6,8-tri-*C*-xylosylflavone was isolated from *Asplenium*.[156]
- Several isomers were detected under the di-*C*-arabinosylflavone patterns: di-*C*-α-L-arabinopyranosyl in *Schnabellia*,[76] *Plagiochila*, and *Plagiochasma*,[318] 6-*C*-α-L-arabinopyranosyl-8-*C*-β-L-arabinopryanosyl in *Viola*,[401] and 6-*C*-β-L-arabinopyranosyl-8-*C*-α-L-arabinopyranosyl in *Schnabellia*.[76]

TABLE 14.2
Naturally Occurring Mono-*C*-Glycosylflavonoids. (Hypothetic Structures Previously Mentioned in the Last Edition Have Been Removed If Not Confirmed in Their True Structure; the New Compounds for the Period 1992 to 2004 Are Indicated by an Asterisk [*])

Compounds	Sources	Ref.
C-GLYCOSYLFLAVONES		
(*) 8-*C*-β-D-Glucosylchrysin		
(5,7-diOH 8-Gl)	*Scutellaria baicalensis* (Lab.)	Miyaichi and Tomimori (1994)
		Zhang et al. (1997)
	Scutellaria amoena (Lab.)	Zhou and Yang (2000)
(*) 6-*C*-β-D-Glucosylchrysin		
(5,7-diOH 6-Gl)	*Scutellaria baicalensis* (Lab.)	Zhang et al. (1997)
(*) Kaplanin		
(5-OH 7-OMe 8-Glc)	*Piper ihotzkyanum* (Pip.)	Moreira et al. (2000)
Bayin		
(7,4′-diOH 8-Glc)	*Castanospermum australe* (Leg.)	Eade et al. (1962, 1966)
Isovitexin (saponaretin)		
(5,7,4′-triOH 6-Glc)	*Vitex lucens* (Verb.)	Horowitz and Gentili (1964)
	Many sources	
(*) Cucumerin B		
8-*C*-(4-OH-1-ethyl benzene) isovitexin	*Cucumis sativus* (Cucur.)	McNally et al. (2003)
(*) 8-*C*-*p*-OH benzyl isovitexin		
(5,7,4′-triOH 6-Glc 8-OH-benzyl)	*Citrullus colocynthis* (Cucur.)	Maatooq et al. (1997)
Vitexin		
(5,7,4′-triOH 8-Glc)	*Vitex lucens* (Verb.)	Horowitz and Gentili (1964)
	Many sources	
(*) Cucumerin A		
6-*C*-(4-OH-1-ethyl benzene) vitexin	*Cucumis sativus* (Cucur.)	McNally et al. (2003)
(*) 6-*C*-*p*-OH benzyl vitexin		
(5,7,4′-triOH 8-Glc 6-OH-benzyl)	*Citrullus colocynthis* (Cucur.)	Maatooq et al. (1996)
8-*C*-β-D-Galactopyranosylapigenin		
(5,7,4′-triOH 8-Gal)	*Briza media* (Gram.)	Castledine and Harborne (1976)
Cerarvensin		
(5,7,4′-triOH 6-Xyl)	*Cerastium arvense* (Caryo.)	Dubois et al. (1982)
Isomollupentin		
(5,7,4′-triOH 6-Ara)	*Spergularia rubra* (Caryo.)	Bouillant et al. (1979)
Mollupentin		
(5,7,4′-triOH 8-Ara)	*Mollugo pentaphylla* (Mollug.)	Chopin et al. (1979)
Isofurcatain		
(5,7,4′-triOH 6-Rha)	*Metzgeria furcata* (Bryo.)	Markham et al. (1982)
	Otanthus maritimus (Comp.)	El-Sayed et al. (1992)
3′-Deoxyderhamnosylmaysin		
(5,7,4′-triOH 6-(6-deoxy-xylo-hexos-4-ulosyl))	*Zea mays* (Gram.)	Elliger et al. (1980)
6-*C*-β-D-Galactopyranosylapigenin		
(5,7,4′-triOH 6-Gal)	*Semecarpus kurzii* (Anacard.)	Jain et al. (1990)
Torosaflavone A		
(5,7,4′-triOH 6-Olio)	*Cassia torosa* (Leg.)	Kitanaka et al. (1989)
Swertisin		
(5,4′-diOH 7-OMe 6-Glc)	*Swertia japonica* (Gent.)	Komatsu and Tomimori (1966)
	Iris germanica (Irid.)	Ali et al. (1993)

TABLE 14.2
Naturally Occurring Mono-*C*-Glycosylflavonoids. (Hypothetic Structures Previously Mentioned in the Last Edition Have Been Removed If Not Confirmed in Their True Structure; the New Compounds for the Period 1992 to 2004 Are Indicated by an Asterisk [*]) — *continued*

Compounds	Sources	Ref.
	Passiflora incarnata (Passifl.)	Rahman et al. (1997)
	Zizyphus jujuba (Rham.)	Cheng et al. (2000)
Isoswertisin		
(5,4′-diOH 7-OMe 8-Glc)	*Centaurea cyanus* (Comp.)	Asen and Jurd (1967)
	Deschampsia antarctica (Gram.)	Webby and Markham (1994)
8-*C*-Rhamnosylgenkwanin		
(5,4′-diOH 7-OMe 8-Rha)	*Adina cordifolia* (Rub.)	Srivastava and Srivastava (1986)
Isomolludistin		
(5,4′-diOH 7-OMe 6-Ara)	*Mollugo distica* (Mollug.)	Chopin et al. (1978)
Molludistin		
(5,4′-diOH 7-OMe 8-Ara)	*Mollugo distica* (Mollug.)	Chopin et al. (1978)
Isocytisoside		
(5,7-diOH 4′-OMe 6-Glc)	*Fortunella margarita* (Rut.)	Horowitz et al. (1974)
	Combretum quadrangulare (Comb.)	Banskota et al. (2000)
Cytisoside		
(5,7-diOH 4′-OMe 8-Glc)	*Cytisus laburnum* (Leg.)	Paris (1957)
8-*C*-Rhamnosyl-5-*O*-methylacacetin		
(7-OH 5,4′-diOMe 8-Rha)	*Adina cordifolia* (Rub.)	Srivastava and Srivastava (1986)
Embigenin		
(5-OH 7,4′-diOMe 6-Glc)	*Iris tectorum* (Irid.)	Hirose et al. (1962)
Isoembigenin		
(5-OH 7,4′-diOMe 8-Glc)	*Siphonoglossa sessilis* (Acan.)	Hilsenbeck and Mabry (1983)
	Ornithocephalinae (Orch.)	Williams et al. (1994a)
7,4-Di-*O*-methylisomollupentin		
(5-OH 7,4′-diOMe 6-Ara)	*Asterostigma riedelianum* (Arac.)	Markham and Williams (1980)
8-*O*-methylswertisin (precatorin I)		
(5,4′-diOH 7,8-diOMe 6-Glc)	*Siphonoglossa* sp. (Acan.)	Hilsenbeck and Mabry (1990)
	Abrus precatorius (Leg.)	Ma et al. (1998)
6-*C*-Galactosylisoscutellarein		
(5,7,8,4′-tetraOH 6-Gal)	*Stellaria dichotoma* (Caryo.)	Yasukawa et al. (1982)
8-*C*-Glucosyl-6,7-di-*O*-methyl-scutellarein	*Siphonoglossa* sp. (Acan.)	Hilsenbeck and Mabry (1990)
(abrusin)		
(5,4′-diOH 6,7-diOMe 8-Glc)	*Abrus precatorius* (Leg.)	Markham et al. (1989)
Isoorientin		
(5,7,3′,4′-tetraOH 6-Glc)	*Polygonum Orientale* (Polyg.)	Hörhammer et al. (1958)
	Many sources	
Orientin		
(5,7,3′,4′-tetraOH 8-Glc)	*Polygonum Orientale* (Polyg.)	Hörhammer et al. (1958)
	Many sources	
6-*C*-β-ᴅ-Xylopyranosylluteolin		
(5,7,3′,4′-tetraOH 6-Xyl)	*Phlox drummondii* (Polem.)	Mabry et al. (1971)
Derhamnosylmaysin		
(5,7,3′,4′-tetraOH 6-(6-deoxyxylo-hexos-4-ulosyl))	*Zea mays* (Gram.)	Elliger et al. (1980)

continued

TABLE 14.2
Naturally Occurring Mono-C-Glycosylflavonoids. (Hypothetic Structures Previously Mentioned in the Last Edition Have Been Removed If Not Confirmed in Their True Structure; the New Compounds for the Period 1992 to 2004 Are Indicated by an Asterisk [*]) — continued

Compounds	Sources	Ref.
6-C-Galactosylluteolin		
(5,7,3',4'-tetraOH 6-Gal)	*Muhlenbergia* sp. (Gram.)	Peterson and Rieseberg (1987)
	Zea mays (Gram.)	Snook et al. (1994)
8-C-Galactosylluteolin		
(5,7,3',4'-tetraOH 8-Gal)	*Parkinsonia aculeata* (Leg.)	El-Sayed et al. (1990)
6-C-α-L-Arabinosylluteolin		
(5,7,3'4'-tetraOH 6-Ara)	*Muhlenbergia* sp. (Gram.)	Herrera and Bain (1991)
	Sasa borealis (Gram.)	Yoon et al. (2000)
8-C-α-L-Arabinosylluteolin		
(5,7,3',4'-tetraOH 8-Ara)	*Mucuna sempervirens* (Leg.)	Ishikura and Yoshitama (1988)
6-C-Quinovopyranosylluteolin		
(5,7,3',4'-tetraOH 6-Chino)	*Passiflora edulis* (Pass.)	Mareck et al. (1991)
6-C-Fucopyranosylluteolin		
(5,7,3',4'-tetraOH 6-Fuco)	*Passiflora edulis* (Pass.)	Mareck et al. (1991)
(*) Demethyltorosaflavone C	*Cassia nomame* (Leg.)	Kitanaka and Takido (1992)
Parkinsonin A		
(7,3',4'-triOH 5-OMe 8-Glc)	*Parkinsonia aculeata* (Leg.)	Bhatia et al. (1966)
Swertiajaponin		
(5,3'4'-triOH 7-OMe 6-Glc)	*Swertia japonica* (Gent.)	Komatsu and Tomimori (1966)
	Deschampsia antarctica (Gram.)	Webby and Markham (1994)
Isoswertiajaponin		
(5,3',4'-triOH 7-OMe 8-Glc)	*Gnetum gnemon* (Gnet.)	Wallace and Morris (1978)
	Deschampsia antarctica (Gram.)	Webby and Markham (1994)
(+)Isoscoparin		
(5,7,4'-triOH 3'-OMe 6-Glc)	*Hordeum vulgare* (Gram.)	Seikel et al. (1962)
	Barbarea vulgaris (Brass.)	Senatore et al. (2000)
	Centaurea triumfetti (Comp.)	Gonnet (1993)
	Citrullus colocynthis (Cucur.)	Maatooq et al. (1997)
	Vellozia sp. (Velloz.)	Williams et al. (1994b)
(−)Isoscoparin		
(5,7,4'-triOH 3'-OMe 6-Glc)	*Arenaria kansuensis* (Caryo.)	Wu et al. (1990)
Scoparin		
(5,7,4'-triOH 3'-OMe 8-Glc)	*Sarothamnus scoparius* (Leg.)	Chopin et al. (1968)
(*) 6-C-β-Fucopyranosylchrysoeriol		
(5,7,4'-triOH 3'-OMe 6-Fuc)	*Zea mays* (Gram.)	Suzuki et al. (2003b)
6-C-β-L-Boivinopyranosyl-Chrysoeriol: alternanthin		
(5,7,4'-triOH 3'-OMe 6-Boiv)	*Alternanthera philoxeroides* (Amar.)	Zhou et al. (1988)
	Zea mays (Gram.)	Suzuki et al. (2003a)
3'-O-Methylderhamnosylmaysin		
(5,7,4'-triOH 3'-OMe 6-(6-deoxyxylo-hexos-4-ulosyl))	*Zea mays* (Gram.)	Elliger et al. (1980)
6-C-β-D-Glucopyranosyldiosmetin		

TABLE 14.2
Naturally Occurring Mono-*C*-Glycosylflavonoids. (Hypothetic Structures Previously Mentioned in the Last Edition Have Been Removed If Not Confirmed in Their True Structure; the New Compounds for the Period 1992 to 2004 Are Indicated by an Asterisk [*]) — *continued*

Compounds	Sources	Ref.
(5,7,3'-triOH 4'-OMe 6-Glc)	*Citrus limon* (Rut.)	Gentili and Horowitz (1968)
8-*C*-β-D-Glucopyranosyldiosmetin		
(5,7,3'-triOH 4'-OMe 8-Glc)	*Citrus limon* (Rut.)	Gentili and Horowitz (1968)
Torosaflavone B		
(5,7,3'-triOH 4'-OMe 6-Olio)	*Cassia torosa* (Leg.)	Kitanaka et al. (1989)
	Cassia occidentalis (Leg.)	Hatano et al. (1999)
Parkinsonin B		
(3',4'-diOH 5,7-diOMe 8-Glc)	*Parkinsonia aculeata* (Leg.)	Bhatia et al. (1966)
7,3'-Di-*O*-methylisoorientin		
(5,4'-diOH 7,3'-diOMe 6-Glc)	*Achillea cretica* (Comp.)	Valant et al. (1980)
7,3'-Di-*O*-methylorientin		
(5,4'-diOH 7,3'-diOMe 8-Glc)	*Saccharum* sp. (Gram.)	Mabry et al. (1984)
6-*C*-β-D-Glucopyranosylpilloin		
(5,3'-diOH 7,4'-diOMe 6-Glc)	*Parkinsonia aculeata* (Leg.)	El-Sayed et al. (1991)
7,3',4'-Tri-*O*-methylisoorientin		
(5-OH 7,3',4'-triOMe 6-Glc)	*Linum maritimum* (Lin.)	Volk and Sinn (1968)
Isoaffinetin		
(5,7,3',4',5'-pentaOH 6-Glc)	*Polygonum affine* (Polyg.)	Krause (1976a,b)
	Plumbago zeylanica (Plumb.)	Lin and Chou (2003)
	Frullania sp. (Bryo.)	Kraut et al. (1993)
Affinetin		
(5,7,3',4',5'-pentaOH 8-Glc)	*Trichomanes venosum* (Pterido.)	Markham and Wallace (1980)
Isopyrenin		
(5,7,4'-triOH 3',5'-diOMe 6-Glc)	*Gentiana pyrenaica* (Gent.)	Marston et al. (1976)
6-*C*-Glucosyl-5,7-dihydroxy-8,3',4',5'-tetramethoxyflavone	*Vitex negundo* (Verb.)	Subramanian and Misra (1979)
(*) 6-*C*-β-D-Glucopyranosyl 5,7,2',4',5'-Pentahydroxyflavone	*Hordeum vulgare* (Gram.)	Norbaek et al. (2000)
(*) Telephioidin		
(5,7,2',3',5',6'-hexahydroxy-6-*C*-β-D-Glc)	*Polygala telephioides* (Poly.)	Kumar et al. (1999)
C-GLYCOSYLFLAVONOLS		
8-*C*-Glucosyl-5-deoxykæmpferol		
(3,7,4'-triOH 8-Glc)	*Pterocarpus marsupium* (Leg.)	Bezuidenhoudt et al. (1987)
8-*C*-Glucosylfisetin		
(3,7,3',4'-tetraOH 8-Glc)	*Pterocarpus marsupium* (Leg.)	Bezuidenhoudt et al. (1987)
6-*C*-Glucosylkæmpferol		
(3,5,7,4'-tetraOH 6-Glc)	*Zelkova* sp. (Ulm.)	Hayashi et al. (1987)
	Cyclopia intermedia (Leg.)	Kamara et al. (2003)
(*) 8-*C*-β-D-Glucopyranosylkaempferol		
(3,5,7,4'-tetraOH 8-Glc)	*Cyclopia intermedia* (Leg.)	Kamara et al. (2003)
Keyakinin		
(3,5,4'-triOH 7-OMe 6-Glc)	*Zelkova serrata* (Ulm.)	Funaoka (1956)
6-*C*-Glucosylquercetin		

continued

TABLE 14.2
Naturally Occurring Mono-*C*-Glycosylflavonoids. (Hypothetic Structures Previously Mentioned in the Last Edition Have Been Removed If Not Confirmed in Their True Structure; the New Compounds for the Period 1992 to 2004 Are Indicated by an Asterisk [*]) — *continued*

Compounds	Sources	Ref.
(3,5,7,3′,4′-pentaOH 6-Glc) Keyakinin B	*Ageratina calophylla* (Comp.)	Fang et al. (1986)
(3,5,3′,4′-tetraOH 7-OMe 6-Glc)	*Zelkova serrata* (Ulm.)	Hillis and Horn (1966)
8-*C*-Rhamnosyl-5,7,3′-trihydroxy-3,4′-dimethoxy flavone	*Adina cordifolia* (Rub.)	Srivastava and Srivastava (1986)
8-*C*-Rhamnosyleuropetin		
(3,5,3′,4′,5′-pentaOH 7-OMe 8-Rha) (*) Moghanin A	*Cassia sophera* (Leg.)	Tiwari and Bajpai (1981)
(3,5,7,4′-tetraOH 3′,5′-diOMe 6-*C*-β-D-Glc) (*) Shamimin	*Moghania macrophylla* (Leg.)	Wu et al. (1997)
3,5,7,2′,4′,5′-hexaOH 6-*C*-Glc	*Bombax ceiba* (Bomb.)	Faizi and Ali (1999)
C-GLYCOSYLFLAVANONES		
Aervanone		
(7,4-diOH 8-Gal)	*Aervia persica* (Amar.)	Garg et al. (1980)
Hemiphloin		
(5,7,4′-triOH 6-Glc)	*Eucalyptus hemiphloia* (Myrt.) *Acacia saligna* (Leg.) *Betula platyphylla* (Bet.)	Hillis and Carle (1963) El-Shafae et al. (1998) Lee (1994)
Isohemiphloin		
(5,7,4′-triOH 8-Glc)	*Eucalyptus hemiphloia* (Myrt.)	Hillis and Horn (1965)
Palodulcin B		
(7,3′,4′-triOH 8-Gal)	*Eysenhardtia polystachia* (Leg.)	Vita-Finzi et al. (1980)
C-GLYCOSYLFLAVANONOLS		
(*) 8-*C*-Glucosyldihydrokaempferol		
(3,5,7,4′-tetraOH 8-Glc)	*Acacia saligna* (Leg.)	El-Sawi (2001)
6-*C*-Glucosyldihydrokæmpferol		
(3,5,7,4′-tetraOH 6-Glc)	*Zelkova* sp. (Ulm.) *Betula platytphylla* (Bet.)	Hayashi et al. (1987) Lee (1994)
Keyakinol		
(3,5,4′-triOH 7-OMe 6-Glc)	*Zelkova serrata* (Ulm.)	Funaoka (1956)
6-*C*-Glucosyldihydroquercetin		
(3,5,7,3′,4′-pentaOH 6-Glc)	*Zelkova* sp. (Ulm.)	Hayashi et al. (1987)
C-GLYCOSYLCHALCONES		
3′-*C*-Glucosylisoliquiritigenin		
(2′,4,4′-triOH 3′-Glc)	*Cladrastis platycarpa* (Leg.)	Ohashi et al. (1977)
C-GLYCOSYLDIHYDROCHALCONES		
Nothofagin		
(2′,4′,6′,4-tetraOH *C*-Gly)	*Nothofagus fusca* (Fag.)	Hillis and Inoue (1967)
Konnanin		
(2′,4′,6′,3,4-pentaOH *C*-Gly)	*Nothofagus fusca* (Fag.)	Hillis and Inoue (1967)
Aspalathin		
(2′,4′,6′,3,4-pentaOH 3′-Glc)	*Aspalathus linearis* (Leg.)	Dahlgren (1963)

TABLE 14.2
Naturally Occurring Mono-*C*-Glycosylflavonoids. (Hypothetic Structures Previously Mentioned in the Last Edition Have Been Removed If Not Confirmed in Their True Structure; the New Compounds for the Period 1992 to 2004 Are Indicated by an Asterisk [*])
— *continued*

Compounds	Sources	Ref.
C-GLYCOSYL-α-HYDROXYDIHYDROCHALCONES		
Coatline A		
(α-2′,4′,4-tetraOH 3′-Glc)	*Eysenhardtia polystachya* (Leg.)	Beltrami et al. (1982)
Coatline B		
(α-2′,4′,3,4-pentaOH 3′-Glc)	*Eysenhardtia polystachya* (Leg.)	Beltrami et al. (1982)
C-GLYCOSYL-β-HYDROXYDIHYDROCHALCONES		
Pterosupin		
(β-2,2′,4′,4-tetraOH 3′-Glc)	*Pterocarpus marsupium* (Leg.)	Adinarayana et al. (1982)
C-GLYCOSYLFLAVANOLS		
6-*C*-Glucosyl-(−)-epicatechin		
(3,5,7,3′,4′-pentaOH 6-Glc)	*Cinnamomum cassia* (Laur.)	Morimoto et al. (1986a)
8-*C*-Glucosyl-(−)-epicatechin		
(3,5,7,3′,4′-pentaOH 8-Glc)	*Cinnamomum cassia* (Laur.)	Morimoto et al. (1986a)
C-GLYCOSYLPROANTHOCYANIDINS		
6-*C*-Glucosylprocyanidin B2	*Cinnamomum cassia* (Laur.)	Morimoto et al. (1986b)
8-*C*-Glucosylprocyanidin B2	*Cinnamomum cassia* (Laur.)	Morimoto et al. (1986b)
C-GLYCOSYLQUINOCHALCONE		
Carthamin	*Carthamus tinctorius* (Comp.)	Takahashi et al. (1982)
C-GLYCOSYLISOFLAVONES		
Puerarin		
(7,4′-diOH 8-Glc)	*Pueraria thunbergiana* (Leg.)	Murakami et al. (1960)
	Zizyphus jujuba (Rham.)	Cheng et al. (2000)
8-*C*-β-D-Glucopyranosylgenistein		
(5,7,4′-triOH 8-Glc)	*Lupinus luteus* (Leg.)	Zapesochnaya and Laman (1977)
	Lupinus luteus (Leg.)	Zavodnik et al. (2000)
8-*C*-Glucosylprunetin		
(5,4′-diOH 7-OMe 8-Glc)	*Dalbergia paniculata* (Leg.)	Parthasarathy et al. (1974)
Isovolubilin		
(5-OH 7,4′-diOMe 6-Rha)	*Dalbergia volubilis* (Leg.)	Chawla et al. (1974)
Volubilin		
(5-OH 7,4′-diOMe 8-Rha)	*Dalbergia volubilis* (Leg.)	Chawla et al. (1974)
6-*C*-Glucosylorobol		
(5,7,3′,4′-tetraOH 6-Glc)	*Dalbergia monetaria* (Leg.)	Nunes et al. (1989)
8-*C*-Glucosylorobol		
(5,7,3′,4′-tetraOH 8-Glc)	*Lupinus luteus* (Leg.)	Zapesochnaya and Laman (1977)
Dalpanitin		
(5,7,4′-triOH 3′-OMe 8-Glc)	*Dalbergia paniculata* (Leg.)	Adinarayana and Rao (1972)
Volubilinin		
(5,7-diOH 6,4′-diOMe 8-Glc)	*Dalbergia volubilis* (Leg.)	Chawla et al. (1976)
(*) Dalpaniculin		

| (5,7-diOH 2′,4′,5′,6-tetraOMe 8-Glc) | *Dalbergia paniculata* (Leg.) | Rao and Rao (1991) |

TABLE 14.2
Naturally Occurring Mono-*C*-Glycosylflavonoids. (Hypothetic Structures Previously Mentioned in the Last Edition Have Been Removed If Not Confirmed in Their True Structure; the New Compounds for the Period 1992 to 2004 Are Indicated by an Asterisk [*]) — continued

Compounds	Sources	Ref.
C-GLYCOSYLISOFLAVANONES		
Dalpanin	*Dalbergia paniculata* (Leg.)	Adinarayana and Rao (1975)
Macrocarposide		
6-*C*-Glucosyldalbergioidin		
(5,7,2′,4′-tetraOH 6-Glc)	*Pterocarpus macrocarpus* (Leg.)	Verma et al. (1986)
C-GLYCOSYLCHROMONES		
Aloeresin B		
(7-OH 5-Me 2-Acetonyl 8-Glc)	*Aloe* sp. (Lil.)	Haynes and Holsworth (1970)
7-*O*-Methyl-5-methyl 2-(2-hydroxy) propyl 8-*C*-glucoside	*Aloe* sp. (Lil.)	Speranza et al. (1986)
C-GLYCOSYLFLAVONE-XANTHONES		
(*) Swertifrancheside		
5,7,3′,4′-tetraOH 6-*C*-β-D		
Glucopyranosyl 8-*C*-(1‴,5‴,8‴-triOH 3‴-OMe Xanthonyl) flavone	*Swertia franchetiana* (Gent.)	Wang et al. (1994)

14.2.3 *O*-GLYCOSYL-*C*-GLYCOSYLFLAVONOIDS

Forty-six new compounds were described during the period for review.

A large number of new flavonoid aglycones in mono-*C*-glycosides have been described as supports for *O*-glycosidic bonds:

– On apigenin base: 8-*C*-*p*-hydroxybenzylvitexin in *Citrullus*,[223] apigenin-6-*C*-(6″-*O*-galactosyl) galactoside in *Cecropia*,[285] mollupentin in *Allophylus*,[143] 8-methoxyswertisin in *Abrus*,[222] and 6-*C*-(6-deoxy-ribo-hexos-3-ulosyl)apigenin in *Cassia*.[130]
– On luteolin base: 6-*C*-fucosylluteolin and 6-*C*-quinovosylluteolin in *Zea*,[335] 6-*C*-(6-deoxy-ribo-hexos-3-ulosyl)luteolin in *Cassia*,[130] and 8-quinovosylluteolin in *Turnera*.[297]
– On 7-*O*-methylluteolin base: isoswertiajaponin in *Deschampsia*.[383]
– On chrysoeriol base: 6-*C*-boivinosylchrysoeriol, 6-*C*-fucosylchrysoeriol, and 6-*C*-(6-deoxy-ribo-hexos-4-ulosyl)chrysoeriol in *Zea*[333,335,345,346] and 8-*C*-xylosylchrysoeriol in *Scleranthus*.[403]
– On diosmetin base: torosaflavone B and 6-*C*-(6-deoxy-ribo-hexos-3-ulosyl)diosmetin in *Cassia*.[130]
– On flavonol base: 6-*C*-glucosyl-3-*O*-glucosyl kaempferol in *Cyclopia*,[168] 6-*C*-rhamnosylrhamnetin in *Frullania*,[329] and 8-*C*-glucosylquercetin in *Eucalyptus*.[232]

In addition, four groups of closely related molecules were described:

– Series of cassiaoccidentalins built on a 6-*C*-(6-deoxy-ribo-hexos-3-ulosyl)flavone general structure, and isolated from *Cassia*[130] — cassiaoccidentalin A: 6-*C*-glycosyl (2″-*O*-rhamnosyl)apigenin; B: 6-*C*-glycosyl (2″-*O*-rhamnosyl)luteolin; and C: 6-*C*-glycosyl (2″-*O*-rhamnosyl)diosmetin.

TABLE 14.3
Naturally Occurring Di-*C*-Glycosylflavonoids and Tri-*C*-Glycosylflavonoids. (Hypothetic Structures Previously Mentioned in the Last Edition Have Been Removed If Not Confirmed in Their True Structure; the New Compounds for the Period 1992 to 2004 Are Indicated by an Asterisk [*])

Compounds	Sources	Ref.
DI-*C*-GLYCOSYLFLAVONES		
6-*C*-Glucopyranosyl-8-*C*-arabinopyranosylchrysin		
(5,7-diOH 6-Glc 8-Ara)	*Scutellaria baicalensis* (Lab.)	Takagi et al. (1981)
	Scutellaria amoena (Lab.)	Zhou and Yang (2000)
6-*C*-Arabinopyranosyl-8-*C*-glucopyranosylchrysin		
(5,7-diOH 6-Ara 8-Glc)	*Scutellaria baicalensis* (Lab.)	Takagi et al. (1981)
Vicenin-2		
(5,7,4′-triOH 6,8-diGlc)	*Citrus lemon* (Rut.)	Chopin et al. (1964)
	Many sources	
3,6-Di-*C*-glucosylapigenin		
(5,7,4′-triOH 3,6-diGlc)	*Citrus unshiu* (Rut.)	Matsubara et al. (1985a)
3,8-Di-*C*-glucosylapigenin		
(5,7,4′-triOH 3,8-diGlc)	*Citrus sudachi* (Rut.)	Matsubara et al. (1985b)
6,8-Di-*C*-galactopyranosylapigenin		
(5,7,4′-triOH 6,8-diGal)	*Stellaria dichotoma* (Caryo.)	Yasukawa et al. (1982)
6-*C*-Glucopyranosyl-8-*C*-galactopyranosylapigenin		
(5,7,4′-triOH 6-Glc-8-Gal)	*Cerastium arvense* (Caryo.)	Dubois et al. (1984)
Vicenin-3		
(5,7,4′-triOH 6-Glc-8-Xyl)	*Vitex lucens* (Verb.)	Seikel et al. (1966)
Vicenin-1		
(5,7,4′-triOH 6-Xyl-8-Glc)	*Vitex lucens* (Verb.)	Seikel et al. (1966)
	Athyriaceae/Aspleniaceae (Pterid.)	Umikalson et al. (1994)
	Cotoneaster wilsonii (Ros.)	Chang and Jeon (2003)
	Prosopis chilensis (Leg.)	Elrady and Saad (1994)
Violanthin		
(5,7,4′-triOH 6-Glc-8-Rha)	*Viola tricolor* (Viol.)	Hörhammer et al. (1965)
	Viola arvensis (Viol.)	Carnat et al. (1998)
Isoviolanthin		
(5,7,4′-triOH 6-Rha-8-Glc)	*Angiopteris evecta* (Pterid.)	Wallace et al. (1979)
	Athyriaceae/Aspleniaceae (Pterid.)	Umikalson et al. (1994)
(*) 6-*C*-β-D-Glucopyranosyl-8-*C*-β-D-apiofuranosylapigenin		
(5,7,4′-triOH 6-Glc-8-Apio)	*Xanthosoma violaceum* (Arac.)	Picerno et al. (2003)
(*) 6-*C*-β-D-Glucopyranosyl-8-*C*-β-D-ribopyranosylapigenin		
(5,7,4′-triOH 6-Glc-8-Rib)	*Passiflora incarnata* (Passif.)	Chimichi et al. (1998)
Schaftoside		
(5,7,4′-triOH 6-Glc-8-Ara)	*Silene schafta* (Caryo.)	Chopin et al. (1974)
	Many sources	
Isoschaftoside		
(5,7,4′-triOH 6-Ara-8-Glc)	*Flourensia cernua* (Comp.)	Dillon et al. (1976)
	Many sources	
Neoschaftoside		
(5,7,4′-triOH 6-Glc-8-Ara)	*Catananche coerulea* (Comp.)	Proliac et al. (1973)

continued

TABLE 14.3
Naturally Occurring Di-*C*-Glycosylflavonoids and Tri-*C*-Glycosylflavonoids. (Hypothetic Structures Previously Mentioned in the Last Edition Have Been Removed If Not Confirmed in Their True Structure; the New Compounds for the Period 1992 to 2004 Are Indicated by an Asterisk [*]) — *continued*

Compounds	Sources	Ref.
	Atractylis carduus (Comp.)	Melek et al. (1992)
	Capsicum annuum (Solan.)	Materska et al. (2003)
	Viola yedoensis (Viol.)	Xie et al. (2003)
Neoisoschaftoside		
(5,7,4'-triOH 6-Ara-8-Glc)	*Minum undulatum* (Bryo.)	Osterdahl (1979)
Isocorymboside		
(5,7,4'-triOH 6-Gal-8-Ara)	*Polygonatum multiflorum* (Lil.)	Chopin et al. (1977b)
Corymboside		
(5,7,4'-triOH 6-Ara-8-Glc)	*Carlina corymbosa* (Comp.)	Besson et al. (1979)
6,8-Di-*C*-arabinosylapigenin		
(5,7,4'-triOH 6,8-diAra)	*Melilotus alba* (Leg.)	Specht et al. (1976)
(*) 6,8-Di-*C*-α-L-arabinopyranosylapigenin		
(5,7,4'-triOH 6,8-diAra)	*Schnabelia tetradonta* (Lab.)	Dou et al. (2002)
(*) 6-*C*-α-L-Arabinopyranosyl-8-*C*-β-L-arabinopyranosylapigenin		
(5,7,4'-triOH 6,8-diAra)	*Viola yedoensis* (Viol.)	Xie et al. (2003)
(*) 6-*C*-β-L-Arabinopyranosyl-8-*C*-α-L-arabinopyranosylapigenin		
(5,7,4'-triOH 6,8-diAra)	*Schnabelia tetradonta* (Lab.)	Dou et al. (2002)
6-*C*-β-D-Xylopyranosyl-8-*C*-α-L-arabinopyranosylapigenin		
(5,7,4'-triOH 6-Xyl-8-Ara)	*Mollugo pentaphylla* (Mollug.)	Chopin et al. (1982)
	Viola yedoensis (Viol.)	Xie et al. (2003)
6-*C*-Arabinosyl-8-*C*-xylosylapigenin		
(5,7,4'-triOH 6-Ara-8-Xyl)	*Mollugo pentaphylla* (Mollug.)	Chopin et al. (1982)
	Viola yedoensis (Viol.)	Xie et al. (2003)
Neocorymboside		
(5,7,4'-triOH 6-Ara-8-Gal)	*Atractylis gummifera* (Comp.)	Chaboud et al. (1988)
(*) 6-*C*-β-D-Glucopyranosyl-8-*C*-β-D-6-deoxygulopyranosylapigenin		
(5,7,4'-triOH 6-Glc-8-Deoxygul)	*Viola arvensis* (Viol.)	Carnat et al. (1998)
6-*C*-Glucosyl-8-*C*-galactosylgenkwanin		
(5,4'-diOH 7-OMe 6-Glc-8-Gal)	*Glycine max* (Leg.)	Jay et al. (1984)
6-*C*-Glucosyl-8-*C*-arabinosylgenkwanin		
(5,4'-diOH 7-OMe 6-Glc-8-Ara)	*Almeidea guyanensis* (Rut.)	Wirasutisna et al. (1986)
Almeidein		
(5,4'-diOH 7-OMe 6,8-diAra)	*Almeida guyanensis* (Rut.)	Jay et al. (1979)
6,8-Di-*C*-glucosylgenkwanin		
(5,4'-diOH 7-OMe 6,8-diGlc)	*Galipea trifoliata* (Rut.)	Baktiar et al. (1990)
3,6-Di-*C*-glucosylacacetin		
(5,7-diOH 4'-OMe 3,6-diGlc)	*Fortunella japonica* (Rut.)	Kumamoto et al. (1985b)
7,4'-Di-*O*-methyl-6,8-di-*C*-arabinosylapigenin		

(5-OH 7,4′-diOMe 6,8-diAra)   *Asterostigma riedelianum* (Arac.)   Markham and Williams (1980)

TABLE 14.3
Naturally Occurring Di-*C*-Glycosylflavonoids and Tri-*C*-Glycosylflavonoids. (Hypothetic Structures Previously Mentioned in the Last Edition Have Been Removed If Not Confirmed in Their True Structure; the New Compounds for the Period 1992 to 2004 Are Indicated by an Asterisk [*]) — *continued*

Compounds	Sources	Ref.
Lucenin-2		
(5,7,3′,4′-tetraOH 6,8-diGlc)	*Vitex lucens* (Verb.)	Seikel et al. (1966)
	Many sources	
Lucenin-3		
(5,7,3′,4′-tetraOH 6-Glc-8-Xyl)	*Vitex lucens* (Verb.)	Seikel et al. (1966)
Lucenin-1		
(5,7,3′,4′-tetraOH 6-Xyl-8-Glc)	*Vitex lucens* (Verb.)	Seikel et al. (1966)
Carlinoside		
(5,7,3′4′-tetraOH 6-Glc-8-Ara)	*Carlina vulgaris* (Comp.)	Raynaud and Rasolojaona (1976)
	Hordeum vulgare (Gram.)	Norbaek et al. (2000)
Isocarlinoside		
(5,7,3′,4′-tetraOH 6-Ara-8-Glc)	*Lespedeza capitata* (Leg.)	Linard et al. (1982)
	Capsicum annuum (Solan.)	Materska et al. (2003)
	Viola yedoensis (Viol.)	Xie et al. (2003)
Neocarlinoside		
(5,7,3′,4′-tetraOH 6-Glc-8-Ara)	*Lespedeza capitata* (Leg.)	Linard et al. (1982)
(*) 6-*C*-β-D-Glucopyranosyl-8-*C*-α-L-rhamnopyranosylluteolin (Elatin)		
(5,7,3′,4′-tetraOH 6-Glc-8-Rha)	*Plagiomnium elatum* (Bryo.)	Anhut et al. (1992)
(*) 6,8-Di-*C*-α-L-arabinosylluteolin		
(5,7,4′-triOH 6,8-diAra)	*Plagiochasma rupestre* (Bryo.)	Schoeneborn and Mues (1993)
(*) 6-*C*-hexosyl-8-*C*-rhamnosylchrysoeriol		
(5,7,4′-triOH 3′-OMe 6-hex-8-rha)	*Plagiomnium elatum* (Bryo.)	Anhut et al. (1992)
6,8-Di-*C*-Glucosylchrysoeriol (Stellarin-2)		
(5,7,4′-triOH 3′-OMe 6,8-diGlc)	*Stellaria holostea* (Caryo.)	Zoll and Nouvel (1974)
	Cydonia oblonga (Ros.)	Ferreres et al. (2003)
	Citrus limon (Rut.)	Gil Izquierdo et al. (2004)
6-*C*-Glucosyl-8-*C*-arabinosylchrysoeriol		
(5,7,4′-triOH 3′-OMe 6-Glc-8-Ara)	*Trichophorum cespitorum* (Cyp.)	Salmenkallio et al. (1982)
(*) 6-*C*-β-D-Glucosyl-8-*C*-β-D-xylosylchrysoeriol		
(5,7,4′-triOH 3′-OMe 6-Glc-8-Xyl)	*Raputia paraensis* (Rut.)	Bakhtiar et al. (1991)
(*) 6-*C*-Xylosyl-8-*C*-glucosylchrysoeriol		
(5,7,4′-triOH 3′-OMe 6-Xyl-8-Glc)	*Raputia paraensis* (Rut.)	Bakhtiar et al. (1991)
6-*C*-Arabinosyl-8-*C*-glucosylchrysoeriol		
(5,7,4′-triOH 3′-OMe 6-Ara-8-Glc)	*Trichophorum cespitorum* (Cyp.)	Salmenkallio et al. (1982)
6,8-Di-*C*-glucosyldiosmetin		
(5,7,3′-triOH 4′-OMe 6,8-diGlc)	*Citrus limon* (Rut.)	Kumamoto et al. (1985a)
	Citrus sp. (Rut.)	Tsiklauri and Shalashvili (1995)
3,8-Di-*C*-glucosyldiosmetin		
(5,7,3′-triOH 4′-OMe 3,8-diGlc)	*Citrus sudachi* (Rut.)	Matsubara et al. (1985b)
6,8-Di-*C*-glucosyltricetin		
(5,7,3′,4′,5′-pentaOH 6,8-diGlc)	*Plagiochila asplenoides* (Bryo.)	Mues and Zinsmeister (1976)
	Frullania polysticha (Bryo.)	Kraut et al. (1993)

continued

TABLE 14.3
Naturally Occurring Di-*C*-Glycosylflavonoids and Tri-*C*-Glycosylflavonoids. (Hypothetic Structures Previously Mentioned in the Last Edition Have Been Removed If Not Confirmed in Their True Structure; the New Compounds for the Period 1992 to 2004 Are Indicated by an Asterisk [*]) — *continued*

Compounds	Sources	Ref.
	Plagiochila jamesonii (Bryo.)	Schoeneborn and Mues (1993)
6-*C*-β-D-Glucosyl-8-*C*-α-L-arabinosyltricetin		
(5,7,3′,4′,5′-pentaOH 6-Glc-8-Ara)	*Radula complanata* (Bryo.)	Markham and Mues (1984)
	Plagiochila jamesonii (Bryo.)	Schoeneborn and Mues (1993)
6-*C*-Arabinosyl-8-*C*-glucosyltricetin		
(5,7,3′,4′,5′-pentaOH 6-Ara-8-Glc)	*Apometzgeria pubescens* (Bryo.)	Theodor et al. (1980, 1981a)
(*) 6,8-Di-*C*-α-L-arabinopyranosyltricetin		
(5,7,3′,4′,5′-pentaOH 6,8-diAra)	*Plagiochila jamesonii* (Bryo.)	Schoeneborn and Mues (1993)
6,8-Di-*C*-glucosyltricin		
(5,7,4′-triOH 3′,5′-diOMe 6,8-diGlc)	*Apometzgeria pubescens* (Bryo.)	Theodor et al. (1980, 1981a)
6-*C*-Glucosyl-8-*C*-arabinosyltricin		
(5,7,4′-triOH 3′,5′-diOMe 6-Glc-8-Ara)	*Apometzgeria pubescens* (Bryo.)	Theodor et al. (1980, 1981a)
6-*C*-Arabinosyl-8-*C*-glucosyltricin		
(5,7,4′-triOH 3′,5′-diOMe 6-Ara-8-Glc)	*Apometzgeria pubescens* (Bryo.)	Theodor et al. (1980, 1981a)
6,8-Di-*C*-arabinosyltricin		
(5,7,4′-triOH 3′,5′-diOMe 6,8-diAra)	*Apometzgeria pubescens* (Bryo.)	Theodor et al. (1980, 1981a)
6,8-Di-*C*-arabinosylapometzgerin		
(5,7,5′-triOH 3′,4′-diOMe 6,8-diAra)	*Apometzgeria pubescens* (Bryo.)	Theodor et al. (1980, 1981a)
DI-*C*-GLYCOSYLFLAVANONES		
6,8-Di-*C*-glucosylnaringenin		
(5,7,4′-triOH 6,8-diGlc)	*Zizyphus jujuba* (Rham.)	Okamura et al. (1981)
DI-*C*-GLYCOSYLISOFLAVONES		
Paniculatin		
(5,7,4′-triOH 6,8-diGlc)	*Dalbergia paniculata* (Leg.)	Narayanan and Seshadri (1971)
6,8-Di-*C*-glucosylorobol		
(5,7,3′,4′-tetraOH 6,8-diGlc)	*Dalbergia nitidula* (Leg.)	Van Heerden et al. (1980)
DI-*C*-GLYCOSYLQUINOCHALCONES		
Safflor yellow A	*Carthamus tinctoria* (Comp.)	Takahashi et al. (1982)
Safflor yellow B	*Carthamus tinctoria* (Comp.)	Takahashi et al. (1982)
TRI-*C*-GLYCOSYLFLAVONES		
(*) 3,6,8-Tri-*C*-xylosylapigenin		
(5,7,4′-triOH 3,6,8-triXyl)	*Asplenium viviparum* (Pteri.)	Imperato (1993)

- Series of rhamnellaflavosides based on 6-*C*-glycosyl (4′-*O*-glucosyl)apigenin, with three new compounds characterized in *Rhamnella*[350] — ramnellaflavoside A: 6-*C*-β-D-oliopyranosyl; B: 6-*C*-β-D-boivinopryanosyl; and C: 6-*C*-β-D-4-epioliosyl.
- Series of maysins (6-*C*-glycosyl, 2″-*O*-rhamnosylflavone) isolated from *Zea*,[333,335,345,346] with the following basic structures: 6-*C*-(6-deoxy-ribo-hexos-4-ulosyl)apigenin, 6-*C*-fucosylluteolin, 6-*C*-quinovosylluteolin, 6-*C*-β-fucosylchrysoeriol, 6-*C*-β-boivinosylchrysoeriol, and 6-*C*-(6-deoxy-ribo-hexos-4-ulosyl)chrysoeriol.
- Series of precatorins isolated from *Abrus*[222] with precatorin II: 2″-*O*-apiofuranosyl 8-*O*-methylswertisin; III: 2″-*O*-apiofuranosylswertisin (precatorin I: 8-*O*-methylswertisin).

TABLE 14.4
Naturally Occurring *C*-Glycosylflavonoids *O*-Glycosides. (Hypothetic Structures Previously Mentioned in the Last Edition Have Been Removed If Not Confirmed in Their True Structure; the New Compounds for the Period 1992 to 2004 Are Indicated by an Asterisk [*])

Compounds	Sources	Ref.
MONO-*C*-GLYCOSYLFLAVONES		
Bayin		
2″-*O*-Rhamnoside (sophoraflavone A)	*Sophora subprostata* (Leg.)	Shirataki et al. (1986)
Isovitexin		
X‴-Rhamnoside	*Bocoa* sp. (Leg.)	Kite and Ireland (2002)
X″-Xyloside	*Itea/Pterostemom* (Saxif.)	Bohm et al. (1999)
7-*O*-Glucoside (saponarin)	*Saponaria officinalis* (Caryo.)	Barger (1906)
	Many sources	
7-*O*-Galactoside (neosaponarin)	*Melandrium album* (Caryo.)	Wagner et al. (1979)
7-*O*-Rhamnoside	*Yeatesia viridiflora* (Acan.)	Hilsenbeck et al. (1984)
7-*O*-Xyloside	*Melandrium album* (Caryo.)	Van Brederode and Nigtevecht (1972)
7-*O*-Rhamnosylglucoside	*Passiflora platyloba* (Pass.)	Ayanoglu et al. (1982)
7,2″-Di-*O*-glucoside	*Melandrium album* (Caryo.)	Van Brederode and Nigtevecht (1974)
7,2″-Di-*O*-galactoside	*Gentiana depressa* (Gent.)	Chulia (1984)
7-*O*-Glucoside 2″-*O*-arabinoside	*Melandrium album* (Caryo.)	Van Brederode and Nigtevecht (1972)
7-*O*-Glucoside 2″-*O*-rhamnoside	*Melandrium album* (Caryo.)	Van Brederode and Nigtevecht (1972)
7-*O*-Galactoside 2″-*O*-glucoside	*Melandrium album* (Caryo.)	Wagner et al. (1979)
7-*O*-Galactoside 2″-*O*-rhamnoside	*Melandrium album* (Caryo.)	Wagner et al. (1979)
7-*O*-Galactoside 2″-*O*-arabinoside	*Silene pratensis* (Caryo.)	Steyns et al. (1983)
7-*O*-Xyloside 2″-*O*-glucoside	*Melandrium album* (Caryo.)	Van Brederode and Nigtvecht (1974)
7-*O*-Xyloside 2″-*O*-rhamnoside	*Melandrium album* (Caryo.)	Van Brederode and Nigtvecht (1974)
7-*O*-Xyloside 2″-*O*-arabinoside	*Melandrium album* (Caryo.)	Van Brederode and Nigtvecht (1974)
7-*O*-Arabinoside 2″-*O*-glucoside	*Silene dioica* (Caryo.)	Mastenbroek et al. (1983)
4′-*O*-Glucoside (isosaponarin)	*Spirodela oligorrhiza* (Lemn.)	Jurd et al. (1957)
4′-*O*-Arabinoside	*Leptodactylon* sp. (Polem.)	Smith et al. (1982)
4′-*X*-*O*-Diglucoside	*Cucumis sativus* (Cucurb.)	Krauze and Cisowski (2001)
4′,2″-Di-*O*-glucoside	*Gentiana asclepiadea* (Gent.)	Goetz and Jacot-Guillarmod (1977)
4′-*O*-Glucoside 2″-*O*-arabinoside	*Vaccaria segetalis* (Caryo.)	Baeva et al. (1974)
(*) 4′-*O*-Rhamnoside	*Xanthosoma violaceum* (Arac.)	Picerno et al. (2003)
2″-*O*-Arabinoside	*Avena sativa* (Gram.)	Chopin et al. (1977a)
	Centaurea triumfetti (Comp.)	Gonnet (1993)
2″-*O*-Galactoside	*Secale cereale* (Gram.)	Dellamonica et al. (1983)
2″-*O*-Glucoside	*Oxalis acetosella* (Oxal.)	Tschesche and Struckmeyer (1976)
	Centaurea triumfetti (Comp.)	Gonnet (1993)
	Cucumis sativus (Cucurb.)	Krauze and Cisowski (2001)
	Passiflora incarnata (Passif.)	Li et al. (1991), Rahman et al. (1997)
	Zizyphus jujuba (Rham.)	Cheng et al. (2000)
2″-*O*-Ramnoside	*Crataegus monogyna* (Ros.)	Nikolov et al. (1976)
	Allophylus edulis (Sapin.)	Hoffmann et al. (1992)
	Biophytum sensitivum (Oxal.)	Bucar et al. (1998)
	Tripterospermum japonicum (Gent.)	Otsuka and Kijima (2001)
2″-*O*-Xyloside	*Desmodium canadense* (Leg.)	Chernobrovaya (1973)
	Tripterospermum japonicum (Gent.)	Otsuka and Kijima (2001)
6″-*O*-Arabinoside	*Swertia perennis* (Gent.)	Hostettmann and Jacot-Guillarmod (1976)

continued

TABLE 14.4
Naturally Occurring *C*-Glycosylflavonoids *O*-Glycosides. (Hypothetic Structures Previously Mentioned in the Last Edition Have Been Removed If Not Confirmed in Their True Structure; the New Compounds for the Period 1992 to 2004 Are Indicated by an Asterisk [*]) — *continued*

Compounds	Sources	Ref.
(*) 6″-*O*-Glucoside	*Gentiana arisanensis* (Gent.)	Lin et al. (1997)
(*) 6″-*O*-Rhamnoside	*Vitis* sp. (Vitid.)	Moore and Giannasi (1994)
	Vigna radiata (Legum.)	Larsen et al. (1995)
8-*C-p*-Hydroxybenzylisovitexin		
(*) 4′-*O*-Glucoside	*Citrullus colocynthis* (Cucurb.)	Maatooq et al. (1997)
Vitexin		
X″-*O*-Rhamnoside	*Bocoa* sp. (Leg.)	Kite and Ireland (2002)
X″-*O*-Xyloside	*Itea/Pterostemom* (Saxif.)	Bohm et al. (1999)
7-*O*-Glucoside	*Trigonella fœnumgreacum* (Leg.)	Adamska and Lutomski (1971)
7-*O*-Rhamnosylglucoside	*Phoenix* sp. (Palm.)	Williams et al. (1973)
4′-*O*-Glucoside	*Brisa media* (Gram.)	Williams and Murray (1972)
4′-*O*-Galactoside	*Crotalaria retusa* (Leg.)	Srinivasan and Subramanian (1983)
4′-*O*-Rhamnoglucoside	*Crataegus oxyacantha* (Ros.)	Lewak (1966)
4′-*O*-Glucoside 2″-*O*-rhamnoside	*Passiflora coactilis* (Pass.)	Escobar et al. (1983)
(*) 2″-*O*-α-D-Arabinofuranoside	*Cotoneaster thymaefolia* (Ros.)	Palme et al. (1994)
2″-*O*-Glucoside	*Polygonatum odoratum* (Lil.)	Morita et al. (1976)
	Alternanthera maritima (Amaranth.)	Salvador and Dias (2004)
2″-*O*-Rhamnoside	*Crataegus monogyna* (Ros.)	Nikolov et al. (1976)
	Alternanthera maritima (Amaranth.)	Salvador and Dias (2004)
	Allophylus edulis (Sapind.)	Hoffmann et al. (1992)
	Clusia sandiensis (Gutt.)	Delle Monache (1991)
	Cotoneaster thymaefolia (Ros.)	Palme et al. (1994)
	Crataegus monogyna (Ros.)	Rehwald et al. (1994)
	Crataegus sinaica (Ros.)	Kim and Kim (1993)
	Crataegus pinnatifida (Ros.)	Zhang and Xu (2002)
	Rhamnella inaequilatera (Rham.)	Takeda et al. (2003)
	Turnera diffusa (Turn.)	Piacente et al. (2002)
2″-*O*-Xyloside	*Vitex lucens* (Verb.)	Seikel et al. (1959)
	Beta vulgaris (Chenop.)	Gil et al. (1998)
	Citrus sp. (Rut.)	Tsiklauri and Shalashvili (1995), Manthey et al. (2001)
2″-*O*-Sophoroside	*Polygonatum odoratum* (Lil.)	Morita et al. (1976)
4″-*O*-Rhamnoside	*Crataegus curvisepala* (Ros.)	Batyuk et al. (1966)
	Silene conoidea (Caryo.)	Ali et al. (1999)
6″-*O*-Gentiobioside (marginatoside)	*Piper marginatum* (Pip.)	Tillequin et al. (1978)
(*) 6″-*O*-Glucoside	*Xanthosoma violaceum* (Arac.)	Picerno et al. (2003)
6″-*O*-Rhamnoside	*Larix sibirica* (Gymno.)	Medvedeva et al. (1974)
6″-*O*-Xyloside	*Gypsophila paniculata* (Caryo.)	Darmograi et al. (1968)
(*) 6″-*O*-Rhamnosyl-4‴-*O*-glucosyl-2‴-*O*-galactosyl (panzhihuacycaside)	*Cycas panzhihuaensis* (Cyc.)	Zhou et al. (2002)
6-*C*-Xylosylapigenin (cerarvensin)		
7-*O*-Glucoside	*Cerastium arvense* (Caryo.)	Dubois et al. (1980)

2″-O-Rhamnoside   *Phlox drummondii* (Polem.)   Bouillant et al. (1978)

TABLE 14.4
Naturally Occurring C-Glycosylflavonoids O-Glycosides. (Hypothetic Structures Previously Mentioned in the Last Edition Have Been Removed If Not Confirmed in Their True Structure; the New Compounds for the Period 1992 to 2004 Are Indicated by an Asterisk [*]) — *continued*

Compounds	Sources	Ref.
	Allophylus edulis (Sapind.)	Hoffmann et al. (1992)
	Rhamnella inaequilatera (Rham.)	Takeda et al. (2004)
6-C-Arabinosylapigenin (isomollupentin)		
7-O-Glucoside	*Cerastium arvense* (Caryo.)	Dubois et al. (1985)
	Climacanthus nutans (Acant.)	Teshima et al. (1997)
7-O-Rhamnosylglucoside	*Passiflora platyloba* (Pass.)	Ayanoglu et al. (1982)
7,2″-Di-O-glucoside	*Spergularia rubra* (Caryo.)	Bouillant et al. (1979)
7-O-Glucoside 2″-O-arabinoside	*Cerastium arvense* (Caryo.)	Dubois et al. (1983)
7-O-Glucoside 2″-O-xyloside	*Cerastium arvense* (Caryo.)	Dubois et al. (1983)
4′-O-Glucoside	*Cerastium arvense* (Caryo.)	Dubois et al. (1985)
2″-O-Glucoside	*Cerastium arvense* (Caryo.)	Dubois et al. (1985)
4″-O-Rhamnoside (hemsleyanoside)	*Tetrastigma hemsleyanum* (Vitid.)	Liu et al. (2002)
6-C-β-D-Oliopyranosylapigenin		
(*) 4′-O-β-D-Glucopyranoside (rhamnellaflavoside A)	*Rhamnella inaequilatera* (Rham.)	Takeda et al. (2004)
6-C-β-D-Boivinopyranosylapigenin		
(*) 4′-O-β-D-Glucopyranoside (rhamnellaflavoside B)	*Rhamnella inaequilatera* (Rham.)	Takeda et al. (2004)
6-C-β-D-4-Epioliosylapigenin		
(*) 4′-O-β-D-Glucopyranoside (rhamnellaflavoside C)	*Rhamnella inaequilatera* (Rham.)	Takeda et al. (2004)
6-C -Rhamnosylapigenin (isofurcatain)		
7-O-Glucoside	*Metzgeria furcata* (Bryo.)	Markham et al. (1982)
6-C-Galactosylapigenin		
(*) 6″-O-Galactoside	*Cecropia lyratifolia* (Urt.)	Oliveira et al. (2003)
6-C-(6-Deoxy-xylo-hexos-4-ulosyl)apigenin		
2″-O-Rhamnoside (apimaysin)	*Zea mays* (Gram.)	Elliger et al. (1980)
6-C-(6-Deoxy-ribo-hexos-3-ulosyl)apigenin		
(*) 2″-O-Rhamnoside (cassiaoccidentalin A)	*Cassia occidentalis* (Leg.)	Hatano et al. (1999)
8-C-Arabinosylapigenin (mollupentin)		
(*) 2″-O-Rhamnoside	*Allophylus edulis* (Sapind.)	Hoffmann et al. (1992)
(*) 4″-O-Rhamnoside (isohemsleyanoside)	*Tetrastigma hemsleyanum* (Vitid.)	Liu et al. (2002)
Swertisin		
5-O-Glucoside	*Enicostemma hyssopifolium* (Gent.)	Ghosal and Jaiswal (1980)
4′-O-Glucoside		
4′-O-Glucoside (flavocommelinin)	*Commelina* sp.(Comm.)	Komatsu et al. (1968)
4′-O-Rhamnoside	*Passiflora biflora* (Pass.)	McCormick and Mabry (1983)
(*) 4′-O- Glucoside 2″-O-rhamnoside	*Felicia amelloides* (Comp.)	Bloor (1999)
(*) 2″-O-Apiofuranoside (precatorin III)	*Abrus precatorius* (Leg.)	Ma et al. (1998)
2″-O-Arabinoside	*Achillea fragrantissima* (Comp.)	Ahmed et al. (1988)
2″-O-Glucoside (spinosin)	*Zizyphus vulgaris* (Rham.)	Woo et al. (1979)
	Desmodium tortuosum (Leg.)	Lewis et al. (2000)
	Zizyphus jujuba (Rham.)	Cheng et al. (2000)
2″-O-Rhamnoside	*Gemmingia chinensis* (Irid.)	Shirane et al. (1982)
	Aleurites moluccana (Euphorb.)	Meyre-Silva et al. (1999)

continued

TABLE 14.4
Naturally Occurring *C*-Glycosylflavonoids *O*-Glycosides. (Hypothetic Structures Previously Mentioned in the Last Edition Have Been Removed If Not Confirmed in Their True Structure; the New Compounds for the Period 1992 to 2004 Are Indicated by an Asterisk [*]) — *continued*

Compounds	Sources	Ref.
4″-*O*-Glucoside (zivulgarin)	*Zizyphus spinosus* (Rham.)	Zeng et al. (1987)
6″-*O*-Rhamnoside (fagovatin)	*Fagraea obovata* (Logan.)	Qasim et al. (1987)
8-*O*-Methylswertisin		
(*) 2″-*O*-Apiofuranoside (precatorin II)	*Abrus precatorius* (Leg.)	Ma et al. (1998)
Isoswertisin		
5-*O*-Glucoside	*Enicostemma hyssopifolium* (Gent.)	Ghosal and Jaiswal (1980)
4′-*O*-Glucoside	*Triticum aestivum* (Gram.)	Julian et al. (1971)
(*) 2″-*O*-β-Arabinoside	*Deschampsia antarctica* (Gram.)	Webby and Markham (1994)
2″-*O*-Glucoside (isospinosin)	*Gnetum* sp. (Gymn.)	Ouabonzi et al. (1983)
	Zizyphus jujuba (Rham.)	Cheng et al. (2000)
2″-*O*-Rhamnoside	*Avena sativa* (Gram.)	Chopin et al. (1977a)
2″-*O*-Xyloside	*Gnetum* sp. (Gymn.)	Ouabonzi et al. (1983)
Isomolludistin		
2″-*O*-Glucoside	*Asterostigma riedelianum* (Arac.)	Markham and Williams (1980)
Molludistin		
2″-*O*-Glucoside	*Almeidea guyanensis* (Rut.)	Jay et al. (1979)
2″-*O*-Xyloside	*Almeidea guyanensis* (Rut.)	Wirasutisna et al. (1986)
2″-*O*-Rhamnoside	*Mollugo distica* (Mollug.)	Chopin et al. (1978)
Isocytisoside		
7-*O*-Glucoside	*Gentiana pyrenaica* (Gent.)	Marston et al. (1976)
2″-*O*-Glucoside	*Securigera coronilla* (Leg.)	Jay et al. (1980)
2″-*O*-Rhamnoside (isomargariten)	*Fortunella margarita* (Rut.)	Horowitz et al. (1974)
	Ornothicephalinae (Orch.)	Williams et al. (1994 a)
(*) 3″-*O*-α-L-Rhamnopyranoside	*Anthurium versicolor* (Arac.)	Aquino et al. (2001)
(*) 3″-*O*-β-D-Xylopyranoside	*Anthurium versicolor* (Arac.)	Aquino et al. (2001)
(*) 6″-*O*-β-D-Apiofuranoside	*Anthurium versicolor* (Arac.)	Aquino et al. (2001)
Cytisoside		
7-*O*-Glucoside	*Trema aspera* (Ulm.)	Oelrichs et al. (1968)
2″-*O*-Rhamnoside (margariten)	*Fortunella margarita* (Rut.)	Horowitz et al. (1974)
(*) 3″-*O*-β-D-Rhamnopyranoside	*Anthurium versicolor* (Arac.)	Aquino et al. (2001)
Embigenin		
2″-*O*-Glucoside (embinoidin)	*Siphonoglossa sessilis* (Acan.)	Hilsenbeck and Mabry (1983)
2″-*O*-Rhamnoside (embinin)	*Iris tectorum* (Irid.)	Hirose et al. (1962)
	Ornithocephalinae (Orch.)	Williams et al. (1994a)
7-4′-Di-*O*-methylisomollupentin		
2″-*O*-Glucoside	*Asterostigma riedelianum* (Arac.)	Markham and Williams (1980)
Abrusin		
2″-*O*-β-apiofuranoside	*Abrus precatorius* (Leg.)	Markham et al. (1989)
	Abrus precatorius (Leg.)	Ma et al. (1998)
Isoorientin		
X″-*O*-Xyloside	*Itea/Pterostemom* (Saxif.)	Bohm et al. (1999)
7-*O*-Apioside	*Vellozia* sp. (Vell.)	Williams et al. (1991)
	Vellozia sp. (Velloz.)	Harborne et al. (1993)
	(Velloz.)(Velloz.)	Williams et al. (1994b)

TABLE 14.4
Naturally Occurring C-Glycosylflavonoids O-Glycosides. (Hypothetic Structures Previously Mentioned in the Last Edition Have Been Removed If Not Confirmed in Their True Structure; the New Compounds for the Period 1992 to 2004 Are Indicated by an Asterisk [*]) — continued

Compounds	Sources	Ref.
7-O-Galactoside	*Eminium spiculatum* (Epacr.)	Shammas and Couladi (1988)
7-O-Glucoside (lutonarin)	*Hordeum vulgare* (Gram.)	Seikel and Bushnell (1959)
	Biophytum sensitivum (Oxal.)	Bucar et al. (1998)
	Bryonia sp. (Cucurb.)	Krauze and Cisowski (1994)
	Hordeum vulgare (Gram.)	Markham and Mitchell (2003)
	Lagenaria siceraria (Cucurb.)	Krauze and Cisowski (1994)
	Plagiomnium sp. (Bryo.)	Anhut et al. (1992)
	Vellozia sp. (Velloz.)	Harborne et al. (1993, 1994)
	(Velloz.)	Williams et al. (1994b)
(*) 7-O-Di-glucoside	*Vellozia* sp. (Vell.)	Williams et al. (1991)
(*) 7-O-Rhamnoside	*Hyparrhenia hirta* (Gram.)	Bouaziz et al. (2001)
7-O-Rhamnosylglucoside	*Triticum aestivum* (Gram.)	Julian et al. (1971)
(*) 7-O-Xyloside	*Vellozia* sp. (Velloz.)	Harborne et al. (1993, 1994)
	(Velloz.)	Williams et al. (1994b)
3'-O-Glucoside	*Gentiana nivalis* (Gent.)	Hostettmann andet Jacot-Guillarmod (1974)
3'-O-Glucuronide	*Rynchospora eximia* (Cyp.)	Williams and Harborne (1977)
3'-O-Neohesperidoside	*Plagiomnium affine* (Bryo.)	Freitag et al. (1986)
3'-O-Sophoroside	*Plagiomnium affine* (Bryo.)	Freitag et al. (1986)
3',6''-Di-O-glucoside	*Gentiana pedicellata* (Gent.)	Chulia and Mariotte (1985)
4'-O-Glucoside	*Briza media* (Gram.)	Williams and Murray (1972)
4',2''-Di-O-glucoside	*Gentiana asclepiadea* (Gent.)	Goetz and Jacot-Guillarmod (1977)
(*) 2''-O-Apiofuranoside	*Achillea nobilis* (Comp.)	Marchart et al. (2003), Krenn et al. (2003)
2''-O-Arabinoside	*Avena sativa* (Gram.)	Chopin et al. (1977a)
	Centaurea triumfetti (Comp.)	Gonnet (1993)
2''-O-β-L-Arabinofuranoside	*Trichomanes venosum* (Pterido.)	Markham and Wallace (1980)
2''-O-Glucoside	*Gentiana verna* (Gent.)	Hostettmann and Jacot-Guillarmod (1975)
	Centaurea triumfetti (Comp.)	Gonnet (1993)
	Passiflora incarnata (Passif.)	Qimin et al. (1991)
		Rahman et al. (1997)
2''-O-Mannoside	*Poa annua* (Gram.)	Rofi and Pomilio (1987)
2''-O-Rhamnoside	*Coronilla varia* (Leg.)	Sherwood et al. (1973)
	Biophytum sensitivum (Oxal.)	Bucar et al. (1998)
2''-O-Xyloside	*Desmodium canadense* (Leg.)	Chernobrovaya (1973)
6''-O-Arabinoside	*Swertia perennis* (Gent.)	Hostettmann and Jacot-Guillarmod (1976)
6''-O-Glucoside	*Gentiana pedicellata* (Gent.)	Chulia and Mariotte (1985), Chulia et al. (1986)
	Gentiana arisanensis (Gent.)	Lin et al. (1997)
6''-O-Di-glucoside	*Triticum* sp. (Gram.)	Harborne et al. (1986a)
6''-O-Rhamnoside	*Triticum* sp. (Gram.)	Harborne et al. (1986)
Orientin		
7-O-Glucoside	*Phœnix canariensis* (Palm.)	Harborne et al. (1974)
	Vellozia sp. (Vell.)	Williams et al. (1994b)
7-O-Rhamnoside	*Linum usitatissimum* (Lin.)	Ibrahim and Shaw (1970)
4'-O-Glucoside	*Briza media* (Gram.)	Williams and Murray (1972)
4'-O-Glucoside 2''-O-rhamnoside	*Passiflora coactilis* (Pass.)	Escobar et al. (1983)

continued

TABLE 14.4
Naturally Occurring *C*-Glycosylflavonoids *O*-Glycosides. (Hypothetic Structures Previously Mentioned in the Last Edition Have Been Removed If Not Confirmed in Their True Structure; the New Compounds for the Period 1992 to 2004 Are Indicated by an Asterisk [*]) — *continued*

Compounds	Sources	Ref.
2″-*O*-β-L-Arabinofuranoside	*Trichomanes venosum* (Pterido.)	Markham and Wallace (1980)
(*) 2″-*O*-α-L-Arabinopyranoside	*Deschampsia antarctica* (Gram.)	Webby and Markham (1994)
2″-*O*-Glucoside	*Cannabis sativa* (Cannab.)	Segelman et al. (1978)
2″-*O*-Rhamnoside	*Crataegus monogyna* (Ros.)	Nikolov et al. (1976)
	Allophylus edulis (Sapind.)	Hoffmann et al. (1992)
	Turnera diffusa (Turn.)	Piacente et al. (2002)
2″-*O*-Xyloside (adonivernith)	*Adonis vernalis* (Ranun.)	Hörhammer et al. (1960)
6-*C*-(6-Deoxy-xylo-hexos-4-ulosyl)luteolin		
2″-*O*-α-L-Rhamnoside (maysin)	*Zea mays* (Gram.)	Waisse et al. (1979)
6-*C*-Fucosylluteolin		
(*) 2″-*O*-α-L-Rhamnoside	*Zea mays* (Gram.)	Snook et al. (1995)
(ax-4″-OH maysin)	*Mimosa pudica* (Mim.)	Lobstein et al. (2002)
6-*C*-Quinovopyranosylluteolin		
(*) 2″-*O*-α-L-Rhamnoside	*Zea mays* (Gram.)	Snook et al. (1995)
(eq-4″-OH maysin)		
6-*C*-(6-Deoxy-ribo-hexos-3-ulosyl)luteolin		
(*) 2″-*O*-Rhamnoside (cassiaoccidentalin B)	*Cassia occidentalis* (Leg.)	Hatano et al. (1999)
	Mimosa pudica (Leg.)	Lobstein et al. (2002)
6-*C*-Xylosylluteolin		
2″-*O*-Rhamnoside	*Phlox drummondii* (Polem.)	Bouillant et al. (1984)
8-*C*-Quinovopyranosylluteolin		
(*) 2″-*O*-α-L-Rhamnopyranoside	*Turnera diffusa* (Turn.)	Piacente et al. (2002)
Swertiajaponin		
3′-*O*-Gentiobioside	*Phragmites australis* (Gram.)	Nawwar et al. (1980)
3′-*O*-Glucoside	*Phragmites australis* (Gram.)	Nawwar et al. (1980)
(*) 4′-*O*-Di-glucoside	*Cucumis sativus* (Cucurb.)	Krauze and Cisowski (2001)
4′-*O*-Rhamnoside	*Passiflora biflora* (Pass.)	McCormick and Mabry (1983)
2″-*O*-Glucoside (luteoayamenin)	*Iris nertshinskia* (Irid.)	Hirose et al. (1981)
2″-*O*-Rhamnoside	*Securigera coronilla* (Leg.)	Jay et al. (1980)
Isoswertiajaponin		
(*) 2″-*O*-Arabinopyranoside	*Deschampsia antarctica* (Gram.)	Webby and Markham (1994)
6-*C*-β-Boivinopyranosylchrysoeriol		
(*) 7-*O*-β-Glucopyranoside	*Zea mays* (Gram.)	Suzuki et al. (2003b)
6-*C*-β-Fucosylchrysoeriol		
(*) 2″-*O*-Rhamnoside		
(ax-4″-OH 3′-OMe maysin)	*Zea mays* (Gram.)	Snook et al. (1995)
6-*C*-(6-Deoxy-xylo-hexos-4-ulosyl)chrysoeriol		
(*) 2″-*O*-α-L-Rhamnoside	*Zea mays* (Gram.)	Snook et al. (1993)
(3′-OMe maysin)		
8-*C*-Xylopyranosylchrysoeriol		
(*) 2″-*O*-Glucoside	*Scleranthus uncinatus* (Caryo.)	Yayli et al. (2001)
8-*C*-β-L-Xylofuranosylchrysoeriol		

(*) 2″-O-Glucoside   *Scleranthus uncinatus* (Caryo.)   Yayli et al. (2001)

TABLE 14.4
Naturally Occurring *C*-Glycosylflavonoids *O*-Glycosides. (Hypothetic Structures Previously Mentioned in the Last Edition Have Been Removed If Not Confirmed in Their True Structure; the New Compounds for the Period 1992 to 2004 Are Indicated by an Asterisk [*]) — *continued*

Compounds	Sources	Ref.
Isoscoparin		
7-O-Glucoside	*Hordeum vulgare* (Gram.)	Seikel et al. (1962)
	Hordeum vulgare (Gram.)	Norbaek et al. (2000)
	Plagiomnium sp. (Bryo.)	Anhut et al. (1992)
2″-O-Glucoside	*Oryza sativa* (Gram.)	Besson et al. (1985)
	Centaurea trimufetti (Comp.)	Gonnet (1993)
	Passiflora incarnata (Pass.)	Rahman et al. (1997)
2″-O-Rhamnoside	*Silene alba* (Caryo.)	Van Brederode and Kamps-Heinsbroek (1981)
Scoparin		
2″-O-Glucoside	*Setaria italica* (Gram.)	Gluchoff-Fiasson et al. (1989)
2″-O-Rhamnoside	*Passiflora coactilis* (Pass.)	Escobar et al. (1983)
2″-O-Xyloside	*Setaria italica* (Gram.)	Gluchoff-Fiasson et al. (1989)
(*) 6″-O-Glucoside	*Turnera diffusa* (Turn.)	Piacente et al. (2002)
Episcoparin		
7-O-Glucoside (knautoside)	*Knautia montana* (Dips.)	Zemtsova and Bandyukova (1974)
Torosaflavone B		
(*) 3′-O-β-D-Glucoside	*Cassia torosa* (Leg.)	Kitanaka and Takido (1992)
	Cassia occidentalis (Leg.)	Hatano et al. (1999)
8-*C*-Glucosyldiosmetin		
2″-O-Rhamnoside	*Fortunella japonica* (Rut.)	Kumamoto et al. (1985b)
(*) 4″-O-Rhamnopyranoside	*Silene conoidea* (Caryo.)	Ali et al. (1999)
6-*C*-(6-Deoxy-ribo-hexos-3-ulosyl)diomestin		
(*) 2″-O-Rhamnoside (cassiaoccidentalin C)	*Cassia occidentalis* (Leg.)	Hatano et al. (1999)
7,3′,4′-Tri-*O*-methylisoorientin		
2″-O-Rhamnoside (linoside B)	*Linum maritimum* (Lin.)	Volk and Sinn (1968)
Isopyrenin		
7-O-Glucoside	*Gentiana pyrenaica* (Gent.)	Marston et al. (1976)
6-*C*-Glucosyl-5,7-dihydroxy-8,3′,4′,5′-tetramethoxyflavone		
5-O-Rhamnoside	*Vitex negundo* (Verb.)	Subramanian and Misra (1979)
MONO-*C*-GLYCOSYLFLAVONOLS		
6-*C*-β-D-Glucopyranosylkaempferol		
(*) 3-O-β-D-Glucopyranoside	*Cyclopia intermedia* (Leg.)	Kamara et al. (2003)
6-*C*-Rhamnopyranosylrhamnetin		
(*) 3-O-Glucopyranoside	*Frullania tamarisei* (Bryo.)	Singh and Singh (1991)
8-*C*-Glucosylquercetin		
(*) 2″-O-Rhamnoside	*Eucalyptus globulus* (Myrt.)	Manguro et al. (1995)
MONO-*C*-GLYCOSYLISOFLAVONES		
Puerarin		
6″-O-β-Apiofuranoside (mirificin)	*Pueraria mirifica* (Leg.)	Ingham et al. (1986)
6″-O-Xyloside	*Pueraria lobata* (Leg.)	Kinjo et al. (1987)

continued

TABLE 14.4
Naturally Occurring *C*-Glycosylflavonoids *O*-Glycosides. (Hypothetic Structures Previously Mentioned in the Last Edition Have Been Removed If Not Confirmed in Their True Structure; the New Compounds for the Period 1992 to 2004 Are Indicated by an Asterisk [*]) — *continued*

Compounds	Sources	Ref.
8-*C*-Glucosylgenistein		
6''-*O*-Apiosyl	*Pueraria lobata* (Leg.)	Kinjo et al. (1987)
MONO-*C*-GLYCOSYLCHROMONES		
Chromones		
Aloeresin B		
7-*O*-Glucoside	*Aloe* sp. (Lil.)	Speranza et al. (1985)
7-Hydroxy-5-methyl-2-acetonyl 6-*C*-glucopyranosylchromone 2''-*O*-glucoside	*Chrozophora prostrata* (Euph.)	Agrawal and Singh (1988)
DI-*C*-GLYCOSYLFLAVONES		
Vicenin-2		
6''-*O*-Glucoside	*Stellaria holostea* (Caryo.)	Bouillant et al. (1984)
(*) 2''-2'''-Di-*O*-β-Glucopyranoside	*Ephedra aphylla* (Ephed.)	Hussein et al. (1997)
Schaftoside		
6''-*O*-Glucoside	*Stellaria holostea* (Caryo.)	Bouillant et al. (1984)
6,8-Di-*C*-glucosylgenkwanin		
2'''-*O*-Xyloside	*Galipea trifoliata* (Rut.)	Bakhtiar et al. (1990)
Lucenin-2		
3'-*O*-Glucoside	*Pleurozia* sp. (Bryo.)	Mues et al. (1991)

Finally, in two molecules the flavone structure was linked to four sugar residues: 2'',2'''-di-*O*-β-glucopyranosylvicenin-2 with two *C*-glycosides and two *O*-glycosides, isolated from *Ephedra*,[153] and 6''-*O*-rhamnosyl 4'''-*O*-glucosyl 2'''-*O*-galactosylvitexin with one *C*-glycoside and three *O*-glycosides, named panzhihuacycaside, and isolated from *Cycas*[416] (Figure 14.2).

14.2.4 *O*-ACYL-*C*-GLYCOSYLFLAVONOIDS

Forty new compounds have been reported in this class during the last 12 years. Many references concerned the well-known *O*-acetyl, *O*-caffeoyl, *O*-*p*-coumaroyl, and *O*-feruloyl esters. The *p*-hydroxybenzoyl and *p*-methoxybenzoyl groups gave rise to two new natural 6''-*O*-acylated isoorientin derivatives in *Polygonum*[418]: perfoliatumin A and B.

However, the true novelty comes with four *O*-galloyl esters identified in *Terminalia*[213] and *Pelargonium*,[200] six *O*-methylbutyryl esters in *Trollius*,[420] one *O*-malonyl ester in *Beta*,[109] and one *O*-(3-hydroxy-3-methyl)glutaroyl ester in *Glycyrrhiza*.[216]

14.3 SEPARATION AND IDENTIFICATION OF *C*-GLYCOSYLFLAVONOIDS

14.3.1 CAPILLARY ZONE ELECTROPHORESIS SEPARATION

For the separation of *C*-glycosylflavones, high-performance liquid chromatography (HPLC) has been mainly used in recent works. However, we have to underline a few interesting contributions on the performance of capillary zone electrophoresis (CZE) in this field.

TABLE 14.5
Naturally Occurring Acyl-C-Glycosylflavonoids. (Hypothetic Structures Previously Mentioned in the Last Edition Have Been Removed If Not Confirmed in Their True Structure; the New Compounds for the Period 1992 to 2004 Are Indicated by an Asterisk [*])

Compounds	Sources	Ref.
MONO-C-GLYCOSYLFLAVONES		
Isovitexin		
(*) 7-O-(6‴-Caffeoyl)-β-D-glucopyranoside	*Bryonia dioica* (Cucurb.)	Krauze and Cisowski (1995a)
(*) 7-O-(6‴-O-E-p-Coumaroyl)glucoside	*Hordeum vulgare* (Gram.)	Norbaek et al. (2003)
7-O-Feruloylglucoside	*Silene pratense* (Caryo.)	Niemann (1981, 1982)
	Hordeum vulgare (Gram.)	Norbaek et al. (2003)
(*) 7-O-(6‴-O-E-Feruloyl)glucoside	*Hordeum vulgare* (Gram.)	Norbaek et al. (2003)
7-Sulfate	(Palm.)	Williams et al. (1973)
(*) 2″,6″-Di-O-acetyl	*Crotalaria thebaica* (Leg.)	Ibraheim (1994)
(*) 2″-O-(6‴-(E)-p-Coumaroyl)glucoside	*Cucumis sativus* (Cucurb.)	Aboud-Zaid et al. (2001)
(*) 2″-O-(6‴-(E)-p-Coumaroyl)		
Glucoside 4′-O-glucoside	*Cucumis sativus* (Cucurb.)	Aboud-Zaid et al. (2001)
2″-O-Feruloyl	*Gentiana punctata* (Gent.)	Luong and Jacot-Guillarmod (1977)
(*) 2″-O-Feruloylglucoside	*Bryonia* sp. (Cucurb.)	Krauze and Cisowski (1994)
	Cucumis sativus (Cucurb.)	Aboud-Zaid et al. (2001)
2″-O-(E)-Feruloyl 4′-O-glucoside	*Gentiana punctata* (Gent.)	Luong and Jacot-Guillarmod (1977)
(*) 2″-O-(6‴-(E)-Feruloyl)glucoside		
4′-O-Glucoside	*Cucumis sativus* (Cucurb.)	Aboud-Zaid et al. (2001)
(*) 2″-O-Galloyl	*Terminalia catappa* (Combr.)	Lin et al. (2000)
	Pelargonium reniforme (Ger.)	Latte et al. (2002)
(*) 6″-O-Acetyl	*Crotalaria thebaica* (Leg.)	Ibraheim (1994)
Vitexin		
X-O-Acetyl 2″-O-rhamnoside	*Crataegus sinaica* (Ros.)	El-Mousallamy (1998)
7-Sulfate	(Palm.)	Williams et al. (1973)
7-O-Rutinoside sulfate	(Palm.)	Williams et al. (1973)
2″-O-Acetyl	*Crateagus sanguinea* (Ros.)	Kashnikova et al. (1984)
	Crataegus pinnatifida (Ros.)	Zang and Xu (2002)
(*) 2″-O-Acetyl 4″-O-rhamnoside	*Crataegus monogyna* (Ros.)	Rehwald et al. (1994)
(*) 2″-O-(2‴-Methylbutyryl)	*Trollius ledebouri* (Ranun.)	Zou et al. (2004)
(*) 2″-O-(3‴,4‴-Dimethylbutyryl)	*Trollius ledebouri* (Ranun.)	Zou et al. (2004)
2″-O-p-Coumaroyl	*Trigonella fœnum-grœcum* (Leg.)	Sood et al. (1976)
2″-O-p-Coumaroyl 7-O-glucoside	*Mollugo oppositifolia* (Mollug.)	Chopin et al. (1984)
(*) 2″-O-Galloyl	*Terminalia catappa* (Combr.)	Lin et al. (2000)
	Pelargonium reniforme (Ger.)	Latte et al. (2002)
2″-O-p-Hydroxybenzoyl	*Vitex lucens* (Verb.)	Horowitz and Gentili (1966)
(*) 3″-O-Acetyl	*Crataegus pinnatifida* (Ros.)	Zang and Xu (2002)
(*) 6″-O-Acetyl	*Crataegus pinnatifida* (Ros.)	Zang and Xu (2002)
6″-O-Acetyl 4′-O-rhamnoside		
(Cratenacin)	*Crataegus curvisepala* (Ros.)	Batyuk et al. (1966)
(*) 6″-O-Acetyl 2″-O-rhamnoside	*Clusia sandiensis* (Gutt.)	Delle Monache (1991)
(*) 6″-O-Malonyl 2″-O-xyloside	*Beta vulgaris* (Chenop.)	Gil et al. (1998)

continued

TABLE 14.5
Naturally Occurring Acyl-*C*-Glycosylflavonoids. (Hypothetic Structures Previously Mentioned in the Last Edition Have Been Removed If Not Confirmed in Their True Structure; the New Compounds for the Period 1992 to 2004 Are Indicated by an Asterisk [*]) — *continued*

Compounds	Sources	Ref.
(*) 2‴-*O*-Acetyl		
2″-*O*-α-L-Rhamnopyranosyl	*Tripterospermum japonicum* (Gent.)	Otsuka and Kijima (2001)
(*) 3″,4‴-Di-*O*-acetyl 2″-*O*-rhamnoside	*Crataegus sinaica* (Ros.)	El-Mousallamy (1998)
4‴-*O*-Acetyl 2″-*O*-rhamnoside	*Crataegus monogyna* (Ros.)	Nikolov et al. (1976)
	Crataegus sinaica (Ros.)	El-Mousallamy (1998)
	Tripterospermum japonicum (Gent.)	Otsuka and Kijima (2001)
8-*C*-β-D-Glucofuranosylapigenin		
(*) 2″-*O*-Acetyl	*Crataegus pinnatifida* (Ros.)	Zang and Xu (2002)
8-*C*-Galactosylapigenin		
6″-*O*-Acetyl	*Briza media* (Gram.)	Chari et al. (1980)
Swertisin		
6‴-*O*-*p*-Coumaroyl 2″-*O*-glucoside	*Zizyphus jujuba* (Rham.)	Woo et al. (1980)
6‴-*O*-Feruloyl 2″-*O*-glucoside	*Zizyphus jujuba* (Rham.)	Woo et al. (1980)
6‴-*O*-Sinapoyl 2″-*O*-glucoside	*Zizyphus jujuba* (Rham.)	Woo et al. (1980)
Isoswertisin		
2″-*O*-Acetyl	*Brackenridgea zanguebaria* (Ochn.)	Bombardelli et al. (1974)
(*) 2″-*O*-(2‴-Methylbutyryl)	*Trollius ledebouri* (Ranun.)	Zou et al. (2004)
(*) 3″-*O*-(2‴-Methylbutyryl)	*Trollius ledebouri* (Ranun.)	Zou et al. (2004)
(*) 6‴-*O*-Feruloyl 2″-*O*-glucoside	*Ziziphus jujuba* (Rham.)	Cheng et al. (2000)
Cytisoside		
O-Acetyl 7-*O*-glucoside		
O-Acetyl 7-*O*-glucoside (tremasperin)	*Trema aspera* (Ulm.)	Oelrichs et al. (1968)
Embigenin		
2‴-*O*-Acetyl 2″-*O*-rhamnoside	*Iris lactea* (Irid.)	Pryakhina et al. (1984)
7,4′-Di-*O*-methylisomollupentin		
2″-*O*-Caffeoylglucoside	*Asterostigma riedelianum* (Arac.)	Markham and Williams (1980)
Isoorientin		
(*) 7-*O*-(6‴-*O*-*E*-Feruloyl)glucoside	*Hordeum vulgare* (Gram.)	Norbaek et al. (2003)
2″-*O*-Acetyl	*Rumex acetosa* (Polyg.)	Kato and Morita (1990)
2″,6″-Di-*O*-acetyl	*Rumex acetosa* (Polyg.)	Kato and Morita (1990)
2″-*O*-(*E*)-Caffeoyl	*Gentiana burseri* (Gent.)	Jacot-Guillarmod et al. (1975)
2″-*O*-(*E*)-Caffeoyl glucoside	*Cucumis melo* (Cucurb.)	Monties et al. (1976)
2″-*O*-(*E*)-Caffeoyl 4′-*O*-glucoside	*Gentiana punctata* (Gent.)	Luong and Jacot-Guillarmod (1977)
2″-*O*-(*E*)-*p*-Coumaroyl	*Gentiana* sp. (Gent.)	Luong et al. (1980)
2″-*O*-(*E*)-Feruloyl	*Gentiana burseri* (Gent.)	Jacot-Guillarmod et al. (1975)
2″-*O*-(*E*)-Feruloyl 4′-*O*-glucoside	*Gentiana burseri* (Gent.)	Jacot-Guillarmod et al. (1975)
(*) 2″-*O*-Galloyl	*Pelargonium reniforme* (Ger.)	Latte et al. (2002)
2″-*O*-β-Glucosyl-(*E*)-caffeoyl 4′-*O*-glucoside	*Gentiana burseri* (Gent.)	Jacot-Guillarmod et al. (1975)
2″-*O*-(4-*O*-Glucosyl-2,4,5-trihydroxy-(*E*)-cinnamoyl) 4′-*O*-glucoside	*Gentiana* sp. (Gent.)	Luong et al. (1981)
2″-*O*-*p*-Hydroxylbenzoyl	*Gentiana asclepiadea* (Gent.)	Goetz and Jacot-Guillarmod (1978)
2″-*O*-*p*-Hydroxybenzoyl 4′-*O*-glucoside	*Gentiana asclepiadea* (Gent.)	Goetz and Jacot-Guillarmod (1978)

TABLE 14.5
Naturally Occurring Acyl-*C*-Glycosylflavonoids. (Hypothetic Structures Previously Mentioned in the Last Edition Have Been Removed If Not Confirmed in Their True Structure; the New Compounds for the Period 1992 to 2004 Are Indicated by an Asterisk [*]) — *continued*

Compounds	Sources	Ref.
(*) 6″-*O*-Acetyl	*Crotalaria thebaica* (Leg.)	Ibraheim (1994)
(*) 6″-*O*-Caffeoyl	*Gentiana arisensis* (Gent.)	Kuo et al. (1996)
6″-*O*-Caffeoyl 7-sulfate	(Palm.)	Williams et al. (1973)
(*) 6″-*O*-*p*-Hydroxybenzoyl (perfoliatumin A)	*Polygonum perfoliatum* (Polyg.)	Zhu et al. (2000)
(*) 6″-*O*-*p*-Methoxybenzoyl (perfoliatumin B)	*Polygonum perfoliatum* (Polyg.)	Zhu et al. (2000)
Orientin		
(*) 7-*O*-Caffeoyl	*Vellozia* sp. (Velloz.)	Harborne et al. (1994)
	(Velloz.)	Williams et al. (1994b)
7-*O*-Glucoside sulfate	(Palm.)	Williams et al. (1973)
7-Sulfate	(Palm.)	Williams et al. (1973)
2″-*O*-Acetyl	*Hypericum hirsutum* (Hyper.)	Kitanov et al. (1979)
2″,6″-Di-*O*-acetyl	*Rumex acetosa* (Polyg.)	Kato and Morita (1990)
(*) 2″-*O*-Caffeoyl	*Vitex polygama* (Verb.)	Leitoa and Delle Monache (1998)
(*) 2″-*O*-Galloyl	*Pelargonium reniforme* (Ger.)	Latte et al. (2002)
(*) 2″-*O*-(2‴-Methylbutyryl)	*Trollius ledebouri* (Ranun.)	Zou et al. (2004)
(*) 2″-*O*-(3‴,4‴-Dimethylbutyryl)	*Trollius ledebouri* (Ranun.)	Zou et al. (2004)
6″-*O*-Feruloyl 2″-*O*-xyloside	*Setaria italica* (Gram.)	Gluchoff-Fiasson et al. (1989)
Isoscoparin		
(*) 2″-*O*-Feruloyl	*Bryonia* sp. (Cucurb.)	Krauze and Cisowski (1994)
6‴-*O*-*p*-Coumaroyl 2″-*O*-glucoside	*Oryza sativa* (Gram.)	Besson et al. (1985)
	Cucumis sativus (Cucurb.)	Aboud-Zaid et al. (2001)
(*) 6‴-*O*-*p*-Coumaroyl 2″-*O*-glucosyl 4′-*O*-Glucoside	*Cucumis sativus* (Cucurb.)	Aboud-Zaid et al. (2001)
6‴-*O*-Feruloyl 2″-*O*-glucoside	*Oryza sativa* (Gram.)	Besson et al. (1985)
Scoparin		
6″-*O*-Acetyl	*Sarothamnus scoparius* (Leg.)	Prum-Bousquet et al. (1977)
7,3′,4′-Tri-*O*-methylisoorientin		
6″-*O*-Acetyl 2″-*O*-rhamnoside (linoside A)	*Linum maritimum* (Lin.)	Chari et al. (1978)
8-*C*-β-D-Xylopyranosylchrysoeriol		
(*) 4′-*O*-Acetyl 2″-*O*-glucoside	*Scleranthus unicinatus* (Caryo.)	Yayli et al. (2001)
MONO-*C*-GLYCOSYLISOFLAVONES		
Puerarin		
4′,6″-Di-*O*-acetyl	*Pueraria tuberosa* (Leg.)	Bhutani et al. (1969)
8-*C*-Glucosylorobol		
6″-*O*-Acetyl	*Dalbergia monetaria* (Leg.)	Nunes et al. (1989)
8-*C*-GLUCOSYLCHROMONES		
2′-*O*-*p*-Coumaroyl 5-methyl-7-hydroxy 2-acetonyl		
Aloeresin A	*Aloe* sp. (Lil.)	Gramatica et al. (1982)
2′-*O*-*p*-Coumaroyl 5-methyl-7-*O*-glucosyl 2-acetonyl		

continued

TABLE 14.5
**Naturally Occurring Acyl-C-Glycosylflavonoids. (Hypothetic Structures Previously
Mentioned in the Last Edition Have Been Removed If Not Confirmed in Their True
Structure; the New Compounds for the Period 1992 to 2004 Are Indicated by an
Asterisk [*]) — continued**

Compounds	Sources	Ref.
Aloeresin C	*Aloe* sp. (Lil.)	Speranza et al. (1985)
2'-O-p-Coumaroyl 5-methyl-7-methoxyl 2-(2-hydroxy)propyl	*Aloe ferox* (Lil.)	Speranza et al. (1986)
Aloeresin D		
2'-O-p-Methoxycoumaroyl 5-methyl-7- hydroxy 2-acetonyl	*Aloe excelsa* (Lil.)	Mebe (1987)
DI-C-GLYCOSYLFLAVONES		
Vicenin-1		
(*) 6''-O-Acetyl	*Prosopis chilensis* (Leg.)	Elrady and Saad (1994)
6-C-α-L-Rhamnosylapigenin		
(*) 8-C-(6'''-3-Hydroxy-3-methylglutaroyl)glucoside	*Glycyrrhiza eurycarpa* (Leg.)	Liu et al. (1994)
Isoschaftoside		
2'''-O-Feruloyl	*Metzegeria* sp. (Bryo.)	Theodor et al. (1981b)

14.3.1.1 Qualitative Aspects

The first application of CZE concerns a mixture of C-glycosylflavones from different market
samples of *Passiflora incarnata*.[373] Factors mainly responsible for differences in electrophoretic mobility (EM) of C-glycosylflavones are identical to those defined from the separation of
flavonoid-O-glycosides: number and position of the free hydroxy groups on the flavone
skeleton, number and type of attached sugar groups. The best technical compromise between
resolution efficiency and retention time was obtained using an operating buffer of 5%
methanol in 75 nM borate buffer at pH 10, with the capillary temperature maintained at
50°C, the applied voltage at 18 kV, and an operating current of 40 µA. From *Passiflora*

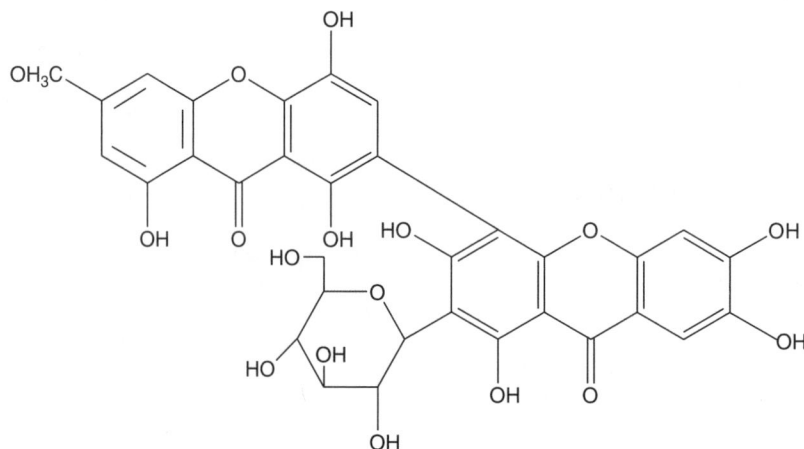

FIGURE 14.1 The structure of the flavone–xanthone dimer, swertifrancheside, isolated from *Swertia
franchetiana*. (From Wang, J. et al., *J. Nat. Prod.*, 57, 211, 1994. With permission.)

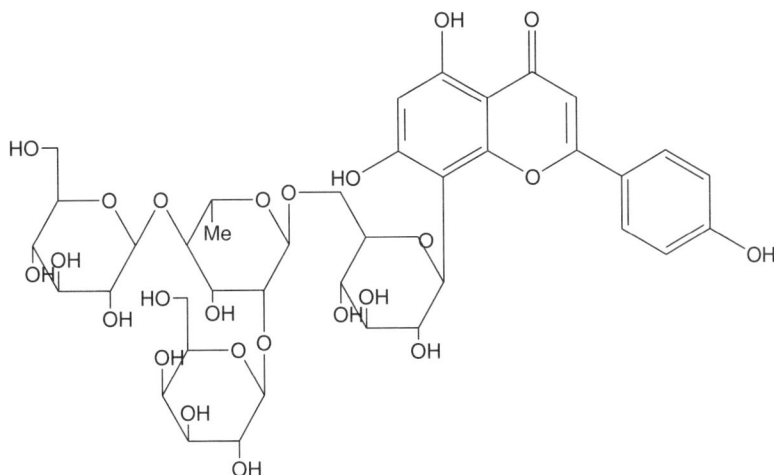

FIGURE 14.2 The structure of panzhihuacycaside isolated from *Cycas panzhihuaensis*. (From Zhou, Y. et al., *Zhiwu Xuebao*, 44, 101, 2002. With permission.)

incarnata extracts, 11 different *C*-glycosylflavones were separated on the electrophoregram. This technical approach permitted qualitative comparison of the chemical contents of different geographic samples of *Passiflora*.

Moreover, typical features of the electrophoretic behavior of *C*-glycosylflavones were discussed in relation to their structural characteristics. The EM values of flavone 6-*C*-glycosides are always lower than for flavone 8-*C*-glycoside isomers. For the flavone diglycoside derivatives of the same molecular weight, lucenin-2 and isoorientin 2″-*O*-glucoside, vicenin-2 and isovitexin 2″-*O*-glucoside, the EM values of *O*-glycosyl derivatives are higher than those of corresponding di-*C*-glycosyl flavones. The *C*-7 substitution on ring A of the flavone has a strong influence on EM values. Thus, glycosylation or methylation results in the loss of the most acidic hydroxyl group and consequently in a lower EM. Similarly, the increase of molecular size also results in a lower EM; e.g., isovitexin is higher than isoorientin, vicenin-2 is higher than lucenin-2, and schaftoside and isoshaftoside are higher than vicenin-2.

14.3.1.2 Quantitative Aspects

The CZE technique was also applied to complex mixtures of *C*-glycosylflavones from leaves of *Achillea setacea*, in order to quantify each compound.[235] The optimum separation was obtained using 25 mM sodium tetraborate with 20% methanol (pH 9.3). Six *C*-glycosylflavones, five *O*-glycosylflavonoids, and three caffeoyl derivatives were efficiently separated. For the quantification, calibration curves were established. A linear correlation from 20 to 200 µg/ml was found for vitexin with a coefficient R^2 very close to 1.0. The detection limit of the CZE method was about 9.3 µg/ml, corresponding to 0.023% in the drug. Concerning the reproducibility of extractions, the relative standard deviation was less than 5%, with the migration time coefficients of variation ranging from 0.34 to 0.45%. The accuracy of the method was examined by recovery studies: the results are around 98 to 99% with a relative standard deviation of about 3 to 5%.

14.3.2 High-Speed Countercurrent Chromatography Separation

Separation of *C*-glycosylflavones by using high-speed countercurrent chromatography (HSCCC) was described by Oliveira et al.[285] using an ethylacetate extract of *Cecropia lyratifolia.* The different components were separated in two steps: initially the mixture chloroform–methanol–water (46:25:29) was used as solvent system with the upper phase as the stationary phase. The flavonoid fraction was purified and concentrated in this stationary phase. Thereafter, the flavonoid block was submitted to a further HSCCC procedure using ethylacetate–butanol–methanol–water mixture (35:10:11:44) with the upper phase as the stationary phase. HPLC analysis of the collected fractions revealed the separation of 6-*C*-(6″-*O*-galactoside)-galactoside apigenin (20 mg), isoorientin (45 mg), and a mixture of orientin and isovitexin (74 mg). The HSCCC technique has been found to be the most effective method to resolve this flavonoid mixture, since it does not display the undesirable adsorption phenomenon and sample loss that are associated with other chromatographic techniques.

14.3.3 NEW POLYMERS FOR COLUMN CHROMATOGRAPHY

An unusual chromatographic support, chitin, was tried for separation of swertisin and 2″-*O*-rhamnosylswertisin in a methanolic extract of leaves of *Aleurites moluccana.*[268] The chitin is a straight copolymer composed of β-(1,4)-linked GlcNAc units (85%) and β-(1,4)2-amino-2-deoxy-D-glucose units (15%) with a three-dimensional α-helical configuration stabilized by intramolecular hydrogen bonding. Chitin is an important natural polysaccharide with a great variety of functional groups: —OH, —NH$_2$, —NHCOCH$_3$. From chitin, a new polymer was prepared, chitin-100 or fully *N*-acetylated chitin, under heterogeneous conditions using a binary mixture of methanol–formamide and acetic anhydride. After processing, the *N*-acetylation degree was 99.2%. The ethyl acetate fraction of *Aleurites moluccana* (150 mg) was chromatographed on chitin, chitin-100, and silica gel; quantitative results showed that chitin-100 is the best sorbent for separation of swertisin and 2″-*O*-rhamnosylswertisin (12.8 and 15.8 mg, respectively) from the ethyl acetate extract. In comparison, on silica gel only 3.0 mg of swertisin and 9.2 mg of 2″-*O*-rhamnosylswertisin were eluted.

14.3.4 MASS SPECTROMETRY AND STRUCTURAL IDENTIFICATION

The development of mass spectrometry (MS) procedures for the study of *C*-glycosylflavonoids is particularly significant because of the resistance of the *C*-glycosidic bond to hydrolysis. It is important, from the structural point of view, to determine the molecular mass and to localize the sugar substituents on the aglycone moiety. This has been achieved in the past by electron impact EI-MS of permethylated derivatives.

14.3.4.1 Fast Atom Bombardment and Tandem Mass Spectrometry

More recently, fast atom bombardment (FAB), which enables direct analysis on underivatized *O*- and *C*-glycosylflavones, in combination with collisionally activated dissociation (CAD) and tandem MS, has been shown to be useful for structural characterization of various kinds of compounds. The advantages of using FAB in combination with CAD and tandem MS have been documented for *O*- and *C*-glycosylflavones. The characteristic fragment ions of (M − H)⁻ ions allow the differentiation between *C*-glycosylation at the 6- and 8-positions.[27,69] An application of liquid secondary ion mass spectrometry, which is a variant of FAB, in combination with CAD and linked scanning at constant B/E, to obtain daughter ion spectra of (M + H)⁺ and (M − H)⁻ ions was investigated for di-*C*-glycosylflavones and *O*-glycosyl-*C*-glycosylflavones.[208,209] Low-energy collisions allow the determination of the type of the terminal *O*-linked sugar residues, while high-energy collisions enable identification of the type and the linkage position of the *C*-linked sugars. The results illustrate that the daughter ion spectra of both (M + H)⁺ and (M − H)⁻ ions show diagnostic ions, which allow differentiation between the isomeric 6,8-di-*C*-glycosides and provide proof of sugar linkage in *O*-glycosyl-*C*-glycosylflavones.

14.3.4.2 Liquid Chromatography–Mass Spectrometry: Thermospray and Atmospheric Pressure Ionization

Much data on the structure of flavonoids in crude or semipurified plant extracts have been obtained by HPLC coupled with MS, in order to obtain information on sugar and acyl moieties not revealed by ultraviolet spectrum, without the need to isolate and hydrolyze the compounds. In the last decade, soft ionization MS techniques have been used in this respect, e.g., thermospray (TSP) and atmospheric pressure ionization (API). However, the most used methods for the determination of phenols in crude plant extracts were the coupling of liquid chromatography (LC) and MS with API techniques such as electrospray ionization (ESI) MS and atmospheric pressure chemical ionization (APCI) MS. ESI and APCI are soft ionization techniques that generate mainly protonated molecules for relatively small metabolites such as flavonoids.

TSP was used in a so-called buffer ionization mode requiring the addition of a volatile buffer before vaporizing the HPLC eluate, and the discharge ionization mode that is based on a discharge electrode that produces a plasma of the HPLC eluent. LC–TSP-MS allowed the identification of several compounds from *Tea C*-glycosylflavone extracts[174] and *Citrus* di-*C*-glycosylflavone mixtures.[23]

The coupling of LC and MS with API techniques was applied to the structural characterization of *Ocimum* mono-*C* and di-*C*-glycosylflavones[121] by LC–APCI-MS and *Theobroma* mono-*C*-glycosylflavones by LC–ESI-MS.[317]

14.3.4.3 LC–API-MS–MS

Molecular mass information alone, however, is not sufficient for the online (LC–MS) structural elucidation of natural compounds, and fragment information generated by collision-induced dissociation (CID) MS–MS becomes necessary for partial online identification. An interesting application of the low-energy CID-MS–MS spectra on various *C*-mono and *C*-diglycosylflavones was made in LC–APCI-MS–MS and LC–ESI-MS–MS modes.[382] These procedures were studied on two types of MS instruments: a quadrupole time-of-flight one and an ion trap one. It has been demonstrated that fragment ions provide important structural information for the nature and site of attachment of sugars in *O*-glycosyl *C*-glycosylflavones, and on the distribution of substituents between the A and B rings of the flavone skeleton.

14.3.5 NMR Spectroscopy and Fine Structure Elucidation

14.3.5.1 Improved NMR Techniques

^{1}H and ^{13}C NMR spectroscopies were systematically used to elucidate the structure of *C*-glycosylflavonoids; the assignment of signals was based upon various experiments, HMBC, HMQC, COSY, etc., as shown in the following examples:

In the case of isoorientin 6″-*O*-caffeate isolated from *Gentiana arisanensis*,[198] the ^{13}C NMR spectrum was assigned by ^{1}H-decoupled spectra, DEPT pulse sequence, ^{1}H–^{13}C COSY spectrum, long-range ^{13}C–^{1}H COSY, and NOESY experiments; the ^{1}H NMR spectrum was analyzed with the aid of ^{1}H–^{1}H COSY and ^{1}H–^{13}C COSY.

For the new *C*-hydroxybenzyl glycoflavones from *Citrullus colocynthis*,[223] the complete ^{13}C NMR data were assigned and additionally confirmed by 2D-COSY, HETCOR, and selective INEPT experiments.

The structure of shamimin[99,100] isolated from the leaves of *Bombax ceiba* was determined by studies of the one-bond ^{1}H–^{13}C connectivities from the HMQC/HeteroCOSY spectrum and by long-range correlations from HMBC or COLOC experiments. The ^{1}H COSY-45 and ^{1}H–^{1}H COSY spectrum revealed the vicinal couplings in the sugar part; the *J*-resolved spectrum confirmed the assignment of the sugar protons.

In cassiaoccidentalins A, B, and C,[130] ^{1}H–^{1}H COSY spectrum and HMBC correlations substantiated the 3-keto structures and indicated that the rhamnose residue is attached to *O*-2″ of the 3-keto sugar residue. NOESY spectrum was used to assign the structure of 6-deoxyribo-hexos-3-ulose.

In kaplanin, the β configuration of the sugar in the 5-hydroxy-7-methoxy-8-*C*-β-glucosyl-flavone[265] was indicated by ^{1}H NMR and supported by ^{1}H–^{1}H COSY. ^{1}H and ^{13}C NMR and HETCOR spectra gave the precise location of the glucose moiety.

The complex structure isolated from *Scleranthus uncinatus*,[403,404] 5,7-dihydroxy-3′-meth-oxy-4′-acetoxyflavone-8-*C*-β-D-xylopyranoside-2″-*O*-glucoside, was determined by using 1D NMR (^{1}H, ^{13}C, DEPT) and 2D NMR (H-COSY, TOCSY, HMQC, HMBC, NOESY) data; sequence and linkage of the sugar chain and acylation site were confirmed by observation of inter-residue NOEs in the NOESY spectrum.

For the five new acetylated flavone *C*-glycosides (such as isovitexin 2″-*O*-(6‴-(*E*)-*p*-coumaroyl)glucoside-4′-*O*-glucoside) isolated from *Cucumis sativus*,[3] the structures were elucidated using 1D NMR (^{1}H, XSROESY, ^{13}C) and DEPT, DQF-COSY, HSQC, and HMBC experiments. The precise nature of linkages between the residues was determined using ^{1}H–^{1}H dipolar connectivities detected in 1D XSROESY experiments.

4-Hydroxy-1-ethylbenzene derivatives of vitexin and isovitexin (cucumerin A and B, two regioisomers)[228] were studied by ^{1}H and ^{13}C NMR analysis with DEPT, TOCSY, and COSY experiments; HMBC and NOESY correlations were also used for assignment of the hydroxyethylbenzene residue.

14.3.5.2 NMR and Rotational Isomers

Another aspect often underlined in the NMR studies on *C*-glycosylflavonoids is the broadening or the doubling of the signals due to the existence of rotational isomers. In the 1960s, the broadening of NMR signals was reported in *C*-glycosylflavones and was explained by the existence of rotational isomers. Kato and Morita[173] observed the doubling of signals in ^{1}H and ^{13}C NMR of peracetylated *C*-glycosylflavones due to the restricted rotation of the acetylated glucosyl moieties; the conformations of rotational isomers of hepta-*O*-acetylvitexin and octa-*O*-acetylorientin were assigned as +*sp* (major) and −*sc* (minor) for both compounds after COSY and NOESY experiments on NMR spectra in CDCl$_3$.

During the structural elucidation of the first flavone–xanthone dimer (swertifranche-side)[381] through a series of 1D or 2D NMR techniques including COSY, phase-sensitive ROESY, reversed-detected HMQC, HMBC, and selective INEPT experiments, ^{1}H and ^{13}C signals appeared as two peaks indicating two conformers; and classically, the coalescence of the split signal was observed as the temperature was increased.

The fine structure of the three cassiaoccidentalins[130] with a 6-*C* (2″-*O*-rhamnosyl) 6-deoxy-ribo-hexos-3-ulosyl moiety was established on the basis of spectroscopic evidence, as already mentioned; the ^{1}H and ^{13}C NMR spectra showed signals of two conformers due to hindered rotation around the *C*-6-glycosidic linkage, with duplication or broadening of the signals remaining even at elevated temperatures. The NOESY spectrum showed a cross-peak between H-8 of the flavone nucleus and H-6‴ of the rhamnose for the major conformer and a cross-peak between OH-5 and rhamnose H-2‴ for the minor conformer.

An initial study of the ^{1}H NMR of spinosin,[54,207] 6-C-(2''-O-glucosyl)-glucosyl-7-O-methyl apigenin, revealed a doubling of many of the signals and it was noted that the paired signals were in a nearly 1:1 ratio; the ^{13}C NMR data showed similar duplication of signals. HMBC data permitted the matching of each signal of a duplicated pair of carbon signals of the flavone nucleus to the corresponding signal of the pair given by its attached proton. A HSBC experiment was necessary in order to achieve the carbon and proton chemical shifts assignments for the sugar moiety. The 2D NMR experiments allowed the complete assignment of the ^{1}H and ^{13}C data to each rotamer for the disaccharide part of the molecule.

The pattern of duplication of signals suggested that at 303 K there were two rotamers of about equal energies. At 338 K, the signals were still detected independently although the two signals had broadened and moved closer together. The two signals coalesced at 363 K for the entire ^{1}H spectrum and the free energy of activation for rotation was calculated using the Eyring equation: 75 kJ mol^{-1}. On the basis of these observations, it was proposed that spinosin showed rotational isomers separated by an energy barrier about the C-glycosidic bond sufficiently high to prevent fast exchange between the two conformers at room temperature. The data from T-ROESY experiments were in accordance with this conclusion. Hence, the H-1'' proton is oriented *syn* to the methoxy group in one rotamer while in the other the H-2'' proton is *syn* to it. Grid Scan procedure enables systematic exploration of conformations by stepwise alteration of selected torsion angles over specified range of values.

It was noteworthy that the ^{1}H NMR spectra of compounds 2''-O-galloyl-vitexin and 2''-O-galloyl-orientin at room temperatures displayed severe line broadening of the well-separated 2''-H signal of the glycosyl unit in each instance.[200] Such a phenomenon presumably reflected restricted conformational flexibility at the C6–C1'' bond associated with the presence of the 2''-O-galloyl group, and a limitation in the dynamic rotational isomerism. Under the same conditions, ^{1}H NMR spectra of 2''-O-galloyl-isovitexin and 2''-O-galloyl-isoorientin displayed a typical duplication of most of the signals; the two rotamers coexisted in the ratio of 8:1 in each case. Interestingly, this ratio was solvent-dependent: with acetone-d6 the signal intensity of the minor rotamer was enhanced, culminating in a 3:1 relative abundance of the two conformers. The authors underlined that when the 2''-position is occupied by a rhamnosyl instead of a galloyl residue, no doubling of signals is found, suggesting that the presence of an additional sugar at the 2''-position locks the 8-glucosyl unit in a position precluding its interaction with the aromatic B-ring. In the C-6 glycosylated compounds the preferred conformations are those that are stabilized via attractive *p*-stacking between the A-ring and the galloyl aromatic ring. This conformation is obviously stable enough to retard free rotation about the C-6-glycosyl bond to such an extent that broadened ^{1}H NMR resonances are observed. In the C-8 glycosylated analogs the conformation is stabilized by *p*-stacking between the aromatic ring of the ester and the B-ring of the flavone. This interaction produces a more pronounced stabilizing effect with the rotation being slowed down to such an extent that two distinct rotamers are observed, hence resulting in duplication of ^{1}H and ^{13}C NMR resonances.

The ability to show several conformers was used for the 2'-hydroxyisoorientin isolated from barley leaves, as a new method to measure NOE signals of C-glycoside flavonoids.[279] When NMR measurements were performed very close to the freezing temperature of the dimethyl sulfoxide (DMSO) solvent, the molecular rotation and exchange among the protons were highly hindered so that the rotamers could be observed separately because of the slow correlation time. Therefore, it became possible to observe clear NOE signals between spatially close protons; by preirradiation of a signal for a longer time than the correlation time, but shorter than the chemical exchange time for difference NOE. The residue signals differed from each other. Also, NOE signals between protons of the hydroxyl groups and the chromophore could be observed due to slow proton exchange among phenolic hydoxyl groups in DMSO at low temperature. A negative NOE produced by irradiation of H-1'' was observed at OH-5 and OH-7, and by irradiation of H-3 of the chromophore, a negative NOE was produced in the signals of H-6′ and OH-2′.

14.4 SYNTHESIS OF *C*-GLYCOSYLFLAVONOIDS

The synthesis of *C*-glycosylflavones has been achieved mainly by Friedel–Crafts-type reactions or Koenigs–Knorr glycosylations. However, yields were poor due to steric hindrance of the substrates. Several other procedures have been proposed in the last decade.

14.4.1   *O–C* Glycoside Rearrangement

An aryl *C*-glycosylation method via *O–C* glycoside rearrangement was developed by Kumazawa et al.[196] The glycosylation using phloroglucinol diacetate as an acceptor and 2,3,4,6-tetra-*O*-benzyl-α-D-glucopyranosyl fluoride as donor afforded the *O*-glucoside with a yield of 95%. When phloroglucinol dimethyl ether replaced the phloroglucinol diacetate, the reaction afforded both *O*- and *C*-glucosides in a total yield of 84% (30 and 54, respectively). Using the same glycosyl donor and several 2-acetyl phloroglucinol derivatives (di-methyl, di-benzyl, methyl benzyl so that there was a free hydroxyl group *ortho* to the acetyl group) as glycosyl acceptors, the glycosylation reaction gave the *C*-glucoside predominantly. This method was successfully applied to the synthesis of *C*-glucosyl chalcone.

More recently, Kumazawa et al.[197] reported on the synthesis of several 8-*C*-glucosylflavones from phloracetophenone derivatives via the same stereoselective *O–C* glycosyl rearrangement.

By selective protection of the phloracetophenone with methoxymethyl chloride, and then regioselective benzylation with benzyl chloride, the 2,4-dibenzylphloracetophenone was synthesized. By reacting with benzyl-protected glucopyranosyl fluoride in the presence of boron trifluoride diethyl etherate, this compound led to a *C*-glucosyl derivative via a highly regio- and stereoselective *O–C* glycosyl rearrangement. Its aldol condensation with 3,4-*bis*-benzyloxybenzaldehyde afforded a *C*-glucosyl chalcone derivative. This chalcone was oxidized in the presence of selenium dioxide to give the partially benzyl-protected orientin. All benzyl-protected groups on this flavone were removed by hydrogenolysis to afford orientin. If the partial benzyl-protection orientin was previously methylated with Me$_2$SO$_4$, the deprotection by hydrogenolysis led to parkinsonin A (5-methyl orientin). Alternatively, if the 2-benzyl-4-methyl phloracetophenone is used as the *C*-glucosylation acceptor, the aldol condensation produces isoswertiajaponin, and after the methylation process parkinsonin B (5,7-dimethyl orientin) is obtained.

14.4.2   Fries-Type Rearrangement

Another strategy was developed[95,230] for the synthesis of phenolic *C*-glycosyl derivatives via glycosylation of partially *O*-unprotected phenols followed by Fries-type rearrangement of the *O*-glycoside intermediate. Thus, with *O*-glycosyl trichloroacetimidates as donors, only catalytic amounts of trimethylsilyl triflate as promoters are required for reaction with phenol derivatives as acceptors. The studies have involved mono-*C*-glycosylation and di-*C*-glycosylation of partially *O*-protected phenols.

For the mono-*C*-glycosylphenol, the commercially available 2,4,6-trihydroxyacetophenone was chosen and selectively methylated at *C*-2 and *C*-4. The partially protected phenol was glycosylated with the *O*-benzyl-protected glucosyl trichloroacetimidate in the presence of trimethylsilyl triflate as promoter to give directly a *C*-(benzyl protected)glycosylphenol. The unprotected hydroxyl group of this compound was converted with benzoyl chloride into a fully protected *C*-glycoside phenol. Treatment of the benzoate derivative with sodium hydroxide in

DMSO resulted in an intramolecular ester condensation (Baker–Venkataraman-type re-arrangement) to give a C15 phenol the cyclization of which furnished the *O*-benzyl-protected flavone-*C*-glycoside. After deprotection, isoembigenin was obtained.

For a di-*C*-glycosyl compound, the *O*-benzyl-protected *O*-glucosyl trichloroacetimidate was reacted with 3,5-dimethoxyphloroglucinol in the presence of trimethylsilyl triflate to lead to 2-*C*-(tetra-*O*-benzylglucosyl) 3,5-dimethoxyphenol. Further reaction of this last compound with a second donor molecule under similar reaction conditions afforded the 2,6-di-*C*-glucosyl 3,5-dimethoxyphloroglucinol in very high yield (93%). Similarly, when the first intermediate 3,5-dimethoxy 2-*C*-(tetra-*O*-benzylglucosyl)phenol was reacted with *O*-(acetyl)-galactosyl trichloroacetimidate, another di-*C*-glycosyl phenol was obtained in good yield: 3,5-dimethoxy-6-*C*-(tetra-*O*-acetyl-galactosyl)-2-*C*-(tetra-*O*-benzyglucosyl)phenol.

14.4.3 VIA *C*-β-D-GLUCOPYRANOSYL 2,6-DIMETHOXYBENZENE

Very recently, a new strategy was developed[201] for the synthesis of a *C*-glycosylisoflavone, puerarin. The key intermediate is the β-D-glucopyranosyl-2,6-dimethoxybenzene. This was obtained by coupling a lithiated aromatic reagent (2,6-dimethoxybenzene, for example) with a benzyl protected glycopyranolactone, followed by hydride reduction. The *C*-(benzyl protected) glucoside was obtained in 56% overall yield. Hydrogenolysis of this compound proceeded smoothly to give the *C*-glucosylated 2,6-dimethoxybenzene, the acetylation of which gave a tetraacetate derivative. Reaction of this last compound in anhydrous $AlCl_3$ gave *C*-(tetraacetylglucosyl) 6-methoxy 2-hydroxy-3-acetyl benzene. After removal of acetyl groups on the glucose, the compound was condensed with *p*-methoxybenzaldehyde to give a chalcone. Oxidative rearrangement of the acetyl-protected chalcone in a mixed solvent of methanol and trimethylorthofolate followed by refluxing in methanol with HCl gave 7,4′-di-*O*-methylpuerarin in 84% yield. After demethylation, puerarin was obtained in 35% yield. The overall yield of the synthesis of puerarin was about 10%.

14.5 *C*-GLYCOSYLFLAVONOIDS AND VACUOLAR STORAGE

A number of different functions have been attributed to plant secondary substances, especially *C*-glycosylflavonoids; in many cases these functions require rather high local concentration (millimolar range) and many of these compounds are harmful to the plant producing these compounds. Therefore, the presence and synthesis require a strict compartmentalization of the sites of production and storage. For many metabolites, it has been shown that they are efficiently stored within the vacuoles. Consequently, transport mechanisms for vacuole depos-ition of glycosylated compounds are important. This raises the question of which structural features determine the specificity of glycosylated compounds for their recognition either by a glucoside pump or by a secondary energized glucoside transporter. This question was consid-ered by Frangne et al.[103] and Klein et al.[181]: in *Hordeum vulgare* the twofold glucosylated saponarin (apigenin 6-*C*-glucosyl-7-*O*-glucoside) accumulates as the major compound while apigenin 6-*C*-glucoside, the precursor, is present only in trace amounts. Saponarin is taken up by a proton antiport system into barley vacuoles and the transport activity is strongly reduced in a barley mutant impaired in flavonoid biosynthesis. The isovitexin was a competitive inhibitor of saponarin uptake, arguing for the fact that the same vacuolar transporter accepts both flavone glycosides. Interestingly, the saponarin was taken up by an ABC-type transporter into vacuoles of another species (*Arabidopsis*) that does not synthesize this class of flavonoids. Thus, specific proton antiport systems would be responsible for the vacuolar transport of endogenous glucosides, whereas vacuolar ABC-type transporters would be involved for exogenous compounds, which would be considered as foreign substances to be detoxified.

14.6 BIOLOGICAL PROPERTIES OF *C*-GLYCOSYLFLAVONOIDS

14.6.1 ANTIOXIDANT ACTIVITY OF *C*-GLYCOSYLFLAVONOIDS

Antioxidants are scavengers of oxygen radicals or hydroxy radicals that attack polyunsaturated fatty acids in cell membranes giving rise to lipid peroxidation. Lipid peroxidation is strongly associated with aging and carcinogenesis. Synthetic antioxidants such as butylated hydroxyanisole and butylated hydroxytoluene have been widely used as antioxidants for foods, but the uses have begun to be restricted because of their toxicity. Therefore, the interest for natural antioxidants has increased greatly. Many natural resources have been tested for their antioxidant effect and some of them accumulated large amounts of *C*-glycosylflavones.

Xanthosoma violaceum, Araceae, widely distributed in Dominican Republic, Puerto Rico, Guatemala, and Equator, was investigated for *in vitro* antioxidant and free-radical scavenger activities.[298] The fraction rich in *C*-glycosylflavones showed with 1,1-diphenyl-2-picrylhydrazyl radical (DPPH) test an EC_{50} of 11.6 µg/ml, compared with quercetin (2.3 µg/ml) and α-tocopherol (10.1 µg/ml).

Terminalia catappa, Combretaceae, commonly used in the folk medicine in Taiwan, contains several *C*-glycosylflavones, two of them galloyl esters of vitexin and isovitexin. The isolated compounds were tested for antioxidative activity on Cu^{2+}/O_2-induced low-density lipoprotein peroxidation.[213] IC_{50} values were 2.1 µ*M* for vitexin derivatives and 4.5 µ*M* for isovitexin derivatives compared to 4.0 µ*M* for probucol chosen as positive control.

Genistein 8-*C*-glucoside, isolated from *Lupinus luteus*, develops a clear-cut antioxidant effect in liver homogenates and microsomes preventing the destruction of cytochrome P450 and its conversion to an inactive form cytochrome P420. Moreover, oxidative damage caused by tertbutyl hydroperoxide and hypochloric acid on red blood cells was inhibited by about 50% on pretreatment with this *C*-glucosyl isoflavone in a concentration range of 3 to 5 m*M*.[44,406]

Other reports have concerned the antioxidant effect of *C*-glycosylflavones from *Carthamus tinctorius*,[203] *Anthurium versicolor*,[17] Lemon fruits,[259,260] and *Hordeum vulgare*.[239,258,286]

14.6.2 *C*-GLYCOSYLFLAVONOIDS AND PLANT–INSECT RELATIONSHIPS

Natural resistance in crops is an important aspect of any integrated pest management program, and it has been a major target in response to environmental concerns and increased resistance of insects to traditional insecticides.

14.6.2.1 *Zea mays* and *Helicoverpa zea*

The development of natural resistance in corn (*Zea mays*) to the corn earworm (CEW) (*Helicoverpa zea*) received many contributions. The CEW is a major insect pest of maize and other crops (cotton, soybeans, etc.); the eggs are laid on the silks, and the larvae access the ear by feeding through the silk channel.

Host-plant resistance to CEW by antibiosis is caused by the presence of several *C*-glycoflavones: maysin, apimaysin, methoxymaysin[333]; maysin seemed to be one of the most important natural resistance factors in corn silk.[334] Maysin is the predominant compound in most genotypes and a highly significant relationship was found between silk maysin concentration in fresh organs and weights of corn earworm larvae fed diets containing silk extracts. It has been shown that corn silk with maysin levels higher than 0.2% fresh weight reduced larval weights by 50% of controls, while silk maysin higher than 0.4% reduced weights by about 70%. Upon ingestion by CEW, the flavone molecules are oxidized to quinines, which bind amino acids making them unavailable and thus inhibiting larval growth.

The results of flavonoid analyses[70,384] showed that there is a wide range in silk maysin levels, from 0.01 to 0.5% fresh weight. A significant proportion (one fifth) of these sources presented silk maysin levels above the 0.2% fresh weight threshold, considered to be significant for cornworm antibiosis.

The presence of *o*-dihydroxy substituent on the B-ring of the *C*-glycosylflavones is important for antibiosis activity, while the nature of the sugar is not important. Consequently, the discovery of other *o*-dihydroxyflavonoids in corn silks would be advantageous in breeding resistance with a broad and diverse chemical basis. In the course of the survey of many corn inbreds, populations, and lines for flavonoid contents, several reduced maysin analogs were found, e.g., equatorial 4″-OH-maysin, axial 4″-OH-maysin, axial 4″-OH-3′-methoxymaysin.[335] The laboratory bioassays for CEW larval growth inhibition showed that the 4″-OH-maysins are almost as active as maysin in reducing larval growth. Genetic control of these particular *C*-glycosylflavones received much attention during the last 10 years.[45,202,227,347,410]

14.6.2.2 *Oryza sativa* and *Nilaparvata lugens*

Another example concerns the Brown planthopper (*Nilaparvata lugens*), a serious rice pest in Asia. It has been shown that the resistance factors to the insect are located in the phloem.[119] The chemical analysis of the sap revealed the presence of three main *C*-glycosyl flavones: schaftoside, isoshaftoside, and probably neoschaftoside. When sucrose was fed together with these *C*-glycosylflavones in concentrations similar to those in the stems, the mean duration of the ingestion pattern was reduced by 75% as compared with sucrose alone, evidencing the deterrent role of these compounds in the phloem of resistant rice varieties. Simultaneously, a large increase in the mean probing frequency was recorded from 2.3 to 6.6 over 2 h. It is likely that whenever the insects encounter a deterrent substance, they react by increasing their probing frequency in order to find a more acceptable food source.

14.6.3 C-GLYCOSYLFLAVONOIDS AND ANTIMOLLUSCIDAL ACTIVITY

El-Sawi[89] reported the molluscicidal activity of alcoholic extracts from aerial parts of *Acacia saligna*, which exhibited strong activity against *Biomphalaria alexandrina*, the intermediate host of *Schistosoma mansoni*. These extracts were particularly rich in flavonoids and more especially in a new compound identified as 8-*C*-glucosyldihydrokaempferol.

14.6.4 C-GLYCOSYLFLAVONOIDS AND PLANT–MICROORGANISM RELATIONSHIPS

14.6.4.1 Arbuscular Mycorrhiza: *Cucumis melo* and *Glomus caledonium*

Arbuscular mycorrhiza is the most widespread form of symbiotic association between soil-borne fungi and plant roots. This symbiosis confers benefits to the host-plant growth and development through the acquisition of phosphate and minerals from the soil by the fungi. Moreover, it enhances the plant resistance to pathogens and environmental stresses. It has been calculated that the arbuscular mycorrhizal fungi are able to colonize the roots of 80% of terrestrial plants. Despite the ubiquitous occurrence of this symbiosis and its importance in sustainable agriculture, the mechanisms for the formation of a functional symbiosis are almost entirely unknown. The role of *C*-glycosylflavones has been underlined[13] in the case of the arbuscular mycorrhizal colonization by *Glomus caledonium* in melon roots (*Cucumis melo*). The accumulation of 2″-*O*-glucosylvitexin was caused by a phosphate deficiency. This compound stimulated arbuscular mycorrhiza formation in melon roots. At low phosphate levels, the degree of colonization in control roots (22%) was much greater than when grown under high phosphate conditions (8.8%). With a treatment of roots by $2''$-O-glucosylvitexin, the degree of colonization was markedly increased by up to 35% on low phosphate and 25% on high phosphate levels.

14.6.4.2 Disease: *Cucumis sativus* and *Podosphaera xanthii*

The phenolic compositions of cucumber leaves (*Cucumis sativus* var. *corona*) were analyzed under different selective pressures of powdery mildew fungi.[228] Cucumber plants were tested under two treatments: plants infected with *Podosphaera xanthii* and rendered resistant with bioprotectant treatment elicitor of phytoalexin production, and control plants that consisted of infected, nonelicited susceptible plants. The leaf tissue of cucumber expressing induced resistance against powdery mildew fungi was characterized by high accumulation of two major *C*-glycosylflavones, cucumerin A and cucumerin B, which are new compounds with an original substitution, 4-hydroxy-1-ethylbenzene, and six known *C*-glycosylflavones. All these compounds were found in much higher concentrations within leaves of elicited cucumber plants while they were near absent within susceptible control plants. A role for these compounds in this species was suspected as phytoalexins.

14.6.5 ANTIBACTERIAL ROLE OF *C*-GLYCOSYLFLAVONOIDS

Due to the antimicrobial activity of many of the phenolic compounds against different bacterial and fungal strains, several reports about the antibacterial effects of *C*-glycosylflavones have appeared.

The extracts of *Triticum aestivum* were studied on six different bacteria. The activity was measured by using the agar-diffusion method. The acetone extracts rich in di-*C*-glycosyl and *C*-diglycosylflavones showed antibacterial effects on *Sarcina lutea* and *Bacillus subtilis* (two Gram-positive bacteria), while the ethanolic extracts rich in mono-*C*-glycosylflavones showed an antibacterial effect on *Escherichia coli* (a Gram-negative bacteria). The extracts did not have any effect on *Bacillus pumilus*, *Brodetella brochiseptica*, and *Staphylococcus aureus*.[255]

C-Glycosylflavones of *Arum palaestinum* showed antimicrobial activity against *E. coli* and *Staphylococcus aureus*, which were the most susceptible bacteria to vitexin and isoorientin.[9]

Antimicrobial activity was also described for *C*-glycosylflavones from *Otostegia fruticosa*: vicenin-2.[2,99]

14.6.6 *C*-GLYCOSYLFLAVONOIDS AND GENERAL ANIMAL PHYSIOLOGY

14.6.6.1 Antinociceptive Activity

The leaves of *Aleurites moluccana* contain $2''$-O-rhamnosylswertisin and swertisin. The antinociceptive effect of both compounds was evaluated by the writhing test in mice.[254] The results indicated that the first derivative inhibits, dose dependently, the abdominal constrictions caused by acetic acid with an ID_{50} value of 6.9 to 10.2 μM/kg and maximal inhibition of 92%. When compared with aspirin $ID_{50} = 133$ μM/kg, the *C*-glycosylflavone was about 16-fold more potent. On the other hand, the swertisin alone did not show any effect.

14.6.6.2 Antispasmodic Activity

Arum palaestinum contains isoorientin and vitexin, which were studied for their antispasmodic activity on isolated organs.[8] Isoorientin in concentrations ranging from 10^{-7} to 6×10^{-4} M decreased the frequency and the amplitude of the phasic contractions of uterine segments isolated from rats and guinea pigs; the EC_{50} values on the amplitude of contractions were 2.05×10^{-4} M in rats and 5.66×10^{-5} M in guinea pigs. The myolytic activity of isoorientin on uterine smooth muscle could be explained as due to inhibition of phosphodiesterases and consequently to an increase in the cellular concentration of cyclic nucleotides.

14.6.6.3 Sedative Activity

The seeds of *Ziziphus jujuba* var. *spinosa* (Bunge) are used as a sedative in Chinese medicine. They accumulate eight *C*-glycosylflavonoids based on the aglycones apigenin and genistein. Among them, spinosin and swertisin possess significant sedative activities. The oral administration of these compounds (4×10^{-5} M/kg) prolonged pentobarbital-induced sleeping time by about 30% compared to the control group.[54]

14.6.6.4 Antihepatotoxic Activity

The leaves of *Allophyllus edulis* are used for the treatment of liver ailments, such as jaundice, in the traditional medicine of Paraguay. Ten *C*-glycosylflavones were isolated from this plant: schaftoside, vicenin-2, lucenin-2, isovitexin 2″-*O*-rhamnoside, cerarvensin 2″-*O*-rhamnoside, mollupentin 2″-*O*-rhamnoside, etc. Monitoring of antihepatotoxic activity of these compounds was achieved against CCl_4 toxicity in primary cultured rat hepatocytes. The protection against CCl_4 toxicity was 35% with schaftoside, 50% with isovitexin 2″-*O*-rhamnoside, and 45% for cerarvensin 2″-*O*-rhamnoside. The *C*-8 isomers are less efficient than *C*-6 isomers and if a rhamnosyl residue is attached to the *C*-bound sugar, an enhancement in antihepatotoxic activity can be registered. If the free hydroxy group at position 7 of 6-*C*-glycosylflavones is replaced by *O*-glycosyl or *O*-methyl, the hepatoprotective activity is markedly enhanced.[143]

Green tea has a preventive effect on D-galactosamine (D-GalN)-induced liver injury in rats. The rats were given free access to the experimental diets for 10 days; at the 11th day, the D-galactosamine was injected intraperitoneally, and the activities of alanine aminotransferase and aspartate aminotransferase were measured from the plasma 22 h after the beginning of the treatment. When the experimental diet contained isoschaftoside isolated from green tea, the increase of plasma enzyme activities was restricted to about 30%.[375]

Extracts of leaves of *Combretum quadrangulare* showed promising hepatoprotective effect on D-GalN or tumor necrosis factor-α (TNF-α)-induced cell death in primary cultured mouse hepatocytes. The hepatoprotective effect of *C*-glycosylflavones from *C. quadrangulare* was determined by evaluation of the serum enzyme level in comparison with the D-GalN or TNF-α treated control.[24] At the concentration of 200 μg/ml, orientin and isoscoparin possessed an inhibitory effect on TNF-α-induced cell death of about 20%, while vitexin at 100 μg/ml produces a 95% inhibitory effect.

14.6.6.5 Antiinflammatory Activity

Much of the activity of *Citrus* flavonoids (flavanone glycosides, flavone *O*- and *C*-glycosides, polymethoxyflavones) appears to impact blood leukocytes and microvascular endothelial cells, and it is not surprising that two of the main areas of research on the biological actions of *Citrus* flavonoids have been inflammation and cancer. Inflammation is typically characterized by increased permeability of endothelial tissue and influxes of blood leukocytes into the interstitium, resulting in edema. Many properties of *Citrus* flavonoids could be linked to the abilities of these compounds to inhibit enzymes involved in cell activation: phosphodiesterases, kinases, topoisomerases, etc. Signal transduction in early stages of inflammation involves phosphodiesterases and among them, phosphodiesterase-4 isozyme influences the expression of the TNF-α and other proinflammatory protein cytokines. So, inhibition of this enzyme is a target for the potential treatment of inflammatory diseases. Vitexin, isovitexin, and 2″-O-rhamnosylvitexin appeared as inhibitors of this enzyme at IC_{50} of 7.1, 25, and 25 μM, respectively. Comparatively, the assays with polymethoxyflavones such as nobiletin or with the flavanone naringin gave IC_{50} of 3.2 and 10 μM, respectively. These results could be attributed to the abilities of *Citrus* flavonoids to interact with the nucleotide binding sites of phosphodiesterases because of similarities in the ring structures of flavonoids and the adenosine of ATP.[233]

Others reports describe the antiinflammatory activity of *C*-glycosylflavones from *Swertia franchetiana*[381] and *Pinelliae tuber*.[244]

14.6.6.6 Antidiabetic Activity

Related to the common uses of corn silk in Chinese folk medicine, *C*-glycosylflavones from corn silk (chrysoeriol 6-*C*-fucopyranoside, chrysoeriol 6-*C*-boivinopyranoside, chrysoeriol 6-*C*-boivinopyranosyl-7-*O*-glucopyranoside, 4″-OH-3′-methoxymaysin) were tested for the prevention of diabetic complications. These complications arise when reducing sugars react nonenzymatically with amino groups in proteins, lipids, etc., through a series of reactions forming Schiff bases and Amadori products. This leads to the accumulation of some kinds of aggregate or advanced glycation end products (AGEs) such as pentosidine or *N-E-*(carboxymethyl)lysine (CML). This process is known as glycation and the formation of these AGEs in the human body is associated with the inducement of diabetic complications.

The CML is the most characterized AGE and is referred to as a glycoxidation product. The inhibitory effects of *C*-glycosylflavones on the CML formation were tested by enzyme-linked immunosorbent assay in kidney diabetic subjects.[345,346] The results showed that the percent inhibition was about 53% for chrysoeriol 6-*C*-boivinosyl 7-*O*-glucoside, 64% for chrysoeriol 6-*C*-boivinosyl, 80% chrysoeriol 6-*C*-fucosyl, and only 2% for 4″-OH-3′-methoxymaysin versus 60% for the standard glycation inhibitor, aminoguanidine.

14.6.6.7 Antihypertensive Activity

The effects of crude flavonoids of lemon juice on blood pressure were examined using spontaneously hypertensive rats. The systolic blood pressure of the animals fed a diet with lemon crude flavonoids for 16 weeks was significantly lower than the control group. When the animals were fed a diet containing a purified fraction of *Citrus* flavonoids (6,8-di-*C*-glycosylflavones), the same result was reached after 4 weeks. The flavonoid glycosides had an inhibitory effect on angiotensin I converting enzyme.[261]

Use of *Bombax ceiba* leaves to reduce blood sugar levels is common in North Pakistan; Saleem et al.[313] tried an evaluation of the hypotensive activities (and hypoglycemic effect) of the shamimin, a new *C*-glycosylflavone isolated from this plant. This compound caused 81, 67, and 51% falls in blood pressure at the doses of 15, 3, and 1 mg/kg, respectively. The effect lasted for 2 to 4 min. Shamimin also produced a significant hypoglycemic effect in Sprague–Dawley rats at the dose of 500 mg/kg; the decrease remained constant for the next 4 h. Plasma sugar levels showed a decline of 26%.

14.7 MANUFACTURE AND CULTIVATION PROCESS AND *C*-GLYCOSYLFLAVONOIDS

The influence of a process on the flavonoid content was studied with *Pennisetum americanium*, the pearl millet that is largely grown as a food grain in Africa and Asia. Traditional processing of millet for food involves the removal of some of the outer layers of the kernel by pounding the grain in a mortar. The decorticated grain is steeped overnight in water containing tamarind bean extract or sour milk to bleach pigments and finally the bleached flour is then cooked. According to the variety, the content of *C*-glycosylflavones is 137 to 275 mg/100 g with a mean at 157 mg/100 g. A decortication level of 20% alters the *C*-glycosylflavone content to a range of 31 to 95 mg/100 g with a mean at 64. The cooking significantly reduced the concentration of *C*-glycosylflavones by about 30%. This result is important since millet diets were linked with the high incidence of endemic goiter in millet-consuming populations. Animal feeding studies[12] showed that the process in which 25% of the grain was removed as bran successfully removed the antithyroid proprieties of pearl millet as demonstrated by patterns of serum thyroid hormones and thyroid histopathology.

Among Mediterranean fruits orange juice must be highlighted, as this product is a major source of flavonoid intake in the diet of developed countries. *Citrus* flavonoids possess health-promoting properties thanks to flavanones, polymethoxyflavones, and *C*-glycosylflavones. An important question is the process applied to natural resources and its role on natural compounds. The flavonoid content of orange juices[110] produced by different processing techniques (hand squeezed, pasteurized, mildly pasteurized, etc.) was compared. For *C*-glycosylflavones, freshly prepared hand-squeezed navel orange juice contained 80 mg/l of vicenin-2, the pasteurized juice only 27 to 51 mg/l, and the mild pasteurized form 40 to 80 mg/l, indicating that the pasteurization process at higher temperature leads to a strong decrease in juice *C*-glycosylflavones.

A very recent report[111] concerns the cultivation process with the effect of grafting on the chemical quality of *Citrus* lemon juice. *Citrus aurantium* and *C. macrophylla* trees were selected as rootstock, and seven interstocks were also used (from different cultivars of *C. sinensis*, *C. aurantifolia*, and *C. reticulata*). The rootstock grafting is an agronomical technique able to improve production and quality of fruit, while the interstock grafting increases longevity of trees and decreases the thickness of the trunk at the grafting point. The rootstock was a more important factor than interstock on the total flavonoid content of lemon juice. Regarding the individual flavonoids, di-*C*-glucosyldiosmetin was the flavonoid most affected by the type of rootstock used (range: 400 to 1000 mg/l according to the variety).

REFERENCES

(+) Before a reference means citation obtained from the previous edition of *The Flavonoids*, Harborne J.B., Ed., Chapman & Hall, London, 1992, chap. 3.
1. Abourashed, E.A., Vanderplank, J.R., and Khan, I.A., High-speed extraction and HPLC fingerprinting of medicinal plants-application to *Passiflora* flavonoids, *Pharm. Biol.*, 40, 81, 2002.
2. Aboutabl, E.A., Sokkar, N.M., and Awaad, A.S., Phytochemical study of *Otostegia fruticosa* (Forsk.) and biological activity, *Bull. Fac. Pharm. (Cairo Univ.)*, 38, 115, 2000.
3. Abou-Zaid, M.M. et al., Acylated flavone *C*-glycosides from *Cucumis sativus*, *Phytochemistry*, 58, 167, 2001.
4. +Adamska, M. and Lutomski, J., *Planta Med.*, 20, 224, 1971.
5. +Adinarayana, D. and Rao, J.R., *Tetrahedron*, 28, 5377, 1972.
6. +Adinarayana, D. and Rao, J.R., *Proc. Indian Acad. Sci.*, 81, 23, 1975.
7. +Adinarayana, D. et al., *Z. Naturforsch.*, 37, 145, 1982.
8. Afifi, F.U., Khalil, E., and Abdalla, S., Effect of isoorientin isolated from *Arum palaestinum* on uterine smooth muscle of rats and guinea pigs, *J. Ethnopharmacol.*, 65, 173, 1999.
9. Afifi, F.U., Shervington, A., and Darwish, R., Phytochemical and biological evaluation of *Arum palaestinum*. Part 1: flavone *C*-glycosides, *Act. Tec. Leg. Med.*, 8, 105, 1997.
10. +Agrawal, A. and Singh, J., *Phytochemistry*, 27, 3692, 1988.

11. +Ahmed, A.A. et al., *J. Nat. Prod.*, 51, 971, 1988.

12. Akingbala, J.O., Effect of processing on flavonoids in millet (*Pennisetum americanum*) flour, *Cereal Chem.*, 68, 180, 1991.

13. Akiyama, K., Matsuoka, H., and Hayashi, H., Isolation and identification of a phosphate deficiency-induced *C*-glycosylflavonoid that stimulates arbuscular mycorrhiza formation in melon roots, *Mol. Plant–Microbe Interact.*, 15, 334, 2002.

14. Ali, A.A., El-Emary, N.A., and Darwish, F.M., Studies on the constituents of two *Iris* species, *Bull. Pharm. Sci. Assiut Univ.*, 16, 159, 1993.

15. Ali, Z. et al., Two new *C*-glycosylflavones from *Silene conoidea*, *Nat. Prod. Lett.*, 13, 121, 1999.

16. Anhut, S. et al., Flavone-*C*-glycosides from the mosses *Plagiomnium elatum* and *Plagiomnium cuspidatum*, *Z. Naturforsch. C: Biosci*, 47, 654, 1992.

17. Aquino, R. et al., Phenolic constituents and antioxidant activity of an extract of *Anthurium versicolor* leaves, *J. Nat. Prod.*, 64, 1019, 2001.

18. +Asen, S. and Jurd, L., *Phytochemistry*, 6, 577, 1967.

19. +Ayanoglu, E. et al., *Phytochemistry*, 21, 799, 1982.

20. +Baeva, R.T. et al., *Khim. Prir. Soedin.*, 171, 1974.

21. +Bakhtiar, A. et al., *Phytochemistry*, 29, 3840, 1990.

22. Bakhtiar, A. et al., *C*-Glycosylflavones from *Raputia paraensis*, *Phytochemistry*, 30, 1339, 1991.

23. Baldi, A. et al., Identification of nonvolatile components in lemon peel by high-performance liquid chromatography with confirmation by mass spectrometry and diode-array detection, *J. Chromatogr. A*, 718, 89, 1995.

24. Banskota, A.H. et al., Hepatoprotective effect of *Combretum quadrangulare* and its constituents, *Biol. Pharm. Bull.*, 23, 456, 2000.

25. +Barger, G., *J. Chem. Soc.*, 89, 1210, 1906.

26. +Batyuk, V.S., Chernobrovaya, N.V., and Prokopenko, A.P., *Khim. Prir. Soedin.*, 2, 288, 1966.

27. +Becchi, M. and Fraysse, D., *Biomed. Environ. Mass Spectrom.*, 18, 122, 1989.

28. +Beltrami, E. et al., *Phytochemistry*, 21, 2931, 1982.

29. +Besson, E. et al., *Phytochemistry*, 18, 1899, 1979.

30. +Besson, E. et al, *Phytochemistry*, 24, 1061, 1985.

31. +Bezuidenhoud, B.C.B., Brandt, E.V., and Ferreira, D., *Phytochemistry*, 26, 531, 1987.

32. +Bhatia, V.K., Gupta, S.R., and Seshadri, T.R., *Tetrahedron*, 22, 1147, 1966.

33. +Bhutani, S.P., Chibber, S.S., and Seshadri, T.R., *Indian J. Chem.*, 7, 210, 1969.

34. Bloor, S.J., Novel pigments and copigmentation in the blue marguerite daisy, *Phytochemistry*, 50, 1395, 1999.

35. Bohm, B.A. et al., Flavonoids, DNA and relationships of *Itea* and *Pterostemon*, *Biochem. Syst. Ecol.*, 27, 79, 1999.

36. +Bombardelli, E. et al., *Phytochemistry*, 13, 295, 1974.

37. Bouaziz, M. et al., Flavonoids from *Hyparrhenia hirta* Stapf (Poaceae) growing in Tunisia, *Biochem. Syst. Ecol.*, 29, 849, 2001.

38. +Bouillant, M.L. et al., *Phytochemistry*, 17, 527, 1978.

39. +Bouillant, M.L. et al., *Phytochemistry*, 18, 1043, 1979.

40. +Bouillant, M.L. et al., *Phytochemistry*, 23, 2653, 1984.

41. +Boyet, C. and Jay, M., *Biochem. Syst. Ecol.*, 17, 443, 1989.

42. Bucar, F. et al., Phenolic compounds from *Biophytum sensitivum*, *Pharmazie*, 53, 651, 1998.

43. Budzianowski, J., Budzianowska, A., and Kromer, K., Naphtalene glucoside and other phenolics from the shoot and callus cultures of *Drosophyllum lusitanicum*, *Phytochemistry*, 61, 421, 2002.

44. Buko, V. et al., Inhibition of oxidative damage of red blood cells and liver tissue by genistein-8-*C*-glucoside, *Adv. Exp. Med. Biol.*, 500, 271, 2001.

45. Byrne, P.F. et al., Quantitative trait loci and metabolic pathways: genetic control of the concentration of maysin, a corn earworm resistance factor, in maize silks, *Proc. Natl. Acad. Sci. USA*, 93, 8820, 1996.

46. Carnat, A.P. et al.,Violarvensin, a new flavone di-*C*-glycoside from *Viola arvensis*, *J. Nat. Prod.*, 61, 272, 1998.

47. +Castledine, R.M. and Harborne, J.B., *Phytochemistry*, 15, 803, 1976.

48. +Chaboud, A., Dellamonica, G., and Raynaud, J., *Phytochemistry*, 27, 2360, 1988.

49. Chang, C.S. and Jeon, J., Leaf flavonoids in *Cotoneaster wilsonii* (Rosaceae) from the island Ulleung-do Korea, *Biochem. Syst. Ecol.*, 31, 171, 2003.

50. +Chari, V.M. et al., 11th IUPAC Int. Symp. on the Chemistry of Natural Products, Symposium papers, 2, 279, 1978.

51. +Chari, V.M., Harborne, J.B., and Williams, C.A., *Phytochemistry*, 19, 983, 1980.

52. +Chawla, H., Chibber, S.S., and Seshadri, T.R., *Phytochemistry*, 13, 2301, 1974.

53. +Chawla, H., Chibber, S.S., and Seshadri, T.R., *Phytochemistry*, 15, 235, 1976.

54. Cheng, G. et al., Flavonoids from *Ziziphus jujuba* Mil var. *spinosa*, *Tetrahedron*, 56, 8915, 2000.

55. +Chernobrovaya, N.V., *Khim. Prir. Soedin.*, 801, 1973.

56. Chimichi, S. et al., Isolation and characterisation of an unknown flavonoid in dry extracts from *Passiflora incarnata*, *Nat. Prod. Lett.*, 11, 225, 1998.

57. +Chopin, J. et al., *Phytochemistry*, 16, 2041, 1977a.

58. +Chopin, J. et al., *Phytochemistry*, 16, 1999, 1977b.

59. +Chopin, J et al., *Phytochemistry*, 17, 299, 1978.

60. +Chopin, J. et al., *Phytochemistry*, 21, 2367, 1982.

61. +Chopin, J. et al., *Phytochemistry*, 23, 2106, 1984.

62. +Chopin, J., Besson, E., and Nair, A.G.R., *Phytochemistry*, 18, 2059, 1979.

63. +Chopin, J., Bouillant, M.L., Wagner, H., and Galle, K., *Phytochemistry*, 13, 2583, 1974.

64. +Chopin, J., Durix, A., and Bouillant, M.L., *C.R. Acad. Sci. Paris, Ser. C*, 266, 1334, 1968.

65. +Chopin, J., Roux, B., and Durix, A., *C.R. Acad. Sci. Paris, Ser. C*, 259, 3111, 1964.

66. +Chulia, A.J., Thèse Doc. Pharm. Université Grenoble, France, 1984.

67. +Chulia, A.J. et al., in *C.R.J. Int. Groupe Polyphénols*, Montpellier Fr., INRA Ed., Versailles Fr., 13, 50, 1986.

68. +Chulia, A.J. and Mariotte, A.M., *J. Nat. Prod.*, 48, 480, 1985.

69. Claeys, M. et al., Mass spectrometric studies on flavonoid glycosides, *Proc. Phytochem. Soc. Eur.*, 40, 182, 1996.

70. Cortes-Cruz, M., Snook, M., and Mac Mullen, M.D., The genetic basis of *C*-glycosyl flavone B-ring modification in maize (*Zea mays* L.) silks, *Genome*, 46, 182, 2003.

71. +Dahlgren, R., *Opera Botanika*, 9, 1, 1963.

72. +Darmograi, V.N., Litvinenko, V.I., and Krivenchuk, P.E., *Khim. Prir. Soedin.*, 248, 1968.

73. +Dellamonica, G. et al., *Phytochemistry*, 22, 2627, 1983.

74. Delle Monache, F., Chemistry of the *Clusia* genus. Part 7: flavonoid-*C*-glycosides from *Clusia sandiensis*, *Rev. Latinoam. Quim.*, 22, 27, 1991.

75. +Dillon, M.O. et al., *Phytochemistry*, 15, 1085, 1976.

76. Dou, H. et al., Chemical constituents of the aerial parts of *Schnabelia tetradonta*, *J. Nat. Prod.*, 65, 1777, 2002.

77. +Dubois, M.A. et al., *Groupe Polyphenols Neuchâtel.*, 1980.

78. +Dubois, M.A. et al., *Phytochemistry*, 21, 1141, 1982.

79. +Dubois, M.A. et al., *Phytochemistry*, 23, 706, 1984.

80. +Dubois, M.A., Zoll, A., and Chopin, J., *Phytochemistry*, 22, 2879, 1983.

81. +Dubois, M.A., Zoll, A., and Chopin, J., *Phytochemistry*, 24, 1077, 1985.

82. +Eade, R.A., Salasoo, I., and Simes, J.J.H., *Chem. Ind.*, 1720, 1962.

83. +Eade, R.A., Salasoo, I., and Simes, J.J.H., *Aust. J. Chem.*, 19, 1717, 1966.

84. El-Fishawi, A.M., Phytochemical study of *Feronia elephantum* Correa, *Zag. J. Pharm. Sci.*, 3, 76, 1994.

85. El-Hela, A.A., Singab, A.B., and El-Azizi, M.M.J., Flavone-*C*-glucosides from *Verbena bipinnatifida* Nutt growing in Egypt, *J. Pharm. Sci.*, 25, 232, 2000.

86. +Elliger, C.A. et al., *Phytochemistry*, 19, 293, 1980.

87. El-Mousallamy, A.M.D., Chemical investigation of the constitutive flavonoid glycosides of the leaves of *Crataegus sinaica*, *Nat. Prod. Sci.*, 4, 53, 1998.

88. Elrady, H. and Saad, A., *C*-glycosylflavones from *Prosopis chilensis*, *Mansoura J. Pharm. Sci.*, 10, 254, 1994.
89. El-Sawi, S.A., A new rare 8-*C*-glucosylflavonoid and other eight flavonoids from the molluscicidal plant *Acacia saligna* Wendl., *Pharm. Pharm. Lett.*, 11, 30, 2001.
90. +El-Sayed, N.H. et al., in *C.R.J. Int. Groupe Polyphénols*, Strasbourg Fr., INRA Ed., Versailles Fr., 15, 356, 1990.
91. +El-Sayed, N.H. et al., *Phytochemistry*, 30, 2442, 1991.
92. El-Sayed, N.H. et al., Flavonoids of *Otanthus maritimus*, *Rev. Latinoam. Quim.*, 23, 1, 1992.
93. El-Sayed, N.H. et al., Flavonoid glycosides from *Lagonychium farcatum* and *Herminiera elaphroxylon*, *Asian J. Chem.*, 9, 549, 1997.
94. El-Shafae, A., Azza, M., and El-Domiaty, M.M., Flavonoidal constituents of the flowers of *Acacia saligna*, *Zagazig J. Pharm. Sci.*, 7, 48, 1998.
95. El-Telbani, E. et al., Synthesis of bis(*C*-glycosyl)flavonoid precursors, *Carbohydr. Res.*, 306, 463, 1998.
96. Engelhardt, U.H., Finger, A., and Kuhr, S., Determination of *C*-glycosides in tea, *Z. Lebensm. Unters. Forsch.*, 197, 239, 1993.
97. Engelhardt, U.H., Lakenbrink, C., and Lapczynski, S., Antioxidative phenolic compounds in green/black tea and other methylxanthine containing beverages, *ACS Symp. Ser.*, 754, 111, 2000.
98. +Escobar, L.K., Liu, Y.L., and Mabry, T.J., *Phytochemistry*, 22, 796, 1983.
99. Faizi, S. and Ali, M., Shamimin: a new flavonol *C*-glycoside from leaves of *Bombax ceiba*, *Planta Med.*, 65, 383, 1999.
100. Faizi, S. and Ali, M., 1H and 13C NMR of acetyl derivatives of shamimin, a flavonol-*C*-glycoside, *Magn. Reson. Chem.*, 38, 701, 2000.
101. +Fang, N., Leidig, M., and Mabry, T.J., *Phytochemistry*, 25, 927, 1986.
102. Ferreres, F. et al., Approach to the study of *C*-glycosylflavones by ion trap HPLC-PAD–ESI/MS/MS: application to seeds of quince (*Cydonia oblonga*), *Phytochem. Anal.*, 14, 352, 2003.
103. Frangne, N. et al., Flavone glucoside uptake into barley mesophyll and *Arabidopsis* cell culture vacuoles. Energization occurs by H^+-antiport and ATP-binding cassette-type, *Plant Physiol.*, 128, 726, 2002.
104. +Freitag, P. et al., *Phytochemistry*, 25, 669, 1986.
105. +Funaoka, K., *Chem. Abstr.*, 50, 14279, 1956.
106. +Garg, S.P., Bhushan, R., and Kapoor, R.C., *Phytochemistry*, 19, 1265, 1980.
107. +Gentili, B. and Horowitz, R.M., *J. Org. Chem.*, 33, 1571, 1968.
108. +Ghosal, S. and Jaiswal, D.K., *J. Pharm. Sci.*, 69, 53, 1980.
109. Gil, M.I., Ferreres, F., and Tomas-Barberan, F.A., Effect of modified atmosphere packaging on the flavonoids and vitamin C content of minimally processed Swiss chard (*Beta vulgaris* Subspecies *cycla*), *J. Agric. Food Chem.*, 46, 2007, 1998.
110. Gil-Izquierdo, A. et al., *In vitro* availability of flavonoids and other phenolics in orange juice, *J. Agric. Food Chem.*, 49, 1035, 2001.
111. Gil-Izquierdo, A. et al., Effect of the rootstock and interstock grafted in lemon tree (*Citrus limon* (L.) Burm.) on the flavonoid content of lemon juice, *J. Agric. Food Chem.*, 52, 324, 2004.
112. +Gluchoff-Fiasson, K., Jay, M., and Viricel, M.R., *Phytochemistry*, 28, 2471, 1989.
113. +Goetz, M. and Jacot-Guillarmod, A., *Helv. Chim. Acta*, 60, 1322, 1977.
114. +Goetz, M. and Jacot-Guillarmod, A., *Helv. Chim. Acta*, 61, 1373, 1978.
115. Gonnet, J.F., Flavonoid glycoside variation in wild specimens of *Centaurea montana* (Compositae), *Biochem. Syst. Ecol.*, 20, 149, 1992.
116. Gonnet, J.F., Flavonoid glycoside variation in wild specimens of *Centaurea triumfetti* (Compositae) and comments on its relationships with *Centaurea montana* based on flavonoid fingerprints, *Biochem. Syst. Ecol.*, 21, 389, 1993.
117. Gonnet, J.F., Flavonoid glycoside variation in the progeny of wild specimens of *Centaurea montana* and comments on the origin of their natural diversity, *Biochem. Syst. Ecol.*, 24, 447, 1996.
118. +Gramatica, P. et al., *Tetrahedron Lett.*, 2423, 1982.

119. Grayer, R.J. et al., Phenolics in rice phloem sap as sucking deterrents to the brown planthopper, *Nilaparvata lugens*, Int. Symp. on Nat. Phenols in Plant Resistance, 2, 691, 1994.

120. Grayer, R.J. et al., The application of atmospheric pressure chemical ionization liquid chromatography–mass spectrometry in the chemotaxonomic study of flavonoids: characterization of flavonoids from *Ocimum gratissimum* var., *Phytochem. Anal.*, 11, 257, 2000.

121. Grayer, R.J. et al., Leaf flavonoid glycosides as chemosystematic characters in *Ocimum*, *Biochem. Syst. Ecol.*, 30, 327, 2002.

122. Grayer, R.J. and De Kok, R.P.J., Flavonoids and verbascoside as chemotaxonomic characters in the genera *Oxera* and *Faradaya* (Labiatae), *Biochem. Syst. Ecol.*, 26, 729, 1998.

123. Grayer-Barkmeijer, R.J. and Tomas-Barberan, F.A., 8-Hydroxylated flavone-*O*-glycosides and other flavonoids in chemotypes of *Gratiola officinalis*, *Phytochemistry*, 34, 205, 1993.

124. Haggag, E.G. et al., Flavonoids from the leaves of *Citrus aurantium* (sour orange) and *Citrus sinensis* (sweet orange), *Asian J. Chem.*, 11, 707, 1999.

125. +Harborne, J.B. et al., *Plant. Syst. Evol.*, 154, 251, 1986a.

126. Harborne, J.B. et al., Ten isoprenylated and *C*-methylated flavonoids from the leaves of three *Vellozia* species, *Phytochemistry*, 34, 219, 1993.

127. Harborne, J.B. et al., Variations in the lipohilic and vacuolar flavonoids of the genus *Vellozia*, *Phytochemistry*, 35, 1475, 1994.

128. +Harborne, J.B., Heywood, V.H., and Chen, X.Y., *Biochem. Syst. Ecol.*, 14, 81, 1986b.

129. Harder, L.H. and Christensen, L.P., A new flavone *O*-glycoside and other constituents from wheat leaves (*Triticum aestivum* L.), *Z. Naturforsch. C: Biosci.*, 55, 337, 2000.

130. Hatano, T. et al., *C*-glycosidic flavonoids from *Cassia occidentalis*, *Phytochemistry*, 52, 1379, 1999.

131. +Hayashi, Y., Ghara, S., and Takahashi, T., *Mokuzai Gakkaish*, 33, 511, 1987.

132. +Haynes, L.J. and Holsworth, D.K., *J. Chem. Soc. C*, 2581, 1970.

133. +Herrera, Y. and Bain, J.F., *Biochem. Syst. Ecol.*, 19, 665, 1991.

134. +Hillis, W.E. and Carle, A., *Aust. J. Chem.*, 16, 147, 1963.

135. +Hillis, W.E. and Horn, D.H.S., *Aust. J. Chem.*, 18, 531, 1965.

136. +Hillis, W.E. and Horn, D.H.S., *Aust. J. Chem.*, 19, 705, 1966.

137. +Hillis, W.E. and Inoue, T., *Phytochemistry*, 6, 59, 1967.

138. +Hilsenbeck, R.A. and Mabry, T.J., *Phytochemistry*, 22, 2215, 1983.

139. +Hilsenbeck, R.A. and Mabry, T.J., *Phytochemistry*, 29, 2181, 1990.

140. +Hilsenbeck, R.A., Wright, S.J., and Mabry, T.J., *J. Nat. Prod.*, 47, 312, 1984.

141. +Hirose, R. et al., *Agric. Biol. Chem.*, 45, 551, 1981.

142. +Hirose, Y. et al., *Kumamoto Pharm. Bull.*, 5, 4, 1962.

143. Hoffmann-Bohm, K. et al., Antihepatotoxic *C*-glycosylflavones from the leaves of *Allophylus edulis* var. *edulis* and *gracilis*, *Plant. Med.*, 58, 544, 1992.

144. +Hörhammer, L. et al., *Tetrahedron Lett.*, 22, 1707, 1965.

145. +Hörhammer, L., Wagner, H., and Gloggengiesser, F., *Arch. Pharm.*, 291, 126, 1958.

146. +Hörhammer, L., Wagner, H., and Leeb, W., *Arch. Pharm.*, 65, 264, 1960.

147. +Horowitz, R.M. and Gentili, B., *Chem. Ind.*, 498, 1964.

148. +Horowitz, R.M. and Gentili, B., *Chem. Ind.*, 625, 1966.

149. +Horowitz, R.M., Gentili, B., and Gaffield, W., ACS Meeting, Los Angeles, USA, 1974.

150. +Hostettmann, K. and Jacot-Guillarmod, A., *Helv. Chim. Acta*, 57, 204, 1974.

151. +Hostettmann, K. and Jacot-Guillarmod, A., *Helv. Chim. Acta*, 58, 130, 1975.

152. +Hostettmann, K. and Jacot-Guillarmod, A., *Helv. Chim. Acta*, 59, 1584, 1976.

153. Hussein, S.A.M. et al., Flavonoids from *Ephedra aphylla*, *Phytochemistry*, 45, 1529, 1997.

154. Ibraheim, Z.Z., Further constituents of *Crotalaria thebaica* (Del.) Dc growing in Egypt, *Bull. Fac. Sci. Assiut Univ. B*, 23, 49, 1994.

155. +Ibrahim, R.K. and Shaw, M.A., *Phytochemistry*, 9, 1855, 1970.

156. Imperato, F., 3,6,8-Tri-*C*-xylosylapigenin from *Asplenium viviparum*, *Phytochemistry*, 33, 729, 1993.

157. +Ingham, J.L. et al., *Phytochemistry*, 25, 1772, 1986.

158. +Ishikura, N. and Yoshitama, K., *Phytochemistry*, 27, 1555, 1988.

159. Ito, H. et al., Polyphenols from *Eriobotrya japonica* and their cytotoxicity against human oral tumor cell lines, *Chem. Pharm. Bull.*, 48, 687, 2000.

160. +Jacot-Guillarmot, A., Luong, M.D., and Hostettmann, K., *Helv. Chim. Acta*, 58, 1477, 1975.

161. +Jain, N. et al., *J. Chem. Res. Synop.*, 3, 80, 1990.

162. +Jay, M. et al., *Phytochemistry*, 18, 184, 1979.

163. +Jay, M. et al., *Biochem. Syst. Ecol.*, 8, 127, 1980.

164. +Jay, M., Lameta-d'Arcy, A., and Viricel, M.R., *Phytochemistry*, 23, 1153, 1984.

165. Jeong, S.J. et al., Flavonoids from the seeds of *Phaseolus radiatus*, *Saengyak Hakhoechi*, 29, 357, 1998.

166. +Julian, E.A. et al., *Phytochemistry*, 10, 3185, 1971.

167. +Jurd, L., Geissman, T.A., and Seikel, M.K., *Arch. Biochem. Biophys.*, 67, 284, 1957.

168. Kamara, B.I. et al., Polyphenols from honeybush tea (*Cyclopia intermedia*), *J. Agric. Food Chem.*, 51, 3874, 2003.

169. Kamel, M.S., Flavone *C*-glycosides from *Lupinus hartwegii*, *Phytochemistry*, 63, 449, 2003.

170. +Kartnig, T., Hiermann, A., and Azzam, S., *Sci. Pharm.*, 55, 95, 1987.

171. +Kashnikova, M.V. et al., *Khim. Prir. Soedin.*, 108, 1984.

172. +Kato, T. and Morita, Y., *Chem. Pharm. Bull.*, 38, 2277, 1990.

173. Kato, T. and Morita Y., The rotational isomers of peracetylated *C*-glycosylflavones, *Heterocycles*, 35, 965, 1993.

174. Kiehne, A. and Engelhardt, U.H., Thermospray-LC–MS analysis of various groups of polyphenols in tea. Part 1. Catechins, flavonol *O*-glycosides, and flavone *C*-glycosides, *Z. Lebensm. Unters. Forsch.*, 202, 48, 1996.

175. Kim, J.S. and Kim, I.H., Pharmacoconstituents of *Crataegus pinnatifida* var. *pubescens* leaves, *Yakhak Hoechi*, 37, 193, 1993.

176. +Kinjo, J. et al., *Pharm. Bull.*, 35, 4846, 1987.

177. +Kitanaka, S., Ogata, K., and Takido, M., *Chem. Pharm. Bull.*, 37, 2441, 1989.

178. Kitanaka, S. and Takido, M., Demethyltorosaflavones C and D from *Cassia nomame*, *Chem. Pharm. Bull.*, 40, 249, 1992.

179. +Kitanov, G., Blinova, K.F., and Akhtardzhiev, K., *Khim. Prir. Soedin.*, 154, 231, 1979.

180. Kite, G.C. and Ireland, H., Non-protein amino acids of *Bocoa* (Leguminosae; Papilionoideae), *Phytochemistry*, 59, 163, 2002.

181. Klein, M. et al., Different energization mechanisms drive the vacuolar uptake of a flavonoid glucoside and a herbicide glucoside, *J. Biol. Chem.*, 271, 29666, 1996.

182. +Komatsu, M. et al., *Chem. Pharm. Bull.*, 16, 1413, 1968.

183. +Komatsu, M. and Tomimori, T., *Tetrahedron Lett.*, 1611, 1966.

184. +Krause, J., *Z. Pflanzenphysiol.*, 79, 372, 1976a.

185. +Krause, J., *Z. Pflanzenphysiol.*, 79, 465, 1976b.

186. Kraut, L., Mues, R., and Sim-Sim, M., Acylated flavone and glycerol glucosides from two *Frullania* species, *Phytochemistry*, 34, 211, 1993.

187. Kraut, L. et al., Flavonoids from some *Frullania* species (Hepaticae), *Z. Naturforsch.*, 50, 345, 1995.

188. Krauze-Baranowska, M. and Cisowski, W., High-performance liquid chromatographic determination of flavone *C*-glycosides in some species of the Cucurbitaceae family, *J. Chromatogr. A*, 675, 240, 1994.

189. Krauze-Baranowska, M. and Cisowski, W., Flavone *C*-glycosides from *Bryonia alba* and *B. dioica*, *Phytochemistry*, 39, 727, 1995a.

190. Krauze-Baranowska, M. and Cisowski, W., Isolation and identification of *C*-glycoside flavones from *Lagenaria siceraria*, *Acta Pol. Pharm.*, 52, 137, 1995b.

191. Krauze-Baranowska, M. and Cisowski, W., Flavonoids from some species of the genus *Cucumis*, *Biochem. Syst. Ecol.*, 29, 321, 2001.

192. Krenn, L. et al., Flavonoids from *Achillea nobilis* L., *Z. Naturforsch. C: Biosci.*, 58, 11, 2003.

193. +Kumamoto, H. et al., *Nogeikagaku Kaishi*, 59, 677, 1985a.

194. +Kumamoto, H. et al., *Agric. Biol. Chem.*, 49, 2613, 1985b.

195. Kumar, J.K. et al., Flavone glycosides from *Polygala telephioides* and *Polygala arvensis*, *Nat. Prod. Lett.*, 14, 35, 1999.

196. Kumazawa, T. et al., Practical synthesis of a *C*-glycosyl flavonoid via *O–C* glycoside rearrangement, *Bull. Chem. Soc. Jap.*, 68, 1379, 1995.

197. Kumazawa, T. et al., Synthesis of 8-*C*-glucosylflavones, *Carbohydr. Res.*, 334, 183, 2001.

198. Kuo, S.H. et al., A flavone *C*-glycoside and an aromatic glucoside from *Gentiana* species, *Phytochemistry*, 41, 309, 1996.

199. Larsen, L.M., Olsen, C.E., and Wieczorkowska, E., The distribution of flavonoids in developing mung bean (*Vigna radiata* L.) organs and in seeds, in *C.R.J. Int. Groupe Polyphénols*, Palma de Mallorca Sp., INRA Ed., Versailles Fr., 69, 319, 1995.

200. Latté, K.P. et al., *O*-Galloyl-*C*-glycosylflavones from *Pelargonium reniforme*, *Phytochemistry*, 59, 419, 2002.

201. Lee, D.Y.W., Zhang, W.Y., and Kamati, V.V.R., Total synthesis of puerarin, an isoflavone *C*-glycoside, *Tetrahedron Lett.*, 44, 6857, 2003.

202. Lee, E.A. et al., Genetic mechanisms underlying apimaysin and maysin synthesis and corn earworm antibiosis in maize (*Zea mays* L.), *Genetics*, 149, 1997, 1998.

203. Lee, J.Y. et al., Antioxidative flavonoids from leaves of *Carthamus tinctorius*, *Arch. Pharm. Res.*, 25, 313, 2002.

204. Lee, M.W., Flavonoids from the leaves of *Betula platyphylla* var. *latifolia*, *Seangyak Hakhoechi*, 25, 199, 1994.

205. Leitao, S.G. and Delle Monache, F., 2″-*O*-caffeoylorientin from *Vitex polygama*, *Phytochemistry*, 49, 2167, 1998.

206. +Lewak, S., *Rocz. Chem.*, 40, 445, 1966.

207. Lewis, K.C. et al., Room-temperature (^{1}H, ^{13}C) and variable-temperature (^{1}H) NMR studies on spinosin., *Magn. Reson. Chem.*, 38, 771, 2000.

208. Li, Q.M. et al., Mass spectral characterization of *C*-glycosidic flavonoids isolated from a medicinal plant (*Passiflora incarnata*), *J. Chromatogr.*, 562, 435, 1991.

209. Li, Q.M. et al., Differentiation of 6-*C*- and 8-*C*-glycosidic flavonoids by positive ion fast atom bombardment and tandem mass spectrometry, *Biol. Mass Spectrom.*, 21, 213, 1992.

210. Li, W., Asada, Y., and Yoshikawa, T., Flavonoid constituents from *Glycyrrhiza glabra* hairy root cultures, *Phytochemistry*, 55, 447, 2000.

211. Lin, L.C. et al., A new flavone *C*-glycoside and antiplatelet and vasorelaxing flavones from *Gentiana arisanensis*, *J. Nat. Prod.*, 60, 851, 1997.

212. Lin, L.C. and Chou, C.J., Meroterpenes and *C*-glucosylflavonoids from the aerial parts of *Plumbago zeylanica*, *Chin. Pharm. J.*, 55, 77, 2003.

213. Lin, Y.L. et al., Flavonoid glycosides from *Terminalia catappa* L., *J. Chin. Chem. Soc.*, 47, 253, 2000.

214. +Linard, A. et al., *Phytochemistry*, 21, 797, 1982.

215. Liu, D. et al., New *C*-glycosylflavones from *Tetrastigma hemsleyanum*, *Zhiwu Xuebo*, 44, 227, 2002.

216. Liu, M. et al., An acylated flavone *C*-glycoside from *Glycyrrhiza eurycarpa*, *Phytochemistry*, 36, 1089, 1994.

217. Lobstein, A. et al., 4″-Hydroxymaysin and cassiaoccidentalin B, two unusual *C*-glycosylflavones from *Mimosa pudica* (Mimosaceae), *Biochem. Syst. Ecol.*, 30, 375, 2002.

218. Lu, Y. and Foo, L.Y., Flavonoid and phenolic glycosides from *Salvia officinalis*, *Phytochemistry*, 55, 263, 2000.

219. +Luong, M.D. et al., *Helv. Chim. Acta*, 64, 2741, 1981.

220. +Luong, M.D., Fombasso, P., and Jacot-Guillarmod, A., *Helv. Chim. Acta*, 63, 244, 1980.

221. +Luong, M.D. and Jacot-Guillarmod, A., *Helv. Chim. Acta*, 60, 2099, 1977.

222. Ma, C.M., Nakamura, N., and Hattori, M., Saponins and *C*-glycosyl flavones from the seeds of *Abrus precatorius*, *Chem. Pharm. Bull.*, 46, 982, 1998.

223. Maatooq, G.T. et al., *C*-*p*-hydroxybenzylglycosylflavones from *Citrullus colocynthis*, *Phytochemistry*, 44, 187, 1997.

224. +Mabry, T.J. et al., *Phytochemistry*, 10, 677, 1971.

225. +Mabry, T.J. et al., *J. Nat. Prod.*, 47, 127, 1984.

226. +Mac Cormick, S. and Mabry, T.J., *Phytochemistry*, 22, 798, 1983.

227. Mac Mullen, M.D. et al., Quantitative trait loci and metabolic pathways, *Proc. Natl. Acad. Sci. USA*, 95, 1996, 1998.

228. Mac Nally, D. et al., Complex *C*-glycosylflavonoid phytoalexins from *Cucumis sativus*, *J. Nat. Prod.*, 66, 1280, 2003.

229. +Maggi, L. et al., *J. Chromatogr.*, 478, 225, 1989.

230. Mahlin, J.A., Jung, K.H., and Schmidt, R.R., Glycosyl imidates, synthesis of flavone *C*-glycosides vitexin, isovitexin and isoembigenin, *Liebigs Ann. Chem.*, 3, 461, 1995.

231. Majumder, P.L. and Sen, R.C., Chemical constituents of the orchid *Tylostylis discolor*: occurrence of flavonoid *C*-glucoside in Orchidaceae plant, *J. Indian Chem. Soc.*, 71, 649, 1994.

232. Manguro, L.O.A., Mukonyl, K.W., and Githiomi, J.K., A new flavonol glycoside from *Eucalyptus globulus* subsp. *Maidenii*, *Nat. Prod. Lett.*, 7, 163, 1995.

233. Manthey, J.A., Guthrie, N., and Grohmann, K., Biological properties of citrus flavonoids pertaining to cancer and inflammation, *Curr. Med. Chem.*, 8, 135, 2001.

234. Marchart, E. et al., Analysis of flavonoids in *Achillea nobilis* L., *Sci. Pharm.*, 71, 133, 2003.

235. Marchart, E. and Kopp, B., Capillary electrophoretic separation and quantification of flavone-*O*- and *C*-glycosides in *Achillea setacea* W. et K., *J. Chromatogr. A*, 792, 363, 2003.

236. +Mareck, U. et al., *Phytochemistry*, 30, 3486, 1991.

237. +Markham, K.R. et al., *Z. Naturforsch.*, 37, 562, 1982.

238. +Markham, K.R. et al., *Phytochemistry*, 28, 299, 1989.

239. Markham, K.R. and Mitchell, K.A., The mis-identification of the major antioxidant flavonoids in young barley (*Hordeum vulgare*) leaves, *Z. Naturforsch. C: Biosci.*, 58, 53, 2003.

240. +Markham, K.R. and Mues, R., *Z. Naturforsch.*, 39, 309, 1984.

241. +Markham, K.R. and Wallace, J.W., *Phytochemistry*, 19, 415, 1980.

242. +Markham, K.R. and Williams, C.A., *Phytochemistry*, 19, 2789, 1980.

243. +Marston, A., Hostettman, K., and Jacot-Guillarmod, A., *Helv. Chim. Acta*, 59, 2596, 1976.

244. Maruno, M., Active principles of Pinelliae tuber and new preparation of crude drug, *WakanIyakugaku Zasshi*, 14, 81, 1997.

245. +Mastenbroek, O. et al., *Plant Syst. Evol.*, 141, 257, 1983.

246. Materska, M. et al., Isolation and structure elucidation of flavonoid and phenolic acid glycosides from pericap of hot pepper fruit *Capsicum annuum* L., *Phytochemistry*, 63, 893, 2003.

247. +Matsubara, Y. et al., *Agric. Biol. Chem.*, 49, 909, 1985a.

248. +Matsubara, Y. et al., *Nippon Nogeikayaku Kaishi*, 59, 405, 1985b.

249. Matsumoto, S. et al., Evidence by flavonoid markers of four natural hybrids among *Asplenium normale* and related species, *Biochem. Syst. Ecol.*, 31, 51, 2003.

250. +Mebe, P.P., *Phytochemistry*, 26, 2646, 1987.

251. +Medvedeva, S.A, Tjukavhina, N.A., and Ivanova, S.Z., *Khim. Drev.*, 15, 144, 1974.

252. Melek, F.R. et al., *Atractylis carduus angustifolia* flavonoids and anti-inflammatory activity, *Egypt. J. Pharm. Sci.*, 33, 11, 1992.

253. Meyre-Silva, C. et al., A triterpene and a flavonoid *C*-glycoside from *Aleurites moluccana* L. Willd., *Act. Farm. Bonaer.*, 16, 169, 1997.

254. Meyre-Silva, C. et al., Isolation of a *C*-glycoside flavonoid with antinociceptive action from *Aleurites moluccana* leaves, *Planta Med.*, 65, 293, 1999.

255. Michael, H.N., Guergues, S.N., and Sandak, R.N., Some polyphenolic constituents of *Titricum aestivum* (wheat bran, Sakha 69) and their antibacterial effect, *Asian J. Chem.*, 10, 256, 1998.

256. +Mihajlovic, N., *Arch. Farm.*, 38, 29, 1988.

257. Miyaichi, Y. and Tomimori, T., Constituents of *Scutellaria* species XVI. On the phenol glycosides of the root of *Scutellaria baicalensis* Georgi, *Nat. Med.*, 48, 215, 1994.

258. Miyake, T., Possible inhibition of atherosclerosis by a flavonoid isolated from young green barley leaves, *ACS Symp. Ser.*, 702, 178, 1998.

259. Miyake, Y. et al., Isolation of *C*-glucosylflavone from lemon peel and antioxidative activity of flavonoid compounds in lemon fruit, *J. Agric. Food Chem.*, 45, 4619, 1997.

260. Miyake, Y. et al., Suppressive effect of components in lemon juice on blood pressure in spontaneously hypertensive rats, *Food Sci. Technol. Int.*, 4, 29, 1998a.

261. Miyake, Y. et al., Characteristics of antioxidative flavonoid glycosides in lemon fruit, *Food Sci. Technol. Int.*, 4, 48, 1998b.

262. +Monties, B., Bouillant, M.L., and Chopin, J., *Phytochemistry*, 15, 1053, 1976.

263. Moore, M.O. and Giannasi, D.E., Foliar flavonoids of eastern North America *Vitis* (Vitaceae) north of Mexico, *Plant Syst. Evol.*, 193, 21, 1994.

264. +Morita, N., Arisawa, M., and Yoshikawa, A., *J. Pharm. Soc.*, 96, 1180, 1976.

265. Moreira, D.L., Guimaraes, E.F., and Kaplan, M.A.C., A *C*-glucosylflavone from leaves of *Piper Ihotzkyanum*, *Phytochemistry*, 55, 783, 2000.

266. +Moritomo, S., Nonaka, G., and Nishioka, I., *Chem. Pharm. Bull.*, 34, 633, 1986a.

267. +Moritomo, S., Nonaka, G., and Nishioka, I., *Chem. Pharm. Bull.*, 34, 643, 1986b.

268. Morsch, M. et al., Separation of *C*-glycoside flavonoids from *Aleurites moluccana* using chitin and full N-acetylated chitin, *Z. Naturforsch. C: Biosci.*, 57, 957, 2002.

269. Mosharrafa, S.A.M., Kawashty, S.A., and Saleh, N.A.M., Flavonoids of *Pistacia atlantica* (Desf.), *Bull. Nat. Res. Centre (Egypt)*, 24, 109, 1999.

270. +Mues, R., Klein, R., and Gradstein, S.R., *J. Hattor. Bot. Lab.*, 70, 79, 1991.

271. +Mues, R. and Zinsmeister, H.D., *Phytochemistry*, 15, 1757, 1976.

272. +Murakami, T., Nishikawa, Y., and Ando, T., *Chem. Pharm. Bull.*, 8, 688, 1960.

273. +Narayanan, V. and Seshadri, T.R., *Indian J. Chem.*, 9, 14, 1971.

274. +Nawwar, M.A.M., El Sissi, H.I., and Barakat, H.H., *Phytochemistry*, 19, 1854, 1980.

275. +Niemann, G.J., *Acta Bot. Neerl.*, 30, 475, 1981.

276. +Niemann, G.J., *Rev. Latinoam. Quim.*, 13, 74, 1982.

277. +Nikolov, N., Horowitz, R.M., and Gentili, B., Int. Congress for Research on Medicinal Plants Munich, Abstracts of papers, Section A, 1976.

278. Norbaek, R. et al., Flavone *C*-glycoside, phenolic acid and nitrogen contents in leaves of barley subject to organic fertilization treatments, *J. Agric. Food Chem.*, 51, 809, 2003.

279. Norbaek, R., Brandt, K., and Kondo T., Identification of flavone *C*-glycosides including a new flavonoid chromophore from barley leaves (*Hordeum vulgare* L.) by improved NMR techniques, *J. Agric. Food Chem.*, 48, 1703, 2000.

280. +Nunes, D.S., Haag, A., and Bestmann, H.J., *Liebigs Ann. Chem.*, 4, 331, 1989.

281. +Oelrichs, P., Marshall, J.T.B., and Williams, D.H., *J. Chem. Soc.*, 941, 1968.

282. Ogihara, Y., Phenolic compounds from *Scutellaria pontica*, *Turk. J. Chem.*, 26, 581, 2002.

283. +Ohashi, H., Goto, M., and Imamura, H., *Phytochemistry*, 16, 1106, 1977.

284. +Okamura, N., Yagi, A., and Nishioka, I., *Chem. Pharm. Bull.*, 29, 3507, 1981.

285. Oliveira, R.R. et al., High-speed countercurrent chromatography as a valuable tool to isolate *C*-glycosylflavones from *Cecropia lyratiloba* Miquel., *Phytochem. Anal.*, 14, 96, 2003.

286. Osawa, T. et al., A novel antioxidant isolated from young green barley leaves, *J. Agric. Food Chem.*, 40, 1135, 1992.

287. +Osterdahl, B.G., *Acta Chem. Scand.*, 33, 400, 1979.

288. Otsuka, H. et al., Glochiflavanosides A–D: flavanol glucosides from the leaves of *Glochidion zeylanicum* (Gaertn) A. Juss, *Chem. Pharm. Bull.*, 49, 921, 2001.

289. Otsuka, H. and Kijima, K., An iridoid gentiobioside, a benzophenone glucoside and acylated flavone *C*-glycosides from *Tripterospermum japonicum* (Sieb. Et Zucc.) Maxim, *Chem. Pharm. Bull.*, 49, 699, 2001.

290. +Ouabonzi, A., Bouillant, M.L., and Chopin, J., *Phytochemistry*, 22, 2632, 1983.

291. Palme, E. et al., Flavonoid glycosides from *Cotoneaster thymaefolia*, *Phytochemistry*, 35, 1381, 1994.

292. +Paris, R.R., *C.R. Acad. Sci. Paris*, 245, 443, 1957.

293. +Parthasarathy, M.R., Seshadri, T.R., and Varma, R.S., *Curr. Sci.*, 43, 74, 1974.

294. Pastene, E.R. et al., Separation by capillary electrophoresis of *C*-glycosylflavonoids in *Passiflora* sp. extracts, *Bol. Soc. Chil. Quim.*, 45, 461, 2000.

295. Pereira, C.A.M. and Vilegas, J.H.Y., Chemical and pharmacological constituents of *Passiflora alata* Dryander, *Passiflora edulis* Sims and *Passiflora incarnata* L., *Rev. Bras. Plant. Med.*, 3, 1, 2000.

296. +Peterson, P.M. and Rieseberg, L.H., *Biochem. Syst. Ecol.*, 15, 647, 1987.

297. Piacente, S. et al., Flavonoids and arbutin from *Turnera diffusa*, *Z. Naturforsch.*, 57, 983, 2002.

298. Picerno, P. et al., Phenolic constituents and antioxidant properties of *Xanthosoma violaceum* leaves, *J. Agric. Food Chem.*, 51, 6423, 2003.

299. +Proliac, A. et al., *C.R. Acad. Sci. Ser. D*, 277, 2813, 1973.

300. Proliac, A, Chaboud, A., and Raynaud, J., Isolation et identification of three *C*-glycosylflavones from leafy stems of *Arum dracunculus* L., *Pharmazie*, 47, 646, 1992.

301. +Prum-Bousquet, M., Tillequin, F., and Paris, R.R., *Lloydia*, 40, 591, 1977.

302. +Pryakhina, N.I., Sheichenko, V.L., and Blinova, K.F., *Khim. Prir. Soedin.*, 589, 1984.

303. +Qasim, M.A. et al., *Phytochemistry*, 26, 2871, 1987.

304. Qiao, S.Y. et al., Rapid discovery and analysis of *C*-glycosylflavones from *Stellaria media* (L.) Cyr. By MS/MS, *Tianran Yaowu*, 1, 120, 2003.

305. Rahman, K. et al., Isoscoparin-2''-*O*-glucoside from *Passiflora incarnata*, *Phytochemistry*, 45, 1093, 1997.

306. Rao, J.R. and Rao, R.S., Dalpaniculatin a *C*-glycosylisoflavone from *Dalbergia paniculata* seeds, *Phytochemistry*, 30, 715, 1991.

307. +Raynaud, J. and Rasolojaona, L., *C.R. Acad. Sci. Paris Ser. D*, 282, 1059, 1976.

308. Rehwald, A., Meier, B., and Sticher, O., Qualitative and quantitative reversed-phase high-performance liquid chromatography of flavonoids in *Crataegus* leaves and flowers, *J. Chromatogr. A*, 677, 25, 1994.

309. Ricci, C.V. et al., Flavonoids of *Syngonanthus* Ruhl. (Eriocaulaceae): taxonomic implications, *Biochem. Syst. Ecol.*, 24, 577, 1996.

310. +Rofi, R.D. and Pomilio, A.B., *Phytochemistry*, 26, 859, 1987.

311. Salatino, A. et al., Distribution and evolution of secondary metabolites in Eriocaulaceae, Lythraceae and Velloziaceae from "campos rupestres", *Gen. Mol. Biol.*, 23, 931, 2000.

312. Sakushima, A. et al., *C*-glycosylflavone and other compounds of *Boreava orientalis*, *Nat. Med.*, 50, 65, 1996.

313. Saleem, R. et al., Hypotensive, hypoglycaemic and toxicological studies on the flavonol *C*-glycoside Shamimin from *Bombax ceiba*, *Plant. Med.*, 65, 331, 1999.

314. Saleh, N.A.M., El-Hadidi, M.N., and Arafa, R.F.M., Flavonoids and anthraquinones of some Egyptian *Rumex* species (Polygonaceae), *Biochem. Syst. Ecol.*, 21, 301, 1993.

315. +Salmenkallio, S. et al., *Phytochemistry*, 21, 2990, 1982.

316. Salvador, M.J. and Dias, D.A., Flavone *C*-glycosides from *Alternanthera maritima* (Mart.) St. Hill (Amaranthaceae), *Biochem. Syst. Ecol.*, 32, 107, 2004.

317. Sanchez-Rabaneda, F. et al., Liquid chromatographic/electrospray ionization tandem mass spectrometric study of the phenolic composition of cocoa (*Theobroma cacao*), *J. Mass Spectrom.*, 38, 35, 2003.

318. Schoeneborn, R. and Mues, R., Flavone di-*C*-glycosides from *Plagiochila jamesonii* and *Plagiochila rupestre*, *Phytochemistry*, 34, 1143, 1993.

319. +Segelman, A.G. et al., *Phytochemistry*, 17, 824, 1978.

320. +Seikel, M.K. and Bushnell, A.J., *J. Org. Chem.*, 24, 1995, 1959.

321. +Seikel, M.K., Bushnell, A.J., and Birzgalis, R., *Arch. Biochem. Biophys.*, 99, 461, 1962.

322. +Seikel, M.K., Chow, J.H.S., and Feldman, L., *Phytochemistry*, 5, 439, 1966.

323. +Seikel, M.K., Holder, D.J., and Birzgalis, R., *Arch. Biochem. Biophys.*, 85, 272, 1959.

324. Senatore, F., D'Agostino, M., and Dini, I., Flavonoid glycosides of *Barbarea vulgaris* L. (Brassicaceae), *J. Agric. Food Chem.*, 48, 2659, 2000.

325. +Shammas, G. and Couladi, M., *Sci. Pharm.*, 56, 277, 1988.

326. Sharaf, M., El-Ansari, M.A., and Saleh, N.A.M., Flavonoids of four *Cleome* and three *Capparis* species, *Biochem. Syst. Ecol.*, 25, 161, 1997.

327. +Sherwood, R.T. et al., *Phytochemistry*, 12, 2275, 1973.

328. +Shirataki, Y., Yokoc, I., and Komatsu, M., *J. Nat. Prod.*, 49, 645, 1986.

329. Singh, M. and Singh, J., New flavonoids from *Frullania tamarisci*, *Fitoterapia*, 62, 187, 1991.

330. Skaltsa, H. et al., Flavonoids from *Scutellaria albida*, *Sci. Pharm.*, 64, 103, 1996.

331. Slimestad, R., Flavonoids in buds and young needles of *Picea*, *Pinus* and *Abies*, *Biochem. Syst. Ecol.*, 31, 1247, 2003.

332. +Smith, D.W., Glennie, C.W., and Harborne, J.B., *Biochem. Syst. Ecol.*, 10, 37, 1982.

333. Snook, M.E. et al., Levels of maysin and maysin analogues in silks of maize germplasm, *J. Agric. Food Chem.*, 41, 1481, 1993.

334. Snook, M.E. et al., New flavone *C*-glycosides from corn (*Zea mays* L.) for the control of the corn earworm (*Helicoverpa zea*), *ACS Symp. Ser.*, 557, 122, 1994.

335. Snook, M.E. et al., New C–4″-hydroxy derivatives of maysin and 3′-methoxymaysin isolated from corn silks (*Zea mays*), *J. Agric. Food Chem.*, 43, 2740, 1995.

336. Snook, M.E. et al., New flavones from corn (*Zea mays* L.) silk including a novel biflavone that contribute resistance to the corn earworm (*Helicoverpa Zea* (Boddie)), Abs. 222nd ACS Nat. Meeting, Chicago, 26, 2001.

337. +Sood, A.R. et al., *Phytochemistry*, 15, 351, 1976.

338. +Specht, J.E., Gorz, H.J., and Haskins, F.A., *Phytochemistry*, 15, 133, 1976.

339. +Speranza, G. et al., *Phytochemistry*, 24, 1571, 1985.

340. +Speranza, G. et al., *Phytochemistry*, 25, 2219, 1986.

341. +Srinivasan, K.K. and Subramanian, S.S., *Arogya* (*Manipal, India*), 9, 89, 1983.

342. +Srivastava, S.D. and Srivastava, S.K., *Curr. Sci.*, 55, 1069, 1986.

343. +Steyns, J.M. et al., *Z. Naturforsch.*, 38, 544, 1983.

344. +Subramanian, P.M. and Misra, G.S., *J. Nat. Prod.*, 42, 540, 1979.

345. Suzuki, R., Okada, Y., and Okuyama, T., Two flavone *C*-glycosides from the style of *Zea mays* with glycation inhibitory activity, *J. Nat. Prod.*, 66, 564, 2003a.

346. Suzuki, R., Okada, Y., and Okuyama, T., A new flavone *C*-glycosides from the style of *Zea mays* with glycation inhibitory activity, *Chem. Pharm. Bull.*, 51, 1186, 2003b.

347. Szalma, S.J. et al., Duplicate loci as QTL: the role of chalcone synthase loci in flavone and phenylpropanoid biosynthesis in maize, *Crop Sci.*, 42, 1679, 2002.

348. +Takagi, S., Yamachi, M., and Inoue, K., *Phytochemistry*, 20, 2443, 1981.

349. +Takahashi, Y. et al., *Tetrahedron Lett.*, 23, 5163, 1982.

350. Takeda, Y. et al., *C,O*-Bisglycosylapigenins from the leaves of *Rhamnella inaequilatera*, *Phytochemistry*, 65, 463, 2004.

351. Tanaka, N. et al., Constituents of bamboos and bamboo grasses, *Yakugaku Zasshi*, 118, 332, 1998.

352. Teshima, K.I. et al., *C*-glycosylflavones from *Climacanthus nutans*, *Nat. Med.*, 51, 557, 1997.

353. Tekelova, D. et al., Determination of flavonoids derivatives of cinnamic acid and phenols in *Ligustrum vulgare* L., *Farm. Obzor.*, 68, 233, 1999.

354. +Theodor, R. et al., *Phytochemistry*, 19, 1965, 1980.

355. +Theodor, R. et al., *Phytochemistry*, 20, 1457, 1981a.

356. +Theodor, R. et al., *Phytochemistry*, 20, 1851, 1981b.

357. +Tillequin, F. et al., *Planta Med.*, 33, 46, 1978.

358. +Tiwari, R.D. and Bajpai, M., *Indian J. Chem.*, 20, 450, 1981.

359. Touati, D. and Fkih-Tetouani, S., Flavonoids of *Daphne laureola* L., *Plant. Med. Phytother.*, 26, 43, 1993.

360. +Tschesche, R. and Struckmeyer, K., *Chem. Ber.*, 109, 2901, 1976.

361. Tsiklauri, G.C. and Shalashvili, A.G., Flavonoid *C*-glucosides of tangerine Iveria leaves, *Soobshcheniya Akad. Nauk Gruzii*, 151, 131, 1995.

362. Umikalsom, Y., Grayer-Barkmeijer, R.J., and Harborne, J.B., A comparison of the flavonoids in Athyriaceae and Aspleniaceae, *Biochem. Syst. Ecol.*, 22, 587, 1994.

363. Umikalsom, Y. and Harborne, J.B., Flavone *C*-glycosides from the pinnae of three *Asplenium* species, *Pertanika*, 14, 143, 1991.

364. +Valant, K., Besson, E., and Chopin, J., *Phytochemistry*, 19, 156, 1980.

365. Valant-Vetschera, K.M., Therapeutic significance of *C*-glycosylflavone accumulation in *Achillea*, *Sci. Pharm.*, 62, 323, 1994.

366. Valsakumari, M.K. and Sulochana, N., Chemical examination of *Acacia leucophoea* willd, *J. Indian Chem. Soc.*, 68, 673, 1991.

367. +Van Brederode, J. and Kamps-Heinsbroek, R., *Z. Naturforsch.*, 36, 486, 1981.

368. +Van Brederode, J. and Nigtevecht, G., *Mol. Gen. Genet.*, 118, 247, 1972.

369. +Van Brederode, J. and Nigtevecht, G., *Biochem. Genet.*, 11, 65, 1974.

370. +Van Heerden, F.R., Brandt, E.V., and Roux, D.G., *J. Chem. Soc. Perkin Trans. 1*, 2463, 1980.
371. +Verma, K.S. et al., *Planta Med.*, 315, 1986.
372. +Vita-Finzi, P. et al., 12th IUPAC Int. Symp. on Chem. of Nat. Prod., Abstracts, 167, 1980.
373. Voirin, B. et al., Separation of flavone *C*-glycosides and qualitative analysis of *Passiflora incarnata* L. by capillary zone electrophoresis, *Phytochem. Anal.*, 11, 90, 2000.
374. +Volk, O. and Sinn, M., *Z. Naturforsch.*, 23, 1017, 1968.
375. Wada, S. et al., Glycosidic flavonoids as rat-liver injury preventing compounds from green tea, *Biosci. Biotechnol. Biochem.*, 64, 2262, 2000.
376. +Wagner, H., *Phytochemistry*, 18, 907, 1979.
377. +Waiss, A.C. et al., *J. Econ. Entomol.*, 72, 256, 1979.
378. +Wallace, J.W. et al., *Phytochemistry*, 18, 1077, 1979.
379. Wallace, J.W., Chemotaxonomy of the Hymenophyllaceae.II. *C*-glycosylflavones and flavone-*O*-glycosides of *Trichomanes* s.I., *Am. J. Bot.*, 83, 1304, 1996.
380. +Wallace, J.W. and Morris, G., *Phytochemistry*, 17, 1809, 1978.
381. Wang, J. et al., Swertifrancheside, an HIV-reverse transcriptase inhibitor and the first flavone–xanthone dimer, from *Swertia franchetiana*, *J. Nat. Prod.*, 57, 211, 1994.
382. Waridel, P. et al., Evaluation of quadrupole time-of-flight tandem mass spectrometry and ion trap multiple-stage mass spectrometry for the differentiation of *C*-glycosidic flavonoid isomers, *J. Chromatogr. A*, 926, 29, 2001.
383. Webby, R. and Markham, K.R., Isoswertiajaponin 2″-*O*-β-arabinopyranoside and other flavone-*C*-glycosides from the Antartic grass *Deschampsia antartica*, *Phytochemistry*, 36, 1323, 1994.
384. Widstrom, N.W. and Snook, M.E., Genetic variation for maysin and its analogues in crosses among corn inbreds, *Crop Sci.*, 38, 372, 1998.
385. Williams, C.A. et al., Methylated *C*-glycosylflavones as taxonomic markers in orchids of the subtribe Ornithocephalinae, *Phytochemistry*, 37, 1045, 1994a.
386. Williams, C.A. et al., Differences in flavonoid patterns between genera within the Velloziaceae, *Phytochemistry*, 36, 931, 1994b.
387. Williams, C.A. et al., Flavonoid patterns and the revised classification of Australian Restionaceae, *Phytochemistry*, 49, 529, 1998.
388. +Williams, C.A. and Harborne, J.B., *Biochem. Syst. Ecol.*, 5, 45, 1977.
389. +Williams, C.A., Harborne, J.B., and Clifford, H.T., *Phytochemistry*, 12, 2417, 1973.
390. +Williams, C.A., Harborne, J.B., and de Menezes, N.L., *Biochem. Syst. Ecol.*, 19, 483, 1991.
391. +Williams, C.A. and Murray, B.G., *Phytochemistry*, 11, 2507, 1972.
392. Williams, C.A., Newman, M., and Gibby, M., The application of leaf phenolic evidence for systematic studies within the genus *Pelargonium* (Geraniaceae), *Biochem. Syst. Ecol.*, 28, 119, 2000.
393. Williams, C.A., Onylagha, J.C., and Harborne, J.B., Flavonoid profiles in leaves, flowers and stems of forty-nine members of the Phaseolineae, *Biochem. Syst. Ecol.*, 23, 655, 1995.
394. +Wirasutisna, K.R. et al., *Phytochemistry*, 25, 558, 1986.
395. +Woo, W.S. et al., *Phytochemistry*, 18, 353, 1979.
396. +Woo, W.S. et al., *Phytochemistry*, 19, 279, 1980.
397. +Wu, F. et al., *Chem. Pharm. Bull.*, 38, 2281, 1990.
398. Wu, J.B. et al., A flavonol *C*-glycoside from *Mohgania macrophylla*, *Phytochemistry*, 45, 1727, 1997.
399. Wyatt, R., Lanes, D.M., and Stoneburner, A., Chemosystematics of Mniaceae II. Flavonoids of *Plagiomnium* section *Rosulata*, *Bryologist*, 94, 443, 1991.
400. Xavier, H.S. and D'Angelo, L.C.A., Flavone *C*-glycosides from the leaves of *Jatropha polhiana molissima*, *Fitoterapia*, 66, 468, 1995.
401. Xie, C., Veitch, N.C., Houghton, P.J., and Simmonds, M.S.J., Flavone *C*-glycosides from *Viola yedoensis* Makino, *Chem. Pharm. Bull.*, 51, 1204, 2003.
402. +Yasukawa, K., Yamamouchi, S., and Tikido, M., *Yakugaku Zasshi*, 102, 292, 1982.
403. Yayli, N., Seymen, H., and Baltaci, C., Flavone *C*-glycosides from *Scleranthus uncinatus*, *Phytochemistry*, 58, 607, 2001.
404. Yayli, N. et al., Phenolic and flavone *C*-glycosides from *Scleranthus uncinatus*, *Pharm. Biol.*, 40, 369, 2002.

405. Yoon, K.D., Kim, C.Y., and Huh, H., The flavone glycosides of *Sasa borealis*, *Saengyak Hakhoechi*, 31, 224, 2000.

406. Zavodnik, L.B. et al., Inhibition by genistein-8-*C*-glycoside of some oxidative processes in liver microsomes and erythrocytes, *Curr. Top. Biochem.*, 24, 241, 2000.

407. +Zapesochnaya, G.G. and Laman, N.A., *Khim. Prir. Soedin.*, 862, 1977.

408. +Zemtsova, G.N. and Bandyukova, V.A., *Khim. Prir. Soedin.*, 107, 1974.

409. +Zeng, L., Zhang, R., and Wang, X., *Yaoxue Xuebao*, 22, 114, 1987.

410. Zhang, P. et al., A maize QTL for silk maysin levels contains duplicated Myb-homologous genes which jointly regulate flavone biosynthesis, *Plant Mol. Biol.*, 52, 1, 2003.

411. Zhang, P.C. and Xu, S.X., Two new *C*-glucoside flavonoids from leaves of *Crataegus pinnatifida* Bge. Var. major N.E. Br., *Chin. Chem. Lett.*, 13, 337, 2002.

412. Zhang, Y. et al., A new flavone *C*-glycoside from *Scutellaria baicalensis*, *Chin. Chem. Lett.*, 5, 849, 1994.

413. Zhang, Y. et al., A new flavone *C*-glycoside from *Scutellaria baicalensis*, *J. Chin. Pharm. Sci.*, 6, 182, 1997.

414. Zhang, Y. and Yang, C., Chemical studies on *Gentianella azurea*, a Tibetan medicinal plant, *Yunnan Zhiwu Yanjiu*, 16, 401, 1994.

415. +Zhou, B.N., Blasko, G., and Cordell, G.A., *Phytochemistry*, 27, 3633, 1988.

416. Zhou, Y. et al., New *C*-glycosylflavone from leaves of *Cycas panzhihuaensis*, *Zhiwu Xuebao*, 44, 101, 2002.

417. Zhou, Z. and Yang, C., Five new flavonoid glycosides from *Scutellaria amoena*, *Yunnan Zhiwu Yanjiu*, 22, 475, 2000.

418. Zhu, G., Wang, D., and Meng, J., New compounds from *Polygonum perfoliatum*, *Indian J. Heterocycl. Chem.*, 10, 41, 2000.

419. +Zoll, A. and Nouvel, G., *Plant. Med. Phytother.*, 8, 134, 1974.

420. Zou, J.H., Yang, J.S., and Zhou, L., Acylated flavone *C*-glycosides from *Trollius ledebouri*, *J. Nat. Prod.*, 67, 664, 2004.

421. Zulu, R.M. et al., Flavonoids from the roots of two *Rhynchosia* species used in the preparation of a Zambian beverage, *J. Sci. Food Agric.*, 65, 347, 1994.

15 Flavanones and Dihydroflavonols

Renée J. Grayer and Nigel C. Veitch

CONTENTS

15.1 GENERAL INTRODUCTION

In three of the volumes of *The Flavonoids, Advances in Research*, published between 1975 and 1993, flavanones and dihydroflavonols were part of the chapter on "Minor Flavonoids," expertly written by Professor Bruce Bohm.[1-4] These "Minor Flavonoid" chapters also included chalcones, dihydrochalcones, and aurones. The term "Minor Flavonoids" was first used by Harborne in 1967 to encompass not only flavanones, chalcones, and aurones, but also isoflavonoids, biflavonyls, and leucoanthocyanidins, because so few compounds belonging to each of these flavonoid classes were known at that time.[5] For example, only about 30 flavanone and dihydroflavonol aglycones, 19 chalcones, and 7 aurones were known in 1967. The number of known "minor flavonoids" increased considerably in the next two decades, so that when the checklist for *The Flavonoids, Advances in Research Since 1980* was published in 1988, 429 known flavanones and dihydroflavonols (including glycosides) were listed, 268 chalcones and dihydrochalcones, and 29 aurones.[6] In the last 15 years, the total number of known compounds in these flavonoid classes has more than doubled, so that the term "minor flavonoids" is no longer appropriate. Consequently, it has been decided that separate chapters should be devoted to the flavanones and dihydroflavonols (this chapter), and chalcones, dihydrochalcones, and aurones (Chapter 16).

The general structures and atom numbering of flavanones and dihydroflavonols are given in Figure 15.1. Flavanones (also called dihydroflavones) and dihydroflavonols (also called 3-hydroxyflavanones or flavanonols) lack the double bond between carbons 2 and 3 in the C-ring of the flavonoid skeleton, which is present in flavones and flavonols. Thus, in flavanones, C-2 bears one hydrogen atom in addition to the phenolic B-ring, and C-3 two hydrogen atoms. Two stereoisomeric forms of each flavanone structure are possible, since C-2 is a center of asymmetry (epimeric center). Consequently, the B-ring can be either in the (2S)- or (2R)-configuration (see Figure 15.1). The great majority of the flavanones isolated from plants are laevorotatory (−)- or (2S)-flavanones, because the enzymatic reaction catalyzing the conversion of chalcones to flavanones is stereospecific. The C-3 atom of dihydroflavonols bears both a hydrogen atom and a hydroxyl group, and is therefore an additional center of asymmetry (see Figure 15.1). Thus, four stereoisomers are possible for each dihydroflavonol structure, (2R,3R), (2R,3S), (2S,3R), and (2S,3S). All four configurations have been found in naturally occurring dihydroflavonols, but the (2R,3R)-configuration is by far the most common.

As in all other flavonoids, there is structural variation in flavanones and dihydroflavonols because of variation in hydroxylation, methoxylation, methylation, prenylation, benzylation, glycosylation, etc. of suitable carbon atoms in the skeleton, i.e., C-5, C-6, C-7, and C-8 of the A-ring, C-2′, C-3′, C-4′, C-5′, and C-6′ of the B-ring, and C-2 of the C-ring in both flavanones

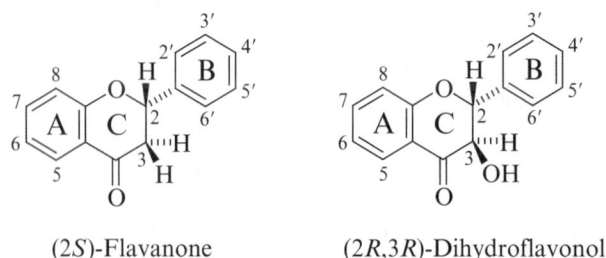

(2S)-Flavanone   (2R,3R)-Dihydroflavonol

FIGURE 15.1 Skeletons of a (2S)-flavanone and a (2R,3R)-dihydroflavonol, showing the numbering of the carbons and naming of the rings.

and dihydroflavonols. In addition, the hydroxyl group at C-3 in dihydroflavonols can be methylated, glycosylated, or esterified.

Biogenetically, chalcones are the immediate precursors of flavanones, and some flavanones isomerize by ring opening into chalcones during isolation from plants or after chemical treatment with alkali. In turn, flavanones are intermediates in the biosynthesis of most other flavonoid groups, including flavones, flavonols, and isoflavonoids. For more information on the biosynthesis of flavonoids and flavanones in particular, the reader is referred to Chapter 3 and reviews by Heller and Forkmann.[7–9]

This chapter deals mainly with the flavanones and dihydroflavonols that have been newly reported from 1992 to 2003. Most of the literature references for these compounds have been obtained by searching *Chemical Abstracts*, *The Combined Chemical Dictionary Database of Natural Products*, and reviews published in *Natural Product Reports*.[10–13] For the compilation of the checklist of all known flavanones and dihydroflavonols (see appendix), the information given in the four volumes of *The Flavonoids, Advances in Research*,[1–4] and that in the two volumes of *The Handbook of Natural Flavonoids*[14,15] were used for the compounds recorded before 1992. Biflavonoids containing one or two flavanone or dihydroflavonol subunits have been omitted from the present chapter and checklist, since so many of these dimers and heterodimers are known that they warrant a separate chapter. However, flavanone–chalcone heterodimers and Diels–Alder adducts of chalcones and flavanones are discussed in Chapter 16 (see Table 16.4 and Table 16.5). The same applies to a number of flavanone–aurone and flavanone–auronol heterodimers (see Table 16.15). In this chapter, the newly reported flavanones and dihydroflavonols are presented and discussed in separate sections on aglycones and glycosides. The aglycones are further subdivided into different groups based on their substituents. The newly reported compounds are arranged in ten tables. The following data are supplied for each compound in the tables: *O*-substitution and *C*-substitution patterns, which provide their semisystematic name; molecular formula and relative molecular mass; trivial name where given; the plant species, family, and organ from which the compound was isolated; and the literature reference from which these details were obtained. In some more complex compounds, further modifications of the flavonoid skeleton or side groups are such that it is difficult or impossible to describe them in terms of their *O*- and *C*-substituents alone; in these cases the trivial name is used and the structure presented in one of the figures. Although most flavanones discussed in this chapter are in the (2*S*)-configuration, it is not always clear from papers describing new compounds whether the stereochemistry has been verified. In cases where the authors have determined the configuration as a (2*S*)-flavanone, the corresponding entry in each table has been annotated with an asterisk after the compound number; in the few flavanones where the configuration has been determined to be (2*R*), this has also been indicated. For dihydroflavonols, an asterisk indicates that the compound is in the (2*R*,3*R*)-configuration.

In the text, the most interesting compounds in each section are discussed with regard to their different substitution patterns, their distribution in the plant kingdom, and their biological activities if these have been studied. Furthermore, some aspects of the biosynthesis of selected compounds or groups of compounds are described. Methods and techniques for the detection, separation, purification, and identification of flavanones and dihydroflavonols are not presented in this chapter, but can be found in Chapters 1 and 2 of this book and elsewhere.[16] However, it is worth pointing out here that these compounds can easily be distinguished from other groups of flavonoids when analyzed by high-performance liquid chromatography with diode array or ultraviolet (UV) detection, since most flavanones and dihydroflavonols exhibit a characteristic maximum at a wavelength of ca. 290 nm, accompanied by a small shoulder at ca. 330 to 360 nm. In the case of a glycoside acylated with a phenolic acid, the UV spectrum of the acid is superimposed on that of the flavanone or dihydroflavonol.

15.2 FLAVANONES

15.2.1 SIMPLE FLAVANONES

15.2.1.1 Simple Flavanones with O-Substitution Only

Flavanones substituted by hydroxy, methoxy, methylenedioxy, and C-methyl or related groups could conveniently be called "simple flavanones," in contrast to flavanones bearing more complex substituents such as prenyl and benzyl groups. Simple flavanones without C-substitution, newly reported between 1992 and 2003, are presented in Table 15.1 (compounds 1–35).[17–49] Table 15.1 has been subdivided into groups according to the number of O-containing substituents, and within the groups the compounds have been arranged according to their number of hydroxy and methoxy substituents and relative molecular mass. As can be seen, no new mono-O- or di-O-substituted simple flavanones have been discovered in the review period and only one new tri-O-substituted flavanone has been isolated in the last 12 years, 5,2′-dihydroxy-7-methoxyflavanone (dihydroechioidin, 1) from *Andrographis echioides* (Acanthaceae).[17] In contrast, 17 new tetra-O-substituted flavanones have been reported in this period, which increases the number of known compounds in this group by 50%. For the new penta-O- and hexa-O-substituted flavanones there are nine and five newly discovered compounds, respectively, and the latter are the first five hexa-oxygenated "simple" flavanones recorded from plants.

An interesting new compound in the tetra-O-substituted group is 2,5-dihydroxy-6,7-dimethoxyflavanone (mosloflavanone, 7, Figure 15.2), because accumulation of 2-hydroxylated flavanones is rather rare in plants. This flavanone was isolated independently from two species of Lamiaceae, *Mosla soochouensis*[22] and *Collinsia canadensis*.[23] Mosloflavanone was shown to have antifungal activity against *Cladosporium cucumerinum* (thin-layer chromatography assay), and radical scavenging activities in the DPPH (1,1-diphenyl-2-picrylhydrazyl radical) spectrophotometric assay.[22] Alyssifolinone (4) from another species of Lamiaceae, *Teucrium alyssifolium*, has a 3′,5′-di-O-substitution pattern in the B-ring, which is also quite rare.[20]

A striking feature of about half of the new flavanones is that they contain a 2′-hydroxyl or 2′-methoxyl group in the B-ring, most notably in the genus *Andrographis* (Acanthaceae), which has yielded many new 2′-oxygenated flavanones. These include compound 1 mentioned above, compound 9 from *A. lineata*,[25] compound 16 from *A. viscosula*,[32] compound 18 from *A. rothii*,[34] and compound 27 from *A. affinis*.[42] The rhizomes of *Iris tenuifolia* (Iridaceae) are the source of the largest number of new 2′-O-substituted simple flavanones in a single species, namely 2, 5, 12, and 19.[18] Additionally, the 2′-oxygenated dihydroflavonols corresponding to flavanones 5 and 12 were found in the same species[18] (see Table 15.9). Two new 2′-O-substituted flavanones have also been reported from a representative of the family Moraceae, *Artocarpus heterophyllus*, flavanones 10[26] and 23, which is 5-hydroxy-7,2′,4′,6′-tetramethoxyflavanone.[39] The 8-prenyl derivative of 23 and related prenylated flavanones also occur in *A. heterophyllus*.[39] Prenylated derivatives of the unmethylated parent compound of 23, 5,7,2′,4′,6′-pentahydroxyflavanone, are abundant in species of *Echinosophora* and *Sophora* (Leguminosae) (see Table 15.3), but the corresponding nonprenylated pentahydroxyflavanone has yet not been recorded. New 2′-O-substituted flavanones have also been reported from species in the families Asteraceae (11),[27] Leguminosae (21),[36,37] and Boraginaceae (22).[38] The hexa-O-substituted (2S)-5,2′-dihydroxy-6,7,8,6′-tetramethoxyflavanone (30) from *Scutellaria oxystegia* (Lamiaceae)[45] has a 2′,6′-dioxygenation pattern in the B-ring, in common with several previously reported flavanones from the genus *Scutellaria*.[15] This oxygenation pattern is rare in other plant taxa.

An interesting new penta-O-substituted flavanone is 8-hydroxyhesperetin (5,7,8,3′-tetrahydroxy-4′-methoxyflavanone, 20), which was produced by the fungus *Aspergillus saitoi*

TABLE 15.1
Flavanones with Hydroxy, Methoxy, and Methylenedioxy Substituents Reported from 1992 to 2003

No.	O-Substitution of Flavanone	Mol. Formula	M_r	Trivial Name	Plant Source	Family	Organ[a]	Ref.
	Tri-O-substituted							
1*	5,2'-DiOH-7-OMe	$C_{16}H_{14}O_5$	286	Dihydroechioidin	Andrographis echioides	Acanthaceae	Whole	17
	Tetra-O-substituted							
2*	5,2'-DiOH-6,7-methylenedioxy	$C_{16}H_{12}O_6$	300		Iris tenuifolia	Iridaceae	Root	18
3	5,7,8-TriOH-4'-OMe	$C_{16}H_{14}O_6$	302		Licania densiflora	Chrysobalanaceae	Leaf	19
4	5,7,3'-TriOH-5'-OMe	$C_{16}H_{14}O_6$	302	Alyssifolinone	Teucrium alyssifolium	Lamiaceae	Aerial	20
5*	5,2',3'-TriOH-7-OMe	$C_{16}H_{14}O_6$	302		Iris tenuifolia	Iridaceae	Root	18
6	6,7,8-TriOH-5-OMe	$C_{16}H_{14}O_6$	302	Oresbiusin	Isodon oresbius	Lamiaceae	Whole	21
7*	2,5-DiOH-6,7-diOMe	$C_{17}H_{16}O_6$	316	Mosloflavanone (Figure 15.2)	Mosla soochouensis	Lamiaceae	Whole	22
					Collinsia canadensis	Lamiaceae	Aerial	23
8	5,7-DiOH-8,4'-diOMe	$C_{17}H_{16}O_6$	316		Chromolaena subscandens	Asteraceae	Leaf	24
9	5,2'-DiOH-7,8-diOMe	$C_{17}H_{16}O_6$	316		Andrographis lineata	Acanthaceae	Whole	25
10	5,2'-DiOH-7,4'-diOMe	$C_{17}H_{16}O_6$	316		Artocarpus heterophyllus	Moraceae	Root	26
11	5,2'-DiOH-7,5'-diOMe	$C_{17}H_{16}O_6$	316		Eupatorium odoratum	Asteraceae	Aerial	27
12*	5,3'-DiOH-7,2'-diOMe	$C_{17}H_{16}O_6$	316		Iris tenuifolia	Iridaceae	Root	18
13	7,8-DiOH-6,4'-diOMe	$C_{17}H_{16}O_6$	316		Tecoma stans	Bignoniaceae	Flower	28
14	7,4'-DiOH-8,3'-diOMe	$C_{17}H_{16}O_6$	316		Wedelia asperrina	Asteraceae	Aerial	29
15*	5,7-DiOMe-3',4'-methylenedioxy	$C_{18}H_{16}O_6$	328		Caesalpinia pulcherrima	Leguminosae	Aerial	30
					Bauhinia variegata	Leguminosae	RootB	31
16	(2R)-5-OH-7,2',3'-triOMe	$C_{18}H_{18}O_6$	330		Andrographis viscosula	Acanthaceae	Whole	32
17	8-OH-5,6,7-triOMe	$C_{18}H_{18}O_6$	330	Kwangsienin A	Fissistigma kwangsiense	Annonaceae	StemB	33
18	5,7,2',5'-TetraOMe	$C_{19}H_{20}O_6$	344		Andrographis rothii	Acanthaceae	Whole	34
	Penta-O-substituted							
19*	5,2',3'-TriOH-6,7-methylenedioxy	$C_{16}H_{12}O_7$	316		Iris tenuifolia	Iridaceae	Root	18
20	5,7,8,3'-TetraOH-4'-OMe	$C_{16}H_{14}O_7$	318	8-Hydroxyhesperetin	Biotransformation product of hesperidin from citrus by Aspergillus saitoi	—	—	35
21	5,2',5'-TriOH-6,7-diOMe	$C_{17}H_{16}O_7$	332	Dioclein	Dioclea grandiflora	Leguminosae	Root	36
					Acacia longifolia	Leguminosae	Root	37

continued

TABLE 15.1
Flavanones with Hydroxy, Methoxy, and Methylenedioxy Substituents Reported from 1992 to 2003 — *continued*

No.	O-Substitution of Flavanone	Mol. Formula	M_r	Trivial Name	Plant Source	Family	Organ[a]	Ref.
22*	5,2'-DiOH-7,4',5'-triOMe	$C_{18}H_{18}O_7$	346		*Onosma hispida*	Boraginaceae		38
23	5-OH-7,2',4',6'-TetraOMe	$C_{19}H_{20}O_7$	360	Heteroflavanone A	*Artocarpus heterophyllus*	Moraceae	RootB	39
24	4'-OH-5,6,7,3'-TetraOMe	$C_{19}H_{20}O_7$	360	Agecorynin E	*Ageratum corymbosum*	Asteraceae	Aerial	40
25*	5,6,7,8,4'-PentaOMe	$C_{20}H_{22}O_7$	374		*Citrus kinokuni*	Rutaceae	Fruit	41
26*	5,6,7,3',4'-PentaOMe	$C_{20}H_{22}O_7$	374		*Citrus kinokuni*	Rutaceae	Fruit	41
27*	5,7,2',3',4'-PentaOMe	$C_{20}H_{22}O_7$	374		*Andrographis affinis*	Acanthaceae	Whole	42
Hexa-O-substituted								
28	5,7-DiOH-6,3'-diOMe-4',5'-methylenedioxy	$C_{18}H_{16}O_8$	360	Agamanone	*Agave americana*	Agavaceae	Aerial	43
29	5,7,3'-TriOH-6,4',5'-triOMe	$C_{18}H_{18}O_8$	362		*Greigia sphacelata*	Bromeliaceae	Aerial	44
30*	5,2'-DiOH-6,7,8,6'-tetraOMe	$C_{19}H_{20}O_8$	376		*Scutellaria oxystegia*	Lamiaceae	Root	45
31	5,3'-DiOH-6,7,4',5'-tetraOMe	$C_{19}H_{20}O_8$	376		*Greigia sphacelata*	Bromeliaceae	Aerial	44
32	5,6,7,3',4',5'-HexaOMe	$C_{21}H_{24}O_8$	404		*Neoraputia magnifica*	Rutaceae	Stem	46
Flavanone esters								
33	5,7-DiOH-flavanone (pinocembrin) 7-O-benzoate	$C_{22}H_{16}O_5$	360		*Lophopappus tarapacanus*	Asteraceae	Aerial	47
34	5,7,4'-TriOH-3'-OMe-flavanone 4'-O-isobutyrate	$C_{20}H_{20}O_7$	372		*Eriodictyon californicum*	Hydrophyllaceae	Leaf	48
35	5,7,4'-TriOH-flavanone (naringenin) 7-sulfate	$C_{15}H_{12}O_8S$	352		Fermentation product of naringenin using *Cunninghamella elegans*	—	—	49

*(2S)-Flavanones.
[a]Whole, whole plant; Aerial, aerial parts; Root, roots or other underground parts; RootB, root bark; StemB, stem bark.

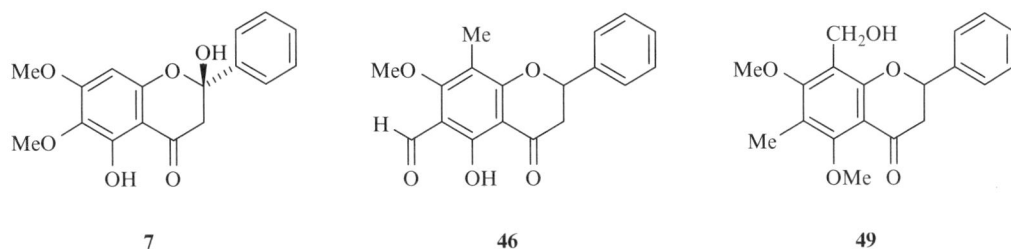

FIGURE 15.2 Flavanones with simple patterns of substitution.

when incubated with the flavanone glycoside hesperidin (hesperetin 7-rutinoside), which is common in *Citrus* fruits. The fungus also produced the known 6- and 8-hydroxylated derivatives of naringenin (carthamidin and isocarthamidin, respectively) from another common citrus flavanone glycoside, naringin (naringenin 7-neohesperidoside).[35] These three flavanones, produced by *A. saitoi*, were found to be potent radical scavengers and antioxidants using the DPPH test and methyllinoleate oxidation system, and had a greater activity than the original glycosides. The activities of the two 8-hydroxyflavanones were comparable to that of α-tocopherol, but that of the 6-hydroxy derivative was weaker.[35]

Two new pentamethoxyflavanones (**25** and **26**) have been reported from *Citrus kinokuni* (Rutaceae)[41] and the new hexamethoxyflavanone (**32**) from another member of the Rutaceae, *Neoraputia magnifica*.[46] The oxygenation pattern of **32** (5,6,7,3′,4′,5′-hexa-*O*) is also found in the new hexa-*O*-substituted flavanones **29** and **31** from *Greigia sphacelata* (Bromeliaceae)[44] and the methylenedioxyflavanone **28** from *Agave americana* (Agavaceae).[43]

Three new esters of flavanone aglycones are listed in Table 15.1. These are the 7-*O*-benzoate of pinocembrin (**33**) from *Lophopappus tarapacanus* (Asteraceae),[47] the 4′-*O*-isobutyrate of eriodictyol 3′-methyl ether (**34**) from *Eriodictyon californicum* (Hydrophyllaceae), which is a potential cancer chemopreventive agent,[48] and the 7-sulfate of naringenin (**35**), which was obtained as a fermentation product of naringenin using the fungus *Cunninghamella elegans* NRRL 1392 in 23% yield.[49]

During the last 12 years, several publications have dealt with the biological activities of some known "simple" flavonoids. For example, the antifungal activity of the previously known mono-oxygenated 7-hydroxyflavanone, isolated from the roots of *Virola surinamensis* (Myristicaceae), was tested against *Cladosporium cladosporoides*, and the compound had a tenfold higher activity than the positive control nystatin.[50] The 7-methyl ether of naringenin, sakuranetin (5,4′-dihydroxy-7-methoxyflavanone), named after the Japanese name for cherry tree, sakura, from which it was first isolated nearly 100 years ago, also has antifungal activity. Sakuranetin was isolated as a phytoalexin (an antifungal compound not present in healthy plants, but produced in large quantities after infection or physical damage of the plant) from UV-irradiated rice leaves[51] and later also from rice leaves infected with the rice blast fungus.[52] The ED$_{50}$ value of this flavanone against spore germination of *Pyricularia oryza*, the rice blast pathogen, was ca. 15 ppm. The sakuranetin content of resistant cultivars of rice after infection with *P. oryzae* was generally much higher than that of susceptible cultivars,[51,52] so that the production of this phytoalexin may be involved in the resistance of rice plants against the blast pathogen. Sakuranetin has also been found as an induced compound in rice plants after infection by the stem nematode *Ditylenchus angustus*.[53] Infection of bark of the Weymouth pine (*Pinus strobus*, Pinaceae) with the pinewood nematode, *Bursphelenchus xylophilus*, induced the production of a related flavanone, pinocembrin (5,7-dihydroxyflavanone).[54] Sakuranetin and the corresponding dihydroflavonol, aromadendrin 7-methyl ether, isolated from the leaves of *Trixis vauthieri* (Asteraceae), were active against the trypomastigote forms of *Trypanosoma cruzi*, the protozoan that causes Chagas' disease, which affects 18 million people in Latin America. The trypanocidal activity of sakuranetin at 0.5 mg/ml was 100%, and that of aromadendrin 7-methyl ether at the same concentration was 86 ± 13%.[55]

15.2.1.2 Simple Flavanones with Both *O*- and *C*-Substitution

Fourteen new simple flavanones with *C*-methyl, *C*-hydroxymethyl, and *C*-formyl substituents reported between 1992 and 2003 are listed in Table 15.2 (compounds **36–49**).[56–66] For each compound the *O*- and *C*-substituents are given in separate columns. Two interesting new compounds are the related *C*-methylated 2-hydroxyflavanones, 2,5-dihydroxy-7-methoxy-6-*C*-methylflavanone (**36**) and 2,5-dihydroxy-7-methoxy-8-*C*-methylflavanone (**37**), which have been reported from the stem-bark of *Friesodielsia enghiana* (Annonaceae).[56] The same compounds were later isolated from the leaves of *Leptospermum polygalifolium* ssp. *polygalifolium* (Myrtaceae), together with the 6-*C*- and 8-*C*-regioisomers of 2,5,7,4′-tetrahydroxydihydroflavonol (see Table 15.9).[67] *C*-Methylflavonoids are quite common in the family Myrtaceae, and a new 8-*C*-methylated flavanone (**40**) was found in another species from this family, *Callistemon coccineus*.[59] The 6-*C*-regioisomer of the latter compound, **39**, has been reported from the conifer *Pseudotsuga wilsoniana* (Pinaceae).[58] From the related *Pseudotsuga sinensis* the 6-*C*-methylated flavanone **41** was reported,[60] with the same unusual 3′,5′-*O*-substitution of the flavanone B-ring as alyssifolinone (**4**). Another unusual B-ring oxygenation pattern, based on 2′,4′-di-*O*-substitution, is present in flavanone **43** from the roots of *Terminalia alata* (Combretaceae),[62] and also in 2′-hydroxymatteucinol (**45**) from the fern *Matteucia orientalis*.[64] The 6,8-di-*C*-methylated flavanone, isomatteucinol (**44**), which bears a sole 3′-methoxy group in the B-ring, was isolated from the same species.[63] The only monocot from which *C*-methylflavanones have been reported in the last 12 years is *Vellozia nanuzae* (Velloziaceae), from which 6-*C*-methyleriodictyol 7,3′,4′-trimethyl ether (**42**) was obtained in addition to a range of isoprenylated flavonoids.[61] The 8-*C*-methyl derivative of naringenin 4′-methyl ether, 8-methylisosakuranetin (**38**), has been reported from *Amaranthus caudatus* (Amaranthaceae), a member of the plant order Caryophyllales.[57] Unusual flavanones and chalcones with both *C*-methyl and *C*-formyl or *C*-hydroxymethyl substituents have been discovered in another member of the Caryophyllales, *Petiveria alliacea* (Phytolaccaceae). These include 5-hydroxy-7-methoxy-6-*C*-formyl-8-*C*-methylflavanone (leridal, **46**, Figure 15.2), 5-hydroxy-7-methoxy-6-*C*-hydroxymethyl-8-*C*-methylflavanone (leridol, **48**), and 5,7-dimethoxy-6-*C*-hydroxymethyl-8-*C*-methylflavanone (5-*O*-methylleridol, **49**, Figure 15.2).[65] However, the related 5,7-dihydroxy-6-*C*-formyl-8-*C*-methylflavanone (7-demethylleridal) reported as new from the same species[68] had already been recorded previously as lawinal from *Unona lawii* (Annonaceae).[15] Lawinal isolated from species of *Desmos* (also Annonaceae) exhibited a potent anti-HIV activity with EC_{50} values of 0.022 µg/ml and a therapeutic index of 489. Thus, the compound was considered to be an excellent lead for the development of anti-HIV drugs.[69] A new *C*-formylflavanone, 7-hydroxy-5-methoxy-6-*C*-methyl-8-*C*-formylflavanone (desmosflavanone II, **47**), has been obtained from another species of this genus, *D. cochinchinensis*.[66]

15.2.2 Isoprenylated Flavanones

15.2.2.1 Introduction

In the first volume of *The Flavonoids*, published in 1975, only eight isoprenylated flavanones were described.[1] Since then many more have been isolated and characterized, and at present the number of known isoprenylated flavanones (excluding dihydroflavonols) is nearly 300. Most of these substances have been obtained from species of Leguminosae, although their distribution is not quite so restricted as thought previously, as isoprenylated flavanones have

TABLE 15.2
Flavanones with C-Methyl, C-Hydroxymethyl, and C-Formyl Substituents Reported from 1992 to 2003

No.	O-Substitution	C-Substitution[a]	Mol. Formula	M_r	Trivial Name	Plant Source	Family	Organ[a]	Ref.
36	2,5-DiOH-7-OMe	6-Me	$C_{17}H_{16}O_5$	300		Friesodielsia enghiana	Annonaceae	StemB	56
37	2,5-DiOH-7-OMe	8-Me	$C_{17}H_{16}O_5$	300		Friesodielsia enghiana	Annonaceae	StemB	56
38	5,7-DiOH-4'-OMe	8-Me	$C_{17}H_{16}O_5$	300	8-Methylisosakuranetin	Amaranthus caudatus	Amaranthaceae	Flower	57
39	5,4'-DiOH-7-OMe	6-Me	$C_{17}H_{16}O_5$	300	6-Methylsakuranetin	Pseudotsuga wilsoniana	Pinaceae	Heartwood	58
40	5,4'-DiOH-7-OMe	8-Me	$C_{17}H_{16}O_5$	300	8-Methylsakuranetin	Callistemon coccineus	Myrtaceae	Leaf wax	59
41	5,7',5'-TetraOH	6-Me	$C_{16}H_{14}O_6$	302		Pseudotsuga sinensis	Pinaceae		60
42	5-OH-7,3',4'-TriOMe	6-Me	$C_{19}H_{20}O_6$	344	6-Methyleriodictyol 7,3',4'-trimethyl ether	Vellozia nanuzae	Velloziaceae	Leaf	61
43	5,7,2',4'-TetraOMe	8-Me	$C_{20}H_{22}O_6$	358		Terminalia alata	Combretaceae	Root	62
44	5,7-DiOH-3'-OMe	6,8-diMe	$C_{18}H_{18}O_5$	314	Isomatteucinol	Matteucia orientalis	Aspleniaceae, Pteridophyta		63
45	5,7,2'-TriOH-4'-OMe	6,8-diMe	$C_{18}H_{18}O_6$	330	2'-Hydroxymatteucinol	Matteucia orientalis	Aspleniaceae, Pteridophyta		64
46	5-OH-7-OMe	6-Formyl, 8-Me	$C_{18}H_{16}O_5$	312	Leridal (Figure 15.2)	Petiveria alliacea	Phytolaccaceae	Leaf	65
47	7-OH-5-OMe	6-Me, 8-formyl	$C_{18}H_{16}O_5$	312	Desmosflavanone II	Desmos cochinchinensis	Annonaceae	Root	66
48	5-OH-7-OMe	6-CH$_2$OH, 8-Me	$C_{18}H_{18}O_5$	314	Leridol	Petiveria alliacea	Phytolaccaceae	Leaf	65
49	5,7-DiOMe	6-CH$_2$OH, 8-Me	$C_{19}H_{20}O_5$	328	5-O-Methylleridol (Figure 15.2)	Petiveria alliacea	Phytolaccaceae	Leaf	65

*(2S)-Flavanones.

[a] Whole, whole plant; Aerial, aerial parts; Root, roots or other underground parts; StemB, stem bark.

been isolated from 14 different plant families in the last 12 years. A useful review on isoprenylated flavonoids up to 1995 is that by Barron and Ibrahim.[70]

Until a decade ago, the mevalonate pathway was assumed to be the general biosynthetic route leading to isoprenoid substituents, but then a new mevalonate-independent route was discovered, initially in bacteria but later also in higher plants, the glyceraldehyde pyruvate pathway.[71] To investigate whether the mevalonate or the glyceraldehyde pyruvate pathway is used for the biosynthesis of isoprenylated flavanones, [1-^{13}C]-glucose was administered to *Glycyrrhiza glabra* roots, and it was found that the prenyl groups of glabrol (7,4'-dihydroxy-8,3'-di-*C*-prenylflavanone) were synthesized via the latter pathway.[72]

A wide range of different isoprenoid substituents have been found conjugated to flavanones from plants. The most common of these is the 3-methylbut-2-enyl group, also described as 3,3-dimethylallyl or isopentenyl, but for convenience often called "prenyl" group, a term which is used in this chapter. In some flavonoids, the monoprenyl group is rearranged into a 1,1-dimethylallyl substituent. When two isoprenoid units are linked, a geranyl, neryl, or lavandulyl moiety may be formed, and linkage of three isoprenoid units may result in a farnesyl group. The isoprenylated side chains may be modified by reduction, oxidation, and dehydration. Furthermore, cyclization of the isoprenoid substituent with an *ortho*-hydroxyl group on the phenolic A- or B-ring may result in furano- or dimethylpyrano ring structures. For convenience, the isoprenylated flavanones in this section have been divided into three groups, (1) those with noncyclic isoprenoid groups, (2) flavanones with furano or dihydro-furano rings, and (3) flavanones with dimethylpyrano or dimethyldihydropyrano rings. This division is rather artificial from a biogenetic point of view, as these three types of moieties are closely related and representatives of all three groups may be found in the same plant species, although the isoprenylated flavanones of the Moraceae and Rutaceae tend to bear cyclic rather than noncyclic isoprenoid substituents.

15.2.2.2 Flavanones with Noncyclic Isoprenoid Substituents

Flavanones with noncyclic isoprenoid substituents reported during 1992 to 2003 are listed in Table 15.3 (compounds **50–140**).[39,61,73–127] They have been divided into six groups according to their isoprenoid side chains: prenyl, 1,1-dimethylallyl, geranyl, lavandulyl, farnesyl, and oxygen-containing isoprenyl substituents. Within these groups the compounds have been arranged according to the number of prenyl groups (mono-, di-, or triprenylated) and then according to the number of oxygens in the molecular formula and by relative molecular mass. For each compound, the *O*- and *C*-substituents are given in separate columns. More than two thirds of the 90 flavanones listed in Table 15.3 have been isolated from representatives of the family Leguminosae, and more than 30 from species of the genus *Sophora* or the related *Echinosophora*, notably by Iinuma and coworkers.[82,85,87,90,102,108,112,114,116] Most of the iso-prenoid flavanones have been obtained from root tissue, but others have been found in a wide variety of other organs, such as stems, leaves, tubers, flowers, and seeds.

Most of the isoprenyl groups are *C*-linked to the flavanone skeleton, but in some compounds they are *O*-linked, for example, in 4'-*O*-prenylsakuranetin (**58**) and 4'-*O*-geranyl-naringenin (**98**) from *Boronia coerulescens* ssp. *spinescens* (Rutaceae),[80] ponganone V (**67**) from *Pongamia pinnata* (Leguminosae),[88] monotesone A (**62**) from *Monotes engleri* (Diptero-carpaceae),[83] and 5-hydroxy-7-neryloxyflavanone (**76**, Figure 15.3) from *Helichrysum rugu-losum* (Asteraceae).[93] The *C*-prenyl groups are usually attached to C-6 or C-8 of the A-ring of the flavonoid skeleton, but may also be attached to C-3', C-5', or C-6' of the B-ring. Examples are the 3',5'-di-*C*-prenylated abyssinone V 4'-methyl ether (**79**) from *Erythrina burttii*[97] and *E. abyssinica* (Leguminosae),[98] the 5',6'-di-*C*-prenylated abyssinin III (**84**) from *E. abyssinica*,[86] and the 6,8,5'-tri-*C*-prenylated isoamoritin (**88**) from *Amorpha fruticosa* (Leguminosae).[103]

TABLE 15.3
Flavanones with Noncyclic Isoprenoid Substituents Reported from 1992 to 2003

No.	O-Substitution	C-Substitution[a]	Mol. Formula	M_r	Trivial Name	Plant Source	Family	Organ[b]	Ref.
PRENYLFLAVANONES									
Di-O-substituted monoprenyl									
—	5,4'-DiOH	6-Pr	$C_{20}H_{20}O_4$	324	Crotaramosmin[c]	Crotalaria ramossisima	Leguminosae		73
50	5,7-DiOH	6-Me-8-Pr	$C_{21}H_{22}O_4$	338		Dalea caerulea	Leguminosae		74
51*	4'-OH-7-OMe	8-Pr	$C_{21}H_{22}O_4$	338	Mundulea flavanone A	Mundulea suberosa	Leguminosae	StemB	75
Tri-O-substituted monoprenyl									
52*	5,7,2'-TriOH	8-Pr	$C_{20}H_{20}O_5$	340	Kushenol S	Sophora flavescens	Leguminosae	Root	76
53*	5,7,4'-TriOH	6-Me-8-Pr	$C_{21}H_{22}O_5$	354		Eysenhardtia texana	Leguminosae	Aerial	77
54*	5,7,4'-TriOH	8-Me-6-Pr	$C_{21}H_{22}O_5$	354		Eysenhardtia texana	Leguminosae	Aerial	77
55	5,7-DiOH-6-OMe	8-Pr	$C_{21}H_{22}O_5$	354	Agrandol	Dioclea grandiflora	Leguminosae	RootB	78
56	(2R)-5,7-DiOH-8-OMe	6-Pr	$C_{21}H_{22}O_5$	354	Microfolione	Cedrelopsis microfoliata	Ptaeroxylaceae	StemB	79
57*	5,4'-DiOH-7-OMe	8-Pr	$C_{21}H_{22}O_5$	354	Mundulea flavanone B	Mundulea suberosa	Leguminosae	StemB	75
58	5-OH-7-OMe-4-OPr		$C_{21}H_{22}O_5$	354	4'-O-Prenylsakuranetin	Boronia coerulescens ssp. spinescens	Rutaceae	Aerial	80
59	5-OH-7,4'-diOMe	8-Pr	$C_{22}H_{24}O_5$	368	Flowerine	Azadirachta indica	Meliaceae	Flower	81
Tetra-O-substituted monoprenyl									
60	5,7,8,4'-TetraOH	3'-(3-Me-but-3-enyl)	$C_{20}H_{20}O_6$	356	Flowerone	Azadirachta indica	Meliaceae	Flower	81
61*	5,7,2',4'-TetraOH	8-Pr	$C_{20}H_{20}O_6$	356	Leachianone G	Sophora leachiana	Leguminosae	Root	82
62	5,7,3'-TriOH-4'-OPr		$C_{20}H_{20}O_6$	356	Monotesone A	Monotes engleri	Dipterocarpaceae		83
63	(2R)-5,7,2'-TriOH-8-OMe	6-Pr	$C_{21}H_{22}O_6$	370	Dioflorin	Dioclea grandiflora	Leguminosae	RootB	84
64	5,7,4'-TriOH-3'-OMe	8-Pr	$C_{21}H_{22}O_6$	370	Exiguaflavanone K	Sophora exigua	Leguminosae	Root	85
65	5,7,4'-TriOH-3'-OMe	5'-Pr	$C_{21}H_{22}O_6$	370	Abyssinin II	Erythrina abyssinica	Leguminosae	StemB	86
66*	5,2',4'-TriOH-7-OMe	8-Pr	$C_{21}H_{22}O_6$	370	Kenusanone I	Echinosophora koreensis	Leguminosae	Stem	87
67	7-OMe-3',4'-methylenedioxy-6-O-Pr		$C_{22}H_{22}O_6$	382	Ponganone V	Pongamia pinnata	Leguminosae	RootB	88
68	5,7-DiOH-3',4'-diOMe	5'-Pr	$C_{22}H_{24}O_6$	384		Melilotus alba	Leguminosae	Leaf, flower	89
Penta-O-substituted monoprenyl									
69*	5,7,2',4'-TetraOH-5'-OMe	6-Pr	$C_{21}H_{22}O_7$	386	Kushenol V	Sophora flavescens	Leguminosae	Root	76

continued

TABLE 15.3
Flavanones with Noncyclic Isoprenoid Substituents Reported from 1992 to 2003 — continued

No.	O-Substitution	C-Substitution[a]	Mol. Formula	M_r	Trivial Name	Plant Source	Family	Organ[b]	Ref.
70	5,7,2',4'-TetraOH-5'-OMe	8-Pr	$C_{21}H_{22}O_7$	386	Kushenol W	*Sophora flavescens*	Leguminosae	Root	76
71*	5,7,2',6'-TetraOH-4'-OMe	8-Pr	$C_{21}H_{22}O_7$	386	Kenusanone D	*Echinosophora koreensis*	Leguminosae	Root	90
72*	5,2',6'-TriOH-7,4'-diOMe	8-Pr	$C_{22}H_{24}O_7$	400	Kenusanone E	*Echinosophora koreensis*	Leguminosae	Root	90
73	5,7-DiOH-2',4',6'-triOMe	8-Pr	$C_{23}H_{26}O_7$	414	Heteroflavanone C	*Artocarpus heterophyllus*	Moraceae	RootB	91
74	5-OH-7,2',4',6'-tetraOMe	8-Pr	$C_{24}H_{28}O_7$	428	Heteroflavanone B	*Artocarpus heterophyllus*	Moraceae	RootB	39
75*	5,7,4'-TriOH-3'-OMe	6-(β-Hydroxyethyl)-8-Pr	$C_{23}H_{26}O_7$	414	Laxiflorin	*Derris laxiflora*	Leguminosae	Aerial	92
Di-O-substituted diprenyl									
76	5-OH-7-O-neryl		$C_{25}H_{28}O_4$	392	(Figure 15.3)	*Helichrysum rugulosum*	Asteraceae		93
Tri-O-substituted diprenyl									
77	5,7,4'-TriOH	6,3'-diPr	$C_{25}H_{28}O_5$	408	Paratocarpin L or Macaranga flavanone B	*Paratocarpus venenosa* / *Macaranga pleiostemma*	Moraceae / Euphorbiaceae	Leaf / StemB	94 / 95
78	5,7-DiOH-4'-OMe	3'-(3-Me-but-1,3-dienyl)-5'-Pr	$C_{26}H_{28}O_5$	420	Burttinonedehydrate	*Erythrina burttii*	Leguminosae	StemB	96
79	5,7-DiOH-4'-OMe	3',5'-diPr	$C_{26}H_{30}O_5$	422	Abyssinone V 4'-methyl ether	*Erythrina burttii*	Leguminosae	StemB	97
79	5,7-DiOH-4'-OMe	3',5'-diPr	$C_{26}H_{30}O_5$	422	Abyssinone V 4'-methyl ether	*Erythrina abyssinica*	Leguminosae	StemB	98
80*	5,7-DiOH-4'-OMe	6-Me-8,3'-diPr	$C_{27}H_{32}O_5$	436	Lespedezaflavanone F	*Lespedeza formosa*	Leguminosae	RootB	99
81*	5,7-DiOH-4'-OMe	8-Me-6,3'-diPr	$C_{27}H_{32}O_5$	436	Lespedezaflavanone G	*Lespedeza formosa*	Leguminosae	RootB	99
Tetra-O-substituted diprenyl									
82	5,7,3',4'-TetraOH	6,8-diPr	$C_{25}H_{28}O_6$	424	6,8-Diprenyleriodictyol	*Vellozia nanuzae*	Velloziaceae	Leaf	61
83	5,7,3',4'-TetraOH	6,5'-diPr	$C_{25}H_{28}O_6$	424		*Schoenus nigricans*	Cyperaceae	Tuber	100
84	5,7,3',4'-TetraOH	5',6'-diPr	$C_{25}H_{28}O_6$	424	Abyssinin III	*Erythrina abyssinica*	Leguminosae	StemB	86
85	5,7,3',5'-TetraOH	6,8-diPr	$C_{25}H_{28}O_6$	424	Monotesone B	*Monotes engleri*	Dipterocarpaceae		83
86	5,2',4'-TriOH-7-OMe	8,5'-diPr	$C_{26}H_{30}O_6$	438	Maackiaflavanone	*Maackia amurensis* ssp. *buergeri*	Leguminosae	Root	101
Penta-O-substituted diprenyl									
87*	5,7,2',4',6'-PentaOH	6,8-diPr	$C_{25}H_{28}O_7$	440	Kenusanone B	*Echinosophora koreensis*	Leguminosae	Root	102

Tetra-O-substituted triprenyl

No.	Substitution	Prenyl	Formula	MW	Name	Species	Family	Part	Ref
88	5,7,3'-TriOH-4'-OMe	6,8,5'-triPr	$C_{31}H_{38}O_6$	506	Isoamoritin	*Amorpha fruticosa*	Leguminosae	Root	103
	1,1-DIMETHYLALLYL-FLAVANONES								
89*	5,7,4'-TriOH	6-(1,1-DMA)	$C_{20}H_{20}O_5$	340	(Figure 15.3)	*Monotes engleri*	Dipterocarpaceae	Leaf	104
90	5,7,4'-TriOH	8-(1,1-DMA)	$C_{20}H_{20}O_5$	340	Ugonin E	*Helminthostachys zeylanica*	Ophioglossaceae/ Pteridophyta	Rhiz	105
91*	5,7,3',4'-TetraOH	6-(1,1-DMA)	$C_{20}H_{20}O_6$	356	(Figure 15.3)	*Monotes engleri*	Dipterocarpaceae	Leaf	104
92*	5,7,4'-TriOH-3'-OMe	6-(1,1-DMA)	$C_{21}H_{22}O_6$	370	(Figure 15.3)	*Monotes engleri*	Dipterocarpaceae	Leaf	104
93	5,7,2',4'-TetraOH	6-Pr-5'-(1,1-DMA)	$C_{25}H_{28}O_6$	424		*Dalea elegans*	Leguminosae	Root	106
94*	5,7,2',4'-TetraOH	8-Pr-5'-(1,1-DMA)	$C_{25}H_{28}O_6$	424		*Dalea scandens* var. *paucifolia*	Leguminosae	Root	107
95*	5,7,4'-TriOH-2'-OMe	8-Pr-5'-(1,1-DMA)	$C_{26}H_{30}O_6$	438		*Dalea scandens* var. *paucifolia*	Leguminosae	Root	107
	GERANYL-FLAVANONES								
96*	7,4'-DiOH	8-Ger	$C_{25}H_{28}O_4$	392	Prostratol F	*Sophora prostrata*	Leguminosae	Root	108
97	5,7,4'-TriOH	3'-Ger	$C_{25}H_{28}O_5$	408	Macaranga flavanone A	*Macaranga pleiostemona*	Euphorbiaceae	Leaf	95
98	5,7-DiOH-4'-OGer		$C_{25}H_{28}O_5$	408	4-O-Geranylnaringenin	*Boronia :aerulescens* ssp. *spinescens*	Rutaceae	Aerial	80
99	5,7,2',4'-TetraOH	8-Ger	$C_{25}H_{28}O_6$	424	Sophorallavanone C	*Echinosophora koreensis*	Leguminosae	Stem	87
100	5,7,2',4'-TetraOH	3'-Ger	$C_{25}H_{28}O_6$	424	Sanggenol A	*Morus cathayana*	Moraceae	RootB	109
101	5,7,2',4'-TetraOH	6-Ger-8-Pr	$C_{30}H_{36}O_6$	492	Macrourone C	*Morus macroura*	Moraceae	StemB	110
102*	5,7,2',4',6'-PentaOH	6-Ger	$C_{25}H_{28}O_7$	440	Sophorallavanone D	*Echinosophora koreensis*	Leguminosae	Root	102
103*	5,7,2',4',6'-PentaOH	8-Ger	$C_{25}H_{28}O_7$	440	Sophorallavanone E	*Echinosophora koreensis*	Leguminosae	Root	102
104	5,7,2',4',6'-PentaOH	6-Ger-8-Pr	$C_{30}H_{36}O_7$	508	Tomentosanol E (Figure 15.3)	*Sophora tomentosa*	Leguminosae	Root	111
	LAVANDULYL-FLAVANONES								
105	5,7,4'-TriOH	8-Lav	$C_{25}H_{28}O_5$	408	Leachianone E	*Sophora leachiana*	Leguminosae	Root	112
106	5,7,4'-TriOH	8-(2-isopropyl-5-Me-5-hexenyl)	$C_{25}H_{28}O_5$	408	Remangiflavanone A	*Physena madagascariensis*	Capparaceae	Leaf	113
107*	7,2'-DiOH-5-OMe	8-Lav	$C_{26}H_{30}O_5$	422	Kushenol R (Figure 15.3)	*Sophora flavescens*	Leguminosae	Root	76
108*	7,4'-DiOH-5-OMe	8-Lav	$C_{26}H_{30}O_5$	422	Kushenol U	*Sophora flavescens*	Leguminosae	Root	76
109	7,4'-DiOH-2'-OMe	8-Lav	$C_{26}H_{30}O_5$	422	Alopecurone G	*Sophora alopecuroides*	Leguminosae	Root	114
110	5,7,2',4'-TetraOH	8-(2-isopropyl-5-Me-5-hexenyl)	$C_{25}H_{28}O_6$	424	Remangiflavanone B	*Physena madagascariensis*	Capparaceae	Leaf	113

continued

TABLE 15.3
Flavanones with Noncyclic Isoprenoid Substituents Reported from 1992 to 2003 — *continued*

No.	O-Substitution[a]	C-Substitution[a]	Mol. Formula	M_r	Trivial Name	Plant Source	Family	Organ[b]	Ref.
111	5,7,2',6'-TetraOH	8-Lav	$C_{25}H_{28}O_6$	424	Exiguaflavanone A	*Sophora exigua*	Leguminosae	Root	115
112*	5,2',5'-TriOH-7-OMe	8-Lav	$C_{26}H_{30}O_6$	438	Exiguaflavanone F	*Sophora exigua*	Leguminosae	Root	116
113	5,2',6'-TriOH-7-OMe	8-Lav	$C_{26}H_{30}O_6$	438	Exiguaflavanone B	*Sophora exigua*	Leguminosae	Root	115
114	7,2',4'-TriOH-5-OMe	8-Lav	$C_{26}H_{30}O_6$	438	Kurarinone	*Sophora flavescens*	Leguminosae	Root	117
						Gentiana macrophylla	Gentianaceae	Root	118
115	7,4'-DiOH-5,2'-diOMe	8-Lav	$C_{27}H_{32}O_6$	452	Kurarinone 2'-methyl ether	*Sophora flavescens*	Leguminosae	Root	117
116*	5,7,2',4',6'-PentaOH	6-Lav	$C_{25}H_{28}O_7$	440	Exiguaflavanone C	*Sophora exigua*	Leguminosae	Root	116
117*	5,7,2',4',6'-PentaOH	8-Lav	$C_{25}H_{28}O_7$	440	Exiguaflavanone G	*Sophora exigua*	Leguminosae	Root	85
118*	5,2',4'-TriOH-7,5'-diOMe	8-Lav	$C_{27}H_{32}O_7$	468	Exiguaflavanone E	*Sophora exigua*	Leguminosae	Root	116
119	5,7,2',4',6'-PentaOH	6-Pr-8-Lav	$C_{30}H_{36}O_7$	508	Exiguaflavanone J	*Sophora exigua*	Leguminosae	Root	85
120*	5,7,2',6'-TetraOH-4'-OMe	6-Pr-8-Lav	$C_{31}H_{38}O_7$	522	Exiguaflavanone D	*Sophora exigua*	Leguminosae	Root	116

FARNESYL-FLAVANONES

No.	O-Substitution[a]	C-Substitution[a]	Mol. Formula	M_r	Trivial Name	Plant Source	Family	Organ[b]	Ref.
121	5,7,3',4'-TetraOH	6-Farnesyl	$C_{30}H_{36}O_6$	492	(Figure 15.3)	*Boronia ramosa*	Rutaceae	Aerial	119

FLAVANONES BEARING HYDROXY- OR EPOXY-PRENYL GROUPS

No.	O-Substitution[a]	C-Substitution[a]	Mol. Formula	M_r	Trivial Name	Plant Source	Family	Organ[b]	Ref.
122	5,7,4'-TriOH	8-(2-OH-3-Me-but-3-enyl)	$C_{20}H_{20}O_6$	356	Tomentosanol D	*Sophora tomentosa*	Leguminosae	Root	111
123	5,7-DiOH-4'-OMe	8-(2-OH-3-Me-but-3-enyl)	$C_{21}H_{22}O_6$	370		*Bosistoa brassii*	Rutaceae	Leaf	120
124	(2R)-7,4'-DiOH-5-OMe	8-[4-OH-3-Me-(2Z)-butenyl]	$C_{21}H_{22}O_6$	370		d d	—	—	121
125	(2R)-7,4'-DiOH-5-OMe	8-[5-OH-3-Me-(2E)-butenyl]	$C_{21}H_{22}O_6$	370		dd	—	—	121
126	5,7,4'-TriOH	8-Pr-6-(2-OH-3-Me-but-3-enyl)	$C_{25}H_{28}O_6$	424	Lupiniols A1 and A2	*Lupinus luteus*	Leguminosae	Root	122

No.	Substitution	Substituent	Formula	MW	Name	Species	Family	Part	Ref.
127	5,7,4'-TriOH	8-Pr-3'-(2-OH-3-Me-but-3-enyl)	$C_{25}H_{28}O_6$	424	Lupiniol B	*Lupinus luteus*	Leguminosae	Root	122
128*	5,7,2'-TriOH	8-(5-OH-2-isopropenyl-5-Me-hexyl)	$C_{25}H_{30}O_6$	426	Kushenol T	*Sophora flavescens*	Leguminosae	Root	76
129	5,7-DiOH-4'-OMe	3'-Pr-5'-(3-OH-3-Me-buten-1-yl)	$C_{26}H_{30}O_6$	438	Burttinone	*Erythrina burttii*	Leguminosae	StemB	97
130	5,7-DiOH-4'-OMe	8-Pr-3'-(3-OH-3-Me-butyl)	$C_{26}H_{32}O_6$	440		*Azadirachta indica*	Meliaceae	Aerial	123
131*	5,7,3',4'-TetraOH	8-[(E)-3-Hydroxymethyl-2-butenyl]	$C_{20}H_{20}O_7$	372	Licoleafol (Figure 15.3)	*Glycyrrhiza uralensis*	Leguminosae	Leaf	124
132	5,7,3',4'-TetraOH	6-(3-OH-3-methylbutyl)	$C_{20}H_{22}O_7$	374	6-(3-Hydroxy-isopentanyl)-eriodictyol	*Vellozia nanuzae*	Velloziaceae	Leaf	61
133*	5,7-DiOH-2',4'-diOMe	8-(2,3-epoxy-3-Me-butyl)	$C_{22}H_{24}O_7$	400		*Atylosia scarabaeoides*	Leguminosae	Root	125
134	5,7,3',4'-TetraOH	6-Pr-8-(2-OH-3-Me-but-3-enyl)	$C_{25}H_{28}O_7$	440	Dorsmanin H	*Dorstenia mannii*	Moraceae	Twig	126
135	5,7,3',4'-TetraOH	8-Pr-6-(2-OH-3-Me-but-3-enyl)	$C_{25}H_{28}O_7$	440	(Figure 15.3)	*Monotes engleri*	Dipterocarpaceae	Leaf	83
136*	5,7,2',4'-TetraOH	8-[2-(2-OH-isopropyl)-5-Me-4-hexenyl]	$C_{25}H_{30}O_7$	442	Kushenol Q	*Sophora flavescens*	Leguminosae	Root	76
137	5,7,2',6'-TetraOH	8-[2-(2-OH-isopropyl)-5-Me-4-hexenyl]	$C_{25}H_{30}O_7$	442	Exiguaflavanone M	*Sophora exigua*	Leguminosae	Root	85
138	5,7,4'-TriOH-2'-OMe	8-(5-OH-5-Me-2-isopropenyl-trans-hex-3-enyl)	$C_{26}H_{30}O_7$	454	Leachianone D	*Sophora leachiana*	Leguminosae	Root	112
139*	5,7,4'-TriOH-2'-OMe	8-(5-OH-5-Me-2-isopropenyl-hexyl)	$C_{26}H_{32}O_7$	456	Kushenol P1	*Sophora flavescens*	Leguminosae	Root	76
140	5,7,3',4'-TetraOH	8-(4-OAc-3-Me-but-2-enyl)	$C_{22}H_{22}O_8$	414	Kanzonol S	*Glycyrriza eurycarpa*	Leguminosae	Aerial	127

*(2S)-Flavanones.

aPr, prenyl (3,3-dimethylallyl = 3-methylbut-2-enyl); Ger, geranyl; Lav, lavandulyl; DMA, dimethylallyl.

bWhole, whole plant; Root, roots or other underground parts; RootB, root bark; Aerial, aerial parts; Bark, stem bark.

cStructure has been revised to a dihydrochalcone (see Chapter 16).

dMicrobial metabolite of xanthohumol using culture broth of *Cunninghamella echinulata* NRRL 3655.

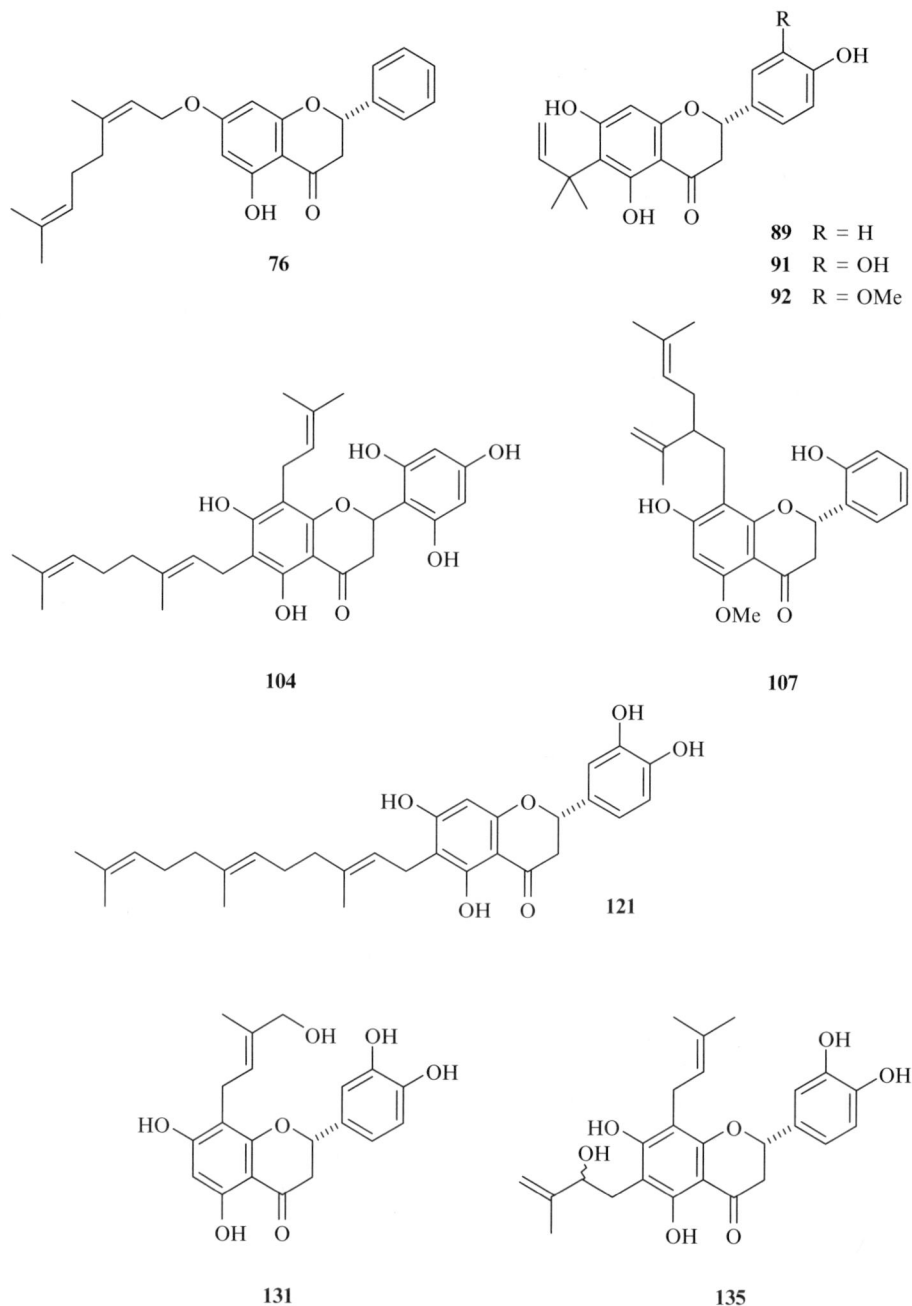

FIGURE 15.3 Examples of flavanones bearing noncyclic isoprenoid substituents.

A related compound to **79**, but containing two double bonds in one of the prenyl groups, is 5,7-dihydroxy-4′-methoxy-3′-*C*-(3-methylbut-1,5-dienyl)-5′-*C*-prenylflavanone (burttinone-dehydrate, **78**), also isolated from *E. burttii*.[96]

Some of the compounds listed in Table 15.3 not only bear prenyl substituents, but also *C*-methyl or related groups, e.g., flavanone **50** from *Dalea caerulea* (Leguminosae),[74] the regioisomers **53** and **54** from *Eysenhardtia texana* (Leguminosae),[77] and the regioisomers lespedezaflavanones F (**80**) and G (**81**), from *Lespedeza formosa* (Leguminosae).[99] Flavanones **53** and **54** from *E. texana* showed antibacterial activity as inhibitors of the growth of *Staphylococcus aureus* at a concentration of 0.1 mg/ml. Compound **54** also inhibited the growth of *Candida albicans* in an agar-gel diffusion assay[77] and thus showed antifungal activity. An unusual *C*-hydroxyethyl group is present in laxiflorin (5,7,4′-trihydroxy-3′-methoxy-6-*C*-(β-hydroxyethyl)-8-*C*-prenylflavanone, **75**) from *Derris laxiflora* (Leguminosae).[92] Laxiflorin showed significant inhibitory activity against protein-tyrosine kinase.[92]

As mentioned above, more than 30 of the newly reported noncyclic isoprenylated flavanones have been isolated from the genus *Sophora* (six species) and the related *Echinosophora koreensis* (Leguminosae), almost all from roots. Most of these isoprenylated flavanones are characterized by 2′-substitution of the B-ring. The 2′,4′- and 2′,4′,6′-oxygenation patterns are found most frequently, but 2′-*O*-, 2′,6′-di-*O*-, and 2′,4′,5′-tri-*O*-substitution also occurs. Examples of isoprenylated 2′,4′,6′-tri-oxygenated flavanones are kenusanone D (5,7,2′,6′-tetrahydroxy-4′-methoxy-8-*C*-prenylflavanone, **71**) and its 7-methyl ether, kenusanone E (**72**) from *Echinosophora koreensis*.[90] In 2′,6′-dioxygenated flavanones, such as **71** and **72**, the resonances of H-2 and H-3 show unusual chemical shift values in the ^{1}H nuclear magnetic resonance (NMR) spectrum. Oxygenation at C-2′ causes a slight downfield shift of H-2, while oxygenation at both C-2′ and C-6′ causes a conspicuous downfield shift of H-2 by ca. 0.3 ppm in most solvents. When flavanones are oxygenated at both C-2′ and C-6′, H-3$_{eq}$ is shifted upfield (ca. 0.3 to 0.4 ppm), whereas H-3$_{ax}$ is shifted downfield (0.5 to 0.8 ppm) in acetone-d_6 and DMSO-d_6, but not in CDCl$_3$. These shifts can be used as a diagnostic tool for the structural determination of both 2′,6′-dioxygenated flavanones and dihydroflavonols.[90]

Another feature of the *Sophora* flavanones is substitution with a ten-carbon geranyl or lavandulyl side group, which is sometimes hydroxylated. Eight new flavanones from *S. exigua* are lavandulylated, exiguaflavanones A–G, and J (**111–113** and **116–120**).[85,116] Lavandulyl groups are also present in the newly reported leachianone E (**105**) from *S. leachiana*[112] and kushenols R (**107**, Figure 15.3) and U (**108**) from *S. flavescens*.[76] In all but one compound, the lavandulyl substituent is attached at C-8 (see Table 15.3). Geranyl rather than lavandulyl side chains are found in the newly reported sophoraflavanones C–E (**99, 102, 103**) from *Echinosophora koreensis*,[87,102] prostratol F (**96**) from *S. prostrata*,[108] and tomentosanol E (**104**, Figure 15.3) from *S. tomentosa*.[111] A hydroxylated prenyl group is present in tomentosanol D (**122**) from the same species,[111] and hydroxylated lavandulyl groups (5-hydroxy-2-isopropenyl-5-methylhexyl) in kushenols P$_1$ (**139**), Q (**136**), and T (**128**) from *S. flavescens*[76] and leachianone D (**138**) from *S. leachiana*.[112] Again, the modified lavandulyl group was attached at C-8 of the flavanone skeleton. Biosynthetic studies of the ten-carbon lavandulyl side chain were carried out using cell suspension cultures of *S. flavescens* in which 8-prenyltransferase activity had been detected.[128] When naringenin was used as a prenyl acceptor, only 8-prenylnaringenin was formed. However, when 2′-hydroxynaringenin was added as a prenyl acceptor, the 6-prenyl- and 8-lavandulylflavanones were the main products and very little of the 8-prenyl derivative was produced. These results suggest that the 2′-hydroxy group of naringenin may play an important role in the formation of the lavandulyl group,[128] and that it may not be a coincidence that the majority of the lavandulylflavanones have a 2′-oxygenated B-ring and that the lavandulyl group is usually attached to C-8.

Some of the prenylated and lavandulylated flavanones from *S. flavescens*, kushenols P–S, exhibited significant antibacterial activities against the Gram-positive bacteria, *Staphylococcus aureus*, *S. epidermidis*, *Bacillus subtilis*, and *Propionibacterium acnes*. They also exhibited antiandrogenic activities.[76] Kurarinone (**114**), 2′-*O*-methylkurarinone (**115**), and the known sophoraflavanone G and leachianone A from the roots of *S. flavescens*, which all have 8-lavandulyl substitution and 2′,4′-di-*O*-substitution of the B-ring, exhibited cytotoxic activity against human myeloid leukemia HC-60 cells with IC_{50} values of 13.7, 18.5, 12.5, and 11.3 μM, respectively.[117]

As is clear from Table 15.3, most of the newly reported flavanones with noncyclic isoprenoid groups have been isolated from members of the Leguminosae. However, one or more new prenylated flavanones have also been reported from the families Ptaeroxylaceae, Rutaceae, Meliaceae, Asteraceae, Euphorbiaceae, Velloziaceae, Cyperaceae, Ophioglossaceae, Capparaceae, Gentianaceae, and especially Moraceae and Dipterocarpaceae. For example, two highly methoxylated 5,7,2′,4′,6′-penta-O-substituted prenylated flavanones, showing the same unusual B-ring substitution as found in many prenylflavanones from *Sophora*, have been reported from the root bark of *Artocarpus heterophyllus* (Moraceae) as heteroflavanones C (**73**)[91] and B (**74**).[39]

From a chemosystematic point of view, it is interesting to note that prenylated flavonoids such as microfolione (**56**) have been found in a species of the family Ptaeroxylaceae,[79] because the relationships of this family with other families were disputed in the past. Most taxonomists considered the Ptaeroxylaceae closely related to families in the order Rutales to which the Rutaceae and Meliaceae belong, whereas others considered it related to the Sapindaceae. Flavonoid chemistry supports a close relationship to the Rutaceae and Meliaceae, as isoprenylated flavanones also occur in these families, e.g., **58** and **98** in *Boronia coerulescens* ssp. *spinescens*,[80] the farnesyl-bearing **121** (Figure 15.3) in *B. ramosa* (Rutaceae),[119] and flowerine (**59**) and flowerone (**60**) in *Azadirachta indica* (Meliaceae).[81] Microfolione is one of the few new flavanones for which the (2R)-configuration has been determined.

The occurrence of prenyloxyflavanones in *Monotes engleri* (Dipterocarpaceae) has already been discussed, but it is worth adding here that five further new isoprenylated flavanones have been reported from this species, including three 6-C-1,1-dimethylallyl (1,1-DMA)-substituted flavanones (**89**, **91**, and **92**, Figure 15.3).[83] These compounds and the known 6,8-diprenyleriodictyol and its 3′-methyl ether, hirvanone, from the same plant displayed cytotoxic activity against several human cancer cell lines, whereas another novel isoprenylated flavanone from *M. engleri*, bearing a hydroxylated prenyl group (**135**, Figure 15.3), was nontoxic.[83] Two further 1,1-DMA-substituted flavanones, **94** and **95** from *Dalea scandens* var. *paucifolia* (Leguminosae), showed significant activity against methicillin-susceptible and -resistant strains of *Staphylococcus aureus*.[107]

A species of the family Euphorbiaceae, *Macaranga pleiostemma*, also yielded biologically active isoprenylated flavanones, macaranga flavanones A (**97**) and B (**77**), which showed significant antibacterial activity against *Escherichia coli*.[95] Flavanone **77** was also reported as paratocarpin L from *Parartocarpus* (incorrectly spelled as *Paratocarpus*) *venenosa* (Moraceae).[94] Two more antibacterial flavanones bearing an 8-C-(2-isopropyl-5-methyl-5-hexenyl) side chain, which is similar to a lavandulyl group, but which has a double bond between C-5 and C-6 of the isoprenoid chain instead of between C-4 and C-5, are remangiflavanones A (**106**) and B (**110**). These were isolated from *Physena madagascariensis*, family Capparaceae. The compounds were bacteriocidal at 9 and 4 μM, respectively, against *Staphylococcus aureus*, and also bacteriocidal or antibacterial against seven other bacteria, with minimum effective concentrations of 0.3 to 10 μM. This species also produces a flavanone dimer, remangiflavanone C, which was not active.[113]

Vellozia and *Schoenus* are the only genera of Monocotyledoneae from which new isoprenylated flavanones have been reported, 6,8-diprenyleriodictyol (**82**) from the leaves of *Vellozia nanuzae* (Velloziaceae)[61] and 5,7,3′,4′-tetrahydroxy-6,5′-di-C-prenylflavanone (**83**) from the tubers of *Schoenus nigricans* (Cyperaceae).[100] The known regioisomer of **83**, 5,7,3′,4′-tetrahydroxy-8,5′-di-C-prenylflavanone, was also reported from the latter species.[100]

Several new flavanones were produced as biotransformation products of the prenylchalcone, xanthohumol, when this compound was incubated with the microorganism

Cunninghamella echinulata NRRL 3655. Xanthohumol is an important constituent of beer, as it is a major constituent of hops (*Humulus lupulus*, Cannabinaceae) used in the brewing industry. Microbial transformation studies of xanthohumol were carried out to generate compounds similar to mammalian metabolites of the constituents of hops. Two flavanones bearing the same hydroxylated prenyl side chain, the *E,Z*-isomers **124** and **125**, were some of the transformation products obtained. These and other transformation compounds from xanthohumol were tested for anticancer activity, but did not show cytotoxicity at 25 μl/ml against a range of human cancer cell lines as well as noncancerous Vero cells. However, antimalarial activity was exhibited by **125** (but not by **124**) against D6 (chloroquinine-sensitive) strains of the malaria parasite, *Plasmodium falciparium* (IC_{50}, 2 μg/ml).[121] The known flavanone 8-prenylnaringenin is present in low concentrations in beer as an isomerization product of the chalcone desmethylxanthohumol (another hop constituent) and is the most potent phytoestrogen so far discovered.[129] 8-Prenylnaringenin was also tested for antifungal activity together with semisynthetic 6- and 3'-prenylnaringenin, using *Cladosporium herbarum* as the test organism. The 8-prenylflavanone was more fungitoxic than the corresponding 3'-prenylated isomer, whereas 6-prenylnaringenin was inactive, indicating a strong regiospecific influence of the prenyl side chain on fungitoxicity.[122]

During a survey of species and hybrids of *Glycyrrhiza* (Leguminosae) from Central Asia, licoleafol (**131**, Figure 15.3), a new flavanone that contains a hydroxylated prenyl side chain, was found to be a chemotaxonomic marker to distinguish *G. uralensis*, in which the compound occurred in high levels, from *G. glabra*, in which it could not be detected. A morphologically intermediate plant, which was thought to be a hybrid between *G. uralensis* and *G. glabra*, produced small amounts of licoleafol.[124]

The compound listed in Table 15.3 as crotaramosmin from *Crotalaria ramosissima* (Leguminosae) was reported as the new 5,4'-dihydroxy-6-*C*-prenylflavanone.[73] This structure was reassigned on the basis of detailed spectroscopic studies as 2',4-dihydroxy-6'',6''-dimethylpyrano[2'',3'':4',3']-dihydrochalcone (compound **286** in Chapter 16).[130]

15.2.2.3   Flavanones with Furano Rings

When a *C*-prenyl group attached to a phenolic ring of a flavonoid forms a linkage with an *ortho*-hydroxyl, the result may be a heterocyclic isoprop(en)ylfurano or dimethylpyrano ring, or dihydro derivatives of these. Dimethylpyranoflavanones are discussed in the next section. Further modification of the furano ring may lead to hydroxylation or loss of the isopropyl side chain. Figure 15.4 shows representatives of flavanones with various types of furano substituents and gives the numbering of the atoms used in this chapter, since various different numbering conventions are presented in the literature.

More than 20 furanoflavanones have been reported during the period under review, whereas only one furanoflavanone was known previously. The newly reported structures are listed in Table 15.4 (compounds **141–161**).[61,98,104,122,126,131–140] The compounds are arranged according to the number of oxygens in the molecular formula and the relative molecular masses. More than half of the furanoflavanones contain two isoprenoid substituents as they also bear an additional *C*-prenyl group.

In the new compounds, the furano ring is usually attached to C-7 and C-8 (or C-6) of the flavanone, and incorporates the oxygen at C-7. However, in abyssinoflavanone IV (**150**) from *Erythrina abyssinica* (Leguminosae), C-5' and the oxygen at C-4' are involved in the ring structure,[98] and in flavanones **151** and **161** from *Paramignya griffithii* (Rutaceae), C-6 and the oxygen at C-5 are part of the ring.[139]

Flavanones bearing a simple furano ring without an isopropyl side chain include compounds **141** from *Millettia erythrocalyx*,[131] **143** from *Lonchocarpus latifolius*,[133] **144** from

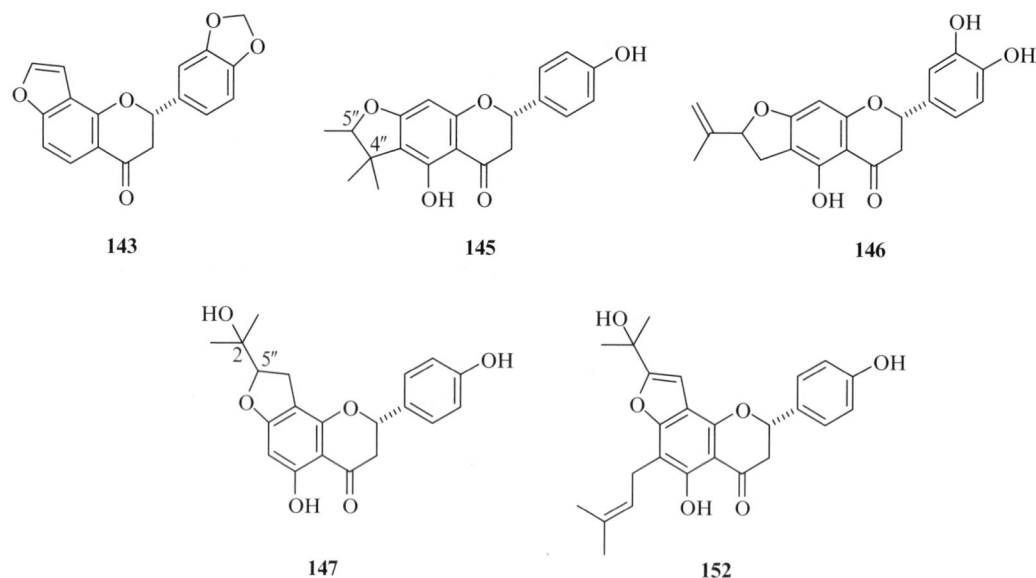

FIGURE 15.4 Examples of furanoflavanones.

L. subglaucescens,[134] and **150** from *Erythrina abyssinica*.[98] All these species belong to the Leguminosae. Although the majority of the furanoflavanones have been found in members of the Leguminosae, as is the case with most other classes of prenylated flavonoids, the families Moraceae and Rutaceae are also good sources. From another family rich in isoprenylated flavonoids, Velloziaceae, a flavanone having an isopropenyl group at C-5″ of the dihydrofurano ring has been isolated, velloeriodictyol (**146**).[61] The same 5″-isopropenyldihydrofurano substituent is present in emoroidenone (**142**) from *Tephrosia emoroides* (Leguminosae). This flavanone has a strong antifeedant activity against the larvae of the insect *Chilo portellus*.[132] The furano or dihydrofurano rings of most of the remaining flavanones in Table 15.4 bear hydroxylated isopropyl groups, e.g., compound **147** found in *Macaranga conifera* (Euphorbiaceae),[135] phellodensin D (**148**) from both *Phellodendron chinense* var. *glabriusculum* (Rutaceae)[137] and *Broussonetia papyrifera* (Moraceae),[136] flavanones **152–156** from *Lupinus luteus* (Leguminosae),[122] and compounds **157–160** from *Dorstenia mannii* (Moraceae).[126,140] Compound **152**, lupinenol, is distinguished from the other new flavanones from *L. luteus* in having a furano rather than a dihydrofurano ring. It should be noted that C-5″ of the dihydrofurano ring is an asymmetric center, and therefore two stereoisomers are possible. Although the dihydrofuranoflavanones lonchocarpols C and D were known compounds, the two epimers of each had not been reported previously. They have now been isolated from *L. luteus* and separated by high-performance liquid chromatography to yield lonchocarpol C_1 (**153**) and its 5″-epimer lonchocarpol C_2 (**154**), and lonchocarpol D_1 (**155**) and its 5″-epimer lonchocarpol D_2 (**156**).[122] The 5″-epimers of dorsmanins F (**157**) and G (**158**) from *Dorstenia mannii* have also been reported, as epidorsmanins F (**159**) and G (**160**), respectively.[140]

Furanoflavanones **152–156** from *L. luteus* were tested for antifungal activity using *Cladosporium herbarum* as the test organism. Lonchocarpol D_1 showed the highest fungitoxicity and there was evidence for stereospecific activity, as lonchocarpol D_2 was much less active. Lonchocarpols C_1 and C_2 were very weakly antifungal.[122] Flavanone **148** (discussed above) and the known (2S)-abyssinone II from *Broussonetia papyrifera* (Moraceae) are very potent inhibitors of aromatase, which catalyzes the final step in estrogen biosynthesis, with IC_{50} values of 0.1 and 0.4 μ*M*, respectively.[136]

TABLE 15.4
Furanoflavanones Reported from 1992 to 2003

No.	OH-, OMe-, and Methylenedioxy Substituents[a]	Furano and Prenyl Substituents[a]	Mol. Formula	M_r	Trivial Name	Plant Source	Family	Organ[b]	Ref.
141	6-OMe	Furano[2'',3'':7,8]	$C_{17}H_{14}O_4$	282		Millettia erythrocalyx	Leguminosae	Root	131
142	5-OMe	5''-Isopropenyldihydrofurano[2'',3'':7,8]	$C_{21}H_{20}O_4$	336	Emoroidenone	Tephrosia emoroides	Leguminosae		132
143	3',4'-Methylenedioxy	Furano[2'',3'':7,8]	$C_{18}H_{12}O_5$	308	Figure 15.4	Lonchocarpus latifolius	Leguminosae	Root	133
144*	5,6-DiOMe	Furano[2'',3'':7,8]	$C_{19}H_{16}O_5$	324		Lonchocarpus subglaucescens	Leguminosae		134
145	5,4'-DiOH	4'',4''-Dimethyl-5''-methyldihydrofurano[2'',3'':7,6]	$C_{20}H_{20}O_5$	340	Figure 15.4	Monotes engleri	Dipterocarpaceae	Leaf	104
146	5,3',4'-TriOH	5''-Isopropenyldihydrofurano[2'',3'':7,6]	$C_{20}H_{18}O_6$	354	Velloeriodictyol, (Figure 15.4)	Vellozia glabra	Velloziaceae	Leaf	61
147	5,4'-DiOH	5''-(2-OH-isopropyl)dihydrofurano[2'',3'':7,8]	$C_{20}H_{20}O_6$	356	Figure 15.4	Macaranga conifera	Euphorbiaceae	Leaf	135
148	2,4'-DiOH	5''-(2-OH-isopropyl)dihydrofurano[2'',3'':7,8]	$C_{20}H_{20}O_6$	356		Broussonetia papyrifera	Moraceae	Whole	136
					Phellodensin D	Phellodendron chinense var. glabriusculum	Rutaceae	Leaf	137
149	4'-OH-5-OMe	(E)-5''-(2-OH-isopropyl)dihydrofurano[2'',3'':7,8]	$C_{21}H_{22}O_5$	370		[c]			138
150	5,7,3'-TriOH	2-Pr-furano[2'',3'':4,5]	$C_{22}H_{22}O_5$	382	Abyssinoflavanone IV	Erythrina abyssinica	Leguminosae		98
151	3',4'-DiOH-7-OMe	8-Pr-furano[2'',3'':5,6]	$C_{23}H_{22}O_6$	394		Paramignya griffithii	Rutaceae	Stem	139
152	5,4'-DiOH	6-Pr-5''-(2-OH-isopropyl)furano[2'',3'':7,8]	$C_{25}H_{26}O_6$	422	Lupinenol (Figure 15.4)	Lupinus luteus	Leguminosae	Root	122
153	5,4'-DiOH	6-Pr-5''-(2-OH-isopropyl)dihydrofurano[2'',3'':7,8]	$C_{25}H_{28}O_6$	424	Lonchocarpol C₁	Lupinus luteus	Leguminosae	Root	122
154	5,4'-DiOH	5''-Epimer of 153	$C_{25}H_{28}O_6$	424	Lonchocarpol C₂	Lupinus luteus	Leguminosae	Root	122
155	5,4'-DiOH	8-Pr-5''-(2-OH-isopropyl)dihydrofurano[2'',3'':7,6]	$C_{25}H_{28}O_6$	424	Lonchocarpol D₁	Lupinus luteus	Leguminosae	Root	122
156	5,4'-DiOH	5''-Epimer of 155	$C_{25}H_{28}O_6$	424	Lonchocarpol D₂	Lupinus luteus	Leguminosae	Root	122
157	5,3',4'-TriOH	6-Pr-5''-(2-OH-isopropyl)dihydrofurano[2'',3'':7,8]	$C_{25}H_{28}O_7$	440	Dorsmanin F	Dorstenia mannii	Moraceae	Twig	126
158	5,3',4'-TriOH	8-Pr-5''-(2-OH-isopropyl)dihydrofurano[2'',3'':7,6]	$C_{25}H_{28}O_7$	440	Dorsmanin G	Dorstenia mannii	Moraceae	Twig	126
159	5,3',4'-TriOH	5''-Epimer of 157	$C_{25}H_{28}O_7$	440	Epidorsmanin F	Dorstenia mannii	Moraceae	Twig	140
160	5,3',4'-TriOH	5''-Epimer of 158	$C_{25}H_{28}O_7$	440	Epidorsmanin G	Dorstenia mannii	Moraceae	Twig	140
161	3',4'-DiOH-7-OMe	8-Pr-5''-(2-OH-isopropyl)furano[2'',3'':5,6]	$C_{26}H_{28}O_7$	452		Paramignya griffithii	Rutaceae	Stem	139

[a] Pr, prenyl.
[b] Whole, whole plant.
[c] Microbial metabolite of xanthohumol using a culture broth of Pichia membranifaciens.

In the previous section, the microbial transformation of the isoprenylated chalcone, xanthohumol, was discussed. When a culture broth of a different fungus, *Pichia membranifaciens*, was used, the new furanoflavonone **149** was produced from xanthohumol.[138]

15.2.2.4 Flavanones with Pyrano Rings

Some 50 new dimethylpyrano (DMP), dimethyldihydropyrano (DMDHP), and related pyranoflavanones have been reported from 1992 to 2003, more than doubling the number of known pyranoflavanones. The newly reported compounds are presented in Table 15.5 (compounds **162–211**).[82,85–88,94,98,126,134,135,140–162] They are firstly arranged according to the number of oxygen atoms in the compounds, and within these groups according to their relative molecular masses. There is no free hydroxyl group in maximaflavanone A (**162**) from *Tephrosia maxima* (Leguminosae),[141] but there is of course an oxygen atom incorporated in the DMP group. Similarly, there is no free hydroxyl in dorsmanin B (**168**) from *Dorstenia mannii* (Moraceae),[147] although it is di-*O*-substituted, since two *O*-containing DMDHP groups are attached to this flavanone. Figure 15.5 shows representatives of DMP and DMDHP flavanones, and the numbering system used in this chapter for pyranoflavanones. The place of attachment of the pyrano ring structure to the flavonoid is more varied than that of the furanoflavanones discussed above. Most commonly the structures are characterized by ring closure of a prenyl group at C-8 (or C-6) with the oxygen at C-7; less commonly C-3′ (or C-5′) and an oxygen at C-4′ are involved in the pyrano ring, and rarely C-6 and the oxygen at C-5. The newly reported DMP flavanones have been isolated from species belonging to the families Leguminosae, Moraceae, Rutaceae, Asteraceae, and Euphorbiaceae, and the new DMDHP flavanones from the same families apart from the Rutaceae. There does not seem to be a chemosystematic difference between the occurrence of these two groups, since all genera in which new DMDHP flavanones have been recorded also produced new DMP flavanones, except *Echinosophora koreensis*, but this species is closely related to the genus *Sophora* in which both types were present.

Many of the new pyranoflavanones show the common 4′-*O*- or 3′,4′-di-*O*-substitution patterns of the B-ring, but in most of the compounds from *Sophora* and *Echinosophora* (Leguminosae) C-2′ is oxygenated in addition to or instead of C-4′ and sometimes C-6′, e.g., in kenusanone J (**181**) from *E. koreensis*,[87] leachianone F (**193**) from *S. leachiana*,[82] and exiguaflavanones L (**194**), I (**208**), and H (**209**) from *S. exigua*.[85] The latter two compounds are also special because they additionally contain a lavandulyl group. These *Sophora* and *Echinosophora* DMP-flavanones are closely related to the noncyclic isoprenoid-containing kenusanones, leachianones, and exiguaflavanones discussed previously.[82,85,87,90,102,112,115,116]

2′,4′-Di-*O*-substitution of the B-ring is also found in three new pyranoflavanones from *Morus* species (Moraceae), sanggenol L (**189**, Figure 15.5) from *M. mongolica*,[156] and sanggenols N (**188**) and O (**187**) from *M. australis*.[155] In the latter compound, both the 2′-*O*- and 4′-*O*-functionalities are part of DMP rings. In sanggenol L (**189**), the pyrano ring seems to have been formed from a geranyl unit rather than the usual prenyl unit, since C-6″ bears only one methyl group, and an additional 6-carbon isohexenyl chain (4-methylpent-3-enyl) (Figure 15.5).

More than half of the compounds in this section bear a second *C*-isoprenoid group in addition to the dimethylpyrano group. This is either a prenyl, geranyl, or lavandulyl group, or a second dimethylpyrano group. A 2′-*C*-geranyl group is found in tanariflavanone A (**201**, Figure 15.5), isolated from *Macaranga tanarius* (Euphorbiaceae).[158] In the related tanariflavanone B (**200**, Figure 15.5) from *M. tanarius*, the 2′-*C*-geranyl has formed a ring with the hydroxyl on C-3′ to form the same 6″-methyl-6″-(4-methylpent-3-enyl)pyrano ring as found

TABLE 15.5
Dimethylpyrano- and Dimethyldihydropyranoflavanones Reported from 1992 to 2003

No.	OH-, OMe-, OPr-, and Methylenedioxy-Substituents[a]	Dimethylpyrano-, Dimethyldihydropyrano-, and Other C- substituents[a]	Mol. Formula	M_r	Trivial Name	Plant Source	Family	Organ[b]	Ref.
162	—	6-Pr-6''-DMP[2'',3'':7,8]	$C_{25}H_{26}O_3$	374	Maximaflavanone A	*Tephrosia maxima*	Leguminosae	Root	141
163	4'-OMe	6',6''-DMDHP[2'',3'':7,8]	$C_{21}H_{20}O_4$	336	Dorspoinsettifolin	*Dorstenia poinsettifolia*	Moraceae	Twig	142
164*	5'-OH	6-(3-Me-but-1,3-dienyl)- 6',6''- DMP[2'',3'':7,8]	$C_{25}H_{24}O_4$	388	Spinoflavanone A	*Tephrosia spinosa*	Leguminosae	Whole	143
165	7-OH	6-Pr-6'',6''-DMP[2'',3'':4,3]	$C_{25}H_{26}O_4$	390	Dinklagin A	*Dorstenia dinklagei*	Moraceae	Twig	144
166	4'-OH	3'-Pr-6'',6''-DMP[2'',3'':7,8]	$C_{25}H_{26}O_4$	390	Shinflavanone	*Glycyrrhiza glabra*	Leguminosae	Root	145
167*	7-OH	8-Pr-6'',6''-DMDHP[2'',3'':4,3]	$C_{25}H_{28}O_4$	392	Euchrenone a17	*Euchresta formosana*	Asteraceae	Root	146
168*	—	Bis(6'',6''-DMDHP[2'',3'':7,6][2'',3'':4,3])	$C_{25}H_{28}O_4$	392	Dorsmanin B	*Dorstenia mannii*	Moraceae	Twig	147
169	5,4-DiOH	6'',6''-DMP[2'',3'':7,6]	$C_{20}H_{18}O_5$	338	Paratocarpin K	*Parartocarpus venenosa*	Moraceae		94
170	5-OH-4-OMe	6'',6''-DMP-[2'',3'':7,8]	$C_{21}H_{20}O_5$	352	(Figure 15.5)	*Macaranga conifera*	Euphorbiaceae	Leaf	135
171	5,4-DiOMe	6'',6''-DMP[2'',3'':7,8]	$C_{22}H_{22}O_5$	366	Glyllavanone A	*Glycosmis citrifolia*	Rutaceae	Leaf	148
172*	3',4'-DiOMe	6'',6''-DMP[2'',3'':7,8]	$C_{22}H_{22}O_5$	366	Ponganone III	*Pongamia pinnata*	Leguminosae	RootB	88
						Lonchocarpus subglaucescens	Leguminosae	Root	134
173	5,7-DiOH	6-Pr-6'',6''-DMP[2'',3'':4,3]	$C_{25}H_{26}O_5$	406	Paratocarpin H	*Parartocarpus venenosa*	Moraceae		94
174	5,4-DiOH	3'-Pr-6'',6''-DMP[2'',3'':7,6]	$C_{25}H_{26}O_5$	406	Paratocarpin I	*Parartocarpus venenosa*	Moraceae		94
175	5,4-DiOH	6''-Me,6''-(4-methylpent-3-enyl)-pyrano[2'',3'':7,8]	$C_{25}H_{26}O_5$	406	Cycloaltilisin 7	*Artocarpus altilis*	Moraceae	Bud covers	149
176	5-OH	Bis(6',6''-DMDHP[2'',3'':7,6][2'',3'':4,3])	$C_{25}H_{28}O_5$	408	Paratocarpin J (Figure 15.5)	*Parartocarpus venenosa*	Moraceae		94
177	5,7-DiOH	8-Pr-6',6''-DMDHP[2'',3'':4,3]	$C_{25}H_{28}O_5$	408	Euchrenone a16	*Euchresta formosana*	Leguminosae	Root	150
178	3',4'-DiOMe	6-Pr-6',6''-DMP[2'',3'':7,8]	$C_{27}H_{30}O_5$	434		*Cubé resin*[c]	Leguminosae	Root	151
179	5-OH	6-Pr-bis(6',6''-DMP [2'',3'':7,8][2'',3'':4,3])	$C_{30}H_{32}O_5$	472	Euchrenone a15	*Euchresta tubulosa*	Leguminosae	Root	152
180	5-OH	8-Pr-bis(6',6''-DMP [2'',3'':7,6][2'',3'':4,3])	$C_{30}H_{32}O_5$	472	Euchrenone a14	*Euchresta tubulosa*	Leguminosae	Root	152
181	5,2',4'-TriOH	6',6''-DMDHP[2'',3'':7,8]	$C_{20}H_{20}O_6$	356	Kenusanone J	*Echinosophora koreensis*	Leguminosae	Stem	87
182*	5,7-DiOH-3'-OMe	6',6''-DMP[2'',3'':4,5]	$C_{21}H_{20}O_6$	368	Abyssinin I	*Erythrina abyssinica*	Leguminosae	StemB	86

continued

TABLE 15.5
Dimethylpyrano- and Dimethyldihydropyranoflavanones Reported from 1992 to 2003 — *continued*

No.	OH-, OMe-, OPr-, and Methylenedioxy-Substituents[a]	Dimethylpyrano-, Dimethyldihydropyrano-, and Other C- substituents[a]	Trivial Name	Mol. Formula	M_r	Plant Source	Family	Organ[b]	Ref.
183*	5,3'-DiOH-4-OMe	6'',6''-DMP[2'',3'':7,8]		$C_{21}H_{20}O_6$	368	*Feronia limonia*	Rutaceae	StemB	153
184*	5,4-DiOH	8-Hydroxymethyl-6'',6''-DMP[2'',3'':7,6]	Figure 15.5	$C_{21}H_{20}O_6$	368	*Derris reticulata*	Leguminosae	Stem	154
185	5,3',4'-TriOMe	6'',6''-DMP[2'',3'':7,8]	Glyflavanone B	$C_{23}H_{24}O_6$	396	*Glycosmis citrifolia*	Rutaceae	Leaf	148
186	6,3',4'-TriOMe	6'',6''-DMP[2'',3'':7,8]	Ponganone IV	$C_{23}H_{24}O_6$	396	*Pongamia pinnata*	Leguminosae	RootB	88
187	5,7-DiOH	Bis(6'',6''-DMP [2'',3'':2',3''][2'',3'':4',5'])	Sanggenol O	$C_{25}H_{24}O_6$	420	*Morus australis*	Moraceae	RootB	155
188	5,7,2'-TriOH	5''-Pr-6'',6''-DMP[2'',3'':4',3']	Sanggenol N	$C_{25}H_{26}O_6$	422	*Morus australis*	Moraceae	RootB	155
189	5,2',4'-TriOH	6''-Me-6''-(4-methylpent-3-enyl)-pyrano[2'',3'':7,8]	Sanggenol L (Figure 15.5)	$C_{25}H_{26}O_6$	422	*Morus mongolica*	Moraceae	StemB	156
190	5,3',4'-TriOH	8-Pr-6'',6''-DMP[2'',3'':7,6]	Dorsmanin I	$C_{25}H_{26}O_6$	422	*Dorstenia mannii*	Moraceae	Twig	140
191	5,4-DiOH	8-(2,3-Epoxy-3-Me-butyl)-6'',6''-DMP[2'',3'':7,6]	2''',3'''-Epoxylupinifolin	$C_{25}H_{26}O_6$	422	*Derris reticulata*	Leguminosae	Stem	157
192	5,4-DiOH	8-(2-OH-3-Me-but-3-enyl)-6'',6''-DMP[2'',3'':7,6]	Dereticulatin	$C_{25}H_{26}O_6$	422	*Derris reticulata*	Leguminosae	Stem	157
193	5,2',4'-TriOH	5''-Pr-6'',6''-DMDHP[2'',3'':7,8]	Leachianone F	$C_{25}H_{28}O_6$	424	*Sophora leachiana*	Leguminosae	Root	82
194	5,2',6'-TriOH	5''-Pr-6'',6''-DMDHP[2'',3'':7,8]	Exiguaflavanone L	$C_{25}H_{28}O_6$	424	*Sophora exigua*	Leguminosae	Root	85
195	5,3',4'- TriOH	8-Pr-6'',6''-DMDHP[2'',3'':7,6]	Dorsmanin J	$C_{25}H_{28}O_6$	424	*Dorstenia mannii*	Moraceae	Twig	140
196	3',4'-DiOH	Bis(6'',6''-DMDHP [2',3'':5,6][2'',3'':7,8])	Dorsmanin E	$C_{25}H_{28}O_6$	424	*Dorstenia mannii*	Moraceae	Twig	126
197	5,3'-DiOH-4-OMe	6-Pr-6'',6''-DMP[2'',3'':7,8]		$C_{26}H_{28}O_6$	436	*Cubé resin*[c]	Leguminosae	Root	151
198	5,4-DiOH-3'-OMe	6-Pr-6'',6''-DMP[2'',3'':7,8]		$C_{26}H_{28}O_6$	436	*Cubé resin*[c]	Leguminosae	Root	151

No.	Substitution	Substituent[a]	Formula	MW	Name	Species	Family	Tissue[b]	Ref.
199	5-OH-3',4'-DiOMe	6-Pr-6'',6''-DMP[2'',3'':7,8]	$C_{27}H_{30}O_6$	450		Cubé resin[c]	Leguminosae	Root	151
200	5,7,4'-TriOH	6-Pr-6''-Me,6''-(4-methylpent-3-enyl)-pyrano[2'',3'':3;2]	$C_{30}H_{34}O_6$	490	Tanariflavanone B (Figure 15.5)	Macaranga tanarius	Euphorbiaceae	Leaf (fallen)	158
201	5,3',4'-TriOH	2'-Ger-(5''-OH-6'',6''-DMDHP[2'',3'':7,6]	$C_{30}H_{36}O_6$	492	Tanariflavanone A (Figure 15.5)	Macaranga tanarius	Euphorbiaceae	Leaf (fallen)	158
202	5,4'-DiOH	8-(1-OH-2,3-epoxy-3-Me-butyl)-6'',6''-DMP[2'',3'':7,6]	$C_{25}H_{26}O_7$	438	1'''-OH-2''',3'''-Epoxy-lupinifolin (Figure 15.5)	Derris reticulata	Leguminosae	Stem	159
203	5,7,3'-TriOH	2-Pr-(5''-OH-6'',6''-DMDHP[2'',3'':4,5']	$C_{25}H_{28}O_7$	440	Abyssinoflavanone V	Erythrina abyssinica	Leguminosae		98
204	5,7-DiOH	(5''-OH-6'',6''-DMP[2'',3'':7,6])-6'',6'''-DMDHP[2''',3''':4,3']	$C_{25}H_{28}O_7$	440	Abyssinoflavanone VI	Erythrina abyssinica	Leguminosae		98
205*	5,4'-DiOH	8-(2,3-diOH-3-Me-butyl)-6'',6''-DMP[2'',3'':7,6]	$C_{25}H_{28}O_7$	440	2'',3''-Dihydroxylupinifolin (Figure 15.5)	Derris reticulata	Leguminosae	Stem	154
206	5,4'-DiOH	See Figure 15.5	$C_{26}H_{28}O_7$	452	Derriflavanone	Derris laxiflora	Leguminosae	Root	160
207	5,4'-DiOH	See Figure 15.5	$C_{26}H_{28}O_7$	452	Epi-derriflavanone	Derris laxiflora	Leguminosae	Root	160
208	5,2',4',6'-TetraOH	6-Lav-6'',6''-DMP[2'',3'':7,8]	$C_{30}H_{34}O_7$	506	Exiguaflavanone I	Sophora exigua	Leguminosae	Root	85
209	5,2',4',6'-TetraOH	8-Lav-6'',6''-DMP[2'',3'':7,6]	$C_{30}H_{34}O_7$	506	Exiguaflavanone H	Sophora exigua	Leguminosae	Root	85
210	5,7,3'-TriOH	(4'',5''-diOH-6'',6''-DMDHP[2'',3'':4',5'])	$C_{20}H_{20}O_8$	388	Sigmoidin G	Erythrina sigmoidea	Leguminosae	StemB	161
211	2',3',6'-TriOH	8-Pr-5'-(2,4-diOH-phenyl)-6'',6''-DMP[2'',3'':7,6]	$C_{32}H_{30}H_8$	528	Eriosemaone C	Eriosema tuberosum	Leguminosae	Root	162

*(2S)-Flavanones.

[a] Pr, prenyl; Ger, geranyl; Lav, lavandulyl; DMP, dimethylpyrano; DMDHP, dimethyldihydropyrano.

[b] Rhiz, rhizomes; Tissue, plant tissue culture; RootB, root bark; Aerial, aerial parts; StemB, stem bark.

[c] Cubé resin is an extract of the roots of Lonchocarpus utilis and L. urucu (Leguminosae).

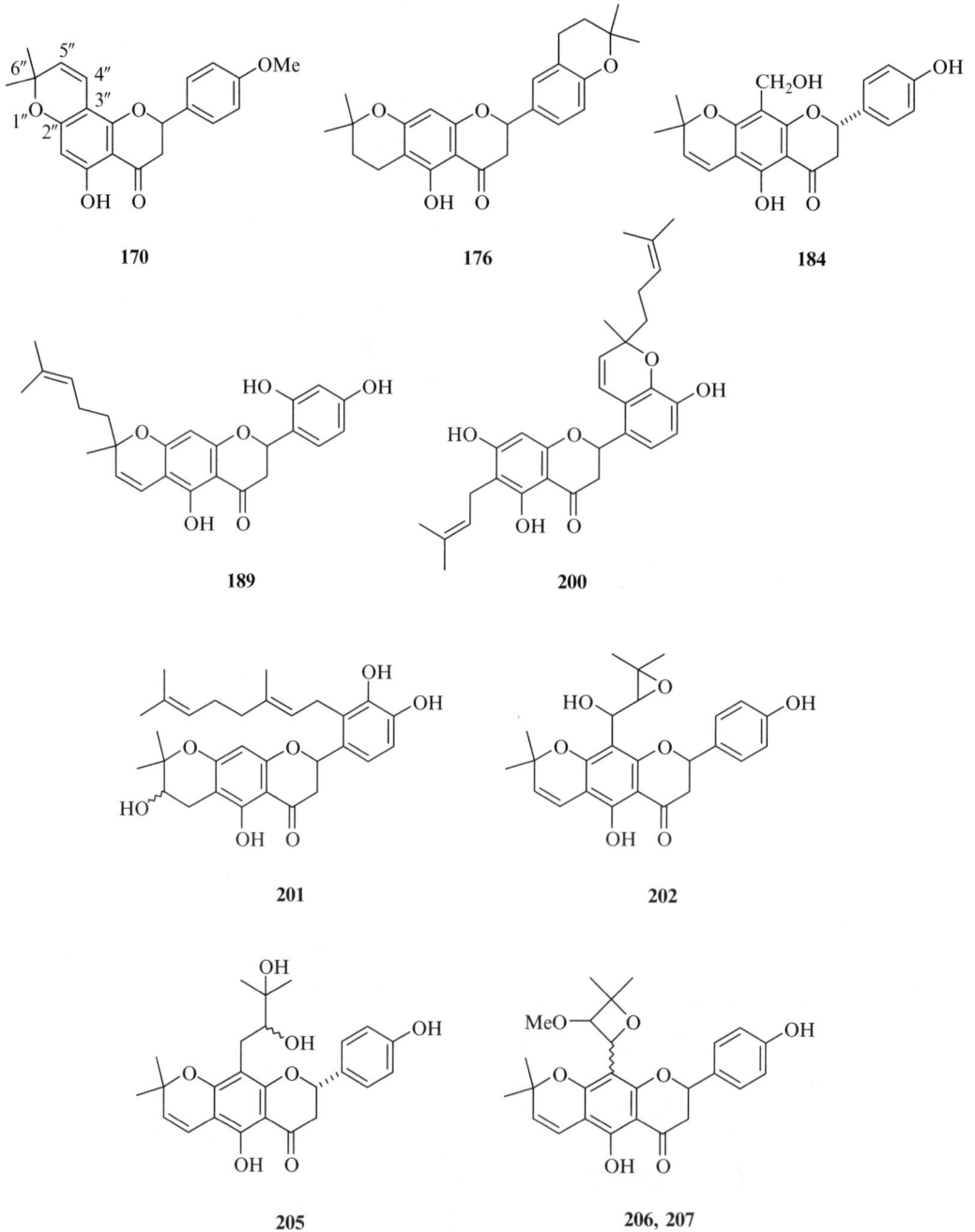

170

176

184

189

200

201

202

205

206, 207

FIGURE 15.5 Examples of pyranoflavanones.

in sanggenol L (**189**, see above). Flavanones **200** and **201** were obtained from fallen leaves and showed allelopathic activity (inhibition of radicle growth of lettuce seedlings at 200 ppm).[158] Two DMP rings are present in euchrenones a$_{14}$ and a$_{15}$ (**180** and **179**) from *Euchresta tubulosa* (Leguminosae)[152] and in sanggenol O (**187**) from *Morus australis* (Moraceae),[155] whereas two DMDHP rings are present in dorsmanins B (**168**)[147] and E (**196**)[126] from *Dorstenia mannii*

(Moraceae) and paratocarpin J (**176**, Figure 15.5) from *Parartocarpus venenosa* (Moraceae).[94] Abyssinoflavanone VI (**204**) from *Erythrina abyssinica* (Leguminosae) contains both DMP and DMDHP ring structures.[98]

Seven new flavanones have been reported from species of *Derris* (Leguminosae), which in addition to a DMP ring have unusual substituents. For example, flavanone **184** (Figure 15.5) from *D. reticulata* bears an 8-*C*-hydroxymethyl substituent, and 2″,3″-dihydroxylupinifolin (**205**, Figure 15.5) from the same species bears an 8-*C*-isopentyl group hydroxylated in the 2- and 3-positions.[154] In the related 2‴,3‴-epoxylupinifolin (**191**), also from *D. reticulata*, there is an epoxy group between C-2 and C-3 of the 8-*C*-isopentyl chain;[157] 1‴-hydroxy-2‴,3‴-epoxylupinifolin (**202**, Figure 15.5) is similar to **191**, but there is an additional hydroxyl at C-1 in the 8-*C*-isopentyl group.[159] In dereticulatin (**192**), there is an 8-*C*-pentenyl group that is hydroxylated at C-2.[157] The close biogenetic relationships among compounds **191**, **192**, **202**, and **205** are obvious, but **184** seem different because there is no isoprenyl group attached to C-8. However, it was suggested that **184** could be derived from the co-occurring 1‴-OH-2‴,3‴-epoxylupinifolin by acid-catalyzed opening of the epoxide ring, followed by carbon–carbon bond cleavage to produce a precursor of **184**.[154] *In vitro* bioassay evaluation of **184** and **191** revealed cytotoxic activity in the P-388 cell line with IC_{50} values of 6.4 and 1.3 μm/ml, respectively, but they were inactive against the KB cell line. Even more unusual are the diastereoisomeric flavanones derriflavanone (**206**) and epi-derriflavanone (**207**) from *Derris laxiflora*. These contain a four-membered heterocyclic 3-methoxy-2,2-dimethyl-oxetane ring, which is C–C linked at C-4 to C-8 of 5,4′-dihydroxy-6″,6″-DMP[2″,3″:7,6]flavanone (Figure 15.5).[160] The two compounds are C-4-epimers. Four new bioactive pyranoflavanones, **178** and **197–199**, have been reported from cubé resin, which is an extract of the roots of *Lonchocarpus utilis* and *L. urucu* (Leguminosae) used as an insecticide and piscicide.[151] The active principles of cubé resin are rotenone (44%) and deguelin (22%), but many other flavonoids are present as minor components. A number of these minor constituents, including the four pyranoflavanones, were tested for other activities, including the following: (1) inhibition of NADH:ubiquinone oxidoreductase *in vitro*; (2) inhibition of phorbol ester-induced ornithine decarboxylase activity in cultured MCF-7 cells; and (3) cytotoxicity in MCF-7 and Hepa 1c1c7 cells. All four pyranoflavanones were active in these tests, but compound **198** was the most potent. This flavanone differs from the other three in having a free 4′-hydroxyl substituent, a feature that may be important for the activity. The well-known isoflavone genistein and stilbene *trans*-resveratrol, also obtained from the same plant, were much less active in these bioassays than the pyranoflavanones.[151]

Pyranoflavanone **183** from *Feronia limonia* (Rutaceae) was active against both Gram-positive and Gram-negative bacteria (at 100 μg/ml against *Staphylococcus aureus*, *Escherichia coli*, and *Klebsiella erogenes*), but did not show any antifungal activity.[153]

15.2.3 Complex Flavanones

15.2.3.1 Benzylated Flavanones

Six new flavanones having one or more *C*-benzyl groups attached to C-6 or C-8, or both, have been reported from members of the Annonaceae, a family from which many *C*-benzylated flavanones and dihydrochalcones had previously been described.[15] The new compounds (**212–217**) are presented in Table 15.6.[163–166] Macrophyllol (**212**) and macrophyllol A (**213**, Figure 15.6) have been isolated from the roots of *Uvaria macrophylla*. These consist of 5-hydroxy-6,7-dimethoxyflavanone with one 2-hydroxy-5-methoxybenzyl unit attached to C-8 and C-6, respectively.[163,164] The roots of *Xylopia africana* yielded isouvarinol (**214**, Figure 15.6), in which one 2-hydroxybenzyl unit is attached to C-6 and two units to C-8. Its

TABLE 15.6
C-Benzylated Flavanones Reported from 1992 to 2003

No.	OH- and OMe- Substitution	C-Benzyl (Bn) Substitution	Formula	M_r	Trivial Name	Plant Source	Family	Organ	Ref.
212	5-OH-6,7-diOMe	8-(2-OH-5-OMe-Bn)	$C_{25}H_{24}O_7$	436	Macrophyllol	*Uvaria macrophylla*	Annonaceae	Root	163
213	5-OH-7,8-diOMe	6-(2-OH-5-OMe-Bn)	$C_{25}H_{24}O_7$	436	Macrophyllol A (Figure 15.6)	*Uvaria macrophylla*	Annonaceae	Root	164
214	5,7-DiOH	6-(2-OH-Bn)-8-(2 × 2-OH-Bn)	$C_{36}H_{30}O_7$	574	Isouvarinol (Figure 15.6)	*Xylopia africana*	Annonaceae	Root	165
215	5,7-DiOH	6-(3 × 2-OH-Bn)-8-(2-OH-Bn)	$C_{43}H_{36}O_8$	680	2''''-OH-3''''-Benzyluvarinol	*Xylopia africana*	Annonaceae	Root	166
216	5,7-DiOH	6-(2-OH-Bn)-8-(3 × 2-OH-Bn)	$C_{43}H_{36}O_8$	680	2''''-OH-5''''-Benzylisouvarinol A	*Xylopia africana*	Annonaceae	Root	166
217	5,7-DiOH	6-(2 × 2-OH-Bn)-8-(2 × 2-OH-Bn)	$C_{43}H_{36}O_8$	680	2''-OH-5''-Benzylisouvarinol B	*Xylopia africana*	Annonaceae	Root	166

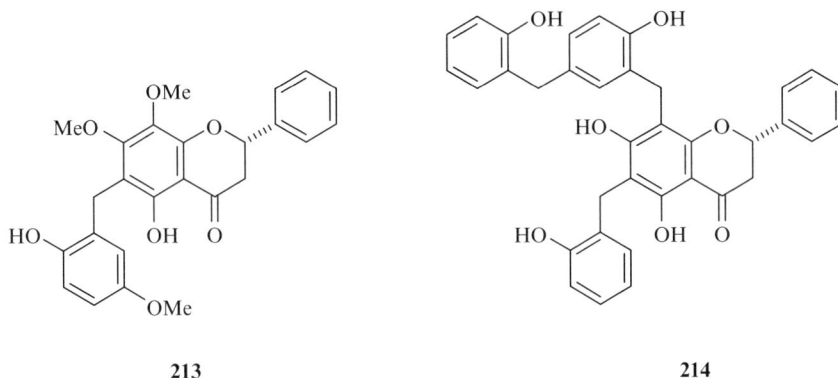

213   **214**

FIGURE 15.6 *C*-Benzyl-substituted flavanones.

known regioisomer, the cytotoxic and antibacterial uvarinol, was also produced by the same plant.[165] In addition, three tetra-(2-hydroxybenzyl)flavanones have been reported from *X. africana*, with the cumbersome names 2′′′′′-hydroxy-3′′′-benzyluvarinol (**215**), in which C-6 bears three 2-hydroxybenzyl units and C-8 one such unit; 2′′′′′-hydroxy-5′′′′-benzylisou-varinol (**216**), with one 2-hydroxybenzyl unit at C-6 and three units at C-8; and 2′′′-hydroxy-5′′-benzylisouvarinol (**217**) with two units at both C-6 and C-8.[166]

15.2.3.2 Flavanone-Stilbenes

The roots of several species of *Sophora* (Leguminosae) produce flavanones that have a stilbene, usually resveratrol, condensed to one of the phenolic rings. Six such flavanone-stilbenes, alopecurones A–F, have been reported from *S. alopecuroides* (**218–223**, Table 15.7).[114] In these compounds, resveratrol is condensed to the A-ring of the flavanone (Figure 15.7). In alopecurones A, B, D, and E, the A-ring additionally bears a 8-*C*-lavandulyl substituent, and the B-ring is 2′,4′-dioxygenated, so that they have a similar substitution in this respect to alopecurone G (**109**), a compound discussed previously (see Table 15.3). Alopecurones C and F are characterized by an 8-*C*-prenyl side chain.[114] In another new flavonostilbene, leachianone I (**224**, Figure 15.7) from *S. leachiana*, the flavanone part is also 5,7,2′,4′-tetraoxygenated, but it bears a hydroxylated prenyl group at C-8 and the resveratrol unit is condensed to the B-ring of the flavanone.[167]

15.2.3.3 Anastatins

Flavanones with a novel carbon skeleton, anastatins A (**225**) and B (**226**, Table 15.7), have been isolated from *Anastatica hierochuntica* (Brassicaceae), an Egyptian medicinal herb used to treat fatigue and uterine haemorrhage.[168] In these flavanones, a 3,4-dihydroxyphenyl group can be considered to form a benzofuran ring with the 7-hydroxyl of naringenin. In anastatin A, C-7 and C-6 of naringenin are part of the furan ring, whereas in anastatin B the furan ring involves C-7 and C-8 (Figure 15.8). The hepatoprotective effects of anastatins A and B on D-galactosamine-induced cytotoxicity in primary cultured mouse hepatocytes have been determined. The activities found were compared with those of the common flavanones and dihydroflavonols, naringenin, eriodictyol, aromadendrin, and taxifolin, from the same species, and with a commercial sample of the flavanone mixture silybin, which is a known hepatoprotective agent. The hepatoprotective activities of anastatins A and B appeared to be stronger than those of all the other compounds tested, including silybin.[168]

TABLE 15.7
Complex Flavanones Reported from 1992 to 2003

No.	Flavanone	Formula	M_r	Plant Source	Family	Organ[a]	Ref.
	Flavanone-stilbenes (Figure 15.7)						
218	Alopecurone A	$C_{39}H_{38}O_9$	650	*Sophora alopecuroides*	Leguminosae	Root	114
219	Alopecurone B	$C_{39}H_{38}O_9$	650	*Sophora alopecuroides*	Leguminosae	Root	114
220	Alopecurone C	$C_{34}H_{30}O_8$	566	*Sophora alopecuroides*	Leguminosae	Root	114
221	Alopecurone D	$C_{40}H_{40}O_9$	664	*Sophora alopecuroides*	Leguminosae	Root	114
222	Alopecurone E	$C_{40}H_{40}O_9$	664	*Sophora alopecuroides*	Leguminosae	Root	114
223	Alopecurone F	$C_{34}H_{30}O_9$	582	*Sophora alopecuroides*	Leguminosae	Root	114
224	Leachianone I	$C_{34}H_{30}O_{10}$	598	*Sophora leachiana*	Leguminosae	Root	167
	Anastatins (Figure 15.8)						
225	Anastatin A	$C_{21}H_{14}O_7$	378	*Anastatica hierochuntica*	Brassicaceae	Whole	168
226	Anastatin B	$C_{21}H_{14}O_7$	378	*Anastatica hierochuntica*	Brassicaceae	Whole	168
	Complex Myrtaceae flavanones (Figure 15.9)						
227	Baeckea flavanone	$C_{30}H_{34}O_7$	506	*Baeckea frutescens*	Myrtaceae	Aerial	169
228	BF-4	$C_{30}H_{32}O_6$	488	*Baeckea frutescens*	Myrtaceae	Leaf	170
229	BF-5	$C_{30}H_{32}O_6$	488	*Baeckea frutescens*	Myrtaceae	Leaf	170
230	BF-6	$C_{33}H_{30}O_6$	522	*Baeckea frutescens*	Myrtaceae	Leaf	170
231	Leucadenone A	$C_{33}H_{32}O_7$	540	*Melaleuca leucadendron*	Myrtaceae	Leaf	171
232	Leucadenone B	$C_{33}H_{32}O_7$	540	*Melaleuca leucadendron*	Myrtaceae	Leaf	171
233	Leucadenone C	$C_{33}H_{32}O_7$	540	*Melaleuca leucadendron*	Myrtaceae	Leaf	171
234	Leucadenone D	$C_{33}H_{32}O_7$	540	*Melaleuca leucadendron*	Myrtaceae	Leaf	171
235	Lumaflavanone A (2*S*)	$C_{30}H_{32}O_6$	488	*Luma checken*	Myrtaceae	Leaf	172
236	Lumaflavanone B (2*R*)	$C_{30}H_{32}O_6$	488	*Luma checken*	Myrtaceae	Leaf	172
237	Lumaflavanone C	$C_{30}H_{34}O_7$	506	*Luma checken*	Myrtaceae	Leaf	172
	Calomelanols (Figure 15.10)						
238	Calomelanol G	$C_{25}H_{20}O_7$	432	*Pityrogramma calomelanos*	Adiantaceae, Pteridophyta	Frond	173
239	Calomelanol H	$C_{24}H_{18}O_6$	402	*Pityrogramma calomelanos*	Adiantaceae, Pteridophyta	Frond	173
240	Calomelanol I	$C_{24}H_{18}O_6$	402	*Pityrogramma calomelanos*	Adiantaceae, Pteridophyta	Frond	173
241	Calomelanol J	$C_{24}H_{18}O_5$	386	*Pityrogramma calomelanos*	Adiantaceae, Pteridophyta	Frond	173
	Diarylheptanoids (Figure 15.11)						
242	Calyxin C	$C_{35}H_{34}O_8$	582	*Alpinia blepharocalyx*	Zingiberaceae	Seed	174
243	Calyxin D	$C_{35}H_{34}O_8$	582	*Alpinia blepharocalyx*	Zingiberaceae	Seed	174
244	Calyxin G	$C_{35}H_{34}O_8$	582	*Alpinia blepharocalyx*	Zingiberaceae	Seed	174
245	Calyxin J	$C_{42}H_{38}O_9$	686	*Alpinia blepharocalyx*	Zingiberaceae	Seed	174
246	Calyxin K	$C_{35}H_{34}O_8$	582	*Alpinia blepharocalyx*	Zingiberaceae	Seed	174
247	Calyxin M	$C_{35}H_{34}O_8$	582	*Alpinia blepharocalyx*	Zingiberaceae	Seed	174
248	Epicalyxin C	$C_{35}H_{34}O_8$	582	*Alpinia blepharocalyx*	Zingiberaceae	Seed	174
249	Epicalyxin D	$C_{35}H_{34}O_8$	582	*Alpinia blepharocalyx*	Zingiberaceae	Seed	174
250	Epicalyxin G	$C_{35}H_{34}O_8$	582	*Alpinia blepharocalyx*	Zingiberaceae	Seed	174
251	Epicalyxin J	$C_{42}H_{38}O_9$	686	*Alpinia blepharocalyx*	Zingiberaceae	Seed	174
252	Epicalyxin K	$C_{35}H_{34}O_8$	582	*Alpinia blepharocalyx*	Zingiberaceae	Seed	174
253	Epicalyxin M	$C_{35}H_{34}O_8$	582	*Alpinia blepharocalyx*	Zingiberaceae	Seed	174

TABLE 15.7
Complex Flavanones Reported from 1992 to 2003 — *continued*

No.	Flavanone	Formula	M_r	Plant Source	Family	Organ[a]	Ref.
	Miscellaneous (Figure 15.12)						
254	Kurziflavolactone A (2*R*)	$C_{32}H_{30}O_7$	526	*Cryptocarya kurzii*	Lauraceae	Leaf	175
255	Kurziflavolactone B (2*S*)	$C_{32}H_{30}O_7$	526	*Cryptocarya kurzii*	Lauraceae	Leaf	175
256	Kurziflavolactone C (2*S*)	$C_{32}H_{30}O_7$	526	*Cryptocarya kurzii*	Lauraceae	Leaf	175
257	Kurziflavolactone D (2*R*)	$C_{32}H_{30}O_7$	526	*Cryptocarya kurzii*	Lauraceae	Leaf	175
258	Tephrorin A	$C_{24}H_{26}O_7$	426	*Tephrosia purpurea*	Leguminosae		176
259	Tephrorin B	$C_{30}H_{28}O_6$	484	*Tephrosia purpurea*	Leguminosae		176

[a]RootB, root bark; Aerial, aerial parts; Whole, whole plant.

15.2.3.4 Complex Myrtaceae Flavanones

A series of very unusual flavanones has been detected in the genera *Baeckea*, *Luma*, and *Melaleuca*, all belonging to the family Myrtaceae (Table 15.7).[169–172] The compounds are based on 6-*C*-methylpinocembrin with an unusual substituent at C-8, including a methylated phloroglucinol-based ring structure fused by a 6-carbon heterocyclic ring (which bears either

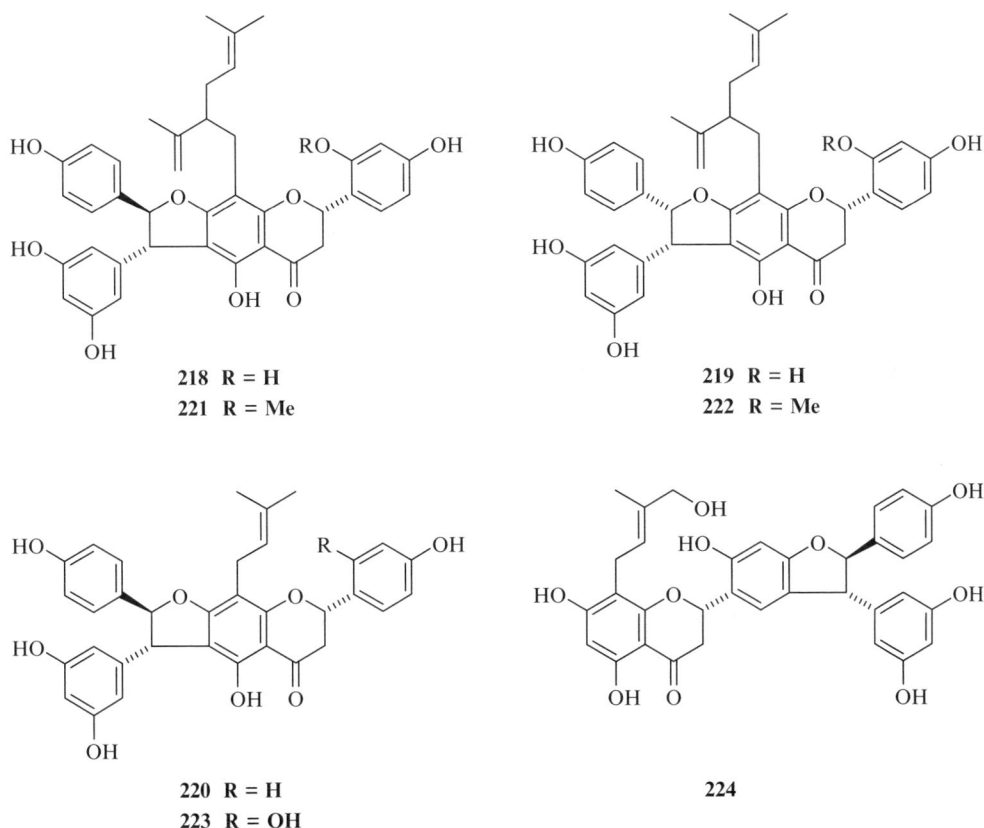

218 R = H
221 R = Me

219 R = H
222 R = Me

220 R = H
223 R = OH

224

FIGURE 15.7 Flavanone-stilbenes.

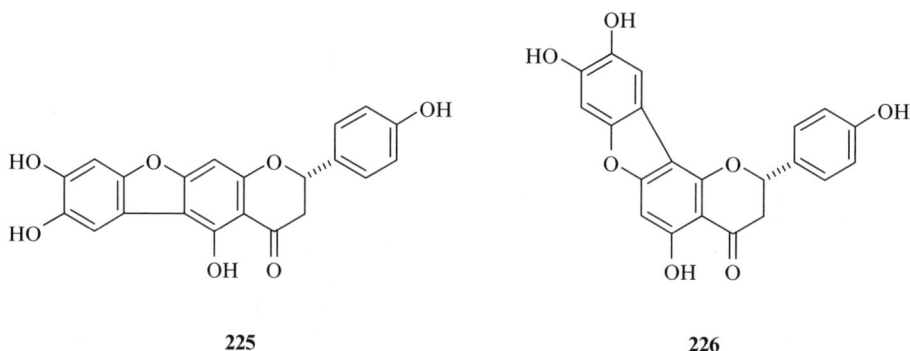

225   226

FIGURE 15.8 Anastatins.

an isopropyl or a phenyl group) to the A-ring of the flavanone (Figure 15.9). Four such compounds have been isolated from the leaves of *Baeckea frutescens*, a southeastern-Asian shrub used for treating rheumatism, fever, and snake bites. In the flavanone **227**, the phloroglucinol-like ring is connected via an isobutyl chain to C-8 of the flavanone.[169] Two more flavanones, BF-4 (**228**) and BF-5 (**229**), are C-2 epimers and contain an isopropyl side chain, whereas a third, BF-6 (**230**), is slightly different since it bears a phenyl side group and has two ketone functions in the phloroglucinol-like ring.[170] BF-4 and BF-5 showed strong cytotoxic activity against leukemia cells (L1210) in tissue culture (IC$_{50}$ = 0.2 to 0.5 µg/ml).[170] Structures very similar to BF-6 (bearing two ketones and a phenyl rather than an isopropyl side group) were obtained from the leaves of *Melaleuca leucadendron*. There are four epimeric centers in the molecule, so that many different stereoisomers are possible. The four obtained so far are known as leucadenones A–D (**231–234**).[171] For the absolute configurations of these compounds, see Figure 15.9.

Bioassay-guided fractionation of an extract of *Luma checken* leaves led to the isolation of three further related compounds, lumaflavanones A (**235**), B (**236**), and C (**237**),[172] which are structurally similar to both the *Baeckea* and *Melaleuca* flavanones (Figure 15.9). The luma-flavanones were active in the brine shrimp test, they showed insect-antifeedant activity against *Spodoptera littoralis*, and exhibited antifungal activity against *Botrytis cinerea*.[172] All the active flavanones of *Baeckea* and *Luma* bear an isopropyl rather than phenyl side group, so that the isopropyl chain may be important for these bioactivities. These unusual flavanones may also have chemosystematic importance, as they are all found in genera from the same family.

15.2.3.5 Calomelanols

The farinose exudate of the frond of the fern *Pityrogramma calomelanos* (Adiantaceae) has been the source of complex flavonoids characterized by a novel C$_6$–C$_3$–C$_6$–C$_3$–C$_6$ skeleton (Table 15.7).[173] This group includes four flavanones, calomelanols G (**238**), H (**239**), I (**240**), and J (**241**) (Figure 15.10). In these compounds, a molecule of *p*-coumaric or cinnamic acid appears to be fused with the A-ring of the flavanone. Biosynthetic pathways for these complex flavanones and related flavones, chalcones, and dihydrochalcones in *P. calomelanos* and other *Pityrogramma* species have been proposed by the authors.[173]

15.2.3.6 Diarylheptanoid Flavanones

Diarylheptanoids are characteristic phenolics found in the family Zingiberaceae, e.g., curcumin, the bioactive yellow pigment from the spice turmeric (roots of *Curcuma longa*) and similar compounds from ginger (*Zingiber officinalis*). A series of novel diarylheptanoids

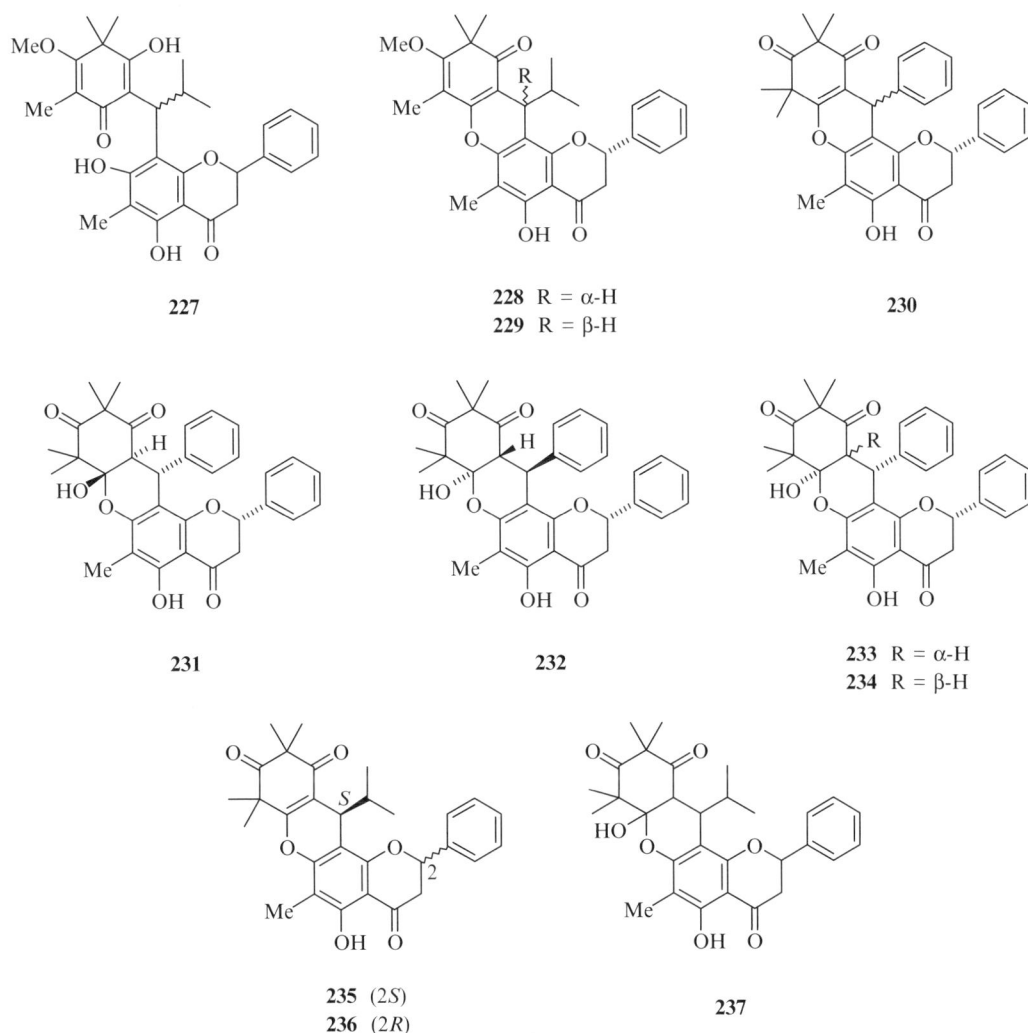

227

228 R = α-H
229 R = β-H

230

231

232

233 R = α-H
234 R = β-H

235 (2*S*)
236 (2*R*)

237

FIGURE 15.9 Complex Myrtaceae flavanones.

conjugated with chalcones and flavanones has been discovered in the seeds of another Zingiberaceae species, *Alpinia blepharocalyx*, which is used in Chinese traditional medicine for the treatment of stomach disorders (Table 15.7).[174] The series includes 12 diarylheptanoid flavanones, calyxins C, D, G, J, K, and M (**242–247**) and the corresponding epicalyxins (**248–253**). In all these compounds the flavonoid moiety is 5-*O*-methylnaringenin, with C-5 or C-7 of the diarylheptanoid attached to C-8 of the flavanone via a C–C linkage (Figure 15.11).[174] The calyxins and epicalyxins differ from each other in their configuration of the 2-position of the flavanone, but the absolute configurations in this position have not been determined. The compounds can be arranged into four groups according to structural features of the diarylheptanoid moiety. In calyxins and epicalyxins C and D, the heptanoid chain is acyclic and the flavanone is attached to C-7 of the diarylheptanoid (Figure 15.11). This C-7 atom is an asymmetric center, and calyxins C (**242**) and D (**243**) are thus 7-epimers. Epicalyxins C (**248**) and D (**249**) are also 7-epimers, and so are the pairs calyxins G (**244**) and K (**246**), and epicalyxins G (**250**) and K (**252**). However, in calyxins and epicalyxins

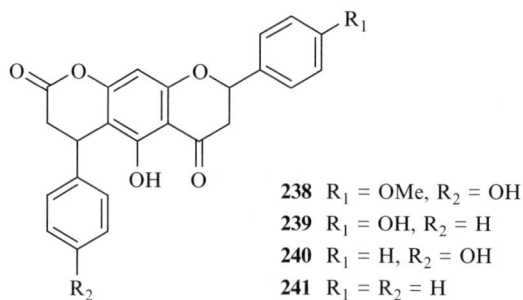

FIGURE 15.10 Calomelanols.

238 R$_1$ = OMe, R$_2$ = OH
239 R$_1$ = OH, R$_2$ = H
240 R$_1$ = H, R$_2$ = OH
241 R$_1$ = R$_2$ = H

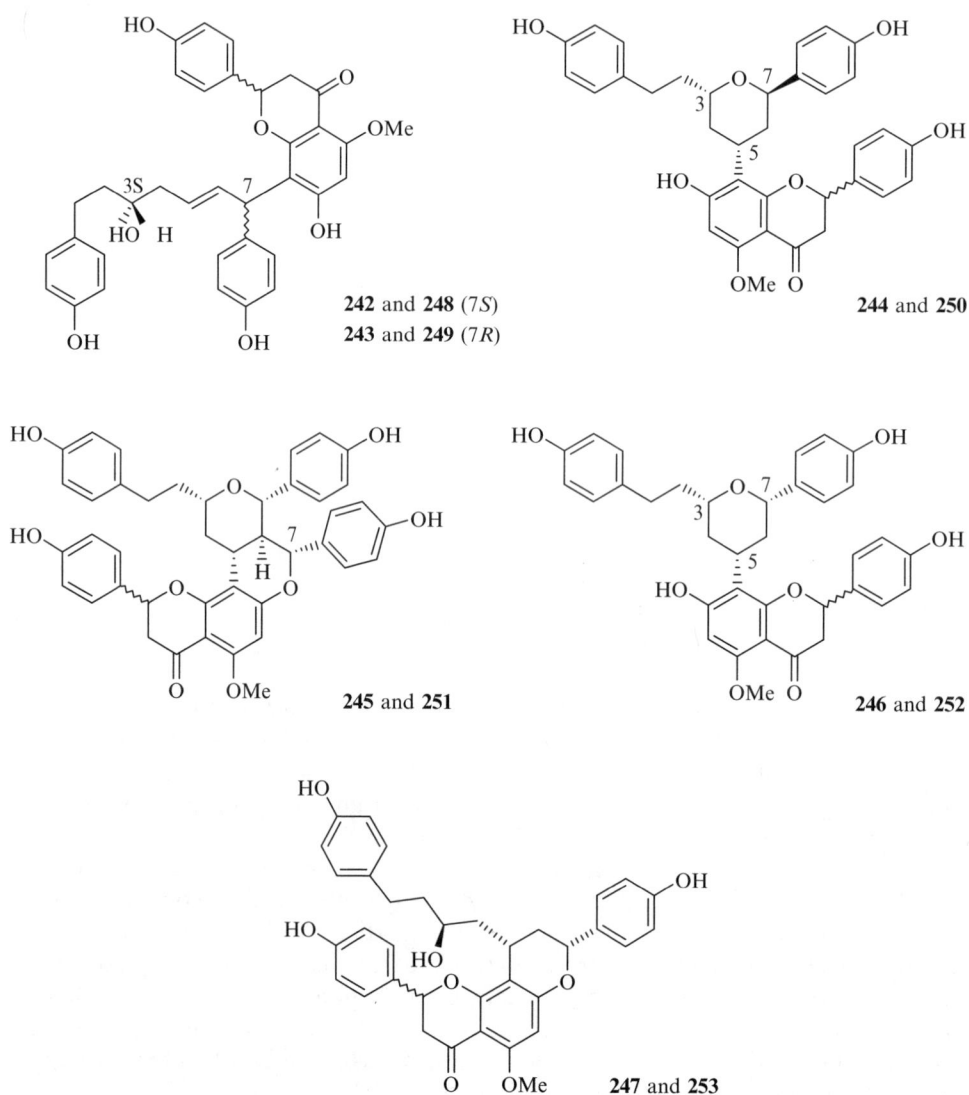

242 and 248 (7S)
243 and 249 (7R)

244 and 250

245 and 251

246 and 252

247 and 253

FIGURE 15.11 Diarylheptanoid flavanones.

G and K the heptanoid chain has formed a ring and the flavanone is attached to C-5 of the diarylheptanoid. Calyxin J (**245**) and epicalyxin J (**251**) also contain a cyclic heptanoid, which forms a second tetrahydropyrano ring with the 7-hydroxyl group of the flavanone, and in addition they contain a *p*-hydroxybenzyl group (Figure 15.11). Finally, in calyxin M (**247**) and epicalyxin M (**253**), C-5 and C-7 of the heptanoid chain are attached to C-8 and the 7-hydroxyl of the flavanone, respectively, to form a tetrahydropyrano ring. In a range of bioactivity tests, it was found that the epimeric mixture of calyxin and epicalyxin J (**245** and **251**) exhibited a strong antiproliferative activity against human HT-1080 fibrosarcoma (IC$_{50}$ 0.3 μM) and had some activity against murine colon 26-L5 carcinoma (IC$_{50}$ 13.7). The other flavanone-diaryl heptanoids were much less active in these tests[174] (see also Section 16.2.6).

15.2.3.7 Miscellaneous Complex Flavanones

Flavonoids containing an unusual 17-carbon substituent fused to the A-ring have been found in *Cryptocarya kurzii* (Lauraceae), including four flavanones, kurziflavolactones A–D (**254**–**257**, Table 15.7).[175] The substituent consists of a phenylpropenyl part and an aliphatic part including a tetrahydropyran ring. In kurziflavanones A (**254**) and B (**255**), which are C-2 epimers, the side chain forms a ring with the 7-hydroxyl and C-8 of the flavanone to form an additional ring. In the C-2 epimers **256** and **257**, the side chain is attached to the 7-hydroxyl and C-6 of the flavanone (Figure 15.12). Kurziflavanone B showed slight cytotoxicity against KB cells, with an IC$_{50}$ value of 4 μg/ml.[175]

Tephrosia purpurea (Leguminosae) is the source of two flavanones bearing a novel tetrahydrofurano ring as a side chain at the C-8 position, (+)-tephrorins A (**258**) and B (**259**).[176] The compounds differ from each other in the substitution of the flavanone A-ring and that of

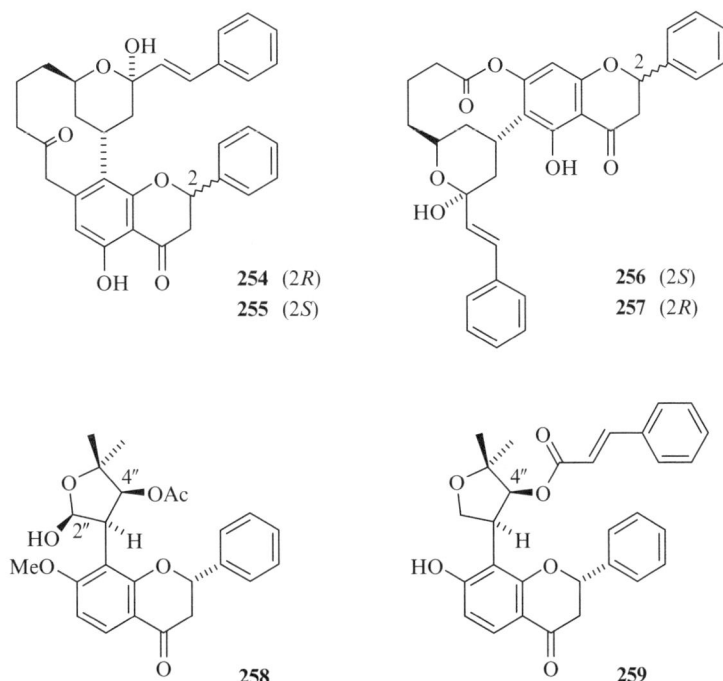

FIGURE 15.12 Kurziflavolactones (**254–257**) and tephrorins (**258** and **259**).

C-2″ and C-4″ of the side group; **258** bears a methoxy group at C-7, a hydroxyl at C-2″ and the C-4″ hydroxyl is acylated with acetic acid, whereas **259** bears a hydroxyl at C-7, C-2″ is not hydroxylated and the C-4″ hydroxyl is acylated with (E)-cinnamic acid (Figure 15.12). The two flavanones were evaluated for their potential as quinone reductase inducers in cultured mouse Hepa 1c1c7 cells. Tephrorin A significantly reduced quinone reductase activity, whereas tephrorin B was inactive. It was thought that the presence of the bulky cinnamic acid group at C-4″ may affect its biological activity.[176]

15.2.4 FLAVANONE GLYCOSIDES

Table 15.8 presents a list of 69 flavanone glycosides reported between 1992 and 2003 (compounds **260–328**).[137,145,177–228] The new glycosides are based on 35 different aglycones, most of which are "simple flavanones" bearing only hydroxy and methoxy substituents, and these are arranged according to their O-substitution pattern in the same manner as in Table 15.1. Four of the aglycones are C-methylated and four are prenylated. In Table 15.8, glycosides of the same aglycone are grouped together, and glycosides based on C-methyl and prenylated aglycones follow those with only O-substituents. The linkages between aglycones and sugars, and between two sugars or a sugar and an acyl group are C–O–C in all flavanones, the only exception being a C–C linkage in (2R)- and (2S)-eriodictyol 6-C-glucosides (**298** and **299**) from rooibos tea (prepared from *Aspalathus linearis*, Leguminosae), but these are artifacts and do not occur in unprocessed plant material. Rooibos leaves are processed, like those of *Camellia sinensis* tea, by a fermentation and drying process, which may cause the natural dihydrochalcone C-glycosides present in the plant to be converted into flavanone C-glycosides.[209]

In approximately one third of the new glycosides only one sugar is attached to the flavanone (monosides) and in a further third two sugars are attached as a disaccharide to one hydroxyl of the flavanone, e.g., homoesperetin 7-rhamnosyl(1 → 6)glucoside (**311**) from *Vernonia diffusa* (Asteraceae).[217] The correct term for this type of glycoside is bioside and not diglycoside, because there is only one glycosidic linkage with the flavonoid.[5] Five compounds are diglycosides, i.e., that they have one or more sugars attached to two different hydroxyls (two glycosidic linkages with the flavanone),[5] e.g., carthamidin 6,7-diglucoside (**293**) from *Carthamus tinctorius* (Asteraceae).[204] There are six triosides (a combination of three sugars attached to one hydroxyl),[5] and in two of those compounds the sugar chains are branched. These are the 7-(2,6-dirhamnosyl)glucosides of naringenin (**285**) and hesperetin (**307**). Both compounds were obtained from the fruits of *Citrus junos* (Rutaceae)[196] and **308** was first described from leaves of *Buddleja madagascariensis* (Buddlejaceae).[215] The hesperetin glycoside inhibits the influenza A virus.[196]

Glucose is by far the most common sugar in the new flavanone glycosides, either as a monoside or as one or more of the sugars in the biosides, triosides, diglycosides, or acylated glycosides; glucose is lacking in only eight of the compounds. The second most common sugar in the newly reported glycosides is apiose, which was found in 15 different glycosides. This is a little surprising, since apiose is not considered to be a common sugar in flavonoid glycosides in general. However, most of the new apiose-containing glycosides have been reported from only a few genera, especially *Glycyrrhiza*, in which nine such compounds were found, e.g., the diglycoside liquiritigenin 7-apiofuranoside-4′-glucoside (**265**) from *Glycyrrhiza inflata* (Leguminosae).[184] Rhamnose is the third most common sugar (in 14 glycosides), glucuronic acid is present in three of the glycosides, and arabinose and xylose in two. Galactose is only present in alhagidin from *Alhagi pseudalhagi* (Leguminosae), which is hesperetin 7-galactosyl(1 → 2)[rhamnosyl(1 → 6)]glucoside (**309**)[199] and the unusual sugar fucose has been reported to be present in isosakuranetin 7-fucopyranosyl(1 → 6)glucoside (longitin, **286**)

TABLE 15.8
New Flavanone Glycosides Reported from 1992 to 2003

No.	Flavanone Glycoside	Formula	M_r	Trivial Name	Plant Source	Family	Organ[a]	Ref.
	7-OH-Flavanone							
260	7-Glucoside	$C_{21}H_{22}O_8$	402		Clerodendrum phlomides	Lamiaceae	Leaf	177
					Cochlospermum regium	Bixaceae	Leaf	178
	5,7-DiOH-Flavanone (pinocembrin)							
261	7-Glucoside (2S)	$C_{21}H_{22}O_9$	418	Pinocembroside	*Glycyrrhiza glabra*	Leguminosae	Aerial	179
					Penthorum chinense	Saxifragaceae		180
262	7-Rhamnosylglucoside	$C_{27}H_{32}O_{13}$	564		*Onychium japonicum*	Adianthaceae, Pteridophyta	Aerial	181
263	7-Apiosyl(1 → 5)apiosyl(1 → 2)glucoside	$C_{31}H_{38}O_{17}$	682		*Viscum angulatum*	Viscaceae	Whole	182
	5-OH-7-OMe-Flavanone (pinostrobin)							
264	5-Glucoside	$C_{22}H_{24}O_9$	432		*Pyracantha coccinea*	Rosaceae	Root	183
	7,4'-diOH-Flavanone (liquiritigenin)							
265	7-Apiofuranoside-4'-glucoside	$C_{26}H_{30}O_{13}$	550		*Glycyrrhiza inflata*	Leguminosae	Root	184
266	7-(3-Acetylapioside)-4'-glucoside	$C_{28}H_{32}O_{14}$	592		*Glycyrrhiza inflata*	Leguminosae	Root	184
267	4'-[3-Acetylapiosyl(1 → 2glucoside]	$C_{28}H_{32}O_{14}$	592		*Glycyrrhiza uralensis*	Leguminosae	Root	185
268	7-Glucoside-4-apiosyl(1 → 2)glucoside	$C_{32}H_{40}O_{18}$	712	Glucoliquiritin apioside	*Glycyrrhiza glabra*	Leguminosae	Root	145
269	Licorice glycoside E	$C_{35}H_{35}O_{14}N$	693	Figure. 15.13	*Glycyrrhiza uralensis*	Leguminosae	Root	186
270	4'-[4-*p*-Coumaroylapiosyl(1 → 2) glucoside] (2R)	$C_{35}H_{36}O_{15}$	696		*Glycyrrhiza uralensis*	Leguminosae	Root	186
271	4'-[4-*p*-Coumaroylapiosyl(1 → 2) glucoside] (2S)	$C_{35}H_{36}O_{15}$	696		*Glycyrrhiza uralensis*	Leguminosae	Root	186
272	4'-[4-Feruloylapiosyl(1 → 2)glucoside] (2R)	$C_{36}H_{38}O_{16}$	726		*Glycyrrhiza uralensis*	Leguminosae	Root	186
273	4'-[4-Feruloylapiosyl(1 → 2)glucoside] (2S)	$C_{36}H_{38}O_{16}$	726		*Glycyrrhiza uralensis*	Leguminosae	Root	186
	5,6,7-TriOH-Flavanone (dihydrobaicalein)							
274	7-Glucoside (2S)	$C_{21}H_{22}O_{10}$	434		*Cephalocereus senilis*	Cactaceae	Cells[b]	187
	5,7,8-TriOH-Flavanone (dihydronorwogonin)							
275	5-Glucoside	$C_{21}H_{22}O_{10}$	434		*Pyracantha coccinea*	Rosaceae	Root	183
	5,7,2'-TriOH-Flavanone							
276	7-Glucoside (2S)	$C_{21}H_{22}O_{10}$	434		*Scutellaria ramosissima*	Lamiaceae	Aerial	188
277	7-O-(Methyl-β-D-glucuronate) (2S)	$C_{22}H_{22}O_{11}$	462		*Scutellaria ramosissima*	Lamiaceae	Aerial	189

continued

TABLE 15.8
New Flavanone Glycosides Reported from 1992 to 2003 — _continued_

No.	Flavanone Glycoside	Formula	M_r	Trivial Name	Plant Source	Family	Organ[a]	Ref.
278	7-O-(Ethyl-β-D-glucuronate) (2S) _5,7,4'-TriOH-Flavanone (naringenin)_	$C_{23}H_{24}O_{11}$	476		_Scutellaria ramosissima_	Lamiaceae	Aerial	189
279	4'-Rhamnoside	$C_{21}H_{22}O_9$	418		_Crotalaria striata_	Leguminosae	Stem	190
280	7-(6-Acetylglucoside)	$C_{23}H_{24}O_{11}$	476		_Acacia saligna_	Leguminosae	Flower	191
281	7-(2-_p_-Coumaroylglucoside)	$C_{30}H_{28}O_{12}$	580		_Ricinus communis_	Euphorbiaceae	Seed	192
282	7-[3-Acetyl-6-_p_-coumaroylglucoside]	$C_{32}H_{30}O_{13}$	622		_Blepharis ciliaris_	Acanthaceae	Aerial	193
283	7-(4,6-Digalloylglucoside)	$C_{35}H_{30}O_{18}$	738		_Acacia farnesiana_	Leguminosae	Fruit	194
284	7-Rhamnosyl(1 → 2)(4-O-methyl-β-D-glucoside)	$C_{28}H_{34}O_{14}$	594	Fumotonaringenin	_Microlepia marginata_	Dennstaedtiaceae, Pteridophytae	Fruit	195
285	7-(2,6-Dirhamnosylglucoside) _5,7-DiOH-4'-OMe-Flavanone (isosakuranetin)_	$C_{33}H_{42}O_{18}$	726		_Citrus junos_	Rutaceae	Fruit	196
286	7-Fucopyranosyl(1 → 6)glucoside	$C_{27}H_{32}O_{14}$	580	Longitin	_Mentha longifolia_	Lamiaceae	Aerial	197
287	7-α-L-Arabinofuranosyl(1 → 6)glucoside _7,4'-DiOH-5-OMe-Flavanone (5-O-methylnaringenin)_	$C_{27}H_{32}O_{14}$	580		_Punica granatum_	Punicaceae	StemB	198
288	4'-Glucoside	$C_{22}H_{24}O_{10}$	448	Alhagitin	_Alhagi pseudalhagi_	Leguminosae	Whole	199
289	7-Neohesperidoside-4'-glucoside _5-OH-7,8-diOMe_	$C_{34}H_{44}O_{19}$	756		_Clerodendrum phlomoides_	Lamiaceae	Root	200
290	5-Rhamnoside _4'-OH-5,7-diOMe-Flavanone_	$C_{23}H_{26}O_9$	446		_Albizzia procera_	Leguminosae	Stem	201
291	4'-[2-(5-Cinnamoyl)-β-D-apiofuranosyl]glucoside _2,5,7,4'-TetraOH-Flavanone (2-hydroxynaringenin)_	$C_{37}H_{40}O_{15}$	724		_Viscum album_ ssp. _album_	Viscaceae		202
292	7-Glucoside _5,6,7,4'-TetraOH-Flavanone (carthamidin)_	$C_{21}H_{22}O_{11}$	450		_Chaenomeles sinensis_	Rosaceae	Fruit	203
293	6,7-Diglucoside _5,7,8,4'-TetraOH-Flavanone (isocarthamidin)_	$C_{27}H_{32}O_{16}$	612		_Carthamus tinctorius_	Asteraceae		204
294	7-Rhamnoside	$C_{21}H_{22}O_{10}$	434		_Spartium junceum_	Leguminosae	Aerial	205
295	8-Glucoside _5,7,2',4'-TetraOH-Flavanone (steppogenin)_	$C_{21}H_{22}O_{11}$	450	3-Desoxycallunin	_Calluna vulgaris_	Ericaceae	Flower	206
296	4'-Glucoside _5,7,2',5'-TetraOH-Flavanone_	$C_{21}H_{22}O_{11}$	450		_Maclura tinctoria_	Moraceae	StemB	207
297	7-Glucoside	$C_{21}H_{22}O_{11}$	450	Coccinoside B	_Pyracantha coccinea_	Rosaceae	Leaf	208

No.	Compound	Trivial name	Formula	MW	Species	Family	Part	Ref
	5,7,3',4'-TetraOH-Flavanone (eriodictyol)							
298	6-C-Glucoside (2R)		$C_{21}H_{22}O_{11}$	450	*Aspalathus linearis*	Leguminosae	Leaf, stem[c]	209
299	6-C-Glucoside (2S)		$C_{21}H_{22}O_{11}$	450	*Aspalathus linearis*	Leguminosae	Leaf, stem[c]	209
300	7-Glucuronide (2S)		$C_{21}H_{20}O_{12}$	464	*Chrysanthemum indicum*	Asteraceae	Flower	210
301	7-(6-Acetylglucoside)		$C_{23}H_{24}O_{12}$	492	*Pyracantha coccinea*	Rosaceae	Leaf	208
302	7-α-L-Arabinofuranosyl(1 → 6)glucoside	Coccinoside A	$C_{26}H_{30}O_{15}$	582	*Punica granatum*	Punicaceae	StemB	198
303	7-(6-p-Coumaroylglucoside) (2S)		$C_{30}H_{28}O_{13}$	596	*Phyllanthus emblica*	Euphorbiaceae	Aerial	211
304	3'-(6-p-Coumaroylglucoside)		$C_{30}H_{28}O_{13}$	596	*Malus × domestica*	Rosaceae	Leaf[d]	212
305	7-(6-Galloylglucoside) (2S)		$C_{28}H_{26}O_{15}$	602	*Phyllanthus emblica*	Euphorbiaceae	Aerial	211
	7,8,3',4'-TetraOH-Flavanone (iso-okanin)							
306	7-(2,4,6-Triacetylglucoside)		$C_{27}H_{28}O_{14}$	576	*Bidens pilosa*	Asteraceae	Aerial	213
	5,6,7-TriOH-4'-OMe-Flavanone (4'-methylcarthamidin)							
307	7-(2-p-Coumaroylglucoside)		$C_{31}H_{30}O_{13}$	610	*Crotalaria prostrata*	Leguminosae	Leaf	214
	5,7,3'-TriOH-4'-OMe-Flavanone (hesperetin)							
308	7-(2,6-Dirhamnosylglucoside)		$C_{34}H_{44}O_{19}$	756	*Buddleja madagascariensis*	Buddlejaceae	Leaf	215
308	7-(2,6-Dirhamnosylglucoside)		$C_{34}H_{44}O_{19}$	756	*Citrus junos*	Rutaceae	Fruit	196
309	7-Galactosyl(1 → 2][rhamnosyl(1 → 6)] glucoside	Alhagidin	$C_{34}H_{44}O_{20}$	810	*Alhagi pseudalhagi*	Leguminosae	Whole	199
	5,7,4'-TriOH-3'-OMe-Flavanone							
310	7-β-L-Rhamnosyl(1 → 6)glucoside		$C_{28}H_{34}O_{15}$	610	*Clematis armandii*	Ranunculaceae	Aerial	216
	5,7-DiOH-3',4'-diOMe-Flavanone (homoesperetin)							
311	7-Rhamnosyl(1 → 6)glucoside		$C_{29}H_{36}O_{15}$	624	*Vernonia diffusa*	Asteraceae	Wood	217
	5,3'-DiOH-7,4'-diOMe-Flavanone (persicogenin)							
312	3'-Glucoside		$C_{23}H_{26}O_{11}$	478	*Prunus amygdalus*	Rosaceae	StemB	218
	5,4'-DiOH-7,3'-diOMe-Flavanone							
313	4'-Apiosyl(1 → 2)glucoside		$C_{28}H_{34}O_{15}$	610	*Viscum ahiformosanae*	Viscaceae	Aerial	219
	4-OH-5,7,2'-triOMe-Flavanone							
314	4'-Rhamnosyl(1 → 6)glucoside		$C_{30}H_{38}O_{15}$	638	*Terminalia alata*	Combretaceae	Root	220
	5,7,4'-TriOH-3',5'-OMe-Flavanone							
315	5-Glucoside (2R)	Peruvianoside I	$C_{23}H_{26}O_{12}$	494	*Thevetia peruviana*	Apocynaceae	Leaf	221
316	5-Glucoside (2S)	Peruvianoside II	$C_{23}H_{26}O_{12}$	494	*Thevetia peruviana*	Apocynaceae	Leaf	221
	5,7-diOH-6-C-Me-Flavanone (strobopinin)							
317	7-Xylosyl(1 → 3)xyloside		$C_{26}H_{30}O_{12}$	534	*Mosla chinensis*	Lamiaceae	Leaf	222
	5,7-diOH-6,8-di-C-Me-Flavanone							

continued

TABLE 15.8
New Flavanone Glycosides Reported from 1992 to 2003 — continued

No.	Flavanone Glycoside	Formula	M_r	Trivial Name	Plant Source	Family	Organ[a]	Ref.
318	7-[6-(3-OH-3-Methylglutaryl)glucoside]	$C_{29}H_{34}O_{13}$	590	Matteuorienate B (Figure 15.13)	*Matteucia orientalis*	Aspleniaceae, Pteridophyta	Rhiz	223
	5,7,4'-TriOH-6,8-di-C-Me-Flavanone (farrerol)							
319	7-β-D-Apiofuranosyl(1 → 6)glucoside	$C_{28}H_{34}O_{14}$	594	Miconioside B	*Miconia traillii*	Melastomataceae	Leaf, stem	224
	5,7-DiOH-4'-OMe-6,8-di-C-Me-Flavanone (matteucinol)							
320	7-Apiosyl(1 → 6)glucoside	$C_{29}H_{36}H_{14}$	608		*Rhododendron simsii*	Ericaceae	Leaf	225
321	7-α-L-Arabinopyranosyl(1 → 6)glucoside	$C_{29}H_{36}H_{14}$	608	Miconioside A	*Miconia traillii*	Melastomataceae	Aerial	224
322	7-(4,6-(S)-Hexahydroxydiphenylglucoside)	$C_{38}H_{34}O_{18}$	778		*Miconia myriantha*	Melastomataceae	Aerial	226
323	7-(4,6-Digalloylglucoside)	$C_{38}H_{36}O_{18}$	780		*Miconia myriantha*	Melastomataceae	Aerial	226
324	7-[6-(3-OH-3-Methylglutaryl)glucoside]	$C_{30}H_{36}O_{14}$	620	Matteuorienate A (Figure 15.13)	*Matteucia orientalis*	Aspleniaceae, Pteridophyta	Rhiz	223
	5,4'-DiOH-6-C-Prenylflavanone							
325	4'-Xylosyl(1 → 2)rhamnoside	$C_{31}H_{38}O_{12}$	602		*Gliricidia maculata*	Leguminosae	Seed	227
	5,7,4'-TriOH-8-C-Prenylflavanone							
326	7-Glucoside	$C_{26}H_{30}O_{10}$	502	Phellodensin F	*Phellodendron chinense var. glabriusculum*	Rutaceae	Leaf	137
	5,7,4'-TriOH-8-C-(3-Hydroxymethyl-2-butenyl)flavanone							
327	7-Glucoside	$C_{26}H_{30}O_{11}$	518	Phellodensin E	*Phellodendron chinense var. glabriusculum*	Rutaceae	Leaf	137
	Sophora flavanone I							
328	7-Glucoside	$C_{45}H_{48}O_{14}$	812	Figure. 15.13	*Sophora stenophylla*	Leguminosae	Root	228

Notes: If the configuration and form of the sugars has not been specified, rhamnoside, α-L-rhamnopyranoside; apioside, β-apiofuranoside; glucoside, β-D-glucopyranoside; galactoside, β-D-galactopyranoside; xyloside, β-D-xylopyranoside; glucuronide, β-D-glucuronopyranoside. Cinnamic acids are assumed in the *E*-form.
[a] Aerial, aerial parts; Whole, whole plant; StemB, stem bark; Rhiz, rhizome.
[b] Chitin-treated cell suspension cultures.
[c] Fermented leaves and stem.
[d] Leaves treated with prohexadione-Ca.

from *Mentha longifolia* (Lamiaceae).[197] However, the identification of fucose was not fully supported by a complete set of ^{1}H and ^{13}C resonance assignments for the sugar in the NMR spectra of this compound and should therefore be regarded as tentative.

More than 20 of the new glycosides are acylated, *p*-coumaroyl and acetyl being the most common acyl groups. Diacylation is present in naringenin 7-[3-acetyl-6-(*E*)-*p*-coumaroylglucoside] (**282**) from *Blepharis ciliaris* (Acanthaceae),[193] and in the 7-(4,6-digalloylglucosides) of naringenin (**283**) from *Acacia farnesiana* (Leguminosae)[195] and matteucinol (**323**) from *Miconia myriantha* (Melastomataceae).[226] Another glycoside from *M. myriantha* (**322**) is very similar to **323**, but here the two gallic acid molecules are *C–C* linked to form one acyl group, the two carboxyl groups of which are esterified with the 4- and 6-hydroxyls of the glucose, resulting in matteucinol 7-(4,6-hexahydroxydiphenylglucoside), in which the sugar and acyl groups form a ring.[226] Triacylation is found in only one new glycoside, iso-okanin 7-(2,4,6-triacetylglucoside) (**306**) from *Bidens pilosa* (Asteraceae).[213]

In two thirds of the newly reported flavanone glycosides the sugars are attached to the 7-hydroxyl of the flavanone moiety. Linkage to the 4′-hydroxyl is also very common, however, as this is found in 17 structures listed in Table 15.8. Examples include an interesting group of new flavanone glycosides reported from *Glycyrrhiza uralensis*, which all contain liquiritigenin as the aglycone and apiose and glucose in the sugar moiety.[186] In licorice glycosides C_1 (**272**) and C_2 (**273**), which are epimers at C-2, the apiose is acylated with ferulic acid, whereas in licorice glycosides D_1 (**270**) and D_2 (**271**) (also epimers at C-2), the acyl group is *p*-coumaric acid. Licorice glycoside E (**269**) has the same basic structure, but the hydroxycinnamoyl has been replaced by an indole-2-carboxyl group (Figure 15.13). It is the only nitrogen-containing flavanone glycoside.[186]

There are five new 5-*O*-glycosides, flavanones **264**, **275**, **290**, **315**, and **316**, but only one 8-*O*-glycoside, isocarthamidin 8-glucoside (3-desoxycallunin, **294**) from *Calluna vulgaris* (Ericaceae).[206]

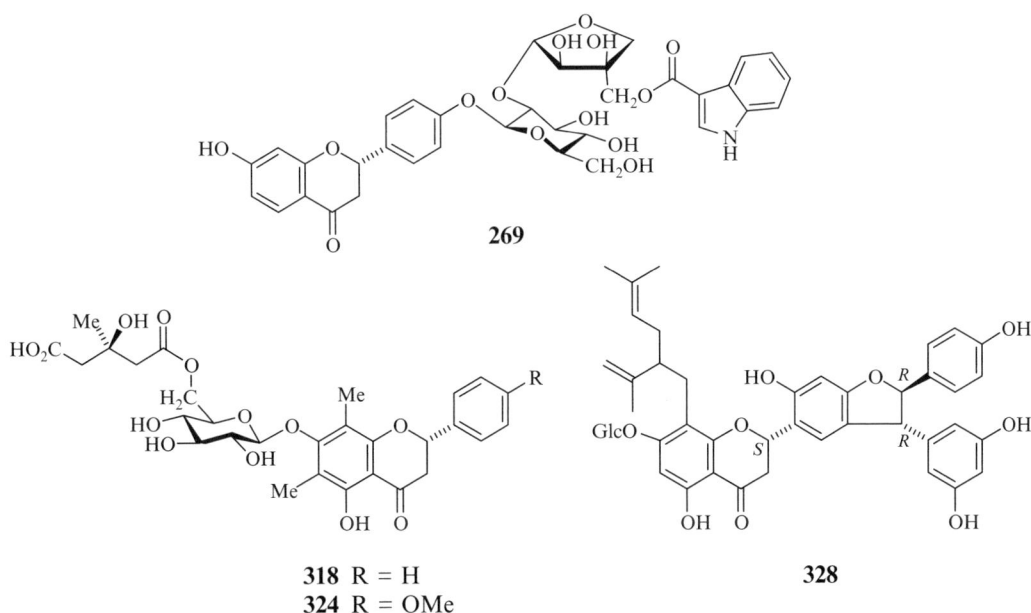

269

318 R = H
324 R = OMe

328

FIGURE 15.13 Examples of flavanone glycosides.

In Table 15.8, the configurations of the sugars are assumed to be β-D for glucose and glucuronic acid, and α-L for rhamnose, but the 7-β-L-rhamnosyl(1 → 6)glucoside of 5,7,4′-trihydroxy-3′-methoxyflavanone (**310**) has been reported from *Clematis armandi* (Ranunculaceae).[216] The β-configuration of the rhamnose is not supported by NMR data, which indicates that the α-rhamnosyl sugar is present as expected.

New glycosides of *C*-methylflavanones have been isolated from species of just four different families, the Aspleniaceae, Ericaceae, Lamiaceae, and Melastomataceae. The most unusual glycosides of these are the acylated matteuorienates A (**324**) and B (**318**) from the fern *Matteucia orientalis* (Aspleniaceae). They are the 3-methylglutaric acid esters of matteucinol and farrerol 7-glucoside, respectively (Figure 15.13). Both matteuorienates have been found to be potent lens aldose reductase inhibitors.[223] Glycosides of prenylated flavanones have been found in species of Leguminosae and Rutaceae, which are well-known sources of isoprenylated flavanones in general. The aglycone of glycoside **328**, sophoraflavanone I, is a flavanonostilbene very similar to the alopecurones discussed in Section 15.2.3.2 (see Figure 15.13).

Some new flavanone glycosides that are interesting from an ecological or bioactivity point of view and have not yet been mentioned include (2*S*)-dihydrobaicalein 7-*O*-glucoside (**274**), a compound induced by chitin in cell suspension cultures of "old man" cactus, *Cephalocereus senilis* (Cactaceae).[187] A rhamnosylglucoside of pinocembrin (**262**) has been isolated from aerial parts of the fern *Onichium japonicum* (Adiantaceae), in which it occurs in high levels (0.13% of the dried weight). This glycoside showed moderate activity against P-388 lymphocytic leukemia with an IC$_{50}$ of 2.58 μg/ml, but did not exhibit any activity against Hela cells up to 20 μg/ml.[181] Unfortunately, the authors did not specify the linkage between rhamnose and glucose in the sugar moiety. (2*S*)-Eriodictyol 7-glucuronide (**300**) and its known (2*R*)-epimer from *Chrysanthemum indicum* (Asteraceae) showed inhibitory activity toward rat lens aldose reductase,[210] and finally, 4′-hydroxy-5,7,2′-trimethoxyflavanone 4′-rhamnosyl(1 → 6)glucoside (**314**) from *Terminalia alata* (Combretaceae) displayed antifungal activity.[220]

15.3 DIHYDROFLAVONOLS

15.3.1 DIHYDROFLAVONOL AGLYCONES

The number of new dihydroflavonols or flavanonols reported during the 1992 to 2003 period is much smaller than that of the respective flavanones, and they are collated in a single table (Table 15.9, compounds **329–396**).[18,57,67,76,78,80,90,109,133,134,161,229–264] Within this table, the compounds are arranged in the same manner as the flavanones, into "simple dihydroflavonols," *C*-methylated and isoprenylated dihydroflavonols, etc.

15.3.1.1 Simple Dihydroflavonols with *O*-Substitution Only

Only 15 newly reported "simple dihydroflavonols" are listed in Table 15.9. Several of the compounds are just simply the C-3 epimers of known (2*R*,3*R*)-dihydroflavonols, e.g., (2*R*,3*S*)-3-hydroxy-5,7-dimethoxyflavanone (**329**) from the fern *Woodsia scopulina* (Dryopteridaceae),[229] (2*R*,3*S*)-3,5,5′,4′-tetrahydroxy-7-methoxyflavanone (3-epipadmatin, **333**) from *Inula graveolens* (Asteraceae),[232] (2*R*,3*S*)-3,5,3′-trihydroxy-7,4′-dimethoxyflavanone (**337**) from *Lannea coromandelica* (Anacardiaceae),[231] and (2*R*,3*S*)-3,5,7,3′,4′,5′-hexahydroxyflavanone (hovenitin III, **338**) from the seeds and fruits of *Hovenia dulcis* (Rhamnaceae).[235] The previously known (2*R*,3*R*)-stereoisomer of hovenitin III has the trivial name ampelopsin.[15] The same plant source has yielded the (2*R*,3*R*)- and (2*R*,3*S*)-epimers of 3,5,7,4′,5′-pentahydroxy-3′-methoxyflavanone, which are known as hovenitins I (**340**) and II (**341**), respectively.[235]

TABLE 15.9
Dihydroflavonols (Flavanonols) Reported from 1992 to 2003

No.	O-Substitution	Other Substituents[a]	Formula	M_r	Trivial Name	Plant Source	Family	Organ[b]	Ref.
Simple dihydroflavonols									
Tri-O-substituted									
329 (2R,3S)	3-OH-5,7-diOMe		$C_{17}H_{16}O_5$	300		*Woodsia scopulina*	Dryopteridaceae, Pteridophyta	Frond	229
Tetra-O-substituted									
330*	3,5,8-TriOH-7-OMe		$C_{16}H_{14}O_6$	302		*Muntingia calabura*	Elaeocarpaceae	Leaf	230
331*	3-OH-5,7,4'-triOMe		$C_{18}H_{18}O_6$	330		*Lannea coromandelica*	Anacardiaceae	StemB	231
Penta-O-substituted									
332	3,5,2',3'-TetraOH-7-OMe		$C_{16}H_{14}O_7$	318		*Iris tenuifolia*	Iridaceae	Root	18
333 (2R,3S)	3,5,3',4'-TetraOH-7-OMe		$C_{16}H_{14}O_7$	318	3-Epipadmatin	*Inula graveolens*	Asteraceae	Aerial	232
334	3,5,7-TriOH-6,4'-diOMe		$C_{17}H_{16}O_7$	332		*Prunus domestica*	Rosaceae	Wood	233
335*	3,5,2'-TriOH-7,5'-diOMe		$C_{17}H_{16}O_7$	332		*Blumea balsamifera*	Asteraceae	Aerial	234
336	3,5,3'-TriOH-7,2'-diOMe		$C_{17}H_{16}O_7$	332		*Iris tenuifolia*	Iridaceae	Root	18
337 (2R,3S)	3,5,3'-TriOH-7,4'-diOMe		$C_{17}H_{16}O_7$	332		*Lannea coromandelica*	Anacardiaceae	StemB	231
Hexa-O-substituted									
338 (2R,3S)	3,5,7,3',4',5'-HexaOH		$C_{15}H_{12}O_8$	320	Hovenitin III	*Hovenia dulcis*	Rhamnaceae	Seed, fruit	235
339	3,5,7,2',5'-PentaOH-6-OMe		$C_{16}H_{14}O_8$	334	Diosalol	*Dioclea grandiflora*	Leguminosae	RootB	78
340*	3,5,7,4',5'-PentaOH-3'-OMe		$C_{16}H_{14}O_8$	334	Hovenitin I	*Hovenia dulcis*	Rhamnaceae	Seed, fruit	235
341 (2R,3S)	3,5,7,4',5'-PentaOH-3'-OMe		$C_{16}H_{14}O_8$	334	Hovenitin II	*Hovenia dulcis*	Rhamnaceae	Seed, fruit	235
342*	3,5,3',4'-TetraOH-7,8-diOMe		$C_{17}H_{16}O_8$	348		*Erica cinerea*	Ericaceae	Flower	236
Hepta-O-substituted									
343*	3,5,3'-TriOH-8,5'-diOMe-6,7-methylenedioxy		$C_{18}H_{16}O_9$	376	Plumbaginol	*Plumbago indica*	Plumbaginaceae	Aerial	237
Esters									
344 (2R,3S)	5,7-diOH-3-O-acetate		$C_{17}H_{14}O_6$	314	cis-Pinobanksin 3-O-acetate	*Woodsia scopulina*	Dryopteridaceae, Pteridophyta	Frond	229
345	5,7,3'-TriOH-3-isobutyrate		$C_{19}H_{18}O_7$	358		*Flourensia retinophylla*	Asteraceae	Aerial	238
346	5,7-DiOH-4'-OMe-3-O-acetate		$C_{18}H_{16}O_7$	344		*Afromomum hanburyi*	Zingiberaceae	Aerial	239
347 (2R,3S)	5,4'-DiOH-7-OMe-3-O-acetate		$C_{18}H_{16}O_7$	344		*Inula graveolens*	Asteraceae	Aerial	232
348 (2R,3R)	5,2'-DiOH-7,8-diOMe-3-O-acetate		$C_{19}H_{18}O_8$	374		*Notholaena sulphurea*	Pteridaceae, Pteridophyta	Frond	240

continued

TABLE 15.9
Dihydroflavonols (Flavononols) Reported from 1992 to 2003 — continued

No.	O-Substitution	Other Substituents[a]	Formula	M_r	Trivial Name	Plant Source	Family	Organ[b]	Ref.
	C-Methyl-substituted								
349	3,5,7-TriOH-4'-OMe	6-Me	$C_{17}H_{16}O_6$	316		*Amaranthus caudatus*	Amaranthaceae	Flower	57
350	2,3,5,7,4'-PentaOH	6-Me	$C_{16}H_{14}O_7$	318		*Leptospermum polygalifolium* ssp. *polygalifolium*	Myrtaceae	Leaf	67
351	2,3,5,7,4'-PentaOH	8-Me	$C_{16}H_{14}O_7$	318		*Leptospermum polygalifolium* ssp. *polygalifolium*	Myrtaceae	Leaf	67
352*	3,5,7,3',4'-PentaOH	6'-CH2OH	$C_{16}H_{14}O_8$	334		*Trifolium alexandrium*	Leguminosae	Seed	241
	Prenyl-, geranyl-, and lavandulyl-substituted dihydroflavonols								
353*	3,5,7-TriOH-8-OMe	6-Pr	$C_{21}H_{22}O_6$	370	Dioclenol	*Dioclea grandiflora*	Leguminosae	RootB	242
354*	3,5,3',4'-TetraOH-7-OPr		$C_{20}H_{20}O_7$	372		*Pterocaulon alopecuroides*	Asteraceae	Aerial	243
355*	3,5,7,4'-TetraOH-6-OMe	8-Pr	$C_{21}H_{22}O_7$	386	Floranol	*Dioclea grandiflora*	Leguminosae	Root	244
356*	3,5,4'-TriOH-7,3'-diOMe	6-Pr	$C_{22}H_{24}O_7$	400		*Rhynchosia densiflora*	Leguminosae	Leaf	245
357*	3,5,4'-TriOH-7,3'-diOMe	8-Pr	$C_{22}H_{24}O_7$	400	Scariosin	*Paracalyx scariosa*	Leguminosae	Leaf	246
358	3,5,7,2',5'-PentaOH-6-OMe	8-Pr	$C_{21}H_{22}O_8$	402	Paraibanol	*Dioclea grandiflora*	Leguminosae	RootB	78
359	3,5,7,4'-TetraOH	6,8-diPr	$C_{25}H_{28}O_6$	424	6,8-Diprenyl-aromadendrin	*Monotes africanus*	Dipterocarpaceae	Leaf	247
360*	3,5,4'-TriOH-7-O-Ger		$C_{25}H_{28}O_6$	424		*Boronia caerulescens* ssp. *spicata*	Rutaceae	Aerial	80
361	3,5,7,4'-TetraOH	8-Pr-3'-Ger	$C_{30}H_{36}O_6$	492	Sanggenol C	*Morus cathayana*	Moraceae	RootB	109
362*	3,5,7,2',4'-PentaOH	8-Lav	$C_{25}H_{28}O_7$	440	Kushenol X	*Sophora flavescens*	Leguminosae	Root	76
363	3,5,7,3',4'-PentaOH	8,2',6'-triPr	$C_{30}H_{36}O_7$	508	Petalostemumol	*Petalostemum purpureum*	Leguminosae	Whole	248
364	3,5,7,2',4'-PentaOH	6-Pr-3'-Ger	$C_{30}H_{36}O_7$	508	Sanggenol K	*Morus cathayana*	Moraceae	RootB	249
365	3,5,7,2',4'-PentaOH	3'-Ger-5-Pr	$C_{30}H_{36}O_7$	508	Sanggenol D	*Morus cathayana*	Moraceae	RootB	109
366	3,5,7,2',4'-PentaOH	8,5'-diPr-3'-Ger	$C_{35}H_{44}O_7$	576	Sanggenol E	*Morus cathayana*	Moraceae	RootB	109
367	3,5,7,2',4'-PentaOH	6-(3-OH-3-Me-butyl)-8-Lav	$C_{30}H_{36}O_8$	524	Kosamol A	*Sophora flavescens*	Leguminosae	Root	250
368*	3,5,7,2',6'-PentaOH-4'-OMe	6-Ger-8-Pr	$C_{31}H_{38}O_8$	538	Kenusanone C	*Echinosophora koreensis*	Leguminosae	Root	90
	Furano-dihydroflavonols								
369	3-OMe	[2'',3'':7,8]Furanoflavanone	$C_{18}H_{14}O_4$	294		*Lonchocarpus latifolius*	Leguminosae	Root	133
370*	3,5,6-TriOMe	[2'',3'':7,8]Furanoflavanone	$C_{20}H_{18}O_6$	354		*Lonchocarpus subglaucescens*	Leguminosae		134
	Dimethylpyrano-dihydroflavonols								
371	3-OH	Bis(6'',6'-DMP[2'',3'':5,6][2'',3'':7,8])	$C_{25}H_{24}O_5$	404	MS II	*Mundulea suberosa*	Leguminosae	Root	251
372*	3,4'-DiOH	3'-Pr-6'',6'-DMP[2'',3'':7,8]	$C_{25}H_{26}O_5$	406	Kanzonol Z	*Glycyrrhiza glabra*	Leguminosae	Root	252
373	3,5,2'-TriOH	8-Pr-6'',6'-DMP[2'',3'':7,6]	$C_{25}H_{26}O_6$	422	Jayacanol	*Lonchocarpus oaxacensis*	Leguminosae	Root	253

No.	Substitution		Trivial name	Molecular formula	MW	Species	Family	Plant part	Ref.
374	3,4'-DiOH-5-OMe	8-Pr-6'',6''-DMP[2'',3'':7,6]		$C_{26}H_{28}O_6$	436	*Lonchocarpus atropurpureus*	Leguminosae	Root	254
375	5,2'-DiOH-3-OMe	8-Pr-6'',6''-DMP[2'',3'':7,6]		$C_{26}H_{28}O_6$	436	*Mundulea suberosa*	Leguminosae	StemB	255
376*	3,5,4'-TriOH-3'-OMe	6'',6''-DMP[2'',3'':7,6]	Eriotrinol	$C_{21}H_{20}O_7$	384	*Lonchocarpus atropurpureus*	Leguminosae	Root	254
						Erythrina eriotricha	Leguminosae	StemB	161
Isoprenylated C-3'/C-2'-ether linked dihydroflavonols (Figure 15.14)									
377	2,6-diPr	3,5,7,4'-TetraOH	Sanggenol F	$C_{25}H_{26}O_7$	438	*Morus cathayana*	Moraceae	RootB	249
378	2,3'-diPr-6'',6''-DMP[2'',3'':7,6]	3,5,4'-TriOH	Sorocein D	$C_{30}H_{32}O_7$	504	*Sorocea bonplandii*	Moraceae	RootB	256
379	2,3'-diPr-6'',6''-DMP[2'',3'':7,8]	3,5,4'-TriOH	Sorocein E	$C_{30}H_{32}O_7$	504	*Sorocea ilicifolia*	Moraceae	RootB	257
380	2,6,8-triPr	3,5,7,4'-TetraOH	Sorocein G	$C_{30}H_{34}O_7$	506	*Sorocea ilicifolia*	Moraceae	RootB	257
381	2,6,3'-triPr	3,5,7,4'-TetraOH	Sorocein F	$C_{30}H_{34}O_7$	506	*Sorocea ilicifolia*	Moraceae	RootB	257
382	2-Pr-8-Ger	3,5,7,4'-TetraOH	Sanggenol I	$C_{30}H_{34}O_7$	506	*Morus cathayana*	Moraceae	RootB	249
383	2-Pr-3'-Ger	3,5,7,4'-TetraOH	Sanggenol G	$C_{30}H_{34}O_7$	506	*Morus cathayana*	Moraceae	RootB	249
384	2-Farnesyl	3,5,7,4'-TetraOH	Sanggenol H	$C_{30}H_{34}O_7$	506	*Morus cathayana*	Moraceae	RootB	249
C-Benzyl-substituted (Figure 15.15)									
385	6-(4-OH-Benzyl)	3,5,7,4'-TetraOH	Gericudranin E	$C_{22}H_{18}O_7$	394	*Cudrania tricuspidata*	Moraceae	StemB	258
386	6-(4-OH-Benzyl)	3,5,7,3',4'-PentaOH	Gericudranin B	$C_{22}H_{18}O_8$	410	*Cudrania tricuspidata*	Moraceae	StemB	259
387	8-(4-OH-Benzyl)	3,5,7,3',4'-PentaOH	Gericudranin C	$C_{22}H_{18}O_8$	410	*Cudrania tricuspidata*	Moraceae	StemB	259
388	6,8-Di-(4-OH-benzyl)	3,5,7,4'-TetraOH	Gericudranin D	$C_{29}H_{24}O_8$	500	*Cudrania tricuspidata*	Moraceae	StemB	258
389	6,8-Di-(4-OH-benzyl)	3,5,7,3',4'-PentaOH	Gericudranin A	$C_{29}H_{24}O_9$	516	*Cudrania tricuspidata*	Moraceae	StemB	259
Miscellaneous substitutions (Figure 15.16)									
390	6-(3-Oxobutyl)	3,5,7,3',4'-PentaOH		$C_{19}H_{18}O_8$	374	*Bauhinia purpurea*	Leguminosae	Heart-wood	260
391	3'-(4-OH-Benzaldehyde)	3,5,7,4'-TetraOH	Hypnogenol F	$C_{22}H_{16}O_8$	408	*Hypnum cupressiforme*	Hypnaceae, Musci	Gam.	261
392	3'-(4-OH-Benzoic acid)	3,5,7,4'-TetraOH	Hypnum acid	$C_{22}H_{16}O_9$	424	*Hypnum cupressiforme*	Hypnaceae, Musci	Gam.	262
393	3'-(4-OH-Benzoic acid methyl ester)	3,5,7,4'-TetraOH	Hypnum acid methyl ester	$C_{23}H_{18}O_9$	438	*Hypnum cupressiforme*	Hypnaceae, Musci	Gam.	262
394	*rel*-5-Hydroxy-7,4'-dimethoxy-2''S-(2,4,5-trimethoxy-*E*-styryl)tetrahydrofuro[4''R,5''R:2,3]flavanonol			$C_{30}H_{30}O_{10}$	550	*Alpinia flabellata*	Zingiberaceae	Leaf	263
395	*rel*-5-Hydroxy-7,4'-dimethoxy-3''S-(2,4,5-trimethoxy-*E*-styryl)tetrahydrofuro[4''R,5''R:2,3]flavanonol			$C_{30}H_{30}O_{10}$	550	*Alpinia flabellata*	Zingiberaceae	Leaf	263
396	5,7,4'-TriOH-3-*O*-(1,8,14-trimethylhexadecanyl)		Muscanone	$C_{34}H_{50}O_6$	554	*Commiphora wightii*	Burseraceae	Trunk	264

*Dihydroflavonols for which the (2*R*,3*R*)-configuration has been confirmed.

[a]Pr, prenyl; Ger, geranyl; Lav, lavandulyl; DMP, dimethylpyrano; DMDHP, dimethyldihydropyrano.

[b]StemB, stem bark; RootB, root bark; Whole, whole plant; Gam, gametophyte.

Dihydroflavonols **332** and **336** from *Iris tenuifolia* (Iridaceae) show 2′,3′-dioxygenation of the B-ring.[18] Flavanones with this unusual B-ring substitution pattern have also been found in this species (see Section 15.2.1.1). An equally unusual 2′,5′-dioxygenation pattern for the B-ring is present in flavanonol **335** from *Blumea balsamifera* (Asteraceae).[234] A bioactive dihydroflavonol with 8-*O*-substitution of the A-ring, (2*R*,3*R*)-3,5,8-dihydroxy-7-methoxyfla-vanone (**330**), has been isolated from the leaves of *Muntingia calabura* (Elaeocarpaceae).[230] This compound and chalcones and isoflavones from the same species were active in a quinone reductase induction assay with cultured Hepa lclc (mouse hepatoma cells).[230] Flowers of the heather *Erica cinerea* also contain a novel 8-oxygenated dihydroflavonol, 3,5,3′,4′-tetra-hydroxy-7,8-dimethoxyflavanone (**342**).[236] In plumbaginol (**343**) from *Plumbago indica* (Plumbaginaceae), all the carbons of the A-ring are *O*-substituted and the B-ring has the unusual 3′,5′-dioxygenation pattern. The compound also bears a methylenedioxy substituent at C-6 and C-7.[237]

Two known dihydroflavonols, 3-hydroxy-5-methoxy-6,7-methylenedioxyflavanone and 3,5,7,4′-tetrahydroxy-3′-methoxyflavanone, are produced as phytoalexins by sugarbeet roots (*Beta vulgaris*, Chenopodiaceae) when inoculated with *Rhizoctonia solani*. They had not been previously reported for *B. vulgaris*.[265]

The 3-hydroxyl of dihydroflavonols frequently forms an ester with various acids and previously 18 dihydroflavonols-esters have been described, notably acetates.[15] In the last decade or so, several new acetates have been added to this list (compounds **344–348**). (2*R*,3*R*)-3,5,2′-Trihydroxy-7,8-dimethoxyflavanone 3-*O*-acetate (**348**), which has an unusual oxygenation pattern in both the A- and B-rings, has been isolated from the frond of the fern *Notholaena sulphurea* (Pteridaceae),[240] where it is present in the yellow farinose coating on the lower leaf surface. This acetylated dihydroflavonol is characteristic of the chemotype of *N. sulphurea* that has a yellow exudate. In contrast, the white form of this species mainly produces methylated dihydrochalcones.[240] Isobutyric acid represents the acyl group in the dihydroflavonol **345** from *Flourensia retinophylla* (Asteraceae).[238]

15.3.1.2 Simple Dihydroflavonols with *O*- and *C*-Substitution

Three new *C*-methyl-substituted dihydroflavonols are listed in Table 15.9 (compounds **349–351**), together with one *C*-hydroxymethylflavanonol (**352**) from *Trifolium alexandrinum* (Leguminosae), in which the hydroxymethyl group is attached to C-6′.[241] The flowers of *Amaranthus caudatus* (Amaranthaceae) were the source of the 6-methyl-substituted dihydroflavonol **349**.[57] The 2-hydroxylated methylflavanonol, 2,3,5,7,4′-pentahydroxy-6-*C*-methylflavanone (**350**), has been isolated from *Leptospermum polygalifolium* ssp. *polygali-folium*, together with its 8-*C*-methyl regioisomer **351**.[67] Both *A. caudatus* and *L. polygalifolium* ssp. *polygalifolium* also contain related *C*-methylflavanones[57,67] (see Section 15.2.1.2).

15.3.1.3 Dihydroflavonols with Noncyclic Isoprenoid Substituents

Table 15.9 also presents the newly reported dihydroflavonols with prenyl, geranyl, and lavandulyl side chains (compounds **353–368**), which have been mainly found in species of the family Leguminosae. For example, the root bark of *Dioclea grandiflora* has been the source of three new prenylated dihydroflavonols, dioclenol (**353**),[242] floranol (**355**),[244] and paraibanol (**358**).[78] Paraibanol is the 8-prenyl derivative of the dihydroflavonol diosalol (3,5,7,2′,5′-pentahydroxy-6-methoxyflavanone, **339**), which was isolated from the same plant.[78] Other newly reported prenylated dihydroflavonols from this family include the monoprenylated flavanones **356** from *Rhynchosia densiflora*[245] and scariosin (**357**) from

Paracalyx scariosa.[246] Lavandulyl groups are present in both kushenol X (**362**)[76] and kosamol A (**367**)[250] from *Sophora flavescens.* Petalostemumol (**364**) from *Petalostemum purpureum* is 8,2′,6′-triprenylated; its structure was determined by a single-crystal x-ray diffraction analysis.[248] The compound showed good activity against the Gram-positive bacteria *Staphyllococcus aureus* and *Bacillus subtilis* ATCC 6633, with MIC values of 3.12 and 0.78 μg/ml, respectively. Activity against the Gram-negative bacterium *Escherichia coli* and against the yeast *Candida albicans* was only moderate in comparison (MIC 6.25 and 12.5 μg/ml, respectively).[248]

The family Moraceae is another good source of dihydroflavonols bearing noncyclic isoprenylated substituents. From the root bark of *Morus cathayana* four flavanonols with both prenyl and geranyl groups have been isolated, sanggenols C–E and K (**361, 365, 366**, and **364**).[109,249] Only a few new prenylated dihydroflavonols have been reported from families other than the Leguminosae and Moraceae, e.g., the 7-*O*-geranyl derivative of aromadendrin (**360**) from *Boronia caerulescens* ssp. *spicata* (Rutaceae).[80] From a species of Dipterocarpaceae, *Monotes africanus*, the new 6,8-diprenylaromadendrin (**359**) has been obtained together with a range of prenylated flavonoids belonging to other classes. These compounds were tested for HIV-inhibitory activity in the XTT-based whole cell screen[247] and all 6,8-diprenylated flavonoids isolated, including the dihydroflavonol **359**, the flavanone 6,8-diprenylnaringenin (lonchocarpol A), and the flavanonol 6,8-diprenylkaempferol, showed activity, with IC_{50} values of 4.7, 2.7, and 5.8 μg/ml, respectively.[247]

The known dihydroflavonol kushenol I (3,7,2′,4′-tetrahydroxy-5-methoxy-8-*C*-lavandulylflavanone) was isolated, together with the corresponding flavanone kurarinone, from the roots of *Gentiana macrophylla* (Gentianaceae) and shown to be active against the plant pathogenic fungus *Cladosporium cucumerinum.* Kurarinone also inhibited the growth of the human pathogenic yeast *Candida albicans.*[118] It is interesting that a dihydroflavonol and flavanone with substitutions typical for root flavonoids in species of the Leguminosae (lavandulyl side chain and 2′,4′-di-*O*-substitution of the B-ring) have been found in the roots of a totally unrelated species. The flavonoids in plant roots have not been studied nearly as well as those of leaves. Perhaps these unusually substituted flavonoids, which are often associated with antifungal activity and may protect the roots against attacks by fungi and parasites, are much more widespread in plant roots than are presently realized.

15.3.1.4 Dihydroflavonols with Furano or Pyrano Rings

Two new dihydroflavonols with furano rings and six with pyrano rings are listed in Table 15.9 (compounds **369–376**). They were all isolated from members of the Leguminosae. Both furanoflavanonols have been obtained from the genus *Lonchocarpus*, which is also a source of furanoflavanones. Dihydroflavonol **369** was found in the roots of *L. latifolius*[133] and **370** in *L. subglaucescens.*[134] In these compounds, the furano ring does not bear an isopropenyl side chain and no free hydroxyl groups are present, only methoxyl groups.

Two of the new dimethylpyranoflavanonols have also been found in the genus *Lonchocarpus*; jayacanol (**373**), in the roots of *L. oaxacensis*,[253] and compound **375** in *L. atropurpureus*,[254] which also contained jayacanol. These substances are characterized by 2′-hydroxylation of the B-ring. Jayacanol (**373**) and its known 2′-deoxy derivative, mundulinol, were tested for their activity against the wood-rotting fungus *Postia placenta.* Only the 2′-hydroxylated jayacanol was active.[253] Two further dimethylpyranoflavanonols are constituents of *Mundulea suberosa*, MS II (**371**)[251] and compound **374**.[255] The remaining two new dihydroflavonols in this class are kanzonol Z (**372**) from *Glycyrrhiza glabra*[252] and eriotrinol (**376**) from *Erythrina eriotricha.*[161]

15.3.1.5 Isoprenylated Dihydroflavonols with a C-3-C-2′ Ether Linkage

In the 1980s, several unusual dihydroflavonoids were reported from the root bark of *Morus mongolica* (Moraceae), characterized by an ether linkage between C-3 and C-2′ of the flavonoid, e.g., sanggenons A and M.[266] They were thought to bear a prenyl group at C-3 and a hydroxyl group at C-2. More recently, similar compounds, soroceins D–G, were discovered in the root bark of species belonging to the related genus *Sorocea* (*S. bonplandii* and *S. ilicifolia*, Moraceae), but these were identified as bearing a prenyl group at C-2 and hydroxyl at C-3.[256,257] This led to a structure revision of some of the sanggenons, so that they are now also described as 2-prenylated dihydroflavonols rather than 2-hydroxylated 3-prenylflavanones.[267] Several related dihydroflavonols have recently been discovered in root bark extracts of *Morus cathayana* (sanggenols F–I).[249] These soroceins and sanggenols are hydroxylated at C-3, C-5, C-7, and C-4′ (see Table 15.9, compounds **377–384**, and Figure 15.14). In sanggenol F (**377**), there is an additional prenyl group at C-8,[249] whereas in soroceins F (**381**) and G (**380**) there are two additional prenyl groups, at C-6 and C-3′ in **381** and C-6 and C-8 in **380**.[257] In soroceins D (**378**) and E (**379**), a dimethylpyrano group is attached to the A-ring and an additional prenyl group to C-3′.[256,257] In sanggenol G (**383**), there is a geranyl substituent at C-3′ in addition to the prenyl at C-2, whereas in sanggenol I (**382**), the geranyl group is attached to C-8. In sanggenol H (**384**), there is a 15-carbon farnesyl chain at C-2 instead of a prenyl group and the molecule bears no further isoprenoid substituents.[249]

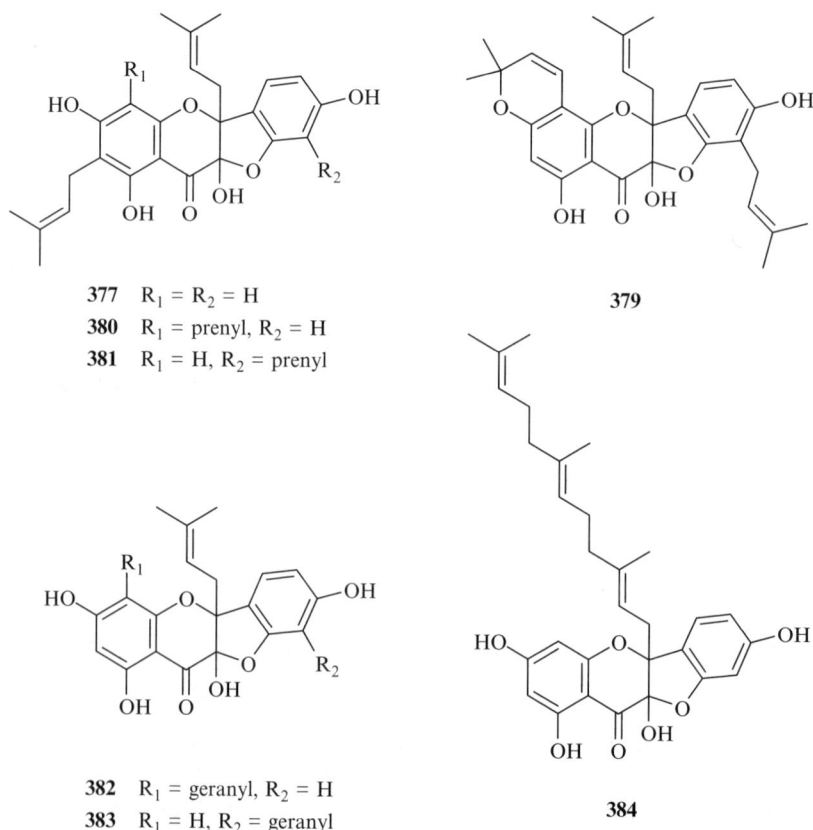

377 R₁ = R₂ = H	
380 R₁ = prenyl, R₂ = H	**379**
381 R₁ = H, R₂ = prenyl	

382 R₁ = geranyl, R₂ = H	
383 R₁ = H, R₂ = geranyl	**384**

FIGURE 15.14 C-2-Isoprenylated C-3–C-2′-ether-linked dihydroflavonols.

15.3.1.6 Benzylated Dihydroflavonols

The first *C*-benzylated dihydroflavonols, gericudranins A (**389**), B (**386**), C (**387**), D (**388**), and E (**385**), have been discovered in stem bark extracts of *Cudrania tricuspidata* (Moraceae) (Table 15.9; Figure 15.15).[258,259] Gericudranins A–C are based on taxifolin (5,7,3′,4′-tetrahydroxydihydroflavonol), whereas gericudranins D and E are based on aromadendrin (5,7,4′-trihydroxydihydroflavonol), and they all contain a 4-hydroxybenzyl moiety that is attached to C-6 or C-8 (or both) of the flavanonol. The antitumor activity of gericudranins A–C was tested, indicating cytotoxicity to the human tumor cell lines CRL 1579 (skin), LOX-IMVI (skin), MOLT-4F (leukemia), KM 12 (colon), and UO-31 (renal) in culture, with ED_{50} values of 2.7 to 31.3 μg/ml.[259]

15.3.1.7 Dihydroflavonols with Miscellaneous Substituents

There are a few newly reported dihydroflavonols with unusual side groups or skeletons that cannot easily be arranged into any of the groups discussed so far and are therefore treated together in a "miscellaneous" group (Table 15.9). These include 6-(3-oxobutyl)taxifolin (**390**, Figure 15.16) from *Bauhinia purpurea* (Leguminosae), which bears an uncommon C–C-linked side chain.[260] The gametophyte of the moss *Hypnum cupressiforme* (Hypnaceae) produces a range of dihydroflavonols with C–C-linked attachments, all at C-3′. In hypnogenol F (**391**, Figure 15.16), C-3′ of aromadendrin (3,5,7,4′-tetrahydroxyflavanone) is linked to C-3 of 4-hydroxybenzaldehyde,[261] whereas in hypnum acid (**392**) and hypnum acid methyl ester (**393**), C-3′ of aromadendrin is linked to C-3 of 4-hydroxybenzoic acid and its methyl ester, respectively.[262] This species also produces a wide variety of biflavonoids.[261,262] From the rare Japanese species *Alpinia flabellata* (Zingiberaceae), two flavonol–phenylbutadiene adducts have been isolated, **394** and **395** (Figure 15.16). In each of these compounds, the substitution of the phenolic ring in the phenylbutanoid moiety is 2,4,5-trimethoxy, and the butenoid chain forms a tetrahydrofurano ring with the dihydroflavonol.[263] An antifungal aromadendrin derivative, muscanone (**396**, Figure 15.16), was obtained from the trunk of *Commiphora wightii* (Burseraceae). This dihydroflavonol has a 1,8,14-trimethylhexadecanyl side chain ether-linked to the 3-hydroxyl of aromadendrin. The compound was active against *Candida albicans* at 250 μg/ml.[264]

15.3.2 DIHYDROFLAVONOL GLYCOSIDES

Relatively few dihydroflavonol glycosides have been reported during the period between 1992 and 2003; 16 are listed in Table 15.10 (compounds **397–412**) based on ten different aglycones,

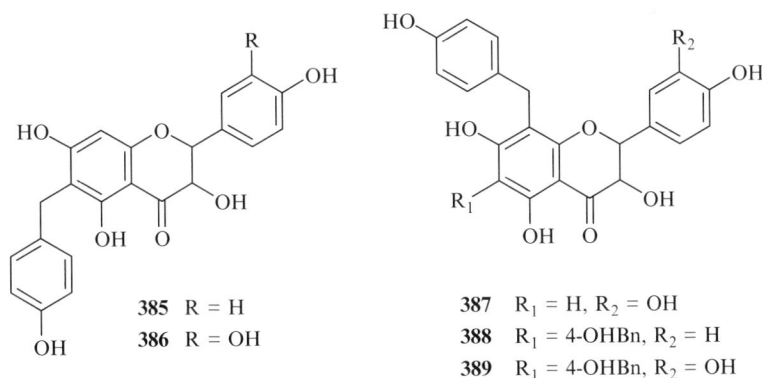

FIGURE 15.15 *C*-Benzyl-substituted dihydroflavonols (Bn, benzyl).

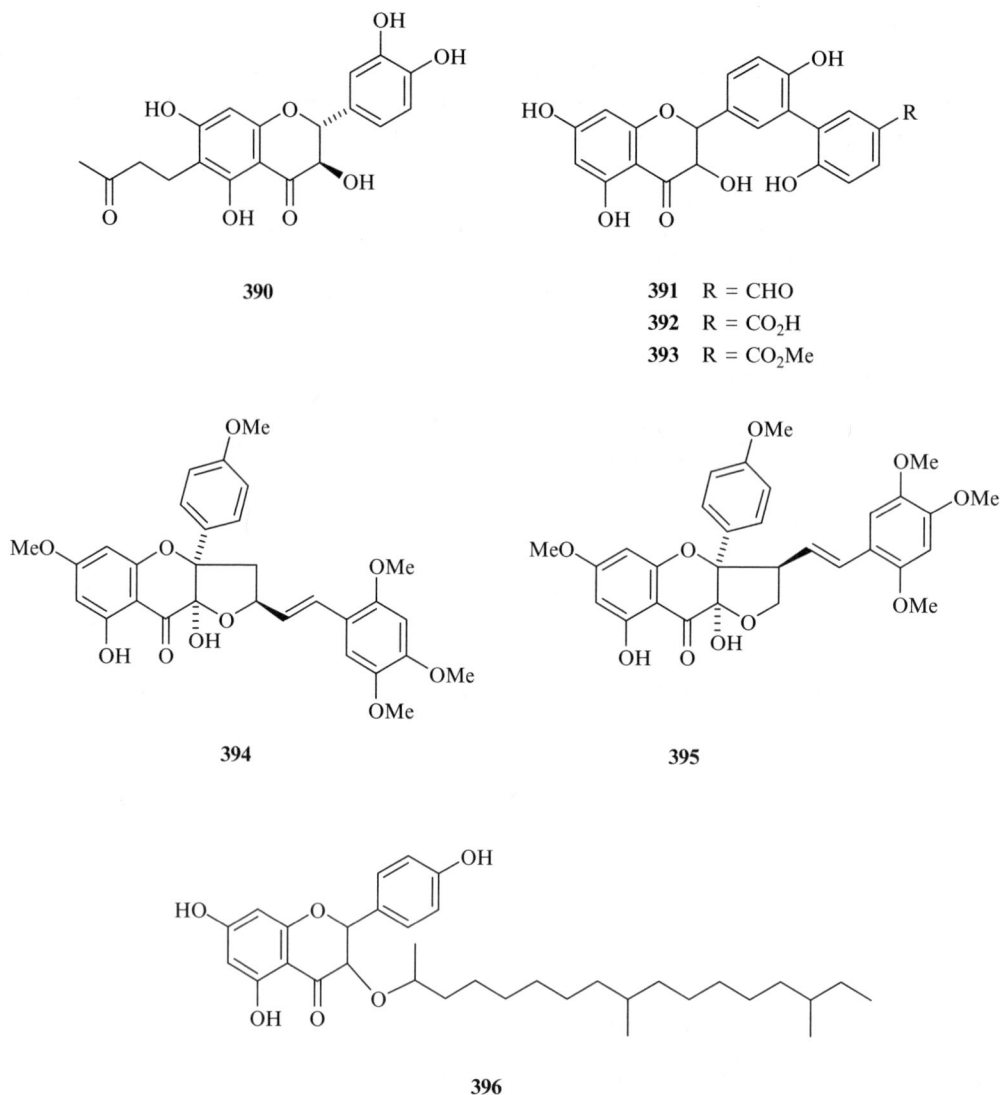

390

391 R = CHO
392 R = CO$_2$H
393 R = CO$_2$Me

394

395

396

FIGURE 15.16 Dihydroflavonols with miscellaneous substituents.

including two that are prenylated.[206,268–281] The conventions used in this table are the same as those described for the flavanone glycosides in Section 15.2.4. Rhamnose and glucose are the most frequent sugar moieties, appearing in seven and six of the dihydroflavonol glycosides, respectively. Most are monosides; only two biosides, aromadendrin 7-rhamnosyl(1 → 4)galactoside (**397**) from *Crotalaria laburnifolia* (Leguminosae)[268] and taxifolin 7-rhamnosyl(1 → 6)glucoside (**406**) from *Platycodon grandiflorum* (Campanulaceae),[275] are among the new compounds, and one diglycoside (taxifolin 3,7-dirhamnoside, **405**) from *Hypericum japonicum* (Clusiaceae = Guttiferae).[274] Some of the new dihydroflavonol glycosides are stereoisomers, e.g., the 3-rhamnosides of (2R,3R)-, (2R,3S)-, and (2S,3S)-3,5,7,3′,5′-pentahydroxyflavanone (**408**, **409**, and **410**, respectively), **408** from *Excoecaria agallocha* (Euphorbiaceae),[277] and **409** (smitilbin) and **410** (neosmitilbin) from *Smilax glabra* (Smilacaceae).[278,279]

The new glycosides include arabinose in different forms and configurations, e.g., as β-D-arabinopyranose in dihydrorhamnetin 3-β-D-arabinopyranoside (**407**) from *Pyrola*

TABLE 15.10
Dihydroflavonol Glycosides Reported from 1992 to 2003

No.	Dihydroflavonol Glycoside[a]	Formula	M_r	Plant Source	Family	Organ[b]	Ref.
	3,5,7,4'-TetraOH-Flavanone (aromadendrin)						
397	7-Rhamnosyl(1 → 4)galactoside	$C_{27}H_{32}O_{15}$	596	*Crotalaria laburnifolia*	Leguminosae	Seed	268
398 (2R,3R)	7-(6-[4-Hydroxy-2-methylenebutanoyl]glucoside) (Figure 15.17)	$C_{26}H_{28}O_{13}$	548	*Afzelia bella*	Leguminosae	StemB	269
	3,5,7,8,4'-PentaOH-Flavanone						
399	8-(2-Acetylglucoside) (2''-acetylcallunin)	$C_{23}H_{24}O_{13}$	508	*Calluna vulgaris*	Ericaceae	Flower	206
	3,5,7,2',5'-PentaOH-Flavanone						
400	3-Rhamnoside	$C_{21}H_{22}O_{11}$	450	*Plinia pinnata*	Myrtaceae	Aerial	270
	3,7,3',4'-PentaOH-Flavanone (taxifolin)						
401 (2R,3R)	3-α-Arabinopyranoside	$C_{20}H_{20}O_{11}$	436	*Rhododendron ferrugineum*	Ericaceae	Leaf, flower	271
402 (2R,3S)	3-α-Arabinopyranoside	$C_{20}H_{20}O_{11}$	436	*Rhododendron ferrugineum*	Ericaceae	Leaf, flower	271
403	3-α-L-Arabinofuranoside	$C_{20}H_{20}O_{11}$	436	*Fragaria × ananassa*	Rosaceae	Root	272
404	3-(3-Cinnamoylrhamnoside)	$C_{30}H_{28}O_{12}$	580	*Andira inermis*	Leguminosae	Leaf	273
405	3,7-Dirhamnoside	$C_{27}H_{32}O_{15}$	596	*Hypericum japonicum*	Clusiacea	Aerial	274
406 (2R,3R)	7-Rhamnosyl(1 → 6)glucoside (flavoplatycoside)	$C_{27}H_{32}O_{16}$	612	*Platycodon grandiflorum*	Campanulaceae	Seed	275
	3,5,3',4'-Tetra-7-OMe-Flavanone (padmatin, dihydrorhamnetin)						
407	3-β-D-Arabinopyranoside	$C_{21}H_{22}O_{11}$	450	*Pyrola elliptica*	Pyrolaceae	Whole	276
	3,5,7,3',5'-PentaOH-Flavanone						
408 (2R,3R)	3-Rhamnoside	$C_{21}H_{22}O_{11}$	450	*Excoecaria agallocha*	Euphorbiaceae	Stem	277
409 (2R,3S)	3-Rhamnoside (smitilbin)	$C_{21}H_{22}O_{11}$	450	*Smilax glabra*	Smilacaceae	Rhiz	278
410 (2S,3S)	3-Rhamnoside (neosmitilbin)	$C_{21}H_{22}O_{11}$	450	*Smilax glabra*	Smilacaceae	Rhiz	279
	3,5,7,4'-TetraOH-3'-Pr-Flavanone						
411	7-Glucoside (phellochinin)	$C_{26}H_{30}O_{11}$	518	*Phellodendron chinense*	Rutaceae	Leaf	280
	3,5,7,3',4'-PentaOH-6-Pr-Flavanone (6-prenyltaxifolin)						
412	7-Glucoside	$C_{26}H_{30}O_{12}$	534	*Ochna integerrima*	Ochnaceae	Leaf	281

[a] All sugars are *O*-substituted unless specified as *C*-glycosides; rhamnoside, α-L-rhamnopyranoside; glucoside, β-D-glucopyranoside; arabinose is as specified. Pr, prenyl.
[b] StemB, stem bark; Aerial, aerial parts; Whole, whole plant; Rhiz, rhizome.

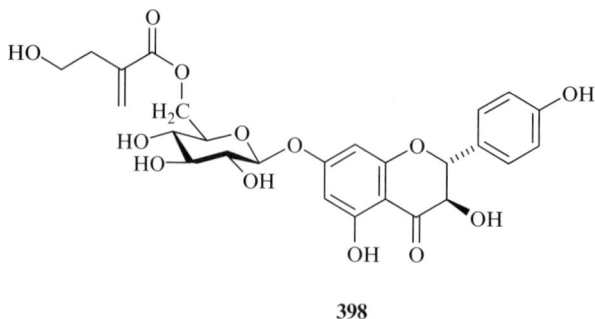

398

FIGURE 15.17 Acylated dihydroflavonol glycoside.

elliptica (Pyrolaceae),[276] and as α-L-arabinofuranose in taxifolin 3-α-L-arabinofuranoside (**403**) from strawberry roots (Rosaceae).[272] Furthermore, the 3-α-arabinopyranosides of the epimers (2*R*,3*R*)- and (2*R*,3*S*)-taxifolin (**401** and **402**) have been reported from leaves and flowers of *Rhododendron ferrugineum* (Ericaceae).[271] The Ericaceae are a well-known source of dihydroflavonols, and the new acylated 2″-acetylcallunin (**399**) has been reported from the flowers of the common heather, *Calluna vulgaris*.[206] In this compound, the sugar moiety is attached to the 8-hydroxyl of the dihydroflavanol; in all other new glycosides it is attached to the 3- or 7-hydroxyl. Two more new acylated dihydroflavonols have been described, both from legumes, (2*R*,3*R*)-aromadendrin 7-(6-[4-hydroxy-2-methylenebutanoyl]glucoside) (**398**, Figure 15.17) from the stem bark of *Afzelia bella*[269] and taxifolin 3-(3-cinnamoylrhamnoside) (**404**, Figure 15.17) from the leaves of *Andira inermis*.[273] The latter compound exhibited antiplasmodial activity with an IC$_{50}$ value of 10.4 μg/ml against the *Plasmodium falciparum* chloroquinone-sensitive strain PoW, and an IC$_{50}$ of 4.2 μg/ml against the resistant strain Dd2.[273]

Two new prenylated dihydroflavonol glycosides are listed in Table 15.10, 6-prenyltaxifolin 7-glucoside (**412**) from *Ochna integerrima* (Ochnaceae)[281] and phellochinin (**411**) from *Phellodendron chinense* (Rutaceae),[280] which is the 7-glucoside of 3′-prenylaromadendrin. The presence of prenylated dihydroflavonol glycosides appears to be a chemosystematic marker for *Phellodendron*, since previously four related compounds have been isolated from four different species of this genus, *P. amurense*, *P. japonicum*, *P. lavallei*, and *P. sachalinense*.[15]

15.4 ACKNOWLEDGMENTS

The authors are very grateful to Dr Christine A. Williams for supplying many literature references and data on flavanones and dihydroflavonols for the period 2001 to 2003.

REFERENCES

1. Bohm, B.A., Flavanones and dihydroflavonols, in *The Flavonoids*, Harborne, J.B., Mabry, T.J., and Mabry, H., Eds, Chapman & Hall, London, 1975, Chap. 11.
2. Bohm, B.A., The minor flavonoids, in *The Flavonoids: Advances in Research*, Harborne, J.B. and Mabry, T.J., Eds, Chapman & Hall, London, 1982, Chap. 6.
3. Bohm, B.A., The minor flavonoids, in *The Flavonoids: Advances in Research Since 1980*, Harborne, J.B., Ed., Chapman & Hall, London, 1988, Chap. 9.
4. Bohm, B.A., The minor flavonoids, in *The Flavonoids: Advances in Research Since 1986*, Harborne, J.B., Ed., Chapman & Hall, London, 1993, Chap. 9.

5. Harborne, J.B., *Comparative Biochemistry of the Flavonoids*, Academic Press, New York, 1967, Chap. 3.
6. Harborne, J.B., *The Flavonoids: Advances in Research Since 1980*, Chapman & Hall, London, 1988, 581.
7. Heller, W. and Forkmann, G., Biosynthesis, in *The Flavonoids: Advances in Research Since 1980*, Harborne, J.B., Ed., Chapman & Hall, London, 1988, Chap. 11.
8. Heller, W. and Forkmann, G., Biosynthesis of flavonoids, in *The Flavonoids: Advances in Research Since 1986*, Harborne, J.B., Ed., Chapman & Hall, London, 1993, Chap. 11.
9. Forkmann, G. and Heller, W., Biosynthesis of flavonoids, in *Comprehensive Natural Products Chemistry*, Vol. 1, Barton, D., Nakanishi, K., and Meth-Cohn, O., Eds, Elsevier, Amsterdam, 1999, Chap. 1.26.
10. Harborne, J.B. and Williams, C.A., Anthocyanins and other flavonoids, *Nat. Prod. Rep.*, 12, 639, 1995.
11. Harborne, J.B. and Williams, C.A., Anthocyanins and other flavonoids, *Nat. Prod. Rep.*, 15, 631, 1998.
12. Harborne, J.B. and Williams, C.A., Anthocyanins and other flavonoids, *Nat. Prod. Rep.*, 18, 310, 2001.
13. Williams, C.A. and Grayer, R.J., Anthocyanins and other flavonoids, *Nat. Prod. Rep.*, 21, 539, 2004.
14. Harborne, J.B. and Baxter, H., *The Handbook of Natural Flavonoids*, Vol. 1, John Wiley and Sons, Chichester, 1999.
15. Harborne, J.B. and Baxter, H., *The Handbook of Natural Flavonoids*, Vol. 2, John Wiley and Sons, Chichester, 1999.
16. Grayer, R.J., Flavanoids, in *Methods in Plant Biochemistry, Vol. 1, Plant Phenolics*, Harborne, J.B., Ed., Academic Press, London, 1989, Chap. 8.
17. Jayaprakasam, B. et al., Dihydroechioidinin, a flavanone from *Andrographis echioides*, *Phytochemistry*, 52, 935, 1999.
18. Kojima, K. et al., Flavanones from *Iris tenuifolia*, *Phytochemistry*, 44, 711, 1997.
19. Braca, A. et al., Three flavonoids from *Licania densiflora*, *Phytochemistry*, 51, 1125, 1999.
20. Topcu, G. et al., A new flavanone from *Teucrium alyssifolium*, *Turk. J. Chem.*, 20, 265, 1996.
21. Huang, H., Sun, H., and Zhao, S., Flavonoids from *Isodon oresbius*, *Phytochemistry*, 42, 1247, 1996.
22. Wang, Q. et al., A new 2-hydroxyflavanone from *Mosla soochouensis*, *Planta Med.*, 65, 729, 1999.
23. Stevens, J.F. et al., A novel 2-hydroxyflavanone from *Collinsia canadensis*, *J. Nat. Prod.*, 62, 392, 1999.
24. Amaro-Luis, J.M. and Mendez, P.D., Flavonoids from the leaves of *Chromolaena subscandens*, *J. Nat. Prod.*, 56, 610, 1993.
25. Kishore, P.H. et al., Flavonoids from *Andrographis lineata*, *Phytochemistry*, 63, 457, 2003.
26. Lin, C.-N., Lu, C.-M., and Huang, P.-L., Flavonoids from *Artocarpus heterophyllus*, *Phytochemistry*, 39, 1447, 1995.
27. Hai, M.A., Koushik, S., and Ahmad, M.U., Chemical constituents of *Eupatorium odoratum* Linn. (Compositae), *J. Bangladesh Chem. Soc.*, 8, 139, 1995.
28. Srivastava, B.K. and Reddy, M.V.R.K., A new flavanone from the flowers of *Tecoma stans*, *Oriental J. Chem.*, 10, 81, 1994.
29. Calanasan, C.A. and MacLeod, J.K., A diterpenoid sulphate and flavonoids from *Wedelia asperrima*, *Phytochemistry*, 47, 1093, 1998.
30. Srinivas, K.V.N.S. et al., Flavonoids from *Caesalpinia pulcherrima*, *Phytochemistry*, 63, 789, 2003.
31. Reddy, M.V.B. et al., A flavanone and a dihydrodibenzoxepin from *Bauhinia variegata*, *Phytochemistry*, 64, 879, 2003.
32. Rao, Y.K. et al., Flavonoids from *Andrographis visculosa*, *Chem. Pharm. Bull.*, 51, 1374, 2003.
33. Shang, L., Zhao, B., and Hao, X., New flavonoid from *Fissistigma kwangsiense*, *Yunnan Zhiwu Yanjiu*, 16, 191, 1994.
34. Reddy, M.K. et al., Two new flavonoids from *Andrographis rothii*, *Chem. Pharm. Bull.*, 51, 191, 2003.

35. Miyake, Y. et al., New potent antioxidative hydroxyflavanones produced with *Aspergillus saitoi* from flavanone glycoside in citrus fruit, *Biosci. Biotechnol. Biochem.*, 67, 1443, 2003.
36. Bhattacharyya, J., Batista, J.S., and Almeida, R.N., Dioclein, a flavanone from the roots of *Dioclea grandiflora*, *Phytochemistry*, 38, 277, 1995.
37. Anam, E.M., A flavanone from the roots of *Acacia longifolia* (Leguminosae), *Indian J. Heterocycl. Chem.*, 7, 63, 1997.
38. Ahmad, I. et al., Cholinesterase inhibitory constituents from *Onosma hispida*, *Chem. Pharm. Bull.*, 51, 412, 2003.
39. Lu, C.-M. and Lin, C.-N., Two 2′,4′,6′-trioxygenated flavanones from *Artocarpus heterophyllus*, *Phytochemistry*, 33, 909, 1993.
40. Quijano, L. et al., Flavonoids from *Ageratum corymbosum*, *Phytochemistry*, 31, 2859, 1992.
41. Iwase, Y. et al., Isolation and identification of two new flavanones and a chalcone from *Citrus kinokuni*, *Chem. Pharm. Bull.*, 49, 1356, 2001.
42. Reddy, M.V.B. et al., New 2′-oxygenated flavonoids from *Andrographis affinis*, *J. Nat. Prod.*, 66, 295, 2003.
43. Parmar, V.S. et al., Agamanone, a flavanone from *Agave americana*, *Phytochemistry*, 31, 2567, 1992.
44. Flagg, M.L. et al., Two novel flavanones from *Greigia sphacelata*, *J. Nat. Prod.*, 63, 1689, 2000.
45. Chemesova, I.I., Iinuma, M., and Budantsev, A.L., Flavonoids of *Scutellaria oxystegia* Juz., *Rastit. Resur.*, 29, 75, 1993.
46. Passador, E.A.P. et al., A pyrano chalcone and a flavanone from *Neoraputia magnifica*, *Phytochemistry*, 45, 1533, 1997.
47. Hoeneisen, M. et al., Flavanones of *Lophopappus tarapacanus* and triterpenoids of *Pachylaena atriplicifolia*, *Phytochemistry*, 34, 1653, 1993.
48. Liu, Y.L. et al., Isolation of potential cancer chemopreventive agents from *Eriodictyon californicum*, *J. Nat. Prod.*, 55, 357, 1992.
49. Ibrahim, A.-R.S., Sulfation of naringenin by *Cunninghamella elegans*, *Phytochemistry*, 53, 209, 2000.
50. Lopes, N.P., Kato, M.J., and Yoshida, M., Antifungal constituents from roots of *Virola surinamensis*, *Phytochemistry*, 51, 29, 1999.
51. Kodama, O. et al., Sakuranetin, a flavanone phytoalexin from ultraviolet-irradiated rice leaves, *Phytochemistry*, 31, 3807, 1992.
52. Dillon, V.M. et al., Differences in phytoalexin response among rice cultivars resistance to blast, *Phytochemistry*, 44, 599, 1997.
53. Plowright, R.A. et al., The induction of phenolic compounds in rice after infection by the stem nematode *Ditylenchus angustus*, *Nematologica*, 42, 564, 1996.
54. Hanawa, F., Yamada, T., and Nakashima, T., Phytoalexins from *Pinus strobus* bark infected with pinewood nematode, *Bursaphelenchus xylophilus*, *Phytochemistry*, 57, 223, 2001.
55. Ribeiro, A. et al., Trypanocidal flavonoids from *Trixis vauthieri*, *J. Nat. Prod.*, 60, 836, 1997.
56. Fleischer, T.C., Waigh, R.D., and Waterman, P.G., Bisabolene sesquiterpenes and flavonoids from *Friesodielsia enghiana*, *Phytochemistry*, 44, 315, 1997.
57. Srivastava, B.K. and Reddy, M.V.R.K., New flavonoids from the flowers of *Amaranthus caudatus*, *Oriental J. Chem.*, 10, 294, 1994.
58. Hsieh, Y.-L., Fang, J.-M., and Cheng, Y.-S., Terpenoids and flavonoids from *Pseudotsuga wilsoniana*, *Phytochemistry*, 47, 845, 1998.
59. Wollenweber, E. et al., *C*-Methylflavonoids from the leaf waxes of some Myrtaceae, *Phytochemistry*, 55, 965, 2000.
60. Yi, J., Zhang, G., and Li, B., Studies on the chemical consituents of *Pseudotsuga sinensis*, *Yaoxue Xuebao*, 37, 352, 2002.
61. Harborne, J.B. et al., Ten isoprenylated and *C*-methylated flavonoids from the leaves of three *Vellozia* species, *Phytochemistry*, 34, 219, 1993.
62. Srivastava, S.K., Srivastava, S.D., and Chouksey, B.K., New constituents from *Terminalia alata*, *Fitoterapia*, 70, 390, 1999.

63. Jiang, J. et al., A new flavanone from *Matteucia orientalis*, *Zhongguo Yaoke Dexue Xuebao*, 25, 199, 1994.

64. Basnet, P. et al., 2′-Hydroxymatteucinol, a new *C*-methyl flavanone derivative from *Matteucia orientalis*; potent hypoglycemic activity in streptozotocin (STZ)-induced diabetic rat, *Chem. Pharm. Bull.*, 41, 1790, 1993.

65. Delle Monache, F. and Cuca Suarez, L.E., 6-*C*-Formyl and 6-*C*-hydroxymethylflavanones from *Petiveria alliacea*, *Phytochemistry*, 31, 2481, 1992.

66. Wu, J.-H. et al., Desmosflavanone II: a new flavanone from *Desmos cochinchinensis* Lour., *J. Chin. Pharm. Sci.*, 6, 119, 1997.

67. Mustafa, K.A., Perry, N.B., and Weavers, R.T., 2-Hydroxyflavanones from *Leptospermum polygalifolium* subsp. *polygalifolium*. Equilibrating sets of hemiacetal isomers, *Phytochemistry*, 64, 1285, 2003.

68. Delle Monache, F., Menichini, F., and Suarez, L.E.C., *Petiveria alliacea*. II. Further flavonoids and triterpenes, *Gazz. Chim. Ital.*, 126, 275, 1996.

69. Wu, J.-H. et al., Anti-AIDS agents 54. A potent anti-HIV chalcone and flavonoids from the genus *Desmos*, *Bioorg. Med. Chem. Lett.*, 13, 1813, 2003.

70. Barron, D. and Ibrahim, R.K., Isoprenylated flavonoids — a survey, *Phytochemistry*, 43, 921, 1996.

71. Rohmer, M. et al., Isoprenoid biosynthesis in bacteria: a novel pathway for the early steps leading to isopentenyl diphosphate, *Biochem. J.*, 295, 517, 1993.

72. Asada, Y., Li, W., and Yoshikawa, T., Biosynthesis of the dimethylallyl moiety of glabrol in *Glycyrrhiza glabra* hairy root cultures via a non-mevalonate pathway, *Phytochemistry*, 55, 323, 2000.

73. Khalilullah, M. et al., Crotaramosmin, a new prenylated flavanone from *Crotalaria ramosissima*, *J. Nat. Prod.*, 55, 229, 1992.

74. Aranguo, A.I. and Gonzales, G.J., Prenyl flavanones of *Dalea caerulea*, *Rev. Columb. Quim.*, 23, 1, 1994.

75. Rao, E.V., Sridhar, P., and Prasad, Y.R., Two prenylated flavanones from *Mundulea suberosa*, *Phytochemistry*, 46, 1271, 1997.

76. Kuroyanagi, M. et al., Antibacterial and antiandrogen flavonoids from *Sophora flavescens*, *J. Nat. Prod.*, 62, 1595, 1999.

77. Wächter, G.A. et al., Antibacterial and antifungal flavanones from *Eysenhardtia texana*, *Phytochemistry*, 52, 1469, 1999.

78. Jenkins, T. et al., Flavonoids from the root-bark of *Dioclea grandiflora*, *Phytochemistry*, 52, 723, 1999.

79. Koorbanally, N.A. et al., Bioactive constituents of *Cedrelopsis microfoliata*, *J. Nat. Prod.*, 65, 1349, 2002.

80. Ahsan, M. et al., Novel *O*-prenylated flavonoids from two varieties of *Boronia coerulescens*, *Phytochemistry*, 37, 259, 1994.

81. Siddiqui, B.S. et al., Chemical constituents of the flowers of *Azadirachta indica*, *Helv. Chim. Acta*, 86, 2787, 2003.

82. Iinuma, M., Ohyama, M., and Toshiyuki, T., Three new phenolic compounds from the roots of *Sophora leachiana*, *J. Nat. Prod.*, 56, 2212, 1993.

83. Garo, E., Wolfender, J.-L., and Hostettmann, K., Prenylated flavanones from *Monotes engleri*: on-line structure elucidation by LC/UV/NMR, *Helv. Chim. Acta*, 81, 754, 1998.

84. Bhattacharyya, J. et al., Dioflorin, a minor flavonoid from *Dioclea grandiflora*, *J. Nat. Prod.*, 61, 413, 1998.

85. Iinuma, M. et al., Eight phenolic compounds in roots of *Sophora exigua*, *Phytochemistry*, 35, 785, 1994.

86. Ichimaru, M. et al., Structural elucidation of new flavanones isolated from *Erythrina abyssinica*, *J. Nat. Prod.*, 59, 1113, 1996.

87. Iinuma, M. et al., Five flavonoid compounds from *Echinosophora koreensis*, *Phytochemistry*, 33, 1241, 1993.

88. Tanaka, T. et al., Flavonoids in the root bark of *Pongamia pinnata*, *Phytochemistry*, 31, 993, 1992.

89. Souleles, C., A new prenylated flavanone from *Melilotus alba*, *Sci. Pharm.*, 60, 10, 1992.

90. Iinuma, M. et al., Three 2′,4′,6′-trioxygenated flavanones in roots of *Echinosophora koreensis*, *Phytochemistry*, 31, 2855, 1992.

91. Lu, C.-M. and Lin, C.-N., Flavonoids and 9-hydroxytridecyldocosanoate from *Artocarpus heterophyllus*, *Phytochemistry*, 35, 781, 1994.

92. Kim, Y.H. et al., Prenylated flavanones from *Derris laxiflora*, *Nat. Prod. Lett.*, 6, 223, 1995.

93. Randriaminaly, M. et al., Lipophilic constituents from *Helichrysum* species endemic to Madagascar, *Z. Naturforsch.*, 47C, 10, 1992.

94. Hano, Y. et al., Constituents of the Moraceae plants. 25. Paratocarpins F–L, seven new isoprenoid substituted flavonoids from *Paratocarpus venenosa* Zoll., *Heterocycles*, 41, 2313, 1995.

95. Schütz, B.A. et al., Prenylated flavanones from the leaves of *Macaranga pleiostemona*, *Phytochemistry*, 40, 1273, 1995.

96. Yenesew, A. et al., Two prenylated flavonoids from the stem bark of *Erythrina burtii*, *Phytochemistry*, 63, 445, 2003.

97. Yenesew, A. et al., Two prenylated flavanones from stem bark of *Erythrina burttii*, *Phytochemistry*, 48, 1439, 1998.

98. Moriyasu, M. et al., Minor flavonoids from *Erythrina abyssinica*, *J. Nat. Prod.*, 61, 185, 1998.

99. Li, J., Yuan, H., and Wang, M., Two flavanones from the root bark of *Lespedeza formosa*, *Phytochemistry*, 31, 3664, 1992.

100. Dawidar, A.M. et al., Prenylstilbenes and prenylflavanones from *Schoenus nigricans*, *Phytochemistry*, 36, 803, 1994.

101. Matsuura, N. et al., A prenylated flavanone from roots of *Maackia amurensis* subsp. *buergeri*, *Phytochemistry*, 36, 255, 1994.

102. Iinuma, M. et al., Three 2′,4′,6′-trioxygenated flavanones in roots of *Echinosophora koreensis*, *Phytochemistry*, 31, 665, 1992.

103. Ohyama, M., Tanaka, T., and Iinuma, M., A prenylated flavanone from roots of *Amorpha fruticosa*, *Phytochemistry*, 48, 907, 1998.

104. Seo, E.-K. et al., Cytotoxic prenylated flavanones from *Monotes engleri*, *Phytochemistry*, 45, 509, 1997.

105. Huang, Y.-L., Antioxidant flavonoids from the rhizomes of *Helminthostachys zeylanica*, *Phytochemistry*, 64, 1277, 2003.

106. Caffaratti, M. et al., Prenylated flavanones from *Dalea elegans*, *Phytochemistry*, 36, 1083, 1994.

107. Nanayakkara, N.P., Burardt, C.L. Jr., and Jacob, M.R., Flavonoids with activity against methicillin-resistant *Staphylococcus aureus* from *Dalea scandens* var. *paucifolia*, *Planta Med.*, 68, 519, 2002.

108. Iinuma, M., Ohyama, M., and Tanaka, T., Flavonoids in roots of *Sophora prostrata*, *Phytochemistry*, 38, 539, 1995.

109. Fukai, T. et al., Components of the root bark of *Morus cathayana*. 1. Structures of five new isoprenylated flavonoids, sanggenols A–E and a diprenyl-2-arylbenzofuran, mulberrofuran V, *Heterocycles*, 43, 425, 1996.

110. Sun, S.G., Chen, R.Y., and Yu, D.Q., A new flavanone from the bark of *Morus macroura* Miq., *Chin. Chem. Lett.*, 12, 233, 2001.

111. Tanaka, T. et al., Flavonoids from the root and stem of *Sophora tomentosa*, *Phytochemistry*, 46, 1431, 1997.

112. Iinuma, M. et al., Two flavanones from roots of *Sophora leachiana*, *Phytochemistry*, 31, 721, 1992.

113. Deug, Y. et al., New antimicrobial flavanones from *Physena madagascariensis*, *J. Nat. Prod.*, 63, 1082, 2000.

114. Iinuma, M., Ohyama, M., and Tanaka, T., Six flavonostilbenes and a flavanone in roots of *Sophora alopecuroides*, *Phytochemistry*, 38, 519, 1995.

115. Ruangrungsi, N. et al., Three flavanones with a lavandulyl group in the roots of *Sophora exigua*, *Phytochemistry*, 31, 999, 1992.

116. Iinuma, M. et al., Seven phenolic compounds in the roots of *Sophora exigua*, *Phytochemistry*, 33, 203, 1993.

117. Kang, T.-H. et al., Cytotoxic lavandulyl flavanones from *Sophora flavescens*, *J. Nat. Prod.*, 63, 680, 2000.
118. Tan, R.X. et al., Acyl secoiridoids and antifungal constituents from *Gentiana macrophylla*, *Phytochemistry*, 42, 1305, 1996.
119. Ahsan, M. et al., Farnesyl acetophenone and flavanone compounds from the aerial parts of *Boronia ramosa*, *J. Nat. Prod.*, 57, 673, 1994.
120. Parsons, I.C., Gray, A.I., and Waterman, P.G., New triterpenes and flavonoids from the leaves of *Bosistoa brassii*, *J. Nat. Prod.*, 56, 46, 1993.
121. Herath, W. et al., Identification and biological activity of microbial metabolites of xanthohumol, *Chem. Pharm. Bull.*, 51, 1237, 2003.
122. Tahara, S. et al., Prenylated flavonoids in the roots of yellow lupin, *Phytochemistry*, 36, 1261, 1994.
123. Balasubramanian, C. et al., Flavonoid from resin glands of *Azadirachta indica*, *Phytochemistry*, 34, 1194, 1993.
124. Hayashi, H. et al., Field survey of *Glycyrrhiza* plants in Central Asia (2). Characterization of phenolics and their variation in the leaves of *Glycyrrhiza* plants collected in Kazakhstan, *Chem. Pharm. Bull.*, 51, 1147, 2003.
125. Yadava, R.N. and Singh, R.K., A novel epoxyflavanone from *Atylosia scarabaeoides* roots, *Fitoterapia*, 69, 122, 1998.
126. Ngadjui, B. et al., Prenylated flavanones from the twigs of *Dorstenia mannii*, *Phytochemistry*, 50, 1401, 1999.
127. Fukai, T. et al., Phenolic constituents of *Glycyrrhiza* species. 19. Phenolic constituents of aerial parts of *Glycyrrhiza eurycarpa*, *Nat. Med.*, 50, 247, 1996.
128. Yamamoto, H., Senda, M., and Inoue, K., Flavanone 8-dimethylallyltransferase in *Sophora flavescens* cell suspension cultures, *Phytochemistry*, 54, 649, 2000.
129. Milligan, S.R. et al., Identification of a potent phytoestrogen in hops (*Humulus lupulus* L.) and beer, *J. Clin. Endocrinol. Metabol.*, 84, 2249, 1999.
130. Rao, M.S. et al., A revised structure for crotaramosmin from *Crotolaria ramosissima*, *J. Nat. Prod.*, 61, 1148, 1998.
131. Sritularak, B. et al., Flavonoids from the roots of *Millettia erythrocalyx*, *Phytochemistry*, 61, 943, 2002.
132. Machocho, A.K. et al., Three new flavonoids from the root of *Tephrosia emoroides* and their antifeedant activity against the larvae of the spotted stalk borer *Chilo partellus* Swinhoe, *Int. J. Pharmacogn.*, 33, 222, 1995.
133. Magalhães, A.F. et al., Flavonoids from *Lonchocarpus latifolius* roots, *Phytochemistry*, 55, 787, 2000.
134. Magalhães, A.F. et al., Twenty-three flavonoids from *Lonchocarpus subglaucescens*, *Phytochemistry*, 42, 1459, 1996.
135. Jang, D.S. et al., Prenylated flavonoids of the leaves of *Macaranga conifera* with inhibitory activity against cyclooxygenase-2, *Phytochemistry*, 61, 867, 2002.
136. Lee, D. et al., Aromatase inhibitors from *Broussonetia papyrifera*, *J. Nat. Prod.*, 64, 1286, 2001.
137. Wu, T.S. et al., Constituents of leaves of *Phellodendron chinense* var. *glabriusculum*, *Heterocycles*, 60, 397, 2003.
138. Herath, W.H.M.W., Ferreira, D., and Khan, I.A., Microbial transformation of xanthohumol, *Phytochemistry*, 62, 673, 2003.
139. Wattanapiromsakul, C. and Waterman, P.G., Flavanone, triterpene and chromene derivatives from the stems of *Paramignya griffithii*, *Phytochemistry*, 55, 269, 2000.
140. Ngadjui, B.T. et al., Prenylated flavonoids from the aerial parts of *Dorstenia mannii*, *Phytochemistry*, 55, 915, 2000.
141. Rao, E.V., Prasad, Y.R., and Murthy, M.S.R., A prenylated flavanone from *Tephrosia maxima*, *Phytochemistry*, 37, 111, 1994.
142. Ngadjui, B.T. et al., Prenylated flavonoids and a dihydro-4-phenylcoumarin from *Dorstenia poinsettifolia*, *Phytochemistry*, 51, 119, 1999.
143. Rao, E.V. and Prasad, Y.R., Prenylated flavonoids from *Tephrosia spinosa*, *Phytochemistry*, 32, 183, 1993.

144. Ngadjui, B.T. et al., Dinklagins A, B and C: three prenylated flavonoids and other constituents from the twigs of *Dorstenia dinklagei*, *Phytochemistry*, 61, 99, 2002.
145. Kitagawa, I. et al., Chemical studies of Chinese licorice roots. I. Elucidation of five new flavonoid constituents from the roots of *Glycyrrhiza glabra* L. collected in Xinjiang, *Chem. Pharm. Bull.*, 42, 1056, 1994.
146. Lo, W.-L. et al., The constituents of *Euchresta formosana*, *J. Chin. Chem. Soc. (Taipei, Taiwan)*, 49, 421, 2002.
147. Ngadjui, B.T. et al., Geranylated and prenylated flavonoids from the twigs of *Dorstenia mannii*, *Phytochemistry*, 48, 349, 1998.
148. Wu, T.-S., Chang, F.-C., and Wu, P.-L., Flavonoids, amidosulfoxides and an alkaloid from the leaves of *Glycosmis citrifolia*, *Phytochemistry*, 39, 1453, 1995.
149. Patil, A.D. et al., A new dimeric dihydrochalcone and a new prenylated flavone from the bud covers of *Artocarpus altilis*: inhibitors of cathepsin K, *J. Nat. Prod.*, 65, 624, 2002.
150. Lo, W.-L. et al., Coumaronochromones and flavanones from *Euchresta formosana* roots, *Phytochemistry*, 60, 839, 2002.
151. Fang, N. and Casida, J.E., New bioactive flavonoids and stilbenes in cubé resin insecticide, *J. Nat. Prod.*, 62, 205, 1999.
152. Matsuura, N. et al., Flavonoids and a benzofuran in roots of *Euchresta tubulosa*, *Phytochemistry*, 33, 701, 1993.
153. Rahman, M.M. and Gray, A.I., Antimicrobial constituents from the stem bark of *Feronia limonia*, *Phytochemistry*, 59, 73, 2002.
154. Mahidol, C. et al., Two new pyranoflavanones from the stems of *Derris reticulata*, *Heterocycles*, 57, 1287, 2002.
155. Shi, Y.-Q. et al., Phenolic constituents of the root bark of Chinese *Morus australis*, *Nat. Med.*, 55, 143, 2001.
156. Shi, Y.-Q. et al., Cytotoxic flavonoids with isoprene groups from *Morus mongolica*, *J. Nat. Prod.*, 64, 181, 2001.
157. Mahidol, C. et al., Prenylated flavanones from *Derris reticulata*, *Phytochemistry*, 45, 825, 1997.
158. Tseng, M.-H. et al., Allelopathic prenylflavanones from the fallen leaves of *Macaranga tanarius*, *J. Nat. Prod.*, 64, 827, 2001.
159. Prawat, H., Mahidol, C., and Ruchirawat, S., Reinvestigation of *Derris reticulata*, *Pharm. Biol.*, 38, 63, 2000.
160. Lin, Y.L., Chen, Y.L., and Kuo, Y.H., Two new flavanones and two new chalcones from the root of *Derris laxiflora* Benth., *Chem. Pharm. Bull.*, 40, 2295, 1992.
161. Nkengfack, A.E. et al., Further flavonoids from *Erythrina* species, *Phytochemistry*, 32, 1305, 1993.
162. Ma, W.G. et al., Polyphenols from *Eriosema tuberosum*, *Phytochemistry*, 39, 1049, 1995.
163. Zhang, H.L. and Chen, R.Y., A new flavanone from the roots of *Uvaria macrophylla*, *Chin. Chem. Lett.*, 12, 791, 2001.
164. Wang, S. et al., A novel dihydroflavone from the roots of *Uvaria macrophylla*, *Chin. Chem. Lett.*, 13, 857, 2002.
165. Ekpa, O.D., Anam, E.M., and Gariboldi, P.V., Uvarinol and novel isouvarinol: two *C*-benzylated flavanones from *Xylopia africana* (Annonaceae), *Indian J. Chem.*, 32B, 1295, 1993.
166. Anam, E.M., 2″″-Hydroxy-3‴-benzyluvarinol, 2″″-hydroxy-5″″-benzylisouvarinol-A and 2‴-hydroxy-5″-benzylisouvarinol-B: three novel tetra-*C*-benzylated flavanones from the root extract of *Xylopia africana* (Benth.) Oliver (Annonaceae), *Indian J. Chem.*, 33B, 1009, 1994.
167. Iinuma, M. et al., A flavonostilbene and two stilbene oligomers in roots of *Sophora leachiana*, *Phytochemistry*, 37, 1157, 1994.
168. Yoshikawa, M. et al., Anastatins A and B, new skeletal flavonoids with hepatoprotective activities from the desert plant *Anastatica hierochuntica*, *Bioorg. Med. Chem. Lett.*, 13, 1045, 2003.
169. Tsui, W.-Y. and Brown, G.D., Unusual metabolites of *Baeckea frutescens*, *Tetrahedron*, 52, 9735, 1996.
170. Makino, M. and Fujimoto, Y., Flavanones from *Baeckea frutescens*, *Phytochemistry*, 50, 273, 1999.

171. Lee, C.-K., Leucadenone A–D, the novel class flavanone from the leaves of *Melaleuca leucadendron* L., *Tetrahedron Lett.*, 40, 7255, 1999.

172. Labbe, C. et al., Bioactive flavanones from *Luma checken*, *Coll. Czech. Chem. Commun.*, 67, 115, 2002.

173. Asai, F. et al., Five complex flavonoids in the farinose exudate of *Pityrogramma calomelanos*, *Phytochemistry*, 31, 2487, 1992.

174. Kadota, S. et al., Novel diarylheptanoids of *Alpinia blepharocalyx*, *Curr. Top. Med. Chem.*, 3, 203, 2003.

175. Fu, X. et al., Flavanone and chalcone derivatives from *Cryptocarya kurzii*, *J. Nat. Prod.*, 56, 1153, 1993.

176. Chang, L.C. et al., Absolute configuration of novel bioactive flavonoids from *Tephrosia purpurea*, *Org. Lett.*, 2, 515, 2000.

177. Roy, R. and Pandey, V.B., A chalcone glycoside from *Clerodendron phlomidis*, *Indian J. Nat. Prod.*, 11, 13, 1995.

178. De Lima, D.P. et al., A flavanone glycoside from *Cochlospermum regium*, *Fitoterapia*, 66, 545, 1995.

179. Yuldashev, M.P., New flavanone glucoside from *Glycyrrhiza glabra*, *Chem. Nat. Comp. (Eng. Trans. of Khim. Prir. Soedin.)*, 37, 224, 2001.

180. Feng, H. et al., Studies on chemical constituents from *Penthorum chinense* Pursh., *Zhongguo Zhongyao Zazhi*, 26, 260, 2001.

181. Xu, Y., Kubo, I., and Ma, Y., A cytotoxic flavanone glycoside from *Onychium japonicum*: structure of onychin, *Phytochemistry*, 33, 510, 1993.

182. Lin, J.-H., Chiou, Y.-N., and Lin, Y.-L., Phenolic glycosides from *Viscum angulatum*, *J. Nat. Prod.*, 65, 638, 2002.

183. Bilia, A.R. et al., Flavonoids from *Pyracantha coccinea* roots, *Phytochemistry*, 33, 1449, 1993.

184. Wang, B. et al., Two new flavanone glycosides from *Glycyrrhiza inflata*, *Yaoxue Xuebao*, 32, 199, 1997.

185. Yin, S., A novel flavanone glycoside from *Glycyrrhiza uralensis* Fisch., *Zhangguo Yaoke Daxue Xuebao*, 50, 19, 1999.

186. Hatano, T. et al., Acylated flavonoid glycosides and accompanying phenolics from licorice, *Phytochemistry*, 47, 287, 1998.

187. Liu, Q., Dixon, R.A., and Mabry, T.J., Additional flavonoids from elicitor-treated cell cultures of *Cephalocereus senilis*, *Phytochemistry*, 34, 167, 1993.

188. Yuldashev, M.P. et al., Structure of two new flavonoids from *Scutellaria ramosissima*, *Khim. Prir. Soedin.*, 355, 1994.

189. Yuldashev, M.P., Batirov, E.Kh., and Malikov, V.M., Flavonoids of the epigeal part of *Scutellaria ramosissima*, *Khim. Prir. Soedin.*, 178, 1992.

190. Yadava, R.N., Mathews, S.R., and Jain, N., Naringenin 4'-*O*-α-L-rhamnopyranoside, a novel flavanone glycoside from the stem of *Crotalaria striata* D.C., *J. Indian Chem. Soc.*, 74, 426, 1997.

191. El-Shafae, A.M. and El-Domiaty, M.M., Flavonoidal constituents of the flowers of *Acacia saligna*, *Zagazig J. Pharm. Sci.*, 7, 48, 1998.

192. Yuldashev, M.P. et al., Acylated flavanone glycosides from *Ricinus communis*, *Khim. Prir. Soedin.*, 362, 1993.

193. Harraz, F.M. et al., Acylated flavonoids from *Blepharis ciliaris*, *Phytochemistry*, 43, 521, 1996.

194. Barakat, H.H. et al., Flavonoid galloyl glucosides from the pods of *Acacia farnesiana*, *Phytochemistry*, 51, 139, 1999.

195. Tanaka, N. et al., Chemical and chemotaxonomical studies of ferns. LXXXIII. Constituent variation of *Microlepia marginata* (1), *Yakugaku Zasshi*, 113, 70, 1993.

196. Kim, H.K., Jeon, W.K., and Ko, B.S., Flavanone glycosides from *Citrus junos* and their anti-influenza activity, *Planta Med.*, 67, 548, 2001.

197. Ali, M.S. et al., A chlorinated monoterpene ketone, acylated β-sitosterol glycosides and a flavanone glycoside from *Mentha longifolia* (Lamiaceae), *Phytochemistry*, 59, 889, 2002.

198. Srivastava, R., Chauhan, D., and Chauhan, J.S., Flavonoid diglycosides from *Punica granatum*, *Indian J. Chem.*, 40B, 170, 2001.

199. Singh, V.P., Yadav, B., and Pandey, V.B., Flavanone glycosides from *Alhagi pseudoalhagi*, *Phytochemistry*, 51, 587, 1999.

200. Anam, E.M., Novel flavone and chalcone glycosides from *Clerodendron phlomidis* (Verbenaceae), *Indian J. Chem.*, 36B, 897, 1997.

201. Yadava, R.N. and Tripathi, P., A novel flavanone glycoside from *Albizzia procera* Benth., *J. Inst. Chem. (India)*, 71, 202, 1999.

202. Orhan, D.D., Calis, I., and Ergun, F., Two new flavonoid glycosides from *Viscum album* ssp. *album*, *Pharm. Biol.*, 40, 380, 2002.

203. Kim, H.K., Jeon, W.K., and Ko, B.S., Flavanone glycosides from the fruits of *Chaenomeles sinensis*, *Nat. Prod. Sci.*, 6, 79, 2000.

204. Li, F., He, Z.-S., and Ye, Y., Flavonoids from *Carthamus tinctorius*, *Chin. J. Chem.*, 20, 699, 2002.

205. Bilia, A.R. et al., Flavonoids and a saponin from *Spartium junceum*, *Phytochemistry*, 34, 847, 1993.

206. Allais, D.P. et al., 3-Desoxycallunin and 2″-acetylcallunin, two minor 2,3-dihydroflavonoid glucosides from *Calluna vulgaris*, *Phytochemistry*, 39, 427, 1995.

207. El-Sohly, H.N. et al., Flavonoids from *Maclura tinctoria*, *Phytochemistry*, 52, 141, 1999.

208. Bilia, A.R. et al., New constituents from *Pyracantha coccinea* leaves, *J. Nat. Prod.*, 55, 1741, 1992.

209. Marais, C. et al., (*S*)- and (*R*)-eriodictyol-6-*C*-β-D-glucopyranoside, novel keys to the fermentation of rooibos (*Aspalathus linearis*), *Phytochemistry*, 55, 43, 2000.

210. Matsuda, H. et al., Medicinal flowers. VI. Absolute stereostructures of two new flavanone glycosides and a phenylbutanoid glycoside from the flowers of *Chrysanthemum indicum* L.: their inhibitory activities for rat lens aldose reductase, *Chem. Pharm. Bull.*, 50, 972, 2002.

211. Zhang, Y.-J. et al., Two new acylated flavanone glycosides from the leaves and branches of *Phyllanthus emblica*, *Chem. Pharm. Bull.*, 50, 841, 2002.

212. Roemmelt, S. et al., Formation of novel flavonoids in apple (*Malus × domestica*) treated with the 2-oxoglutarate-dependent dioxygenase inhibitor prohexadione-Ca, *Phytochemistry*, 64, 709, 2003.

213. Wang, J. et al., Flavonoids from *Bidens pilosa* var. *radiata*, *Phytochemistry*, 46, 1275, 1997.

214. Yadava, R.N., Singh, A., and Reddy, K.I.S., Flavanone glycoside from seeds of *Crotalaria prostrata*, *J. Inst. Chem. (India)*, 71, 231, 1999.

215. Emam, A.M. et al., Two flavonoid triglycosides from *Buddleja madagascariensis*, *Phytochemistry*, 48, 739, 1998.

216. Chen, Y. et al., Isolation and structure of clematine, a new flavanone glycoside from *Clematis armandii* Franch., *Tetrahedron*, 49, 5169, 1993.

217. De Carvalho, M.G., Da Costa, P.M., and Dos Santos Abreu, H., Flavanones from *Vernonia diffusa*, *J. Braz. Chem. Soc.*, 10, 163, 1999.

218. Rawat, M.S.M. et al., A persicogenin 3′-glucoside from the stem bark of *Prunus amygdalus*, *Phytochemistry*, 38, 1519, 1995.

219. Chou, C.J., Ko, H.C., and Lin, L.C., Flavonoid glycosides from *Viscum alniformosanae*. *J. Nat. Prod.*, 62, 1421, 1999.

220. Srivastava, S.K., Srivastava, S.D., and Chouksey, B.K., New antifungal constituents from *Terminalia alata*, *Fitoterapia*, 72, 106, 2001.

221. Tewtrakul, S. et al., Flavanone and flavonol glycosides from the leaves of *Thevetia peruviana* and their HIV-1 transcriptase and HIV-1 integrase inhibitory activities, *Chem. Pharm. Bull.*, 50, 630, 2002.

222. Zheng, S. et al., Flavonoid constituents from *Mosla chinensis* Maxim., *Indian J. Chem.*, 35B, 392, 1996.

223. Kadota, S. et al., Matteuorienate A and B, two new and potent aldose reductase inhibitors from *Matteucia orientalis* (Hook.) Trev., *Chem. Pharm. Bull.*, 42, 1712, 1994.

224. Zhang, Z. et al., Flavanone glycosides from *Miconia traillii*, *J. Nat. Prod.*, 66, 39, 2003.

225. Takahashi, H. et al., Triterpene and flavanone glycoside from *Rhododendron simsii*, *Phytochemistry*, 56, 875, 2001.

226. Li, X.-C. et al., Phenolic compounds from *Miconia myriantha* inhibiting *Candida* aspartic proteases, *J. Nat. Prod.*, 64, 1282, 2001.

227. Sreedevi, E. and Rao, J.T., A new prenylated flavanone glycoside from the seeds of *Gliricidia maculata*, *Fitoterapia*, 71, 392, 2000.
228. Masayoshi, O. et al., Phenolic compounds isolated from the roots of *Sophora stenophylla*, *Chem. Pharm. Bull.*, 46, 663, 1998.
229. Economides, C. and Adam, K.-P., Lipophilic flavonoids from the fern *Woodsia scopulina*, *Phytochemistry*, 49, 859, 1998.
230. Su, B.-N. et al. Activity-guided isolation of the chemical constituents of *Muntingia calabura* using a quinone reductase induction assay, *Phytochemistry*, 63, 335, 2003.
231. Islam, M.T. and Tahara, S., Dihydroflavonols from *Lannea coromandelica*, *Phytochemistry*, 54, 901, 2000.
232. Öksüz, S. and Topcu, G., A eudesmanolide and other constituents from *Inula graveolens*, *Phytochemistry*, 31, 195, 1992.
233. Parmar, V.S. et al., Dihydroflavonols from *Prunus domestica*, *Phytochemistry*, 31, 2185, 1992.
234. Barua, N.C. and Sharma, R.P., (2*R*,3*R*)-7,5'-Dimethoxy-3,5,2'-trihydroxyflavanone from *Blumea balsamifera*, *Phytochemistry*, 31, 4040, 1992.
235. Yoshikawa, M. et al., Bioactive constituents of Chinese natural medicines. III. Absolute stereostructures of new dihydroflavonols, hovenitins I, II and III, isolated from Hoveniae Semen Seu Fructus, the seed and fruit of *Hovenia dulcis* Thumb. (Rhamnaceae): inhibitory effect on alcohol-induced muscular relaxation and hepatoprotective activity, *Yakugaku Zasshi*, 117, 108, 1997.
236. Bennini, B. et al., (2*R*,3*R*)-Dihydroflavonol aglycone and glycosides from *Erica cinerea*, *Phytochemistry*, 33, 1233, 1993.
237. Dinda, B., Chel, G., and Achari, B., A dihydroflavonol from *Plumbago indica*, *Phytochemistry*, 35, 1083, 1994.
238. Stuppner, H. and Müller, E.P., Rare flavonoid aglycones from *Flourensia retinophylla*, *Phytochemistry*, 37, 1185, 1994.
239. Tsopmo, A., Tchuendem, M.H.K., and Ayafor, J.F., 3-Acetoxy-5,7-dihydroxy-4'-methoxyflavanone, from *Afromomum hanburyi*, *Nat. Prod. Lett.*, 9, 33, 1996.
240. Wollenweber, E., Dörr, M., and Stevens, J.F., A dihydroflavonol with taxonomic significance from the fern *Notholaena sulphurea*, *Z. Naturforsch.*, 56C, 499, 2001.
241. Maatooq, G.T., Trifolexin: a new flavanonol derivative from *Trifolium alexandrium* seeds, *Mansoura J. Pharm. Sci.*, 13, 70, 1997.
242. Bhattacharyya, J. et al., Dioclenol, a minor flavanonol from the root-bark of *Dioclea grandiflora*, *Phytochemistry*, 46, 385, 1997.
243. Vilegas, W. et al., Coumarins and a flavonoid from *Pterocaulon alopecuroides*, *Phytochemistry*, 38, 1017, 1994.
244. Lemos, V.S. et al., Total assignments of ^{1}H and ^{13}C NMR spectra of a new prenylated flavanone from *Dioclea grandiflora*, *Magn. Reson. Chem.*, 40, 793, 2002.
245. Rao, K.V. and Gunasekar, D., *C*-Prenylated dihydroflavonol from *Rhynchosia densiflora*, *Phytochemistry*, 48, 1453, 1998.
246. Nia, M.A. et al., Two new prenylated flavonoids from *Paracalyx scariosa*, *J. Nat. Prod.*, 55, 1152, 1992.
247. Meragelman, K.M., McKee, T.C., and Boyd, M.R., Anti-HIV prenylated flavonoids from *Monotes africanus*, *J. Nat. Prod.*, 64, 546, 2001.
248. Hufford, C.D. et al., Antimicrobial compounds from *Petalostemum purpureum*, *J. Nat. Prod.*, 56, 1878, 1993.
249. Fukai, T. et al., Isoprenylated flavanones from *Morus cathayana*, *Phytochemistry*, 47, 273, 1998.
250. Ryu, S.Y. et al., A novel flavonoid from *Sophora flavescens*, *Planta Med.*, 62, 361, 1996.
251. Satyanarayana, P., Anjaneyulu, V., and Viswanatham, K.N., Two dihydroflavonols from *Mundulea suberosa* Benth., *Indian J. Chem.*, 35B, 1235, 1996.
252. Fukai, T. et al., An isoprenylated flavanone from *Glycyrrhiza glabra* and rec-assay of licorice phenols, *Phytochemistry*, 49, 2005, 1998.
253. Alavez-Solano, D. et al., Flavanones and 3-hydroxyflavanones from *Lonchocarpus oaxacensis*, *Phytochemistry*, 55, 953, 2000.

254. Magalhães, A.F. et al., Dihydroflavonols and flavanones from *Lonchocarpus atropurpureus* roots, *Phytochemistry*, 52, 1681, 1999.

255. Rao, E.V. et al., A prenylated dihydroflavonol from *Mundulea suberosa*, *Phytochemistry*, 50, 1417, 1999.

256. Messana, I. et al., Three new flavanone derivatives from the root bark of *Sorocea bonplandii* Baillon., *Heterocycles*, 38, 1287, 1994.

257. Ferrari, F. and Messana, I., Prenylated flavanones from *Sorocea ilicifolia*, *Phytochemistry*, 38, 251, 1995.

258. Lee, I.-K. et al., Two benzylated dihydroflavonols from *Cudrania tricuspidata*, *J. Nat. Prod.*, 58, 1614, 1995.

259. Lee, I.-K. et al., Cytotoxic benzyl dihydroflavonols from *Cudrania tricuspidata*, *Phytochemistry*, 41, 213, 1996.

260. Kuo, Y.-H., Yeh, M.-H., and Huang, S.-L., A novel 6-butyl-3-hydroxyflavanone from heartwood of *Bauhinia purpurea*, *Phytochemistry*, 49, 2529, 1998.

261. Sievers, H. et al., Hypnogenols and other dihydroflavonols from the moss *Hypnum cupressiforme*, *Phytochemistry*, 31, 3233, 1992.

262. Sievers, H. et al., Further biflavonoids and 3'-phenylflavonoids from *Hypnum cupressiforme*, *Phytochemistry*, 35, 795, 1994.

263. Kikuzaki, H. and Tesaki, S., New flavonol-phenylbutadiene adducts from the leaves of *Alpinia flabellata*, *J. Nat. Prod.*, 65, 389, 2002.

264. Fatope, M.O. et al., Muscanone: a 3-*O*-(1'',8'',14''-trimethylhexadecanyl)naringenin from *Commiphora wightii*, *Phytochemistry*, 62, 1251, 2003.

265. Elliger, C.A. and Halloin, J.M., Phenolics induced in *Beta vulgaris* by *Rhizoctonia solani* infection, *Phytochemistry*, 37, 691, 1994.

266. Nomura, T., Phenolic compounds of the mulberry tree and related plants, *Prog. Chem. Org. Nat. Prod.*, 53, 87, 1988.

267. Hano, Y. et al., Revised structure of sanggenol A, *Heterocycles*, 45, 867, 1997.

268. Yadava, R.N. and Singh, A., A novel flavanone glycoside from the seeds of *Crotalaria laburnifolia*, *Fitoterapia*, 64, 276, 1993.

269. Binitu, O.A. and Cordell, G.A., Constituents of *Afzelia bella* stem bark, *Phytochemistry*, 56, 827, 2001.

270. Mendez, J. et al., 5,7,2',5'-Tetrahydroxydihydroflavonol 3-rhamnoside from *Plinia pinnata*, *Phytochemistry*, 36, 1087, 1994.

271. Chosson, E. et al., Dihydroflavonol glycosides from *Rhododendron ferrugineum*, *Phytochemistry*, 49, 1431, 1998.

272. Ishimaru, K. et al., Taxifolin 3-arabinoside from *Fragaria × ananassa*, *Phytochemistry*, 40, 345, 1995.

273. Kraft, C. et al., Andinermals A–C, antiplasmodial constituents from *Andira inermis*, *Phytochemistry*, 58, 769, 2001.

274. Wu, Q.-L. et al., Chromone glycosides and flavonoids from *Hypericum japonicum*, *Phytochemistry*, 49, 1417, 1998.

275. Inada, A. et al., Phytochemical studies of seeds of medicinal plants. II. A new dihydroflavonol glycoside and a new 3-methyl-1-butanol glycoside from seeds of *Platycodon grandiflorum* A. De Candolle, *Chem. Pharm. Bull.*, 40, 3081, 1992.

276. Bergeron, C. et al., Flavonoids from *Pyrola elliptica*, *Phytochemistry*, 49, 233, 1998.

277. Konishi, T. et al., Three diterpenoids (excoecarins V1–V3) and a flavanone glycoside from the fresh stem of *Excoecaria agallocha*, *Chem. Pharm. Bull.*, 51, 1142, 2003.

278. Chen, T. et al., A new flavanone isolated from Rhizoma Smilacis Glabrae and the structural requirements of its derivatives for preventing immunological hepatocyte damage, *Planta Med.*, 65, 56, 1999.

279. Chen, T. et al., A flavanonol glycoside from *Smilax glabra*, *Chin. Chem. Lett.*, 13, 537, 2002.

280. Guo, S. et al., Studies on flavonoids in leaves of *Phellodendron chinense*, *Jinan Daxue Xuebo Ziran Kexue Yu Yixueban*, 19, 68, 1998.

281. Likhitwitayawuid, K. et al., Flavonoids from *Ochna integerrima*, *Phytochemistry*, 56, 353, 2001.

APPENDIX

Checklist of Known Flavanones and Dihydroflavonols [a]

FLAVANONES
Flavanone aglycones bearing hydroxy, methoxy, and methylenedioxy substituents only
Mono-*O*-substituted
1. 7-OH (H2/1011)
2. 7-OMe (H2/1012)

Di-*O*-substituted
3. 5,7-DiOH (pinocembrin, H2/1022)
4. 7,4'-DiOH (liquiritigenin, H2/1082)
5. 5-OH-7-OMe (pinostrobin, H2/1031)
6. 7-OH-5-OMe (alpinetin, H2/1030)
7. 7-OH-8-OMe (isolarrein, H2/1013)
8. 7-OH-4'-OMe (H2/1342)
9. 4'-OH-7-OMe (H2/1090)
10. 5,7-DiOMe (H2/1032)
11. 7,8-DiOMe (H2/1015)

Tri-*O*-substituted
12. 2,5,7-TriOH (H2/1319)
13. 5,6,7-TriOH (dihydrobaicalein, H2/1033)
14. 5,7,8-TriOH (dihydronorwogonin, H2/1038)
15. 5,7,2'-TriOH (H2/1392)
16. 5,7,3'-TriOH (H2/1393)
17. 5,7,4'-TriOH (naringenin, H2/1098)
18. 5,8,4'-TriOH (H2/1304)
19. 7,8,4'-TriOH (H2/1081)
20. 7,3',4'-TriOH (butin, H2/1092)
21. 5,6-DiOH-4'-OMe (H2/1303)
22. 5,7-DiOH-6-OMe (dihydrooroxylin, H2/1035)
23. 5,7-DiOH-8-OMe (dihydrowogonin, H2/1043)
24. 5,7-DiOH-4'-OMe (isosakuranetin, H2/1119)
25. 5,8-DiOH-7-OMe (H2/1357)
26. 5,2'-DiOH-7-OMe (dihydroechioidin, **1**)
27. 5,4'-DiOH-7-OMe (sakuranetin, H2/1117)

[a]This checklist of flavanones and dihydroflavonols contains compounds of these flavonoid classes reported in the literature as natural products to the end of 2003. Compounds recorded before 1992 are cross-referenced to numbered entries in volume 2 of the *Handbook of Natural Flavonoids*, using the prefix "H2."[15] Compounds reported from 1992 to 2003 are shown by numbers in bold typeface, corresponding to the entries in Table 15.1 to Table 15.10. The compounds are arranged into flavanone aglycones, flavanone glycosides, dihydroflavonol aglycones, and dihydroflavonol glycosides, and within these classes as (i) derivatives with hydroxy, methoxy, and methylenedioxy substituents; (ii) derivatives with methyl, hydroxymethyl, and formyl substituents; (iii) derivatives with prenyl, geranyl, and lavandulyl substituents; (iv) derivatives with isoprenyl ring structures; (v) benzylated derivatives; (vi) miscellaneous structures. Within these groups the compounds are generally arranged according to the number of *O*-substituents on the flavonoid skeleton and compounds with the same number of *O*-substituents are grouped numerically according to the numbering of the A- and B-rings. The unprimed numbers (A-ring) precede the primed numbers (B-ring), and fully hydroxylated compounds come before those with methoxy groups. Only the (2*R*)-configuration of flavanones and not the more common (2*S*)-configuration has been indicated in the checklist. Similarly, the common (2*R*,3*R*)-configuration has generally not been given in the checklist for dihydroflavonols, and only in cases where several different configurations of the same dihydroflavonol have been found is this indicated in the table. Sugars are assumed to be *O*-linked unless specified as *C*-linked.

28. 7,4'-DiOH-5-OMe (H2/1126)
29. 5-OH-6,7-diOMe (onysilin, H2/1036)
30. 5-OH-7,8-diOMe (H2/1041)
31. 5-OH-7,4'-diOMe (H2/1127)
32. 6-OH-5,7-diOMe (H2/1037)
33. 7-OH-5,4'-diOMe (tsugafolin, H2/1349)
34. 8-OH-5,7-diOMe (H2/1042)
35. 3'-OH-7,4'-diOMe (tithonin, H2/1093)
36. 4'-OH-5,7-diOMe (H2/1128)
37. 5,7,8-TriOMe (H2/1045)
38. 5,7,4'-TriOMe (H2/1131)
39. 6,3',4'-TriOMe (H2/1178)
Tetra-*O*-substituted
40. 5,6,7,4'-TetraOH (carthamidin, H2/1132)
41. 5,7,8,4'-TetraOH (isocarthamidin, H2/1134)
42. 5,7,2',4'-TetraOH (steppogenin, H2/1277)
43. 5,7,2',5'-TetraOH (H2/1264)
44. 5,7,2',6'-TetraOH (H2/1248)
45. 5,7,3',4'-TetraOH (eriodictyol, H2/1193)
46. 6,7,3',4'-TetraOH (plathymenin, H2/1184)
47. 7,8,3',4'-TetraOH (isookanin, H2/1185)
48. 5,7,8-TriOH-4'-OMe (**3**)
49. 5,7,3'-TriOH-4'-OMe (hesperetin, H2/1203)
50. 5,7,3'-TriOH-5'-OMe (alyssifolinone, **4**)
51. 5,7,4'-TriOH-6-OMe (H2/1135)
52. 5,7,4'-TriOH-8-OMe (H2/1358)
53. 5,7,4'-TriOH-3'-OMe (homoeriodictyol, H2/1202)
54. 5,8,2'-TriOH-7-OMe (H2/1359)
55. 5,2',3'-TriOH-7-OMe (**5**)
56. 5,2',4'-TriOH-7-OMe (artocarpanone, H2/1272)
57. 5,2',6'-TriOH-7-OMe (scutamoenin, H2/1324)
58. 5,3',4'-TriOH-7-OMe (sternbin, H2/1214)
59. 6,7,8-TriOH-5-OMe (oresbiusin, **6**)
60. 7,2',6'-TriOH-5-OMe (H2/1249)
61. 7,3',4'-TriOH-8-OMe (8-methoxybutin, H2/1188)
62. 2,5-DiOH-6,7-diOMe (mosloflavanone, **7**)
63. 5,7-DiOH-8,2'-diOMe (H2/1162)
64. 5,7-DiOH-8,4'-diOMe (**8**)
65. 5,8-DiOH-6,7-diOMe (didymocarpin A, H2/1046)
66. 5,2'-DiOH-7,8-diOMe (dihydroskullcap flavone I, **9**)
67. 5,2'-DiOH-7,4'-diOMe (**10**)
68. 5,2'-DiOH-7,5'-diOMe (**11**)
69. 5,3'-DiOH-7,2'-diOMe (**12**)
70. 5,3'-DiOH-7,4'-diOMe (persicogenin, H2/1210)
71. 5,4'-DiOH-6,7-diOMe (H2/1155)
72. 5,4'-DiOH-7,8-diOMe (H2/1157)
73. 5,4'-DiOH-7,3'-diOMe (H2/1209)
74. 7,8-DiOH-6,4'-diOMe (**13**)
75. 7,4'-DiOH-8,3'-diOMe (**14**)
76. 5-OH-6,7,8-triOMe (H2/1048)

77. 5-OH-6,7,4′-triOMe (H2/1156)
78. 5-OH-7,8,4′-triOMe (H2/1158)
79. 5-OH-7,2′,3′-triOMe (**16**)
80. 5-OH-7,3′,4′-triOMe (H2/1259)
81. 6-OH-5,7,8-triOMe (isopedicin, H2/1047)
82. 7-OH-5,8,2′-triOMe (H2/1163)
83. 7-OH-5,2′,4′-triOMe (cerasinone, H2/1265)
84. 8-OH-5,6,7-triOMe (kwangsienin A, **17**; H2/1400)
85. 4′-OH-5,6,7-triOMe (H2/1180)
86. 5,6,7,8-TetraOMe (kanakugin, H2/1049)
87. 5,6,7,4′-TetraOMe (H2/1136)
88. 5,7,2′,4′-TetraOMe (arjunone, H2/1266)
89. 5,7,2′,5′-TetraOMe (**18**)

Penta-*O*-substituted

90. 5,7,3′,4′,5′-PentaOH (H2/1252)
91. 5,7,8,3′-TetraOH-4′-OMe (**20**)
92. 5,7,2′,5′-TetraOH-6-OMe (H2/1247)
93. 5,7,3′,4′-TetraOH-6-OMe (H2/1327)
94. 5,7,3′,4′-TetraOH-8-OMe (H2/1360)
95. 5,7,4′-TriOH-8,3′-diOMe (H2/1216)
96. 5,2′,5′-TriOH-6,7-diOMe (dioclein, **21**)
97. 5,2′,5′-TriOH-7,8-diOMe (H2/1323)
98. 5,3′,4′-TriOH-7,5′-diOMe (H2/1254)
99. 5,6-DiOH-7,8,4′-triOMe (H2/1159)
100. 5,2′-DiOH-6,7,6′-triOMe (H2/1250)
101. 5,2′-DiOH-7,8,6′-triOMe (H2/1251)
102. 5,2′-DiOH-7,4′,5′-triOMe (**22**)
103. 5,3′-DiOH-6,7,4′-triOMe (H2/1217)
104. 5,3′-DiOH-7,8,4′-triOMe (H2/1218)
105. 5,4′-DiOH-6,7,8-triOMe (H2/1160)
106. 6,4′-DiOH-5,7,3′-triOMe (agestricin D, H2/1219)
107. 5-OH-6,7,8,4′-tetraOMe (H2/1179)
108. 5-OH-6,7,3′,4′-tetraOMe (H2/1328)
109. 5-OH-7,2′,4′,6′-tetraOMe (heteroflavanone A, **23**; H2/1448)
110. 5-OH-7,3′,4′,5′-tetraOMe (H2/1270)
111. 6-OH-5,7,3′,4′-tetraOMe (agestricin C, H2/1220)
112. 4′-OH-5,6,7,3′-tetraOMe (agecorynin A, **24**)
113. 5,6,7,8,4′-PentaOMe (**25**)
114. 5,6,7,3′,4′-PentaOMe (**26**)
115. 5,7,2′,3′,4′-PentaOMe (**27**)

Hexa-*O*-substituted

116. 5,7,3′-TriOH-6,4′,5′-triOMe (**29**)
117. 5,2′-DiOH-6,7,8,6′-tetraOMe (**30**; H2/1403)
118. 5,3′-DiOH-6,7,4′,5′-tetraOMe (**31**)
119. 5,6,7,3′,4′,5′-HexaOMe (**32**)

Hepta-*O*-substituted

120. 5,6,7,8,3′,4′,5′-HeptaOMe (H2/1329)
121. 5,6,7,2′,3′,4′,5′-HeptaOMe (H2/1258)

Methylene dioxy-substituted

122. 5,2′-DiOH-6,7-methylenedioxy (**2**)

123. 5,7-DiOMe-3′,4′-methylenedioxy (**15**)
124. 5,2′-DiOMe-6,7-methylenedioxy (betagarin, H2/1301)
125. 5,4′-DiOMe-6,7-methylenedioxy (H2/1161)
126. 6,7-DiOMe-3′,4′-methylenedioxy (H2/1310)
127. 7,2′-DiOMe-4′,5′-methylenedioxy (H2/1307)
128. 5,2′,3′-TriOH-6,7-methylenedioxy (**19**)
129. 6-OH-5,7-diOMe-3′,4′-methylenedioxy (agestricin B, H2/1221)
130. 5,7-DiOH-6,3′-diOMe-4′,5′-methylenedioxy (agamanone, **28**; H2/1404)
131. 5,6,7,3′-TetraOMe-4′,5′-methylenedioxy (agecorynin A, H2/1256)
132. 5,6,7,8,3′-PentaOH-4′,5′-methylenedioxy (agecorynin B, H2/1257)
Esters
133. 5,7-DiOH-7-*O*-benzoate (**33**; H2/1394)
134. 5,8-DiOH-7-OMe-8-*O*-acetate (H2/1040)
135. 5,7,4′-TriOH-3′-OMe-4′-*O*-isobutyrate (**34**)
Flavanone aglycones with *C*-methyl, *C*-hydroxymethyl, and *C*-formyl substituents
Di-*O*-substituted
136. 5,7-DiOH-6-Me (strobopinin, H2/1050)
137. 5,7-DiOH-8-Me (cryptostrobin, H2/1054)
138. 5-OH-7-OMe-6-Me (H2/1053)
139. 5-OH-7-OMe-8-Me (H2/1055)
140. 7-OH-5-OMe-6-Me (comptonin, H2/1052)
141. 7,4′-DiOMe-6-Me (H2/1091)
142. 5,7-DiOMe-6-Me (H2/1332)
143. 5,7-DiOH-6,8-diMe (desmethoxymatteucinol, H2/1056)
144. 5-OH-7-OMe-6,8-diMe (H2/1057)
145. 7-OH-5-OMe-6,8-diMe (H2/1069)
Tri-*O*-substituted
146. 5,7,4′-TriOH-3-Me (3-methylnaringenin, H2/1320)
147. 5,7,4′-TriOH-3-Me-8-Cl (H2/1321)
148. 5,7,4′-TriOH-6-Me (poriol, H2/1139)
149. 2,5-DiOH-7-OMe-6-Me (**36**)
150. 2,5-DiOH-7-OMe-8-Me (**37**)
151. 5,7-DiOH-4′-OMe-8-Me (**38**)
152. 5,4′-DiOH-7-OMe-6-Me (**39**)
153. 5,4′-DiOH-7-OMe-8-Me (**40**)
154. 5,7-DiOH-6-formyl-8-Me (lawinal, H2/1058)
155. 5-OH-7-OMe-6-formyl-8-Me (leridal, **46**; H2/1395)
156. 7-OH-5-OMe-6-Me-8-formyl (desmosflavanone II, **47**)
157. 5-OH-7-OMe-6-hydroxyMe-8-Me (leridol, **48**; H2/1396)
158. 5,7-DiOMe-6-hydroxyMe-8-Me (**49**; H2/1397)
159. 5,7,4′-TriOH-6,8-diMe (farrerol, H2/1141)
160. 5,7-DiOH-3′-OMe-6,8-diMe (isomatteucinol, **44**)
161. 5,7-DiOH-4′-OMe-6,8-diMe (matteucinol, H2/1145)
162. 5,4′-DiOH-7-OMe-6,8-diMe (angophorol, H2/1144)
Tetra-*O*-substituted
163. 5,7,3′,5′-TetraOH-6-Me (**41**)
164. 5-OH-7,3′,4′-triOMe-6-Me (**42**; H2/1402)
165. 3′-OH-5,7,4′-triOMe-8-Me (H2/1224)
166. 5,7,2′,4′-TetraOMe-8-Me (**43**)
167. 5,7,3′,4′-TetraOH-6,8-diMe (cyrtominetin, H2/1262)

168. 5,7,2′-TriOH-4′-OMe-6,8-diMe (**45**)
169. 5,7,4′-TriOH-3′-OMe-6,8-diMe (H2/1225)
Penta-*O*-substituted
170. 5,7,3′-TriOH-4′,5′-diOMe-6,8-diMe (H2/1255)

Flavanone aglycones bearing noncyclic isoprenyl substituents[a]
Mono-*O*-substituted
171. 7-OH-8-Pr (ovaliflavanone B, H2/1021)
172. 7-OH-6,8-diPr (ovaliflavanone A, H2/1020)
173. 6-OMe-5-(1,1-diMe-2-OH-prop-2-enyl) (falciformin, H2/1014)
174. 7-OMe-8-Pr (isoderricin A, H2/1305)
175. 7-OPr (isoderricidin, H2/1016)
176. 7-OPr-8-(3-OH-3-Me-*trans*-buten-1-yl) (H2/1074)
Di-*O*-substituted
177. 5,7-DiOH-6-Me-8-Pr (**50**)
178. 5,7-DiOH-6-Pr (H2/1059)
179. 5,7-DiOH-6,8-diPr (H2/1077)
180. 5,7-DiOH-8-Ger (H2/1065)
181. 5,7-DiOH-8-Me-6-Pr (H2/1076)
182. 5,7-DiOH-8-Pr (glabranin, H2/1060)
183. 5,7-DiOH-8-(γ-Me-γ-formylallyl) (H2/1064)
184. 5,7-DiOH-8-(4-OH-3-Me-2-butenyl) (H2/1063)
185. 7,4′-DiOH-6-Pr (bavachin, H2/1150)
186. 7,4′-DiOH-8-Pr (isobavachin, H2/1088)
187. 7,4′-DiOH-3′-Pr (abyssinone II, H2/1094)
188. 7,4′-DiOH-6,8-diPr (H2/1087)
189. 7,4′-DiOH-8,3′-diPr (glabrol, H2/1372)
190. 7,4′-DiOH-8-Ger (prostratol F, **96**)
191. 7,4′-DiOH-3′,5′-diPr (abyssinone IV, H2/1095)
192. 7-4′-DiOH-8,3′, 5′-tripr((-)-sophoranone (H2/1089)
193. 5-OH-7-OMe-6-Pr (H2/1061)
194. 5-OH-7-OMe-8-Pr (tephrinone, H2/1066)
195. 5-OH-7-OMe-8-(3-OH-3-Me-*trans*-but-1-enyl) (tephroleocarpin A, H2/1440)
196. 5-OH-7-OMe-8-(3-OH-3-Me-butyl) (tephrowatsin C, H2/1075)
197. 5-OH-7-O-(3-Me-2,3-epoxybutoxy) (H2/1051)
198. 5-OH-7-OPr (7-*O*-prenylpinocembrin, H2/1078)
199. 5-OH-7-OPr-8-Me (7-*O*-prenylcryptostrobin, H2/1079)
200. 5-OH-7-O-neryl (**76**)
201. 5-OH-7-OPr-8-Pr (H2/1071)
202. 4′-OH-7-OMe-6-Pr (bavachinin, H2/1151)
203. 4′-OH-7-OMe-8-Pr (mundulea flavanone A, **51**)
204. 5-OMe-7-OPr (H2/1072)
205. 5-OMe-7-OPr-8-Pr (H2/1073)
206. 5,7-DiOMe-8-Pr (candidone, H2/1296)
207. 5,7-DiOMe-8-(2,3-epoxy-3-Me-butyl) (epoxycandidone, H2/1368)
208. 5,7-DiOMe-8-(3-OH-3-Me-buten-1-yl) (quercetol C, H2/1335)
Tri-*O*-substituted
209. 5,7,2′-TriOH-8-Pr (kushenol S, **52**)
210. 5,7,2′-TriOH-8-Lav (kushenol A, H2/1173)
211. 5,7,2′-TriOH-8-(5-OH-2-isopropenyl-5-Me-hexyl) (kushenol T, **128**)

212. 5,7,4′-TriOH-6-(1,1-DMA) (**89**)
213. 5,7,4′-TriOH-6-Pr (H2/1147)
214. 5,7,4′-TriOH-6-Pr-8-Me (**54**)
215. 5,7,4′-TriOH-6-Me-8-Pr (**53**)
216. 5,7,4′-TriOH-6-Ger (bonannione A, H2/1176)
217. 5,7,4′-TriOH-6-Pr-8-(2,3-diOH-3-Me-butyl) (lonchocarpol B, H2/1337)
218. 5,7,4′-TriOH-6,8-diPr (lonchocarpol A, H2/1336)
219. 5,7,4′-TriOH-6,3′-diPr (paratocarpin L; macaranga flavanone B, **77**)
220. 5,7,4′-TriOH-6,8,3′-triPr (amorilin, H2/1172)
221. 5,7,4′-TriOH-8-(1,1-DMA) (ugonin E, **90**)
222. 5,7,4′-TriOH-8-Pr (sophoraflavanone B, H2/1148)
223. 5,7,4′-TriOH-8-Ger (sophoraflavanone A, H2/1177)
224. 5,7,4′-TriOH-8-Lav (leachianone E, **105**; H2/1435)
225. 5,7,4′-TriOH-8-(2-isopropyl-5-Me-5-hexenyl) (remangiflavanone A, **106**)
226. 5,7,4′-TriOH-8-(2-OH-3-Me-but-3-enyl) (tomentosanol D, **122**)
227. 5,7,4′-TriOH-8-Pr-6-(2-OH-3-Me-but-3-enyl) (lupiniols A1 and A2, **126**; H2/1424)
228. 5,7,4′-TriOH-8-Pr-3′-(2-OH-3-Me-but-3-enyl) (lupiniol B, **127**; H2/1425)
229. 5,7,4′-TriOH-8-Pr-3′-(3-Me-2,3-epoxybutenyl) (flemiflavanone D, H2/1175)
230. 5,7,4′-TriOH-8,3′-diPr (euchrestaflavanone A, H2/1169)
231. 5,7,4′-TriOH-8,3′,5′-triPr (hydroxysophoranone, H2/1171)
232. 5,7,4′-TriOH-3′-Pr (licoflavanone H2/1364)
233. 5,7,4′-TriOH-3′-Ger (macaranga flavanone A, **97**)
234. 5,7,4′-TriOH-3′,5′-diPr (abyssinone V, H2/1375)
235. 7,2′,4′-TriOH-8-Pr (euchrenone a$_7$, H2/1365)
236. 7,2′,4′-TriOH-8-Lav (lehmannin, H2/1312)
237. 5,7-DiOH-6-OMe-8-Pr (agrandol, **55**)
238. 5,7-DiOH-8-OMe-6-Pr (microfolione, **56**)
239. 5,7-DiOH-4′-OPr (selinone, H2/1137)
240. 5,7-DiOH-4′-OGer (**98**; H2/1406)
241. 5,7-DiOH-4′-OMe-8-Pr (H2/1366)
242. 5,7-DiOH-4′-OMe-8-(2-OH-3-Me-3-butenyl) (**123**; H2/1420)
243. 5,7-DiOH-4′-OMe-8-Pr-3′-(3-OH-3-Me-butyl) (**130**; H2/1421)
244. 5,7-DiOH-4′-OMe-8,3′-diPr (H2/1170)
245. 5,7-DiOH-4′-OMe-6-Me-8,3′-diPr (lespedezaflavanone G, **80**; H2/1409)
246. 5,7-DiOH-4′-OMe-8-Me-6,3′-diPr (lespedezaflavanone F, **81**; H2/1408)
247. 5,7-DiOH-4′-OMe-3′-(3-Me-but-1,3-dienyl)-5′-Pr (burttinonedehydrate, **78**)
248. 5,7-DiOH-4′-OMe-3′-Pr-5′-(3-OH-3-Me-buten-1-yl) (burttinone, **114**)
249. 5,7-DiOH-4′-OMe-3′,5′-diPr (**79**)
250. 5,4′-DiOH-7-OMe-6-Pr (H2/1166)
251. 5,4′-DiOH-7-OMe-8-Pr (mundulea flavanone B, **57**)
252. 5,4′-DiOH-7-OPr (H2/1168)
253. 7,2′-DiOH-5-OMe-8-Lav (kushenol R, **107**)
254. 7,2′-DiOH-4′-OMe-8,5′-diPr (euchrenone a$_8$, H2/1376)
255. 7,4′-DiOH-5-OMe-8-Pr (isoxanthohumol, H2/1149)
256. 7,4′-DiOH-5-OMe-8-Lav (kushenol U, **108**)
257. (2R)-7,4′-DiOH-5-OMe-8-[4-OH-3-Me-(2Z)-butenyl] (**124**)
258. (2R)-7,4′-DiOH-5-OMe-8-[5-OH-3-Me-(2E)-butenyl] (**125**)
259. 7,4′-DiOH-2′-OMe-8-Lav (alopecurone G, **109**)
260. 5-OH-7-OMe-4′-OPr (**58**; H2/1407)
261. 5-OH-7,4′-diOMe-8-Pr (flowerine, **59**)

262. 5-OH-8-OMe-7-OPr (H2/1361)

263. 7-OH-3',4'-methylenedioxy-8-Pr (ovaliflavanone C, H2/1189)

264. 7-OH-3',4'-methylenedioxy-6,8-diPr (ovaliflavanone D, H2/1190)

265. 7-OMe-3',4'-methylenedioxy-8-Pr (H2/1306)

Tetra-O-substituted

266. 5,7,8,4'-TetraOH-3'-(3-methylbut-3-enyl) (flowerone, **60**)

267. 5,7,2',4'-TetraOH-6-Lav (kushenol F, H2/1244)

268. 5,7,2',4'-TetraOH-6,8-diPr (kushenol E, H2/1380)

269. 5,7,2',4'-TetraOH-6-Pr-8-Lav (kushenol B, H2/1246)

270. 5,7,2',4'-TetraOH-6-Ger-8-Pr (**101**)

271. 5,7,2',4'-TetraOH-6-Pr-5'-(1,1-DMA) (**93**; H2/1458)

272. 5,7,2',4'-TetraOH-6,8,3'-triPr (lespedezaflavanone E, H2/1391)

273. 5,7,2',4'-TetraOH-6,8,5'-triPr (amorisin, H2/1233)

274. 5,7,2',4,-TetraOH-8-Pr (leachianone G, **61**)

275. 5,7,2',4'-TetraOH-8-Ger (sophoraflavanone C, **99**; H2/1412)

276. 5,7,2',4'-TetraOH-8-Lav (sophoraflavanone G, H2/1274)

277. 5,7,2',4'-TetraOH-8-(2-isopropyl-5-Me-5-hexenyl) (remangiflavanone B, **110**)

278. 5,7,2',4'-TetraOH-8-(2-(2-OH-isopropyl)-5-Me-4-hexenyl) (kushenol Q, **136**)

279. 5,7,2',4'-TetraOH-8,5'-diPr (lespedezaflavanone D, H2/1381)

280. 5,7,2',4'-TetraOH-8-Pr-5'-(1,1-DMA) (**94**)

281. 5,7,2',4'-TetraOH-3'-Ger (sanggenol A, **100**)

282. 5,7,2',4'-TetraOH-5'-Ger (kuwanon E, H2/1238)

283. 5,7,2',6'-TetraOH-8-Lav (exiguaflavanone A, **111**; H2/1432)

284. 5,7,2',6'-TetraOH-8-[2-(2-OH-isopropyl)-5-Me-4-hexenyl] (exiguaflavanone M, **137**; H2/1455)

285. 5,7,3',4'-TetraOH-6-Pr (6-prenyleriodictyol, H2/1226)

286. 5,7,3',4'-TetraOH-6-(1,1-DMA) (**91**)

287. 5,7,3',4'-TetraOH-6-Ger (diplacone, H2/1236)

288. 5,7,3',4'-TetraOH-6-farnesyl (**121**)

289. 5,7,3',4'-TetraOH-6-(2-OH-3-Me-but-3-enyl)-8-Pr (**135**)

290. 5,7,3',4'-TetraOH-6-(3-OH-3-Me-butyl) (**132**; H2/1422)

291. 5,7,3',4'-TetraOH-6-Pr-8-(2-OH-3-Me-but-3-enyl) (dorsmanin H, **134**)

292. 5,7,3',4'-TetraOH-6,8-diPr (**82**; H2/1410)

293. 5,7,3',4'-TetraOH-6,5'-diPr (**83**; H2/1457)

294. 5,7,3',4'-TetraOH-8-Pr (8-prenyleriodictyol, H2/1227)

295. 5,7,3',4'-TetraOH-8-[(E)-3-hydroxymethyl-2-butenyl] (licoleafol, **131**)

296. 5,7,3',4'-TetraOH-8-(4-OAc-3-Me-but-2-enyl) (kanzonol S, **140**)

297. 5,7,3',4'-TetraOH-8,5'-diPr (gancaonin E, H2/1382)

298. 5,7,3',4'-TetraOH-2'-Pr (2'-prenyleriodictyol, H2/1367)

299. 5,7,3',4'-TetraOH-2',5'-diPr (sigmoidin A, H2/1229)

300. 5,7,3',4'-TetraOH-5'-Pr (sigmoidin B, H2/1228)

301. 5,7,3',4'-TetraOH-5',6'-diPr (abyssinin III, **84**)

302. 5,7,3',5'-TetraOH-6,8-diPr (monotesone B, **85**)

303. (2R)-5,7,2'-TriOH-8-OMe-6-Pr (dioflorin, **63**)

304. 5,7,2'-TriOH-4'-OMe-6,8-diPr (flemiflavanone A, H2/1268)

305. 5,7,3'-TriOH-4'-OMe-6-Ger (4'-O-methyldiplacone, H2/1237)

306. 5,7,3'-TriOH-4'-OMe-6,8,5'-triPr (isoamoritin, **88**)

307. 5,7,3'-TriOH-4'-OMe-5'-Pr (4'-methylsigmoidin B, H2/1369)

308. 5,7,3'-TriOH-4'-OPr (monotesone A, **62**)

309. 5,7,4′-TriOH-2′-OMe-8-(5-OH-5-Me-2-isopropenyl-*trans*-hex-3-enyl) (leachianone D, **138**; H2/1434)
310. 5,7,4′-TriOH-2′-OMe-8-(5-OH-5-Me-2-isopropenyl-hexyl) (kushenol P1, **139**)
311. 5,7,4′-TriOH-3′-OMe-6-(1,1-DMA) (**92**)
312. 5,7,4′-TriOH-3′-OMe-6-(β-OH-ethyl)-8-Pr (laxiflorin, **75**)
313. 5,7,4′-TriOH-3′-OMe-6,8,5′-triPr (amoritin, H2/1234)
314. 5,7,4′-TriOH-3′-OMe-8-Pr (exiguaflavanone K, **64**; H2/1453)
315. 5,7,4′-TriOH-3′-OMe-5′-Pr (abyssinin II, **65**)
316. 5,7,4′-TriOH-2′-OMe-6-Lav (isokurarinone, H2/1245)
317. 5,7,4′-TriOH-2′-OMe-8-Lav (leachianone A, H2/1346)
318. 5,7,4′-TriOH-2′-OMe-8-Pr-5′-(1,1-DMA) (**95**)
319. 5,7,4′-TriOH-3′-OMe-6,8-diPr (hiravanone, H2/1383)
320. 5,2′,4′-TriOH-7-OMe-8-Pr (kenusanone I, **66**; H2/1411)
321. 5,2′,4′-TriOH-7-OMe-8,5′-diPr (maackiaflavanone, **86**; H2/1456)
322. 5,2′,5′-TriOH-7-OMe-8-Lav (exiguaflavanone F, **112**; H2/1447)
323. 5,2′,6′-TriOH-7-OMe-8-Lav (exiguaflavanone B, **113**; H2/1433)
324. 5,3′,4′-TriOH-7-OMe-8-Pr (H2/1267)
325. 5,3′,4′-TriOH-7-OMe-6,8-diPr (amoradicin, H2/1231)
326. 7,2′,4′-TriOH-5-OMe-8-Lav (kurarinone, **114**)
327. 5,7-DiOH-2′,4′-diOMe-8-(2,3-epoxy-3-Me-butyl) (**133**)
328. 5,7-DiOH-3′,4′-diOMe-5′-Pr (**68**)
329. 5,4′-DiOH-7,3′-diOMe-6,8-diPr (amoradinin, H2/1232)
330. 7,4′-DiOH-5,2′-diOMe-8-Lav (**115**)
331. 5-OH-7,3′-diOMe-4′-OPr (H2/1260)
332. 7-OMe-3′,4′-Methylenedioxy-6-OPr (ponganone V, **67**)
Penta-*O*-substituted
333. 5,7,2′,4′,6′-PentaOH-6-Ger (sophoraflavanone D, **102**; H2/1413)
334. 5,7,2′,4′,6′-PentaOH-6-Lav (exiguaflavanone C, **116**; H2/1444)
335. 5,7,2′,4′,6′-PentaOH-6,8-diPr (kenusanone B, **87**; H2/1415)
336. 5,7,2′,4′,6′-PentaOH-6-Ger-8-Pr (tomentosanol E, **104**)
337. 5,7,2′,4′,6′-PentaOH-6-Pr-8-Lav (exiguaflavanone J, **119**; H2/1452)
338. 5,7,2′,4′,6′-PentaOH-8-Ger (sophoraflavanone E, **103**; H2/1414)
339. 5,7,2′,4′,6′-PentaOH-8-Lav (exiguaflavanone G, **117**; H2/1449)
340. 5,7,2′,4′-TetraOH-5′-OMe-6-Pr (kushenol V, **69**)
341. 5,7,2′,4′-TetraOH-5′-OMe-8-Pr (kushenol W, **70**)
342. 5,7,2′,6′-TetraOH-4′-OMe-6-Pr-8-Lav (exiguaflavanone D, **120**; H2/1445)
343. 5,7,2′,6′-TetraOH-4′-OMe-8-Pr (kenusanone D, **71**; H2/1417)
344. 5,2′,4′-TriOH-7,5′-diOMe-8-Lav (exiguaflavanone E, **118**; H2/1446)
345. 5,2′,6′-TriOH-7,4′-diOMe-8-Pr (kenusanone E, **72**; H2/1416)
346. 5,7-DiOH-2′,4′,6′-triOMe-8-Pr (heteroflavanone C, **73**; H2/1418)
347. 5-OH-7,2′,4′,6′-tetraOMe-8-Pr (heteroflavanone B, **74**; H2/1419)
Flavanone aglycones bearing furano substituents[a]
Di-*O*-substituted
348. 5-OMe-5″-isoprenyldihydrofurano[2″,3″:7,8] (emoroidenone, **142**)
349. 6-OMe-furano[2″,3″:7,8] (**141**)
Tri-*O*-substituted
350. 5,6-DiOMe-furano[2″,3″:7,8] (**144**)
351. 5,4′-DiOH-6-Pr-5″-(2-OH-isopropyl)furano[2″,3″:7,8] (lupinenol, **152**; H2/1426)
352. 5,4′-DiOH-6-Pr-5″-(2-OH-isopropyl)dihydrofurano[2″,3″:7,8] (lonchocarpol C$_1$, **153**)

353. 5″-Epimer of lonchocarpol C_1 (lonchocarpol C_2, **154**)
354. 5,4′-DiOH-8-Pr-5″-(2-OH-isopropyl)furano[2″,3″:7,6] (lonchocarpol D_1, **155**)
355. 5″-Epimer of lonchocarpol D_1 (lonchocarpol D_2, **156**)
356. 5,4′-DiOH-4″,4″-dimethyl-5″-methyldihydrofurano[2″,3″:7,6] (**145**)
357. 5,4′-DiOH-[5″-(2-OH-isopropyl)dihydrofurano][2″,3″:7,8] (**147**)
358. 2′,4′-DiOH-[5″-(2-OH-isopropyl)dihydrofurano][2″,3″:7,8] (phellodensin D, **148**)
359. 4′-OH-5-OMe-(E)-5″-(2-OH-isopropyl)dihydrofurano[2″,3″:7,8] (**149**)
360. 4′-OH-bis(5″-(2-OH-isopropyl)dihydrofurano[2″,3″:7,8][2″,3″:5,6]) (lonchocarpol E, H2/1340)
361. 3′,4′-Methylenedioxyfurano[2″,3″:7,8] (**143**)

Tetra-O-substituted
362. 5,7,3′-TriOH-2′-Pr-furano[2″,3″:4′,5′] (abyssinoflavanone IV, **150**)
363. 5,3′,4′-TriOH-6-Pr-(2-OH-isopropyl)furano[2″,3″:7,8] (dorsmanin F, **157**)
364. 5″-Epimer of dorsmanin F (epidorsmanin F, **159**)
365. 5,3′,4′-TriOH-8-Pr-(2-OH-isopropyl)furano[2″,3″:7,6] (dorsmanin G, **158**)
366. 5″-Epimer of dorsmanin G (epidorsmanin G, **160**)
367. 5,3′,4′-TriOH-5″-isopropenyldihydrofurano[2″,3″:7,6] (velloeriodictyol, **146**; H2/1423)
368. 3′,4′-DiOH-7-OMe-8-Pr-furano[2″,3″:5,6] (**151**)
369. 3′,4′-DiOH-7-OMe-8-Pr-5″-(2-OH-isopropyl)furano[2″,3″:5,6] (**161**)

Penta-O-substituted
370. 5-OH-6-OMe-3′,4′-methylenedioxyfurano[2″,3″:7,8] (H2/1222)

Flavanone aglycones bearing pyrano substituents[a]
Mono-O-substituted
371. 6″,6″-DMP[2″,3″:7,8] (isolonchocarpin, H2/1017)
372. 6-Pr-6″,6″-DMP[2″,3″:7,8] (maximaflavanone A, **162**)
373. 8-Pr-6″,6″-DMP[2″,3″:7,6] (H2/1018)

Di-O-substituted
374. 5-OH-6-(3-Me-but-1,3-dienyl)-6″,6″-DMP[2″,3″:7,8] (spinoflavanone A, **164**)
375. 5-OH-6-Pr-6″,6″-DMP[2″,3″:7,8] (fulvinervin A, H2/1080)
376. 5-OH-6″,6″-DMP[2″,3″:7,8] (obovatin, H2/1344)
377. 7-OH-6-Pr-6″,6″-DMP[2″,3″:4′,3′] (dinklagin A, **165**)
378. 7-OH-8-Pr-6″,6″-DMP[2″,3″:4′,3′] (euchrenone a_5, H2/1371)
379. 7-OH-8-Pr-6″,6″-DMDHP[2″,3″:4′,3′] (euchrenone a_{17}, **167**)
380. 7-OH-8,5′-diPr-6″,6″-DMP[2″,3″:4′,3′] (sophoranochromene, H2/1154)
381. 7-OH-5′-Pr-6″,6″-DMP[2″,3″:4′,3′] (abyssinone III, H2/1097)
382. 7-OH-6″,6″-DMP[2″,3″:4′,3′] (abyssinone I, H2/1096)
383. 4′-OH-3′-Pr-6″,6″-DMP[2″,3″:7,8] (shinflavanone, **166**)
384. 4′-OH-6″,6″-DMP[2″,3″:7,8] (4′-hydroxyisolonchocarpin, H2/1311)
385. 5-OMe-6″,6″-DMP[2″,3″:7,8] (obovatin methyl ether, H2/1062)
386. 6-OMe-6″,6″-DMP[2″,3″:7,8] (ovalichromene, H2/1019)
387. 4′-OMe-6″,6″-DMP[2″,3″:7,8] (dorspoinsettifolin, **163**)
388. Bis(6″,6″-DMP[2″,3″:7,8][2″,3″:4′,3′]) (xambioona, H2/1345)
389. Bis(6″,6″-DMDHP[2″,3″:7,6][2″,3″:4′,3′]) (dormanin B, **168**)

Tri-O-substituted
390. 5,7-DiOH-6-Pr-6″,6″-DMP[2″,3″:4′,3′] (paratocarpin H, **173**)
391. 5,7-DiOH-6,8-diPr-6″,6″-DMP[2″,3″:4′,3′] (euchrenone a_4, H2/1384)
392. 5,7-DiOH-8-Pr-6″,6″-DMDHP[2″,3″:4′,3′] (euchrenone a_{16}, **177**)
393. 5,2′-DiOH-8-Pr-6″,6″-DMP[2″,3″:7,6] (minimiflorin, H2/1164)
394. 5,4′-DiOH-6-Pr-6″,6″-DMP[2″,3″:7,8] (cajaflavanone, H2/1152)
395. 5,4′-DiOH-8-Pr-6″,6″-DMP[2″,3″:7,6] (lupinifolin, H2/1165)

396. 5,4'-DiOH-8-hydroxymethyl-6'',6''-DMP[2'',3'':7,6] (**184**)
397. 5,4'-DiOH-8-(1-OH-2,3-epoxy-3-Me-butyl)-6'',6''-DMP[2'',3'':7,6] (1'''-OH-2''',3'''-epoxylupinifolin, **202**)
398. 5,4'-DiOH-8-(2,3-diOH-3-Me-butyl)-6'',6''-DMP[2'',3'':7,6] (2'',3''-dihydroxylupinifolin, **205**)
399. 5,4'-DiOH-8-(2,3-epoxy-3-Me-butyl)-6'',6''-DMP[2'',3'':7,6] (2''',3'''-epoxylupinifolin, **191**)
400. 5,4'-DiOH-8-(2-OH-3-Me-but-3-enyl)-6'',6''-DMP[2'',3'':7,6] (dereticulatin, **192**)
401. 5,4'-DiOH-3'-Pr-6'',6''-DMP[2'',3'':7,6] (paratocarpin I, **174**)
402. 5,4'-DiOH-3'-Pr-6'',6''-DMP[2'',3'':7,8] (euchrenone a₂, H2/1374)
403. 5,4'-DiOH-6'',6''-DMP[2'',3'':7,6] (paratocarpin K, **169**)
404. 5,4'-DiOH-6'',6''-DMP[2'',3'':7,8] (citflavanone, H2/1363)
405. 5,4'-DiOH-6''-Me,6'',(4-methylpent-3-enyl)pyrano[2'',3'':7,8] (cycloaltilisin 7, **175**)
406. 5-OH-4'-OMe-6'',6''-DMP-[2'',3'':7,8] (**170**)
407. 5,4'-DiOMe-6'',6''-DMP[2'',3'':7,8] (glyflavanone A, **171**)
408. 3',4'-DiOMe-6-Pr-6'',6''-DMP[2'',3'':7,8] (**178**)
409. 3',4'-DiOMe-6'',6''-DMP[2'',3'':7,8] (ponganone III, **172**)
410. 3',4'-Methylenedioxy-6'',6''-DMP[2'',3'':7,8] (ovalichromene B, H2/1191)
411. 5-OH-bis(6'',6''-DMP[2'',3'':7,8][2''',3''':4',3']) (euchrenone a₁, H2/1373)
412. 5-OH-bis(6'',6''-DMDHP[2'',3'':7,6][2''',3''':4',3']) (paratocarpin J, **176**)
413. 5-OH-6-Pr-bis(6'',6''-DMP[2'',3'':7,8][2''',3''':4',3']) (euchrenone a₁₅, **179**)
414. 5-OH-8-Pr-bis(6'',6''-DMP[2'',3'':7,6][2''',3''':4',3']) (euchrenone a₁₄, **180**)

Tetra-*O*-substituted

415. 5,7,2'-TriOH-6-Pr-6'',6''-DMP[2'',3'':4',5'] (cudraflavanone A, H2/1242)
416. 5,7,2'-TriOH-6,8-diPr-6'',6''-DMP[2'',3'':4',5'] (euchrenone a₆, H2/1388)
417. 5,7,2'-TriOH-8-Pr-6'',6''-DMP[2'',3'':4',5'] (euchrestaflavanone C, H2/1241)
418. 5,7,2'-TriOH-5'-Pr-6'',6''-DMP[2'',3'':4',3'] (sanggenol N, **188**)
419. 5,7,2'-TriOH-6'',6''-DMP[2'',3'':4',3'] (sanggenon F, H2/1239)
420. 5,7,3'-TriOH-6,8-diPr-6'',6''-DMP[2'',3'':4',5'] (amorinin, H2/1235)
421. 5,7,3'-TriOH-2'-Pr-(5''-OH-6'',6''-DMDHP[2'',3'':4',5']) (abyssinoflavanone V, **203**)
422. 5,7,3'-TriOH-2'-Pr-6'',6''-DMP[2'',3'':4',5'] (sigmoidin F, H2/1379)
423. 5,7,3'-TriOH-(4'',5''-diOH-6'',6''-DMDHP[2'',3'':4',5']) (sigmoidin G, **210**, H2/1429)
424. 5,7,3'-TriOH-(5''-OH-6'',6''-DMDHP[2'',3'':4',5']) (sigmoidin D, H2/1370)
425. 5,7,3'-TriOH-6'',6''-DMP[2'',3'':4',5'] (sigmoidin C, H2/1230)
426. 5,7,4'-TriOH-6-Pr-6''-Me,6''-(4-Me-pent-3-enyl)pyrano[2'',3'':3',2'] (tanariflavanone B, **200**)
427. 5,7,4'-TriOH-6'',6''-DMP[2'',3'':2',3'] (sanggenon H, H2/1240)
428. 5,2',4'-TriOH-6-Pr-6'',6''-DMP[2'',3'':7,8] (euchrenone a₉, H2/1377)
429. 5,2',4'-TriOH-8-Pr-6'',6''-DMP[2'',3'':7,6] (flemichin D, H2/1269)
430. 5,2',4'-TriOH-(5''-Pr-6'',6''-DMDHP[2'',3'':7,8]) (leachianone F, **193**)
431. 5,2',4'-TriOH-6'',6''-DMDHP[2'',3'':7,8] (kenusanone J, **181**)
432. 5,2',4'-TriOH-6''-Me,6''-(4''-Me-pent-3-enyl)pyrano[2'',3'':7,6] (kuwanol C, H2/1341)
433. 5,2',4'-TriOH-6''-Me,6''-(4''-Me-pent-3-enyl)pyrano[2'',3'':7,8] (sanggenol L, **189**)
434. 5,2',6'-TriOH-8-Pr-6'',6''-DMP[2'',3'':7,6] (orotinin, H2/1378)
435. 5,2',6'-TriOH-5''-Pr-6'',6''-DMP[2'',3'':7,8] (exiguaflavanone L, **194**)
436. 5,3',4'-TriOH-8-Pr-6'',6''-DMP[2'',3'':7,6] (dorsmanin I, **190**)
437. 5,3',4'-TriOH-8-Pr-6'',6''-DMDHP[2'',3'':7,6] (dorsmanin J, **195**)
438. 5,3',4'-TriOH-8,5'-diPr-6'',6''-DMP[2'',3'':7,6] (amoridin, H2/1389)
439. 5,3',4'-TriOH-2'-Ger-(5''-OH-6'',6''-DMDHP[2'',3'':7,6]) (tanariflavanone A, **201**)
440. 2',3',6'-TriOH-8-Pr-5'-(2,4-diOH-phenyl)-6'',6''-DMP[2'',3'':7,6] (eriosemaone C, **211**)

441. 5,7-DiOH-3'-OMe-6'',6''-DMP[2'',3'':4',5'] (abyssinin I, **182**)
442. 5,7-DiOH-bis(6'',6''-DMP[2'',3'':2',3'][2'',3'':4',5']) (sanggenol O, **187**)
443. 5,7-DiOH-(5''-OH-6'',6''-DMP[2'',3'':7,6])-6'',6''-DMDHP[2'',3'':4',3'] (abyssinoflavanone VI, **204**)
444. 5,2'-DiOH-6-Pr-bis(6'',6''-DMP[2'',3'':7,8][2'',3'':4',5']) (euchrenone a$_{12}$, H2/1386)
445. 5,2'-DiOH-8-Pr-bis(6'',6''-DMP[2'',3'':7,6][2'',3'':4',5']) (euchrenone a$_{11}$, H2/1385)
446. 5,2'-DiOH-8-Pr-6''DMP6'[2'',3'':7,8],(5''-OH-6'',6''-DMDHP[2'',3'':4',3']) (flemichin E, H2/1280)
447. 5,2'-DiOH-4'-OMe-6-Pr-6'',6''-DMP[2'',3'':7,8] (fleminone, H2/1243)
448. 5,2'-DiOH-6''-Me,6''-(4-Me-pent-3-enyl)pyrano[2'',3'':7,8],6'',6''-DMP[2'',3'':4',3'] (flemichin A, H2/1279)
449. 5,3'-DiOH-8-Pr-bis(6'',6''-DMP[2'',3'':7,6][2'',3'':4',5']) (amorin, H2/1387)
450. 5,3'-DiOH-4'-OMe-6'',6''-DMP[2'',3'':7,8] (**183**)
451. 5,3'-DiOH-4'-OMe-6-Pr-6'',6''-DMP[2'',3'':7,8] (**197**)
452. 5,4'-DiOH-2'-OMe-5''-Pr-6'',6''-DMDHP[2'',3'':7,8] (leachianone B, H2/1436)
453. 5,4'-DiOH-3'-OMe-6-Pr-6'',6''-DMP[2'',3'':7,8] (**198**)
454. 5,4'-DiOH-3'-OMe-8-Pr-6'',6''-DMP[2'',3'':7,6] (3'-methyoxylupinifolin, H2/1427)
455. 5,4'-DiOH-3'-OMe-8,5'-diPr-6'',6''-DMP[2'',3'':7,6] (amoricin, H2/1390)
456. 3',4'-DiOH-bis(6'',6''-DMDHP[2'',3'':5,6][2'',3'':7,8]) (dorsmanin E, **196**)
457. 5-OH-3',4'-DiOMe-6-Pr-6'',6''-DMP[2'',3'':7,8] (**199**)
458. 5,3',4'-TriOMe-6'',6''-DMP[2'',3'':7,8] (glyflavanone B, **185**)
459. 6,3',4'-TriOMe-6'',6''-DMP[2'',3'':7,8] (ponganone IV, **186**)
460. 5-OMe-3',4'-methylenedioxy-6'',6''-DMP[2'',3'':7,8] (isoglabrachromene, H2/1442)
461. 6-OMe-3',4'-methylenedioxy-6'',6''-DMP[2'',3'':7,8] (ovalichromene A, H2/1192)
Penta-*O*-substituted
462. 5,2',4',6'-TetraOH-6-Lav-6'',6''-DMP[2'',3'':7,8] (exiguaflavanone I, **208**)
463. 5,2',4',6'-TetraOH-8-Lav-6'',6''-DMP[2'',3'':7,6] (exiguaflavanone H, **209**)

Flavanone aglycones bearing benzyl substituents
Mono-*C*-benzyl
464. 5,7-DiOH-6-(2-OH-benzyl) (isochamanetin, H2/1067)
465. 5,7-DiOH-8-(2-OH-benzyl) (chamanetin, H2/1068)
466. 7-OH-5-OMe-8-(2-OH-benzyl) (5-*O*-methylchamanetin, H2/1070)
467. 5-OH-6,7-diOMe-8-(2-OH-5-OMe-benzyl) (macrophyllol, **212**)
468. 5-OH-7,8-diOMe-6-(2-OH-5-OMe-benzyl) (macrophyllol A, **213**)
Di-*C*-benzyl
469. 5,7-DiOH-6,8-di(2-OH-benzyl) (dichamanetin, H2/1302)
Tri-*C*-benzyl
470. 5,7-DiOH-6-(2 × 2-OH-benzyl)-8-(2-OH-benzyl) (uvarinol, H2/1284)
471. 5,7-DiOH-6-(2-OH-benzyl)-8-(2 × 2-OH-benzyl) (isouvarinol, **214**)
Tetra-*C*-benzyl
472. 5,7-DiOH-6-(3 × 2-OH-benzyl)-8-(2-OH-benzyl) (2'''''-OH-3'''-benzyluvarinol, **215**)
473. 5,7-DiOH-6-(2-OH-benzyl)-8-(2 × 2-OH-benzyl) (2'''''-OH-5''''-benzyluvarinol A, **216**)
474. 5,7-DiOH-6-(2 × 2-OH-benzyl)-8-(2 × 2-OH-benzyl) (2'''-OH-5''-benzyluvarinol B, **217**)
Complex flavanone aglycones (arranged alphabetically)
475. Alopecurone A (**218**)
476. Alopecurone B (**219**)
477. Alopecurone C (**220**)
478. Alopecurone D (**221**)
479. Alopecurone E (**222**)
480. Alopecurone F (**223**)

481. Anastatin A (**225**)
482. Anastatin B (**226**)
483. Baeckea flavanone (**227**)
484. BF-4 (**228**)
485. BF-5 (**229**)
486. BF-6 (**230**)
487. Breverin (H2/1273)
488. Calomelanol G (**238**)
489. Calomelanol H (**239**)
490. Calomelanol I (**240**)
491. Calomelanol J (**241**)
492. Calyxin C (**242**)
493. Calyxin D (**243**)
494. Calyxin G (**244**)
495. Calyxin J (**245**)
496. Calyxin K (**246**)
497. Calyxin M (**247**)
498. Derriflavanone (**206**)
499. Epicalyxin C (**248**)
500. Epicalyxin D (**249**)
501. Epicalyxin G (**250**)
502. Epicalyxin J (**251**)
503. Epicalyxin K (**252**)
504. Epicalyxin M (**253**)
505. Epiderriflavanone (**207**)
506. Kurziflavolactone A (*2R*) (**254**)
507. Kurziflavolactone B (*2S*) (**255**)
508. Kurziflavolactone C (*2S*) (**256**)
509. Kurziflavolactone D (*2R*) (**257**)
510. Kuwanon D (H2/1291)
511. Kuwanon F (H2/1290)
512. Kuwanon L (H2/1295)
513. Leachianone C (H2/1437)
514. Leachianone I (**224**)
515. Lepidissipyrone (H2/1317)
516. Leucadenone A (**231**)
517. Leucadenone B (**232**)
518. Leucadenone C (**233**)
519. Leucadenone D (**234**)
520. Linderatone (H2/1315)
521. Louisfieserone A (H2/1281)
522. Louisfieserone B (H2/1282)
523. Lumaflavanone A (**235**)
524. Lumaflavanone B (**236**)
525. Lumaflavanone C (**237**)
526. Neolinderatone (H2/1316)
527. Neosilyhermin A (H2/1288)
528. Neosilyhermin B (H2/1289)
529. 8-Prenyllepidissipyrone (H2/1318)
530. Protofarrerol (H2/1334)

531. Purpurin (H2/1283)
532. Remerin (H2/1275)
533. Scaberin (H2/1276)
534. Silandrin (H2/1285)
535. Silymonin (H2/1286)
536. Silyhermin (H2/1287)
537. Sophoraflavanone I (H2/1438)
538. Tephroleocarpin B (H2/1441)
539. Tephrorin A (**258**)
540. Tephrorin B (**259**)

Flavanone glycosides[a]

7-OH-flavanone

541. 7-Glucoside (**260**)

5,7-DiOH-flavanone (pinocembrin)

542. 5-Glucoside (H2/1023)
543. 7-Apiosyl(1 → 5)apiosyl(1 → 2)glucoside (**263**)
544. 7-Glucoside (**261**)
545. 7-Neohesperidoside 2‴-acetate (H2/1026)
546. 7-Neohesperidoside 3‴-acetate (H2/1027)
547. 7-Neohesperidoside 4‴-acetate (H2/1028)
548. 7-Neohesperidoside 6″-acetate (H2/1029)
549. 7-Rhamnoside (H2/1024)
550. 7-Rhamnosyl(1 → 2)glucoside (sarotanoside, onychin, H2/1025)
551. 7-Rhamnosyl(1 → 6)glucoside (**262**)

5-OH-7-OMe-flavanone (pinostrobin)

552. 5-Glucoside (**264**; H2/1398)

7,4′-DiOH-flavanone (liquiritigenin)

553. 7-(3-Acetylapioside)-4′-glucoside (**266**)
554. 7-Apioside-4′-glucoside (**265**)
555. 7-Glucoside (neoliquiritin, H2/1084)
556. 7-Glucoside-4′-apiosyl(1 → 2)glucoside (glucoliquiritin apioside, **268**)
557. 7-Rhamnosylglucoside (rhamnoliquiritin, H2/1086)
558. 4′-[3-Acetylapiosyl(1 → 2)glucoside] (**267**)
559. 4′-Apiosyl(1 → 2)glucoside (H2/1085)
560. 4′-[4-*p*-Coumaroylapiosyl(1 → 2)glucoside] (2*S*) (licorice glycoside D$_2$, **271**)
561. 4′-[4-*p*-Coumaroylapiosyl(1 → 2)glucoside] (2*R*) (licorice glycoside D$_1$, **270**)
562. 4′-[4-Feruloylapiosyl(1 → 2)glucoside] (2*S*) (licorice glycoside C$_2$, **272**)
563. 4′-[4-Feruloylapiosyl(1 → 2)glucoside] (2*R*) (licorice glycoside C$_1$, **273**)
564. 4′-Glucoside (liquiritin, H2/1083)
565. Licorice glycoside E (**269**)

5,6,7-TriOH-flavanone (dihydrobaicalein)

566. 7-Glucoside (**274**)
567. 7-Glucuronide (H2/1034)

5,7,8-TriOH-flavanone (dihydronorwogonin)

568. 5-Glucoside (**275**)
569. 7-Glucuronide (H2/1039)

5,7,2′-TriOH-flavanone

570. 7-Glucoside (**276**, H2/1326)
571. 7-*O*-(Ethyl-β-D-glucopyranosiduronate) (**278**)
572. 7-*O*-(Methyl-β-D-glucopyranosiduronate) (**277**)

5,7,4′-TriOH-flavanone (naringenin)

573. 5-Glucoside (floribundoside, salipurposide, H2/1099)
574. 5-Rhamnoside (H2/1100)
575. 5-Rhamnosyl(1 → 2)glucoside (= 5-neohesperidoside) (H2/1101)
576. 5,7-Diglucoside (H2/1110)
577. 7-[3-Acetyl-6-(*E*)-*p*-coumaroylglucoside] (**282**)
578. 7-(6-Acetylglucoside) (**280**)
579. 7-Arabinopyranosyl(1 → 6)glucoside (H2/1113)
580. 7-(2-*p*-Coumaroylglucoside) (**281**)
581. 7-(3-*p*-Coumaroylglucoside) (H2/1108)
582. 7-(6-*p*-Coumaroylglucoside) (H2/1107)
583. 7-(2,6-Dirhamnosylglucoside) (**285**)
584. 7-(3,6-Di-*p*-coumaroylglucoside) (H2/1109)
585. 7-(4,6-Digalloylglucoside) (**283**)
586. 7-Galactosyl(1→4)glucoside (H2/1114)
587. 7-(6-Galloylglucoside) (H2/1106)
588. 7-Glucoside (prunin, H2/1102)
589. 7-Neohesperidoside (naringin) (H2/1112)
590. 7-Neohesperidoside-4′-glucoside (H2/1298)
591. 7-Neohesperidoside-6″-malonate (H2/1443)
592. 7-Rhamnoside (naringerin, H2/1103)
593. 7-Rhamnosyl(1 → 2)(4-*O*-methyl-β-D-glucoside) (fumotonaringenin) (**284**)
594. 7-Rhamnosyl(1 → 6)glucoside (=7-rutinoside) (narirutin, H2/1111)
595. 7-Rutinoside-4′-glucoside (H2/1297)
596. 7-Xylosylglucoside (H2/1116)
597. 4′-Galactoside (H2/1105)
598. 4′-Glucoside (H2/1104)
599. 4′-Rhamnoside (**279**)
600. 4′-Rutinoside (H2/1115)

7,3′,4′-TriOH-flavanone (butin)

601. 7-Glucoside (isocoreopsin 2, H2/1182)
602. 7,3′-Diglucoside (butrin, H2/1183)
603. 3′-Glucoside (isomonospermoside, H2/1181)

5,7-DiOH-8-OMe-flavanone (dihydrowogonin)

604. 7-Glucoside (H2/1044)

5,7-DiOH-4′-OMe-flavanone (isosakuranetin)

605. 5-Glucoside (H2/1120)
606. 7-α-L-Arabinofuranosyl(1 → 6)glucoside (**287**)
607. 7-Fucopyranosyl(1 → 6)glucoside (longitin, **286**)
608. 7-Galactoside (puddumin B, H2/1348)
609. 7-Glucoside (H2/1122)
610. 7-Glucosyl(1 → 4)rhamnoside (acinoside, H2/1125)
611. 7-Neohesperidoside (poncirin, citrifolioside, H2/1124)
612. 7-Rhamnoside (isosakuranin, H2/1347)
613. 7-Rutinoside (didymin, neoponcirin, H2/1123)
614. 7-Xyloside (H2/1121)

5,2′-DiOH-7-OMe-flavanone

615. 2′-Glucoside (haplanthin, H2/1138)

5,4′-DiOH-7-OMe-flavanone (sakuranetin)

616. 5-Glucoside (sakuranin, H2/1118)

6,7-DiOH-5-OMe-flavanone

617. 7-Glucoside (H2/1399)

7,4'-DiOH-5-OMe-flavanone (5-O-methyl-naringenin)

618. 7-Glucoside (puddumin A, H2/1314)

619. 7-Neohesperidoside-4'-glucoside (**289**)

620. 4'-Glucoside (alhagitin, **288**)

621. 4'-Rhamnosylglucoside (H2/1130)

622. 4'-Xylosyl(1 → 4)arabinoside (H2/1129)

5-OH-7,8-diOMe

623. 5-Rhamnoside (**290**)

624. 5-Glucoside (andrographidin, H2/1362)

4'-OH-5,7-diOMe-flavanone (5,7-di-O-methyl-naringenin)

625. 4'-[(5-Cinnamoyl)-β-D-apiofuranosyl(1 → 2)glucoside] (**291**)

2,5,7,4'-TetraOH-flavanone (2-hydroxynaringenin)

626. 7-Glucoside (**292**)

5,6,7,4'-TetraOH-flavanone (carthamidin)

627. 5-Glucoside (H2/1133)

628. 6,7-Diglucoside (**293**)

629. 7-Rhamnoside (H21401/)

5,7,8,4'-TetraOH-flavanone (isocarthamidin)

630. 7-Rhamnoside (**294**)

631. 8-Glucoside (3-desoxycallunin, **295**)

5,7,2',4'-TetraOH-flavanone (steppogenin)

632. 7-Glucoside (stepposide, H2/1278)

633. 4'-Glucoside (**296**)

5,7,2',5'-TetraOH-flavanone

634. 7-Glucoside (coccinoside B, **297**)

635. 7-Rutinoside (H2/1322)

5,7,3',4'-TetraOH-flavanone (eriodictyol)

636. 5-Glucoside (H2/1195)

637. 5-Rhamnoside (H2/1194)

638. 6-C-Glucoside (2S) (**298**)

639. 6-C-Glucoside (2R) (**299**)

640. 7-(6-Acetylglucoside) (coccinoside A, **301**)

641. 7-α-L-Arabinofuranosyl(1 → 6)glucoside (**302**)

642. 7-(6-p-Coumaroylglucoside) (**303**)

643. 7-(6-Galloylglucoside) (**305**)

644. 7-Glucoside (H2/1196)

645. 7-Glucoside 3',4',2'',3'',4'',6''-hexaacetate (hexaacetylpyracanthoside, H2/1439)

646. 7-Glucuronide (2S) (**300**)

647. 7-Glucuronide (2R) (H2/1350)

648. 7-Neohesperidoside (neoeriocitrin, H2/1201)

649. 7-Rhamnoside (eriodictin, H2/1197)

650. 7-Rutinoside (eriocitrin, H2/1200)

651. 3'-(6-p-Coumaroylglucoside) (**304**)

652. 3'-Glucoside (H2/1198)

653. 5,3'-Diglucoside (H2/1199)

7,8,3',4'-TetraOH-flavanone (iso-okanin)

654. 7-Glucoside (H2/1186)

655. 7-Rhamnoside (H2/1187)

656. 7-(2,4,6-Triacetylglucoside) (**306**)
5,6,7-TriOH-4'-OMe-flavanone (*4'-methylcarthamidin*)
657. 7-(2-*p*-Coumaroylglucoside) **307**
5,7,3'-TriOH-4'-OMe-flavanone (*hesperetin*)
658. 5-Glucoside (H2/1204)
659. 7-(2,6-Dirhamnosylglucoside) (**308**)
660. 7-Galactosyl(1 → 3)[rhamnosyl(1 → 6)]glucoside] (alhagidin, **309**)
661. 7-Glucoside (H2/1205)
662. 7-Neohesperidoside (neohesperidin, H2/1208)
663. 7-Rhamnoside (H2/1206)
664. 7-Rutinoside (hesperidin, H2/1207)
5,7,4'-TriOH-3'-OMe-flavanone
665. 7-(6-Acetylglucoside) (viscumneoside VI, H2/1355)
666. 7-Apiosyl(1 → 5)apiosyl(1 → 2)glucoside (viscumneoside V, H2/1356)
667. 7-Glucoside (viscumside A, H2/1354)
668. 7-Glucoside-4'-apioside (viscumneoside I, H2/1353)
669. 7-Rutinoside (**310**)
5,3',4'-TriOH-7-OMe-flavanone
670. 3'-Glucoside (H2/1352)
7,3',4'-TriOH-5-OMe-flavanone
671. 7-Xylosyl(1 → 4)arabinoside (H2/1213)
5,7-DiOH-3',4'-diOMe-flavanone (*homoesperetin*)
672. 5-Glucoside (H2/1212)
673. 7-Rutinoside (**311**)
5,3'-DiOH-7,4'-diOMe-flavanone (*persicogenin*)
674. 5-Glucoside (persicoside, H2/1211)
675. 3'-Glucoside (**312**)
5,4'-DiOH-7,3'-diOMe-flavanone
676. 4'-Apiosyl(1 → 2)glucoside (**313**)
7,3'-DiOH-5,4'-diOMe-flavanone
677. 7-Glucoside (H2/1215)
4'-OH-5,7,2'-triOMe-flavanone
678. 4'-Rhamnosyl(1 → 6)glucoside (**314**)
2,5,7,3',4'-PentaOH-flavanone
679. 5-Glucoside (H2/1263)
5,7,3',4',5'-PentaOH-flavanone
680. 3'-Glucoside (plantagoside, H2/1253)
5,7,4'-TriOH-3',5'-OMe-flavanone
681. 5-Glucoside (2*R*) (peruvianoside I, **315**)
682. 5-Glucoside (2*S*) (peruvianoside II, **316**)
7,8,4'-TriOH-3',5'-OMe-flavanone
683. 4'-Glucoside (H2/1308)
5,2'-DiOH-7,8,6'-triOMe-flavanone
684. 2'-Glucuronide (H2/1325)
5-OH-6,7,3',4',5'-OMe-flavanone
685. 5-Rhamnoside (H2/1271)
7-OH-6,8-diMe-flavanone
686. 7-Arabinoside (H2/1309)
5,7-diOH-6-Me-flavanone (*strobopinin*)
687. 7-Galactoside (H2/1330)

688. 7-Glucoside (H2/1331)
689. 7-Xylosyl(1 → 3)xyloside (**317**)
5,7-diOH-6,8-diMe-flavanone
690. 7-[6-(3-OH-3-Methylglutaryl)glucoside] (matteuorienate B, **318**)
5,7,4'-TriOH-6-Me-flavanone (poriol)
691. 7-Glucoside (poriolin, H2/1140)
5,7,4'-TriOH-6,8-diMe-flavanone (farrerol)
692. 5,7-Diglucoside (H2/1143)
693. 7-β-D-Apiosyl(1 → 6)glucoside (miconioside B, **319**)
694. 7-Glucoside (cyrtopterin, H2/1142)
5,7-diOH-4'-OMe-6,8-diMe-flavanone (matteucinol)
695. 7-β-L-Apiosyl(1 → 6)glucoside (**320**)
696. 7-α-L-Arabinopyranosyl(1 → 6)glucoside (miconioside A, **321**)
697. 7-(4,6-Digalloylglucoside) (**323**)
698. 7-Glucoside (H2/1146)
699. 7-[4,6-(S)-Hexahydroxydiphenylglucoside] (**322**)
700. 7-[6-(3-OH-3-Methylglutaryl)glucoside] (matteuorienate A, **324**)
5,7,3',4'-Tetra-OH-6-Me-flavanone
701. 7-Glucoside (H2/1261)
5,7,3',4'-Tetra-OH-5'-Me-flavanone
702. 3'-Galactosyl(1 → 4)rhamnoside (mesuein, H2/1333)
7,3',4'-TriOH-5-OMe-6-Me-flavanone
703. 7-Glucoside (H2/1223)
5,4'-DiOH-6-Pr-flavanone
704. 4'-Xylosyl(1 → 2)rhamnoside (**325**)
5,7,4'-TriOH-8-Pr-flavanone
705. 7-Glucoside (phellodensin F, **326**)
706. 7,4'-Diglucoside (flavaprenin 7,4'-diglucoside)
5,6,7,4'-TetraOH-8-Pr-flavanone
707. 5-Rutinoside (nirurin, H2/1174)
5,7,4'-TriOH-8-(3-hydroxymethyl-2-butenyl)flavanone
708. 7-Glucoside (phellodensin E, **327**)
5-OH-7,4'-diOMe-6,8-diMe-flavanone
709. 5-Galactoside (H2/1167)
Sophora flavanone I
710. 7-Glucoside (**328**)

DIHYDROFLAVONOLS
Dihydroflavonol aglycones bearing hydroxy, methoxy, and methylenedioxy substituents only
Di-*O*-substituted
711. 3,7-DiOH-flavanone (7-OH-flavanonol, H2/1459)
Tri-*O*-substituted
712. 3,5,7-TriOH-flavanone (pinobanksin, H2/1460)
713. 3,7,4'-TriOH-flavanone (garbanzol, H2/1466)
714. 3,5-DiOH-7-OMe-flavanone (alpinone, H2/1463)
715. 3,7-DiOH-5-OMe-flavanone (H2/1462)
716. 3,7-DiOH-6-OMe-flavanone (H2/1465)
717. 5,7-DiOH-3-OMe-flavanone (H2/1461)
718. 3-OH-5,7-diOMe-(2*R*,3*R*)-flavanone (H2/1464)
719. 3-OH-5,7-diOMe-(2*R*,3*S*)-flavanone (**329**)

Tetra-*O*-substituted

720. 3,5,7,2′-TetraOH-flavanone (H2/1468)
721. 3,5,7,4′-TetraOH-flavanone (aromadendrin, H2/1469)
722. 3,7,3′,4′-TetraOH-(2*R*,3*R*)-flavanone ((2*R*,3*R*)-fustin, H2/1477)
723. 3,7,3′,4′-TetraOH-(2*S*,3*S*)-flavanone ((2*S*,3*S*)-fustin, H2/1478)
724. 3,5,7-TriOH-6-OMe-flavanone (alnustinol, H2/1467)
725. 3,5,7-TriOH-4′-OMe-flavanone (H2/1473)
726. 3,5,8-TriOH-7-OMe-flavanone (**330**)
727. 3,5,4′-TriOH-7-OMe-flavanone (H2/1472)
728. 3,7,4′-TriOH-5-OMe-flavanone (H2/1471)
729. 5,7,4′-TriOH-3-OMe-flavanone (H2/1470)
730. 7,3′,4′-TriOH-3-OMe-(2*R*,3*R*)-flavanone (H2/1479)
731. 3,5-DiOH-6,7-methylenedioxy-flavanone (H2/1519)
732. 3,5-DiOH-7,4′-diOMe-flavanone (H2/1475)
733. 3,4′-DiOH-5,7-diOMe-flavanone (H2/1474)
734. 3′,4′-DiOH-3,7-diOMe-flavanone (H2/1480)
735. 3-OH-5-OMe-6,7-methylenedioxy-flavanone (H2/1520)
736. 3-OH-5,7,4′-triOMe-flavanone (**331**)
737. 7-OH-3,5,4′-triOMe-flavanone (H2/1476)

Penta-*O*-substituted

738. 3,5,7,2′,4′-PentaOH-flavanone (dihydromorin, H2/1487)
739. 3,5,7,2′,5′-PentaOH-flavanone (H2/1488)
740. 3,5,7,2′,6′-PentaOH-flavanone (H2/1491)
741. 3,5,7,3′,4′-PentaOH-(2*R*,3*R*)-flavanone ((2*R*,3*R*)-taxifolin, H2/1492)
742. 3,5,7,3′,4′-PentaOH-(2*S*,3*S*)-flavanone ((2*S*,3*S*)-taxifolin, H2/1493)
743. 3,7,8,3′,4′-PentaOH-(2*R*,3*R*)-flavanone ((2*R*,3*R*)-8-hydroxyfustin, H2/1502)
744. 3,7,8,3′,4′-PentaOH-(2*R*,3*S*)-flavanone ((2*R*,3*S*)-8-hydroxyfustin, H2/1503)
745. 3,7,3′,4′,5′-PentaOH-flavanone (dihydrorobinetin, H2/1505)
746. 3,5,7,2′-TetraOH-5′-OMe-flavanone (H2/1489)
747. 3,5,7,3′-TetraOH-4′-OMe-flavanone (H2/1499)
748. 3,5,7,4′-TetraOH-6-OMe-flavanone (H2/1481)
749. 3,5,7,4′-TetraOH-8-OMe-flavanone (H2/1485)
750. 3,5,7,4′-TetraOH-3′-OMe-flavanone (dihydroisorhamnetin, H2/1498)
751. 3,5,2′,3′-TetraOH-7-OMe-flavanone (**332**)
752. 3,5,3′,4′-TetraOH-7-OMe-(2*R*,3*R*)-flavanone (padmatin, H2/1496)
753. 3,5,3′,4′-TetraOH-7-OMe-(2*R*,3*S*)-flavanone (epipadmatin, **333**; H2/1497)
754. 3,7,3′,4′-TetraOH-5-OMe-flavanone (H2/1495)
755. 3,7,3′,4′-TetraOH-8-OMe-flavanone (8-methoxyfustin, H2/1504)
756. 3,7,3′,5′-TetraOH-4′-OMe-flavanone (sepinol, H2/1506)
757. 5,7,3′,4′-TetraOH-3-OMe-flavanone (H2/1494)
758. 3,5,7-TriOH-6,4′-diOMe-flavanone (**334**; H2/1483)
759. 3,5,7-TriOH-8,4′-diOMe-flavanone (H2/1486)
760. 3,5,2′-TriOH-7,8-diOMe-flavanone (H2/1484)
761. 3,5,2′-TriOH-7,5′-diOMe-flavanone (**335**, H2/1490)
762. 3,5,3′-TriOH-7,2′-diOMe-flavanone (**336**)
763. 3,5,3′-TriOH-7,4′-diOMe-(2*R*,3*R*)-flavanone (H2/1501)
764. 3,5,3′-TriOH-7,4′-diOMe-(2*R*,3*S*)-flavanone (**337**)
765. 3,5,4′-TriOH-6,7-diOMe-flavanone (H2/1482)
766. 3,5,4′-TriOH-7,3′-diOMe-flavanone (H2/1500)

Hexa-*O*-substituted

767. 3,5,7,3′,4′,5′-HexaOH-(2*R*,3*R*)-flavanone (ampelopsin, H2/1510)
768. 3,5,7,3′,4′,5′-HexaOH-(2*R*,3*S*)-flavanone (hovenitin III, **338**)
769. 3,5,7,2′,5′-PentaOH-6-OMe-flavanone (diosalol, **339**)
770. 3,5,7,3′,4′-PentaOH-6-OMe-flavanone (H2/1507)
771. 3,5,7,3′,5′-PentaOH-4′-OMe-flavanone (pallasiin, H2/1511)
772. 3,5,7,4′,5′-PentaOH-3′-OMe-(2*R*,3*R*)-flavanone (hovenitin I, **340**)
773. 3,5,7,4′,5′-PentaOH-3′-OMe-(2*R*,3*S*)-flavanone (hovenitin II, **341**)
774. 3,5,7,3′-TetraOH-8,4′-diOMe-flavanone (H2/1509)
775. 3,5,7,4′-TetraOH-3′,5′-diOMe-flavanone (dihydrosyringetin, H2/1512)
776. 3,5,3′,4′-TetraOH-7,8-diOMe-flavanone (**342**; H2/1508)
Hepta-*O*-substituted
777. 3,5,3′-TriOH-8,5′-diOMe-6,7-methylenedioxy-flavanone (plumbaginol, **343**)
Dihydroflavonol aglycones with *C*-methyl and *C*-hydroxymethyl substituents
Tri-*O*-substituted
778. 3,5,7-TriOH-6-Me-flavanone (H2/1513)
Tetra-*O*-substituted
779. 3,5,7-TriOH-4′-OMe-6-Me-flavanone (**349**)
780. 3,7,4′-TriOH-5-OMe-6-Me-flavanone (H2/1514)
Penta-*O*-substituted
781. 2,3,5,7,4′-PentaOH-6-Me-flavanone (**350**)
782. 2,3,5,7,4′-PentaOH-8-Me-flavanone (**351**)
783. 3,5,7,3′,4′-PentaOH-6-Me-flavanone (cedeodarin, H2/1515)
784. 3,5,7,3′,4′-PentaOH-8-Me-flavanone (deodarin, H2/1516)
785. 3,5,7,3′,4′-PentaOH-6′-hydroxymethyl-flavanone (**352**)
Hexa-*O*-substituted
786. 3,5,7,3′,4′,5′-HexaOH-6-Me-flavanone (cedrin, H2/1517)
787. 3,5,7,3′,4′,5′-HexaOH-6,8-diMe-flavanone (H2/1518)
Dihydroflavonol aglycones with prenyl, geranyl, and lavandulyl substituents[a]
Tri-*O*-substituted
788. 3,5,7-TriOH-6-Ger-(2*R*,3*S*)-flavanone (H2/1567)
789. 3,5,7-TriOH-6-Pr-(2*R*,3*R*)-flavanone (glepidotin B, H2/1528)
790. 3,5,7-TriOH-6-Pr-(2*R*,3*S*)-flavanone (H2/1529)
791. 3,7,4′-TriOH-8,3′-diPr-flavanone (3-hydroxyglabrol, H2/1532)
792. 3,7,4′-TriOH-8,3′,5′-triPr (H2/1533)
793. 3,5-DiOH-7-OMe-8-Pr-flavanone (leaserone, H2/1531)
Tetra-*O*-substituted
794. 3,5,7,4′-TetraOH-6-Ger-flavanone (bonanniol A, H2/1568)
795. 3,5,7,4′-TetraOH-6-Pr-flavanone (shuterin, H2/1534)
796. 3,5,7,4′-TetraOH-6,8-diPr-flavanone (6,8-diprenylaromadendrin, **359**)
797. 3,5,7,4′-TetraOH-8-Pr-flavanone (H2/1537)
798. 3,5,7,4′-TetraOH-8-(3-OH-3-Me-butyl)-flavanone (H2/1538)
799. 3,5,7,4′-TetraOH-8,3′-diPr-flavanone (lespedezaflavanone C, H2/1543)
800. 3,5,7,4′-TetraOH-8-Pr-3′-Ger-flavanone (sanggenol C, **361**)
801. 3,5,7-TriOH-8-OMe-6-Pr-flavanone (dioclenol, **353**)
802. 3,5,4′-TriOH-7-OGer-flavanone (**360**; H2/1569)
803. 3,5,4′-TriOH-7-OPr-flavanone (H2/1536)
804. 3,7,4′-TriOH-5-OMe-6-Ger-flavanone (bonanniol B, H2/1570)
805. 3,7,4′-TriOH-5-OMe-8-Pr-flavanone (H2/1545)
Penta-*O*-substituted
806. 3,5,7,2′,4′-PentaOH-6-Pr-8-Lav (kushenol M, H2/1565)

807. 3,5,7,2′,4′-PentaOH-6-Pr-3′-Ger-flavanone (sanggenol K, **364**)
808. 3,5,7,2′,4′-PentaOH-6-(3-OH-3-Me-butyl)-8-Lav-flavanone (kosamol A, **367**)
809. 3,5,7,2′,4′-PentaOH-6,8-diPr-flavanone (kushenol L, H2/1548)
810. 3,5,7,2′,4′-PentaOH-8-Lav-flavanone (kushenol X, **362**)
811. 3,5,7,2′,4′-PentaOH-8-(2-isopropenyl-5-OH-5-Me-hexyl)flavanone (kushenol G, H2/1571)
812. 3,5,7,2′,4′-PentaOH-8,5′-diPr-3′-Ger-flavanone (sanggenol E, **366**)
813. 3,5,7,2′,4′-PentaOH-3′-Ger-5′-Pr-flavanone (sanggenol D, **365**)
814. 3,5,7,3′,4′-PentaOH-8,2′,6′-triPr-flavanone (petalostemumol, **363**; H2/1559)
815. 3,5,7,3′-TetraOH-4′-OMe-6-Ger-flavanone (diplacol, H2/1576)
816. 3,5,7,3′-TetraOH-4′-OMe-6-Ger-flavanone (diplacol 4′-methyl ether, H2/1577)
817. 3,5,7,4′-TetraOH-6-OMe-8-Pr-flavanone (floranol, **355**)
818. 3,5,7,4′-TetraOH-3′-OMe-8-Pr-flavanone (H2/1561)
819. 3,5,3′,4′-TetraOH-7-OPr-flavanone (**354**)
820. 3,7,2′,4′-TetraOH-5-OMe-8-(2-isopropenyl-5-OH-5-Me-hexyl)(2R,3R)-flavanone (kushenol H, H2/1574)
821. 3,7,2′,4′-TetraOH-5-OMe-8-(2-isopropenyl-5-OH-5-Me-hexyl)(2R,3S)-flavanone (kushenol K, H2/1575)
822. 3,7,2′,4′-TetraOH-5-OMe-8-Lav-(2R,3R)-flavanone (kushenol I, H2/1572)
823. 3,7,2′,4′-TetraOH-5-OMe-8-Lav-(2R,3S)-flavanone (kushenol N, H2/1573)
824. 3,5,3′-TriOH-7,4′-diOMe-6-Pr-flavanone (isotirumalin, H2/1563)
825. 3,5,3′-TriOH-7,4′-diOMe-8-Pr-flavanone (tirumalin, H2/1564)
826. 3,5,4′-TriOH-7,3′-diOMe-6-Pr-flavanone (**356**)
827. 3,5,4′-TriOH-7,3′-diOMe-8-Pr-flavanone (scariosin, **357**; H2/1562)

Hexa-*O*-substituted

828. 3,5,7,2′,5′-PentaOH-6-OMe-8-Pr-flavanone (paraibanol, **358**)
829. 3,5,7,2′,6′-PentaOH-4′-OMe-6-Ger-8-Pr-flavanone (kenusanone C, **368**; H2/1566)

Dihydroflavonol aglycones bearing furano substituents

830. 3-OMe-[2″,3″:7,8]furanoflavanone (**369**)
831. 3,5,6-TriOMe-[2″,3″:7,8]furanoflavanone (**370**)
832. 3,5,4′-TriOH-5″-isopropenyldihydrofurano[2″,3″:7,6]-(2R,3R)-flavanone (shuterol, H2/1535)
833. 3,5,4′-TriOH-5″-isopropenyldihydrofurano[2″,3″:7,8]flavanone (H2/1539)
834. 3,5,2′,4′-TetraOH-5″-isopropenyldihydrofurano[2″,3″:7,8]-(2R,3R)flavanone (shuterone A, H2/1546)
835. 3,5,2′,4′-TetraOH-5″-isopropenyldihydrofurano[2″,3″:7,8]-(2S,3R)-flavanone (shuterone B, H2/1547)
836. 3,5,6-TriOMe-3′,4′-methylenedioxyfurano[2″,3″:7,8]flavanone (H2/1521)

Dihydroflavonol aglycones bearing pyrano substituents[a]

837. 3-OH-6″,6″-DMP[2″,3″:7,8]flavanone (3-hydroxyisolonchocarpin, H2/1527)
838. 3-OH-bis(6″,6″-DMP[2″,3″:5,6][2″,3″:7,8])flavanone (MS II, **371**)
839. 3,5-DiOH-8-Pr-6″,6″-DMP[2″,3″:7,6]flavanone (mundulinol, H2/1530)
840. 3,4′-DiOH-3′-Pr-6″,6″-DMP[2″,3″:7,8]flavanone (kanzonol Z, **372**)
841. 3,5,2′-TriOH-8-Pr-6″,6″-DMP[2″,3″:7,6]flavanone (jayacanol, **373**)
842. 3,5,4′-TriOH-8-Pr-6″,6″-DMP[2″,3″:7,6]flavanone (lupinifolinol, H2/1542)
843. 3,5,4′-TriOH-6″,6″-DMP[2″,3″:7,8]flavanone (H2/1540)
844. 3,5,4′-TriOH-6″,6″-DHDMP[2″,3″:7,8]flavanone (H2/1541)
845. 3,4′-DiOH-5-OMe-8-Pr-6″,6″-DMP[2″,3″:7,6]flavanone (**374**)
846. 5,2′-DiOH-3-OMe-8-Pr-6″,6″-DMP[2″,3″:7,6]flavanone (**375**)

847. 5,4'-DiOH-3-OMe-8-Pr-6'',6''-DMP[2'',3'':7,6]flavanone (3-*O*-methyllupinifolinol, H2/1544)
848. 3,5,7-TriOH-2'-OMe-6'',6''-DMP[2'',3'':4',3'](2*S*,3*S*)-flavanone (H2/1556)
849. 3,5,4'-TriOH-3'-OMe-6'',6''-DMP[2'',3'':7,6]flavanone (eriotrinol, **376**; H2/1560)

Dihydroflavonol aglycones having a C-3–C-2' ether link and an isoprenylated C-2

850. Sanggenol F (**377**)
851. Sanggenol G (**383**)
852. Sanggenol H (**384**)
853. Sanggenol I (**382**)
854. Sorocein D (**378**)
855. Sorocein E (**379**)
856. Sorocein F (**381**)
857. Sorocein G (**380**)

Dihydroflavonol aglycones bearing benzyl substituents

858. 3,5,7,4'-TetraOH-6-(4-OH-benzyl)-flavanone (gericudranin E, **385**, H2/1579)
859. 3,5,7,4'-TetraOH-6,8-di(4-OH-benzyl)-flavanone (gericudranin D, **388**, H2/1580)
860. 3,5,7,3',4'-PentaOH-6-(4-OH-benzyl)-flavanone (gericudranin C, **386**, H2/1581)
861. 3,5,7,3',4'-PentaOH-6,8-di(4-OH-benzyl)flavanone (gericudranin A, **389**, H2/1583)
862. 3,5,7,3',4'-PentaOH-8-(4-OH-benzyl)-flavanone (gericudranin B, **387**, H2/1582)

Dihydroflavonol aglycones bearing miscellaneous substituents

863. Crombeone (H2/1524)
864. 7,8-Dihydrooxepinodihydroquercetin (H2/1558)
865. *rel*-5-Hydroxy-7,4'-dimethoxy-2''*S*-(2,4,5-trimethoxy-*E*-styryl)tetrahydrofuro[4''*R*,5''*R*:2,3] (**394**)
866. *rel*-5-Hydroxy-7,4'-dimethoxy-3''*S*-(2,4,5-trimethoxy-*E*-styryl)tetrahydrofuro[4''*R*,5''*R*:2,3] (**395**)
867. Isosilybin (H2/1585)
868. Isosilychristin (H2/1587)
869. 5-Methoxypeltogynone (H2/1525)
870. 5-Methoxymopanone (H2/1526)
871. Mopanone (H2/1523)
872. Peltogynone (H2/1522)
873. 3,5,7,3',4'-PentaOH-6-(3-oxobutyl)-flavanone (**390**)
874. Pseudotsuganol (H2/1590)
875. Silybin (H2/1584)
876. Silychristin (H2/1586)
877. Silydianin (H2/1588)
878. 3,5,7,4'-TetraOH-3'-(4-OH-benzaldehyde) (hypnogenol F, **391**)
879. 3,5,7,4'-TetraOH-3'-(4-OH-benzoic acid) (hypnum acid, **392**)
880. 3,5,7,4'-TetraOH-3'-(4-OH-benzoic acid methyl ester) (hypnum acid methyl ester, **393**)
881. 3,5,7,4'-TetraOH-3'-(5''-formyl-2''-hydroxyphenyl)flavanone (H2/1578)
882. 5,7,4'-TriOH-3-*O*-(1,8,14-trimethylhexadecanyl)flavanone (muscanone, **396**)

Dihydroflavonol esters

883. 3,5,7-TriOH-(2*R*,3*R*)-flavanone 3-acetate ((2*R*,3*R*)-pinobanksin 3-acetate, H2/1658)
884. 3,5,7-TriOH-(2*R*,3*S*)-flavanone 3-acetate ((2*R*,3*S*)-pinobanksin 3-acetate, **344**)
885. 3,5,7-TriOH-flavanone 3-propionate (pinobanksin 3-propionate, H2/1660)
886. 3,5,7-TriOH-flavanone 3-benzoate (pinobanksin 3-benzoate, H2/1661)
887. 3,5,7-TriOH-flavanone 3-cinnamate (pinobanksin 3-cinnamate, H2/1662)
888. 3,5-DiOH-7-OMe-flavanone 3-acetate (3-acetylalpinone, H2/1663)
889. 3,5,7,3'-TetraOH-flavanone 3-isobutyrate (**345**)

890. 3,5,7,4'-TetraOH-flavanone 3-acetate (aromadendrin 3-acetate, H2/1664)
891. 3,5,7-TriOH-4'-OMe-flavanone 3-acetate (**346**)
892. 3,5,4'-TriOH-7-OMe-(2*R*,3*R*)-flavanone 3-acetate (H2/1665)
893. 3,5,4'-TriOH-7-OMe-(2*R*,3*S*)-flavanone 3-acetate (**347**)
894. 3,5,7,3',4'-PentaOH-flavanone 3-acetate (taxifolin 3-acetate, H2/1670)
895. 3,5,7,4'-TetraOH-6-OMe-flavanone 3-acetate (H2/1666)
896. 3,5,7,4'-TetraOH-8-OMe-flavanone 3-acetate (H2/1668)
897. 3,5,7,4'-TetraOH-8-OMe-flavanone 3-angelate (H2/1669)
898. 3,5,7,4'-TetraOH-3'-OMe-flavanone 3-acetate (H2/1672)
899. 3,5,3',4'-TetraOH-7-OMe-flavanone 3-acetate (padmatin 3-acetate, H2/1671)
900. 3,5,2'-TriOH-7,8-diOMe-flavanone 3-acetate (**348**)
901. 3,5,2'-TriOH-7,8-diOMe-flavanone 2'-acetate (H2/1667)
902. 3,5,7,3',4'-PentaOH-6-OMe-flavanone 3-acetate (H2/1673)
903. 3,5,7,3',4',5'-HexaOH-flavanone 3-gallate-3'-sulfate (myricatin, H2/1674)
904. 3,5,7-TriOH-6-Me 3-acetate (H2/1675)
905. 3,5,7-TriOH-6-Pr 3-acetate (H2/1676)

Dihydroflavonol glycosides[a]
3,5,7-TriOH-flavanone (pinobanksin)
906. 5-Galactosyl(1 → 4)glucoside (H2/1591)
3,7,4'-TriOH-flavanone (garbanzol)
907. 3-Glucoside (lecontin, H2/1592)
3,7-DiOH-4'-OMe-flavanone
908. 7-Xylosyl(1→ 6)glucoside (kushenol J, H2/1593)
3,5,7,4'-TetraOH-flavanone (aromadendrin)
909. 3-β-L-Arabinopyranoside (H2/1594)
910. 3-Galactoside (H2/1595)
911. 3-Glucoside (H2/1596)
912. 3-Rhamnoside (2*R*,3*R*) (engeletin, H2/1597)
913. 3-Rhamnoside (2*R*,3*S*) (isoengeletin, H2/1598)
914. 3-Rhamnoside-5-glucoside (H2/1603)
915. 7-Glucoside (sinensin, H2/1599)
916. 7-[6-(4-Hydroxy-2-methylenebutanoyl)glucoside] (**398**)
917. 7-Rhamnoside (H2/1600)
918. 7-Rhamnosyl(1 → 4)galactoside (**397**; H2/1604)
919. 4'-Glucoside (H2/1602)
920. 4'-Xyloside (H2/1601)
3,7,3',4'-TetraOH-flavanone (fustin)
921. 3-Glucoside (H2/1611)
922. 3,7-Diglucoside (H2/1613)
923. 7-Rhamnoside (H2/1612)
3,5,7-TriOH-4'-OMe-flavanone (dihydrokaempferide)
924. 3-Glucuronide (H2/1607)
925. 7-Rhamnoside (H2/1608)
3,5,4'-TriOH-7-OMe-flavanone
926. 5-Glucoside (H2/1605)
927. 5,4'-Diglucoside (micrantoside, H2/1606)
3,5-DiOH-7,4'-diOMe-flavanone
928. 3-Rhamnoside (aurapin, H2/1609)
929. 5-Glucoside (H2/1610)

3,5,7,8,4'-PentaOH-flavanone
930. 8-Glucoside (callunin, H2/1614)
931. 8-(2-Acetylglucoside) (2''-acetylcallunin, **399**; H2/1615)
3,5,7,2',5'-PentaOH-flavanone
932. 3-Rhamnoside (**400**; H2/1617)
3,5,7,3',4'-PentaOH-flavanone (taxifolin)
933. 3-Apioside (H2/1618)
934. 3-α-Arabinofuranoside (**403**; H2/1619)
935. 3-α-Arabinopyranoside (2*R*,3*R*) (**401**)
936. 3-α-Arabinopyranoside (2*R*,3*S*) (**402**)
937. 3-Galactoside (dihydrohyperin, H2/1624)
938. 3-Galactosyl(1 → 6)glucoside (H2/1640)
939. 3-Glucoside (2*R*,3*R*) (glucodistylin, H2/1625)
940. 3-Glucoside (2*R*,3*S*) (isoglucodistylin, H2/1626)
941. 3-Glucoside (2*S*,3*S*) (H2/1627)
942. 3-Glucosyl(1 → 3)rhamnoside (huangqioside E, H2/1641)
943. 3-Glucosyl(1 → 4)rhamnoside (H2/1642)
944. 3-Rhamnoside (2*R*,3*R*) (astilbin, H2/1629)
945. 3-Rhamnoside (2*R*,3*S*) (isoastilbin, H2/1630)
946. 3-Rhamnoside (2*S*,3*R*) (neoisoastilbin, H2/1631)
947. 3-Rhamnoside (2*S*,3*S*) (neoastilbin, H2/1632)
948. 3-Xyloside (2*R*,3*R*) (H2/1620)
949. 3-Xyloside (2*R*,3*S*) (H2/1621)
950. 3-Xyloside (2*S*,3*R*) (H2/1622)
951. 3-Xyloside (2*S*,3*S*) (H2/1623)
952. 3,5-Dirhamnoside (H2/1643)
953. 3,7-Dirhamnoside (**405**)
954. 3-(3-Cinnamoylrhamnoside) (**404**)
955. 3-(6-Galloylglucoside) (taxillusin, H2/1628)
956. 5-Galactoside (H2/1633)
957. 7-Galactoside (H2/1634)
958. 7-Glucoside (H2/1635)
959. 7-Rhamnoside (H2/1636)
960. 7-Rhamnosyl(1 → 6)glucoside (flavoplatycoside, **406**; H2/1644)
961. 3'-Glucoside (H2/1637)
962. 3'-(6-Phenylacetylglucoside) (H2/1638)
963. 4'-Glucoside (H2/1639)
3,5,7,3',5'-PentaOH-flavanone
964. 3-Rhamnoside (2*R*,3*R*) (**408**)
965. 3-Rhamnoside (2*R*,3*S*) (smitilbin, **409**)
966. 3-Rhamnoside (2*S*,3*S*) (neosmitilbin, **410**)
3,5,7,3'-TetraOH-4'-OMe-flavanone
967. 3-Rhamnosyl(1 → 6)glucoside (anacheiloside, H2/1647)
3,5,3',4'-TetraOH-7-OMe-flavanone (padmatin, dihydrorhamnetin)
968. 3β-(-D-Arabinopyranoside (**407**)
969. 3-Glucoside (H2/1645)
970. 5-Glucoside (H2/1646)
3,5,7-TriOH-8,4'-diOMe-flavanone
971. 7-Glucoside (dihydroprudomenin, H2/1616)

3,5,7,3',4',5'-HexaOH-flavanone (*ampelopsin*)

972. 3-Rhamnoside (H2/1648)

973. 7-Glucoside (dihydroprudomenin, 2/1649)

974. 3'-Glucoside (H2/1650)

3,5,7,3',4'-PentaOH-5'-OMe-flavanone (*5'-O-methylampelopsin*)

975. 4'-Rhamnoside (H2/1651)

3,5,7,4'-TetraOH-6-Me-flavanone

976. 7-Glucoside (H2/1652)

3,5,7,3',4',5'-HexaOH-6-Me-flavanone

977. 3'-Glucoside (cedrinoside, H2/1653)

3,5,7,4'-TetraOH-8-Pr-flavanone

978. 7-Glucoside (phellamurin, H2/1655)

3,5,7,4'-TetraOH-3'-Pr-flavanone

979. 7-Glucoside (phellochinin, **411**)

3,5,4'-TriOH-6'',6''-DMDHP[2'',3'':7,8]flavanone

980. 3-Glucoside (phellodendroside, H2/1657)

3,5,7,4'-TetraOH-6-(3-OH-3-Me-butyl)-flavanone

981. 7-Glucoside (phellavin, H2/1654)

3,5,7,4'-TetraOH-8-(3-OH-3-Me-butyl)-flavanone

982. 7,3''-Diglucoside (dihydrophelloside, H2/1656)

3,5,7,3',4'-PentaOH-6-Pr-flavanone

983. 7-Glucoside (**412**)

---

[a]Pr, prenyl; Ger, geranyl; Lav, lavandulyl; DMP, dimethylpyrano; DMDHP, dimethyldihydropyrano.
Rhamnoside, α-L-rhamnopyranoside; apioside, β-apiofuranoside; glucoside, β-D-glucopyranoside; galactoside, β-D-galactopyranoside; xyloside, β-D-xylopyranoside; glucuronide, β-D-glucuronopyranoside. Cinnamic acids are assumed to be in the *E*-form.

Nigel C. Veitch and Renée J. Grayer

CONTENTS

16.1   GENERAL INTRODUCTION

The chalcones, dihydrochalcones, and aurones are three distinctive classes of compound that comprise more than 900 of all the naturally occurring flavonoids reported in the literature to the end of 2003. Viewed from a historical perspective, the chalcones and aurones are best known as the yellow to orange colored flower pigments of some species of *Coreopsis* and other Asteraceae taxa. The distribution of these compounds is not restricted to flowers, however, and examples of all three classes can be found in many different plant tissues. The chalcones are structurally one of most diverse groups of flavonoids, as witnessed by the formation of a wide range of dimers, oligomers, Diels–Alder adducts, and conjugates of various kinds. At the same time, they are of great significance biosynthetically as the immediate precursors of all other classes of flavonoid. Underlying these important attributes is the unique feature that distinguishes chalcones and dihydrochalcones from other flavonoids, the open-chain three-carbon structure linking the A- and B-rings in place of a heterocyclic C-ring (Figure 16.1). In plants, chalcones are converted to the corresponding (2*S*)-flavanones in a stereospecific reaction catalyzed by the enzyme chalcone isomerase. This close structural and biogenetic relationship between chalcones and flavanones explains why they often co-occur as natural products. It is also the reason why chalcones, dihydrochalcones, and aurones are sometimes described together with flavanones and dihydroflavonols.[1–3] Whether this group should continue to be known as the "minor flavonoids" is a matter for debate; however, the significant increase in the number of new examples of each of these flavonoid classes in the recent literature suggests that this title may no longer be appropriate. For this reason, the chalcones, dihydrochalcones, and aurones are treated separately from the flavanones and dihydroflavonols (Chapter 15) in this volume. The main purpose of this chapter is to review the scientific literature on all chalcones, dihydrochalcones, and aurones reported as new natural products during the period from 1992 to 2003. This continues the tradition set by the four volumes of the *Advances in Flavonoids* series,[1–4] which provide the most comprehensive treatment available of the scientific literature relating to flavonoids up to the end of 1991. It is not the purpose of this review to describe general methods for the detection, isolation, and characterization of chalcones, dihydrochalcones, and aurones, as these have been discussed extensively not only in the "Advances" series[1–4] but also in more general texts on flavonoids.[5–10] However, points of interest relating to the source, identification, and biological activity of new compounds are covered, as well as wider issues such as biosynthesis and chemosystematic or ecological significance. The chemical synthesis of these three flavonoid classes (a major subject in its own right) is not treated here, although synthetic proced-

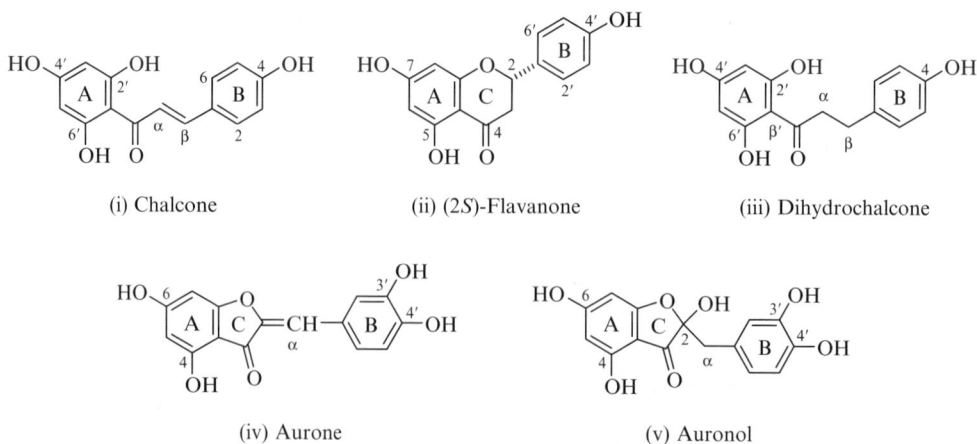

(i) Chalcone   (ii) (2*S*)-Flavanone   (iii) Dihydrochalcone

(iv) Aurone   (v) Auronol

FIGURE 16.1 Guide to the structures, ring labeling, and atom numbering of chalcones, dihydrochalcones, aurones, and auronols. The corresponding flavanone structure is shown for reference.

ures that are specific to new compounds will be highlighted. The total number of flavonoids presented in this chapter as new natural products is 377, comprising 248 chalcones, 91 dihydrochalcones, and 38 aurones. Among these are a small number of known compounds for which new or revised structures have been proposed. It should be noted that new sources of well-known compounds have not been documented in this survey.

16.1.1 NOMENCLATURE

The nomenclature and in particular the atom numbering of chalcones, dihydrochalcones, and aurones remain a potential source of confusion when compared to that of other classes of flavonoid. The A- and B-rings of all the flavonoids have the same origin in biosynthetic terms, with the A-rings derived from the acetate pathway and the B-rings from the shikimate pathway. Similarly, all the structures are written by convention with the A-ring to the left (although this convention may break down for the more complex chalcones, such as dimers, oligomers, and Diels–Alder adducts). The crucial difference is in the style of atom numbering, in which primed numbers are used to refer to the A-ring of chalcones and dihydrochalcones, but to the B-ring of other flavonoid classes, including the aurones. Similarly, the B-rings of chalcones and dihydrochalcones carry the nonprimed numbers instead of the A-ring. The numbering scheme followed for chalcones and dihydrochalcones is also different, because the C_3 unit linking the A- and B-rings is referred to only in terms of carbonyl (β'), α- and β-carbons, whereas the equivalent carbon atoms of the heterocyclic C-rings of other flavonoids are numbered together with the rest of the molecule. An example of the two systems is shown in Figure 16.1 for clarification. Notice that the aurone numbering system is anomalous because of the five-membered C-ring. The result is that the A-ring positions equivalent to other flavonoids (excluding chalcones and dihydrochalcones) bear a number one less in value. Aurones are also referred to as 2-benzylidenecoumaranones, although the correct systematic name is 2-benzylidene-3(2H)-benzofuranone. Similarly, auronols may also be found described as 2-hydroxy-2-benzylcoumaranones.

The IUPAC-approved systematic name for chalcone of 1,3-diphenyl-2-propen-1-one is generally thought too cumbersome for routine use, even for simple naturally occurring derivatives such as the commonly found 2′,4′,4-trihydroxychalcone (isoliquiritigenin), which bears the systematic name 1-(2,4-dihydroxyphenyl)-3-(4-hydroxyphenyl)-2-propen-1-one. For this reason, the use of semi-systematic (2′,4′,4-trihydroxychalcone) and trivial (isoliquiritigenin) names is widespread. So far as this chapter is concerned, semi-systematic names are given where possible, although the complexity of some of the compounds precludes the use of any but trivial names. The new compounds are arranged sequentially in 15 tables following the order chalcones (Table 16.1–Table 16.7), dihydrochalcones (Table 16.8–Table 16.12), and aurones (Table 16.13–Table 16.15). Within each class, the compounds are introduced as (i) simple derivatives (i.e., hydroxy, methoxy, methylenedioxy, and methyl substituted), (ii) isoprenylated derivatives, (iii) glycosides, and (iv) dimers, oligomers, adducts, and other conjugates or special groups. In tables of types (i) to (iii), the compounds are arranged in order of increasing O-substitution. Within each group the precedence of A- over B-ring over α- and β-carbons is observed for the individual substitution patterns, thus (2′,4′,6′)-tri-O-substituted chalcones will be found listed before their (2′,4′,4)-tri-O-substituted analogs. Within the subgroups compounds with the greatest number of free hydroxyl groups take precedence, thus 2′,4′,2-trihydroxy-6′-methoxychalcone **6**) appears before 2′-hydroxy-4′,6′, 2-trimethoxychalcone (**7**), although both are examples of (2′,4′,6′,2)-tetra-O-substituted chalcones. The purpose of this system is to allow groups of compounds with a particular substitution pattern to be identified easily for comparison with previous listings.[1–4] Furthermore, this arrangement of compounds on a structural basis enables trends of potential

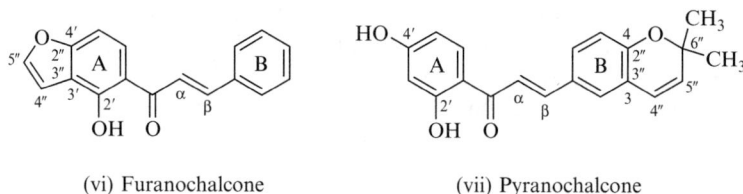

(vi) Furanochalcone   (vii) Pyranochalcone

FIGURE 16.2 Atom numbering in furano- and pyranochalcones.

biosynthetic or chemosystematic interest to be recognized more easily. With regard to isoprenylated derivatives, the isoprenoid-derived groups are treated as substituents to the flavonoid skeleton, for example, as prenyl-, geranyl-, furano-, and pyranochalcones. The numbering system used to identify the location of the fused-ring substituents is indicated in Figure 16.2. In tables listing chalcone, dihydrochalcone, and aurone glycosides, the sugars can be assumed to be pyranosides unless otherwise stated. The absolute configurations of D and L for the sugars have been omitted as most were not determined experimentally. A checklist of all new chalcones, dihydrochalcones, and aurones (**1–377**) arranged by molecular formula is given in Appendix A. This is followed by a list containing entries for all known compounds of these flavonoid classes published to the end of 2003 (Appendix B).

16.1.2 OVERVIEW OF CHALCONE BIOSYNTHESIS

Much has been written about the biosynthesis of flavonoids, and several recent review articles are available.[11–13] Only a few essential points relating to chalcone biosynthesis will be mentioned here. An outline of the most important enzyme-catalyzed reactions leading to the production of chalcones and 6'-deoxychalcones is given in Figure 16.3. Central to this scheme is the biosynthesis of 2',4',6',4-tetrahydroxychalcone, which is also known by the trivial names of chalconaringenin or naringenin chalcone. This compound is formed by sequential condensation of three molecules of malonyl coenzyme A (malonyl-CoA) with one of p-coumaroyl-CoA, a reaction catalyzed by chalcone synthase.[14] The final step is believed to be the cyclization of a tetraketide precursor (Figure 16.3). The production of chalcones as natural products represents the convergence of two biosynthetic pathways, the acetate (leading to the A-ring) and the shikimate (leading to the B-ring), respectively, as mentioned previously. It might be supposed that the extent of hydroxyl substitution in the B-ring could be controlled by introduction of different shikimate-derived cinnamoyl-CoA precursors. Thus, cinnamoyl-, p-coumaroyl-, caffeoyl-, and 3,4,5-trihydroxycinnamoyl-CoA would give an unsubstituted B-ring, 4-hydroxy, 3,4-dihydroxy, and 3,4,5-trihydroxy substitution patterns, respectively. However, many studies indicate that the extent of B-ring hydroxyl substitution in flavonoids is controlled at the C_{15} level, rather than by incorporation of specific cinnamoyl-CoA derivatives.[11] Thus, further elaboration of the B-ring resulting from p-coumaroyl-CoA (4-hydroxyl) appears to be achieved through specific hydroxylase and methyl transferase enzymes. Some evidence has been presented for the existence of a specific 3-hydroxylase for chalcones that is distinct from the general flavonoid 3'-hydroxylase,[15] although both enzymes are known to be cytochrome P450-dependent monooxygenases. How the number of hydroxyl groups in the A-ring of chalcones is controlled is also an interesting question because of the existence of many 6'-deoxychalcone derivatives. A typical example is isoliquiritigenin (2',4',4-trihydroxychalcone), a common constituent of

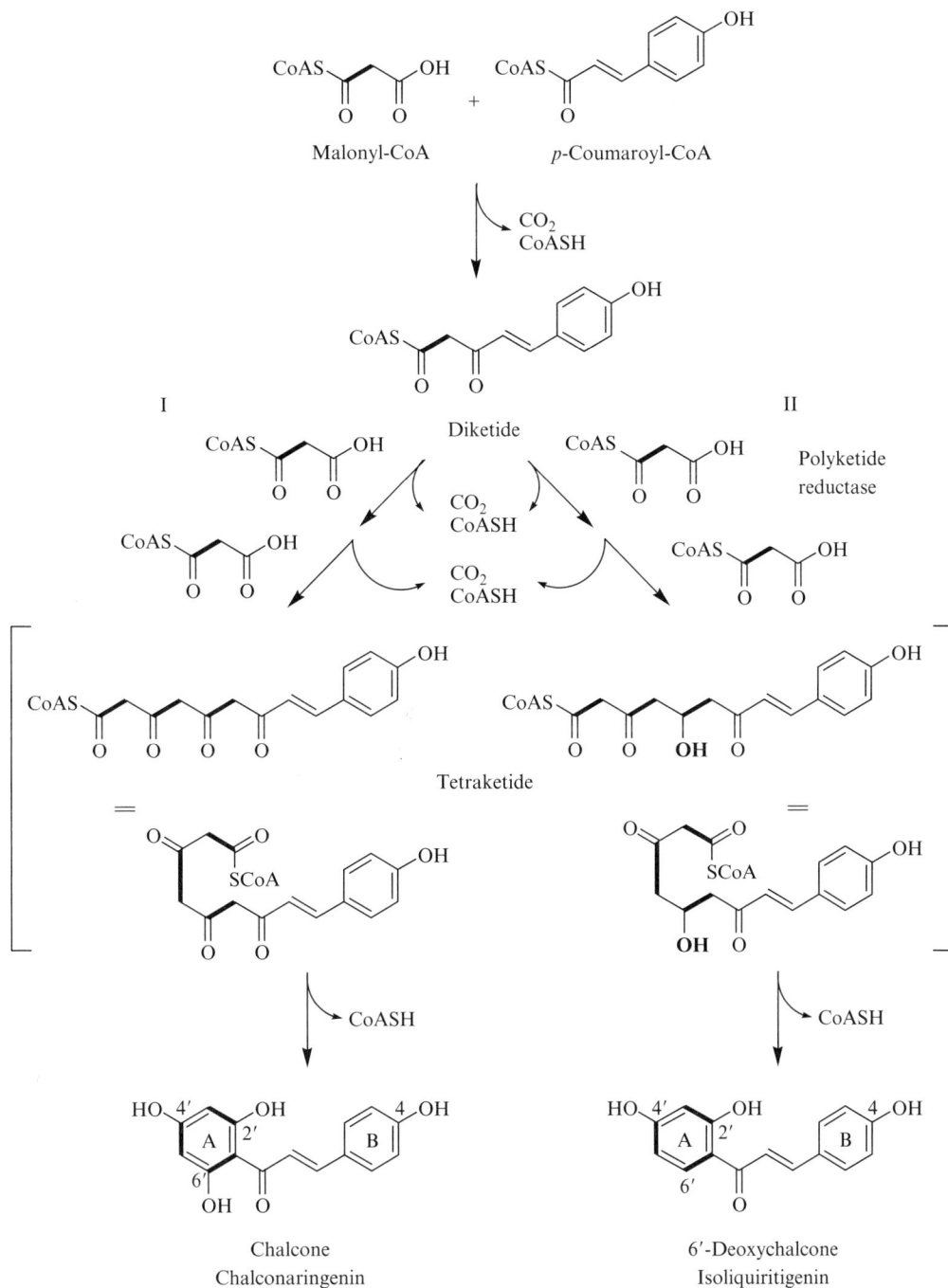

FIGURE 16.3 Overview of the biosynthesis of (I) chalcones and (II) 6'-deoxychalcones. The sequential condensation of three molecules of malonyl-CoA (acetate pathway) and p-coumaroyl-CoA (shikimate pathway) is catalyzed by the enzyme chalcone synthase.[11] The production of 6'-deoxychalcones is thought to involve an additional reduction step at the tri- or tetraketide level, catalyzed by polyketide reductase.[14,16] The origin of the A-ring carbons derived from the acetate pathway is indicated in bold. CoA, coenzyme A.

the Leguminosae. The most likely explanation for their formation is a reduction step at the polyketide level, as shown in Figure 16.3. This type of reaction is catalyzed by a NADPH-dependent monomeric enzyme referred to in the literature as either chalcone ketide reductase or polyketide reductase.[14,16] An insight into the biosynthesis of the less common C-methyl chalcones has resulted from the identification of a chalcone synthase-related protein in *Pinus strobus* seedlings that catalyzes a chain extension reaction with methylmalonyl-CoA.[17] Incorporation is predicted to occur at the diketide stage in order to produce 2′,4′,6′-trihydroxy-3′-methylchalcone, the immediate precursor of the flavanones cryptostrobin (5,7-dihydroxy-8-methylflavanone) and strobopinin (5,7-dihydroxy-6-methylflavanone) found in this species.[17]

A small group of chalcone derivatives for which the typical O-substitution patterns of the A- and B-rings are apparently reversed are known as retrochalcones.[2–4] A typical example is echinatin (4′,4-dihydroxy-2-methoxychalcone), a compound isolated from tissue culture of *Glycyrrhiza echinata* (Leguminosae),[18] which is the retrochalcone equivalent of isoliquiritigenin 2′-methyl ether (4′,4-dihydroxy-2′-methoxychalcone). The first studies on the biosynthesis of echinatin by Saitoh and colleagues show that in contrast to normal chalcones, the A-ring is shikimate-derived and the B-ring acetate-derived.[19] The co-occurrence of echinatin, licodione, and licodione 2′-methyl ether in *G. echinata* cell cultures suggested that retrochalcones might be formed from chalcones with dibenzoylmethanes (the keto tautomers of β-hydroxychalcones) as intermediates.[20–22] The possibility of 2-hydroxyflavanones as precursors to dibenzoylmethanes was also considered.[22] Additional support for this hypothesis came from the observation that isoliquiritigenin, liquiritigenin (7,4′-dihydroxyflavanone), 7,4′-dihydroxyflavone, licodione, licodione 2′-methyl ether, and echinatin were all constituents of *G. pallidiflora*.[23] More recent work on the enzymology of cultured licorice cells (*G. echinata*) confirms the intermediacy of both 2-hydroxyflavanones and dibenzoylmethanes in retrochalcone synthesis,[24,25] as summarized in Figure 16.4. Some new examples of both retrochalcones and retrodihydrochalcones can be found in Table 16.1 and Table 16.8, respectively.

Little is known about the biosynthesis of dihydrochalcones from chalcones, but an enzyme involved in aurone biosynthesis has recently been identified for the first time.[26] This important breakthrough came from the work of Nakayama and colleagues on the origins of the yellow flower color of the snapdragon, *Antirrhinum majus* (Scrophulariaceae). This is due to aurone glycosides located in vacuoles of the epidermal cells of the flowers, such as the 6-O-glucosides of aureusidin (4,6,3′,4′-tetrahydroxyaurone) and bracteatin (4,6,3′,4′,5′-pentahydroxyaurone). The aglycones of these aurones are now known to be formed through the action of aureusidin synthase on the chalcone precursors, chalconaringenin (2′,4′,6′,4-tetrahydroxychalcone) and 2′,4′,6′,3,4-pentahydroxychalcone (Figure 16.5).[27–29] This enzyme has been characterized as a monomeric 39-kDa glycoprotein with a binuclear copper center. Sequence homology analysis indicates that it belongs to the family of plant polyphenol oxidases,[26] enzymes that catalyze the conversion of monophenols to *ortho*-diphenols and *ortho*-quinones. Two important features of the mechanism involving aureusidin synthase (Figure 16.5) are the *ortho*-hydroxylation of the chalcone B-ring and the oxidative cyclization step to give the aurone C-ring.[28,29] However, aureusidin can also be formed from 2′,4′,6′,3,4-pentahydroxychalcone, presumably by the initial transformation of the o-dihydroxy B-ring of the latter to an o-diquinone. The biosynthetic origin of aurones with only one or no hydroxyl groups in the B-ring is unknown at present.

For details of the enzymes that further modify chalcone and other flavonoid structures by hydroxylation, methylation, glycosylation, and other processes, the review by Forkmann and Heller can be recommended.[11]

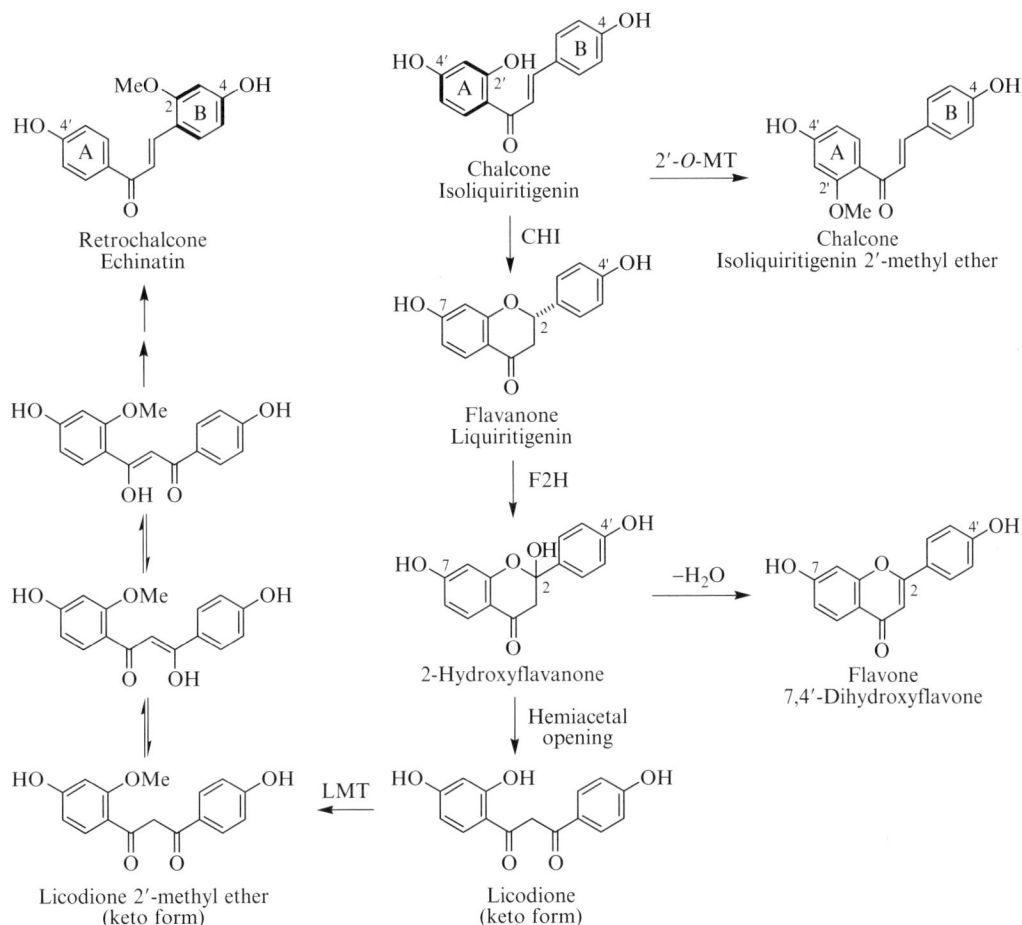

FIGURE 16.4 An outline of the biosynthesis of retrochalcones based on the production of echinatin in cell cultures of *Glycyrrhiza echinata* (Leguminosae).[19–22,24,25] The transformation of isoliquiritigenin to its 2'-methyl ether is a reaction catalyzed by 2'-*O*-methyltransferase (2'-*O*-MT) in *Medicago sativa* (Leguminosae) cell cultures.[11] Enzymes involved in the biosynthesis of echinatin include chalcone isomerase (CHI), flavanone 2-hydroxylase (F2H), and licodione methyl transferase (LMT). The order of the final sequence of reactions leading to echinatin remains to be confirmed.

16.2 CHALCONES

16.2.1 Chalcones with Simple Patterns of *O*-Substitution

16.2.1.1 Structures and Synthetic Derivatives

New chalcone aglycones reported in the literature between 1992 and 2003 are shown in Table 16.1, with some examples illustrated in Figure 16.6. This listing excludes the larger group of isoprenylated chalcones that are found in Table 16.2, but does include a small number of examples with *C*-methyl, formyl, and halogen substituents. Most of the patterns of *O*-substitution (hydroxy, methoxy, and methylenedioxy) represented by the chalcones in Table 16.1 are well known,[1–4,10] although both new and rare combinations continue to be discovered. Among these are a di-*O*-substituted chalcone (**1**) isolated from the whole plant of *Primula macrophylla* (= *P. stuartii*) (Primulaceae) and known previously only as a synthetic

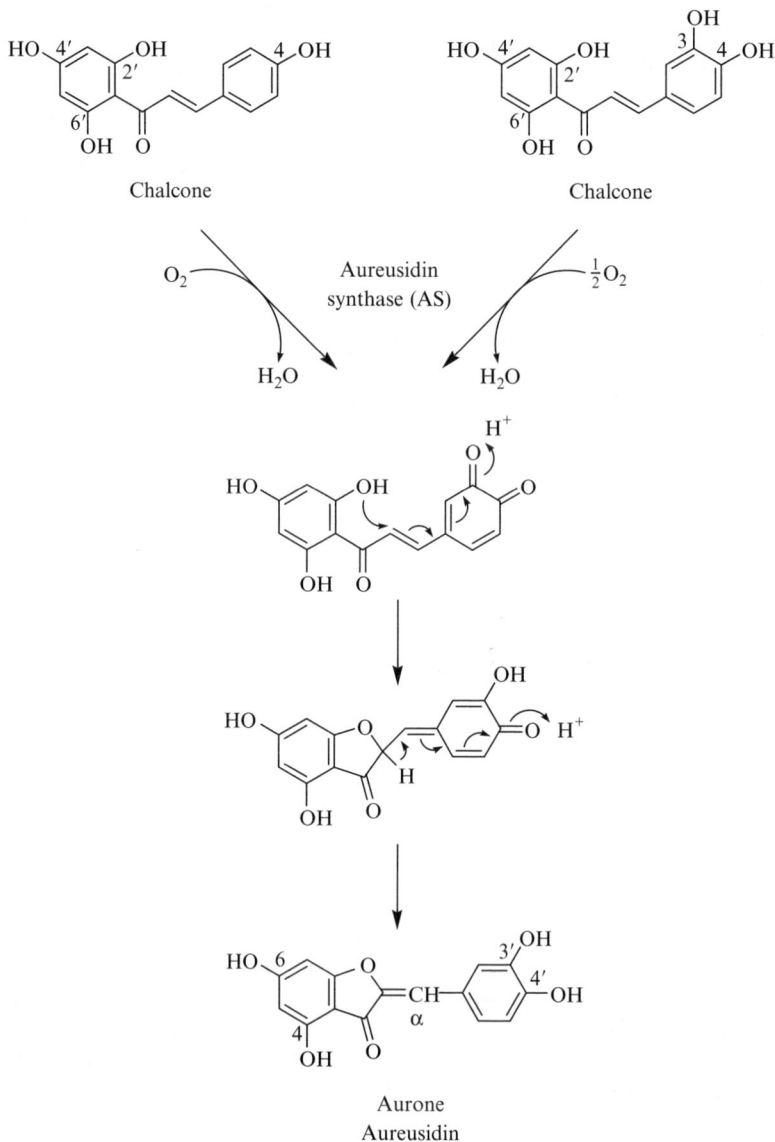

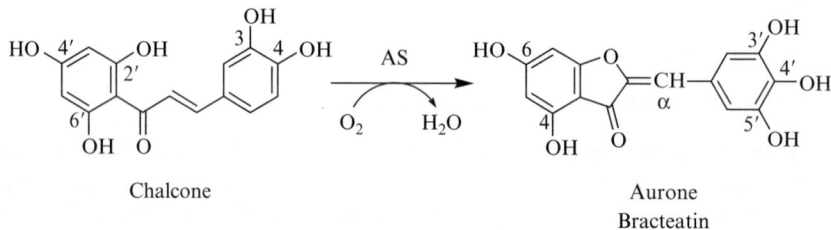

FIGURE 16.5 Reactions catalyzed by the plant polyphenol oxidase, aureusidin synthase (AS), in the transformation of chalcones to aurones.[26–29] (A) The biosynthesis of aureusidin from either tetra- or pentahydroxychalcone precursors through an *ortho*-diquinone intermediate. (B) The aureusidin synthase-catalyzed formation of bracteatin from a pentahydroxychalcone precursor.

TABLE 16.1
New Chalcones with Simple Patterns of O-Substitution Reported in the Literature from 1992 to 2003

No.	OH	OMe	Other	Mol. Formula	Trivial Name	Source	Family	Ref.
Di O-substituted								
	(3',3)							
1	3,3	—	—	$C_{15}H_{12}O_3$		*Primula macrophylla*	Primulaceae	30
Tri O-substituted								
	(2',4',6')							
2	2',6'	4'	3'-CHO, 5'-Me	$C_{18}H_{16}O_5$	Leridalchalcone	*Petiveria alliacea*	Phytolaccaceae	31
	(2',5',4)							
3	2',5',4	—	—	$C_{15}H_{12}O_4$		*Platymiscium yucatanum*	Leguminosae	32
	(4',2,4)							
4	4	4',2	—	$C_{17}H_{16}O_4$	Glypallichalcone	*Glycyrrhiza pallidiflora*	Leguminosae	33
Tetra O-substituted								
	(2',3',4,4)							
5	2,4	3',4'	—	$C_{17}H_{16}O_5$	Heliannone A	*Helianthus annuus*	Asteraceae	34
	(2',4',6',2)							
6	2',4,2	6'	—	$C_{16}H_{14}O_5$		*Scutellaria strigillosa*	Lamiaceae	35
7	2'	4',6',2	—	$C_{18}H_{18}O_5$		*Andrographis lineata*	Acanthaceae	36
	(2',4',6',4)							
8	2',4	6',4	—	$C_{17}H_{16}O_5$		*Vitex leptobotrys*	Verbenaceae	37
9	2'	4',6',4	5-Br	$C_{18}H_{17}O_5Br$		*Garcinia nervosa*	Guttiferae	38
10	4'	2',6',4	—	$C_{18}H_{18}O_5$		*Vitex leptobotrys*	Verbenaceae	37
	(2',4',3,4)							
11	2',3,4	4'	—	$C_{16}H_{14}O_5$	Calythropsin	*Calythropsis aurea*	Myrtaceae	39
12	—	2',4'	3,4-OCH$_2$O—	$C_{18}H_{16}O_5$		*Millettia erythrocalyx*	Leguminosae	40
	(2',3,4,5)							
13	2'	3,4,5	—	$C_{18}H_{18}O_5$	Crotaoprostrin	*Crotalaria prostrata*	Leguminosae	41
	(2,3,4,6)							
14	2	3,4,6	—	$C_{18}H_{18}O_5$	Tepanone	*Ellipeia cuneifolia*	Annonaceae	42

continued

TABLE 16.1
New Chalcones with Simple Patterns of *O*-Substitution Reported in the Literature from 1992 to 2003 — *continued*

No.	OH	OMe	Other	Mol. Formula	Trivial Name	Source	Family	Ref.
Penta *O*-substituted								
15	2',4,4 *(2',3',4',3,4)*	3',3	—	$C_{17}H_{16}O_6$		*Wedelia asperrima*	Asteraceae	43
16	2' *(2',4',6',2,3)*	4',6',2,3	—	$C_{19}H_{20}O_6$		*Caesalpinia pulcherrima*	Leguminosae	44
17	— *(2',4',6',3,4)*	2',4',6'	3,4-OCH$_2$O-	$C_{19}H_{18}O_6$		*Millettia leucantha*	Leguminosae	45
18	3',4',3,4 *(3',4',2,3,4)*	2	—	$C_{16}H_{14}O_6$		*Glycyrrhiza uralensis*	Leguminosae	46
Hexa *O*-substituted								
19	3,4 *(2',3',4',5',6',4)*	2',4',5',6'	—	$C_{19}H_{20}O_7$		*Didymocarpus leucocalyx*	Gesneriaceae	47
20	6,3,4 *(2',3',4',6',3,4)*	2',3',4'	—	$C_{18}H_{18}O_7$	Hamilcone	*Uvaria hamiltonii*	Annonaceae	48
21	2'	3',4',6',3,4	—	$C_{20}H_{22}O_7$		*Citrus kinokuni*	Rutaceae	49
22	— *(2',4',6',2,3,4)*	2',4',6',2,3,4	—	$C_{21}H_{24}O_7$		*Andrographis neesiana*	Acanthaceae	50
β-Hydroxy substituted								
23	2',β *(2',4',6',β)*	4',6'	3'-Me	$C_{18}H_{18}O_5$		*Leptospermum scoparium*	Myrtaceae	51
24	β *(3',4',3,4,β)*	—	3',4':3,4-bis (-OCH$_2$O-)	$C_{17}H_{12}O_6$	Galiposin	*Galipea granulosa*	Rutaceae	52
25	β *(2',4',6',3,4,β)*	2',4',6'	3,4-OCH$_2$O-	$C_{19}H_{18}O_7$	Ponganone X	*Pongamia pinnata*	Leguminosae	53
26	— *(3',4',2,4,6,β)*	2,4,6,β	3',4'-OCH$_2$O-	$C_{20}H_{20}O_7$		*Millettia leucantha*	Leguminosae	45

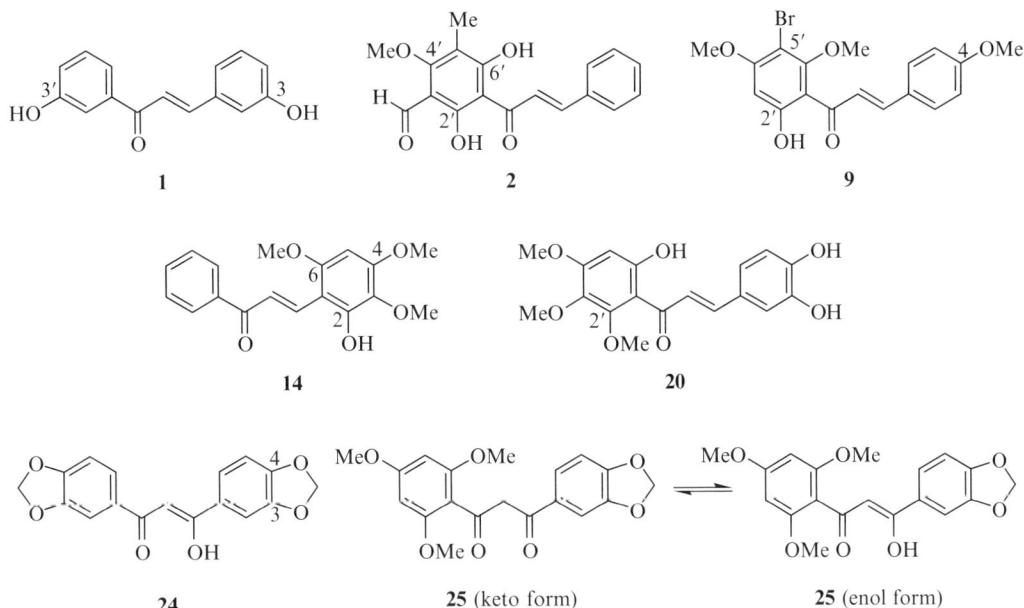

FIGURE 16.6 Chalcones with simple patterns of *O*-substitution (see Table 16.1).

product.[30] Other di-*O*-substituted chalcones found previously in this genus are 2′,2-dihydroxychalcone from *P. denticulata* and 2′,β-dihydroxychalcone from *P. pulverulenta*.[54,55] An extract of the heartwood of *Platymiscium yucatanum* (Leguminosae), a tropical wood that is highly resistant to the fungi *Coriolus versicolor* and *Lenzites trabea*, yielded an antifungal chalcone (**3**) with the rare (2′,5′,4)-*O*-substitution pattern.[32] Several chalcones produced by synthesis have now been found as natural products, including **12**, **13**, and **16**.[40,41,44] The structure of 2′,4′,4-trihydroxy-3,3′-dimethoxychalcone (**15**), a constituent of the aerial parts of *Wedelia asperrima* (Asteraceae), was confirmed by chemical synthesis as well as by spectroscopic methods.[43] Compounds **13** and **14** are both examples of retrochalcones, in which the typical patterns of *O*-substitution in the A- and B-rings appear to be reversed. The methyl ether of tepanone (**14**), 2,3,4,6-tetramethoxychalcone, prepared either by methylation of the natural product or by Claisen–Schmidt condensation of acetophenone with 2,3,4,6-tetramethoxybenzaldehyde, showed *trans–cis* isomerization.[42] The instability of the parent compound (tepanone) may be due to a similar isomerization step, as the *cis*-isomer readily converts to a 2-hydroxyflav-3-ene derivative (or a colored flavylium salt in the presence of acid). Among the more highly substituted chalcones listed in Table 16.1 is a new example from leaves of *Didymocarpus leucocalyx* (Gesneriaceae) with a fully substituted A-ring (**19**).[47] Several other compounds of this type were obtained previously from *D. pedicellata*, although their distribution is not exclusive to this genus.[10]

Leridalchalcone (**2**) is a rare example of a chalcone with both *C*-methyl and formyl substituents isolated from the aerial parts of *Petiveria alliacea* (Phytolaccaceae).[31] Perhaps more striking is a report from Ilyas et al. describing the first halogenated chalcone to be obtained as a natural product.[38] This compound, 5′-bromo-2′-hydroxy-4′,6′,4-trimethoxychalcone (**9**), was obtained from the leaves of *Garcinia nervosa* (Guttiferae) together with the known derivatives 2′-hydroxy-4′,4-dimethoxychalcone and 2′-hydroxy-4′,6′,3,4-tetramethoxychalcone. The presence of bromine as a substituent was confirmed by mass spectrometry and chemical analysis. The plant material used for the isolation work was collected

TABLE 16.2
New Isoprenylated Chalcones Reported in the Literature from 1992 to 2003

No.	O-Substituents	Other Substituents	Mol. Formula.	Trivial Name	Source	Family	Ref.
	Di O-substituted						
	(2',4')						
27	2',4'-diOH	3',5'-Diprenyl	$C_{25}H_{28}O_3$	Spinochalcone A	*Tephrosia spinosa*	Leguminosae	63
28	2'-OH	6''-(4-Methylpent-3-enyl)-6''-methylpyrano[2'',3'':4',3']	$C_{25}H_{26}O_3$	Spinochalcone B	*Tephrosia spinosa*	Leguminosae	63
29	2'-OH	3'-Prenyl, 6'',6''-dimethylpyrano[2'',3'':4',5']	$C_{25}H_{26}O_3$	Spinochalcone C	*Tephrosia spinosa*	Leguminosae	64
30	2'-OH	Complex (FIGURE 16.7)	$C_{21}H_{20}O_5$	(+)-Tephrosone	*Tephrosia purpurea*	Leguminosae	65
	Tri O-substituted						
	(2',4',6')						
31	2',4',6'-triOH	3'-Neryl	$C_{25}H_{28}O_4$		*Helichrysum retrosum*	Asteraceae	66
32	2',4'-diOH	3'-Prenyl, 5''-(2-hydroxyisopropyl)-4'',5''-dihydrofurano[2'',3'':6',5']	$C_{25}H_{28}O_5$	Cedrediprenone	*Cedrelopsis grevei*	Ptaeroxylaceae	67
33	4',6'-diOH	6'',6''-Dimethyl-5''-hydroxy-4'',5''-dihydropyrano[2'',3'':2',3']	$C_{20}H_{20}O_5$		*Helichrysum aphelexiodes*	Asteraceae	66
34	2'-OH, 6'-OMe	5''-Isopropenyl-4'',5''-dihydrofurano[2'',3'':4',3']	$C_{21}H_{20}O_4$	Crassichalcone	*Tephrosia crassifolia*	Leguminosae	68
35	2'-OH, 6'-OMe	4'',5''-Epoxy(6'',6''-dimethyl-4'',5''-dihydropyrano[2'',3'':4',3'])	$C_{21}H_{20}O_5$	Epoxyobovatachalcone	*Tephrosia carrollii*	Leguminosae	69
36	2'-OH, 6'-OMe	6'',6''-Dimethyl-5''-hydroxy-4'',5''-dihydropyrano[2'',3'':4',3']	$C_{21}H_{22}O_5$	6'-Methoxy-helikrausichalcone	*Cedrelopsis grevei*	Ptaeroxylaceae	67
37	2'-OH, 6'-OMe	Complex (FIGURE 16.7)	$C_{24}H_{24}O_7$	(+)-Tephropurpurin	*Tephrosia purpurea*	Leguminosae	70
38	4'-OH, 2'-OMe	6'',6''-Dimethylpyrano[2'',3'':6',5]	$C_{21}H_{20}O_4$	Cedreprenone	*Cedrelopsis grevei*	Ptaeroxylaceae	67
39	4'-OH, 6'-OMe	6'',6''-Dimethyl-5''-hydroxy-4'',5''-dihydropyrano[2'',3'':2',3']	$C_{21}H_{22}O_5$		*Helichrysum aphelexiodes*	Asteraceae	66
40	6'-OH, 4'-OMe	Complex-C_{10} (FIGURE 16.7)	$C_{26}H_{30}O_5$	Linderol A	*Lindera umbellata*	Lauraceae	71
	(2',4',4)						
41	2',4',4-triOH	3'-(2-Hydroxy-3-methylbut-3-enyl)	$C_{20}H_{20}O_5$		*Maclura tinctoria*	Moraceae	72
42	2',4',4-triOH	3'-(4-Coumaroyloxy-3-methyl-but-2(*E*)-enyl)	$C_{29}H_{26}O_7$	Isogemichalcone B	*Hypericum geminiflorum*	Guttiferae	73
43	2',4',4-triOH	3'-(4-Coumaroyloxy-3-methyl-but-2(*Z*)-enyl)	$C_{29}H_{26}O_7$	Gemichalcone B	*Hypericum geminiflorum*	Guttiferae	73
44	2',4',4-triOH	3'-(4-Feruloyloxy-3-methyl-but-2(*Z*)-enyl)	$C_{30}H_{28}O_8$	Gemichalcone A	*Hypericum geminiflorum*	Guttiferae	73
45	2',4',4-triOH	3',3-Diprenyl	$C_{25}H_{28}O_4$	Kanzonol C	*Glycyrrhiza* sp. *Glycyrrhiza eurycarpa*	Leguminosae	74 75
46	2',4',4-triOH	3'-Prenyl, 3-(2-hydroxy-3-methylbut-3-enyl)	$C_{25}H_{28}O_5$	Paratocarpin D	*Parartocarpus venenosa*	Moraceae	76
47	2',4',4-triOH	3'-(2-Hydroxy-3-methylbut-3-enyl), 3-prenyl	$C_{25}H_{28}O_5$	Paratocarpin E	*Parartocarpus venenosa*	Moraceae	76
48	2',4',4-triOH	5',3-Diprenyl	$C_{25}H_{28}O_4$	Stipulin	*Dalbergia stipulacea*	Leguminosae	77

No.	Substitution	Substituent	Compound	Formula	Species	Family	Ref.
49	2',4,4'-triOH	3-Prenyl	Licoagrochalcone A	$C_{20}H_{20}O_4$	*Glycyrrhiza glabra*	Leguminosae	78
50	2',4,4'-triOH	3-(2-Hydroxy-3-methylbut-3-enyl), 5-prenyl	Anthyllin	$C_{25}H_{28}O_5$	*Anthyllis hermanniae*	Leguminosae	79
51	2',4-diOH	6',6''-Dimethylpyrano[2'',3'':4,3]	Kanzonol B	$C_{20}H_{18}O_4$	*Glycyrrhiza eurycarpa*	Leguminosae	75
52	2',4-diOH	3'-Prenyl, 6',6''-dimethylpyrano[2'',3'':4,3]	Paratocarpin C	$C_{25}H_{26}O_4$	*Parartocarpus venenosa*	Moraceae	76
53	2',4-diOH	3'-Prenyl, 5''-(2-hydroxyisopropyl)-4''-hydroxy-4'',5''-dihydrofurano[2'',3'':4,3]	Paratocarpin G	$C_{25}H_{28}O_6$	*Parartocarpus venenosa*	Moraceae	80
54	2',4-diOH	5'-Prenyl, 6',6''-dimethylpyrano[2'',3'':4,3]	Anthyllisone	$C_{25}H_{26}O_4$	*Anthyllis hermanniae*	Leguminosae	79
55	2',4-diOH, 4-OMe	3'-Geranyl	Xanthoangelol F	$C_{26}H_{30}O_4$	*Angelica keiskei*	Umbelliferae	81
56	2',4-diOH, 4-OMe	3'-(3,7-Dimethyl-6-hydroxyocta-2,7-dienyl)	Xanthoangelol G	$C_{26}H_{30}O_5$	*Angelica keiskei*	Umbelliferae	81
57	2',4-diOH	6',6''-Dimethyl-4',5''-dihydropyrano[2'',3'':4,3']	Dorsmanin A	$C_{20}H_{20}O_4$	*Dorstenia mannii*	Moraceae	82
58	2',4-diOH	6',6''-Dimethyl-5''-hydroxy-4'',5''-dihydropyrano[2'',3'':4,3']		$C_{20}H_{20}O_5$	*Dorstenia zenkeri*	Moraceae	83
59	2',4-diOH	3'-Prenyl, 6',6''-dimethylpyrano[2'',3'':4,3']	Paratocarpin B	$C_{25}H_{26}O_4$	*Parartocarpus venenosa*	Moraceae	76
60	2'-OH	Bis(6',6''-dimethylpyrano[2'',3'':4,3][2'',3'':4,3]	Paratocarpin A	$C_{25}H_{24}O_4$	*Parartocarpus venenosa*	Moraceae	76
61	2'-OH	5''-(2-Hydroxyisopropyl)-4'',5''-dihydrofurano[2'',3'':4,3'], 6',6''-dimethylpyrano[2'',3'':4,3]	Paratocarpin F	$C_{25}H_{26}O_5$	*Parartocarpus venenosa*	Moraceae	80
62	2'-OH	Bis(6',6''-dimethyl-4',5''-dihydropyrano[2'',3'':4,3][2'',3'':4,3]	Artoindonesianin J	$C_{25}H_{28}O_4$	*Artocarpus bracteata*	Moraceae	84
63	2'-OH	Bis(6',6''-dimethyl-4',5''-dihydropyrano[2'',3'':4,5][2'',3'':4,3]		$C_{25}H_{28}O_4$	*Dorstenia kameruniana*	Moraceae	85
64	4-OH, 4'-OMe	6',6''-Dimethyl-5''-hydroxy-4'',5''-dihydropyrano[2'',3'':2,3]	Xanthoangelol H	$C_{21}H_{22}O_5$	*Angelica keiskei*	Umbelliferae	81
(4',2,4)							
65	4,4'-diOH, 2-OMe	3-Prenyl	Licochalcone C	$C_{21}H_{22}O_4$	*Glycyrrhiza inflata*	Leguminosae	86
66	2,4-diOH	6',6''-Dimethylpyrano[2'',3'':4,3']	Munsericin	$C_{20}H_{18}O_4$	*Mundulea sericea*	Leguminosae	87
67	4'-OH, 2-OMe	6',6''-Dimethylpyrano[2'',3'':4,3]	Licoagrochalcone B	$C_{21}H_{20}O_4$	*Glycyrrhiza glabra*	Leguminosae	88
68	4'-OH, 2-OMe	5''-(2-Hydroxyisopropyl)-4'',5''-dihydrofurano[2'',3'':4,3]	Licoagrochalcone D	$C_{21}H_{22}O_5$	*Glycyrrhiza glabra*	Leguminosae	88
Tetra *O*-substituted (2',4',6,4)							
69	2',4,6,4-tetraOH	3'-Geranyl	3'-Geranylchalconaringenin	$C_{25}H_{28}O_5$	*Humulus lupulus*	Cannabinaceae	89
70	2',4,6'-triOH, 6-OMe	3'-(2-Hydroxy-3-methylbut-3-enyl)	Xanthohumol D	$C_{21}H_{22}O_6$	*Humulus lupulus*	Cannabinaceae	90
71	2',4,4'-triOH, 6-OMe	3',5'-Diprenyl	5'-Prenylxanthohumol	$C_{26}H_{30}O_5$	*Humulus lupulus*	Cannabinaceae	89
72	2',4,4'-triOH	3'-Prenyl, 6',6''-dimethylpyrano[2'',3'':6,5]	Xanthohumol E	$C_{25}H_{26}O_5$	*Humulus lupulus*	Cannabinaceae	90

continued

TABLE 16.2
New Isoprenylated Chalcones Reported in the Literature from 1992 to 2003 — continued

No.	O-substituents	Other Substituents	Mol. Formula.	Trivial Name	Source	Family	Ref.
73	2',6',4-triOH, 4'-OMe	3'-Prenyl	$C_{21}H_{22}O_5$	Xanthogalenol	*Humulus lupulus*	Cannabinaceae	90
74	2,4-diOH, 6'-OMe	6'',6''-Dimethylpyrano[2'',3'':4',3']	$C_{21}H_{20}O_5$	Xanthohumol C, dehydrocycloxanthohumol	*Humulus lupulus*	Cannabinaceae	89
75	2,4-diOH, 6'-OMe	6'',6''-Dimethyl-4''-hydroxy-4'', 5''- dihydropyrano[2'',3'':4',3']	$C_{21}H_{22}O_6$	Isodehydrocycloxanthohumol hydrate	*Humulus lupulus*	Cannabinaceae	91
76	2,4-diOH, 6'-OMe	6'',6''-Dimethyl-5''-hydroxy-4'',5''- dihydropyrano[2'',3'':4',3']	$C_{21}H_{22}O_6$	Xanthohumol B, dehydrocycloxanthohumol hydrate	*Humulus lupulus*	Cannabinaceae	89
77	2,4-diOH	Bis(6'',6''-dimethylpyrano)[2'',3'':4',3'],[2'',3'':6',5']	$C_{25}H_{24}O_5$	Laxichalcone	*Derris laxiflora*	Leguminosae	92
78	2,4-diOH	6'',6''-Dimethylpyrano[2'',3'':4',3], 6'', 6''-dimethyl-4''-hydroxy-5''-methoxy-4'', 5''-dihydropyrano[2'',3'':6',5']	$C_{26}H_{28}O_7$	Derrichalcone	*Derris laxiflora*	Leguminosae	92
79	2'-OH, 4',4-diOMe	6'',6''-Dimethylpyrano[2'',3'':6',5']	$C_{22}H_{22}O_5$		*Neoraputia magnifica*	Rutaceae	93
80	2'-OH, 6',4-diOMe (2',4',2,4)	6'',6''-Dimethylpyrano[2'',3'':4',3']	$C_{22}H_{22}O_5$	Glychalcone A	*Glycosmis citrifolia*	Rutaceae	94
81	2',4',2,4-tetraOH	3'-(4-Coumaroyloxy-3-methyl-but-2(E)-enyl)	$C_{29}H_{26}O_8$	Demethoxyisogemichalcone C	*Broussonetia papyrifera*	Moraceae	95
82	2',4',2,4-tetraOH	3'-(4-Feruloyloxy-3-methyl-but-2(E)-enyl)	$C_{30}H_{28}O_9$	Isogemichalcone C	*Broussonetia papyrifera*	Moraceae	95
83	2',4',2,4-tetraOH (2',4',3,4)	3'-(4-Feruloyloxy-3-methyl-but-2(Z)-enyl)	$C_{30}H_{28}O_9$	Gemichalcone C	*Hypericum geminiflorum*	Guttiferae	96
84	2',4',3,4-tetraOH	3'-Geranyl	$C_{25}H_{28}O_5$		*Artocarpus incisus*	Moraceae	97
85	2',4',3,4-tetraOH	3',5-Digeranyl	$C_{35}H_{44}O_5$		*Dorstenia prorepens*	Moraceae	83
86	2',4',3-triOH	3'-Prenyl, 6'',6''-dimethylpyrano[2'',3'':4,5]	$C_{25}H_{26}O_5$		*Glycyrrhiza* sp. *Glycyrrhiza glabra*	Leguminosae	74 98
87	2',4,4-triOH	3'-Prenyl, 6''-(4-methylpent-3-enyl), 6''-methylpyrano[2'',3'':3,2]	$C_{30}H_{34}O_5$	Poinsettifolin B	*Dorstenia poinsettifolia*	Moraceae	99
88	2,3-diOH	Bis(6'',6''-dimethylpyrano)[2'',3'':4',3'],[2'',3'':4,5]	$C_{25}H_{24}O_5$	Glyinflanin G	*Glycyrrhiza inflata*	Leguminosae	100
89	2'-OH, 3,4-diOMe	6'',6''-Dimethylpyrano[2'',3'':4',3']	$C_{22}H_{22}O_5$	3,4-Dimethoxylonchocarpin	*Lonchocarpus subglaucescens*	Leguminosae	101

No.	Substitution	Substituent	Name	Formula	Species	Family	Ref.
	(3',4',2,4)						
90	3',4',4-triOH, 2-OMe	3-Prenyl	Licoagrochalcone C	$C_{21}H_{22}O_5$	*Glycyrrhiza glabra*	Leguminosae	88
	(4',2,3,4)						
91	4',3,4-triOH, 2-OMe	3'-Prenyl	Licochalcone D	$C_{21}H_{22}O_5$	*Glycyrrhiza inflata*	Leguminosae	86
	Penta O-substituted						
	(2',4',5',3,4)						
92	2'-OH, 5',3,4-triOMe	6'',6''-Dimethylpyrano[2'',3'':4',3']	Ponganone VI	$C_{23}H_{24}O_6$	*Pongamia pinnata*	Leguminosae	53
	(2',4',6',3,4)						
93	2'-OH, 4',3,4-triOMe	6'',6''-Dimethylpyrano[2'',3'':6',5']		$C_{23}H_{24}O_6$	*Neoraputia magnifica*	Rutaceae	93
	(2',4',2,4,5)						
94	2'-OH, 6',3,4-triOMe	6'',6''-Dimethylpyrano[2'',3'':4',3']	Glychalcone B	$C_{23}H_{24}O_6$	*Glycosmis citrifolia*	Rutaceae	94
	(2',4',2,4,5)						
95	2',4',2,4,5-pentaOH	3'-Prenyl	Ramosismin	$C_{20}H_{20}O_6$	*Crotalaria ramosissima*	Leguminosae	102
	Hexa O-substituted						
	(2',3',4',6',3,4)						
96	2'-OH, 3',6'-diOMe, 3,4-OCH$_2$O-	Furano[2'',3'':4',5']		$C_{20}H_{16}O_7$	*Lonchocarpus subglaucescens*	Leguminosae	101
	(2',4',6',3,4,5)						
97	6-OH, 4',3,4,5-tetraOMe	6'',6''-Dimethylpyrano[2'',3'':2',3']		$C_{24}H_{26}O_7$	*Neoraputia magnifica*	Rutaceae	103
	β-Hydroxy substituted						
	(2',4',6',β)						
98	2',6',β-triOH, 4'-OMe	3'-Prenyl		$C_{21}H_{22}O_5$	*Tephrosia major*	Leguminosae	104
99	2',6',β-triOMe	6'',6''-Dimethylpyrano[2'',3'':4',3']	7-Methoxypraecansone B	$C_{23}H_{24}O_5$	*Pongamia pinnata*	Leguminosae	105
	(2',4,4,β)						
100	2',4',4,β-tetraOH	5',3-Diprenyl	Glycyrdione A (glyinflanin A)	$C_{25}H_{28}O_5$	*Glycyrrhiza inflata*	Leguminosae	106
101	2',4',4,β-tetraOH	5'-Prenyl, 3-(2-hydroxy-3-methylbut-3-enyl)	Glyinflanin E	$C_{25}H_{28}O_6$	*Glycyrrhiza inflata*	Leguminosae	100
102	2',4',4,β-tetraOH	3-Prenyl	Kanzonol A	$C_{20}H_{20}O_5$	*Glycyrrhiza eurycarpa*	Leguminosae	75
103	2',4',β-triOH	5'-Prenyl, 6'',6''-dimethylpyrano[2'',3'':4,3]	Glycyrdione B	$C_{25}H_{26}O_5$	*Glycyrrhiza inflata*	Leguminosae	106
104	2',4,β-triOH	6'',6''-Dimethylpyrano[2'',3'':4',5']	Glyinflanin B	$C_{20}H_{18}O_5$	*Glycyrrhiza inflata*	Leguminosae	107
105	2',4,β-triOH	3-Prenyl, 6'',6''-dimethylpyrano[2'',3'':4',5']	Glyinflanin C (glycyrdione C)	$C_{25}H_{26}O_5$	*Glycyrrhiza inflata*	Leguminosae	107, 108
106	2',4,β-triOH	3-Prenyl, 5''-(2-hydroxyisopropyl)-4'',5''-dihydrofurano[2'',3'':4',5]	Glyinflanin F	$C_{25}H_{28}O_6$	*Glycyrrhiza inflata*	Leguminosae	100
107	2',β-diOH	Bis(6'',6''-dimethylpyrano[2'',3'':4',5'][2'',3'':4,3]	Glyinflanin D	$C_{25}H_{24}O_5$	*Glycyrrhiza inflata*	Leguminosae	107
	(2',4',6',3,4,β)						
108	β-OH, 2',6'-diOMe, 3,4-OCH$_2$O-	6'',6''-Dimethylpyrano[2'',3'':4',3']	Pongapinone A	$C_{23}H_{22}O_7$	*Pongamia pinnata*	Leguminosae	109

from a site on contaminated land, which may explain the unexpected occurrence of the bromochalcone.

16.2.1.2 Biological Activity

Bioassay data are available for some of the new chalcones listed in Table 16.1. For example, Macías et al. assessed the allelopathic activity of heliannone A (**5**) and other flavonoids from *Helianthus annuus* cultivars, including the known chalcone, kukulkanin B (2′,4′,4-trihydroxy-3′-methoxychalcone).[34] Despite the structural similarity of these compounds, heliannone A inhibited germination of both tomato and barley and shoot growth of barley, whereas kukulkanin B inhibited only the shoot growth of tomato. Calythropsin (**11**) was obtained together with its dihydrochalcone analog (**263**) during cytotoxicity-guided fractionation of an extract of the roots of *Calythropsis aurea* (Myrtaceae).[39] It showed weak cytotoxicity in the United States National Cancer Institute panel of 60 cell lines, a small effect on tubulin polymerization and weak antimitotic activity. Dihydrocalythropsin (**263**) was 10-fold less potent in the cell line panel compared to calythropsin, but is more abundant as a constituent of the original extract.[39] Moderate cytotoxicity was also reported for 2′,4′,6′-trimethoxy-3,4-methylenedioxychalcone (**17**), one of several chalcones and dihydrochalcones isolated from the stem bark of *Millettia leucantha* (Leguminosae).[45] Hamilcone (**20**) was tested for both cytotoxicity and DNA strand-scission activity but was only weakly active in these assays.[48] The penta-*O*-substituted chalcone **18** was one of several phenolic constituents isolated in a study of a commercial licorice sourced from *Glycyrrhiza uralensis* (Leguminosae) and used in northeastern China.[46] This compound shows a potent scavenging effect on the 1,1-diphenyl-2-picrylhydrazyl radical and forms a stable radical in solution that can be detected using electron paramagnetic resonance (EPR) spectroscopy. The authors speculate that the stability of this radical may be responsible for the good activity shown by the chalcone in the 1,1-diphenyl-2-picryhydrazyl (DPPH) assay for antioxidant activity.[46]

16.2.1.3 β-Hydroxychalcones

The β-hydroxychalcones are a relatively small group of chalcones that occur as the enol-tautomers of dibenzoylmethane derivatives (see Figure 16.6, **25**). Four new examples are listed in Table 16.1, and some additional isoprenylated derivatives will be found in Table 16.2. The extent of keto–enol tautomerism is largely solvent dependent, and nuclear magnetic resonance (NMR) spectroscopy provides one of the best methods to determine the ratio of the tautomers present. In ^{1}H NMR spectra recorded in $CDCl_3$, the exchangeable proton of the β-OH of the enol tautomer appears as a 1H singlet at ca. 16 ppm, whereas the α-CH_2 protons of the keto tautomer appear as a 2H singlet at ca. 4.50 ppm. Another diagnostic resonance is the 1H methine singlet of the enol tautomer (α-CH), which is found at ca. 6.50 ppm, with its corresponding C-α resonance at 90 to 92 ppm in ^{13}C NMR spectra. For example, galiposin (**24**), a constituent of the bark of *Galipea granulosa* (Rutaceae), exists entirely in the enol form in solution; a sharp 1H singlet appears at 16.94 ppm (β-OH) and no resonance is found in the spectral region corresponding to the α-CH_2 protons of the keto form.[52] The ^{1}H and ^{13}C resonances of the enol methine (α-CH) are found at 6.62 and 91.6 ppm, respectively. This compound is also remarkable as the first bis(methylenedioxy)chalcone to be reported in the literature. The ^{1}H NMR spectrum of ponganone X (**25**) in $CDCl_3$ indicated a keto–enol tautomer ratio of 1:4, based on the relative intensities of the corresponding α-CH_2 and β-OH resonances at 4.31 ppm (2H, s) and 16.35 ppm (1H, s), respectively.[53] Similarly, the *C*-methylated derivative, 2′,β-dihydroxy-4′,6′-dimethoxy-3′-methylchalcone (**23**), from the aerial parts of *Leptospermum scoparium* (a plant from the

Myrtaceae used in Australian traditional medicine), exists in a keto–enol tautomer ratio of 3:2.[51] This compound was known previously as a synthetic product but only limited physical (melting point) and no spectroscopic data were available.[56] The β-hydroxychalcone constituent obtained from *Millettia leucantha* (**26**) is a further example of a retrochalcone.[45]

16.2.2 ISOPRENYLATED CHALCONES

16.2.2.1 Structural Diversity and Chemosystematic Trends

Table 16.2 lists more than 80 examples of new isoprenylated chalcones reported in the period 1992 to 2003 (see also Figure 16.7). Almost half of the compounds described here are from the Leguminosae, a trend that is also evident in earlier surveys.[1–4] Other plant families that are well represented in Table 16.2 are the Moraceae and the Cannabinaceae. The literature on isoprenylated flavonoids in general has been reviewed by Barron and Ibrahim to the end of 1994.[57] The phenolic constituents of *Glycyrrhiza* species (licorice), among which are many isoprenylated chalcones, were the subject of an extensive review that includes literature published up to the end of 1996.[58] Nomura and Hano have reviewed the literature on isoprenylated phenolic compounds of the Moraceae to the end of 1993.[59] More recent descriptions of isoprenylated flavonoids are available for the hop plant, *Humulus lupulus* (Cannabinaceae),[60] and the Moraceae genera *Artocarpus*[61] and *Dorstenia*.[62]

The arrangement of the Table 16.2 entries follows the scheme adopted in Table 16.1, in which compounds are listed in order of increasing *O*-substitution. Within each heading, specific patterns of A- and B-ring substitution are noted. This system of structure classification allows types of chalcone that are specific to a particular species, genus, or family to be recognized more clearly. For example, isoprenylated (2′,4′)-di-*O*-substituted (**27–30**)[63–65] and (4′,2,4)-tri-*O*-substituted (**65–68**)[86–88] chalcones reported in Table 16.2 are restricted to the Leguminosae. The compounds within these groupings are often closely related, such as spinochalcones A–C (**27–29**) obtained from the roots of *Tephrosia spinosa* (Leguminosae).[63,64] In contrast, (+)-tephrosone (**30**), a constituent of *T. purpurea*, has the unusual feature of two fused furan rings,[65] as was also found in (+)-tephropurpurin (**37**), isolated from the same species (Figure 16.7).[70] The absolute configurations of the stereogenic centers of **30** were determined on the basis of ¹H NMR analysis of Mosher esters and by comparison with the recently determined absolute configuration of the related compound, (+)-purpurin, a flavanone first isolated in 1980 from seeds of *T. purpurea*.[65,110,111] Isoprenylated chalcones of the Moraceae are predominantly isoliquiritigenin (2′,4′,4-trihydroxychalcone) derivatives, such as paratocarpins A–G (**46, 47, 52, 53, 59–61**) from *Parartocarpus venenosa* (cited incorrectly as *Paratocarpus venenosa*).[76,80] Other typical substitution patterns for the Moraceae are represented in Table 16.2 by isoprenylated 2′,4′,2,4- (**81, 82**)[95] and 2′,4′,3,4-tetrahydroxychalcones (**84, 85, 87**).[83,97,99] Compounds **81** and **82** represent a particularly unusual type of chalcone in which a prenyl side-chain is esterified by a hydroxycinnamic acid.[95] Four similar compounds (**42–44, 83**) have been isolated from the heartwood and root of *Hypericum geminiflorum* (Guttiferae).[73,96]

Further investigation of the chemical constituents of the hop plant, *Humulus lupulus* (Cannabinaceae), has revealed a series of isoprenylated chalconaringenin derivatives (**69–76**), most of which are minor components related to xanthohumol (2′,4′,4-trihydroxy-6′-methoxy-3′-prenylchalcone), the major chalcone of this species.[89–91] The new compounds, which are also components of the resin secreted by glandular trichomes (lupulin glands) on female inflorescences ("hop cones") and the undersides of young leaves, occur at concentrations that are 10- to 100-fold less than xanthohumol.[60] Liquid chromatography coupled to mass spectrometry was used to detect the presence of 3′,5′-diprenylchalconaringenin in hop extracts, a compound identified by comparison with a synthetic sample.[90] This isoprenylated

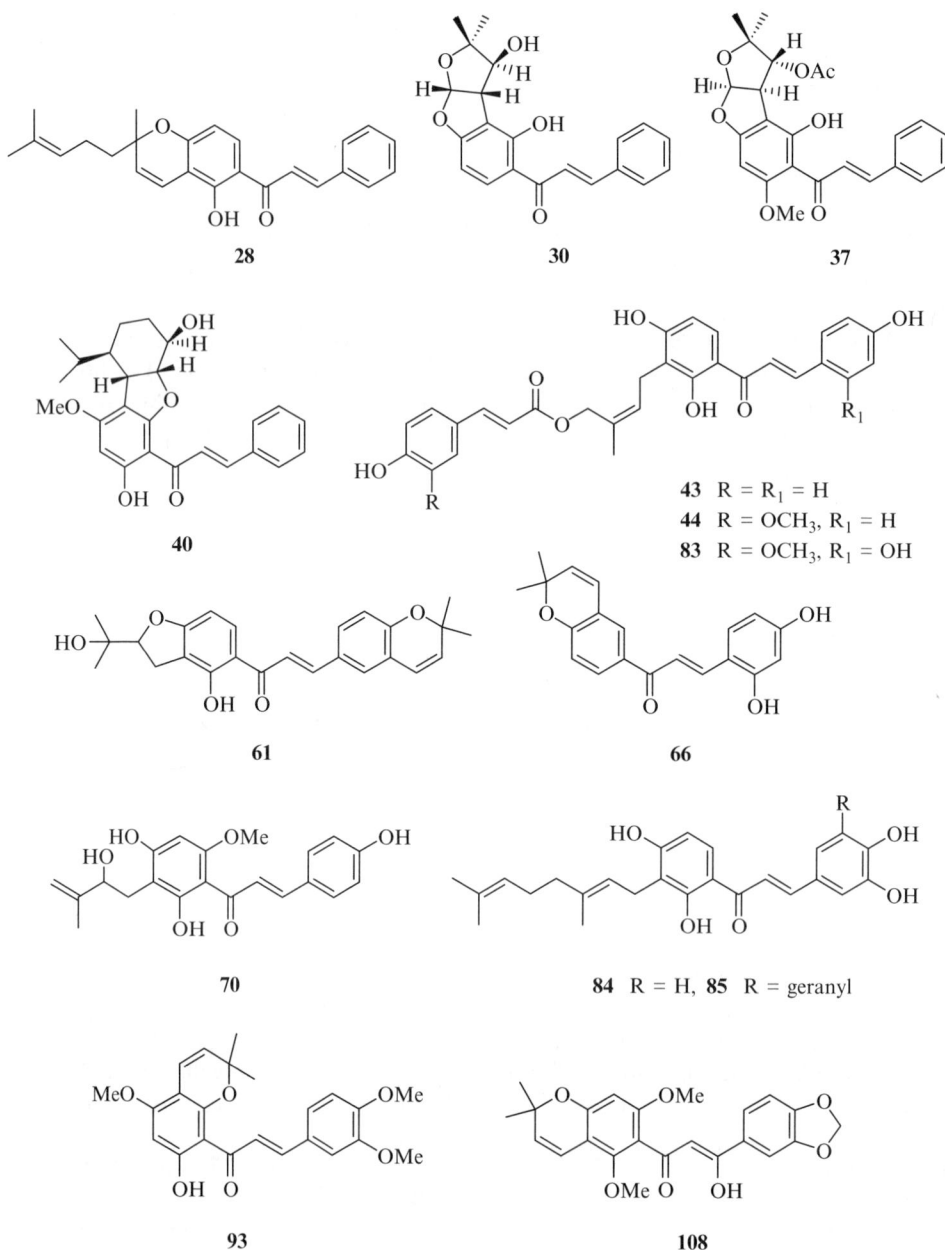

FIGURE 16.7 Isoprenylated chalcones (see Table 16.2).

chalcone still awaits isolation and characterization from a natural source. An extensive survey of the distribution of chalcones in 120 accessions of *H. lupulus* varieties and cultivars from three main geographical regions (Europe, Japan, and North America) has been made by Stevens and colleagues.[90] The compounds with the greatest diagnostic value for distinguishing regional chemotypes and possible evolutionary lineages are the 4′-*O*-methylchalcones, xanthogalenol (**73**), xanthohumol 4′-methyl ether, and chalconaringenin 4′,6′-dimethyl ether. The characteristic isoprenylated chalcones of *H. lupulus* were found to be absent from *H. japonicus*, a species from eastern Asia.[90]

A survey of lipophilic phenolic compounds of *Helichrysum* endemics (Asteraceae) of Madagascar reveals that the constituent chalcones are largely characterized by unsubstituted B-rings and either *C*- or *O*-isoprenylation in the A-ring.[66] Three new chalcones (**31, 33, 39**) were obtained in the study, including one with the less commonly observed neryl substituent (**31**). It is interesting to note from a taxonomic viewpoint that this profile of chalcones and other phenolic substituents was not substantially different from that recorded for other African species of *Helichrysum*, indicating that these compounds represent chemical characters that were conserved during the origin of the endemic taxa of Madagascar.[66] For a comprehensive treatment of the flavonoids of the Asteraceae and their distribution, the monograph by Bohm and Stuessy is recommended.[112]

The Lauraceae is a family from which relatively few isoprenylated chalcones have been reported, and only one new example, linderol A (**40**),[71] from the bark of *Lindera umbellata*, is listed in Table 16.2. In this compound, the B-ring is unsubstituted (Figure 16.7), a common feature of Lauraceae chalcones.[10] Several other unusual monoterpene-substituted chalcones have been reported previously from the Lauraceae genera *Aniba* and *Lindera*.[10] The unusual structure of linderol A (**40**), in which the A-ring is functionalized by a cyclized monoterpenyl substituent, prompted the first total synthesis of this molecule.[113]

16.2.2.2 Biological Activity

The cancer chemopreventive properties of (+)-tephrosone (**30**) and (+)-tephropurpurin (**37**) from *Tephrosia purpurea* (Leguminosae) have been assessed in a cell-based quinone reductase induction assay.[65,70] The induction of quinone reductase, a phase II drug-metabolizing enzyme, is believed to be an important protection mechanism against tumor initiation. Chang et al. demonstrated that (+)-tephropurpurin (**37**) was three times more active than sulforaphane, the positive control used in the assay, and had low cytotoxicity.[70] This compound may thus be a useful lead for development as a cancer chemopreventive agent. In a similar search for cancer chemopreventive agents from natural sources, munsericin (**66**) was found to inhibit phorbol ester-induced ornithine decarboxylase activity in cell culture.[87] The potential health-promoting effects of xanthohumol and other prenylated flavonoids from hops (*Humulus lupulus*) have been of great interest due to the presence of these compounds in beer. Stevens and Page characterize xanthohumol as a "broad spectrum" cancer chemopreventive agent with three important properties: (i) inhibition of the metabolic activation of procarcinogens, (ii) induction of carcinogen-detoxifying enzymes, and (iii) early stage inhibition of tumor growth.[60] Much of the literature on the biological activity of xanthohumol and other hop flavonoids, including the new isoprenylated chalcones listed in Table 16.2 (**69–76**), has been reviewed recently.[60]

Kanzonol C (**45**), obtained by bioassay-guided fractionation of licorice roots (*Glycyrrhiza* sp.), showed potent antileishmanial activity *in vitro*, using an assay based on the inhibition of thymidine uptake in proliferating promastigotes of *Leishmania donovani*.[74] This compound had been synthesized previously as a potential antiulcer drug,[114] but was not known as a natural product. A second chalcone (**86**) obtained from licorice roots showed relatively poor activity in the same assay.[74] The biological activity of many chalcones and other flavonoids from *Glycyrrhiza* species was reviewed in 1998 by Nomura and Fukai.[58]

The biological activity of isoprenylated chalcones from the Moraceae has been described both in original papers and reviews.[61,62] Of particular interest is the potent 5-α-reductase inhibition shown by a geranylated chalcone (**84**) isolated from leaves of *Artocarpus incisus*.[97] The inhibitory effect is decreased by a factor of 2 when the geranyl substituent is lacking, as in butein (2′,4′,3,4-tetrahydroxychalcone). Compound **41**, which was obtained from the leaves of *Maclura tinctoria* together with four known isoprenylated flavonoids, showed inhibitory activity toward the AIDS-related opportunistic pathogens, *Candida albicans* and *Cryptococcus neoformans*.[72] Among the reports of biological activity of chalcones from plant families less well represented in Table 16.2 are the superoxide scavenging properties of cedrediprenone (**32**), a compound obtained from an extract of the fruits and seeds of *Cedrelopsis grevei* (Ptaeroxylaceae),[67] and the potent inhibitory activity of linderol A (**40**) on melanin biosynthesis in cultured B-16 melanoma cells.[71] The fruits of *Neoraputia magnifica* (Rutaceae) yielded a pyranochalcone (**93**), with weak inhibitory activity toward glycosomal glyceraldehyde-3-phosphate dehydrogenase from *Trypanosoma cruzi*, the causative agent of Chagas' disease.[93] Two polymethoxylated flavones isolated from the same source showed potent inhibition in this assay.

16.2.2.3   β-Hydroxychalcones

The 11 new isoprenylated β-hydroxychalcones (**98–108**) listed in Table 16.2 are restricted to three genera of the Leguminosae, *Glycyrrhiza*, *Pongamia*, and *Tephrosia*. The distribution of the tautomeric dibenzoylmethane (keto) and β-hydroxychalcone (enol) forms of the compounds can be assessed under solution conditions by NMR spectroscopy, as outlined in Section 16.2.1.3. For example, kanzonol A (**102**) was found to adopt an equilibrium mixture of keto–enol tautomers of ca. 2:3,[75] whereas the corresponding ratio for glycyrdiones A (**100**) and B (**103**) was close to 2:1.[106] Pongapinone A (**108**), which was obtained by bioassay-guided fractionation of extracts of the bark of *Pongamia pinnata*, inhibited the production of interleukin-1 with an IC_{50} value of 2.5 μg/ml.[109] In a related study, activity-guided fractionation of extracts of stem bark of the same species yielded 7-methoxypraecansone B (**99**), a compound active in the quinone reductase induction assay.[105]

16.2.3   Chalcone Glycosides

The structures of the 25 new chalcone glycosides reported from 1992 to 2003 (Table 16.3 and Figure 16.8) reflect trends already evident from analysis of ca. 60 such compounds described prior to 1992.[10] Thus, isoliquiritigenin glycosides (**110–112**) appear to be typical of the Leguminosae (although not exclusive to this family), while glycosides of okanin and its methyl ethers (**125–131**) are characteristic of the Asteraceae and, in particular, the genus *Bidens*. The glycosidic profile of chalcone glycosides is relatively conservative compared to that of the glycosides of flavonols and flavones. Almost all chalcone monoglycosides are β-glucopyranosides, and only a few disaccharides are encountered with any frequency. No tri- or higher oligosaccharides have yet been reported as glycosidic components of chalcones. Only one new disaccharide is listed among the new chalcone glycosides in Table 16.3, the 4′-*O*-β-xylopyranosyl(1‴ → 6″)-β-glucopyranoside (primveroside) of okanin 4-methyl ether (**130**) isolated from aerial parts of *Bidens campylotheca*.[129] The new bisdesmosidic triglycoside **110**, isolated from the roots of *Glycyrrhiza aspera* (Leguminosae), is only the second chalcone glycoside with three sugars to be reported in the literature.[116] Its structure was determined by spectroscopic methods and confirmed by partial enzymatic hydrolysis to the known compound, isoliquiritin apioside. The disaccharide component of **110**, β-apiofuranosyl(1 → 2)-β-glucopyranoside, is found in several chalcone glycosides of the Leguminosae (**110–112**)[116,117] and Loranthaceae (**123**).[126] A further bisdesmosidic chalcone triglycoside (**120**) has been reported from the leaves of *Asarum canadense* (Aristolochiaceae).[122] Variation in the types of organic acids attached to the sugars of chalcone glycosides by ester linkages is limited compared to other acylated flavonoid glycosides; for example, prior to 1992 only acetic, malonic, and *p*-coumaric acids had been reported.[10] Among the new chalcone glycosides in Table 16.3 are the first examples with cinnamic (**123**),[126] caffeic (**126, 127, 129**),[128] and ferulic (**112**)[117] acids as acylating groups.

The majority of the chalcone glycosides listed in Table 16.3 are based on well-known aglycones such as isoliquiritigenin, chalconaringenin, and okanin. However, the aglycone of 2′,3′,4-trihydroxychalcone 4-*O*-glucoside (**109**)[115] is unknown as a natural product. The first isoprenylated chalcone glycoside (**113**) has now been obtained from the stem bark of *Maclura tinctoria* (Moraceae) together with the acylated chalcone glycoside **117** and several known chalcone glycosides and flavanones.[118] Measurement of the free radical scavenging potential of these compounds in two different antioxidant assays showed **113** to have the greatest activity. The characterization of a new chalcone glycoside isolated from the flowers of *Clerodendron phlomidis* (Verbenaceae) as a di-*O*-α-glucopyranoside (**122**)[125] calls for close scrutiny as it is the β-anomer of this sugar that is invariably found in flavonoid glycosides.[132] The determination of the configuration of the α-glucopyranose sugars was supported by the measurement of $^3J_{H-1,H-2}$ coupling constants of 3 Hz for the anomeric protons in the ^1H NMR spectrum of this compound (the corresponding coupling constant for the β-anomer is typically 7 Hz);[132] however, this structure deserves further investigation.

16.2.4 CHALCONE DIMERS AND OLIGOMERS

The number of biflavonoid and oligomeric flavonoid structures in which chalcones are incorporated has increased significantly in the period 1992 to 2003, as Table 16.4 indicates. Not only have both dimers and heterodimers of chalcones been reported, but also oligomers comprising up to six chalcone-derived structural units (Figure 16.9; see also Chapter 17). The dimers and oligomers are found most commonly in the Ochnaceae, and in particular from species in the genera *Lophira* and *Ochna*. Chalcone dimers are also well represented in the Anacardiaceae. For example, rhuschalcone VI (**143**) is one of six bichalcones obtained from either twigs or stem bark of *Rhus pyroides*, a shrub commonly found in eastern Botswana.[134,135] It is the only example described to date of a dimer in which two chalcones are linked by a single C–C bond, in this case between C-5′ of the A-ring and C-3 of the B-ring of two molecules of isoliquiritigenin (2′,4′,4-trihydroxychalcone). A second unsymmetrical dimer, rhuschalcone V (**142**), is characterized by a C–C bond between C-5′ of the A-ring of isoliquiritigenin and C-3 of the B-ring of the equivalent dihydrochalcone, davidigenin (2′,4′,4-trihydroxydihydrochalcone).[135] Rhuschalcones I–IV (**138–141**) are unsymmetrical *O*-linked chalcone dimers comprising combinations of isoliquiritigenin and its 4′-methyl ether.[135] Three of these (**138–140**, Figure 16.9) are characterized by C–O–C bonds from C-5′ (A-ring) to C-4 (B-ring OH of the second molecule) and one (**141**) by a similar linkage from C-5′ (A-ring of isoliquiritigenin 4′-methyl ether) to C-4′ (A-ring OH of isoliquiritigenin). The structures of rhuschalcones I–III (**138–140**) were confirmed by total synthesis based on Ullmann coupling of 4-hydroxy-2′,4′-dimethoxychalcone and 5′-bromo-2′,4′,4-trimethoxychalcone.[135] Rhuschalones I–VI (**138–143**) were screened for cytotoxicity in the United States National Cancer Institute panel of 60 different human tumor cell lines. The most potent activity was shown by rhuschalcone IV on melanoma cell lines. A general observation was that rhuschalcones as a group of compounds show activity on colon cancer cell lines.[135] The four chalcone dimers (**134–137**) obtained from *Myracrodruon urundeuva* (Anacardiaceae)[133] are structurally more complex than the rhuschalcones, but similar to licobichalcone (**146**), a novel biflavonoid from the roots of *Glycyrrhiza uralensis* (Leguminosae).[137] A biosynthetic pathway proposed for the formation of licobichalcone is based on the coupling of radicals of the constituent monomer, licochalcone B (4′,3,4-trihydroxy-2-methoxychalcone). Subsequent rearrangements result in a condensed bichalcone structure with an additional six-membered ring. On this basis, the monomeric precursors of the condensed structures of urundeuvine A (**134**), B (**135**), and matosine (**137**) are revealed as butein (2′,4′,3,4-tetrahydroxychalcone) and isoli-

TABLE 16.3
New Chalcone Glycosides Reported from 1992 to 2003

No.	Compound	Mol. Formula	Source	Family	Ref.
	2′,3′,4-Trihydroxychalcone				
109	4-O-Glucoside	$C_{21}H_{22}O_9$	*Ammi majus*	Umbelliferae	115
	2′,4′,4-Trihydroxychalcone (isoliquiritigenin)				
110	4′-O-Glucoside 4-O-apiofuranosyl-(1‴ → 2″)-glucoside	$C_{32}H_{40}O_{18}$	*Glycyrrhiza aspera*	Leguminosae	116
111	4-O-(5‴-O-p-Coumaroyl)-apiofuranosyl(1‴ → 2″)-glucoside	$C_{35}H_{36}O_{15}$	*Glycyrrhiza uralensis*	Leguminosae	117
112	4-O-(5‴-O-Feruloyl)-apiofuranosyl(1‴ → 2″)-glucoside	$C_{36}H_{38}O_{16}$	*Glycyrrhiza uralensis*	Leguminosae	117
	2′,4′,4-Trihydroxy-3′-prenylchalcone				
113	4′-O-Glucoside	$C_{26}H_{30}O_9$	*Maclura tinctoria*	Moraceae	118
	2′,3′,4′,4-Tetrahydroxychalcone				
114	4′-O-(2″-O-p-Coumaroyl)glucoside	$C_{30}H_{28}O_{12}$	*Maclura tinctoria*	Moraceae	118
115	4′-O-(6″-O-p-Coumaroyl)glucoside	$C_{30}H_{28}O_{12}$	*Bidens leucantha*	Asteraceae	119
116	4′-O-(2″-O-Acetyl-6″-O-cinnamoyl)glucoside	$C_{32}H_{30}O_{12}$	*Bidens andicola*	Asteraceae	120
117	4′-O-(2″-O-p-Coumaroyl-6″-O-acetyl)glucoside	$C_{32}H_{30}O_{13}$	*Maclura tinctoria*	Moraceae	118
	2′,6′,2-Trihydroxy-4′-methoxychalcone				
118	2′-O-Glucoside (androechin)	$C_{22}H_{24}O_{10}$	*Andrographis echiodes*	Acanthaceae	121
	2′,4′,6′,4-Tetrahydroxychalcone (chalconaringenin)				
119	2′,4′-Di-O-glucoside	$C_{27}H_{32}O_{15}$	*Asarum canadense*	Aristolochiaceae	122
			Asarum macranthum		123
120	2′-O-Glucoside 4′-O-gentobioside	$C_{33}H_{42}O_{20}$	*Asarum canadense*	Aristolochiaceae	122
	2′,4′,4-Trihydroxy-6′-methoxychalcone (helichrysetin)				
121	4-O-Glucoside	$C_{22}H_{24}O_{10}$	*Pyracantha coccinea*	Rosaceae	124
122	4′,4-Di-O-α-glucoside	$C_{28}H_{34}O_{15}$	*Clerodendron phlomidis*	Verbenaceae	125
	2′,4-Dihydroxy-4′,6′-dimethoxychalcone (flavokawin C)				
123	4-O-(5‴-O-p-Cinnamoyl)-apiofuranosyl(1‴ → 2″)-glucoside	$C_{37}H_{40}O_{15}$	*Viscum album* ssp. *album*	Loranthaceae	126
	2′,4′,4-Trihydroxy-3-methoxychalcone (homobutein)				
124	4′-O-Glucoside	$C_{22}H_{24}O_{10}$	*Wedelia asperrima*	Asteraceae	43
	2′,3′,4′,3,4-Pentahydroxychalcone (okanin)				
125	4′-O-(4″,6″-Di-O-acetylglucoside)	$C_{25}H_{26}O_{13}$	*Bidens pilosa* var. *radiata*	Asteraceae	127
126	4′-O-(2″-O-Caffeoyl-6″-O-acetylglucoside)	$C_{32}H_{30}O_{15}$	*Bidens frondosa*	Asteraceae	128
127	4′-O-(2″-O-Caffeoyl-6″-O-p-coumaroylglucoside)	$C_{39}H_{34}O_{16}$	*Bidens frondosa*	Asteraceae	128
	2′,3′,4′,3-Tetrahydroxy-4-methoxychalcone (okanin 4-methyl ether)				
128	4′-O-(6″-O-p-Coumaroylglucoside)	$C_{31}H_{30}O_{13}$	*Bidens frondosa*	Asteraceae	128
129	4′-O-(2″-O-Caffeoyl-6″-O-acetylglucoside)	$C_{33}H_{32}O_{15}$	*Bidens frondosa*	Asteraceae	128
130	4′-O-Primveroside	$C_{27}H_{32}O_{15}$	*Bidens campylotheca*	Asteraceae	129
	2′,4′,4-Trihydroxy-3′,3-dimethoxychalcone				
131	4′-O-Glucoside	$C_{23}H_{26}O_{11}$	*Wedelia asperrima*	Asteraceae	43
	2′-Hydroxy-4′,6′,2,4-tetramethoxychalcone				
132	2′-O-Glucoside	$C_{25}H_{30}O_{11}$	*Terminalia alata*	Combretaceae	130
	4′,6′,3-Trihydroxy-2′,4-dimethoxychalcone				
133	4′-O-Rutinoside	$C_{29}H_{36}O_{15}$	*Myrtus communis*	Myrtaceae	131

quiritigenin (2′,4′,4-trihydroxychalcone), while urundeuvine C (**136**) is a heterodimer based on okanin (2′,3′,4′,3,4-pentahydroxychalcone) and isoliquiritigenin.[137]

The chalcone dimers of the Ochnaceae differ structurally to those of the other plant families represented in Table 16.4. Compounds **147–149** and **151** are dimers in which combinations of isoliquiritigenin and davidigenin (2′,4′,4-trihydroxydihydrochalcone) are coupled through a dihydrofurano ring formed between the B-ring of one molecule (at C-3 and 4-OH) with the α- and β-carbons of the other (Figure 16.9).[138–140] Similarly, the structures of

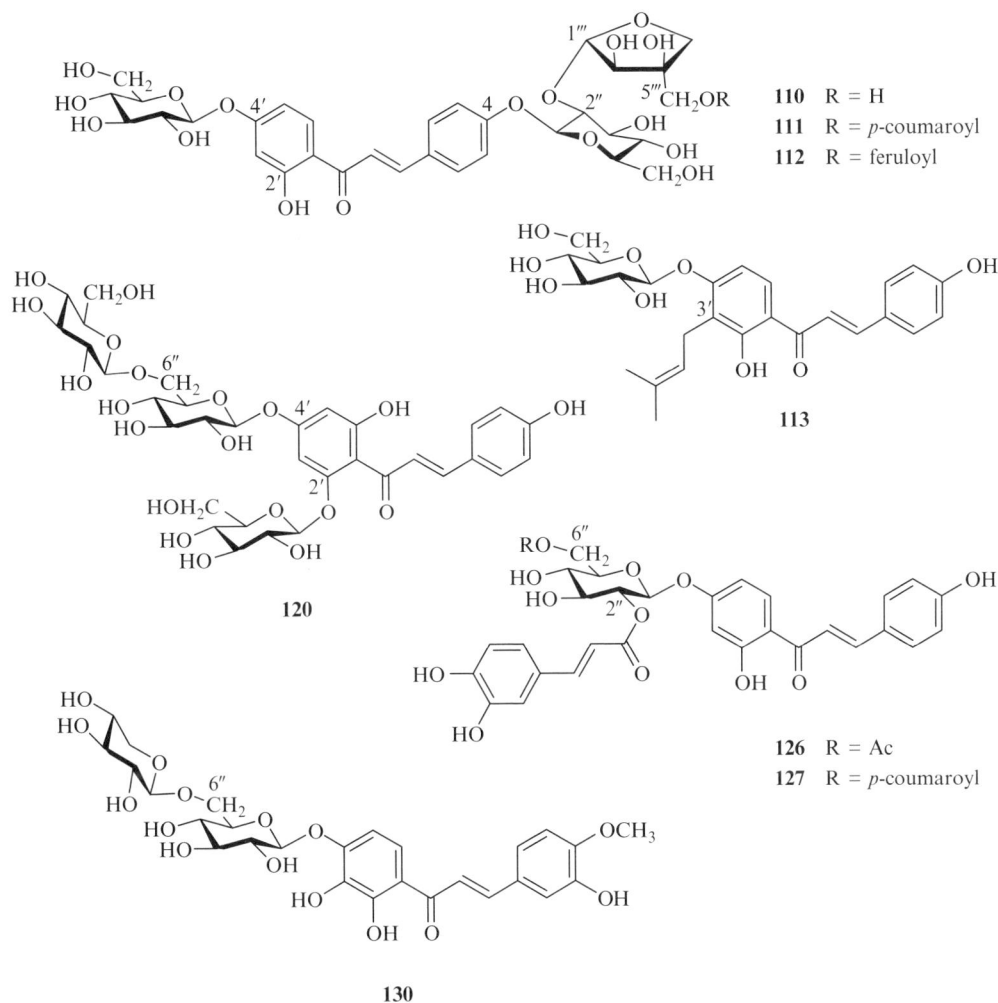

FIGURE 16.8 Chalcone glycosides (see Table 16.3).

calodenin A (**150**) and flavumone A (**152**) appear to result from condensation of the A-ring of phloretin (2′,4′,6′,4-tetrahydroxydihydrochalcone) and 2′,3′,4′,6′,4-pentahydroxychalcone, respectively, with the α,β-carbons of isoliquiritigenin.[140,141] Azobechalcone A (**147**) showed potent inhibition of Epstein–Barr virus early antigen induction caused by the tumor promoter teleocidin B-4.[138] Perhaps the most unusual of the chalcone dimer structures are two compounds (**144, 145**) isolated from kamala, a red colored exudate found on glandular trichomes of the surface of the fruits of *Mallotus philippensis* (Euphorbiaceae).[136] Kamalachalcone A (**144**) is formed by the condensation of two molecules of 2′,4′-dihydroxy-3′-methyl-6″,6″-dimethylpyrano[2″,3″:6′,5′]chalcone, a known constituent of the same plant species.[158] Similarly, kamalachalcone B (**145**) is a dimer comprised of the same chalcone and rottlerin, a benzylated pyranochalcone from *Rottlera tinctoria* (Euphorbiaceae).[10,159] A cyclobutane chalcone dimer based on the 2 + 2 cycloaddition of two molecules of chalconaringenin 6′,4-dimethyl ether (**8**) was isolated during a study of the constituents of *Goniothalamus gardneri* (Annonaceae), but is considered to be an artifact.[160]

TABLE 16.4
Chalcone Dimers, Heterodimers, and Oligomers Reported from 1992 to 2003

No.	Compound	Mol. Formula	Source	Family	Ref.
	Dimers (chalcone)				
134	Urundeuvine A	$C_{30}H_{22}O_9$	*Myracrodruon urundeuva*	Anacardiaceae	133
135	Urundeuvine B	$C_{30}H_{20}O_9$	*Myracrodruon urundeuva*	Anacardiaceae	133
136	Urundeuvine C	$C_{30}H_{22}O_{10}$	*Myracrodruon urundeuva*	Anacardiaceae	133
137	Matosine	$C_{30}H_{24}O_{10}$	*Myracrodruon urundeuva*	Anacardiaceae	133
138	Rhuschalcone I	$C_{32}H_{26}O_8$	*Rhus pyroides*	Anacardiaceae	134
139	Rhuschalcone II	$C_{30}H_{22}O_8$	*Rhus pyroides*	Anacardiaceae	135
140	Rhuschalcone III	$C_{31}H_{24}O_8$	*Rhus pyroides*	Anacardiaceae	135
141	Rhuschalcone IV	$C_{31}H_{24}O_8$	*Rhus pyroides*	Anacardiaceae	135
142	Rhuschalcone V	$C_{30}H_{24}O_8$	*Rhus pyroides*	Anacardiaceae	135
143	Rhuschalcone VI	$C_{30}H_{22}O_8$	*Rhus pyroides*	Anacardiaceae	135
144	Kamalachalcone A	$C_{42}H_{40}O_8$	*Mallotus philippensis*	Euphorbiaceae	136
145	Kamalachalcone B	$C_{51}H_{48}O_{12}$	*Mallotus philippensis*	Euphorbiaceae	136
146	Licobichalcone	$C_{32}H_{26}O_{10}$	*Glycyrrhiza uralensis*	Leguminosae	137
147	Azobechalcone A	$C_{31}H_{26}O_8$	*Lophira alata*	Ochnaceae	138
148	Isolophirone C	$C_{30}H_{22}O_8$	*Ochna afzelii*	Ochnaceae	139
149	Dihydrolophirone C	$C_{30}H_{24}O_8$	*Ochna afzelii*	Ochnaceae	139
150	Calodenin A	$C_{30}H_{22}O_9$	*Ochna calodendron*	Ochnaceae	140
151	Lophirone K	$C_{30}H_{22}O_9$	*Ochna calodendron*	Ochnaceae	140
152	Flavumone A	$C_{30}H_{20}O_{10}$	*Ouratea flava*	Ochnaceae	141
	Dimers (chalcone–flavan)				
153	Daphnodorin J	$C_{30}H_{24}O_9$	*Daphne odora*	Thymelaeaceae	142
154	Daphnodorin M	$C_{30}H_{22}O_{10}$	*Daphne acutiloba*	Thymelaeaceae	143
155	Daphnodorin N	$C_{30}H_{22}O_{10}$	*Daphne acutiloba*	Thymelaeaceae	143
	Dimers (chalcone–flavan-3-ol)				
156	Daphnodorin I	$C_{30}H_{22}O_{10}$	*Daphne odora*	Thymelaeaceae	144
157	Genkwanol B	$C_{30}H_{22}O_{11}$	*Daphne genkwa*	Thymelaeaceae	145
158	Genkwanol C	$C_{30}H_{22}O_{11}$	*Daphne genkwa*	Thymelaeaceae	146
	Dimers (chalcone–flavanone)				
159–162	Chalcocaryanones A–D	$C_{34}H_{28}O_8$	*Cryptocarya infectoria*	Lauraceae	147
163	6‴-Hydroxylophirone B	$C_{30}H_{22}O_9$	*Ochna integerrima*	Ochnaceae	148
164	6‴-Hydroxylophirone B 4‴-*O*-glucoside	$C_{36}H_{32}O_{14}$	*Ochna integerrima*	Ochnaceae	148
165	Flavumone B	$C_{30}H_{20}O_9$	*Ouretea flava*	Ochnaceae	141
	Dimers (chalcone–flavone)				
166–169	*Aristolochia* dimers (isomers)	$C_{33}H_{24}O_{11}$	*Aristolochia ridicula*	Aristolochiaceae	149
170	Cissampeloflavone	$C_{34}H_{26}O_{11}$	*Cissampelos pareira*	Menispermaceae	150
171	Calodenone	$C_{31}H_{24}O_8$	*Ochna calodendron*	Ochnaceae	151
	Trimers				
172	Caloflavan A	$C_{45}H_{38}O_{13}$	*Ochna calodendron*	Ochnaceae	152
173	Caloflavan B	$C_{45}H_{38}O_{13}$	*Ochna calodendron*	Ochnaceae	152
	Tetramers				
174	*Aristolochia* tetraflavonoid	$C_{66}H_{46}O_{21}$	*Aristolochia ridicula*	Aristolochiaceae	149
175	Alatachalcone	$C_{60}H_{48}O_{15}$	*Lophira alata*	Ochnaceae	153
176	Isolophirachalcone	$C_{60}H_{48}O_{15}$	*Lophira alata*	Ochnaceae	138
177	Lophiroflavan A	$C_{60}H_{48}O_{15}$	*Lophira alata*	Ochnaceae	154
178	Lophiroflavan B	$C_{60}H_{50}O_{15}$	*Lophira alata*	Ochnaceae	155
179	Lophiroflavan C	$C_{60}H_{50}O_{15}$	*Lophira alata*	Ochnaceae	155
	Pentamer				
180	Ochnachalcone	$C_{75}H_{62}O_{21}$	*Ochna calodendron*	Ochnaceae	156
	Hexamer				
181	Azobechalcone	$C_{90}H_{70}O_{22}$	*Lophira alata*	Ochnaceae	157

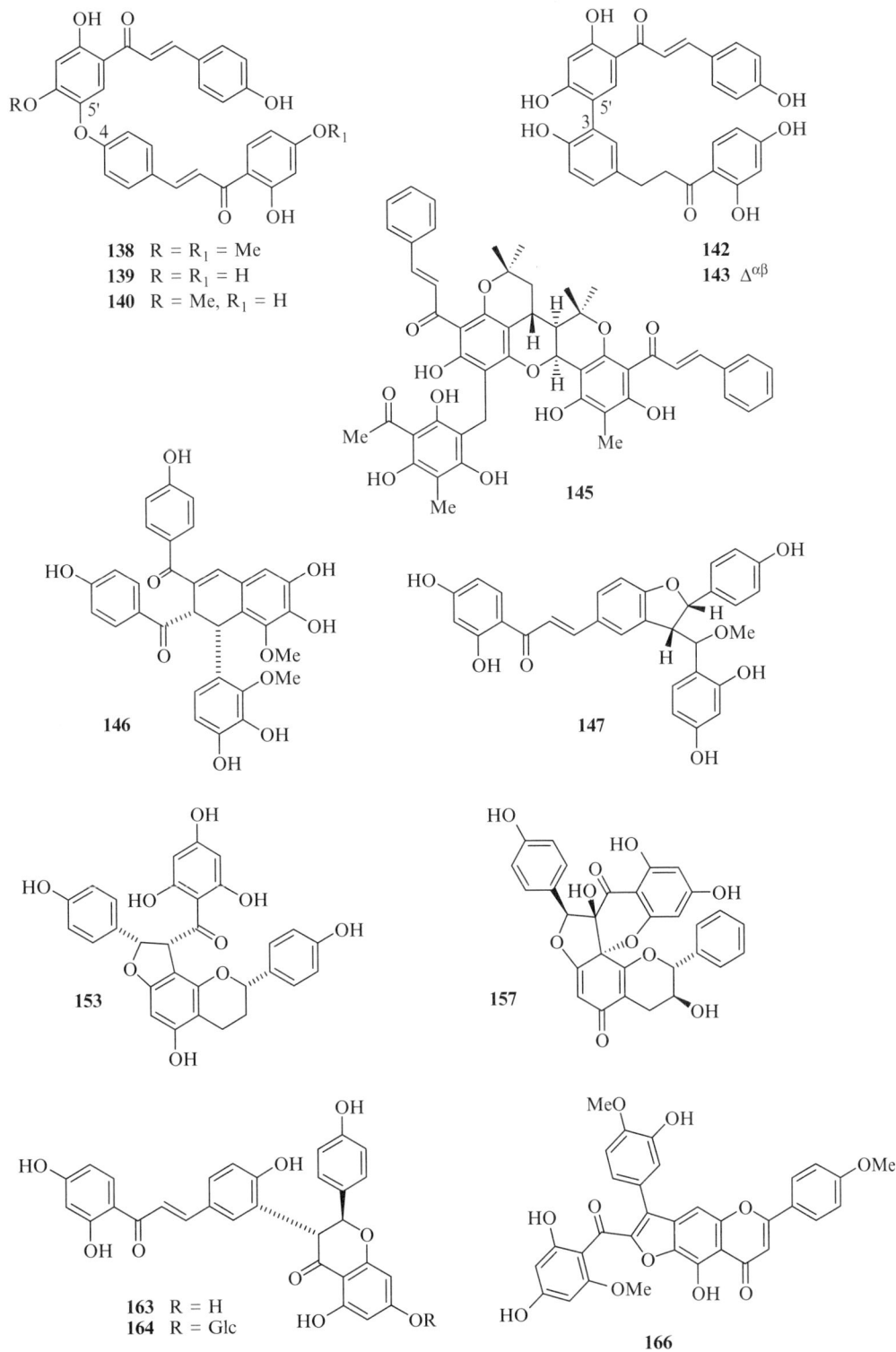

138 R = R$_1$ = Me
139 R = R$_1$ = H
140 R = Me, R$_1$ = H

142
143 $\Delta^{\alpha\beta}$

145

146

147

153

157

163 R = H
164 R = Glc

166

FIGURE 16.9 Chalcone dimers, heterodimers, and oligomers (see Table 16.4). — *continued*

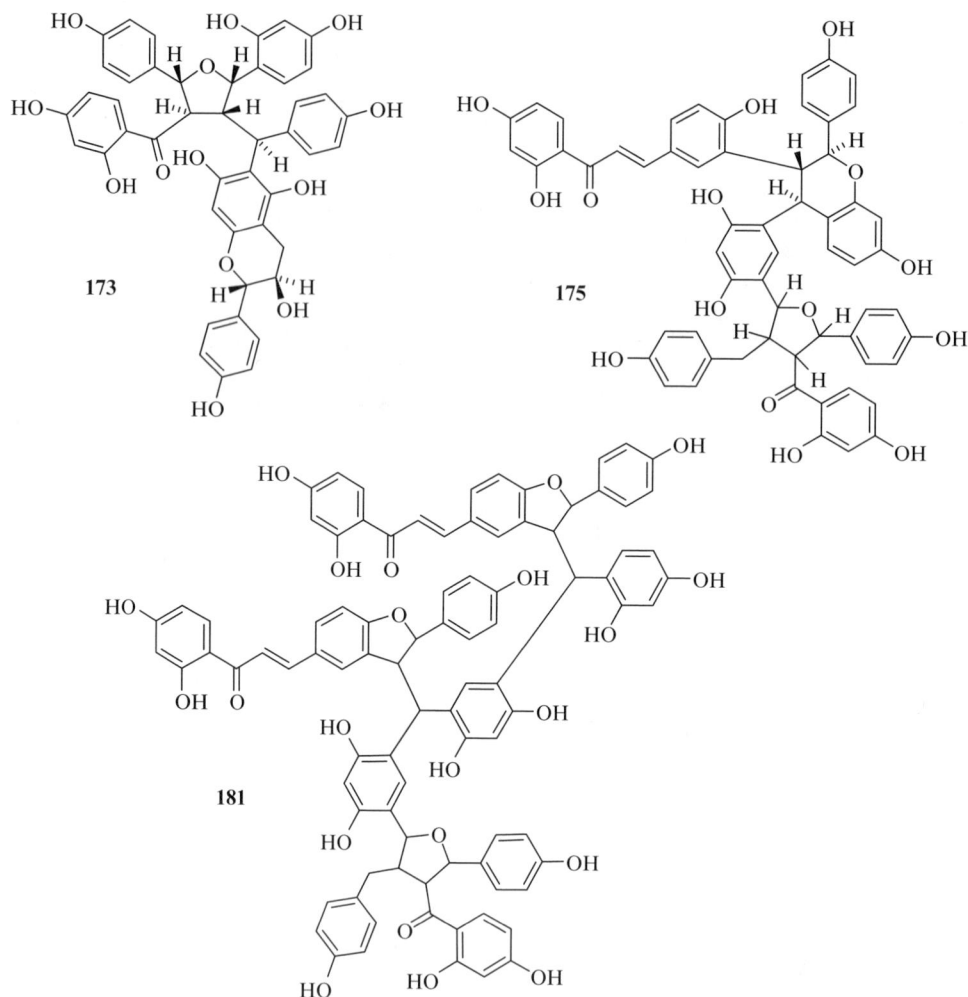

FIGURE 16.9 Chalcone dimers, heterodimers, and oligomers (see Table 16.4).

Rare dimers comprising a chalcone and a second class of flavonoid continue to be reported from a small number of plant species, as indicated in Table 16.4 and Figure 16.9. These "mixed" dimers include two series of compounds derived from the condensation of chalconaringenin with either apigeniflavan (5,7,4′-trihydroxyflavan) (**153–155**)[142,143] or afzelechin (3,5,7,4′-tetrahydroxyflavan) (**156–158**),[144–146] followed by internal cyclization reactions. Dimers of flavanones or flavones with flavans or flavan-3-ols may also be produced as an outcome of similar cyclization and rearrangement processes.[144,161,162] The compounds are characteristic constituents of the roots and bark of some species of *Daphne* (Thymelaeaceae),[142–146] although genkwanol B (**157**) has also been reported from the roots of *Wikstroemia sikokiana* (Thymelaeaceae), together with many other related biflavonoids.[161] The absolute configuration of genkwanol B was obtained by Mosher's method and x-ray crystallography of a permethylated derivative.[163] The stem bark of *Ochna integerrima* (Ochnaceae) yielded the chalcone–flavanone dimer, 6‴-hydroxylophirone C (**163**), characterized by a C–C bond between C-3 of the B-ring of isoliquiritigenin to C-3 of the C-ring of naringenin.[148] A second derivative (**164**) represents a similarly linked dimer of isoliquiritigenin and naringenin

7-*O*-β-glucopyranoside (prunin). The absolute configuration of the flavanone components of these compounds was determined to be (2*S*,3*R*) from circular dichroism spectra.[148] The remarkable chalcone–flavanone dimers, chalcocaryanones A–D (**159–162**), were obtained from an extract of the trunk bark of *Cryptocarya infectoria* (Lauraceae).[147] Both the chalcone and flavanone components of these dimers show the unusual feature of reduced A-rings (See Section 17.3.1.27). The constituent chalcone monomer is cryptocaryone, whose absolute configuration was determined by x-ray crystallographic analysis of a synthetic bromo deriva-tive.[147] Four chalcone–flavone dimers (**166–169**) have been reported from the stems of *Aristolochia ridicula* (Aristolochiaceae), together with a novel tetramer (**174**).[149] The chalcone component is the rare chalconaringenin 6′,4-dimethyl ether (**8**), 2′,4′,3-trihydroxy-6′,4-dimethoxychalcone, or 2′,4′,4-trihydroxy-6′,3-dimethoxychalcone. The latter two compounds have not been reported as natural products. The flavone components of these dimers, 5,6-dihydroxy-4′-methoxyflavone and 5,6,4′-trihydroxy-3′-methoxyflavone, are also notable for the unusual substitution pattern of the A-ring. In structural terms, the monomers are linked through the creation of an additional furan ring based on 6-OH and C-7 of the flavone A-ring and the α,β-carbons of the chalcone (Figure 16.9 and Section 17.3.1.19). The tetramer **174** comprises a biflavone *O*-linked to a chalcone–flavone dimer based on chalconaringenin 6′-methyl ether and 5,6-dihydroxy-3′,4′-dimethoxyflavone, and is probably the result of oxida-tive coupling of the biflavonoid units see Section 17.3.3.[149] Cissampeloflavone (**170**) is a chalcone–flavone dimer with a structure similar to those of the *Aristolochia* biflavonoids (**166–169**).[150] It showed good activity in antiprotozoal assays against *Trypanosoma cruzi* and *T. brucei rhodiense* and low cytotoxicity in the human KB cell line. The constituent monomers of cissampeloflavone are the more common flavokawin A (2′-hydroxy-4′,6′,4-trimethoxychal-cone) and chrysoeriol (5,7,4′-trihydroxy-3′-methoxyflavone). Calodenone (**171**), isolated from the stem bark of *Ochna calodendron* (Ochnaceae), appears to be a rearranged biflavonoid based on chalcone (isoliquiritigenin 4′-methyl ether) and flavone components, but nothing is known of its biosynthesis.[151] A closely related derivative (lophirone A) with isoliquiritigenin as the chalcone component is a known constituent of *Lophira lanceolata* (Ochnaceae).[164]

Caloflavans A and B are trimers isolated from the leaves of *Ochna calodendron* that differ only in the attachment of the chalcone dimer, isombamichalcone,[165] to either C-8 or C-6 of the A-ring of afzelechin, respectively.[152] The chalcone tetramers and higher oligomers listed in Table 16.4 are based on the condensation of isoliquiritigenin, with the exception of the *Aristolochia* tetramer (**174**) discussed above. Some are of interest as antitumor promoters, such as alatachalcone (**175**)[153] and isolophirachalcone (**176**),[148] constituents of the bark of *Lophira alata* (Ochnaceae), a rich source of bi- and tetraflavonoids. Tih et al. consider that lophiroflavans A and B (**177, 178**) arise from coupling of the biflavonoids, isombamichalcone, and lophirone H, whereas the immediate precursors of lophiroflavan C (**179**) appear to be lophirone H and mbamichalcone.[154,155] Ochnachalcone (**180**), a pentaflavonoid from the stem bark of *Ochna calodendron*, has a most unusual structure (see Section 17.3.4) in which two molecules of the chalcone dimer isombamichalcone are linked by C–C bonds to C-6 and C-8 of the A-ring of (2*R*,3*S*)-3,5,7,4′-tetrahydroxyflavan (afzelechin).[156] The extraordinary hexa-flavonoid, azobechalcone (**181**), which should not be confused with the chalcone dimer azobechalcone A (**147**), has a structure based on the condensation of six molecules of isoliquiritigenin (Figure 16.9).[157] In biosynthetic terms, its immediate precursors are consid-ered to be the bi- and tetraflavonoids, lophirone C and lophirachalcone,[165] respectively. The structural determination and NMR spectral assignment of these oligomers is a challenging task requiring extensive use of two-dimensional NMR (COSY, HMQC, HMBC) and meas-urement of "through-space" proton–proton connectivities by NOE and ROE experiments. These methods have also been used to investigate the relative configuration and stereochem-istry of other chalcone oligomer substructures such as those of ochnachalcone (**180**).[156]

16.2.5 DIELS–ALDER ADDUCTS OF CHALCONES

One of the most characteristic features of the structural chemistry of chalcones is their ability to act as dienophiles in enzyme-catalyzed Diels–Alder reactions (Figure 16.10). The dienes that participate in the formation of the so-called Diels–Alder adducts range from simple isoprene and monoterpene compounds to coumarins and other classes of flavonoid. A full list of Diels–Alder adducts of chalcones reported in the literature from 1992 to 2003 is given in Table 16.5, which is arranged according to the diene component of the compounds. New or revised structures for some adducts described in the literature before 1992 are also included here. An important point that is evident from Table 16.5 is the predominance of Diels–Alder adducts of chalcones in the Moraceae, and in particular, the genus *Morus* (mulberry trees). Other plant families are represented only within the chalcone–monoterpene group of adducts, which are found mainly in the Annonaceae. A commentary on some aspects of the chemistry and biosynthesis of Diels–Alder adducts from *Morus* was published in 1999;[185] however, this continues to be an area of active research interest. Some representative examples of the Diels–Alder adducts listed in Table 16.5 are illustrated in Figure 16.11.

Sanggenon R (**182**) is a good example of a Diels–Alder adduct whose chalcone origins may not be immediately obvious (Figure 16.10). According to the biosynthetic scheme proposed by Hano et al., the initial adduct formed by Diels–Alder reaction of 2′,4′,2,4-tetrahydroxychalcone and isoprene is subject to several further rearrangement and oxidation steps to produce sanggenon R.[166] This compound is one of many Diels–Alder adducts and isoprenylated flavonoids obtained from the root bark of *Morus* sp., the source of the Chinese herbal medicine "Sang-Bai-Pi." Four pairs of regioisomeric Diels–Alder adducts formed by reaction of chalcones with the monoterpene myrcene (7-methyl-3-methylene-1,6-octadiene)

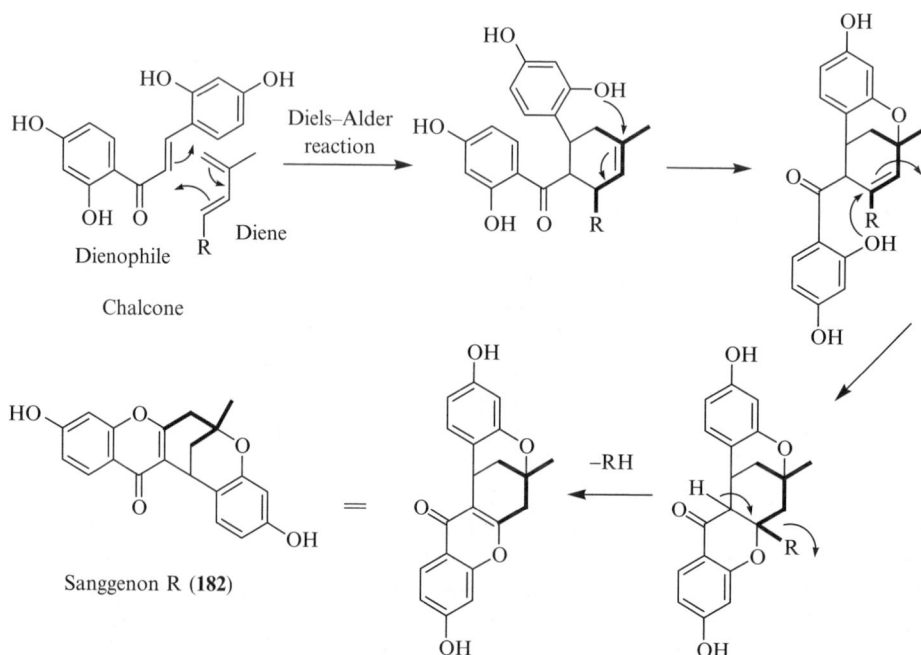

FIGURE 16.10 A biosynthetic pathway proposed for the formation of the chalcone–isoprene Diels–Alder adduct, sanggenon R (**182**), after Hano et al.[166] The fate of the diene is indicated in bold type.

have been characterized from the Annonaceae taxa, *Cyathocalyx crinatus* and *Fissistigma lanuginosum*.[167,168] Of these, crinatusins A$_1$ (**183**) and A$_2$ (**184**) originate from the reaction of 2′,4′-dihydroxy-6′-methoxy-3′,5′-dimethylchalcone and myrcene. The remaining pairs of regioisomers, crinatusins B$_1$ (**185**) and B$_2$ (**186**), C$_1$ (**187**) and C$_2$ (**188**), and fissistin (**189**) and isofissistin (**190**), are based on stercurensin (2′,4′-dihydroxy-6′-methoxy-3′-methylchalcone), cardamonin (2′,4′-dihydroxy-6′-methoxychalcone), and 2′,5′-dihydroxy-3′,4′,6′-trimethoxychalcone, respectively. The crinatusins showed lethal toxicity in the brine shrimp bioassay.[167] Cytotoxicity toward KB cells was recorded for both fissistin and isofissistin (IC$_{50}$ = 0.15 µg/ml).[168] Two adducts (**192**, **193**) isolated from the rhizomes of *Kaempferia pandurata* (Zingiberaceae) are regioisomers resulting from the Diels–Alder reaction of cardamonin with β-ocimene (3,7-dimethyl-1,3,6-octatriene).[170] A closely related Diels–Alder adduct (**191**) derived from 2′,4′,6′-trihydroxychalcone and β-ocimene and found in the rhizomes of *Boesenbergia pandurata* (Zingiberaceae) was reported first by Trakoontivakorn et al.,[169] who noted its similarity to panduratin A, a known Diels–Alder adduct of 2′,6′-dihydroxy-4′-methoxychalcone and β-ocimene.[186] These authors assigned the incorrect trivial name of "4-hydroxypanduratin A" to **191**. In a subsequent paper, Tuchinda et al. described an anti-inflammatory compound identical to **191** and from the same source, as a new chalcone derivative, giving it the incorrect trivial name "(−)-hydroxypanduratin A."[187] The correct trivial name for this Diels–Alder adduct treated as a derivative of panduratin A is 4′-demethylpanduratin A. The racemic nicolaioidesin A (**194**) is diastereoisomeric to panduratin A, to which (±)-nicolaioidesin B (**195**) is a positional isomer.[171]

Palodesangrens A–E (**197**–**201**), obtained from the bark of the Amazonian tree, *Brosimum rubescens* (Moraceae), are Diels–Alder adducts of chalcones and an isoprenylated coumarin.[172] Three of the compounds (**199**–**201**) were effective inhibitors toward the formation of a 5α-dihydrotestosterone–androgen receptor complex implicated in androgen-dependent diseases. Palodesangretins A and B (**202**, **203**), isolated from the same source, are considered to originate from the Diels–Alder reaction of chalcones and 8-hydroxy-7-methoxy-6-(3-methyl-but-1,3-dienyl)coumarin followed by a five-membered ring closure.[173] The leaves of *Morus insignis* yielded mulberrofuran U (**204**), a new example of a Diels–Alder adduct of 2′,4′,2,4-tetrahydroxychalcone (a typical chalcone of the Moraceae) and an isoprenylated arylbenzofuran.[174] Several related compounds have been reported prior to 1992, also from species of *Morus*.[10] It is perhaps not surprising that Diels–Alder adducts can be formed solely from isoprenylated chalcones and several new examples (**205**–**207**) are listed in Table 16.5. For example, the formation of dorstenone (**206**) proceeds by Diels–Alder reaction of 2′,4′,4-trihydroxy-3′-prenylchalcone (dienophile) and the corresponding dehydro derivative, 2′,4′,4-trihydroxy-3′-(3-methyl-but-1,3-dienyl)chalcone (diene).[176] Similarly, artonin X (**205**), a constituent of the bark of *Artocarpus heterophylla* (cited incorrectly as *A. heterophyllus*), is an adduct of 2′,4′,2,4-tetrahydroxy-3′-prenylchalcone (dienophile) and 2′,4′,4-trihydroxy-3′-(3-methyl-but-1,3-dienyl)chalcone (diene).[175] A more complex bichalcone (**207**) isolated from twigs of *Dorstenia zenkeri* is considered to be an elaborated form of the Diels–Alder adduct, **206**, modified by subsequent cyclization reactions (Figure 16.10).[83]

Among the most intensively investigated of all the chalcone Diels–Alder adducts are a group obtained solely from *Morus* species in which the diene component of the reaction is a dehydroprenylflavanone. The structures of several such compounds published prior to 1992 have now been revised on the basis of new spectroscopic and chemical data. Among the most important of the techniques used were two-dimensional NMR and circular dichroism spectroscopy. The revised structures listed in Table 16.5 are those of sanggenons C (**210**), D (**211**), E (**212**), and O (**213**).[179,180] In these compounds, the flavanones show the common feature of 3-hydroxy-2-prenyl substitution with an ether linkage between C-3 and C-2′ of the B-ring. A method for determining the absolute configurations at C-2 and C-3 has now been

TABLE 16.5
Diels–Alder Adducts of Chalcones Reported from 1992 to 2003

No.	Compound	Mol. Formula	Source	Family	Ref.
	Chalcone–isoprene				
182	Sanggenon R	$C_{20}H_{16}O_5$	*Morus* sp.	Moraceae	166
	Chalcone–monoterpene				
183	Crinatusin A_1	$C_{28}H_{34}O_4$	*Cyathocalyx crinatus*	Annonaceae	167
184	Crinatusin A_2	$C_{28}H_{34}O_4$	*Cyathocalyx crinatus*	Annonaceae	167
185	Crinatusin B_1	$C_{27}H_{32}O_4$	*Cyathocalyx crinatus*	Annonaceae	167
186	Crinatusin B_2	$C_{27}H_{32}O_4$	*Cyathocalyx crinatus*	Annonaceae	167
187	Crinatusin C_1	$C_{26}H_{30}O_4$	*Cyathocalyx crinatus*	Annonaceae	167
188	Crinatusin C_2	$C_{26}H_{30}O_4$	*Cyathocalyx crinatus*	Annonaceae	167
189	Fissistin	$C_{28}H_{34}O_6$	*Fissistigma lanuginosum*	Annonaceae	168
190	Isofissistin	$C_{28}H_{34}O_6$	*Fissistigma lanuginosum*	Annonaceae	168
191	*Boesenbergia* adduct	$C_{25}H_{28}O_4$	*Boesenbergia pandurata*	Zingiberaceae	169
192–193	*Kaempferia* adducts	$C_{26}H_{30}O_4$	*Kaempferia pandurata*	Zingiberaceae	170
194–196	($\pm$)-Nicolaioidesins A–C	$C_{26}H_{30}O_4$	*Renealmia nicolaioides*	Zingiberaceae	171
	Chalcone–coumarin				
197	Palodesangren A	$C_{30}H_{26}O_7$	*Brosimum rubescens*	Moraceae	172
198	Palodesangren B	$C_{30}H_{26}O_7$	*Brosimum rubescens*	Moraceae	172
199	Palodesangren C	$C_{30}H_{26}O_6$	*Brosimum rubescens*	Moraceae	172
200	Palodesangren D	$C_{30}H_{26}O_6$	*Brosimum rubescens*	Moraceae	172
201	Palodesangren E	$C_{31}H_{28}O_7$	*Brosimum rubescens*	Moraceae	172
202	Palodesangretin A	$C_{31}H_{28}O_8$	*Brosimum rubescens*	Moraceae	173
203	Palodesangretin B	$C_{31}H_{28}O_8$	*Brosimum rubescens*	Moraceae	173
	Chalcone–arylbenzofuran				
204	Mulberrofuran U	$C_{39}H_{36}O_9$	*Morus insignis*	Moraceae	174
	Chalcone–chalcone				
205	Artonin X	$C_{40}H_{38}O_9$	*Artocarpus heterophylla*	Moraceae	175
206	Dorstenone	$C_{40}H_{38}O_8$	*Dorstenia barteri*	Moraceae	176
207	*Dorstenia* chalcone dimer	$C_{40}H_{38}O_8$	*Dorstenia zenkeri*	Moraceae	83
	Chalcone–flavanone				
208	Sanggenol J	$C_{45}H_{44}O_{12}$	*Morus cathayana*	Moraceae	177
209	Sanggenol M	$C_{45}H_{46}O_{11}$	*Morus mongolica*	Moraceae	178
210	Sanggenon C [a]	$C_{40}H_{36}O_{12}$	*Morus mongolica*	Moraceae	179
211	Sanggenon D [a]	$C_{40}H_{36}O_{12}$	*Morus mongolica*	Moraceae	179
212	Sanggenon E [a]	$C_{45}H_{44}O_{12}$	*Morus mongolica*	Moraceae	179
213	Sanggenon O [a]	$C_{40}H_{36}O_{12}$	*Morus mongolica*	Moraceae	180
214	Sanggenon S	$C_{40}H_{34}O_{12}$	*Morus* sp.	Moraceae	166
215	Sanggenon T	$C_{40}H_{40}O_{12}$	*Morus* sp.	Moraceae	166
216	Cathayanon A	$C_{40}H_{36}O_{12}$	*Morus cathayana*	Moraceae	181
217	Cathayanon B	$C_{40}H_{36}O_{12}$	*Morus cathayana*	Moraceae	181
	Chalcone–flavone				
218	Artonin I	$C_{40}H_{36}O_{11}$	*Artocarpus heterophylla*	Moraceae	182
219	Multicaulisin	$C_{40}H_{36}O_{11}$	*Morus multicaulis*	Moraceae	183
	Miscellaneous				
220	Sorocenol B	$C_{31}H_{28}O_7$	*Sorocea bonplandii*	Moraceae	184

[a] Indicates revised structure.

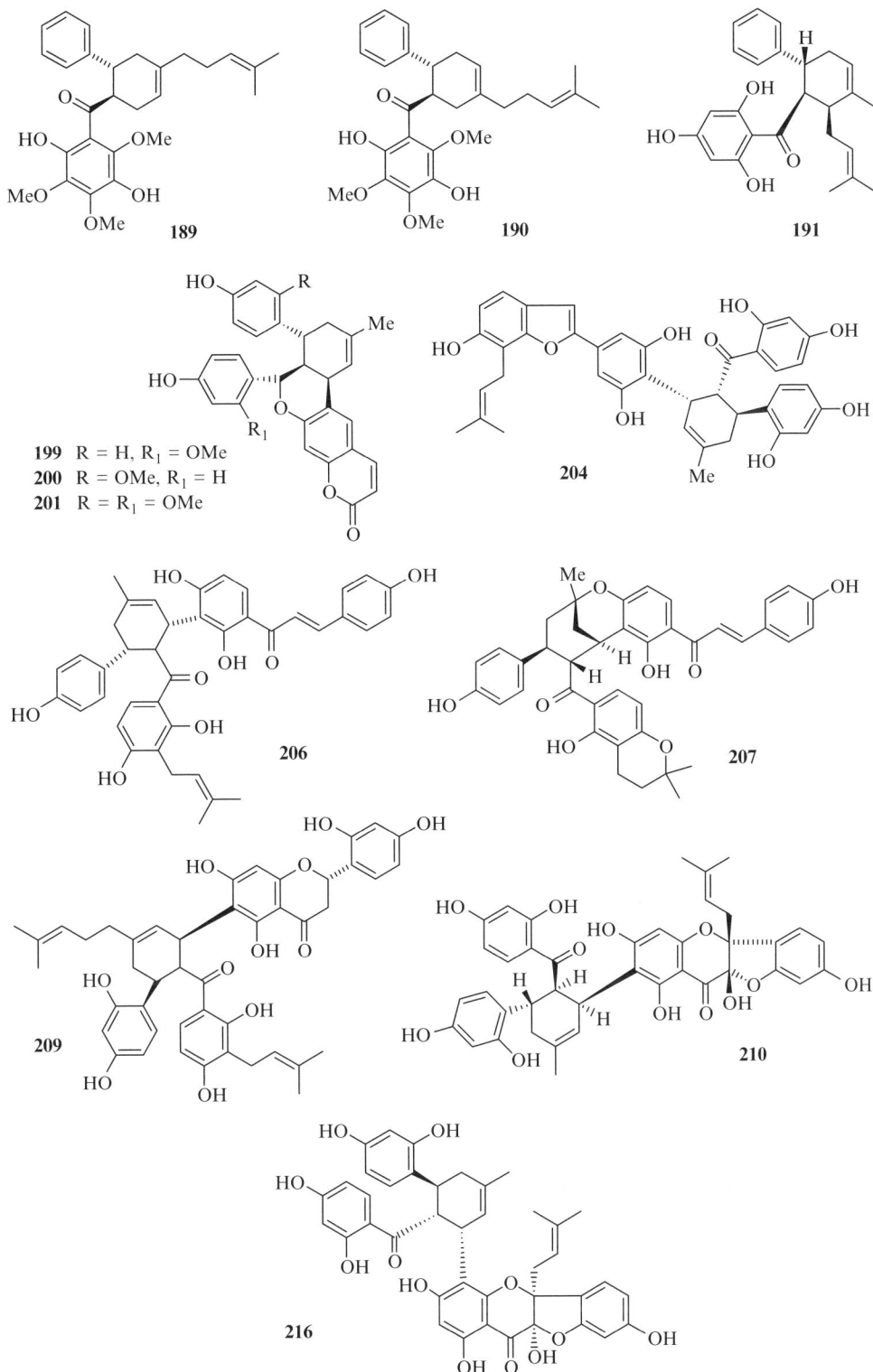

FIGURE 16.11 Diels–Alder adducts of chalcones (see Table 16.5).

described.[188] New Diels–Alder adducts incorporating this type of flavanone include sanggenol J (**208**)[177] and cathayanons A and B (**216, 217**).[181] The structure of cathayanon A was confirmed by x-ray crystallography. In sanggenons C (**210**), D (**211**), and O (**213**) and cathayanons A (**216**) and B (**217**), the chalcone dienophile is the commonly found 2′,4′,2,4-tetrahydroxychalcone, whereas in sanggenon E it is the 3′-prenyl derivative of the same chalcone. In contrast, the more unusual 2′,2,4-trihydroxy-6″,6″-dimethyl-4″,5″-dihydropyrano[2″,3″:4′,3′]chalcone is the dienophile component of sanggenol J (**208**).[177] Sanggenol M (**209**) is a Diels–Alder adduct of 2′,4′,2,4-tetrahydroxy-3′-prenylchalcone and 6-dehydrogeranyl-5,7,2′,4′-tetrahydroxyflavanone.[178] It showed the greatest cytotoxicity toward human oral tumor cell lines of a series of isoprenylated flavanones and chalcone–flavanone Diels–Alder adducts isolated from *Morus mongolica*.[178] Sanggenon T (**215**) is also an example of a Diels–Alder adduct in which the diene component is a dehydrogeranylflavanone derivative.[166] In this respect both sanggenol M (**209**) and sanggenon T (**215**) resemble the known derivative sanggenon G.[189] The structural origins of sanggenon S (**214**) are less easy to discern, but it appears to be an adduct of the sanggenon C-type that has undergone further cyclization and rearrangement.[166] Two Diels–Alder adducts in which the diene is an isoprenylated flavone have also been characterized, artonin I (**218**), from *Artocarpus heterophylla*,[182] and multicaulisin (**219**),[183] a constituent of the roots of *Morus multicaulis*. Partial enzymatic synthesis of artonin I by introduction of artocarpesin (5,7,2′,4′-tetrahydroxy-6-prenylflavone), a known constituent of *A. heterophylla*, into a *Morus alba* cell culture was used to confirm that the diene component was 6- rather than 8-isoprenylated.[182]

16.2.6 Chalcone Conjugates

The seeds of *Alpinia blepharocalyx* (Zingiberaceae) contain a novel series of diarylheptanoids, some of which are characterized by conjugation to either chalcones (Table 16.6) or flavanones. This plant is used in Chinese traditional medicine to treat stomach disorders, and extracts of the seeds show both hepatoprotective and antiproliferative activities.[200] The structural determination, stereochemical assignment, biosynthesis, and biological activity of these *Alpinia* diarylheptanoids have been described in detail in a recent review,[200] where the following classification of chalcone-bearing diarylheptanoids was proposed (see Figure 16.12 for examples):

(i) Acyclic diarylheptanoids with a chalcone at C-7: calyxin B (**223**), epicalyxin B (**224**), calyxin H (**228**), and epicalyxin H (**229**)
(ii) Acyclic diarylheptanoids with a chalcone at C-5: calyxin A (**221**) and deoxycalyxin A (**222**)
(iii) Cyclic diarylheptanoids with a chalcone at C-5: calyxin F (**225**), epicalyxin F (**226**), and 6-hydroxycalyxin F (**227**)
(iv) Cyclic diarylheptanoids with a chalcone and an additional *p*-hydroxybenzyl group: calyxin I (**230**) and epicalyxin I (**231**)
(v) Diarylheptanoids with an ether bond between C-7 and the C-5-linked chalcone: calyxin L (**232**)
(vi) Dimeric diarylheptanoids conjugated to a chalcone: blepharocalyxins A, B, and E (**233–235**)

Many of the chalcone–diarylheptanoid conjugates occur as pairs of epimers, as illustrated by calyxin B and epicalyxin B (Figure 16.12). These can be separated by high-performance liquid chromatography (HPLC) using chiral column packings.[192] The absolute configurations of chiral carbons with secondary hydroxyl functions were determined using Mosher's method.[192] Similarly, the relative stereochemistry of other chiral centers, for example, in ring structures, was assigned on the basis of correlations observed in ROE experiments. The chalcone component of all the conjugates (221–235) is 2′,4′,4-trihydroxy-6′-methoxychalcone (helichrysetin). In several of the compounds (221–229, 235), the chalcone is linked only by a C–C bond from C-3′ of the A-ring. A shared tetrahydropyran ring based on C-3′ and O-2′ of the chalcone A-ring to C-5 and C-7 of the diarylheptanoid, respectively, characterizes calyxin I (230), epicalyxin I (231), and calyxin L (232).[191] In blepharocalyxins A (233) and B (234), the chalcone is linked to two diarylheptanoid units, one by a C–C bond from C-3′ of the A-ring and the other at the α,β-carbons through a tetrahydropyran ring.[194] Some potential pathways for the biosynthesis of these molecules have been proposed.[200] The most promising biological activities uncovered in a series of assays are the potent inhibition of nitric oxide (NO) formation by blepharocalyxin B (234)[201] and the antiproliferative effect of calyxin B (223) and epicalyxin F (226) toward the human HT-1080 fibrosarcoma and colon 26-L5 carcinoma cell lines, respectively.[202]

Table 16.6 also includes several chalcone conjugates that do not share an obvious structural relationship with other more clearly defined groups of compounds. Some of these conjugates are illustrated in Figure 16.12. Among the most interesting is a labdane diterpenoid linked by a C–C bond to C-3′ of the A-ring of 2′,4′-dihydroxy-6′-methoxychalcone (cardamonin) (237),[198] and didymocalyxin B (239), a compound thought to be formed by the oxidative coupling of an enolic tautomer of cinnamoylacetic acid with 2′,6′-dihydroxy-3′,4′-dimethoxychalcone (pashanone).[47] The genus *Cryptocarya* (Lauraceae) is the source of several unusual chalcone derivatives including the chalcone–flavone dimers 159–162.[147] To their number must be added the unique conjugate, kurzichalcolactone (238),[199] and infectocaryone (240),[147] a chalcone with a reduced A-ring that is related structurally to cryptocaryone.

16.2.7 Quinochalcones

A small group of chalcones characterized as either chalcoquinones or chalcoquinone glycosides are shown in Table 16.7 and Figure 16.13. Desmosdumotin C (241) is a constituent of the roots of *Desmos dumosus* (Asteraceae), which showed significant and selective cytotoxicity when evaluated against a panel of cancer cell lines.[203] The structure of this compound was confirmed by x-ray crystallography.[203] The *gem*-diprenylquinochalcone, tunicatachalcone (243), was obtained from the roots of *Tephrosia tunicata* (Leguminosae) together with several other isoprenylated flavonoids.[205] It is closely related in structure to munchiwarin (242), an orange pigment isolated from the roots of *Crotalaria trifoliastrum* (Leguminosae) for which the crystal structure has been determined.[204] This indicates that the conjugated quinochalcone skeleton is essentially planar, with the *gem*-diprenyl substituents directed above and below the plane of the molecule and perpendicular to it. The third prenyl substituent of the quinochalcone A-ring lies in the plane of the molecule.

Of all the quinochalcones studied to date, the red and yellow glycosidic pigments obtained from the flowers of *Carthamus tinctorius* (the safflower) are probably the most fascinating. Their use in coloring and flavoring foods, as medicines, and for making dyes, has been known since ancient times. The results of the first studies on the chemical properties of the pigments were published by Johann Beckmann in 1773,[212] while professor at the University of Göttingen. During the first half of the nineteenth century a number of scientists continued to investigate and debate the chemical composition of the pigments.[213–216] It was not until 1979 that the structure of the red pigment (carthamin) was solved by Obara and Onodera.[217] The absolute configuration of this unusual *C*-glycosylquinochalcone dimer has been obtained more recently from the analysis of circular dichroism spectra of synthetic model

TABLE 16.6
Chalcone Conjugates Reported from 1992 to 2003

No.	Compound	Mol. Formula	Source	Family	Ref.
	Chalcone–diarylheptanoid				
221	Calyxin A	$C_{35}H_{34}O_9$	*Alpinia blepharocalyx*	Zingiberaceae	190
222	Deoxycalyxin A	$C_{35}H_{34}O_8$	*Alpinia blepharocalyx*	Zingiberaceae	191
223	Calyxin B	$C_{35}H_{34}O_8$	*Alpinia blepharocalyx*	Zingiberaceae	192
224	Epicalyxin B	$C_{35}H_{34}O_8$	*Alpinia blepharocalyx*	Zingiberaceae	192
225	Calyxin F	$C_{35}H_{34}O_8$	*Alpinia blepharocalyx*	Zingiberaceae	190
226	Epicalyxin F	$C_{35}H_{34}O_8$	*Alpinia blepharocalyx*	Zingiberaceae	191
227	6-Hydroxycalyxin F	$C_{35}H_{34}O_9$	*Alpinia blepharocalyx*	Zingiberaceae	190
228	Calyxin H	$C_{35}H_{34}O_7$	*Alpinia blepharocalyx*	Zingiberaceae	193
229	Epicalyxin H	$C_{35}H_{34}O_7$	*Alpinia blepharocalyx*	Zingiberaceae	193
230	Calyxin I	$C_{42}H_{38}O_9$	*Alpinia blepharocalyx*	Zingiberaceae	191
231	Epicalyxin I	$C_{42}H_{38}O_9$	*Alpinia blepharocalyx*	Zingiberaceae	191
232	Calyxin L	$C_{35}H_{34}O_8$	*Alpinia blepharocalyx*	Zingiberaceae	191
	Chalcone–Bis(diarylheptanoid)				
233	Blepharocalyxin A	$C_{54}H_{54}O_{11}$	*Alpinia blepharocalyx*	Zingiberaceae	193, 194
234	Blepharocalyxin B	$C_{54}H_{54}O_{11}$	*Alpinia blepharocalyx*	Zingiberaceae	193, 194
235	Blepharocalyxin E	$C_{54}H_{54}O_{11}$	*Alpinia blepharocalyx*	Zingiberaceae	195, 196
	Miscellaneous				
236	2′,4′-Dihydroxy-3′-C-(2,6-dihydroxybenzyl)-6′-methoxychalcone	$C_{23}H_{20}O_6$	*Desmos chinensis*	Annonaceae	197
237	2′,4′-Dihydroxy-6′-methoxy-3′-(8,17-epoxy-16-oxo-12,14-labdadien-15-yl)chalcone	$C_{36}H_{42}O_6$	*Alpinia katsumadai*	Zingiberaceae	198
238	Kurzichalcolactone	$C_{32}H_{30}O_7$	*Cryptocarya kurzii*	Lauraceae	199
239	Didymocalyxin B	$C_{28}H_{22}O_7$	*Didymocarpus leucocalyx*	Gesneriaceae	47
240	Infectocaryone	$C_{18}H_{18}O_4$	*Cryptocarya infectoria*	Lauraceae	147

compounds.[218] The flowers of safflower are yellow on first opening and turn red after a few days. Two groups have now solved the structure of one of the yellow pigment precursors of carthamin, naming it precarthamin (**247**).[209,210] It differs from carthamin only by the presence of an additional carboxyl group at the bridging carbon between the two C-glycosylquino-chalcone monomers. Analysis of the flavonoid pigments of the orange flowers of *C. tinctorius* cv. Ken-ba revealed the presence of another yellow pigment and carthamin precursor, anhydrosafflor yellow B (**244**).[206] The techniques used to obtain the structure of this compound included circular dichroism spectroscopy, NOE and ROE measurements by ^{1}H NMR, and molecular mechanics calculations. In the same study, the flavonoid and quinochalcone constituents of three distinct *C. tinctorius* cultivars were compared by HPLC and a scheme for the biosynthesis of quinochalcone pigments from chalcone precursors was proposed.[206] Several other yellow pigments that are C-glycosylquinochalcone monomers have been reported, including hydroxysafflor yellow A (**246**),[208] tinctormine (**248**),[208,211] and

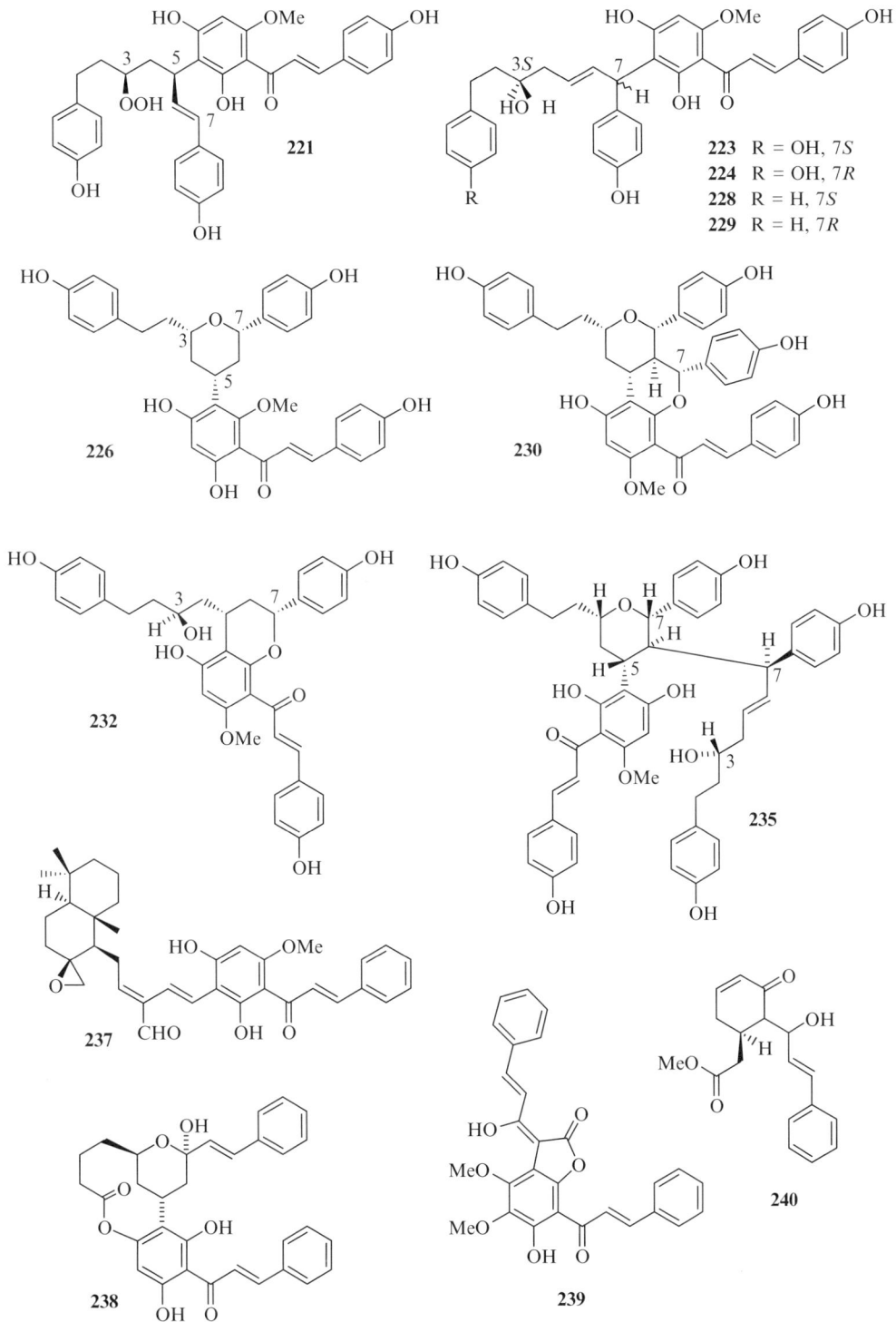

FIGURE 16.12 Chalcone conjugates (see Table 16.6).

TABLE 16.7
Quinochalcones Reported from 1992 to 2003

No.	Compound	Mol. Formula	Source	Family	Ref.
	Aglycones				
241	Desmosdumotin C	$C_{19}H_{20}O_4$	*Desmos dumosus*	Asteraceae	203
242	Munchiwarin	$C_{30}H_{36}O_4$	*Crotalaria trifoliastrum*	Leguminosae	204
243	Tunicatachalcone	$C_{26}H_{30}O_4$	*Tephrosia tunicata*	Leguminosae	205
	C-Glycosides				
244	Anhydrosafflor yellow B	$C_{48}H_{52}O_{26}$	*Carthamus tinctorius*	Asteraceae	206
245	Cartormin	$C_{27}H_{29}O_{13}N$	*Carthamus tinctorius*	Asteraceae	207
246	Hydroxysafflor yellow A	$C_{27}H_{32}O_{16}$	*Carthamus tinctorius*	Asteraceae	208
247	Precarthamin	$C_{44}H_{44}O_{24}$	*Carthamus tinctorius*	Asteraceae	209, 210
248	Tinctormine	$C_{27}H_{31}O_{14}N$	*Carthamus tinctorius*	Asteraceae	208, 211

cartormin (**245**).[207] Tinctormine (**248**), a potent Ca^{2+} antagonist, is structurally remarkable for its unusual *C*-linked polyhydroxylated pyrrolidine ring (Figure 16.13). The structure of cartormin (**245**), a quinochalcone *C*-glucoside with a novel *C*-glycosylpyrrole substituent, was confirmed by x-ray crystallography.[207]

16.3 DIHYDROCHALCONES

16.3.1 DIHYDROCHALCONES WITH SIMPLE PATTERNS OF O-SUBSTITUTION

16.3.1.1 Structure, Chemosystematic Trends, and Biological Activity

New dihydrochalcones reported in the literature from 1992 to 2003 are listed in Table 16.8 and some examples are shown in Figure 16.14. The two methyl ethers (**249, 250**) of the simplest of the *O*-substituted dihydrochalcone found in plants, 2′,4′-dihydroxydihydrochalcone, have now been isolated from leaf surface extracts of *Empetrum nigrum* (Empetraceae) together with the parent compound.[219] The latter was cited as new, although it had been reported previously from *Ceratiola ericoides*,[235] another species of the same family. The three-dimensional structure of the 2′-methyl ether (**250**) was solved using x-ray crystallography by Krasnov et al.[236] As expected, the molecule is essentially planar apart from the unsubstituted B-ring, and the dihedral angle between the planes of the two phenyl rings is close to 80°. A di-*O*-substituted retrodihydrochalcone obtained from the red-colored resin of *Dracaena cinnabari*, a tree endemic to the island of Socotra, was characterized as 4-hydroxy-2-methoxydihydrochalcone (**251**).[220] This pattern of substitution has not been reported previously. The resin, which is more commonly known as dragon's blood, can also be obtained from *D. draco*, the source of 4′,2,4-trihydroxydihydrochalcone (**256**).[224] Three further retrodihydrochalcones obtained from species of *Dracaena* are **257, 265**, and **266** (Table 16.8). One of these (**266**), a constituent of the stem bark of *D. loureiri*, showed estrogenic activity comparable to that of the isoflavones daidzein and genistein.[230] It is interesting to note that most of the retro-dihydrochalcones cited in the pre-1992 literature were also found in *Dracaena*.[10] However, compounds **258** and **268**, which are examples of the (2′,3′,4′,6′)-*O*-substitution pattern and its retrodihydrochalcone equivalent, respectively, were both obtained from species of *Uvaria* (Annonaceae).[226,231] Lusianin (**260**) and two known chalcones isolated from the orchid *Lusia volucris* are cited by Majumder et al.[228] as the first examples of this class of compound to be

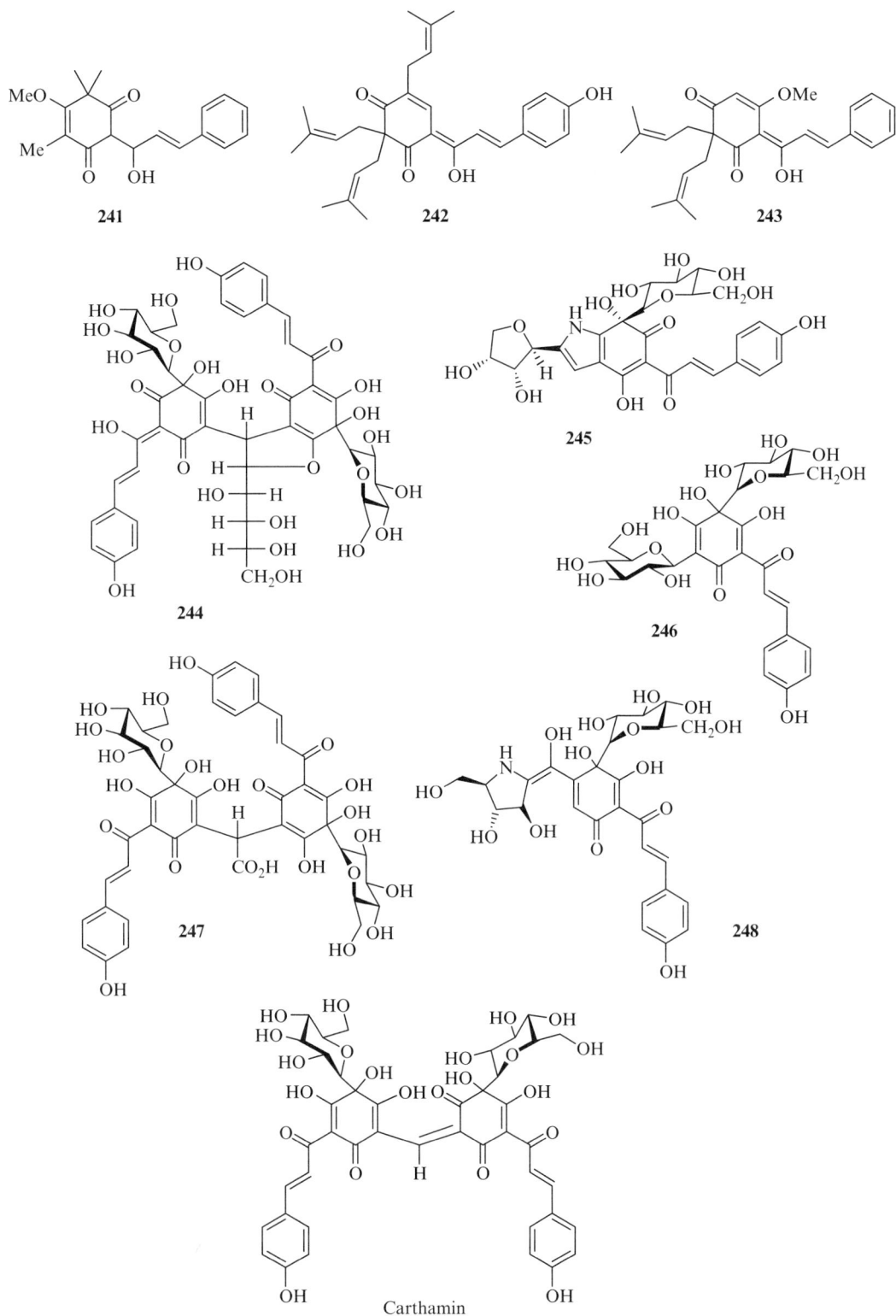

FIGURE 16.13 Quinochalcones and quinochalcone glycosides (see Table 16.7).

TABLE 16.8
New Dihydrochalcones with Simple Patterns of *O*-Substitution Reported in the Literature from 1992 to 2003

No.	OH	OMe	Other	Mol. Formula	Trivial Name	Source	Family	Ref.
Di *O*-substituted	(2′,4′)							
249	2′	4′	—	$C_{16}H_{16}O_3$		*Empetrum nigrum*	Empetraceae	219
250	4′	2′	—	$C_{16}H_{16}O_3$		*Empetrum nigrum*	Empetraceae	219
	(2,4)							
251	4	2	—	$C_{16}H_{16}O_3$		*Dracaena cinnabari*	Dracaenaceae	220
Tri *O*-substituted	(2′,4′,6′)							
252	2′,4′,6′	—	3′-Me	$C_{16}H_{16}O_4$		*Leptospermum recurvum*	Myrtaceae	221
253	2′,4′	6′	3′-Me	$C_{17}H_{18}O_4$	Myrigalone H	*Myrica gale*	Myricaceae	222
254	2′,6′	4′	3′-Me	$C_{17}H_{18}O_4$	Myrigalone G	*Myrica gale*	Myricaceae	222
	(2′,4′,4)							
255	4	2′,4′	—	$C_{17}H_{18}O_4$		*Crinum bulbispermum*	Amaryllidaceae	223
	(4′,2,4)							
256	4′,2,4	—	—	$C_{15}H_{14}O_4$		*Dracaena draco*	Dracaenaceae	224
	(4′,2,6)							
257	4′	2,6	—	$C_{17}H_{18}O_4$		*Dracaena cochinchinensis*	Dracaenaceae	225
Tetra *O*-substituted	(2′,3′,4′,6′)							
258	2′,3′	4′,6′	—	$C_{17}H_{18}O_5$		*Uvaria dulcis*	Annonaceae	226
259	2′	3′,4′,6′	—	$C_{18}H_{20}O_5$		*Fissistigma bracteolatum*	Annonaceae	227
	(2′,3′,4′,4)							
260	2′,4′	3′,4	—	$C_{17}H_{18}O_5$	Lusianin	*Lusia volucris*	Orchidaceae	228
	(2′,4′,6′,4)							
261	2′,4	4′,6′	—	$C_{17}H_{18}O_5$		*Iryanthera lancifolia*	Myristicaceae	229
262	2′	4′,6′,4	—	$C_{18}H_{20}O_5$		*Goniothalamus gardneri*	Annonaceae	160

No.	OH	OMe	OCH₂O	Name	Formula	Species	Family	Ref.
	(2',4',3,4)							
263	2',3,4	4'	—		$C_{16}H_{16}O_5$	*Calythropsis aurea*	Myrtaceae	39
264	—	2',4'	3,4-OCH₂O-	Dihydrocalythropsin Ponganone VII	$C_{18}H_{18}O_5$	*Pongamia pinnata*	Leguminosae	53
	(4',2,4,6)							
265	4',2	4,6	—		$C_{17}H_{18}O_5$	*Dracaena loureiri*	Dracaenaceae	230
266	4,4	2,6	—		$C_{17}H_{18}O_5$	*Dracaena loureiri*	Dracaenaceae	230
	(2,3,4,6)							
267	2	3,4,6	—		$C_{18}H_{20}O_5$	*Fissistigma bracteolatum*	Annonaceae	227
268	6	2,3,4	—		$C_{18}H_{20}O_5$	*Uvaria mocoli*	Annonaceae	231
Penta *O*-substituted								
	(2',3',4',5',6')							
269	2',5'	3',4',6'	—	Dihydropedicin	$C_{18}H_{20}O_6$	*Fissistigma lanuginosum*	Annonaceae	168
270	3',5'	2',4',6'	—		$C_{18}H_{20}O_6$	*Lindera lucida*	Lauraceae	232
	(2',4',6',3,4)							
271	2',6'	4',3,4	—		$C_{18}H_{20}O_6$	*Pityrogramma tartarea*	Hemionitidaceae	233
272	—	2',4',6'	3,4-OCH₂O-		$C_{19}H_{20}O_6$	*Millettia leucantha*	Leguminosae	45
α-Hydroxy-substituted								
	(2',4',3,4,α)							
273	2,4',3,4,α	—	—		$C_{15}H_{14}O_6$	*Eysenhardtia polystachya*	Leguminosae	234
β-Hydroxy-substituted								
	(2',4',6',4,β)							
274	2',4',4,β	6'	—		$C_{16}H_{16}O_6$	*Vitex leptobotrys*	Verbenaceae	37

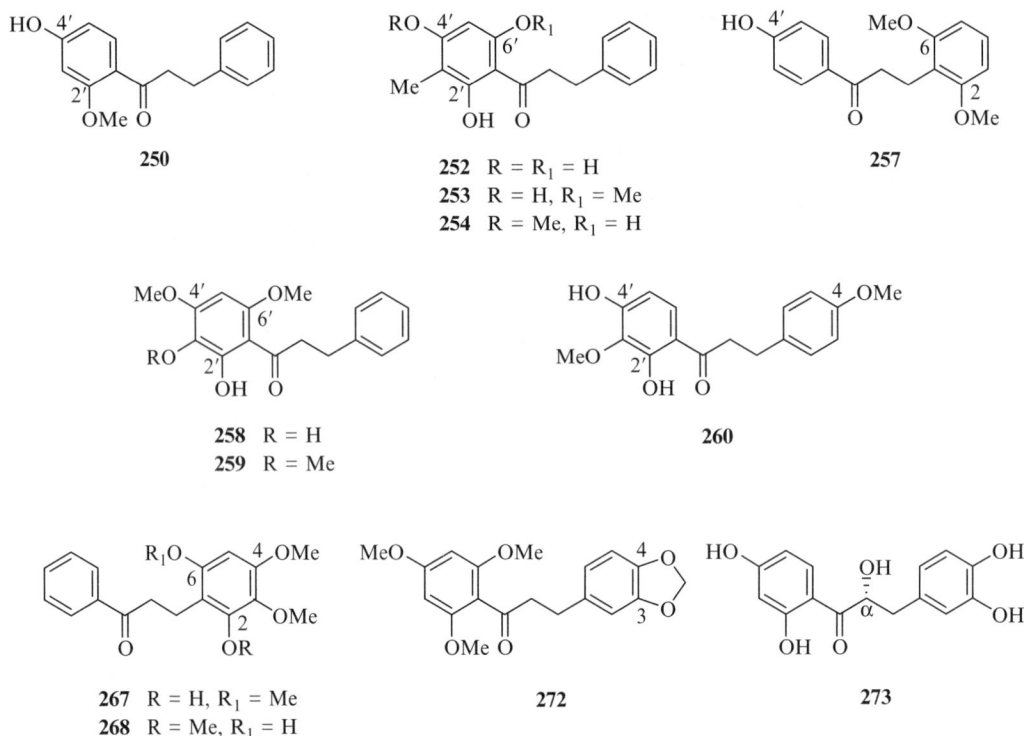

FIGURE 16.14 Dihydrochalcones with simple patterns of *O*-substitution (see Table 16.8).

found in the Orchidaceae. The (2′,3′,4′,4)-*O*-substitution pattern of lusianin has not been found previously in dihydrochalcones although several chalcone equivalents are known from the Leguminosae.[10] Two new phloretin (2′,4′,6′,4-tetrahydroxydihydrochalcone) methyl ethers (**261, 262**) were isolated from extracts of the pericarps of *Iryanthera lancifolia* (Myristicaceae) and the aerial parts of *Goniothalamus gardneri* (Annonaceae), respectively.[160,229] Phloretin 4′,6′-dimethyl ether (**261**) was obtained both as the free dihydrochalcone and as the diastereoisomeric lignan conjugates, iryantherins K (**338**) and L (**339**) (see Table 16.12).[229] The compounds showed greater antioxidant activity than either α-tocopherol or quercetin in an assay based on the inhibition of lipid peroxidation. The phloretin trimethyl ether derivative, **262**, was previously known as a synthetic product.[160] Moderate antiherpes simplex virus activity has been measured for 2′,4′,6′-trimethoxy-3,4-methylenedioxydihydrochalcone (**272**) and the known *O*-methyldihydromilletenone. These compounds were obtained from the stem bark of *Millettia leucantha* together with the chalcones **17** and **26** (Table 16.1).[45]

Myrigalones G (**254**) and H (**253**), constituents of the fruits of *Myrica gale* (Myricaceae),[222] are two examples of a small group of *C*-methylated dihydrochalcones all of which show 3′-methyl substitution of the A-ring (Table 16.8). Myrigalone G (**254**) was also obtained together with 2′,4′,6′-trihydroxy-3′-methyldihydrochalcone (**252**) from the dried foliage of *Leptospermum recurvum* (Myrtaceae),[221] a plant endemic to Mt. Kinabalu in Sabah, Malaysia. In assays for antiviral activity (herpes-simplex virus) only myrigalone G (**254**) yielded positive results; however, **252** showed both cytotoxicity and antimicrobial activity. Synthesis of **252** by a standard three-step procedure from phloroglucinol yielded 2′,4′,6′-trihydroxy-3′-formyldihydrochalcone as an intermediate and precursor to the final product.[221] This formyl derivative was reported previously as a natural product from *Psidium acutangulum*

(Myrtaceae),[237] from which the related 2',4',6'-trihydroxy-3'-formyl-5'-methyldihydrochalcone was also obtained.[238] The NMR spectra of the synthetic formyl derivative show evidence for exchange broadening, which was attributed to conformational exchange between two stable hydrogen-bonded rotamers from molecular mechanics and *ab initio* calculations.[221]

16.3.1.2 α- and β-Hydroxydihydrochalcones

The α-hydroxydihydrochalcones are a small subclass of compounds found mainly in the Leguminosae. To their number can be added the new derivative **273**, a constituent of extracts of the bark and wood of *Eysenhardtia polystachya*.[234] This species is already known as the source of the α-hydroxydihydrochalcone *C*-glucosides, coatline A and B.[239] Comparison of the circular dichroism spectrum of **273** with that of related compounds allowed the configuration at C-α to be deduced as *R* (see Figure 16.14). Most known examples of β-hydroxydihydrochalcones are also constituents of the Leguminosae, thus the isolation of 2',4',4, β-tetrahydroxy-6'-methoxydihydrochalcone (**274**) from the aerial parts of *Vitex leptobotrys* represents the first report of this type of compound in the Verbenaceae.[37] Nel et al. have reported an improved method for the enantioselective synthesis of β-hydroxydihydrochalcones.[312]

16.3.2 ISOPRENYLATED DIHYDROCHALCONES

A comparison of Table 16.2 and Table 16.9 indicates that the number of new isoprenylated dihydrochalcones reported in the literature from 1992 to 2003 is far fewer than the equivalent number of chalcones; nevertheless, 28 examples are listed in the latter. Some of the most interesting compounds are illustrated in Figure 16.15. Many of the dihydrochalcones are characterized by prenyl, geranyl, furano, and pyrano substituents, which are unexceptional in structural terms. However, a group of isoprenylated (2',4',6')-*O*-substituted dihydrochalcones (**275–285**) found in the Annonaceae and Piperaceae calls for special comment. Two derivatives (**275, 276**)[240] isolated from extracts of the stem bark of *Mitrella kentii* (Annonaceae) were characterized as enantiomers of the known compounds, neolinderatin and linderatin, obtained previously from *Lindera umbellata* (Lauraceae).[251] For example, the ^{1}H and ^{13}C NMR spectra of **276** were superimposable with those of (+)-neolinderatin, but the optical rotation values of the compounds were of opposite sign and their circular dichroism spectra showed an inverse relationship.[240] Compounds **275** and **276** were, therefore, assigned as (−)-linderatin and (−)-neolinderatin, respectively. Both are unusual dihydrochalcone derivatives substituted by *p*-menthenyl groups at either C-3' (**275**) or C-3' and C-5' (**276**) of the A-ring. *In vitro* testing of each enantiomer of the two pairs against a non-small-cell bronchopulmonary lung carcinoma cell line uncovered promising activity for (−)-linderatin ($IC_{50} = 3.8 \, \mu g/ml$), whereas the other enantiomers proved to be inactive ($IC_{50} > 30 \, \mu g/ml$).[240] A compound isolated from leaves of *Piper aduncum* (Piperaceae) with identical physical and spectroscopic data to (+)-methyllinderatin,[252] but with the opposite sign of optical rotation, was characterized as (−)-methyllinderatin (**277**).[241] Five 6'-*O*-linked *p*-menthenyl derivatives of 2',6'-dihydroxy-4'-methoxydihydrochalcone, aductins A–E (**281–285**), were also obtained from the same source.[241] The *p*-menthenyl substituent is cyclized to the dihydrochalcone A-ring through an additional C–C bond at C-5' in all but adunctin A (**281**). Adunctins B–D (**282–284**) and (−)-methyllinderatin (**277**) showed antibacterial activity toward *Micrococcus luteus*. Cytotoxicity toward a KB nasopharyngeal carcinoma cell line was shown by (−)-methyllinderatin (**277**), whereas adunctins A–E were inactive (**281–285**).[241] Two isoprenylated derivatives of 2',6'-dihydroxy-4'-methoxydihydrochalcone with additional methyl *p*-hydroxybenzoate substitution were obtained by bioassay-guided fractionation in a later

TABLE 16.9
New Isoprenylated Dihydrochalcones Reported in the Literature from 1992 to 2003

No.	O-Substituents (2',4',6')	Other Substituents	Mol. Formula	Trivial Name	Source	Family	Ref.
Tri O-substituted							
(2',4',6')							
275	2',4',6'-triOH	3'-C$_{10}$	C$_{25}$H$_{30}$O$_4$	(−)-Linderatin	*Mitrella kentii*	Annonaceae	240
276	2',4',6'-triOH	3',5'-diC$_{10}$	C$_{35}$H$_{46}$O$_4$	(−)-Neolinderatin	*Mitrella kentii*	Annonaceae	240
277	2',6'-diOH, 4'-OMe	3'-C$_{10}$	C$_{26}$H$_{32}$O$_4$	(−)-Methyllinderatin	*Piper aduncum*	Piperaceae	241
278	2',6'-diOH, 4'-OMe	5'-(1''-Aryl)prenyl	C$_{29}$H$_{30}$O$_7$	Piperaduncin A	*Piper aduncum*	Piperaceae	242
279	2'-OH, 4'-OMe	5''-Arylfurano[2'',3'':6,5]	C$_{26}$H$_{22}$O$_7$	Longicaudatin	*Piper longicaudatum*	Piperaceae	243
280	2'-OH, 4'-OMe	4''-Aryl-5''-(2-hydroxyisopropyl) dihydrofurano[2'',3'':6,5']	C$_{29}$H$_{30}$O$_8$	Piperaduncin B	*Piper aduncum*	Piperaceae	242
281	2'-OH, 4'-OMe, 6'-O-C$_{10}$	—	C$_{26}$H$_{32}$O$_4$	Adunctin A	*Piper aduncum*	Piperaceae	241
282	2'-OH, 4'-OMe	[5',6']-C$_{10}$	C$_{26}$H$_{30}$O$_4$	Adunctin B	*Piper aduncum*	Piperaceae	241
283	2'-OH, 4'-OMe	[5',6']-C$_{10}$	C$_{26}$H$_{30}$O$_4$	Adunctin C	*Piper aduncum*	Piperaceae	241
284	2'-OH, 4'-OMe	[5',6']-C$_{10}$	C$_{26}$H$_{30}$O$_4$	Adunctin D	*Piper aduncum*	Piperaceae	241
285	2'-OH, 4'-OMe	[5',6']-C$_{10}$	C$_{26}$H$_{32}$O$_5$	Adunctin E	*Piper aduncum*	Piperaceae	241
(2',4,4)							
286	2',4-diOH	6',6''-Dimethylpyrano[2'',3'':4,3]	C$_{20}$H$_{20}$O$_4$	Crotaramosmin	*Crotalaria ramosissima*	Leguminosae	244
287	2'-OH, 4-OMe	6',6''-Dimethylpyrano[2'',3'':4,3]	C$_{21}$H$_{22}$O$_4$	Crotaramin	*Crotalaria ramosissima*	Leguminosae	245
Tetra O-substituted							
(2',3',4',6')							
288	4',6'-diOH, 3'-OMe, 2'-oxo	3'-Prenyl	C$_{21}$H$_{24}$O$_5$		*Helichrysum aphelexiodes*	Asteraceae	66
(2',4',6,4)							
289	2',4',6,4-tetraOH	3,5-Diprenyl	C$_{25}$H$_{30}$O$_5$		*Boronia inconspicua*	Rutaceae	246
290	2',4',6,4-tetraOH	3-Geranyl, 5-prenyl	C$_{30}$H$_{38}$O$_5$		*Boronia inconspicua*	Rutaceae	246
291	2',4-triOH, 6'-OMe	3'-Prenyl	C$_{21}$H$_{24}$O$_5$	α,β-Dihydroxanthohumol	*Humulus lupulus*	Cannabinaceae	91
(2',4,3,4)							
292	2',3,4-triOH	6',6''-Dimethylpyrano[2'',3'':4,3]	C$_{20}$H$_{20}$O$_5$	Crotin	*Crotalaria ramosissima*	Leguminosae	245
293	2'-OMe, 3,4-OCH$_2$O-	Furano[2'',3'':4,3]	C$_{19}$H$_{16}$O$_5$		*Lonchocarpus subglaucescens*	Leguminosae	101
(2',4,4,α)							
294	2',4,4,α-tetraOH	5',3-Diprenyl	C$_{25}$H$_{30}$O$_5$	Kanzonol Y	*Glycyrrhiza glabra*	Leguminosae	247

No.	Substitution	Substituent	Trivial name	Formula	Species	Family	Ref.
	(2',4',4',α) continued						
295	2',4,α-triOH, 4'-geranyloxy	—		$C_{25}H_{30}O_5$	*Millettia usaramensis* ssp. *usaramensis*	Leguminosae	248
	Penta O-substituted						
	(2',4',6',3,4)						
296	2',4,6,3,4-pentaOH	3',5-Diprenyl		$C_{25}H_{30}O_6$	*Esenbeckia grandiflora* ssp. *grandiflora*	Rutaceae	249
297	2',4,6,3,4-pentaOH	3'-Geranyl, 5-prenyl		$C_{30}H_{38}O_6$	*Esenbeckia grandiflora* ssp. *grandiflora*	Rutaceae	249
298	2',4,6,3-tetraOH	3'-Geranyl, 6',6''-dimethylpyrano[2'',3'':4,5]		$C_{30}H_{36}O_6$	*Esenbeckia grandiflora* ssp. *grandiflora*	Rutaceae	249
299	2',4,6,3-tetraOH, 4-OMe	3',5-Diprenyl		$C_{26}H_{32}O_6$	*Metrodorea nigra*	Rutaceae	250
300	2',6',3-triOH, 4-OMe	5-Prenyl, 6'',6''-dimethyl-4'',5''-dihydropyrano[2'',3'':4,3']		$C_{26}H_{32}O_6$	*Metrodorea nigra*	Rutaceae	250
	(2',4',3,4,β)						
301	2',β-diOMe, 3,4-OCH₂O-	Furano[2'',3'':4',3']	Ponganone IX	$C_{20}H_{18}O_6$	*Pongamia pinnata*	Leguminosae	53
	Hexa O-substituted						
	(2',4',5',3,4,β)						
302	2',5',β-triOMe, 3,4-OCH₂O-	6'',6''-Dimethylpyrano[2'',3'':4',3']	Ponganone VIII	$C_{24}H_{26}O_7$	*Pongamia pinnata*	Leguminosae	53

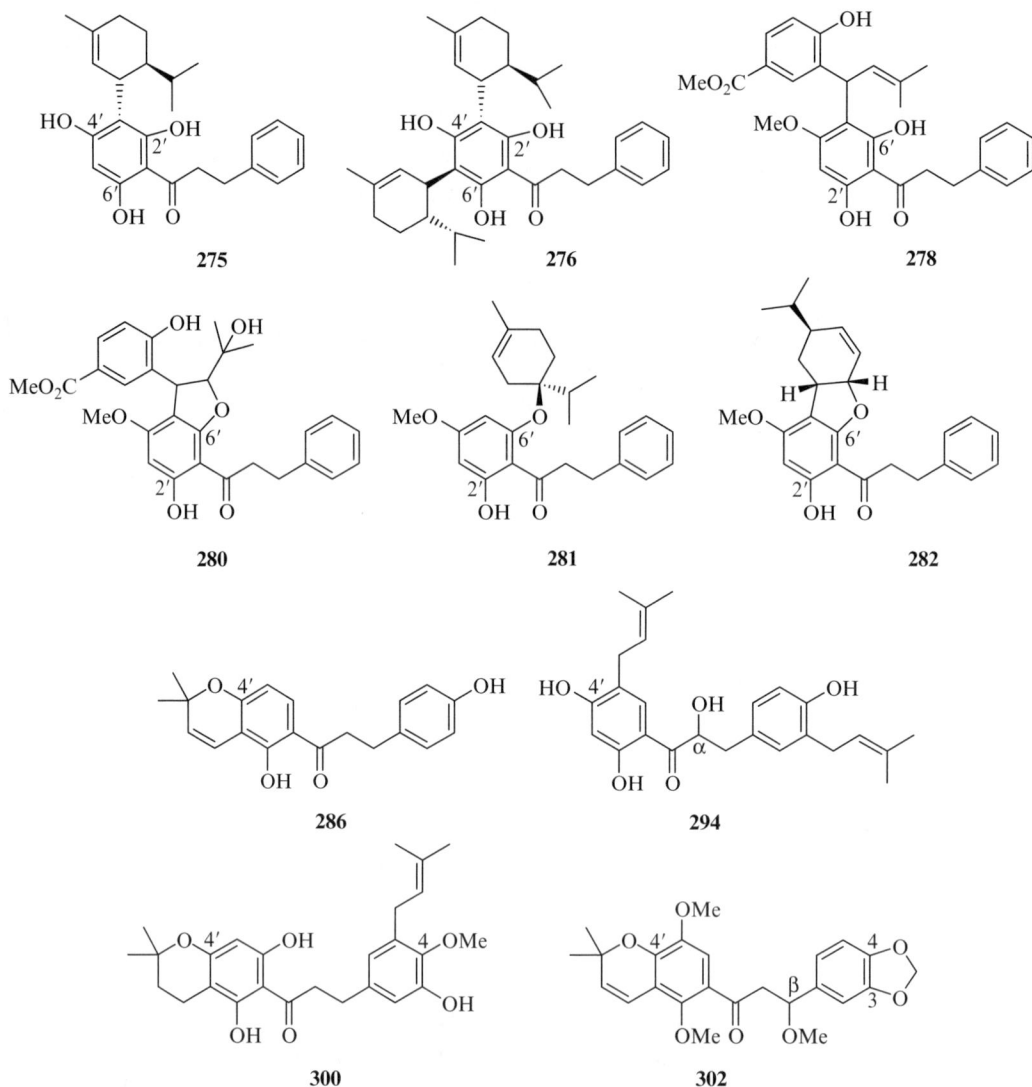

FIGURE 16.15 Isoprenylated dihydrochalcones (see Table 16.9).

study of *P. aduncum* constituents.[242] The compounds piperaduncins A (**278**) and B (**280**) showed antibacterial activity toward both *Bacillus subtilis* and *Micrococcus luteus* and were cytotoxic toward a KB nasopharyngeal carcinoma cell line. Bioassay-guided fractionation of an extract of dried leaves and twigs of *Piper longicaudatum* (Piperaceae) gave piperaduncin B (**280**) and a new constituent, longicaudatin (**279**), a furanodihydrochalcone with the furan ring substituted by methyl *p*-hydroxybenzoate.[243]

Rao et al. revised the structure of crotaramosmin (**286**), formerly described as a prenyl-flavanone, to 2′,4-dihydroxy-6″,6″-dimethylpyrano[2″,3″:4′,3′]dihydrochalcone.[244] Two derivatives of crotaramosmin, crotaramin (**287**) and crotin (**292**), were later isolated from the same source, *Crotalaria ramosissima* (cited incorrectly as *Crotolaria ramosissima*).[245] Kanzonol Y (**294**) is an uncommon diprenylated α-hydroxydihydrochalcone found in cultivated licorice (*Glycyrrhiza glabra*)[247] for which the absolute configuration was determined by

Mosher's method as $\alpha(R)$.[253] All the new isoprenylated α- and β-hydroxydihydrochalcones reported in Table 16.9 are constituents of the Leguminosae, a general trend observed both in Table 16.8 and the earlier literature.[10] Another apparent trend is the predominance of isoprenylated (2′,4′,6′,3,4)-penta-O-substituted dihydrochalcones in species of the Rutaceae (**296–300**).[249,250] Only one other example of this type of compound has been recorded in the literature, a geranylated derivative from *Helichrysum monticola* (Asteraceae).[254]

16.3.3 DIHYDROCHALCONE GLYCOSIDES

Relatively few dihydrochalcone glycosides have been reported in the literature, with only 11 O-glycosides and 3 C-glycosides published prior to 1992, according to the *Handbook of Natural Flavonoids*.[9,10] The O-glycosides are mainly 2′- and 4′-O-glucosides of simple dihydrochalcones while only one higher glycoside, the incompletely characterized 2′-O-xylosyl-glucoside of phloretin (2′,4′,6′,4-tetrahydroxydihydrochalcone), has been described.[255] Table 16.10 lists new dihydrochalcone glycosides published in the period 1992 to 2003, and some examples are shown in Figure 16.16. Among the new compounds are the first acylated dihydrochalcone glycosides, including two acetylated glucosides of phloretin (**306**, **307**).[258,259] Of particular interest are thonningianins A and B (**303**, **304**), unusual galloylated glucosides of 2′,4′,6′-trihydroxydihydrochalcone isolated from the roots of the African medicinal herb *Thonningia sanguinea* (Balanophoraceae).[256] Both structures feature a 4′-O-glucopyranosyl moiety acylated at 4-OH and 6-OH by a C–C linked digallic acid. The compounds are effective scavengers of the DPPH radical, according to analysis by EPR spectroscopy.[256] In contrast, salicifolioside A, a 4′-O-gentobioside of 2′,4′-dihydroxy-3′,6′-dimethoxydihydroxy-chalcone isolated from aerial parts of *Polygonum salicifolium* (Polygonaceae), was not active in the DPPH assay.[257] The β-Glc-(1 → 6)-β-Glc interglycosidic linkage of this compound was confirmed from HMBC data.

The first di-C-glycoside of a dihydrochalcone has been reported as a characteristic constituent of species of *Fortunella* (Rutaceae).[260] The compound, phloretin 3′,5′-di-C-glucoside (**308**), accumulates in the fruits and leaves. Only two of 28 species of *Citrus* (*C. halimii* and *C. madurensis*) surveyed contain this glycoside and it was absent from *Poncirus trifoliata*. In contrast, large amounts were detected in many *Fortunella–Citrus* hybrids. The authors suggest that accumulation of phloretin 3′,5′-di-C-glucoside is a generic trait of *Fortunella*, and that inheritance of the trait among intergeneric hybrids is under the control of a dominant allele.[260] A 3′-C-xyloside of 2′,4′,3,4,$\alpha(R)$-pentahydroxydihydrochalcone (**314**), the corresponding aglycone (**273**), and a compound described as the 3′-O-xyloside of the same dihydrochalcone were obtained from *Eysenhardtia polystachya* (Leguminosae).[234] However, the NMR data presented for the latter compound do not support its identification as an O-glycoside. Two 3′-C-glucosides of α-hydroxydihydrochalcones were isolated previously from *E. polystachya* (coatline A and B)[239] and C-glucosides of both α-hydroxy- (coatline A)[266] and β-hydroxydihydrochalcones have been reported from *Pterocarpus marsupium* (Leguminosae).[267,268]

16.3.4 DIHYDROCHALCONE DIMERS

The only example of a dihydrochalcone dimer reported in the literature prior to 1992 is brackenin, a Cα–Cα linked dimer of davidigenin from *Brackenridgea zanguebarica* (Ochnaceae).[269] Several new examples of this rare class of dihydrochalcones cited between 1992 and 2003 are shown in Table 16.11 and Figure 16.17 (note that the "mixed" chalcone–dihydrochalcone dimers **142** and **148–150** are described in Table 16.4 and Section 16.2.4). The leaves and inflorescences of *Iryanthera sagotiana* (Myristicaceae) yielded a dimer (**317**) comprising

TABLE 16.10
New Dihydrochalcone Glycosides Reported from 1992 to 2003

No.	Compound	Mol. Formula	Source	Family	Ref.
	2′,4′,6′-Trihydroxydihydrochalcone				
303	4′-O-(3″-O-Galloyl-4″,6″-O,O-hexahydroxydiphenoylglucoside) (thonningianin A)	$C_{42}H_{34}O_{21}$	*Thonningia sanguinea*	Balanophoraceae	256
304	4′-O-(4″,6″-O,O-Hexahydroxydiphenoylglucoside) (thonningianin B)	$C_{35}H_{30}O_{17}$	*Thonningia sanguinea*	Balanophoraceae	256
	2′,4′-Dihydroxy-3′,6′-dimethoxydihydrochalcone				
305	4′-O-Glucosyl-(1‴ → 6″)glucoside (salicifolioside A)	$C_{29}H_{38}O_{15}$	*Polygonum salicifolium*	Polygonaceae	257
	2′,4′,6′,4-Tetrahydroxydihydrochalcone (phloretin)				
306	2′-O-(6″-O-Acetylglucoside)	$C_{23}H_{26}O_{11}$	*Loiseleuria procumbens*	Ericaceae	258
307	4′-O-(2″-O-Acetylglucoside)	$C_{23}H_{26}O_{11}$	*Lithocarpus pachyphyllus*	Fagaceae	259
308	3′,5′-Di-C-glucoside	$C_{27}H_{34}O_{15}$	*Fortunella* spp.	Rutaceae	260
	2′,4′,6′-Trihydroxy-4-methoxydihydrochalcone				
309	2′-O-Glucoside	$C_{22}H_{26}O_{10}$	*Iryanthera sagotiana*	Myristicaceae	261
	2′,4-Dihydroxy-4′,6′-diacetoxydihydrochalcone				
310	2′-O-Glucoside (zosterin)	$C_{25}H_{28}O_{12}$	*Zostera* sp.	Zosteraceae	262
	4-Hydroxy-2′,4′,6′-trimethoxydihydrochalcone				
311	4-O-Glucoside (bidenoside B)	$C_{24}H_{30}O_{10}$	*Bidens bipinnata*	Asteraceae	263
	2′,4′,4,α-Tetrahydroxydihydrochalcone				
312	α-O-Glucoside (licoagroside F)	$C_{21}H_{24}O_{10}$	*Glycyrrhiza pallidiflora*	Leguminosae	264
	2′,4′,4,β-Tetrahydroxydihydrochalcone				
313	2′-O-Glucoside (rocymosin B)	$C_{21}H_{24}O_{10}$	*Rosa cymosa*	Rosaceae	265
	2′,4′,3,4,α-Pentahydroxydihydrochalcone				
314	3′-C-Xyloside	$C_{20}H_{22}O_{10}$	*Eysenhardtia polystachya*	Leguminosae	234

two molecules of 2′,4′,6′-trihydroxy-4-methoxydihydrochalcone linked by a C–C bond between C-3′ of each A-ring.[261] Cycloaltilisin 6 (**319**) is the only known example of an isoprenylated dihydrochalcone dimer and a constituent of the bud covers of *Artocarpus altilis* (Moraceae).[272] This compound, which comprises two molecules of 2′,4′,3,4-tetrahydroxy-2′-geranyldihydrochalcone linked by a C–C bond between the B-rings (Figure 16.17), is a potent inhibitor of cathepsin K, a novel cysteine protease implicated in osteoporosis. A novel feature of the structure of piperaduncin C (**318**) is a methylene bridge connecting C-3′ of the A-rings of two molecules of 2′,6′-dihydroxy-4′-methoxydihydrochalcone.[242] Littorachalcone (**315**)[270] and verbenachalcone (**316**)[271] are the first examples of ether-linked dihydrochalcone dimers and were isolated from the aerial parts of *Verbena littoralis* (Verbenaceae). Both compounds enhanced the effect of neural growth factor on stimulating neurite outgrowth in PC12D cells, with littorachalcone (**315**) showing greater potency. A characteristic structural feature of both littorachalcone (**315**) and verbenachalcone (**316**) is an ether bridge between the

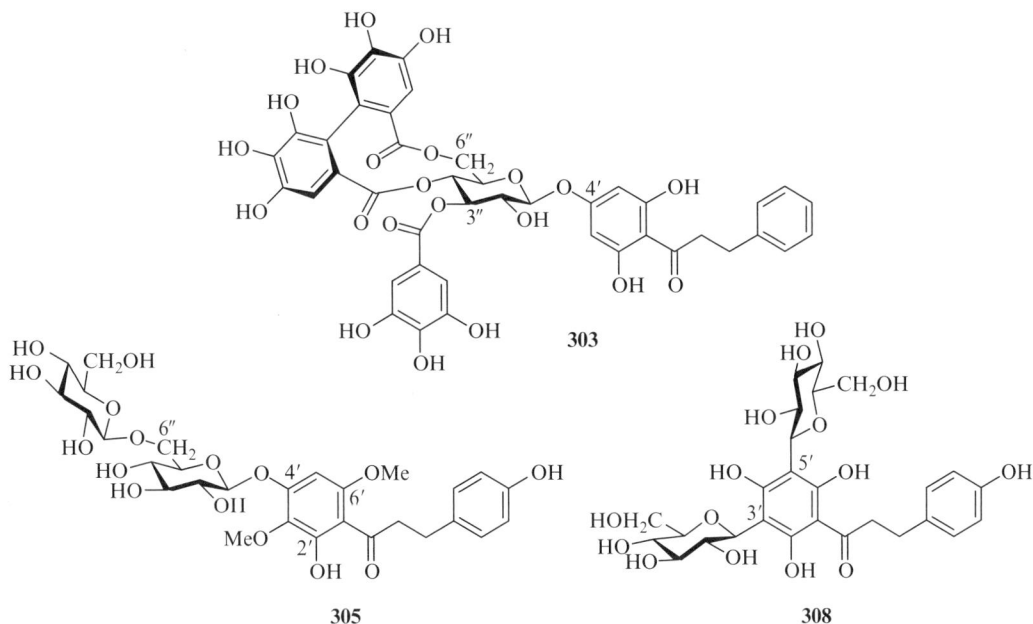

FIGURE 16.16 Dihydrochalcone glycosides (see Table 16.10).

B-rings of their constituent dihydrochalcone monomers. A concise synthesis of verbenachalcone (**316**) by catalytic copper-mediated coupling of phenol and aryl halides has been reported by Xing et al., who also prepared two further derivatives for preliminary structure–activity studies.[277] One of the latter, the corresponding bichalcane (deoxo) derivative of verbenachalcone showed no activity in the neural outgrowth stimulation bioassay mentioned above.

A resin from *Dracaena cinnabari* known as "dragon's blood," mentioned previously as the source of the dihydrochalcone **251** (Section 16.3.1), also yielded cinnabarone (**321**), a dihydrochalcone linked by a C–C bond from C-5 of its B-ring to the deoxo-carbon of a deoxotetrahydrochalcone.[220] The closely related compound cochinchinenin (**320**) is characterized by the same linkage, and was isolated from "Chinese dragon's blood," the resin of *D. cochinchinensis*.[273] It is interesting to note that both cochinchinenin and cinnabarone are dimers derived from retrodihydrochalcones, given the prevalence of this type of compound in species of *Dracaena* (Table 16.8).

Trianguletin (**322**) is a novel dihydrochalcone–flavonol dimer in which the A-rings of the two flavonoids are linked by a methylene bridge similar to that of piperaduncin C (**318**).[274] In the case of trianguletin, this bridge connects C-3′ of 2′,6′-dihydroxy-4′-methoxy-5′-methyldihydrochalcone with C-8 of 3,5,7-trihydroxy-4′-methoxyflavone (kaempferide). The structure of this compound, which is a constituent of the farinose exudate of the fronds of the fern, *Pentagramma triangularis* ssp. *triangularis*, was confirmed by x-ray crystallography.[274] Four structurally related dimers also obtained from farinose exudates of *P. triangularis* comprise slightly different combinations of dihydrochalcones and flavonols; for example, trianguletin "B" (**323**) contains 5,7-dihydroxy-3-methoxyflavone (galangin 3-methyl ether) instead of kaempferide.[275] Similarly, trianguletins "C" to "E" (**324–326**) comprise 2′,6′,4-trihydroxy-4′-methoxy-5′-methyldihydrochalcone with kaempferide, ermanin (kaempferol 3,4′-dimethyl ether), and galangin 3-methyl ether, respectively.[276]

TABLE 16.11
Dihydrochalcone Dimers and Heterodimers Reported from 1992 to 2003

No.	Compound	Mol. Formula	Source	Family	Ref.
	Dihydrochalcone–dihydrochalcone				
315	Littorachalcone	$C_{30}H_{26}O_8$	*Verbena littoralis*	Verbenaceae	270
316	Verbenachalcone	$C_{31}H_{28}O_9$	*Verbena littoralis*	Verbenaceae	271
317	3',3'-Bis(2',4',6'-trihydroxy-4-methoxydihydrochalcone)	$C_{32}H_{30}O_{10}$	*Iryanthera sagotiana*	Myristicaceae	261
318	Piperaduncin C	$C_{33}H_{32}O_8$	*Piper aduncum*	Piperaceae	242
319	Cycloaltilisin 6	$C_{50}H_{58}O_{10}$	*Artocarpus altilis*	Moraceae	272
	Dihydrochalcone–deoxotetrahydrochalcone				
320	Cochinchinenin	$C_{31}H_{30}O_7$	*Dracaena cochinchinensis*	Liliaceae	273
321	Cinnabarone	$C_{32}H_{32}O_7$	*Dracaena cinnabari*	Liliaceae	220
	Dihydrochalcone–flavonol				
322	Trianguletin	$C_{34}H_{30}O_{10}$	*Pentagramma triangularis* ssp. *triangularis*	Adiantaceae	274
323	Trianguletin "B"	$C_{34}H_{30}O_9$	*Pentagramma triangularis*	Adiantaceae	275
324	Trianguletin "C"	$C_{34}H_{30}O_{11}$	*Pentagramma triangularis*	Adiantaceae	276
325	Trianguletin "D"	$C_{35}H_{32}O_{11}$	*Pentagramma triangularis*	Adiantaceae	276
326	Trianguletin "E"	$C_{34}H_{30}O_{10}$	*Pentagramma triangularis*	Adiantaceae	276

16.3.5 BENZYLATED DIHYDROCHALCONES

New reports of *C*-benzylated dihydrochalcones are summarized in Table 16.12. Examples of this unique group of dihydrochalcones were known previously only from *Xylopia africana* and species of *Uvaria* (Annonaceae),[10,278] and these taxa continue to be the only recorded sources of the compounds. Those listed in Table 16.12 were obtained from either the root bark of *Uvaria leptocladon* or the dried roots of *Xylopia africana*, and are based on uvangoletin (2',4'-dihydroxy-6'-methoxydihydrochalcone), with the exception of **327**, which is a derivative of 2',6'-dihydroxy-4'-methoxydihydrochalcone.[279] Some examples of this group of dihydrochalcones are shown in Figure 16.18. The 2-hydroxybenzyl substituents (2-OHBn) can be at either C-3' or C-5' of the A-ring or at both positions. Many of the compounds have chains of 2-hydroxybenzyl unit linked successively from C-5_n to C-1_{n+1}, which may also incorporate a C-3_n to C-1_{n+1} linkage. For example, compounds **332** and **333** feature chains of four 2-hydroxybenzyl groups as substituents to the dihydrochalcone A-ring.[283] Preliminary data on the antibacterial activity of some of the compounds (**330–333**) have been published.[282,283]

16.3.6 DIHYDROCHALCONE–LIGNAN CONJUGATES

Iryantherins G–L (**334–339**) are flavonolignans thought to arise from the oxidative coupling of dihydrochalcones and lignans (Figure 16.18). They represent further examples of a series of compounds reported exclusively from species of *Iryanthera* (Myristicaceae). The diastereo-isomeric pairs iryantherins G and H (**334, 335**) and I and J (**336, 337**) were isolated from the fruits of *I. grandis*, a known source of dihydrochalcones and lignans.[284] The relative configurations of the lignan components of iryantherins G–J were deduced from NOE

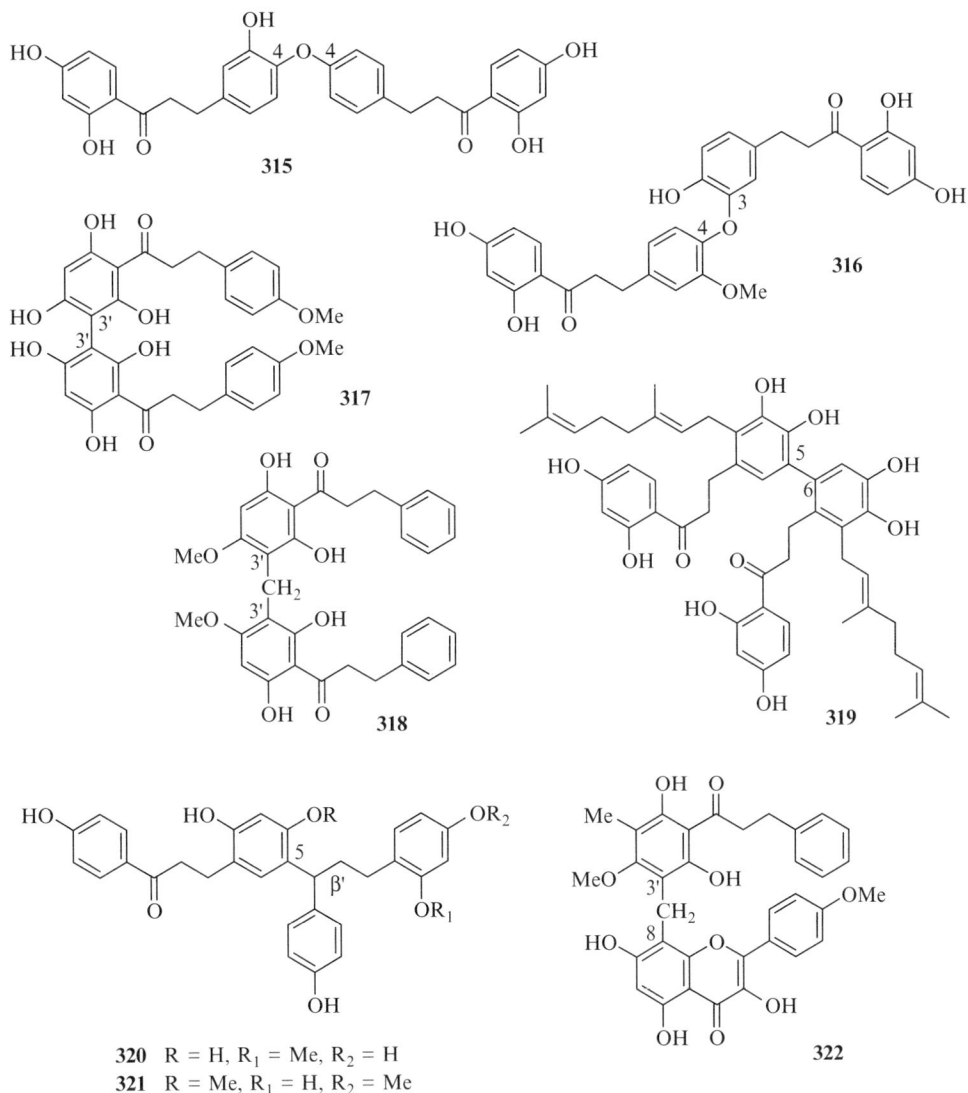

FIGURE 16.17 Dihydrochalcone dimers and heterodimers (see Table 16.11).

measurements and coupling constant data. The lignans are linked to C-3′ of the A-rings of either 2′,4′,6′-trihydroxy-4-methoxy- (G and H) or 2′,4′,6′-trihydroxy-3,4-methylenedioxydi-hydrochalcone (I and J). An additional pair of flavonolignan diastereoisomers (iryantherins K and L) was isolated subsequently by Silva et al. from the pericarps of *I. lancifolia*.[229] Both compounds showed moderate antioxidant activity. Iryantherin L (**339**) appears to be syn-onymous with iryantherin B, a compound found previously in both *I. laevis* and *I. ulei* for which the relative stereochemistry was not defined.[285] Iryantherins K (**338**) and L (**339**) are also known as constituents of the stem bark of *I. megistophylla*, and were assayed for antibacterial, antifungal, antiviral, and antiacetylcholinesterase activity.[286] Iryantherin K (**338**) showed high levels of inhibition against potato virus, but only moderate inhibition of acetylcholinesterase.

TABLE 16.12
Miscellaneous Dihydrochalcones Reported from 1992 to 2003

No.	Compound	Mol. Formula	Source	Family	Ref.
	***C*-Benzylated Dihydrochalcones**[a]				
327	2′,6′-DiOH, 4′-OMe, 3′-(2-OHBn)	$C_{23}H_{22}O_5$	*Xylopia africana*	Annonaceae	279
328	2′,4′-DiOH, 6′-OMe, 3′-(2-OHBn), 5′-(2 × 2-OHBn) (triuvaretin)	$C_{37}H_{34}O_7$	*Uvaria leptocladon*	Annonaceae	280
			Xylopia africana	Annonaceae	281
329	2′,4′-DiOH, 6′-OMe, 3′-(2 × 2-OHBn), 5′-(2-OHBn) (isotriuvaretin)	$C_{37}H_{34}O_7$	*Uvaria leptocladon*	Annonaceae	280
			Xylopia africana	Annonaceae	281
330	2′,4′-DiOH, 6′-OMe, 3′-(2-OHBn), 5′-(3 × 2-OHBn)	$C_{44}H_{40}O_8$	*Xylopia africana*	Annonaceae	282
331	2′,4′-DiOH, 6′-OMe, 3′-(3 × 2-OHBn), 5′-(2-OHBn)	$C_{44}H_{40}O_8$	*Xylopia africana*	Annonaceae	282
332	2′,4′-DiOH, 6′-OMe, 3′-(2-OHBn), 5′-(4 × 2-OHBn)	$C_{51}H_{46}O_9$	*Xylopia africana*	Annonaceae	283
333	2′,4′-DiOH, 6′-OMe, 3′-(4 × 2-OHBn), 5′-(2-OHBn)	$C_{51}H_{46}O_9$	*Xylopia africana*	Annonaceae	283
	Dihydrochalcone–lignans				
334	Iryantherin G	$C_{34}H_{36}O_7$	*Iryanthera grandis*	Myristicaeae	284
335	Iryantherin H	$C_{34}H_{36}O_7$	*Iryanthera grandis*	Myristicaeae	284
336	Iryantherin I	$C_{34}H_{34}O_8$	*Iryanthera grandis*	Myristicaeae	284
337	Iryantherin J	$C_{34}H_{34}O_8$	*Iryanthera grandis*	Myristicaeae	284
338	Iryantherin K	$C_{35}H_{38}O_7$	*Iryanthera lancifolia (I. ulei)*	Myristicaeae	229
339	Iryantherin L	$C_{35}H_{38}O_7$	*Iryanthera lancifolia (I. ulei)*	Myristicaeae	229

[a] Bn, benzyl.

16.4 AURONES

16.4.1 AURONES AND AURONOLS

The number of new aurone and auronol aglycones reported in the literature between 1992 and 2003 is relatively small, as Table 16.13 indicates. Some examples of these compounds are illustrated in Figure 16.19. Only six of the compounds listed in Table 16.13 represent new combinations of hydroxy and methoxy substituents. Among these are two examples from Asteraceae taxa, 5-hydroxy-4,6,4′-trimethoxyaurone (**342**) from the flowers of cultivated sunflowers (*Helianthus annuus*)[288] and 4,6,7,4′-tetrahydroxyaurone (**343**) from *Helminthia echioides* (= *Picris echioides*).[289,290] Hamiltrone (3′,4′-dihydroxy-4,5,6-trimethoxyaurone) (**350**) showed the highest DNA strand-scission activity of several constituents isolated from a combined leaf and stem extract of *Uvaria hamiltonii* (Annonaceae).[48] The structural similarity of this aurone to combretastatin A-4, a tumor vascular targeting agent, prompted Lawrence et al. to develop a total synthesis for the compound.[295] This was achieved in four steps and 37% overall yield from benzofuranone and benzaldehyde precursors and facilitated the preparation of a series of substituted aurone analogs. Synthetic hamiltrone showed only poor cell-growth inhibition against a K562 cell line ($IC_{50} = 12 \mu M$), but the activity of 3′-hydroxy-5,6,7,4′-tetramethoxyaurone was more than 200-fold greater ($IC_{50} = 50$ nM).[295] A general observation was that 5,6,7-trimethoxyaurones showed greater cell-growth inhibition

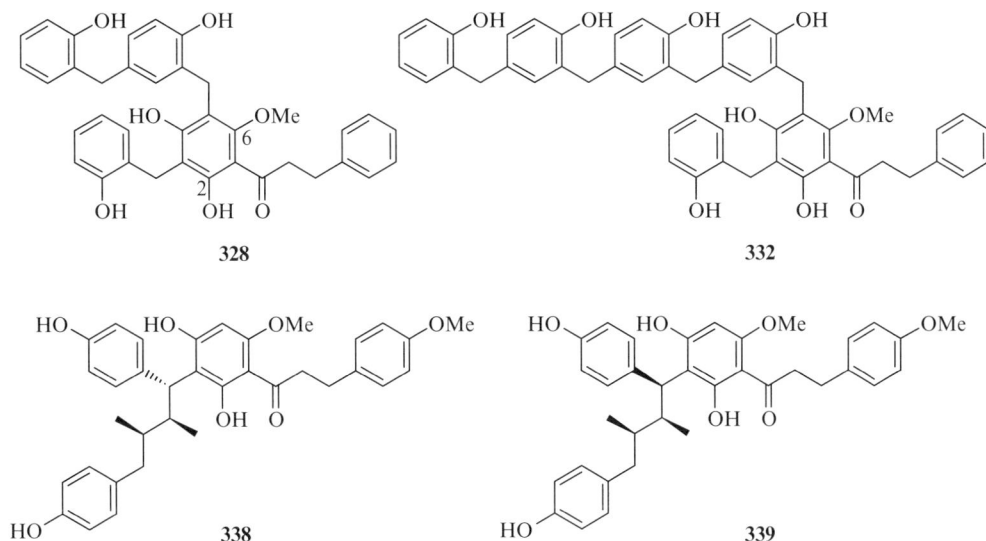

FIGURE 16.18 *C*-Benzylated dihydrochalcones and dihydrochalcone–lignan conjugates (see Table 16.12).

than the corresponding 4,5,6-trimethoxyaurones. The two new auronols from the bark of *Pseudolarix amabilis* (Pinaceae), amaronols A (**351**) and B (**352**), were considered to be enantiomeric pairs because of the reversible hemiketal at C-2.[293] These compounds were inactive in a series of antifungal and antibacterial assays against test organisms.

The ground rhizomes and roots of *Cyperus capitatus* (Cyperaceae) are the only known source of aurone aglycones with *C*-methyl substituents (**344–348**),[291,292] although two aurone glycosides with C-7 methyl groups were described in 1989 from *Pterocarpus marsupium* (Leguminosae).[296,297] Two new examples of isoprenylated aurones based on sulfuretin (6,3′,4′-trihydroxyaurone) have been found in the cortex of *Broussonetia papyrifera* (Moraceae) and hairy root cultures of *Glycyrrhiza glabra* (Leguminosae), and assigned the trivial names of broussoaurone A (**340**)[287] and licoagroaurone, respectively (**341**).[88] Licoagroaurone is the first aurone to be isolated from species of *Glycyrrhiza* (licorice), from which more than 100 flavonoids have already been described.[253] Perhaps the most unusual of the aurones in Table 16.13 are the chlorinated derivatives **353** and **354**, obtained from the marine brown alga *Spatoglossum variabile*.[294] These are the first halogenated aurones to be described from a natural source. The auronol 4′-chloro-2-hydroxyaurone (**354**) was found as a racemic mixture of (*R*)- and (*S*)-epimers at C-2.

16.4.2 Aurone Glycosides

Nine aurone and auronol glycosides reported in the literature between 1992 and 2003 are listed in Table 16.14. Structures of some of these compounds are given in Figure 16.20. Dalmaisione D (**355**) from the roots of *Polygala dalmaisiana* (Polygalaceae) is an unusual example of an aurone glycoside for which the corresponding aglycone, 2′-hydroxyaurone, is unknown.[298] Indeed, no other 2′-*O*-substituted aurones have been reported in the literature. Species from the genus *Bidens* (Asteraceae) are a well-documented source of glucosides and acylated glucosides of maritimetin (6,7,3′,4′-tetrahydroxyaurone)[10] and two new di- and tri-acetylated 6-*O*-glucosides (**358**, **359**) have now been described from the aerial parts of *B. bipinnata* and *B. pilosa* var. *radiata*, respectively.[263,127] Caulesauroneside (**356**), an aurone

TABLE 16.13
New Aurones and Auronols Reported in the Literature from 1992 to 2003

No.	OH	OMe	Other	Mol. Formula	Trivial Name	Source	Family	Ref.
Tri O-substituted								
	(6,3',4')							
340	6,3',4'	—	5-Pr	$C_{20}H_{18}O_5$	Broussoaurone A	*Broussonetia papyrifera*	Moraceae	287
341	6,3',4'	—	7-Pr	$C_{20}H_{18}O_5$	Licoagroaurone	*Glycyrrhiza glabra*	Leguminosae	88
Tetra O-substituted								
	(4,5,6,4')							
342	5	4,6,4'	—	$C_{18}H_{16}O_6$		*Helianthus annuus*	Asteraceae	288
	(4,6,7,4')							
343	4,6,7,4'	—	—	$C_{15}H_{10}O_6$		*Helminthia echioides*	Asteraceae	289, 290
	(4,6,3',4')							
344	4,6,3',4'	—	5-Me	$C_{16}H_{12}O_6$		*Cyperus capitatus*	Cyperaceae	291
345	4,6,3',4'	—	7-Me	$C_{16}H_{12}O_6$		*Cyperus capitatus*	Cyperaceae	291
346	6,3',4'	4	5-Me	$C_{17}H_{14}O_6$		*Cyperus capitatus*	Cyperaceae	291
347	6,3',4'	4	7-Me	$C_{17}H_{14}O_6$		*Cyperus capitatus*	Cyperaceae	291
348	6,3'	4,4'	5-Me	$C_{18}H_{16}O_6$		*Cyperus capitatus*	Cyperaceae	291
349	—	4,6,3',4'	—	$C_{19}H_{18}O_6$		*Cyperus capitatus*	Cyperaceae	292
Penta O-substituted								
	(4,5,6,3',4')							
350	3',4'	4,5,6	—	$C_{18}H_{16}O_7$	Hamiltrone	*Uvaria hamiltonii*	Annonaceae	48
Auronols								
	(2,4,6,3',4',5')							
351	2,4,6,3',4',5'	—	—	$C_{15}H_{12}O_8$	Amaronol A	*Pseudolarix amabilis*	Pinaceae	293
352	2,4,6,3',5'	4'	—	$C_{16}H_{14}O_8$	Amaronol B	*Pseudolarix amabilis*	Pinaceae	293
Halogenated								
353	—	—	4'-Cl	$C_{15}H_9ClO_2$		*Spatoglossum variabile*	Dictyotaceae	294
354	2	—	4'-Cl	$C_{15}H_{11}ClO_3$		*Spatoglossum variabile*	Dictyotaceae	294

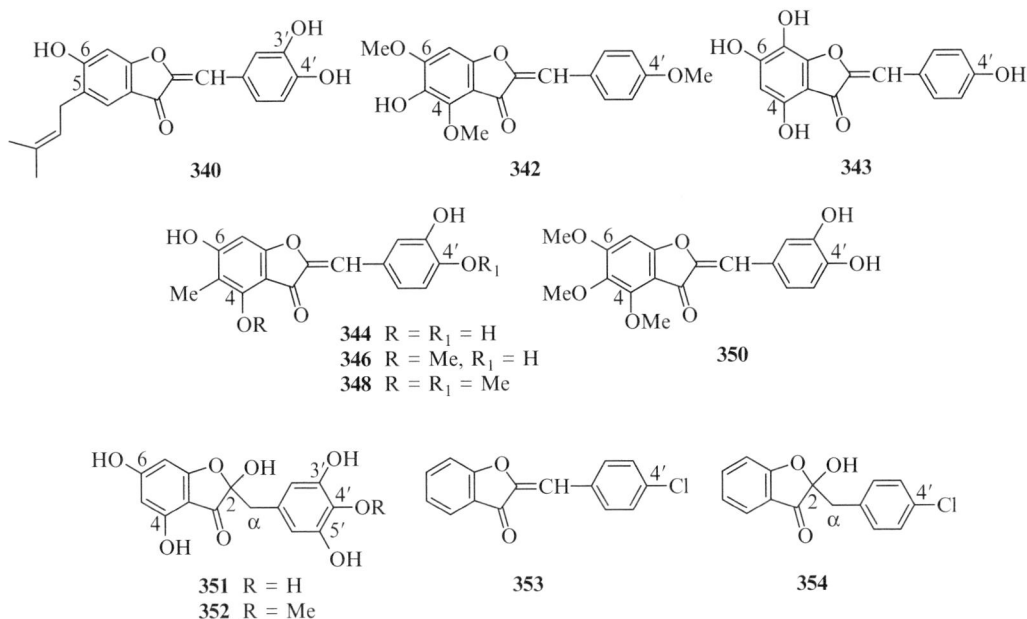

FIGURE 16.19 Aurones and auronols (see Table 16.13).

diglucoside from the roots and rhizomes of *Asarum longirhizomatosum*, is the first aurone to be found in the Aristolochiaceae.[299] A *C*-methylated aurone rhamnoside (360)[300] from the heartwood of *Pterocarpus santalinus* (Leguminosae) is an analog of similar structures described previously from *P. marsupium*.[296,297] The characterization of the glycosidic component of a second aurone glycoside (357) from *P. santalinus* as neohesperidoside must be regarded as preliminary due to the incomplete nature of the supporting spectroscopic data.[300]

The first auronol glycosides to be described are all from genera of the Rhamnaceae and based on maesopsin (2,4,6,4′-tetrahydroxy-2-benzylcoumaranone). Hovetrichosides C (361) and D (363) were obtained from the bark of *Hovenia trichocarea* together with maesopsin and neolignan and phenylpropanoid glycosides.[301] Similarly, maesopsin 6-*O*-glucoside (362) was obtained together with maesopsin from root bark of *Ceanothus americanus*.[302] These compounds showed poor activity as inhibitors of the growth of both Gram-negative, anaerobic periodontal pathogens and Gram-positive carcinogenic bacteria.[302]

16.4.3 AURONE AND AURONOL DIMERS

The first biaurone to be reported in the literature has been isolated from the gametophytes of two species of moss, *Aulacomnium androgynum* and *A. palustre*, and named aulacomnium-biaureusidin (364).[303] As the trivial name suggests, it is a dimer of the well-known compound aureusidin (4,6,3′,4′-tetrahydroxyaurone), and is characterized by a C–C bond from C-5′ of the B-ring of one aurone monomer to C-5 of the A-ring of the other (Figure 16.21). Two further biaurones have now been discovered as indicated in Table 16.15. One is a dimer of sulfuretin (6,3′,4′-trihydroxyaurone) based on a C–C linkage between Cα atoms (365)[304] while licoagrone (366), the first isoprenylated biaurone, features a C–C bond from C-2′ of 7-prenylsulfuretin to C-2 of a 5,3′-diprenylhispidol analog.[305] Geiger and Markham have described the first aurone heterodimer, campylopusaurone (367), from the mosses *Campylopus clavatus* and *C. holomitrium*.[306] In this unique compound, a C–C bond links C-5′ of the

TABLE 16.14
New Aurone and Auronol Glycosides Reported from 1992 to 2003

No.	Compound	Mol. Formula	Source	Family	Ref.
	2'-Hydroxyaurone				
355	2'-*O*-Glucosyl-(1''' → 6'')-glucoside (dalmaisione D)	$C_{27}H_{30}O_{13}$	*Polygala dalmaisiana*	Polygalaceae	298
	4,6,4'-Trihydroxyaurone				
356	4,6-Di-*O*-glucoside (caulesauroneside)	$C_{27}H_{30}O_{15}$	*Asarum longirhizomatosum*	Aristolochiaceae	299
357	4-*O*-Rhamnosyl-(1''' → 2'')-glucoside	$C_{27}H_{30}O_{14}$	*Pterocarpus santalinus*	Leguminosae	300
	6,7,3',4'-Tetrahydroxyaurone (maritimetin)				
358	6-*O*-(3'',6''-Di-*O*-acetylglucoside) (bidenoside A)	$C_{25}H_{24}O_{13}$	*Bidens bipinnata*	Asteraceae	263
359	6-*O*-(3'',4'',6''-Tri-*O*-acetylglucoside)	$C_{27}H_{26}O_{14}$	*Bidens pilosa* var. *radiata*	Asteraceae	127
	4,6-Dihydroxy,3',4',5'-Trimethoxy-7-methylaurone				
360	4-*O*-Rhamnoside	$C_{25}H_{28}O_{12}$	*Pterocarpus santalinus*	Leguminosae	300
	2,4,6,4'-Tetrahydroxy-2-benzylcoumaranone (maesopsin)				
361	4-*O*-Glucoside (hovetrichoside C)	$C_{21}H_{22}O_{11}$	*Hovenia trichocarea*	Rhamnaceae	301
362	6-*O*-Glucoside	$C_{21}H_{22}O_{11}$	*Ceanothus americanus*	Rhamnaceae	302
363	4-*O*-Glucoside 4'-*O*-rhamnoside (hovetrichoside D)	$C_{27}H_{32}O_{15}$	*Hovenia trichocarea*	Rhamnaceae	301

B-ring of the aurone, aureusidin, to C-6 of the A-ring of the flavanone, eriodictyol (5,7,3',4'-tetrahydroxyflavanone).

An extraordinary series of rare auronol dimers and heterodimers (**368–377**) has been described in several papers by Bekker and colleagues.[307–310] Some examples of the structures of these compounds are shown in Figure 16.21. The source of the dimers is the heartwood of

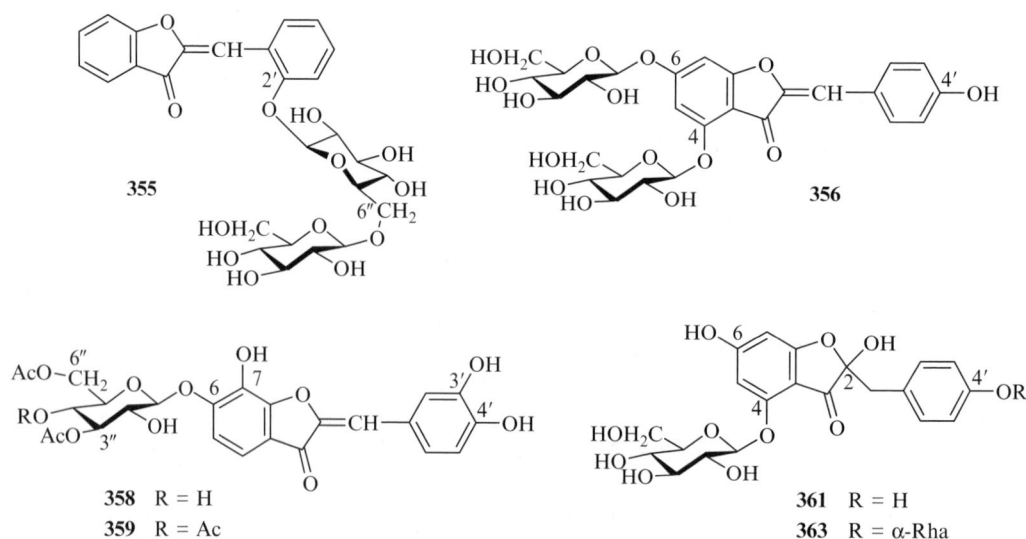

FIGURE 16.20 Aurone and auronol glycosides (see Table 16.14).

TABLE 16.15
Aurone and Auronol Dimers and Heterodimers Reported from 1992 to 2003

No.	Compound	Mol. Formula	Source	Family	Ref.
	Aurone–aurone				
364	Aulacomniumbiaureusidin	$C_{30}H_{18}O_{12}$	*Aulacomnium androgynum* *Aulacomnium palustre*	Aulacomniaceae	303
365	Disulfuretin	$C_{30}H_{18}O_{10}$	*Cotinus coggygria*	Anacardiaceae	304
366	Licoagrone	$C_{45}H_{42}O_{10}$	*Glycyrrhiza glabra*	Leguminosae	305
	Aurone–flavanone				
367	Campylopusaurone	$C_{30}H_{20}O_{12}$	*Campylopus clavatus* *Campylopus holomitrum*	Dicranaceae	306
	Auronol–auronol				
368	(2S)-2-Deoxymaesopsin-(2 → 7)-(2R)-maesopsin	$C_{30}H_{22}O_{11}$	*Berchemia zeyheri*	Rhamnaceae	307
369	(2R)-2-Deoxymaesopsin-(2 → 7)-(2S)-maesopsin	$C_{30}H_{22}O_{11}$	*Berchemia zeyheri*	Rhamnaceae	307
370	(2R)-2-Deoxymaesopsin-(2 → 7)-(2R)-maesopsin	$C_{30}H_{22}O_{11}$	*Berchemia zeyheri*	Rhamnaceae	307
371	(2S)-2-Deoxymaesopsin-(2 → 7)-(2S)-maesopsin	$C_{30}H_{22}O_{11}$	*Berchemia zeyheri*	Rhamnaceae	307
	Auronol–flavanone				
372	(2R,3S)-Naringenin-(3α → 5)-(2R)-maesopsin	$C_{30}H_{22}O_{11}$	*Berchemia zeyheri*	Rhamnaceae	308
373	(2R,3S)-Naringenin-(3α → 5)-(2S)-maesopsin	$C_{30}H_{22}O_{11}$	*Berchemia zeyheri*	Rhamnaceae	308
374	(2R,3S)-Naringenin-(3α → 7)-(2R)-maesopsin (zeyherin)[a]	$C_{30}H_{22}O_{11}$	*Berchemia zeyheri*	Rhamnaceae	309
375	(2R,3S)-Naringenin-(3α → 7)-(2S)-maesopsin[a]	$C_{30}H_{22}O_{11}$	*Berchemia zeyheri*	Rhamnaceae	309
	Auronol–isoflavanone				
376	(2S,3R)-Dihydrogenistein-(2α → 7)-(2R)-maesopsin	$C_{30}H_{22}O_{11}$	*Berchemia zeyheri*	Rhamnaceae	308
377	(2S,3R)-Dihydrogenistein-(2α → 7)-(2S)-maesopsin	$C_{30}H_{22}O_{11}$	*Berchemia zeyheri*	Rhamnaceae	308

[a] Indicates revised structure

Berchemia zeyheri (Rhamnaceae), a tree native to southern Africa which is prized for its beautiful wood, known as "pink ivory" or "red ivory." The complexity of the phenolic compounds present in heartwood extracts prompted their analysis as permethylated derivatives. Stereochemical features were determined by using both NMR and circular dichroism spectroscopy of the parent compounds and their degradation products. These methods were used successfully to obtain a full stereochemical description of the zeyherin epimers **374** and **375**,[309] which were first isolated in 1971 but not fully characterized at that time.[311] Subsequent work has led to the discovery of further auronol dimers and novel heterodimers with flavanone or isoflavanone constituents as summarized in Table 16.15.[307,308,310]

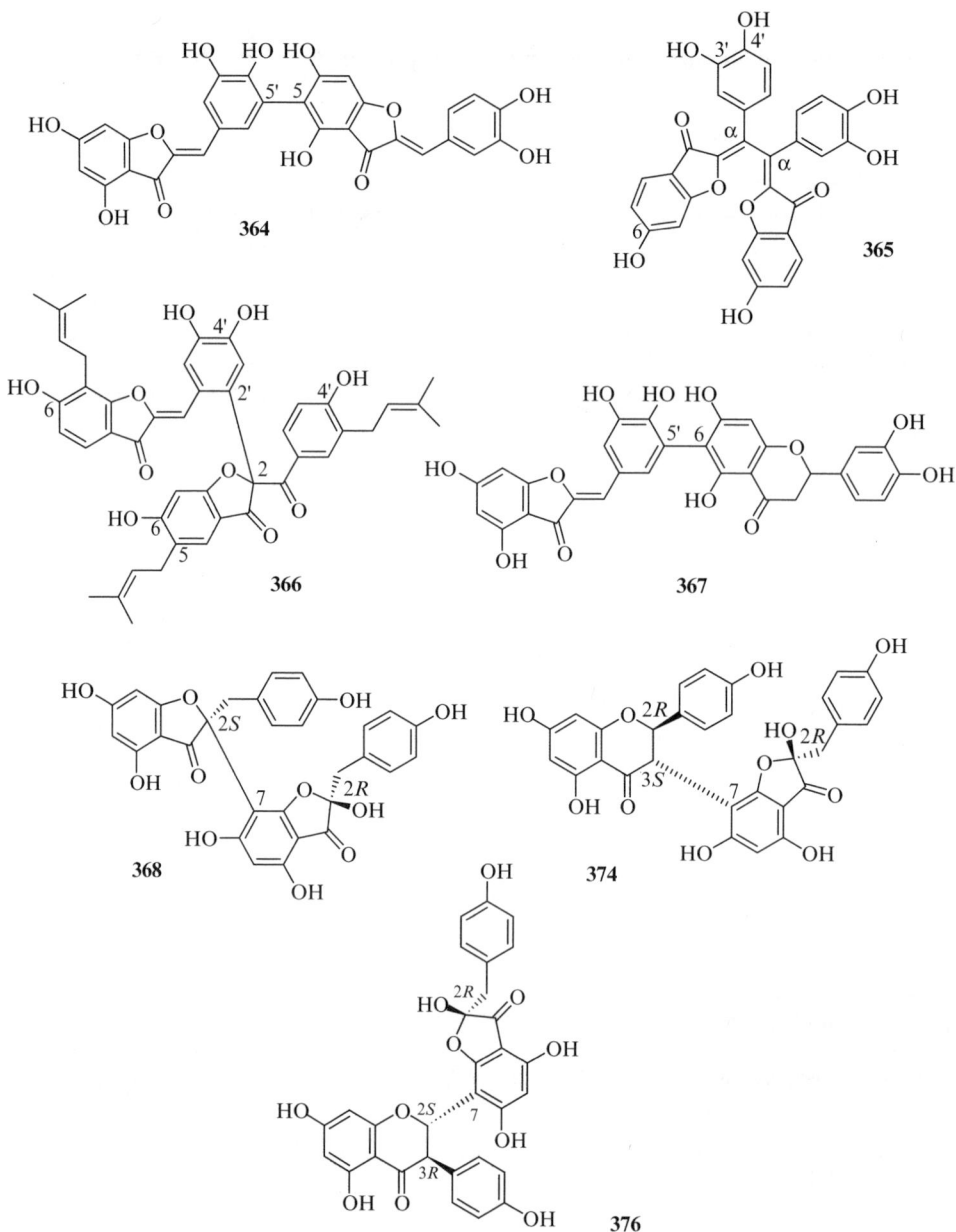

FIGURE 16.21 Aurone and auronol dimers and heterodimers (see Table 16.15).

REFERENCES

1. Bohm, B.A., The minor flavonoids, in *The Flavonoids: Advances in Research Since 1986*, Harborne, J.B., Ed., Chapman & Hall, London, 1993, chap. 9.

2. Bohm, B.A., The minor flavonoids, in *The Flavonoids: Advances in Research Since 1980*, Harborne, J.B., Ed., Chapman & Hall, London, 1988, chap. 9.

3. Bohm, B.A., The minor flavonoids, in *The Flavonoids: Advances in Research*, Harborne, J.B. and Mabry, T.J., Eds., Chapman & Hall, London, 1982, chap. 6.

4. Bohm, B.A., Chalcones, aurones and dihydrochalcones, in *The Flavonoids*, Harborne, J.B., Mabry, T.J., and Mabry, H., Eds., Chapman & Hall, London, 1975, chap. 9.

5. Mabry, T.J., Markham, K.R., and Thomas, M.B., *The Systematic Identification of Flavonoids*, Springer-Verlag, Berlin, 1970.

6. Markham, K.R., *Techniques of Flavonoid Identification*, Academic Press, London, 1982.

7. Bohm, B.A., Chalcones and aurones, in *Methods in Plant Biochemistry*, Vol. 1, Dey, P.M. and Harborne, J.B., Eds., Academic Press, London, 1989, chap. 7.

8. Bohm, B.A., *Introduction to Flavonoids*, Harwood Academic, Amsterdam, 1998.

9. Harborne, J.B. and Baxter, H., Eds., *The Handbook of Natural Flavonoids*, Vol. 1, John Wiley & Sons, Chichester, 1999.

10. Harborne, J.B. and Baxter, H., Eds., *The Handbook of Natural Flavonoids*, Vol. 2, John Wiley & Sons, Chichester, 1999.

11. Forkmann, G. and Heller, W., Biosynthesis of flavonoids, in *Comprehensive Natural Products Chemistry*, Vol. 1, Barton, D., Nakanishi, K., and Meth-Cohn, O., Eds., Elsevier, Amsterdam, 1999, chap. 1.26.

12. Heath, R.J. and Rock, C.O., The Claisen condensation in biology, *Nat. Prod. Rep.*, 19, 581, 2002.

13. Springob, K. et al., Recent advances in the biosynthesis and accumulation of anthocyanins, *Nat. Prod. Rep.*, 20, 288, 2003.

14. Schröder, J., The chalcone/stilbene synthase-type family of condensing enzymes, in *Comprehensive Natural Products Chemistry*, Vol. 1, Barton, D., Nakanishi, K., and Meth-Cohn, O., Eds., Elsevier, Amsterdam, 1999, chap. 1.27.

15. Wimmer, G. et al., Enzymatic hydroxylation of 6'-deoxychalcones with protein preparations from petals of *Dahlia variabilis*, *Phytochemistry*, 47, 1013, 1998.

16. Schröder, J., A family of plant-specific polyketide synthases: facts and predictions, *Trends Plant Sci.*, 2, 373, 1997.

17. Schröder, J., Plant polyketide synthases: a chalcone synthase-type enzyme which performs a condensation reaction with methylmalonyl-CoA in the biosynthesis of C-methylated chalcones, *Biochemistry*, 37, 8417, 1998.

18. Furuya, T., Matsumoto, K., and Hikichi, M., Echinatin, a new chalcone from tissue culture of *Glycyrrhiza echinata*, *Tetrahedron Lett.*, 12, 2567, 1971.

19. Saitoh, T. et al., Biosynthesis of echinatin. A new biosynthetical scheme of retrochalcone, *Tetrahedron Lett.*, 16, 4463, 1975.

20. Ayabe, S. et al., Flavonoids from the cultured cells of *Glycyrrhiza echinata*, *Phytochemistry*, 19, 2179, 1980.

21. Ayabe, S. et al., Biosynthesis of a retrochalcone, echinatin: involvement of *O*-methyltransferase to licodione, *Phytochemistry*, 19, 2332, 1980.

22. Ayabe, S. and Furuya, T., Studies on plant tissue cultures. Part 36. Biosynthesis of a retrochalcone, echinatin, and other flavonoids in the culture cells of *Glycyrrhiza echinata*. A new route to a chalcone with transposed A- and B-rings, *J. Chem. Soc. Perkin Trans. 1*, 2725, 1982.

23. Kajiyama, K. et al., Flavonoids and isoflavonoids of chemotaxonomic significance from *Glycyrrhiza pallidiflora* (Leguminosae), *Biochem. Syst. Ecol.*, 21, 785, 1993.

24. Otani, K. et al., Licodione synthase, a cytochrome P450 monooxygenase catalyzing 2-hydroxylation of 5-deoxyflavanone, in cultured *Glycyrrhiza echinata* L. cells, *Plant Physiol.*, 105, 1427, 1994.

25. Akashi, T., Aoki, T., and Ayabe, S., Identification of a cytochrome P450 cDNA encoding (2*S*)-flavanone 2-hydroxylase of licorice (*Glycyrrhiza echinata* L.) which represents licodione synthase and flavone synthase II, *FEBS Lett.*, 431, 287, 1998.

26. Nakayama, T. et al., Aureusidin synthase: a polyphenol oxidase homolog responsible for flower coloration, *Science*, 290, 1163, 2000.

27. Sata, T. et al., Enzymatic formation of aurones in the extracts of yellow snapdragon flowers, *Plant Sci.*, 160, 229, 2001.

28. Nakayama, T. et al., Specificity analysis and mechanism of aurone synthesis catalyzed by aureusidin synthase, a polyphenol oxidase homolog responsible for flower coloration, *FEBS Lett.*, 499, 107, 2001.

29. Nakayama, T., Enzymology of aurone biosynthesis, *J. Biosci. Bioeng.*, 94, 487, 2002.

30. Ahmad, V.U. et al., Isolation of 3,3'-dihydroxychalcone from *Primula macrophylla*, *J. Nat. Prod.*, 55, 956, 1992.

31. Delle Monache, F., Menichini, F., and Suarez, L.E.C., *Petiveria alliacea*: II. Further flavonoids and triterpenes, *Gazz. Chim. Ital.*, 126, 275, 1996.
32. Reyes-Chilpa, R. et al., Flavonoids and isoflavonoids with antifungal properties from *Platymiscium yucatanum* heartwood, *Holzforschung*, 52, 459, 1998.
33. Cai, L.-N. et al., Chemical constituents of *Glycyrrhiza pallidiflora* Maxim., *Yaoxue Xuebao*, 27, 748, 1992 (*Chem. Abstr.*, 118, 165182d, 1993).
34. Maćias, F.A. et al., Bioactive flavonoids from *Helianthus annuus* cultivars, *Phytochemistry*, 45, 683, 1997.
35. Miyaichi, Y. et al., Studies on the constituents of *Scutellaria* species XX. Constituents of roots of *Scutellaria strigillosa* Hemsl., *Nat. Med. (Tokyo)*, 53, 237, 1999.
36. Kishore, P.H. et al., Flavonoids from *Andrographis lineata*, *Phytochemistry*, 63, 457, 2003.
37. Thuy, T.T. et al., Chalcones and ecdysteroids from *Vitex leptobotrys*, *Phytochemistry*, 49, 2603, 1998.
38. Ilyas, M. et al., A novel chalcone from *Garcinia nervosa*, *J. Chem. Res. (S)*, 231, 2002.
39. Beutler, J.A. et al., Two new cytotoxic chalcones from *Calythropsis aurea*, *J. Nat. Prod.*, 56, 1718, 1993.
40. Sritularak, B. et al., Flavonoids from the roots of *Millettia erythrocalyx*, *Phytochemistry*, 61, 943, 2002.
41. Krohn, K., Steingrover, K., and Rao, M.S., Isolation and synthesis of chalcones with different degrees of saturation, *Phytochemistry*, 61, 931, 2002.
42. Colegate, S.M. et al., Tepanone, a retrochalcone from *Ellipeia cuneifolia*, *Phytochemistry*, 31, 2123, 1992.
43. Calanasan, C.A. and MacLeod, J.K., A diterpenoid sulphate and flavonoids from *Wedelia asperrima*, *Phytochemistry*, 47, 1093, 1998.
44. Srinivas, K.V.N.S. et al., Flavanoids from *Caesalpinia pulcherrima*, *Phytochemistry*, 63, 789, 2003.
45. Phrutivorapongkul, A. et al., Studies on the chemical constituents of stem bark of *Millettia leucantha*: isolation of new chalcones with cytotoxic, anti-herpes simplex virus and anti-inflammatory activities, *Chem. Pharm Bull.*, 51, 187, 746 (Erratum), 2003.
46. Hatano, T. et al., Phenolic constituents of liquorice. VII. A new chalcone with a potent radical scavenging activity and accompanying phenolics from liquorice, *Chem. Pharm. Bull.*, 45, 1485, 1997.
47. Segawa, A. et al., Studies on Nepalese crude drugs. XXV. Phenolic constituents of the leaves of *Didymocarpus leucocalyx* C.B. CLARKE (Gesneriaceae), *Chem. Pharm. Bull.*, 47, 1404, 1999.
48. Huang, L. et al., New compounds with DNA strand-scission activity from the combined leaf and stem of *Uvaria hamiltonii*, *J. Nat. Prod.*, 61, 446, 1998.
49. Iwase, Y. et al., Isolation and identification of two new flavanones and a chalcone from *Citrus kinokuni*, *Chem. Pharm. Bull.*, 49, 1356, 2001.
50. Reddy, M.K. et al., A new chalcone and a flavone from *Andrographis neesiana*, *Chem. Pharm. Bull.*, 51, 854, 2003.
51. Mayer, R., A β-hydroxychalcone from *Leptospermum scoparium*, *Planta Med.*, 59, 269, 1993.
52. López, J.A. et al., Galiposin: a new β-hydroxychalcone from *Galipea granulosa*, *Planta Med.*, 64, 76, 1998.
53. Tanaka, T. et al., Flavonoids in root bark of *Pongamia pinnata*, *Phytochemistry*, 31, 993, 1992.
54. Wollenweber, E. and Mann, K., New flavonoids from farinose *Primula* exudates, *Biochem. Physiol. Pflanz.*, 181, 667, 1986.
55. Wollenweber, E. et al., 5,2,5′-Trihydroxyflavone and 2′,β-dihydroxychalcone from *Primula pulverulenta*, *Phytochemistry*, 28, 295, 1989.
56. Matsuura, S., The structure of cryptostrobin and strobopinin, the flavanones from the heartwood of *Pinus strobus*, *Pharm. Bull. (Tokyo)*, 5, 195, 1957.
57. Barron, D. and Ibrahim, R.K., Isoprenylated flavonoids — a survey, *Phytochemistry*, 43, 921, 1996.
58. Nomura, T. and Fukai, T., Phenolic constituents of licorice, in *Progress in the Chemistry of Organic Natural Products*, Vol. 73, Herz, W., Kirby, G.W., Moore, R.E., Steglich, W., and Tamm, Ch., Eds., Springer-Verlag, Wien, 1998, 1–158.

59. Nomura, T. and Hano, Y., Isoprenoid-substituted phenolic-compounds of Moraceous plants, *Nat. Prod. Rep.*, 11, 205, 1994.

60. Stevens, J.F. and Page, J.E., Xanthohumol and related prenylflavonoids from hops and beer: to your good health! *Phytochemistry*, 65, 1317, 2004.

61. Nomura, T., Hano, Y., and Aida, M., Isoprenoid-substituted flavonoids from *Artocarpus* plants (Moraceae), *Heterocycles*, 47, 1179, 1998.

62. Ngadjui, B.T. and Abegaz, B.M., The chemistry and pharmacology of the genus *Dorstenia* (Moraceae), in *Studies in Natural Products Chemistry*, Vol. 29, Atta-ur-Rahman, Ed., Elsevier, Amsterdam, 2003, 761.

63. Rao, E.V. and Prasad, Y.R., Two chalcones from *Tephrosia spinosa*, *Phytochemistry*, 31, 2121, 1992.

64. Rao, E.V. and Prasad, Y.R., Prenylated flavonoids from *Tephrosia spinosa*, *Phytochemistry*, 32, 183, 1993.

65. Chang, L.C. et al., Absolute configuration of novel bioactive flavonoids from *Tephrosia purpurea*, *Org. Lett.*, 2, 515, 2000.

66. Randriaminahy, M. et al., Lipophilic phenolic constituents from *Helichrysum* species endemic to Madagascar, *Z. Naturforsch. C*, 47, 10, 1992.

67. Koorbanally, N.A. et al., Chalcones from the seed of *Cedrelopsis grevei* (Ptaeroxylaceae), *Phytochemistry*, 62, 1225, 2003.

68. Gómez-Garibay, F. et al., An unusual isopropenyldihydrofuran biflavanol from *Tephrosia crassifolia*, *Phytochemistry*, 52, 1159, 1999.

69. Gómez-Garibay, F. et al., Chromene chalcones from *Tephrosia carrollii* and the revised structure of oaxacacin, *Z. Naturforsch. C*, 56, 969, 2001.

70. Chang, L.C. et al., Activity-guided isolation of constituents of *Tephrosia purpurea* with the potential to induce the phase II enzyme, quinone reductase, *J. Nat. Prod.*, 60, 869, 1997.

71. Mimaki, Y. et al., A novel hexahydrodibenzofuran derivative with potent inhibitory activity on melanin biosynthesis of cultured B-16 melanoma cells from *Lindera umbellata* bark, *Chem. Pharm. Bull.*, 43, 893, 1995.

72. El-Sohly, H.N. et al., Antifungal chalcones from *Maclura tinctoria*, *Planta Med.*, 67, 87, 2001.

73. Chung, M.-I. et al., Phenolics from *Hypericum geminiflorum*, *Phytochemistry*, 44, 943, 1997.

74. Christensen, S.B. et al., An antileishmanial chalcone from Chinese licorice roots, *Planta Med.*, 60, 121, 1994.

75. Fukai, T., Nishizawa, J., and Nomura, T., Five isoprenoid-substituted flavonoids from *Glycyrrhiza eurycarpa*, *Phytochemistry*, 35, 515, 1994.

76. Hano, Y. et al., Constituents of the Moraceae plants. 22. Paratocarpins A–E, five new isoprenoid-substituted chalcones from *Paratocarpus venenosa* Zoll., *Heterocycles*, 41, 191, 1995.

77. Bhatt, P. and Dayal, R., Stipulin, a prenylated chalcone from *Dalbergia stipulacea*, *Phytochemistry*, 31, 719, 1992.

78. Asada, Y., Li, W., and Yoshikawa, T., Isoprenylated flavonoids from hairy root cultures of *Glycyrrhiza glabra*, *Phytochemistry*, 47, 389, 1998.

79. Pistelli, L. et al., Isoflavonoids and chalcones from *Anthyllis hermanniae*, *Phytochemistry*, 42, 1455, 1996.

80. Hano, Y. et al., Paratocarpins F–L, seven new isoprenoid-substituted flavonoids from *Paratocarpus venenosa* Zoll., *Heterocycles*, 41, 2313, 1995.

81. Nakata, K., Taniguchi, M., and Baba, K., Three chalcones from *Angelica keiskei*, *Nat. Med. (Tokyo)*, 53, 329, 1999.

82. Ngadjui, B.T. et al., Geranylated and prenylated flavonoids from the twigs of *Dorstenia mannii*, *Phytochemistry*, 48, 349, 1998.

83. Abegaz, B.M. et al., Chalcones and other constituents of *Dorstenia prorepens* and *Dorstenia zenkeri*, *Phytochemistry*, 59, 877, 2002.

84. Ersam, T. et al., A new isoprenylated chalcone, artoindonesianin J, from the root and tree bark of *Artocarpus bracteata*, *J. Chem. Res. (S)*, 187, 2002.

85. Abegaz, B.M. et al., Prenylated chalcones and flavones from the leaves of *Dorstenia kameruniana*, *Phytochemistry*, 49, 1147, 1998.

86. Kajiyama, K. et al., Two prenylated retrochalcones from *Glycyrrhiza inflata*, *Phytochemistry*, 31, 3229, 1992.

87. Luyengi, L. et al., Rotenoids and chalcones from *Mundulea sericea* that inhibit phorbol ester-induced ornithine decarboxylase activity, *Phytochemistry*, 36, 1523, 1994.

88. Li, W., Asada, Y., and Yoshikawa, T., Flavonoid constituents from *Glycyrrhiza glabra* hairy root cultures, *Phytochemistry*, 55, 447, 2000.

89. Stevens, J.F. et al., Prenylflavonoids from *Humulus lupulus*, *Phytochemistry*, 44, 1575, 1997.

90. Stevens, J.F. et al., Prenylflavonoid variation in *Humulus lupulus*: distribution and taxonomic significance of xanthogalenol and 4'-*O*-methylxanthohumol, *Phytochemistry*, 53, 759, 2000.

91. Etteldorf, N., Etteldorf, N., and Becker, H., New chalcones from hop *Humulus lupulus* L., *Z. Naturforsch. C*, 54, 610, 1999.

92. Lin, Y.-L., Chen, Y.-L., and Kuo, Y.-H., Two new flavanones and two new chalcones from the roots of *Derris laxiflora* Benth, *Chem. Pharm. Bull.*, 40, 2295, 1992.

93. Tomazela, D.M. et al., Pyrano chalcones and a flavone from *Neoraputia magnifica* and their *Trypanosoma cruzi* glycosomal glyceraldehyde-3-phosphate dehydrogenase-inhibitory activities, *Phytochemistry*, 55, 643, 2000.

94. Wu, T.-S., Chang, F.-C., and Wu, P.-L., Flavonoids, amidosulfoxides and an alkaloid from the leaves of *Glycosmis citrifolia*, *Phytochemistry*, 39, 1453, 1995.

95. Lee, D. et al., Aromatase inhibitors from *Broussonetia papyrifera*, *J. Nat. Prod.*, 64, 1286, 2001.

96. Chung, M.-I. et al., A new chalcone, xanthones, and a xanthonolignoid from *Hypericum geminiflorum*, *J. Nat. Prod.*, 62, 1033, 1999.

97. Shimizu, K. et al., A geranylated chalcone with 5α-reductase inhibitory properties from *Artocarpus incisus*, *Phytochemistry*, 54, 737, 2000.

98. Kinoshita, T. et al., The isolation of new pyrano-2-arylbenzofuran derivatives from the root of *Glycyrrhiza glabra*, *Chem. Pharm. Bull.*, 44, 1218, 1996.

99. Tsopmo, A. et al., Geranylated flavonoids from *Dorstenia poinsettifolia*, *Phytochemistry*, 48, 345, 1998.

100. Fukai, T. and Nomura, T., Isoprenoid-substituted flavonoids from roots of *Glycyrrhiza inflata*, *Phytochemistry*, 38, 759, 1995.

101. Magalhães, A.F. et al., Twenty-three flavonoids from *Lonchocarpus subglaucescens*, *Phytochemistry*, 42, 1459, 1996.

102. Khalilullah, M.D., Sharma, V.M., and Rao, P.S., Ramosismin, a new prenylated chalcone from *Crotalaria ramosissima*, *Fitoterapia*, 64, 232, 1993.

103. Passador, E.A.P. et al., A pyrano chalcone and a flavanone from *Neoraputia magnifica*, *Phytochemistry*, 45, 1533, 1997.

104. Gómez-Garibay, F. et al., Flavonoids from *Tephrosia major*. A new prenyl-beta-hydroxychalcone, *Z. Naturforsch. C*, 57, 579, 2002.

105. Carcache-Blanco, E.J. et al., Constituents of the stem bark of *Pongamia pinnata* with the potential to induce quinone reductase, *J. Nat. Prod.*, 66, 1197, 2003.

106. Demizu, S. et al., Prenylated dibenzoylmethane derivatives from the root of *Glycyrrhiza inflata* (Xinjiang licorice), *Chem. Pharm. Bull.*, 40, 392, 1992.

107. Zeng, L. et al., Four new isoprenoid-substituted dibenzoylmethane derivatives, glyinflanins A, B, C, and D from the roots of *Glycyrrhiza inflata*, *Heterocycles*, 34, 85, 1992.

108. Kajiyama, K. et al., New prenylflavones and dibenzoylmethane from *Glycyrrhiza inflata*, *J. Nat. Prod.*, 55, 1197, 1992.

109. Kitagawa, I. et al., Indonesian medicinal plants. II. Chemical structures of pongapinones A and B, two new phenyl propanoids from the bark of *Pongamia pinnata* (Papilionaceae), *Chem. Pharm. Bull.*, 40, 2041, 1992.

110. Pirrung, M.C. et al., Revised relative and absolute stereochemistry of (+)-purpurin, *J. Nat. Prod.*, 61, 89, 1998.

111. Gupta, R.K., Krishnamurti, M., and Parthasarathi, J., Purpurin, a new flavanone from *Tephrosia purpurea* seeds, *Phytochemistry*, 19, 1264, 1980.

112. Bohm, B.A. and Stuessy, T.F., *Flavonoids of the Sunflower Family (Asteraceae)*, Springer-Verlag, Wien, 2001.

113. Yamashita, M. et al., First total synthesis of ($\pm$)-linderol A, a tricyclic hexahydrodibenzofuran constituent of *Lindera umbellata* bark, with potent inhibitory activity on melanin biosynthesis of cultured B-16 melanoma cells, *J. Org. Chem.*, 68, 1216, 2003.

114. Kyogoku, K. et al., Anti-ulcer effect of isoprenyl flavonoids. II. Synthesis and anti-ulcer activity of new chalcones related to sophoradin, *Chem. Pharm. Bull.*, 27, 2943, 1979.

115. Elgamal, M.H.A., Shalaby, N.M.M., and Duddeck, H., Isolation of two adjuncts from the fruits of *Ammi majus* L., *Nat. Prod. Lett.*, 3, 209, 1993.

116. Kitagawa, I. et al., Chemical studies of chinese licorice-roots. II. Five new flavonoid constituents from the roots of *Glycyrrhiza aspera* Pall. collected in Xinjiang, *Chem. Pharm. Bull.*, 46, 1511, 1998.

117. Hatano, T. et al., Acylated flavonoid glycosides and accompanying phenolics from licorice, *Phytochemistry*, 47, 287, 1998.

118. Cioffi, G. et al., Antioxidant chalcone glycosides and flavanones from *Maclura* (*Chlorophora*) *tinctoria*, *J. Nat. Prod.*, 66, 1061, 2003.

119. De Tommasi, N. et al., Flavonol and chalcone ester glycosides from *Bidens leucantha*, *J. Nat. Prod.*, 60, 270, 1997.

120. De Tommasi, N., Piacente, S., and Pizza, C., Flavonol and chalcone ester glycosides from *Bidens andicola*, *J. Nat. Prod.*, 61, 973, 1998.

121. Jayaprakasam, B. et al., Androechin, a new chalcone glucoside from *Andrographis echiodes*, *J. Asian Nat. Prod. Res.*, 3, 43, 2001.

122. Iwashina, T. and Kitajima, J., Chalcone and flavonol glycosides from *Asarum canadense* (Aristolochiaceae), *Phytochemistry*, 55, 971, 2000.

123. Lin, L.-C. et al., Studies on the constituents of *Asarum macranthum*, *Chin. Pharm. J. (Taipei)*, 47, 501, 1995.

124. Bilia, A.R., Morelli, I., and Marsili, A., Two glucosides from *Pyracantha coccinea* roots: a new lignan and a new chalcone, *Tetrahedron*, 50, 5181, 1994.

125. Roy, R. and Pandey, V.B., A chalcone glycoside from *Clerodendron phlomidis*, *Phytochemistry*, 37, 1775, 1994.

126. Orhan, D.D., Çalis, I., and Ergun, F., Two new flavonoid glycosides from *Viscum album* ssp. *album*, *Pharm. Biol.*, 40, 380, 2002.

127. Wang, J. et al., Flavonoids from *Bidens pilosa* var. *radiata*, *Phytochemistry*, 46, 1275, 1997.

128. Karikome, H., Ogawa, K., and Sashida, Y., New acylated glucosides of chalcone from the leaves of *Bidens frondosa*, *Chem. Pharm. Bull.*, 40, 689, 1992.

129. Redl, K., Davis, B., and Bauer, R., Chalcone glycosides from *Bidens campylotheca*, *Phytochemistry*, 32, 218, 1993.

130. Srivastava, S.K., Srivastava, S.D., and Chouksey, B.K., New constituents of *Terminalia alata*, *Fitoterapia*, 70, 390, 1999.

131. Martín, T. et al., Polyphenolic compounds from pericarps of *Myrtus communis*, *Pharm. Biol.*, 37, 28, 1999.

132. Markham, K.R. and Geiger, H., ^{1}H nuclear magnetic resonance spectroscopy of flavonoids and their glycosides in hexadeuterodimethylsulfoxide, in *The Flavonoids: Advances in Research Since 1986*, Harborne, J.B., Ed., Chapman & Hall, London, 1993, chap. 10.

133. Bandeira, M.A.M., de Matos, F.J.A., and Braz-Filho, R., Structural elucidation and total assignment of the ^{1}H and ^{13}C NMR spectra of new chalcone dimers, *Magn. Reson. Chem.*, 41, 1009, 2003.

134. Masesane, I.B. et al., A bichalcone from the twigs of *Rhus pyroides*, *Phytochemistry*, 53, 1005, 2000.

135. Mdee, L.K., Yeboah, S.O., and Abegaz, B.M., Rhuschalcones II–VI, five new bichalcones from the root bark of *Rhus pyroides*, *J. Nat. Prod.*, 66, 599, 2003.

136. Tanaka, T et al., Dimeric chalcone derivatives from *Mallotus philippensis*, *Phytochemistry*, 48, 1423, 1998.

137. Bai, H. et al., A novel biflavonoid from roots of *Glycyrrhiza uralensis* cultivated in China, *Chem. Pharm. Bull.*, 51, 1095, 2003.

138. Murakami, A. et al., Possible anti-tumour promoters: bi- and tetraflavonoids from *Lophira alata*, *Phytochemistry*, 31, 2689, 1992.

139. Pegnyemb, D.E. et al., Biflavonoids from *Ochna afzelii*, *Phytochemistry*, 57, 579, 2001.

140. Messanga, B. et al., Biflavonoids from *Ochna calodendron*, *Phytochemistry*, 35, 791, 1994.

141. Mbing, J.N. et al., Two biflavonoids from *Ouratea flava* stem bark, *Phytochemistry*, 63, 427, 2003.

142. Taniguchi, M., Fujiwara, A., and Baba, K., Three flavonoids from *Daphne odora*, *Phytochemistry*, 45, 183, 1997.

143. Taniguchi, M. et al., Two biflavonoids from *Daphne acutiloba*, *Phytochemistry*, 49, 863, 1998.

144. Taniguchi, M. and Baba, K., Three biflavonoids from *Daphne odora*, *Phytochemistry*, 42, 1447, 1996.

145. Baba, K., Taniguchi, M., and Kozawa, M., A spirobiflavonoid genkwanol B from *Daphne genkwa*, *Phytochemistry*, 31, 975, 1992.

146. Baba, K., Taniguchi, M., and Kozawa, M., A third spirobiflavonoid genkwanol C from *Daphne genkwa*, *Phytochemistry*, 33, 913, 1993.

147. Dumontet, V. et al., New cytotoxic flavonoids from *Cryptocarya infectoria*, *Tetrahedron*, 57, 6189, 2001.

148. Kaewamatawong, R. et al., Novel biflavonoids from the stem bark of *Ochna integerrima*, *J. Nat. Prod.*, 65, 1027, 2002.

149. Carneiro, F.J.C. et al., Bi- and tetraflavonoids from *Aristolochia ridicula*, *Phytochemistry*, 55, 823, 2000.

150. Ramírez, I. et al., Cissampeloflavone, a chalcone–flavone dimer from *Cissampelos pareira*, *Phytochemistry*, 64, 654, 1421 (Erratum), 2003.

151. Messanga, B.B. et al., Calodenone, a new isobiflavonoid from *Ochna calodendron*, *J. Nat. Prod.*, 55, 245, 1992.

152. Messanga, B.B. et al., Triflavonoids of *Ochna calodendron*, *Phytochemistry*, 59, 435, 2002.

153. Murakami, A. et al., Chalcone tetramers, lophirachalcone and alatachalcone, from *Lophira alata* as possible anti-tumor promoters, *Biosci. Biotechnol. Biochem.*, 56, 769, 1992.

154. Tih, A. et al., Tetraflavonoids of *Lophira alata*, *Phytochemistry*, 31, 981, 1992.

155. Tih, A. et al., Lophiroflavans B and C, tetraflavonoids of *Lophira alata*, *Phytochemistry*, 31, 3595, 1992.

156. Messanga, B.B. et al., Isolation and structural elucidation of a new pentaflavonoid from *Ochna calodendron*, *New. J. Chem.*, 25, 1098, 2001.

157. Tih, A.E. et al., A novel hexaflavonoid from *Lophira alata*, *Tetrahedron Lett.*, 40, 4721, 1999.

158. Crombie, L. et al., Constituents of kamala. Isolation and structure of 2 new components, *J. Chem. Soc. C*, 2625, 1968.

159. Dean, F.M., *Naturally Occurring Oxygen Ring Compounds*, Butterworths, London, 1963, 228.

160. Seidel, V., Bailleul, F., and Waterman, P.G., (*Rel*)-1β,2α-di-(2,4-dihydroxy-6-methoxybenzoyl)-3β,4α-di-(4-methoxyphenyl)-cyclobutane and other flavonoids from the aerial parts of *Goniothalamus gardneri* and *Goniothalamus thwaitesii*, *Phytochemistry*, 55, 439, 2000.

161. Baba, K., Taniguchi, M., and Kozawa, M., Three biflavonoids from *Wikstroemia sikokiana*, *Phytochemistry*, 37, 879, 1994.

162. Baba, K. et al., Biflavonoids from *Daphne odora*, *Phytochemistry*, 38, 1021, 1995.

163. Baba, K. et al., Stereochemistry of the spirobiflavonoid genkwanol B from *Daphne genkwa*, *Phytochemistry*, 32, 221, 1993.

164. Ghogomu, R. et al., Lophirone A, a biflavonoid with unusual skeleton from *Lophira lanceolata*, *Tetrahedron Lett.*, 28, 2967, 1987.

165. Tih, R.G. et al., Structures of isombamichalcone and lophirochalcone, bi- and tetra-flavonoids from *Lophira lanceolata*, *Tetrahedron Lett.*, 30, 1807, 1989.

166. Hano, Y. et al., Sanggenons R, S and T, three new isoprenylated phenols from the Chinese crude drug "Sang-Bai-Pi" (Morus root bark), *Heterocycles*, 40, 953, 1995.

167. Shibata, K. et al., Crinatusins, bioactive Diels-Alder adducts from *Cyathocalyx crinatus*, *Tetrahedron*, 56, 8821, 2000.

168. Alias, Y. et al., An antimitotic and cytotoxic chalcone from *Fissistigma lanuginosum*, *J. Nat. Prod.*, 58, 1160, 1995.

169. Trakoontivakorn, G. et al., Structural analysis of a novel antimutagenic compound, 4-hydroxy-panduratin A, and the antimutagenic activity of flavonoids in a Thai spice, fingerroot (*Boesenber-*

gia pandurata Schult.) against mutagenic heterocyclic amines, *J. Agric. Food. Chem.*, 49, 3046, 2001.

170. Pandji, C. et al., Insecticidal constituents from four species of the Zingiberaceae, *Phytochemistry*, 34, 415, 1993.

171. Gu, J.-Q. et al., Activity-guided isolation of constituents of *Renealmia nicolaioides* with the potential to induce the phase II enzyme quinone reductase, *J. Nat. Prod.*, 65, 1616, 2002.

172. Shirota, O. et al., Antiandrogenic natural Diels–Alder-type adducts from *Brosimum rubescens*, *J. Nat. Prod.*, 60, 997, 1997.

173. Shirota, O. et al., Two chalcone–prenylcoumarin Diels–Alder adducts from *Brosimum rubescens*, *Phytochemistry*, 47, 1381, 1998.

174. Basnet, P. et al., Two new 2-arylbenzofuran derivatives from hypoglycemic activity-bearing fractions of *Morus insignis*, *Chem. Pharm. Bull.*, 41, 1238, 1993.

175. Shinomiya, K. et al., A Diels–Alder-type adduct from *Artocarpus heterophyllus*, *Phytochemistry*, 40, 1317, 1995.

176. Tsopmo, A. et al., A new Diels–Alder-type adduct flavonoid from *Dorstenia barteri*, *J. Nat. Prod.*, 62, 1432, 1999.

177. Fukai, T. et al., Isoprenylated flavanones from *Morus cathayana*, *Phytochemistry*, 47, 273, 1998.

178. Shi, Y.-Q. et al., Cytotoxic flavonoids with isoprenoid groups from *Morus mongolica*, *J. Nat. Prod.*, 64, 181, 2001.

179. Hano, Y. et al., Revised structure of sanggenon A, *Heterocycles*, 45, 867, 1997.

180. Shi, Y.-Q., Fukai, T., and Nomura, T., Structure of sanggenon O, a Diels–Alder type adduct derived from a chalcone and a dehydroprenylated sanggenon-type flavanone from *Morus cathayana*, *Heterocycles*, 54, 639, 2001.

181. Shen, R.-C. and Lin, M., Diels–Alder type adducts from *Morus cathayana*, *Phytochemistry*, 57, 1231, 2001.

182. Hano, Y. et al., A novel way of determining the structure of artonin I, an optically active Diels–Alder type adduct, with the aid of an enzyme system of *Morus alba* cell cultures, *J. Chem. Soc. Chem. Commun.*, 1177, 1992.

183. Ferrari, F. et al., Multicaulisin, a new Diels–Alder type adduct from *Morus multicaulis*, *Fitoterapia*, 71, 213, 2000.

184. Hano, Y. et al., Constituents of the Moraceae plants. 23. Sorocenols A and B, two new isoprenylated phenols from the root bark of *Sorocea bonplandii* Baillon, *Heterocycles*, 41, 1035, 1995.

185. Nomura, T., The chemistry and biosynthesis of isoprenylated flavonoids from moraceous plants, *Pure Appl. Chem.*, 71, 1115, 1999.

186. Tuntiwachwuttikul, P. et al., (1′*RS*,2′*SR*,6′*RS*)-(2,6-Dihydroxy-4-methoxyphenyl)-[3′-methyl-2′-(3″-methylbut-2″-enyl)-6′-phenylcyclohex-3′-enyl] methanone (Panduratin A) — a constituent of the red rhizomes of a variety of *Boesenbergia pandurata*, *Aust. J. Chem.*, 37, 449, 1984.

187. Tuchinda, P. et al., Anti-inflammatory cyclohexenyl chalcone derivatives in *Boesenbergia pandurata*, *Phytochemistry*, 59, 169, 2002.

188. Shi, Y.-Q. et al., Absolute structures of 3-hydroxy-2-prenylflavanones with an ether linkage between the 2′- and 3-positions from Moraceous plants, *Heterocycles*, 55, 13, 2001.

189. Fukai, T. et al., Structure of sanggenon G, a new Diels–Alder adduct from Chinese crude drug "Sang-Bai-Pi" (*Morus* root bark), *Heterocycles*, 20, 611, 1983.

190. Prasain, J.K. et al., Novel diarylheptanoids from the seeds of *Alpinia blepharocalyx*: revised structure of calyxin A, *J. Chem. Res. (S)*, 22, 1998.

191. Tezuka, Y. et al., Eleven novel diarylheptanoids and two unusual diarylheptanoid derivatives from the seeds of *Alpinia blepharocalyx*, *J. Nat. Prod.*, 64, 208, 2001.

192. Prasain, J.K. et al., Six novel diarylheptanoids bearing chalcone or flavanone moiety from the seeds of *Alpinia blepharocalyx*, *Tetrahedron*, 53, 7833, 1997.

193. Prasain, J.K. et al., Calyxin H, epicalyxin H and blepharocalyxins A and B, novel diarylheptanoids from the seeds of *Alpinia blepharocalyx*, *J. Nat. Prod.*, 61, 212, 1998.

194. Kadota, S. et al., Blepharocalyxins A and B, novel diarylheptanoids from *Alpinia blepharocalyx*, and their inhibitory effect on NO formation in murine macrophages, *Tetrahedron Lett.*, 37, 7283, 1996.

195. Tezuka, Y. et al., Blepharocalyxins C–E: three novel antiproliferative diarylheptanoids from the seeds of *Alpinia blepharocalyx*, *Tetrahedron Lett.*, 41, 5903, 2000.

196. Ali, M.S. et al., Blepharocalyxins C–E, three new dimeric diarylheptanoids, and related compounds from the seeds of *Alpinia blepharocalyx*, *J. Nat. Prod.*, 64, 491, 2001.

197. Rahman, M.M., Qais, N., and Rashid, M.A., A new C-benzylated chalcone from *Desmos chinensis*, *Fitoterapia*, 74, 511, 2003.

198. Ngo, K.-S. and Brown, G.D., Stilbenes, monoterpenes, diarylheptanoids, labdanes and chalcones from *Alpinia katsumadai*, *Phytochemistry*, 47, 1117, 1998.

199. Fu, X. et al., Flavanone and chalcone derivatives from *Cryptocarya kurzii*, *J. Nat. Prod.*, 56, 1153, 1993.

200. Kadota, S. et al., Novel diarylheptanoids of *Alpinia blepharocalyx*, *Curr. Top. Med. Chem.*, 3, 203, 2003.

201. Prasain, J.K. et al., Inhibitory effect of diarylheptanoids on nitric oxide production in activated murine macrophages, *Biol. Pharm. Bull.*, 21, 371, 1998.

202. Ali, M.S. et al., Antiproliferative activity of diarylheptanoids from the seeds of *Alpinia blepharocalyx*, *Biol. Pharm. Bull.*, 24, 525, 2001.

203. Wu, J.-H. et al., Desmosdumotin, a novel cytotoxic principle from *Desmos dumosus*, *Tetrahedron Lett.*, 43, 1391, 2002.

204. Yang, S.-W. et al., Munchiwarin, a prenylated chalcone from *Crotalaria trifoliastrum*, *J. Nat. Prod.*, 61, 1274, 1998.

205. Andrei, C.C. et al., C-prenylflavonoids from roots of *Tephrosia tunicata*, *Phytochemistry*, 55, 799, 2000.

206. Kazuma, K. et al., Quinochalcones and flavonoids from fresh florets in different cultivars of *Carthamus tinctorius* L., *Biosci. Biotechnol. Biochem.*, 64, 1588, 2000.

207. Yin, H.-B. and He, Z.-S., A novel semi-quinone chalcone sharing a pyrrole ring C-glycoside from *Carthamus tinctorius*, *Tetrahedron Lett.*, 41, 1955, 2000.

208. Meselhy, M.R. et al., Two new quinochalcone yellow pigments from *Carthamus tinctorius* and Ca^{2+} antagonistic activity of tinctormine, *Chem. Pharm. Bull.*, 41, 1796, 1993.

209. Kumazawa, T. et al., Precursor of carthamin, a constituent of safflower, *Chem. Lett.*, 2343, 1994.

210. Kazuma, K. et al., Structure of precarthamin, a biosynthetic precursor of carthamin, *Biosci. Biotechnol. Biochem.*, 59, 1588, 1995.

211. Meselhy, M.R. et al., Tinctormine, a novel Ca^{2+} antagonist N-containing quinochalcone C-glycoside from *Carthamus tinctorius* L., *Chem. Pharm. Bull.*, 40, 3355, 1992.

212. Beckmann, J., Experimenta lanas inficiendi floribus Carthami tinctorii, *Novi Commentarii Societatis Regiae Scientiarum Gottingensis*, 4, 89, 1773.

213. Dufour, J.B., Sur la composition chimique de la fleur du carthame (*Cartamus tinctorius* Linnaei), *Ann. Chim.*, 48, 283, 1803.

214. Marchais, Observations sur un mémoire publié dans le cahier des Annales de chimie du mois de frimaire, page 283, par M. Dufour, pharmacien, Sur la composition chimique de la fleur du carthame, *Ann. Chim.*, 50, 73, 1804.

215. Preisser, F., Dissertation sur l'origine et la nature des matières colorantes organiques, et étude spéciale de l'action de l'oxygène sur ces principes immédiats, *J. Pharm. Chim.*, Ser. 3, 5, 191, 1844.

216. Preisser, F., Ueber den Ursprung and die Beschaffenheit der organische Farbstoffe und besonders über die Einwirkung des Sauerstoffes auf dieselben, *J. Prakt. Chem.*, 32, 129, 1844.

217. Obara, H. and Onodera, J.-I., Structure of carthamin, *Chem. Lett.*, 201, 1979.

218. Sata, S. et al., Synthesis of (+),(−)-model compounds and absolute configuration of carthamin; a red pigment in the flower petals of safflower, *Chem. Lett.*, 833, 1996.

219. Wollenweber, E. et al., Lipophilic phenolics from the leaves of *Empetrum nigrum* — chemical structures and exudate localization, *Bot. Acta*, 105, 300, 1992.

220. Masaoud, M. et al., Flavonoids of dragon's blood from *Dracaena cinnabari*, *Phytochemistry*, 38, 745, 1995.

221. Mustafa, K.A. et al., Hydrogen-bonded rotamers of 2′,4′,6′-trihydroxy-3′-formyldihydrochalcone, an intermediate in the synthesis of a dihydrochalcone from *Leptospermum recurvum*, *Tetrahedron*, 59, 6113, 2003.

222. Malterud, K.E., C-Methylated dihydrochalcones from *Myrica gale* fruit extract, *Acta Pharm. Nord.*, 4, 65, 1992.

223. Ramadan, M.A. et al., Minor phenolics from *Crinum bulbispermum* bulbs, *Phytochemistry*, 54, 891, 2000.

224. González, A.G. et al., Phenolic compounds of Dragon's blood from *Dracaena draco*, *J. Nat. Prod.*, 63, 1297, 2000.

225. Lu, W. et al., Studies on chemical constituents of chloroform extract of *Dracaena cochinchinensis*, *Yaoxue Xuebao*, 33, 755, 1998.

226. Sinz, A. et al., Phenolic compounds from *Anomianthus dulcis*, *Phytochemistry*, 50, 1069, 1999.

227. Lien, T.P. et al., Chalconoids from *Fissistigma bracteolatum*, *Phytochemistry*, 53, 991, 2000.

228. Majumder, P.L., Lahiri, S., and Mukhoti, N., Chalcone and dihydrochalcone derivatives from the orchid *Lusia volucris*, *Phytochemistry*, 40, 271, 1995.

229. Silva, D.H.S. et al., Dihydrochalcones and flavonolignans from *Iryanthera lancifolia*, *J. Nat. Prod.*, 62, 1475, 1999.

230. Ichikawa, K. et al., Retrodihydrochalcones and homoisoflavones isolated from Thai medicinal plant *Dracaena loureiri* and their estrogen agonist activity, *Planta Med.*, 63, 540, 1997.

231. Fleischer, T.C., Waigh, R.D., and Waterman, P.G., A novel retrodihydrochalcone from the stem bark of *Uvaria mocoli*, *Phytochemistry*, 47, 1387, 1998.

232. Leong, Y.-W. et al., A dihydrochalcone from *Lindera lucida*, *Phytochemistry*, 47, 891, 1998.

233. Iinuma, M., Tanaka, T., and Asai, F., Flavonoids in frond exudates of *Pityrogramma tartarea*, *Phytochemistry*, 36, 941, 1994.

234. Alvarez, L. and Delgado, G., *C*- and *O*-glycosyl-α-hydroxydihydrochalcones from *Eysenhardtia polystachya*, *Phytochemistry*, 50, 681, 1999.

235. Tanrisever, N. et al., Ceratiolin and other flavonoids from *Ceratiola ericoides*, *Phytochemistry*, 26, 175, 1987.

236. Krasnov, E.A. et al., Phenolic components of *Empetrum nigrum* extract and the crystal structure of one of them, *Chem. Nat. Compd.*, 36, 493, 2000.

237. Miles, D.H. et al., 3'-Formyl-2',4',6'-trihydroxydihydrochalcone from *Psidium acutangulum*, *Phytochemistry*, 30, 1131, 1991.

238. Miles, D.H. et al., 3'-Formyl-2',4',6'-trihydroxy-5'-methyldihydrochalcone, a prospective new agrochemical from *Psidium acutangulum*, *J. Nat. Prod.*, 53, 1548, 1990.

239. Beltrami, E. et al., Coatline A and B, two *C*-glucosyl-α-hydroxydihydrochalcones from *Eysenhardtia polystachya*, *Phytochemistry*, 21, 2931, 1982.

240. Benosman, A. et al., New terpenylated dihydrochalcone derivatives isolated from *Mitrella kentii*, *J. Nat. Prod.*, 60, 921, 1997.

241. Orjala, J. et al., New monoterpene-substituted dihydrochalcones from *Piper aduncum*, *Helv. Chim. Acta*, 76, 1481, 1993.

242. Orjala, J. et al., Cytotoxic and antibacterial dihydrochalcones from *Piper aduncum*, *J. Nat. Prod.*, 57, 18, 1994.

243. Joshi, A.S. et al., Dihydrochalcones from *Piper longicaudatum*, *Planta Med.*, 67, 186, 2001.

244. Rao, M.S. et al., A revised structure for crotaramosmin from *Crotolaria ramosissima*, *J. Nat. Prod.*, 61, 1148, 1998.

245. Kumar, J.K. et al., Further dihydrochalcones from *Crotolaria ramosissima*, *J. Braz. Chem. Soc.*, 10, 278, 1999.

246. Ahsan, M., Armstrong, J.A., and Waterman, P.G., Dihydrochalcones from the aerial parts of *Boronia inconspicua*, *Phytochemistry*, 36, 799, 1994.

247. Fukai, T. et al., Isoprenylated flavonoids from underground parts of *Glycyrrhiza glabra*, *Phytochemistry*, 43, 1119, 1996.

248. Yenesew, A., Midiwo, J.O., and Waterman, P.G., Rotenoids, isoflavones and chalcones from the stem bark of *Millettia usaramensis* subspecies *usaramensis*, *Phytochemistry*, 47, 295, 1998.

249. Trani, M. et al., Dihydrochalcones and coumarins of *Esembeckia grandiflora* subsp. *grandiflora*, *Gazz. Chim. Ital.*, 127, 415, 1997.

250. Müller, A.H. et al., Dihydrochalcones, coumarins and alkaloids from *Metrodorea nigra*, *Phytochemistry*, 40, 1797, 1995.

251. Ichino, K., Tanaka, H., and Ito, K., Isolation and structure of 2 new flavonoids from *Lindera umbellata*, *Chem. Lett.*, 363, 1989.

252. Ichino, K., Two flavonoids from two *Lindera umbellata* varieties, *Phytochemistry*, 28, 955, 1989.

253. Fukai, T. et al., An isoprenylated flavanone from *Glycyrrhiza glabra* and rec-assay of licorice phenols, *Phytochemistry*, 49, 2005, 1998.

254. Jakupovic, J. et al., Twenty-one acylphloroglucinol derivatives and further constituents from South African *Helichrysum* species, *Phytochemistry*, 28, 1119, 1989.

255. Williams, A.H., Dihydrochalcones, in *Comparative Phytochemistry*, Swain, T., Ed., Academic Press, London, 1966, chap. 17.

256. Ohtani, I.I. et al., Thonningianins A and B, new antioxidants from the African medicinal herb *Thonningia sanguinea*, *J. Nat. Prod.*, 63, 676, 2000.

257. Calis, I. et al., Phenylvaleric acid and flavonoid glycosides from *Polygonum salicifolium*, *J. Nat. Prod.*, 62, 1101, 1999.

258. Cuendet, M. et al., A stilbene and dihydrochalcones with radical scavenging activities from *Loiseleuria procumbens*, *Phytochemistry*, 54, 871, 2000.

259. Qin, X.-D. and Liu, J.-K., A new sweet dihydrochalcone-glucoside from leaves of *Lithocarpus pachyphyllus* (Kurz) Rehd. (Fagaceae), *Z. Naturforsch. C*, 58, 759, 2003.

260. Ogawa, K. et al., 3′,5′-Di-C-β-glucopyranosylphloretin, a flavonoid characteristic of the genus *Fortunella*, *Phytochemistry*, 57, 737, 2001.

261. Silva, D.H.S., Yoshida, M., and Kato, M.J., Flavonoids from *Iryanthera sagotiana*, *Phytochemistry*, 46, 579, 1997.

262. Yang, Z. et al., Chemical studies on the marine algae *Zostera* sp., *Zhongguo Haiyang Yaowu*, 11, 1, 1992 (*Chem. Abstr.* 118, 18925v, 1993).

263. Li, S. et al., A new aurone glucoside and a new chalcone glucoside from *Bidens bipinnata* Linne, *Heterocycles*, 61, 557, 2003.

264. Li, W. et al., Flavonoids from *Glycyrrhiza pallidiflora* hairy root cultures, *Phytochemistry*, 60, 351, 2002.

265. Yoshida, T., Feng, W.-S., and Okuda, T., Two polyphenol glycosides and tannins from *Rosa cymosa*, *Phytochemistry*, 32, 1033, 1993.

266. Bezuidenhoudt, B.C.B., Brandt, E.V., and Ferreira, D., Flavonoid analogues from *Pterocarpus* species, *Phytochemistry*, 26, 531, 1987.

267. Adinarayana, D. et al., Structure elucidation of pterosupin from *Pterocarpus marsupium*, the first naturally occurring C-glycosyl-β-hydroxy-dihydrochalcone, *Z. Naturforsch. C*, 37, 145, 1982.

268. Maurya, R. et al., Constituents of *Pterocarpus marsupium*, *J. Nat. Prod.*, 47, 179, 1984.

269. Drewes, S.E. and Hudson, N.A., Brackenin, a dimeric dihydrochalcone from *Brackenridgea zanguebarica*, *Phytochemistry*, 22, 2823, 1983.

270. Li, Y. et al., Littorachalcone, a new enhancer of NGF-mediated neurite outgrowth, from *Verbena littoralis*, *Chem. Pharm. Bull.*, 51, 872, 2003.

271. Li, Y.-S. et al., Verbenachalcone, a novel dimeric dihydrochalcone with potentiating activity on nerve growth factor-action from *Verbena littoralis*, *J. Nat. Prod.*, 64, 806, 2001.

272. Patil, A.D. et al., A new dimeric dihydrochalcone and a new prenylated flavone from the bud covers of *Artocarpus altilis*: potent inhibitors of cathepsin K, *J. Nat. Prod.*, 65, 624, 2002.

273. Zhou, Z., Wang, J., and Yang, C., Cochinchinenin-new chalcone dimer from Chinese dragon blood, *Yaoxue Xuebao*, 36, 200, 2001.

274. Roitman, J.N., Wong, R.Y., and Wollenweber, E., Methylene bisflavonoids from frond exudate of *Pentagramma triangularis* ssp *triangularis*, *Phytochemistry*, 34, 297, 1993.

275. Iinuma, M. et al., Two biflavonoids in the farinose exudate of *Pentagramma triangularis*, *Phytochemistry*, 35, 1043, 1994.

276. Iinuma, M. et al., Unusual biflavonoids in the farinose exudate of *Pentagramma triangularis*, *Phytochemistry*, 44, 705, 1997.

277. Xing, X. et al., Utilization of a copper-catalyzed diaryl ether synthesis for the preparation of verbenachalcone, *Tetrahedron*, 58, 7903, 2002.

278. Parmar, V.S. et al., Novel constituents of *Uvaria* species, *Nat. Prod. Rep.*, 11, 219, 1994.

279. Anam, E.M., 7-*O*-Methylbenzylflavanones and 4′-*O*-methylbenzyldihydrochalcone from *Xylopia africana* (Benth.) Oliver, *Indian J. Chem.*, 33B, 870, 1994.

280. Nkunya, M.H.H. et al., Benzylated dihydrochalcones from *Uvaria leptocladon*, *Phytochemistry*, 32, 1297, 1993.

281. Anam, E.M., 6″′′-Hydroxybenzyldiuvaretins and related compounds from *Xylopia africana* (Benth.) Oliver, *Indian J. Chem.*, 32B, 1051, 1993.

282. Anam, E.M., 5″,5″′′-(2″′′,2″′′′-Dihydroxybenzyl) diuvaretin and 5″′,3″′′-(2″′′,2″′′′-dihydroxydibenzyl)iso-diuvaretin: two novel C-dibenzylated dihydrochalcones from the root extract of *Xylopia africana* (Benth.) Oliver (Annonaceae), *Indian J. Chem.*, 36B, 57, 1997.

283. Anam, E.M., 2″′′,2″′′′,2″′′′′-Trihydroxy-5″,3″′′,5″′′′-tribenzyldiuvaretin & 2″′′,2″′′′,2″′′′′-trihydroxy-5″′,3″′′,5″′′′-tribenzyl isodiuvaretin: two novel tri-*C*-benzylated dihydrochalcones from the root extract of *Xylopia africana* (Benth.) Oliver (Annonaceae), *Indian J. Chem.*, 33B, 204, 1994.

284. Silva, D.H.S. et al., Flavonolignoids from fruits of *Iryanthera grandis*, *Phytochemistry*, 38, 1013, 1995.

285. Conserva, L.M. et al., Iryantherins, lignoflavonoids of novel structural types from the Myristicaceae, *Phytochemistry*, 29, 3911, 1990.

286. Ming, D.S. et al., Bioactive constituents from *Iryanthera megistophylla*, *J. Nat. Prod.*, 65, 1412, 2002.

287. Fang, S.-C., Shieh, B.-J., and Lin, C.-N., Phenolic constituents of Formosan *Broussonetia papyrifera*, *Phytochemistry*, 37, 851, 1994.

288. Alfatafta, A.A. and Mullin, C.A., Epicuticular terpenoids and an aurone from flowers of *Helianthus annuus*, *Phytochemistry*, 31, 4109, 1992.

289. Milovanović, M. and Djermanović, M., Constituents of *Helminthia echoides*, *Fitoterapia*, 65, 377, 1994.

290. Milovanović, M. et al., Antioxidant activities of the constituents of *Picris echoides*, *J. Serbian Chem. Soc.*, 67, 7, 2002.

291. Seabra, R.M. et al., Methylaurones from *Cyperus capitatus*, *Phytochemistry*, 48, 1429, 1998.

292. Seabra, R.M. et al., Methoxylated aurones from *Cyperus capitatus*, *Phytochemistry*, 45, 839, 1997.

293. Li, X.-C. et al., Two auronols from *Pseudolarix amabalis*, *J. Nat. Prod.*, 62, 767, 1999.

294. Atta-ur-Rahman et al., Two new aurones from marine brown alga *Spatoglossum variabile*, *Chem. Pharm. Bull.*, 49, 105, 2001.

295. Lawrence, N.J. et al., The total synthesis of an aurone isolated from *Uvaria hamiltonii*: aurones and flavones as anticancer agents, *Bioorg. Med. Chem. Lett.*, 13, 3759, 2003.

296. Mohan, P. and Joshi, T., Two aurone glycosides from the flowers of *Pterocarpus marsupium*, *Phytochemistry*, 28, 1287, 1989.

297. Mohan, P. and Joshi, T., Two anthochlor pigments from heartwood of *Pterocarpus marsupium*, *Phytochemistry*, 28, 2529, 1989.

298. Kobayashi, S., Miyase, T., and Noguchi, H., Polyphenolic glycosides and oligosaccharide multiesters from the roots of *Polygala dalmaisiana*, *J. Nat. Prod.*, 65, 319, 2002.

299. Zhang, S.-X. et al., Glycosyl flavonoids from the roots and rhizomes of *Asarum longerhizomatosum*, *J. Asian Nat. Prod. Res.*, 5, 25, 2003.

300. Singh, J., Aurone glycosides from *Pterocarpus santalinus*, *J. Indian Chem. Soc.*, 80, 190, 2003.

301. Yoshikawa, K. et al., Hovetrichosides C–G, five new glycosides of two auronols, two neolignans, and a phenylpropanoid from the bark of *Hovenia trichocarea*, *J. Nat. Prod.*, 61, 786, 1998.

302. Li, X.-C., Cai, L., and Wu, C.D., Antimicrobial compounds from *Ceanothus americanus* against oral pathogens, *Phytochemistry*, 46, 97, 1997.

303. Hahn, H. et al., The first biaurone, a triflavone and biflavonoids from two *Aulacomnium* species, *Phytochemistry*, 40, 573, 1995.

304. Westenburg, H.E. et al., Activity-guided isolation of antioxidative constituents of *Cotinus coggygria*, *J. Nat. Prod.*, 63, 1696, 2000.

305. Asada, Y., Li, W., and Yoshikawa, T., The first prenylated biaurone, licoagrone from hairy root cultures of *Glycyrrhiza glabra*, *Phytochemistry* 50, 1015, 1999.

306. Geiger, H. and Markham, K.R., Campylopusaurone, an auronoflavanone biflavonoid from the mosses *Campylopus clavatus* and *Campylopus holomitrum*, *Phytochemistry*, 31, 4325, 1992.

307. Bekker, R. et al., Biflavonoids. Part 5: Structure and stereochemistry of the first bibenzofuranoids, *Tetrahedron*, 56, 5297, 2000.
308. Bekker, R., Brandt, E.V., and Ferreira, D., Biflavonoids. Part 4. Structure and stereochemistry of novel flavanone- and the first isoflavanone–benzofuranone biflavonoids, *Tetrahedron*, 55, 10005, 1999.
309. Bekker, R., Brandt, E.V., and Ferreira, D., Absolute configuration of flavanone–benzofuranone-type biflavonoids and 2-benzyl-2-hydroxybenzofuranones, *J. Chem. Soc., Perkins Trans. 1*, 2535, 1996.
310. Bekker, R., Brandt, E.V., and Ferreira, D., Structure and stereochemistry of the first isoflavanone–benzofuranone biflavonoids, *Tetrahedron Lett.*, 39, 6407, 1998.
311. du Volsteedt, F.R. and Roux, D.G., Zeyherin, a natural 3,8-coumaranonylflavanone from *Phyllogeiton zeyheri* Sond., *Tetrahedron Lett.*, 12, 1647, 1971.
312. Nel, R.J.J. et al., Enantioselective synthesis of flavonoids. Part 5. Poly-oxygenated β-hydroxydi-hydrochalcones, *Tetrahedron Lett.*, 39, 5623, 1998.

APPENDIX A   Index of Molecular Formulae

Chalcones, dihydrochalcones, and aurones described in the literature from 1992 to 2003

Formula	Compound Number
$C_{15}H_9ClO_2$	353
$C_{15}H_{10}O_6$	343
$C_{15}H_{11}ClO_3$	354
$C_{15}H_{12}O_3$	1
$C_{15}H_{12}O_4$	3
$C_{15}H_{12}O_8$	351
$C_{15}H_{14}O_4$	256
$C_{15}H_{14}O_6$	273
$C_{16}H_{12}O_6$	344, 345
$C_{16}H_{14}O_5$	6, 11
$C_{16}H_{14}O_6$	18
$C_{16}H_{14}O_8$	352
$C_{16}H_{16}O_3$	249, 250, 251
$C_{16}H_{16}O_4$	252
$C_{16}H_{16}O_5$	263
$C_{16}H_{16}O_6$	274
$C_{17}H_{12}O_6$	24
$C_{17}H_{14}O_6$	346, 347
$C_{17}H_{16}O_4$	4
$C_{17}H_{16}O_5$	5, 8
$C_{17}H_{16}O_6$	15
$C_{17}H_{18}O_4$	253, 254, 255, 257
$C_{17}H_{18}O_5$	258, 260, 261, 265, 266
$C_{18}H_{16}O_5$	2, 12
$C_{18}H_{16}O_6$	342, 348
$C_{18}H_{16}O_7$	350
$C_{18}H_{17}BrO_5$	9
$C_{18}H_{18}O_4$	240
$C_{18}H_{18}O_5$	7, 10, 13, 14, 23, 264
$C_{18}H_{18}O_7$	20
$C_{18}H_{20}O_5$	259, 262, 267, 268

APPENDIX A   Index of Molecular Formulae — *continued*

Formula	Compound Number
$C_{18}H_{20}O_6$	269, 270, 271
$C_{19}H_{16}O_5$	293
$C_{19}H_{18}O_6$	17, 349
$C_{19}H_{18}O_7$	25
$C_{19}H_{20}O_4$	241
$C_{19}H_{20}O_6$	16, 272
$C_{19}H_{20}O_7$	19
$C_{20}H_{16}O_5$	182
$C_{20}H_{16}O_7$	96
$C_{20}H_{18}O_4$	51, 66
$C_{20}H_{18}O_5$	104, 340, 341
$C_{20}H_{18}O_6$	301
$C_{20}H_{20}O_4$	49, 57, 286
$C_{20}H_{20}O_5$	33, 41, 58, 102, 292
$C_{20}H_{20}O_6$	95
$C_{20}H_{20}O_7$	26
$C_{20}H_{22}O_7$	21
$C_{20}H_{22}O_{10}$	314
$C_{21}H_{20}O_4$	34, 38, 67
$C_{21}H_{20}O_5$	30, 35, 74
$C_{21}H_{22}O_4$	65, 287
$C_{21}H_{22}O_5$	36, 39, 64, 68, 73, 90, 91, 98
$C_{21}H_{22}O_6$	70, 75, 76
$C_{21}H_{22}O_9$	109
$C_{21}H_{22}O_{11}$	361, 362
$C_{21}H_{24}O_5$	288, 291
$C_{21}H_{24}O_7$	22
$C_{21}H_{24}O_{10}$	312, 313
$C_{22}H_{22}O_5$	79, 80, 89
$C_{22}H_{24}O_{10}$	118, 121, 124
$C_{22}H_{26}O_{10}$	309
$C_{23}H_{20}O_6$	236
$C_{23}H_{22}O_5$	327
$C_{23}H_{22}O_7$	108
$C_{23}H_{24}O_5$	99
$C_{23}H_{24}O_6$	92, 93, 94
$C_{23}H_{26}O_{11}$	131, 306, 307
$C_{24}H_{24}O_7$	37
$C_{24}H_{26}O_7$	97, 302
$C_{24}H_{30}O_{10}$	311
$C_{25}H_{24}O_4$	60
$C_{25}H_{24}O_5$	77, 88, 107
$C_{25}H_{24}O_{13}$	358
$C_{25}H_{26}O_3$	28, 29

APPENDIX A   Index of Molecular Formulae — *continued*

Formula	Compound Number
$C_{25}H_{26}O_4$	52, 54, 59
$C_{25}H_{26}O_5$	61, 72, 86, 103, 105
$C_{25}H_{26}O_{13}$	125
$C_{25}H_{28}O_3$	27
$C_{25}H_{28}O_4$	31, 45, 48, 62, 63, 191
$C_{25}H_{28}O_5$	32, 46, 47, 50, 69, 84, 100
$C_{25}H_{28}O_6$	53, 101, 106
$C_{25}H_{28}O_{12}$	310, 360
$C_{25}H_{30}O_4$	275
$C_{25}H_{30}O_5$	289, 294, 295
$C_{25}H_{30}O_6$	296
$C_{25}H_{30}O_{11}$	132
$C_{26}H_{22}O_7$	279
$C_{26}H_{28}O_7$	78
$C_{26}H_{30}O_4$	55, 187, 188, 192, 193, 194, 195, 196, 243, 282, 283, 284
$C_{26}H_{30}O_5$	40, 56, 71
$C_{26}H_{30}O_9$	113
$C_{26}H_{32}O_4$	277, 281
$C_{26}H_{32}O_5$	285
$C_{26}H_{32}O_6$	299, 300
$C_{27}H_{26}O_{14}$	359
$C_{27}H_{29}O_{13}N$	245
$C_{27}H_{30}O_{13}$	355
$C_{27}H_{30}O_{14}$	357
$C_{27}H_{30}O_{15}$	356
$C_{27}H_{31}O_{14}N$	248
$C_{27}H_{32}O_4$	185, 186
$C_{27}H_{32}O_{15}$	119, 130, 363
$C_{27}H_{32}O_{16}$	246
$C_{27}H_{34}O_{15}$	308
$C_{28}H_{22}O_7$	239
$C_{28}H_{34}O_4$	183, 184
$C_{28}H_{34}O_6$	189, 190
$C_{28}H_{34}O_{15}$	122
$C_{29}H_{26}O_7$	42, 43
$C_{29}H_{26}O_8$	81
$C_{29}H_{30}O_7$	278
$C_{29}H_{30}O_8$	280
$C_{29}H_{36}O_{15}$	133
$C_{29}H_{38}O_{15}$	305
$C_{30}H_{18}O_{10}$	365
$C_{30}H_{18}O_{12}$	364
$C_{30}H_{20}O_9$	135, 165
$C_{30}H_{20}O_{10}$	152

APPENDIX A   Index of Molecular Formulae — *continued*

APPENDIX A   Index of Molecular Formulae — *continued*

Formula	Compound Number
$C_{34}H_{34}O_8$	336, 337
$C_{34}H_{36}O_7$	334, 335
$C_{35}H_{30}O_{17}$	304
$C_{35}H_{32}O_{11}$	325
$C_{35}H_{34}O_7$	228, 229
$C_{35}H_{34}O_8$	222–226, 232
$C_{35}H_{34}O_9$	221, 227
$C_{35}H_{36}O_{15}$	111
$C_{35}H_{38}O_7$	338, 339
$C_{35}H_{44}O_5$	85
$C_{35}H_{46}O_4$	276
$C_{36}H_{32}O_{14}$	164
$C_{36}H_{38}O_{16}$	112
$C_{36}H_{42}O_6$	237
$C_{37}H_{34}O_7$	328, 329
$C_{37}H_{40}O_{15}$	123
$C_{39}H_{34}O_{16}$	127
$C_{39}H_{36}O_9$	204
$C_{40}H_{34}O_{12}$	214
$C_{40}H_{36}O_{11}$	218, 219
$C_{40}H_{36}O_{12}$	210, 211, 213, 216, 217
$C_{40}H_{38}O_8$	206, 207
$C_{40}H_{38}O_9$	205
$C_{40}H_{40}O_{12}$	215
$C_{42}H_{34}O_{21}$	303
$C_{42}H_{38}O_9$	230, 231
$C_{42}H_{40}O_8$	144
$C_{44}H_{40}O_8$	330, 331
$C_{44}H_{44}O_{24}$	247
$C_{45}H_{38}O_{13}$	172, 173
$C_{45}H_{42}O_{10}$	366
$C_{45}H_{44}O_{12}$	208, 212
$C_{45}H_{46}O_{11}$	209
$C_{48}H_{52}O_{26}$	244
$C_{50}H_{58}O_{10}$	319
$C_{51}H_{46}O_9$	332, 333
$C_{51}H_{48}O_{12}$	145
$C_{54}H_{54}O_{11}$	233–235
$C_{60}H_{48}O_{15}$	175, 176, 177
$C_{60}H_{50}O_{15}$	178, 179
$C_{66}H_{46}O_{21}$	174
$C_{75}H_{62}O_{21}$	180
$C_{90}H_{70}O_{22}$	181

APPENDIX B   Checklist of Known Chalcones, Dihydrochalcones, and Aurones[a]

CHALCONES
Chalcone aglycones
Mono-*O*-substituted
   (4')
1.   4'-OMe (H2/454)
Di-*O*-substituted
   (2',4')
2.   2',4'-diOH (H2/457)
3.   2',4'-diOH, 3'-prenyl (isocordoin, H2/582)
4.   2',4'-diOH, 3'-(1,1-dimethylallyl) (ψ-isocordoin, H2/583)
5.   2',4'-diOH, 3',5'-diprenyl (spinochalcone A, **27**)
6.   2'-OH, 4'-OMe, 3'-prenyl (derricin, H2/589)
7.   2'-OH, 4'-prenyloxy (derricidin, cordoin, H2/590)
8.   2'-OH, furano[2″,3″:4',3'] (H2/556)
9.   2'-OH, 5″-(2-hydroxyisopropyl)-4″,5″-dihydrofurano[2″,3″:4',3'] (flemistrictin B, H2/586)
10.   2'-OH, 6″,6″-dimethylpyrano[2″,3″:4',3'] (lonchocarpin, H2/587)
11.   2'-OH, 6″,6″-dimethyl-5″-hydroxy-4″,5″-dihydropyrano[2″,3″:4',3'] (flemistrictin C, H2/588)
12.   2'-OH, 6″-(4-methylpent-3-enyl),6″-methylpyrano[2″,3″:4',3'] (spinochalcone B, **28**)
13.   2'-OH, 3'-prenyl, 6″,6″-dimethylpyrano[2″,3″:4',5'] (spinochalcone C, **29**)
14.   2'-OH, complex substituent ((+)-tephrosone, **30**)
15.   4'-OH, 2'-OMe (H2/458)
16.   4'-OH, 5″-(2-hydroxyisopropyl)-4″,5″-dihydrofurano[2″,3″:2',3'] (flemistrictin E, H2/584)
17.   4'-OH, 6″,6″-dimethyl-5″-hydroxy-4″,5″-dihydropyrano[2″,3″:2',3'] (flemistrictin F, H2/585)
18.   2'-OMe, furano[2″,3″:4',3'] (ovalitenin A, H2/557)
19.   2'-OMe, 4-Me, furano[2″,3″:4',3'] (purpuritenin A, H2/559)
20.   2'-OMe, α-Me, furano[2″,3″:4',3'] (purpuritenin B, H2/560)
   (2',2)
21.   2',2-diOH (H2/455)
   (2',β)
22.   2',β-diOH (H2/530)
   (3',3)
23.   3',3-diOH (**1**)
   (4',4)
24.   4',4-diOH (H2/456)
Tri-*O*-substituted
   (2',3',4')

---

[a] This checklist of chalcones, dihydrochalcones, and aurones contains compounds of these classes reported in the literature as natural products to the end of 2003. Compounds published before 1992 are cross-referenced to numbered entries in volumes 1 and 2 of the *Handbook of Natural Flavonoids*[9,10] using the abbreviations H1 and H2, respectively. Compounds published from 1992 to 2003 are cross-referenced to Table 16.1– Table 16.15 using numbers in bold type. The compounds are listed according to the system outlined in Section 16.1.1, with the exception that isoprenylated derivatives are included under the heading of aglycones. Bn, benzyl.

APPENDIX B   Checklist of Known Chalcones, Dihydrochalcones, and Aurones — *continued*

25. 2′,4′-diOH, 3′-OMe (larrein, H2/468)
    (2′,3′,2)
26. 2′,3′,2-triOH, 4′-geranyl (flemiwallichin E, H2/661)
    (2′,4′,5′)
27. 2′,4′-diOH, 5′-OMe (flemichapparin, H2/469)
28. 2′,5′-diOH, 6″,6″-dimethylpyrano[2″,3″:4′,3′] (flemichapparin A, H2/616)
29. 2′,5′-diOH, (6″-(4-hydroxy-4-methylpent-2-enyl),6″-methyl)pyrano[2″,3″:4′,3′] (flemiwallichin F, H2/666)
    (2′,4′,6′)
30. 2′,4′,6′-triOH (H2/470)
31. 2′,4′,6′-triOH, 3′-prenyl (desmethylisoxanthohumol, H2/617)
32. 2′,4′,6′-triOH, 3′-geranyl (H2/667)
33. 2′,4′,6′-triOH, 3′-neryl (**31**)
34. 2′,4′,6′-triOH, 3′-C$_{10}$ (linderachalcone, H2/668)
35. 2′,4′,6′-triOH, 3′,5′-diC$_{10}$ (neolinderachalcone, H2/674)
36. 2′,4′-diOH, 6′-OMe (cardamonin, H2/472)
37. 2′,4′-diOH, 6′-OMe, 3′-Me (stercurensin, H2/579)
38. 2′,4′-diOH, 6′-OMe, 3′,5′-diMe (H2/581)
39. 2′,4′-diOH, 6′-OMe, 3′-prenyl (H2/622)
40. 2′,4′-diOH, 3′-Me, 6″,6″-dimethylpyrano[2″,3″:6′,5′] (H2/659)
41. 2′,4′-diOH, 3′-prenyl, 5″-(2-hydroxyisopropyl)-4″,5″-dihydrofurano[2″,3″:6′,5′] (cedrediprenone, **32**)
42. 2′,4′-diOH, 6″,6″-dimethylpyrano[2″,3″:6′,5′], 3′-(3-acetyl-5-methyl-2,4,6-trihydroxybenzyl) (rottlerin, H2/699)
43. 2′,6′-diOH, 4′-OMe (H2/471)
44. 2′,6′-diOH, 4′-OMe, 3′-CHO, 5′-Me (leridalchalcone, **2**)
45. 2′,6′-diOH, 4′-OMe, 3′-C$_{10}$ (H2/669)
46. 2′,6′-diOH, 4′-prenyloxy (H2/625)
47. 2′,6′-diOH, 4′-OMe, 3′-Me (triangularin, H2/578)
48. 2′,6′-diOH, 4′-OMe, 3′-prenyl (isoxanthohumol, H2/620)
49. 2′,6′-diOH, 6″,6″-dimethyl-5″-hydroxy-4″,5″-dihydropyrano[2″,3″:4′,3′] (helikrausichalcone, H2/618)
50. 4′,6′-diOH, 2′-OMe, 3′-Me (aurentiacin A, H2/577)
51. 4′,6′-diOH, 6″,6″-dimethyl-5″-hydroxy-4″,5″-dihydropyrano[2″,3″:2′,3′] (**33**)
52. 2′-OH, 4′,6′-diOMe (flavokawin B, H2/473)
53. 2′-OH, 4′,6′-diOMe, 3′-Me (aurentiacin, H2/580)
54. 2′-OH, 4′,6′-diOMe, 3′-prenyl (ovalichalcone, H2/624)
55. 2′-OH, 6′-OMe, 4′-prenyloxy (H2/626)
56. 2′-OH, 6′-OMe, 5″-isopropenyl-4″,5″-dihydrofurano[2″,3″:4′,3′] (crassichalcone, **34**)
57. 2′-OH, 6′-OMe, 6″,6″-dimethylpyrano[2″,3″:4′,3′] (pongachalcone I, oaxacacin (revised structure)[69] H2/623)
58. 2′-OH, 6′-OMe, 4″,5″-epoxy(6″,6″-dimethyl-4″,5″-dihydropyrano[2″,3″:4′,3′]) (epoxyobovatachalcone, **35**)
59. 2′-OH, 6′-OMe, 6″,6″-dimethyl-5″-hydroxy-4″,5″-dihydropyrano[2″,3″:4′,3′] (6′-methoxyhelikrausichalcone, **36**)
60. 2′-OH, 6′-OMe, (6″-(4-methylpent-3-enyl),6″-methyl)pyrano[2″,3″:4′,3′] (boesenbergin B, H2/671)

APPENDIX B   Checklist of Known Chalcones, Dihydrochalcones, and Aurones — *continued*

61. 2′-OH, 6′-OMe, complex substituent ((+)-tephropurpurin, **37**)
62. 2′-OH, bis(6″,6″-dimethylpyrano)[2″,3″:4′,3′],[2″,3″:6′,5′] (flemiculosin, H2/627)
63. 2′-OH, complex (4′,6′-di-*O*-,5′-*C*-) substituent, 3′-prenyl (3′-prenylrubranine, H2/697)
64. 4′-OH, 2′-OMe, 6″,6″-dimethylpyrano[2″,3″:6′,5′] (cedreprenone, **38**)
65. 4′-OH, 6′-OMe, 6″,6″-dimethyl-5″-hydroxy-4″,5″-dihydropyrano[2″,3″:2′,3′] (**39**)
66. 6′-OH, 4′-OMe, 6″,6″-dimethyl-5″-hydroxy-4″,5″-dihydropyrano[2″,3″:2′,3′] (helichromanochalcone, H2/621)
67. 6′-OH, 4′-OMe, (6″-(4-methylpent-3-enyl),6″-methyl)pyrano[2″,3″:2′,3′] (boesenbergin A, H2/670)
68. 6′-OH, 4′-OMe, complex substituent (linderol A, **40**)
69. 6′-OH, complex (2′,4′-di-*O*-,3′-*C*-) substituent (rubranine, H2/672)
     (2′,4′,4)
70. 2′,4′,4-triOH (isoliquiritigenin, H2/462)
71. 2′,4′,4-triOH, 3′-prenyl (isobavachalcone, H2/597)
72. 2′,4′,4-triOH, 3′-(2-hydroxy-3-methylbut-3-enyl) (**41**)
73. 2′,4′,4-triOH, 3′-geranyl (xanthoangelol, H2/663)
74. 2′,4′,4-triOH, 3′-(3,7-dimethyl-6-hydroxy-2,7-octadienyl) (xanthoangelol B, H2/664-structure drawing incorrect)
75. 2′,4′,4-triOH, 3′-(3-methyl-6-oxo-2-hexenyl) (xanthoangelol C, H2/698)
76. 2′,4′,4-triOH, 3′-(4-coumaroyloxy-3-methyl-but-2(*E*)-enyl) (isogemichalcone B, **42**)
77. 2′,4′,4-triOH, 3′-(4-coumaroyloxy-3-methyl-but-2(*Z*)-enyl) (gemichalcone B, **43**)
78. 2′,4′,4-triOH, 3′-(4-feruloyloxy-3-methyl-but-2(*Z*)-enyl) (gemichalcone A, **44**)
79. 2′,4′,4-triOH, 3′,5′-diprenyl (H2/612)
80. 2′,4′,4-triOH, 3′,3-diprenyl (kanzonol C, **45**)
81. 2′,4′,4-triOH, 3′-prenyl, 3-(2-hydroxy-3-methylbut-3-enyl) (paratocarpin D, **46**)
82. 2′,4′,4-triOH, 3′-(2-hydroxy-3-methylbut-3-enyl), 3-prenyl (paratocarpin E, **47**)
83. 2′,4′,4-triOH, 3′,3,5-triprenyl (sophoradin, H2/613)
84. 2′,4′,4-triOH, 5′-prenyl (broussochalcone B, H2/606)
85. 2′,4′,4-triOH, 5′,3-diprenyl (stipulin, **48**)
86. 2′,4′,4-triOH, 3-prenyl (licoagrochalcone A, **49**)
87. 2′,4′,4-triOH, 3,5-diprenyl (abyssinone VI, H2/610)
88. 2′,4′,4-triOH, 3-(2-hydroxy-3-methylbut-3-enyl), 5-prenyl (anthyllin, **50**)
89. 2′,4′-diOH, 4-OMe (H2/463)
90. 2′,4′-diOH, 6″,6″-dimethylpyrano[2″,3″:4,3] (kanzonol B, **51**)
91. 2′,4′-diOH, 3′-prenyl, 6″,6″-dimethylpyrano[2″,3″:4,3] (paratocarpin C, **52**)
92. 2′,4′-diOH, 3′-prenyl, 5″-(2-hydroxyisopropyl)-4″-hydroxy-4″,5″-dihydrofurano[2″,3″:4,3] (paratocarpin G, **53**)
93. 2′,4′-diOH,3′,5-diprenyl,6″,6″-dimethylpyrano[2″,3″:4,3](sophoradochromene,H2/614)
94. 2′,4′-diOH, 5-prenyl, 6″,6″-dimethylpyrano[2″,3″:4,3] (anthyllisone, **54**)
95. 2′,4-diOH, 4′-OMe (H2/465)
96. 2′,4-diOH, 4′-OMe, 5′-CHO (neobavachalcone, H2/576)
97. 2′,4-diOH, 4′-OMe, 3′-prenyl (H2/602)
98. 2′,4-diOH, 4′-OMe, 3′-(2-hydroxy-3-methylbut-3-enyl) (xanthoangelol D, H2/603)
99. 2′,4-diOH, 4′-OMe, 3′-(2-hydroperoxy-3-methylbut-3-enyl) (xanthoangelol E, H2/604)
100. 2′,4-diOH, 4′-OMe, 3′-geranyl (xanthoangelol F, **55**)
101. 2′,4-diOH, 4′-OMe, 3′-(3,7-dimethyl-6-hydroxy-2,7-octadienyl) (xanthoangelol G, **56**)
102. 2′,4-diOH, 4′-OMe, 5′-prenyl (bavachalcone, H2/608)

APPENDIX B Checklist of Known Chalcones, Dihydrochalcones, and Aurones — *continued*

103. 2′,4-diOH, 4′-prenyloxy (H2/605)
104. 2′,4-diOH, 4′-geranyloxy (H2/662)
105. 2′,4-diOH, 5″-(2-hydroxyisopropyl)-4″,5″-dihydrofurano[2″,3″,4′,3′] (bakuchalcone, H2/598)
106. 2′,4-diOH, 6″,6″-dimethylpyrano[2″,3″:4′,3′] (isobavachromene, H2/599)
107. 2′,4-diOH, 6″,6″-dimethyl-4″,5″-dihydropyrano[2″,3″:4′,3′] (dorsmanin A, **57**)
108. 2′,4-diOH, 6″,6″-dimethyl-5″-hydroxy-4″,5″-dihydropyrano[2″,3″:4′,3′] (**58**)
109. 2′,4-diOH, 3-prenyl, 6″,6″-dimethylpyrano[2″,3″:4′,3′] (paratocarpin B, **59**)
110. 2′,4-diOH, 6″,6″-dimethylpyrano[2″,3″:4′,5′] (bavachromene, H2/607)
111. 2′,4-diOH, (6″-(4-methylpent-3-enyl),6″-methyl)pyrano[2″,3″:4′,3′] (lespeol, H2/665)
112. 4′,4-diOH, 2′-OMe (H2/464)
113. 4′,4-diOH, 2′-OMe, 5′-CHO (isoneobavachalcone, H2/575)
114. 4′,4-diOH, 6″,6″-dimethyl-5″-hydroxy-4″,5″-dihydropyrano[2″,3″:2′,3′] (bavachromanol, H2/600)
115. 2′-OH, 4′,4-diOMe (H2/466)
116. 2′-OH, 4′,4-diOMe, 5′-prenyl (H2/609)
117. 2′-OH, 4-OMe, 6″,6″-dimethylpyrano[2″,3″:4′,3′] (H2/601)
118. 2′-OH, bis(6″,6″-dimethylpyrano)[2″,3″:4′,3′],[2′′,3′′:4,3] (paratocarpin A, **60**)
119. 2′-OH, 5″-(2-hydroxyisopropyl)-4″,5″-dihydrofurano[2″,3″:4′,3′], 6″,6″-dimethylpyrano[2″,3″:4,3] (paratocarpin F, **61**)
120. 2′-OH, bis(6″,6″-dimethyl-4″,5″-dihydropyrano)[2″,3″:4′,3′],[2′′,3′′:4,3] (artoindonesianin J, **62**)
121. 2′-OH, bis(6″,6″-dimethyl-4″,5″-dihydropyrano)[2″,3″:4′,5′],[2′′,3′′:4,3] (**63**)
122. 4-OH, 4′-OMe, 6″,6″-dimethyl-5″-hydroxy-4″,5″-dihydropyrano[2″,3″:2′,3′] (xanthoangelol H, **64**)
    (2′,4′,β)
123. β-OH, 2′,4-diOMe, 5′-prenyl (pongagallone A, H2/540)
124. β-OH, 2′-OMe, 6″,6″-dimethylpyrano[2″,3″:4′,3′] (H2/539)
125. β-OH, 2′-OMe, furano[2″,3″:4′,3′] (pongamol, H2/555)
126. 2′,β-diOMe, furano[2″,3″:4′,3′] (H2/558)
    (2′,5′,4)
127. 2′,5′,4-triOH (**3**)
128. 2′,5′-diOH, 4-OMe (H2/467)
    (2′,3,4)
129. 2′,3,4-triOH (H2/461)
    (3′,4′,4)
130. 4′,4-diOH, 6″,6″-dimethyl-4″,5″-dihydropyrano[2″,3″:3′,2′] (crotmadine, H2/615)
    (4′,2,4)
131. 4′,4-diOH, 2-OMe (echinatin, H2/459)
132. 4′,4-diOH, 2-OMe, 3-prenyl (licochalcone C, **65**)
133. 4′,4-diOH, 2-OMe, 5-(1,1-dimethylallyl) (licochalcone A, H2/594)
134. 2,4-diOH, 6″,6″-dimethylpyrano[2″,3″:4′,3′] (munsericin, **66**)
135. 4′-OH, 2-OMe, 6″,6″-dimethylpyrano[2″,3″:4,3] (licoagrochalcone B, **67**)
136. 4′-OH, 2-OMe, 5″-(2-hydroxyisopropyl)-4″,5″-dihydrofurano[2″,3″:4,3] (licoagrochalcone D, **68**)
137. 4-OH, 4′,2-diOMe (glypallichalcone, **4**)
    (2,4,6)

APPENDIX B Checklist of Known Chalcones, Dihydrochalcones, and Aurones — *continued*

138. 4,6-diOH, 2-OMe, 3-CHO, 5-Me (H2/574)

Tetra-*O*-substituted

(*2′,3′,4′,6′*)

139. 2′,3′,6′-triOH, 6″,6″-dimethylpyrano[2″,3″:4′,5′] (mallotus A, H2/647)

140. 2′,3′-diOH, 4′,6′-diOMe (H2/498)

141. 2′,4′-diOH, 3′,6′-diOMe (H2/497)

142. 2′,4′-diOH, 6′-OMe, 3′-angeloyloxy (H2/644)

143. 2′,4′-diOH, 6′-OMe, 3′-(2-methylbutyryloxy) (H2/645)

144. 2′,4′-diOH, 6′-OMe, 3′-isovaleryloxy (H2/646)

145. 2′,6′-diOH, 3′,4′-diOMe (pashanone, H2/496)

146. 2′-OH, 3′,4′,6′-triOMe (H2/500)

147. 6′-OH, 2′,3′,4′-triOMe (helilandin B, H2/499)

148. 2′,6′-diOMe, 3′,4′-OCH$_2$O- (helilandin A, H2/563)

149. 2′,3′,4′,6′-tetraOMe (H2/501)

(*2′,3′,4′,4*)

150. 2′,3′,4′,4-tetraOH (H2/485)

151. 2′,4′,4-triOH, 3′-OMe (kukulkanin B, H2/486)

152. 2′,4′-diOH, 3′,4-diOMe (kukulkanin A, H2/487)

153. 2′,4-diOH, 3′,4′-diOMe (heliannone A, **5**)

(*2′,4′,5′,2*)

154. 2′,4′,2-triOH, 5′-OMe, 3′-geranyl (H2/681)

155. 2′,5′,2-triOH, (6″-(4-methylpent-3-enyl),6″-methyl)pyrano[2″,3″:4′,3′] (flemingin A, H2/680)

(*2′,4′,5′,4*)

156. 2′,5′,4-triOH, (6″-(4-methylpent-3-enyl),6″-methyl)pyrano[2″,3″:4′,3′] (flemingin D, H2/682)

157. 2′,5′,4-triOH, (6″-(4-hydroxy-4-methylpent-2-enyl),6″-methyl)pyrano[2″,3″:4′,3′] (flemingin E, H2/683)

158. 2′,5′,4-triOH, (6″-(3-hydroxy-4-methylpent-4-enyl),6″-methyl)pyrano[2″,3″:4′,3′] (flemingin F, H2/684)

(*2′,4′,6′,2*)

159. 2′,4′,2-triOH, 6′-OMe (**6**)

160. 2′-OH, 4′,6′,2-triOMe (**7**)

(*2′,4′,6′,4*)

161. 2′,4′,6′,4-tetraOH (chalconaringenin, H2/488)

162. 2′,4′,6′,4-tetraOH, 3′-geranyl (**69**)

163. 2′,4′,6′-triOH, 4-OMe (H2/489)

164. 2′,4′,4-triOH, 6′-OMe (helichrysetin, H2/491)

165. 2′,4′,4-triOH, 6′-OMe, 3′-prenyl (xanthohumol, H2/638)

166. 2′,4′,4-triOH, 6′-OMe, 3′-(2-hydroxy-3-methylbut-3-enyl) (xanthohumol D, **70**)

167. 2′,4′,4-triOH, 6′-OMe, 3′,5′-diprenyl (5′-prenylxanthohumol, **71**)

168. 2′,4′,4-triOH, 3′-prenyl, 6″,6″-dimethylpyrano[2″,3″:6′,5′] (xanthohumol E, **72**)

169. 2′,4′,4-triOH, 3′-(3-acetyl-5-methyl-2,4,6-trihydroxybenzyl), 6″,6″-dimethylpyrano[2″,3″:6′,5′] (4-hydroxyrottlerin, H2/701)

170. 2′,6′,4-triOH, 4′-OMe (neosakuranetin, H2/490)

171. 2′,6′,4-triOH, 4′-OMe, 3′-prenyl (xanthogalenol, **73**)

172. 2′,6′,4-triOH, 4′-prenyloxy (H2/640)

APPENDIX B   Checklist of Known Chalcones, Dihydrochalcones, and Aurones — *continued*

173. 2',6',4-triOH, 3'-prenyl, 6'',6''-dimethylpyrano[2'',3'':4',5'] (sericone, H2/641)
174. 2',4'-diOH, 6',4-diOMe (**8**)
175. 2',6'-diOH, 4',4-diOMe (gymnogrammene, H2/492)
176. 2',4-diOH, 4',6'-diOMe (flavokawin C, H2/494)
177. 2',4-diOH, 4',6'-diOMe, 3'-prenyl (H2/639)
178. 2',4-diOH, 6'-OMe, 6'',6''-dimethylpyrano[2'',3'':4',3'] (xanthohumol C, **74**)
179. 2',4-diOH, 6'-OMe, 6'',6''-dimethyl-4''-hydroxy-4'',5''-dihydropyrano[2'',3'':4',3']
    (isodehydrocycloxanthohumol hydrate, **75**)
180. 2',4-diOH, 6'-OMe, 6'',6''-dimethyl-5''-hydroxy-4'',5''-dihydropyrano[2'',3'':4',3']
    (xanthohumol B, **76**)
181. 2',4-diOH, bis(6'',6''-dimethylpyrano)[2'',3'':4',3'],[2'',3'':6',5'] (laxichalcone, **77**)
182. 2',4-diOH, 6'',6''-dimethylpyrano[2'',3'':4',3'], 6'',6''-dimethyl-4''-hydroxy-5''-methoxy-
    4'',5''-dihydropyrano[2'',3'':6',5'] (derrichalcone, **78**)
183. 4',4-diOH, 2',6'-diOMe (H2/493)
184. 6',4-diOH, 4'-OMe, 6'',6''-dimethylpyrano[2'',3'':2',3'] (citrunobin, H2/637)
185. 2'-OH, 4',6',4-triOMe (flavokawin A, H2/495)
186. 2'-OH, 4',6',4-triOMe, 5'-Br (**9**)
187. 2'-OH, 4',4-diOMe, 6'',6''-dimethylpyrano[2'',3'':6',5'] (**79**)
188. 2'-OH, 6',4-diOMe, 6'',6''-dimethylpyrano[2'',3'':4',3'] (glychalcone A, **80**)
189. 4'-OH, 2',6',4-triOMe (**10**)
     (2',4',6',β)
190. 2',4',6',β-tetraOH, 3'-CHO, 5'-Me (H2/537)
191. 2',4',β-triOH, 6'-OMe (H2/534)
192. 2',6',β-triOH, 4'-OMe (H2/533)
193. 2',6',β-triOH, 4'-OMe, 3'-CHO, 5'-Me (H2/536)
194. 2',6',β-triOH, 4'-OMe, 3'-prenyl (**98**)
195. 2',β-diOH, 4',6'-diOMe, 3'-Me (**23**)
196. 2',β-diOH, 6'-OMe, 6'',6''-dimethylpyrano[2'',3'':4',3'] (demethylpraecansone B, H2/538)
197. β-OH, 2',6'-diOMe, 6'',6''-dimethylpyrano[2'',3'':4',3'] (praecansone B, H2/549)
198. 2',6',β-triOMe, 6'',6''-dimethylpyrano[2'',3'':4',3'] (**99**)
     (2',4',2,4)
199. 2',4',2,4-tetraOH (H2/477)
200. 2',4',2,4-tetraOH, 3'-prenyl (morachalcone A, H2/630)
201. 2',4',2,4-tetraOH, 3'-lavandulyl (ammothamnidin, H2/677)
202. 2',4',2,4-tetraOH, 3'-(4-coumaroyloxy-3-methyl-but-2(*E*)-enyl)
    (demethoxyisogemichalcone C, **81**)
203. 2',4',2,4-tetraOH, 3'-(4-feruloyloxy-3-methyl-but-2(*E*)-enyl) (isogemichalcone C, **82**)
204. 2',4',2,4-tetraOH, 3'-(4-feruloyloxy-3-methyl-but-2(*Z*)-enyl) (gemichalcone C, **83**)
205. 2',4',2,4-tetraOH, 5-geranyl (kuwanol D, H2/676)
206. 2',2,4-triOH, 3'-prenyl, 6'',6''-dimethylpyrano[2'',3'':4',5'] (H2/631)
     (2',4',2,5)
207. 2',2,5-triOH (6''-(4-methylpent-3-enyl),6''-methyl)pyrano[2'',3'':4',3'] (flemiwallichin D,
    H2/679)
     (2',4',3,4)
208. 2',4',3,4-tetraOH (butein, H2/478)
209. 2',4',3,4-tetraOH, 3'-geranyl (**84**)
210. 2',4',3,4-tetraOH, 3',5-digeranyl (**85**)

APPENDIX B   Checklist of Known Chalcones, Dihydrochalcones, and Aurones — *continued*

211. 2',4',3,4-tetraOH, 5'-prenyl (broussochalcone A, H2/634)
212. 2',4',3-triOH, 4-OMe (H2/480)
213. 2',4',3-triOH, 3'-prenyl 6'',6''-dimethylpyrano[2'',3'':4,5] (**86**)
214. 2',4',4-triOH, 3-OMe (homobutein, H2/479)
215. 2',4',4-triOH, 3'-prenyl, (6''-(4-methylpent-3-enyl),6''-methyl)pyrano[2'',3'':3,2] (poinsettifolin B, **87**)
216. 2',3,4-triOH, 4'-OMe (calythropsin, **11**)
217. 2',3,4-triOH, 6'',6''-dimethylpyrano[2'',3'':4',3'] (H2/632)
218. 4',3,4-triOH, 2'-OMe (sappanchalcone, H2/481)
219. 2',4'-diOH, 3,4-diOMe (H2/483)
220. 2',3-diOH, bis(6'',6''-dimethylpyrano)[2'',3'':4',3'],[2'',3'':4,5] (glyinflanin G, **88**)
221. 2',4-diOH, 3'-OMe, 6'',6''-dimethylpyrano[2'',3'':4',3'] (pongachalcone II, H2/633)
222. 2'-OH, 3,4-diOMe, 6'',6''-dimethylpyrano[2'',3'':4',3'] (3,4-dimethoxylonchocarpin, **89**)
223. 2'-OH, 3,4-OCH$_2$O-, 6'',6''-dimethylpyrano[2'',3'':4',3'] (glabrachromene II, H2/571)
224. 2'-OMe, 3,4-OCH$_2$O-, furano[2'',3'':4',3'] (ovalitenin C, H2/561)
225. 2',4'-diOMe, 3,4-OCH$_2$O- (**12**)
     (2',4',3,5)
226. 2',4',3,5-tetraOH (pseudosindorin, H2/484)
     (2',4',4,β)
227. 2',4',4,β-tetraOH (licodione, H2/531)
228. 2',4',4,β-tetraOH, 5'-prenyl (H2/541)
229. 2',4',4,β-tetraOH, 5',3-diprenyl (glycyrdione A, **100**)
230. 2',4',4,β-tetraOH, 5'-prenyl, 3-(2-hydroxy-3-methylbut-3-enyl) (glyinflanin E, **101**)
231. 2',4',4,β-tetraOH, 3-prenyl (kanzonol A, **102**)
232. 2',4',β-triOH, 5'-prenyl, 6'',6''-dimethylpyrano[2'',3'':4,3] (glycyrdione B, **103**)
233. 2',4,β-triOH, 6'',6''-dimethylpyrano[2'',3'':4,5] (glyinflanin B, **104**)
234. 2',4,β-triOH, 3-prenyl, 6'',6''-dimethylpyrano[2'',3'':4,5] (glyinflanin C, **105**)
235. 2',4,β-triOH, 3-prenyl, 5''-(2-hydroxyisopropyl)-4'',5''-dihydrofurano[2'',3'':4,5] (glyinflanin F, **106**)
236. 4',4,β-triOH, 2'-OMe (H2/532)
237. 2',β-diOH, bis(6'',6''-dimethylpyrano)[2'',3'':4,5],[2'',3'':4,3] (glyinflanin D, **107**)
     (2',3,4,5)
238. 2'-OH, 3,4,5-triOMe (crotaoprostrin, **13**)
     (3',4',2,4)
239. 3',4',4-triOH, 2-OMe, 3-prenyl (licoagrochalcone C, **90**)
     (4',2,3,4)
240. 4',3,4-triOH, 2-OMe (licochalcone B, H2/475)
241. 4',3,4-triOH, 2-OMe, 3'-prenyl (licochalcone D, **91**)
     (4',2,4,6)
242. 2,4-diOH, 4',6-diOMe (H2/476)
     (2,3,4,6)
243. 2-OH, 3,4,6-triOMe (tepanone, **14**)
     (2,4,6,β)
244. 2,6,β-triOMe, 6'',6''-dimethylpyrano[2'',3'':4,3] (praecansone A, H2/629)
Penta-*O*-substituted
     (2',3',4',5',6')
245. 2',4'-diOH, 3',5',6'-triOMe (isodidymocarpin, H2/520)

APPENDIX B   Checklist of Known Chalcones, Dihydrochalcones, and Aurones — *continued*

246. 2',5'-diOH, 3',4',6'-triOMe (pedicin, H2/519)
247. 4',6'-diOH, 2',5'-diOMe, 3'-angeloyloxy (H2/656)
248. 4',6'-diOH, 2',5'-diOMe, 3'-(2-methylbutyryloxy) (melafolone, H2/657)
249. 4',6'-diOH, 2',5'-diOMe, 3'-isovaleryloxy (valafolone, H2/658)
250. 2'-OH, 3',4',5',6'-tetraOMe (kanakugiol, H2/522)
251. 3'-OH, 2',4',5',6'-tetraOMe (H2/521)
252. 2',3',4',5',6'-pentaOMe (pedicellin, H2/523)
       (2',3',4',6',2)
253. 2',4'-diOH, 3',6',2-triOMe (H2/504)
       (2',3',4',6',4)
254. 2',4-diOH, 3',4',6'-triOMe (H2/517)
255. 2',4-diOH, 6'-OMe, 3',4'-OCH₂O- (H2/564)
256. 6',4-diOH, 2',3',4'-triOMe (H2/516)
257. 6'-OH, 2',3',4',4-tetraOMe (H2/518)
       (2',3',4',3,4)
258. 2',3',4',3,4-pentaOH (okanin, H2/506)
259. 2',4',3,4-tetraOH, 3'-OMe (lanceoletin, H2/507)
260. 2',4',4-triOH, 3',3-diOMe (**15**)
261. 2'-OH, 3',4',3,4-tetraOMe (H2/508)
       (2',4',5',2,5)
262. 2',4',2,5-tetraOH, 5'-OMe, 3'-geranyl (homoflemingin, H2/694)
263. 2',5',2,5-tetraOH (6''-(4-methylpent-3-enyl),6''-methyl)pyrano[2'',3'':4',3'] (flemingin C, H2/691)
       (2',4',5',2,6)
264. 2',5',2,6-tetraOH (6''-(4-methylpent-3-enyl),6''-methyl)pyrano[2'',3'':4',3'] (flemingin B, H2/692)
       (2',4',5',3,4)
265. 2',4',5',3,4-pentaOH (neoplathymenin, H2/509)
266. 2',4'-diOH, 5'-OMe, 3,4-OCH₂O- (prosogerin B, H2/565)
267. 2'-OH, 5',3,4-triOMe, 6',6''-dimethylpyrano[2'',3'':4',3'] (ponganone VI, **92**)
268. 5'-OH, 2'-OMe, 3,4-OCH₂O-, furano[2'',3'':4',3'] (H2/562)
       (2',4',6',2,3)
269. 2'-OH, 4',6',2,3-tetraOMe (**16**)
       (2',4',6',2,4)
270. 2',4',2,4-tetraOH, 6'-OMe, 3'-lavandulyl (kuraridin, H2/687)
271. 2',4',2,4-tetraOH, 6'-OMe, 3'-(5-hydroxy-2-isopropenyl-5-methylhexyl) (kuraridinol, H2/688)
272. 2',4',4-triOH, 6',2-diOMe 3'-lavandulyl (kushenol D, H2/678)
273. 2',4'-diOH, 6',2,4-triOMe (cerasin, H2/502)
274. 2'-OH, 4',6',2,4-tetraOMe (cerasidin, H2/503)
       (2',4',6',2,5)
275. 2',4',2,5-tetraOH, 6'-OMe, 3'-geranyl (flemiwallichin C, H2/690)
276. 2',6',2,5-tetraOH (6''-(4-methylpent-3-enyl),6''-methyl)pyrano[2'',3'':4',3'] (flemiwallichin B, H2/689)
       (2',4',6',2,6)
277. 2',6',2,6-tetraOH (6''-(4-methylpent-3-enyl),6''-methyl)pyrano[2'',3'':4',3'] (flemiwallichin A, H2/693)

APPENDIX B   Checklist of Known Chalcones, Dihydrochalcones, and Aurones — *continued*

278. 2′,2,6-triOH, 6′-OMe, 3′-prenyl, 6″,6″-dimethylpyrano[2″,3″:4′,5′] (orotinichalcone, H2/649)

(2′,4′,6′,3,4)

279. 2′,4′,6′,3,4-pentaOH (H2/510)

280. 2′,4′,6′,3-tetraOH, 4-OMe, 3′,2-diprenyl (antiarone C, H2/655)

281. 2′,4′,6′,3-tetraOH, 4-OMe, 2,5-diprenyl (antiarone E, H2/652)

282. 2′,4′,6′,3-tetraOH, 4-OMe, (2,β)-C_5 (antiarone J, H2/653)

283. 2′,4′,6′,4-tetraOH, 3-OMe (H2/511)

284. 2′,4′,6′,4-tetraOH, 3-OMe, 3′,2-diprenyl (antiarone D, H2/654)

285. 2′,4′,3,4-tetraOH, 6″,6″-dimethylpyrano[2″,3″:6′,5′], 3′-(3-acetyl-5-methyl-2,4,6-trihydroxybenzyl) (3,4-dihydroxyrottlerin, H2/700)

286. 2′,6′,3,4-tetraOH, 3′,4′-dihydrooxepino (H2/651)

287. 2′,4′,6′-triOH, 3,4-diOMe, (2,β)-C_5 (antiarone K, H2/650)

288. 2′,6′,4-triOH, 4′,3-diOMe (H2/512)

289. 2′,3-diOH, 4′,6′,4-triOMe (H2/513)

290. 2′-OH, 4′,6′,3,4-tetraOMe (H2/514)

291. 2′-OH, 4′,3,4-triOMe, 6″,6″-dimethylpyrano[2″,3″:6′,5′] (**93**)

292. 2′-OH, 6′,3,4-triOMe, 6″,6″-dimethylpyrano[2″,3″:4′,3′] (glychalcone B, **94**)

293. 2′-OH, 4′,6′-diOMe, 3,4-OCH$_2$O- (tephrone, H2/566)

294. 2′-OH, 4′,6′-diOMe, 3,4-OCH$_2$O-, 3′-prenyl (ovalichalcone A, H2/572)

295. 2′-OH, 6′-OMe, 3,4-OCH$_2$O-, 6″,6″-dimethylpyrano[2″,3″:4′,3′] (glabrachromene I, H2/573)

296. 2′,4′,6′-triOMe, 3,4-OCH$_2$O- (**17**)

(2′,4′,2,4,5)

297. 2′,4′,2,4,5-pentaOH, 3′-prenyl (ramosismin, **95**)

298. 2′-OH, 2,4,5-triOMe, 6″,6″-dimethylpyrano[2″,3″:4′,3′] (glabrachalcone, H2/648)

(2′,4′,3,4,5)

299. 2′,4′,3,4,5-pentaOH (robtein, H2/505)

(2′,4′,3,4,α)

300. 2′,4′,3,4,α-pentaOH (H2/515)

301. 2′,4′,3,4-tetraOH, *cyclo*[α-OCH$_2$-2] (mopachalcone, H2/720)

(2′,4′,3,4,β)

302. β-OH, 2′,4′-diOMe, 3,4-OCH$_2$O- (milletenone, H2/552)

303. β-OH, 2′,4′-diOMe, 3,4-OCH$_2$O-, 5′-prenyl (pongagallone B, H2/550)

304. β-OH, 2′,4′-diOMe, 3,4-OCH$_2$O-, α-Me (tinosporinone, H2/553)

305. β-OH, 2′-OMe, 3,4-OCH$_2$O-, furano[2″,3″:4′,3′] (ovalitenone, H2/554)

(2′,4′,4,5,α)

306. 2′,4′,4,5-tetraOH, *cyclo*[α-OCH$_2$-2] (peltochalcone, H2/719)

(3′,4′,2,3,4)

307. 3′,4′,3,4-tetraOH, 2-OMe (**18**)

(3′,4′,3,4,β)

308. β-OH, 3′,4′:3,4-bis(-OCH$_2$O-) (galiposin, **24**)

Hexa-*O*-substituted

(2′,3′,4′,5′,6′,4)

309. 2′,5′-diOH, 3′,4′,6′,4-tetraOMe (H2/527)

310. 3′,4-diOH, 2′,4′,5′,6′-tetraOMe (**19**)

(2′,3′,4′,6′,3,4)

APPENDIX B Checklist of Known Chalcones, Dihydrochalcones, and Aurones — *continued*

311. 6',3,4-triOH, 2',3',4'-triOMe (hamilcone, **20**)
312. 3',6'-diOH, 2',4'-diOMe, 3,4-OCH$_2$O- (agestricin, H2/567)
313. 2'-OH, 3',4',6',3,4-pentaOMe (**21**)
314. 2'-OH, 3',6'-diOMe, 3,4-OCH$_2$O-, furano[2'',3'':4',5'] (**96**)
315. 6'-OH, 2',3',4',3,4-pentaOMe (H2/526)
    (2',4',6',2,3,4)
316. 2',4',6',2,3,4-hexaOMe (**22**)
    (2',4',6',2,4,5)
317. 2'-OH, 4',6',2,4,5-pentaOMe (rubone, H2/524)
    (2',4',6',3,4,5)
318. 2'-OH, 4',6',3,4,5-pentaOMe (H2/525)
319. 6'-OH, 4',3,4,5-tetraOMe, 6'',6''-dimethylpyrano[2'',3'':2',3'] (**97**)
    (2',4',6',3,4,β)
320. β-OH, 2',4',6'-triOMe, 3,4-OCH$_2$O- (ponganone X, **25**)
321. β-OH, 2',6'-diOMe, 3,4-OCH$_2$O-, 6'',6''-dimethylpyrano[2'',3'':4',3'] (pongapinone A, **108**)
    (3',4',2,4,6,β)
322. 2,4,6,β-tetraOMe, 3',4'-OCH$_2$O- (**26**)
Octa-*O*-substituted
    (2',3',4',6',2,3,4,5)
323. 2',2-diOH, 6',3,4,5-tetraOMe, 3',4'-OCH$_2$O- (H2/568)
324. 6'-OH, 2',3',4',2,3,4,5-heptaOMe (H2/528)
325. 6'-OH, 2',2,3,4,5-pentaOMe, 3',4'-OCH$_2$O- (H2/569)
326. 2',3',4',6',2,3,4,5-octaOMe (H2/529)
327. 2',6',2,3,4,5-hexaOMe, 3',4'-OCH$_2$O- (H2/570)
    (2',4',5',2,3,4,5,β)
328. β-OH, 2',4',5',2,3,4,5-heptaOMe (H2/535)

Chalcone glycosides
4'-Hydroxychalcone
329. 4'-*O*-Glucoside (H2/721)
2',4-Dihydroxychalcone
330. 4-*O*-Glucoside (H2/723)
3,4-Dihydroxychalcone
331. 4-*O*-β-Arabinosyl-(1''' → 4'')-galactoside (H2/722)
2',3',4-Trihydroxychalcone
332. 4-*O*-Glucoside (**109**)
2',4',4-Trihydroxychalcone (isoliquiritigenin)
333. 2'-*O*-Glucosyl-(1''' → 4'')-rhamnoside (H2/729)
334. 4'-*O*-Glucoside (neoisoliquiritin, H2/725)
335. 4'-*O*-Glucosylglucoside (H2/730)
336. 4',4-Di-*O*-glucoside (H2/728)
337. 4'-*O*-Glucoside 4-*O*-apiofuranosyl-(1''' → 2'')-glucoside (**110**)
338. 4'-*O*-Glucosylglucoside 4-*O*-glucoside (H2/731)
339. 4-*O*-Glucoside (isoliquiritin, H2/724)
340. 4-*O*-Apiofuranosyl-(1''' → 2'')-glucoside (licuroside, H2/726)
341. 4-*O*-(5'''-*O*-*p*-Coumaroyl)apiofuranosyl-(1''' → 2'')-glucoside (**111**)
342. 4-*O*-(5'''-*O*-Feruloyl)apiofuranosyl-(1''' → 2'')-glucoside (**112**)
343. 4-*O*-Rhamnosylglucoside (H2/727)

APPENDIX B Checklist of Known Chalcones, Dihydrochalcones, and Aurones — *continued*

344. 3'-*C*-Glucoside (H1/2597)

2',4',4-Trihydroxy-3'-prenylchalcone (3'-prenylisoliquiritigenin)

345. 4'-*O*-Glucoside (**113**)

2',3',4',4-Tetrahydroxychalcone

346. 4'-*O*-(2''-*O*-*p*-Coumaroyl)glucoside (**114**)

347. 4'-*O*-(6''-*O*-*p*-Coumaroyl)glucoside (**115**)

348. 4'-*O*-(2''-*O*-Acetyl-6''-*O*-cinnamoyl)glucoside (**116**)

349. 4'-*O*-(2'''-*O*-*p*-Coumaroyl-6''-*O*-acetyl)glucoside (**117**)

2',6',2-Trihydroxy-4'-methoxychalcone

350. 2'-*O*-Glucoside (androechin, **118**)

2',4',6',4-Tetrahydroxychalcone (chalconaringenin)

351. 2'-*O*-Glucoside (isosalipurposide, H2/744)

352. 2'-*O*-(6''-*O*-*p*-Coumaroyl)glucoside (H2/742)

353. 2'-*O*-Xyloside (H2/743)

354. 2'-*O*-Rhamnosyl-(1''' → 4'')-glucoside (H2/747)

355. 2'-*O*-Rhamnosyl-(1''' → 4'')-xyloside (H2/746)

356. 2',4'-Di-*O*-glucoside (**119**)

357. 2'-*O*-Glucoside 4'-*O*-gentobioside (**120**)

358. 4'-*O*-Glucoside (H2/745)

359. 4-*O*-Glucoside (H2/741)

2',4',4-Trihydroxy-6'-methoxychalcone (helichrysetin)

360. 4'-*O*-Glucoside (helichrysin, H2/749)

361. 4-*O*-Glucoside (**121**)

362. 4',4-Di-*O*-α-glucoside (**122**)

2',6',4-Trihydroxy-4'-methoxychalcone (neosakuranetin)

363. 2'-*O*-Glucoside (neosakuranin, H2/748)

2',4-Dihydroxy-4',6'-dimethoxychalcone (flavokawin C)

364. 4-*O*-Glucoside (H2/751)

365. 4-*O*-Apiosyl-(1''' → 2'')-glucoside (H2/752)

366. 4-*O*-(5'''-*O*-*p*-Cinnamoyl)apiofuranosyl-(1''' → 2'')-glucoside (**123**)

2',4',6',β-Tetrahydroxychalcone

367. 2'-*O*-Glucoside (H2/753)

368. 4'-*O*-Glucoside (H2/754)

2',4',3,4-Tetrahydroxychalcone (butein)

369. 2',3-Di-*O*-glucoside (H2/735)

370. 4'-*O*-Glucoside (coreopsin, H2/733)

371. 4'-*O*-Malonylglucoside (H2/734)

372. 4'-*O*-Arabinosyl-(1''' → 4'')-galactoside (H2/737)

373. 4'-*O*-Glucosylglucoside (H2/738)

374. 4'-*O*-Malonylsophoroside (H2/739)

375. 4',3-Di-*O*-glucoside (isobutrin, H2/736)

376. 3-*O*-Glucoside (monospermoside, H2/732)

2',4',4-Trihydroxy-3-methoxychalcone (homobutein)

377. 4'-*O*-Glucoside (**124**)

378. 4-*O*-Glucoside (H2/740)

2',4-Dihydroxy-4',6',3-trimethoxychalcone

APPENDIX B   Checklist of Known Chalcones, Dihydrochalcones, and Aurones — *continued*

379.   4-*O*-Glucoside (H2/780)
2′,3′,4′,3,4-Pentahydroxychalcone (okanin)
380.   3′-*O*-Glucoside (H2/755)
381.   3′,4′-Di-*O*-glucoside (H2/764)
382.   4′-*O*-Glucoside (marein, H2/756)
383.   4′-*O*-(6″-*O*-Acetyl)glucoside (H2/757)
384.   4′-*O*-(6″-*O*-*p*-Coumaroyl)glucoside (H2/758)
385.   4′-*O*-(4″,6″-Di-*O*-acetylglucoside) (**125**)
386.   4′-*O*-(4″-*O*-Acetyl-6″-*O*-*p*-coumaroyl)glucoside (H2/759)
387.   4′-*O*-(2″-*O*-Caffeoyl-6″-*O*-acetylglucoside) (**126**)
388.   4′-*O*-(2″-*O*-Caffeoyl-6″-*O*-*p*-coumaroylglucoside) (**127**)
389.   4′-*O*-(2″,4″,6″-Tri-*O*-acetyl)glucoside (H2/760)
390.   4′-*O*-(2″,4″-Di-*O*-acetyl-6″-*O*-*p*-coumaroyl)glucoside (H2/762)
391.   4′-*O*-(3″,4″,6″-Tri-*O*-acetyl)glucoside (H2/761)
392.   4′-*O*-(3″,4″-Di-*O*-acetyl-6″-*O*-*p*-coumaroyl)glucoside (H2/763)
393.   4′-*O*-α-Arabinofuranosyl-(1‴ → 4″)-glucoside (H2/765)
394.   4′-*O*-Glucosyl-(1‴ → 6″)-glucoside (H2/766)
2′,3′,4′,3-Tetrahydroxy-4-methoxychalcone (okanin 4-methyl ether)
395.   3′-*O*-Glucoside (H2/767)
396.   3′-*O*-(6″-*O*-Acetyl)glucoside (H2/768)
397.   4′-*O*-Glucoside (H2/769)
398.   4′-*O*-(6″-*O*-Acetyl)glucoside (H2/770)
399.   4′-*O*-(6″-*O*-*p*-Coumaroylglucoside) (**128**)
400.   4′-*O*-(2″-*O*-Caffeoyl-6″-*O*-acetylglucoside) (**129**)
401.   4′-*O*-Primveroside (**130**)
2′,4′,3,4-Tetrahydroxy-3′-methoxychalcone (lanceoletin)
402.   4′-*O*-Glucoside (lanceolin, H2/772)
2′,3′,4′-Trihydroxy-3,4-dimethoxychalcone (okanin 3,4-dimethyl ether)
403.   4′-*O*-Glucoside (H2/773)
2′,4′,4-Trihydroxy-3′,3-dimethoxychalcone (okanin 3′,3-dimethyl ether)
404.   4′-*O*-Glucoside (**131**)
2′,4′-Dihydroxy-3′,3,4-trimethoxychalcone (okanin 3′,3,4-trimethyl ether)
405.   4′-*O*-Glucoside (H2/774)
2′,4′,5′,3,4-Pentahydroxychalcone
406.   4′-*O*-Glucoside (stillopsin, H2/776)
2′-Hydroxy-4′,6′,2,4-tetramethoxychalcone
407.   2′-*O*-Glucoside (**132**)
2′,4′,6′,3,4-Pentahydroxychalcone
408.   2′-*O*-Glucoside (H2/777)
409.   4′-*O*-Glucoside (H2/778)
2′,4′,6′,4-Tetrahydroxy-3-methoxychalcone
410.   2′-*O*-Glucoside (H2/779)
4′,6′,3-Trihydroxy-2′,4-dimethoxychalcone
411.   4′-*O*-Rutinoside (**133**)
2′,4′,6′,3,4,5-Hexahydroxychalcone
412.   2′-*O*-Glucoside (H2/781)
2′,4′,6′,3,4,β-Hexahydroxychalcone

APPENDIX B Checklist of Known Chalcones, Dihydrochalcones, and Aurones — *continued*

413. 2'-*O*-Glucoside (H2/782)

2',3',4',6',3,4,α-Heptahydroxychalcone

414. 2'-*O*-Glucoside (H2/783)

Chalcone dimers and oligomers

Dimers (chalcone)

415. Azobechalcone A (**147**)
416. *Brackenridgea* orange pigment (H1/3090)
417. Calodenin A (**150**)
418. Calodenin B (H1/3089)
419. α-Diceroptene (H1/3145)
420. Dihydrolophirone C (**149**)
421. α-Diohobanin (H1/3144)
422. Flavumone A (**152**)
423. Isolophirone C (**148**)
424. Isombamichalcone (H1/3097)
425. Kamalachalcone A (**144**)
426. Kamalachalcone B (**145**)
427. Licobichalcone (**146**)
428. Lophirone C (H1/3092)
429. Lophirone F (H1/3098)
430. Lophirone G (H1/3099)
431. Lophirone K (**151**)
432. Matosine (**137**)
433. Mbamichalcone (H1/3096)
434. Rhuschalcone I (**138**)
435. Rhuschalcone II (**139**)
436. Rhuschalcone III (**140**)
437. Rhuschalcone IV (**141**)
438. Rhuschalcone V (**142**)
439. Rhuschalcone VI (**143**)
440. Urundeuvine A (**134**)
441. Urundeuvine B (**135**)
442. Urundeuvine C (**136**)

Dimers (chalcone–flavan)

443. Daphnodorin A (H2/1727)
444. Daphnodorin C (H2/1728)
445. Daphnodorin J (**153**)
446. Daphnodorin M (**154**)
447. Daphnodorin N (**155**)

Dimers (chalcone–flavan-3-ol)

448. Daphnodorin B (H2/1895)
449. Daphnodorin I (**156**)
450. Dihydrodaphnodorin B (H2/1896)
451. Genkwanol A[145]
452. Genkwanol B (**157**)
453. Genkwanol C (**158**)
454. Larixinol (H2/1897)

APPENDIX B   Checklist of Known Chalcones, Dihydrochalcones, and Aurones — *continued*

Dimers (chalcone–flavanone)
455.   Chalcocaryanone A (**159**)
456.   Chalcocaryanone B (**160**)
457.   Chalcocaryanone C (**161**)
458.   Chalcocaryanone D (**162**)
459.   Flavumone B (**165**)
460.   6‴-Hydroxylophirone B (**163**)
461.   6‴-Hydroxylophirone B 4‴-*O*-glucoside (**164**)
462.   Lophirone B (H1/3086)
463.   Lophirone H (H1/3087)
464.   Occidentoside (H1/2946)
Dimers (chalcone–flavene)
465.   Bongosin (H1/3085)
Dimers (chalcone–flavone)
466.   *Aristolochia* dimer "A" (**166**)
467.   *Aristolochia* dimer "B" (**167**)
468.   *Aristolochia* dimer "C" (**168**)
469.   *Aristolochia* dimer "D" (**169**)
470.   Calodenone (**171**)
471.   Chamaechromone (H1/3127)
472.   Cissampeloflavone (**170**)
473.   Lophirone A (H1/3125)
Trimers
474.   Caloflavan A (**172**)
475.   Caloflavan B (**173**)
Tetramers
476.   Alatachalcone (**175**)
477.   *Aristolochia* tetraflavonoid (**174**)
478.   Isolophirachalcone (**176**)
479.   Lophirochalcone (H1/3133)
480.   Lophiroflavan A (**177**)
481.   Lophiroflavan B (**178**)
482.   Lophiroflavan C (**179**)
Pentamer
483.   Ochnachalcone (**180**)
Hexamer
484.   Azobechalcone (**181**)
Chalcone Diels–Alder adducts
Chalcone–isoprene
485.   Sanggenon R (**182**)
Chalcone–monoterpene
486.   *Boesenbergia* adduct (**191**)
487.   Crinatusin A_1 (**183**)
488.   Crinatusin A_2 (**184**)
489.   Crinatusin B_1 (**185**)
490.   Crinatusin B_2 (**186**)
491.   Crinatusin C_1 (**187**)

APPENDIX B   Checklist of Known Chalcones, Dihydrochalcones, and Aurones — *continued*

492.   Crinatusin C$_2$ (**188**)
493.   Fissistin (**189**)
494.   Isofissistin (**190**)
495.   Isoschefflerin (H2/685)
496.   *Kaempferia* adduct "A" (**192**)
497.   *Kaempferia* adduct "B" (**193**)
498.   ($\pm$)-Nicolaioidesin A (**194**)
499.   ($\pm$)-Nicolaioidesin B (**195**)
500.   ($\pm$)-Nicolaioidesin C (**196**)
501.   Panduratin A (H2/673)
502.   Panduratin B (H2/675)
503.   Schefflerin (H2/686)

Chalcone–coumarin
504.   Palodesangren A (**197**)
505.   Palodesangren B (**198**)
506.   Palodesangren C (**199**)
507.   Palodesangren D (**200**)
508.   Palodesangren E (**201**)
509.   Palodesangretin A (**202**)
510.   Palodesangretin B (**203**)

Chalcone–arylbenzofuran
511.   Albafuran C (H2/954)
512.   Albanol A (mulberrofuran G) (H2/957)
513.   Albanol B (H2/958, possible artifact)
514.   Chalcomoracin (H2/966)
515.   Mulberrofuran C (H2/983)
516.   Mulberrofuran E (H2/984)
517.   Mulberrofuran J (H2/985)
518.   Mulberrofuran O (H2/986)
519.   Mulberrofuran T (H2/987)
520.   Mulberrofuran U (**204**)

Chalcone–chalcone
521.   Artonin C (H2/959)
522.   Artonin D (H2/960)
523.   Artonin X (**205**)
524.   Brosimone A (H2/963)
525.   Brosimone D (H2/964)
526.   *Dorstenia* chalcone dimer (**207**)
527.   Dorstenone (**206**)
528.   Kuwanon I (H2/968)
529.   Kuwanon J (H2/969)
530.   Kuwanon Q (H2/974)
531.   Kuwanon R (H2/975)
532.   Kuwanon V (H2/976)
533.   Sorocein B (H2/995)

Chalcone–flavanone
534.   Kuwanon K (H2/970)

APPENDIX B   Checklist of Known Chalcones, Dihydrochalcones, and Aurones — *continued*

535.   Kuwanon N (H2/971)
536.   Kuwanon O (H2/972)
537.   Sanggenol J (**208**)
538.   Sanggenol M (**209**)
539.   Sanggenon C (**210**)
540.   Sanggenon D (**211**)
541.   Sanggenon E (**212**)
542.   Sanggenon O (**213**)
543.   Sanggenon P (H2/993)
544.   Sanggenon Q (H2/994)
545.   Sanggenon S (**214**)
546.   Sanggenon T (**215**)
547.   Cathayanon A (**216**)
548.   Cathayanon B (**217**)

Chalcone–flavone
549.   Albanin F (kuwanon G, moracenin B) (H2/955)
550.   Albanin G (kuwanon H, moracenin A) (H2/956)
551.   Artonin I (**218**)
552.   Brosimone B (H2/965)
553.   Kuwanon W (H2/977)
554.   Moracenin C (H2/981)
555.   Moracenin D (H2/982)
556.   Multicaulisin (**219**)

Chalcone–stilbene
557.   Kuwanol E (H2/967)
558.   Kuwanon P (H2/973)
559.   Kuwanon X (H2/978)
560.   Kuwanon Y (H2/979)
561.   Kuwanon Z (H2/980)

Miscellaneous
562.   Sorocenol B (**220**)

Chalcone conjugates
Chalcone–diarylheptanoid
563.   Calyxin A (**221**)
564.   Calyxin B (**223**)
565.   Calyxin F (**225**)
566.   Calyxin H (**228**)
567.   Calyxin I (**230**)
568.   Calyxin L (**232**)
569.   Deoxycalyxin A (**222**)
570.   Epicalyxin B (**224**)
571.   Epicalyxin F (**226**)
572.   Epicalyxin H (**229**)
573.   Epicalyxin I (**231**)
574.   6-Hydroxycalyxin F (**227**)
Chalcone–bis(diarylheptanoid)
575.   Blepharocalyxin A (**233**)

APPENDIX B   Checklist of Known Chalcones, Dihydrochalcones, and Aurones — *continued*

576. Blepharocalyxin B (**234**)
577. Blepharocalyxin E (**235**)
Miscellaneous
578. 8-Caffeoyl-3,4-dihydro-5,7-dihydroxy-4-phenylcoumarin (H2/704)
579. 8-Cinnamoyl-3,4-dihydro-5,7-dihydroxy-4-phenylcoumarin (H2/702)
580. 8-*p*-Coumaroyl-3,4-dihydro-5,7-dihydroxy-4-phenylcoumarin (H2/703)
581. Cryptocaryone (H2/716)
582. Didymocalyxin B (**239**)
583. 2′,4′-Dihydroxy-3′-*C*-(2,6-dihydroxybenzyl)-6′-methoxychalcone (**236**)
584. 2′,4′-Dihydroxy-6′-methoxy-3′-(8,17-epoxy-16-oxo-12,14-labdadien-15-yl)chalcone (**237**)
585. Lophirone D (H2/705)
586. Lophirone E (H2/706)
587. Infectocaryone (**240**)
588. Kurzichalcolactone (**238**)

Quinochalcones
Aglycones
589. Ceroptene (H2/713)
590. Desmosdumotin C (**241**)
591. 2-Hydroxy-7,8-dehydrograndiflorone (H2/712)
592. Methylpedicinin (H2/715)
593. Munchiwarin (**242**)
594. Pedicinin (H2/714)
595. Ohobanin (H2/711)
596. Tunicatachalcone (**243**)
C-Glycosides
597. Anhydrosafflor yellow B (**244**)
598. Carthamin (H1/2845)
599. Cartormin (**245**)
600. Hydroxysafflor yellow A (**246**)
601. Precarthamin (**247**)
602. Safflomin A (H1/2944)
603. Safflomin C (H1/2945)
604. Safflor yellow A (H1/2734)
605. Safflor yellow B (H1/2846)
606. Tinctormine (**248**)

DIHYDROCHALCONES
Dihydrochalcone aglycones
607. Dihydrochalcone (H2/784)
Di-*O*-substituted
   (2′,4′)
608. 2′,4′-diOH (H2/785)
609. 2′-OH, 4′-OMe (**249**)
610. 2′-OH, 4′-prenyloxy (dihydrocordoin, H2/875)
611. 4′-OH, 2′-OMe (**250**)
   (2,4)

APPENDIX B Checklist of Known Chalcones, Dihydrochalcones, and Aurones — *continued*

612. 4-OH, 2-OMe (**251**)

Tri-*O*-substituted

  (2′,4′,6′)

613. 2′,4′,6′-triOH (H2/791)
614. 2′,4′,6′-triOH, 3′-Me (**252**)
615. 2′,4′,6′-triOH, 3′-formyl (H2/868)
616. 2′,4′,6′-triOH, 3′-prenyl (H2/877)
617. 2′,4′,6′-triOH, 3′-C_{10} ((+)-linderatin, H2/889)
618. 2′,4′,6′-triOH, 3′-C_{10} ((−)-linderatin, **275**)
619. 2′,4′,6′-triOH, 3′,5′-diprenyl (H2/882)
620. 2′,4′,6′-triOH, 3′,5′-diC_{10} ((+)-neolinderatin, H2/891)
621. 2′,4′,6′-triOH, 3′,5′-diC_{10} ((−)-neolinderatin, **276**)
622. 2′,4′-diOH, 6′-OMe (uvangoletin, H2/793)
623. 2′,4′-diOH, 6′-OMe, 3′-Me (myrigalone H, **253**)
624. 2′,4′-diOH, 6′-OMe, 3′,5′-diMe (angoletin, H2/872)
625. 2′,4′-diOH, 3′-Me, 6″,6″-dimethyl-4″-hydroxy-4″,5″-dihydropyrano[2″,3″:6′,5′] (H2/886)
626. 2′,4′-diOH, 3′-Me, 6″,6″-dimethyl-5″-hydroxy-4″,5″-dihydropyrano [2″,3″:6′,5′] (H2/887)
627. 2′,6′-diOH, 4′-OMe (H2/792)
628. 2′,6′-diOH, 4′-prenyloxy (H2/880)
629. 2′,6′-diOH, 3′,4′-dihydrooxepino (H2/878)
630. 2′,6′-diOH, 4′-OMe, 3′-Me (myrigalone G, **254**)
631. 2′,6′-diOH, 4′-OMe, 3′,5′-diMe (myrigalon B, H2/871)
632. 2′,6′-diOH, 4′-OMe, 3′-prenyl (H2/879)
633. 2′,6′-diOH, 4′-OMe, 3′-C_{10} ((+)-methyllinderatin, H2/890)
634. 2′,6′-diOH, 4′-OMe, 3′-C_{10} ((−)-methyllinderatin, **277**)
635. 2′,6′-diOH, 4′-OMe, 5′-(1″-aryl)prenyl (piperaduncin A, **278**)
636. 2′-OH, 4′,6′-diOMe (dihydroflavokawin B, H2/794)
637. 2′-OH, 4′,6′-diOMe, 3′-Me (H2/869)
638. 2′-OH, 4′,6′-diOMe, 3′-formyl, 5′-Me (H2/870)
639. 2′-OH, 4′-OMe, 5″-arylfurano[2″,3″:6′,5′] (longicaudatin, **279**)
640. 2′-OH, 4′-OMe, 4″-aryl-5″-(2-hydroxy isopropyl)dihydrofurano[2″,3″:6′,5′] (piperaduncin B, **280**)
641. 2′-OH, 4′-OMe, 6′-*O*-C_{10} (adunctin A, **281**)
642. 2′-OH, 4′-OMe, [5′,6′]-C_{10} (adunctin B, **282**)
643. 2′-OH, 4′-OMe, [5′,6′]-C_{10} (adunctin C, **283**)
644. 2′-OH, 4′-OMe, [5′,6′]-C_{10} (adunctin D, **284**)
645. 2′-OH, 4′-OMe, [5′,6′]-C_{10} (adunctin E, **285**)
646. 2′-OH, 6′-OMe, 4′-prenyloxy (H2/881)

  (2′,4′,4)

647. 2′,4′,4-triOH (davidigenin, H2/789)
648. 2′,4′,4-triOH, 3′,5′-diprenyl (gancaonin J, H2/876)
649. 2′,4′,4-triOH, 3-geranyl (H2/888)
650. 2′,4-diOH, 4′-OMe (H2/790)
651. 2′,4-diOH, 6″,6″-dimethylpyrano[2″,3″:4′,3′] (crotaramosmin, **286**)
652. 2′-OH, 4-OMe, 6″,6″-dimethylpyrano[2″,3″:4′,3′] (crotaramin, **287**)

APPENDIX B   Checklist of Known Chalcones, Dihydrochalcones, and Aurones — *continued*

653.   4-OH, 2',4'-diOMe (**255**)

   (2',4',α)

654.   2',α-diOH, furano[2'',3'':4',3'] (castillene E, H2/826)

   (2',4',β)

655.   2',β-diOMe, furano[2'',3'':4',3'] (ovalitenin B, H2/827)

   (4',2,4)

656.   4',2,4-triOH (**256**)

657.   4',2-diOH, 4-OMe (H2/787)

658.   4',4-diOH, 2-OMe (loureirin C, H2/786)

659.   4'-OH, 2,4-diOMe (loureirin A, H2/788)

   (4',2,6)

660.   4'-OH, 2,6-diOMe (**257**)

Tetra-*O*-substituted

   (2',3',4',6')

661.   2',4',6'-triOH, 3'-OMe, 5'-prenyl (H2/885)

662.   2',3'-diOH, 4',6'-diOMe (**258**)

663.   2',6'-diOH, 3',4'-diOMe (dihydropashanone, H2/805)

664.   2',6'-diOMe, 3',4'-OCH$_2$O- (H2/820)

665.   3',6'-diOMe, 2'-OMe, 6'',6''-dimethylpyrano[2'',3'':4',5'] (flemistrictin D, H2/884)

666.   2'-OH, 3',4',6'-triOMe (**259**)

   (2',3',4',4)

667.   2',4'-diOH, 3',4-diOMe (lusianin, **260**)

   (2',4',6',4)

668.   2',4',6',4-tetraOH (phloretin, H2/799)

669.   2',4',6',4-tetraOH, 3,5-diprenyl (**289**)

670.   2',4',6',4-tetraOH, 3-geranyl, 5-prenyl (**290**)

671.   2',4',6'-triOH, 4-OMe (H2/800)

672.   2',4',4-triOH, 6'-OMe (H2/802)

673.   2',4',4-triOH, 6'-OMe, 3'-prenyl (α,β-dihydroxanthohumol, **291**)

674.   2',6',4-triOH, 4'-OMe (asebogenin, H2/801)

675.   2',6',4-triOH, 4'-OMe, 3'-Me (H2/873)

676.   2',6',4-triOH, 4'-OMe, 3',5'-diMe (H2/874)

677.   2',4'-diOH, 6',4-diOMe (H2/804)

678.   2',6'-diOH, 4',4-diOMe (calomelanone, H2/803)

679.   2',4-diOH, 4',6'-diOMe (**261**)

680.   2'-OH, 4',6',4-triOMe (**262**)

   (2',4',3,4)

681.   2',4',3,4-tetraOH, 2-geranyl (H2/892)

682.   2',3,4-triOH, 4'-OMe (dihydrocalythropsin, **263**)

683.   2',4',4-triOH, 3-OMe (H2/797)

684.   2',4',4-triOH, (6''-(4-methylpent-3-enyl),6''-methyl)pyrano[2'',3'':3,2] (H2/893)

685.   2',3,4-triOH, 6'',6''-dimethylpyrano[2'',3'':4',3'] (crotin, **292**)

686.   2',4'-diOH, 3,4-OCH$_2$O- (H2/819)

687.   2'-OMe, 3,4-OCH$_2$O-, furano[2'',3'':4',3'] (**293**)

688.   2',4'-diOMe, 3,4-OCH$_2$O- (ponganone VII, **264**)

   (2',4',4,α)

APPENDIX B   Checklist of Known Chalcones, Dihydrochalcones, and Aurones — *continued*

689. 2′,4′,4,α-tetraOH (H2/811)
690. 2′,4′,4,α-tetraOH, 5′,3-diprenyl (kanzonol Y, **294**)
691. 2′,4′,α-triOH, 4-OMe (H2/812)
692. 2′,4,α-triOH, 4′-geranyloxy (**295**)
693. 4′,4,α-triOH, 2′-OMe (H2/813)
694. 2′,α-diOH, 4′,4-diOMe (odoratol, H2/814)
695. α-OH, 2′,4′,4-triOMe (H2/815)
      (4′,2,4,6)
696. 4′,2,4-triOH, 6-OMe (loureirin D, H2/795)
697. 4′,2-diOH, 4,6-diOMe (**265**)
698. 4′,4-diOH, 2,6-diOMe (**266**)
699. 4′-OH, 2,4,6-triOMe (loureirin B, H2/796)
      (2,3,4,6)
700. 2-OH, 3,4,6-triOMe (**267**)
701. 6-OH, 2,3,4-triOMe (**268**)
Penta-*O*-substituted
      (2′,3′,4′,5′,6′)
702. 2′,5′-diOH, 3′,4′,6′-triOMe (dihydropedicin, **269**)
703. 3′,5′-diOH, 2′,4′,6′-triOMe (**270**)
704. 2′-OH, 3′,4′,5′,6′-tetraOMe (dihydrokanakugiol, H2/810)
      (2′,3′,4′,6′,4)
705. 2′,3′,4′,6′,4-pentaOH, 5′-geranyl (H2/896)
706. 2′,3′,4′,6′,4-pentaOH, 5′-neryl (H2/897)
      (2′,3′,4′,6′,β)
707. β-OH, 2′,6′-diOMe, 3′,4′-OCH₂O- (H2/825)
      (2′,4′,6′,3,4)
708. 2′,4′,6′,3,4-pentaOH (H2/806)
709. 2′,4′,6′,3,4-pentaOH, 3′,5-diprenyl (**296**)
710. 2′,4′,6′,3,4-pentaOH, 3′-geranyl, 5-prenyl (**297**)
711. 2′,4′,6′,3-tetraOH, 3′-geranyl, 6″,6″-dimethylpyrano[2″,3″:4,5] (**298**)
712. 2′,4′,6′,3-tetraOH, 4-OMe, 3′,5-diprenyl (**299**)
713. 2′,4′,3,4-tetraOH, 6′-OMe, 3′-geranyl (H2/895)
714. 2′,6′,3-triOH, 4-OMe, 5-prenyl, 6″,6″-dimethyl-4″,5″-dihydropyrano[2″,3″:4,5′] (**300**)
715. 2′,4′-diOH, 6′,3,4-triOMe (H2/808)
716. 2′,4′-diOH, 6′-OMe, 3,4-OCH₂O- (H2/821)
717. 2′,6′-diOH, 4′,3,4-triOMe (**271**)
718. 2′,4′,6′-triOMe, 3,4-OCH₂O- (**272**)
      (2′,4′,6′,4,α)
719. 2′,4′,6′,4,α-pentaOH (nubigenol, H2/817)
      (2′,4′,6′,4,β)
720. 2′,4′,4,β-tetraOH, 6′-OMe (**274**)
      (2′,4′,3,4,α)
721. 2′,4′,3,4,α-pentaOH (**273**)
722. 4′,3,4,α-tetraOH, 2′-OMe (H2/816)
      (2′,4′,3,4,β)
723. 2′,4′,β-triOMe, 3,4-OCH₂O- (dihydromilletenone methyl ether, H2/824)

APPENDIX B   Checklist of Known Chalcones, Dihydrochalcones, and Aurones — *continued*

724. 2′,β-diOMe, 3,4-OCH$_2$O-, furano[2″,3″:4′,3′] (ponganone IX, **301**)
    (3′,4′,2,4,β)

725. 2,4,β-triOMe, 3′,4′-OCH$_2$O- (dihydroisomilletenone methyl ether, H2/823)

Hexa-*O*-substituted
    (2′,3′,4′,5′,6′,4)

726. 2′,6′-diOH, 3′,4-diOMe, 4′,5′-OCH$_2$O- (H2/822)
    (2′,3′,4′,6′,3,4)

727. 6′-OH, 2′,3′-diOMe, 3,4-OCH$_2$O-, furano[2″,3″:4′,5′] (H2/828)
    (2′,4′,5′,3,4,β)

728. 2′,5′,β-triOMe, 3,4-OCH$_2$O-, 6″,6″-dimethylpyrano[2″,3″,4′,3′]
    (ponganone VIII, **302**)
    (2′,4′,3,4,5,β)

729. 2′,4′,3,5,β-pentaOH, 4-OMe (gliricidol, H2/818)

Dihydrochalcone glycosides

2′,4′,6′-Trihydroxydihydrochalcone

730. 4′-*O*-(3″-*O*-Galloyl-4″,6″-*O,O*-hexahydroxydiphenoylglucoside) (thonningianin A, **303**)

731. 4′-*O*-(4″,6″-*O,O*-Hexahydroxydiphenoylglucoside) (thonningianin B, **304**)

2′,4′,2-Trihydroxydihydrochalcone (davidigenin)

732. 2′-*O*-Glucoside (davidioside, H2/857)

733. 4′-*O*-Glucoside (confusoside, H2/858)

2′,4-Dihydroxy-4′-methoxydihydrochalcone

734. 2′-*O*-Glucoside (H2/859)

2′,4′-Dihydroxy-3′,6′-dimethoxydihydrochalcone

735. 4′-*O*-Glucosyl-(1‴ → 6″)-glucoside (salicifolioside A, **305**)

2′,4′,6′,4-Tetrahydroxydihydrochalcone (phloretin)

736. 2′-*O*-Glucoside (phloridzin, H2/860)

737. 2′-*O*-(6″-*O*-Acetyl)glucoside (**306**)

738. 2′-*O*-Rhamnoside (glycyphyllin, H2/861)

739. 2′-*O*-Xylosylglucoside (H2/863)

740. 4′-*O*-Glucoside (trilobatin, H2/862)

741. 4′-*O*-(2″-*O*-Acetyl)glucoside (**307**)

742. 3′,5′-Di-*C*-glucoside (**308**)

2′,4′,6′-Trihydroxy-4-methoxydihydrochalcone

743. 2′-*O*-Glucoside (**309**)

2′,6′,4-Trihydroxy-4′-methoxydihydrochalcone (asebogenin)

744. 2′-*O*-Glucoside (asebotin, H2/864)

2′,4′-Dihydroxy-4′,6′-diacetoxydihydrochalcone

745. 2′-*O*-Glucoside (zosterin, **310**)

4-Hydroxy-2′,4′,6′-trimethoxydihydrochalcone

746. 4-*O*-Glucoside (bidenoside B, **311**)

2′,4′,4,α-Tetrahydroxydihydrochalcone

747. α-*O*-Glucoside (licoagroside F, **312**)

748. 3′-*C*-Glucoside (coatline A, H1/2598)

2′,4′,4,β-Tetrahydroxydihydrochalcone

749. 2′-*O*-Glucoside (rocymosin B, **313**)

750. 3′-*C*-Glucoside (pterosupin, H1/2600)

2′,4′,6′,3,4-Pentahydroxydihydrochalcone

APPENDIX B   Checklist of Known Chalcones, Dihydrochalcones, and Aurones — *continued*

751. 2′-*O*-Galactoside (H2/865)
752. 2′-*O*-Glucoside (H2/866)
753. 4′-*O*-Glucoside (sieboldin, H2/867)
754. 3′-*C*-Glucoside (aspalathin, H1/2579)
2′,4′,3,4,α-Pentahydroxydihydrochalcone
755. 3′-*C*-Glucoside (coatline B, H1/2599)
756. 3′-*C*-Xyloside (**314**)

Dihydrochalcone dimers
Dihydrochalcone–dihydrochalcone
757. Brackenin (H1/3095)
758. Cycloaltilisin 6 (**319**)
759. Iryantherin F (H2/846)
760. Littorachalcone (**315**)
761. Piperaduncin C (**318**)
762. 3′,3′-bis(2′,4′,6′-Trihydroxy-4-methoxydihydrochalcone) (**317**)
763. Verbenachalcone (**316**)
Dihydrochalcone–deoxotetrahydrochalcone
764. Cinnabarone (**321**)
765. Cochinchinenin (**320**)
Dihydrochalcone–flavonol
766. Trianguletin (**322**)
767. Trianguletin 'B' (**323**)
768. Trianguletin 'C' (**324**)
769. Trianguletin 'D' (**325**)
770. Trianguletin 'E' (**326**)

Dihydrochalcone conjugates
C-Benzyl derivatives
771. 2′,4′-diOH, 6′-OMe, 3′-(2-OHBn) (uvaretin, H2/830)
772. 2′,4′-diOH, 6′-OMe, 3′-(2 × 2-OHBn) (angoluvarin, H2/832)
773. 2′,4′-diOH, 6′-OMe, 3′-(2-OHBn), 5′-Me (anguvetin, H2/831)
774. 2′,4′-diOH, 6′-OMe, 3′,5′-di(2-OHBn) (diuvaretin, H2/833)
775. 2′,4′-DiOH, 6′-OMe, 3′-(2-OHBn), 5′-(2 × 2-OHBn) (triuvaretin, **328**)
776. 2′,4′-DiOH, 6′-OMe, 3′-(2-OHBn), 5′-(3 × 2-OHBn) (**330**)
777. 2′,4′-DiOH, 6′-OMe, 3′-(2-OHBn), 5′-(4 × 2-OHBn) (**332**)
778. 2′,4′-DiOH, 6′-OMe, 3′-(2 × 2-OHBn), 5′-(2-OHBn) (isotriuvaretin, **329**)
779. 2′,4′-DiOH, 6′-OMe, 3′-(3 × 2-OHBn), 5′-(2-OHBn) (**331**)
780. 2′,4′-DiOH, 6′-OMe, 3′-(4 × 2-OHBn), 5′-(2-OHBn) (**333**)
781. 2′,6′-diOH, 4′-OMe, 3′-(2-OHBn) (**327**)
782. 4′,6′-diOH, 2′-OMe, 3′-(2-OHBn) (isouvaretin, H2/829)
783. Chamuvaretin (H2/834)
Dihydrochalcone–lignans
784. Iryantherin A (H2/841)
785. Iryantherin B (H2/842)
786. Iryantherin C (H2/843)
787. Iryantherin D (H2/844)
788. Iryantherin E (H2/845)
789. Iryantherin G (**334**)

APPENDIX B   Checklist of Known Chalcones, Dihydrochalcones, and Aurones — *continued*

790.   Iryantherin H (**335**)
791.   Iryantherin I (**336**)
792.   Iryantherin J (**337**)
793.   Iryantherin K (**338**)
794.   Iryantherin L (**339**)

Miscellaneous

795.   Calomelanol A (H2/836)
796.   Calomelanol B (H2/837)
797.   Calomelanol C (H2/838)
798.   Calomelanol D-1 (H2/835)

Quinodihydrochalcones

799.   Ceratiolin (H2/850)
800.   Grandiflorone (H2/849)
801.   Grenoblone (H2/854)
802.   *Helichrysum aphelexiodes* quinodihydrochalcone (**288**)
803.   *Helichrysum forskahlii* quinodihydrochalcone (H2/852)
804.   Helihumulone (H2/853)
805.   Helilupolone (H2/856)
806.   4-Hydroxygrenoblone (H2/855)
807.   *Myrica gale* quinodihydrochalcone (H2/848)
808.   Syzygiol (H2/851)

AURONES

Aurone aglycones

Mono-*O*-substituted
   (*6*)
809.   Furano[2″,3″:6,7] (H2/910)

Di-*O*-substituted
   (*4,6*)
810.   4-OH, Furano[2″,3″:6,7] (H2/911)
811.   4-OMe, Furano[2″,3″:6,7] (H2/912)
   (*6,4′*)
812.   6,4′-diOH (hispidol, H2/898)

Tri-*O*-substituted
   (*4,5,6*)
813.   6-OH, 4,5-OCH$_2$O- (H2/909)
   (*4,6,4′*)
814.   4,6,4′-triOH (H2/899)
   (*6,3′,4′*)
815.   6,3′,4′-triOH (sulfuretin, H2/900)
816.   6,3′,4′-triOH, 5-prenyl (broussoaurone A, **340**)
817.   6,3′,4′-triOH, 7-prenyl (licoagroaurone, **341**)
818.   3′,4′-OCH$_2$O-, furano[2″,3″:6,7] (H2/913)

Tetra-*O*-substituted
   (*4,5,6,4′*)
819.   5-OH, 4,6,4′-triOMe (**342**)
   (*4,6,7,4′*)

APPENDIX B   Checklist of Known Chalcones, Dihydrochalcones, and Aurones — *continued*

820. 4,6,7,4'-tetraOH (**343**)
    (4,6,3',4')
821. 4,6,3',4'-tetraOH (aureusidin, H2/901)
822. 4,6,3',4'-tetraOH, 5-Me (**344**)
823. 4,6,3',4'-tetraOH, 7-Me (**345**)
824. 4,6,3',4'-tetraOH, 5,2'-diprenyl (antiarone A, H2/907)
825. 4,6,3',4'-tetraOH, 2',5'-diprenyl (antiarone B, H2/908)
826. 6,3',4'-triOH, 4-OMe (rengasin, H2/902)
827. 6,3',4'-triOH, 4-OMe, 5-Me (**346**)
828. 6,3',4'-triOH, 4-OMe, 7-Me (**347**)
829. 6,3'-diOH, 4,4'-diOMe, 5-Me (**348**)
830. 4,6,3',4'-tetraOMe (**349**)
    (6,7,3',4')
831. 6,7,3',4'-tetraOH (maritimetin, H2/903)
832. 6,3',4'-triOH, 7-OMe (leptosidin, H2/904)
Penta-*O*-substituted
    (4,5,6,3',4')
833. 3',4'-diOH, 4,5,6-triOMe (hamiltrone, **350**)
    (4,6,3',4',5')
834. 4,6,3',4',5'-pentaOH (bracteatin, H2/905)
Halogenated
835. 4'-Cl (**353**)
Miscellaneous
836. Derriobtusone A (H2/914)
837. Derriobtusone B (H2/915)
Auronol aglycones
Di-*O*-substituted
    (2,6)
838. 2-OMe, furano[2'',3'':6,7] (castillene A, H2/949)
Tetra-*O*-substituted
    (2,4,6,4')
839. 2,4,6,4'-tetraOH (maesopsin, H2/943)
840. 2,6,4'-triOH, 4-OMe (carpusin, H2/944)
    (2,6,3',4')
841. 2,6,3',4'-tetraOH (H2/945)
842. 2,6,3'-triOH, 4'-OMe (H2/946)
843. 2-OMe, 3',4'-OCH$_2$O-, furano[2'',3'':6,7] (castillene D, H2/952)
Penta-*O*-substituted
    (2,4,6,3',4')
844. 2,4,6,3',4'-pentaOH (alphitonin, H2/947)
    (2,6,7,3',4')
845. 2,6,7,3',4'-pentaOH (nigrescin, H2/948)
Hexa-*O*-substituted
    (2,4,6,3',4',5')
846. 2,4,6,3',4',5'-hexaOH (amaronol A, **351**)
847. 2,4,6,3',5'-pentaOH, 4'-OMe (amaronol B, **352**)
Halogenated

APPENDIX B   Checklist of Known Chalcones, Dihydrochalcones, and Aurones — *continued*

848.   2-OH, 4'-Cl (**354**)

Miscellaneous

849.   Castillene B (H2/950)

850.   Castillene C (H2/951)

851.   Crombenin (H2/953)

Aurone and auronol glycosides

2'-Hydroxyaurone

852.   2'-*O*-Glucosyl-(1''' → 6'')-glucoside (dalmaisione D, **355**)

6,4'-Dihydroxyaurone (hispidol)

853.   6-*O*-Glucoside (H2/916)

6,4'-Dihydroxy-7-methylaurone

854.   6-*O*-Rhamnoside (H2/941)

4,6,4'-Trihydroxyaurone

855.   4-*O*-Rhamnosyl-(1''' → 2'')-glucoside (**357**)

856.   4,6-di-*O*-Glucoside (caulesauroneside, **356**)

857.   6-*O*-Rhamnoside (H2/917)

4,6,4'-Trihydroxy-7-methylaurone

858.   4-*O*-Rhamnoside (H2/942)

6,3',4'-Trihydroxyaurone (sulfuretin)

859.   6-*O*-Glucoside (sulfurein, H2/918)

860.   6-*O*-Glucosylglucoside (H2/919)

861.   6,3'-Di-*O*-glucoside (palasitrin, H2/920)

4,6,3',4'-Tetrahydroxyaurone (aureusidin)

862.   4-*O*-Glucoside (cernuoside, H2/921)

863.   4,6-Di-*O*-glucoside (H2/925)

864.   6-*O*-Glucoside (aureusin, H2/922)

865.   6-*O*-Glucuronide (H2/924)

866.   6-*O*-Rhamnoside (H2/923)

6,7,3',4'-Tetrahydroxyaurone (maritimetin)

867.   6-*O*-Glucoside (maritimein, H2/926)

868.   6-*O*-(6''-*O*-Acetyl)glucoside (H2/927)

869.   6-*O*-(6''-*O*-*p*-Coumaroyl)glucoside (H2/928)

870.   6-*O*-(3'',6''-Di-*O*-acetyl)glucoside (bidenoside A, **358**)

871.   6-*O*-(4'',6''-Di-*O*-acetyl)glucoside (H2/929)

872.   6-*O*-(2'',4'',6''-Tri-*O*-acetyl)glucoside (H2/930)

873.   6-*O*-(3'',4'',6''-Tri-*O*-acetyl)glucoside (**359**)

874.   6-*O*-Glucosylglucoside (H2/932)

875.   7-*O*-Glucoside (H2/931)

6,3',4'-Trihydroxy-7-methoxyaurone (leptosidin)

876.   6-*O*-Glucoside (leptosin, H2/933)

877.   6-*O*-Glucosyl-(1''' → 4'')-rhamnoside (H2/935)

878.   6-*O*-Xylosyl-(1''' → 4'')-arabinoside (H2/934)

4,6,3',4',5'-Pentahydroxyaurone (bracteatin)

879.   4-*O*-Glucoside (bractein, H2/936)

880.   6-*O*-Glucoside (H2/937)

4,6-Dihydroxy-3',4',5'-trimethoxy-7-methylaurone

APPENDIX B   Checklist of Known Chalcones, Dihydrochalcones, and Aurones — *continued*

881.   4-*O*-Rhamnoside (**360**)

6,3′,4′,5′-Tetrahydroxy-4-methoxyaurone (bracteatin 4-methyl ether)

882.   6-*O*-Rhamnosyl-(1‴ → 4″)-glucoside (subulin, H2/938)

2,4,6,4′-Tetrahydroxy-2-benzylcoumaranone (maesopsin)

883.   4-*O*-Glucoside (hovetrichoside C, **361**)

884.   6-*O*-Glucoside (**362**)

885.   4-*O*-Glucoside 4′-*O*-rhamnoside (hovetrichoside D, **363**)

Miscellaneous

886.   Neoraufuracin (H2/939)

887.   Ambofuracin (H2/940)

Aurone and auronol dimers

Aurone–aurone

888.   Aulacomniumbiaureusidin (**364**)

889.   Disulfuretin (**365**)

890.   Licoagrone (**366**)

Aurone–flavanone

891.   Campylopusaurone (**367**)

Auronol–auronol

892.   (2*S*)-2-Deoxymaesopsin-(2 → 7)-(2*R*)-maesopsin (**368**)

893.   (2*R*)-2-Deoxymaesopsin-(2 → 7)-(2*S*)-maesopsin (**369**)

894.   (2*R*)-2-Deoxymaesopsin-(2 → 7)-(2*R*)-maesopsin (**370**)

895.   (2*S*)-2-Deoxymaesopsin-(2 → 7)-(2*S*)-maesopsin (**371**)

Auronol–flavanone

896.   (2*R*,3*S*)-Naringenin-(3α → 5)-(2*R*)-maesopsin (**372**)

897.   (2*R*,3*S*)-Naringenin-(3α → 5)-(2*S*)-maesopsin (**373**)

898.   (2*R*,3*S*)-Naringenin-(3α → 7)-(2*R*)-maesopsin (zeyherin epimer, **374**)

899.   (2*R*,3*S*)-Naringenin-(3α → 7)-(2*S*)-maesopsin (zeyherin epimer, **375**)

Auronol–isoflavanone

900.   (2*S*,3*R*)-Dihydrogenistein-(2α → 7)-(2*R*)-maesopsin (**376**)

901.   (2*S*,3*R*)-Dihydrogenistein-(2α → 7)-(2*S*)-maesopsin (**377**)

17 Bi-, Tri-, Tetra-, Penta-, and Hexaflavonoids

Daneel Ferreira, Desmond Slade, and Jannie P.J. Marais

CONTENTS

17.1 INTRODUCTION

Several comprehensive reviews dealing with bi- and triflavonoids have been published.[1–7] This chapter focuses on compounds that have been reported since the last review by Geiger was published in 1994.[7] Since that time tetra-, penta-, and hexaflavonoids have also been identified, hence necessitating a change in the title of this chapter compared to the previous review.[7]

Together with the proanthocyanidins, the bi- and triflavonoids constitute the two major classes of complex C6–C3–C6 secondary metabolites. These compounds represent products of phenol oxidative coupling of flavones, flavonols, dihydroflavonols, flavanones, isoflavones, aurones, auronols, and chalcones, and thus predominantly possess a carbonyl group at C-4 or its equivalent in every constituent unit. However, there are now several examples where the presence of a C-4 carbonyl group is not evident, e.g., the flavanone–isoflavan, licoagrodin (**91**),[88] where the carbonyl group was subject to secondary modifications, e.g., the calycopterones (**127–129**),[102,104] or where the carbon framework is obscured by, e.g., a dienone–phenol rearrangement as in licobichalcone (**35**).[68] It should also be emphasized that the terms bi- and triflavonoids are used loosely. A multitude of compounds that "do not arise" via phenol oxidative coupling of the aforementioned classes of monomeric flavonoids possessing C-4 carbonyl functional groups are also being reported as "bi- or triflavonoids." This is exemplified by bichalcone (**139**), a "biflavonoid" generated via an intermolecular Diels–Alder process.[105] Several additional examples can be found in Chapter 11 or by simply entering "biflavonoid" in one of the several powerful electronic search engines that cover the primary literature.

Essentially all the biflavonoid classes covered in the Geiger review[7] were supplemented since 1992. In addition, several new classes have been reported, e.g., the bi-4-aryldihydrocoumarins (neoflavones) (**16 and 17**),[29] the biauronols (**19–22**),[38] the isoflavanone–auronols (**100 and 101**),[35,36] a number of O-linked bichalcones (see Section 17.3.1.5), and the first (I-6,O,II-8)- (**42**),[89] (I-2′,II-8)- (**43**),[87] and (I-3,II-6)- (**49–52**)[45,84] biflavones. Such a rapid growth not only in the number of new compounds but also in the extended diversity in the location of the interflavonoid bond is discernable in terms of their genesis via phenol oxidative coupling reactions.[8] Notable in these radical couplings is the exclusive formation of carbon–carbon and carbon–oxygen bonds but the conspicuous absence of the generation of oxygen–oxygen bonds.

17.2 NOMENCLATURE

There is no commonly accepted trivial nomenclature for the bi- and triflavonoids and higher oligomeric forms. Their full systematic names, not to mention their often complex common names, are extremely cumbersome. Geiger and Quinn have proposed a system[3–5,7] that requires frequent reference for understanding and has become increasingly difficult to implement as the number of new compounds has grown. No doubt, the nomenclature of this class of compounds is in disarray and it depends on active researchers in the field to select a simple but logical system and then to use it consistently.

Locksley[2] has proposed a system in which the position of substitution of the upper (I) and lower (II) units in biflavonoids is defined from the usual numbering of the parent flavonoid, and each unit is also described from the parent monomer, e.g., apigenin, luteolin, naringenin, genistein, etc. Thus, the well-known biflavones, amentoflavone is named apigeninyl-(I-3′, II-8)-apigenin, the tri-O-methyl derivative (**11**) 7,4′-di-O-methylnaringeninyl-(I-3′,II-8)-4′-O-methylnaringenin, and the flavanone–auronol (**74**) (2S)-naringeninyl-(I-3α,II-5)-(2R)-maesopsin. The consistent implementation of this method for naming the many straightforward examples should be strongly encouraged.

17.3 STRUCTURE AND DISTRIBUTION

The naturally occurring bi- and triflavonoids, together with their plant sources are listed in the following sections. The entries listed are confined to new compounds reported in the post-1992 period or those that have not been dealt with in Geiger's 1994 review.[7] In order to be comprehensive, the listed compounds must be considered in conjunction with the tables in the Geiger and Quinn[3–5,7] and Hemingway[6] reviews. Since many of the analogs have been reported under trivial names these will be retained to facilitate future electronic literature searches.

The first compounds belonging to the tetra-, penta-, and hexaflavonoid classes have also been reported. These complex structures are covered in Sections 17.3.3–17.3.5. As with all published complex structures the reader must avail himself or herself of the correctness of the proposed structure based upon the supporting experimental data.

With a few exceptions the issue of absolute configuration of optically active analogs is often completely ignored. In a large number of cases absolute configuration is readily accessible from chiroptical data. The utility of the circular dichroism method in this regard has been amply demonstrated in several reports. Useful information may be extracted from papers dealing with the chiroptical properties of monomeric constituent units like flavanones and 3-hydroxyflavanones (dihydroflavonols),[110] auronols,[114] a summary of the various classes of stereogenic monomers (Ref. 113 and references cited therein), dimeric compounds like the flavanone– and isoflavanone–auronol- and bis-auronol-type biflavonoids,[35–38] and several other classes of biflavonoids.[23,74,111,112]

In addition to the multitude of biological activities reported in the references listed with the compound or natural source, supplementary information can also be found in Refs. 90 and 97. Representative contributions regarding synthesis of the biflavonoids are available in Refs. 91–93, 96, and 99, while useful NMR and x-ray data may be retrieved from Refs. 94 and 98, respectively.

A comprehensive listing of chalcone dimers and oligomers, dihydrochalcone dimers, and aurone and auronal dimers may also be found in Chapter 16.

17.3.1 BIFLAVONOIDS

17.3.1.1 Agathisflavones [(I-6,II-8)-Coupling]

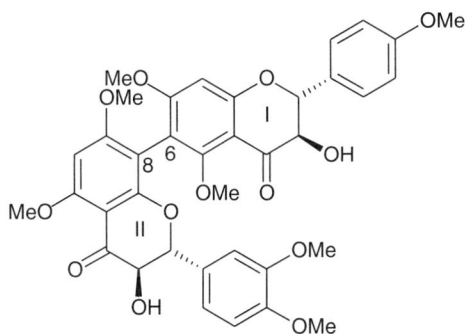

(1) *Ouratea multiflora*, relative configuration, Ref. 25

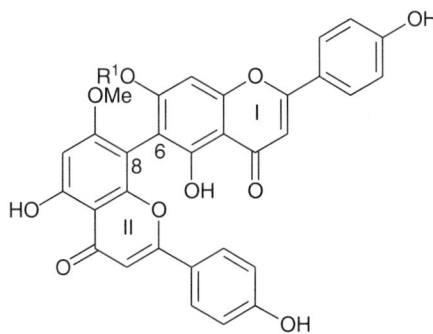

(2) R[1] = Me — *Ouratea spectabilis*, Ref. 28
(3) R[1] = H — *Ouratea hexasperma*, Ref. 33

17.3.1.2 Amentoflavones [(I-3′,II-8)-Coupling]

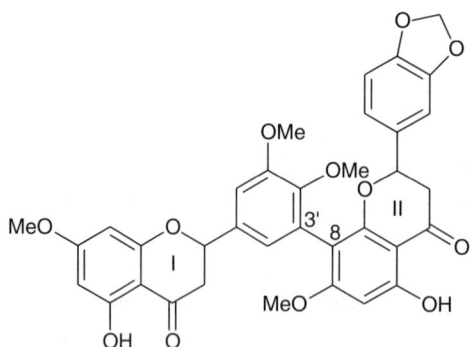

(**4**) Anacarduflavanone — *Semecarpus anacardium* Linn., rel. config., Ref. 18

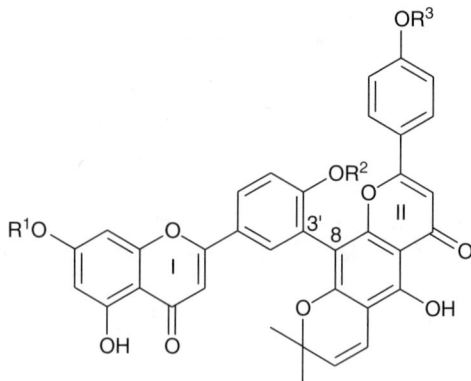

(**5**) R¹ = R² = R³ = H — Pyranoamentoflavone — *Calophyllum inophylloide*, Ref. 19
(**6**) R¹ = Me, R² = R³ = H
(**7**) R¹ = R³ = H, R² = Me
(**8**) R¹ = R³ = Me, R² = H — *Calophyllum venulosum*, Refs. 20,21
(**9**) R¹ = R² = Me, R³ = H

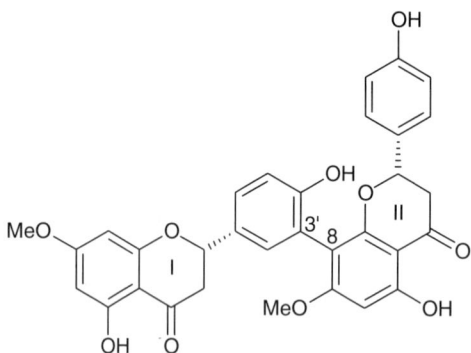

(**12**) (I-2S,II-2S)-I-7,II-7-Di-O-methyltetrahydroamentoflavone — *Rhus retinorrhoea*, abs. config., Ref. 23

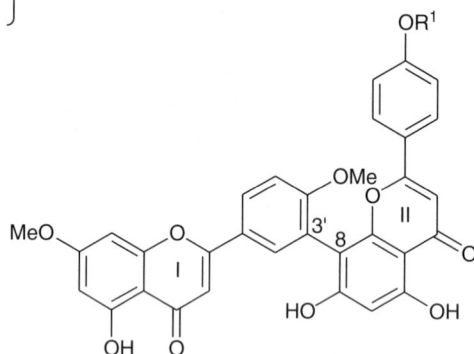

(**10**) R¹ = H
(**11**) R¹ = Me — *Taxus baccata*, Ref. 22
Amentoflavone is the free phenolic form of compound (**11**)

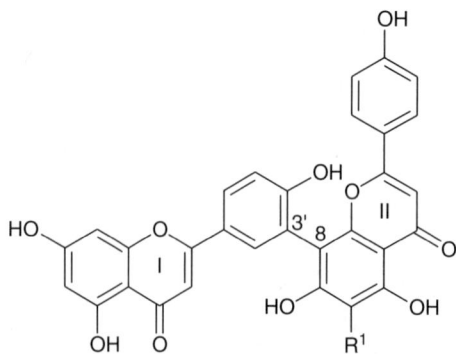

(**13**) R¹ =
(**14**) R¹ = — *Calophyllum venulosum*, Ref. 21

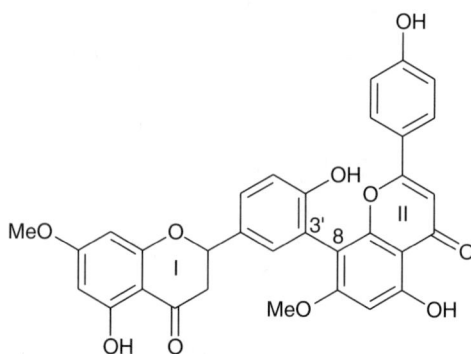

(**15**) *Amentotaxus yunnanensis*, rel. config., Ref. 44

17.3.1.3 Bi-4-aryldihydrocoumarins (Bineoflavones)

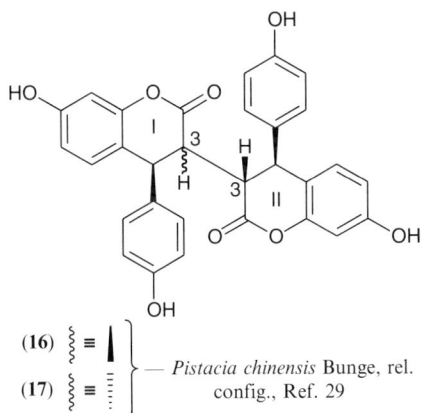

(16) $\lessgtr \equiv$ ▮
(17) $\lessgtr \equiv$ ⋮ } — *Pistacia chinensis* Bunge, rel. config., Ref. 29

17.3.1.4 Biauronols and Biaurones

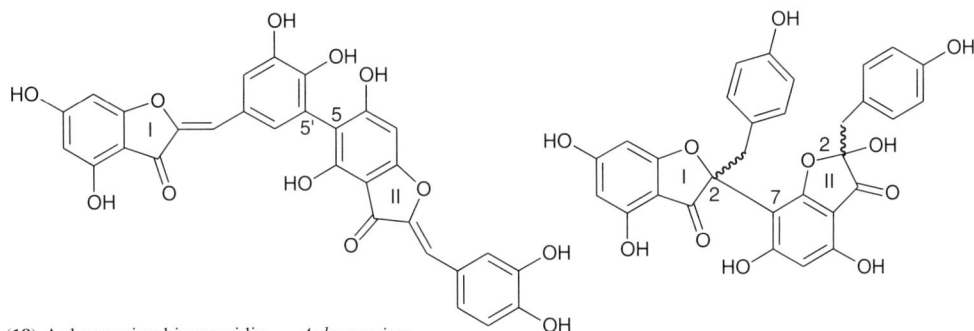

(18) Aulacomniumbiaureusidin — *Aulacomnium palustre* and *A. androgynum*, Ref. 24

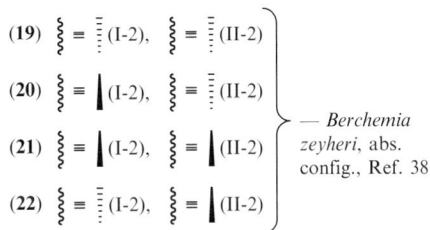

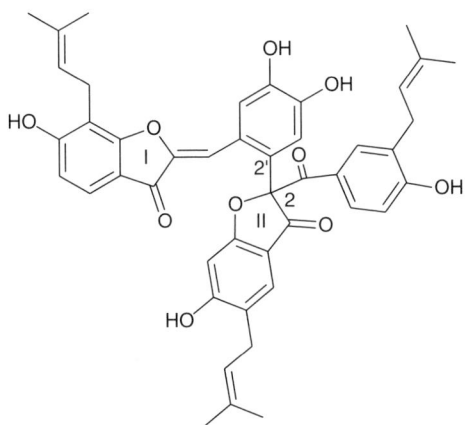

(19) $\lessgtr \equiv$ ⋮ (I-2), $\lessgtr \equiv$ ⋮ (II-2)

(20) $\lessgtr \equiv$ ▮ (I-2), $\lessgtr \equiv$ ⋮ (II-2)

(21) $\lessgtr \equiv$ ▮ (I-2), $\lessgtr \equiv$ ▮ (II-2)

(22) $\lessgtr \equiv$ ⋮ (I-2), $\lessgtr \equiv$ ▮ (II-2)

} — *Berchemia zeyheri*, abs. config., Ref. 38

(23) Licoagrone — *Glycyrrhiza glabra*, Ref. 56

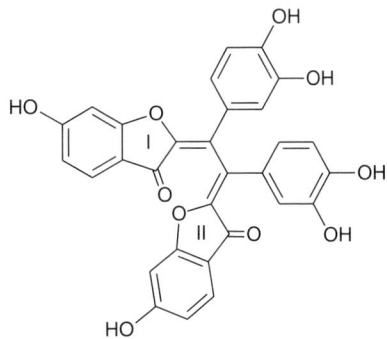

(24) Disulfuretin — *Cotinus coggygria*, Ref. 57

17.3.1.5   Bichalcones

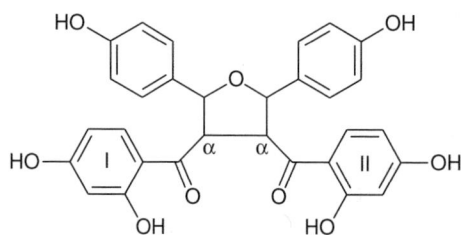

(**25**) Cordigone — *Cordia goetzei*, no config. indicated, Ref. 31

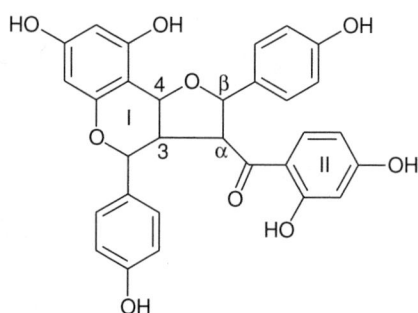

(**26**) Cordigol — *Cordia goetzei*, no config. indicated, Ref. 31

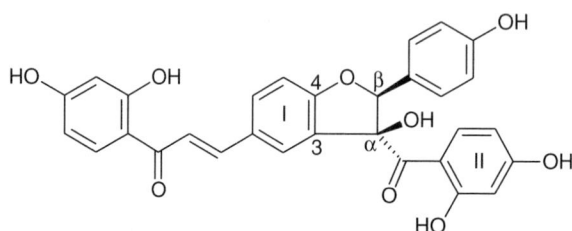

(**27**) Lophirone K — *Ochna calodendron*, rel. config., Ref. 32

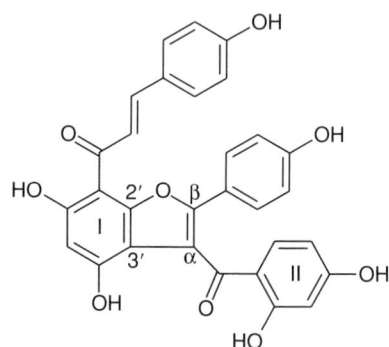

(**28**) Calodenin A — *Ochna calodendron*, Ref. 32

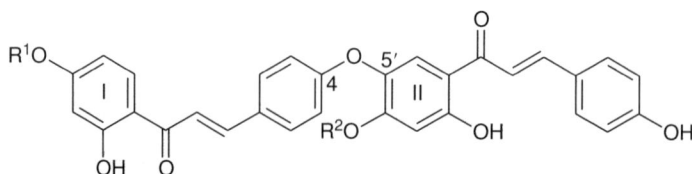

(**29**) $R^1 = R^2 = Me$ — Rhuschalcone I
(**30**) $R^1 = R^2 = H$ — Rhuschalcone II   — *Rhus pyroides* Burch.   Ref. 58
(**31**) $R^1 = Me, R^2 = H$ — Rhuschalcone III   Ref. 67

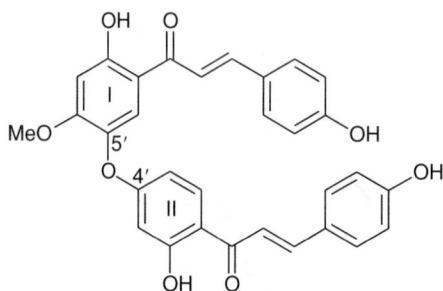

(**32**) Rhuschalcone IV — *Rhus pyroides* Burch., Ref. 67

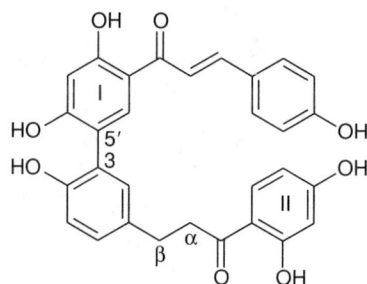

(**33**) Rhuschalcone V
(**34**) α,β-Double bond, rhuschalcone VI   — *Rhus pyroides* Burch., Ref. 67

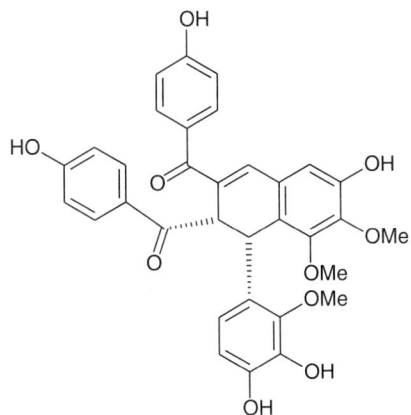

(35) Licobichalcone — *Glycyrrhiza uralensis*, rel. config., Ref. 68

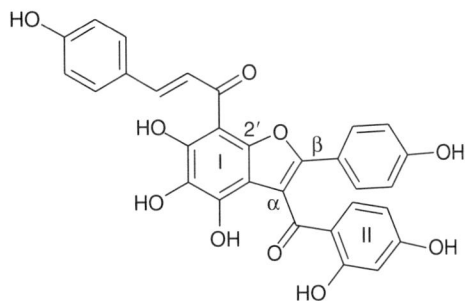

(36) Flavumone A — *Ouratea flava*, Ref. 64

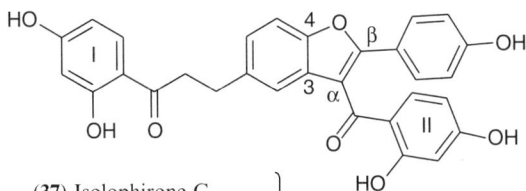

(37) Isolophirone C
(38) Dihydrolophirone C, α,β-dihydro analog of isolophirone C
— *Ochna afzelii*, Ref. 66

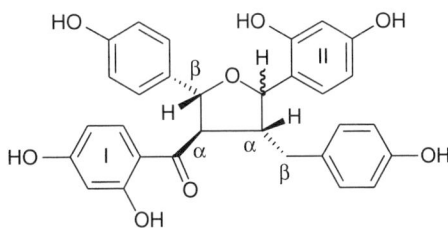

(39) ≡ , Mbamichalcone — *Lophira alata*, rel. config., Ref. 61

(40) ≡ , Isombamichalcone — *Lophira lanceolata*, rel. config., Ref. 62

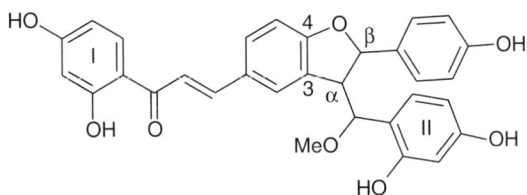

(41) Azobechalcone — *Lophira alata*, rel. config., Ref. 50

17.3.1.6   (I-6,*O*,II-8)-Biflavones

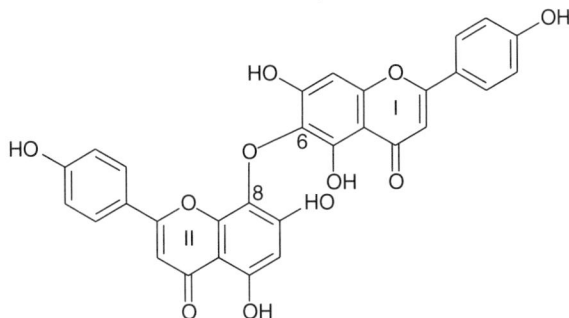

(42) *Viburnum cotinifolium*, Ref. 89

17.3.1.7 (I-2′,II-8)-Biflavones

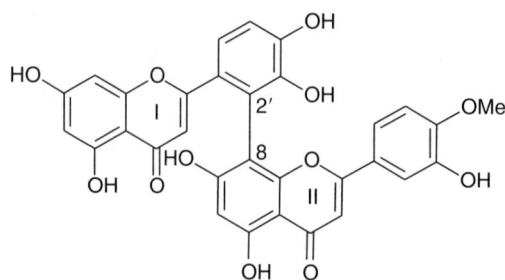

(**43**) Philonotisflavone-II-4′-methyl ether — *Mnium hornum*, Ref. 87

17.3.1.8 (I-2′,II-6)-Biflavones and Aurone–Flavones

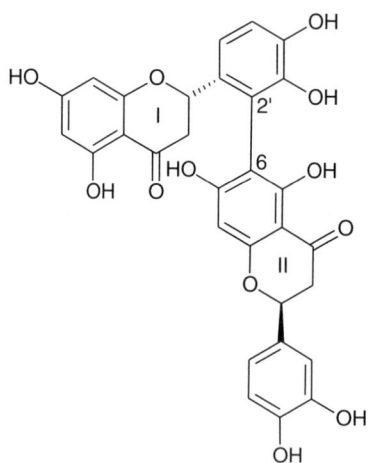

(**44**) Tetrahydrodicranolomin — *Pilotrichella flexilis*, abs. config., Ref. 71

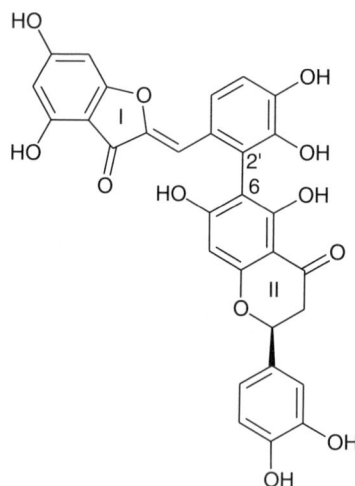

(**45**) Pilotrichellaaurone — *Pilotrichella flexilis*, abs. config., Ref. 71

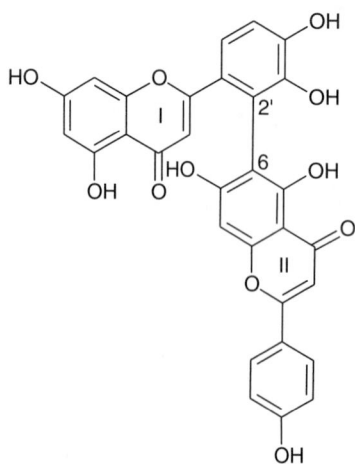

(**46**) 3‴-Desoxydicranolomin ⎤
(**47**) I-2,3-Dihydro derivative ⎦ — *Plagiomnium undulatum*, Ref. 85

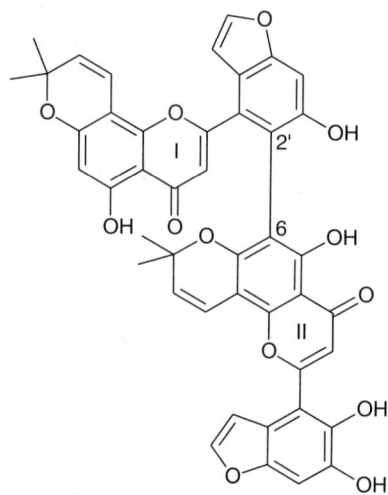

(**48**) Leucaediflavone — *Leucaena diversifolia*, Ref. 107

17.3.1.9 (I-3,II-6)-Biflavones

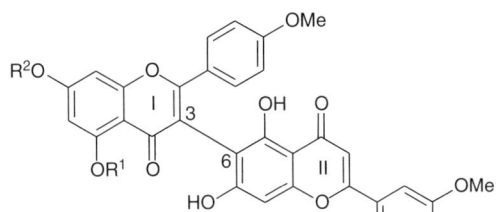

(49) R^1 = H, R^2 = Me
(50) R^1 = Me, R^2 = H } — *Aristolochia ridicula*, Ref. 45

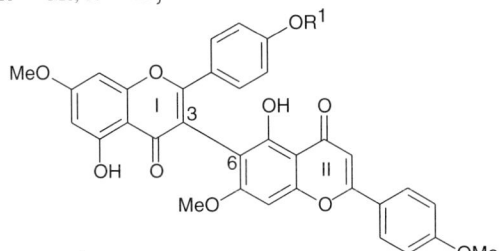

(51) R^1 = H, Stephaflavone A
(52) R^1 = Me, Stephaflavone B } — *Stephania tetrandra*, Ref. 84

17.3.1.10 (I-4′,*O*,II-8)-Biflavones

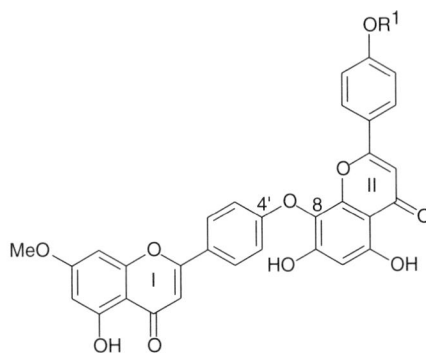

(53) R^1 = H
(54) R^1 = Me } — *Ouratea semiserrata*, Ref. 26

17.3.1.11 (I-3,*O*,II-4′)-Biflavones

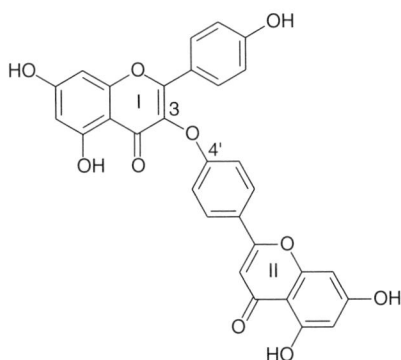

(55) Delicaflavone — *Selaginella delicatula*, Ref. 10

17.3.1.12 (I-3′,II-3)-Biflavones

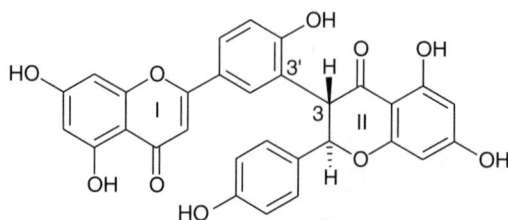

(**56**) Lanceolatin A — *Lophira lanceolatum*, rel. config., Ref. 83

17.3.1.13 (I-2′,II-2′)-Biflavonols

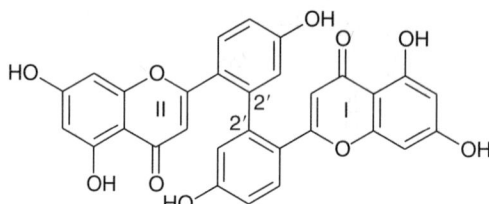

(**57**) *Garcinia nervosa*, Ref. 108

17.3.1.14 (I-3′,II-3′)-Biflavones

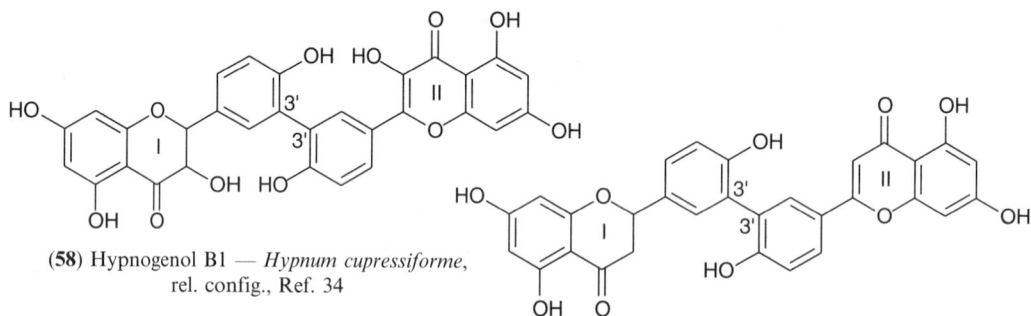

(**58**) Hypnogenol B1 — *Hypnum cupressiforme*, rel. config., Ref. 34

(**59**) 2,3-Dihydroapigeninyl-(I-3′,II-3′)-apigenin — *Homalothecium lutescens*, rel. config., Ref. 40

17.3.1.15 Bi-Isoflavonoids

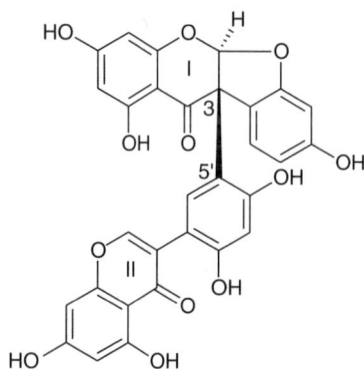

(**60**) *Lupinus albus* L., rel. config., Ref. 12

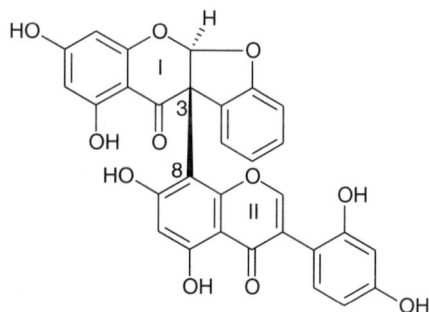

(**61**) *Lupinus albus* L., rel. config., Ref. 12

(65) Dehydroxyhexaspermone C
— *Ochna macrocalyx*, rel. config., Ref. 13

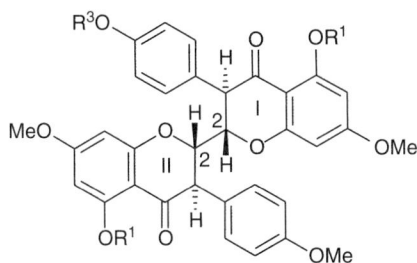

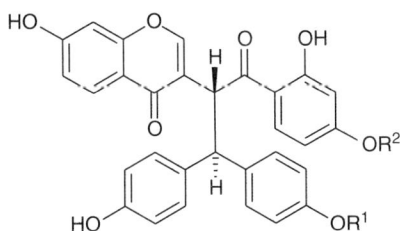

(62) R^1 = R^2 = H, R^3 = Me, Hexaspermone A
(63) R^1 = R^3 = H, R^2 = Me, Hexaspermone B } — *Ouratea hexasperma*, rel. config., Ref. 30
(64) R^1 = R^2 = R^3= H, Hexaspermone C

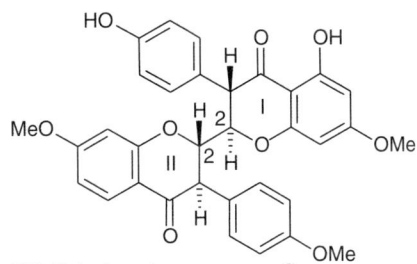

(66) R^1 = R^2 = Me, Afzelone D — *Ochna afzelii*, rel. config., Ref. 16
(67) R^1 = H, R^2 = Me, Calodenone — *Ochna calodendron*, rel. config., Ref. 42
(68) R^1 = R^2 = H, Lophirone A — *Lophira lanceolata*, rel. config., Ref. 65

17.3.1.16 Chamaejasmins [(I-3,II-3)-Coupling]

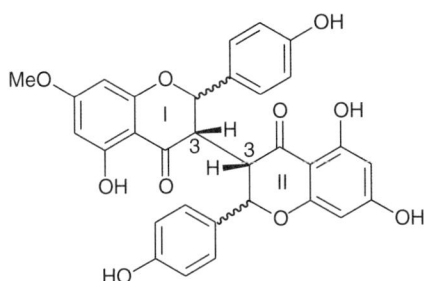

(69) *Stellera chamaejasme* L., rel. config., Ref. 27

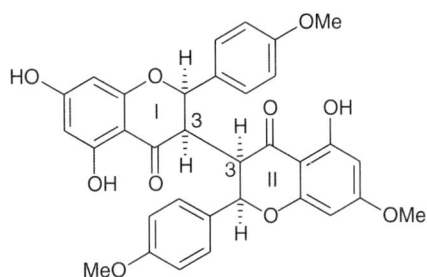

(70) Ruixianglangdusu A — *Stellera chamaejasme*, rel. config., Ref. 81

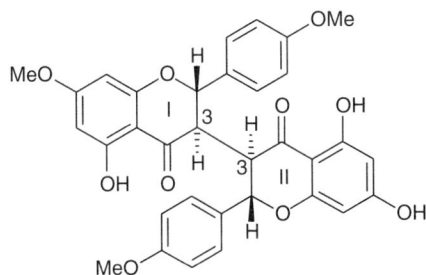

(71) Ruixianglangdusu B — *Stellera chamaejasme*, rel. config., Ref. 81

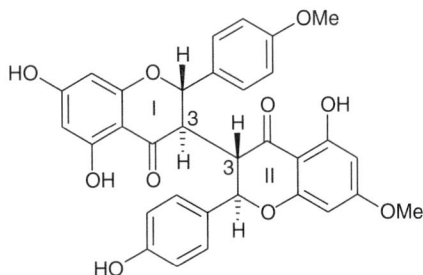

(72) Sikokianin — *Wikstroemia sikokiana*, rel. config., Ref. 82
Also named Sikokianin C — *Wikstroemia indica*, Ref. 109

17.3.1.17   Cupressuflavones [(I-8,II-8)-Coupling]

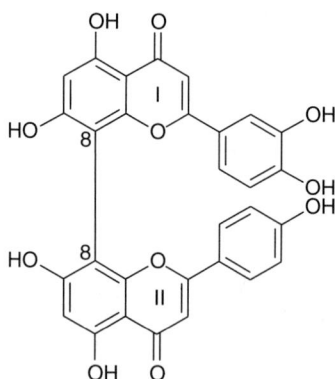

(73) Mogathin (I-3'-hydroxycupressuflavone)
— *Glossostemon bruguieri* (Desf.), Ref. 9

17.3.1.18   Flavanone–Auronols

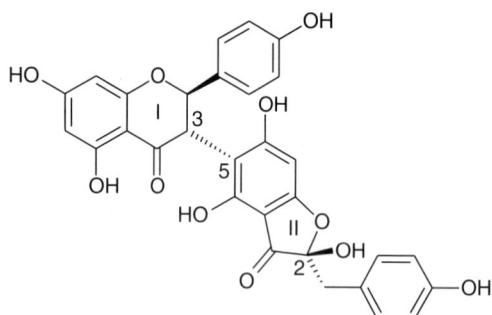

(74) Diastereoisomer shown ⎫
(75) (II-2)-*S*-epimer   ⎬ — *Berchemia zeyheri*, abs. config., Refs. 35, 36

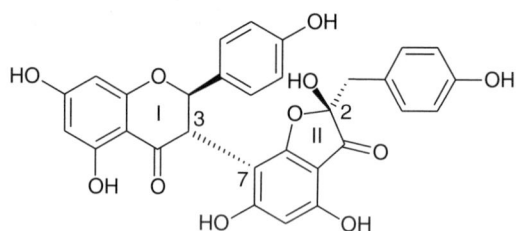

(76) Diastereoisomer shown ⎫ —*Berchemia zeyheri*,
(77) (II-2)-*S*-epimer   ⎬ abs. config., Ref. 37

17.3.1.19   Flavanone and Flavone–Chalcones

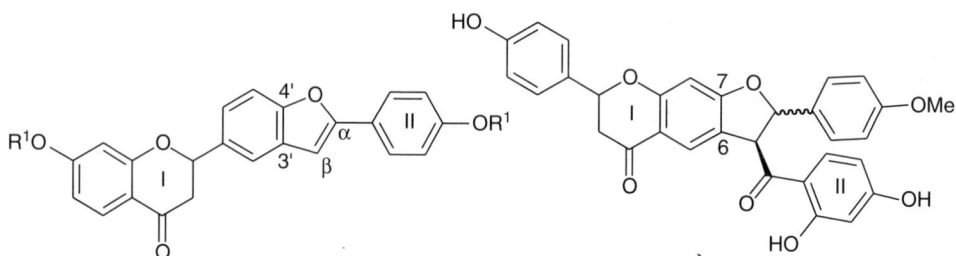

(78) R¹ = H, Lophirone I ⎫ — "Cleaved" chalcones — *Lophira*
(79) R¹ = Me, Lophirone J ⎬ — *lanceolata*, rel. config., Ref. 48

(80) —Afzelone A⎤ — *C*-β-epimers, *Ochna afzelii*,
(81) Afzelone B   ⎦ — rel. config., Ref.

(**83**) $R^1 = R^2 =$ H, 6'''-Hydroxylophirone B   — *Ochna integerrima*,
(**84**) $R^1 =$ H, $R^2 = \beta$-D-Glc   — abs. config., Ref. 52

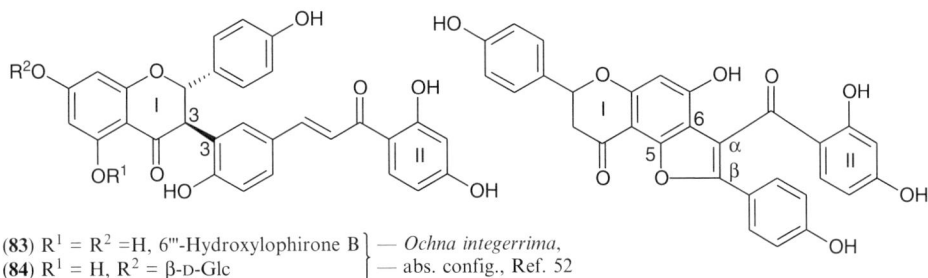

(**82**) Flavumone B — *Ouratea flava*, rel.
config., Ref. 64

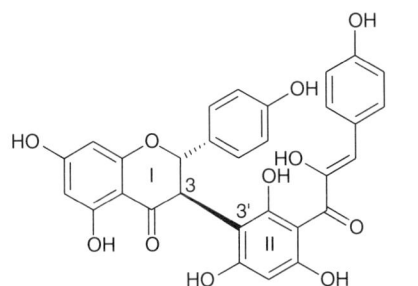

(**85**) Flavanone-α-hydroxychalcone — *Berchemia zeyheri*, abs. config., Ref. 37

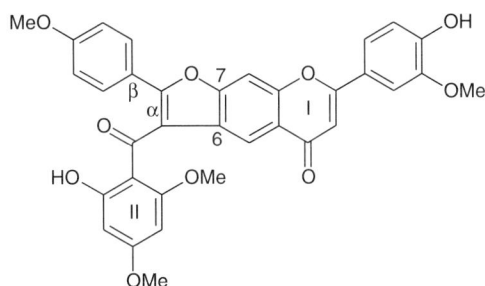

(**86**) Cissampeloflavone — *Cissampelos pareira*, Ref. 53

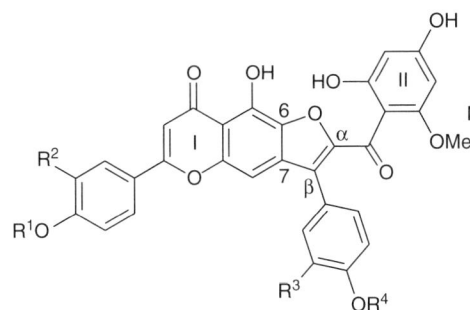

(**87**) $R^1 =$ Me, $R^2 = R^4 =$ H, $R^3 =$ OMe
(**88**) $R^1 = R^3 =$ H, $R^2 =$ OMe, $R^4 =$ Me   — *Aristolochia ridicula*, Ref. 45
(**89**) $R^1 = R^4 =$ Me, $R^2 =$ H, $R^3 =$ OH
Note the odd I-A hydroxylation pattern.

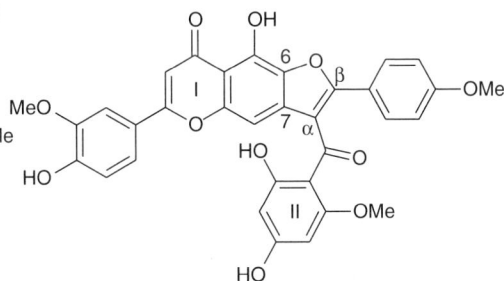

(**90**) *Aristolochia ridicula*, Ref. 45
Note the odd I-A hydroxylation pattern.

17.3.1.20   Flavanone–Isoflavans

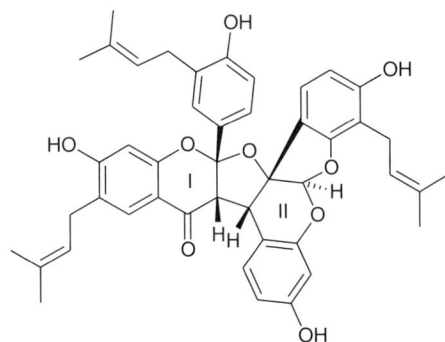

(**91**) Licoagrodin — *Glycyrrhiza glabra*, rel. config., Ref. 88

17.3.1.21   GB-flavones [(I-3,II-8)-Coupling]

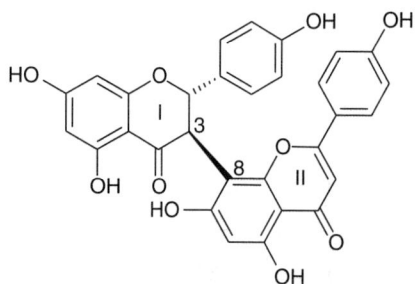

(**92**) Garcinianin atropisomers — *Garcinia kola*,
rel. config., Ref. 69

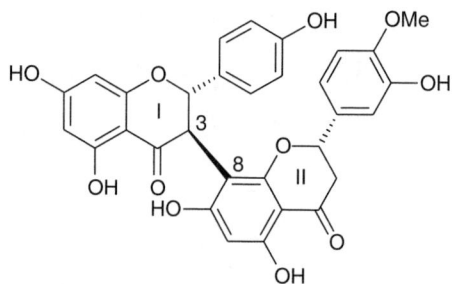

(**93**) GB-2a-II-4'-OMe — *Rheedia gardneriana*,
rel. config., Ref. 70

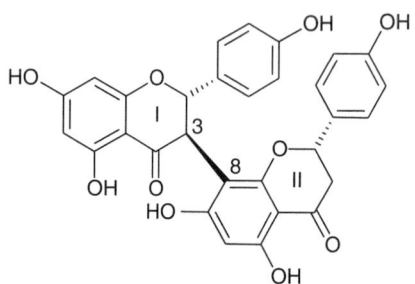

(**94**) (+)-GB-lb — *Garcinia kola*, abs. config., Ref. 72

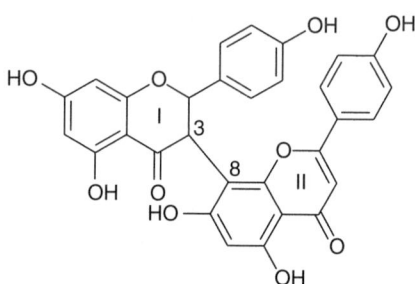

(**95**) Pancibiflavonol — *Callophyllum panciflorum*,
rel. config., Ref. 73

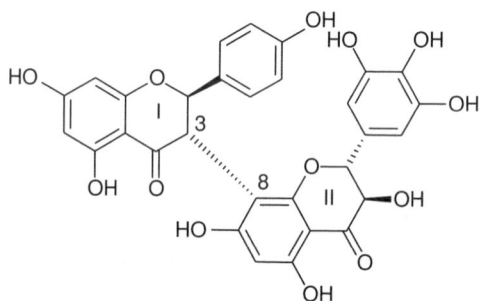

(**96**) GB-4, (I-2*R*,3*S*; II-2*R*,3*R*) ⎤ — *Gnidia involucrata*,
(**97**) GB-4a, (I-2*S*,3*R*; II-2*R*,3*R*) ⎦   abs. config., Ref. 74

17.3.1.22   Hinokiflavones [(I-4′,*O*,II-6)-Coupling]

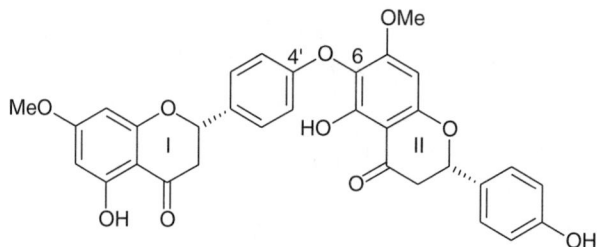

(**98**) I-7,II-7-Di-*O*-methyltetrahydrohinokiflavone — *Cycas beddomei*, abs. config., Ref. 80

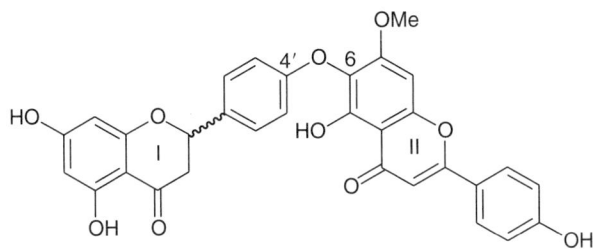

(**99**) 2,3-Dihydroisocryptomerin — *Selaginella delicatula*, rel. config., Ref. 10

17.3.1.23   Isoflavanone–Auronols [(I-2,II-7)-Coupling]

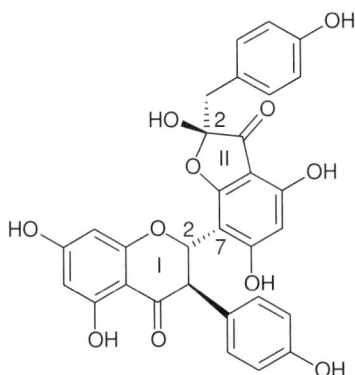

(**100**) Diastereoisomer shown } — *Berchemia zeyheri*, abs.
(**101**) (II-2)-*S*-epimer   }   config., Refs. 35, 36

17.3.1.24   Ochnaflavones [(I-3′,*O*,II-4′)-Coupling]

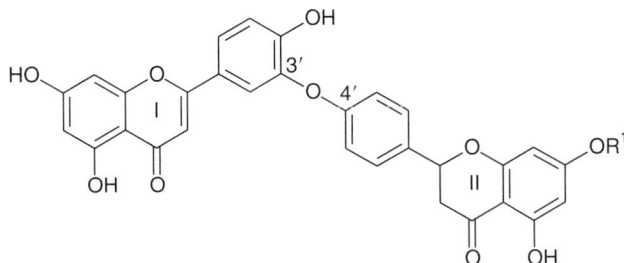

(**102**) R^1 = H — *Luxemburgia nobilis* (EICHL), rel. config., Refs. 14, 77
(**103**) R^1 = Me — *Ochna integerrima*, rel. config., Ref. 77

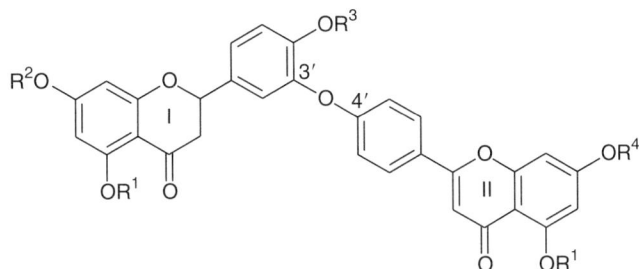

(**104**) R^1 = R^2 = R^3 = R^4 = H }
(**105**) R^1 = R^3 = R^4 = H, R^2 = Me } — *Ochna obtusata*, rel. config., Ref. 75
(**106**) R^1 = H, R^2 = R^3 = R^4 = Me — *Ochna beddomei*, rel. config., Ref. 78

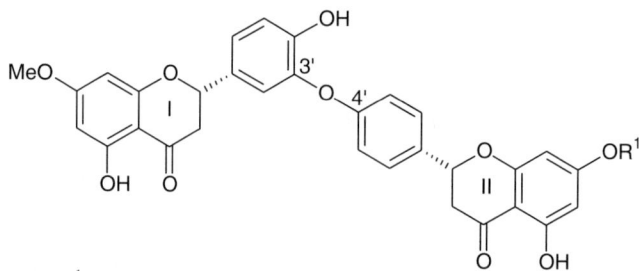

(107) R^1 = H — *Ochna beddomei*, abs. config., Ref. 76
(108) R^1 = Me — *Quntinia acutifolia*, rel. config., Ref. 79

17.3.1.25 Robustaflavones [(I-3′,II-6)-Coupling]

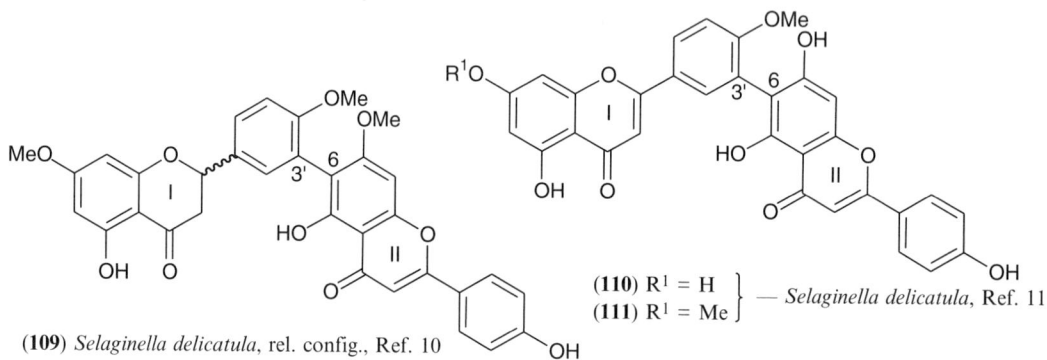

(109) *Selaginella delicatula*, rel. config., Ref. 10

(110) R^1 = H
(111) R^1 = Me } — *Selaginella delicatula*, Ref. 11

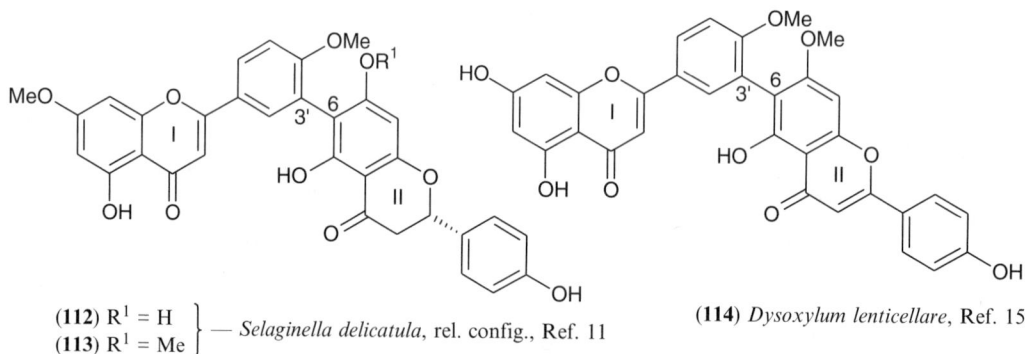

(112) R^1 = H
(113) R^1 = Me } — *Selaginella delicatula*, rel. config., Ref. 11

(114) *Dysoxylum lenticellare*, Ref. 15

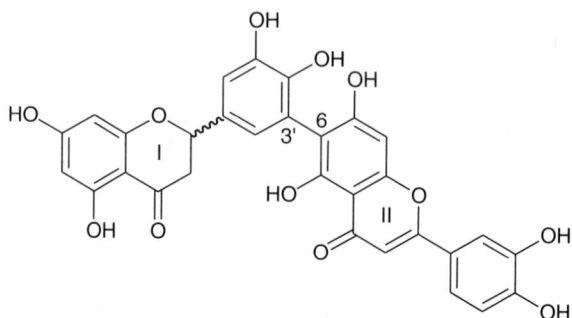

(115) *Plagiomnium undulatum*, rel. config., Ref. 85

17.3.1.26   Succedaneaflavones [(I-6,II-6)-Coupling]

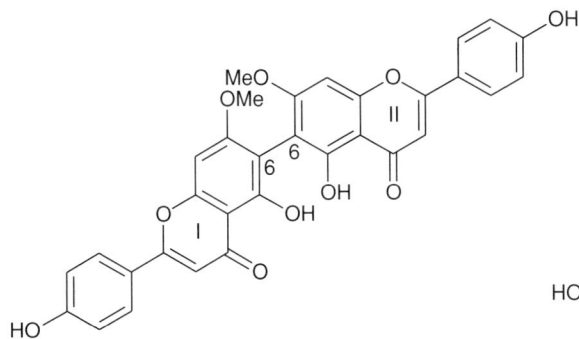

(**116**) 6,6″-Bigenkwanin — *Ouratea spectabilis*, Ref. 28

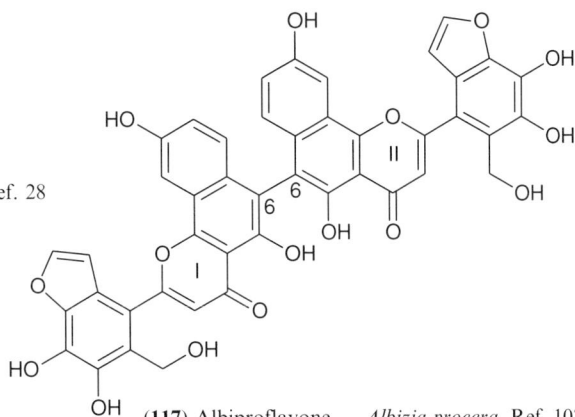

(**117**) Albiproflavone — *Albizia procera*, Ref. 107
Note the naphthopyrano functionalities.

17.3.1.27   "Unusual" Biflavonoids

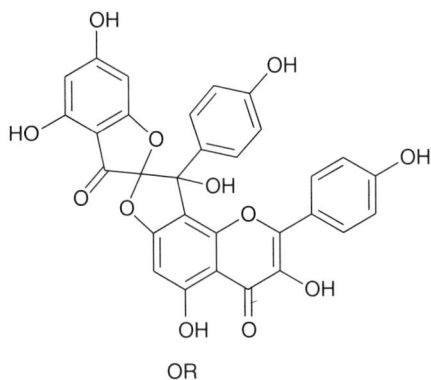

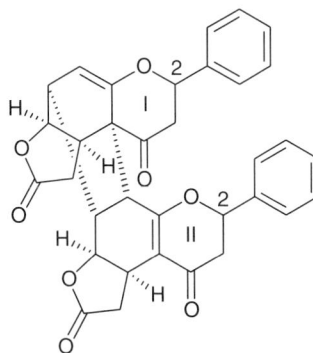

(**119**) Bicaryone A (I-2*S*, II-2*S*)
(**120**) Bicaryone B (I-2*S*, II-2*R*)
(**121**) Bicaryone C (I-2*R*, II-2*S*)
(**122**) Bicaryone D (I-2*R*, II-2*R*)
— *Cryptocarya infectoria*, Ref. 101

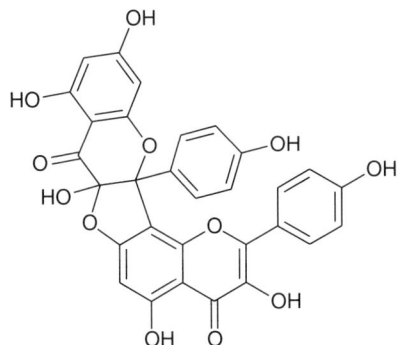

(**118**) VC-15B (vahlia biflavone) — *Vahlia capensis*, Ref. 100

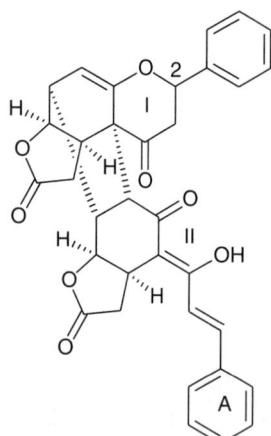

(**123**) Chalcocaryanone A (I-2*R*)
(**124**) Chalcocaryone B (I-2*S*)
Cryptocarya infectoria, Ref. 101

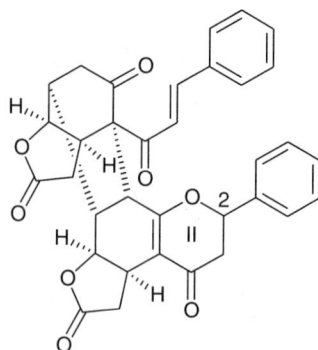

(**125**) Chalcocaryanone C (I-2*S*)
(**126**) Chalcocaryone D (I-2*R*)
Cryptocarya infectoria, Ref. 101

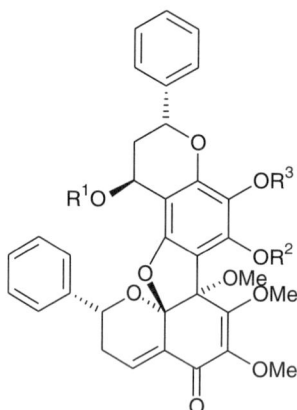

(**127**) R^1 = R^3 = Me, R^2 = H, Calycopterone
(**128**) R^1 = R^2 = Me, R^3 = H, Isocalycopterone
(**129**) R^1 = R^2 = H, R^3 = Me, 4-Demethylcalycopterone

Calycopteris floribunda Lamk., Ref. 102

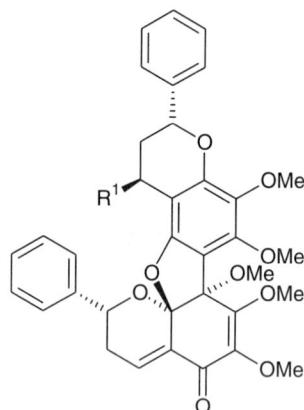

(**130**) R^1 = OH, Neocalycopterone
(**131**) R^1 = OMe, Neocalycopterone-4-Me

Calycopteris floribunda, Ref. 102

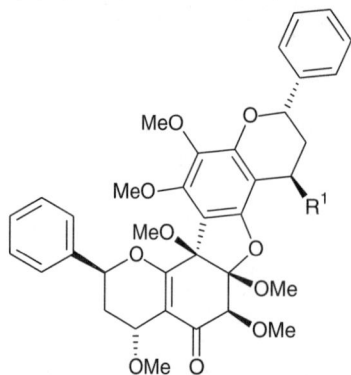

(**132**) R^1 = OMe, Calyflorenone A
(**133**) R^1 = OH, Calyflorenone B

Calycopteris floribunda, Ref. 103

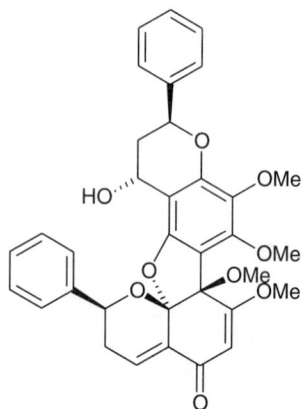

(**134**) 6"-Demethoxyneocalycopterone
Calycopteris floribunda, abs. config., Ref. 104

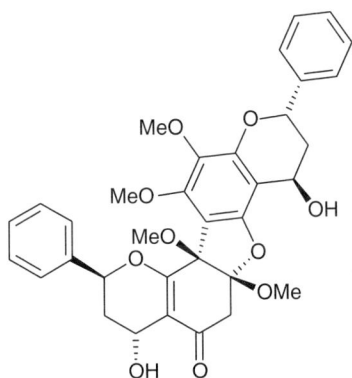

(135) Calyflorenone D — *Calycopteris floribunda*, abs. config., Ref. 104

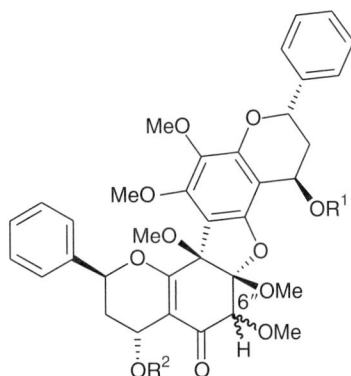

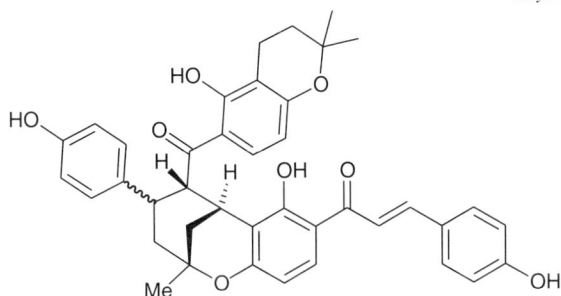

(136) R^1 = R^2 = H, 6"β-OMe, Calyflorenone C
(137) R^1 = H, R^2 = Me, 6"α-OMe, 6-*epi*-Calyflorenone B
(138) R^1 = R^2 = H, 6"α-OMe, 6-*epi*-Calyflorenone C

Calycopteris floribunda, abs. config., Ref. 104

(139) *Dorstenia zenkeri*, rel. config., Ref. 105

17.3.1.28 Miscellaneous Biflavonoids

17.3.1.28.1 Methylene-Bridged Analogs

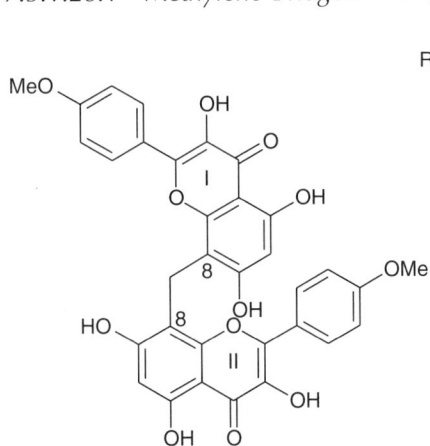

(140) Pentagrametin — *Pentagramma triangularis* spp. *triangularis*, Ref. 41

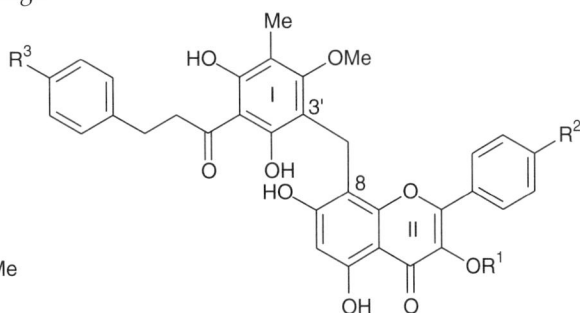

(141) R^1 = R^3 = H, R^2 = OMe, Trianguletin
— *Pentagramma triangularis* spp. *triangularis*, Ref. 41

(142) R^1 = Me, R^2 = R^3 = H
— *Pentagramma triangularis*, Ref. 43

(143) R^1 = H, R^2 = OMe, R^3 = OH
(144) R^1 = Me, R^2 = OMe, R^3 = OH
(145) R^1 = Me, R^2 = H, R^3 = OH
— *Pentagramma triangularis*, Ref.

17.3.1.28.2 Hypnumbiflavonoid A

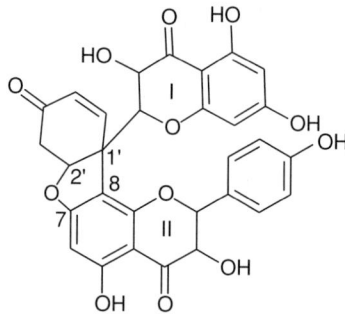

(**146**) Hypnumbiflavonoid A — *Hypnum cupressiforme*, rel. config., Ref. 34
See also the structures of two phenylacetic acid-substituted
aromadendrin analogs from the same source.

17.3.1.28.3 Bongosin

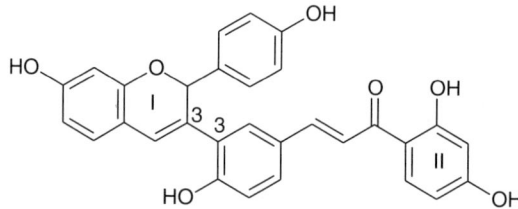

(**147**) Bongosin — *Lophira alata*, Ref. 59

17.3.2 TRIFLAVONOIDS

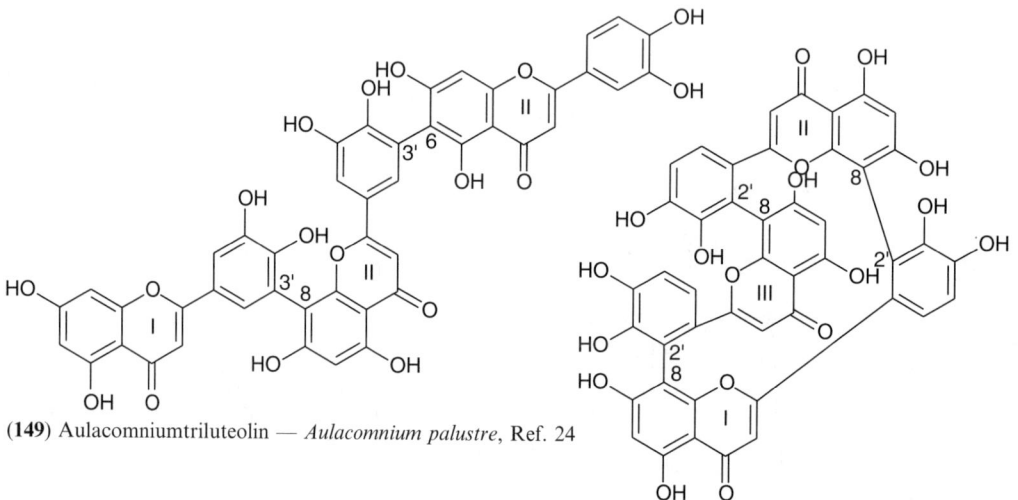

(**149**) Aulacomniumtriluteolin — *Aulacomnium palustre*, Ref. 24

(**148**) Cyclobartramiatriluteolin — *Bartramia stricta*, Ref. 17

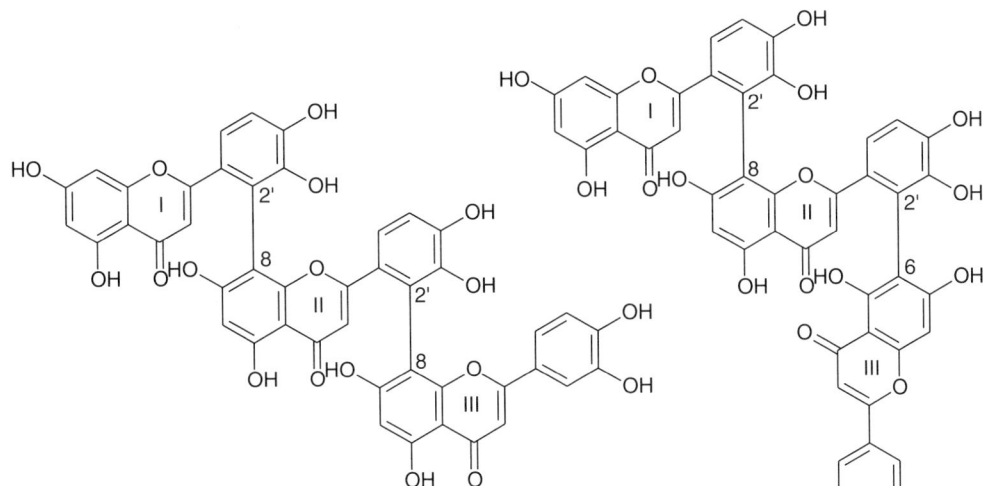

(**150**) Epibartamiatriluteolin atropisomers
Bartramia pomiformis and *B. stricta*, Ref. 55

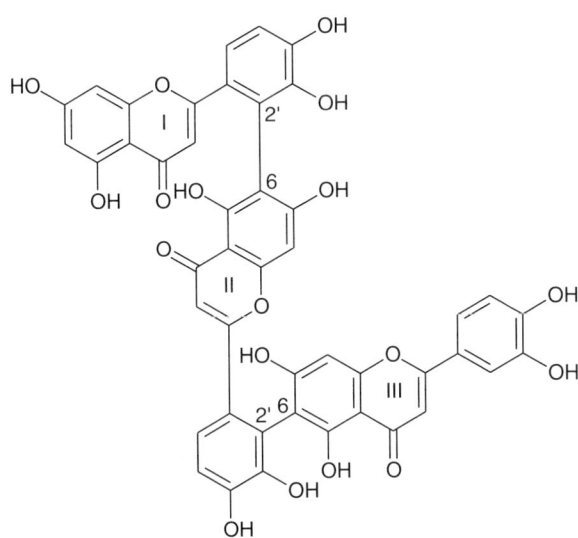

(**151**) Strictatriluteolin atropisomers — *Bartramia pomiformis* and *B. stricta*, Ref. 55

(**152**) Distichumtriluteolin — *Rhizogonium distichum*, Ref. 86

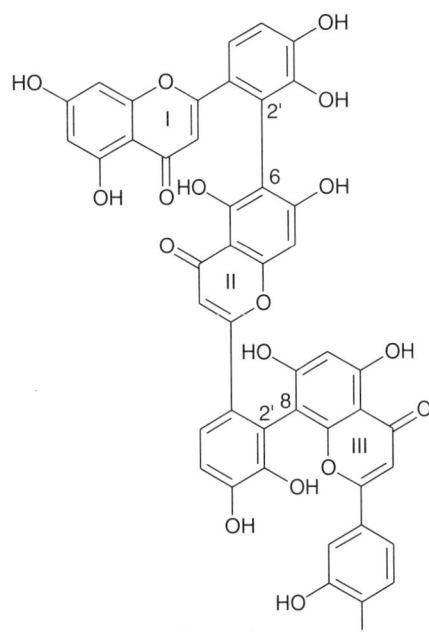

(**153**) Rhizogoniumtriluteolin — *Rhizogonium disticum*, Ref. 86

17.3.3 TETRAFLAVONOIDS

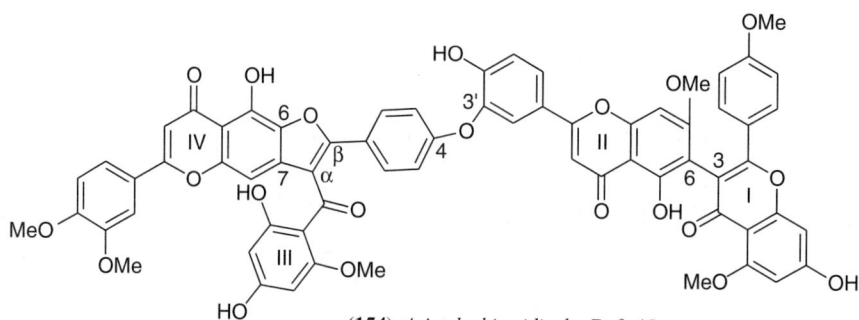

(154) *Aristolochia ridicula*, Ref. 45
Note the odd hydroxylation pattern of the IV-A-ring.

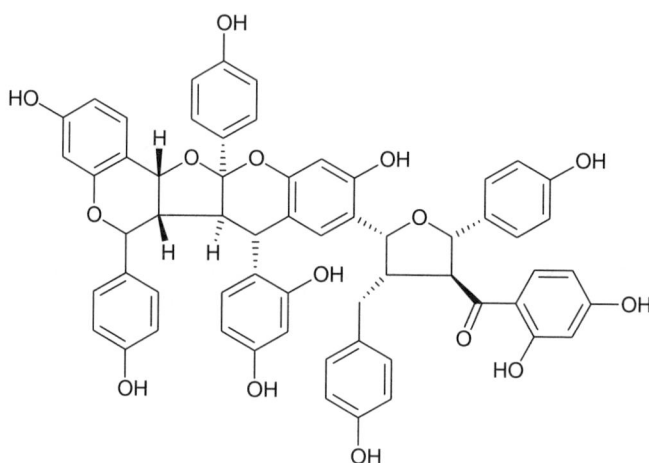

(155) Lophiroflavan A — *Lophira alata*, rel. config., Ref. 49

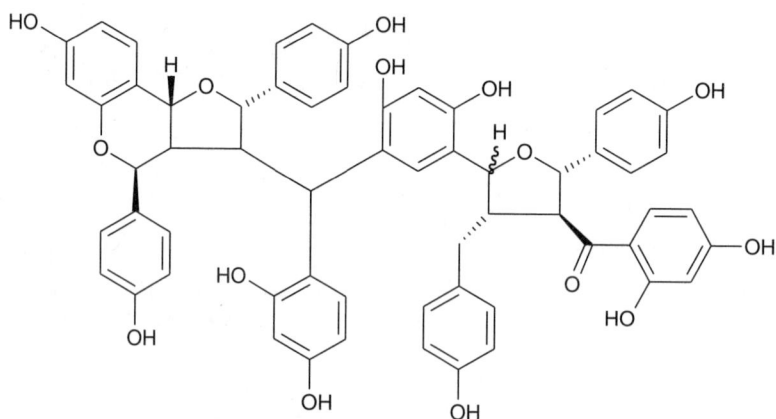

(156) = , Lophiroflavan B
(157) = , Lophiroflavan C
— *Lophira alata*, rel. config., Ref. 47

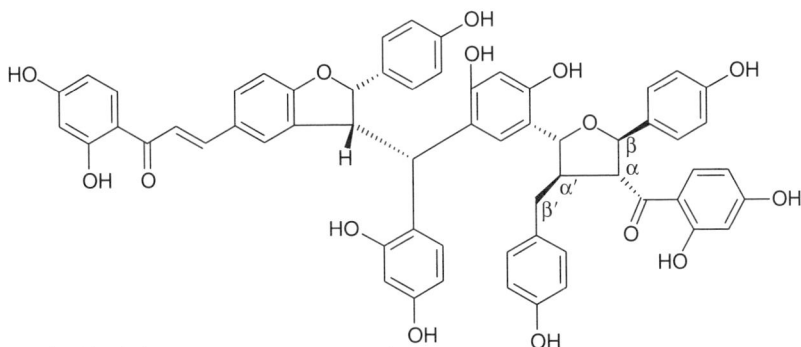

(**158**) Isolophirachalcone
(**159**) (C-α,α′,β-enantiomer) Lophirachalcone   } — *Lophira alata*, rel. config., Refs. 50, 51

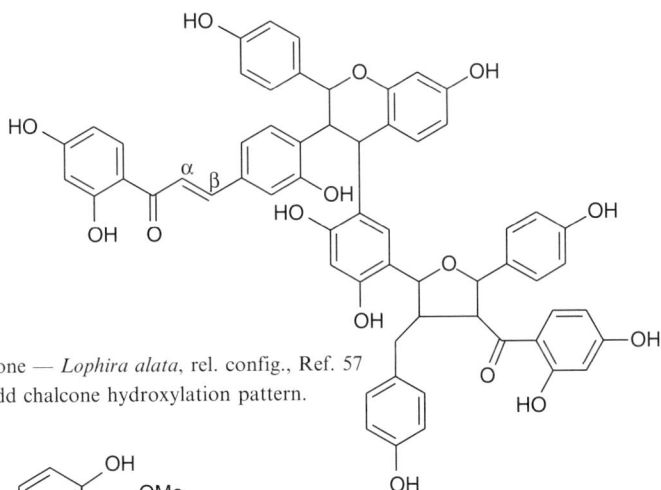

(**160**) Alatachalcone — *Lophira alata*, rel. config., Ref. 57
Note the odd chalcone hydroxylation pattern.

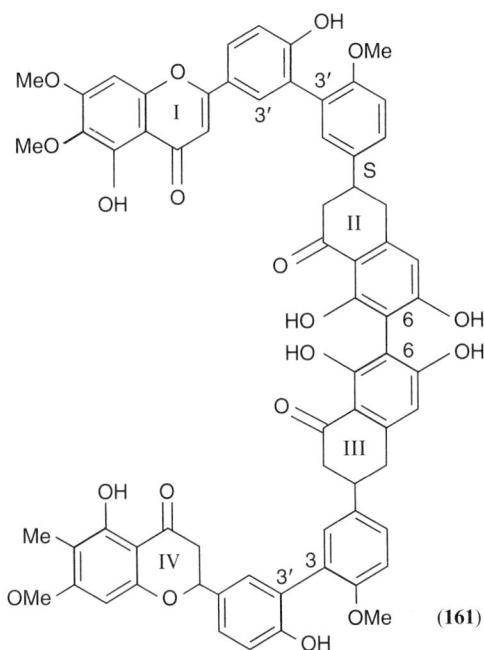

(**161**) Lophirochalcone — *Lophira lanceolata*, rel. config., Ref. 62

(**162**) Taiwanhomoflavone C — *Cephalotaxus wilsoniana*, Ref. 60

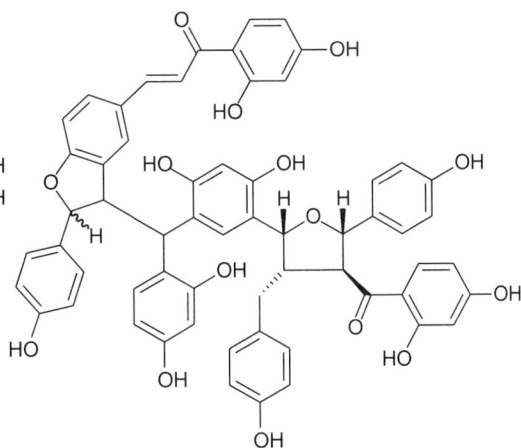

17.3.4   Pentaflavonoids

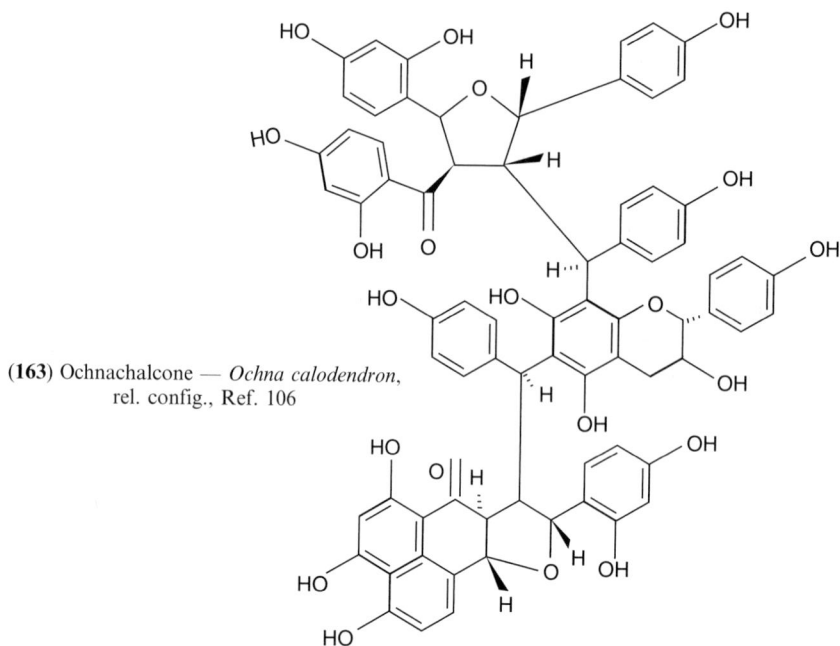

(163) Ochnachalcone — *Ochna calodendron*,
rel. config., Ref. 106

17.3.5   Hexaflavonoids

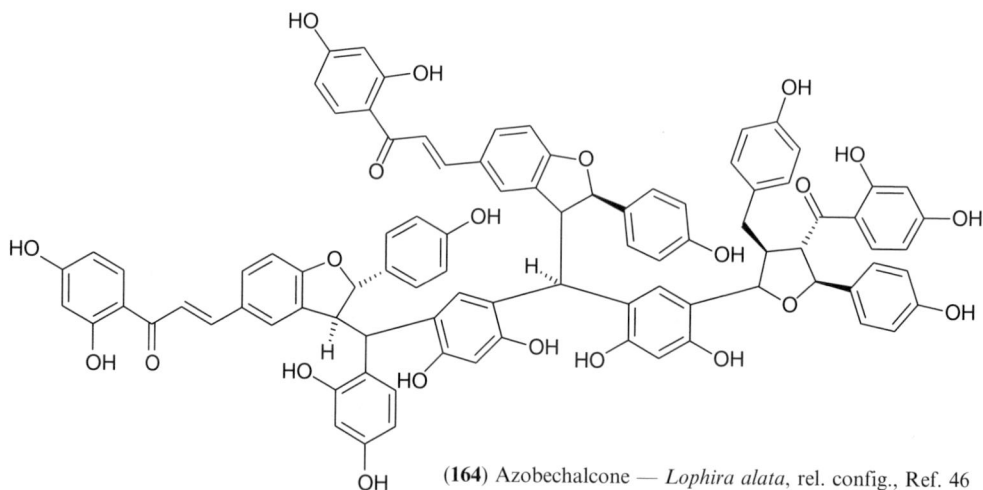

(164) Azobechalcone — *Lophira alata*, rel. config., Ref. 46

REFERENCES

1. Baker, W. and Ollis, W.D., Biflavonyls, in *Recent Developments in the Chemistry of Natural Phenolic Compounds*, Ollis, W.D., Ed., Pergamon, London, 1961, 152–184.
2. Locksley, H.D., The chemistry of biflavonoid compounds, *Fortschr. Chem. Org. Naturst.*, 30, 207, 1973.
3. Geiger, H. and Quinn, C., Biflavonoids, in *The Flavonoids*, Harborne, J.B., Mabry, T.J., and Mabry, H., Eds., Chapman & Hall, London, 1975, 692–742.

4. Geiger, H. and Quinn, C., Biflavonoids, in *The Flavonoids*, Harborne, J.B. and Mabry, T.J., Eds., Chapman & Hall, London, 1982, 505–534.

5. Geiger, H. and Quinn, C., Biflavonoids, in *The Flavonoids: Advances in Research Since 1980*, Harborne, J.B. and Mabry, T.J., Eds., Chapman & Hall, London, 1988, 99–124.

6. Hemingway, R.W., Biflavonoids and proanthocyanidins, in *Natural Products of Woody Plants 1: Chemicals Extraneous to the Lignocellulosic Cell Wall*, Rowe, J.W., Ed., Springer Verlag, New York, 1989, 571–651.

7. Geiger, H., Biflavonoids and triflavonoids, in *The Flavonoids: Advances in Research Since 1986*, Harborne, J.B., Ed., Chapman & Hall, London, 1994, 95–115.

8. Taylor, W.I. and Battersby, A.R., *Oxidative Coupling of Phenols*, Marcel Dekker, New York, 1967.

9. Meselhy, M.R., Constituents from Moghat, the roots of *Glossostemon bruguieri* (Desf.), *Molecules*, 8, 614, 2003.

10. Lin, L-C. and Chou, C-J., Three new biflavonoids from *Selaginella delicatula*, *Chin. Pharm. J.*, 52, 211, 2000.

11. Lin, L-C., Kuo, Y-C., and Chou, C-J., Cytotoxic biflavonoids from *Selaginella delicatula*, *J. Nat. Prod.*, 63, 627, 2000.

12. Mitsuyoshi, S. et al., A new class of biflavonoids: 2′-hydroxygenistein dimers from the roots of white lupin, *Z. Naturforsch.*, 55, 165, 2000.

13. Tang, S. et al., Biflavonoids with cytotoxic and antibacterial activity from *Ochna macrocalyx*, *Planta Med.*, 69, 247, 2003.

14. De Oliveira, M.C.C. et al., New biflavonoid and other constituents from *Luxemburgia nobilis* (EICHL), *J. Braz. Chem. Soc.*, 13, 119, 2002.

15. He., K. et al., A biflavonoid from *Dysoxylum lenticellare*, *Phytochemistry*, 42, 1199, 1996.

16. Pegnyemb, D.E. et al., Isolation and structure elucidation of a new isobiflavonoid from *Ochna afzelii*, *Pharm. Biol.*, 41, 92, 2003.

17. Geiger, H. et al., Cyclobartramiatriluteolin, a unique triflavonoid from *Bartramia stricta*, *Phytochemistry*, 39, 465, 1995.

18. Murthy, S.S.S., Naturally occurring biflavonoid derivatives. 11. New biflavonoid from *Semecarpus anacardium* Linn., *Chim. Act. Turc.*, 20, 33, 1992.

19. Goh, S.H. et al., Neoflavonoid and biflavonoid constituents from *Calophyllum inophylloide*, *J. Nat. Prod.*, 55, 1415, 1992.

20. Cao, S-G., Sim, K-Y., and Goh, S-H., Minor methylated pyranoamentoflavones from *Calophyllum venulosum*, *Nat. Prod. Lett.*, 15, 291, 2001.

21. Cao, S-G., Sim, K-Y., and Goh, S-H., Biflavonoids of *Calophyllum venulosum*, *J. Nat. Prod.*, 60, 1245, 1997.

22. Reddy, B.P. and Krupadanam, G.L.D., Chemical constituents of the leaves of Himalayan *Taxus baccata*: use of DQF-COSY in the structure elucidation of biflavones, *Indian J. Chem.*, 35B, 283, 1996.

23. Ahmed, M.S. et al., A weakly antimalarial biflavanone from *Rhus retinorrhoea*, *Phytochemistry*, 58, 599, 2001.

24. Hahn, H. et al., The first biaurone, a triflavone and biflavonoids from two *Aulacomnium* species, *Phytochemistry*, 40, 573, 1995.

25. D'arc Felicio, J. et al., Biflavonoids from *Ouratea multiflora*, *Fitoterapia*, 72, 453, 2001.

26. Velandia, J.R. et al., Biflavonoids and a glucopyranoside derivative from *Ouratea semiserrata*, *Phytochem. Anal.*, 13, 283, 2002.

27. Jiang, Z-H. et al., Biflavanones, diterpenes, and coumarins from the roots of *Stellera chamaejasme* L., *Chem. Pharm. Bull.*, 50, 137, 2002.

28. D'arc Felicio, J. et al., Inhibition of lens aldose reductase by biflavones from *Ouratea spectabilis*, *Planta Med.*, 61, 217, 1995.

29. Nishimura, S. et al., Structures of 4-arylcoumarin (neoflavone) dimers isolated from *Pistacia chinensis* Bunge and their estrogen-like activity, *Chem. Pharm. Bull.*, 48, 505, 2000.

30. Moreira, I.C. et al., Isoflavanone dimers hexaspermone A, B and C from *Ouratea hexasperma*, *Phytochemistry*, 35, 1567, 1994.

31. Marston, A., Slacamin, I., and Hostettmann, K., Antifungal polyphenols from *Cordia goetzei* Guerke, *Helv. Chim. Acta*, 71, 1210, 1988.

32. Messanga, B. et al., Biflavonoids from *Ochna calodendron*, *Phytochemistry*, 35, 791, 1994.

33. Moreira, I.C. et al., A flavone dimer from *Ouratea hexasperma*, *Phytochemistry*, 51, 833, 1999.

34. Sievers, H. et al., Further biflavonoids and 3′-phenylflavonoids from *Hypnum cupressiforme*, *Phytochemistry*, 35, 795, 1994.

35. Bekker, R., Brandt, E.V., and Ferreira, D., Structure and stereochemistry of the first isoflavanone–benzofuranone biflavonoids, *Tetrahedron Lett.*, 39, 6407, 1998.

36. Bekker, R., Brandt, E.V., and Ferreira, D., Biflavonoids. Part 4. Structure and stereochemistry of novel flavanone– and the first isoflavanone–benzofuranone biflavonoids, *Tetrahedron*, 55, 10005, 1999.

37. Bekker, R., Brandt, E.V., and Ferreira, D., Absolute configuration of flavanone–benzofuranone-type biflavonoids and 2-benzyl-2-hydroxybenzofuranones, *J. Chem. Soc. Perkin Trans.*, 1, 2535, 1996.

38. Bekker, R. et al., Biflavonoids. Part 5: structure and stereochemistry of the first bibenzofuranoids, *Tetrahedron*, 56, 5297, 2000.

39. Rani, M.S. et al., A biflavonoid from *Cycas beddomei*, *Phytochemistry*, 47, 319, 1998.

40. Seeger, T. et al., Biflavonoids from the moss, *Homalothecium lutescens*, *Phytochemistry*, 34, 295, 1993.

41. Roitman, J.N., Wong, R.Y., and Wollenweber, E., Methylene bisflavonoids from frond exudates of *Pentagramma triangularis* ssp *triangularis*, *Phytochemistry*, 34, 297, 1993.

42. Messanga, B.B. et al., Calodenone, a new isobiflavonoid from *Ochna calodendron*, *J. Nat Prod.*, 55, 245, 1992.

43. Iinuma, M. et al., Two biflavonoids in the farinose exudate of *Pentagramma triangularis*, *Phytochemistry*, 35, 1043, 1994.

44. Li, S-H. et al., Chemical constituents from *Amentotaxus yunnanensis* and *Torreya yunnanensis*, *J. Nat. Prod.*, 66, 1002, 2003.

45. Carneiro, F.J.C. et al., Bi- and tetraflavonoids from *Aristolochia ridicula*, *Phytochemistry*, 55, 823, 2000.

46. Tih, A.E. et al., A novel hexaflavonoid from *Lophira alata*, *Tetrahedron Lett.*, 40, 4721, 1999.

47. Tih, A.E. et al., Lophiroflavans B and C, tetraflavonoids of *Lophira alata*, *Phytochemistry*, 31, 3595, 1992.

48. Tih, R.G., Tih, A.E., and Sondengam, B.L., Structures of lophirones I and J, minor cleaved chalcone dimers of *Lophira lanceolata*, *J. Nat. Prod.*, 57, 142, 1994.

49. Tih, A. et al., Tetraflavonoids of *Lophira alata*, *Phytochemistry*, 31, 981, 1992.

50. Murakami, A. et al., Possible antitumor promoters: bi- and tetraflavonoids from *Lophira alata*, *Phytochemistry*, 31, 2689, 1992.

51. Murakami, A. et al., Chalcone tetramers, lophirachalcone and alatachalcone from *Lophira alata* as possible antitumor promoters, *Biosci. Biotechnol. Biochem.*, 56, 769, 1992.

52. Kaewamatawong, R. et al., Novel biflavonoids from the stem bark of *Ochna integerrima*, *J. Nat. Prod.*, 65, 1027, 2002.

53. Ramírez, I. et al., Cissampeloflavone, a chalcone–flavone dimer from *Cissampelos pareira*, *Phytochemistry*, 64, 645, 2003.

54. Iinuma, M. et al., Unusual biflavonoids in the farinose exudate of *Pentagramma triangularis*, *Phytochemistry*, 44, 705, 1997.

55. Seeger, T. et al., Isomeric triluteolins from *Bartramia stricta* and *Bartramia pomiformis*, *Phytochemistry*, 40, 1531, 1995.

56. Asada, Y., Li, W., and Yoshikawa, T., The first prenylated biaurone, licoagrone from hairy root cultures of *Glycyrrhiza glabra*, *Phytochemistry*, 50, 1015, 1999.

57. Westenburg, H.E. et al., Activity-guided isolation of antioxidative constituents of *Cotinus coggygria*, *J. Nat. Prod.*, 63, 1696, 2000.

58. Masesane, I.B. et al., A bichalcone from the twigs of *Rhus pyroides*, *Phytochemistry*, 53, 1005, 2000.

59. Tih, A.E. et al., Bongosin: a new chalcone-dimer from *Lophira alata*, *J. Nat. Prod.*, 53, 964, 1990.

60. Messanga, B.L. et al., Isolation and structure elucidation of a new pentaflavonoid from *Ochna calodendron*, *New J. Chem.*, 25, 1098, 2001.

61. Tih, A.E. et al., A new chalcone dimer from *Lophira alata, Tetrahedron Lett.*, 29, 5797, 1988.

62. Tih, R.G. et al., Structures of isombamichalcone and lophirochalcone, bi- and tetraflavonoids from *Lophira lanceolata, Tetrahedron Lett.*, 30, 1807, 1989.

63. Pegnyemb, D.E. et al., Flavonoids of *Ochna afzelii, Phytochemistry*, 64, 661, 2003.

64. Mbing, J.N. et al., Two biflavonoids from *Ouratea flava* stem bark, *Phytochemistry*, 63, 427, 2003.

65. Ghogomu, R. et al., Lopirone A, a biflavonoid with unusual skeleton from *Lophira lanceolata, Tetrahedron Lett.*, 28, 2967, 1987.

66. Pegnyemb, D.E. et al., Biflavonoids from *Ochna afzelii, Phytochemistry*, 57, 579, 2001.

67. Mdee, L.K., Yeboah, S.O., and Abegaz, B.M., Rhuschalcones II–VI, five new bichalcones from the root bark of *Rhus pyroides, J. Nat. Prod.*, 66, 599, 2003.

68. Bai, H. et al., A novel biflavonoid from roots of *Glycyrrhiza uralensis* cultivated in China, *Chem. Pharm. Bull.*, 51, 1095, 2003.

69. Terashima, K., Aqil, M., and Niwa, M., Garcinianin, a novel biflavonoid from the roots of *Garcinia kola, Heterocycles*, 41, 2245, 1995.

70. Cechinel Filho, V. et al., I3-naringenin-II8-4′-OMe-eriodyctol: a new potential analgesic agent isolated from *Rheedia gardneriana* leaves, *Z. Naturforsch.*, 55, 820, 2000.

71. Brinkmeier, E. et al., The cooccurrence of different biflavonoid types in *Pilotrichella flexilis, Z. Naturforsch.*, 55, 866, 2000.

72. Terashima, K., A study of biflavanones from the stems of *Garcinia kola* (Guttiferae), *Heterocycles*, 50, 283, 1999.

73. Ito, C. et al., A new biflavonoid from *Calophyllum panciflorum* with antitumor-promoting activity, *J. Nat. Prod.*, 62, 1668, 1999.

74. Ferrari, J. et al., Isolation and on-line LC/CD analysis of 3,8″-linked biflavonoids from *Gnidia involucrate, Helv. Chim. Acta*, 86, 2768, 2003.

75. Rao, K.V. et al., Two new biflavonoids from *Ochna obtusata, J. Nat. Prod.*, 60, 632, 1997.

76. Jayaprakasam, B. et al., 7-*O*-methyltetrahydroochnaflavone, a new biflavanone from *Ochna beddomei, J. Nat. Prod.*, 63, 507, 2000.

77. Likhitwitayawuid, K. et al., Flavonoids from *Ochna integerrima, Phytochemistry*, 56, 353, 2001.

78. Jayakrishna, G. et al., A new biflavonoid from *Ochna beddomei, J. Asian Nat. Prod. Res.*, 5, 83, 2003.

79. Ariyasena, J. et al., Ether-linked biflavonoids from *Quintinia acutifolia, J. Nat. Prod.*, 67, 693, 2004.

80. Jayaprakasam, B. et al., A biflavanone from *Cycas beddomei, Phytochemistry*, 53, 515, 2000.

81. Xu, Z. et al., New biflavanones and bioactive compounds from *Stellera chamaejasme, Yaoxue Xuebao*, 36, 668, 2001.

82. Baba, K., Taniguchi, M., and Kozawa, M., Three biflavonoids from *Wikstroemia sikokiana, Phytochemistry*, 37, 879, 1994.

83. Pegnyemb, D.E., Ghogomu-Tih, R., and Sondengam, B.L., Minor biflavonoids from *Lophira lanceolata, J. Nat. Prod.*, 57, 1275, 1994.

84. Si, D. et al., Biflavonoids from the aerial parts of *Stephania tetrandra, Phytochemistry*, 58, 563, 2001.

85. Rampendahl, C. et al., The biflavonoids of *Plagiomnium undulatum, Phytochemistry*, 41, 1621, 1996.

86. Geiger, H. and Seeger, T., Triflavones and a biflavone from the moss *Rhizogonium distichum, Z. Naturforsch.*, 55, 870, 2000.

87. Brinkmeier, E., Geiger, H., and Zinsmeister, D., Biflavonoids and 4,2′-epoxy-3-phenylcoumarins from the moss *Mnium hornum, Phytochemistry*, 52, 297, 1999.

88. Li, W., Asada, Y., and Yoshikawa, T., Flavonoid constituents from *Glycyrrhiza glabra* hairy root cultures, *Phytochemistry*, 55, 447, 2000.

89. Muhaisen, H.M. et al., Flavonoid from *Viburnum cotinifolium, J. Chem. Res. (S)*, 480, 2002.

90. Lin, Y-M. et al., *In vitro* anti-HIV activity of biflavonoids isolated from *Rhus succedanea* and *Garcinia multiflora, J. Nat. Prod.*, 60, 884, 1997.

91. Donnelly, D.M.X., Fitzpatrick, B.M., and Finet, J-P., Aryllead triacetates as synthons for the synthesis of biflavonoids. Part 1. Synthesis and reactivity of a flavonyllead triacetate, *J. Chem. Soc., Perkin Trans. 1*, 1791, 1994.

92. Donnelly, D.M.X. et al., Aryllead triacetates as synthons for the synthesis of biflavonoids. Part 2. Synthesis of a garcinia-type biflavonoid, *J. Chem. Soc., Perkin Trans. 1*, 1797, 1994.
93. Shimamura, T. et al., Biogenetic synthesis of biflavonoids, lophirones B and C, from *Lophira lanceolata*, *Heterocycles*, 43, 2223, 1996.
94. Geiger, H. and Markham, K.R., The ^{1}H-NMR spectra of the biflavones isocryptomerin and cryptomerin B. A critical comment on two recent publications on the biflavone patterns of *Selaginella selaginoides* and *S. denticulate*, *Z. Naturforsch.*, 51, 757, 1996.
95. Geiger, H. et al., ^{1}H-NMR assignments in biflavonoid spectra by proton-detected C–H correlation, *Z. Naturforsch.*, 48, 821, 1993.
96. Zhao, L-Y. et al., Total synthesis of several 8,8″-biflavones, *Chin. Chem. Lett.*, 9, 351, 1998.
97. Nunome, S. et al., *In vitro* antimalarial activity of biflavonoids from *Wikstroemia indica*, *Planta Med.*, 70, 76, 2004.
98. Jiang, R-W. et al., Molecular structures and π–π interactions in some flavonoids and biflavonoids, *J. Mol. Struct.*, 642, 77, 2002.
99. Zembower, D.E. and Zhang, H., Total synthesis of robustaflavone, a potential anti-hepatitis B agent, *J. Org. Chem.*, 63, 9300, 1998.
100. Majinda, R.R.T. et al., Phenolic and antibacterial constituents of *Vahlia capensis*, *Planta Med.*, 63, 268, 1997.
101. Dumontet, V. et al., New cytotoxic flavonoids from *Cryptocarya infectoria*, *Tetrahedron*, 57, 6189, 2001.
102. Wall, M.E. et al., Plant antitumor agents. 31. The calycopterones, a new class of biflavonoids with novel cytotoxicity in a diverse panel of human tumor cell lines, *J. Med. Chem.*, 37, 1465, 1994.
103. Mayer, R., Calycopterones and calyflorenones, novel biflavonoids from *Calycopteris floribunda*, *J. Nat. Prod.*, 62, 1274, 1999.
104. Mayer, R., Five biflavonoids from *Calycopteris floribunda* (Combretaceae), *Phytochemistry*, 65, 593, 2004.
105. Abegaz, B.N. et al., Chalcones and other constituents of *Dorstenia prorepens* and *Dorstenia zenkeri*, *Phytochemistry*, 59, 877, 2002.
106. Wang, L-W. et al., New alkaloids and a tetraflavonoid from *Cephalotaxus wilsoniana*, *J. Nat. Prod.*, 67, 1182, 2004.
107. Yadav, S. and Bhadoria, B.K., Novel biflavonoids from the leaves of *Leucaena diversifolia* and *Albizia procera* and their protein binding efficiency, *J. Indian Chem. Soc.*, 81, 392, 2004.
108. Parveem, M. et al., A new biflavonoid from leaves of *Garcinia nervosa*, *Nat. Prod. Res.*, 18, 269, 2004.
109. Nunome, S. et al., *In vitro* antimalarial activity of biflavonoids from *Wikstroemia indica*, *Planta Med.*, 70, 76, 2004.
110. Gaffield, W., Circular dichroism, optically rotatory dispersion and absolute configuration of flavanones, 3-hydroxyflavanones and their glycosides. Determination of agylcone chirality in flavanone glycosides, *Tetrahedron*, 26, 4093, 1970.
111. Duddeck, H., Snatzke, G., and Yemul, S.S., ^{13}C-NMR and CD of some 3,8″-biflavonoids from *Garcinia* species and of related flavanones, *Phytochemistry*, 17, 1369, 1978.
112. Li, X-C. et al., Absolute configuration, conformation, and chiral properties of flavanone-(3 → 8″)-flavone biflavonoids from *Rheedia acuminata*, *Tetrahedron*, 58, 8709, 2002.
113. Slade, D., Ferreira, D., and Marais, J.P.J., Circular dichroism, a powerful tool for the assessment of absolute configuration of flavonoids, *Phytochemistry*, 2005, in press.
114. Bekker, R. et al., Resolution and absolute configuration of naturally occurring auronols, *J. Nat. Prod.*, 64, 345, 2001.

APPENDIX
Checklist for Isoflavonoids Described in the Literature During the Period 1991–2004

	Name	Trivial Name	Mass	Formula	Ref.
	ISOFLAVANES				
1	7,4'-Dihydroxyisoflavan	Equol	242.28	C15H14O3	2719
2	7,2',4'-Trihydroxyisoflavan	Demethylvestitol	258.28	C15H14O4	2720
3	2',4'-Dihydroxy-7-methoxyisoflavan	Neovestitol	272.30	C16H16O4	2723
4	7,2'-Dihydroxy-4'-methoxyisoflavan	Vestitol	272.30	C16H16O4	2721
5	7,4'-Dihydroxy-2'-methoxyisoflavan	Isovestitol	272.30	C16H16O4	2722
6	7,4'-Dihydroxy-3'-methoxyisoflavan		272.30	C16H16O4	F90
7	2',5'-Diketo-7-hydroxy-4'-methoxyisoflavan	Claussequinone	286.29	C16H14O5	2758
8	2'-Hydroxy-7,4'-dimethoxyisoflavan	Isosativan	286.33	C17H18O4	2725
9	4'-Hydroxy-7,2'-dimethoxyisoflavan	Arvensan	286.33	C17H18O4	2726
10	7-Hydroxy-2',4'-dimethoxyisoflavan	Sativan	286.33	C17H18O4	2724
11	3,7,2'-Trihydroxy-4'-methoxyisoflavan		288.30	C16H16O5	F28
12	(3R)-7,2',3-Trihydroxy-4'-methoxyisoflavan	Arizonicanol A	288.30	C16H16O5	F255
13	7,2',4'-Trihydroxy-5'-methoxyisoflavan	Lespedezol G1	288.30	C16H16O5	F171
14	7-Hydroxy-2'-methoxy-4',5'-methylenedioxyisoflavan	Astraciceran	300.31	C17H16O5	2727
15	2',4'-Dihydroxy-5,7-dimethoxyisoflavan	Lotisoflavan	302.33	C17H18O5	2732
16	(3R)-7,2'-Dihydroxy-4',5'-dimethoxyisoflavan	Methoxyvestitol	302.33	C17H18O5	3059
17	(3R)-8,2'-Dihydroxy-7,4'-dimethoxyisoflavan		302.33	C17H18O5	F255
18	(3S)-7,2'-Dihydroxy-3',4'-dimethoxyisoflavan	Isomucronulatol	302.33	C17H18O5	2728
19	(3S)-7,2'-Dihydroxy-8,4'-dimethoxyisoflavan	8-Methoxyvestitol	302.33	C17H18O5	F63
20	7,2'-Dihydroxy-5,4'-dimethoxyisoflavan	5-Methoxyvestitol	302.33	C17H18O5	2731
21	7,3'-Dihydroxy-2',4'-dimethoxyisoflavan	Mucronulatol	302.33	C17H18O5	2729
22	7,4'-Dihydroxy-2',3'-dimethoxyisoflavan	Sphaerosin	302.33	C17H18O5	2730
23	(3R)-7,8,2',3'-Tetrahydroxy-4'-methoxyisoflavan	(3R)-8,3'-Dihydroxyvestitol	304.30	C16H16O6	3062
24	(3R)-2',5'-Diketo-3-hydroxy-7,8-dimethoxyisoflavan	Astragaluquinone	316.31	C17H16O6	F63
25	2',5'-Diketo-7-hydroxy-3',4'-dimethoxyisoflavan	Pendulone	316.31	C17H16O6	2759
26	(3R)-2',5'-Diketo-7-hydroxy-8,4'-dimethoxyisoflavan	Mucroquinone	316.31	C17H16O6	2760a
27	(3S)-2',5'-Diketo-7-hydroxy-8,4'-dimethoxyisoflavan	Mucroquinone	316.31	C17H16O6	2760b
28	(3R)-2'-Hydroxy-7,3',4'-trimethoxyisoflavan	7-O-Methylisomucronulatol	316.35	C18H20O5	F261

continued

APPENDIX
Checklist for Isoflavonoids Described in the Literature During the Period 1991–2004 — *continued*

	Name	Trivial Name	Mass	Formula	Ref.
29	(3R)-7-Hydroxy-2',3',4'-trimethoxyisoflavan		316.35	C18H20O5	F257
30	(3R,4S)-4,2'3',4'-Tetrahydroxy-6,7-methylenedioxyisoflavan		318.28	C16H14O7	F30
31	6,7,3'-Trihydroxy-2',4'-dimethoxyisoflavan		318.33	C17H18O6	2733
32	7,8,3'-Trihydroxy-2',4'-dimethoxyisoflavan	Bolusanthol A	318.33	C17H18O6	2734
33	2,4'-Dihydroxy-6'',6''-dimethylpyrano[2'',3'':7,8]isoflavan	Bryaflavan	324.37	C20H20O4	2740
34	(3R)-2',4'-Dihydroxy-6'',6''-dimethylpyrano[2'',3'':7,6]isoflavan	8-Demetylduartin	324.37	C20H20O4	F282
35	(3R)-7,4'-Dihydroxy-6'',6''-dimethylpyrano[2'',3'':2',3']isoflavan	Glabridin	324.37	C20H20O4	F273
36	7,2'-Dihydroxy-5''-(1-methylethenyl)-4'',5''-dihydrofurano[2'',3'':4',5']isoflavan	Eryvarin C	324.37	C20H20O4	3061
37	7,2'-Dihydroxy-6'',6''-dimethylpyrano[2'',3'':4',3']isoflavan	Erythbidin A	324.37	C20H20O4	2739
38	3'-Hydroxy-2',4'-dimethoxyfurano[2'',3'':8,7]isoflavan	Crotmarine	326.34	C19H18O5	F220
39	(3R)-2',4'-Dihydroxy-5'',4'',4'''-trimethyl-4'',5''-dihydrofurano[2'',3'':7,6]isoflavan	Phaseollinisoflavan	326.39	C20H22O4	3064
40	(3R)-7,2',4'-Trihydroxy-5''-(1,1-dimethyl-2-propenyl)isoflavan	Smiranicin	326.39	C20H22O4	F347
41	(3R)-7,2',4'-Trihydroxy-6-prenylisoflavan	Cyclomillinol	326.39	C20H22O4	F347
42	7,2',4'-Trihydroxy-6-(1,2-dimethyl-2-propenyl)isoflavan	Manuifolin K	326.39	C20H22O4	F204
43	(3R)-2',5'-Diketo-7,3',4'-trimethoxyisoflavan	Manuifolin H	330.34	C18H18O6	F82
44	(3R)-2',5'-Diketo-7,4',6'-trimethoxyisoflavan	Neomillinol	330.34	C18H18O6	F83
45	8-Carboxyaldehyde-7,3'-dihydroxy-2',4'-dimethoxyisoflavan	Colutequinone	330.34	C18H18O6	2729
46	(3R)-2',5'-Dihydroxy-7,3',4'-trimethoxyisoflavan	Colutequinone B	332.36	C18H20O6	F82
47	(3R)-3',5'-Dihydroxy-7,2',4'-trimethoxyisoflavan	Sphaerosin s3	332.36	C18H20O6	F83
48	(3R)-7,2'-Dihydroxy-8,3',4'-trimethoxyisoflavan	Colutehydroquinone	332.36	C18H20O6	3060
49	(3R)-7,5'-Dihydroxy-2',3',4'-trimethoxyisoflavan	Coluteol	332.36	C18H20O6	F257
50	7,3'-Dihydroxy-8,2',4'-trimethoxyisoflavan	Isoduartin	332.36	C18H20O6	2736
51	7,4'-Dihydroxy-2',3',6'-trimethoxyisoflavan	Duartin	332.36	C18H20O6	2735
52	2'-Hydroxy-4'-methoxy-6'',6''-dimethylpyrano[2'',3'':7,6]isoflavan	Lonchocarpan	338.40	C21H22O4	F77
53	2'-Hydroxy-4'-methoxy-6'',6''-dimethylpyrano[2'',3'':7,8]isoflavan	Gancaonin X	338.40	C21H22O4	2742
54	(6aR,11aR)-9-Hydroxy-3-methoxy-10-prenylpterocarpan	4-O-Methylglabridin	338.40	C21H22O4	F168
55	7-Hydroxy-2'-methoxy-6'',6''-dimethylpyrano[2'',3'':4',3']isoflavan	2'-O-Methylphaseollinisoflavan	338.40	C21H22O4	2741
56	7-Hydroxy-4'-methoxy-6'',6''-dimethylpyrano[2'',3'':2',3']isoflavan	Gancaonin Y	338.40	C21H22O4	F77
57	(3R)-7,2'-Dihydroxy-4-methoxy-3'-prenylisoflavan	Gancaonin Z	340.42	C21H24O4	F77
58	(3R)-7,2'-Dihydroxy-4-methoxy-6-(1,1-dimethyl-2-propenyl)isoflavan	(R)-Isomillinol B	340.42	C21H24O4	F212

59	(3R)-7,4'-Dihydroxy-2'-methoxy-6-(1,1-dimethyl-2-propenyl)isoflavan	Millinol B	340.42	C21H24O4	3063
60	(3S)-7,2'-Dihydroxy-4'-methoxy-8-prenylisoflavan	4'-O-Methylpreglabridin	340.42	C21H24O4	3065
61	7,4'-Dihydroxy-2'-methoxy-3'-prenylisoflavan	2'-O-Methylphaseollidinisoflavan	340.42	C21H24O4	2743
62	7,2',4'-Trihydroxy-6-(1-hydroxymethyl-1-methyl-2-propenyl)isoflavan	Millinolol	342.39	C20H22O5	F204
63	2',5'-Diketo-7-hydroxy-8,3',4'-trimethoxyisoflavan	Amorphaquinone	346.34	C18H18O7	2761
64	3',6-Diketo-7-hydroxy-8,2',4'-trimethoxyisoflavan	Laurentiquinone	346.34	C18H18O7	F117
65	6,8,2'-Trihydroxy-7,3',4'-trimethoxyisoflavan	Machaerol C	348.35	C18H20O7	2737
66	2'-Hydroxy-4',5'-methylenedioxy-6'',6''-dimethylpyrano[2'',3'':7,8]isoflavan	Leiocin	352.38	C21H20O5	2744
67	2',3'-Dihydroxy-4'-methoxy-6'',6''-dimethylpyrano[2'',3'':7,6]isoflavan	Arizonicanol B	354.40	C21H22O5	F289
68	2',3'-Dihydroxy-4'-methoxy-6'',6''-dimethylpyrano[2'',3'':7,8]isoflavan	3'-Hydroxy-4'-O-methylglabridin	354.40	C21H22O5	F121
69	2',4'-Dihydroxy-3'-methoxy-6'',6''-dimethylpyrano[2'',3'':7,8]isoflavan	3'-Methoxyglabridin	354.40	C21H22O5	2746
70	2',4'-Dihydroxy-5-methoxy-6'',6''-dimethylpyrano[2'',3'':7,6]isoflavan	Neorauflavane	354.40	C21H22O5	2747
71	7,4'-Dihydroxy-5-methoxy-6'',6''-dimethylpyrano[2'',3'':7,6]isoflavan	Gancaonol C	354.40	C21H22O5	F70
72	(3R)-7,2',4'-Trihydroxy-5-methoxy-6-prenylisoflavan	Glyasperin C	356.41	C21H24O5	F350
73	7,2',3'-Trihydroxy-4'-methoxy-5'-(1,1-dimethyl-2-propenyl)isoflavan	Secundiflorol G	356.41	C21H24O5	F289
74	7,3',4'-Trihydroxy-2'-methoxy-5'-(1,1-dimethyl-2-propenyl)isoflavan	a,a-Dimethylallylcyclolobin	356.41	C21H24O5	2748
75	2',5'-Diketo-6,7,3',4'-tetramethoxyisoflavan	Abruquinone A	360.37	C19H20O7	2762
76	(3S)-7,3'-Dihydroxy-8,2',4',5'-tetramethoxyisoflavan		362.38	C29H22O7	F10
77	6,2'-Dihydroxy-7,8,3',4'-tetramethoxyisoflavan	Machaerol B	362.38	C19H22O7	2738
78	6,2'-Dihydroxy-4',5'-methylenedioxy-6'',6''-dimethylpyrano[2'',3'':7,8]isoflavan	Leiocinol	368.39	C21H20O6	2745
79	(3R)-3'-Hydroxy-2',4'-dimethoxy-6'',6''-dimethylpyrano[2'',3'':7,8]isoflavan	Glyasperin H	368.43	C22H24O5	F151
80	4'-Hydroxy-2',3'-dimethoxy-6'',6''-dimethylpyrano[2'',3'':7,6]isoflavan	Sphaerosinin	368.43	C22H24O5	2749
81	(3R)-2',4'-Dihydroxy-5,7-dimethoxy-6-prenylisoflavan	Glyasperin D	370.45	C22H26O5	F350
82	(3R)-7,2'-Dihydroxy-5,4-dimethoxy-3'-prenylisoflavan	Kanzonol R	370.45	C22H26O5	F76
83	7,3'-Dihydroxy-2',4'-dimethoxy-5'-(1,1-dimethyl-2-propenyl)isoflavan	Unanisoflavan	370.45	C22H26O5	2750
84	7,3'-Dihydroxy-2',4'-dimethoxy-8-prenylisoflavan	Sphaerosin s1	370.45	C22H26O5	F151
85	(3S)-7-Hydroxy-8,2',3',4',5'-pentamethoxyisoflavan		376.40	C20H24O7	F10
86	2',5'-Diketo-6-hydroxy-7,8,3',4'-tetramethoxyisoflavan	Abruquinone C	376.37	C19H20O8	2763
87	(3S)-2',5'-Diketo-7-hydroxy-6,8,3',4'-tetramethoxyisoflavan	Abruquinone D	376.37	C19H20O8	F133
88	(3S)-2',5'-Diketo-8-hydroxy-6,7,3',4'-tetramethoxyisoflavan	Abruquinone F	376.37	C19H20O8	F133
89	(6aR,11aR)-9-Acetoxy-3-methoxy-2-prenylpterocarpan		380.43	C23H24O5	F168
90	(3R)-8-Carboxyaldehyde-7,4'-dihydroxy-5-methoxy-6'',6''-dimethylpyrano[2'',3'':2',3']isoflavan	Kanzonol O	382.41	C22H22O6	F76
91	(3R)-8-Carboxyaldehyde-7,2',4'-trihydroxy-5-methoxy-3'-prenylisoflavan	Kanzonol N	384.42	C22H24O6	F76
92	2',5'-Diketo-6,7,8,3',4'-pentamethoxyisoflavan	Abruquinone B	390.39	C20H22O8	2764

continued

APPENDIX
Checklist for Isoflavonoids Described in the Literature During the Period 1991–2004 — continued

	Name	Trivial Name	Mass	Formula	Ref.
93	2'-Hydroxy-(2'',3'':7,8),(2''',3''':4',3')-bis(6,6-dimethylpyrano)isoflavan	Hispaglabridin B	390.48	C25H26O4	2751
94	4'-Hydroxy-(2'',3'':7,6),(2''',3''':2',3')-bis(6,6-dimethylpyrano)isoflavan	Glyinflanin J	390.48	C25H26O4	F75
95	4'-Hydroxy-(2'',3'':7,8),(2''',3''':2',3')-bis(6,6-dimethylpyrano)isoflavan	Glyinflanin K	390.48	C25H26O4	F75
96	2',4'-Dihydroxy-3'-prenyl-6'',6''-dimethylpyrano[2'',3'':7,8]isoflavan	Hispaglabridin A	392.50	C25H28O4	2752
97	2',4'-Dihydroxy-8-prenyl-6'',6''-dimethylpyrano[2'',3'':7,6]isoflavan	Eryzerin D	392.50	C25H28O4	F277
98	(3R)-7,4'-Dihydroxy-6-prenyl-6'',6''-dimethylpyrano[2'',3'':2',3']isoflavan	Glyinflanin I	392.50	C25H28O4	F75
99	7,2'-Dihydroxy-8-prenyl-6'',6''-dimethylpyrano[2'',3'':4',3']isoflavan		392.50	C25H28O4	F121
100	(3R)-7,2'-Dihydroxy-5'-(1,1-dimethyl-2-propenyl)-4'-prenyloxyisoflavan	Manuifolin D	394.50	C25H30O4	F346
101	(3R)-7,2',4'-Trihydroxy-5'-(1-isopropylethenyl)-8-prenylisoflavan	Manuifolin F	394.50	C25H30O4	F346
102	(3R)-7,2',4'-Trihydroxy-6,5-bis(1,1-dimethyl-2-propenyl)isoflavan	Manuifolin E	394.50	C25H30O4	F346
103	(3R)-7,2',4'-Trihydroxy-8,3-diprenyloxyisoflavan	Kanzonol X	394.50	C25H30O4	F73
104	(3R)-8-Carboxyaldehyde-7,2-dihydroxy-5,4-dimethoxy-3'-prenylisoflavan	Kanzonol M	398.46	C23H26O6	F76
105	2'-Hydroxy-4-methoxy-6''-methyl-6''-(4-methyl-3-pentenyl)pyrano[2'',3'':7,8]isoflavan	Heminitidulan	406.51	C26H30O4	2753
106	(3R)-7,2',4'-Trihydroxy-6,8-diprenylisoflavan	Eryzerin C	406.51	C26H30O4	F277
107	(3R)-7,4'-Dihydroxy-2'-methoxy-5'-(1,1-dimethyl-2-propenyl)-8-prenyloxyisoflavon	Tenuifolin A	408.53	C26H32O4	F348
108	(3R)-7,2',4'-Trihydroxy-5'-(1,1-dimethyl-2-propenyl)-8-(3-hydroxy-3-methylbutyl)isoflavan	Manuifolin G	412.52	C25H32O5	F347
109	(3S)-2',5'-Diketo-6,7,8,3',4',6-hexamethoxyisoflavan	Abruquinone E	420.41	C21H24O9	F133
110	2'-Hydroxy-5',6-methylenedioxy-6''-methyl-6''-(4-methyl-3-pentenyl)pyrano[2'',3'':7,8]isoflavan	Nitidulan	420.50	C26H28O5	2754
111	2',3'-Dihydroxy-4'-methoxy-6''-methyl-6''-(4-methyl-3-pentenyl)pyrano[2'',3'':7,8]isoflavan	Nitidulin	422.51	C26H30O5	2755
112	(3R)-4'-Hydroxy-5-methoxy-6'',6'',6''',6'''-tetramethylpyrano[2'',3'':2',3']-4''',5''-dihydropyrano[2''',3''':7,6]isoflavan	Kanzonol J	422.51	C26H30O5	F74
113	(3R)-2',4'-Dihydroxy-5-methoxy-6'',6''-dimethyl-3'-prenyl-4',5''-dihydropyrano[2'',3'':7,6]isoflavan	Kanzonol H	424.54	C26H32O5	F74
114	5,2',4'-Trihydroxy-7-methoxy-6,3-diprenylisoflavan	Licoricidin	424.54	C26H32O5	2756
115	7,2',4'-Trihydroxy-5-methoxy-6,3'-diprenylisoflavan		424.54	C26H32O5	F168
116	(3R)-4'-Hydroxy-5,7-dimethoxy-6-prenyl-6'',6''-dimethylpyrano[2'',3'':2',3']isoflavan	Kanzonol I	436.54	C27H32O5	F74
117	(+)-2',4'-Dihydroxy-5,7-dimethoxy-6,3-diprenylisoflavan	7-O-Methyllicoricidin	438.57	C27H34O5	3066
118	(3R,4R)-7,2'-Dihydroxy-4'-methoxy-4-(2,4-dihydroxy-5'-(1,1-dimethyl-2-propenyl)phenyl)isoflavan	Manuifolin Q	448.51	C27H28O6	F349
119	2'-Hydroxy-4'-methoxy-4''-phenylpyrano[2'',3'':7,6]-6''-ylidene-5'''-hydroxy-2'''-methoxy-2''',5'''-cyclohexadien-1'''-one-isoflavan	Neocandenaton	522.54	C32H26O7	F21

Isoflavan glycosides

No.	Compound	Common name	MW	Formula	Ref
120	7,2'-Dihydroxy-3',4'-dimethoxyisoflavan 7-O-glucoside	Isomucronulatol 7-O-glucoside	464.47	$C_{23}H_{28}O_{10}$	3236
121	(3R)-7,2'-Dihydroxy-3',4'-dimethoxyisoflavan 7,2'-O-diglucoside	Isomucronulatol 7,2'-di-O-glucoside	626.60	$C_{29}H_{38}O_{15}$	F261
122	(3R)-7,2',5'-Trihydroxy-3',4'-dimethoxyisoflavan 2',5'-O-diglucoside	5'-Hydroxyisomucronulatol 2',5'-di-O-glucoside	642.60	$C_{29}H_{38}O_{16}$	F261

ISOFLAVENES

No.	Compound	Common name	MW	Formula	Ref
123	7,2',4'-Trihydroxyisoflav-2-ene	Lespedezol F1	256.25	$C_{15}H_{12}O_4$	F171
124	7,4'-Dihydroxyisoflav-3-ene	Haginin E	240.26	$C_{15}H_{12}O_3$	F170
125	7,2',4'-Trihydroxyisoflav-3-ene	Hagenin D	256.26	$C_{15}H_{12}O_4$	2765
126	7,2'-Dihydroxy-4'-methoxyisoflav-3-ene	Bolusanthin III	270.29	$C_{16}H_{14}O_4$	F116
127	7,4'-Dihydroxy-2'-methoxyisoflav-3-ene	Hagenin B	270.29	$C_{16}H_{14}O_4$	2766
128	7,2',3'-Trihydroxy-4'-methoxyisoflav-3-ene	Sepiol	286.29	$C_{16}H_{14}O_5$	2767
129	7,2',4'-Trihydroxy-3'-methoxyisoflav-3-ene	Hagenin C	286.29	$C_{16}H_{14}O_5$	2768
130	7-Hydroxy-2'-methoxy-4',5'-methylenedioxyisoflav-3-ene	Judaicin	298.30	$C_{17}H_{14}O_5$	F260
131	7,2'-Dihydroxy-3',4'-dimethoxyisoflav-3-ene	Odoriflavene	300.31	$C_{17}H_{16}O_5$	3067
132	7,2'-Dihydroxy-8,4'-dimethoxyisoflav-3-ene		300.31	$C_{17}H_{16}O_5$	3068
133	7,3'-Dihydroxy-2',4'-dimethoxyisoflav-3-ene	2'-O-Methylsepiol	300.31	$C_{17}H_{16}O_5$	2769
134	7,4'-Dihydroxy-2',3'-dimethoxyisoflav-3-ene	Hagenin A	300.31	$C_{17}H_{16}O_5$	2770
135	7,4'-Dihydroxy-2',5'-dimethoxyisoflav-3-ene	Eryvarin H	300.31	$C_{17}H_{16}O_5$	F276
136	6,7,3'-Trihydroxy-2',4'-dimethoxyisoflav-3-ene		316.31	$C_{17}H_{16}O_6$	2771
137	7,4'-Dihydroxy-6'',6''-dimethylpyrano[2'',3'':2',3']isoflav-3-ene	Glabrene	322.36	$C_{20}H_{18}O_4$	F73
138	7-Hydroxy-2,2'-dimethoxy-4',5'-methylenedioxyisoflav-3-ene	2-Methoxyjudaicin	328.32	$C_{18}H_{16}O_6$	F300
139	3-Hydroxy-2'-methoxy-6'',6''-dimethylpyrano[2'',3'':4',3']isoflav-3-ene	Eryvarin I	336.40	$C_{21}H_{20}O_4$	F276
140	4'-Hydroxy-2'-methoxy-6'',6''-dimethylpyrano[2'',3'':7,8]isoflav-3-ene	Erypoegin B	336.40	$C_{21}H_{20}O_4$	F278
141	7,4'-Dihydroxy-2'-methoxy-3'-prenylisoflav-3-ene	Bidwillol A	338.40	$C_{21}H_{22}O_4$	F104
142	7,4'-Dihydroxy-2'-methoxy-6-(1,1-dimethyl-2-propenyl)isoflav-3-ene	Burttinol A	338.40	$C_{21}H_{22}O_4$	F337
143	7,4'-Dihydroxy-2'-methoxy-8-prenylisoflav-3-ene	Erypoegin A	338.40	$C_{21}H_{22}O_4$	F278
144	2',4'-Dihydroxy-5-methoxy-6'',6''-dimethylpyrano[2'',3'':7,6]isoflavene	Neorauflavene	352.38	$C_{21}H_{20}O_5$	2773
145	7,2',4'-Trihydroxy-5-methoxy-6-prenylisoflav-3-ene	Dehydroglyasperin C	354.40	$C_{21}H_{22}O_5$	F124

Isoflavene glycosides

No.	Compound	Common name	MW	Formula	Ref
146	7-Hydroxy-2'-methoxy-4',5'-methylenedioxyisoflav-3-ene 7-O-glucoside	Judaicin 7-O-glucoside	460.44	$C_{23}H_{24}O_{10}$	F260
147	7-Hydroxy-2'-methoxy-4',5'-methylenedioxyisoflav-3-ene 7-O-(6''-malonylglucoside)	Judaicin 7-O-(6''-malonylglucoside)	546.48	$C_{26}H_{26}O_{13}$	F260

Appendix —
Checklist for Isoflavonoids

Øyvind M. Andersen

The following checklist of isoflavonoids contains compounds reported in the literature as natural products to the end of 2004. Compounds published before 1991 are referenced to numbered entries in Volume 2 of the *Handbook of Natural Flavonoids* (J.B. Harborne and H. Baxter), John Wiley & Sons, Chichester, 1999, using a number consisting of four digits. Compounds published in the period 1991–2004 are referenced with numbers having F as prefix before the number of the publication found in the reference list. The various isoflavonoid classes are shown in Figure A1.

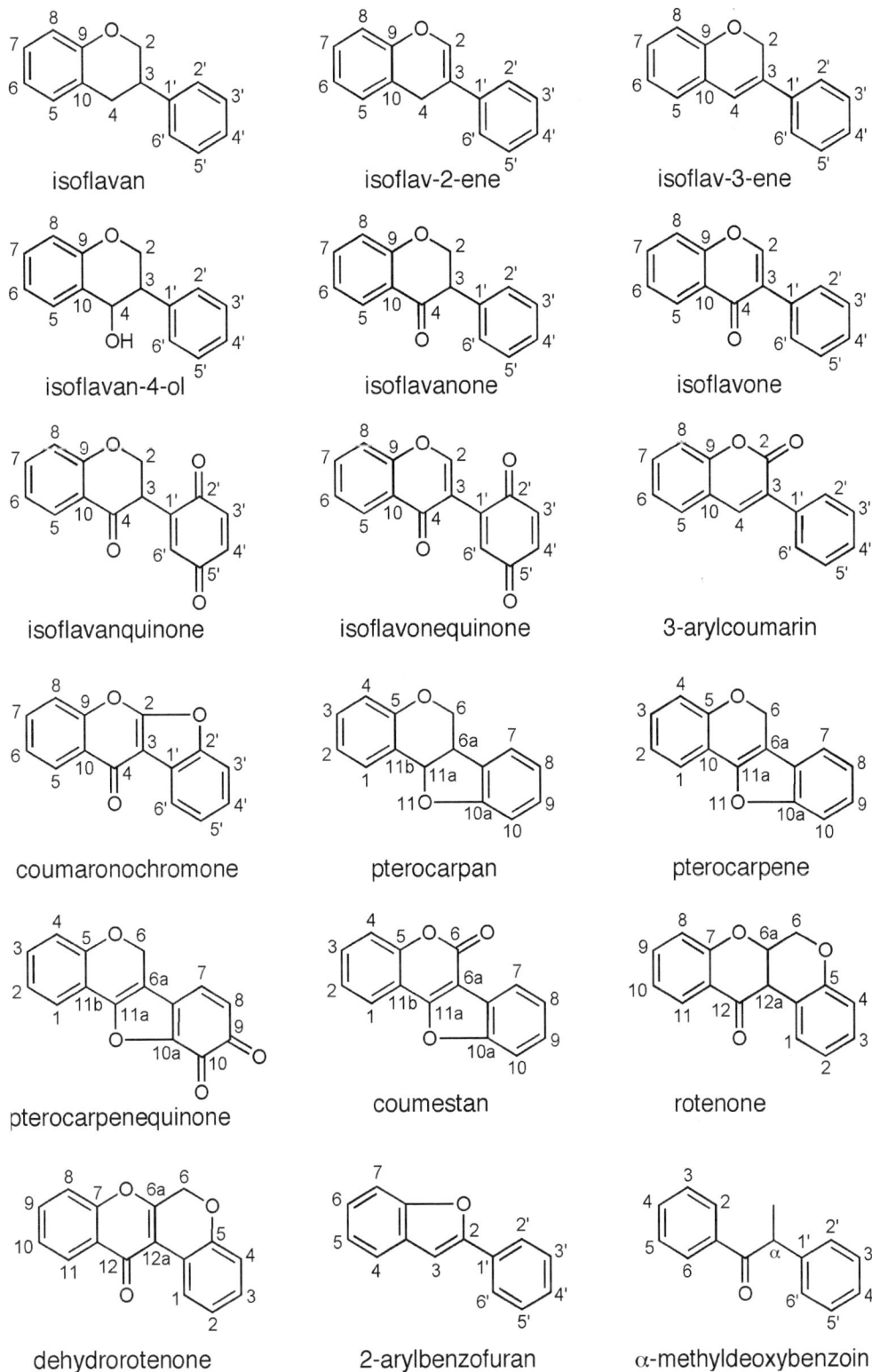

isoflavan

isoflav-2-ene

isoflav-3-ene

isoflavan-4-ol

isoflavanone

isoflavone

isoflavanquinone

isoflavonequinone

3-arylcoumarin

coumaronochromone

pterocarpan

pterocarpene

pterocarpenequinone

coumestan

rotenone

dehydrorotenone

2-arylbenzofuran

α-methyldeoxybenzoin

FIGURE A1 Isoflavonoid classes.

APPENDIX
Checklist for Isoflavonoids Described in the Literature During the Period 1991–2004 — *continued*

	Name	Trivial Name	Mass	Formula	Ref.
	ISOFLAVAN-4-OLS				
148	5-Hydroxy-8-methoxy-6,7-methylenedioxyisoflavan-4-ol	Laphatinol	316.31	C17H16O6	3069
149	2′-Methoxy-4′,5′-methylenedioxyfurano[2″,3″:7,6]isoflavan-4-ol	Ambanol	340.33	C19H16O6	2757
	ISOFLAVANONES				
150	7,4′-Dihydroxyisoflavanone	Dihydrodaidzein	256.26	C15H12O4	2514
151	7-Hydroxy-4′-methoxyisoflavanone	Dihydroformononetin	270.29	C16H14O4	2515
152	7,2′,4′-Trihydroxyisoflavanone		272.26	C15H12O5	2516
153	5,7-Dihydroxy-4′-methoxyisoflavanone	Dihydrobiochanin A	286.29	C16H14O5	2536
154	7,2′-Dihydroxy-4′-methoxyisoflavanone	Vestitone	286.29	C16H14O5	2517
155	7,3′-Dihydroxy-4′-methoxyisoflavanone	3′-Hydroxydihydroformononetin	286.29	C16H14O5	2518
156	5,7,2′,4′-Tetrahydroxy-4′-isoflavanone	Dalbergioidin	288.26	C15H12O6	2537
157	7,2′-Dihydroxy-4′,5′-methylenedioxyisoflavanone	Sophorol	300.27	C16H12O6	2521
158	2′-Hydroxy-7,4′-dimethoxyisoflavanone	Isosativanone	300.31	C17H16O5	2522
159	7-Hydroxy-2′,4′-dimethoxyisoflavanone	Sativanone	300.31	C17H16O5	2519
160	7-Hydroxy-3′,4′-dimethoxyisoflavanone	3′-Methoxydihydroformononetin	300.31	C17H16O5	2520
161	3,7,2′-Trihydroxy-4′-methoxyisoflavanone		302.29	C16H14O6	F132
162	(3R)-4′-Methoxy-3,7,2′-trihydroxyisoflavanone		302.29	C16H14O6	F35
163	(3R)-7,2′,3′-Trihydroxy-4′-methoxyisoflavanone		302.29	C16H14O6	2986
164	5,2′,4′-Trihydroxy-7-methoxyisoflavanone	Dihydrocajanin	302.29	C16H14O6	F192
165	5,7,2′-Trihydroxy-4′-methoxyisoflavanone	Ferreirin	302.29	C16H14O6	2538
166	5,7,3′-Trihydroxy-4′-methoxyisoflavanone	Kenusanone G	302.29	C16H14O6	F100
167	5,7,4′-Trihydroxy-2′-methoxyisoflavanone	Isoferreirin	302.29	C16H14O6	2539
168	7,2′,4′-Trihydroxy-5′-methoxyisoflavanone	Lespedol D	302.29	C16H14O6	F171
169	7-Hydroxy-2′-methoxy-4′,5′-methylenedioxyisoflavanone	Onogenin	314.29	C17H14O6	2523
170	5,2′,4′-Trihydroxy-7-methoxy-6-methylisoflavanone	Ougenin	316.31	C17H16O6	2543
171	5,4′-Dihydroxy-5,2′-dimethoxyisoflavanone		316.31	C17H16O6	2980
172	5,4′-Dihydroxy-7,2′-dimethoxyisoflavanone	Cajanol	316.31	C17H16O6	2541
173	5,7-Dihydroxy-2′,4′-dimethoxyisoflavanone	Homoferreirin	316.31	C17H16O6	2540
174	7,3′-Dihydroxy-2′,4′-dimethoxyisoflavanone	Violanone	316.31	C17H16O6	2524

No.	Systematic name	Common name	Formula	MW	Ref.
175	7,4'-Dihydroxy-2',3'-dimethoxyisoflavanone	Lespedeol C	C17H16O6	316.31	2525
176	7,4'-Dihydroxy-2',5'-dimethoxyisoflavanone	Eryvarin M	C17H16O6	316.31	F280
177	7,4'-Dihydroxy-3',5'-dimethoxyisoflavanone		C17H16O6	316.31	2987
178	3,5,3,2'-Tetrahydroxy-4-methoxyisoflavanone	Ferreirino	C16H14O7	318.28	F60
179	3,5,7,3'-Tetrahydroxy-4-methoxyisoflavanone	Bolusanthin	C16H14O7	318.28	2981
180	7-Hydroxy-6'',6''-dimethylpyrano[2'',3'':4',3']isoflavanone	5-Deoxylicoisoflavanone	C20H18O4	322.36	F168
181	2',3'-Dimethoxyfurano[2'',3'':7,6]isoflavanone		C19H16O5	324.34	2529
182	6,7-Dimethoxy-3',4'-methylenedioxyisoflavanone	Neoraunone	C18H16O6	328.32	2526
183	2'-Hydroxy-7,3',4'-trimethoxyisoflavanone	3'-Methoxyisosativanone	C18H18O6	330.34	2528
184	7-Hydroxy-2',3',4'-trimethoxyisoflavanone	3'-O-Methylviolanone	C18H18O6	330.34	2527
185	5,7,2',4'-Tetrahydroxy-5'-methoxy-6-methylisoflavanone	Lespedol E	C17H16O7	332.30	F171
186	5,7,3'-Trihydroxy-2',4'-dimethoxyisoflavanone	Secundiflorol H	C17H16O7	332.30	F289
187	5,7,4'-Trihydroxy-2',3'-dimethoxyisoflavanone	Parvisflavanone	C17H16O7	332.30	2542
188	2'-Methoxy-4',5'-methylenedioxyfurano[2'',3'':7,6]isoflavanone	Neotenone	C19H14O6	338.32	2530
189	7,4'-Dihydroxy-6'',6''-dimethylpyrano[2'',3'':2',3']isoflavanone	5-Deoxyglyasperin F	C20H18O5	338.36	F168
190	(+)-5,7,4'-Trihydroxy-6-prenylisoflavanone	(+)-Dihydrowighteone	C20H20O5	340.38	2992
191	7,2',4'-Trihydroxy-3'-prenylisoflavanone		C20H20O5	340.38	F168
192	7,2',4'-Trihydroxy-8-prenylisoflavanone	2,3-Dihydro-2'-hydroxyneobavaisoflavanone	C20H20O5	340.38	2531
193	2',4-Dihydroxy-7-methoxy-6-prenylisoflavanone	4-Deoxykievitone	C18H18O7	346.34	F317
194	5,7-Dihydroxy-2',4',5'-trimethoxyisoflavanone	Eryvellutinone	C18H18O7	346.34	F48
195	7-Hydroxy-2'-methoxy-6'',6''-dimethylpyrano[2'',3'':4',5']isoflavanone	Sigmoidin H	C21H20O5	352.38	F182
196	2'-Hydroxy-6'-methoxy-4',5'-methylenedioxyfurano[2'',3'':7,6]isoflavanone	Erosenone	C19H14O7	354.32	2534
197	2',3',4'-Trimethoxyfurano[2'',3'':7,6]isoflavanone	Nepseudin	C20H18O6	354.36	2532
198	2',4',5'-Trimethoxyfurano[2'',3'':7,6]isoflavanone	Ambonone	C20H18O6	354.36	2533
199	5,2',4'-Trihydroxy-5'-(1-methylethenyl)-4',5''-dihydrofurano(2'',3'':7,6)isoflavanone	Uncinanone B	C20H18O6	354.36	F297
200	5,2',4'-Trihydroxy-6'',6''-dimethylpyrano[2'',3'':7,8]isoflavanone	Cyclokievitone 1'',2''-dehydrocyclokievitone	C20H18O6	354.36	2545
201	5,7,2'-Trihydroxy-6'',6''-dimethylpyrano[2'',3'':4',3']isoflavanone	Licoisoflavanone	C20H18O6	354.36	2544
202	5,7,4'-Trihydroxy-6'',6''-dimethylpyrano(2'',3'':2',3')isoflavanone	Glyasperin F	C20H18O6	354.36	F168
203	7,2'-Dihydroxy-4'-methoxy-5'-prenylisoflavanone	Prostratol C	C21H22O5	354.40	F106
204	7,3'-Dihydroxy-4'-methoxy-2'-prenylisoflavanone	Secundiflorol F	C21H22O5	354.40	F242
205	7,4'-Dihydroxy-2'-methoxy-5'-prenylisoflavanone	Sigmoidin I	C21H22O5	354.40	F187
206	5,7,2',4'-Tetrahydroxy-6-prenylisoflavanone	Diphysolone	C20H20O6	356.37	F297
207	5,7,2',4'-Tetrahydroxy-8-prenylisoflavanone	Kievitone	C20H20O6	356.37	2546
208	5,7,3',4'-Tetrahydroxy-5-prenylisoflavanone	Bolusanthol B	C20H20O6	356.37	F28

continued

APPENDIX
Checklist for Isoflavonoids Described in the Literature During the Period 1991–2004 — continued

	Name	Trivial Name	Mass	Formula	Ref.
209	7,2',4'-Trihydroxy-8-(4-hydroxy-3-dimethyl-2-butenyl)isoflavanone	5-Deoxykievitol	356.37	C20H20O6	2991
210	(S)-5,7,2',4'-Tetrahydroxy-3'-prenylisoflavanone	Dihydrolicoisoflavone	356.37	C20H20O6	F59
211	7,2',4'-Trihydroxy-8-(3-hydroxy-3-methylbutyl)isoflavanone	5-Deoxykievitone hydrate	358.39	C20H22O6	2982
212	5,7,3'-Trihydroxy-6,4',5'-trimethoxyisoflavanone	2,3-Dihydroirigenin	362.33	C18H18O8	F112
213	5,4'-Dihydroxy-2'-methoxy-5''-(1-methylethenyl)-4'',5''-dihydrofurano-(2'',3'';7,6)-isoflavanone	Uncinanone C	368.39	C21H20O6	F297
214	5,7-Dihydroxy-2'-methoxy-6'',6''-dimethylpyrano[2'',3'':4',3]isoflavanone	Isosophoronol	368.39	C21H20O6	2547
215	7,2'-Dihydroxy-5-methoxy-6'',6''-dimethylpyrano[2'',3'':4',3]isoflavanone	Glyasperin M	368.39	C21H20O6	F72
216	7,3'-Dihydroxy-5-methoxy-6'',6''-dimethylpyrano[2'',3'':4',5]isoflavanone	Glycyrrhisoflavanone	368.39	C21H20O6	2990
217	(3R)-7,2',3'-Trihydroxy-4'-methoxy-5'-(1,1-dimethyl-2-propenyl)isoflavanone	3'-O-Demethylpervilleanone	370.40	C21H22O6	F80
218	(3R)-7,2',3'-Trihydroxy-4'-methoxy-5'-prenylisoflavanone		370.40	C21H22O6	F303
219	(3S)-5,7,3'-Trihydroxy-4'-methoxy-5'-prenylisoflavanone		370.40	C21H22O6	2989
220	5,2',4'-Trihydroxy-7-methoxy-6-prenylisoflavanone	Glyasperin B	370.40	C21H22O6	F350
221	5,7,2'-Trihydroxy-4'-methoxy-5'-prenylisoflavanone	Vogelin A	370.40	C21H22O6	F14
222	5,7,2'-Trihydroxy-4'-methoxy-6-prenylisoflavanone	Diphysolidone	370.40	C21H22O6	2984
223	5,7,2'-Trihydroxy-4'-methoxy-8-prenylisoflavanone	4-O-Methylkievitone	370.40	C21H22O6	2985
224	5,7,3'-Trihydroxy-4'-methoxy-2'-prenylisoflavanone	Arizonicanol D	370.40	C21H22O6	F289
225	5,7,4'-Trihydroxy-2'-methoxy-3'-prenylisoflavanone	Sophoraisoflavanone A	370.40	C21H22O6	2548
226	5,7,4'-Trihydroxy-2'-methoxy-5'-(1,1-dimethyl-2-propenyl)isoflavanone	Fraserinone A	370.40	C21H22O6	F102
227	5,7,4'-Trihydroxy-2'-methoxy-5'-prenylisoflavanone	Erypoegin C	370.40	C21H22O6	F278
228	5,7,2',4'-Tetrahydroxy-8-(4-hydroxy-3-dimethyl-2-butenyl)isoflavanone	Kievitol	372.37	C20H20O7	2993
229	5,7,3',4'-Tetrahydroxy-5'-(2-epoxy-3-methylbutyl)isoflavanone		372.37	C20H20O7	F28
230	5-Hydroxy-7,2'-dimethoxy-6'',6''-dimethylpyrano[2'',3'':4',5]isoflavanone	Erypoegin G	382.41	C22H22O6	F272
231	3,5,7-Trihydroxy-2'-methoxy-6'',6''-dimethylpyrano[2'',3'':4',3]isoflavanone	Sophoronol	384.39	C21H20O7	2999
232	(3R)-7,2'-Dihydroxy-3',4'-dimethoxy-5'-(1,1-dimethyl-2-propenyl)isoflavanone	Pervilleanone	384.42	C22H24O6	F80
233	5,4'-Dihydroxy-7,2'-dimethoxy-5'-(1,1-dimethyl-2-propenyl)isoflavanone	Echinoisosophoranone	384.42	C22H24O6	2988
234	5,4'-Dihydroxy-7,2'-dimethoxy-5'-prenylisoflavanone	Erypoegin D	384.42	C22H24O6	F278
235	7,4'-Dihydroxy-2',5'-dimethoxy-6-prenylisoflavanone	Sigmoidin J	384.42	C22H24O6	F181
236	7,4'-Dihydroxy-2',5'-dimethoxy-8-prenylisoflavanone	Eryvarin N	384.42	C22H24O6	F280
237	(R)-5,2',4'-Trihydroxy-7-methoxy-6-methyl-8-prenylisoflavanone	Desmodianone B	384.42	C22H24O6	F51
238	(S)-5,2'-Dihydroxy-7,4'-dimethoxy-6-prenylisoflavanone	Glyasperin K	384.42	C22H24O6	F72
239	3,5,7,4'-Tetrahydroxy-2'-methoxy-3'-prenylisoflavanone	Kenusanone F	386.40	C21H22O7	F100

No.	Systematic name	Common name	Formula	MW	Ref.
240	3,7,2',3'-Tetrahydroxy-4'-methoxy-5'-(1,1-dimethyl-2-propenyl)isoflavanone	Secundifloran	C21H22O7	386.40	2535
241	(3R)-5,7,2',3'-Tetrahydroxy-4'-methoxy-5'-prenylisoflavanone		C21H22O7	386.40	F303
242	(3S)-3,7,2',3'-Tetrahydroxy-4'-methoxy-5'-prenylisoflavanone		C21H22O7	386.40	F303
243	7,4'-Dihydroxy-3'-((2E)-3,7-dimethyl-2,6-octadienyl)isoflavanone	Tetrapterol E	C25H28O4	392.50	F101
244	7,4'-Dihydroxy-4'-methoxy-6,3'-diprenylisoflavanone	Tetrapterol I	C25H28O4	392.50	F241
245	(+)-3,5,7-Trihydroxy-2',4'-dimethoxy-3'-prenylisoflavanone	(+)-Echinoisoflavanone	C22H24O7	400.43	2998
246	5,7,3'-Trihydroxy-2',4'-dimethoxy-5'-(1,1-dimethyl-2-propenyl)isoflavanone	Secundiflorol E	C22H24O7	400.43	F242
247	5,7,3'-Trihydroxy-2',4'-dimethoxy-6-prenylisoflavanone	Arizonicanol C	C22H24O7	400.43	F289
248	3,5,7,2',3'-Tetrahydroxy-4'-methoxy-3'-(1,1-dimethyl-2-propenyl)isoflavanone	Secundiflorol A	C21H22O8	402.39	F103
249	2',4'-Dihydroxy-6-prenyl-6'',6''-dimethylpyrano[2'',3'':7,8]isoflavanone	Orientanol F	C25H26O5	406.47	F281
250	2',4'-Dihydroxy-6''-(4-methyl-3-pentenyl)-6''-methylpyrano[2'',3'':7,6]isoflavanone		C25H26O5	406.47	F167
251	2',4'-Dihydroxy-8-prenyl-6'',6''-dimethylpyrano[2'',3'':7,6]isoflavanone	Bidwillon B	C25H26O5	406.47	F97
252	5,7,4'-Trihydroxy-3'-((2E)-3,7-dimethyl-2,6-octadienyl)isoflavanone	Tetrapterol D	C25H28O5	408.50	F101
253	5,7,4'-Trihydroxy-6,3'-diprenylisoflavanone	Bolusanthol C	C25H28O5	408.50	F28
254	7,2',4'-Trihydroxy-5'-((2E)-3,7-dimethyl-2,6-octadienyl)isoflavanone	Prostratol A	C25H28O5	408.50	F106
255	7,2',4'-Trihydroxy-6,3'-diprenylisoflavanone		C25H28O5	408.50	F168
256	7,2',4'-Trihydroxy-6,8-diprenylisoflavanone	Eriotrichin B	C25H28O5	408.50	F186
257	7,2',4'-Trihydroxy-6,8-diprenylisoflavanone	Bidwillon A	C25H28O5	408.50	F97
258	(±)-7,2',4'-Trihydroxy-8,3'-diprenylisoflavanone	Eryzerin A	C25H28O5	408.50	F277
259	(3R)-5-Hydroxy-2',4',5'-trimethoxy-6'',6''-dimethylpyrano[2'',3'':4',5'][1'''-methylphenyl][3''',4''',4'',5'']isoflavanone	Saclenone	C23H24O7	412.44	F338
260	5,7,2'-Trihydroxy-6''-methyl-6'-(4-methyl-3-pentenyl)pyrano[2'',3'':7,6]isoflavanone	Tetrapterol A	C25H26O6	418.45	F101
261	5,2',4'-Trihydroxy-3'-prenyl-6''-methyl-6''-(4-methyl-3-pentenyl)pyrano[2'',3'':7,6]isoflavanone	Lespedeol A	C25H26O6	422.48	2550
262	5,2',4'-Trihydroxy-3'-prenyl-6'',6''-dimethylpyrano[2'',3'':7,6]isoflavanone	Cajanone	C25H26O6	422.48	2549
263	5,7,4'-Trihydroxy-8-prenyl-6'',6''-dimethylpyrano[2'',3'':7,6]isoflavanone	2,3-Dihydroauriculatin	C25H26O6	422.48	2997
264	(3R)-5,7-Dihydroxy-4'-methoxy-6,3'-diprenylisoflavanone	Vogelin D	C26H30O5	422.51	F209
265	(3R)-7,4'-Dihydroxy-2'-methoxy-6,8-diprenylisoflavanone	Eryzerin B	C26H30O5	422.51	F277
266	7,4'-Dihydroxy-2'-methoxy-6-((2E)-3,7-dimethyl-2,6-octadienyl)isoflavanone		C26H30O5	422.51	F167
267	3,7,2',4'-Tetrahydroxy-6,8-diprenylisoflavanone	Orientanol D	C25H28O6	424.50	F281
268	5,7,2',4'-Tetrahydroxy-5'-[(2E)-3,7-dimethyl-2,6-octadienyl]isoflavanone	Kenusanone A	C25H28O6	424.50	F98
269	5,7,2',4'-Tetrahydroxy-6-geranylisoflavanone	Lespedeol B	C25H28O6	424.50	2551
270	5,7,2',4'-Tetrahydroxy-6,3'-diprenylisoflavanone	Glisoflavanone	C25H28O6	424.50	F88
271	5,7,2',4'-Tetrahydroxy-6,5'-diprenylisoflavanone	Tetrapterol G	C25H28O6	424.50	F241
272	5,7,2',4'-Tetrahydroxy-6,8-diprenylisoflavanone	Orientanol E	C25H28O6	424.50	F281
273	5,7,2',4'-Tetrahydroxy-8-[(2E)-3,7-dimethyl-2,6-octadienyl]isoflavanone	Kenusanone H	C25H28O6	424.50	F100

continued

APPENDIX
Checklist for Isoflavonoids Described in the Literature During the Period 1991–2004 — continued

	Name	Trivial Name	Mass	Formula	Ref.
274	5,7,2',4'-Tetrahydroxy-8-prenyl-3'-(1,1-dimethyl-2-propenyl)isoflavanone	Dalversinol A	424.50	C25H28O6	F26
275	5,7,2',4'-Tetrahydroxy-8,3'-diprenylisoflavanone	3'-Dimethylallylkievitone	424.50	C25H28O6	2996
276	5,7,2'-Trihydroxy-6-methyl-6'',6''-dimethyl-4'-5'-dihydropyrano[2'',3'':4',5']-3'''-methylbenzo[1''',6''':5'',4'']isoflavanone	Methyltetrapterol A	432.47	C26H24O6	F31
277	5,4'-Dihydroxy-2'-methoxy-3'-prenyl-6'',6''-dimethylpyrano[2'',3'':7,6]isoflavanone	2'-O-Methylcajanone	436.50	C26H28O6	2552
278	5,7-Dihydroxy-6-methyl-3-(1a,2,3,3a,8b,8c-hexahydro-6-hydroxy-1,1,3a-trimethyl-1H-4-oxabenzo[f]cyclobut[cd]inden-7-yl)-2,3-dihydro 4H-1-benzopyran-4-one	Desmodianone D	436.50	C26H28O6	F31
279	5,7,2'-Trihydroxy-6-methyl-6'',6''-dimethyl-4',5'-dihydropyrano[2',3':4',5']-2'''-methylcyclohexene[5''',6''':5'',4'']isoflavanone	Desmodianone E	436.50	C26H28O6	F31
280	(R)-5,7,2'-Trihydroxy-6,6''-dimethyl-6''-(4-methylpent-3-enyl)pyrano(2'',3'':4',5')isoflavanone	Desmodianone A	436.50	C26H28O6	F51
281	(3S)-5,2',4'-Trihydroxy-7-methoxy-6,3'-diprenylisoflavanone	Kanzonol G	438.51	C26H30O6	F74
282	5,7,2'-Trihydroxy-4'-methoxy-6,3'-diprenylisoflavanone	Sophoraisoflavanone B	438.51	C26H30O6	2554
283	5,7,2'-Trihydroxy-4'-methoxy-8,5'-diprenylisoflavanone	Tetrapterol H	438.51	C26H30O6	F241
284	5,7,4'-Trihydroxy-2'-methoxy-6,3'-diprenyisoflavanone	Isosophoranone	438.51	C26H30O6	2553
285	5,7,4'-Trihydroxy-2'-methoxy-8-((2E,6E)-3,7-dimethyl-2,6-octadienyl)isoflavanone	Tetrapterol C	438.51	C26H30O6	F101
286	5,7,4'-Trihydroxy-2'-methoxy-8,3'-diprenylisoflavanone	Phyllanone B	438.51	C26H30O6	F219
287	(R)-5,7,2',4'-Tetrahydroxy-6-methyl-5'-(3,7-dimethylocta-2(E),6-dienyl)isoflavanone	Desmodianone C	438.51	C26H30O6	F51
288	3,5,7,2',4'-Pentahydroxy-8,3'-diprenylisoflavanone	Bolusanthin II	440.49	C25H28O7	F29
289	5,7,2',4',5'-Pentahydroxy-6,8-diprenylisoflavanone	Erysengalensein B	440.49	C25H28O7	F306
290	3,5,7-Trihydroxy-2'-methoxy-8-prenyl-6'',6''-dimethylpyrano[2'',3'':4',3']isoflavanone	Phyllanone A	452.50	C26H28O7	F219
291	5,2',4'-Trihydroxy-5'-methoxy-8-prenyl-6'',6''-dimethylpyrano[2'',3'':7,6]isoflavanone	Erysengalensein B	452.50	C26H28O7	F306
292	5,7,4'-Trihydroxy-3'-geranyl-6-prenylisoflavanone	Sophoraisoflavanone C	476.62	C30H36O5	2994
293	5,7,2',4'-Tetrahydroxy-5'-geranyl-6-prenylisoflavanone	Sophoraisoflavanone D	492.62	C30H36O5	2995

Isoflavanon glycosides

	Name	Trivial Name	Mass	Formula	Ref.
294	7,4'-Dihydroxyisoflavanone 7-O-glucoside	Dihydrodaidzin	418.39	C21H22O9	F92
295	7-Hydroxy-4'-methoxyisoflavanone 7-O-glucoside	Dihydroformononetin 7-O-glucoside	432.43	C22H24O9	3219
296	5,7,4'-Trihydroxyisoflavanone 7-O-glucoside	Dihydrogenistin	434.39	C21H22O10	F92
297	5,2',4'-Tetrahydroxyisoflavanone 4'-O-glucoside	Dalbergioidin 4'-O-glucoside	450.39	C21H22O11	F99
298	7-Hydroxy-2'-methoxy-4',5'-methylenedioxyisoflavanone 7-O-glucoside	Onogenin 7-O-glucoside	476.44	C23H24O11	3220

ISOFLAVONES

No.	Name	Trivial name	MW	Formula	Code
299	5,8-Dihydroxy-7-methoxyisoflavone		238.24	C15H10O3	F243
300	7-Hydroxy-2-methylisoflavone		252.27	C16H12O3	2339
301	5,7-Dihydroxyisoflavone		254.24	C15H10O4	2839
302	7,4'-Dihydroxyisoflavone	Daidzein	254.24	C15H10O4	2340
303	8,4'-Dihydroxyisoflavone		254.24	C15H10O4	2870
304	7-Methoxy-2-methylisoflavone		266.30	C17H14O3	2341
305	4'-Hydroxy-7-methoxyisoflavone	Isoformononetin	268.27	C16H12O4	2343
306	5-Hydroxy-4'-methoxyisoflavone	Pallidiflorin	268.27	C16H12O4	2867
307	5-Hydroxy-7-methoxyisoflavone		268.27	C16H12O4	2416
308	7-Hydroxy-4'-methoxyisoflavone	Formononetin	268.27	C16H12O4	2342
309	5,7,3'-Trihydroxyisoflavone		270.24	C15H10O5	2840
310	5,7,4'-Trihydroxyisoflavone	Genistein	270.24	C15H10O5	2417
311	6,7,4'-Trihydroxyisoflavone	Demethyltexasin	270.24	C15H10O5	2346
312	7,2',4'-Trihydroxyisoflavone	2'-Hydroxydaidzein	270.24	C15H10O5	2344
313	7,3',4'-Trihydroxyisoflavone	3'-Hydroxydaidzein	270.24	C15H10O5	2345
314	7,8,4'-Trihydroxyisoflavone	8-Hydroxydaidzein	270.24	C15H10O5	2877
315	3'-Carboxyaldehyde-7,4'-dihydroxyisoflavone	Corylinal	282.25	C16H10O5	2347
316	7-Hydroxy-3',4'-methylenedioxyisoflavone	Pseudobaptigenin	282.25	C16H10O5	2348
317	5,7-Dimethoxyisoflavone		282.30	C17H14O4	2418
318	7,4'-Dimethoxyisoflavone	7,4'-Di-O-methyldaidzein	282.30	C17H14O4	2349
319	2',4'-Dihydroxy-7-methoxyisoflavone	2'-Methoxyformononetin	284.27	C16H12O5	F113
320	5,4'-Dihydroxy-7-methoxyisoflavone	Prunetin	284.27	C16H12O5	2420
321	5,7-Dihydroxy-3'-methoxyisoflavone	Mutabilein	284.27	C16H12O5	F55
322	5,7-Dihydroxy-4'-methoxyisoflavone	Biochanin A	284.27	C16H12O5	2419
323	5,7,2'-Trihydroxy-6-methylisoflavone	Abronisoflavone	284.27	C16H12O5	F321
324	6,4'-Dihydroxy-7-methoxyisoflavone	Kakkatin	284.27	C16H12O5	2355
325	6,7-Dihydroxy-4'-methoxyisoflavone	Texasin	284.27	C16H12O5	2354
326	6,7-Dihydroxy 3'-methoxyisoflavone		284.27	C16H12O5	F344
327	7,2'-Dihydroxy-4'-methoxyisoflavone	2'-Hydroxyformononetin	284.27	C16H12O5	2350
328	7,2'-Dihydroxy-5'-methoxyisoflavone		284.27	C16H12O5	F343
329	7,2'-Dihydroxy-6-methoxyisoflavone		284.27	C16H12O5	2843
330	7,3'-Dihydroxy-4'-methoxyisoflavone	3'-Hydroxyformononetin	284.27	C16H12O5	2352
331	7,3',4',5'-Tetrahydroxyisoflavone	Baptigenin	284.27	C15H10O6	2358

continued

APPENDIX
Checklist for Isoflavonoids Described in the Literature During the Period 1991–2004 — *continued*

Name	Trivial Name	Mass	Formula	Ref.
332 7,4'-Dihydroxy-2'-methoxyisoflavone	Theralin	284.27	C16H12O5	2351
333 7,4'-Dihydroxy-3'-methoxyisoflavone	3-Methoxydaidzein	284.27	C16H12O5	2353
334 7,4'-Dihydroxy-5-methoxyisoflavone	5-O-Methylgenistein	284.27	C16H12O5	2421
335 7,4'-Dihydroxy-6-methoxyisoflavone	Glycitein	284.27	C16H12O5	2356
336 7,8-Dihydroxy-4'-methoxyisoflavone	Retusin	284.27	C16H12O5	2357
337 8,4'-Dihydroxy-7-methoxyisoflavone		284.27	C16H12O5	F95
338 5,6,7,4'-Tetrahydroxyisoflavone	6-Hydroxygenistein	286.24	C15H10O6	2424
339 5,7,2',4'-Tetrahydroxyisoflavone	2'-Hydroxygenistein	286.24	C15H10O6	2422
340 5,7,3',4'-Tetrahydroxyisoflavone	Orobol	286.24	C15H10O6	2423
341 7,8,2',4'-Tetrahydroxyisoflavone		286.24	C15H10O6	2878
342 7-Acetoxy-2-methylisoflavone		294.31	C18H14O4	2359
343 8-Acetyl-7-hydroxy-2-methylisoflavone	Glyzarin	294.31	C18H14O4	2360
344 4'-Methoxy-7,8-methylenedioxyisoflavone	Maximaisoflavone H	296.28	C17H12O5	2845
345 7-Methoxy-3',4'-methylenedioxyisoflavone	Pseudobaptigenin methyl ether	296.28	C17H12O5	2361
346 (4''S,5''R)-5,7,4',5''-Tetrahydroxy-6'',6''-dimethyl-4'',5''-dihydropyrano[2'',3'':4',3']isoflavone	Dihydroisoderrondiol	296.32	C20H18O7	F201
347 7,2'-Dimethoxy-8-methylisoflavone		296.32	C18H16O4	F27
348 5,2'-Dihydroxy-6,7-methylenedioxyisoflavone	Irisone B	298.25	C16H10O6	2881
349 5,4'-Dihydroxy-6,7-methylenedioxyisoflavone	Irilone	298.25	C16H10O6	2427
350 5,7-Dihydroxy-3',4'-methylenedioxyisoflavone	5-Hydroxypseudobaptigenin	298.25	C16H10O6	2425
351 7,2'-Dihydroxy-3',4'-methylenedioxyisoflavone	Glyzaglabrin	298.25	C16H10O6	2363
352 2',5'-Diketo-7-hydroxy-4'-methoxyisoflavone	Bowdichione	298.25	C16H10O6	2511
353 2'-Hydroxy-7,4'-dimethoxyisoflavone		298.30	C17H14O5	F191
354 4'-Hydroxy-6,7-methoxyisoflavone		298.30	C17H14O5	F159
355 4'-Hydroxy-7,3'-dimethoxyisoflavone	Sayanedine	298.30	C17H14O5	2364
356 5-Hydroxy-7,4'-dimethoxyisoflavone		298.30	C17H14O5	F352
357 5,7,4'-Trihydroxy-6,8-dimethylisoflavone		298.30	C17H14O5	F33
358 6-Hydroxy-7,4'-dimethoxyisoflavone	Alfalone	298.30	C17H14O5	2875
359 7-Hydroxy-2',4'-dimethoxyisoflavone	2'-Methoxyformononetin	298.30	C17H14O5	F113
360 7-Hydroxy-3',4'-dimethoxyisoflavone	Cladrin	298.30	C17H14O5	2362
361 7-Hydroxy-3',5'-dimethoxyisoflavone		298.30	C17H14O5	F1
362 7-Hydroxy-5,4'-dimethoxyisoflavone	5-O-Methylbiochanin A	298.30	C17H14O5	2426

No.	Name	Common name	Formula	MW	Ref.
363	7-Hydroxy-6,4'-dimethoxyisoflavone	Afrormosin	C17H14O5	298.30	2365
364	7-Hydroxy-8,4'-dimethoxyisoflavone	8-O-Methylretusin	C17H14O5	298.30	2366
365	5,2',4'-Trihydroxy-7-methoxyisoflavone	Cajanin	C16H12O6	300.27	2431
366	5,3',4'-Trihydroxy-7-methoxyisoflavone	Santal	C16H12O6	300.27	2432
367	5,6,4'-Trihydroxy-7-methoxyisoflavone		C16H12O6	300.27	F91
368	5,7,2'-Trihydroxy-4'-methoxyisoflavone	2'-Hydroxybiochanin A	C16H12O6	300.27	2428
369	5,7,3'-Trihydroxy-4'-methoxyisoflavone	Pratensein	C16H12O6	300.27	2429
370	5,7,4'-Trihydroxy-2'-methoxyisoflavone		C16H12O6	300.27	F125
371	5,7,4'-Trihydroxy-3'-methoxyisoflavone	3'-O-Methylorobol	C16H12O6	300.27	2430
372	5,7,4'-Trihydroxy-6-methoxyisoflavone	Tectorigenin	C16H12O6	300.27	2433
373	5,7,4'-Trihydroxy-8-methoxyisoflavone	Isotectorigenin	C16H12O6	300.27	2434
374	7,2',3'-Trihydroxy-4'-methoxyisoflavone	Koparin	C16H12O6	300.27	2367
375	7,2',4'-Trihydroxy-3'-methoxyisoflavone		C16H12O6	300.27	2838
376	7,3',5'-Trihydroxy-4'-methoxyisoflavone	Gliricidin	C16H12O6	300.27	2368
377	7,4',5'-Trihydroxy-2'-methoxyisoflavone		C16H12O6	300.27	F25
378	7,8,3'-Trihydroxy-4'-methoxyisoflavone		C16H12O6	300.27	2879
379	7,8,4'-Trihydroxy-6-methoxyisoflavone		C16H12O6	300.27	2893
380	5,7,8,2',4'-Tetrahydroxyisoflavone		C15H10O5	302.24	2892
381	(7,8:3',4')-Bis(methylenedioxy)isoflavone	Maximaisoflavone A	C17H10O6	310.26	2369
382	2'-Hydroxy-5-methoxy-6,7-methylenedioxyisoflavone	Betavulgarin	C17H12O6	312.28	2435
383	4'-Hydroxy-5-methoxy-6,7-methylenedioxyisoflavone	Irisolone	C17H12O6	312.28	2436
384	5-Hydroxy-2'-methoxy-6,7-methylenedioxyisoflavone	Irisone A	C17H12O6	312.28	2882
385	5-Hydroxy-7-methoxy-3',4'-methylenedioxyisoflavone		C17H12O6	312.28	F352
386	6-Hydroxy-7-methoxy-3',4'-methylenedioxyisoflavone	Acicerone	C17H12O6	312.28	F139
387	7-Hydroxy-2'-methoxy-4',5'-methylenedioxyisoflavone	Cuneatin	C17H12O6	312.28	2370
388	7-Hydroxy-6-methoxy-3',4'-methylenedioxyisoflavone	Fujikinetin	C17H12O6	312.28	2372
389	7-Hydroxy-8-methoxy-3',4'-methylenedioxyisoflavone	Maximaisoflavone E	C17H12O6	312.28	2846
390	7,3',4'-Trimethoxyisoflavone	Cabreuvin	C18H16O5	312.32	2371
391	2',5'-Diketo-5,7-dihydroxy-4'-methoxyisoflavone	5-Hydroxybowdichione	C16H10O7	314.25	2979
392	2,4'-Dihydroxy-5,7-dimethoxyisoflavone		C17H14O6	314.29	2871
393	5,2'-Dihydroxy-6,7-dimethoxyisoflavone	Irilin A	C17H14O6	314.29	F18
394	5,2'-Dihydroxy-7,8-dimethoxyisoflavone		C17H14O6	314.29	2890
395	5,3'-Dihydroxy-7,4'-dimethoxyisoflavone		C17H14O6	314.29	2842
396	5,4'-Dihydroxy-6,7-dimethoxyisoflavone	7-O-Methyltectorigenin	C17H14O6	314.29	2439

continued

APPENDIX
Checklist for Isoflavonoids Described in the Literature During the Period 1991–2004 — *continued*

	Name	Trivial Name	Mass	Formula	Ref.
397	5,4'-Dihydroxy-7,3'-dimethoxyisoflavone	7,3'-Dimethylorobol	314.29	C17H14O6	2872
398	5,7-Dihydroxy-2',4'-dimethoxyisoflavone	2'-Methoxybiochanin A	314.29	C17H14O6	2437
399	5,7-Dihydroxy-3',4'-dimethoxyisoflavone	Pratensein 3'-O-methyl ether	314.29	C17H14O6	2841
400	5,7-Dihydroxy-6,2'-dimethoxyisoflavone		314.29	C17H14O6	2850
401	5,7-Dihydroxy-6,4'-dimethoxyisoflavone	Irisolidone	314.29	C17H14O6	2438
402	5,7-Dihydroxy-8,4'-dimethoxyisoflavone		314.29	C17H14O6	F23
403	6,4'-Dihydroxy-5,7-dimethoxyisoflavone	Muningin	314.29	C17H14O6	2440
404	7,2'-Dihydroxy-4',5'-dimethoxyisoflavone		314.29	C17H14O6	F35
405	7,2'-Dihydroxy-6,4-dimethoxyisoflavone		314.29	C17H14O6	2876
406	7,3'-Dihydroxy-6,4-dimethoxyisoflavone	Odoratin	314.29	C17H14O6	2373
407	7,3'-Dihydroxy-8,4-dimethoxyisoflavone	3'-Hydroxy-8-O-methylretusin	314.29	C17H14O6	2374
408	7,4'-Dihydroxy-3',5'-dimethoxyisoflavone		314.29	C17H14O6	F37
409	7,4'-Dihydroxy-5,3'-dimethoxyisoflavone	Gerontoisoflavone A	314.29	C17H14O6	F36
410	7,8-Dihydroxy-6,4'-dimethoxyisoflavone	Dipteryxin	314.29	C17H14O6	2375
411	5,6,7,2'-Tetrahydroxy-3'-methoxyisoflavone		316.27	C16H12O7	2852
412	5,6,7,3'-Tetrahydroxy-4'-methoxyisoflavone		316.27	C16H12O7	F112
413	5,6,7,4'-Tetrahydroxy-8-methoxyisoflavone		316.27	C16H12O7	2855
414	5,7,2',4'-Tetrahydroxy-5'-methoxyisoflavone	Piscerygenin	316.27	C16H12O7	F176
415	5,7,3',4'-Tetrahydroxy-6-methoxyisoflavone	Irilin D	316.27	C16H12O7	F42
416	5,7,3',5'-Tetrahydroxy-4'-methoxyisoflavone	Junipegenin A	316.27	C16H12O7	2441
417	4'-Hydroxy-6'',6''-dimethylpyrano[2'',3'':7,6]isoflavone		320.34	C20H16O4	2397
418	4'-Hydroxy-6'',6''-dimethylpyrano[2'',3'':7,8]isoflavone	Erythrinin A	320.34	C20H16O4	F104
419	6-Hydroxy-6'',6''-dimethylpyrano[2'',3'':4',3']isoflavone	Bidwillon C	320.34	C20H16O4	F4
420	7-Hydroxy-6'',6''-dimethylpyrano[2'',3'':4',3']isoflavone	Corylin	320.34	C20H16O4	2396
421	7-Hydroxy-4'-prenyloxyisoflavone	Nordurlettone	322.36	C20H18O4	2868
422	7,4'-Dihydroxy-3'-prenylisoflavone	Neobavaisoflavone	322.36	C20H18O4	2398
423	7,4'-Dihydroxy-8-prenylisoflavone	8-Prenyldaidzein	322.36	C20H18O4	F85
424	7-Acetoxy-2'-methoxy-8-methylisoflavone		324.34	C19H16O5	F27
425	3',4'-Dimethoxy-6,7-methylenedioxyisoflavone		326.31	C18H14O6	2378
426	3',4'-Dimethoxy-7,8-methylenedioxyisoflavone	Maximaisoflavone D	326.31	C18H14O6	2848
427	5,2'-Dimethoxy-6,7-methylenedioxyisoflavone	Tlatlancuayin	326.31	C18H14O6	2443

Appendix

No.	Isoflavone name	Trivial name	Formula	MW	Ref.
428	5,4'-Dimethoxy-6,7-methylenedioxyisoflavone	Irisolone methyl ether	C18H14O6	326.31	2444
429	5,7-Dimethoxy-3',4'-methylenedioxyisoflavone	Derrustone	C18H14O6	326.31	2442
430	6,7-Dimethoxy-3',4'-methylenedioxyisoflavone	Fujikinetin methyl ether	C18H14O6	326.31	2377
431	7,2'-Dimethoxy-4',5'-methylenedioxyisoflavone	Cuneatin methyl ether	C18H14O6	326.31	2376
432	7,3'-Dimethoxy-4',5'-methylenedioxyisoflavone		C18H14O6	326.31	2880
433	7,8-Dimethoxy-3',4'-methylenedioxyisoflavone		C18H14O6	326.31	F336
434	5-Carboxyaldehyde-5,7,3'-trihydroxy-4'-methoxyisoflavone	5'-Formylpratensein	C17H12O7	328.28	2883
435	5,3'-Dihydroxy-2'-methoxy-6,7-methylenedioxyisoflavone		C17H12O7	328.28	F15
436	5,3'-Dihydroxy-4'-methoxy-6,7-methylenedioxyisoflavone	Soforanarin A	C17H12O7	328.28	2853
437	5,7-Dihydroxy-6-methoxy-3',4'-methylenedioxyisoflavone	Dalspinin	C17H12O7	328.28	F356
438	6,3'-Dihydroxy-7-methoxy-4',5'-methylenedioxyisoflavone		C17H12O7	328.28	F312
439	6,7-Dihydroxy-3'-methoxy-4',5'-methylenedioxyisoflavone		C17H12O7	328.28	2380
440	2'-Hydroxy-7,3',4'-trimethoxyisoflavone		C18H16O6	328.32	F217
441	3'-Hydroxy-6,7,4'-trimethoxyisoflavone	Isocladrastin	C18H16O6	328.32	2445
442	5-Hydroxy-6,7,4'-trimethoxyisoflavone	7,4'-Di-O-methyltectorigenin	C18H16O6	328.32	F299
443	5-Hydroxy-7,3',4'-trimethoxyisoflavone		C18H16O6	328.32	F313
444	6-Hydroxy-7,2',4'-trimethoxyisoflavone		C18H16O6	328.32	2379
445	7-Hydroxy-2',4',5'-trimethoxyisoflavone		C18H16O6	328.32	2446
446	7-Hydroxy-5,6,4'-trimethoxyisoflavone	5-Methoxyafrormosin	C18H16O6	328.32	2447
447	7-Hydroxy-5,8,4'-trimethoxyisoflavone	Iso-5-methoxyafrormosin	C18H16O6	328.32	2381
448	7-Hydroxy-6,3',4'-trimethoxyisoflavone	Cladrastin	C18H16O6	328.32	2382
449	7-Hydroxy-8,3',4'-trimethoxyisoflavone		C18H16O6	328.32	2884
450	5,3',4'-Trihydroxy-6,7-dimethoxyisoflavone		C17H14O7	330.29	F3
451	5,6,3'-Trihydroxy-7,2'-dimethoxyisoflavone		C17H14O7	330.29	F91
452	5,6,4'-Trihydroxy-7,3'-dimethoxyisoflavone		C17H14O7	330.29	2873
453	5,7,2'-Trihydroxy-4',5'-dimethoxyisoflavone	2'-Hydroxy-5'-methoxybiochanin A	C17H14O7	330.29	2448
454	5,7,2'-Trihydroxy-6,5'-dimethoxyisoflavone	Podospicatin	C17H14O7	330.29	2449
455	5,7,3'-Trihydroxy-6,4'-dimethoxyisoflavone	Iristectorigenin A	C17H14O7	330.29	2451
456	5,7,3'-Trihydroxy-8,4'-dimethoxyisoflavone		C17H14O7	330.29	F111
457	5,7,4'-Trihydroxy-2',5'-dimethoxyisoflavone	Olibergin A	C17H14O7	330.29	2450
458	5,7,4'-Trihydroxy-6,3'-dimethoxyisoflavone	Iristectorigenin B	C17H14O7	330.29	F23
459	5,7,4'-Trihydroxy-6,8-dimethoxyisoflavone		C17H14O7	330.29	F3
460	5,8,3'-Trihydroxy-7,2'-dimethoxyisoflavone		C17H14O7	330.29	F3
461	6,7,3'-Trihydroxy-5,2'-dimethoxyisoflavone		C17H14O7	330.29	F3

continued

APPENDIX
Checklist for Isoflavonoids Described in the Literature During the Period 1991–2004 — continued

	Name	Trivial Name	Mass	Formula	Ref.
462	7,8,3'-Trihydroxy-6,4'-dimethoxyisoflavone		330.29	C17H14O7	F253
463	4'-Methoxy-6'',6''-dimethylpyrano[2'',3'':7,8]isoflavone	Calopogoniumisoflavone A	334.37	C21H18O4	2399
464	2'-Methoxy-4',5'-methylenedioxyfurano[2'',3'':7,6]isoflavone	Dehydroneotenone	336.30	C19H12O6	2402
465	3',4'-Methyleneoxyfurano[2'',3'':7,8]isoflavone	Garhwalin	336.30	C19H12O6	2934
466	5,4'-Dihydroxy-5''-(1-methylethenyl)-4'',5''-dihydrofurano[2'',3'':7,6]isoflavone	Licoagroisoflavone	336.34	C20H16O5	F141
467	5,4'-Dihydroxy-6'',6''-dimethylpyrano[2'',3'':7,6]isoflavone	Alpinumisoflavone 5	336.34	C20H16O5	2468
468	5,4'-Dihydroxy-6'',6''-dimethylpyrano[2'',3'':7,8]isoflavone	Derrone	336.34	C20H16O5	2469
469	5,7-Dihydroxy-6'',6''-dimethylpyrano[2'',3'':4',3']isoflavone	Isoderrone	336.34	C20H16O5	2898
470	7,2-Dihydroxy-6'',6''-dimethylpyrano(2'',3'':4',3')isoflavone	Eurycarpin B	336.34	C20H16O5	F353
471	7,2-Dihydroxy-6'',6''-dimethylpyrano[2'',3'':4',5']isoflavone	Puerarone	336.34	C20H16O5	2896
472	7,2'-Dihydroxy-6'',6''-dimethylpyrano[2'',3'':4',3']isoflavone	Glabron	336.34	C20H16O5	2400
473	7,4'-Methoxy-4'-prenyloxyisoflavone	Durlettone	336.38	C21H20O4	2401
474	4'-Methoxy-7-prenyloxyisoflavone	Maximaisoflavone J	336.39	C21H20O4	2835
475	5,4'-Dihydroxy-6'',6''-dimethyl-4'',5''-dihydropyrano[2'',3'':7,6]isoflavone	Erythrivarone A	338.36	C20H18O5	F94
476	5,4'-Dihydroxy-6'',6''-dimethyl-4'',5''-dihydropyrano[2'',3'':7,8]isoflavone	Crotalarin	338.36	C20H18O5	2933
477	5,7,4'-Trihydroxy-3'-prenylisoflavone	Isowighteone	338.36	C20H18O5	2856
478	5,7,4'-Trihydroxy-6-(1,1-dimethyl-2-propenyl)isoflavone	6-(1,1-Dimethylallyl)genistein	338.36	C20H18O5	2470
479	5,7,4'-Trihydroxy-6-prenylisoflavone	Wighteone	338.36	C20H18O5	2471
480	5,7,4'-Trihydroxy-8-(1,1-dimethylprop-2-enyl)isoflavone		338.36	C20H18O5	F210
481	5,7,4'-Trihydroxy-8-prenylisoflavone	Lupiwighteone	338.36	C20H18O5	2921
482	7,2',4'-Trihydroxy-3'-prenylisoflavone	Eurycarpin A	338.36	C20H18O5	F354
483	7,5''-Dihydroxy-6'',6''-dimethyl-4'',5''-dihydropyrano[2'',3'':4',3']isoflavone	Psoralenol	338.36	C20H18O5	2403
484	3'-Methoxy-6,7:4',5'-bis(methylenedioxy)isoflavone		340.28	C18H12O7	F299
485	5-Methoxy-6,7:3',4'-bis(methylenedioxy)isoflavone		340.28	C18H12O7	F299
486	5,7,8-Trimethoxy-3',4'-methylenedioxyisoflavone		340.33	C19H16O6	F180
487	3'-Hydroxy-5,4'-dimethoxy-6,7-methylenedioxyisoflavone	Iriskumaonin	342.30	C18H14O7	2453
488	3'-Hydroxy-5,5'-dimethoxy-6,7-methylenedioxyisoflavone	Isoiriskashmirianin	342.30	C18H14O7	2888
489	4'-Hydroxy-3',5'-dimethoxy-6,7-methylenedioxyisoflavone	Kashmigenin	342.30	C18H14O7	F217
490	4'-Hydroxy-5,3'-dimethoxy-6,7-methylenedioxyisoflavone	Iriskashmirianin	342.30	C18H14O7	2887
491	5-Hydroxy-3',4'-dimethoxy-6,7-me¹hylenedioxyisoflavone	Squarrosin	342.30	C18H14O7	2886
492	5-Hydroxy-7,3'-dimethoxy-4,5-methylenedioxyisoflavone	Sanchemarroquin	342.30	C18H14O7	F56

No.	Name	Common name	Formula	MW	Ref.
493	6-Hydroxy-7,2'-dimethoxy-4',5'-methylenedioxyisoflavone	Dipteryxine	C18H14O7	342.30	2384
494	7-Hydroxy-5,6-dimethoxy-3',4'-methylenedioxyisoflavone	Platycarpanetin	C18H14O7	342.30	2452
495	7-Hydroxy-5,8-dimethoxy-3',4'-methylenedioxyisoflavone	Dalpatein	C18H14O7	342.30	2454
496	7-Hydroxy-6,2'-dimethoxy-4',5'-methylenedioxyisoflavone		C18H14O7	342.30	2385
497	7-Hydroxy-8,2'-dimethoxy-4',5'-methylenedioxyisoflavone	Maximaisoflavone F	C18H14O7	342.30	2849
498	6,7,3',4'-Tetramethoxyisoflavone		C19H18O6	342.35	2386
499	7,2',4',5'-Tetramethoxyisoflavone		C19H18O6	342.35	2383
500	5,3'-Dihydroxy-6,7,2'-trimethoxyisoflavone		C18H16O7	344.32	F3
501	5,3'-Dihydroxy-7,4',5'-trimethoxyisoflavone	Vavain	C18H16O7	344.32	F189
502	5,3'-Dihydroxy-7,8,2'-trimethoxyisoflavone		C18H16O7	344.32	2891
503	5,4'-Dihydroxy-7,2',5'-trimethoxyisoflavone	Derrugenin	C18H16O7	344.32	2455
504	5,7-Dihydroxy-2',4',5'-trimethoxyisoflavone	7-Demethylrobustigenin	C18H16O7	344.32	F336
505	5,7-Dihydroxy-3',4',5'-trimethoxyisoflavone	Panchovillin	C18H16O7	344.32	2874
506	5,7-Dihydroxy-6,3',4'-trimethoxyisoflavone	Junipegenin B	C18H16O7	344.32	2456
507	5,7-Dihydroxy-6,8,4'-trimethoxyisoflavone		C18H16O7	344.32	F23
508	5,7-Dihydroxy-8,3',4'-trimethoxyisoflavone		C18H16O7	344.32	2457
509	6,3'-Dihydroxy-5,7,2'-trimethoxyisoflavone		C18H16O7	344.32	F3
510	7,3'-Dihydroxy-5,6,2'-trimethoxyisoflavone		C18H16O7	344.32	F3
511	8,3'-Dihydroxy-5,7,2'-trimethoxyisoflavone		C18H16O7	344.32	F3
512	3',4'-Methylenedioxy-6'',6''-dimethylpyrano[2'',3'':7,8]isoflavone	Calopogoniumisoflavone B	C21H16O5	348.36	2404
513	3',4'-Methylenedioxy-7-prenyloxyisoflavone	Maximaisoflavone B	C21H18O5	350.37	2405
514	4'-Hydroxy-5-methoxy-6'',6''-dimethylpyrano[2'',3'':7,8]isoflavone	Lanceolone	C21H18O5	350.37	F199
515	5-Hydroxy-4'-methoxy-6'',6''-dimethylpyrano[2'',3'':7,6]isoflavone	4'-O-Methylalpinumisoflavone	C21H18O5	350.37	2472
516	5-Hydroxy-4'-methoxy-6'',6''-dimethylpyrano[2'',3'':7,8]isoflavone	4'-O-Methylderrone	C21H18O5	350.37	2473
517	5-Methoxy-6'',6''-dimethylpyrano[2'',3'':7,6]isoflavone	Indicanine C	C21H18O5	350.37	F305
518	7-Hydroxy-3'-methoxy-6'',6''-dimethylpyrano[2'',3'':4',5']isoflavone	Piscisoflavone C	C21H18O5	350.37	F176
519	7,8-Methylenedioxy-4'-prenyloxyisoflavone	Auricularin	C21H18O5	350.37	F213
520	5,2',4'-Trihydroxy-5''-(1-methylethenyl)-4'',5''-dihydrofurano[2'',3'':7,6]isoflavone	Lupinisoflavone A	C20H16O6	352.34	2862
521	5,2',4'-Trihydroxy-6'',6''-dimethylpyrano[2'',3'':7,6]isoflavone	Parvisoflavone B	C20H16O6	352.34	2475
522	5,2',4'-Trihydroxy-6'',6''-dimethylpyrano[2'',3'':7,8]isoflavone	Parvisoflavone A	C20H16O6	352.34	2476
523	5,4-Dihydroxy-4'',5''-epoxy-6'',6''-dimethyl-4'',5''-dihydropyrano[2'',3'':7,6]isoflavone	Anagyroidisoflavone B	C20H16O6	352.34	F222
524	5,4'-Dihydroxy-5''-(1-hydroxy-1-methylethyl)furano[2'',3'':7,6]isoflavone	Erysubin A	C20H16O6	352.34	F283
525	5,4'-Dihydroxy-6''-hydroxymethyl-6''-methylpyrano[2'',3'':6,7]isoflavone	Hydroxyalpinumisoflavone	C20H16O6	352.34	F202

continued

APPENDIX
Checklist for Isoflavonoids Described in the Literature During the Period 1991–2004 — continued

	Name	Trivial Name	Mass	Formula	Ref.
526	5,4'-Dihydroxy-6''-hydroxymethyl-6''-methylpyrano[2'',3'':7,6]isoflavone	Erysubin B	352.34	C20H16O6	F283
527	5,7,2'-Trihydroxy-6'',6''-dimethylpyrano[2'',3'':4',3']isoflavone	Licoisoflavone B	352.34	C20H16O6	2474
528	5,7,3'-Trihydroxy-6'',6''-dimethylpyrano[2'',3'':4',5]isoflavone	Semilicoisoflavone B	352.34	C20H16O6	2905
529	5,7,4'-Trihydroxy-5''-(1-methylethenyl)-4'',5''-dihydrofurano[2'',3'':2',3']isoflavone	Crotarin	352.34	C20H16O6	2900
530	5,7,4'-Trihydroxy-6'',6''-dimethylpyrano[2'',3'':2',3]isoflavone	Sophoraisoflavone A	352.34	C20H16O6	2899
531	3'-Hydroxy-4'-methoxy-7-prenyloxyisoflavone		352.38	C21H20O5	2836
532	5,4'-Dihydroxy-7'-methoxy-6-prenylisoflavone	Gancaonin G	352.38	C21H20O5	2913
533	5,7-Dihydroxy-4'-methoxy-3'-prenylisoflavone		352.38	C21H20O5	F336
534	5,7-Dihydroxy-4'-methoxy-6-prenylisoflavone	Gancaonin A	352.38	C21H20O5	2912
535	5,7-Dihydroxy-4'-methoxy-8-prenylisoflavone	Gancaonin M	352.38	C21H20O5	2922
536	7,4'-Dihydroxy-5-methoxy-8-prenylisoflavone	5-Methyllupiwighteone	352.38	C21H20O5	2923
537	(2E)-5,7,4'-Trihydroxy-3'-(4-hydroxy-3-methyl-2-butenyl)isoflavone	Vogelin E	354.36	C20H18O6	F209
538	(2E)-5,7,4'-Trihydroxy-6-(4-hydroxy-3-methyl-2-butenyl)isoflavone	Glabrisoflavone	354.36	C20H18O6	F345
539	5,2',4'-Trihydroxy-4'',4'',5''-trimethyl-4',5''-dihydrofurano(2'',3'':7,6)isoflavone		354.36	C20H18O6	F210
540	5,2',4'-Trihydroxy-7-prenyloxyisoflavone		354.36	C20H18O6	F210
541	5,4'-Dihydroxy-5''-(1-hydroxy-1-methylethyl)-4'',5''-dihydrofurano[2'',3'':7,6]isoflavanone	Erythrinin C	354.36	C20H18O6	2479
542	5,7-Dihydroxy-5''-(1-hydroxy-1-methylethyl)-4'',5''-dihydrofurano[2'',3'':4',3']isoflavone	Lupinisoflavone C	354.36	C20H18O6	2857
543	5,7,2',4'-Tetrahydroxy-3'-prenylisoflavone	Licoisoflavone A	354.36	C20H18O6	2477
544	5,7,2',4'-Tetrahydroxy-5'-(1,1-dimethyl-2-propenyl)isoflavone	Fremontin	354.36	C20H18O6	2901
545	5,7,2',4'-Tetrahydroxy-5'-prenylisoflavone	Allolicoisoflavone A	354.36	C20H18O6	F176
546	5,7,2',4'-Tetrahydroxy-6-prenylisoflavone	Luteone	354.36	C20H18O6	2478
547	5,7,2',4'-Tetrahydroxy-8-(1,1-dimethylprop-2-enyl)isoflavone		354.36	C20H18O6	F210
548	5,7,2',4'-Tetrahydroxy-8-prenylisoflavone	2,3-Dehydrokievitone	354.36	C20H18O6	2480
549	5,7,3',4'-Tetrahydroxy-6-prenylisoflavone		354.36	C20H18O6	2915
550	5,7,3',4'-Tetrahydroxy-8-prenylisoflavone		354.36	C20H18O6	2925
551	5,7,4'-Trihydroxy-6-(2-hydroxy-3-methyl-3-butenyl)isoflavone	Gancaonin L	354.36	C20H18O6	F222
552	5,7,4'-Trihydroxy-6-(4-hydroxy-3-methyl-2-butenyl)isoflavone	Laburnetin	354.36	C20H18O6	2917
553	5,7,4'-Trihydroxy-8-(4-hydroxy-3-methyl-2-butenyl)isoflavone	Hydroxywighteone	354.36	C20H18O6	2927
554	5,7,5'(S)-Trihydroxy-6'',dimethyl-4'',5''-dihydropyrano[2'',3'':4',3']isoflavone	Gancaonin C	354.36	C20H18O6	F142
555	5,2',5'-Trimethoxy-6,7-methylenedioxyisoflavone	Ficuisoflavone	354.36	C20H18O6	2844
556	5,3',4'-Trimethoxy-6,7-methylenedioxyisoflavone	Hemerocallone	356.33	C19H16O7	2459
		Iriskumaonin methyl ether	356.33	C19H16O7	

557	5,6,7-Trimethoxy-3',4'-methylenedioxyisoflavone	Odoratine	356.33	C19H16O7	2458
558	6,7,2'-Trimethoxy-4',5'-methylenedioxyisoflavone		356.33	C19H16O7	2387
559	6,7,3'-Trimethoxy-4',5'-methylenedioxyisoflavone		356.33	C19H16O7	2388
560	6,7,8-Trimethoxy-3',4'-methylenedioxyisoflavone	Petalostetin	356.33	C19H16O7	2390
561	7,8,2'-Trimethoxy-3',4'-methylenedioxyisoflavone	Maximaisoflavone L	356.33	C19H16O7	F205
562	7,8,2'-Trimethoxy-4',5'-methylenedioxyisoflavone		356.33	C19H16O7	2389
563	8,3',4'-Trimethoxy-6,7-methylenedioxyisoflavone		356.33	C19H16O7	2391
564	5,3'-Dihydroxy-4',5'-dimethoxy-6,7-methylenedioxyisoflavone		358.30	C18H14O8	2854
565	5,7-Dihydroxy-6,2'-dimethoxy-4',5'-methylenedioxyisoflavone		358.30	C18H14O8	2889
566	3'-Hydroxy-5,6,7,2'-tetramethoxyisoflavone		358.35	C19H18O7	F3
567	5-Hydroxy-6,7,3',4'-tetramethoxyisoflavone	Belamcandin	358.35	C19H18O7	2885
568	5-Hydroxy-7,2',4',5'-tetramethoxyisoflavone	Robustigenin	358.35	C19H18O7	2460
569	6-Hydroxy-2',4',5',7-tetramethoxyisoflavone		358.35	C19H18O7	F130
570	7-Hydroxy-6,2',4',5'-trimethoxyisoflavone		358.35	C19H18O7	2392
571	7,4'-Dihydroxy-5,3',5'-trimethoxy-6-methylisoflavone	Brachyrachisin	358.35	C19H18O7	F221
572	5,7-Dihydroxy-4'-(p-methylbenzyl)isoflavone		358.39	C23H18O4	F178
573	5,7,3'-Trihydroxy-6,4',5'-trimethoxyisoflavone	Irigenin	360.32	C18H16O8	2461
574	5,7,4'-Trihydroxy-2',3',6'-trimethoxyisoflavone	Nervosin	360.32	C18H16O8	F107
575	6,4',5'-Trihydroxy-5,7,3'-trimethoxyisoflavone	Soforanarin B	360.32	C18H16O8	F15
576	7,8,5'-Trihydroxy-6,3',4'-trimethoxyisoflavone		360.32	C18H16O8	F208
577	3-Hydroxy-4',5'-methylenedioxy-6'',6''-dimethylpyrano[2'',3'':7,8]isoflavone	Norisojamicin	364.35	C21H16O6	F333
578	3'-Carboxyaldehyde-5,4'-dihydroxy-6'',6''-dimethylpyrano[2'',3'':7,6]isoflavone	Scandenal	364.35	C21H16O6	F43
579	3'-Hydroxy-4,5-methylenedioxy-6'',6''-dimethylpyrano[2'',3'':7,8]isoflavone	Norisojamicin	364.35	C21H16O6	F342
580	5-Hydroxy-3',4'-methylenedioxy-6'',6''-dimethylpyrano[2'',3'':7,6]isoflavone	Robustone	364.35	C21H16O6	2481
581	5-Hydroxy-3',4'-methylenedioxy-6'',6''-dimethylpyrano[2'',3'':7,8]isoflavone		364.35	C21H16O6	2932
582	5,4'-Dimethoxy-5''-(1-methylethenyl)-4'',5''-dihydrofurano[2'',3'':7,6]isoflavone	Thonninginisoflavone	364.40	C22H20O5	F13
583	5,4'-Dimethoxy-6'',6''-dimethylpyrano[2'',3'':7,6]isoflavone	Alpinumisoflavone dimethyl ether	364.40	C22H20O5	2482
584	6,4'-Dimethoxy-6'',6''-dimethylpyrano[2'',3'':7,8]isoflavone	6-Methoxycalopogonium	364.40	C22H20O5	F340
585	2'-Hydroxy-3',4'-methylenedioxy-7-prenyloxyisoflavone	Maximaisoflavone K	366.36	C21H18O6	F205
586	2',4'-Dihydroxy-5-methoxy-6'',6''-dimethylpyrano[2'',3'':7,8]isoflavone	Barpisoflavone C	366.36	C21H18O6	2931
587	5,3'-Dihydroxy-4'-methoxy-6'',6''-dimethylpyrano[2'',3'':7,6]isoflavone	3'-Hydroxyalpinumisoflavone 4'-methyl ether	366.36	C21H18O6	2484
588	5,7-Dihydroxy-3',4'-methylenedioxy-6-prenylisoflavone	Derrubone	366.36	C21H18O6	2483
589	7,2'-Dihydroxy-5-methoxy-6'',6''-dimethylpyrano[2'',3'':4',5']isoflavone	Glycyrrhiza isoflavone B	366.36	C21H18O6	F89

continued

APPENDIX
Checklist for Isoflavonoids Described in the Literature During the Period 1991–2004 — *continued*

	Name	Trivial Name	Mass	Formula	Ref.
590	7,2'-Dihydroxy-5'-methoxy-6'',6''-dimethylpyrano[2'',3'':4',3']isoflavone	Piscisoflavone B	366.36	C21H18O6	F176
591	7,4'-Dihydroxy-5'-methoxy-6'',6''-dimethylpyrano[2'',3'':2']isoflavone	Piscisoflavone D	366.36	C21H18O6	F264
592	3',4'-Dihydroxy-7-prenyloxyisoflavone		366.41	C22H22O5	2837
593	5,2',4'-Trihydroxy-7-methoxy-6-prenylisoflavone	7-O-Methylluteone	368.39	C21H20O6	F339
594	5,7-Dihydroxy-6-methoxy-4'-prenyloxyisoflavone	Isoaurmillone	368.39	C21H20O6	2851
595	5,7-Dihydroxy-8-methoxy-4'-prenyloxyisoflavone	Aurmillone	368.39	C21H20O6	2485
596	5,7,2'-Trihydroxy-4'-methoxy-5'-prenylisoflavone	Lysisteisoflavanone	368.39	C21H20O6	F62
597	5,7,2'-Trihydroxy-4'-methoxy-6-prenylisoflavone	Gancaonin N	368.39	C21H20O6	2914
598	5,7,3'-Trihydroxy-4'-methoxy-5'-prenylisoflavone		368.39	C21H20O6	F336
599	5,7,3'-Trihydroxy-4'-methoxy-6-prenylisoflavone	Gancaonin B	368.39	C21H20O6	2916
600	5,7,4'-Trihydroxy-2-methoxy-5'-prenylisoflavone	Vogelin F	368.39	C21H20O6	F209
601	5,7,4'-Trihydroxy-3'-methoxy-5'-prenylisoflavone	2'-Deoxypiscerythrone	368.39	C21H20O6	2859
602	5,7,4'-Trihydroxy-3'-methoxy-8-prenylisoflavone		368.39	C21H20O6	2926
603	5,7,4'-Trihydroxy-5'-methoxy-3'-prenylisoflavone	2'-Deoxypiscerythrone	368.39	C21H20O6	F264
604	7,2'-Dihydroxy-6'-methoxy-6'',6''-dimethyl-4'',5''-dihydropyrano[2'',3'':4',5']isoflavone		368.39	C21H20O6	F89
605	7,2',3'-Trihydroxy-4'-methoxy-3'-(1,1-dimethyl-2-propenyl)isoflavone	Secundiflorol C	368.39	C21H20O6	F103
606	7,2',4'-Trihydroxy-5-methoxy-8-prenylisoflavone	Barpisoflavone B	368.39	C21H20O6	2924
607	7,2',4'-Trihydroxy-5'-methoxy-3'-prenylisoflavone	Piscisoflavone A	368.39	C21H20O6	F176
608	7,3',4'-Trihydroxy-5-methoxy-5'-prenylisoflavone	Glisoflavone	368.39	C21H20O6	2902
609	7,4',6'-Trihydroxy-3'-methoxy-2'-prenylisoflavone	Kwakhurin	368.39	C21H20O6	2897
610	5,2'-Dimethoxy-6,7:4',5'-bis(methylenedioxy)isoflavone		370.32	C19H14O8	F299
611	5,3'-Dimethoxy-6,7:4',5'-bis(methylenedioxy)isoflavone		370.32	C19H14O8	F299
612	5,2',4'-Trihydroxy-5'-(1-hydroxy-1-methylethyl)-4'',5''-dihydrofurano[2'',3'':7,6]isoflavone	Lupinisoflavone B	370.36	C20H18O7	2863
613	5,2',4'-Trihydroxy-5'-(1-hydroxy-1-methylethyl)-4'',5''-dihydrofurano[2'',3'':7,8]isoflavone	Lunatone	370.36	C20H18O7	2935
614	5,7,2'-Trihydroxy-5'-(1-hydroxy-1-methylethyl)-4'',5''-dihydrofurano[2'',3'':4',3']isoflavone	Lupinisoflavone D	370.36	C20H18O7	2858
615	5,7,2',4'-Tetrahydroxy-8-(4-hydroxy-3-methyl-2-butenyl)isoflavone	2,3-Dehydrokievitol	370.36	C20H18O7	2928
616	5-Hydroxy-3',4',5'-trimethoxy-6,7-methylenedioxyisoflavone		372.33	C19H16O8	F322
617	5,6,7,3',4'-Pentamethoxyisoflavone		372.37	C20H20O7	F299
618	5,7,2',4'-Tetrahydroxy-8-(3-hydroxy-3-methyl-2-butyl)isoflavone	2,3-Dehydrokievitol hydrate	372.37	C20H20O7	2930
619	5,7,2',4',5'-Pentamethoxyisoflavone	Robustigenin methyl ether	372.37	C20H20O7	2462
620	6,7,2',3',4'-Tetramethoxyisoflavone		372.37	C20H20O7	2393

621	6,7,2′,4′,5′-Tetramethoxyisoflavone		C20H20O7	372.37	2394
622	6,7,3′,4′,5′-Tetramethoxyisoflavone		C20H20O7	372.37	2395
623	7,8,3′,4′,5′-Pentamethoxyisoflavone		C20H20O7	372.37	F40
624	5,3′-Dihydroxy-6,7,8,2′-tetramethoxyisoflavone		C19H18O8	374.35	2894
625	5,7-Dihydroxy-6,2′,3′,4′-tetramethoxyisoflavone	Irisjaponin B	C19H18O8	374.35	F169
626	5,7-Dihydroxy-6,2′,4′,5′-tetramethoxyisoflavone	Caviunin	C19H18O8	374.35	2463
627	5,7-Dihydroxy-6,3′,4′,5′-tetramethoxyisoflavone	Junipegenin C	C19H18O8	374.35	2464
628	5,7-Dihydroxy-8,2′,4′,5′-tetramethoxyisoflavone	Isocaviunin	C19H18O8	374.35	2465
629	5-Hydroxy-5′-methoxy-2′-prenyloxazol[2″,3″:4′,3′]isoflavone	Piscerythoxazole	C22H19O5N1	377.39	F177
630	7-Hydroxy-5′-methoxy-2′-prenyloxazol[2″,3″:4′,3′]isoflavone	Piscerythoxazole	C22H19O5N1	377.39	F264
631	2′-Methoxy-4′,5′-methylenedioxy-6″,6″-dimethylpyrano[2″,3″:7,8]isoflavone	Jamaicin	C22H18O6	378.38	2406
632	3′-Methoxy-4′,5′-methylenedioxy-6″,6″-dimethylpyrano[2″,3″:7,8]isoflavone	Isojamaicin	C22H18O6	378.38	2911
633	5-Methoxy-3′,4′-methylenedioxy-5″-(1-methylethenyl)-4″,5″-dihydrofurano[2″,3″:7,6]isoflavone	Glabrescione A	C22H18O6	378.38	2487
634	5-Methoxy-3′,4′-methylenedioxy-6″,6″-dimethylpyrano[2″,3″:7,6]isoflavone	Robustone methyl ether	C22H18O6	378.38	2486
635	6-Methoxy-3′,4′-methylenedioxy-6″,6″-dimethylpyrano[2″,3″:7,8]isoflavone	Durmillone	C22H18O6	378.38	2407
636	2′-Hydroxy-5,4-dimethoxy-6″,6″-dimethylpyrano[2″,3″:7,6]isoflavone	Indicanin E	C22H20O6	380.40	F184
637	2′-Methoxy-4′,5′-methylenedioxy-7-prenyloxyisoflavone	Maximaisoflavone C	C22H20O6	380.40	2408
638	6-Hydroxy-3′,4′-dimethoxy-6″,6″-dimethylpyrano[2″,3″:7,8]isoflavone		C22H20O6	380.40	F341
639	7-Hydroxy-6-methoxy-3′,4′-methylenedioxy-8-prenyloxyisoflavone	Predurmillone	C22H20O6	380.40	2936
640	8-Methoxy-3′,4′-methylenedioxy-7-prenyloxyisoflavone		C22H20O6	380.40	2847
641	5,5″-Dihydroxy-3′,4′-dimethylenedioxy-6″,6″-dimethyl-4″,5″-dihydropyrano[2″,3″:7,6]isoflavone	Harpalycin	C21H18O7	382.27	F47
642	5,7,3′,5′-Tetrahydroxy-5-methoxy-6″,6″-dimethyl-4″,5″-dihydropyrano[2″,3″:4′,5′]isoflavone	Glycyrrhiza isoflavone A	C20H18O7	382.37	F89
643	5,7-Dihydroxy-3′,4′-dimethoxy-5′-prenylisoflavone	Piscerythrinetin	C22H22O6	382.41	2903
644	7,3′-Dihydroxy-5,4′-dimethoxy-5′-prenylisoflavone		C22H22O6	382.41	2904
645	7,6′-Dihydroxy-2′,4′-dimethoxy-3′-prenylisoflavone	Licoricone	C22H22O6	382.41	2409
646	5,4-Dihydroxy-5″-(1-hydroxy-1-methylethyl)-4″-methoxyfurano[2″,3″:7,6]isoflavone		C21H20O7	384.39	F138
647	5,4′,5″-Trihydroxy-4″-methoxy-6″,6″-dimethyl-4″,5″-dihydropyrano[2″,3″:7,6]isoflavone	Anagyroidisoflavone A	C21H20O7	384.39	F222
648	5,7-Dihydroxy-3′-methoxy-5″-(1-hydroxy-1-methylethyl)-4″,5″-dihydrofurano[2″,3″:4′,5′]isoflavone	Piscerynetol	C21H20O7	384.39	F264
649	5,7,2′-Trihydroxy-4′-methoxy-8-(4-hydroxy-3-methyl-2-butenyl)isoflavone	Gancaonin D	C21H20O7	384.39	2929
650	5,7,2′,3′-Tetrahydroxy-4′-methoxy-3′-(1,1-dimethyl-2-propenyl)isoflavone	Secundiflorol B	C21H20O7	384.39	F103
651	5,7,2′,4′-Tetrahydroxy-5′-methoxy-3′-prenylisoflavone	Piscerythrone	C21H20O7	384.39	2488
652	5,7,2′,4′-Tetrahydroxy-5′-methoxy-6-prenylisoflavone	Isopiscerythrone	C21H20O7	384.39	F176
653	5,7,3′,4′-Tetrahydroxy-5′-methoxy-2′-prenylisoflavone	Piscidone	C21H20O7	384.39	F268

continued

APPENDIX

Checklist for Isoflavonoids Described in the Literature During the Period 1991–2004 — *continued*

	Name	Trivial Name	Mass	Formula	Ref.
654	5,7,4'-Trihydroxy-3'-methoxy-6'',6''-dimethylpyrano[2'',3'':5',6']isoflavone	Piscidanone	384.39	C21H20O7	F176
655	5,2',3',6'-Tetrahydroxy-6,7-methylenedioxyisoflavone	Madhushazone	386.36	C20H18O8	F245
656	5,3',4',5'-Tetramethoxy-6,7-methylenedioxyisoflavone	Irisflorentin	386.36	C20H18O8	2466
657	5,6,7,8-Tetramethoxy-3',4'-methylenedioxyisoflavone		386.36	C20H18O8	2467
658	7,4',6'-Trihydroxy-3'-methoxy-2'-(3-hydroxy-3-methylbutyl)isoflavone		386.40	C21H22O7	2895
659	6,8,3'-Trichloro-5,4'-dihydroxy-7-methoxyisoflavone	Kwakhurin hydrate	387.60	C16H9O5Cl3	F301
660	4'-Prenyloxy-6'',6''-dimethylpyrano[2'',3'':7,8]isoflavone		388.46	C25H24O4	F341
661	7,3'-Dihydroxy-8,2'-diprenylisoflavone	Erysubin F	390.48	C25H26O4	F275
662	5-Hydroxy-8-methoxy-3',4'-methylenedioxy-6'',6''-dimethylpyrano[2'',3'':7,6]isoflavone		394.38	C22H18O7	F156
663	6-Hydroxy-5-methoxy-3',4'-methylenedioxy-6'',6''-dimethylpyrano[2'',3'':7,8]isoflavone		394.38	C22H18O7	F331
664	2',4',5'-Trimethoxy-6'',6''-dimethylpyrano[2'',3'':7,6]isoflavone	Griffonianone B	394.42	C23H22O6	F118
665	2',4',5'-Trimethoxy-6'',6''-dimethylpyrano[2'',3'':7,8]isoflavone		394.42	C23H22O6	2410
666	6,3',4'-Trimethoxy-6'',6''-dimethylpyrano[2'',3'':7,8]isoflavone	Barbigerone	394.42	C23H22O6	F341
667	4'-Hydroxy-2',5'-dimethoxy-6'',6''-dimethylpyrano[2'',3'':7,6]isoflavone	Durallone	396.40	C22H20O7	2490
668	5,4'-Dihydroxy-2',6'-dimethoxy-6'',6''-dimethylpyrano[2'',3'':7,8]isoflavone	Elongatin	396.40	C22H20O7	F49
669	5,4'-Dihydroxy-3',5'-dimethoxy-6'',6''-dimethylpyrano[2'',3'':7,6]isoflavone	4'-Demethyltoxicarol	396.40	C22H20O7	2918
670	7-Hydroxy-6,3',4'-trimethoxy-8-prenylisoflavone	Pumilaisoflavone D	396.44	C23H24O6	F341
671	5,7,4'-Trihydroxy-5'-methoxy-5''-(1-hydroxy-1-methylethyl)-4'',5''-dihydrofurano[2'',3'':2',3']isoflavone	Predurallone	398.36	C21H18O8	F264
672	5,7-Dihydroxy-4',5'-dimethoxy-3'-((1E)-3-hydroxy-3-methyl-1-butenyl)isoflavone	Piscerisoflavone E	398.41	C22H22O7	F264
673	5,7,2'-Trihydroxy-4',5'-dimethoxy-3'-prenylisoflavone	Piscerynetin	398.41	C22H22O7	2906
674	5,7,4'-Trihydroxy-2',5'-dimethoxy-6-prenylisoflavone	2'-Hydroxypiscerythrinetin	398.41	C22H22O7	2860
675	5,7,2'-Trihydroxy-5'-methoxy-5''-(1-hydroxy-1-methylethyl)-4'',5''-dihydrofurano[2'',3'':4',3']isoflavone	Viridiflorin	400.38	C21H20O8	F264
676	5,7,2',4'-Tetrahydroxy-5'-methoxy-3'-((2E)-4-hydroxy-3-methyl-2-butenyl)isoflavone	Piscerisoflavone B	400.38	C21H20O8	F264
677	5,7,2',5'-Tetrahydroxy-5'-methoxy-6'',6''-dimethyl-4'',5''-dihydropyrano[2'',3'':4',3']isoflavone	Piscerisoflavone D	400.38	C21H20O8	F264
678	5,7,4'-Trihydroxy-5'-methoxy-5''-(1-hydroxy-1-methylethyl)-4'',5''-dihydrofurano[2'',3'':2',3']isoflavone	Piscerisoflavone A	400.38	C21H20O8	F264
679	5,7,4',5'-Tetrahydroxy-5'-methoxy-6'',6''-dimethyl-4'',5''-dihydropyrano[2'',3'':2',3']isoflavone	Piscerisoflavone C	400.38	C21H20O8	F264
680	5,7,4',5'-Tetrahydroxy-5'-methoxy-6'',6''-dimethyl-4'',5''-dihydropyrano[2'',3'':3',2']isoflavone	Piscidanol	400.38	C21H20O8	F264
681	6,7,8,3',4',5'-Hexamethoxyisoflavone		402.39	C21H22O8	F40
682	5-Hydroxy-(2'',3'':7,6),(2''',3''':4',3')-bis(6,6-dimethylpyrano)isoflavone	Ulexin B	402.45	C25H22O5	F165
683	5-Hydroxy-(2'',3'':7,8),(2''',3''':4',3')-bis(6,6-dimethylpyrano)isoflavone	Ulexone B	402.45	C25H22O5	2962
684	4'-(6''-Methyl)salicylate-5,7-dihydroxyisoflavone	Genistein-4'-(6''-methyl)salicylate	404.37	C23H16O7	F159

685	5,7-Dihydroxy-6,2',3',4',5'-pentamethoxyisoflavone	Irisjaponin A	404.37	C20H20O9	F169
686	5-Hydroxy-4'-prenyloxy-6'',6''-dimethylpyrano[2'',3'':7,6]isoflavone	4'-O-prenylalpinumisoflavone	404.46	C25H24O5	2861
687	5,4'-Dihydroxy-3'-prenyl-6'',6''-dimethylpyrano[2'',3'':7,6]isoflavone	Chandalone	404.46	C25H24O5	2491
688	5,4'-Dihydroxy-3'-prenyl-6'',6''-dimethylpyrano[2'',3'':7,8]isoflavone	Scanderone	404.46	C25H24O5	F157
689	5,4'-Dihydroxy-6-prenyl-6'',6''-dimethylpyrano[2'',3'':7,6]isoflavone	Osajin	404.46	C25H24O5	2492
690	5,4'-Dihydroxy-8-prenyl-6'',6''-dimethylpyrano[2'',3'':7,6]isoflavone	Warangalone	404.46	C25H24O5	2493
691	5,7-Dihydroxy-6-prenyl-6'',6''-dimethylpyrano[2'',3'':4',3']isoflavone	Isochandalone	404.46	C25H24O5	2940
692	5,7-Dihydroxy-8-prenyl-6'',6''-dimethylpyrano[2'',3'':4',3']isoflavone	Ulexone A	404.46	C25H24O5	2960
693	5,7,4'-Trihydroxy-8-((1R,6R)-3-methyl-6-(1-methylethenyl)-2-cyclohexen-1-yl)isoflavone	Ficusin A	404.46	C25H24O5	F5
694	4'-Methoxy-7-geranyloxyisoflavone	7-Geranylformononetin	404.51	C26H28O4	2869
695	3',4'-Dihydroxy-7-(3,7-dimethyl-2(E),6-octadienyloxy)isoflavone		406.47	C25H26O5	F332
696	4'-Hydroxy-(2''',3''':5,6),(2''',3''':7,8)-bis(4,5-dihydro-6,6-dimethylpyrano)isoflavone	Erythrivarone B	406.47	C25H26O5	F94
697	5,7,2',4'-Tetrahydroxy-3'-prenyl-5'-(1,1-dimethyl-2-propenyl)isoflavone	Fremontone	406.47	C25H26O5	2908
698	5,7,4'-Trihydroxy-3',5'-diprenylisoflavone	3',5'-Diprenylgenistein	406.47	C25H26O5	2907
699	5,7,4'-Trihydroxy-6,3'-diprenylisoflavone	Lupalbigenin	406.47	C25H26O5	2494
700	5,7,4'-Trihydroxy-6,8-diprenylisoflavone	6,8-Diprenylgenistein	406.47	C25H26O5	2495
701	5,7,4'-Trihydroxy-8-(3,7-dimethyl-2,6-octadienyl)isoflavone	Lespedezol E1	406.47	C25H26O5	F171
702	5,7,4'-Trihydroxy-8,3'-diprenylisoflavone	Isolupalbigenin	406.47	C25H26O5	F267
703	7-Hydroxy-4'-geranyloxyisoflavone	Conrauinone D	406.47	C25H26O5	F69
704	2',6'-Dimethylenedioxy-6'',6''-dimethylpyrano[2'',3'':7,8]isoflavone	Ferrugone	408.41	C23H20O7	2411
705	5,6-Dimethoxy-3',4'-methylenedioxy-6'',6''-dimethylpyrano[2'',3'':7,8]isoflavone	5-Methoxydurmillone	408.41	C23H20O7	2938
706	6,2'-Dimethoxy-4',5'-methylenedioxy-6'',6''-dimethylpyrano[2'',3'':7,8]isoflavone	Ichthynone	408.41	C23H20O7	2412
707	5-Hydroxy-2',4',5'-trimethoxy-6'',6''-dimethylpyrano[2'',3'':7,8]isoflavone	Toxicarolisoflavone	410.42	C23H22O7	2496
708	7-Hydroxy-2',5'-dimethoxy-3',4'-methylenedioxy-8-prenylisoflavone	Preferrugone	410.42	C23H22O7	2910
709	7-Hydroxy-5,6-dimethoxy-3',4'-methylenedioxy-8-prenylisoflavone	Pre-5-methoxydurmillone	410.42	C23H22O7	2937
710	7,2',4',5'-Tetramethoxy-8-prenylisoflavone	Prebarbigerone	410.47	C24H26O6	2909
711	5,7-Dihydroxy-4',5'-dimethoxy-5''-(1-hydroxy-1-methylethyl)-4'',5''-dihydrofurano[2''',3''':2':3']isoflavone	Piscerisoflavone F	414.41	C22H22O8	F264
712	5,7,5'-Trihydroxy-4',5'-dimethoxy-6'',6''-dimethyl-4'',5''-dihydropyrano[2''',3''':2':3']isoflavone	Piscerisoflavone G	414.41	C22H22O8	F264
713	5,7,2'-Trihydroxy-4',5'-dimethoxy-3-(3-hydroxy-3-methyl-butyl)isoflavone		416.42	C22H24O8	F264
714	2'-Methoxy-(2'',3'':7,6),(2''',3''':4',3')-bis(6,6-dimethylpyrano)isoflavone	Munetone	416.48	C26H24O5	2413
715	5-Hydroxy-5'''-(1-hydroxy-1-methylethyl)furano[2''',3''':7,6]-6'',6''-dimethylpyrano[2'',3'':4',3']isoflavone	Ulexin C	418.45	C25H22O6	F164
716	5,2-Dihydroxy-(2'':3'':7,6),(2''',3''':4',3)-bis(6,6-dimethylpyrano)isoflavone	Kraussianone 1	418.45	C25H22O6	F58
717	5,7-Dihydroxy-6''-hydroxymethyl-6'',6''-trimethyl-(2'',3'':7,8),(2''',3''':4,3)-bis(pyrano)isoflavone	Ulexone D	418.45	C25H22O6	2963

continued

APPENDIX
Checklist for Isoflavonoids Described in the Literature During the Period 1991–2004 — *continued*

	Name	Trivial Name	Mass	Formula	Ref.
718	3′,4′-Methylenedioxy-7-((2E)-3,7-dimethyl-2,6-octadienyloxy)isoflavone	7-O-Geranylpseudobaptigenin	418.49	C26H26O5	F333
719	4′-Hydroxy-5-methoxy-6-prenyl-6″,6″-dimethylpyrano[2″,3″:7,8]isoflavone	Scandinone	418.49	C26H26O5	2497
720	5-Methoxy-4′-prenyloxy-6″,6″-dimethylpyrano[2″,3″:7,6]isoflavone		418.49	C26H26O5	F13
721	5-Hydroxy-5″-(1-hydroxy-1-methylethyl)-4″,5″-dihydrofurano[2″,3″:7,8]-6″,6″-dimethylpyrano[2‴,3‴:4′,3′]isoflavone	Ulexone C	420.46	C25H24O6	2964
722	5-Hydroxy-5‴-(1-hydroxy-1-methylethyl)-4‴,5‴-dihydrofurano[2‴,3‴:7,6]-6″,6″-dimethylpyrano[2″,3″:4′,3′]isoflavone	Ulexin D	420.46	C25H24O6	F164
723	5,2′-Dihydroxy-4′-prenyloxy-6″,6″-dimethylpyrano[2″,3″:7,6]isoflavone	Isoauriculatin	420.46	C25H24O6	2498
724	5,2′,4′-Trihydroxy-3′-prenyl-6″,6″-dimethylpyrano[2″,3″:7,6]isoflavone	Angustone C	420.46	C25H24O6	2954
725	5,2′,4′-Trihydroxy-8-prenyl-6″,6″-dimethylpyrano[2″,3″:7,6]isoflavone	Auriculatin	420.46	C25H24O6	2501
726	5,3′-Dihydroxy-4′-prenyloxy-6″,6″-dimethylpyrano[2″,3″:7,6]isoflavone	Isoauriculasin	420.46	C25H24O6	2499
727	5,3′,4′-Trihydroxy-6-prenyl-6″,6″-dimethylpyrano[2″,3″:7,8]isoflavone	Pomiferin	420.46	C25H24O6	2500
728	5,3′,4′-Trihydroxy-8-prenyl-6″,6″-dimethylpyrano[2″,3″:7,6]isoflavone	Auriculasin	420.46	C25H24O6	2502
729	5,4′-Dihydroxy-6-(2-hydroxy-3-methyl-3-butenyl)-6″,6″-dimethylpyrano[2″,3″:7,8]isoflavone	Euchrenone b9	420.46	C25H24O6	2973
730	5,4′-Dihydroxy-8-(2-hydroxy-3-methyl-3-butenyl)-6″,6″-dimethylpyrano[2″,3″:7,6]isoflavone	Erysenegalensein M	420.46	C25H24O6	F309
731	5,4′-Dihydroxy-8-(2-hydroxy-3-methyl-3-butenyl)-6″,6″-dimethylpyrano[2″,3″:7,6]isoflavone	Euchrenone b8	420.46	C25H24O6	2968
732	5,4′-Dihydroxy-8-(2,3-epoxy-3-methylbutyl)-6″,6″-dimethylpyrano[2″,3″:7,6]isoflavone	Erysenegalensein G	420.46	C25H24O6	F310
733	5,4′,7″-Trihydroxy-7″-methyl-4″-(1-methylethenyl)-3″a,4″,5″,6″,7″,7″a-hexahydrobenzofurano[2″,3″:7,8]isoflavone	Ficusin B	420.46	C25H24O6	F5
734	5,7-Dihydroxy-6-(2-hydroxy-3-methyl-3-butenyl)-6″,6″-dimethylpyrano[2″,3″:4′,3′]isoflavone	Ulexin A	420.46	C25H24O6	F165
735	5,7,2′-Trihydroxy-6-prenyl-6″,6″-dimethylpyrano[2″,3″:4′,3′]isoflavone	Angustone B	420.46	C25H24O6	2944
736	5,7,2′-Trihydroxy-6-prenyl-6″,6″-dimethylpyrano[2″,3″:4′,5′]isoflavone	Kraussianone 2	420.46	C25H24O6	F58
737	5,7,3′-Trihydroxy-6-prenyl-6″,6″-dimethylpyrano[2″,3″:4′,5′]isoflavone	Gancaonin H	420.46	C25H24O6	2948
737b	5,7,3′-Trihydroxy-2′-prenyl-6″,6″-dimethylpyrano[2″,3″:4′,5′]isoflavone		420.46	C25H24O6	F4b
738	5,7,4′-Trihydroxy-8-prenyl-6″,6″-dimethylpyrano[2″,3″:2′,3′]isoflavone	Glyasperin N	420.46	C25H24O6	F72
739	4′-Methoxy-7-(8-hydroxy-3,7-dimethyl-2(E),6(Z)-octadienyloxy)isoflavone		420.50	C26H28O5	F332
740	5,4′-Dihydroxy-7-methoxy-8,3′-diprenylisoflavone	7-O-Methylisolupalbigenin	420.50	C26H28O5	F164
741	5,7-Dihydroxy-4′-methoxy-8-geranylisoflavone	Olibergin A	420.50	C26H28O5	F111
742	7-Hydroxy-5-methoxy-4′-geranyloxyisoflavone	Conrauinone C	420.50	C26H28O5	F69
743	7,4′-Dihydroxy-5-methoxy-6,8-diprenylisoflavone	Derrisisoflavone A	420.50	C26H28O5	F232
744	3,5,7,4′-Tetrahydroxy-6,3′-diprenylisoflavone	Glyasperin A	422.48	C25H26O6	F350

No.	Name	Trivial name	Formula	MW	Code
745	4″,5″-Dihydro-5,7-dihydroxy-3′-prenyl-6″,6″-dimethylpyrano[2″,3″:4′,5′]isoflavone	Vogelin G	C25H26O6	422.48	F209
746	5,4′-Dihydroxy-5″-(1-hydroxy-1-methylethyl)-prenyl-4″,5″-dihydrofurano[2″,3″:7,6]isoflavone	Lupinisoflavone G	C25H26O6	422.48	2957
747	5,4′-Dihydroxy-5″-(1-hydroxy-1-methylethyl)-6-prenyl-4″,5″-dihydrofurano[2″,3″:7,8]isoflavone	Isosenegalensin	C25H26O6	422.48	F62
748	5,4′-Dihydroxy-5″-(1-hydroxy-1-methylethyl)-8-prenyl-4″,5″-dihydrofurano[2″,3″:7,6]isoflavone	Euchrenone b10	C25H26O6	422.48	F279
749	5,4′-Dihydroxy-6-(3-hydroxy-3-methylbutyl)-6″,6″-dimethylpyrano[2″,3″:7,8]isoflavone	Euchrenone b7	C25H26O6	422.48	2974
750	5,4′-Dihydroxy-8-(3-hydroxy-3-methylbutyl)-6″,6″-dimethylpyrano[2″,3″:7,6]isoflavone	Euchrenone b6	C25H26O6	422.48	2969
751	5,4′,5″-Trihydroxy-3′-prenyl-4″,5″-dihydropyrano[2″,3″:7,6]isoflavone	Lupinisolone A	C25H26O6	422.48	2956
752	5,7,2′,4′-Tetrahydroxy-6,3′-diprenylisoflavone	Angustone A	C25H26O6	422.48	2503
753	5,7,2′,4′-Tetrahydroxy-6,8-diprenylisoflavone	8-Prenylluteone	C25H26O6	422.48	2965
754	5,7,2′,4′-Tetrahydroxy-8,3′-diprenylisoflavone	2′-Hydroxyisolupalbigenin	C25H26O6	422.48	2961
755	5,7,2′,4′-Tetrahydroxy-8,5′-diprenylisoflavanone	Vogelin C	C25H26O6	422.48	F14
756	5,7,3′,4′-Tetrahydroxy-6,5′-diprenylisoflavone	Isoangustone A	C25H26O6	422.48	2947
757	5,7,3′,4′-Tetrahydroxy-6,8-diprenylisoflavone	6,8-Diprenyllorobol	C25H26O6	422.48	2504
757b	5,7,3′,4′-Tetrahydroxy-2′,5′-diprenylisoflavone		C25H26O6	422.48	F4b
758	5,7,4′-Trihydroxy-3-(2-hydroxy-3-methyl-3-butenyl)-6-prenylisoflavone	Derrisisoflavone B	C25H26O6	422.48	F232
759	5,7,4′-Trihydroxy-6-(2-hydroxy-3-methyl-3-butenyl)-3′-prenylisoflavone	Lupinisol A	C25H26O6	422.48	2950
760	5,7,4′-Trihydroxy-6-(2-hydroxy-3-methyl-3-butenyl)-8-prenylisoflavone	Isoerysenegalensein E	C25H26O6	422.48	F62
761	5,7,4′-Trihydroxy-8-(2-hydroxy-3-methyl-3-butenyl)-6-prenylisoflavone	Erysenegalensein E	C25H26O6	422.48	F307
762	5,7,5′-Trihydroxy-6-prenyl-6″,6″-dimethyl-4″,5″-dihydropyrano[2″,3″:4′,3′]isoflavone	Derrisisoflavone F	C25H26O6	422.48	F232
763	7,2′,4′,5′-Tetrahydroxy-3′-(3,7-dimethyl-2,6-octadienyl)isoflavone	Lespedezol E2	C25H26O6	422.48	F171
764	7,4′-Dihydroxy-5″-(1-hydroxy-1-methylethyl)-8′-prenyl-6″,6″-dimethyl-4″,5″-dihydropyrano[2″,3″:5,6]isoflavone	Eryvarin B	C25H26O6	422.48	F284
765	6,8,3′,5′-Tetrachloro-5,4′-dihydroxy-7-methoxyisoflavone		C16H9O5Cl3	423.05	F301
766	5-Hydroxy-2′-methoxy-6″,6″-dimethylpyrano[2″,3″:7,6]-4′-prenyloxyisoflavone		C26H26O6	434.49	F213
767	5,2′-Dihydroxy-4′-methoxy-8-prenyl-6″,6″-dimethylpyrano[2″,3″:7,6]isoflavone	Auriculin	C26H26O6	434.49	2505
768	5″-Hydroxy-2′-methoxy-4′,5″-dihydro-(2″,3″:7,6),(2″,3″:4′,3)-bis(6,6-dimethylpyrano)isoflavone	Mundulone	C26H26O6	434.49	2415
769	5-Hydroxy-5″-(1-hydroxy-1-methylethyl)-4″,5″-dihydrofurano[2″,3″:4,3]-6′-dimethylpyrano[2″,3″:7,6]isoflavone		C25H24O7	436.46	F43
770	5,2′-Dihydroxy-5″-(1-hydroxy-1-methylethyl)-4″,5″-dihydrofurano[2″,3″:7,6]-6″,6″-dimethylpyrano[2″,3″:4′,3′]isoflavone	Lupinisoflavone K	C25H24O7	436.46	2959
771	5,2′-Dihydroxy-5″-(1-hydroxy-1-methylethyl)-4″,5″-dihydrofurano[2″,3″:4′,3]-6″,6′-dimethylpyrano[2″,3″:7,6]isoflavone	Lupinisoflavone L	C25H24O7	436.46	2955
772	5,2′,4′-Trihydroxy-8-(2-hydroxy-3-methyl-3-butenyl)-6″,6″-dimethylpyrano[2″,3″:7,6]isoflavone	Erysenegalensein L	C25H24O7	436.46	F309
773	5,2′,4′-Trihydroxy-8-(2,3-epoxy-3-methylbutyl)-6″,6″-dimethylpyrano[2″,3″:7,6]isoflavone	Erysenegalensein F	C25H24O7	436.46	F310

continued

APPENDIX
Checklist for Isoflavonoids Described in the Literature During the Period 1991–2004 — continued

	Name	Trivial Name	Mass	Formula	Ref.
774	5,2',5''-Trihydroxy-4',5''-dihydro-(2''',3'':7,6),(2''',3''':4',3')-bis(6,6-dimethylpyrano)isoflavone	Kraussianone 6	436.46	C25H24O7	F57
775	5,7,2'-Trihydroxy-6-(2-hydroxy-3-methyl-3-butenyl)-6'',6''-dimethylpyrano[2'',3'':4',3']isoflavone	Kraussianone 7	436.46	C25H24O7	F57
776	2'-Hydroxy-4'-methoxy-(2'',3'':5,6),(2''',3''':7,8)-bis(4,5-dihydro-6,6-dimethylpyrano)isoflavone	1'',2'-Dihydro-2'-hydroxycycloosajin	436.50	C26H28O6	2506
777	4'-Hydroxy-3'-methoxy-(2'',3'':5,6),(2''',3''':7,8)-bis(4,5-dihydro-6,6-dimethylpyrano)isoflavone	1'',2'-Dihydro-O-methylcyclopomiferin	436.50	C26H28O6	2507
778	4'-Hydroxy-5-methoxy-5''-(1-hydroxy-1-methylethyl)-6-prenyl-4',5''-dihydrofurano[2'',3'':7,8]isoflavone	Derrisisoflavone C	436.50	C26H28O6	F232
779	5,4'-Dihydroxy-8-(3-methoxy-3-methylbutyl)-6'',6''-dimethylpyrano(2'',3'':7,6)isoflavone	Eturunagarone	436.50	C26H28O6	F215
780	5,7,3'-Trihydroxy-4'-methoxy-6,8-diprenylisoflavone	6,8-Diprenylpratensein	436.50	C26H28O6	2866
781	5,7,4'-Trihydroxy-2'-methoxy-6,3'-diprenylisoflavone	2'-Methoxylupalbigenin	436.50	C26H28O6	2864
782	5,7,4'-Trihydroxy-2'-methoxy-6,8-diprenylisoflavone	Euchrenone b15	436.50	C26H28O6	F175
783	5,7,4'-Trihydroxy-3'-methoxy-2',5'-diprenylisoflavone	Millewanin A	436.50	C26H28O6	F110
784	7,4'-Dihydroxy-5-methoxy-6-(2-hydroxy-3-methyl-3-butenyl)-8-prenylisoflavone	Derrisisoflavone D	436.50	C26H28O6	F232
785	7,4'-Dihydroxy-5-methoxy-8-(2-hydroxy-3-methyl-3-butenyl)-6-prenylisoflavone	Derrisisoflavone E	436.50	C26H28O6	F232
786	4'-Amino-5,7,3'-trihydroxy-5'-methoxy-2',6'-diprenylisoflavone	Piscerythramine	437.52	C26H29O6N1	2978
787	5,6,2'-Trimethoxy-4',5'-methylenedioxy-6'',6''-dimethylpyrano[2'',3'':7,8]isoflavone	Conrauinone A	438.44	C24H22O8	F68
788	5-Hydroxy-(2'',3'':7,6),(2''',3''':4',3')-bis(5-(1-hydroxy-1-methylethyl))-4,5-dihydrofurano)isoflavone	Isolupinisoflavone E	438.48	C25H26O7	F142
789	5,2',4'-Trihydroxy-5''-(1-hydroxy-1-methylethyl)-3'-prenyl-4',5''-dihydrofurano[2'',3'':7,6]isoflavone	Lupinisoflavone H	438.48	C25H26O7	2958
790	5,2',4'-Trihydroxy-5''-(1-hydroxy-1-methylethyl)-8-prenyl-4',5''-dihydrofurano[2'',3'':7,6]isoflavone	Erysenegalensein H	438.48	C25H26O7	F308
791	5,2',4'-Trihydroxy-8-prenyl-6'',6''-dimethyl-4',5''-dihydropyrano[2'',3'':7,6]isoflavone	Erysenegalensein I	438.48	C25H26O7	F308
792	5,2',4,5''-Tetrahydroxy-6-prenyl-6'',6''-dimethyl-4',5''-dihydropyrano[2'',3'':7,8]isoflavone	Erysenegalensein O	438.48	C25H26O7	F190
793	5,4'-Dihydroxy-8-(2-hydroperoxy-3-methyl-3-butenyl)-6'',6''-dimethyl-4',5''-dihydropyrano[2'',3'':7,6]isoflavone	Millewanin E	438.48	C25H26O7	F110
794	5,7,2'-Trihydroxy-5''-(1-hydroxy-1-methylethyl)-6-prenyl-4',5''-dihydrofurano[2'',3'':4',3']isoflavone	Lupinisoflavone I	438.48	C25H26O7	2946
795	5,7,2'-Trihydroxy-6-(3-hydroxy-3-methyl-3-butyl)-6'',6''-dimethylpyrano[2'',3'':4',3']isoflavone	Kraussianone 3	438.48	C25H26O7	F58
796	5,7,2'-Trihydroxy-6-(3-hydroxy-3-methylbutyl)-6'',6''-dimethylpyrano[2'',3'':4',3']isoflavone	Kanzonol T	438.48	C25H26O7	F79
797	5,7,2',4'-Tetrahydroxy-3'-(2-hydroxy-3-methyl-3-butenyl)-6-prenylisoflavone	Lupinisol C	438.48	C25H26O7	2941
798	5,7,2',4'-Tetrahydroxy-6-(2-hydroxy-3-methyl-3-butenyl)-3'-prenylisoflavone	Lupinisol B	438.48	C25H26O7	2951
799	5,7,2',4'-Tetrahydroxy-6-(2-hydroxy-3-methyl-3-butenyl)-8-prenylisoflavone	Erysenegalensein N	438.48	C25H26O7	F190
800	5,7,2',4'-Tetrahydroxy-8-(2-hydroxy-3-methyl-3-butenyl)-6-prenylisoflavone	Erysenegalensein D	438.48	C25H26O7	F307
801	5,7,2',5''-Tetrahydroxy-6-prenyl-6'',6''-dimethyl-4',5''-dihydropyrano[2''',3''':4',3']isoflavone	Lupinisolone B	438.48	C25H26O7	2945
802	5,7,4'-Trihydroxy-5''-(1-hydroxy-1-methylethyl)-6-prenyl-4',5''-dihydrofurano[2'',3'':2',3']isoflavone	Lupinisoflavone J	438.48	C25H26O7	2943

803	5,7,4',5''-Tetrahydroxy-6-prenyl-6'',6''-dimethyl-4'',5''-dihydropyrano[2'',3'':2':2',3']isoflavone	Lupinisolone C	438.48	C25H26O7	2942
804	4'-Methoxy-7-((2E)-6,7-dihydroxy-3,7-dimethyl-2-octaenyloxy)isoflavanone	Griffonianone D	438.51	C26H30O6	F330
805	6-Prenylisocaviunin	6-Prenylisocaviunin	442.47	C24H26O8	2939
806	5,7-Dihydroxy-8,2,4',5'-tetramethoxy-8-prenylisoflavone	Griffonianone C	448.51	C27H28O6	F331
807	7-Keto-6-methoxy-3',4'-methylenedioxy-8,8-diprenylisoflavone	Cajaisoflavone	450.49	C26H26O7	2414
808	2'-Hydroxy-4',6'-dimethoxy-3'-prenyl-6'',6''-dimethylpyrano[2'',3'':7,6]isoflavone	Euchrenone b5	450.49	C26H26O7	2966
809	5,7,2'-Trihydroxy-4',5'-methylenedioxy-6,8-diprenylisoflavone	Flemiphilippinin B	450.52	C27H30O6	F41
810	5,7-Dihydroxy-3',4'-dimethoxy-6,8-diprenylisoflavone	Glabrescione B	450.52	C27H30O6	2508
811	5,7-Dimethoxy-3',4'-diprenyloxyisoflavone	Isopiscerythramine	451.51	C26H29O6N1	F177
811b	4'-Amino-5,7,3'-trihydroxy-5'-methoxy-8,2'-diprenylisoflavone	Torvanol A	452.43	C20H20O10S1	F11b
812	4-Sulfate-2'-hydroxy-6,3'-dimethoxy-5'-(2-propenoic acid)isoflavone	Erythbigenin	452.50	C26H28O7	F263
813	5,7,3',4'-Tetrahydroxy-5'-methoxy-2',6'-diprenylisoflavone	Lupinisoflavone M	456.49	C25H28O8	2952
814	5,7-Dihydroxy-6-(2,3-dihydroxy-3-methylbutyl)-5''-(1-hydroxy-1-methylethyl)-4'',5''-dihydrofurano[2'',3'':4',3']isoflavone				
814	5-Hydroxy-2'-methoxy-4',5-methylenedioxy-6-prenyl-6'',6''-dimethylpyrano[2'',3'':7,8]isoflavone	Euchrenone b3	462.50	C27H26O7	2972
815	5-Hydroxy-3',5'-dimethoxy-4-(1,1-dimethyl-2-propenyloxy)-5''-(1-methylethenyl)-4'',5''-dihydrofurano[2'',3'':7,6]isoflavone	Pumilaisoflavone B	464.52	C27H28O7	2920
816	5-Hydroxy-3',5'-dimethoxy-4-(1,1-dimethyl-2-propenyloxy)-6'',6''-dimethylpyrano[2'',3'':7,6]isoflavone	Pumilaisoflavone A	464.52	C27H28O7	2919
817	5,7-Dihydroxy-2'-methoxy-4',5'-methylenedioxy-6,8-diprenylisoflavone	Euchrenone b4	464.52	C27H28O7	2967
818	6-Methoxy-3',4'-methylenedioxy-7-(7-hydroxy-3,7-dimethyl-2(E),5'-octadienyloxy)isoflavone	Conrauinone A	464.52	C27H28O7	F68
819	5,4'-Dihydroxy-5''-(1-hydroxyethyloxy-1-methylethyl)-8-prenyl-4'',5''-dihydrofurano[2'',3'':7,6]isoflavone	Eriotriochin	466.53	C27H30O7	2971
820	5,7,4'-Trihydroxy-3',5'-dimethoxy-6,2-diprenylisoflavone	Pumilaisoflavone C	466.53	C27H30O7	2949
821	5,7,4'-Trihydroxy-5'-methoxy-5''-(1-hydroxy-1-methylethyl)-6-prenyl-4',5''-dihydrofurano[2'',3'':3',2']isoflavone	Erythbigenol A	468.50	C26H28O8	F264
822	5,7,4'-Trihydroxy-5'-methoxy-5''-(1-hydroxy-1-methylethyl)-6-prenyl-4',5''-dihydrofurano[2'',3'':3',2']isoflavone	Erythbigenol B	468.50	C26H28O8	F264
823	5,7,4'-Trihydroxy-5'-methoxy-6-prenyl-6'',6''-dimethyl-4'',5''-dihydropyrano[2'',3'':2']isoflavone	Erythbigenone B	468.50	C26H28O8	F264
824	5,7,4'-Trihydroxy-5'-methoxy-6-prenyl-6'',6''-dimethyl-4'',5''-dihydropyrano[2'',3'':2']isoflavone	Erythbigenone A	468.50	C26H28O8	F264
825	5-Hydroxy-2',4'-dimethoxy-8-(3-hydroxy-3-methylbutyl)-6'',6''-dimethyl-4'',5''-dihydropyrano[2'',3'':7,6]isoflavone		468.55	C27H32O7	2509
826	5-Hydroxy-3',4'-dimethoxy-8-(3-hydroxy-3-methylbutyl)-6'',6''-dimethyl-4'',5''-dihydropyrano[2'',3'':7,6]isoflavone		468.55	C27H32O7	2510
827	5,7,2'-Trihydroxy-6-(2,3-dihydroxy-3-methylbutyl)-5''-(1-hydroxy-1-methylethyl)-4'',5''-dihydrofurano[2'',3'':4',3']isoflavone	Lupinisoflavone N	472.49	C25H28O9	2953
828	5,7,4'-Trihydroxy-5''-((2E)-3,7-dimethyl-2,6-octadienyl)-2'-prenylisoflavone	Millewanin D	474.60	C30H34O5	F110
829	5,7,4'-Trihydroxy-6,8,3'-triprenylisoflavone	Euchrenone b1	474.60	C30H34O5	2976
830	5,7,4'-Trihydroxy-8,3',5'-triprenylisoflavone	Flemiphyllin	474.60	C30H34O5	2865

continued

APPENDIX
Checklist for Isoflavonoids Described in the Literature During the Period 1991–2004 — *continued*

	Name	Trivial Name	Mass	Formula	Ref.
831	5,2',4'-Trihydroxy-6,3'-diprenyl-6'',6''-dimethylpyrano[2'',3'':7,8]isoflavone	Euchrenone b16	488.57	C30H32O6	F175
832	5,3',4'-Trihydroxy-8-prenyl-6'',6''-dimethyl-5''-(1,1-dimethyl-2-propenyl)pyrano[2'',3'':7,6]isoflavone	Flemiphilippinin A	488.57	C30H32O6	F41
833	5,7,3',4'-Tetrahydroxy-5'-((2E)-3,7-dimethyl-2,6-octadienyl)-2'-prenylisoflavone	Millewanin C	490.59	C30H34O6	F110
834	5,7,2',4'-Tetrahydroxy-6,8,3'-triprenylisoflavone	Euchrenone b2	490.60	C30H34O6	2977
835	5,4'-Dihydroxy-2'-methoxy-6,3'-diprenyl-6'',6''-dimethylpyrano[2'',3'':7,8]isoflavone	Euchrenone b13	502.60	C31H34O6	F175
836	5,4'-Dihydroxy-2'-methoxy-8,3'-diprenyl-6'',6''-dimethylpyrano[2'',3'':7,6]isoflavone	Euchrenone b12	502.60	C31H34O6	F175
837	5,7,4'-Trihydroxy-2'-methoxy-6,8,3-triprenylisoflavone	Euchrenone b14	504.61	C31H36O6	F175
838	5,7,4'-Trihydroxy-3'-methoxy-5'-((2E)-3,7-dimethyl-2,6-octadienyl)-2'-prenylisoflavone	Millewanin B	504.61	C31H36O6	F110
839	5,5'-Dihydroxy-2'-methoxy-6-prenyl-(2'',3'':7,8),(2'',3'':4',3)-bis(6,6-dimethylpyrano)isoflavone	Euchrenone b11	516.48	C31H32O7	F175
840	5,5'-Dihydroxy-2'-methoxy-8-prenyl-(2'',3'':7,6),(2'',3'':4',3)-bis(6,6-dimethylpyrano)isoflavone	Euchrenone b10	516.48	C31H32O7	F175
841	5,7-Dihydroxy-8-(2-hydroxy-3-methyl-3-butenyl)-4'-(1-(hydroxymethyl)pentacosyloxy)-6-prenylisoflavone	Indicanin D	803.16	C51H78O7	F184

Isoflavone glycosides

	Name	Trivial Name	Mass	Formula	Ref.
842	7,4'-Dihydroxyisoflavone 7-O-(2-O-methylrhamnoside)	Daidzein G 3	414.41	C22H22O8	F93
843	5,7,4'-Trihydroxyisoflavone 5-O-rhamnoside	Genestein G 1	416.39	C21H20O9	F93
844	5,7,4'-Trihydroxyisoflavone 7-O-rhamnoside	Genistein 4'-O-rhamnoside	416.39	C21H20O9	F224
845	7,4'-Dihydroxyisoflavone 4'-O-glucoside	Daidzein 4'-O-glucoside	416.39	C21H20O9	3110
846	7,4'-Dihydroxyisoflavone 7-O-glucoside	Daidzin 7-O-glucoside	416.39	C21H20O9	3108
847	7,8-Dihydroxy-4'-methoxyisoflavone 7-O-arabinoside	Retusin 7-O-arabinoside	416.39	C21H20O9	3215
848	4'-Hydroxy-7-methoxyisoflavone-4'-O-glucoside	Isononin	430.41	C22H22O9	F351
849	5,7-Dihydroxy-6-methoxyisoflavone 7-O-rhamnoside		430.41	C22H22O9	3191
850	7-Hydroxy-4'-methoxyisoflavone 7-O-glucoside	Formononetin 7-O-glucoside	430.41	C22H22O9	3113
851	5,7,2'-Trihydroxyisoflavone 7-O-glucoside	Isogenistein 7-O-glucoside	432.39	C21H20O10	3148
852	5,7,4'-Trihydroxyisoflavone 4'-O-glucoside	Genistein 4'-O-glucoside	432.39	C21H20O10	3141
853	5,7,4'-Trihydroxyisoflavone 5-O-glucoside	Genistein 5-O-glucoside	432.39	C21H20O10	3207
854	5,7,4'-Trihydroxyisoflavone 7-O-glucoside	Genistin 7-O-glucoside	432.39	C21H20O10	3139
855	6,7,4'-Trihydroxyisoflavone 4'-O-glucoside	Demethyltexasin 4'-O-glucoside	432.39	C21H20O10	3117
856	7,2',4'-Trihydroxyisoflavone-4'-O-glucoside		432.39	C21H20O10	F179
857	7-Hydroxy-3',4'-methylenedioxyisoflavone 7-O-glucoside	Pseudobaptigenin 7-O-glucoside	444.40	C22H20O10	3118

No.	Name	Systematic name	Formula	MW	Code
858	Daidzein G 2	7,4'-Dihydroxy-6-methoxyisoflavone 7-O-(2-O-methylrhamnoside)	C23H24O9	444.43	F93
859	Prunetin 4'-O-galactoside	5,4'-Dihydroxy-7-methoxyisoflavone 4'-O-galactoside	C22H22O10	446.41	3213
860	Prunetin 4'-O-glucoside	5,4'-Dihydroxy-7-methoxyisoflavone 4'-O-glucoside	C22H22O10	446.41	3155
861	Prunetin 5-O-glucoside	5,4'-Dihydroxy-7-methoxyisoflavone 5-O-glucoside	C22H22O10	446.41	3212
862	Mutabilin	5,7-Dihydroxy-3'-methoxyisoflavone 7-O-glucoside	C22H22O10	446.41	F55
863	Biochanin A 7-O-glucoside	5,7-Dihydroxy-4'-methoxyisoflavone 7-O-glucoside	C22H22O10	446.41	3149
864	4'-O-Methyl-genistein-8-C-glucoside	5,7-Dihydroxy-4'-methoxyisoflavone 8-C-glucoside	C22H22O10	446.41	F2
865	Texasin 7-O-glucoside	6,7-Dihydroxy-4'-methoxyisoflavone 7-O-glucoside	C22H22O10	446.41	3123
866	Calycosin 7-O-galactoside	7,3'-Dihydroxy-4'-methoxyisoflavone 7-O-galactoside	C22H22O10	446.41	3205
867	Calycosin 7-O-glucoside	7,3'-Dihydroxy-4'-methoxyisoflavone 7-O-glucoside	C22H22O10	446.41	3121
868	Genistein 5-methyl ether 4'-glucoside	7,4'-Dihydroxy-5-methoxyisoflavone 4'-O-glucoside	C22H22O10	446.41	F194
869	7-O-Methylgenistein 7-O-glucoside	7,4'-Dihydroxy-5-methoxyisoflavone 7-O-glucoside	C22H22O10	446.41	3211
870	Glycetein 7-O-glucoside	7,4'-Dihydroxy-6-methoxyisoflavone 7-O-glucoside	C22H22O10	446.41	3124
871		7,5'-Hydroxy-3'-methoxyisoflavone-7-O-glucoside	C22H22O10	446.41	F34
872	Retusin 7-O-glucoside	7,8-Dihydroxy-4'-methoxyisoflavone 7-O-glucoside	C22H22O10	446.41	3190
873	Derriscandenoside A	7,8-Dihydroxy-4'-methoxyisoflavone 8-O-glucoside	C22H22O10	446.41	F218
874	2'-Hydroxygenistein 4'-O-glucoside	5,7,2',4'-Tetrahydroxyisoflavone 4'-O-glucoside	C21H20O11	448.39	F240
875	2'-Hydroxygenistein 7-O-glucoside	5,7,2',4'-Tetrahydroxyisoflavone 7-O-glucoside	C21H20O11	448.39	F240
876		5,7,2',4'-Tetrahydroxyisoflavone 8-C-glucoside	C21H20O11	448.39	F234
877	Orobol 7-O-glucoside	5,7,3',4'-Tetrahydroxyisoflavone 7-O-glucoside	C21H20O11	448.39	3156
878	Daidzin 6''-O-acetate	7,4'-Dihydroxyisoflavone 7-O-(6''-acetylglucoside)	C23H22O10	458.42	3109
879	Irilone 4'-O-glucoside	7-Hydroxy-5,4'-dimethoxy-8-methylisoflavone 7-O-rhamnoside	C24H26O9	458.47	3197
880	6-Hydroxypseudobaptigenin 7-O-glucoside	5,4'-Dihydroxy-6,7-methylenedioxyisoflavone 4'-O-glucoside	C22H20O11	460.39	3161
881		5,7-Dihydroxy-3',4'-methylenedioxyisoflavone 7-O-glucoside	C22H20O11	460.39	3159
882		5,7-Dihydroxy-3',4'-dimethoxyisoflavone 7-O-rhamnoside	C23H24O10	460.44	F61
883	Irisolidone 7-O-rhamnoside	5,7-Dihydroxy-6,2'-dimethoxyisoflavone 7-O-rhamnoside	C23H24O10	460.44	F226
884	Cladrin 7-O-glucoside	5,7-Dihydroxy-6,4'-dimethoxyisoflavone 7-O-rhamnoside	C23H24O10	460.44	3192
885	Afrormosin 7-O-galactoside	7-Hydroxy-3',4'-dimethoxyisoflavone 7-O-glucoside	C23H24O10	460.44	3126
886	Afrormosin 7-O-glucoside	7-Hydroxy-6,4'-dimethoxyisoflavone 7-O-galactoside	C23H24O10	460.44	F198
887	8-O-Methylretusin 7-O-glucoside	7-Hydroxy-6,4'-dimethoxyisoflavone 7-O-glucoside	C23H24O10	460.44	3128
888		7-Hydroxy-8,4'-dimethoxyisoflavone 7-O-glucoside	C23H24O10	460.44	3131
889		5,3',4'-Trihydroxy-7-methoxyisoflavone 3'-O-glucoside	C22H22O11	462.41	F129
890	Pratensein 7-O-glucoside	5,7,3'-Trihydroxy-4'-methoxyisoflavone 7-O-glucoside	C22H22O11	462.41	3162
891		5,7,3',4'-Tetrahydroxy-8-methylisoflavone 7-O-glucoside	C22H22O11	462.41	F61

continued

APPENDIX
Checklist for Isoflavonoids Described in the Literature During the Period 1991–2004 — *continued*

	Name	Trivial Name	Mass	Formula	Ref.
892	5,7,4'-Trihydroxy-3'-methoxyisoflavone 7-O-glucoside	3'-O-Methylorobol 7-O-glucoside	462.41	C22H22O11	3163
893	5,7,4'-Trihydroxy-6-methoxyisoflavone 4'-O-glucoside	Tectorigenin 4'-glucoside	462.41	C22H22O11	F236
894	5,7,4'-Trihydroxy-6-methoxyisoflavone 7-O-glucoside	Tectorigenin 7-O-glucoside	462.41	C22H22O11	3164
895	5,7,4'-Trihydroxy-8-methoxyisoflavone 7-O-glucoside	Isotectorigenin 7-O-glucoside	462.41	C22H22O11	3217
896	7-Hydroxy-4'-methoxyisoflavone 7-O-(6''-acetylglucoside)	Formononetin 7-O-(6''-acetylglucoside)	472.45	C24H24O10	3201
897	4'-Hydroxy-5-methoxy-6,7-methylenedioxyisoflavone 4'-O-glucoside	Germanaism B	474.42	C23H22O11	F17
898	5,7,4'-Trihydroxyisoflavone 7-O-(6''-acetylglucoside)	Genistin 6''-O-acetate	474.42	C23H22O11	3140
899	6,3'-Dihydroxy-7-methoxy-4',5'-methylenedioxyisoflavone 6-O-rhamnoside		474.42	C23H22O11	F356
900	6,7-Dihydroxy-3'-methoxy-4',5'-methylenedioxyisoflavone 6-O-rhamnoside		474.42	C23H22O11	F312
901	7-Hydroxy-6-methoxy-3',4'-methylenedioxyisoflavone 7-O-glucoside	Fujikinetin 7-O-glucoside	474.42	C23H22O11	3133
902	7-Hydroxy-6'-methoxy-3',4'-methylenedioxyisoflavone 7-O-glucoside	6'-Methoxypseudobaptigenin 7-O-glucoside	474.42	C23H22O11	F137
903	5,7,2'-Trihydroxy-4',5'-methylenedioxyisoflavone 2'-O-glucoside		476.39	C22H20O12	F197
904	5,4'-Dihydroxy-6,7-dimethoxyisoflavone 4'-O-galactoside	7-O-Methyltectorigenin 4'-O-galactoside	476.44	C23H24O11	3216
905	5,4'-Dihydroxy-6,7-dimethoxyisoflavone 4'-O-glucoside	7-O-Methyltectorigenin 4'-O-glucoside	476.44	C23H24O11	3169
906	5,7-Dihydroxy-6,4'-dimethoxyisoflavone 7-O-galactoside		476.44	C23H24O11	F231
907	5,7-Dihydroxy-6,4'-dimethoxyisoflavone 7-O-glucoside	Irisolidone 7-O-glucoside	476.44	C23H24O11	3167
908	5,7-Dihydroxy-8,4'-dimethoxyisoflavone 7-O-glucoside		476.44	C23H24O11	F7
909	7,2'-Dihydroxy-3',4'-dimethoxyisoflavone 7-O-glucoside		476.44	C23H24O11	3206
910	7,3'-Dihydroxy-6,4'-dimethoxyisoflavone 7-O-glucoside	Odoratin-7-O-glucoside	476.44	C23H24O11	F302
911	5,6,7,4'-Tetrahydroxy-8-methoxyisoflavone 7-O-glucoside		478.41	C22H22O12	F225
912	7,4'-Dihydroxy-6-methoxyisoflavone 7-O-(6''-acetylglucoside)	Glycitein 7-O-(6''-acetylglucoside)	488.45	C24H24O11	F131
913	5,4'-Dihydroxy-2'-methoxy-6,7-methylenedioxyisoflavone 4'-O-glucoside	Iriflogenin 4'-O-glucoside	490.32	C23H22O12	3172
914	5,7-Hydroxy-6-methoxy-3',4'-methylenedioxyisoflavone 7-O-galactoside	Dalspinin 7-O-galactoside	490.32	C23H22O12	3195
915	6,3'-Dihydroxy-7-methoxy-4',5'-methylenedioxyisoflavone 6-O-glucoside		490.32	C23H22O12	F356
916	6,7-Dihydroxy-3'-methoxy-4',5'-methylenedioxyisoflavone 6-O-glucoside		490.32	C23H22O12	F312
917	7-Hydroxy-2',4',5'-trimethoxyisoflavone 7-O-glucoside		490.47	C24H26O11	3135
918	7-Hydroxy-5,6,4'-trimethoxyisoflavone 7-O-glucoside	5-Methoxyafrormosin 7-O-glucoside	490.47	C24H26O11	3173
919	7-Hydroxy-6,3',5'-trimethoxyisoflavone 7-O-glucoside	Cladrastin 7-O-glucoside	490.47	C24H26O11	3136
920	5,7,2'-Trihydroxy-6,4'-dimethoxyisoflavone 7-O-glucoside	Iristectorigenin A 7-O-glucoside	492.44	C23H24O12	3175
921	5,7,4'-Trihydroxy-6,3'-dimethoxyisoflavone 7-O-glucoside	Iristectorigenin B 7-O-glucoside	492.44	C23H24O12	3176
922	5,7,4'-Trihydroxy-8,3'-dimethoxyisoflavone 7-O-glucoside	Homotectorigenin 7-O-glucoside	492.44	C23H24O12	3177

No.	Name	Trivial name	Formula	MW	Code
922b	7,2',5'-Trihydroxy-6,4'-dimethoxyisoflavone 2'-O-glucoside	Licoagroside A	C23H24O12	492.44	F141b
923	5,7,4'-Trihydroxy-6-prenylisoflavone 7-O-glucoside	Genisteone	C26H28O10	500.49	F202
924	7,4'-Dihydroxyisoflavone 7-O-(6''-malonylglucoside)	Daidzin 6''-O-malonate	C24H22O12	502.43	F135
925	7,4'-Dihydroxyisoflavone 8-C-(6''-malonylglucoside)	Puerarin 6''-O-malonate	C24H22O12	502.43	F135
926	4'-Hydroxy-5,3'-dimethoxy-6,7-methylenedioxyisoflavone 4'-O-glucoside	Germanaism A	C24H24O12	504.45	F17
927	7-Hydroxy-2',6-dimethoxy-4',5'-methylenedioxyisoflavone 7-O-glucoside	Dalpatein 7-O-glucoside	C24H24O12	504.45	3138
928	7-Hydroxy-5,6-dimethoxy-3',4'-methylenedioxyisoflavone 7-O-glucoside	Isoplatycarpanetin 7-O-glucoside	C24H24O12	504.45	3178
929	7-Hydroxy-5,8-dimethoxy-3',4'-methylenedioxyisoflavone 7-O-glucoside	Platycarpanetin 7-O-glucoside	C24H24O12	504.45	3179
930	5,2',3'-Trihydroxy-7-methoxy-2-hydroxymethyl-6-methylisoflavone 3'-O-glucoside	Mirabijalone C	C24H26O12	506.46	F316
931	5,3'-Dihydroxy-7,4',5'-trimethoxyisoflavone 3'-O-glucoside	Vavain 3'-O-glucoside	C24H26O12	506.46	F189
932	5,7-Dihydroxy-6,3',4'-trimethoxyisoflavone 7-O-glucoside	Junipegenin 7-O-glucoside	C24H26O12	506.46	3194
933	6,8,3',5'-Tetrahydroxy-7,4'-dimethoxyisoflavone 6-O-glucoside		C23H24O13	508.43	F208
934	7-Hydroxy-4'-methoxyisoflavone 7-O-(2'',6''-diacetylglucoside)	2'',6''-O-Diacetylonimin	C26H26O11	514.48	F92
935	5,7,4'-Trihydroxyisoflavone 7-O-(6''-succinylglucoside)	Genistein 7-O-(6''-succinylglucoside)	C25H24O12	516.45	F294
936	7-Hydroxy-4'-methoxyisoflavone 7-O-(6''-malonylglucoside)	Formononetin 7-O-(6''-malonylglucoside)	C25H24O12	516.46	3116
937	5,7,2',4'-Tetrahydroxy-3'-prenylisoflavone 4'-O-glucoside	licoisoflavone A 4'-O-glucoside	C26H28O11	516.49	F240
938	5,7,2',4'-Tetrahydroxy-3'-prenylisoflavone 7-O-glucoside	licoisoflavone A 7-O-glucoside	C26H28O11	516.49	F240
939	5,7,2',4'-Tetrahydroxy-6-prenylisoflavone 7-O-glucoside	luteone 7-O-glucoside	C26H28O11	516.49	F240
940	5,7,4'-Trihydroxyisoflavone 4'-O-(6''-malonylglucoside)	Genistein 7-O-(6''-O-malonylglucoside)	C24H22O13	518.43	F240
941	5,7,4'-Trihydroxyisoflavone 7-O-(6''-malonylglucoside)	Genistein 7-O-(6''-malonylglucoside)	C24H22O13	518.43	3147
942	5,7,4'-Trihydroxyisoflavone 8-C-(6''-malonylglucoside)	Genistein 8-C-glucoside 6''-O-malonate	C24H22O13	518.43	F135
943	5,7,3'-Trihydroxy-6,4',5'-trimethoxyisoflavone 7-O-glucoside	Irigenin 7-O-glucoside	C24H26O13	522.46	3181
944	5,7,4'-Trihydroxy-6,3',5'-trimethoxyisoflavone 7-O-glucoside		C24H26O13	522.46	3196
945	5,7-Dihydroxy-4'-methoxyisoflavone 7-O-(6''-malonylglucoside)	Biochanin A 7-O-(6''-malonylglucoside)	C25H24O13	532.46	3154
946	7,4'-Dihydroxyisoflavone 7-O-(6''-succinylglucoside)	Daidzein 7-O-(6''-succinylglucoside)	C25H24O13	532.46	F294
947	5,7,2',4'-Tetrahydroxyisoflavone 4'-O-(6''-malonylglucoside)	2'-Hydroxygenisteine 7-O-(6''-malonylglucoside)	C24H22O14	534.43	F240
948	5,7,2',4'-Tetrahydroxyisoflavone 7-O-(6''-malonyl)glucoside	2'-Hydroxygenistein 7-O-(6''-malonylglucoside)	C24H22O14	534.43	F240
949	5,7,3',4'-Trihydroxyisoflavone 7-O-(6''-malonylglucoside)	Orobol 7-O-(6''-malonylglucoside)	C24H22O14	534.43	3187
950	5,7-Dihydroxy-6,2',4',5'-tetramethoxyisoflavone 7-O-glucoside	Caviunin 7-O-glucoside	C25H28O13	536.49	3182
951	5,7-Dihydroxy-6,2',4',5'-tetramethoxyisoflavone 8-C-glucoside	Dalpaniculin	C25H28O13	536.49	F214
952	5,7-Dihydroxy-8,2',4',5'-tetramethoxyisoflavone 7-O-glucoside	Isocaviunin 7-O-glucoside	C25H28O13	536.49	3184
953	7,5'-Dihydroxy-5,4'-dimethoxy-3'-prenylisoflavone 7-O-galactoside		C28H32O11	544.55	F328
954	7-Hydroxy-6,4'-dimethoxyisoflavone 7-O-(6''-malonylglucoside)	Afrormosin 7-O-(6''-malonylglucoside)	C26H26O13	546.48	3214
955	7,4'-Dihydroxy-6-methoxyisoflavone 7-O-(6''-succinylglucoside)	Glycitein 7-O-(6''-succinylglucoside)	C26H26O13	546.48	F294

continued

APPENDIX
Checklist for Isoflavonoids Described in the Literature During the Period 1991–2004 — continued

	Name	Trivial Name	Mass	Formula	Ref.
956	5,7,3'-Trihydroxy-4'-methoxyisoflavone 7-O-(6''-malonylglucoside)	Pratensein 7-O-(6''-malonylglucoside)	548.46	C25H24O14	3189
957	7,4'-Dihydroxyisoflavone 7-O-(6''-apiosylglucoside)	Daidzein 7-O-(6''-apiosylglucoside)	548.50	C26H28O13	3200
958	7,4'-Dihydroxyisoflavone 7-O-glucoside-4'-O-apioside	Daidzein 7-O-glucoside-4'-O-apioside	548.50	C26H28O13	3199
959	7-Hydroxy-4'-methoxyisoflavone 7-O-(2''-p-hydroxybenzoylglucoside)	Formononetin 7-O-(2''-p-hydroxybenzoylglucoside)	550.52	C29H26O11	3204
960	5,7,4'-Trihydroxyisoflavone 7,4'-bis(3-cymaroside)	Genistein 4',7-O-bis(3-cymaroside)	558.57	C29H34O11	F159
961	7-Hydroxy-4'-methoxyisoflavone 7-O-(2''-apiosylglucoside)	Formononetin 7-O-(6''-apiosylglucoside)	562.53	C27H30O13	3202
962	7-Hydroxy-4'-methoxyisoflavone 7-O-(6''-xylosylglucoside)	Formononetin 7-O-(6''-xylosylglucoside)	562.53	C27H30O13	3203
963	7-Hydroxy-4'-methoxyisoflavone 7-O-(rhamnosylglucoside)	Daidzein 7-O-rhamnosylglucoside	562.53	C27H30O13	3112
964	5,7,4'-Trihydroxyisoflavone 7-O-(6''-apiosylglucoside)	Genistein 7-O-(6''-apiosylglucoside)	564.50	C26H28O14	3208
965	5,7,4'-Trihydroxyisoflavone 7-O-(6''-arabinosylglucoside)	Genistin 6''-O-arabinosyl	564.50	C26H28O14	F323
966	5,7,4'-Trihydroxyisoflavone 7-O-(6''-xylosylglucoside)	Genistin 6''-O-xylosyl	564.50	C26H28O14	F323
967	5,7,4'-Trihydroxyisoflavone 7-O-glucoside-4'-O-apioside	Genistein 7-O-glucoside-4'-O-apioside	564.50	C26H28O14	3210
968	5,7,4'-Trihydroxyisoflavone 7,4'-O-di(2-O-methylrhamnoside)	Daidzein G 1	574.58	C29H34O12	F93
969	5,7,4'-Trihydroxyisoflavone 5-O-rhamnoside-7-O-(2-O-methylrhamnoside)	Genestein G 2	576.56	C28H32O13	F93
970	7-Hydroxy-4'-methoxyisoflavone 7-O-(6''-rhamnosylglucoside)	Formononetin 7-O-rutinoside	576.56	C28H32O13	3114
971	4',7-Dihydroxy-6-methoxyisoflavone 7-O-(6''-xylosylglucoside)		578.53	C27H30O14	F195
972	5,7-Dihydroxy-4'-methoxyisoflavone 7-O-(2''-apiosylglucoside)	Coromandelin	578.53	C27H30O14	F211
973	5,7-Dihydroxy-4'-methoxyisoflavone 7-O-(6''-apiosylglucoside)	Biochanin A 7-O-(6''-apiosylglucoside)	578.53	C27H30O14	3153
974	5,7-Dihydroxy-4'-methoxyisoflavone 7-O-(6''-xylosylglucoside)	Biochanin A 7-O-(6''-xylosylglucoside)	578.53	C27H30O14	3152
975	5,7,4'-Trihydroxyisoflavone 4'-O-(2''-rhamnosylglucoside)	Genistein 7-O-neohesperidoside	578.53	C27H30O14	3145
976	5,7,4'-Trihydroxyisoflavone 7-O-(2''-p-coumaroylglucoside)	Genistein 7-O-(2''-p-coumaroylglucoside)	578.53	C30H26O12	3209
977	5,7,4'-Trihydroxyisoflavone 7-O-(6''-rhamnosylglucoside)	Genistein 7-O-rutinoside	578.53	C27H30O14	3144
978	7,4'-Dihydroxy-6-methoxyisoflavone 7-(6''-xylosylglucoside)	Tectorigenin 7-O-(6''-xylosylglucoside)	578.53	C27H30O14	F188
979	7,4'-Dihydroxyisoflavone 7-O-glucoside-4'-O-glucoside	Daidzein 7,4'-O-diglucoside	578.53	C27H30O14	3111
980	5,2'-Dihydroxy-8-prenyl-6'',6''-dimethylpyrano[2',3'':7,6]isoflavone 4'-O-glucoside	Auriculatin 4'-O-glucoside	582.60	C31H34O11	F183
981	7-Hydroxy-3',4'-methylenedioxyisoflavone 7-O-(rhamnosylglucoside)	Pseudobaptigenin 7-O-rhamnosylglucoside	590.54	C28H30O14	3119
982	4',5-Dihydroxy-7-methoxyisoflavone 3'-O-(3''-E-cinnamoylglucoside)		592.55	C31H28O12	F128
983	5,7-Dihydroxy-4'-methoxyisoflavone 7-O-(6''-rhamnosylglucoside)	Biochanin A 7-O-rutinoside	592.56	C28H32O14	3150
984	7-Hydroxy-4'-methoxyisoflavone 7-O-(3''-glucosylglucoside)	Formononetin 7-O-laminariobioside	592.56	C28H32O14	3115
985	7-Hydroxy-4'-methoxyisoflavone 7-O-(6''-glucosylglucoside)	Formononetin 7-O-gentiobioside	592.56	C28H32O14	F96
986	7-Hydroxy-6,4'-dimethoxyisoflavone 7-O-(2''-apiosylglucoside)	Afromosin 7-O-(2''-apiosylglucoside)	592.56	C28H32O14	F296

No.	Name	Trivial name	MW	Formula	Code
987	7,3'-Dihydroxy-4'-methoxyisoflavone 7-O-(rhamnosylglucoside)	Calycosin 7-O-rhamnosylglucoside	592.56	C28H32O14	3122
988	7,8-Dihydroxy-4'-methoxyisoflavone 7-O-(2''-rhamnosylglucoside)	Retusin 7-O-neohesperidoside	592.56	C28H32O14	3125
989	7,8-Dihydroxy-4'-methoxyisoflavone 7-O-(6''-rhamnosylglucoside)	Derriscandenoside B	592.56	C28H32O14	F218
990	4',7-Dihydroxy-6-methoxyisoflavone 7-O-(6''-xylosylglucoside)		594.53	C27H30O15	F195
991	5,6,7,4'-Tetrahydroxyisoflavone 7-O-(rhamnosylglucoside)	6-Hydroxygenistein 7-O-rhamnosylglucoside	594.53	C27H30O15	3158
992	5,7,2',4'-Tetrahydroxyisoflavone 6-C-(2''-rhamnosylglucoside)	Nodosin	594.53	C27H30O15	F108
993	5,7,3',4'-Tetrahydroxyisoflavone 7-O-(rhamnosylglucoside)	Orobol 7-O-rhamnosylglucoside	594.53	C27H30O15	3157
994	5,7,4'-Trihydroxyisoflavone 7-O-(6''-apiosylglucoside)	Tectorigenin 7-O-(6''-apiosylglucoside)	594.53	C27H30O15	F65
995	5,7,4'-Trihydroxyisoflavone 7-O-glucosylglucoside	Genistein 7-O-glucosylglucoside	594.53	C27H30O15	3142
996	5,7,4'-Trihydroxyisoflavone 7-O-glucoside-4'-O-glucoside	Genistein 7,4'-O-glucoside	594.53	C27H30O15	3143
997	5,7,4'-Trihydroxyisoflavone 8-C-glucoside-4'-O-glucoside	Genistein 8-C-glucoside-4'-O-glucoside	594.53	C27H30O15	F319
998	7,2',4'-Trihydroxyisoflavone 7-O-glucoside-4'-O-glucoside	2'-Hydroxydaidzein 7,4'-O-diglucoside	594.53	C27H30O15	3186
999	7-Hydroxy-3',4'-methylenedioxyisoflavone 7-O-(3''-glucosylglucoside)	Pseudobaptigenin 7-O-laminariobioside	606.54	C28H30O15	3120
1000	7-Hydroxy-6,4'-dimethoxyisoflavone 7-O-(6''-rhamnosylglucoside)	Derriscandenoside D	606.58	C29H34O14	F218
1001	7-Hydroxy-8,4'-dimethoxyisoflavone 7-O-(6''-rhamnosylglucoside)	Derriscanoside B	606.58	C29H34O14	F140
1002	8-Hydroxy-7,4'-dimethoxyisoflavone 8-O-(6''-rhamnosylglucoside)	Derriscandenoside C	606.58	C29H34O14	F218
1003	5,4'-Dihydroxy-6,7-dimethoxyisoflavone 4'-O-(6''-apiosylglucoside)		608.54	C28H32O15	F160
1004	5,7-Dihydroxy-4'-methoxyisoflavone 7-O-(6''-glucosylglucoside)	Biochanin A 7-O-gentiobioside	608.54	C28H32O15	3151
1005	5,7-Dihydroxy-6,4'-dimethoxyisoflavone 7-O-(2''-apiosylglucoside)	Pubescidin	608.54	C28H32O15	F295
1006	5,7-Dihydroxy-6,4'-dimethoxyisoflavone 7-O-(xylosylglucoside)	Irisolidone 7-O-xylosylglucoside	608.54	C28H32O15	3168
1007	7,4'-Dihydroxy-5-methoxyisoflavone 7,4'-O-diglucoside	Isoprunetin 7,4'-O-diglucoside	608.54	C28H32O15	F319
1008	5,6,7,4'-Tetrahydroxyisoflavone 4'-O-(6''-glucosylglucoside)	Germanaism D	610.53	C27H30O16	F16
1009	5,7,2',4'-Tetrahydroxyisoflavone 7,4'-O-diglucoside	2'-Hydroxygenistein 7,4'-O-diglucoside	610.53	C27H30O16	F319
1010	5,7,3',4'-Tetrahydroxyisoflavone 7-O-(2''-glucosylglucoside)	Orobol 7-O-sophoroside	610.53	C27H30O16	3188
1011	5,7-Dihydroxy-3',4'-methylenedioxyisoflavone 7-O-(glucosylglucoside)		622.54	C28H30O16	3160
1012	6,3'-Dihydroxy-7-methoxy-4',5'-methylenedioxyisoflavone 6-O-(6''-xylosylglucoside)		622.54	C28H30O16	F356
1013	6,7-Dihydroxy-3'-methoxy-4',5'-methylenedioxyisoflavone 6-O-(6''-xylosylglucoside)		622.54	C28H30O16	F312
1014	5,4'-Dihydroxy-6,7-dimethoxyisoflavone 4'-O-(rhamnosylglucoside)	7-O-Methyltectorigenin 4'-O-rhamnosylglucoside	622.58	C29H34O15	3170
1015	5,7-Dihydroxy-6,4'-dimethoxyisoflavone 7-O-(6''-O-rhamnosylglucoside)	Derriscandenoside E	622.58	C29H34O15	F218
1016	7-Hydroxy-3',4'-dimethoxyisoflavone 7-O-(3''-glucosylglucoside)	Cladrin 7-O-laminariobioside	622.58	C29H34O15	3127
1017	7-Hydroxy-6,4'-dimethoxyisoflavone 7-O-(3''-glucosylglucoside)	Afrormosin 7-O-laminariobioside	622.58	C29H34O15	3130
1018	7-Hydroxy-8,4'-dimethoxyisoflavone 7-O-(3''-glucosylglucoside)	8-O-Methylretusin 7-O-laminariobioside	622.58	C29H34O15	3132
1019	5,7,4'-Trihydroxy-6-methoxyisoflavone 4'-O-(6''-glucosylglucoside)	Tectorigenin 4'-O-(6''-glucosylglucoside)	624.56	C28H32O16	F66
1020	5,7,4'-Trihydroxy-6-methoxyisoflavone 7-O-(6''-glucosylglucoside)	Tectorigenin 7-O-gentiobioside	624.56	C28H32O16	3165

continued

APPENDIX
Checklist for Isoflavonoids Described in the Literature During the Period 1991–2004 — *continued*

Name	Trivial Name	Mass	Formula	Ref.
1021	5,7,4′-Trihydroxy-6-methoxyisoflavone 7-O-glucoside-4′-O-glucoside	Tectorigenin 7,4′-O-diglucoside	624.56	C28H32O16	F250
1022	4′-Hydroxy-5-methoxy-6,7-methylenedioxyisoflavone 4′-O-(6″-glucosylglucoside)	Germanaism E	636.56	C29H32O16	F16
1023	7-Hydroxy-6-methoxy-3′,4′-methylenedioxyisoflavone 7-O-(3″-glucosylglucoside)	Fujikinetin 7-O-laminariobioside	636.56	C29H32O16	3134
1024	5,3′-Dihydroxy-7,4′,5′-trimethoxyisoflavone 3′-O-(6″-arabinosylglucoside)		638.58	C29H34O16	F298
1025	5,4′-Dihydroxy-6,7-dimethoxyisoflavone 4′-O-(6″-glucosylglucoside)	7-O-Methyltectorigenin 4′-O-gentiobioside	638.58	C29H34O16	3171
1026	5,7,3′-Trihydroxy-6,4′,5′-trimethoxyisoflavone 7-(6″-(4-hydroxybenzoyl)glucoside)	6″-O-p-Hydroxybenzoyliridin	642.56	C31H30O15	F112
1027	7-Hydroxy-4′-methoxy-3′-prenylisoflavone 7-O-(2″-p-coumaroylglucoside)	4′-O-Methylneobavaisoflavone 7-O-(2″-p-coumaroylglucoside)	644.68	C36H36O11	3218
1028	5-Hydroxy-7,3′,4′-trimethoxy-8-methylisoflavone 5-O-(2″-rhamnosylglucoside)		650.64	C31H38O15	3198
1029	5,7-Dihydroxy-6,3′,4′-trimethoxyisoflavone 7-O-(6″-rhamnosylglucoside)		652.61	C30H36O16	F326
1030	7-Hydroxy-5,6,4′-trimethoxyisoflavone 7-O-(3″-glucosylglucoside)	5-Methoxyafrormosin 7-O-laminariobioside	652.61	C30H36O16	3174
1031	7-Hydroxy-6,3′,4′-trimethoxyisoflavone 7-O-(3″-glucosylglucoside)	Cladrastin 7-O-laminariobioside	652.61	C30H36O16	3137
1032	5,7,2′-Trihydroxy-6,4′-dimethoxyisoflavone 7-O-(6″-glucosylglucoside)	Iristectorigenin 7-O-gentiobioside	654.58	C29H34O17	3193
1033	5,7,4′-Trihydroxy-6,3′-dimethoxyisoflavone 7-O-(6″-glucosylglucoside)	Iristectorigenin B 7-O-(6″-glucosylglucoside)	654.58	C29H34O17	F66
1034	7-Hydroxy-5,8-dimethoxy-3′,4′-methylenedioxyisoflavone 7-O-(3″-glucosylglucoside)	Platycarpanetin 7-O-laminariobioside	666.59	C30H34O17	3180
1035	5,7,3′-Trihydroxy-6,4′,5′-trimethoxyisoflavone 7-(6″-O-(4-hydroxy-3-methoxybenzoyl)glucoside)	Shegansu A	672.59	C32H32O16	F357
1036	7-Hydroxy-5,6,2′-trimethoxyisoflavone 7-O-(2″-p-coumaroylglucoside)		680.61	C34H32O15	F329
1037	5,7-Dihydroxy-6,2′,4′,5′-tetramethoxyisoflavone 7-O-(rhamnosylglucoside)	Caviunin 7-O-rhamnosylglucoside	682.64	C31H38O17	3183
1038	5,7,3′-Trihydroxy-6,4′,5′-trimethoxyisoflavone 7-O-(6″-glucosylglucoside)	Germanaism C	684.60	C30H36O18	F16
1039	5,7-Dihydroxy-8,2′,4′,5′-tetramethoxyisoflavone 7-O-(6″-glucosylglucoside)	Isocaviunin 7-O-gentiobioside	698.64	C31H38O18	3185
1040	5,7-Dihydroxy-4′-methoxyisoflavone 7-O-(6″-(5‴-(apiosyl)-apiosyl)-glucoside)	Biochanin A 7-O-(6″-(5‴-(apiosyl)apiosyl)glucoside)	710.63	C32H38O18	F45
1041	5,7,2′,4′-Tetrahydroxy-6,3′-diprenylisoflavone 5-O-(4″-rhamnosylrhamnoside)		714.75	C37H46O14	F327
1042	7-Hydroxy-4′-methoxyisoflavone 7-O-(2″,6″-di-O-E-p-coumaroyl)-glucoside	Formononetin 7-O-(2″,6″-di-O-E-p-coumaroyl)glucoside	722.69	C40H34O13	F251
1043	5,7,4′-Trihydroxyisoflavone 7-O-rhamnoside-4′-O-(2″-rhamnosylglucoside)		724.66	C33H40O18	F290
1044	5,7,4′-Trihydroxyisoflavone 7-O-glucoside-4′-O-(2″-rhamnosylglucoside)		740.66	C33H40O19	F290
1045	5,7,4′-Trihydroxyisoflavone 7-O-rhamnoside-4′-O-(2″-glucosylglucoside)		740.66	C33H40O19	F290

No.	Name	Trivial name	Formula	MW	Ref.
1046	5,7,4'-Trihydroxyisoflavone 7-O-glucoside-4'-O-(2''-glucosyl)glucoside)		C33H40O20	756.66	F290
1047	4'-Hydroxy-5-methoxy-6,7-methylenedioxyisoflavone 4'-O-(2''-glucosyl-6''-rhamnosylglucoside)	Germanaism F	C35H42O20	782.70	F16
1048	3',4'-Dihydroxy-5-methoxy-6,7-methylenedioxyisoflavone 3'-O-glucoside-4'-O-(2''-O-(4''''-acetyl-2''''-methoxyphenyl)glucoside)	Germanaism G	C38H40O19	800.71	F16
1049	5,7,4'-Trihydroxyisoflavone 7-O-(4''-glucosylapioside)-4'-O-(4''-glucosylapioside)	Genistein 7,4'-O-bis(glucosylapioside)	C37H46O23	858.76	3146

3-ARYLCOUMARINS

No.	Name	Trivial name	Formula	MW	Ref.
1050	7,2'-Dihydroxy-4'-methoxy-3-phenylcoumarin		C16H12O5	284.27	2802
1051	7,2'-Dihydroxy-4',5'-methylenedioxy-3-phenylcoumarin		C16H10O6	298.25	2803
1052	4'-Hydroxy-4,7,2'-trimethoxy-3-phenylcoumarin	Melimessanol B	C18H16O6	328.32	F153
1053	2'-Methoxy-4',5'-methylenedioxy-3-phenylcoumarin	Pachyrrhizin	C19H12O6	336.30	2804
1054	2',4-Dihydroxy-6'',6''-dimethylpyrano[2'',3'':7,6]-3-phenylcoumarin	Kanzonol W	C20H16O5	336.34	F73
1055	2',4,5'-Trimethoxyfurano[2'',3'':7,6]-3-phenylcoumarin		C20H16O6	352.34	F154
1056	7,4'-Dihydroxy-2'-methoxy-3'-prenyl-3-phenylcoumarin	Eryvarin O	C21H20O5	352.38	F280
1057	4,5,7-Trimethoxy-4',5'-methylenedioxy-3-phenylcoumarin	Derrusnin	C19H16O7	356.33	2807
1058	2',4-Dimethoxy-6'',6''-dimethylpyrano[2'',3'':7,8]-3-phenylcoumarin		C22H22O5	364.40	F122
1059	8,2'-Dimethoxy-4',5'-methylenedioxyfurano[2'',3'':7,6]-3-phenylcoumarin	Neofolin	C20H14O7	366.33	2805
1060	4,4'-Dihydroxy-5-methoxy-6'',6''-dimethylpyrano[2'',3'':7,6]-3-phenylcoumarin	Indicanine B	C21H18O6	366.36	F305
1061	7,2'-Dihydroxy-5-methoxy-6'',6''-dimethylpyrano[2'',3'':4,3]-3-phenylcoumarin	Glyasperin L	C21H18O6	366.36	F72
1062	2',4-Dihydroxy-5-methoxy-6'',6''-dimethyl-4'',5''-dihydropyrano[2'',3'':7,6]-3-phenylcoumarin	Isoglycycoumarin	C21H20O6	368.39	3071
1063	7,2',4'-Trihydroxy-5-methoxy-3'-prenyl-3-phenylcoumarin	Gancaonin W	C21H20O6	368.39	F78
1064	7,2',4'-Trihydroxy-5-methoxy-6-prenyl-3-phenylcoumarin	Glycycoumarin	C21H20O6	368.39	3070
1065	7,2',4'-Trihydroxy-5-methoxy-8-(1,1-dimethyl-2-propenyl)-3-phenylcoumarin	Licoarylcoumarin	C21H20O6	368.39	3073
1066	4-Hydroxy-5,6,7-trimethoxy-4',5'-methylenedioxy-3-phenylcoumarin		C19H16O8	372.33	F332
1067	2'-Methoxy-3',4-methylenedioxy-6'',6''-dimethylpyrano[2'',3'':7,6]-3-phenylcoumarin	Pervilleanin	C22H18O6	378.38	F193
1068	4-Hydroxy-5,4'-dimethoxy-6'',6''-dimethylpyrano[2'',3'':7,6]-3-phenylcoumarin	Robustic acid	C22H20O6	380.40	2808
1069	2',4-Dihydroxy-5,7-dimethoxy-6-prenyl-3-phenylcoumarin	Glycyrin	C22H22O6	382.41	2806
1070	4-Hydroxy-5,4'-dimethoxy-5''-(1-methylethenyl)-4'',5''-dihydrofurano[2'',3'':7,6]-3-phenylcoumarin	Indicanine A	C22H20O6	382.41	F185
1071	7-Hydroxy-5,2-dimethoxy-6'',6''-dimethyl-4',5''-dihydropyrano[2'',3':4,3]-3-phenylcoumarin	Gancaonol A	C22H22O6	382.41	F70
1072	2',4-Dihydroxy-5-methoxy-5''-(1-hydroxy-1-methylethyl)-4'',5''-dihydrofurano[2'',3'':7,6]-3-phenylcoumarin	Licofuranocoumarin	C21H20O7	384.39	F88
1073	2',4-Dihydroxy-5-methoxy-6''-(hydroxymethyl)-6''-methyl-4'',5''-dihydropyrano[2'',3'':7,6]-3-phenylcoumarin	Licopyranocoumarin	C21H20O7	384.39	3072

continued

APPENDIX
Checklist for Isoflavonoids Described in the Literature During the Period 1991–2004 — continued

	Name	Trivial Name	Mass	Formula	Ref.
1074	4-Hydroxy-5-methoxy-3',4'-methylenedioxy-6'',6''-dimethylpyrano[2'',3'':7,6]-3-phenylcoumarin	Robustin	394.38	C22H18O7	2810
1075	4-Hydroxy-5-methoxy-3',4'-methylenedioxy-6'',6''-dimethylpyrano[2'',3'':7,8]-3-phenylcoumarin	Isorobustin	394.38	C22H18O7	3076
1076	4,5,4'-Trimethoxy-6'',6''-dimethylpyrano[2'',3'':7,6]-3-phenylcoumarin	Robustic acid methyl ether	394.42	C23H22O6	2809
1077	7,2'4'-Trihydroxy-8,3'-diprenyl-3-phenylcoumarin	Licocoumarin A	406.47	C25H26O5	F19
1078	4-Hydroxy-5,8,4'-trimethoxy-5''-(1-methylethenyl)furano[2'',3'':7,6]-3-phenylcoumarin	Thonningine B	408.41	C23H20O7	3075
1079	4,5-Dimethoxy-3',4'-methylenedioxy-5''-(1-methylethenyl)-4'',5''-dihydrofurano[2'',3'':7,6]-3-phenylcoumarin	Glabrescin	408.41	C23H20O7	2812
1080	4,5-Dimethoxy-3',4'-methylenedioxy-6'',6''-dimethylpyrano[2'',3'':7,6]-3-phenylcoumarin	Robustin methyl ether	408.41	C23H20O7	2811
1081	4,5-Dimethoxy-3',4'-methylenedioxy-6'',6''-dimethylpyrano[2'',3'':7,8]-3-phenylcoumarin	Isorobustin 4-methyl ether	408.41	C23H20O7	3077
1082	4-Hydroxy-5,8-dimethoxy-3',4'-methylenedioxy-5''-(1-methylethenyl)furano[2'',3'':7,6]-3-phenylcoumarin	Thonningine A	422.39	C23H18O8	3074
1083	4,4'-Dihydroxy-5-methoxy-6-prenyl-6'',6''-dimethylpyrano[2'',3'':7,8]-3-phenylcoumarin	Scandenin	434.49	C26H26O6	2813
1084	4,4'-Dihydroxy-5-methoxy-8-prenyl-6'',6''-dimethylpyrano[2'',3'':7,6]-3-phenylcoumarin	Lonchocarpic acid	434.49	C26H26O6	2814
1085	4-Hydroxy-5,3',4'-trimethoxy-5''-(1-methylethenyl)-4'',5''-dihydrofurano[2'',3'':7,6]-3-phenylcoumarin	Thonningin C	438.43	C24H22O8	F12
1086	4-Hydroxy-5,4'-dimethoxy-8-prenyl-6'',6''-dimethylpyrano[2'',3'':7,6]-3-phenylcoumarin	Lonchocarpenin	448.51	C27H28O6	2815
1087	4-Keto-4'-hydroxy-5-methoxy-6-prenyl-3-(2-oxopropyl)-6',6'-dimethylpyrano[2'',3'':7,8]-3-phenylcoumarin	Lonchocarpin	490.54	C29H30O7	F155

COUMARONOCHROMONES

	Name	Trivial Name	Mass	Formula	Ref.
1088	Coumaronochromone	Wrightiadione	248.23	C16H8O3	F144
1089	5,7,3'-Trihydroxycoumaronochromone	Coccineone A	284.23	C15H8O6	F67
1090	5,7,4'-Trihydroxycoumaronochromone	Lupinalbin A	284.23	C15H8O6	3093
1091	5,4'-Dihydroxy-7-methoxycoumaronochromone	Sophorophenolone	298.25	C16H10O6	F291
1092	7,4'-Dihydroxy-5-methoxycoumaronochromone	Desmoxyphyllin B	298.25	C16H10O6	F172
1093	5,7,4'-Trihydroxy-5'-methoxycoumaronochromone	Desmoxyphyllin A	314.25	C16H10O7	F172
1094	5,7,5'-Trihydroxy-4'-methoxycoumaronochromone	Oblonginol	314.25	C16H10O7	F145
1095	5,4'-Dihydroxy-6'',6''-dimethylpyrano[2'',3'':7,6]coumaronochromone	Lupinalbin H	350.32	C20H14O6	F267
1096	5,7,4'-Trihydroxy-3'-prenylcoumaronochromone	Lupinalbin D	352.34	C20H16O6	3094
1097	5,7,4'-Trihydroxy-6-prenylcoumaronochromone	Lupinalbin B	352.34	C20H16O6	3096
1098	5,4'-Dihydroxy-5''-(1-hydroxy-1-methylethyl)-4'',5''-dihydrofurano[2'',3'':7,6]coumaronochromone	Lupinalbin C	368.34	C20H16O7	3097
1099	5,7-Dihydroxy-5''-(1-hydroxy-1-methylethyl)-4'',5''-dihydrofurano[2'',3'':4,3]coumaronochromone	Lupinalbin E	368.34	C20H16O7	3095
1100	3,2',4'-Trihydroxy-8-prenylfurano[2'',3'':7,6]coumaronochromone	Erysenegalensein K	378.38	C22H18O6	F311

No.	Name	Trivial name	MW	Formula	Ref
1101	5,7,4'-Trihydroxy-3'-methoxy-5'-prenylcoumaronochromone	Lisetin	382.27	C21H18O7	2512
1102	5,7,5''-Trihydroxy-4'-methoxy-6'',6''-dimethyl-4'',5''-dihydropyrano[2'',3'':3',2']coumaronochromone	Lisetinone	398.36	C21H18O8	F264
1103	5,4-Dihydroxy-8-prenyl-6'',6''-dimethylpyrano[2'',3'':7,6]coumaronochromone	Millettin	418.45	C25H22O6	2513
1104	5,7,4'-Trihydroxy-6,3'-diprenylcoumaronochromone	Lupinalbin F	420.46	C25H24O6	3098
1105	5,7,5'-Trihydroxy-6,8-diprenyl-6'',6''-dimethylpyrano[2'',3'':4',3']coumaronochromone	Euchretin A	420.46	C25H24O6	3102
1106	7,4',5'-Trihydroxy-3-(3,7-dimethyl-2,6-octadienyl)coumaronochromone	Lespedezol C1	420.46	C25H24O6	F170
1107	5,4',5'-Trihydroxy-6-prenyl-6'',6''-dimethylpyrano[2'',3'':7,8]coumaronochromone	Euchretin G	434.44	C25H22O7	F173
1108	5,5'-Dihydroxy-6'',6'',6''',6'''-tetramethyl-4'',5''-dihydropyrano[2'',3'':7,8]pyrano[2''',3''':4',3]coumaronochromone	Formosanatin D	434.44	C25H22O7	F148
1109	5,7,5'-Trihydroxy-8-prenyl-6'',6''-dimethylpyrano[2'',3'':4',3]coumaronochromone	Euchretin H	434.44	C25H22O7	F173
1110	3,5,6'-Trihydroxy-8-prenyl-6'',6''-dimethylpyrano[2'',3'':7,6]coumaronochromone	Erysenegalensein J	436.46	C25H24O7	F311
1111	5,7,4',5'-Tetrahydroxy-6,8-diprenylcoumaronochromone	Euchretin F	436.46	C25H24O7	F173
1112	5,7,4',5'-Tetrahydroxy-8,3'-diprenylcoumaronochromone	Euchretin B	436.46	C25H24O7	3099
1113	5,7,5'-Trihydroxy-8-prenyl-6'',6''-dimethyl-4'',5''-dihydropyrano[2'',3'':4',3']coumaronochromone	Euchretin L	436.46	C25H24O7	F149
1114	5,7,4'-Trihydroxy-5'-methoxy-8,3'-diprenylcoumaronochromone	8-Dimethylallyllilisetin	450.49	C26H26O7	3100
1115	7,4'-Dihydroxy-5'-methoxycoumaronochromone 7-O-glucoside	Desmoxyphyllin B 7-O-glucoside	460.39	C22H20O11	F172
1116	5,7,4'-Trihydroxy-5'-methoxycoumaronochromone 7-O-glucoside	Desmoxyphyllin A 7-O-glucoside	476.39	C22H20O12	F172
1117	5,5'-Dihydroxy-6-prenyl-(2'',3'':7,8),(2''',3''':4',3)-bis(6,6-dimethylpyrano)coumaronochromone	Euchretin E	500.54	C30H28O7	F174
1118	5,5'-Dihydroxy-8-prenyl-(2'',3'':7,6),(2''',3''':4',3)-bis(6,6-dimethylpyrano)coumaronochromone	Euchretin D	500.54	C30H28O7	F174
1119	5,5'-Dihydroxy-8-prenyl-6'',6'',6''',6'''-tetramethyl-4'',5''-dihydropyrano[2'',3'':7,6]pyrano[2''',3''':4',3']coumaronochromone	Formosanatin C	502.56	C30H30O7	F148
1120	7,5'-Dihydroxy-8-prenyl-6'',6'',6''',6'''-tetramethyl-4'',5''-dihydropyrano[2'',3'':5,6]pyrano[2''',3''':4',3]coumaronochromone	Euchretin J	502.56	C30H30O7	F149
1121	5,7,4',5'-Tetrahydroxy-6,8,3'-triprenylcoumaronochromone	Euchretin C	504.58	C30H32O7	3101
1122	7,5'-Dihydroxy-8-prenyl-(2'',3'':5,6),(2''',3''':4',3)-bis(6,6-dimethyl-4,5-dihydropyrano)coumaronochromone	Euchretin K	504.58	C30H32O7	F149
1123	5,7,4'-Trihydroxy-5'-methoxy-6,8,3'-triprenylcoumaronochromone	Euchretin N	518.60	C31H34O7	F149
1124	5,7,5'-Trihydroxy-8-prenyl-6-(3-hydroxy-3-methylbutyl)-6'',6''-dimethylpyrano[2'',3'':7,8]pyrano[2''',3''':4',3]coumaronochromone	Formosanatin A	520.57	C30H32O8	F148
1125	5,7,4',5'-Tetrahydroxy-6-(3-hydroxy-3-methylbutyl)-8,3'-diprenylcoumaronochromone	Euchretin M	522.59	C30H34O8	F149
1126	5,7,5',4''(S),5''(R)-Pentahydroxy-6'',6''-dimethyl-4'',5''-dihydropyrano[2'',3'':4',3]coumaronochromone	Formosanatin B	536.57	C30H32O9	F148

COUMARANOCHROMANONES

No.	Name	Trivial name	MW	Formula	Ref
1127	(2R,3S)-3,5,7,4'-Tetrahydroxy-6-prenylcoumaranochroman-4-one	Lupino. C	370.35	C20H18O7	F266
1128	(2S,3S)-3,5,7,4'-Tetrahydroxy-5'-methoxy-3'-prenylcoumaranochroman-4-one	Piscerythrol	400.38	C21H20O8	F266

continued

APPENDIX
Checklist for Isoflavonoids Described in the Literature During the Period 1991–2004 — *continued*

	Name	Trivial Name	Mass	Formula	Ref.
1129	(2*S*,3*S*)-3,5,7,4′-Tetrahydroxy-6,3′-diprenylcoumaranochroman-4-one	Lupinol A	438.48	C25H26O7	F266
1130	(2*S*,3*S*)-5,7,4′-Trihydroxy-3-methoxy-6,3′-diprenylcoumaranochroman-4-one	Lupinol B	452.50	C26H28O7	F266
	PTEROCARPANES				
1131	(6a*S*,11a*S*)-4,9-Dihydroxy-3,8-dimethoxypterocarpan	Derricarpin	316.31	C17H16O6	F147
1132	3,9-Dihydroxypterocarpan	Demethylmedicarpin	256.26	C15H12O4	2608
1133	3-Hydroxy-9-methoxypterocarpan	(−)Medicarpin	270.29	C16H14O4	2609
1134	9-Hydroxy-3-methoxypterocarpan	Isomedicarpin	270.29	C16H14O4	2610
1135	3,4,9-Trihydroxypterocarpan	4-Hydroxydemethylmedicarpin	272.26	C15H12O5	2611
1136	3,6a,9-Trihydroxypterocarpan	Glycinol	272.26	C15H12O5	2680
1137	9-Hydroxyfurano[2′,3′:3,2]pterocarpan	Neodunol	280.28	C17H12O4	2640
1138	3-Hydroxy-8,9-methylenedioxypterocarpan	(−)Maackiain	284.27	C16H12O5	2612
1139	3,9-Dimethoxypterocarpan	(−)Homopterocarpin	284.31	C17H16O4	2613
1140	1,9-Dihydroxy-3-methoxypterocarpan	Desmocarpin	286.29	C16H14O5	3021
1141	3,10-Dihydroxy-9-methoxypterocarpan	Vesticarpan	286.29	C16H14O5	2615
1142	3,4-Dihydroxy-9-methoxypterocarpan	4-Hydroxymedicarpin	286.29	C16H14O5	2616
1143	3,6a-Dihydroxy-9-methoxypterocarpan	6a-Hydroxymedicarpin	286.29	C16H14O5	2681
1144	3,7-Dihydroxy-9-methoxypterocarpan	Nissicarpin	286.29	C16H14O5	3018
1145	3,9-Dihydroxy-10-methoxypterocarpan	Nissolin	286.29	C16H14O5	2614
1146	3,9-Dihydroxy-8-methoxypterocarpan	Kushenin	286.29	C16H14O5	3020
1147	4,9-Dihydroxy-3-methoxypterocarpan	Melilotocarpan B	286.29	C16H14O5	3023
1148	6a,9-Dihydroxy-3-methoxypterocarpan	6a-Hydroxyisomedicarpin	286.29	C16H14O5	2682
1149	(6a*R*,11a*R*)-3,8-Dihydroxy-9-methoxypterocarpan		286.29	C16H14O5	3034
1150	(6a*R*,11a*R*)-3,9-Dihydroxy-8-methoxypterocarpan	Lespedezol D1	286.29	C16H14O5	F171
1151	(6a*R*,11a*R*,11b*S*)-3-Keto-11b-hydroxy-9-methoxy-1,2-dihydropterocarpan	11b-Hydroxydihydromedicarpin	288.30	C16H16O5	F22
1152	9-Methoxyfurano[2′,3′:3,2]pterocarpan	9-*O*-Methylneodunol	294.31	C18H14O4	3030
1153	(6a*R*,11a*R*)-5′,5′-Dimethyl-4′,5′-dihydrofurano[2′,3′:3,2]pterocarpan	Methyl-oroxylopterocarpan	294.34	C19H18O3	F8
1154	3-Methoxy-8,9-methylenedioxypterocarpan	Pterocarpin	298.30	C17H14O5	2617
1155	(6a*R*,11a*R*)-2-Carboxyaldehyde-9-hydroxy-3-methoxypterocarpan	Erysubin C	298.30	C17H14O5	F275
1156	2,3-Dihydroxy-8,9-methylenedioxypterocarpan	2-Hydroxymaackiain	300.27	C16H12O6	3036

1157	3,4-Dihydroxy-8,9-methylenedioxypterocarpan	4-Hydroxymaackiain	300.27	C16H12O6	2621
1158	3,6a-Dihydroxy-8,9-methylenedioxypterocarpan	6a-Hydroxymaackiain	300.27	C16H12O6	2683
1159	(6aS,11aS)-3,4-Dihydroxy-8,9-methylenedioxypterocarpan		300.27	C16H12O6	F39
1160	3-Hydroxy-2,9-dimethoxypterocarpan	2-Methoxymedicarpin	300.31	C17H16O5	2619
1161	3-Hydroxy-4,9-dimethoxypterocarpan	4-Methoxymedicarpin	300.31	C17H16O5	2622
1162	3-Hydroxy-8,9-dimethoxypterocarpan	Secundiflorol I	300.31	C17H16O5	F289
1163	4-Hydroxy-3,9-dimethoxypterocarpan	4-Hydroxyhomopterocarpin	300.31	C17H16O5	2623
1164	6a-Hydroxy-3,9-dimethoxypterocarpan	Variabilin	300.31	C17H16O5	2684
1165	(6aR,11aR)-10-Hydroxy-3,9-dimethoxypterocarpan		300.31	C17H16O5	F256
1166	(6aR,11aR)-3-Hydroxy-8,9-dimethoxypterocarpan		300.31	C17H16O5	F6
1167	(6aR,11aR)-3-Hydroxy-9,10-dimethoxypterocarpan	Methylnissolin	300.31	C17H16O5	2618
1168	(6aR,11aR)-8-Hydroxy-3,9-dimethoxypterocarpan		300.31	C17H16O5	3035
1169	(6R,6aS,11aR)-3-Hydroxy-6,9-dimethoxypterocarpan	Sophoracarpan A	300.31	C17H16O5	3038
1170	7-Hydroxy-3,9-dimethoxypterocarpan	Fruticarpin	300.31	C17H16O5	3019
1171	9-Hydroxy-2,3-dimethoxypterocarpan	Sparticarpin	300.31	C17H16O5	2620
1172	3,6a,7-Trihydroxy-9-methoxypterocarpan	6a,7-Dihydroxymedicarpin	302.29	C16H14O6	2685
1173	(6aR,11aR,11bS)-3-Keto-11b-hydroxy-8,9-methylenedioxy-1,2-dihydropterocarpan	11b-Hydroxydihydromaackiain	302.29	C16H14O6	F22
1174	(6aS,11aS)-4-Dihydroxy-3-methoxy-8,9-methylenedioxypterocarpan		302.29	C16H14O6	F39
1175	8,9-Methyleneoxyfurano[2′,3′:3,2]pterocarpan	Neodulin	308.29	C18H12O5	2641
1176	(6aR,11aS)-(3,4:8,9)-Bis(methylenedioxy)pterocarpan		312.28	C17H12O6	F292
1177	2-Hydroxy-3-methoxy-8,9-methylenedioxypterocarpan	2-Hydroxypterocarpin	314.29	C17H14O6	2624
1178	3-Hydroxy-4-methoxy-8,9-methylenedioxypterocarpan	4-Methoxymaackiain	314.29	C17H14O6	2626
1179	3-Methoxy-6a-hydroxy-8,9-methylenedioxypterocarpan	Pisatin	314.29	C17H14O6	2686
1180	4-Hydroxy-3-methoxy-8,9-methylenedioxypterocarpan	4-Hydroxypterocarpin	314.29	C17H14O6	2627
1181	(6R,6aS,11aR)-3-Hydroxy-6-methoxy-8,9-methylenedioxypterocarpan	Sophoracarpan B	314.29	C17H14O6	3039
1182	2,3,9-Trimethoxypterocarpan	2-Methoxyhomopterocarpin	314.34	C18H18O5	2625
1183	3,4,9-Trimethoxypterocarpan	4-Methoxyhomopterocarpin	314.34	C18H18O5	2628
1184	(6aR,11aR)-3,9,10-Trimethoxypterocarpan	3,9-di-O-Methylnissolin	314.34	C18H18O5	F261
1185	(6S,6aS,11aR)-3,6,9-Trimethoxypterocarpan	6-Methoxyhomopterocarpin	314.34	C18H18O5	3032
1186	3,6a,7-Trihydroxy-8,9-methylenedioxypterocarpan	6a,7-Dihydroxymaackiain	316.27	C16H12O7	2687
1187	(6S,6aR,11aR)-3,6,6a-Trihydroxy-8,9-methylenedioxypterocarpan	6,6a-Dihydroxymaackiain	316.27	C16H12O7	F252
1188	2,10-Dihydroxy-3,9-dimethoxypterocarpan	Mucronucarpan	316.31	C17H16O6	2631
1189	2,8-Dihydroxy-3,9-dimethoxypterocarpan		316.31	C17H16O6	2630
1190	3,7-Dihydroxy-3,9-dimethoxypterocarpan	Nissolicarpin	316.31	C17H16O6	3022

continued

APPENDIX

Checklist for Isoflavonoids Described in the Literature During the Period 1991–2004 — continued

	Name	Trivial Name	Mass	Formula	Ref.
1191	3,9-Dihydroxy-7,10-dimethoxypterocarpan	Philenopteran	316.31	C17H16O6	2629
1192	4,10-Dihydroxy-3,9-dimethoxypterocarpan	Melilotocarpan D	316.31	C17H16O6	3024
1193	(6aR,11aR)-4,9-Dihydroxy-3,10-dimethoxypterocarpan	Melilotocarpan E	316.31	C17H16O6	3025
1194	(6S,6aS,11aR,11bS)-3-Keto-11b-hydroxy-6,9-dimethoxy-1,2-dihydropterocarpan	Kushecarpin A	318.33	C17H18O6	F134
1195	(6aR,11aR)-3-Hydroxy-5'-(1-methylethenyl)furano[2',3':9,10]pterocarpan	Crotafuran A	320.34	C20H16O4	F320
1196	(6aR,11aR)-5'-Acetyl-3-hydroxyfurano[2',3':9,10]pterocarpan	Crotafuran B	322.31	C19H14O5	F320
1197	3-Hydroxy-6',6'-dimethylpyrano[2',3':9,10]pterocarpan	Phaseollin	322.36	C20H18O4	2642
1198	3-Hydroxy-6',6'-dimethylpyrano[2',3':9,8]pterocarpan	Isoneorautenol	322.36	C20H18O4	3028
1199	(6aR,11aR)-3-Hydroxy-5'-(1-methylethenyl)furano[2',3':9,10]pterocarpan	Barbacarpan	322.36	C20H18O4	F20
1200	9-Hydroxy-6',6'-dimethylpyrano[2',3':3,2]pterocarpan	Neorautenol	322.36	C20H18O4	2643
1201	6a-Hydroxy-8,9-methyleneoxyfurano[2',3':3,2]pterocarpan	Neobanol	324.29	C18H12O6	2691
1202	9,10-Dimethoxyfurano[2',3':3,2]pterocarpan	Ambonane	324.34	C19H16O5	2646
1203	3,9-Dihydroxy-10-prenylpterocarpan	Phaseollidin	324.37	C20H20O4	2645
1204	3,9-Dihydroxy-2-prenylpterocarpan	Calocarpin	324.37	C20H20O4	2647
1205	3,9-Dihydroxy-8-prenylpterocarpan	Sophorapterocarpan A	324.37	C20H20O4	2644
1206	(6aR,11aR)-3,9-Dihydroxy-6a-prenylpterocarpan	Lespedezol D2	324.37	C20H20O4	F171
1207	6a-Hydroxy-(3,4:8,9)-bis(methylenedioxy)pterocarpan	Acanthocarpan	328.28	C17H12O7	2688
1208	2,3-Dimethoxy-8,9-methylenedioxypterocarpan	2-Methoxypterocarpin	328.32	C18H16O6	2632
1209	3,4-Dimethoxy-8,9-methylenedioxypterocarpan	4-Methoxypterocarpin	328.32	C18H16O6	2633
1210	(6S,6aS,11aR)-3,6-Dimethoxy-8,9-methylenedioxypterocarpan	6-Methoxypterocarpin	328.32	C18H16O6	3033
1211	1,3-Dihydroxy-2-methoxy-8,9-methylenedioxypterocarpan	Trifolian	330.29	C17H14O7	2635
1212	3,6a-Dihydroxy-2-methoxy-8,9-methylenedioxypterocarpan	Hildecarpin	330.29	C17H14O7	3047
1213	3,6a-Dihydroxy-4-methoxy-8,9-methylenedioxypterocarpan	Tephrocarpin	330.29	C17H14O7	2689
1214	(6aR,11aR)-2,6a-Dihydroxy-3-methoxy-8,9-methylenedioxypterocarpan	2-Hydroxypisatin	330.29	C17H14O7	F126
1215	3-Hydroxy-7,9,10-trimethoxypterocarpan	9-O-Methylphilenopteran	330.34	C18H18O6	2634
1216	4-Hydroxy-2,3,9-trimethoxypterocarpan		330.34	C18H18O6	2636
1217	4-Hydroxy-3,9,10-trimethoxypterocarpan	Melilotocarpan C	330.34	C18H18O6	3026
1218	(6aR,11aR)-10-Hydroxy-3,4,9-trimethoxypterocarpan	Odoricarpan	330.34	C18H18O6	3027
1219	(6S,6aS,11aR,11bS)-3-Keto-11b-hydroxy-6-methoxy-8,9-methylenedioxy-1,2-dihydropterocarpan	Kushecarpin C	332.30	C17H16O7	F134
1220	6a-Hydroxy-5'-(1-methylethenyl)furano[2',3':3,4]pterocarpan	Clandestacarpin	336.34	C20H16O5	2692
1221	6a,9-Dihydroxy-5'-(1-methylethenyl)furano[2',3':3,4]pterocarpan	Clandestacarpin	336.34	C20H16O5	3048

No.	Systematic name	Name	Formula	MW	Ref
1222	(6aS,11aS)-3,6a-Dihydroxy-5'-(1-methylethenyl)furano[2',3':9,10]pterocarpan	Crotafuran C	C20H16O5	336.34	F320
1223	3-Hydroxy-9-methoxy-10-prenylpterocarpan	Eryvarin D	C21H20O4	336.38	F282
1224	9-Methoxy-6',6'-dimethylpyrano[2',3':3,4]pterocarpan	Hemileiocarpin	C21H20O4	336.38	2648
1225	(6aR,11aR)-5'-Acetyl-3,6a-dihydroxyfurano[2',3':9,10]pterocarpan	Crotafuran D	C19H14O6	338.31	F320
1226	4-Methoxy-8,9-methyleneoxyfurano[2',3':3,2]pterocarpan	Ficinin	C19H14O6	338.32	2651
1227	1,9-Dihydroxy-5'-(1-methylethenyl)-4',5'-dihydrofurano[2',3':3,4]pterocarpan	Apiocarpin	C20H18O5	338.36	2650
1228	3,6a-Dihydroxy-6',6'-dimethylpyrano[2',3':9,10]pterocarpan	Tuberosin	C20H18O5	338.36	2693
1229	6a,9-Dihydroxy-5'(R)-(1-methylethenyl)furano[2',3':3,2]pterocarpan	Glyceollin III	C20H18O5	338.36	2695
1230	6a,9-Dihydroxy-5'(S)-(1-methylethenyl)furano[2',3':3,2]pterocarpan	Canescacarpin	C20H18O5	338.36	2696
1231	6a,9-Dihydroxy-6',6'-dimethylpyrano[2',3':3,2]pterocarpan	Glyceollin II	C20H18O5	338.36	2694
1232	6a,9-Dihydroxy-6',6'-dimethylpyrano[2',3':3,4]pterocarpan	Glyceollin I	C20H18O5	338.36	2697
1233	3-Hydroxy-9-methoxy-10-prenylpterocarpene	Sandwicensin	C21H22O4	338.40	2649
1234	(6aR,11aR)-3-Hydroxy-9-methoxy-8-prenylpterocarpan	Prostratol D	C21H22O4	338.40	F105
1235	(6aR,11aR)-9-Hydroxy-3-methoxy-2-prenylpterocarpan	Orientanol B	C21H22O4	338.40	F288
1236	3,6a,9-Trihydroxy-10-prenylpterocarpan	Sandwicarpin	C20H20O5	340.38	2698
1237	3,6a,9-Trihydroxy-2-prenylpterocarpan	Glyceocarpin	C20H20O5	340.38	2699
1238	3,6a,9-Trihydroxy-4-prenylpterocarpan	Glyceollidin I	C20H20O5	340.38	2700
1239	3,9-Dihydroxy-10-(2(R)-hydroxy-3-methyl-3-butenyl)pterocarpan	Dolichinin A	C20H20O5	340.38	2652
1240	3,9-Dihydroxy-10-(2(S)-hydroxy-3-methyl-3-butenyl)pterocarpan	Dolichinin B	C20H20O5	340.38	2653
1241	(6aR,11aR)-3-Hydroxy-5'-(1-hydroxy-1-methylethyl)-4',5'-dihydrofurano[2',3':9,10]pterocarpan	Erystagallin C	C20H20O5	340.38	F287
1242	(6aR,11aR)-3,8,9-Trihydroxy-6a-prenylpterocarpan	Lespedezol D3	C20H20O5	340.38	F171
1243	(6aR,11aR)-3,9,10-Trihydroxy-4-prenylpterocarpan	Bitucarpin B	C20H20O5	340.38	F203
1244	2-Hydroxy-3,4-dimethoxy-8,9-methylenedioxypterocarpan	2-Hydroxy-4-methoxypterocarpin	C18H16O7	344.32	3037
1245	6a-Hydroxy-2,3-dimethoxy-8,9-methylenedioxypterocarpan	Lathycarpin	C18H16O7	344.32	2690
1246	2,8-Dihydroxy-3,9,10-trimethoxypterocarpan		C18H18O7	346.34	2637
1247	(6S,6aS,11aR,11bR)-3-Keto-6,11b-dimethoxy-8,9-methylenedioxy-1,2-dihydropterocarpan	Kushecarpin B	C18H18O7	346.34	F134
1248	9-Hydroxy-10-prenylfurano[2',3':3,2]pterocarpan	Erybraedin E	C22H20O4	348.40	3041
1249	(6aR,11aR)-8,9-Methylenedioxy-5'-(1-methylethenyl)-4',5'-dihydrofurano[2',3':3,2]pterocarpan	Emoroidocarpan	C21H18O5	350.37	F152
1250	8,9-Methylenedioxy-6',6'-dimethylpyrano[2',3':3,2]pterocarpan	Neorautenane	C21H18O5	350.37	2654
1251	8,9-Methylenedioxy-6',6'-dimethylpyrano[2',3':3,4]pterocarpan	Leiocarpin	C21H18O5	350.37	2655
1252	3-Hydroxy-2-prenyl-8,9-methylenedioxypterocarpan	Edunol	C21H20O5	352.38	2656
1253	3-Hydroxy-9-methoxy-6',6'-dimethylpyrano[2',3':1,2]pterocarpan	Bolucarpan D	C21H20O5	352.38	F29
1254	(6aR,11aR)-10-Hydroxy-9-methoxy-5'-(1-methylethenyl)-4',5'-dihydrofurano[2',3':3,2]pterocarpan	Pervillin	C21H20O5	352.38	F193
1255	(6aR,11aR)-3-Hydroxy-1-methoxy-6',6'-dimethylpyrano[2',3':9,10]pterocarpan	1-Methoxyphaseollidin	C21H20O5	352.38	F124

continued

APPENDIX
Checklist for Isoflavonoids Described in the Literature During the Period 1991–2004 — *continued*

	Name	Trivial Name	Mass	Formula	Ref.
1256	(6a*R*,11a*R*)-3-Hydroxy-8,9-methylenedioxy-4-prenylpterocarpan	4'-Dehydroxycabenegrin A-I	352.38	C21H20O5	F46
1257	(6a*R*,11a*R*)-8-Hydroxy-9-methoxy-5'-(1-methylethenyl)-4',5'-dihydrofurano[2',3':3,2]pterocarpan	Pervillinin	352.38	C21H20O5	F193
1258	8,9-Methylenedioxy-4',5'-dihydro-6',6'-dimethylpyrano[2',3':3,2]pterocarpan	Neorautane	352.38	C21H20O5	2657
1259	9-Hydroxy-1-methoxy-6',6'-dimethylpyrano[2',3':3,2]pterocarpan	Edulenanol	352.38	C21H20O5	2658
1260	(6a*R*,11a*R*)-3,9-Dimethoxy-4-prenylpterocarpan	Bitucarpin A	352.42	C22H24O4	F203
1261	6a,9-Dihydroxy-5'-(hydroxymethylethyl)furano[2',3':3,2]pterocarpan	Glyceofuran	354.36	C20H18O6	2703
1262	3-Hydroxy-9-methoxy-6',6'-dimethyl-4',5'-dihydropyrano[2',3':1,2]pterocarpan	Bolucarpan C	354.40	C21H22O5	F29
1263	3,10-Dihydroxy-9-methoxy-8-(1,1-dimethyl-2-propenyl)pterocarpan	Arizonicanol E	354.40	C21H22O5	F289
1264	3,6a-Dihydroxy-9-methoxy-10-prenylpterocarpan	Cristacarpin	354.40	C21H22O5	2701
1265	3,9-Dihydroxy-1-methoxy-10-prenylpterocarpan	1-Methoxyphaseollidin	354.40	C21H22O5	2659
1266	3,9-Dihydroxy-1-methoxy-2-prenylpterocarpan	Edudiol	354.40	C21H22O5	2660
1267	3,9-Dihydroxy-8-methoxy-4-prenylpterocarpan	Eryvarin K	354.40	C21H22O5	F276
1268	6a,9-Dihydroxy-3-methoxy-2-prenylpterocarpan	Glyceollin IV	354.40	C21H22O5	2702
1269	(6a*R*,11a*R*)-1,9-Dihydroxy-3-methoxy-2-prenylpterocarpan		354.40	C21H22O5	F229
1270	(6a*R*,11a*R*)-3,8-Dihydroxy-9-methoxy-6a-prenylpterocarpan		354.40	C21H22O5	F171
1271	(6a*R*,11a*R*)-9-Hydroxy-1-methoxy-3-prenyloxypterocarpan	Lespedezol D4	354.40	C21H22O5	F124
1272	3-Hydroxy-9-methoxy-10-(3-hydroxy-3-methylbutyl)pterocarpan		354.40	C21H24O5	F120
1273	(6a*R*,11a*R*)-2,3,4-Trimethoxy-9,8-methylenedioxypterocarpan	Asperopterocarpin	356.41	C19H18O7	F166
1274	8-Hydroxy-3,4,9,10-tetramethoxypterocarpan		360.37	C19H20O7	2638
1275	3-Sulfate-8,9-methylenedioxypterocarpan	(–)-Maackiain sulfate	364.33	C16H12O8S	F196
1276	1-Hydroxy-8,9-methylenedioxy-6',6'-dimethylpyrano[2',3':3,2]pterocarpan	Neorautenanol	366.36	C21H18O6	2661
1277	3-Hydroxy-8,9-methylenedioxy-6',6'-dimethylpyrano[2',3':1,2]pterocarpan	Bolucarpan B	366.36	C21H18O6	F29
1278	1,9-Dimethoxy-6',6'-dimethylpyrano[2',3':3,2]pterocarpan	Edulenane	366.41	C22H22O5	2662
1279	(6a*R*,11a*R*)-8-Hydroxy-5'-hexylethyl-4',5'-dihydrofurano[2',3':3,2]pterocarpan	Hexyl-oroxylopterocarpan	366.45	C23H26O4	F8
1280	3-Hydroxy-8,9-methylenedioxy-4-(4-hydroxy-3-methyl-2-butenyl)pterocarpan	Cabenegrin A-II	368.39	C21H20O6	3031
1281	3-Hydroxy-8,9-methylenedioxy-6',6'-dimethyl-3',4'-dihydropyrano[2',3':1,2]pterocarpan	Bolucarpan A	368.39	C21H20O6	F29
1282	5'-Hydroxy-8,9-methylenedioxy-6',6'-dimethyl-4',5'-dihydropyrano[2',3':3,2]pterocarpan	Neorautanol	368.39	C21H20O6	2663
1283	6a-Hydroxy-9-methoxy-5'-(hydroxymethylethyl)furano[2',3':3,2]pterocarpan	9-*O*-Methylglyceofuran	368.39	C21H20O6	2704
1284	1,9-Dimethoxy-6',6'-dimethyl-4',5'-dihydropyrano[2',3':3,2]pterocarpan	Edulane	368.43	C22H24O5	2665
1285	3-Hydroxy-1,9-dimethoxy-2-prenylpterocarpan	Edulenol	368.43	C22H24O5	2664

No.	Name	Trivial name	Formula	MW	Ref
1286	(6aR,11aR)-9-Hydroxy-1,3-dimethoxy-2-prenylpterocarpan	Kanzonol P	C22H24O5	368.43	F76
1287	3-Hydroxy-8,9-methylenedioxy-2-(4-hydroxy-3-methylbutyl)pterocarpan	Cabenegrin A-II	C21H22O6	370.40	3029
1288	(6aS,11aR,11bS)-3-Keto-6a,11b-dihydroxy-9-methoxy-10-prenylpterocarpan	Hydroxycristacarpone	C21H22O6	370.40	F286
1289	(6aS,11aS,5'S)-3,6a,5'-Trihydroxy-6',6'-dimethyl-4',5'-dihydropyrano[2',3':9,10]pterocarpan	Eryvarin A	C21H22O6	370.40	F284
1290	(6aS,11aS)-3,6a-Dihydroxy-9-methoxy-10-(2-keto-3-methylbutyl)-pterocarpan	Erypoegin I	C21H22O6	370.40	F272
1291	(6aS,11aS)-3,6a,8-Trihydroxy-9-methoxy-10-prenylpterocarpan	Sphenostylin B	C21H22O6	370.40	3049
1292	2,8-Dihydroxy-3,4,9,10-tetramethoxypterocarpan		C19H20O8	376.37	2639
1293	(6aR,11aR)-8-Hydroxy-5'-heptylethyl-4',5'-dihydrofurano[2',3':3,2]pterocarpan	Heptyl-oroxylopterocarpan	C24H28O4	380.48	F8
1294	(6aS,11aS)-6a-Hydroxy-8,9-methylenedioxy-5'-(1-hydroxymethylethenyl)-4',5'-dihydrofurano[2',3':3,2]pterocarpan	Hildecarpidin	C21H18O7	382.37	3053
1295	1-Methoxy-8,9-methylenedioxy-4',5'-dihydro-6',6'-dimethylpyrano[2',3':3,2]pterocarpan	Neorautanin	C22H22O6	382.41	2667
1296	3-Hydroxy-4-methoxy-8,9-methylenedioxy-2-prenylpterocarpan	Neoraucarpanol	C22H22O6	382.41	2668
1297	9-Hydroxy-1,8-dimethoxy-6',6'-dimethylpyrano[2',3':3,2]pterocarpan	Desmodin	C22H22O6	382.41	2666
1298	(6aS,11aS)-3,6a-Dihydroxy-9-methoxy-10-prenylpterocarpan	Orientanol A	C21H24O7	388.42	F285
1299	(6aS,11aS)-3,6a,8-Trihydroxy-9-methoxy-10-(3-hydroxy-3-methylbutyl)pterocarpan	Sphenostylin C	C21H24O7	388.42	3051
1300	3-Hydroxy-2-prenyl-6',6'-dimethylpyrano[2',3':9,10]pterocarpan	Folitenol	C25H26O4	390.48	2669
1301	6',6'-Dimethyl-4',5'-dihydropyrano[2',3':3,2]-6'',6''-dimethylpyrano[2'',3'':9,10]pterocarpan	Folinin	C25H26O4	390.48	2670
1302	(6aR,11aR)-3-Hydroxy-4-prenyl-6',6'-dimethylpyrano[2',3':9,10]pterocarpan	Erybraedin B	C25H26O4	390.48	3045
1303	(6aR,11aR)-3-Hydroxy-4-prenyl-6',6'-dimethylpyrano[2',3':9,8]pterocarpan	Erybraedin D	C25H26O4	390.48	3043
1304	(6aR,11aR)-9-Hydroxy-10-prenyl-6',6'-dimethylpyrano[2',3':3,2]pterocarpan	Orientanol C	C25H26O4	390.48	F288
1305	3,9-Dihydroxy-10-geranylpterocarpan	Lespedezin	C25H28O4	392.50	2672
1306	3,9-Dihydroxy-2,10-diprenylpterocarpan	Erythrabyssin II	C25H28O4	392.50	2674
1307	3,9-Dihydroxy-2,4-diprenylpterocarpan	Eryvarin J	C25H28O4	392.50	F276
1308	3,9-Dihydroxy-2,8-diprenylpterocarpan	Ficifolinol	C25H28O4	392.50	2673
1309	3,9-Dihydroxy-6a,10-diprenylpterocarpan	Lespein	C25H28O4	392.50	2671
1310	(6aR,11aR)-3,9-Dihydroxy-2-(1,1-dimethyl-2-propenyl)-10-prenylpterocarpan	Striatin	C25H28O4	392.50	F158
1311	(6aR,11aR)-3,9-Dihydroxy-4,10-diprenylpterocarpan	Erybraedin A	C25H28O4	392.50	3044
1312	(6aR,11aR)-3,9-Dihydroxy-4,8-diprenylpterocarpan	Erybraedin C	C25H28O4	392.50	3042
1313	3,4-Dimethoxy-8,9-methylenedioxy-2-prenylpterocarpan	Neoraucarpanol	C23H24O6	396.44	2675
1314	(6aS,11aS)-6a-Hydroxy-3,8,9-trimethoxy-10-prenylpterocarpan	Sphenostylin A	C23H26O6	398.46	3050
1315	1,3-Dihydroxy-6',6'-dimethylpyrano[2',3':9,8][1''-methylphenyl][3'',4'':4',5']pterocarpan	Tetrapterol B	C25H22O5	402.45	F101
1316	(6aS,11aS)-3,6a-Dihydroxy-8,9-dimethoxy-10-(3-hydroxy-3-methylbutyl)pterocarpan	Sphenostylin D	C22H26O7	402.45	3052
1317	3-Hydroxy-9-methoxy-2,10-diprenylpterocarpan	Eryvarin E	C26H28O4	404.51	F282

continued

APPENDIX
Checklist for Isoflavonoids Described in the Literature During the Period 1991–2004 — continued

	Name	Trivial Name	Mass	Formula	Ref.
1318	(6aR,11aR)-1-Hydroxy-2-prenyl-6'',6''-dimethyl-4'',5''-dihydropyrano[2'',3'':8,9]-6',6'-dimethylpyrano[2',3':3,2]pterocarpan		406.47	C25H26O5	F230
1319	(6aS,11aS)-3,6a-Dihydroxy-2-prenyl-6',6'-dimethylpyrano[2',3':9,10]pterocarpan	Erysubin E	406.47	C25H26O5	F275
1320	(6aR,11aR)-3-Hydroxy-9-methoxy-2,10-diprenylpterocarpan	Erycristin	406.51	C26H30O4	3040
1321	(6aR,11aR)-3-Hydroxy-9-methoxy-8-((2E)-3,7-dimethyl-2,6-octadienyl)pterocarpan	Prostratol E	406.51	C26H30O4	F105
1322	3,6a,9-Trihydroxy-2,10-diprenylpterocarpan	2-Prenyl-6a-hydroxyphaseollidin	408.50	C25H28O5	3054
1323	(6aR,11aR)-3,5'-Dihydroxy-2-prenyl-6',6'-dimethyl-4',5'-dihydropyrano[2',3':9,10]pterocarpan	Erysubin D	408.50	C25H28O5	F275
1324	1-Methoxy-(2':3:3,2),(2'':3'':9,10)-bis(6,6-dimethylpyrano)pterocarpan	Gangetinin	418.49	C26H26O5	2677
1325	8,9-Methylenedioxy-6'-methyl-6-prenylpyrano[2',3':3,2]pterocarpan	Nitiducarpin	418.49	C26H26O5	2676
1326	3-Hydroxy-8,9-methylenedioxy-4-geranylpterocarpan	Nitiducol	420.50	C26H28O5	2678
1327	(6aR,11aR)-3-Hydroxy-1-methoxy-2-prenyl-6',6'-dimethylpyrano[2',3':9,8]pterocarpan	Kanzonol F	420.50	C26H28O5	F74
1328	(6aS,11aS)-6a-Hydroxy-9-methoxy-10-prenyl-6',6'-dimethylpyrano[2',3':3,2]pterocarpan	Erypoegin J	420.50	C26H28O5	F272
1329	9-Hydroxy-1-methoxy-10-prenyl-6',6'-dimethylpyrano[2',3':3,2]pterocarpan	Gangetin	420.50	C26H28O5	2679
1330	3,9-Dihydroxy-1-methoxy-2,8-diprenylpterocarpan	1-Methoxyficifolinol	422.51	C26H30O5	3046
1331	(6aS,11aS)-3,11a-Dihydroxy-9-methoxy-2,10-diprenylpterocarpan	Erystagallin A	422.51	C26H30O5	F287
1332	(6aS,11aS)-3,6a-Dihydroxy-9-methoxy-4,10-diprenylpterocarpan	Eryzerin E	422.51	C26H30O5	F277
1333	(6aR,11aR)-3,8,5'-Trihydroxy-6'-methyl-6'-(4-methyl-3-pentenyl)-4',5'-dihydropyrano[2',3':9,10]pterocarpan	Lespedezol D5	424.50	C25H28O6	F171
1334	(6aS,11aS)-3,9,6a-Trihydroxy-1-methoxy-2,10-diprenylpterocarpan	Erystagallin B	438.51	C26H30O6	F287
1335	(6aR,11aR)-5'-Dodecanyl-5'-methyl-4',5'-dihydrofurano[2',3':3,2]pterocarpan	Dodecanyl-oroxylopterocarpan	448.64	C30H40O3	F8
	Pterocarpan glycosides				
1336	(6aR,11aR)-3-Hydroxy-8,9-methylenedioxypterocarpan 3-O-rhamnoside		430.41	C22H22O9	F228
1337	3-Hydroxy-9-methoxypterocarpan 3-O-glucoside	Medicarpin 3-O-glucoside	432.43	C22H24O9	3229
1338	3-Hydroxy-8,9-methylenedioxypterocarpan 3-O-galactoside	Maackiain 3-O-galactoside	446.41	C22H22O10	3233
1339	3-Hydroxy-8,9-methylenedioxypterocarpan 3-O-glucoside	(+)-Maackiain 3-O-glucoside	446.41	C22H22O10	3231
1340	3-Hydroxy-8,9-methylenedioxypterocarpan 3-O-glucoside	(−)-Maackiain 3-O-glucoside	446.41	C22H22O10	3232
1341	3-Hydroxy-9,10-dimethoxypterocarpan 3-O-glucoside	Methylnissolin 3-O-glucoside	462.45	C23H26O10	3235
1342	3-Hydroxy-9,10-methylenedioxypterocarpan 3-O-(6'-acetylglucoside)	Trifolirhizin 6'-monoacetate	488.45	C24H24O11	3234
1343	3-Hydroxy-9-methoxypterocarpan 3-O-(6'-malonylglucoside)	Medicarpin 3-O-(6'-malonylglucoside)	518.48	C25H26O12	3230
1344	3,4-Dihydroxy-9-methoxypterocarpan 3,4-di-O-glucoside	4-Hydroxymedicarpin 3,4-O-diglucoside	610.56	C28H34O15	F235

PTEROCARPENE

No.	Trivial name	Systematic name	MW	Mol. formula	Ref.
1345	Anhydroglycinol	3,9-Dihydroxypterocarpene	254.24	C15H10O4	2705
1346	Lespedezol A1	9-Hydroxy-3-methoxypterocarpene	268.26	C16H12O4	F170
1347	Neorauteen	9-Hydroxyfurano[2',3':3,2]pterocarpene	278.27	C17H10O4	2715
1348		3-Hydroxy-8,9-methylenedioxypterocarpene	282.25	C16H10O5	3058
1349	Anhydroglycinol	3,9-Dimethoxypterocarpene	282.30	C17H14O4	2706
1350	Lespedezol A4	8,9-Dihydroxy-3-methoxypterocarpene	284.26	C16H12O5	F171
1351	Anhydropisatin	3-Methoxy-8,9-methylenedioxypterocarpene	296.28	C17H12O5	2707
1352	Neoduleen	8,9-Methyleneoxyfurano[2',3':3,2]pterocarpene	306.28	C18H10O5	2716
1353		3-Hydroxy-4-methoxy-8,9-methylenedioxypterocarpene	312.28	C17H12O6	2709
1354	Andirol B	3-Hydroxy-8,9-methylenedioxy-6-hydroxymethylpterocarpene	312.28	C17H12O6	F127
1355		3,9,10-Trimethoxypterocarpene	312.32	C18H16O5	2708
1356	Anhydroglycinol	1,7-Dihydroxy-3,9-dimethoxypterocarpene	314.29	C17H14O6	F314
1357	Anhydrotuberosin	3-Hydroxy-6',6'-dimethylpyrano[2',3':9,8]pterocarpene	320.34	C20H16O4	3055
1358	Erypoegin H	3,9-Dihydroxy-10-prenylpterocarpene	322.36	C20H18O4	F272
1359	Erypoegin E	3,9-Dihydroxy-4-prenylpterocarpene	322.36	C20H18O4	F278
1360	Bryacarpene 2	10-Hydroxy-3,8,9-trimethoxypterocarpene	328.32	C18H16O6	2710
1361	Bryacarpene 4	4-Hydroxy-3,9,10-trimethoxypterocarpene	328.32	C18H16O6	2711
1362	3-O-Methylanhydrotuberosin	3-Methoxy-6',6'-dimethylpyrano[2',3':9,8]pterocarpene	334.37	C21H18O4	3056
1363	Leiocalycin	2-Hydroxy-1,3-dimethoxy-8,9-methylenedioxypterocarpene	342.30	C18H14O7	2713
1364	Andirol A	3-Hydroxy-7-methoxy-8,9-methylenedioxy-6-hydroxymethylpterocarpene	342.30	C18H14O7	F127
1365	Bryacarpene 3	3,8,9,10-Tetramethoxypterocarpene	342.35	C19H18O6	2712
1366	Bryacarpene 1	4,10-Dihydroxy-3,8,9-trimethoxypterocarpene	344.32	C18H16O7	2714
1367	Hedysarimpterocarpene B	1,3-Dihydroxy-9-methoxy-10-prenylpterocarpene	352.38	C21H20O5	F315
1368	Puemiricarpene	3,9-Dihydroxy-8-methoxy-7-prenylpterocarpene	352.38	C21H20O5	F38
1369	Erycristagallin	3,9-Dihydroxy-2,10-diprenylpterocarpene	390.48	C25H26O4	3057
1370	Lespedezol A3	3,8-Dihydroxy-6'-methyl-6-(4-methyl-3-pentenyl)pyrano[2',3':9,10]pterocarpene	404.46	C25H24O5	F170
1371	Lespedezol A2	3,8,9-Trihydroxy-10-(3,7-dimethyl-2,6-octadienyl)pterocarpene	406.47	C25H26O5	F170
1372	Lespedezol A5	3,8,9-Trihydroxy-7-(methylmethanoate)-10-(3,7-dimethyl-2,6-octadienyl)pterocarpene	464.52	C27H28O7	F171

PTEROCARPENEQUINONES

No.	Trivial name	Systematic name	MW	Mol. formula	Ref.
1373	4-Deoxybryaquinone	9,10-Diketo-3,7-dimethoxypterocarpene	312.28	C17H12O6	2717
1374	Bryaquinone	9,10-Diketo-4-hydroxy-3,7-dimethoxypterocarpene	328.28	C17H12O7	2718

continued

APPENDIX
Checklist for Isoflavonoids Described in the Literature During the Period 1991–2004 — continued

COUMESTANES

	Name	Trivial Name	Mass	Formula	Ref.
1375	3,9-Dihydroxycoumestan	Coumestrol	268.23	C15H8O5	2774
1376	3-Hydroxy-9-methoxycoumestan	9-O-Methylcoumestrol	282.25	C16H10O5	2775
1377	8,9-Dihydroxy-1-methylcoumestan	Mutisifurocoumarin	282.25	C16H10O5	3090
1378	1,3,9-Trihydroxycoumestan	Aureol	284.23	C15H8O6	3078
1379	2,3,9-Trihydroxycoumestan	Lucernol	284.23	C15H8O6	2777
1380	3,4,9-Trihydroxycoumestan	4-Hydroxycoumestrol	284.23	C15H8O6	F182
1381	3,7,9-Trihydroxycoumestan	Repensol	284.23	C15H8O6	2776
1382	3-Hydroxy-8,9-methyledioxycoumestan	Medicagol	296.24	C16H8O6	2778
1383	3,9-Dimethoxycoumestan	Coumestrol dimethyl ether	296.28	C17H12O5	2779
1384	2,9-Dihydroxy-3-methoxycoumestan	Melimessanol A	298.25	C16H10O6	F153
1385	3,7-Dihydroxy-9-methoxycoumestan	Trifoliol	298.25	C16H10O6	2780
1386	3,9-Dihydroxy-1-methoxycoumestan	Isotrifoliol	298.25	C16H10O6	F88
1387	3,9-Dihydroxy-4-methoxycoumestan	Pongacoumestan	298.25	C16H10O6	F324
1388	3,9-Dihydroxy-8-methoxycoumestan	8-Methoxycoumestrol	298.25	C16H10O6	2781
1389	4,9-Dihydroxy-3-methoxycoumestan	Sativol	298.25	C16H10O6	2782
1390	1,3,8,9-Tetrahydroxy-2-prenylcoumestan		300.23	C15H8O7	F32
1391	1,3,8,9-Tetrahydroxycoumestan	Demethylwedelolactone	300.23	C15H8O7	2783
1392	3-Methoxy-8,9-methyledioxycoumestan	Flemichapparin C	310.26	C17H10O6	2784
1393	3-Hydroxy-7,9-dimethoxycoumestan	Wairol	312.28	C17H12O6	2785
1394	3-Hydroxy-8,9-dimethoxycoumestan	3-hydroxy-8,9-Dimethoxycoumestan	312.28	C17H12O6	2786
1395	1,8,9-Trihydroxy-3-methoxycoumestan	Wedelolactone	314.25	C16H10O7	2787
1396	3-Carboxylic acid-5,6-dihydroxy-2-(2',4',6'-trihydroxyphenyl)benzofuran	Norwedelic acid	318.24	C15H10O8	3082
1397	8,9-Methyledioxyfurano[2',3':3,2]coumestan	Erosnin	320.26	C18H8O6	2792
1398	2-Hydroxy-3-methoxy-8,9-methyledioxycoumestan	2-Hydroxyflemichapparin C	326.26	C17H10O7	2788
1399	3-Hydroxy-2-methoxy-8,9-methyledioxycoumestan	Tephrosol	326.26	C17H10O7	2789
1400	3-Hydroxy-4-methoxy-8,9-methyledioxycoumestan	Sophoracoumestan B	326.26	C17H10O7	2790
1401	1,9-Dihydroxy-3,8-dimethoxy-2-prenylcoumestan		328.28	C17H12O7	F32
1402	3,9-Dihydroxy-1,8-dimethoxy-2-prenylcoumestan		328.28	C17H12O7	F32
1403	3-Hydroxy-6',6'-dimethylpyrano[2',3':9,8]coumestan	Sophoracoumestan A	334.33	C20H14O5	2793

No.	Systematic name	Trivial name	Formula	MW	Code
1404	9-Hydroxy-6',6'-dimethylpyrano[2',3':3,4]coumestan	Plicadin	$C_{20}H_{14}O_5$	334.33	F216
1405	3-Hydroxy-6',6'-dimethyl-4',5'-dihydropyran[2',3':9,8]coumestan	Sojagol	$C_{20}H_{16}O_5$	336.34	2794
1406	3,9-Dihydroxy-10-prenylcoumestan	Isosojagol	$C_{20}H_{16}O_5$	336.34	3080
1407	3,9-Dihydroxy-2-prenylcoumestan	Psoralidin	$C_{20}H_{16}O_5$	336.34	2795
1408	3,9-Dihydroxy-4-prenylcoumestan	Phaseol	$C_{20}H_{16}O_5$	336.34	3081
1409	9-Hydroxy-6',6'-dimethyl-4',5'-dihydropyran[2',3':3,2]coumestan	Isopsoralidin	$C_{20}H_{16}O_5$	336.34	2796
1410	9-Hydroxy-1,3,8-trimethoxy-2-prenylcoumestan		$C_{18}H_{14}O_7$	342.30	F32
1411	3-Methoxy-6',6'-dimethylpyrano[2',3':9,8]coumestan	Tuberostan	$C_{21}H_{16}O_5$	348.36	3079
1412	3-Hydroxy-5'-(1-hydroxy-1-methylethyl)-4',5'-dihydrofurano[2',3':9,8]coumestan	Bavacoumestan B	$C_{20}H_{16}O_6$	352.34	3084
1413	3,3'-Dihydroxy-6',6'-dimethyl-4',5'-dihydropyrano[2',3':9,8]coumestan	Bavacoumestan A	$C_{20}H_{16}O_6$	352.34	3083
1414	3,9-Dihydroxy-2-(3-methyl-2,3-epoxybutyl)coumestan	Psoralidin oxide	$C_{20}H_{16}O_6$	352.34	2797
1415	2-Hydroxy-1,3-dimethoxy-8,9-methylenedioxycoumestan		$C_{18}H_{12}O_8$	356.29	2791
1416	1,3,8,9-Tetramethoxy-2-prenylcoumestan		$C_{19}H_{16}O_7$	356.33	F32
1417	8,9-Methylenedioxy-5'-(1-methylethenyl)-4',5'-dihydrofurano[2',3':3,2]coumestan	Tephcalostan	$C_{21}H_{14}O_6$	362.33	F123
1418	7-Hydroxy-9-methoxy-6',6'-dimethylpyrano[2',3':3,4]coumestan	Gancaonol B	$C_{21}H_{16}O_6$	364.36	F70
1419	9-Hydroxy-1-methoxy-6',6'-dimethylpyrano[2',3':9,8]coumestan	Sojagol	$C_{21}H_{16}O_6$	364.36	3092
1420	1,9-Dihydroxy-3-methoxy-2-prenylcoumestan	Glycyrol	$C_{21}H_{18}O_6$	366.36	2798
1421	3,9-Dihydroxy-1-methoxy-8-prenylcoumestan		$C_{21}H_{18}O_6$	366.36	3088
1422	3,9-Dihydroxy-4-methoxy-8-prenylcoumestan	Puerarostan	$C_{21}H_{18}O_6$	366.36	3089
1423	3,9-Dihydroxy-8-methoxy-7-prenylcoumestan	Mirificoumestan	$C_{21}H_{18}O_6$	366.36	3086
1424	9-Hydroxy-3-methoxy-6',6'-dimethyl-4',5'-dihydropyrano[2',3':1,2]coumestan	Isoglycyrol	$C_{21}H_{18}O_6$	366.36	2799
1425	9,4',5'-Trihydroxy-6',6'-dimethyl-4',5'-dihydropyrano[2',3':3,2]coumestan	Corylidin	$C_{20}H_{16}O_7$	368.34	2800
1426	9-Hydroxy-1,3-dimethoxy-2-prenylcoumestan	1-O-Methylglycyrol	$C_{22}H_{20}O_6$	380.40	2801
1427	3,9-Dihydroxy-8-methoxy-7-(3-hydroxy-3-methylbutyl)coumestan	Mirificoumestan hydrate	$C_{21}H_{20}O_7$	384.39	3085
1428	3,9-Dihydroxy-8-methoxy-7-(2,3-dihydroxy-3-methylbutyl)coumestan	Mirificoumestan glycol	$C_{21}H_{20}O_8$	400.39	3087
1429	3,9-Dihydroxy-2-geranylcoumestan	Puerarol	$C_{25}H_{24}O_5$	404.46	3090
1430	3,9-Dihydroxy-2,10-diprenylcoumestan	Sigmoidin K	$C_{25}H_{24}O_5$	404.46	F181
1431	3,8,9-Trihydroxy-10-(3,7-dimethyl-2,6-octadienyl)coumestan	Lespedezol A6	$C_{25}H_{24}O_6$	420.46	F171

Coumestan glycosides

No.	Systematic name	Trivial name	Formula	MW	Code
1432	3-Hydroxy-9-methoxycoumestan 3-O-glucoside	Licoagroside C	$C_{22}H_{20}O_{10}$	444.39	F141
1433	1,3,8,9-Tetrahydroxycoumestan 3-O-glucoside	3-Demethylwedelolactone-O-glucoside	$C_{21}H_{18}O_{12}$	462.37	3238
1434	3,9-Dihydroxycoumestan 3-O-glucoside	Coumestrol 3-O-glucoside	$C_{21}H_{18}O_{10}$	430.37	3237

continued

APPENDIX
Checklist for Isoflavonoids Described in the Literature During the Period 1991–2004 — *continued*

ROTENONES

	Name	Trivial Name	Mass	Formula	Ref.
1435	6,9,11-Trihydroxyrotenone	Coccineone B	298.25	C16H10O6	F67
1436	(−)-4,9,11,12a-Tetrahydroxyrotenone		316.27	C16H12O7	3008
1437	(−)-4,11,12a-Trihydroxy-9-methoxyrotenone		330.29	C17H14O7	3009
1438	2,3-Methylenedioxyfurano[2′,3′:9,10]rotenone	Dolineone	336.30	C19H12O6	2557
1439	2,3,9-Trimethoxyrotenone	Munduserone	342.35	C19H18O6	2555
1440	(−)-4,11,12a-Trihydroxy-9-methoxy-10-methylrotenone	Boeravinone C	344.32	C18H16O7	3010
1441	(6aR,12aS)-11,12a-Dihydroxy-9,10-dimethoxyrotenone	9-Methoxyirispurinol	344.32	C18H16O7	F236
1442	12a-Hydroxy-2,3-methylenedioxyfurano[2′,3′:9,10]rotenone	12a-Hydroxydolineone	352.30	C19H12O7	2577
1443	2,3-Dimethoxyrfurano[2′,3′:9,10]rotenone	Erosone	352.34	C20H16O6	2559
1444	2,3-Dimethoxyrfurano[2′,3′:9,8]rotenone	Elliptone	352.34	C20H16O6	2558
1445	12-Hydroxy-2,3-dimethoxyfurano[2′,3′:9,8]rotenone	Elliptinol	354.36	C20H18O6	3005
1446	(6aR,12aR)-12a-Hydroxy-2,3-dimethoxyfurano[2′,3′:9,8]rotenone	12a-Hydroxyelliptone	354.36	C20H18O6	F293
1447	(6aR,12aS)-12a-Hydroxy-2,3,8,9-bis(methylenedioxy)rotenone	Usararotenoid A	356.29	C18H12O8	F342
1448	(6aR,12R,12aR)-12,12a-Dihydroxy-12-dihydro-2,3:8,9-bis(methylenedioxy)rotenone	12-Dihydrousararotenoid A	358.30	C18H14O8	F342
1449	11-Hydroxy-2,3,9-trimethoxyrotenone	Sermundone	358.35	C19H18O7	2556
1450	12a-Hydroxy-2,3,9-trimethoxyrotenone	12a-Hydroxymunduserone	358.35	C19H18O7	2575
1451	(6aR,12aS)-4,11,12a-Trihydroxy-9-methoxy-8,10-dimethylrotenone	Mirabijalone A	358.35	C19H18O7	F316
1452	12a-Methoxy-2,3-methylenedioxyfurano[2′,3′:9,10]rotenone	12a-Methoxydolineone	366.33	C20H14O7	2578
1453	8-Methoxy-2,3-methylenedioxyfurano[2′,3′:9,10]rotenone	Pachyrrizone	366.33	C20H14O7	2560
1454	11-Hydroxy-2,3-dimethoxyrfurano[2′,3′:9,8]rotenone	Malaccol	368.34	C20H16O7	2561
1455	12a-Hydroxy-2,3-dimethoxyfurano[2′,3′:9,10]rotenone	12a-Hydroxyerosone	368.34	C20H16O7	2579
1456	(6aS,12aS)-12a-Hydroxy-2,3-dimethoxyfurano[2′,3′:9,8]rotenone	12a-Hydroxyelliptone	368.34	C20H16O7	F109
1457	(6aR,12aS)-12a-Hydroxy-8,9-dimethoxy-2,3-methylenedioxyrotenone	Usararotenoid B	372.33	C19H16O8	F342
1458	6,11-Dihydroxy-2,3,9-trimethoxyrotenone	Dihydrostemonal	374.35	C19H18O8	3001
1459	(6aR,12aR)-11,12a-Dihydroxy-2,3,9-trimethoxyrotenone	6-Deoxyclitoriacetal	374.35	C19H18O8	F143
1460	(6aS,12aR)-6,12a-Dihydroxy-2,3,9-trimethoxyrotenone	11-Deoxyclitoriacetal	374.35	C19H18O8	F162
1461	(6R,6aS,12aR)-6,9,11,12a-Tetrahydroxy-2,3-dimethoxyrotenone	9-Demethylclitoriacetal	376.31	C18H16O9	F246
1462	2,3-Methylenedioxy-5′-(1-methylethenyl)-4′,5′-dihydrofurano[2′,3′:9,8]rotenone	Isomillettone	378.38	C22H18O6	2563
1463	2,3-Methylenedioxy-6′,6′-dimethylpyrano[2′,3′:9,8]rotenone	Millettone	378.38	C22H18O6	2562
1464	12a-Hydroxy-8-methoxy-2,3-methylenedioxyfurano[2′,3′:9,10]rotenone	12a-Hydroxypachyrrhizone	382.33	C20H14O8	2581

1465	2,3,12a-Trimethoxyfurano[2′,3′:9,10]rotenone	Neobanone	382.37	C20H18O7	2580
1466	6,11,12a-Trihydroxy-2,3,9-trimethoxyrotenone	Clitoriacetal	390.35	C19H18O9	2576
1467	12a-Hydroxy-2,3-methylenedioxy-5′-(1-methylethenyl)-4′,5′-dihydrofurano[2′,3′:9,8]rotenone	12a-Hydroxymillettone	394.38	C22H18O7	2583
1468	12a-Hydroxy-2,3-methylenedioxy-6′,6′-dimethylpyrano[2′,3′:9,8]rotenone	Millettosin	394.38	C22H18O7	2582
1469	(6aR,12aS)-12a-Hydroxy-2,3-methylenedioxy-6′,6′-dimethoxypyr[2′,3′:9,8]rotenone	12a-Epimillettosin	394.38	C22H18O7	F342
1470	2,3-Dimethoxy-5′-(1-methylethenyl)-4′,5′-dihydrofurano[2′,3′:9,8]rotenone	Rotenone	394.42	C23H22O6	2565
1471	2,3-Dimethoxy-6′,6′-dimethylpyrano[2′,3′:9,8]rotenone	Deguelin	394.42	C23H22O6	2564
1472	9-Hydroxy-2,3-dimethoxy-10-(1,4-pentadienyl)rotenone	Myriconol	394.42	C23H22O6	2566
1473	3-Hydroxy-2-methoxy-5′-(1-hydroxymethyl-1-ethenyl)-4′,5′-dihydrofurano[2′,3′:9,8]rotenone	3-O-Demethylamorphigenin	396.40	C22H20O7	3000
1474	9-Hydroxy-2,3-dimethoxy-8-prenylrotenone	Rotenonic acid	396.44	C23H24O6	2567
1475	11-Hydroxy-2,3-dimethoxy-5′-(1-methylethenyl)-4′,5′-dihydrofurano[2′,3′:9,8]rotenone	Sumatrol	410.42	C23H22O7	2570
1475b	(12R)-12-Hydroxy-2,3-dimethoxy-5′-(1″-hydroxymethylethenyl)-4′,5′-dihydrofurano[2′,3′:9,8]rotenone	Dalcochinin	412.44	C23H24O7	F261b
1476	11-Hydroxy-2,3-dimethoxy-6′,6′-dimethylpyrano[2′,3′:9,8]rotenone	Toxicarol	410.42	C23H22O7	2569
1477	12a-Hydroxy-2,3-dimethoxy-5′-(1-methylethenyl)-4′,5′-dihydrofurano[2′,3′:9,8]rotenone	12a-Hydroxyrotenone	410.42	C23H22O7	2585
1478	12a-Hydroxy-2,3-dimethoxy-6′,6′-dimethylpyrano[2′,3′:9,8]rotenone	Tephrosin	410.42	C23H22O7	2584
1479	2,3-Dimethoxy-5′-(1-hydroxymethylethenyl)-4′,5′-dihydrofurano[2′,3′:9,8]rotenone	Amorphigenin	410.42	C23H22O7	2568
1480	6-Hydroxy-2,3-dimethoxy-5′-(1-methyl-1-ethenyl)-4′,5′-dihydrofurano[2′,3′:9,8]rotenone	6-Hydroxyrotenone	410.42	C23H22O7	3003
1481	(6aR,12aS)-12a-Hydroxy-9-methoxy-2,3-methylenedioxy-8-prenylrotenone	Usararotenoid C	410.42	C23H22O7	F334
1482	(6R,6aS,12aS)-6-Hydroxy-2,3-dimethoxy-6′,6′-dimethylpyrano[2′,3′:9,8]rotenone	Hydroxydeguelin	410.42	C23H22O7	F150
1483	2,3-Dimethoxy-5′-(1-hydroxy-1-methylethyl)-4′,5′-dihydrofurano[2′,3′:9,8]rotenone	Dalpanol	412.44	C23H24O7	2572
1484	2,3-Dimethoxy-5′-(1-hydroxymethylethyl)-4′,5′-dihydrofurano[2′,3′:9,8]rotenone	Dihydroamorphigenin	412.44	C23H24O7	2571
1485	cis-9,12a-Dihydroxy-2,3-dimethoxy-8-prenylrotenone	cis-12a-Hydroxyrot-2′-enonic acid	412.44	C23H24O7	3006
1486	3,12a-Dihydroxy-2-methoxy-5′-(2-hydroxy-1-methylethyl)-4′,5′-dihydrofurano[2′,3′:9,8]rotenone	Volubinol	414.41	C22H22O8	3007
1487	6-Acetoxy-11-hydroxy-2,3,9-trimethoxyrotenone	6-Acetyldihydrostemonal	416.39	C21H20O9	3002
1488	12a-Methoxy-2,3-dimethoxy-5′-(1-methylethenyl)-4′,5′-dihydrofurano[2′,3′:9,8]rotenone	12a-Methoxyrotenone	424.45	C24H24O7	2586
1489	(+)-2,3,10-Trimethoxy-6′,6′-dimethylpyrano[2′,3′:9,8]rotenone	Erythynone	424.45	C24H24O7	3004
1490	11,12a-Dihydroxy-2,3-dimethoxy-5′-(1-methylethenyl)-4′,5′-dihydrofurano[2′,3′:9,8]rotenone	Villosinol	426.42	C23H22O8	2589
1491	11,12a-Dihydroxy-2,3-dimethoxy-6′,6′-dimethylpyrano[2′,3′:9,8]rotenone	12a-Hydroxytephrosin	426.42	C23H22O8	2588
1492	12a-Hydroxy-2,3-dimethoxy-5′-(1-hydroxymethylethenyl)-4′,5′-dihydrofurano[2′,3′:9,8]rotenone	Dalbinol	426.42	C23H22O8	2587
1493	6,11-Dihydroxy-2,3-dimethoxy-5′-(1-methylethenyl)-4′,5′-dihydrofurano[2′,3′:9,10]rotenone	Villosin	426.42	C23H22O8	2573
1494	(6aR,12aR)-11,12a-Dihydroxy-2,3-dimethoxy-6′,6′-dimethylpyrano[2′,3′:9,8]rotenone	12a-hydroxy-b-toxicarol	426.42	C23H22O8	F11
1495	(6R,6aS,12aR)-6,12a-Dihydroxy-2,3-dimethoxy-6′,6′-dimethylpyrano[2′,3′:9,8]rotenone	Hydroxytephrosin	426.42	C23H22O8	F150
1496	12,12a-Dihydroxy-2,3-dimethoxy-5′-(1-hydroxymethylethenyl)-4′,5′-dihydrofurano[2′,3′:9,8]rotenone	12-Dihydrodalbinol	428.44	C23H24O8	2590
1497	2,3-Dimethoxy-5′-(1-hydroxy-1-hydroxymethylethyl)-4′,5′-dihydrofurano[2′,3′:9,8]rotenone	Amorphigenol	428.44	C23H24O8	2574

continued

APPENDIX
Checklist for Isoflavonoids Described in the Literature During the Period 1991–2004 — *continued*

	Name	Trivial Name	Mass	Formula	Ref.
1498	(6aS,12aS,4′S,5′R)-4′,5′-Dihydroxy-2,3,4′-dimethoxy-6′,6′-dimethyl-4′,5-dihydropyrano[2′,3′:9,8]rotenone	Derrisin	428.44	C23H24O8	F269
1499	(+)-12a-Hydroxy-2,3,10-trimethoxy-6′,6′-dimethylpyrano[2′,3′:9,8]rotenone	(+)-12a-Hydroxyerythynone	440.45	C24H24O8	3011
1500	6,11,12a-Trihydroxy-2,3-dimethoxy-5′-(1-methylethenyl)-4′,5′-dihydrofurano[2′,3′:9,8]rotenone	Villol	442.42	C23H22O9	2591
1501	(6aR,12aR)-11,12a-Dihydroxy-2,3,4′-trimethoxy-6′,6′-dimethyl-4′,5-dihydropyrano(2′,3′:9,8]rotenone		474.46	C24H26O10	F115
1502	(6aS,12aS)-12-Keto-12a-hydroxy-3,4-dimethoxy-9-((2E,5E)-7-hydroxy-3,7-dimethyl-2,5-octadienyloxy)) rotenone	Griffonianone A	496.55	C28H32O8	F331
	Rotenone glycosides				
1503	(6aR,12aR)-11,12a-Dihydroxy-2,3,9-trimethoxyrotenone 11-O-glucoside	6-Deoxyclitoriacetal 11-O-glucoside	536.49	C25H28O13	F200
1504	(6S,6aS,12aR)-6,12a-Dihydroxy-2,3,9-trimethoxyrotenone 11-O-glucoside	Rotenoid 11a-O-glucoside	536.49	C25H28O13	F161
1505	(6R,6aS,12aR)-6,11,12a-Trihydroxy-2,3,9-trimethoxyrotenone 11-O-glucoside	Clitoriacetal 11-O-glucoside	552.48	C25H28O14	F44
1506	2,3-Dimethoxy-5′-(1″-hydroxymethylethenyl)-4′,5′-dihydrofurano[2′,3′:9,8]rotenone 1″-O-glucoside	Amorphigenin 7-O-glucoside	572.57	C29H32O12	3221
1507	(6aR,12aR)-11,12a-Dihydroxy-2,3-dimethoxy 5′-(1-hydroxy-1-methylethyl)furano[2′,3′:9,8]rotenone 1″-O-glucoside	Dehydrodalpanol O-glucoside	572.57	C29H32O12	F214
1507b	(12R)-12-Hydroxy-2,3-dimethoxy-5′-(1″-hydroxymethylethenyl)-4′,5′-dihydrofurano[2′,3′:9,8]rotenone 1″-O-glucoside	Dalcochinin O-glucoside	574.58	C29H34O12	F261b
1508	2,3-Dimethoxy-5′-(1″-hydroxy-1″-methylethyl)-4′,5′-dihydrofurano[2′,3′:9,8]rotenone 1″-O-glucoside	Dalpanol O-glucoside	574.58	C29H34O12	3223
1509	12a-Hydroxy-2,3-dimethoxy-5′-(1″-hydroxymethylethenyl)-4′,5′-dihydrofurano[2′,3′:9,8]rotenone 1″-O-glucoside	Dalbinol O-glucoside	588.57	C29H32O13	3226
1510	12-Hydroxy-2,3-dimethoxy-5′-(1″-hydroxymethylethenyl)-4′,5′-dihydrofurano[2′,3′:9,8]-12-dihydrorotenone 1″-O-glucoside	12-Dihydrodalbinol O-glucoside	590.58	C29H34O13	3228
1511	2,3-Dimethoxy-5′-(1″-hydroxy-1″-hydroxymethylethyl)-4′,5′-dihydrofurano[2′,3′:9,8]rotenone 1″-O-glucoside	Amorphigenol O-glucoside	590.58	C29H34O13	3224
1512	2,3-Dimethoxy-5′-(1″-hydroxymethylethenyl)-4′,5′-dihydrofurano[2′,3′:9,8]rotenone 1″-O-(6‴-arabinosylglucoside)	Amorphigenin 7-O-vicianoside	704.69	C34H40O16	3222
1513	2,3-Dimethoxy-5′-(1″,2″-dihydroxy-1″-methylethyl)-4′,5′-dihydrofurano[2′,3′:9,8]rotenone 1″-O-(6‴-arabinosylglucoside)	Amorphigenol O-vicianoside	704.69	C34H40O16	3225
1514	12a-Hydroxy-2,3-dimethoxy-5′-(1″-hydroxymethylethenyl)-4′,5′-dihydrofurano[2′,3′:9,8]rotenone 1″-O-(6‴-arabinosylglucoside)	Dalbinol O-vicianoside	720.68	C34H40O17	3227

DEHYDROROTENONES

No.	Name	Systematic name	MW	Formula	Code
1515	Pongarotene	Furano[2′,3′:9,8]-6a,12a-didehydrororotenone	290.27	C18H10O4	F249
1516	Boeravinone B	6,9,11-Trihydroxy-10-methyl-6a,12a-didehydrororotenone	312.28	C17H12O6	3015
1517		1,11-Dihydroxy-9,10-methylenedioxy-6a,12a-didehydrororotenone	326.26	C17H10O7	F250
1518	Boeravinone F	6-Oxo-3,9,11-Trihydroxy-10-methyl-6a,12a-didehydrororotenone	326.26	C17H10O7	F136
1519	Boeravinone A	9,11-Dihydroxy-6-methoxy-10-methyl-6a,12a-didehydrororotenone	326.31	C18H14O6	3016
1520	Boeravinone E	3,6,11-Tetrahydroxy-10-methyl-6a,12a-didehydrororotenone	328.28	C17H12O7	F136
1521	Dehydrodolineone	2,3-Methylenedioxyfurano[2′,3′:9,10]-6a,12a-didehydrororotenone	334.29	C19H10O6	2595
1522	Mirabijalone D	3,6,11-Trihydroxy-9-methoxy-10-methyl-6a,12a-didehydrororotenone	342.30	C18H14O7	F316
1523	Boeravinone D	3,9,11-Trihydroxy-6-methoxy-10-methyl-6a,12a-didehydrororotenone	342.30	C18H14O7	F136
1524	Mirabijalone B	4,6,9,11-Tetrahydroxy-8,10-dimethyl-6a,12a-didehydrororotenone	342.30	C18H14O7	F316
1525	Repenone	6-Acetoxy-9,10,11-trihydroxy-6a,12a-didehydrororotenone	356.29	C18H12O8	3013
1526	Dehydropacchyrrhizone	8-Methoxy-2,3-methylenedioxyfurano[2′,3′:9,10]-6a,12a-didehydrororotenone	364.31	C20H12O7	2596
1527	Stemonone	6-Oxo-11-hydroxy-2,3,9-trimethoxy-6a,12a-didehydrororotenone	370.32	C19H14O8	2592
1528	Repenol	6-Acetoxy-3,9,10,11-tetrahydroxy-6a,12a-didehydrororotenone	372.29	C18H12O9	3014
1529	Stemonal	6,11-Dihydroxy-2,3,9-trimethoxy-6a,12a-didehydrororotenone	372.33	C19H16O8	2593
1530	Dehydromillettone	2,3-Methylenedioxy-6′,6′-dimethylpyrano[2′,3′:9,8]-6a,12a-didehydrororotenone	376.37	C22H16O6	2597
1531	Dehydrorotenone	2,3-Dimethoxy-5′-(1-methylethenyl)-4′,5′-dihydrofurano[2′,3′:9,8]-6a,12a-didehydrororotenone	392.41	C23H20O6	2599
1532	Dehydroguelin	2,3-Dimethoxy-6′,6′-dimethylpyrano[2′,3′:9,8]-6a,12a-didehydrororotenone	392.41	C23H20O6	2598
1533	Dehydrodihydrororotenone	2,3-Dimethoxy-5′-(1-methylethyl)-4′,5′-dihydrofurano[2′,3′:9,8]-6a,12a-didehydrororotenone	394.42	C23H22O6	3017
1534	Diffusarotenoid	6-Pentanoate-4,9-dihydroxy-10-methyl-6a,12a-didehydrororotenone	396.40	C22H20O7	F84
1535	Stemonalacetal	11-Dihydroxy-6-ethoxy-2,3,9-trimethoxy-6a,12a-didehydrororotenone	400.38	C21H20O8	2594
1536	Rotenonone	6-Oxo-2,3-dimethoxy-5′-(1-methylethenyl)-4′,5′-dihydrofurano[2′,3′:9,8]-6a,12a-didehydrororotenone	406.39	C23H18O7	2600
1537	Villosol	11-Hydroxy-2,3-dimethoxy-5′-(1-methylethenyl)-4′,5′-dihydrofurano[2′,3′:9,8]-6a,12a-didehydrororotenone	408.41	C23H20O7	2604
1538	6a,12a-Dehydro-a-toxicarol	11-Hydroxy-2,3-dimethoxy-6′,6′-dimethylpyrano[2′,3′:9,10]-6a,12a-didehydrororotenone	408.41	C23H20O7	F147
1539	Dehydrotoxicarol	11-Hydroxy-2,3-dimethoxy-6′,6′-dimethylpyrano[2′,3′:9,8]-6a,12a-didehydrororotenone	408.41	C23H20O7	2603
1540	Dehydroamorphigenin	2,3-Dimethoxy-5′-(1-hydroxymethylethenyl)-4′,5′-dihydrofurano[2′,3′:9,8]-6a,12a-didehydrororotenone	408.41	C23H20O7	2602
1541	Amorpholone	6-Hydroxy-2,3-dimethoxy-5′-(1-methylethenyl)-4′,5′-dihydrofurano[2′,3′:9,8]-6a,12a-didehydrororotenone	408.41	C23H20O7	2601
1542	Dehydroalpanol	2,3-Dimethoxy-5′-(1-hydroxy-1-methylethenyl)furan[2′,3′:9,8]-6a,12a-didehydrororotenone	410.42	C23H22O7	2605
1543	6-Ketodehydroamorphigenin	6-Oxo-11-hydroxy-2,3-dimethoxy-5′-(1-hydroxymethylethenyl)furan[2′,3′:9,8]-6a,12a-didehydrororotenone	422.39	C23H18O8	F258
1544	Villosone	6-Oxo-11-hydroxy-2,3-dimethoxy-5′-(1-methylethenyl)-4′,5′-dihydrofuran[2′,3′:9,8]-6a,12a-didehydrororotenone	422.39	C23H18O8	2606
1545	6-Oxo-6a,12a-didehydro-a-toxicarol	6-Oxo-11-hydroxy-2,3-dimethoxy-6′,6′-dimethoxypyran[2′,3′:9,8]-6a,12a-didehydrororotenone	422.39	C23H18O8	F146
1546	6a,12a-Dehydrovillosin	(5′R,6S)-6,11-Dihydroxy-2,3-dimethoxy-5′-(1-methylethenyl)furano[2′,3′:9,8]-6a,12a-didehydrororotenone	424.41	C23H20O8	F207

continued

APPENDIX
Checklist for Isoflavonoids Described in the Literature During the Period 1991–2004 — *continued*

	Name	Trivial Name	Mass	Formula	Ref.
1547	(+)-6,11-Dihydroxy-2,3-dimethoxy-6′,6′-dimethylpyrano[2′,3′:9,8]-6a,12a-didehydrorotenone	6-Hydroxy-6a,12a-dehydro-α-toxicarol	424.41	C23H20O8	3012x
1548	2,3-Dimethoxy-6-methoxy-5′-(1-methylethenyl)-4′,5′-dihydrofurano[2′,3′:9,8]-6a,12a-didehydrorotenone	Villinol	438.44	C24H22O8	2607
1549	9,11-Dihydroxy-2,3,6-trimethoxy-8-prenyl-6a,12a-didehydrorotenone		440.45	C24H24O8	F206
1550	11,5′-Dihydroxy-2,3,4′-trimethoxy-6′,6′-dimethyl-4′,5′-dihydropyrano[2′,3′:9,8]-6a,12a-didehydrorotenone		456.45	C24H24O9	F114
	Dehydrorotenone glycosides				
1551	6,9-Dihydroxy-2,3,10-trimethoxy-6a,12a-didehydrorotenone 9-*O*-glucoside		534.47	C25H26O13	F248
	2-ARYLBENZOFURAN				
1552	6,2′,4′-Trihydroxy-2-arylbenzofuran		242.23	C14H10O4	2821
1553	5-Hydroxy-4′-methoxy-2,3-dihydro-2-arylbenzofuran	Corsifuran B	242.27	C15H14O3	F304
1554	5-Hydroxy-6-methoxy-3-methyl-2-arylbenzofuran	Parvifuran	254.29	C16H14O3	2822
1555	5,4′-Dimethoxy-2-arylbenzofuran	Corsifuran C	254.29	C16H14O3	F304
1556	6,2′-Dihydroxy-4′-methoxy-2-arylbenzofuran	Centrolobofuran	256.26	C15H12O4	3103
1557	6,4′-Dihydroxy-2′-methoxy-2-arylbenzofuran	6-Demethylvignafuran	256.26	C15H12O4	2823
1558	5,4′-Dimethoxy-2,3-dihydro-2-arylbenzofuran	Corsifuran A	256.30	C16H16O3	F304
1559	2′,4′-Dihydroxy-5,6-methylenedioxy-2-arylbenzofuran		270.24	C15H10O5	2825
1560	4′-Hydroxy-2′,6-dimethoxy-2-arylbenzofuran	Vignafuran	270.29	C16H14O4	2824
1561	2′-Hydroxy-4′-methoxy-5,6-methylenedioxy-2-arylbenzofuran		284.27	C16H12O5	2826
1562	6-Hydroxy-2′-methoxy-4′,5′-methylenedioxy-2-arylbenzofuran	Cicerfuran	284.27	C16H12O5	F259
1563	5,2′-Dihydroxy-6,4′-dimethoxy-2-arylbenzofuran	Sainfuran	286.29	C16H14O5	3104
1564	5,6-Dihydroxy-2′,4′-dimethoxy-2-arylbenzofuran		286.29	C16H14O5	2829
1565	6,3′-Dihydroxy-2′,4′-dimethoxy-2-arylbenzofuran	Pterofuran	286.29	C16H14O5	2827
1566	6,4′-Dihydroxy-2′,3′-dimethoxy-2-arylbenzofuran	Isopterofuran	286.29	C16H14O5	2828
1567	6,4′-Dihydroxy-2′,5′-dimethoxy-2-arylbenzofuran	Eryvarin L	286.29	C16H14O5	F276
1568	3-Carboxyaldehyde-6′-hydroxy-6,4′-dimethoxy-2-arylbenzofuran	Melimessanol C	298.29	C17H15O5	F153
1569	2′,4′-Dihydroxy-3′-methoxy-5,6-methylenedioxy-2-arylbenzofuran	Sophorafuran A	300.27	C16H12O6	2830
1570	5-Hydroxy-6,2′,4′-trimethoxy-2-arylbenzofuran	Methylsainfuran	300.31	C17H16O5	3105
1571	5,4′-Dihydroxy-6″,6″-dimethylpyrano[2″,3″:2′,3′]-2-arylbenzofuran	Kanzonol U	308.33	C19H16O4	F73

1572	6,2'-Dihydroxy-6'',6''-dimethylpyrano[2'',3'':4',3']-2-arylbenzofuran	Glyinflanin H	308.33	C19H16O4	F75
1573	3-Carboxyaldehyde-6,4'-dihydroxy-2',5'-dimethoxy-2-arylbenzofuran	Eryvarin P	314.29	C17H14O6	F280
1574	3-Carboxylic acid methyl ester-4,6,3',4'-tetrahydroxy-2-arylbenzofuran	Oryzafuran	316.27	C16H12O7	F87
1575	6,3'-Dihydroxy-5'-methoxy-4'-prenyl-2-arylbenzofuran	Artoindonesianin O	324.37	C20H20O4	F86
1576	6,4'-Dihydroxy-2'-methoxy-3'-prenyl-2-arylbenzofuran	Bidwillol B	324.37	C20H20O4	F104
1577	6,4'-Dihydroxy-2'-methoxy-6-(1,1-dimethyl-2-propenyl)-2-arylbenzofuran	Burttinol D	324.37	C20H20O4	F337
1578	5,5',5'-Trihydroxy-6-(4-hydroxy-3-methyl-2(E)-butenyl)-2-arylbenzofuran	w-Hydroxymoracin N	326.34	C19H18O5	F163
1578b	3-Carboxyaldehyde-4,3',4'-trihydroxy-6,2-dimethoxy-2-arylbenzofuran	Andinermal C	330.29	C17H14O7	F127b
1579	2',4'-Dihydroxy-4-methoxy-6',6'-dimethylpyrano[2'',3'':6,5]-2-arylbenzofuran	Bryebinal	338.36	C20H18O5	2832
1580	6,2',4'-Trihydroxy-4-methoxy-5-prenyl-2-arylbenzofuran	Licocoumarone	340.38	C20H20O5	3106
1580b	3-Carboxyaldehyde-4,3'-dihydroxy-6,2',4'-trimethoxy-2-arylbenzofuran	Andinermal A	344.32	C18H16O7	F127b
1581	3-Carboxyaldehyde-2',4'-dihydroxy-6-methoxy-2-arylbenzofuran	Erypoegin F	352.38	C21H20O5	F272
1582	2',4'-Dihydroxy-4,6-dimethoxy-5-prenyl-2-arylbenzofuran	Gancaonin I	354.40	C21H22O5	3107
1583	6,2'-Dihydroxy-3,4'-dimethoxy-5-prenyl-2-arylbenzofuran	Ambofuranol	354.40	C21H22O5	2833
1583b	3-Carboxyaldehyde-4,3'-dihydroxy-2'',5''-dihydrofurano[3'',4'':5,6]-2-arylbenzofuran	Andinermal B	356.33	C19H16O7	F127b
1584	3'-Hydroxy-4,6,2',4'-tetramethoxy-3-hydroxymethyl-2-arylbenzofuran	Andinermol	360.37	C19H20O7	F127
1585	4,3'-Dihydroxy-5,7,2',4'-tetramethoxy-2-arylbenzofuran	Bryebinal	374.35	C19H18O8	2831
1586	5,4'-Dihydroxy-6-prenyl-6'',6''-dimethylpyrano[2'',3'':2',3']-2-arylbenzofuran	Kanzonol V	376.44	C24H24O4	F73
1587	6,3',5'-Trihydroxy-2',6'-diprenyl-2-arylbenzofuran	Mulberrofuran V	378.46	C24H26O4	F71
1588	3-Carboxyaldehyde-6,2',4'-trihydroxy-7,5'-diprenyl-2-arylbenzofuran	Eryvarin Q	406.47	C25H26O5	F280
1589	5,6,2',4'-Tetrahydroxy-3-methyl-7-(3,7-dimethyl-2,6-octadienyl)-2-arylbenzofuran	Lespedezol B1	408.50	C25H28O5	F170
1590	6,3',5'-Trihydroxy-5-methoxy-4-((2E)-3,7-dimethyl-2,6-octadienyl)-2-arylbenzofuran	Mulberrofuran Y	408.50	C25H28O5	F238
1591	6,3',5'-Trihydroxy-5-methoxy-2-arylbenzofuran 3'-O-glucoside	Schoenoside	434.39	C21H22O10	F119
1592	3',5'-Dihydroxy-2',6'-diprenyl-6'',6''-dimethylpyrano[2'',3'':6,7]-2-arylbenzofuran	Artoindonesianin Y	444.56	C29H32O4	F254
1593	6,3',5'-Trihydroxy-2'-((2E,6E)-3,7,11-trimethyl-2,6,10-dodecatrienyl)-2-arylbenzofuran	Mulberrofuran W	446.58	C29H34O4	F238
1594	6,3',5'-Trihydroxy-7,2',6'-triprenyl-2-arylbenzofuran	Artoindonesianin X	446.58	C29H34O4	F254
1595	6,5'-Dihydroxy-3'-methoxy-7-[(2E)-3,7-dimethyl-2,6-octadienyl]-2'-prenyl-2-arylbenzofuran	Sanggenofuran A	460.60	C30H36O4	F239
1596	(2''R)-6,5'-Dihydroxy-7-((2E)-3,7-dimethyl-2,6-octadienyl)-5''-(1-hydroxy-1-methylethyl)-4'',5''-dihydrofurano[2'',3'':3,2]-2-arylbenzofuran	Mulberrofuran X	462.58	C29H34O5	F238
1597	6,3',5'-Trihydroxy-5-methoxy-4-((2E)-3,7-dimethyl-2,6-octadienyl)-2'-prenyl-2-arylbenzofuran	Mulberrofuran Z	476.62	C30H36O5	F238
1598	6,3',5'-Trihydroxy-7-prenyl-4'-((1S,5S,6S)-6-(2,4-dihydroxyphenyl)-5-(2,4-dihydroxyphenylmethanonyl)-3-methyl-2-cyclohexen-1-yl)-2-arylbenzofuran	Mulberroside U	648.70	C39H36O9	F24

2-Arylbenzofuran glycosides

1599	6,3',5'-Trihydroxy-2-arylbenzofuran 3'-O-glucoside	Moracin M 3'-O-glucoside	404.37	C20H20O9	F24
1600	3',5',5''-Trihydroxy-6'',6''-dimethyl-4'',5''-dihydropyrano[2'',3'':6,5]-2-arylbenzofuran 3'-O-glucoside	Mulberroside C	458.47	C24H26O9	F262

continued

APPENDIX
Checklist for Isoflavonoids Described in the Literature During the Period 1991–2004 — continued

	Name	Trivial Name	Mass	Formula	Ref.
	2,5-DIARYLBENZOFURAN				
1601	(2R,3S,3″S,5′S)-3,3″-Dihydroxy-4′-methoxy-2,3-dihydro-2,5″-diaryl-(cyclopentan[1″,2″:2,3]benzofuran)	Ferrugin	434.49	C26H26O6	F50
	3-ARYL-2,3-DIHYDROBENZOFURANS				
1602	3-Carboxyaldehyde-6,4′-dihydroxy-2′,5′-dimethoxy-3-aryl-2,3-dihydrobenzofuran	Eryvarin R	316.31	C17H16O6	F280
1603	(2S,3S)-7,4′-Dihydroxy-3′-methoxy-2-hydroxypropyl-5-hydroxyethyl-2-aryl-2,3-dihydrobenzofuran 4′-O-rhamnoside	Mulberrofuran Z	476.62	C30H36O5	F237
	2-ARYLBENZOFURANQUINONE				
1604	3-Carboxyaldehyde-4,7-diketo-3′-hydroxy-6,2′,4′trimethoxy-5-prenyl-2-arylbenzofuranquinone	Bryebinalquinone	358.30	C18H14O8	2834
	α-METHYLDEOXYBENZOINS				
1605	2,4-Dihydroxy-4′-methoxy-α-methyldeoxybenzoin	Angolensin	272.30	C16H16O4	2816
1606	2-Hydroxy-4,4′-dimethoxy-″α -methyldeoxybenzoin	4-O-Methylangolensin	286.33	C17H18O4	2818
1607	4-Hydroxy-2,4′-dimethoxy-″α -methyldeoxybenzoin	2-O-Methylangolensin	286.33	C17H18O4	2817
1608	2-Hydroxy-4′-methoxy-4-[[1,2,3,4,4a,7,8,8a-octahydro-1,6-dimethyl-4-(1-methylethyl)-1-naphthalenyl]oxy]α-methyldeoxybenzoin	4-O-a-Cadinylangolensin	476.66	C31H40O4	2819
1609	6-Hydroxy-4′-methoxy-4-[[1,2,3,4,4a,7,8,8a-octahydro-1,6-dimethyl-4-(1-methylethyl)-1-naphthalenyl]oxy]-α-methyldeoxybenzoin	4-O-T-Cadinylangolensin	476.66	C31H40O4	2820

This checklist of isoflavonoids contains various compounds reported in the literature as natural products to the end of 2004. Compounds published before 1991 are referenced to numbered entries in Volume 2 of the Handbook of Natural Flavonoids (J.B. Harborne and H. Baxter), John Wiley & Sons, Chichester 1999, using a number consisting of four digits. Compounds published in the period 1991–2004 are referenced with numbers having F as prefix before the number of the publication found in the reference list.

REFERENCES

Reference list for natural isoflavonoids reported in the period 1991–2004.

1. Abd El-Latif, R.R. et al., A new isoflavone from *Astragalus peregrinus*, *Chemistry of Natural Compounds* (Translation of *Khimiya Prirodnykh Soedinenii*), 39, 536, 2003.

2. Abdel-Halim, O.B. et al., Isoflavonoids and alkaloids from *Spartidium saharae*, *Natural Product Sciences*, 6, 189, 2000.

3. Abegaz, B.M. and Woldu, Y., Isoflavonoids from the roots of *Salsola somalensis*, *Phytochemistry*, 30, 1281, 1991.

4. Agarwal, D., Sah, P., and Garg, S.P., A new isoflavone from the seeds of *Psoralea corylifolia* Linn, *Oriental Journal of Chemistry*, 16, 541, 2000.

4b. Ahn, E.-M. et al., Prenylated flavonoids from *Moghania philippinensis*, *Phytochemistry*, 64, 1389, 2003.

5. Aida, M., Hano, Y., and Nomura, T., Constituents of the Moraceae plants. 26. Ficusins A and B, two new cyclic-monoterpene-substituted isoflavones from *Ficus septica* Barm. F., *Heterocycles*, 41, 2761, 1995.

6. Al-Hazimi, H.M. and Al-Andis, N.M., Minor pterocarpanoids from *Melilotus alba*, *Journal of Saudi Chemical Society*, 4, 215, 2000.

7. Ali, A.A., El-Emary, N.A., and Darwish, F.M., Studies on the constituents of two *Iris* species, *Bulletin of Pharmaceutical Sciences, Assiut University*, 16, 159, 1993.

8. Ali, M., Chaudhary, A., and Ramachandram, R., New pterocarpans from *Oroxylum indicum* stem bark, *Indian Journal of Chemistry*, 38B, 950, 1999.

10. Alvarez, L. et al., Cytotoxic isoflavans from *Eysenhardtia polystachya*, *Journal of Natural Products*, 61, 767, 1998.

11. Andrei, C.C. et al., Dimethylchromene rotenoids from *Tephrosia candida*, *Phytochemistry*, 46, 1081, 1997.

11b. Arthan, D. et al., Antiviral isoflavonoid sulfate and steroidal glycosides from the fruits of *Solanum torvum*, *Phytochemistry*, 59, 459, 2002.

12. Asomaning, W.A. et al., Isoflavones and coumarins from *Milletia thonningii*, *Phytochemistry*, 51, 937, 1999.

13. Asomaning, W.A. et al., Pyrano- and dihydrofurano-isoflavones from *Milletia thonningii*, *Phytochemistry*, 39, 1215, 1995.

14. Atindehou, K.K. et al., Three new prenylated isoflavonoids from the root bark of *Erythrina vogelii*, *Planta Medica*, 68, 181, 2002.

15. Atta-Ur-Rahman et al., Two new isoflavanoids from the rhizomes of *Iris soforana*, *Natural Product Research*, 18, 465, 2004.

16. Atta-ur-Rahman, et al., Isoflavonoid glycosides from the rhizomes of *Iris germanica*, *Helvetica Chimica Acta*, 86, 3354, 2003.

17. Atta-ur-Rahman, et al., Isoflavonoid glycosides from the rhizomes of *Iris germanica*, *Chemical & Pharmaceutical Bulletin*, 50, 1100, 2002.

18. Ayatollahi, S.M. et al., Two isoflavones from *Iris songarica* Schrenk. Daru, *Journal of Faculty of Pharmacy, Tehran University of Medical Sciences*, 12, 54, 2004.

19. Baba, M. et al., Studies of the Egyptian traditional folk medicines. III. A new diprenylated 3-arylcoumarin from Egyptian licorice, *Heterocycles*, 51, 387, 1999.

20. Babu, U.V., Bhandari, S.P.S., and Garg, H.S., Barbacarpan, a pterocarpan from *Crotalaria barbata*, *Phytochemistry*, 48, 1457, 1998.

21. Barragan-Huerta, B.E. et al., Neocandenatone, an isoflavan–cinnamylphenol quinone methide pigment from *Dalbergia congestiflora*, *Phytochemistry*, 65, 925, 2004.

22. Barrero, A.F., Cabrera, E., and Garcia, I.R., Pterocarpans from *Ononis viscosa* subsp. *breviflora*, *Phytochemistry*, 48, 187, 1998.

23. Bashir, A. et al., Isoflavones and xanthones from *Polygala virgata*, *Phytochemistry*, 31, 309, 1992.

24. Basnet, P. et al., Two new 2-arylbenzofuran derivatives from hypoglycemic activity-bearing fractions of *Morus insignis*, *Chemical & Pharmaceutical Bulletin*, 41, 1238, 1993.

25. Bekker, M. et al., An isoflavanoid–neoflavonoid and an *O*-methylated isoflavone from the heartwood of *Dalbergia nitidula*, *Phytochemistry*, 59, 415, 2002.
26. Belofsky, G. et al., Phenolic metabolites of *Dalea versicolor* that enhance antibiotic activity against model pathogenic bacteria, *Journal of Natural Products*, 67, 481, 2004.
27. Bhatnagar, R. and Kapoor, R.C., Phytochemical investigation of *Tephrosia purpurea* seeds, *Indian Journal of Chemistry*, 39B, 879, 2000.
28. Bojase, G. et al., Two new isoflavanoids from *Bolusanthus speciosus*, *Bulletin of the Chemical Society of Ethiopia*, 15, 131, 2001.
29. Bojase, G., Antimicrobial flavonoids from *Bolusanthus speciosus*, *Planta Medica*, 68, 615, 2002.
30. Bojase, G., Wanjala, C.C.W., and Majinda, R.R.T., Flavonoids from the stem bark of *Bolusanthus speciosus*, *Phytochemistry*, 56, 837, 2001.
31. Botta, B. et al., Three isoflavanones with cannabinoid-like moieties from *Desmodium canum*, *Phytochemistry*, 64, 599, 2003.
32. Brinkmeier, E., Geiger, H., and Zinsmeister, H.D., Biflavonoids and 4,2'-epoxy-3-phenylcoumarins from the moss *Mnium hornum*, *Phytochemistry*, 52, 297, 1999.
33. Calderon, A.I. et al., Isolation and structure elucidation of an isoflavone and a sesterterpenoic acid from *Henriettella fascicularis*, *Journal of Natural Products*, 65, 1749, 2002.
34. Cao, Z.Z. et al., A new isoflavone glucoside from *Astragalus membranaceus*, *Chinese Chemical Letters*, 9, 537, 1998.
35. Chan, S.C. et al., Three new flavonoids and antiallergic, anti-inflammatory constituents from the heartwood of *Dalbergia odorifera*, *Planta Medica*, 64, 153, 1998.
36. Chang, C.-H. et al., Flavonoids and a prenylated xanthone from *Cufrania cochinchinensis* var. *gerontogea*, *Phytochemistry*, 40, 945, 1995.
37. Chang, L.C. et al., Activity-guided isolation of constituents of *Tephrosia purpurea* with the potential to induce the phase II enzyme, quinone reductase, *Journal of Natural Products*, 60, 869, 1997.
38. Chansakaow, S. et al., Isoflavonoids from *Pueraria mirifica* and their estrogenic activity, *Planta Medica*, 66, 572, 2000.
39. Chaudhuri, S.K. et al., Isolation and structure identification of an active DNA strand-scission agent, (+)-3,4-dihydroxy-8,9-methylenedioxypterocarpan, *Journal of Natural Products*, 58, 1966, 1995.
40. Chaudhuri, S.K. et al., Two isoflavones from the bark of *Petalostemon purpureus*, *Phytochemistry*, 41, 1625, 1996.
41. Chen, M., Lou, S., and Chen, J., Two isoflavones from *Flemingia philippinensis*, *Phytochemistry*, 30, 3842, 1991.
42. Choudhary, M.I. et al., Four new flavones and a new isoflavone from *Iris bungei*, *Journal of Natural Products*, 64, 857, 2001.
43. Chuankamnerdkarn, M. et al., Prenylated isoflavones from *Derris scandens*, *Heterocycles*, 57, 1901, 2002.
44. Da Silva, B.P., Bernardo, R.R., and Parente, J.P., Clitoriacetal 11-*O*-β-D-glucopyranoside from *Clitoria fairchildiana*, *Phytochemistry*, 47, 121, 1998.
45. da Silva, B.P., Velozo, L.S.M., and Parente, J.P., Biochanin A triglycoside from *Andira inermis*, *Fitoterapia*, 71, 663, 2000.
46. Da Silva, G.L. et al., 4'-Dehydroxycabenegrin A–I from roots of *Harpalyce brasiliana*, *Phytochemistry*, 46, 1059, 1997.
47. Da Silva, G.L., A new isoflavone isolated from *Harpalyce brasiliana*, *Journal of the Brazilian Chemical Society*, 10, 438, 1999.
48. Da-Cunha, E.V.L. et al., Eryvellutinone, an isoflavanone from the stem bark of *Erythrina vellutina*, *Phytochemistry*, 43, 1371, 1996.
49. Dagne, E., Mammo, W., and Sterner, O., Flavonoids of *Tephrosia polyphylla*, *Phytochemistry*, 31, 3662, 1992.
50. Dean, F.M. et al., An isoflavanoid from *Aglaia ferruginaea*, an Australian member of the Meliaceae, *Phytochemistry*, 34, 1537, 1993.

51. Delle Monache, G. et al., Antimicrobial isoflavanones from *Desmodium canum*, *Phytochemistry*, 41, 537, 1996.

53. Derese, S. et al., A new isoflavone from stem bark of *Millettia dura*, *Bulletin of the Chemical Society of Ethiopia*, 17, 113, 2003.

55. Dini, I., Schettino, O., and Dini, A., Studies on the constituents of *Lupinus mutabilis* (Fabaceae). Isolation and characterization of two new isoflavonoid derivatives, *Journal of Agricultural and Food Chemistry*, 46, 5089, 1998.

56. Dominguez, X.A. et al., Bioactive isoflavonoids from the bark of "K'anawte" (*Piscidia priscipula* Sarg.), *Revista Latinoamericana de Quimica*, 22, 94, 1991.

57. Drewes, S.E. et al., Minor pyrano-isoflavones from *Eriosema kraussianum*: activity-, structure-, and chemical reaction studies, *Phytochemistry*, 65, 1955, 2004.

58. Drewes, S.E. et al., Pyrano-isoflavones with erectile-dysfunction activity from *Eriosema kraussianum*, *Phytochemistry*, 59, 739, 2002.

59. DuBois, J.L. and Sneden, A.T., Dihydrolicoisoflavone, a new isoflavanone from *Swartzia polyphylla*, *Journal of Natural Products*, 58, 629, 1995.

60. DuBois, J.L. and Sneden, A.T., Ferreirinol, a new 3-hydroxyisoflavanone from *Swartzia polyphylla*, *Journal of Natural Products*, 59, 902, 1996.

61. Elgindi, M.R. et al., Isoflavones and a phenylethanoid from *Verbascum sinaiticum*, *Asian Journal of Chemistry*, 11, 1534, 1999.

62. El-Masry, S. et al., Prenylated flavonoids of *Erythrina lysistemon* grown in Egypt, *Phytochemistry*, 60, 783, 2002.

63. El-Sebakhy, N.A.A. et al., Antimicrobial isoflavans from *Astragalus* species, *Phytochemistry*, 36, 1387, 1994.

65. Farag, S.F. et al., Isoflavonoid glycosides from *Dalbergia sissoo*, *Phytochemistry*, 57, 1263, 2001.

66. Farag, S.F. et al., Isoflavonoids and flavone glycosides from rhizomes of *Iris carthaliniae*, *Phytochemistry*, 50, 1407, 1999.

67. Ferrari, F. et al., Two new isoflavonoids from *Boerhaavia coccinea*, *Journal of Natural Products*, 54, 597, 1991.

68. Fuendjiep, V. et al., Conrauinones A and B, two new isoflavones from stem bark of *Millettia conraui*, *Journal of Natural Products*, 61, 380, 1998.

69. Fuendjiep, V. et al., The *Millettia* of Cameroon. Part 9. Conrauinones C and D, two isoflavones from stem bark of *Millettia conraui*, *Phytochemistry*, 47, 113, 1998.

70. Fukai, T. et al., Anti-*Helicobacter pylori* flavonoids from licorice extract, *Life Sciences*, 71, 1449, 2002.

71. Fukai, T. et al., Components of the root bark of *Morus cathayana*. 1. Structures of five new isoprenylated flavonoids, sanggenols A–E and a diprenyl-2-arylbenzofuran, mulberrofuran V, *Heterocycles*, 43, 425, 1996.

72. Fukai, T. et al., Four isoprenoid-substituted flavonoids from *Glycyrrhiza aspera*, *Phytochemistry*, 36, 233, 1994.

73. Fukai, T. et al., Isoprenylated flavonoids from underground parts of *Glycyrrhiza glabra*, *Phytochemistry*, 43, 1119, 1996.

74. Fukai, T. et al., Phenolic constituents of *Glycyrrhiza* species. Five new isoprenoid-substituted flavonoids, kanzonols F–J, from *Glycyrrhiza uralensis*, *Heterocycles*, 36, 2565, 1993.

75. Fukai, T. and Nomura, T., Isoprenoid-substituted flavonoids from roots of *Glycyrrhiza inflata*, *Phytochemistry*, 38, 759, 1995.

76. Fukai, T. et al., Phenolic constituents of *Glycyrrhiza* species. 16. Five new isoprenoid-substituted flavonoids, kanzonols M–P and R, from two *Glycyrrhiza* species, *Heterocycles*, 38, 1089, 1994.

77. Fukai, T. et al., Phenolic constituents of *Glycyrrhiza* species. 17. Three isoprenoid-substituted isoflavans, gancaonins X–Z, from Chinese folk medicine Tiexin Gancao (root xylems of *Glycyrrhiza* species), *Natural Medicines*, 48, 203, 2004.

78. Fukai, T., Kato, H., and Nomura, T., Phenolic constituents of *Glycyrrhiza* species. 11. A new prenylated 3-arylcoumarin, gancaonin W, from Licorice, *Shoyakugaku Zasshi*, 47, 326, 1993.

79. Fukai, T., Tantai, L., and Nomura, T., Isoprenoid-substituted flavonoids from *Glycyrrhiza glabra*, *Phytochemistry*, 43, 531, 1996.

80. Galeffi, C. et al., Two prenylated isoflavanones from *Millettia pervilleana*, *Phytochemistry*, 45, 189, 1997.

82. Grosvenor, P.W. and Gray, D.O., Colutequinone and colutehydroquinone, antifungal isoflavonoids from *Colutea arborescens*, *Phytochemistry*, 43, 377, 1996.

83. Grosvenor, P.W. and Gray, D.O., Coluteol and colutequinone B, more antifungal isoflavonoids from *Colutea arborescens*, *Journal of Natural Products*, 61, 99, 1998.

84. Gupta, J. and Ali, M., Chemical constituents of *Boerhaavia diffusa* Linn. roots, *Indian Journal of Chemistry*, 37B, 912, 1998.

85. Hakamatsuka, T., Ebizuka, Y., and Sankawa, U., Induced isoflavonoids from copper chloride-treated stems of *Pueraria lobata*, *Phytochemistry*, 30, 1481, 1991.

86. Hakim, E.H. et al., Artoindonesianins N and O, new prenylated stilbene and prenylated aryl-benzofuran derivatives from *Artocarpus gomezianus*, *Fitoterapia*, 73, 597, 2002.

87. Han, S.J., Ryu, S.N., and Kang, S.S., A new 2-arylbenzofuran with antioxidant activity from the black colored rice (*Oryza sativa* L.) bran, *Chemical & Pharmaceutical Bulletin*, 52, 1365, 2004.

88. Hatano, T. et al., Minor flavonoids from licorice, *Phytochemistry*, 55, 959, 2000.

89. Hatano, T. et al., Phenolic constituents of liquorice. VII. A new chalcone with a potent radical scavenging activity and accompanying phenolics from liquorice, *Chemical & Pharmaceutical Bulletin*, 45, 1485, 1997.

90. Herath, H.M.T.B. et al., Isoflavonoids and a pterocarpan from *Gliricidia sepium*, *Phytochemistry*, 47, 117, 1998.

91. Herz, W., Pethtel, K.D., and Raulais, D. Isoflavones, a sesquiterpene lactone–monoterpene adduct and other constituents of *Gaillardia species*, *Phytochemistry*, 30, 1273, 1991.

92. Hosny, M. and Rosazza, J.P.N., New isoflavone and triterpene glycosides from soybeans, *Journal of Natural Products*, 65, 805, 2002.

93. Hu, J.-F. et al., New 2-*O*-methylrhamno-isoflavones from *Streptomyces* sp., *Natural Product Research*, 17, 451, 2003.

94. Huang, K.-F. and Yen, Y.-F., Three prenylated isoflavones from *Erythrina variegata*, *Journal of the Chinese Chemical Society*, 43, 515, 1996.

95. Huang, W., Duan, J., and Li, Z., Studies on chemical constituents of *Maackia amurensis*, *Zhongguo Zhongyao Zazhi*, 26, 403, 2001.

96. Hwang, M.-H., Kwon, Y.-S., and Kim, C.-M., A new isoflavone glycoside from heartwood of *Maackia fauriei*, *Natural Medicines*, 52, 527, 1998.

97. Iinuma M. et al., Phenolic constituents in *Erythrina* × *bidwilli* and their activity against oral microbial organisms, *Chemical & Pharmaceutical Bulletin*, 40, 2749, 1992.

98. Iinuma, M. et al., An isoflavanone from roots of *Echinosophora koreensis*, *Phytochemistry*, 30, 3153, 1991.

99. Iinuma, M. et al., Anti-oral microbial activity of isoflavonoids in root bark of *Ormosia monosperma*, *Phytochemistry*, 37, 889, 1994.

100. Iinuma, M. et al., Five flavonoid compounds from *Echinosophora koreensis*, *Phytochemistry*, 33, 1241, 1993.

101. Iinuma, M. et al., Flavonoid compounds in roots of *Sophora tetraptera*, *Phytochemistry*, 39, 667, 1995.

102. Iinuma, M. et al., Isoflavoniods in roots of *Sophora fraseri*, *Phytochemistry*, 34, 1654, 1993.

103. Iinuma, M. et al., Isoflavonoids in roots of *Sophora secundiflora*, *Phytochemistry*, 39, 907, 1995.

104. Iinuma, M. et al., Phenolic compounds in *Erythrina* × *bidwillii* and their activity against oral microbial organisms, *Heterocycles*, 39, 687, 1994.

105. Iinuma, M., Ohyama, M., and Tanaka, T., Flavonoids in roots of *Sophora prostrata*, *Phytochemistry*, 38, 539, 1995.

106. Iinuma, M., Ohyama, M., and Tanaka, T., Three isoflavanones from roots of *Sophora prostrata*, *Phytochemistry*, 37, 1713, 1994.

107. Ilyas, M. et al., Isoflavones from *Garcinia nervosa*, *Phytochemistry*, 36, 807, 1994.

108. Ilyas, M. et al., Nodosin, a novel *C*-glycosylisoflavone from *Cassia nodosa*, *Journal of Chemical Research*, 3, 88, 1994.

109. Ito, C. et al., Cancer chemopreventive activity of rotenoids from *Derris trifoliate*, *Planta Medica*, 70, 8, 2004.

110. Ito, C. et al., Chemical constituents of *Millettia taiwaniana*: structure elucidation of five new isoflavonoids and their cancer chemopreventive activity, *Journal of Natural Products*, 67, 1125, 2004.

111. Ito, C. et al., Isoflavonoids from *Dalbergia olivari*, *Phytochemistry*, 64, 1265, 2003.

112. Ito, H., Onoue, S., and Yoshida, T., Isoflavonoids from *Belamcanda chinensis*, *Chemical & Pharmaceutical Bulletin*, 49, 1229, 2001.

113. Jain, L. et al., Flavonoids from *Eschscholzia californica*, *Phytochemistry*, 41, 661, 1996.

114. Jang, D.S. et al., Potential cancer chemopreventive constituents of the leaves of *Macaranga triloba*, *Phytochemistry*, 65, 345, 2004.

115. Jang, D.S. et al., Potential cancer chemopreventive flavonoids from the stems of *Tephrosia toxicaria*, *Journal of Natural Products*, 66, 1166, 2003.

116. Kajiyama, K. et al., Flavonoids and isoflavonoids of chemotaxonomic significance from *Glycyrrhiza pallidiflora* (Leguminosae), *Biochemical Systematics and Ecology*, 21, 785, 1993.

117. Kamnaing, P. et al., An isoflavan-quinone and a flavonol from *Millettia laurentii*, *Phytochemistry*, 51, 829, 1999.

118. Kamperdick, C. et al., Flavones and isoflavones from *Millettia ichthyochtona*, *Phytochemistry*, 48, 577, 1998.

119. Kanchanapoom, T. et al., Stilbene and 2-arylbenzofuran glucosides from the rhizomes of *Schoenocaulon officinale*, *Chemical & Pharmaceutical Bulletin*, 50, 863, 2002.

120. Khaomek, P. et al., A new pterocarpan from *Erythrina fusca*, *Heterocycles*, 63, 879, 2004.

121. Kinoshita, T. et al., Isoflavan derivatives from *Glycyrrhiza glabra* (licorice), *Heterocycles*, 43, 581, 1996.

122. Kinoshita, T., Tamura, Y., and Mizutani, K., Isolation and synthesis of two new 3-arylcoumarin derivatives from the root of *Glycyrrhiza glabra* (licorice), and structure revision of an antioxidant isoflavonoid glabrene, *Natural Product Letters*, 9, 289, 1997.

123. Kishore, P.H. et al., A new coumestan from *Tephrosia calophylla*, *Chemical & Pharmaceutical Bulletin*, 51, 194, 2003.

124. Kitagawa, I. et al., Chemical studies of Chinese licorice-roots. II. Five new flavonoid constituents from the roots of *Glycyrrhiza aspera* PALL. collected in Xinjiang, *Chemical & Pharmaceutical Bulletin*, 46, 1511, 1998.

125. Ko, H.-H. et al., Anti-inflammatory flavonoids and pterocarpanoid from *Crotalaria pallida* and *C. assamica*, *Bioorganic & Medicinal Chemistry Letters*, 14, 1011, 2004.

126. Kobayashi, A., Akiyama, K., and Kawazu, K., A pterocarpan, (+)-2-hydroxypisatin from *Pisum sativum*, *Phytochemistry*, 32, 77, 1992.

127. Kraft, C. et al., Andirol A and B, two unique 6-hydroxymethylpterocarpenes from *Andira inermis*, *Zeitschrift fuer Naturforschung C*, 57, 785, 2002.

127b. Kraft, C. et al., Andinermals A–C, antiplasmodial constituents from *Andira inermis, Phytochemistry*, 58, 769, 2001.

128. Krishnaveni, K.S. and Rao, J.V.S., A new acylated isoflavone glucoside from *Pterocarpus santalinus*, *Chemical & Pharmaceutical Bulletin*, 48, 1373, 2000.

129. Krishnaveni, K.S. and Rao, J.V.S., A new isoflavone glucoside from *Pterocarpus santalinus*, *Journal of Asian Natural Products Research*, 2, 219, 2000.

130. Krishnaveni, K.S. and Rao, J.V.S., An isoflavone from *Pterocarpus santalinus*, *Phytochemistry*, 53, 605, 2000.

131. Kudou, S. et al., A new isoflavone glycoside in soybean seeds (*Glycine max* Merrill), glycitein 7-*O*-β-D-(6″-*O*-acetyl)-glucopyranoside, *Agricultural and Biological Chemistry*, 55, 859, 1991.

132. Kulesh, N.I. et al., Isoflavonoids from heartwood of *Maackia amurensis* Rupr. et Maxim. *Chemistry of Natural Compounds* (Translation of *Khimiya Prirodnykh Soedinenii*), 37, 29, 2001.

133. Kuo, S.-C. et al., Potent antiplatelet, anti-inflammatory and antiallergic isoflavanquinones from the roots of *Abrus precatorius*, *Planta Medica*, 61, 307, 1995.

134. Kuroyanagi, M. et al., Antibacterial and antiandrogen flavonoids from *Sophora flavescens*, *Journal of Natural Products*, 62, 1595, 1999.

135. Kwon, I.-B. and Park, H.-H., Isoflavonoids of kudzu (*Pueraria lobata*) and bioconversion of exogenous compounds into their malonylglucosides by its cell cultures, *Foods & Food Ingredients Journal of Japan*, 163, 86, 1995.

136. Lami, N., Kadota, S., and Kikuchi, T., Constituents of the roots of *Boerhaavia diffusa* L. IV. Isolation and structure determination of boeravinones D, E, and F, *Chemical & Pharmaceutical Bulletin*, 39, 1863, 1991.

137. Lazaro, M.L. et al., An isoflavone glucoside from *Retama sphaerocarpa* Boissier, *Phytochemistry*, 48, 401, 1998.

138. Lee, S.-J. et al., Prenylated flavonoids from *Maclura pomifera*, *Phytochemistry*, 49, 2573, 1998.

139. Lenssen, A.W. et al., Acicerone: an isoflavone from *Astragalus cicer*, *Phytochemistry*, 36, 1185, 1994.

140. Li, D. et al., Two isoflavonoid glycosides from *Derris scandens*, *Yaoxue Xuebao*, 34, 43, 1999.

141. Li, W. et al., Flavonoids from *Glycyrrhiza pallidiflora* hairy root cultures, *Phytochemistry*, 58, 595, 2001.

141b. Li, W., Asada, Y., and Yoshikawa, T., Flavonoid constituents from *Glycyrrhiza glabra* hairy root cultures, *Phytochemistry*, 55, 447, 2000.

142. Li, Y.-C. and Kuo, Y.-H., Two new isoflavones from the bark of *Ficus microcarpa*, *Journal of Natural Products*, 60, 292, 1997.

143. Lin, L.J. et al., Traditional medicinal plants of Thailand. Part 18. 6-Deoxyclitoriacetal from *Clitoria macrophylla*, *Phytochemistry*, 31, 4329, 1992.

144. Lin, L.J. et al., Traditional medicinal plants of Thailand. Part 21. Wrightiadione from *Wrightia tomentosa*, *Phytochemistry*, 31, 4333, 1992.

145. Lin, Y.L. and Kuo, ueh H., Two new coumaronochromone derivatives, oblongin and oblonginol from the roots of *Derris oblonga* benth., *Heterocycles*, 36, 1501, 1993.

146. Lin, Y.-L. and Kuo, Y.-H, 6-Oxo-6a,12a-dehydro-a-toxicarol, a 6-oxodehydrorotenone from the roots of *Derris oblonga* Benth, *Heterocycles*, 41, 1959, 1995.

147. Lin, Y.L. and Kuo, Y.H., 6a,12a-Dehydro-b-toxicarol and derricarpin, two new isoflavonoids, from the roots of *Derris oblonga* Benth, *Chemical & Pharmaceutical Bulletin*, 41, 1456, 1993.

148. Lo, W.-L. et al., Coumaronochromones and flavanones from *Euchresta formosana* roots, *Phytochemistry*, 60, 839, 2002.

149. Lo, W.L. et al., Cytotoxic coumaronochromones from the roots of *Euchresta formosana*, *Planta Medica*, 68, 146, 2002.

150. Luyengi, L. et al., Rotenoids and chalcones from *Mundulea sericea* that inhibit phorbol ester-induced ornithine decarboxylase activity, *Phytochemistry*, 36, 1523, 1994.

151. Ma, Z.-J. et al., Isoflavans from *Sphaerophysa salsula*, *Pharmazie*, 57, 75, 2002.

152. Machocho, A.K. et al., Three new flavonoids from the root of *Tephrosia emoroides* and their antifeedant activity against the larvae of the spotted stalk borer *Chilo partellus* Swinhoe, *International Journal of Pharmacognosy*, 33, 222, 1995.

153. Macias, F.A. et al., Natural products as allelochemicals. Bioactive phenolics and polar compounds from *Melilotus messanensis*, *Phytochemistry*, 50, 35, 1999.

154. Magalhaes, A.F. et al., Flavonoids and 3-phenylcoumarins from the seeds of *Pachyrhizus tuberosus*, *Phytochemistry*, 31, 1831, 1992.

155. Magalhaes, A.F. et al., Prenylated flavonoids from *Deguelia hatschbachii* and their systematic significance in *Deguelia*, *Phytochemistry*, 57, 77, 2001.

156. Magalhaes, A.F. et al., Twenty-three flavonoids from *Lonchocarpus subglaucescens*, *Phytochemistry*, 42, 1459, 1996.

157. Mahabusarakam, W. et al., A benzil and isoflavone derivatives from *Derris scandens* Benth. *Phytochemistry*, 65, 1185, 2004.

158. Manjary, F. et al., A prenylated pterocarpan from *Mundulea striata*, *Phytochemistry*, 33, 515, 1993.

159. Maskey, R.P. et al., Flavones and new isoflavone derivatives from microorganisms: isolation and structure elucidation, *Zeitschrift fuer Naturforschung B*, 58, 686, 2003.

160. Mathias, L. et al., A new isoflavone glycoside from *Dalbergia nigra*, *Journal of Natural Products*, 61, 1158, 1998.

161. Mathias, L. et al., Novel 11a-*O*-β-ᴅ-glucopyranosylrotenoid isolated from *Clitoria fairchildiana*, *Natural Product Letters*, 11, 119, 1998.

162. Mathias, L., Mors, W.B., and Parente, J.P., Rotenoids from seeds of *Clitoria fairchildiana*, *Phytochemistry*, 48, 1449, 1998.

163. Matsuyama, S., Kuwahara, Y., and Suzuki, T., A new 2-arylbenzofuran, w-hydroxymoracin N, from mulberry leaves, *Agricultural and Biological Chemistry*, 55, 1409, 1991.

164. Maximo, P. et al., Flavonoids from *Ulex airensis* and *Ulex europaeus* ssp. *europaeus*, *Journal of Natural Products*, 65, 175, 2002.

165. Maximo, P. et al., Flavonoids from *Ulex species*, *Zeitschrift fuer Naturforschung C*, 55, 506, 2000.

166. Maximo, P. and Lourenco, A., A pterocarpan from *Ulex parviflorus*, *Phytochemistry*, 48, 359, 1998.

167. Maximov, O.B. et al., New prenylated isoflavanones and other constituents of *Lespedeza bicolor*, *Fitoterapia*, 75, 96, 2004.

168. McKee, T.C. et al., Isolation and characterization of new anti-HIV and cytotoxic leads from plants, marine, and microbial organisms, *Journal of Natural Products*, 60, 431, 1997.

169. Minami, H. et al., Highly oxygenated isoflavones from *Iris japonica*, *Phytochemistry*, 41, 1219, 1996.

170. Miyase, T. et al., Antioxidants from *Lespedeza homoloba* (I), *Phytochemistry*, 52, 303, 1999.

171. Miyase, T. et al., Antioxidants from *Lespedeza homoloba* (II), *Phytochemistry*, 52, 311, 1999.

172. Mizuno, M. et al., Coumaronochromones from leaves of *Desmodium oxyphyllum*, *Phytochemistry*, 31, 361, 1992.

173. Mizuno, M. et al., Coumaronochromones in the roots of *Euchresta japonica*, *Phytochemistry*, 31, 643, 1992.

174. Mizuno, M. et al., Four flavonoids in the roots of *Euchresta formosana*, *Phytochemistry*, 30, 3095, 1991.

175. Mizuno, M. et al., Isoflavones from roots of *Euchresta japonica*, *Phytochemistry*, 31, 675, 1992.

176. Moriyama, M. et al., Isoflavones from the root bark of *Piscidia erythrina*, *Phytochemistry*, 31, 683, 1992.

177. Moriyama, M. et al., Isoflavonoid alkaloids from *Piscidia erythrina*, *Phytochemistry*, 32, 1317, 1993.

178. Muhaisen, H.M.H. et al., Flavonoids from *Acacia tortilis*, *Journal of Chemical Research*, 6, 276, 2002.

179. Mun'im, A. et al., Estrogenic and acetylcholinesterase-enhancement activity of a new isoflavone, 7,2′,4′-trihydroxyisoflavone-4′-*O*-β-ᴅ-glucopyranoside from *Crotalaria sessiliflora*, *Cytotechnology*, 43, 127, 2004.

180. Nia, M. and Gunasekar, D., A new isoflavone from the root bark of *Ochna squarrosa*, *Fitoterapia*, 63, 249, 1992.

181. Nkengfack, A. et al., *Erythrina* studies. 30. Sigmoidins J and K, two new prenylated isoflavonoids from *Erythrina sigmoidea*, *Journal of Natural Products*, 57, 1172, 1994.

182. Nkengfack, A.E. et al., An isoflavanone and a coumestan from *Erythrina sigmoidea*, *Phytochemistry*, 35, 521, 1994.

183. Nkengfack, A.E. et al., Auriculatin 4′-*O*-glucoside: a new prenylated isoflavone glycoside from *Erythrina eriotricha*, *Planta Medica*, 57, 488, 1991.

184. Nkengfack, A.E. et al., Cytotoxic isoflavones from *Erythrina indica*, *Phytochemistry*, 58, 1113, 2001.

185. Nkengfack, A.E. et al., Indicanine A, a new 3-phenylcoumarin from root bark of *Erythrina indica*, *Journal of Natural Products*, 63, 855, 2000.

186. Nkengfack, A.E. et al., Prenylated isoflavanone from *Erythrina eriotricha*, *Phytochemistry*, 40, 1803, 1995.

187. Nkengfack, A.E. et al., Prenylated isoflavanone from the roots of *Erythrina sigmoidea*, *Phytochemistry*, 36, 1047, 1994.

188. Nohara, T. et al., The new isoflavonoid of *Pueraria lobata*, *Japan Kokai Tokkyo Koho*, Patent 2000.

189. Noreen, Y., Two new isoflavones from *Ceiba pentandra* and their effect on cyclooxygenase-catalyzed prostaglandin biosynthesis, *Journal of Natural Products*, 61, 8, 1998.
190. Oh, W.K. et al., Prenylated isoflavonoids from *Erythrina senegalensis*, *Phytochemistry*, 51, 1147, 1999.
191. Omobuwajo, O.R., Adesanya, S.A., and Babalola, G.O., Isoflavonoids from *Pycnanthus angolensis* and *Baphia nitida*, *Phytochemistry*, 31, 1013, 1992.
192. Osawa, K. et al., Isoflavanones from the heartwood of *Swartzia polyphylla* and their antibacterial activity against cariogenic bacteria, *Chemical & Pharmaceutical Bulletin*, 40, 2970, 1992.
193. Palazzino, G. et al., Prenylated isoflavonoids from *Millettia pervilleana*, *Phytochemistry*, 63, 471, 2003.
194. Palme, E., Bilia, A.R., and Morelli, I., Flavonols and isoflavones from *Cotoneaster simonsii*, *Phytochemistry*, 42, 903, 1996.
195. Park, H.-J. et al., Isoflavone glycosides from the flowers of *Pueraria thunbergiana*, *Phytochemistry*, 51, 147, 1999.
196. Park, J.A. et al., A new pterocarpan, (−)-maackiain sulfate, from the roots of *Sophora subprostrata*, *Archives of Pharmacal Research*, 26, 1009, 2003.
197. Park, Y. et al., New isoflavone glycoside from the woods of *Sophora japonica*, *Bulletin of the Korean Chemical Society*, 25, 147, 2004.
198. Parvez, M. and Rahman, A., A novel antimicrobial isoflavone galactoside from *Cnestis ferruginea* (Connaraceae), *Journal of the Chemical Society of Pakistan*, 14, 221, 1992.
199. Pegnyemb, D.E. et al., A new benzoylglucoside and a new prenylated isoflavone from *Lophira lanceolata*, *Journal of Natural Products*, 61, 801, 1998.
200. Pereira da Silva, B., Bernardo, R.R., and Paz Parente, J., A new rotenoid glucoside, 6-deoxyclitoriacetal 11-O-β-D-glucopyranoside, from the roots of *Clitoria fairchildiana*, *Planta Medica*, 64, 285, 1998.
201. Pistelli, L. et al., A new isoflavone from Genista corsica, *Journal of Natural Products*, 63, 504, 2000.
202. Pistelli, L. et al., Flavonoids from *Genista ephedroides*, *Journal of Natural Products*, 61, 1404, 1998.
203. Pistelli, L. et al., Pterocarpans from *Bituminaria morisiana* and *Bituminaria bituminosa*, *Phytochemistry*, 64, 595, 2003.
204. Prakash Rao, C., Prashant, A., and Krupadanam, G.L.D., Two prenylated isoflavans from *Millettia racemosa*, *Phytochemistry*, 41, 1223, 1996.
205. Prasad, Y.R. and Chakradhar, V., Two new isoflavones from *Tephrosia maxima*, *International Journal of Chemical Sciences*, 2, 223, 2004.
206. Prashant, A. and Krupadanam, G.L.D., A new prenylated dehydrorotenoid from *Tephrosia villosa* seeds, *Journal of Natural Products*, 56, 765, 1993.
207. Prashant, A. and Krupadanam, G.L.D., Dehydro-6-hydroxyrotenoid and lupenone from *Tephrosia villosa*, *Phytochemistry*, 32, 484, 1993.
208. Purev, O. et al., New isoflavones and flavanol from *Iris potaninii*, *Chemical & Pharmaceutical Bulletin*, 50, 1367, 2002.
209. Queiroz, E.F. et al., Prenylated isoflavonoids from the root bark of *Erythrina vogelii*, *Journal of Natural Products*, 65, 403, 2002.
210. Rahman, M.M. et al., New salicylic acid and isoflavone derivatives from *Flemingia paniculata*, *Journal of Natural Products*, 67, 402, 2004.
211. Ramesh, P. and Yuvarajan, C.R., Coromandelin, a new isoflavone apioglucoside from the leaves of *Dalbergia coromandeliana*, *Journal of Natural Products*, 58, 1240, 1995.
212. Rao, C.P. and Krupadanam, G.L.D., An isoflavan from *Millettia racemosa*, *Phytochemistry*, 35, 1597, 1994.
213. Rao, E.V., Prasad, Y.R., and Ganapaty, S., Three prenylated isoflavones from *Milletia auriculata*, *Phytochemistry*, 31, 1015, 1992.
214. Rao, J.R. and Rao, R.S., Dalpaniculin, a C-glycosylisoflavone from *Dalbergia paniculata* seeds, *Phytochemistry*, 30, 715, 1991.
215. Rao, M.N., Krupadanam, G.L.D., and Srimannarayana, G., Four isoflavones and two 3-arylcoumarins from stems of *Derris scandens*, *Phytochemistry*, 37, 267, 1994.

216. Rasool, N. et al., A benzoquinone and a coumestan from *Psoralea plicata*, *Phytochemistry*, 30, 2800, 1991.

217. Razdan, T.K. et al., Isocladrastin and kashmigenin — two isoflavones from *Iris kashmiriana*, *Phytochemistry*, 41, 947, 1996.

218. Rukachaisirikul, V. et al., Isoflavone glycosides from *Derris scandens*, *Phytochemistry*, 60, 827, 2002.

219. Russell, G.B., New isoflavonoids from root bark of kowhai (*Sophora microphylla*), *Australian Journal of Chemistry*, 50, 333, 1997.

220. Sairafianpour, M. et al., Leishmanicidal and antiplasmodial activity of constituents of *Smirnowia iranica*, *Journal of Natural Products*, 65, 1754, 2002.

221. Sanchez, C.S. et al., A new isoflavone from *Swartzia* (Leguminosae) Brachyrachisin, *Acta Amazonica*, 29, 419, 1999.

222. Sato, H. et al., Isoflavones from pods of *Laburnum anagyroides*, *Phytochemistry*, 39, 673, 1995.

224. Saxena, V.K. and Singal, M., Genistein 4′-O-α-L-rhamnopyranoside from *Pithecellobium dulce*, *Fitoterapia*, 69, 305, 1998.

225. Saxena, V.K. and Chourasia, S., A new isoflavone from the roots of *Asparagus racemosus*, *Fitoterapia*, 72, 307, 2001.

226. Saxena, V.K. and Mishra, L.N., A new isoflavone glycoside from the roots of *Sesbania grandiflora*, *Research Journal of Chemistry and Environment*, 3, 69, 1999.

228. Saxena, V.K. and Nigam, S., A methylenedioxypterocarpan from *Melilotus indica*, *Fitoterapia*, 68, 343, 1997.

229. Saxena, V.K. and Nigam, S., A novel prenylated pterocarpan from *Melilotus indica*, *Fitoterapia*, 68, 403, 1997.

230. Saxena, V.K. and Nigam, S., Novel antifungal pterocarpan from *Crotalaria mucronata*, *Journal of the Institution of Chemists*, 68, 137, 1996.

231. Saxena, V.K. and Sharma, D.N., A new isoflavone from the roots of *Abrus precatorius*, *Fitoterapia*, 70, 328, 1999.

232. Sekine, T. et al., Six diprenylisoflavones, derrisisoflavones A–F, from *Derris scandens*, *Phytochemistry*, 52, 87, 1999.

234. Shafiullah, M. et al., A new isoflavone C-glycoside from *Cassia siamea*, *Fitoterapia*, 66, 439, 1995.

235. Shaker, K.H. et al., A new triterpenoid saponin from *Ononis spinosa* and two new flavonoid glycosides from *Ononis vaginalis*, *Zeitschrift fuer Naturforschung B*, 59, 124, 2004.

236. Shawl, A.S. and Kumar, T., Isoflavonoids from *Iris crocea*, *Phytochemistry*, 31, 1399, 1992.

237. Shen, Y., Kojima, Y., and Terazawa, M., Two lignan rhamnosides from birch leaves, *Journal of Wood Science*, 45, 326, 1999.

238. Shi, Y.-Q. et al., Cytotoxic flavonoids with isoprenoid groups from *Morus mongolica*, *Journal of Natural Products*, 64, 181, 2001.

239. Shi, Y.-Q. et al., Phenolic constituents of the root bark of Chinese *Morus australis*, *Natural Medicines*, 55, 143, 2001.

240. Shibuya, Y. et al., New isoflavone glucosides from white lupine (*Lupinus albus* L.), *Zeitschrift fuer Naturforschung C*, 46, 513, 1991.

241. Shirataki, Y. et al., Studies on the constituents of *Sophora* species, *Phytochemistry*, 50, 695, 1999.

242. Shirataki, Y., Isoflavanones in roots of *Sophora secundiflora*, *Phytochemistry*, 44, 715, 1997.

243. Shirwaikar, A. and Srinivasan, K.K., New flavonoids from the flowers of *Thespesia populnea*, *Journal of Medicinal and Aromatic Plant Sciences*, 18, 266, 1996.

245. Siddiqui, B.S. et al., Chemical constituents from the fruits of *Madhuca latifolia*, *Helvetica Chimica Acta*, 87, 1194, 2004.

246. Silva, B.P., Bernardo, R.R., and Parente, J.P. Rotenoids from roots of *Clitoria fairchildiana*, *Phytochemistry*, 49, 1787, 1998.

248. Silva, B.P., Bernardo, R.R., and Parente, J.P., A novel rotenoid glucoside from the roots of *Clitoria fairchildiana*, *Fitoterapia*, 69, 49, 1998.

249. Simin, K. et al., Structure and biological activity of a new rotenoid from *Pongamia pinnata*, *Natural Product Letters*, 16, 351, 2002.

250. Singab, A.N.B., Flavonoids from *Iris spuria* (Zeal) cultivated in Egypt, *Archives of Pharmacal Research*, 27, 1023, 2004.

251. Singh, R.K. and Choubey, A., Acylated formononetin monoglucoside from *Pterocarpus marsupium*, *Journal of the Institution of Chemists*, 71, 103, 1999.

252. Soby, S., Bates, R., and Van Etten, H., Oxidation of the phytoalexin maackiain to 6,6a-dihydroxy-maackiain by *Colletotrichum gloeosporioides*, *Phytochemistry*, 45, 925, 1997.

253. Socorro, M.P., Pinto, A., Kaiser, C.R., New isoflavonoid from *Dipterix odorata*, *Zeitschrift fuer Naturforschung B*, 58, 1206, 2003.

254. Soekamto, N.H. et al., Artoindonesianins X and Y, two isoprenylated 2-arylbenzofurans, from *Artocarpus fretessi* (Moraceae), *Phytochemistry*, 64, 831, 2003.

255. Song, C. et al., Antimicrobial isoflavans from *Astragalus membranaceus* (Fisch.) Bunge, *Zhiwu Xuebao*, 39, 486, 1997.

256. Song, C. et al., Pterocarpans and isoflavans from *Astragalus membranaceus* bunge, *Zhiwu Xuebao*, 39, 1169, 1997.

257. Spencer, G.F. et al., A pterocarpan and two isoflavans from alfalfa, *Phytochemistry*, 30, 4147, 1991.

258. Sripathi, S.K. et al., First occurrence of a xanthone and isolation of a 6-ketodehydroprotenoid from *Dalbergia sissoides*, *Phytochemistry*, 37, 911, 1994.

259. Stevenson, P.C. and Veitch, N.C., A 2-arylbenzofuran from roots of *Cicer bijugum* associated with *Fusarium* wilt resistance, *Phytochemistry*, 48, 947, 1998.

260. Stevenson, P.C. and Veitch, N.C., Isoflavenes from the roots of *Cicer judaicum*, *Phytochemistry*, 43, 695, 1996.

261. Subarnas, A., Oshima, Y., and Hikino, H., Isoflavans and a pterocarpan from *Astragalus mongholicus*, *Phytochemistry*, 30, 2777, 1991.

261b. Svasti, J. et al., Dalcochinin-8′-*O*-β-D-glucoside and its β-glucosidase enzyme from *Dalbergia cochinchinensis*, *Phytochemistry*, 50, 739, 1999.

262. Syah, Y.M. et al., Andalasin A, a new stilbene dimer from *Morus macroura*, *Fitoterapia*, 71, 630, 2000.

263. Tahara, S. et al., 5,7,3′,4′,5′-Pentaoxygenated and 2′,6′-diprenylated isoflavones from *Piscidia erythrina*, *Phytochemistry*, 30, 2769, 1991.

264. Tahara, S. et al., Diverse oxygenated isoflavonoids from *Piscidia erythrina*, *Phytochemistry*, 34, 303, 1993.

266. Tahara, S. et al., Naturally occurring coumaranochroman-4-ones: a new class of isoflavonoids from lupines and Jamaican dogwood. *Zeitschrift fuer Naturforschung C*, 46, 331, 1991.

267. Tahara, S. et al., Prenylated flavonoids in the roots of yellow lupin, *Phytochemistry*, 36, 1261, 1994.

268. Tahara, S. et al., Structure revision of piscidone, a major isoflavonoid in the root bark of *Piscidia erythrina*, *Phytochemistry*, 31, 679, 1992.

269. Takashima, J. et al., Derrisin, a new rotenoid from *Derris malaccensis* plain and anti-*Helicobacter pylori* activity of its related constituents, *Journal of Natural Products*, 65, 611, 2002.

272. Tanaka, H. et al., An arylbenzofuran and four isoflavonoids from the roots of *Erythrina poeppigiana*, *Phytochemistry*, 63, 597, 2003.

273. Tanaka, H. et al., An isoflavan from *Erythrina* × *bidwillii*, *Phytochemistry*, 47, 1397, 1998.

275. Tanaka, H. et al., Erysubins C–F, four isoflavonoids from *Erythrina suberosa* var. *glabrescences*, *Phytochemistry*, 56, 769, 2001.

276. Tanaka, H. et al., Four new isoflavonoids and a new 2-arylbenzofuran from the roots of *Erythrina variegata*, *Heterocycles*, 60, 2767, 2003.

277. Tanaka, H. et al., Isoflavonoids from roots of *Erythrina zeyheri*, *Phytochemistry*, 64, 753, 2003.

278. Tanaka, H. et al., Isoflavonoids from the roots of *Erythrina poeppigiana*, *Phytochemistry*, 60, 789, 2002.

279. Tanaka, H. et al., Revised structures for Senegalensin and Euchrenone b10, *Journal of Natural Products*, 64, 1336, 2001.

280. Tanaka, H. et al., Six new constituents from the roots of *Erythrina variegata*, *Chemistry & Biodiversity*, 1, 1101, 2004.

281. Tanaka, H. et al., Three isoflavanones from *Erythrina orientalis, Phytochemistry*, 48, 355, 1998.
282. Tanaka, H. et al., Three new isoflavonoids from *Erythrina variegata, Heterocycles*, 55, 2341, 2001.
283. Tanaka, H. et al., Two new isoflavones from *Erythrina suberosa* var. *glabrescences, Heterocycles*, 48, 2661, 1998.
284. Tanaka, H. et al., Two new isoflavonoids from *Erythrina variegata, Planta Medica*, 66, 578, 2000.
285. Tanaka, H., Tanaka, T., and Etoh, H., A pterocarpan from *Erythrina orientalis, Phytochemistry*, 45, 205, 1997.
286. Tanaka, H., Tanaka, T., and Etoh, H., A pterocarpan from *Erythrina orientalis, Phytochemistry*, 42, 1473, 1996.
287. Tanaka, H., Tanaka, T., and Etoh, H., Three pterocarpans from *Erythrina crista-galli, Phytochemistry*, 45, 835, 1997.
288. Tanaka, H., Tanaka, T., and Etoh, H., Two pterocarpans from *Erythrina orientalis, Phytochemistry*, 47, 475, 1998.
289. Tanaka, T. et al., Isoflavonoids from *Sophora secundiflora, S. arizonica* and *S. gypsophila, Phytochemistry*, 48, 1187, 1998.
290. Tang, Y. et al., Four new isoflavone triglycosides from *Sophora japonica, Journal of Natural Products*, 64, 1107, 2001.
291. Tang, Y.-P. et al., A new coumaronochromone from *Sophora japonica, Journal of Asian Natural Products Research*, 4, 1, 2002.
292. Tarus, P.K. et al., Flavonoids from *Tephrosia aequilata, Phytochemistry*, 60, 375, 2002.
293. Thasana, N., Chuankamnerdkarn, M., and Ruchirawat, S., A new 12α-hydroxyelliptone from the stems of *Derris malaccensis, Heterocycles*, 55, 1121, 2001.
294. Toda, T. et al., New 6-*O*-acyl isoflavone glycosides from soybeans fermented with *Bacillus subtilis* (natto). I. 6-*O*-succinylated isoflavone glycosides and their preventive effects on bone loss in ovariectomized rats fed a calcium-deficient diet, *Biological & Pharmaceutical Bulletin*, 22, 1193, 1999.
295. Tostes, J.B.F., Da Silva, A.J.R., and Parente, J.P., Pubescidin, an isoflavone glycoside from *Centrosema pubescens, Phytochemistry*, 45, 1069, 1997.
296. Tostes, J.B.F., Silva, A.J.R., and Parente, J.P., Isoflavone glycosides from *Centrosema pubescens, Phytochemistry*, 50, 1087, 1999.
297. Tsanuo, M.K. et al., Isoflavanones from the allelopathic aqueous root exudate of *Desmodium uncinatum, Phytochemistry*, 64, 265, 2003.
298. Ueda, H. et al., A new isoflavone glycoside from *Ceiba pentandra* (L.) Gaertner, *Chemical & Pharmaceutical Bulletin*, 50, 403, 2003.
299. Veitch, N.C. et al., Six new isoflavones and a 5-deoxyflavonol glycoside from the leaves of *Ateleia herbert-smithii, Journal of Natural Products*, 66, 210, 2003.
300. Veitch, N.C. and Stevenson, P.C., 2-Methoxyjudaicin, an isoflavene from the roots of *Cicer bijugum, Phytochemistry*, 44, 1587, 1997.
301. Velandia, J.R. et al., Novel trichloro- and tetrachloroisoflavone isolated from *Ouratea semiserrata, Natural Product Letters*, 12, 191, 1998.
302. Velozo, L.S.M. et al., Odoratin 7-*O*-β-D-glucopyranoside from *Bowdichia virgilioides, Phytochemistry*, 52, 1473, 1999.
303. Vila, J. et al., Prenylisoflavanones from *Geoffroea decorticans, Phytochemistry*, 49, 2525, 1998.
304. von Reuss, S.H. and Koenig, W.A., Corsifurans A–C, 2-arylbenzofurans of presumed stilbenoid origin from *Corsinia coriandrina* (Hepaticae), *Phytochemistry*, 65, 3113, 2004.
305. Waffo A.K. et al., Indicanines B and C, two isoflavonoid derivatives from the root bark of *Erythrina indica, Phytochemistry*, 53, 981, 2000.
306. Wandji, J. et al., Erysenegalenseins B and C, two new prenylated isoflavanones from *Erythrina senegalensis, Journal of Natural Products*, 58, 105, 1995.
307. Wandji, J. et al., *Erythrina* studies. Part 24. Two isoflavones from *Erythrina senegalensis, Phytochemistry*, 35, 245, 1994.
308. Wandji, J. et al., *Erythrina* studies. Part 26. Erysenegalenseins H and I: two new isoflavones from *Erythrina senegalensis, Planta Medica*, 60, 178, 1994.

309. Wandji, J. et al., Isoflavones and alkaloids from the stem bark and seeds of *Erythrina senegalensis*, *Phytochemistry*, 39, 677, 1995.

310. Wandji, J. et al., Part 27. Epoxyisoflavones from *Erythrina senegalensis*, *Phytochemistry*, 35, 1573, 1994.

311. Wandji, J. et al., Prenylated isoflavonoids from *Erythrina sensegalensis*, *Phytochemistry*, 38, 1309, 1995.

312. Wang, D.-Y. et al., Four new isoflavones from *Ampelopsis grossedentata*, *Journal of Asian Natural Products Research*, 4, 303, 2002.

313. Wang, S., Ghisalberti, E.L., and Ridsdill-Smith, J., Bioactive isoflavonols and other components from *Trifolium subterraneum*, *Journal of Natural Products*, 61, 508, 1998.

314. Wang, W. et al., Studies on flavonoid constituents of *Hedysarum multijugum*, *Yaoxue Xuebao*, 37, 196, 2002.

315. Wang, W. et al., Two new pterocarpenes from *Hedysarum multijugum*, *Journal of Asian Natural Products Research*, 5, 31, 2003.

316. Wang, Y.-F. et al., New rotenoids from roots of *Mirabilis jalapa*, *Helvetica Chimica Acta*, 85, 2342, 2002.

317. Wanjala, C.C.W. and Majinda, R.R.T., A new isoflavanone from the stem bark of *Erythrina latissima*, *Fitoterapia*, 71, 400, 2000.

319. Watanabe, K., Kinjo, J., and Nohara, T., Leguminous plants. XXXIX. Three new isoflavonoid glycosides from *Lupinus luteus* and L. *polyphyllus arboreus*, *Chemical & Pharmaceutical Bulletin*, 41, 394, 1993.

320. Weng, J.-R., Yen, M.-H., and Lin, C.-N., New pterocarpanoids of *Crotalaria pallida* and *Crotalaria assamica*, *Helvetica Chimica Acta*, 85, 847, 2002.

321. Wollenweber, E., Papendieck, S., and Schilling, G., A novel *C*-methyl isoflavone from *Abronia latifolia*, *Natural Product Letters*, 3, 119, 1993.

322. Woo, W.S. and Woo, E.H., An isoflavone noririsflorentin from *Belamcanda chinensis*, *Phytochemistry*, 33, 939, 1993.

323. Xu, D. et al., Isolation of a new isoflavone from soybean germ, *Zhongcaoyao*, 34, 1065, 2003.

324. Yadav, P.P., Ahmad, G., and Maurya, R., Furanoflavonoids from *Pongamia pinnata* fruits, *Phytochemistry*, 65, 439, 2004.

326. Yadava, R.N. and Barsainya, D., A new isoflavone glycoside from seeds of *Cotula anthemoids* Linn., *Journal of the Institution of Chemists*, 70, 128, 1998.

327. Yadava, R.N. and Kumar, S., A novel isoflavone from the stems of *Ageratum conyzoides*, *Fitoterapia*, 70, 475, 1999.

328. Yadava, R.N. and Kumar, S., A novel isoflavone glycoside from the leaves of *Sphaeranthus indicus*, *Fitoterapia*, 70, 127, 1999.

329. Yadava, R.N. and Syeda, Y. An isoflavone glycoside from the seeds of *Trichosanthes anguina*, *Phytochemistry*, 36, 1519, 1994.

330. Yankep, E. et al., Griffonianone D, an isoflavone with anti-inflammatory activity from the root bark of *Millettia griffoniana*, *Journal of Natural Products*, 66, 1288, 2003.

331. Yankep, E. et al., Further isoflavonoid metabolites from *Millettia griffoniana* (Bail), *Phytochemistry*, 56, 363, 2001.

332. Yankep, E. et al., The Millettia of Cameroon. *O*-Geranylated isoflavones and a 3-phenylcoumarin from *Millettia griffoniana*, *Phytochemistry*, 49, 2521, 1998.

333. Yankep, E., Fomum, Z.T., and Dagne, E., An *O*-geranylated isoflavone from *Millettia griffoniana*, *Phytochemistry*, 46, 591, 1997.

334. Yenesew, A. et al., Anti-plasmodial activities and x-ray crystal structures of rotenoids from *Millettia usaramensis* subspecies *usaramensis*, *Phytochemistry*, 64, 773, 2003.

336. Yenesew, A. et al., Four isoflavones from the stem bark of *Erythrina sacleuxii*, *Phytochemistry*, 49, 247, 1998.

337. Yenesew, A. et al., Three isoflav-3-enes and a 2-arylbenzofuran from the root bark of *Erythrina burttii*, *Phytochemistry*, 59, 337, 2002.

338. Yenesew, A. et al., Two isoflavanones from the stem bark of *Erythrina sacleuxii*, *Phytochemistry*, 55, 457, 2000.

339. Yenesew, A. et al., Two prenylated flavonoids from the stem bark of *Erythrina burttii*, *Phytochemistry*, 63, 445, 2003.

340. Yenesew, A., Midiwo, J.O., and Waterman, P.G., 6-Methoxycalopogonium isoflavone A: a new isoflavone from the seed pods of *Millettia dura*, *Journal of Natural Products*, 60, 806, 1997.

341. Yenesew, A., Midiwo, J.O., and Waterman, P.G., Four isoflavones from seed pods of *Millettia dura*, *Phytochemistry*, 41, 951, 1996.

342. Yenesew, A., Midiwo, J.O., and Waterman, P.G., Rotenoids, isoflavones and chalcones from the stem bark of *Millettia usaramensis* subspecies *usaramensis*, *Phytochemistry*, 47, 295, 1998.

343. Yi, Y. et al., A new isoflavone from *Smilax glabra*, *Molecules* (electronic publication), 3, 145, 1998.

344. Yi, Y. et al., Studies on chemical constituents of *Smilax glabra* Roxb. (IV), *Yaoxue Xuebao*, 33, 873, 1998.

345. Yuldashev, M.P. et al., Structural study of glabrisoflavone, a novel isoflavone from *Glycyrrhiza glabra* L., *Russian Journal of Bioorganic Chemistry* (Translation of *Bioorganicheskaya Khimiya*), 26, 784, 2000.

346. Zeng, J.-F. et al., Manuifolins D, E, and F: new isoflavonoids from *Maackia tenuifolia*, *Journal of Natural Products*, 60, 918, 1997.

347. Zeng, J.-F. et al., Three prenylated isoflavans from *Maackia tenuifolia*, *Phytochemistry*, 47, 903, 1998.

348. Zeng, J.-F. et al., Two isoprenoid-substituted isoflavans from roots of *Maackia tenuifolia*, *Phytochemistry*, 43, 893, 1996.

349. Zeng, J.-F., Tan, C.-H., and Zhu, D.-Y., Manuifolin Q, an unusual 4-aryl-substituted isoflavan from *Maackia tenuifolia*, *Journal of Asian Natural Products Research*, 6, 45, 2004.

350. Zeng, L. et al., Phenolic constituents of *Glycyrrhiza* species. 8. Four new prenylated flavonoids, glyasperins A, B, C, and D from the roots of *Glycyrrhiza* aspera, *Heterocycles*, 34, 575, 1992.

351. Zhang, H.J., Liu, Y., and Zhang, R.Y., Chemical studies of flavonoid compounds from *Glycyrrhiza uralensis* Fisch., *Yaoxue Xuebao*, 29, 471, 1994.

352. Zhang, L., Ju, M., and Hu, C., Five isoflavonoid compounds from roots of *Caragana sinica*, *Journal of Chinese Pharmaceutical Sciences*, 6, 122, 1997.

353. Zhang, Y. et al., Isoflavones from *Glycyrrhiza eurycarpa*, *Yaoxue Xuebao*, 32, 301, 1997.

354. Zhang, Y.M. et al., Eurycarpin A. A new isoflavone from *Glycyrrhiza eurycarpa*, *Chinese Chemical Letters*, 6, 477, 1995.

356. Zhong, C. and Wang, D., Four new isoflavones from *Premna microphylla*, *Indian Journal of Heterocyclic Chemistry*, 12, 143, 2002.

357. Zhou, L.X. and Lin, M., Studies on chemical constituents of *Belamcanda chinensis* (L.) DC. II, *Chinese Chemical Letters*, 8, 133, 1997.

Index